Solar System Data

Body	Mass (kg)	Mean Radius (m)	Period (s)	Distance from the Sun (m)
Mercury	3.18×10^{23}	2.43×10^{6}	7.60×10^{6}	5.79×10^{10}
Venus	4.88×10^{24}	6.06×10^{6}	1.94×10^{7}	1.08×10^{11}
Earth	5.98×10^{24}	6.37×10^{6}	3.156×10^{7}	1.496×10^{11}
Mars	6.42×10^{23}	3.37×10^{6}	5.94×10^{7}	2.28×10^{11}
Jupiter	1.90×10^{27}	6.99×10^{7}	3.74×10^{8}	7.78×10^{11}
Saturn	5.68×10^{26}	5.85×10^{7}	9.35×10^{8}	1.43×10^{12}
Uranus	8.68×10^{25}	2.33×10^{7}	2.64×10^{9}	2.87×10^{12}
Neptune	1.03×10^{26}	2.21×10^{7}	5.22×10^{9}	4.50×10^{12}
Pluto	$\approx 1.4 \times 10^{22}$	$\approx 1.5 \times 10^{6}$	7.82×10^{9}	5.91×10^{12}
Moon	7.36×10^{22}	1.74×10^{6}	—	—
Sun	1.991×10^{30}	6.96×10^{8}	—	—

Physical Data Often Used[a]

Average Earth–Moon distance	3.84×10^{8} m
Average Earth–Sun distance	1.496×10^{11} m
Average radius of the Earth	6.37×10^{6} m
Density of air (20°C and 1 atm)	1.20 kg/m^3
Density of water (20°C and 1 atm)	1.00×10^{3} kg/m^3
Free-fall acceleration	9.80 m/s^2
Mass of the Earth	5.98×10^{24} kg
Mass of the Moon	7.36×10^{22} kg
Mass of the Sun	1.99×10^{30} kg
Standard atmospheric pressure	1.013×10^{5} Pa

[a] These are the values of the constants as used in the text.

Some Prefixes for Powers of Ten

Power	Prefix	Abbreviation	Power	Prefix	Abbreviation
10^{-24}	yocto	y	10^{1}	deka	da
10^{-21}	zepto	z	10^{2}	hecto	h
10^{-18}	atto	a	10^{3}	kilo	k
10^{-15}	femto	f	10^{6}	mega	M
10^{-12}	pico	p	10^{9}	giga	G
10^{-9}	nano	n	10^{12}	tera	T
10^{-6}	micro	μ	10^{15}	peta	P
10^{-3}	milli	m	10^{18}	exa	E
10^{-2}	centi	c	10^{21}	zetta	Z
10^{-1}	deci	d	10^{24}	yotta	Y

Pedagogical Color Chart

Part 1 (Chapters 1–15) : Mechanics

Displacement and position vectors

Linear ($\mathbf{v}$) and angular ($\boldsymbol{\omega}$) velocity vectors

Velocity component vectors

Force vectors ($\mathbf{F}$)

Force component vectors

Acceleration vectors ($\mathbf{a}$)

Acceleration component vectors

Linear ($\mathbf{p}$) and angular ($\mathbf{L}$) momentum vectors

Torque vectors ($\boldsymbol{\tau}$)

Linear or rotational motion directions

Springs

Pulleys

Part 4 (Chapters 23–34) : Electricity and Magnetism

Electric fields

Magnetic fields

Positive charges

Negative charges

Resistors

Batteries and other DC power supplies

Switches

Capacitors

Inductors (coils)

Voltmeters

Ammeters

AC Generators

Ground symbol

Part 5 (Chapters 35–38) : Light and Optics

Light rays

Lenses and prisms

Mirrors

Objects

Images

Open the door to the fascinating world of physics

Physics, the most fundamental of all natural sciences, will reveal to you the basic principles of the Universe. And while physics can seem challenging, its true beauty lies in the sheer simplicity of fundamental physical theories—theories and concepts that can alter and expand your view of the world around you. Other courses that follow will use the same principles, so it is important that you understand and are able to apply the various concepts and theories discussed in the text. **Physics for Scientists and Engineers, Sixth Edition** is your guide to this fascinating science.

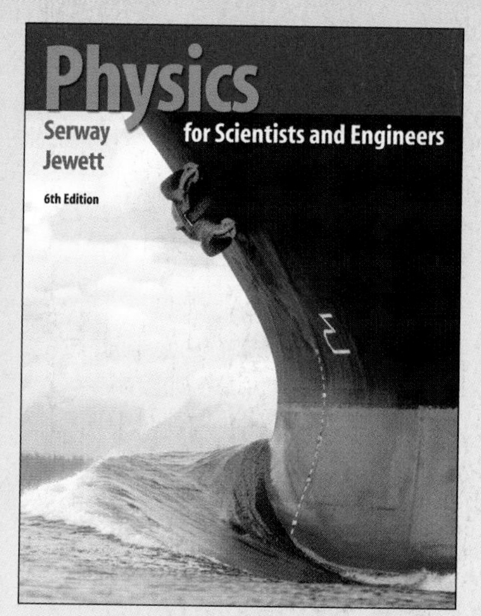

Your quick start for studying smart

Achieve success in your physics course by making the most of what **Physics for Scientists and Engineers, Sixth Edition** has to offer you. From a host of in-text features to a range of Web resources, you'll have everything you need to understand the natural forces and principles of physics:

- **Dynamic built-in study aids.** Throughout every chapter the authors have built in a wide range of examples, exercises, and illustrations that will help you understand and appreciate the laws of physics. *See pages 2 and 3 for more information.*
- **A powerful Web-based learning system.** The text is fully integrated with **PhysicsNow**, an interactive learning system that tailors itself to your needs in the course. It's like having a personal tutor available whenever you need it! *See pages 4–7 to explore* **PhysicsNow**.

Physics Now

Your *Quick Start for Studying Smart* begins with this special tour through the book. On the following pages you'll discover how **Physics for Scientists and Engineers, Sixth Edition** and **PhysicsNow** not only enhance your experience in this course, but help you to succeed!

Quick Start for Studying Smart!

Take charge of your success

Everything you need to succeed in your course is available to you in Physics for Scientists and Engineers, Sixth Edition. Authors Serway and Jewett have filled their text with learning tools and study aids that will clarify concepts and help you build a solid base of knowledge. The end result: confidence in the classroom, in your study sessions, and in your exams.

THE RIGHT APPROACH

Start out right! Early on in the text the authors outline a general problem-solving strategy that will enable you to increase your accuracy in solving problems, enhance your understanding of physical concepts, eliminate initial worry or lack of direction in approaching a problem, and organize your work. The problem-solving strategy is integrated into the *Coached Problems* found on **PhysicsNow** to reinforce this key skill. (See pages 4–7 for more information about the **PhysicsNow** Web-based and student-centered learning system.)

GENERAL PROBLEM-SOLVING STRATEGY

Conceptualize

- The first thing to do when approaching a problem is to *think about* and *understand* the situation. Study carefully any diagrams, graphs, tables, or photographs that accompany the problem. Imagine a movie, running in your mind, of what happens in the problem.
- If a diagram is not provided, you should almost always make a quick drawing of the situation. Indicate any known values, perhaps in a table or directly on your sketch.
- Now focus on what algebraic or numerical information is given in the problem. Carefully read the problem statement, looking for key phrases such as "starts from rest" ($v_i = 0$), "stops" ($v_f = 0$), or "freely falls" ($a_y = -g = -9.80\ \mathrm{m/s^2}$).
- Now focus on the expected result of solving the problem. Exactly what is the question asking? Will the final result be numerical or algebraic? Do you know what units to expect?
- Don't forget to incorporate information from your own experiences and common sense. What should a reasonable answer look like? You wouldn't expect to calculate the speed of an automobile to be

Analyze

- Now you must analyze the problem and strive for a mathematical solution. Because you have already categorized the problem, it should not be too difficult to select relevant equations that apply to the type of situation in the problem. For example, if the problem involves a particle moving under constant acceleration, Equations 2.9 to 2.13 are relevant.
- Use algebra (and calculus, if necessary) to solve symbolically for the unknown variable in terms of what is given. Substitute in the appropriate numbers, calculate the result, and round it to the proper number of significant figures.

Finalize

- This is the most important part. Examine your numerical answer. Does it have the correct units? Does it meet your expectations from your conceptualization of the problem? What about the algebraic form of the result — before you substituted numerical values? Does it make sense? Examine the variables in the problem to see whether the answer would change in a physically meaningful way if they were drastically increased or decreased or even became zero. Looking at limiting cases to see whether they yield expected values is a very useful way to make sure that you are obtaining reasonable results.

Think about how this problem compares with others you have done. How was it similar? In what critical ways did it differ? Why was this problem assigned? You should have learned something by doing it. Can you figure out what? If it is a new category of problem, be sure you understand it so that you can use it as a model for solving future problems in the same category.

hen solving complex problems, you may need to idenfy a series of sub-problems and apply the problem-solv-

PROBLEM-SOLVING HINTS

Problem-Solving Hints help you approach homework assignments with greater confidence. General strategies and suggestions are included for solving the types of problems featured in the worked examples, end-of-chapter problems, and **PhysicsNow**. This feature helps you identify the essential steps in solving problems and increases your skills as a problem solver.

PROBLEM-SOLVING HINTS

Applying Newton's Laws

The following procedure is recommended when dealing with problems involving Newton's laws:

- Draw a simple, neat diagram of the system to help *conceptualize* the problem.
- *Categorize* the problem: if any acceleration component is zero, the particle is in equilibrium in this direction and $\Sigma F = 0$. If not, the particle is undergoing an acceleration, the problem is one of nonequilibrium in this direction, and $\Sigma F = ma$.
- *Analyze* the problem by isolating the object whose motion is being analyzed. Draw a free-body diagram for this object. For systems containing more than one object, draw *separate* free-body diagrams for each object. *Do not* include in the free-body diagram forces exerted by the object on its surroundings.
- Establish convenient coordinate axes for each object and find the components of the forces along these axes. Apply Newton's second law, $\Sigma \mathbf{F} = m\mathbf{a}$, in component form. Check your dimensions to make sure that all terms have unit
- Solve the comp have as many ir complete soluti
- *Finalize* by maki Also check the variables. By do

WORKED EXAMPLES

Reinforce your understanding of essential problem-solving techniques using a large number of realistic *Worked Examples*. In many cases, these examples serve as models for solving the end-of-chapter problems. Numerous *Worked Examples* include specific references to the general problem-solving strategy to illustrate the underlying concepts and methodology used in arriving at a correct solution. This will help you understand the logic behind the solution and the advantage of using a particular approach to solve the problem. **PhysicsNow** also features a number of worked examples to further enhance your understanding of problem solving and to give you even more practice solving problems.

Example 4.3 The Long Jump

A long-jumper (Fig. 4.12) leaves the ground at an angle of 20.0° above the horizontal and at a speed of 11.0 m/s.

(A) How far does he jump in the horizontal direction? (Assume his motion is equivalent to that of a particle.)

Solution We *conceptualize* the motion of the long-jumper as equivalent to that of a simple projectile such as the ball in Example 4.2, and *categorize* this problem as a projectile motion problem. Because the initial speed and launch angle are given, and because the final height is the same as the initial height, we further categorize this problem as satisfying the conditions for which Equations 4.13 and 4.14 can be used. This is the most direct way to *analyze* this problem describi the gen

provides a graphical representation of the flight of the long-jumper. As before, we set our origin of coordinates at the takeoff point and label the peak as Ⓐ and the landing point as Ⓑ. The horizontal motion is described by Equation 4.11:

$$x_f = x_{\mathrm{B}} = (v_i \cos \theta_i) t_{\mathrm{B}} = (11.0\ \mathrm{m/s})(\cos 20.0°) t_{\mathrm{B}}$$

The value of x_{B} can be found if the time of landing t_{B} is known. We can find t_{B} by remembering that $a_y = -g$ and by using the y part of Equation 4.8a. We also note that at the top of the jump the vertical component of velocity $v_{y\mathrm{A}}$ is zero:

$$v_{yf} = v_{y\mathrm{A}} = v_i \sin \theta_i - g t_{\mathrm{A}}$$

Figure 4.12 (Example 4.3) Mike Powell, current holder of the world long jump record of 8.95 m.

Mike Powell/Allsport/Getty Images

This is the time at which the long-jumper is at the *top* of the jump. Because of the symmetry of the vertical motion, another 0.384 s passes before the jumper returns to the ground. Therefore, the time at which the jumper lands is $t_{\mathrm{B}} = 2t_{\mathrm{A}} = 0.768$ s. Substituting this value into the above expression for x_f gives

$$x_f = x_{\mathrm{B}} = (11.0\ \mathrm{m/s})(\cos 20.0°)(0.768\ \mathrm{s}) = \boxed{7.94\ \mathrm{m}}$$

This is a reasonable distance for a world-class athlete.

(B) What is the maximum height reached?

Solution We find the maximum height reached by using Equation 4.12:

$$y_{\max} = y_{\mathrm{A}} = (v_i \sin \theta_i) t_{\mathrm{A}} - \tfrac{1}{2} g t_{\mathrm{A}}^2$$
$$= (11.0\ \mathrm{m/s})(\sin 20.0°)(0.384\ \mathrm{s})$$
$$-\tfrac{1}{2}(9.80\ \mathrm{m/s^2})(0.384\ \mathrm{s})^2 = \boxed{0.722\ \mathrm{m}}$$

To *finalize* this problem, find the answers to parts (a) and (b) using Equations 4.13 and 4.14. The results should agree. Treating the long-jumper as a particle is an oversimplification. Nevertheless, the values obtained are consistent with experience in sports. We learn that we can model a complicated system such as a long-jumper as a particle and still obtain results that are reasonable.

Quick Start for Studying Smart!

… = −1, and we can write the acceleration expression as

where k is a dimensionless constant of proportionality. Knowing the dimensions of a, r, and v, we see that the dimensional equation must be

$$\frac{\mathrm{L}}{\mathrm{T}^2} = \mathrm{L}^n\left(\frac{\mathrm{L}}{\mathrm{T}}\right)^m = \frac{\mathrm{L}^{n+m}}{\mathrm{T}^m}$$

$$a = kr^{-1}v^2 = k\frac{v^2}{r}$$

When we discuss uniform circular motion later, we shall see that $k = 1$ if a consistent set of units is used. The constant k would not equal 1 if, for example, v were in km/h and you wanted a in m/s^2.

▲ PITFALL PREVENTION

1.5 Always Include Units

When performing calculations, include the units for every quantity and carry the units through the entire calculation. Avoid the temptation to drop the units early and then attach the expected units once you have an answer. By including the units in every step, you can detect errors if the units for the answer turn out to be incorrect.

1.5 Conversion of Units

Sometimes it is necessary to convert units from one measurement system to another, or to convert within a system, for example, from kilometers to meters. Equalities between SI and U.S. customary units of length are as follows:

$$1 \text{ mile} = 1\,609 \text{ m} = 1.609 \text{ km} \qquad 1 \text{ ft} = 0.304\,8 \text{ m} = 30.48 \text{ cm}$$

$$1 \text{ m} = 39.37 \text{ in.} = 3.281 \text{ ft} \qquad 1 \text{ in.} = 0.025\,4 \text{ m} = 2.54 \text{ cm (exactly)}$$

A more complete list of conversion factors can be found in Appendix A.

Units can be treated as algebraic quantities that can cancel each other. For example, suppose we wish to convert 15.0 in. to centimeters. Because 1 in. is defined as exactly 2.54 cm, we find that

PITFALL PREVENTIONS

An easy way to make sure you avoid common mistakes—while studying *and* while taking exams! These helpful notes, found in the margins, allow you to correct misconceptions before applying them to new material.

Example 4.5 That's Quite an Arm!

A stone is thrown from the top of a building upward at an angle of 30.0° to the horizontal with an initial speed of 20.0 m/s, as shown in Figure 4.14. If the height of the building is 45.0 m,

(A) how long before the stone hits the ground?

Solution We *conceptualize* the problem by studying Figure 4.14, in which we have indicated the various parameters. By now, it should be natural to *categorize* this as a projectile motion problem.

To *analyze* the problem, let us once again separate motion into two components. The initial x and y components of the stone's velocity are

$$v_{xi} = v_i \cos\theta_i = (20.0 \text{ m/s})\cos 30.0° = 17.3 \text{ m/s}$$

$$v_{yi} = v_i \sin\theta_i = (20.0 \text{ m/s})\sin 30.0° = 10.0 \text{ m/s}$$

To find t, we can use $y_f = y_i + v_{yi}t + \frac{1}{2}a_y t^2$ (Eq. 4.9a) with $y_i = 0$, $y_f = -45.0$ m, $a_y = -g$, and $v_{yi} = 10.0$ m/s (there is a negative sign on the numerical value of y_f because we have chosen the top of the building as the origin):

$$-45.0 \text{ m} = (10.0 \text{ m/s})t - \tfrac{1}{2}(9.80 \text{ m/s}^2)t^2$$

Solving the quadratic equation for t gives, for the positive root, $t = 4.22$ s. To *finalize* this part, think: Does the negative root have any physical meaning?

(B) What is the speed of the stone just before it strikes the ground?

Solution We can use Equation 4.8a, $v_{yf} = v_{yi} + a_y t$, with $t = 4.22$ s to obtain the y component of the velocity just before the stone strikes the ground:

$$v_{yf} = 10.0 \text{ m/s} - (9.80 \text{ m/s}^2)(4.22 \text{ s}) = -31.4 \text{ m/s}$$

Because $v_{xf} = v_{xi} = 17.3$ m/s, the required speed is

$$v_f = \sqrt{v_{xf}^2 + v_{yf}^2} = \sqrt{(17.3)^2 + (-31.4)^2} \text{ m/s} = 35.9 \text{ m/s}$$

What If? **What if a horizontal wind is blowing in the same direction as the ball is thrown and it causes the ball to have a horizontal acceleration component $a_x = 0.500$ m/s^2. Which part of this example, (a) or (b), will have a different answer?**

Answer Recall that the motions in the x and y directions are independent. Thus, the horizontal wind cannot affect the vertical motion. The vertical motion determines the time of the projectile in the air, so the answer to (a) does not change. The wind will cause the horizontal velocity component to increase with time, so that the final speed will change in part (b).

We can find the new final horizontal velocity component by using Equation 4.8a:

$$v_{xf} = v_{xi} + a_x t = 17.3 \text{ m/s} + (0.500 \text{ m/s}^2)(4.22 \text{ s}) = 19.4 \text{ m/s}$$

and the new final speed:

$$v_f = \sqrt{v_{xf}^2 + v_{yf}^2} = \sqrt{(19.4)^2 + (-31.4)^2} \text{ m/s} = 36.9 \text{ m/s}$$

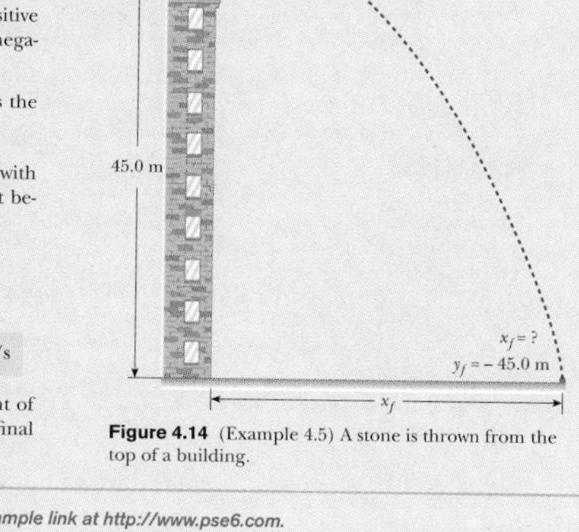

Figure 4.14 (Example 4.5) A stone is thrown from the top of a building.

…ht of … final

…ample link at http://www.pse6.com.

WHAT IF?

What if you could get even more out of the text's worked examples? You can with this feature. The authors change some of the data or assumptions in about one-third of the *Worked Examples* and then explore the consequences. This allows you to apply concepts and problem-solving skills to new situations as well as test the final result to see if it's realistic.

Quick Quiz 5.2 An object experiences no acceleration. Which of the following *cannot* be true for the object? (a) A single force acts on the object. (b) No forces act on the object. (c) Forces act on the object, but the forces cancel.

Quick Quiz 5.3 An object experiences a net force and exhibits an acceleration in response. Which of the following statements is *always* true? (a) The object moves in the direction of the force. (b) The … velocity. (c) The acceleration is in the same … the object increases.

Quick Quiz 5.4 You push an objec… with a constant force for a time interval Δ… object. You repeat the experiment, but with … terval is now required to reach the same fina… (e) $\Delta t/4$.

Answers to Quick Quizzes

5.1 (d). Choice (a) is true. Newton's first law tells us that motion requires no force: an object in motion continues to move at constant velocity in the absence of external forces. Choice (b) is also true. A stationary object can have several forces acting on it, but if the vector sum of all these external forces is zero, there is no net force and the object remains stationary.

5.2 (a). If a single force acts, this force constitutes the net force and there is an acceleration according to Newton's second law.

5.3 (c). Newton's second law relates only the force and the acceleration. Direction of motion is part of an object's *velocity*, and force determines the direction of acceleration, not that of velocity.

5.4 (d). With twice the force, the object will experience twice the acceleration. Because the force is constant, the acceler-

QUICK QUIZZES

Test your understanding of the concepts before investigating the worked examples. Found throughout each chapter, these quizzes give you an opportunity to test your conceptual understanding while also making the material more interactive. Comprehensive answers are found at the end of each chapter. A supplementary set of *Quick Quizzes* is available on the text's Web site, accessible through **www.pse6.com**.

"You do not know anything until you have practiced."

R. P. Feynman, Nobel Laureate in Physics

What do you need to learn now?

Take charge of your learning with **PhysicsNow**, a powerful, interactive study tool that will help you manage and maximize your study time. This collection of dynamic technology resources will assess your unique study needs, giving you an individualized learning plan that will enhance your conceptual and computational skills. Designed to maximize your time investment, **PhysicsNow** helps you succeed by focusing your study time on the concepts you need to master. And best of all, **PhysicsNow** is FREE with every new copy of this text!

INTERACTIVE LEARNING BEGINS WITH THIS BOOK

PhysicsNow directly links with **Physics for Scientists and Engineers, Sixth Edition**. **PhysicsNow** and this text were built in concert to enhance each other and provide you with a seamless, integrated learning system. As you work through the chapters you'll see notes that direct you to the media-enhanced activities on **PhysicsNow (www.pse6.com)**. This precise page-by-page integration means you'll spend less time flipping through pages and navigating Web sites for useful exercises. It is far easier to understand physics if you see it in action, and **PhysicsNow** enables you to *become* a part of the action!

Take a practice test for this chapter by clicking on the Practice Test link at http://www.pse6.com.

SUMMARY

Scalar quantities are those that have only magnitude and no associated direction. **Vector quantities** have both magnitude and direction and obey the laws of vector addition. The magnitude of a vector is *always* a positive number.

When two or more vectors are added together, all of them must have the same units and all of them must be the same type of quantity. We can add two vectors **A** and **B** graphically. In this method (Fig. 3.6), the resultant vector **R** = **A** + **B** runs from the tail of **A** to the tip of **B**.

A second method of adding vectors involves **components** of the vectors. The x component A_x of the vector **A** is equal to the projection of **A** along the x axis of a coordinate system, as shown in Figure 3.13, where $A_x = A\cos\theta$. The y component A_y of **A** is the projection of **A** along the y axis, where $A_y = A\sin\theta$. Be sure you can determine which trigonometric functions you should use in all situations, especially when θ is defined as something other than the counterclockwise angle from the positive x axis.

If a vector **A** has an x component A_x and a y component A_y, the vector can be expressed in unit – vector form as $\mathbf{A} = A_x\hat{\mathbf{i}} + A_y\hat{\mathbf{j}}$. In this is notation, $\hat{\mathbf{i}}$ is a unit vector pointing in the positive x direction, and $\hat{\mathbf{j}}$ is a unit vector pointing in the positive y direction. Because $\hat{\mathbf{i}}$ and $\hat{\mathbf{j}}$ are unit vectors, $|\hat{\mathbf{i}}| = |\hat{\mathbf{j}}| = 1$.

We can find the resultant of two or more vectors by resolving all vectors into their x and y components, adding their resultant x and y components, and then using the Pythagorean theorem to find the magnitude of the resultant vector. We can find the angle that the resultant vector makes with respect to the x axis by using a suitable trigonometric function.

QUESTIONS

1. Two vectors have unequal magnitudes. Can their sum be zero? Explain.
2. Can the magnitude of a particle's displacement be greater

4. Which of the following are vectors and which are not: force, temperature, the volume of water in a can, the ratings of a TV show, the height of a building, the velocity of ... he Universe?

... y plane. For what orientations of **A** ... ents be negative? For what orienta... s have opposite signs?

Example 4.5 That's Quite an Arm!

A stone is thrown from the top of a building upward at an angle of 30.0° to the horizontal with an initial speed of 20.0 m/s, as shown in Figure 4.14. If the height of the building is 45.0 m,

(A) how long before the stone hits the ground?

Solution We *conceptualize* the problem by studying Figure 4.14, in which we have indicated the various parameters. By now, it should be natural to *categorize* this as a projectile motion problem.

To *analyze* the problem, let us once again separate motion into two components. The initial x and y components of the stone's velocity are

$$v_{xi} = v_i\cos\theta_i = (20.0\text{ m/s})\cos 30.0° = 17.3\text{ m/s}$$

$$v_{yi} = v_i\sin\theta_i = (20.0\text{ m/s})\sin 30.0° = 10.0\text{ m/s}$$

To find t, we can use $y_f = y_i + v_{yi}t + \frac{1}{2}a_yt^2$ (Eq. 4.9a) with $y_i = 0$, $y_f = -45.0$ m, $a_y = -g$, and $v_{yi} = 10.0$ m/s (there is a negative sign on the numerical value of y_f because we have chosen the top of the building as the origin):

$$-45.0\text{ m} = (10.0\text{ m/s})t - \tfrac{1}{2}(9.80\text{ m/s}^2)t^2$$

Solving the quadratic equation for t gives, for the positive root, $t =$ 4.22 s . To *finalize* this part, think: Does the negative root have any physical meaning?

(B) What is the speed of the stone just before it strikes the ground?

Solution We can use Equation 4.8a, $v_{yf} = v_{yi} + a_yt$, with $t = 4.22$ s to obtain the y component of the velocity just before the stone strikes the ground:

$$v_{yf} = 10.0\text{ m/s} - (9.80\text{ m/s}^2)(4.22\text{ s}) = -31.4\text{ m/s}$$

Because $v_{xf} = v_{xi} = 17.3$ m/s, the required speed is

$$v_f = \sqrt{v_{xf}^2 + v_{yf}^2} = \sqrt{(17.3)^2 + (-31.4)^2}\text{ m/s} = 35.9\text{ m/s}$$

To *finalize* this part, is it reasonable that the y component of the final velocity is negative? Is it reasonable that the final speed is larger than the initial speed of 20.0 m/s?

What If? What if a horizontal wind is blowing in the same direction as the ball is thrown and it causes the ball to have a horizontal acceleration component $a_x = 0.500$ m/s². Which part of this example, (a) or (b), will have a different answer?

Answer Recall that the motions in the x and y directions are independent. Thus, the horizontal wind cannot affect the vertical motion. The vertical motion determines the time of the projectile in the air, so the answer to (a) does not change. The wind will cause the horizontal velocity component to increase with time, so that the final speed will change in part (b).

We can find the new final horizontal velocity component by using Equation 4.8a:

$$v_{xf} = v_{xi} + a_xt = 17.3\text{ m/s} + (0.500\text{ m/s}^2)(4.22\text{ s}) = 19.4\text{ m/s}$$

and the new final speed:

$$v_f = \sqrt{v_{xf}^2 + v_{yf}^2} = \sqrt{(19.4)^2 + (-31.4)^2}\text{ m/s} = 36.9\text{ m/s}$$

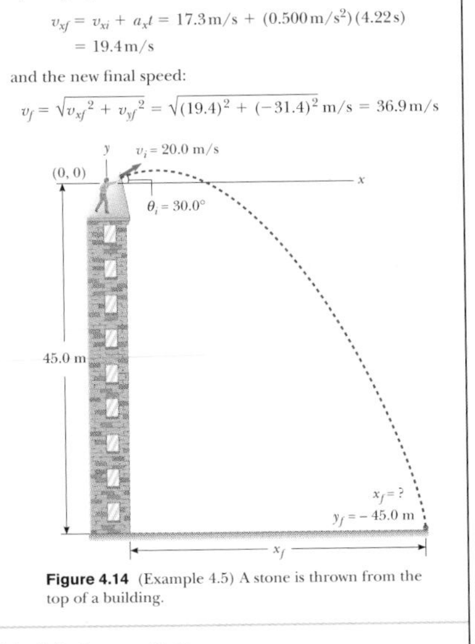

Figure 4.14 (Example 4.5) A stone is thrown from the top of a building.

Investigate this situation at the Interactive Worked Example link at http://www.pse6.com.

Quick Start for Studying Smart!

GO ONLINE AT www.pse6.com

Log on to **PhysicsNow** at **www.pse6.com** by using the free pincode packaged with this text.* You'll immediately notice the system's easy-to-use, browser-based format. Getting to where you need to go is as easy as a click of the mouse. The **PhysicsNow** system is made up of three interrelated parts:

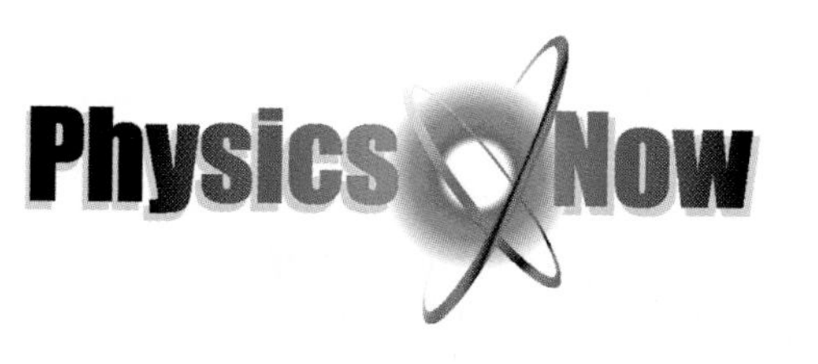

- **How Much Do I Know?**
- **What Do I Need to Learn?**
- **What Have I Learned?**

These three interrelated elements work together, but are distinct enough to allow you the freedom to explore only those assets that meet your personal needs. You can use **PhysicsNow** like a traditional Web site, accessing all assets of a particular chapter and exploring on your own. The best way to maximize the system and *your* time is to start by taking the *Pre-Test.*

* Free PIN codes are only available with new copies of **Physics for Scientists and Engineers, Sixth Edition.**

HOW MUCH DO I KNOW?

The Pre-Test is the first step in creating your *Personalized Learning Plan.* Each *Pre-Test* is based on the end-of-chapter homework problems and includes approximately 15 questions.

Once you've completed the *Pre-Test* you'll be presented with a detailed *Learning Plan* that outlines the elements you need to review to master the chapter's most essential concepts.

At each stage, the text is referenced to reinforce its value as a learning tool.

Turn the page to view problems from a sample *Personalized Learning Plan.*

WHAT DO I NEED TO LEARN?

Once you've completed the *Pre-Test* you're ready to work the problems in your *Personalized Learning Plan*—problems that will help you master concepts essential to your success in this course.

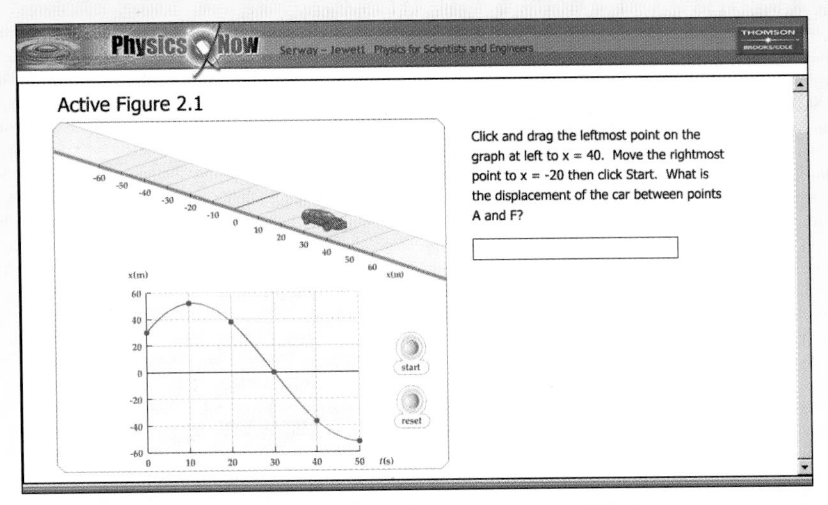

More than 200 *Active Figures* are taken from the text and animated to help you visualize physics in action. Each figure is paired with a question to help you focus on physics at work, and a brief quiz ensures that you understand the concept played out in the animations.

PhysicsNow Serway – Jewett Physics for Scientists and Engineers

THOMSON BROOKS/COLE

Coached Problem 2.3

Section Score: 0%

Current Item - Unanswered

A baseball is hit so that it travels straight upward after being struck by the bat. A fan observes that it requires 3.00 s for the ball to reach its maximum height. Find (a) its initial velocity and (b) the maximum height it reaches.

Type your answer

Try Another | See Example | Work in Steps | Explain

Use the **Arrow Buttons** and **Item Buttons** to navigate through the tutorial. Select your answer(s) for each question and click the **Submit** button to enter your answer. Click the **New Set** button to restart.

You'll continue to master the concepts though *Coached Problems.* These engaging problems reinforce the lessons in the text by taking a step-by-step approach to problem-solving methodology. Each *Coached Problem* gives you the option of working a question and receiving feedback, or seeing a solution worked for you. You'll find approximately five *Coached Problems* per chapter.

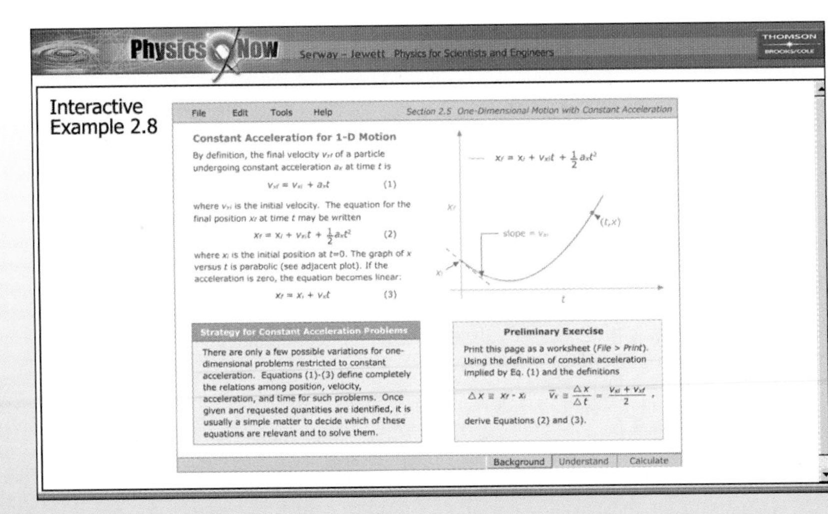

You'll strengthen your problem-solving and visualization skills by working through the *Interactive Examples.* Each step in the examples uses the authors' problem-solving methodology that is introduced in the text (see page 2 of this Visual Preface). You'll find *Interactive Examples* for each chapter of the text.

WHAT HAVE I LEARNED?

After working through the problems highlighted in your personal *Learning Plan* you'll move on to a *Chapter Quiz.* These multiple-choice quizzes present you with questions that are similar to those you might find in an exam. You can even e-mail your quiz results to your instructor.

Once you've completed the quiz you'll receive your results in the form of a percentage. If you need to improve your score, **PhysicsNow** will take you back through the system, beginning with *What Do I Know?*, and work with you as you continue to build your knowledge and skills and master concepts.

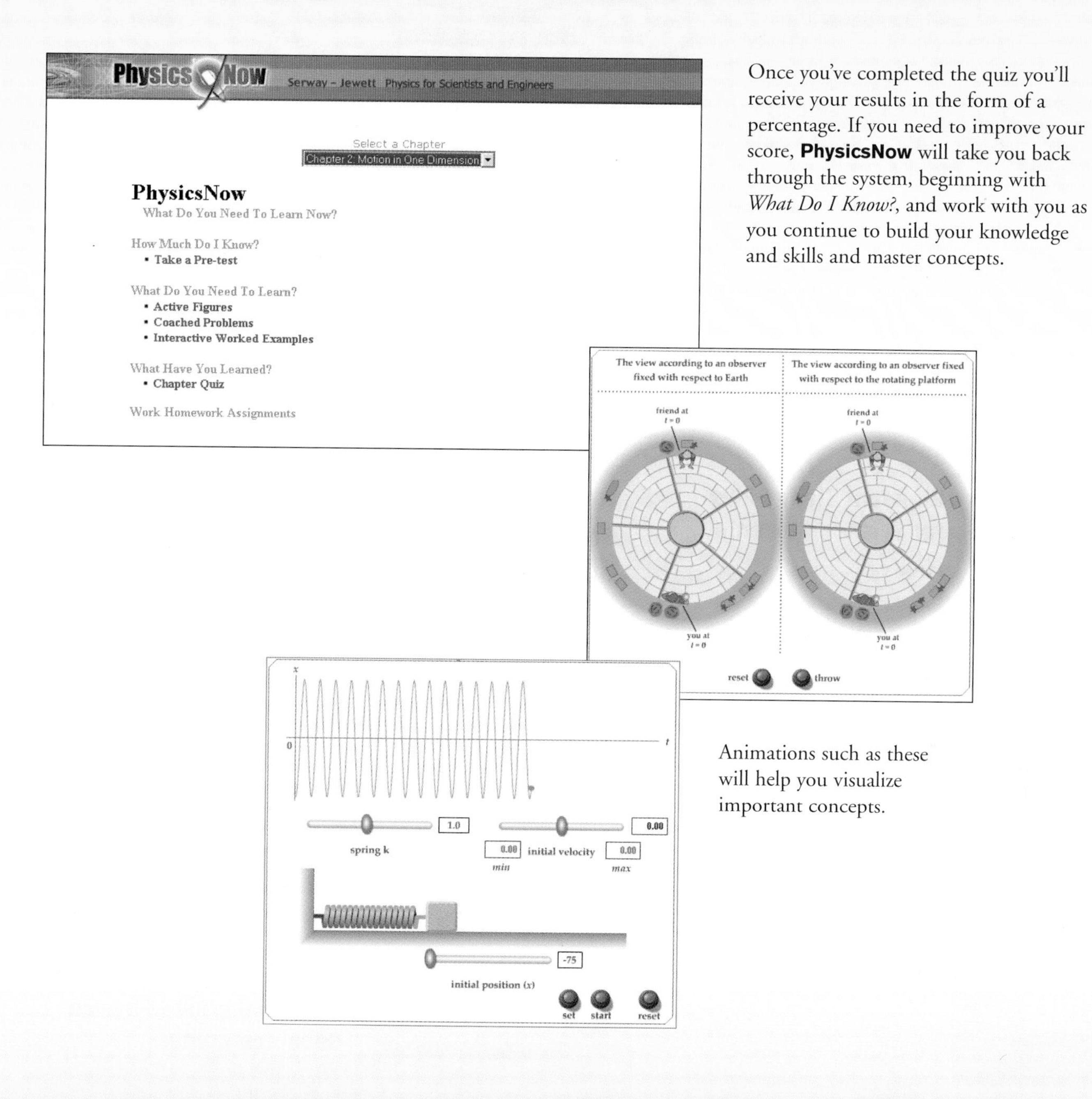

Animations such as these will help you visualize important concepts.

Chart your own course for success . . .

Log on to **www.pse6.com** to take advantage of **PhysicsNow!**

Quick Start for Studying Smart!

Make the most of the course and your time with these exclusive study tools

Enrich your experience outside of the classroom with a host of resources designed to help you excel in the course. To purchase any of these supplements, contact your campus bookstore or visit our online BookStore at www.brookscole.com.

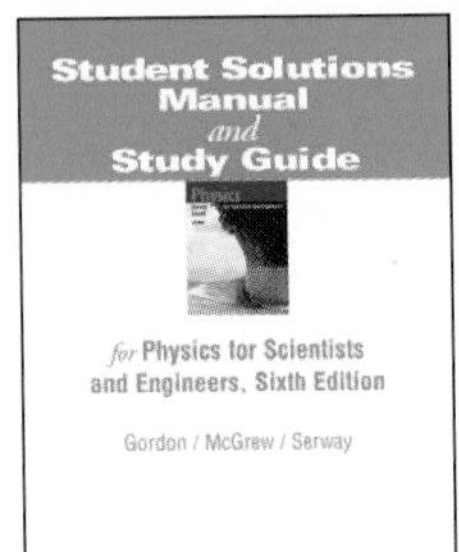

Student Solutions Manual with Study Guide

Volume I ISBN: 0-534-40855-9
Volume II ISBN: 0-534-40856-7

by John R. Gordon, Ralph McGrew, and Raymond Serway
This two-volume manual features detailed solutions to 20% of the end-of-chapter problems from the text. The manual also features a list of important equations, concepts, and answers to selected end-of-chapter questions.

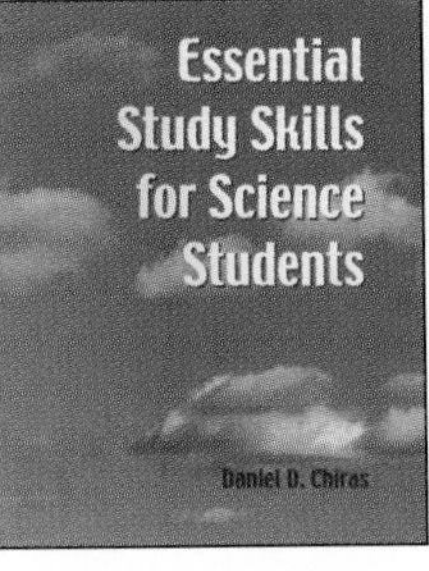

Essential Study Skills for Science Students

ISBN: 0-534-37595-2

by Daniel D. Chiras
Written specifically for science students, this book discusses how to develop good study habits, sharpen memory, learn more quickly, get the most out of lectures, prepare for tests, produce excellent term papers, and improve critical-thinking skills.

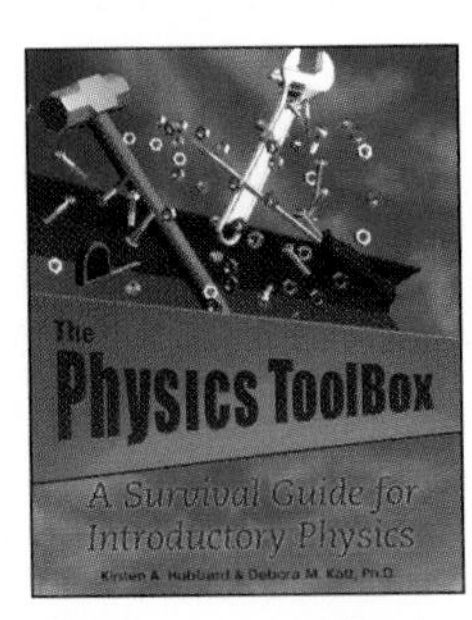

The Physics Toolbox: A Survival Guide for Introductory Physics

ISBN: 0-03-034652-5

by Kirsten A. Hubbard and Debora M. Katz
This "paperback mentor" gives you the material critical for success in physics, including an introduction to the nature of physics and science, a look at what to expect and how to succeed, a verbal overview of the concepts you'll encounter, and an extensive review of the math you'll need to solve the problems.

WebTUTOR™ Advantage

WebTutor™ Advantage on WebCT and Blackboard

WebCT ISBN: 0-534-40859-1
Blackboard ISBN: 0-534-40950-4

WebTutor Advantage offers real-time access to a full array of study tools, including chapter outlines, summaries, learning objectives, glossary flashcards (with audio), practice quizzes, **InfoTrac® College Edition** exercises, and Web links. **WebTutor Advantage** also provides robust communication tools, such as a course calendar, asynchronous discussion, real-time chat, a whiteboard, and an integrated e-mail system. Also new to **WebTutor Advantage** is access to *NewsEdge,* an online news service that brings the latest news to the **WebTutor Advantage** site daily.
Contact your instructor for more information.

Additional Web resources . . .

available FREE to you with each new copy of this text:

InfoTrac® College Edition

When you purchase a new copy of **Physics for Scientists and Engineers, Sixth Edition**, you automatically receive a FREE four-month subscription to **InfoTrac College Edition!** Newly improved, this extensive online library opens the door to the full text (not just abstracts) of countless articles from thousands of publications including *American Scientist, Physical Review, Science, Science Weekly*, and more! Use the passcode included with the new copy of this text and log on to **www.infotrac-college.com** to explore the wealth of resources available to you—24 hours a day and seven days a week!
Available only to college and university students. Journals subject to change.

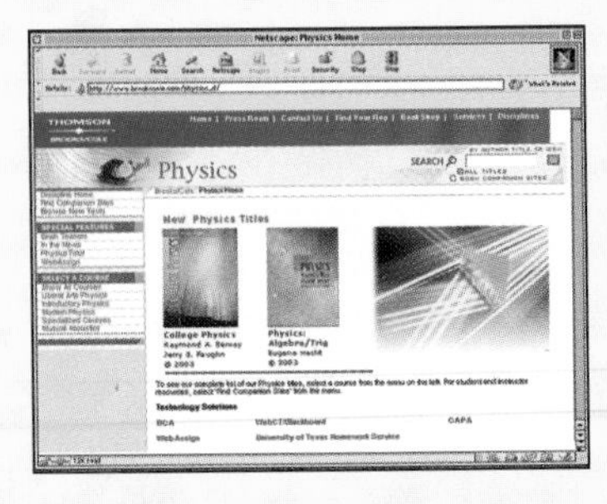

The Brooks/Cole Physics Resource Center

http://physics.brookscole.com
Here you'll find even more opportunities to hone your skills and expand your knowledge. **The Brooks/Cole Physics Resource Center** is filled with helpful content that will engage you while you master the material. You'll find additional online quizzes, Web links, animations, and *NewEdge*—an online news service that brings the latest news to this site daily.

6th Edition • Volume 1

PHYSICS

for Scientists and Engineers

Raymond A. Serway

John W. Jewett, Jr.

California State Polytechnic University–Pomona

THOMSON
BROOKS/COLE

Australia • Canada • Mexico • Singapore • Spain
United Kingdom • United States

THOMSON

BROOKS/COLE

Editor-in-Chief: Michelle Julet
Publisher: David Harris
Physics Editor: Chris Hall
Development Editor: Susan Dust Pashos
Assistant Editor: Rebecca Heider, Alyssa White
Editorial Assistant: Seth Dobrin, Jessica Howard
Technology Project Manager: Sam Subity
Marketing Manager: Kelley McAllister
Marketing Assistant: Sandra Perin
Advertising Project Manager: Stacey Purviance
Project Manager, Editorial Production: Teri Hyde
Print/Media Buyer: Barbara Britton
Permissions Editor: Joohee Lee

Production Service: Sparkpoint Communications, a division of J. B. Woolsey Associates
Text Designer: Lisa Devenish
Photo Researcher: Terri Wright
Copy Editor: Andrew Potter
Illustrator: Rolin Graphics
Cover Designer: Lisa Devenish
Cover Image: Water Displaced by Oil Tanker, © Stuart Westmorland/CORBIS
Compositor: Progressive Information Technologies
Cover Printer: Quebecor World, Versailles
Printer: Quebecor World, Versailles

Printed in the United States
4 5 6 7 07 06

For more information about our products, contact us at:
Thomson Learning Academic Resource Center
1-800-423-0563

For permission to use material from this text, contact us by:
Phone: 1-800-730-2214
Fax: 1-800-730-2215
Web: http://www.thomsonrights.com

Library of Congress Control Number: 2003100126

Student Edition: ISBN 0-534-40956-3

Instructor's Edition: ISBN 0-534-40843-5

Brooks/Cole—Thomson Learning
10 Davis Drive
Belmont, CA 94002
USA

Asia
Thomson Learning
5 Shenton Way #01-01
UIC Building
Singapore 068808

Australia
Nelson Thomson Learning
102 Dodds Street
South Melbourne, Victoria 3205
Australia

Canada
Nelson Thomson Learning
1120 Birchmount Road
Toronto, Ontario M1K 5G4
Canada

Europe/Middle East/Africa
Thomson Learning
High Holborn House
50/51 Bedford Row
London WC1R 4LR
United Kingdom

Latin America
Thomson Learning
Seneca, 53
Colonia Polanco
11560 Mexico D.F.
Mexico

Spain
Paraninfo Thomson Learning
Calle/Magallanes, 25
28015 Madrid, Spain

We dedicate this book to the courageous astronauts who died on the space shuttle Columbia *on February 1, 2003. The women and men of the international team lost their lives not in a contest between countries or a struggle for necessities but in advancing one of humankind's noblest creations—science.*

Contents Overview • Volume 1

Gerard Vandystadt/ Photo Researchers, Inc.

PART 1 Mechanics 1

PART 2 Oscillations and Mechanical Waves 451

Steve Niedorf/Getty Images

PART 3 Thermodynamics 579

Table of Contents • Volume 1

The Telegraph Colour Library/Getty Images

elektraVision/Index Stock Imagery

Billy Hustace/Getty Images

Courtesy Tourism Malaysia

Richard Megna/Fundamental Photographs

Andy Sachs/Getty Images

Don Bonsey/Getty Images

©1973 Kim Vandiver & Harold E. Edgerton/Courtesy of Palm Press, Inc.

PART 3 Thermodynamics 579

Lowell Georgia/CORBIS

About the Authors

Raymond A. Serway received his doctorate at Illinois Institute of Technology and is Professor Emeritus at James Madison University. Dr. Serway began his teaching career at Clarkson University, where he conducted research and taught from 1967 to 1980. His second academic appointment was at James Madison University as Professor of Physics and Head of the Physics Department from 1980 to 1986. He remained at James Madison University until his retirement in 1997. He was the recipient of the Madison Scholar Award at James Madison University in 1990, the Distinguished Tea ing Award at Clarkson University in 1977, and the Alumni Achievement Award Utica College in 1985. As Guest Scientist at the IBM Research Laboratory in Switzerland, he worked with K. Alex Müller, 1987 Nobel Prize recipient. Dr. S In held research appointments at Rome Air Development Center from 196 high- IIT Research Institute from 1963 to 1967, and as a visiting scientist ston and tional Laboratory, where he collaborated with his mentor and friend sixth edition addition to earlier editions of this textbook, Dr. Serway is the co school textbook *Physics* with Jerry Faughn, published by Holt, R *ysics* with Clem co-author of the third edition of *Principles of Physics* with Joh than 40 research of *College Physics* with Jerry Faughn, and the second edition ore than 70 presenta- Moses and Curt Moyer. In addition, Dr. Serway has pu eth enjoy traveling, golf- papers in the field of condensed matter physics and children and five grandchil- tions at professional meetings. Dr. Serway and his ing, gardening, and spending quality time with dren.

John W. Jewett, Jr. earned hi ate at Ohio State University, specializing in optical and magnetic properties ensed matter. Dr. Jewett began his academic career at Richard Stockton Colle ew Jersey, where he taught from 1974 to 1984. He is currently Professor of Ph California State Polytechnic University, Pomona. Throughout his teaching ca Dr. Jewett has been active in promoting science edu- cation. In addition to re g four National Science Foundation grants, he helped found and direct the S ern California Area Modern Physics Institute (SCAMPI). He also directed Scie IMPACT (Institute for Modern Pedagogy and Creative Teach- ing), which works with teachers and schools to develop effective science curricula. Dr. Jewett's honors include the Stockton Merit Award at Richard Stockton College, the Outstanding Professor Award at California State Polytechnic University for 1991–1992, and the Excellence in Undergraduate Physics Teaching Award from the American Association of Physics Teachers (AAPT) in 1998. He has given over 80 presentations at professional meetings, including presentations at international conferences in China and Japan. In addition to his work on this textbook, he is co-author of the third edition of *Principles of Physics* with Ray Serway and author of *The World of Physics . . . Mysteries, Magic, and Myth*. Dr. Jewett enjoys playing piano, traveling, and collecting antiques that can be used as demonstration apparatus in physics lectures, as well as spending time with his wife Lisa and their children and grandchildren.

Preface

Mark Cooper/Corbis Stock

In writing this sixth edition of *Physics for Scientists and Engineers*, we continue our ongoing efforts to improve the clarity of presentation and we again include new pedagogical features that help support the learning and teaching processes. ...wing on positive feedback from users of the fifth edition and reviewers' suggestions, ...e refined the text in order to better meet the needs of students and teachers. We ... the first time integrated a powerful collection of media resources into many of ...tions, examples, and end-of-chapter problems in the text. These resources ... Web-based learning system ***PhysicsNow*** and are flagged by the media icon ... tails are described below.

... is intended for a course in introductory physics for students major-ing ... gineering. The entire contents of the text in its extended version could b... three-semester course, but it is possible to use the material in shorter seq... omission of selected chapters and sections. The mathemati-cal backgroun... ent taking this course should ideally include one semester of calculus. If t... ssible, the student should be enrolled in a concurrent course in introduct...

Objectives

This introductory physics text... two main objectives: to provide the student with a clear and logical presentation ... c concepts and principles of physics, and to strengthen an understanding of th... and principles through a broad range of interesting applications to the real wo... et these objectives, we have placed emphasis on sound physical arguments an... m-solving methodology. At the same time, we have attempted to motivate the ... through practical examples that demonstrate the role of physics in other discip... ncluding engineering, chemistry, and medicine.

Changes in the Sixth Edition

A large number of changes and improvements have been m... in preparing the sixth edition of this text. Some of the new features are based on our experiences and on current trends in science education. Other changes have been incorporated in response to comments and suggestions offered by users of the fifth edition and reviewers of the manuscript. The following represent the major changes in the sixth edition:

Active Figures Many diagrams from the text have been animated to form **Active Figures,** part of the ***PhysicsNow*** integrated Web-based learning system. By visualizing phenomena and processes that cannot be fully represented on a static page, students greatly increase their conceptual understanding. **Active Figures** are identified with the media icon. An addition to the figure caption in **blue type** describes briefly the nature and contents of the animation.

Interactive Worked Examples Approximately 40 of the worked examples in Volume 1 have been identified as interactive, labeled with the media icon. As part of the ***PhysicsNow*** Web-based learning system, students can engage in an extension of the problem solved in the example. This often includes elements of both visualization and calculation, and may also involve prediction and intuition building. Often the interactivity is inspired by the **"What If?"** question we posed in the example text.

What If? Approximately one-third of the worked examples in the text contain this new feature. At the completion of the example solution, a **What If?** question offers a variation on the situation posed in the text of the example. For instance, this feature might explore the effects of changing the conditions of the situation, determine what happens when a quantity is taken to a particular limiting value, or question whether additional information can be determined about the problem situation. The answer to the question generally includes both a conceptual response and a mathematical response. This feature encourages students to think about the results of the example and assists in conceptual understanding of the principles. It also prepares students to encounter novel problems featured on exams. Some of the end-of-chapter problems also carry the **"What If?"** feature.

Quick Quizzes The number of Quick Quiz questions in each chapter has been increased. Quick Quizzes provide students with opportunities to test their understanding of the physical concepts presented. The questions require students to make decisions on the basis of sound reasoning, and some of them have been written to help students overcome common misconceptions. Quick Quizzes have been cast in an objective format, including multiple choice, true–false, and ranking. Answers to all Quick Quiz questions are found at the end of each chapter. Additional Quick Quizzes that can be used in classroom teaching are available on the instructor's companion Web site. Many instructors choose to use such questions in a "peer instruction" teaching style, but they can be used in standard quiz format as well.

Pitfall Preventions These new features are placed in the margins of the text and address common student misconceptions and situations in which students often follow unproductive paths. Over 200 Pitfall Preventions are provided to help students avoid common mistakes and misunderstandings.

General Problem-Solving Strategy A general strategy to be followed by the student is outlined at the end of Chapter 2 and provides students with a structured process for solving problems. In Chapters 3 through 5, the strategy is employed explicitly in every example so that students learn how it is applied. In the remaining chapters, the strategy appears explicitly in one example per chapter so that students are encouraged throughout the course to follow the procedure.

Line-by-Line Revision The entire text has been carefully edited to improve clarity of presentation and precision of language. We hope that the result is a book that is both accurate and enjoyable to read.

Problems A substantial revision of the end-of-chapter problems was made in an effort to improve their variety and interest, while maintaining their clarity and quality. Approximately 17% of the problems (about 550) are new. All problems have been carefully edited. Solutions to approximately 20% of the end-of-chapter problems are included in the *Student Solutions Manual and Study Guide.* These problems are identified by boxes around their numbers. A smaller subset of solutions, identified by the media icon, are available on the World Wide Web **(http://www.pse6.com)** as coached solutions with hints. Targeted feedback is provided for students whose instructors adopt *Physics for Scientists and Engineers,* sixth edition. See the next section for a complete description of other features of the problem set.

Content Changes The content and organization of the textbook is essentially the same as that of the fifth edition. An exception is that Chapter 13 (Oscillatory Motion) in the fifth edition has been moved to the Chapter 15 position in the sixth edition, in order to form a cohesive four-chapter Part 2 on oscillations and waves. Many sections in various chapters have been streamlined, deleted, or combined with other sections to allow for a more balanced presentation. The chapters on Modern Physics, Chapters 39–46, have been extensively rewritten to provide more up-to-date material as well as modern applications. A more detailed list of content changes can be found on the instructor's companion Web site.

Content

The material in this book covers fundamental topics in classical physics and provides an introduction to modern physics. The book is divided into six parts. Part 1 (Chapters 1 to 14) deals with the fundamentals of Newtonian mechanics and the physics of fluids, Part 2 (Chapters 15 to 18) covers oscillations, mechanical waves, and sound, Part 3 (Chapters 19 to 22) addresses heat and thermodynamics, Part 4 (Chapters 23 to 34) treats electricity and magnetism, Part 5 (Chapters 35 to 38) covers light and optics, and Part 6 (Chapters 39 to 46) deals with relativity and modern physics. Each part opener includes an overview of the subject matter covered in that part, as well as some historical perspectives.

Text Features

Most instructors would agree that the textbook selected for a course should be the student's primary guide for understanding and learning the subject matter. Furthermore, the textbook should be easily accessible and should be styled and written to facilitate instruction and learning. With these points in mind, we have included many pedagogical features in the textbook that are intended to enhance its usefulness to both students and instructors. These features are as follows:

Style To facilitate rapid comprehension, we have attempted to write the book in a style that is clear, logical, and engaging. We have chosen a writing style that is somewhat informal and relaxed so that students will find the text appealing and enjoyable to read. New terms are carefully defined, and we have avoided the use of jargon.

Previews All chapters begin with a brief preview that includes a discussion of the chapter's objectives and content.

Important Statements and Equations Most important statements and definitions are set in **boldface** type or are highlighted with a background screen for added emphasis and ease of review. Similarly, important equations are highlighted with a background screen to facilitate location.

Bruce Ayers/Getty Images

Problem-Solving Hints In several chapters, we have included general strategies for solving the types of problems featured both in the examples and in the end-of-chapter problems. This feature helps students to identify necessary steps in problem solving and to eliminate any uncertainty they might have. Problem-solving strategies are highlighted with a light red background screen for emphasis and ease of location.

Marginal Notes Comments and notes appearing in blue type in the margin can be used to locate important statements, equations, and concepts in the text.

Pedagogical Use of Color Readers should consult the **pedagogical color chart** (second page inside the front cover) for a listing of the color-coded symbols used in the text diagrams, Web-based **Active Figures,** and diagrams within **Interactive Worked Examples.** This system is followed consistently whenever possible, with slight variations made necessary by the complexity of physical situations depicted in Volume 2.

Mathematical Level We have introduced calculus gradually, keeping in mind that students often take introductory courses in calculus and physics concurrently. Most steps are shown when basic equations are developed, and reference is often made to mathematical appendices at the end of the textbook. Vector products are introduced later in the text, where they are needed in physical applications. The dot product is introduced in Chapter 7, which addresses energy and energy transfer; the cross product is introduced in Chapter 11, which deals with angular momentum.

Worked Examples A large number of worked examples of varying difficulty are presented to promote students' understanding of concepts. In many cases, the examples serve as models for solving the end-of-chapter problems. Because of the increased emphasis on understanding physical concepts, many examples are conceptual in nature and are labeled as such. The examples are set off in boxes, and the answers to examples with numerical solutions are highlighted with a background screen. We have already mentioned that a number of examples are designated as interactive and are part of the ***PhysicsNow*** Web-based learning system.

Questions Questions of a conceptual nature requiring verbal or written responses are provided at the end of each chapter. Over 1 000 questions are included in this edition. Some questions provide the student with a means of self-testing the concepts presented in the chapter. Others could serve as a basis for initiating classroom discussions. Answers to selected questions are included in the *Student Solutions Manual and Study Guide,* and answers to all questions are found in the *Instructor's Solutions Manual.*

Significant Figures Significant figures in both worked examples and end-of-chapter problems have been handled with care. Most numerical examples are worked out to either two or three significant figures, depending on the precision of the data provided. End-of-chapter problems regularly state data and answers to three-digit precision.

Problems An extensive set of problems is included at the end of each chapter; in all, over 3 000 problems are given throughout the text. Answers to odd-numbered problems are provided at the end of the book in a section whose pages have colored edges for ease of location. For the convenience of both the student and the instructor, about two thirds of the problems are keyed to specific sections of the chapter. The remaining problems, labeled "Additional Problems," are not keyed to specific sections.

Usually, the problems within a given section are presented so that the straightforward problems (those with black problem numbers) appear first. For ease of identification, the numbers of intermediate-level problems are printed in blue, and those of challenging problems are printed in magenta.

- **Review Problems** Many chapters include review problems requiring the student to combine concepts covered in the chapter with those discussed in previous chapters. These problems reflect the cohesive nature of the principles in the text and verify that physics is not a scattered set of ideas. When facing real-world issues such as global warming or nuclear weapons, it may be necessary to call on ideas in physics from several parts of a textbook such as this one.
- **Paired Problems** To allow focused practice in solving problems stated in symbolic terms, some end-of-chapter numerical problems are paired with the same problems in symbolic form. Paired problems are identified by a common light red background screen.
- **Computer- and Calculator-Based Problems** Many chapters include one or more problems whose solution requires the use of a computer or graphing calculator. Computer modeling of physical phenomena enables students to obtain graphical representations of variables and to perform numerical analyses.
- **Coached Problems with Hints** These have been described above as part of the ***PhysicsNow*** Web-based learning system. These problems are identified by the media icon and targeted feedback is provided to students of instructors adopting the sixth edition.

Units The international system of units (SI) is used throughout the text. The U.S. customary system of units is used only to a limited extent in the chapters on mechanics, heat, and thermodynamics.

Summaries Each chapter contains a summary that reviews the important concepts and equations discussed in that chapter. A marginal note in blue type next to each chapter summary directs students to a practice test (Post-Test) for the chapter.

Courtesy NASA

Appendices and Endpapers Several appendices are provided at the end of the textbook. Most of the appendix material represents a review of mathematical concepts and techniques used in the text, including scientific notation, algebra, geometry, trigonometry, differential calculus, and integral calculus. Reference to these appendices is made throughout the text. Most mathematical review sections in the appendices include worked examples and exercises with answers. In addition to the mathematical reviews, the appendices contain tables of physical data, conversion factors, atomic masses, and the SI units of physical quantities, as well as a periodic table of the elements. Other useful information, including fundamental constants and physical data, planetary data, a list of standard prefixes, mathematical symbols, the Greek alphabet, and standard abbreviations of units of measure, appears on the endpapers.

Student Ancillaries

Student Solutions Manual and Study Guide by John R. Gordon, Ralph McGrew, and Raymond Serway. This two-volume manual features detailed solutions to 20% of the end-of-chapter problems from the text. The manual also features a list of important equations, concepts, and notes from key sections of the text, in addition to answers to selected end-of-chapter questions. Volume 1 contains Chapters 1 through 22 and Volume 2 contains Chapters 23 through 46.

WebTutor™ on WebCT and Blackboard **WebTutor** offers students real-time access to a full array of study tools, including chapter outlines, summaries, learning objectives, glossary flashcards (with audio), practice quizzes, **InfoTrac® College Edition** exercises, and Web links.

InfoTrac® College Edition Adopters and their students automatically receive a four-month subscription to **InfoTrac® College Edition** with every new copy of this book. Newly improved, this extensive online library opens the door to the full text (not just abstracts) of countless articles from thousands of publications including *American Scientist, Physical Review, Science, Science Weekly,* and more! Available only to college and university students. Journals subject to change.

The Brooks/Cole Physics Resource Center You will find additional online quizzes, Web links and animations at **http://physics.brookscole.com.**

Ancillaries for Instructors

The first four ancillaries below are available to qualified adopters. Please consult your local sales representative for details.

Instructor's Solutions Manual by Ralph McGrew and James A. Currie. This two-volume manual contains complete worked solutions to all of the end-of-chapter problems in the textbook as well as answers to even-numbered problems. The solutions to problems new to the sixth edition are marked for easy identification by the instructor. New to this edition are complete answers to the conceptual questions in the main text. Volume 1 contains Chapters 1 through 22 and Volume 2 contains Chapters 23 through 46.

Printed Test Bank by Edward Adelson. This two-volume test bank contains approximately 2 300 multiple-choice questions. These questions are also available in electronic format with complete answers and solutions in the Brooks/Cole Assessment test program. Volume 1 contains Chapters 1 through 22 and Volume 2 contains Chapters 23 through 46.

Multimedia Manager This easy-to-use multimedia lecture tool allows you to quickly assemble art and database files with notes to create fluid lectures. The CD-ROM set (Volume 1, Chapters 1–22; Volume 2, Chapters 23–46) includes a database of animations, video clips, and digital art from the text as well as electronic files of the *Instructor's Solutions Manual and Test Bank.* The simple interface makes it easy for you to incorporate graphics, digital video, animations, and audio clips into your lectures.

Transparency Acetates Each volume contains approximately 100 acetates featuring art from the text. Volume 1 contains Chapters 1 through 22 and Volume 2 contains Chapters 23 through 46.

George Semple

Brooks/Cole Assessment With a balance of efficiency, high performance, simplicity and versatility, **Brooks/Cole Assessment (BCA)** gives you the power to transform the learning and teaching experience. **BCA** is fully integrated testing, tutorial, and course management software accessible by instructors and students anytime, anywhere. Delivered for FREE in a browser-based format without the need for any proprietary software or plug-ins, **BCA** uses correct scientific notation to provide the drill of basic skills that students need, enabling the instructor to focus more time in higher-level learning activities (i.e., concepts and applications). Students can have unlimited practice in questions and problems, building their own confidence and skills. Results flow automatically to a grade book for tracking so that instructors will be better able to assess student understanding of the material, even prior to class or an actual test.

WebTutor™ on WebCT and Blackboard With **WebTutor's** text-specific, preformatted content and total flexibility, instructors can easily create and manage their own personal Web site. **WebTutor's** course management tool gives instructors the ability to provide virtual office hours, post syllabi, set up threaded discussions, track student progress with the quizzing material, and much more. **WebTutor** also provides robust communication tools, such as a course calendar, asynchronous discussion, real-time chat, a whiteboard, and an integrated e-mail system.

Additional Options for Online Homework For detailed information and demonstrations, contact your Thomson • Brooks/Cole representative or visit the following:

- WebAssign: A Web-based Homework System
  http://www.webassign.net or contact WebAssign at *webassign@ncsu.edu*
- Homework Service
  http://hw.ph.utexas.edu/hw.html or contact *moore@physics.utexas.edu*
- CAPA: A Computer-Assisted Personalized Approach
  http://capa4.lite.msu.edu/homepage/

Instructor's Companion Web Site Consult the instructor's site at *http://www.pse6.com* for additional Quick Quiz questions, a detailed list of content changes since the fifth edition, a problem correlation guide, images from the text, and sample PowerPoint lectures. Instructors adopting the sixth edition of *Physics for Scientists and Engineers* may download these materials after securing the appropriate password from their local Thomson • Brooks/Cole sales representative.

Teaching Options

The topics in this textbook are presented in the following sequence: classical mechanics, oscillations and mechanical waves, and heat and thermodynamics (Volume 1); followed by electricity and magnetism, electromagnetic waves, optics, relativity, and modern physics (Volume 2). This presentation represents a traditional sequence, with the subject of mechanical waves being presented before electricity and magnetism. Some instructors may prefer to cover this material after completing electricity and magnetism (i.e., after Chapter 34). The chapter on relativity is placed near the end of the text because this topic often is treated as an introduction to the era of "modern physics." If time permits, instructors may choose to cover Chapter 39 in Volume 1 after completing Chapter 13, as it concludes the material on Newtonian mechanics.

For those instructors teaching a two-semester sequence, some sections and chapters could be deleted without any loss of continuity. The following sections in Volume 1 can be considered optional for this purpose:

2.7 Kinematic Equations Derived from Calculus

4.6 Relative Velocity and Relative Acceleration

6.3 Motion in Accelerated Frames

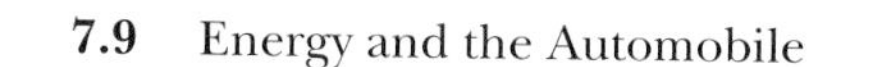

6.4 Motion in the Presence of Resistive Forces

6.5 Numerical Modeling in Particle Dynamics

7.9 Energy and the Automobile

- **8.6** Energy Diagrams and Equilibrium of a System
- **9.7** Rocket Propulsion
- **11.5** The Motion of Gyroscopes and Tops
- **11.6** Angular Momentum as a Fundamental Quantity
- **14.7** Other Applications of Fluid Dynamics
- **15.6** Damped Oscillations
- **15.7** Forced Oscillations
- **16.6** The Linear Wave Equation
- **17.5** Digital Sound Recording
- **17.6** Motion Picture Sound
- **18.6** Standing Waves in Rods and Membranes
- **18.8** Nonsinusoidal Wave Patterns
- **21.7** Mean Free Path
- **22.8** Entropy on a Microscopic Scale

Topham Picturepoint/The Image Works

Acknowledgments

The sixth edition of this textbook was prepared with the guidance and assistance of many professors who reviewed selections of the manuscript, the pre-revision text, or both. We wish to acknowledge the following scholars and express our sincere appreciation for their suggestions, criticisms, and encouragement:

Edward Adelson, *Ohio State University*
Michael R. Cohen, *Shippensburg University*
Jerry D. Cook, *Eastern Kentucky University*
J. William Dawicke, *Milwaukee School of Engineering*
N. John DiNardo, *Drexel University*
Andrew Duffy, *Boston University*
Robert J. Endorf, *University of Cincinnati*
F. Paul Esposito, *University of Cincinnati*
Joe L. Ferguson, *Mississippi State University*
Perry Ganas, *California State University, Los Angeles*
John C. Hardy, *Texas A&M University*
Michael Hayes, *University of Pretoria (South Africa)*
John T. Ho, *The State University of New York, Buffalo*
Joseph W. Howard, *Salisbury University*
Robert Hunt, *Johnson County Community College*
Walter S. Jaronski, *Radford University*
L. R. Jordan, *Palm Beach Community College*
Teruki Kamon, *Texas A & M University*
Louis E. Keiner, *Coastal Carolina University*
Edwin H. Lo, *American University*
James G. McLean, *The State University of New York, Geneseo*
Richard E. Miers, *Indiana University–Purdue University, Fort Wayne*
Oscar Romulo Ochoa, *The College of New Jersey*
Paul S. Ormsby, *Moraine Valley Community College*
Didarul I. Qadir, *Central Michigan University*
Judith D. Redling, *New Jersey Institute of Technology*
Richard W. Robinett, *Pennsylvania State University*
Om P. Rustgi, *SUNY College at Buffalo*
Mesgun Sebhatu, *Winthrop University*
Natalia Semushkina, *Shippensburg University*
Uwe C. Täuber, *Virginia Polytechnic Institute*
Perry A. Tompkins, *Samford University*
Doug Welch, *McMaster University, Ontario*
Augden Windelborn, *Northern Illinois University*
Jerzy M. Wrobel, *University of Missouri, Kansas City*
Jianshi Wu, *Fayetteville State University*
Michael Zincani, *University of Dallas*

Volume 1 was carefully checked for accuracy by Michael Kotlarchyk *(Rochester Institute of Technology)*, Chris Vuille *(Embry-Riddle Aeronautical University)*, Laurencin Dunbar *(St. Louis Community College)*, and William Dawicke *(Milwaukee School of Engineering.)* We thank them for their diligent efforts under schedule pressure!

We are grateful to Ralph McGrew for organizing the end-of-chapter problems, writing many new problems, and his excellent suggestions for improving the content of the textbook. Problems new to this edition were written by Edward Adelson, Ronald Bieniek, Michael Browne, Andrew Duffy, Robert Forsythe, Perry Ganas, Michael Hones, John Jewett, Boris Korsunsky, Edwin Lo, Ralph McGrew, Raymond Serway, and Jerzy Wrobel, with the help of Bennett Simpson and JoAnne Maniago. Students Alexander Coto, Karl Payne, and Eric Peterman made corrections to problems taken from previous editions, as did teachers David Aspnes, Robert Beichner, Joseph Biegen, Vasili Haralambous, Frank Hayes, Erika Hermon, Ken Menningen, Henry Nebel, and Charles Teague. We are grateful to authors John R. Gordon and Ralph McGrew and compositor Michael Rudmin for preparing the *Student Solutions Manual and Study Guide.* Authors Ralph McGrew and James Currie and compositor Mary Toscano have prepared an excellent *Instructor's Solutions Manual,* and we thank them. Edward Adelson has carefully edited and improved the Test Bank for the sixth edition. Kurt Vandervoort prepared extra Quick Quiz questions for the instructor's companion Web site.

Special thanks and recognition go to the professional staff at the Brooks/Cole Publishing Company—in particular Susan Pashos, Rebecca Heider and Alyssa White (who managed the ancillary program and so much more), Jessica Howard, Peter McGahey, Teri Hyde, Michelle Julet, David Harris, and Chris Hall—for their fine work during the development and production of this textbook. We are most appreciative of Sam Subity's masterful management of the ***PhysicsNow*** media program. Kelley McAllister is our energetic Marketing Manager, and Stacey Purviance coordinates our marketing communications. We recognize the skilled production service provided by the staff at Sparkpoint Communications, the excellent artwork produced by Rolin Graphics, and the dedicated photo research efforts of Terri Wright.

Finally, we are deeply indebted to our wives and children for their love, support, and long-term sacrifices.

Frank Oberle/Getty Images

Raymond A. Serway
Leesburg, Virginia

John W. Jewett, Jr.
Pomona, California

To the Student

It is appropriate to offer some words of advice that should be of benefit to you, the student. Before doing so, we assume that you have read the Preface, which describes the various features of the text that will help you through the course.

How to Study

Very often instructors are asked, "How should I study physics and prepare for examinations?" There is no simple answer to this question, but we would like to offer some suggestions that are based on our own experiences in learning and teaching over the years.

First and foremost, maintain a positive attitude toward the subject matter, keeping in mind that physics is the most fundamental of all natural sciences. Other science courses that follow will use the same physical principles, so it is important that you understand and are able to apply the various concepts and theories discussed in the text.

Concepts and Principles

It is essential that you understand the basic concepts and principles before attempting to solve assigned problems. You can best accomplish this goal by carefully reading the textbook before you attend your lecture on the covered material. When reading the text, you should jot down those points that are not clear to you. We've purposely left wide margins in the text to give you space for making notes. Also be sure to make a diligent attempt at answering the questions in the Quick Quizzes as you come to them in your reading. We have worked hard to prepare questions that help you judge for yourself how well you understand the material. Study carefully the **What If?** features that appear with many of the worked examples. These will help you to extend your understanding beyond the simple act of arriving at a numerical result. The Pitfall Preventions will also help guide you away from common misunderstandings about physics. During class, take careful notes and ask questions about those ideas that are unclear to you. Keep in mind that few people are able to absorb the full meaning of scientific material after only one reading. Several readings of the text and your notes may be necessary. Your lectures and laboratory work supplement reading of the textbook and should clarify some of the more difficult material. You should minimize your memorization of material. Successful memorization of passages from the text, equations, and derivations does not necessarily indicate that you understand the material. Your understanding of the material will be enhanced through a combination of efficient study habits, discussions with other students and with instructors, and your ability to solve the problems presented in the textbook. Ask questions whenever you feel clarification of a concept is necessary.

Study Schedule

It is important that you set up a regular study schedule, preferably a daily one. Make sure that you read the syllabus for the course and adhere to the schedule set by your instructor. The lectures will make much more sense if you read the corresponding text material before attending them. As a general rule, you should devote about two hours of study time for every hour you are in class. If you are having trouble with the course, seek the advice of the instructor or other students who have taken the course. You may find it necessary to seek further instruction from experienced students. Very often, instructors offer review sessions in addition to regular class periods. It is important that you avoid the practice of delaying study until a day or two before an exam. More often than not, this approach has disastrous results. Rather than undertake an all-night study session, briefly review the basic concepts and equations, and get a good night's rest. If you feel you need additional help in understanding the concepts, in preparing for exams, or in problem solving, we suggest that you acquire a copy of the *Student Solutions Manual and Study Guide* that accompanies this textbook; this manual should be available at your college bookstore.

George Semple

Use the Features

You should make full use of the various features of the text discussed in the Preface. For example, marginal notes are useful for locating and describing important equations and concepts, and **boldfaced** type indicates important statements and definitions. Many useful tables are contained in the Appendices, but most are incorporated in the text where they are most often referenced. Appendix B is a convenient review of mathematical techniques.

Answers to odd-numbered problems are given at the end of the textbook, answers to Quick Quizzes are located at the end of each chapter, and answers to selected end-of-chapter questions are provided in the *Student Solutions Manual and Study Guide.* Problem-Solving Strategies and Hints are included in selected chapters throughout the text and give you additional information about how you should solve problems. The Table of Contents provides an overview of the entire text, while the Index enables you to locate specific material quickly. Footnotes sometimes are used to supplement the text or to cite other references on the subject discussed.

After reading a chapter, you should be able to define any new quantities introduced in that chapter and to discuss the principles and assumptions that were used to arrive at certain key relations. The chapter summaries and the review sections of the *Student Solutions Manual and Study Guide* should help you in this regard. In some cases, it may be necessary for you to refer to the index of the text to locate certain topics. You should be able to associate with each physical quantity the correct symbol used to represent that quantity and the unit in which the quantity is specified. Furthermore, you should be able to express each important equation in a concise and accurate prose statement.

Problem Solving

R. P. Feynman, Nobel laureate in physics, once said, "You do not know anything until you have practiced." In keeping with this statement, we strongly advise that you develop the skills necessary to solve a wide range of problems. Your ability to solve problems will be one of the main tests of your knowledge of physics, and therefore you should try to solve as many problems as possible. It is essential that you understand basic concepts and principles before attempting to solve problems. It is good practice to try to find alternate solutions to the same problem. For example, you can solve problems in mechanics using Newton's laws, but very often an alternative method that draws on energy considerations is more direct. You should not deceive yourself into thinking that you understand a problem merely because you have seen it solved in class. You must be able to solve the problem and similar problems on your own.

The approach to solving problems should be carefully planned. A systematic plan is especially important when a problem involves several concepts. First, read the problem several times until you are confident you understand what is being asked. Look for any key words that will help you interpret the problem and perhaps allow you to make certain assumptions. Your ability to interpret a question properly is an integral part of problem solving. Second, you should acquire the habit of writing down the information given in a problem and those quantities that need to be found; for example, you might construct a table listing both the quantities given and the quantities to be found. This procedure is sometimes used in the worked examples of the textbook. Finally, after you have decided on the method you feel is appropriate for a given problem, proceed with your solution. Specific problem-solving strategies (Hints) of this type are included in the text and are highlighted with a light red screen. We have also developed a General Problem-Solving Strategy to help guide you through complex problems. If you follow the steps of this procedure *(Conceptualize, Categorize, Analyze, Finalize)*, you will not only find it easier to come up with a solution, but you will also gain more from your efforts. This Strategy is located at the end of Chapter 2 (page 47) and is used in all worked examples in Chapters 3 through 5 so that you can learn how to apply it. In the remaining chapters, the Strategy is used in one example per chapter as a reminder of its usefulness.

Often, students fail to recognize the limitations of certain equations or physical laws in a particular situation. It is very important that you understand and remember the assumptions that underlie a particular theory or formalism. For example, certain equations in kinematics apply only to a particle moving with constant acceleration. These equations are not valid for describing motion whose acceleration is not constant, such as the motion of an object connected to a spring or the motion of an object through a fluid.

© Phil Degginger/Stone/Getty

Experiments

Physics is a science based on experimental observations. In view of this fact, we recommend that you try to supplement the text by performing various types of "hands-on" experiments, either at home or in the laboratory. These can be used to test ideas and models discussed in class or in the textbook. For example, the common Slinky™ toy is excellent for studying traveling waves; a ball swinging on the end of a long string can be used to investigate pendulum motion; various masses attached to the end of a vertical spring or rubber band can be used to determine their elastic nature; an old pair of Polaroid sunglasses and some discarded lenses and a magnifying glass are the components of various experiments in optics; and an approximate measure of the free-fall acceleration can be determined simply by measuring with a stopwatch the time it takes for a ball to drop from a known height. The list of such experiments is endless. When physical models are not available, be imaginative and try to develop models of your own.

New Media

We strongly encourage you to use the ***PhysicsNow*** Web-based learning system that accompanies this textbook. It is far easier to understand physics if you see it in action, and these new materials will enable you to become a part of that action. ***PhysicsNow*** media described in the Preface are accessed at the URL *http://www.pse6.com,* and feature a three-step learning process consisting of a Pre-Test, a personalized learning plan, and a Post-Test.

In addition to other elements, ***PhysicsNow*** includes the following Active Figures and Interactive Worked Examples from Volume 1:

Chapter 2
Active Figures 2.1, 2.3, 2.9, 2.10, and 2.13
Examples 2.8 and 2.12

Chapter 3
Active Figures 3.3 and 3.16

Chapter 4
Active Figures 4.5, 4.7, and 4.11
Examples 4.4 and 4.5

Chapter 5
Active Figure 5.16
Examples 5.9 and 5.10

Chapter 6
Active Figures 6.2, 6.12, and 6.15
Examples 6.4, 6.5, and 6.7

Chapter 7
Active Figure 7.10
Examples 7.9 and 7.11

Chapter 8
Active Figures 8.4 and 8.16
Examples 8.2 and 8.4

Chapter 9
Active Figures 9.8, 9.9, 9.13, 9.16, and 9.17
Examples 9.1, 9.5, and 9.8

Chapter 10
Active Figures 10.4, 10.14, and 10.30
Examples 10.12, 10.13, and 10.14

Chapter 11
Active Figures 11.1, 11.3, and 11.4
Examples 11.6 and 11.10

Chapter 12
Active Figures 12.14, 12.16, 12.17, and 12.18
Examples 12.3 and 12.4

Chapter 13
Active Figures 13.1, 13.5, and 13.7
Example 13.1

Chapter 14
Active Figures 14.9 and 14.10
Example 14.10

Chapter 15
Active Figures 15.1, 15.2, 15.7, 15.9, 15.10, 15.14, 15.17, and 15.22

Chapter 16
Active Figures 16.7, 16.8, 16.10, and 16.17
Example 16.5

Chapter 17
Active Figures 17.2, 17.8, and 17.9
Example 17.6

Chapter 18
Active Figures 18.2, 18.4, 18.9, 18.10, 18.22, and 18.25
Example 18.5

Chapter 19
Active Figure 19.12

Chapter 20
Active Figures 20.4, 20.5, and 20.7
Example 20.10

Chapter 21
Active Figures 21.2, 21.4, 21.11, and 21.12
Example 21.4

Chapter 22
Active Figures 22.2, 22.5, 22.10, 22.11, 22.12, 22.13, and 22.19
Example 22.10

An Invitation to Physics

It is our sincere hope that you too will find physics an exciting and enjoyable experience and that you will profit from this experience, regardless of your chosen profession. Welcome to the exciting world of physics!

The scientist does not study nature because it is useful; he studies it because he delights in it, and he delights in it because it is beautiful. If nature were not beautiful, it would not be worth knowing, and if nature were not worth knowing, life would not be worth living.

—Henri Poincaré

USA

Mechanics

Physics, the most fundamental physical science, is concerned with the basic principles of the Universe. It is the foundation upon which the other sciences—astronomy, biology, chemistry, and geology—are based. The beauty of physics lies in the simplicity of the fundamental physical theories and in the manner in which just a small number of fundamental concepts, equations, and assumptions can alter and expand our view of the world around us.

The study of physics can be divided into six main areas:

1. *classical mechanics,* which is concerned with the motion of objects that are large relative to atoms and move at speeds much slower than the speed of light;
2. *relativity,* which is a theory describing objects moving at any speed, even speeds approaching the speed of light;
3. *thermodynamics,* which deals with heat, work, temperature, and the statistical behavior of systems with large numbers of particles;
4. *electromagnetism,* which is concerned with electricity, magnetism, and electromagnetic fields;
5. *optics,* which is the study of the behavior of light and its interaction with materials;
6. *quantum mechanics,* a collection of theories connecting the behavior of matter at the submicroscopic level to macroscopic observations.

The disciplines of mechanics and electromagnetism are basic to all other branches of classical physics (developed before 1900) and modern physics (c. 1900–present). The first part of this textbook deals with classical mechanics, sometimes referred to as *Newtonian mechanics* or simply *mechanics.* This is an appropriate place to begin an introductory text because many of the basic principles used to understand mechanical systems can later be used to describe such natural phenomena as waves and the transfer of energy by heat. Furthermore, the laws of conservation of energy and momentum introduced in mechanics retain their importance in the fundamental theories of other areas of physics.

Today, classical mechanics is of vital importance to students from all disciplines. It is highly successful in describing the motions of different objects, such as planets, rockets, and baseballs. In the first part of the text, we shall describe the laws of classical mechanics and examine a wide range of phenomena that can be understood with these fundamental ideas. ■

◀ *Liftoff of the space shuttle* Columbia. *The tragic accident of February 1, 2003 that took the lives of all seven astronauts aboard happened just before Volume 1 of this book went to press. The launch and operation of a space shuttle involves many fundamental principles of classical mechanics, thermodynamics, and electromagnetism. We study the principles of classical mechanics in Part 1 of this text, and apply these principles to rocket propulsion in Chapter 9. (NASA)*

Chapter 1

Physics and Measurement

CHAPTER OUTLINE

▲ *The workings of a mechanical clock. Complicated timepieces have been built for centuries in an effort to measure time accurately. Time is one of the basic quantities that we use in studying the motion of objects. (elektraVision/Index Stock Imagery)*

Like all other sciences, physics is based on experimental observations and quantitative measurements. The main objective of physics is to find the limited number of fundamental laws that govern natural phenomena and to use them to develop theories that can predict the results of future experiments. The fundamental laws used in developing theories are expressed in the language of mathematics, the tool that provides a bridge between theory and experiment.

When a discrepancy between theory and experiment arises, new theories must be formulated to remove the discrepancy. Many times a theory is satisfactory only under limited conditions; a more general theory might be satisfactory without such limitations. For example, the laws of motion discovered by Isaac Newton (1642–1727) in the 17th century accurately describe the motion of objects moving at normal speeds but do not apply to objects moving at speeds comparable with the speed of light. In contrast, the special theory of relativity developed by Albert Einstein (1879–1955) in the early 1900s gives the same results as Newton's laws at low speeds but also correctly describes motion at speeds approaching the speed of light. Hence, Einstein's special theory of relativity is a more general theory of motion.

Classical physics includes the theories, concepts, laws, and experiments in classical mechanics, thermodynamics, optics, and electromagnetism developed before 1900. Important contributions to classical physics were provided by Newton, who developed classical mechanics as a systematic theory and was one of the originators of calculus as a mathematical tool. Major developments in mechanics continued in the 18th century, but the fields of thermodynamics and electricity and magnetism were not developed until the latter part of the 19th century, principally because before that time the apparatus for controlled experiments was either too crude or unavailable.

A major revolution in physics, usually referred to as *modern physics,* began near the end of the 19th century. Modern physics developed mainly because of the discovery that many physical phenomena could not be explained by classical physics. The two most important developments in this modern era were the theories of relativity and quantum mechanics. Einstein's theory of relativity not only correctly described the motion of objects moving at speeds comparable to the speed of light but also completely revolutionized the traditional concepts of space, time, and energy. The theory of relativity also shows that the speed of light is the upper limit of the speed of an object and that mass and energy are related. Quantum mechanics was formulated by a number of distinguished scientists to provide descriptions of physical phenomena at the atomic level.

Scientists continually work at improving our understanding of fundamental laws, and new discoveries are made every day. In many research areas there is a great deal of overlap among physics, chemistry, and biology. Evidence for this overlap is seen in the names of some subspecialties in science—biophysics, biochemistry, chemical physics, biotechnology, and so on. Numerous technological advances in recent times are the result of the efforts of many scientists, engineers, and technicians. Some of the most notable developments in the latter half of the 20th century were (1) unmanned planetary explorations and manned moon landings, (2) microcircuitry and high-speed computers, (3) sophisticated imaging techniques used in scientific research and medicine, and

(4) several remarkable results in genetic engineering. The impacts of such developments and discoveries on our society have indeed been great, and it is very likely that future discoveries and developments will be exciting, challenging, and of great benefit to humanity.

1.1 Standards of Length, Mass, and Time

The laws of physics are expressed as mathematical relationships among physical quantities that we will introduce and discuss throughout the book. Most of these quantities are *derived quantities,* in that they can be expressed as combinations of a small number of *basic quantities.* In mechanics, the three basic quantities are length, mass, and time. All other quantities in mechanics can be expressed in terms of these three.

If we are to report the results of a measurement to someone who wishes to reproduce this measurement, a *standard* must be defined. It would be meaningless if a visitor from another planet were to talk to us about a length of 8 "glitches" if we do not know the meaning of the unit glitch. On the other hand, if someone familiar with our system of measurement reports that a wall is 2 meters high and our unit of length is defined to be 1 meter, we know that the height of the wall is twice our basic length unit. Likewise, if we are told that a person has a mass of 75 kilograms and our unit of mass is defined to be 1 kilogram, then that person is 75 times as massive as our basic unit.[1] Whatever is chosen as a standard must be readily accessible and possess some property that can be measured reliably. Measurements taken by different people in different places must yield the same result.

In 1960, an international committee established a set of standards for the fundamental quantities of science. It is called the **SI** (Système International), and its units of length, mass, and time are the *meter, kilogram,* and *second,* respectively. Other SI standards established by the committee are those for temperature (the *kelvin*), electric current (the *ampere*), luminous intensity (the *candela*), and the amount of substance (the *mole*).

Length

In A.D. 1120 the king of England decreed that the standard of length in his country would be named the *yard* and would be precisely equal to the distance from the tip of his nose to the end of his outstretched arm. Similarly, the original standard for the foot adopted by the French was the length of the royal foot of King Louis XIV. This standard prevailed until 1799, when the legal standard of length in France became the *meter,* defined as one ten-millionth the distance from the equator to the North Pole along one particular longitudinal line that passes through Paris.

Many other systems for measuring length have been developed over the years, but the advantages of the French system have caused it to prevail in almost all countries and in scientific circles everywhere. As recently as 1960, the length of the meter was defined as the distance between two lines on a specific platinum–iridium bar stored under controlled conditions in France. This standard was abandoned for several reasons, a principal one being that the limited accuracy with which the separation between the lines on the bar can be determined does not meet the current requirements of science and technology. In the 1960s and 1970s, the meter was defined as 1 650 763.73 wavelengths of orange-red light emitted from a krypton-86 lamp. However, in October 1983, **the meter (m) was redefined as the distance traveled by light in vacuum during a time of 1/299 792 458 second.** In effect, this

[1] The need for assigning numerical values to various measured physical quantities was expressed by Lord Kelvin (William Thomson) as follows: "I often say that when you can measure what you are speaking about, and express it in numbers, you should know something about it, but when you cannot express it in numbers, your knowledge is of a meager and unsatisfactory kind. It may be the beginning of knowledge but you have scarcely in your thoughts advanced to the state of science."

Table 1.1

Approximate Values of Some Measured Lengths

	Length (m)
Distance from the Earth to the most remote known quasar	1.4×10^{26}
Distance from the Earth to the most remote normal galaxies	9×10^{25}
Distance from the Earth to the nearest large galaxy (M 31, the Andromeda galaxy)	2×10^{22}
Distance from the Sun to the nearest star (Proxima Centauri)	4×10^{16}
One lightyear	9.46×10^{15}
Mean orbit radius of the Earth about the Sun	1.50×10^{11}
Mean distance from the Earth to the Moon	3.84×10^{8}
Distance from the equator to the North Pole	1.00×10^{7}
Mean radius of the Earth	6.37×10^{6}
Typical altitude (above the surface) of a satellite orbiting the Earth	2×10^{5}
Length of a football field	9.1×10^{1}
Length of a housefly	5×10^{-3}
Size of smallest dust particles	$\sim 10^{-4}$
Size of cells of most living organisms	$\sim 10^{-5}$
Diameter of a hydrogen atom	$\sim 10^{-10}$
Diameter of an atomic nucleus	$\sim 10^{-14}$
Diameter of a proton	$\sim 10^{-15}$

latest definition establishes that the speed of light in vacuum is precisely 299 792 458 meters per second.

Table 1.1 lists approximate values of some measured lengths. You should study this table as well as the next two tables and begin to generate an intuition for what is meant by a length of 20 centimeters, for example, or a mass of 100 kilograms or a time interval of 3.2×10^7 seconds.

Mass

The SI unit of mass, **the kilogram (kg), is defined as the mass of a specific platinum–iridium alloy cylinder kept at the International Bureau of Weights and Measures at Sèvres, France.** This mass standard was established in 1887 and has not been changed since that time because platinum–iridium is an unusually stable alloy. A duplicate of the Sèvres cylinder is kept at the National Institute of Standards and Technology (NIST) in Gaithersburg, Maryland (Fig. 1.1a).

Table 1.2 lists approximate values of the masses of various objects.

Time

Before 1960, the standard of time was defined in terms of the *mean solar day* for the year 1900. (A solar day is the time interval between successive appearances of the Sun at the highest point it reaches in the sky each day.) The *second* was defined as $\left(\frac{1}{60}\right)\left(\frac{1}{60}\right)\left(\frac{1}{24}\right)$ of a mean solar day. The rotation of the Earth is now known to vary slightly with time, however, and therefore this motion is not a good one to use for defining a time standard.

In 1967, the second was redefined to take advantage of the high precision attainable in a device known as an *atomic clock* (Fig. 1.1b), which uses the characteristic frequency of the cesium-133 atom as the "reference clock." **The second (s) is now defined as 9 192 631 770 times the period of vibration of radiation from the cesium atom.**[2]

[2] *Period* is defined as the time interval needed for one complete vibration.

PITFALL PREVENTION

1.1 No Commas in Numbers with Many Digits

We will use the standard scientific notation for numbers with more than three digits, in which groups of three digits are separated by spaces rather than commas. Thus, 10 000 is the same as the common American notation of 10,000. Similarly, $\pi = 3.14159265$ is written as 3.141 592 65.

Table 1.2

Masses of Various Objects (Approximate Values)

	Mass (kg)
Observable Universe	$\sim 10^{52}$
Milky Way galaxy	$\sim 10^{42}$
Sun	1.99×10^{30}
Earth	5.98×10^{24}
Moon	7.36×10^{22}
Shark	$\sim 10^{3}$
Human	$\sim 10^{2}$
Frog	$\sim 10^{-1}$
Mosquito	$\sim 10^{-5}$
Bacterium	$\sim 1 \times 10^{-15}$
Hydrogen atom	1.67×10^{-27}
Electron	9.11×10^{-31}

PITFALL PREVENTION

1.2 Reasonable Values

Generating intuition about typical values of quantities is important because when solving problems you must think about your end result and determine if it seems reasonable. If you are calculating the mass of a housefly and arrive at a value of 100 kg, this is *unreasonable*—there is an error somewhere.

(a) (b)

Figure 1.1 (a) The National Standard Kilogram No. 20, an accurate copy of the International Standard Kilogram kept at Sèvres, France, is housed under a double bell jar in a vault at the National Institute of Standards and Technology. (b) The nation's primary time standard is a cesium fountain atomic clock developed at the National Institute of Standards and Technology laboratories in Boulder, Colorado. The clock will neither gain nor lose a second in 20 million years.

To keep these atomic clocks—and therefore all common clocks and watches that are set to them—synchronized, it has sometimes been necessary to add leap seconds to our clocks.

Since Einstein's discovery of the linkage between space and time, precise measurement of time intervals requires that we know both the state of motion of the clock used to measure the interval and, in some cases, the location of the clock as well. Otherwise, for example, global positioning system satellites might be unable to pinpoint your location with sufficient accuracy, should you need to be rescued.

Approximate values of time intervals are presented in Table 1.3.

Table 1.3

Approximate Values of Some Time Intervals

	Time Interval (s)
Age of the Universe	5×10^{17}
Age of the Earth	1.3×10^{17}
Average age of a college student	6.3×10^{8}
One year	3.2×10^{7}
One day (time interval for one revolution of the Earth about its axis)	8.6×10^{4}
One class period	3.0×10^{3}
Time interval between normal heartbeats	8×10^{-1}
Period of audible sound waves	$\sim 10^{-3}$
Period of typical radio waves	$\sim 10^{-6}$
Period of vibration of an atom in a solid	$\sim 10^{-13}$
Period of visible light waves	$\sim 10^{-15}$
Duration of a nuclear collision	$\sim 10^{-22}$
Time interval for light to cross a proton	$\sim 10^{-24}$

Table 1.4

Prefixes for Powers of Ten

Power	Prefix	Abbreviation
10^{-24}	yocto	y
10^{-21}	zepto	z
10^{-18}	atto	a
10^{-15}	femto	f
10^{-12}	pico	p
10^{-9}	nano	n
10^{-6}	micro	μ
10^{-3}	milli	m
10^{-2}	centi	c
10^{-1}	deci	d
10^{3}	kilo	k
10^{6}	mega	M
10^{9}	giga	G
10^{12}	tera	T
10^{15}	peta	P
10^{18}	exa	E
10^{21}	zetta	Z
10^{24}	yotta	Y

In addition to SI, another system of units, the *U.S. customary system,* is still used in the United States despite acceptance of SI by the rest of the world. In this system, the units of length, mass, and time are the foot (ft), slug, and second, respectively. In this text we shall use SI units because they are almost universally accepted in science and industry. We shall make some limited use of U.S. customary units in the study of classical mechanics.

In addition to the basic SI units of meter, kilogram, and second, we can also use other units, such as millimeters and nanoseconds, where the prefixes *milli-* and *nano-* denote multipliers of the basic units based on various powers of ten. Prefixes for the various powers of ten and their abbreviations are listed in Table 1.4. For example, 10^{-3} m is equivalent to 1 millimeter (mm), and 10^3 m corresponds to 1 kilometer (km). Likewise, 1 kilogram (kg) is 10^3 grams (g), and 1 megavolt (MV) is 10^6 volts (V).

1.2 Matter and Model Building

If physicists cannot interact with some phenomenon directly, they often imagine a **model** for a physical system that is related to the phenomenon. In this context, a model is a system of physical components, such as electrons and protons in an atom. Once we have identified the physical components, we make predictions about the behavior of the system, based on the interactions among the components of the system and/or the interaction between the system and the environment outside the system.

As an example, consider the behavior of *matter.* A 1-kg cube of solid gold, such as that at the left of Figure 1.2, has a length of 3.73 cm on a side. Is this cube nothing but wall-to-wall gold, with no empty space? If the cube is cut in half, the two pieces still retain their chemical identity as solid gold. But what if the pieces are cut again and again, indefinitely? Will the smaller and smaller pieces always be gold? Questions such as these can be traced back to early Greek philosophers. Two of them—Leucippus and his student Democritus—could not accept the idea that such cuttings could go on forever. They speculated that the process ultimately must end when it produces a particle

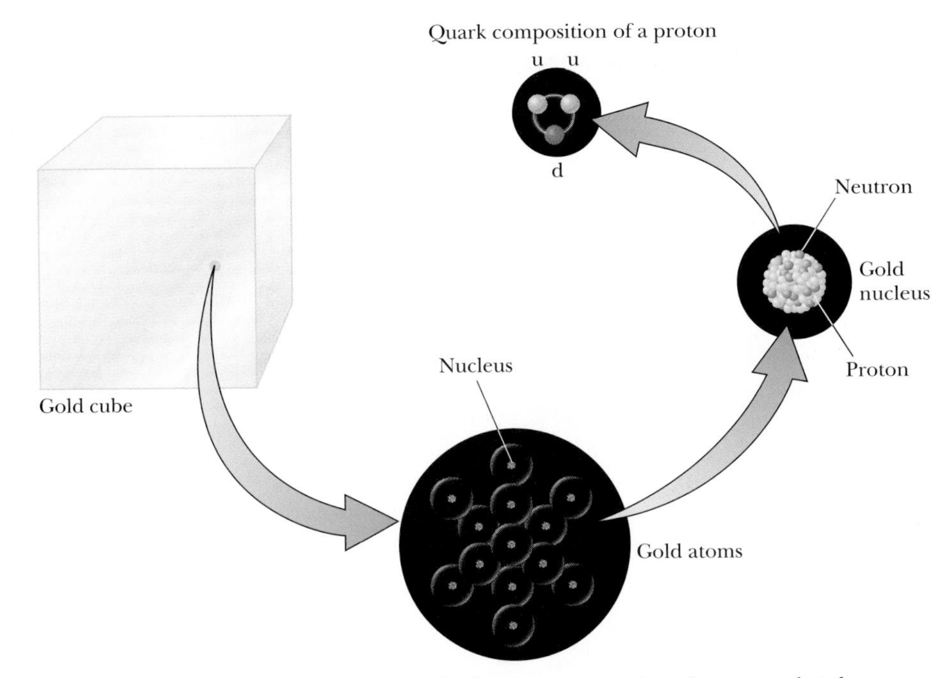

Figure 1.2 Levels of organization in matter. Ordinary matter consists of atoms, and at the center of each atom is a compact nucleus consisting of protons and neutrons. Protons and neutrons are composed of quarks. The quark composition of a proton is shown.

that can no longer be cut. In Greek, *atomos* means "not sliceable." From this comes our English word *atom.*

Let us review briefly a number of historical models of the structure of matter. The Greek model of the structure of matter was that all ordinary matter consists of atoms, as suggested to the lower right of the cube in Figure 1.2. Beyond that, no additional structure was specified in the model—atoms acted as small particles that interacted with each other, but internal structure of the atom was not a part of the model.

In 1897, J. J. Thomson identified the electron as a charged particle and as a constituent of the atom. This led to the first model of the atom that contained internal structure. We shall discuss this model in Chapter 42.

Following the discovery of the nucleus in 1911, a model was developed in which each atom is made up of electrons surrounding a central nucleus. A nucleus is shown in Figure 1.2. This model leads, however, to a new question—does the nucleus have structure? That is, is the nucleus a single particle or a collection of particles? The exact composition of the nucleus is not known completely even today, but by the early 1930s a model evolved that helped us understand how the nucleus behaves. Specifically, scientists determined that occupying the nucleus are two basic entities, protons and neutrons. The proton carries a positive electric charge, and a specific chemical element is identified by the number of protons in its nucleus. This number is called the **atomic number** of the element. For instance, the nucleus of a hydrogen atom contains one proton (and so the atomic number of hydrogen is 1), the nucleus of a helium atom contains two protons (atomic number 2), and the nucleus of a uranium atom contains 92 protons (atomic number 92). In addition to atomic number, there is a second number characterizing atoms—**mass number,** defined as the number of protons plus neutrons in a nucleus. The atomic number of an element never varies (i.e., the number of protons does not vary) but the mass number can vary (i.e., the number of neutrons varies).

The existence of neutrons was verified conclusively in 1932. A neutron has no charge and a mass that is about equal to that of a proton. One of its primary purposes is to act as a "glue" that holds the nucleus together. If neutrons were not present in the nucleus, the repulsive force between the positively charged particles would cause the nucleus to come apart.

But is this where the process of breaking down stops? Protons, neutrons, and a host of other exotic particles are now known to be composed of six different varieties of particles called **quarks,** which have been given the names of *up, down, strange, charmed, bottom,* and *top*. The up, charmed, and top quarks have electric charges of $+\frac{2}{3}$ that of the proton, whereas the down, strange, and bottom quarks have charges of $-\frac{1}{3}$ that of the proton. The proton consists of two up quarks and one down quark, as shown at the top in Figure 1.2. You can easily show that this structure predicts the correct charge for the proton. Likewise, the neutron consists of two down quarks and one up quark, giving a net charge of zero.

This process of building models is one that you should develop as you study physics. You will be challenged with many mathematical problems to solve in this study. One of the most important techniques is to build a model for the problem—identify a system of physical components for the problem, and make predictions of the behavior of the system based on the interactions among the components of the system and/or the interaction between the system and its surrounding environment.

1.3 Density and Atomic Mass

In Section 1.1, we explored three basic quantities in mechanics. Let us look now at an example of a derived quantity—**density.** The density ρ (Greek letter rho) of any substance is defined as its *mass per unit volume:*

A table of the letters in the Greek alphabet is provided on the back endsheet of the textbook.

$$\rho \equiv \frac{m}{V} \tag{1.1}$$

For example, aluminum has a density of 2.70 g/cm^3, and lead has a density of 11.3 g/cm^3. Therefore, a piece of aluminum of volume 10.0 cm^3 has a mass of 27.0 g, whereas an equivalent volume of lead has a mass of 113 g. A list of densities for various substances is given in Table 1.5.

The numbers of protons and neutrons in the nucleus of an atom of an element are related to the **atomic mass** of the element, which is defined as the mass of a single atom of the element measured in **atomic mass units** (u) where 1 u = $1.660\,538\,7 \times 10^{-27}$ kg.

Table 1.5

Densities of Various Substances

Substance	Density ρ (10^3 kg/m^3)
Platinum	21.45
Gold	19.3
Uranium	18.7
Lead	11.3
Copper	8.92
Iron	7.86
Aluminum	2.70
Magnesium	1.75
Water	1.00
Air at atmospheric pressure	0.0012

The atomic mass of lead is 207 u and that of aluminum is 27.0 u. However, the ratio of atomic masses, 207 u/27.0 u = 7.67, does not correspond to the ratio of densities, $(11.3 \times 10^3 \text{ kg/m}^3)/(2.70 \times 10^3 \text{ kg/m}^3) = 4.19$. This discrepancy is due to the difference in atomic spacings and atomic arrangements in the crystal structures of the two elements.

Quick Quiz 1.1 In a machine shop, two cams are produced, one of aluminum and one of iron. Both cams have the same mass. Which cam is larger? (a) the aluminum cam (b) the iron cam (c) Both cams have the same size.

Example 1.1 How Many Atoms in the Cube?

A solid cube of aluminum (density 2.70 g/cm³) has a volume of 0.200 cm³. It is known that 27.0 g of aluminum contains 6.02×10^{23} atoms. How many aluminum atoms are contained in the cube?

Solution Because density equals mass per unit volume, the mass of the cube is

$$m = \rho V = (2.70 \text{ g/cm}^3)(0.200 \text{ cm}^3) = 0.540 \text{ g}$$

To solve this problem, we will set up a ratio based on the fact that the mass of a sample of material is proportional to the number of atoms contained in the sample. This technique of solving by ratios is very powerful and should be studied and understood so that it can be applied in future problem solving. Let us express our proportionality as $m = kN$, where m is the mass of the sample, N is the number of atoms in the sample, and k is an unknown proportionality constant. We write this relationship twice, once for the actual sample of aluminum in the problem and once for a 27.0-g sample, and then we divide the first equation by the second:

$$\begin{aligned} m_{\text{sample}} &= kN_{\text{sample}} \\ m_{27.0 \text{ g}} &= kN_{27.0 \text{ g}} \end{aligned} \quad \longrightarrow \quad \frac{m_{\text{sample}}}{m_{27.0 \text{ g}}} = \frac{N_{\text{sample}}}{N_{27.0 \text{ g}}}$$

Notice that the unknown proportionality constant k cancels, so we do not need to know its value. We now substitute the values:

$$\frac{0.540 \text{ g}}{27.0 \text{ g}} = \frac{N_{\text{sample}}}{6.02 \times 10^{23} \text{ atoms}}$$

$$N_{\text{sample}} = \frac{(0.540 \text{ g})(6.02 \times 10^{23} \text{ atoms})}{27.0 \text{ g}}$$

$$= \boxed{1.20 \times 10^{22} \text{ atoms}}$$

PITFALL PREVENTION

1.3 Setting Up Ratios

When using ratios to solve a problem, keep in mind that *ratios come from equations.* If you start from equations known to be correct and can divide one equation by the other as in Example 1.1 to obtain a useful ratio, you will avoid reasoning errors. So write the known equations first!

1.4 Dimensional Analysis

The word *dimension* has a special meaning in physics. It denotes the physical nature of a quantity. Whether a distance is measured in units of feet or meters or fathoms, it is still a distance. We say its dimension is *length.*

The symbols we use in this book to specify the dimensions of length, mass, and time are L, M, and T, respectively.[3] We shall often use brackets [ ] to denote the dimensions of a physical quantity. For example, the symbol we use for speed in this book is v, and in our notation the dimensions of speed are written $[v] = \text{L/T}$. As another example, the dimensions of area A are $[A] = \text{L}^2$. The dimensions and units of area, volume, speed, and acceleration are listed in Table 1.6. The dimensions of other quantities, such as force and energy, will be described as they are introduced in the text.

In many situations, you may have to derive or check a specific equation. A useful and powerful procedure called *dimensional analysis* can be used to assist in the derivation or to check your final expression. Dimensional analysis makes use of the fact that

[3] The *dimensions* of a quantity will be symbolized by a capitalized, non-italic letter, such as L. The *symbol* for the quantity itself will be italicized, such as L for the length of an object, or t for time.

Table 1.6

Units of Area, Volume, Velocity, Speed, and Acceleration

System	Area (L^2)	Volume (L^3)	Speed (L/T)	Acceleration (L/T^2)
SI	m^2	m^3	m/s	m/s^2
U.S. customary	ft^2	ft^3	ft/s	ft/s^2

dimensions can be treated as algebraic quantities. For example, quantities can be added or subtracted only if they have the same dimensions. Furthermore, the terms on both sides of an equation must have the same dimensions. By following these simple rules, you can use dimensional analysis to help determine whether an expression has the correct form. The relationship can be correct only if the dimensions on both sides of the equation are the same.

To illustrate this procedure, suppose you wish to derive an equation for the position x of a car at a time t if the car starts from rest and moves with constant acceleration a. In Chapter 2, we shall find that the correct expression is $x = \frac{1}{2}at^2$. Let us use dimensional analysis to check the validity of this expression. The quantity x on the left side has the dimension of length. For the equation to be dimensionally correct, the quantity on the right side must also have the dimension of length. We can perform a dimensional check by substituting the dimensions for acceleration, L/T^2 (Table 1.6), and time, T, into the equation. That is, the dimensional form of the equation $x = \frac{1}{2}at^2$ is

$$L = \frac{L}{\cancel{T^2}} \cdot \cancel{T^2} = L$$

The dimensions of time cancel as shown, leaving the dimension of length on the right-hand side.

A more general procedure using dimensional analysis is to set up an expression of the form

$$x \propto a^n t^m$$

where n and m are exponents that must be determined and the symbol $\propto$ indicates a proportionality. This relationship is correct only if the dimensions of both sides are the same. Because the dimension of the left side is length, the dimension of the right side must also be length. That is,

$$[a^n t^m] = L = L^1T^0$$

Because the dimensions of acceleration are L/T^2 and the dimension of time is T, we have

$$(L/T^2)^n T^m = L^1T^0$$

$$(L^n T^{m-2n}) = L^1T^0$$

The exponents of L and T must be the same on both sides of the equation. From the exponents of L, we see immediately that $n = 1$. From the exponents of T, we see that $m - 2n = 0$, which, once we substitute for n, gives us $m = 2$. Returning to our original expression $x \propto a^n t^m$, we conclude that $x \propto at^2$. This result differs by a factor of $\frac{1}{2}$ from the correct expression, which is $x = \frac{1}{2}at^2$.

Quick Quiz 1.2 True or False: Dimensional analysis can give you the numerical value of constants of proportionality that may appear in an algebraic expression.

PITFALL PREVENTION

1.4 Symbols for Quantities

Some quantities have a small number of symbols that represent them. For example, the symbol for time is almost always t. Others quantities might have various symbols depending on the usage. Length may be described with symbols such as x, y, and z (for position), r (for radius), a, b, and c (for the legs of a right triangle), ℓ (for the length of an object), d (for a distance), h (for a height), etc.

Example 1.2 Analysis of an Equation

Show that the expression $v = at$ is dimensionally correct, where v represents speed, a acceleration, and t an instant of time.

Solution For the speed term, we have from Table 1.6

$$[v] = \frac{\mathrm{L}}{\mathrm{T}}$$

The same table gives us $\mathrm{L/T^2}$ for the dimensions of acceleration, and so the dimensions of at are

$$[at] = \frac{\mathrm{L}}{\mathrm{T}^{\not{2}}}\not{\mathrm{T}} = \frac{\mathrm{L}}{\mathrm{T}}$$

Therefore, the expression is dimensionally correct. (If the expression were given as $v = at^2$ it would be dimensionally *incorrect*. Try it and see!)

Example 1.3 Analysis of a Power Law

Suppose we are told that the acceleration a of a particle moving with uniform speed v in a circle of radius r is proportional to some power of r, say r^n, and some power of v, say v^m. Determine the values of n and m and write the simplest form of an equation for the acceleration.

Solution Let us take a to be

$$a = kr^n v^m$$

where k is a dimensionless constant of proportionality. Knowing the dimensions of a, r, and v, we see that the dimensional equation must be

$$\frac{\mathrm{L}}{\mathrm{T}^2} = \mathrm{L}^n\left(\frac{\mathrm{L}}{\mathrm{T}}\right)^m = \frac{\mathrm{L}^{n+m}}{\mathrm{T}^m}$$

This dimensional equation is balanced under the conditions

$$n + m = 1 \qquad \text{and} \qquad m = 2$$

Therefore $n = -1$, and we can write the acceleration expression as

$$a = kr^{-1}v^2 = k\frac{v^2}{r}$$

When we discuss uniform circular motion later, we shall see that $k = 1$ if a consistent set of units is used. The constant k would not equal 1 if, for example, v were in km/h and you wanted a in $\mathrm{m/s^2}$.

PITFALL PREVENTION

1.5 Always Include Units

When performing calculations, include the units for every quantity and carry the units through the entire calculation. Avoid the temptation to drop the units early and then attach the expected units once you have an answer. By including the units in every step, you can detect errors if the units for the answer turn out to be incorrect.

1.5 Conversion of Units

Sometimes it is necessary to convert units from one measurement system to another, or to convert within a system, for example, from kilometers to meters. Equalities between SI and U.S. customary units of length are as follows:

$$1 \text{ mile} = 1\,609 \text{ m} = 1.609 \text{ km} \qquad 1 \text{ ft} = 0.304\,8 \text{ m} = 30.48 \text{ cm}$$

$$1 \text{ m} = 39.37 \text{ in.} = 3.281 \text{ ft} \qquad 1 \text{ in.} = 0.025\,4 \text{ m} = 2.54 \text{ cm (exactly)}$$

A more complete list of conversion factors can be found in Appendix A.

Units can be treated as algebraic quantities that can cancel each other. For example, suppose we wish to convert 15.0 in. to centimeters. Because 1 in. is defined as exactly 2.54 cm, we find that

$$15.0 \text{ in.} = (15.0 \not{\text{in.}})\left(\frac{2.54 \text{ cm}}{1 \not{\text{in.}}}\right) = 38.1 \text{ cm}$$

where the ratio in parentheses is equal to 1. Notice that we choose to put the unit of an inch in the denominator and it cancels with the unit in the original quantity. The remaining unit is the centimeter, which is our desired result.

Quick Quiz 1.3 The distance between two cities is 100 mi. The number of kilometers between the two cities is (a) smaller than 100 (b) larger than 100 (c) equal to 100.

Example 1.4 Is He Speeding?

On an interstate highway in a rural region of Wyoming, a car is traveling at a speed of 38.0 m/s. Is this car exceeding the speed limit of 75.0 mi/h?

Solution We first convert meters to miles:

$$(38.0\ \cancel{\text{m}}/\text{s})\left(\frac{1\ \text{mi}}{1\,609\ \cancel{\text{m}}}\right) = 2.36 \times 10^{-2}\ \text{mi/s}$$

Now we convert seconds to hours:

$$(2.36 \times 10^{-2}\ \text{mi}/\cancel{\text{s}})\left(\frac{60\ \cancel{\text{s}}}{1\ \cancel{\text{min}}}\right)\left(\frac{60\ \cancel{\text{min}}}{1\ \text{h}}\right) = 85.0\ \text{mi/h}$$

Thus, the car is exceeding the speed limit and should slow down.

What If? **What if the driver is from outside the U.S. and is familiar with speeds measured in km/h? What is the speed of the car in km/h?**

Answer We can convert our final answer to the appropriate units:

$$(85.0\ \cancel{\text{mi}}/\text{h})\left(\frac{1.609\ \text{km}}{1\ \cancel{\text{mi}}}\right) = 137\ \text{km/h}$$

Figure 1.3 shows the speedometer of an automobile, with speeds in both mi/h and km/h. Can you check the conversion we just performed using this photograph?

Figure 1.3 The speedometer of a vehicle that shows speeds in both miles per hour and kilometers per hour.

1.6 Estimates and Order-of-Magnitude Calculations

It is often useful to compute an approximate answer to a given physical problem even when little information is available. This answer can then be used to determine whether or not a more precise calculation is necessary. Such an approximation is usually based on certain assumptions, which must be modified if greater precision is needed. We will sometimes refer to an *order of magnitude* of a certain quantity as the power of ten of the number that describes that quantity. Usually, when an order-of-magnitude calculation is made, the results are reliable to within about a factor of 10. If a quantity increases in value by three orders of magnitude, this means that its value increases by a factor of about $10^3 = 1\,000$. We use the symbol $\sim$ for "is on the order of." Thus,

$$0.008\,6 \sim 10^{-2} \qquad 0.002\,1 \sim 10^{-3} \qquad 720 \sim 10^{3}$$

The spirit of order-of-magnitude calculations, sometimes referred to as "guesstimates" or "ball-park figures," is given in the following quotation: "Make an estimate before every calculation, try a simple physical argument . . . before every derivation, guess the answer to every puzzle."[4] Inaccuracies caused by guessing too low for one number are often canceled out by other guesses that are too high. You will find that with practice your guesstimates become better and better. Estimation problems can be fun to work as you freely drop digits, venture reasonable approximations for

[4] E. Taylor and J. A. Wheeler, *Spacetime Physics: Introduction to Special Relativity,* 2nd ed., San Francisco, W. H. Freeman & Company, Publishers, 1992, p. 20.

unknown numbers, make simplifying assumptions, and turn the question around into something you can answer in your head or with minimal mathematical manipulation on paper. Because of the simplicity of these types of calculations, they can be performed on a *small* piece of paper, so these estimates are often called "back-of-the-envelope calculations."

Example 1.5 Breaths in a Lifetime

Estimate the number of breaths taken during an average life span.

Solution We start by guessing that the typical life span is about 70 years. The only other estimate we must make in this example is the average number of breaths that a person takes in 1 min. This number varies, depending on whether the person is exercising, sleeping, angry, serene, and so forth. To the nearest order of magnitude, we shall choose 10 breaths per minute as our estimate of the average. (This is certainly closer to the true value than 1 breath per minute or 100 breaths per minute.) The number of minutes in a year is approximately

$$1 \text{ yr}\left(\frac{400 \text{ days}}{1 \text{ yr}}\right)\left(\frac{25 \text{ h}}{1 \text{ day}}\right)\left(\frac{60 \text{ min}}{1 \text{ h}}\right) = 6 \times 10^5 \text{ min}$$

Notice how much simpler it is in the expression above to multiply 400×25 than it is to work with the more accurate 365×24. These approximate values for the number of days in a year and the number of hours in a day are close enough for our purposes. Thus, in 70 years there will be $(70 \text{ yr})(6 \times 10^5 \text{ min/yr}) = 4 \times 10^7$ min. At a rate of 10 breaths/min, an individual would take 4×10^8 breaths in a lifetime, or on the order of 10^9 breaths.

What If? What if the average life span were estimated as 80 years instead of 70? Would this change our final estimate?

Answer We could claim that $(80 \text{ yr})(6 \times 10^5 \text{ min/yr}) = 5 \times 10^7$ min, so that our final estimate should be 5×10^8 breaths. This is still on the order of 10^9 breaths, so an order-of-magnitude estimate would be unchanged. Furthermore, 80 years is 14% larger than 70 years, but we have overestimated the total time interval by using 400 days in a year instead of 365 and 25 hours in a day instead of 24. These two numbers together result in an overestimate of 14%, which cancels the effect of the increased life span!

Example 1.6 It's a Long Way to San Jose

Estimate the number of steps a person would take walking from New York to Los Angeles.

Solution Without looking up the distance between these two cities, you might remember from a geography class that they are about 3 000 mi apart. The next approximation we must make is the length of one step. Of course, this length depends on the person doing the walking, but we can estimate that each step covers about 2 ft. With our estimated step size, we can determine the number of steps in 1 mi. Because this is a rough calculation, we round 5 280 ft/mi to 5 000 ft/mi. (What percentage error does this introduce?) This conversion factor gives us

$$\frac{5\,000 \text{ ft/mi}}{2 \text{ ft/step}} = 2\,500 \text{ steps/mi}$$

Now we switch to scientific notation so that we can do the calculation mentally:

$$(3 \times 10^3 \text{ mi})(2.5 \times 10^3 \text{ steps/mi}) = 7.5 \times 10^6 \text{ steps} \sim 10^7 \text{ steps}$$

So if we intend to walk across the United States, it will take us on the order of ten million steps. This estimate is almost certainly too small because we have not accounted for curving roads and going up and down hills and mountains. Nonetheless, it is probably within an order of magnitude of the correct answer.

Example 1.7 How Much Gas Do We Use?

Estimate the number of gallons of gasoline used each year by all the cars in the United States.

Solution Because there are about 280 million people in the United States, an estimate of the number of cars in the country is 100 million (guessing that there are between two and three people per car). We also estimate that the average distance each car travels per year is 10 000 mi. If we assume a gasoline consumption of 20 mi/gal or 0.05 gal/mi, then each car uses about 500 gal/yr. Multiplying this by the total number of cars in the United States gives an estimated total consumption of $5 \times 10^{10} \text{ gal} \sim 10^{11}$ gal.

1.7 Significant Figures

When certain quantities are measured, the measured values are known only to within the limits of the experimental uncertainty. The value of this uncertainty can depend on various factors, such as the quality of the apparatus, the skill of the experimenter, and the number of measurements performed. The number of **significant figures** in a measurement can be used to express something about the uncertainty.

As an example of significant figures, suppose that we are asked in a laboratory experiment to measure the area of a computer disk label using a meter stick as a measuring instrument. Let us assume that the accuracy to which we can measure the length of the label is ± 0.1 cm. If the length is measured to be 5.5 cm, we can claim only that its length lies somewhere between 5.4 cm and 5.6 cm. In this case, we say that the measured value has two significant figures. Note that the significant figures include the first estimated digit. Likewise, if the label's width is measured to be 6.4 cm, the actual value lies between 6.3 cm and 6.5 cm. Thus we could write the measured values as (5.5 ± 0.1) cm and (6.4 ± 0.1) cm.

Now suppose we want to find the area of the label by multiplying the two measured values. If we were to claim the area is $(5.5 \text{ cm})(6.4 \text{ cm}) = 35.2 \text{ cm}^2$, our answer would be unjustifiable because it contains three significant figures, which is greater than the number of significant figures in either of the measured quantities. A good rule of thumb to use in determining the number of significant figures that can be claimed in a multiplication or a division is as follows:

When multiplying several quantities, the number of significant figures in the final answer is the same as the number of significant figures in the quantity having the lowest number of significant figures. The same rule applies to division.

Applying this rule to the previous multiplication example, we see that the answer for the area can have only two significant figures because our measured quantities have only two significant figures. Thus, all we can claim is that the area is 35 cm^2, realizing that the value can range between $(5.4 \text{ cm})(6.3 \text{ cm}) = 34 \text{ cm}^2$ and $(5.6 \text{ cm})(6.5 \text{ cm}) = 36 \text{ cm}^2$.

Zeros may or may not be significant figures. Those used to position the decimal point in such numbers as 0.03 and 0.007 5 are not significant. Thus, there are one and two significant figures, respectively, in these two values. When the zeros come after other digits, however, there is the possibility of misinterpretation. For example, suppose the mass of an object is given as 1 500 g. This value is ambiguous because we do not know whether the last two zeros are being used to locate the decimal point or whether they represent significant figures in the measurement. To remove this ambiguity, it is common to use scientific notation to indicate the number of significant figures. In this case, we would express the mass as 1.5×10^3 g if there are two significant figures in the measured value, 1.50×10^3 g if there are three significant figures, and 1.500×10^3 g if there are four. The same rule holds for numbers less than 1, so that 2.3×10^{-4} has two significant figures (and so could be written 0.000 23) and 2.30×10^{-4} has three significant figures (also written 0.000 230). In general, **a significant figure in a measurement is a reliably known digit (other than a zero used to locate the decimal point) or the first estimated digit.**

For addition and subtraction, you must consider the number of decimal places when you are determining how many significant figures to report:

When numbers are added or subtracted, the number of decimal places in the result should equal the smallest number of decimal places of any term in the sum.

PITFALL PREVENTION

1.6 Read Carefully

Notice that the rule for addition and subtraction is different from that for multiplication and division. For addition and subtraction, the important consideration is the number of *decimal places*, not the number of *significant figures*.

For example, if we wish to compute 123 + 5.35, the answer is 128 and not 128.35. If we compute the sum 1.000 1 + 0.000 3 = 1.000 4, the result has five significant figures, even though one of the terms in the sum, 0.000 3, has only one significant figure. Likewise, if we perform the subtraction 1.002 − 0.998 = 0.004, the result has only one significant figure even though one term has four significant figures and the other has three. In this book, **most of the numerical examples and end-of-chapter problems will yield answers having three significant figures.** When carrying out estimates we shall typically work with a single significant figure.

If the number of significant figures in the result of an addition or subtraction must be reduced, there is a general rule for rounding off numbers, which states that the last digit retained is to be increased by 1 if the last digit dropped is greater than 5. If the last digit dropped is less than 5, the last digit retained remains as it is. If the last digit dropped is equal to 5, the remaining digit should be rounded to the nearest even number. (This helps avoid accumulation of errors in long arithmetic processes.)

A technique for avoiding error accumulation is to delay rounding of numbers in a long calculation until you have the final result. Wait until you are ready to copy the final answer from your calculator before rounding to the correct number of significant figures.

Quick Quiz 1.4 Suppose you measure the position of a chair with a meter stick and record that the center of the seat is 1.043 860 564 2 m from a wall. What would a reader conclude from this recorded measurement?

Example 1.8 Installing a Carpet

A carpet is to be installed in a room whose length is measured to be 12.71 m and whose width is measured to be 3.46 m. Find the area of the room.

Solution If you multiply 12.71 m by 3.46 m on your calculator, you will see an answer of 43.976 6 m^2. How many of these numbers should you claim? Our rule of thumb for multiplication tells us that you can claim only the number of significant figures in your answer as are present in the measured quantity having the lowest number of significant figures. In this example, the lowest number of significant figures is three in 3.46 m, so we should express our final answer as 44.0 m^2.

SUMMARY

Take a practice test for this chapter by clicking on the Practice Test link at http://www.pse6.com.

The three fundamental physical quantities of mechanics are length, mass, and time, which in the SI system have the units meters (m), kilograms (kg), and seconds (s), respectively. Prefixes indicating various powers of ten are used with these three basic units.

The **density** of a substance is defined as its *mass per unit volume.* Different substances have different densities mainly because of differences in their atomic masses and atomic arrangements.

The method of **dimensional analysis** is very powerful in solving physics problems. Dimensions can be treated as algebraic quantities. By making estimates and performing order-of-magnitude calculations, you should be able to approximate the answer to a problem when there is not enough information available to completely specify an exact solution.

When you compute a result from several measured numbers, each of which has a certain accuracy, you should give the result with the correct number of **significant figures.** When multiplying several quantities, the number of significant figures in the final answer is the same as the number of significant figures in the quantity having the lowest number of significant figures. The same rule applies to division. When numbers are added or subtracted, the number of decimal places in the result should equal the smallest number of decimal places of any term in the sum.

QUESTIONS

1. What types of natural phenomena could serve as time standards?
2. Suppose that the three fundamental standards of the metric system were length, *density*, and time rather than length, *mass*, and time. The standard of density in this system is to be defined as that of water. What considerations about water would you need to address to make sure that the standard of density is as accurate as possible?
3. The height of a horse is sometimes given in units of "hands." Why is this a poor standard of length?
4. Express the following quantities using the prefixes given in Table 1.4: (a) 3×10^{-4} m (b) 5×10^{-5} s (c) 72×10^{2} g.
5. Suppose that two quantities A and B have different dimensions. Determine which of the following arithmetic operations *could* be physically meaningful: (a) $A + B$ (b) A/B (c) $B - A$ (d) AB.
6. If an equation is dimensionally correct, does this mean that the equation must be true? If an equation is not dimensionally correct, does this mean that the equation cannot be true?
7. Do an order-of-magnitude calculation for an everyday situation you encounter. For example, how far do you walk or drive each day?
8. Find the order of magnitude of your age in seconds.
9. What level of precision is implied in an order-of-magnitude calculation?
10. Estimate the mass of this textbook in kilograms. If a scale is available, check your estimate.
11. In reply to a student's question, a guard in a natural history museum says of the fossils near his station, "When I started work here twenty-four years ago, they were eighty million years old, so you can add it up." What should the student conclude about the age of the fossils?

PROBLEMS

1, 2, 3 = straightforward, intermediate, challenging □ = full solution available in the *Student Solutions Manual and Study Guide*

= coached solution with hints available at http://www.pse6.com = computer useful in solving problem

= paired numerical and symbolic problems

Section 1.2 Matter and Model Building

Note: Consult the endpapers, appendices, and tables in the text whenever necessary in solving problems. For this chapter, Appendix B.3 may be particularly useful. Answers to odd-numbered problems appear in the back of the book.

1. A crystalline solid consists of atoms stacked up in a repeating lattice structure. Consider a crystal as shown in Figure P1.1a. The atoms reside at the corners of cubes of side $L = 0.200$ nm. One piece of evidence for the regular arrangement of atoms comes from the flat surfaces along which a crystal separates, or cleaves, when it is broken. Suppose this crystal cleaves along a face diagonal, as shown in Figure P1.1b. Calculate the spacing d between two adjacent atomic planes that separate when the crystal cleaves.

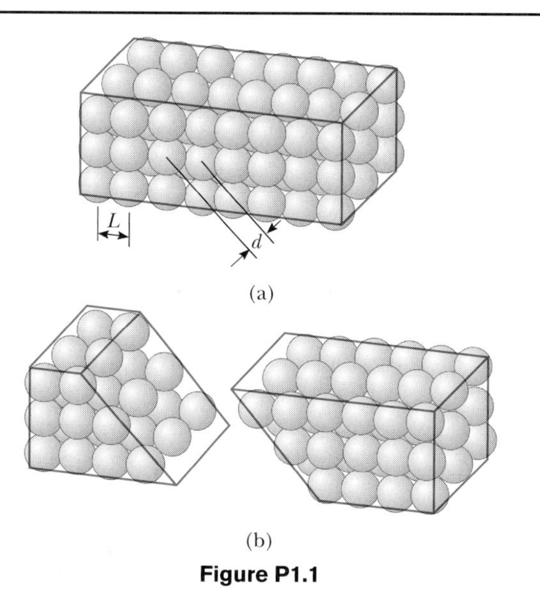

Figure P1.1

Section 1.3 Density and Atomic Mass

2. Use information on the endpapers of this book to calculate the average density of the Earth. Where does the value fit among those listed in Tables 1.5 and 14.1? Look up the density of a typical surface rock like granite in another source and compare the density of the Earth to it.

3. The standard kilogram is a platinum–iridium cylinder 39.0 mm in height and 39.0 mm in diameter. What is the density of the material?

4. A major motor company displays a die-cast model of its first automobile, made from 9.35 kg of iron. To celebrate its hundredth year in business, a worker will recast the model in gold from the original dies. What mass of gold is needed to make the new model?

5. What mass of a material with density ρ is required to make a hollow spherical shell having inner radius r_1 and outer radius r_2?

6. Two spheres are cut from a certain uniform rock. One has radius 4.50 cm. The mass of the other is five times greater. Find its radius.

7. Calculate the mass of an atom of (a) helium, (b) iron, and (c) lead. Give your answers in grams. The atomic masses of these atoms are 4.00 u, 55.9 u, and 207 u, respectively.

8. The paragraph preceding Example 1.1 in the text mentions that the atomic mass of aluminum is $27.0\,\text{u} = 27.0 \times 1.66 \times 10^{-27}$ kg. Example 1.1 says that 27.0 g of aluminum contains 6.02×10^{23} atoms. (a) Prove that each one of these two statements implies the other. (b) **What If?** What if it's not aluminum? Let M represent the numerical value of the mass of one atom of any chemical element in atomic mass units. Prove that M grams of the substance contains a particular number of atoms, the same number for all elements. Calculate this number precisely from the value for u quoted in the text. The number of atoms in M grams of an element is called *Avogadro's number* N_A. The idea can be extended: Avogadro's number of molecules of a chemical compound has a mass of M grams, where M atomic mass units is the mass of one molecule. Avogadro's number of atoms or molecules is called one *mole,* symbolized as 1 mol. A periodic table of the elements, as in Appendix C, and the chemical formula for a compound contain enough information to find the molar mass of the compound. (c) Calculate the mass of one mole of water, H_2O. (d) Find the molar mass of CO_2.

9. On your wedding day your lover gives you a gold ring of mass 3.80 g. Fifty years later its mass is 3.35 g. On the average, how many atoms were abraded from the ring during each second of your marriage? The atomic mass of gold is 197 u.

10. A small cube of iron is observed under a microscope. The edge of the cube is 5.00×10^{-6} cm long. Find (a) the mass of the cube and (b) the number of iron atoms in the cube. The atomic mass of iron is 55.9 u, and its density is $7.86\ \text{g/cm}^3$.

11. A structural I beam is made of steel. A view of its cross-section and its dimensions are shown in Figure P1.11. The density of the steel is $7.56 \times 10^3\ \text{kg/m}^3$. (a) What is the mass of a section 1.50 m long? (b) Assume that the atoms are predominantly iron, with atomic mass 55.9 u. How many atoms are in this section?

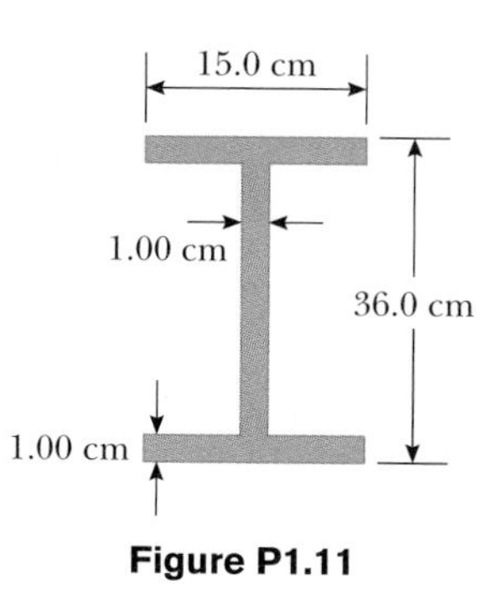

Figure P1.11

12. A child at the beach digs a hole in the sand and uses a pail to fill it with water having a mass of 1.20 kg. The mass of one molecule of water is 18.0 u. (a) Find the number of water molecules in this pail of water. (b) Suppose the quantity of water on Earth is constant at 1.32×10^{21} kg. How many of the water molecules in this pail of water are likely to have been in an equal quantity of water that once filled one particular claw print left by a Tyrannosaur hunting on a similar beach?

Section 1.4 Dimensional Analysis

13. The position of a particle moving under uniform acceleration is some function of time and the acceleration. Suppose we write this position $s = ka^m t^n$, where k is a dimensionless constant. Show by dimensional analysis that this expression is satisfied if $m = 1$ and $n = 2$. Can this analysis give the value of k?

14. Figure P1.14 shows a *frustrum of a cone.* Of the following mensuration (geometrical) expressions, which describes (a) the total circumference of the flat circular faces (b) the volume (c) the area of the curved surface? (i) $\pi(r_1 + r_2)[h^2 + (r_1 - r_2)^2]^{1/2}$ (ii) $2\pi(r_1 + r_2)$ (iii) $\pi h(r_1^2 + r_1 r_2 + r_2^2)$.

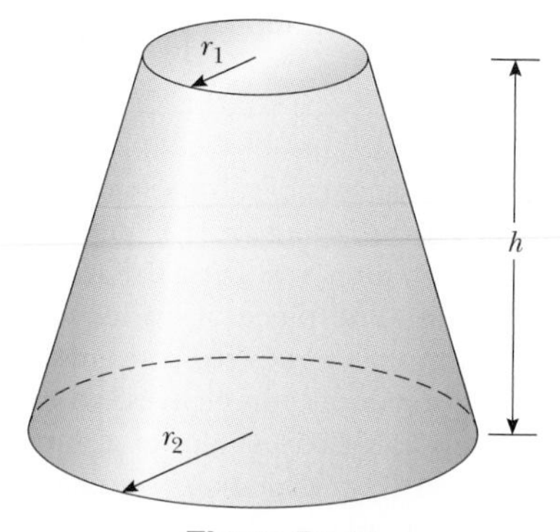

Figure P1.14

15. Which of the following equations are dimensionally correct?
(a) $v_f = v_i + ax$
(b) $y = (2\ \text{m})\cos(kx)$, where $k = 2\ \text{m}^{-1}$.

16. (a) A fundamental law of motion states that the acceleration of an object is directly proportional to the resultant force exerted on the object and inversely proportional to its mass. If the proportionality constant is defined to have no dimensions, determine the dimensions of force. (b) The newton is the SI unit of force. According to the results for (a), how can you express a force having units of newtons using the fundamental units of mass, length, and time?

17. Newton's law of universal gravitation is represented by

$$F = \frac{GMm}{r^2}$$

Here F is the magnitude of the gravitational force exerted by one small object on another, M and m are the masses of the objects, and r is a distance. Force has the SI units $\text{kg}\cdot\text{m/s}^2$. What are the SI units of the proportionality constant G?

Section 1.5 Conversion of Units

18. A worker is to paint the walls of a square room 8.00 ft high and 12.0 ft along each side. What surface area in square meters must she cover?

19. Suppose your hair grows at the rate 1/32 in. per day. Find the rate at which it grows in nanometers per second. Because the distance between atoms in a molecule is on the order of 0.1 nm, your answer suggests how rapidly layers of atoms are assembled in this protein synthesis.

20. The volume of a wallet is 8.50 in.^{3} Convert this value to m^3, using the definition 1 in. = 2.54 cm.

21. A rectangular building lot is 100 ft by 150 ft. Determine the area of this lot in m^2.

22. An auditorium measures 40.0 m × 20.0 m × 12.0 m. The density of air is 1.20 kg/m^3. What are (a) the volume of the room in cubic feet and (b) the weight of air in the room in pounds?

23. Assume that it takes 7.00 minutes to fill a 30.0-gal gasoline tank. (a) Calculate the rate at which the tank is filled in gallons per second. (b) Calculate the rate at which the tank is filled in cubic meters per second. (c) Determine the time interval, in hours, required to fill a 1-m^3 volume at the same rate. (1 U.S. gal = 231 in.^{3})

24. Find the height or length of these natural wonders in kilometers, meters and centimeters. (a) The longest cave system in the world is the Mammoth Cave system in central Kentucky. It has a mapped length of 348 mi. (b) In the United States, the waterfall with the greatest single drop is Ribbon Falls, which falls 1 612 ft. (c) Mount McKinley in Denali National Park, Alaska, is America's highest mountain at a height of 20 320 ft. (d) The deepest canyon in the United States is King's Canyon in California with a depth of 8 200 ft.

25. A solid piece of lead has a mass of 23.94 g and a volume of 2.10 cm^3. From these data, calculate the density of lead in SI units (kg/m^3).

26. A *section* of land has an area of 1 square mile and contains 640 acres. Determine the number of square meters in 1 acre.

27. An ore loader moves 1 200 tons/h from a mine to the surface. Convert this rate to lb/s, using 1 ton = 2 000 lb.

28. (a) Find a conversion factor to convert from miles per hour to kilometers per hour. (b) In the past, a federal law mandated that highway speed limits would be 55 mi/h. Use the conversion factor of part (a) to find this speed in kilometers per hour. (c) The maximum highway speed is now 65 mi/h in some places. In kilometers per hour, how much increase is this over the 55 mi/h limit?

29. At the time of this book's printing, the U.S. national debt is about \$6 trillion. (a) If payments were made at the rate of \$1 000 per second, how many years would it take to pay off the debt, assuming no interest were charged? (b) A dollar bill is about 15.5 cm long. If six trillion dollar bills were laid end to end around the Earth's equator, how many times would they encircle the planet? Take the radius of the Earth at the equator to be 6 378 km. (*Note:* Before doing any of these calculations, try to guess at the answers. You may be very surprised.)

30. The mass of the Sun is 1.99×10^{30} kg, and the mass of an atom of hydrogen, of which the Sun is mostly composed, is 1.67×10^{-27} kg. How many atoms are in the Sun?

31. One gallon of paint (volume = 3.78×10^{-3} m^3) covers an area of 25.0 m^2. What is the thickness of the paint on the wall?

32. A pyramid has a height of 481 ft and its base covers an area of 13.0 acres (Fig. P1.32). If the volume of a pyramid is given by the expression $V = \frac{1}{3}Bh$, where B is the area of the base and h is the height, find the volume of this pyramid in cubic meters. (1 acre = 43 560 ft^2)

Sylvain Grandadam/Photo Researchers, Inc.

Figure P1.32 Problems 32 and 33.

33. The pyramid described in Problem 32 contains approximately 2 million stone blocks that average 2.50 tons each. Find the weight of this pyramid in pounds.

34. Assuming that 70% of the Earth's surface is covered with water at an average depth of 2.3 mi, estimate the mass of the water on the Earth in kilograms.

35. A hydrogen atom has a diameter of approximately 1.06×10^{-10} m, as defined by the diameter of the spherical electron cloud around the nucleus. The hydrogen nucleus has a diameter of approximately 2.40×10^{-15} m. (a) For a scale model, represent the diameter of the hydrogen atom by the length of an American football field

(100 yd = 300 ft), and determine the diameter of the nucleus in millimeters. (b) The atom is how many times larger in volume than its nucleus?

36. The nearest stars to the Sun are in the Alpha Centauri multiple-star system, about 4.0×10^{13} km away. If the Sun, with a diameter of 1.4×10^9 m, and Alpha Centauri A are both represented by cherry pits 7.0 mm in diameter, how far apart should the pits be placed to represent the Sun and its neighbor to scale?

37. The diameter of our disk-shaped galaxy, the Milky Way, is about 1.0×10^5 lightyears (ly). The distance to Messier 31, which is Andromeda, the spiral galaxy nearest to the Milky Way, is about 2.0 million ly. If a scale model represents the Milky Way and Andromeda galaxies as dinner plates 25 cm in diameter, determine the distance between the two plates.

38. The mean radius of the Earth is 6.37×10^6 m, and that of the Moon is 1.74×10^8 cm. From these data calculate (a) the ratio of the Earth's surface area to that of the Moon and (b) the ratio of the Earth's volume to that of the Moon. Recall that the surface area of a sphere is $4\pi r^2$ and the volume of a sphere is $\frac{4}{3}\pi r^3$.

39. One cubic meter (1.00 m^3) of aluminum has a mass of 2.70×10^3 kg, and 1.00 m^3 of iron has a mass of 7.86×10^3 kg. Find the radius of a solid aluminum sphere that will balance a solid iron sphere of radius 2.00 cm on an equal-arm balance.

40. Let ρ_{Al} represent the density of aluminum and ρ_{Fe} that of iron. Find the radius of a solid aluminum sphere that balances a solid iron sphere of radius r_{Fe} on an equal-arm balance.

Section 1.6 Estimates and Order-of-Magnitude Calculations

41. Estimate the number of Ping-Pong balls that would fit into a typical-size room (without being crushed). In your solution state the quantities you measure or estimate and the values you take for them.

42. An automobile tire is rated to last for 50 000 miles. To an order of magnitude, through how many revolutions will it turn? In your solution state the quantities you measure or estimate and the values you take for them.

43. Grass grows densely everywhere on a quarter-acre plot of land. What is the order of magnitude of the number of blades of grass on this plot? Explain your reasoning. Note that 1 acre = 43 560 ft^2.

44. Approximately how many raindrops fall on a one-acre lot during a one-inch rainfall? Explain your reasoning.

45. Compute the order of magnitude of the mass of a bathtub half full of water. Compute the order of magnitude of the mass of a bathtub half full of pennies. In your solution list the quantities you take as data and the value you measure or estimate for each.

46. Soft drinks are commonly sold in aluminum containers. To an order of magnitude, how many such containers are thrown away or recycled each year by U.S. consumers? How many tons of aluminum does this represent? In your solution state the quantities you measure or estimate and the values you take for them.

47. To an order of magnitude, how many piano tuners are in New York City? The physicist Enrico Fermi was famous for asking questions like this on oral Ph.D. qualifying examinations. His own facility in making order-of-magnitude calculations is exemplified in Problem 45.48.

Section 1.7 Significant Figures

Note: Appendix B.8 on propagation of uncertainty may be useful in solving some problems in this section.

48. A rectangular plate has a length of (21.3 ± 0.2) cm and a width of (9.8 ± 0.1) cm. Calculate the area of the plate, including its uncertainty.

49. The radius of a circle is measured to be (10.5 ± 0.2) m. Calculate the (a) area and (b) circumference of the circle and give the uncertainty in each value.

50. How many significant figures are in the following numbers? (a) 78.9 ± 0.2 (b) 3.788×10^9 (c) 2.46×10^{-6} (d) 0.005 3.

51. The radius of a solid sphere is measured to be (6.50 ± 0.20) cm, and its mass is measured to be (1.85 ± 0.02) kg. Determine the density of the sphere in kilograms per cubic meter and the uncertainty in the density.

52. Carry out the following arithmetic operations: (a) the sum of the measured values 756, 37.2, 0.83, and 2.5; (b) the product $0.003\,2 \times 356.3$; (c) the product $5.620 \times \pi$.

53. The *tropical year,* the time from vernal equinox to the next vernal equinox, is the basis for our calendar. It contains 365.242 199 days. Find the number of seconds in a tropical year.

54. A farmer measures the distance around a rectangular field. The length of the long sides of the rectangle is found to be 38.44 m, and the length of the short sides is found to be 19.5 m. What is the total distance around the field?

55. A sidewalk is to be constructed around a swimming pool that measures (10.0 ± 0.1) m by (17.0 ± 0.1) m. If the sidewalk is to measure (1.00 ± 0.01) m wide by (9.0 ± 0.1) cm thick, what volume of concrete is needed, and what is the approximate uncertainty of this volume?

Additional Problems

56. In a situation where data are known to three significant digits, we write 6.379 m = 6.38 m and 6.374 m = 6.37 m. When a number ends in 5, we arbitrarily choose to write 6.375 m = 6.38 m. We could equally well write 6.375 m = 6.37 m, "rounding down" instead of "rounding up," because we would change the number 6.375 by equal increments in both cases. Now consider an order-of-magnitude estimate, in which we consider factors rather than increments. We write $500 \text{ m} \sim 10^3 \text{ m}$ because 500 differs from 100 by a factor of 5 while it differs from 1 000 by only a factor of 2. We write $437 \text{ m} \sim 10^3 \text{ m}$ and $305 \text{ m} \sim 10^2 \text{ m}$. What distance differs from 100 m and from 1 000 m by equal factors, so that we could equally well choose to represent its order of magnitude either as $\sim 10^2$ m or as $\sim 10^3$ m?

57. For many electronic applications, such as in computer chips, it is desirable to make components as small as possible to keep the temperature of the components low and to increase the speed of the device. Thin metallic coatings (films) can be used instead of wires to make electrical connections. Gold is especially useful because it does not oxidize readily. Its atomic mass is 197 u. A gold film can be no thinner than the size of a gold atom. Calculate the minimum coating thickness, assuming that a gold atom occupies a cubical volume in the film that is equal to the volume it occupies in a large piece of metal. This geometric model yields a result of the correct order of magnitude.

58. The basic function of the carburetor of an automobile is to "atomize" the gasoline and mix it with air to promote rapid combustion. As an example, assume that 30.0 cm^3 of gasoline is atomized into N spherical droplets, each with a radius of 2.00×10^{-5} m. What is the total surface area of these N spherical droplets?

59. The consumption of natural gas by a company satisfies the empirical equation $V = 1.50t + 0.008\,00t^2$, where V is the volume in millions of cubic feet and t the time in months. Express this equation in units of cubic feet and seconds. Assign proper units to the coefficients. Assume a month is equal to 30.0 days.

60. In physics it is important to use mathematical approximations. Demonstrate that for small angles ($< 20°$)

$$\tan \alpha \approx \sin \alpha \approx \alpha = \pi\alpha'/180°$$

where α is in radians and α' is in degrees. Use a calculator to find the largest angle for which $\tan \alpha$ may be approximated by $\sin \alpha$ if the error is to be less than 10.0%.

61. A high fountain of water is located at the center of a circular pool as in Figure P1.61. Not wishing to get his feet wet, a student walks around the pool and measures its circumference to be 15.0 m. Next, the student stands at the edge of the pool and uses a protractor to gauge the angle of elevation of the top of the fountain to be 55.0°. How high is the fountain?

Figure P1.61

62. Collectible coins are sometimes plated with gold to enhance their beauty and value. Consider a commemorative quarter-dollar advertised for sale at \$4.98. It has a diameter of 24.1 mm, a thickness of 1.78 mm, and is completely covered with a layer of pure gold 0.180 μm thick. The volume of the plating is equal to the thickness of the layer times the area to which it is applied. The patterns on the faces of the coin and the grooves on its edge have a negligible effect on its area. Assume that the price of gold is \$10.0 per gram. Find the cost of the gold added to the coin. Does the cost of the gold significantly enhance the value of the coin?

63. There are nearly $\pi \times 10^7$ s in one year. Find the percentage error in this approximation, where "percentage error" is defined as

$$\text{Percentage error} = \frac{|\text{assumed value} - \text{true value}|}{\text{true value}} \times 100\%$$

64. Assume that an object covers an area A and has a uniform height h. If its cross-sectional area is uniform over its height, then its volume is given by $V = Ah$. (a) Show that $V = Ah$ is dimensionally correct. (b) Show that the volumes of a cylinder and of a rectangular box can be written in the form $V = Ah$, identifying A in each case. (Note that A, sometimes called the "footprint" of the object, can have any shape and the height can be replaced by average thickness in general.)

65. A child loves to watch as you fill a transparent plastic bottle with shampoo. Every horizontal cross-section is a circle, but the diameters of the circles have different values, so that the bottle is much wider in some places than others. You pour in bright green shampoo with constant volume flow rate 16.5 cm^3/s. At what rate is its level in the bottle rising (a) at a point where the diameter of the bottle is 6.30 cm and (b) at a point where the diameter is 1.35 cm?

66. One cubic centimeter of water has a mass of 1.00×10^{-3} kg. (a) Determine the mass of 1.00 m^3 of water. (b) Biological substances are 98% water. Assume that they have the same density as water to estimate the masses of a cell that has a diameter of 1.0 μm, a human kidney, and a fly. Model the kidney as a sphere with a radius of 4.0 cm and the fly as a cylinder 4.0 mm long and 2.0 mm in diameter.

67. Assume there are 100 million passenger cars in the United States and that the average fuel consumption is 20 mi/gal of gasoline. If the average distance traveled by each car is 10 000 mi/yr, how much gasoline would be saved per year if average fuel consumption could be increased to 25 mi/gal?

68. A creature moves at a speed of 5.00 furlongs per fortnight (not a very common unit of speed). Given that 1 furlong = 220 yards and 1 fortnight = 14 days, determine the speed of the creature in m/s. What kind of creature do you think it might be?

69. The distance from the Sun to the nearest star is about 4×10^{16} m. The Milky Way galaxy is roughly a disk of diameter $\sim 10^{21}$ m and thickness $\sim 10^{19}$ m. Find the order of magnitude of the number of stars in the Milky Way. Assume the distance between the Sun and our nearest neighbor is typical.

70. The data in the following table represent measurements of the masses and dimensions of solid cylinders of aluminum, copper, brass, tin, and iron. Use these data to calculate the densities of these substances. Compare your results for aluminum, copper, and iron with those given in Table 1.5.

Substance	Mass (g)	Diameter (cm)	Length (cm)
Aluminum	51.5	2.52	3.75
Copper	56.3	1.23	5.06
Brass	94.4	1.54	5.69
Tin	69.1	1.75	3.74
Iron	216.1	1.89	9.77

71. (a) How many seconds are in a year? (b) If one micrometeorite (a sphere with a diameter of 1.00×10^{-6} m) strikes each square meter of the Moon each second, how many years will it take to cover the Moon to a depth of 1.00 m? To solve this problem, you can consider a cubic box on the Moon 1.00 m on each edge, and find how long it will take to fill the box.

Answers to Quick Quizzes

1.1 (a). Because the density of aluminum is smaller than that of iron, a larger volume of aluminum is required for a given mass than iron.

1.2 False. Dimensional analysis gives the units of the proportionality constant but provides no information about its numerical value. To determine its numerical value requires either experimental data or geometrical reasoning. For example, in the generation of the equation $x = \frac{1}{2}at^2$, because the factor $\frac{1}{2}$ is dimensionless, there is no way of determining it using dimensional analysis.

1.3 (b). Because kilometers are shorter than miles, a larger number of kilometers is required for a given distance than miles.

1.4 Reporting all these digits implies you have determined the location of the center of the chair's seat to the nearest $\pm 0.000\ 000\ 000\ 1$ m. This roughly corresponds to being able to count the atoms in your meter stick because each of them is about that size! It would be better to record the measurement as 1.044 m: this indicates that you know the position to the nearest millimeter, assuming the meter stick has millimeter markings on its scale.

Chapter 2

Motion in One Dimension

▲ One of the physical quantities we will study in this chapter is the velocity of an object moving in a straight line. Downhill skiers can reach velocities with a magnitude greater than 100 km/h. (Jean Y. Ruszniewski/Getty Images)

CHAPTER OUTLINE

Jean Y. Rusznieski/Stone/Getty

As a first step in studying classical mechanics, we describe motion in terms of space and time while ignoring the agents that caused that motion. This portion of classical mechanics is called *kinematics.* (The word *kinematics* has the same root as *cinema.* Can you see why?) In this chapter we consider only motion in one dimension, that is, motion along a straight line. We first define position, displacement, velocity, and acceleration. Then, using these concepts, we study the motion of objects traveling in one dimension with a constant acceleration.

From everyday experience we recognize that motion represents a continuous change in the position of an object. In physics we can categorize motion into three types: translational, rotational, and vibrational. A car moving down a highway is an example of translational motion, the Earth's spin on its axis is an example of rotational motion, and the back-and-forth movement of a pendulum is an example of vibrational motion. In this and the next few chapters, we are concerned only with translational motion. (Later in the book we shall discuss rotational and vibrational motions.)

In our study of translational motion, we use what is called the **particle model**—we describe the moving object as a *particle* regardless of its size. In general, **a particle is a point-like object—that is, an object with mass but having infinitesimal size.** For example, if we wish to describe the motion of the Earth around the Sun, we can treat the Earth as a particle and obtain reasonably accurate data about its orbit. This approximation is justified because the radius of the Earth's orbit is large compared with the dimensions of the Earth and the Sun. As an example on a much smaller scale, it is possible to explain the pressure exerted by a gas on the walls of a container by treating the gas molecules as particles, without regard for the internal structure of the molecules.

2.1 Position, Velocity, and Speed

Position

The motion of a particle is completely known if the particle's position in space is known at all times. A particle's **position** is the location of the particle with respect to a chosen reference point that we can consider to be the origin of a coordinate system.

Consider a car moving back and forth along the x axis as in Figure 2.1a. When we begin collecting position data, the car is 30 m to the right of a road sign, which we will use to identify the reference position $x = 0$. (Let us assume that all data in this example are known to two significant figures. To convey this information, we should report the initial position as 3.0×10^1 m. We have written this value in the simpler form 30 m to make the discussion easier to follow.) We will use the particle model by identifying some point on the car, perhaps the front door handle, as a particle representing the entire car.

We start our clock and once every 10 s note the car's position relative to the sign at $x = 0$. As you can see from Table 2.1, the car moves to the right (which we have

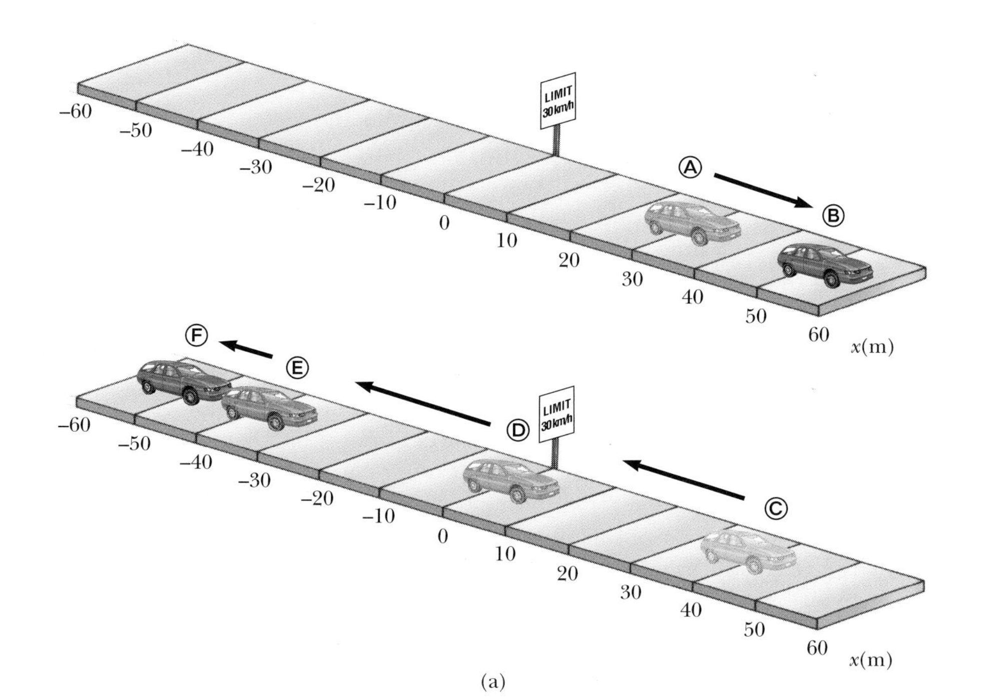

Table 2.1

Position of the Car at Various Times

Position	t(s)	x(m)
Ⓐ	0	30
Ⓑ	10	52
Ⓒ	20	38
Ⓓ	30	0
Ⓔ	40	−37
Ⓕ	50	−53

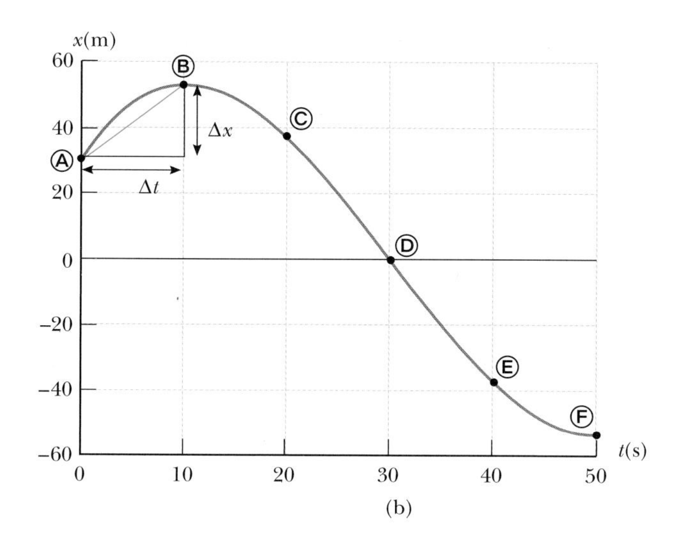

Active Figure 2.1 (a) A car moves back and forth along a straight line taken to be the x axis. Because we are interested only in the car's translational motion, we can model it as a particle. (b) Position–time graph for the motion of the "particle."

At the Active Figures link at* http://www.pse6.com, *you can move each of the six points Ⓐ through Ⓕ and observe the motion of the car pictorially and graphically as it follows a smooth path through the six points.

defined as the positive direction) during the first 10 s of motion, from position Ⓐ to position Ⓑ. After Ⓑ, the position values begin to decrease, suggesting that the car is backing up from position Ⓑ through position Ⓕ. In fact, at Ⓓ, 30 s after we start measuring, the car is alongside the road sign (see Figure 2.1a) that we are using to mark our origin of coordinates. It continues moving to the left and is more than 50 m to the left of the sign when we stop recording information after our sixth data point. A graphical representation of this information is presented in Figure 2.1b. Such a plot is called a *position–time graph.*

Displacement

Given the data in Table 2.1, we can easily determine the change in position of the car for various time intervals. The **displacement** of a particle is defined as its change in position in some time interval. As it moves from an initial position x_i to a final position x_f, the displacement of the particle is given by $x_f - x_i$. We use the Greek letter delta (Δ) to denote the *change* in a quantity. Therefore, we write the displacement, or change in position, of the particle as

$$\Delta x \equiv x_f - x_i \tag{2.1}$$

From this definition we see that Δx is positive if x_f is greater than x_i and negative if x_f is less than x_i.

It is very important to recognize the difference between displacement and distance traveled. **Distance** is the length of a path followed by a particle. Consider, for example, the basketball players in Figure 2.2. If a player runs from his own basket down the court to the other team's basket and then returns to his own basket, the *displacement* of the player during this time interval is zero, because he ended up at the same point as he started. During this time interval, however, he covered a *distance* of twice the length of the basketball court.

Displacement is an example of a vector quantity. Many other physical quantities, including position, velocity, and acceleration, also are vectors. In general, **a vector quantity requires the specification of both direction and magnitude.** By contrast, **a scalar quantity has a numerical value and no direction.** In this chapter, we use positive (+) and negative (−) signs to indicate vector direction. We can do this because the chapter deals with one-dimensional motion only; this means that any object we study can be moving only along a straight line. For example, for horizontal motion let us arbitrarily specify to the right as being the positive direction. It follows that any object always moving to the right undergoes a positive displacement $\Delta x > 0$, and any object moving to the left undergoes a negative displacement, so that $\Delta x < 0$. We shall treat vector quantities in greater detail in Chapter 3.

For our basketball player in Figure 2.2, if the trip from his own basket to the opposing basket is described by a displacement of + 28 m, the trip in the reverse direction represents a displacement of − 28 m. Each trip, however, represents a distance of 28 m, because distance is a scalar quantity. The total distance for the trip down the court and back is 56 m. Distance, therefore, is always represented as a positive number, while displacement can be either positive or negative.

There is one very important point that has not yet been mentioned. Note that the data in Table 2.1 results only in the six data points in the graph in Figure 2.1b. The smooth curve drawn through the six points in the graph is only a *possibility* of the actual motion of the car. We only have information about six instants of time—we have no idea what happened in between the data points. The smooth curve is a *guess* as to what happened, but keep in mind that it is *only* a guess.

If the smooth curve does represent the actual motion of the car, the graph contains information about the entire 50-s interval during which we watch the car move. It is much easier to see changes in position from the graph than from a verbal description or even a table of numbers. For example, it is clear that the car was covering more ground during the middle of the 50-s interval than at the end. Between positions Ⓒ and Ⓓ, the car traveled almost 40 m, but during the last 10 s, between positions Ⓔ and Ⓕ, it moved less than half that far. A common way of comparing these different motions is to divide the displacement Δx that occurs between two clock readings by the length of that particular time interval Δt. This turns out to be a very useful ratio, one that we shall use many times. This ratio has been given a special name—*average velocity*. **The average velocity $\bar{v}_x$ of a particle is defined as the**

Ken White/Allsport/Getty Images

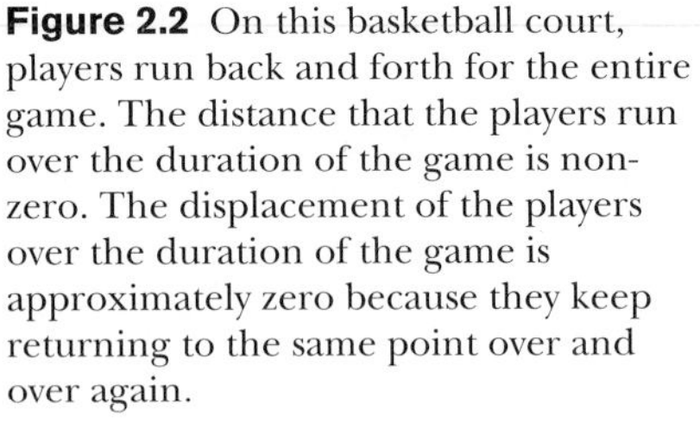

Figure 2.2 On this basketball court, players run back and forth for the entire game. The distance that the players run over the duration of the game is nonzero. The displacement of the players over the duration of the game is approximately zero because they keep returning to the same point over and over again.

particle's displacement Δx divided by the time interval Δt during which that displacement occurs:

$$\overline{v}_x \equiv \frac{\Delta x}{\Delta t} \tag{2.2}$$

Average velocity

where the subscript x indicates motion along the x axis. From this definition we see that average velocity has dimensions of length divided by time (L/T)—meters per second in SI units.

The average velocity of a particle moving in one dimension can be positive or negative, depending on the sign of the displacement. (The time interval Δt is always positive.) If the coordinate of the particle increases in time (that is, if $x_f > x_i$), then Δx is positive and $\overline{v}_x = \Delta x/\Delta t$ is positive. This case corresponds to a particle moving in the positive x direction, that is, toward larger values of x. If the coordinate decreases in time (that is, if $x_f < x_i$) then Δx is negative and hence $\overline{v}_x$ is negative. This case corresponds to a particle moving in the negative x direction.

We can interpret average velocity geometrically by drawing a straight line between any two points on the position–time graph in Figure 2.1b. This line forms the hypotenuse of a right triangle of height Δx and base Δt. The slope of this line is the ratio $\Delta x/\Delta t$, which is what we have defined as average velocity in Equation 2.2. For example, the line between positions Ⓐ and Ⓑ in Figure 2.1b has a slope equal to the average velocity of the car between those two times, $(52\text{ m} - 30\text{ m})/(10\text{ s} - 0) = 2.2\text{ m/s}$.

In everyday usage, the terms *speed* and *velocity* are interchangeable. In physics, however, there is a clear distinction between these two quantities. Consider a marathon runner who runs more than 40 km, yet ends up at his starting point. His total displacement is zero, so his average velocity is zero! Nonetheless, we need to be able to quantify how fast he was running. A slightly different ratio accomplishes this for us. The **average speed** of a particle, a scalar quantity, is defined as **the total distance traveled divided by the total time interval required to travel that distance:**

$$\text{Average speed} = \frac{\text{total distance}}{\text{total time}} \tag{2.3}$$

Average speed

The SI unit of average speed is the same as the unit of average velocity: meters per second. However, unlike average velocity, average speed has no direction and hence carries no algebraic sign. Notice the distinction between average velocity and average speed—average velocity (Eq. 2.2) is the *displacement* divided by the time interval, while average speed (Eq. 2.3) is the *distance* divided by the time interval.

Knowledge of the average velocity or average speed of a particle does not provide information about the details of the trip. For example, suppose it takes you 45.0 s to travel 100 m down a long straight hallway toward your departure gate at an airport. At the 100-m mark, you realize you missed the rest room, and you return back 25.0 m along the same hallway, taking 10.0 s to make the return trip. The magnitude of the average *velocity* for your trip is $+75.0\text{ m}/55.0\text{ s} = +1.36\text{ m/s}$. The average *speed* for your trip is $125\text{ m}/55.0\text{ s} = 2.27\text{ m/s}$. You may have traveled at various speeds during the walk. Neither average velocity nor average speed provides information about these details.

2.1 Average Speed and Average Velocity

The magnitude of the average velocity is *not* the average speed. For example, consider the marathon runner discussed here. The magnitude of the average velocity is zero, but the average speed is clearly not zero.

Quick Quiz 2.1 Under which of the following conditions is the magnitude of the average velocity of a particle moving in one dimension smaller than the average speed over some time interval? (a) A particle moves in the $+x$ direction without reversing. (b) A particle moves in the $-x$ direction without reversing. (c) A particle moves in the $+x$ direction and then reverses the direction of its motion. (d) There are no conditions for which this is true.

Example 2.1 Calculating the Average Velocity and Speed

Find the displacement, average velocity, and average speed of the car in Figure 2.1a between positions Ⓐ and Ⓕ.

Solution From the position–time graph given in Figure 2.1b, note that $x_A = 30$ m at $t_A = 0$ s and that $x_F = -53$ m at $t_F = 50$ s. Using these values along with the definition of displacement, Equation 2.1, we find that

$$\Delta x = x_F - x_A = -53 \text{ m} - 30 \text{ m} = -83 \text{ m}$$

This result means that the car ends up 83 m in the negative direction (to the left, in this case) from where it started. This number has the correct units and is of the same order of magnitude as the supplied data. A quick look at Figure 2.1a indicates that this is the correct answer.

It is difficult to estimate the average velocity without completing the calculation, but we expect the units to be meters per second. Because the car ends up to the left of where we started taking data, we know the average velocity must be negative. From Equation 2.2,

$$\bar{v}_x = \frac{\Delta x}{\Delta t} = \frac{x_f - x_i}{t_f - t_i} = \frac{x_F - x_A}{t_F - t_A}$$

$$= \frac{-53 \text{ m} - 30 \text{ m}}{50 \text{ s} - 0 \text{ s}} = \frac{-83 \text{ m}}{50 \text{ s}}$$

$$= -1.7 \text{ m/s}$$

We cannot unambiguously find the average speed of the car from the data in Table 2.1, because we do not have information about the positions of the car between the data points. If we adopt the assumption that the details of the car's position are described by the curve in Figure 2.1b, then the distance traveled is 22 m (from Ⓐ to Ⓑ) plus 105 m (from Ⓑ to Ⓕ) for a total of 127 m. We find the car's average speed for this trip by dividing the distance by the total time (Eq. 2.3):

$$\text{Average speed} = \frac{127 \text{ m}}{50 \text{ s}} = 2.5 \text{ m/s}$$

2.2 Instantaneous Velocity and Speed

Often we need to know the velocity of a particle at a particular instant in time, rather than the average velocity over a finite time interval. For example, even though you might want to calculate your average velocity during a long automobile trip, you would be especially interested in knowing your velocity at the *instant* you noticed the police car parked alongside the road ahead of you. In other words, you would like to be able to specify your velocity just as precisely as you can specify your position by noting what is happening at a specific clock reading—that is, at some specific instant. It may not be immediately obvious how to do this. What does it mean to talk about how fast something is moving if we "freeze time" and talk only about an individual instant? This is a subtle point not thoroughly understood until the late 1600s. At that time, with the invention of calculus, scientists began to understand how to describe an object's motion at any moment in time.

To see how this is done, consider Figure 2.3a, which is a reproduction of the graph in Figure 2.1b. We have already discussed the average velocity for the interval during which the car moved from position Ⓐ to position Ⓑ (given by the slope of the dark blue line) and for the interval during which it moved from Ⓐ to Ⓕ (represented by the slope of the light blue line and calculated in Example 2.1). Which of these two lines do you think is a closer approximation of the initial velocity of the car? The car starts out by moving to the right, which we defined to be the positive direction. Therefore, being positive, the value of the average velocity during the Ⓐ to Ⓑ interval is more representative of the initial value than is the value of the average velocity during the Ⓐ to Ⓕ interval, which we determined to be negative in Example 2.1. Now let us focus on the dark blue line and slide point Ⓑ to the left along the curve, toward point Ⓐ, as in Figure 2.3b. The line between the points becomes steeper and steeper, and as the two points become extremely close together, the line becomes a tangent line to the curve, indicated by the green line in Figure 2.3b. The slope of this tangent line

2.2 Slopes of Graphs

In any graph of physical data, the *slope* represents the ratio of the change in the quantity represented on the vertical axis to the change in the quantity represented on the horizontal axis. Remember that *a slope has units* (unless both axes have the same units). The units of slope in Figure 2.1b and Figure 2.3 are m/s, the units of velocity.

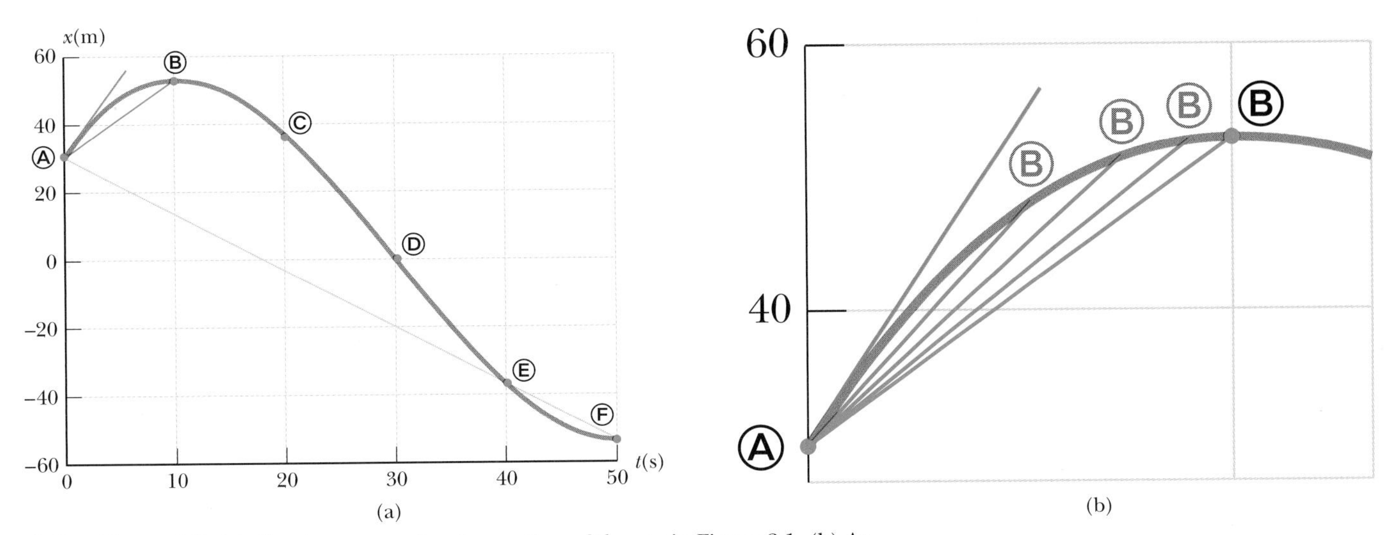

Active Figure 2.3 (a) Graph representing the motion of the car in Figure 2.1. (b) An enlargement of the upper-left-hand corner of the graph shows how the blue line between positions Ⓐ and Ⓑ approaches the green tangent line as point Ⓑ is moved closer to point Ⓐ.

At the Active Figures link at* http://www.pse6.com, *you can move point* Ⓑ *as suggested in (b) and observe the blue line approaching the green tangent line.

represents the velocity of the car at the moment we started taking data, at point Ⓐ. What we have done is determine the *instantaneous velocity* at that moment. In other words, **the instantaneous velocity v_x equals the limiting value of the ratio $\Delta x/\Delta t$ as Δt approaches zero:**[1]

$$v_x \equiv \lim_{\Delta t \to 0} \frac{\Delta x}{\Delta t} \tag{2.4}$$

In calculus notation, this limit is called the *derivative* of x with respect to t, written dx/dt:

Instantaneous velocity

$$v_x \equiv \lim_{\Delta t \to 0} \frac{\Delta x}{\Delta t} = \frac{dx}{dt} \tag{2.5}$$

The instantaneous velocity can be positive, negative, or zero. When the slope of the position–time graph is positive, such as at any time during the first 10 s in Figure 2.3, v_x is positive—the car is moving toward larger values of x. After point Ⓑ, v_x is negative because the slope is negative—the car is moving toward smaller values of x. At point Ⓑ, the slope and the instantaneous velocity are zero—the car is momentarily at rest.

From here on, we use the word *velocity* to designate instantaneous velocity. When it is *average velocity* we are interested in, we shall always use the adjective *average*.

The **instantaneous speed** of a particle is defined as the magnitude of its instantaneous velocity. As with average speed, instantaneous speed has no direction associated with it and hence carries no algebraic sign. For example, if one particle has an instantaneous velocity of $+25$ m/s along a given line and another particle has an instantaneous velocity of -25 m/s along the same line, both have a speed[2] of 25 m/s.

2.3 Instantaneous Speed and Instantaneous Velocity

In Pitfall Prevention 2.1, we argued that the magnitude of the average velocity is not the average speed. Notice the difference when discussing instantaneous values. The magnitude of the instantaneous velocity *is* the instantaneous speed. In an infinitesimal time interval, the magnitude of the displacement is equal to the distance traveled by the particle.

[1] Note that the displacement Δx also approaches zero as Δt approaches zero, so that the ratio looks like 0/0. As Δx and Δt become smaller and smaller, the ratio $\Delta x/\Delta t$ approaches a value equal to the slope of the line tangent to the x-versus-t curve.

[2] As with velocity, we drop the adjective for instantaneous speed: "Speed" means instantaneous speed.

Conceptual Example 2.2 The Velocity of Different Objects

Consider the following one-dimensional motions: **(A)** A ball thrown directly upward rises to a highest point and falls back into the thrower's hand. **(B)** A race car starts from rest and speeds up to 100 m/s. **(C)** A spacecraft drifts through space at constant velocity. Are there any points in the motion of these objects at which the instantaneous velocity has the same value as the average velocity over the entire motion? If so, identify the point(s).

Solution (A) The average velocity for the thrown ball is zero because the ball returns to the starting point; thus its displacement is zero. (Remember that average velocity is defined as $\Delta x/\Delta t$.) There is one point at which the instantaneous velocity is zero—at the top of the motion.

(B) The car's average velocity cannot be evaluated unambiguously with the information given, but it must be some value between 0 and 100 m/s. Because the car will have every instantaneous velocity between 0 and 100 m/s at some time during the interval, there must be some instant at which the instantaneous velocity is equal to the average velocity.

(C) Because the spacecraft's instantaneous velocity is constant, its instantaneous velocity at *any* time and its average velocity over *any* time interval are the same.

Example 2.3 Average and Instantaneous Velocity

A particle moves along the x axis. Its position varies with time according to the expression $x = -4t + 2t^2$ where x is in meters and t is in seconds.[3] The position–time graph for this motion is shown in Figure 2.4. Note that the particle moves in the negative x direction for the first second of motion, is momentarily at rest at the moment $t = 1$ s, and moves in the positive x direction at times $t > 1$ s.

(A) Determine the displacement of the particle in the time intervals $t = 0$ to $t = 1$ s and $t = 1$ s to $t = 3$ s.

Solution During the first time interval, the slope is negative and hence the average velocity is negative. Thus, we know that the displacement between Ⓐ and Ⓑ must be a negative number having units of meters. Similarly, we expect the displacement between Ⓑ and Ⓓ to be positive.

In the first time interval, we set $t_i = t_A = 0$ and $t_f = t_B = 1$ s. Using Equation 2.1, with $x = -4t + 2t^2$, we obtain for the displacement between $t = 0$ and $t = 1$ s,

$$\Delta x_{A\to B} = x_f - x_i = x_B - x_A$$
$$= [-4(1) + 2(1)^2] - [-4(0) + 2(0)^2]$$
$$= -2 \text{ m}$$

To calculate the displacement during the second time interval ($t = 1$ s to $t = 3$ s), we set $t_i = t_B = 1$ s and $t_f = t_D = 3$ s:

$$\Delta x_{B\to D} = x_f - x_i = x_D - x_B$$
$$= [-4(3) + 2(3)^2] - [-4(1) + 2(1)^2]$$
$$= +8 \text{ m}$$

These displacements can also be read directly from the position–time graph.

(B) Calculate the average velocity during these two time intervals.

Solution In the first time interval, $\Delta t = t_f - t_i = t_B - t_A = 1$ s. Therefore, using Equation 2.2 and the displacement calculated in (a), we find that

$$\bar{v}_{x(A\to B)} = \frac{\Delta x_{A\to B}}{\Delta t} = \frac{-2 \text{ m}}{1 \text{ s}} = -2 \text{ m/s}$$

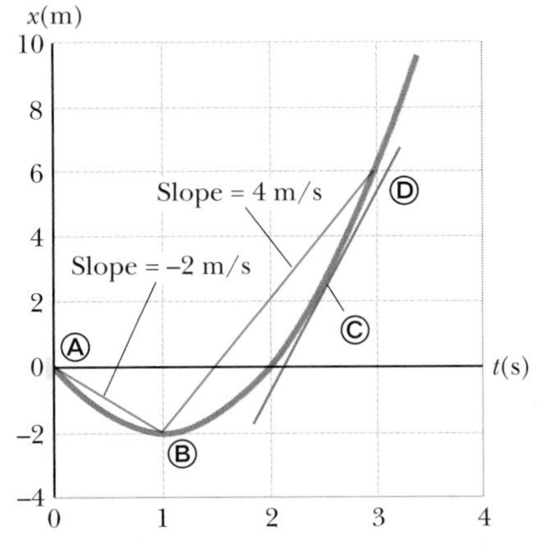

Figure 2.4 (Example 2.3) Position–time graph for a particle having an x coordinate that varies in time according to the expression $x = -4t + 2t^2$.

In the second time interval, $\Delta t = 2$ s; therefore,

$$\bar{v}_{x(B\to D)} = \frac{\Delta x_{B\to D}}{\Delta t} = \frac{8 \text{ m}}{2 \text{ s}} = +4 \text{ m/s}$$

These values are the same as the slopes of the lines joining these points in Figure 2.4.

(C) Find the instantaneous velocity of the particle at $t = 2.5$ s.

Solution We can guess that this instantaneous velocity must be of the same order of magnitude as our previous results, that is, a few meters per second. By measuring the slope of the green line at $t = 2.5$ s in Figure 2.4, we find that

$$v_x = +6 \text{ m/s}$$

[3] Simply to make it easier to read, we write the expression as $x = -4t + 2t^2$ rather than as $x = (-4.00 \text{ m/s})t + (2.00 \text{ m/s}^2)t^{2.00}$. When an equation summarizes measurements, consider its coefficients to have as many significant digits as other data quoted in a problem. Consider its coefficients to have the units required for dimensional consistency. When we start our clocks at $t = 0$, we usually do not mean to limit the precision to a single digit. Consider any zero value in this book to have as many significant figures as you need.

2.3 Acceleration

In the last example, we worked with a situation in which the velocity of a particle changes while the particle is moving. This is an extremely common occurrence. (How constant is your velocity as you ride a city bus or drive on city streets?) It is possible to quantify changes in velocity as a function of time similarly to the way in which we quantify changes in position as a function of time. When the velocity of a particle changes with time, the particle is said to be *accelerating*. For example, the magnitude of the velocity of a car increases when you step on the gas and decreases when you apply the brakes. Let us see how to quantify acceleration.

Suppose an object that can be modeled as a particle moving along the x axis has an initial velocity v_{xi} at time t_i and a final velocity v_{xf} at time t_f, as in Figure 2.5a.

The average acceleration $\bar{a}_x$ of the particle is defined as the *change* in velocity Δv_x divided by the time interval Δt during which that change occurs:

$$\bar{a}_x \equiv \frac{\Delta v_x}{\Delta t} = \frac{v_{xf} - v_{xi}}{t_f - t_i} \qquad (2.6)$$

Average acceleration

As with velocity, when the motion being analyzed is one-dimensional, we can use positive and negative signs to indicate the direction of the acceleration. Because the dimensions of velocity are L/T and the dimension of time is T, acceleration has dimensions of length divided by time squared, or L/T^2. The SI unit of acceleration is meters per second squared (m/s^2). It might be easier to interpret these units if you think of them as meters per second per second. For example, suppose an object has an acceleration of $+2$ m/s^2. You should form a mental image of the object having a velocity that is along a straight line and is increasing by 2 m/s during every interval of 1 s. If the object starts from rest, you should be able to picture it moving at a velocity of $+2$ m/s after 1 s, at $+4$ m/s after 2 s, and so on.

In some situations, the value of the average acceleration may be different over different time intervals. It is therefore useful to define the *instantaneous acceleration* as the limit of the average acceleration as Δt approaches zero. This concept is analogous to the definition of instantaneous velocity discussed in the previous section. If we imagine that point Ⓐ is brought closer and closer to point Ⓑ in Figure 2.5a and we take the limit of $\Delta v_x/\Delta t$ as Δt approaches zero, we obtain the instantaneous acceleration:

$$a_x \equiv \lim_{\Delta t \to 0} \frac{\Delta v_x}{\Delta t} = \frac{dv_x}{dt} \qquad (2.7)$$

Instantaneous acceleration

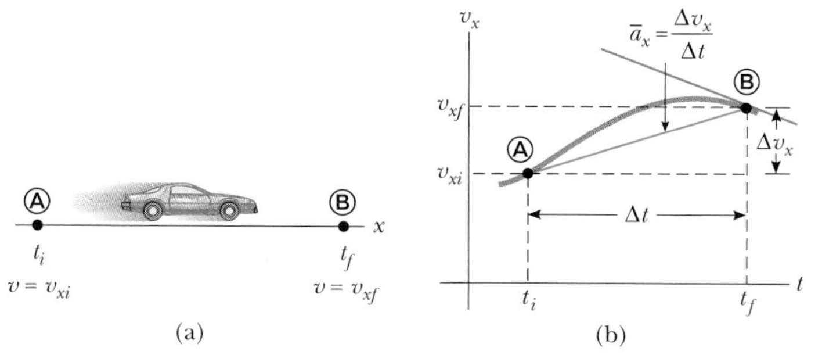

Figure 2.5 (a) A car, modeled as a particle, moving along the x axis from Ⓐ to Ⓑ has velocity v_{xi} at $t = t_i$ and velocity v_{xf} at $t = t_f$. (b) Velocity–time graph (rust) for the particle moving in a straight line. The slope of the blue straight line connecting Ⓐ and Ⓑ is the average acceleration in the time interval $\Delta t = t_f - t_i$.

PITFALL PREVENTION

2.4 Negative Acceleration

Keep in mind that *negative acceleration does not necessarily mean that an object is slowing down.* If the acceleration is negative, and the velocity is negative, the object is speeding up!

PITFALL PREVENTION

2.5 Deceleration

The word *deceleration* has the common popular connotation of *slowing down.* We will not use this word in this text, because it further confuses the definition we have given for negative acceleration.

That is, **the instantaneous acceleration equals the derivative of the velocity with respect to time,** which by definition is the slope of the velocity–time graph. The slope of the green line in Figure 2.5b is equal to the instantaneous acceleration at point Ⓑ. Thus, we see that just as the velocity of a moving particle is the slope at a point on the particle's x-t graph, the acceleration of a particle is the slope at a point on the particle's v_x-t graph. One can interpret the derivative of the velocity with respect to time as the time rate of change of velocity. If a_x is positive, the acceleration is in the positive x direction; if a_x is negative, the acceleration is in the negative x direction.

For the case of motion in a straight line, the direction of the velocity of an object and the direction of its acceleration are related as follows. **When the object's velocity and acceleration are in the same direction, the object is speeding up. On the other hand, when the object's velocity and acceleration are in opposite directions, the object is slowing down.**

To help with this discussion of the signs of velocity and acceleration, we can relate the acceleration of an object to the *force* exerted on the object. In Chapter 5 we formally establish that **force is proportional to acceleration:**

$$F \propto a$$

This proportionality indicates that acceleration is caused by force. Furthermore, force and acceleration are both vectors and the vectors act in the same direction. Thus, let us think about the signs of velocity and acceleration by imagining a force applied to an object and causing it to accelerate. Let us assume that the velocity and acceleration are in the same direction. This situation corresponds to an object moving in some direction that experiences a force acting in the same direction. In this case, the object speeds up! Now suppose the velocity and acceleration are in opposite directions. In this situation, the object moves in some direction and experiences a force acting in the opposite direction. Thus, the object slows down! It is very useful to equate the direction of the acceleration to the direction of a force, because it is easier from our everyday experience to think about what effect a force will have on an object than to think only in terms of the direction of the acceleration.

Quick Quiz 2.2 If a car is traveling eastward and slowing down, what is the direction of the force on the car that causes it to slow down? (a) eastward (b) westward (c) neither of these.

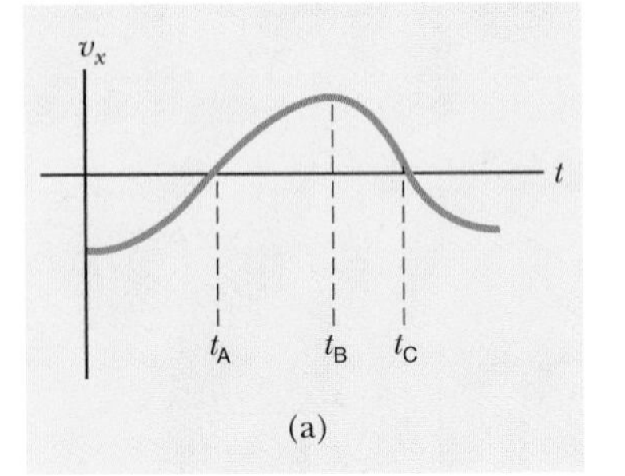

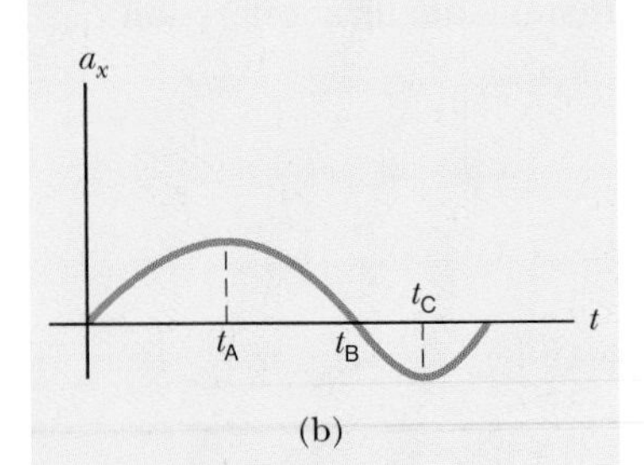

Figure 2.6 The instantaneous acceleration can be obtained from the velocity–time graph (a). At each instant, the acceleration in the a_x versus t graph (b) equals the slope of the line tangent to the v_x versus t curve (a).

From now on we shall use the term *acceleration* to mean instantaneous acceleration. When we mean average acceleration, we shall always use the adjective *average.*

Because $v_x = dx/dt$, the acceleration can also be written

$$a_x = \frac{dv_x}{dt} = \frac{d}{dt}\left(\frac{dx}{dt}\right) = \frac{d^2x}{dt^2} \tag{2.8}$$

That is, in one-dimensional motion, the acceleration equals the *second derivative* of x with respect to time.

Figure 2.6 illustrates how an acceleration–time graph is related to a velocity–time graph. The acceleration at any time is the slope of the velocity–time graph at that time. Positive values of acceleration correspond to those points in Figure 2.6a where the velocity is increasing in the positive x direction. The acceleration reaches a maximum at time t_A, when the slope of the velocity–time graph is a maximum. The acceleration then goes to zero at time t_B, when the velocity is a maximum (that is, when the slope of the v_x-t graph is zero). The acceleration is negative when the velocity is decreasing in the positive x direction, and it reaches its most negative value at time t_C.

Quick Quiz 2.3 Make a velocity–time graph for the car in Figure 2.1a. The speed limit posted on the road sign is 30 km/h. True or false? The car exceeds the speed limit at some time within the interval.

Conceptual Example 2.4 Graphical Relationships between x, v_x, and a_x

The position of an object moving along the x axis varies with time as in Figure 2.7a. Graph the velocity versus time and the acceleration versus time for the object.

Solution The velocity at any instant is the slope of the tangent to the x-t graph at that instant. Between $t = 0$ and $t = t_A$, the slope of the x-t graph increases uniformly, and so the velocity increases linearly, as shown in Figure 2.7b. Between t_A and t_B, the slope of the x-t graph is constant, and so the velocity remains constant. At t_D, the slope of the x-t graph is zero, so the velocity is zero at that instant. Between t_D and t_E, the slope of the x-t graph and thus the velocity are negative and decrease uniformly in this interval. In the interval t_E to t_F, the slope of the x-t graph is still negative, and at t_F it goes to zero. Finally, after t_F, the slope of the x-t graph is zero, meaning that the object is at rest for $t > t_F$.

The acceleration at any instant is the slope of the tangent to the v_x-t graph at that instant. The graph of acceleration versus time for this object is shown in Figure 2.7c. The acceleration is constant and positive between 0 and t_A, where the slope of the v_x-t graph is positive. It is zero between t_A and t_B and for $t > t_F$ because the slope of the v_x-t graph is zero at these times. It is negative between t_B and t_E because the slope of the v_x-t graph is negative during this interval.

Note that the sudden changes in acceleration shown in Figure 2.7c are unphysical. Such instantaneous changes cannot occur in reality.

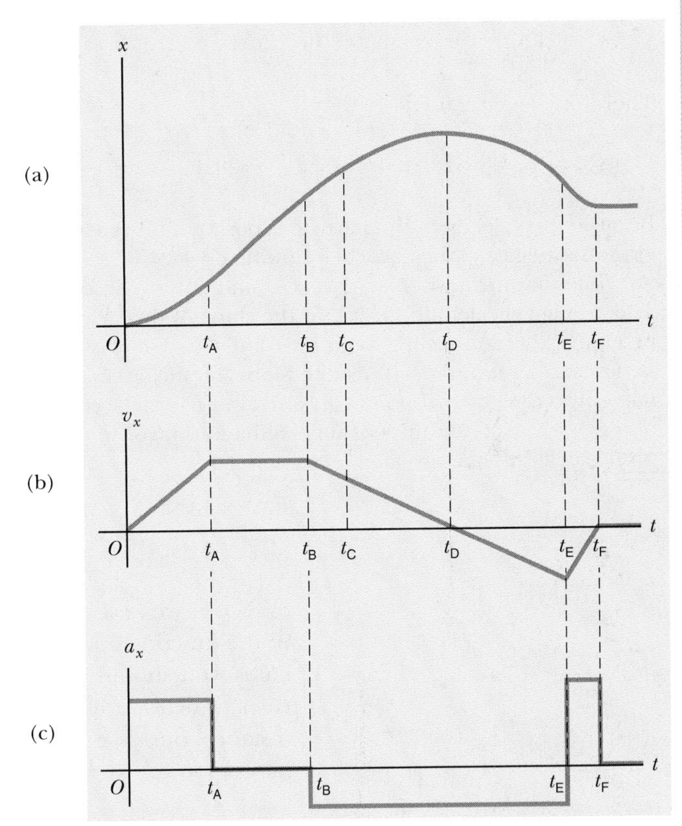

Figure 2.7 (Example 2.4) (a) Position–time graph for an object moving along the x axis. (b) The velocity–time graph for the object is obtained by measuring the slope of the position–time graph at each instant. (c) The acceleration–time graph for the object is obtained by measuring the slope of the velocity–time graph at each instant.

Example 2.5 Average and Instantaneous Acceleration

The velocity of a particle moving along the x axis varies in time according to the expression $v_x = (40 - 5t^2)$ m/s, where t is in seconds.

(A) Find the average acceleration in the time interval $t = 0$ to $t = 2.0$ s.

Solution Figure 2.8 is a v_x-t graph that was created from the velocity versus time expression given in the problem statement. Because the slope of the entire v_x-t curve is negative, we expect the acceleration to be negative.

We find the velocities at $t_i = t_A = 0$ and $t_f = t_B = 2.0$ s by substituting these values of t into the expression for the velocity:

$$v_{xA} = (40 - 5t_A{}^2) \text{ m/s} = [40 - 5(0)^2] \text{ m/s} = +40 \text{ m/s}$$

$$v_{xB} = (40 - 5t_B{}^2) \text{ m/s} = [40 - 5(2.0)^2] \text{ m/s} = +20 \text{ m/s}$$

Therefore, the average acceleration in the specified time interval $\Delta t = t_B - t_A = 2.0$ s is

$$\bar{a}_x = \frac{v_{xf} - v_{xi}}{t_f - t_i} = \frac{v_{xB} - v_{xA}}{t_B - t_A} = \frac{(20 - 40) \text{ m/s}}{(2.0 - 0) \text{ s}}$$

$$= \boxed{-10 \text{ m/s}^2}$$

The negative sign is consistent with our expectations—namely, that the average acceleration, which is represented by the slope of the line joining the initial and final points on the velocity–time graph, is negative.

(B) Determine the acceleration at $t = 2.0$ s.

Solution The velocity at any time t is $v_{xi} = (40 - 5t^2)$ m/s and the velocity at any later time $t + \Delta t$ is

$$v_{xf} = 40 - 5(t + \Delta t)^2 = 40 - 5t^2 - 10t\,\Delta t - 5(\Delta t)^2$$

Therefore, the change in velocity over the time interval Δt is

$$\Delta v_x = v_{xf} - v_{xi} = [-10t\,\Delta t - 5(\Delta t)^2] \text{ m/s}$$

Dividing this expression by Δt and taking the limit of the result as Δt approaches zero gives the acceleration at *any* time t:

$$a_x = \lim_{\Delta t \to 0} \frac{\Delta v_x}{\Delta t} = \lim_{\Delta t \to 0} (-10t - 5\Delta t) = -10t \text{ m/s}^2$$

Therefore, at $t = 2.0$ s,

$$a_x = (-10)(2.0) \text{ m/s}^2 = -20 \text{ m/s}^2$$

Because the velocity of the particle is positive and the acceleration is negative, the particle is slowing down.

Note that the answers to parts (A) and (B) are different. The average acceleration in (A) is the slope of the blue line in Figure 2.8 connecting points Ⓐ and Ⓑ. The instantaneous acceleration in (B) is the slope of the green line tangent to the curve at point Ⓑ. Note also that the acceleration is *not* constant in this example. Situations involving constant acceleration are treated in Section 2.5.

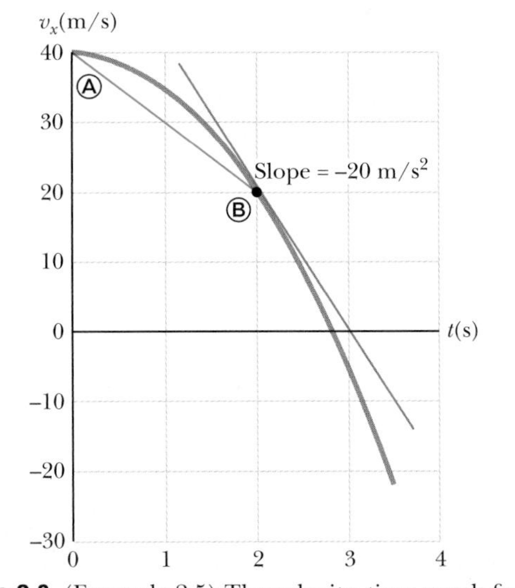

Figure 2.8 (Example 2.5) The velocity–time graph for a particle moving along the x axis according to the expression $v_x = (40 - 5t^2)$ m/s. The acceleration at $t = 2$ s is equal to the slope of the green tangent line at that time.

So far we have evaluated the derivatives of a function by starting with the definition of the function and then taking the limit of a specific ratio. If you are familiar with calculus, you should recognize that there are specific rules for taking derivatives. These rules, which are listed in Appendix B.6, enable us to evaluate derivatives quickly. For instance, one rule tells us that the derivative of any constant is zero. As another example, suppose x is proportional to some power of t, such as in the expression

$$x = At^n$$

where A and n are constants. (This is a very common functional form.) The derivative of x with respect to t is

$$\frac{dx}{dt} = nAt^{n-1}$$

Applying this rule to Example 2.5, in which $v_x = 40 - 5t^2$, we find that the acceleration is $a_x = dv_x/dt = -10t$.

2.4 Motion Diagrams

The concepts of velocity and acceleration are often confused with each other, but in fact they are quite different quantities. It is instructive to use motion diagrams to describe the velocity and acceleration while an object is in motion.

A *stroboscopic* photograph of a moving object shows several images of the object, taken as the strobe light flashes at a constant rate. Figure 2.9 represents three sets of strobe photographs of cars moving along a straight roadway in a single direction, from left to right. The time intervals between flashes of the stroboscope are equal in each part of the diagram. In order not to confuse the two vector quantities, we use red for velocity vectors and violet for acceleration vectors in Figure 2.9. The vectors are

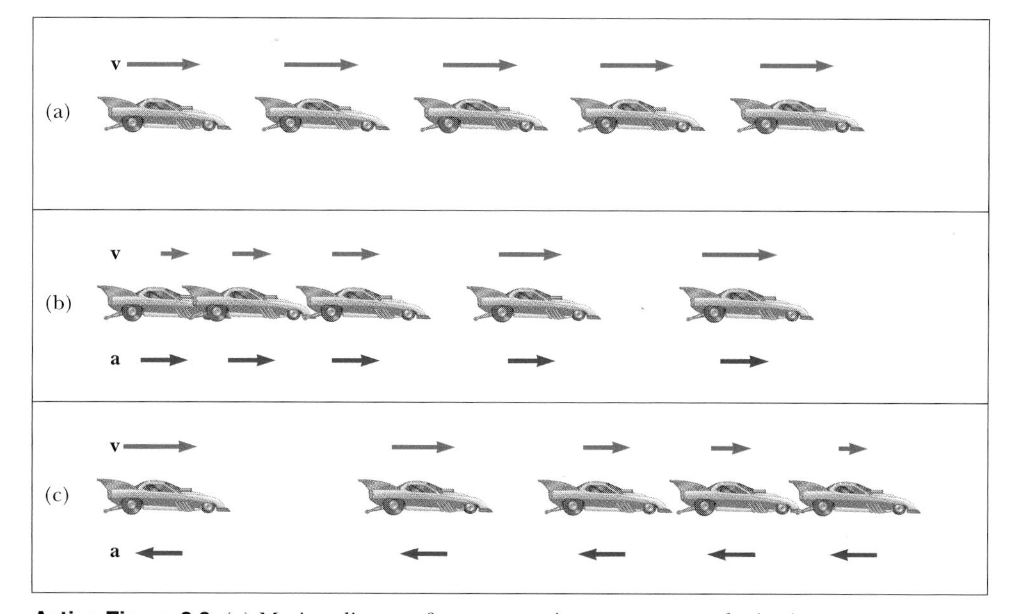

Active Figure 2.9 (a) Motion diagram for a car moving at constant velocity (zero acceleration). (b) Motion diagram for a car whose constant acceleration is in the direction of its velocity. The velocity vector at each instant is indicated by a red arrow, and the constant acceleration by a violet arrow. (c) Motion diagram for a car whose constant acceleration is in the direction *opposite* the velocity at each instant.

At the Active Figures link at http://www.pse6.com, you can select the constant acceleration and initial velocity of the car and observe pictorial and graphical representations of its motion.

sketched at several instants during the motion of the object. Let us describe the motion of the car in each diagram.

In Figure 2.9a, the images of the car are equally spaced, showing us that the car moves through the same displacement in each time interval. This is consistent with the car moving with *constant positive velocity* and *zero acceleration.* We could model the car as a particle and describe it as a particle moving with constant velocity.

In Figure 2.9b, the images become farther apart as time progresses. In this case, the velocity vector increases in time because the car's displacement between adjacent positions increases in time. This suggests that the car is moving with a *positive velocity* and a *positive acceleration.* The velocity and acceleration are in the same direction. In terms of our earlier force discussion, imagine a force pulling on the car in the same direction it is moving—it speeds up.

In Figure 2.9c, we can tell that the car slows as it moves to the right because its displacement between adjacent images decreases with time. In this case, this suggests that the car moves to the right with a constant negative acceleration. The velocity vector decreases in time and eventually reaches zero. From this diagram we see that the acceleration and velocity vectors are *not* in the same direction. The car is moving with a *positive velocity* but with a *negative acceleration.* (This type of motion is exhibited by a car that skids to a stop after applying its brakes.) The velocity and acceleration are in opposite directions. In terms of our earlier force discussion, imagine a force pulling on the car opposite to the direction it is moving—it slows down.

The violet acceleration vectors in Figures 2.9b and 2.9c are all of the same length. Thus, these diagrams represent motion with constant acceleration. This is an important type of motion that will be discussed in the next section.

Quick Quiz 2.4 Which of the following is true? (a) If a car is traveling eastward, its acceleration is eastward. (b) If a car is slowing down, its acceleration must be negative. (c) A particle with constant acceleration can never stop and stay stopped.

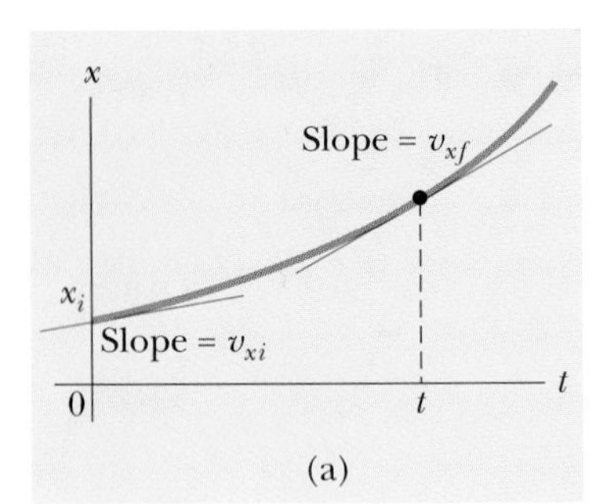

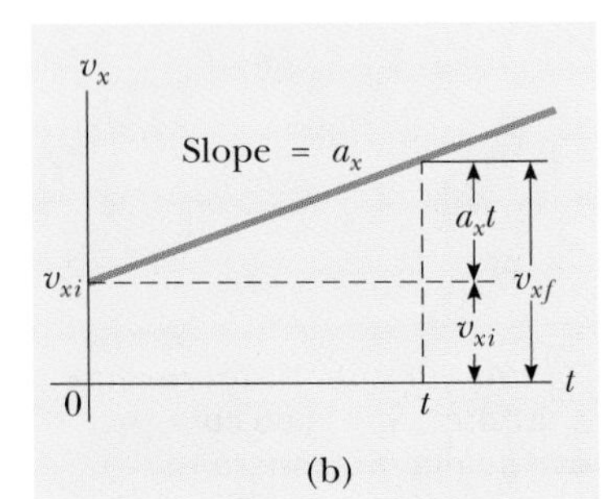

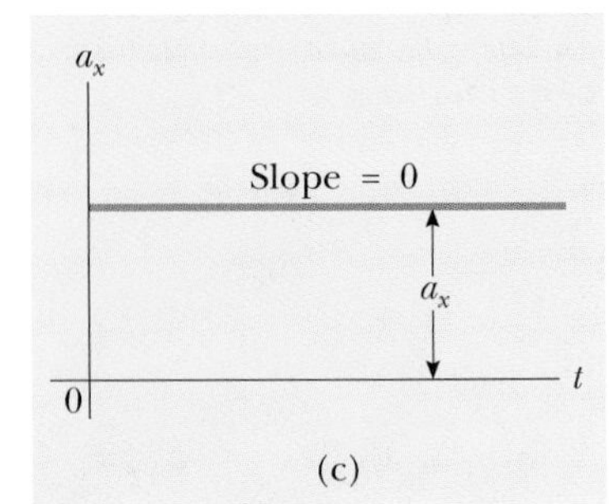

Active Figure 2.10 A particle moving along the x axis with constant acceleration a_x; (a) the position–time graph, (b) the velocity–time graph, and (c) the acceleration–time graph.

At the Active Figures link at http://www.pse6.com, you can adjust the constant acceleration and observe the effect on the position and velocity graphs.

2.5 One-Dimensional Motion with Constant Acceleration

If the acceleration of a particle varies in time, its motion can be complex and difficult to analyze. However, a very common and simple type of one-dimensional motion is that in which the acceleration is constant. When this is the case, the average acceleration $\overline{a}_x$ over any time interval is numerically equal to the instantaneous acceleration a_x at any instant within the interval, and the velocity changes at the same rate throughout the motion.

If we replace $\overline{a}_x$ by a_x in Equation 2.6 and take $t_i = 0$ and t_f to be any later time t, we find that

$$a_x = \frac{v_{xf} - v_{xi}}{t - 0}$$

or

$$v_{xf} = v_{xi} + a_x t \qquad \text{(for constant } a_x\text{)} \tag{2.9}$$

This powerful expression enables us to determine an object's velocity at *any* time t if we know the object's initial velocity v_{xi} and its (constant) acceleration a_x. A velocity–time graph for this constant-acceleration motion is shown in Figure 2.10b. The graph is a straight line, the (constant) slope of which is the acceleration a_x; this is consistent with the fact that $a_x = dv_x/dt$ is a constant. Note that the slope is positive; this indicates a positive acceleration. If the acceleration were negative, then the slope of the line in Figure 2.10b would be negative.

When the acceleration is constant, the graph of acceleration versus time (Fig. 2.10c) is a straight line having a slope of zero.

Because velocity at constant acceleration varies linearly in time according to Equation 2.9, we can express the average velocity in any time interval as the arithmetic mean of the initial velocity v_{xi} and the final velocity v_{xf}:

$$\overline{v}_x = \frac{v_{xi} + v_{xf}}{2} \qquad \text{(for constant } a_x\text{)} \tag{2.10}$$

Note that this expression for average velocity applies *only* in situations in which the acceleration is constant.

We can now use Equations 2.1, 2.2, and 2.10 to obtain the position of an object as a function of time. Recalling that Δx in Equation 2.2 represents $x_f - x_i$, and recognizing that $\Delta t = t_f - t_i = t - 0 = t$, we find

$$x_f - x_i = \overline{v}t = \tfrac{1}{2}(v_{xi} + v_{xf})t$$

Position as a function of velocity and time

$$x_f = x_i + \tfrac{1}{2}(v_{xi} + v_{xf})t \qquad \text{(for constant } a_x\text{)} \tag{2.11}$$

This equation provides the final position of the particle at time t in terms of the initial and final velocities.

We can obtain another useful expression for the position of a particle moving with constant acceleration by substituting Equation 2.9 into Equation 2.11:

$$x_f = x_i + \tfrac{1}{2}[v_{xi} + (v_{xi} + a_x t)]t$$

Position as a function of time

$$x_f = x_i + v_{xi}t + \tfrac{1}{2}a_x t^2 \qquad \text{(for constant } a_x\text{)} \tag{2.12}$$

This equation provides the final position of the particle at time t in terms of the initial velocity and the acceleration.

The position–time graph for motion at constant (positive) acceleration shown in Figure 2.10a is obtained from Equation 2.12. Note that the curve is a parabola.

The slope of the tangent line to this curve at $t = 0$ equals the initial velocity v_{xi}, and the slope of the tangent line at any later time t equals the velocity v_{xf} at that time.

Finally, we can obtain an expression for the final velocity that does not contain time as a variable by substituting the value of t from Equation 2.9 into Equation 2.11:

$$x_f = x_i + \frac{1}{2}(v_{xi} + v_{xf})\left(\frac{v_{xf} - v_{xi}}{a_x}\right) = \frac{v_{xf}^2 - v_{xi}^2}{2a_x}$$

$$v_{xf}^2 = v_{xi}^2 + 2a_x(x_f - x_i) \qquad \text{(for constant } a_x\text{)} \tag{2.13}$$

Velocity as a function of position

This equation provides the final velocity in terms of the acceleration and the displacement of the particle.

For motion at *zero* acceleration, we see from Equations 2.9 and 2.12 that

$$\left.\begin{aligned} v_{xf} &= v_{xi} = v_x \\ x_f &= x_i + v_x t \end{aligned}\right\} \quad \text{when } a_x = 0$$

That is, when the acceleration of a particle is zero, its velocity is constant and its position changes linearly with time.

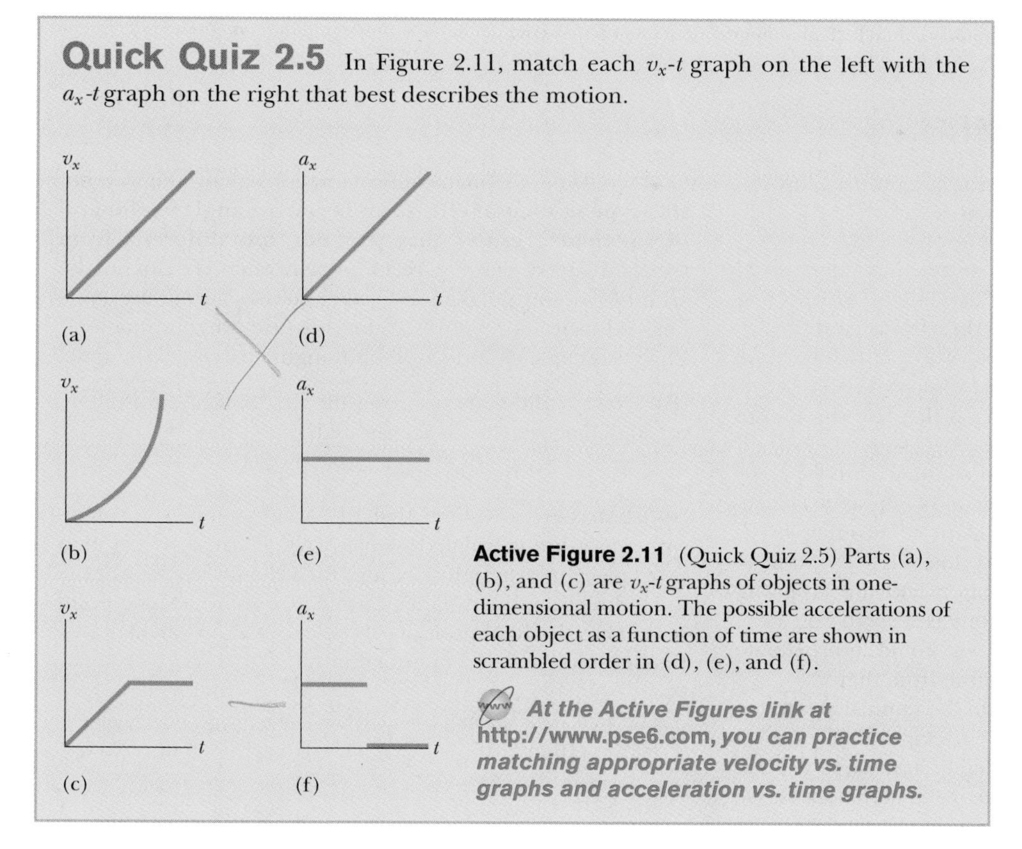

Quick Quiz 2.5 In Figure 2.11, match each v_x-t graph on the left with the a_x-t graph on the right that best describes the motion.

Active Figure 2.11 (Quick Quiz 2.5) Parts (a), (b), and (c) are v_x-t graphs of objects in one-dimensional motion. The possible accelerations of each object as a function of time are shown in scrambled order in (d), (e), and (f).

At the Active Figures link at http://www.pse6.com, you can practice matching appropriate velocity vs. time graphs and acceleration vs. time graphs.

Equations 2.9 through 2.13 are **kinematic equations that may be used to solve any problem involving one-dimensional motion at constant acceleration.** Keep in mind that these relationships were derived from the definitions of velocity and

Table 2.2

Kinematic Equations for Motion of a Particle Under Constant Acceleration

Equation	Information Given by Equation
$v_{xf} = v_{xi} + a_x t$	Velocity as a function of time
$x_f = x_i + \frac{1}{2}(v_{xi} + v_{xf})t$	Position as a function of velocity and time
$x_f = x_i + v_{xi}t + \frac{1}{2}a_x t^2$	Position as a function of time
$v_{xf}^2 = v_{xi}^2 + 2a_x(x_f - x_i)$	Velocity as a function of position

Note: Motion is along the x axis.

acceleration, together with some simple algebraic manipulations and the requirement that the acceleration be constant.

The four kinematic equations used most often are listed in Table 2.2 for convenience. The choice of which equation you use in a given situation depends on what you know beforehand. Sometimes it is necessary to use two of these equations to solve for two unknowns. For example, suppose initial velocity v_{xi} and acceleration a_x are given. You can then find (1) the velocity at time t, using $v_{xf} = v_{xi} + a_x t$ and (2) the position at time t, using $x_f = x_i + v_{xi}t + \frac{1}{2}a_x t^2$. You should recognize that the quantities that vary during the motion are position, velocity, and time.

You will gain a great deal of experience in the use of these equations by solving a number of exercises and problems. Many times you will discover that more than one method can be used to obtain a solution. Remember that these equations of kinematics *cannot* be used in a situation in which the acceleration varies with time. They can be used only when the acceleration is constant.

Example 2.6 Entering the Traffic Flow

(A) Estimate your average acceleration as you drive up the entrance ramp to an interstate highway.

Solution This problem involves more than our usual amount of estimating! We are trying to come up with a value of a_x, but that value is hard to guess directly. The other variables involved in kinematics are position, velocity, and time. Velocity is probably the easiest one to approximate. Let us assume a final velocity of 100 km/h, so that you can merge with traffic. We multiply this value by (1 000 m/1 km) to convert kilometers to meters and then multiply by (1 h/3 600 s) to convert hours to seconds. These two calculations together are roughly equivalent to dividing by 3. In fact, let us just say that the final velocity is $v_{xf} \approx 30$ m/s. (Remember, this type of approximation and the dropping of digits when performing estimations is okay. If you were starting with U.S. customary units, you could approximate 1 mi/h as roughly 0.5 m/s and continue from there.)

Now we assume that you started up the ramp at about one third your final velocity, so that $v_{xi} \approx 10$ m/s. Finally, we assume that it takes about 10 s to accelerate from v_{xi} to v_{xf}, basing this guess on our previous experience in automobiles. We can then find the average acceleration, using Equation 2.6:

$$\bar{a}_x = \frac{v_{xf} - v_{xi}}{t} \approx \frac{30\text{ m/s} - 10\text{ m/s}}{10\text{ s}}$$

$$= 2\text{ m/s}^2$$

Granted, we made many approximations along the way, but **this type of mental effort can be surprisingly useful and often yields results that are not too different from those derived from careful measurements.** Do not be afraid to attempt making educated guesses and doing some fairly drastic number rounding to simplify estimations. Physicists engage in this type of thought analysis all the time.

(B) How far did you go during the first half of the time interval during which you accelerated?

Solution Let us assume that the acceleration is constant, with the value calculated in part (A). Because the motion takes place in a straight line and the velocity is always in the same direction, the distance traveled from the starting point is equal to the final position of the car. We can calculate the final position at 5 s from Equation 2.12:

$$x_f = x_i + v_{xi}t + \frac{1}{2}a_x t^2$$

$$\approx 0 + (10\text{ m/s})(5\text{ s}) + \frac{1}{2}(2\text{ m/s}^2)(5\text{ s})^2 = 50\text{ m} + 25\text{ m}$$

$$= 75\text{ m}$$

This result indicates that if you had not accelerated, your initial velocity of 10 m/s would have resulted in a 50-m movement up the ramp during the first 5 s. The additional 25 m is the result of your increasing velocity during that interval.

Example 2.7 Carrier Landing

A jet lands on an aircraft carrier at 140 mi/h ($\approx$63 m/s).

(A) What is its acceleration (assumed constant) if it stops in 2.0 s due to an arresting cable that snags the airplane and brings it to a stop?

Solution We define our x axis as the direction of motion of the jet. A careful reading of the problem reveals that in addition to being given the initial speed of 63 m/s, we also know that the final speed is zero. We also note that we have no information about the change in position of the jet while it is slowing down. Equation 2.9 is the only equation in Table 2.2 that does not involve position, and so we use it to find the acceleration of the jet, modeled as a particle:

$$a_x = \frac{v_{xf} - v_{xi}}{t} \approx \frac{0 - 63 \text{ m/s}}{2.0 \text{ s}}$$

$$= -31 \text{ m/s}^2$$

(B) If the plane touches down at position $x_i = 0$, what is the final position of the plane?

Solution We can now use any of the other three equations in Table 2.2 to solve for the final position. Let us choose Equation 2.11:

$$x_f = x_i + \tfrac{1}{2}(v_{xi} + v_{xf})t = 0 + \tfrac{1}{2}(63 \text{ m/s} + 0)(2.0 \text{ s})$$

$$= 63 \text{ m}$$

If the plane travels much farther than this, it might fall into the ocean. The idea of using arresting cables to slow down landing aircraft and enable them to land safely on ships originated at about the time of the first World War. The cables are still a vital part of the operation of modern aircraft carriers.

What If? Suppose the plane lands on the deck of the aircraft carrier with a speed higher than 63 m/s but with the same acceleration as that calculated in part (A). How will that change the answer to part (B)?

Answer If the plane is traveling faster at the beginning, it will stop farther away from its starting point, so the answer to part (B) should be larger. Mathematically, we see in Equation 2.11 that if v_{xi} is larger, then x_f will be larger.

If the landing deck has a length of 75 m, we can find the maximum initial speed with which the plane can land and still come to rest on the deck at the given acceleration from Equation 2.13:

$$v_{xf}^2 = v_{xi}^2 + 2a_x(x_f - x_i)$$

$$\rightarrow v_{xi} = \sqrt{v_{xf}^2 - 2a_x(x_f - x_i)}$$

$$= \sqrt{0 - 2(-31 \text{ m/s}^2)(75 \text{ m} - 0)}$$

$$= 68 \text{ m/s}$$

Example 2.8 Watch Out for the Speed Limit! Interactive

A car traveling at a constant speed of 45.0 m/s passes a trooper hidden behind a billboard. One second after the speeding car passes the billboard, the trooper sets out from the billboard to catch it, accelerating at a constant rate of 3.00 m/s^2. How long does it take her to overtake the car?

Solution Let us model the car and the trooper as particles. A sketch (Fig. 2.12) helps clarify the sequence of events.

First, we write expressions for the position of each vehicle as a function of time. It is convenient to choose the position of the billboard as the origin and to set $t_B = 0$ as the time the trooper begins moving. At that instant, the car has already traveled a distance of 45.0 m because it has traveled at a constant speed of $v_x = 45.0$ m/s for 1 s. Thus, the initial position of the speeding car is $x_B = 45.0$ m.

Because the car moves with constant speed, its acceleration is zero. Applying Equation 2.12 (with $a_x = 0$) gives for the car's position at any time t:

$$x_{\text{car}} = x_B + v_{x\,\text{car}}t = 45.0 \text{ m} + (45.0 \text{ m/s})t$$

A quick check shows that at $t = 0$, this expression gives the car's correct initial position when the trooper begins to move: $x_{\text{car}} = x_B = 45.0$ m.

The trooper starts from rest at $t_B = 0$ and accelerates at 3.00 m/s^2 away from the origin. Hence, her position at any time t can be found from Equation 2.12:

$$x_f = x_i + v_{xi}t + \tfrac{1}{2}a_x t^2$$

$$x_{\text{trooper}} = 0 + (0)t + \tfrac{1}{2}a_x t^2 = \tfrac{1}{2}(3.00 \text{ m/s}^2)t^2$$

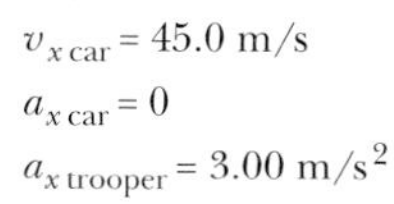

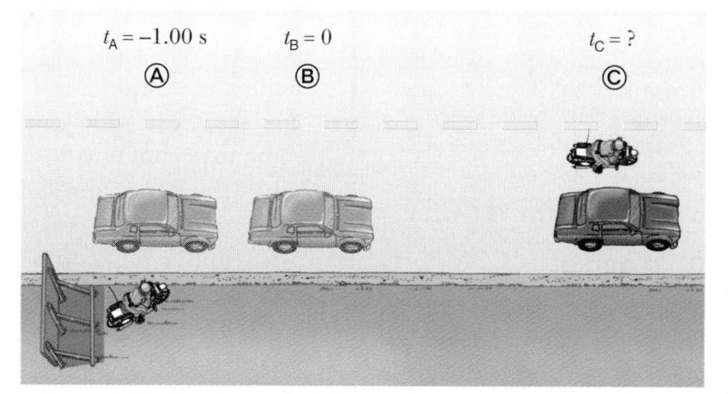

Figure 2.12 (Example 2.8) A speeding car passes a hidden trooper.

The trooper overtakes the car at the instant her position matches that of the car, which is position ©:

$$x_{\text{trooper}} = x_{\text{car}}$$

$$\tfrac{1}{2}(3.00 \text{ m/s}^2)t^2 = 45.0 \text{ m} + (45.0 \text{ m/s})t$$

This gives the quadratic equation

$$1.50t^2 - 45.0t - 45.0 = 0$$

The positive solution of this equation is $t =$ 31.0 s.

(For help in solving quadratic equations, see Appendix B.2.)

What If? What if the trooper had a more powerful motorcycle with a larger acceleration? How would that change the time at which the trooper catches the car?

Answer If the motorcycle has a larger acceleration, the trooper will catch up to the car sooner, so the answer for the time will be less than 31 s. Mathematically, let us cast the final quadratic equation above in terms of the parameters in the problem:

$$\tfrac{1}{2}a_x t^2 - v_{x\,\text{car}}t - x_{\text{B}} = 0$$

The solution to this quadratic equation is,

$$t = \frac{v_{x\,\text{car}} \pm \sqrt{v_{x\,\text{car}}^2 + 2a_x x_{\text{B}}}}{a_x}$$

$$= \frac{v_{x\,\text{car}}}{a_x} + \sqrt{\frac{v_{x\,\text{car}}^2}{a_x^{\,2}} + \frac{2x_{\text{B}}}{a_x}}$$

where we have chosen the positive sign because that is the only choice consistent with a time $t > 0$. Because all terms on the right side of the equation have the acceleration a_x in the denominator, increasing the acceleration will decrease the time at which the trooper catches the car.

You can study the motion of the car and trooper for various velocities of the car at the Interactive Worked Example link at http://www.pse6.com.

Galileo Galilei

Italian physicist and astronomer (1564–1642)

Galileo formulated the laws that govern the motion of objects in free fall and made many other significant discoveries in physics and astronomy. Galileo publicly defended Nicholaus Copernicus's assertion that the Sun is at the center of the Universe (the heliocentric system). He published *Dialogue Concerning Two New World Systems* to support the Copernican model, a view which the Church declared to be heretical. *(North Wind)*

PITFALL PREVENTION

2.6 *g* and g

Be sure not to confuse the italicized symbol *g* for free-fall acceleration with the nonitalicized symbol g used as the abbreviation for "gram."

2.6 Freely Falling Objects

It is well known that, in the absence of air resistance, all objects dropped near the Earth's surface fall toward the Earth with the same constant acceleration under the influence of the Earth's gravity. It was not until about 1600 that this conclusion was accepted. Before that time, the teachings of the great philosopher Aristotle (384–322 B.C.) had held that heavier objects fall faster than lighter ones.

The Italian Galileo Galilei (1564–1642) originated our present-day ideas concerning falling objects. There is a legend that he demonstrated the behavior of falling objects by observing that two different weights dropped simultaneously from the Leaning Tower of Pisa hit the ground at approximately the same time. Although there is some doubt that he carried out this particular experiment, it is well established that Galileo performed many experiments on objects moving on inclined planes. In his experiments he rolled balls down a slight incline and measured the distances they covered in successive time intervals. The purpose of the incline was to reduce the acceleration; with the acceleration reduced, Galileo was able to make accurate measurements of the time intervals. By gradually increasing the slope of the incline, he was finally able to draw conclusions about freely falling objects because a freely falling ball is equivalent to a ball moving down a vertical incline.

You might want to try the following experiment. Simultaneously drop a coin and a crumpled-up piece of paper from the same height. If the effects of air resistance are negligible, both will have the same motion and will hit the floor at the same time. In the idealized case, in which air resistance is absent, such motion is referred to as *free-fall.* If this same experiment could be conducted in a vacuum, in which air resistance is truly negligible, the paper and coin would fall with the same acceleration even when the paper is not crumpled. On August 2, 1971, such a demonstration was conducted on the Moon by astronaut David Scott. He simultaneously released a hammer and a feather, and they fell together to the lunar surface. This demonstration surely would have pleased Galileo!

When we use the expression *freely falling object,* we do not necessarily refer to an object dropped from rest. **A freely falling object is any object moving freely under the influence of gravity alone, regardless of its initial motion. Objects thrown upward or downward and those released from rest are all falling freely once they**

are released. Any freely falling object experiences an acceleration directed *downward*, regardless of its initial motion.

We shall denote the magnitude of the *free-fall acceleration* by the symbol g. The value of g near the Earth's surface decreases with increasing altitude. Furthermore, slight variations in g occur with changes in latitude. It is common to define "up" as the $+y$ direction and to use y as the position variable in the kinematic equations. At the Earth's surface, the value of g is approximately 9.80 m/s^2. Unless stated otherwise, we shall use this value for g when performing calculations. For making quick estimates, use $g = 10$ m/s^2.

If we neglect air resistance and assume that the free-fall acceleration does not vary with altitude over short vertical distances, then the motion of a freely falling object moving vertically is equivalent to motion in one dimension under constant acceleration. Therefore, the equations developed in Section 2.5 for objects moving with constant acceleration can be applied. The only modification that we need to make in these equations for freely falling objects is to note that the motion is in the vertical direction (the y direction) rather than in the horizontal direction (x) and that the acceleration is downward and has a magnitude of 9.80 m/s^2. Thus, we always choose $a_y = -g = -9.80$ m/s^2, where the negative sign means that the acceleration of a freely falling object is downward. In Chapter 13 we shall study how to deal with variations in g with altitude.

Quick Quiz 2.6 A ball is thrown upward. While the ball is in free fall, does its acceleration (a) increase (b) decrease (c) increase and then decrease (d) decrease and then increase (e) remain constant?

Quick Quiz 2.7 After a ball is thrown upward and is in the air, its speed (a) increases (b) decreases (c) increases and then decreases (d) decreases and then increases (e) remains the same.

PITFALL PREVENTION

2.7 The Sign of g

Keep in mind that g is a *positive number*—it is tempting to substitute -9.80 m/s^2 for g, but resist the temptation. Downward gravitational acceleration is indicated explicitly by stating the acceleration as $a_y = -g$.

2.8 Acceleration at the Top of The Motion

It is a common misconception that the acceleration of a projectile at the top of its trajectory is zero. While the velocity at the top of the motion of an object thrown upward momentarily goes to zero, *the acceleration is still that due to gravity* at this point. If the velocity and acceleration were both zero, the projectile would stay at the top!

Conceptual Example 2.9 The Daring Sky Divers

A sky diver jumps out of a hovering helicopter. A few seconds later, another sky diver jumps out, and they both fall along the same vertical line. Ignore air resistance, so that both sky divers fall with the same acceleration. Does the difference in their speeds stay the same throughout the fall? Does the vertical distance between them stay the same throughout the fall?

Solution At any given instant, the speeds of the divers are different because one had a head start. In any time interval Δt after this instant, however, the two divers increase their speeds by the same amount because they have the same acceleration. Thus, the difference in their speeds remains the same throughout the fall.

The first jumper always has a greater speed than the second. Thus, in a given time interval, the first diver covers a greater distance than the second. Consequently, the separation distance between them increases.

Example 2.10 Describing the Motion of a Tossed Ball

A ball is tossed straight up at 25 m/s. Estimate its velocity at 1-s intervals.

Solution Let us choose the upward direction to be positive. Regardless of whether the ball is moving upward or downward, its vertical velocity changes by approximately -10 m/s for every second it remains in the air. It starts out at 25 m/s. After 1 s has elapsed, it is still moving upward but at 15 m/s because its acceleration is downward (downward acceleration causes its velocity to decrease). After another second, its upward velocity has dropped to 5 m/s. Now comes the tricky part—after another half second, its velocity is zero. The ball has gone as high as it will go. After the last half of this 1-s interval, the ball is moving at -5 m/s. (The negative sign tells us that the ball is now moving in the negative direction, that is, *downward.* Its velocity has changed from $+5$ m/s to -5 m/s during that 1-s interval. The change in velocity is still -5 m/s $-$ $(+5$ m/s$)$ $= -10$ m/s in that second.) It continues downward, and after another 1 s has elapsed, it is falling at a velocity of -15 m/s. Finally, after another 1 s, it has reached its original starting point and is moving downward at -25 m/s.

Conceptual Example 2.11 Follow the Bouncing Ball

A tennis ball is dropped from shoulder height (about 1.5 m) and bounces three times before it is caught. Sketch graphs of its position, velocity, and acceleration as functions of time, with the $+y$ direction defined as upward.

Solution For our sketch let us stretch things out horizontally so that we can see what is going on. (Even if the ball were moving horizontally, this motion would not affect its vertical motion.)

From Figure 2.13a we see that the ball is in contact with the floor at points Ⓑ, Ⓓ, and Ⓕ. Because the velocity of the ball changes from negative to positive three times during these bounces (Fig. 2.13b), the slope of the position–time graph must change in the same way. Note that the time interval between bounces decreases. Why is that?

During the rest of the ball's motion, the slope of the velocity–time graph in Fig. 2.13b should be $-9.80\ \text{m/s}^2$. The acceleration–time graph is a horizontal line at these times because the acceleration does not change when the ball is in free fall. When the ball is in contact with the floor, the velocity changes substantially during a very short time

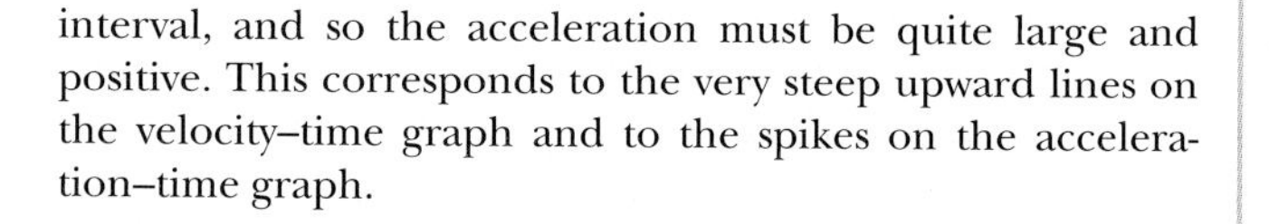

interval, and so the acceleration must be quite large and positive. This corresponds to the very steep upward lines on the velocity–time graph and to the spikes on the acceleration–time graph.

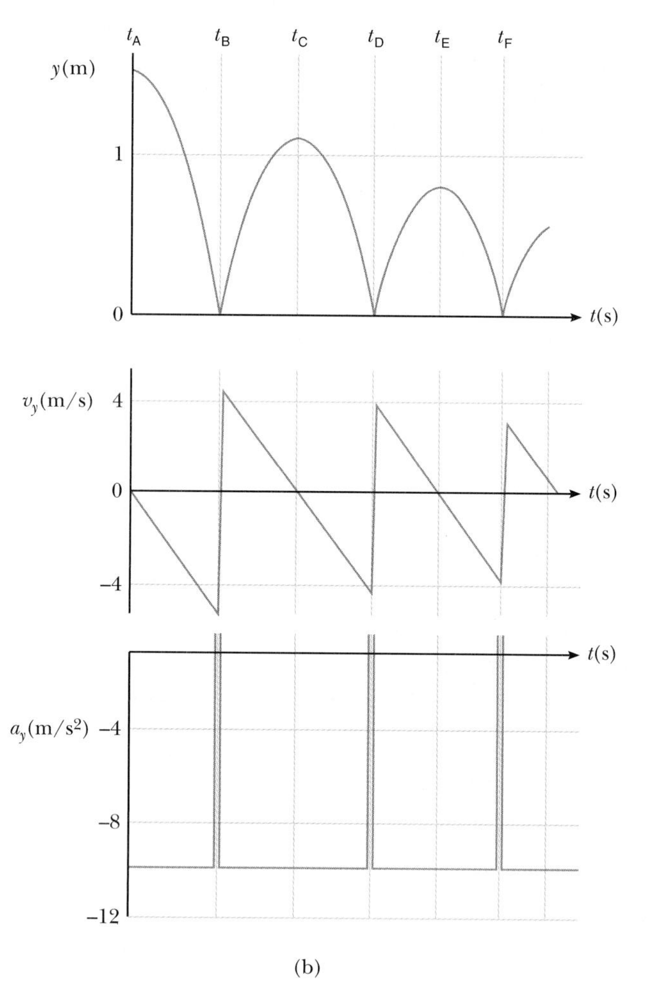

Active Figure 2.13 (Conceptual Example 2.11) (a) A ball is dropped from a height of 1.5 m and bounces from the floor. (The horizontal motion is not considered here because it does not affect the vertical motion.) (b) Graphs of position, velocity, and acceleration versus time.

At the Active Figures link at* http://www.pse6.com, *you can adjust both the value for g and the amount of "bounce" of the ball, and observe the resulting motion of the ball both pictorially and graphically.

Quick Quiz 2.8 Which values represent the ball's vertical velocity and acceleration at points Ⓐ, Ⓒ, and Ⓔ in Figure 2.13a?

(a) $v_y = 0,\ a_y = -9.80\ \text{m/s}^2$

(b) $v_y = 0,\ a_y = 9.80\ \text{m/s}^2$

(c) $v_y = 0,\ a_y = 0$

(d) $v_y = -9.80\ \text{m/s},\ a_y = 0$

Example 2.12 Not a Bad Throw for a Rookie!

Interactive

A stone thrown from the top of a building is given an initial velocity of 20.0 m/s straight upward. The building is 50.0 m high, and the stone just misses the edge of the roof on its way down, as shown in Figure 2.14. Using $t_A = 0$ as the time the stone leaves the thrower's hand at position Ⓐ, determine **(A)** the time at which the stone reaches its maximum height, **(B)** the maximum height, **(C)** the time at which the stone returns to the height from which it was thrown, **(D)** the velocity of the stone at this instant, and **(E)** the velocity and position of the stone at $t = 5.00$ s.

Solution (A) As the stone travels from Ⓐ to Ⓑ, its velocity must change by 20 m/s because it stops at Ⓑ. Because gravity causes vertical velocities to change by about 10 m/s for every second of free fall, it should take the stone about 2 s to go from Ⓐ to Ⓑ in our drawing. To calculate the exact time t_B at which the stone reaches maximum height, we use Equation 2.9, $v_{yB} = v_{yA} + a_y t$, noting that $v_{yB} = 0$ and setting the start of our clock readings at $t_A = 0$:

$$0 = 20.0 \text{ m/s} + (-9.80 \text{ m/s}^2)t$$

$$t = t_B = \frac{20.0 \text{ m/s}}{9.80 \text{ m/s}^2} = 2.04 \text{ s}$$

Our estimate was pretty close.

(B) Because the average velocity for this first interval is 10 m/s (the average of 20 m/s and 0 m/s) and because it travels for about 2 s, we expect the stone to travel about 20 m. By substituting our time into Equation 2.12, we can find the maximum height as measured from the position of the thrower, where we set $y_A = 0$:

$$y_{\text{max}} = y_B = y_A + v_{xA}t + \tfrac{1}{2}a_y t^2$$

$$y_B = 0 + (20.0 \text{ m/s})(2.04 \text{ s}) + \tfrac{1}{2}(-9.80 \text{ m/s}^2)(2.04 \text{ s})^2$$

$$= 20.4 \text{ m}$$

Our free-fall estimates are very accurate.

(C) There is no reason to believe that the stone's motion from Ⓑ to Ⓒ is anything other than the reverse of its motion from Ⓐ to Ⓑ. The motion from Ⓐ to Ⓒ is symmetric. Thus, the time needed for it to go from Ⓐ to Ⓒ should be twice the time needed for it to go from Ⓐ to Ⓑ. When the stone is back at the height from which it was thrown (position Ⓒ), the y coordinate is again zero. Using Equation 2.12, with $y_C = 0$, we obtain

$$y_C = y_A + v_{yA}t + \tfrac{1}{2}a_y t^2$$

$$0 = 0 + 20.0t - 4.90t^2$$

This is a quadratic equation and so has two solutions for $t = t_C$. The equation can be factored to give

$$t(20.0 - 4.90t) = 0$$

One solution is $t = 0$, corresponding to the time the stone starts its motion. The other solution is $t = 4.08$ s, which

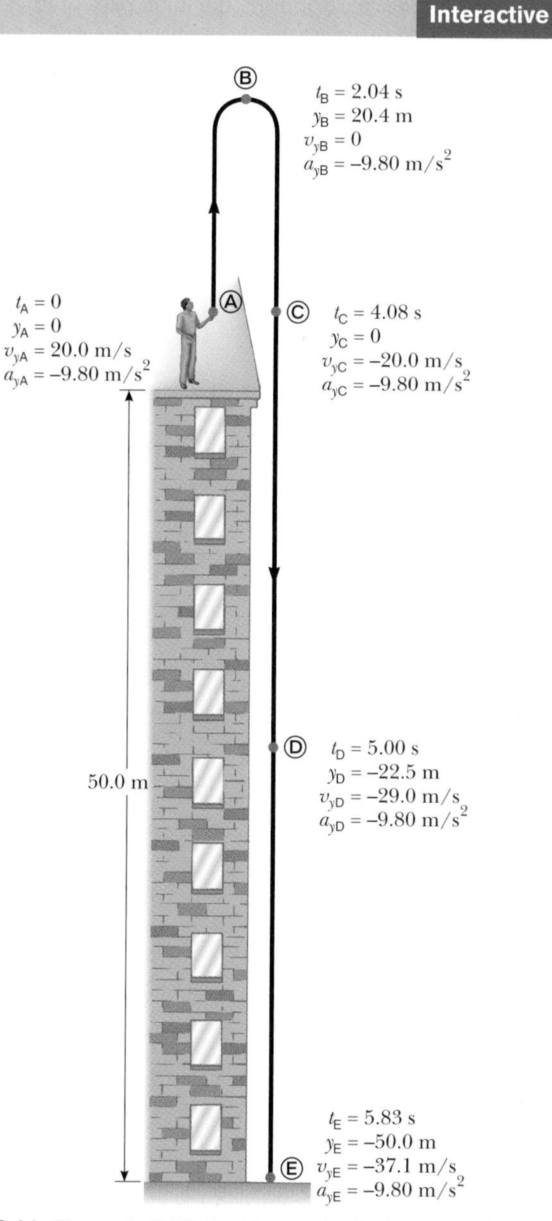

Figure 2.14 (Example 2.12) Position and velocity versus time for a freely falling stone thrown initially upward with a velocity $v_{yi} = 20.0$ m/s.

is the solution we are after. Notice that it is double the value we calculated for t_B.

(D) Again, we expect everything at Ⓒ to be the same as it is at Ⓐ, except that the velocity is now in the opposite direction. The value for t found in (c) can be inserted into Equation 2.9 to give

$$v_{yC} = v_{yA} + a_y t = 20.0 \text{ m/s} + (-9.80 \text{ m/s}^2)(4.08 \text{ s})$$

$$= -20.0 \text{ m/s}$$

The velocity of the stone when it arrives back at its original height is equal in magnitude to its initial velocity but opposite in direction.

(E) For this part we ignore the first part of the motion (Ⓐ→Ⓑ) and consider what happens as the stone falls from position Ⓑ, where it has zero vertical velocity, to position Ⓓ. We define the initial time as $t_B = 0$. Because the given time for this part of the motion relative to our new zero of time is 5.00 s − 2.04 s = 2.96 s, we estimate that the acceleration due to gravity will have changed the speed by about 30 m/s. We can calculate this from Equation 2.9, where we take $t = 2.96$ s:

$$v_{yD} = v_{yB} + a_y t = 0 \text{ m/s} + (-9.80 \text{ m/s}^2)(2.96 \text{ s})$$
$$= -29.0 \text{ m/s}$$

We could just as easily have made our calculation between positions Ⓐ (where we return to our original initial time $t_A = 0$) and Ⓓ:

$$v_{yD} = v_{yA} + a_y t = 20.0 \text{ m/s} + (-9.80 \text{ m/s}^2)(5.00 \text{ s})$$
$$= -29.0 \text{ m/s}$$

To further demonstrate that we can choose different initial instants of time, let us use Equation 2.12 to find the position of the stone at $t_D = 5.00$ s (with respect to $t_A = 0$) by defining a new initial instant, $t_C = 0$:

$$y_D = y_C + v_{yC}t + \tfrac{1}{2}a_y t^2$$
$$= 0 + (-20.0 \text{ m/s})(5.00 \text{ s} - 4.08 \text{ s})$$
$$+ \tfrac{1}{2}(-9.80 \text{ m/s}^2)(5.00 \text{ s} - 4.08 \text{ s})^2$$
$$= -22.5 \text{ m}$$

What If? **What if the building were 30.0 m tall instead of 50.0 m tall? Which answers in parts (A) to (E) would change?**

Answer None of the answers would change. All of the motion takes place in the air, and the stone does not interact with the ground during the first 5.00 s. (Notice that even for a 30.0-m tall building, the stone is above the ground at $t = 5.00$ s.) Thus, the height of the building is not an issue. Mathematically, if we look back over our calculations, we see that we never entered the height of the building into any equation.

You can study the motion of the thrown ball at the Interactive Worked Example link at http://www.pse6.com.

2.7 Kinematic Equations Derived from Calculus

This section assumes the reader is familiar with the techniques of integral calculus. If you have not yet studied integration in your calculus course, you should skip this section or cover it after you become familiar with integration.

The velocity of a particle moving in a straight line can be obtained if its position as a function of time is known. Mathematically, the velocity equals the derivative of the position with respect to time. It is also possible to find the position of a particle if its velocity is known as a function of time. In calculus, the procedure used to perform this task is referred to either as *integration* or as finding the *antiderivative.* Graphically, it is equivalent to finding the area under a curve.

Suppose the v_x-t graph for a particle moving along the x axis is as shown in Figure 2.15. Let us divide the time interval $t_f - t_i$ into many small intervals, each of

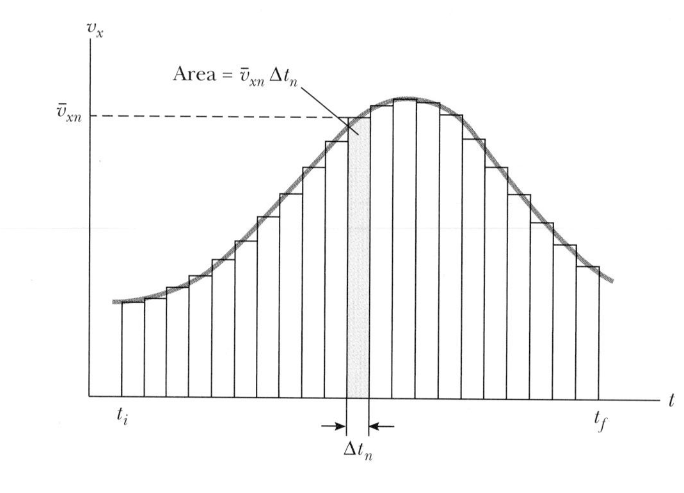

Figure 2.15 Velocity versus time for a particle moving along the x axis. The area of the shaded rectangle is equal to the displacement Δx in the time interval Δt_n, while the total area under the curve is the total displacement of the particle.

duration Δt_n. From the definition of average velocity we see that the displacement during any small interval, such as the one shaded in Figure 2.15, is given by $\Delta x_n = \overline{v}_{xn}\,\Delta t_n$ where $\overline{v}_{xn}$ is the average velocity in that interval. Therefore, the displacement during this small interval is simply the area of the shaded rectangle. The total displacement for the interval $t_f - t_i$ is the sum of the areas of all the rectangles:

$$\Delta x = \sum_n \overline{v}_{xn}\,\Delta t_n$$

where the symbol Σ (upper case Greek sigma) signifies a sum over all terms, that is, over all values of n. In this case, the sum is taken over all the rectangles from t_i to t_f. Now, as the intervals are made smaller and smaller, the number of terms in the sum increases and the sum approaches a value equal to the area under the velocity–time graph. Therefore, in the limit $n \rightarrow \infty$, or $\Delta t_n \rightarrow 0$, the displacement is

$$\Delta x = \lim_{\Delta t_n \to 0} \sum_n v_{xn}\,\Delta t_n \tag{2.14}$$

or

Displacement = area under the v_x-t graph

Note that we have replaced the average velocity $\overline{v}_{xn}$ with the instantaneous velocity v_{xn} in the sum. As you can see from Figure 2.15, this approximation is valid in the limit of very small intervals. Therefore if we know the v_x-t graph for motion along a straight line, we can obtain the displacement during any time interval by measuring the area under the curve corresponding to that time interval.

The limit of the sum shown in Equation 2.14 is called a **definite integral** and is written

Definite integral

$$\lim_{\Delta t_n \to 0} \sum_n v_{xn}\,\Delta t_n = \int_{t_i}^{t_f} v_x(t)\,dt \tag{2.15}$$

where $v_x(t)$ denotes the velocity at any time t. If the explicit functional form of $v_x(t)$ is known and the limits are given, then the integral can be evaluated. Sometimes the v_x-t graph for a moving particle has a shape much simpler than that shown in Figure 2.15. For example, suppose a particle moves at a constant velocity v_{xi}. In this case, the v_x-t graph is a horizontal line, as in Figure 2.16, and the displacement of the particle during the time interval Δt is simply the area of the shaded rectangle:

$$\Delta x = v_{xi}\,\Delta t \qquad (\text{when } v_x = v_{xi} = \text{constant})$$

As another example, consider a particle moving with a velocity that is proportional to t, as in Figure 2.17. Taking $v_x = a_x t$, where a_x is the constant of proportionality (the

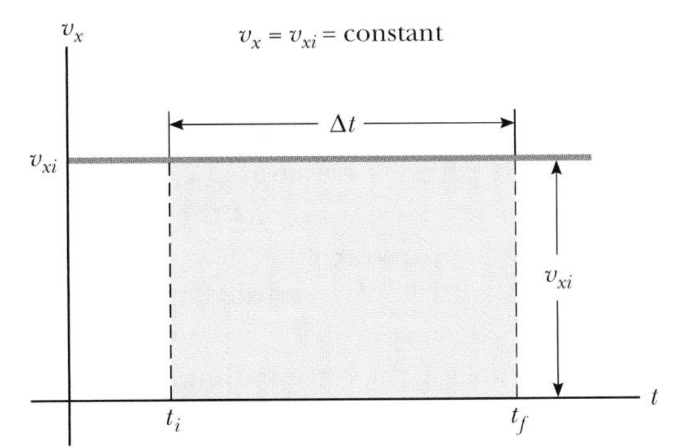

Figure 2.16 The velocity–time curve for a particle moving with constant velocity v_{xi}. The displacement of the particle during the time interval $t_f - t_i$ is equal to the area of the shaded rectangle.

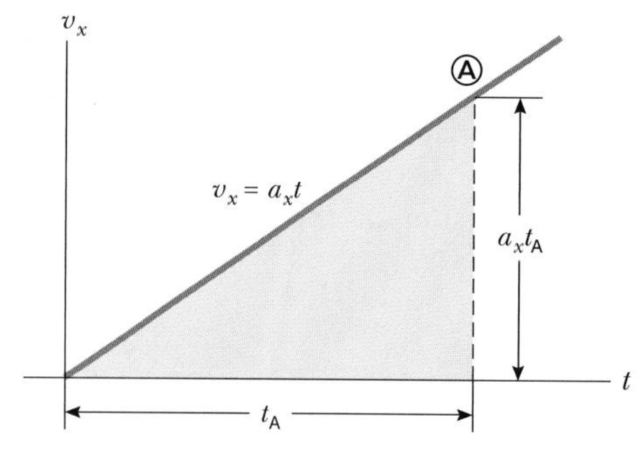

Figure 2.17 The velocity–time curve for a particle moving with a velocity that is proportional to the time.

acceleration), we find that the displacement of the particle during the time interval $t = 0$ to $t = t_A$ is equal to the area of the shaded triangle in Figure 2.17:

$$\Delta x = \tfrac{1}{2}(t_A)(a_x t_A) = \tfrac{1}{2} a_x t_A{}^2$$

Kinematic Equations

We now use the defining equations for acceleration and velocity to derive two of our kinematic equations, Equations 2.9 and 2.12.

The defining equation for acceleration (Eq. 2.7),

$$a_x = \frac{dv_x}{dt}$$

may be written as $dv_x = a_x dt$ or, in terms of an integral (or antiderivative), as

$$v_{xf} - v_{xi} = \int_0^t a_x dt$$

For the special case in which the acceleration is constant, a_x can be removed from the integral to give

$$v_{xf} - v_{xi} = a_x \int_0^t dt = a_x(t - 0) = a_x t \qquad (2.16)$$

which is Equation 2.9.

Now let us consider the defining equation for velocity (Eq. 2.5):

$$v_x = \frac{dx}{dt}$$

We can write this as $dx = v_x\, dt$, or in integral form as

$$x_f - x_i = \int_0^t v_x dt$$

Because $v_x = v_{xf} = v_{xi} + a_x t$, this expression becomes

$$\begin{aligned} x_f - x_i &= \int_0^t (v_{xi} + a_x t)\, dt = \int_0^t v_{xi}\, dt + a_x \int_0^t t dt = v_{xi}(t - 0) + a_x\left(\frac{t^2}{2} - 0\right) \\ &= v_{xi} t + \tfrac{1}{2} a_x t^2 \end{aligned}$$

which is Equation 2.12.

Besides what you might expect to learn about physics concepts, a very valuable skill you should hope to take away from your physics course is the ability to solve complicated problems. The way physicists approach complex situations and break them down into manageable pieces is extremely useful. On the next page is a general problem-solving strategy that will help guide you through the steps. To help you remember the steps of the strategy, they are called *Conceptualize, Categorize, Analyze, and Finalize.*

GENERAL PROBLEM-SOLVING STRATEGY

Conceptualize

- The first thing to do when approaching a problem is to *think about* and *understand* the situation. Study carefully any diagrams, graphs, tables, or photographs that accompany the problem. Imagine a movie, running in your mind, of what happens in the problem.
- If a diagram is not provided, you should almost always make a quick drawing of the situation. Indicate any known values, perhaps in a table or directly on your sketch.
- Now focus on what algebraic or numerical information is given in the problem. Carefully read the problem statement, looking for key phrases such as "starts from rest" ($v_i = 0$), "stops" ($v_f = 0$), or "freely falls" ($a_y = -g = -9.80 \text{ m/s}^2$).
- Now focus on the expected result of solving the problem. Exactly what is the question asking? Will the final result be numerical or algebraic? Do you know what units to expect?
- Don't forget to incorporate information from your own experiences and common sense. What should a reasonable answer look like? For example, you wouldn't expect to calculate the speed of an automobile to be 5×10^6 m/s.

Categorize

- Once you have a good idea of what the problem is about, you need to *simplify* the problem. Remove the details that are not important to the solution. For example, model a moving object as a particle. If appropriate, ignore air resistance or friction between a sliding object and a surface.
- Once the problem is simplified, it is important to *categorize* the problem. Is it a simple *plug-in problem,* such that numbers can be simply substituted into a definition? If so, the problem is likely to be finished when this substitution is done. If not, you face what we can call an *analysis problem*—the situation must be analyzed more deeply to reach a solution.
- If it is an analysis problem, it needs to be categorized further. Have you seen this type of problem before? Does it fall into the growing list of types of problems that you have solved previously? Being able to classify a problem can make it much easier to lay out a plan to solve it. For example, if your simplification shows that the problem can be treated as a particle moving under constant acceleration and you have already solved such a problem (such as the examples in Section 2.5), the solution to the present problem follows a similar pattern.

Analyze

- Now you must analyze the problem and strive for a mathematical solution. Because you have already categorized the problem, it should not be too difficult to select relevant equations that apply to the type of situation in the problem. For example, if the problem involves a particle moving under constant acceleration, Equations 2.9 to 2.13 are relevant.
- Use algebra (and calculus, if necessary) to solve symbolically for the unknown variable in terms of what is given. Substitute in the appropriate numbers, calculate the result, and round it to the proper number of significant figures.

Finalize

- This is the most important part. Examine your numerical answer. Does it have the correct units? Does it meet your expectations from your conceptualization of the problem? What about the algebraic form of the result—before you substituted numerical values? Does it make sense? Examine the variables in the problem to see whether the answer would change in a physically meaningful way if they were drastically increased or decreased or even became zero. Looking at limiting cases to see whether they yield expected values is a very useful way to make sure that you are obtaining reasonable results.
- Think about how this problem compares with others you have solved. How was it similar? In what critical ways did it differ? Why was this problem assigned? You should have learned something by doing it. Can you figure out what? If it is a new category of problem, be sure you understand it so that you can use it as a model for solving future problems in the same category.

When solving complex problems, you may need to identify a series of sub-problems and apply the problem-solving strategy to each. For very simple problems, you probably don't need this strategy at all. But when you are looking at a problem and you don't know what to do next, remember the steps in the strategy and use them as a guide.

For practice, it would be useful for you to go back over the examples in this chapter and identify the *Conceptualize, Categorize, Analyze,* and *Finalize* steps. In the next chapter, we will begin to show these steps explicitly in the examples.

SUMMARY

Take a practice test for this chapter by clicking the Practice Test link at **http://www.pse6.com.**

After a particle moves along the x axis from some initial position x_i to some final position x_f, its **displacement** is

$$\Delta x \equiv x_f - x_i \tag{2.1}$$

The **average velocity** of a particle during some time interval is the displacement Δx divided by the time interval Δt during which that displacement occurs:

$$\bar{v}_x \equiv \frac{\Delta x}{\Delta t} \tag{2.2}$$

The **average speed** of a particle is equal to the ratio of the total distance it travels to the total time interval during which it travels that distance:

$$\text{Average speed} = \frac{\text{total distance}}{\text{total time}} \tag{2.3}$$

The **instantaneous velocity** of a particle is defined as the limit of the ratio $\Delta x/\Delta t$ as Δt approaches zero. By definition, this limit equals the derivative of x with respect to t, or the time rate of change of the position:

$$v_x \equiv \lim_{\Delta t \to 0} \frac{\Delta x}{\Delta t} = \frac{dx}{dt} \tag{2.5}$$

The **instantaneous speed** of a particle is equal to the magnitude of its instantaneous velocity.

The **average acceleration** of a particle is defined as the ratio of the change in its velocity Δv_x divided by the time interval Δt during which that change occurs:

$$\bar{a}_x \equiv \frac{\Delta v_x}{\Delta t} = \frac{v_{xf} - v_{xi}}{t_f - t_i} \tag{2.6}$$

The **instantaneous acceleration** is equal to the limit of the ratio $\Delta v_x/\Delta t$ as Δt approaches 0. By definition, this limit equals the derivative of v_x with respect to t, or the time rate of change of the velocity:

$$a_x \equiv \lim_{\Delta t \to 0} \frac{\Delta v_x}{\Delta t} = \frac{dv_x}{dt} \tag{2.7}$$

When the object's velocity and acceleration are in the same direction, the object is speeding up. On the other hand, when the object's velocity and acceleration are in opposite directions, the object is slowing down. Remembering that $F \propto a$ is a useful way to identify the direction of the acceleration.

The **equations of kinematics** for a particle moving along the x axis with uniform acceleration a_x (constant in magnitude and direction) are

$$v_{xf} = v_{xi} + a_x t \tag{2.9}$$

$$x_f = x_i + \bar{v}_x t = x_i + \tfrac{1}{2}(v_{xi} + v_{xf})t \tag{2.11}$$

$$x_f = x_i + v_{xi}t + \tfrac{1}{2}a_x t^2 \tag{2.12}$$

$$v_{xf}{}^2 = v_{xi}{}^2 + 2a_x(x_f - x_i) \tag{2.13}$$

An object falling freely in the presence of the Earth's gravity experiences a free-fall acceleration directed toward the center of the Earth. If air resistance is neglected, if the motion occurs near the surface of the Earth, and if the range of the motion is small compared with the Earth's radius, then the free-fall acceleration g is constant over the range of motion, where g is equal to 9.80 m/s^2.

Complicated problems are best approached in an organized manner. You should be able to recall and apply the *Conceptualize, Categorize, Analyze,* and *Finalize* steps of the General Problem-Solving Strategy when you need them.

QUESTIONS

1. The speed of sound in air is 331 m/s. During the next thunderstorm, try to estimate your distance from a lightning bolt by measuring the time lag between the flash and the thunderclap. You can ignore the time it takes for the light flash to reach you. Why?

2. The average velocity of a particle moving in one dimension has a positive value. Is it possible for the instantaneous velocity to have been negative at any time in the interval? Suppose the particle started at the origin $x = 0$. If its average velocity is positive, could the particle ever have been in the $-x$ region of the axis?

3. If the average velocity of an object is zero in some time interval, what can you say about the displacement of the object for that interval?

4. Can the instantaneous velocity of an object at an instant of time ever be greater in magnitude than the average velocity over a time interval containing the instant? Can it ever be less?

5. If an object's average velocity is nonzero over some time interval, does this mean that its instantaneous velocity is never zero during the interval? Explain your answer.

6. If an object's average velocity is zero over some time interval, show that its instantaneous velocity must be zero at some time during the interval. It may be useful in your proof to sketch a graph of x versus t and to note that $v_x(t)$ is a continuous function.

7. If the velocity of a particle is nonzero, can its acceleration be zero? Explain.

8. If the velocity of a particle is zero, can its acceleration be nonzero? Explain.

9. Two cars are moving in the same direction in parallel lanes along a highway. At some instant, the velocity of car A exceeds the velocity of car B. Does this mean that the acceleration of A is greater than that of B? Explain.

10. Is it possible for the velocity and the acceleration of an object to have opposite signs? If not, state a proof. If so, give an example of such a situation and sketch a velocity–time graph to prove your point.

11. Consider the following combinations of signs and values for velocity and acceleration of a particle with respect to a one-dimensional x axis:

Velocity	*Acceleration*
a. Positive	Positive
b. Positive	Negative
c. Positive	Zero
d. Negative	Positive
e. Negative	Negative
f. Negative	Zero
g. Zero	Positive
h. Zero	Negative

Describe what a particle is doing in each case, and give a real life example for an automobile on an east-west one-dimensional axis, with east considered the positive direction.

12. Can the equations of kinematics (Eqs. 2.9–2.13) be used in a situation where the acceleration varies in time? Can they be used when the acceleration is zero?

13. A stone is thrown vertically upward from the roof of a building. Does the position of the stone depend on the location chosen for the origin of the coordinate system? Does the stone's velocity depend on the choice of origin? Explain your answers.

14. A child throws a marble into the air with an initial speed v_i. Another child drops a ball at the same instant. Compare the accelerations of the two objects while they are in flight.

15. A student at the top of a building of height h throws one ball upward with a speed of v_i and then throws a second ball downward with the same initial speed $|v_i|$. How do the final velocities of the balls compare when they reach the ground?

16. An object falls freely from height h. It is released at time zero and strikes the ground at time t. (a) When the object is at height $0.5h$, is the time earlier than $0.5t$, equal to $0.5t$, or later than $0.5t$? (b) When the time is $0.5t$, is the height of the object greater than $0.5h$, equal to $0.5h$, or less than $0.5h$? Give reasons for your answers.

17. You drop a ball from a window on an upper floor of a building. It strikes the ground with speed v. You now repeat the drop, but you have a friend down on the street who throws another ball upward at speed v. Your friend throws the ball upward at exactly the same time that you drop yours from the window. At some location, the balls pass each other. Is this location *at* the halfway point between window and ground, *above* this point, or *below* this point?

PROBLEMS

1, 2, 3 = straightforward, intermediate, challenging □ = full solution available in the *Student Solutions Manual and Study Guide*

= coached solution with hints available at http://www.pse6.com = computer useful in solving problem

= paired numerical and symbolic problems

Section 2.1 Position, Velocity, and Speed

1. The position of a pinewood derby car was observed at various times; the results are summarized in the following table. Find the average velocity of the car for (a) the first second, (b) the last 3 s, and (c) the entire period of observation.

t(s)	0	1.0	2.0	3.0	4.0	5.0
x(m)	0	2.3	9.2	20.7	36.8	57.5

2. (a) Sand dunes in a desert move over time as sand is swept up the windward side to settle in the lee side. Such "walking" dunes have been known to walk 20 feet in a year and can travel as much as 100 feet per year in particularly windy times. Calculate the average speed in each case in m/s. (b) Fingernails grow at the rate of drifting continents, on the order of 10 mm/yr. Approximately how long did it take for North America to separate from Europe, a distance of about 3 000 mi?

3. The position versus time for a certain particle moving along the x axis is shown in Figure P2.3. Find the average velocity in the time intervals (a) 0 to 2 s, (b) 0 to 4 s, (c) 2 s to 4 s, (d) 4 s to 7 s, (e) 0 to 8 s.

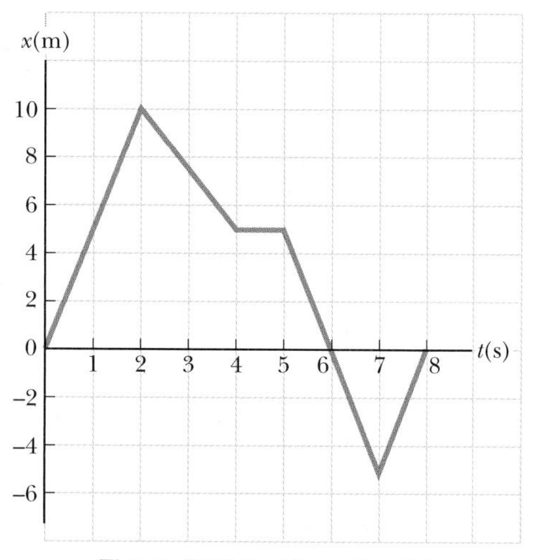

Figure P2.3 Problems 3 and 9

4. A particle moves according to the equation $x = 10t^2$ where x is in meters and t is in seconds. (a) Find the average velocity for the time interval from 2.00 s to 3.00 s. (b) Find the average velocity for the time interval from 2.00 to 2.10 s.

5. A person walks first at a constant speed of 5.00 m/s along a straight line from point A to point B and then back along the line from B to A at a constant speed of 3.00 m/s. What is (a) her average speed over the entire trip? (b) her average velocity over the entire trip?

Section 2.2 Instantaneous Velocity and Speed

6. The position of a particle moving along the x axis varies in time according to the expression $x = 3t^2$, where x is in meters and t is in seconds. Evaluate its position (a) at $t = 3.00$ s and (b) at $3.00 \text{ s} + \Delta t$. (c) Evaluate the limit of $\Delta x/\Delta t$ as Δt approaches zero, to find the velocity at $t = 3.00$ s.

7. A position-time graph for a particle moving along the x axis is shown in Figure P2.7. (a) Find the average velocity in the time interval $t = 1.50$ s to $t = 4.00$ s. (b) Determine the instantaneous velocity at $t = 2.00$ s by measuring the slope of the tangent line shown in the graph. (c) At what value of t is the velocity zero?

8. (a) Use the data in Problem 1 to construct a smooth graph of position versus time. (b) By constructing tangents to the $x(t)$ curve, find the instantaneous velocity of the car at several instants. (c) Plot the instantaneous velocity versus time and, from this, determine the average acceleration of the car. (d) What was the initial velocity of the car?

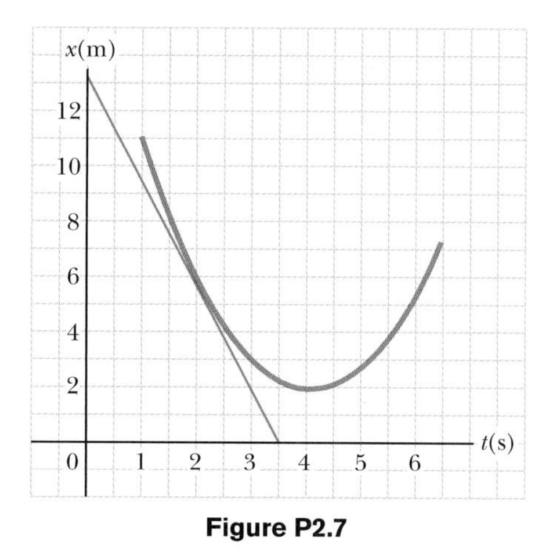

Figure P2.7

9. Find the instantaneous velocity of the particle described in Figure P2.3 at the following times: (a) $t = 1.0$ s, (b) $t = 3.0$ s, (c) $t = 4.5$ s, and (d) $t = 7.5$ s.

10. A hare and a tortoise compete in a race over a course 1.00 km long. The tortoise crawls straight and steadily at its maximum speed of 0.200 m/s toward the finish line. The hare runs at its maximum speed of 8.00 m/s toward the goal for 0.800 km and then stops to tease the tortoise. How close to the goal can the hare let the tortoise approach before resuming the race, which the tortoise wins in a photo finish? Assume that, when moving, both animals move steadily at their respective maximum speeds.

Section 2.3 Acceleration

11. A 50.0-g superball traveling at 25.0 m/s bounces off a brick wall and rebounds at 22.0 m/s. A high-speed camera records this event. If the ball is in contact with the wall for 3.50 ms, what is the magnitude of the average acceleration of the ball during this time interval? (*Note:* 1 ms $= 10^{-3}$ s.)

12. A particle starts from rest and accelerates as shown in Figure P2.12. Determine (a) the particle's speed at $t = 10.0$ s and at $t = 20.0$ s, and (b) the distance traveled in the first 20.0 s.

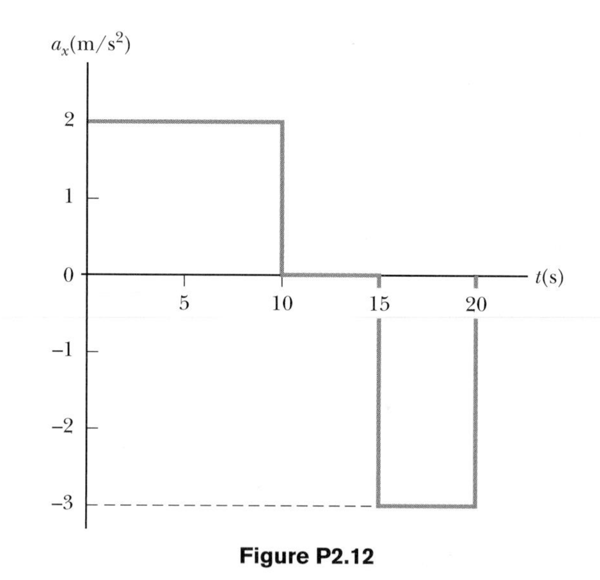

Figure P2.12

13. Secretariat won the Kentucky Derby with times for successive quarter-mile segments of 25.2 s, 24.0 s, 23.8 s, and 23.0 s. (a) Find his average speed during each quarter-mile segment. (b) Assuming that Secretariat's instantaneous speed at the finish line was the same as the average speed during the final quarter mile, find his average acceleration for the entire race. (Horses in the Derby start from rest.)

14. A velocity–time graph for an object moving along the x axis is shown in Figure P2.14. (a) Plot a graph of the acceleration versus time. (b) Determine the average acceleration of the object in the time intervals $t = 5.00$ s to $t = 15.0$ s and $t = 0$ to $t = 20.0$ s.

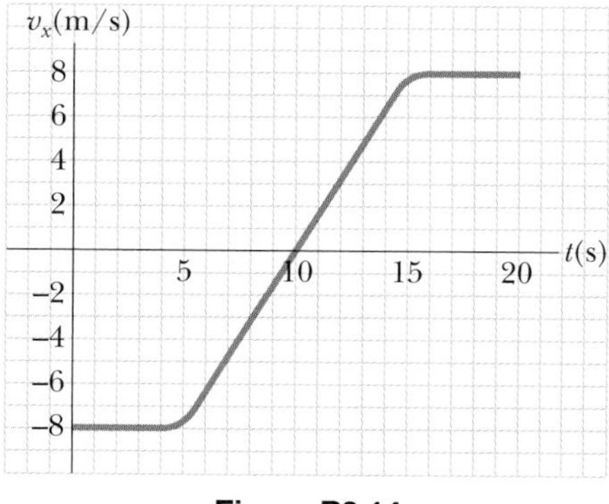

Figure P2.14

15. A particle moves along the x axis according to the equation $x = 2.00 + 3.00t - 1.00t^2$, where x is in meters and t is in seconds. At $t = 3.00$ s, find (a) the position of the particle, (b) its velocity, and (c) its acceleration.

16. An object moves along the x axis according to the equation $x(t) = (3.00t^2 - 2.00t + 3.00)$ m. Determine (a) the average speed between $t = 2.00$ s and $t = 3.00$ s, (b) the instantaneous speed at $t = 2.00$ s and at $t = 3.00$ s, (c) the average acceleration between $t = 2.00$ s and $t = 3.00$ s, and (d) the instantaneous acceleration at $t = 2.00$ s and $t = 3.00$ s.

17. Figure P2.17 shows a graph of v_x versus t for the motion of a motorcyclist as he starts from rest and moves along the road in a straight line. (a) Find the average acceleration for the time interval $t = 0$ to $t = 6.00$ s. (b) Estimate the time at which the acceleration has its greatest positive value and the value of the acceleration at that instant. (c) When is the acceleration zero? (d) Estimate the maximum negative value of the acceleration and the time at which it occurs.

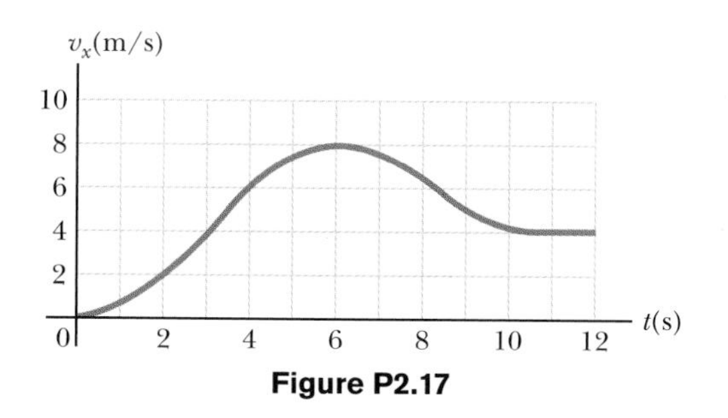

Figure P2.17

Section 2.4 Motion Diagrams

18. Draw motion diagrams for (a) an object moving to the right at constant speed, (b) an object moving to the right and speeding up at a constant rate, (c) an object moving to the right and slowing down at a constant rate, (d) an object moving to the left and speeding up at a constant rate, and (e) an object moving to the left and slowing down at a constant rate. (f) How would your drawings change if the changes in speed were not uniform; that is, if the speed were not changing at a constant rate?

Section 2.5 One-Dimensional Motion with Constant Acceleration

19. Jules Verne in 1865 suggested sending people to the Moon by firing a space capsule from a 220-m-long cannon with a launch speed of 10.97 km/s. What would have been the unrealistically large acceleration experienced by the space travelers during launch? Compare your answer with the free-fall acceleration 9.80 m/s^2.

20. A truck covers 40.0 m in 8.50 s while smoothly slowing down to a final speed of 2.80 m/s. (a) Find its original speed. (b) Find its acceleration.

21. An object moving with uniform acceleration has a velocity of 12.0 cm/s in the positive x direction when its x coordinate is 3.00 cm. If its x coordinate 2.00 s later is -5.00 cm, what is its acceleration?

22. A 745i BMW car can brake to a stop in a distance of 121 ft. from a speed of 60.0 mi/h. To brake to a stop from a speed of 80.0 mi/h requires a stopping distance of 211 ft. What is the average braking acceleration for (a) 60 mi/h to rest, (b) 80 mi/h to rest, (c) 80 mi/h to 60 mi/h? Express the answers in mi/h/s and in m/s^2.

23. A speedboat moving at 30.0 m/s approaches a no-wake buoy marker 100 m ahead. The pilot slows the boat with a constant acceleration of -3.50 m/s^2 by reducing the throttle. (a) How long does it take the boat to reach the buoy? (b) What is the velocity of the boat when it reaches the buoy?

24. Figure P2.24 represents part of the performance data of a car owned by a proud physics student. (a) Calculate from the graph the total distance traveled. (b) What distance does the car travel between the times $t = 10$ s and $t = 40$ s? (c) Draw a graph of its acceleration versus time between $t = 0$ and $t = 50$ s. (d) Write an equation for x as a function of time for each phase of the motion, represented by (i) $0a$, (ii) ab, (iii) bc. (e) What is the average velocity of the car between $t = 0$ and $t = 50$ s?

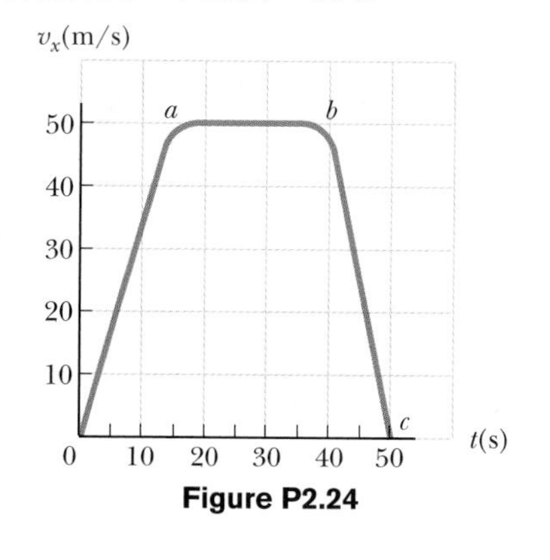

Figure P2.24

25. A particle moves along the x axis. Its position is given by the equation $x = 2 + 3t - 4t^2$ with x in meters and t in seconds. Determine (a) its position when it changes direction and (b) its velocity when it returns to the position it had at $t = 0$.

26. In the Daytona 500 auto race, a Ford Thunderbird and a Mercedes Benz are moving side by side down a straightaway at 71.5 m/s. The driver of the Thunderbird realizes he must make a pit stop, and he smoothly slows to a stop over a distance of 250 m. He spends 5.00 s in the pit and then accelerates out, reaching his previous speed of 71.5 m/s after a distance of 350 m. At this point, how far has the Thunderbird fallen behind the Mercedes Benz, which has continued at a constant speed?

27. A jet plane lands with a speed of 100 m/s and can accelerate at a maximum rate of -5.00 m/s^2 as it comes to rest. (a) From the instant the plane touches the runway, what is the minimum time interval needed before it can come to rest? (b) Can this plane land on a small tropical island airport where the runway is 0.800 km long?

28. A car is approaching a hill at 30.0 m/s when its engine suddenly fails just at the bottom of the hill. The car moves with a constant acceleration of -2.00 m/s^2 while coasting up the hill. (a) Write equations for the position along the slope and for the velocity as functions of time, taking $x = 0$ at the bottom of the hill, where $v_i = 30.0$ m/s. (b) Determine the maximum distance the car rolls up the hill.

29. The driver of a car slams on the brakes when he sees a tree blocking the road. The car slows uniformly with an acceleration of -5.60 m/s^2 for 4.20 s, making straight skid marks 62.4 m long ending at the tree. With what speed does the car then strike the tree?

30. *Help! One of our equations is missing!* We describe constant-acceleration motion with the variables and parameters v_{xi}, v_{xf}, a_x, t, and $x_f - x_i$. Of the equations in Table 2.2, the first does not involve $x_f - x_i$. The second does not contain a_x; the third omits v_{xf} and the last leaves out t. So to complete the set there should be an equation *not* involving v_{xi}. Derive it from the others. Use it to solve Problem 29 in one step.

31. For many years Colonel John P. Stapp, USAF, held the world's land speed record. On March 19, 1954, he rode a rocket-propelled sled that moved down a track at a speed of 632 mi/h. He and the sled were safely brought to rest in 1.40 s (Fig. P2.31). Determine (a) the negative acceleration he experienced and (b) the distance he traveled during this negative acceleration.

32. A truck on a straight road starts from rest, accelerating at 2.00 m/s^2 until it reaches a speed of 20.0 m/s. Then the truck travels for 20.0 s at constant speed until the brakes are applied, stopping the truck in a uniform manner in an additional 5.00 s. (a) How long is the truck in motion? (b) What is the average velocity of the truck for the motion described?

33. An electron in a cathode ray tube (CRT) accelerates from 2.00×10^4 m/s to 6.00×10^6 m/s over 1.50 cm. (a) How long does the electron take to travel this 1.50 cm? (b) What is its acceleration?

34. In a 100-m linear accelerator, an electron is accelerated to 1.00% of the speed of light in 40.0 m before it coasts for 60.0 m to a target. (a) What is the electron's acceleration during the first 40.0 m? (b) How long does the total flight take?

35. Within a complex machine such as a robotic assembly line, suppose that one particular part glides along a straight track. A control system measures the average velocity of the part during each successive interval of time $\Delta t_0 = t_0 - 0$, compares it with the value v_c it should be, and switches a servo motor on and off to give the part a correcting pulse of acceleration. The pulse consists of a constant acceleration a_m applied for time interval $\Delta t_m = t_m - 0$ within the next control time interval Δt_0. As shown in Fig. P2.35, the part may be modeled as having zero acceleration when the motor is off (between t_m and t_0). A computer in the control system chooses the size of the acceleration so that the final velocity of the part will have the correct value v_c. Assume the part is initially at rest and is to have instantaneous velocity v_c at time t_0. (a) Find the required value of a_m in terms of v_c and t_m. (b) Show that the displacement Δx of the part during the time interval Δt_0 is given by $\Delta x = v_c\,(t_0 - 0.5t_m)$. For specified values of v_c and t_0, (c) what is the minimum displacement of the part? (d) What is the maximum displacement of the part? (e) Are both the minimum and maximum displacements physically attainable?

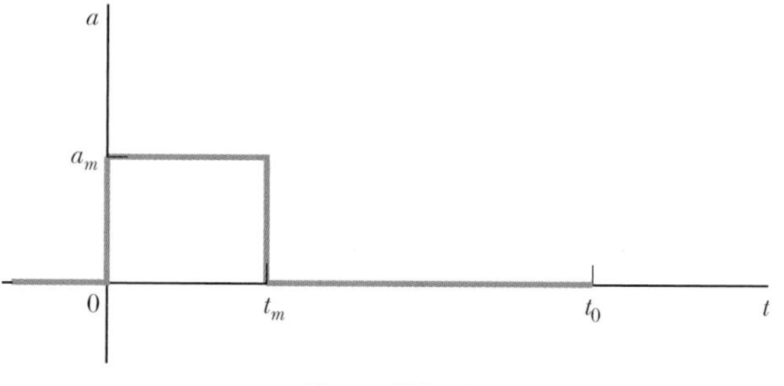

Figure P2.35

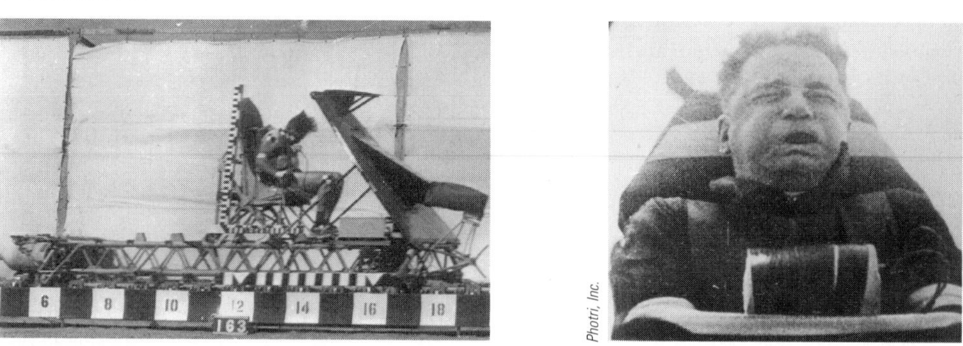

Figure P2.31 *(Left)* Col. John Stapp on rocket sled. *(Right)* Col. Stapp's face is contorted by the stress of rapid negative acceleration.

36. A glider on an air track carries a flag of length ℓ through a stationary photogate, which measures the time interval Δt_d during which the flag blocks a beam of infrared light passing across the photogate. The ratio $v_d = \ell/\Delta t_d$ is the average velocity of the glider over this part of its motion. Suppose the glider moves with constant acceleration. (a) Argue for or against the idea that v_d is equal to the instantaneous velocity of the glider when it is halfway through the photogate in space. (b) Argue for or against the idea that v_d is equal to the instantaneous velocity of the glider when it is halfway through the photogate in time.

37. A ball starts from rest and accelerates at 0.500 m/s^2 while moving down an inclined plane 9.00 m long. When it reaches the bottom, the ball rolls up another plane, where, after moving 15.0 m, it comes to rest. (a) What is the speed of the ball at the bottom of the first plane? (b) How long does it take to roll down the first plane? (c) What is the acceleration along the second plane? (d) What is the ball's speed 8.00 m along the second plane?

38. Speedy Sue, driving at 30.0 m/s, enters a one-lane tunnel. She then observes a slow-moving van 155 m ahead traveling at 5.00 m/s. Sue applies her brakes but can accelerate only at -2.00 m/s^2 because the road is wet. Will there be a collision? If yes, determine how far into the tunnel and at what time the collision occurs. If no, determine the distance of closest approach between Sue's car and the van.

39. Solve Example 2.8, "Watch out for the Speed Limit!" by a graphical method. On the same graph plot position versus time for the car and the police officer. From the intersection of the two curves read the time at which the trooper overtakes the car.

Section 2.6 Freely Falling Objects

> *Note:* In all problems in this section, ignore the effects of air resistance.

40. A golf ball is released from rest from the top of a very tall building. Neglecting air resistance, calculate (a) the position and (b) the velocity of the ball after 1.00, 2.00, and 3.00 s.

41. *Every morning at seven o'clock*
There's twenty terriers drilling on the rock.
The boss comes around and he says, "Keep still
And bear down heavy on the cast-iron drill

And drill, ye terriers, drill." And drill, ye terriers, drill.
It's work all day for sugar in your tea
Down beyond the railway. And drill, ye terriers, drill.

The foreman's name was John McAnn.
By God, he was a blamed mean man.
One day a premature blast went off
And a mile in the air went big Jim Goff. And drill ...

Then when next payday came around
Jim Goff a dollar short was found.
When he asked what for, came this reply:
"You were docked for the time you were up in the sky." And drill...
—American folksong

What was Goff's hourly wage? State the assumptions you make in computing it.

42. A ball is thrown directly downward, with an initial speed of 8.00 m/s, from a height of 30.0 m. After what time interval does the ball strike the ground?

43. A student throws a set of keys vertically upward to her sorority sister, who is in a window 4.00 m above. The keys are caught 1.50 s later by the sister's outstretched hand. (a) With what initial velocity were the keys thrown? (b) What was the velocity of the keys just before they were caught?

44. Emily challenges her friend David to catch a dollar bill as follows. She holds the bill vertically, as in Figure P2.44, with the center of the bill between David's index finger and thumb. David must catch the bill after Emily releases it without moving his hand downward. If his reaction time is 0.2 s, will he succeed? Explain your reasoning.

George Semple

Figure P2.44

45. In Mostar, Bosnia, the ultimate test of a young man's courage once was to jump off a 400-year-old bridge (now destroyed) into the River Neretva, 23.0 m below the bridge. (a) How long did the jump last? (b) How fast was the diver traveling upon impact with the water? (c) If the speed of sound in air is 340 m/s, how long after the diver took off did a spectator on the bridge hear the splash?

46. A ball is dropped from rest from a height h above the ground. Another ball is thrown vertically upwards from the ground at the instant the first ball is released. Determine the speed of the second ball if the two balls are to meet at a height $h/2$ above the ground.

47. A baseball is hit so that it travels straight upward after being struck by the bat. A fan observes that it takes 3.00 s for the ball to reach its maximum height. Find (a) its initial velocity and (b) the height it reaches.

48. It is possible to shoot an arrow at a speed as high as 100 m/s. (a) If friction is neglected, how high would an arrow launched at this speed rise if shot straight up? (b) How long would the arrow be in the air?

49. A daring ranch hand sitting on a tree limb wishes to drop vertically onto a horse galloping under the tree. The constant speed of the horse is 10.0 m/s, and the distance from the limb to the level of the saddle is 3.00 m. (a) What must be the horizontal distance between the saddle and limb when the ranch hand makes his move? (b) How long is he in the air?

50. A woman is reported to have fallen 144 ft from the 17th floor of a building, landing on a metal ventilator box, which she crushed to a depth of 18.0 in. She suffered only minor injuries. Neglecting air resistance, calculate (a) the speed of the woman just before she collided with the ventilator, (b) her average acceleration while in contact with the box, and (c) the time it took to crush the box.

51. The height of a helicopter above the ground is given by $h = 3.00t^3$, where h is in meters and t is in seconds. After 2.00 s, the helicopter releases a small mailbag. How long after its release does the mailbag reach the ground?

52. A freely falling object requires 1.50 s to travel the last 30.0 m before it hits the ground. From what height above the ground did it fall?

Section 2.7 Kinematic Equations Derived from Calculus

53. Automotive engineers refer to the time rate of change of acceleration as the "jerk." If an object moves in one dimension such that its jerk J is constant, (a) determine expressions for its acceleration $a_x(t)$, velocity $v_x(t)$, and position $x(t)$, given that its initial acceleration, velocity, and position are a_{xi}, v_{xi}, and x_i, respectively. (b) Show that $a_x^2 = a_{xi}^2 + 2J(v_x - v_{xi})$.

54. A student drives a moped along a straight road as described by the velocity-versus-time graph in Figure P2.54. Sketch this graph in the middle of a sheet of graph paper. (a) Directly above your graph, sketch a graph of the position versus time, aligning the time coordinates of the two graphs. (b) Sketch a graph of the acceleration versus time directly below the v_x-t graph, again aligning the time coordinates. On each graph, show the numerical values of x and a_x for all points of inflection. (c) What is the acceleration at $t = 6$ s? (d) Find the position (relative to the starting point) at $t = 6$ s. (e) What is the moped's final position at $t = 9$ s?

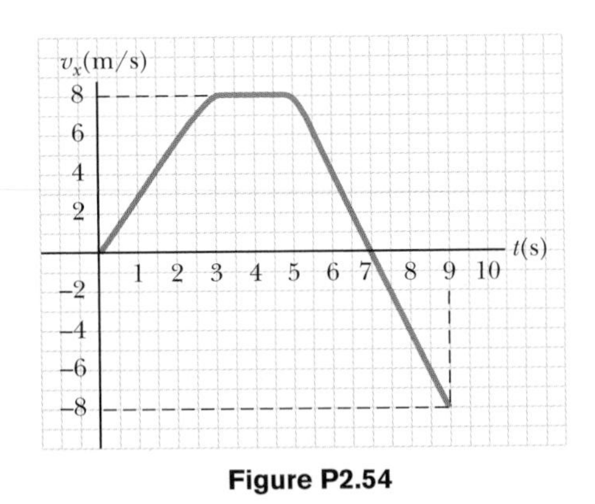

Figure P2.54

55. The speed of a bullet as it travels down the barrel of a rifle toward the opening is given by $v = (-5.00 \times 10^7)t^2 + (3.00 \times 10^5)t$, where v is in meters per second and t is in seconds. The acceleration of the bullet just as it leaves the barrel is zero. (a) Determine the acceleration and position of the bullet as a function of time when the bullet is in the barrel. (b) Determine the length of time the bullet is accelerated. (c) Find the speed at which the bullet leaves the barrel. (d) What is the length of the barrel?

56. The acceleration of a marble in a certain fluid is proportional to the speed of the marble squared, and is given (in SI units) by $a = -3.00v^2$ for $v > 0$. If the marble enters this fluid with a speed of 1.50 m/s, how long will it take before the marble's speed is reduced to half of its initial value?

Additional Problems

57. A car has an initial velocity v_0 when the driver sees an obstacle in the road in front of him. His reaction time is Δt_r, and the braking acceleration of the car is a. Show that the total stopping distance is

$$s_{\text{stop}} = v_0\,\Delta t_r - v_0^2/2a.$$

Remember that a is a negative number.

58. The yellow caution light on a traffic signal should stay on long enough to allow a driver to either pass through the intersection or safely stop before reaching the intersection. A car can stop if its distance from the intersection is greater than the stopping distance found in the previous problem. If the car is less than this stopping distance from the intersection, the yellow light should stay on long enough to allow the car to pass entirely through the intersection. (a) Show that the yellow light should stay on for a time interval

$$\Delta t_{\text{light}} = \Delta t_r - (v_0/2a) + (s_i/v_0)$$

where Δt_r is the driver's reaction time, v_0 is the velocity of the car approaching the light at the speed limit, a is the braking acceleration, and s_i is the width of the intersection. (b) As city traffic planner, you expect cars to approach an intersection 16.0 m wide with a speed of 60.0 km/h. Be cautious and assume a reaction time of 1.10 s to allow for a driver's indecision. Find the length of time the yellow light should remain on. Use a braking acceleration of -2.00 m/s^2.

59. The Acela is the Porsche of American trains. Shown in Figure P2.59a, the electric train whose name is pronounced ah-SELL-ah is in service on the Washington-New York-Boston run. With two power cars and six coaches, it can carry 304 passengers at 170 mi/h. The carriages tilt as much as 6° from the vertical to prevent passengers from feeling pushed to the side as they go around curves. Its braking mechanism uses electric generators to recover its energy of motion. A velocity-time graph for the Acela is shown in Figure P2.59b. (a) Describe the motion of the train in each successive time interval. (b) Find the peak positive acceleration of the train in the motion graphed. (c) Find the train's displacement in miles between $t = 0$ and $t = 200$ s.

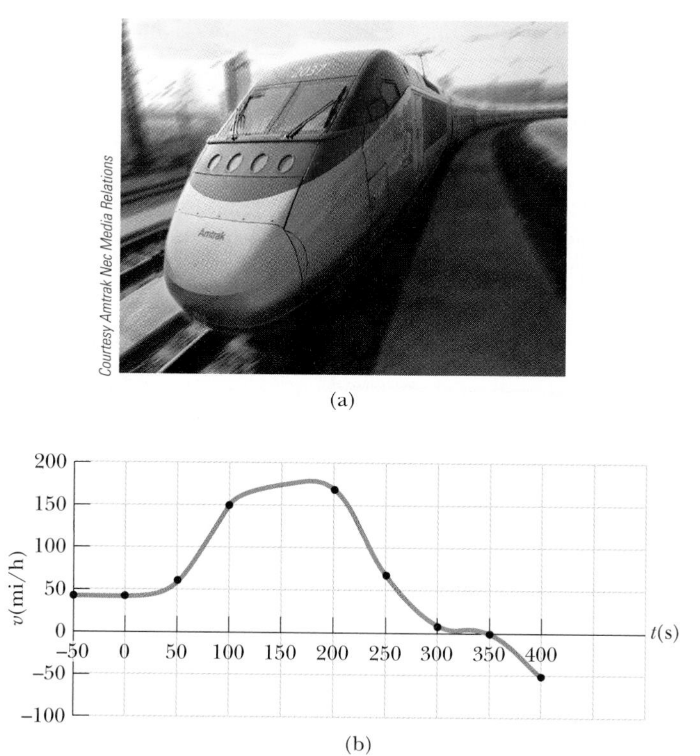

Figure P2.59 (a) The Acela—1 171 000 lb of cold steel thundering along at 150 mi/h. (b) Velocity-versus-time graph for the Acela.

60. Liz rushes down onto a subway platform to find her train already departing. She stops and watches the cars go by. Each car is 8.60 m long. The first moves past her in 1.50 s and the second in 1.10 s. Find the constant acceleration of the train.

61. A dog's hair has been cut and is now getting 1.04 mm longer each day. With winter coming on, this rate of hair growth is steadily increasing, by 0.132 mm/day every week. By how much will the dog's hair grow during 5 weeks?

62. A test rocket is fired vertically upward from a well. A catapult gives it an initial speed of 80.0 m/s at ground level. Its engines then fire and it accelerates upward at 4.00 m/s^2 until it reaches an altitude of 1 000 m. At that point its engines fail and the rocket goes into free fall, with an acceleration of -9.80 m/s^2. (a) How long is the rocket in motion above the ground? (b) What is its maximum altitude? (c) What is its velocity just before it collides with the Earth? (You will need to consider the motion while the engine is operating separate from the free-fall motion.)

63. A motorist drives along a straight road at a constant speed of 15.0 m/s. Just as she passes a parked motorcycle police officer, the officer starts to accelerate at 2.00 m/s^2 to overtake her. Assuming the officer maintains this acceleration, (a) determine the time it takes the police officer to reach the motorist. Find (b) the speed and (c) the total displacement of the officer as he overtakes the motorist.

64. In Figure 2.10b, the area under the velocity versus time curve and between the vertical axis and time t (vertical dashed line) represents the displacement. As shown, this area consists of a rectangle and a triangle. Compute their areas and compare the sum of the two areas with the expression on the right-hand side of Equation 2.12.

65. Setting a new world record in a 100-m race, Maggie and Judy cross the finish line in a dead heat, both taking 10.2 s. Accelerating uniformly, Maggie took 2.00 s and Judy 3.00 s to attain maximum speed, which they maintained for the rest of the race. (a) What was the acceleration of each sprinter? (b) What were their respective maximum speeds? (c) Which sprinter was ahead at the 6.00-s mark, and by how much?

66. A commuter train travels between two downtown stations. Because the stations are only 1.00 km apart, the train never reaches its maximum possible cruising speed. During rush hour the engineer minimizes the time interval Δt between two stations by accelerating for a time interval Δt_1 at a rate $a_1 = 0.100$ m/s^2 and then immediately braking with acceleration $a_2 = -0.500$ m/s^2 for a time interval Δt_2. Find the minimum time interval of travel Δt and the time interval Δt_1.

67. A hard rubber ball, released at chest height, falls to the pavement and bounces back to nearly the same height. When it is in contact with the pavement, the lower side of the ball is temporarily flattened. Suppose that the maximum depth of the dent is on the order of 1 cm. Compute an order-of-magnitude estimate for the maximum acceleration of the ball while it is in contact with the pavement. State your assumptions, the quantities you estimate, and the values you estimate for them.

68. At NASA's John H. Glenn research center in Cleveland, Ohio, free-fall research is performed by dropping experiment packages from the top of an evacuated shaft 145 m high. Free fall imitates the so-called microgravity environment of a satellite in orbit. (a) What is the maximum time interval for free fall if an experiment package were to fall the entire 145 m? (b) Actual NASA specifications allow for a 5.18-s drop time interval. How far do the packages drop and (c) what is their speed at 5.18 s? (d) What constant acceleration would be required to stop an experiment package in the distance remaining in the shaft after its 5.18-s fall?

69. An inquisitive physics student and mountain climber climbs a 50.0-m cliff that overhangs a calm pool of water. He throws two stones vertically downward, 1.00 s apart, and observes that they cause a single splash. The first stone has an initial speed of 2.00 m/s. (a) How long after release of the first stone do the two stones hit the water? (b) What initial velocity must the second stone have if they are to hit simultaneously? (c) What is the speed of each stone at the instant the two hit the water?

70. A rock is dropped from rest into a well. The well is not really 16 seconds deep, as in Figure P2.70. (a) The sound of the splash is actually heard 2.40 s after the rock is released from rest. How far below the top of the well is the surface of the water? The speed of sound in air (at the ambient temperature) is 336 m/s. (b) **What If?** If the travel time for the sound is neglected, what percentage error is introduced when the depth of the well is calculated?

71. To protect his food from hungry bears, a boy scout raises his food pack with a rope that is thrown over a tree limb at height h above his hands. He walks away from the vertical rope with constant velocity v_{boy}, holding the free end of the rope in his hands (Fig. P2.71). (a) Show that the speed v of the food pack is given by $x(x^2 + h^2)^{-1/2}\, v_{\text{boy}}$ where x

By permission of John Hart and Creators Syndicate, Inc.

Figure P2.70

is the distance he has walked away from the vertical rope. (b) Show that the acceleration a of the food pack is $h^2(x^2 + h^2)^{-3/2}\, v_{\text{boy}}^2$. (c) What values do the acceleration a and velocity v have shortly after he leaves the point under the pack ($x = 0$)? (d) What values do the pack's velocity and acceleration approach as the distance x continues to increase?

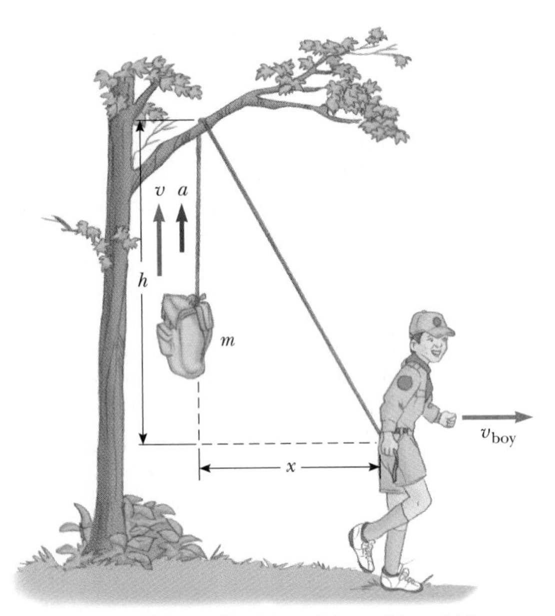

Figure P2.71 Problems 71 and 72.

72. In Problem 71, let the height h equal 6.00 m and the speed v_{boy} equal 2.00 m/s. Assume that the food pack starts from rest. (a) Tabulate and graph the speed–time graph. (b) Tabulate and graph the acceleration-time graph. Let the range of time be from 0 s to 5.00 s and the time intervals be 0.500 s.

73. Kathy Kool buys a sports car that can accelerate at the rate of 4.90 m/s^2. She decides to test the car by racing with another speedster, Stan Speedy. Both start from rest, but experienced Stan leaves the starting line 1.00 s before Kathy. If Stan moves with a constant acceleration of 3.50 m/s^2 and Kathy maintains an acceleration of 4.90 m/s^2, find (a) the time at which Kathy overtakes Stan, (b) the distance she travels before she catches him, and (c) the speeds of both cars at the instant she overtakes him.

74. Astronauts on a distant planet toss a rock into the air. With the aid of a camera that takes pictures at a steady rate, they record the height of the rock as a function of time as given in Table P2.74. (a) Find the average velocity of the rock in the time interval between each measurement and the next. (b) Using these average velocities to approximate instantaneous velocities at the midpoints of the time intervals, make a graph of velocity as a function of time. Does the rock move with constant acceleration? If so, plot a straight line of best fit on the graph and calculate its slope to find the acceleration.

Table P2.74

Height of a Rock versus Time

Time (s)	Height (m)	Time (s)	Height (m)
0.00	5.00	2.75	7.62
0.25	5.75	3.00	7.25
0.50	6.40	3.25	6.77
0.75	6.94	3.50	6.20
1.00	7.38	3.75	5.52
1.25	7.72	4.00	4.73
1.50	7.96	4.25	3.85
1.75	8.10	4.50	2.86
2.00	8.13	4.75	1.77
2.25	8.07	5.00	0.58
2.50	7.90		

75. Two objects, A and B, are connected by a rigid rod that has a length L. The objects slide along perpendicular guide rails, as shown in Figure P2.75. If A slides to the left with a constant speed v, find the velocity of B when $\alpha = 60.0°$.

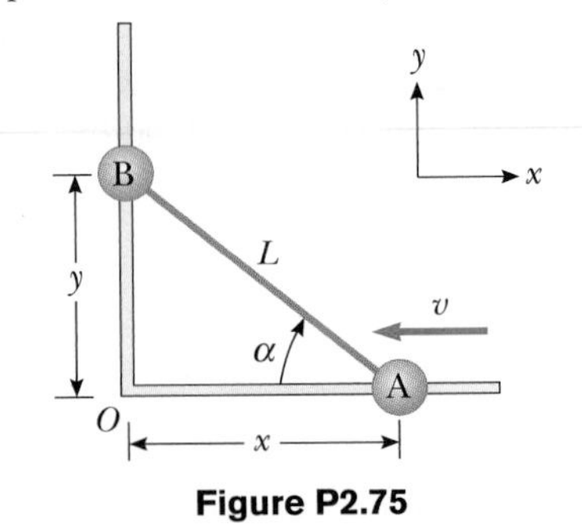

Figure P2.75

Answers to Quick Quizzes

2.1 (c). If the particle moves along a line without changing direction, the displacement and distance traveled over any time interval will be the same. As a result, the magnitude of the average velocity and the average speed will be the same. If the particle reverses direction, however, the displacement will be less than the distance traveled. In turn, the magnitude of the average velocity will be smaller than the average speed.

2.2 (b). If the car is slowing down, a force must be pulling in the direction opposite to its velocity.

2.3 False. Your graph should look something like the following. This v_x-t graph shows that the maximum speed is about 5.0 m/s, which is 18 km/h (= 11 mi/h), and so the driver was not speeding.

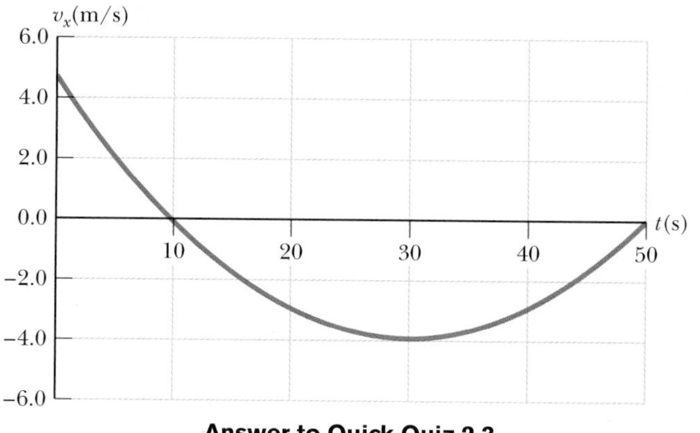

Answer to Quick Quiz 2.3

2.4 (c). If a particle with constant acceleration stops and its acceleration remains constant, it must begin to move again in the opposite direction. If it did not, the acceleration would change from its original constant value to zero. Choice (a) is not correct because the direction of acceleration is not specified by the direction of the velocity. Choice (b) is also not correct by counterexample—a car moving in the $-x$ direction and slowing down has a positive acceleration.

2.5 Graph (a) has a constant slope, indicating a constant acceleration; this is represented by graph (e).

Graph (b) represents a speed that is increasing constantly but not at a uniform rate. Thus, the acceleration must be increasing, and the graph that best indicates this is (d).

Graph (c) depicts a velocity that first increases at a constant rate, indicating constant acceleration. Then the velocity stops increasing and becomes constant, indicating zero acceleration. The best match to this situation is graph (f).

2.6 (e). For the entire time interval that the ball is in free fall, the acceleration is that due to gravity.

2.7 (d). While the ball is rising, it is slowing down. After reaching the highest point, the ball begins to fall and its speed increases.

2.8 (a). At the highest point, the ball is momentarily at rest, but still accelerating at $-g$.

Chapter 3

Vectors

CHAPTER OUTLINE

▲ *These controls in the cockpit of a commercial aircraft assist the pilot in maintaining control over the* velocity *of the aircraft—how fast it is traveling and in what direction it is traveling—allowing it to land safely. Quantities that are defined by both a magnitude and a direction, such as velocity, are called* vector quantities. (Mark Wagner/Getty Images)

In our study of physics, we often need to work with physical quantities that have both numerical and directional properties. As noted in Section 2.1, quantities of this nature are vector quantities. This chapter is primarily concerned with vector algebra and with some general properties of vector quantities. We discuss the addition and subtraction of vector quantities, together with some common applications to physical situations.

Vector quantities are used throughout this text, and it is therefore imperative that you master both their graphical and their algebraic properties.

3.1 Coordinate Systems

Many aspects of physics involve a description of a location in space. In Chapter 2, for example, we saw that the mathematical description of an object's motion requires a method for describing the object's position at various times. This description is accomplished with the use of coordinates, and in Chapter 2 we used the Cartesian coordinate system, in which horizontal and vertical axes intersect at a point defined as the origin (Fig. 3.1). Cartesian coordinates are also called *rectangular coordinates.*

Sometimes it is more convenient to represent a point in a plane by its *plane polar coordinates* (r, θ), as shown in Figure 3.2a. In this *polar coordinate system,* r is the distance from the origin to the point having Cartesian coordinates (x, y), and θ is the angle between a line drawn from the origin to the point and a fixed axis. This fixed axis is usually the positive x axis, and θ is usually measured counterclockwise from it. From the right triangle in Figure 3.2b, we find that $\sin\theta = y/r$ and that $\cos\theta = x/r$. (A review of trigonometric functions is given in Appendix B.4.) Therefore, starting with the plane polar coordinates of any point, we can obtain the Cartesian coordinates by using the equations

$$x = r\cos\theta \tag{3.1}$$

$$y = r\sin\theta \tag{3.2}$$

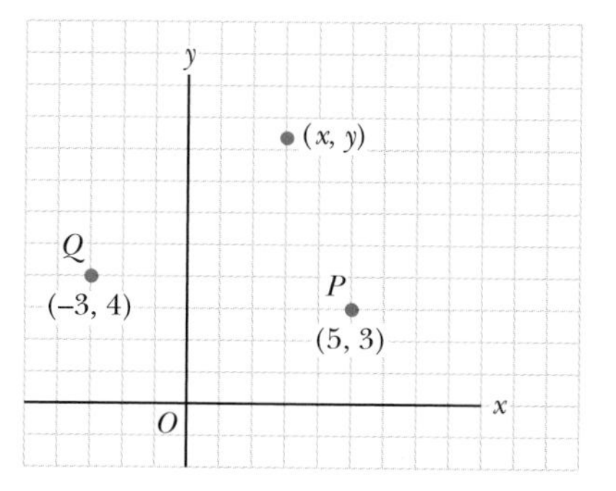

Figure 3.1 Designation of points in a Cartesian coordinate system. Every point is labeled with coordinates (x, y).

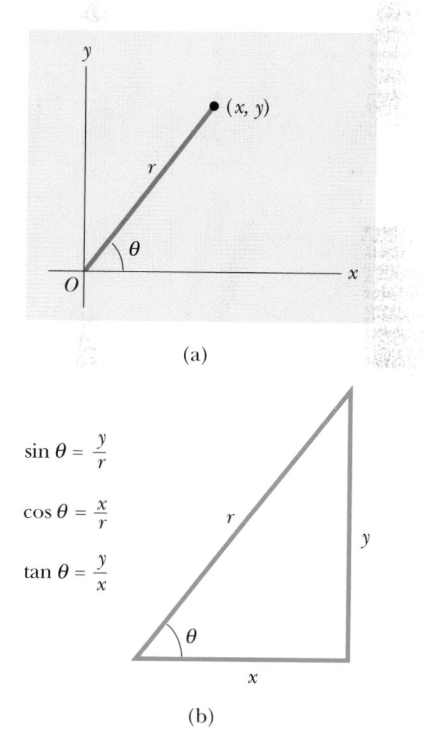

Active Figure 3.2 (a) The plane polar coordinates of a point are represented by the distance r and the angle θ, where θ is measured counterclockwise from the positive x axis. (b) The right triangle used to relate (x, y) to (r, θ).

At the Active Figures link at http://www.pse6.com, you can move the point and see the changes to the rectangular and polar coordinates as well as to the sine, cosine, and tangent of angle θ.

Furthermore, the definitions of trigonometry tell us that

$$\tan \theta = \frac{y}{x} \tag{3.3}$$

$$r = \sqrt{x^2 + y^2} \tag{3.4}$$

Equation 3.4 is the familiar Pythagorean theorem.

These four expressions relating the coordinates (x, y) to the coordinates (r, θ) apply only when θ is defined as shown in Figure 3.2a—in other words, when positive θ is an angle measured counterclockwise from the positive x axis. (Some scientific calculators perform conversions between Cartesian and polar coordinates based on these standard conventions.) If the reference axis for the polar angle θ is chosen to be one other than the positive x axis or if the sense of increasing θ is chosen differently, then the expressions relating the two sets of coordinates will change.

Example 3.1 Polar Coordinates

The Cartesian coordinates of a point in the xy plane are $(x, y) = (-3.50, -2.50)$ m, as shown in Figure 3.3. Find the polar coordinates of this point.

Solution For the examples in this and the next two chapters we will illustrate the use of the General Problem-Solving Strategy outlined at the end of Chapter 2. In subsequent chapters, we will make fewer explicit references to this strategy, as you will have become familiar with it and should be applying it on your own. The drawing in Figure 3.3 helps us to *conceptualize* the problem. We can *categorize* this as a plug-in problem. From Equation 3.4,

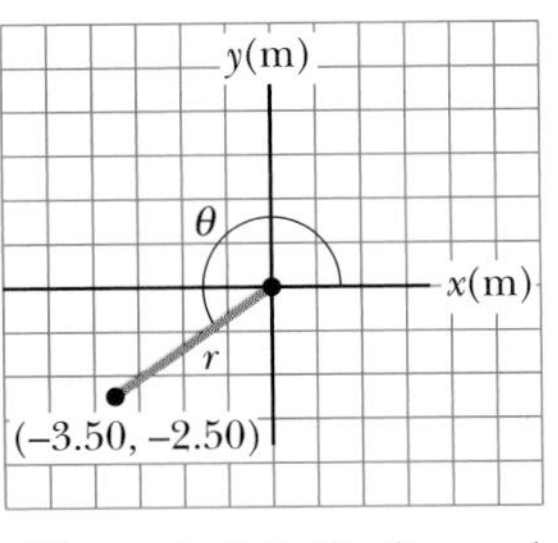

Active Figure 3.3 (Example 3.1) Finding polar coordinates when Cartesian coordinates are given.

At the Active Figures link at http://www.pse6.com, you can move the point in the xy plane and see how its Cartesian and polar coordinates change.

$$r = \sqrt{x^2 + y^2} = \sqrt{(-3.50 \text{ m})^2 + (-2.50 \text{ m})^2} = \boxed{4.30 \text{ m}}$$

and from Equation 3.3,

$$\tan \theta = \frac{y}{x} = \frac{-2.50 \text{ m}}{-3.50 \text{ m}} = 0.714$$

$$\theta = \boxed{216°}$$

Note that you must use the signs of x and y to find that the point lies in the third quadrant of the coordinate system. That is, $\theta = 216°$ and not $35.5°$.

3.2 Vector and Scalar Quantities

As noted in Chapter 2, some physical quantities are scalar quantities whereas others are vector quantities. When you want to know the temperature outside so that you will know how to dress, the only information you need is a number and the unit "degrees C" or "degrees F." Temperature is therefore an example of a *scalar quantity*:

> A **scalar quantity** is completely specified by a single value with an appropriate unit and has no direction.

Other examples of scalar quantities are volume, mass, speed, and time intervals. The rules of ordinary arithmetic are used to manipulate scalar quantities.

If you are preparing to pilot a small plane and need to know the wind velocity, you must know both the speed of the wind and its direction. Because direction is important for its complete specification, velocity is a *vector quantity*:

A **vector quantity** is completely specified by a number and appropriate units plus a direction.

Another example of a vector quantity is displacement, as you know from Chapter 2. Suppose a particle moves from some point Ⓐ to some point Ⓑ along a straight path, as shown in Figure 3.4. We represent this displacement by drawing an arrow from Ⓐ to Ⓑ, with the tip of the arrow pointing away from the starting point. The direction of the arrowhead represents the direction of the displacement, and the length of the arrow represents the magnitude of the displacement. If the particle travels along some other path from Ⓐ to Ⓑ, such as the broken line in Figure 3.4, its displacement is still the arrow drawn from Ⓐ to Ⓑ. Displacement depends only on the initial and final positions, so the displacement vector is independent of the path taken between these two points.

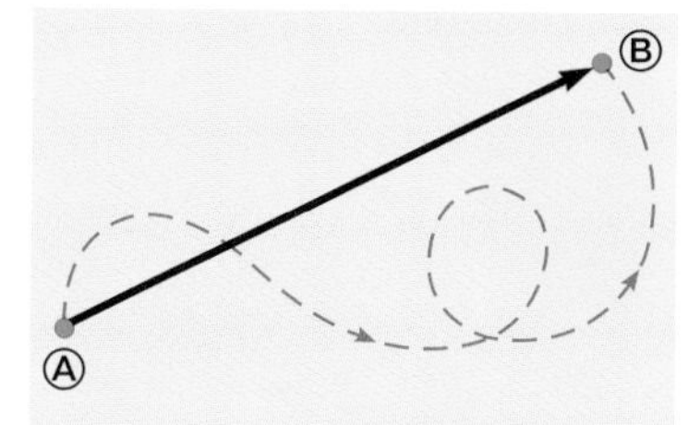

Figure 3.4 As a particle moves from Ⓐ to Ⓑ along an arbitrary path represented by the broken line, its displacement is a vector quantity shown by the arrow drawn from Ⓐ to Ⓑ.

In this text, we use a boldface letter, such as **A**, to represent a vector quantity. Another notation is useful when boldface notation is difficult, such as when writing on paper or on a chalkboard—an arrow is written over the symbol for the vector: $\vec{A}$. The magnitude of the vector **A** is written either A or $|\mathbf{A}|$. The magnitude of a vector has physical units, such as meters for displacement or meters per second for velocity. The magnitude of a vector is *always* a positive number.

Quick Quiz 3.1 Which of the following are vector quantities and which are scalar quantities? (a) your age (b) acceleration (c) velocity (d) speed (e) mass

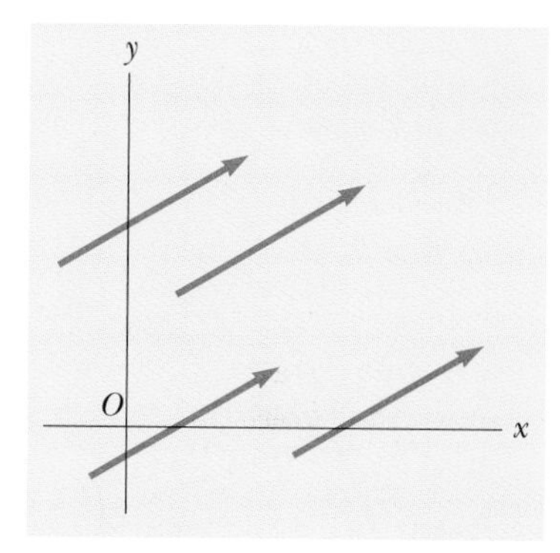

Figure 3.5 These four vectors are equal because they have equal lengths and point in the same direction.

3.3 Some Properties of Vectors

Equality of Two Vectors

For many purposes, two vectors **A** and **B** may be defined to be equal if they have the same magnitude and point in the same direction. That is, $\mathbf{A} = \mathbf{B}$ only if $A = B$ and if **A** and **B** point in the same direction along parallel lines. For example, all the vectors in Figure 3.5 are equal even though they have different starting points. This property allows us to move a vector to a position parallel to itself in a diagram without affecting the vector.

Adding Vectors

The rules for adding vectors are conveniently described by graphical methods. To add vector **B** to vector **A**, first draw vector **A** on graph paper, with its magnitude represented by a convenient length scale, and then draw vector **B** to the same scale with its tail starting from the tip of **A**, as shown in Figure 3.6. The **resultant vector** $\mathbf{R} = \mathbf{A} + \mathbf{B}$ is the vector drawn from the tail of **A** to the tip of **B**.

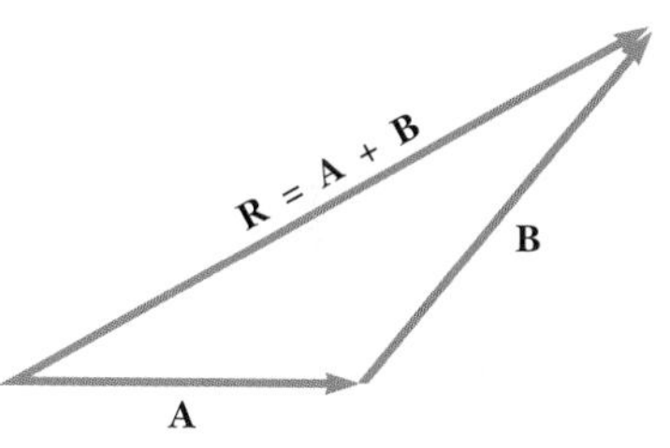

Active Figure 3.6 When vector **B** is added to vector **A**, the resultant **R** is the vector that runs from the tail of **A** to the tip of **B**.

Go to the Active Figures link at http://www.pse6.com.

For example, if you walked 3.0 m toward the east and then 4.0 m toward the north, as shown in Figure 3.7, you would find yourself 5.0 m from where you started, measured at an angle of 53° north of east. Your total displacement is the vector sum of the individual displacements.

A geometric construction can also be used to add more than two vectors. This is shown in Figure 3.8 for the case of four vectors. The resultant vector $\mathbf{R} = \mathbf{A} + \mathbf{B} + \mathbf{C} + \mathbf{D}$ is the vector that completes the polygon. In other words, **R is the vector drawn from the tail of the first vector to the tip of the last vector**.

When two vectors are added, the sum is independent of the order of the addition. (This fact may seem trivial, but as you will see in Chapter 11, the order is important

PITFALL PREVENTION

3.1 Vector Addition versus Scalar Addition

Keep in mind that $\mathbf{A} + \mathbf{B} = \mathbf{C}$ is very different from $A + B = C$. The first is a vector sum, which must be handled carefully, such as with the graphical method described here. The second is a simple algebraic addition of numbers that is handled with the normal rules of arithmetic.

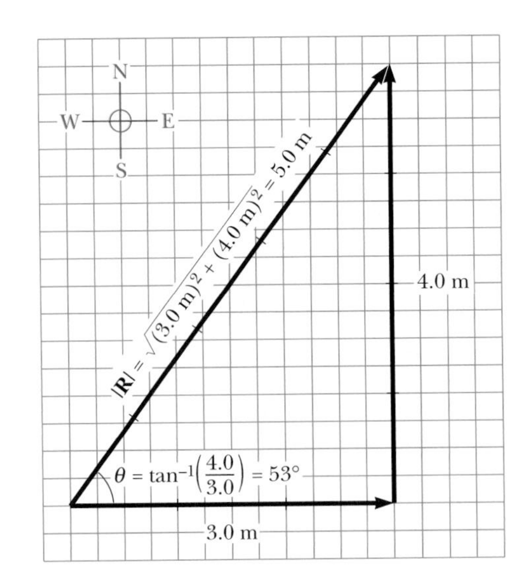

Figure 3.7 Vector addition. Walking first 3.0 m due east and then 4.0 m due north leaves you 5.0 m from your starting point.

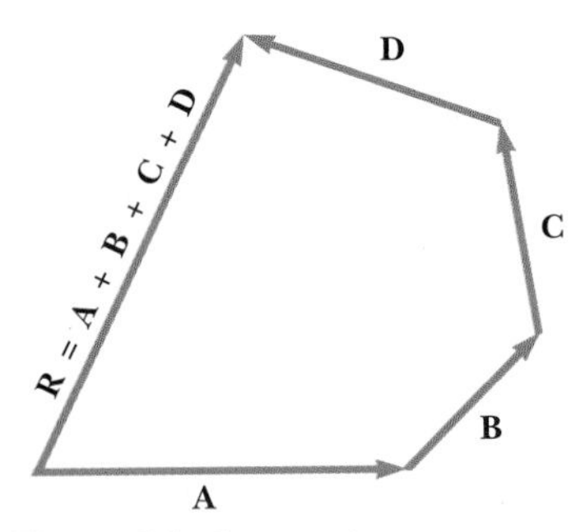

Figure 3.8 Geometric construction for summing four vectors. The resultant vector **R** is by definition the one that completes the polygon.

when vectors are multiplied). This can be seen from the geometric construction in Figure 3.9 and is known as the **commutative law of addition:**

$$\mathbf{A} + \mathbf{B} = \mathbf{B} + \mathbf{A} \tag{3.5}$$

When three or more vectors are added, their sum is independent of the way in which the individual vectors are grouped together. A geometric proof of this rule for three vectors is given in Figure 3.10. This is called the **associative law of addition:**

$$\mathbf{A} + (\mathbf{B} + \mathbf{C}) = (\mathbf{A} + \mathbf{B}) + \mathbf{C} \tag{3.6}$$

In summary, **a vector quantity has both magnitude and direction and also obeys the laws of vector addition** as described in Figures 3.6 to 3.10. When two or more vectors are added together, all of them must have the same units and all of them must be the same type of quantity. It would be meaningless to add a velocity vector (for example, 60 km/h to the east) to a displacement vector (for example, 200 km to the north) because they represent different physical quantities. The same rule also applies to scalars. For example, it would be meaningless to add time intervals to temperatures.

Negative of a Vector

The negative of the vector **A** is defined as the vector that when added to **A** gives zero for the vector sum. That is, $\mathbf{A} + (-\mathbf{A}) = 0$. The vectors **A** and $-\mathbf{A}$ have the same magnitude but point in opposite directions.

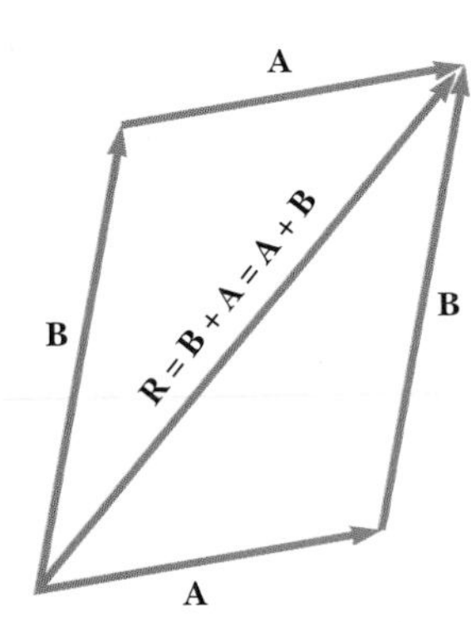

Figure 3.9 This construction shows that $\mathbf{A} + \mathbf{B} = \mathbf{B} + \mathbf{A}$—in other words, that vector addition is commutative.

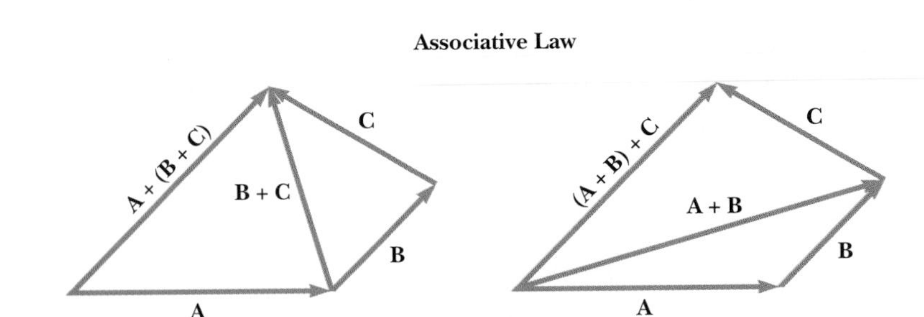

Figure 3.10 Geometric constructions for verifying the associative law of addition.

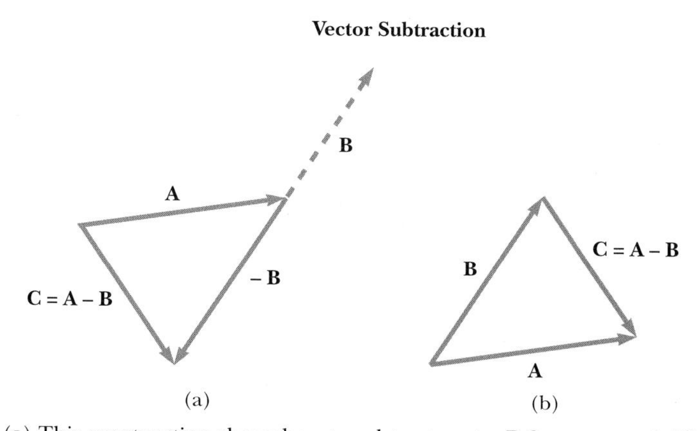

Figure 3.11 (a) This construction shows how to subtract vector **B** from vector **A**. The vector $-$**B** is equal in magnitude to vector **B** and points in the opposite direction. To subtract **B** from **A**, apply the rule of vector addition to the combination of **A** and $-$**B**: Draw **A** along some convenient axis, place the tail of $-$**B** at the tip of **A**, and **C** is the difference **A** $-$ **B**. (b) A second way of looking at vector subtraction. The difference vector **C** = **A** $-$ **B** is the vector that we must add to **B** to obtain **A**.

Subtracting Vectors

The operation of vector subtraction makes use of the definition of the negative of a vector. We define the operation **A** $-$ **B** as vector $-$**B** added to vector **A**:

$$\mathbf{A} - \mathbf{B} = \mathbf{A} + (-\mathbf{B}) \tag{3.7}$$

The geometric construction for subtracting two vectors in this way is illustrated in Figure 3.11a.

Another way of looking at vector subtraction is to note that the difference **A** $-$ **B** between two vectors **A** and **B** is what you have to add to the second vector to obtain the first. In this case, the vector **A** $-$ **B** points from the tip of the second vector to the tip of the first, as Figure 3.11b shows.

Quick Quiz 3.2 The magnitudes of two vectors **A** and **B** are $A = 12$ units and $B = 8$ units. Which of the following pairs of numbers represents the *largest* and *smallest* possible values for the magnitude of the resultant vector **R** = **A** + **B**? (a) 14.4 units, 4 units (b) 12 units, 8 units (c) 20 units, 4 units (d) none of these answers.

Quick Quiz 3.3 If vector **B** is added to vector **A**, under what condition does the resultant vector **A** + **B** have magnitude $A + B$? (a) **A** and **B** are parallel and in the same direction. (b) **A** and **B** are parallel and in opposite directions. (c) **A** and **B** are perpendicular.

Quick Quiz 3.4 If vector **B** is added to vector **A**, which *two* of the following choices must be true in order for the resultant vector to be equal to zero? (a) **A** and **B** are parallel and in the same direction. (b) **A** and **B** are parallel and in opposite directions. (c) **A** and **B** have the same magnitude. (d) **A** and **B** are perpendicular.

Example 3.2 A Vacation Trip

A car travels 20.0 km due north and then 35.0 km in a direction 60.0° west of north, as shown in Figure 3.12a. Find the magnitude and direction of the car's resultant displacement.

Solution The vectors **A** and **B** drawn in Figure 3.12a help us to *conceptualize* the problem. We can *categorize* this as a relatively simple analysis problem in vector addition. The displacement **R** is the resultant when the two individual displacements **A** and **B** are added. We can further categorize this as a problem about the analysis of triangles, so we appeal to our expertise in geometry and trigonometry.

In this example, we show two ways to *analyze* the problem of finding the resultant of two vectors. The first way is to solve the problem geometrically, using graph paper and a protractor to measure the magnitude of **R** and its direction in Figure 3.12a. (In fact, even when you know you are going to be carrying out a calculation, you should sketch the vectors to check your results.) With an ordinary ruler and protractor, a large diagram typically gives answers to two-digit but not to three-digit precision.

The second way to solve the problem is to analyze it algebraically. The magnitude of **R** can be obtained from the law of cosines as applied to the triangle (see Appendix B.4). With $\theta = 180° - 60° = 120°$ and $R^2 = A^2 + B^2 - 2AB\cos\theta$, we find that

$$R = \sqrt{A^2 + B^2 - 2AB\cos\theta}$$

$$= \sqrt{(20.0\text{ km})^2 + (35.0\text{ km})^2 - 2(20.0\text{ km})(35.0\text{ km})\cos 120°}$$

$$= \boxed{48.2\text{ km}}$$

The direction of **R** measured from the northerly direction can be obtained from the law of sines (Appendix B.4):

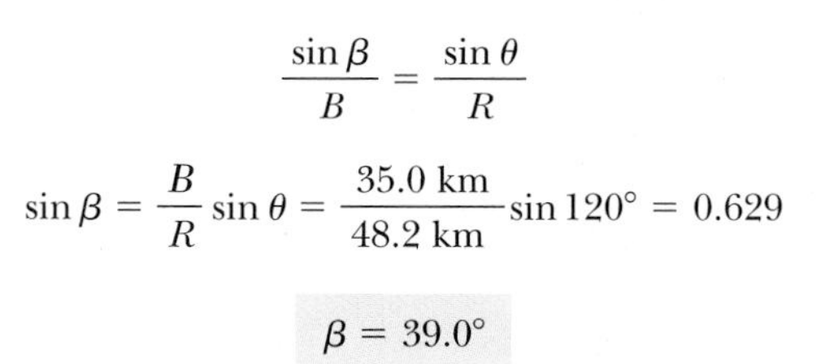

$$\frac{\sin\beta}{B} = \frac{\sin\theta}{R}$$

$$\sin\beta = \frac{B}{R}\sin\theta = \frac{35.0\text{ km}}{48.2\text{ km}}\sin 120° = 0.629$$

$$\boxed{\beta = 39.0°}$$

The resultant displacement of the car is 48.2 km in a direction 39.0° west of north.

We now *finalize* the problem. Does the angle β that we calculated agree with an estimate made by looking at Figure 3.12a or with an actual angle measured from the diagram using the graphical method? Is it reasonable that the magnitude of **R** is larger than that of both **A** and **B**? Are the units of **R** correct?

While the graphical method of adding vectors works well, it suffers from two disadvantages. First, some individuals find using the laws of cosines and sines to be awkward. Second, a triangle only results if you are adding two vectors. If you are adding three or more vectors, the resulting geometric shape is not a triangle. In Section 3.4, we explore a new method of adding vectors that will address both of these disadvantages.

What If? Suppose the trip were taken with the two vectors in reverse order: 35.0 km at 60.0° west of north first, and then 20.0 km due north. How would the magnitude and the direction of the resultant vector change?

Answer They would not change. The commutative law for vector addition tells us that the order of vectors in an addition is irrelevant. Graphically, Figure 3.12b shows that the vectors added in the reverse order give us the same resultant vector.

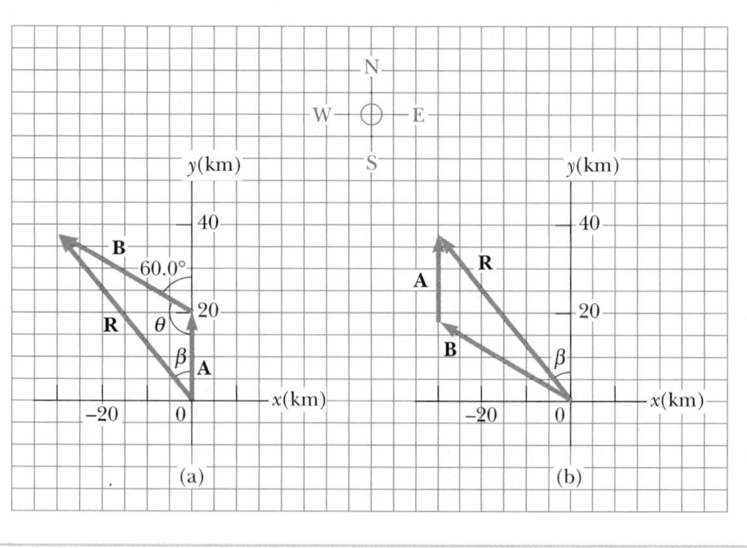

Figure 3.12 (Example 3.2) (a) Graphical method for finding the resultant displacement vector **R** = **A** + **B**. (b) Adding the vectors in reverse order (**B** + **A**) gives the same result for **R**.

Multiplying a Vector by a Scalar

If vector **A** is multiplied by a positive scalar quantity m, then the product $m\mathbf{A}$ is a vector that has the same direction as **A** and magnitude mA. If vector **A** is multiplied by a negative scalar quantity $-m$, then the product $-m\mathbf{A}$ is directed opposite **A**. For example, the vector $5\mathbf{A}$ is five times as long as **A** and points in the same direction as **A**; the vector $-\frac{1}{3}\mathbf{A}$ is one-third the length of **A** and points in the direction opposite **A**.

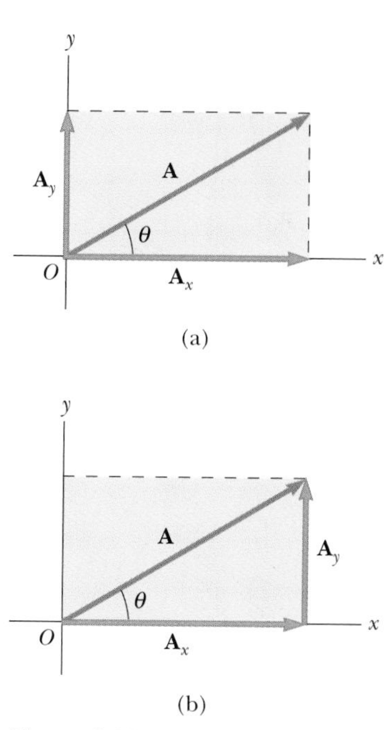

Figure 3.13 (a) A vector **A** lying in the xy plane can be represented by its component vectors $\mathbf{A}_x$ and $\mathbf{A}_y$. (b) The y component vector $\mathbf{A}_y$ can be moved to the right so that it adds to $\mathbf{A}_x$. The vector sum of the component vectors is **A**. These three vectors form a right triangle.

3.4 Components of a Vector and Unit Vectors

The graphical method of adding vectors is not recommended whenever high accuracy is required or in three-dimensional problems. In this section, we describe a method of adding vectors that makes use of the projections of vectors along coordinate axes. These projections are called the **components** of the vector. Any vector can be completely described by its components.

Consider a vector **A** lying in the xy plane and making an arbitrary angle θ with the positive x axis, as shown in Figure 3.13a. This vector can be expressed as the sum of two other vectors $\mathbf{A}_x$ and $\mathbf{A}_y$. From Figure 3.13b, we see that the three vectors form a right triangle and that $\mathbf{A} = \mathbf{A}_x + \mathbf{A}_y$. We shall often refer to the "components of a vector **A**," written A_x and A_y (without the boldface notation). The component A_x represents the projection of **A** along the x axis, and the component A_y represents the projection of **A** along the y axis. These components can be positive or negative. The component A_x is positive if $\mathbf{A}_x$ points in the positive x direction and is negative if $\mathbf{A}_x$ points in the negative x direction. The same is true for the component A_y.

From Figure 3.13 and the definition of sine and cosine, we see that $\cos\theta = A_x/A$ and that $\sin\theta = A_y/A$. Hence, the components of **A** are

Components of the vector A

$$A_x = A\cos\theta \tag{3.8}$$

$$A_y = A\sin\theta \tag{3.9}$$

These components form two sides of a right triangle with a hypotenuse of length A. Thus, it follows that the magnitude and direction of **A** are related to its components through the expressions

$$A = \sqrt{A_x^{\,2} + A_y^{\,2}} \tag{3.10}$$

$$\theta = \tan^{-1}\left(\frac{A_y}{A_x}\right) \tag{3.11}$$

Note that **the signs of the components A_x and A_y depend on the angle θ**. For example, if $\theta = 120°$, then A_x is negative and A_y is positive. If $\theta = 225°$, then both A_x and A_y are negative. Figure 3.14 summarizes the signs of the components when **A** lies in the various quadrants.

When solving problems, you can specify a vector **A** either with its components A_x and A_y or with its magnitude and direction A and θ.

PITFALL PREVENTION

3.2 Component Vectors versus Components

The vectors $\mathbf{A}_x$ and $\mathbf{A}_y$ are the *component vectors* of **A**. These should not be confused with the scalars A_x and A_y, which we shall always refer to as the *components* of **A**.

Quick Quiz 3.5 Choose the correct response to make the sentence true: A component of a vector is (a) always, (b) never, or (c) sometimes larger than the magnitude of the vector.

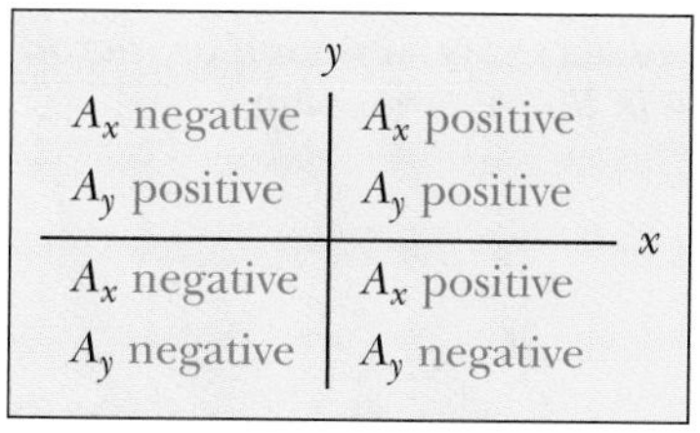

A_x negative, A_y positive	A_x positive, A_y positive
A_x negative, A_y negative	A_x positive, A_y negative

Figure 3.14 The signs of the components of a vector **A** depend on the quadrant in which the vector is located.

Suppose you are working a physics problem that requires resolving a vector into its components. In many applications it is convenient to express the components in a coordinate system having axes that are not horizontal and vertical but are still perpendicular to each other. If you choose reference axes or an angle other than the axes and angle shown in Figure 3.13, the components must be modified accordingly. Suppose a

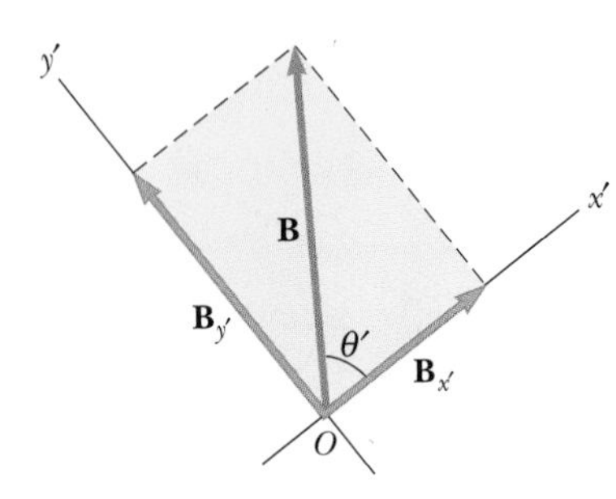

Figure 3.15 The component vectors of **B** in a coordinate system that is tilted.

vector **B** makes an angle θ' with the x' axis defined in Figure 3.15. The components of **B** along the x' and y' axes are $B_{x'} = B \cos \theta'$ and $B_{y'} = B \sin \theta'$, as specified by Equations 3.8 and 3.9. The magnitude and direction of **B** are obtained from expressions equivalent to Equations 3.10 and 3.11. Thus, we can express the components of a vector in any coordinate system that is convenient for a particular situation.

Unit Vectors

Vector quantities often are expressed in terms of unit vectors. **A unit vector is a dimensionless vector having a magnitude of exactly 1.** Unit vectors are used to specify a given direction and have no other physical significance. They are used solely as a convenience in describing a direction in space. We shall use the symbols $\hat{\mathbf{i}}$, $\hat{\mathbf{j}}$, and $\hat{\mathbf{k}}$ to represent unit vectors pointing in the positive *x*, *y*, and *z* directions, respectively. (The "hats" on the symbols are a standard notation for unit vectors.) The unit vectors $\hat{\mathbf{i}}$, $\hat{\mathbf{j}}$, and $\hat{\mathbf{k}}$ form a set of mutually perpendicular vectors in a right-handed coordinate system, as shown in Figure 3.16a. The magnitude of each unit vector equals 1; that is, $|\hat{\mathbf{i}}| = |\hat{\mathbf{j}}| = |\hat{\mathbf{k}}| = 1$.

Consider a vector **A** lying in the *xy* plane, as shown in Figure 3.16b. The product of the component A_x and the unit vector $\hat{\mathbf{i}}$ is the vector $A_x\hat{\mathbf{i}}$, which lies on the *x* axis and has magnitude $|A_x|$. (The vector $A_x\hat{\mathbf{i}}$ is an alternative representation of vector $\mathbf{A}_x$.) Likewise, $A_y\hat{\mathbf{j}}$ is a vector of magnitude $|A_y|$ lying on the *y* axis. (Again, vector $A_y\hat{\mathbf{j}}$ is an alternative representation of vector $\mathbf{A}_y$.) Thus, the unit–vector notation for the vector **A** is

$$\mathbf{A} = A_x\hat{\mathbf{i}} + A_y\hat{\mathbf{j}} \tag{3.12}$$

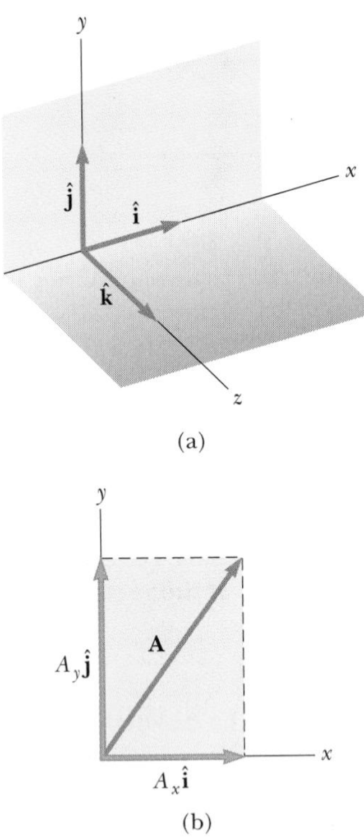

Active Figure 3.16 (a) The unit vectors $\hat{\mathbf{i}}$, $\hat{\mathbf{j}}$, and $\hat{\mathbf{k}}$ are directed along the *x*, *y*, and *z* axes, respectively. (b) Vector $\mathbf{A} = A_x\hat{\mathbf{i}} + A_y\hat{\mathbf{j}}$ lying in the *xy* plane has components A_x and A_y.

At the Active Figures link at http://www.pse6.com you can rotate the coordinate axes in 3-dimensional space and view a representation of vector A in three dimensions.

For example, consider a point lying in the *xy* plane and having Cartesian coordinates (x, y), as in Figure 3.17. The point can be specified by the **position vector r**, which in unit–vector form is given by

$$\mathbf{r} = x\hat{\mathbf{i}} + y\hat{\mathbf{j}} \tag{3.13}$$

This notation tells us that the components of **r** are the lengths *x* and *y*.

Now let us see how to use components to add vectors when the graphical method is not sufficiently accurate. Suppose we wish to add vector **B** to vector **A** in Equation 3.12, where vector **B** has components B_x and B_y. All we do is add the *x* and *y* components separately. The resultant vector $\mathbf{R} = \mathbf{A} + \mathbf{B}$ is therefore

$$\mathbf{R} = (A_x\hat{\mathbf{i}} + A_y\hat{\mathbf{j}}) + (B_x\hat{\mathbf{i}} + B_y\hat{\mathbf{j}})$$

or

$$\mathbf{R} = (A_x + B_x)\hat{\mathbf{i}} + (A_y + B_y)\hat{\mathbf{j}} \tag{3.14}$$

Because $\mathbf{R} = R_x\hat{\mathbf{i}} + R_y\hat{\mathbf{j}}$, we see that the components of the resultant vector are

$$R_x = A_x + B_x \qquad R_y = A_y + B_y \tag{3.15}$$

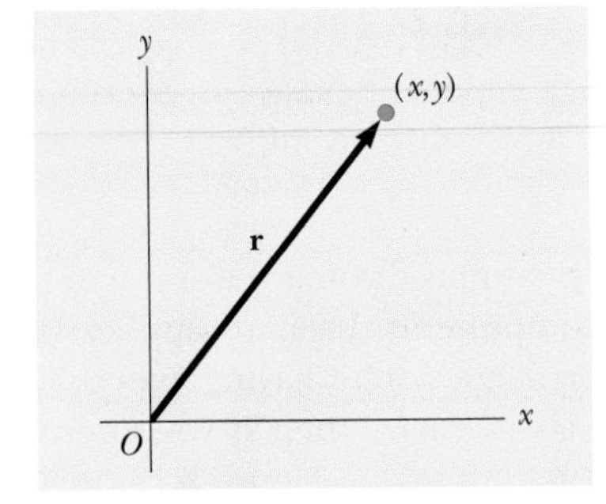

Figure 3.17 The point whose Cartesian coordinates are (x, y) can be represented by the position vector $\mathbf{r} = x\hat{\mathbf{i}} + y\hat{\mathbf{j}}$.

We obtain the magnitude of **R** and the angle it makes with the x axis from its components, using the relationships

$$R = \sqrt{R_x^2 + R_y^2} = \sqrt{(A_x + B_x)^2 + (A_y + B_y)^2} \tag{3.16}$$

$$\tan \theta = \frac{R_y}{R_x} = \frac{A_y + B_y}{A_x + B_x} \tag{3.17}$$

We can check this addition by components with a geometric construction, as shown in Figure 3.18. Remember that you must note the signs of the components when using either the algebraic or the graphical method.

At times, we need to consider situations involving motion in three component directions. The extension of our methods to three-dimensional vectors is straightforward. If **A** and **B** both have x, y, and z components, we express them in the form

$$\mathbf{A} = A_x\hat{\mathbf{i}} + A_y\hat{\mathbf{j}} + A_z\hat{\mathbf{k}} \tag{3.18}$$

$$\mathbf{B} = B_x\hat{\mathbf{i}} + B_y\hat{\mathbf{j}} + B_z\hat{\mathbf{k}} \tag{3.19}$$

The sum of **A** and **B** is

$$\mathbf{R} = (A_x + B_x)\hat{\mathbf{i}} + (A_y + B_y)\hat{\mathbf{j}} + (A_z + B_z)\hat{\mathbf{k}} \tag{3.20}$$

Note that Equation 3.20 differs from Equation 3.14: in Equation 3.20, the resultant vector also has a z component $R_z = A_z + B_z$. If a vector **R** has x, y, and z components, the magnitude of the vector is $R = \sqrt{R_x^2 + R_y^2 + R_z^2}$. The angle θ_x that **R** makes with the x axis is found from the expression $\cos \theta_x = R_x/R$, with similar expressions for the angles with respect to the y and z axes.

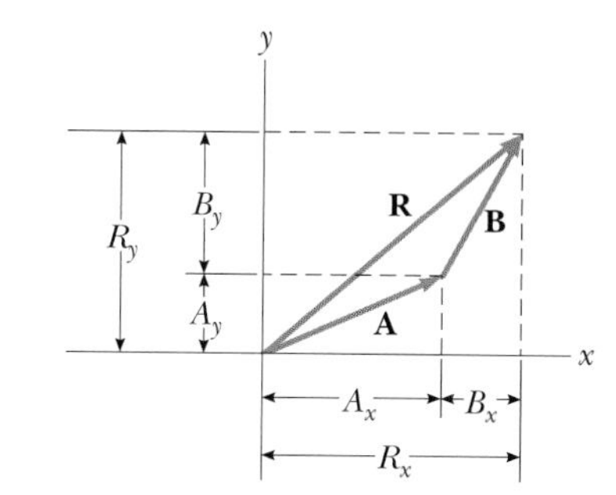

Figure 3.18 This geometric construction for the sum of two vectors shows the relationship between the components of the resultant **R** and the components of the individual vectors.

Quick Quiz 3.6 If at least one component of a vector is a positive number, the vector cannot (a) have any component that is negative (b) be zero (c) have three dimensions.

Quick Quiz 3.7 If $\mathbf{A} + \mathbf{B} = 0$, the corresponding components of the two vectors **A** and **B** must be (a) equal (b) positive (c) negative (d) of opposite sign.

Quick Quiz 3.8 For which of the following vectors is the magnitude of the vector equal to one of the components of the vector? (a) $\mathbf{A} = 2\hat{\mathbf{i}} + 5\hat{\mathbf{j}}$ (b) $\mathbf{B} = -3\hat{\mathbf{j}}$ (c) $\mathbf{C} = +5\hat{\mathbf{k}}$

PITFALL PREVENTION

3.3 *x* and *y* Components

Equations 3.8 and 3.9 associate the cosine of the angle with the x component and the sine of the angle with the y component. This is true *only* because we measured the angle θ with respect to the x axis, so don't memorize these equations. If θ is measured with respect to the y axis (as in some problems), these equations will be incorrect. Think about which side of the triangle containing the components is adjacent to the angle and which side is opposite, and assign the cosine and sine accordingly.

PROBLEM-SOLVING HINTS

Adding Vectors

When you need to add two or more vectors, use this step-by-step procedure:

- Select a coordinate system that is convenient. (Try to reduce the number of components you need to calculate by choosing axes that line up with as many vectors as possible.)
- Draw a labeled sketch of the vectors described in the problem.
- Find the x and y components of all vectors and the resultant components (the algebraic sum of the components) in the x and y directions.
- If necessary, use the Pythagorean theorem to find the magnitude of the resultant vector and select a suitable trigonometric function to find the angle that the resultant vector makes with the x axis.

PITFALL PREVENTION

3.4 Tangents on Calculators

Generally, the inverse tangent function on calculators provides an angle between $-90°$ and $+90°$. As a consequence, if the vector you are studying lies in the second or third quadrant, the angle measured from the positive x axis will be the angle your calculator returns plus $180°$.

Example 3.3 The Sum of Two Vectors

Find the sum of two vectors **A** and **B** lying in the *xy* plane and given by

$$\mathbf{A} = (2.0\hat{\mathbf{i}} + 2.0\hat{\mathbf{j}})\text{ m} \qquad \text{and} \qquad \mathbf{B} = (2.0\hat{\mathbf{i}} - 4.0\hat{\mathbf{j}})\text{ m}$$

Solution You may wish to draw the vectors to *conceptualize* the situation. We *categorize* this as a simple plug-in problem. Comparing this expression for **A** with the general expression $\mathbf{A} = A_x\hat{\mathbf{i}} + A_y\hat{\mathbf{j}}$, we see that $A_x = 2.0$ m and $A_y = 2.0$ m. Likewise, $B_x = 2.0$ m and $B_y = -4.0$ m. We obtain the resultant vector **R**, using Equation 3.14:

$$\mathbf{R} = \mathbf{A} + \mathbf{B} = (2.0 + 2.0)\hat{\mathbf{i}}\text{ m} + (2.0 - 4.0)\hat{\mathbf{j}}\text{ m}$$
$$= (4.0\hat{\mathbf{i}} - 2.0\hat{\mathbf{j}})\text{ m}$$

or

$$R_x = 4.0\text{ m} \qquad R_y = -2.0\text{ m}$$

The magnitude of **R** is found using Equation 3.16:

$$R = \sqrt{R_x^2 + R_y^2} = \sqrt{(4.0\text{ m})^2 + (-2.0\text{ m})^2} = \sqrt{20}\text{ m}$$
$$= 4.5\text{ m}$$

We can find the direction of **R** from Equation 3.17:

$$\tan\theta = \frac{R_y}{R_x} = \frac{-2.0\text{ m}}{4.0\text{ m}} = -0.50$$

Your calculator likely gives the answer $-27°$ for $\theta = \tan^{-1}(-0.50)$. This answer is correct if we interpret it to mean 27° clockwise from the *x* axis. Our standard form has been to quote the angles measured counterclockwise from the $+x$ axis, and that angle for this vector is $\theta = 333°$.

Example 3.4 The Resultant Displacement

A particle undergoes three consecutive displacements: $\mathbf{d}_1 = (15\hat{\mathbf{i}} + 30\hat{\mathbf{j}} + 12\hat{\mathbf{k}})$ cm, $\mathbf{d}_2 = (23\hat{\mathbf{i}} - 14\hat{\mathbf{j}} - 5.0\hat{\mathbf{k}})$ cm and $\mathbf{d}_3 = (-13\hat{\mathbf{i}} + 15\hat{\mathbf{j}})$ cm. Find the components of the resultant displacement and its magnitude.

Solution Three-dimensional displacements are more difficult to imagine than those in two dimensions, because the latter can be drawn on paper. For this problem, let us *conceptualize* that you start with your pencil at the origin of a piece of graph paper on which you have drawn *x* and *y* axes. Move your pencil 15 cm to the right along the *x* axis, then 30 cm upward along the *y* axis, and then 12 cm *vertically away* from the graph paper. This provides the displacement described by $\mathbf{d}_1$. From this point, move your pencil 23 cm to the right parallel to the *x* axis, 14 cm parallel to the graph paper in the $-y$ direction, and then 5.0 cm vertically downward toward the graph paper. You are now at the displacement from the origin described by $\mathbf{d}_1 + \mathbf{d}_2$. From this point, move your pencil 13 cm to the left in the $-x$ direction, and (finally!) 15 cm parallel to the graph paper along the *y* axis. Your final position is at a displacement $\mathbf{d}_1 + \mathbf{d}_2 + \mathbf{d}_3$ from the origin.

Despite the difficulty in conceptualizing in three dimensions, we can *categorize* this problem as a plug-in problem due to the careful bookkeeping methods that we have developed for vectors. The mathematical manipulation keeps track of this motion along the three perpendicular axes in an organized, compact way:

$$\mathbf{R} = \mathbf{d}_1 + \mathbf{d}_2 + \mathbf{d}_3$$
$$= (15 + 23 - 13)\hat{\mathbf{i}}\text{ cm} + (30 - 14 + 15)\hat{\mathbf{j}}\text{ cm} + (12 - 5.0 + 0)\hat{\mathbf{k}}\text{ cm}$$
$$= (25\hat{\mathbf{i}} + 31\hat{\mathbf{j}} + 7.0\hat{\mathbf{k}})\text{ cm}$$

The resultant displacement has components $R_x = 25$ cm, $R_y = 31$ cm, and $R_z = 7.0$ cm. Its magnitude is

$$R = \sqrt{R_x^2 + R_y^2 + R_z^2}$$
$$= \sqrt{(25\text{ cm})^2 + (31\text{ cm})^2 + (7.0\text{ cm})^2} = 40\text{ cm}$$

Example 3.5 Taking a Hike

Interactive

A hiker begins a trip by first walking 25.0 km southeast from her car. She stops and sets up her tent for the night. On the second day, she walks 40.0 km in a direction 60.0° north of east, at which point she discovers a forest ranger's tower.

(A) Determine the components of the hiker's displacement for each day.

Solution We *conceptualize* the problem by drawing a sketch as in Figure 3.19. If we denote the displacement vectors on the first and second days by **A** and **B**, respectively, and use the car as the origin of coordinates, we obtain the vectors shown in Figure 3.19. Drawing the resultant **R**, we can now *categorize* this as a problem we've solved before—an addition of two vectors. This should give you a hint of the power of categorization—many new problems are very similar to problems that we have already solved if we are careful to conceptualize them.

We will *analyze* this problem by using our new knowledge of vector components. Displacement **A** has a magnitude of 25.0 km and is directed 45.0° below the positive *x* axis. From Equations 3.8 and 3.9, its components are

$$A_x = A\cos(-45.0°) = (25.0\text{ km})(0.707) = 17.7\text{ km}$$

$$A_y = A\sin(-45.0°) = (25.0\text{ km})(-0.707) = -17.7\text{ km}$$

The negative value of A_y indicates that the hiker walks in the negative *y* direction on the first day. The signs of A_x and A_y also are evident from Figure 3.19.

The second displacement **B** has a magnitude of 40.0 km and is 60.0° north of east. Its components are

$$B_x = B\cos 60.0° = (40.0\text{ km})(0.500) = \boxed{20.0\text{ km}}$$

$$B_y = B\sin 60.0° = (40.0\text{ km})(0.866) = \boxed{34.6\text{ km}}$$

(B) Determine the components of the hiker's resultant displacement **R** for the trip. Find an expression for **R** in terms of unit vectors.

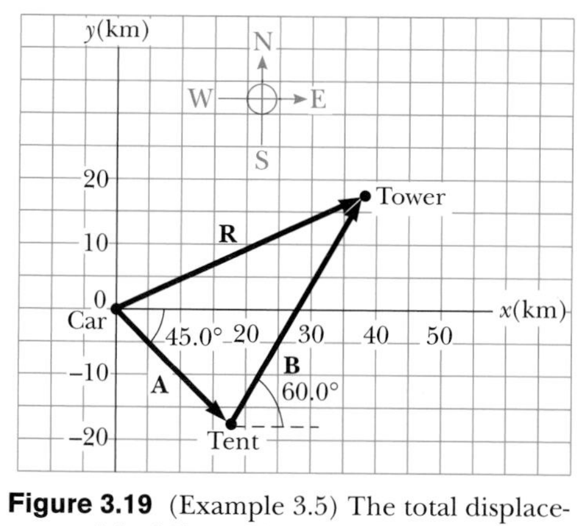

Figure 3.19 (Example 3.5) The total displacement of the hiker is the vector **R** = **A** + **B**.

Solution The resultant displacement for the trip **R** = **A** + **B** has components given by Equation 3.15:

$$R_x = A_x + B_x = 17.7\text{ km} + 20.0\text{ km} = \boxed{37.7\text{ km}}$$

$$R_y = A_y + B_y = -17.7\text{ km} + 34.6\text{ km} = \boxed{16.9\text{ km}}$$

In unit–vector form, we can write the total displacement as

$$\mathbf{R} = \boxed{(37.7\hat{\mathbf{i}} + 16.9\hat{\mathbf{j}})\text{ km}}$$

Using Equations 3.16 and 3.17, we find that the vector **R** has a magnitude of 41.3 km and is directed 24.1° north of east.

Let us *finalize.* The units of **R** are km, which is reasonable for a displacement. Looking at the graphical representation in Figure 3.19, we estimate that the final position of the hiker is at about (38 km, 17 km) which is consistent with the components of **R** in our final result. Also, both components of **R** are positive, putting the final position in the first quadrant of the coordinate system, which is also consistent with Figure 3.19.

Investigate this situation at the Interactive Worked Example link at **http://www.pse6.com.**

Example 3.6 Let's Fly Away!

A commuter airplane takes the route shown in Figure 3.20. First, it flies from the origin of the coordinate system shown to city A, located 175 km in a direction 30.0° north of east. Next, it flies 153 km 20.0° west of north to city B. Finally, it flies 195 km due west to city C. Find the location of city C relative to the origin.

Solution Once again, a drawing such as Figure 3.20 allows us to *conceptualize* the problem. It is convenient to choose the coordinate system shown in Figure 3.20, where the x axis points to the east and the y axis points to the north. Let us denote the three consecutive displacements by the vectors **a**, **b**, and **c**.

We can now *categorize* this problem as being similar to Example 3.5 that we have already solved. There are two primary differences. First, we are adding three vectors instead of two. Second, Example 3.5 guided us by first asking for the components in part (A). The current Example has no such guidance and simply asks for a result. We need to *analyze* the situation and choose a path. We will follow the same pattern that we did in Example 3.5, beginning with finding the components of the three vectors **a**, **b**, and **c**. Displacement **a** has a magnitude of 175 km and the components

$$a_x = a\cos(30.0°) = (175\text{ km})(0.866) = 152\text{ km}$$

$$a_y = a\sin(30.0°) = (175\text{ km})(0.500) = 87.5\text{ km}$$

Displacement **b**, whose magnitude is 153 km, has the components

$$b_x = b\cos(110°) = (153\text{ km})(-0.342) = -52.3\text{ km}$$

$$b_y = b\sin(110°) = (153\text{ km})(0.940) = 144\text{ km}$$

Finally, displacement **c,** whose magnitude is 195 km, has the components

$$c_x = c\cos(180°) = (195\text{ km})(-1) = -195\text{ km}$$

$$c_y = c\sin(180°) = 0$$

Therefore, the components of the position vector **R** from the starting point to city C are

$$R_x = a_x + b_x + c_x = 152\text{ km} - 52.3\text{ km} - 195\text{ km}$$

$$= \boxed{-95.3\text{ km}}$$

$$R_y = a_y + b_y + c_y = 87.5\text{ km} + 144\text{ km} + 0 = \boxed{232\text{ km}}$$

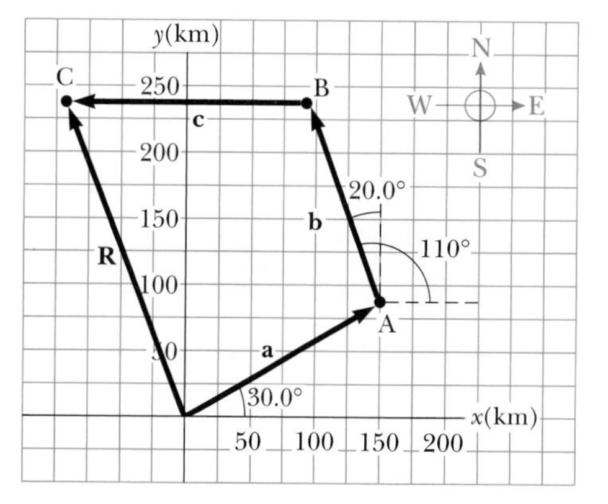

Figure 3.20 (Example 3.6) The airplane starts at the origin, flies first to city A, then to city B, and finally to city C.

In unit–vector notation, $\mathbf{R} = (-95.3\hat{\mathbf{i}} + 232\hat{\mathbf{j}})$ km. Using Equations 3.16 and 3.17, we find that the vector **R** has a magnitude of 251 km and is directed 22.3° west of north.

To *finalize* the problem, note that the airplane can reach city C from the starting point by first traveling 95.3 km due west and then by traveling 232 km due north. Or it could follow a straight-line path to C by flying a distance $R =$ 251 km in a direction 22.3° west of north.

What If? After landing in city C, the pilot wishes to return to the origin along a single straight line. What are the components of the vector representing this displacement? What should the heading of the plane be?

Answer The desired vector **H** (for Home!) is simply the negative of vector **R**:

$$\mathbf{H} = -\mathbf{R} = (+95.3\hat{\mathbf{i}} - 232\hat{\mathbf{j}}) \text{ km}$$

The heading is found by calculating the angle that the vector makes with the x axis:

$$\tan\theta = \frac{R_y}{R_x} = \frac{-232 \text{ m}}{95.3 \text{ m}} = -2.43$$

This gives a heading angle of $\theta = -67.7°$, or 67.7° south of east.

Take a practice test for this chapter by clicking on the Practice Test link at **http://www.pse6.com.**

SUMMARY

Scalar quantities are those that have only a numerical value and no associated direction. **Vector quantities** have both magnitude and direction and obey the laws of vector addition. The magnitude of a vector is *always* a positive number.

When two or more vectors are added together, all of them must have the same units and all of them must be the same type of quantity. We can add two vectors **A** and **B** graphically. In this method (Fig. 3.6), the resultant vector $\mathbf{R} = \mathbf{A} + \mathbf{B}$ runs from the tail of **A** to the tip of **B**.

A second method of adding vectors involves **components** of the vectors. The x component A_x of the vector **A** is equal to the projection of **A** along the x axis of a coordinate system, as shown in Figure 3.13, where $A_x = A\cos\theta$. The y component A_y of **A** is the projection of **A** along the y axis, where $A_y = A\sin\theta$. Be sure you can determine which trigonometric functions you should use in all situations, especially when θ is defined as something other than the counterclockwise angle from the positive x axis.

If a vector **A** has an x component A_x and a y component A_y, the vector can be expressed in unit–vector form as $\mathbf{A} = A_x\hat{\mathbf{i}} + A_y\hat{\mathbf{j}}$. In this notation, $\hat{\mathbf{i}}$ is a unit vector pointing in the positive x direction, and $\hat{\mathbf{j}}$ is a unit vector pointing in the positive y direction. Because $\hat{\mathbf{i}}$ and $\hat{\mathbf{j}}$ are unit vectors, $|\hat{\mathbf{i}}| = |\hat{\mathbf{j}}| = 1$.

We can find the resultant of two or more vectors by resolving all vectors into their x and y components, adding their resultant x and y components, and then using the Pythagorean theorem to find the magnitude of the resultant vector. We can find the angle that the resultant vector makes with respect to the x axis by using a suitable trigonometric function.

QUESTIONS

1. Two vectors have unequal magnitudes. Can their sum be zero? Explain.

2. Can the magnitude of a particle's displacement be greater than the distance traveled? Explain.

3. The magnitudes of two vectors **A** and **B** are $A = 5$ units and $B = 2$ units. Find the largest and smallest values possible for the magnitude of the resultant vector $\mathbf{R} = \mathbf{A} + \mathbf{B}$.

4. Which of the following are vectors and which are not: force, temperature, the volume of water in a can, the ratings of a TV show, the height of a building, the velocity of a sports car, the age of the Universe?

5. A vector **A** lies in the xy plane. For what orientations of **A** will both of its components be negative? For what orientations will its components have opposite signs?

6. A book is moved once around the perimeter of a tabletop with the dimensions 1.0 m × 2.0 m. If the book ends up at its initial position, what is its displacement? What is the distance traveled?

7. While traveling along a straight interstate highway you notice that the mile marker reads 260. You travel until you reach mile marker 150 and then retrace your path to the mile marker 175. What is the magnitude of your resultant displacement from mile marker 260?

8. If the component of vector **A** along the direction of vector **B** is zero, what can you conclude about the two vectors?

9. Can the magnitude of a vector have a negative value? Explain.

10. Under what circumstances would a nonzero vector lying in the *xy* plane have components that are equal in magnitude?

11. If **A** = **B**, what can you conclude about the components of **A** and **B**?

12. Is it possible to add a vector quantity to a scalar quantity? Explain.

13. The resolution of vectors into components is equivalent to replacing the original vector with the sum of two vectors, whose sum is the same as the original vector. There are an infinite number of pairs of vectors that will satisfy this condition; we choose that pair with one vector parallel to the *x* axis and the second parallel to the *y* axis. What difficulties would be introduced by defining components relative to axes that are not perpendicular—for example, the *x* axis and a *y* axis oriented at 45° to the *x* axis?

14. In what circumstance is the *x* component of a vector given by the magnitude of the vector times the sine of its direction angle?

PROBLEMS

1, 2, 3 = straightforward, intermediate, challenging □ = full solution available in the *Student Solutions Manual and Study Guide*

= coached solution with hints available at http://www.pse6.com = computer useful in solving problem

= paired numerical and symbolic problems

Section 3.1 Coordinate Systems

1. The polar coordinates of a point are $r = 5.50$ m and $\theta = 240°$. What are the Cartesian coordinates of this point?

2. Two points in a plane have polar coordinates (2.50 m, 30.0°) and (3.80 m, 120.0°). Determine (a) the Cartesian coordinates of these points and (b) the distance between them.

3. A fly lands on one wall of a room. The lower left-hand corner of the wall is selected as the origin of a two-dimensional Cartesian coordinate system. If the fly is located at the point having coordinates (2.00, 1.00) m, (a) how far is it from the corner of the room? (b) What is its location in polar coordinates?

4. Two points in the *xy* plane have Cartesian coordinates (2.00, − 4.00) m and (− 3.00, 3.00) m. Determine (a) the distance between these points and (b) their polar coordinates.

5. If the rectangular coordinates of a point are given by (2, *y*) and its polar coordinates are (r, 30°), determine *y* and *r*.

6. If the polar coordinates of the point (x, y) are (r, θ), determine the polar coordinates for the points: (a) $(-x, y)$, (b) $(-2x, -2y)$, and (c) $(3x, -3y)$.

Section 3.2 Vector and Scalar Quantities

Section 3.3 Some Properties of Vectors

7. A surveyor measures the distance across a straight river by the following method: starting directly across from a tree on the opposite bank, she walks 100 m along the riverbank to establish a baseline. Then she sights across to the tree. The angle from her baseline to the tree is 35.0°. How wide is the river?

8. A pedestrian moves 6.00 km east and then 13.0 km north. Find the magnitude and direction of the resultant displacement vector using the graphical method.

9. A plane flies from base camp to lake A, 280 km away, in a direction of 20.0° north of east. After dropping off supplies it flies to lake B, which is 190 km at 30.0° west of north from lake A. Graphically determine the distance and direction from lake B to the base camp.

10. Vector **A** has a magnitude of 8.00 units and makes an angle of 45.0° with the positive *x* axis. Vector **B** also has a magnitude of 8.00 units and is directed along the negative *x* axis. Using graphical methods, find (a) the vector sum **A** + **B** and (b) the vector difference **A** − **B**.

11. A skater glides along a circular path of radius 5.00 m. If he coasts around one half of the circle, find (a) the magnitude of the displacement vector and (b) how far the person skated. (c) What is the magnitude of the displacement if he skates all the way around the circle?

12. A force $\mathbf{F}_1$ of magnitude 6.00 units acts at the origin in a direction 30.0° above the positive *x* axis. A second force $\mathbf{F}_2$ of magnitude 5.00 units acts at the origin in the direction of the positive *y* axis. Find graphically the magnitude and direction of the resultant force $\mathbf{F}_1 + \mathbf{F}_2$.

13. Arbitrarily define the "instantaneous vector height" of a person as the displacement vector from the point halfway

between his or her feet to the top of the head. Make an order-of-magnitude estimate of the total vector height of all the people in a city of population 100 000 (a) at 10 o'clock on a Tuesday morning, and (b) at 5 o'clock on a Saturday morning. Explain your reasoning.

14. A dog searching for a bone walks 3.50 m south, then runs 8.20 m at an angle 30.0° north of east, and finally walks 15.0 m west. Find the dog's resultant displacement vector using graphical techniques.

15. Each of the displacement vectors **A** and **B** shown in Fig. P3.15 has a magnitude of 3.00 m. Find graphically (a) **A** + **B**, (b) **A** − **B**, (c) **B** − **A**, (d) **A** − 2**B**. Report all angles counterclockwise from the positive x axis.

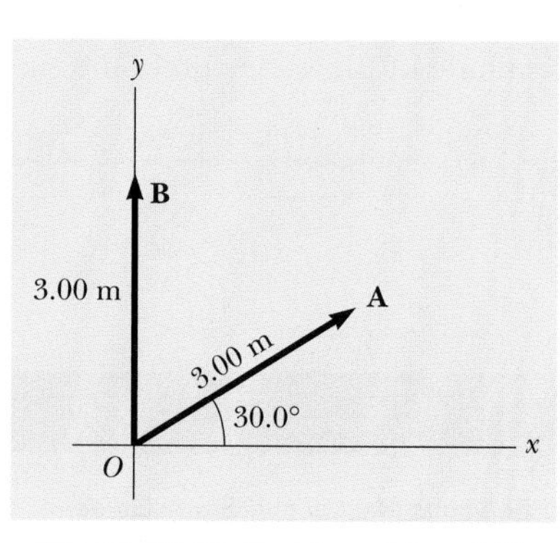

Figure P3.15 Problems 15 and 37.

16. Three displacements are **A** = 200 m, due south; **B** = 250 m, due west; **C** = 150 m, 30.0° east of north. Construct a separate diagram for each of the following possible ways of adding these vectors: $\mathbf{R}_1 = \mathbf{A} + \mathbf{B} + \mathbf{C}$; $\mathbf{R}_2 = \mathbf{B} + \mathbf{C} + \mathbf{A}$; $\mathbf{R}_3 = \mathbf{C} + \mathbf{B} + \mathbf{A}$.

17. A roller coaster car moves 200 ft horizontally, and then rises 135 ft at an angle of 30.0° above the horizontal. It then travels 135 ft at an angle of 40.0° downward. What is its displacement from its starting point? Use graphical techniques.

Section 3.4 Components of a Vector and Unit Vectors

18. Find the horizontal and vertical components of the 100-m displacement of a superhero who flies from the top of a tall building following the path shown in Fig. P3.18.

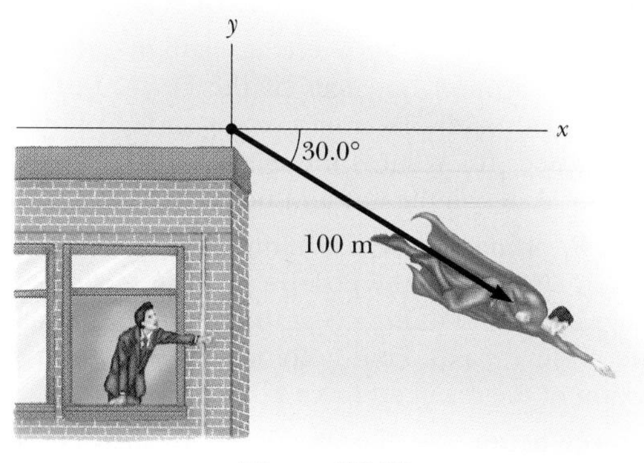

Figure P3.18

19. A vector has an x component of − 25.0 units and a y component of 40.0 units. Find the magnitude and direction of this vector.

20. A person walks 25.0° north of east for 3.10 km. How far would she have to walk due north and due east to arrive at the same location?

21. Obtain expressions in component form for the position vectors having the following polar coordinates: (a) 12.8 m, 150° (b) 3.30 cm, 60.0° (c) 22.0 in., 215°.

22. A displacement vector lying in the xy plane has a magnitude of 50.0 m and is directed at an angle of 120° to the positive x axis. What are the rectangular components of this vector?

23. A girl delivering newspapers covers her route by traveling 3.00 blocks west, 4.00 blocks north, and then 6.00 blocks east. (a) What is her resultant displacement? (b) What is the total distance she travels?

24. In 1992, Akira Matsushima, from Japan, rode a unicycle across the United States, covering about 4 800 km in six weeks. Suppose that, during that trip, he had to find his way through a city with plenty of one-way streets. In the city center, Matsushima had to travel in sequence 280 m north, 220 m east, 360 m north, 300 m west, 120 m south, 60.0 m east, 40.0 m south, 90.0 m west (road construction) and then 70.0 m north. At that point, he stopped to rest. Meanwhile, a curious crow decided to fly the distance from his starting point to the rest location directly ("as the crow flies"). It took the crow 40.0 s to cover that distance. Assuming the velocity of the crow was constant, find its magnitude and direction.

25. While exploring a cave, a spelunker starts at the entrance and moves the following distances. She goes 75.0 m north, 250 m east, 125 m at an angle 30.0° north of east, and 150 m south. Find the resultant displacement from the cave entrance.

26. A map suggests that Atlanta is 730 miles in a direction of 5.00° north of east from Dallas. The same map shows that Chicago is 560 miles in a direction of 21.0° west of north from Atlanta. Modeling the Earth as flat, use this information to find the displacement from Dallas to Chicago.

27. Given the vectors $\mathbf{A} = 2.00\hat{\mathbf{i}} + 6.00\hat{\mathbf{j}}$ and $\mathbf{B} = 3.00\hat{\mathbf{i}} - 2.00\hat{\mathbf{j}}$, (a) draw the vector sum $\mathbf{C} = \mathbf{A} + \mathbf{B}$ and the vector difference $\mathbf{D} = \mathbf{A} - \mathbf{B}$. (b) Calculate **C** and **D**, first in terms of unit vectors and then in terms of polar coordinates, with angles measured with respect to the + x axis.

28. Find the magnitude and direction of the resultant of three displacements having rectangular components (3.00, 2.00) m, (− 5.00, 3.00) m, and (6.00, 1.00) m.

29. A man pushing a mop across a floor causes it to undergo two displacements. The first has a magnitude of 150 cm and makes an angle of 120° with the positive x axis. The resultant displacement has a magnitude of 140 cm and is directed at an angle of 35.0° to the positive x axis. Find the magnitude and direction of the second displacement.

30. Vector **A** has x and y components of − 8.70 cm and 15.0 cm, respectively; vector **B** has x and y components of 13.2 cm and − 6.60 cm, respectively. If $\mathbf{A} - \mathbf{B} + 3\mathbf{C} = 0$, what are the components of **C**?

31. Consider the two vectors $\mathbf{A} = 3\hat{\mathbf{i}} - 2\hat{\mathbf{j}}$ and $\mathbf{B} = -\hat{\mathbf{i}} - 4\hat{\mathbf{j}}$. Calculate (a) $\mathbf{A} + \mathbf{B}$, (b) $\mathbf{A} - \mathbf{B}$, (c) $|\mathbf{A} + \mathbf{B}|$, (d) $|\mathbf{A} - \mathbf{B}|$, and (e) the directions of $\mathbf{A} + \mathbf{B}$ and $\mathbf{A} - \mathbf{B}$.

32. Consider the three displacement vectors $\mathbf{A} = (3\hat{\mathbf{i}} - 3\hat{\mathbf{j}})$ m, $\mathbf{B} = (\hat{\mathbf{i}} - 4\hat{\mathbf{j}})$ m, and $\mathbf{C} = (-2\hat{\mathbf{i}} + 5\hat{\mathbf{j}})$ m. Use the component method to determine (a) the magnitude and direction of the vector $\mathbf{D} = \mathbf{A} + \mathbf{B} + \mathbf{C}$, (b) the magnitude and direction of $\mathbf{E} = -\mathbf{A} - \mathbf{B} + \mathbf{C}$.

33. A particle undergoes the following consecutive displacements: 3.50 m south, 8.20 m northeast, and 15.0 m west. What is the resultant displacement?

34. In a game of American football, a quarterback takes the ball from the line of scrimmage, runs backward a distance of 10.0 yards, and then sideways parallel to the line of scrimmage for 15.0 yards. At this point, he throws a forward pass 50.0 yards straight downfield perpendicular to the line of scrimmage. What is the magnitude of the football's resultant displacement?

35. The helicopter view in Fig. P3.35 shows two people pulling on a stubborn mule. Find (a) the single force that is equivalent to the two forces shown, and (b) the force that a third person would have to exert on the mule to make the resultant force equal to zero. The forces are measured in units of newtons (abbreviated N).

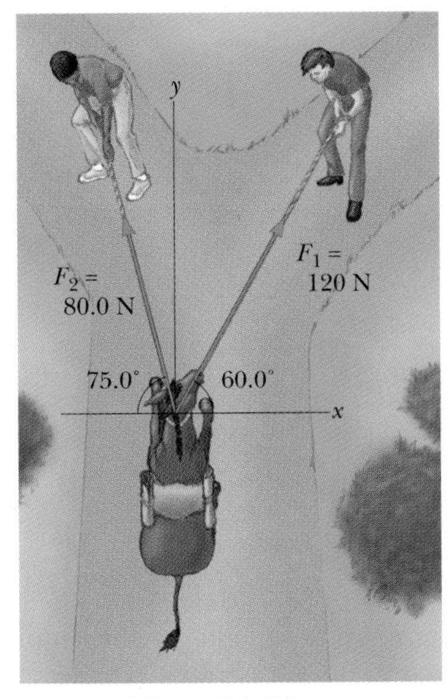

Figure P3.35

36. A novice golfer on the green takes three strokes to sink the ball. The successive displacements are 4.00 m to the north, 2.00 m northeast, and 1.00 m at 30.0° west of south. Starting at the same initial point, an expert golfer could make the hole in what single displacement?

37. Use the component method to add the vectors **A** and **B** shown in Figure P3.15. Express the resultant $\mathbf{A} + \mathbf{B}$ in unit–vector notation.

38. In an assembly operation illustrated in Figure P3.38, a robot moves an object first straight upward and then also to the east, around an arc forming one quarter of a circle of radius 4.80 cm that lies in an east-west vertical plane. The robot then moves the object upward and to the north, through a quarter of a circle of radius 3.70 cm that lies in a north-south vertical plane. Find (a) the magnitude of the total displacement of the object, and (b) the angle the total displacement makes with the vertical.

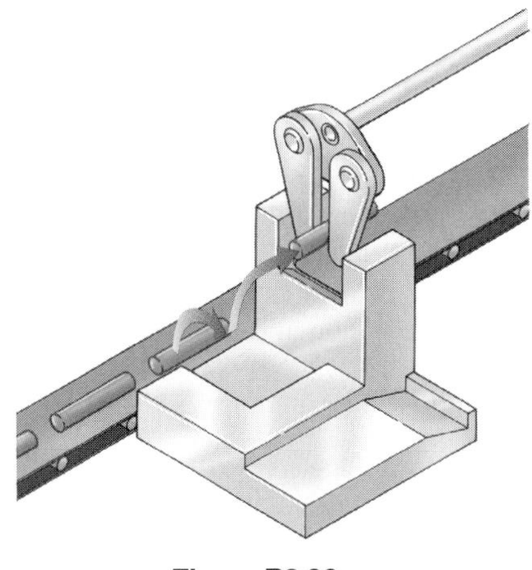

Figure P3.38

39. Vector **B** has *x*, *y*, and *z* components of 4.00, 6.00, and 3.00 units, respectively. Calculate the magnitude of **B** and the angles that **B** makes with the coordinate axes.

40. You are standing on the ground at the origin of a coordinate system. An airplane flies over you with constant velocity parallel to the *x* axis and at a fixed height of 7.60×10^3 m. At time $t = 0$ the airplane is directly above you, so that the vector leading from you to it is $\mathbf{P}_0 = (7.60 \times 10^3\ \text{m})\hat{\mathbf{j}}$. At $t = 30.0$ s the position vector leading from you to the airplane is $\mathbf{P}_{30} = (8.04 \times 10^3\ \text{m})\hat{\mathbf{i}} + (7.60 \times 10^3\ \text{m})\hat{\mathbf{j}}$. Determine the magnitude and orientation of the airplane's position vector at $t = 45.0$ s.

41. The vector **A** has *x*, *y*, and *z* components of 8.00, 12.0, and −4.00 units, respectively. (a) Write a vector expression for **A** in unit–vector notation. (b) Obtain a unit–vector expression for a vector **B** one fourth the length of **A** pointing in the same direction as **A**. (c) Obtain a unit–vector expression for a vector **C** three times the length of **A** pointing in the direction opposite the direction of **A**.

42. Instructions for finding a buried treasure include the following: Go 75.0 paces at 240°, turn to 135° and walk 125 paces, then travel 100 paces at 160°. The angles are measured counterclockwise from an axis pointing to the east, the $+x$ direction. Determine the resultant displacement from the starting point.

43. Given the displacement vectors $\mathbf{A} = (3\hat{\mathbf{i}} - 4\hat{\mathbf{j}} + 4\hat{\mathbf{k}})$ m and $\mathbf{B} = (2\hat{\mathbf{i}} + 3\hat{\mathbf{j}} - 7\hat{\mathbf{k}})$ m, find the magnitudes of the vectors (a) $\mathbf{C} = \mathbf{A} + \mathbf{B}$ and (b) $\mathbf{D} = 2\mathbf{A} - \mathbf{B}$, also expressing each in terms of its rectangular components.

44. A radar station locates a sinking ship at range 17.3 km and bearing 136° clockwise from north. From the same station a rescue plane is at horizontal range 19.6 km, 153° clockwise from north, with elevation 2.20 km. (a) Write the position vector for the ship relative to the plane, letting $\hat{\mathbf{i}}$ represent east, $\hat{\mathbf{j}}$ north, and $\hat{\mathbf{k}}$ up. (b) How far apart are the plane and ship?

45. As it passes over Grand Bahama Island, the eye of a hurricane is moving in a direction 60.0° north of west with a speed of 41.0 km/h. Three hours later, the course of the hurricane suddenly shifts due north, and its speed slows to 25.0 km/h. How far from Grand Bahama is the eye 4.50 h after it passes over the island?

46. (a) Vector **E** has magnitude 17.0 cm and is directed 27.0° counterclockwise from the $+x$ axis. Express it in unit–vector notation. (b) Vector **F** has magnitude 17.0 cm and is directed 27.0° counterclockwise from the $+y$ axis. Express it in unit–vector notation. (c) Vector **G** has magnitude 17.0 cm and is directed 27.0° clockwise from the $-y$ axis. Express it in unit–vector notation.

47. Vector **A** has a negative x component 3.00 units in length and a positive y component 2.00 units in length. (a) Determine an expression for **A** in unit–vector notation. (b) Determine the magnitude and direction of **A**. (c) What vector **B** when added to **A** gives a resultant vector with no x component and a negative y component 4.00 units in length?

48. An airplane starting from airport A flies 300 km east, then 350 km at 30.0° west of north, and then 150 km north to arrive finally at airport B. (a) The next day, another plane flies directly from A to B in a straight line. In what direction should the pilot travel in this direct flight? (b) How far will the pilot travel in this direct flight? Assume there is no wind during these flights.

49. Three displacement vectors of a croquet ball are shown in Figure P3.49, where $|\mathbf{A}| = 20.0$ units, $|\mathbf{B}| = 40.0$ units, and $|\mathbf{C}| = 30.0$ units. Find (a) the resultant in unit–vector notation and (b) the magnitude and direction of the resultant displacement.

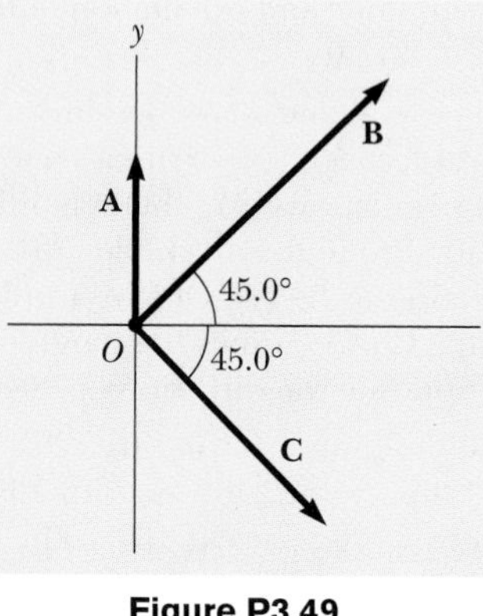

Figure P3.49

50. If $\mathbf{A} = (6.00\hat{\mathbf{i}} - 8.00\hat{\mathbf{j}})$ units, $\mathbf{B} = (-8.00\hat{\mathbf{i}} + 3.00\hat{\mathbf{j}})$ units, and $\mathbf{C} = (26.0\hat{\mathbf{i}} + 19.0\hat{\mathbf{j}})$ units, determine a and b such that $a\mathbf{A} + b\mathbf{B} + \mathbf{C} = 0$.

Additional Problems

51. Two vectors **A** and **B** have precisely equal magnitudes. In order for the magnitude of **A** + **B** to be one hundred times larger than the magnitude of **A** – **B**, what must be the angle between them?

52. Two vectors **A** and **B** have precisely equal magnitudes. In order for the magnitude of **A** + **B** to be larger than the magnitude of **A** – **B** by the factor n, what must be the angle between them?

53. A vector is given by $\mathbf{R} = 2\hat{\mathbf{i}} + \hat{\mathbf{j}} + 3\hat{\mathbf{k}}$. Find (a) the magnitudes of the x, y, and z components, (b) the magnitude of **R**, and (c) the angles between **R** and the x, y, and z axes.

54. The biggest stuffed animal in the world is a snake 420 m long, constructed by Norwegian children. Suppose the snake is laid out in a park as shown in Figure P3.54, forming two straight sides of a 105° angle, with one side 240 m long. Olaf and Inge run a race they invent. Inge runs directly from the tail of the snake to its head and Olaf starts from the same place at the same time but runs along the snake. If both children run steadily at 12.0 km/h, Inge reaches the head of the snake how much earlier than Olaf?

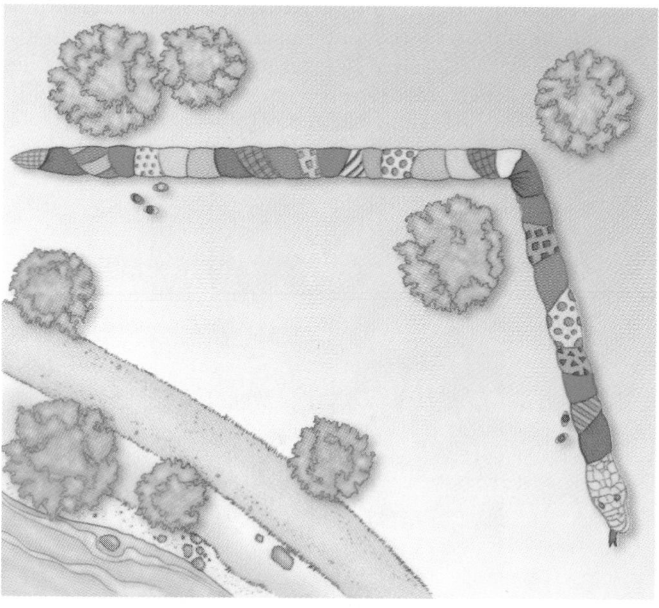
Figure P3.54

55. An air-traffic controller observes two aircraft on his radar screen. The first is at altitude 800 m, horizontal distance 19.2 km, and 25.0° south of west. The second aircraft is at altitude 1 100 m, horizontal distance 17.6 km, and 20.0° south of west. What is the distance between the two aircraft? (Place the x axis west, the y axis south, and the z axis vertical.)

56. A ferry boat transports tourists among three islands. It sails from the first island to the second island, 4.76 km away, in a direction 37.0° north of east. It then sails from the second island to the third island in a direction 69.0° west of north. Finally it returns to the first island, sailing in a direction 28.0° east of south. Calculate the distance between (a) the second and third islands (b) the first and third islands.

57. The rectangle shown in Figure P3.57 has sides parallel to the x and y axes. The position vectors of two corners are **A** = 10.0 m at 50.0° and **B** = 12.0 m at 30.0°. (a) Find the

perimeter of the rectangle. (b) Find the magnitude and direction of the vector from the origin to the upper right corner of the rectangle.

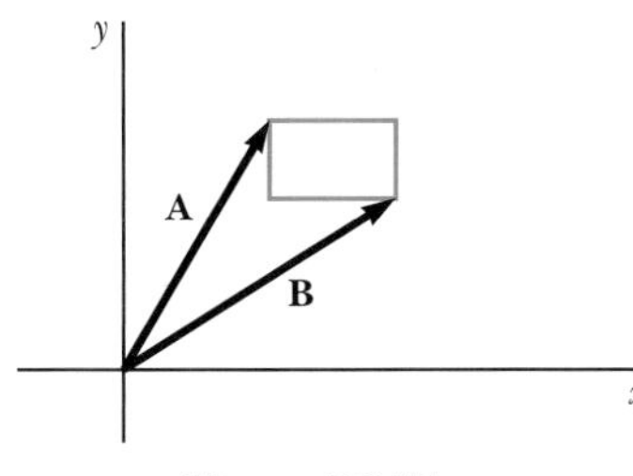

Figure P3.57

58. Find the sum of these four vector forces: 12.0 N to the right at 35.0° above the horizontal, 31.0 N to the left at 55.0° above the horizontal, 8.40 N to the left at 35.0° below the horizontal, and 24.0 N to the right at 55.0° below the horizontal. Follow these steps: Make a drawing of this situation and select the best axes for x and y so you have the least number of components. Then add the vectors by the component method.

59. A person going for a walk follows the path shown in Fig. P3.59. The total trip consists of four straight-line paths. At the end of the walk, what is the person's resultant displacement measured from the starting point?

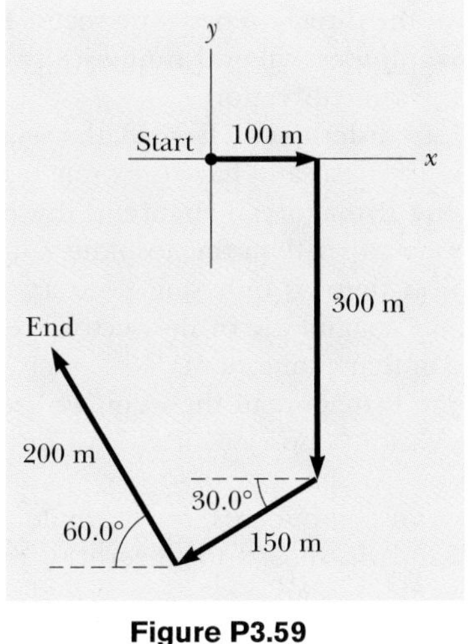

Figure P3.59

60. The instantaneous position of an object is specified by its position vector **r** leading from a fixed origin to the location of the point object. Suppose that for a certain object the position vector is a function of time, given by $\mathbf{r} = 4\hat{\mathbf{i}} + 3\hat{\mathbf{j}} - 2t\hat{\mathbf{k}}$, where r is in meters and t is in seconds. Evaluate $d\mathbf{r}/dt$. What does it represent about the object?

61. A jet airliner, moving initially at 300 mi/h to the east, suddenly enters a region where the wind is blowing at 100 mi/h toward the direction 30.0° north of east. What are the new speed and direction of the aircraft relative to the ground?

62. Long John Silver, a pirate, has buried his treasure on an island with five trees, located at the following points: (30.0 m, − 20.0 m), (60.0 m, 80.0 m), (− 10.0 m, − 10.0 m), (40.0 m, − 30.0 m), and (− 70.0 m, 60.0 m), all measured relative to some origin, as in Figure P3.62. His ship's log instructs you to start at tree A and move toward tree B, but to cover only one half the distance between A and B. Then move toward tree C, covering one third the distance between your current location and C. Next move toward D, covering one fourth the distance between where you are and D. Finally move towards E, covering one fifth the distance between you and E, stop, and dig. (a) Assume that you have correctly determined the order in which the pirate labeled the trees as A, B, C, D, and E, as shown in the figure. What are the coordinates of the point where his treasure is buried? (b) **What if** you do not really know the way the pirate labeled the trees? Rearrange the order of the trees [for instance, B(30 m, − 20 m), A(60 m, 80 m), E(− 10 m, − 10 m), C(40 m, − 30 m), and D(− 70 m, 60 m)] and repeat the calculation to show that the answer does not depend on the order in which the trees are labeled.

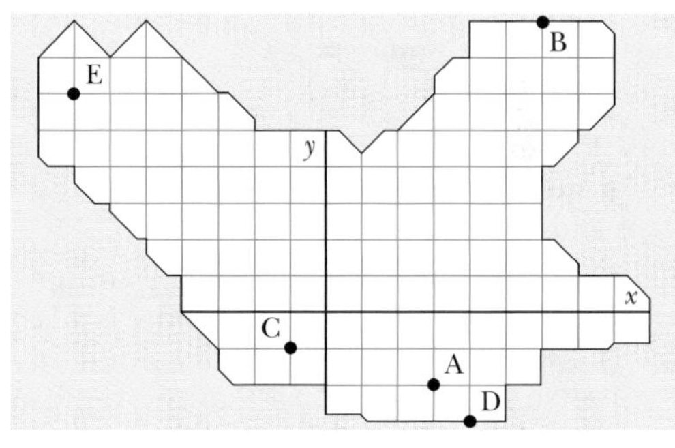

Figure P3.62

63. Consider a game in which N children position themselves at equal distances around the circumference of a circle. At the center of the circle is a rubber tire. Each child holds a rope attached to the tire and, at a signal, pulls on his rope. All children exert forces of the same magnitude F. In the case $N = 2$, it is easy to see that the net force on the tire will be zero, because the two oppositely directed force vectors add to zero. Similarly, if $N = 4$, 6, or any even integer, the resultant force on the tire must be zero, because the forces exerted by each pair of oppositely positioned children will cancel. When an odd number of children are around the circle, it is not so obvious whether the total force on the central tire will be zero. (a) Calculate the net force on the tire in the case $N = 3$, by adding the components of the three force vectors. Choose the x axis to lie along one of the ropes. (b) **What If?** Determine the net force for the general case where N is any integer, odd or even, greater than one. Proceed as follows: Assume that the total force is not zero. Then it must point in some particular direction. Let every child move one position clockwise. Give a reason that the total force must then have a direction turned clockwise by $360°/N$. Argue that the total force must nevertheless be the same as before. Explain that the contradiction proves that the magnitude of the force is zero. This problem illustrates a widely useful technique of proving a result "by symmetry"—by using a bit of the mathematics of *group theory*. The particular situation

is actually encountered in physics and chemistry when an array of electric charges (ions) exerts electric forces on an atom at a central position in a molecule or in a crystal.

64. A rectangular parallelepiped has dimensions a, b, and c, as in Figure P3.64. (a) Obtain a vector expression for the face diagonal vector $\mathbf{R}_1$. What is the magnitude of this vector? (b) Obtain a vector expression for the body diagonal vector $\mathbf{R}_2$. Note that $\mathbf{R}_1$, $c\hat{\mathbf{k}}$, and $\mathbf{R}_2$ make a right triangle and prove that the magnitude of $\mathbf{R}_2$ is $\sqrt{a^2 + b^2 + c^2}$.

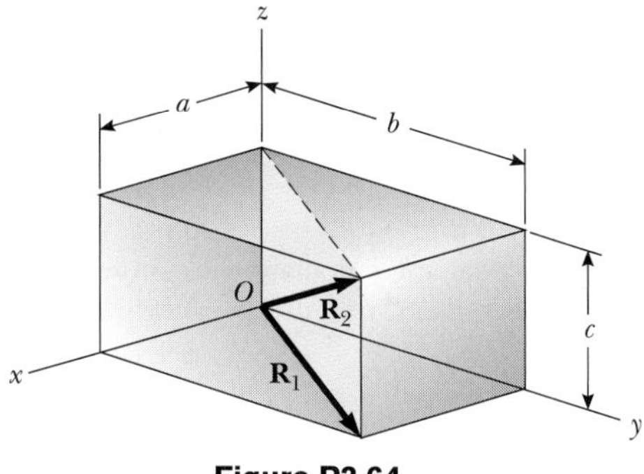

Figure P3.64

65. Vectors **A** and **B** have equal magnitudes of 5.00. If the sum of **A** and **B** is the vector $6.00\hat{\mathbf{j}}$, determine the angle between **A** and **B**.

66. In Figure P3.66 a spider is resting after starting to spin its web. The gravitational force on the spider is 0.150 newton down. The spider is supported by different tension forces in the two strands above it, so that the resultant vector force on the spider is zero. The two strands are perpendicular to each other, so we have chosen the x and y directions to be along them. The tension T_x is 0.127 newton. Find (a) the tension T_y, (b) the angle the x axis makes with the horizontal, and (c) the angle the y axis makes with the horizontal.

Figure P3.66

67. A point P is described by the coordinates (x, y) with respect to the normal Cartesian coordinate system shown in Fig. P3.67. Show that (x', y'), the coordinates of this point in the rotated coordinate system, are related to (x, y) and the rotation angle α by the expressions

$$x' = x \cos \alpha + y \sin \alpha$$

$$y' = -x \sin \alpha + y \cos \alpha$$

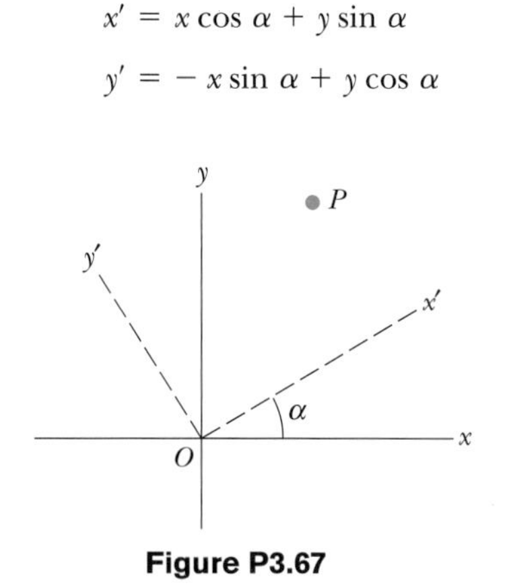

Figure P3.67

Answers to Quick Quizzes

3.1 Scalars: (a), (d), (e). None of these quantities has a direction. Vectors: (b), (c). For these quantities, the direction is necessary to specify the quantity completely.

3.2 (c). The resultant has its maximum magnitude $A + B = 12 + 8 = 20$ units when vector **A** is oriented in the same direction as vector **B**. The resultant vector has its minimum magnitude $A - B = 12 - 8 = 4$ units when vector **A** is oriented in the direction opposite vector **B**.

3.3 (a). The magnitudes will add numerically only if the vectors are in the same direction.

3.4 (b) and (c). In order to add to zero, the vectors must point in opposite directions and have the same magnitude.

3.5 (b). From the Pythagorean theorem, the magnitude of a vector is always larger than the absolute value of each component, unless there is only one nonzero component, in which case the magnitude of the vector is equal to the absolute value of that component.

3.6 (b). From the Pythagorean theorem, we see that the magnitude of a vector is nonzero if at least one component is nonzero.

3.7 (d). Each set of components, for example, the two x components A_x and B_x, must add to zero, so the components must be of opposite sign.

3.8 (c). The magnitude of **C** is 5 units, the same as the z component. Answer (b) is not correct because the magnitude of any vector is always a positive number while the y component of **B** is negative.

Chapter 4

Motion in Two Dimensions

▲ Lava spews from a volcanic eruption. Notice the parabolic paths of embers projected into the air. We will find in this chapter that all projectiles follow a parabolic path in the absence of air resistance. (© Arndt/Premium Stock/PictureQuest)

CHAPTER OUTLINE

In this chapter we explore the kinematics of a particle moving in two dimensions. Knowing the basics of two-dimensional motion will allow us to examine—in future chapters—a wide variety of motions, ranging from the motion of satellites in orbit to the motion of electrons in a uniform electric field. We begin by studying in greater detail the vector nature of position, velocity, and acceleration. As in the case of one-dimensional motion, we derive the kinematic equations for two-dimensional motion from the fundamental definitions of these three quantities. We then treat projectile motion and uniform circular motion as special cases of motion in two dimensions. We also discuss the concept of relative motion, which shows why observers in different frames of reference may measure different positions, velocities, and accelerations for a given particle.

4.1 The Position, Velocity, and Acceleration Vectors

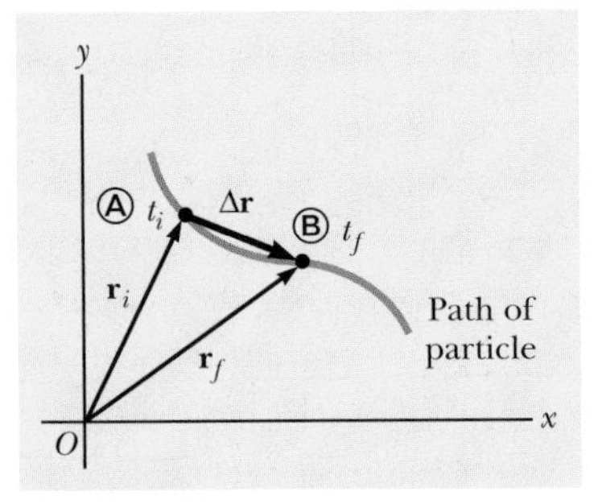

Figure 4.1 A particle moving in the xy plane is located with the position vector $\mathbf{r}$ drawn from the origin to the particle. The displacement of the particle as it moves from Ⓐ to Ⓑ in the time interval $\Delta t = t_f - t_i$ is equal to the vector $\Delta\mathbf{r} = \mathbf{r}_f - \mathbf{r}_i$.

In Chapter 2 we found that the motion of a particle moving along a straight line is completely known if its position is known as a function of time. Now let us extend this idea to motion in the xy plane. We begin by describing the position of a particle by its **position vector r**, drawn from the origin of some coordinate system to the particle located in the xy plane, as in Figure 4.1. At time t_i the particle is at point Ⓐ, described by position vector $\mathbf{r}_i$. At some later time t_f it is at point Ⓑ, described by position vector $\mathbf{r}_f$. The path from Ⓐ to Ⓑ is not necessarily a straight line. As the particle moves from Ⓐ to Ⓑ in the time interval $\Delta t = t_f - t_i$, its position vector changes from $\mathbf{r}_i$ to $\mathbf{r}_f$. As we learned in Chapter 2, displacement is a vector, and the displacement of the particle is the difference between its final position and its initial position. We now define the **displacement vector Δr** for the particle of Figure 4.1 as being the difference between its final position vector and its initial position vector:

Displacement vector

$$\Delta\mathbf{r} \equiv \mathbf{r}_f - \mathbf{r}_i \tag{4.1}$$

The direction of $\Delta\mathbf{r}$ is indicated in Figure 4.1. As we see from the figure, the magnitude of $\Delta\mathbf{r}$ is *less* than the distance traveled along the curved path followed by the particle.

As we saw in Chapter 2, it is often useful to quantify motion by looking at the ratio of a displacement divided by the time interval during which that displacement occurs, which gives the rate of change of position. In two-dimensional (or three-dimensional) kinematics, everything is the same as in one-dimensional kinematics except that we must now use full vector notation rather than positive and negative signs to indicate the direction of motion.

We define the **average velocity** of a particle during the time interval Δt as the displacement of the particle divided by the time interval:

Average velocity

$$\overline{\mathbf{v}} \equiv \frac{\Delta\mathbf{r}}{\Delta t} \tag{4.2}$$

Mark C. Burnett/Photo Researchers, Inc.

Figure 4.2 Bird's-eye view of a baseball diamond. A batter who hits a home run travels around the bases, ending up where he began. Thus, his average velocity for the entire trip is zero. His average speed, however, is not zero and is equal to the distance around the bases divided by the time interval during which he runs around the bases.

Multiplying or dividing a vector quantity by a positive scalar quantity such as Δt changes only the magnitude of the vector, not its direction. Because displacement is a vector quantity and the time interval is a positive scalar quantity, we conclude that the average velocity is a vector quantity directed along $\Delta\mathbf{r}$.

Note that the average velocity between points is *independent of the path* taken. This is because average velocity is proportional to displacement, which depends only on the initial and final position vectors and not on the path taken. As with one-dimensional motion, we conclude that if a particle starts its motion at some point and returns to this point via any path, its average velocity is zero for this trip because its displacement is zero. Figure 4.2 suggests such a situation in a baseball park. When a batter hits a home run, he runs around the bases and returns to home plate. Thus, his average velocity is zero during this trip. His average speed, however, is not zero.

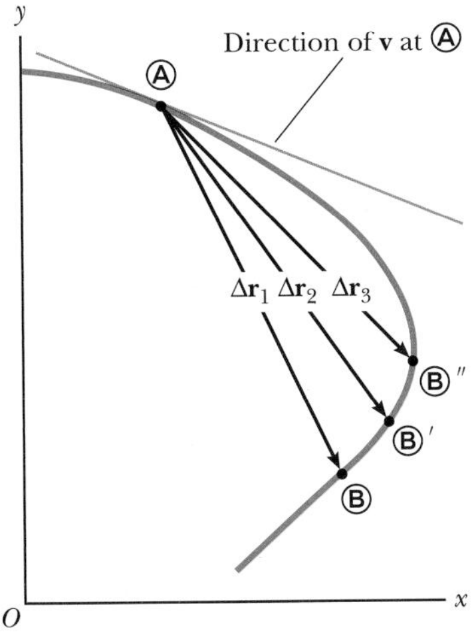

Figure 4.3 As a particle moves between two points, its average velocity is in the direction of the displacement vector $\Delta\mathbf{r}$. As the end point of the path is moved from Ⓑ to Ⓑ′ to Ⓑ″, the respective displacements and corresponding time intervals become smaller and smaller. In the limit that the end point approaches Ⓐ, Δt approaches zero, and the direction of $\Delta\mathbf{r}$ approaches that of the line tangent to the curve at Ⓐ. By definition, the instantaneous velocity at Ⓐ is directed along this tangent line.

Consider again the motion of a particle between two points in the xy plane, as shown in Figure 4.3. As the time interval over which we observe the motion becomes smaller and smaller, the direction of the displacement approaches that of the line tangent to the path at Ⓐ. The **instantaneous velocity v** is defined as the limit of the average velocity $\Delta\mathbf{r}/\Delta t$ as Δt approaches zero:

$$\mathbf{v} \equiv \lim_{\Delta t \to 0} \frac{\Delta\mathbf{r}}{\Delta t} = \frac{d\mathbf{r}}{dt} \tag{4.3}$$

That is, the instantaneous velocity equals the derivative of the position vector with respect to time. The direction of the instantaneous velocity vector at any point in a particle's path is along a line tangent to the path at that point and in the direction of motion.

The magnitude of the instantaneous velocity vector $v = |\mathbf{v}|$ is called the *speed,* which is a scalar quantity.

As a particle moves from one point to another along some path, its instantaneous velocity vector changes from $\mathbf{v}_i$ at time t_i to $\mathbf{v}_f$ at time t_f. Knowing the velocity at these points allows us to determine the average acceleration of the particle—the **average acceleration $\bar{\mathbf{a}}$** of a particle as it moves is defined as the change in the instantaneous velocity vector $\Delta\mathbf{v}$ divided by the time interval Δt during which that change occurs:

$$\bar{\mathbf{a}} \equiv \frac{\mathbf{v}_f - \mathbf{v}_i}{t_f - t_i} = \frac{\Delta\mathbf{v}}{\Delta t} \tag{4.4}$$

PITFALL PREVENTION

4.1 Vector Addition

While the vector addition discussed in Chapter 3 involves *displacement* vectors, vector addition can be applied to *any* type of vector quantity. Figure 4.4, for example, shows the addition of *velocity* vectors using the graphical approach.

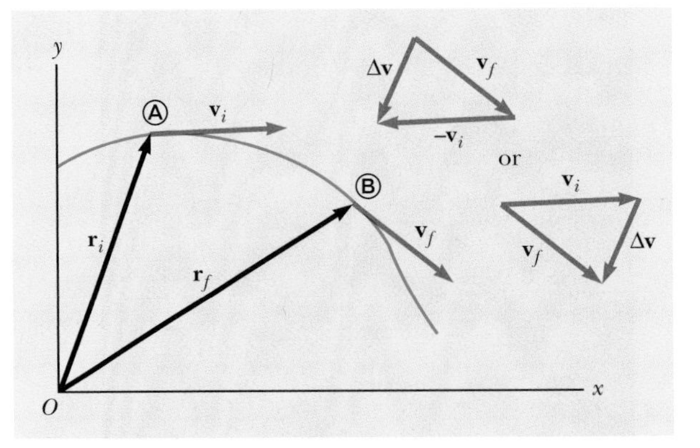

Figure 4.4 A particle moves from position Ⓐ to position Ⓑ. Its velocity vector changes from $\mathbf{v}_i$ to $\mathbf{v}_f$. The vector diagrams at the upper right show two ways of determining the vector $\Delta\mathbf{v}$ from the initial and final velocities.

Because $\bar{\mathbf{a}}$ is the ratio of a vector quantity $\Delta\mathbf{v}$ and a positive scalar quantity Δt, we conclude that average acceleration is a vector quantity directed along $\Delta\mathbf{v}$. As indicated in Figure 4.4, the direction of $\Delta\mathbf{v}$ is found by adding the vector $-\mathbf{v}_i$ (the negative of $\mathbf{v}_i$) to the vector $\mathbf{v}_f$, because by definition $\Delta\mathbf{v} = \mathbf{v}_f - \mathbf{v}_i$.

When the average acceleration of a particle changes during different time intervals, it is useful to define its instantaneous acceleration. The **instantaneous acceleration a** is defined as the limiting value of the ratio $\Delta\mathbf{v}/\Delta t$ as Δt approaches zero:

Instantaneous acceleration

$$\mathbf{a} \equiv \lim_{\Delta t \to 0} \frac{\Delta\mathbf{v}}{\Delta t} = \frac{d\mathbf{v}}{dt} \tag{4.5}$$

In other words, the instantaneous acceleration equals the derivative of the velocity vector with respect to time.

It is important to recognize that various changes can occur when a particle accelerates. First, the magnitude of the velocity vector (the speed) may change with time as in straight-line (one-dimensional) motion. Second, the direction of the velocity vector may change with time even if its magnitude (speed) remains constant, as in curved-path (two-dimensional) motion. Finally, both the magnitude and the direction of the velocity vector may change simultaneously.

Quick Quiz 4.1 Which of the following cannot *possibly* be accelerating? (a) An object moving with a constant speed (b) An object moving with a constant velocity (c) An object moving along a curve.

Quick Quiz 4.2 Consider the following controls in an automobile: gas pedal, brake, steering wheel. The controls in this list that cause an acceleration of the car are (a) all three controls (b) the gas pedal and the brake (c) only the brake (d) only the gas pedal.

4.2 Two-Dimensional Motion with Constant Acceleration

In Section 2.5, we investigated one-dimensional motion in which the acceleration is constant because this type of motion is common. Let us consider now two-dimensional motion during which the acceleration remains constant in both magnitude and direction. This will also be useful for analyzing some common types of motion.

The position vector for a particle moving in the xy plane can be written

$$\mathbf{r} = x\hat{\mathbf{i}} + y\hat{\mathbf{j}} \tag{4.6}$$

where x, y, and $\mathbf{r}$ change with time as the particle moves while the unit vectors $\hat{\mathbf{i}}$ and $\hat{\mathbf{j}}$ remain constant. If the position vector is known, the velocity of the particle can be obtained from Equations 4.3 and 4.6, which give

$$\mathbf{v} = \frac{d\mathbf{r}}{dt} = \frac{dx}{dt}\hat{\mathbf{i}} + \frac{dy}{dt}\hat{\mathbf{j}} = v_x\hat{\mathbf{i}} + v_y\hat{\mathbf{j}} \quad (4.7)$$

Because $\mathbf{a}$ is assumed constant, its components a_x and a_y also are constants. Therefore, we can apply the equations of kinematics to the x and y components of the velocity vector. Substituting, from Equation 2.9, $v_{xf} = v_{xi} + a_xt$ and $v_{yf} = v_{yi} + a_yt$ into Equation 4.7 to determine the final velocity at any time t, we obtain

$$\mathbf{v}_f = (v_{xi} + a_xt)\hat{\mathbf{i}} + (v_{yi} + a_yt)\hat{\mathbf{j}}$$
$$= (v_{xi}\hat{\mathbf{i}} + v_{yi}\hat{\mathbf{j}}) + (a_x\hat{\mathbf{i}} + a_y\hat{\mathbf{j}})t$$
$$\mathbf{v}_f = \mathbf{v}_i + \mathbf{a}t \quad (4.8)$$

Velocity vector as a function of time

This result states that the velocity of a particle at some time t equals the vector sum of its initial velocity $\mathbf{v}_i$ and the additional velocity $\mathbf{a}t$ acquired at time t as a result of constant acceleration. It is the vector version of Equation 2.9.

Similarly, from Equation 2.12 we know that the x and y coordinates of a particle moving with constant acceleration are

$$x_f = x_i + v_{xi}t + \tfrac{1}{2}a_xt^2 \qquad y_f = y_i + v_{yi}t + \tfrac{1}{2}a_yt^2$$

Substituting these expressions into Equation 4.6 (and labeling the final position vector $\mathbf{r}_f$) gives

$$\mathbf{r}_f = (x_i + v_{xi}t + \tfrac{1}{2}a_xt^2)\hat{\mathbf{i}} + (y_i + v_{yi}t + \tfrac{1}{2}a_yt^2)\hat{\mathbf{j}}$$
$$= (x_i\hat{\mathbf{i}} + y_i\hat{\mathbf{j}}) + (v_{xi}\hat{\mathbf{i}} + v_{yi}\hat{\mathbf{j}})t + \tfrac{1}{2}(a_x\hat{\mathbf{i}} + a_y\hat{\mathbf{j}})t^2$$
$$\mathbf{r}_f = \mathbf{r}_i + \mathbf{v}_it + \tfrac{1}{2}\mathbf{a}t^2 \quad (4.9)$$

Position vector as a function of time

which is the vector version of Equation 2.12. This equation tells us that the position vector $\mathbf{r}_f$ is the vector sum of the original position $\mathbf{r}_i$, a displacement $\mathbf{v}_it$ arising from the initial velocity of the particle and a displacement $\frac{1}{2}\mathbf{a}t^2$ resulting from the constant acceleration of the particle.

Graphical representations of Equations 4.8 and 4.9 are shown in Figure 4.5. Note from Figure 4.5a that $\mathbf{v}_f$ is generally not along the direction of either $\mathbf{v}_i$ or $\mathbf{a}$ because the relationship between these quantities is a vector expression. For the same reason,

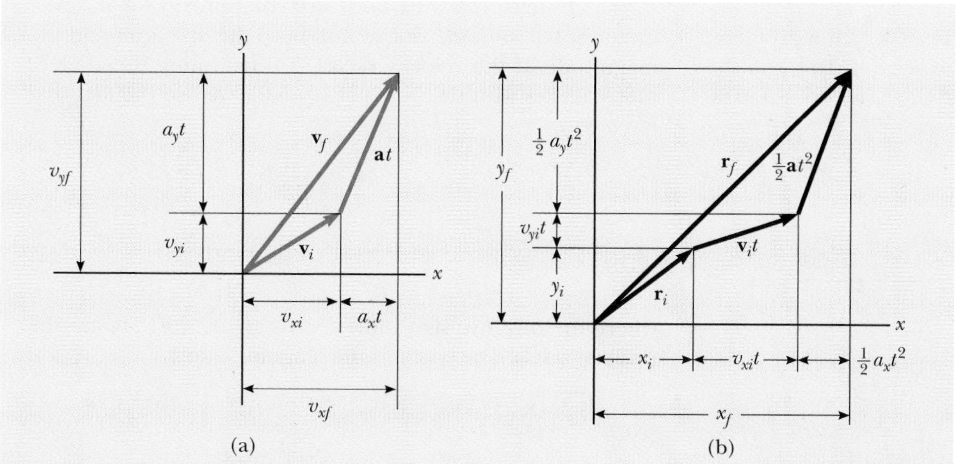

Active Figure 4.5 Vector representations and components of (a) the velocity and (b) the position of a particle moving with a constant acceleration $\mathbf{a}$.

At the Active Figures link at http://www.pse6.com, you can investigate the effect of different initial positions and velocities on the final position and velocity (for constant acceleration).

from Figure 4.5b we see that $\mathbf{r}_f$ is generally not along the direction of $\mathbf{v}_i$ or $\mathbf{a}$. Finally, note that $\mathbf{v}_f$ and $\mathbf{r}_f$ are generally not in the same direction.

Because Equations 4.8 and 4.9 are vector expressions, we may write them in component form:

$$\mathbf{v}_f = \mathbf{v}_i + \mathbf{a}t \qquad \begin{cases} v_{xf} = v_{xi} + a_x t \\ v_{yf} = v_{yi} + a_y t \end{cases} \tag{4.8a}$$

$$\mathbf{r}_f = \mathbf{r}_i + \mathbf{v}_i t + \tfrac{1}{2}\mathbf{a}t^2 \qquad \begin{cases} x_f = x_i + v_{xi}t + \tfrac{1}{2}a_x t^2 \\ y_f = y_i + v_{yi}t + \tfrac{1}{2}a_y t^2 \end{cases} \tag{4.9a}$$

These components are illustrated in Figure 4.5. The component form of the equations for $\mathbf{v}_f$ and $\mathbf{r}_f$ show us that two-dimensional motion at constant acceleration is equivalent to two *independent* motions—one in the x direction and one in the y direction—having constant accelerations a_x and a_y.

Example 4.1 Motion in a Plane

A particle starts from the origin at $t = 0$ with an initial velocity having an x component of 20 m/s and a y component of -15 m/s. The particle moves in the xy plane with an x component of acceleration only, given by $a_x = 4.0\ \text{m/s}^2$.

(A) Determine the components of the velocity vector at any time and the total velocity vector at any time.

Solution After carefully reading the problem, we *conceptualize* what is happening to the particle. The components of the initial velocity tell us that the particle starts by moving toward the right and downward. The x component of velocity starts at 20 m/s and increases by 4.0 m/s every second. The y component of velocity never changes from its initial value of -15 m/s. We sketch a rough motion diagram of the situation in Figure 4.6. Because the particle is accelerating in the $+x$ direction, its velocity component in this direction will increase, so that the path will curve as shown in the diagram. Note that the spacing between successive images increases as time goes on because the speed is increasing. The placement of the acceleration and velocity vectors in Figure 4.6 helps us to further conceptualize the situation.

Because the acceleration is constant, we *categorize* this problem as one involving a particle moving in two dimensions with constant acceleration. To *analyze* such a problem, we use the equations developed in this section. To begin the mathematical analysis, we set $v_{xi} = 20$ m/s, $v_{yi} = -15$ m/s, $a_x = 4.0\ \text{m/s}^2$, and $a_y = 0$.

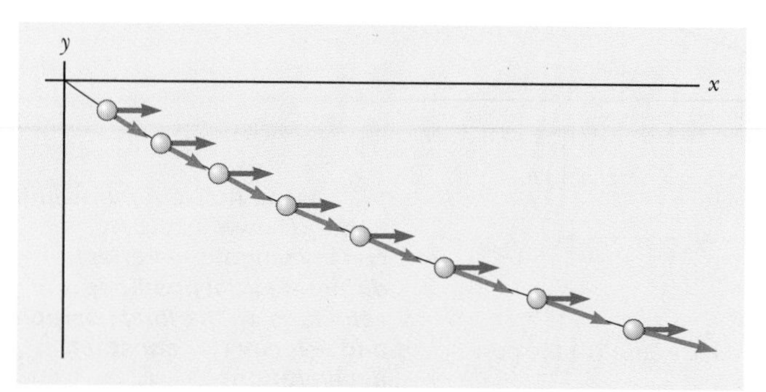

Figure 4.6 (Example 4.1) Motion diagram for the particle.

Equations 4.8a give

(1) $\quad v_{xf} = v_{xi} + a_x t = (20 + 4.0t)\ \text{m/s}$

(2) $\quad v_{yf} = v_{yi} + a_y t = -15\ \text{m/s} + 0 = -15\ \text{m/s}$

Therefore

$$\mathbf{v}_f = v_{xi}\hat{\mathbf{i}} + v_{yi}\hat{\mathbf{j}} = [(20 + 4.0t)\hat{\mathbf{i}} - 15\hat{\mathbf{j}}]\ \text{m/s}$$

We could also obtain this result using Equation 4.8 directly, noting that $\mathbf{a} = 4.0\hat{\mathbf{i}}\ \text{m/s}^2$ and $\mathbf{v}_i = [20\hat{\mathbf{i}} - 15\hat{\mathbf{j}}]$ m/s. To *finalize* this part, notice that the x component of velocity increases in time while the y component remains constant; this is consistent with what we predicted.

(B) Calculate the velocity and speed of the particle at $t = 5.0$ s.

Solution With $t = 5.0$ s, the result from part (A) gives

$$\mathbf{v}_f = [(20 + 4.0(5.0))\hat{\mathbf{i}} - 15\hat{\mathbf{j}}]\ \text{m/s} = (40\hat{\mathbf{i}} - 15\hat{\mathbf{j}})\ \text{m/s}$$

This result tells us that at $t = 5.0$ s, $v_{xf} = 40$ m/s and $v_{yf} = -15$ m/s. Knowing these two components for this two-dimensional motion, we can find both the direction and the magnitude of the velocity vector. To determine the angle θ that $\mathbf{v}$ makes with the x axis at $t = 5.0$ s, we use the fact that $\tan\theta = v_{yf}/v_{xf}$:

(3) $$\theta = \tan^{-1}\left(\frac{v_{yf}}{v_{xf}}\right) = \tan^{-1}\left(\frac{-15\ \text{m/s}}{40\ \text{m/s}}\right) = -21^\circ$$

where the negative sign indicates an angle of 21° below the positive x axis. The speed is the magnitude of $\mathbf{v}_f$:

$$v_f = |\mathbf{v}_f| = \sqrt{v_{xf}^2 + v_{yf}^2} = \sqrt{(40)^2 + (-15)^2}\ \text{m/s} = 43\ \text{m/s}$$

To *finalize* this part, we notice that if we calculate v_i from the x and y components of $\mathbf{v}_i$, we find that $v_f > v_i$. Is this consistent with our prediction?

(C) Determine the x and y coordinates of the particle at any time t and the position vector at this time.

Solution Because $x_i = y_i = 0$ at $t = 0$, Equation 4.9a gives

$$x_f = v_{xi}t + \tfrac{1}{2}a_x t^2 = (20t + 2.0t^2)\text{ m}$$

$$y_f = v_{yi}t = (-15t)\text{ m}$$

Therefore, the position vector at any time t is

$$(4) \qquad \mathbf{r}_f = x_f\hat{\mathbf{i}} + y_f\hat{\mathbf{j}} = [(20t + 2.0t^2)\hat{\mathbf{i}} - 15t\hat{\mathbf{j}}]\text{ m}$$

(Alternatively, we could obtain $\mathbf{r}_f$ by applying Equation 4.9 directly, with $\mathbf{v}_f = (20\hat{\mathbf{i}} - 15\hat{\mathbf{j}})$ m/s and $\mathbf{a} = 4.0\hat{\mathbf{i}}$ m/s^2. Try it!) Thus, for example, at $t = 5.0$ s, $x = 150$ m, $y = -75$ m, and $\mathbf{r}_f = (150\hat{\mathbf{i}} - 75\hat{\mathbf{j}})$ m. The magnitude of the displacement of the particle from the origin at $t = 5.0$ s is the magnitude of $\mathbf{r}_f$ at this time:

$$r_f = |\mathbf{r}_f| = \sqrt{(150)^2 + (-75)^2}\text{ m} = 170\text{ m}$$

Note that this is *not* the distance that the particle travels in this time! Can you determine this distance from the available data?

To *finalize* this problem, let us consider a limiting case for very large values of t in the following **What If?**

What If? **What if we wait a very long time and then observe the motion of the particle? How would we describe the motion of the particle for large values of the time?**

Answer Looking at Figure 4.6, we see the path of the particle curving toward the x axis. There is no reason to assume that this tendency will change, so this suggests that the path will become more and more parallel to the x axis as time grows large. Mathematically, let us consider Equations (1) and (2). These show that the y component of the velocity remains constant while the x component grows linearly with t. Thus, when t is very large, the x component of the velocity will be much larger than the y component, suggesting that the velocity vector becomes more and more parallel to the x axis.

Equation (3) gives the angle that the velocity vector makes with the x axis. Notice that $\theta \to 0$ as the denominator (v_{xf}) becomes much larger than the numerator (v_{yf}).

Despite the fact that the velocity vector becomes more and more parallel to the x axis, the particle does not approach a limiting value of y. Equation (4) shows that both x_f and y_f continue to grow with time, although x_f grows much faster.

4.3 Projectile Motion

Assumptions of projectile motion

Anyone who has observed a baseball in motion has observed projectile motion. The ball moves in a curved path, and its motion is simple to analyze if we make two assumptions: (1) the free-fall acceleration $\mathbf{g}$ is constant over the range of motion and is directed downward,[1] and (2) the effect of air resistance is negligible.[2] With these assumptions, we find that the path of a projectile, which we call its *trajectory*, is *always* a parabola. **We use these assumptions throughout this chapter.**

To show that the trajectory of a projectile is a parabola, let us choose our reference frame such that the y direction is vertical and positive is upward. Because air resistance is neglected, we know that $a_y = -g$ (as in one-dimensional free fall) and that $a_x = 0$. Furthermore, let us assume that at $t = 0$, the projectile leaves the origin ($x_i = y_i = 0$) with speed v_i, as shown in Figure 4.7. The vector $\mathbf{v}_i$ makes an angle θ_i with the horizontal. From the definitions of the cosine and sine functions we have

$$\cos\theta_i = v_{xi}/v_i \qquad \sin\theta_i = v_{yi}/v_i$$

Therefore, the initial x and y components of velocity are

$$v_{xi} = v_i\cos\theta_i \qquad v_{yi} = v_i\sin\theta_i \tag{4.10}$$

Substituting the x component into Equation 4.9a with $x_i = 0$ and $a_x = 0$, we find that

$$x_f = v_{xi}t = (v_i\cos\theta_i)t \tag{4.11}$$

[1] This assumption is reasonable as long as the range of motion is small compared with the radius of the Earth (6.4×10^6 m). In effect, this assumption is equivalent to assuming that the Earth is flat over the range of motion considered.

[2] This assumption is generally *not* justified, especially at high velocities. In addition, any spin imparted to a projectile, such as that applied when a pitcher throws a curve ball, can give rise to some very interesting effects associated with aerodynamic forces, which will be discussed in Chapter 14.

At the Active Figures link at http://www.pse6.com, you can change launch angle and initial speed. You can also observe the changing components of velocity along the trajectory of the projectile.

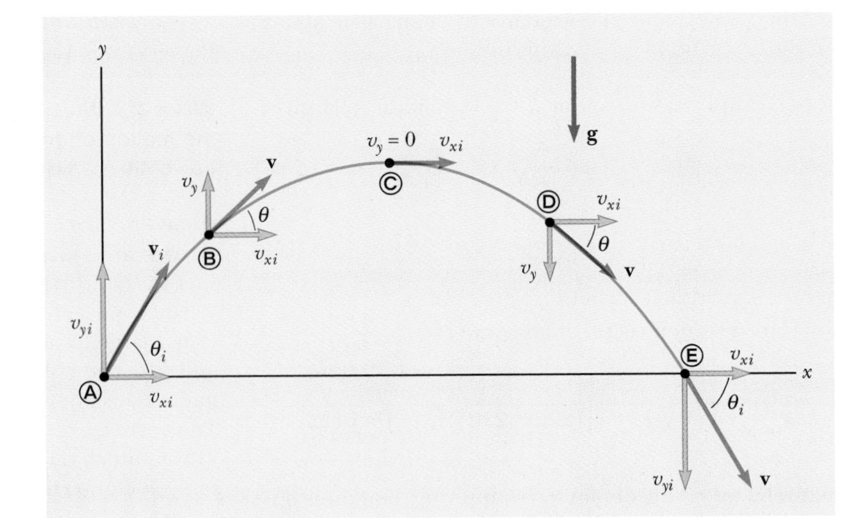

Active Figure 4.7 The parabolic path of a projectile that leaves the origin with a velocity $\mathbf{v}_i$. The velocity vector $\mathbf{v}$ changes with time in both magnitude and direction. This change is the result of acceleration in the negative y direction. The x component of velocity remains constant in time because there is no acceleration along the horizontal direction. The y component of velocity is zero at the peak of the path.

PITFALL PREVENTION

4.2 Acceleration at the Highest Point

As discussed in Pitfall Prevention 2.8, many people claim that the acceleration of a projectile at the topmost point of its trajectory is zero. This mistake arises from confusion between zero vertical velocity and zero acceleration. If the projectile were to experience zero acceleration at the highest point, then its velocity at that point would not change—the projectile would move horizontally at constant speed from then on! This does not happen, because the acceleration is NOT zero anywhere along the trajectory.

The Telegraph Colour Library/Getty Images

A welder cuts holes through a heavy metal construction beam with a hot torch. The sparks generated in the process follow parabolic paths.

Repeating with the y component and using $y_i = 0$ and $a_y = -g$, we obtain

$$y_f = v_{yi}t + \tfrac{1}{2}a_y t^2 = (v_i \sin\theta_i)\,t - \tfrac{1}{2}gt^2 \qquad (4.12)$$

Next, from Equation 4.11 we find $t = x_f/(v_i \cos\theta_i)$ and substitute this expression for t into Equation 4.12; this gives

$$y = (\tan\theta_i)\,x - \left(\frac{g}{2{v_i}^2\cos^2\theta_i}\right)x^2$$

This equation is valid for launch angles in the range $0 < \theta_i < \pi/2$. We have left the subscripts off the x and y because the equation is valid for any point (x, y) along the path of the projectile. The equation is of the form $y = ax - bx^2$, which is the equation of a parabola that passes through the origin. Thus, we have shown that the trajectory of a projectile is a parabola. Note that the trajectory is completely specified if both the initial speed v_i and the launch angle θ_i are known.

The vector expression for the position vector of the projectile as a function of time follows directly from Equation 4.9, with $\mathbf{a} = \mathbf{g}$:

$$\mathbf{r}_f = \mathbf{r}_i + \mathbf{v}_i t + \tfrac{1}{2}\mathbf{g}t^2$$

This expression is plotted in Figure 4.8, for a projectile launched from the origin, so that $\mathbf{r}_i = 0$.

The final position of a particle can be considered to be the superposition of the initial position $\mathbf{r}_i$, the term $\mathbf{v}_i t$, which is the displacement if no acceleration were present, and the term $\frac{1}{2}\mathbf{g}t^2$ that arises from the acceleration due to gravity. In other words, if there were no gravitational acceleration, the particle would continue to move along a straight path in the direction of $\mathbf{v}_i$. Therefore, the vertical distance $\frac{1}{2}\mathbf{g}t^2$ through which the particle "falls" off the straight-line path is the same distance that a freely falling object would fall during the same time interval.

In Section 4.2, we stated that two-dimensional motion with constant acceleration can be analyzed as a combination of two independent motions in the x and y directions, with accelerations a_x and a_y. Projectile motion is a special case of two-dimensional motion with constant acceleration and can be handled in this way, with zero acceleration in the x direction and $a_y = -g$ in the y direction. Thus, **when analyzing projectile motion, consider it to be the superposition of two motions:**

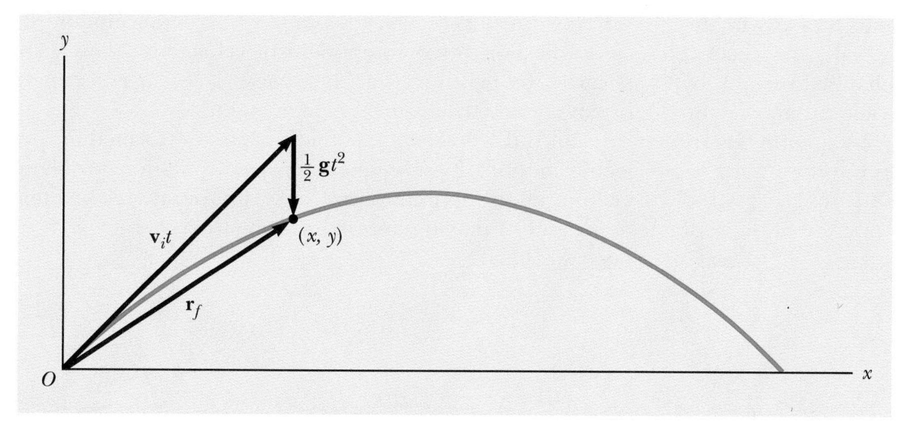

Figure 4.8 The position vector $\mathbf{r}_f$ of a projectile launched from the origin whose initial velocity at the origin is $\mathbf{v}_i$. The vector $\mathbf{v}_i t$ would be the displacement of the projectile if gravity were absent, and the vector $\frac{1}{2}\mathbf{g}t^2$ is its vertical displacement due to its downward gravitational acceleration.

(1) **constant-velocity motion in the horizontal direction** and (2) **free-fall motion in the vertical direction.** The horizontal and vertical components of a projectile's motion are completely independent of each other and can be handled separately, with time t as the common variable for both components.

Quick Quiz 4.3 Suppose you are running at constant velocity and you wish to throw a ball such that you will catch it as it comes back down. In what direction should you throw the ball relative to you? (a) straight up (b) at an angle to the ground that depends on your running speed (c) in the forward direction.

Quick Quiz 4.4 As a projectile thrown upward moves in its parabolic path (such as in Figure 4.8), at what point along its path are the velocity and acceleration vectors for the projectile perpendicular to each other? (a) nowhere (b) the highest point (c) the launch point.

Quick Quiz 4.5 As the projectile in Quick Quiz 4.4 moves along its path, at what point are the velocity and acceleration vectors for the projectile parallel to each other? (a) nowhere (b) the highest point (c) the launch point.

Example 4.2 Approximating Projectile Motion

A ball is thrown in such a way that its initial vertical and horizontal components of velocity are 40 m/s and 20 m/s, respectively. Estimate the total time of flight and the distance the ball is from its starting point when it lands.

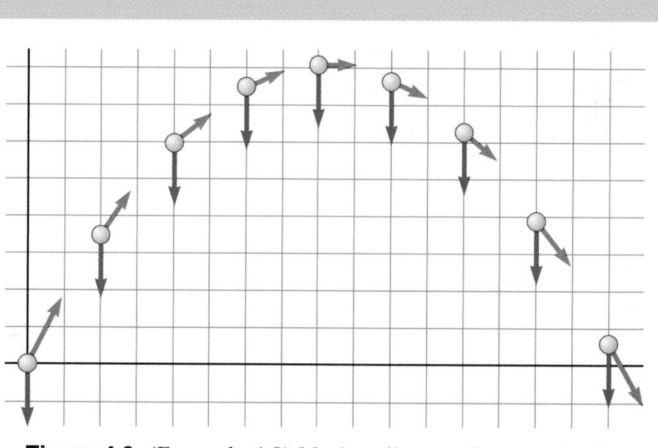

Figure 4.9 (Example 4.2) Motion diagram for a projectile.

Solution A motion diagram like Figure 4.9 helps us *conceptualize* the problem. The phrase "A ball is thrown" allows us to *categorize* this as a projectile motion problem, which we *analyze* by continuing to study Figure 4.9. The acceleration vectors are all the same, pointing downward with a magnitude of nearly 10 m/s^2. The velocity vectors change direction. Their horizontal components are all the same: 20 m/s.

Remember that the two velocity components are independent of each other. By considering the vertical motion first, we can determine how long the ball remains in the air. Because the vertical motion is free-fall, the vertical components of the velocity vectors change, second by second, from 40 m/s to roughly 30, 20, and 10 m/s in the upward direction, and then to 0 m/s. Subsequently, its velocity becomes 10, 20, 30, and 40 m/s in the downward direction. Thus it takes the ball about 4 s to go up and another 4 s to come back down, for a total time of flight of approximately 8 s.

Now we shift our analysis to the horizontal motion. Because the horizontal component of velocity is 20 m/s, and because the ball travels at this speed for 8 s, it ends up approximately 160 m from its starting point.

This is the first example that we have performed for projectile motion. In subsequent projectile motion problems, keep in mind the importance of separating the two components and of making approximations to give you rough expected results.

Horizontal Range and Maximum Height of a Projectile

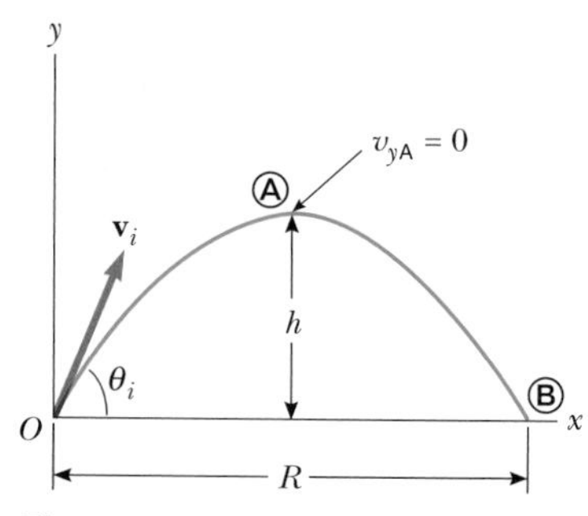

Figure 4.10 A projectile launched from the origin at $t_i = 0$ with an initial velocity $\mathbf{v}_i$. The maximum height of the projectile is h, and the horizontal range is R. At Ⓐ, the peak of the trajectory, the particle has coordinates $(R/2, h)$.

Let us assume that a projectile is launched from the origin at $t_i = 0$ with a positive v_{yi} component, as shown in Figure 4.10. Two points are especially interesting to analyze: the peak point Ⓐ, which has Cartesian coordinates $(R/2, h)$, and the point Ⓑ, which has coordinates $(R, 0)$. The distance R is called the *horizontal range* of the projectile, and the distance h is its *maximum height.* Let us find h and R in terms of v_i, θ_i, and g.

We can determine h by noting that at the peak, $v_{yA} = 0$. Therefore, we can use Equation 4.8a to determine the time t_A at which the projectile reaches the peak:

$$v_{yf} = v_{yi} + a_y t$$

$$0 = v_i \sin\theta_i - g t_A$$

$$t_A = \frac{v_i \sin\theta_i}{g}$$

Substituting this expression for t_A into the y part of Equation 4.9a and replacing $y = y_A$ with h, we obtain an expression for h in terms of the magnitude and direction of the initial velocity vector:

$$h = (v_i \sin\theta_i)\frac{v_i \sin\theta_i}{g} - \tfrac{1}{2}g\left(\frac{v_i \sin\theta_i}{g}\right)^2$$

$$h = \frac{v_i^2 \sin^2\theta_i}{2g} \qquad (4.13)$$

The range R is the horizontal position of the projectile at a time that is twice the time at which it reaches its peak, that is, at time $t_B = 2t_A$. Using the x part of Equation 4.9a, noting that $v_{xi} = v_{xB} = v_i \cos\theta_i$ and setting $x_B = R$ at $t = 2t_A$, we find that

$$R = v_{xi} t_B = (v_i \cos\theta_i) 2t_A$$

$$= (v_i \cos\theta_i)\frac{2v_i \sin\theta_i}{g} = \frac{2v_i^2 \sin\theta_i \cos\theta_i}{g}$$

Using the identity $\sin 2\theta = 2\sin\theta\cos\theta$ (see Appendix B.4), we write R in the more compact form

$$R = \frac{v_i^2 \sin 2\theta_i}{g} \qquad (4.14)$$

The maximum value of R from Equation 4.14 is $R_{max} = v_i^2/g$. This result follows from the fact that the maximum value of $\sin 2\theta_i$ is 1, which occurs when $2\theta_i = 90°$. Therefore, R is a maximum when $\theta_i = 45°$.

Figure 4.11 illustrates various trajectories for a projectile having a given initial speed but launched at different angles. As you can see, the range is a maximum for $\theta_i = 45°$. In addition, for any θ_i other than 45°, a point having Cartesian coordinates $(R, 0)$ can be reached by using either one of two complementary values of θ_i, such as 75° and 15°. Of course, the maximum height and time of flight for one of these values of θ_i are different from the maximum height and time of flight for the complementary value.

PITFALL PREVENTION

4.3 The Height and Range Equations

Equation 4.14 is useful for calculating R only for a symmetric path, as shown in Figure 4.10. If the path is not symmetric, *do not use this equation.* The general expressions given by Equations 4.8 and 4.9 are the *more important* results, because they give the position and velocity components of *any* particle moving in two dimensions at *any* time t.

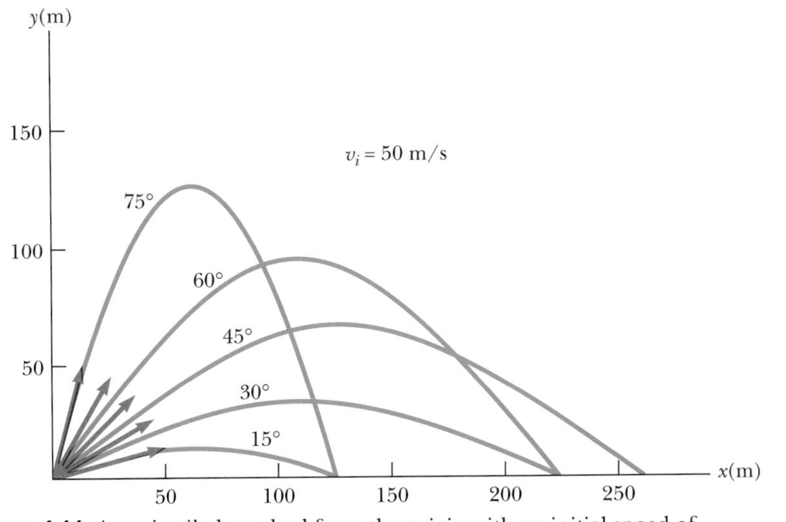

Active Figure 4.11 A projectile launched from the origin with an initial speed of 50 m/s at various angles of projection. Note that complementary values of θ_i result in the same value of R (range of the projectile).

At the Active Figures link at http://www.pse6.com, you can vary the projection angle to observe the effect on the trajectory and measure the flight time.

Quick Quiz 4.6 Rank the launch angles for the five paths in Figure 4.11 with respect to time of flight, from the shortest time of flight to the longest.

PROBLEM-SOLVING HINTS

Projectile Motion

We suggest that you use the following approach to solving projectile motion problems:

- Select a coordinate system and resolve the initial velocity vector into x and y components.
- Follow the techniques for solving constant-velocity problems to analyze the horizontal motion. Follow the techniques for solving constant-acceleration problems to analyze the vertical motion. The x and y motions share the same time t.

Example 4.3 The Long Jump

A long-jumper (Fig. 4.12) leaves the ground at an angle of 20.0° above the horizontal and at a speed of 11.0 m/s.

(A) How far does he jump in the horizontal direction? (Assume his motion is equivalent to that of a particle.)

Solution We *conceptualize* the motion of the long-jumper as equivalent to that of a simple projectile such as the ball in Example 4.2, and *categorize* this problem as a projectile motion problem. Because the initial speed and launch angle are given, and because the final height is the same as the initial height, we further categorize this problem as satisfying the conditions for which Equations 4.13 and 4.14 can be used. This is the most direct way to *analyze* this problem, although the general methods that we have been describing will always give the correct answer. We will take the general approach and use components. Figure 4.10 provides a graphical representation of the flight of the long-jumper. As before, we set our origin of coordinates at the takeoff point and label the peak as Ⓐ and the landing point as Ⓑ. The horizontal motion is described by Equation 4.11:

$$x_f = x_{\text{B}} = (v_i \cos \theta_i) t_{\text{B}} = (11.0 \text{ m/s})(\cos 20.0°) t_{\text{B}}$$

The value of x_{B} can be found if the time of landing t_{B} is known. We can find t_{B} by remembering that $a_y = -g$ and by using the y part of Equation 4.8a. We also note that at the top of the jump the vertical component of velocity $v_{y\text{A}}$ is zero:

$$v_{yf} = v_{y\text{A}} = v_i \sin \theta_i - g t_{\text{A}}$$

$$0 = (11.0 \text{ m/s}) \sin 20.0° - (9.80 \text{ m/s}^2) t_{\text{A}}$$

$$t_{\text{A}} = 0.384 \text{ s}$$

Figure 4.12 (Example 4.3) Mike Powell, current holder of the world long jump record of 8.95 m.

This is the time at which the long-jumper is at the *top* of the jump. Because of the symmetry of the vertical motion, another 0.384 s passes before the jumper returns to the ground. Therefore, the time at which the jumper lands is $t_B = 2t_A = 0.768$ s. Substituting this value into the above expression for x_f gives

$$x_f = x_B = (11.0 \text{ m/s})(\cos 20.0°)(0.768 \text{ s}) = 7.94 \text{ m}$$

This is a reasonable distance for a world-class athlete.

(B) What is the maximum height reached?

Solution We find the maximum height reached by using Equation 4.12:

$$y_{max} = y_A = (v_i \sin \theta_i)t_A - \tfrac{1}{2}gt_A^2$$
$$= (11.0 \text{ m/s})(\sin 20.0°)(0.384 \text{ s})$$
$$-\tfrac{1}{2}(9.80 \text{ m/s}^2)(0.384 \text{ s})^2 = 0.722 \text{ m}$$

To *finalize* this problem, find the answers to parts (A) and (B) using Equations 4.13 and 4.14. The results should agree. Treating the long-jumper as a particle is an oversimplification. Nevertheless, the values obtained are consistent with experience in sports. We learn that we can model a complicated system such as a long-jumper as a particle and still obtain results that are reasonable.

Example 4.4 A Bull's-Eye Every Time

Interactive

In a popular lecture demonstration, a projectile is fired at a target T in such a way that the projectile leaves the gun at the same time the target is dropped from rest, as shown in Figure 4.13. Show that if the gun is initially aimed at the stationary target, the projectile hits the target.

Solution *Conceptualize* the problem by studying Figure 4.13. Notice that the problem asks for no numbers. The expected result must involve an algebraic argument. Because both objects are subject only to gravity, we *categorize* this problem as one involving two objects in free-fall, one moving in one dimension and one moving in two. Let us now *analyze* the problem. A collision results under the conditions stated by noting that, as soon as they are released, the projectile and the target experience the same acceleration, $a_y = -g$. Figure 4.13b shows that the initial y coordinate of the target is $x_T \tan \theta_i$ and that it falls to a position $\frac{1}{2}gt^2$ below this coordinate at time t. Therefore, the y coordinate of the target at any moment after release is

$$y_T = x_T \tan \theta_i - \tfrac{1}{2}gt^2$$

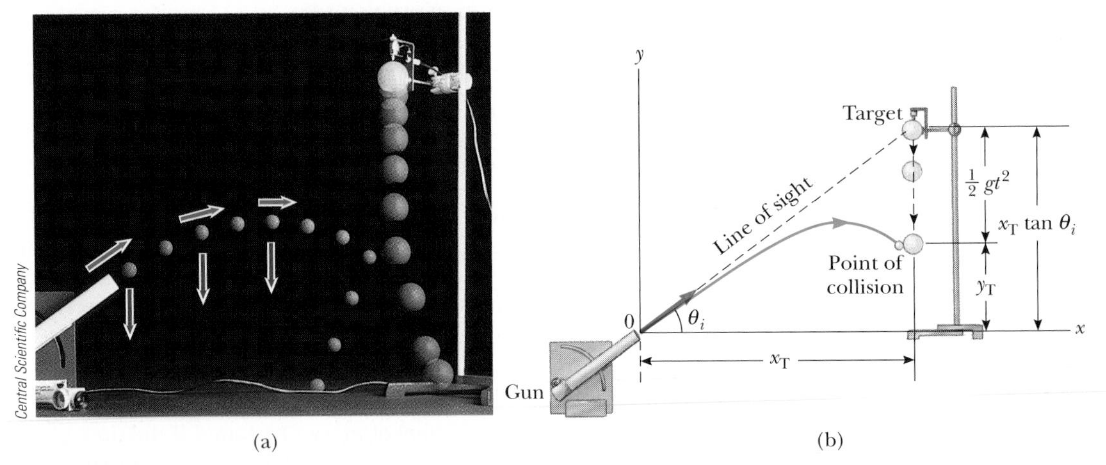

Figure 4.13 (Example 4.4) (a) Multiflash photograph of projectile–target demonstration. If the gun is aimed directly at the target and is fired at the same instant the target begins to fall, the projectile will hit the target. Note that the velocity of the projectile (red arrows) changes in direction and magnitude, while its downward acceleration (violet arrows) remains constant. (b) Schematic diagram of the projectile–target demonstration. Both projectile and target have fallen through the same vertical distance at time t, because both experience the same acceleration $a_y = -g$.

Now if we use Equation 4.9a to write an expression for the y coordinate of the projectile at any moment, we obtain

$$y_P = x_P \tan \theta_i - \tfrac{1}{2}gt^2$$

Thus, by comparing the two previous equations, we see that when the y coordinates of the projectile and target are the same, their x coordinates are the same and a collision results. That is, when $y_P = y_T$, $x_P = x_T$. You can obtain the same result, using expressions for the position vectors for the projectile and target.

To *finalize* this problem, note that a collision can result only when $v_i \sin \theta_i \geq \sqrt{gd/2}$ where d is the initial elevation of the target above the floor. If $v_i \sin \theta_i$ is less than this value, the projectile will strike the floor before reaching the target.

Investigate this situation at the Interactive Worked Example link at http://www.pse6.com.

Example 4.5 That's Quite an Arm! — Interactive

A stone is thrown from the top of a building upward at an angle of 30.0° to the horizontal with an initial speed of 20.0 m/s, as shown in Figure 4.14. If the height of the building is 45.0 m,

(A) how long does it take the stone to reach the ground?

Solution We *conceptualize* the problem by studying Figure 4.14, in which we have indicated the various parameters. By now, it should be natural to *categorize* this as a projectile motion problem.

To *analyze* the problem, let us once again separate motion into two components. The initial x and y components of the stone's velocity are

$$v_{xi} = v_i \cos \theta_i = (20.0 \text{ m/s})\cos 30.0° = 17.3 \text{ m/s}$$

$$v_{yi} = v_i \sin \theta_i = (20.0 \text{ m/s})\sin 30.0° = 10.0 \text{ m/s}$$

To find t, we can use $y_f = y_i + v_{yi}t + \frac{1}{2}a_y t^2$(Eq. 4.9a) with $y_i = 0$, $y_f = -45.0$ m, $a_y = -g$, and $v_{yi} = 10.0$ m/s (there is a negative sign on the numerical value of y_f because we have chosen the top of the building as the origin):

$$-45.0 \text{ m} = (10.0 \text{ m/s})t - \tfrac{1}{2}(9.80 \text{ m/s}^2)t^2$$

Solving the quadratic equation for t gives, for the positive root, $t =$ 4.22 s. To *finalize* this part, think: Does the negative root have any physical meaning?

(B) What is the speed of the stone just before it strikes the ground?

Solution We can use Equation 4.8a, $v_{yf} = v_{yi} + a_y t$, with $t = 4.22$ s to obtain the y component of the velocity just before the stone strikes the ground:

$$v_{yf} = 10.0 \text{ m/s} - (9.80 \text{ m/s}^2)(4.22 \text{ s}) = -31.4 \text{ m/s}$$

Because $v_{xf} = v_{xi} = 17.3$ m/s, the required speed is

$$v_f = \sqrt{v_{xf}^2 + v_{yf}^2} = \sqrt{(17.3)^2 + (-31.4)^2} \text{ m/s} = 35.9 \text{ m/s}$$

To *finalize* this part, is it reasonable that the y component of the final velocity is negative? Is it reasonable that the final speed is larger than the initial speed of 20.0 m/s?

What If? **What if a horizontal wind is blowing in the same direction as the ball is thrown and it causes the ball to have a horizontal acceleration component $a_x = 0.500$ m/s². Which part of this example, (A) or (B), will have a different answer?**

Answer Recall that the motions in the x and y directions are independent. Thus, the horizontal wind cannot affect the vertical motion. The vertical motion determines the time of the projectile in the air, so the answer to (A) does not change. The wind will cause the horizontal velocity component to increase with time, so that the final speed will change in part (B).

We can find the new final horizontal velocity component by using Equation 4.8a:

$$v_{xf} = v_{xi} + a_x t = 17.3 \text{m/s} + (0.500 \text{m/s}^2)(4.22 \text{s})$$
$$= 19.4 \text{m/s}$$

and the new final speed:

$$v_f = \sqrt{v_{xf}^2 + v_{yf}^2} = \sqrt{(19.4)^2 + (-31.4)^2} \text{ m/s} = 36.9 \text{m/s}$$

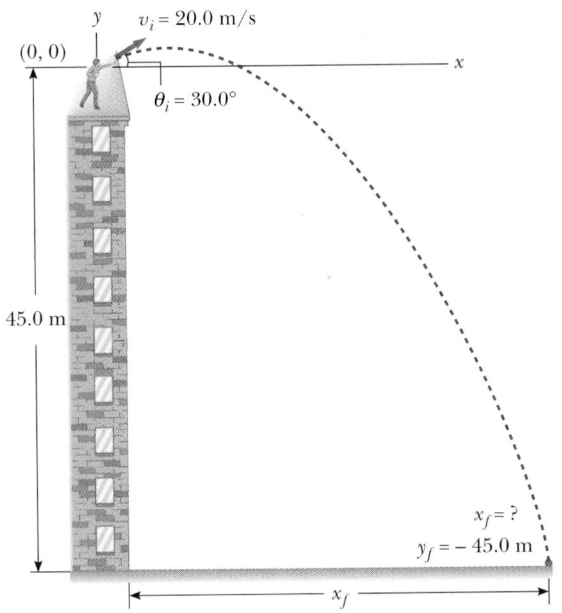

Figure 4.14 (Example 4.5) A stone is thrown from the top of a building.

Investigate this situation at the Interactive Worked Example link at http://www.pse6.com.

Example 4.6 The Stranded Explorers

A plane drops a package of supplies to a party of explorers, as shown in Figure 4.15. If the plane is traveling horizontally at 40.0 m/s and is 100 m above the ground, where does the package strike the ground relative to the point at which it is released?

Solution *Conceptualize* what is happening with the assistance of Figure 4.15. The plane is traveling horizontally when it drops the package. Because the package is in free-fall while moving in the horizontal direction, we *categorize* this as a projectile motion problem. To *analyze* the problem, we choose the coordinate system shown in Figure 4.15, in which the origin is at the point of release of the package. Consider first its horizontal motion. The only equation available for finding the position along the horizontal direction is $x_f = x_i + v_{xi}t$ (Eq. 4.9a). The initial x component of the package velocity is the same as that of the plane when the package is released: 40.0 m/s. Thus, we have

$$x_f = (40.0 \text{ m/s})t$$

If we know t, the time at which the package strikes the ground, then we can determine x_f, the distance the package travels in the horizontal direction. To find t, we use the equations that describe the vertical motion of the package. We know that, at the instant the package hits the ground, its y coordinate is $y_f = -100$ m. We also know that the initial vertical component of the package velocity v_{yi} is zero because at the moment of release, the package has only a horizontal component of velocity.

From Equation 4.9a, we have

$$y_f = -\tfrac{1}{2}gt^2$$

$$-100 \text{ m} = -\tfrac{1}{2}(9.80 \text{ m/s}^2)t^2$$

$$t = 4.52 \text{ s}$$

Substitution of this value for the time into the equation for the x coordinate gives

$$x_f = (40.0 \text{ m/s})(4.52 \text{ s}) = 181 \text{ m}$$

The package hits the ground 181 m to the right of the drop point. To *finalize* this problem, we learn that an object dropped from a moving airplane does not fall straight down. It hits the ground at a point different from the one right below the plane where it was released. This was an important consideration for free-fall bombs such as those used in World War II.

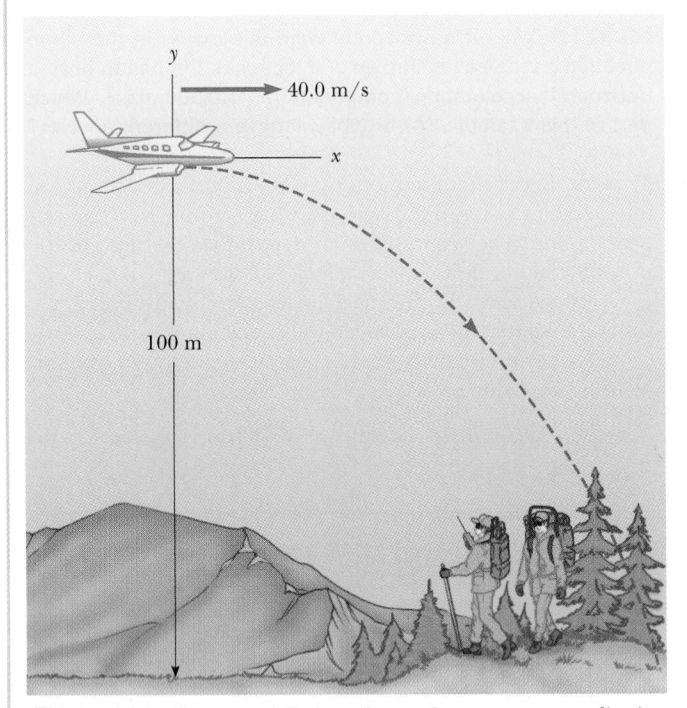

Figure 4.15 (Example 4.6) A package of emergency supplies is dropped from a plane to stranded explorers.

Example 4.7 The End of the Ski Jump

A ski-jumper leaves the ski track moving in the horizontal direction with a speed of 25.0 m/s, as shown in Figure 4.16. The landing incline below him falls off with a slope of 35.0°. Where does he land on the incline?

Solution We can *conceptualize* this problem based on observations of winter Olympic ski competitions. We observe the skier to be airborne for perhaps 4 s and go a distance of about 100 m horizontally. We should expect the value of d, the distance traveled along the incline, to be of the same order of magnitude. We *categorize* the problem as that of a particle in projectile motion.

To *analyze* the problem, it is convenient to select the beginning of the jump as the origin. Because $v_{xi} = 25.0$ m/s and $v_{yi} = 0$, the x and y component forms of Equation 4.9a are

$$(1) \qquad x_f = v_{xi}t = (25.0 \text{ m/s})t$$

$$(2) \qquad y_f = v_{yi}t + \tfrac{1}{2}a_yt^2 = -\tfrac{1}{2}(9.80 \text{ m/s}^2)t^2$$

From the right triangle in Figure 4.16, we see that the jumper's x and y coordinates at the landing point are $x_f = d\cos 35.0°$ and $y_f = -d\sin 35.0°$. Substituting these relationships into (1) and (2), we obtain

$$(3) \qquad d\cos 35.0° = (25.0 \text{ m/s})t$$

$$(4) \qquad -d\sin 35.0° = -\tfrac{1}{2}(9.80 \text{ m/s}^2)t^2$$

Solving (3) for t and substituting the result into (4), we find that $d = 109$ m. Hence, the x and y coordinates of the point at which the skier lands are

$$x_f = d\cos 35.0° = (109 \text{ m})\cos 35.0° = 89.3 \text{ m}$$

$$y_f = -d\sin 35.0° = -(109 \text{ m})\sin 35.0° = -62.5 \text{ m}$$

To *finalize* the problem, let us compare these results to our expectations. We expected the horizontal distance to be on the order of 100 m, and our result of 89.3 m is indeed on

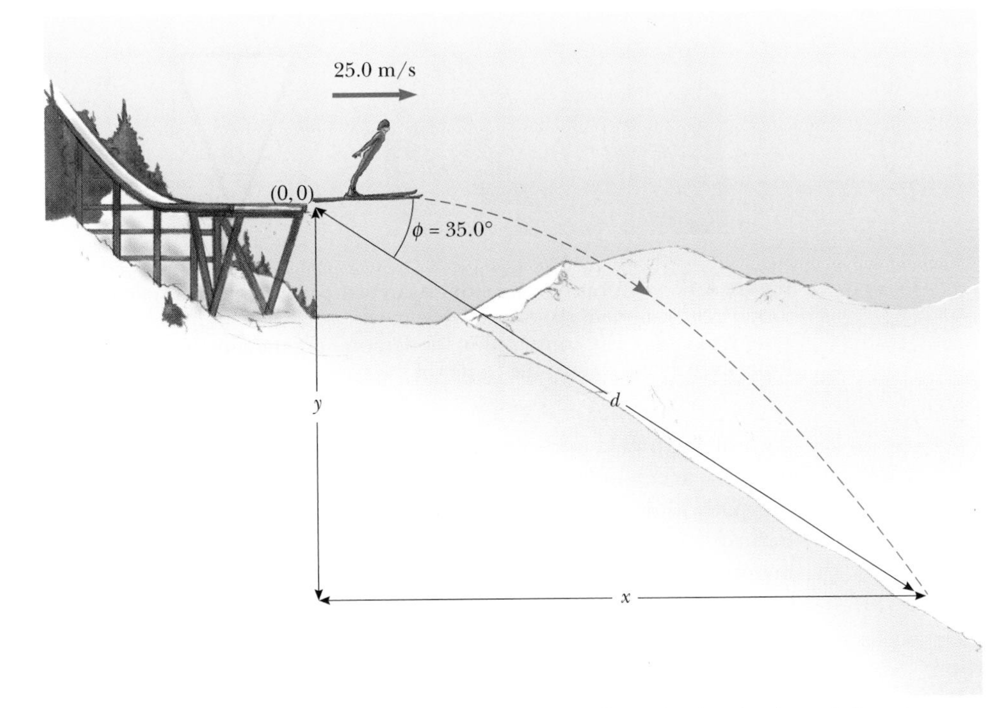

Figure 4.16 (Example 4.7) A ski jumper leaves the track moving in a horizontal direction.

this order of magnitude. It might be useful to calculate the time interval that the jumper is in the air and compare it to our estimate of about 4 s.

What If? **Suppose everything in this example is the same except that the ski jump is curved so that the jumper is projected upward at an angle from the end of the track. Is this a better design in terms of maximizing the length of the jump?**

Answer If the initial velocity has an upward component, the skier will be in the air longer, and should therefore travel further. However, tilting the initial velocity vector upward will reduce the horizontal component of the initial velocity. Thus, angling the end of the ski track upward at a *large* angle may actually *reduce* the distance. Consider the extreme case. The skier is projected at 90° to the horizontal, and simply goes up and comes back down at the end of the ski track! This argument suggests that there must be an optimal angle between 0 and 90° that represents a balance between making the flight time longer and the horizontal velocity component smaller.

We can find this optimal angle mathematically. We modify equations (1) through (4) in the following way, assuming that the skier is projected at an angle θ with respect to the horizontal:

$$\text{(1) and (3)} \quad \rightarrow \quad x_f = (v_i \cos\theta)t = d\cos\phi$$

$$\text{(2) and (4)} \quad \rightarrow \quad y_f = (v_i \sin\theta)t - \tfrac{1}{2}gt^2 = -d\sin\phi$$

If we eliminate the time t between these equations and then use differentiation to maximize d in terms of θ, we arrive (after several steps—see Problem 72!) at the following equation for the angle θ that gives the maximum value of d:

$$\theta = 45° - \frac{\phi}{2}$$

For the slope angle in Figure 4.16, $\phi = 35.0°$; this equation results in an optimal launch angle of $\theta = 27.5°$. Notice that for a slope angle of $\phi = 0°$, which represents a horizontal plane, this equation gives an optimal launch angle of $\theta = 45°$, as we would expect (see Figure 4.11).

4.4 Uniform Circular Motion

Figure 4.17a shows a car moving in a circular path with *constant speed v*. Such motion is called **uniform circular motion,** and occurs in many situations. It is often surprising to students to find that **even though an object moves at a constant speed in a circular path, it still has an acceleration.** To see why, consider the defining equation for average acceleration, $\bar{\mathbf{a}} = \Delta\mathbf{v}/\Delta t$ (Eq. 4.4).

Note that the acceleration depends on *the change in the velocity vector*. Because velocity is a vector quantity, there are two ways in which an acceleration can occur, as mentioned in Section 4.1: by a change in the *magnitude* of the velocity and/or by a change in the *direction* of the velocity. The latter situation occurs for an object moving with constant speed in a circular path. The velocity vector is always tangent to the

PITFALL PREVENTION

4.4 Acceleration of a Particle in Uniform Circular Motion

Remember that acceleration in physics is defined as a change in the *velocity*, not a change in the *speed* (contrary to the everyday interpretation). In circular motion, the velocity vector is changing in direction, so there is indeed an acceleration.

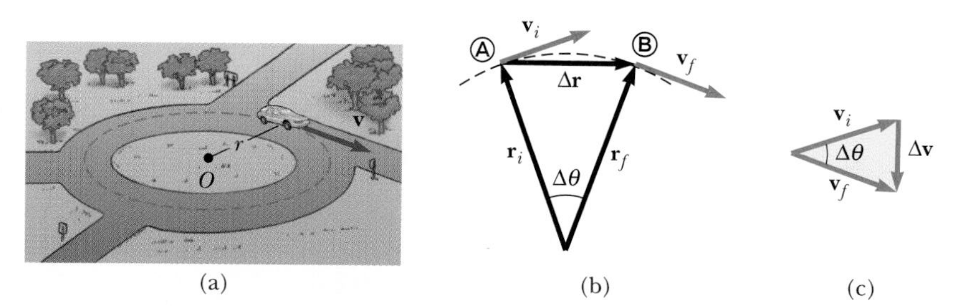

Figure 4.17 (a) A car moving along a circular path at constant speed experiences uniform circular motion. (b) As a particle moves from Ⓐ to Ⓑ, its velocity vector changes from $\mathbf{v}_i$ to $\mathbf{v}_f$. (c) The construction for determining the direction of the change in velocity $\Delta\mathbf{v}$, which is toward the center of the circle for small $\Delta\mathbf{r}$.

path of the object and perpendicular to the radius of the circular path. We now show that the acceleration vector in uniform circular motion is always perpendicular to the path and always points toward the center of the circle. An acceleration of this nature is called a **centripetal acceleration** (*centripetal* means *center-seeking*), and its magnitude is

Centripetal acceleration

$$a_c = \frac{v^2}{r} \tag{4.15}$$

where r is the radius of the circle. The subscript on the acceleration symbol reminds us that the acceleration is centripetal.

First note that the acceleration must be perpendicular to the path followed by the object, which we will model as a particle. If this were not true, there would be a component of the acceleration parallel to the path and, therefore, parallel to the velocity vector. Such an acceleration component would lead to a change in the speed of the particle along the path. But this is inconsistent with our setup of the situation—the particle moves with constant speed along the path. Thus, for *uniform* circular motion, the acceleration vector can only have a component perpendicular to the path, which is toward the center of the circle.

To derive Equation 4.15, consider the diagram of the position and velocity vectors in Figure 4.17b. In addition, the figure shows the vector representing the change in position $\Delta\mathbf{r}$. The particle follows a circular path, part of which is shown by the dotted curve. The particle is at Ⓐ at time t_i, and its velocity at that time is $\mathbf{v}_i$; it is at Ⓑ at some later time t_f, and its velocity at that time is $\mathbf{v}_f$. Let us also assume that $\mathbf{v}_i$ and $\mathbf{v}_f$ differ only in direction; their magnitudes are the same (that is, $v_i = v_f = v$, because it is *uniform* circular motion). In order to calculate the acceleration of the particle, let us begin with the defining equation for average acceleration (Eq. 4.4):

$$\overline{\mathbf{a}} \equiv \frac{\mathbf{v}_f - \mathbf{v}_i}{t_f - t_i} = \frac{\Delta\mathbf{v}}{\Delta t}$$

In Figure 4.17c, the velocity vectors in Figure 4.17b have been redrawn tail to tail. The vector $\Delta\mathbf{v}$ connects the tips of the vectors, representing the vector addition $\mathbf{v}_f = \mathbf{v}_i + \Delta\mathbf{v}$. In both Figures 4.17b and 4.17c, we can identify triangles that help us analyze the motion. The angle $\Delta\theta$ between the two position vectors in Figure 4.17b is the same as the angle between the velocity vectors in Figure 4.17c, because the velocity vector $\mathbf{v}$ is always perpendicular to the position vector $\mathbf{r}$. Thus, the two triangles are *similar.* (Two triangles are similar if the angle between any two sides is the same for both triangles and if the ratio of the lengths of these sides is the same.) This enables us to write a relationship between the lengths of the sides for the two triangles:

$$\frac{|\Delta\mathbf{v}|}{v} = \frac{|\Delta\mathbf{r}|}{r}$$

where $v = v_i = v_f$ and $r = r_i = r_f$. This equation can be solved for $|\Delta\mathbf{v}|$ and the expression so obtained can be substituted into $\bar{\mathbf{a}} = \Delta\mathbf{v}/\Delta t$ to give the magnitude of the average acceleration over the time interval for the particle to move from Ⓐ to Ⓑ:

$$|\bar{\mathbf{a}}| = \frac{|\Delta\mathbf{v}|}{\Delta t} = \frac{v}{r}\frac{|\Delta\mathbf{r}|}{\Delta t}$$

Now imagine that points Ⓐ and Ⓑ in Figure 4.17b become extremely close together. As Ⓐ and Ⓑ approach each other, Δt approaches zero, and the ratio $|\Delta\mathbf{r}|/\Delta t$ approaches the speed v. In addition, the average acceleration becomes the instantaneous acceleration at point Ⓐ. Hence, in the limit $\Delta t \rightarrow 0$, the magnitude of the acceleration is

$$a_c = \frac{v^2}{r}$$

Thus, in uniform circular motion the acceleration is directed inward toward the center of the circle and has magnitude v^2/r.

In many situations it is convenient to describe the motion of a particle moving with constant speed in a circle of radius r in terms of the **period** T, which is defined as the time required for one complete revolution. In the time interval T the particle moves a distance of $2\pi r$, which is equal to the circumference of the particle's circular path. Therefore, because its speed is equal to the circumference of the circular path divided by the period, or $v = 2\pi r/T$, it follows that

Period of circular motion

$$T \equiv \frac{2\pi r}{v} \tag{4.16}$$

PITFALL PREVENTION

4.5 Centripetal Acceleration is not Constant

We derived the magnitude of the centripetal acceleration vector and found it to be constant for uniform circular motion. But *the centripetal acceleration vector is not constant.* It always points toward the center of the circle, but continuously changes direction as the object moves around the circular path.

Quick Quiz 4.7 Which of the following correctly describes the centripetal acceleration vector for a particle moving in a circular path? (a) constant and always perpendicular to the velocity vector for the particle (b) constant and always parallel to the velocity vector for the particle (c) of constant magnitude and always perpendicular to the velocity vector for the particle (d) of constant magnitude and always parallel to the velocity vector for the particle.

Quick Quiz 4.8 A particle moves in a circular path of radius r with speed v. It then increases its speed to $2v$ while traveling along the same circular path. The centripetal acceleration of the particle has changed by a factor of (a) 0.25 (b) 0.5 (c) 2 (d) 4 (e) impossible to determine

Example 4.8 The Centripetal Acceleration of the Earth

What is the centripetal acceleration of the Earth as it moves in its orbit around the Sun?

Solution We *conceptualize* this problem by bringing forth our familiar mental image of the Earth in a circular orbit around the Sun. We will simplify the problem by modeling the Earth as a particle and approximating the Earth's orbit as circular (it's actually slightly elliptical). This allows us to *categorize* this problem as that of a particle in uniform circular motion. When we begin to *analyze* this problem, we realize that we do not know the orbital speed of the Earth in Equation 4.15. With the help of Equation 4.16, however, we can recast Equation 4.15 in terms of the period of the Earth's orbit, which we know is one year:

$$a_c = \frac{v^2}{r} = \frac{\left(\frac{2\pi r}{T}\right)^2}{r} = \frac{4\pi^2 r}{T^2}$$

$$= \frac{4\pi^2(1.496 \times 10^{11}\text{ m})}{(1\text{ yr})^2}\left(\frac{1\text{ yr}}{3.156 \times 10^7\text{ s}}\right)^2$$

$$= 5.93 \times 10^{-3}\text{ m/s}^2$$

To *finalize* this problem, note that this acceleration is much smaller than the free-fall acceleration on the surface of the Earth. An important thing we learned here is the technique of replacing the speed v in terms of the period T of the motion.

4.5 Tangential and Radial Acceleration

Let us consider the motion of a particle along a smooth curved path where the velocity changes both in direction and in magnitude, as described in Figure 4.18. In this situation, the velocity vector is always tangent to the path; however, the acceleration vector **a** is at some angle to the path. At each of three points Ⓐ, Ⓑ, and Ⓒ in Figure 4.18, we draw dashed circles that represent a portion of the actual path at each point. The radius of the circles is equal to the radius of curvature of the path at each point.

As the particle moves along the curved path in Figure 4.18, the direction of the total acceleration vector **a** changes from point to point. This vector can be resolved into two components, based on an origin at the center of the dashed circle: a radial component a_r along the radius of the model circle, and a tangential component a_t perpendicular to this radius. The *total* acceleration vector **a** can be written as the vector sum of the component vectors:

Total acceleration

$$\mathbf{a} = \mathbf{a}_r + \mathbf{a}_t \tag{4.17}$$

The tangential acceleration component causes the change in the speed of the particle. This component is parallel to the instantaneous velocity, and is given by

Tangential acceleration

$$a_t = \frac{d|\mathbf{v}|}{dt} \tag{4.18}$$

The radial acceleration component arises from the change in direction of the velocity vector and is given by

Radial acceleration

$$a_r = -a_c = -\frac{v^2}{r} \tag{4.19}$$

where r is the radius of curvature of the path at the point in question. We recognize the radial component of the acceleration as the centripetal acceleration discussed in Section 4.4. The negative sign indicates that the direction of the centripetal acceleration is toward the center of the circle representing the radius of curvature, which is opposite the direction of the radial unit vector $\hat{\mathbf{r}}$, which always points away from the center of the circle.

Because $\mathbf{a}_r$ and $\mathbf{a}_t$ are perpendicular component vectors of **a**, it follows that the magnitude of **a** is $a = \sqrt{a_r^2 + a_t^2}$. At a given speed, a_r is large when the radius of curvature is small (as at points Ⓐ and Ⓑ in Fig. 4.18) and small when r is large (such as at point Ⓒ). The direction of $\mathbf{a}_t$ is either in the same direction as **v** (if v is increasing) or opposite **v** (if v is decreasing).

In uniform circular motion, where v is constant, $a_t = 0$ and the acceleration is always completely radial, as we described in Section 4.4. In other words, uniform circular motion is a special case of motion along a general curved path. Furthermore, if the direction of **v** does not change, then there is no radial acceleration and the motion is one-dimensional (in this case, $a_r = 0$, but a_t may not be zero).

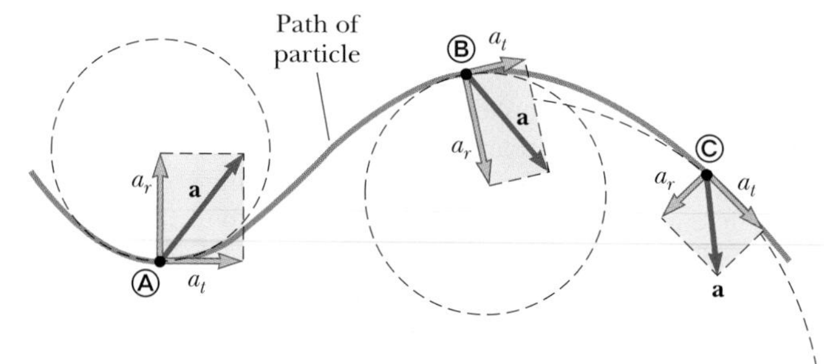
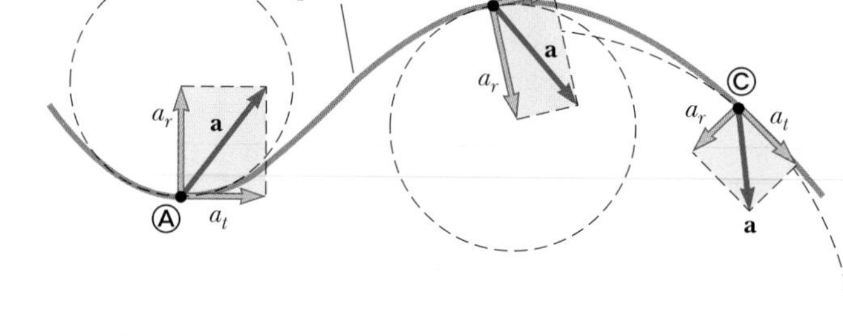

At the Active Figures link at http://www.pse6.com, you can study the acceleration components of a roller coaster car.

Active Figure 4.18 The motion of a particle along an arbitrary curved path lying in the xy plane. If the velocity vector **v** (always tangent to the path) changes in direction and magnitude, the components of the acceleration **a** are a tangential component a_t and a radial component a_r.

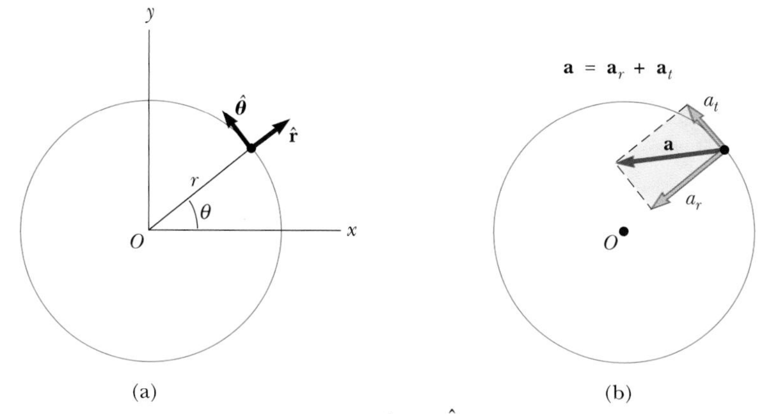

Figure 4.19 (a) Descriptions of the unit vectors $\hat{\mathbf{r}}$ and $\hat{\boldsymbol{\theta}}$. (b) The total acceleration **a** of a particle moving along a curved path (which at any instant is part of a circle of radius r) is the sum of radial and tangential component vectors. The radial component vector is directed toward the center of curvature. If the tangential component of acceleration becomes zero, the particle follows uniform circular motion.

It is convenient to write the acceleration of a particle moving in a circular path in terms of unit vectors. We do this by defining the unit vectors $\hat{\mathbf{r}}$ and $\hat{\boldsymbol{\theta}}$ shown in Figure 4.19a, where $\hat{\mathbf{r}}$ is a unit vector lying along the radius vector and directed radially outward from the center of the circle and $\hat{\boldsymbol{\theta}}$ is a unit vector tangent to the circle. The direction of $\hat{\boldsymbol{\theta}}$ is in the direction of increasing θ, where θ is measured counterclockwise from the positive x axis. Note that both $\hat{\mathbf{r}}$ and $\hat{\boldsymbol{\theta}}$ "move along with the particle" and so vary in time. Using this notation, we can express the total acceleration as

$$\mathbf{a} = \mathbf{a}_t + \mathbf{a}_r = \frac{d|\mathbf{v}|}{dt}\hat{\boldsymbol{\theta}} - \frac{v^2}{r}\hat{\mathbf{r}} \tag{4.20}$$

These vectors are described in Figure 4.19b.

Quick Quiz 4.9 A particle moves along a path and its speed increases with time. In which of the following cases are its acceleration and velocity vectors parallel? (a) the path is circular (b) the path is straight (c) the path is a parabola (d) never.

Quick Quiz 4.10 A particle moves along a path and its speed increases with time. In which of the following cases are its acceleration and velocity vectors perpendicular everywhere along the path? (a) the path is circular (b) the path is straight (c) the path is a parabola (d) never.

Example 4.9 Over the Rise

A car exhibits a constant acceleration of 0.300 m/s^2 parallel to the roadway. The car passes over a rise in the roadway such that the top of the rise is shaped like a circle of radius 500 m. At the moment the car is at the top of the rise, its velocity vector is horizontal and has a magnitude of 6.00 m/s. What is the direction of the total acceleration vector for the car at this instant?

Solution *Conceptualize* the situation using Figure 4.20a. Because the car is moving along a curved path, we can *categorize* this as a problem involving a particle experiencing both tangential and radial acceleration. Now we recognize that this is a relatively simple plug-in problem. The radial acceleration is given by Equation 4.19. With $v = 6.00$ m/s and $r = 500$ m, we find that

$$a_r = -\frac{v^2}{r} = -\frac{(6.00\text{ m/s})^2}{500\text{ m}} = -0.0720\text{ m/s}^2$$

The radial acceleration vector is directed straight downward

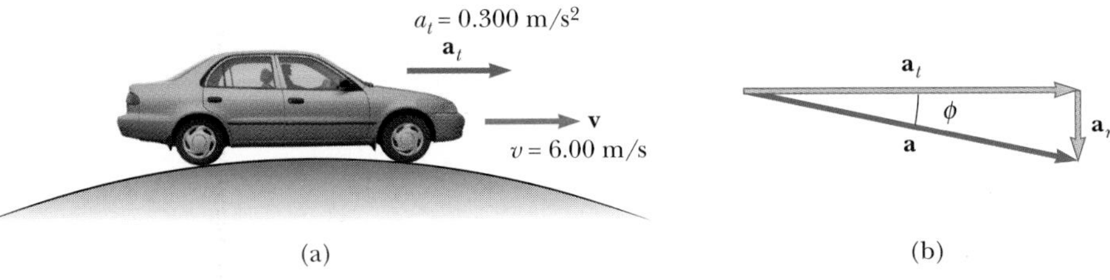

Figure 4.20 (Example 4.9) (a) A car passes over a rise that is shaped like a circle. (b) The total acceleration vector **a** is the sum of the tangential and radial acceleration vectors $\mathbf{a}_t$ and $\mathbf{a}_r$.

while the tangential acceleration vector has magnitude 0.300 m/s^2 and is horizontal. Because $\mathbf{a} = \mathbf{a}_r + \mathbf{a}_t$, the magnitude of **a** is

$$a = \sqrt{a_r^2 + a_t^2} = \sqrt{(-0.0720)^2 + (0.300)^2} \text{ m/s}^2$$

$$= \boxed{0.309 \text{ m/s}^2}$$

If ϕ is the angle between **a** and the horizontal, then

$$\phi = \tan^{-1}\frac{a_r}{a_t} = \tan^{-1}\left(\frac{-0.0720 \text{ m/s}^2}{0.300 \text{ m/s}^2}\right) = \boxed{-13.5°}$$

This angle is measured downward from the horizontal. (See Figure 4.20b.)

4.6 Relative Velocity and Relative Acceleration

In this section, we describe how observations made by different observers in different frames of reference are related to each other. We find that observers in different frames of reference may measure different positions, velocities, and accelerations for a given particle. That is, two observers moving relative to each other generally do not agree on the outcome of a measurement.

As an example, consider two observers watching a man walking on a moving beltway at an airport in Figure 4.21. The woman standing on the moving beltway will see the man moving at a normal walking speed. The woman observing from the stationary floor will see the man moving with a higher speed, because the beltway speed combines with his walking speed. Both observers look at the same man and arrive at different values for his speed. Both are correct; the difference in their measurements is due to the relative velocity of their frames of reference.

Suppose a person riding on a skateboard (observer A) throws a ball in such a way that it appears in this person's frame of reference to move first straight upward and

Figure 4.21 Two observers measure the speed of a man walking on a moving beltway. The woman standing on the beltway sees the man moving with a slower speed than the woman observing from the stationary floor.

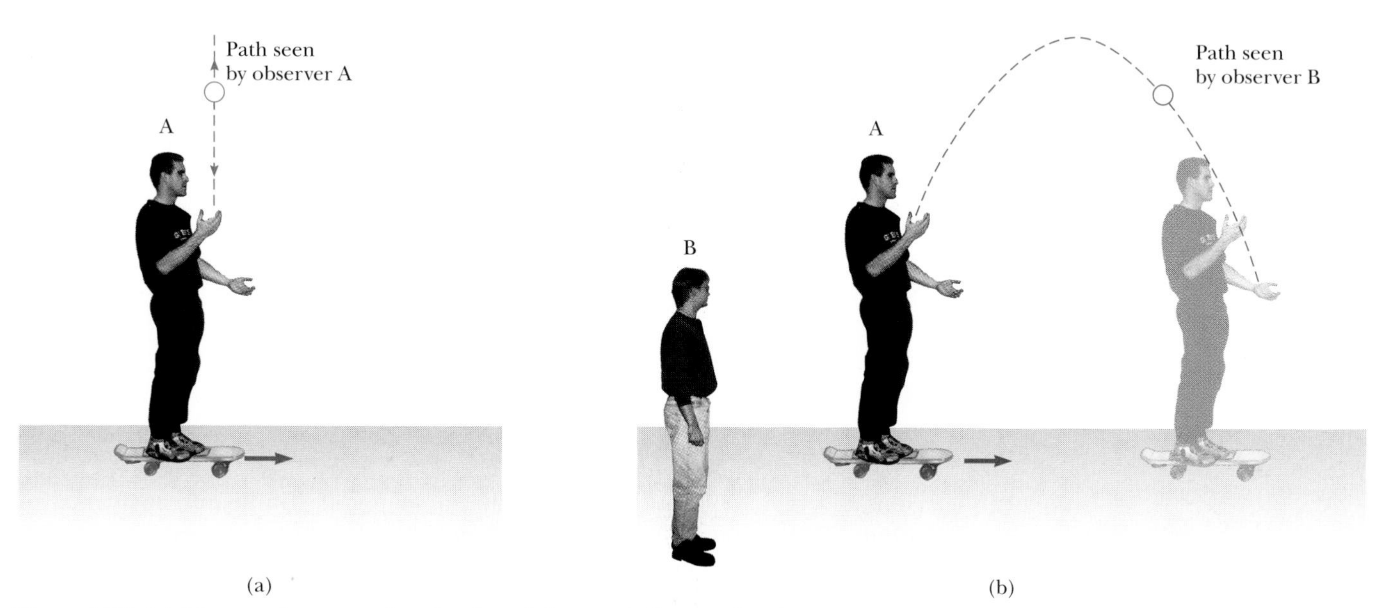

Figure 4.22 (a) Observer A on a moving skateboard throws a ball upward and sees it rise and fall in a straight-line path. (b) Stationary observer B sees a parabolic path for the same ball.

then straight downward along the same vertical line, as shown in Figure 4.22a. An observer B on the ground sees the path of the ball as a parabola, as illustrated in Figure 4.22b. Relative to observer B, the ball has a vertical component of velocity (resulting from the initial upward velocity and the downward acceleration due to gravity) *and* a horizontal component.

Another example of this concept is the motion of a package dropped from an airplane flying with a constant velocity—a situation we studied in Example 4.6. An observer on the airplane sees the motion of the package as a straight line downward toward Earth. The stranded explorer on the ground, however, sees the trajectory of the package as a parabola. Once the package is dropped, and the airplane continues to move horizontally with the same velocity, the package hits the ground directly beneath the airplane (if we assume that air resistance is neglected)!

In a more general situation, consider a particle located at point Ⓐ in Figure 4.23. Imagine that the motion of this particle is being described by two observers, one in reference frame S, fixed relative to Earth, and another in reference frame S', moving to the right relative to S (and therefore relative to Earth) with a constant velocity $\mathbf{v}_0$. (Relative to an observer in S', S moves to the left with a velocity $-\mathbf{v}_0$.) Where an observer stands in a reference frame is irrelevant in this discussion, but for purposes of this discussion let us place each observer at her or his respective origin.

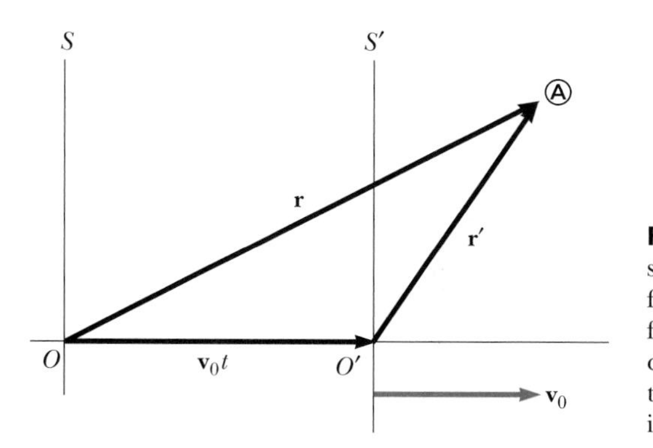

Figure 4.23 A particle located at Ⓐ is described by two observers, one in the fixed frame of reference S, and the other in the frame S', which moves to the right with a constant velocity $\mathbf{v}_0$. The vector $\mathbf{r}$ is the particle's position vector relative to S, and $\mathbf{r}'$ is its position vector relative to S'.

We define the time $t = 0$ as that instant at which the origins of the two reference frames coincide in space. Thus, at time t, the origins of the reference frames will be separated by a distance v_0t. We label the position of the particle relative to the S frame with the position vector $\mathbf{r}$ and that relative to the S' frame with the position vector $\mathbf{r}'$, both at time t. The vectors $\mathbf{r}$ and $\mathbf{r}'$ are related to each other through the expression $\mathbf{r} = \mathbf{r}' + \mathbf{v}_0t$, or

Galilean coordinate transformation

$$\mathbf{r}' = \mathbf{r} - \mathbf{v}_0t \qquad (4.21)$$

If we differentiate Equation 4.21 with respect to time and note that $\mathbf{v}_0$ is constant, we obtain

$$\frac{d\mathbf{r}'}{dt} = \frac{d\mathbf{r}}{dt} - \mathbf{v}_0$$

Galilean velocity transformation

$$\mathbf{v}' = \mathbf{v} - \mathbf{v}_0 \qquad (4.22)$$

where $\mathbf{v}'$ is the velocity of the particle observed in the S' frame and $\mathbf{v}$ is its velocity observed in the S frame. Equations 4.21 and 4.22 are known as **Galilean transformation equations.** They relate the position and velocity of a particle as measured by observers in relative motion.

Although observers in two frames measure different velocities for the particle, they measure the *same acceleration* when $\mathbf{v}_0$ is constant. We can verify this by taking the time derivative of Equation 4.22:

$$\frac{d\mathbf{v}'}{dt} = \frac{d\mathbf{v}}{dt} - \frac{d\mathbf{v}_0}{dt}$$

Because $\mathbf{v}_0$ is constant, $d\mathbf{v}_0/dt = 0$. Therefore, we conclude that $\mathbf{a}' = \mathbf{a}$ because $\mathbf{a}' = d\mathbf{v}'/dt$ and $\mathbf{a} = d\mathbf{v}/dt$. That is, **the acceleration of the particle measured by an observer in one frame of reference is the same as that measured by any other observer moving with constant velocity relative to the first frame.**

Quick Quiz 4.11 A passenger, observer A, in a car traveling at a constant horizontal velocity of magnitude 60 mi/h pours a cup of coffee for the tired driver. Observer B stands on the side of the road and watches the pouring process through the window of the car as it passes. Which observer(s) sees a parabolic path for the coffee as it moves through the air? (a) A (b) B (c) both A and B (d) neither A nor B.

Example 4.10 A Boat Crossing a River

A boat heading due north crosses a wide river with a speed of 10.0 km/h relative to the water. The water in the river has a uniform speed of 5.00 km/h due east relative to the Earth. Determine the velocity of the boat relative to an observer standing on either bank.

Solution To *conceptualize* this problem, imagine moving across a river while the current pushes you along the river. You will not be able to move directly across the river, but will end up downstream, as suggested in Figure 4.24. Because of the separate velocities of you and the river, we can *categorize* this as a problem involving relative velocities. We will *analyze* this problem with the techniques discussed in this section. We know $\mathbf{v}_{br}$, the velocity of the *boat* relative to the *river*, and $\mathbf{v}_{rE}$, the velocity of the *river* relative to *Earth*. What we must find is $\mathbf{v}_{bE}$, the velocity of the *boat* relative to *Earth*. The relationship between these three quantities is

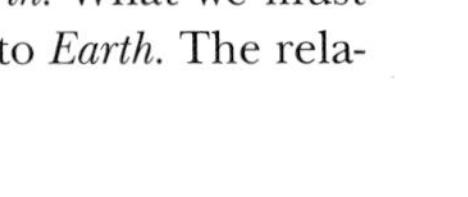

$$\mathbf{v}_{bE} = \mathbf{v}_{br} + \mathbf{v}_{rE}$$

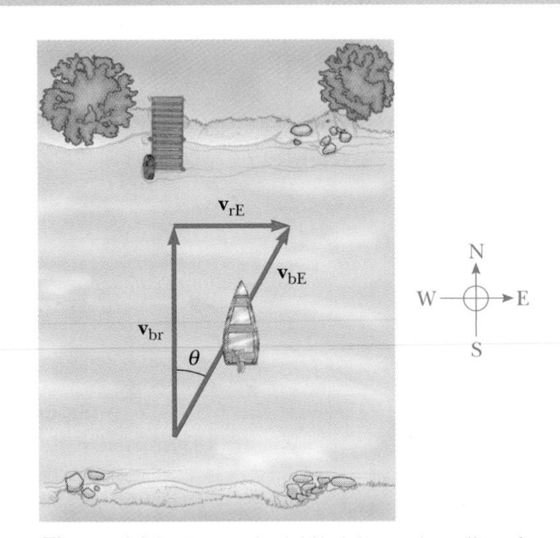

Figure 4.24 (Example 4.10) A boat aims directly across a river and ends up downstream.

The terms in the equation must be manipulated as vector quantities; the vectors are shown in Figure 4.24. The quantity $\mathbf{v}_{br}$ is due north, $\mathbf{v}_{rE}$ is due east, and the vector sum of the two, $\mathbf{v}_{bE}$, is at an angle θ, as defined in Figure 4.24. Thus, we can find the speed v_{bE} of the boat relative to Earth by using the Pythagorean theorem:

$$v_{bE} = \sqrt{v_{br}^2 + v_{rE}^2} = \sqrt{(10.0)^2 + (5.00)^2} \text{ km/h}$$

$$= 11.2 \text{ km/h}$$

The direction of $\mathbf{v}_{bE}$ is

$$\theta = \tan^{-1}\left(\frac{v_{rE}}{v_{br}}\right) = \tan^{-1}\left(\frac{5.00}{10.0}\right) = 26.6°$$

The boat is moving at a speed of 11.2 km/h in the direction 26.6° east of north relative to Earth. To *finalize* the problem, note that the speed of 11.2 km/h is faster than your boat speed of 10.0 km/h. The current velocity adds to yours to give you a larger speed. Notice in Figure 4.24 that your resultant velocity is at an angle to the direction straight across the river, so you will end up downstream, as we predicted.

Example 4.11 Which Way Should We Head?

If the boat of the preceding example travels with the same speed of 10.0 km/h relative to the river and is to travel due north, as shown in Figure 4.25, what should its heading be?

Solution This example is an extension of the previous one, so we have already *conceptualized* and *categorized* the problem. The *analysis* now involves the new triangle shown in Figure 4.25. As in the previous example, we know $\mathbf{v}_{rE}$ and the magnitude of the vector $\mathbf{v}_{br}$, and we want $\mathbf{v}_{bE}$ to be directed across the river. Note the difference between the triangle in Figure 4.24 and the one in Figure 4.25—the hypotenuse in Figure 4.25 is no longer $\mathbf{v}_{bE}$. Therefore, when we use the Pythagorean theorem to find $\mathbf{v}_{bE}$ in this situation, we obtain

$$v_{bE} = \sqrt{v_{br}^2 - v_{rE}^2} = \sqrt{(10.0)^2 - (5.00)^2} \text{ km/h} = 8.66 \text{ km/h}$$

Now that we know the magnitude of $\mathbf{v}_{bE}$, we can find the direction in which the boat is heading:

$$\theta = \tan^{-1}\left(\frac{v_{rE}}{v_{bE}}\right) = \tan^{-1}\left(\frac{5.00}{8.66}\right) = 30.0°$$

To *finalize* this problem, we learn that the boat must head upstream in order to travel directly northward across the river. For the given situation, the boat must steer a course 30.0° west of north.

What If? **Imagine that the two boats in Examples 4.10 and 4.11 are racing across the river. Which boat arrives at the opposite bank first?**

Answer In Example 4.10, the velocity of 10 km/h is aimed directly across the river. In Example 4.11, the velocity that is directed across the river has a magnitude of only 8.66 km/h. Thus, the boat in Example 4.10 has a larger velocity component directly across the river and will arrive first.

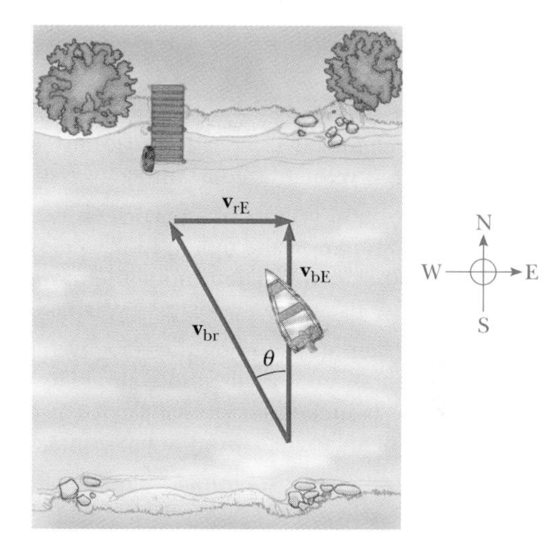

Figure 4.25 (Example 4.11) To move directly across the river, the boat must aim upstream.

SUMMARY

If a particle moves with *constant* acceleration $\mathbf{a}$ and has velocity $\mathbf{v}_i$ and position $\mathbf{r}_i$ at $t = 0$, its velocity and position vectors at some later time t are

$$\mathbf{v}_f = \mathbf{v}_i + \mathbf{a}t \quad (4.8)$$

$$\mathbf{r}_f = \mathbf{r}_i + \mathbf{v}_i t + \tfrac{1}{2}\mathbf{a}t^2 \quad (4.9)$$

Take a practice test for this chapter by clicking on the Practice Test link at **http://www.pse6.com.**

For two-dimensional motion in the *xy* plane under constant acceleration, each of these vector expressions is equivalent to two component expressions—one for the motion in the *x* direction and one for the motion in the *y* direction.

Projectile motion is one type of two-dimensional motion under constant acceleration, where $a_x = 0$ and $a_y = -g$. It is useful to think of projectile motion as the superposition of two motions: (1) constant-velocity motion in the x direction and

(2) free-fall motion in the vertical direction subject to a constant downward acceleration of magnitude $g = 9.80 \text{ m/s}^2$.

A particle moving in a circle of radius r with constant speed v is in **uniform circular motion.** It undergoes a radial acceleration $\mathbf{a}_r$ because the direction of $\mathbf{v}$ changes in time. The magnitude of $\mathbf{a}_r$ is the **centripetal acceleration** a_c:

$$a_c = \frac{v^2}{r} \quad (4.19)$$

and its direction is always toward the center of the circle.

If a particle moves along a curved path in such a way that both the magnitude and the direction of $\mathbf{v}$ change in time, then the particle has an acceleration vector that can be described by two component vectors: (1) a radial component vector $\mathbf{a}_r$ that causes the change in direction of $\mathbf{v}$ and (2) a tangential component vector $\mathbf{a}_t$ that causes the change in magnitude of $\mathbf{v}$. The magnitude of $\mathbf{a}_r$ is v^2/r, and the magnitude of $\mathbf{a}_t$ is $d|\mathbf{v}|/dt$.

The velocity $\mathbf{v}$ of a particle measured in a fixed frame of reference S can be related to the velocity $\mathbf{v}'$ of the same particle measured in a moving frame of reference S' by

$$\mathbf{v}' = \mathbf{v} - \mathbf{v}_0 \quad (4.22)$$

where $\mathbf{v}_0$ is the velocity of S' relative to S.

QUESTIONS

1. Can an object accelerate if its speed is constant? Can an object accelerate if its velocity is constant?
2. If you know the position vectors of a particle at two points along its path and also know the time it took to move from one point to the other, can you determine the particle's instantaneous velocity? Its average velocity? Explain.
3. Construct motion diagrams showing the velocity and acceleration of a projectile at several points along its path if (a) the projectile is launched horizontally and (b) the projectile is launched at an angle θ with the horizontal.
4. A baseball is thrown with an initial velocity of $(10\hat{\mathbf{i}} + 15\hat{\mathbf{j}})$ m/s. When it reaches the top of its trajectory, what are (a) its velocity and (b) its acceleration? Neglect the effect of air resistance.
5. A baseball is thrown such that its initial x and y components of velocity are known. Neglecting air resistance, describe how you would calculate, at the instant the ball reaches the top of its trajectory, (a) its position, (b) its velocity, and (c) its acceleration. How would these results change if air resistance were taken into account?
6. A spacecraft drifts through space at a constant velocity. Suddenly a gas leak in the side of the spacecraft gives it a constant acceleration in a direction perpendicular to the initial velocity. The orientation of the spacecraft does not change, so that the acceleration remains perpendicular to the original direction of the velocity. What is the shape of the path followed by the spacecraft in this situation?
7. A ball is projected horizontally from the top of a building. One second later another ball is projected horizontally from the same point with the same velocity. At what point in the motion will the balls be closest to each other? Will the first ball always be traveling faster than the second ball? What will be the time interval between when the balls hit the ground? Can the horizontal projection velocity of the second ball be changed so that the balls arrive at the ground at the same time?
8. A rock is dropped at the same instant that a ball, at the same initial elevation, is thrown horizontally. Which will have the greater speed when it reaches ground level?
9. Determine which of the following moving objects obey the equations of projectile motion developed in this chapter. (a) A ball is thrown in an arbitrary direction. (b) A jet airplane crosses the sky with its engines thrusting the plane forward. (c) A rocket leaves the launch pad. (d) A rocket moving through the sky after its engines have failed. (e) A stone is thrown under water.
10. How can you throw a projectile so that it has zero speed at the top of its trajectory? So that it has nonzero speed at the top of its trajectory?
11. Two projectiles are thrown with the same magnitude of initial velocity, one at an angle θ with respect to the level ground and the other at angle $90° - \theta$. Both projectiles will strike the ground at the same distance from the projection point. Will both projectiles be in the air for the same time interval?
12. A projectile is launched at some angle to the horizontal with some initial speed v_i, and air resistance is negligible. Is the projectile a freely falling body? What is its acceleration in the vertical direction? What is its acceleration in the horizontal direction?
13. State which of the following quantities, if any, remain constant as a projectile moves through its parabolic trajectory: (a) speed, (b) acceleration, (c) horizontal component of velocity, (d) vertical component of velocity.

14. A projectile is fired at an angle of 30° from the horizontal with some initial speed. Firing the projectile at what other angle results in the same horizontal range if the initial speed is the same in both cases? Neglect air resistance.

15. The maximum range of a projectile occurs when it is launched at an angle of 45.0° with the horizontal, if air resistance is neglected. If air resistance is not neglected, will the optimum angle be greater or less than 45.0°? Explain.

16. A projectile is launched on the Earth with some initial velocity. Another projectile is launched on the Moon with the same initial velocity. Neglecting air resistance, which projectile has the greater range? Which reaches the greater altitude? (Note that the free-fall acceleration on the Moon is about 1.6 m/s^2.)

17. A coin on a table is given an initial horizontal velocity such that it ultimately leaves the end of the table and hits the floor. At the instant the coin leaves the end of the table, a ball is released from the same height and falls to the floor. Explain why the two objects hit the floor simultaneously, even though the coin has an initial velocity.

18. Explain whether or not the following particles have an acceleration: (a) a particle moving in a straight line with constant speed and (b) a particle moving around a curve with constant speed.

19. Correct the following statement: "The racing car rounds the turn at a constant velocity of 90 miles per hour."

20. At the end of a pendulum's arc, its velocity is zero. Is its acceleration also zero at that point?

21. An object moves in a circular path with constant speed v. (a) Is the velocity of the object constant? (b) Is its acceleration constant? Explain.

22. Describe how a driver can steer a car traveling at constant speed so that (a) the acceleration is zero or (b) the magnitude of the acceleration remains constant.

23. An ice skater is executing a figure eight, consisting of two equal, tangent circular paths. Throughout the first loop she increases her speed uniformly, and during the second loop she moves at a constant speed. Draw a motion diagram showing her velocity and acceleration vectors at several points along the path of motion.

24. Based on your observation and experience, draw a motion diagram showing the position, velocity, and acceleration vectors for a pendulum that swings in an arc carrying it from an initial position 45° to the right of the central vertical line to a final position 45° to the left of the central vertical line. The arc is a quadrant of a circle, and you should use the center of the circle as the origin for the position vectors.

25. What is the fundamental difference between the unit vectors $\hat{\mathbf{r}}$ and $\hat{\boldsymbol{\theta}}$ and the unit vectors $\hat{\mathbf{i}}$ and $\hat{\mathbf{j}}$?

26. A sailor drops a wrench from the top of a sailboat's mast while the boat is moving rapidly and steadily in a straight line. Where will the wrench hit the deck? (Galileo posed this question.)

27. A ball is thrown upward in the air by a passenger on a train that is moving with constant velocity. (a) Describe the path of the ball as seen by the passenger. Describe the path as seen by an observer standing by the tracks outside the train. (b) How would these observations change if the train were accelerating along the track?

28. A passenger on a train that is moving with constant velocity drops a spoon. What is the acceleration of the spoon relative to (a) the train and (b) the Earth?

PROBLEMS

1, 2, 3 = straightforward, intermediate, challenging ☐ = full solution available in the *Student Solutions Manual and Study Guide*

= coached solution with hints available at http://www.pse6.com = computer useful in solving problem

= paired numerical and symbolic problems

Section 4.1 The Position, Velocity, and Acceleration Vectors

1. A motorist drives south at 20.0 m/s for 3.00 min, then turns west and travels at 25.0 m/s for 2.00 min, and finally travels northwest at 30.0 m/s for 1.00 min. For this 6.00-min trip, find (a) the total vector displacement, (b) the average speed, and (c) the average velocity. Let the positive x axis point east.

2. A golf ball is hit off a tee at the edge of a cliff. Its x and y coordinates as functions of time are given by the following expressions:

$$x = (18.0 \text{ m/s})t$$

$$\text{and} \quad y = (4.00 \text{ m/s})t - (4.90 \text{ m/s}^2)t^2$$

(a) Write a vector expression for the ball's position as a function of time, using the unit vectors $\hat{\mathbf{i}}$ and $\hat{\mathbf{j}}$. By taking derivatives, obtain expressions for (b) the velocity vector $\mathbf{v}$ as a function of time and (c) the acceleration vector $\mathbf{a}$ as a function of time. Next use unit-vector notation to write expressions for (d) the position, (e) the velocity, and (f) the acceleration of the golf ball, all at $t = 3.00$ s.

3. When the Sun is directly overhead, a hawk dives toward the ground with a constant velocity of 5.00 m/s at 60.0° below the horizontal. Calculate the speed of her shadow on the level ground.

4. The coordinates of an object moving in the xy plane vary with time according to the equations $x = -(5.00 \text{ m}) \sin(\omega t)$ and $y = (4.00 \text{ m}) - (5.00 \text{ m}) \cos(\omega t)$, where ω is a constant and t is in seconds. (a) Determine the components of velocity and components of acceleration at $t = 0$. (b) Write expressions for the position vector, the velocity vector, and the acceleration vector at any time $t > 0$. (c) Describe the path of the object in an xy plot.

Section 4.2 Two-Dimensional Motion with Constant Acceleration

5. At $t = 0$, a particle moving in the xy plane with constant acceleration has a velocity of $\mathbf{v}_i = (3.00\hat{\mathbf{i}} - 2.00\hat{\mathbf{j}})$ m/s and is at the origin. At $t = 3.00$ s, the particle's velocity is $\mathbf{v} = (9.00\hat{\mathbf{i}} + 7.00\hat{\mathbf{j}})$ m/s. Find (a) the acceleration of the particle and (b) its coordinates at any time t.

6. The vector position of a particle varies in time according to the expression $\mathbf{r} = (3.00\hat{\mathbf{i}} - 6.00t^2\hat{\mathbf{j}})$ m. (a) Find expressions for the velocity and acceleration as functions of time. (b) Determine the particle's position and velocity at $t = 1.00$ s.

7. A fish swimming in a horizontal plane has velocity $\mathbf{v}_i = (4.00\hat{\mathbf{i}} + 1.00\hat{\mathbf{j}})$ m/s at a point in the ocean where the position relative to a certain rock is $\mathbf{r}_i = (10.0\hat{\mathbf{i}} - 4.00\hat{\mathbf{j}})$ m. After the fish swims with constant acceleration for 20.0 s, its velocity is $\mathbf{v} = (20.0\hat{\mathbf{i}} - 5.00\hat{\mathbf{j}})$ m/s. (a) What are the components of the acceleration? (b) What is the direction of the acceleration with respect to unit vector $\hat{\mathbf{i}}$? (c) If the fish maintains constant acceleration, where is it at $t = 25.0$ s, and in what direction is it moving?

8. A particle initially located at the origin has an acceleration of $\mathbf{a} = 3.00\hat{\mathbf{j}}$ m/s^2 and an initial velocity of $\mathbf{v}_i = 500\hat{\mathbf{i}}$ m/s. Find (a) the vector position and velocity at any time t and (b) the coordinates and speed of the particle at $t = 2.00$ s.

9. It is not possible to see very small objects, such as viruses, using an ordinary light microscope. An electron microscope can view such objects using an electron beam instead of a light beam. Electron microscopy has proved invaluable for investigations of viruses, cell membranes and subcellular structures, bacterial surfaces, visual receptors, chloroplasts, and the contractile properties of muscles. The "lenses" of an electron microscope consist of electric and magnetic fields that control the electron beam. As an example of the manipulation of an electron beam, consider an electron traveling away from the origin along the x axis in the xy plane with initial velocity $\mathbf{v}_i = v_i\hat{\mathbf{i}}$. As it passes through the region $x = 0$ to $x = d$, the electron experiences acceleration $\mathbf{a} = a_x\hat{\mathbf{i}} + a_y\hat{\mathbf{j}}$, where a_x and a_y are constants. For the case $v_i = 1.80 \times 10^7$ m/s, $a_x = 8.00 \times 10^{14}$ m/s^2 and $a_y = 1.60 \times 10^{15}$ m/s^2, determine at $x = d = 0.0100$ m (a) the position of the electron, (b) the velocity of the electron, (c) the speed of the electron, and (d) the direction of travel of the electron (i.e., the angle between its velocity and the x axis).

Section 4.3 Projectile Motion

Note: Ignore air resistance in all problems and take $g = 9.80$ m/s^2 at the Earth's surface.

10. To start an avalanche on a mountain slope, an artillery shell is fired with an initial velocity of 300 m/s at 55.0° above the horizontal. It explodes on the mountainside 42.0 s after firing. What are the x and y coordinates of the shell where it explodes, relative to its firing point?

11. In a local bar, a customer slides an empty beer mug down the counter for a refill. The bartender is momentarily distracted and does not see the mug, which slides off the counter and strikes the floor 1.40 m from the base of the counter. If the height of the counter is 0.860 m, (a) with what velocity did the mug leave the counter, and (b) what was the direction of the mug's velocity just before it hit the floor?

12. In a local bar, a customer slides an empty beer mug down the counter for a refill. The bartender is momentarily distracted and does not see the mug, which slides off the counter and strikes the floor at distance d from the base of the counter. The height of the counter is h. (a) With what velocity did the mug leave the counter, and (b) what was the direction of the mug's velocity just before it hit the floor?

13. One strategy in a snowball fight is to throw a snowball at a high angle over level ground. While your opponent is watching the first one, a second snowball is thrown at a low angle timed to arrive before or at the same time as the first one. Assume both snowballs are thrown with a speed of 25.0 m/s. The first one is thrown at an angle of 70.0° with respect to the horizontal. (a) At what angle should the second snowball be thrown to arrive at the same point as the first? (b) How many seconds later should the second snowball be thrown after the first to arrive at the same time?

14. An astronaut on a strange planet finds that she can jump a maximum horizontal distance of 15.0 m if her initial speed is 3.00 m/s. What is the free-fall acceleration on the planet?

15. A projectile is fired in such a way that its horizontal range is equal to three times its maximum height. What is the angle of projection?

16. A rock is thrown upward from the level ground in such a way that the maximum height of its flight is equal to its horizontal range d. (a) At what angle θ is the rock thrown? (b) **What If?** Would your answer to part (a) be different on a different planet? (c) What is the range d_{max} the rock can attain if it is launched at the same speed but at the optimal angle for maximum range?

17. A ball is tossed from an upper-story window of a building. The ball is given an initial velocity of 8.00 m/s at an angle of 20.0° below the horizontal. It strikes the ground 3.00 s later. (a) How far horizontally from the base of the building does the ball strike the ground? (b) Find the height from which the ball was thrown. (c) How long does it take the ball to reach a point 10.0 m below the level of launching?

18. The small archerfish (length 20 to 25 cm) lives in brackish waters of southeast Asia from India to the Philippines. This aptly named creature captures its prey by shooting a stream of water drops at an insect, either flying or at rest. The bug falls into the water and the fish gobbles it up. The archerfish has high accuracy at distances of 1.2 m to 1.5 m, and it sometimes makes hits at distances up to 3.5 m. A groove in the roof of its mouth, along with a curled tongue, forms a tube that enables the fish to impart high velocity to the water in its mouth when it suddenly closes its gill flaps. Suppose the archerfish shoots at a target

2.00 m away, at an angle of 30.0° above the horizontal. With what velocity must the water stream be launched if it is not to drop more than 3.00 cm vertically on its path to the target?

19. A place-kicker must kick a football from a point 36.0 m (about 40 yards) from the goal, and half the crowd hopes the ball will clear the crossbar, which is 3.05 m high. When kicked, the ball leaves the ground with a speed of 20.0 m/s at an angle of 53.0° to the horizontal. (a) By how much does the ball clear or fall short of clearing the crossbar? (b) Does the ball approach the crossbar while still rising or while falling?

20. A firefighter, a distance d from a burning building, directs a stream of water from a fire hose at angle θ_i above the horizontal as in Figure P4.20. If the initial speed of the stream is v_i, at what height h does the water strike the building?

Figure P4.20

21. A playground is on the flat roof of a city school, 6.00 m above the street below. The vertical wall of the building is 7.00 m high, to form a meter-high railing around the playground. A ball has fallen to the street below, and a passerby returns it by launching it at an angle of 53.0° above the horizontal at a point 24.0 meters from the base of the building wall. The ball takes 2.20 s to reach a point vertically above the wall. (a) Find the speed at which the ball was launched. (b) Find the vertical distance by which the ball clears the wall. (c) Find the distance from the wall to the point on the roof where the ball lands.

22. A dive bomber has a velocity of 280 m/s at an angle θ below the horizontal. When the altitude of the aircraft is 2.15 km, it releases a bomb, which subsequently hits a target on the ground. The magnitude of the displacement from the point of release of the bomb to the target is 3.25 km. Find the angle θ.

23. A soccer player kicks a rock horizontally off a 40.0-m high cliff into a pool of water. If the player hears the sound of the splash 3.00 s later, what was the initial speed given to the rock? Assume the speed of sound in air to be 343 m/s.

24. A basketball star covers 2.80 m horizontally in a jump to dunk the ball (Fig. P4.24). His motion through space can be modeled precisely as that of a particle at his *center of mass*, which we will define in Chapter 9. His center of mass is at elevation 1.02 m when he leaves the floor. It reaches a maximum height of 1.85 m above the floor, and is at elevation 0.900 m when he touches down again. Determine (a) his time of

Figure P4.24

flight (his "hang time"), (b) his horizontal and (c) vertical velocity components at the instant of takeoff, and (d) his take-off angle. (e) For comparison, determine the hang time of a whitetail deer making a jump with center-of-mass elevations $y_i = 1.20$ m, $y_{max} = 2.50$ m, $y_f = 0.700$ m.

25. An archer shoots an arrow with a velocity of 45.0 m/s at an angle of 50.0° with the horizontal. An assistant standing on the level ground 150 m downrange from the launch point throws an apple straight up with the minimum initial speed necessary to meet the path of the arrow. (a) What is the initial speed of the apple? (b) At what time after the arrow launch should the apple be thrown so that the arrow hits the apple?

26. A fireworks rocket explodes at height h, the peak of its vertical trajectory. It throws out burning fragments in all directions, but all at the same speed v. Pellets of solidified metal fall to the ground without air resistance. Find the smallest angle that the final velocity of an impacting fragment makes with the horizontal.

Section 4.4 Uniform Circular Motion

Note: Problems 8, 10, 12, and 16 in Chapter 6 can also be assigned with this section.

27. The athlete shown in Figure P4.27 rotates a 1.00-kg discus along a circular path of radius 1.06 m. The maximum speed of the discus is 20.0 m/s. Determine the magnitude of the maximum radial acceleration of the discus.

Sam Sargent/Liaison International

Figure P4.27

28. From information on the endsheets of this book, compute the radial acceleration of a point on the surface of the Earth at the equator, due to the rotation of the Earth about its axis.

29. A tire 0.500 m in radius rotates at a constant rate of 200 rev/min. Find the speed and acceleration of a small stone lodged in the tread of the tire (on its outer edge).

30. As their booster rockets separate, Space Shuttle astronauts typically feel accelerations up to $3g$, where $g = 9.80$ m/s^2. In their training, astronauts ride in a device where they experience such an acceleration as a centripetal acceleration. Specifically, the astronaut is fastened securely at the end of a mechanical arm that then turns at constant speed in a horizontal circle. Determine the rotation rate, in revolutions per second, required to give an astronaut a centripetal acceleration of $3.00g$ while in circular motion with radius 9.45 m.

31. Young David who slew Goliath experimented with slings before tackling the giant. He found that he could revolve a sling of length 0.600 m at the rate of 8.00 rev/s. If he increased the length to 0.900 m, he could revolve the sling only 6.00 times per second. (a) Which rate of rotation gives the greater speed for the stone at the end of the sling? (b) What is the centripetal acceleration of the stone at 8.00 rev/s? (c) What is the centripetal acceleration at 6.00 rev/s?

32. The astronaut orbiting the Earth in Figure P4.32 is preparing to dock with a Westar VI satellite. The satellite is in a circular orbit 600 km above the Earth's surface, where the free-fall acceleration is 8.21 m/s^2. Take the radius of the Earth as 6 400 km. Determine the speed of the satellite and the time interval required to complete one orbit around the Earth.

Courtesy of NASA

Figure P4.32

Section 4.5 Tangential and Radial Acceleration

33. A train slows down as it rounds a sharp horizontal turn, slowing from 90.0 km/h to 50.0 km/h in the 15.0 s that it takes to round the bend. The radius of the curve is 150 m. Compute the acceleration at the moment the train speed reaches 50.0 km/h. Assume it continues to slow down at this time at the same rate.

34. An automobile whose speed is increasing at a rate of 0.600 m/s^2 travels along a circular road of radius 20.0 m. When the instantaneous speed of the automobile is 4.00 m/s, find (a) the tangential acceleration component, (b) the centripetal acceleration component, and (c) the magnitude and direction of the total acceleration.

35. Figure P4.35 represents the total acceleration of a particle moving clockwise in a circle of radius 2.50 m at a certain instant of time. At this instant, find (a) the radial acceleration, (b) the speed of the particle, and (c) its tangential acceleration.

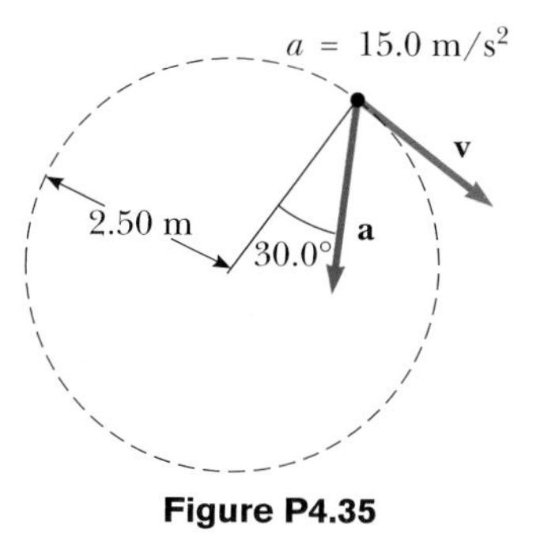

Figure P4.35

36. A ball swings in a vertical circle at the end of a rope 1.50 m long. When the ball is 36.9° past the lowest point on its way up, its total acceleration is $(-22.5\hat{\mathbf{i}} + 20.2\hat{\mathbf{j}})$ m/s². At that instant, (a) sketch a vector diagram showing the components of its acceleration, (b) determine the magnitude of its radial acceleration, and (c) determine the speed and velocity of the ball.

37. A race car starts from rest on a circular track. The car increases its speed at a constant rate a_t as it goes once around the track. Find the angle that the total acceleration of the car makes—with the radius connecting the center of the track and the car—at the moment the car completes the circle.

Section 4.6 Relative Velocity and Relative Acceleration

38. Heather in her Corvette accelerates at the rate of $(3.00\hat{\mathbf{i}} - 2.00\hat{\mathbf{j}})$ m/s², while Jill in her Jaguar accelerates at $(1.00\hat{\mathbf{i}} + 3.00\hat{\mathbf{j}})$ m/s². They both start from rest at the origin of an xy coordinate system. After 5.00 s, (a) what is Heather's speed with respect to Jill, (b) how far apart are they, and (c) what is Heather's acceleration relative to Jill?

39. A car travels due east with a speed of 50.0 km/h. Raindrops are falling at a constant speed vertically with respect to the Earth. The traces of the rain on the side windows of the car make an angle of 60.0° with the vertical. Find the velocity of the rain with respect to (a) the car and (b) the Earth.

40. How long does it take an automobile traveling in the left lane at 60.0 km/h to pull alongside a car traveling in the same direction in the right lane at 40.0 km/h if the cars' front bumpers are initially 100 m apart?

41. A river has a steady speed of 0.500 m/s. A student swims upstream a distance of 1.00 km and swims back to the starting point. If the student can swim at a speed of 1.20 m/s in still water, how long does the trip take? Compare this with the time the trip would take if the water were still.

42. The pilot of an airplane notes that the compass indicates a heading due west. The airplane's speed relative to the air is 150 km/h. If there is a wind of 30.0 km/h toward the north, find the velocity of the airplane relative to the ground.

43. Two swimmers, Alan and Beth, start together at the same point on the bank of a wide stream that flows with a speed v. Both move at the same speed $c(c > v)$, relative to the water. Alan swims downstream a distance L and then upstream the same distance. Beth swims so that her motion relative to the Earth is perpendicular to the banks of the stream. She swims the distance L and then back the same distance, so that both swimmers return to the starting point. Which swimmer returns first? (*Note:* First guess the answer.)

44. A bolt drops from the ceiling of a train car that is accelerating northward at a rate of 2.50 m/s². What is the acceleration of the bolt relative to (a) the train car? (b) the Earth?

45. A science student is riding on a flatcar of a train traveling along a straight horizontal track at a constant speed of 10.0 m/s. The student throws a ball into the air along a path that he judges to make an initial angle of 60.0° with the horizontal and to be in line with the track. The student's professor, who is standing on the ground nearby, observes the ball to rise vertically. How high does she see the ball rise?

46. A Coast Guard cutter detects an unidentified ship at a distance of 20.0 km in the direction 15.0° east of north. The ship is traveling at 26.0 km/h on a course at 40.0° east of north. The Coast Guard wishes to send a speedboat to intercept the vessel and investigate it. If the speedboat travels 50.0 km/h, in what direction should it head? Express the direction as a compass bearing with respect to due north.

Additional Problems

47. *The "Vomit Comet."* In zero-gravity astronaut training and equipment testing, NASA flies a KC135A aircraft along a parabolic flight path. As shown in Figure P4.47, the aircraft climbs from 24 000 ft to 31 000 ft, where it enters the zero-g parabola with a velocity of 143 m/s nose-high at 45.0° and exits with velocity 143 m/s at 45.0° nose-low. During this portion of the flight the aircraft and objects inside its padded cabin are in free fall—they have gone ballistic. The aircraft then pulls out of the dive with an upward acceleration of $0.800g$, moving in a vertical circle with radius 4.13 km. (During this portion of the flight, occupants of the plane perceive an acceleration of $1.8g$.) What are the aircraft (a) speed and (b) altitude at the top of the maneuver? (c) What is the time spent in zero gravity? (d) What is the speed of the aircraft at the bottom of the flight path?

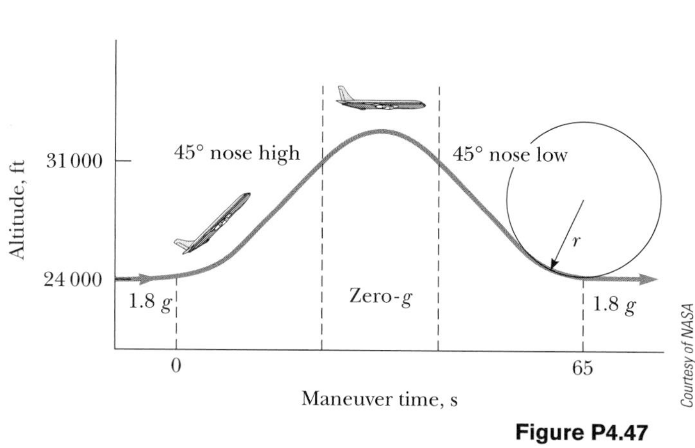

Figure P4.47

48. As some molten metal splashes, one droplet flies off to the east with initial velocity v_i at angle θ_i above the horizontal, and another droplet to the west with the same speed at the same angle above the horizontal, as in Figure P4.48. In terms of v_i and θ_i, find the distance between them as a function of time.

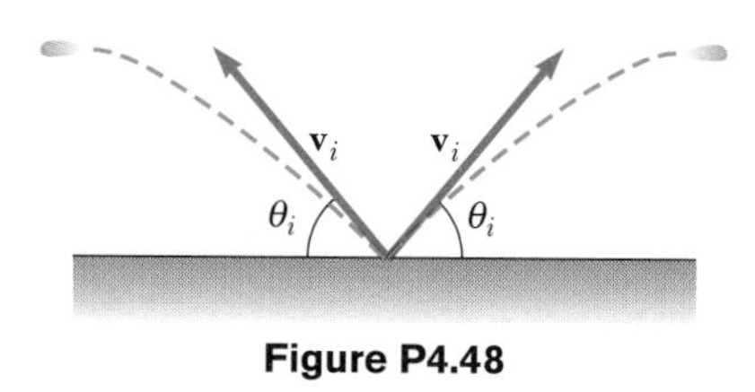

Figure P4.48

49. A ball on the end of a string is whirled around in a horizontal circle of radius 0.300 m. The plane of the circle is 1.20 m above the ground. The string breaks and the ball lands 2.00 m (horizontally) away from the point on the ground directly beneath the ball's location when the string breaks. Find the radial acceleration of the ball during its circular motion.

50. A projectile is fired up an incline (incline angle ϕ) with an initial speed v_i at an angle θ_i with respect to the horizontal ($\theta_i > \phi$), as shown in Figure P4.50. (a) Show that the projectile travels a distance d up the incline, where

$$d = \frac{2v_i^2 \cos\theta_i \sin(\theta_i - \phi)}{g\cos^2\phi}$$

(b) For what value of θ_i is d a maximum, and what is that maximum value?

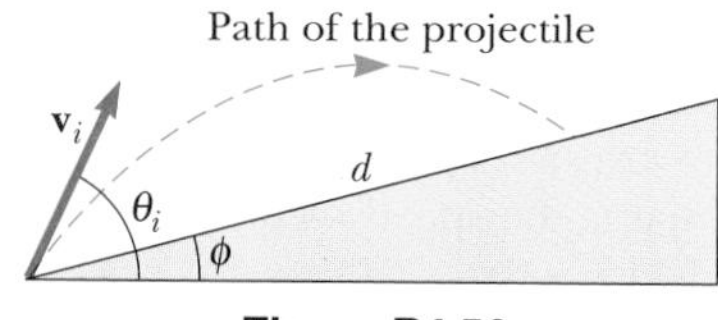

Figure P4.50

51. Barry Bonds hits a home run so that the baseball just clears the top row of bleachers, 21.0 m high, located 130 m from home plate. The ball is hit at an angle of 35.0° to the horizontal, and air resistance is negligible. Find (a) the initial speed of the ball, (b) the time at which the ball reaches the cheap seats, and (c) the velocity components and the speed of the ball when it passes over the top row. Assume the ball is hit at a height of 1.00 m above the ground.

52. An astronaut on the surface of the Moon fires a cannon to launch an experiment package, which leaves the barrel moving horizontally. (a) What must be the muzzle speed of the package so that it travels completely around the Moon and returns to its original location? (b) How long does this trip around the Moon take? Assume that the free-fall acceleration on the Moon is one-sixth that on the Earth.

53. A pendulum with a cord of length r = 1.00 m swings in a vertical plane (Fig. P4.53). When the pendulum is in the two horizontal positions θ = 90.0° and θ = 270°, its speed is 5.00 m/s. (a) Find the magnitude of the radial acceleration and tangential acceleration for these positions. (b) Draw vector diagrams to determine the direc-

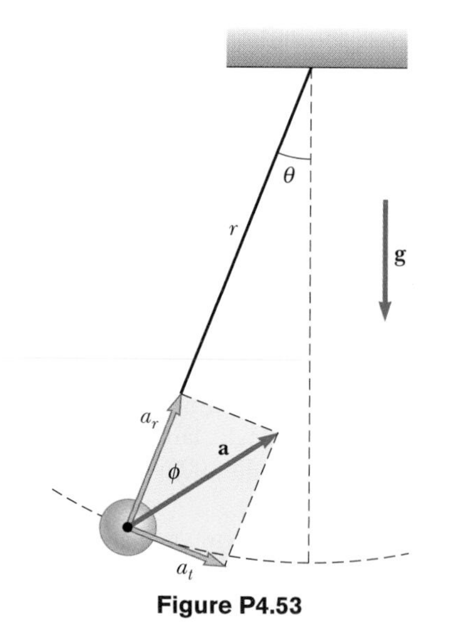

Figure P4.53

tion of the total acceleration for these two positions. (c) Calculate the magnitude and direction of the total acceleration.

54. A basketball player who is 2.00 m tall is standing on the floor 10.0 m from the basket, as in Figure P4.54. If he shoots the ball at a 40.0° angle with the horizontal, at what initial speed must he throw so that it goes through the hoop without striking the backboard? The basket height is 3.05 m.

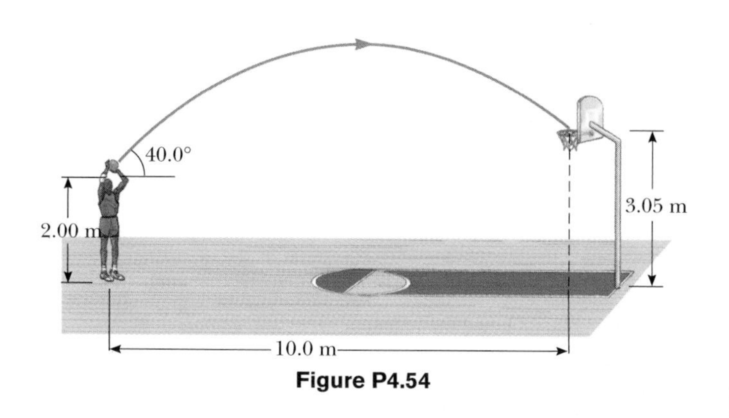

Figure P4.54

55. When baseball players throw the ball in from the outfield, they usually allow it to take one bounce before it reaches the infield, on the theory that the ball arrives sooner that way. Suppose that the angle at which a bounced ball leaves the ground is the same as the angle at which the outfielder threw it, as in Figure P4.55, but that the ball's speed after the bounce is one half of what it was before the bounce. (a) Assuming the ball is always thrown with the same initial speed, at what angle θ should the fielder throw the ball to make it go the same distance D with one bounce (blue path) as a ball thrown upward at 45.0° with no bounce (green path)? (b) Determine the ratio of the times for the one-bounce and no-bounce throws.

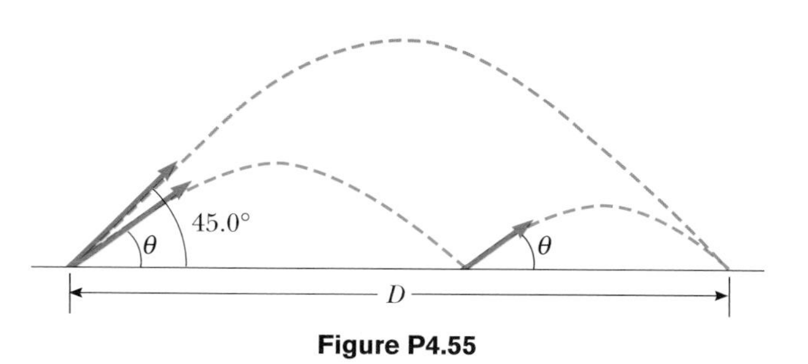

Figure P4.55

56. A boy can throw a ball a maximum horizontal distance of R on a level field. How far can he throw the same ball vertically upward? Assume that his muscles give the ball the same speed in each case.

57. A stone at the end of a sling is whirled in a vertical circle of radius 1.20 m at a constant speed $v_0 = 1.50$ m/s as in Figure P4.57. The center of the sling is 1.50 m above the ground. What is the range of the stone if it is released when the sling is inclined at 30.0° with the horizontal (a) at Ⓐ? (b) at Ⓑ? What is the acceleration of the stone (c) just before it is released at Ⓐ? (d) just after it is released at Ⓐ?

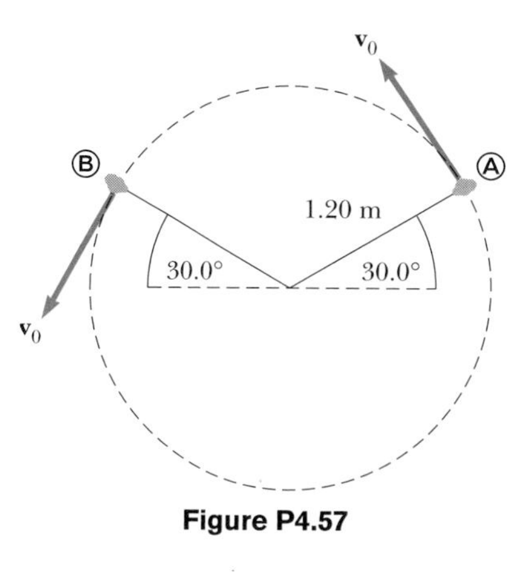

Figure P4.57

58. A quarterback throws a football straight toward a receiver with an initial speed of 20.0 m/s, at an angle of 30.0° above the horizontal. At that instant, the receiver is 20.0 m from the quarterback. In what direction and with what constant speed should the receiver run in order to catch the football at the level at which it was thrown?

59. Your grandfather is copilot of a bomber, flying horizontally over level terrain, with a speed of 275 m/s relative to the ground, at an altitude of 3 000 m. (a) The bombardier releases one bomb. How far will it travel horizontally between its release and its impact on the ground? Neglect the effects of air resistance. (b) Firing from the people on the ground suddenly incapacitates the bombardier before he can call, "Bombs away!" Consequently, the pilot maintains the plane's original course, altitude, and speed through a storm of flak. Where will the plane be when the bomb hits the ground? (c) The plane has a telescopic bomb sight set so that the bomb hits the target seen in the sight at the time of release. At what angle from the vertical was the bomb sight set?

60. A high-powered rifle fires a bullet with a muzzle speed of 1.00 km/s. The gun is pointed horizontally at a large bull's eye target—a set of concentric rings—200 m away. (a) How far below the extended axis of the rifle barrel does a bullet hit the target? The rifle is equipped with a telescopic sight. It is "sighted in" by adjusting the axis of the telescope so that it points precisely at the location where the bullet hits the target at 200 m. (b) Find the angle between the telescope axis and the rifle barrel axis. When shooting at a target at a distance other than 200 m, the marksman uses the telescopic sight, placing its crosshairs to "aim high" or "aim low" to compensate for the different range. Should she aim high or low, and approximately how far from the bull's eye, when the target is at a distance of (c) 50.0 m, (d) 150 m, or (e) 250 m? *Note:* The trajectory of the bullet is everywhere so nearly horizontal that it is a good approximation to model the bullet as fired horizontally in each case. **What if** the target is uphill or downhill? (f) Suppose the target is 200 m away, but the sight line to the target is above the horizontal by 30°. Should the marksman aim high, low, or right on? (g) Suppose the target is downhill by 30°. Should the marksman aim high, low, or right on? Explain your answers.

61. A hawk is flying horizontally at 10.0 m/s in a straight line, 200 m above the ground. A mouse it has been carrying struggles free from its grasp. The hawk continues on its path at the same speed for 2.00 seconds before attempting to retrieve its prey. To accomplish the retrieval, it dives in a straight line at constant speed and recaptures the mouse 3.00 m above the ground. (a) Assuming no air resistance, find the diving speed of the hawk. (b) What angle did the hawk make with the horizontal during its descent? (c) For how long did the mouse "enjoy" free fall?

62. A person standing at the top of a hemispherical rock of radius R kicks a ball (initially at rest on the top of the rock) to give it horizontal velocity $\mathbf{v}_i$ as in Figure P4.62. (a) What must be its minimum initial speed if the ball is never to hit the rock after it is kicked? (b) With this initial speed, how far from the base of the rock does the ball hit the ground?

Figure P4.62

63. A car is parked on a steep incline overlooking the ocean, where the incline makes an angle of 37.0° below the horizontal. The negligent driver leaves the car in neutral, and the parking brakes are defective. Starting from rest at $t = 0$, the car rolls down the incline with a constant acceleration of 4.00 m/s^2, traveling 50.0 m to the edge of a vertical cliff. The cliff is 30.0 m above the ocean. Find (a) the speed of the car when it reaches the edge of the cliff and the time at which it arrives there, (b) the velocity of the car when it lands in the ocean, (c) the total time interval that the car is in motion, and (d) the position of the car when it lands in the ocean, relative to the base of the cliff.

64. A truck loaded with cannonball watermelons stops suddenly to avoid running over the edge of a washed-out bridge (Fig. P4.64). The quick stop causes a number of melons to fly off the truck. One melon rolls over the edge with an initial speed $v_i = 10.0$ m/s in the horizontal direction. A cross-section of the bank has the shape of the bottom half of a parabola with its vertex at the edge of the road, and with the equation $y^2 = 16x$, where x and y are measured in meters. What are the x and y coordinates of the melon when it splatters on the bank?

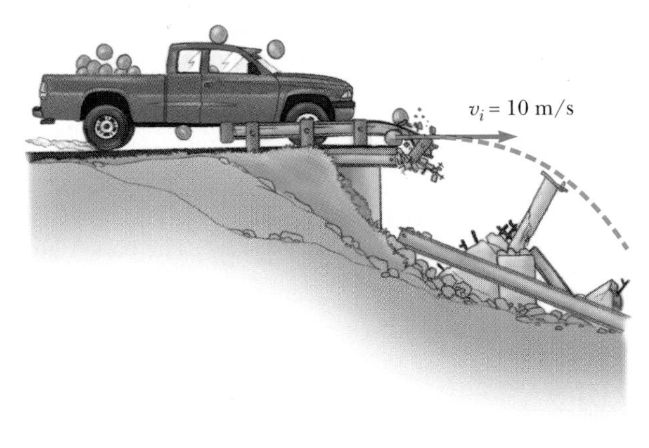

Figure P4.64

65. The determined coyote is out once more in pursuit of the elusive roadrunner. The coyote wears a pair of Acme jet-powered roller skates, which provide a constant horizontal acceleration of 15.0 m/s^2 (Fig. P4.65). The coyote starts at rest 70.0 m from the brink of a cliff at the instant the roadrunner zips past him in the direction of the cliff. (a) If the roadrunner moves with constant speed, determine the minimum speed he must have in order to reach the cliff before the coyote. At the edge of the cliff, the roadrunner escapes by making a sudden turn, while the coyote continues straight ahead. His skates remain horizontal and continue to operate while he is in flight, so that the coyote's acceleration while in the air is $(15.0\hat{\mathbf{i}} - 9.80\hat{\mathbf{j}})$ m/s^2. (b) If the cliff is 100 m above the flat floor of a canyon, determine where the coyote lands in the canyon. (c) Determine the components of the coyote's impact velocity.

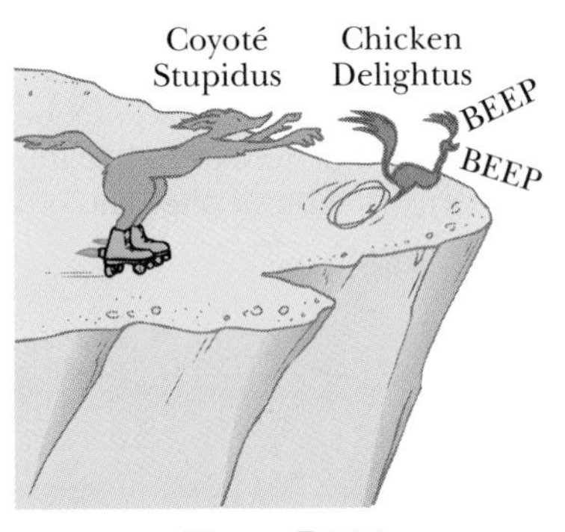

Figure P4.65

66. Do not hurt yourself; do not strike your hand against anything. Within these limitations, describe what you do to give your hand a large acceleration. Compute an order-of-magnitude estimate of this acceleration, stating the quantities you measure or estimate and their values.

67. A skier leaves the ramp of a ski jump with a velocity of 10.0 m/s, 15.0° above the horizontal, as in Figure P4.67. The slope is inclined at 50.0°, and air resistance is negligible. Find (a) the distance from the ramp to where the jumper lands and (b) the velocity components just before the landing. (How do you think the results might be affected if air resistance were included? Note that jumpers lean forward in the shape of an airfoil, with their hands at their sides, to increase their distance. Why does this work?)

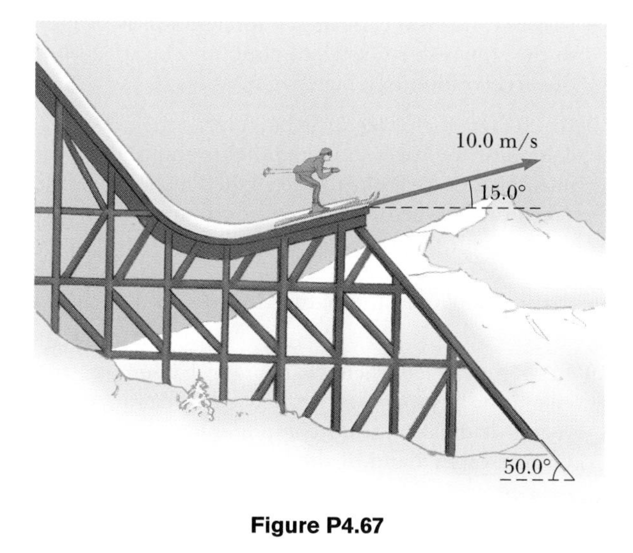

Figure P4.67

68. In a television picture tube (a cathode ray tube) electrons are emitted with velocity $\mathbf{v}_i$ from a source at the origin of coordinates. The initial velocities of different electrons make different angles θ with the x axis. As they move a distance D along the x axis, the electrons are acted on by a constant electric field, giving each a constant acceleration $\mathbf{a}$ in the x direction. At $x = D$ the electrons pass through a circular aperture, oriented perpendicular to the x axis. At the aperture, the velocity imparted to the electrons by the electric field is much larger than $\mathbf{v}_i$ in magnitude. Show that velocities of the electrons going through the aperture radiate from a certain point on the x axis, which is not the origin. Determine the location of this point. This point is called a *virtual source,* and it is important in determining where the electron beam hits the screen of the tube.

69. A fisherman sets out upstream from Metaline Falls on the Pend Oreille River in northwestern Washington State. His small boat, powered by an outboard motor, travels at a constant speed v in still water. The water flows at a lower constant speed v_w. He has traveled upstream for 2.00 km when his ice chest falls out of the boat. He notices that the chest is missing only after he has gone upstream for another 15.0 minutes. At that point he turns around and heads back downstream, all the time traveling at the same speed relative to the water. He catches up with the floating ice chest just as it is about to go over the falls at his starting point. How fast is the river flowing? Solve this problem in two ways. (a) First, use the Earth as a reference frame. With respect to the Earth, the boat travels upstream at speed $v - v_w$ and downstream at $v + v_w$. (b) A second much simpler and more elegant solution is obtained by using the water as the reference frame. This approach has important applications in many more complicated problems; examples are calculating the motion of rockets and satellites and analyzing the scattering of subatomic particles from massive targets.

70. The water in a river flows uniformly at a constant speed of 2.50 m/s between parallel banks 80.0 m apart. You are to deliver a package directly across the river, but you can swim only at 1.50 m/s. (a) If you choose to minimize the time you spend in the water, in what direction should you head? (b) How far downstream will you be carried? (c) **What If?** If you choose to minimize the distance downstream that the river carries you, in what direction should you head? (d) How far downstream will you be carried?

71. An enemy ship is on the east side of a mountain island, as shown in Figure P4.71. The enemy ship has maneuvered to within 2 500 m of the 1 800-m-high mountain peak and can shoot projectiles with an initial speed of 250 m/s. If the western shoreline is horizontally 300 m from the peak, what are the distances from the western shore at which a ship can be safe from the bombardment of the enemy ship?

72. In the **What If?** section of Example 4.7, it was claimed that the maximum range of a ski-jumper occurs for a launch angle θ given by

$$\theta = 45^\circ - \frac{\phi}{2}$$

where ϕ is the angle that the hill makes with the horizontal in Figure 4.16. Prove this claim by deriving the equation above.

Answers to Quick Quizzes

4.1 (b). An object moving with constant velocity has $\Delta\mathbf{v} = 0$, so, according to the definition of acceleration, $\mathbf{a} = \Delta\mathbf{v}/\Delta t = 0$. Choice (a) is not correct because a particle can move at a constant speed and change direction. This possibility also makes (c) an incorrect choice.

4.2 (a). Because acceleration occurs whenever the velocity changes in any way—with an increase or decrease in

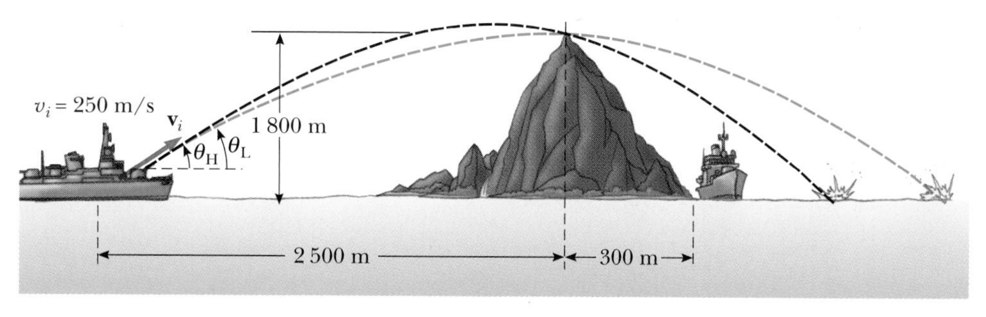

Figure P4.71

speed, a change in direction, or both—all three controls are accelerators. The gas pedal causes the car to speed up; the brake pedal causes the car to slow down. The steering wheel changes the direction of the velocity vector.

4.3 (a). You should simply throw it straight up in the air. Because the ball is moving along with you, it will follow a parabolic trajectory with a horizontal component of velocity that is the same as yours.

4.4 (b). At only one point—the peak of the trajectory—are the velocity and acceleration vectors perpendicular to each other. The velocity vector is horizontal at that point and the acceleration vector is downward.

4.5 (a). The acceleration vector is always directed downward. The velocity vector is never vertical if the object follows a path such as that in Figure 4.8.

4.6 15°, 30°, 45°, 60°, 75°. The greater the maximum height, the longer it takes the projectile to reach that altitude and then fall back down from it. So, as the launch angle increases, the time of flight increases.

4.7 (c). We cannot choose (a) or (b) because the centripetal acceleration vector is not constant—it continuously changes in direction. Of the remaining choices, only (c) gives the correct perpendicular relationship between $\mathbf{a}_c$ and $\mathbf{v}$.

4.8 (d). Because the centripetal acceleration is proportional to the square of the speed, doubling the speed increases the acceleration by a factor of 4.

4.9 (b). The velocity vector is tangent to the path. If the acceleration vector is to be parallel to the velocity vector, it must also be tangent to the path. This requires that the acceleration vector have no component perpendicular to the path. If the path were to change direction, the acceleration vector would have a radial component, perpendicular to the path. Thus, the path must remain straight.

4.10 (d). The velocity vector is tangent to the path. If the acceleration vector is to be perpendicular to the velocity vector, it must have no component tangent to the path. On the other hand, if the speed is changing, there *must* be a component of the acceleration tangent to the path. Thus, the velocity and acceleration vectors are never perpendicular in this situation. They can only be perpendicular if there is no change in the speed.

4.11 (c). Passenger A sees the coffee pouring in a "normal" parabolic path, just as if she were standing on the ground pouring it. The stationary observer B sees the coffee moving in a parabolic path that is extended horizontally due to the constant horizontal velocity of 60 mi/h.

Chapter 5

The Laws of Motion

▲ *A small tugboat exerts a force on a large ship, causing it to move. How can such a small boat move such a large object? (Steve Raymer/CORBIS)*

CHAPTER OUTLINE

In Chapters 2 and 4, we described motion in terms of position, velocity, and acceleration without considering what might cause that motion. Now we consider the cause—what might cause one object to remain at rest and another object to accelerate? The two main factors we need to consider are the forces acting on an object and the mass of the object. We discuss the three basic laws of motion, which deal with forces and masses and were formulated more than three centuries ago by Isaac Newton. Once we understand these laws, we can answer such questions as "What mechanism changes motion?" and "Why do some objects accelerate more than others?"

5.1 The Concept of Force

Everyone has a basic understanding of the concept of force from everyday experience. When you push your empty dinner plate away, you exert a force on it. Similarly, you exert a force on a ball when you throw or kick it. In these examples, the word *force* is associated with muscular activity and some change in the velocity of an object. Forces do not always cause motion, however. For example, as you sit reading this book, a gravitational force acts on your body and yet you remain stationary. As a second example, you can push (in other words, exert a force) on a large boulder and not be able to move it.

What force (if any) causes the Moon to orbit the Earth? Newton answered this and related questions by stating that forces are what cause any change in the velocity of an object. The Moon's velocity is not constant because it moves in a nearly circular orbit around the Earth. We now know that this change in velocity is caused by the gravitational force exerted by the Earth on the Moon. Because only a force can cause a change in velocity, we can think of force as *that which causes an object to accelerate.* In this chapter, we are concerned with the relationship between the force exerted on an object and the acceleration of that object.

An object accelerates due to an external force

What happens when several forces act simultaneously on an object? In this case, the object accelerates only if the net force acting on it is not equal to zero. The **net force** acting on an object is defined as the vector sum of all forces acting on the object. (We sometimes refer to the net force as the *total force*, the *resultant force*, or the *unbalanced force.*) **If the net force exerted on an object is zero, the acceleration of the object is zero and its velocity remains constant.** That is, if the net force acting on the object is zero, the object either remains at rest or continues to move with constant velocity. When the velocity of an object is constant (including when the object is at rest), the object is said to be in **equilibrium.**

Definition of equilibrium

When a coiled spring is pulled, as in Figure 5.1a, the spring stretches. When a stationary cart is pulled sufficiently hard that friction is overcome, as in Figure 5.1b, the cart moves. When a football is kicked, as in Figure 5.1c, it is both deformed and set in motion. These situations are all examples of a class of forces called *contact forces.* That is, they involve physical contact between two objects. Other examples of contact forces are the force exerted by gas molecules on the walls of a container and the force exerted by your feet on the floor.

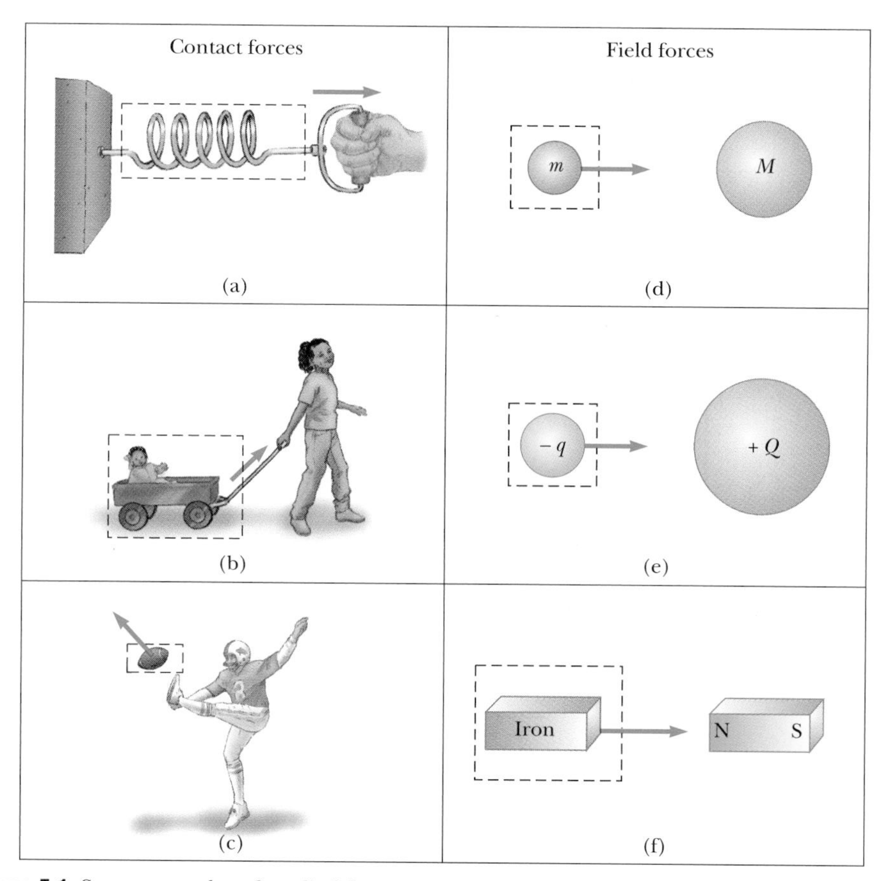

Figure 5.1 Some examples of applied forces. In each case a force is exerted on the object within the boxed area. Some agent in the environment external to the boxed area exerts a force on the object.

Another class of forces, known as *field forces,* do not involve physical contact between two objects but instead act through empty space. The gravitational force of attraction between two objects, illustrated in Figure 5.1d, is an example of this class of force. This gravitational force keeps objects bound to the Earth and the planets in orbit around the Sun. Another common example of a field force is the electric force that one electric charge exerts on another (Fig. 5.1e). These charges might be those of the electron and proton that form a hydrogen atom. A third example of a field force is the force a bar magnet exerts on a piece of iron (Fig. 5.1f).

The distinction between contact forces and field forces is not as sharp as you may have been led to believe by the previous discussion. When examined at the atomic level, all the forces we classify as contact forces turn out to be caused by electric (field) forces of the type illustrated in Figure 5.1e. Nevertheless, in developing models for macroscopic phenomena, it is convenient to use both classifications of forces. The only known *fundamental* forces in nature are all field forces: (1) *gravitational forces* between objects, (2) *electromagnetic forces* between electric charges, (3) *nuclear forces* between subatomic particles, and (4) *weak forces* that arise in certain radioactive decay processes. In classical physics, we are concerned only with gravitational and electromagnetic forces.

Measuring the Strength of a Force

It is convenient to use the deformation of a spring to measure force. Suppose we apply a vertical force to a spring scale that has a fixed upper end, as shown in Figure 5.2a. The spring elongates when the force is applied, and a pointer on the scale reads the value of the applied force. We can calibrate the spring by defining a reference force $\mathbf{F}_1$ as the force that produces a pointer reading of 1.00 cm. (Because force is a vector

Isaac Newton,
English physicist and mathematician (1642–1727)

Isaac Newton was one of the most brilliant scientists in history. Before the age of 30, he formulated the basic concepts and laws of mechanics, discovered the law of universal gravitation, and invented the mathematical methods of calculus. As a consequence of his theories, Newton was able to explain the motions of the planets, the ebb and flow of the tides, and many special features of the motions of the Moon and the Earth. He also interpreted many fundamental observations concerning the nature of light. His contributions to physical theories dominated scientific thought for two centuries and remain important today.
(Giraudon/Art Resource)

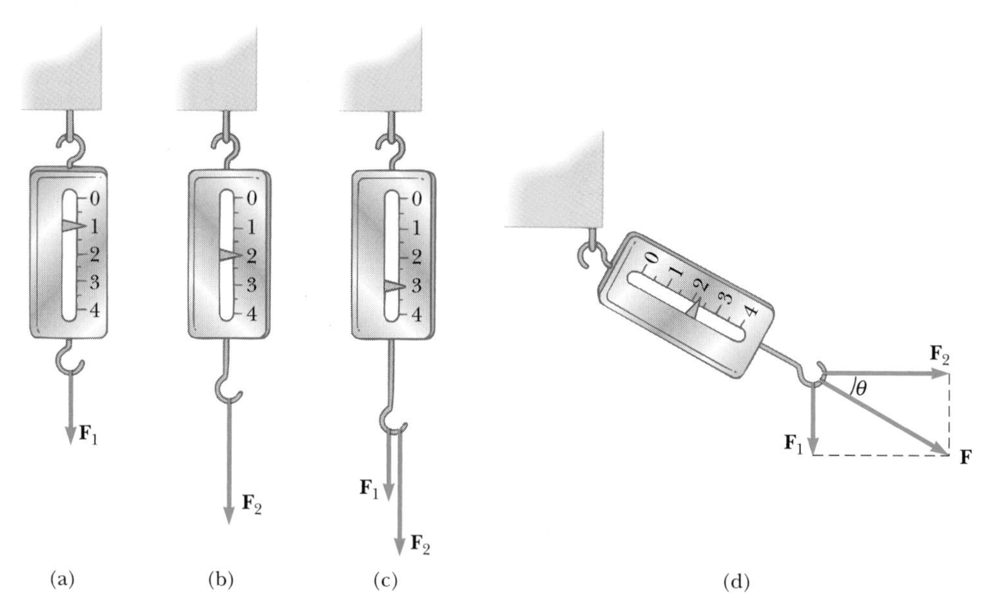

Figure 5.2 The vector nature of a force is tested with a spring scale. (a) A downward force $\mathbf{F}_1$ elongates the spring 1.00 cm. (b) A downward force $\mathbf{F}_2$ elongates the spring 2.00 cm. (c) When $\mathbf{F}_1$ and $\mathbf{F}_2$ are applied simultaneously, the spring elongates by 3.00 cm. (d) When $\mathbf{F}_1$ is downward and $\mathbf{F}_2$ is horizontal, the combination of the two forces elongates the spring $\sqrt{(1.00 \text{ cm})^2 + (2.00 \text{ cm})^2} = 2.24$ cm.

quantity, we use the bold-faced symbol $\mathbf{F}$.) If we now apply a different downward force $\mathbf{F}_2$ whose magnitude is twice that of the reference force $\mathbf{F}_1$, as seen in Figure 5.2b, the pointer moves to 2.00 cm. Figure 5.2c shows that the combined effect of the two collinear forces is the sum of the effects of the individual forces.

Now suppose the two forces are applied simultaneously with $\mathbf{F}_1$ downward and $\mathbf{F}_2$ horizontal, as illustrated in Figure 5.2d. In this case, the pointer reads $\sqrt{5.00 \text{ cm}^2} = 2.24$ cm. The single force $\mathbf{F}$ that would produce this same reading is the sum of the two vectors $\mathbf{F}_1$ and $\mathbf{F}_2$, as described in Figure 5.2d. That is, $|\mathbf{F}| = \sqrt{F_1^2 + F_2^2} = 2.24$ units, and its direction is $\theta = \tan^{-1}(-0.500) = -26.6°$. **Because forces have been experimentally verified to behave as vectors, you must use the rules of vector addition to obtain the net force on an object.**

5.2 Newton's First Law and Inertial Frames

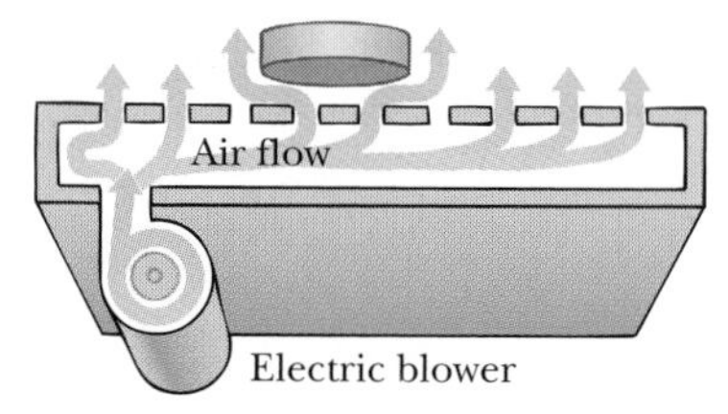

Figure 5.3 On an air hockey table, air blown through holes in the surface allow the puck to move almost without friction. If the table is not accelerating, a puck placed on the table will remain at rest.

We begin our study of forces by imagining some situations. Imagine placing a puck on a perfectly level air hockey table (Fig. 5.3). You expect that it will remain where it is placed. Now imagine your air hockey table is located on a train moving with constant velocity. If the puck is placed on the table, the puck again remains where it is placed. If the train were to accelerate, however, the puck would start moving along the table, just as a set of papers on your dashboard falls onto the front seat of your car when you step on the gas.

As we saw in Section 4.6, a moving object can be observed from any number of reference frames. **Newton's first law of motion,** sometimes called the *law of inertia,* defines a special set of reference frames called *inertial frames.* This law can be stated as follows:

Newton's first law

If an object does not interact with other objects, it is possible to identify a reference frame in which the object has zero acceleration.

Such a reference frame is called an **inertial frame of reference**. When the puck is on the air hockey table located on the ground, you are observing it from an inertial reference frame—there are no horizontal interactions of the puck with any other objects and you observe it to have zero acceleration in that direction. When you are on the train moving at constant velocity, you are also observing the puck from an inertial reference frame. **Any reference frame that moves with constant velocity relative to an inertial frame is itself an inertial frame**. When the train accelerates, however, you are observing the puck from a **noninertial reference frame** because you and the train are accelerating relative to the inertial reference frame of the surface of the Earth. While the puck appears to be accelerating according to your observations, we can identify a reference frame in which the puck has zero acceleration. For example, an observer standing outside the train on the ground sees the puck moving with the same velocity as the train had before it started to accelerate (because there is almost no friction to "tie" the puck and the train together). Thus, Newton's first law is still satisfied even though your observations say otherwise.

Inertial frame of reference

A reference frame that moves with constant velocity relative to the distant stars is the best approximation of an inertial frame, and for our purposes we can consider the Earth as being such a frame. The Earth is not really an inertial frame because of its orbital motion around the Sun and its rotational motion about its own axis, both of which result in centripetal accelerations. However, these accelerations are small compared with g and can often be neglected. For this reason, we assume that the Earth is an inertial frame, as is any other frame attached to it.

Let us assume that we are observing an object from an inertial reference frame. (We will return to observations made in noninertial reference frames in Section 6.3.) Before about 1600, scientists believed that the natural state of matter was the state of rest. Observations showed that moving objects eventually stopped moving. Galileo was the first to take a different approach to motion and the natural state of matter. He devised thought experiments and concluded that it is not the nature of an object to stop once set in motion: rather, it is its nature to *resist changes in its motion*. In his words, "Any velocity once imparted to a moving body will be rigidly maintained as long as the external causes of retardation are removed." For example, a spacecraft drifting through empty space with its engine turned off will keep moving forever—it would *not* seek a "natural state" of rest.

5.1 Newton's First Law

Newton's first law does *not* say what happens for an object with *zero net force,* that is, multiple forces that cancel; it says what happens *in the absence of a force.* This is a subtle but important difference that allows us to define force as that which causes a change in the motion. The description of an object under the effect of forces that balance is covered by Newton's second law.

Given our assumption of observations made from inertial reference frames, we can pose a more practical statement of Newton's first law of motion:

> In the absence of external forces, when viewed from an inertial reference frame, an object at rest remains at rest and an object in motion continues in motion with a constant velocity (that is, with a constant speed in a straight line).

Another statement of Newton's first law

In simpler terms, we can say that **when no force acts on an object, the acceleration of the object is zero.** If nothing acts to change the object's motion, then its velocity does not change. From the first law, we conclude that any *isolated object* (one that does not interact with its environment) is either at rest or moving with constant velocity. The tendency of an object to resist any attempt to change its velocity is called **inertia.**

Quick Quiz 5.1 Which of the following statements is most correct? (a) It is possible for an object to have motion in the absence of forces on the object. (b) It is possible to have forces on an object in the absence of motion of the object. (c) Neither (a) nor (b) is correct. (d) Both (a) and (b) are correct.

5.3 Mass

Imagine playing catch with either a basketball or a bowling ball. Which ball is more likely to keep moving when you try to catch it? Which ball has the greater tendency to remain motionless when you try to throw it? The bowling ball is more resistant to changes in its velocity than the basketball—how can we quantify this concept?

Definition of mass

Mass is that property of an object that specifies how much resistance an object exhibits to changes in its velocity, and as we learned in Section 1.1, the SI unit of mass is the kilogram. The greater the mass of an object, the less that object accelerates under the action of a given applied force.

To describe mass quantitatively, we begin by experimentally comparing the accelerations a given force produces on different objects. Suppose a force acting on an object of mass m_1 produces an acceleration $\mathbf{a}_1$, and the *same force* acting on an object of mass m_2 produces an acceleration $\mathbf{a}_2$. The ratio of the two masses is defined as the *inverse* ratio of the magnitudes of the accelerations produced by the force:

$$\frac{m_1}{m_2} \equiv \frac{a_2}{a_1} \tag{5.1}$$

For example, if a given force acting on a 3-kg object produces an acceleration of 4 m/s^2, the same force applied to a 6-kg object produces an acceleration of 2 m/s^2. If one object has a known mass, the mass of the other object can be obtained from acceleration measurements.

Mass is an inherent property of an object and is independent of the object's surroundings and of the method used to measure it. Also, **mass is a scalar quantity** and thus obeys the rules of ordinary arithmetic. That is, several masses can be combined in simple numerical fashion. For example, if you combine a 3-kg mass with a 5-kg mass, the total mass is 8 kg. We can verify this result experimentally by comparing the accelerations that a known force gives to several objects separately with the acceleration that the same force gives to the same objects combined as one unit.

Mass and weight are different quantities

Mass should not be confused with weight. **Mass and weight are two different quantities.** The weight of an object is equal to the magnitude of the gravitational force exerted on the object and varies with location (see Section 5.5). For example, a person who weighs 180 lb on the Earth weighs only about 30 lb on the Moon. On the other hand, the mass of an object is the same everywhere: an object having a mass of 2 kg on the Earth also has a mass of 2 kg on the Moon.

5.4 Newton's Second Law

Newton's first law explains what happens to an object when no forces act on it. It either remains at rest or moves in a straight line with constant speed. Newton's second law answers the question of what happens to an object that has a nonzero resultant force acting on it.

Imagine performing an experiment in which you push a block of ice across a frictionless horizontal surface. When you exert some horizontal force $\mathbf{F}$ on the block, it moves with some acceleration $\mathbf{a}$. If you apply a force twice as great, you find that the acceleration of the block doubles. If you increase the applied force to $3\mathbf{F}$, the acceleration triples, and so on. From such observations, we conclude that **the acceleration of an object is directly proportional to the force acting on it.**

The acceleration of an object also depends on its mass, as stated in the preceding section. We can understand this by considering the following experiment. If you apply a force $\mathbf{F}$ to a block of ice on a frictionless surface, the block undergoes some acceleration $\mathbf{a}$. If the mass of the block is doubled, the same applied force produces an acceleration $\mathbf{a}/2$. If the mass is tripled, the same applied force produces an acceleration $\mathbf{a}/3$, and so on. According to this observation, we conclude that **the magnitude of the acceleration of an object is inversely proportional to its mass.**

These observations are summarized in **Newton's second law:**

When viewed from an inertial reference frame, the acceleration of an object is directly proportional to the net force acting on it and inversely proportional to its mass.

▲ PITFALL PREVENTION

5.2 Force is the Cause of Changes in Motion

Force *does not* cause motion. We can have motion in the absence of forces, as described in Newton's first law. Force is the cause of *changes* in motion, as measured by acceleration.

Thus, we can relate mass, acceleration, and force through the following mathematical statement of Newton's second law:[1]

$$\sum \mathbf{F} = m\mathbf{a} \tag{5.2}$$

Newton's second law

In both the textual and mathematical statements of Newton's second law above, we have indicated that the acceleration is due to the *net force* $\sum \mathbf{F}$ acting on an object. The **net force** on an object is the vector sum of all forces acting on the object. In solving a problem using Newton's second law, it is imperative to determine the correct net force on an object. There may be many forces acting on an object, but there is only one acceleration.

Note that Equation 5.2 is a vector expression and hence is equivalent to three component equations:

$$\sum F_x = ma_x \qquad \sum F_y = ma_y \qquad \sum F_z = ma_z \tag{5.3}$$

Newton's second law—component form

PITFALL PREVENTION

5.3 *ma* is Not a Force

Equation 5.2 does *not* say that the product $m\mathbf{a}$ is a force. All forces on an object are added vectorially to generate the net force on the left side of the equation. This net force is then equated to the product of the mass of the object and the acceleration that results from the net force. Do *not* include an "$m\mathbf{a}$ force" in your analysis of the forces on an object.

Quick Quiz 5.2 An object experiences no acceleration. Which of the following *cannot* be true for the object? (a) A single force acts on the object. (b) No forces act on the object. (c) Forces act on the object, but the forces cancel.

Quick Quiz 5.3 An object experiences a net force and exhibits an acceleration in response. Which of the following statements is *always* true? (a) The object moves in the direction of the force. (b) The acceleration is in the same direction as the velocity. (c) The acceleration is in the same direction as the force. (d) The velocity of the object increases.

Quick Quiz 5.4 You push an object, initially at rest, across a frictionless floor with a constant force for a time interval Δt, resulting in a final speed of v for the object. You repeat the experiment, but with a force that is twice as large. What time interval is now required to reach the same final speed v? (a) $4\,\Delta t$ (b) $2\,\Delta t$ (c) Δt (d) $\Delta t/2$ (e) $\Delta t/4$.

Unit of Force

The SI unit of force is the **newton**, which is defined as the force that, when acting on an object of mass 1 kg, produces an acceleration of 1 m/s^2. From this definition and Newton's second law, we see that the newton can be expressed in terms of the following fundamental units of mass, length, and time:

$$1\text{ N} \equiv 1\text{ kg}\cdot\text{m/s}^2 \tag{5.4}$$

Definition of the newton

[1] Equation 5.2 is valid only when the speed of the object is much less than the speed of light. We treat the relativistic situation in Chapter 39.

Table 5.1

Units of Mass, Acceleration, and Force[a]

System of Units	Mass	Acceleration	Force
SI	kg	m/s^2	$N = kg \cdot m/s^2$
U.S. customary	slug	ft/s^2	$lb = slug \cdot ft/s^2$

[a] 1 N = 0.225 lb.

In the U.S. customary system, the unit of force is the **pound**, which is defined as the force that, when acting on a 1-slug mass,[2] produces an acceleration of 1 ft/s^2:

$$1\,\text{lb} \equiv 1\,\text{slug} \cdot \text{ft/s}^2 \qquad (5.5)$$

A convenient approximation is that $1\,\text{N} \approx \frac{1}{4}\,\text{lb}$.

The units of mass, acceleration, and force are summarized in Table 5.1.

Example 5.1 An Accelerating Hockey Puck

A hockey puck having a mass of 0.30 kg slides on the horizontal, frictionless surface of an ice rink. Two hockey sticks strike the puck simultaneously, exerting the forces on the puck shown in Figure 5.4. The force $\mathbf{F}_1$ has a magnitude of 5.0 N, and the force $\mathbf{F}_2$ has a magnitude of 8.0 N. Determine both the magnitude and the direction of the puck's acceleration.

Solution *Conceptualize* this problem by studying Figure 5.4. Because we can determine a net force and we want an acceleration, we *categorize* this problem as one that may be solved using Newton's second law. To *analyze* the problem, we resolve the force vectors into components. The net force acting on the puck in the x direction is

$$\sum F_x = F_{1x} + F_{2x} = F_1 \cos(-20°) + F_2 \cos 60°$$
$$= (5.0\text{ N})(0.940) + (8.0\text{ N})(0.500) = 8.7\text{ N}$$

The net force acting on the puck in the y direction is

$$\sum F_y = F_{1y} + F_{2y} = F_1 \sin(-20°) + F_2 \sin 60°$$
$$= (5.0\text{ N})(-0.342) + (8.0\text{ N})(0.866) = 5.2\text{ N}$$

Now we use Newton's second law in component form to find the x and y components of the puck's acceleration:

$$a_x = \frac{\Sigma F_x}{m} = \frac{8.7\text{ N}}{0.30\text{ kg}} = 29\text{ m/s}^2$$

$$a_y = \frac{\Sigma F_y}{m} = \frac{5.2\text{ N}}{0.30\text{ kg}} = 17\text{ m/s}^2$$

The acceleration has a magnitude of

$$a = \sqrt{(29)^2 + (17)^2}\text{ m/s}^2 = 34\text{ m/s}^2$$

and its direction relative to the positive x axis is

$$\theta = \tan^{-1}\left(\frac{a_y}{a_x}\right) = \tan^{-1}\left(\frac{17}{29}\right) = 30°$$

To *finalize* the problem, we can graphically add the vectors in Figure 5.4 to check the reasonableness of our answer. Because the acceleration vector is along the direction of the resultant force, a drawing showing the resultant force helps us check the validity of the answer. (Try it!)

What If? Suppose three hockey sticks strike the puck simultaneously, with two of them exerting the forces shown in Figure 5.4. The result of the three forces is that the hockey puck shows *no* acceleration. What must be the components of the third force?

Answer If there is zero acceleration, the net force acting on the puck must be zero. Thus, the three forces must cancel. We have found the components of the combination of the first two forces. The components of the third force must be of equal magnitude and opposite sign in order that all of the components add to zero. Thus, $F_{3x} = -8.7$ N, $F_{3y} = -5.2$ N.

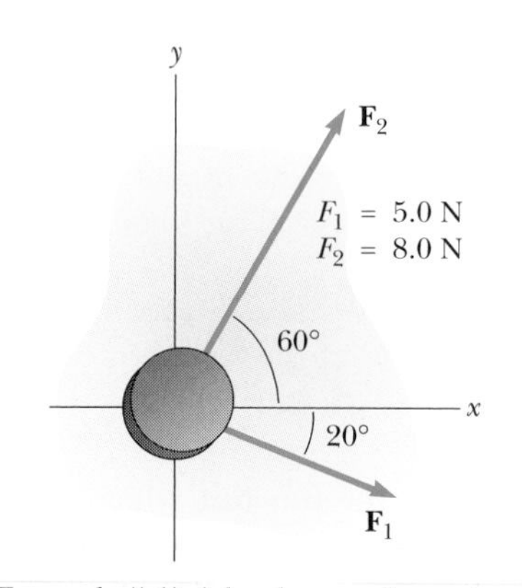

Figure 5.4 (Example 5.1) A hockey puck moving on a frictionless surface accelerates in the direction of the resultant force $\mathbf{F}_1 + \mathbf{F}_2$.

[2] The *slug* is the unit of mass in the U.S. customary system and is that system's counterpart of the SI unit the *kilogram*. Because most of the calculations in our study of classical mechanics are in SI units, the slug is seldom used in this text.

5.5 The Gravitational Force and Weight

We are well aware that all objects are attracted to the Earth. The attractive force exerted by the Earth on an object is called the **gravitational force $\mathbf{F}_g$**. This force is directed toward the center of the Earth,[3] and its magnitude is called the **weight** of the object.

We saw in Section 2.6 that a freely falling object experiences an acceleration $\mathbf{g}$ acting toward the center of the Earth. Applying Newton's second law $\Sigma\mathbf{F} = m\mathbf{a}$ to a freely falling object of mass m, with $\mathbf{a} = \mathbf{g}$ and $\Sigma\mathbf{F} = \mathbf{F}_g$, we obtain

$$\mathbf{F}_g = m\mathbf{g} \tag{5.6}$$

Thus, the weight of an object, being defined as the magnitude of $\mathbf{F}_g$, is equal to mg.

Because it depends on g, weight varies with geographic location. Because g decreases with increasing distance from the center of the Earth, objects weigh less at higher altitudes than at sea level. For example, a 1 000-kg palette of bricks used in the construction of the Empire State Building in New York City weighed 9 800 N at street level, but weighed about 1 N less by the time it was lifted from sidewalk level to the top of the building. As another example, suppose a student has a mass of 70.0 kg. The student's weight in a location where $g = 9.80\ \text{m/s}^2$ is $F_g = mg = 686$ N (about 150 lb). At the top of a mountain, however, where $g = 9.77\ \text{m/s}^2$, the student's weight is only 684 N. Therefore, if you want to lose weight without going on a diet, climb a mountain or weigh yourself at 30 000 ft during an airplane flight!

Because weight is proportional to mass, we can compare the masses of two objects by measuring their weights on a spring scale. At a given location (at which two objects are subject to the same value of g), the ratio of the weights of two objects equals the ratio of their masses.

Equation 5.6 quantifies the gravitational force on the object, but notice that this equation does not require the object to be moving. Even for a stationary object, or an object on which several forces act, Equation 5.6 can be used to calculate the magnitude of the gravitational force. This results in a subtle shift in the interpretation of m in the equation. The mass m in Equation 5.6 is playing the role of determining the strength of the gravitational attraction between the object and the Earth. This is a completely different role from that previously described for mass, that of measuring the resistance to changes in motion in response to an external force. Thus, we call m in this type of equation the **gravitational mass.** Despite this quantity being different in behavior from inertial mass, it is one of the experimental conclusions in Newtonian dynamics that gravitational mass and inertial mass have the same value.

5.4 "Weight of an Object"

We are familiar with the everyday phrase, the "weight of an object." However, weight is not an inherent property of an object, but rather a measure of the gravitational force between the object and the Earth. Thus, weight is a property of a *system* of items—the object and the Earth.

5.5 Kilogram is Not a Unit of Weight

You may have seen the "conversion" 1 kg = 2.2 lb. Despite popular statements of weights expressed in kilograms, the kilogram is not a unit of *weight,* it is a unit of *mass.* The conversion statement is not an equality; it is an *equivalence* that is only valid on the surface of the Earth.

Quick Quiz 5.5 A baseball of mass m is thrown upward with some initial speed. A gravitational force is exerted on the ball (a) at all points in its motion (b) at all points in its motion except at the highest point (c) at no points in its motion.

Quick Quiz 5.6 Suppose you are talking by interplanetary telephone to your friend, who lives on the Moon. He tells you that he has just won a newton of gold in a contest. Excitedly, you tell him that you entered the Earth version of the same contest and also won a newton of gold! Who is richer? (a) You (b) Your friend (c) You are equally rich.

Courtesy of NASA

The life-support unit strapped to the back of astronaut Edwin Aldrin weighed 300 lb on the Earth. During his training, a 50-lb mock-up was used. Although this effectively simulated the reduced weight the unit would have on the Moon, it did not correctly mimic the unchanging mass. It was just as difficult to accelerate the unit (perhaps by jumping or twisting suddenly) on the Moon as on the Earth.

[3] This statement ignores the fact that the mass distribution of the Earth is not perfectly spherical.

Conceptual Example 5.2 How Much Do You Weigh in an Elevator?

You have most likely had the experience of standing in an elevator that accelerates upward as it moves toward a higher floor. In this case, you feel heavier. In fact, if you are standing on a bathroom scale at the time, the scale measures a force having a magnitude that is greater than your weight. Thus, you have tactile and measured evidence that leads you to believe you are heavier in this situation. *Are* you heavier?

Solution No, your weight is unchanged. To provide the acceleration upward, the floor or scale must exert on your feet an upward force that is greater in magnitude than your weight. It is this greater force that you feel, which you interpret as feeling heavier. The scale reads this upward force, not your weight, and so its reading increases.

5.6 Newton's Third Law

If you press against a corner of this textbook with your fingertip, the book pushes back and makes a small dent in your skin. If you push harder, the book does the same and the dent in your skin is a little larger. This simple experiment illustrates a general principle of critical importance known as **Newton's third law:**

Newton's third law

If two objects interact, the force $\mathbf{F}_{12}$ exerted by object 1 on object 2 is equal in magnitude and opposite in direction to the force $\mathbf{F}_{21}$ exerted by object 2 on object 1:

$$\mathbf{F}_{12} = -\mathbf{F}_{21} \tag{5.7}$$

When it is important to designate forces as interactions between two objects, we will use this subscript notation, where $\mathbf{F}_{ab}$ means "the force exerted *by* a *on* b." The third law, which is illustrated in Figure 5.5a, is equivalent to stating that **forces always occur in pairs,** or that **a single isolated force cannot exist.** The force that object 1 exerts on object 2 may be called the *action force* and the force of object 2 on object 1 the *reaction force.* In reality, either force can be labeled the action or reaction force. **The action force is equal in magnitude to the reaction force and opposite in direction. In all cases, the action and reaction forces act on different objects and must be of the same type.** For example, the force acting on a freely falling projectile is the gravitational force exerted by the Earth on the projectile $\mathbf{F}_g = \mathbf{F}_{Ep}$ (E = Earth, p = projectile), and the magnitude of this force is mg. The reaction to this force is the gravitational force exerted by the projectile on the Earth $\mathbf{F}_{pE} = -\mathbf{F}_{Ep}$. The reaction force $\mathbf{F}_{pE}$ must accelerate the Earth toward the projectile just as the action force $\mathbf{F}_{Ep}$ accelerates the projectile toward the Earth. However, because the Earth has such a large mass, its acceleration due to this reaction force is negligibly small.

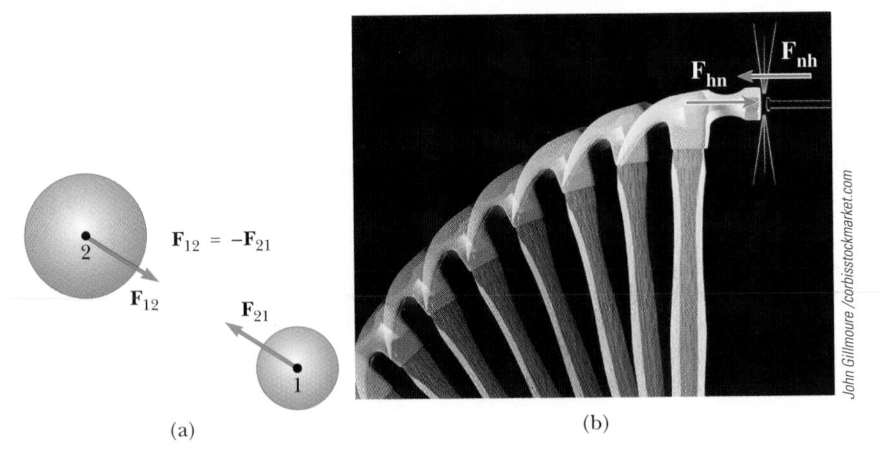

Figure 5.5 Newton's third law. (a) The force $\mathbf{F}_{12}$ exerted by object 1 on object 2 is equal in magnitude and opposite in direction to the force $\mathbf{F}_{21}$ exerted by object 2 on object 1. (b) The force $\mathbf{F}_{hn}$ exerted by the hammer on the nail is equal in magnitude and opposite to the force $\mathbf{F}_{nh}$ exerted by the nail on the hammer.

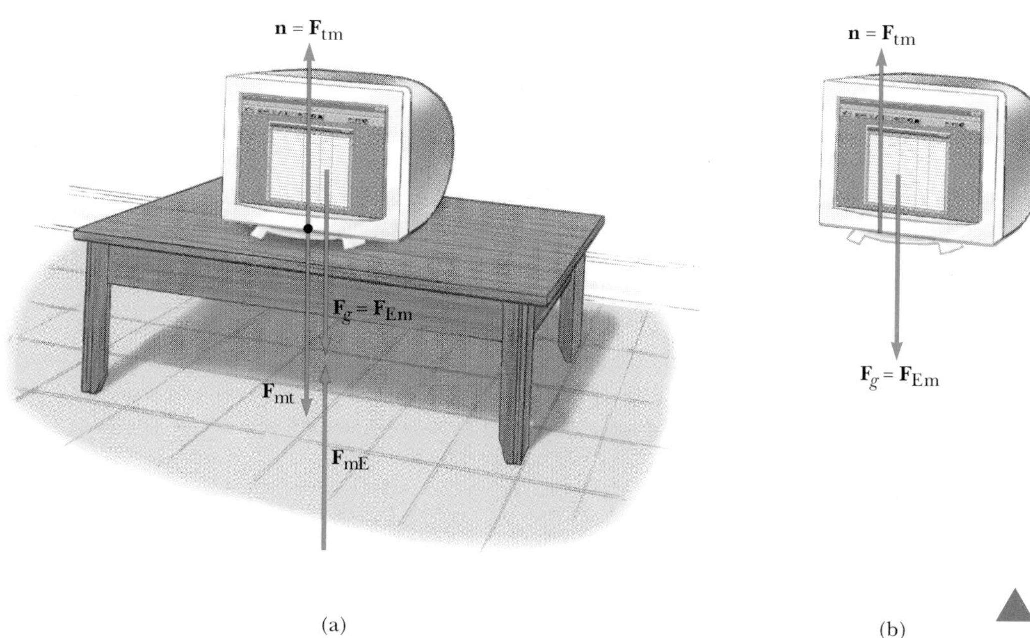

Figure 5.6 (a) When a computer monitor is at rest on a table, the forces acting on the monitor are the normal force $\mathbf{n}$ and the gravitational force $\mathbf{F}_g$. The reaction to $\mathbf{n}$ is the force $\mathbf{F}_{mt}$ exerted by the monitor on the table. The reaction to $\mathbf{F}_g$ is the force $\mathbf{F}_{mE}$ exerted by the monitor on the Earth. (b) The free-body diagram for the monitor.

Another example of Newton's third law is shown in Figure 5.5b. The force $\mathbf{F}_{hn}$ exerted by the hammer on the nail (the action) is equal in magnitude and opposite the force $\mathbf{F}_{nh}$ exerted by the nail on the hammer (the reaction). This latter force stops the forward motion of the hammer when it strikes the nail.

You experience the third law directly if you slam your fist against a wall or kick a football with your bare foot. You can feel the force back on your fist or your foot. You should be able to identify the action and reaction forces in these cases.

The Earth exerts a gravitational force $\mathbf{F}_g$ on any object. If the object is a computer monitor at rest on a table, as in Figure 5.6a, the reaction force to $\mathbf{F}_g = \mathbf{F}_{Em}$ is the force exerted by the monitor on the Earth $\mathbf{F}_{mE} = -\mathbf{F}_{Em}$. The monitor does not accelerate because it is held up by the table. The table exerts on the monitor an upward force $\mathbf{n} = \mathbf{F}_{tm}$, called the **normal force**.[4] This is the force that prevents the monitor from falling through the table; it can have any value needed, up to the point of breaking the table. From Newton's second law, we see that, because the monitor has zero acceleration, it follows that $\Sigma\mathbf{F} = \mathbf{n} - m\mathbf{g} = 0$, or $n = mg$. The normal force balances the gravitational force on the monitor, so that the net force on the monitor is zero. The reaction to $\mathbf{n}$ is the force exerted by the monitor downward on the table, $\mathbf{F}_{mt} = -\mathbf{F}_{tm} = -\mathbf{n}$.

Definition of normal force

Note that the forces acting on the monitor are $\mathbf{F}_g$ and $\mathbf{n}$, as shown in Figure 5.6b. The two reaction forces $\mathbf{F}_{mE}$ and $\mathbf{F}_{mt}$ are exerted on objects other than the monitor. Remember, **the two forces in an action–reaction pair always act on two different objects.**

Figure 5.6 illustrates an extremely important step in solving problems involving forces. Figure 5.6a shows many of the forces in the situation—those acting on the monitor, one acting on the table, and one acting on the Earth. Figure 5.6b, by contrast, shows only the forces acting on *one object*, the monitor. This is a critical drawing called a **free-body diagram**. When analyzing an object subject to forces, we are interested in the net force acting on one object, which we will model as a particle. Thus, a free-body diagram helps us to isolate only those forces on the object and eliminate the other forces from our analysis. The free-body diagram can be simplified further by representing the object (such as the monitor) as a particle, by simply drawing a dot.

PITFALL PREVENTION

5.6 *n* Does Not Always Equal *mg*

In the situation shown in Figure 5.6 and in many others, we find that $n = mg$ (the normal force has the same magnitude as the gravitational force). However, this is *not* generally true. If an object is on an incline, if there are applied forces with vertical components, or if there is a vertical acceleration of the system, then $n \neq mg$. *Always* apply Newton's second law to find the relationship between n and mg.

5.7 Newton's Third Law

This is such an important and often misunderstood concept that it will be repeated here in a Pitfall Prevention. Newton's third law action and reaction forces act on *different* objects. Two forces acting on the same object, even if they are equal in magnitude and opposite in direction, *cannot* be an action–reaction pair.

[4] *Normal* in this context means *perpendicular*.

PITFALL PREVENTION

5.8 Free-body Diagrams

The *most important* step in solving a problem using Newton's laws is to draw a proper sketch—the free-body diagram. Be sure to draw only those forces that act on the object that you are isolating. Be sure to draw *all* forces acting on the object, including any field forces, such as the gravitational force.

Quick Quiz 5.7 If a fly collides with the windshield of a fast-moving bus, which object experiences an impact force with a larger magnitude? (a) the fly (b) the bus (c) the same force is experienced by both.

Quick Quiz 5.8 If a fly collides with the windshield of a fast-moving bus, which object experiences the greater acceleration: (a) the fly (b) the bus (c) the same acceleration is experienced by both.

Quick Quiz 5.9 Which of the following is the reaction force to the gravitational force acting on your body as you sit in your desk chair? (a) The normal force exerted by the chair (b) The force you exert downward on the seat of the chair (c) Neither of these forces.

Quick Quiz 5.10 In a free-body diagram for a single object, you draw (a) the forces acting on the object and the forces the object exerts on other objects, or (b) only the forces acting on the object.

Conceptual Example 5.3 You Push Me and I'll Push You

A large man and a small boy stand facing each other on frictionless ice. They put their hands together and push against each other so that they move apart.

(A) Who moves away with the higher speed?

Solution This situation is similar to what we saw in Quick Quizzes 5.7 and 5.8. According to Newton's third law, the force exerted by the man on the boy and the force exerted by the boy on the man are an action–reaction pair, and so they must be equal in magnitude. (A bathroom scale placed between their hands would read the same, regardless of which way it faced.) Therefore, the boy, having the smaller mass, experiences the greater acceleration. Both individuals accelerate for the same amount of time, but the greater acceleration of the boy over this time interval results in his moving away from the interaction with the higher speed.

(B) Who moves farther while their hands are in contact?

Solution Because the boy has the greater acceleration and, therefore, the greater average velocity, he moves farther during the time interval in which the hands are in contact.

© John Elk III/Stock, Boston Inc./PictureQuest

Rock climbers at rest are in equilibrium and depend on the tension forces in ropes for their safety.

5.7 Some Applications of Newton's Laws

In this section we apply Newton's laws to objects that are either in equilibrium ($\mathbf{a} = 0$) or accelerating along a straight line under the action of constant external forces. Remember that **when we apply Newton's laws to an object, we are interested only in external forces that act on the object.** We assume that the objects can be modeled as particles so that we need not worry about rotational motion. For now, we also neglect the effects of friction in those problems involving motion; this is equivalent to stating that the surfaces are *frictionless*. (We will incorporate the friction force in problems in Section 5.8.)

We usually neglect the mass of any ropes, strings, or cables involved. In this approximation, the magnitude of the force exerted at any point along a rope is the same at all points along the rope. In problem statements, the synonymous terms *light* and *of negligible mass* are used to indicate that a mass is to be ignored when you work the problems. When a rope attached to an object is pulling on the object, the rope exerts a force $\mathbf{T}$ on the object, and the magnitude T of that force is called the **tension** in the rope. Because it is the magnitude of a vector quantity, tension is a scalar quantity.

Objects in Equilibrium

If the acceleration of an object that can be modeled as a particle is zero, the particle is in **equilibrium**. Consider a lamp suspended from a light chain fastened to the ceiling, as in Figure 5.7a. The free-body diagram for the lamp (Figure 5.7b) shows that the forces acting on the lamp are the downward gravitational force $\mathbf{F}_g$ and the upward force $\mathbf{T}$ exerted by the chain. If we apply the second law to the lamp, noting that $\mathbf{a} = 0$, we see that because there are no forces in the x direction, $\Sigma F_x = 0$ provides no helpful information. The condition $\Sigma F_y = ma_y = 0$ gives

$$\sum F_y = T - F_g = 0 \qquad \text{or} \qquad T = F_g$$

Again, note that $\mathbf{T}$ and $\mathbf{F}_g$ are *not* an action–reaction pair because they act on the same object—the lamp. The reaction force to $\mathbf{T}$ is $\mathbf{T}'$, the downward force exerted by the lamp on the chain, as shown in Figure 5.7c. The ceiling exerts on the chain a force $\mathbf{T}''$ that is equal in magnitude to the magnitude of $\mathbf{T}'$ and points in the opposite direction.

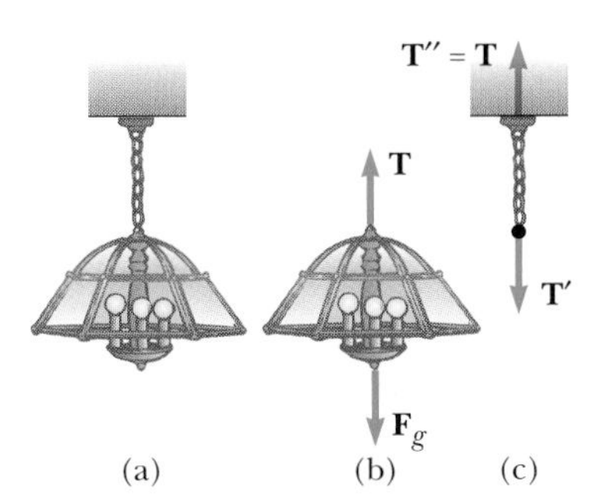

Figure 5.7 (a) A lamp suspended from a ceiling by a chain of negligible mass. (b) The forces acting on the lamp are the gravitational force $\mathbf{F}_g$ and the force $\mathbf{T}$ exerted by the chain. (c) The forces acting on the chain are the force $\mathbf{T}'$ exerted by the lamp and the force $\mathbf{T}''$ exerted by the ceiling.

Objects Experiencing a Net Force

If an object that can be modeled as a particle experiences an acceleration, then there must be a nonzero net force acting on the object. Consider a crate being pulled to the right on a frictionless, horizontal surface, as in Figure 5.8a. Suppose you are asked to find the acceleration of the crate and the force the floor exerts on it. First, note that the horizontal force $\mathbf{T}$ being applied to the crate acts through the rope. The magnitude of $\mathbf{T}$ is equal to the tension in the rope. The forces acting on the crate are illustrated in the free-body diagram in Figure 5.8b. In addition to the force $\mathbf{T}$, the free-body diagram for the crate includes the gravitational force $\mathbf{F}_g$ and the normal force $\mathbf{n}$ exerted by the floor on the crate.

We can now apply Newton's second law in component form to the crate. The only force acting in the x direction is $\mathbf{T}$. Applying $\Sigma F_x = ma_x$ to the horizontal motion gives

$$\sum F_x = T = ma_x \qquad \text{or} \qquad a_x = \frac{T}{m}$$

No acceleration occurs in the y direction. Applying $\Sigma F_y = ma_y$ with $a_y = 0$ yields

$$n + (-F_g) = 0 \qquad \text{or} \qquad n = F_g$$

That is, the normal force has the same magnitude as the gravitational force but acts in the opposite direction.

If $\mathbf{T}$ is a constant force, then the acceleration $a_x = T/m$ also is constant. Hence, the constant-acceleration equations of kinematics from Chapter 2 can be used to obtain the crate's position x and velocity v_x as functions of time. Because $a_x = T/m =$ constant, Equations 2.9 and 2.12 can be written as

$$v_{xf} = v_{xi} + \left(\frac{T}{m}\right)t$$

$$x_f = x_i + v_{xi}t + \frac{1}{2}\left(\frac{T}{m}\right)t^2$$

In the situation just described, the magnitude of the normal force $\mathbf{n}$ is equal to the magnitude of $\mathbf{F}_g$, but this is not always the case. For example, suppose a book is lying on a table and you push down on the book with a force $\mathbf{F}$, as in Figure 5.9. Because the book is at rest and therefore not accelerating, $\Sigma F_y = 0$, which gives $n - F_g - F = 0$, or $n = F_g + F$. In this situation, the normal force is *greater* than the force of gravity. Other examples in which $n \neq F_g$ are presented later.

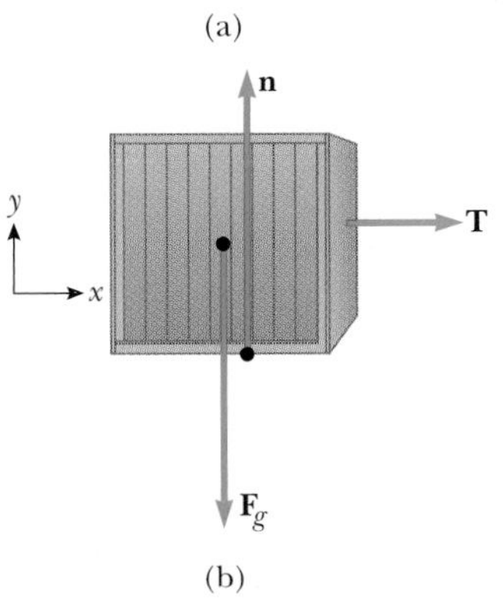

Figure 5.8 (a) A crate being pulled to the right on a frictionless surface. (b) The free-body diagram representing the external forces acting on the crate.

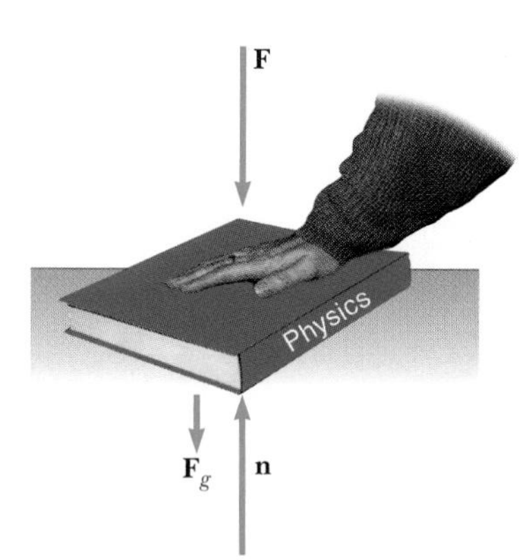

Figure 5.9 When one object pushes downward on another object with a force **F**, the normal force **n** is greater than the gravitational force: $n = F_g + F$.

PROBLEM-SOLVING HINTS

Applying Newton's Laws

The following procedure is recommended when dealing with problems involving Newton's laws:

- Draw a simple, neat diagram of the system to help *conceptualize* the problem.
- *Categorize* the problem: if any acceleration component is zero, the particle is in equilibrium in this direction and $\Sigma F = 0$. If not, the particle is undergoing an acceleration, the problem is one of nonequilibrium in this direction, and $\Sigma F = ma$.
- *Analyze* the problem by isolating the object whose motion is being analyzed. Draw a free-body diagram for this object. For systems containing more than one object, draw *separate* free-body diagrams for each object. *Do not* include in the free-body diagram forces exerted by the object on its surroundings.
- Establish convenient coordinate axes for each object and find the components of the forces along these axes. Apply Newton's second law, $\Sigma \mathbf{F} = m\mathbf{a}$, in component form. Check your dimensions to make sure that all terms have units of force.
- Solve the component equations for the unknowns. Remember that you must have as many independent equations as you have unknowns to obtain a complete solution.
- *Finalize* by making sure your results are consistent with the free-body diagram. Also check the predictions of your solutions for extreme values of the variables. By doing so, you can often detect errors in your results.

Example 5.4 A Traffic Light at Rest

A traffic light weighing 122 N hangs from a cable tied to two other cables fastened to a support, as in Figure 5.10a. The upper cables make angles of 37.0° and 53.0° with the horizontal. These upper cables are not as strong as the vertical cable, and will break if the tension in them exceeds 100 N. Will the traffic light remain hanging in this situation, or will one of the cables break?

Solution We *conceptualize* the problem by inspecting the drawing in Figure 5.10a. Let us assume that the cables do not break so that there is no acceleration of any sort in this problem in any direction. This allows us to *categorize* the problem as one of equilibrium. Because the acceleration of the system is zero, we know that the net force on the light and the net force on the knot are both zero. To *analyze* the

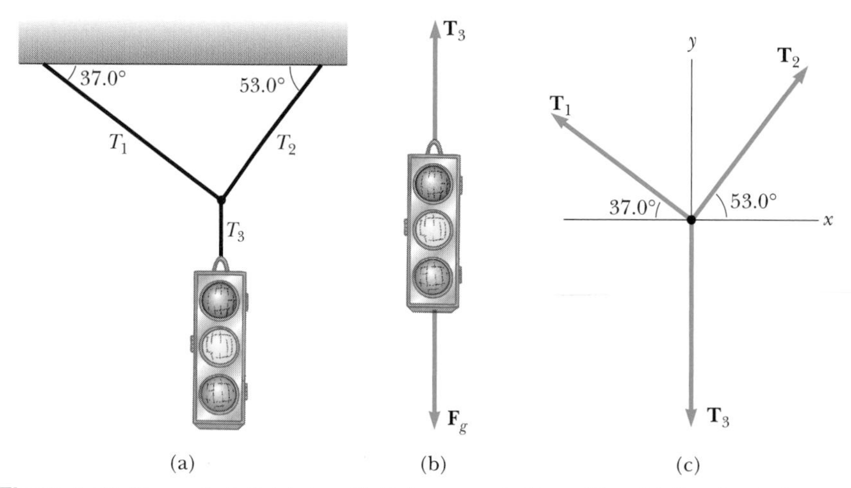

Figure 5.10 (Example 5.4) (a) A traffic light suspended by cables. (b) Free-body diagram for the traffic light. (c) Free-body diagram for the knot where the three cables are joined.

problem, we construct two free-body diagrams—one for the traffic light, shown in Figure 5.10b, and one for the knot that holds the three cables together, as in Figure 5.10c. This knot is a convenient object to choose because all the forces of interest act along lines passing through the knot.

With reference to Figure 5.10b, we apply the equilibrium condition in the y direction, $\Sigma F_y = 0 \rightarrow T_3 - F_g = 0$. This leads to $T_3 = F_g = 122$ N. Thus, the upward force $\mathbf{T}_3$ exerted by the vertical cable on the light balances the gravitational force.

Next, we choose the coordinate axes shown in Figure 5.10c and resolve the forces acting on the knot into their components:

Force	***x* Component**	***y* Component**
$\mathbf{T}_1$	$-T_1 \cos 37.0°$	$T_1 \sin 37.0°$
$\mathbf{T}_2$	$T_2 \cos 53.0°$	$T_2 \sin 53.0°$
$\mathbf{T}_3$	0	-122 N

Knowing that the knot is in equilibrium ($\mathbf{a} = 0$) allows us to write

$$(1) \qquad \sum F_x = -T_1 \cos 37.0° + T_2 \cos 53.0° = 0$$

$$(2) \qquad \sum F_y = T_1 \sin 37.0° + T_2 \sin 53.0° + (-122 \text{ N}) = 0$$

From (1) we see that the horizontal components of $\mathbf{T}_1$ and $\mathbf{T}_2$ must be equal in magnitude, and from (2) we see that the sum of the vertical components of $\mathbf{T}_1$ and $\mathbf{T}_2$ must balance the downward force $\mathbf{T}_3$, which is equal in magnitude to the weight of the light. We solve (1) for T_2 in terms of T_1 to obtain

$$(3) \qquad T_2 = T_1\left(\frac{\cos 37.0°}{\cos 53.0°}\right) = 1.33T_1$$

This value for T_2 is substituted into (2) to yield

$$T_1 \sin 37.0° + (1.33T_1)(\sin 53.0°) - 122 \text{ N} = 0$$

$$T_1 = 73.4 \text{ N}$$

$$T_2 = 1.33T_1 = 97.4 \text{ N}$$

Both of these values are less than 100 N (just barely for T_2), so the cables will not break. Let us *finalize* this problem by imagining a change in the system, as in the following **What If?**

What If? Suppose the two angles in Figure 5.10a are equal. What would be the relationship between T_1 and T_2?

Answer We can argue from the symmetry of the problem that the two tensions T_1 and T_2 would be equal to each other. Mathematically, if the equal angles are called θ, Equation (3) becomes

$$T_2 = T_1\left(\frac{\cos \theta}{\cos \theta}\right) = T_1$$

which also tells us that the tensions are equal. Without knowing the specific value of θ, we cannot find the values of T_1 and T_2. However, the tensions will be equal to each other, regardless of the value of θ.

Conceptual Example 5.5 Forces Between Cars in a Train

Train cars are connected by *couplers*, which are under tension as the locomotive pulls the train. As you move through the train from the locomotive to the last car, does the tension in the couplers increase, decrease, or stay the same as the train speeds up? When the engineer applies the brakes, the couplers are under compression. How does this compression force vary from the locomotive to the last car? (Assume that only the brakes on the wheels of the engine are applied.)

Solution As the train speeds up, tension decreases from front to back. The coupler between the locomotive and the first car must apply enough force to accelerate the rest of the cars. As you move back along the train, each coupler is accelerating less mass behind it. The last coupler has to accelerate only the last car, and so it is under the least tension.

When the brakes are applied, the force again decreases from front to back. The coupler connecting the locomotive to the first car must apply a large force to slow down the rest of the cars, but the final coupler must apply a force large enough to slow down *only* the last car.

Example 5.6 The Runaway Car

A car of mass m is on an icy driveway inclined at an angle θ, as in Figure 5.11a.

(A) Find the acceleration of the car, assuming that the driveway is frictionless.

Solution *Conceptualize* the situation using Figure 5.11a. From everyday experience, we know that a car on an icy incline will accelerate down the incline. (It will do the same thing as a car on a hill with its brakes not set.) This allows us to *categorize* the situation as a nonequilibrium problem—that is, one in which an object accelerates. Figure 5.11b shows the free-body diagram for the car, which we can use to *analyze* the problem. The only forces acting on the car are the normal force $\mathbf{n}$ exerted by the inclined plane, which acts perpendicular to the plane, and the gravitational force $\mathbf{F}_g = m\mathbf{g}$, which acts vertically downward. For problems involving inclined planes, it is convenient to choose the coordinate axes with x along the incline and y perpendicular to it, as in Figure 5.11b. (It is possible to solve the problem with "standard" horizontal and vertical axes. You may want to try this, just for practice.) Then, we replace the gravitational force by a component of magnitude $mg \sin \theta$ along the positive x axis and one of magnitude $mg \cos \theta$ along the negative y axis.

Now we apply Newton's second law in component form, noting that $a_y = 0$:

$$(1) \qquad \sum F_x = mg \sin \theta = ma_x$$

$$(2) \qquad \sum F_y = n - mg \cos \theta = 0$$

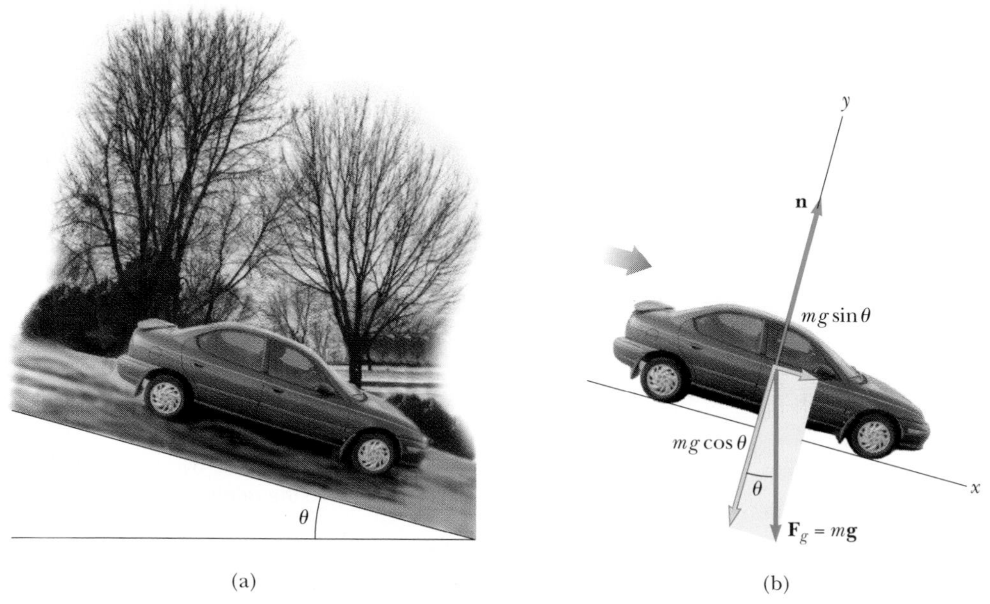

Figure 5.11 (Example 5.6) (a) A car of mass m sliding down a frictionless incline. (b) The free-body diagram for the car. Note that its acceleration along the incline is $g \sin\theta$.

Solving (1) for a_x, we see that the acceleration along the incline is caused by the component of $\mathbf{F}_g$ directed down the incline:

$$(3) \qquad a_x = g \sin\theta$$

To *finalize* this part, note that this acceleration component is independent of the mass of the car! It depends only on the angle of inclination and on g.

From (2) we conclude that the component of $\mathbf{F}_g$ perpendicular to the incline is balanced by the normal force; that is, $n = mg\cos\theta$. This is another example of a situation in which the normal force is *not* equal in magnitude to the weight of the object.

(B) Suppose the car is released from rest at the top of the incline, and the distance from the car's front bumper to the bottom of the incline is d. How long does it take the front bumper to reach the bottom, and what is the car's speed as it arrives there?

Solution *Conceptualize* by imagining that the car is sliding down the hill and you are operating a stop watch to measure the entire time interval until it reaches the bottom. This part of the problem belongs to kinematics rather than to dynamics, and Equation (3) shows that the acceleration a_x is constant. Therefore you should *categorize* this problem as that of a particle undergoing constant acceleration. Apply Equation 2.12, $x_f = x_i + v_{xi}t + \frac{1}{2}a_x t^2$, to *analyze* the car's motion. Defining the initial position of the front bumper as $x_i = 0$ and its final position as $x_f = d$, and recognizing that $v_{xi} = 0$, we obtain

$$d = \tfrac{1}{2}a_x t^2$$

$$(4) \qquad t = \sqrt{\frac{2d}{a_x}} = \sqrt{\frac{2d}{g\sin\theta}}$$

Using Equation 2.13, with $v_{xi} = 0$, we find that

$$v_{xf}^{\ 2} = 2a_x d$$

$$(5) \qquad v_{xf} = \sqrt{2a_x d} = \sqrt{2gd\sin\theta}$$

To *finalize* this part of the problem, we see from Equations (4) and (5) that the time t at which the car reaches the bottom and its final speed v_{xf} are independent of the car's mass, as was its acceleration. Note that we have combined techniques from Chapter 2 with new techniques from the present chapter in this example. As we learn more and more techniques in later chapters, this process of combining information from several parts of the book will occur more often. In these cases, use the General Problem-Solving Strategy to help you identify what techniques you will need.

What If? (A) What previously solved problem does this become if $\theta = 90°$? (B) What problem does this become if $\theta = 0$?

Answer (A) Imagine θ going to 90° in Figure 5.11. The inclined plane becomes vertical, and the car is an object in free-fall! Equation (3) becomes

$$a_x = g\sin\theta = g\sin 90° = g$$

which is indeed the free-fall acceleration. (We find $a_x = g$ rather than $a_x = -g$ because we have chosen positive x to be downward in Figure 5.11.) Notice also that the condition

$n = mg \cos \theta$ gives us $n = mg \cos 90° = 0$. This is consistent with the fact that the car is falling downward *next to* the vertical plane but there is no interaction force between the car and the plane. Equations (4) and (5) give us $t = \sqrt{\dfrac{2d}{g \sin 90°}} = \sqrt{\dfrac{2d}{g}}$ and $v_{xf} = \sqrt{2gd \sin 90°} = \sqrt{2gd}$, both of which are consistent with a falling object.

(B) Imagine θ going to 0 in Figure 5.11. In this case, the inclined plane becomes horizontal, and the car is on a horizontal surface. Equation (3) becomes

$$a_x = g \sin \theta = g \sin 0 = 0$$

which is consistent with the fact that the car is at rest in equilibrium. Notice also that the condition $n = mg \cos \theta$ gives us $n = mg \cos 0 = mg$, which is consistent with our expectation.

Equations (4) and (5) give us $t = \sqrt{\dfrac{2d}{g \sin 0}} \rightarrow \infty$ and $v_{xf} = \sqrt{2gd \sin 0} = 0$. These results agree with the fact that the car does not accelerate, so it will never achieve a nonzero final velocity, and it will take an infinite amount of time to reach the bottom of the "hill"!

Example 5.7 One Block Pushes Another

Two blocks of masses m_1 and m_2, with $m_1 > m_2$, are placed in contact with each other on a frictionless, horizontal surface, as in Figure 5.12a. A constant horizontal force **F** is applied to m_1 as shown. **(A)** Find the magnitude of the acceleration of the system.

Solution *Conceptualize* the situation using Figure 5.12a and realizing that both blocks must experience the *same* acceleration because they are in contact with each other and remain in contact throughout the motion. We *categorize* this as a Newton's second law problem because we have a force applied to a system and we are looking for an acceleration. To *analyze* the problem, we first address the combination of two blocks as a system. Because **F** is the only external horizontal force acting on the system, we have

$$\sum F_x(\text{system}) = F = (m_1 + m_2)a_x$$

$$(1) \qquad a_x = \frac{F}{m_1 + m_2}$$

To *finalize* this part, note that this would be the same acceleration as that of a single object of mass equal to the combined masses of the two blocks in Figure 5.12a and subject to the same force.

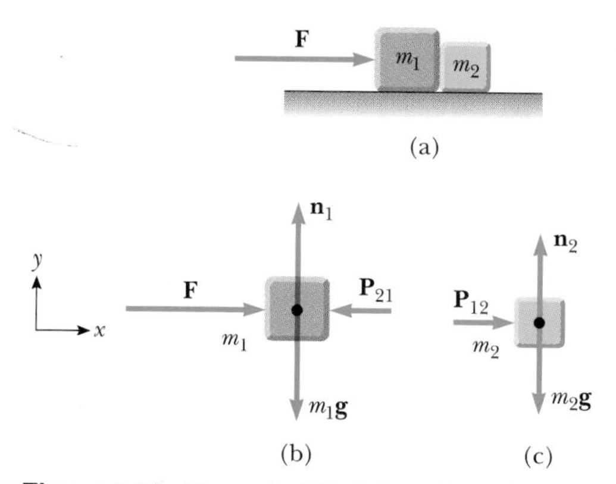

Active Figure 5.12 (Example 5.7) A force is applied to a block of mass m_1, which pushes on a second block of mass m_2. (b) The free-body diagram for m_1. (c) The free-body diagram for m_2.

At the Active Figures link at http://www.pse6.com, you can study the forces involved in this two-block system.

(B) Determine the magnitude of the contact force between the two blocks.

Solution *Conceptualize* by noting that the contact force is internal to the system of two blocks. Thus, we cannot find this force by modeling the whole system (the two blocks) as a single particle. We must now treat each of the two blocks individually by *categorizing* each as a particle subject to a net force. To *analyze* the situation, we first construct a free-body diagram for each block, as shown in Figures 5.12b and 5.12c, where the contact force is denoted by **P**. From Figure 5.12c we see that the only horizontal force acting on m_2 is the contact force $\mathbf{P}_{12}$ (the force exerted by m_1 on m_2), which is directed to the right. Applying Newton's second law to m_2 gives

$$(2) \qquad \sum F_x = P_{12} = m_2 a_x$$

Substituting the value of the acceleration a_x given by (1) into (2) gives

$$(3) \qquad P_{12} = m_2 a_x = \left(\frac{m_2}{m_1 + m_2}\right) F$$

To *finalize* the problem, we see from this result that the contact force P_{12} is *less* than the applied force F. This is consistent with the fact that the force required to accelerate block 2 alone must be less than the force required to produce the same acceleration for the two-block system.

To *finalize* further, it is instructive to check this expression for P_{12} by considering the forces acting on m_1, shown in Figure 5.12b. The horizontal forces acting on m_1 are the applied force **F** to the right and the contact force $\mathbf{P}_{21}$ to the left (the force exerted by m_2 on m_1). From Newton's third law, $\mathbf{P}_{21}$ is the reaction to $\mathbf{P}_{12}$, so $P_{21} = P_{12}$. Applying Newton's second law to m_1 gives

$$(4) \qquad \sum F_x = F - P_{21} = F - P_{12} = m_1 a_x$$

Substituting into (4) the value of a_x from (1), we obtain

$$P_{12} = F - m_1 a_x = F - m_1\left(\frac{F}{m_1 + m_2}\right) = \left(\frac{m_2}{m_1 + m_2}\right) F$$

This agrees with (3), as it must.

What If? **Imagine that the force F in Figure 5.12 is applied toward the left on the right-hand block of mass m_2. Is the magnitude of the force P_{12} the same as it was when the force was applied toward the right on m_1?**

Answer With the force applied toward the left on m_2, the contact force must accelerate m_1. In the original situation, the contact force accelerates m_2. Because $m_1 > m_2$, this will require more force, so the magnitude of $\mathbf{P}_{12}$ is greater than in the original situation.

Example 5.8 Weighing a Fish in an Elevator

A person weighs a fish of mass m on a spring scale attached to the ceiling of an elevator, as illustrated in Figure 5.13. Show that if the elevator accelerates either upward or downward, the spring scale gives a reading that is different from the weight of the fish.

Solution *Conceptualize* by noting that the reading on the scale is related to the extension of the spring in the scale, which is related to the force on the end of the spring as in Figure 5.2. Imagine that a string is hanging from the end of the spring, so that the magnitude of the force exerted on the spring is equal to the tension T in the string. Thus, we are looking for T. The force $\mathbf{T}$ pulls down on the string and pulls up on the fish. Thus, we can *categorize* this problem as one of analyzing the forces and acceleration associated with the fish by means of Newton's second law. To *analyze* the problem, we inspect the free-body diagrams for the fish in Figure 5.13 and note that the external forces acting on the fish are the downward gravitational force $\mathbf{F}_g = m\mathbf{g}$ and the force $\mathbf{T}$ exerted by the scale. If the elevator is either at rest or moving at constant velocity, the fish does not accelerate, and so $\sum F_y = T - F_g = 0$ or $T = F_g = mg$. (Remember that the scalar mg is the weight of the fish.)

If the elevator moves with an acceleration $\mathbf{a}$ relative to an observer standing outside the elevator in an inertial frame (see Fig. 5.13), Newton's second law applied to the fish gives the net force on the fish:

$$(1) \qquad \sum F_y = T - mg = ma_y$$

where we have chosen upward as the positive y direction. Thus, we conclude from (1) that the scale reading T is greater than the fish's weight mg if $\mathbf{a}$ is upward, so that a_y is positive, and that the reading is less than mg if $\mathbf{a}$ is downward, so that a_y is negative.

For example, if the weight of the fish is 40.0 N and $\mathbf{a}$ is upward, so that $a_y = +\ 2.00\ \text{m/s}^2$, the scale reading from (1) is

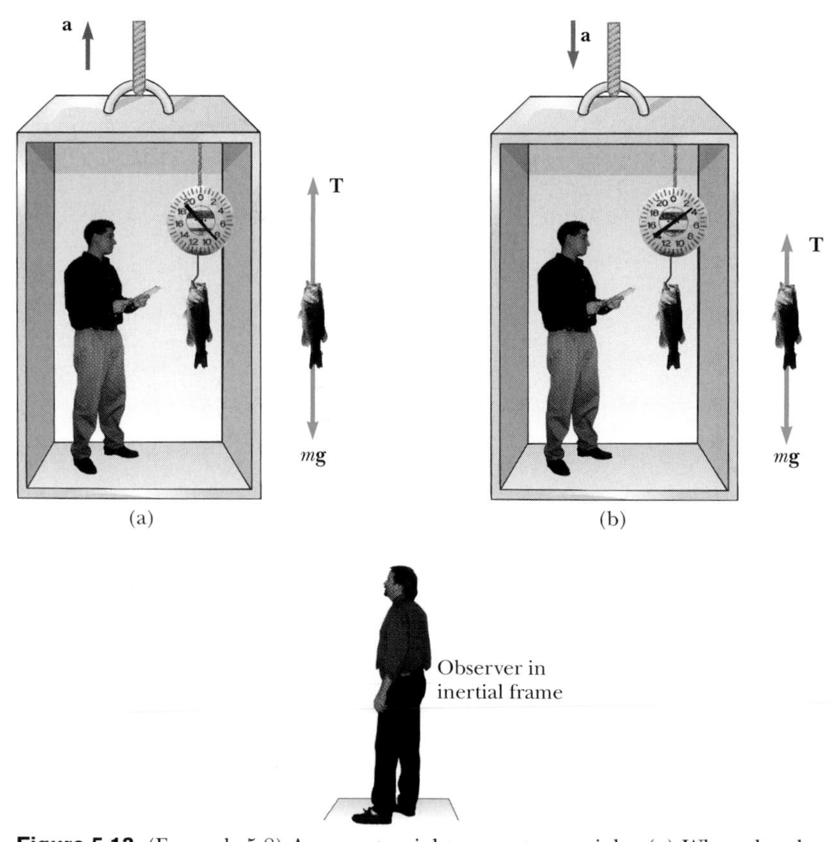

Figure 5.13 (Example 5.8) Apparent weight versus true weight. (a) When the elevator accelerates upward, the spring scale reads a value greater than the weight of the fish. (b) When the elevator accelerates downward, the spring scale reads a value less than the weight of the fish.

$$(2)\quad T = ma_y + mg = mg\left(\frac{a_y}{g} + 1\right)$$

$$= F_g\left(\frac{a_y}{g} + 1\right) = (40.0\text{ N})\left(\frac{2.00\text{ m/s}^2}{9.80\text{ m/s}^2} + 1\right)$$

$$= 48.2\text{ N}$$

If **a** is downward so that $a_y = -2.00\text{ m/s}^2$, then (2) gives us

$$T = F_g\left(\frac{a_y}{g} + 1\right) = (40.0\text{ N})\left(\frac{-2.00\text{ m/s}^2}{9.80\text{ m/s}^2} + 1\right)$$

$$= 31.8\text{ N}$$

To *finalize* this problem, take this advice—if you buy a fish in an elevator, make sure the fish is weighed while the elevator is either at rest or accelerating downward! Furthermore, note that from the information given here, one cannot determine the direction of motion of the elevator.

What If? Suppose the elevator cable breaks, so that the elevator and its contents are in free-fall. What happens to the reading on the scale?

Answer If the elevator falls freely, its acceleration is $a_y = -g$. We see from (2) that the scale reading T is zero in this case; that is, the fish *appears* to be weightless.

Example 5.9 The Atwood Machine

Interactive

When two objects of unequal mass are hung vertically over a frictionless pulley of negligible mass, as in Figure 5.14a, the arrangement is called an *Atwood machine*. The device is sometimes used in the laboratory to measure the free-fall acceleration. Determine the magnitude of the acceleration of the two objects and the tension in the lightweight cord.

Solution *Conceptualize* the situation pictured in Figure 5.14a—as one object moves upward, the other object moves downward. Because the objects are connected by an inextensible string, their accelerations must be of equal magnitude. The objects in the Atwood machine are subject to the gravitational force as well as to the forces exerted by the strings connected to them—thus, we can *categorize* this as a Newton's second law problem. To *analyze* the situation, the free-body diagrams for the two objects are shown in Figure 5.14b. Two forces act on each object: the upward force **T** exerted by the string and the downward gravitational force. In problems such as this in which the pulley is modeled as massless and frictionless, the tension in the string on both sides of the pulley is the same. If the pulley has mass and/or is subject to friction, the tensions on either side are not the same and the situation requires techniques we will learn in Chapter 10.

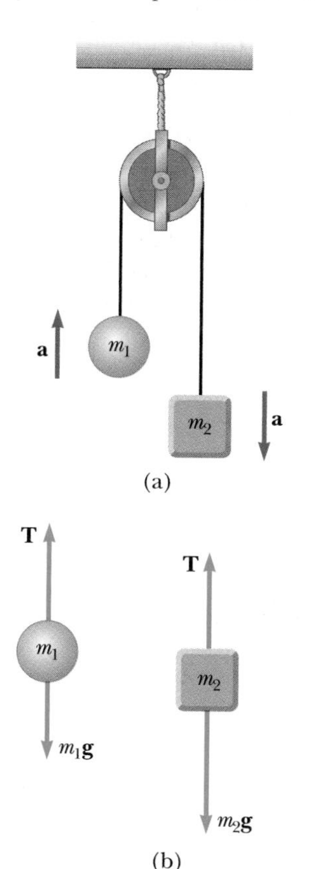

Active Figure 5.14 (Example 5.9) The Atwood machine. (a) Two objects ($m_2 > m_1$) connected by a massless inextensible cord over a frictionless pulley. (b) Free-body diagrams for the two objects.

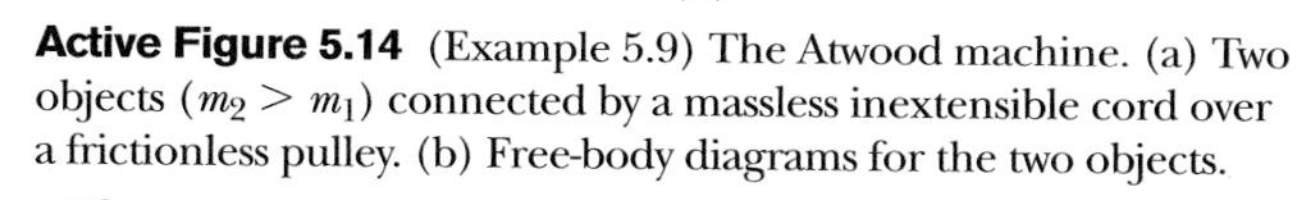

At the Active Figures link at http://www.pse6.com, you can adjust the masses of the objects on the Atwood machine and observe the motion.

We must be very careful with signs in problems such as this. In Figure 5.14a, notice that if object 1 accelerates upward, then object 2 accelerates downward. Thus, for consistency with signs, if we define the upward direction as positive for object 1, we must define the downward direction as positive for object 2. With this sign convention, both objects accelerate in the same direction as defined by the choice of sign. Furthermore, according to this sign convention, the y component of the net force exerted on object 1 is $T - m_1g$, and the y component of the net force exerted on object 2 is $m_2g - T$. Notice that we have chosen the signs of the forces to be consistent with the choices of signs for up and down for each object. If we assume that $m_2 > m_1$, then m_1 must accelerate upward, while m_2 must accelerate downward.

When Newton's second law is applied to object 1, we obtain

$$(1)\qquad \sum F_y = T - m_1g = m_1a_y$$

Similarly, for object 2 we find

$$(2)\qquad \sum F_y = m_2g - T = m_2a_y$$

When (2) is added to (1), T cancels and we have

$$-m_1g + m_2g = m_1a_y + m_2a_y$$

$$(3)\qquad a_y = \left(\frac{m_2 - m_1}{m_1 + m_2}\right)g$$

The acceleration given by (3) can be interpreted as the ratio of the magnitude of the unbalanced force on the system $(m_2 - m_1)g$, to the total mass of the system $(m_1 + m_2)$, as expected from Newton's second law.

When (3) is substituted into (1), we obtain

$$(4) \qquad T = \left(\frac{2m_1 m_2}{m_1 + m_2}\right)g$$

Finalize this problem with the following **What If?**

What If? (A) Describe the motion of the system if the objects have equal masses, that is, $m_1 = m_2$.

(B) Describe the motion of the system if one of the masses is much larger than the other, $m_1 \gg m_2$.

Answer (A) If we have the same mass on both sides, the system is balanced and it should not accelerate. Mathematically, we see that if $m_1 = m_2$, Equation (3) gives us $a_y = 0$. (B) In the case in which one mass is infinitely larger than the other, we can ignore the effect of the smaller mass. Thus, the larger mass should simply fall as if the smaller mass were not there. We see that if $m_1 \gg m_2$, Equation (3) gives us $a_y = -g$.

Investigate these limiting cases at the Interactive Worked Example link at http://www.pse6.com.

Example 5.10 Acceleration of Two Objects Connected by a Cord — Interactive

A ball of mass m_1 and a block of mass m_2 are attached by a lightweight cord that passes over a frictionless pulley of negligible mass, as in Figure 5.15a. The block lies on a frictionless incline of angle θ. Find the magnitude of the acceleration of the two objects and the tension in the cord.

Solution *Conceptualize* the motion in Figure 5.15. If m_2 moves down the incline, m_1 moves upward. Because the objects are connected by a cord (which we assume does not stretch), their accelerations have the same magnitude. We can identify forces on each of the two objects and we are looking for an acceleration, so we *categorize* this as a Newton's second-law problem. To *analyze* the problem, consider the free-body diagrams shown in Figures 5.15b and 5.15c. Applying Newton's second law in component form to the ball, choosing the upward direction as positive, yields

$$(1) \qquad \sum F_x = 0$$

$$(2) \qquad \sum F_y = T - m_1 g = m_1 a_y = m_1 a$$

Note that in order for the ball to accelerate upward, it is necessary that $T > m_1 g$. In (2), we replaced a_y with a because the acceleration has only a y component.

For the block it is convenient to choose the positive x' axis along the incline, as in Figure 5.15c. For consistency with our choice for the ball, we choose the positive direction to be down the incline. Applying Newton's second law in component form to the block gives

$$(3) \qquad \sum F_{x'} = m_2 g \sin\theta - T = m_2 a_{x'} = m_2 a$$

$$(4) \qquad \sum F_{y'} = n - m_2 g \cos\theta = 0$$

In (3) we replaced $a_{x'}$ with a because the two objects have accelerations of equal magnitude a. Equations (1) and (4) provide no information regarding the acceleration. However, if we solve (2) for T and then substitute this value for T into (3) and solve for a, we obtain

$$(5) \qquad a = \frac{m_2 g \sin\theta - m_1 g}{m_1 + m_2}$$

When this expression for a is substituted into (2), we find

$$(6) \qquad T = \frac{m_1 m_2 g(\sin\theta + 1)}{m_1 + m_2}$$

To *finalize* the problem, note that the block accelerates down the incline only if $m_2 \sin\theta > m_1$. If $m_1 > m_2 \sin\theta$,

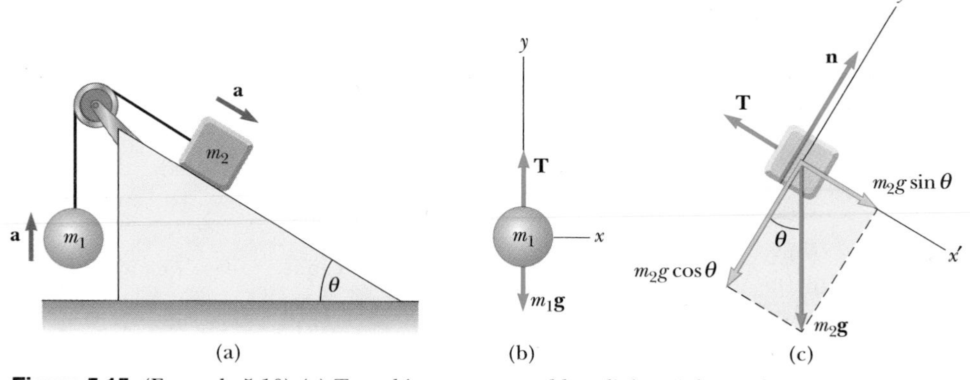

Figure 5.15 (Example 5.10) (a) Two objects connected by a lightweight cord strung over a frictionless pulley. (b) Free-body diagram for the ball. (c) Free-body diagram for the block. (The incline is frictionless.)

then the acceleration is up the incline for the block and downward for the ball. Also note that the result for the acceleration (5) can be interpreted as the magnitude of the net force acting on the system divided by the total mass of the system; this is consistent with Newton's second law.

What If? (A) What happens in this situation if the angle $\theta = 90°$?

(B) What happens if the mass $m_1 = 0$?

Answer (A) If $\theta = 90°$, the inclined plane becomes vertical and there is no interaction between its surface and m_2. Thus, this problem becomes the Atwood machine of Example 5.9. Letting $\theta \rightarrow 90°$ in Equations (5) and (6) causes them to reduce to Equations (3) and (4) of Example 5.9! (B) If $m_1 = 0$, then m_2 is simply sliding down an inclined plane without interacting with m_1 through the string. Thus, this problem becomes the sliding car problem in Example 5.6. Letting $m_1 \rightarrow 0$ in Equation (5) causes it to reduce to Equation (3) of Example 5.6!

Investigate these limiting cases at the Interactive Worked Example link at http://www.pse6.com.

5.8 Forces of Friction

When an object is in motion either on a surface or in a viscous medium such as air or water, there is resistance to the motion because the object interacts with its surroundings. We call such resistance a **force of friction.** Forces of friction are very important in our everyday lives. They allow us to walk or run and are necessary for the motion of wheeled vehicles.

Imagine that you are working in your garden and have filled a trash can with yard clippings. You then try to drag the trash can across the surface of your concrete patio, as in Figure 5.16a. This is a *real* surface, not an idealized, frictionless surface. If we apply an external horizontal force **F** to the trash can, acting to the right, the trash can remains stationary if **F** is small. The force that counteracts **F** and keeps the trash can from moving acts to the left and is called the **force of static friction f**$_s$. As long as the trash can is not moving, $f_s = F$. Thus, if **F** is increased, **f**$_s$ also increases. Likewise, if **F** decreases, **f**$_s$ also

Force of static friction

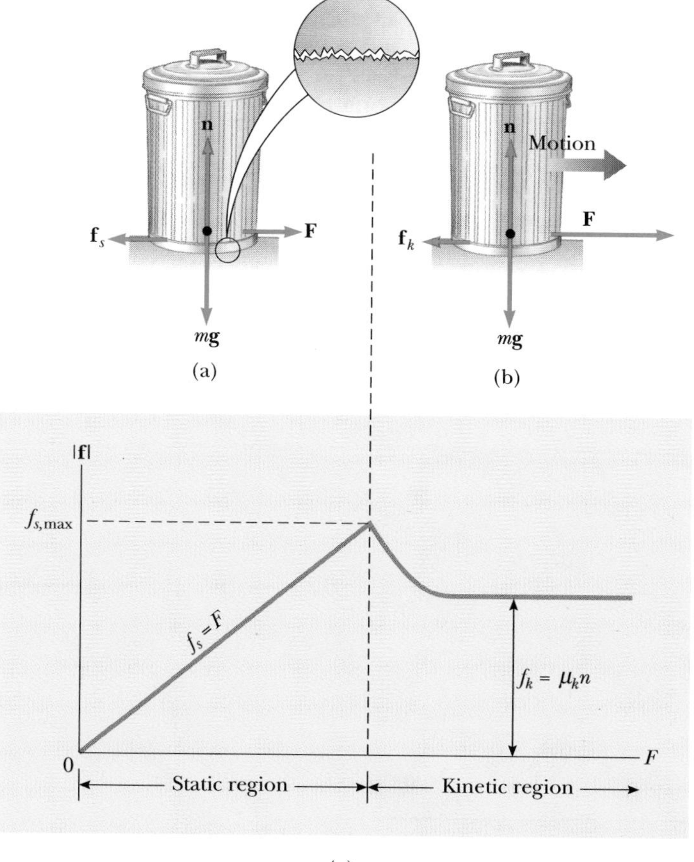

Active Figure 5.16 The direction of the force of friction **f** between a trash can and a rough surface is opposite the direction of the applied force **F**. Because both surfaces are rough, contact is made only at a few points, as illustrated in the "magnified" view. (a) For small applied forces, the magnitude of the force of static friction equals the magnitude of the applied force. (b) When the magnitude of the applied force exceeds the magnitude of the maximum force of static friction, the trash can breaks free. The applied force is now larger than the force of kinetic friction and the trash can accelerates to the right. (c) A graph of friction force versus applied force. Note that $f_{s,max} > f_k$.

At the Active Figures link at http://www.pse6.com *you can vary the applied force on the trash can and practice sliding it on surfaces of varying roughness. Note the effect on the trash can's motion and the corresponding behavior of the graph in (c).*

decreases. Experiments show that the friction force arises from the nature of the two surfaces: because of their roughness, contact is made only at a few locations where peaks of the material touch, as shown in the magnified view of the surface in Figure 5.16a.

At these locations, the friction force arises in part because one peak physically blocks the motion of a peak from the opposing surface, and in part from chemical bonding ("spot welds") of opposing peaks as they come into contact. If the surfaces are rough, bouncing is likely to occur, further complicating the analysis. Although the details of friction are quite complex at the atomic level, this force ultimately involves an electrical interaction between atoms or molecules.

If we increase the magnitude of **F**, as in Figure 5.16b, the trash can eventually slips. When the trash can is on the verge of slipping, f_s has its maximum value $f_{s,\text{max}}$, as shown in Figure 5.16c. When F exceeds $f_{s,\text{max}}$, the trash can moves and accelerates to the right. When the trash can is in motion, the friction force is less than $f_{s,\text{max}}$ (Fig. 5.16c). We call the friction force for an object in motion the **force of kinetic friction** $\mathbf{f}_k$. The net force $F - f_k$ in the x direction produces an acceleration to the right, according to Newton's second law. If $F = f_k$, the acceleration is zero, and the trash can moves to the right with constant speed. If the applied force is removed, the friction force acting to the left provides an acceleration of the trash can in the $-x$ direction and eventually brings it to rest, again consistent with Newton's second law.

Force of kinetic friction

Experimentally, we find that, to a good approximation, both $f_{s,\text{max}}$ and f_k are proportional to the magnitude of the normal force. The following empirical laws of friction summarize the experimental observations:

- The magnitude of the force of static friction between any two surfaces in contact can have the values

$$f_s \leq \mu_s n \tag{5.8}$$

where the dimensionless constant μ_s is called the **coefficient of static friction** and n is the magnitude of the normal force exerted by one surface on the other. The equality in Equation 5.8 holds when the surfaces are on the verge of slipping, that is, when $f_s = f_{s,\text{max}} \equiv \mu_s n$. This situation is called *impending motion.* The inequality holds when the surfaces are not on the verge of slipping.

- The magnitude of the force of kinetic friction acting between two surfaces is

$$f_k = \mu_k n \tag{5.9}$$

where μ_k is the **coefficient of kinetic friction.** Although the coefficient of kinetic friction can vary with speed, we shall usually neglect any such variations in this text.

PITFALL PREVENTION

5.9 The Equal Sign is Used in Limited Situations

In Equation 5.8, the equal sign is used *only* in the case in which the surfaces are just about to break free and begin sliding. Do not fall into the common trap of using $f_s = \mu_s n$ in *any* static situation.

PITFALL PREVENTION

5.10 Friction Equations

Equations 5.8 and 5.9 are *not* vector equations. They are relationships between the *magnitudes* of the vectors representing the friction and normal forces. Because the friction and normal forces are perpendicular to each other, the vectors cannot be related by a multiplicative constant.

Table 5.2

Coefficients of Friction[a]

	μ_s	μ_k
Steel on steel	0.74	0.57
Aluminum on steel	0.61	0.47
Copper on steel	0.53	0.36
Rubber on concrete	1.0	0.8
Wood on wood	0.25–0.5	0.2
Glass on glass	0.94	0.4
Waxed wood on wet snow	0.14	0.1
Waxed wood on dry snow	—	0.04
Metal on metal (lubricated)	0.15	0.06
Ice on ice	0.1	0.03
Teflon on Teflon	0.04	0.04
Synovial joints in humans	0.01	0.003

[a] All values are approximate. In some cases, the coefficient of friction can exceed 1.0.

- The values of μ_k and μ_s depend on the nature of the surfaces, but μ_k is generally less than μ_s. Typical values range from around 0.03 to 1.0. Table 5.2 lists some reported values.
- The direction of the friction force on an object is parallel to the surface with which the object is in contact and opposite to the actual motion (kinetic friction) or the impending motion (static friction) of the object relative to the surface.
- The coefficients of friction are nearly independent of the area of contact between the surfaces. We might expect that placing an object on the side having the most area might increase the friction force. While this provides more points in contact, as in Figure 5.16a, the weight of the object is spread out over a larger area, so that the individual points are not pressed so tightly together. These effects approximately compensate for each other, so that the friction force is independent of the area.

PITFALL PREVENTION

5.11 The Direction of the Friction Force

Sometimes, an incorrect statement about the friction force between an object and a surface is made—"the friction force on an object is opposite to its motion or impending motion"—rather than the correct phrasing, "the friction force on an object is opposite to its motion or impending motion *relative to the surface*." Think carefully about Quick Quiz 5.12.

Quick Quiz 5.11 You press your physics textbook flat against a vertical wall with your hand. What is the direction of the friction force exerted by the wall on the book? (a) downward (b) upward (c) out from the wall (d) into the wall.

Quick Quiz 5.12 A crate is located in the center of a flatbed truck. The truck accelerates to the east, and the crate moves with it, not sliding at all. What is the direction of the friction force exerted by the truck on the crate? (a) to the west (b) to the east (c) No friction force exists because the crate is not sliding.

Quick Quiz 5.13 You place your physics book on a wooden board. You raise one end of the board so that the angle of the incline increases. Eventually, the book starts sliding on the board. If you maintain the angle of the board at this value, the book (a) moves at constant speed (b) speeds up (c) slows down (d) none of these.

Quick Quiz 5.14 You are playing with your daughter in the snow. She sits on a sled and asks you to slide her across a flat, horizontal field. You have a choice of (a) pushing her from behind, by applying a force downward on her shoulders at 30° below the horizontal (Fig. 5.17a), or (b) attaching a rope to the front of the sled and pulling with a force at 30° above the horizontal (Fig 5.17b). Which would be easier for you and why?

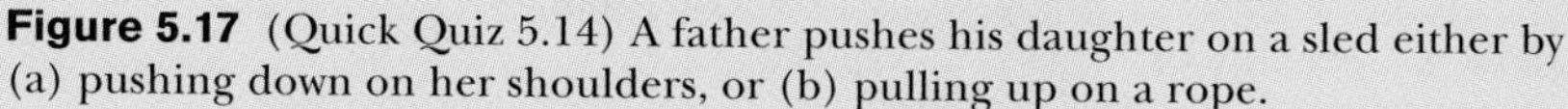

Figure 5.17 (Quick Quiz 5.14) A father pushes his daughter on a sled either by (a) pushing down on her shoulders, or (b) pulling up on a rope.

Conceptual Example 5.11 Why Does the Sled Accelerate?

A horse pulls a sled along a level, snow-covered road, causing the sled to accelerate, as shown in Figure 5.18a. Newton's third law states that the sled exerts a force of equal magnitude and opposite direction on the horse. In view of this, how can the sled accelerate—don't the forces cancel? Under what condition does the system (horse plus sled) move with constant velocity?

Solution Remember that the forces described in Newton's third law act on *different* objects—the horse exerts a force on the sled, and the sled exerts an equal-magnitude and oppositely directed force on the horse. Because we are interested only in the motion of the sled, we do not consider the forces it exerts on the horse. When determining the motion of an object, you must add only the forces on that object. (This is the principle behind drawing a free-body diagram.) The horizontal forces exerted on the sled are the forward force **T** exerted by the horse and the backward force of friction $\mathbf{f}_{\text{sled}}$ between sled and snow (see Fig. 5.18b). When the forward force on the sled exceeds the backward force, the sled accelerates to the right.

The horizontal forces exerted on the horse are the forward force $\mathbf{f}_{\text{horse}}$ exerted by the Earth and the backward tension force **T** exerted by the sled (Fig. 5.18c). The resultant of these two forces causes the horse to accelerate.

The force that accelerates the system (horse plus sled) is the net force $\mathbf{f}_{\text{horse}} - \mathbf{f}_{\text{sled}}$. When $\mathbf{f}_{\text{horse}}$ balances $\mathbf{f}_{\text{sled}}$, the system moves with constant velocity.

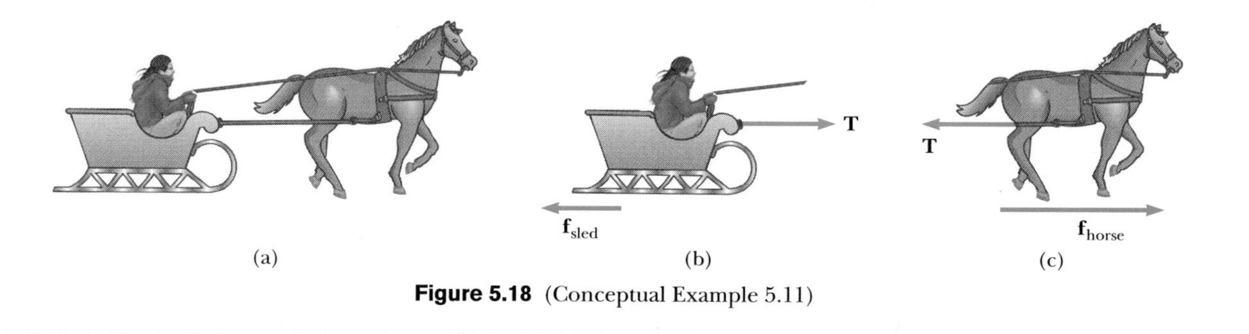

Figure 5.18 (Conceptual Example 5.11)

Example 5.12 Experimental Determination of μ_s and μ_k

The following is a simple method of measuring coefficients of friction: Suppose a block is placed on a rough surface inclined relative to the horizontal, as shown in Figure 5.19. The incline angle is increased until the block starts to move. Show that by measuring the critical angle θ_c at which this slipping just occurs, we can obtain μ_s.

Solution *Conceptualizing* from the free body diagram in Figure 5.19, we see that we can *categorize* this as a Newton's second law problem. To *analyze* the problem, note that the only forces acting on the block are the gravitational force $m\mathbf{g}$, the normal force **n**, and the force of static friction $\mathbf{f}_s$. These forces balance when the block is not moving. When we choose x to be parallel to the plane and y perpendicular to it, Newton's second law applied to the block for this balanced situation gives

$$(1) \qquad \sum F_x = mg \sin \theta - f_s = ma_x = 0$$

$$(2) \qquad \sum F_y = n - mg \cos \theta = ma_y = 0$$

We can eliminate mg by substituting $mg = n/\cos \theta$ from (2) into (1) to find

$$(3) \qquad f_s = mg \sin \theta = \left(\frac{n}{\cos \theta}\right) \sin \theta = n \tan \theta$$

When the incline angle is increased until the block is on the verge of slipping, the force of static friction has reached its maximum value $\mu_s n$. The angle θ in this situation is the critical angle θ_c, and (3) becomes

$$\mu_s n = n \tan \theta_c$$

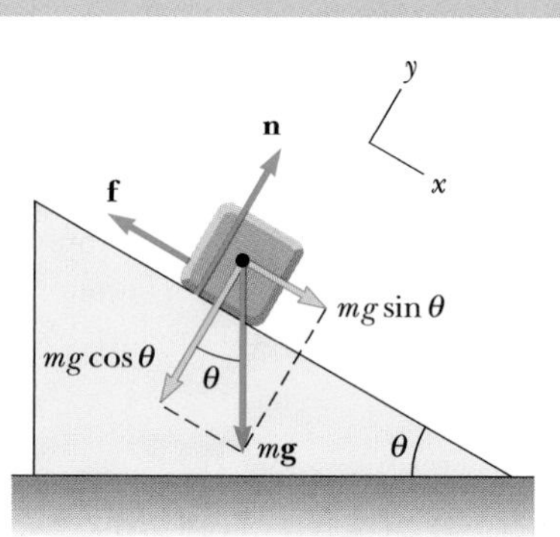

Figure 5.19 (Example 5.12) The external forces exerted on a block lying on a rough incline are the gravitational force $m\mathbf{g}$, the normal force **n**, and the force of friction **f**. For convenience, the gravitational force is resolved into a component along the incline $mg \sin \theta$ and a component perpendicular to the incline $mg \cos \theta$.

$$\mu_s = \tan \theta_c$$

For example, if the block just slips at $\theta_c = 20.0°$, then we find that $\mu_s = \tan 20.0° = 0.364$.

To *finalize* the problem, note that once the block starts to move at $\theta \geq \theta_c$, it accelerates down the incline and the force of friction is $f_k = \mu_k n$. However, if θ is reduced to a value less than θ_c, it may be possible to find an angle θ_c' such that the block moves down the incline with constant speed ($a_x = 0$). In this case, using (1) and (2) with f_s replaced by f_k gives

$$\mu_k = \tan \theta_c'$$

where $\theta_c' < \theta_c$.

Example 5.13 The Sliding Hockey Puck

A hockey puck on a frozen pond is given an initial speed of 20.0 m/s. If the puck always remains on the ice and slides 115 m before coming to rest, determine the coefficient of kinetic friction between the puck and ice.

Solution *Conceptualize* the problem by imagining that the puck in Figure 5.20 slides to the right and eventually comes to rest. To *categorize* the problem, note that we have forces identified in Figure 5.20, but that kinematic variables are provided in the text of the problem. Thus, we must combine the techniques of Chapter 2 with those of this chapter. (We assume that the friction force is constant, which will result in a constant horizontal acceleration.) To *analyze* the situation, note that the forces acting on the puck after it is in motion are shown in Figure 5.20. First, we find the acceleration algebraically in terms of the coefficient of kinetic friction, using Newton's second law. Knowing the acceleration of the puck and the distance it travels, we can then use the equations of kinematics to find the numerical value of the coefficient of kinetic friction.

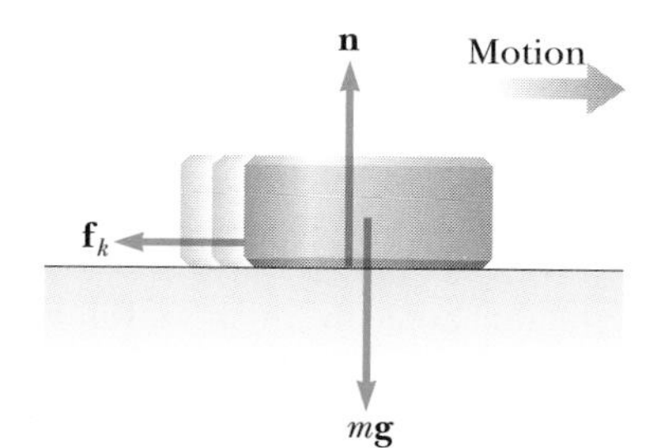

Figure 5.20 (Example 5.13) *After* the puck is given an initial velocity to the right, the only external forces acting on it are the gravitational force $m\mathbf{g}$, the normal force $\mathbf{n}$, and the force of kinetic friction $\mathbf{f}_k$.

Defining rightward and upward as our positive directions, we apply Newton's second law in component form to the puck and obtain

$$(1) \qquad \sum F_x = -f_k = ma_x$$

$$(2) \qquad \sum F_y = n - mg = 0 \qquad (a_y = 0)$$

But $f_k = \mu_k n$, and from (2) we see that $n = mg$. Therefore, (1) becomes

$$-\mu_k n = -\mu_k mg = ma_x$$

$$a_x = -\mu_k g$$

The negative sign means the acceleration is to the left in Figure 5.20; because the velocity of the puck is to the right, this means that the puck is slowing down. The acceleration is independent of the mass of the puck and is constant because we assume that μ_k remains constant.

Because the acceleration is constant, we can use Equation 2.13, $v_{xf}^2 = v_{xi}^2 + 2a_x(x_f - x_i)$, with $x_i = 0$ and $v_f = 0$:

$$0 = v_{xi}^2 + 2a_x x_f = v_{xi}^2 - 2\mu_k g x_f$$

$$\mu_k = \frac{v_{xi}^2}{2gx_f}$$

$$\mu_k = \frac{(20.0 \text{ m/s})^2}{2(9.80 \text{ m/s}^2)(115 \text{ m})} = 0.117$$

To *finalize* the problem, note that μ_k is dimensionless, as it should be, and that it has a low value, consistent with an object sliding on ice.

Example 5.14 Acceleration of Two Connected Objects When Friction Is Present

A block of mass m_1 on a rough, horizontal surface is connected to a ball of mass m_2 by a lightweight cord over a lightweight, frictionless pulley, as shown in Figure 5.21a. A force of magnitude F at an angle θ with the horizontal is applied to the block as shown. The coefficient of kinetic friction between the block and surface is μ_k. Determine the magnitude of the acceleration of the two objects.

Solution *Conceptualize* the problem by imagining what happens as $\mathbf{F}$ is applied to the block. Assuming that $\mathbf{F}$ is not large enough to lift the block, the block will slide to the right and the ball will rise. We can identify forces and we want an acceleration, so we *categorize* this as a Newton's second law problem, one that includes the friction force. To *analyze* the problem, we begin by drawing free-body diagrams for the two objects, as shown in Figures 5.21b and 5.21c. Next, we apply Newton's second law in component form to each object and use Equation 5.9, $f_k = \mu_k n$. Then we can solve for the acceleration in terms of the parameters given.

The applied force $\mathbf{F}$ has x and y components $F\cos\theta$ and $F\sin\theta$, respectively. Applying Newton's second law to both objects and assuming the motion of the block is to the right, we obtain

Motion of block: $(1) \quad \sum F_x = F\cos\theta - f_k - T = m_1 a_x = m_1 a$

$(2) \quad \sum F_y = n + F\sin\theta - m_1 g = m_1 a_y = 0$

Motion of ball: $\sum F_x = m_2 a_x = 0$

$(3) \quad \sum F_y = T - m_2 g = m_2 a_y = m_2 a$

Because the two objects are connected, we can equate the magnitudes of the x component of the acceleration of the block and the y component of the acceleration of the ball. From Equation 5.9 we know that $f_k = \mu_k n$, and from (2) we know that $n = m_1 g - F\sin\theta$ (in this case n is *not* equal to $m_1 g$); therefore,

$$(4) \qquad f_k = \mu_k(m_1 g - F\sin\theta)$$

That is, the friction force is reduced because of the positive y component of $\mathbf{F}$. Substituting (4) and the value of T from (3) into (1) gives

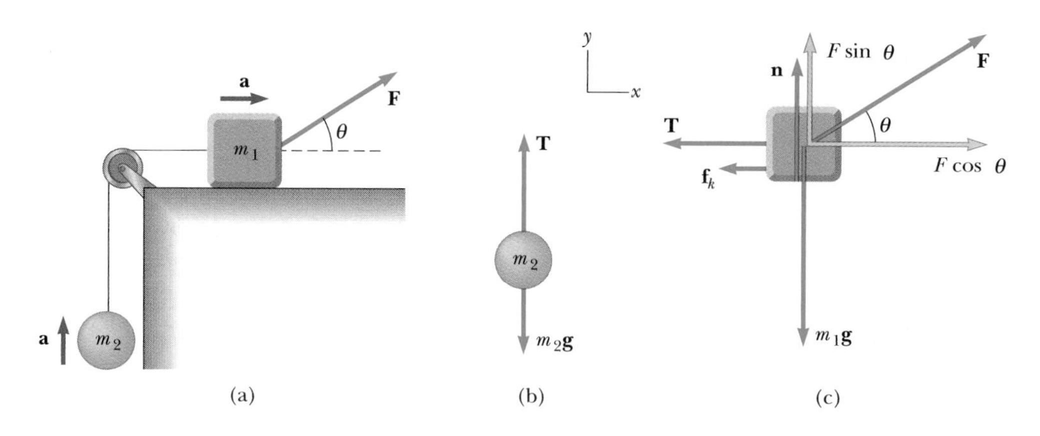

Figure 5.21 (Example 5.14) (a) The external force **F** applied as shown can cause the block to accelerate to the right. (b) and (c) The free-body diagrams assuming that the block accelerates to the right and the ball accelerates upward. The magnitude of the force of kinetic friction in this case is given by $f_k = \mu_k n = \mu_k(m_1 g - F\sin\theta)$.

$$F\cos\theta - \mu_k(m_1 g - F\sin\theta) - m_2(a + g) = m_1 a$$

Solving for a, we obtain

$$(5) \qquad a = \frac{F(\cos\theta + \mu_k \sin\theta) - g(m_2 + \mu_k m_1)}{m_1 + m_2}$$

To *finalize* the problem, note that the acceleration of the block can be either to the right or to the left, depending on the sign of the numerator in (5). If the motion is to the left, then we must reverse the sign of f_k in (1) because the force of kinetic friction must oppose the motion of the block relative to the surface. In this case, the value of a is the same as in (5), with the two plus signs in the numerator changed to minus signs.

This is the final chapter in which we will explicitly show the steps of the General Problem-Solving Strategy in all worked examples. We will refer to them explicitly in occasional examples in future chapters, but you should use the steps in *all* of your problem solving.

Application Automobile Antilock Braking Systems (ABS)

If an automobile tire is rolling and not slipping on a road surface, then the maximum friction force that the road can exert on the tire is the force of static friction $\mu_s n$. One must use static friction in this situation because at the point of contact between the tire and the road, no sliding of one surface over the other occurs if the tire is not skidding. However, if the tire starts to skid, the friction force exerted on it is reduced to the force of kinetic friction $\mu_k n$. Thus, to maximize the friction force and minimize stopping distance, the wheels must maintain pure rolling motion and not skid. An additional benefit of maintaining wheel rotation is that directional control is not lost as it is in skidding. Unfortunately, in emergency situations drivers typically press down as hard as they can on the brake pedal, "locking the brakes." This stops the wheels from rotating, ensuring a skid and reducing the friction force from the static to the kinetic value. To address this problem, automotive engineers have developed antilock braking systems (ABS). The purpose of the ABS is to help typical drivers (whose tendency is to lock the wheels in an emergency) to better maintain control of their automobiles and minimize stopping distance. The system briefly releases the brakes when a wheel is just about to stop turning. This maintains rolling contact between the tire and the pavement. When the brakes are released momentarily, the stopping distance is greater than it would be if the brakes were being applied continuously. However, through the use of computer control, the "brake-off" time is kept to a minimum. As a result, the stopping distance is much less than what it would be if the wheels were to skid.

Let us model the stopping of a car by examining real data. In an issue of *AutoWeek*,[5] the braking performance for a Toyota Corolla was measured. These data correspond to the braking force acquired by a highly trained, professional driver. We begin by assuming constant acceleration. (Why do we need to make this assumption?) The magazine provided the initial speed and stopping distance in non-SI units, which we show in the left and middle sections of Table 5.3. After converting these values to SI, we use $v_f^2 = v_i^2 + 2ax$ to determine the acceleration at different speeds, shown in the right section. These do not vary greatly, and so our assumption of constant acceleration is reasonable.

[5] *AutoWeek* magazine, **48:**22–23, 1998.

Table 5.3

Data for a Toyota Corolla:				
Initial Speed		Stopping Distance		Acceleration
(mi/h)	(m/s)	(ft)	(m)	(m/s^2)
30	13.4	34	10.4	−8.63
60	26.8	143	43.6	−8.24
80	35.8	251	76.5	−8.38

Table 5.4

Stopping Distance With and Without Skidding		
Initial Speed	Stopping Distance	
(mi/h)	no skid (m)	skidding (m)
30	10.4	13.9
60	43.6	55.5
80	76.5	98.9

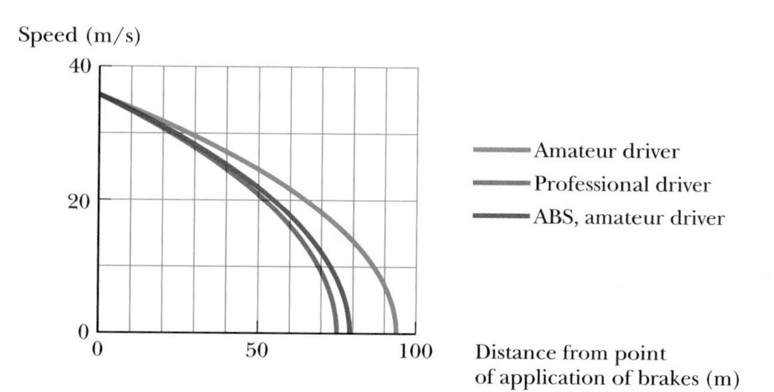

Figure 5.22 This plot of vehicle speed versus distance from the location at which the brakes were applied shows that an antilock braking system (ABS) approaches the performance of a trained professional driver.

We take an average value of acceleration of $-8.4\ m/s^2$, which is approximately $0.86g$. We then calculate the coefficient of friction from $\Sigma F = \mu_s mg = ma$, which gives $\mu_s = 0.86$ for the Toyota. This is lower than the rubber-on-concrete value given in Table 5.2. Can you think of any reasons for this?

We now estimate the stopping distance of the car if the wheels were skidding. From Table 5.2, we see that the difference between the coefficients of static and kinetic friction for rubber against concrete is about 0.2. Let us assume that our coefficients differ by the same amount, so that $\mu_k \approx 0.66$. This allows us to estimate the stopping distances when the wheels are locked and the car skids across the pavement, as shown in the third column of Table 5.4. The results illustrate the advantage of not allowing the wheels to skid.

Because an ABS keeps the wheels rotating, the higher coefficient of static friction is maintained between the tires and road. This approximates the technique of a professional driver who is able to maintain the wheels at the point of maximum friction force. Let us estimate the ABS performance by assuming that the magnitude of the acceleration is not quite as good as that achieved by the professional driver but instead is reduced by 5%.

Figure 5.22 is a plot of vehicle speed versus distance from where the brakes were applied (at an initial speed of 80.0 mi/h = 35.8 m/s) for the three cases of amateur driver, professional driver, and estimated ABS performance (amateur driver). This shows that a markedly shorter distance is necessary for stopping without locking the wheels compared to skidding. In addition a satisfactory value of stopping distance is achieved when the ABS computer maintains tire rotation.

SUMMARY

Take a practice test for this chapter by clicking the Practice Test link at http://www.pse6.com.

An **inertial frame of reference** is one we can identify in which an object that does not interact with other objects experiences zero acceleration. Any frame moving with constant velocity relative to an inertial frame is also an inertial frame. **Newton's first law** states that it is possible to find such a frame, or, equivalently, in the absence of an external force, when viewed from an inertial frame, an object at rest remains at rest and an object in uniform motion in a straight line maintains that motion.

Newton's second law states that the acceleration of an object is directly proportional to the net force acting on it and inversely proportional to its mass. The net force acting on an object equals the product of its mass and its acceleration: $\Sigma \mathbf{F} = m\mathbf{a}$. If the object is either stationary or moving with constant velocity, then the object is in equilibrium and the force vectors must cancel each other.

The **gravitational force** exerted on an object is equal to the product of its mass (a scalar quantity) and the free-fall acceleration: $\mathbf{F}_g = m\mathbf{g}$. The **weight** of an object is the magnitude of the gravitational force acting on the object.

Newton's third law states that if two objects interact, the force exerted by object 1 on object 2 is equal in magnitude and opposite in direction to the force exerted by object 2 on object 1. Thus, an isolated force cannot exist in nature.

The **maximum force of static friction $\mathbf{f}_{s,\text{max}}$** between an object and a surface is proportional to the normal force acting on the object. In general, $f_s \leq \mu_s n$, where μ_s is the **coefficient of static friction** and n is the magnitude of the normal force. When an object slides over a surface, the direction of the **force of kinetic friction $\mathbf{f}_k$** is opposite the direction of motion of the object relative to the surface and is also proportional to the magnitude of the normal force. The magnitude of this force is given by $f_k = \mu_k n$, where μ_k is the **coefficient of kinetic friction.**

QUESTIONS

1. A ball is held in a person's hand. (a) Identify all the external forces acting on the ball and the reaction to each. (b) If the ball is dropped, what force is exerted on it while it is falling? Identify the reaction force in this case. (Neglect air resistance.)

2. If a car is traveling westward with a constant speed of 20 m/s, what is the resultant force acting on it?

3. What is wrong with the statement "Because the car is at rest, there are no forces acting on it"? How would you correct this sentence?

4. In the motion picture *It Happened One Night* (Columbia Pictures, 1934), Clark Gable is standing inside a stationary bus in front of Claudette Colbert, who is seated. The bus suddenly starts moving forward and Clark falls into Claudette's lap. Why did this happen?

5. A passenger sitting in the rear of a bus claims that she was injured as the driver slammed on the brakes, causing a suitcase to come flying toward her from the front of the bus. If you were the judge in this case, what disposition would you make? Why?

6. A space explorer is moving through space far from any planet or star. She notices a large rock, taken as a specimen from an alien planet, floating around the cabin of the ship. Should she push it gently or kick it toward the storage compartment? Why?

7. A rubber ball is dropped onto the floor. What force causes the ball to bounce?

8. While a football is in flight, what forces act on it? What are the action–reaction pairs while the football is being kicked and while it is in flight?

9. The mayor of a city decides to fire some city employees because they will not remove the obvious sags from the cables that support the city traffic lights. If you were a lawyer, what defense would you give on behalf of the employees? Who do you think would win the case in court?

10. A weightlifter stands on a bathroom scale. He pumps a barbell up and down. What happens to the reading on the bathroom scale as this is done? **What if** he is strong enough to actually *throw* the barbell upward? How does the reading on the scale vary now?

11. Suppose a truck loaded with sand accelerates along a highway. If the driving force on the truck remains constant, what happens to the truck's acceleration if its trailer leaks sand at a constant rate through a hole in its bottom?

12. As a rocket is fired from a launching pad, its speed and acceleration increase with time as its engines continue to op-

erate. Explain why this occurs even though the thrust of the engines remains constant.

13. What force causes an automobile to move? A propeller-driven airplane? A rowboat?

14. Identify the action–reaction pairs in the following situations: a man takes a step; a snowball hits a girl in the back; a baseball player catches a ball; a gust of wind strikes a window.

15. In a contest of National Football League behemoths, teams from the Rams and the 49ers engage in a tug-of-war, pulling in opposite directions on a strong rope. The Rams exert a force of 9 200 N and they are winning, making the center of the rope move steadily toward themselves. Is it possible to know the tension in the rope from the information stated? Is it larger or smaller than 9 200 N? How hard are the 49ers pulling on the rope? Would it change your answer if the 49ers were winning or if the contest were even? The stronger team wins by exerting a larger force—on what? Explain your answers.

16. Twenty people participate in a tug-of-war. The two teams of ten people are so evenly matched that neither team wins. After the game they notice that a car is stuck in the mud. They attach the tug-of-war rope to the bumper of the car, and all the people pull on the rope. The heavy car has just moved a couple of decimeters when the rope breaks. Why did the rope break in this situation when it did not break when the same twenty people pulled on it in a tug-of-war?

17. "When the locomotive in Figure Q5.17 broke through the wall of the train station, the force exerted by the locomotive on the wall was greater than the force the wall could exert on the locomotive." Is this statement true or in need of correction? Explain your answer.

18. An athlete grips a light rope that passes over a low-friction pulley attached to the ceiling of a gym. A sack of sand precisely equal in weight to the athlete is tied to the other end of the rope. Both the sand and the athlete are initially at rest. The athlete climbs the rope, sometimes speeding up and slowing down as he does so. What happens to the sack of sand? Explain.

19. If the action and reaction forces are always equal in magnitude and opposite in direction to each other, then doesn't the net vector force on any object necessarily add up to zero? Explain your answer.

20. Can an object exert a force on itself? Argue for your answer.

21. If you push on a heavy box that is at rest, you must exert some force to start its motion. However, once the box is sliding, you can apply a smaller force to maintain that motion. Why?

Roger Viollet, Mill Valley, CA, University Science Books, 1982

Figure Q5.17

22. The driver of a speeding empty truck slams on the brakes and skids to a stop through a distance *d*. (a) If the truck carried a load that doubled its mass, what would be the truck's "skidding distance"? (b) If the initial speed of the truck were halved, what would be the truck's skidding distance?

23. Suppose you are driving a classic car. Why should you avoid slamming on your brakes when you want to stop in the shortest possible distance? (Many cars have antilock brakes that avoid this problem.)

24. A book is given a brief push to make it slide up a rough incline. It comes to a stop and slides back down to the starting point. Does it take the same time to go up as to come down? **What if** the incline is frictionless?

25. A large crate is placed on the bed of a truck but not tied down. (a) As the truck accelerates forward, the crate remains at rest relative to the truck. What force causes the crate to accelerate forward? (b) If the driver slammed on the brakes, what could happen to the crate?

26. Describe a few examples in which the force of friction exerted on an object is in the direction of motion of the object.

PROBLEMS

1, 2, 3 = straightforward, intermediate, challenging □ = full solution available in the *Student Solutions Manual and Study Guide*

= coached solution with hints available at http://www.pse6.com = computer useful in solving problem

= paired numerical and symbolic problems

Sections 5.1 through 5.6

1. A force **F** applied to an object of mass m_1 produces an acceleration of 3.00 m/s^2. The same force applied to a second object of mass m_2 produces an acceleration of 1.00 m/s^2. (a) What is the value of the ratio m_1/m_2? (b) If m_1 and m_2 are combined, find their acceleration under the action of the force **F**.

2. The largest-caliber antiaircraft gun operated by the German air force during World War II was the 12.8-cm Flak 40. This weapon fired a 25.8-kg shell with a muzzle speed of 880 m/s. What propulsive force was necessary to attain the muzzle speed within the 6.00-m barrel? (Assume the shell moves horizontally with constant acceleration and neglect friction.)

3. A 3.00-kg object undergoes an acceleration given by $\mathbf{a} = (2.00\hat{\mathbf{i}} + 5.00\hat{\mathbf{j}})$ m/s^2. Find the resultant force acting on it and the magnitude of the resultant force.

4. The gravitational force on a baseball is $-F_g\hat{\mathbf{j}}$. A pitcher throws the baseball with velocity $v\hat{\mathbf{i}}$ by uniformly accelerating it straight forward horizontally for a time interval $\Delta t = t - 0 = t$. If the ball starts from rest, (a) through what distance does it accelerate before its release? (b) What force does the pitcher exert on the ball?

5. To model a spacecraft, a toy rocket engine is securely fastened to a large puck, which can glide with negligible friction over a horizontal surface, taken as the xy plane. The 4.00-kg puck has a velocity of $300\hat{\mathbf{i}}$ m/s at one instant. Eight seconds later, its velocity is to be $(800\hat{\mathbf{i}} + 10.0\hat{\mathbf{j}})$ m/s. Assuming the rocket engine exerts a constant horizontal force, find (a) the components of the force and (b) its magnitude.

6. The average speed of a nitrogen molecule in air is about 6.70×10^2 m/s, and its mass is 4.68×10^{-26} kg. (a) If it takes 3.00×10^{-13} s for a nitrogen molecule to hit a wall and rebound with the same speed but moving in the opposite direction, what is the average acceleration of the molecule during this time interval? (b) What average force does the molecule exert on the wall?

7. An electron of mass 9.11×10^{-31} kg has an initial speed of 3.00×10^5 m/s. It travels in a straight line, and its speed increases to 7.00×10^5 m/s in a distance of 5.00 cm. Assuming its acceleration is constant, (a) determine the force exerted on the electron and (b) compare this force with the weight of the electron, which we neglected.

8. A woman weighs 120 lb. Determine (a) her weight in newtons (N) and (b) her mass in kilograms (kg).

9. If a man weighs 900 N on the Earth, what would he weigh on Jupiter, where the acceleration due to gravity is 25.9 m/s^2?

10. The distinction between mass and weight was discovered after Jean Richer transported pendulum clocks from Paris to French Guyana in 1671. He found that they ran slower there quite systematically. The effect was reversed when the clocks returned to Paris. How much weight would you personally lose in traveling from Paris, where $g = 9.809\,5$ m/s^2, to Cayenne, where $g = 9.780\,8$ m/s^2? [We will consider how the free-fall acceleration influences the period of a pendulum in Section 15.5.]

11. Two forces $\mathbf{F}_1$ and $\mathbf{F}_2$ act on a 5.00-kg object. If $F_1 = 20.0$ N and $F_2 = 15.0$ N, find the accelerations in (a) and (b) of Figure P5.11.

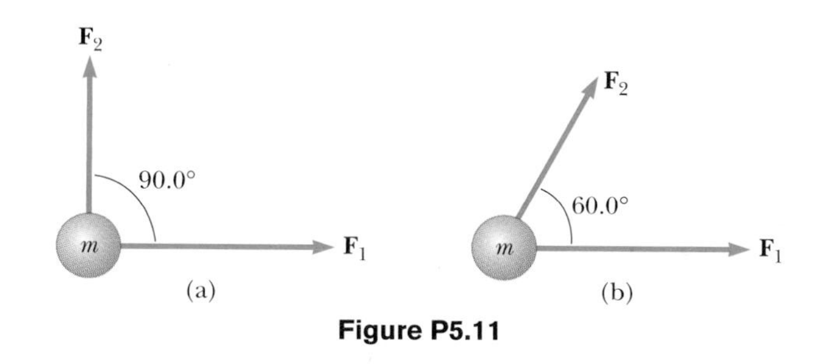

Figure P5.11

12. Besides its weight, a 2.80-kg object is subjected to one other constant force. The object starts from rest and in 1.20 s experiences a displacement of $(4.20\hat{\mathbf{i}} - 3.30\hat{\mathbf{j}})$ m, where the direction of $\hat{\mathbf{j}}$ is the upward vertical direction. Determine the other force.

13. You stand on the seat of a chair and then hop off. (a) During the time you are in flight down to the floor, the Earth is lurching up toward you with an acceleration of what order of magnitude? In your solution explain your logic. Model the Earth as a perfectly solid object. (b) The Earth moves up through a distance of what order of magnitude?

14. Three forces acting on an object are given by $\mathbf{F}_1 = (-2.00\hat{\mathbf{i}} + 2.00\hat{\mathbf{j}})$ N, $\mathbf{F}_2 = (5.00\hat{\mathbf{i}} - 3.00\hat{\mathbf{j}})$ N, and $\mathbf{F}_3 = (-45.0\hat{\mathbf{i}})$ N. The object experiences an acceleration of magnitude 3.75 m/s^2. (a) What is the direction of the acceleration? (b) What is the mass of the object? (c) If the object is initially at rest, what is its speed after 10.0 s? (d) What are the velocity components of the object after 10.0 s?

15. A 15.0-lb block rests on the floor. (a) What force does the floor exert on the block? (b) If a rope is tied to the block and run vertically over a pulley, and the other end is attached to a free-hanging 10.0-lb weight, what is the force exerted by the floor on the 15.0-lb block? (c) If we replace the 10.0-lb weight in part (b) with a 20.0-lb weight, what is the force exerted by the floor on the 15.0-lb block?

Section 5.7 Some Applications of Newton's Laws

16. A 3.00-kg object is moving in a plane, with its x and y coordinates given by $x = 5t^2 - 1$ and $y = 3t^3 + 2$, where x and y are in meters and t is in seconds. Find the magnitude of the net force acting on this object at $t = 2.00$ s.

17. The distance between two telephone poles is 50.0 m. When a 1.00-kg bird lands on the telephone wire midway between the poles, the wire sags 0.200 m. Draw a free-body diagram of the bird. How much tension does the bird produce in the wire? Ignore the weight of the wire.

18. A bag of cement of weight 325 N hangs from three wires as suggested in Figure P5.18. Two of the wires make angles $\theta_1 = 60.0°$ and $\theta_2 = 25.0°$ with the horizontal. If the system is in equilibrium, find the tensions T_1, T_2, and T_3 in the wires.

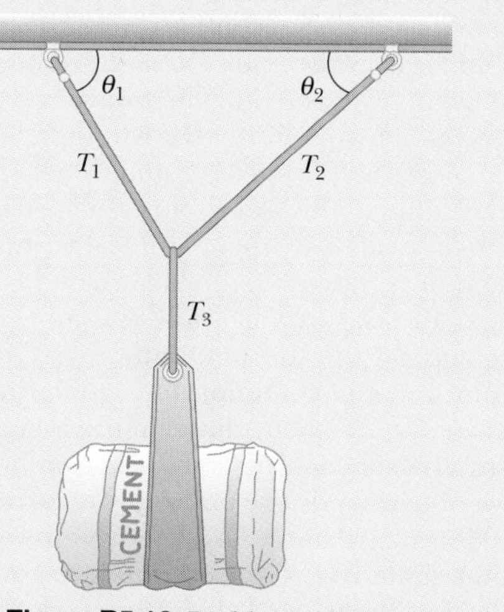

Figure P5.18 Problems 18 and 19.

19. A bag of cement of weight F_g hangs from three wires as shown in Figure P5.18. Two of the wires make angles θ_1 and θ_2 with the horizontal. If the system is in equilibrium, show that the tension in the left-hand wire is

$$T_1 = F_g \cos \theta_2 / \sin (\theta_1 + \theta_2)$$

20. You are a judge in a children's kite-flying contest, and two children will win prizes for the kites that pull most strongly and least strongly on their strings. To measure string tensions, you borrow a weight hanger, some slotted weights, and a protractor from your physics teacher, and use the following protocol, illustrated in Figure P5.20: Wait for a child to get her kite well controlled, hook the hanger onto the kite string about 30 cm from her hand, pile on weight until that section of string is horizontal, record the mass required, and record the angle between the horizontal and the string running up to the kite. (a) Explain how this method works. As you construct your explanation, imagine that the children's parents ask you about your method, that they might make false assumptions about your ability without concrete evidence, and that your explanation is an opportunity to give them confidence in your evaluation technique. (b) Find the string tension if the mass is 132 g and the angle of the kite string is 46.3°.

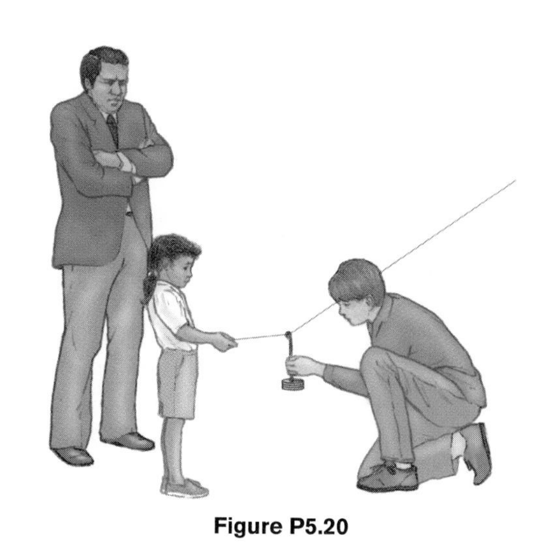

Figure P5.20

21. The systems shown in Figure P5.21 are in equilibrium. If the spring scales are calibrated in newtons, what do they read? (Neglect the masses of the pulleys and strings, and assume the incline in part (c) is frictionless.)

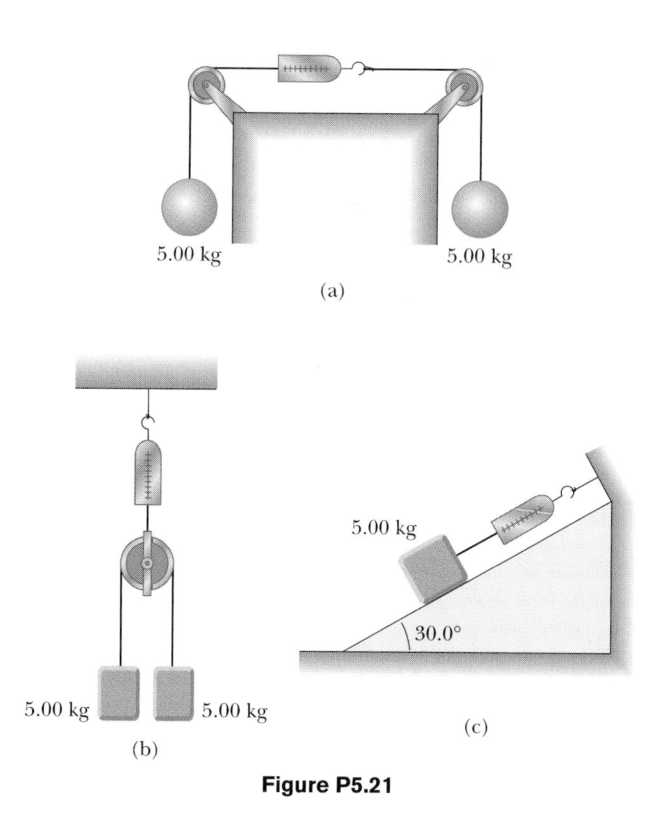

Figure P5.21

22. Draw a free-body diagram of a block which slides down a frictionless plane having an inclination of $\theta = 15.0°$ (Fig. P5.22). The block starts from rest at the top and the length of the incline is 2.00 m. Find (a) the acceleration of

the block and (b) its speed when it reaches the bottom of the incline.

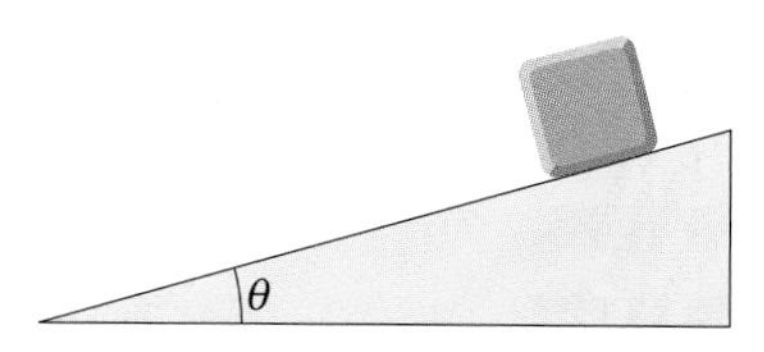

Figure P5.22 Problems 22 and 25.

23. A 1.00-kg object is observed to have an acceleration of 10.0 m/s^2 in a direction 30.0° north of east (Fig. P5.23). The force $\mathbf{F}_2$ acting on the object has a magnitude of 5.00 N and is directed north. Determine the magnitude and direction of the force $\mathbf{F}_1$ acting on the object.

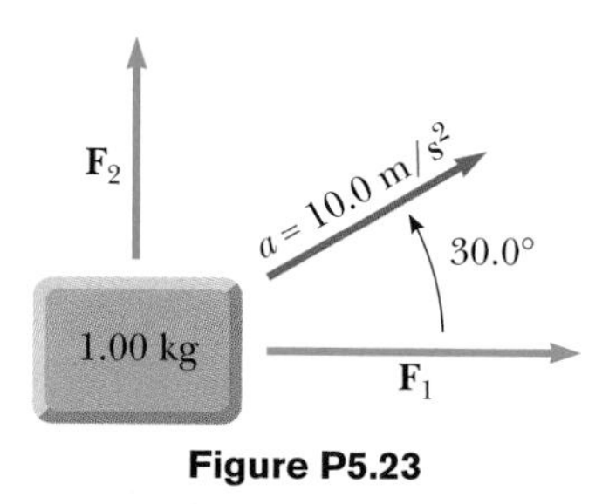

Figure P5.23

24. A 5.00-kg object placed on a frictionless, horizontal table is connected to a string that passes over a pulley and then is fastened to a hanging 9.00-kg object, as in Figure P5.24. Draw free-body diagrams of both objects. Find the acceleration of the two objects and the tension in the string.

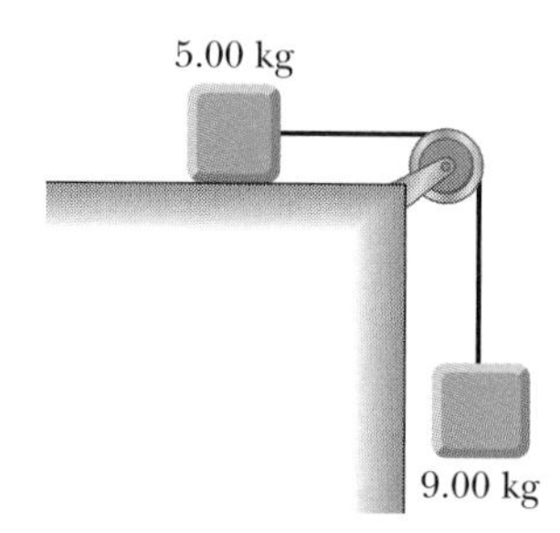

Figure P5.24 Problems 24 and 43.

25. A block is given an initial velocity of 5.00 m/s up a frictionless 20.0° incline (Fig. P5.22). How far up the incline does the block slide before coming to rest?

26. Two objects are connected by a light string that passes over a frictionless pulley, as in Figure P5.26. Draw free-body diagrams of both objects. If the incline is frictionless and if $m_1 = 2.00$ kg, $m_2 = 6.00$ kg, and $\theta = 55.0°$, find (a) the accelerations of the objects, (b) the tension in the string, and (c) the speed of each object 2.00 s after being released from rest.

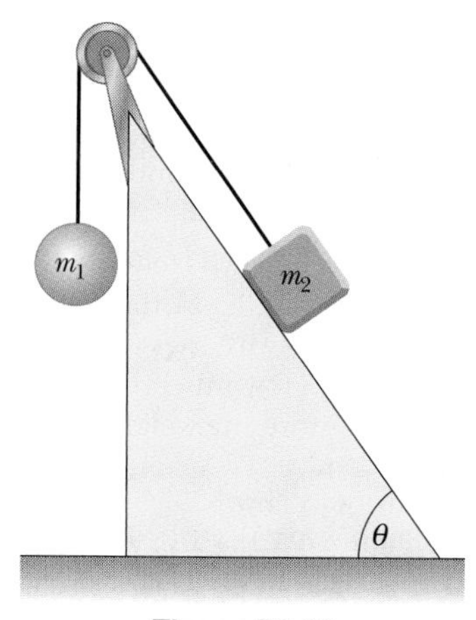

Figure P5.26

27. A tow truck pulls a car that is stuck in the mud, with a force of 2 500 N as in Fig. P5.27. The tow cable is under tension and therefore pulls downward and to the left on the pin at its upper end. The light pin is held in equilibrium by forces exerted by the two bars A and B. Each bar is a *strut*: that is, each is a bar whose weight is small compared to the forces it exerts, and which exerts forces only through hinge pins at its ends. Each strut exerts a force directed parallel to its length. Determine the force of tension or compression in each strut. Proceed as follows: Make a guess as to which way (pushing or pulling) each force acts on the top pin. Draw a free-body diagram of the pin. Use the condition for equilibrium of the pin to translate the free-body diagram into equations. From the equations calculate the forces exerted by struts A and B. If you obtain a positive answer, you correctly guessed the direction of the force. A negative answer means the direction should be reversed, but the absolute value correctly gives the magnitude of the force. If a strut pulls on a pin, it is in tension. If it pushes, the strut is in compression. Identify whether each strut is in tension or in compression.

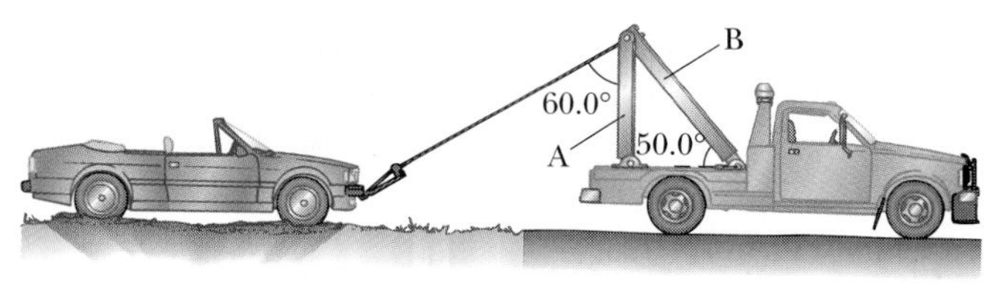

Figure P5.27

28. Two objects with masses of 3.00 kg and 5.00 kg are connected by a light string that passes over a light frictionless pulley to form an Atwood machine, as in Figure 5.14a. Determine (a) the tension in the string, (b) the acceleration of each object, and (c) the distance each object will move in the first second of motion if they start from rest.

29. In Figure P5.29, the man and the platform together weigh 950 N. The pulley can be modeled as frictionless. Determine how hard the man has to pull on the rope to lift himself steadily upward above the ground. (Or is it impossible? If so, explain why.)

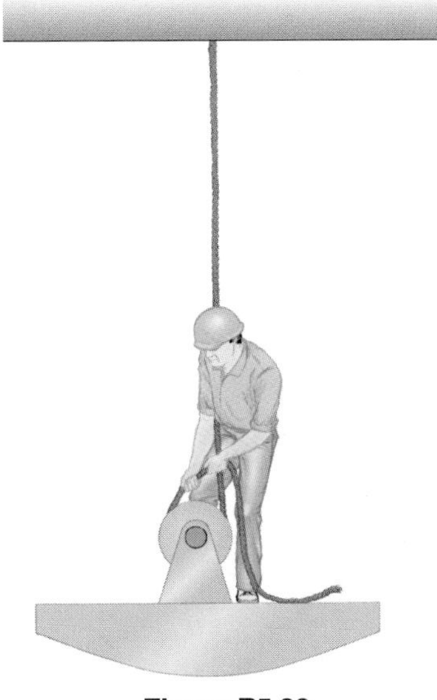

Figure P5.29

30. In the Atwood machine shown in Figure 5.14a, $m_1 = 2.00$ kg and $m_2 = 7.00$ kg. The masses of the pulley and string are negligible by comparison. The pulley turns without friction and the string does not stretch. The lighter object is released with a sharp push that sets it into motion at $v_i = 2.40$ m/s downward. (a) How far will m_1 descend below its initial level? (b) Find the velocity of m_1 after 1.80 seconds.

31. In the system shown in Figure P5.31, a horizontal force $\mathbf{F}_x$ acts on the 8.00-kg object. The horizontal surface is frictionless. (a) For what values of F_x does the 2.00-kg object accelerate upward? (b) For what values of F_x is the tension in the cord zero? (c) Plot the acceleration of the 8.00-kg object versus F_x. Include values of F_x from -100 N to $+100$ N.

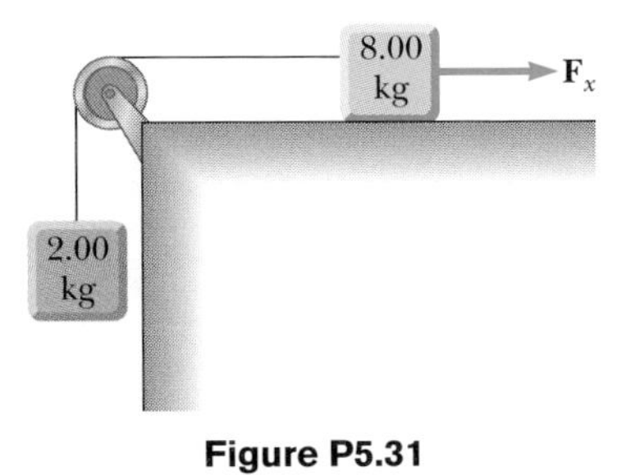

Figure P5.31

32. A frictionless plane is 10.0 m long and inclined at 35.0°. A sled starts at the bottom with an initial speed of 5.00 m/s up the incline. When it reaches the point at which it momentarily stops, a second sled is released from the top of this incline with an initial speed v_i. Both sleds reach the bottom of the incline at the same moment. (a) Determine the distance that the first sled traveled up the incline. (b) Determine the initial speed of the second sled.

33. A 72.0-kg man stands on a spring scale in an elevator. Starting from rest, the elevator ascends, attaining its maximum speed of 1.20 m/s in 0.800 s. It travels with this constant speed for the next 5.00 s. The elevator then undergoes a uniform acceleration in the negative y direction for 1.50 s and comes to rest. What does the spring scale register (a) before the elevator starts to move? (b) during the first 0.800 s? (c) while the elevator is traveling at constant speed? (d) during the time it is slowing down?

34. An object of mass m_1 on a frictionless horizontal table is connected to an object of mass m_2 through a very light pulley P_1 and a light fixed pulley P_2 as shown in Figure P5.34. (a) If a_1 and a_2 are the accelerations of m_1 and m_2, respectively, what is the relation between these accelerations? Express (b) the tensions in the strings and (c) the accelerations a_1 and a_2 in terms of the masses m_1 and m_2, and g.

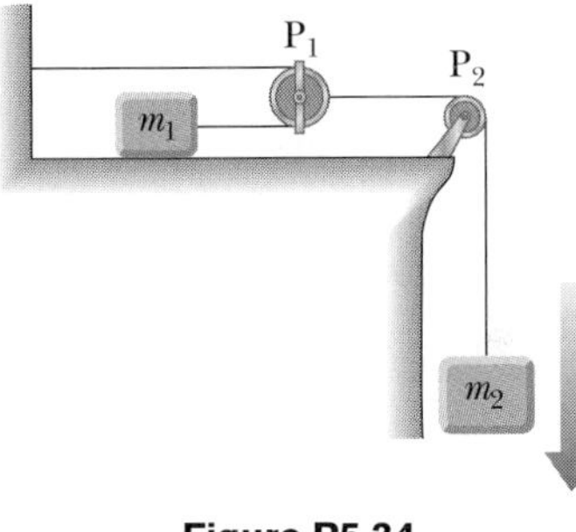

Figure P5.34

Section 5.8 Forces of Friction

35. The person in Figure P5.35 weighs 170 lb. As seen from the front, each light crutch makes an angle of 22.0° with the vertical. Half of the person's weight is supported by the crutches. The other half is supported by the vertical forces of the ground on his feet. Assuming the person is moving

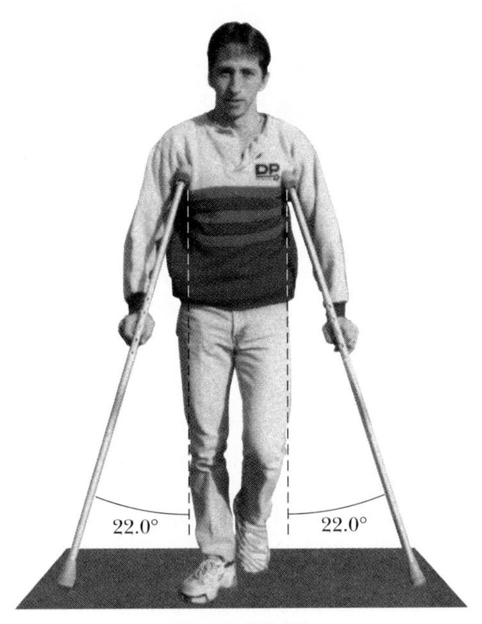

Figure P5.35

with constant velocity and the force exerted by the ground on the crutches acts along the crutches, determine (a) the smallest possible coefficient of friction between crutches and ground and (b) the magnitude of the compression force in each crutch.

36. A 25.0-kg block is initially at rest on a horizontal surface. A horizontal force of 75.0 N is required to set the block in motion. After it is in motion, a horizontal force of 60.0 N is required to keep the block moving with constant speed. Find the coefficients of static and kinetic friction from this information.

37. A car is traveling at 50.0 mi/h on a horizontal highway. (a) If the coefficient of static friction between road and tires on a rainy day is 0.100, what is the minimum distance in which the car will stop? (b) What is the stopping distance when the surface is dry and $\mu_s = 0.600$?

38. Before 1960 it was believed that the maximum attainable coefficient of static friction for an automobile tire was less than 1. Then, about 1962, three companies independently developed racing tires with coefficients of 1.6. Since then, tires have improved, as illustrated in this problem. According to the 1990 Guinness Book of Records, the shortest time in which a piston-engine car initially at rest has covered a distance of one-quarter mile is 4.96 s. This record was set by Shirley Muldowney in September 1989. (a) Assume that, as in Figure P5.38, the rear wheels lifted the front wheels off the pavement. What minimum value of μ_s is necessary to achieve the record time? (b) Suppose Muldowney were able to double her engine power, keeping other things equal. How would this change affect the elapsed time?

Mike Powell/Allsport USA/Getty

Figure P5.38

39. To meet a U.S. Postal Service requirement, footwear must have a coefficient of static friction of 0.5 or more on a specified tile surface. A typical athletic shoe has a coefficient of 0.800. In an emergency, what is the minimum time interval in which a person starting from rest can move 3.00 m on a tile surface if she is wearing (a) footwear meeting the Postal Service minimum? (b) a typical athletic shoe?

40. A woman at an airport is towing her 20.0-kg suitcase at constant speed by pulling on a strap at an angle θ above the horizontal (Fig. P5.40). She pulls on the strap with a 35.0-N force, and the friction force on the suitcase is 20.0 N. Draw a free-body diagram of the suitcase. (a) What angle does the strap make with the horizontal? (b) What normal force does the ground exert on the suitcase?

Figure P5.40

41. A 3.00-kg block starts from rest at the top of a 30.0° incline and slides a distance of 2.00 m down the incline in 1.50 s. Find (a) the magnitude of the acceleration of the block, (b) the coefficient of kinetic friction between block and plane, (c) the friction force acting on the block, and (d) the speed of the block after it has slid 2.00 m.

42. A Chevrolet Corvette convertible can brake to a stop from a speed of 60.0 mi/h in a distance of 123 ft on a level roadway. What is its stopping distance on a roadway sloping downward at an angle of 10.0°?

43. A 9.00-kg hanging weight is connected by a string over a pulley to a 5.00-kg block that is sliding on a flat table (Fig. P5.24). If the coefficient of kinetic friction is 0.200, find the tension in the string.

44. Three objects are connected on the table as shown in Figure P5.44. The table is rough and has a coefficient of kinetic friction of 0.350. The objects have masses of 4.00 kg, 1.00 kg, and 2.00 kg, as shown, and the pulleys are frictionless. Draw free-body diagrams of each of the objects. (a) Determine the acceleration of each object and their directions. (b) Determine the tensions in the two cords.

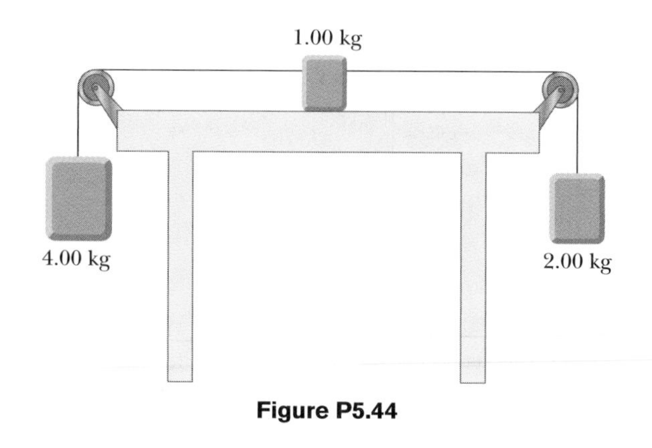

Figure P5.44

45. Two blocks connected by a rope of negligible mass are being dragged by a horizontal force **F** (Fig. P5.45). Suppose that $F = 68.0$ N, $m_1 = 12.0$ kg, $m_2 = 18.0$ kg, and the coefficient of kinetic friction between each block and the surface is 0.100. (a) Draw a free-body diagram for each block.

(b) Determine the tension T and the magnitude of the acceleration of the system.

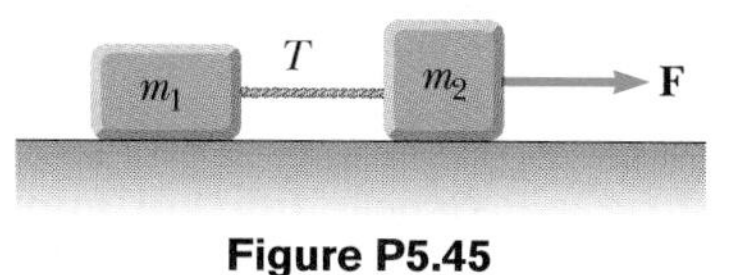

Figure P5.45

46. A block of mass 3.00 kg is pushed up against a wall by a force **P** that makes a 50.0° angle with the horizontal as shown in Figure P5.46. The coefficient of static friction between the block and the wall is 0.250. Determine the possible values for the magnitude of **P** that allow the block to remain stationary.

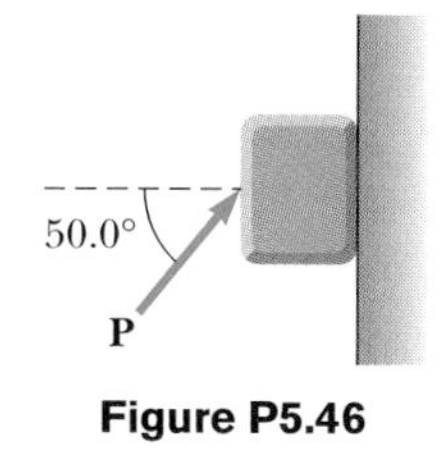

Figure P5.46

47. You and your friend go sledding. Out of curiosity, you measure the constant angle θ that the snow-covered slope makes with the horizontal. Next, you use the following method to determine the coefficient of friction μ_k between the snow and the sled. You give the sled a quick push up so that it will slide up the slope away from you. You wait for it to slide back down, timing the motion. It turns out that the sled takes twice as long to slide down as it does to reach the top point in the round trip. In terms of θ, what is the coefficient of friction?

48. The board sandwiched between two other boards in Figure P5.48 weighs 95.5 N. If the coefficient of friction between the boards is 0.663, what must be the magnitude of the compression forces (assume horizontal) acting on both sides of the center board to keep it from slipping?

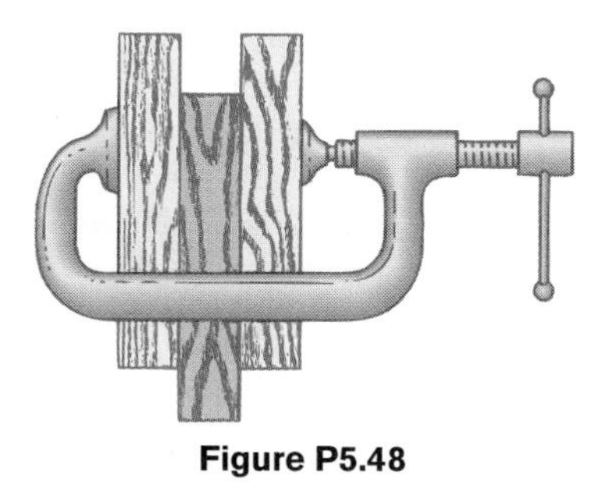

Figure P5.48

49. A block weighing 75.0 N rests on a plane inclined at 25.0° to the horizontal. A force F is applied to the object at 40.0° to the horizontal, pushing it upward on the plane. The coefficients of static and kinetic friction between the block and the plane are, respectively, 0.363 and 0.156. (a) What is the minimum value of F that will prevent the block from slipping down the plane? (b) What is the minimum value of F that will start the block moving up the plane? (c) What value of F will move the block up the plane with constant velocity?

50. **Review problem.** One side of the roof of a building slopes up at 37.0°. A student throws a Frisbee onto the roof. It strikes with a speed of 15.0 m/s and does not bounce, but slides straight up the incline. The coefficient of kinetic friction between the plastic and the roof is 0.400. The Frisbee slides 10.0 m up the roof to its peak, where it goes into free fall, following a parabolic trajectory with negligible air resistance. Determine the maximum height the Frisbee reaches above the point where it struck the roof.

Additional Problems

51. An inventive child named Pat wants to reach an apple in a tree without climbing the tree. Sitting in a chair connected to a rope that passes over a frictionless pulley (Fig. P5.51), Pat pulls on the loose end of the rope with such a force that the spring scale reads 250 N. Pat's true weight is 320 N, and the chair weighs 160 N. (a) Draw free-body diagrams for Pat and the chair considered as separate systems, and another diagram for Pat and the chair considered as one system. (b) Show that the acceleration of the system is *upward* and find its magnitude. (c) Find the force Pat exerts on the chair.

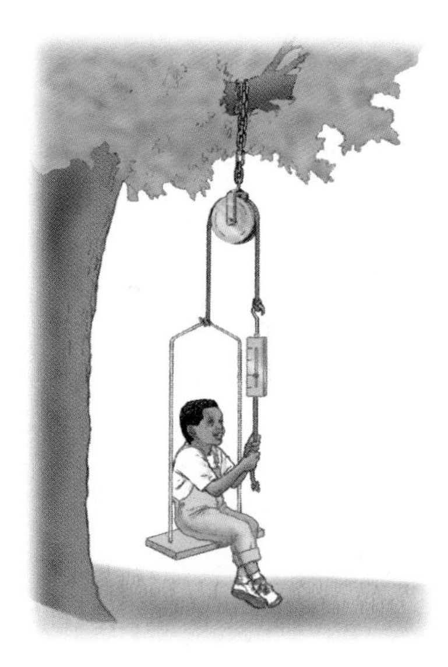

Figure P5.51

52. A time-dependent force, $\mathbf{F} = (8.00\hat{\mathbf{i}} - 4.00t\hat{\mathbf{j}})$ N, where t is in seconds, is exerted on a 2.00-kg object initially at rest. (a) At what time will the object be moving with a speed of 15.0 m/s? (b) How far is the object from its initial position when its speed is 15.0 m/s? (c) Through what total displacement has the object traveled at this time?

53. To prevent a box from sliding down an inclined plane, student A pushes on the box in the direction parallel to the incline, just hard enough to hold the box stationary. In an identical situation student B pushes on the box horizontally. Regard as known the mass m of the box, the coefficient of static friction μ_s between box and incline, and the inclination angle θ. (a) Determine the force A

has to exert. (b) Determine the force B has to exert. (c) If $m = 2.00$ kg, $\theta = 25.0°$, and $\mu_s = 0.160$, who has the easier job? (d) **What if** $\mu_s = 0.380$? Whose job is easier?

54. Three blocks are in contact with each other on a frictionless, horizontal surface, as in Figure P5.54. A horizontal force **F** is applied to m_1. Take $m_1 = 2.00$ kg, $m_2 = 3.00$ kg, $m_3 = 4.00$ kg, and $F = 18.0$ N. Draw a separate free-body diagram for each block and find (a) the acceleration of the blocks, (b) the *resultant* force on each block, and (c) the magnitudes of the contact forces between the blocks. (d) You are working on a construction project. A coworker is nailing up plasterboard on one side of a light partition, and you are on the opposite side, providing "backing" by leaning against the wall with your back pushing on it. Every blow makes your back sting. The supervisor helps you to put a heavy block of wood between the wall and your back. Using the situation analyzed in parts (a), (b), and (c) as a model, explain how this works to make your job more comfortable.

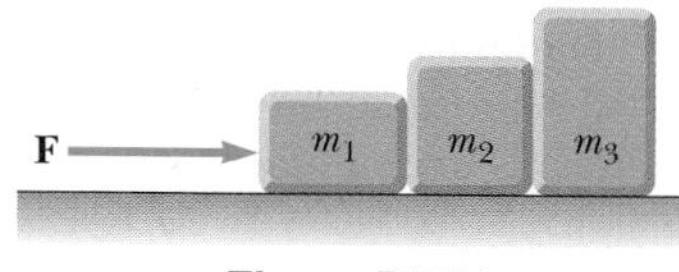

Figure P5.54

55. An object of mass M is held in place by an applied force **F** and a pulley system as shown in Figure P5.55. The pulleys are massless and frictionless. Find (a) the tension in each section of rope, T_1, T_2, T_3, T_4, and T_5 and (b) the magnitude of **F**. *Suggestion:* Draw a free-body diagram for each pulley.

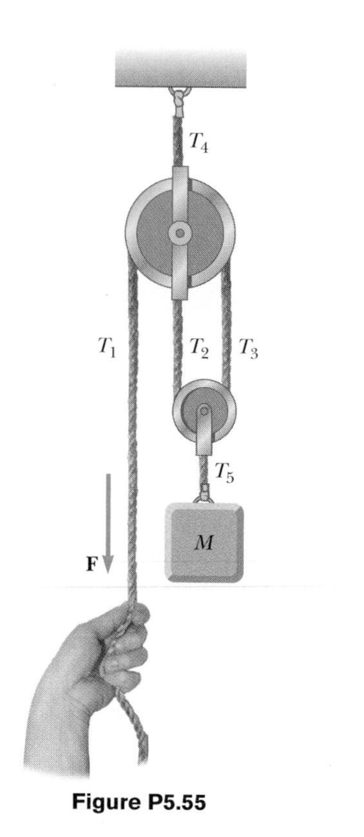

Figure P5.55

56. A high diver of mass 70.0 kg jumps off a board 10.0 m above the water. If his downward motion is stopped 2.00 s after he enters the water, what average upward force did the water exert on him?

57. A crate of weight F_g is pushed by a force **P** on a horizontal floor. (a) If the coefficient of static friction is μ_s and **P** is directed at angle θ below the horizontal, show that the minimum value of P that will move the crate is given by

$$P = \frac{\mu_s F_g \sec\theta}{1 - \mu_s \tan\theta}$$

(b) Find the minimum value of P that can produce motion when $\mu_s = 0.400$, $F_g = 100$ N, and $\theta = 0°$, 15.0°, 30.0°, 45.0°, and 60.0°.

58. **Review problem.** A block of mass $m = 2.00$ kg is released from rest at $h = 0.500$ m above the surface of a table, at the top of a $\theta = 30.0°$ incline as shown in Figure P5.58. The frictionless incline is fixed on a table of height $H = 2.00$ m. (a) Determine the acceleration of the block as it slides down the incline. (b) What is the velocity of the block as it leaves the incline? (c) How far from the table will the block hit the floor? (d) How much time has elapsed between when the block is released and when it hits the floor? (e) Does the mass of the block affect any of the above calculations?

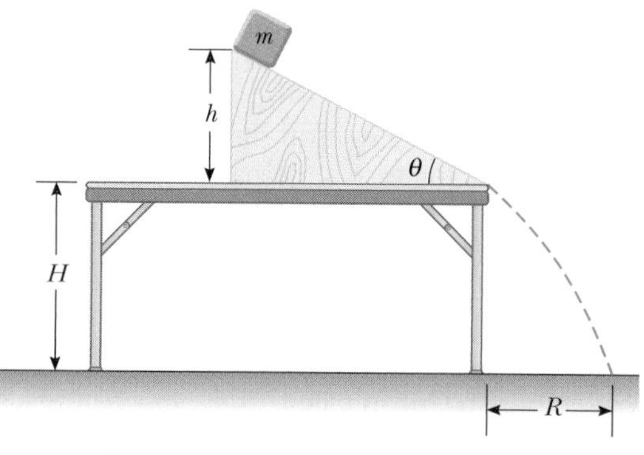

Figure P5.58 Problems 58 and 70.

59. A 1.30-kg toaster is not plugged in. The coefficient of static friction between the toaster and a horizontal countertop is 0.350. To make the toaster start moving, you carelessly pull on its electric cord. (a) For the cord tension to be as small as possible, you should pull at what angle above the horizontal? (b) With this angle, how large must the tension be?

60. Materials such as automobile tire rubber and shoe soles are tested for coefficients of static friction with an apparatus called a James tester. The pair of surfaces for which μ_s is to be measured are labeled B and C in Figure P5.60. Sample C is attached to a foot D at the lower end of a pivoting arm E, which makes angle θ with the vertical. The upper end of the arm is hinged at F to a vertical rod G, which slides freely in a guide H fixed to the frame of the apparatus and supports a load I of mass 36.4 kg. The hinge pin at F is also the axle of a wheel that can roll vertically on the frame. All of the moving parts have masses negligible in comparison to the 36.4-kg load. The pivots are nearly frictionless. The test surface B is attached to a

rolling platform A. The operator slowly moves the platform to the left in the picture until the sample C suddenly slips over surface B. At the critical point where sliding motion is ready to begin, the operator notes the angle θ_s of the pivoting arm. (a) Make a free-body diagram of the pin at F. It is in equilibrium under three forces. These forces are the gravitational force on the load I, a horizontal normal force exerted by the frame, and a force of compression directed upward along the arm E. (b) Draw a free-body diagram of the foot D and sample C, considered as one system. (c) Determine the normal force that the test surface B exerts on the sample for any angle θ. (d) Show that $\mu_s = \tan \theta_s$. (e) The protractor on the tester can record angles as large as 50.2°. What is the greatest coefficient of friction it can measure?

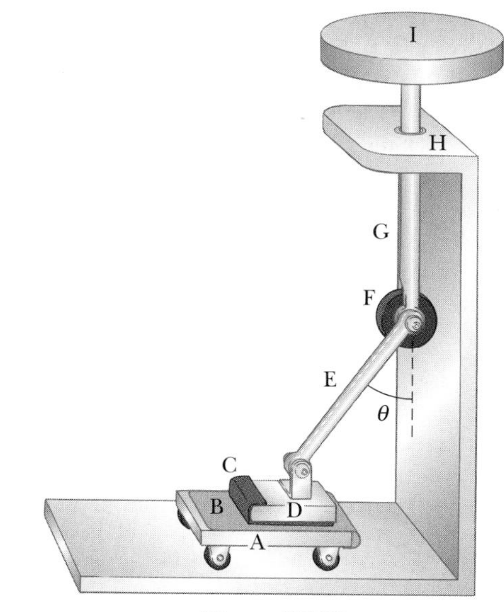

Figure P5.60

61. What horizontal force must be applied to the cart shown in Figure P5.61 in order that the blocks remain stationary relative to the cart? Assume all surfaces, wheels, and pulley are frictionless. (*Hint:* Note that the force exerted by the string accelerates m_1.)

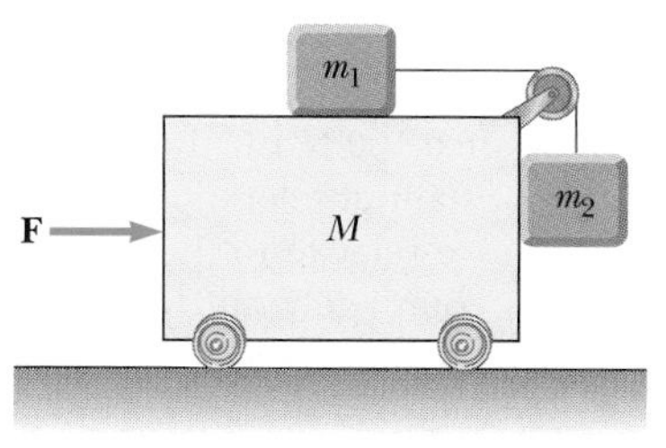

Figure P5.61 Problems 61 and 63.

62. A student is asked to measure the acceleration of a cart on a "frictionless" inclined plane as in Figure 5.11, using an air track, a stopwatch, and a meter stick. The height of the incline is measured to be 1.774 cm, and the total length of the incline is measured to be $d = 127.1$ cm. Hence, the angle of inclination θ is determined from the relation $\sin \theta = 1.774/127.1$. The cart is released from rest at the top of the incline, and its position x along the incline is measured as a function of time, where $x = 0$ refers to the initial position of the cart. For x values of 10.0 cm, 20.0 cm, 35.0 cm, 50.0 cm, 75.0 cm, and 100 cm, the measured times at which these positions are reached (averaged over five runs) are 1.02 s, 1.53 s, 2.01 s, 2.64 s, 3.30 s, and 3.75 s, respectively. Construct a graph of x versus t^2, and perform a linear least-squares fit to the data. Determine the acceleration of the cart from the slope of this graph, and compare it with the value you would get using $a' = g \sin \theta$, where $g = 9.80 \text{ m/s}^2$.

63. Initially the system of objects shown in Figure P5.61 is held motionless. All surfaces, pulley, and wheels are frictionless. Let the force **F** be zero and assume that m_2 can move only vertically. At the instant after the system of objects is released, find (a) the tension T in the string, (b) the acceleration of m_2, (c) the acceleration of M, and (d) the acceleration of m_1. (*Note:* The pulley accelerates along with the cart.)

64. One block of mass 5.00 kg sits on top of a second rectangular block of mass 15.0 kg, which in turn is on a horizontal table. The coefficients of friction between the two blocks are $\mu_s = 0.300$ and $\mu_k = 0.100$. The coefficients of friction between the lower block and the rough table are $\mu_s = 0.500$ and $\mu_k = 0.400$. You apply a constant horizontal force to the lower block, just large enough to make this block start sliding out from between the upper block and the table. (a) Draw a free-body diagram of each block, naming the forces on each. (b) Determine the magnitude of each force on each block at the instant when you have started pushing but motion has not yet started. In particular, what force must you apply? (c) Determine the acceleration you measure for each block.

65. A 1.00-kg glider on a horizontal air track is pulled by a string at an angle θ. The taut string runs over a pulley and is attached to a hanging object of mass 0.500 kg as in Fig. P5.65. (a) Show that the speed v_x of the glider and the

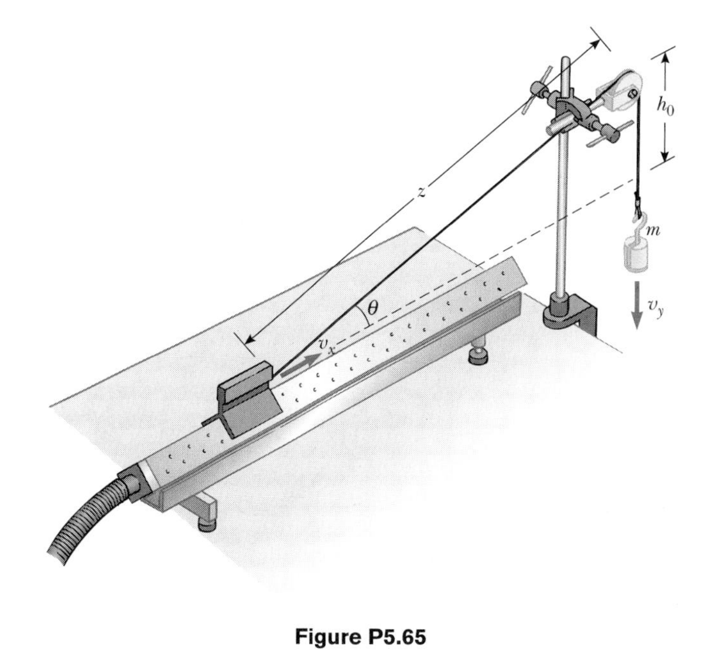

Figure P5.65

speed v_y of the hanging object are related by $v_x = uv_y$, where $u = z(z^2 - h_0^2)^{-1/2}$. (b) The glider is released from rest. Show that at that instant the acceleration a_x of the glider and the acceleration a_y of the hanging object are related by $a_x = ua_y$. (c) Find the tension in the string at the instant the glider is released for $h_0 = 80.0$ cm and $\theta = 30.0°$.

66. Cam mechanisms are used in many machines. For example, cams open and close the valves in your car engine to admit gasoline vapor to each cylinder and to allow the escape of exhaust. The principle is illustrated in Figure P5.66, showing a follower rod (also called a pushrod) of mass m resting on a wedge of mass M. The sliding wedge duplicates the function of a rotating eccentric disk on a camshaft in your car. Assume that there is no friction between the wedge and the base, between the pushrod and the wedge, or between the rod and the guide through which it slides. When the wedge is pushed to the left by the force F, the rod moves upward and does something, such as opening a valve. By varying the shape of the wedge, the motion of the follower rod could be made quite complex, but assume that the wedge makes a constant angle of $\theta = 15.0°$. Suppose you want the wedge and the rod to start from rest and move with constant acceleration, with the rod moving upward 1.00 mm in 8.00 ms. Take $m = 0.250$ kg and $M = 0.500$ kg. What force F must be applied to the wedge?

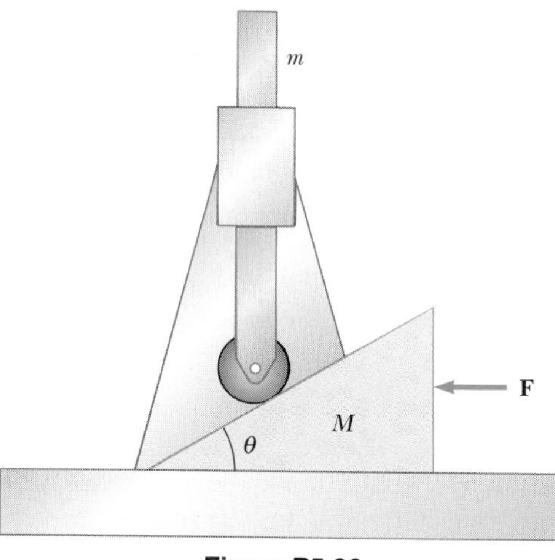

Figure P5.66

67. Any device that allows you to increase the force you exert is a kind of *machine*. Some machines, such as the prybar or the inclined plane, are very simple. Some machines do not even look like machines. An example is the following: Your car is stuck in the mud, and you can't pull hard enough to get it out. However, you have a long cable which you connect taut between your front bumper and the trunk of a stout tree. You now pull sideways on the cable at its midpoint, exerting a force f. Each half of the cable is displaced through a small angle θ from the straight line between the ends of the cable. (a) Deduce an expression for the force exerted on the car. (b) Evaluate the cable tension for the case where $\theta = 7.00°$ and $f = 100$ N.

68. Two blocks of mass 3.50 kg and 8.00 kg are connected by a massless string that passes over a frictionless pulley (Fig. P5.68). The inclines are frictionless. Find (a) the magnitude of the acceleration of each block and (b) the tension in the string.

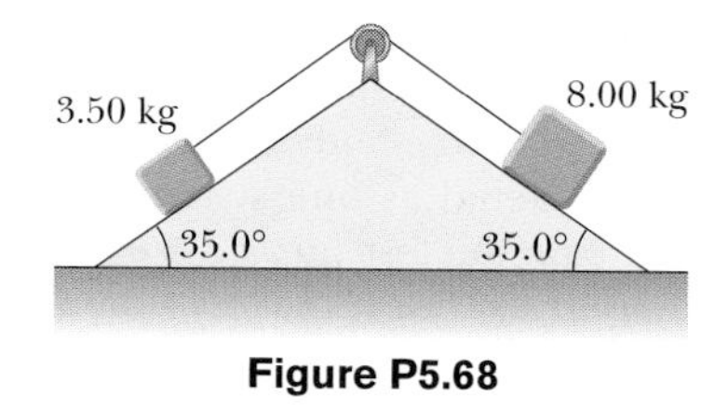

Figure P5.68

69. A van accelerates down a hill (Fig. P5.69), going from rest to 30.0 m/s in 6.00 s. During the acceleration, a toy ($m = 0.100$ kg) hangs by a string from the van's ceiling. The acceleration is such that the string remains perpendicular to the ceiling. Determine (a) the angle θ and (b) the tension in the string.

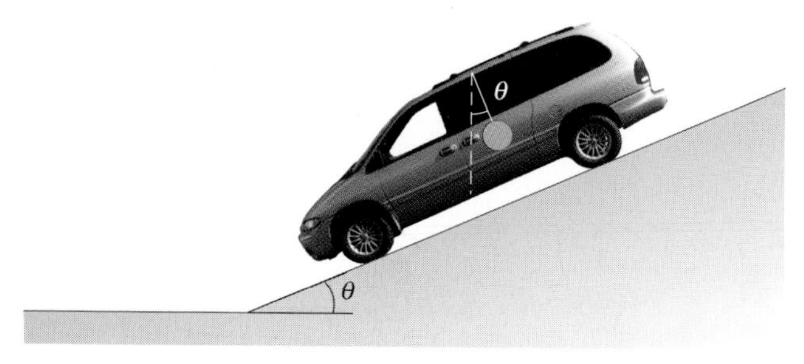

Figure P5.69

70. In Figure P5.58 the incline has mass M and is fastened to the stationary horizontal tabletop. The block of mass m is placed near the bottom of the incline and is released with a quick push that sets it sliding upward. It stops near the top of the incline, as shown in the figure, and then slides down again, always without friction. Find the force that the tabletop exerts on the incline throughout this motion.

71. A magician pulls a tablecloth from under a 200-g mug located 30.0 cm from the edge of the cloth. The cloth exerts a friction force of 0.100 N on the mug, and the cloth is pulled with a constant acceleration of 3.00 m/s^2. How far does the mug move relative to the horizontal tabletop before the cloth is completely out from under it? Note that the cloth must move more than 30 cm relative to the tabletop during the process.

72. An 8.40-kg object slides down a fixed, frictionless inclined plane. Use a computer to determine and tabulate the normal force exerted on the object and its acceleration for a series of incline angles (measured from the horizontal) ranging from 0° to 90° in 5° increments. Plot a graph of the normal force and the acceleration as functions of the incline angle. In the limiting cases of 0° and 90°, are your results consistent with the known behavior?

73. A mobile is formed by supporting four metal butterflies of equal mass m from a string of length L. The points of support are evenly spaced a distance ℓ apart as shown in Figure P5.73. The string forms an angle θ_1 with the ceiling at each end point. The center section of string is horizontal. (a) Find the tension in each section of string in terms of θ_1, m, and g. (b) Find the angle θ_2, in terms of θ_1, that the sections of string between the outside butterflies and the inside butterflies form with the horizontal. (c) Show that the distance D between the end points of the string is

$$D = \frac{L}{5}(2 \cos \theta_1 + 2 \cos [\tan^{-1}(\tfrac{1}{2} \tan \theta_1)] + 1)$$

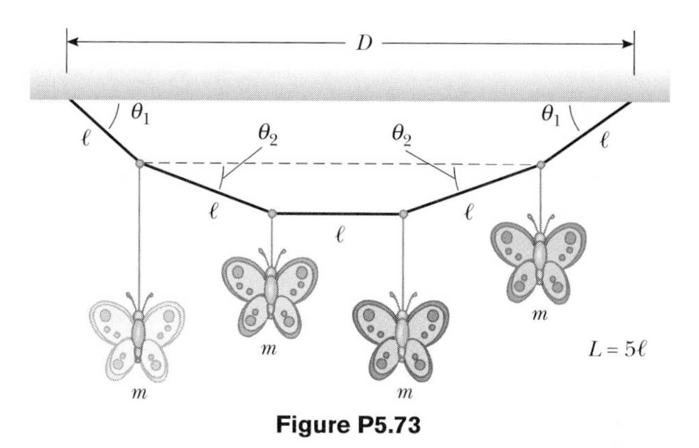

Figure P5.73

Answers to Quick Quizzes

5.1 (d). Choice (a) is true. Newton's first law tells us that motion requires no force: an object in motion continues to move at constant velocity in the absence of external forces. Choice (b) is also true. A stationary object can have several forces acting on it, but if the vector sum of all these external forces is zero, there is no net force and the object remains stationary.

5.2 (a). If a single force acts, this force constitutes the net force and there is an acceleration according to Newton's second law.

5.3 (c). Newton's second law relates only the force and the acceleration. Direction of motion is part of an object's *velocity*, and force determines the direction of acceleration, not that of velocity.

5.4 (d). With twice the force, the object will experience twice the acceleration. Because the force is constant, the acceleration is constant, and the speed of the object (starting from rest) is given by $v = at$. With twice the acceleration, the object will arrive at speed v at half the time.

5.5 (a). The gravitational force acts on the ball at *all* points in its trajectory.

5.6 (b). Because the value of g is smaller on the Moon than on the Earth, more mass of gold would be required to represent 1 newton of weight on the Moon. Thus, your friend on the Moon is richer, by about a factor of 6!

5.7 (c). In accordance with Newton's third law, the fly and bus experience forces that are equal in magnitude but opposite in direction.

5.8 (a). Because the fly has such a small mass, Newton's second law tells us that it undergoes a very large acceleration. The huge mass of the bus means that it more effectively resists any change in its motion and exhibits a small acceleration.

5.9 (c). The reaction force to your weight is an upward gravitational force on the Earth due to you.

5.10 (b). Remember the phrase "free-body." You draw *one* body (one object), free of all the others that may be interacting, and draw only the forces exerted on that object.

5.11 (b). The friction force acts opposite to the gravitational force on the book to keep the book in equilibrium. Because the gravitational force is downward, the friction force must be upward.

5.12 (b). The crate accelerates to the east. Because the only horizontal force acting on it is the force of static friction between its bottom surface and the truck bed, that force must also be directed to the east.

5.13 (b). At the angle at which the book breaks free, the component of the gravitational force parallel to the board is approximately equal to the maximum static friction force. Because the kinetic coefficient of friction is smaller than the static coefficient, at this angle, the component of the gravitational force parallel to the board is larger than the kinetic friction force. Thus, there is a net downhill force parallel to the board and the book speeds up.

5.14 (b). When pulling with the rope, there is a component of your applied force that is upward. This reduces the normal force between the sled and the snow. In turn, this reduces the friction force between the sled and the snow, making it easier to move. If you push from behind, with a force with a downward component, the normal force is larger, the friction force is larger, and the sled is harder to move.

Chapter 6

Circular Motion and Other Applications of Newton's Laws

CHAPTER OUTLINE

▲ *The London Eye, a ride on the River Thames in downtown London. Riders travel in a large vertical circle for a breathtaking view of the city. In this chapter, we will study the forces involved in circular motion. (© Paul Hardy/CORBIS)*

In the preceding chapter we introduced Newton's laws of motion and applied them to situations involving linear motion. Now we discuss motion that is slightly more complicated. For example, we shall apply Newton's laws to objects traveling in circular paths. Also, we shall discuss motion observed from an accelerating frame of reference and motion of an object through a viscous medium. For the most part, this chapter consists of a series of examples selected to illustrate the application of Newton's laws to a wide variety of circumstances.

6.1 Newton's Second Law Applied to Uniform Circular Motion

In Section 4.4 we found that a particle moving with uniform speed v in a circular path of radius r experiences an acceleration that has a magnitude

$$a_c = \frac{v^2}{r}$$

The acceleration is called *centripetal acceleration* because $\mathbf{a}_c$ is directed toward the center of the circle. Furthermore, $\mathbf{a}_c$ is *always* perpendicular to $\mathbf{v}$. (If there were a component of acceleration parallel to $\mathbf{v}$, the particle's speed would be changing.)

Consider a ball of mass m that is tied to a string of length r and is being whirled at constant speed in a horizontal circular path, as illustrated in Figure 6.1. Its weight is supported by a frictionless table. Why does the ball move in a circle? According to Newton's first law, the ball tends to move in a straight line; however, the string prevents

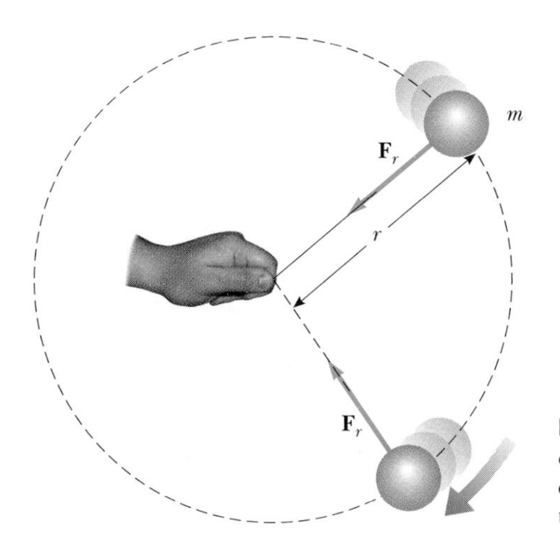

Figure 6.1 Overhead view of a ball moving in a circular path in a horizontal plane. A force $\mathbf{F}_r$ directed toward the center of the circle keeps the ball moving in its circular path.

Mike Powell / Allsport / Getty Images

An athlete in the process of throwing the hammer at the 1996 Olympic Games in Atlanta, Georgia. The force exerted by the chain causes the centripetal acceleration of the hammer. Only when the athlete releases the hammer will it move along a straight-line path tangent to the circle.

motion along a straight line by exerting on the ball a radial force $\mathbf{F}_r$ that makes it follow the circular path. This force is directed along the string toward the center of the circle, as shown in Figure 6.1.

If we apply Newton's second law along the radial direction, we find that the net force causing the centripetal acceleration can be evaluated:

Force causing centripetal acceleration

$$\sum F = ma_c = m\frac{v^2}{r} \tag{6.1}$$

A force causing a centripetal acceleration acts toward the center of the circular path and causes a change in the direction of the velocity vector. If that force should vanish, the object would no longer move in its circular path; instead, it would move along a straight-line path tangent to the circle. This idea is illustrated in Figure 6.2 for the ball whirling at the end of a string in a horizontal plane. If the string breaks at some instant, the ball moves along the straight-line path tangent to the circle at the point where the string breaks.

© Tom Carroll/Index Stock Imagery/PictureQuest

Figure 6.3 (Quick Quiz 6.1 and 6.2) A Ferris wheel located on the Navy Pier in Chicago, Illinois.

Quick Quiz 6.1 You are riding on a Ferris wheel (Fig. 6.3) that is rotating with constant speed. The car in which you are riding always maintains its correct upward orientation—it does not invert. What is the direction of your centripetal acceleration when you are at the *top* of the wheel? (a) upward (b) downward (c) impossible to determine. What is the direction of your centripetal acceleration when you are at the *bottom* of the wheel? (d) upward (e) downward (f) impossible to determine.

Quick Quiz 6.2 You are riding on the Ferris wheel of Quick Quiz 6.1. What is the direction of the normal force exerted by the seat on you when you are at the *top* of the wheel? (a) upward (b) downward (c) impossible to determine. What is the direction of the normal force exerted by the seat on you when you are at the *bottom* of the wheel? (d) upward (e) downward (f) impossible to determine.

PITFALL PREVENTION

6.1 Direction of Travel When the String is Cut

Study Figure 6.2 very carefully. Many students (wrongly) think that the ball will move *radially* away from the center of the circle when the string is cut. The velocity of the ball is *tangent* to the circle. By Newton's first law, the ball continues to move in the direction that it is moving just as the force from the string disappears.

At the Active Figures link at http://www.pse6.com, you can "break" the string yourself and observe the effect on the ball's motion.

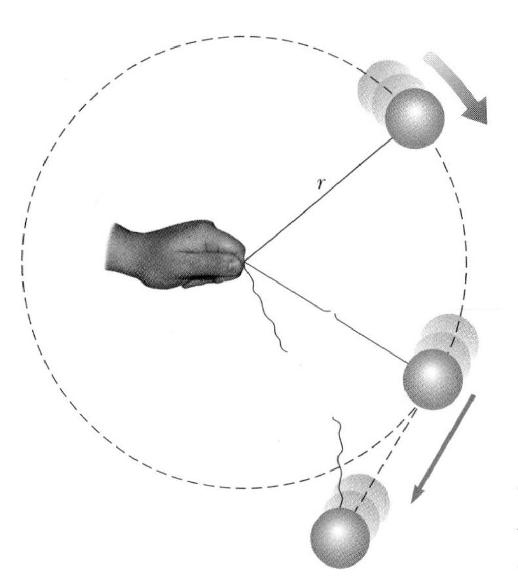

Active Figure 6.2 An overhead view of a ball moving in a circular path in a horizontal plane. When the string breaks, the ball moves in the direction tangent to the circle.

Conceptual Example 6.1 Forces That Cause Centripetal Acceleration

The force causing centripetal acceleration is sometimes called a *centripetal force.* We are familiar with a variety of forces in nature—friction, gravity, normal forces, tension, and so forth. Should we add *centripetal* force to this list?

Solution No; centripetal force *should not* be added to this list. This is a pitfall for many students. Giving the force causing circular motion a name—centripetal force—leads many students to consider it as a new *kind* of force rather than a new *role* for force. A common mistake in force diagrams is to draw all the usual forces and then to add another vector for the centripetal force. But it is not a separate force—it is simply one or more of our familiar forces *acting in the role of a force that causes a circular motion.*

Consider some examples. For the motion of the Earth around the Sun, the centripetal force is *gravity.* For an object sitting on a rotating turntable, the centripetal force is *friction.* For a rock whirled horizontally on the end of a string, the magnitude of the centripetal force is the *tension* in the string. For an amusement-park patron pressed against the inner wall of a rapidly rotating circular room, the centripetal force is the *normal force* exerted by the wall. Furthermore, the centripetal force could be a combination of two or more forces. For example, as you pass through the lowest point of the Ferris wheel in Quick Quiz 6.1, the centripetal force on you is the difference between the normal force exerted by the seat and the gravitational force. We will not use the term *centripetal force* in this book after this discussion.

Example 6.2 The Conical Pendulum

A small object of mass m is suspended from a string of length L. The object revolves with constant speed v in a horizontal circle of radius r, as shown in Figure 6.4. (Because the string sweeps out the surface of a cone, the system is known as a *conical pendulum.*) Find an expression for v.

Solution Conceptualize the problem with the help of Figure 6.4. We categorize this as a problem that combines equilibrium for the ball in the vertical direction with uniform circular motion in the horizontal direction. To analyze the problem, begin by letting θ represent the angle between the string and the vertical. In the free-body diagram shown, the force $\mathbf{T}$ exerted by the string is resolved into a vertical component $T\cos\theta$ and a horizontal component $T\sin\theta$ acting toward the center of revolution. Because the object does not accelerate in the vertical direction, $\Sigma F_y = ma_y = 0$ and the upward vertical component of $\mathbf{T}$ must balance the downward gravitational force. Therefore,

$$(1) \qquad T\cos\theta = mg$$

Because the force providing the centripetal acceleration in this example is the component $T\sin\theta$, we can use Equation 6.1 to obtain

$$(2) \qquad \sum F = T\sin\theta = ma_c = \frac{mv^2}{r}$$

Dividing (2) by (1) and using $\sin\theta/\cos\theta = \tan\theta$, we eliminate T and find that

$$\tan\theta = \frac{v^2}{rg}$$

$$v = \sqrt{rg\tan\theta}$$

From the geometry in Figure 6.4, we see that $r = L\sin\theta$; therefore,

$$v = \sqrt{Lg\sin\theta\tan\theta}$$

Note that the speed is independent of the mass of the object.

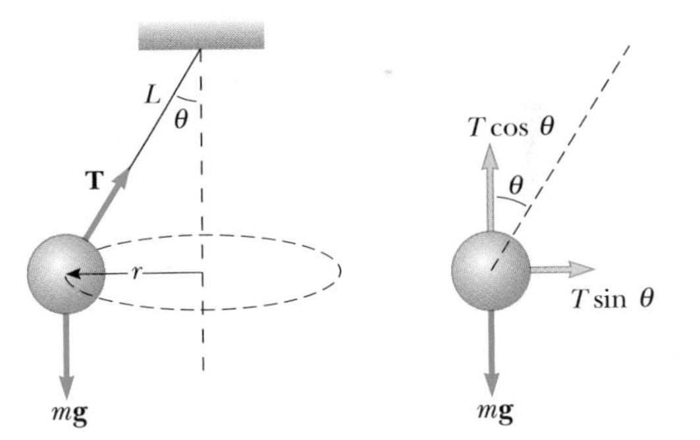

Figure 6.4 (Example 6.2) The conical pendulum and its free-body diagram.

Example 6.3 How Fast Can It Spin?

A ball of mass 0.500 kg is attached to the end of a cord 1.50 m long. The ball is whirled in a horizontal circle as shown in Figure 6.1. If the cord can withstand a maximum tension of 50.0 N, what is the maximum speed at which the ball can be whirled before the cord breaks? Assume that the string remains horizontal during the motion.

Solution It makes sense that the stronger the cord, the faster the ball can twirl before the cord breaks. Also, we expect a more massive ball to break the cord at a lower speed. (Imagine whirling a bowling ball on the cord!)

Because the force causing the centripetal acceleration in this case is the force $\mathbf{T}$ exerted by the cord on the ball, Equation 6.1 yields

$$(1) \qquad T = m\frac{v^2}{r}$$

Solving for v, we have

$$v = \sqrt{\frac{Tr}{m}}$$

This shows that v increases with T and decreases with larger m, as we expect to see—for a given v, a large mass requires a large tension and a small mass needs only a small tension. The maximum speed the ball can have corresponds to the maximum tension. Hence, we find

$$v_{\max} = \sqrt{\frac{T_{\max} r}{m}} = \sqrt{\frac{(50.0 \text{ N})(1.50 \text{ m})}{0.500 \text{ kg}}} = 12.2 \text{ m/s}$$

What If? **Suppose that the ball is whirled in a circle of larger radius at the same speed *v*. Is the cord more likely to break or less likely?**

Answer The larger radius means that the change in the direction of the velocity vector will be smaller for a given time interval. Thus, the acceleration is smaller and the required force from the string is smaller. As a result, the string is less likely to break when the ball travels in a circle of larger radius. To understand this argument better, let us write Equation (1) twice, once for each radius:

$$T_1 = \frac{mv^2}{r_1} \qquad T_2 = \frac{mv^2}{r_2}$$

Dividing the two equations gives us,

$$\frac{T_2}{T_1} = \frac{\left(\frac{mv^2}{r_2}\right)}{\left(\frac{mv^2}{r_1}\right)} = \frac{r_1}{r_2}$$

If we choose $r_2 > r_1$, we see that $T_2 < T_1$. Thus, less tension is required to whirl the ball in the larger circle and the string is less likely to break.

Example 6.4 What Is the Maximum Speed of the Car? — Interactive

A 1 500-kg car moving on a flat, horizontal road negotiates a curve, as shown in Figure 6.5. If the radius of the curve is 35.0 m and the coefficient of static friction between the tires and dry pavement is 0.500, find the maximum speed the car can have and still make the turn successfully.

Solution In this case, the force that enables the car to remain in its circular path is the force of static friction. (*Static* because no slipping occurs at the point of contact between road and tires. If this force of static friction were zero—for example, if the car were on an icy road—the car would continue in a straight line and slide off the road.) Hence, from Equation 6.1 we have

$$(1) \qquad f_s = m\frac{v^2}{r}$$

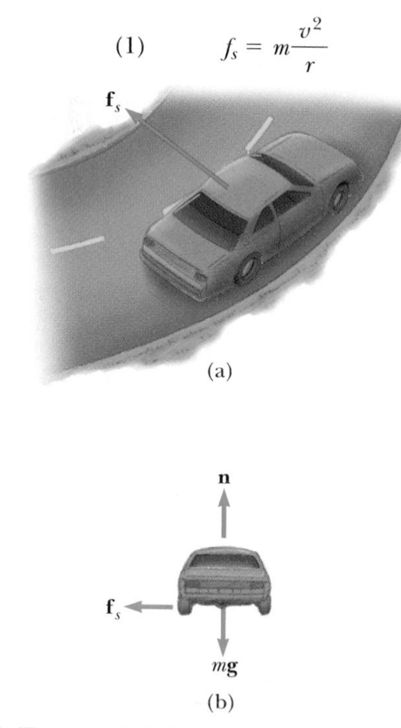

Figure 6.5 (Example 6.4) (a) The force of static friction directed toward the center of the curve keeps the car moving in a circular path. (b) The free-body diagram for the car.

The maximum speed the car can have around the curve is the speed at which it is on the verge of skidding outward. At this point, the friction force has its maximum value $f_{s,\max} = \mu_s n$. Because the car shown in Figure 6.5b is in equilibrium in the vertical direction, the magnitude of the normal force equals the weight ($n = mg$) and thus $f_{s,\max} = \mu_s mg$. Substituting this value for f_s into (1), we find that the maximum speed is

$$(2) \qquad v_{\max} = \sqrt{\frac{f_{s,\max}\, r}{m}} = \sqrt{\frac{\mu_s mgr}{m}} = \sqrt{\mu_s gr}$$

$$= \sqrt{(0.500)(9.80 \text{ m/s}^2)(35.0 \text{ m})}$$

$$= 13.1 \text{ m/s}$$

Note that the maximum speed does not depend on the mass of the car. That is why curved highways do not need multiple speed limit signs to cover the various masses of vehicles using the road.

What If? **Suppose that a car travels this curve on a wet day and begins to skid on the curve when its speed reaches only 8.00 m/s. What can we say about the coefficient of static friction in this case?**

Answer The coefficient of friction between tires and a wet road should be smaller than that between tires and a dry road. This expectation is consistent with experience with driving, because a skid is more likely on a wet road than a dry road.

To check our suspicion, we can solve (2) for the coefficient of friction:

$$\mu_s = \frac{v_{\max}^2}{gr}$$

Substituting the numerical values,

$$\mu_s = \frac{v_{\max}^2}{gr} = \frac{(8.00 \text{ m/s})^2}{(9.80 \text{ m/s}^2)(35.0 \text{ m})} = 0.187$$

This is indeed smaller than the coefficient of 0.500 for the dry road.

Study the relationship between the car's speed, radius of the turn, and the coefficient of static friction between road and tires at the Interactive Worked Example link at **http://www.pse6.com.**

Example 6.5 The Banked Exit Ramp

Interactive

A civil engineer wishes to design a curved exit ramp for a highway in such a way that a car will not have to rely on friction to round the curve without skidding. In other words, a car moving at the designated speed can negotiate the curve even when the road is covered with ice. Such a ramp is usually *banked*; this means the roadway is tilted toward the inside of the curve. Suppose the designated speed for the ramp is to be 13.4 m/s (30.0 mi/h) and the radius of the curve is 50.0 m. At what angle should the curve be banked?

Solution On a level (unbanked) road, the force that causes the centripetal acceleration is the force of static friction between car and road, as we saw in the previous example. However, if the road is banked at an angle θ, as in Figure 6.6, the normal force **n** has a horizontal component $n \sin \theta$ pointing toward the center of the curve. Because the ramp is to be designed so that the force of static friction is zero, only the component $n_x = n \sin \theta$ causes the centripetal acceleration. Hence, Newton's second law for the radial direction gives

$$(1) \qquad \sum F_r = n \sin \theta = \frac{mv^2}{r}$$

The car is in equilibrium in the vertical direction. Thus, from $\Sigma F_y = 0$ we have

$$(2) \qquad n \cos \theta = mg$$

Dividing (1) by (2) gives

$$(3) \qquad \tan \theta = \frac{v^2}{rg}$$

$$\theta = \tan^{-1}\left(\frac{(13.4 \text{ m/s})^2}{(50.0 \text{ m})(9.80 \text{ m/s}^2)}\right) = 20.1°$$

If a car rounds the curve at a speed less than 13.4 m/s, friction is needed to keep it from sliding down the bank (to the left in Fig. 6.6). A driver who attempts to negotiate the curve at a speed greater than 13.4 m/s has to depend on friction to keep from sliding up the bank (to the right in Fig. 6.6). The banking angle is independent of the mass of the vehicle negotiating the curve.

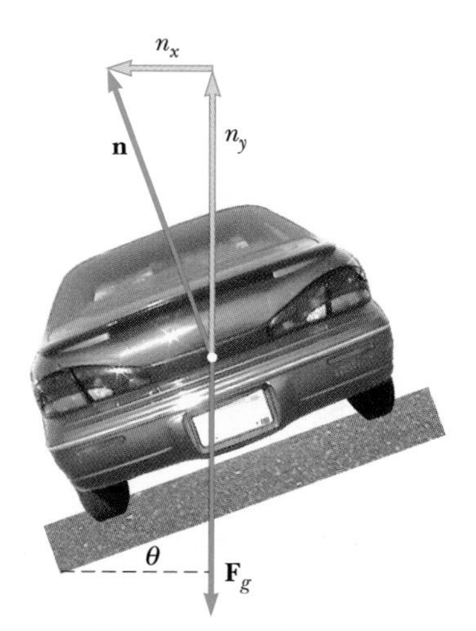

Figure 6.6 (Example 6.5) A car rounding a curve on a road banked at an angle θ to the horizontal. When friction is neglected, the force that causes the centripetal acceleration and keeps the car moving in its circular path is the horizontal component of the normal force.

What If? What if this same roadway were built on Mars in the future to connect different colony centers; could it be traveled at the same speed?

Answer The reduced gravitational force on Mars would mean that the car is not pressed so tightly to the roadway. The reduced normal force results in a smaller component of the normal force toward the center of the circle. This smaller component will not be sufficient to provide the centripetal acceleration associated with the original speed. The centripetal acceleration must be reduced, which can be done by reducing the speed v.

Equation (3) shows that the speed v is proportional to the square root of g for a roadway of fixed radius r banked at a fixed angle θ. Thus, if g is smaller, as it is on Mars, the speed v with which the roadway can be safely traveled is also smaller.

You can adjust the turn radius and banking angle at the Interactive Worked Example link at **http://www.pse6.com.**

Example 6.6 Let's Go Loop-the-Loop!

A pilot of mass m in a jet aircraft executes a loop-the-loop, as shown in Figure 6.7a. In this maneuver, the aircraft moves in a vertical circle of radius 2.70 km at a constant speed of 225 m/s. Determine the force exerted by the seat on the pilot **(A)** at the bottom of the loop and **(B)** at the top of the loop. Express your answers in terms of the weight of the pilot mg.

Solution To conceptualize this problem, look carefully at Figure 6.7. Based on experiences with driving over small hills in a roadway, or riding over the top of a Ferris wheel, you would expect to feel lighter at the top of the path. Similarly, you would expect to feel heavier at the bottom of the path. By looking at Figure 6.7, we expect the answer for (A) to be greater than that for (B) because at the bottom of the loop the normal and gravitational forces act in *opposite* directions, whereas at the top of the loop these two forces act in the *same* direction. The vector sum of these two forces gives the force of constant magnitude that keeps the pilot moving in a circular path at a constant speed. To yield net

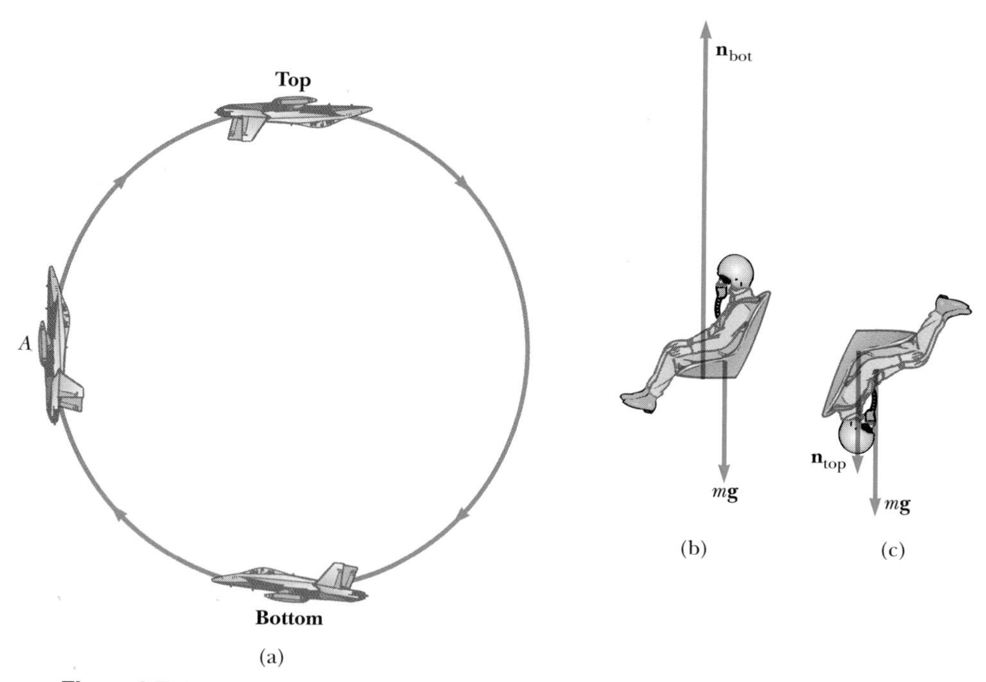

Figure 6.7 (Example 6.6) (a) An aircraft executes a loop-the-loop maneuver as it moves in a vertical circle at constant speed. (b) Free-body diagram for the pilot at the bottom of the loop. In this position the pilot experiences an apparent weight greater than his true weight. (c) Free-body diagram for the pilot at the top of the loop.

force vectors with the same magnitude, the normal force at the bottom must be greater than that at the top. Because the speed of the aircraft is constant (how likely is this?), we can categorize this as a uniform circular motion problem, complicated by the fact that the gravitational force acts at all times on the aircraft.

(A) Analyze the situation by drawing a free-body diagram for the pilot at the bottom of the loop, as shown in Figure 6.7b. The only forces acting on him are the downward gravitational force $\mathbf{F}_g = m\mathbf{g}$ and the upward force $\mathbf{n}_{bot}$ exerted by the seat. Because the net upward force that provides the centripetal acceleration has a magnitude $n_{bot} - mg$, Newton's second law for the radial direction gives

$$\sum F = n_{bot} - mg = m\frac{v^2}{r}$$

$$n_{bot} = mg + m\frac{v^2}{r} = mg\left(1 + \frac{v^2}{rg}\right)$$

Substituting the values given for the speed and radius gives

$$n_{bot} = mg\left(1 + \frac{(225 \text{ m/s})^2}{(2.70 \times 10^3 \text{ m})(9.80 \text{ m/s}^2)}\right) = \boxed{2.91mg}$$

Hence, the magnitude of the force $\mathbf{n}_{bot}$ exerted by the seat on the pilot is *greater* than the weight of the pilot by a factor of 2.91. This means that the pilot experiences an apparent weight that is greater than his true weight by a factor of 2.91.

(B) The free-body diagram for the pilot at the top of the loop is shown in Figure 6.7c. As we noted earlier, both the gravitational force exerted by the Earth and the force $\mathbf{n}_{top}$ exerted by the seat on the pilot act downward, and so the net downward force that provides the centripetal acceleration has a magnitude $n_{top} + mg$. Applying Newton's second law yields

$$\sum F = n_{top} + mg = m\frac{v^2}{r}$$

$$n_{top} = m\frac{v^2}{r} - mg = mg\left(\frac{v^2}{rg} - 1\right)$$

$$n_{top} = mg\left(\frac{(225 \text{ m/s})^2}{(2.70 \times 10^3 \text{ m})(9.80 \text{ m/s}^2)} - 1\right)$$

$$= \boxed{0.913mg}$$

In this case, the magnitude of the force exerted by the seat on the pilot is *less* than his true weight by a factor of 0.913, and the pilot feels lighter. To finalize the problem, note that this is consistent with our prediction at the beginning of the solution.

6.2 Nonuniform Circular Motion

In Chapter 4 we found that if a particle moves with varying speed in a circular path, there is, in addition to the radial component of acceleration, a tangential component having magnitude dv/dt. Therefore, the force acting on the particle must also have a tangential and a radial component. Because the total acceleration is $\mathbf{a} = \mathbf{a}_r + \mathbf{a}_t$, the total force exerted on the particle is $\Sigma\mathbf{F} = \Sigma\mathbf{F}_r + \Sigma\mathbf{F}_t$, as shown in Figure 6.8. The vector $\Sigma\mathbf{F}_r$ is directed toward the center of the circle and is responsible for the centripetal acceleration. The vector $\Sigma\mathbf{F}_t$ tangent to the circle is responsible for the tangential acceleration, which represents a change in the speed of the particle with time.

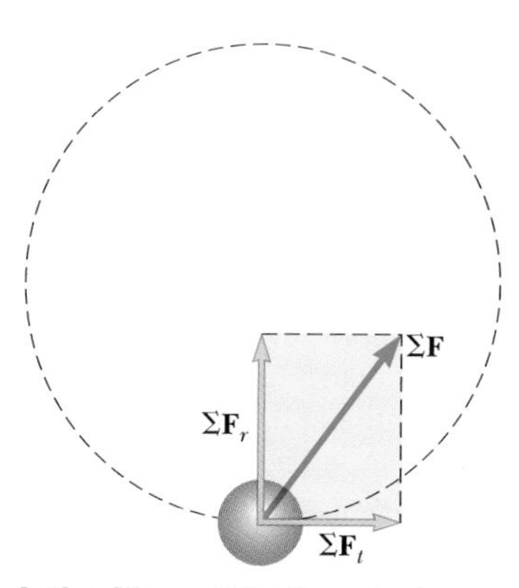

Active Figure 6.8 When the force acting on a particle moving in a circular path has a tangential component ΣF_t, the particle's speed changes. The total force exerted on the particle in this case is the vector sum of the radial force and the tangential force. That is, $\Sigma\mathbf{F} = \Sigma\mathbf{F}_r + \Sigma\mathbf{F}_t$.

At the Active Figures link at http://www.pse6.com, you can adjust the initial position of the particle and compare the component forces acting on the particle to those for a child swinging on a swing set.

Quick Quiz 6.3 Which of the following is *impossible* for a car moving in a circular path? (a) the car has tangential acceleration but no centripetal acceleration. (b) the car has centripetal acceleration but no tangential acceleration. (c) the car has both centripetal acceleration and tangential acceleration.

Quick Quiz 6.4 A bead slides freely along a *horizontal*, curved wire at constant speed, as shown in Figure 6.9. Draw the vectors representing the force exerted by the wire on the bead at points Ⓐ, Ⓑ, and Ⓒ.

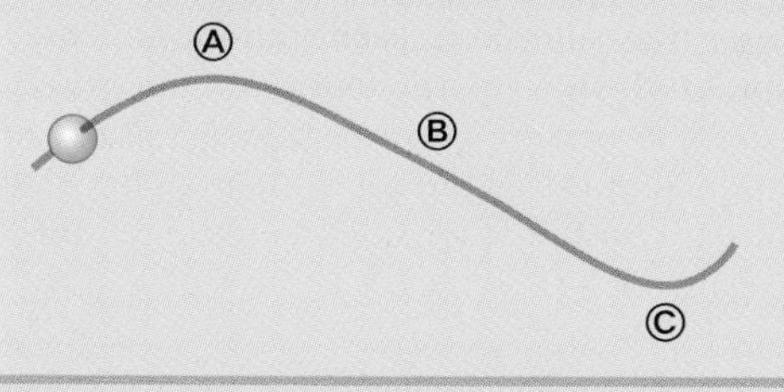

Figure 6.9 (Quick Quiz 6.4 and 6.5) A bead slides along a curved wire.

Quick Quiz 6.5 In Figure 6.9, the bead speeds up with constant tangential acceleration as it moves toward the right. Draw the vectors representing the force on the bead at points Ⓐ, Ⓑ, and Ⓒ.

Robin Smith / Getty Images

Passengers on a "corkscrew" roller coaster experience a radial force toward the center of the circular track and a tangential force due to gravity.

Example 6.7 Keep Your Eye on the Ball

Interactive

A small sphere of mass m is attached to the end of a cord of length R and set into motion in a *vertical* circle about a fixed point O, as illustrated in Figure 6.10a. Determine the tension in the cord at any instant when the speed of the sphere is v and the cord makes an angle θ with the vertical.

Solution Unlike the situation in Example 6.6, the speed is *not* uniform in this example because, at most points along the path, a tangential component of acceleration arises from the gravitational force exerted on the sphere. From the free-body diagram in Figure 6.10a, we see that the only forces acting on the sphere are the gravitational force $\mathbf{F}_g = m\mathbf{g}$ exerted by the Earth and the force $\mathbf{T}$ exerted by the cord. Now we resolve $\mathbf{F}_g$ into a tangential component $mg \sin \theta$ and a radial component $mg \cos \theta$. Applying Newton's second law to the forces acting on the sphere in the tangential direction yields

$$\sum F_t = mg \sin \theta = ma_t$$

$$a_t = g \sin \theta$$

This tangential component of the acceleration causes v to change in time because $a_t = dv/dt$.

Applying Newton's second law to the forces acting on the sphere in the radial direction and noting that both $\mathbf{T}$ and $\mathbf{a}_r$ are directed toward O, we obtain

$$\sum F_r = T - mg \cos \theta = \frac{mv^2}{R}$$

$$T = m\left(\frac{v^2}{R} + g \cos \theta\right)$$

What If? **What if we set the ball in motion with a slower speed? (A) What speed would the ball have as it passes over the top of the circle if the tension in the cord goes to zero instantaneously at this point?**

Answer At the top of the path (Fig. 6.10b), where $\theta = 180°$, we have $\cos 180° = -1$, and the tension equation becomes

$$T_{\text{top}} = m\left(\frac{v_{\text{top}}^2}{R} - g\right)$$

Let us set $T_{\text{top}} = 0$. Then,

$$0 = m\left(\frac{v_{\text{top}}^2}{R} - g\right)$$

$$v_{\text{top}} = \sqrt{gR}$$

(B) What if we set the ball in motion such that the speed at the top is less than this value? What happens?

Answer In this case, the ball never reaches the top of the circle. At some point on the way up, the tension in the string goes to zero and the ball becomes a projectile. It follows a segment of a parabolic path over the top of its motion, rejoining the circular path on the other side when the tension becomes nonzero again.

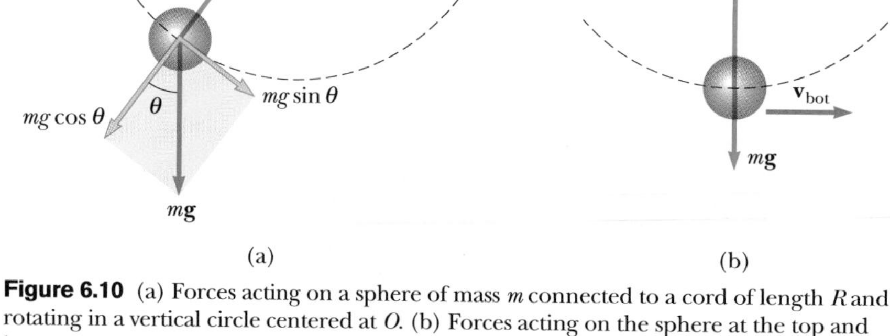

Figure 6.10 (a) Forces acting on a sphere of mass m connected to a cord of length R and rotating in a vertical circle centered at O. (b) Forces acting on the sphere at the top and bottom of the circle. The tension is a maximum at the bottom and a minimum at the top.

Investigate these alternatives at the Interactive Worked Example link at **http://www.pse6.com.**

6.3 Motion in Accelerated Frames

When Newton's laws of motion were introduced in Chapter 5, we emphasized that they are valid only when observations are made in an inertial frame of reference. In this section, we analyze how Newton's second law is applied by an observer in a noninertial frame of reference, that is, one that is accelerating. For example, recall the discussion of the air hockey table on a train in Section 5.2. The train moving at constant velocity represents an inertial frame. The puck at rest remains at rest, and Newton's first law is obeyed. The accelerating train is not an inertial frame. According to you as the observer on the train, there appears to be no visible force on the puck, yet it accelerates from rest toward the back of the train, violating Newton's first law.

As an observer on the accelerating train, if you apply Newton's second law to the puck as it accelerates toward the back of the train, you might conclude that a force has acted on the puck to cause it to accelerate. We call an apparent force such as this a **fictitious force,** because it is due to an accelerated reference frame. Remember that real forces are always due to interactions between two objects. A fictitious force appears to act on an object in the same way as a real force, but you cannot identify a second object for a fictitious force.

The train example above describes a fictitious force due to a change in the speed of the train. Another fictitious force is due to the change in the *direction* of the velocity vector. To understand the motion of a system that is noninertial because of a change in direction, consider a car traveling along a highway at a high speed and approaching a curved exit ramp, as shown in Figure 6.11a. As the car takes the sharp left turn onto the ramp, a person sitting in the passenger seat slides to the right and hits the door. At that point, the force exerted by the door on the passenger keeps her from being ejected from the car. What causes her to move toward the door? A popular but incorrect explanation is that a force acting toward the right in Figure 6.11b pushes her outward. This is often called the "centrifugal force," but it is a fictitious force due to the acceleration associated with the changing direction of the car's velocity vector. (The driver also experiences this effect but wisely holds on to the steering wheel to keep from sliding to the right.)

The phenomenon is correctly explained as follows. Before the car enters the ramp, the passenger is moving in a straight-line path. As the car enters the ramp and travels a curved path, the passenger tends to move along the original straight-line path. This is in accordance with Newton's first law: the natural tendency of an object is to continue moving in a straight line. However, if a sufficiently large force (toward the center of curvature) acts on the passenger, as in Figure 6.11c, she moves in a curved path along with the car. This force is the force of friction between her and the car seat. If this friction force is not large enough, she slides to the right as the seat turns to the left under her. Eventually, she encounters the door, which provides a force large enough to enable her to follow the same curved path as the car. She slides toward the door not because of an outward force but because **the force of friction is not sufficiently great to allow her to travel along the circular path followed by the car.**

Another interesting fictitious force is the "Coriolis force." This is an apparent force caused by changing the radial position of an object in a rotating coordinate system. For example, suppose you and a friend are on opposite sides of a rotating circular platform and you decide to throw a baseball to your friend. As Figure 6.12a shows, at $t = 0$ you throw the ball toward your friend, but by the time t_f when the ball has crossed the platform, your friend has moved to a new position.

Figure 6.12a represents what an observer would see if the ball is viewed while the observer is hovering at rest above the rotating platform. According to this observer, who is in an inertial frame, the ball follows a straight line, as it must according to Newton's first law. Now, however, consider the situation from your friend's viewpoint. Your friend is in a noninertial reference frame because he is undergoing a centripetal acceleration relative to the inertial frame of the Earth's surface. He starts off seeing the baseball coming toward him, but as it crosses the platform, it veers to one side, as shown in Figure 6.12b. Thus, your friend on the rotating platform claims that the ball

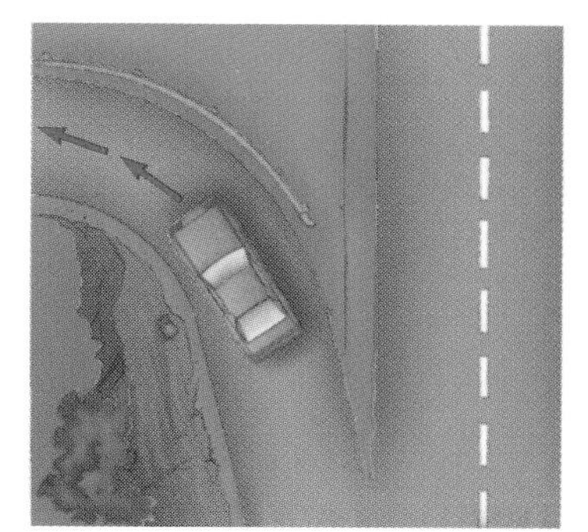

(a)

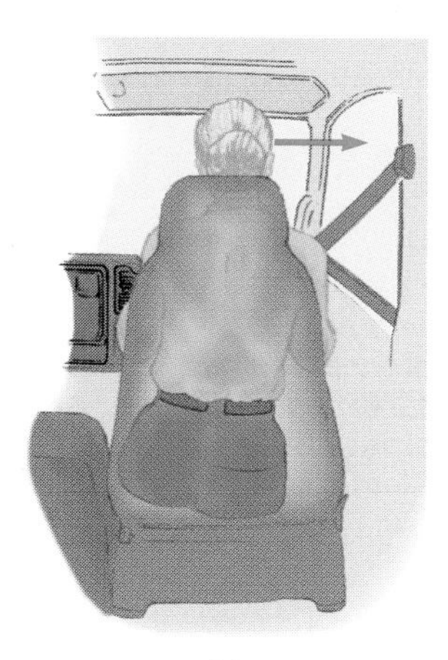

(b)

(c)

Figure 6.11 (a) A car approaching a curved exit ramp. What causes a front-seat passenger to move toward the right-hand door? (b) From the frame of reference of the passenger, a force appears to push her toward the right door, but this is a fictitious force. (c) Relative to the reference frame of the Earth, the car seat applies a leftward force to the passenger, causing her to change direction along with the rest of the car.

PITFALL PREVENTION

6.2 Centrifugal Force

The commonly heard phrase "centrifugal force" is described as a force pulling *outward* on an object moving in a circular path. If you are feeling a "centrifugal force" on a rotating carnival ride, what is the other object with which you are interacting? You cannot identify another object because this is a fictitious force that occurs as a result of your being in a noninertial reference frame.

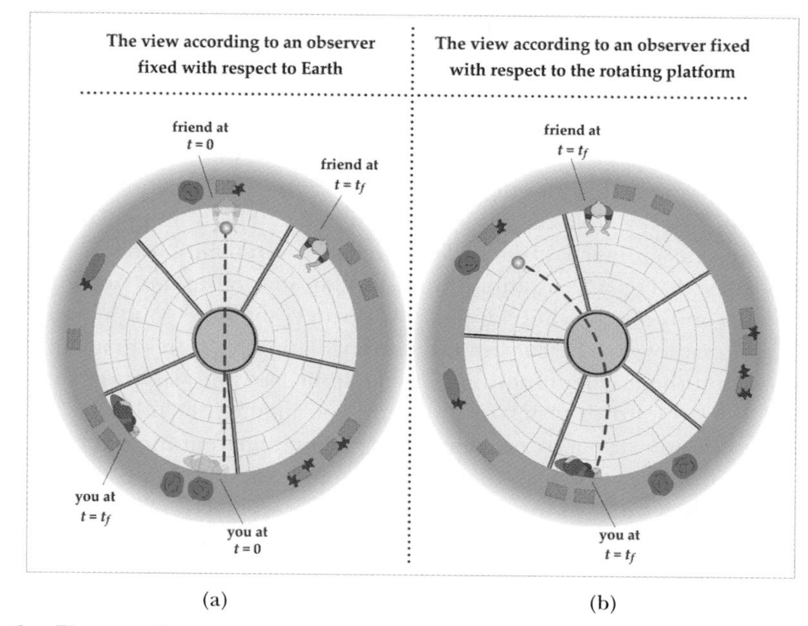

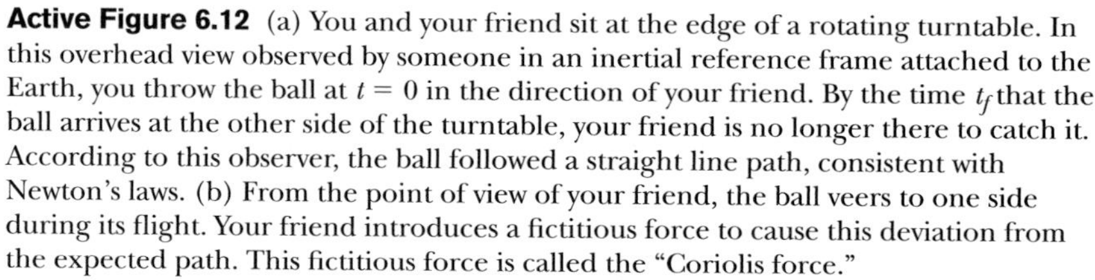

Active Figure 6.12 (a) You and your friend sit at the edge of a rotating turntable. In this overhead view observed by someone in an inertial reference frame attached to the Earth, you throw the ball at $t = 0$ in the direction of your friend. By the time t_f that the ball arrives at the other side of the turntable, your friend is no longer there to catch it. According to this observer, the ball followed a straight line path, consistent with Newton's laws. (b) From the point of view of your friend, the ball veers to one side during its flight. Your friend introduces a fictitious force to cause this deviation from the expected path. This fictitious force is called the "Coriolis force."

At the Active Figures link at http://www.pse6.com, you can observe the ball's path simultaneously from the reference frame of an inertial observer and from the reference frame of the rotating turntable.

does not obey Newton's first law and claims that a force is causing the ball to follow a curved path. This fictitious force is called the Coriolis force.

Fictitious forces may not be real forces, but they can have real effects. An object on your dashboard *really* slides off if you press the accelerator of your car. As you ride on a merry-go-round, you feel pushed toward the outside as if due to the fictitious "centrifugal force." You are likely to fall over and injure yourself if you walk along a radial line while the merry-go-round rotates. The Coriolis force due to the rotation of the Earth is responsible for rotations of hurricanes and for large-scale ocean currents.

Quick Quiz 6.6 Consider the passenger in the car making a left turn in Figure 6.11. Which of the following is correct about forces in the horizontal direction if the person is making contact with the right-hand door? (a) The passenger is in equilibrium between real forces acting to the right and real forces acting to the left. (b) The passenger is subject only to real forces acting to the right. (c) The passenger is subject only to real forces acting to the left. (d) None of these is true.

Example 6.8 Fictitious Forces in Linear Motion

A small sphere of mass m is hung by a cord from the ceiling of a boxcar that is accelerating to the right, as shown in Figure 6.13. The noninertial observer in Figure 6.13b claims that a force, which we know to be fictitious, must act in order to cause the observed deviation of the cord from the vertical. How is the magnitude of this force related to the acceleration of the boxcar measured by the inertial observer in Figure 6.13a?

Solution According to the inertial observer at rest (Fig. 6.13a), the forces on the sphere are the force $\mathbf{T}$ exerted by the cord and the gravitational force. The inertial observer concludes that the acceleration of the sphere is the same as that of the boxcar and that this acceleration is provided by the horizontal component of $\mathbf{T}$. Also, the vertical component of $\mathbf{T}$ balances the gravitational force because the sphere is in equilibrium in the vertical direction. Therefore,

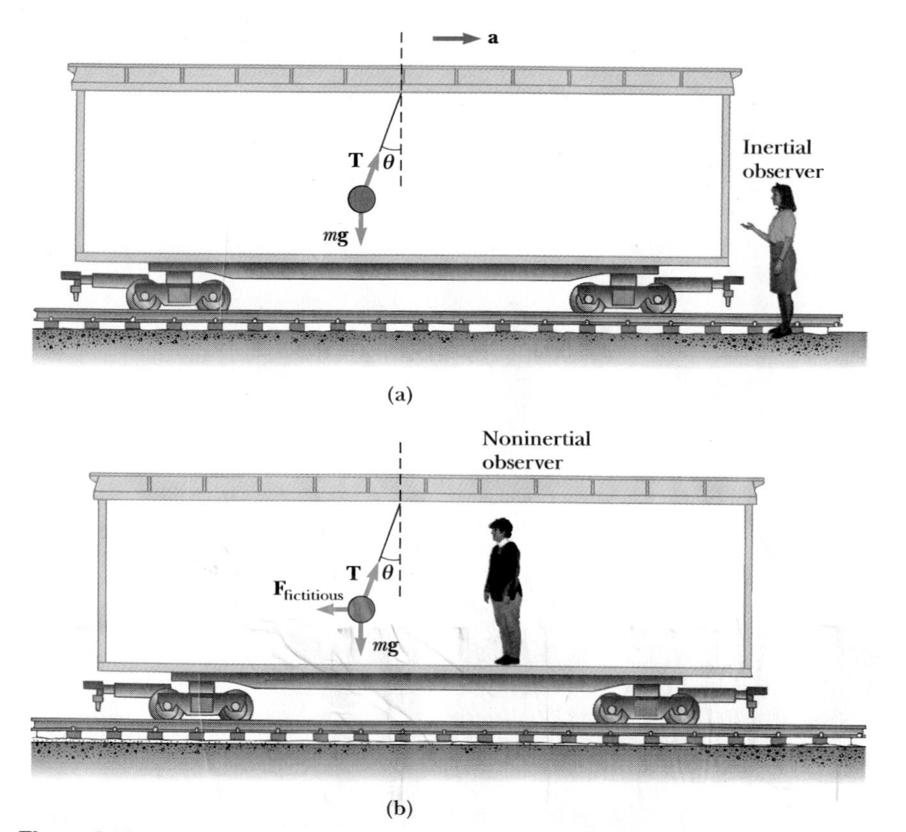

Figure 6.13 (Example 6.8) A small sphere suspended from the ceiling of a boxcar accelerating to the right is deflected as shown. (a) An inertial observer at rest outside the car claims that the acceleration of the sphere is provided by the horizontal component of **T**. (b) A noninertial observer riding in the car says that the net force on the sphere is zero and that the deflection of the cord from the vertical is due to a fictitious force $\mathbf{F}_{\text{fictitious}}$ that balances the horizontal component of **T**.

she writes Newton's second law as $\Sigma\mathbf{F} = \mathbf{T} + m\mathbf{g} = m\mathbf{a}$, which in component form becomes

Inertial observer
$$\begin{cases} (1) \quad \sum F_x = T\sin\theta = ma \\ (2) \quad \sum F_y = T\cos\theta - mg = 0 \end{cases}$$

According to the noninertial observer riding in the car (Fig. 6.13b), the cord also makes an angle θ with the vertical; however, to him the sphere is at rest and so its acceleration is zero. Therefore, he introduces a fictitious force in the horizontal direction to balance the horizontal component of **T** and claims that the net force on the sphere is *zero!* In this noninertial frame of reference, Newton's second law in component form yields

Noninertial observer
$$\begin{cases} \sum F'_x = T\sin\theta - F_{\text{fictitious}} = 0 \\ \sum F'_y = T\cos\theta - mg = 0 \end{cases}$$

We see that these expressions are equivalent to (1) and (2) if $F_{\text{fictitious}} = ma$, where a is the acceleration according to the inertial observer. If we were to make this substitution in the equation for F'_x above, the noninertial observer obtains the same mathematical results as the inertial observer. However, the physical interpretation of the deflection of the cord differs in the two frames of reference.

What If? Suppose the inertial observer wants to measure the acceleration of the train by means of the pendulum (the sphere hanging from the cord). How could she do this?

Answer Our intuition tells us that the angle θ that the cord makes with the vertical should increase as the acceleration increases. By solving (1) and (2) simultaneously for a, the inertial observer can determine the magnitude of the car's acceleration by measuring the angle θ and using the relationship $a = g\tan\theta$. Because the deflection of the cord from the vertical serves as a measure of acceleration, *a simple pendulum can be used as an accelerometer.*

Example 6.9 Fictitious Force in a Rotating System

Suppose a block of mass m lying on a horizontal, frictionless turntable is connected to a string attached to the center of the turntable, as shown in Figure 6.14. How would each of the observers write Newton's second law for the block?

Solution According to an inertial observer (Fig. 6.14a), if the block rotates uniformly, it undergoes an acceleration of magnitude v^2/r, where v is its linear speed. The inertial observer concludes that this centripetal acceleration is provided by the force **T** exerted by the string and writes Newton's second law as $T = mv^2/r$.

According to a noninertial observer attached to the turntable (Fig 6.14b), the block is at rest and its acceleration is zero. Therefore, she must introduce a fictitious outward force of magnitude mv^2/r to balance the inward force exerted by the string. According to her, the net force on the block is zero, and she writes Newton's second law as $T - mv^2/r = 0$.

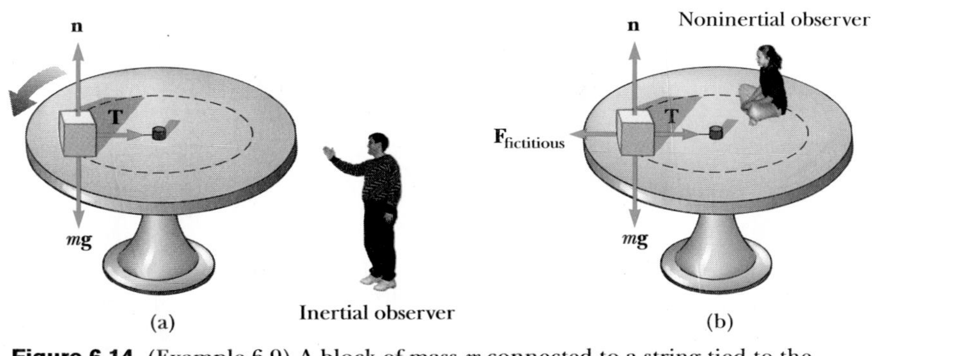

Figure 6.14 (Example 6.9) A block of mass m connected to a string tied to the center of a rotating turntable. (a) The inertial observer claims that the force causing the circular motion is provided by the force **T** exerted by the string on the block. (b) The noninertial observer claims that the block is not accelerating, and therefore she introduces a fictitious force of magnitude mv^2/r that acts outward and balances the force **T**.

6.4 Motion in the Presence of Resistive Forces

In the preceding chapter we described the force of kinetic friction exerted on an object moving on some surface. We completely ignored any interaction between the object and the medium through which it moves. Now let us consider the effect of that medium, which can be either a liquid or a gas. The medium exerts a **resistive force R** on the object moving through it. Some examples are the air resistance associated with moving vehicles (sometimes called *air drag*) and the viscous forces that act on objects moving through a liquid. The magnitude of **R** depends on factors such as the speed of the object, and the direction of **R** is always opposite the direction of motion of the object relative to the medium. Furthermore, the magnitude of **R** nearly always increases with increasing speed.

The magnitude of the resistive force can depend on speed in a complex way, and here we consider only two situations. In the first situation, we assume the resistive force is proportional to the speed of the moving object; this assumption is valid for objects falling slowly through a liquid and for very small objects, such as dust particles, moving through air. In the second situation, we assume a resistive force that is proportional to the square of the speed of the moving object; large objects, such as a skydiver moving through air in free fall, experience such a force.

Resistive Force Proportional to Object Speed

If we assume that the resistive force acting on an object moving through a liquid or gas is proportional to the object's speed, then the resistive force can be expressed as

$$\mathbf{R} = -b\mathbf{v} \tag{6.2}$$

where **v** is the velocity of the object and b is a constant whose value depends on the properties of the medium and on the shape and dimensions of the object. If the object is a sphere of radius r, then b is proportional to r. The negative sign indicates that **R** is in the opposite direction to **v**.

Consider a small sphere of mass m released from rest in a liquid, as in Figure 6.15a. Assuming that the only forces acting on the sphere are the resistive force $\mathbf{R} = -b\mathbf{v}$ and

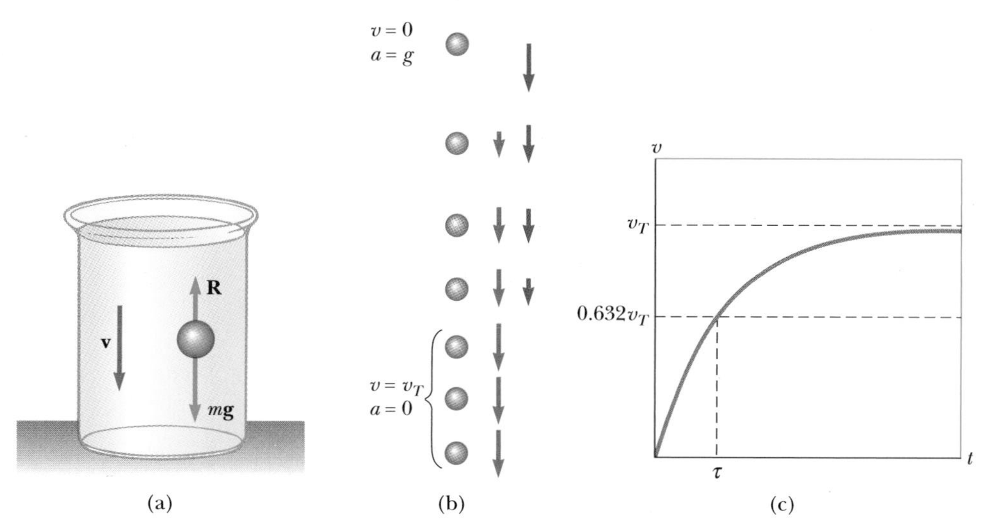

Active Figure 6.15 (a) A small sphere falling through a liquid. (b) Motion diagram of the sphere as it falls. (c) Speed–time graph for the sphere. The sphere reaches a maximum (or terminal) speed v_T, and the time constant τ is the time interval during which it reaches a speed of $0.632v_T$.

At the Active Figures link at http://www.pse6.com, you can vary the size and mass of the sphere and the viscosity (resistance to flow) of the surrounding medium, then observe the effects on the sphere's motion and its speed–time graph.

the gravitational force $\mathbf{F}_g$, let us describe its motion.[1] Applying Newton's second law to the vertical motion, choosing the downward direction to be positive, and noting that $\Sigma F_y = mg - bv$, we obtain

$$mg - bv = ma = m\frac{dv}{dt} \tag{6.3}$$

where the acceleration dv/dt is downward. Solving this expression for the acceleration gives

$$\frac{dv}{dt} = g - \frac{b}{m}v \tag{6.4}$$

This equation is called a *differential equation,* and the methods of solving it may not be familiar to you as yet. However, note that initially when $v = 0$, the magnitude of the resistive force bv is also zero, and the acceleration dv/dt is simply g. As t increases, the magnitude of the resistive force increases and the acceleration decreases. The acceleration approaches zero when the magnitude of the resistive force approaches the sphere's weight. In this situation, the speed of the sphere approaches its **terminal speed** v_T. In reality, the sphere only *approaches* terminal speed but never *reaches* terminal speed.

Terminal speed

We can obtain the terminal speed from Equation 6.3 by setting $a = dv/dt = 0$. This gives

$$mg - bv_T = 0 \qquad \text{or} \qquad v_T = \frac{mg}{b}$$

The expression for v that satisfies Equation 6.4 with $v = 0$ at $t = 0$ is

$$v = \frac{mg}{b}(1 - e^{-bt/m}) = v_T(1 - e^{-t/\tau}) \tag{6.5}$$

This function is plotted in Figure 6.15c. The symbol e represents the base of the natural logarithm, and is also called *Euler's number:* $e = 2.718\ 28$. The **time constant** $\tau = m/b$ (Greek letter tau) is the time at which the sphere released from rest reaches 63.2% of its terminal speed. This can be seen by noting that when $t = \tau$, Equation 6.5 yields $v = 0.632v_T$.

[1] There is also a *buoyant force* acting on the submerged object. This force is constant, and its magnitude is equal to the weight of the displaced liquid. This force changes the apparent weight of the sphere by a constant factor, so we will ignore the force here. We discuss buoyant forces in Chapter 14.

We can check that Equation 6.5 is a solution to Equation 6.4 by direct differentiation:

$$\frac{dv}{dt} = \frac{d}{dt}\left(\frac{mg}{b} - \frac{mg}{b}e^{-bt/m}\right) = -\frac{mg}{b}\frac{d}{dt}e^{-bt/m} = ge^{-bt/m}$$

(See Appendix Table B.4 for the derivative of e raised to some power.) Substituting into Equation 6.4 both this expression for dv/dt and the expression for v given by Equation 6.5 shows that our solution satisfies the differential equation.

Example 6.10 Sphere Falling in Oil

A small sphere of mass 2.00 g is released from rest in a large vessel filled with oil, where it experiences a resistive force proportional to its speed. The sphere reaches a terminal speed of 5.00 cm/s. Determine the time constant τ and the time at which the sphere reaches 90.0% of its terminal speed.

Solution Because the terminal speed is given by $v_T = mg/b$, the coefficient b is

$$b = \frac{mg}{v_T} = \frac{(2.00\text{ g})(980\text{ cm/s}^2)}{5.00\text{ cm/s}} = 392\text{ g/s}$$

Therefore, the time constant τ is

$$\tau = \frac{m}{b} = \frac{2.00\text{ g}}{392\text{ g/s}} = 5.10 \times 10^{-3}\text{ s}$$

The speed of the sphere as a function of time is given by Equation 6.5. To find the time t at which the sphere reaches a speed of $0.900v_T$, we set $v = 0.900v_T$ in Equation 6.5 and solve for t:

$$0.900v_T = v_T(1 - e^{-t/\tau})$$

$$1 - e^{-t/\tau} = 0.900$$

$$e^{-t/\tau} = 0.100$$

$$-\frac{t}{\tau} = \ln(0.100) = -2.30$$

$$t = 2.30\tau = 2.30(5.10 \times 10^{-3}\text{ s})$$

$$= 11.7 \times 10^{-3}\text{ s} = 11.7\text{ ms}$$

Thus, the sphere reaches 90.0% of its terminal speed in a very short time interval.

Air Drag at High Speeds

For objects moving at high speeds through air, such as airplanes, sky divers, cars, and baseballs, the resistive force is approximately proportional to the square of the speed. In these situations, the magnitude of the resistive force can be expressed as

$$R = \tfrac{1}{2}D\rho Av^2 \tag{6.6}$$

where ρ is the density of air, A is the cross-sectional area of the moving object measured in a plane perpendicular to its velocity, and D is a dimensionless empirical quantity called the *drag coefficient*. The drag coefficient has a value of about 0.5 for spherical objects but can have a value as great as 2 for irregularly shaped objects.

Let us analyze the motion of an object in free-fall subject to an upward air resistive force of magnitude $R = \frac{1}{2}D\rho Av^2$. Suppose an object of mass m is released from rest. As Figure 6.16 shows, the object experiences two external forces:[2] the downward gravitational force $\mathbf{F}_g = m\mathbf{g}$ and the upward resistive force $\mathbf{R}$. Hence, the magnitude of the net force is

$$\sum F = mg - \tfrac{1}{2}D\rho Av^2 \tag{6.7}$$

where we have taken downward to be the positive vertical direction. Combining $\sum F = ma$ with Equation 6.7, we find that the object has a downward acceleration of magnitude

$$a = g - \left(\frac{D\rho A}{2m}\right)v^2 \tag{6.8}$$

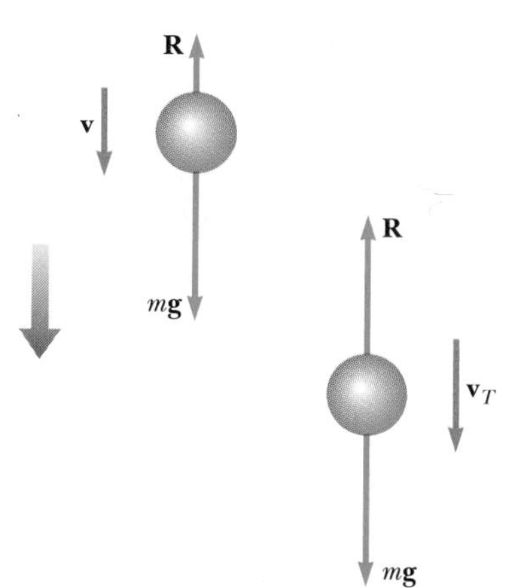

Figure 6.16 An object falling through air experiences a resistive force $\mathbf{R}$ and a gravitational force $\mathbf{F}_g = m\mathbf{g}$. The object reaches terminal speed (on the right) when the net force acting on it is zero, that is, when $\mathbf{R} = -\mathbf{F}_g$ or $R = mg$. Before this occurs, the acceleration varies with speed according to Equation 6.8.

We can calculate the terminal speed v_T by using the fact that when the gravitational force is balanced by the resistive force, the net force on the object is zero and therefore its acceleration is zero. Setting $a = 0$ in Equation 6.8 gives

$$g - \left(\frac{D\rho A}{2m}\right)v_T^2 = 0$$

[2] There is also an upward buoyant force that we neglect.

Table 6.1

Terminal Speed for Various Objects Falling Through Air

Object	Mass (kg)	Cross-Sectional Area (m^2)	v_T (m/s)
Sky diver	75	0.70	60
Baseball (radius 3.7 cm)	0.145	4.2×10^{-3}	43
Golf ball (radius 2.1 cm)	0.046	1.4×10^{-3}	44
Hailstone (radius 0.50 cm)	4.8×10^{-4}	7.9×10^{-5}	14
Raindrop (radius 0.20 cm)	3.4×10^{-5}	1.3×10^{-5}	9.0

so that,

$$v_T = \sqrt{\frac{2mg}{D\rho A}} \tag{6.9}$$

Using this expression, we can determine how the terminal speed depends on the dimensions of the object. Suppose the object is a sphere of radius r. In this case, $A \propto r^2$ (from $A = \pi r^2$) and $m \propto r^3$ (because the mass is proportional to the volume of the sphere, which is $V = \frac{4}{3}\pi r^3$). Therefore, $v_T \propto \sqrt{r}$.

Table 6.1 lists the terminal speeds for several objects falling through air.

Quick Quiz 6.7 A baseball and a basketball, having the same mass, are dropped through air from rest such that their bottoms are initially at the same height above the ground, on the order of 1 m or more. Which one strikes the ground first? (a) the baseball (b) the basketball (c) both strike the ground at the same time.

Conceptual Example 6.11 The Sky Surfer

Consider a sky surfer (Fig. 6.17) who jumps from a plane with her feet attached firmly to her surfboard, does some tricks, and then opens her parachute. Describe the forces acting on her during these maneuvers.

Solution When the surfer first steps out of the plane, she has no vertical velocity. The downward gravitational force causes her to accelerate toward the ground. As her downward speed increases, so does the upward resistive force exerted by the air on her body and the board. This upward force reduces their acceleration, and so their speed increases more slowly. Eventually, they are going so fast that the upward resistive force matches the downward gravitational force. Now the net force is zero and they no longer accelerate, but reach their terminal speed. At some point after reaching terminal speed, she opens her parachute, resulting in a drastic increase in the upward resistive force. The net force (and thus the acceleration) is now upward, in the direction opposite the direction of the velocity. This causes the downward velocity to decrease rapidly; this means the resistive force on the chute also decreases. Eventually the upward resistive force and the downward gravitational force balance each other and a much smaller terminal speed is reached, permitting a safe landing.

(Contrary to popular belief, the velocity vector of a sky diver never points upward. You may have seen a videotape in which a sky diver appears to "rocket" upward once the chute opens. In fact, what happens is that the diver slows down while the person holding the camera continues falling at high speed.)

Figure 6.17 (Conceptual Example 6.11) A sky surfer.

Example 6.12 Falling Coffee Filters

The dependence of resistive force on speed is an empirical relationship. In other words, it is based on observation rather than on a theoretical model. Imagine an experiment in which we drop a series of stacked coffee filters, and measure their terminal speeds. Table 6.2 presents data for these coffee filters as they fall through the air. The time constant τ is small, so that a dropped filter quickly reaches terminal speed. Each filter has a mass of 1.64 g. When the filters are nested together, they stack in such a way that the front-facing surface area does not increase. Determine the relationship between the resistive force exerted by the air and the speed of the falling filters.

Pleated coffee filters can be nested together so that the force of air resistance can be studied.

Solution At terminal speed, the upward resistive force balances the downward gravitational force. So, a single filter falling at its terminal speed experiences a resistive force of

$$R = mg = (1.64 \text{ g})\left(\frac{1 \text{ kg}}{100\,0 \text{ g}}\right)(9.80 \text{ m/s}^2) = 0.016\,1 \text{ N}$$

Two filters nested together experience 0.032 2 N of resistive force, and so forth. A graph of the resistive force on the filters as a function of terminal speed is shown in Figure 6.18a. A straight line would not be a good fit, indicating that the resistive force is *not* proportional to the speed. The behavior is more clearly seen in Figure 6.18b, in which the resistive force is plotted as a function of the square of the terminal speed. This indicates a proportionality of the resistive force to the *square* of the speed, as suggested by Equation 6.6.

Table 6.2

Terminal Speed for Stacked Coffee Filters

Number of Filters	v_T (m/s)[a]
1	1.01
2	1.40
3	1.63
4	2.00
5	2.25
6	2.40
7	2.57
8	2.80
9	3.05
10	3.22

[a] All values of v_T are approximate.

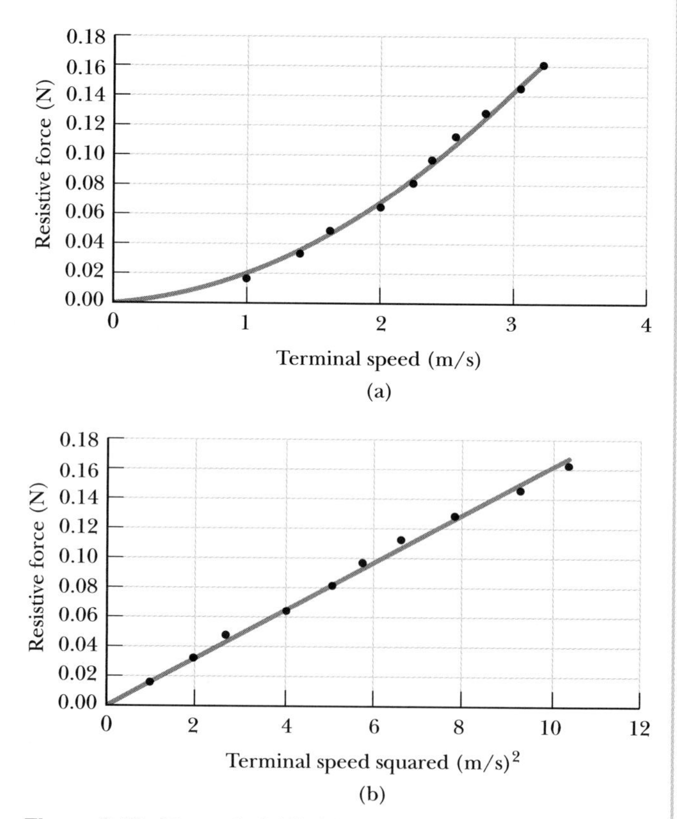

Figure 6.18 (Example 6.12) (a) Relationship between the resistive force acting on falling coffee filters and their terminal speed. The curved line is a second-order polynomial fit. (b) Graph relating the resistive force to the square of the terminal speed. The fit of the straight line to the data points indicates that the resistive force is proportional to the terminal speed squared. Can you find the proportionality constant?

Example 6.13 Resistive Force Exerted on a Baseball

A pitcher hurls a 0.145-kg baseball past a batter at 40.2 m/s (= 90 mi/h). Find the resistive force acting on the ball at this speed.

Solution We do not expect the air to exert a huge force on the ball, and so the resistive force we calculate from Equation 6.6 should not be more than a few newtons.

First, we must determine the drag coefficient D. We do this by imagining that we drop the baseball and allow it to reach terminal speed. We solve Equation 6.9 for D and substitute the appropriate values for m, v_T, and A from Table 6.1. Taking the density of air as 1.20 kg/m^3, we obtain

$$D = \frac{2mg}{v_T{}^2 \rho A} = \frac{2(0.145 \text{ kg})(9.80 \text{ m/s}^2)}{(43 \text{ m/s})^2(1.20 \text{ kg/m}^3)(4.2 \times 10^{-3} \text{ m}^2)}$$
$$= 0.305$$

This number has no dimensions. We have kept an extra digit beyond the two that are significant and will drop it at the end of our calculation.

We can now use this value for D in Equation 6.6 to find the magnitude of the resistive force:

$$R = \tfrac{1}{2} D\rho A v^2$$
$$= \tfrac{1}{2}(0.305)(1.20 \text{ kg/m}^3)(4.2 \times 10^{-3} \text{ m}^2)(40.2 \text{ m/s})^2$$
$$= \boxed{1.2 \text{ N}}$$

6.5 Numerical Modeling in Particle Dynamics[3]

As we have seen in this and the preceding chapter, the study of the dynamics of a particle focuses on describing the position, velocity, and acceleration as functions of time. Cause-and-effect relationships exist among these quantities: Velocity causes position to change, and acceleration causes velocity to change. Because acceleration is the direct result of applied forces, any analysis of the dynamics of a particle usually begins with an evaluation of the net force acting on the particle.

Until now, we have used what is called the *analytical method* to investigate the position, velocity, and acceleration of a moving particle. This method involves the identification of well-behaved functional expressions for the position of a particle (such as the kinematic equations of Chapter 2), generated from algebraic manipulations or the techniques of calculus. Let us review this method briefly before learning about a second way of approaching problems in dynamics. (Because we confine our discussion to one-dimensional motion in this section, boldface notation will not be used for vector quantities.)

If a particle of mass m moves under the influence of a net force ΣF, Newton's second law tells us that the acceleration of the particle is $a = \Sigma F/m$. In general, we apply the analytical method to a dynamics problem using the following procedure:

1. Sum all the forces acting on the particle to find the net force ΣF.
2. Use this net force to determine the acceleration from the relationship $a = \Sigma F/m$.
3. Use this acceleration to determine the velocity from the relationship $dv/dt = a$.
4. Use this velocity to determine the position from the relationship $dx/dt = v$.

The following straightforward example illustrates this method.

Example 6.14 An Object Falling in a Vacuum–Analytical Method

Consider a particle falling in a vacuum under the influence of the gravitational force, as shown in Figure 6.19. Use the analytical method to find the acceleration, velocity, and position of the particle.

Solution The only force acting on the particle is the downward gravitational force of magnitude F_g, which is also the net force. Applying Newton's second law, we set the net force acting on the particle equal to the mass of the particle times its acceleration (taking upward to be the positive y direction):

$$F_g = ma_y = -mg$$

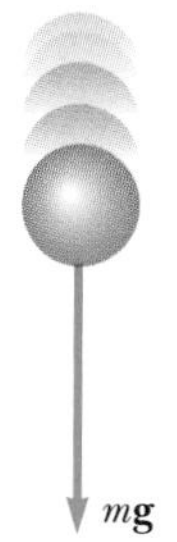

Figure 6.19 (Example 6.14) An object falling in vacuum under the influence of gravity.

[3] The authors are most grateful to Colonel James Head of the U.S. Air Force Academy for preparing this section.

Thus, $a_y = -g$, which means the acceleration is constant. Because $dv_y/dt = a_y$, we see that $dv_y/dt = -g$, which may be integrated to yield

$$v_y(t) = v_{yi} - gt$$

Then, because $v_y = dy/dt$, the position of the particle is obtained from another integration, which yields the well-known result

$$y(t) = y_i + v_{yi}t - \tfrac{1}{2}gt^2$$

In these expressions, y_i and v_{yi} represent the position and speed of the particle at $t_i = 0$.

The analytical method is straightforward for many physical situations. In the "real world," however, complications often arise that make analytical solutions difficult and perhaps beyond the mathematical abilities of most students taking introductory physics. For example, the net force acting on a particle may depend on the particle's position, as in cases where the gravitational acceleration varies with height. Or the force may vary with velocity, as in cases of resistive forces caused by motion through a liquid or gas.

Another complication arises because the expressions relating acceleration, velocity, position, and time are differential equations rather than algebraic ones. Differential equations are usually solved using integral calculus and other special techniques that introductory students may not have mastered.

When such situations arise, scientists often use a procedure called *numerical modeling* to study motion. The simplest numerical model is called the Euler method, after the Swiss mathematician Leonhard Euler (1707–1783).

The Euler Method

In the **Euler method** for solving differential equations, derivatives are approximated as ratios of finite differences. Considering a small increment of time Δt, we can approximate the relationship between a particle's speed and the magnitude of its acceleration as

$$a(t) \approx \frac{\Delta v}{\Delta t} = \frac{v(t + \Delta t) - v(t)}{\Delta t}$$

Then the speed $v(t + \Delta t)$ of the particle at the end of the time interval Δt is approximately equal to the speed $v(t)$ at the beginning of the time interval plus the magnitude of the acceleration during the interval multiplied by Δt:

$$v(t + \Delta t) \approx v(t) + a(t)\,\Delta t \qquad (6.10)$$

Because the acceleration is a function of time, this estimate of $v(t + \Delta t)$ is accurate only if the time interval Δt is short enough such that the change in acceleration during the interval is very small (as is discussed later). Of course, Equation 6.10 is exact if the acceleration is constant.

The position $x(t + \Delta t)$ of the particle at the end of the interval Δt can be found in the same manner:

$$v(t) \approx \frac{\Delta x}{\Delta t} = \frac{x(t + \Delta t) - x(t)}{\Delta t}$$

$$x(t + \Delta t) \approx x(t) + v(t)\,\Delta t \qquad (6.11)$$

You may be tempted to add the term $\frac{1}{2}a(\Delta t)^2$ to this result to make it look like the familiar kinematics equation, but this term is not included in the Euler method because Δt is assumed to be so small that $(\Delta t)^2$ is nearly zero.

If the acceleration at any instant t is known, the particle's velocity and position at a time $t + \Delta t$ can be calculated from Equations 6.10 and 6.11. The calculation then proceeds in a series of finite steps to determine the velocity and position at any later time.

Table 6.3

The Euler Method for Solving Dynamics Problems

Step	Time	Position	Velocity	Acceleration
0	t_0	x_0	v_0	$a_0 = F(x_0, v_0, t_0)/m$
1	$t_1 = t_0 + \Delta t$	$x_1 = x_0 + v_0 \Delta t$	$v_1 = v_0 + a_0 \Delta t$	$a_1 = F(x_1, v_1, t_1)/m$
2	$t_2 = t_1 + \Delta t$	$x_2 = x_1 + v_1 \Delta t$	$v_2 = v_1 + a_1 \Delta t$	$a_2 = F(x_2, v_2, t_2)/m$
3	$t_3 = t_2 + \Delta t$	$x_3 = x_2 + v_2 \Delta t$	$v_3 = v_2 + a_2 \Delta t$	$a_3 = F(x_3, v_3, t_3)/m$
$\vdots$	$\vdots$	$\vdots$	$\vdots$	$\vdots$
n	t_n	x_n	v_n	a_n

The acceleration is determined from the net force acting on the particle, and this force may depend on position, velocity, or time:

$$a(x, v, t) = \frac{\sum F(x, v, t)}{m} \tag{6.12}$$

It is convenient to set up the numerical solution to this kind of problem by numbering the steps and entering the calculations in a table. Table 6.3 illustrates how to do this in an orderly way. Many small increments can be taken, and accurate results can usually be obtained with the help of a computer. The equations provided in the table can be entered into a spreadsheet and the calculations performed row by row to determine the velocity, position, and acceleration as functions of time. The calculations can also be carried out using a programming language, or with commercially available mathematics packages for personal computers. Graphs of velocity versus time or position versus time can be displayed to help you visualize the motion.

One advantage of the Euler method is that the dynamics is not obscured—the fundamental relationships between acceleration and force, velocity and acceleration, and position and velocity are clearly evident. Indeed, these relationships form the heart of the calculations. There is no need to use advanced mathematics, and the basic physics governs the dynamics.

The Euler method is completely reliable for infinitesimally small time increments, but for practical reasons a finite increment size must be chosen. For the finite difference approximation of Equation 6.10 to be valid, the time increment must be small enough that the acceleration can be approximated as being constant during the increment. We can determine an appropriate size for the time increment by examining the particular problem being investigated. The criterion for the size of the time increment may need to be changed during the course of the motion. In practice, however, we usually choose a time increment appropriate to the initial conditions and use the same value throughout the calculations.

The size of the time increment influences the accuracy of the result, but unfortunately it is not easy to determine the accuracy of an Euler-method solution without a knowledge of the correct analytical solution. One method of determining the accuracy of the numerical solution is to repeat the calculations with a smaller time increment and compare results. If the two calculations agree to a certain number of significant figures, you can assume that the results are correct to that precision.

Example 6.15 Euler and the Sphere in Oil Revisited

Consider the sphere falling in oil in Example 6.10. Using the Euler method, find the position and the acceleration of the sphere at the instant that the speed reaches 90.0% of terminal speed.

Solution The net force on the sphere is

$$\Sigma F = -mg + bv$$

Thus, the acceleration values in the last column of Table 6.3 are

$$a = \frac{\Sigma F(x, v, t)}{m} = \frac{-mg + bv}{m} = -g + \frac{bv}{m}$$

Choosing a time increment of 0.1 ms, the first few lines of the spreadsheet modeled after Table 6.3 look like Table 6.4. We see that the speed is increasing while the magnitude of the acceleration is decreasing due to the resistive force. We also see that the sphere does not fall very far in the first millisecond.

Further down the spreadsheet, as shown in Table 6.5, we find the instant at which the sphere reaches the speed $0.900v_T$, which is 0.900×5.00 cm/s = 4.50 cm/s. This calculation shows that this occurs at $t = 11.6$ ms, which agrees within its uncertainty with the value obtained in Example 6.10. The 0.1-ms difference in the two values is due to the approximate nature of the Euler method. If a smaller time increment were used, the instant at which the speed reaches $0.900v_T$ approaches the value calculated in Example 6.10.

From Table 6.5, we see that the position and acceleration of the sphere when it reaches a speed of $0.900v_T$ are

$$y = -0.035 \text{ cm} \qquad \text{and} \qquad a = -99 \text{ cm/s}^2$$

Table 6.4

The Sphere Begins to Fall in Oil

Step	Time (ms)	Position (cm)	Velocity (cm/s)	Acceleration (cm/s²)
0	0.0	0.0000	0.0	−980.0
1	0.1	0.0000	−0.10	−960.8
2	0.2	0.0000	−0.19	−942.0
3	0.3	0.0000	−0.29	−923.5
4	0.4	−0.0001	−0.38	−905.4
5	0.5	−0.0001	−0.47	−887.7
6	0.6	−0.0001	−0.56	−870.3
7	0.7	−0.0002	−0.65	−853.2
8	0.8	−0.0003	−0.73	−836.5
9	0.9	−0.0003	−0.82	−820.1
10	1.0	−0.0004	−0.90	−804.0

Table 6.5

The Sphere Reaches 0.900 v_T

Step	Time (ms)	Position (cm)	Velocity (cm/s)	Acceleration (cm/s²)
110	11.0	−0.0324	−4.43	−111.1
111	11.1	−0.0328	−4.44	−108.9
112	11.2	−0.0333	−4.46	−106.8
113	11.3	−0.0337	−4.47	−104.7
114	11.4	−0.0342	−4.48	−102.6
115	11.5	−0.0346	−4.49	−100.6
116	11.6	−0.0351	−4.50	−98.6
117	11.7	−0.0355	−4.51	−96.7
118	11.8	−0.0360	−4.52	−94.8
119	11.9	−0.0364	−4.53	−92.9
120	12.0	−0.0369	−4.54	−91.1

SUMMARY

Take a practice test for this chapter by clicking the Practice Test link at http://www.pse6.com.

Newton's second law applied to a particle moving in uniform circular motion states that the net force causing the particle to undergo a centripetal acceleration is

$$\sum F = ma_c = \frac{mv^2}{r} \tag{6.1}$$

A particle moving in nonuniform circular motion has both a radial component of acceleration and a nonzero tangential component of acceleration. In the case of a particle rotating in a vertical circle, the gravitational force provides the tangential component of acceleration and part or all of the radial component of acceleration.

An observer in a noninertial (accelerating) frame of reference must introduce **fictitious forces** when applying Newton's second law in that frame. If these fictitious forces are properly defined, the description of motion in the noninertial frame is equivalent to that made by an observer in an inertial frame. However, the observers in the two frames do not agree on the causes of the motion.

An object moving through a liquid or gas experiences a speed-dependent **resistive force.** This resistive force, which opposes the motion relative to the medium, generally increases with speed. The magnitude of the resistive force depends on the size and shape of the object and on the properties of the medium through which the object is moving. In the limiting case for a falling object, when the magnitude of the resistive force equals the object's weight, the object reaches its **terminal speed.** **Euler's method** provides a means for analyzing the motion of a particle under the action of a force that is not simple.

QUESTIONS

1. Why does mud fly off a rapidly turning automobile tire?
2. Imagine that you attach a heavy object to one end of a spring, hold onto the other end of the spring, and then whirl the object in a horizontal circle. Does the spring stretch? If so, why? Discuss this in terms of the force causing the motion to be circular.
3. Describe a situation in which the driver of a car can have a centripetal acceleration but no tangential acceleration.
4. Describe the path of a moving body in the event that its acceleration is constant in magnitude at all times and (a) perpendicular to the velocity; (b) parallel to the velocity.
5. An object executes circular motion with constant speed whenever a net force of constant magnitude acts perpendicular to the velocity. What happens to the speed if the force is not perpendicular to the velocity?
6. Explain why the Earth is not spherical in shape and bulges at the equator.
7. Because the Earth rotates about its axis, it is a noninertial frame of reference. Assume the Earth is a uniform sphere. Why would the apparent weight of an object be greater at the poles than at the equator?
8. What causes a rotary lawn sprinkler to turn?
9. If someone told you that astronauts are weightless in orbit because they are beyond the pull of gravity, would you accept the statement? Explain.
10. It has been suggested that rotating cylinders about 10 mi in length and 5 mi in diameter be placed in space and used as colonies. The purpose of the rotation is to simulate gravity for the inhabitants. Explain this concept for producing an effective imitation of gravity.
11. Consider a rotating space station, spinning with just the right speed such that the centripetal acceleration on the inner surface is g. Thus, astronauts standing on this inner surface would feel pressed to the surface as if they were pressed into the floor because of the Earth's gravitational force. Suppose an astronaut in this station holds a ball above her head and "drops" it to the floor. Will the ball fall just like it would on the Earth?
12. A pail of water can be whirled in a vertical path such that none is spilled. Why does the water stay in the pail, even when the pail is above your head?
13. How would you explain the force that pushes a rider toward the side of a car as the car rounds a corner?
14. Why does a pilot tend to black out when pulling out of a steep dive?
15. The observer in the accelerating elevator of Example 5.8 would claim that the "weight" of the fish is T, the scale reading. This is obviously wrong. Why does this observation differ from that of a person outside the elevator, at rest with respect to the Earth?
16. If you have ever taken a ride in an express elevator of a high-rise building, you may have experienced a nauseating sensation of heaviness or lightness depending on the direction of the acceleration. Explain these sensations. Are we truly weightless in free-fall?
17. A falling sky diver reaches terminal speed with her parachute closed. After the parachute is opened, what parameters change to decrease this terminal speed?
18. Consider a small raindrop and a large raindrop falling through the atmosphere. Compare their terminal speeds. What are their accelerations when they reach terminal speed?
19. On long journeys, jet aircraft usually fly at high altitudes of about 30 000 ft. What is the main advantage of flying at these altitudes from an economic viewpoint?
20. Analyze the motion of a rock falling through water in terms of its speed and acceleration as it falls. Assume that the resistive force acting on the rock increases as the speed increases.
21. "If the current position and velocity of every particle in the Universe were known, together with the laws describing the forces that particles exert on one another, then the whole future of the Universe could be calculated. The future is determinate and preordained. Free will is an illusion." Do you agree with this thesis? Argue for or against it.

PROBLEMS

1, 2, 3 = straightforward, intermediate, challenging □ = full solution available in the *Student Solutions Manual and Study Guide*

= coached solution with hints available at http://www.pse6.com = computer useful in solving problem

= paired numerical and symbolic problems

Section 6.1 Newton's Second Law Applied to Uniform Circular Motion

1. A light string can support a stationary hanging load of 25.0 kg before breaking. A 3.00-kg object attached to the string rotates on a horizontal, frictionless table in a circle of radius 0.800 m, while the other end of the string is held fixed. What range of speeds can the object have before the string breaks?

2. A curve in a road forms part of a horizontal circle. As a car goes around it at constant speed 14.0 m/s, the total force on the driver has magnitude 130 N. What is the total vector force on the driver if the speed is 18.0 m/s instead?

3. In the Bohr model of the hydrogen atom, the speed of the electron is approximately 2.20×10^6 m/s. Find (a) the force acting on the electron as it revolves in a circular orbit of radius 0.530×10^{-10} m and (b) the centripetal acceleration of the electron.

4. In a cyclotron (one type of particle accelerator), a deuteron (of atomic mass 2.00 u) reaches a final speed of 10.0% of the speed of light while moving in a circular path of radius 0.480 m. The deuteron is maintained in the circular path by a magnetic force. What magnitude of force is required?

5. A coin placed 30.0 cm from the center of a rotating, horizontal turntable slips when its speed is 50.0 cm/s. (a) What force causes the centripetal acceleration when the coin is stationary relative to the turntable? (b) What is the coefficient of static friction between coin and turntable?

6. Whenever two *Apollo* astronauts were on the surface of the Moon, a third astronaut orbited the Moon. Assume the orbit to be circular and 100 km above the surface of the Moon, where the acceleration due to gravity is 1.52 m/s^2. The radius of the Moon is 1.70×10^6 m. Determine (a) the astronaut's orbital speed, and (b) the period of the orbit.

7. A crate of eggs is located in the middle of the flat bed of a pickup truck as the truck negotiates an unbanked curve in the road. The curve may be regarded as an arc of a circle of radius 35.0 m. If the coefficient of static friction between crate and truck is 0.600, how fast can the truck be moving without the crate sliding?

8. The cornering performance of an automobile is evaluated on a skidpad, where the maximum speed that a car can maintain around a circular path on a dry, flat surface is measured. Then the centripetal acceleration, also called the lateral acceleration, is calculated as a multiple of the free-fall acceleration g. The main factors affecting the performance are the tire characteristics and the suspension system of the car. A Dodge Viper GTS can negotiate a skidpad of radius 61.0 m at 86.5 km/h. Calculate its maximum lateral acceleration.

9. Consider a conical pendulum with an 80.0-kg bob on a 10.0-m wire making an angle of 5.00° with the vertical (Fig. P6.9). Determine (a) the horizontal and vertical components of the force exerted by the wire on the pendulum and (b) the radial acceleration of the bob.

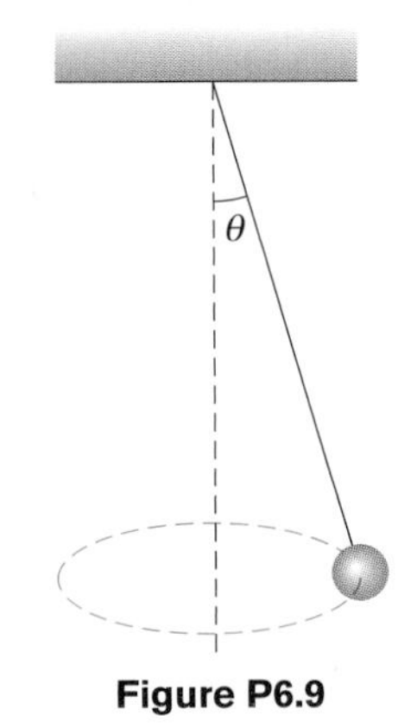

Figure P6.9

10. A car initially traveling eastward turns north by traveling in a circular path at uniform speed as in Figure P6.10. The length of the arc ABC is 235 m, and the car completes the turn in 36.0 s. (a) What is the acceleration when the car is at B located at an angle of 35.0°? Express your answer in terms of the unit vectors $\hat{\mathbf{i}}$ and $\hat{\mathbf{j}}$. Determine (b) the car's average speed and (c) its average acceleration during the 36.0-s interval.

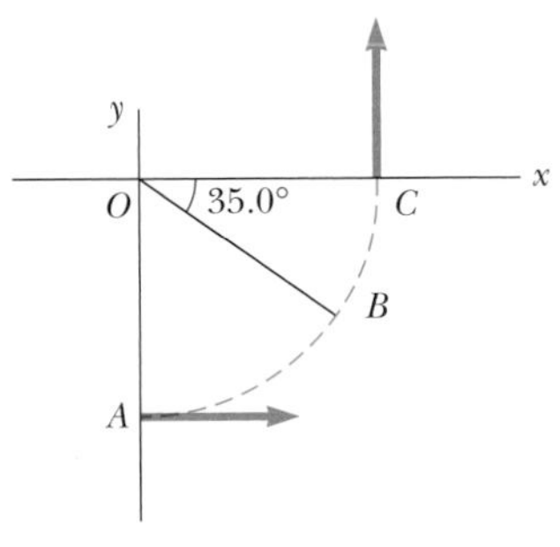

Figure P6.10

11. A 4.00-kg object is attached to a vertical rod by two strings, as in Figure P6.11. The object rotates in a horizontal circle at constant speed 6.00 m/s. Find the tension in (a) the upper string and (b) the lower string.

12. Casting of molten metal is important in many industrial processes. *Centrifugal casting* is used for manufacturing pipes, bearings and many other structures. A variety of sophisticated techniques have been invented, but the basic idea is as illustrated in Figure P6.12. A cylindrical enclosure is rotated rapidly and steadily about a horizontal axis. Molten metal is poured into the rotating cylinder and then cooled, forming the finished product. Turning the cylin-

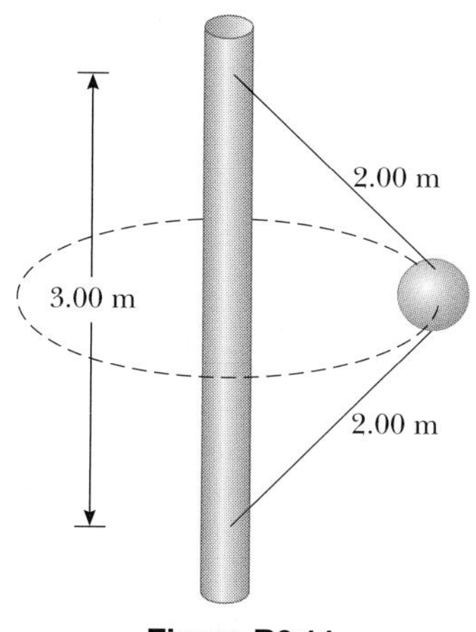

Figure P6.11

der at a high rotation rate forces the solidifying metal strongly to the outside. Any bubbles are displaced toward the axis, so unwanted voids will not be present in the casting. Sometimes it is desirable to form a composite casting, such as for a bearing. Here a strong steel outer surface is poured, followed by an inner lining of special low-friction metal. In some applications a very strong metal is given a coating of corrosion-resistant metal. Centrifugal casting results in strong bonding between the layers.

Suppose that a copper sleeve of inner radius 2.10 cm and outer radius 2.20 cm is to be cast. To eliminate bubbles and give high structural integrity, the centripetal acceleration of each bit of metal should be $100g$. What rate of rotation is required? State the answer in revolutions per minute.

Section 6.2 Nonuniform Circular Motion

13. A 40.0-kg child swings in a swing supported by two chains, each 3.00 m long. If the tension in each chain at the lowest point is 350 N, find (a) the child's speed at the lowest point and (b) the force exerted by the seat on the child at the lowest point. (Neglect the mass of the seat.)

14. A child of mass m swings in a swing supported by two chains, each of length R. If the tension in each chain at the lowest point is T, find (a) the child's speed at the lowest point and (b) the force exerted by the seat on the child at the lowest point. (Neglect the mass of the seat.)

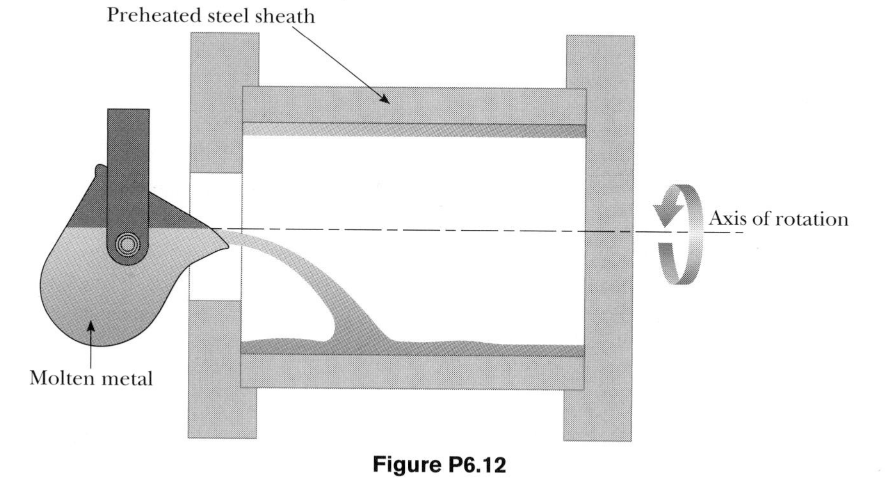

Figure P6.12

15. Tarzan (m = 85.0 kg) tries to cross a river by swinging from a vine. The vine is 10.0 m long, and his speed at the bottom of the swing (as he just clears the water) will be 8.00 m/s. Tarzan doesn't know that the vine has a breaking strength of 1 000 N. Does he make it safely across the river?

16. A hawk flies in a horizontal arc of radius 12.0 m at a constant speed of 4.00 m/s. (a) Find its centripetal acceleration. (b) It continues to fly along the same horizontal arc but increases its speed at the rate of 1.20 m/s^2. Find the acceleration (magnitude and direction) under these conditions.

17. A pail of water is rotated in a vertical circle of radius 1.00 m. What is the minimum speed of the pail at the top of the circle if no water is to spill out?

18. A 0.400-kg object is swung in a vertical circular path on a string 0.500 m long. If its speed is 4.00 m/s at the top of the circle, what is the tension in the string there?

19. A roller coaster car (Fig. P6.19) has a mass of 500 kg when fully loaded with passengers. (a) If the vehicle has a speed of 20.0 m/s at point Ⓐ, what is the force exerted by the track on the car at this point? (b) What is the maximum speed the vehicle can have at Ⓑ and still remain on the track?

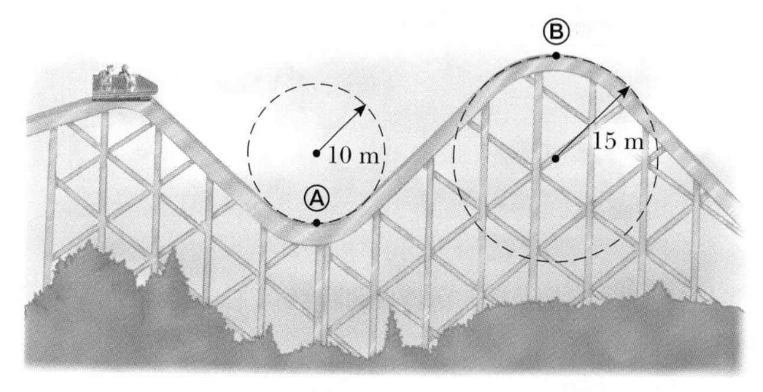

Figure P6.19

20. A roller coaster at the Six Flags Great America amusement park in Gurnee, IL, incorporates some clever design technology and some basic physics. Each vertical loop, instead of being circular, is shaped like a teardrop (Fig. P6.20). The cars ride on the inside of the loop at the top, and the speeds are high enough to ensure that the cars remain on the track. The biggest loop is 40.0 m high, with a maximum speed of 31.0 m/s (nearly 70 mi/h) at the bottom. Suppose

the speed at the top is 13.0 m/s and the corresponding centripetal acceleration is $2g$. (a) What is the radius of the arc of the teardrop at the top? (b) If the total mass of a car plus the riders is M, what force does the rail exert on the car at the top? (c) Suppose the roller coaster had a circular loop of radius 20.0 m. If the cars have the same speed, 13.0 m/s at the top, what is the centripetal acceleration at the top? Comment on the normal force at the top in this situation.

Frank Cezus / Getty Images

Figure P6.20

Section 6.3 Motion in Accelerated Frames

21. An object of mass 5.00 kg, attached to a spring scale, rests on a frictionless, horizontal surface as in Figure P6.21. The spring scale, attached to the front end of a boxcar, has a constant reading of 18.0 N when the car is in motion. (a) If the spring scale reads zero when the car is at rest, determine the acceleration of the car. (b) What constant reading will the spring scale show if the car moves with constant velocity? (c) Describe the forces on the object as observed by someone in the car and by someone at rest outside the car.

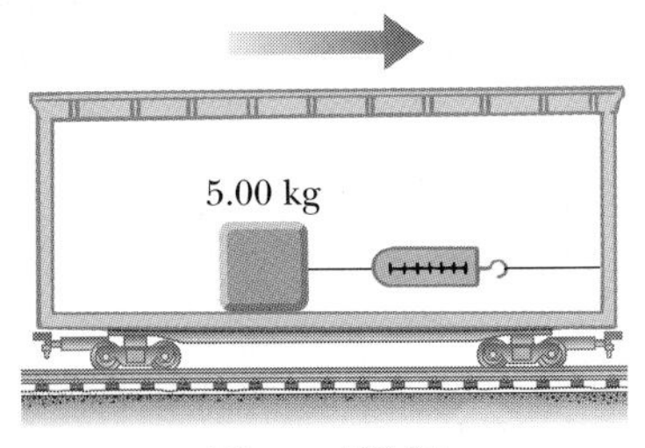

Figure P6.21

22. If the coefficient of static friction between your coffee cup and the horizontal dashboard of your car is $\mu_s = 0.800$, how fast can you drive on a horizontal roadway around a right turn of radius 30.0 m before the cup starts to slide? If you go too fast, in what direction will the cup slide relative to the dashboard?

23. A 0.500-kg object is suspended from the ceiling of an accelerating boxcar as in Figure 6.13. If $a = 3.00$ m/s^2, find (a) the angle that the string makes with the vertical and (b) the tension in the string.

24. A small container of water is placed on a carousel inside a microwave oven, at a radius of 12.0 cm from the center. The turntable rotates steadily, turning through one revolution in each 7.25 s. What angle does the water surface make with the horizontal?

25. A person stands on a scale in an elevator. As the elevator starts, the scale has a constant reading of 591 N. As the elevator later stops, the scale reading is 391 N. Assume the magnitude of the acceleration is the same during starting and stopping, and determine (a) the weight of the person, (b) the person's mass, and (c) the acceleration of the elevator.

26. The Earth rotates about its axis with a period of 24.0 h. Imagine that the rotational speed can be increased. If an object at the equator is to have zero apparent weight, (a) what must the new period be? (b) By what factor would the speed of the object be increased when the planet is rotating at the higher speed? Note that the apparent weight of the object becomes zero when the normal force exerted on it is zero.

27. A small block is at rest on the floor at the front of a railroad boxcar that has length ℓ. The coefficient of kinetic friction between the floor of the car and the block is μ_k. The car, originally at rest, begins to move with acceleration a. The block slides back horizontally until it hits the back wall of the car. At that moment, what is its speed (a) relative to the car? (b) relative to Earth?

28. A student stands in an elevator that is continuously accelerating upward with acceleration a. Her backpack is sitting on the floor next to the wall. The width of the elevator car is L. The student gives her backpack a quick kick at $t = 0$, imparting to it speed v, and making it slide across the elevator floor. At time t, the backpack hits the opposite wall. Find the coefficient of kinetic friction μ_k between the backpack and the elevator floor.

29. A child on vacation wakes up. She is lying on her back. The tension in the muscles on both sides of her neck is 55.0 N as she raises her head to look past her toes and out the motel window. Finally it is not raining! Ten minutes later she is screaming feet first down a water slide at terminal speed 5.70 m/s, riding high on the outside wall of a horizontal curve of radius 2.40 m (Figure P6.29). She raises her head to look forward past her toes; find the tension in the muscles on both sides of her neck.

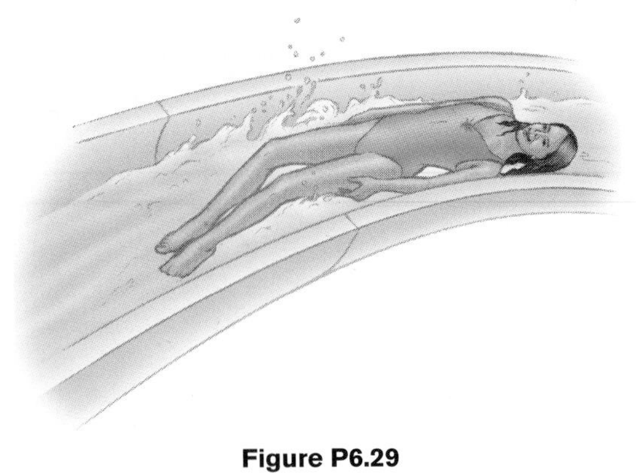

Figure P6.29

30. One popular design of a household juice machine is a conical, perforated stainless steel basket 3.30 cm high with a closed bottom of diameter 8.00 cm and open top of diameter 13.70 cm that spins at 20 000 revolutions per minute about a vertical axis (Figure P6.30). Solid pieces of fruit are chopped into granules by cutters at the bottom of the spinning cone. Then the fruit granules rapidly make their way to the sloping surface where the juice is extracted to the outside of the cone through the mesh perforations. The dry pulp spirals upward along the slope to be ejected from the top of the cone. The juice is collected in an enclosure immediately surrounding the sloped surface of the cone. (a) What centripetal acceleration does a bit of fruit experience when it is spinning with the basket at a point midway between the top and bottom? Express the answer as a multiple of g. (b) Observe that the weight of the fruit is a negligible force. What is the normal force on 2.00 g of fruit at that point? (c) If the effective coefficient of kinetic friction between the fruit and the cone is 0.600, with what acceleration relative to the cone will the bit of fruit start to slide up the wall of the cone at that point, after being temporarily stuck?

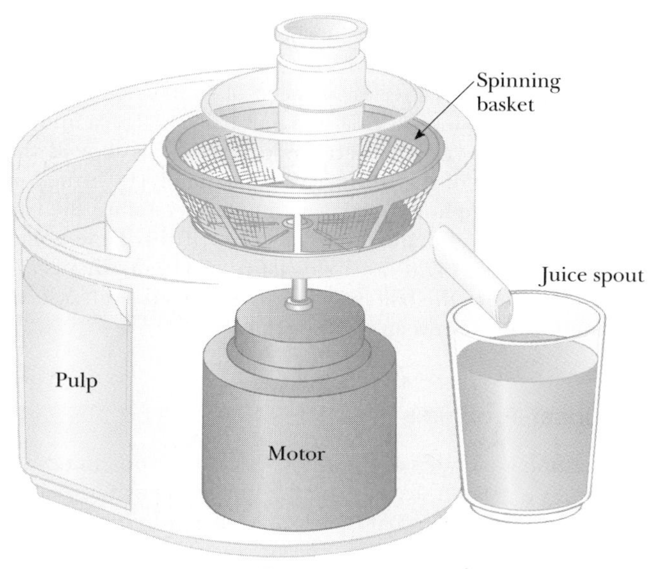

Figure P6.30

31. A plumb bob does not hang exactly along a line directed to the center of the Earth's rotation. How much does the plumb bob deviate from a radial line at 35.0° north latitude? Assume that the Earth is spherical.

Section 6.4 Motion in the Presence of Resistive Forces

32. A sky diver of mass 80.0 kg jumps from a slow-moving aircraft and reaches a terminal speed of 50.0 m/s. (a) What is the acceleration of the sky diver when her speed is 30.0 m/s? What is the drag force on the diver when her speed is (b) 50.0 m/s? (c) 30.0 m/s?

33. A small piece of Styrofoam packing material is dropped from a height of 2.00 m above the ground. Until it reaches terminal speed, the magnitude of its acceleration is given by $a = g - bv$. After falling 0.500 m, the Styrofoam effectively reaches terminal speed, and then takes 5.00 s more to reach the ground. (a) What is the value of the constant b? (b) What is the acceleration at $t = 0$? (c) What is the acceleration when the speed is 0.150 m/s?

34. (a) Estimate the terminal speed of a wooden sphere (density 0.830 g/cm^3) falling through air if its radius is 8.00 cm and its drag coefficient is 0.500. (b) From what height would a freely falling object reach this speed in the absence of air resistance?

35. Calculate the force required to pull a copper ball of radius 2.00 cm upward through a fluid at the constant speed 9.00 cm/s. Take the drag force to be proportional to the speed, with proportionality constant 0.950 kg/s. Ignore the buoyant force.

36. A fire helicopter carries a 620-kg bucket at the end of a cable 20.0 m long as in Figure P6.36. As the helicopter flies to a fire at a constant speed of 40.0 m/s, the cable makes an angle of 40.0° with respect to the vertical. The bucket presents a cross-sectional area of 3.80 m^2 in a plane perpendicular to the air moving past it. Determine the drag coefficient assuming that the resistive force is proportional to the square of the bucket's speed.

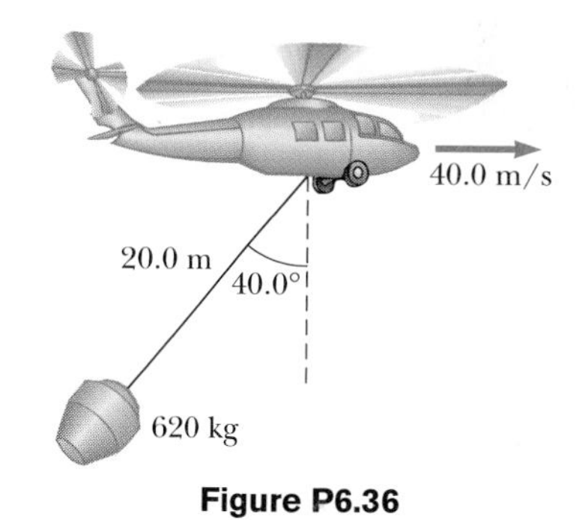

Figure P6.36

37. A small, spherical bead of mass 3.00 g is released from rest at $t = 0$ in a bottle of liquid shampoo. The terminal speed is observed to be $v_T = 2.00$ cm/s. Find (a) the value of the constant b in Equation 6.2, (b) the time τ at which the bead reaches $0.632v_T$, and (c) the value of the resistive force when the bead reaches terminal speed.

38. The mass of a sports car is 1 200 kg. The shape of the body is such that the aerodynamic drag coefficient is 0.250 and the frontal area is 2.20 m^2. Neglecting all other sources of friction, calculate the initial acceleration of the car if it has been traveling at 100 km/h and is now shifted into neutral and allowed to coast.

39. A motorboat cuts its engine when its speed is 10.0 m/s and coasts to rest. The equation describing the motion of the motorboat during this period is $v = v_i e^{-ct}$, where v is the speed at time t, v_i is the initial speed, and c is a constant. At $t = 20.0$ s, the speed is 5.00 m/s. (a) Find the constant c. (b) What is the speed at $t = 40.0$ s? (c) Differentiate the expression for $v(t)$ and thus show that the acceleration of the boat is proportional to the speed at any time.

40. Consider an object on which the net force is a resistive force proportional to the square of its speed. For example, assume that the resistive force acting on a speed skater is $f = -kmv^2$, where k is a constant and m is the skater's mass. The skater crosses the finish line of a straight-line race with

speed v_0 and then slows down by coasting on his skates. Show that the skater's speed at any time t after crossing the finish line is $v(t) = v_0/(1 + ktv_0)$. This problem also provides the background for the two following problems.

41. (a) Use the result of Problem 40 to find the position x as a function of time for an object of mass m, located at $x = 0$ and moving with velocity $v_0\hat{\mathbf{i}}$ at time $t = 0$ and thereafter experiencing a net force $-kmv^2\hat{\mathbf{i}}$. (b) Find the object's velocity as a function of position.

42. At major league baseball games it is commonplace to flash on the scoreboard a speed for each pitch. This speed is determined with a radar gun aimed by an operator positioned behind home plate. The gun uses the Doppler shift of microwaves reflected from the baseball, as we will study in Chapter 39. The gun determines the speed at some particular point on the baseball's path, depending on when the operator pulls the trigger. Because the ball is subject to a drag force due to air, it slows as it travels 18.3 m toward the plate. Use the result of Problem 41(b) to find how much its speed decreases. Suppose the ball leaves the pitcher's hand at 90.0 mi/h = 40.2 m/s. Ignore its vertical motion. Use data on baseballs from Example 6.13 to determine the speed of the pitch when it crosses the plate.

43. You can feel a force of air drag on your hand if you stretch your arm out of the open window of a speeding car. [*Note:* Do not endanger yourself.] What is the order of magnitude of this force? In your solution state the quantities you measure or estimate and their values.

Section 6.5 Numerical Modeling in Particle Dynamics

44. A 3.00-g leaf is dropped from a height of 2.00 m above the ground. Assume the net downward force exerted on the leaf is $F = mg - bv$, where the drag factor is $b = 0.030\,0$ kg/s. (a) Calculate the terminal speed of the leaf. (b) Use Euler's method of numerical analysis to find the speed and position of the leaf, as functions of time, from the instant it is released until 99% of terminal speed is reached. (*Suggestion:* Try $\Delta t = 0.005$ s.)

45. A hailstone of mass 4.80×10^{-4} kg falls through the air and experiences a net force given by

$$F = -mg + Cv^2$$

where $C = 2.50 \times 10^{-5}$ kg/m. (a) Calculate the terminal speed of the hailstone. (b) Use Euler's method of numerical analysis to find the speed and position of the hailstone at 0.2-s intervals, taking the initial speed to be zero. Continue the calculation until the hailstone reaches 99% of terminal speed.

46. A 0.142-kg baseball has a terminal speed of 42.5 m/s (95 mi/h). (a) If a baseball experiences a drag force of magnitude $R = Cv^2$, what is the value of the constant C? (b) What is the magnitude of the drag force when the speed of the baseball is 36.0 m/s? (c) Use a computer to determine the motion of a baseball thrown vertically upward at an initial speed of 36 m/s. What maximum height does the ball reach? How long is it in the air? What is its speed just before it hits the ground?

47. A 50.0-kg parachutist jumps from an airplane and falls to Earth with a drag force proportional to the square of the speed, $R = Cv^2$. Take $C = 0.200$ kg/m (with the parachute closed) and $C = 20.0$ kg/m (with the chute open). (a) Determine the terminal speed of the parachutist in both configurations, before and after the chute is opened. (b) Set up a numerical analysis of the motion and compute the speed and position as functions of time, assuming the jumper begins the descent at 1 000 m above the ground and is in free fall for 10.0 s before opening the parachute. (*Suggestion:* When the parachute opens, a sudden large acceleration takes place; a smaller time step may be necessary in this region.)

48. Consider a 10.0-kg projectile launched with an initial speed of 100 m/s, at an elevation angle of 35.0°. The resistive force is $\mathbf{R} = -b\mathbf{v}$, where $b = 10.0$ kg/s. (a) Use a numerical method to determine the horizontal and vertical coordinates of the projectile as functions of time. (b) What is the range of this projectile? (c) Determine the elevation angle that gives the maximum range for the projectile. (*Suggestion:* Adjust the elevation angle by trial and error to find the greatest range.)

49. A professional golfer hits her 5-iron 155 m (170 yd). A 46.0-g golf ball experiences a drag force of magnitude $R = Cv^2$, and has a terminal speed of 44.0 m/s. (a) Calculate the drag constant C for the golf ball. (b) Use a numerical method to calculate the trajectory of this shot. If the initial velocity of the ball makes an angle of 31.0° (the loft angle) with the horizontal, what initial speed must the ball have to reach the 155-m distance? (c) If this same golfer hits her 9-iron (47.0° loft) a distance of 119 m, what is the initial speed of the ball in this case? Discuss the differences in trajectories between the two shots.

Additional Problems

50. In a home laundry dryer, a cylindrical tub containing wet clothes is rotated steadily about a horizontal axis, as shown in Figure P6.50. So that the clothes will dry uniformly, they are made to tumble. The rate of rotation of the smooth-walled tub is chosen so that a small piece of cloth will lose contact with the tub when the cloth is at an angle of 68.0°

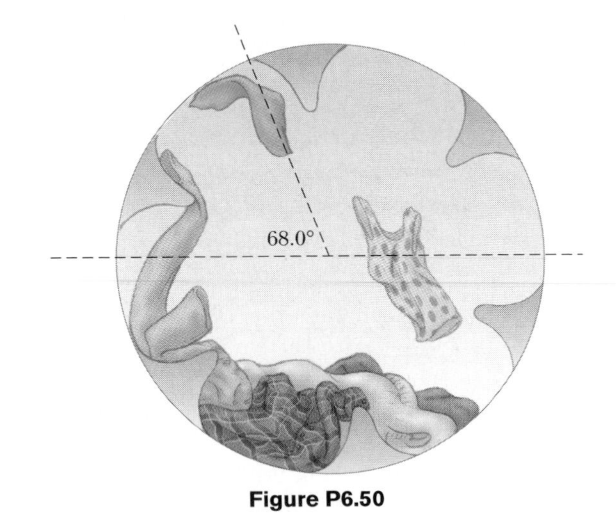

Figure P6.50

above the horizontal. If the radius of the tub is 0.330 m, what rate of revolution is needed?

51. We will study the most important work of Nobel laureate Arthur Compton in Chapter 40. Disturbed by speeding cars outside the physics building at Washington University in St. Louis, Compton designed a speed bump and had it installed. Suppose that a 1 800-kg car passes over a bump in a roadway that follows the arc of a circle of radius 20.4 m as in Figure P6.51. (a) What force does the road exert on the car as the car passes the highest point of the bump if the car travels at 30.0 km/h? (b) **What If?** What is the maximum speed the car can have as it passes this highest point without losing contact with the road?

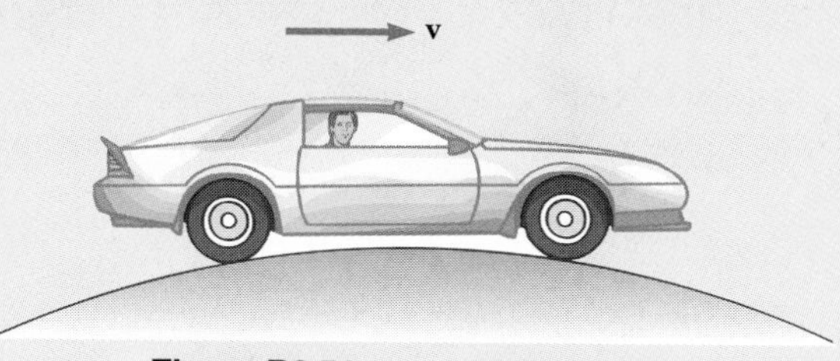

Figure P6.51 Problems 51 and 52.

52. A car of mass m passes over a bump in a road that follows the arc of a circle of radius R as in Figure P6.51. (a) What force does the road exert on the car as the car passes the highest point of the bump if the car travels at a speed v? (b) **What If?** What is the maximum speed the car can have as it passes this highest point without losing contact with the road?

53. Interpret the graph in Figure 6.18(b). Proceed as follows: (a) Find the slope of the straight line, including its units. (b) From Equation 6.6, $R = \frac{1}{2}D\rho Av^2$, identify the theoretical slope of a graph of resistive force versus squared speed. (c) Set the experimental and theoretical slopes equal to each other and proceed to calculate the drag coefficient of the filters. Use the value for the density of air listed on the book's endpapers. Model the cross-sectional area of the filters as that of a circle of radius 10.5 cm. (d) Arbitrarily choose the eighth data point on the graph and find its vertical separation from the line of best fit. Express this scatter as a percentage. (e) In a short paragraph state what the graph demonstrates and compare it to the theoretical prediction. You will need to make reference to the quantities plotted on the axes, to the shape of the graph line, to the data points, and to the results of parts (c) and (d).

54. A student builds and calibrates an accelerometer, which she uses to determine the speed of her car around a certain unbanked highway curve. The accelerometer is a plumb bob with a protractor that she attaches to the roof of her car. A friend riding in the car with her observes that the plumb bob hangs at an angle of 15.0° from the vertical when the car has a speed of 23.0 m/s. (a) What is the centripetal acceleration of the car rounding the curve? (b) What is the radius of the curve? (c) What is the speed of the car if the plumb bob deflection is 9.00° while rounding the same curve?

55. Suppose the boxcar of Figure 6.13 is moving with constant acceleration a up a hill that makes an angle ϕ with the horizontal. If the pendulum makes a constant angle θ with the perpendicular to the ceiling, what is a?

56. (a) A luggage carousel at an airport has the form of a section of a large cone, steadily rotating about its vertical axis. Its metallic surface slopes downward toward the outside, making an angle of 20.0° with the horizontal. A piece of luggage having mass 30.0 kg is placed on the carousel, 7.46 m from the axis of rotation. The travel bag goes around once in 38.0 s. Calculate the force of static friction between the bag and the carousel. (b) The drive motor is shifted to turn the carousel at a higher constant rate of rotation, and the piece of luggage is bumped to another position, 7.94 m from the axis of rotation. Now going around once in every 34.0 s, the bag is on the verge of slipping. Calculate the coefficient of static friction between the bag and the carousel.

57. Because the Earth rotates about its axis, a point on the equator experiences a centripetal acceleration of 0.033 7 m/s², while a point at the poles experiences no centripetal acceleration. (a) Show that at the equator the gravitational force on an object must exceed the normal force required to support the object. That is, show that the object's true weight exceeds its apparent weight. (b) What is the apparent weight at the equator and at the poles of a person having a mass of 75.0 kg? (Assume the Earth is a uniform sphere and take $g = 9.800$ m/s².)

58. An air puck of mass m_1 is tied to a string and allowed to revolve in a circle of radius R on a frictionless horizontal table. The other end of the string passes through a hole in the center of the table, and a counterweight of mass m_2 is tied to it (Fig. P6.58). The suspended object remains in equilibrium while the puck on the tabletop revolves. What is (a) the tension in the string? (b) the radial force acting on the puck? (c) the speed of the puck?

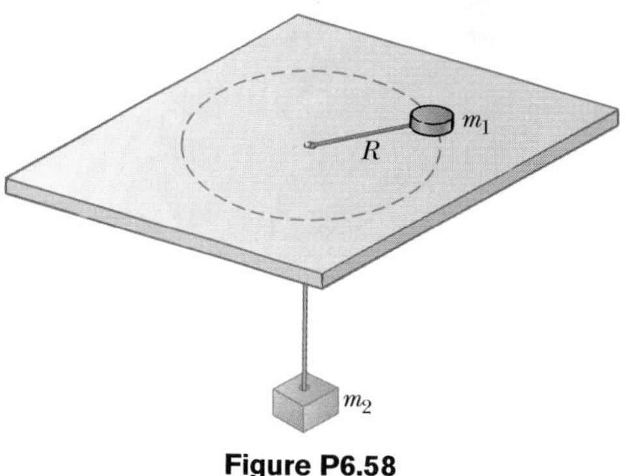

Figure P6.58

59. The pilot of an airplane executes a constant-speed loop-the-loop maneuver in a vertical circle. The speed of the airplane is 300 mi/h, and the radius of the circle is 1 200 ft. (a) What is the pilot's apparent weight at the lowest point if his true weight is 160 lb? (b) What is his apparent weight at the highest point? (c) **What If?** Describe how the pilot could experience weightlessness if both the radius and the speed can be varied. (*Note:* His apparent weight is equal to the magnitude of the force exerted by the seat on his body.)

60. A penny of mass 3.10 g rests on a small 20.0-g block supported by a spinning disk (Fig. P6.60). The coefficients of friction between block and disk are 0.750 (static) and

0.640 (kinetic) while those for the penny and block are 0.520 (static) and 0.450 (kinetic). What is the maximum rate of rotation in revolutions per minute that the disk can have, without the block or penny sliding on the disk?

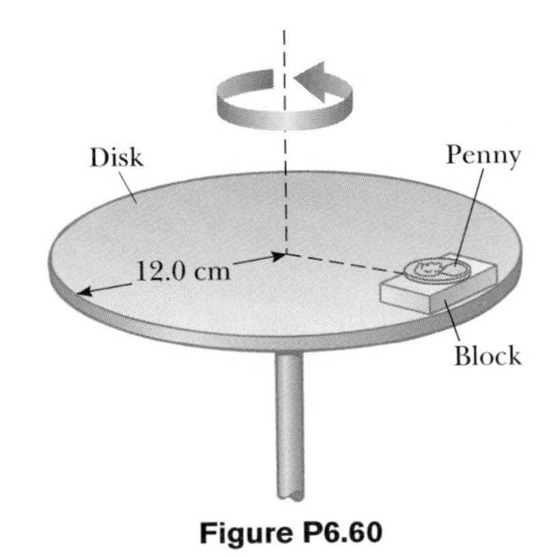

Figure P6.60

61. Figure P6.61 shows a Ferris wheel that rotates four times each minute. It carries each car around a circle of diameter 18.0 m. (a) What is the centripetal acceleration of a rider? What force does the seat exert on a 40.0-kg rider (b) at the lowest point of the ride and (c) at the highest point of the ride? (d) What force (magnitude and direction) does the seat exert on a rider when the rider is halfway between top and bottom?

Figure P6.61

62. A space station, in the form of a wheel 120 m in diameter, rotates to provide an "artificial gravity" of 3.00 m/s^2 for persons who walk around on the inner wall of the outer rim. Find the rate of rotation of the wheel (in revolutions per minute) that will produce this effect.

63. An amusement park ride consists of a rotating circular platform 8.00 m in diameter from which 10.0-kg seats are suspended at the end of 2.50-m massless chains (Fig. P6.63). When the system rotates, the chains make an angle $\theta = 28.0°$ with the vertical. (a) What is the speed of each seat? (b) Draw a free-body diagram of a 40.0-kg child riding in a seat and find the tension in the chain.

64. A piece of putty is initially located at point A on the rim of a grinding wheel rotating about a horizontal axis. The putty is dislodged from point A when the diameter through A is horizontal. It then rises vertically and returns to A at the instant the wheel completes one revolution.

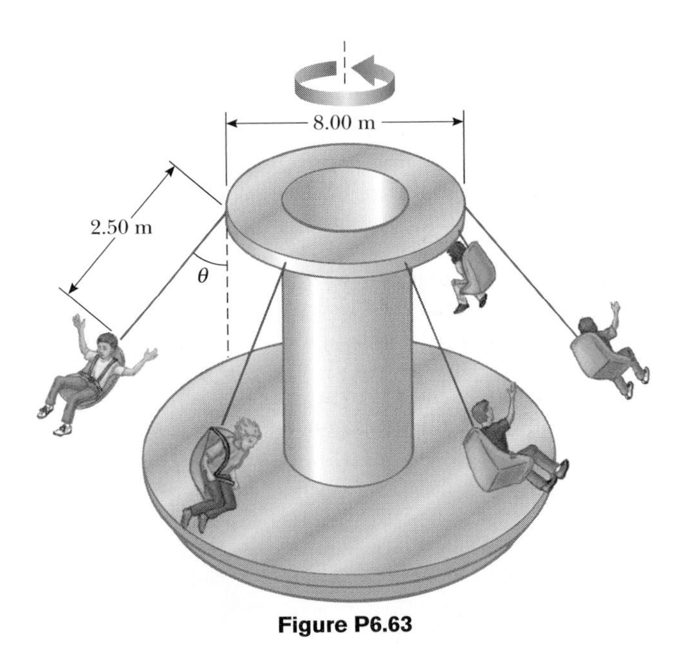

Figure P6.63

(a) Find the speed of a point on the rim of the wheel in terms of the acceleration due to gravity and the radius R of the wheel. (b) If the mass of the putty is m, what is the magnitude of the force that held it to the wheel?

65. An amusement park ride consists of a large vertical cylinder that spins about its axis fast enough such that any person inside is held up against the wall when the floor drops away (Fig. P6.65). The coefficient of static friction between person and wall is μ_s, and the radius of the cylinder is R. (a) Show that the maximum period of revolution necessary to keep the person from falling is $T = (4\pi^2 R\mu_s/g)^{1/2}$. (b) Obtain a numerical value for T if $R = 4.00$ m and $\mu_s = 0.400$. How many revolutions per minute does the cylinder make?

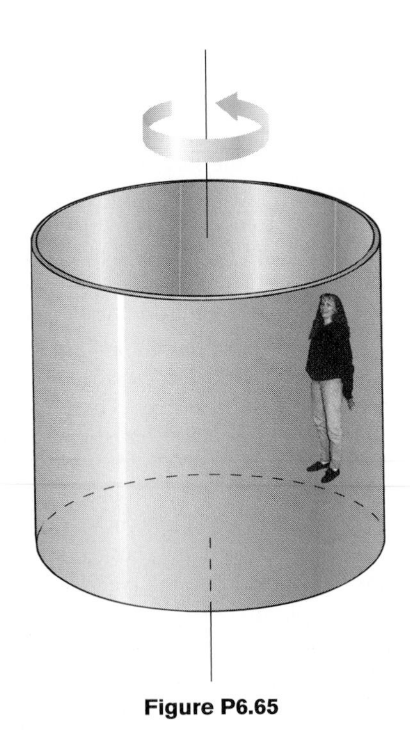

Figure P6.65

66. *An example of the Coriolis effect.* Suppose air resistance is negligible for a golf ball. A golfer tees off from a location precisely at $\phi_i = 35.0°$ north latitude. He hits the ball due south, with range 285 m. The ball's initial velocity is at 48.0° above the horizontal. (a) For how long is the ball in flight? The cup is due south of the golfer's location, and he would have a hole-in-one if the Earth were not rotating. The Earth's rotation makes the tee move in a circle of radius $R_E \cos \phi_i = (6.37 \times 10^6 \text{ m}) \cos 35.0°$, as shown in Figure P6.66. The tee completes one revolution each day. (b) Find the eastward speed of the tee, relative to the stars. The hole is also moving east, but it is 285 m farther south, and thus at a slightly lower latitude ϕ_f. Because the hole moves in a slightly larger circle, its speed must be greater than that of the tee. (c) By how much does the hole's speed exceed that of the tee? During the time the ball is in flight, it moves upward and downward as well as southward with the projectile motion you studied in Chapter 4, but it also moves eastward with the speed you found in part (b). The hole moves to the east at a faster speed, however, pulling ahead of the ball with the relative speed you found in part (c). (d) How far to the west of the hole does the ball land?

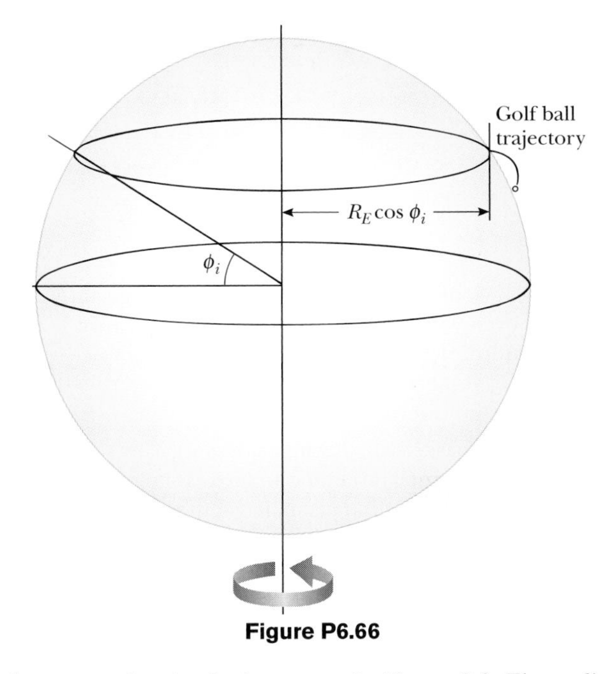

Figure P6.66

67. A car rounds a banked curve as in Figure 6.6. The radius of curvature of the road is R, the banking angle is θ, and the coefficient of static friction is μ_s. (a) Determine the range of speeds the car can have without slipping up or down the road. (b) Find the minimum value for μ_s such that the minimum speed is zero. (c) What is the range of speeds possible if $R = 100$ m, $\theta = 10.0°$, and $\mu_s = 0.100$ (slippery conditions)?

68. A single bead can slide with negligible friction on a wire that is bent into a circular loop of radius 15.0 cm, as in Figure P6.68. The circle is always in a vertical plane and rotates steadily about its vertical diameter with (a) a period of 0.450 s. The position of the bead is described by the angle θ that the radial line, from the center of the loop to the bead, makes with the vertical. At what angle up from the bottom of the circle can the bead stay motionless relative to the turning circle? (b) **What If?** Repeat the problem if the period of the circle's rotation is 0.850 s.

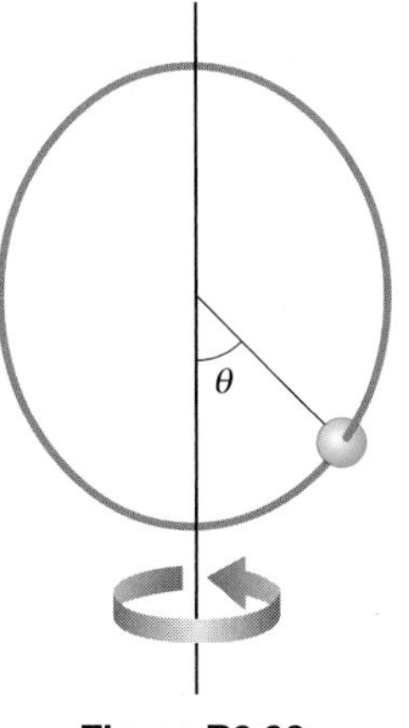

Figure P6.68

69. The expression $F = arv + br^2v^2$ gives the magnitude of the resistive force (in newtons) exerted on a sphere of radius r (in meters) by a stream of air moving at speed v (in meters per second), where a and b are constants with appropriate SI units. Their numerical values are $a = 3.10 \times 10^{-4}$ and $b = 0.870$. Using this expression, find the terminal speed for water droplets falling under their own weight in air, taking the following values for the drop radii: (a) 10.0 μm, (b) 100 μm, (c) 1.00 mm. Note that for (a) and (c) you can obtain accurate answers without solving a quadratic equation, by considering which of the two contributions to the air resistance is dominant and ignoring the lesser contribution.

70. A 9.00-kg object starting from rest falls through a viscous medium and experiences a resistive force $\mathbf{R} = -b\mathbf{v}$, where $\mathbf{v}$ is the velocity of the object. If the object reaches one-half its terminal speed in 5.54 s, (a) determine the terminal speed. (b) At what time is the speed of the object three-fourths the terminal speed? (c) How far has the object traveled in the first 5.54 s of motion?

71. A model airplane of mass 0.750 kg flies in a horizontal circle at the end of a 60.0-m control wire, with a speed of 35.0 m/s. Compute the tension in the wire if it makes a constant angle of 20.0° with the horizontal. The forces exerted on the airplane are the pull of the control wire, the gravitational force, and aerodynamic lift, which acts at 20.0° inward from the vertical as shown in Figure P6.71.

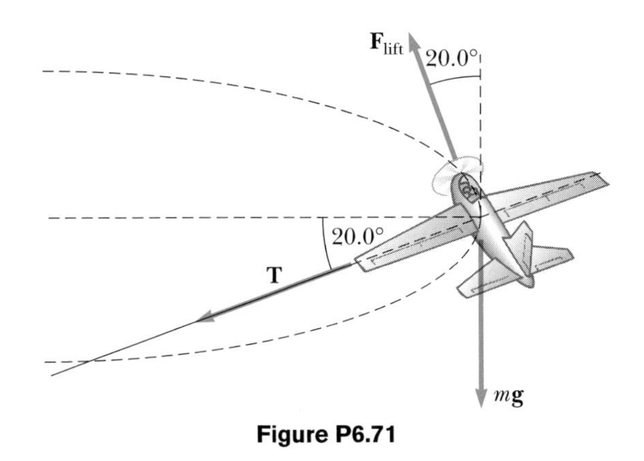

Figure P6.71

72. Members of a skydiving club were given the following data to use in planning their jumps. In the table, d is the distance fallen from rest by a sky diver in a "free-fall stable spread position," versus the time of fall t. (a) Convert the distances in feet into meters. (b) Graph d (in meters) versus t. (c) Determine the value of the terminal speed v_T by finding the slope of the straight portion of the curve. Use a least-squares fit to determine this slope.

t (s)	d (ft)	t (s)	d (ft)
1	16	11	1 309
2	62	12	1 483
3	138	13	1 657
4	242	14	1 831
5	366	15	2 005
6	504	16	2 179
7	652	17	2 353
8	808	18	2 527
9	971	19	2 701
10	1 138	20	2 875

73. If a single constant force acts on an object that moves on a straight line, the object's velocity is a linear function of time. The equation $v = v_i + at$ gives its velocity v as a function of time, where a is its constant acceleration. **What if** velocity is instead a linear function of position? Assume that as a particular object moves through a resistive medium, its speed decreases as described by the equation $v = v_i - kx$, where k is a constant coefficient and x is the position of the object. Find the law describing the total force acting on this object.

Answers to Quick Quizzes

6.1 (b), (d). The centripetal acceleration is always toward the center of the circular path.

6.2 (a), (d). The normal force is always perpendicular to the surface that applies the force. Because your car maintains its orientation at all points on the ride, the normal force is always upward.

6.3 (a). If the car is moving in a circular path, it must have centripetal acceleration given by Equation 4.15.

6.4 Because the speed is constant, the only direction the force can have is that of the centripetal acceleration. The force is larger at Ⓒ than at Ⓐ because the radius at Ⓒ is smaller. There is no force at Ⓑ because the wire is straight.

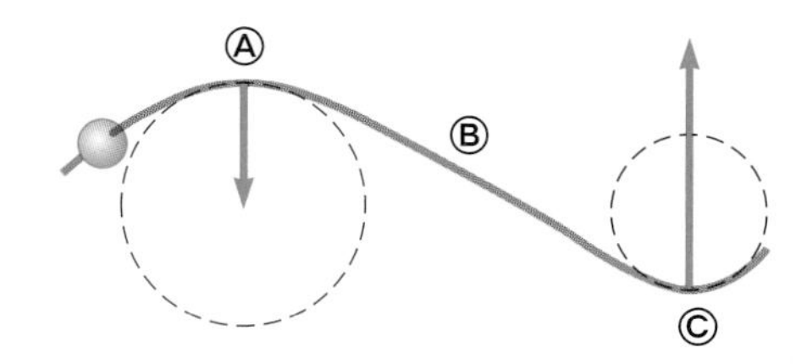

6.5 In addition to the forces in the centripetal direction in Quick Quiz 6.4, there are now tangential forces to provide the tangential acceleration. The tangential force is the same at all three points because the tangential acceleration is constant.

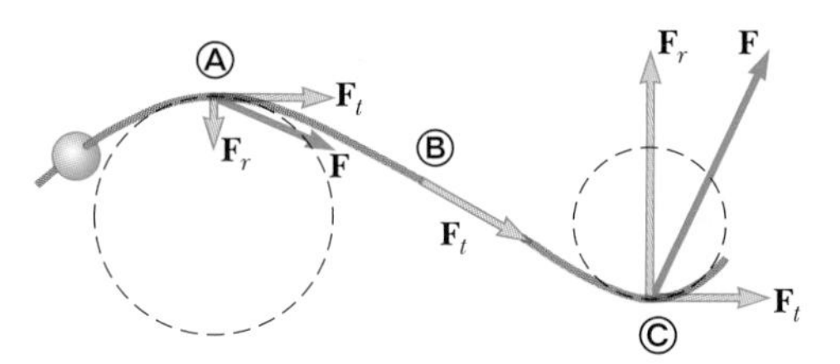

6.6 (c). The only forces acting on the passenger are the contact force with the door and the friction force from the seat. Both of these are real forces and both act to the left in Figure 6.11. Fictitious forces should never be drawn in a force diagram.

6.7 (a). The basketball, having a larger cross-sectional area, will have a larger force due to air resistance than the baseball. This will result in a smaller net force in the downward direction and a smaller downward acceleration.

Chapter 7

Energy and Energy Transfer

▲ On a wind farm, the moving air does work on the blades of the windmills, causing the blades and the rotor of an electrical generator to rotate. Energy is transferred out of the system of the windmill by means of electricity. (Billy Hustace/Getty Images)

CHAPTER OUTLINE

181

The concept of energy is one of the most important topics in science and engineering. In everyday life, we think of energy in terms of fuel for transportation and heating, electricity for lights and appliances, and foods for consumption. However, these ideas do not really define energy. They merely tell us that fuels are needed to do a job and that those fuels provide us with something we call energy.

The definitions of quantities such as position, velocity, acceleration, and force and associated principles such as Newton's second law have allowed us to solve a variety of problems. Some problems that could theoretically be solved with Newton's laws, however, are very difficult in practice. These problems can be made much simpler with a different approach. In this and the following chapters, we will investigate this new approach, which will include definitions of quantities that may not be familiar to you. Other quantities may sound familiar, but they may have more specific meanings in physics than in everyday life. We begin this discussion by exploring the notion of energy.

Energy is present in the Universe in various forms. *Every* physical process that occurs in the Universe involves energy and energy transfers or transformations. Unfortunately, despite its extreme importance, energy cannot be easily defined. The variables in previous chapters were relatively concrete; we have everyday experience with velocities and forces, for example. The notion of energy is more abstract, although we do have *experiences* with energy, such as running out of gasoline, or losing our electrical service if we forget to pay the utility bill.

The concept of energy can be applied to the dynamics of a mechanical system without resorting to Newton's laws. This "energy approach" to describing motion is especially useful when the force acting on a particle is not constant; in such a case, the acceleration is not constant, and we cannot apply the constant acceleration equations that were developed in Chapter 2. Particles in nature are often subject to forces that vary with the particles' positions. These forces include gravitational forces and the force exerted on an object attached to a spring. We shall describe techniques for treating such situations with the help of an important concept called *conservation of energy*. This approach extends well beyond physics, and can be applied to biological organisms, technological systems, and engineering situations.

Our problem-solving techniques presented in earlier chapters were based on the motion of a particle or an object that could be modeled as a particle. This was called the *particle model*. We begin our new approach by focusing our attention on a *system* and developing techniques to be used in a *system model*.

7.1 Systems and Environments

In the system model mentioned above, we focus our attention on a small portion of the Universe—the **system**—and ignore details of the rest of the Universe outside of the system. A critical skill in applying the system model to problems is *identifying the system*.

A valid system may

- be a single object or particle
- be a collection of objects or particles
- be a region of space (such as the interior of an automobile engine combustion cylinder)
- vary in size and shape (such as a rubber ball, which deforms upon striking a wall)

7.1 Identify the System

The most important step to take in solving a problem using the energy approach is to identify the appropriate system of interest. Make sure this is the *first* step you take in solving a problem.

Identifying the *need* for a system approach to solving a problem (as opposed to a particle approach) is part of the "categorize" step in the General Problem-Solving Strategy outlined in Chapter 2. Identifying the particular system and its nature is part of the "analyze" step.

No matter what the particular system is in a given problem, there is a **system boundary,** an imaginary surface (not necessarily coinciding with a physical surface) that divides the Universe into the system and the **environment** surrounding the system.

As an example, imagine a force applied to an object in empty space. We can define the object as the system. The force applied to it is an influence on the system from the environment that acts across the system boundary. We will see how to analyze this situation from a system approach in a subsequent section of this chapter.

Another example is seen in Example 5.10 (page 130). Here the system can be defined as the combination of the ball, the cube, and the string. The influence from the environment includes the gravitational forces on the ball and the cube, the normal and friction forces on the cube, and the force exerted by the pulley on the string. The forces exerted by the string on the ball and the cube are internal to the system and, therefore, are not included as an influence from the environment.

We shall find that there are a number of mechanisms by which a system can be influenced by its environment. The first of these that we shall investigate is *work.*

7.2 Work Done by a Constant Force

Almost all the terms we have used thus far—velocity, acceleration, force, and so on—convey a similar meaning in physics as they do in everyday life. Now, however, we encounter a term whose meaning in physics is distinctly different from its everyday meaning—*work.*

To understand what work means to the physicist, consider the situation illustrated in Figure 7.1. A force is applied to a chalkboard eraser, and the eraser slides

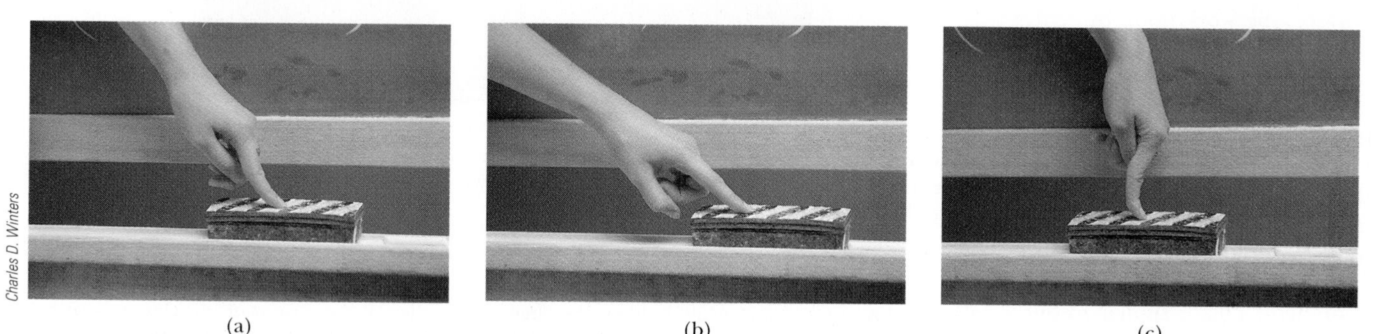

Charles D. Winters

(a) (b) (c)

Figure 7.1 An eraser being pushed along a chalkboard tray.

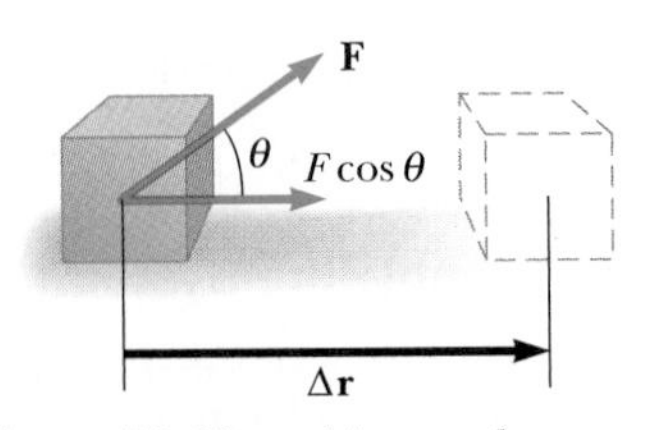

Figure 7.2 If an object undergoes a displacement **Δr** under the action of a constant force **F**, the work done by the force is $F\Delta r \cos\theta$.

along the tray. If we want to know how effective the force is in moving the eraser, we must consider not only the magnitude of the force but also its direction. Assuming that the magnitude of the applied force is the same in all three photographs, the push applied in Figure 7.1b does more to move the eraser than the push in Figure 7.1a. On the other hand, Figure 7.1c shows a situation in which the applied force does not move the eraser at all, regardless of how hard it is pushed. (Unless, of course, we apply a force so great that we break the chalkboard tray.) So, in analyzing forces to determine the work they do, we must consider the vector nature of forces. We must also know how far the eraser moves along the tray if we want to determine the work associated with that displacement. Moving the eraser 3 m requires more work than moving it 2 cm.

Let us examine the situation in Figure 7.2, where an object undergoes a displacement along a straight line while acted on by a constant force **F** that makes an angle θ with the direction of the displacement.

Work done by a constant force

The **work** W done on a system by an agent exerting a constant force on the system is the product of the magnitude F of the force, the magnitude Δr of the displacement of the point of application of the force, and cos θ, where θ is the angle between the force and displacement vectors:

$$W \equiv F\,\Delta r \cos\theta \qquad (7.1)$$

PITFALL PREVENTION

7.2 What is being Displaced?

The displacement in Equation 7.1 is that of *the point of application of the force.* If the force is applied to a particle or a non-deformable, non-rotating system, this displacement is the same as the displacement of the particle or system. For deformable systems, however, these two displacements are often not the same.

PITFALL PREVENTION

7.3 Work is Done by . . . on . . .

Not only must you identify the system, you must also identify the interaction of the system with the environment. When discussing work, always use the phrase, "the work done by . . . on . . . " After "by," insert the part of the environment that is interacting directly with the system. After "on," insert the system. For example, "the work done by the hammer on the nail" identifies the nail as the system and the force from the hammer represents the interaction with the environment. This is similar to our use in Chapter 5 of "the force exerted by . . . on . . ."

As an example of the distinction between this definition of work and our everyday understanding of the word, consider holding a heavy chair at arm's length for 3 min. At the end of this time interval, your tired arms may lead you to think that you have done a considerable amount of work on the chair. According to our definition, however, you have done no work on it whatsoever.[1] You exert a force to support the chair, but you do not move it. A force does no work on an object if the force does not move through a displacement. This can be seen by noting that if $\Delta r = 0$, Equation 7.1 gives $W = 0$—the situation depicted in Figure 7.1c.

Also note from Equation 7.1 that the work done by a force on a moving object is zero when the force applied is perpendicular to the displacement of its point of

Gerard Vandystadt/Photo Researchers, Inc.

The weightlifter does no work on the weights as he holds them on his shoulders. (If he could rest the bar on his shoulders and lock his knees, he would be able to support the weights for quite some time.) Did he do any work when he raised the weights to this height?

[1] Actually, you do work while holding the chair at arm's length because your muscles are continuously contracting and relaxing; this means that they are exerting internal forces on your arm. Thus, work is being done by your body—but internally on itself rather than on the chair.

application. That is, if $\theta = 90°$, then $W = 0$ because $\cos 90° = 0$. For example, in Figure 7.3, the work done by the normal force on the object and the work done by the gravitational force on the object are both zero because both forces are perpendicular to the displacement and have zero components along an axis in the direction of $\Delta\mathbf{r}$.

The sign of the work also depends on the direction of $\mathbf{F}$ relative to $\Delta\mathbf{r}$. The work done by the applied force is positive when the projection of $\mathbf{F}$ onto $\Delta\mathbf{r}$ is in the same direction as the displacement. For example, when an object is lifted, the work done by the applied force is positive because the direction of that force is upward, in the same direction as the displacement of its point of application. When the projection of $\mathbf{F}$ onto $\Delta\mathbf{r}$ is in the direction opposite the displacement, W is negative. For example, as an object is lifted, the work done by the gravitational force on the object is negative. The factor $\cos\theta$ in the definition of W (Eq. 7.1) automatically takes care of the sign.

If an applied force $\mathbf{F}$ is in the same direction as the displacement $\Delta\mathbf{r}$, then $\theta = 0$ and $\cos 0 = 1$. In this case, Equation 7.1 gives

$$W = F\Delta r$$

Work is a scalar quantity, and its units are force multiplied by length. Therefore, the SI unit of work is the **newton·meter** (N·m). This combination of units is used so frequently that it has been given a name of its own: the **joule** (J).

An important consideration for a system approach to problems is to note that **work is an energy transfer.** If W is the work done on a system and W is positive, energy is transferred *to* the system; if W is negative, energy is transferred *from* the system. Thus, if a system interacts with its environment, this interaction can be described as a transfer of energy across the system boundary. This will result in a change in the energy stored in the system. We will learn about the first type of energy storage in Section 7.5, after we investigate more aspects of work.

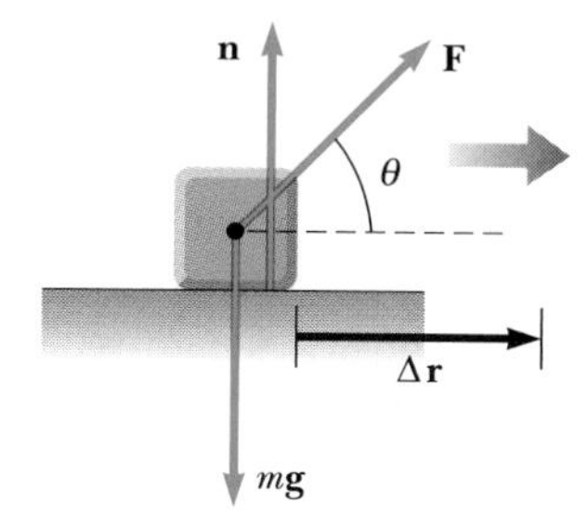

Figure 7.3 When an object is displaced on a frictionless, horizontal surface, the normal force $\mathbf{n}$ and the gravitational force $m\mathbf{g}$ do no work on the object. In the situation shown here, $\mathbf{F}$ is the only force doing work on the object.

PITFALL PREVENTION

7.4 Cause of the Displacement

We can calculate the work done by a force on an object, but that force is *not* necessarily the cause of the object's displacement. For example, if you lift an object, work is done by the gravitational force, although gravity is not the cause of the object moving upward!

Quick Quiz 7.1 The gravitational force exerted by the Sun on the Earth holds the Earth in an orbit around the Sun. Let us assume that the orbit is perfectly circular. The work done by this gravitational force during a short time interval in which the Earth moves through a displacement in its orbital path is (a) zero (b) positive (c) negative (d) impossible to determine.

Quick Quiz 7.2 Figure 7.4 shows four situations in which a force is applied to an object. In all four cases, the force has the same magnitude, and the displacement of the object is to the right and of the same magnitude. Rank the situations in order of the work done by the force on the object, from most positive to most negative.

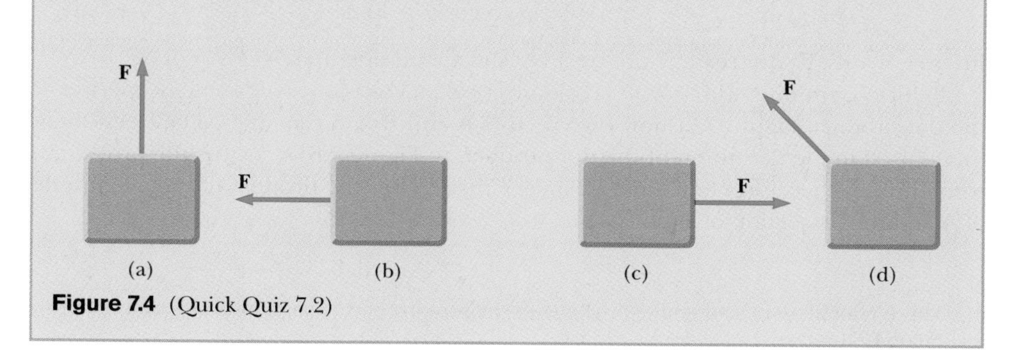

Figure 7.4 (Quick Quiz 7.2)

Example 7.1 Mr. Clean

A man cleaning a floor pulls a vacuum cleaner with a force of magnitude $F = 50.0$ N at an angle of 30.0° with the horizontal (Fig. 7.5a). Calculate the work done by the force on the vacuum cleaner as the vacuum cleaner is displaced 3.00 m to the right.

Solution Figure 7.5a helps conceptualize the situation. We are given a force, a displacement, and the angle between the two vectors, so we can categorize this as a simple problem that will need minimal analysis. To analyze the situation, we identify the vacuum cleaner as the system and draw a free-body diagram as shown in Figure 7.5b. Using the definition of work (Eq. 7.1),

$$W = F\,\Delta r\cos\theta = (50.0\text{ N})(3.00\text{ m})(\cos 30.0°)$$

$$= 130\text{ N}\cdot\text{m} = \boxed{130\text{ J}}$$

To finalize this problem, notice in this situation that the normal force $\mathbf{n}$ and the gravitational $\mathbf{F}_g = m\mathbf{g}$ do no work on the vacuum cleaner because these forces are perpendicular to its displacement.

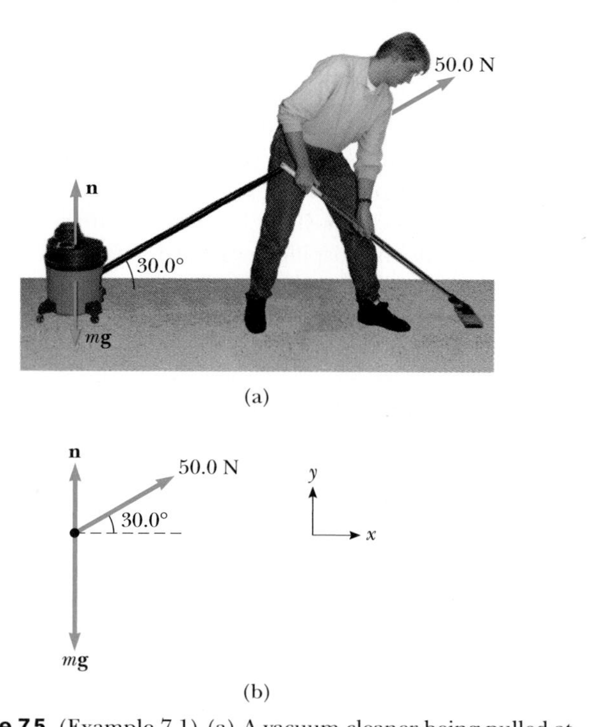

Figure 7.5 (Example 7.1) (a) A vacuum cleaner being pulled at an angle of 30.0° from the horizontal. (b) Free-body diagram of the forces acting on the vacuum cleaner.

7.3 The Scalar Product of Two Vectors

Because of the way the force and displacement vectors are combined in Equation 7.1, it is helpful to use a convenient mathematical tool called the **scalar product** of two vectors. We write this scalar product of vectors $\mathbf{A}$ and $\mathbf{B}$ as $\mathbf{A}\cdot\mathbf{B}$. (Because of the dot symbol, the scalar product is often called the **dot product.**)

In general, the scalar product of any two vectors $\mathbf{A}$ and $\mathbf{B}$ is a scalar quantity equal to the product of the magnitudes of the two vectors and the cosine of the angle θ between them:

Scalar product of any two vectors A and B

$$\mathbf{A}\cdot\mathbf{B} \equiv AB\cos\theta \qquad (7.2)$$

Note that $\mathbf{A}$ and $\mathbf{B}$ need not have the same units, as is the case with any multiplication.

Comparing this definition to Equation 7.1, we see that we can express Equation 7.1 as a scalar product:

$$W = F\,\Delta r\cos\theta = \mathbf{F}\cdot\Delta\mathbf{r} \qquad (7.3)$$

In other words, $\mathbf{F}\cdot\Delta\mathbf{r}$ (read "$\mathbf{F}$ dot $\Delta\mathbf{r}$") is a shorthand notation for $F\,\Delta r\cos\theta$.

Before continuing with our discussion of work, let us investigate some properties of the dot product. Figure 7.6 shows two vectors $\mathbf{A}$ and $\mathbf{B}$ and the angle θ between them that is used in the definition of the dot product. In Figure 7.6, $B\cos\theta$ is the projection of $\mathbf{B}$ onto $\mathbf{A}$. Therefore, Equation 7.2 means that $\mathbf{A}\cdot\mathbf{B}$ is the product of the magnitude of $\mathbf{A}$ and the projection of $\mathbf{B}$ onto $\mathbf{A}$.[2]

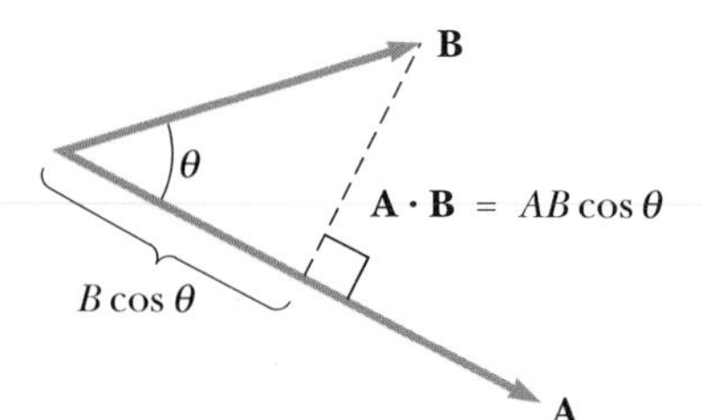

Figure 7.6 The scalar product $\mathbf{A}\cdot\mathbf{B}$ equals the magnitude of $\mathbf{A}$ multiplied by $B\cos\theta$, which is the projection of $\mathbf{B}$ onto $\mathbf{A}$.

[2] This is equivalent to stating that $\mathbf{A}\cdot\mathbf{B}$ equals the product of the magnitude of $\mathbf{B}$ and the projection of $\mathbf{A}$ onto $\mathbf{B}$.

From the right-hand side of Equation 7.2 we also see that the scalar product is **commutative**.[3] That is,

$$\mathbf{A}\cdot\mathbf{B} = \mathbf{B}\cdot\mathbf{A}$$

Finally, the scalar product obeys the **distributive law of multiplication,** so that

$$\mathbf{A}\cdot(\mathbf{B} + \mathbf{C}) = \mathbf{A}\cdot\mathbf{B} + \mathbf{A}\cdot\mathbf{C}$$

The dot product is simple to evaluate from Equation 7.2 when **A** is either perpendicular or parallel to **B**. If **A** is perpendicular to **B** ($\theta = 90°$), then $\mathbf{A}\cdot\mathbf{B} = 0$. (The equality $\mathbf{A}\cdot\mathbf{B} = 0$ also holds in the more trivial case in which either **A** or **B** is zero.) If vector **A** is parallel to vector **B** and the two point in the same direction ($\theta = 0$), then $\mathbf{A}\cdot\mathbf{B} = AB$. If vector **A** is parallel to vector **B** but the two point in opposite directions ($\theta = 180°$), then $\mathbf{A}\cdot\mathbf{B} = -AB$. The scalar product is negative when $90° < \theta \leq 180°$.

The unit vectors $\hat{\mathbf{i}}$, $\hat{\mathbf{j}}$, and $\hat{\mathbf{k}}$, which were defined in Chapter 3, lie in the positive x, y, and z directions, respectively, of a right-handed coordinate system. Therefore, it follows from the definition of $\mathbf{A}\cdot\mathbf{B}$ that the scalar products of these unit vectors are

Dot products of unit vectors

$$\hat{\mathbf{i}}\cdot\hat{\mathbf{i}} = \hat{\mathbf{j}}\cdot\hat{\mathbf{j}} = \hat{\mathbf{k}}\cdot\hat{\mathbf{k}} = 1 \quad (7.4)$$

$$\hat{\mathbf{i}}\cdot\hat{\mathbf{j}} = \hat{\mathbf{i}}\cdot\hat{\mathbf{k}} = \hat{\mathbf{j}}\cdot\hat{\mathbf{k}} = 0 \quad (7.5)$$

Equations 3.18 and 3.19 state that two vectors **A** and **B** can be expressed in component vector form as

$$\mathbf{A} = A_x\hat{\mathbf{i}} + A_y\hat{\mathbf{j}} + A_z\hat{\mathbf{k}}$$

$$\mathbf{B} = B_x\hat{\mathbf{i}} + B_y\hat{\mathbf{j}} + B_z\hat{\mathbf{k}}$$

Using the information given in Equations 7.4 and 7.5 shows that the scalar product of **A** and **B** reduces to

$$\mathbf{A}\cdot\mathbf{B} = A_xB_x + A_yB_y + A_zB_z \quad (7.6)$$

(Details of the derivation are left for you in Problem 6.) In the special case in which $\mathbf{A} = \mathbf{B}$, we see that

$$\mathbf{A}\cdot\mathbf{A} = A_x^2 + A_y^2 + A_z^2 = A^2$$

PITFALL PREVENTION

7.5 Work is a Scalar

Although Equation 7.3 defines the work in terms of two vectors, *work is a scalar*—there is no direction associated with it. *All* types of energy and energy transfer are scalars. This is a major advantage of the energy approach—we don't need vector calculations!

Quick Quiz 7.3 Which of the following statements is true about the relationship between $\mathbf{A}\cdot\mathbf{B}$ and $(-\mathbf{A})\cdot(-\mathbf{B})$? (a) $\mathbf{A}\cdot\mathbf{B} = -[(-\mathbf{A})\cdot(-\mathbf{B})]$; (b) If $\mathbf{A}\cdot\mathbf{B} = AB\cos\theta$, then $(-\mathbf{A})\cdot(-\mathbf{B}) = AB\cos(\theta + 180°)$; (c) Both (a) and (b) are true. (d) Neither (a) nor (b) is true.

Quick Quiz 7.4 Which of the following statements is true about the relationship between the dot product of two vectors and the product of the magnitudes of the vectors? (a) $\mathbf{A}\cdot\mathbf{B}$ is larger than AB; (b) $\mathbf{A}\cdot\mathbf{B}$ is smaller than AB; (c) $\mathbf{A}\cdot\mathbf{B}$ could be larger or smaller than AB, depending on the angle between the vectors; (d) $\mathbf{A}\cdot\mathbf{B}$ could be equal to AB.

[3] This may seem obvious, but in Chapter 11 you will see another way of combining vectors that proves useful in physics and is not commutative.

Example 7.2 The Scalar Product

The vectors **A** and **B** are given by $\mathbf{A} = 2\hat{\mathbf{i}} + 3\hat{\mathbf{j}}$ and $\mathbf{B} = -\hat{\mathbf{i}} + 2\hat{\mathbf{j}}$.

(A) Determine the scalar product $\mathbf{A}\cdot\mathbf{B}$.

Solution Substituting the specific vector expressions for **A** and **B**, we find,

$$\begin{aligned}\mathbf{A}\cdot\mathbf{B} &= (2\hat{\mathbf{i}} + 3\hat{\mathbf{j}})\cdot(-\hat{\mathbf{i}} + 2\hat{\mathbf{j}})\\ &= -2\hat{\mathbf{i}}\cdot\hat{\mathbf{i}} + 2\hat{\mathbf{i}}\cdot 2\hat{\mathbf{j}} - 3\hat{\mathbf{j}}\cdot\hat{\mathbf{i}} + 3\hat{\mathbf{j}}\cdot 2\hat{\mathbf{j}}\\ &= -2(1) + 4(0) - 3(0) + 6(1)\\ &= -2 + 6 = \boxed{4}\end{aligned}$$

where we have used the facts that $\hat{\mathbf{i}}\cdot\hat{\mathbf{i}} = \hat{\mathbf{j}}\cdot\hat{\mathbf{j}} = 1$ and $\hat{\mathbf{i}}\cdot\hat{\mathbf{j}} = \hat{\mathbf{j}}\cdot\hat{\mathbf{i}} = 0$. The same result is obtained when we use Equation 7.6 directly, where $A_x = 2$, $A_y = 3$, $B_x = -1$, and $B_y = 2$.

(B) Find the angle θ between **A** and **B**.

Solution The magnitudes of **A** and **B** are

$$A = \sqrt{A_x^2 + A_y^2} = \sqrt{(2)^2 + (3)^2} = \sqrt{13}$$

$$B = \sqrt{B_x^2 + B_y^2} = \sqrt{(-1)^2 + (2)^2} = \sqrt{5}$$

Using Equation 7.2 and the result from part (a) we find that

$$\cos\theta = \frac{\mathbf{A}\cdot\mathbf{B}}{AB} = \frac{4}{\sqrt{13}\sqrt{5}} = \frac{4}{\sqrt{65}}$$

$$\theta = \cos^{-1}\frac{4}{8.06} = \boxed{60.2^\circ}$$

Example 7.3 Work Done by a Constant Force

A particle moving in the xy plane undergoes a displacement $\Delta\mathbf{r} = (2.0\hat{\mathbf{i}} + 3.0\hat{\mathbf{j}})$ m as a constant force $\mathbf{F} = (5.0\hat{\mathbf{i}} + 2.0\hat{\mathbf{j}})$ N acts on the particle.

(A) Calculate the magnitudes of the displacement and the force.

Solution We use the Pythagorean theorem:

$$\Delta r = \sqrt{(\Delta x)^2 + (\Delta y)^2} = \sqrt{(2.0)^2 + (3.0)^2} = \boxed{3.6\text{ m}}$$

$$F = \sqrt{F_x^2 + F_y^2} = \sqrt{(5.0)^2 + (2.0)^2} = \boxed{5.4\text{ N}}$$

(B) Calculate the work done by **F**.

Solution Substituting the expressions for **F** and $\Delta\mathbf{r}$ into Equation 7.3 and using Equations 7.4 and 7.5, we obtain

$$\begin{aligned}W = \mathbf{F}\cdot\Delta\mathbf{r} &= [(5.0\hat{\mathbf{i}} + 2.0\hat{\mathbf{j}})\text{ N}]\cdot[(2.0\hat{\mathbf{i}} + 3.0\hat{\mathbf{j}})\text{ m}]\\ &= (5.0\hat{\mathbf{i}}\cdot 2.0\hat{\mathbf{i}} + 5.0\hat{\mathbf{i}}\cdot 3.0\hat{\mathbf{j}} + 2.0\hat{\mathbf{j}}\cdot 2.0\hat{\mathbf{i}} + 2.0\hat{\mathbf{j}}\cdot 3.0\hat{\mathbf{j}})\text{ N}\cdot\text{m}\\ &= [10 + 0 + 0 + 6]\text{ N}\cdot\text{m} = \boxed{16\text{ J}}\end{aligned}$$

7.4 Work Done by a Varying Force

Consider a particle being displaced along the x axis under the action of a force that varies with position. The particle is displaced in the direction of increasing x from $x = x_i$ to $x = x_f$. In such a situation, we cannot use $W = F\Delta r\cos\theta$ to calculate the work done by the force because this relationship applies only when **F** is constant in magnitude and direction. However, if we imagine that the particle undergoes a very small displacement Δx, shown in Figure 7.7a, the x component F_x of the force is approximately constant over this small interval; for this small displacement, we can approximate the work done by the force as

$$W \approx F_x\Delta x$$

This is just the area of the shaded rectangle in Figure 7.7a. If we imagine that the F_x versus x curve is divided into a large number of such intervals, the total work done for the displacement from x_i to x_f is approximately equal to the sum of a large number of such terms:

$$W \approx \sum_{x_i}^{x_f} F_x\Delta x$$

If the size of the displacements is allowed to approach zero, the number of terms in the sum increases without limit but the value of the sum approaches a definite value equal to the area bounded by the F_x curve and the x axis:

$$\lim_{\Delta x \to 0} \sum_{x_i}^{x_f} F_x \Delta x = \int_{x_i}^{x_f} F_x dx$$

Therefore, we can express the work done by F_x as the particle moves from x_i to x_f as

$$W = \int_{x_i}^{x_f} F_x dx \tag{7.7}$$

This equation reduces to Equation 7.1 when the component $F_x = F \cos \theta$ is constant.

If more than one force acts on a system *and the system can be modeled as a particle*, the total work done on the system is just the work done by the net force. If we express the net force in the x direction as ΣF_x, then the total work, or *net work*, done as the particle moves from x_i to x_f is

$$\sum W = W_{\text{net}} = \int_{x_i}^{x_f} \left(\sum F_x \right) dx \tag{7.8}$$

If the system cannot be modeled as a particle (for example, if the system consists of multiple particles that can move with respect to each other), we cannot use Equation 7.8. This is because different forces on the system may move through different displacements. In this case, we must evaluate the work done by each force separately and then add the works algebraically.

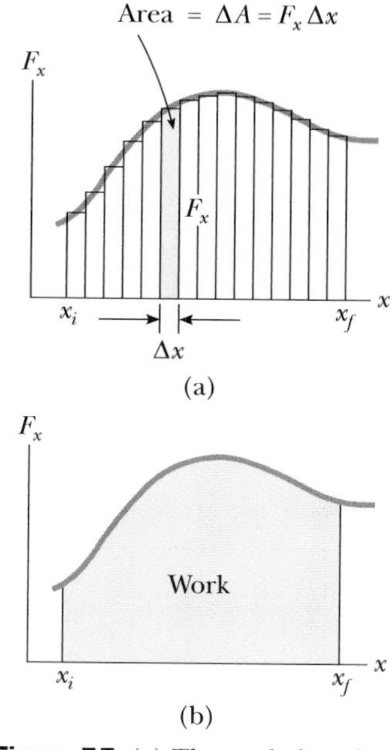

Figure 7.7 (a) The work done by the force component F_x for the small displacement Δx is $F_x \Delta x$, which equals the area of the shaded rectangle. The total work done for the displacement from x_i to x_f is approximately equal to the sum of the areas of all the rectangles. (b) The work done by the component F_x of the varying force as the particle moves from x_i to x_f is *exactly* equal to the area under this curve.

Example 7.4 Calculating Total Work Done from a Graph

A force acting on a particle varies with x, as shown in Figure 7.8. Calculate the work done by the force as the particle moves from $x = 0$ to $x = 6.0$ m.

Solution The work done by the force is equal to the area under the curve from $x_Ⓐ = 0$ to $x_Ⓒ = 6.0$ m. This area is equal to the area of the rectangular section from Ⓐ to Ⓑ plus the area of the triangular section from Ⓑ to Ⓒ. The area of the rectangle is (5.0 N)(4.0 m) = 20 J, and the area of the triangle is $\frac{1}{2}$(5.0 N)(2.0 m) = 5.0 J. Therefore, the total work done by the force on the particle is 25 J.

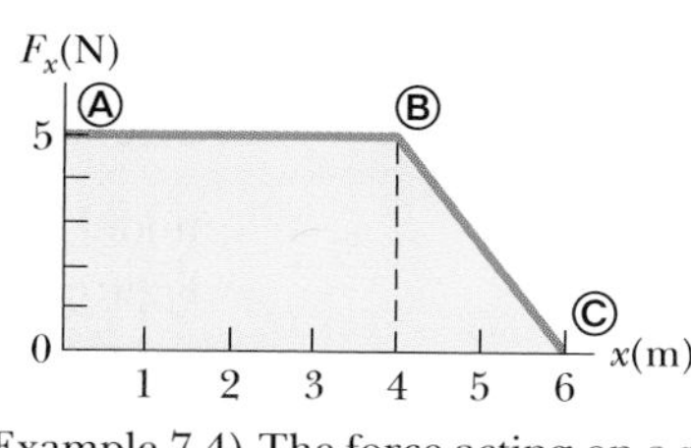

Figure 7.8 (Example 7.4) The force acting on a particle is constant for the first 4.0 m of motion and then decreases linearly with x from $x_Ⓑ = 4.0$ m to $x_Ⓒ = 6.0$ m. The net work done by this force is the area under the curve.

Example 7.5 Work Done by the Sun on a Probe

The interplanetary probe shown in Figure 7.9a is attracted to the Sun by a force given by

$$F = -\frac{1.3 \times 10^{22}}{x^2}$$

in SI units, where x is the Sun-probe separation distance. Graphically and analytically determine how much work is done by the Sun on the probe as the probe–Sun separation changes from 1.5×10^{11} m to 2.3×10^{11} m.

Graphical Solution The negative sign in the equation for the force indicates that the probe is attracted to the Sun. Because the probe is moving away from the Sun, we expect to obtain a negative value for the work done on it. A spreadsheet or other numerical means can be used to generate a graph like that in Figure 7.9b. Each small square of the grid corresponds to an area $(0.05 \text{ N})(0.1 \times 10^{11} \text{ m}) = 5 \times 10^8$ J. The work done is equal to the shaded area in Figure 7.9b. Because there are approximately 60 squares shaded, the total

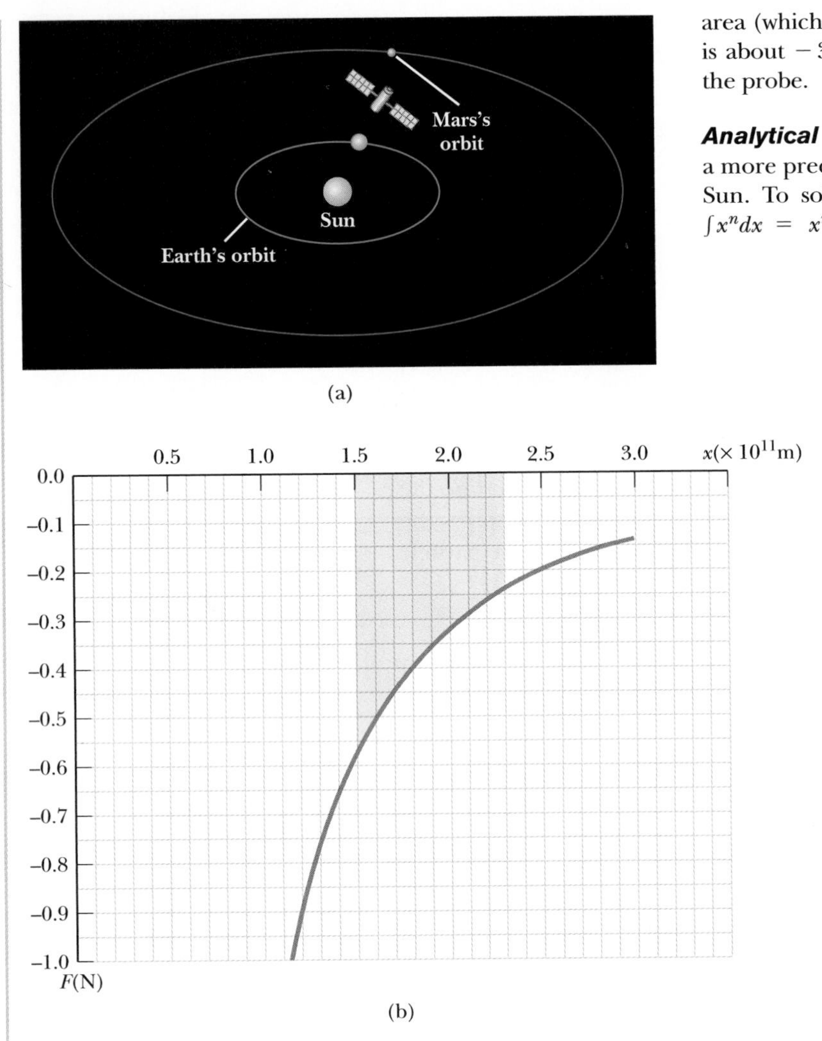

area (which is negative because the curve is below the x axis) is about -3×10^{10} J. This is the work done by the Sun on the probe.

Analytical Solution We can use Equation 7.7 to calculate a more precise value for the work done on the probe by the Sun. To solve this integral, we make use of the integral $\int x^n dx = x^{n+1}/(n+1)$ with $n = -2$:

$$W = \int_{1.5\times 10^{11}}^{2.3\times 10^{11}} \left(\frac{-1.3 \times 10^{22}}{x^2}\right) dx$$

$$= (-1.3 \times 10^{22}) \int_{1.5\times 10^{11}}^{2.3\times 10^{11}} x^{-2} dx$$

$$= (-1.3 \times 10^{22}) \left(\frac{x^{-1}}{-1}\right)\Bigg|_{1.5\times 10^{11}}^{2.3\times 10^{11}}$$

$$= (-1.3 \times 10^{22}) \left(\frac{-1}{2.3 \times 10^{11}} - \frac{-1}{1.5 \times 10^{11}}\right)$$

$$= -3.0 \times 10^{10} \text{ J}$$

Figure 7.9 (Example 7.5) (a) An interplanetary probe moves from a position near the Earth's orbit radially outward from the Sun, ending up near the orbit of Mars. (b) Attractive force versus distance for the interplanetary probe.

Work Done by a Spring

A model of a common physical system for which the force varies with position is shown in Figure 7.10. A block on a horizontal, frictionless surface is connected to a spring. If the spring is either stretched or compressed a small distance from its unstretched (equilibrium) configuration, it exerts on the block a force that can be expressed as

Spring force

$$F_s = -kx \tag{7.9}$$

where x is the position of the block relative to its equilibrium ($x = 0$) position and k is a positive constant called the **force constant** or the **spring constant** of the spring. In other words, the force required to stretch or compress a spring is proportional to the amount of stretch or compression x. This force law for springs is known as **Hooke's law**. The value of k is a measure of the *stiffness* of the spring. Stiff springs have large k values, and soft springs have small k values. As can be seen from Equation 7.9, the units of k are N/m.

The negative sign in Equation 7.9 signifies that the force exerted by the spring is always directed *opposite* to the displacement from equilibrium. When $x > 0$ as in Figure 7.10a, so that the block is to the right of the equilibrium position, the spring force is directed to the left, in the negative x direction. When $x < 0$ as in Figure 7.10c, the block is to the left of equilibrium and the spring force is directed to the right, in the positive x direction. When $x = 0$ as in Figure 7.10b, the spring is unstretched and $F_s = 0$.

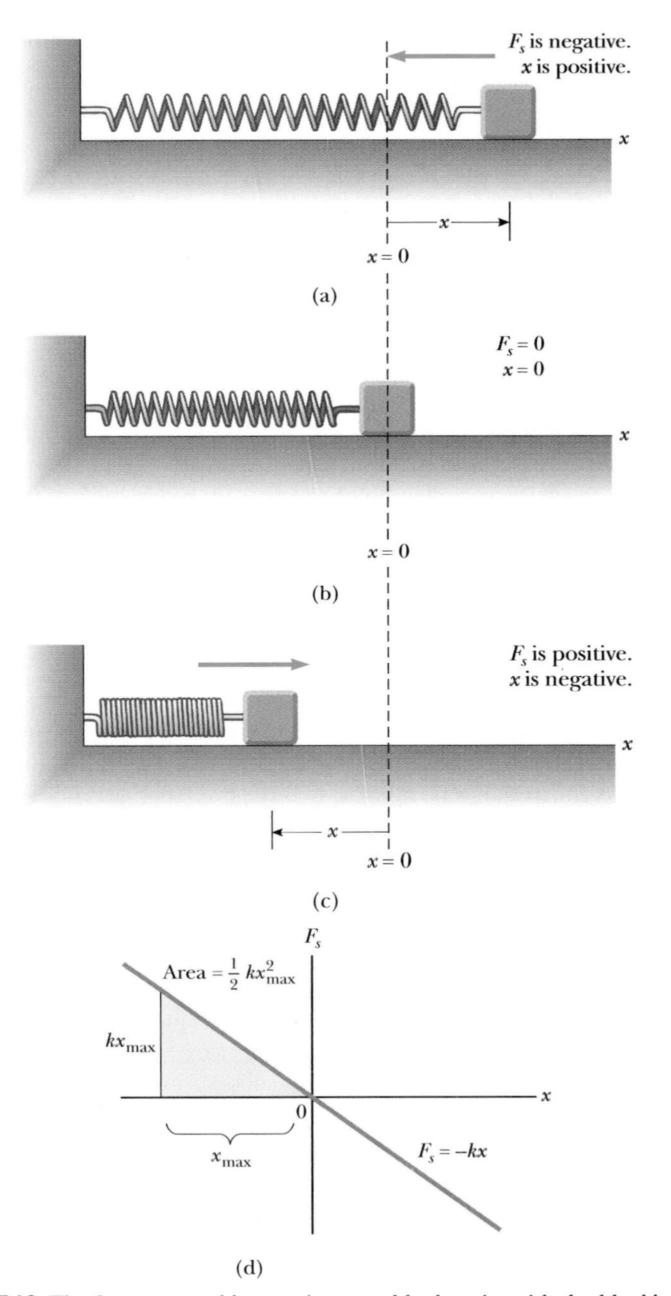

Active Figure 7.10 The force exerted by a spring on a block varies with the block's position x relative to the equilibrium position $x = 0$. (a) When x is positive (stretched spring), the spring force is directed to the left. (b) When x is zero (natural length of the spring), the spring force is zero. (c) When x is negative (compressed spring), the spring force is directed to the right. (d) Graph of F_s versus x for the block–spring system. The work done by the spring force as the block moves from $-x_{max}$ to 0 is the area of the shaded triangle, $\frac{1}{2}kx_{max}^2$.

At the Active Figures link at http://www.pse6.com, you can observe the block's motion for various maximum displacements and spring constants.

Because the spring force always acts toward the equilibrium position ($x = 0$), it is sometimes called a *restoring force.* If the spring is compressed until the block is at the point $-x_{max}$ and is then released, the block moves from $-x_{max}$ through zero to $+x_{max}$. If the spring is instead stretched until the block is at the point $+x_{max}$ and is then released, the block moves from $+x_{max}$ through zero to $-x_{max}$. It then reverses direction, returns to $+x_{max}$, and continues oscillating back and forth.

Suppose the block has been pushed to the left to a position $-x_{max}$ and is then released. Let us identify the block as our system and calculate the work W_s done by the spring force on the block as the block moves from $x_i = -x_{max}$ to $x_f = 0$. Applying

Equation 7.7 and assuming the block may be treated as a particle, we obtain

$$W_s = \int_{x_i}^{x_f} F_s dx = \int_{-x_{\max}}^{0} (-kx)\, dx = \tfrac{1}{2} kx_{\max}^2 \qquad (7.10)$$

where we have used the integral $\int x^n dx = x^{n+1}/(n+1)$ with $n = 1$. The work done by the spring force is positive because the force is in the same direction as the displacement of the block (both are to the right). Because the block arrives at $x = 0$ with some speed, it will continue moving, until it reaches a position $+x_{\max}$. When we consider the work done by the spring force as the block moves from $x_i = 0$ to $x_f = x_{\max}$, we find that $W_s = -\frac{1}{2}kx_{\max}^2$ because for this part of the motion the displacement is to the right and the spring force is to the left. Therefore, the *net* work done by the spring force as the block moves from $x_i = -x_{\max}$ to $x_f = x_{\max}$ is *zero.*

Figure 7.10d is a plot of F_s versus x. The work calculated in Equation 7.10 is the area of the shaded triangle, corresponding to the displacement from $-x_{\max}$ to 0. Because the triangle has base $x_{\max}$ and height $kx_{\max}$, its area is $\frac{1}{2}kx_{\max}^2$, the work done by the spring as given by Equation 7.10.

If the block undergoes an arbitrary displacement from $x = x_i$ to $x = x_f$, the work done by the spring force on the block is

Work done by a spring

$$W_s = \int_{x_i}^{x_f} (-kx)\, dx = \tfrac{1}{2}kx_i^2 - \tfrac{1}{2}kx_f^2 \qquad (7.11)$$

For example, if the spring has a force constant of 80 N/m and is compressed 3.0 cm from equilibrium, the work done by the spring force as the block moves from $x_i = -3.0$ cm to its unstretched position $x_f = 0$ is 3.6×10^{-2} J. From Equation 7.11 we also see that the work done by the spring force is zero for any motion that ends where it began ($x_i = x_f$). We shall make use of this important result in Chapter 8, in which we describe the motion of this system in greater detail.

Equations 7.10 and 7.11 describe the work done by the spring on the block. Now let us consider the work done on the spring by an *external agent* that stretches the spring very slowly from $x_i = 0$ to $x_f = x_{\max}$, as in Figure 7.11. We can calculate this work by noting that at any value of the position, the *applied force* $\mathbf{F}_{\text{app}}$ is equal in magnitude and opposite in direction to the spring force $\mathbf{F}_s$, so that $F_{\text{app}} = -(-kx) = kx$. Therefore, the work done by this applied force (the external agent) on the block–spring system is

$$W_{F_{\text{app}}} = \int_0^{x_{\max}} F_{\text{app}}\, dx = \int_0^{x_{\max}} kx\, dx = \tfrac{1}{2}kx_{\max}^2$$

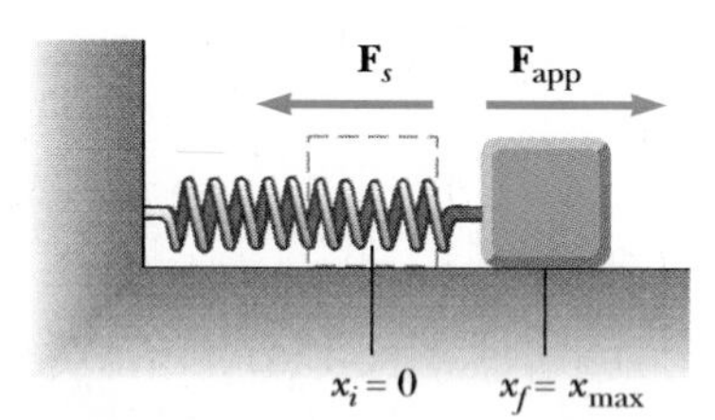

Figure 7.11 A block being pulled from $x_i = 0$ to $x_f = x_{\max}$ on a frictionless surface by a force $\mathbf{F}_{\text{app}}$. If the process is carried out very slowly, the applied force is equal in magnitude and opposite in direction to the spring force at all times.

This work is equal to the negative of the work done by the spring force for this displacement.

The work done by an applied force on a block–spring system between arbitrary positions of the block is

$$W_{F_{\text{app}}} = \int_{x_i}^{x_f} F_{\text{app}}\, dx = \int_{x_i}^{x_f} kx\, dx = \tfrac{1}{2}kx_f^2 - \tfrac{1}{2}kx_i^2 \qquad (7.12)$$

Notice that this is the negative of the work done by the spring as expressed by Equation 7.11. This is consistent with the fact that the spring force and the applied force are of equal magnitude but in opposite directions.

Quick Quiz 7.5 A dart is loaded into a spring-loaded toy dart gun by pushing the spring in by a distance *d*. For the next loading, the spring is compressed a distance 2*d*. How much work is required to load the second dart compared to that required to load the first? (a) four times as much (b) two times as much (c) the same (d) half as much (e) one-fourth as much.

Example 7.6 Measuring k for a Spring

A common technique used to measure the force constant of a spring is demonstrated by the setup in Figure 7.12. The spring is hung vertically, and an object of mass m is attached to its lower end. Under the action of the "load" mg, the spring stretches a distance d from its equilibrium position.

(A) If a spring is stretched 2.0 cm by a suspended object having a mass of 0.55 kg, what is the force constant of the spring?

Solution Because the object (the system) is at rest, the upward spring force balances the downward gravitational force $m\mathbf{g}$. In this case, we apply Hooke's law to give $|\mathbf{F}_s| = kd = mg$, or

$$k = \frac{mg}{d} = \frac{(0.55 \text{ kg})(9.80 \text{ m/s}^2)}{2.0 \times 10^{-2} \text{ m}} = 2.7 \times 10^2 \text{ N/m}$$

(B) How much work is done by the spring as it stretches through this distance?

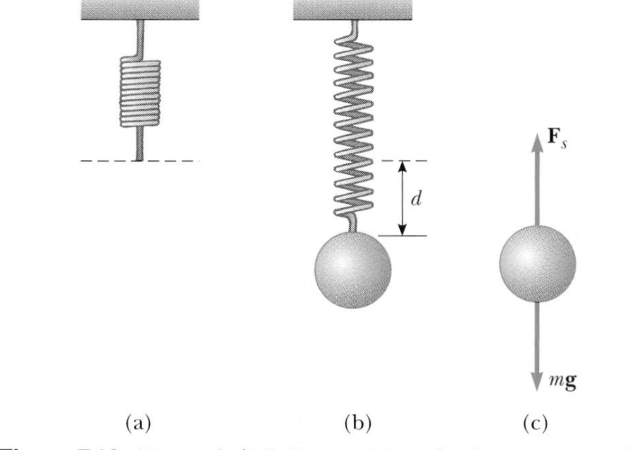

Figure 7.12 (Example 7.6) Determining the force constant k of a spring. The elongation d is caused by the attached object, which has a weight mg. Because the spring force balances the gravitational force, it follows that $k = mg/d$.

Solution Using Equation 7.11,

$$W_s = 0 - \tfrac{1}{2}kd^2 = -\tfrac{1}{2}(2.7 \times 10^2 \text{ N/m})(2.0 \times 10^{-2} \text{ m})^2$$

$$= -5.4 \times 10^{-2} \text{ J}$$

What If? **Suppose this measurement is made on an elevator with an upward vertical acceleration a. Will the unaware experimenter arrive at the same value of the spring constant?**

Answer The force $\mathbf{F}_s$ in Figure 7.12 must be larger than $m\mathbf{g}$ to produce an upward acceleration of the object. Because $\mathbf{F}_s$ must increase in magnitude, and $|\mathbf{F}_s| = kd$, the spring must extend farther. The experimenter sees a larger extension for the same hanging weight and therefore measures the spring constant to be smaller than the value found in part (A) for $\mathbf{a} = 0$.

Newton's second law applied to the hanging object gives

$$\sum F_y = |\mathbf{F}_s| - mg = ma_y$$

$$kd - mg = ma_y$$

$$d = \frac{m(g + a_y)}{k}$$

where k is the *actual* spring constant. Now, the experimenter is unaware of the acceleration, so she claims that $|\mathbf{F}_s| = k'd = mg$ where k' is the spring constant as measured by the experimenter. Thus,

$$k' = \frac{mg}{d} = \frac{mg}{\left(\dfrac{m(g + a_y)}{k}\right)} = \frac{g}{g + a_y}k$$

If the acceleration of the elevator is upward so that a_y is positive, this result shows that the measured spring constant will be smaller, consistent with our conceptual argument.

7.5 Kinetic Energy and the Work–Kinetic Energy Theorem

We have investigated work and identified it as a mechanism for transferring energy into a system. One of the possible outcomes of doing work on a system is that the system changes its speed. In this section, we investigate this situation and introduce our first type of energy that a system can possess, called *kinetic energy*.

Consider a system consisting of a single object. Figure 7.13 shows a block of mass m moving through a displacement directed to the right under the action of a net force $\Sigma\mathbf{F}$, also directed to the right. We know from Newton's second law that the block moves with an acceleration $\mathbf{a}$. If the block moves through a displacement $\Delta\mathbf{r} = \Delta x\hat{\mathbf{i}} = (x_f - x_i)\hat{\mathbf{i}}$, the work done by the net force $\Sigma\mathbf{F}$ is

$$\sum W = \int_{x_i}^{x_f} \sum F\,dx \qquad (7.13)$$

Using Newton's second law, we can substitute for the magnitude of the net force $\Sigma F = ma$, and then perform the following chain-rule manipulations on the integrand:

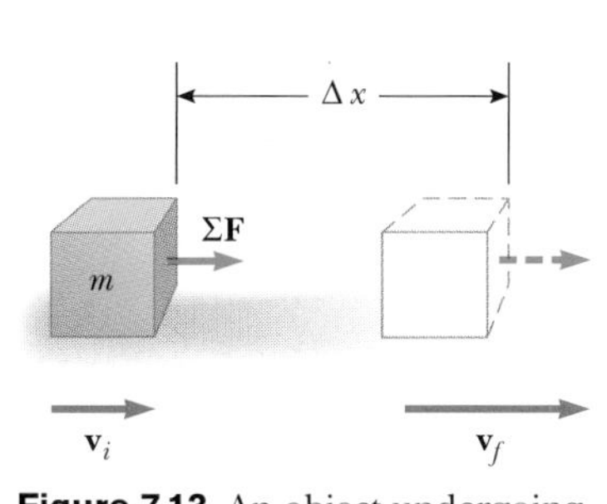

Figure 7.13 An object undergoing a displacement $\Delta\mathbf{r} = \Delta x\hat{\mathbf{i}}$ and a change in velocity under the action of a constant net force $\Sigma\mathbf{F}$.

$$\sum W = \int_{x_i}^{x_f} madx = \int_{x_i}^{x_f} m \frac{dv}{dt} dx = \int_{x_i}^{x_f} m \frac{dv}{dx} \frac{dx}{dt} dx = \int_{v_i}^{v_f} mv\, dv$$

$$\sum W = \tfrac{1}{2}mv_f^2 - \tfrac{1}{2}mv_i^2 \qquad (7.14)$$

where v_i is the speed of the block when it is at $x = x_i$ and v_f is its speed at x_f.

This equation was generated for the specific situation of one-dimensional motion, but it is a general result. It tells us that the work done by the net force on a particle of mass m is equal to the difference between the initial and final values of a quantity $\frac{1}{2}mv^2$. The quantity $\frac{1}{2}mv^2$ represents the energy associated with the motion of the particle. This quantity is so important that it has been given a special name—**kinetic energy.** Equation 7.14 states that the net work done on a particle by a net force $\Sigma\mathbf{F}$ acting on it equals the change in kinetic energy of the particle.

In general, the kinetic energy K of a particle of mass m moving with a speed v is defined as

Kinetic energy

$$K \equiv \tfrac{1}{2}mv^2 \qquad (7.15)$$

Kinetic energy is a scalar quantity and has the same units as work. For example, a 2.0 kg object moving with a speed of 4.0 m/s has a kinetic energy of 16 J. Table 7.1 lists the kinetic energies for various objects.

It is often convenient to write Equation 7.14 in the form

Work–kinetic energy theorem

$$\sum W = K_f - K_i = \Delta K \qquad (7.16)$$

Another way to write this is $K_f = K_i + \Sigma W$, which tells us that the final kinetic energy is equal to the initial kinetic energy plus the change due to the work done.

Equation 7.16 is an important result known as the **work–kinetic energy theorem:**

> In the case in which work is done on a system and the only change in the system is in its speed, the work done by the net force equals the change in kinetic energy of the system.

The work–kinetic energy theorem indicates that the speed of a particle will *increase* if the net work done on it is *positive,* because the final kinetic energy will be greater than the initial kinetic energy. The speed will *decrease* if the net work is *negative,* because the final kinetic energy will be less than the initial kinetic energy.

PITFALL PREVENTION

7.6 Conditions for the Work–Kinetic Energy Theorem

The work–kinetic energy theorem is important, but limited in its application—it is not a general principle. There are many situations in which other changes in the system occur besides its speed, and there are other interactions with the environment besides work. A more general principle involving energy is conservation of energy in Section 7.6.

Table 7.1

Kinetic Energies for Various Objects

Object	Mass (kg)	Speed (m/s)	Kinetic Energy (J)
Earth orbiting the Sun	5.98×10^{24}	2.98×10^{4}	2.66×10^{33}
Moon orbiting the Earth	7.35×10^{22}	1.02×10^{3}	3.82×10^{28}
Rocket moving at escape speed[a]	500	1.12×10^{4}	3.14×10^{10}
Automobile at 65 mi/h	2 000	29	8.4×10^{5}
Running athlete	70	10	3 500
Stone dropped from 10 m	1.0	14	98
Golf ball at terminal speed	0.046	44	45
Raindrop at terminal speed	3.5×10^{-5}	9.0	1.4×10^{-3}
Oxygen molecule in air	5.3×10^{-26}	500	6.6×10^{-21}

[a] Escape speed is the minimum speed an object must reach near the Earth's surface in order to move infinitely far away from the Earth.

Because we have only investigated translational motion through space so far, we arrived at the work–kinetic energy theorem by analyzing situations involving translational motion. Another type of motion is *rotational motion,* in which an object spins about an axis. We will study this type of motion in Chapter 10. The work–kinetic energy theorem is also valid for systems that undergo a change in the rotational speed due to work done on the system. The windmill in the chapter opening photograph is an example of work causing rotational motion.

The work–kinetic energy theorem will clarify a result that we have seen earlier in this chapter that may have seemed odd. In Section 7.4, we arrived at a result of zero net work done when we let a spring push a block from $x_i = -x_{max}$ to $x_f = x_{max}$. Notice that the speed of the block is continually changing during this process, so it may seem complicated to analyze this process. The quantity ΔK in the work–kinetic energy theorem, however, only refers to the initial and final points for the speeds—it does not depend on details of the path followed between these points. Thus, because the speed is zero at both the initial and final points of the motion, the net work done on the block is zero. We will see this concept of path independence often in similar approaches to problems.

Earlier, we indicated that work can be considered as a mechanism for transferring energy into a system. Equation 7.16 is a mathematical statement of this concept. We do work ΣW on a system and the result is a transfer of energy across the boundary of the system. The result on the system, in the case of Equation 7.16, is a change ΔK in kinetic energy. We will explore this idea more fully in the next section.

PITFALL PREVENTION

7.7 The Work-Kinetic Energy Theorem: Speed, not Velocity

The work–kinetic energy theorem relates work to a change in the *speed* of an object, not a change in its velocity. For example, if an object is in uniform circular motion, the speed is constant. Even though the velocity is changing, no work is done by the force causing the circular motion.

Quick Quiz 7.6 A dart is loaded into a spring-loaded toy dart gun by pushing the spring in by a distance *d*. For the next loading, the spring is compressed a distance 2*d*. How much faster does the second dart leave the gun compared to the first? (a) four times as fast (b) two times as fast (c) the same (d) half as fast (e) one-fourth as fast.

Example 7.7 A Block Pulled on a Frictionless Surface

A 6.0-kg block initially at rest is pulled to the right along a horizontal, frictionless surface by a constant horizontal force of 12 N. Find the speed of the block after it has moved 3.0 m.

Solution We have made a drawing of this situation in Figure 7.14. We could apply the equations of kinematics to determine the answer, but let us practice the energy approach. The block is the system, and there are three external forces acting on the system. The normal force balances the gravitational force on the block, and neither of these vertically acting forces does work on the block because their points of application are horizontally displaced. Thus, the net external force acting on the block is the 12-N force. The work done by this force is

$$W = F\Delta x = (12\,\text{N})(3.0\,\text{m}) = 36\,\text{J}$$

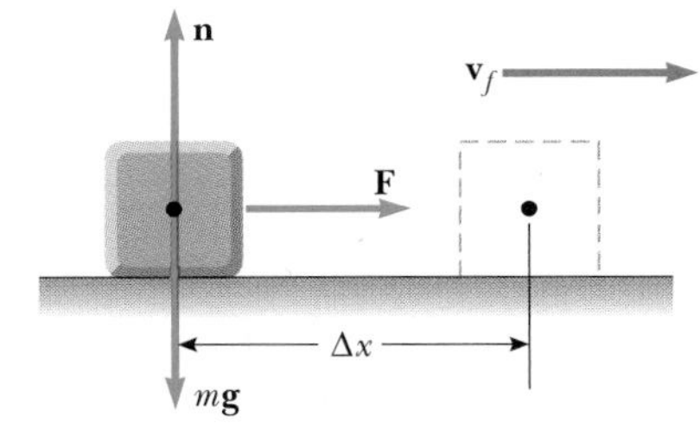

Figure 7.14 (Example 7.7) A block pulled to the right on a frictionless surface by a constant horizontal force.

Using the work–kinetic energy theorem and noting that the initial kinetic energy is zero, we obtain

$$W = K_f - K_i = \tfrac{1}{2}mv_f^{\,2} - 0$$

$$v_f = \sqrt{\frac{2W}{m}} = \sqrt{\frac{2(36\,\text{J})}{6.0\,\text{kg}}} = 3.5\,\text{m/s}$$

What If? **Suppose the magnitude of the force in this example is doubled to $F' = 2F$. The 6.0-kg block accelerates to 3.5 m/s due to this applied force while moving through a displacement $\Delta x'$. (A) How does the displacement $\Delta x'$ compare to the original displacement Δx? (B) How does the time interval $\Delta t'$ for the block to accelerate from rest to 3.5 m/s compare to the original interval Δt?**

Answer (A) If we pull harder, the block should accelerate to a given speed in a shorter distance, so we expect $\Delta x' < \Delta x$. Mathematically, from the work–kinetic energy theorem $W = \Delta K$, we find

$$F'\Delta x' = \Delta K = F\Delta x$$

$$\Delta x' = \frac{F}{F'}\Delta x = \frac{F}{2F}\Delta x = \tfrac{1}{2}\Delta x$$

and the distance is shorter as suggested by our conceptual argument.

(B) If we pull harder, the block should accelerate to a given speed in a shorter time interval, so we expect $\Delta t' < \Delta t$. Mathematically, from the definition of average velocity,

$$\bar{v} = \frac{\Delta x}{\Delta t} \quad \rightarrow \quad \Delta t = \frac{\Delta x}{\bar{v}}$$

Because both the original force and the doubled force cause the same change in velocity, the average velocity $\bar{v}$ is the same in both cases. Thus,

$$\Delta t' = \frac{\Delta x'}{\bar{v}} = \frac{\frac{1}{2}\Delta x}{\bar{v}} = \frac{1}{2}\Delta t$$

and the time interval is shorter, consistent with our conceptual argument.

Conceptual Example 7.8 Does the Ramp Lessen the Work Required?

A man wishes to load a refrigerator onto a truck using a ramp, as shown in Figure 7.15. He claims that less work would be required to load the truck if the length L of the ramp were increased. Is his statement valid?

Solution No. Suppose the refrigerator is wheeled on a dolly up the ramp at constant speed. Thus, $\Delta K = 0$. The normal force exerted by the ramp on the refrigerator is directed at 90° to the displacement and so does no work on the refrigerator. Because $\Delta K = 0$, the work–kinetic energy theorem gives

$$W_{\text{net}} = W_{\text{by man}} + W_{\text{by gravity}} = 0$$

The work done by the gravitational force equals the product of the weight mg of the refrigerator, the height h through which it is displaced, and cos 180°, or $W_{\text{by gravity}} = -mgh$. (The negative sign arises because the downward gravitational force is opposite the displacement.) Thus, the man must do the same amount of work mgh on the refrigerator, *regardless* of the length of the ramp. Although less force is required with a longer ramp, that force must act over a greater distance.

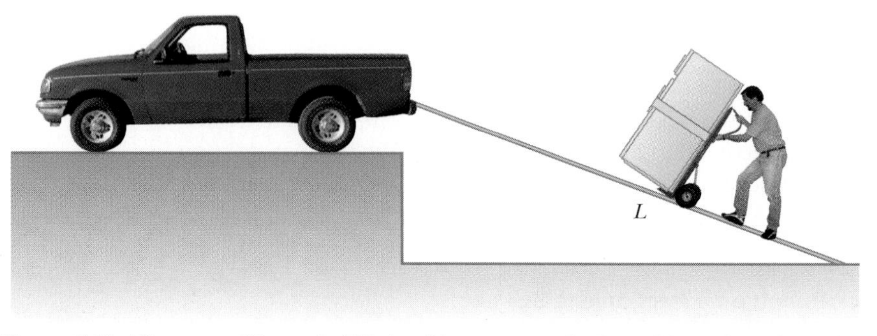

Figure 7.15 (Conceptual Example 7.8) A refrigerator attached to a frictionless wheeled dolly is moved up a ramp at constant speed.

7.6 The Nonisolated System—Conservation of Energy

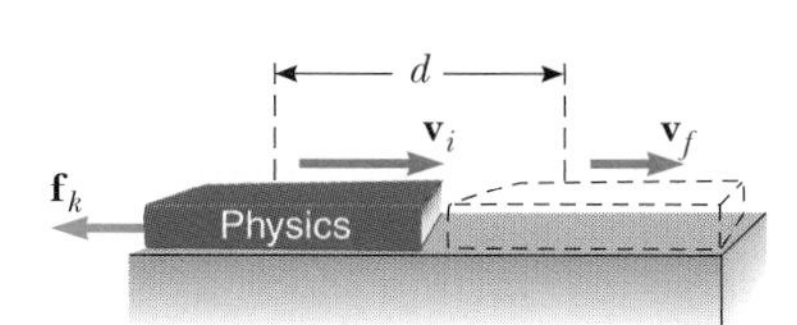

Figure 7.16 A book sliding to the right on a horizontal surface slows down in the presence of a force of kinetic friction acting to the left. The initial velocity of the book is $\mathbf{v}_i$, and its final velocity is $\mathbf{v}_f$. The normal force and the gravitational force are not included in the diagram because they are perpendicular to the direction of motion and therefore do not influence the book's speed.

We have seen examples in which an object, modeled as a particle, is acted on by various forces, resulting in a change in its kinetic energy. This very simple situation is the first example of the **nonisolated system**—a common scenario in physics problems. Physical problems for which this scenario is appropriate involve systems that interact with or are influenced by their environment, causing some kind of change in the system. If a system does not interact with its environment it is an **isolated system**, which we will study in Chapter 8.

The work–kinetic energy theorem is our first example of an energy equation appropriate for a nonisolated system. In the case of the work–kinetic energy theorem, the interaction is the work done by the external force, and the quantity in the system that changes is the kinetic energy.

In addition to kinetic energy, we now introduce a second type of energy that a system can possess. Let us imagine the book in Figure 7.16 sliding to the right on the surface of a heavy table and slowing down due to the friction force. Suppose the *surface* is the system. Then the friction force from the sliding book does work on the surface. The force on the surface is to the right and the displacement of the point of application of the force is to the right—the work is positive. But the surface is not moving after the book has stopped. Positive work has been done on the surface, yet there is no increase in the surface's kinetic energy. Is this a violation of the work–kinetic energy theorem?

It is not really a violation, because this situation does not fit the description of the conditions given for the work–kinetic energy theorem. Work is done on the system of the surface, but the result of that work is *not* an increase in kinetic energy. From your everyday experience with sliding over surfaces with friction, you can probably guess that the surface will be *warmer* after the book slides over it. (Rub your hands together briskly to experience this!) Thus, the work that was done on the surface has gone into warming the surface rather than increasing its speed. We call the energy associated with an object's temperature its **internal energy,** symbolized E_{int}. (We will define internal energy more generally in Chapter 20.) In this case, the work done on the surface does indeed represent energy transferred into the system, but it appears in the system as internal energy rather than kinetic energy.

We have now seen two methods of storing energy in a system—kinetic energy, related to motion of the system, and internal energy, related to its temperature. A third method, which we cover in Chapter 8, is *potential energy.* This is energy related to the configuration of a system in which the components of the system interact by forces. For example, when a spring is stretched, *elastic potential energy* is stored in the spring due to the force of interaction between the spring coils. Other types of potential energy include gravitational and electric.

We have seen only one way to transfer energy into a system so far—work. We mention below a few other ways to transfer energy into or out of a system. The details of these processes will be studied in other sections of the book. We illustrate these in Figure 7.17 and summarize them as follows:

Work, as we have learned in this chapter, is a method of transferring energy to a system by applying a force to the system and causing a displacement of the point of application of the force (Fig. 7.17a).

Mechanical waves (Chapters 16–18) are a means of transferring energy by allowing a disturbance to propagate through air or another medium. This is the method by which energy (which you detect as sound) leaves your clock radio through the loudspeaker and enters your ears to stimulate the hearing process (Fig. 7.17b). Other examples of mechanical waves are seismic waves and ocean waves.

Heat (Chapter 20) is a mechanism of energy transfer that is driven by a temperature difference between two regions in space. One clear example is thermal conduction, a mechanism of transferring energy by microscopic collisions. For example, a metal spoon in a cup of coffee becomes hot because fast-moving electrons and atoms in the submerged portion of the spoon bump into slower ones in the nearby part of the handle (Fig. 7.17c). These particles move faster because of the collisions and bump into the next group of slow particles. Thus, the internal energy of the spoon handle rises from energy transfer due to this bumping process.[4]

Matter transfer (Chapter 20) involves situations in which matter physically crosses the boundary of a system, carrying energy with it. Examples include filling your automobile tank with gasoline (Fig. 7.17d), and carrying energy to the rooms of your home by circulating warm air from the furnace, a process called *convection.*

[4] The process we call heat can also proceed by convection and radiation, as well as conduction. Convection and radiation, described in Chapter 20, overlap with other types of energy transfer in our list of six.

PITFALL PREVENTION

7.8 Heat is not a Form of Energy

The word *heat* is one of the most misused words in our popular language. In this text, heat is a method of *transferring* energy, *not* a form of storing energy. Thus, phrases such as "heat content," "the heat of the summer," and "the heat escaped" all represent uses of this word that are inconsistent with our physics definition. See Chapter 20.

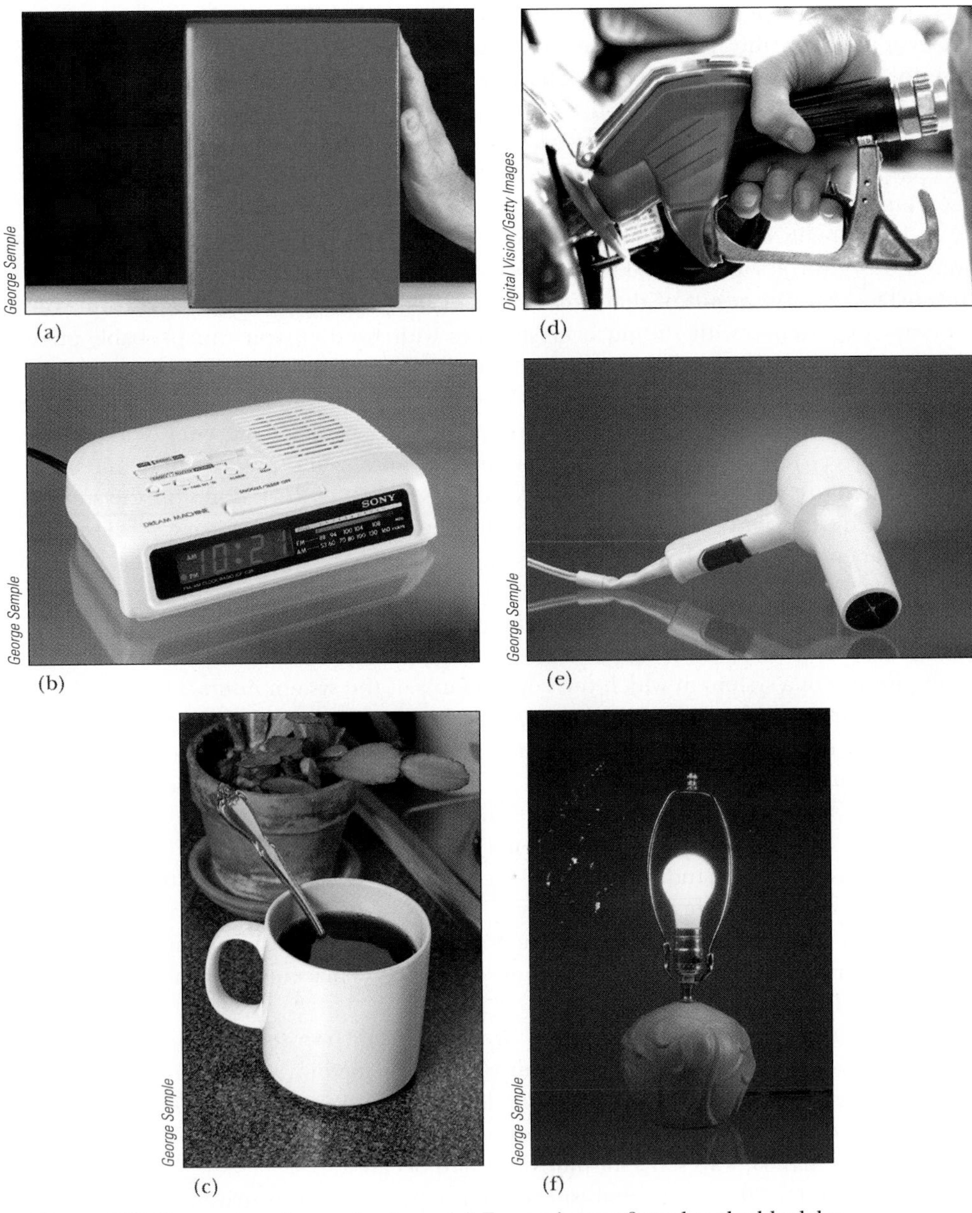

Figure 7.17 Energy transfer mechanisms. (a) Energy is transferred to the block by *work*; (b) energy leaves the radio from the speaker by *mechanical waves*; (c) energy transfers up the handle of the spoon by *heat*; (d) energy enters the automobile gas tank by *matter transfer*; (e) energy enters the hair dryer by *electrical transmission*; and (f) energy leaves the light bulb by *electromagnetic radiation*.

Electrical Transmission (Chapters 27–28) involves energy transfer by means of electric currents. This is how energy transfers into your hair dryer (Fig. 7.17e), stereo system, or any other electrical device.

Electromagnetic radiation (Chapter 34) refers to electromagnetic waves such as light, microwaves, radio waves, and so on (Fig. 7.17f). Examples of this method of transfer include cooking a baked potato in your microwave oven and light energy traveling from the Sun to the Earth through space.[5]

[5] Electromagnetic radiation and work done by field forces are the only energy transfer mechanisms that do not require molecules of the environment to be available at the system boundary. Thus, systems surrounded by a vacuum (such as planets) can only exchange energy with the environment by means of these two possibilities.

One of the central features of the energy approach is the notion that **we can neither create nor destroy energy—energy is always *conserved*.** Thus, **if the total amount of energy in a system changes, it can *only* be due to the fact that energy has crossed the boundary of the system by a transfer mechanism such as one of the methods listed above.** This is a general statement of the principle of **conservation of energy.** We can describe this idea mathematically as follows:

$$\Delta E_{\text{system}} = \sum T \qquad (7.17)$$

Conservation of energy

where E_{system} is the total energy of the system, including all methods of energy storage (kinetic, internal, and potential, as discussed in Chapter 8) and T is the amount of energy transferred across the system boundary by some mechanism. Two of our transfer mechanisms have well-established symbolic notations. For work, $T_{\text{work}} = W$, as we have seen in the current chapter, and for heat, $T_{\text{heat}} = Q$, as defined in Chapter 20. The other four members of our list do not have established symbols.

This is no more complicated in theory than is balancing your checking account statement. If your account is the system, the change in the account balance for a given month is the sum of all the transfers—deposits, withdrawals, fees, interest, and checks written. It may be useful for you to think of energy as the *currency of nature*!

Suppose a force is applied to a nonisolated system and the point of application of the force moves through a displacement. Suppose further that the only effect on the system is to change its speed. Then the only transfer mechanism is work (so that ΣT in Equation 7.17 reduces to just W) and the only kind of energy in the system that changes is the kinetic energy (so that ΔE_{system} reduces to just ΔK). Equation 7.17 then becomes

$$\Delta K = W$$

which is the work–kinetic energy theorem. The work–kinetic energy theorem is a special case of the more general principle of conservation of energy. We shall see several more special cases in future chapters.

Quick Quiz 7.7 By what transfer mechanisms does energy enter and leave (a) your television set; (b) your gasoline-powered lawn mower; (c) your hand-cranked pencil sharpener?

Quick Quiz 7.8 Consider a block sliding over a horizontal surface with friction. Ignore any sound the sliding might make. If we consider the system to be the *block*, this system is (a) isolated (b) nonisolated (c) impossible to determine.

Quick Quiz 7.9 If we consider the system in Quick Quiz 7.8 to be the *surface*, this system is (a) isolated (b) nonisolated (c) impossible to determine.

Quick Quiz 7.10 If we consider the system in Quick Quiz 7.8 to be the *block and the surface*, this system is (a) isolated (b) nonisolated (c) impossible to determine.

7.7 Situations Involving Kinetic Friction

Consider again the book in Figure 7.16 sliding to the right on the surface of a heavy table and slowing down due to the friction force. Work is done by the friction force because there is a force and a displacement. Keep in mind, however, that our equations for work involve the displacement *of the point of application of the force.* The friction force is spread out over the entire contact area of an object sliding on a surface, so the force is not localized at a point. In addition, the magnitudes of the friction forces at various points are constantly changing as spot welds occur, the surface and the book deform locally, and so on. The points of application of the friction force on the book are jumping all over the face of the book in contact with the surface. This means that the displacement of the point of application of the friction force (assuming we could calculate it!) is not the same as the displacement of the book.

The work–kinetic energy theorem is valid for a particle or an object that can be modeled as a particle. When an object cannot be treated as a particle, however, things become more complicated. For these kinds of situations, Newton's second law is still valid for the system, even though the work–kinetic energy theorem is not. In the case of a nondeformable object like our book sliding on the surface,[6] we can handle this in a relatively straightforward way.

Starting from a situation in which a constant force is applied to the book, we can follow a similar procedure to that in developing Equation 7.14. We start by multiplying each side of Newton's second law (x component only) by a displacement Δx of the book:

$$\left(\sum F_x\right)\Delta x = (ma_x)\Delta x \tag{7.18}$$

For a particle under constant acceleration, we know that the following relationships (Eqs. 2.9 and 2.11) are valid:

$$a_x = \frac{v_f - v_i}{t} \qquad \Delta x = \tfrac{1}{2}(v_i + v_f)t$$

where v_i is the speed at $t = 0$ and v_f is the speed at time t. Substituting these expressions into Equation 7.18 gives

$$\left(\sum F_x\right)\Delta x = m\left(\frac{v_f - v_i}{t}\right)\tfrac{1}{2}(v_i + v_f)t$$

$$\left(\sum F_x\right)\Delta x = \tfrac{1}{2}m{v_f}^2 - \tfrac{1}{2}m{v_i}^2$$

This *looks* like the work–kinetic energy theorem, but *the left hand side has not been called work.* The quantity Δx is the displacement of the book—*not* the displacement of the point of application of the friction force.

Let us now apply this equation to a book that has been projected across a surface. We imagine that the book has an initial speed and slows down due to friction, the only force in the horizontal direction. The net force on the book is the kinetic friction force $\mathbf{f}_k$, which is directed opposite to the displacement Δx. Thus,

$$\left(\sum F_x\right)\Delta x = -f_k\Delta x = \tfrac{1}{2}m{v_f}^2 - \tfrac{1}{2}m{v_i}^2 = \Delta K$$

$$-f_k\Delta x = \Delta K \tag{7.19}$$

which mathematically describes the decrease in kinetic energy due to the friction force.

We have generated these results by assuming that a book is moving along a straight line. An object could also slide over a surface with friction and follow a curved path. In this case, Equation 7.19 must be generalized as follows:

Change in kinetic energy due to friction

$$-f_k d = \Delta K \tag{7.20}$$

where d is the length of the path followed by an object.

If there are other forces besides friction acting on an object, the change in kinetic energy is the sum of that due to the other forces from the work–kinetic energy theorem, and that due to friction:

[6] The overall shape of the book remains the same, which is why we are saying it is nondeformable. On a microscopic level, however, there is deformation of the book's face as it slides over the surface.

$$\Delta K = -f_k d + \sum W_{\text{other forces}} \quad (7.21a)$$

or

$$K_f = K_i - f_k d + \sum W_{\text{other forces}} \quad (7.21b)$$

Now consider the larger system of the book *and* the surface as the book slows down under the influence of a friction force alone. There is no work done across the boundary of this system—the system does not interact with the environment. There are no other types of energy transfer occurring across the boundary of the system, assuming we ignore the inevitable sound the sliding book makes! In this case, Equation 7.17 becomes

$$\Delta E_{\text{system}} = \Delta K + \Delta E_{\text{int}} = 0$$

The change in kinetic energy of this book-plus-surface system is the same as the change in kinetic energy of the the book alone in Equation 7.20, because the book is the only part of the book-surface system that is moving. Thus,

$$-f_k d + \Delta E_{\text{int}} = 0$$

$$\Delta E_{\text{int}} = f_k d \quad (7.22)$$

Change in internal energy due to friction

Thus, the increase in internal energy of the system is equal to the product of the friction force and the displacement of the book.

The conclusion of this discussion is that **the result of a friction force is to transform kinetic energy into internal energy, and the increase in internal energy is equal to the decrease in kinetic energy.**

Quick Quiz 7.11 You are traveling along a freeway at 65 mi/h. Your car has kinetic energy. You suddenly skid to a stop because of congestion in traffic. Where is the kinetic energy that your car once had? (a) All of it is in internal energy in the road. (b) All of it is in internal energy in the tires. (c) Some of it has transformed to internal energy and some of it transferred away by mechanical waves. (d) All of it is transferred away from your car by various mechanisms.

Example 7.9 A Block Pulled on a Rough Surface

Interactive

A 6.0-kg block initially at rest is pulled to the right along a horizontal surface by a constant horizontal force of 12 N.

(A) Find the speed of the block after it has moved 3.0 m if the surfaces in contact have a coefficient of kinetic friction of 0.15. (This is Example 7.7, modified so that the surface is no longer frictionless.)

Solution Conceptualize this problem by realizing that the rough surface is going to apply a friction force opposite to the applied force. As a result, we expect the speed to be lower than that found in Example 7.7. The surface is rough and we are given forces and a distance, so we categorize this as a situation involving kinetic friction that must be handled by means of Equation 7.21. To analyze the problem, we have made a drawing of this situation in Figure 7.18a. We identify the block as the system, and there are four external forces interacting with the system. The normal force balances the gravitational force on the block, and neither of these vertically acting forces does work on the block because their points of application are displaced horizontally. The applied force does work just as in Example 7.7:

$$W = F\Delta x = (12\text{ N})(3.0\text{ m}) = 36\text{ J}$$

In this case we must use Equation 7.21a to calculate the kinetic energy change due to friction, $\Delta K_{\text{friction}}$. Because the block is in equilibrium in the vertical direction, the normal force $\mathbf{n}$ counterbalances the gravitational force $m\mathbf{g}$, so we have $n = mg$. Hence, the magnitude of the friction force is

$$f_k = \mu_k n = \mu_k mg = (0.15)(6.0\text{ kg})(9.80\text{ m/s}^2) = 8.82\text{ N}$$

The change in kinetic energy of the block due to friction is

$$\Delta K_{\text{friction}} = -f_k d = -(8.82\text{N})(3.0\text{m}) = -26.5\text{ J}$$

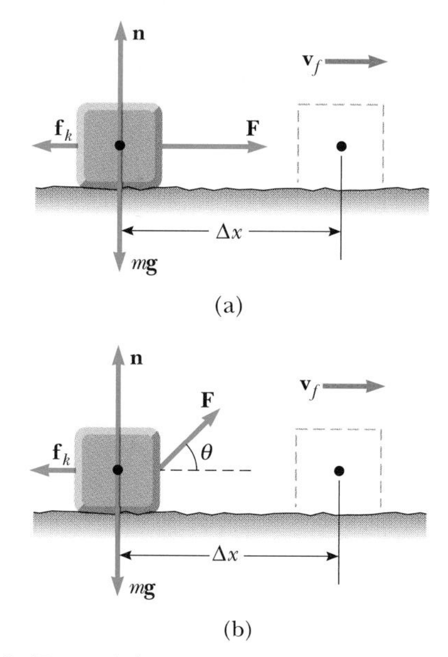

Figure 7.18 (Example 7.9) (a) A block pulled to the right on a rough surface by a constant horizontal force. (b) The applied force is at an angle θ to the horizontal.

The final speed of the block follows from Equation 7.21b:

$$\tfrac{1}{2}mv_f^{\,2} = \tfrac{1}{2}mv_i^{\,2} - f_k d + \sum W_{\text{other forces}}$$

$$v_f = \sqrt{v_i^{\,2} + \frac{2}{m}\left(-f_k d + \sum W_{\text{other forces}}\right)}$$

$$= \sqrt{0 + \frac{2}{6.0\,\text{kg}}(-26.5\,\text{J} + 36\,\text{J})}$$

$$= \boxed{1.8\,\text{m/s}}$$

To finalize this problem note that, after covering the same distance on a frictionless surface (see Example 7.7), the speed of the block was 3.5 m/s.

(B) Suppose the force $\mathbf{F}$ is applied at an angle θ as shown in Figure 7.18b. At what angle should the force be applied to achieve the largest possible speed after the block has moved 3.0 m to the right?

Solution The work done by the applied force is now

$$W = F\Delta x \cos\theta = Fd\cos\theta$$

where $\Delta x = d$ because the path followed by the block is a straight line. The block is in equilibrium in the vertical direction, so

$$\sum F_y = n + F\sin\theta - mg = 0$$

and

$$n = mg - F\sin\theta$$

Because $K_i = 0$, Equation 7.21b can be written,

$$K_f = -f_k d + \sum W_{\text{other forces}}$$

$$= -\mu_k n d + Fd\cos\theta$$

$$= -\mu_k(mg - F\sin\theta)d + Fd\cos\theta$$

Maximizing the speed is equivalent to maximizing the final kinetic energy. Consequently, we differentiate K_f with respect to θ and set the result equal to zero:

$$\frac{d(K_f)}{d\theta} = -\mu_k(0 - F\cos\theta)d - Fd\sin\theta = 0$$

$$\mu_k\cos\theta - \sin\theta = 0$$

$$\tan\theta = \mu_k$$

For $\mu_k = 0.15$, we have,

$$\theta = \tan^{-1}(\mu_k) = \tan^{-1}(0.15) = \boxed{8.5°}$$

Try out the effects of pulling the block at various angles at the Interactive Worked Example link at **http://www.pse6.com.**

Conceptual Example 7.10 Useful Physics for Safer Driving

A car traveling at an initial speed v slides a distance d to a halt after its brakes lock. Assuming that the car's initial speed is instead $2v$ at the moment the brakes lock, estimate the distance it slides.

Solution Let us assume that the force of kinetic friction between the car and the road surface is constant and the same for both speeds. According to Equation 7.20, the friction force multiplied by the distance d is equal to the initial kinetic energy of the car (because $K_f = 0$). If the speed is doubled, as it is in this example, the kinetic energy is quadrupled. For a given friction force, the distance traveled is four times as great when the initial speed is doubled, and so the estimated distance that the car slides is $4d$.

Example 7.11 A Block–Spring System

Interactive

A block of mass 1.6 kg is attached to a horizontal spring that has a force constant of 1.0×10^3 N/m, as shown in Figure 7.10. The spring is compressed 2.0 cm and is then released from rest.

(A) Calculate the speed of the block as it passes through the equilibrium position $x = 0$ if the surface is frictionless.

Solution In this situation, the block starts with $v_i = 0$ at $x_i = -2.0$ cm, and we want to find v_f at $x_f = 0$. We use Equation 7.10 to find the work done by the spring with $x_{\text{max}} = x_i = -2.0\text{ cm} = -2.0 \times 10^{-2}$ m:

$$W_s = \tfrac{1}{2}kx_{\text{max}}^2 = \tfrac{1}{2}(1.0 \times 10^3\,\text{N/m})(-2.0 \times 10^{-2}\,\text{m})^2 = 0.20\,\text{J}$$

Using the work–kinetic energy theorem with $v_i = 0$, we set the change in kinetic energy of the block equal to the work done on it by the spring:

$$W_s = \tfrac{1}{2}mv_f^2 - \tfrac{1}{2}mv_i^2$$

$$v_f = \sqrt{v_i^2 + \frac{2}{m}W_s}$$

$$= \sqrt{0 + \frac{2}{1.6\text{ kg}}(0.20\text{ J})}$$

$$= 0.50\text{ m/s}$$

(B) Calculate the speed of the block as it passes through the equilibrium position if a constant friction force of 4.0 N retards its motion from the moment it is released.

Solution Certainly, the answer has to be less than what we found in part (A) because the friction force retards the motion. We use Equation 7.20 to calculate the kinetic energy lost because of friction and add this negative value to the kinetic energy we calculated in the absence of friction. The kinetic energy lost due to friction is

$$\Delta K = -f_k d = -(4.0\text{ N})(2.0 \times 10^{-2}\text{ m}) = -0.080\text{ J}$$

In part (A), the work done by the spring was found to be 0.20 J. Therefore, the final kinetic energy in the presence of friction is

$$K_f = 0.20\text{ J} - 0.080\text{ J} = 0.12\text{ J} = \tfrac{1}{2}mv_f^2$$

$$v_f = \sqrt{\frac{2K_f}{m}} = \sqrt{\frac{2(0.12\text{ J})}{1.6\text{ kg}}} = 0.39\text{ m/s}$$

As expected, this value is somewhat less than the 0.50 m/s we found in part (A). If the friction force were greater, then the value we obtained as our answer would have been even smaller.

What If? What if the friction force were increased to 10.0 N? What is the block's speed at $x = 0$?

Answer In this case, the loss of kinetic energy as the block moves to $x = 0$ is

$$\Delta K = -f_k d = -(10.0\text{ N})(2.0 \times 10^{-2}\text{ m}) = -0.20\text{ J}$$

which is equal in magnitude to the kinetic energy at $x = 0$ without the loss due to friction. Thus, all of the kinetic energy has been transformed by friction when the block arrives at $x = 0$ and its speed at this point is $v = 0$.

In this situation as well as that in part (B), the speed of the block reaches a maximum at some position other than $x = 0$. Problem 70 asks you to locate these positions.

Investigate the role of the spring constant, amount of spring compression, and surface friction at the Interactive Worked Example link at **http://www.pse6.com.**

7.8 Power

Consider Conceptual Example 7.8 again, which involved rolling a refrigerator up a ramp into a truck. Suppose that the man is not convinced by our argument that the work is the same regardless of the length of the ramp and sets up a long ramp with a gentle rise. Although he will do the same amount of work as someone using a shorter ramp, he will take longer to do the work simply because he has to move the refrigerator over a greater distance. While the work done on both ramps is the same, there is *something* different about the tasks—the *time interval* during which the work is done.

The time rate of energy transfer is called **power.** We will focus on work as the energy transfer method in this discussion, but keep in mind that the notion of power is valid for *any* means of energy transfer. If an external force is applied to an object (which we assume acts as a particle), and if the work done by this force in the time interval Δt is W, then the **average power** during this interval is defined as

$$\overline{\mathcal{P}} \equiv \frac{W}{\Delta t}$$

Thus, while the same work is done in rolling the refrigerator up both ramps, less power is required for the longer ramp.

In a manner similar to the way we approached the definition of velocity and acceleration, we define the **instantaneous power** $\mathcal{P}$ as the limiting value of the average power as Δt approaches zero:

$$\mathcal{P} \equiv \lim_{\Delta t \to 0} \frac{W}{\Delta t} = \frac{dW}{dt}$$

where we have represented the infinitesimal value of the work done by dW. We find from Equation 7.3 that $dW = \mathbf{F} \cdot d\mathbf{r}$. Therefore, the instantaneous power can be written

Instantaneous power

$$\mathcal{P} = \frac{dW}{dt} = \mathbf{F} \cdot \frac{d\mathbf{r}}{dt} = \mathbf{F} \cdot \mathbf{v} \qquad (7.23)$$

where we use the fact that $\mathbf{v} = d\mathbf{r}/dt$.

In general, power is defined for any type of energy transfer. Therefore, the most general expression for power is

$$\mathcal{P} = \frac{dE}{dt} \qquad (7.24)$$

where dE/dt is the rate at which energy is crossing the boundary of the system by a given transfer mechanism.

The SI unit of power is joules per second (J/s), also called the **watt** (W) (after James Watt):

The watt

$$1 \text{ W} = 1 \text{ J/s} = 1 \text{ kg} \cdot \text{m}^2/\text{s}^3$$

A unit of power in the U.S. customary system is the **horsepower** (hp):

$$1 \text{ hp} = 746 \text{ W}$$

A unit of energy (or work) can now be defined in terms of the unit of power. One **kilowatt-hour** (kWh) is the energy transferred in 1 h at the constant rate of 1 kW = 1 000 J/s. The amount of energy represented by 1 kWh is

$$1 \text{ kWh} = (10^3 \text{ W})(3\,600 \text{ s}) = 3.60 \times 10^6 \text{ J}$$

Note that a kilowatt-hour is a unit of energy, not power. When you pay your electric bill, you are buying energy, and the amount of energy transferred by electrical transmission into a home during the period represented by the electric bill is usually expressed in kilowatt-hours. For example, your bill may state that you used 900 kWh of energy during a month, and you are being charged at the rate of 10¢ per kWh. Your obligation is then $90 for this amount of energy. As another example, suppose an electric bulb is rated at 100 W. In 1.00 hour of operation, it would have energy transferred to it by electrical transmission in the amount of $(0.100 \text{ kW})(1.00 \text{ h}) = 0.100 \text{ kWh} = 3.60 \times 10^5 \text{ J}$.

PITFALL PREVENTION

7.9 W, *W*, and watts

Do not confuse the symbol W for the watt with the italic symbol W for work. Also, remember that the watt already represents a rate of energy transfer, so that "watts per second" does not make sense. The watt is *the same as* a joule per second.

Quick Quiz 7.12 An older model car accelerates from rest to speed v in 10 seconds. A newer, more powerful sports car accelerates from rest to $2v$ in the same time period. What is the ratio of the power of the newer car to that of the older car? (a) 0.25 (b) 0.5 (c) 1 (d) 2 (e) 4

Example 7.12 Power Delivered by an Elevator Motor

An elevator car has a mass of 1 600 kg and is carrying passengers having a combined mass of 200 kg. A constant friction force of 4 000 N retards its motion upward, as shown in Figure 7.19a.

(A) What power delivered by the motor is required to lift the elevator car at a constant speed of 3.00 m/s?

Solution The motor must supply the force of magnitude T that pulls the elevator car upward. The problem states that the speed is constant, which provides the hint that $a = 0$. Therefore we know from Newton's second law that $\Sigma F_y = 0$. The free-body diagram in Figure 7.19b specifies the upward direction as positive. From Newton's second law we obtain

$$\sum F_y = T - f - Mg = 0$$

where M is the *total* mass of the system (car plus passengers), equal to 1 800 kg. Therefore,

$$\begin{aligned} T &= f + Mg \\ &= 4.00 \times 10^3 \text{ N} + (1.80 \times 10^3 \text{ kg})(9.80 \text{ m/s}^2) \\ &= 2.16 \times 10^4 \text{ N} \end{aligned}$$

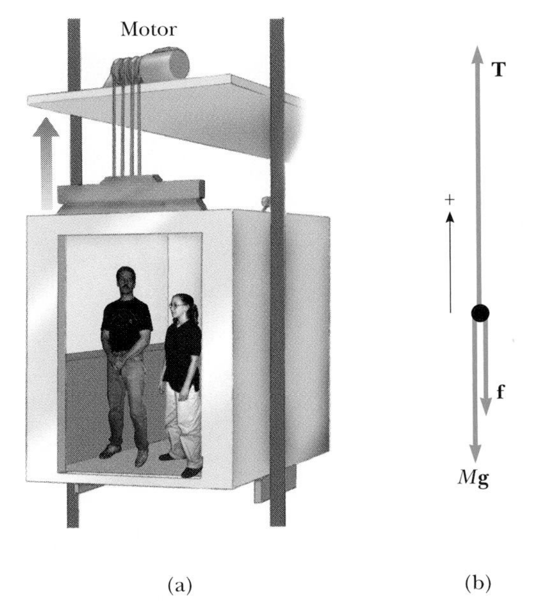

Figure 7.19 (Example 7.12) (a) The motor exerts an upward force **T** on the elevator car. The magnitude of this force is the tension T in the cable connecting the car and motor. The downward forces acting on the car are a friction force **f** and the gravitational force $\mathbf{F}_g = M\mathbf{g}$. (b) The free-body diagram for the elevator car.

Using Equation 7.23 and the fact that **T** is in the same direction as **v**, we find that

$$\mathcal{P} = \mathbf{T}\cdot\mathbf{v} = Tv$$

$$= (2.16 \times 10^4\ \text{N})(3.00\ \text{m/s}) = \boxed{6.48 \times 10^4\ \text{W}}$$

(B) What power must the motor deliver at the instant the speed of the elevator is v if the motor is designed to provide the elevator car with an upward acceleration of $1.00\ \text{m/s}^2$?

Solution We expect to obtain a value greater than we did in part (A), where the speed was constant, because the motor must now perform the additional task of accelerating the car. The only change in the setup of the problem is that in this case, $a > 0$. Applying Newton's second law to the car gives

$$\sum F_y = T - f - Mg = Ma$$

$$T = M(a + g) + f$$

$$= (1.80 \times 10^3\ \text{kg})(1.00\ \text{m/s}^2 + 9.80\ \text{m/s}^2)$$

$$+ 4.00 \times 10^3\ \text{N}$$

$$= 2.34 \times 10^4\ \text{N}$$

Therefore, using Equation 7.23, we obtain for the required power

$$\mathcal{P} = Tv = \boxed{(2.34 \times 10^4\ \text{N})v}$$

where v is the instantaneous speed of the car in meters per second. To compare to part (A), let $v = 3.00$ m/s, giving a power of

$$\mathcal{P} = (2.34 \times 10^4\ \text{N})(3.00\ \text{m/s}) = 7.02 \times 10^4\ \text{W}$$

This is larger than the power found in part (A), as we expect.

7.9 Energy and the Automobile

Automobiles powered by gasoline engines are very inefficient machines. Even under ideal conditions, less than 15% of the chemical energy in the fuel is used to power the vehicle. The situation is much worse than this under stop-and-go driving conditions in a city. In this section, we use the concepts of energy, power, and friction to analyze automobile fuel consumption.

Many mechanisms contribute to energy loss in an automobile. About 67% of the energy available from the fuel is lost in the engine. This energy ends up in the atmosphere, partly via the exhaust system and partly via the cooling system. (As explained in Chapter 22, energy loss from the exhaust and cooling systems is required by a fundamental law of thermodynamics.) Approximately 10% of the available energy is lost to friction in the transmission, drive shaft, wheel and axle bearings, and differential. Friction in other moving parts transforms approximately 6% of the energy to internal energy, and 4% of the energy is used to operate fuel and oil pumps and such accessories as power steering and air conditioning. This leaves a mere 13% of the available energy to propel the automobile! This energy is used mainly to balance the energy loss due to flexing of the tires and the friction caused by the air, which is more commonly referred to as *air resistance.*

Let us examine the power required to provide a force in the forward direction that balances the combination of the two friction forces. The coefficient of rolling friction μ between the tires and the road is about 0.016. For a 1 450-kg car, the weight is 14 200 N and on a horizontal roadway the force of rolling friction has a magnitude of $\mu n = \mu mg = 227$ N. As the car's speed increases, a small reduction in the normal force

Table 7.2

Friction Forces and Power Requirements for a Typical Car[a]

v(mi/h)	v(m/s)	n(N)	f_r(N)	f_a(N)	f_t(N)	$\mathcal{P} = f_t v$(kW)
0	0	14 200	227	0	227	0
20	8.9	14 100	226	48	274	2.4
40	17.9	13 900	222	192	414	7.4
60	26.8	13 600	218	431	649	17.4
80	35.8	13 200	211	767	978	35.0
100	44.7	12 600	202	1 199	1 400	62.6

[a] In this table, n is the normal force, f_r is rolling friction, f_a is air friction, f_t is total friction, and $\mathcal{P}$ is the power delivered to the wheels.

occurs as a result of decreased pressure as air flows over the top of the car. (This phenomenon is discussed in Chapter 14.) This reduction in the normal force causes a reduction in the force of rolling friction f_r with increasing speed, as the data in Table 7.2 indicate.

Now let us consider the effect of the resistive force that results from the movement of air past the car. For large objects, the resistive force f_a associated with air friction is proportional to the square of the speed (see Section 6.4) and is given by Equation 6.6:

$$f_a = \tfrac{1}{2} D \rho A v^2$$

where D is the drag coefficient, ρ is the density of air, and A is the cross-sectional area of the moving object. We can use this expression to calculate the f_a values in Table 7.2, using $D = 0.50$, $\rho = 1.20\ \text{kg/m}^3$, and $A \approx 2\ \text{m}^2$.

The magnitude of the total friction force f_t is the sum of the rolling friction force and the air resistive force:

$$f_t = f_r + f_a$$

At low speeds, rolling friction is the predominant resistive force, but at high speeds air drag predominates, as shown in Table 7.2. Rolling friction can be decreased by a reduction in tire flexing (for example, by an increase in the air pressure slightly above recommended values) and by the use of radial tires. Air drag can be reduced through the use of a smaller cross-sectional area and by streamlining the car. Although driving a car with the windows open increases air drag and thus results in a 3% decrease in mileage, driving with the windows closed and the air conditioner running results in a 12% decrease in mileage.

The total power needed to maintain a constant speed v is $f_t v$, and this is the power that must be delivered to the wheels. For example, from Table 7.2 we see that at $v = 26.8$ m/s (60 mi/h) the required power is

$$\mathcal{P} = f_t v = (649\ \text{N})(26.8\ \text{m/s}) = 17.4\ \text{kW}$$

This power can be broken down into two parts: (1) the power $f_r v$ needed to compensate for rolling friction, and (2) the power $f_a v$ needed to compensate for air drag. At $v = 26.8$ m/s, we obtain the values

$$\mathcal{P}_r = f_r v = (218\ \text{N})(26.8\ \text{m/s}) = 5.84\ \text{kW}$$

$$\mathcal{P}_a = f_a v = (431\ \text{N})(26.8\ \text{m/s}) = 11.6\ \text{kW}$$

Note that $\mathcal{P} = \mathcal{P}_r + \mathcal{P}_a$ and 67% of the power is used to compensate for air drag.

On the other hand, at $v = 44.7$ m/s(100 mi/h), $\mathcal{P}_r = 9.03$ kW, $\mathcal{P}_a = 53.6$ kW, $\mathcal{P} = 62.6$ kW and 86% of the power is associated with air drag. This shows the importance of air drag at high speeds.

Example 7.13 Gas Consumed by a Compact Car

A compact car has a mass of 800 kg, and its efficiency is rated at 18%. (That is, 18% of the available fuel energy is delivered to the wheels.) Find the amount of gasoline used to accelerate the car from rest to 27 m/s (60 mi/h). Use the fact that the energy equivalent of 1 gal of gasoline is 1.3×10^8 J.

Solution The energy required to accelerate the car from rest to a speed v is equal to its final kinetic energy, $\frac{1}{2}mv^2$:

$$K = \tfrac{1}{2}mv^2 = \tfrac{1}{2}(800\text{ kg})(27\text{ m/s})^2 = 2.9 \times 10^5\text{ J}$$

If the engine were 100% efficient, each gallon of gasoline would supply 1.3×10^8 J of energy. Because the engine is only 18% efficient, each gallon delivers an energy of only $(0.18)(1.3 \times 10^8\text{ J}) = 2.3 \times 10^7$ J. Hence, the number of gallons used to accelerate the car is

$$\text{Number of gal} = \frac{2.9 \times 10^5\text{ J}}{2.3 \times 10^7\text{ J/gal}} = \boxed{0.013\text{ gal}}$$

Let us estimate that it takes 10 s to achieve the indicated speed. The distance traveled during this acceleration is

$$\Delta x = \bar{v}\Delta t = \frac{v_{xf} + v_{xi}}{2}(\Delta t) = \frac{27\text{ m/s} + 0}{2}(10\text{ s})$$
$$= 135\text{ m} \approx 0.08\text{ mi}$$

At a constant cruising speed, 0.013 gal of gasoline is sufficient to propel the car nearly 0.5 mi, over six times farther. This demonstrates the extreme energy requirements of stop-and-start driving.

Example 7.14 Power Delivered to the Wheels

Suppose the compact car in Example 7.13 has a gas mileage of 35 mi/gal at 60 mi/h. How much power is delivered to the wheels?

Solution We find the rate of gasoline consumption by dividing the speed by the gas mileage:

$$\frac{60\text{ mi/h}}{35\text{ mi/gal}} = 1.7\text{ gal/h}$$

Using the fact that each gallon is equivalent to 1.3×10^8 J, we find that the total power used is

$$\mathcal{P} = (1.7\text{ gal/h})(1.3 \times 10^8\text{ J/gal})\left(\frac{1\text{ h}}{3.6 \times 10^3\text{ s}}\right)$$
$$= 62\text{ kW}$$

Because 18% of the available power is used to propel the car, the power delivered to the wheels is $(0.18)(62\text{ kW}) = \boxed{11\text{ kW.}}$ This is 37% less than the 17.4-kW value obtained for the 1 450-kg car discussed in the text. Vehicle mass is clearly an important factor in power-loss mechanisms.

Example 7.15 Car Accelerating Up a Hill

Consider a car of mass m that is accelerating up a hill, as shown in Figure 7.20. An automotive engineer measures the magnitude of the total resistive force to be

$$f_t = (218 + 0.70v^2)\text{ N}$$

where v is the speed in meters per second. Determine the power the engine must deliver to the wheels as a function of speed.

Solution The forces on the car are shown in Figure 7.20, in which **F** is the force of friction from the road that propels the car; the remaining forces have their usual meaning.

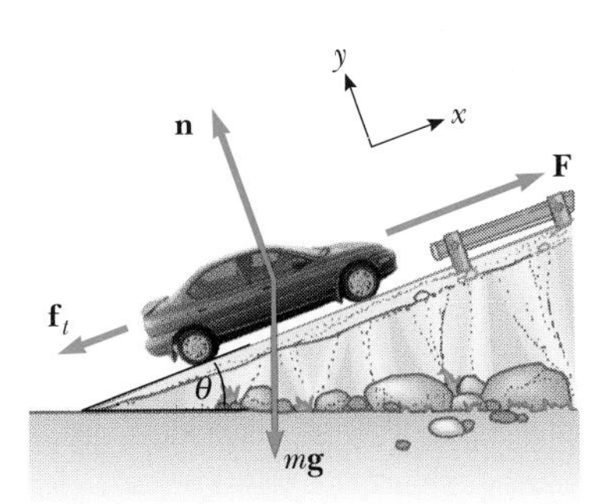

Figure 7.20 (Example 7.15) A car climbs a hill.

Applying Newton's second law to the motion along the road surface, we find that

$$\sum F_x = F - f_t - mg\sin\theta = ma$$
$$F = ma + mg\sin\theta + f_t$$
$$= ma + mg\sin\theta + (218 + 0.70v^2)$$

Therefore, the power required to move the car forward is

$$\mathcal{P} = Fv = mva + mvg\sin\theta + 218v + 0.70v^3$$

The term mva represents the power that the engine must deliver to accelerate the car. If the car moves at constant speed, this term is zero and the total power requirement is reduced. The term $mvg\sin\theta$ is the power required to provide a force to balance a component of the gravitational force as the car moves up the incline. This term would be zero for motion on a horizontal surface. The term $218v$ is the power required to provide a force to balance rolling friction, and the term $0.70v^3$ is the power needed against air drag.

If we take $m = 1\,450$ kg, $v = 27$ m/s (= 60 mi/h), $a = 1.0\text{ m/s}^2$, and $\theta = 10°$, then the various terms in $\mathcal{P}$ are calculated to be

$$mva = (1\,450\text{ kg})(27\text{ m/s})(1.0\text{ m/s}^2)$$
$$= 39\text{ kW} = 52\text{ hp}$$

$$mvg \sin\theta = (1\,450\ \text{kg})(27\ \text{m/s})(9.80\ \text{m/s}^2)(\sin 10°) = 67\ \text{kW} = 89\ \text{hp}$$

$$218v = 218(27\ \text{m/s}) = 5.9\ \text{kW} = 7.9\ \text{hp}$$

$$0.70v^3 = 0.70(27\ \text{m/s})^3 = 14\ \text{kW} = 18\ \text{hp}$$

Hence, the total power required is 126 kW or 167 hp.

Note that the power requirements for traveling at constant speed on a horizontal surface are only 20 kW, or 27 hp (the sum of the last two terms). Furthermore, if the mass were halved (as in the case of a compact car), then the power required also is reduced by almost the same factor.

SUMMARY

Take a practice test for this chapter by clicking on the Practice Test link at http://www.pse6.com.

A **system** is most often a single particle, a collection of particles or a region of space. A **system boundary** separates the system from the **environment**. Many physics problems can be solved by considering the interaction of a system with its environment.

The **work** W done on a system by an agent exerting a constant force $\mathbf{F}$ on the system is the product of the magnitude Δr of the displacement of the point of application of the force and the component $F\cos\theta$ of the force along the direction of the displacement $\Delta\mathbf{r}$:

$$W \equiv F\,\Delta r\cos\theta \tag{7.1}$$

The **scalar product** (dot product) of two vectors $\mathbf{A}$ and $\mathbf{B}$ is defined by the relationship

$$\mathbf{A}\cdot\mathbf{B} \equiv AB\cos\theta \tag{7.2}$$

where the result is a scalar quantity and θ is the angle between the two vectors. The scalar product obeys the commutative and distributive laws.

If a varying force does work on a particle as the particle moves along the x axis from x_i to x_f, the work done by the force on the particle is given by

$$W \equiv \int_{x_i}^{x_f} F_x\,dx \tag{7.7}$$

where F_x is the component of force in the x direction.

The **kinetic energy** of a particle of mass m moving with a speed v is

$$K \equiv \tfrac{1}{2}mv^2 \tag{7.15}$$

The **work–kinetic energy theorem** states that if work is done on a system by external forces and the only change in the system is in its speed, then

$$\sum W = K_f - K_i = \tfrac{1}{2}mv_f^{\,2} - \tfrac{1}{2}mv_i^{\,2} \tag{7.14, 7.16}$$

For a nonisolated system, we can equate the change in the total energy stored in the system to the sum of all the transfers of energy across the system boundary. For an isolated system, the total energy is constant—this is a statement of **conservation of energy.**

If a friction force acts, the kinetic energy of the system is reduced and the appropriate equation to be applied is

$$\Delta K = -f_k d + \sum W_{\text{other forces}} \tag{7.21a}$$

or

$$K_f = K_i - f_k d + \sum W_{\text{other forces}} \tag{7.21b}$$

The **instantaneous power** $\mathcal{P}$ is defined as the time rate of energy transfer. If an agent applies a force $\mathbf{F}$ to an object moving with a velocity $\mathbf{v}$, the power delivered by that agent is

$$\mathcal{P} \equiv \frac{dW}{dt} = \mathbf{F}\cdot\mathbf{v} \tag{7.23}$$

QUESTIONS

1. When a particle rotates in a circle, a force acts on it directed toward the center of rotation. Why is it that this force does no work on the particle?
2. Discuss whether any work is being done by each of the following agents and, if so, whether the work is positive or negative: (a) a chicken scratching the ground, (b) a person studying, (c) a crane lifting a bucket of concrete, (d) the gravitational force on the bucket in part (c), (e) the leg muscles of a person in the act of sitting down.
3. When a punter kicks a football, is he doing any work on the ball while his toe is in contact with it? Is he doing any work on the ball after it loses contact with his toe? Are any forces doing work on the ball while it is in flight?
4. Cite two examples in which a force is exerted on an object without doing any work on the object.
5. As a simple pendulum swings back and forth, the forces acting on the suspended object are the gravitational force, the tension in the supporting cord, and air resistance. (a) Which of these forces, if any, does no work on the pendulum? (b) Which of these forces does negative work at all times during its motion? (c) Describe the work done by the gravitational force while the pendulum is swinging.
6. If the dot product of two vectors is positive, does this imply that the vectors must have positive rectangular components?
7. For what values of θ is the scalar product (a) positive and (b) negative?
8. As the load on a vertically hanging spiral spring is increased, one would not expect the F_s-versus-x graph line to remain straight, as shown in Figure 7.10d. Explain qualitatively what you would expect for the shape of this graph as the load on the spring is increased.
9. A certain uniform spring has spring constant k. Now the spring is cut in half. What is the relationship between k and the spring constant k' of each resulting smaller spring? Explain your reasoning.
10. Can kinetic energy be negative? Explain.
11. Discuss the work done by a pitcher throwing a baseball. What is the approximate distance through which the force acts as the ball is thrown?
12. One bullet has twice the mass of a second bullet. If both are fired so that they have the same speed, which has more kinetic energy? What is the ratio of the kinetic energies of the two bullets?
13. Two sharpshooters fire 0.30-caliber rifles using identical shells. A force exerted by expanding gases in the barrels accelerates the bullets. The barrel of rifle A is 2.00 cm longer than the barrel of rifle B. Which rifle will have the higher muzzle speed?
14. (a) If the speed of a particle is doubled, what happens to its kinetic energy? (b) What can be said about the speed of a particle if the net work done on it is zero?
15. A car salesman claims that a souped-up 300-hp engine is a necessary option in a compact car, in place of the conventional 130-hp engine. Suppose you intend to drive the car within speed limits (≤ 65 mi/h) on flat terrain. How would you counter this sales pitch?
16. Can the average power over a time interval ever be equal to the instantaneous power at an instant within the interval? Explain.
17. In Example 7.15, does the required power increase or decrease as the force of friction is reduced?
18. The kinetic energy of an object depends on the frame of reference in which its motion is measured. Give an example to illustrate this point.
19. Words given precise definitions in physics are sometimes used in popular literature in interesting ways. For example, a rock falling from the top of a cliff is said to be "gathering force as it falls to the beach below." What does the phrase "gathering force" mean, and can you repair this phrase?
20. In most circumstances, the normal force acting on an object and the force of static friction do zero work on the object. However, the reason that the work is zero is different for the two cases. Explain why each does zero work.
21. "A level air track can do no work." Argue for or against this statement.
22. Who first stated the work–kinetic energy theorem? Who showed that it is useful for solving many practical problems? Do some research to answer these questions.

PROBLEMS

1, 2, 3 = straightforward, intermediate, challenging ☐ = full solution available in the *Student Solutions Manual and Study Guide*

= coached solution with hints available at http://www.pse6.com = computer useful in solving problem

= paired numerical and symbolic problems

Section 7.2 Work Done by a Constant Force

1. A block of mass 2.50 kg is pushed 2.20 m along a frictionless horizontal table by a constant 16.0-N force directed 25.0° below the horizontal. Determine the work done on the block by (a) the applied force, (b) the normal force exerted by the table, and (c) the gravitational force. (d) Determine the total work done on the block.
2. A shopper in a supermarket pushes a cart with a force of 35.0 N directed at an angle of 25.0° downward from the horizontal. Find the work done by the shopper on the cart as he moves down an aisle 50.0 m long.
3. Batman, whose mass is 80.0 kg, is dangling on the free end of a 12.0-m rope, the other end of which is fixed to a tree limb above. He is able to get the rope in motion

as only Batman knows how, eventually getting it to swing enough that he can reach a ledge when the rope makes a 60.0° angle with the vertical. How much work was done by the gravitational force on Batman in this maneuver?

4. A raindrop of mass 3.35×10^{-5} kg falls vertically at constant speed under the influence of gravity and air resistance. Model the drop as a particle. As it falls 100 m, what is the work done on the raindrop (a) by the gravitational force and (b) by air resistance?

Section 7.3 The Scalar Product of Two Vectors

5. Vector **A** has a magnitude of 5.00 units, and **B** has a magnitude of 9.00 units. The two vectors make an angle of 50.0° with each other. Find $\mathbf{A}\cdot\mathbf{B}$.

6. For any two vectors **A** and **B**, show that $\mathbf{A}\cdot\mathbf{B} = A_xB_x + A_yB_y + A_zB_z$. (*Suggestion:* Write **A** and **B** in unit vector form and use Equations 7.4 and 7.5.)

Note: In Problems 7 through 10, calculate numerical answers to three significant figures as usual.

7. A force $\mathbf{F} = (6\hat{\mathbf{i}} - 2\hat{\mathbf{j}})$ N acts on a particle that undergoes a displacement $\Delta\mathbf{r} = (3\hat{\mathbf{i}} + \hat{\mathbf{j}})$ m. Find (a) the work done by the force on the particle and (b) the angle between **F** and $\Delta\mathbf{r}$.

8. Find the scalar product of the vectors in Figure P7.8.

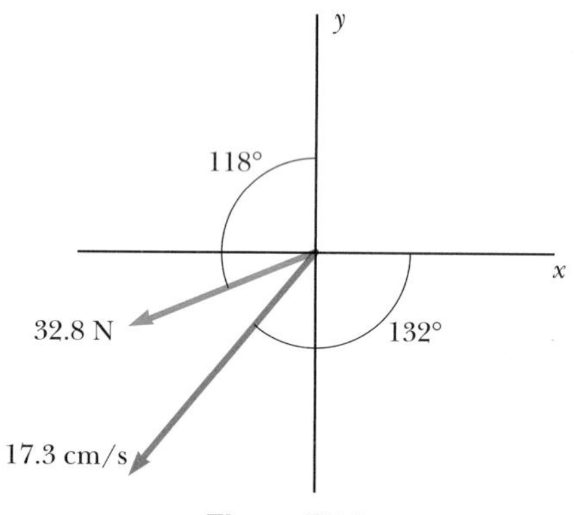

Figure P7.8

9. Using the definition of the scalar product, find the angles between (a) $\mathbf{A} = 3\hat{\mathbf{i}} - 2\hat{\mathbf{j}}$ and $\mathbf{B} = 4\hat{\mathbf{i}} - 4\hat{\mathbf{j}}$; (b) $\mathbf{A} = -2\hat{\mathbf{i}} + 4\hat{\mathbf{j}}$ and $\mathbf{B} = 3\hat{\mathbf{i}} - 4\hat{\mathbf{j}} + 2\hat{\mathbf{k}}$; (c) $\mathbf{A} = \hat{\mathbf{i}} - 2\hat{\mathbf{j}} + 2\hat{\mathbf{k}}$ and $\mathbf{B} = 3\hat{\mathbf{j}} + 4\hat{\mathbf{k}}$.

10. For $\mathbf{A} = 3\hat{\mathbf{i}} + \hat{\mathbf{j}} - \hat{\mathbf{k}}$, $\mathbf{B} = -\hat{\mathbf{i}} + 2\hat{\mathbf{j}} + 5\hat{\mathbf{k}}$, and $\mathbf{C} = 2\hat{\mathbf{j}} - 3\hat{\mathbf{k}}$, find $\mathbf{C}\cdot(\mathbf{A} - \mathbf{B})$.

Section 7.4 Work Done by a Varying Force

11. The force acting on a particle varies as in Figure P7.11. Find the work done by the force on the particle as it moves (a) from $x = 0$ to $x = 8.00$ m, (b) from $x = 8.00$ m to $x = 10.0$ m, and (c) from $x = 0$ to $x = 10.0$ m.

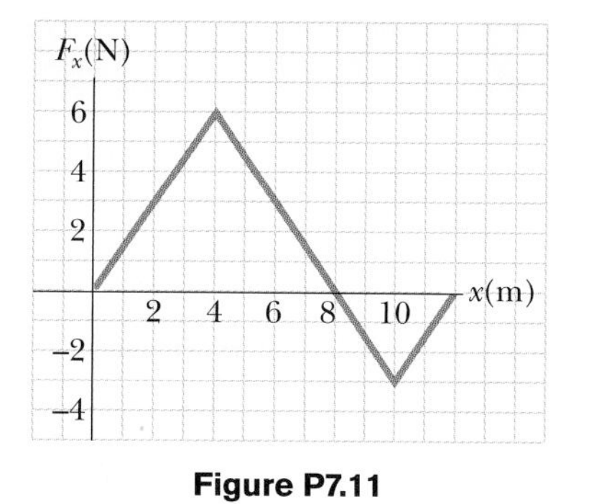

Figure P7.11

12. The force acting on a particle is $F_x = (8x - 16)$ N, where x is in meters. (a) Make a plot of this force versus x from $x = 0$ to $x = 3.00$ m. (b) From your graph, find the net work done by this force on the particle as it moves from $x = 0$ to $x = 3.00$ m.

13. A particle is subject to a force F_x that varies with position as in Figure P7.13. Find the work done by the force on the particle as it moves (a) from $x = 0$ to $x = 5.00$ m, (b) from $x = 5.00$ m to $x = 10.0$ m, and (c) from $x = 10.0$ m to $x = 15.0$ m. (d) What is the total work done by the force over the distance $x = 0$ to $x = 15.0$ m?

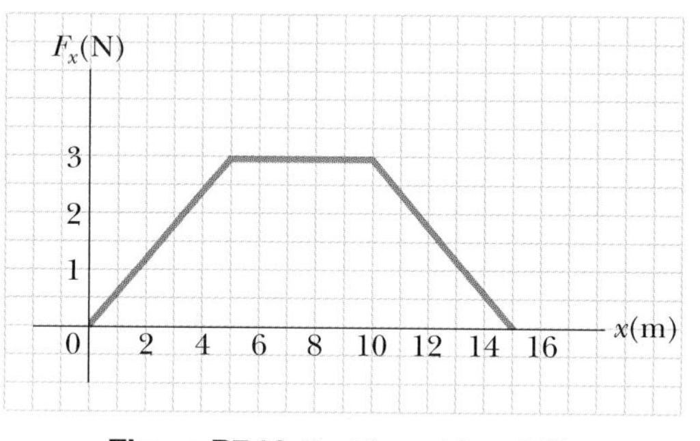

Figure P7.13 Problems 13 and 28.

14. A force $\mathbf{F} = (4x\hat{\mathbf{i}} + 3y\hat{\mathbf{j}})$ N acts on an object as the object moves in the x direction from the origin to $x = 5.00$ m. Find the work $W = \int \mathbf{F}\cdot d\mathbf{r}$ done on the object by the force.

15. When a 4.00-kg object is hung vertically on a certain light spring that obeys Hooke's law, the spring stretches 2.50 cm. If the 4.00-kg object is removed, (a) how far will the spring stretch if a 1.50-kg block is hung on it, and (b) how much work must an external agent do to stretch the same spring 4.00 cm from its unstretched position?

16. An archer pulls her bowstring back 0.400 m by exerting a force that increases uniformly from zero to 230 N. (a) What is the equivalent spring constant of the bow? (b) How much work does the archer do in pulling the bow?

17. Truck suspensions often have "helper springs" that engage at high loads. One such arrangement is a leaf spring with a helper coil spring mounted on the axle, as in Figure P7.17. The helper spring engages when the main leaf spring is compressed by distance y_0, and then helps to support any additional load. Consider a leaf spring constant of 5.25×10^5 N/m, helper spring constant of 3.60×10^5 N/m, and $y_0 = 0.500$ m. (a) What is the

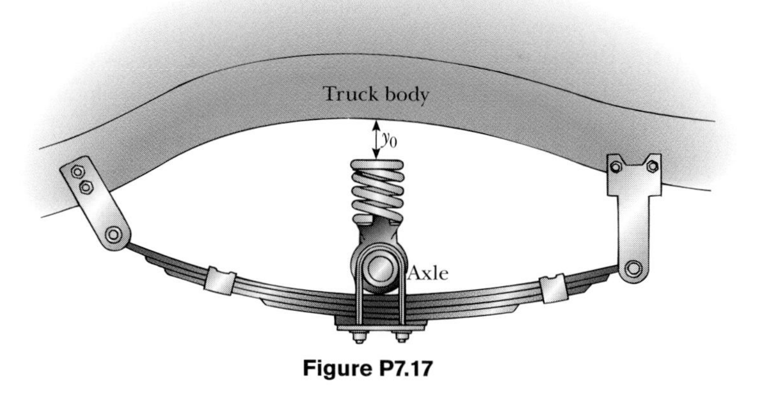

Figure P7.17

compression of the leaf spring for a load of 5.00×10^5 N? (b) How much work is done in compressing the springs?

18. A 100-g bullet is fired from a rifle having a barrel 0.600 m long. Assuming the origin is placed where the bullet begins to move, the force (in newtons) exerted by the expanding gas on the bullet is $15\ 000 + 10\ 000x - 25\ 000x^2$, where x is in meters. (a) Determine the work done by the gas on the bullet as the bullet travels the length of the barrel. (b) **What If?** If the barrel is 1.00 m long, how much work is done, and how does this value compare to the work calculated in (a)?

19. If it takes 4.00 J of work to stretch a Hooke's-law spring 10.0 cm from its unstressed length, determine the extra work required to stretch it an additional 10.0 cm.

20. A small particle of mass m is pulled to the top of a frictionless half-cylinder (of radius R) by a cord that passes over the top of the cylinder, as illustrated in Figure P7.20. (a) If the particle moves at a constant speed, show that $F = mg \cos \theta$. (*Note*: If the particle moves at constant speed, the component of its acceleration tangent to the cylinder must be zero at all times.) (b) By directly integrating $W = \int \mathbf{F} \cdot d\mathbf{r}$, find the work done in moving the particle at constant speed from the bottom to the top of the half-cylinder.

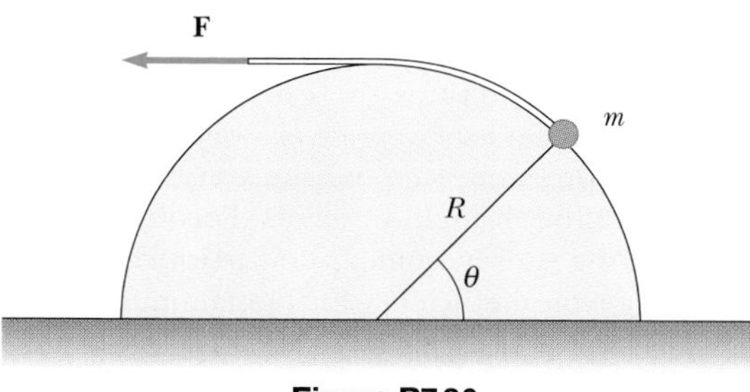

Figure P7.20

21. A light spring with spring constant 1 200 N/m is hung from an elevated support. From its lower end a second light spring is hung, which has spring constant 1 800 N/m. An object of mass 1.50 kg is hung at rest from the lower end of the second spring. (a) Find the total extension distance of the pair of springs. (b) Find the effective spring constant of the pair of springs as a system. We describe these springs as *in series.*

22. A light spring with spring constant k_1 is hung from an elevated support. From its lower end a second light spring is hung, which has spring constant k_2. An object of mass m is hung at rest from the lower end of the second spring. (a) Find the total extension distance of the pair of springs. (b) Find the effective spring constant of the pair of springs as a system. We describe these springs as *in series.*

23. Express the units of the force constant of a spring in SI base units.

Section 7.5 Kinetic Energy and the Work–Kinetic Energy Theorem

Section 7.6 The Nonisolated System–Conservation of Energy

24. A 0.600-kg particle has a speed of 2.00 m/s at point Ⓐ and kinetic energy of 7.50 J at point Ⓑ. What is (a) its kinetic energy at Ⓐ? (b) its speed at Ⓑ? (c) the total work done on the particle as it moves from Ⓐ to Ⓑ?

25. A 0.300-kg ball has a speed of 15.0 m/s. (a) What is its kinetic energy? (b) **What If?** If its speed were doubled, what would be its kinetic energy?

26. A 3.00-kg object has a velocity $(6.00\hat{\mathbf{i}} - 2.00\hat{\mathbf{j}})$ m/s. (a) What is its kinetic energy at this time? (b) Find the total work done on the object if its velocity changes to $(8.00\hat{\mathbf{i}} + 4.00\hat{\mathbf{j}})$ m/s. (*Note:* From the definition of the dot product, $v^2 = \mathbf{v} \cdot \mathbf{v}$.)

27. A 2 100-kg pile driver is used to drive a steel I-beam into the ground. The pile driver falls 5.00 m before coming into contact with the top of the beam, and it drives the beam 12.0 cm farther into the ground before coming to rest. Using energy considerations, calculate the average force the beam exerts on the pile driver while the pile driver is brought to rest.

28. A 4.00-kg particle is subject to a total force that varies with position as shown in Figure P7.13. The particle starts from rest at $x = 0$. What is its speed at (a) $x = 5.00$ m, (b) $x = 10.0$ m, (c) $x = 15.0$ m?

29. You can think of the work–kinetic energy theorem as a second theory of motion, parallel to Newton's laws in describing how outside influences affect the motion of an object. In this problem, solve parts (a) and (b) separately from parts (c) and (d) to compare the predictions of the two

theories. In a rifle barrel, a 15.0-g bullet is accelerated from rest to a speed of 780 m/s. (a) Find the work that is done on the bullet. (b) If the rifle barrel is 72.0 cm long, find the magnitude of the average total force that acted on it, as $F = W/(\Delta r \cos \theta)$. (c) Find the constant acceleration of a bullet that starts from rest and gains a speed of 780 m/s over a distance of 72.0 cm. (d) If the bullet has mass 15.0 g, find the total force that acted on it as $\Sigma F = ma$.

30. In the neck of the picture tube of a certain black-and-white television set, an electron gun contains two charged metallic plates 2.80 cm apart. An electric force accelerates each electron in the beam from rest to 9.60% of the speed of light over this distance. (a) Determine the kinetic energy of the electron as it leaves the electron gun. Electrons carry this energy to a phosphorescent material on the inner surface of the television screen, making it glow. For an electron passing between the plates in the electron gun, determine (b) the magnitude of the constant electric force acting on the electron, (c) the acceleration, and (d) the time of flight.

Section 7.7 Situations Involving Kinetic Friction

31. A 40.0-kg box initially at rest is pushed 5.00 m along a rough, horizontal floor with a constant applied horizontal force of 130 N. If the coefficient of friction between box and floor is 0.300, find (a) the work done by the applied force, (b) the increase in internal energy in the box-floor system due to friction, (c) the work done by the normal force, (d) the work done by the gravitational force, (e) the change in kinetic energy of the box, and (f) the final speed of the box.

32. A 2.00-kg block is attached to a spring of force constant 500 N/m as in Figure 7.10. The block is pulled 5.00 cm to the right of equilibrium and released from rest. Find the speed of the block as it passes through equilibrium if (a) the horizontal surface is frictionless and (b) the coefficient of friction between block and surface is 0.350.

33. A crate of mass 10.0 kg is pulled up a rough incline with an initial speed of 1.50 m/s. The pulling force is 100 N parallel to the incline, which makes an angle of 20.0° with the horizontal. The coefficient of kinetic friction is 0.400, and the crate is pulled 5.00 m. (a) How much work is done by the gravitational force on the crate? (b) Determine the increase in internal energy of the crate–incline system due to friction. (c) How much work is done by the 100-N force on the crate? (d) What is the change in kinetic energy of the crate? (e) What is the speed of the crate after being pulled 5.00 m?

34. A 15.0-kg block is dragged over a rough, horizontal surface by a 70.0-N force acting at 20.0° above the horizontal. The block is displaced 5.00 m, and the coefficient of kinetic friction is 0.300. Find the work done on the block by (a) the 70-N force, (b) the normal force, and (c) the gravitational force. (d) What is the increase in internal energy of the block-surface system due to friction? (e) Find the total change in the block's kinetic energy.

35. A sled of mass m is given a kick on a frozen pond. The kick imparts to it an initial speed of 2.00 m/s. The coefficient of kinetic friction between sled and ice is 0.100. Use energy considerations to find the distance the sled moves before it stops.

Section 7.8 Power

36. The electric motor of a model train accelerates the train from rest to 0.620 m/s in 21.0 ms. The total mass of the train is 875 g. Find the average power delivered to the train during the acceleration.

37. A 700-N Marine in basic training climbs a 10.0-m vertical rope at a constant speed in 8.00 s. What is his power output?

38. Make an order-of-magnitude estimate of the power a car engine contributes to speeding the car up to highway speed. For concreteness, consider your own car if you use one. In your solution state the physical quantities you take as data and the values you measure or estimate for them. The mass of the vehicle is given in the owner's manual. If you do not wish to estimate for a car, consider a bus or truck that you specify.

39. A skier of mass 70.0 kg is pulled up a slope by a motor-driven cable. (a) How much work is required to pull him a distance of 60.0 m up a 30.0° slope (assumed frictionless) at a constant speed of 2.00 m/s? (b) A motor of what power is required to perform this task?

40. A 650-kg elevator starts from rest. It moves upward for 3.00 s with constant acceleration until it reaches its cruising speed of 1.75 m/s. (a) What is the average power of the elevator motor during this period? (b) How does this power compare with the motor power when the elevator moves at its cruising speed?

41. An energy-efficient lightbulb, taking in 28.0 W of power, can produce the same level of brightness as a conventional bulb operating at power 100 W. The lifetime of the energy efficient bulb is 10 000 h and its purchase price is $17.0, whereas the conventional bulb has lifetime 750 h and costs $0.420 per bulb. Determine the total savings obtained by using one energy-efficient bulb over its lifetime, as opposed to using conventional bulbs over the same time period. Assume an energy cost of $0.080 0 per kilowatt-hour.

42. Energy is conventionally measured in Calories as well as in joules. One Calorie in nutrition is one kilocalorie, defined as 1 kcal = 4 186 J. Metabolizing one gram of fat can release 9.00 kcal. A student decides to try to lose weight by exercising. She plans to run up and down the stairs in a football stadium as fast as she can and as many times as necessary. Is this in itself a practical way to lose weight? To evaluate the program, suppose she runs up a flight of 80 steps, each 0.150 m high, in 65.0 s. For simplicity, ignore the energy she uses in coming down (which is small). Assume that a typical efficiency for human muscles is 20.0%. This means that when your body converts 100 J from metabolizing fat, 20 J goes into doing mechanical work (here, climbing stairs). The remainder goes into extra internal energy. Assume the student's mass is 50.0 kg. (a) How many times must she run the flight of stairs to lose one pound of fat? (b) What is her average power output, in watts and in horsepower, as she is running up the stairs?

43. For saving energy, bicycling and walking are far more efficient means of transportation than is travel by automobile. For example, when riding at 10.0 mi/h a cyclist uses food energy at a rate of about 400 kcal/h above what he would use if merely sitting still. (In exercise physiology, power is often measured in kcal/h rather than in watts. Here 1 kcal = 1 nutritionist's Calorie = 4 186 J.) Walking at 3.00 mi/h requires about 220 kcal/h. It is interesting to compare these values with the energy consumption required for travel by car. Gasoline yields about 1.30×10^8 J/gal. Find the fuel economy in equivalent miles per gallon for a person (a) walking, and (b) bicycling.

Section 7.9 Energy and the Automobile

44. Suppose the empty car described in Table 7.2 has a fuel economy of 6.40 km/liter (15 mi/gal) when traveling at 26.8 m/s (60 mi/h). Assuming constant efficiency, determine the fuel economy of the car if the total mass of passengers plus driver is 350 kg.

45. A compact car of mass 900 kg has an overall motor efficiency of 15.0%. (That is, 15% of the energy supplied by the fuel is delivered to the wheels of the car.) (a) If burning one gallon of gasoline supplies 1.34×10^8 J of energy, find the amount of gasoline used in accelerating the car from rest to 55.0 mi/h. Here you may ignore the effects of air resistance and rolling friction. (b) How many such accelerations will one gallon provide? (c) The mileage claimed for the car is 38.0 mi/gal at 55 mi/h. What power is delivered to the wheels (to overcome frictional effects) when the car is driven at this speed?

Additional Problems

46. A baseball outfielder throws a 0.150-kg baseball at a speed of 40.0 m/s and an initial angle of 30.0°. What is the kinetic energy of the baseball at the highest point of its trajectory?

47. While running, a person dissipates about 0.600 J of mechanical energy per step per kilogram of body mass. If a 60.0-kg runner dissipates a power of 70.0 W during a race, how fast is the person running? Assume a running step is 1.50 m long.

48. The direction of any vector **A** in three-dimensional space can be specified by giving the angles α, β, and γ that the vector makes with the x, y, and z axes, respectively. If $\mathbf{A} = A_x\hat{\mathbf{i}} + A_y\hat{\mathbf{j}} + A_z\hat{\mathbf{k}}$, (a) find expressions for $\cos\alpha$, $\cos\beta$, and $\cos\gamma$ (these are known as *direction cosines*), and (b) show that these angles satisfy the relation $\cos^2\alpha + \cos^2\beta + \cos^2\gamma = 1$. (*Hint:* Take the scalar product of **A** with $\hat{\mathbf{i}}$, $\hat{\mathbf{j}}$, and $\hat{\mathbf{k}}$ separately.)

49. A 4.00-kg particle moves along the x axis. Its position varies with time according to $x = t + 2.0t^3$, where x is in meters and t is in seconds. Find (a) the kinetic energy at any time t, (b) the acceleration of the particle and the force acting on it at time t, (c) the power being delivered to the particle at time t, and (d) the work done on the particle in the interval $t = 0$ to $t = 2.00$ s.

50. The spring constant of an automotive suspension spring increases with increasing load due to a spring coil that is widest at the bottom, smoothly tapering to a smaller diameter near the top. The result is a softer ride on normal road surfaces from the narrower coils, but the car does not bottom out on bumps because when the upper coils collapse, they leave the stiffer coils near the bottom to absorb the load. For a tapered spiral spring that compresses 12.9 cm with a 1 000-N load and 31.5 cm with a 5 000-N load, (a) evaluate the constants a and b in the empirical equation $F = ax^b$ and (b) find the work needed to compress the spring 25.0 cm.

51. A bead at the bottom of a bowl is one example of an object in a stable equilibrium position. When a physical system is displaced by an amount x from stable equilibrium, a restoring force acts on it, tending to return the system to its equilibrium configuration. The magnitude of the restoring force can be a complicated function of x. For example, when an ion in a crystal is displaced from its lattice site, the restoring force may not be a simple function of x. In such cases we can generally imagine the function $F(x)$ to be expressed as a power series in x, as $F(x) = -(k_1x + k_2x^2 + k_3x^3 + \ldots)$. The first term here is just Hooke's law, which describes the force exerted by a simple spring for small displacements. For small excursions from equilibrium we generally neglect the higher order terms, but in some cases it may be desirable to keep the second term as well. If we model the restoring force as $F = -(k_1x + k_2x^2)$, how much work is done in displacing the system from $x = 0$ to $x = x_{max}$ by an applied force $-F$?

52. A traveler at an airport takes an escalator up one floor, as in Figure P7.52. The moving staircase would itself carry him upward with vertical velocity component v between entry and exit points separated by height h. However, while the escalator is moving, the hurried traveler climbs the steps of the escalator at a rate of n steps/s. Assume that the height of each step is h_s. (a) Determine the amount of chemical energy converted into mechanical energy by the traveler's leg muscles during his escalator ride, given that

Figure P7.52

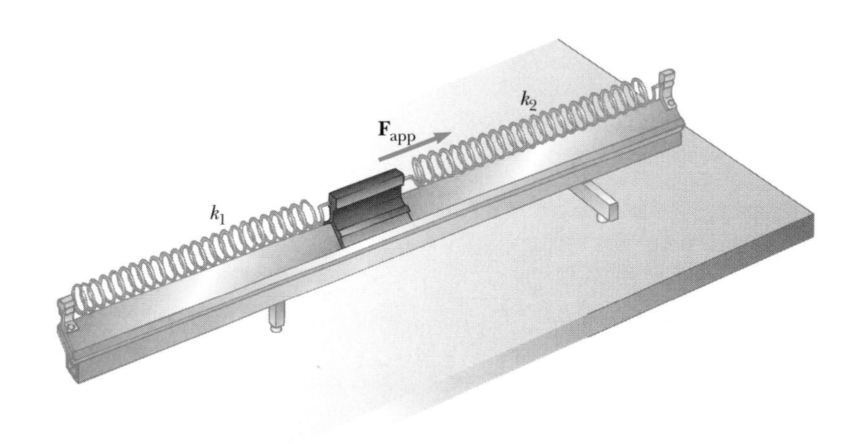

Figure P7.56

his mass is m. (b) Determine the work the escalator motor does on this person.

53. A mechanic pushes a car of mass m, doing work W in making it accelerate from rest. Neglecting friction between car and road, (a) what is the final speed of the car? During this time, the car moves a distance d. (b) What constant horizontal force did the mechanic exert on the car?

54. A 5.00-kg steel ball is dropped onto a copper plate from a height of 10.0 m. If the ball leaves a dent 3.20 mm deep, what is the average force exerted by the plate on the ball during the impact?

55. A single constant force $\mathbf{F}$ acts on a particle of mass m. The particle starts at rest at $t = 0$. (a) Show that the instantaneous power delivered by the force at any time t is $\mathcal{P} = (F^2/m)t$. (b) If $F = 20.0$ N and $m = 5.00$ kg, what is the power delivered at $t = 3.00$ s?

56. Two springs with negligible masses, one with spring constant k_1 and the other with spring constant k_2, are attached to the endstops of a level air track as in Figure P7.56. A glider attached to both springs is located between them. When the glider is in equilibrium, spring 1 is stretched by extension x_{i1} to the right of its unstretched length and spring 2 is stretched by x_{i2} to the left. Now a horizontal force $\mathbf{F}_{app}$ is applied to the glider to move it a distance x_a to the right from its equilibrium position. Show that in this process (a) the work done on spring 1 is $\frac{1}{2}k_1(x_a^2 + 2x_a x_{i1})$, (b) the work done on spring 2 is $\frac{1}{2}k_2(x_a^2 - 2x_a x_{i2})$, (c) x_{i2} is related to x_{i1} by $x_{i2} = k_1 x_{i1}/k_2$, and (d) the total work done by the force F_{app} is $\frac{1}{2}(k_1 + k_2)x_a^2$.

57. As the driver steps on the gas pedal, a car of mass 1 160 kg accelerates from rest. During the first few seconds of motion, the car's acceleration increases with time according to the expression

$$a = (1.16\text{ m/s}^3)t - (0.210\text{ m/s}^4)t^2 + (0.240\text{ m/s}^5)t^3$$

(a) What work is done by the wheels on the car during the interval from $t = 0$ to $t = 2.50$ s? (b) What is the output power of the wheels at the instant $t = 2.50$ s?

58. A particle is attached between two identical springs on a horizontal frictionless table. Both springs have spring constant k and are initially unstressed. (a) If the particle is pulled a distance x along a direction perpendicular to the initial configuration of the springs, as in Figure P7.58, show that the force exerted by the springs on the particle is

$$\mathbf{F} = -2kx\left(1 - \frac{L}{\sqrt{x^2 + L^2}}\right)\hat{\mathbf{i}}$$

(b) Determine the amount of work done by this force in moving the particle from $x = A$ to $x = 0$.

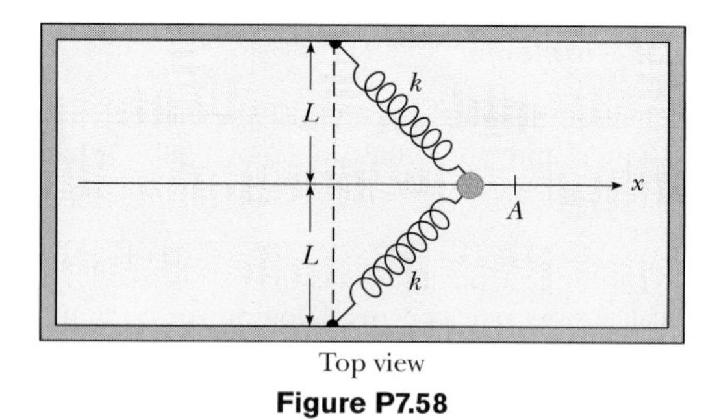

Figure P7.58

59. A rocket body of mass M will fall out of the sky with terminal speed v_T after its fuel is used up. What power output must the rocket engine produce if the rocket is to fly (a) at its terminal speed straight up; (b) at three times the terminal speed straight down? In both cases assume that the mass of the fuel and oxidizer remaining in the rocket is negligible compared to M. Assume that the force of air resistance is proportional to the square of the rocket's speed.

60. **Review problem.** Two constant forces act on a 5.00-kg object moving in the xy plane, as shown in Figure P7.60. Force $\mathbf{F}_1$ is 25.0 N at 35.0°, while $\mathbf{F}_2$ is 42.0 N at 150°. At time $t = 0$, the object is at the origin and has velocity $(4.00\hat{\mathbf{i}} + 2.50\hat{\mathbf{j}})$ m/s. (a) Express the two forces in unit-vector notation. Use unit-vector notation for your other answers. (b) Find the total force on the object. (c) Find the object's acceleration. Now, considering the instant $t = 3.00$ s, (d) find the object's velocity, (e) its location, (f) its kinetic energy from $\frac{1}{2}mv_f^2$, and (g) its kinetic energy from $\frac{1}{2}mv_i^2 + \Sigma\mathbf{F}\cdot\Delta\mathbf{r}$.

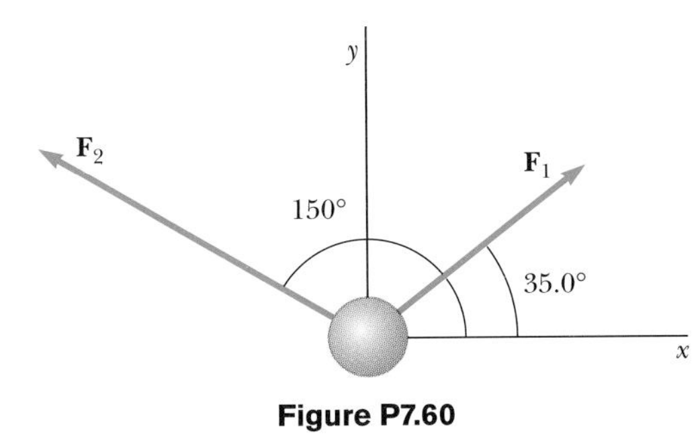

Figure P7.60

61. A 200-g block is pressed against a spring of force constant 1.40 kN/m until the block compresses the spring 10.0 cm. The spring rests at the bottom of a ramp inclined at 60.0° to the horizontal. Using energy considerations, determine how far up the incline the block moves before it stops (a) if there is no friction between the block and the ramp and (b) if the coefficient of kinetic friction is 0.400.

62. When different weights are hung on a spring, the spring stretches to different lengths as shown in the following table. (a) Make a graph of the applied force versus the extension of the spring. By least-squares fitting, determine the straight line that best fits the data. (You may not want to use all the data points.) (b) From the slope of the best-fit line, find the spring constant k. (c) If the spring is extended to 105 mm, what force does it exert on the suspended weight?

F (N)	2.0	4.0	6.0	8.0	10	12	14	16	18	20	22
L (mm)	15	32	49	64	79	98	112	126	149	175	190

63. The ball launcher in a pinball machine has a spring that has a force constant of 1.20 N/cm (Fig. P7.63). The surface on which the ball moves is inclined 10.0° with respect to the horizontal. If the spring is initially compressed 5.00 cm, find the launching speed of a 100-g ball when the plunger is released. Friction and the mass of the plunger are negligible.

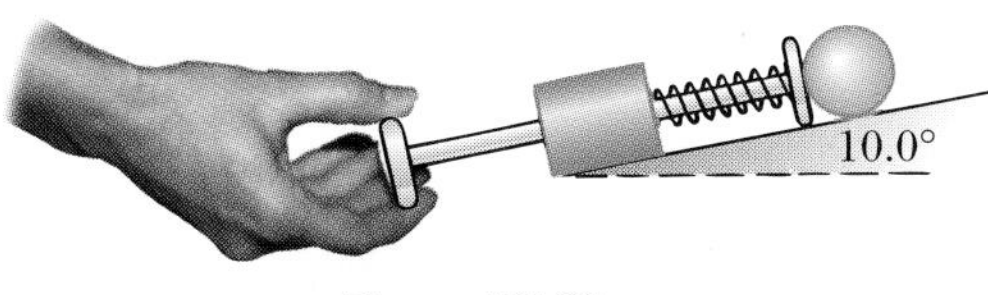

Figure P7.63

64. A 0.400-kg particle slides around a horizontal track. The track has a smooth vertical outer wall forming a circle with a radius of 1.50 m. The particle is given an initial speed of 8.00 m/s. After one revolution, its speed has dropped to 6.00 m/s because of friction with the rough floor of the track. (a) Find the energy converted from mechanical to internal in the system due to friction in one revolution. (b) Calculate the coefficient of kinetic friction. (c) What is the total number of revolutions the particle makes before stopping?

65. In diatomic molecules, the constituent atoms exert attractive forces on each other at large distances and repulsive forces at short distances. For many molecules, the Lennard-Jones law is a good approximation to the magnitude of these forces:

$$F = F_0\left[2\left(\frac{\sigma}{r}\right)^{13} - \left(\frac{\sigma}{r}\right)^{7}\right]$$

where r is the center-to-center distance between the atoms in the molecule, σ is a length parameter, and F_0 is the force when $r = \sigma$. For an oxygen molecule, we find that $F_0 = 9.60 \times 10^{-11}$ N and $\sigma = 3.50 \times 10^{-10}$ m. Determine the work done by this force if the atoms are pulled apart from $r = 4.00 \times 10^{-10}$ m to $r = 9.00 \times 10^{-10}$ m.

66. As it plows a parking lot, a snowplow pushes an ever-growing pile of snow in front of it. Suppose a car moving through the air is similarly modeled as a cylinder pushing a growing plug of air in front of it. The originally stationary air is set into motion at the constant speed v of the cylinder, as in Figure P7.66. In a time interval Δt, a new disk of air of mass Δm must be moved a distance $v\Delta t$ and hence must be given a kinetic energy $\frac{1}{2}(\Delta m)v^2$. Using this model, show that the automobile's power loss due to air resistance is $\frac{1}{2}\rho Av^3$ and that the resistive force acting on the car is $\frac{1}{2}\rho Av^2$, where ρ is the density of air. Compare this result with the empirical expression $\frac{1}{2}D\rho Av^2$ for the resistive force.

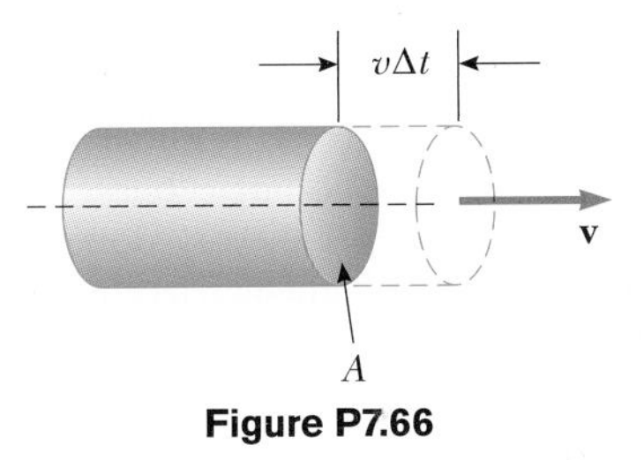

Figure P7.66

67. A particle moves along the x axis from $x = 12.8$ m to $x = 23.7$ m under the influence of a force

$$F = \frac{375}{x^3 + 3.75x}$$

where F is in newtons and x is in meters. Using numerical integration, determine the total work done by this force on the particle during this displacement. Your result should be accurate to within 2%.

68. A windmill, such as that in the opening photograph of this chapter, turns in response to a force of high-speed air resistance, $R = \frac{1}{2}D\rho Av^2$. The power available is $\mathcal{P} = Rv = \frac{1}{2}D\rho\pi r^2v^3$, where v is the wind speed and we have assumed a circular face for the windmill, of radius r. Take the drag coefficient as $D = 1.00$ and the density of air from the front endpaper. For a home windmill with $r = 1.50$ m, calculate the power available if (a) $v = 8.00$ m/s and (b) $v = 24.0$ m/s. The power delivered to the generator is limited by the efficiency of the system, which is about 25%. For comparison, a typical home needs about 3 kW of electric power.

69. More than 2 300 years ago the Greek teacher Aristotle wrote the first book called *Physics*. Put into more precise terminology, this passage is from the end of its Section Eta:

Let $\mathcal{P}$ be the power of an agent causing motion; w, the thing moved; d, the distance covered; and Δt, the time interval required. Then (1) a power equal to $\mathcal{P}$ will in a period of time equal to Δt move $w/2$ a distance $2d$; or (2) it will move $w/2$ the given distance d in the time interval $\Delta t/2$. Also, if (3) the given power $\mathcal{P}$ moves the given object w a distance $d/2$ in time interval $\Delta t/2$, then (4) $\mathcal{P}/2$ will move $w/2$ the given distance d in the given time interval Δt.

(a) Show that Aristotle's proportions are included in the equation $\mathcal{P}\Delta t = bwd$ where b is a proportionality constant. (b) Show that our theory of motion includes this part of Aristotle's theory as one special case. In particular, describe a situation in which it is true, derive the equation representing Aristotle's proportions, and determine the proportionality constant.

70. Consider the block-spring-surface system in part (b) of Example 7.11. (a) At what position x of the block is its speed a maximum? (b) In the **What If?** section of this example, we explored the effects of an increased friction force of 10.0 N. At what position of the block does its maximum speed occur in this situation?

Answers to Quick Quizzes

7.1 (a). The force does no work on the Earth because the force is pointed toward the center of the circle and is therefore perpendicular to the direction of the displacement.

7.2 c, a, d, b. The work in (c) is positive and of the largest possible value because the angle between the force and the displacement is zero. The work done in (a) is zero because the force is perpendicular to the displacement. In (d) and (b), negative work is done by the applied force because in neither case is there a component of the force in the direction of the displacement. Situation (b) is the most negative value because the angle between the force and the displacement is 180°.

7.3 (d). Answer (a) is incorrect because the scalar product $(-\mathbf{A}) \cdot (-\mathbf{B})$ is equal to $\mathbf{A} \cdot \mathbf{B}$. Answer (b) is incorrect because $AB \cos(\theta + 180°)$ gives the negative of the correct value.

7.4 (d). Because of the range of values of the cosine function, $\mathbf{A} \cdot \mathbf{B}$ has values that range from AB to $-AB$.

7.5 (a). Because the work done in compressing a spring is proportional to the square of the compression distance x, doubling the value of x causes the work to increase fourfold.

7.6 (b). Because the work is proportional to the square of the compression distance x and the kinetic energy is proportional to the square of the speed v, doubling the compression distance doubles the speed.

7.7 (a) For the television set, energy enters by electrical transmission (through the power cord) and electromagnetic radiation (the television signal). Energy leaves by heat (from hot surfaces into the air), mechanical waves (sound from the speaker), and electromagnetic radiation (from the screen). (b) For the gasoline-powered lawn mower, energy enters by matter transfer (gasoline). Energy leaves by work (on the blades of grass), mechanical waves (sound), and heat (from hot surfaces into the air). (c) For the hand-cranked pencil sharpener, energy enters by work (from your hand turning the crank). Energy leaves by work (done on the pencil) and mechanical waves (sound).

7.8 (b). The friction force represents an interaction with the environment of the block.

7.9 (b). The friction force represents an interaction with the environment of the surface.

7.10 (a). The friction force is internal to the system, so there are no interactions with the environment.

7.11 (c). The brakes and the roadway are warmer, so their internal energy has increased. In addition, the sound of the skid represents transfer of energy away by mechanical waves.

7.12 (e). Because the speed is doubled, the kinetic energy is four times as large. This kinetic energy was attained for the newer car in the same time interval as the smaller kinetic energy for the older car, so the power is four times as large.

Chapter 8

Potential Energy

▲ *A strobe photograph of a pole vaulter. During this process, several types of energy transformations occur. The two types of potential energy that we study in this chapter are evident in the photograph.* Gravitational potential energy *is associated with the change in vertical position of the vaulter relative to the Earth.* Elastic potential energy *is evident in the bending of the pole. (©Harold E. Edgerton/Courtesy of Palm Press, Inc.)*

CHAPTER OUTLINE

In Chapter 7 we introduced the concepts of kinetic energy associated with the motion of members of a system and internal energy associated with the temperature of a system. In this chapter we introduce *potential energy*, the energy associated with the configuration of a system of objects that exert forces on each other.

The potential energy concept can be used only when dealing with a special class of forces called *conservative forces*. When only conservative forces act within an isolated system, the kinetic energy gained (or lost) by the system as its members change their relative positions is balanced by an equal loss (or gain) in potential energy. This balancing of the two forms of energy is known as the *principle of conservation of mechanical energy*.

Potential energy is present in the Universe in various forms, including gravitational, electromagnetic, chemical, and nuclear. Furthermore, one form of energy in a system can be converted to another. For example, when a system consists of an electric motor connected to a battery, the chemical energy in the battery is converted to kinetic energy as the shaft of the motor turns. The transformation of energy from one form to another is an essential part of the study of physics, engineering, chemistry, biology, geology, and astronomy.

8.1 Potential Energy of a System

In Chapter 7, we defined a system in general, but focused our attention primarily on single particles or objects under the influence of an external force. In this chapter, we consider systems of two or more particles or objects interacting via a force that is *internal* to the system. The kinetic energy of such a system is the algebraic sum of the kinetic energies of all members of the system. There may be systems, however, in which one object is so massive that it can be modeled as stationary and its kinetic energy can be neglected. For example, if we consider a ball–Earth system as the ball falls to the ground, the kinetic energy of the system can be considered as just the kinetic energy of the ball. The Earth moves so slowly in this process that we can ignore its kinetic energy. On the other hand, the kinetic energy of a system of two electrons must include the kinetic energies of both particles.

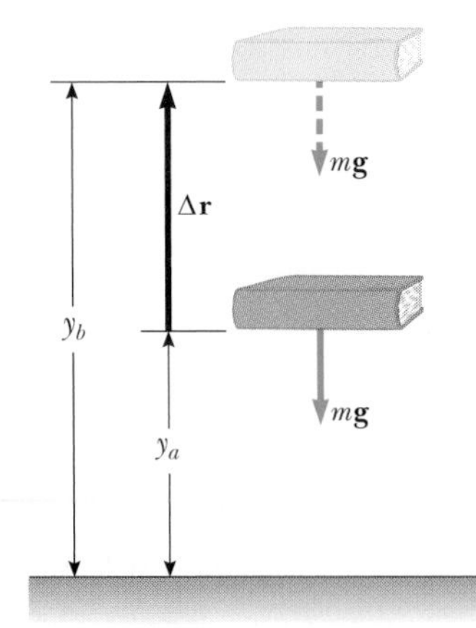

Figure 8.1 The work done by an external agent on the system of the book and the Earth as the book is lifted from a height y_a to a height y_b is equal to $mgy_b - mgy_a$.

Let us imagine a system consisting of a book and the Earth, interacting via the gravitational force. We do some work on the system by lifting the book slowly through a height $\Delta y = y_b - y_a$, as in Figure 8.1. According to our discussion of energy and energy transfer in Chapter 7, this work done on the system must appear as an increase in energy of the system. The book is at rest before we perform the work and is at rest after we perform the work. Thus, there is no change in the kinetic energy of the system. There is no reason why the temperature of the book or the Earth should change, so there is no increase in the internal energy of the system.

Because the energy change of the system is not in the form of kinetic energy or internal energy, it must appear as some other form of energy storage. After lifting the book, we could release it and let it fall back to the position y_a. Notice that the book (and, therefore, the system) will now have kinetic energy, and its source is in the work that was done in lifting the book. While the book was at the highest point, the energy of the system had the *potential* to become kinetic energy, but did not do so until the book was allowed to fall. Thus, we call the energy storage mechanism before we release the book **potential energy.** We will find that a potential energy can only be associated with specific types of forces. In this particular case, we are discussing **gravitational potential energy.**

Let us now derive an expression for the gravitational potential energy associated with an object at a given location above the surface of the Earth. Consider an external agent lifting an object of mass m from an initial height y_a above the ground to a final height y_b, as in Figure 8.1. We assume that the lifting is done slowly, with no acceleration, so that the lifting force can be modeled as being equal in magnitude to the weight of the object—the object is in equilibrium and moving at constant velocity. The work done by the external agent on the system (object and Earth) as the object undergoes this upward displacement is given by the product of the upward applied force $\mathbf{F}_{\text{app}}$ and the upward displacement $\Delta \mathbf{r} = \Delta y \hat{\mathbf{j}}$:

$$W = (\mathbf{F}_{\text{app}}) \cdot \Delta \mathbf{r} = (mg\hat{\mathbf{j}}) \cdot [(y_b - y_a)\hat{\mathbf{j}}] = mgy_b - mgy_a \quad (8.1)$$

Notice how similar this equation is to Equation 7.14 in the preceding chapter. In each equation, the work done on a system equals a difference between the final and initial values of a quantity. In Equation 7.14, the work represents a transfer of energy into the system, and the increase in energy of the system is kinetic in form. In Equation 8.1, the work represents a transfer of energy into the system, and the system energy appears in a different form, which we have called gravitational potential energy.

Thus, we can identify the quantity mgy as the gravitational potential energy U_g:

Gravitational potential energy

$$U_g \equiv mgy \quad (8.2)$$

The units of gravitational potential energy are joules, the same as those of work and kinetic energy. Potential energy, like work and kinetic energy, is a scalar quantity. Note that Equation 8.2 is valid only for objects near the surface of the Earth, where g is approximately constant.[1]

Using our definition of gravitational potential energy, Equation 8.1 can now be rewritten as

$$W = \Delta U_g \quad (8.3)$$

which mathematically describes the fact that the work done on the system in this situation appears as a change in the gravitational potential energy of the system.

The gravitational potential energy depends only on the vertical height of the object above the surface of the Earth. The same amount of work must be done on an object–Earth system whether the object is lifted vertically from the Earth or is pushed starting from the same point up a frictionless incline, ending up at the same height. This can be shown by calculating the work with a displacement having both vertical and horizontal components:

$$W = (\mathbf{F}_{\text{app}}) \cdot \Delta \mathbf{r} = (mg\hat{\mathbf{j}}) \cdot [(x_b - x_a)\hat{\mathbf{i}} + (y_b - y_a)\hat{\mathbf{j}}] = mgy_b - mgy_a$$

where there is no term involving x in the final result because $\hat{\mathbf{j}} \cdot \hat{\mathbf{i}} = 0$.

In solving problems, you must choose a reference configuration for which the gravitational potential energy is set equal to some reference value, which is normally zero. The choice of reference configuration is completely arbitrary because the important quantity is the *difference* in potential energy and this difference is independent of the choice of reference configuration.

It is often convenient to choose as the reference configuration for zero potential energy the configuration in which an object is at the surface of the Earth, but this is not essential. Often, the statement of the problem suggests a convenient configuration to use.

▲ PITFALL PREVENTION

8.1 Potential Energy Belongs to a System

Potential energy is always associated with a *system* of two or more interacting objects. When a small object moves near the surface of the Earth under the influence of gravity, we may sometimes refer to the potential energy "associated with the object" rather than the more proper "associated with the system" because the Earth does not move significantly. We will not, however, refer to the potential energy "of the object" because this clearly ignores the role of the Earth.

[1] The assumption that g is constant is valid as long as the vertical displacement is small compared with the Earth's radius.

Quick Quiz 8.1 Choose the correct answer. The gravitational potential energy of a system (a) is always positive (b) is always negative (c) can be negative or positive.

Quick Quiz 8.2 An object falls off a table to the floor. We wish to analyze the situation in terms of kinetic and potential energy. In discussing the kinetic energy of the system, we (a) must include the kinetic energy of both the object and the Earth (b) can ignore the kinetic energy of the Earth because it is not part of the system (c) can ignore the kinetic energy of the Earth because the Earth is so massive compared to the object.

Quick Quiz 8.3 An object falls off a table to the floor. We wish to analyze the situation in terms of kinetic and potential energy. In discussing the potential energy of the system, we identify the system as (a) both the object and the Earth (b) only the object (c) only the Earth.

Example 8.1 The Bowler and the Sore Toe

A bowling ball held by a careless bowler slips from the bowler's hands and drops on the bowler's toe. Choosing floor level as the $y = 0$ point of your coordinate system, estimate the change in gravitational potential energy of the ball–Earth system as the ball falls. Repeat the calculation, using the top of the bowler's head as the origin of coordinates.

Solution First, we need to estimate a few values. A bowling ball has a mass of approximately 7 kg, and the top of a person's toe is about 0.03 m above the floor. Also, we shall assume the ball falls from a height of 0.5 m. Keeping nonsignificant digits until we finish the problem, we calculate the gravitational potential energy of the ball–Earth system just before the ball is released to be $U_i = mgy_i = (7\text{ kg})(9.80\text{ m/s}^2)(0.5\text{ m}) = 34.3\text{ J}$. A similar calculation for when the ball reaches his toe gives $U_f = mgy_f = (7\text{ kg})(9.80\text{ m/s}^2)(0.03\text{ m}) = 2.06\text{ J}$. So, the change in gravitational potential energy of the ball–Earth system is $\Delta U_g = U_f - U_i = -32.24\text{ J}$. We should probably keep only one digit because of the roughness of our estimates; thus, we estimate that the change in gravitational potential energy is -30 J. The system had 30 J of gravitational potential energy relative to the top of the toe before the ball began its fall.

When we use the bowler's head (which we estimate to be 1.50 m above the floor) as our origin of coordinates, we find that $U_i = mgy_i = (7\text{ kg})(9.80\text{ m/s}^2)(-1\text{ m}) = -68.6\text{ J}$ and $U_f = mgy_f = (7\text{ kg})(9.80\text{ m/s}^2)(-1.47\text{ m}) = -100.8\text{ J}$. The change in gravitational potential energy of the ball–Earth system is $\Delta U_g = U_f - U_i = -32.24\text{ J} \approx -30\text{ J}$. This is the same value as before, as it must be.

8.2 The Isolated System–Conservation of Mechanical Energy

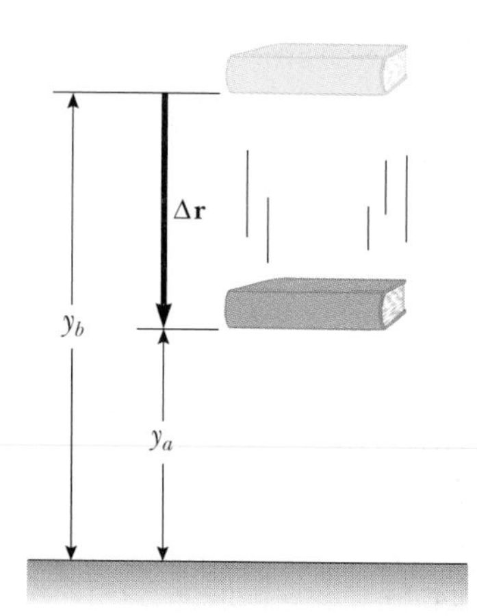

Figure 8.2 The work done by the gravitational force on the book as the book falls from y_b to a height y_a is equal to $mgy_b - mgy_a$.

The introduction of potential energy allows us to generate a powerful and universally applicable principle for solving problems that are difficult to solve with Newton's laws. Let us develop this new principle by thinking about the book–Earth system in Figure 8.1 again. After we have lifted the book, there is gravitational potential energy stored in the system, which we can calculate from the work done by the external agent on the system, using $W = \Delta U_g$.

Let us now shift our focus to the work done on the book alone by the gravitational force (Fig. 8.2) as the book falls back to its original height. As the book falls from y_b to y_a, the work done by the gravitational force on the book is

$$W_{\text{on book}} = (m\mathbf{g}) \cdot \Delta\mathbf{r} = (-mg\hat{\mathbf{j}}) \cdot [(y_a - y_b)\hat{\mathbf{j}}] = mgy_b - mgy_a \qquad (8.4)$$

From the work–kinetic energy theorem of Chapter 7, the work done on the book is equal to the change in the kinetic energy of the book:

$$W_{\text{on book}} = \Delta K_{\text{book}}$$

Therefore, equating these two expressions for the work done on the book,

$$\Delta K_{\text{book}} = mgy_b - mgy_a \tag{8.5}$$

Now, let us relate each side of this equation to the *system* of the book and the Earth. For the right-hand side,

$$mgy_b - mgy_a = -(mgy_a - mgy_b) = -(U_f - U_i) = -\Delta U_g$$

where U_g is the gravitational potential energy of the system. For the left-hand side of Equation 8.5, because the book is the only part of the system that is moving, we see that $\Delta K_{\text{book}} = \Delta K$, where K is the kinetic energy of the system. Thus, with each side of Equation 8.5 replaced with its system equivalent, the equation becomes

$$\Delta K = -\Delta U_g \tag{8.6}$$

▲ PITFALL PREVENTION

8.2 Conditions on Equation 8.6

Equation 8.6 is true for only one of two categories of forces. These forces are called *conservative forces,* as discussed in the next section.

This equation can be manipulated to provide a very important general result for solving problems. First, we bring the change in potential energy to the left side of the equation:

$$\Delta K + \Delta U_g = 0 \tag{8.7}$$

On the left, we have a sum of changes of the energy stored in the system. The right hand is zero because there are no transfers of energy across the boundary of the system—the book–Earth system is *isolated* from the environment.

We define the sum of kinetic and potential energies as **mechanical energy:**

$$E_{\text{mech}} = K + U_g$$

We will encounter other types of potential energy besides gravitational later in the text, so we can write the general form of the definition for mechanical energy without a subscript on U:

Mechanical energy of a system

$$E_{\text{mech}} \equiv K + U \tag{8.8}$$

where U represents the total of *all* types of potential energy.

Let us now write the changes in energy in Equation 8.7 explicitly:

$$(K_f - K_i) + (U_f - U_i) = 0$$

The mechanical energy of an isolated, friction-free system is conserved.

$$K_f + U_f = K_i + U_i \tag{8.9}$$

For the gravitational situation that we have described, Equation 8.9 can be written as

$$\tfrac{1}{2}mv_f^{\,2} + mgy_f = \tfrac{1}{2}mv_i^{\,2} + mgy_i$$

As the book falls to the Earth, the book–Earth system loses potential energy and gains kinetic energy, such that the total of the two types of energy always remains constant.

Equation 8.9 is a statement of **conservation of mechanical energy** for an **isolated system.** An isolated system is one for which there are no energy transfers across the boundary. The energy in such a system is conserved—the sum of the kinetic and potential energies remains constant. (This statement assumes that no *nonconservative forces* act within the system; see Pitfall Prevention 8.2.)

8.3 Mechanical Energy in an Isolated System

Equation 8.9 is not the only statement we can make for an isolated system. This describes conservation of *mechanical energy only* for the isolated system. We will see shortly how to include internal energy. In later chapters, we will generate new conservation statements (and associated equations) related to other conserved quantities.

Quick Quiz 8.4 In an isolated system, which of the following is a correct statement of the quantity that is conserved? (a) kinetic energy (b) potential energy (c) kinetic energy plus potential energy (d) both kinetic energy and potential energy.

Quick Quiz 8.5 A rock of mass m is dropped to the ground from a height h. A second rock, with mass $2m$, is dropped from the same height. When the second rock strikes the ground, its kinetic energy is (a) twice that of the first rock (b) four times that of the first rock (c) the same as that of the first rock (d) half as much as that of the first rock (e) impossible to determine.

Quick Quiz 8.6 Three identical balls are thrown from the top of a building, all with the same initial speed. The first is thrown horizontally, the second at some angle above the horizontal, and the third at some angle below the horizontal, as shown in Figure 8.3. Neglecting air resistance, rank the speeds of the balls at the instant each hits the ground.

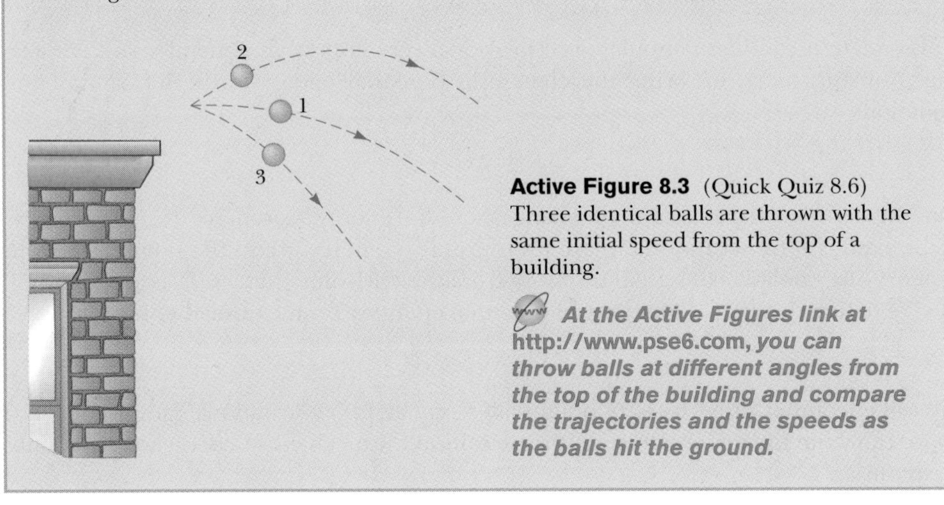

Active Figure 8.3 (Quick Quiz 8.6) Three identical balls are thrown with the same initial speed from the top of a building.

At the Active Figures link at http://www.pse6.com, you can throw balls at different angles from the top of the building and compare the trajectories and the speeds as the balls hit the ground.

Elastic Potential Energy

We are familiar now with gravitational potential energy; let us explore a second type of potential energy. Consider a system consisting of a block plus a spring, as shown in Figure 8.4. The force that the spring exerts on the block is given by $F_s = -kx$. In the previous chapter, we learned that the work done by an external applied force F_{app} on a system consisting of a block connected to the spring is given by Equation 7.12:

$$W_{F_{\text{app}}} = \tfrac{1}{2}kx_f^2 - \tfrac{1}{2}kx_i^2 \tag{8.10}$$

In this situation, the initial and final x coordinates of the block are measured from its equilibrium position, $x = 0$. Again (as in the gravitational case), we see that the work done on the system is equal to the difference between the initial and final values of an expression related to the configuration of the system. The **elastic potential energy** function associated with the block–spring system is defined by

Elastic potential energy stored in a spring

$$U_s \equiv \tfrac{1}{2}kx^2 \tag{8.11}$$

The elastic potential energy of the system can be thought of as the energy stored in the deformed spring (one that is either compressed or stretched from its equilibrium position). To visualize this, consider Figure 8.4, which shows a spring on a frictionless, horizontal surface. When a block is pushed against the spring (Fig. 8.4b) and the spring is compressed a distance x, the elastic potential energy stored in the spring is $\tfrac{1}{2}kx^2$.

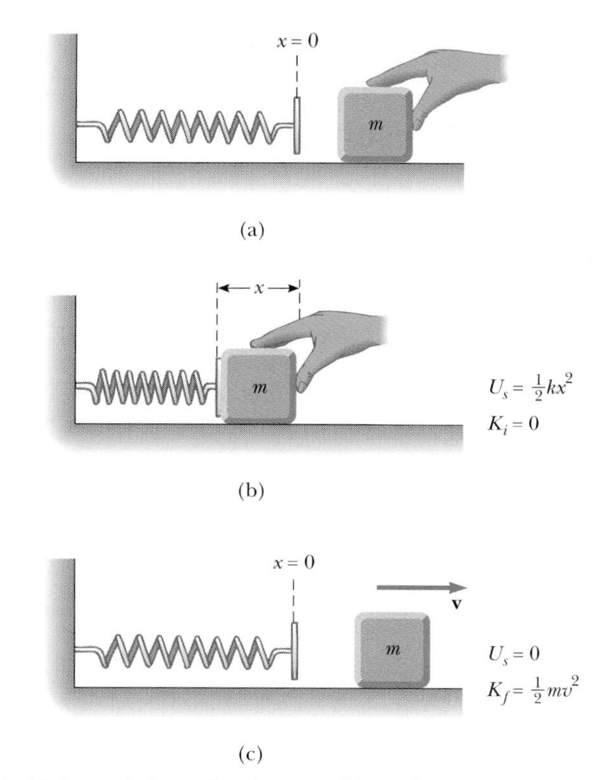

Active Figure 8.4 (a) An undeformed spring on a frictionless horizontal surface. (b) A block of mass m is pushed against the spring, compressing it a distance x. (c) When the block is released from rest, the elastic potential energy stored in the spring is transferred to the block in the form of kinetic energy.

At the Active Figures link at http://www.pse6.com, you can compress the spring by varying amounts and observe the effect on the block's speed.

When the block is released from rest, the spring exerts a force on the block and returns to its original length. The stored elastic potential energy is transformed into kinetic energy of the block (Fig. 8.4c).

The elastic potential energy stored in a spring is zero whenever the spring is undeformed ($x = 0$). Energy is stored in the spring only when the spring is either stretched or compressed. Furthermore, the elastic potential energy is a maximum when the spring has reached its maximum compression or extension (that is, when $|x|$ is a maximum). Finally, because the elastic potential energy is proportional to x^2, we see that U_s is always positive in a deformed spring.

Quick Quiz 8.7 A ball is connected to a light spring suspended vertically, as shown in Figure 8.5. When displaced downward from its equilibrium position and released, the ball oscillates up and down. In the system of *the ball, the spring, and the Earth,* what forms of energy are there during the motion? (a) kinetic and elastic potential (b) kinetic and gravitational potential (c) kinetic, elastic potential, and gravitational potential (d) elastic potential and gravitational potential.

Quick Quiz 8.8 Consider the situation in Quick Quiz 8.7 once again. In the system of *the ball and the spring,* what forms of energy are there during the motion? (a) kinetic and elastic potential (b) kinetic and gravitational potential (c) kinetic, elastic potential, and gravitational potential (d) elastic potential and gravitational potential.

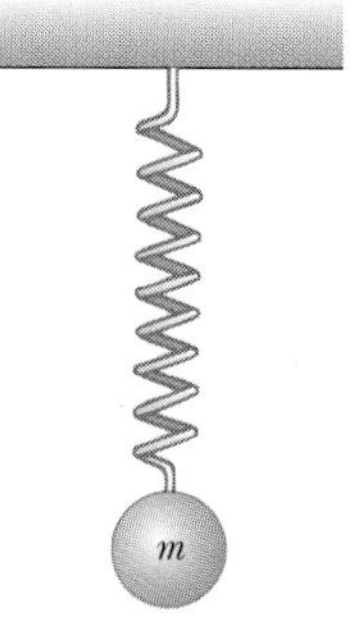

Figure 8.5 (Quick Quizzes 8.7 and 8.8) A ball connected to a massless spring suspended vertically. What forms of potential energy are associated with the system when the ball is displaced downward?

PROBLEM-SOLVING HINTS

Isolated Systems–Conservation of Mechanical Energy

We can solve many problems in physics using the principle of conservation of mechanical energy. You should incorporate the following procedure when you apply this principle:

- Define your isolated system, which may include two or more interacting particles, as well as springs or other structures in which elastic potential energy can be stored. Be sure to include all components of the system that exert forces on each other. Identify the initial and final configurations of the system.
- Identify configurations for zero potential energy (both gravitational and spring). If there is more than one force acting within the system, write an expression for the potential energy associated with each force.
- If friction or air resistance is present, mechanical energy of the system is not conserved and the techniques of Section 8.4 must be employed.
- If mechanical energy of the system is conserved, you can write the total energy $E_i = K_i + U_i$ for the initial configuration. Then, write an expression for the total energy $E_f = K_f + U_f$ for the final configuration that is of interest. Because mechanical energy is conserved, you can equate the two total energies and solve for the quantity that is unknown.

Example 8.2 Ball in Free Fall

Interactive

A ball of mass m is dropped from a height h above the ground, as shown in Figure 8.6.

(A) Neglecting air resistance, determine the speed of the ball when it is at a height y above the ground.

Solution Figure 8.6 and our everyday experience with falling objects allow us to conceptualize the situation. While we can readily solve this problem with the techniques of Chapter 2, let us take an energy approach and categorize this as an energy problem for practice. To analyze the problem, we identify the system as the ball and the Earth. Because there is no air resistance and the system is isolated, we apply the principle of conservation of mechanical energy to the ball–Earth system.

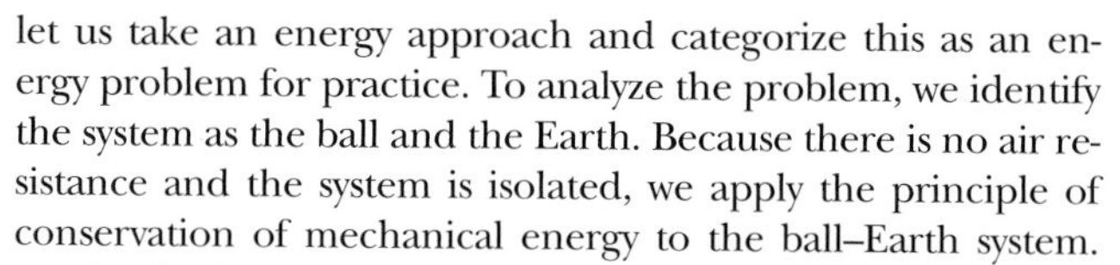

At the instant the ball is released, its kinetic energy is $K_i = 0$ and the potential energy of the system is $U_i = mgh$. When the ball is at a distance y above the ground, its kinetic energy is $K_f = \frac{1}{2}mv_f^2$ and the potential energy relative to the ground is $U_f = mgy$. Applying Equation 8.9, we obtain

$$K_f + U_f = K_i + U_i$$

$$\frac{1}{2}mv_f^2 + mgy = 0 + mgh$$

$$v_f^2 = 2g(h - y)$$

$$v_f = \sqrt{2g(h - y)}$$

The speed is always positive. If we had been asked to find the ball's velocity, we would use the negative value of the square root as the y component to indicate the downward motion.

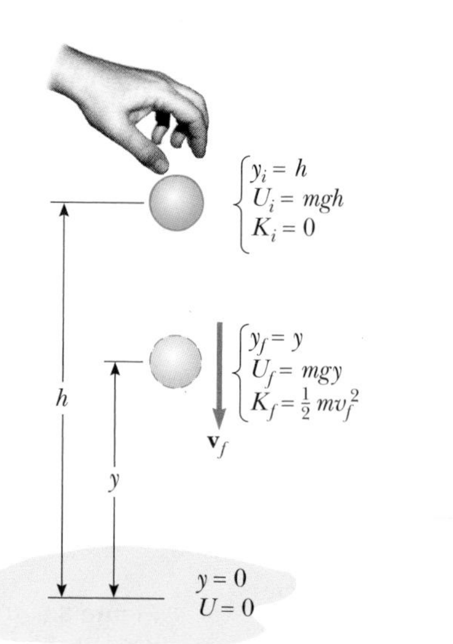

Figure 8.6 (Example 8.2) A ball is dropped from a height h above the ground. Initially, the total energy of the ball–Earth system is potential energy, equal to mgh relative to the ground. At the elevation y, the total energy is the sum of the kinetic and potential energies.

(B) Determine the speed of the ball at y if at the instant of release it already has an initial upward speed v_i at the initial altitude h.

Solution In this case, the initial energy includes kinetic energy equal to $\frac{1}{2}mv_i^2$ and Equation 8.9 gives

$$\frac{1}{2}mv_f^2 + mgy = \frac{1}{2}mv_i^2 + mgh$$

$$v_f^{\,2} = v_i^{\,2} + 2g(h - y)$$

$$v_f = \sqrt{v_i^{\,2} + 2g(h - y)}$$

Note that this result is consistent with the expression $v_{yf}^{\,2} = v_{yi}^{\,2} - 2g(y_f - y_i)$ from kinematics, where $y_i = h$. Furthermore, this result is valid even if the initial velocity is at an angle to the horizontal (Quick Quiz 8.6) for two reasons: (1) energy is a scalar, and the kinetic energy depends only on the magnitude of the velocity; and (2) the change in the gravitational potential energy depends only on the change in position in the vertical direction.

What If? **What if the initial velocity $\mathbf{v}_i$ in part (B) were downward? How would this affect the speed of the ball at position y?**

Answer We might be tempted to claim that throwing it downward would result in it having a higher speed at y than if we threw it upward. Conservation of mechanical energy, however, depends on kinetic and potential energies, which are scalars. Thus, the direction of the initial velocity vector has no bearing on the final speed.

Compare the effect of upward, downward, and zero initial velocities at the Interactive Worked Example link at **http://www.pse6.com.**

Example 8.3 The Pendulum

A pendulum consists of a sphere of mass m attached to a light cord of length L, as shown in Figure 8.7. The sphere is released from rest at point Ⓐ when the cord makes an angle θ_A with the vertical, and the pivot at P is frictionless.

(A) Find the speed of the sphere when it is at the lowest point Ⓑ.

Solution The only force that does work on the sphere is the gravitational force. (The force applied by the cord is always perpendicular to each element of the displacement and hence does no work.) Because the pendulum–Earth system is isolated, the energy of the system is conserved. As the pendulum swings, continuous transformation between potential and kinetic energy occurs. At the instant the pendulum is released, the energy of the system is entirely potential energy. At point Ⓑ the pendulum has kinetic energy, but the system has lost some potential energy. At Ⓒ the system has regained its initial potential energy, and the kinetic energy of the pendulum is again zero.

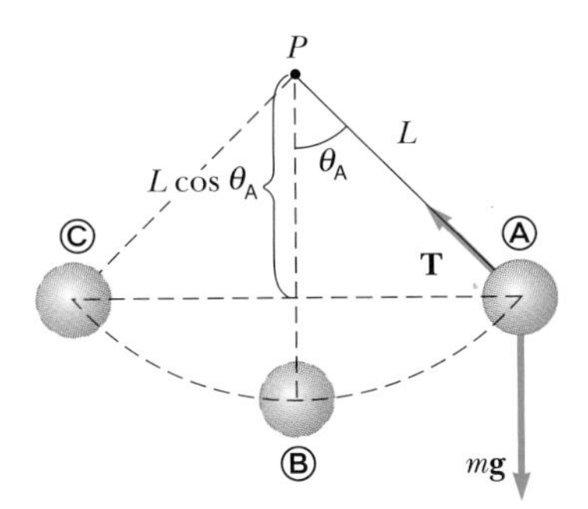

Figure 8.7 (Example 8.3) If the sphere is released from rest at the angle θ_A, it will never swing above this position during its motion. At the start of the motion, when the sphere is at position Ⓐ, the energy of the sphere–Earth system is entirely potential. This initial potential energy is transformed into kinetic energy when the sphere is at the lowest elevation Ⓑ. As the sphere continues to move along the arc, the energy again becomes entirely potential energy when the sphere is at Ⓒ.

If we measure the y coordinates of the sphere from the center of rotation, then $y_A = -L\cos\theta_A$ and $y_B = -L$. Therefore, $U_A = -mgL\cos\theta_A$ and $U_B = -mgL$.

Applying the principle of conservation of mechanical energy to the system gives

$$K_B + U_B = K_A + U_A$$

$$\tfrac{1}{2}mv_B^{\,2} - mgL = 0 - mgL\cos\theta_A$$

$$(1) \qquad v_B = \sqrt{2gL(1 - \cos\theta_A)}$$

(B) What is the tension T_B in the cord at Ⓑ?

Solution Because the tension force does no work, it does not enter into an energy equation, and we cannot determine the tension using the energy method. To find T_B, we can apply Newton's second law to the radial direction. First, recall that the centripetal acceleration of a particle moving in a circle is equal to v^2/r directed toward the center of rotation. Because $r = L$ in this example, Newton's second law gives

$$(2) \qquad \sum F_r = mg - T_B = ma_r = -m\frac{v_B^{\,2}}{L}$$

Substituting Equation (1) into Equation (2) gives the tension at point Ⓑ as a function of θ_A:

$$(3) \qquad T_B = mg + 2mg(1 - \cos\theta_A) = mg(3 - 2\cos\theta_A)$$

From Equation (2) we see that the tension at Ⓑ is greater than the weight of the sphere. Furthermore, Equation (3) gives the expected result that $T_B = mg$ when the initial angle $\theta_A = 0$. Note also that part (A) of this example is categorized as an energy problem while part (B) is categorized as a Newton's second law problem.

Example 8.4 A Grand Entrance

Interactive

You are designing an apparatus to support an actor of mass 65 kg who is to "fly" down to the stage during the performance of a play. You attach the actor's harness to a 130-kg sandbag by means of a lightweight steel cable running smoothly over two frictionless pulleys, as in Figure 8.8a. You need 3.0 m of cable between the harness and the nearest pulley so that the pulley can be hidden behind a curtain. For the apparatus to work successfully, the sandbag must

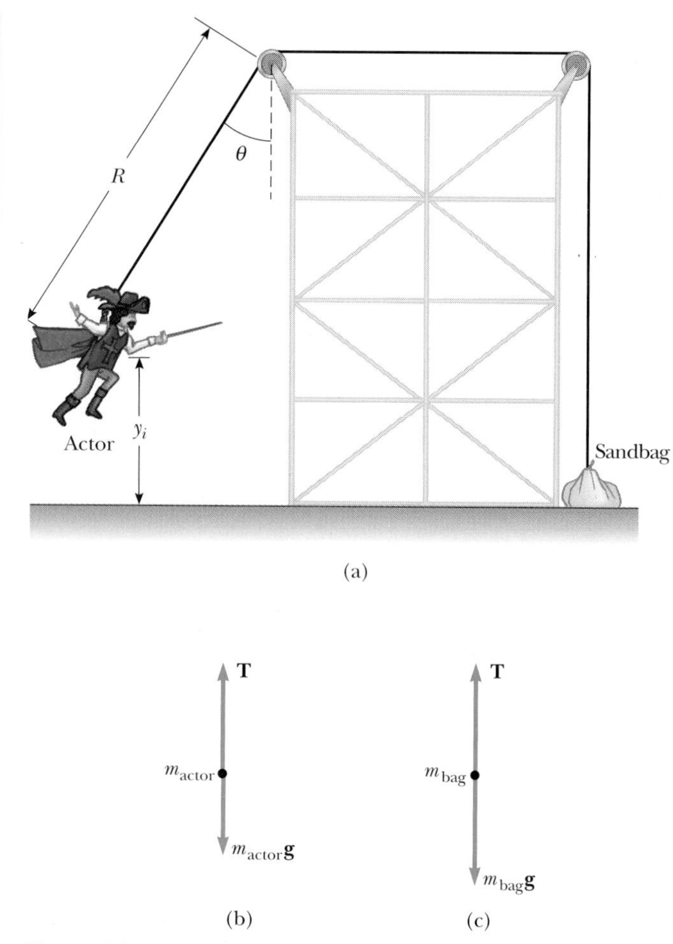

Figure 8.8 (Example 8.4) (a) An actor uses some clever staging to make his entrance. (b) Free-body diagram for the actor at the bottom of the circular path. (c) Free-body diagram for the sandbag.

never lift above the floor as the actor swings from above the stage to the floor. Let us call the initial angle that the actor's cable makes with the vertical θ. What is the maximum value θ can have before the sandbag lifts off the floor?

Solution We must use several concepts to solve this problem. To conceptualize, imagine what happens as the actor approaches the bottom of the swing. At the bottom, the cable is vertical and must support his weight as well as provide centripetal acceleration of his body in the upward direction. At this point, the tension in the cable is the highest and the sandbag is most likely to lift off the floor. Looking first at the swinging of the actor from the initial point to the lowest point, we categorize this as an energy problem involving an isolated system—the actor and the Earth. We use the principle of conservation of mechanical energy for the system to find the actor's speed as he arrives at the floor as a function of the initial angle θ and the radius R of the circular path through which he swings.

Applying conservation of mechanical energy to the actor–Earth system gives

$$K_f + U_f = K_i + U_i$$

$$(1) \qquad \tfrac{1}{2} m_{\text{actor}} v_f^2 + 0 = 0 + m_{\text{actor}} g y_i$$

where y_i is the initial height of the actor above the floor and v_f is the speed of the actor at the instant before he lands. (Note that $K_i = 0$ because he starts from rest and that $U_f = 0$ because we define the configuration of the actor at the floor as having a gravitational potential energy of zero.) From the geometry in Figure 8.8a, and noting that $y_f = 0$, we see that $y_i = R - R\cos\theta = R(1 - \cos\theta)$. Using this relationship in Equation (1), we obtain

$$(2) \qquad v_f^2 = 2gR(1 - \cos\theta)$$

Next, we focus on the instant the actor is at the lowest point. Because the tension in the cable is transferred as a force applied to the sandbag, we categorize the situation at this instant as a Newton's second law problem. We apply Newton's second law to the actor at the bottom of his path, using the free-body diagram in Figure 8.8b as a guide:

$$\sum F_y = T - m_{\text{actor}} g = m_{\text{actor}} \frac{v_f^2}{R}$$

$$(3) \qquad T = m_{\text{actor}} g + m_{\text{actor}} \frac{v_f^2}{R}$$

Finally, we note that the sandbag lifts off the floor when the upward force exerted on it by the cable exceeds the gravitational force acting on it; the normal force is zero when this happens. Thus, when we focus our attention on the sandbag, we categorize this part of the situation as another Newton's second law problem. A force T of the magnitude given by Equation (3) is transmitted by the cable to the sandbag. If the sandbag is to be just lifted off the floor, the normal force on it becomes zero and we require that $T = m_{\text{bag}} g$, as in Figure 8.8c. Using this condition together with Equations (2) and (3), we find that

$$m_{\text{bag}} g = m_{\text{actor}} g + m_{\text{actor}} \frac{2gR(1 - \cos\theta)}{R}$$

Solving for $\cos\theta$ and substituting in the given parameters, we obtain

$$\cos\theta = \frac{3m_{\text{actor}} - m_{\text{bag}}}{2m_{\text{actor}}} = \frac{3(65\text{ kg}) - 130\text{ kg}}{2(65\text{ kg})} = 0.50$$

$$\theta = 60°$$

Note that we had to combine techniques from different areas of our study—energy and Newton's second law. Furthermore, we see that the length R of the cable from the actor's harness to the leftmost pulley did not appear in the final algebraic equation. Thus, the final answer is independent of R.

What If? **What if a stagehand locates the sandbag so that the cable from the sandbag to the right-hand pulley in Figure 8.8a is not vertical but makes an angle ϕ with the vertical? If the actor swings from the angle found in the solution above, will the sandbag lift off the floor? Assume that the length R remains the same.**

Answer In this situation, the gravitational force acting on the sandbag is no longer parallel to the cable. Thus, only a component of the force in the cable acts against the gravitational force, and the vertical resultant of this force component and the gravitational force should be downward. As a result, there should be a nonzero normal force to balance this resultant, and the sandbag should *not* lift off the floor.

If the sandbag is in equilibrium in the y direction and the normal force from the floor goes to zero, Newton's second law gives us $T\cos\phi = m_{\text{bag}}g$. In this case, Equation (3) gives

$$\frac{m_{\text{bag}}g}{\cos\phi} = m_{\text{actor}}g + m_{\text{actor}}\frac{{v_f}^2}{R}$$

Substituting for v_f from Equation (2) gives

$$\frac{m_{\text{bag}}g}{\cos\phi} = m_{\text{actor}}g + m_{\text{actor}}\frac{2gR(1-\cos\theta)}{R}$$

Solving for $\cos\theta$, we have

$$(4)\qquad \cos\theta = \frac{3m_{\text{actor}} - \dfrac{m_{\text{bag}}}{\cos\phi}}{2m_{\text{actor}}}$$

For $\phi = 0$, which is the situation in Figure 8.8a, $\cos\phi = 1$. For nonzero values of ϕ, the term $\cos\phi$ is smaller than 1. This makes the numerator of the fraction in Equation (4) smaller, which makes the angle θ larger. Thus, the sandbag remains on the floor if the actor swings from a larger angle. If he swings from the original angle, the sandbag remains on the floor. For example, suppose $\phi = 10°$. Then, Equation (4) gives

$$\cos\theta = \frac{3(65\text{ kg}) - \dfrac{130\text{ kg}}{\cos 10°}}{2(65\text{ kg})} = 0.48 \longrightarrow \theta = 61°$$

Thus, if he swings from 60°, he is swinging from an angle below the new maximum allowed angle, and the sandbag remains on the floor.

One factor we have not addressed is the friction force between the sandbag and the floor. If this is not large enough, the sandbag may break free and start to slide horizontally as the actor reaches some point in his swing. This will cause the length R to increase, and the actor may have a frightening moment as he begins to drop in addition to swinging!

Let the actor fly or crash without injury to people at the Interactive Worked Example link at http://www.pse6.com. *You may choose to include the effect of friction between the sandbag and the floor.*

Example 8.5 The Spring-Loaded Popgun

The launching mechanism of a toy gun consists of a spring of unknown spring constant (Fig. 8.9a). When the spring is compressed 0.120 m, the gun, when fired vertically, is able to launch a 35.0-g projectile to a maximum height of 20.0 m above the position of the projectile before firing.

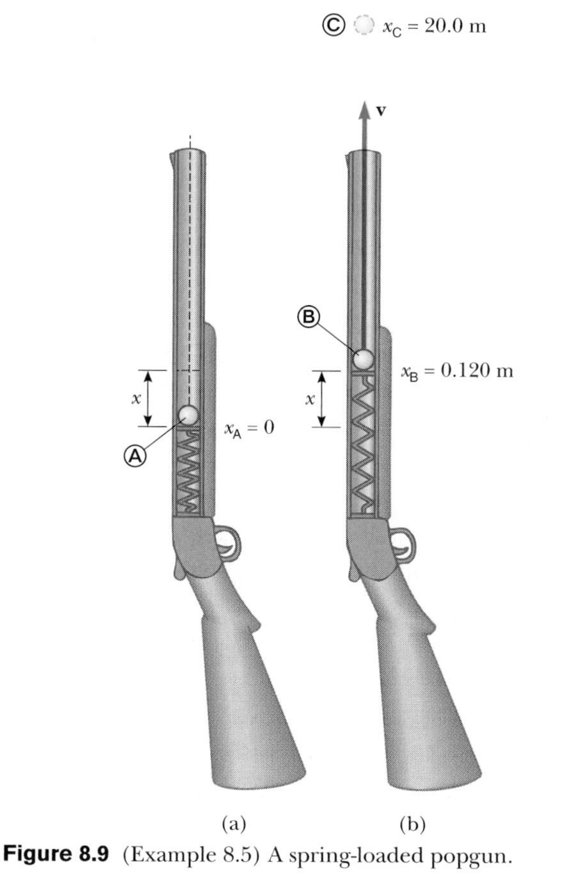

Figure 8.9 (Example 8.5) A spring-loaded popgun.

(A) Neglecting all resistive forces, determine the spring constant.

Solution Because the projectile starts from rest, its initial kinetic energy is zero. If we take the zero configuration for the gravitational potential energy of the projectile–spring–Earth system to be when the projectile is at the lowest position x_A, then the initial gravitational potential energy of the system also is zero. The mechanical energy of this system is conserved because the system is isolated.

Initially, the only mechanical energy in the system is the elastic potential energy stored in the spring of the gun, $U_{sA} = \frac{1}{2}kx^2$, where the compression of the spring is $x = 0.120$ m. The projectile rises to a maximum height $x_C = h = 20.0$ m, and so the final gravitational potential energy of the system when the projectile reaches its peak is mgh. The final kinetic energy of the projectile is zero, and the final elastic potential energy stored in the spring is zero. Because the mechanical energy of the system is conserved, we find that

$$E_C = E_A$$

$$K_C + U_{gC} + U_{sC} = K_A + U_{gA} + U_{sA}$$

$$0 + mgh + 0 = 0 + 0 + \tfrac{1}{2}kx^2$$

$$k = \frac{2mgh}{x^2} = \frac{2(0.035\,0\text{ kg})(9.80\text{ m/s}^2)(20.0\text{ m})}{(0.120\text{ m})^2}$$

$$= \boxed{953\text{ N/m}}$$

(B) Find the speed of the projectile as it moves through the equilibrium position of the spring (where $x_B = 0.120\text{ m}$) as shown in Figure 8.9b.

Solution As already noted, the only mechanical energy in the system at Ⓐ is the elastic potential energy $\frac{1}{2}kx^2$. The total energy of the system as the projectile moves through the equilibrium position of the spring includes the kinetic energy of the projectile $\frac{1}{2}mv_B{}^2$ and the gravitational potential energy mgx_B of the system. Hence, the principle of conservation of mechanical energy in this case gives

$$E_B = E_A$$

$$K_B + U_{gB} + U_{sB} = K_A + U_{gA} + U_{sA}$$

$$\tfrac{1}{2}mv_B{}^2 + mgx_B + 0 = 0 + 0 + \tfrac{1}{2}kx^2$$

Solving for v_B gives

$$v_B = \sqrt{\frac{kx^2}{m} - 2gx_B}$$

$$= \sqrt{\frac{(953\text{ N/m})(0.120\text{ m})^2}{(0.0350\text{ kg})} - 2(9.80\text{ m/s}^2)(0.120\text{ m})}$$

$$= 19.7\text{ m/s}$$

8.3 Conservative and Nonconservative Forces

As an object moves downward near the surface of the Earth, the work done by the gravitational force on the object does not depend on whether it falls vertically or slides down a sloping incline. All that matters is the change in the object's elevation. However, the energy loss due to friction on that incline depends on the distance the object slides. In other words, the path makes no difference when we consider the work done by the gravitational force, but it does make a difference when we consider the energy loss due to friction forces. We can use this varying dependence on path to classify forces as either conservative or nonconservative.

Of the two forces just mentioned, the gravitational force is conservative and the friction force is nonconservative.

Conservative Forces

Conservative forces have these two equivalent properties:

Properties of a conservative force

1. The work done by a conservative force on a particle moving between any two points is independent of the path taken by the particle.
2. The work done by a conservative force on a particle moving through any closed path is zero. (A closed path is one in which the beginning and end points are identical.)

▲ PITFALL PREVENTION

8.4 Similar Equation Warning

Compare Equation 8.12 to Equation 8.3. These equations are similar except for the negative sign, which is a common source of confusion. Equation 8.3 tells us that the work done *by an outside agent* on a system causes an increase in the potential energy of the system (with no change in the kinetic or internal energy). Equation 8.12 states that work done *on a component of a system by a conservative force internal to an isolated system* causes a decrease in the potential energy of the system (with a corresponding increase in kinetic energy).

The gravitational force is one example of a conservative force, and the force that a spring exerts on any object attached to the spring is another. As we learned in the preceding section, the work done by the gravitational force on an object moving between any two points near the Earth's surface is $W_g = mgy_i - mgy_f$. From this equation, we see that W_g depends only on the initial and final y coordinates of the object and hence is independent of the path. Furthermore, W_g is zero when the object moves over any closed path (where $y_i = y_f$).

For the case of the object–spring system, the work W_s done by the spring force is given by $W_s = \frac{1}{2}kx_i{}^2 - \frac{1}{2}kx_f{}^2$ (Eq. 7.11). Again, we see that the spring force is conservative because W_s depends only on the initial and final x coordinates of the object and is zero for any closed path.

We can associate a potential energy for a system with any conservative force acting between members of the system and can do this only for conservative forces. In the previous section, the potential energy associated with the gravitational force was defined as $U_g \equiv mgy$. In general, the work W_c done by a conservative force on an object that is a member of a system as the object moves from one position to another is equal to the initial value of the potential energy of the system minus the final value:

$$W_c = U_i - U_f = -\Delta U \qquad (8.12)$$

This equation should look familiar to you. It is the general form of the equation for work done by the gravitational force (Eq. 8.4) as an object moves relative to the Earth and that for the work done by the spring force (Eq. 7.11) as the extension of the spring changes.

Nonconservative Forces

A force is **nonconservative** if it does not satisfy properties 1 and 2 for conservative forces. Nonconservative forces acting within a system cause a *change* in the mechanical energy E_{mech} of the system. We have defined mechanical energy as the sum of the kinetic and all potential energies. For example, if a book is sent sliding on a horizontal surface that is not frictionless, the force of kinetic friction reduces the book's kinetic energy. As the book slows down, its kinetic energy decreases. As a result of the friction force, the temperatures of the book and surface increase. The type of energy associated with temperature is internal energy, which we introduced in Chapter 7. Only part of the book's kinetic energy is transformed to internal energy in the book. The rest appears as internal energy in the surface. (When you trip and fall while running across a gymnasium floor, not only does the skin on your knees warm up, so does the floor!) Because the force of kinetic friction transforms the mechanical energy of a system into internal energy, it is a nonconservative force.

As an example of the path dependence of the work, consider Figure 8.10. Suppose you displace a book between two points on a table. If the book is displaced in a straight line along the blue path between points Ⓐ and Ⓑ in Figure 8.10, you do a certain amount of work against the kinetic friction force to keep the book moving at a constant speed. Now, imagine that you push the book along the brown semicircular path in Figure 8.10. You perform more work against friction along this longer path than along the straight path. The work done depends on the path, so the friction force cannot be conservative.

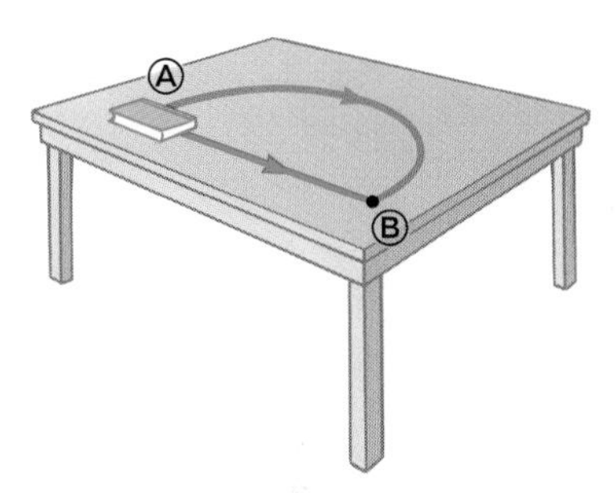

Figure 8.10 The work done against the force of kinetic friction depends on the path taken as the book is moved from Ⓐ to Ⓑ. The work is greater along the red path than along the blue path.

8.4 Changes in Mechanical Energy for Nonconservative Forces

As we have seen, if the forces acting on objects within a system are conservative, then the mechanical energy of the system is conserved. However, if some of the forces acting on objects within the system are not conservative, then the mechanical energy of the system changes.

Consider the book sliding across the surface in the preceding section. As the book moves through a distance d, the only force that does work on it is the force of kinetic friction. This force causes a decrease in the kinetic energy of the book. This decrease was calculated in Chapter 7, leading to Equation 7.20, which we repeat here:

$$\Delta K = -f_k d \quad (8.13)$$

Suppose, however, that the book is part of a system that also exhibits a change in potential energy. In this case, $-f_k d$ is the amount by which the *mechanical* energy of the system changes because of the force of kinetic friction. For example, if the book moves on an incline that is not frictionless, there is a change in both the kinetic energy and the gravitational potential energy of the book–Earth system. Consequently,

$$\Delta E_{\text{mech}} = \Delta K + \Delta U_g = -f_k d$$

In general, if a friction force acts within a system,

$$\Delta E_{\text{mech}} = \Delta K + \Delta U = -f_k d \quad (8.14)$$

Change in mechanical energy of a system due to friction within the system

where ΔU is the change in *all* forms of potential energy. Notice that Equation 8.14 reduces to Equation 8.9 if the friction force is zero.

Quick Quiz 8.9 A block of mass m is projected across a horizontal surface with an initial speed v. It slides until it stops due to the friction force between the block and the surface. The same block is now projected across the horizontal surface with an initial speed $2v$. When the block has come to rest, how does the distance from the projection point compare to that in the first case? (a) It is the same. (b) It is twice as large. (c) It is four times as large. (d) The relationship cannot be determined.

Quick Quiz 8.10 A block of mass m is projected across a horizontal surface with an initial speed v. It slides until it stops due to the friction force between the block and the surface. The surface is now tilted at 30°, and the block is projected up the surface with the same initial speed v. Assume that the friction force remains the same as when the block was sliding on the horizontal surface. When the block comes to rest momentarily, how does the decrease in mechanical energy of the block–surface–Earth system compare to that when the block slid over the horizontal surface? (a) It is the same. (b) It is larger. (c) It is smaller. (d) The relationship cannot be determined.

PROBLEM-SOLVING HINTS

Isolated Systems–Nonconservative Forces

You should incorporate the following procedure when you apply energy methods to a system in which nonconservative forces are acting:

- Follow the procedure in the first three bullets of the Problem-Solving Hints in Section 8.2. If nonconservative forces act within the system, the third bullet should tell you to use the techniques of this section.
- Write expressions for the total initial and total final mechanical energies of the system. The difference between the total final mechanical energy and the total initial mechanical energy equals the change in mechanical energy of the system due to friction.

Example 8.6 Crate Sliding Down a Ramp

A 3.00-kg crate slides down a ramp. The ramp is 1.00 m in length and inclined at an angle of 30.0°, as shown in Figure 8.11. The crate starts from rest at the top, experiences a constant friction force of magnitude 5.00 N, and continues to move a short distance on the horizontal floor after it leaves the ramp. Use energy methods to determine the speed of the crate at the bottom of the ramp.

Solution Because $v_i = 0$, the initial kinetic energy of the crate–Earth system when the crate is at the top of the ramp is zero. If the y coordinate is measured from the bottom of the ramp (the final position of the crate, for which the gravitational potential energy of the system is zero) with the upward direction being positive, then $y_i = 0.500$ m. Therefore, the total mechanical energy of the system when the crate is at the top is all potential energy:

$$E_i = K_i + U_i = 0 + U_i = mgy_i$$
$$= (3.00 \text{ kg})(9.80 \text{ m/s}^2)(0.500 \text{ m}) = 14.7 \text{ J}$$

When the crate reaches the bottom of the ramp, the potential energy of the system is zero because the elevation of

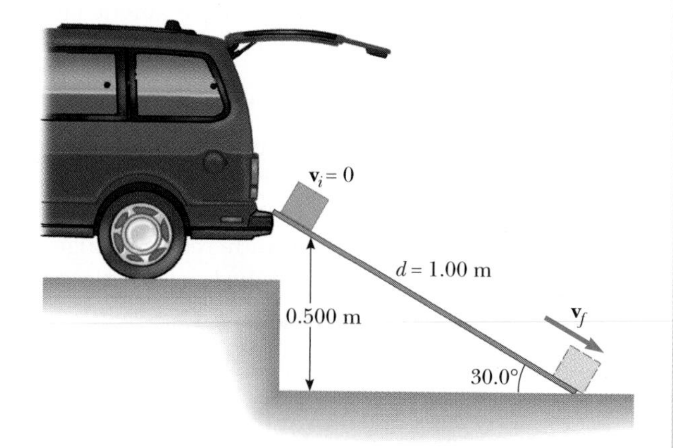

Figure 8.11 (Example 8.6) A crate slides down a ramp under the influence of gravity. The potential energy decreases while the kinetic energy increases.

the crate is $y_f = 0$. Therefore, the total mechanical energy of the system when the crate reaches the bottom is all kinetic energy:

$$E_f = K_f + U_f = \tfrac{1}{2}mv_f^2 + 0$$

We cannot say that $E_i = E_f$ because a nonconservative force reduces the mechanical energy of the system. In this case, Equation 8.14 gives $\Delta E_{\text{mech}} = -f_k d$, where d is the distance the crate moves along the ramp. (Remember that the forces normal to the ramp do no work on the crate because they are perpendicular to the displacement.) With $f_k = 5.00$ N and $d = 1.00$ m, we have

$$(1) \qquad -f_k d = (-5.00 \text{ N})(1.00 \text{ m}) = -5.00 \text{ J}$$

Applying Equation 8.14 gives

$$E_f - E_i = \tfrac{1}{2}mv_f^2 - mgy_i = -f_k d$$

$$(2) \qquad \tfrac{1}{2}mv_f^2 = 14.7 \text{ J} - 5.00 \text{ J} = 9.70 \text{ J}$$

$$v_f^2 = \frac{19.4 \text{ J}}{3.00 \text{ kg}} = 6.47 \text{ m}^2/\text{s}^2$$

$$v_f = 2.54 \text{ m/s}$$

What If? **A cautious worker decides that the speed of the crate when it arrives at the bottom of the ramp may be so large that its contents may be damaged. Therefore, he replaces the ramp with a longer one such that the new ramp makes an angle of 25° with the ground. Does this new ramp reduce the speed of the crate as it reaches the ground?**

Answer Because the ramp is longer, the friction force will act over a longer distance and transform more of the mechanical energy into internal energy. This reduces the kinetic energy of the crate, and we expect a lower speed as it reaches the ground.

We can find the length d of the new ramp as follows:

$$\sin 25° = \frac{0.500 \text{ m}}{d} \longrightarrow d = \frac{0.500 \text{ m}}{\sin 25°} = 1.18 \text{ m}$$

Now, Equation (1) becomes

$$-f_k d = (-5.00 \text{ N})(1.18 \text{ m}) = -5.90 \text{ J}$$

and Equation (2) becomes

$$\tfrac{1}{2}mv_f^2 = 14.7 \text{ J} - 5.90 \text{ J} = 8.80 \text{ J}$$

leading to

$$v_f = 2.42 \text{ m/s}$$

The final speed is indeed lower than in the higher-angle case.

Example 8.7 Motion on a Curved Track

A child of mass m rides on an irregularly curved slide of height $h = 2.00$ m, as shown in Figure 8.12. The child starts from rest at the top.

(A) Determine his speed at the bottom, assuming no friction is present.

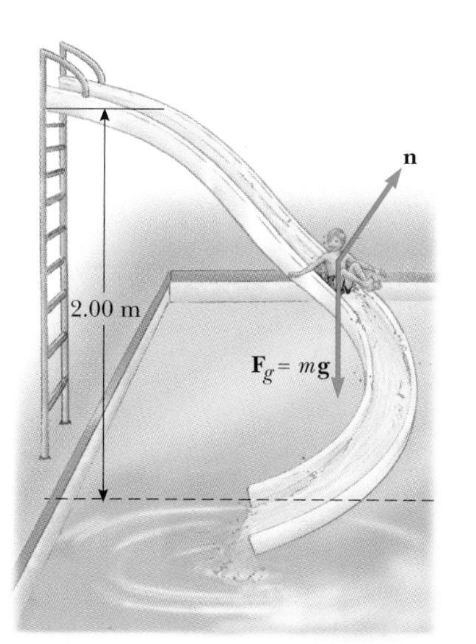

Figure 8.12 (Example 8.7) If the slide is frictionless, the speed of the child at the bottom depends only on the height of the slide.

Solution Although you have no experience on totally frictionless surfaces, you can conceptualize that your speed at the bottom of a frictionless ramp would be greater than in the situation in which friction acts. If we tried to solve this problem with Newton's laws, we would have a difficult time because the acceleration of the child continuously varies in direction due to the irregular shape of the slide. The child–Earth system is isolated and frictionless, however, so we can categorize this as a conservation of energy problem and search for a solution using the energy approach. (Note that the normal force **n** does no work on the child because this force is always perpendicular to each element of the displacement.) To analyze the situation, we measure the y coordinate in the upward direction from the bottom of the slide so that $y_i = h$, $y_f = 0$, and we obtain

$$K_f + U_f = K_i + U_i$$

$$\tfrac{1}{2}mv_f^2 + 0 = 0 + mgh$$

$$v_f = \sqrt{2gh}$$

Note that the result is the same as it would be had the child fallen vertically through a distance h! In this example, $h = 2.00$ m, giving

$$v_f = \sqrt{2gh} = \sqrt{2(9.80 \text{ m/s}^2)(2.00 \text{ m})} = 6.26 \text{ m/s}$$

(B) If a force of kinetic friction acts on the child, how much mechanical energy does the system lose? Assume that $v_f = 3.00$ m/s and $m = 20.0$ kg.

Solution We categorize this case, with friction, as a problem in which a nonconservative force acts. Hence, mechanical energy is not conserved, and we must use Equation 8.14 to find the loss of mechanical energy due to friction:

$$\begin{aligned}\Delta E_{\text{mech}} &= (K_f + U_f) - (K_i + U_i) \\ &= (\tfrac{1}{2}mv_f^2 + 0) - (0 + mgh) = \tfrac{1}{2}mv_f^2 - mgh \\ &= \tfrac{1}{2}(20.0 \text{ kg})(3.00 \text{ m/s})^2 \\ &\quad - (20.0 \text{ kg})(9.80 \text{ m/s}^2)(2.00 \text{ m}) \\ &= -302 \text{ J}\end{aligned}$$

Again, ΔE_{mech} is negative because friction is reducing the mechanical energy of the system. (The final mechanical energy is less than the initial mechanical energy.)

What If? Suppose you were asked to find the coefficient of friction μ_k for the child on the slide. Could you do this?

Answer We can argue that the same final speed could be obtained by having the child travel down a short slide with large friction or a long slide with less friction. Thus, there does not seem to be enough information in the problem to determine the coefficient of friction.

The energy loss of 302 J must be equal to the product of the friction force and the length of the slide:

$$-f_k d = -302 \text{ J}$$

We can also argue that the friction force can be expressed as $\mu_k n$, where n is the magnitude of the normal force. Thus,

$$\mu_k n d = 302 \text{ J}$$

If we try to evaluate the coefficient of friction from this relationship, we run into two problems. First, there is no single value of the normal force n unless the angle of the slide relative to the horizontal remains fixed. Even if the angle were fixed, we do not know its value. The second problem is that we do not have information about the length d of the slide. Thus, we cannot find the coefficient of friction from the information given.

Example 8.8 Let's Go Skiing!

A skier starts from rest at the top of a frictionless incline of height 20.0 m, as shown in Figure 8.13. At the bottom of the incline, she encounters a horizontal surface where the coefficient of kinetic friction between the skis and the snow is 0.210. How far does she travel on the horizontal surface before coming to rest, if she simply coasts to a stop?

Solution The system is the skier plus the Earth, and we choose as our configuration of zero potential energy that in which the skier is at the bottom of the incline. While the skier is on the frictionless incline, the mechanical energy of the system remains constant, and we find, as we did in Example 8.7, that

$$v_{\text{B}} = \sqrt{2gh} = \sqrt{2(9.80 \text{ m/s}^2)(20.0 \text{ m})} = 19.8 \text{ m/s}$$

Now we apply Equation 8.14 as the skier moves along the rough horizontal surface from Ⓑ to Ⓒ. The change in mechanical energy along the horizontal surface is

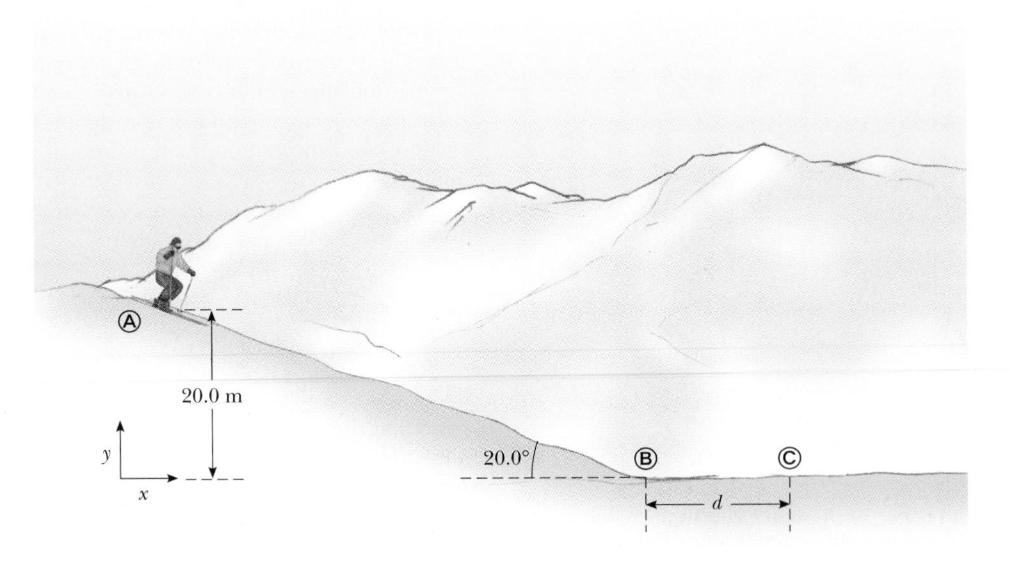

Figure 8.13 (Example 8.8) The skier slides down the slope and onto a level surface, stopping after a distance d from the bottom of the hill.

$\Delta E_{\text{mech}} = -f_k d$, where d is the horizontal distance traveled by the skier.

To find the distance the skier travels before coming to rest, we take $K_C = 0$. With $v_B = 19.8$ m/s and the friction force given by $f_k = \mu_k n = \mu_k mg$, we obtain

$$\Delta E_{\text{mech}} = E_C - E_B = -\mu_k mgd$$

$$(K_C + U_C) - (K_B + U_B) = (0 + 0) - (\tfrac{1}{2}mv_B^2 + 0)$$
$$= -\mu_k mgd$$

$$d = \frac{v_B^2}{2\mu_k g} = \frac{(19.8 \text{ m/s})^2}{2(0.210)(9.80 \text{ m/s}^2)} = 95.2 \text{ m}$$

Example 8.9 Block-Spring Collision

A block having a mass of 0.80 kg is given an initial velocity $v_A = 1.2$ m/s to the right and collides with a spring of negligible mass and force constant $k = 50$ N/m, as shown in Figure 8.14.

(A) Assuming the surface to be frictionless, calculate the maximum compression of the spring after the collision.

Solution Our system in this example consists of the block and spring. All motion takes place in a horizontal plane, so we do not need to consider changes in gravitational potential energy. Before the collision, when the block is at Ⓐ, it has kinetic energy and the spring is uncompressed, so the elastic potential energy stored in the spring is zero. Thus, the total mechanical energy of the system before the collision is just $\frac{1}{2}mv_A^2$. After the collision, when the block is at Ⓒ, the spring is fully compressed; now the block is at rest and so has zero kinetic energy, while the energy stored in the spring has its maximum value $\frac{1}{2}kx^2 = \frac{1}{2}kx_{\text{max}}^2$, where the origin of coordinates $x = 0$ is chosen to be the equilibrium position of the spring and x_{max} is the maximum compression of the spring, which in this case happens to be x_C. The total mechanical energy of the system is conserved because no nonconservative forces act on objects within the system.

Because the mechanical energy of the system is conserved, the kinetic energy of the block before the collision equals the maximum potential energy stored in the fully compressed spring:

$$E_C = E_A$$

$$K_C + U_{sC} = K_A + U_{sA}$$

$$0 + \tfrac{1}{2}kx_{\text{max}}^2 = \tfrac{1}{2}mv_A^2 + 0$$

$$x_{\text{max}} = \sqrt{\frac{m}{k}}\,v_A = \sqrt{\frac{0.80 \text{ kg}}{50 \text{ N/m}}}\,(1.2 \text{ m/s})$$

$$= 0.15 \text{ m}$$

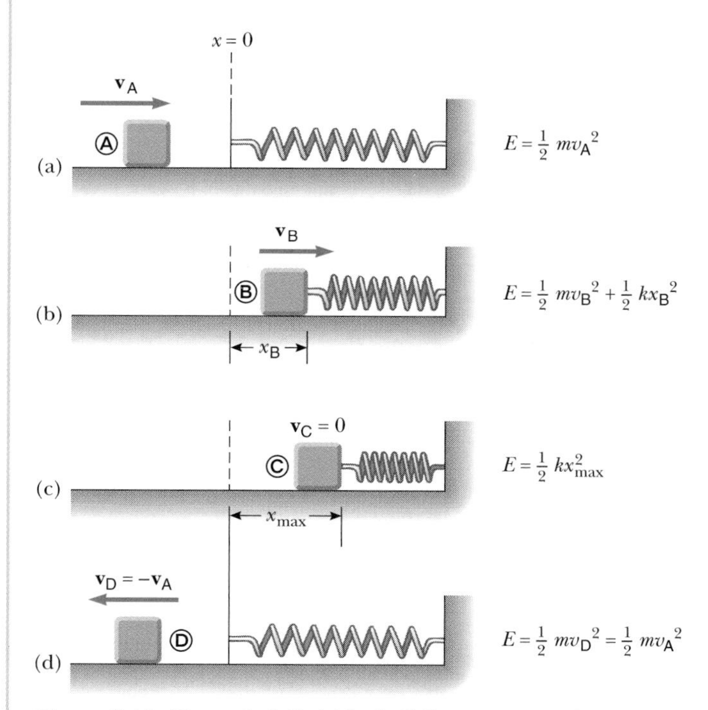

Figure 8.14 (Example 8.9) A block sliding on a smooth, horizontal surface collides with a light spring. (a) Initially the mechanical energy is all kinetic energy. (b) The mechanical energy is the sum of the kinetic energy of the block and the elastic potential energy in the spring. (c) The energy is entirely potential energy. (d) The energy is transformed back to the kinetic energy of the block. The total energy of the system remains constant throughout the motion.

(B) Suppose a constant force of kinetic friction acts between the block and the surface, with $\mu_k = 0.50$. If the speed of the block at the moment it collides with the spring is $v_A = 1.2$ m/s, what is the maximum compression x_C in the spring?

Solution In this case, the mechanical energy of the system is *not* conserved because a friction force acts on the block. The magnitude of the friction force is

$$f_k = \mu_k n = \mu_k mg = 0.50(0.80 \text{ kg})(9.80 \text{ m/s}^2) = 3.92 \text{ N}$$

Therefore, the change in the mechanical energy of the system due to friction as the block is displaced from the equilibrium position of the spring (where we have set our origin) to x_C is

$$\Delta E_{\text{mech}} = -f_k x_C = (-3.92x_C)$$

Substituting this into Equation 8.14 gives

$$\Delta E_{\text{mech}} = E_f - E_i = (0 + \tfrac{1}{2}kx_C^2) - (\tfrac{1}{2}mv_A^2 + 0) = -f_k x_C$$

$$\tfrac{1}{2}(50)x_C^2 - \tfrac{1}{2}(0.80)(1.2)^2 = -3.92x_C$$

$$25x_C^2 + 3.92x_C - 0.576 = 0$$

Solving the quadratic equation for x_C gives $x_C = 0.092$ m and $x_C = -0.25$ m. The physically meaningful root is $x_C = 0.092$ m. The negative root does not apply to this situation because the block must be to the right of the origin (positive value of x) when it comes to rest. Note that the value of 0.092 m is less than the distance obtained in the frictionless case of part (A). This result is what we expect because friction retards the motion of the system.

Example 8.10 Connected Blocks in Motion

Two blocks are connected by a light string that passes over a frictionless pulley, as shown in Figure 8.15. The block of mass m_1 lies on a horizontal surface and is connected to a spring of force constant k. The system is released from rest when the spring is unstretched. If the hanging block of mass m_2 falls a distance h before coming to rest, calculate the coefficient of kinetic friction between the block of mass m_1 and the surface.

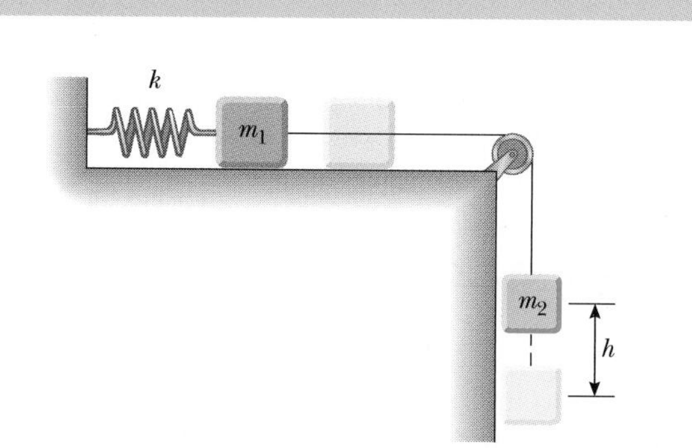

Figure 8.15 (Example 8.10) As the hanging block moves from its highest elevation to its lowest, the system loses gravitational potential energy but gains elastic potential energy in the spring. Some mechanical energy is lost because of friction between the sliding block and the surface.

Solution The key word *rest* appears twice in the problem statement. This suggests that the configurations associated with rest are good candidates for the initial and final configurations because the kinetic energy of the system is zero for these configurations. (Also note that because we are concerned only with the beginning and ending points of the motion, we do not need to label events with circled letters as we did in the previous two examples. Simply using i and f is sufficient to keep track of the situation.) In this situation, the system consists of the two blocks, the spring, and the Earth. We need to consider two forms of potential energy: gravitational and elastic. Because the initial and final kinetic energies of the system are zero, $\Delta K = 0$, and we can write

$$(1) \qquad \Delta E_{\text{mech}} = \Delta U_g + \Delta U_s$$

where $\Delta U_g = U_{gf} - U_{gi}$ is the change in the system's gravitational potential energy and $\Delta U_s = U_{sf} - U_{si}$ is the change in the system's elastic potential energy. As the hanging block falls a distance h, the horizontally moving block moves the same distance h to the right. Therefore, using Equation 8.14, we find that the loss in mechanical energy in the system due to friction between the horizontally sliding block and the surface is

$$(2) \qquad \Delta E_{\text{mech}} = -f_k h = -\mu_k m_1 g h$$

The change in the gravitational potential energy of the system is associated with only the falling block because the vertical coordinate of the horizontally sliding block does not change. Therefore, we obtain

$$(3) \qquad \Delta U_g = U_{gf} - U_{gi} = 0 - m_2 g h$$

where the coordinates have been measured from the lowest position of the falling block.

The change in the elastic potential energy of the system is that stored in the spring:

$$(4) \qquad \Delta U_s = U_{sf} - U_{si} = \tfrac{1}{2}kh^2 - 0$$

Substituting Equations (2), (3), and (4) into Equation (1) gives

$$-\mu_k m_1 g h = -m_2 g h + \tfrac{1}{2}kh^2$$

$$\mu_k = \frac{m_2 g - \frac{1}{2}kh}{m_1 g}$$

This setup represents a way of measuring the coefficient of kinetic friction between an object and some surface. As you can see from the problem, sometimes it is easier to work with the changes in the various types of energy rather than the actual values. For example, if we wanted to calculate the numerical value of the gravitational potential energy associated with the horizontally sliding block, we would need to specify the height of the horizontal surface relative to the lowest position of the falling block. Fortunately, this is not necessary because the gravitational potential energy associated with the first block does not change.

8.5 Relationship Between Conservative Forces and Potential Energy

In an earlier section we found that the work done on a member of a system by a conservative force between the members does not depend on the path taken by the moving member. The work depends only on the initial and final coordinates. As a consequence, we can define a **potential energy function *U*** such that the work done by a conservative force equals the decrease in the potential energy of the system. Let us imagine a system of particles in which the configuration changes due to the motion of one particle along the x axis. The work done by a conservative force $\mathbf{F}$ as a particle moves along the x axis is[2]

[2] For a general displacement, the work done in two or three dimensions also equals $-\Delta U$, where $U = U(x, y, z)$. We write this formally as $W = \int_i^f \mathbf{F} \cdot d\mathbf{r} = U_i - U_f$.

$$W_c = \int_{x_i}^{x_f} F_x \, dx = -\Delta U \tag{8.15}$$

where F_x is the component of **F** in the direction of the displacement. That is, the work done by a conservative force acting between members of a system equals the negative of the change in the potential energy associated with that force when the configuration of the system changes, where the change in the potential energy is defined as $\Delta U = U_f - U_i$. We can also express Equation 8.15 as

$$\Delta U = U_f - U_i = -\int_{x_i}^{x_f} F_x \, dx \tag{8.16}$$

Therefore, ΔU is negative when F_x and dx are in the same direction, as when an object is lowered in a gravitational field or when a spring pushes an object toward equilibrium.

The term *potential energy* implies that the system has the potential, or capability, of either gaining kinetic energy or doing work when it is released under the influence of a conservative force exerted on an object by some other member of the system. It is often convenient to establish some particular location x_i of one member of a system as representing a reference configuration and measure all potential energy differences with respect to it. We can then define the potential energy function as

$$U_f(x) = -\int_{x_i}^{x_f} F_x \, dx + U_i \tag{8.17}$$

The value of U_i is often taken to be zero for the reference configuration. It really does not matter what value we assign to U_i because any nonzero value merely shifts $U_f(x)$ by a constant amount and only the *change* in potential energy is physically meaningful.

If the conservative force is known as a function of position, we can use Equation 8.17 to calculate the change in potential energy of a system as an object within the system moves from x_i to x_f.

If the point of application of the force undergoes an infinitesimal displacement dx, we can express the infinitesimal change in the potential energy of the system dU as

$$dU = -F_x \, dx$$

Therefore, the conservative force is related to the potential energy function through the relationship[3]

$$F_x = -\frac{dU}{dx} \tag{8.18}$$

Relation of force between members of a system to the potential energy of the system

That is, **the x component of a conservative force acting on an object within a system equals the negative derivative of the potential energy of the system with respect to x.**

We can easily check this relationship for the two examples already discussed. In the case of the deformed spring, $U_s = \frac{1}{2}kx^2$, and therefore

$$F_s = -\frac{dU_s}{dx} = -\frac{d}{dx}\left(\tfrac{1}{2}kx^2\right) = -kx$$

which corresponds to the restoring force in the spring (Hooke's law). Because the gravitational potential energy function is $U_g = mgy$, it follows from Equation 8.18 that $F_g = -mg$ when we differentiate U_g with respect to y instead of x.

We now see that U is an important function because a conservative force can be derived from it. Furthermore, Equation 8.18 should clarify the fact that adding a constant to the potential energy is unimportant because the derivative of a constant is zero.

[3] In three dimensions, the expression is, $\mathbf{F} = -\frac{\partial U}{\partial x}\hat{\mathbf{i}} - \frac{\partial U}{\partial y}\hat{\mathbf{j}} - \frac{\partial U}{\partial z}\hat{\mathbf{k}}$ where $\frac{\partial U}{\partial x}$ etc. are partial derivatives. In the language of vector calculus, **F** equals the negative of the *gradient* of the scalar quantity $U(x, y, z)$.

Quick Quiz 8.11 What does the slope of a graph of $U(x)$ versus x represent? (a) the magnitude of the force on the object (b) the negative of the magnitude of the force on the object (c) the x component of the force on the object (d) the negative of the x component of the force on the object.

8.6 Energy Diagrams and Equilibrium of a System

The motion of a system can often be understood qualitatively through a graph of its potential energy versus the position of a member of the system. Consider the potential energy function for a block–spring system, given by $U_s = \frac{1}{2}kx^2$. This function is plotted versus x in Figure 8.16a. (A common mistake is to think that potential energy on the graph represents height. This is clearly not the case here, where the block is only moving horizontally.) The force F_s exerted by the spring on the block is related to U_s through Equation 8.18:

$$F_s = -\frac{dU_s}{dx} = -kx$$

As we saw in Quick Quiz 8.11, the x component of the force is equal to the negative of the slope of the U-versus-x curve. When the block is placed at rest at the equilibrium position of the spring ($x = 0$), where $F_s = 0$, it will remain there unless some external force F_{ext} acts on it. If this external force stretches the spring from equilibrium, x is positive and the slope dU/dx is positive; therefore, the force F_s exerted by the spring is negative and the block accelerates back toward $x = 0$ when released. If the external force compresses the spring, then x is negative and the slope is negative; therefore, F_s is positive and again the mass accelerates toward $x = 0$ upon release.

Stable equilibrium

From this analysis, we conclude that the $x = 0$ position for a block–spring system is one of **stable equilibrium.** That is, any movement away from this position results in a force directed back toward $x = 0$. In general, **configurations of stable equilibrium correspond to those for which $U(x)$ is a minimum.**

From Figure 8.16 we see that if the block is given an initial displacement x_{max} and is released from rest, its total energy initially is the potential energy $\frac{1}{2}kx_{\text{max}}^2$ stored in the

At the Active Figures link at http://www.pse6.com, *you can observe the block oscillate between its turning points and trace the corresponding points on the potential energy curve for varying values of k.*

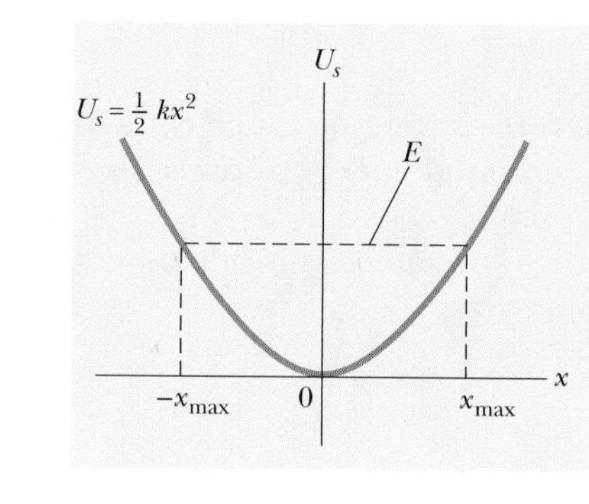

(a)

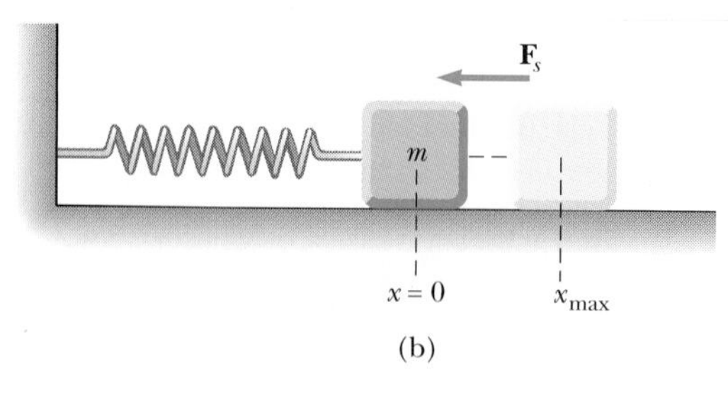

(b)

Active Figure 8.16 (a) Potential energy as a function of x for the frictionless block–spring system shown in (b). The block oscillates between the turning points, which have the coordinates $x = \pm x_{\text{max}}$. Note that the restoring force exerted by the spring always acts toward $x = 0$, the position of stable equilibrium.

spring. As the block starts to move, the system acquires kinetic energy and loses an equal amount of potential energy. Because the total energy of the system must remain constant, the block oscillates (moves back and forth) between the two points $x = -x_{max}$ and $x = +x_{max}$, called the *turning points.* In fact, because no energy is lost (no friction), the block will oscillate between $-x_{max}$ and $+x_{max}$ forever. (We discuss these oscillations further in Chapter 15.) From an energy viewpoint, the energy of the system cannot exceed $\frac{1}{2}kx_{max}^2$; therefore, the block must stop at these points and, because of the spring force, must accelerate toward $x = 0$.

Another simple mechanical system that has a configuration of stable equilibrium is a ball rolling about in the bottom of a bowl. Anytime the ball is displaced from its lowest position, it tends to return to that position when released.

Now consider a particle moving along the x axis under the influence of a conservative force F_x, where the U-versus-x curve is as shown in Figure 8.17. Once again, $F_x = 0$ at $x = 0$, and so the particle is in equilibrium at this point. However, this is a position of **unstable equilibrium** for the following reason: Suppose that the particle is displaced to the right ($x > 0$). Because the slope is negative for $x > 0$, $F_x = -dU/dx$ is positive, and the particle accelerates away from $x = 0$. If instead the particle is at $x = 0$ and is displaced to the left ($x < 0$), the force is negative because the slope is positive for $x < 0$, and the particle again accelerates away from the equilibrium position. The position $x = 0$ in this situation is one of unstable equilibrium because for any displacement from this point, the force pushes the particle farther away from equilibrium. The force pushes the particle toward a position of lower potential energy. A pencil balanced on its point is in a position of unstable equilibrium. If the pencil is displaced slightly from its absolutely vertical position and is then released, it will surely fall over. In general, **configurations of unstable equilibrium correspond to those for which $U(x)$ is a maximum.**

Unstable equilibrium

Finally, a situation may arise where U is constant over some region. This is called a configuration of **neutral equilibrium.** Small displacements from a position in this region produce neither restoring nor disrupting forces. A ball lying on a flat horizontal surface is an example of an object in neutral equilibrium.

Neutral equilibrium

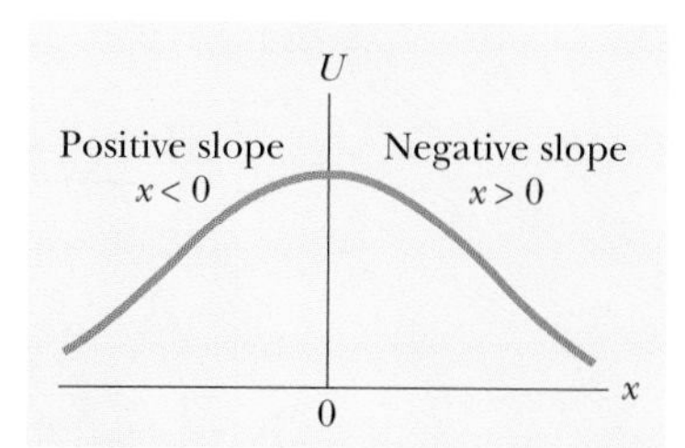

Figure 8.17 A plot of U versus x for a particle that has a position of unstable equilibrium located at $x = 0$. For any finite displacement of the particle, the force on the particle is directed away from $x = 0$.

Example 8.11 Force and Energy on an Atomic Scale

The potential energy associated with the force between two neutral atoms in a molecule can be modeled by the Lennard–Jones potential energy function:

$$U(x) = 4\epsilon\left[\left(\frac{\sigma}{x}\right)^{12} - \left(\frac{\sigma}{x}\right)^{6}\right]$$

where x is the separation of the atoms. The function $U(x)$ contains two parameters σ and ϵ that are determined from experiments. Sample values for the interaction between two atoms in a molecule are $\sigma = 0.263$ nm and $\epsilon = 1.51 \times 10^{-22}$ J.

(A) Using a spreadsheet or similar tool, graph this function and find the most likely distance between the two atoms.

Solution We expect to find stable equilibrium when the two atoms are separated by some equilibrium distance and the potential energy of the system of two atoms (the molecule) is a minimum. One can minimize the function $U(x)$ by taking its derivative and setting it equal to zero:

$$\frac{dU(x)}{dx} = 4\epsilon\frac{d}{dx}\left[\left(\frac{\sigma}{x}\right)^{12} - \left(\frac{\sigma}{x}\right)^{6}\right] = 0$$

$$= 4\epsilon\left[\frac{-12\sigma^{12}}{x^{13}} - \frac{-6\sigma^{6}}{x^{7}}\right] = 0$$

Solving for x—the equilibrium separation of the two atoms in the molecule—and inserting the given information yields

$$x = 2.95 \times 10^{-10}\ \text{m}.$$

We graph the Lennard–Jones function on both sides of this critical value to create our energy diagram, as shown in Figure 8.18a. Notice that $U(x)$ is extremely large when the atoms are very close together, is a minimum when the atoms are at their critical separation, and then increases again as the atoms move apart. When $U(x)$ is a minimum, the atoms are in stable equilibrium; this indicates that this is the most likely separation between them.

(B) Determine $F_x(x)$—the force that one atom exerts on the other in the molecule as a function of separation—and argue that the way this force behaves is physically plausible when the atoms are close together and far apart.

Solution Because the atoms combine to form a molecule, the force must be attractive when the atoms are far apart. On the other hand, the force must be repulsive when the two atoms are very close together. Otherwise, the molecule would collapse in on itself. Thus, the force must change sign at the critical separation, similar to the way spring forces switch sign in the change from extension to compression. Applying Equation 8.18 to the Lennard–Jones potential energy function gives

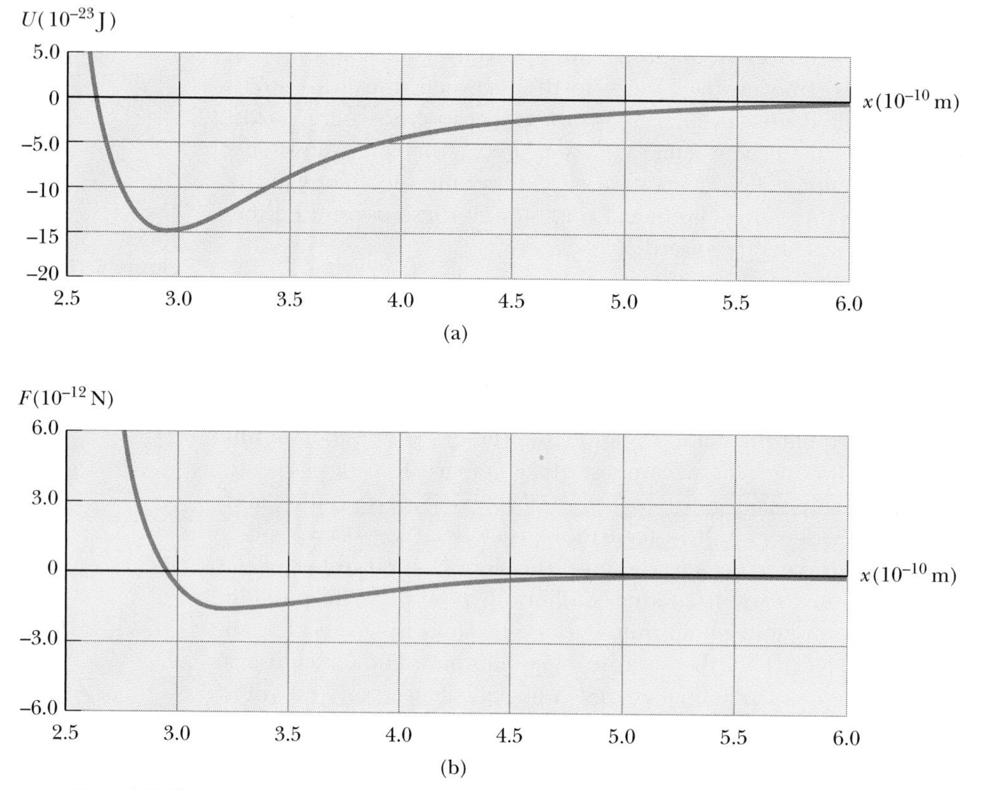

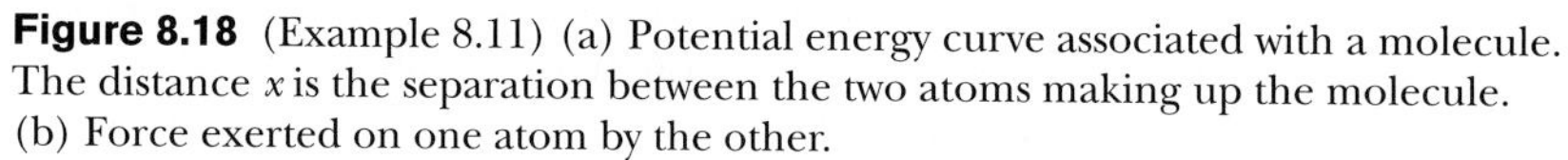

Figure 8.18 (Example 8.11) (a) Potential energy curve associated with a molecule. The distance x is the separation between the two atoms making up the molecule. (b) Force exerted on one atom by the other.

$$F_x = -\frac{dU(x)}{dx} = -4\epsilon \frac{d}{dx}\left[\left(\frac{\sigma}{x}\right)^{12} - \left(\frac{\sigma}{x}\right)^{6}\right]$$

$$= 4\epsilon\left[\frac{12\sigma^{12}}{x^{13}} - \frac{6\sigma^{6}}{x^{7}}\right]$$

This result is graphed in Figure 8.18b. As expected, the force is positive (repulsive) at small atomic separations, zero when the atoms are at the position of stable equilibrium [recall how we found the minimum of $U(x)$], and negative (attractive) at greater separations. Note that the force approaches zero as the separation between the atoms becomes very great.

SUMMARY

Take a Practice Test for this chapter by clicking on the Practice Test link at http://www.pse6.com.

If a particle of mass m is at a distance y above the Earth's surface, the **gravitational potential energy** of the particle–Earth system is

$$U_g \equiv mgy \tag{8.2}$$

The **elastic potential energy** stored in a spring of force constant k is

$$U_s \equiv \tfrac{1}{2}kx^2 \tag{8.11}$$

A reference configuration of the system should be chosen, and this configuration is often assigned a potential energy of zero.

A force is **conservative** if the work it does on a particle moving between two points is independent of the path the particle takes between the two points. Furthermore, a force is conservative if the work it does on a particle is zero when the particle moves through an arbitrary closed path and returns to its initial position. A force that does not meet these criteria is said to be **nonconservative.**

The **total mechanical energy of a system** is defined as the sum of the kinetic energy and the potential energy:

$$E_{\text{mech}} \equiv K + U \tag{8.8}$$

If a system is isolated and if no nonconservative forces are acting on objects inside the system, then the total mechanical energy of the system is constant:

$$K_f + U_f = K_i + U_i \tag{8.9}$$

If nonconservative forces (such as friction) act on objects inside a system, then mechanical energy is not conserved. In these situations, the difference between the total final mechanical energy and the total initial mechanical energy of the system equals the energy transformed to internal energy by the nonconservative forces.

A **potential energy function** U can be associated only with a conservative force. If a conservative force **F** acts between members of a system while one member moves along the x axis from x_i to x_f, then the change in the potential energy of the system equals the negative of the work done by that force:

$$U_f - U_i = -\int_{x_i}^{x_f} F_x\,dx \tag{8.16}$$

Systems can be in three types of equilibrium configurations when the net force on a member of the system is zero. Configurations of **stable equilibrium** correspond to those for which $U(x)$ is a minimum. Configurations of **unstable equilibrium** correspond to those for which $U(x)$ is a maximum. **Neutral equilibrium** arises where U is constant as a member of the system moves over some region.

QUESTIONS

1. If the height of a playground slide is kept constant, will the length of the slide or the presence of bumps make any difference in the final speed of children playing on it? Assume the slide is slick enough to be considered frictionless. Repeat this question assuming friction is present.
2. Explain why the total energy of a system can be either positive or negative, whereas the kinetic energy is always positive.
3. One person drops a ball from the top of a building while another person at the bottom observes its motion. Will these two people agree on the value of the gravitational potential energy of the ball–Earth system? On the change in potential energy? On the kinetic energy?
4. Discuss the changes in mechanical energy of an object–Earth system in (a) lifting the object, (b) holding the object at a fixed position, and (c) lowering the object slowly. Include the muscles in your discussion.
5. In Chapter 7, the work–kinetic energy theorem, $W = \Delta K$, was introduced. This equation states that work done on a system appears as a change in kinetic energy. This is a special-case equation, valid if there are no changes in any other type of energy such as potential or internal. Give some examples in which work is done on a system, but the change in energy of the system is not that of kinetic energy.
6. If three conservative forces and one nonconservative force act within a system, how many potential-energy terms appear in the equation that describes the system?
7. If only one external force acts on a particle, does it necessarily change the particle's (a) kinetic energy? (b) velocity?
8. A driver brings an automobile to a stop. If the brakes lock so that the car skids, where is the original kinetic energy of the car, and in what form is it after the car stops? Answer the same question for the case in which the brakes do not lock, but the wheels continue to turn.
9. You ride a bicycle. In what sense is your bicycle solar-powered?
10. In an earthquake, a large amount of energy is "released" and spreads outward, potentially causing severe damage. In what form does this energy exist before the earthquake, and by what energy transfer mechanism does it travel?
11. A bowling ball is suspended from the ceiling of a lecture hall by a strong cord. The ball is drawn away from its equilibrium position and released from rest at the tip of the demonstrator's nose as in Figure Q8.11. If the demonstrator remains stationary, explain why she is not struck by the ball on its return swing. Would this demonstrator be safe if the ball were given a push from its starting position at her nose?
12. Roads going up mountains are formed into switchbacks, with the road weaving back and forth along the face of the slope such that there is only a gentle rise on any portion of the roadway. Does this require any less work to be done by an automobile climbing the mountain compared to driving on a roadway that is straight up the slope? Why are switchbacks used?
13. As a sled moves across a flat snow-covered field at constant velocity, is any work done? How does air resistance enter into the picture?
14. You are working in a library, reshelving books. You lift a book from the floor to the top shelf. The kinetic energy of the book on the floor was zero, and the kinetic energy of the book on the top shelf is zero, so there is no change

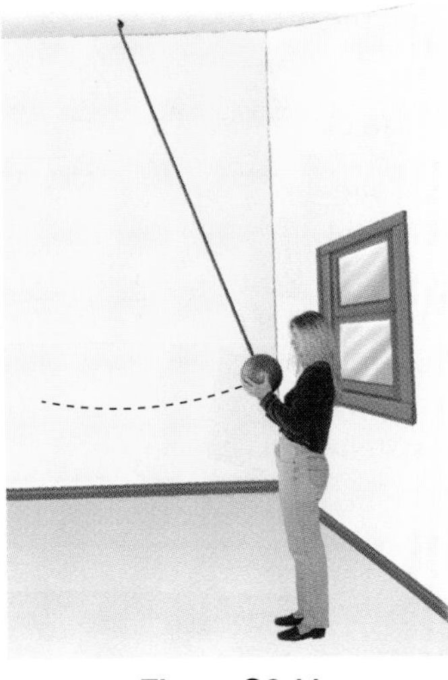

Figure Q8.11

in kinetic energy. Yet you did some work in lifting the book. Is the work–kinetic energy theorem violated?

15. A ball is thrown straight up into the air. At what position is its kinetic energy a maximum? At what position is the gravitational potential energy of the ball–Earth system a maximum?

16. A pile driver is a device used to drive objects into the Earth by repeatedly dropping a heavy weight on them. By how much does the energy of the pile driver–Earth system increase when the weight it drops is doubled? Assume the weight is dropped from the same height each time.

17. Our body muscles exert forces when we lift, push, run, jump, and so forth. Are these forces conservative?

18. A block is connected to a spring that is suspended from the ceiling. If the block is set in motion and air resistance is neglected, describe the energy transformations that occur within the system consisting of the block, Earth, and spring.

19. Describe the energy transformations that occur during (a) the pole vault (b) the shot put (c) the high jump. What is the source of energy in each case?

20. Discuss the energy transformations that occur during the operation of an automobile.

21. What would the curve of U versus x look like if a particle were in a region of neutral equilibrium?

22. A ball rolls on a horizontal surface. Is the ball in stable, unstable, or neutral equilibrium?

23. Consider a ball fixed to one end of a rigid rod whose other end pivots on a horizontal axis so that the rod can rotate in a vertical plane. What are the positions of stable and unstable equilibrium?

PROBLEMS

1, 2, 3 = straightforward, intermediate, challenging □ = full solution available in the *Student Solutions Manual and Study Guide*

www = coached solution with hints available at http://www.pse6.com ▭ = computer useful in solving problem

▭ = paired numerical and symbolic problems

Section 8.1 Potential Energy of a System

1. A 1 000-kg roller coaster train is initially at the top of a rise, at point Ⓐ. It then moves 135 ft, at an angle of 40.0° below the horizontal, to a lower point Ⓑ. (a) Choose point Ⓑ to be the zero level for gravitational potential energy. Find the potential energy of the roller coaster–Earth system at points Ⓐ and Ⓑ, and the change in potential energy as the coaster moves. (b) Repeat part (a), setting the zero reference level at point Ⓐ.

2. A 400-N child is in a swing that is attached to ropes 2.00 m long. Find the gravitational potential energy of the child–Earth system relative to the child's lowest position when (a) the ropes are horizontal, (b) the ropes make a 30.0° angle with the vertical, and (c) the child is at the bottom of the circular arc.

3. A person with a remote mountain cabin plans to install her own hydroelectric plant. A nearby stream is 3.00 m wide and 0.500 m deep. Water flows at 1.20 m/s over the brink of a waterfall 5.00 m high. The manufacturer promises only 25.0% efficiency in converting the potential energy of the water–Earth system into electric energy. Find the power she can generate. (Large-scale hydroelectric plants, with a much larger drop, are more efficient.)

Section 8.2 The Isolated System–Conservation of Mechanical Energy

4. At 11:00 A.M. on September 7, 2001, more than 1 million British school children jumped up and down for one minute. The curriculum focus of the "Giant Jump" was on earthquakes, but it was integrated with many other topics, such as exercise, geography, cooperation, testing hypotheses, and setting world records. Children built their own seismographs, which registered local effects. (a) Find the mechanical energy released in the experiment. Assume that 1 050 000 children of average mass 36.0 kg jump twelve times each, raising their centers of mass by 25.0 cm each time and briefly resting between one jump and the next. The free-fall acceleration in Britain is 9.81 m/s^2. (b) Most of the energy is converted very rapidly into internal energy within the bodies of the children and the floors of the school buildings. Of the energy that propagates into the ground, most produces high-frequency "microtremor" vibrations that are rapidly damped and cannot travel far. Assume that 0.01% of the energy is carried away by a long-range seismic wave. The magnitude of an earthquake on the Richter scale is given by

$$M = \frac{\log E - 4.8}{1.5}$$

where E is the seismic wave energy in joules. According to this model, what is the magnitude of the demonstration quake? (It did not register above background noise overseas or on the seismograph of the Wolverton Seismic Vault, Hampshire.)

5. A bead slides without friction around a loop-the-loop (Fig. P8.5). The bead is released from a height $h = 3.50R$. (a) What is its speed at point Ⓐ? (b) How large is the normal force on it if its mass is 5.00 g?

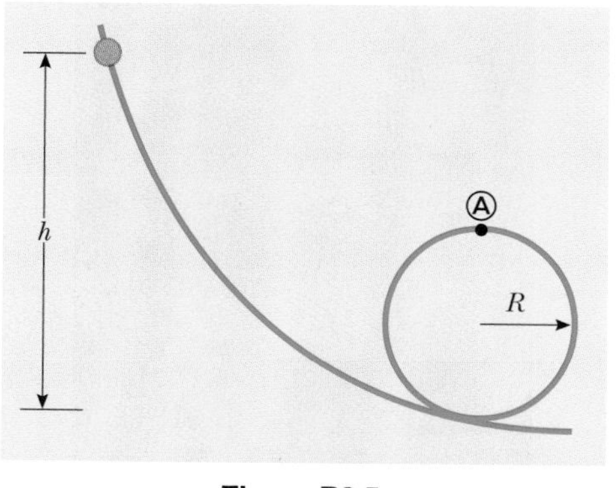

Figure P8.5

6. Dave Johnson, the bronze medalist at the 1992 Olympic decathlon in Barcelona, leaves the ground at the high jump with vertical velocity component 6.00 m/s. How far does his center of mass move up as he makes the jump?

7. A glider of mass 0.150 kg moves on a horizontal frictionless air track. It is permanently attached to one end of a massless horizontal spring, which has a force constant of 10.0 N/m both for extension and for compression. The other end of the spring is fixed. The glider is moved to compress the spring by 0.180 m and then released from rest. Calculate the speed of the glider (a) at the point where it has moved 0.180 m from its starting point, so that the spring is momentarily exerting no force and (b) at the point where it has moved 0.250 m from its starting point.

8. A loaded ore car has a mass of 950 kg and rolls on rails with negligible friction. It starts from rest and is pulled up a mine shaft by a cable connected to a winch. The shaft is inclined at 30.0° above the horizontal. The car accelerates uniformly to a speed of 2.20 m/s in 12.0 s and then continues at constant speed. (a) What power must the winch motor provide when the car is moving at constant speed? (b) What maximum power must the winch motor provide? (c) What total energy transfers out of the motor by work by the time the car moves off the end of the track, which is of length 1 250 m?

9. A simple pendulum, which we will consider in detail in Chapter 15, consists of an object suspended by a string. The object is assumed to be a particle. The string, with its top end fixed, has negligible mass and does not stretch. In the absence of air friction, the system oscillates by swinging back and forth in a vertical plane. If the string is 2.00 m long and makes an initial angle of 30.0° with the vertical, calculate the speed of the particle (a) at the lowest point in its trajectory and (b) when the angle is 15.0°.

10. An object of mass m starts from rest and slides a distance d down a frictionless incline of angle θ. While sliding, it contacts an unstressed spring of negligible mass as shown in Figure P8.10. The object slides an additional distance x as it is brought momentarily to rest by compression of the spring (of force constant k). Find the initial separation d between object and spring.

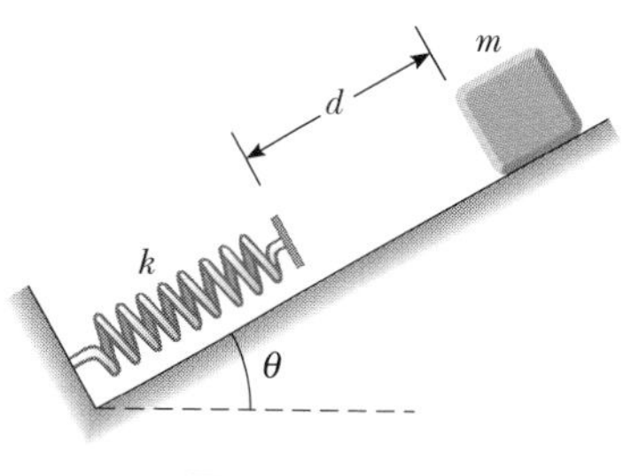

Figure P8.10

11. A block of mass 0.250 kg is placed on top of a light vertical spring of force constant 5 000 N/m and pushed downward so that the spring is compressed by 0.100 m. After the block is released from rest, it travels upward and then leaves the spring. To what maximum height above the point of release does it rise?

12. A circus trapeze consists of a bar suspended by two parallel ropes, each of length ℓ, allowing performers to swing in a vertical circular arc (Figure P8.12). Suppose a performer with mass m holds the bar and steps off an elevated platform, starting from rest with the ropes at an angle θ_i with respect to the vertical. Suppose the size of the performer's body is small compared to the length ℓ, that she does not pump the trapeze to swing higher, and that air resistance is negligible. (a) Show that when the ropes make an angle θ with the vertical, the performer must exert a force

$$mg(3\cos\theta - 2\cos\theta_i)$$

in order to hang on. (b) Determine the angle θ_i for which

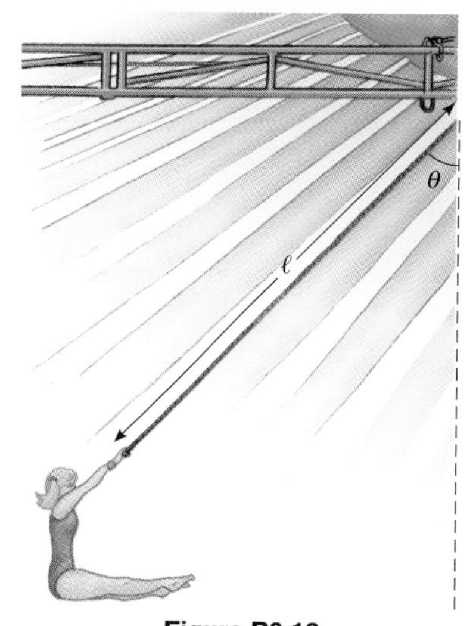

Figure P8.12

the force needed to hang on at the bottom of the swing is twice the performer's weight.

13. Two objects are connected by a light string passing over a light frictionless pulley as shown in Figure P8.13. The object of mass 5.00 kg is released from rest. Using the principle of conservation of energy, (a) determine the speed of the 3.00-kg object just as the 5.00-kg object hits the ground. (b) Find the maximum height to which the 3.00-kg object rises.

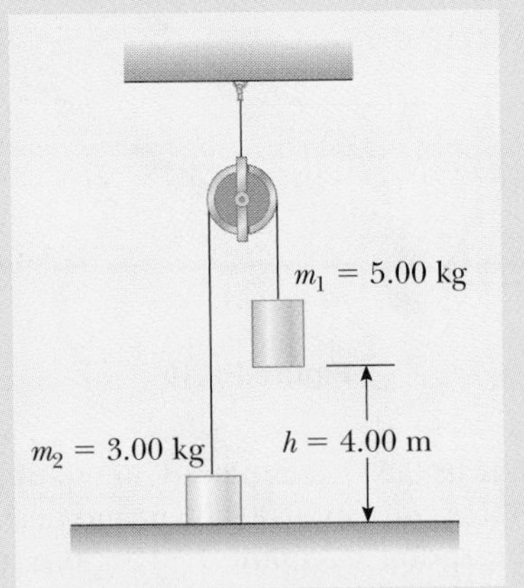

Figure P8.13 Problems 13 and 14.

14. Two objects are connected by a light string passing over a light frictionless pulley as in Figure P8.13. The object of mass m_1 is released from rest at height h. Using the principle of conservation of energy, (a) determine the speed of m_2 just as m_1 hits the ground. (b) Find the maximum height to which m_2 rises.

15. A light rigid rod is 77.0 cm long. Its top end is pivoted on a low-friction horizontal axle. The rod hangs straight down at rest with a small massive ball attached to its bottom end. You strike the ball, suddenly giving it a horizontal velocity so that it swings around in a full circle. What minimum speed at the bottom is required to make the ball go over the top of the circle?

16. Air moving at 11.0 m/s in a steady wind encounters a windmill of diameter 2.30 m and having an efficiency of 27.5%. The energy generated by the windmill is used to pump water from a well 35.0 m deep into a tank 2.30 m above the ground. At what rate in liters per minute can water be pumped into the tank?

17. A 20.0-kg cannon ball is fired from a cannon with muzzle speed of 1 000 m/s at an angle of 37.0° with the horizontal. A second ball is fired at an angle of 90.0°. Use the conservation of energy principle to find (a) the maximum height reached by each ball and (b) the total mechanical energy at the maximum height for each ball. Let $y = 0$ at the cannon.

18. A 2.00-kg ball is attached to the bottom end of a length of fishline with a breaking strength of 10 lb (44.5 N). The top end of the fishline is held stationary. The ball is released from rest with the line taut and horizontal ($\theta = 90.0°$). At what angle θ (measured from the vertical) will the fishline break?

19. A daredevil plans to bungee-jump from a balloon 65.0 m above a carnival midway (Figure P8.19). He will use a uniform elastic cord, tied to a harness around his body, to stop his fall at a point 10.0 m above the ground. Model his body as a particle and the cord as having negligible mass and obeying Hooke's force law. In a preliminary test, hanging at rest from a 5.00-m length of the cord, he finds that his body weight stretches it by 1.50 m. He will drop from rest at the point where the top end of a longer section of the cord is attached to the stationary balloon. (a) What length of cord should he use? (b) What maximum acceleration will he experience?

Figure P8.19

20. **Review problem.** The system shown in Figure P8.20 consists of a light inextensible cord, light frictionless pulleys, and blocks of equal mass. It is initially held at rest so that the blocks are at the same height above the ground. The blocks are then released. Find the speed of block A at the moment when the vertical separation of the blocks is h.

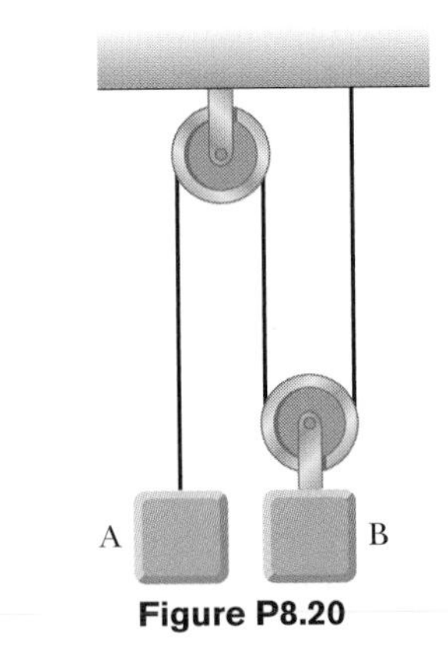

Figure P8.20

Section 8.3 Conservative and Nonconservative Forces

21. A 4.00-kg particle moves from the origin to position C, having coordinates $x = 5.00$ m and $y = 5.00$ m. One force on the particle is the gravitational force acting in the negative y direction (Fig. P8.21). Using Equation 7.3, calculate the

work done by the gravitational force in going from O to C along (a) OAC. (b) OBC. (c) OC. Your results should all be identical. Why?

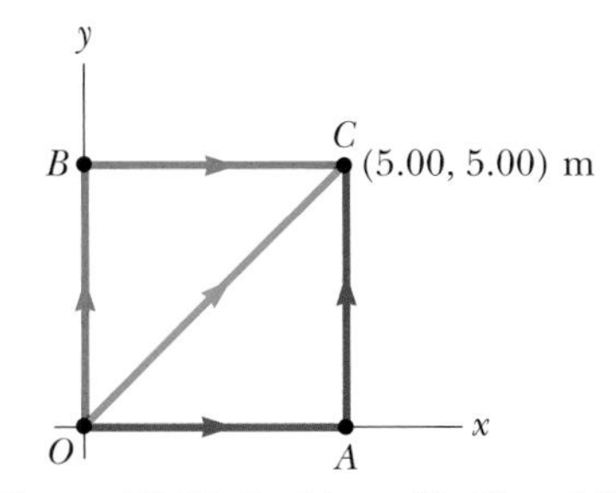

Figure P8.21 Problems 21, 22 and 23.

22. (a) Suppose that a constant force acts on an object. The force does not vary with time, nor with the position or the velocity of the object. Start with the general definition for work done by a force

$$W = \int_i^f \mathbf{F} \cdot d\mathbf{r}$$

and show that the force is conservative. (b) As a special case, suppose that the force $\mathbf{F} = (3\hat{\mathbf{i}} + 4\hat{\mathbf{j}})$ N acts on a particle that moves from O to C in Figure P8.21. Calculate the work done by $\mathbf{F}$ if the particle moves along each one of the three paths OAC, OBC, and OC. (Your three answers should be identical.)

23. A force acting on a particle moving in the xy plane is given by $\mathbf{F} = (2y\hat{\mathbf{i}} + x^2\hat{\mathbf{j}})$ N, where x and y are in meters. The particle moves from the origin to a final position having coordinates $x = 5.00$ m and $y = 5.00$ m, as in Figure P8.21. Calculate the work done by $\mathbf{F}$ along (a) OAC, (b) OBC, (c) OC. (d) Is $\mathbf{F}$ conservative or nonconservative? Explain.

24. A particle of mass $m = 5.00$ kg is released from point Ⓐ and slides on the frictionless track shown in Figure P8.24. Determine (a) the particle's speed at points Ⓑ and Ⓒ and (b) the net work done by the gravitational force in moving the particle from Ⓐ to Ⓒ.

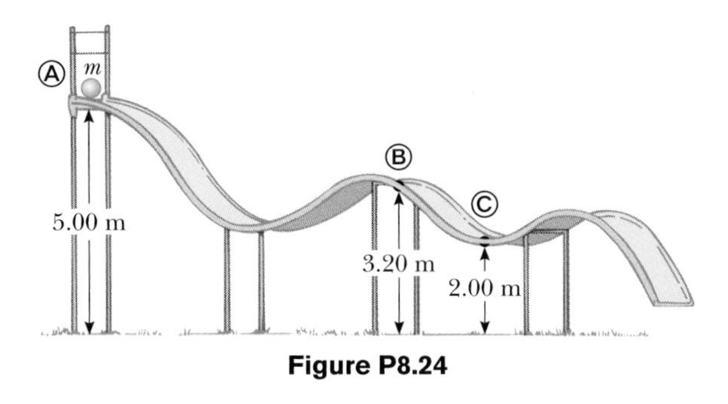

Figure P8.24

25. A single constant force $\mathbf{F} = (3\hat{\mathbf{i}} + 5\hat{\mathbf{j}})$ N acts on a 4.00-kg particle. (a) Calculate the work done by this force if the particle moves from the origin to the point having the vector position $\mathbf{r} = (2\hat{\mathbf{i}} - 3\hat{\mathbf{j}})$ m. Does this result depend on the path? Explain. (b) What is the speed of the particle at $\mathbf{r}$ if its speed at the origin is 4.00 m/s? (c) What is the change in the potential energy?

Section 8.4 Changes in Mechanical Energy for Nonconservative Forces

26. At time t_i, the kinetic energy of a particle is 30.0 J and the potential energy of the system to which it belongs is 10.0 J. At some later time t_f, the kinetic energy of the particle is 18.0 J. (a) If only conservative forces act on the particle, what are the potential energy and the total energy at time t_f? (b) If the potential energy of the system at time t_f is 5.00 J, are there any nonconservative forces acting on the particle? Explain.

27. In her hand a softball pitcher swings a ball of mass 0.250 kg around a vertical circular path of radius 60.0 cm before releasing it from her hand. The pitcher maintains a component of force on the ball of constant magnitude 30.0 N in the direction of motion around the complete path. The speed of the ball at the top of the circle is 15.0 m/s. If she releases the ball at the bottom of the circle, what is its speed upon release?

28. An electric scooter has a battery capable of supplying 120 Wh of energy. If friction forces and other losses account for 60.0% of the energy usage, what altitude change can a rider achieve when driving in hilly terrain, if the rider and scooter have a combined weight of 890 N?

29. The world's biggest locomotive is the MK5000C, a behemoth of mass 160 metric tons driven by the most powerful engine ever used for rail transportation, a Caterpillar diesel capable of 5 000 hp. Such a huge machine can provide a gain in efficiency, but its large mass presents challenges as well. The engineer finds that the locomotive handles differently from conventional units, notably in braking and climbing hills. Consider the locomotive pulling no train, but traveling at 27.0 m/s on a level track while operating with output power 1 000 hp. It comes to a 5.00% grade (a slope that rises 5.00 m for every 100 m along the track). If the throttle is not advanced, so that the power level is held steady, to what value will the speed drop? Assume that friction forces do not depend on the speed.

30. A 70.0-kg diver steps off a 10.0-m tower and drops straight down into the water. If he comes to rest 5.00 m beneath the surface of the water, determine the average resistance force exerted by the water on the diver.

31. The coefficient of friction between the 3.00-kg block and the surface in Figure P8.31 is 0.400. The system starts from rest. What is the speed of the 5.00-kg ball when it has fallen 1.50 m?

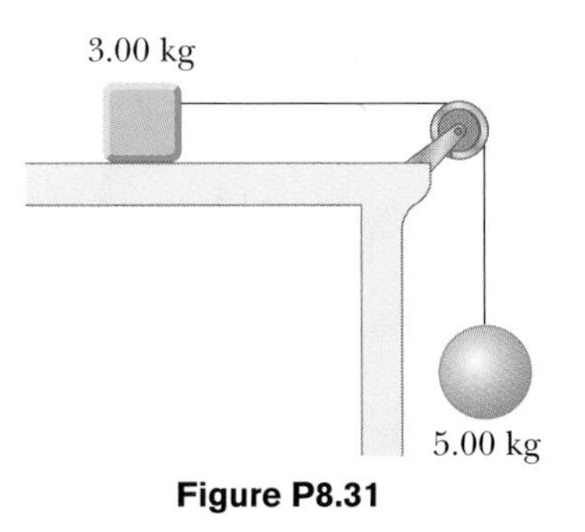

Figure P8.31

32. A boy in a wheelchair (total mass 47.0 kg) wins a race with a skateboarder. The boy has speed 1.40 m/s at the crest of a slope 2.60 m high and 12.4 m long. At the bottom of the slope his speed is 6.20 m/s. If air resistance and rolling resistance can be modeled as a constant friction force of 41.0 N, find the work he did in pushing forward on his wheels during the downhill ride.

33. A 5.00-kg block is set into motion up an inclined plane with an initial speed of 8.00 m/s (Fig. P8.33). The block comes to rest after traveling 3.00 m along the plane, which is inclined at an angle of 30.0° to the horizontal. For this motion determine (a) the change in the block's kinetic energy, (b) the change in the potential energy of the block–Earth system, and (c) the friction force exerted on the block (assumed to be constant). (d) What is the coefficient of kinetic friction?

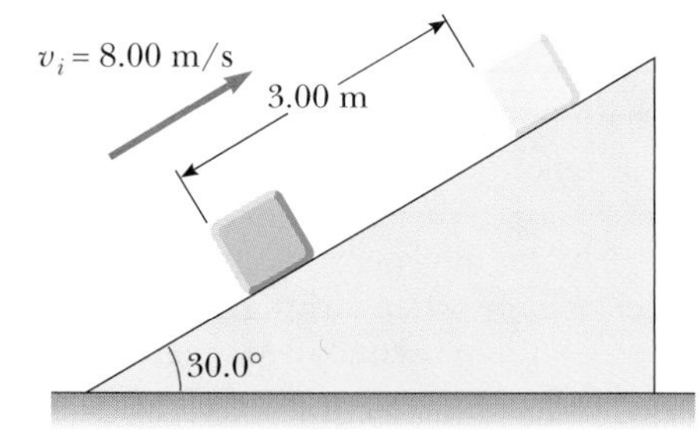

Figure P8.33

34. An 80.0-kg skydiver jumps out of a balloon at an altitude of 1 000 m and opens the parachute at an altitude of 200 m. (a) Assuming that the total retarding force on the diver is constant at 50.0 N with the parachute closed and constant at 3 600 N with the parachute open, what is the speed of the diver when he lands on the ground? (b) Do you think the skydiver will be injured? Explain. (c) At what height should the parachute be opened so that the final speed of the skydiver when he hits the ground is 5.00 m/s? (d) How realistic is the assumption that the total retarding force is constant? Explain.

35. A toy cannon uses a spring to project a 5.30-g soft rubber ball. The spring is originally compressed by 5.00 cm and has a force constant of 8.00 N/m. When the cannon is fired, the ball moves 15.0 cm through the horizontal barrel of the cannon, and there is a constant friction force of 0.032 0 N between the barrel and the ball. (a) With what speed does the projectile leave the barrel of the cannon? (b) At what point does the ball have maximum speed? (c) What is this maximum speed?

36. A 50.0-kg block and a 100-kg block are connected by a string as in Figure P8.36. The pulley is frictionless and of negligible mass. The coefficient of kinetic friction between the 50.0 kg block and incline is 0.250. Determine the change in the kinetic energy of the 50.0-kg block as it moves from Ⓐ to Ⓑ, a distance of 20.0 m.

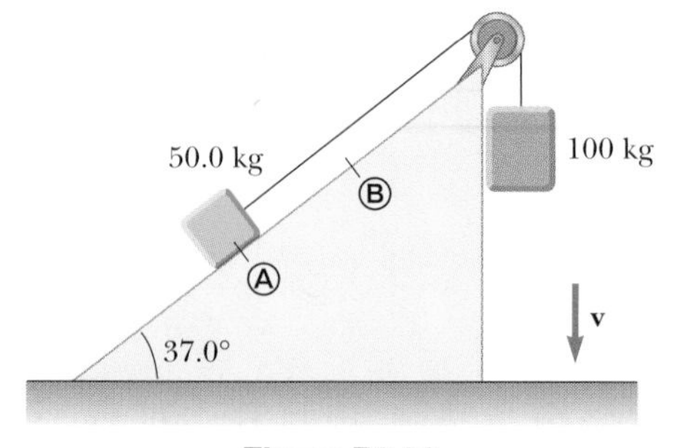

Figure P8.36

37. A 1.50-kg object is held 1.20 m above a relaxed massless vertical spring with a force constant of 320 N/m. The object is dropped onto the spring. (a) How far does it compress the spring? (b) **What If?** How far does it compress the spring if the same experiment is performed on the Moon, where $g = 1.63$ m/s^2? (c) **What If?** Repeat part (a), but this time assume a constant air-resistance force of 0.700 N acts on the object during its motion.

38. A 75.0-kg skysurfer is falling straight down with terminal speed 60.0 m/s. Determine the rate at which the skysurfer–Earth system is losing mechanical energy.

39. A uniform board of length L is sliding along a smooth (frictionless) horizontal plane as in Figure P8.39a. The board then slides across the boundary with a rough horizontal surface. The coefficient of kinetic friction between the board and the second surface is μ_k. (a) Find the acceleration of the board at the moment its front end has traveled a distance x beyond the boundary. (b) The board stops at the moment its back end reaches the boundary, as in Figure P8.39b. Find the initial speed v of the board.

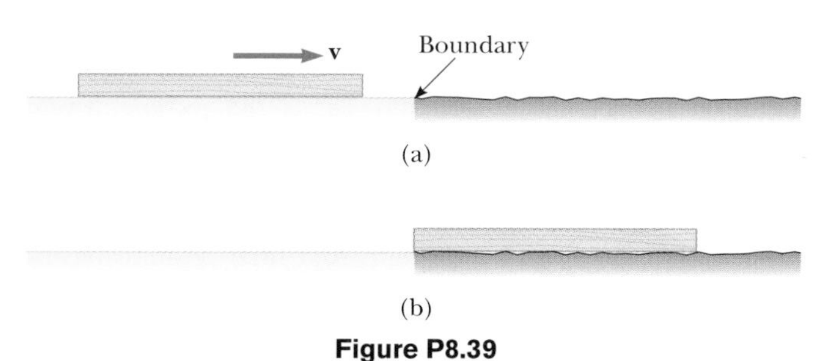

Figure P8.39

Section 8.5 Relationship Between Conservative Forces and Potential Energy

40. A single conservative force acting on a particle varies as $\mathbf{F} = (-Ax + Bx^2)\hat{\mathbf{i}}$ N, where A and B are constants and x is in meters. (a) Calculate the potential-energy function $U(x)$ associated with this force, taking $U = 0$ at $x = 0$. (b) Find the change in potential energy and the change in kinetic energy as the particle moves from $x = 2.00$ m to $x = 3.00$ m.

41. A single conservative force acts on a 5.00-kg particle. The equation $F_x = (2x + 4)$ N describes the force, where x is in meters. As the particle moves along the x axis from $x = 1.00$ m to $x = 5.00$ m, calculate (a) the work done by this force, (b) the change in the potential energy of the system, and (c) the kinetic energy of the particle at $x = 5.00$ m if its speed is 3.00 m/s at $x = 1.00$ m.

42. A potential-energy function for a two-dimensional force is of the form $U = 3x^3y - 7x$. Find the force that acts at the point (x, y).

43. The potential energy of a system of two particles separated by a distance r is given by $U(r) = A/r$, where A is a constant. Find the radial force $\mathbf{F}_r$ that each particle exerts on the other.

Section 8.6 Energy Diagrams and Equilibrium of a System

44. A right circular cone can be balanced on a horizontal surface in three different ways. Sketch these three equilibrium configurations, and identify them as positions of stable, unstable, or neutral equilibrium.

45. For the potential energy curve shown in Figure P8.45, (a) determine whether the force F_x is positive, negative, or zero at the five points indicated. (b) Indicate points of stable, unstable, and neutral equilibrium. (c) Sketch the curve for F_x versus x from $x = 0$ to $x = 9.5$ m.

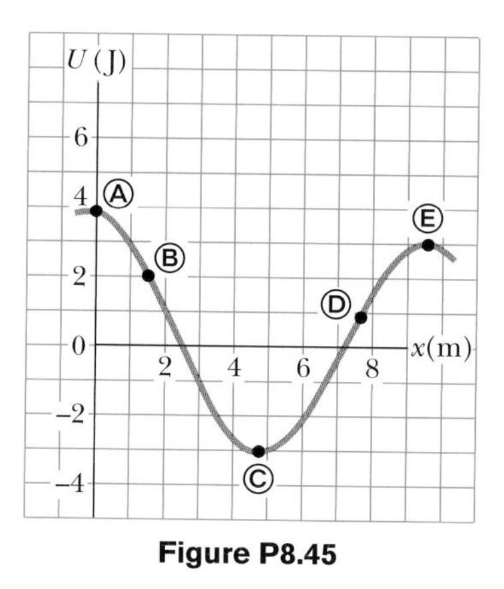

Figure P8.45

46. A particle moves along a line where the potential energy of its system depends on its position r as graphed in Figure P8.46. In the limit as r increases without bound, $U(r)$ approaches +1 J. (a) Identify each equilibrium position for this particle. Indicate whether each is a point of stable, unstable, or neutral equilibrium. (b) The particle will be bound if the total energy of the system is in what range? Now suppose that the system has energy −3 J. Determine (c) the range of positions where the particle can be found, (d) its maximum kinetic energy, (e) the location where it has maximum kinetic energy, and (f) the *binding energy* of the system—that is, the additional energy that it would have to be given in order for the particle to move out to $r \to \infty$.

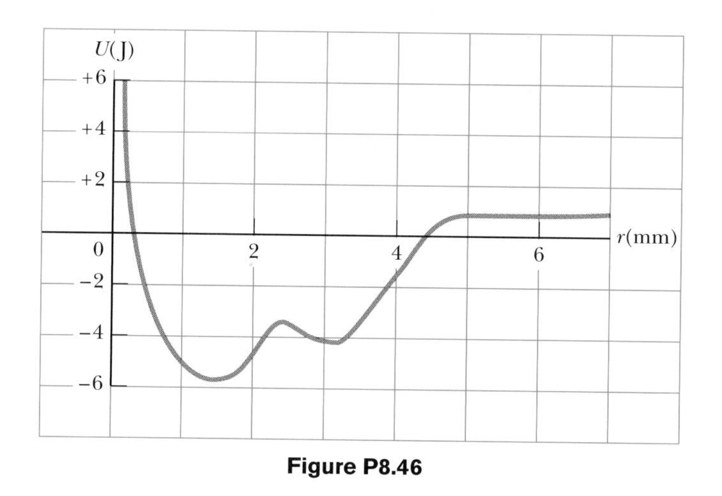

Figure P8.46

47. A particle of mass 1.18 kg is attached between two identical springs on a horizontal frictionless tabletop. The springs have force constant k and each is initially unstressed. (a) If the particle is pulled a distance x along a direction perpendicular to the initial configuration of the springs, as in Figure P8.47, show that the potential energy of the system is

$$U(x) = kx^2 + 2kL\left(L - \sqrt{x^2 + L^2}\right)$$

(*Hint*: See Problem 58 in Chapter 7.) (b) Make a plot of $U(x)$ versus x and identify all equilibrium points. Assume that $L = 1.20$ m and $k = 40.0$ N/m. (c) If the particle is pulled 0.500 m to the right and then released, what is its speed when it reaches the equilibrium point $x = 0$?

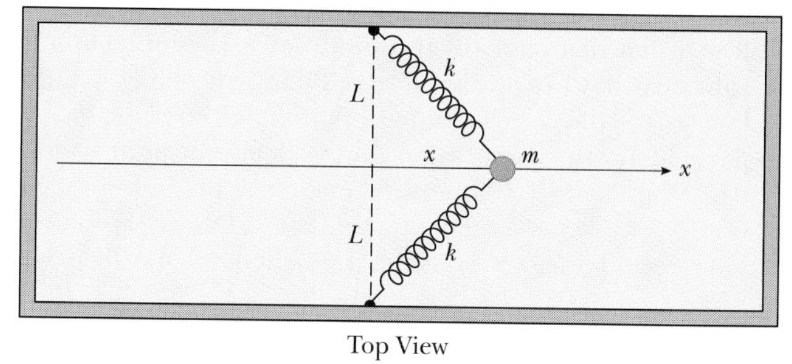

Figure P8.47

Additional Problems

48. A block slides down a curved frictionless track and then up an inclined plane as in Figure P8.48. The coefficient of kinetic friction between block and incline is μ_k. Use energy methods to show that the maximum height reached by the block is

$$y_{max} = \frac{h}{1 + \mu_k \cot \theta}$$

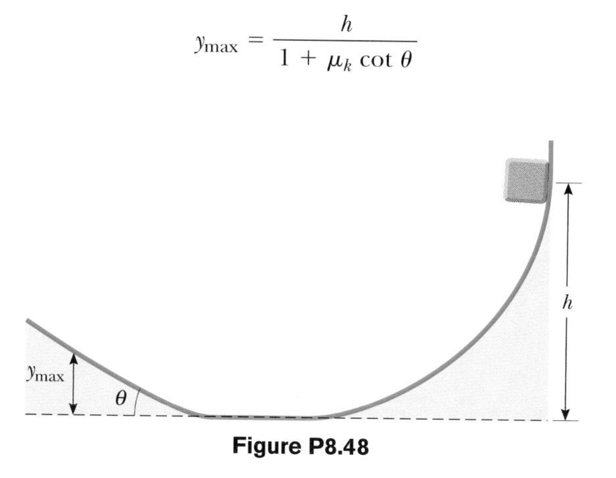

Figure P8.48

49. Make an order-of-magnitude estimate of your power output as you climb stairs. In your solution, state the physical quantities you take as data and the values you measure or estimate for them. Do you consider your peak power or your sustainable power?

50. Review problem. The mass of a car is 1 500 kg. The shape of the body is such that its aerodynamic drag coefficient is $D = 0.330$ and the frontal area is 2.50 m^2. Assuming that the drag force is proportional to v^2 and neglecting other sources of friction, calculate the power required to maintain a speed of 100 km/h as the car climbs a long hill sloping at 3.20°.

51. Assume that you attend a state university that started out as an agricultural college. Close to the center of the campus is a tall silo topped with a hemispherical cap. The cap is frictionless when wet. Someone has somehow balanced a pumpkin at the highest point. The line from the center of curvature of the cap to the pumpkin makes an angle $\theta_i = 0°$ with the vertical. While you happen to be standing nearby in the middle of a rainy night, a breath of wind makes the pumpkin start sliding downward from rest. It loses contact with the cap when the line from the center of the hemisphere to the pumpkin makes a certain angle with the vertical. What is this angle?

52. A 200-g particle is released from rest at point Ⓐ along the horizontal diameter on the inside of a frictionless, hemispherical bowl of radius $R = 30.0$ cm (Fig. P8.52). Calculate (a) the gravitational potential energy of the particle–Earth system when the particle is at point Ⓐ relative to point Ⓑ, (b) the kinetic energy of the particle at point Ⓑ, (c) its speed at point Ⓑ, and (d) its kinetic energy and the potential energy when the particle is at point Ⓒ.

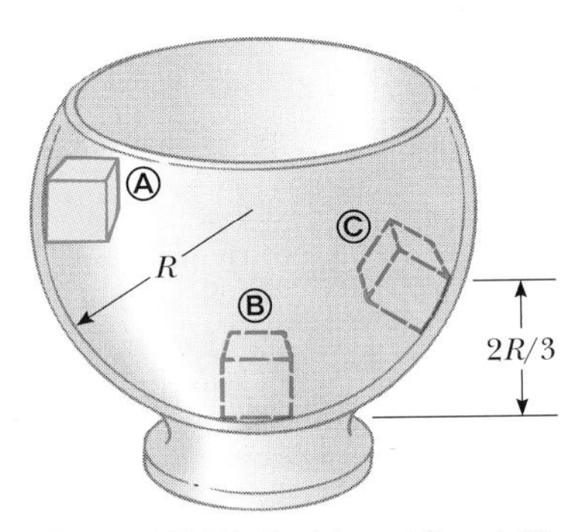

Figure P8.52 Problems 52 and 53.

53. **What If?** The particle described in Problem 52 (Fig. P8.52) is released from rest at Ⓐ, and the surface of the bowl is rough. The speed of the particle at Ⓑ is 1.50 m/s. (a) What is its kinetic energy at Ⓑ? (b) How much mechanical energy is transformed into internal energy as the particle moves from Ⓐ to Ⓑ? (c) Is it possible to determine the coefficient of friction from these results in any simple manner? Explain.

54. A 2.00-kg block situated on a rough incline is connected to a spring of negligible mass having a spring constant of 100 N/m (Fig. P8.54). The pulley is frictionless. The block is released from rest when the spring is unstretched. The block moves 20.0 cm down the incline before coming to rest. Find the coefficient of kinetic friction between block and incline.

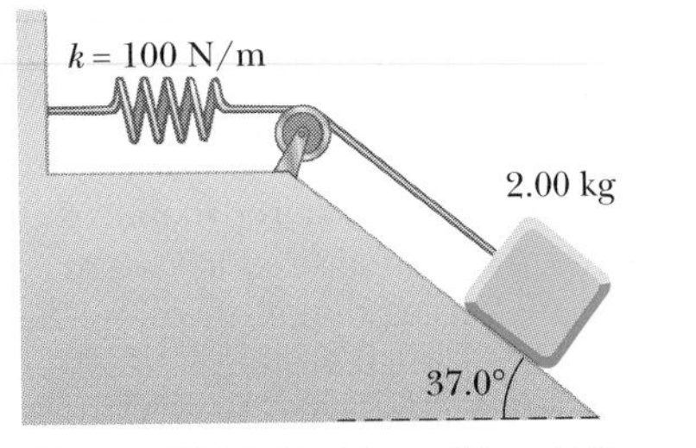

Figure P8.54 Problems 54 and 55.

55. **Review problem.** Suppose the incline is frictionless for the system described in Problem 54 (Fig. P8.54). The block is released from rest with the spring initially unstretched. (a) How far does it move down the incline before coming to rest? (b) What is its acceleration at its lowest point? Is the acceleration constant? (c) Describe the energy transformations that occur during the descent.

56. A child's pogo stick (Fig. P8.56) stores energy in a spring with a force constant of 2.50×10^4 N/m. At position Ⓐ ($x_A = -0.100$ m), the spring compression is a maximum and the child is momentarily at rest. At position Ⓑ ($x_B = 0$), the spring is relaxed and the child is moving upward. At position Ⓒ, the child is again momentarily at rest at the top of the jump. The combined mass of child and pogo stick is 25.0 kg. (a) Calculate the total energy of the child–stick–Earth system if both gravitational and elastic potential energies are zero for $x = 0$. (b) Determine x_C. (c) Calculate the speed of the child at $x = 0$. (d) Determine the value of x for which the kinetic energy of the system is a maximum. (e) Calculate the child's maximum upward speed.

Figure P8.56

57. A 10.0-kg block is released from point Ⓐ in Figure P8.57. The track is frictionless except for the portion between points Ⓑ and Ⓒ, which has a length of 6.00 m. The block travels down the track, hits a spring of force constant 2 250 N/m, and compresses the spring 0.300 m from its equilibrium position before coming to rest momentarily. Determine the coefficient of kinetic friction between the block and the rough surface between Ⓑ and Ⓒ.

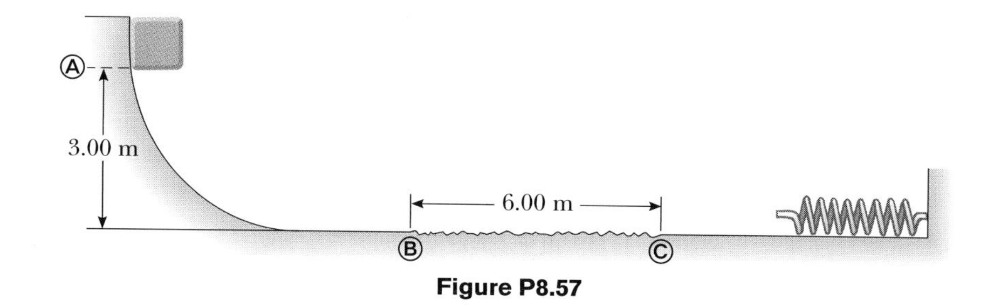

Figure P8.57

58. The potential energy function for a system is given by $U(x) = -x^3 + 2x^2 + 3x$. (a) Determine the force F_x as a function of x. (b) For what values of x is the force equal to zero? (c) Plot $U(x)$ versus x and F_x versus x, and indicate points of stable and unstable equilibrium.

59. A 20.0-kg block is connected to a 30.0-kg block by a string that passes over a light frictionless pulley. The 30.0-kg block is connected to a spring that has negligible mass and a force constant of 250 N/m, as shown in Figure P8.59. The spring is unstretched when the system is as shown in the figure, and the incline is frictionless. The 20.0-kg block is pulled 20.0 cm down the incline (so that the 30.0-kg block is 40.0 cm above the floor) and released from rest. Find the speed of each block when the 30.0-kg block is 20.0 cm above the floor (that is, when the spring is unstretched).

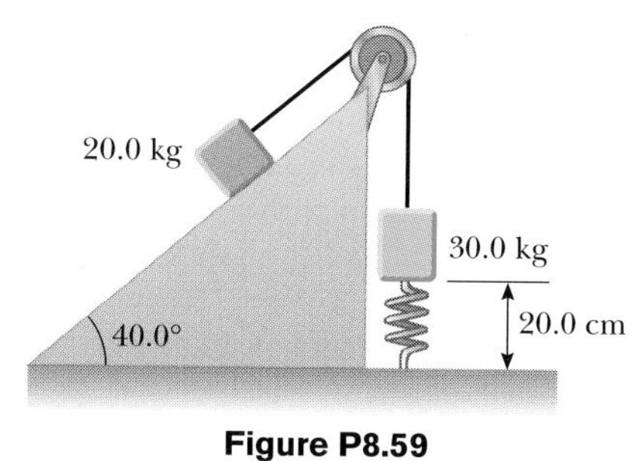

Figure P8.59

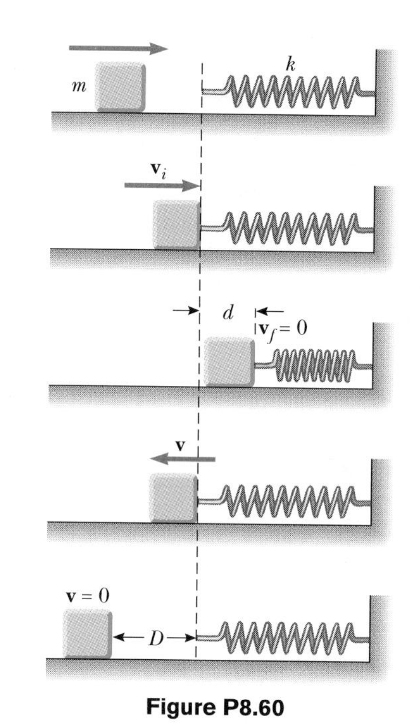

Figure P8.60

60. A 1.00-kg object slides to the right on a surface having a coefficient of kinetic friction 0.250 (Fig. P8.60). The object has a speed of $v_i = 3.00$ m/s when it makes contact with a light spring that has a force constant of 50.0 N/m. The object comes to rest after the spring has been compressed a distance d. The object is then forced toward the left by the spring and continues to move in that direction beyond the spring's unstretched position. Finally, the object comes to rest a distance D to the left of the unstretched spring. Find (a) the distance of compression d, (b) the speed v at the unstretched position when the object is moving to the left, and (c) the distance D where the object comes to rest.

61. A block of mass 0.500 kg is pushed against a horizontal spring of negligible mass until the spring is compressed a distance x (Fig. P8.61). The force constant of the spring is 450 N/m. When it is released, the block travels along a frictionless, horizontal surface to point B, the bottom of a vertical circular track of radius $R = 1.00$ m, and continues to move up the track. The speed of the block at the bottom of the track is $v_B = 12.0$ m/s, and the block experiences an average friction force of 7.00 N while sliding up the track. (a) What is x? (b) What speed do you predict for the block at the top of the track? (c) Does the block actually reach the top of the track, or does it fall off before reaching the top?

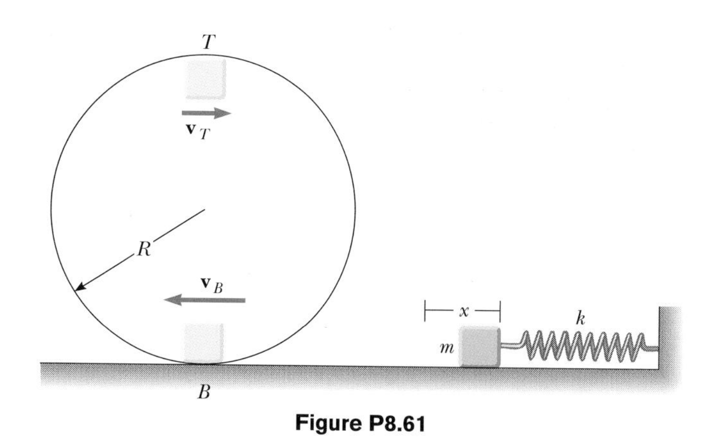

Figure P8.61

62. A uniform chain of length 8.00 m initially lies stretched out on a horizontal table. (a) If the coefficient of static friction between chain and table is 0.600, show that the chain will begin to slide off the table if at least 3.00 m of it hangs over the edge of the table. (b) Determine the speed of the chain

as all of it leaves the table, given that the coefficient of kinetic friction between the chain and the table is 0.400.

63. A child slides without friction from a height h along a curved water slide (Fig. P8.63). She is launched from a height $h/5$ into the pool. Determine her maximum airborne height y in terms of h and θ.

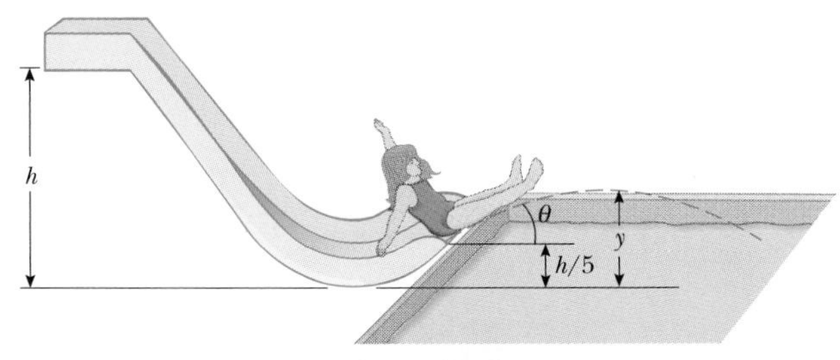

Figure P8.63

64. Refer to the situation described in Chapter 5, Problem 65. A 1.00-kg glider on a horizontal air track is pulled by a string at angle θ. The taut string runs over a light pulley at height $h_0 = 40.0$ cm above the line of motion of the glider. The other end of the string is attached to a hanging mass of 0.500 kg as in Fig. P5.65. (a) Show that the speed of the glider v_x and the speed of the hanging mass v_y are related by $v_y = v_x \cos \theta$. The glider is released from rest when $\theta = 30.0°$. Find (b) v_x and (c) v_y when $\theta = 45.0°$. (d) Explain why the answers to parts (b) and (c) to Chapter 5, Problem 65 do not help to solve parts (b) and (c) of this problem.

65. Jane, whose mass is 50.0 kg, needs to swing across a river (having width D) filled with man-eating crocodiles to save Tarzan from danger. She must swing into a wind exerting constant horizontal force **F**, on a vine having length L and initially making an angle θ with the vertical (Fig. P8.65). Taking $D = 50.0$ m, $F = 110$ N, $L = 40.0$ m, and $\theta = 50.0°$, (a) with what minimum speed must Jane begin her swing in order to just make it to the other side? (b) Once the rescue is complete, Tarzan and Jane must swing back across the river. With what minimum speed must they begin their swing? Assume that Tarzan has a mass of 80.0 kg.

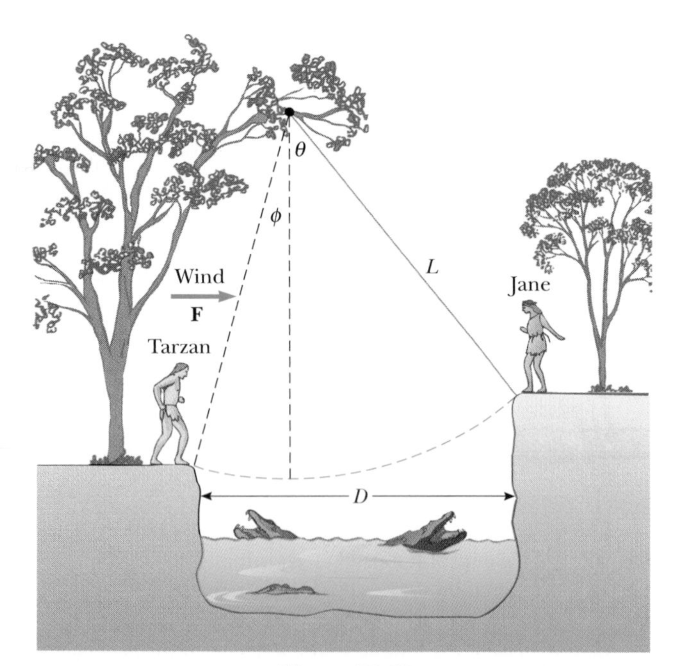

Figure P8.65

66. A 5.00-kg block free to move on a horizontal, frictionless surface is attached to one end of a light horizontal spring. The other end of the spring is held fixed. The spring is compressed 0.100 m from equilibrium and released. The speed of the block is 1.20 m/s when it passes the equilibrium position of the spring. The same experiment is now repeated with the frictionless surface replaced by a surface for which the coefficient of kinetic friction is 0.300. Determine the speed of the block at the equilibrium position of the spring.

67. A skateboarder with his board can be modeled as a particle of mass 76.0 kg, located at his center of mass (which we will study in Chapter 9). As in Figure P8.67, the skateboarder starts from rest in a crouching position at one lip of a half-pipe (point Ⓐ). The half-pipe is a dry water channel, forming one half of a cylinder of radius 6.80 m with its axis horizontal. On his descent, the skateboarder moves without friction so that his center of mass moves through one quarter of a circle of radius 6.30 m. (a) Find his speed at the bottom of the half-pipe (point Ⓑ). (b) Find his centripetal acceleration. (c) Find the normal force n_B acting on the skateboarder at point Ⓑ. Immediately after passing point Ⓑ, he stands up and raises his arms, lifting his center of mass from 0.500 m to 0.950 m above the concrete (point Ⓒ). To account for the conversion of chemical into mechanical energy, model his legs as doing work by pushing him vertically up, with a constant force equal to the normal force n_B, over a distance of 0.450 m. (You will be able to solve this problem with a more accurate model in Chapter 11.) (d) What is the work done on the skateboarder's body in this process? Next, the skateboarder glides upward with his center of mass moving in a quarter circle of radius 5.85 m. His body is horizontal when he passes point Ⓓ, the far lip of the half-pipe. (e) Find his speed at this location. At last he goes ballistic, twisting around while his center of mass moves vertically. (f) How high above point Ⓓ does he rise? (g) Over what time interval is he airborne before he touches down, 2.34 m below the level of point Ⓓ? [*Caution:* Do not try this yourself without the required skill and protective equipment, or in a drainage channel to which you do not have legal access.]

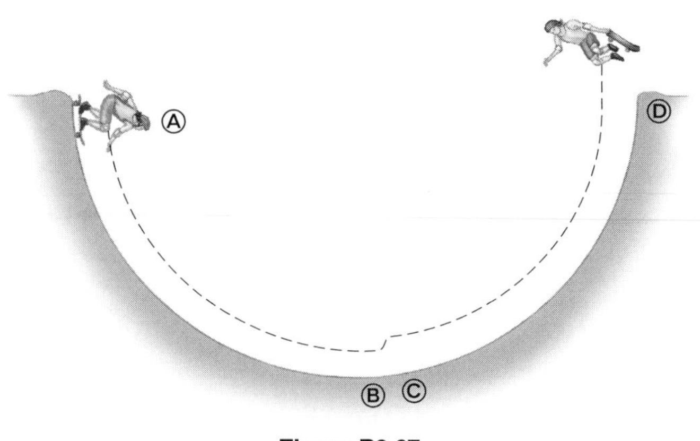

Figure P8.67

68. A block of mass M rests on a table. It is fastened to the lower end of a light vertical spring. The upper end of the spring is fastened to a block of mass m. The upper block is pushed down by an additional force $3mg$, so the spring compression is $4mg/k$. In this configuration the upper block is released from rest. The spring lifts the lower block off the table. In terms of m, what is the greatest possible value for M?

69. A ball having mass m is connected by a strong string of length L to a pivot point and held in place in a vertical position. A wind exerting constant force of magnitude F is blowing from left to right as in Figure P8.69a. (a) If the ball is released from rest, show that the maximum height H reached by the ball, as measured from its initial height, is

$$H = \frac{2L}{1 + (mg/F)^2}$$

Check that the above result is valid both for cases when $0 \leq H \leq L$ and for $L \leq H \leq 2L$. (b) Compute the value of H using the values $m = 2.00$ kg, $L = 2.00$ m, and $F = 14.7$ N. (c) Using these same values, determine the *equilibrium* height of the ball. (d) Could the equilibrium height ever be larger than L? Explain.

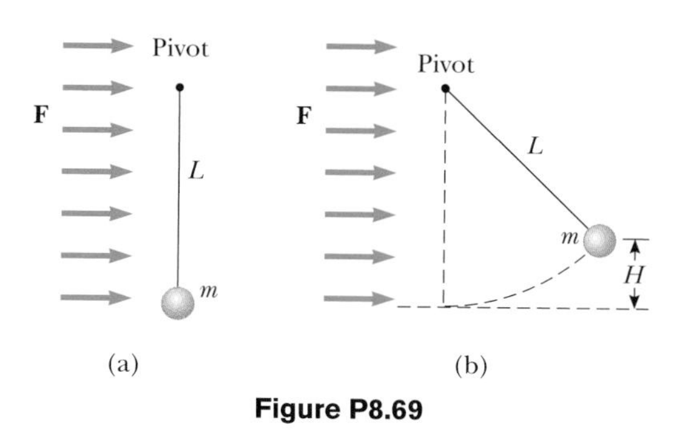

Figure P8.69

70. A ball is tied to one end of a string. The other end of the string is held fixed. The ball is set moving around a vertical circle without friction, and with speed $v_i = \sqrt{Rg}$ at the top of the circle, as in Figure P8.70. At what angle θ should the string be cut so that the ball will then travel through the center of the circle?

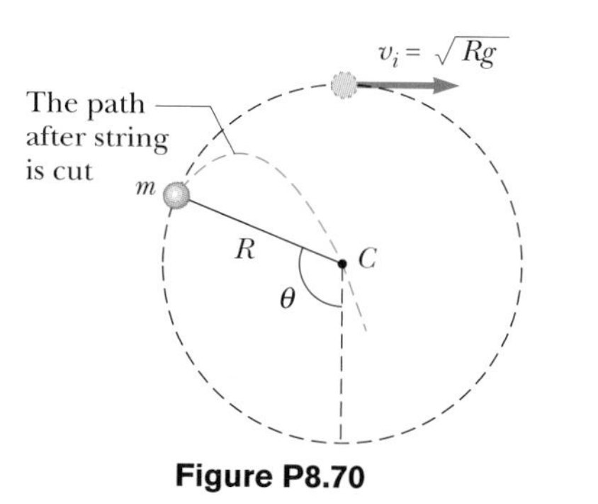

Figure P8.70

71. A ball whirls around in a vertical circle at the end of a string. If the total energy of the ball–Earth system remains constant, show that the tension in the string at the bottom is greater than the tension at the top by six times the weight of the ball.

72. A pendulum, comprising a light string of length L and a small sphere, swings in the vertical plane. The string hits a peg located a distance d below the point of suspension (Fig. P8.72). (a) Show that if the sphere is released from a height below that of the peg, it will return to this height after the string strikes the peg. (b) Show that if the pendulum is released from the horizontal position ($\theta = 90°$) and is to swing in a complete circle centered on the peg, then the minimum value of d must be $3L/5$.

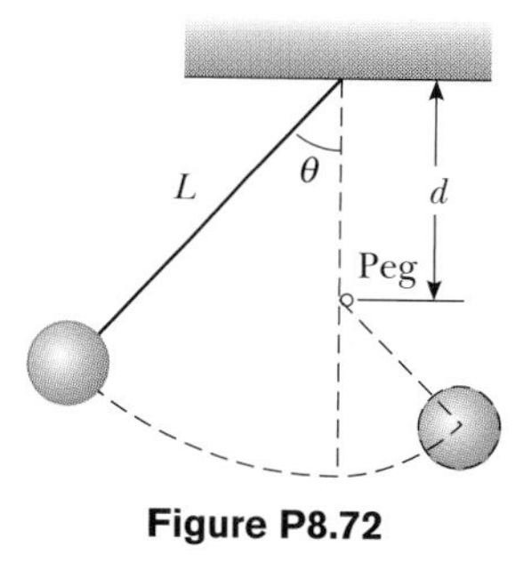

Figure P8.72

73. A roller-coaster car is released from rest at the top of the first rise and then moves freely with negligible friction. The roller coaster shown in Figure P8.73 has a circular loop of radius R in a vertical plane. (a) Suppose first that the car barely makes it around the loop: at the top of the loop the riders are upside down and feel weightless. Find the required height of the release point above the bottom of the loop in terms of R. (b) Now assume that the release point is at or above the minimum required height. Show that the normal force on the car at the bottom of the loop exceeds the normal force at the top of the loop by six times the weight of the car. The normal force on each rider follows the same rule. Such a large normal force is dangerous and very uncomfortable for the riders. Roller coasters are therefore not built with circular loops in vertical planes. Figure P6.20 and the photograph on page 157 show two actual designs.

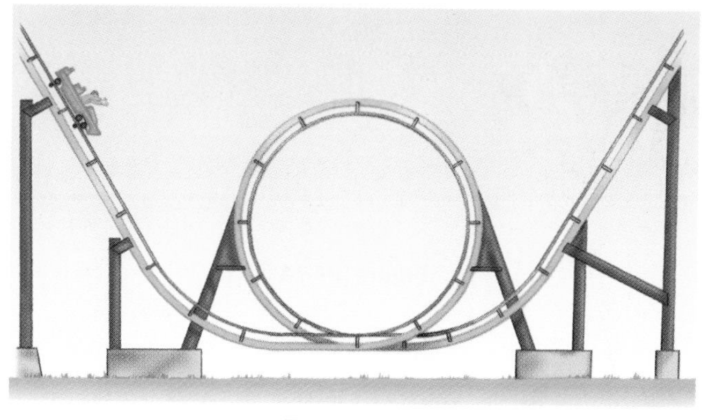

Figure P8.73

74. **Review problem.** In 1887 in Bridgeport, Connecticut, C. J. Belknap built the water slide shown in Figure P8.74. A rider on a small sled, of total mass 80.0 kg, pushed off to start at the top of the slide (point Ⓐ) with a speed of 2.50 m/s. The chute was 9.76 m high at the top, 54.3 m long, and 0.51 m wide. Along its length, 725 wheels made

friction negligible. Upon leaving the chute horizontally at its bottom end (point Ⓒ), the rider skimmed across the water of Long Island Sound for as much as 50 m, "skipping along like a flat pebble," before at last coming to rest and swimming ashore, pulling his sled after him. According to *Scientific American*, "The facial expression of novices taking their first adventurous slide is quite remarkable, and the sensations felt are correspondingly novel and peculiar." (a) Find the speed of the sled and rider at point Ⓒ. (b) Model the force of water friction as a constant retarding force acting on a particle. Find the work done by water friction in stopping the sled and rider. (c) Find the magnitude of the force the water exerts on the sled. (d) Find the magnitude of the force the chute exerts on the sled at point Ⓑ. (e) At point Ⓒ the chute is horizontal but curving in the vertical plane. Assume its radius of curvature is 20.0 m. Find the force the chute exerts on the sled at point Ⓒ.

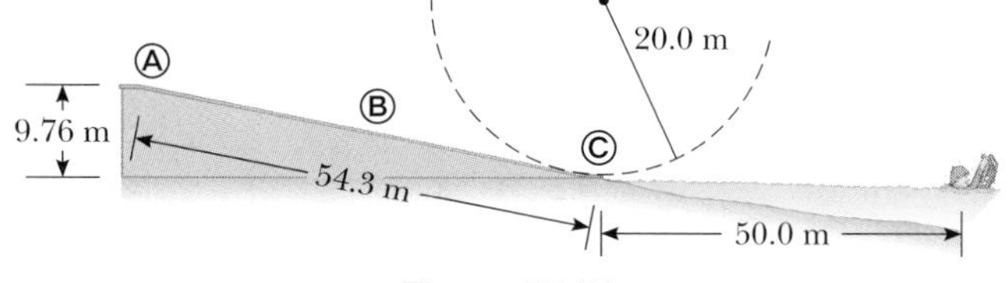

Figure P8.74

Answers to Quick Quizzes

8.1 (c). The sign of the gravitational potential energy depends on your choice of zero configuration. If the two objects in the system are closer together than in the zero configuration, the potential energy is negative. If they are farther apart, the potential energy is positive.

8.2 (c). The reason that we can ignore the kinetic energy of the massive Earth is that this kinetic energy is so small as to be essentially zero.

8.3 (a). We must include the Earth if we are going to work with gravitational potential energy.

8.4 (c). The total mechanical energy, kinetic plus potential, is conserved.

8.5 (a). The more massive rock has twice as much gravitational potential energy associated with it compared to the lighter rock. Because mechanical energy of an isolated system is conserved, the more massive rock will arrive at the ground with twice as much kinetic energy as the lighter rock.

8.6 $v_1 = v_2 = v_3$. The first and third balls speed up after they are thrown, while the second ball initially slows down but then speeds up after reaching its peak. The paths of all three balls are parabolas, and the balls take different times to reach the ground because they have different initial velocities. However, all three balls have the same speed at the moment they hit the ground because all start with the same kinetic energy and the ball–Earth system undergoes the same change in gravitational potential energy in all three cases.

8.7 (c). This system exhibits changes in kinetic energy as well as in both types of potential energy.

8.8 (a). Because the Earth is not included in the system, there is no gravitational potential energy associated with the system.

8.9 (c). The friction force must transform four times as much mechanical energy into internal energy if the speed is doubled, because kinetic energy depends on the square of the speed. Thus, the force must act over four times the distance.

8.10 (c). The decrease in mechanical energy of the system is $f_k d$, where d is the distance the block moves along the incline. While the force of kinetic friction remains the same, the distance d is smaller because a component of the gravitational force is pulling on the block in the direction opposite to its velocity.

8.11 (d). The slope of a $U(x)$-versus-x graph is by definition $dU(x)/dx$. From Equation 8.18, we see that this expression is equal to the negative of the x component of the conservative force acting on an object that is part of the system.

Chapter 9

Linear Momentum and Collisions

▲ *A moving bowling ball carries momentum, the topic of this chapter. In the collision between the ball and the pins, momentum is transferred to the pins. (Mark Cooper/Corbis Stock Market)*

CHAPTER OUTLINE

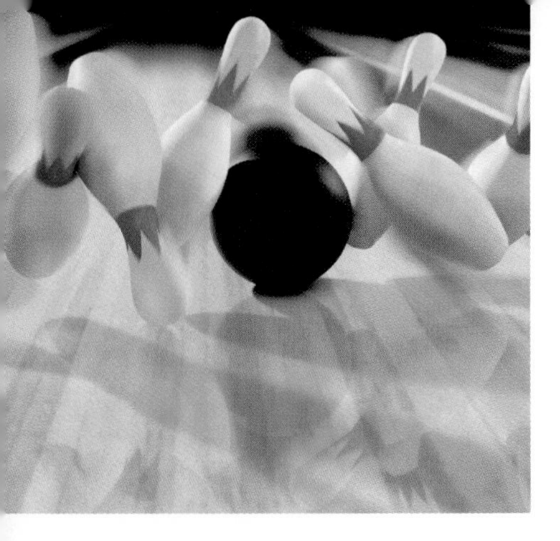

Consider what happens when a bowling ball strikes a pin, as in the opening photograph. The pin is given a large velocity as a result of the collision; consequently, it flies away and hits other pins or is projected toward the backstop. Because the average force exerted on the pin during the collision is large (resulting in a large acceleration), the pin achieves the large velocity very rapidly and experiences the force for a very short time interval. According to Newton's third law, the pin exerts a reaction force on the ball that is equal in magnitude and opposite in direction to the force exerted by the ball on the pin. This reaction force causes the ball to accelerate, but because the ball is so much more massive than the pin, the ball's acceleration is much less than the pin's acceleration.

Although F and a are large for the pin, they vary in time—a complicated situation! One of the main objectives of this chapter is to enable you to understand and analyze such events in a simple way. First, we introduce the concept of *momentum,* which is useful for describing objects in motion. Imagine that you have intercepted a football and see two players from the opposing team approaching you as you run with the ball. One of the players is the 180-lb quarterback who threw the ball; the other is a 300-lb lineman. Both of the players are running toward you at 5 m/s. However, because the two players have different masses, intuitively you know that you would rather collide with the quarterback than with the lineman. The momentum of an object is related to both its mass and its velocity. The concept of momentum leads us to a second conservation law, that of conservation of momentum. This law is especially useful for treating problems that involve collisions between objects and for analyzing rocket propulsion. In this chapter we also introduce the concept of the center of mass of a system of particles. We find that the motion of a system of particles can be described by the motion of one representative particle located at the center of mass.

9.1 Linear Momentum and Its Conservation

In the preceding two chapters we studied situations that are complex to analyze with Newton's laws. We were able to solve problems involving these situations by applying a conservation principle—conservation of energy. Consider another situation—a 60-kg archer stands on frictionless ice and fires a 0.50-kg arrow horizontally at 50 m/s. From Newton's third law, we know that the force that the bow exerts on the arrow will be matched by a force in the opposite direction on the bow (and the archer). This will cause the archer to begin to slide backward on the ice. But with what speed? We cannot answer this question directly using *either* Newton's second law or an energy approach—there is not enough information.

Despite our inability to solve the archer problem using our techniques learned so far, this is a very simple problem to solve if we introduce a new quantity that describes motion, *linear momentum.* Let us apply the General Problem-Solving Strategy and *conceptualize* an isolated system of two particles (Fig. 9.1) with masses m_1 and m_2 and moving with velocities $\mathbf{v}_1$ and $\mathbf{v}_2$ at an instant of time. Because the system is isolated, the only force on one particle is that from the other particle and we can *categorize* this as a situation in which Newton's laws will be useful. If a force from particle 1 (for example, a gravitational force) acts on particle 2, then there must be a second force—equal in magnitude but opposite in direction—that particle 2 exerts on particle 1. That is, they form a Newton's third law action–reaction pair, so that $\mathbf{F}_{12} = -\mathbf{F}_{21}$. We can express this condition as

$$\mathbf{F}_{21} + \mathbf{F}_{12} = 0$$

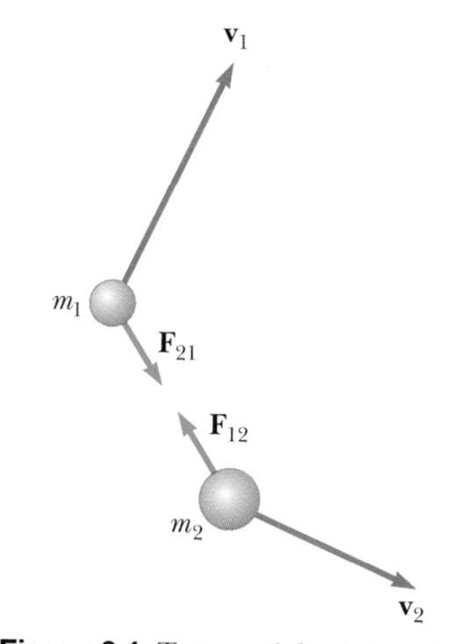

Figure 9.1 Two particles interact with each other. According to Newton's third law, we must have $\mathbf{F}_{12} = -\mathbf{F}_{21}$.

Let us further *analyze* this situation by incorporating Newton's second law. Over some time interval, the interacting particles in the system will accelerate. Thus, replacing each force with $m\mathbf{a}$ gives

$$m_1\mathbf{a}_1 + m_2\mathbf{a}_2 = 0$$

Now we replace the acceleration with its definition from Equation 4.5:

$$m_1\frac{d\mathbf{v}_1}{dt} + m_2\frac{d\mathbf{v}_2}{dt} = 0$$

If the masses m_1 and m_2 are constant, we can bring them into the derivatives, which gives

$$\frac{d(m_1\mathbf{v}_1)}{dt} + \frac{d(m_2\mathbf{v}_2)}{dt} = 0$$

$$\frac{d}{dt}(m_1\mathbf{v}_1 + m_2\mathbf{v}_2) = 0 \qquad (9.1)$$

To *finalize* this discussion, note that the derivative of the sum $m_1\mathbf{v}_1 + m_2\mathbf{v}_2$ with respect to time is zero. Consequently, this sum must be constant. We learn from this discussion that the quantity $m\mathbf{v}$ for a particle is important, in that the sum of these quantities for an isolated system is conserved. We call this quantity *linear momentum*:

Definition of linear momentum of a particle

The **linear momentum** of a particle or an object that can be modeled as a particle of mass m moving with a velocity $\mathbf{v}$ is defined to be the product of the mass and velocity:

$$\mathbf{p} \equiv m\mathbf{v} \qquad (9.2)$$

Linear momentum is a vector quantity because it equals the product of a scalar quantity m and a vector quantity $\mathbf{v}$. Its direction is along $\mathbf{v}$, it has dimensions ML/T, and its SI unit is kg · m/s.

If a particle is moving in an arbitrary direction, $\mathbf{p}$ must have three components, and Equation 9.2 is equivalent to the component equations

$$p_x = mv_x \qquad p_y = mv_y \qquad p_z = mv_z$$

As you can see from its definition, the concept of momentum[1] provides a quantitative distinction between heavy and light particles moving at the same velocity. For example, the momentum of a bowling ball moving at 10 m/s is much greater than that of a tennis ball moving at the same speed. Newton called the product $m\mathbf{v}$ *quantity of motion*; this is perhaps a more graphic description than our present-day word *momentum*, which comes from the Latin word for movement.

Using Newton's second law of motion, we can relate the linear momentum of a particle to the resultant force acting on the particle. We start with Newton's second law and substitute the definition of acceleration:

$$\sum\mathbf{F} = m\mathbf{a} = m\frac{d\mathbf{v}}{dt}$$

[1] In this chapter, the terms *momentum* and *linear momentum* have the same meaning. Later, in Chapter 11, we shall use the term *angular momentum* when dealing with rotational motion.

In Newton's second law, the mass m is assumed to be constant. Thus, we can bring m inside the derivative notation to give us

Newton's second law for a particle

$$\sum \mathbf{F} = \frac{d(m\mathbf{v})}{dt} = \frac{d\mathbf{p}}{dt} \tag{9.3}$$

This shows that **the time rate of change of the linear momentum of a particle is equal to the net force acting on the particle.**

This alternative form of Newton's second law is the form in which Newton presented the law and is actually more general than the form we introduced in Chapter 5. In addition to situations in which the velocity vector varies with time, we can use Equation 9.3 to study phenomena in which the mass changes. For example, the mass of a rocket changes as fuel is burned and ejected from the rocket. We cannot use $\Sigma \mathbf{F} = m\mathbf{a}$ to analyze rocket propulsion; we must use Equation 9.3, as we will show in Section 9.7.

The real value of Equation 9.3 as a tool for analysis, however, arises if we apply it to a *system* of two or more particles. As we have seen, this leads to a law of conservation of momentum for an isolated system. Just as the law of conservation of energy is useful in solving complex motion problems, the law of conservation of momentum can greatly simplify the analysis of other types of complicated motion.

Quick Quiz 9.1 Two objects have equal kinetic energies. How do the magnitudes of their momenta compare? (a) $p_1 < p_2$ (b) $p_1 = p_2$ (c) $p_1 > p_2$ (d) not enough information to tell.

Quick Quiz 9.2 Your physical education teacher throws a baseball to you at a certain speed, and you catch it. The teacher is next going to throw you a medicine ball whose mass is ten times the mass of the baseball. You are given the following choices: You can have the medicine ball thrown with (a) the same speed as the baseball (b) the same momentum (c) the same kinetic energy. Rank these choices from easiest to hardest to catch.

Using the definition of momentum, Equation 9.1 can be written

$$\frac{d}{dt}(\mathbf{p}_1 + \mathbf{p}_2) = 0$$

Because the time derivative of the total momentum $\mathbf{p}_{\text{tot}} = \mathbf{p}_1 + \mathbf{p}_2$ is *zero*, we conclude that the *total* momentum of the system must remain constant:

$$\mathbf{p}_{\text{tot}} = \mathbf{p}_1 + \mathbf{p}_2 = \text{constant} \tag{9.4}$$

or, equivalently,

$$\mathbf{p}_{1i} + \mathbf{p}_{2i} = \mathbf{p}_{1f} + \mathbf{p}_{2f} \tag{9.5}$$

where $\mathbf{p}_{1i}$ and $\mathbf{p}_{2i}$ are the initial values and $\mathbf{p}_{1f}$ and $\mathbf{p}_{2f}$ the final values of the momenta for the two particles for the time interval during which the particles interact. Equation 9.5 in component form demonstrates that the total momenta in the x, y, and z directions are all independently conserved:

$$p_{ix} = p_{fx} \qquad p_{iy} = p_{fy} \qquad p_{iz} = p_{fz} \tag{9.6}$$

This result, known as the **law of conservation of linear momentum**, can be extended to any number of particles in an isolated system. It is considered one of the most important laws of mechanics. We can state it as follows:

PITFALL PREVENTION

9.1 Momentum of a *System* is Conserved

Remember that the momentum of an isolated *system* is conserved. The momentum of one particle within an isolated system is not necessarily conserved, because other particles in the system may be interacting with it. Always apply conservation of momentum to an isolated *system*.

Whenever two or more particles in an isolated system interact, the total momentum of the system remains constant.

Conservation of momentum

This law tells us that **the total momentum of an isolated system at all times equals its initial momentum.**

Notice that we have made no statement concerning the nature of the forces acting on the particles of the system. The only requirement is that the forces must be *internal* to the system.

Quick Quiz 9.3 A ball is released and falls toward the ground with no air resistance. The isolated system for which momentum is conserved is (a) the ball (b) the Earth (c) the ball and the Earth (d) impossible to determine.

Quick Quiz 9.4 A car and a large truck traveling at the same speed make a head-on collision and stick together. Which vehicle experiences the larger change in the magnitude of momentum? (a) the car (b) the truck (c) The change in the magnitude of momentum is the same for both. (d) impossible to determine.

Example 9.1 The Archer — Interactive

Let us consider the situation proposed at the beginning of this section. A 60-kg archer stands at rest on frictionless ice and fires a 0.50-kg arrow horizontally at 50 m/s (Fig. 9.2). With what velocity does the archer move across the ice after firing the arrow?

Solution We *cannot* solve this problem using Newton's second law, $\Sigma\mathbf{F} = m\mathbf{a}$, because we have no information about the force on the arrow or its acceleration. We *cannot* solve this problem using an energy approach because we do not know how much work is done in pulling the bow back or how much potential energy is stored in the bow. However, we *can* solve this problem very easily with conservation of momentum.

Let us take the system to consist of the archer (including the bow) and the arrow. The system is not isolated because the gravitational force and the normal force act on the system. However, these forces are vertical and perpendicular to the motion of the system. Therefore, there are no external forces in the horizontal direction, and we can consider the system to be isolated in terms of momentum components in this direction.

The total horizontal momentum of the system before the arrow is fired is zero ($m_1\mathbf{v}_{1i} + m_2\mathbf{v}_{2i} = 0$), where the archer is particle 1 and the arrow is particle 2. Therefore, the total horizontal momentum after the arrow is fired must be zero; that is,

$$m_1\mathbf{v}_{1f} + m_2\mathbf{v}_{2f} = 0$$

We choose the direction of firing of the arrow as the positive x direction. With $m_1 = 60$ kg, $m_2 = 0.50$ kg, and $\mathbf{v}_{2f} = 50\hat{\mathbf{i}}$ m/s, solving for $\mathbf{v}_{1f}$, we find the recoil velocity of the archer to be

$$\mathbf{v}_{1f} = -\frac{m_2}{m_1}\mathbf{v}_{2f} = -\left(\frac{0.50\text{ kg}}{60\text{ kg}}\right)(50\hat{\mathbf{i}}\text{ m/s}) = \boxed{-0.42\hat{\mathbf{i}}\text{ m/s}}$$

The negative sign for $\mathbf{v}_{1f}$ indicates that the archer is moving to the left after the arrow is fired, in the direction opposite the direction of motion of the arrow, in accordance with Newton's third law. Because the archer is much more massive than the arrow, his acceleration and consequent velocity are much smaller than the acceleration and velocity of the arrow.

What If? **What if the arrow were shot in a direction that makes an angle θ with the horizontal? How will this change the recoil velocity of the archer?**

Answer The recoil velocity should decrease in magnitude because only a component of the velocity is in the x direction.

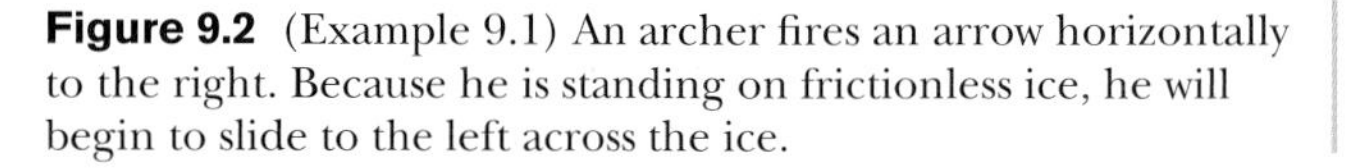

Figure 9.2 (Example 9.1) An archer fires an arrow horizontally to the right. Because he is standing on frictionless ice, he will begin to slide to the left across the ice.

If the arrow were shot straight up, for example, there would be no recoil at all—the archer would just be pressed down into the ice because of the firing of the arrow.

Only the x component of the momentum of the arrow should be used in a conservation of momentum statement, because momentum is only conserved in the x direction. In the y direction, the normal force from the ice and the gravitational force are external influences on the system. Conservation of momentum in the x direction gives us

$$m_1 v_{1f} + m_2 v_{2f} \cos\theta = 0$$

leading to

$$v_{1f} = -\frac{m_2}{m_1} v_{2f} \cos\theta$$

For $\theta = 0$, $\cos\theta = 1$ and this reduces to the value when the arrow is fired horizontally. For nonzero values of θ, the cosine function is less than 1 and the recoil velocity is less than the value calculated for $\theta = 0$. If $\theta = 90°$, $\cos\theta = 0$, and there is no recoil velocity v_{1f}, as we argued conceptually.

At the Interactive Worked Example link at **http://www.pse6.com,** *you can change the mass of the archer and the mass and speed of the arrow.*

Example 9.2 Breakup of a Kaon at Rest

One type of nuclear particle, called the *neutral kaon* (K^0), breaks up into a pair of other particles called *pions* (π^+ and π^-) that are oppositely charged but equal in mass, as illustrated in Figure 9.3. Assuming the kaon is initially at rest, prove that the two pions must have momenta that are equal in magnitude and opposite in direction.

Solution The breakup of the kaon can be written

$$K^0 \longrightarrow \pi^+ + \pi^-$$

If we let $\mathbf{p}^+$ be the final momentum of the positive pion and $\mathbf{p}^-$ the final momentum of the negative pion, the final momentum of the system consisting of the two pions can be written

$$\mathbf{p}_f = \mathbf{p}^+ + \mathbf{p}^-$$

Because the kaon is at rest before the breakup, we know that $\mathbf{p}_i = 0$. Because the momentum of the isolated system (the kaon before the breakup, the two pions afterward) is conserved, $\mathbf{p}_i = \mathbf{p}_f = 0$, so that $\mathbf{p}^+ + \mathbf{p}^- = 0$, or

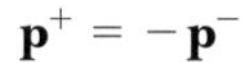

$$\mathbf{p}^+ = -\mathbf{p}^-$$

An important point to learn from this problem is that even though it deals with objects that are very different from those in the preceding example, the physics is identical: *linear momentum is conserved in an isolated system.*

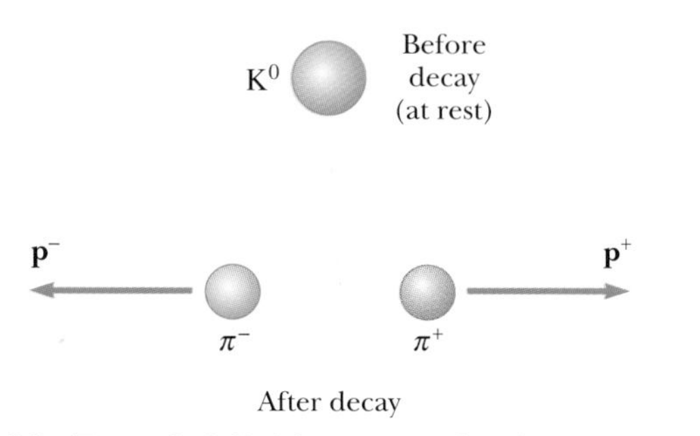

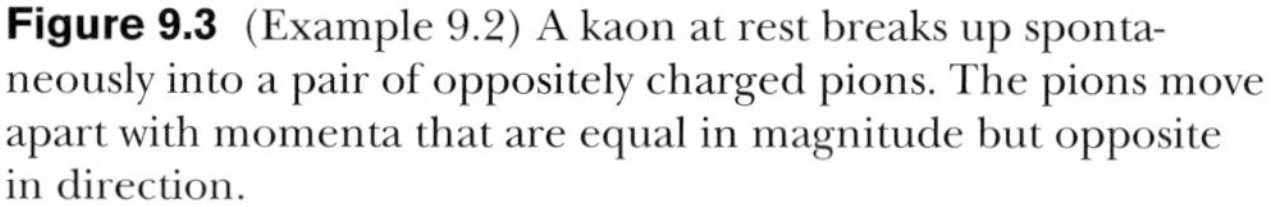

Figure 9.3 (Example 9.2) A kaon at rest breaks up spontaneously into a pair of oppositely charged pions. The pions move apart with momenta that are equal in magnitude but opposite in direction.

9.2 Impulse and Momentum

According to Equation 9.3, the momentum of a particle changes if a net force acts on the particle. Knowing the change in momentum caused by a force is useful in solving some types of problems. To build a better understanding of this important concept, let us assume that a single force $\mathbf{F}$ acts on a particle and that this force may vary with time. According to Newton's second law, $\mathbf{F} = d\mathbf{p}/dt$, or

$$d\mathbf{p} = \mathbf{F}dt \tag{9.7}$$

We can integrate[2] this expression to find the change in the momentum of a particle when the force acts over some time interval. If the momentum of the particle changes from $\mathbf{p}_i$ at time t_i to $\mathbf{p}_f$ at time t_f, integrating Equation 9.7 gives

[2] Note that here we are integrating force with respect to time. Compare this with our efforts in Chapter 7, where we integrated force with respect to position to find the work done by the force.

$$\Delta \mathbf{p} = \mathbf{p}_f - \mathbf{p}_i = \int_{t_i}^{t_f} \mathbf{F}\, dt \tag{9.8}$$

To evaluate the integral, we need to know how the force varies with time. The quantity on the right side of this equation is called the **impulse** of the force **F** acting on a particle over the time interval $\Delta t = t_f - t_i$. Impulse is a vector defined by

$$\mathbf{I} \equiv \int_{t_i}^{t_f} \mathbf{F}\, dt \tag{9.9}$$

Impulse of a force

Equation 9.8 is an important statement known as the **impulse–momentum theorem:**[3]

Impulse–momentum theorem

The impulse of the force **F** acting on a particle equals the change in the momentum of the particle.

This statement is equivalent to Newton's second law. From this definition, we see that impulse is a vector quantity having a magnitude equal to the area under the force–time curve, as described in Figure 9.4a. In this figure, it is assumed that the force varies in time in the general manner shown and is nonzero in the time interval $\Delta t = t_f - t_i$. The direction of the impulse vector is the same as the direction of the change in momentum. Impulse has the dimensions of momentum—that is, ML/T. Note that impulse is *not* a property of a particle; rather, it is a measure of the degree to which an external force changes the momentum of the particle. Therefore, when we say that an impulse is given to a particle, we mean that momentum is transferred from an external agent to that particle.

Because the force imparting an impulse can generally vary in time, it is convenient to define a time-averaged force

$$\overline{\mathbf{F}} \equiv \frac{1}{\Delta t} \int_{t_i}^{t_f} \mathbf{F}\, dt \tag{9.10}$$

where $\Delta t = t_f - t_i$. (This is an application of the mean value theorem of calculus.) Therefore, we can express Equation 9.9 as

$$\mathbf{I} \equiv \overline{\mathbf{F}} \Delta t \tag{9.11}$$

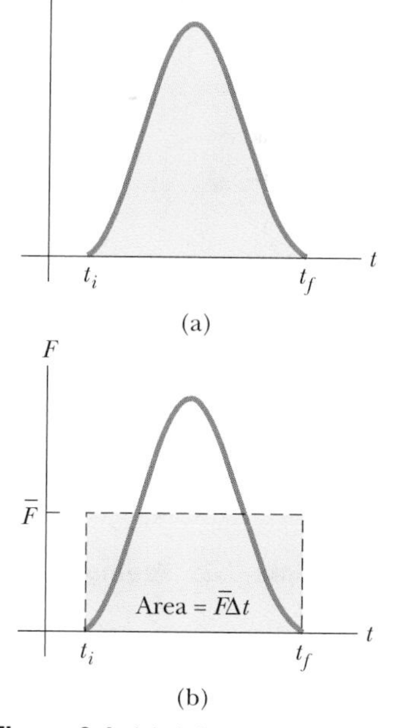

Figure 9.4 (a) A force acting on a particle may vary in time. The impulse imparted to the particle by the force is the area under the force-versus-time curve. (b) In the time interval Δt, the time-averaged force (horizontal dashed line) gives the same impulse to a particle as does the time-varying force described in part (a).

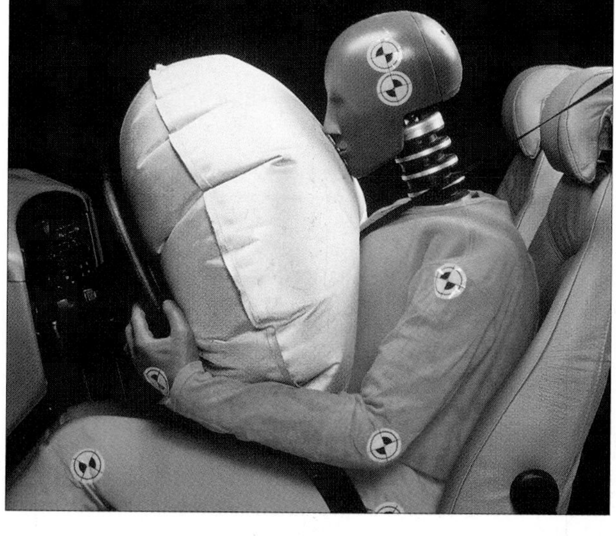

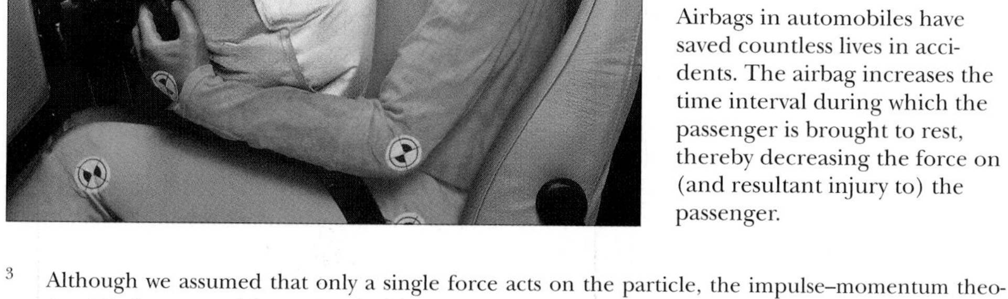

Courtesy of Saab

Airbags in automobiles have saved countless lives in accidents. The airbag increases the time interval during which the passenger is brought to rest, thereby decreasing the force on (and resultant injury to) the passenger.

[3] Although we assumed that only a single force acts on the particle, the impulse–momentum theorem is valid when several forces act; in this case, we replace **F** in Equation 9.8 with $\Sigma\mathbf{F}$.

This time-averaged force, shown in Figure 9.4b, can be interpreted as the constant force that would give to the particle in the time interval Δt the same impulse that the time-varying force gives over this same interval.

In principle, if $\mathbf{F}$ is known as a function of time, the impulse can be calculated from Equation 9.9. The calculation becomes especially simple if the force acting on the particle is constant. In this case, $\overline{\mathbf{F}} = \mathbf{F}$ and Equation 9.11 becomes

$$\mathbf{I} = \mathbf{F}\Delta t \tag{9.12}$$

In many physical situations, we shall use what is called the **impulse approximation, in which we assume that one of the forces exerted on a particle acts for a short time but is much greater than any other force present.** This approximation is especially useful in treating collisions in which the duration of the collision is very short. When this approximation is made, we refer to the force as an *impulsive force.* For example, when a baseball is struck with a bat, the time of the collision is about 0.01 s and the average force that the bat exerts on the ball in this time is typically several thousand newtons. Because this contact force is much greater than the magnitude of the gravitational force, the impulse approximation justifies our ignoring the gravitational forces exerted on the ball and bat. When we use this approximation, it is important to remember that $\mathbf{p}_i$ and $\mathbf{p}_f$ represent the momenta *immediately* before and after the collision, respectively. Therefore, in any situation in which it is proper to use the impulse approximation, the particle moves very little during the collision.

Quick Quiz 9.5 Two objects are at rest on a frictionless surface. Object 1 has a greater mass than object 2. When a constant force is applied to object 1, it accelerates through a distance d. The force is removed from object 1 and is applied to object 2. At the moment when object 2 has accelerated through the same distance d, which statements are true? (a) $p_1 < p_2$ (b) $p_1 = p_2$ (c) $p_1 > p_2$ (d) $K_1 < K_2$ (e) $K_1 = K_2$ (f) $K_1 > K_2$.

Quick Quiz 9.6 Two objects are at rest on a frictionless surface. Object 1 has a greater mass than object 2. When a force is applied to object 1, it accelerates for a time interval Δt. The force is removed from object 1 and is applied to object 2. After object 2 has accelerated for the same time interval Δt, which statements are true? (a) $p_1 < p_2$ (b) $p_1 = p_2$ (c) $p_1 > p_2$ (d) $K_1 < K_2$ (e) $K_1 = K_2$ (f) $K_1 > K_2$.

Quick Quiz 9.7 Rank an automobile dashboard, seatbelt, and airbag in terms of (a) the impulse and (b) the average force they deliver to a front-seat passenger during a collision, from greatest to least.

Example 9.3 Teeing Off

A golf ball of mass 50 g is struck with a club (Fig. 9.5). The force exerted by the club on the ball varies from zero, at the instant before contact, up to some maximum value and then back to zero when the ball leaves the club. Thus, the force–time curve is qualitatively described by Figure 9.4. Assuming that the ball travels 200 m, estimate the magnitude of the impulse caused by the collision.

Solution Let us use Ⓐ to denote the position of the ball when the club first contacts it, Ⓑ to denote the position of the ball when the club loses contact with the ball, and Ⓒ to denote the position of the ball upon landing. Neglecting air resistance, we can use Equation 4.14 for the range of a projectile:

$$R = x_{\text{C}} = \frac{v_{\text{B}}^2}{g}\sin 2\theta_{\text{B}}$$

Let us assume that the launch angle θ_{B} is 45°, the angle that provides the maximum range for any given launch velocity. This assumption gives $\sin 2\theta_{\text{B}} = 1$, and the launch velocity of the ball is

$$v_{\text{B}} = \sqrt{Rg} \approx \sqrt{(200 \text{ m})(9.80 \text{ m/s}^2)} = 44 \text{ m/s}$$

© Harold and Esther Edgerton Foundation 2002, courtesy of Palm Press, Inc.

Figure 9.5 (Example 9.3) A golf ball being struck by a club. Note the deformation of the ball due to the large force from the club.

Considering initial and final values of the ball's velocity for the time interval for the collision, $v_i = v_A = 0$ and $v_f = v_B$. Hence, the magnitude of the impulse imparted to the ball is

$$I = \Delta p = mv_B - mv_A = (50 \times 10^{-3}\ \text{kg})(44\ \text{m/s}) - 0$$

$$= \boxed{2.2\ \text{kg}\cdot\text{m/s}}$$

What If? **What if you were asked to find the average force on the ball during the collision with the club? Can you determine this value?**

Answer With the information given in the problem, we cannot find the average force. Considering Equation 9.11, we would need to know the time interval of the collision in order to calculate the average force. If we *assume* that the time interval is 0.01 s as it was for the baseball in the discussion after Equation 9.12, we can estimate the magnitude of the average force:

$$\overline{F} = \frac{I}{\Delta t} = \frac{2.2\ \text{kg}\cdot\text{m/s}}{0.01\ \text{s}} = 2 \times 10^2\ \text{N}$$

where we have kept only one significant figure due to our rough estimate of the time interval.

Example 9.4 How Good Are the Bumpers?

In a particular crash test, a car of mass 1 500 kg collides with a wall, as shown in Figure 9.6. The initial and final velocities of the car are $\mathbf{v}_i = -15.0\hat{\mathbf{i}}$ m/s and $\mathbf{v}_f = 2.60\hat{\mathbf{i}}$ m/s, respectively. If the collision lasts for 0.150 s, find the impulse caused by the collision and the average force exerted on the car.

Solution Let us assume that the force exerted by the wall on the car is large compared with other forces on the car so that we can apply the impulse approximation. Furthermore, we note that the gravitational force and the normal force exerted by the road on the car are perpendicular to the motion and therefore do not affect the horizontal momentum.

The initial and final momenta of the car are

$$\mathbf{p}_i = m\mathbf{v}_i = (1\,500\ \text{kg})(-15.0\hat{\mathbf{i}}\ \text{m/s})$$
$$= -2.25 \times 10^4\hat{\mathbf{i}}\ \text{kg}\cdot\text{m/s}$$

$$\mathbf{p}_f = m\mathbf{v}_f = (1\,500\ \text{kg})(2.60\hat{\mathbf{i}}\ \text{m/s})$$
$$= 0.39 \times 10^4\hat{\mathbf{i}}\ \text{kg}\cdot\text{m/s}$$

Hence, the impulse is equal to

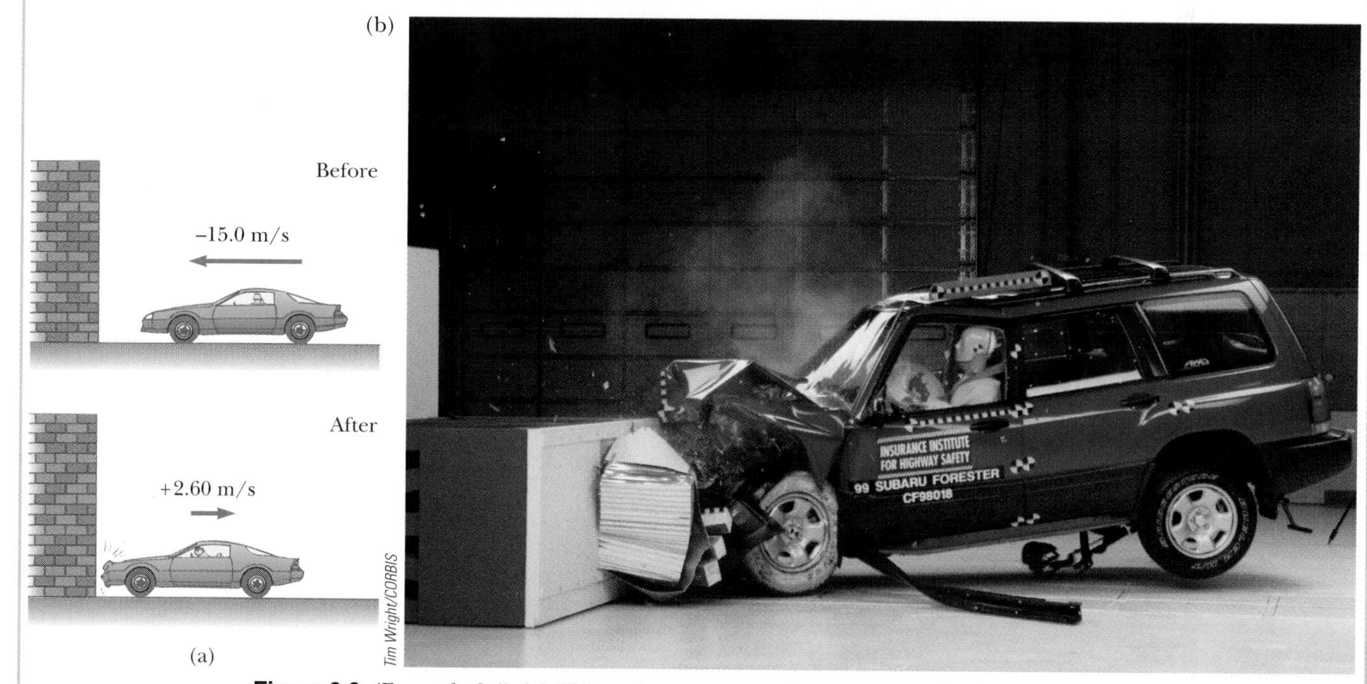

Tim Wright/CORBIS

Figure 9.6 (Example 9.4) (a) This car's momentum changes as a result of its collision with the wall. (b) In a crash test, much of the car's initial kinetic energy is transformed into energy associated with the damage to the car.

$$\mathbf{I} = \Delta\mathbf{p} = \mathbf{p}_f - \mathbf{p}_i = 0.39 \times 10^4\hat{\mathbf{i}}\ \text{kg}\cdot\text{m/s} - (-2.25 \times 10^4\hat{\mathbf{i}}\ \text{kg}\cdot\text{m/s})$$

$$\mathbf{I} = \boxed{2.64 \times 10^4\hat{\mathbf{i}}\ \text{kg}\cdot\text{m/s}}$$

The average force exerted by the wall on the car is

$$\overline{\mathbf{F}} = \frac{\Delta\mathbf{p}}{\Delta t} = \frac{2.64 \times 10^4\hat{\mathbf{i}}\ \text{kg}\cdot\text{m/s}}{0.150\ \text{s}} = \boxed{1.76 \times 10^5\hat{\mathbf{i}}\ \text{N}}$$

In this problem, note that the signs of the velocities indicate the reversal of directions. What would the mathematics be describing if both the initial and final velocities had the same sign?

What If? What if the car did not rebound from the wall? Suppose the final velocity of the car is zero and the time interval of the collision remains at 0.150 s. Would this represent a larger or a smaller force by the wall on the car?

Answer In the original situation in which the car rebounds, the force by the wall on the car does two things in the time interval—it (1) stops the car and (2) causes it to move away from the wall at 2.60 m/s after the collision. If the car does not rebound, the force is only doing the first of these, stopping the car. This will require a *smaller* force.

Mathematically, in the case of the car that does not rebound, the impulse is

$$\mathbf{I} = \Delta\mathbf{p} = \mathbf{p}_f - \mathbf{p}_i = 0 - (-2.25 \times 10^4\hat{\mathbf{i}}\ \text{kg}\cdot\text{m/s}) = 2.25 \times 10^4\hat{\mathbf{i}}\ \text{kg}\cdot\text{m/s}$$

The average force exerted by the wall on the car is

$$\overline{\mathbf{F}} = \frac{\Delta\mathbf{p}}{\Delta t} = \frac{2.25 \times 10^4\hat{\mathbf{i}}\ \text{kg}\cdot\text{m/s}}{0.150\ \text{s}} = 1.50 \times 10^5\hat{\mathbf{i}}\ \text{N}$$

which is indeed smaller than the previously calculated value, as we argued conceptually.

9.3 Collisions in One Dimension

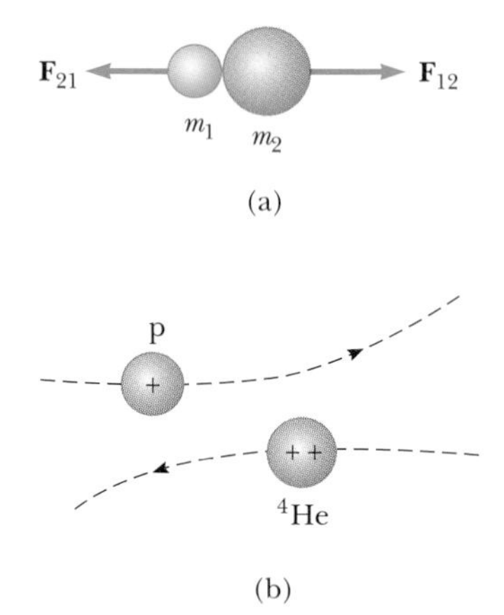

Figure 9.7 (a) The collision between two objects as the result of direct contact. (b) The "collision" between two charged particles.

In this section we use the law of conservation of linear momentum to describe what happens when two particles collide. We use the term **collision** to represent an event during which two particles come close to each other and interact by means of forces. The time interval during which the velocities of the particles change from initial to final values is assumed to be short. The interaction forces are assumed to be much greater than any external forces present, so we can use the impulse approximation.

A collision may involve physical contact between two macroscopic objects, as described in Figure 9.7a, but the notion of what we mean by collision must be generalized because "physical contact" on a submicroscopic scale is ill-defined and hence meaningless. To understand this, consider a collision on an atomic scale (Fig. 9.7b), such as the collision of a proton with an alpha particle (the nucleus of a helium atom). Because the particles are both positively charged, they repel each other due to the strong electrostatic force between them at close separations and never come into "physical contact."

When two particles of masses m_1 and m_2 collide as shown in Figure 9.7, the impulsive forces may vary in time in complicated ways, such as that shown in Figure 9.4. Regardless of the complexity of the time behavior of the force of interaction, however, this force is internal to the system of two particles. Thus, the two particles form an isolated system, and the momentum of the system must be conserved. Therefore, the total momentum of an isolated system just before a collision equals the total momentum of the system just after the collision.

In contrast, the total kinetic energy of the system of particles may or may not be conserved, depending on the type of collision. In fact, whether or not kinetic energy is conserved is used to classify collisions as either *elastic* or *inelastic.*

Elastic collision

An **elastic collision** between two objects is one in which **the total kinetic energy (as well as total momentum) of the system is the same before and after the collision.** Collisions between certain objects in the macroscopic world, such as billiard balls, are only *approximately* elastic because some deformation and loss of kinetic energy take place. For example, you can hear a billiard ball collision, so you know that some of the energy is being transferred away from the system by sound. An elastic collision must be perfectly silent! *Truly* elastic collisions occur between atomic and subatomic particles.

Inelastic collision

An **inelastic collision** is one in which **the total kinetic energy of the system is not the same before and after the collision (even though the momentum of the system is conserved).** Inelastic collisions are of two types. When the colliding objects stick together after the collision, as happens when a meteorite collides with the Earth, the collision is called **perfectly inelastic.** When the colliding objects do not stick together, but some kinetic energy is lost, as in the case of a rubber ball colliding with a hard surface, the collision is called **inelastic** (with no modifying adverb). When the rubber ball collides with the hard surface, some of the kinetic energy of the ball is lost when the ball is deformed while it is in contact with the surface.

In most collisions, the kinetic energy of the system is *not* conserved because some of the energy is converted to internal energy and some of it is transferred away by means of sound. Elastic and perfectly inelastic collisions are limiting cases; most collisions fall somewhere between them.

In the remainder of this section, we treat collisions in one dimension and consider the two extreme cases—perfectly inelastic and elastic collisions. The important distinction between these two types of collisions is that **momentum of the system is conserved in all collisions, but kinetic energy of the system is conserved only in elastic collisions.**

PITFALL PREVENTION

9.2 Inelastic Collisions

Generally, inelastic collisions are hard to analyze unless additional information is provided. This appears in the mathematical representation as having more unknowns than equations.

Perfectly Inelastic Collisions

Consider two particles of masses m_1 and m_2 moving with initial velocities $\mathbf{v}_{1i}$ and $\mathbf{v}_{2i}$ along the same straight line, as shown in Figure 9.8. The two particles collide head-on, stick together, and then move with some common velocity $\mathbf{v}_f$ after the collision. Because the momentum of an isolated system is conserved in *any* collision, we can say that the total momentum before the collision equals the total momentum of the composite system after the collision:

$$m_1\mathbf{v}_{1i} + m_2\mathbf{v}_{2i} = (m_1 + m_2)\mathbf{v}_f \quad (9.13)$$

Solving for the final velocity gives

$$\mathbf{v}_f = \frac{m_1\mathbf{v}_{1i} + m_2\mathbf{v}_{2i}}{m_1 + m_2} \quad (9.14)$$

Before collision

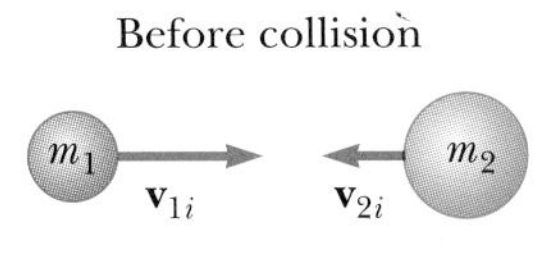

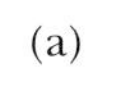

(a)

After collision

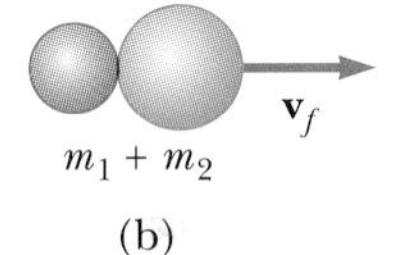

(b)

Active Figure 9.8 Schematic representation of a perfectly inelastic head-on collision between two particles: (a) before collision and (b) after collision.

At the Active Figures link at http://www.pse6.com, you can adjust the masses and velocities of the colliding objects to see the effect on the final velocity.

Elastic Collisions

Consider two particles of masses m_1 and m_2 moving with initial velocities $\mathbf{v}_{1i}$ and $\mathbf{v}_{2i}$ along the same straight line, as shown in Figure 9.9. The two particles collide head-on and then leave the collision site with different velocities, $\mathbf{v}_{1f}$ and $\mathbf{v}_{2f}$. If the collision is elastic, both the momentum and kinetic energy of the system are conserved. Therefore, considering velocities along the horizontal direction in Figure 9.9, we have

$$m_1v_{1i} + m_2v_{2i} = m_1v_{1f} + m_2v_{2f} \quad (9.15)$$

$$\tfrac{1}{2}m_1v_{1i}^2 + \tfrac{1}{2}m_2v_{2i}^2 = \tfrac{1}{2}m_1v_{1f}^2 + \tfrac{1}{2}m_2v_{2f}^2 \quad (9.16)$$

Because all velocities in Figure 9.9 are either to the left or the right, they can be represented by the corresponding speeds along with algebraic signs indicating direction. We shall indicate v as positive if a particle moves to the right and negative if it moves to the left.

In a typical problem involving elastic collisions, there are two unknown quantities, and Equations 9.15 and 9.16 can be solved simultaneously to find these. An alternative approach, however—one that involves a little mathematical manipulation of Equation 9.16—often simplifies this process. To see how, let us cancel the factor $\frac{1}{2}$ in Equation 9.16 and rewrite it as

$$m_1(v_{1i}^2 - v_{1f}^2) = m_2(v_{2f}^2 - v_{2i}^2)$$

and then factor both sides:

$$m_1(v_{1i} - v_{1f})(v_{1i} + v_{1f}) = m_2(v_{2f} - v_{2i})(v_{2f} + v_{2i}) \quad (9.17)$$

Next, let us separate the terms containing m_1 and m_2 in Equation 9.15 to obtain

$$m_1(v_{1i} - v_{1f}) = m_2(v_{2f} - v_{2i}) \quad (9.18)$$

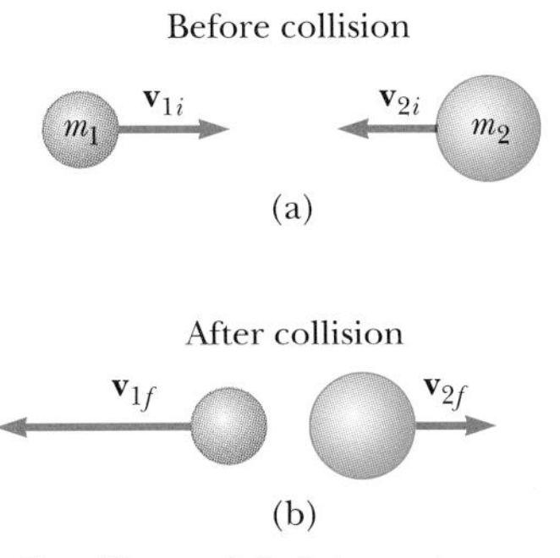

(b)

Active Figure 9.9 Schematic representation of an elastic head-on collision between two particles: (a) before collision and (b) after collision.

At the Active Figures link at http://www.pse6.com, you can adjust the masses and velocities of the colliding objects to see the effect on the final velocities.

PITFALL PREVENTION

9.3 Not a General Equation

We have spent some effort on deriving Equation 9.19, but remember that it can only be used in a very *specific* situation—a one-dimensional, elastic collision between two objects. The *general* concept is conservation of momentum (and conservation of kinetic energy if the collision is elastic) for an isolated system.

To obtain our final result, we divide Equation 9.17 by Equation 9.18 and obtain

$$v_{1i} + v_{1f} = v_{2f} + v_{2i}$$

$$v_{1i} - v_{2i} = -(v_{1f} - v_{2f}) \tag{9.19}$$

This equation, in combination with Equation 9.15, can be used to solve problems dealing with elastic collisions. According to Equation 9.19, the *relative* velocity of the two particles before the collision, $v_{1i} - v_{2i}$, equals the negative of their relative velocity after the collision, $-(v_{1f} - v_{2f})$.

Suppose that the masses and initial velocities of both particles are known. Equations 9.15 and 9.19 can be solved for the final velocities in terms of the initial velocities because there are two equations and two unknowns:

$$v_{1f} = \left(\frac{m_1 - m_2}{m_1 + m_2}\right)v_{1i} + \left(\frac{2m_2}{m_1 + m_2}\right)v_{2i} \tag{9.20}$$

$$v_{2f} = \left(\frac{2m_1}{m_1 + m_2}\right)v_{1i} + \left(\frac{m_2 - m_1}{m_1 + m_2}\right)v_{2i} \tag{9.21}$$

It is important to use the appropriate signs for v_{1i} and v_{2i} in Equations 9.20 and 9.21. For example, if particle 2 is moving to the left initially, then v_{2i} is negative.

Let us consider some special cases. If $m_1 = m_2$, then Equations 9.20 and 9.21 show us that $v_{1f} = v_{2i}$ and $v_{2f} = v_{1i}$. That is, the particles exchange velocities if they have equal masses. This is approximately what one observes in head-on billiard ball collisions—the cue ball stops, and the struck ball moves away from the collision with the same velocity that the cue ball had.

If particle 2 is initially at rest, then $v_{2i} = 0$, and Equations 9.20 and 9.21 become

Elastic collision: particle 2 initially at rest

$$v_{1f} = \left(\frac{m_1 - m_2}{m_1 + m_2}\right)v_{1i} \tag{9.22}$$

$$v_{2f} = \left(\frac{2m_1}{m_1 + m_2}\right)v_{1i} \tag{9.23}$$

If m_1 is much greater than m_2 and $v_{2i} = 0$, we see from Equations 9.22 and 9.23 that $v_{1f} \approx v_{1i}$ and $v_{2f} \approx 2v_{1i}$. That is, when a very heavy particle collides head-on with a very light one that is initially at rest, the heavy particle continues its motion unaltered after the collision and the light particle rebounds with a speed equal to about twice the initial speed of the heavy particle. An example of such a collision would be that of a moving heavy atom, such as uranium, striking a light atom, such as hydrogen.

If m_2 is much greater than m_1 and particle 2 is initially at rest, then $v_{1f} \approx -v_{1i}$ and $v_{2f} \approx 0$. That is, when a very light particle collides head-on with a very heavy particle that is initially at rest, the light particle has its velocity reversed and the heavy one remains approximately at rest.

PITFALL PREVENTION

9.4 Momentum and Kinetic Energy in Collisions

Momentum of an isolated system is conserved in *all* collisions. Kinetic energy of an isolated system is conserved *only* in elastic collisions. Why? Because there are several types of energy into which kinetic energy can transform, or be transferred out of the system (so that the system may *not* be isolated in terms of energy during the collision). However, there is only one type of momentum.

Quick Quiz 9.8 In a perfectly inelastic one-dimensional collision between two objects, what condition alone is necessary so that *all* of the original kinetic energy of the system is gone after the collision? (a) The objects must have momenta with the same magnitude but opposite directions. (b) The objects must have the same mass. (c) The objects must have the same velocity. (d) The objects must have the same speed, with velocity vectors in opposite directions.

Quick Quiz 9.9 A table-tennis ball is thrown at a stationary bowling ball. The table-tennis ball makes a one-dimensional elastic collision and bounces back along the same line. After the collision, compared to the bowling ball, the table-tennis ball has (a) a larger magnitude of momentum and more kinetic energy (b) a smaller magnitude of momentum and more kinetic energy (c) a larger magnitude of momentum and less kinetic energy (d) a smaller magnitude of momentum and less kinetic energy (e) the same magnitude of momentum and the same kinetic energy.

Example 9.5 The Executive Stress Reliever **Interactive**

An ingenious device that illustrates conservation of momentum and kinetic energy is shown in Figure 9.10. It consists of five identical hard balls supported by strings of equal lengths. When ball 1 is pulled out and released, after the almost-elastic collision between it and ball 2, ball 5 moves out, as shown in Figure 9.10b. If balls 1 and 2 are pulled out and released, balls 4 and 5 swing out, and so forth. Is it ever possible that when ball 1 is released, balls 4 and 5 will swing out on the opposite side and travel with half the speed of ball 1, as in Figure 9.10c?

Solution No, such movement can never occur if we assume the collisions are elastic. The momentum of the system before the collision is mv, where m is the mass of ball 1 and v is its speed just before the collision. After the collision, we would have two balls, each of mass m moving with speed $v/2$. The total momentum of the system after the collision would be $m(v/2) + m(v/2) = mv$. Thus, momentum of the system is conserved. However, the kinetic energy just before the collision is $K_i = \frac{1}{2}mv^2$ and that after the collision is $K_f = \frac{1}{2}m(v/2)^2 + \frac{1}{2}m(v/2)^2 = \frac{1}{4}mv^2$. Thus, kinetic energy of the system is *not* conserved. The only way to have both momentum and kinetic energy conserved is for one ball to move out when one ball is released, two balls to move out when two are released, and so on.

What If? **Consider what would happen if balls 4 and 5 are glued together so that they must move together. Now what happens when ball 1 is pulled out and released?**

Answer We are now forcing balls 4 and 5 to come out together. We have argued that we cannot conserve both momentum and energy in this case. However, we assumed that ball 1 stopped after striking ball 2. What if we do not make this assumption? Consider the conservation equations with the assumption that ball 1 moves after the collision. For conservation of momentum,

$$p_i = p_f$$

$$mv_{1i} = mv_{1f} + 2mv_{4,5f}$$

where $v_{4,5f}$ refers to the final speed of the ball 4–ball 5 combination. Conservation of kinetic energy gives us

$$K_i = K_f$$

$$\frac{1}{2}mv_{1i}^2 = \frac{1}{2}mv_{1f}^2 + \frac{1}{2}(2m)v_{4,5f}^2$$

Combining these equations, we find

$$v_{4,5f} = \frac{2}{3}v_{1i} \qquad v_{1f} = -\frac{1}{3}v_{1i}$$

Thus, balls 4 and 5 come out together and ball 1 bounces back from the collision with one third of its original speed.

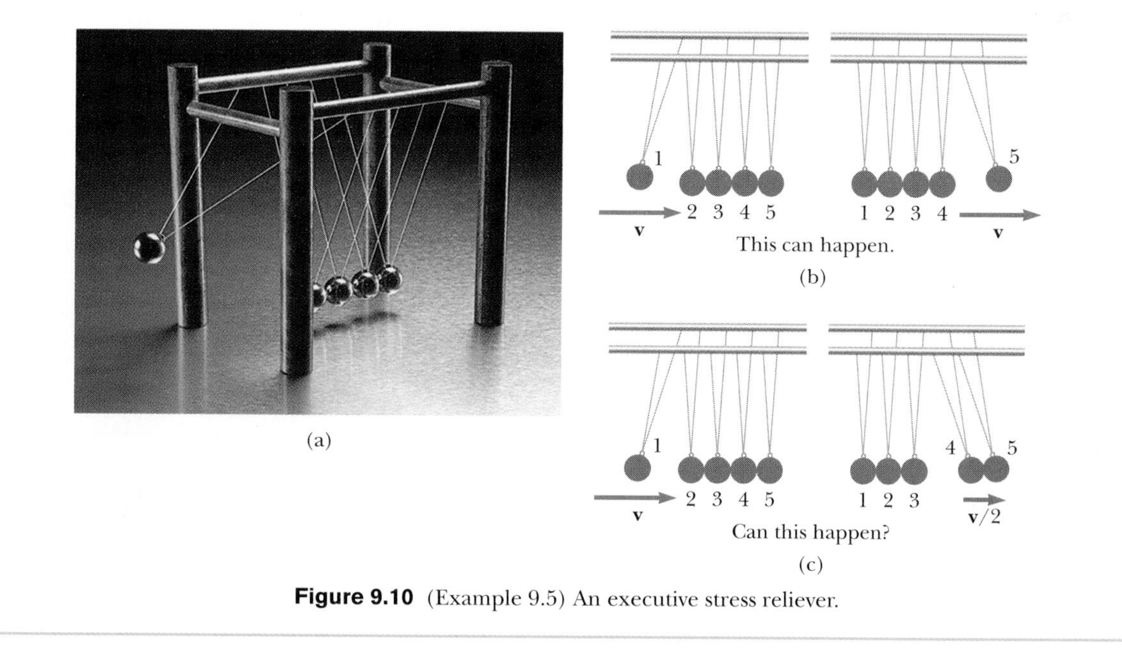

Figure 9.10 (Example 9.5) An executive stress reliever.

At the Interactive Worked Example link at **http://www.pse6.com**, *you can "glue" balls 4 and 5 together to see the situation discussed above.*

Example 9.6 Carry Collision Insurance!

An 1 800-kg car stopped at a traffic light is struck from the rear by a 900-kg car, and the two become entangled, moving along the same path as that of the originally moving car. If the smaller car were moving at 20.0 m/s before the collision, what is the velocity of the entangled cars after the collision?

Solution The phrase "become entangled" tells us that this is a perfectly inelastic collision. We can guess that the final speed is less than 20.0 m/s, the initial speed of the smaller car. The total momentum of the system (the two cars) before the collision must equal the total momentum immediately after the collision because momentum of an isolated system is conserved in any type of collision. The magnitude of the total momentum of the system before the collision is equal to that of the smaller car because the larger car is initially at rest:

$$p_i = m_1 v_i = (900 \text{ kg})(20.0 \text{ m/s}) = 1.80 \times 10^4 \text{ kg}\cdot\text{m/s}$$

After the collision, the magnitude of the momentum of the entangled cars is

$$p_f = (m_1 + m_2)v_f = (2\,700 \text{ kg})v_f$$

Equating the initial and final momenta of the system and solving for v_f, the final velocity of the entangled cars, we have

$$v_f = \frac{p_i}{m_1 + m_2} = \frac{1.80 \times 10^4 \text{ kg}\cdot\text{m/s}}{2\,700 \text{ kg}} = 6.67 \text{ m/s}$$

Because the final velocity is positive, the direction of the final velocity is the same as the velocity of the initially moving car.

What If? Suppose we reverse the masses of the cars—a stationary 900-kg car is struck by a moving 1 800-kg car. Is the final speed the same as before?

Answer Intuitively, we can guess that the final speed will be higher, based on common experiences in driving. Mathematically, this should be the case because the system has a larger momentum if the initially moving car is the more massive one. Solving for the new final velocity, we find

$$v_f = \frac{p_i}{m_1 + m_2} = \frac{(1\,800 \text{ kg})(20.0 \text{ m/s})}{2\,700 \text{ kg}} = 13.3 \text{ m/s}$$

which is indeed higher than the previous final velocity.

Example 9.7 The Ballistic Pendulum

The ballistic pendulum (Fig. 9.11) is an apparatus used to measure the speed of a fast-moving projectile, such as a bullet. A bullet of mass m_1 is fired into a large block of wood of mass m_2 suspended from some light wires. The bullet embeds in the block, and the entire system swings through a height h. How can we determine the speed of the bullet from a measurement of h?

Solution Figure 9.11a helps to conceptualize the situation. Let configuration Ⓐ be the bullet and block before the collision, and configuration Ⓑ be the bullet and block immediately after colliding. The bullet and the block form an isolated system, so we can categorize the collision between them as a conservation of momentum problem. The collision is perfectly inelastic. To analyze the collision, we note that Equation 9.14 gives the speed of the system right after the collision when we assume the impulse approximation. Noting that $v_{2A} = 0$, Equation 9.14 becomes

$$(1) \qquad v_B = \frac{m_1 v_{1A}}{m_1 + m_2}$$

For the process during which the bullet–block combination swings upward to height h (ending at configuration Ⓒ), we focus on a *different* system—the bullet, the block, and the Earth. This is an isolated system for energy, so we categorize this part of the motion as a conservation of mechanical energy problem:

$$K_B + U_B = K_C + U_C$$

We begin to analyze the problem by finding the total kinetic energy of the system right after the collision:

$$(2) \qquad K_B = \tfrac{1}{2}(m_1 + m_2)v_B^2$$

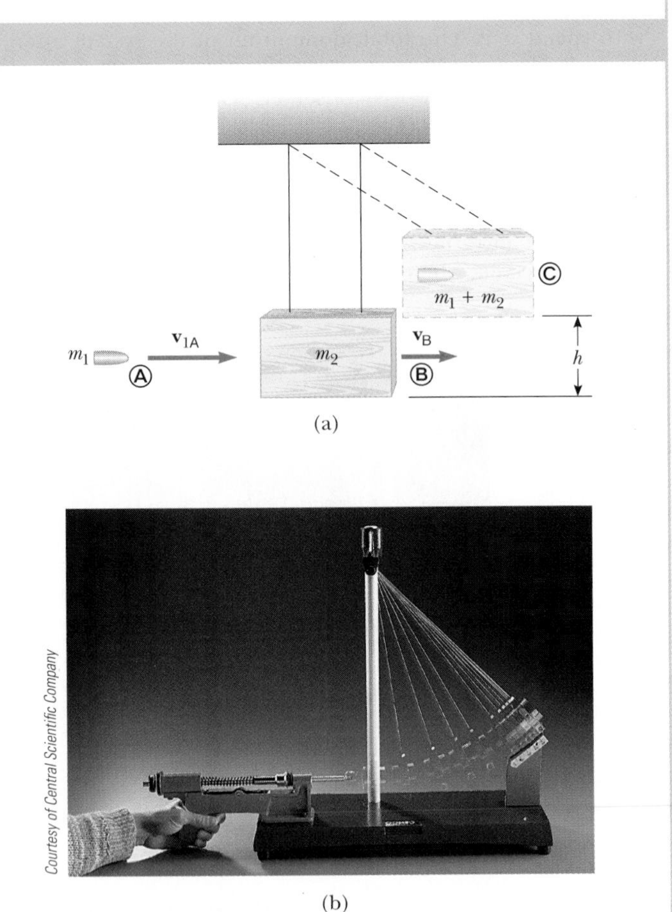

Figure 9.11 (Example 9.7) (a) Diagram of a ballistic pendulum. Note that $\mathbf{v}_{1A}$ is the velocity of the bullet just before the collision and $\mathbf{v}_B$ is the velocity of the bullet-block system just after the perfectly inelastic collision. (b) Multiflash photograph of a ballistic pendulum used in the laboratory.

Substituting the value of v_B from Equation (1) into Equation (2) gives

$$K_B = \frac{m_1^2 v_{1A}^2}{2(m_1 + m_2)}$$

This kinetic energy immediately after the collision is *less* than the initial kinetic energy of the bullet, as expected in an inelastic collision.

We define the gravitational potential energy of the system for configuration Ⓑ to be zero. Thus, $U_B = 0$ while $U_C = (m_1 + m_2)gh$. Conservation of energy now leads to

$$\frac{m_1^2 v_{1A}^2}{2(m_1 + m_2)} + 0 = 0 + (m_1 + m_2)gh$$

Solving for v_{1A}, we obtain

$$v_{1A} = \left(\frac{m_1 + m_2}{m_1}\right)\sqrt{2gh}$$

To finalize this problem, note that we had to solve this problem in two steps. Each step involved a different system and a different conservation principle. Because the collision was assumed to be perfectly inelastic, some mechanical energy was converted to internal energy. It would have been *incorrect* to equate the initial kinetic energy of the incoming bullet to the final gravitational potential energy of the bullet–block–Earth combination.

Example 9.8 A Two-Body Collision with a Spring

Interactive

A block of mass $m_1 = 1.60$ kg initially moving to the right with a speed of 4.00 m/s on a frictionless horizontal track collides with a spring attached to a second block of mass $m_2 = 2.10$ kg initially moving to the left with a speed of 2.50 m/s, as shown in Figure 9.12a. The spring constant is 600 N/m.

(A) Find the velocities of the two blocks after the collision.

Solution Because the spring force is conservative, no kinetic energy is converted to internal energy during the compression of the spring. Ignoring any sound made when the block hits the spring, we can model the collision as being elastic. Equation 9.15 gives us

$$m_1 v_{1i} + m_2 v_{2i} = m_1 v_{1f} + m_2 v_{2f}$$

$$(1.60 \text{ kg})(4.00 \text{ m/s}) + (2.10 \text{ kg})(-2.50 \text{ m/s}) = (1.60 \text{ kg})v_{1f} + (2.10 \text{ kg})v_{2f}$$

$$(1) \qquad 1.15 \text{ kg}\cdot\text{m/s} = (1.60 \text{ kg})v_{1f} + (2.10 \text{ kg})v_{2f}$$

Equation 9.19 gives us

$$v_{1i} - v_{2i} = -(v_{1f} - v_{2f})$$

$$4.00 \text{ m/s} - (-2.50 \text{ m/s}) = -v_{1f} + v_{2f}$$

$$(2) \qquad 6.50 \text{ m/s} = -v_{1f} + v_{2f}$$

Multiplying Equation (2) by 1.60 kg gives us

$$(3) \qquad 10.4 \text{ kg}\cdot\text{m/s} = -(1.60 \text{ kg})v_{1f} + (1.60 \text{ kg})v_{2f}$$

Adding Equations (1) and (3) allows us to find v_{2f}:

$$11.55 \text{ kg}\cdot\text{m/s} = (3.70 \text{ kg})v_{2f}$$

$$v_{2f} = \frac{11.55 \text{ kg}\cdot\text{m/s}}{3.70 \text{ kg}} = 3.12 \text{ m/s}$$

Now, Equation (2) allows us to find v_{1f}:

$$6.50 \text{ m/s} = -v_{1f} + 3.12 \text{ m/s}$$

$$v_{1f} = -3.38 \text{ m/s}$$

(B) During the collision, at the instant block 1 is moving to the right with a velocity of +3.00 m/s, as in Figure 9.12b, determine the velocity of block 2.

Solution Because the momentum of the system of two blocks is conserved *throughout* the collision for the system of two blocks, we have, for *any* instant during the collision,

$$m_1 v_{1i} + m_2 v_{2i} = m_1 v_{1f} + m_2 v_{2f}$$

We choose the final instant to be that at which block 1 is moving with a velocity of + 3.00 m/s:

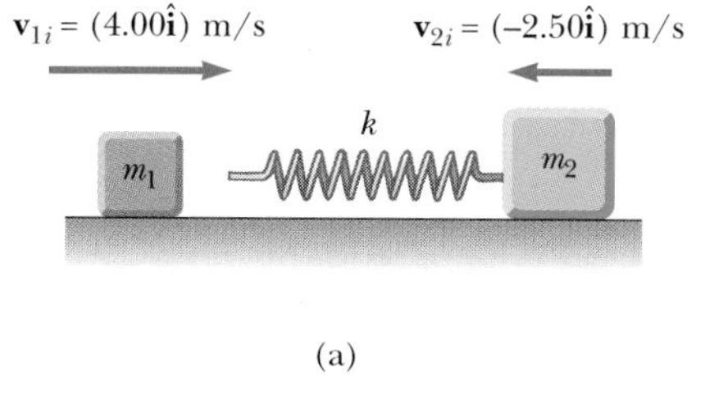

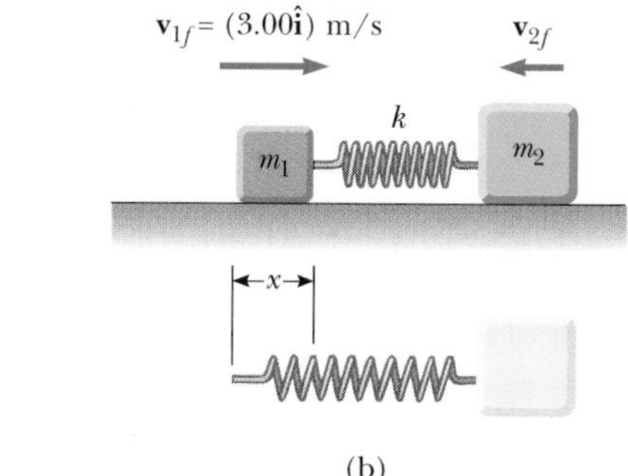

Figure 9.12 (Example 9.8) A moving block approaches a second moving block that is attached to a spring.

$$(1.60 \text{ kg})(4.00 \text{ m/s}) + (2.10 \text{ kg})(-2.50 \text{ m/s}) = (1.60 \text{ kg})(3.00 \text{ m/s}) + (2.10 \text{ kg})v_{2f}$$

$$v_{2f} = -1.74 \text{ m/s}$$

The negative value for v_{2f} means that block 2 is still moving to the left at the instant we are considering.

(C) Determine the distance the spring is compressed at that instant.

Solution To determine the distance that the spring is compressed, shown as x in Figure 9.12b, we can use the principle of conservation of mechanical energy for the system of the spring and two blocks because no friction or other nonconservative forces are acting within the system. We choose the initial configuration of the system to be that existing just before block 1 strikes the spring and the final configuration to be that when block 1 is moving to the right at 3.00 m/s. Thus, we have

$$K_i + U_i = K_f + U_f$$

$$\tfrac{1}{2}m_1{v_{1i}}^2 + \tfrac{1}{2}m_2{v_{2i}}^2 + 0 = \tfrac{1}{2}m_1{v_{1f}}^2 + \tfrac{1}{2}m_2{v_{2f}}^2 + \tfrac{1}{2}kx^2$$

Substituting the given values and the result to part (B) into this expression gives

$$x = 0.173 \text{ m}$$

(D) What is the *maximum* compression of the spring during the collision?

Solution The maximum compression would occur when the two blocks are moving with the same velocity. The conservation of momentum equation for the system can be written

$$m_1v_{1i} + m_2v_{2i} = (m_1 + m_2)v_f$$

where the initial instant is just before the collision and the final instant is when the blocks are moving with the same velocity v_f. Solving for v_f,

$$v_f = \frac{m_1v_{1i} + m_2v_{2i}}{m_1 + m_2}$$

$$= \frac{(1.60 \text{ kg})(4.00 \text{ m/s}) + (2.10 \text{ kg})(-2.50 \text{ m/s})}{1.60 \text{ kg} + 2.10 \text{ kg}}$$

$$= 0.311 \text{ m/s}$$

Now, we apply conservation of mechanical energy between these two instants as in part (C):

$$K_i + U_i = K_f + U_f$$

$$\tfrac{1}{2}m_1{v_{1i}}^2 + \tfrac{1}{2}m_2{v_{2i}}^2 + 0 = \tfrac{1}{2}(m_1 + m_2){v_f}^2 + \tfrac{1}{2}kx^2$$

Substituting the given values into this expression gives

$$x = 0.253 \text{ m}$$

At the Interactive Worked Example link at **http://www.pse6.com,** *you can change the masses and speeds of the blocks and freeze the motion at the maximum compression of the spring.*

Example 9.9 Slowing Down Neutrons by Collisions

In a nuclear reactor, neutrons are produced when an atom splits in a process called *fission.* These neutrons are moving at about 10^7 m/s and must be slowed down to about 10^3 m/s before they take part in another fission event. They are slowed down by passing them through a solid or liquid material called a *moderator.* The slowing-down process involves elastic collisions. Show that a neutron can lose most of its kinetic energy if it collides elastically with a moderator containing light nuclei, such as deuterium (in "heavy water," D_2O) or carbon (in graphite).

Solution Let us assume that the moderator nucleus of mass m_m is at rest initially and that a neutron of mass m_n and initial speed v_{ni} collides with it head-on. Because these are elastic collisions, both momentum and kinetic energy of the neutron–nucleus system are conserved. Therefore, Equations 9.22 and 9.23 can be applied to the head-on collision of a neutron with a moderator nucleus. We can represent this process by a drawing such as Figure 9.9 with $\mathbf{v}_{2i} = 0$.

The initial kinetic energy of the neutron is

$$K_{ni} = \tfrac{1}{2}m_n{v_{ni}}^2$$

After the collision, the neutron has kinetic energy $\tfrac{1}{2}m_n{v_{nf}}^2$, and we can substitute into this the value for v_{nf} given by Equation 9.22:

$$K_{nf} = \tfrac{1}{2}m_n{v_{nf}}^2 = \tfrac{1}{2}m_n\left(\frac{m_n - m_m}{m_n + m_m}\right)^2 {v_{ni}}^2$$

Therefore, the fraction f_n of the initial kinetic energy possessed by the neutron after the collision is

$$(1) \qquad f_n = \frac{K_{nf}}{K_{ni}} = \left(\frac{m_n - m_m}{m_n + m_m}\right)^2$$

From this result, we see that the final kinetic energy of the neutron is small when m_m is close to m_n and zero when $m_n = m_m$.

We can use Equation 9.23, which gives the final speed of the particle that was initially at rest, to calculate the kinetic energy of the moderator nucleus after the collision:

$$K_{mf} = \tfrac{1}{2}m_m{v_{mf}}^2 = \frac{2{m_n}^2m_m}{(m_n + m_m)^2}{v_{ni}}^2$$

Hence, the fraction f_m of the initial kinetic energy transferred to the moderator nucleus is

$$(2) \qquad f_m = \frac{K_{mf}}{K_{ni}} = \frac{4m_nm_m}{(m_n + m_m)^2}$$

Because the total kinetic energy of the system is conserved, Equation (2) can also be obtained from Equation (1) with the condition that $f_n + f_m = 1$, so that $f_m = 1 - f_n$.

Suppose that heavy water is used for the moderator. For collisions of the neutrons with deuterium nuclei in D_2O ($m_m = 2m_n$), $f_n = 1/9$ and $f_m = 8/9$. That is, 89% of the neutron's kinetic energy is transferred to the deuterium nucleus. In practice, the moderator efficiency is reduced because head-on collisions are very unlikely.

How do the results differ when graphite (^{12}C, as found in pencil lead) is used as the moderator?

9.4 Two-Dimensional Collisions

In Section 9.1, we showed that the momentum of a system of two particles is conserved when the system is isolated. For any collision of two particles, this result implies that the momentum in each of the directions x, y, and z is conserved. An important subset of collisions takes place in a plane. The game of billiards is a familiar example involving multiple collisions of objects moving on a two-dimensional surface. For such two-dimensional collisions, we obtain two component equations for conservation of momentum:

$$m_1 v_{1ix} + m_2 v_{2ix} = m_1 v_{1fx} + m_2 v_{2fx}$$

$$m_1 v_{1iy} + m_2 v_{2iy} = m_1 v_{1fy} + m_2 v_{2fy}$$

where we use three subscripts in these equations to represent, respectively, (1) the identification of the object, (2) initial and final values, and (3) the velocity component.

Let us consider a two-dimensional problem in which particle 1 of mass m_1 collides with particle 2 of mass m_2, where particle 2 is initially at rest, as in Figure 9.13. After the collision, particle 1 moves at an angle θ with respect to the horizontal and particle 2 moves at an angle ϕ with respect to the horizontal. This is called a *glancing* collision. Applying the law of conservation of momentum in component form and noting that the initial y component of the momentum of the two-particle system is zero, we obtain

$$m_1 v_{1i} = m_1 v_{1f} \cos\theta + m_2 v_{2f} \cos\phi \qquad (9.24)$$

$$0 = m_1 v_{1f} \sin\theta - m_2 v_{2f} \sin\phi \qquad (9.25)$$

where the minus sign in Equation 9.25 comes from the fact that after the collision, particle 2 has a y component of velocity that is downward. We now have two independent equations. As long as no more than two of the seven quantities in Equations 9.24 and 9.25 are unknown, we can solve the problem.

If the collision is elastic, we can also use Equation 9.16 (conservation of kinetic energy) with $v_{2i} = 0$ to give

$$\tfrac{1}{2}m_1 {v_{1i}}^2 = \tfrac{1}{2}m_1 {v_{1f}}^2 + \tfrac{1}{2}m_2 {v_{2f}}^2 \qquad (9.26)$$

9.5 Don't Use Equation 9.19

Equation 9.19, relating the initial and final relative velocities of two colliding objects, is only valid for one-dimensional elastic collisions. Do not use this equation when analyzing two-dimensional collisions.

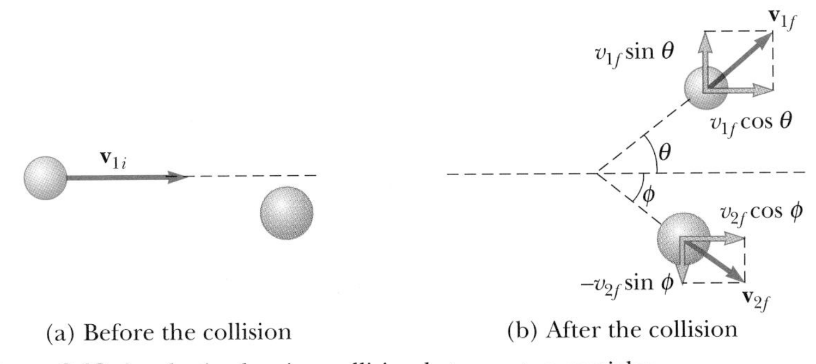

(a) Before the collision (b) After the collision

Active Figure 9.13 An elastic glancing collision between two particles.

At the Active Figures link at http://www.pse6.com, you can adjust the speed and position of the blue particle and the masses of both particles to see the effects.

Knowing the initial speed of particle 1 and both masses, we are left with four unknowns (v_{1f}, v_{2f}, θ, and ϕ). Because we have only three equations, one of the four remaining quantities must be given if we are to determine the motion after the collision from conservation principles alone.

If the collision is inelastic, kinetic energy is *not* conserved and Equation 9.26 does *not* apply.

PROBLEM-SOLVING HINTS

Two-Dimensional Collisions

The following procedure is recommended when dealing with problems involving two-dimensional collisions between two objects:

- Set up a coordinate system and define your velocities with respect to that system. It is usually convenient to have the x axis coincide with one of the initial velocities.
- In your sketch of the coordinate system, draw and label all velocity vectors and include all the given information.
- Write expressions for the x and y components of the momentum of each object before and after the collision. Remember to include the appropriate signs for the components of the velocity vectors.
- Write expressions for the total momentum of the system in the x direction before and after the collision and equate the two. Repeat this procedure for the total momentum of the system in the y direction.
- If the collision is inelastic, kinetic energy of the system is *not* conserved, and additional information is probably required. If the collision is *perfectly* inelastic, the final velocities of the two objects are equal. Solve the momentum equations for the unknown quantities.
- If the collision is *elastic*, kinetic energy of the system is conserved, and you can equate the total kinetic energy before the collision to the total kinetic energy after the collision to obtain an additional relationship between the velocities.

Example 9.10 Collision at an Intersection

A 1 500-kg car traveling east with a speed of 25.0 m/s collides at an intersection with a 2 500-kg van traveling north at a speed of 20.0 m/s, as shown in Figure 9.14. Find the direction and magnitude of the velocity of the wreckage after the collision, assuming that the vehicles undergo a perfectly inelastic collision (that is, they stick together).

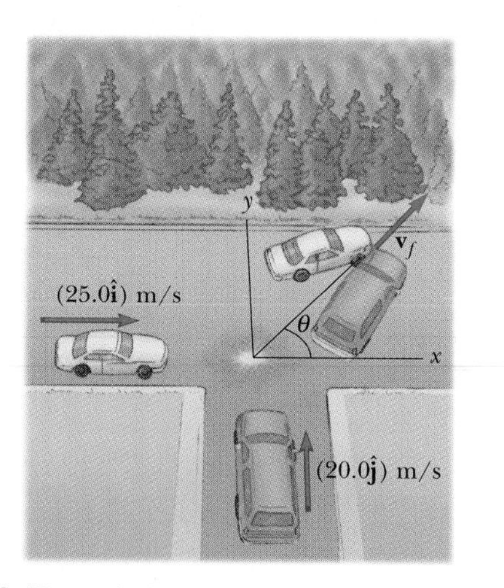

Figure 9.14 (Example 9.10) An eastbound car colliding with a northbound van.

Solution Let us choose east to be along the positive x direction and north to be along the positive y direction. Before the collision, the only object having momentum in the x direction is the car. Thus, the magnitude of the total initial momentum of the system (car plus van) in the x direction is

$$\sum p_{xi} = (1\,500 \text{ kg})(25.0 \text{ m/s}) = 3.75 \times 10^4 \text{ kg}\cdot\text{m/s}$$

Let us assume that the wreckage moves at an angle θ and speed v_f after the collision. The magnitude of the total momentum in the x direction after the collision is

$$\sum p_{xf} = (4\,000 \text{ kg})v_f \cos\theta$$

Because the total momentum in the x direction is conserved, we can equate these two equations to obtain

$$(1) \qquad 3.75 \times 10^4 \text{ kg}\cdot\text{m/s} = (4\,000 \text{ kg})v_f \cos\theta$$

Similarly, the total initial momentum of the system in the y direction is that of the van, and the magnitude of this momentum is $(2\,500 \text{ kg})(20.0 \text{ m/s}) = 5.00 \times 10^4 \text{ kg}\cdot\text{m/s}$. Applying conservation of momentum to the y direction, we have

$$\sum p_{yi} = \sum p_{yf}$$

$$(2) \qquad 5.00 \times 10^4 \text{ kg}\cdot\text{m/s} = (4\,000 \text{ kg})v_f \sin\theta$$

If we divide Equation (2) by Equation (1), we obtain

$$\frac{\sin\theta}{\cos\theta} = \tan\theta = \frac{5.00 \times 10^4}{3.75 \times 10^4} = 1.33$$

$$\theta = 53.1°$$

When this angle is substituted into Equation (2), the value of v_f is

$$v_f = \frac{5.00 \times 10^4 \text{ kg}\cdot\text{m/s}}{(4\,000 \text{ kg})\sin 53.1°} = 15.6 \text{ m/s}$$

It might be instructive for you to draw the momentum vectors of each vehicle before the collision and the two vehicles together after the collision.

Example 9.11 Proton–Proton Collision

A proton collides elastically with another proton that is initially at rest. The incoming proton has an initial speed of 3.50×10^5 m/s and makes a glancing collision with the second proton, as in Figure 9.13. (At close separations, the protons exert a repulsive electrostatic force on each other.) After the collision, one proton moves off at an angle of 37.0° to the original direction of motion, and the second deflects at an angle of ϕ to the same axis. Find the final speeds of the two protons and the angle ϕ.

Solution The pair of protons is an isolated system. Both momentum and kinetic energy of the system are conserved in this glancing elastic collision. Because $m_1 = m_2$, $\theta = 37.0°$, and we are given that $v_{1i} = 3.50 \times 10^5$ m/s, Equations 9.24, 9.25, and 9.26 become

$$(1) \qquad v_{1f}\cos 37° + v_{2f}\cos\phi = 3.50 \times 10^5 \text{ m/s}$$

$$(2) \qquad v_{1f}\sin 37.0° - v_{2f}\sin\phi = 0$$

$$(3) \qquad v_{1f}^2 + v_{2f}^2 = (3.50 \times 10^5 \text{ m/s})^2 = 1.23 \times 10^{11} \text{ m}^2/\text{s}^2$$

We rewrite Equations (1) and (2) as follows:

$$v_{2f}\cos\phi = 3.50 \times 10^5 \text{ m/s} - v_{1f}\cos 37.0°$$

$$v_{2f}\sin\phi = v_{1f}\sin 37.0°$$

Now we square these two equations and add them:

$$v_{2f}^2\cos^2\phi + v_{2f}^2\sin^2\phi$$
$$= 1.23 \times 10^{11} \text{ m}^2/\text{s}^2 - (7.00 \times 10^5 \text{ m/s})v_{1f}\cos 37.0° + v_{1f}^2\cos^2 37.0° + v_{1f}^2\sin^2 37.0°$$

$$v_{2f}^2 = 1.23 \times 10^{11} - (5.59 \times 10^5)v_{1f} + v_{1f}^2$$

Substituting into Equation (3) gives

$$v_{1f}^2 + [1.23 \times 10^{11} - (5.59 \times 10^5)v_{1f} + v_{1f}^2] = 1.23 \times 10^{11}$$

$$2v_{1f}^2 - (5.59 \times 10^5)v_{1f} = (2v_{1f} - 5.59 \times 10^5)v_{1f} = 0$$

One possibility for the solution of this equation is $v_{1f} = 0$, which corresponds to a head-on collision—the first proton stops and the second continues with the same speed in the same direction. This is not what we want. The other possibility is

$$2v_{1f} - 5.59 \times 10^5 = 0 \longrightarrow v_{1f} = 2.80 \times 10^5 \text{ m/s}$$

From Equation (3),

$$v_{2f} = \sqrt{1.23 \times 10^{11} - v_{1f}^2} = \sqrt{1.23 \times 10^{11} - (2.80 \times 10^5)^2} = 2.12 \times 10^5 \text{ m/s}$$

and from Equation (2),

$$\phi = \sin^{-1}\left(\frac{v_{1f}\sin 37.0°}{v_{2f}}\right) = \sin^{-1}\left(\frac{(2.80 \times 10^5)\sin 37.0°}{2.12 \times 10^5}\right) = 53.0°$$

It is interesting to note that $\theta + \phi = 90°$. This result is *not* accidental. Whenever two objects of equal mass collide elastically in a glancing collision and one of them is initially at rest, their final velocities are perpendicular to each other. The next example illustrates this point in more detail.

Example 9.12 Billiard Ball Collision

In a game of billiards, a player wishes to sink a target ball in the corner pocket, as shown in Figure 9.15. If the angle to the corner pocket is 35°, at what angle θ is the cue ball deflected? Assume that friction and rotational motion are unimportant and that the collision is elastic. Also assume that all billiard balls have the same mass m.

Solution Let ball 1 be the cue ball and ball 2 be the target ball. Because the target ball is initially at rest, conservation of kinetic energy (Eq. 9.16) for the two-ball system gives

$$\tfrac{1}{2}m_1 v_{1i}^2 = \tfrac{1}{2}m_1 v_{1f}^2 + \tfrac{1}{2}m_2 v_{2f}^2$$

But $m_1 = m_2 = m$, so that

(1) $$v_{1i}^2 = v_{1f}^2 + v_{2f}^2$$

Applying conservation of momentum to the two-dimensional collision gives

(2) $$m_1\mathbf{v}_{1i} = m_1\mathbf{v}_{1f} + m_2\mathbf{v}_{2f}$$

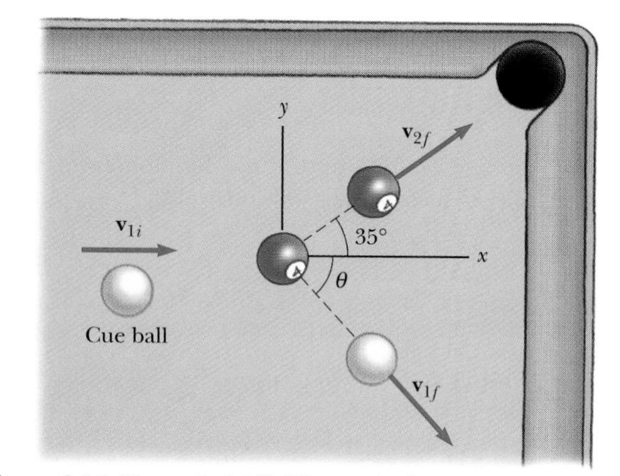

Figure 9.15 (Example 9.12) The cue ball (white) strikes the number 4 ball (blue) and sends it toward the corner pocket.

Note that because $m_1 = m_2 = m$, the masses also cancel in Equation (2). If we square both sides of Equation (2) and use the definition of the dot product of two vectors from Section 7.3, we obtain

$$v_{1i}^2 = (\mathbf{v}_{1f} + \mathbf{v}_{2f}) \cdot (\mathbf{v}_{1f} + \mathbf{v}_{2f}) = v_{1f}^2 + v_{2f}^2 + 2\mathbf{v}_{1f} \cdot \mathbf{v}_{2f}$$

Because the angle between $\mathbf{v}_{1f}$ and $\mathbf{v}_{2f}$ is $\theta + 35°$, $\mathbf{v}_{1f} \cdot \mathbf{v}_{2f} = v_{1f}v_{2f}\cos(\theta + 35°)$, and so

(3) $$v_{1i}^2 = v_{1f}^2 + v_{2f}^2 + 2v_{1f}v_{2f}\cos(\theta + 35°)$$

Subtracting Equation (1) from Equation (3) gives

$$0 = 2v_{1f}v_{2f}\cos(\theta + 35°)$$

$$0 = \cos(\theta + 35°)$$

$$\theta + 35° = 90° \quad \text{or} \quad \theta = 55°$$

This result shows that whenever two equal masses undergo a glancing elastic collision and one of them is initially at rest, they move in perpendicular directions after the collision. The same physics describes two very different situations, protons in Example 9.11 and billiard balls in this example.

9.5 The Center of Mass

In this section we describe the overall motion of a mechanical system in terms of a special point called the **center of mass** of the system. The mechanical system can be either a group of particles, such as a collection of atoms in a container, or an extended object, such as a gymnast leaping through the air. We shall see that the center of mass of the system moves as if all the mass of the system were concentrated at that point. Furthermore, if the resultant external force on the system is $\Sigma\mathbf{F}_{\text{ext}}$ and the total mass of the system is M, the center of mass moves with an acceleration given by $\mathbf{a} = \Sigma\mathbf{F}_{\text{ext}}/M$. That is, the system moves as if the resultant external force were applied to a single particle of mass M located at the center of mass. This behavior is independent of other motion, such as rotation or vibration of the system. This is the *particle model* that was introduced in Chapter 2.

Consider a mechanical system consisting of a pair of particles that have different masses and are connected by a light, rigid rod (Fig. 9.16). The position of the center of mass of a system can be described as being the *average position* of the system's mass. The center of mass of the system is located somewhere on the line joining the two particles and is closer to the particle having the larger mass. If a single force is applied at a point on the rod somewhere between the center of mass and the less massive particle, the system rotates clockwise (see Fig. 9.16a). If the force is applied at a point on the rod somewhere between the center of mass and the more massive particle, the system rotates counterclockwise (see Fig. 9.16b). If the force is applied at the center of mass, the system moves in the direction of $\mathbf{F}$ without rotating (see Fig. 9.16c). Thus, the center of mass can be located with this procedure.

The center of mass of the pair of particles described in Figure 9.17 is located on the x axis and lies somewhere between the particles. Its x coordinate is given by

$$x_{\text{CM}} \equiv \frac{m_1x_1 + m_2x_2}{m_1 + m_2} \tag{9.27}$$

For example, if $x_1 = 0$, $x_2 = d$, and $m_2 = 2m_1$, we find that $x_{CM} = \frac{2}{3}d$. That is, the center of mass lies closer to the more massive particle. If the two masses are equal, the center of mass lies midway between the particles.

We can extend this concept to a system of many particles with masses m_i in three dimensions. The x coordinate of the center of mass of n particles is defined to be

$$x_{CM} \equiv \frac{m_1x_1 + m_2x_2 + m_3x_3 + \cdots + m_nx_n}{m_1 + m_2 + m_3 + \cdots + m_n} = \frac{\sum_i m_ix_i}{\sum_i m_i} = \frac{\sum_i m_ix_i}{M} \tag{9.28}$$

where x_i is the x coordinate of the ith particle. For convenience, we express the total mass as $M \equiv \sum_i m_i$ where the sum runs over all n particles. The y and z coordinates of the center of mass are similarly defined by the equations

$$y_{CM} \equiv \frac{\sum_i m_iy_i}{M} \quad \text{and} \quad z_{CM} \equiv \frac{\sum_i m_iz_i}{M} \tag{9.29}$$

The center of mass can also be located by its position vector $\mathbf{r}_{CM}$. The Cartesian coordinates of this vector are x_{CM}, y_{CM}, and z_{CM}, defined in Equations 9.28 and 9.29. Therefore,

$$\mathbf{r}_{CM} = x_{CM}\hat{\mathbf{i}} + y_{CM}\hat{\mathbf{j}} + z_{CM}\hat{\mathbf{k}} = \frac{\sum_i m_ix_i\hat{\mathbf{i}} + \sum_i m_iy_i\hat{\mathbf{j}} + \sum_i m_iz_i\hat{\mathbf{k}}}{M}$$

$$\mathbf{r}_{CM} \equiv \frac{\sum_i m_i\mathbf{r}_i}{M} \tag{9.30}$$

where $\mathbf{r}_i$ is the position vector of the ith particle, defined by

$$\mathbf{r}_i \equiv x_i\hat{\mathbf{i}} + y_i\hat{\mathbf{j}} + z_i\hat{\mathbf{k}}$$

Although locating the center of mass for an extended object is somewhat more cumbersome than locating the center of mass of a system of particles, the basic ideas we have discussed still apply. We can think of an extended object as a system containing a large number of particles (Fig. 9.18). The particle separation is very small, and so the object can be considered to have a continuous mass distribution. By dividing the object into elements of mass Δm_i with coordinates x_i, y_i, z_i, we see that the x coordinate of the center of mass is approximately

$$x_{CM} \approx \frac{\sum_i x_i\,\Delta m_i}{M}$$

with similar expressions for y_{CM} and z_{CM}. If we let the number of elements n approach infinity, then x_{CM} is given precisely. In this limit, we replace the sum by an integral and Δm_i by the differential element dm:

$$x_{CM} = \lim_{\Delta m_i \to 0} \frac{\sum_i x_i\,\Delta m_i}{M} = \frac{1}{M}\int x\,dm \tag{9.31}$$

Likewise, for y_{CM} and z_{CM} we obtain

$$y_{CM} = \frac{1}{M}\int y\,dm \quad \text{and} \quad z_{CM} = \frac{1}{M}\int z\,dm \tag{9.32}$$

We can express the vector position of the center of mass of an extended object in the form

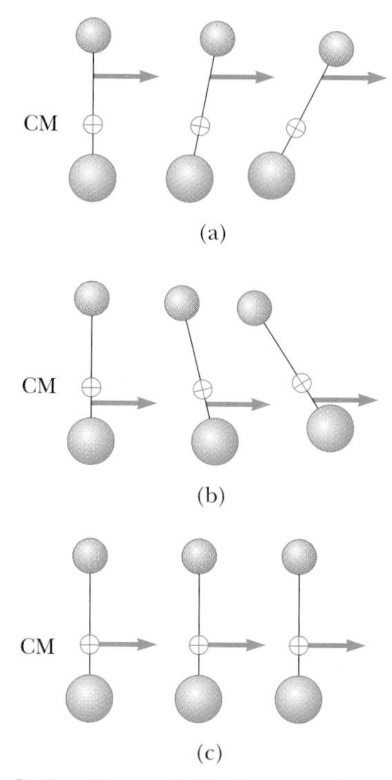

Active Figure 9.16 Two particles of unequal mass are connected by a light, rigid rod. (a) The system rotates clockwise when a force is applied between the less massive particle and the center of mass. (b) The system rotates counterclockwise when a force is applied between the more massive particle and the center of mass. (c) The system moves in the direction of the force without rotating when a force is applied at the center of mass.

At the Active Figures link at http://www.pse6.com, you can choose the point at which to apply the force.

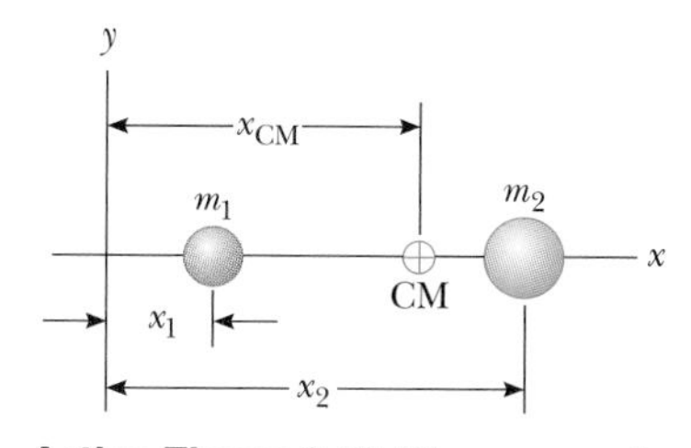

Active Figure 9.17 The center of mass of two particles of unequal mass on the x axis is located at x_{CM}, a point between the particles, closer to the one having the larger mass.

At the Active Figures link at http://www.pse6.com, you can adjust the masses and positions of the particles to see the effect on the location of the center of mass.

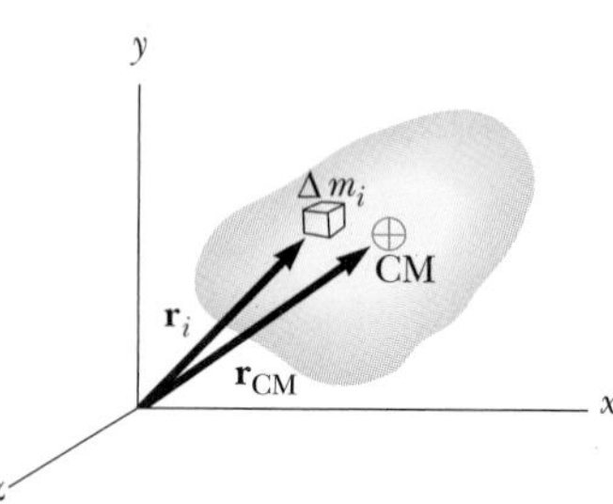

Figure 9.18 An extended object can be considered to be a distribution of small elements of mass Δm_i. The center of mass is located at the vector position $\mathbf{r}_{\text{CM}}$, which has coordinates x_{CM}, y_{CM}, and z_{CM}.

$$\mathbf{r}_{\text{CM}} = \frac{1}{M}\int \mathbf{r}\,dm \qquad (9.33)$$

which is equivalent to the three expressions given by Equations 9.31 and 9.32.

The center of mass of any symmetric object lies on an axis of symmetry and on any plane of symmetry.[4] For example, the center of mass of a uniform rod lies in the rod, midway between its ends. The center of mass of a sphere or a cube lies at its geometric center.

The center of mass of an irregularly shaped object such as a wrench can be determined by suspending the object first from one point and then from another. In Figure 9.19, a wrench is hung from point A, and a vertical line AB (which can be established with a plumb bob) is drawn when the wrench has stopped swinging. The wrench is then hung from point C, and a second vertical line CD is drawn. The center of mass is halfway through the thickness of the wrench, under the intersection of these two lines. In general, if the wrench is hung freely from any point, the vertical line through this point must pass through the center of mass.

Because an extended object is a continuous distribution of mass, each small mass element is acted upon by the gravitational force. The net effect of all these forces is equivalent to the effect of a single force $M\mathbf{g}$ acting through a special point, called the **center of gravity.** If $\mathbf{g}$ is constant over the mass distribution, then the center of gravity coincides with the center of mass. If an extended object is pivoted at its center of gravity, it balances in any orientation.

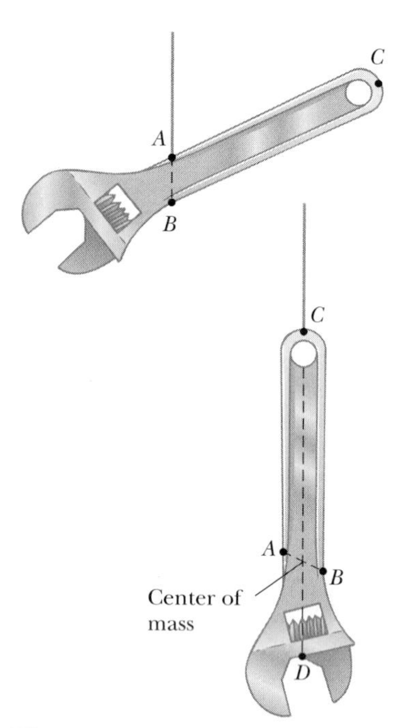

Figure 9.19 An experimental technique for determining the center of mass of a wrench. The wrench is hung freely first from point A and then from point C. The intersection of the two lines AB and CD locates the center of mass.

Quick Quiz 9.10 A baseball bat is cut at the location of its center of mass as shown in Figure 9.20. The piece with the smaller mass is (a) the piece on the right (b) the piece on the left (c) Both pieces have the same mass. (d) impossible to determine.

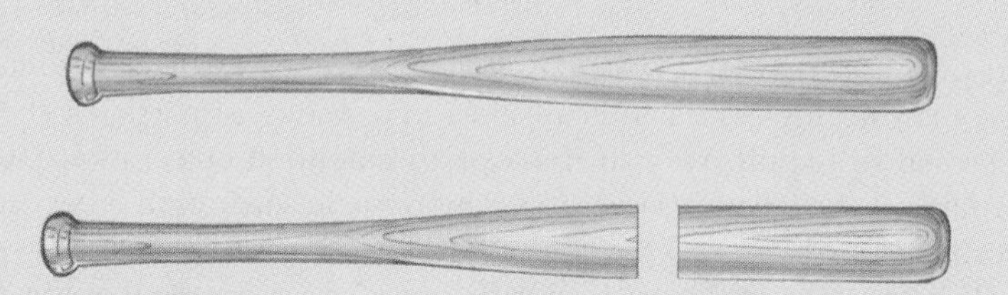

Figure 9.20 (Quick Quiz 9.10) A baseball bat cut at the location of its center of mass.

Example 9.13 The Center of Mass of Three Particles

A system consists of three particles located as shown in Figure 9.21a. Find the center of mass of the system.

Solution We set up the problem by labeling the masses of the particles as shown in the figure, with $m_1 = m_2 = 1.0$ kg and $m_3 = 2.0$ kg. Using the defining equations for the coordinates of the center of mass and noting that $z_{\text{CM}} = 0$, we obtain

$$x_{\text{CM}} = \frac{\sum_i m_i x_i}{M} = \frac{m_1x_1 + m_2x_2 + m_3x_3}{m_1 + m_2 + m_3}$$

$$= \frac{(1.0\text{ kg})(1.0\text{ m}) + (1.0\text{ kg})(2.0\text{ m}) + (2.0\text{ kg})(0)}{1.0\text{ kg} + 1.0\text{ kg} + 2.0\text{ kg}}$$

$$= \frac{3.0\text{ kg}\cdot\text{m}}{4.0\text{ kg}} = 0.75\text{ m}$$

[4] This statement is valid only for objects that have a uniform mass per unit volume.

$$y_{CM} = \frac{\sum_i m_i y_i}{M} = \frac{m_1 y_1 + m_2 y_2 + m_3 y_3}{m_1 + m_2 + m_3}$$

$$= \frac{(1.0 \text{ kg})(0) + (1.0 \text{ kg})(0) + (2.0 \text{ kg})(2.0 \text{ m})}{4.0 \text{ kg}}$$

$$= \frac{4.0 \text{ kg}\cdot\text{m}}{4.0 \text{ kg}} = 1.0 \text{ m}$$

The position vector to the center of mass measured from the origin is therefore

$$\mathbf{r}_{CM} \equiv x_{CM}\hat{\mathbf{i}} + y_{CM}\hat{\mathbf{j}} = (0.75\hat{\mathbf{i}} + 1.0\hat{\mathbf{j}}) \text{ m}$$

We can verify this result graphically by adding together $m_1\mathbf{r}_1 + m_2\mathbf{r}_2 + m_3\mathbf{r}_3$ and dividing the vector sum by M, the total mass. This is shown in Figure 9.21b.

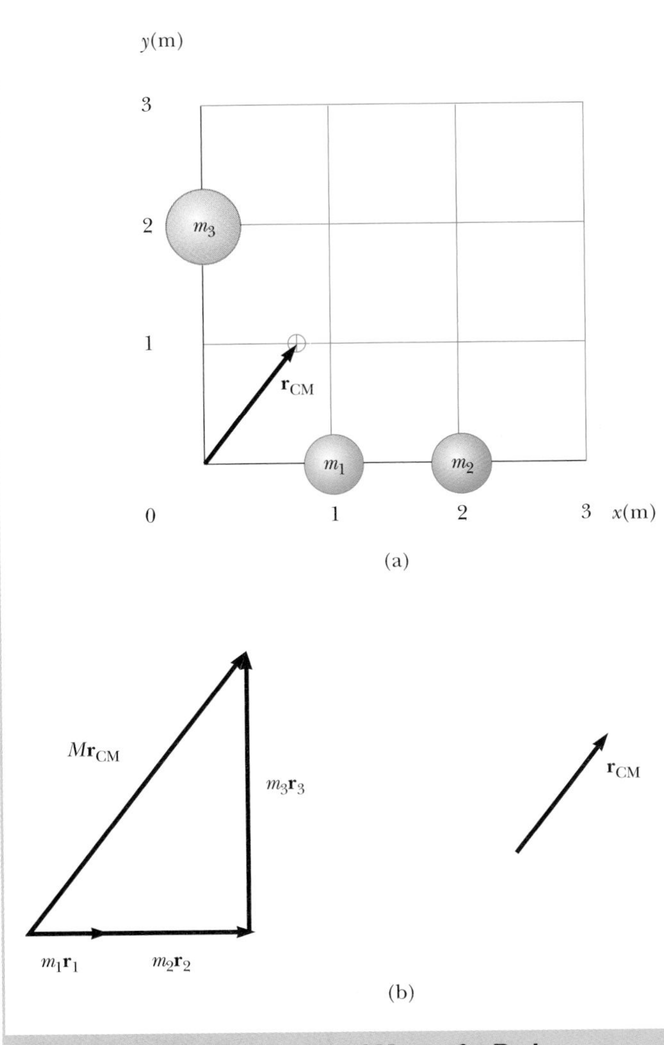

Figure 9.21 (Example 9.13) (a) Two 1.0-kg particles are located on the x axis and a single 2.0-kg particle is located on the y axis as shown. The vector indicates the location of the system's center of mass. (b) The vector sum of $m_i\mathbf{r}_i$ and the resulting vector for $\mathbf{r}_{CM}$.

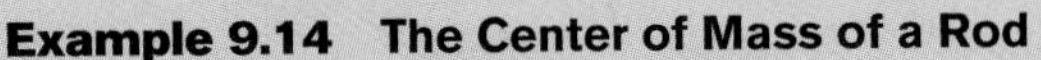

Example 9.14 The Center of Mass of a Rod

(A) Show that the center of mass of a rod of mass M and length L lies midway between its ends, assuming the rod has a uniform mass per unit length.

Solution The rod is shown aligned along the x axis in Figure 9.22, so that $y_{CM} = z_{CM} = 0$. Furthermore, if we call the mass per unit length λ (this quantity is called the *linear mass density*), then $\lambda = M/L$ for the uniform rod we assume here. If we divide the rod into elements of length dx, then the mass of each element is $dm = \lambda\, dx$. Equation 9.31 gives

$$x_{CM} = \frac{1}{M}\int x\,dm = \frac{1}{M}\int_0^L x\lambda\,dx = \frac{\lambda}{M}\left.\frac{x^2}{2}\right|_0^L = \frac{\lambda L^2}{2M}$$

Because $\lambda = M/L$, this reduces to

$$x_{CM} = \frac{L^2}{2M}\left(\frac{M}{L}\right) = \frac{L}{2}$$

One can also use symmetry arguments to obtain the same result.

(B) Suppose a rod is *nonuniform* such that its mass per unit length varies linearly with x according to the expression $\lambda = \alpha x$, where α is a constant. Find the x coordinate of the center of mass as a fraction of L.

Solution In this case, we replace dm by $\lambda\, dx$, where λ is not constant. Therefore, x_{CM} is

$$x_{CM} = \frac{1}{M}\int x\,dm = \frac{1}{M}\int_0^L x\lambda\,dx = \frac{1}{M}\int_0^L x\alpha x\,dx$$

$$= \frac{\alpha}{M}\int_0^L x^2\,dx = \frac{\alpha L^3}{3M}$$

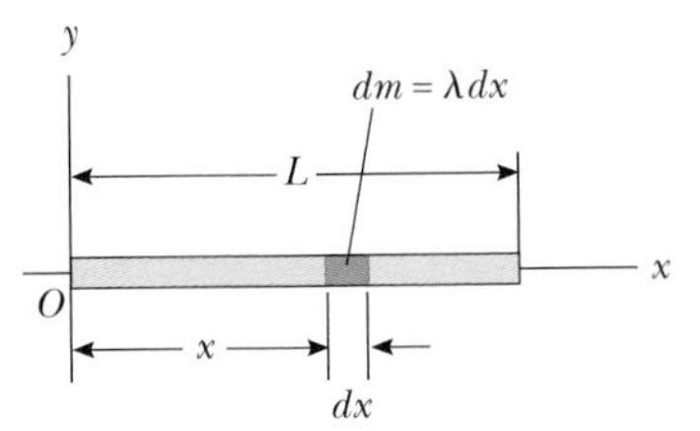

Figure 9.22 (Example 9.14) The geometry used to find the center of mass of a uniform rod.

We can eliminate α by noting that the total mass of the rod is related to α through the relationship

$$M = \int dm = \int_0^L \lambda\, dx = \int_0^L \alpha x\, dx = \frac{\alpha L^2}{2}$$

Substituting this into the expression for x_{CM} gives

$$x_{CM} = \frac{\alpha L^3}{3\alpha L^2/2} = \tfrac{2}{3}L$$

Example 9.15 The Center of Mass of a Right Triangle

You have been asked to hang a metal sign from a single vertical wire. The sign has the triangular shape shown in Figure 9.23a. The bottom of the sign is to be parallel to the ground. At what distance from the left end of the sign should you attach the support wire?

Solution The wire must be attached at a point directly above the center of gravity of the sign, which is the same as its center of mass because it is in a uniform gravitational field. We assume that the triangular sign has a uniform density and total mass M. Because the sign is a continuous distribution of mass, we must use the integral expression in Equation 9.31 to find the x coordinate of the center of mass.

We divide the triangle into narrow strips of width dx and height y as shown in Figure 9.23b, where y is the height of the hypotenuse of the triangle above the x axis for a given value of x. The mass of each strip is the product of the volume of the strip and the density ρ of the material from which the sign is made: $dm = \rho yt\, dx$, where t is the thickness of the metal sign. The density of the material is the total mass of the sign divided by its total volume (area of the triangle times thickness), so

$$dm = \rho yt\, dx = \left(\frac{M}{\frac{1}{2}abt}\right) yt\, dx = \frac{2My}{ab}\, dx$$

Using Equation 9.31 to find the x coordinate of the center of mass gives

$$x_{CM} = \frac{1}{M}\int x\, dm = \frac{1}{M}\int_0^a x\, \frac{2My}{ab}\, dx = \frac{2}{ab}\int_0^a xy\, dx$$

To proceed further and evaluate the integral, we must express y in terms of x. The line representing the hypotenuse of the triangle in Figure 9.23b has a slope of b/a and passes through the origin, so the equation of this line is $y = (b/a)x$. With this substitution for y in the integral, we have

$$x_{CM} = \frac{2}{ab}\int_0^a x\left(\frac{b}{a}x\right) dx = \frac{2}{a^2}\int_0^a x^2\, dx = \frac{2}{a^2}\left[\frac{x^3}{3}\right]_0^a$$

$$= \tfrac{2}{3}a$$

Thus, the wire must be attached to the sign at a distance two thirds of the length of the bottom edge from the left end. We could also find the y coordinate of the center of mass of the sign, but this is not needed in order to determine where the wire should be attached.

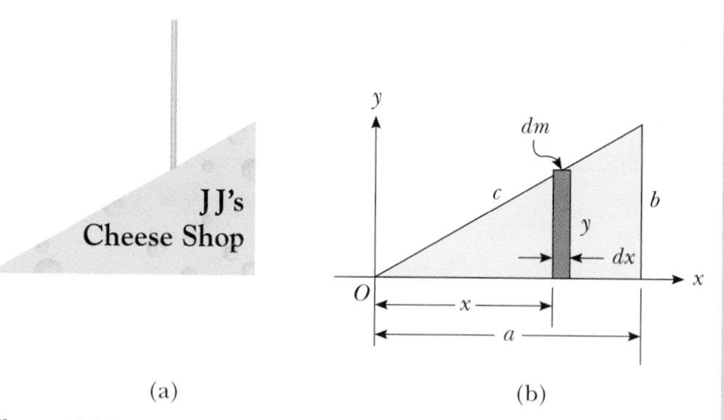

Figure 9.23 (Example 9.15) (a) A triangular sign to be hung from a single wire. (b) Geometric construction for locating the center of mass.

9.6 Motion of a System of Particles

We can begin to understand the physical significance and utility of the center of mass concept by taking the time derivative of the position vector given by Equation 9.30. From Section 4.1 we know that the time derivative of a position vector is by definition a velocity. Assuming M remains constant for a system of particles, that is, no particles enter or leave the system, we obtain the following expression for the **velocity of the center of mass** of the system:

Velocity of the center of mass

$$\mathbf{v}_{CM} = \frac{d\mathbf{r}_{CM}}{dt} = \frac{1}{M}\sum_i m_i \frac{d\mathbf{r}_i}{dt} = \frac{\sum_i m_i \mathbf{v}_i}{M} \quad (9.34)$$

where $\mathbf{v}_i$ is the velocity of the ith particle. Rearranging Equation 9.34 gives

Total momentum of a system of particles

$$M\mathbf{v}_{CM} = \sum_i m_i \mathbf{v}_i = \sum_i \mathbf{p}_i = \mathbf{p}_{tot} \quad (9.35)$$

Therefore, we conclude that the **total linear momentum of the system equals the total mass multiplied by the velocity of the center of mass.** In other words, the total linear momentum of the system is equal to that of a single particle of mass M moving with a velocity $\mathbf{v}_{\text{CM}}$.

If we now differentiate Equation 9.34 with respect to time, we obtain the **acceleration of the center of mass** of the system:

$$\mathbf{a}_{\text{CM}} = \frac{d\mathbf{v}_{\text{CM}}}{dt} = \frac{1}{M}\sum_i m_i \frac{d\mathbf{v}_i}{dt} = \frac{1}{M}\sum_i m_i \mathbf{a}_i \qquad (9.36)$$

Acceleration of the center of mass

Rearranging this expression and using Newton's second law, we obtain

$$M\mathbf{a}_{\text{CM}} = \sum m_i \mathbf{a}_i = \sum_i \mathbf{F}_i \qquad (9.37)$$

where $\mathbf{F}_i$ is the net force on particle i.

The forces on any particle in the system may include both external forces (from outside the system) and internal forces (from within the system). However, by Newton's third law, the internal force exerted by particle 1 on particle 2, for example, is equal in magnitude and opposite in direction to the internal force exerted by particle 2 on particle 1. Thus, when we sum over all internal forces in Equation 9.37, they cancel in pairs and we find that the net force on the system is caused *only* by external forces. Thus, we can write Equation 9.37 in the form

$$\sum \mathbf{F}_{\text{ext}} = M\mathbf{a}_{\text{CM}} \qquad (9.38)$$

Newton's second law for a system of particles

That is, **the net external force on a system of particles equals the total mass of the system multiplied by the acceleration of the center of mass.** If we compare this with Newton's second law for a single particle, we see that the particle model that we have used for several chapters can be described in terms of the center of mass:

The center of mass of a system of particles of combined mass M moves like an equivalent particle of mass M would move under the influence of the net external force on the system.

Finally, we see that if the net external force is zero, then from Equation 9.38 it follows that

$$M\mathbf{a}_{\text{CM}} = M\frac{d\mathbf{v}_{\text{CM}}}{dt} = 0$$

so that

$$M\mathbf{v}_{\text{CM}} = \mathbf{p}_{\text{tot}} = \text{constant} \qquad (\text{when } \sum \mathbf{F}_{\text{ext}} = 0) \qquad (9.39)$$

That is, the total linear momentum of a system of particles is conserved if no net external force is acting on the system. It follows that for an isolated system of particles, both the total momentum and the velocity of the center of mass are constant in time, as shown in Figure 9.24. This is a generalization to a many-particle system of the law of conservation of momentum discussed in Section 9.1 for a two-particle system.

Suppose an isolated system consisting of two or more members is at rest. The center of mass of such a system remains at rest unless acted upon by an external force. For example, consider a system made up of a swimmer standing on a raft, with the system initially at rest. When the swimmer dives horizontally off the raft, the raft moves in the direction opposite to that of the swimmer and the center of mass of the system remains at rest (if we neglect friction between raft and water). Furthermore, the linear momentum of the diver is equal in magnitude to that of the raft, but opposite in direction.

As another example, suppose an unstable atom initially at rest suddenly breaks up into two fragments of masses M_1 and M_2, with velocities $\mathbf{v}_1$ and $\mathbf{v}_2$, respectively. Because the total momentum of the system before the breakup is zero, the total momentum of the

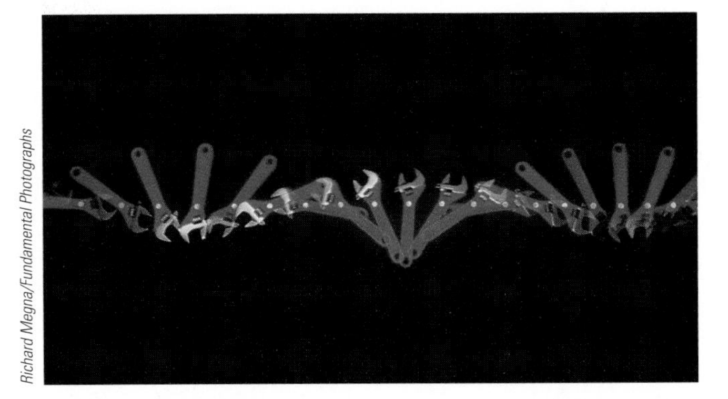

Richard Megna/Fundamental Photographs

Figure 9.24 Multiflash photograph showing an overhead view of a wrench moving on a horizontal surface. The white dots are located at the center of mass of the wrench and show that the center of mass moves in a straight line as the wrench rotates.

system after the breakup must also be zero. Therefore, $M_1\mathbf{v}_1 + M_2\mathbf{v}_2 = 0$. If the velocity of one of the fragments is known, the recoil velocity of the other fragment can be calculated.

Quick Quiz 9.11 The vacationers on a cruise ship are eager to arrive at their next destination. They decide to try to speed up the cruise ship by gathering at the bow (the front) and running all at once toward the stern (the back) of the ship. While they are running toward the stern, the speed of the ship is (a) higher than it was before (b) unchanged (c) lower than it was before (d) impossible to determine.

Quick Quiz 9.12 The vacationers in Quick Quiz 9.11 stop running when they reach the stern of the ship. After they have all stopped running, the speed of the ship is (a) higher than it was before they started running (b) unchanged from what it was before they started running (c) lower than it was before they started running (d) impossible to determine.

Conceptual Example 9.16 The Sliding Bear

Suppose you tranquilize a polar bear on a smooth glacier as part of a research effort. How might you estimate the bear's mass using a measuring tape, a rope, and knowledge of your own mass?

Solution Tie one end of the rope around the bear, and then lay out the tape measure on the ice with one end at the bear's original position, as shown in Figure 9.25. Grab hold of the free end of the rope and position yourself as

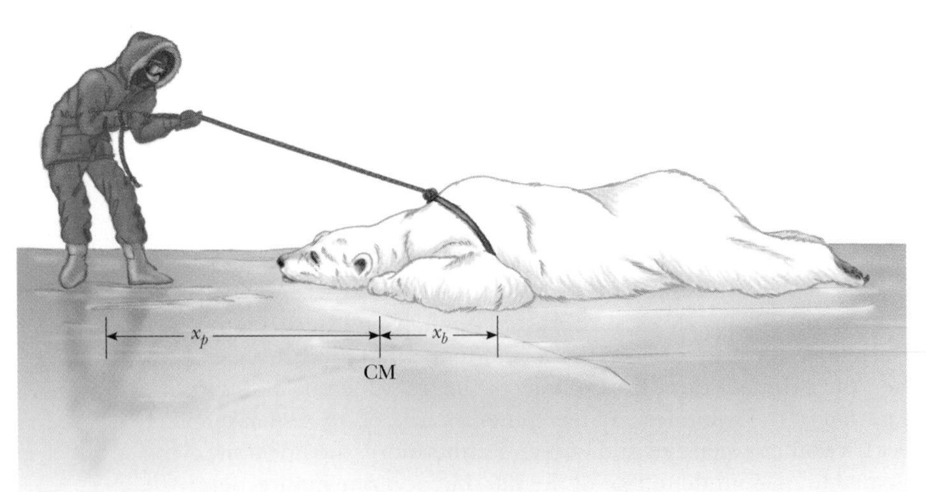

Figure 9.25 (Conceptual Example 9.16) The center of mass of an isolated system remains at rest unless acted on by an external force. How can you determine the mass of the polar bear?

shown, noting your location. Take off your spiked shoes, and pull on the rope hand over hand. Both you and the bear will slide over the ice until you meet. From the tape, observe how far you slide, x_p, and how far the bear slides, x_b. The point where you meet the bear is the fixed location of the center of mass of the system (bear plus you), and so you can determine the mass of the bear from $m_b x_b = m_p x_p$. (Unfortunately, you cannot return to your spiked shoes and so you are in big trouble if the bear wakes up!)

Conceptual Example 9.17 Exploding Projectile

A projectile fired into the air suddenly explodes into several fragments (Fig. 9.26). What can be said about the motion of the center of mass of the system made up of all the fragments after the explosion?

Solution Neglecting air resistance, the only external force on the projectile is the gravitational force. Thus, if the projectile did not explode, it would continue to move along the parabolic path indicated by the dashed line in Figure 9.26. Because the forces caused by the explosion are internal, they do not affect the motion of the center of mass of the system (the fragments). Thus, after the explosion, the center of mass of the fragments follows the same parabolic path the projectile would have followed if there had been no explosion.

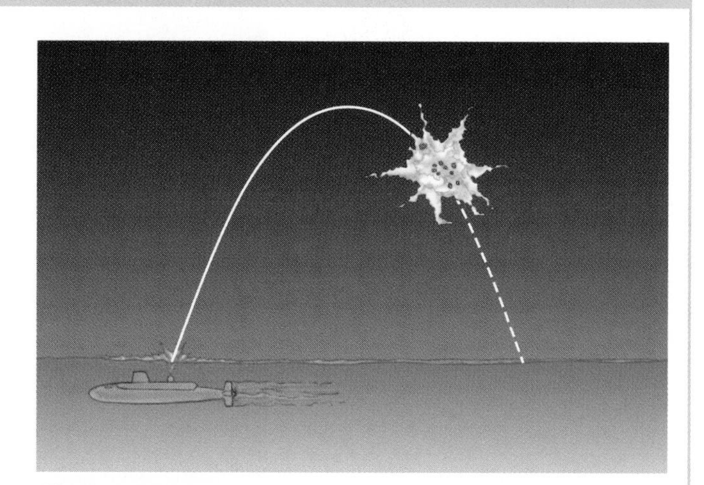

Figure 9.26 (Conceptual Example 9.17) When a projectile explodes into several fragments, the center of mass of the system made up of all the fragments follows the same parabolic path the projectile would have taken had there been no explosion.

Example 9.18 The Exploding Rocket

A rocket is fired vertically upward. At the instant it reaches an altitude of 1 000 m and a speed of 300 m/s, it explodes into three fragments having equal mass. One fragment continues to move upward with a speed of 450 m/s following the explosion. The second fragment has a speed of 240 m/s and is moving east right after the explosion. What is the velocity of the third fragment right after the explosion?

Solution Let us call the total mass of the rocket M; hence, the mass of each fragment is $M/3$. Because the forces of the explosion are internal to the system and cannot affect its total momentum, the total momentum $\mathbf{p}_i$ of the rocket just before the explosion must equal the total momentum $\mathbf{p}_f$ of the fragments right after the explosion.

Before the explosion,

$$\mathbf{p}_i = M\mathbf{v}_i = M(300\hat{\mathbf{j}} \text{ m/s})$$

After the explosion,

$$\mathbf{p}_f = \frac{M}{3}(240\hat{\mathbf{i}} \text{ m/s}) + \frac{M}{3}(450\hat{\mathbf{j}} \text{ m/s}) + \frac{M}{3}\mathbf{v}_f$$

where $\mathbf{v}_f$ is the unknown velocity of the third fragment. Equating these two expressions (because $\mathbf{p}_i = \mathbf{p}_f$) gives

$$\frac{M}{3}\mathbf{v}_f + \frac{M}{3}(240\hat{\mathbf{i}} \text{ m/s}) + \frac{M}{3}(450\hat{\mathbf{j}} \text{ m/s}) = M(300\hat{\mathbf{j}} \text{ m/s})$$

$$\mathbf{v}_f = (-240\hat{\mathbf{i}} + 450\hat{\mathbf{j}}) \text{ m/s}$$

What does the sum of the momentum vectors for all the fragments look like?

9.7 Rocket Propulsion

When ordinary vehicles such as cars and locomotives are propelled, the driving force for the motion is friction. In the case of the car, the driving force is the force exerted by the road on the car. A locomotive "pushes" against the tracks; hence, the driving force is the force exerted by the tracks on the locomotive. However, a rocket moving in space has no road or tracks to push against. Therefore, the source of the propulsion of a rocket must be something other than friction. Figure 9.27 is a dramatic photograph of a spacecraft at liftoff. **The operation of a rocket depends upon the law of conservation of linear momentum as applied to a system of particles, where the system is the rocket plus its ejected fuel.**

Figure 9.27 At liftoff, enormous thrust is generated by the space shuttle's liquid-fuel engines, aided by the two solid-fuel boosters. This photograph shows the liftoff of the space shuttle *Columbia,* which was lost in a tragic accident during its landing attempt on February 1, 2003 (shortly before this volume went to press).

Rocket propulsion can be understood by first considering a mechanical system consisting of a machine gun mounted on a cart on wheels. As the gun is fired, each bullet receives a momentum $m\mathbf{v}$ in some direction, where $\mathbf{v}$ is measured with respect to a stationary Earth frame. The momentum of the system made up of cart, gun, and bullets must be conserved. Hence, for each bullet fired, the gun and cart must receive a compensating momentum in the opposite direction. That is, the reaction force exerted by the bullet on the gun accelerates the cart and gun, and the cart moves in the direction opposite that of the bullets. If n is the number of bullets fired each second, then the average force exerted on the gun is $\overline{\mathbf{F}} = nm\mathbf{v}$.

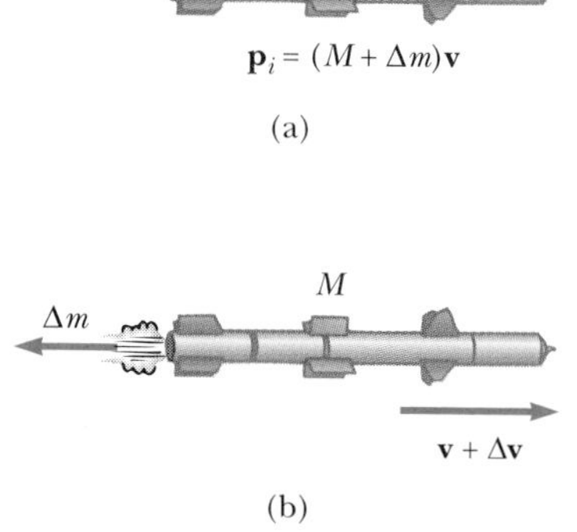

Figure 9.28 Rocket propulsion. (a) The initial mass of the rocket plus all its fuel is $M + \Delta m$ at a time t, and its speed is v. (b) At a time $t + \Delta t$, the rocket's mass has been reduced to M and an amount of fuel Δm has been ejected. The rocket's speed increases by an amount Δv.

In a similar manner, as a rocket moves in free space, its linear momentum changes when some of its mass is released in the form of ejected gases. **Because the gases are given momentum when they are ejected out of the engine, the rocket receives a compensating momentum in the opposite direction.** Therefore, the rocket is accelerated as a result of the "push," or thrust, from the exhaust gases. In free space, the center of mass of the system (rocket plus expelled gases) moves uniformly, independent of the propulsion process.[5]

Suppose that at some time t, the magnitude of the momentum of a rocket plus its fuel is $(M + \Delta m)v$, where v is the speed of the rocket relative to the Earth (Fig. 9.28a). Over a short time interval Δt, the rocket ejects fuel of mass Δm, and so at the end of this interval the rocket's speed is $v + \Delta v$, where Δv is the change in speed of the rocket (Fig. 9.28b). If the fuel is ejected with a speed v_e relative to the rocket (the subscript "e" stands for *exhaust,* and v_e is usually called the *exhaust speed*), the velocity of the fuel relative to a stationary frame of reference is $v - v_e$. Thus, if we equate the total initial mo-

[5] It is interesting to note that the rocket and machine gun represent cases of the reverse of a perfectly inelastic collision: momentum is conserved, but the kinetic energy of the system increases (at the expense of chemical potential energy in the fuel).

mentum of the system to the total final momentum, we obtain

$$(M + \Delta m)v = M(v + \Delta v) + \Delta m(v - v_e)$$

where M represents the mass of the rocket and its remaining fuel after an amount of fuel having mass Δm has been ejected. Simplifying this expression gives

$$M\,\Delta v = v_e\,\Delta m$$

We also could have arrived at this result by considering the system in the center-of-mass frame of reference, which is a frame having the same velocity as the center of mass of the system. In this frame, the total momentum of the system is zero; therefore, if the rocket gains a momentum $M\,\Delta v$ by ejecting some fuel, the exhausted fuel obtains a momentum $v_e\,\Delta m$ in the *opposite* direction, so that $M\Delta v - v_e\Delta m = 0$. If we now take the limit as Δt goes to zero, we let $\Delta v \rightarrow dv$ and $\Delta m \rightarrow dm$. Furthermore, the increase in the exhaust mass dm corresponds to an equal decrease in the rocket mass, so that $dm = -dM$. Note that dM is negative because it represents a decrease in mass, so $-dM$ is a positive number. Using this fact, we obtain

$$M\,dv = v_e\,dm = -v_e\,dM \qquad (9.40)$$

Courtesy of NASA

The force from a nitrogen-propelled hand-controlled device allows an astronaut to move about freely in space without restrictive tethers, using the thrust force from the expelled nitrogen.

We divide the equation by M and integrate, taking the initial mass of the rocket plus fuel to be M_i and the final mass of the rocket plus its remaining fuel to be M_f. This gives

$$\int_{v_i}^{v_f} dv = -v_e \int_{M_i}^{M_f} \frac{dM}{M}$$

$$v_f - v_i = v_e \ln\left(\frac{M_i}{M_f}\right) \qquad (9.41)$$

Expression for rocket propulsion

This is the basic expression for rocket propulsion. First, it tells us that the increase in rocket speed is proportional to the exhaust speed v_e of the ejected gases. Therefore, the exhaust speed should be very high. Second, the increase in rocket speed is proportional to the natural logarithm of the ratio M_i/M_f. Therefore, this ratio should be as large as possible, which means that the mass of the rocket without its fuel should be as small as possible and the rocket should carry as much fuel as possible.

The **thrust** on the rocket is the force exerted on it by the ejected exhaust gases. We can obtain an expression for the thrust from Equation 9.40:

$$\text{Thrust} = M\frac{dv}{dt} = \left| v_e \frac{dM}{dt} \right| \qquad (9.42)$$

This expression shows us that the thrust increases as the exhaust speed increases and as the rate of change of mass (called the *burn rate*) increases.

Example 9.19 A Rocket in Space

A rocket moving in space, far from all other objects, has a speed of 3.0×10^3 m/s relative to the Earth. Its engines are turned on, and fuel is ejected in a direction opposite the rocket's motion at a speed of 5.0×10^3 m/s relative to the rocket.

(A) What is the speed of the rocket relative to the Earth once the rocket's mass is reduced to half its mass before ignition?

Solution We can guess that the speed we are looking for must be greater than the original speed because the rocket is accelerating. Applying Equation 9.41, we obtain

$$v_f = v_i + v_e \ln\left(\frac{M_i}{M_f}\right)$$

$$= 3.0 \times 10^3 \text{ m/s} + (5.0 \times 10^3 \text{ m/s})\ln\left(\frac{M_i}{0.5\,M_i}\right)$$

$$= 6.5 \times 10^3 \text{ m/s}$$

(B) What is the thrust on the rocket if it burns fuel at the rate of 50 kg/s?

Solution Using Equation 9.42,

$$\text{Thrust} = \left| v_e \frac{dM}{dt} \right| = (5.0 \times 10^3 \text{ m/s})(50 \text{ kg/s})$$

$$= 2.5 \times 10^5 \text{ N}$$

Example 9.20 Fighting a Fire

Two firefighters must apply a total force of 600 N to steady a hose that is discharging water at the rate of 3 600 L/min. Estimate the speed of the water as it exits the nozzle.

Solution The water is exiting at 3 600 L/min, which is 60 L/s. Knowing that 1 L of water has a mass of 1 kg, we estimate that about 60 kg of water leaves the nozzle every second. As the water leaves the hose, it exerts on the hose a thrust that must be counteracted by the 600-N force exerted by the firefighters. So, applying Equation 9.42 gives

$$\text{Thrust} = \left| v_e \frac{dM}{dt} \right|$$

$$600 \text{ N} = |v_e(60 \text{ kg/s})|$$

$$v_e = 10 \text{ m/s}$$

Firefighting is dangerous work. If the nozzle should slip from their hands, the movement of the hose due to the thrust it receives from the rapidly exiting water could injure the firefighters.

***Take a practice test for this chapter by clicking on the Practice Test link at* http://www.pse6.com.**

SUMMARY

The **linear momentum p** of a particle of mass m moving with a velocity **v** is

$$\mathbf{p} \equiv m\mathbf{v} \tag{9.2}$$

The law of **conservation of linear momentum** indicates that the total momentum of an isolated system is conserved. If two particles form an isolated system, the momentum of the system is conserved regardless of the nature of the force between them. Therefore, the total momentum of the system at all times equals its initial total momentum, or

$$\mathbf{p}_{1i} + \mathbf{p}_{2i} = \mathbf{p}_{1f} + \mathbf{p}_{2f} \tag{9.5}$$

The **impulse** imparted to a particle by a force **F** is equal to the change in the momentum of the particle:

$$\mathbf{I} \equiv \int_{t_i}^{t_f} \mathbf{F}\,dt = \Delta\mathbf{p} \tag{9.8, 9.9}$$

This is known as the **impulse–momentum theorem.**

Impulsive forces are often very strong compared with other forces on the system and usually act for a very short time, as in the case of collisions.

When two particles collide, the total momentum of the isolated system before the collision always equals the total momentum after the collision, regardless of the nature of the collision. An **inelastic collision** is one for which the total kinetic energy of the system is not conserved. A **perfectly inelastic collision** is one in which the colliding bodies stick together after the collision. An **elastic collision** is one in which the kinetic energy of the system is conserved.

In a two- or three-dimensional collision, the components of momentum of an isolated system in each of the directions (x, y, and z) are conserved independently.

The position vector of the center of mass of a system of particles is defined as

$$\mathbf{r}_{\text{CM}} \equiv \frac{\sum_i m_i \mathbf{r}_i}{M} \tag{9.30}$$

where $M = \sum_i m_i$ is the total mass of the system and $\mathbf{r}_i$ is the position vector of the ith particle.

The position vector of the center of mass of an extended object can be obtained from the integral expression

$$\mathbf{r}_{\text{CM}} = \frac{1}{M}\int \mathbf{r}\, dm \tag{9.33}$$

The velocity of the center of mass for a system of particles is

$$\mathbf{v}_{\text{CM}} = \frac{\sum_i m_i \mathbf{v}_i}{M} \tag{9.34}$$

The total momentum of a system of particles equals the total mass multiplied by the velocity of the center of mass.

Newton's second law applied to a system of particles is

$$\sum \mathbf{F}_{\text{ext}} = M\mathbf{a}_{\text{CM}} \tag{9.38}$$

where $\mathbf{a}_{\text{CM}}$ is the acceleration of the center of mass and the sum is over all external forces. The center of mass moves like an imaginary particle of mass M under the influence of the resultant external force on the system.

QUESTIONS

1. Does a large force always produce a larger impulse on an object than a smaller force does? Explain.
2. If the speed of a particle is doubled, by what factor is its momentum changed? By what factor is its kinetic energy changed?
3. If two particles have equal kinetic energies, are their momenta necessarily equal? Explain.
4. While in motion, a pitched baseball carries kinetic energy and momentum. (a) Can we say that it carries a force that it can exert on any object it strikes? (b) Can the baseball deliver more kinetic energy to the object it strikes than the ball carries initially? (c) Can the baseball deliver to the object it strikes more momentum than the ball carries initially? Explain your answers.
5. An isolated system is initially at rest. Is it possible for parts of the system to be in motion at some later time? If so, explain how this might occur.
6. If two objects collide and one is initially at rest, is it possible for both to be at rest after the collision? Is it possible for one to be at rest after the collision? Explain.
7. Explain how linear momentum is conserved when a ball bounces from a floor.
8. A bomb, initially at rest, explodes into several pieces. (a) Is linear momentum of the system conserved? (b) Is kinetic energy of the system conserved? Explain.
9. A ball of clay is thrown against a brick wall. The clay stops and sticks to the wall. Is the principle of conservation of momentum violated in this example?
10. You are standing perfectly still, and then you take a step forward. Before the step your momentum was zero, but afterward you have some momentum. Is the principle of conservation of momentum violated in this case?
11. When a ball rolls down an incline, its linear momentum increases. Is the principle of conservation of momentum violated in this process?
12. Consider a perfectly inelastic collision between a car and a large truck. Which vehicle experiences a larger change in kinetic energy as a result of the collision?
13. A sharpshooter fires a rifle while standing with the butt of the gun against his shoulder. If the forward momentum of a bullet is the same as the backward momentum of the gun, why isn't it as dangerous to be hit by the gun as by the bullet?
14. A pole-vaulter falls from a height of 6.0 m onto a foam rubber pad. Can you calculate his speed just before he reaches the pad? Can you calculate the force exerted on him by the pad? Explain.
15. Firefighters must apply large forces to hold a fire hose steady (Fig. Q9.15). What factors related to the projection of the water determine the magnitude of the force needed to keep the end of the fire hose stationary?

Figure Q9.15

16. A large bed sheet is held vertically by two students. A third student, who happens to be the star pitcher on the baseball team, throws a raw egg at the sheet. Explain why the egg does not break when it hits the sheet, regardless of its initial speed. (If you try this demonstration, make sure the pitcher hits the sheet near its center, and do not allow the egg to fall on the floor after being caught.)

17. A skater is standing still on a frictionless ice rink. Her friend throws a Frisbee straight at her. In which of the following cases is the largest momentum transferred to the skater? (a) The skater catches the Frisbee and holds onto it. (b) The skater catches the Frisbee momentarily, but then drops it vertically downward. (c) The skater catches the Frisbee, holds it momentarily, and throws it back to her friend.

18. In an elastic collision between two particles, does the kinetic energy of each particle change as a result of the collision?

19. Three balls are thrown into the air simultaneously. What is the acceleration of their center of mass while they are in motion?

20. A person balances a meter stick in a horizontal position on the extended index fingers of her right and left hands. She slowly brings the two fingers together. The stick remains balanced and the two fingers always meet at the 50-cm mark regardless of their original positions. (Try it!) Explain.

21. NASA often uses the gravity of a planet to "slingshot" a probe on its way to a more distant planet. The interaction of the planet and the spacecraft is a collision in which the objects do not touch. How can the probe have its speed increased in this manner?

22. The Moon revolves around the Earth. Model its orbit as circular. Is the Moon's linear momentum conserved? Is its kinetic energy conserved?

23. A raw egg dropped to the floor breaks upon impact. However, a raw egg dropped onto a thick foam rubber cushion from a height of about 1 m rebounds without breaking. Why is this possible? If you try this experiment, be sure to catch the egg after its first bounce.

24. Can the center of mass of an object be located at a position at which there is no mass? If so, give examples.

25. A juggler juggles three balls in a continuous cycle. Any one ball is in contact with his hands for one fifth of the time. Describe the motion of the center of mass of the three balls. What average force does the juggler exert on one ball while he is touching it?

26. Does the center of mass of a rocket in free space accelerate? Explain. Can the speed of a rocket exceed the exhaust speed of the fuel? Explain.

27. Early in the twentieth century, Robert Goddard proposed sending a rocket to the moon. Critics objected that in a vacuum, such as exists between the Earth and the Moon, the gases emitted by the rocket would have nothing to push against to propel the rocket. According to *Scientific American* (January 1975), Goddard placed a gun in a vacuum and fired a blank cartridge from it. (A blank cartridge contains no bullet and fires only the wadding and the hot gases produced by the burning gunpowder.) What happened when the gun was fired?

28. Explain how you could use a balloon to demonstrate the mechanism responsible for rocket propulsion.

29. On the subject of the following positions, state your own view and argue to support it. (a) The best theory of motion is that force causes acceleration. (b) The true measure of a force's effectiveness is the work it does, and the best theory of motion is that work done on an object changes its energy. (c) The true measure of a force's effect is impulse, and the best theory of motion is that impulse injected into an object changes its momentum.

PROBLEMS

1, 2, 3 = straightforward, intermediate, challenging ☐ = full solution available in the *Student Solutions Manual and Study Guide*
= coached solution with hints available at http://www.pse6.com = computer useful in solving problem
= paired numerical and symbolic problems

Section 9.1 Linear Momentum and its Conservation

1. A 3.00-kg particle has a velocity of $(3.00\hat{\mathbf{i}} - 4.00\hat{\mathbf{j}})$ m/s. (a) Find its x and y components of momentum. (b) Find the magnitude and direction of its momentum.

2. A 0.100-kg ball is thrown straight up into the air with an initial speed of 15.0 m/s. Find the momentum of the ball (a) at its maximum height and (b) halfway up to its maximum height.

3. How fast can you set the Earth moving? In particular, when you jump straight up as high as you can, what is the order of magnitude of the maximum recoil speed that you give to the Earth? Model the Earth as a perfectly solid object. In your solution, state the physical quantities you take as data and the values you measure or estimate for them.

4. Two blocks of masses M and $3M$ are placed on a horizontal, frictionless surface. A light spring is attached to one

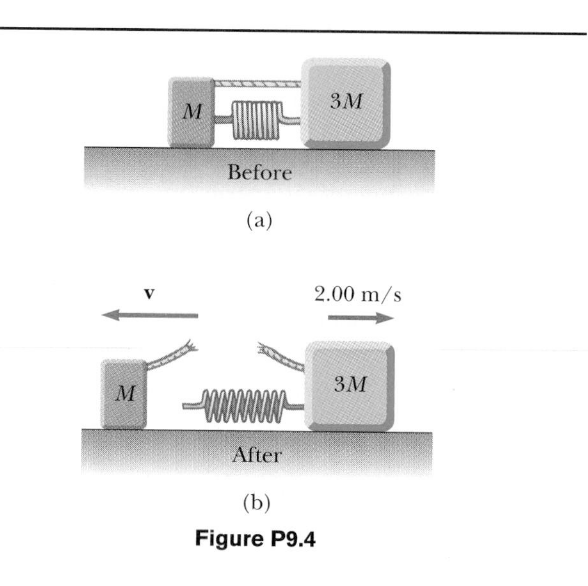

Figure P9.4

of them, and the blocks are pushed together with the spring between them (Fig. P9.4). A cord initially holding the blocks together is burned; after this, the block of mass $3M$ moves to the right with a speed of 2.00 m/s. (a) What is the speed of the block of mass M? (b) Find the original elastic potential energy in the spring if $M = 0.350$ kg.

5. (a) A particle of mass m moves with momentum p. Show that the kinetic energy of the particle is $K = p^2/2m$. (b) Express the magnitude of the particle's momentum in terms of its kinetic energy and mass.

Section 9.2 Impulse and Momentum

6. A friend claims that, as long as he has his seatbelt on, he can hold on to a 12.0-kg child in a 60.0 mi/h head-on collision with a brick wall in which the car passenger compartment comes to a stop in 0.050 0 s. Show that the violent force during the collision will tear the child from his arms. A child should always be in a toddler seat secured with a seat belt in the back seat of a car.

7. An estimated force–time curve for a baseball struck by a bat is shown in Figure P9.7. From this curve, determine (a) the impulse delivered to the ball, (b) the average force exerted on the ball, and (c) the peak force exerted on the ball.

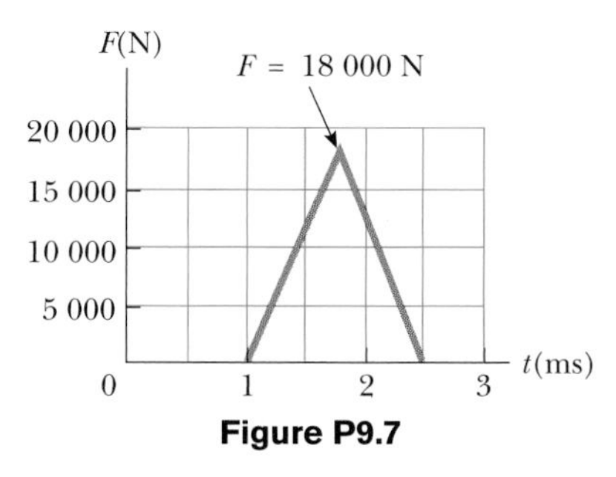

Figure P9.7

8. A ball of mass 0.150 kg is dropped from rest from a height of 1.25 m. It rebounds from the floor to reach a height of 0.960 m. What impulse was given to the ball by the floor?

9. A 3.00-kg steel ball strikes a wall with a speed of 10.0 m/s at an angle of 60.0° with the surface. It bounces off with the same speed and angle (Fig. P9.9). If the ball is in contact with the wall for 0.200 s, what is the average force exerted by the wall on the ball ?

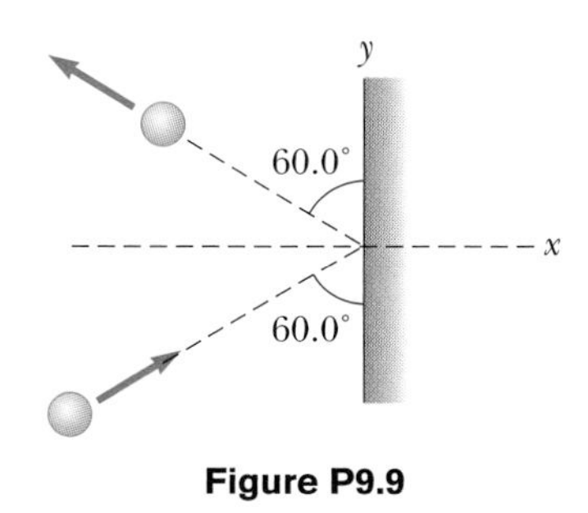

Figure P9.9

10. A tennis player receives a shot with the ball (0.060 0 kg) traveling horizontally at 50.0 m/s and returns the shot with the ball traveling horizontally at 40.0 m/s in the opposite direction. (a) What is the impulse delivered to the ball by the racquet? (b) What work does the racquet do on the ball?

11. In a slow-pitch softball game, a 0.200-kg softball crosses the plate at 15.0 m/s at an angle of 45.0° below the horizontal. The batter hits the ball toward center field, giving it a velocity of 40.0 m/s at 30.0° above the horizontal. (a) Determine the impulse delivered to the ball. (b) If the force on the ball increases linearly for 4.00 ms, holds constant for 20.0 ms, and then decreases to zero linearly in another 4.00 ms, what is the maximum force on the ball?

12. A professional diver performs a dive from a platform 10 m above the water surface. Estimate the order of magnitude of the average impact force she experiences in her collision with the water. State the quantities you take as data and their values.

13. A garden hose is held as shown in Figure P9.13. The hose is originally full of motionless water. What additional force is necessary to hold the nozzle stationary after the water flow is turned on, if the discharge rate is 0.600 kg/s with a speed of 25.0 m/s?

Figure P9.13

14. A glider of mass m is free to slide along a horizontal air track. It is pushed against a launcher at one end of the track. Model the launcher as a light spring of force constant k compressed by a distance x. The glider is released from rest. (a) Show that the glider attains a speed of $v = x(k/m)^{1/2}$. (b) Does a glider of large or of small mass attain a greater speed? (c) Show that the impulse imparted to the glider is given by the expression $x(km)^{1/2}$. (d) Is a greater impulse injected into a large or a small mass? (e) Is more work done on a large or a small mass?

Section 9.3 Collisions in One Dimension

15. High-speed stroboscopic photographs show that the head of a golf club of mass 200 g is traveling at 55.0 m/s just before it strikes a 46.0-g golf ball at rest on a tee. After the collision, the club head travels (in the same direction) at 40.0 m/s. Find the speed of the golf ball just after impact.

16. An archer shoots an arrow toward a target that is sliding toward her with a speed of 2.50 m/s on a smooth, slippery surface. The 22.5-g arrow is shot with a speed of 35.0 m/s and passes through the 300-g target, which is stopped by the impact. What is the speed of the arrow after passing through the target?

17. A 10.0-g bullet is fired into a stationary block of wood (m = 5.00 kg). The relative motion of the bullet stops inside the block. The speed of the bullet-plus-wood combination immediately after the collision is 0.600 m/s. What was the original speed of the bullet?

18. A railroad car of mass 2.50×10^4 kg is moving with a speed of 4.00 m/s. It collides and couples with three other coupled railroad cars, each of the same mass as the single car and moving in the same direction with an initial speed of 2.00 m/s. (a) What is the speed of the four cars after the collision? (b) How much mechanical energy is lost in the collision?

19. Four railroad cars, each of mass 2.50×10^4 kg, are coupled together and coasting along horizontal tracks at speed v_i toward the south. A very strong but foolish movie actor, riding on the second car, uncouples the front car and gives it a big push, increasing its speed to 4.00 m/s southward. The remaining three cars continue moving south, now at 2.00 m/s. (a) Find the initial speed of the cars. (b) How much work did the actor do? (c) State the relationship between the process described here and the process in Problem 18.

20. Two blocks are free to slide along the frictionless wooden track ABC shown in Figure P9.20. The block of mass m_1 = 5.00 kg is released from A. Protruding from its front end is the north pole of a strong magnet, repelling the north pole of an identical magnet embedded in the back end of the block of mass m_2 = 10.0 kg, initially at rest. The two blocks never touch. Calculate the maximum height to which m_1 rises after the elastic collision.

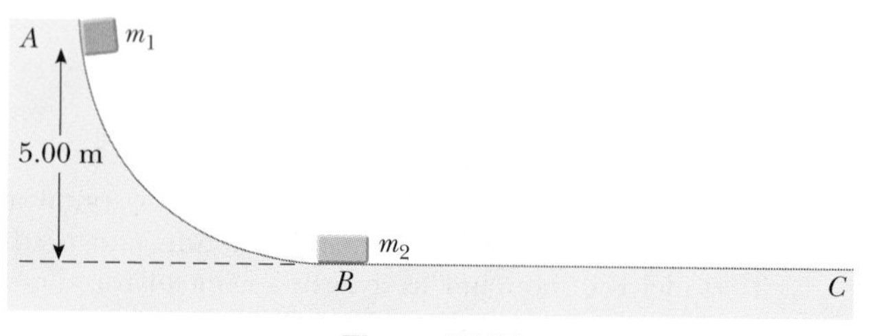

Figure P9.20

21. A 45.0-kg girl is standing on a plank that has a mass of 150 kg. The plank, originally at rest, is free to slide on a frozen lake, which is a flat, frictionless supporting surface. The girl begins to walk along the plank at a constant speed of 1.50 m/s relative to the plank. (a) What is her speed relative to the ice surface? (b) What is the speed of the plank relative to the ice surface?

22. Most of us know intuitively that in a head-on collision between a large dump truck and a subcompact car, you are better off being in the truck than in the car. Why is this? Many people imagine that the collision force exerted on the car is much greater than that experienced by the truck. To substantiate this view, they point out that the car is crushed, whereas the truck is only dented. This idea of unequal forces, of course, is false. Newton's third law tells us that both objects experience forces of the same magnitude. The truck suffers less damage because it is made of stronger metal. But what about the two drivers? Do they experience the same forces? To answer this question, suppose that each vehicle is initially moving at 8.00 m/s and that they undergo a perfectly inelastic head-on collision. Each driver has mass 80.0 kg. Including the drivers, the total vehicle masses are 800 kg for the car and 4 000 kg for the truck. If the collision time is 0.120 s, what force does the seatbelt exert on each driver?

23. A neutron in a nuclear reactor makes an elastic head-on collision with the nucleus of a carbon atom initially at rest. (a) What fraction of the neutron's kinetic energy is transferred to the carbon nucleus? (b) If the initial kinetic energy of the neutron is 1.60×10^{-13} J, find its final kinetic energy and the kinetic energy of the carbon nucleus after the collision. (The mass of the carbon nucleus is nearly 12.0 times the mass of the neutron.)

24. As shown in Figure P9.24, a bullet of mass m and speed v passes completely through a pendulum bob of mass M. The bullet emerges with a speed of $v/2$. The pendulum bob is suspended by a stiff rod of length ℓ and negligible mass. What is the minimum value of v such that the pendulum bob will barely swing through a complete vertical circle?

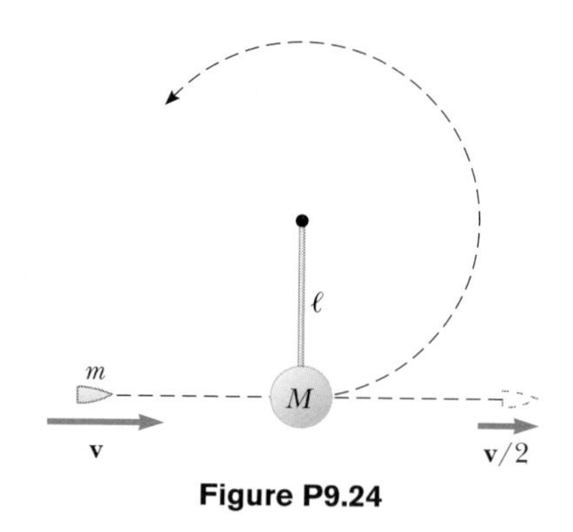

Figure P9.24

25. A 12.0-g wad of sticky clay is hurled horizontally at a 100-g wooden block initially at rest on a horizontal surface. The clay sticks to the block. After impact, the block slides 7.50 m before coming to rest. If the coefficient of friction between the block and the surface is 0.650, what was the speed of the clay immediately before impact?

26. A 7.00-g bullet, when fired from a gun into a 1.00-kg block of wood held in a vise, penetrates the block to a depth of 8.00 cm. **What If?** This block of wood is placed on a frictionless horizontal surface, and a second 7.00-g bullet is fired from the gun into the block. To what depth will the bullet penetrate the block in this case?

27. (a) Three carts of masses 4.00 kg, 10.0 kg, and 3.00 kg move on a frictionless horizontal track with speeds of 5.00 m/s, 3.00 m/s, and 4.00 m/s, as shown in Figure P9.27. Velcro couplers make the carts stick together after colliding. Find the final velocity of the train of three carts. (b) **What If?** Does your answer require that all the carts collide and stick together at the same time? What if they collide in a different order?

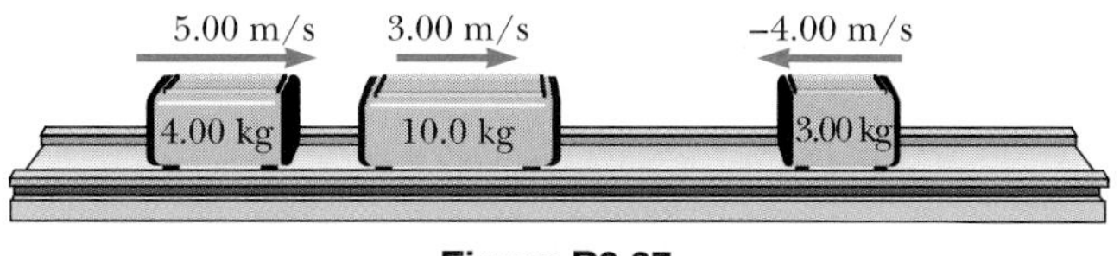

Figure P9.27

Section 9.4 Two-Dimensional Collisions

28. A 90.0-kg fullback running east with a speed of 5.00 m/s is tackled by a 95.0-kg opponent running north with a speed of 3.00 m/s. If the collision is perfectly inelastic, (a) calculate the speed and direction of the players just after the tackle and (b) determine the mechanical energy lost as a result of the collision. Account for the missing energy.

29. Two shuffleboard disks of equal mass, one orange and the other yellow, are involved in an elastic, glancing collision. The yellow disk is initially at rest and is struck by the orange disk moving with a speed of 5.00 m/s. After the collision, the orange disk moves along a direction that makes an angle of 37.0° with its initial direction of motion. The velocities of the two disks are perpendicular after the collision. Determine the final speed of each disk.

30. Two shuffleboard disks of equal mass, one orange and the other yellow, are involved in an elastic, glancing collision. The yellow disk is initially at rest and is struck by the orange disk moving with a speed v_i. After the collision, the orange disk moves along a direction that makes an angle θ with its initial direction of motion. The velocities of the two disks are perpendicular after the collision. Determine the final speed of each disk.

31. The mass of the blue puck in Figure P9.31 is 20.0% greater than the mass of the green one. Before colliding, the pucks approach each other with momenta of equal magnitudes and opposite directions, and the green puck has an initial speed of 10.0 m/s. Find the speeds of the pucks after the collision if half the kinetic energy is lost during the collision.

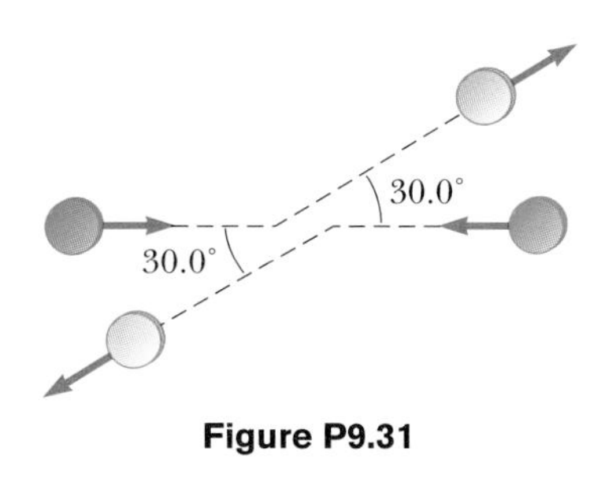

Figure P9.31

32. Two automobiles of equal mass approach an intersection. One vehicle is traveling with velocity 13.0 m/s toward the east, and the other is traveling north with speed v_{2i}. Neither driver sees the other. The vehicles collide in the intersection and stick together, leaving parallel skid marks at an angle of 55.0° north of east. The speed limit for both roads is 35 mi/h, and the driver of the northward-moving vehicle claims he was within the speed limit when the collision occurred. Is he telling the truth?

33. A billiard ball moving at 5.00 m/s strikes a stationary ball of the same mass. After the collision, the first ball moves, at 4.33 m/s, at an angle of 30.0° with respect to the original line of motion. Assuming an elastic collision (and ignoring friction and rotational motion), find the struck ball's velocity after the collision.

34. A proton, moving with a velocity of $v_i \hat{\mathbf{i}}$, collides elastically with another proton that is initially at rest. If the two protons have equal speeds after the collision, find (a) the speed of each proton after the collision in terms of v_i and (b) the direction of the velocity vectors after the collision.

35. An object of mass 3.00 kg, moving with an initial velocity of $5.00\hat{\mathbf{i}}$ m/s, collides with and sticks to an object of mass 2.00 kg with an initial velocity of $-3.00\hat{\mathbf{j}}$ m/s. Find the final velocity of the composite object.

36. Two particles with masses m and $3m$ are moving toward each other along the x axis with the same initial speeds v_i. Particle m is traveling to the left, while particle $3m$ is traveling to the right. They undergo an elastic glancing collision such that particle m is moving downward after the collision at right angles from its initial direction. (a) Find the final speeds of the two particles. (b) What is the angle θ at which the particle $3m$ is scattered?

37. An unstable atomic nucleus of mass 17.0×10^{-27} kg initially at rest disintegrates into three particles. One of the particles, of mass 5.00×10^{-27} kg, moves along the y axis with a speed of 6.00×10^6 m/s. Another particle, of mass 8.40×10^{-27} kg, moves along the x axis with a speed of 4.00×10^6 m/s. Find (a) the velocity of the third particle and (b) the total kinetic energy increase in the process.

Section 9.5 The Center of Mass

38. Four objects are situated along the y axis as follows: a 2.00 kg object is at + 3.00 m, a 3.00-kg object is at + 2.50 m, a 2.50-kg object is at the origin, and a 4.00-kg object is at − 0.500 m. Where is the center of mass of these objects?

39. A water molecule consists of an oxygen atom with two hydrogen atoms bound to it (Fig. P9.39). The angle between the two bonds is 106°. If the bonds are 0.100 nm long, where is the center of mass of the molecule?

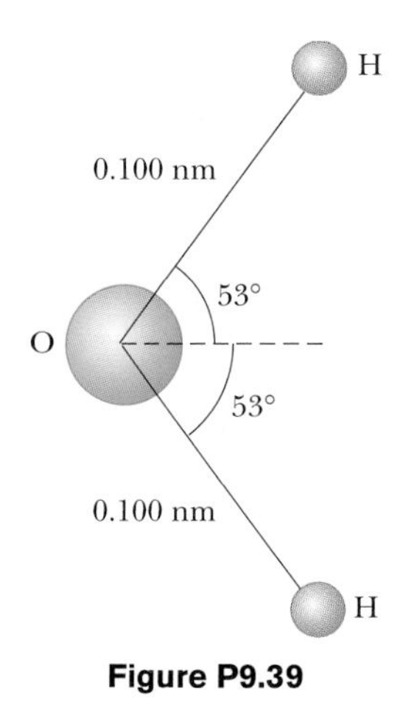

Figure P9.39

40. The mass of the Earth is 5.98×10^{24} kg, and the mass of the Moon is 7.36×10^{22} kg. The distance of separation, measured between their centers, is 3.84×10^{8} m. Locate the center of mass of the Earth–Moon system as measured from the center of the Earth.

41. A uniform piece of sheet steel is shaped as in Figure P9.41. Compute the x and y coordinates of the center of mass of the piece.

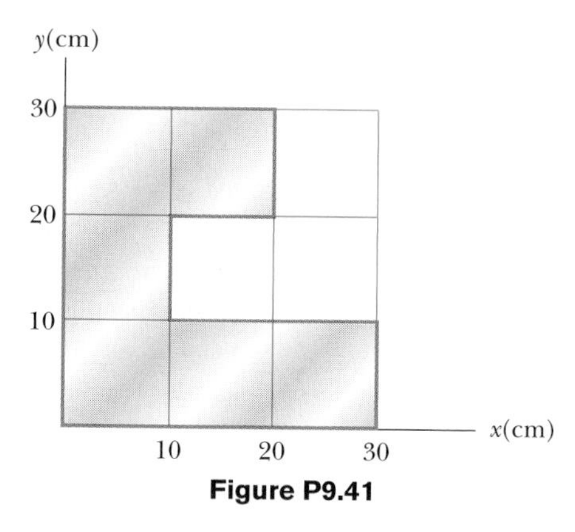

Figure P9.41

42. (a) Consider an extended object whose different portions have different elevations. Assume the free-fall acceleration is uniform over the object. Prove that the gravitational potential energy of the object–Earth system is given by $U_g = Mgy_{CM}$ where M is the total mass of the object and y_{CM} is the elevation of its center of mass above the chosen reference level. (b) Calculate the gravitational potential energy associated with a ramp constructed on level ground with stone with density 3 800 kg/m^3 and everywhere 3.60 m wide. In a side view, the ramp appears as a right triangle with height 15.7 m at the top end and base 64.8 m (Figure P9.42).

Figure P9.42

43. A rod of length 30.0 cm has linear density (mass-per-length) given by

$$\lambda = 50.0 \text{ g/m} + 20.0x \text{ g/m}^2,$$

where x is the distance from one end, measured in meters. (a) What is the mass of the rod? (b) How far from the $x = 0$ end is its center of mass?

44. In the 1968 Olympic Games, University of Oregon jumper Dick Fosbury introduced a new technique of high jumping called the "Fosbury flop." It contributed to raising the world record by about 30 cm and is presently used by nearly every world-class jumper. In this technique, the jumper goes over the bar face up while arching his back as much as possible, as in Figure P9.44a. This action places his center of mass outside his body, below his back. As his body goes over the bar, his center of mass passes below the bar. Because a given energy input implies a certain elevation for his center of mass, the action of arching his back means his body is higher than if his back were straight. As a model, consider the jumper as a thin uniform rod of length L. When the rod is straight, its center of mass is at its center. Now bend the rod in a circular arc so that it subtends an angle of 90.0° at the center of the arc, as shown in Figure P9.44b. In this configuration, how far outside the rod is the center of mass?

(a)

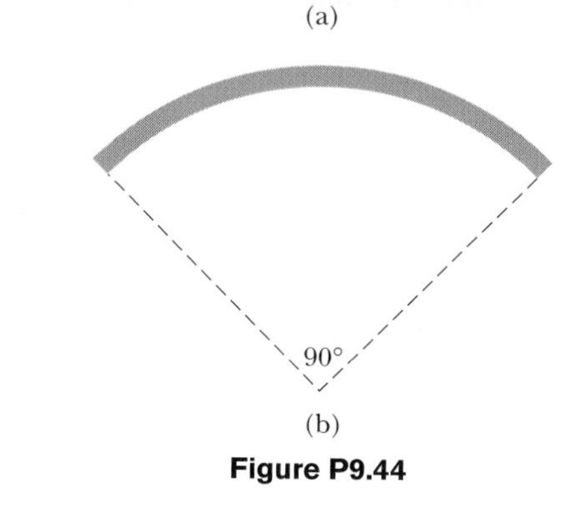

(b)

Figure P9.44

Section 9.6 Motion of a System of Particles

45. A 2.00-kg particle has a velocity $(2.00\hat{\mathbf{i}} - 3.00\hat{\mathbf{j}})$ m/s, and a 3.00-kg particle has a velocity $(1.00\hat{\mathbf{i}} + 6.00\hat{\mathbf{j}})$ m/s. Find (a) the velocity of the center of mass and (b) the total momentum of the system.

46. Consider a system of two particles in the xy plane: $m_1 =$ 2.00 kg is at the location $\mathbf{r}_1 = (1.00\hat{\mathbf{i}} + 2.00\hat{\mathbf{j}})$ m and has a velocity of $(3.00\hat{\mathbf{i}} + 0.500\hat{\mathbf{j}})$ m/s; $m_2 = 3.00$ kg is at $\mathbf{r}_2 =$ $(-4.00\hat{\mathbf{i}} - 3.00\hat{\mathbf{j}})$ m and has velocity $(3.00\hat{\mathbf{i}} - 2.00\hat{\mathbf{j}})$ m/s.

(a) Plot these particles on a grid or graph paper. Draw their position vectors and show their velocities. (b) Find the position of the center of mass of the system and mark it on the grid. (c) Determine the velocity of the center of mass and also show it on the diagram. (d) What is the total linear momentum of the system?

47. Romeo (77.0 kg) entertains Juliet (55.0 kg) by playing his guitar from the rear of their boat at rest in still water, 2.70 m away from Juliet, who is in the front of the boat. After the serenade, Juliet carefully moves to the rear of the boat (away from shore) to plant a kiss on Romeo's cheek. How far does the 80.0-kg boat move toward the shore it is facing?

48. A ball of mass 0.200 kg has a velocity of $150\hat{\mathbf{i}}$ m/s; a ball of mass 0.300 kg has a velocity of $-0.400\hat{\mathbf{i}}$ m/s. They meet in a head-on elastic collision. (a) Find their velocities after the collision. (b) Find the velocity of their center of mass before and after the collision.

Section 9.7 Rocket Propulsion

49. The first stage of a Saturn V space vehicle consumed fuel and oxidizer at the rate of 1.50×10^4 kg/s, with an exhaust speed of 2.60×10^3 m/s. (a) Calculate the thrust produced by these engines. (b) Find the acceleration of the vehicle just as it lifted off the launch pad on the Earth if the vehicle's initial mass was 3.00×10^6 kg. *Note:* You must include the gravitational force to solve part (b).

50. Model rocket engines are sized by thrust, thrust duration, and total impulse, among other characteristics. A size C5 model rocket engine has an average thrust of 5.26 N, a fuel mass of 12.7 g, and an initial mass of 25.5 g. The duration of its burn is 1.90 s. (a) What is the average exhaust speed of the engine? (b) If this engine is placed in a rocket body of mass 53.5 g, what is the final velocity of the rocket if it is fired in outer space? Assume the fuel burns at a constant rate.

51. A rocket for use in deep space is to be capable of boosting a total load (payload plus rocket frame and engine) of 3.00 metric tons to a speed of 10 000 m/s. (a) It has an engine and fuel designed to produce an exhaust speed of 2 000 m/s. How much fuel plus oxidizer is required? (b) If a different fuel and engine design could give an exhaust speed of 5 000 m/s, what amount of fuel and oxidizer would be required for the same task?

52. *Rocket Science.* A rocket has total mass $M_i = 360$ kg, including 330 kg of fuel and oxidizer. In interstellar space it starts from rest, turns on its engine at time $t = 0$, and puts out exhaust with relative speed $v_e = 1\,500$ m/s at the constant rate $k = 2.50$ kg/s. The fuel will last for an actual burn time of 330 kg/(2.5 kg/s) = 132 s, but define a "projected depletion time" as $T_p = M_i/k =$ 144 s. (This would be the burn time if the rocket could use its payload and fuel tanks as fuel, and even the walls of the combustion chamber.) (a) Show that during the burn the velocity of the rocket is given as a function of time by

$$v(t) = -v_e \ln[1 - (t/T_p)]$$

(b) Make a graph of the velocity of the rocket as a function of time for times running from 0 to 132 s. (c) Show that the acceleration of the rocket is

$$a(t) = v_e/(T_p - t)$$

(d) Graph the acceleration as a function of time. (e) Show that the position of the rocket is

$$x(t) = v_e(T_p - t)\ln[1 - (t/T_p)] + v_e t$$

(f) Graph the position during the burn.

53. An orbiting spacecraft is described not as a "zero-g," but rather as a "microgravity" environment for its occupants and for on-board experiments. Astronauts experience slight lurches due to the motions of equipment and other astronauts, and due to venting of materials from the craft. Assume that a 3 500-kg spacecraft undergoes an acceleration of 2.50 $\mu g = 2.45 \times 10^{-5}$ m/s^2 due to a leak from one of its hydraulic control systems. The fluid is known to escape with a speed of 70.0 m/s into the vacuum of space. How much fluid will be lost in 1 h if the leak is not stopped?

Additional Problems

54. Two gliders are set in motion on an air track. A spring of force constant k is attached to the near side of one glider. The first glider, of mass m_1, has velocity $\mathbf{v}_1$, and the second glider, of mass m_2, moves more slowly, with velocity $\mathbf{v}_2$, as in Figure P9.54. When m_1 collides with the spring attached to m_2 and compresses the spring to its maximum compression x_{max}, the velocity of the gliders is $\mathbf{v}$. In terms of $\mathbf{v}_1$, $\mathbf{v}_2$, m_1, m_2, and k, find (a) the velocity $\mathbf{v}$ at maximum compression, (b) the maximum compression x_{max}, and (c) the velocity of each glider after m_1 has lost contact with the spring.

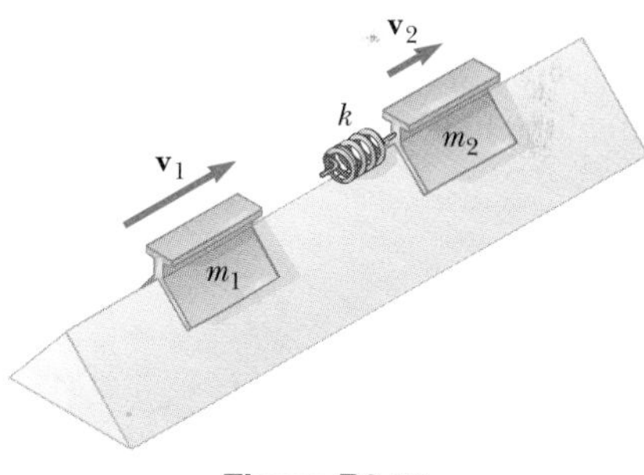

Figure P9.54

55. **Review problem**. A 60.0-kg person running at an initial speed of 4.00 m/s jumps onto a 120-kg cart initially at rest (Figure P9.55). The person slides on the cart's top surface and finally comes to rest relative to the cart. The coefficient of kinetic friction between the person and the cart is 0.400. Friction between the cart and ground can be neglected. (a) Find the final velocity of the person and cart relative to the ground. (b) Find the friction force acting on the person while he is sliding across the top surface of the cart. (c) How long does the friction force act on the person? (d) Find the change in momentum of the person and the change in momentum of the cart. (e) Determine the displacement of the person relative to the ground while he is sliding on the cart. (f) Determine the displacement of the cart relative to the ground while the person is sliding. (g) Find the change in

kinetic energy of the person. (h) Find the change in kinetic energy of the cart. (i) Explain why the answers to (g) and (h) differ. (What kind of collision is this, and what accounts for the loss of mechanical energy?)

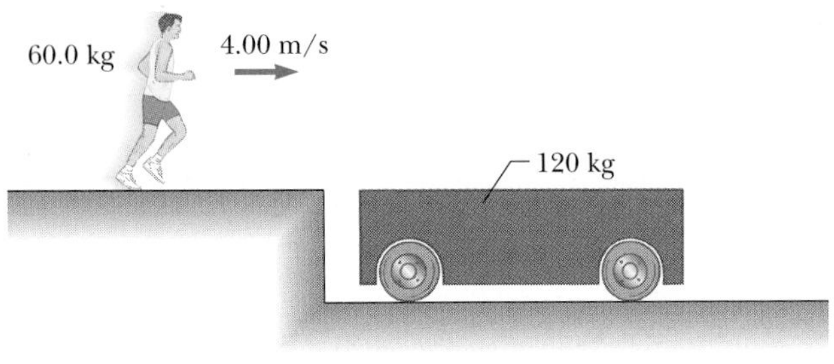

Figure P9.55

56. A golf ball (m = 46.0 g) is struck with a force that makes an angle of 45.0° with the horizontal. The ball lands 200 m away on a flat fairway. If the golf club and ball are in contact for 7.00 ms, what is the average force of impact? (Neglect air resistance.)

57. An 80.0-kg astronaut is working on the engines of his ship, which is drifting through space with a constant velocity. The astronaut, wishing to get a better view of the Universe, pushes against the ship and much later finds himself 30.0 m behind the ship. Without a thruster, the only way to return to the ship is to throw his 0.500-kg wrench directly away from the ship. If he throws the wrench with a speed of 20.0 m/s relative to the ship, how long does it take the astronaut to reach the ship?

58. A bullet of mass m is fired into a block of mass M initially at rest at the edge of a frictionless table of height h (Fig. P9.58). The bullet remains in the block, and after impact the block lands a distance d from the bottom of the table. Determine the initial speed of the bullet.

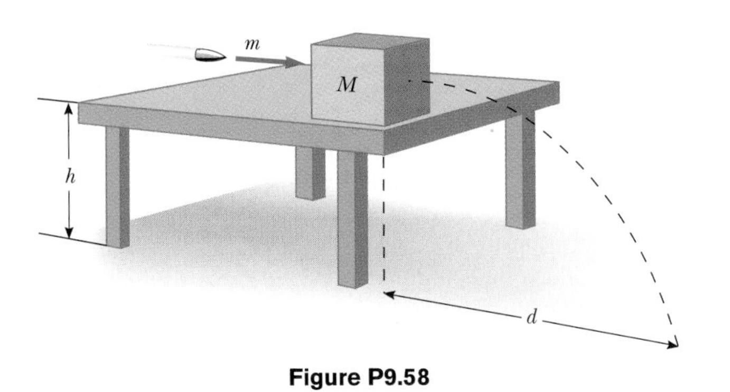

Figure P9.58

59. A 0.500-kg sphere moving with a velocity $(2.00\hat{\mathbf{i}} - 3.00\hat{\mathbf{j}} + 1.00\hat{\mathbf{k}})$ m/s strikes another sphere of mass 1.50 kg moving with a velocity $(-1.00\hat{\mathbf{i}} + 2.00\hat{\mathbf{j}} - 3.00\hat{\mathbf{k}})$ m/s. (a) If the velocity of the 0.500-kg sphere after the collision is $(-1.00\hat{\mathbf{i}} + 3.00\hat{\mathbf{j}} - 8.00\hat{\mathbf{k}})$ m/s, find the final velocity of the 1.50-kg sphere and identify the kind of collision (elastic, inelastic, or perfectly inelastic). (b) If the velocity of the 0.500-kg sphere after the collision is $(-0.250\hat{\mathbf{i}} + 0.750\hat{\mathbf{j}} - 2.00\hat{\mathbf{k}})$ m/s, find the final velocity of the 1.50-kg sphere and identify the kind of collision. (c) **What If?** If the velocity of the 0.500-kg sphere after the collision is $(-1.00\hat{\mathbf{i}} + 3.00\hat{\mathbf{j}} + a\hat{\mathbf{k}})$ m/s, find the value of a and the velocity of the 1.50-kg sphere after an elastic collision.

60. A small block of mass m_1 = 0.500 kg is released from rest at the top of a curve-shaped frictionless wedge of mass m_2 = 3.00 kg, which sits on a frictionless horizontal surface as in Figure P9.60a. When the block leaves the wedge, its velocity is measured to be 4.00 m/s to the right, as in Figure P9.60b. (a) What is the velocity of the wedge after the block reaches the horizontal surface? (b) What is the height h of the wedge?

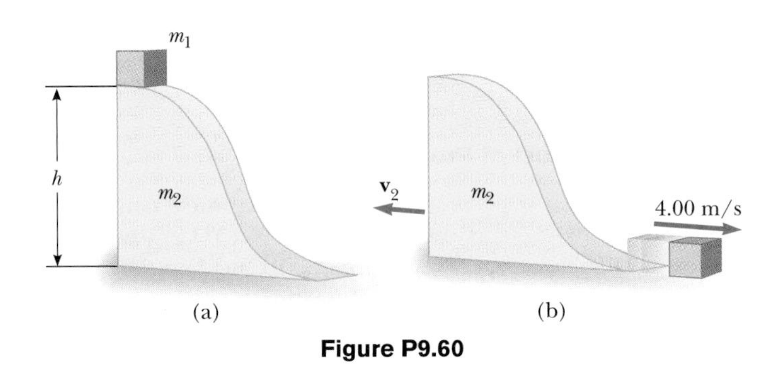

Figure P9.60

61. A bucket of mass m and volume V is attached to a light cart, completely covering its top surface. The cart is given a quick push along a straight, horizontal, smooth road. It is raining, so as the cart cruises along without friction, the bucket gradually fills with water. By the time the bucket is full, its speed is v. (a) What was the initial speed v_i of the cart? Let ρ represent the density of water. (b) **What If?** Assume that when the bucket is half full, it develops a slow leak at the bottom, so that the level of the water remains constant thereafter. Describe qualitatively what happens to the speed of the cart after the leak develops.

62. A 75.0-kg firefighter slides down a pole while a constant friction force of 300 N retards her motion. A horizontal 20.0-kg platform is supported by a spring at the bottom of the pole to cushion the fall. The firefighter starts from rest 4.00 m above the platform, and the spring constant is 4 000 N/m. Find (a) the firefighter's speed just before she collides with the platform and (b) the maximum distance the spring is compressed. (Assume the friction force acts during the entire motion.)

63. George of the Jungle, with mass m, swings on a light vine hanging from a stationary tree branch. A second vine of equal length hangs from the same point, and a gorilla of larger mass M swings in the opposite direction on it. Both vines are horizontal when the primates start from rest at the same moment. George and the gorilla meet at the lowest point of their swings. Each is afraid that one vine will break, so they grab each other and hang on. They swing upward together, reaching a point where the vines make an angle of 35.0° with the vertical. (a) Find the value of the ratio m/M. (b) **What If?** Try this at home. Tie a small magnet and a steel screw to opposite ends of a string. Hold the cen-

ter of the string fixed to represent the tree branch, and reproduce a model of the motions of George and the gorilla. What changes in your analysis will make it apply to this situation? **What If?** Assume the magnet is strong, so that it noticeably attracts the screw over a distance of a few centimeters. Then the screw will be moving faster just before it sticks to the magnet. Does this make a difference?

64. A cannon is rigidly attached to a carriage, which can move along horizontal rails but is connected to a post by a large spring, initially unstretched and with force constant $k = 2.00 \times 10^4$ N/m, as in Figure P9.64. The cannon fires a 200-kg projectile at a velocity of 125 m/s directed 45.0° above the horizontal. (a) If the mass of the cannon and its carriage is 5 000 kg, find the recoil speed of the cannon. (b) Determine the maximum extension of the spring. (c) Find the maximum force the spring exerts on the carriage. (d) Consider the system consisting of the cannon, carriage, and shell. Is the momentum of this system conserved during the firing? Why or why not?

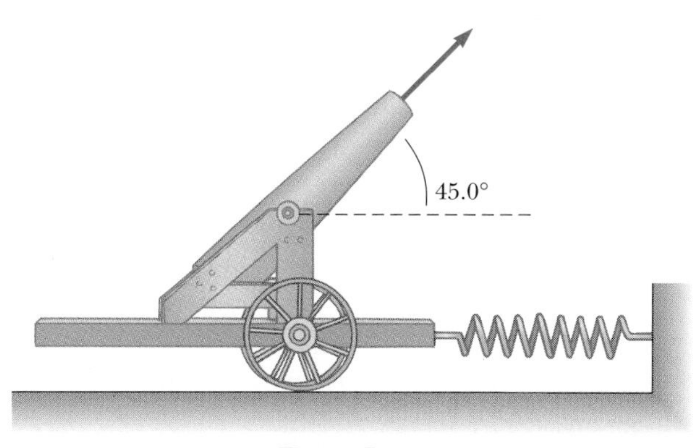

Figure P9.64

65. A student performs a ballistic pendulum experiment using an apparatus similar to that shown in Figure 9.11b. She obtains the following average data: $h = 8.68$ cm, $m_1 = 68.8$ g, and $m_2 = 263$ g. The symbols refer to the quantities in Figure 9.11a. (a) Determine the initial speed v_{1A} of the projectile. (b) The second part of her experiment is to obtain v_{1A} by firing the same projectile horizontally (with the pendulum removed from the path), by measuring its final horizontal position x and distance of fall y (Fig. P9.65). Show that the initial speed of the projectile is related to x and y through the relation

$$v_{1A} = \frac{x}{\sqrt{2y/g}}$$

What numerical value does she obtain for v_{1A} based on her measured values of $x = 257$ cm and $y = 85.3$ cm? What factors might account for the difference in this value compared to that obtained in part (a)?

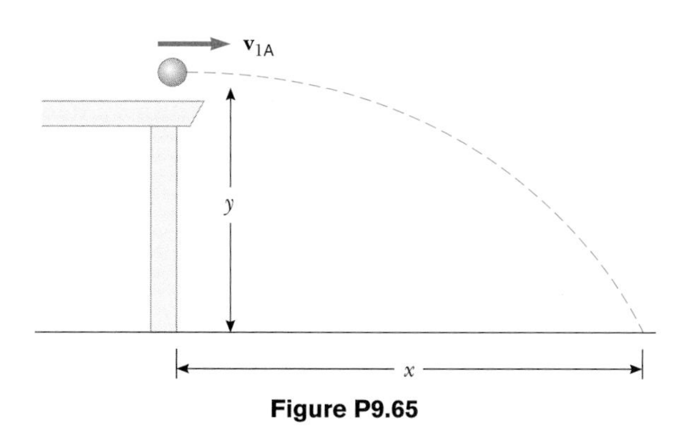

Figure P9.65

66. Small ice cubes, each of mass 5.00 g, slide down a frictionless track in a steady stream, as shown in Figure P9.66. Starting from rest, each cube moves down through a net vertical distance of 1.50 m and leaves the bottom end of the track at an angle of 40.0° above the horizontal. At the highest point of its subsequent trajectory, the cube strikes a vertical wall and rebounds with half the speed it had upon impact. If 10.0 cubes strike the wall per second, what average force is exerted on the wall?

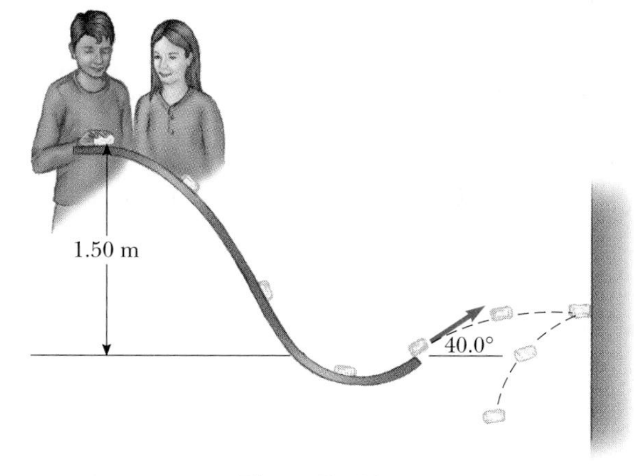

Figure P9.66

67. A 5.00-g bullet moving with an initial speed of 400 m/s is fired into and passes through a 1.00-kg block, as in Figure P9.67. The block, initially at rest on a frictionless, horizontal surface, is connected to a spring with force constant 900 N/m. If the block moves 5.00 cm to the right after impact, find (a) the speed at which the bullet emerges from the block and (b) the mechanical energy converted into internal energy in the collision.

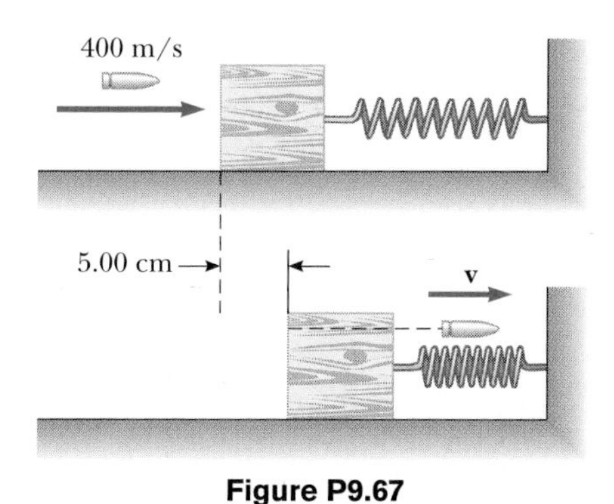

Figure P9.67

68. Consider as a system the Sun with the Earth in a circular orbit around it. Find the magnitude of the change in the velocity of the Sun relative to the center of mass of the

system over a period of 6 months. Neglect the influence of other celestial objects. You may obtain the necessary astronomical data from the endpapers of the book.

69. **Review problem.** There are (one can say) three coequal theories of motion: Newton's second law, stating that the total force on an object causes its acceleration; the work–kinetic energy theorem, stating that the total work on an object causes its change in kinetic energy; and the impulse–momentum theorem, stating that the total impulse on an object causes its change in momentum. In this problem, you compare predictions of the three theories in one particular case. A 3.00-kg object has velocity $7.00\hat{\mathbf{j}}$ m/s. Then, a total force $12.0\hat{\mathbf{i}}$ N acts on the object for 5.00 s. (a) Calculate the object's final velocity, using the impulse–momentum theorem. (b) Calculate its acceleration from $\mathbf{a} = (\mathbf{v}_f - \mathbf{v}_i)/\Delta t$. (c) Calculate its acceleration from $\mathbf{a} = \Sigma\mathbf{F}/m$. (d) Find the object's vector displacement from $\Delta\mathbf{r} = \mathbf{v}_i t + \frac{1}{2}\mathbf{a}t^2$. (e) Find the work done on the object from $W = \mathbf{F}\cdot\Delta\mathbf{r}$. (f) Find the final kinetic energy from $\frac{1}{2}mv_f^2 = \frac{1}{2}m\mathbf{v}_f\cdot\mathbf{v}_f$. (g) Find the final kinetic energy from $\frac{1}{2}mv_i^2 + W$.

70. A rocket has total mass $M_i = 360$ kg, including 330 kg of fuel and oxidizer. In interstellar space it starts from rest. Its engine is turned on at time $t = 0$, and it puts out exhaust with relative speed $v_e = 1\,500$ m/s at the constant rate 2.50 kg/s. The burn lasts until the fuel runs out, at time 330 kg/(2.5 kg/s) = 132 s. Set up and carry out a computer analysis of the motion according to Euler's method. Find (a) the final velocity of the rocket and (b) the distance it travels during the burn.

71. A chain of length L and total mass M is released from rest with its lower end just touching the top of a table, as in Figure P9.71a. Find the force exerted by the table on the chain after the chain has fallen through a distance x, as in Figure P9.71b. (Assume each link comes to rest the instant it reaches the table.)

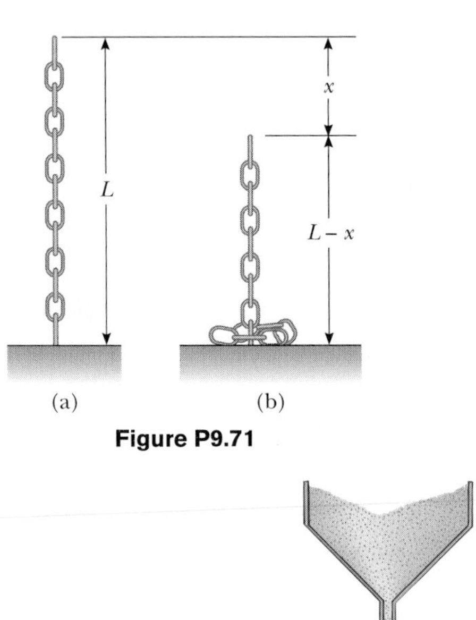

Figure P9.71

72. Sand from a stationary hopper falls onto a moving conveyor belt at the rate of 5.00 kg/s as in Figure P9.72. The conveyor belt is supported by frictionless rollers and moves at a constant speed of 0.750 m/s under the action of a constant horizontal external force $\mathbf{F}_{ext}$ supplied by the motor that drives the belt. Find (a) the sand's rate of change of momentum in the horizontal direction, (b) the force of friction exerted by the belt on the sand, (c) the external force $\mathbf{F}_{ext}$, (d) the work done by $\mathbf{F}_{ext}$ in 1 s, and (e) the kinetic energy acquired by the falling sand each second due to the change in its horizontal motion. (f) Why are the answers to (d) and (e) different?

73. A golf club consists of a shaft connected to a club head. The golf club can be modeled as a uniform rod of length ℓ and mass m_1 extending radially from the surface of a sphere of radius R and mass m_2. Find the location of the club's center of mass, measured from the center of the club head.

Answers to Quick Quizzes

9.1 (d). Two identical objects ($m_1 = m_2$) traveling at the same speed ($v_1 = v_2$) have the same kinetic energies and the same magnitudes of momentum. It also is possible, however, for particular combinations of masses and velocities to satisfy $K_1 = K_2$ but not $p_1 = p_2$. For example, a 1-kg object moving at 2 m/s has the same kinetic energy as a 4-kg object moving at 1 m/s, but the two clearly do not have the same momenta. Because we have no information about masses and speeds, we cannot choose among (a), (b), or (c).

9.2 (b), (c), (a). The slower the ball, the easier it is to catch. If the momentum of the medicine ball is the same as the momentum of the baseball, the speed of the medicine ball must be 1/10 the speed of the baseball because the medicine ball has 10 times the mass. If the kinetic energies are the same, the speed of the medicine ball must be $1/\sqrt{10}$ the speed of the baseball because of the squared speed term in the equation for K. The medicine ball is hardest to catch when it has the same speed as the baseball.

9.3 (c). The ball and the Earth exert forces on each other, so neither is an isolated system. We must include both in the system so that the interaction force is internal to the system.

9.4 (c). From Equation 9.4, if $\mathbf{p}_1 + \mathbf{p}_2 =$ constant, then it follows that $\Delta\mathbf{p}_1 + \Delta\mathbf{p}_2 = 0$ and $\Delta\mathbf{p}_1 = -\Delta\mathbf{p}_2$. While the change in momentum is the same, the change in the *velocity* is a lot larger for the car!

9.5 (c) and (e). Object 2 has a greater acceleration because of its smaller mass. Therefore, it takes less time to travel the distance d. Even though the force applied to objects 1 and 2 is the same, the change in momentum is less for object 2 because Δt is smaller. The work $W = Fd$ done on

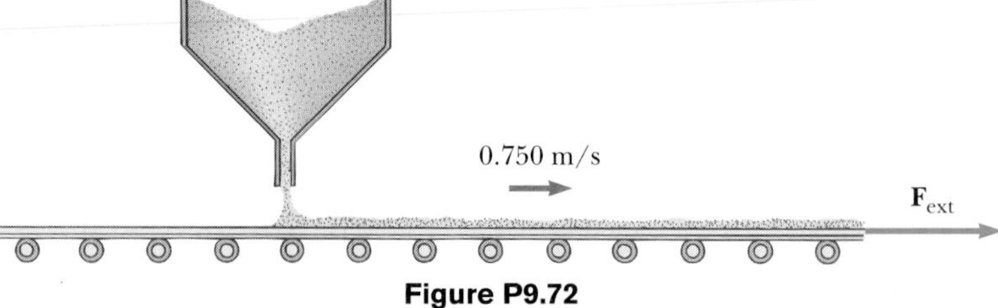

Figure P9.72

both objects is the same because both F and d are the same in the two cases. Therefore, $K_1 = K_2$.

9.6 (b) and (d). The same impulse is applied to both objects, so they experience the same change in momentum. Object 2 has a larger acceleration due to its smaller mass. Thus, the distance that object 2 covers in the time interval Δt is larger than that for object 1. As a result, more work is done on object 2 and $K_2 > K_1$.

9.7 (a) All three are the same. Because the passenger is brought from the car's initial speed to a full stop, the change in momentum (equal to the impulse) is the same regardless of what stops the passenger. (b) Dashboard, seatbelt, airbag. The dashboard stops the passenger very quickly in a front-end collision, resulting in a very large force. The seatbelt takes somewhat more time, so the force is smaller. Used along with the seatbelt, the airbag can extend the passenger's stopping time further, notably for his head, which would otherwise snap forward.

9.8 (a). If all of the initial kinetic energy is transformed, then nothing is moving after the collision. Consequently, the final momentum of the system is necessarily zero and, therefore, the initial momentum of the system must be zero. While (b) and (d) *together* would satisfy the conditions, neither one *alone* does.

9.9 (b). Because momentum of the two-ball system is conserved, $\mathbf{p}_{Ti} + 0 = \mathbf{p}_{Tf} + \mathbf{p}_B$. Because the table-tennis ball bounces back from the much more massive bowling ball with approximately the same speed, $\mathbf{p}_{Tf} = -\mathbf{p}_{Ti}$. As a consequence, $\mathbf{p}_B = 2\mathbf{p}_{Ti}$. Kinetic energy can be expressed as $K = p^2/2m$. Because of the much larger mass of the bowling ball, its kinetic energy is much smaller than that of the table-tennis ball.

9.10 (b). The piece with the handle will have less mass than the piece made up of the end of the bat. To see why this is so, take the origin of coordinates as the center of mass before the bat was cut. Replace each cut piece by a small sphere located at the center of mass for each piece. The sphere representing the handle piece is farther from the origin, but the product of less mass and greater distance balances the product of greater mass and less distance for the end piece:

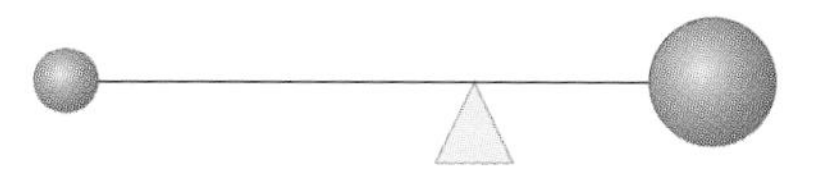

9.11 (a). This is the same effect as the swimmer diving off the raft that we just discussed. The vessel–passengers system is isolated. If the passengers all start running one way, the speed of the vessel increases (a *small* amount!) the other way.

9.12 (b). Once they stop running, the momentum of the system is the same as it was before they started running—you cannot change the momentum of an isolated system by means of internal forces. In case you are thinking that the passengers could do this over and over to take advantage of the speed increase *while* they are running, remember that they will slow the ship down every time they return to the bow!

Chapter 10

Rotation of a Rigid Object About a Fixed Axis

CHAPTER OUTLINE

▲ *The Malaysian pastime of* gasing *involves the spinning of tops that can have masses up to 20 kg. Professional spinners can spin their tops so that they might rotate for hours before stopping. We will study the rotational motion of objects such as these tops in this chapter. (Courtesy Tourism Malaysia)*

When an extended object such as a wheel rotates about its axis, the motion cannot be analyzed by treating the object as a particle because at any given time different parts of the object have different linear velocities and linear accelerations. We can, however, analyze the motion by considering an extended object to be composed of a collection of particles, each of which has its own linear velocity and linear acceleration.

In dealing with a rotating object, analysis is greatly simplified by assuming that the object is rigid. A **rigid object** is one that is nondeformable—that is, the relative locations of all particles of which the object is composed remain constant. All real objects are deformable to some extent; however, our rigid-object model is useful in many situations in which deformation is negligible.

Rigid object

10.1 Angular Position, Velocity, and Acceleration

Figure 10.1 illustrates an overhead view of a rotating compact disc. The disc is rotating about a fixed axis through O. The axis is perpendicular to the plane of the figure. Let us investigate the motion of only one of the millions of "particles" making up the disc. A particle at P is at a fixed distance r from the origin and rotates about it in a circle of radius r. (In fact, *every* particle on the disc undergoes circular motion about O.) It is convenient to represent the position of P with its polar coordinates (r, θ), where r is the distance from the origin to P and θ is measured *counterclockwise* from some reference line as shown in Figure 10.1a. In this representation, the only coordinate for the particle that changes in time is the angle θ; r remains constant. As the particle moves along the circle from the reference line $(\theta = 0)$, it moves through an arc of length s, as in Figure 10.1b. The arc length s is related to the angle θ through the relationship

$$s = r\theta \quad (10.1a)$$

$$\theta = \frac{s}{r} \quad (10.1b)$$

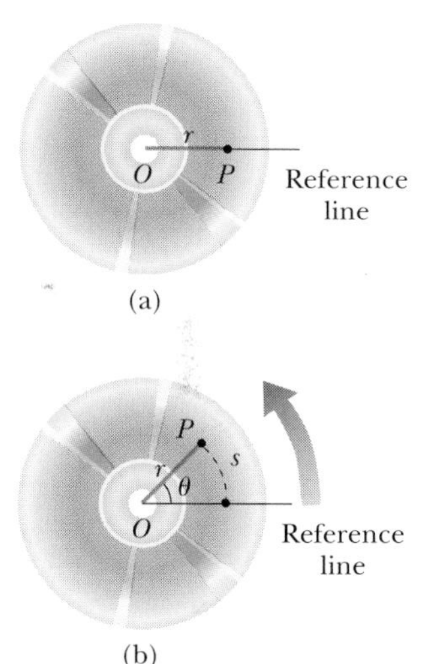

Figure 10.1 A compact disc rotating about a fixed axis through O perpendicular to the plane of the figure. (a) In order to define angular position for the disc, a fixed reference line is chosen. A particle at P is located at a distance r from the rotation axis at O. (b) As the disc rotates, point P moves through an arc length s on a circular path of radius r.

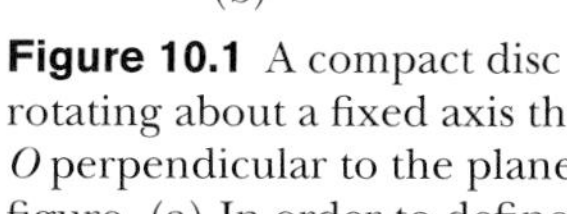

Note the dimensions of θ in Equation 10.1b. Because θ is the ratio of an arc length and the radius of the circle, it is a pure number. However, we commonly give θ the artificial unit **radian** (rad), where

one radian is the angle subtended by an arc length equal to the radius of the arc.

Because the circumference of a circle is $2\pi r$, it follows from Equation 10.1b that 360° corresponds to an angle of $(2\pi r/r)$ rad $= 2\pi$ rad. (Also note that 2π rad corresponds

PITFALL PREVENTION

10.1 Remember the Radian

In rotational equations, we must use angles expressed in *radians*. Don't fall into the trap of using angles measured in degrees in rotational equations.

to one complete revolution.) Hence, 1 rad $= 360°/2\pi \approx 57.3°$. To convert an angle in degrees to an angle in radians, we use the fact that π rad $= 180°$, or

$$\theta\,(\text{rad}) = \frac{\pi}{180°}\,\theta\,(\text{deg})$$

For example, 60° equals $\pi/3$ rad and 45° equals $\pi/4$ rad.

Because the disc in Figure 10.1 is a rigid object, as the particle moves along the circle from the reference line, every other particle on the object rotates through the same angle θ. Thus, **we can associate the angle θ with the entire rigid object as well as with an individual particle.** This allows us to define the *angular position* of a rigid object in its rotational motion. We choose a reference line on the object, such as a line connecting O and a chosen particle on the object. The **angular position** of the rigid object is the angle θ between this reference line on the object and the fixed reference line in space, which is often chosen as the x axis. This is similar to the way we identify the position of an object in translational motion—the distance x between the object and the reference position, which is the origin, $x = 0$.

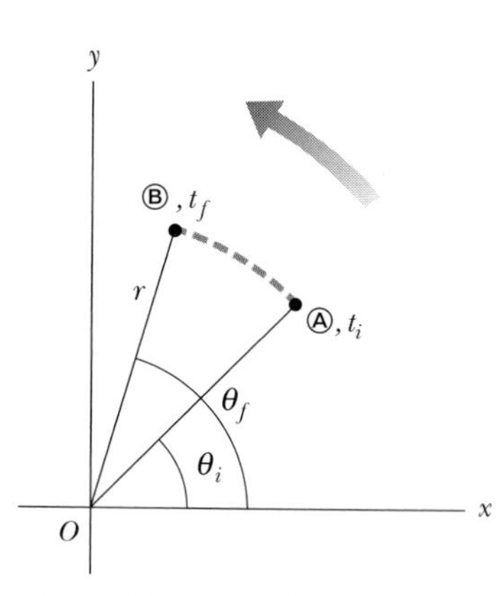

Figure 10.2 A particle on a rotating rigid object moves from Ⓐ to Ⓑ along the arc of a circle. In the time interval $\Delta t = t_f - t_i$, the radius vector moves through an angular displacement $\Delta\theta = \theta_f - \theta_i$.

As the particle in question on our rigid object travels from position Ⓐ to position Ⓑ in a time interval Δt as in Figure 10.2, the reference line of length r sweeps out an angle $\Delta\theta = \theta_f - \theta_i$. This quantity $\Delta\theta$ is defined as the **angular displacement** of the rigid object:

$$\Delta\theta \equiv \theta_f - \theta_i$$

The rate at which this angular displacement occurs can vary. If the rigid object spins rapidly, this displacement can occur in a short time interval. If it rotates slowly, this displacement occurs in a longer time interval. These different rotation rates can be quantified by introducing *angular speed*. We define the **average angular speed** $\overline{\omega}$ (Greek omega) as the ratio of the angular displacement of a rigid object to the time interval Δt during which the displacement occurs:

Average angular speed

$$\overline{\omega} \equiv \frac{\theta_f - \theta_i}{t_f - t_i} = \frac{\Delta\theta}{\Delta t} \tag{10.2}$$

In analogy to linear speed, the **instantaneous angular speed** ω is defined as the limit of the ratio $\Delta\theta/\Delta t$ as Δt approaches zero:

Instantaneous angular speed

$$\omega \equiv \lim_{\Delta t \to 0} \frac{\Delta\theta}{\Delta t} = \frac{d\theta}{dt} \tag{10.3}$$

Angular speed has units of radians per second (rad/s), which can be written as second^{-1} (s^{-1}) because radians are not dimensional. We take ω to be positive when θ is increasing (counterclockwise motion in Figure 10.2) and negative when θ is decreasing (clockwise motion in Figure 10.2).

Quick Quiz 10.1 A rigid object is rotating in a counterclockwise sense around a fixed axis. Each of the following pairs of quantities represents an initial angular position and a final angular position of the rigid object. Which of the sets can *only* occur if the rigid object rotates through more than 180°? (a) 3 rad, 6 rad (b) −1 rad, 1 rad (c) 1 rad, 5 rad.

Quick Quiz 10.2 Suppose that the change in angular position for each of the pairs of values in Quick Quiz 10.1 occurs in 1 s. Which choice represents the lowest average angular speed?

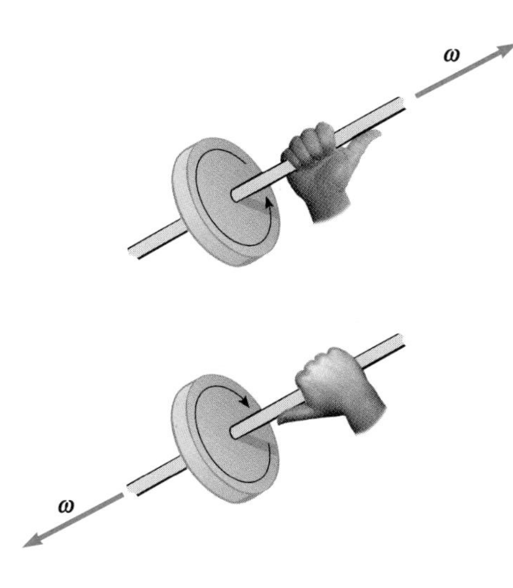

Figure 10.3 The right-hand rule for determining the direction of the angular velocity vector.

If the instantaneous angular speed of an object changes from ω_i to ω_f in the time interval Δt, the object has an angular acceleration. The **average angular acceleration** $\overline{\alpha}$ (Greek alpha) of a rotating rigid object is defined as the ratio of the change in the angular speed to the time interval Δt during which the change in the angular speed occurs:

$$\overline{\alpha} \equiv \frac{\omega_f - \omega_i}{t_f - t_i} = \frac{\Delta\omega}{\Delta t} \qquad (10.4)$$

Average angular acceleration

In analogy to linear acceleration, the **instantaneous angular acceleration** is defined as the limit of the ratio $\Delta\omega/\Delta t$ as Δt approaches zero:

$$\alpha \equiv \lim_{\Delta t \to 0} \frac{\Delta\omega}{\Delta t} = \frac{d\omega}{dt} \qquad (10.5)$$

Instantaneous angular acceleration

Angular acceleration has units of radians per second squared (rad/s^2), or just second^{-2} (s^{-2}). Note that α is positive when a rigid object rotating counterclockwise is speeding up or when a rigid object rotating clockwise is slowing down during some time interval.

When a rigid object is rotating about a *fixed* axis, **every particle on the object rotates through the same angle in a given time interval and has the same angular speed and the same angular acceleration.** That is, the quantities θ, ω, and α characterize the rotational motion of the entire rigid object as well as individual particles in the object. Using these quantities, we can greatly simplify the analysis of rigid-object rotation.

Angular position (θ), angular speed (ω), and angular acceleration (α) are analogous to linear position (x), linear speed (v), and linear acceleration (a). The variables θ, ω, and α differ dimensionally from the variables x, v, and a only by a factor having the unit of length. (See Section 10.3.)

We have not specified any direction for angular speed and angular acceleration. Strictly speaking, ω and α are the magnitudes of the angular velocity and the angular acceleration vectors[1] $\boldsymbol{\omega}$ and $\boldsymbol{\alpha}$, respectively, and they should always be positive. Because we are considering rotation about a fixed axis, however, we can use nonvector notation and indicate the directions of the vectors by assigning a positive or negative sign to ω and α, as discussed earlier with regard to Equations 10.3 and 10.5. For rotation about a fixed axis, the only direction that uniquely specifies the rotational motion is the direction along the axis of rotation. Therefore, the directions of $\boldsymbol{\omega}$ and $\boldsymbol{\alpha}$ are along this axis. If an object rotates in the xy plane as in Figure 10.1, the direction of $\boldsymbol{\omega}$ is out of the plane of the diagram when the rotation is counterclockwise and into the plane of the diagram when the rotation is clockwise. To illustrate this convention, it is convenient to use the *right-hand rule* demonstrated in Figure 10.3. When the four fingers of the right

[1] Although we do not verify it here, the instantaneous angular velocity and instantaneous angular acceleration are vector quantities, but the corresponding average values are not. This is because angular displacements do not add as vector quantities for finite rotations.

PITFALL PREVENTION

10.2 Specify Your Axis

In solving rotation problems, you must specify an axis of rotation. This is a new feature not found in our study of translational motion. The choice is arbitrary, but once you make it, you must maintain that choice consistently throughout the problem. In some problems, the physical situation suggests a natural axis, such as the center of an automobile wheel. In other problems, there may not be an obvious choice, and you must exercise judgement.

hand are wrapped in the direction of rotation, the extended right thumb points in the direction of $\boldsymbol{\omega}$. The direction of $\boldsymbol{\alpha}$ follows from its definition $\boldsymbol{\alpha} \equiv d\boldsymbol{\omega}/dt$. It is in the same direction as $\boldsymbol{\omega}$ if the angular speed is increasing in time, and it is antiparallel to $\boldsymbol{\omega}$ if the angular speed is decreasing in time.

Quick Quiz 10.3 A rigid object is rotating with an angular speed $\omega < 0$. The angular velocity vector $\boldsymbol{\omega}$ and the angular acceleration vector $\boldsymbol{\alpha}$ are antiparallel. The angular speed of the rigid object is (a) clockwise and increasing (b) clockwise and decreasing (c) counterclockwise and increasing (d) counterclockwise and decreasing.

10.2 Rotational Kinematics: Rotational Motion with Constant Angular Acceleration

In our study of linear motion, we found that the simplest form of accelerated motion to analyze is motion under constant linear acceleration. Likewise, for rotational motion about a fixed axis, the simplest accelerated motion to analyze is motion under constant angular acceleration. Therefore, we next develop kinematic relationships for this type of motion. If we write Equation 10.5 in the form $d\omega = \alpha\, dt$, and let $t_i = 0$ and $t_f = t$, integrating this expression directly gives

Rotational kinematic equations

$$\omega_f = \omega_i + \alpha t \qquad \text{(for constant } \alpha\text{)} \tag{10.6}$$

where ω_i is the angular speed of the rigid object at time $t = 0$. Equation 10.6 allows us to find the angular speed ω_f of the object at any later time t. Substituting Equation 10.6 into Equation 10.3 and integrating once more, we obtain

$$\theta_f = \theta_i + \omega_i t + \tfrac{1}{2}\alpha t^2 \qquad \text{(for constant } \alpha\text{)} \tag{10.7}$$

where θ_i is the angular position of the rigid object at time $t = 0$. Equation 10.7 allows us to find the angular position θ_f of the object at any later time t. If we eliminate t from Equations 10.6 and 10.7, we obtain

$$\omega_f^2 = \omega_i^2 + 2\alpha(\theta_f - \theta_i) \qquad \text{(for constant } \alpha\text{)} \tag{10.8}$$

This equation allows us to find the angular speed ω_f of the rigid object for any value of its angular position θ_f. If we eliminate α between Equations 10.6 and 10.7, we obtain

$$\theta_f = \theta_i + \tfrac{1}{2}(\omega_i + \omega_f)t \qquad \text{(for constant } \alpha\text{)} \tag{10.9}$$

Notice that these kinematic expressions for rotational motion under constant angular acceleration are of the same mathematical form as those for linear motion under constant linear acceleration. They can be generated from the equations for linear motion by making the substitutions $x \rightarrow \theta$, $v \rightarrow \omega$, and $a \rightarrow \alpha$. Table 10.1 compares the kinematic equations for rotational and linear motion.

PITFALL PREVENTION

10.3 Just Like Translation?

Equations 10.6 to 10.9 and Table 10.1 suggest that rotational kinematics is just like translational kinematics. That is almost true, with two key differences: (1) in rotational kinematics, you must specify a rotation axis (per Pitfall Prevention 10.2); (2) in rotational motion, the object keeps returning to its original orientation—thus, you may be asked for the number of revolutions made by a rigid object. This concept has no meaning in translational motion, but is specified by $\Delta\theta$, which is analogous to Δx.

Table 10.1

Kinematic Equations for Rotational and Linear Motion Under Constant Acceleration

Rotational Motion About Fixed Axis	Linear Motion
$\omega_f = \omega_i + \alpha t$	$v_f = v_i + at$
$\theta_f = \theta_i + \omega_i t + \frac{1}{2}\alpha t^2$	$x_f = x_i + v_i t + \frac{1}{2}at^2$
$\omega_f^2 = \omega_i^2 + 2\alpha(\theta_f - \theta_i)$	$v_f^2 = v_i^2 + 2a(x_f - x_i)$
$\theta_f = \theta_i + \frac{1}{2}(\omega_i + \omega_f)t$	$x_f = x_i + \frac{1}{2}(v_i + v_f)t$

Quick Quiz 10.4 Consider again the pairs of angular positions for the rigid object in Quick Quiz 10.1. If the object starts from rest at the initial angular position, moves counterclockwise with constant angular acceleration, and arrives at the final angular position with the same angular speed in all three cases, for which choice is the angular acceleration the highest?

Example 10.1 Rotating Wheel

A wheel rotates with a constant angular acceleration of 3.50 rad/s^2.

(A) If the angular speed of the wheel is 2.00 rad/s at $t_i = 0$, through what angular displacement does the wheel rotate in 2.00 s?

Solution We can use Figure 10.2 to represent the wheel. We arrange Equation 10.7 so that it gives us angular displacement:

$$\Delta\theta = \theta_f - \theta_i = \omega_i t + \tfrac{1}{2}\alpha t^2$$
$$= (2.00 \text{ rad/s})(2.00 \text{ s}) + \tfrac{1}{2}(3.50 \text{ rad/s}^2)(2.00 \text{ s})^2$$
$$= 11.0 \text{ rad} = (11.0 \text{ rad})(57.3^\circ/\text{rad}) = 630^\circ$$

(B) Through how many revolutions has the wheel turned during this time interval?

Solution We multiply the angular displacement found in part (A) by a conversion factor to find the number of revolutions:

$$\Delta\theta = 630^\circ \left(\frac{1 \text{ rev}}{360^\circ}\right) = 1.75 \text{ rev}$$

(C) What is the angular speed of the wheel at $t = 2.00$ s?

Solution Because the angular acceleration and the angular speed are both positive, our answer must be greater than 2.00 rad/s. Using Equation 10.6, we find

$$\omega_f = \omega_i + \alpha t = 2.00 \text{ rad/s} + (3.50 \text{ rad/s}^2)(2.00 \text{ s})$$
$$= 9.00 \text{ rad/s}$$

We could also obtain this result using Equation 10.8 and the results of part (A). Try it!

What If? Suppose a particle moves along a straight line with a constant acceleration of 3.50 m/s^2. If the velocity of the particle is 2.00 m/s at $t_i = 0$, through what displacement does the particle move in 2.00 s? What is the velocity of the particle at $t = 2.00$ s?

Answer Notice that these questions are translational analogs to parts (A) and (C) of the original problem. The mathematical solution follows exactly the same form. For the displacement,

$$\Delta x = x_f - x_i = v_i t + \tfrac{1}{2}at^2$$
$$= (2.00 \text{ m/s})(2.00 \text{ s}) + \tfrac{1}{2}(3.50 \text{ m/s}^2)(2.00 \text{ s})^2$$
$$= 11.0 \text{ m}$$

and for the velocity,

$$v_f = v_i + at = 2.00 \text{ m/s} + (3.50 \text{ m/s}^2)(2.00 \text{ s}) = 9.00 \text{ m/s}$$

Note that there is no translational analog to part (B) because translational motion is not repetitive like rotational motion.

10.3 Angular and Linear Quantities

In this section we derive some useful relationships between the angular speed and acceleration of a rotating rigid object and the linear speed and acceleration of a point in the object. To do so, we must keep in mind that when a rigid object rotates about a fixed axis, as in Figure 10.4, **every particle of the object moves in a circle whose center is the axis of rotation.**

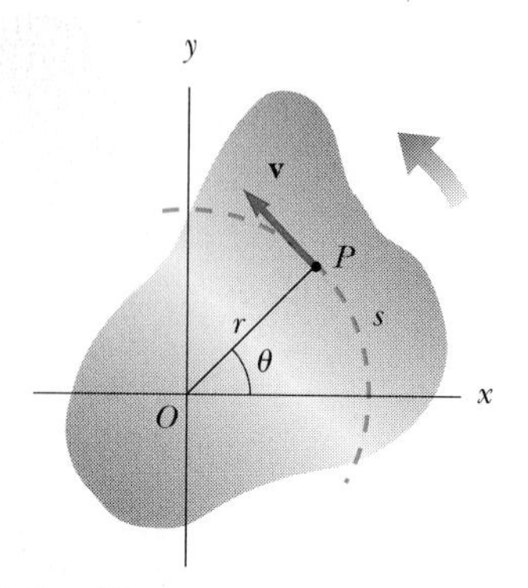

Active Figure 10.4 As a rigid object rotates about the fixed axis through O, the point P has a tangential velocity $\mathbf{v}$ that is always tangent to the circular path of radius r.

At the Active Figures link at http://www.pse6.com, you can move point P and observe the tangential velocity as the object rotates.

Because point P in Figure 10.4 moves in a circle, the linear velocity vector $\mathbf{v}$ is always tangent to the circular path and hence is called *tangential velocity.* The magnitude of the tangential velocity of the point P is by definition the tangential speed $v = ds/dt$, where s is the distance traveled by this point measured along the circular path. Recalling that $s = r\theta$ (Eq. 10.1a) and noting that r is constant, we obtain

$$v = \frac{ds}{dt} = r\frac{d\theta}{dt}$$

Because $d\theta/dt = \omega$ (see Eq. 10.3), we see that

$$v = r\omega \tag{10.10}$$

That is, the tangential speed of a point on a rotating rigid object equals the perpendicular distance of that point from the axis of rotation multiplied by the angular speed. Therefore, although every point on the rigid object has the same *angular* speed, not every point has the same *tangential* speed because r is not the same for all points on the object. Equation 10.10 shows that the tangential speed of a point on the rotating object increases as one moves outward from the center of rotation, as we would intuitively expect. The outer end of a swinging baseball bat moves much faster than the handle.

We can relate the angular acceleration of the rotating rigid object to the tangential acceleration of the point P by taking the time derivative of v:

$$a_t = \frac{dv}{dt} = r\frac{d\omega}{dt}$$

Relation between tangential and angular acceleration

$$a_t = r\alpha \tag{10.11}$$

That is, the tangential component of the linear acceleration of a point on a rotating rigid object equals the point's distance from the axis of rotation multiplied by the angular acceleration.

In Section 4.4 we found that a point moving in a circular path undergoes a radial acceleration a_r of magnitude v^2/r directed toward the center of rotation (Fig. 10.5). Because $v = r\omega$ for a point P on a rotating object, we can express the centripetal acceleration at that point in terms of angular speed as

$$a_c = \frac{v^2}{r} = r\omega^2 \tag{10.12}$$

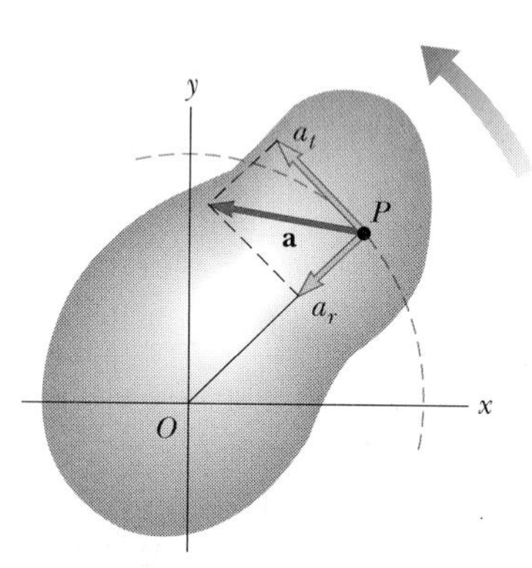

Figure 10.5 As a rigid object rotates about a fixed axis through O, the point P experiences a tangential component of linear acceleration a_t and a radial component of linear acceleration a_r. The total linear acceleration of this point is $\mathbf{a} = \mathbf{a}_t + \mathbf{a}_r$.

The total linear acceleration vector at the point is $\mathbf{a} = \mathbf{a}_t + \mathbf{a}_r$, where the magnitude of $\mathbf{a}_r$ is the centripetal acceleration a_c. Because $\mathbf{a}$ is a vector having a radial and a tangential component, the magnitude of $\mathbf{a}$ at the point P on the rotating rigid object is

$$a = \sqrt{a_t^2 + a_r^2} = \sqrt{r^2\alpha^2 + r^2\omega^4} = r\sqrt{\alpha^2 + \omega^4} \tag{10.13}$$

Quick Quiz 10.5 Andy and Charlie are riding on a merry-go-round. Andy rides on a horse at the outer rim of the circular platform, twice as far from the center of the circular platform as Charlie, who rides on an inner horse. When the merry-go-round is rotating at a constant angular speed, Andy's angular speed is (a) twice Charlie's (b) the same as Charlie's (c) half of Charlie's (d) impossible to determine.

Quick Quiz 10.6 Consider again the merry-go-round situation in Quick Quiz 10.5. When the merry-go-round is rotating at a constant angular speed, Andy's tangential speed is (a) twice Charlie's (b) the same as Charlie's (c) half of Charlie's (d) impossible to determine.

Example 10.2 CD Player

On a compact disc (Fig. 10.6), audio information is stored in a series of pits and flat areas on the surface of the disc. The information is stored digitally, and the alternations between pits and flat areas on the surface represent binary ones and zeroes to be read by the compact disc player and converted back to sound waves. The pits and flat areas are detected by a system consisting of a laser and lenses. The length of a string of ones and zeroes representing one piece of information is the same everywhere on the disc, whether the information is near the center of the disc or near its outer edge. In order that this length of ones and zeroes always passes by the laser–lens system in the same time period, the tangential speed of the disc surface at the location of the lens must be constant. This requires, according to Equation 10.10, that the angular speed vary as the laser–lens system moves radially along the disc. In a typical compact disc player, the constant speed of the surface at the point of the laser–lens system is 1.3 m/s.

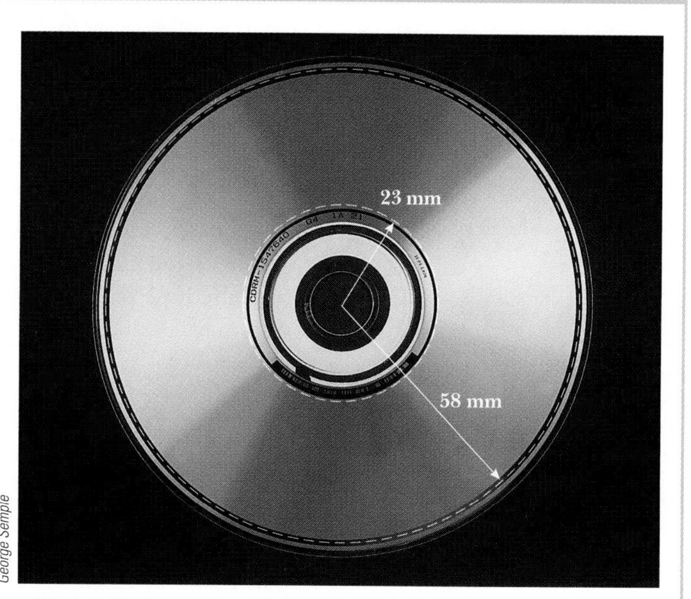

Figure 10.6 (Example 10.2) A compact disc.

(A) Find the angular speed of the disc in revolutions per minute when information is being read from the innermost first track ($r = 23$ mm) and the outermost final track ($r = 58$ mm).

Solution Using Equation 10.10, we can find the angular speed that will give us the required tangential speed at the position of the inner track,

$$\omega_i = \frac{v}{r_i} = \frac{1.3\text{ m/s}}{2.3 \times 10^{-2}\text{ m}} = 57\text{ rad/s}$$

$$= (57\text{ rad/s})\left(\frac{1\text{ rev}}{2\pi\text{ rad}}\right)\left(\frac{60\text{ s}}{1\text{ min}}\right)$$

$$= \boxed{5.4 \times 10^2\text{ rev/min}}$$

For the outer track,

$$\omega_f = \frac{v}{r_f} = \frac{1.3\text{ m/s}}{5.8 \times 10^{-2}\text{ m}} = 22\text{ rad/s}$$

$$= \boxed{2.1 \times 10^2\text{ rev/min}}$$

The player adjusts the angular speed ω of the disc within this range so that information moves past the objective lens at a constant rate.

(B) The maximum playing time of a standard music CD is 74 min and 33 s. How many revolutions does the disc make during that time?

Solution We know that the angular speed is always decreasing, and we assume that it is decreasing steadily, with α constant. If $t = 0$ is the instant that the disc begins, with angular speed of 57 rad/s, then the final value of the time t is (74 min)(60 s/min) + 33 s = 4 473 s. We are looking for the angular displacement $\Delta\theta$ during this time interval. We use Equation 10.9:

$$\Delta\theta = \theta_f - \theta_i = \tfrac{1}{2}(\omega_i + \omega_f)t$$

$$= \tfrac{1}{2}(57\text{ rad/s} + 22\text{ rad/s})(4\,473\text{ s})$$

$$= 1.8 \times 10^5\text{ rad}$$

We convert this angular displacement to revolutions:

$$\Delta\theta = 1.8 \times 10^5\text{ rad}\left(\frac{1\text{ rev}}{2\pi\text{ rad}}\right) = \boxed{2.8 \times 10^4\text{ rev}}$$

(C) What total length of track moves past the objective lens during this time?

Solution Because we know the (constant) linear velocity and the time interval, this is a straightforward calculation:

$$x_f = v_i t = (1.3\text{ m/s})(4\,473\text{ s}) = \boxed{5.8 \times 10^3\text{ m}}$$

More than 5.8 km of track spins past the objective lens!

(D) What is the angular acceleration of the CD over the 4 473-s time interval? Assume that α is constant.

Solution The most direct approach to solving this problem is to use Equation 10.6 and the results to part (A). We should obtain a negative number for the angular acceleration because the disc spins more and more slowly in the positive direction as time goes on. Our answer should also be relatively small because it takes such a long time—more than an hour—for the change in angular speed to be accomplished:

$$\alpha = \frac{\omega_f - \omega_i}{t} = \frac{22\text{ rad/s} - 57\text{ rad/s}}{4\,473\text{ s}}$$

$$= \boxed{-7.8 \times 10^{-3}\text{ rad/s}^2}$$

The disc experiences a very gradual decrease in its rotation rate, as expected.

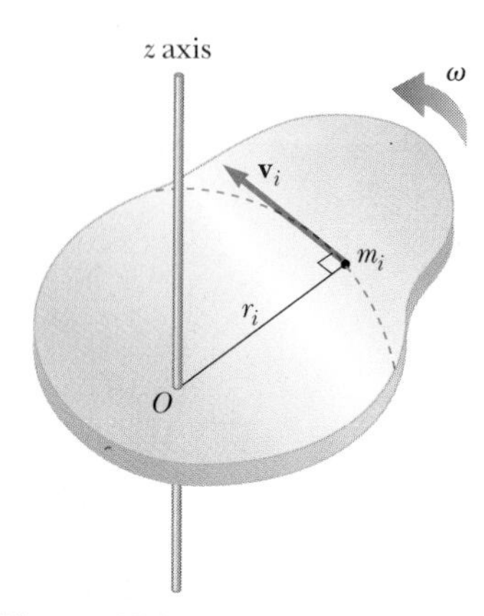

Figure 10.7 A rigid object rotating about the z axis with angular speed ω. The kinetic energy of the particle of mass m_i is $\frac{1}{2}m_i v_i^2$. The total kinetic energy of the object is called its rotational kinetic energy.

10.4 Rotational Kinetic Energy

In Chapter 7, we defined the kinetic energy of an object as the energy associated with its motion through space. An object rotating about a fixed axis remains stationary in space, so there is no kinetic energy associated with translational motion. The individual particles making up the rotating object, however, are moving through space—they follow circular paths. Consequently, there should be kinetic energy associated with rotational motion.

Let us consider an object as a collection of particles and assume that it rotates about a fixed z axis with an angular speed ω. Figure 10.7 shows the rotating object and identifies one particle on the object located at a distance r_i from the rotation axis. Each such particle has kinetic energy determined by its mass and tangential speed. If the mass of the ith particle is m_i and its tangential speed is v_i, its kinetic energy is

$$K_i = \tfrac{1}{2}m_i v_i^2$$

To proceed further, recall that although every particle in the rigid object has the same angular speed ω, the individual tangential speeds depend on the distance r_i from the axis of rotation according to the expression $v_i = r_i\omega$ (see Eq. 10.10). The *total* kinetic energy of the rotating rigid object is the sum of the kinetic energies of the individual particles:

$$K_R = \sum_i K_i = \sum_i \tfrac{1}{2}m_i v_i^2 = \tfrac{1}{2}\sum_i m_i r_i^2\omega^2$$

We can write this expression in the form

$$K_R = \tfrac{1}{2}\left(\sum_i m_i r_i^2\right)\omega^2 \qquad (10.14)$$

where we have factored ω^2 from the sum because it is common to every particle. We simplify this expression by defining the quantity in parentheses as the **moment of inertia *I*:**

Moment of inertia

$$I \equiv \sum_i m_i r_i^2 \qquad (10.15)$$

From the definition of moment of inertia, we see that it has dimensions of ML^2 (kg · m^2 in SI units).[2] With this notation, Equation 10.14 becomes

Rotational kinetic energy

$$K_R = \tfrac{1}{2}I\omega^2 \qquad (10.16)$$

Although we commonly refer to the quantity $\frac{1}{2}I\omega^2$ as **rotational kinetic energy,** it is not a new form of energy. It is ordinary kinetic energy because it is derived from a sum over individual kinetic energies of the particles contained in the rigid object. However, the mathematical form of the kinetic energy given by Equation 10.16 is convenient when we are dealing with rotational motion, provided we know how to calculate I.

It is important that you recognize the analogy between kinetic energy associated with linear motion $\frac{1}{2}mv^2$ and rotational kinetic energy $\frac{1}{2}I\omega^2$. The quantities I and ω in rotational motion are analogous to m and v in linear motion, respectively. (In fact, I takes the place of m and ω takes the place of v every time we compare a linear-motion equation with its rotational counterpart.) The moment of inertia is a measure of the resistance of an object to changes in its rotational motion, just as mass is a measure of the tendency of an object to resist changes in its linear motion.

PITFALL PREVENTION

10.4 No Single Moment of Inertia

There is one major difference between mass and moment of inertia. Mass is an inherent property of an object. The moment of inertia of an object depends on your choice of rotation axis. Thus, there is no single value of the moment of inertia for an object. There is a minimum value of the moment of inertia, which is that calculated about an axis passing through the center of mass of the object.

[2] Civil engineers use moment of inertia to characterize the elastic properties (rigidity) of such structures as loaded beams. Hence, it is often useful even in a nonrotational context.

Quick Quiz 10.7 A section of hollow pipe and a solid cylinder have the same radius, mass, and length. They both rotate about their long central axes with the same angular speed. Which object has the higher rotational kinetic energy? (a) the hollow pipe (b) the solid cylinder (c) they have the same rotational kinetic energy (d) impossible to determine.

Example 10.3 The Oxygen Molecule

Consider an oxygen molecule (O_2) rotating in the xy plane about the z axis. The rotation axis passes through the center of the molecule, perpendicular to its length. The mass of each oxygen atom is 2.66×10^{-26} kg, and at room temperature the average separation between the two atoms is $d = 1.21 \times 10^{-10}$ m. (The atoms are modeled as particles.)

(A) Calculate the moment of inertia of the molecule about the z axis.

Solution This is a straightforward application of the definition of I. Because each atom is a distance $d/2$ from the z axis, the moment of inertia about the axis is

$$I = \sum_i m_i r_i^2 = m\left(\frac{d}{2}\right)^2 + m\left(\frac{d}{2}\right)^2 = \frac{md^2}{2}$$

$$= \frac{(2.66 \times 10^{-26}\text{ kg})(1.21 \times 10^{-10}\text{ m})^2}{2}$$

$$= \boxed{1.95 \times 10^{-46}\text{ kg}\cdot\text{m}^2}$$

This is a very small number, consistent with the minuscule masses and distances involved.

(B) If the angular speed of the molecule about the z axis is 4.60×10^{12} rad/s, what is its rotational kinetic energy?

Solution We apply the result we just calculated for the moment of inertia in the equation for K_R:

$$K_R = \tfrac{1}{2} I\omega^2$$

$$= \tfrac{1}{2}(1.95 \times 10^{-46}\text{ kg}\cdot\text{m}^2)(4.60 \times 10^{12}\text{ rad/s})^2$$

$$= \boxed{2.06 \times 10^{-21}\text{ J}}$$

Example 10.4 Four Rotating Objects

Four tiny spheres are fastened to the ends of two rods of negligible mass lying in the xy plane (Fig. 10.8). We shall assume that the radii of the spheres are small compared with the dimensions of the rods.

(A) If the system rotates about the y axis (Fig. 10.8a) with an angular speed ω, find the moment of inertia and the rotational kinetic energy about this axis.

Solution First, note that the two spheres of mass m, which lie on the y axis, do not contribute to I_y (that is, $r_i = 0$ for these spheres about this axis). Applying Equation 10.15, we obtain

$$I_y = \sum_i m_i r_i^2 = Ma^2 + Ma^2 = \boxed{2Ma^2}$$

Therefore, the rotational kinetic energy about the y axis is

$$K_R = \tfrac{1}{2} I_y\omega^2 = \tfrac{1}{2}(2Ma^2)\omega^2 = \boxed{Ma^2\omega^2}$$

The fact that the two spheres of mass m do not enter into this result makes sense because they have no motion about the axis of rotation; hence, they have no rotational kinetic energy. By similar logic, we expect the moment of inertia about the x axis to be $I_x = 2mb^2$ with a rotational kinetic energy about that axis of $K_R = mb^2\omega^2$.

(B) Suppose the system rotates in the xy plane about an axis (the z axis) through O (Fig. 10.8b). Calculate the moment of inertia and rotational kinetic energy about this axis.

Solution Because r_i in Equation 10.15 is the distance between a sphere and the axis of rotation, we obtain

$$I_z = \sum_i m_i r_i^2 = Ma^2 + Ma^2 + mb^2 + mb^2 = \boxed{2Ma^2 + 2mb^2}$$

$$K_R = \tfrac{1}{2} I_z\omega^2 = \tfrac{1}{2}(2Ma^2 + 2mb^2)\omega^2 = \boxed{(Ma^2 + mb^2)\omega^2}$$

Comparing the results for parts (A) and (B), we conclude that the moment of inertia and therefore the rotational kinetic energy associated with a given angular speed depend on the axis of rotation. In part (B), we expect the result to include all four spheres and distances because all four spheres are rotating in the xy plane. Furthermore, the fact that the rotational kinetic energy in part (A) is smaller than that in part (B) indicates, based on the work–kinetic energy theorem, that it would require less work to set the system into rotation about the y axis than about the z axis.

What If? What if the mass M is much larger than m? How do the answers to parts (A) and (B) compare?

Answer If $M >> m$, then m can be neglected and the moment of inertia and rotational kinetic energy in part (B) become

$$I_z = 2Ma^2 \quad \text{and} \quad K_R = Ma^2\omega^2$$

which are the same as the answers in part (A). If the masses m of the two red spheres in Figure 10.8 are negligible, then these spheres can be removed from the figure and rotations about the y and z axes are equivalent.

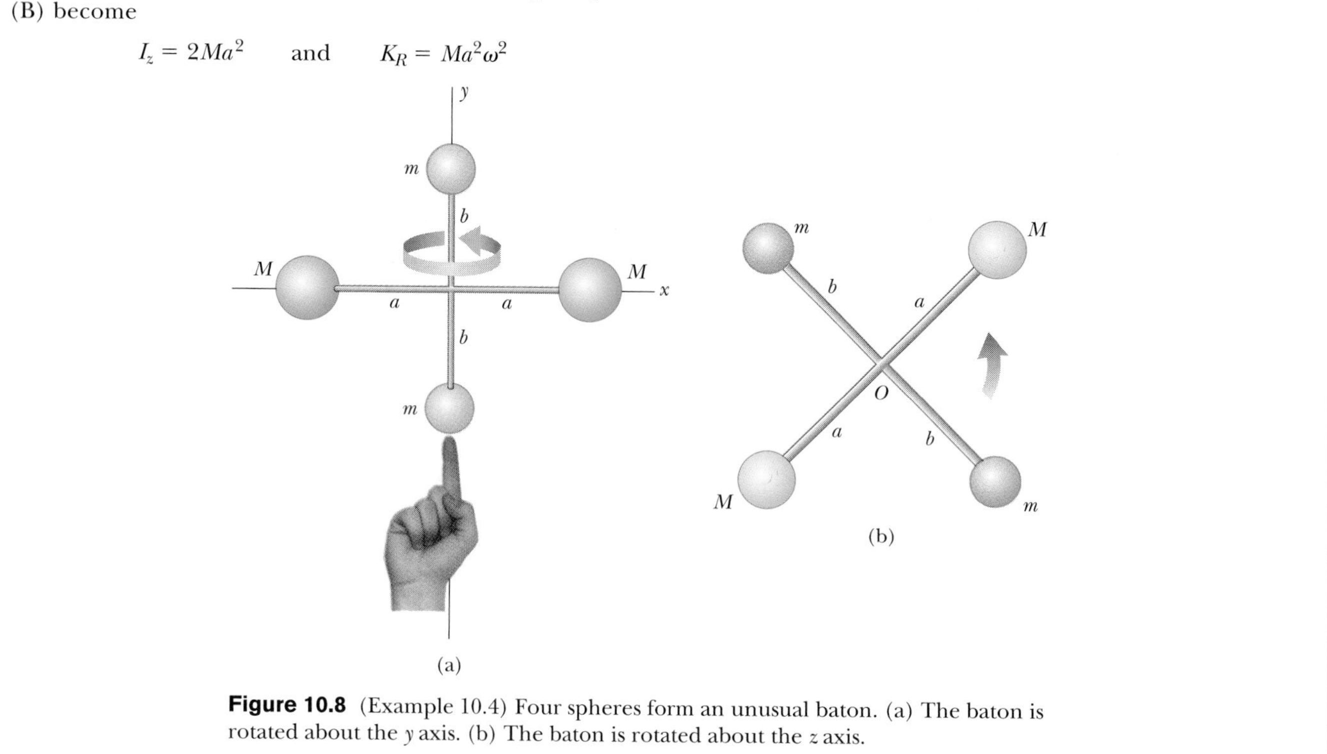

Figure 10.8 (Example 10.4) Four spheres form an unusual baton. (a) The baton is rotated about the y axis. (b) The baton is rotated about the z axis.

10.5 Calculation of Moments of Inertia

We can evaluate the moment of inertia of an extended rigid object by imagining the object to be divided into many small volume elements, each of which has mass Δm_i. We use the definition $I = \sum_i r_i^2 \Delta m_i$ and take the limit of this sum as $\Delta m_i \rightarrow 0$. In this limit, the sum becomes an integral over the volume of the object:

Moment of inertia of a rigid object

$$I = \lim_{\Delta m_i \rightarrow 0} \sum_i r_i^2 \Delta m_i = \int r^2\, dm \qquad (10.17)$$

It is usually easier to calculate moments of inertia in terms of the volume of the elements rather than their mass, and we can easily make that change by using Equation 1.1, $\rho = m/V$, where ρ is the density of the object and V is its volume. From this equation, the mass of a small element is $dm = \rho\, dV$. Substituting this result into Equation 10.17 gives

$$I = \int \rho r^2\, dV$$

If the object is homogeneous, then ρ is constant and the integral can be evaluated for a known geometry. If ρ is not constant, then its variation with position must be known to complete the integration.

The density given by $\rho = m/V$ sometimes is referred to as *volumetric mass density* because it represents mass per unit volume. Often we use other ways of expressing density. For instance, when dealing with a sheet of uniform thickness t, we can define a *surface mass density* $\sigma = \rho t$, which represents *mass per unit area*. Finally, when mass is distributed along a rod of uniform cross-sectional area A, we sometimes use *linear mass density* $\lambda = M/L = \rho A$, which is the *mass per unit length*.

Example 10.5 Uniform Thin Hoop

Find the moment of inertia of a uniform thin hoop of mass M and radius R about an axis perpendicular to the plane of the hoop and passing through its center (Fig. 10.9).

Solution Because the hoop is thin, all mass elements dm are the same distance $r = R$ from the axis, and so, applying Equation 10.17, we obtain for the moment of inertia about the z axis through O:

$$I_z = \int r^2\,dm = R^2 \int dm = MR^2$$

Note that this moment of inertia is the same as that of a single particle of mass M located a distance R from the axis of rotation.

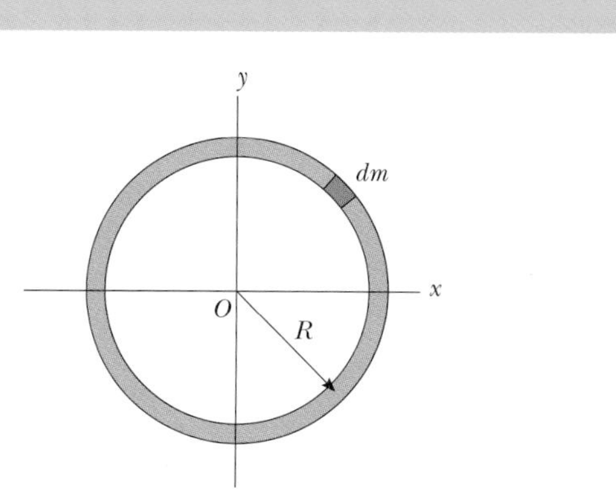

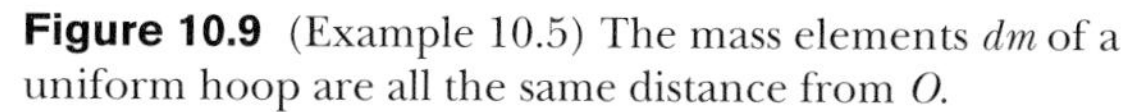
Figure 10.9 (Example 10.5) The mass elements dm of a uniform hoop are all the same distance from O.

Example 10.6 Uniform Rigid Rod

Calculate the moment of inertia of a uniform rigid rod of length L and mass M (Fig. 10.10) about an axis perpendicular to the rod (the y axis) and passing through its center of mass.

Solution The shaded length element dx in Figure 10.10 has a mass dm equal to the mass per unit length λ multiplied by dx:

$$dm = \lambda\,dx = \frac{M}{L}\,dx$$

Substituting this expression for dm into Equation 10.17, with $r^2 = x^2$, we obtain

$$I_y = \int r^2\,dm = \int_{-L/2}^{L/2} x^2 \frac{M}{L}\,dx = \frac{M}{L}\int_{-L/2}^{L/2} x^2\,dx$$

$$= \frac{M}{L}\left[\frac{x^3}{3}\right]_{-L/2}^{L/2} = \frac{1}{12}ML^2$$

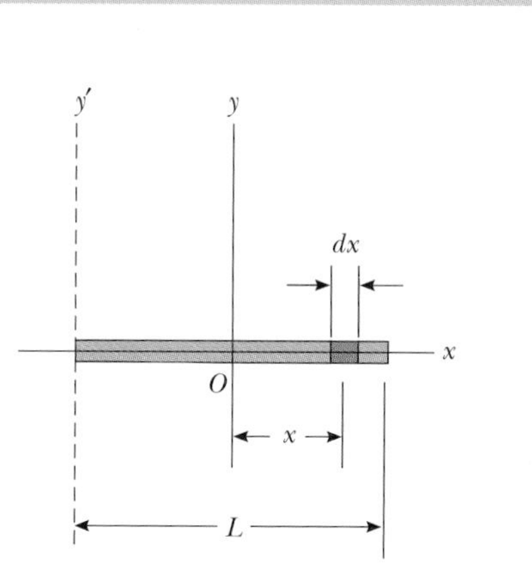

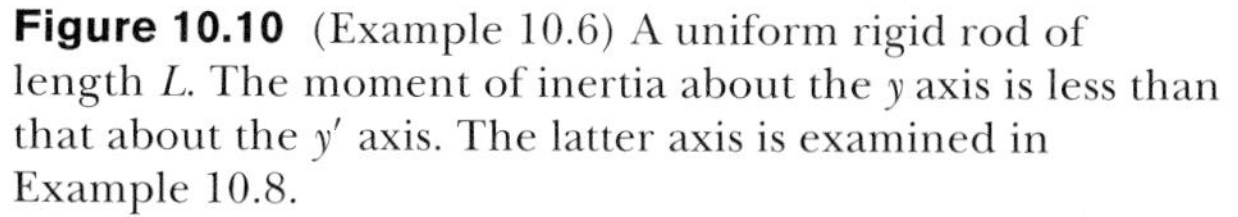
Figure 10.10 (Example 10.6) A uniform rigid rod of length L. The moment of inertia about the y axis is less than that about the y' axis. The latter axis is examined in Example 10.8.

Example 10.7 Uniform Solid Cylinder

A uniform solid cylinder has a radius R, mass M, and length L. Calculate its moment of inertia about its central axis (the z axis in Fig. 10.11).

Solution It is convenient to divide the cylinder into many cylindrical shells, each of which has radius r, thickness dr, and length L, as shown in Figure 10.11. The volume dV of each shell is its cross-sectional area multiplied by its length: $dV = L\,dA = L(2\pi r)\,dr$. If the mass per unit volume is ρ, then the mass of this differential volume element is $dm = \rho\,dV = 2\pi\rho L r\,dr$. Substituting this expression for dm into Equation 10.17, we obtain

$$I_z = \int r^2\,dm = \int r^2(2\pi\rho L r\,dr) = 2\pi\rho L\int_0^R r^3 dr = \tfrac{1}{2}\pi\rho L R^4$$

Because the total volume of the cylinder is $\pi R^2 L$, we see that $\rho = M/V = M/\pi R^2 L$. Substituting this value for ρ into the above result gives

$$I_z = \tfrac{1}{2}MR^2$$

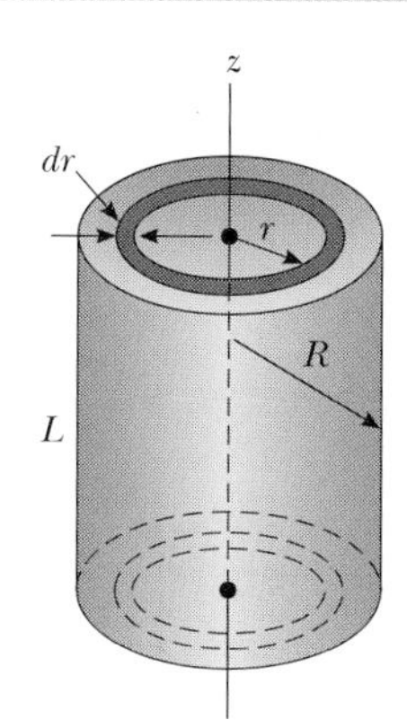

Figure 10.11 (Example 10.7) Calculating I about the z axis for a uniform solid cylinder.

What If? What if the length of the cylinder in Figure 10.11 is increased to $2L$, while the mass M and radius R are held fixed? How does this change the moment of inertia of the cylinder?

Answer Note that the result for the moment of inertia of a cylinder does not depend on L, the length of the cylinder. In other words, it applies equally well to a long cylinder and a flat disk having the same mass M and radius R. Thus, the moment of inertia of the cylinder would not be affected by changing its length.

Table 10.2 gives the moments of inertia for a number of objects about specific axes. The moments of inertia of rigid objects with simple geometry (high symmetry) are relatively easy to calculate provided the rotation axis coincides with an axis of symmetry. The calculation of moments of inertia about an arbitrary axis can be cumbersome, however, even for a highly symmetric object. Fortunately, use of an important theorem, called the **parallel-axis theorem,** often simplifies the calculation. Suppose the moment of inertia about an axis through the center of mass of an object is I_{CM}. The parallel-axis theorem states that the moment of inertia about any axis parallel to and a distance D away from this axis is

Parallel-axis theorem

$$I = I_{CM} + MD^2 \tag{10.18}$$

To prove the parallel-axis theorem, suppose that an object rotates in the xy plane about the z axis, as shown in Figure 10.12, and that the coordinates of the center of mass are x_{CM}, y_{CM}. Let the mass element dm have coordinates x, y. Because this

Table 10.2

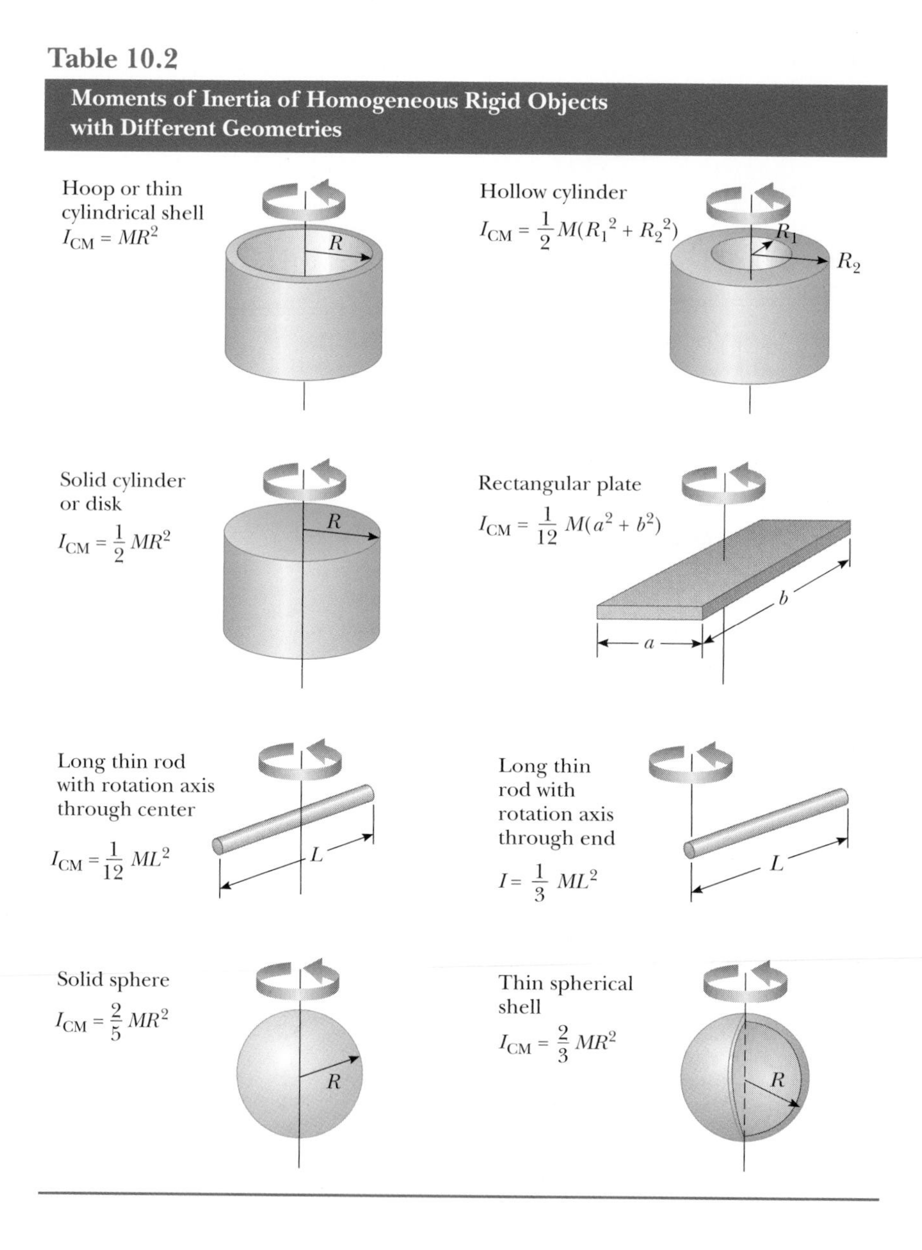

Moments of Inertia of Homogeneous Rigid Objects with Different Geometries
Hoop or thin cylindrical shell
$I_{CM} = MR^2$
R
Hollow cylinder
$I_{CM} = \frac{1}{2}M(R_1^2 + R_2^2)$
R_1
R_2
Solid cylinder or disk
$I_{CM} = \frac{1}{2}MR^2$
R
Rectangular plate
$I_{CM} = \frac{1}{12}M(a^2 + b^2)$
a
b
Long thin rod with rotation axis through center
$I_{CM} = \frac{1}{12}ML^2$
L
Long thin rod with rotation axis through end
$I = \frac{1}{3}ML^2$
L
Solid sphere
$I_{CM} = \frac{2}{5}MR^2$
R
Thin spherical shell
$I_{CM} = \frac{2}{3}MR^2$
R

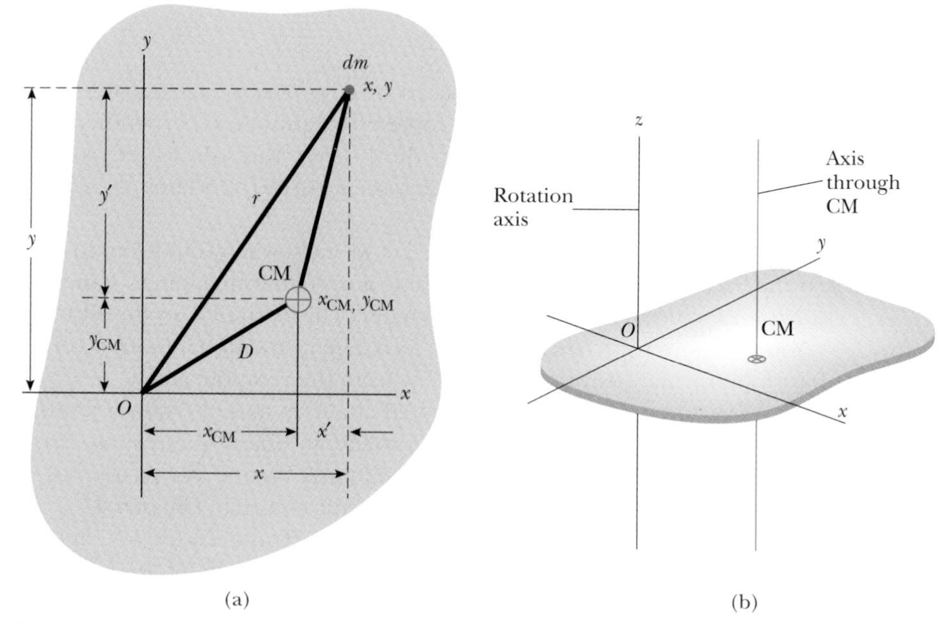

Figure 10.12 (a) The parallel-axis theorem: if the moment of inertia about an axis perpendicular to the figure through the center of mass is I_{CM}, then the moment of inertia about the z axis is $I_z = I_{CM} + MD^2$. (b) Perspective drawing showing the z axis (the axis of rotation) and the parallel axis through the CM.

element is a distance $r = \sqrt{x^2 + y^2}$ from the z axis, the moment of inertia about the z axis is

$$I = \int r^2\, dm = \int (x^2 + y^2)\, dm$$

However, we can relate the coordinates x, y of the mass element dm to the coordinates of this same element located in a coordinate system having the object's center of mass as its origin. If the coordinates of the center of mass are x_{CM}, y_{CM} in the original coordinate system centered on O, then from Figure 10.12a we see that the relationships between the unprimed and primed coordinates are $x = x' + x_{CM}$ and $y = y' + y_{CM}$. Therefore,

$$I = \int [(x' + x_{CM})^2 + (y' + y_{CM})^2]\, dm$$

$$= \int [(x')^2 + (y')^2]\, dm + 2x_{CM} \int x'dm + 2y_{CM} \int y'dm + (x_{CM}^2 + y_{CM}^2) \int dm$$

The first integral is, by definition, the moment of inertia about an axis that is parallel to the z axis and passes through the center of mass. The second two integrals are zero because, by definition of the center of mass, $\int x'dm = \int y'dm = 0$. The last integral is simply MD^2 because $\int dm = M$ and $D^2 = x_{CM}^2 + y_{CM}^2$. Therefore, we conclude that

$$I = I_{CM} + MD^2$$

Example 10.8 Applying the Parallel-Axis Theorem

Consider once again the uniform rigid rod of mass M and length L shown in Figure 10.10. Find the moment of inertia of the rod about an axis perpendicular to the rod through one end (the y' axis in Fig. 10.10).

Solution Intuitively, we expect the moment of inertia to be greater than $I_{CM} = \frac{1}{12}ML^2$ because there is mass up to a distance of L away from the rotation axis, while the farthest distance in Example 10.6 was only $L/2$. Because the distance between the center-of-mass axis and the y' axis is $D = L/2$, the parallel-axis theorem gives

$$I = I_{CM} + MD^2 = \tfrac{1}{12}ML^2 + M\left(\frac{L}{2}\right)^2 = \tfrac{1}{3}ML^2$$

So, it is four times more difficult to change the rotation of a rod spinning about its end than it is to change the motion of one spinning about its center.

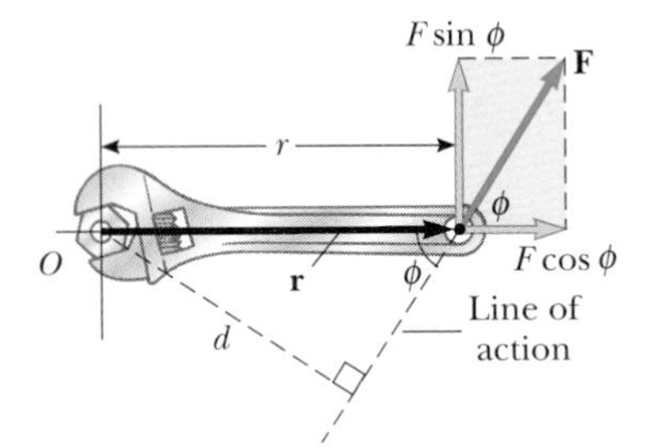

Figure 10.13 The force **F** has a greater rotating tendency about O as F increases and as the moment arm d increases. The component $F \sin \phi$ tends to rotate the wrench about O.

PITFALL PREVENTION

10.5 Torque Depends on Your Choice of Axis

Like moment of inertia, there is no unique value of the torque—its value depends on your choice of rotation axis.

10.6 Torque

Why are a door's hinges and its doorknob placed near opposite edges of the door? Imagine trying to rotate a door by applying a force of magnitude F perpendicular to the door surface but at various distances from the hinges. You will achieve a more rapid rate of rotation for the door by applying the force near the doorknob than by applying it near the hinges.

If you cannot loosen a stubborn bolt with a socket wrench, what would you do in an effort to loosen the bolt? You may intuitively try using a wrench with a longer handle or slip a pipe over the existing wrench to make it longer. This is similar to the situation with the door. You are more successful at causing a change in rotational motion (of the door or the bolt) by applying the force farther away from the rotation axis.

When a force is exerted on a rigid object pivoted about an axis, the object tends to rotate about that axis. The tendency of a force to rotate an object about some axis is measured by a vector quantity called **torque** τ (Greek tau). Torque is a vector, but we will consider only its magnitude here and explore its vector nature in Chapter 11.

Consider the wrench pivoted on the axis through O in Figure 10.13. The applied force **F** acts at an angle ϕ to the horizontal. We define the magnitude of the torque associated with the force **F** by the expression

$$\tau \equiv rF \sin \phi = Fd \qquad (10.19)$$

Moment arm

where r is the distance between the pivot point and the point of application of **F** and d is the perpendicular distance from the pivot point to the line of action of **F**. (The *line of action* of a force is an imaginary line extending out both ends of the vector representing the force. The dashed line extending from the tail of **F** in Figure 10.13 is part of the line of action of **F**.) From the right triangle in Figure 10.13 that has the wrench as its hypotenuse, we see that $d = r \sin \phi$. The quantity d is called the **moment arm** (or *lever arm*) of **F**.

In Figure 10.13, the only component of **F** that tends to cause rotation is $F \sin \phi$, the component perpendicular to a line drawn from the rotation axis to the point of application of the force. The horizontal component $F \cos \phi$, because its line of action passes through O, has no tendency to produce rotation about an axis passing through O. From the definition of torque, we see that the rotating tendency increases as F increases and as d increases. This explains the observation that it is easier to rotate a door if we push at the doorknob rather than at a point close to the hinge. We also want to apply our push as closely perpendicular to the door as we can. Pushing sideways on the doorknob will not cause the door to rotate.

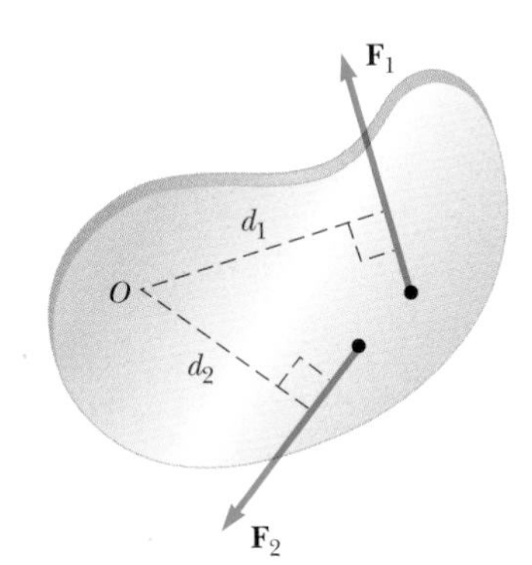

Active Figure 10.14 The force $\mathbf{F}_1$ tends to rotate the object counterclockwise about O, and $\mathbf{F}_2$ tends to rotate it clockwise.

At the Active Figures link at http://www.pse6.com, you can change the magnitudes, directions, and points of application of forces F_1 and F_2 to see how the object accelerates under the action of the two forces.

If two or more forces are acting on a rigid object, as in Figure 10.14, each tends to produce rotation about the axis at O. In this example, $\mathbf{F}_2$ tends to rotate the object clockwise and $\mathbf{F}_1$ tends to rotate it counterclockwise. We use the convention that the sign of the torque resulting from a force is positive if the turning tendency of the force is counterclockwise and is negative if the turning tendency is clockwise. For example, in Figure 10.14, the torque resulting from $\mathbf{F}_1$, which has a moment arm d_1, is positive and equal to $+F_1 d_1$; the torque from $\mathbf{F}_2$ is negative and equal to $-F_2 d_2$. Hence, the *net* torque about O is

$$\sum \tau = \tau_1 + \tau_2 = F_1 d_1 - F_2 d_2$$

Torque should not be confused with force. Forces can cause a change in linear motion, as described by Newton's second law. Forces can also cause a change in rotational motion, but the effectiveness of the forces in causing this change depends on both the forces and the moment arms of the forces, in the combination that we call *torque.* Torque has units of force times length—newton · meters in SI units—and should be reported in these units. Do not confuse torque and work, which have the same units but are very different concepts.

Quick Quiz 10.8 If you are trying to loosen a stubborn screw from a piece of wood with a screwdriver and fail, should you find a screwdriver for which the handle is (a) longer or (b) fatter?

Quick Quiz 10.9 If you are trying to loosen a stubborn bolt from a piece of metal with a wrench and fail, should you find a wrench for which the handle is (a) longer (b) fatter?

Example 10.9 The Net Torque on a Cylinder

A one-piece cylinder is shaped as shown in Figure 10.15, with a core section protruding from the larger drum. The cylinder is free to rotate about the central axis shown in the drawing. A rope wrapped around the drum, which has radius R_1, exerts a force $\mathbf{T}_1$ to the right on the cylinder. A rope wrapped around the core, which has radius R_2, exerts a force $\mathbf{T}_2$ downward on the cylinder.

(A) What is the net torque acting on the cylinder about the rotation axis (which is the z axis in Figure 10.15)?

Solution The torque due to $\mathbf{T}_1$ is $-R_1T_1$. (The sign is negative because the torque tends to produce clockwise rotation.) The torque due to $\mathbf{T}_2$ is $+R_2T_2$. (The sign is positive because the torque tends to produce counterclockwise rotation.) Therefore, the net torque about the rotation axis is

$$\sum \tau = \tau_1 + \tau_2 = R_2T_2 - R_1T_1$$

We can make a quick check by noting that if the two forces are of equal magnitude, the net torque is negative because $R_1 > R_2$. Starting from rest with both forces of equal magnitude acting on it, the cylinder would rotate clockwise because $\mathbf{T}_1$ would be more effective at turning it than would $\mathbf{T}_2$.

(B) Suppose $T_1 = 5.0$ N, $R_1 = 1.0$ m, $T_2 = 15.0$ N, and $R_2 = 0.50$ m. What is the net torque about the rotation axis, and which way does the cylinder rotate starting from rest?

Solution Evaluating the net torque,

$$\sum \tau = (15\ \text{N})(0.50\ \text{m}) - (5.0\ \text{N})(1.0\ \text{m}) = 2.5\ \text{N}\cdot\text{m}$$

Because this torque is positive, the cylinder will begin to rotate in the counterclockwise direction.

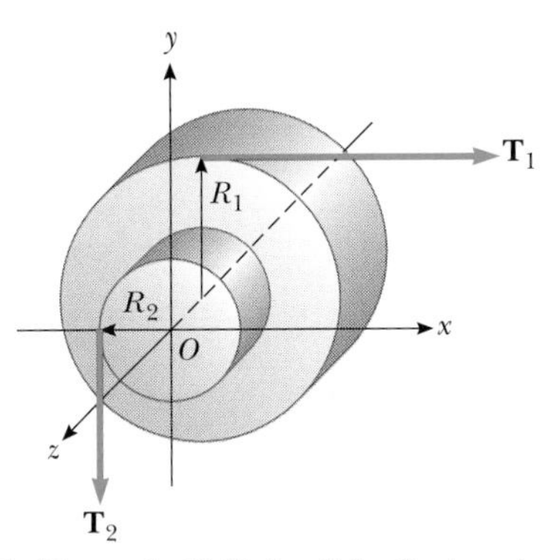

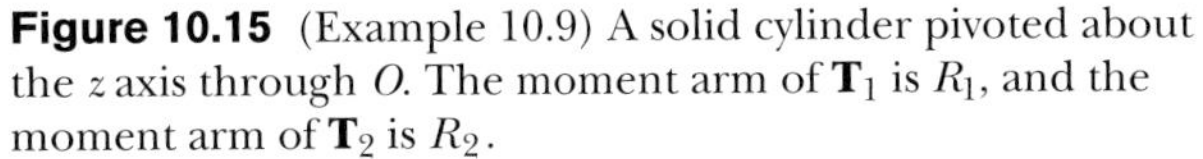

Figure 10.15 (Example 10.9) A solid cylinder pivoted about the z axis through O. The moment arm of $\mathbf{T}_1$ is R_1, and the moment arm of $\mathbf{T}_2$ is R_2.

10.7 Relationship Between Torque and Angular Acceleration

In Chapter 4, we learned that a net force on an object causes an acceleration of the object and that the acceleration is proportional to the net force (Newton's second law). In this section we show the rotational analog of Newton's second law—the angular acceleration of a rigid object rotating about a fixed axis is proportional to the net torque acting about that axis. Before discussing the more complex case of rigid-object rotation, however, it is instructive first to discuss the case of a particle moving in a circular path about some fixed point under the influence of an external force.

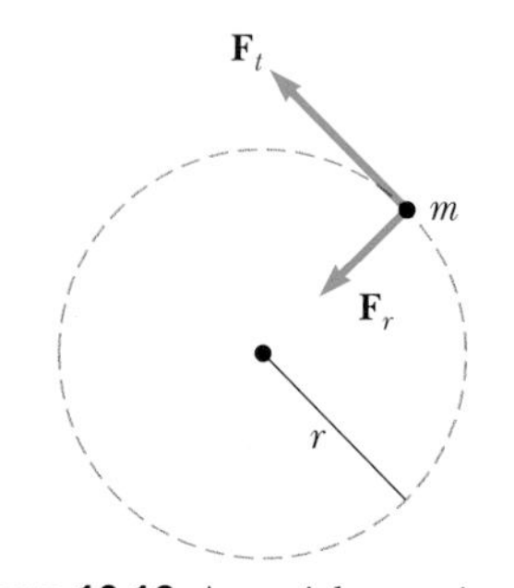

Figure 10.16 A particle rotating in a circle under the influence of a tangential force $\mathbf{F}_t$. A force $\mathbf{F}_r$ in the radial direction also must be present to maintain the circular motion.

Consider a particle of mass m rotating in a circle of radius r under the influence of a tangential force $\mathbf{F}_t$ and a radial force $\mathbf{F}_r$, as shown in Figure 10.16. The tangential force provides a tangential acceleration $\mathbf{a}_t$, and

$$F_t = ma_t$$

The magnitude of the torque about the center of the circle due to $\mathbf{F}_t$ is

$$\tau = F_t r = (ma_t) r$$

Because the tangential acceleration is related to the angular acceleration through the relationship $a_t = r\alpha$ (see Eq. 10.11), the torque can be expressed as

$$\tau = (mr\alpha) r = (mr^2) \alpha$$

Recall from Equation 10.15 that mr^2 is the moment of inertia of the particle about the z axis passing through the origin, so that

$$\tau = I\alpha \tag{10.20}$$

That is, **the torque acting on the particle is proportional to its angular acceleration,** and the proportionality constant is the moment of inertia. Note that $\tau = I\alpha$ is the rotational analog of Newton's second law of motion, $F = ma$.

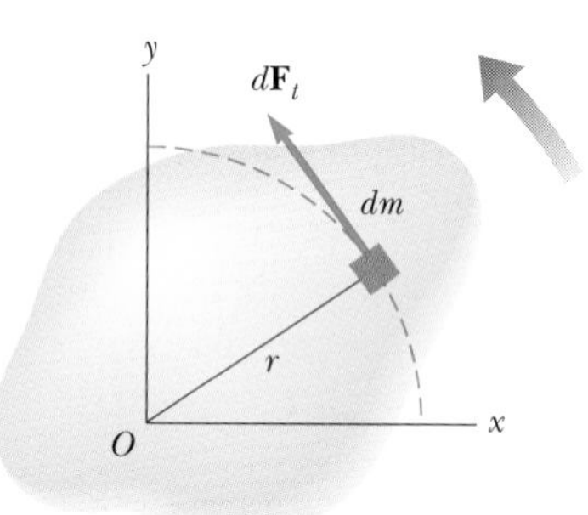

Figure 10.17 A rigid object rotating about an axis through O. Each mass element dm rotates about O with the same angular acceleration α, and the net torque on the object is proportional to α.

Now let us extend this discussion to a rigid object of arbitrary shape rotating about a fixed axis, as in Figure 10.17. The object can be regarded as an infinite number of mass elements dm of infinitesimal size. If we impose a Cartesian coordinate system on the object, then each mass element rotates in a circle about the origin, and each has a tangential acceleration $\mathbf{a}_t$ produced by an external tangential force $d\mathbf{F}_t$. For any given element, we know from Newton's second law that

$$dF_t = (dm)\, a_t$$

The torque $d\tau$ associated with the force $d\mathbf{F}_t$ acts about the origin and is given by

$$d\tau = r\, dF_t = a_t r\, dm$$

Because $a_t = r\alpha$, the expression for $d\tau$ becomes

$$d\tau = \alpha r^2\, dm$$

Although each mass element of the rigid object may have a different linear acceleration $\mathbf{a}_t$, they all have the *same* angular acceleration α. With this in mind, we can integrate the above expression to obtain the net torque $\Sigma\tau$ about O due to the external forces:

$$\sum \tau = \int \alpha r^2\, dm = \alpha \int r^2\, dm$$

where α can be taken outside the integral because it is common to all mass elements. From Equation 10.17, we know that $\int r^2\, dm$ is the moment of inertia of the object about the rotation axis through O, and so the expression for $\Sigma\tau$ becomes

Torque is proportional to angular acceleration

$$\sum \tau = I\alpha \tag{10.21}$$

Note that this is the same relationship we found for a particle moving in a circular path (see Eq. 10.20). So, again we see that the net torque about the rotation axis is proportional to the angular acceleration of the object, with the proportionality factor being I, a quantity that depends upon the axis of rotation and upon the size and shape of the object. In view of the complex nature of the system, the relationship $\Sigma\tau = I\alpha$ is strikingly simple and in complete agreement with experimental observations.

Finally, note that the result $\Sigma\tau = I\alpha$ also applies when the forces acting on the mass elements have radial components as well as tangential components. This is because the line of action of all radial components must pass through the axis of rotation, and hence all radial components produce zero torque about that axis.

Quick Quiz 10.10 You turn off your electric drill and find that the time interval for the rotating bit to come to rest due to frictional torque in the drill is Δt. You replace the bit with a larger one that results in a doubling of the moment of inertia of the entire rotating mechanism of the drill. When this larger bit is rotated at the same angular speed as the first and the drill is turned off, the frictional torque remains the same as that for the previous situation. The time for this second bit to come to rest is (a) $4\Delta t$ (b) $2\Delta t$ (c) Δt (d) $0.5\Delta t$ (e) $0.25\Delta t$ (f) impossible to determine.

Example 10.10 Rotating Rod

A uniform rod of length L and mass M is attached at one end to a frictionless pivot and is free to rotate about the pivot in the vertical plane, as in Figure 10.18. The rod is released from rest in the horizontal position. What is the initial angular acceleration of the rod and the initial linear acceleration of its right end?

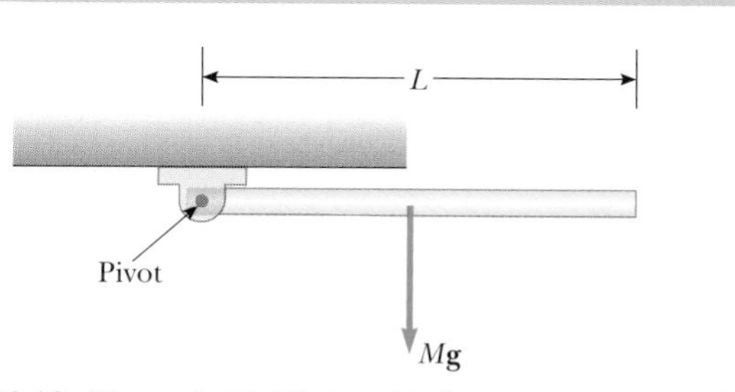

Figure 10.18 (Example 10.10) A rod is free to rotate around a pivot at the left end.

Solution We cannot use our kinematic equations to find α or a because the torque exerted on the rod varies with its angular position and so neither acceleration is constant. We have enough information to find the torque, however, which we can then use in Equation 10.21 to find the initial α and then the initial a.

The only force contributing to the torque about an axis through the pivot is the gravitational force $M\mathbf{g}$ exerted on the rod. (The force exerted by the pivot on the rod has zero torque about the pivot because its moment arm is zero.) To compute the torque on the rod, we assume that the gravitational force acts at the center of mass of the rod, as shown in Figure 10.18. The magnitude of the torque due to this force about an axis through the pivot is

$$\tau = Mg\left(\frac{L}{2}\right)$$

With $\Sigma\tau = I\alpha$, and $I = \frac{1}{3}ML^2$ for this axis of rotation (see Table 10.2), we obtain

$$(1) \qquad \alpha = \frac{\tau}{I} = \frac{Mg(L/2)}{\frac{1}{3}ML^2} = \frac{3g}{2L}$$

All points on the rod have this initial angular acceleration.

To find the initial linear acceleration of the right end of the rod, we use the relationship $a_t = r\alpha$ (Eq. 10.11), with $r = L$:

$$a_t = L\alpha = \tfrac{3}{2}g$$

What If? **What if we were to place a penny on the end of the rod and release the rod? Would the penny stay in contact with the rod?**

Answer The result for the initial acceleration of a point on the end of the rod shows that $a_t > g$. A penny will fall at acceleration g. This means that if we place a penny at the end of the rod and then release the rod, the end of the rod falls faster than the penny does! The penny does not stay in contact with the rod. (Try this with a penny and a meter stick!)

This raises the question as to the location on the rod at which we can place a penny that *will* stay in contact as both begin to fall. To find the linear acceleration of an arbitrary point on the rod at a distance $r < L$ from the pivot point, we combine (1) with Equation 10.11:

$$a_t = r\alpha = \frac{3g}{2L}r$$

For the penny to stay in contact with the rod, the limiting case is that the linear acceleration must be equal to that due to gravity:

$$a_t = g = \frac{3g}{2L}r$$

$$r = \tfrac{2}{3}L$$

Thus, a penny placed closer to the pivot than two thirds of the length of the rod will stay in contact with the falling rod while a penny farther out than this point will lose contact.

Conceptual Example 10.11 Falling Smokestacks and Tumbling Blocks

When a tall smokestack falls over, it often breaks somewhere along its length before it hits the ground, as shown in Figure 10.19. The same thing happens with a tall tower of children's toy blocks. Why does this happen?

Solution As the smokestack rotates around its base, each higher portion of the smokestack falls with a larger tangential acceleration than the portion below it. (The tangential acceleration of a given point on the smokestack is proportional to the distance of that portion from the base.) As the angular acceleration increases as the smokestack tips farther, higher portions of the smokestack experience an acceleration greater than that which could result from gravity alone; this is similar to the situation described in Example 10.10. This can happen only if these portions are being pulled downward by a force in addition to the gravitational force. The force that causes this to occur is the shear force from lower portions of the smokestack. Eventually the shear force that provides this acceleration is greater than the smokestack can withstand, and the smokestack breaks.

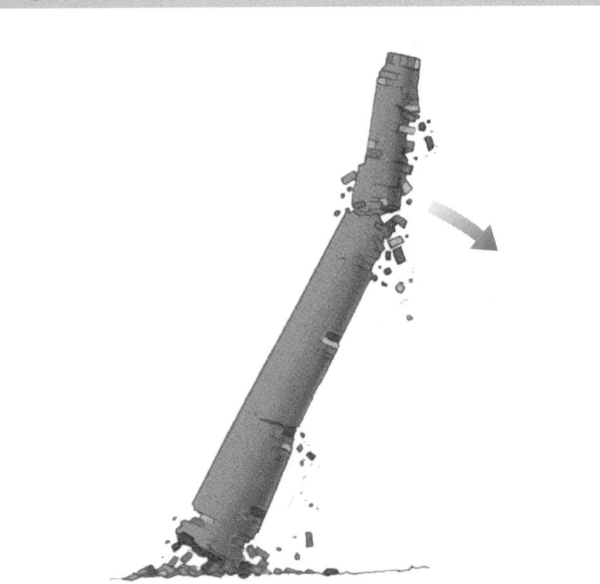

Figure 10.19 (Conceptual Example 10.11) A falling smokestack breaks at some point along its length.

Example 10.12 Angular Acceleration of a Wheel

Interactive

A wheel of radius R, mass M, and moment of inertia I is mounted on a frictionless horizontal axle, as in Figure 10.20. A light cord wrapped around the wheel supports an object of mass m. Calculate the angular acceleration of the wheel, the linear acceleration of the object, and the tension in the cord.

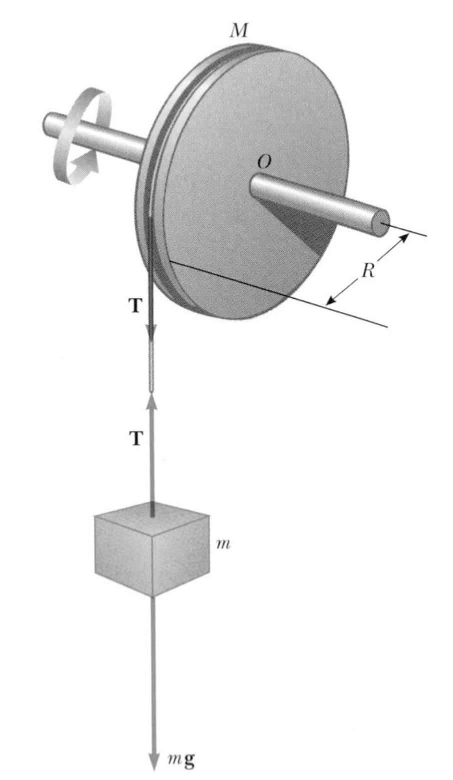

Figure 10.20 (Example 10.12) An object hangs from a cord wrapped around a wheel.

Solution The magnitude of the torque acting on the wheel about its axis of rotation is $\tau = TR$, where T is the force exerted by the cord on the rim of the wheel. (The gravitational force exerted by the Earth on the wheel and the normal force exerted by the axle on the wheel both pass through the axis of rotation and thus produce no torque.) Because $\sum \tau = I\alpha$, we obtain

$$\sum \tau = I\alpha = TR$$

$$\text{(1)} \qquad \alpha = \frac{TR}{I}$$

Now let us apply Newton's second law to the motion of the object, taking the downward direction to be positive:

$$\sum F_y = mg - T = ma$$

$$\text{(2)} \qquad a = \frac{mg - T}{m}$$

Equations (1) and (2) have three unknowns: α, a, and T. Because the object and wheel are connected by a cord that does not slip, the linear acceleration of the suspended object is equal to the tangential acceleration of a point on the rim of the wheel. Therefore, the angular acceleration α of the wheel and the linear acceleration of the object are related by $a = R\alpha$. Using this fact together with Equations (1) and (2), we obtain

$$\text{(3)} \qquad a = R\alpha = \frac{TR^2}{I} = \frac{mg - T}{m}$$

$$\text{(4)} \qquad T = \frac{mg}{1 + (mR^2/I)}$$

Substituting Equation (4) into Equation (2) and solving for a and α, we find that

$$(5) \qquad a = \frac{g}{1 + (I/mR^2)}$$

$$\alpha = \frac{a}{R} = \frac{g}{R + (I/mR)}$$

What If? **What if the wheel were to become very massive so that *I* becomes very large? What happens to the acceleration *a* of the object and the tension *T*?**

Answer If the wheel becomes infinitely massive, we can imagine that the object of mass m will simply hang from the cord without causing the wheel to rotate.

We can show this mathematically by taking the limit $I \rightarrow \infty$, so that Equation (5) becomes

$$a = \frac{g}{1 + (I/mR^2)} \longrightarrow 0$$

This agrees with our conceptual conclusion that the object will hang at rest. We also find that Equation (4) becomes

$$T = \frac{mg}{1 + (mR^2/I)} \longrightarrow \frac{mg}{1 + 0} = mg$$

This is consistent with the fact that the object simply hangs at rest in equilibrium between the gravitational force and the tension in the string.

At the Interactive Worked Example link at **http://www.pse6.com,** *you can change the masses of the object and the wheel as well as the radius of the wheel to see the effect on how the system moves.*

Example 10.13 Atwood's Machine Revisited

Interactive

Two blocks having masses m_1 and m_2 are connected to each other by a light cord that passes over two identical frictionless pulleys, each having a moment of inertia I and radius R, as shown in Figure 10.21a. Find the acceleration of each block and the tensions T_1, T_2, and T_3 in the cord. (Assume no slipping between cord and pulleys.)

Solution Compare this situation with the Atwood machine of Example 5.9 (p. 129). The motion of m_1 and m_2 is similar to the motion of the two blocks in that example. The primary differences are that in the present example we have two pulleys and each of the pulleys has mass. Despite these differences, the apparatus in the present example is indeed an Atwood machine.

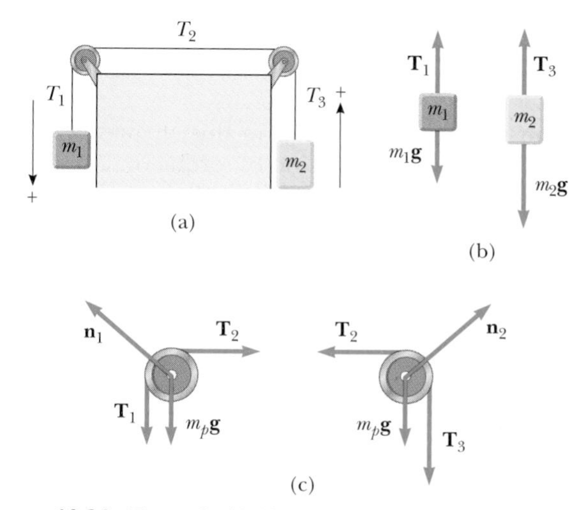

Figure 10.21 (Example 10.13) (a) Another look at Atwood's machine. (b) Free-body diagrams for the blocks. (c) Free-body diagrams for the pulleys, where $m_p\mathbf{g}$ represents the gravitational force acting on each pulley.

We shall define the downward direction as positive for m_1 and upward as the positive direction for m_2. This allows us to represent the acceleration of both masses by a single variable a and also enables us to relate a positive a to a positive (counterclockwise) angular acceleration α of the pulleys. Let us write Newton's second law of motion for each block, using the free-body diagrams for the two blocks as shown in Figure 10.21b:

$$(1) \qquad m_1 g - T_1 = m_1 a$$

$$(2) \qquad T_3 - m_2 g = m_2 a$$

Next, we must include the effect of the pulleys on the motion. Free-body diagrams for the pulleys are shown in Figure 10.21c. The net torque about the axle for the pulley on the left is $(T_1 - T_2)R$, while the net torque for the pulley on the right is $(T_2 - T_3)R$. Using the relation $\Sigma\tau = I\alpha$ for each pulley and noting that each pulley has the same angular acceleration α, we obtain

$$(3) \qquad (T_1 - T_2)R = I\alpha$$

$$(4) \qquad (T_2 - T_3)R = I\alpha$$

We now have four equations with five unknowns: α, a, T_1, T_2, and T_3. We also have a fifth equation that relates the accelerations, $a = R\alpha$. These equations can be solved simultaneously. Adding Equations (3) and (4) gives

$$(5) \qquad (T_1 - T_3)R = 2I\alpha$$

Adding Equations (1) and (2) gives

$$T_3 - T_1 + m_1 g - m_2 g = (m_1 + m_2)a$$

$$(6) \qquad T_1 - T_3 = (m_1 - m_2)g - (m_1 + m_2)a$$

Substituting Equation (6) into Equation (5), we have

$$[(m_1 - m_2)g - (m_1 + m_2)a]R = 2I\alpha$$

Because $\alpha = a/R$, this expression can be simplified to

$$(m_1 - m_2)g - (m_1 + m_2)a = 2I\frac{a}{R^2}$$

$$(7) \qquad a = \frac{(m_1 - m_2)g}{m_1 + m_2 + 2(I/R^2)}$$

Note that if $m_1 > m_2$, the acceleration is positive; this means that the left block accelerates downward, the right block accelerates upward, and both pulleys accelerate counterclockwise. If $m_1 < m_2$, the acceleration is negative and the motions are reversed. If $m_1 = m_2$, no acceleration occurs at all. You should compare these results with those found in Example 5.9.

The expression for a can be substituted into Equations (1) and (2) to give T_1 and T_3. From Equation (1),

$$T_1 = m_1g - m_1a = m_1(g - a)$$

$$= m_1\left(g - \frac{(m_1 - m_2)g}{m_1 + m_2 + 2(I/R^2)}\right)$$

$$= 2m_1g\left(\frac{m_2 + (I/R^2)}{m_1 + m_2 + 2(I/R^2)}\right)$$

Similarly, from Equation (2),

$$T_3 = m_2g + m_2a = 2m_2g\left(\frac{m_1 + (I/R^2)}{m_1 + m_2 + 2(I/R^2)}\right)$$

Finally, T_2 can be found from Equation (3):

$$T_2 = T_1 - \frac{I\alpha}{R} = T_1 - \frac{Ia}{R^2}$$

$$= 2m_1g\left(\frac{m_2 + (I/R^2)}{m_1 + m_2 + 2(I/R^2)}\right) - \frac{I}{R^2}\left(\frac{(m_1 - m_2)g}{m_1 + m_2 + 2(I/R^2)}\right)$$

$$= \frac{2m_1m_2 + (m_1 + m_2)(I/R^2)}{m_1 + m_2 + 2(I/R^2)}\, g$$

What If? **What if the pulleys become massless? Does this reduce to a previously solved problem?**

Answer If the pulleys become massless, the system should behave in the same way as the massless-pulley Atwood machine that we investigated in Example 5.9. The only difference is the existence of two pulleys instead of one.

Mathematically, if $I \rightarrow 0$, Equation (7) becomes

$$a = \frac{(m_1 - m_2)g}{m_1 + m_2 + 2(I/R^2)} \longrightarrow a = \left(\frac{m_1 - m_2}{m_1 + m_2}\right)g$$

which is the same result as Equation (3) in Example 5.9. Although the expressions for the three tensions in the present example are different from each other, all three expressions become, in the limit $I \rightarrow 0$,

$$T = \left(\frac{2m_1m_2}{m_1 + m_2}\right)g$$

which is the same as Equation (4) in Example 5.9.

At the Interactive Worked Example link at **http://www.pse6.com,** *you can change the masses of the blocks and the pulleys to see the effect on the motion of the system.*

10.8 Work, Power, and Energy in Rotational Motion

Up to this point in our discussion of rotational motion in this chapter, we focused on an approach involving force, leading to a description of torque on a rigid object. We now see how an energy approach can be useful to us in solving rotational problems.

We begin by considering the relationship between the torque acting on a rigid object and its resulting rotational motion in order to generate expressions for power and a rotational analog to the work–kinetic energy theorem. Consider the rigid object pivoted at O in Figure 10.22. Suppose a single external force **F** is applied at P, where **F** lies in the plane of the page. The work done by **F** on the object as it rotates through an infinitesimal distance $ds = r\,d\theta$ is

$$dW = \mathbf{F}\cdot d\mathbf{s} = (F\sin\phi)\,r\,d\theta$$

where $F\sin\phi$ is the tangential component of **F**, or, in other words, the component of the force along the displacement. Note that *the radial component of* **F** *does no work because it is perpendicular to the displacement.*

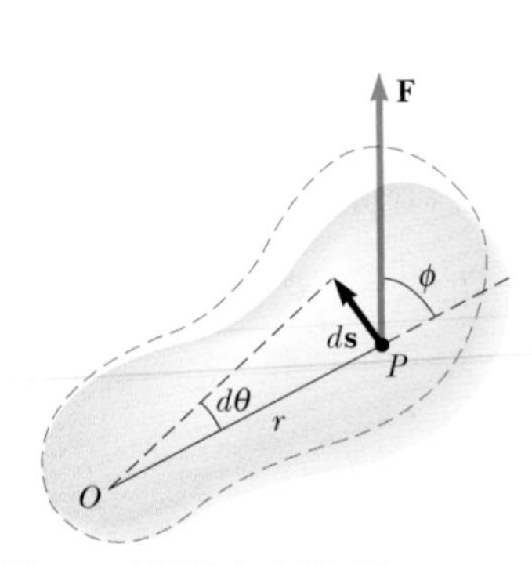

Figure 10.22 A rigid object rotates about an axis through O under the action of an external force **F** applied at P.

Because the magnitude of the torque due to **F** about O is defined as $rF \sin \phi$ by Equation 10.19, we can write the work done for the infinitesimal rotation as

$$dW = \tau \, d\theta \tag{10.22}$$

The rate at which work is being done by **F** as the object rotates about the fixed axis through the angle $d\theta$ in a time interval dt is

$$\frac{dW}{dt} = \tau \frac{d\theta}{dt}$$

Because dW/dt is the instantaneous power $\mathcal{P}$ (see Section 7.8) delivered by the force and $d\theta/dt = \omega$, this expression reduces to

$$\mathcal{P} = \frac{dW}{dt} = \tau\omega \tag{10.23}$$

Power delivered to a rotating rigid object

This expression is analogous to $\mathcal{P} = Fv$ in the case of linear motion, and the expression $dW = \tau \, d\theta$ is analogous to $dW = F_x \, dx$.

In studying linear motion, we found the energy approach extremely useful in describing the motion of a system. From what we learned of linear motion, we expect that when a symmetric object rotates about a fixed axis, the work done by external forces equals the change in the rotational energy.

To show that this is in fact the case, let us begin with $\Sigma\tau = I\alpha$. Using the chain rule from calculus, we can express the resultant torque as

$$\sum \tau = I\alpha = I \frac{d\omega}{dt} = I \frac{d\omega}{d\theta}\frac{d\theta}{dt} = I \frac{d\omega}{d\theta}\omega$$

Rearranging this expression and noting that $\Sigma\tau \, d\theta = dW$, we obtain

$$\sum \tau \, d\theta = dW = I\omega \, d\omega$$

Integrating this expression, we obtain for the total work done by the net external force acting on a rotating system

$$\sum W = \int_{\omega_i}^{\omega_f} I\omega \, d\omega = \tfrac{1}{2}I\omega_f^{\,2} - \tfrac{1}{2}I\omega_i^{\,2} \tag{10.24}$$

Work–kinetic energy theorem for rotational motion

where the angular speed changes from ω_i to ω_f. That is, the **work–kinetic energy theorem for rotational motion** states that

> the net work done by external forces in rotating a symmetric rigid object about a fixed axis equals the change in the object's rotational energy.

In general, then, combining this with the translational form of the work–kinetic energy theorem from Chapter 7, the net work done by external forces on an object is the change in its *total* kinetic energy, which is the sum of the translational and rotational kinetic energies. For example, when a pitcher throws a baseball, the work done by the pitcher's hands appears as kinetic energy associated with the ball moving through space as well as rotational kinetic energy associated with the spinning of the ball.

In addition to the work–kinetic energy theorem, other energy principles can also be applied to rotational situations. For example, if a system involving rotating objects is isolated, the principle of conservation of energy can be used to analyze the system, as in Example 10.14 below.

Table 10.3 lists the various equations we have discussed pertaining to rotational motion, together with the analogous expressions for linear motion. The last two equations in Table 10.3, involving angular momentum L, are discussed in Chapter 11 and are included here only for the sake of completeness.

Table 10.3

Useful Equations in Rotational and Linear Motion

Rotational Motion About a Fixed Axis	Linear Motion
Angular speed $\omega = d\theta/dt$	Linear speed $v = dx/dt$
Angular acceleration $\alpha = d\omega/dt$	Linear acceleration $a = dv/dt$
Net torque $\Sigma\tau = I\alpha$	Net force $\Sigma F = ma$
If $\alpha =$ constant $\begin{cases} \omega_f = \omega_i + \alpha t \\ \theta_f = \theta_i + \omega_i t + \frac{1}{2}\alpha t^2 \\ \omega_f^2 = \omega_i^2 + 2\alpha(\theta_f - \theta_i) \end{cases}$	If $a =$ constant $\begin{cases} v_f = v_i + at \\ x_f = x_i + v_i t + \frac{1}{2}at^2 \\ v_f^2 = v_i^2 + 2a(x_f - x_i) \end{cases}$
Work $W = \int_{\theta_i}^{\theta_f} \tau \, d\theta$	Work $W = \int_{x_i}^{x_f} F_x \, dx$
Rotational kinetic energy $K_R = \frac{1}{2}I\omega^2$	Kinetic energy $K = \frac{1}{2}mv^2$
Power $\mathcal{P} = \tau\omega$	Power $\mathcal{P} = Fv$
Angular momentum $L = I\omega$	Linear momentum $p = mv$
Net torque $\Sigma\tau = dL/dt$	Net force $\Sigma F = dp/dt$

Quick Quiz 10.11 A rod is attached to the shaft of a motor at the center of the rod so that the rod is perpendicular to the shaft, as in Figure 10.23a. The motor is turned on and performs work W on the rod, accelerating it to an angular speed ω. The system is brought to rest, and the rod is attached to the shaft of the motor at one end of the rod as in Figure 10.23b. The motor is turned on and performs work W on the rod. The angular speed of the rod in the second situation is (a) 4ω (b) 2ω (c) ω (d) 0.5ω (e) 0.25ω (f) impossible to determine.

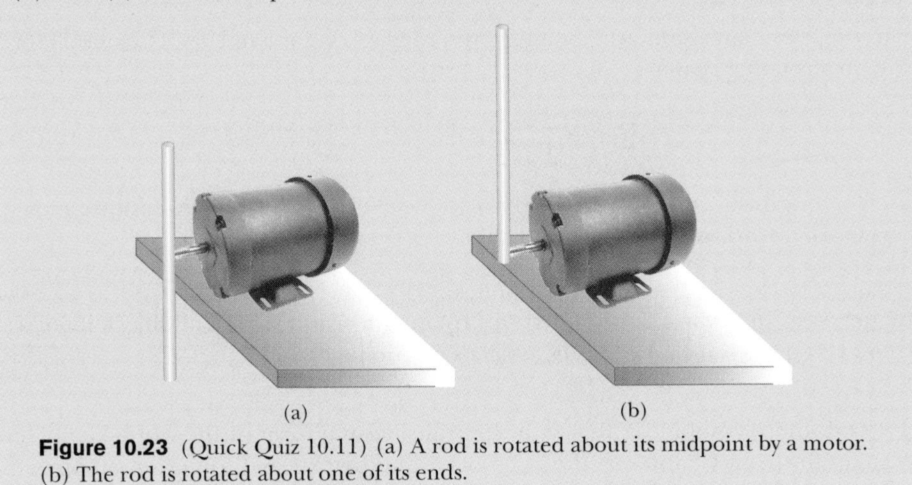

Figure 10.23 (Quick Quiz 10.11) (a) A rod is rotated about its midpoint by a motor. (b) The rod is rotated about one of its ends.

Example 10.14 Rotating Rod Revisited

Interactive

A uniform rod of length L and mass M is free to rotate on a frictionless pin passing through one end (Fig 10.24). The rod is released from rest in the horizontal position.

(A) What is its angular speed when it reaches its lowest position?

Solution To conceptualize this problem, consider Figure 10.24 and imagine the rod rotating downward through a quarter turn about the pivot at the left end. In this situation, the angular acceleration of the rod is not constant. Thus, the kinematic equations for rotation (Section 10.2) cannot be used to solve this problem. As we found with translational motion, however, an energy approach can make such a seemingly insoluble problem relatively easy. We categorize this as a conservation of energy problem.

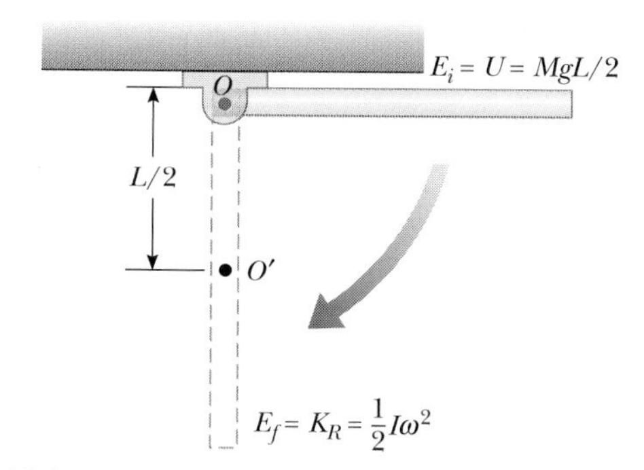

Figure 10.24 (Example 10.14) A uniform rigid rod pivoted at O rotates in a vertical plane under the action of the gravitational force.

To analyze the problem, we consider the mechanical energy of the system of the rod and the Earth. We choose the configuration in which the rod is hanging straight down as the reference configuration for gravitational potential energy and assign a value of zero for this configuration. When the rod is in the horizontal position, it has no rotational kinetic energy. The potential energy of the system in this configuration relative to the reference configuration is $MgL/2$ because the center of mass of the rod is at a height $L/2$ higher than its position in the reference configuration. When the rod reaches its lowest position, the energy is entirely rotational energy $\frac{1}{2}I\omega^2$, where I is the moment of inertia about the pivot, and the potential energy of the system is zero. Because $I = \frac{1}{3}ML^2$ (see Table 10.2) and because the system is isolated with no nonconservative forces acting, we apply conservation of mechanical energy for the system:

$$K_f + U_f = K_i + U_i$$

$$\frac{1}{2}I\omega^2 + 0 = \frac{1}{2}(\frac{1}{3}ML^2)\omega^2 = 0 + \frac{1}{2}MgL$$

$$\omega = \sqrt{\frac{3g}{L}}$$

(B) Determine the tangential speed of the center of mass and the tangential speed of the lowest point on the rod when it is in the vertical position.

Solution These two values can be determined from the relationship between tangential and angular speeds. We know ω from part (A), and so the tangential speed of the center of mass is

$$v_{\text{CM}} = r\omega = \frac{L}{2}\omega = \frac{1}{2}\sqrt{3gL}$$

Because r for the lowest point on the rod is twice what it is for the center of mass, the lowest point has a tangential speed v equal to

$$v = 2v_{\text{CM}} = \sqrt{3gL}$$

To finalize this problem, note that the initial configuration in this example is the same as that in Example 10.10. In Example 10.10, however, we could only find the initial angular acceleration of the rod. We cannot use this and the kinematic equations to find the angular speed of the rod at its lowest point because the angular acceleration is not constant. Applying an energy approach in the current example allows us to find something that we cannot in Example 10.10.

At the Interactive Worked Example link at **http://www.pse6.com,** *you can alter the mass and length of the rod and see the effect on the velocity at the lowest point.*

Example 10.15 Energy and the Atwood Machine

Consider two cylinders having different masses m_1 and m_2, connected by a string passing over a pulley, as shown in Figure 10.25. The pulley has a radius R and moment of inertia I about its axis of rotation. The string does not slip on the pulley, and the system is released from rest. Find the linear speeds of the cylinders after cylinder 2 descends through a distance h, and the angular speed of the pulley at this time.

Solution We will solve this problem by applying energy methods to an Atwood machine with a massive pulley. Because the string does not slip, the pulley rotates about the axle. We can neglect friction in the axle because the axle's radius is small relative to that of the pulley, so the frictional torque is much smaller than the torque applied by the two cylinders, provided that their masses are quite different. Consequently, the system consisting of the two cylinders, the pulley, and the Earth is isolated with no nonconservative forces acting; thus, the mechanical energy of the system is conserved.

We define the zero configuration for gravitational potential energy as that which exists when the system is released. From Figure 10.25, we see that the descent of cylinder 2 is associated with a decrease in system potential energy and the rise of cylinder 1 represents an increase in

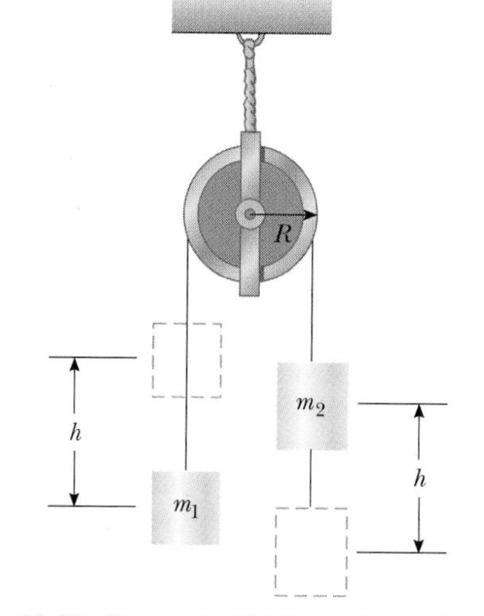

Figure 10.25 (Example 10.15) An Atwood machine.

potential energy. Because $K_i = 0$ (the system is initially at rest), we have

$$K_f + U_f = K_i + U_i$$

$$(\tfrac{1}{2}m_1{v_f}^2 + \tfrac{1}{2}m_2{v_f}^2 + \tfrac{1}{2}I{\omega_f}^2) + (m_1gh - m_2gh) = 0 + 0$$

where v_f is the same for both blocks. Because $v_f = R\omega_f$, this expression becomes

$$\left(\tfrac{1}{2}m_1{v_f}^2 + \tfrac{1}{2}m_2{v_f}^2 + \tfrac{1}{2}\frac{I}{R^2}{v_f}^2\right) = (m_2gh - m_1gh)$$

$$\tfrac{1}{2}\left(m_1 + m_2 + \frac{I}{R^2}\right){v_f}^2 = (m_2gh - m_1gh)$$

Solving for v_f, we find

$$v_f = \left[\frac{2(m_2 - m_1)gh}{[m_1 + m_2 + (I/R^2)]}\right]^{1/2}$$

The angular speed of the pulley at this instant is

$$\omega_f = \frac{v_f}{R} = \frac{1}{R}\left[\frac{2(m_2 - m_1)gh}{(m_1 + m_2 + (I/R^2))}\right]^{1/2}$$

10.9 Rolling Motion of a Rigid Object

In this section we treat the motion of a rigid object rolling along a flat surface. In general, such motion is very complex. Suppose, for example, that a cylinder is rolling on a straight path such that the axis of rotation remains parallel to its initial orientation in space. As Figure 10.26 shows, a point on the rim of the cylinder moves in a complex path called a *cycloid.* However, we can simplify matters by focusing on the center of mass rather than on a point on the rim of the rolling object. As we see in Figure 10.26, the center of mass moves in a straight line. If an object such as a cylinder rolls without slipping on the surface (we call this *pure rolling motion*), we can show that a simple relationship exists between its rotational and translational motions.

Consider a uniform cylinder of radius R rolling without slipping on a horizontal surface (Fig. 10.27). As the cylinder rotates through an angle θ, its center of mass moves a linear distance $s = R\theta$ (see Eq. 10.1a). Therefore, the linear speed of the center of mass for pure rolling motion is given by

$$v_{\text{CM}} = \frac{ds}{dt} = R\frac{d\theta}{dt} = R\omega \qquad (10.25)$$

where ω is the angular speed of the cylinder. Equation 10.25 holds whenever a cylinder or sphere rolls without slipping and is the **condition for pure rolling motion.**

PITFALL PREVENTION

10.6 Equation 10.25 Looks Familiar

Equation 10.25 looks very similar to Equation 10.10, so be sure that you are clear on the difference. Equation 10.10 gives the *tangential* speed of a point on a *rotating* object located a distance r from the rotation axis if the object is rotating with angular speed ω. Equation 10.25 gives the *translational* speed of the center of mass of a *rolling* object of radius R rotating with angular speed ω.

Figure 10.26 One light source at the center of a rolling cylinder and another at one point on the rim illustrate the different paths these two points take. The center moves in a straight line (green line), while the point on the rim moves in the path called a cycloid (red curve).

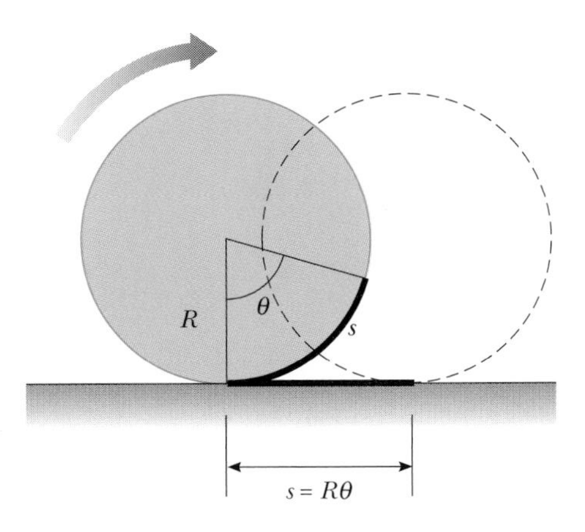

Figure 10.27 For pure rolling motion, as the cylinder rotates through an angle θ, its center moves a linear distance $s = R\theta$.

The magnitude of the linear acceleration of the center of mass for pure rolling motion is

$$a_{CM} = \frac{dv_{CM}}{dt} = R\frac{d\omega}{dt} = R\alpha \tag{10.26}$$

where α is the angular acceleration of the cylinder.

The linear velocities of the center of mass and of various points on and within the cylinder are illustrated in Figure 10.28. A short time after the moment shown in the drawing, the rim point labeled P might rotate from the six o'clock position to, say, the seven o'clock position, while the point Q would rotate from the ten o'clock position to the eleven o'clock position, and so on. Note that the linear velocity of any point is in a direction perpendicular to the line from that point to the contact point P. At any instant, the part of the rim that is at point P is at rest relative to the surface because slipping does not occur.

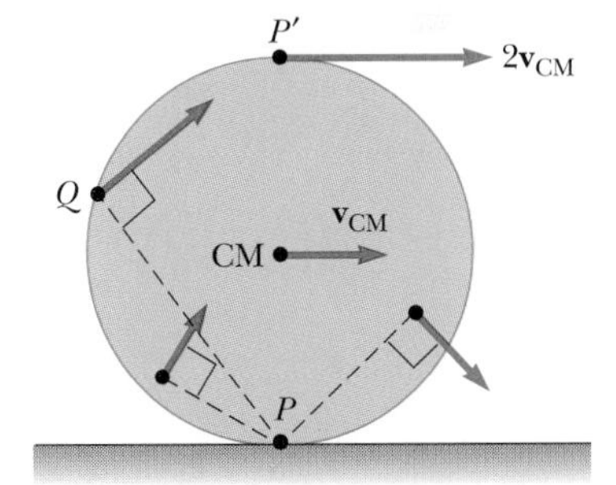

Figure 10.28 All points on a rolling object move in a direction perpendicular to an axis through the instantaneous point of contact P. In other words, all points rotate about P. The center of mass of the object moves with a velocity $\mathbf{v}_{CM}$, and the point P' moves with a velocity $2\mathbf{v}_{CM}$.

All points on the cylinder have the same angular speed. Therefore, because the distance from P' to P is twice the distance from P to the center of mass, P' has a speed $2v_{CM} = 2R\omega$. To see why this is so, let us model the rolling motion of the cylinder in Figure 10.29 as a combination of translational (linear) motion and rotational motion. For the pure translational motion shown in Figure 10.29a, imagine that the cylinder does not rotate, so that each point on it moves to the right with speed v_{CM}. For the pure rotational motion shown in Figure 10.29b, imagine that a rotation axis through the center of mass is stationary, so that each point on the cylinder has the same angular speed ω. The combination of these two motions represents the rolling motion shown in Figure 10.29c. Note in Figure 10.29c that the top of the cylinder has linear speed $v_{CM} + R\omega = v_{CM} + v_{CM} = 2v_{CM}$, which is greater than the linear speed of any other point on the cylinder. As mentioned earlier, the center of mass moves with linear speed v_{CM} while the contact point between the surface and cylinder has a linear speed of zero.

We can express the total kinetic energy of the rolling cylinder as

$$K = \tfrac{1}{2}I_P\omega^2 \tag{10.27}$$

where I_P is the moment of inertia about a rotation axis through P. Applying the parallel-axis theorem, we can substitute $I_P = I_{CM} + MR^2$ into Equation 10.27 to obtain

$$K = \tfrac{1}{2}I_{CM}\omega^2 + \tfrac{1}{2}MR^2\omega^2$$

or, because $v_{CM} = R\omega$,

$$K = \tfrac{1}{2}I_{CM}\omega^2 + \tfrac{1}{2}Mv_{CM}^2 \tag{10.28}$$

Total kinetic energy of a rolling object

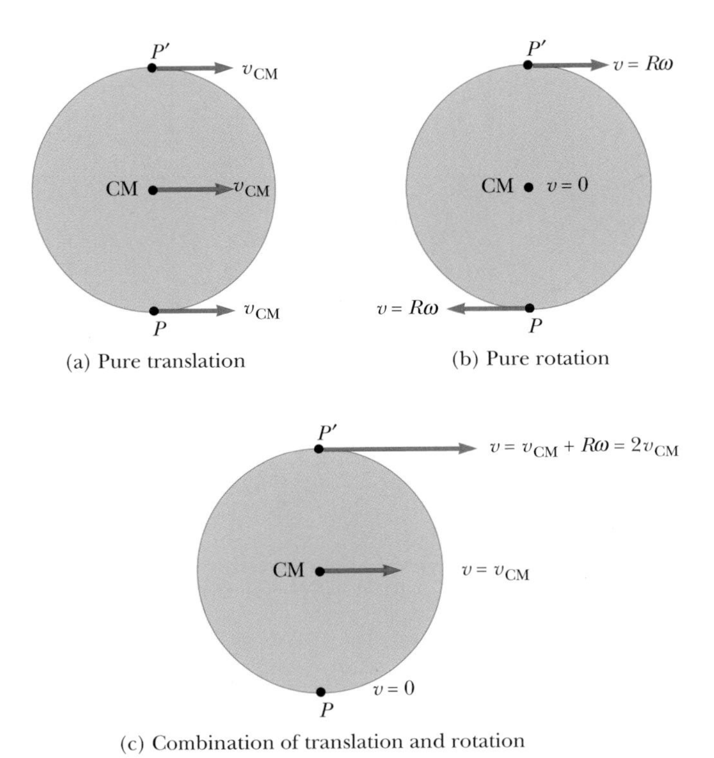

Figure 10.29 The motion of a rolling object can be modeled as a combination of pure translation and pure rotation.

The term $\frac{1}{2}I_{CM}\omega^2$ represents the rotational kinetic energy of the cylinder about its center of mass, and the term $\frac{1}{2}Mv_{CM}{}^2$ represents the kinetic energy the cylinder would have if it were just translating through space without rotating. Thus, we can say that the **total kinetic energy of a rolling object is the sum of the rotational kinetic energy about the center of mass and the translational kinetic energy of the center of mass.**

We can use energy methods to treat a class of problems concerning the rolling motion of an object down a rough incline. For example, consider Figure 10.30, which shows a sphere rolling without slipping after being released from rest at the top of the incline. Note that accelerated rolling motion is possible only if a friction force is present between the sphere and the incline to produce a net torque about the center of mass. Despite the presence of friction, no loss of mechanical energy occurs because the contact point is at rest relative to the surface at any instant. (On the other hand, if the sphere were to slip, mechanical energy of the sphere–incline–Earth system would be lost due to the nonconservative force of kinetic friction.)

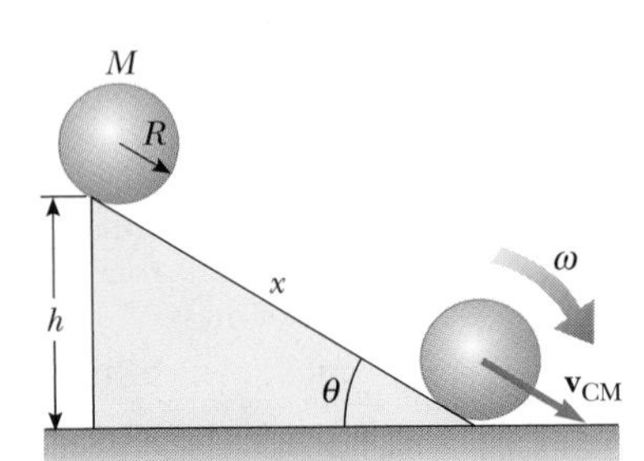

Active Figure 10.30 A sphere rolling down an incline. Mechanical energy of the sphere–incline–Earth system is conserved if no slipping occurs.

At the Active Figures link at http://www.pse6.com, you can roll several objects down the hill and see how the final speed depends on the type of object.

Using the fact that $v_{CM} = R\omega$ for pure rolling motion, we can express Equation 10.28 as

$$K = \tfrac{1}{2}I_{CM}\left(\frac{v_{CM}}{R}\right)^2 + \tfrac{1}{2}Mv_{CM}{}^2$$

$$K = \tfrac{1}{2}\left(\frac{I_{CM}}{R^2} + M\right)v_{CM}{}^2 \qquad (10.29)$$

For the system of the sphere and the Earth, we define the zero configuration of gravitational potential energy to be when the sphere is at the bottom of the incline. Thus, conservation of mechanical energy gives us

$$K_f + U_f = K_i + U_i$$

$$\tfrac{1}{2}\left(\frac{I_{CM}}{R^2} + M\right)v_{CM}{}^2 + 0 = 0 + Mgh$$

$$v_{CM} = \left(\frac{2gh}{1 + (I_{CM}/MR^2)}\right)^{1/2} \qquad (10.30)$$

Quick Quiz 10.12 A ball rolls without slipping down incline A, starting from rest. At the same time, a box starts from rest and slides down incline B, which is identical to incline A except that it is frictionless. Which arrives at the bottom first? (a) the ball (b) the box (c) Both arrive at the same time. (d) impossible to determine

Quick Quiz 10.13 Two solid spheres roll down an incline, starting from rest. Sphere A has twice the mass and twice the radius of sphere B. Which arrives at the bottom first? (a) sphere A (b) sphere B (c) Both arrive at the same time. (d) impossible to determine

Quick Quiz 10.14 Two spheres roll down an incline, starting from rest. Sphere A has the same mass and radius as sphere B, but sphere A is solid while sphere B is hollow. Which arrives at the bottom first? (a) sphere A (b) sphere B (c) Both arrive at the same time. (d) impossible to determine

Example 10.16 Sphere Rolling Down an Incline

For the solid sphere shown in Figure 10.30, calculate the linear speed of the center of mass at the bottom of the incline and the magnitude of the linear acceleration of the center of mass.

Solution For a uniform solid sphere, $I_{CM} = \frac{2}{5}MR^2$ (see Table 10.2), and therefore Equation 10.30 gives

$$v_{CM} = \left(\frac{2gh}{1 + (\frac{2}{5}MR^2/MR^2)}\right)^{1/2} = (\tfrac{10}{7}gh)^{1/2}$$

Notice that this is less than $\sqrt{2gh}$, which is the speed an object would have if it simply slid down the incline without rotating (see Example 8.7).

To calculate the linear acceleration of the center of mass, we note that the vertical displacement is related to the displacement x along the incline through the relationship $h = x \sin\theta$. Hence, after squaring both sides, we can express the equation above as

$$v_{CM}{}^2 = \tfrac{10}{7}gx \sin\theta$$

Comparing this with the expression from kinematics, $v_{CM}{}^2 = 2a_{CM}x$ (see Eq. 2.13), we see that the acceleration of the center of mass is

$$a_{CM} = \tfrac{5}{7}g \sin\theta$$

These results are interesting because both the speed and the acceleration of the center of mass are *independent* of the mass and the radius of the sphere! That is, **all homogeneous solid spheres experience the same speed and acceleration on a given incline,** as we argued in the answer to Quick Quiz 10.13.

If we were to repeat the acceleration calculation for a hollow sphere, a solid cylinder, or a hoop, we would obtain similar results in which only the factor in front of $g \sin\theta$ would differ. The constant factors that appear in the expressions for v_{CM} and a_{CM} depend only on the moment of inertia about the center of mass for the specific object. In all cases, the acceleration of the center of mass is *less* than $g \sin\theta$, the value the acceleration would have if the incline were frictionless and no rolling occurred.

SUMMARY

If a particle moves in a circular path of radius r through an angle θ (measured in radians), the arc length it moves through is $s = r\theta$.

The **angular position** of a rigid object is defined as the angle θ between a reference line attached to the object and a reference line fixed in space. The **angular displacement** of a particle moving in a circular path or a rigid object rotating about a fixed axis is $\Delta\theta \equiv \theta_f - \theta_i$.

Take a practice test for this chapter by clicking on the Practice Test link at http://www.pse6.com.

The **instantaneous angular speed** of a particle moving in a circular path or of a rigid object rotating about a fixed axis is

$$\omega \equiv \frac{d\theta}{dt} \qquad (10.3)$$

The **instantaneous angular acceleration** of a particle moving in a circular path or a rotating rigid object is

$$\alpha \equiv \frac{d\omega}{dt} \qquad (10.5)$$

When a rigid object rotates about a fixed axis, every part of the object has the same angular speed and the same angular acceleration.

If an object rotates about a fixed axis under constant angular acceleration, one can apply equations of kinematics that are analogous to those for linear motion under constant linear acceleration:

$$\omega_f = \omega_i + \alpha t \qquad (10.6)$$

$$\theta_f = \theta_i + \omega_i t + \tfrac{1}{2}\alpha t^2 \qquad (10.7)$$

$$\omega_f^2 = \omega_i^2 + 2\alpha(\theta_f - \theta_i) \qquad (10.8)$$

$$\theta_f = \theta_i + \tfrac{1}{2}(\omega_i + \omega_f)t \qquad (10.9)$$

A useful technique in solving problems dealing with rotation is to visualize a linear version of the same problem.

When a rigid object rotates about a fixed axis, the angular position, angular speed, and angular acceleration are related to the linear position, linear speed, and linear acceleration through the relationships

$$s = r\theta \qquad (10.1a)$$

$$v = r\omega \qquad (10.10)$$

$$a_t = r\alpha \qquad (10.11)$$

The **moment of inertia of a system of particles** is defined as

$$I \equiv \sum_i m_i r_i^2 \qquad (10.15)$$

If a rigid object rotates about a fixed axis with angular speed ω, its **rotational kinetic energy** can be written

$$K_R = \tfrac{1}{2} I\omega^2 \qquad (10.16)$$

where I is the moment of inertia about the axis of rotation.

The **moment of inertia of a rigid object** is

$$I = \int r^2\, dm \qquad (10.17)$$

where r is the distance from the mass element dm to the axis of rotation.

The magnitude of the **torque** associated with a force $\mathbf{F}$ acting on an object is

$$\tau = Fd \qquad (10.19)$$

where d is the moment arm of the force, which is the perpendicular distance from the rotation axis to the line of action of the force. Torque is a measure of the tendency of the force to change the rotation of the object about some axis.

If a rigid object free to rotate about a fixed axis has a **net external torque** acting on it, the object undergoes an angular acceleration α, where

$$\sum \tau = I\alpha \qquad (10.21)$$

The rate at which work is done by an external force in rotating a rigid object about a fixed axis, or the **power** delivered, is

$$\mathcal{P} = \tau\omega \tag{10.23}$$

If work is done on a rigid object and the only result of the work is rotation about a fixed axis, the net work done by external forces in rotating the object equals the change in the rotational kinetic energy of the object:

$$\sum W = \tfrac{1}{2}I\omega_f^2 - \tfrac{1}{2}I\omega_i^2 \tag{10.24}$$

The **total kinetic energy** of a rigid object rolling on a rough surface without slipping equals the rotational kinetic energy about its center of mass, $\frac{1}{2}I_{\text{CM}}\omega^2$, plus the translational kinetic energy of the center of mass, $\frac{1}{2}Mv_{\text{CM}}^2$:

$$K = \tfrac{1}{2}I_{\text{CM}}\omega^2 + \tfrac{1}{2}Mv_{\text{CM}}^2 \tag{10.28}$$

QUESTIONS

1. What is the angular speed of the second hand of a clock? What is the direction of $\boldsymbol{\omega}$ as you view a clock hanging on a vertical wall? What is the magnitude of the angular acceleration vector $\boldsymbol{\alpha}$ of the second hand?

2. One blade of a pair of scissors rotates counterclockwise in the xy plane. What is the direction of $\boldsymbol{\omega}$? What is the direction of $\boldsymbol{\alpha}$ if the magnitude of the angular velocity is decreasing in time?

3. Are the kinematic expressions for θ, ω, and α valid when the angular position is measured in degrees instead of in radians?

4. If a car's standard tires are replaced with tires of larger outside diameter, will the reading of the speedometer change? Explain.

5. Suppose $a = b$ and $M > m$ for the system of particles described in Figure 10.8. About which axis (x, y, or z) does the moment of inertia have the smallest value? the largest value?

6. Suppose that the rod in Figure 10.10 has a nonuniform mass distribution. In general, would the moment of inertia about the y axis still be equal to $ML^2/12$? If not, could the moment of inertia be calculated without knowledge of the manner in which the mass is distributed?

7. Suppose that just two external forces act on a stationary rigid object and the two forces are equal in magnitude and opposite in direction. Under what condition does the object start to rotate?

8. Suppose a pencil is balanced on a perfectly frictionless table. If it falls over, what is the path followed by the center of mass of the pencil?

9. Explain how you might use the apparatus described in Example 10.12 to determine the moment of inertia of the wheel. (If the wheel does not have a uniform mass density, the moment of inertia is not necessarily equal to $\frac{1}{2}MR^2$.)

10. Using the results from Example 10.12, how would you calculate the angular speed of the wheel and the linear speed of the suspended counterweight at $t = 2$ s, if the system is released from rest at $t = 0$? Is the expression $v = R\omega$ valid in this situation?

11. If a small sphere of mass M were placed at the end of the rod in Figure 10.24, would the result for ω be greater than, less than, or equal to the value obtained in Example 10.14?

12. Explain why changing the axis of rotation of an object changes its moment of inertia.

13. The moment of inertia of an object depends on the choice of rotation axis, as suggested by the parallel-axis theorem. Argue that an axis passing through the center of mass of an object must be the axis with the smallest moment of inertia.

14. Suppose you remove two eggs from the refrigerator, one hard-boiled and the other uncooked. You wish to determine which is the hard-boiled egg without breaking the eggs. This can be done by spinning the two eggs on the floor and comparing the rotational motions. Which egg spins faster? Which rotates more uniformly? Explain.

15. Which of the entries in Table 10.2 applies to finding the moment of inertia of a long straight sewer pipe rotating about its axis of symmetry? Of an embroidery hoop rotating about an axis through its center and perpendicular to its plane? Of a uniform door turning on its hinges? Of a coin turning about an axis through its center and perpendicular to its faces?

16. Is it possible to change the translational kinetic energy of an object without changing its rotational energy?

17. Must an object be rotating to have a nonzero moment of inertia?

18. If you see an object rotating, is there necessarily a net torque acting on it?

19. Can a (momentarily) stationary object have a nonzero angular acceleration?

20. In a tape recorder, the tape is pulled past the read-and-write heads at a constant speed by the drive mechanism. Consider the reel from which the tape is pulled. As the tape is pulled from it, the radius of the roll of remaining tape decreases. How does the torque on the reel change with time? How does the angular speed of the reel change in time? If the drive mechanism is switched on so that the tape is suddenly jerked with a large force, is the tape more likely to break when it is being pulled from a nearly full reel or from a nearly empty reel?

21. The polar diameter of the Earth is slightly less than the equatorial diameter. How would the moment of inertia of the Earth about its axis of rotation change if some mass from near the equator were removed and transferred to the polar regions to make the Earth a perfect sphere?

22. Suppose you set your textbook sliding across a gymnasium floor with a certain initial speed. It quickly stops moving because of a friction force exerted on it by the floor. Next, you start a basketball rolling with the same initial speed. It keeps rolling from one end of the gym to the other. Why does the basketball roll so far? Does friction significantly affect its motion?

23. When a cylinder rolls on a horizontal surface as in Figure 10.28, do any points on the cylinder have only a vertical component of velocity at some instant? If so, where are they?

24. Three objects of uniform density—a solid sphere, a solid cylinder, and a hollow cylinder—are placed at the top of an incline (Fig. Q10.24). They are all released from rest at the same elevation and roll without slipping. Which object reaches the bottom first? Which reaches it last? Try this at home and note that the result is independent of the masses and the radii of the objects.

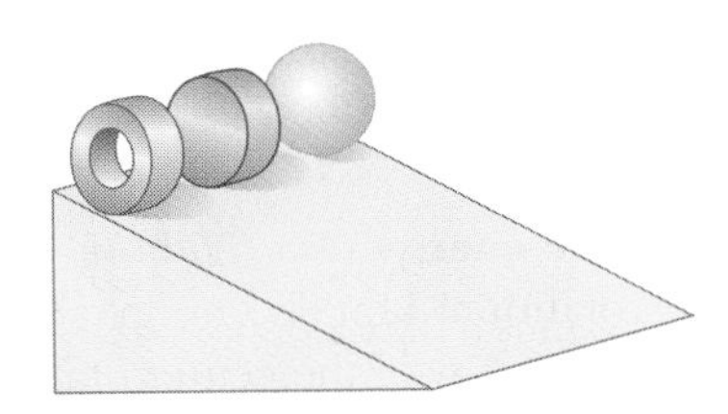

Figure Q10.24 Which object wins the race?

25. In a soap-box derby race, the cars have no engines; they simply coast down a hill to race with one another. Suppose you are designing a car for a coasting race. Do you want to use large wheels or small wheels? Do you want to use solid disk-like wheels or hoop-like wheels? Should the wheels be heavy or light?

PROBLEMS

1, 2, 3 = straightforward, intermediate, challenging □ = full solution available in the *Student Solutions Manual and Study Guide*

= coached solution with hints available at http://www.pse6.com = computer useful in solving problem

= paired numerical and symbolic problems

Section 10.1 Angular Position, Velocity, and Acceleration

1. During a certain period of time, the angular position of a swinging door is described by $\theta = 5.00 + 10.0t + 2.00t^2$, where θ is in radians and t is in seconds. Determine the angular position, angular speed, and angular acceleration of the door (a) at $t = 0$ (b) at $t = 3.00$ s.

Section 10.2 Rotational Kinematics: Rotational Motion with Constant Angular Acceleration

2. A dentist's drill starts from rest. After 3.20 s of constant angular acceleration, it turns at a rate of 2.51×10^4 rev/min. (a) Find the drill's angular acceleration. (b) Determine the angle (in radians) through which the drill rotates during this period.

3. A wheel starts from rest and rotates with constant angular acceleration to reach an angular speed of 12.0 rad/s in 3.00 s. Find (a) the magnitude of the angular acceleration of the wheel and (b) the angle in radians through which it rotates in this time.

4. An airliner arrives at the terminal, and the engines are shut off. The rotor of one of the engines has an initial clockwise angular speed of 2 000 rad/s. The engine's rotation slows with an angular acceleration of magnitude 80.0 rad/s^2. (a) Determine the angular speed after 10.0 s. (b) How long does it take the rotor to come to rest?

5. An electric motor rotating a grinding wheel at 100 rev/min is switched off. With constant negative angular acceleration of magnitude 2.00 rad/s^2, (a) how long does it take the wheel to stop? (b) Through how many radians does it turn while it is slowing down?

6. A centrifuge in a medical laboratory rotates at an angular speed of 3 600 rev/min. When switched off, it rotates 50.0 times before coming to rest. Find the constant angular acceleration of the centrifuge.

7. The tub of a washer goes into its spin cycle, starting from rest and gaining angular speed steadily for 8.00 s, at which time it is turning at 5.00 rev/s. At this point the person doing the laundry opens the lid, and a safety switch turns off the washer. The tub smoothly slows to rest in 12.0 s. Through how many revolutions does the tub turn while it is in motion?

8. A rotating wheel requires 3.00 s to rotate through 37.0 revolutions. Its angular speed at the end of the 3.00-s interval is 98.0 rad/s. What is the constant angular acceleration of the wheel?

9. (a) Find the angular speed of the Earth's rotation on its axis. As the Earth turns toward the east, we see the sky turning toward the west at this same rate.
(b) *The rainy Pleiads wester*
And seek beyond the sea
The head that I shall dream of
That shall not dream of me.
–A. E. Housman (© Robert E. Symons)

Cambridge, England, is at longitude 0°, and Saskatoon, Saskatchewan, is at longitude 107° west. How much time elapses after the Pleiades set in Cambridge until these stars fall below the western horizon in Saskatoon?

10. A merry-go-round is stationary. A dog is running on the ground just outside its circumference, moving with a constant angular speed of 0.750 rad/s. The dog does not change his pace when he sees what he has been looking for: a bone resting on the edge of the merry-go-round one third of a revolution in front of him. At the instant the dog sees the bone ($t = 0$), the merry-go-round begins to move in the direction the dog is running, with a constant angular acceleration of 0.015 0 rad/s^2. (a) At what time will the dog reach the bone? (b) The confused dog keeps running and passes the bone. How long after the merry-go-round starts to turn do the dog and the bone draw even with each other for the second time?

Section 10.3 Angular and Linear Quantities

11. Make an order-of-magnitude estimate of the number of revolutions through which a typical automobile tire turns in 1 yr. State the quantities you measure or estimate and their values.

12. A racing car travels on a circular track of radius 250 m. If the car moves with a constant linear speed of 45.0 m/s, find (a) its angular speed and (b) the magnitude and direction of its acceleration.

13. A wheel 2.00 m in diameter lies in a vertical plane and rotates with a constant angular acceleration of 4.00 rad/s^2. The wheel starts at rest at $t = 0$, and the radius vector of a certain point P on the rim makes an angle of 57.3° with the horizontal at this time. At $t = 2.00$ s, find (a) the angular speed of the wheel, (b) the tangential speed and the total acceleration of the point P, and (c) the angular position of the point P.

14. Figure P10.14 shows the drive train of a bicycle that has wheels 67.3 cm in diameter and pedal cranks 17.5 cm long. The cyclist pedals at a steady angular rate of 76.0 rev/min. The chain engages with a front sprocket 15.2 cm in diameter and a rear sprocket 7.00 cm in diameter. (a) Calculate the speed of a link of the chain relative to the bicycle frame. (b) Calculate the angular speed of the bicycle wheels. (c) Calculate the speed of the bicycle relative to the road. (d) What pieces of data, if any, are not necessary for the calculations?

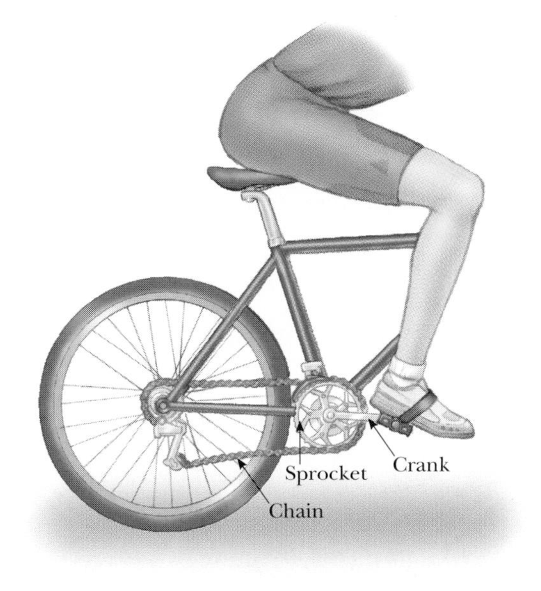

Figure P10.14

15. A discus thrower (Fig. P10.15) accelerates a discus from rest to a speed of 25.0 m/s by whirling it through 1.25 rev. Assume the discus moves on the arc of a circle 1.00 m in radius. (a) Calculate the final angular speed of the discus. (b) Determine the magnitude of the angular acceleration of the discus, assuming it to be constant. (c) Calculate the time interval required for the discus to accelerate from rest to 25.0 m/s.

Figure P10.15

16. A car accelerates uniformly from rest and reaches a speed of 22.0 m/s in 9.00 s. If the diameter of a tire is 58.0 cm, find (a) the number of revolutions the tire makes during this motion, assuming that no slipping occurs. (b) What is the final angular speed of a tire in revolutions per second?

17. A disk 8.00 cm in radius rotates at a constant rate of 1 200 rev/min about its central axis. Determine (a) its angular speed, (b) the tangential speed at a point 3.00 cm from its center, (c) the radial acceleration of a point on the rim, and (d) the total distance a point on the rim moves in 2.00 s.

18. A car traveling on a flat (unbanked) circular track accelerates uniformly from rest with a tangential acceleration of 1.70 m/s^2. The car makes it one quarter of the way around the circle before it skids off the track. Determine the coefficient of static friction between the car and track from these data.

19. Consider a tall building located on the Earth's equator. As the Earth rotates, a person on the top floor of the building moves faster than someone on the ground with respect to an inertial reference frame, because the latter person is closer to the Earth's axis. Consequently, if an object is dropped from the top floor to the ground a distance h below, it lands east of the point vertically below where it was dropped. (a) How far to the east will the object land? Express your answer in terms of h, g, and the angular speed ω of the Earth. Neglect air resistance, and assume that the free-fall acceleration is constant over this range of heights. (b) Evaluate the eastward displacement for $h = 50.0$ m. (c) In your judgment, were we justified in ignoring this aspect of the *Coriolis effect* in our previous study of free fall?

Section 10.4 Rotational Kinetic Energy

20. Rigid rods of negligible mass lying along the y axis connect three particles (Fig. P10.20). If the system rotates about the x axis with an angular speed of 2.00 rad/s, find (a) the moment of inertia about the x axis and the total rotational kinetic energy evaluated from $\frac{1}{2}I\omega^2$ and (b) the tangential speed of each particle and the total kinetic energy evaluated from $\Sigma\frac{1}{2}m_iv_i^2$.

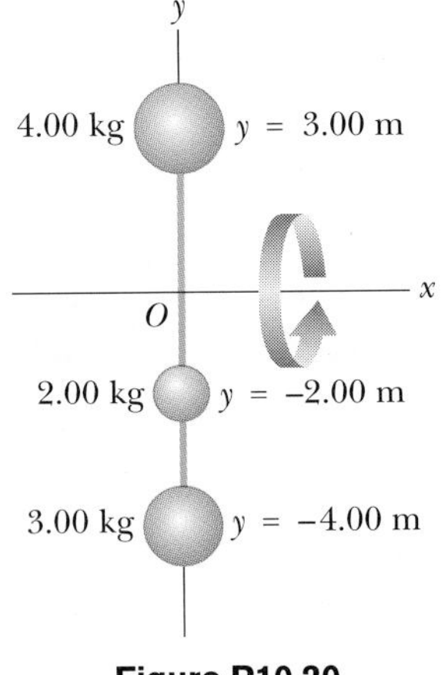

Figure P10.20

21. The four particles in Figure P10.21 are connected by rigid rods of negligible mass. The origin is at the center of the rectangle. If the system rotates in the xy plane about the z axis with an angular speed of 6.00 rad/s, calculate (a) the moment of inertia of the system about the z axis and (b) the rotational kinetic energy of the system.

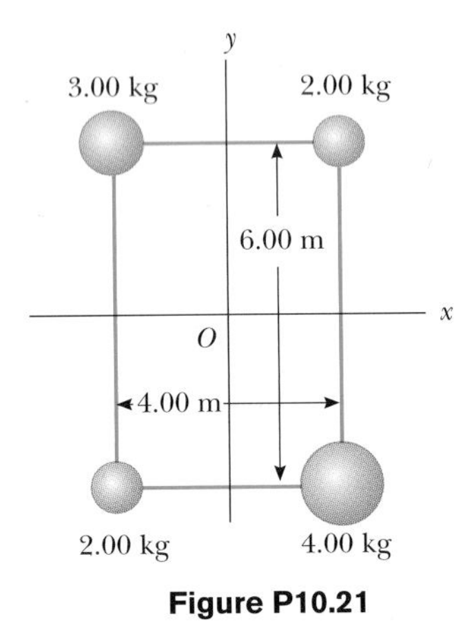

Figure P10.21

22. Two balls with masses M and m are connected by a rigid rod of length L and negligible mass as in Figure P10.22. For an axis perpendicular to the rod, show that the system has the minimum moment of inertia when the axis passes through the center of mass. Show that this moment of inertia is $I = \mu L^2$, where $\mu = mM/(m + M)$.

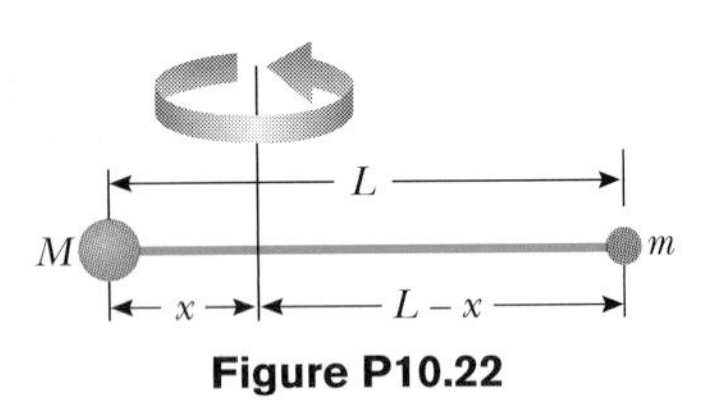

Figure P10.22

Section 10.5 Calculation of Moments of Inertia

23. Three identical thin rods, each of length L and mass m, are welded perpendicular to one another as shown in Figure P10.23. The assembly is rotated about an axis that passes through the end of one rod and is parallel to another. Determine the moment of inertia of this structure.

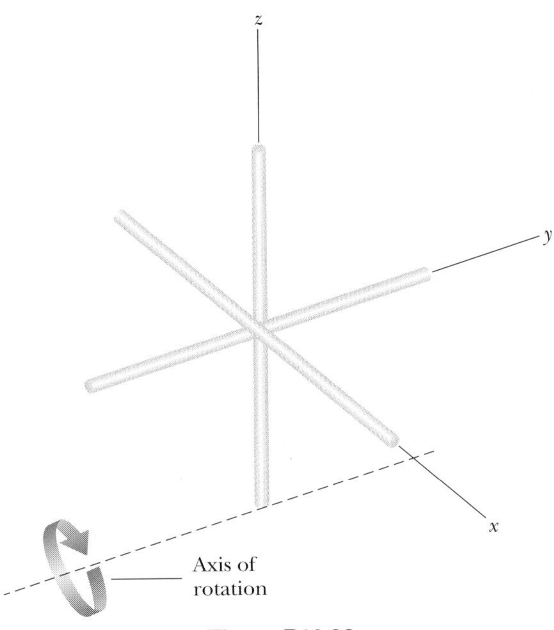

Figure P10.23

24. Figure P10.24 shows a side view of a car tire. Model it as having two sidewalls of uniform thickness 0.635 cm and a tread wall of uniform thickness 2.50 cm and width 20.0 cm. Assume the rubber has uniform density 1.10×10^3 kg/m^3. Find its moment of inertia about an axis through its center.

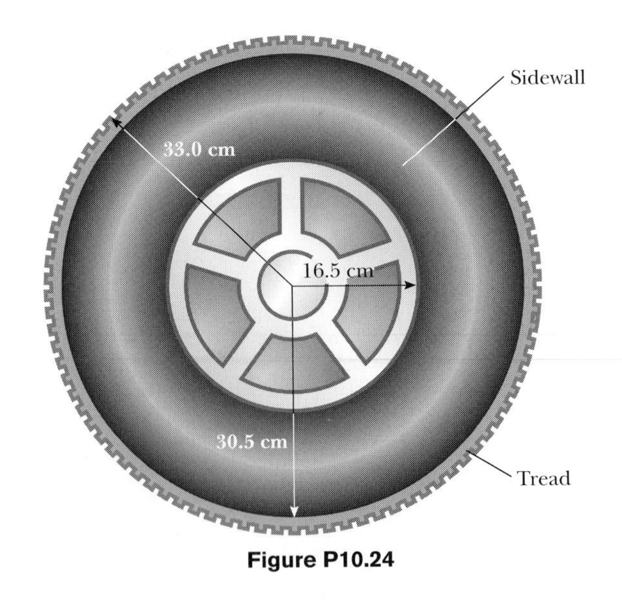

Figure P10.24

25. A uniform thin solid door has height 2.20 m, width 0.870 m, and mass 23.0 kg. Find its moment of inertia for rotation on its hinges. Is any piece of data unnecessary?

26. *Attention! About face!* Compute an order-of-magnitude estimate for the moment of inertia of your body as you stand tall and turn about a vertical axis through the top of your head and the point halfway between your ankles. In your solution state the quantities you measure or estimate and their values.

27. The density of the Earth, at any distance r from its center, is approximately

$$\rho = [14.2 - 11.6(r/R)] \times 10^3 \text{ kg/m}^3$$

where R is the radius of the Earth. Show that this density leads to a moment of inertia $I = 0.330MR^2$ about an axis through the center, where M is the mass of the Earth.

28. Calculate the moment of inertia of a thin plate, in the shape of a right triangle, about an axis that passes through one end of the hypotenuse and is parallel to the opposite leg of the triangle, as in Figure P10.28a. Let M represent the mass of the triangle and L the length of the base of the triangle perpendicular to the axis of rotation. Let h represent the height of the triangle and w the thickness of the plate, much smaller than L or h. Do the calculation in either or both of the following ways, as your instructor assigns:

(a) Use Equation 10.17. Let an element of mass consist of a vertical ribbon within the triangle, of width dx, height y, and thickness w. With x representing the location of the ribbon, show that $y = hx/L$. Show that the density of the material is given by $\rho = 2M/Lwh$. Show that the mass of the ribbon is $dm = \rho yw\, dx = 2Mx\, dx/L^2$. Proceed to use Equation 10.17 to calculate the moment of inertia.

(b) Let I represent the unknown moment of inertia about an axis through the corner of the triangle. Note that Example 9.15 demonstrates that the center of mass of the triangle is two thirds of the way along the length L, from the corner toward the side of height h. Let I_{CM} represent the moment of inertia of the triangle about an axis through the center of mass and parallel to side h. Demonstrate that $I = I_{CM} + 4ML^2/9$. Figure P10.28b shows the same object in a different orientation. Demonstrate that the moment of inertia of the triangular plate, about the y axis is $I_h = I_{CM} + ML^2/9$. Demonstrate that the sum of the moments of inertia of the triangles shown in parts (a) and (b) of the figure must be the moment of inertia of a rectangular sheet of mass $2M$ and length L, rotating like a door about an axis along its edge of height h. Use information in Table 10.2 to write down the moment of inertia of the rectangle, and set it equal to the sum of the moments of inertia of the two triangles. Solve the equation to find the moment of inertia of a triangle about an axis through its center of mass, in terms of M and L. Proceed to find the original unknown I.

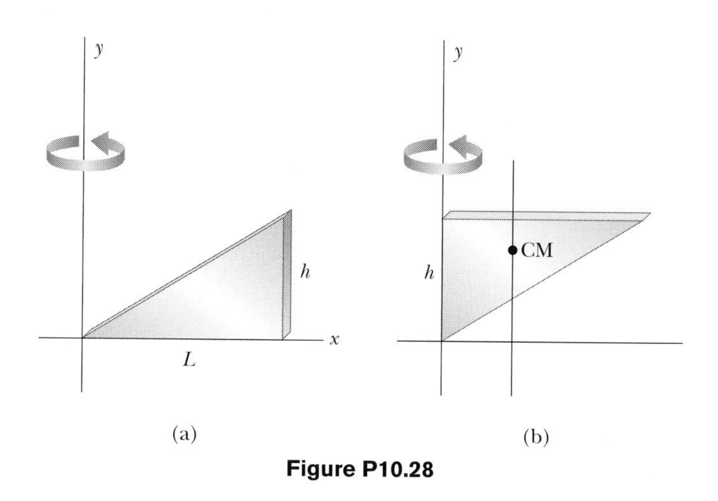

Figure P10.28

29. Many machines employ cams for various purposes, such as opening and closing valves. In Figure P10.29, the cam is a circular disk rotating on a shaft that does not pass through the center of the disk. In the manufacture of the cam, a uniform solid cylinder of radius R is first machined. Then an off-center hole of radius $R/2$ is drilled, parallel to the axis of the cylinder, and centered at a point a distance $R/2$ from the center of the cylinder. The cam, of mass M, is then slipped onto the circular shaft and welded into place. What is the kinetic energy of the cam when it is rotating with angular speed ω about the axis of the shaft?

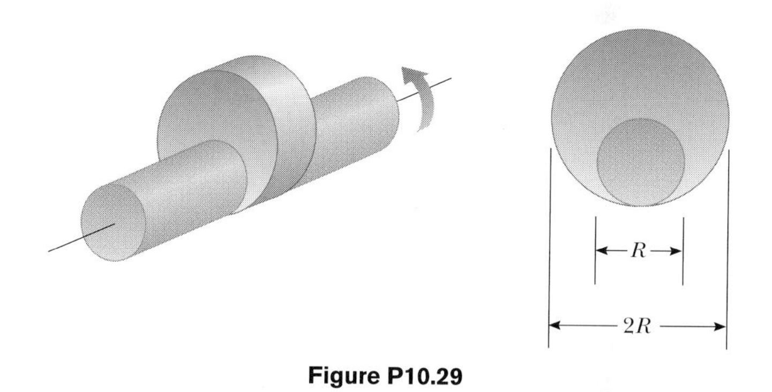

Figure P10.29

Section 10.6 Torque

30. The fishing pole in Figure P10.30 makes an angle of 20.0° with the horizontal. What is the torque exerted by the fish about an axis perpendicular to the page and passing through the fisher's hand?

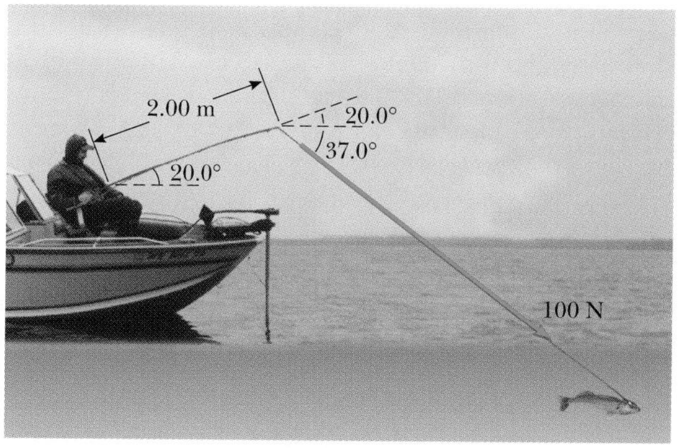

Figure P10.30

31. Find the net torque on the wheel in Figure P10.31 about the axle through O if $a = 10.0$ cm and $b = 25.0$ cm.

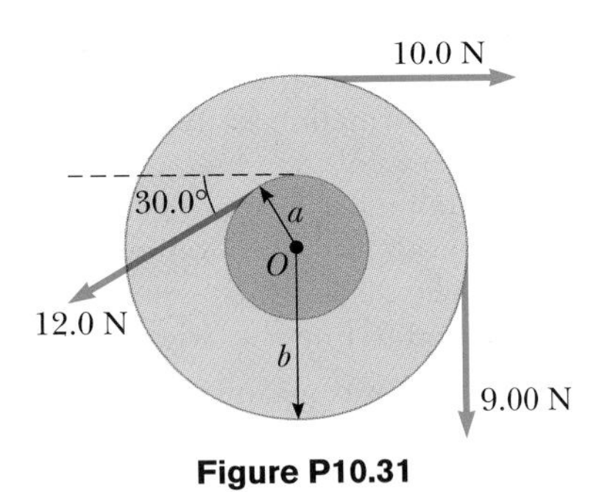

Figure P10.31

32. The tires of a 1 500-kg car are 0.600 m in diameter, and the coefficients of friction with the road surface are $\mu_s = 0.800$ and $\mu_k = 0.600$. Assuming that the weight is evenly distributed on the four wheels, calculate the maximum torque that can be exerted by the engine on a driving wheel without spinning the wheel. If you wish, you may assume the car is at rest.

33. Suppose the car in Problem 32 has a disk brake system. Each wheel is slowed by the friction force between a single brake pad and the disk-shaped rotor. On this particular car, the brake pad contacts the rotor at an average distance of 22.0 cm from the axis. The coefficients of friction between the brake pad and the disk are $\mu_s = 0.600$ and $\mu_k = 0.500$. Calculate the normal force that the pad must apply to the rotor in order to slow the car as quickly as possible.

Section 10.7 Relationship between Torque and Angular Acceleration

34. A grinding wheel is in the form of a uniform solid disk of radius 7.00 cm and mass 2.00 kg. It starts from rest and accelerates uniformly under the action of the constant torque of 0.600 N · m that the motor exerts on the wheel. (a) How long does the wheel take to reach its final operating speed of 1 200 rev/min? (b) Through how many revolutions does it turn while accelerating?

35. A model airplane with mass 0.750 kg is tethered by a wire so that it flies in a circle 30.0 m in radius. The airplane engine provides a net thrust of 0.800 N perpendicular to the tethering wire. (a) Find the torque the net thrust produces about the center of the circle. (b) Find the angular acceleration of the airplane when it is in level flight. (c) Find the linear acceleration of the airplane tangent to its flight path.

36. The combination of an applied force and a friction force produces a constant total torque of 36.0 N · m on a wheel rotating about a fixed axis. The applied force acts for 6.00 s. During this time the angular speed of the wheel increases from 0 to 10.0 rad/s. The applied force is then removed, and the wheel comes to rest in 60.0 s. Find (a) the moment of inertia of the wheel, (b) the magnitude of the frictional torque, and (c) the total number of revolutions of the wheel.

37. A block of mass $m_1 = 2.00$ kg and a block of mass $m_2 = 6.00$ kg are connected by a massless string over a pulley in the shape of a solid disk having radius $R = 0.250$ m and mass $M = 10.0$ kg. These blocks are allowed to move on a fixed block-wedge of angle $\theta = 30.0°$ as in Figure P10.37. The coefficient of kinetic friction is 0.360 for both blocks. Draw free-body diagrams of both blocks and of the pulley. Determine (a) the acceleration of the two blocks and (b) the tensions in the string on both sides of the pulley.

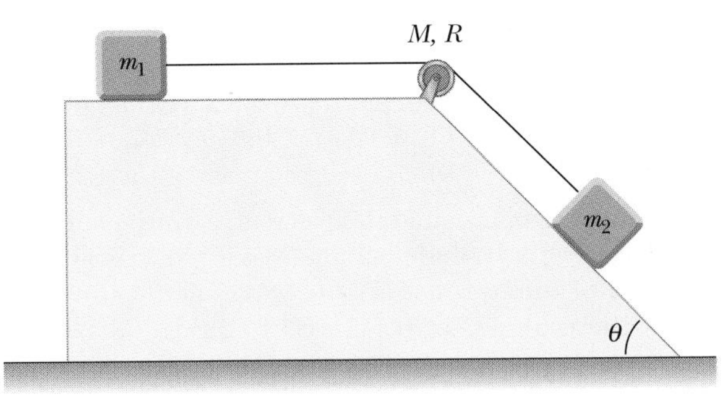

Figure P10.37

38. A potter's wheel—a thick stone disk of radius 0.500 m and mass 100 kg—is freely rotating at 50.0 rev/min. The potter can stop the wheel in 6.00 s by pressing a wet rag against the rim and exerting a radially inward force of 70.0 N. Find the effective coefficient of kinetic friction between wheel and rag.

39. An electric motor turns a flywheel through a drive belt that joins a pulley on the motor and a pulley that is rigidly attached to the flywheel, as shown in Figure P10.39. The flywheel is a solid disk with a mass of 80.0 kg and a diameter of 1.25 m. It turns on a frictionless axle. Its pulley has much smaller mass and a radius of 0.230 m. If the tension in the upper (taut) segment of the belt is 135 N and the flywheel has a clockwise angular acceleration of 1.67 rad/s^2, find the tension in the lower (slack) segment of the belt.

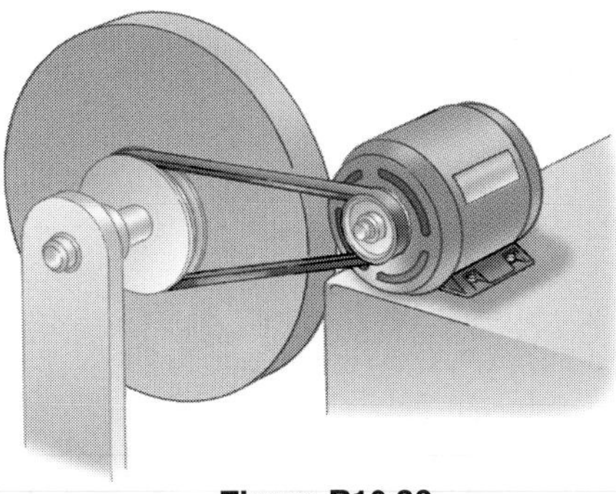

Figure P10.39

Section 10.8 Work, Power, and Energy in Rotational Motion

40. Big Ben, the Parliament tower clock in London, has an hour hand 2.70 m long with a mass of 60.0 kg, and

a minute hand 4.50 m long with a mass of 100 kg (Fig. P10.40). Calculate the total rotational kinetic energy of the two hands about the axis of rotation. (You may model the hands as long, thin rods.)

Figure P10.40 Problems 40 and 74.

41. In a city with an air-pollution problem, a bus has no combustion engine. It runs on energy drawn from a large, rapidly rotating flywheel under the floor of the bus. The flywheel is spun up to its maximum rotation rate of 4 000 rev/min by an electric motor at the bus terminal. Every time the bus speeds up, the flywheel slows down slightly. The bus is equipped with regenerative braking so that the flywheel can speed up when the bus slows down. The flywheel is a uniform solid cylinder with mass 1 600 kg and radius 0.650 m. The bus body does work against air resistance and rolling resistance at the average rate of 18.0 hp as it travels with an average speed of 40.0 km/h. How far can the bus travel before the flywheel has to be spun up to speed again?

42. The top in Figure P10.42 has a moment of inertia of $4.00 \times 10^{-4}\ \text{kg} \cdot \text{m}^2$ and is initially at rest. It is free to rotate about the stationary axis AA'. A string, wrapped around a peg along the axis of the top, is pulled in such a manner as to maintain a constant tension of 5.57 N. If the string does not slip while it is unwound from the peg, what is the angular speed of the top after 80.0 cm of string has been pulled off the peg?

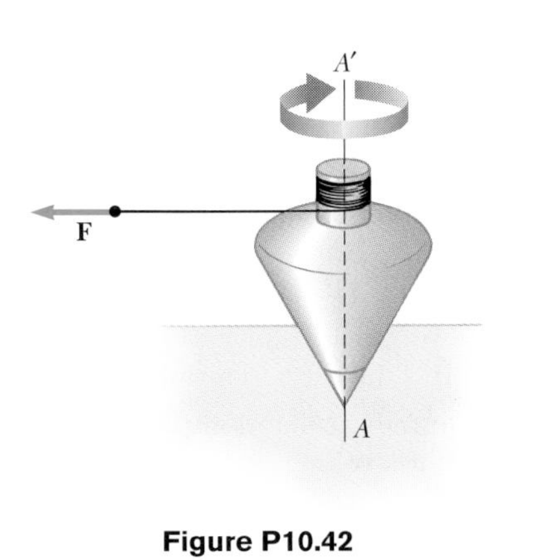

Figure P10.42

43. In Figure P10.43 the sliding block has a mass of 0.850 kg, the counterweight has a mass of 0.420 kg, and the pulley is a hollow cylinder with a mass of 0.350 kg, an inner radius of 0.020 0 m, and an outer radius of 0.030 0 m. The coefficient of kinetic friction between the block and the horizontal surface is 0.250. The pulley turns without friction on its axle. The light cord does not stretch and does not slip on the pulley. The block has a velocity of 0.820 m/s toward the pulley when it passes through a photogate. (a) Use energy methods to predict its speed after it has moved to a second photogate, 0.700 m away. (b) Find the angular speed of the pulley at the same moment.

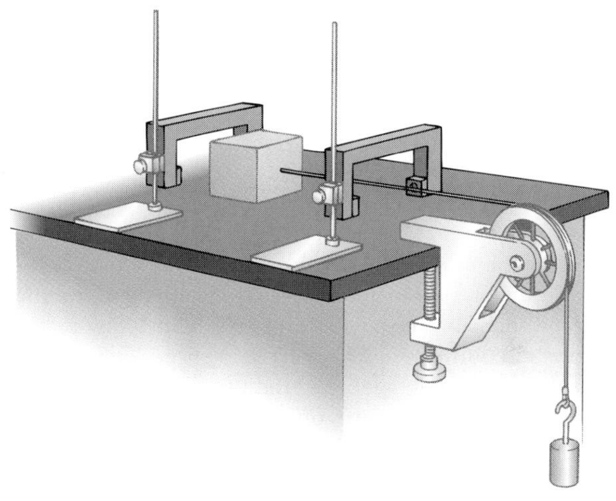

Figure P10.43

44. A cylindrical rod 24.0 cm long with mass 1.20 kg and radius 1.50 cm has a ball of diameter 8.00 cm and mass 2.00 kg attached to one end. The arrangement is originally vertical and stationary, with the ball at the top. The system is free to pivot about the bottom end of the rod after being given a slight nudge. (a) After the rod rotates through ninety degrees, what is its rotational kinetic energy? (b) What is the angular speed of the rod and ball? (c) What is the linear speed of the ball? (d) How does this compare to the speed if the ball had fallen freely through the same distance of 28 cm?

45. An object with a weight of 50.0 N is attached to the free end of a light string wrapped around a reel of radius 0.250 m and mass 3.00 kg. The reel is a solid disk, free to rotate in a vertical plane about the horizontal axis passing through its center. The suspended object is released 6.00 m above the floor. (a) Determine the tension in the string, the acceleration of the object, and the speed with which the object hits the floor. (b) Verify your last answer by using the principle of conservation of energy to find the speed with which the object hits the floor.

46. A 15.0-kg object and a 10.0-kg object are suspended, joined by a cord that passes over a pulley with a radius of 10.0 cm and a mass of 3.00 kg (Fig. P10.46). The cord has a negligible mass and does not slip on the pulley. The pulley rotates on its axis without friction. The objects start from rest 3.00 m apart. Treat the pulley as a uniform disk, and determine the speeds of the two objects as they pass each other.

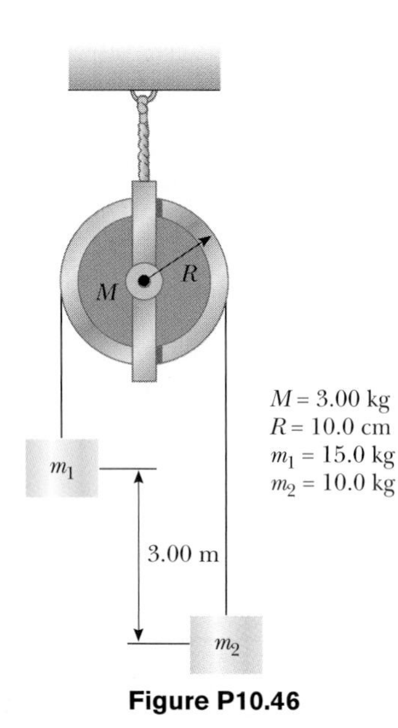

Figure P10.46

47. This problem describes one experimental method for determining the moment of inertia of an irregularly shaped object such as the payload for a satellite. Figure P10.47 shows a counterweight of mass m suspended by a cord wound around a spool of radius r, forming part of a turntable supporting the object. The turntable can rotate without friction. When the counterweight is released from rest, it descends through a distance h, acquiring a speed v. Show that the moment of inertia I of the rotating apparatus (including the turntable) is $mr^2(2gh/v^2 - 1)$.

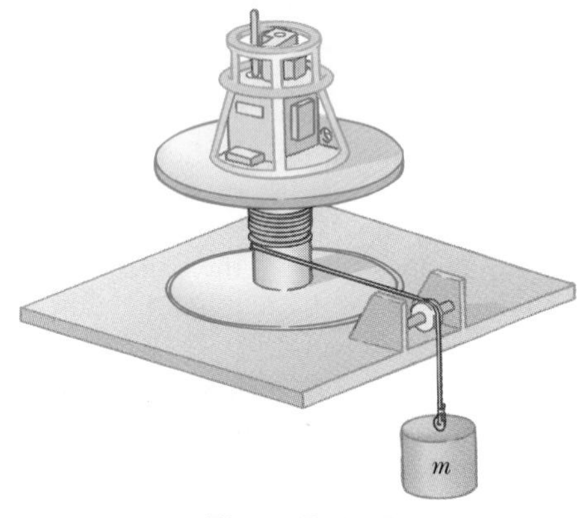

Figure P10.47

48. A horizontal 800-N merry-go-round is a solid disk of radius 1.50 m, started from rest by a constant horizontal force of 50.0 N applied tangentially to the edge of the disk. Find the kinetic energy of the disk after 3.00 s.

49. (a) A uniform solid disk of radius R and mass M is free to rotate on a frictionless pivot through a point on its rim (Fig. P10.49). If the disk is released from rest in the position shown by the blue circle, what is the speed of its center of mass when the disk reaches the position indicated by the dashed circle? (b) What is the speed of the lowest point on the disk in the dashed position? (c) **What If?** Repeat part (a) using a uniform hoop.

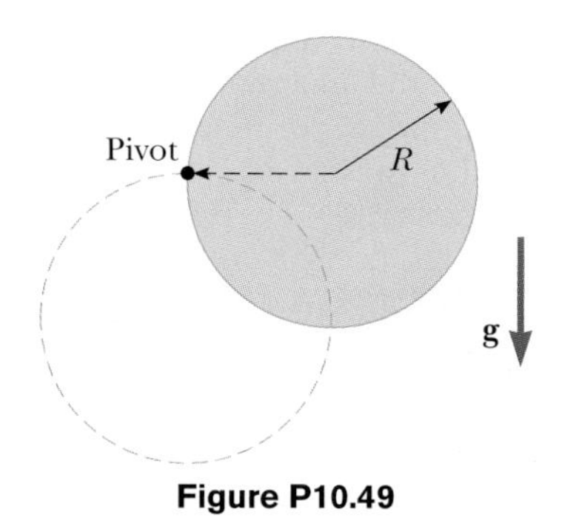

Figure P10.49

50. The head of a grass string trimmer has 100 g of cord wound in a light cylindrical spool with inside diameter 3.00 cm and outside diameter 18.0 cm, as in Figure P10.50. The cord has a linear density of 10.0 g/m. A single strand of the cord extends 16.0 cm from the outer edge of the spool. (a) When switched on, the trimmer speeds up from 0 to 2 500 rev/min in 0.215 s. (a) What average power is delivered to the head by the trimmer motor while it is accelerating? (b) When the trimmer is cutting grass, it spins at 2 000 rev/min and the grass exerts an average tangential force of 7.65 N on the outer end of the cord, which is still at a radial distance of 16.0 cm from the outer edge of the spool. What is the power delivered to the head under load?

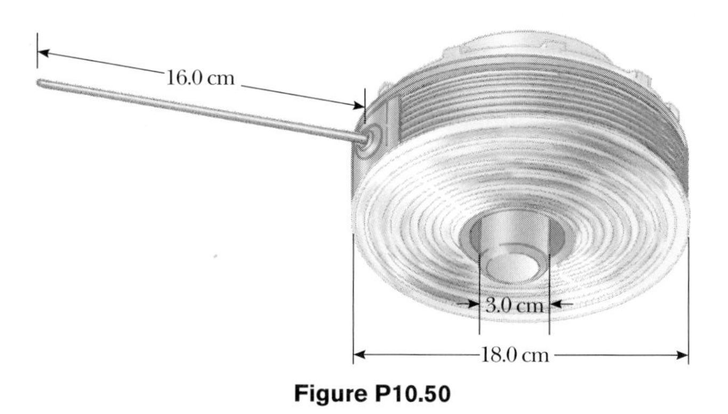

Figure P10.50

Section 10.9 Rolling Motion of a Rigid Object

51. A cylinder of mass 10.0 kg rolls without slipping on a horizontal surface. At the instant its center of mass has a speed of 10.0 m/s, determine (a) the translational kinetic energy of its center of mass, (b) the rotational kinetic energy about its center of mass, and (c) its total energy.

52. A bowling ball has mass M, radius R, and a moment of inertia of $\frac{2}{5}MR^2$. If it starts from rest, how much work must be done on it to set it rolling without slipping at a linear speed v? Express the work in terms of M and v.

53. (a) Determine the acceleration of the center of mass of a uniform solid disk rolling down an incline making angle θ with the horizontal. Compare this acceleration with that of a uniform hoop. (b) What is the minimum coefficient of friction required to maintain pure rolling motion for the disk?

54. A uniform solid disk and a uniform hoop are placed side by side at the top of an incline of height h. If they are released from rest and roll without slipping, which object reaches the bottom first? Verify your answer by calculating their speeds when they reach the bottom in terms of h.

55. A metal can containing condensed mushroom soup has mass 215 g, height 10.8 cm, and diameter 6.38 cm. It is placed at rest on its side at the top of a 3.00-m-long incline that is at 25.0° to the horizontal, and it is then released to roll straight down. Assuming mechanical energy conservation, calculate the moment of inertia of the can if it takes 1.50 s to reach the bottom of the incline. Which pieces of data, if any, are unnecessary for calculating the solution?

56. A tennis ball is a hollow sphere with a thin wall. It is set rolling without slipping at 4.03 m/s on a horizontal section of a track, as shown in Figure P10.56. It rolls around the inside of a vertical circular loop 90.0 cm in diameter and finally leaves the track at a point 20.0 cm below the horizontal section. (a) Find the speed of the ball at the top of the loop. Demonstrate that it will not fall from the track. (b) Find its speed as it leaves the track. **What If?** (c) Suppose that static friction between ball and track were negligible, so that the ball slid instead of rolling. Would its speed then be higher, lower, or the same at the top of the loop? Explain.

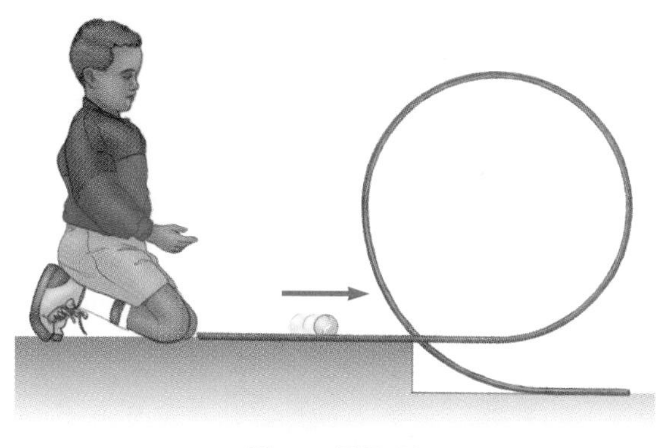

Figure P10.56

Additional Problems

57. As in Figure P10.57, toppling chimneys often break apart in mid-fall because the mortar between the bricks cannot withstand much shear stress. As the chimney begins to fall, shear forces must act on the topmost sections to accelerate them tangentially so that they can keep up with the rotation of the lower part of the stack. For simplicity, let us model the chimney as a uniform rod of length ℓ pivoted at the lower end. The rod starts at rest in a vertical position (with the frictionless pivot at the bottom) and falls over under the influence of gravity. What fraction of the length of the rod has a tangential acceleration greater than $g \sin \theta$, where θ is the angle the chimney makes with the vertical axis?

Figure P10.57 A building demolition site in Baltimore, MD. At the left is a chimney, mostly concealed by the building, that has broken apart on its way down. Compare with Figure 10.19.

58. **Review problem.** A mixing beater consists of three thin rods, each 10.0 cm long. The rods diverge from a central hub, separated from each other by 120°, and all turn in the same plane. A ball is attached to the end of each rod. Each ball has cross-sectional area 4.00 cm^2 and is so shaped that it has a drag coefficient of 0.600. Calculate the power input required to spin the beater at 1 000 rev/min (a) in air and (b) in water.

59. A 4.00-m length of light nylon cord is wound around a uniform cylindrical spool of radius 0.500 m and mass 1.00 kg. The spool is mounted on a frictionless axle and is initially at rest. The cord is pulled from the spool with a constant acceleration of magnitude 2.50 m/s^2. (a) How much work has been done on the spool when it reaches an angular speed of 8.00 rad/s? (b) Assuming there is enough cord on the spool, how long does it take the spool to reach this angular speed? (c) Is there enough cord on the spool?

60. A videotape cassette contains two spools, each of radius r_s, on which the tape is wound. As the tape unwinds from the first spool, it winds around the second spool. The tape moves at constant linear speed v past the heads between the spools. When all the tape is on the first spool, the tape has an outer radius r_t. Let r represent the outer radius of the tape on the first spool at any instant while the tape is being played. (a) Show that at any instant the angular speeds of the two spools are

$$\omega_1 = v/r \qquad \text{and} \qquad \omega_2 = v/(r_s^2 + r_t^2 - r^2)^{1/2}$$

(b) Show that these expressions predict the correct maximum and minimum values for the angular speeds of the two spools.

61. A long uniform rod of length L and mass M is pivoted about a horizontal, frictionless pin through one end. The rod is released from rest in a vertical position, as shown in Figure P10.61. At the instant the rod is horizontal, find (a) its angular speed, (b) the magnitude of its angular acceleration, (c) the x and y components of the acceleration of its center of mass, and (d) the components of the reaction force at the pivot.

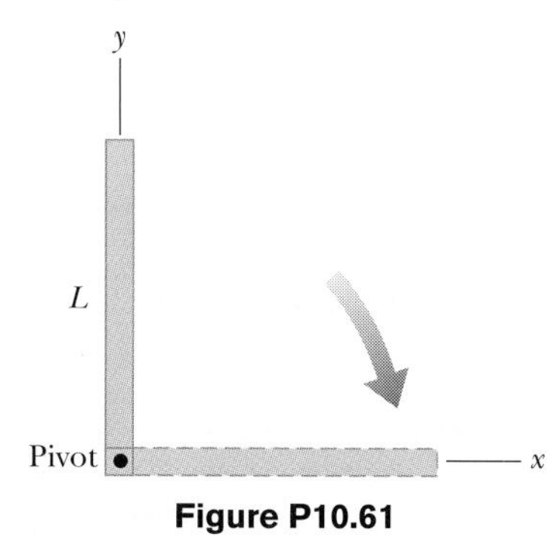

Figure P10.61

62. A shaft is turning at 65.0 rad/s at time $t = 0$. Thereafter, its angular acceleration is given by

$$\alpha = -10.0 \text{ rad/s}^2 - 5.00t \text{ rad/s}^3,$$

where t is the elapsed time. (a) Find its angular speed at $t = 3.00$ s. (b) How far does it turn in these 3 s?

63. A bicycle is turned upside down while its owner repairs a flat tire. A friend spins the other wheel, of radius 0.381 m, and observes that drops of water fly off tangentially. She measures the height reached by drops moving vertically (Fig. P10.63). A drop that breaks loose from the tire on one turn rises $h = 54.0$ cm above the tangent point. A drop that breaks loose on the next turn rises 51.0 cm above the tangent point. The height to which the drops rise decreases because the angular speed of the wheel decreases. From this information, determine the magnitude of the average angular acceleration of the wheel.

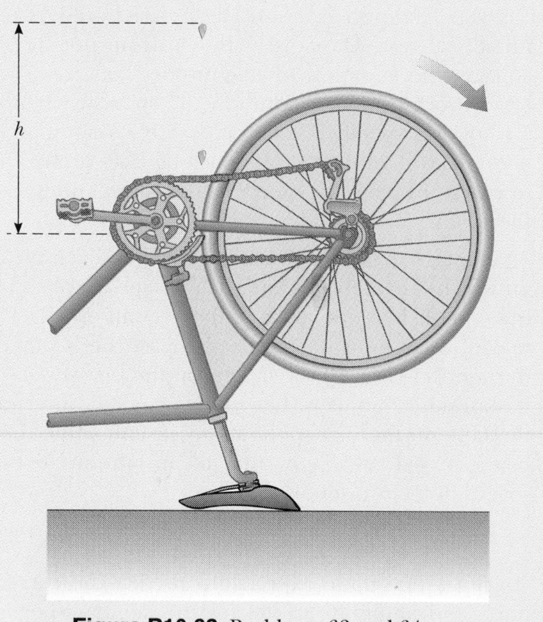

Figure P10.63 Problems 63 and 64.

64. A bicycle is turned upside down while its owner repairs a flat tire. A friend spins the other wheel, of radius R, and observes that drops of water fly off tangentially. She measures the height reached by drops moving vertically (Fig. P10.63). A drop that breaks loose from the tire on one turn rises a distance h_1 above the tangent point. A drop that breaks loose on the next turn rises a distance $h_2 < h_1$ above the tangent point. The height to which the drops rise decreases because the angular speed of the wheel decreases. From this information, determine the magnitude of the average angular acceleration of the wheel.

65. A cord is wrapped around a pulley of mass m and radius r. The free end of the cord is connected to a block of mass M. The block starts from rest and then slides down an incline that makes an angle θ with the horizontal. The coefficient of kinetic friction between block and incline is μ. (a) Use energy methods to show that the block's speed as a function of position d down the incline is

$$v = \sqrt{\frac{4gdM(\sin\theta - \mu\cos\theta)}{m + 2M}}$$

(b) Find the magnitude of the acceleration of the block in terms of μ, m, M, g, and θ.

66. (a) What is the rotational kinetic energy of the Earth about its spin axis? Model the Earth as a uniform sphere and use data from the endpapers. (b) The rotational kinetic energy of the Earth is decreasing steadily because of tidal friction. Find the change in one day, assuming that the rotational period decreases by 10.0 μs each year.

67. Due to a gravitational torque exerted by the Moon on the Earth, our planet's rotation period slows at a rate on the order of 1 ms/century. (a) Determine the order of magnitude of the Earth's angular acceleration. (b) Find the order of magnitude of the torque. (c) Find the order of magnitude of the size of the wrench an ordinary person would need to exert such a torque, as in Figure P10.67. Assume the person can brace his feet against a solid firmament.

Figure P10.67

68. The speed of a moving bullet can be determined by allowing the bullet to pass through two rotating paper disks mounted a distance d apart on the same axle (Fig. P10.68). From the angular displacement $\Delta\theta$ of the two bullet holes in the disks and the rotational speed of the disks, we can determine the speed v of the bullet. Find the bullet speed for the following data: $d = 80$ cm, $\omega = 900$ rev/min, and $\Delta\theta = 31.0°$.

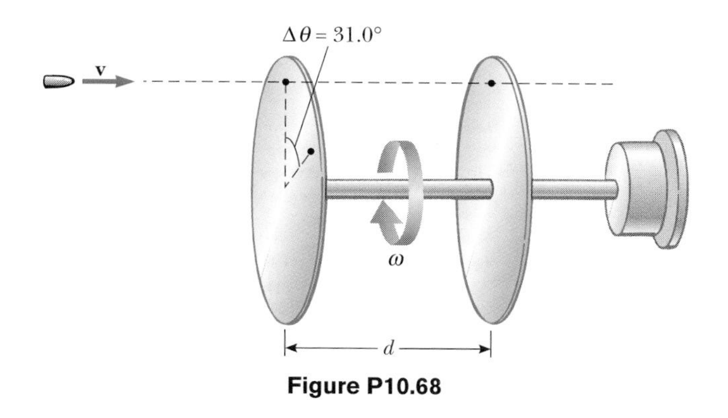

Figure P10.68

69. A uniform, hollow, cylindrical spool has inside radius $R/2$, outside radius R, and mass M (Fig. P10.69). It is mounted so that it rotates on a fixed horizontal axle. A counterweight of mass m is connected to the end of a string wound around the spool. The counterweight falls from rest at $t = 0$ to a position y at time t. Show that the torque due to the friction forces between spool and axle is

$$\tau_f = R\left[m\left(g - \frac{2y}{t^2}\right) - M\frac{5y}{4t^2}\right]$$

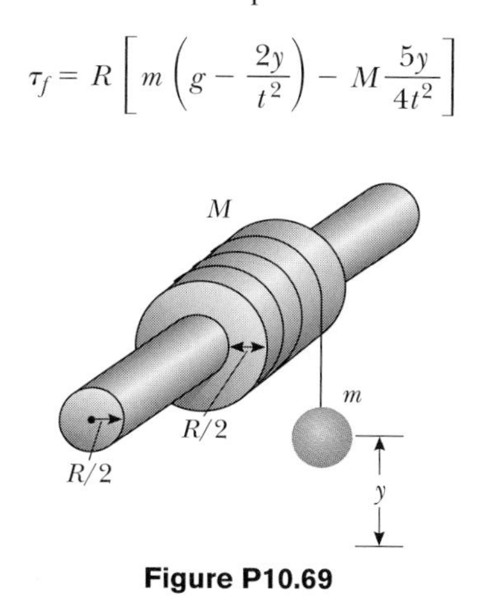

Figure P10.69

70. The reel shown in Figure P10.70 has radius R and moment of inertia I. One end of the block of mass m is connected to a spring of force constant k, and the other end is fastened to a cord wrapped around the reel. The reel axle and the incline are frictionless. The reel is wound counterclockwise so that the spring stretches a distance d from its unstretched position and is then released from rest. (a) Find the angular speed of the reel when the spring is again unstretched. (b) Evaluate the angular speed numerically at this point if $I = 1.00$ kg · m^2, $R = 0.300$ m, $k = 50.0$ N/m, $m = 0.500$ kg, $d = 0.200$ m, and $\theta = 37.0°$.

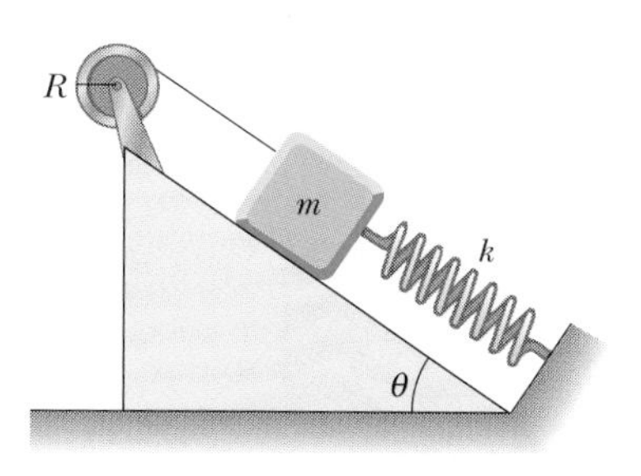

Figure P10.70

71. Two blocks, as shown in Figure P10.71, are connected by a string of negligible mass passing over a pulley of radius 0.250 m and moment of inertia I. The block on the frictionless incline is moving up with a constant acceleration of 2.00 m/s^2. (a) Determine T_1 and T_2, the tensions in the two parts of the string. (b) Find the moment of inertia of the pulley.

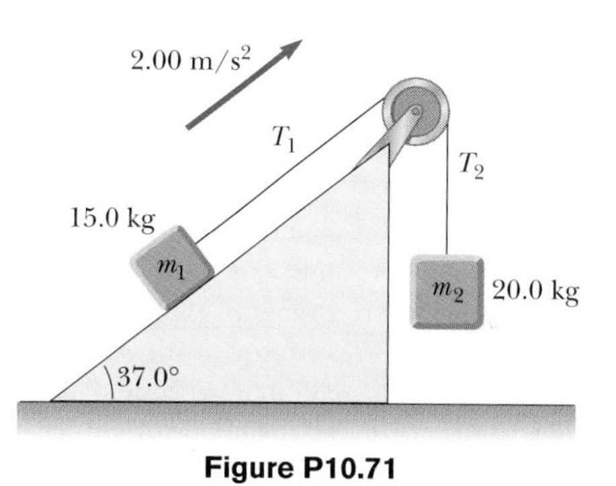

Figure P10.71

72. A common demonstration, illustrated in Figure P10.72, consists of a ball resting at one end of a uniform board of length ℓ, hinged at the other end, and elevated at an angle θ. A light cup is attached to the board at r_c so that it will catch the ball when the support stick is suddenly removed. (a) Show that the ball will lag behind the falling board when θ is less than 35.3°. (b) If the board is 1.00 m long and is supported at this limiting angle, show that the cup must be 18.4 cm from the moving end.

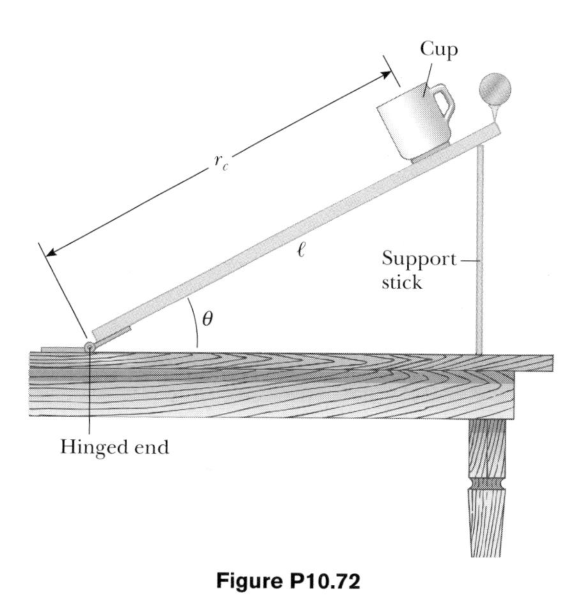

Figure P10.72

73. As a result of friction, the angular speed of a wheel changes with time according to

$$\frac{d\theta}{dt} = \omega_0 e^{-\sigma t}$$

where ω_0 and σ are constants. The angular speed changes from 3.50 rad/s at $t = 0$ to 2.00 rad/s at $t = 9.30$ s. Use this information to determine σ and ω_0. Then determine (a) the magnitude of the angular acceleration at $t = 3.00$ s, (b) the number of revolutions the wheel makes in the first 2.50 s, and (c) the number of revolutions it makes before coming to rest.

74. The hour hand and the minute hand of Big Ben, the Parliament tower clock in London, are 2.70 m and 4.50 m long and have masses of 60.0 kg and 100 kg, respectively (see Figure P10.40). (a) Determine the total torque due to the weight of these hands about the axis of rotation when the time reads (i) 3:00 (ii) 5:15 (iii) 6:00 (iv) 8:20 (v) 9:45. (You may model the hands as long, thin uniform rods.) (b) Determine all times when the total torque about the axis of rotation is zero. Determine the times to the nearest second, solving a transcendental equation numerically.

75. (a) Without the wheels, a bicycle frame has a mass of 8.44 kg. Each of the wheels can be roughly modeled as a uniform solid disk with a mass of 0.820 kg and a radius of 0.343 m. Find the kinetic energy of the whole bicycle when it is moving forward at 3.35 m/s. (b) Before the invention of a wheel turning on an axle, ancient people moved heavy loads by placing rollers under them. (Modern people use rollers too. Any hardware store will sell you a roller bearing for a lazy susan.) A stone block of mass 844 kg moves forward at 0.335 m/s, supported by two uniform cylindrical tree trunks, each of mass 82.0 kg and radius 0.343 m. No slipping occurs between the block and the rollers or between the rollers and the ground. Find the total kinetic energy of the moving objects.

76. A uniform solid sphere of radius r is placed on the inside surface of a hemispherical bowl with much larger radius R. The sphere is released from rest at an angle θ to the vertical and rolls without slipping (Fig. P10.76). Determine the angular speed of the sphere when it reaches the bottom of the bowl.

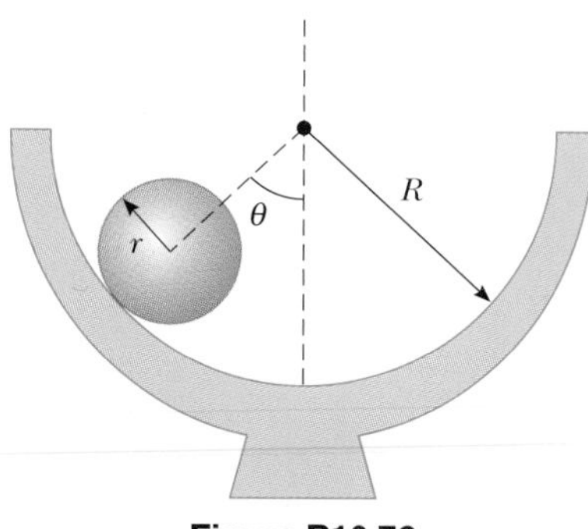

Figure P10.76

77. A string is wound around a uniform disk of radius R and mass M. The disk is released from rest with the string vertical and its top end tied to a fixed bar (Fig. P10.77). Show that (a) the tension in the string is one third of the weight of the disk, (b) the magnitude of the acceleration of the center of mass is $2g/3$, and (c) the speed of the center of mass is $(4gh/3)^{1/2}$ after the disk has descended through distance h. Verify your answer to (c) using the energy approach.

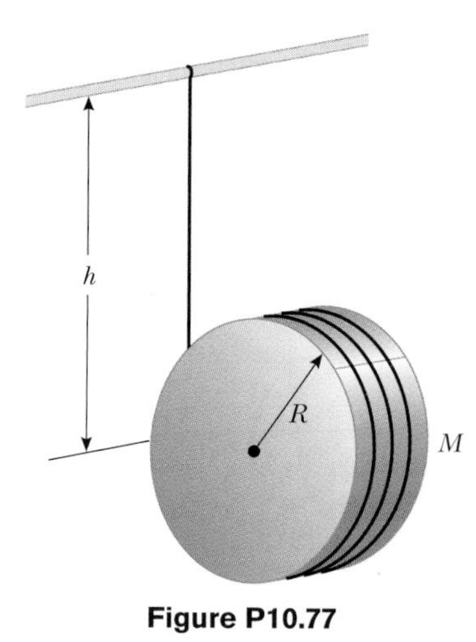

Figure P10.77

78. A constant horizontal force $\mathbf{F}$ is applied to a lawn roller in the form of a uniform solid cylinder of radius R and mass M (Fig. P10.78). If the roller rolls without slipping on the horizontal surface, show that (a) the acceleration of the center of mass is $2\mathbf{F}/3M$ and (b) the minimum coefficient of friction necessary to prevent slipping is $F/3Mg$. (*Hint:* Take the torque with respect to the center of mass.)

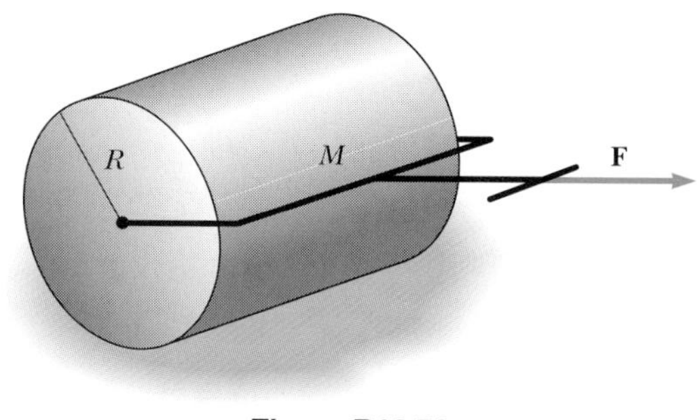

Figure P10.78

79. A solid sphere of mass m and radius r rolls without slipping along the track shown in Figure P10.79. It starts from rest with the lowest point of the sphere at height h above the bottom of the loop of radius R, much larger than r. (a) What is the minimum value of h (in terms of R) such that the sphere completes the loop? (b) What are the force components on the sphere at the point P if $h = 3R$?

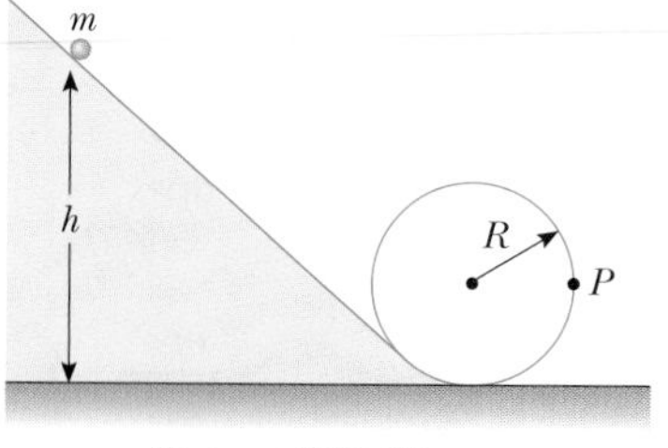

Figure P10.79

80. A thin rod of mass 0.630 kg and length 1.24 m is at rest, hanging vertically from a strong fixed hinge at its top end. Suddenly a horizontal impulsive force $(14.7\hat{\mathbf{i}})$ N is applied to it. (a) Suppose the force acts at the bottom end of the rod. Find the acceleration of its center of mass and the horizontal force the hinge exerts. (b) Suppose the force acts at the midpoint of the rod. Find the acceleration of this point and the horizontal hinge reaction. (c) Where can the impulse be applied so that the hinge will exert no horizontal force? This point is called the *center of percussion*.

81. A bowler releases a bowling ball with no spin, sending it sliding straight down the alley toward the pins. The ball continues to slide for a distance of what order of magnitude, before its motion becomes rolling without slipping? State the quantities you take as data, the values you measure or estimate for them, and your reasoning.

82. Following Thanksgiving dinner your uncle falls into a deep sleep, sitting straight up facing the television set. A naughty grandchild balances a small spherical grape at the top of his bald head, which itself has the shape of a sphere. After all the children have had time to giggle, the grape starts from rest and rolls down without slipping. It will leave contact with your uncle's scalp when the radial line joining it to the center of curvature makes what angle with the vertical?

83. (a) A thin rod of length h and mass M is held vertically with its lower end resting on a frictionless horizontal surface. The rod is then released to fall freely. Determine the speed of its center of mass just before it hits the horizontal surface. (b) **What If?** Now suppose the rod has a fixed pivot at its lower end. Determine the speed of the rod's center of mass just before it hits the surface.

84. A large, cylindrical roll of tissue paper of initial radius R lies on a long, horizontal surface with the outside end of the paper nailed to the surface. The roll is given a slight shove ($v_i \approx 0$) and commences to unroll. Assume the roll has a uniform density and that mechanical energy is conserved in the process. (a) Determine the speed of the center of mass of the roll when its radius has diminished to r. (b) Calculate a numerical value for this speed at $r = 1.00$ mm, assuming $R = 6.00$ m. (c) **What If?** What happens to the energy of the system when the paper is completely unrolled?

85. A spool of wire of mass M and radius R is unwound under a constant force $\mathbf{F}$ (Fig. P10.85). Assuming the spool is a uniform solid cylinder that doesn't slip, show that (a) the acceleration of the center of mass is $4\mathbf{F}/3M$ and (b) the force of friction is to the *right* and equal in magnitude to $F/3$. (c) If the cylinder starts from rest and rolls without slipping, what is the speed of its center of mass after it has rolled through a distance d?

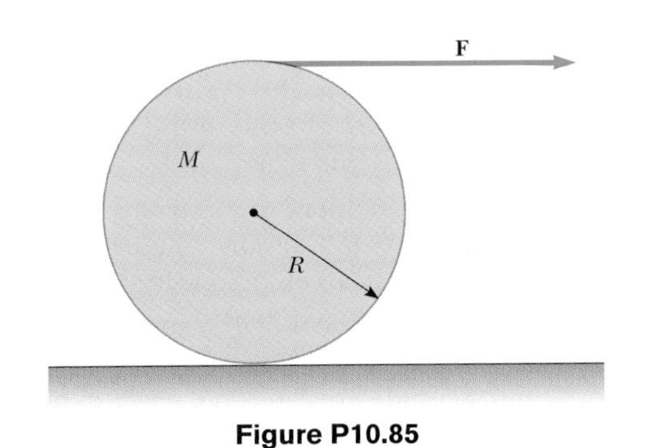

Figure P10.85

86. A plank with a mass $M = 6.00$ kg rides on top of two identical solid cylindrical rollers that have $R = 5.00$ cm and $m = 2.00$ kg (Fig. P10.86). The plank is pulled by a constant horizontal force $\mathbf{F}$ of magnitude 6.00 N applied to the end of the plank and perpendicular to the axes of the cylinders (which are parallel). The cylinders roll without slipping on a flat surface. There is also no slipping between the cylinders and the plank. (a) Find the acceleration of the plank and of the rollers. (b) What friction forces are acting?

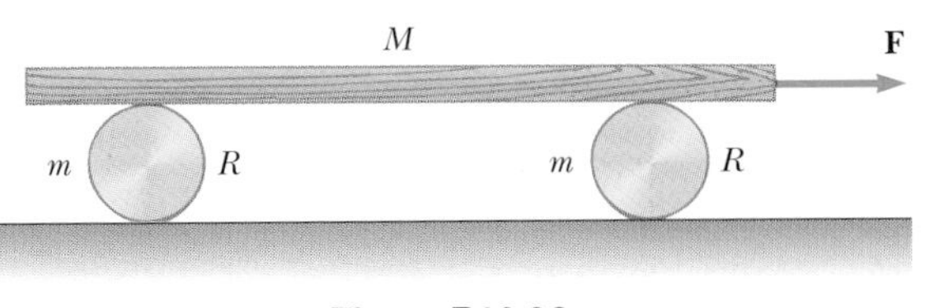

Figure P10.86

87. A spool of wire rests on a horizontal surface as in Figure P10.87. As the wire is pulled, the spool does not slip at the contact point P. On separate trials, each one of the forces $\mathbf{F}_1$, $\mathbf{F}_2$, $\mathbf{F}_3$, and $\mathbf{F}_4$ is applied to the spool. For each one of these forces, determine the direction the spool will roll. Note that the line of action of $\mathbf{F}_2$ passes through P.

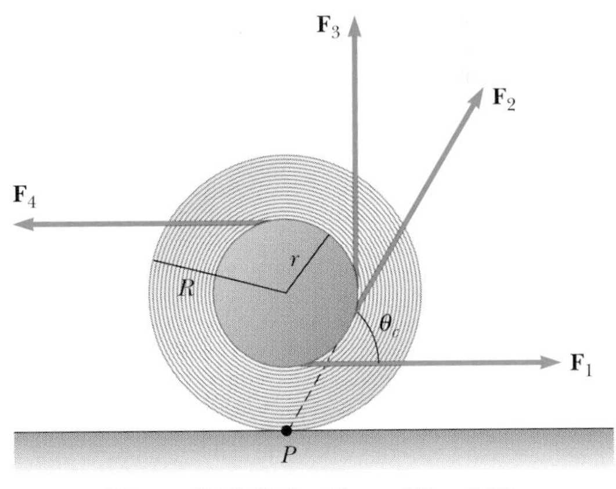

Figure P10.87 Problems 87 and 88.

88. Refer to Problem 87 and Figure P10.87. The spool of wire has an inner radius r and an outer radius R. The angle θ between the applied force and the horizontal can be varied. Show that the critical angle for which the spool does not roll is given by

$$\cos \theta_c = \frac{r}{R}$$

If the wire is held at this angle and the force increased, the spool will remain stationary until it slips along the floor.

89. In a demonstration known as the ballistics cart, a ball is projected vertically upward from a cart moving with constant velocity along the horizontal direction. The ball lands in the catching cup of the cart because both the cart a ball have the same horizontal component of velocity. **What If?** Now consider a ballistics cart on an incline making an angle θ with the horizontal as in Figure P10.89. The cart (including wheels) has a mass M and the moment of inertia of each of the two wheels is $mR^2/2$. (a) Using conservation of energy (assuming no friction between cart and axles) and assuming pure rolling motion (no slipping), show that the acceleration of the cart along the incline is

$$a_x = \left(\frac{M}{M + 2m}\right) g \sin\theta$$

(b) Note that the x component of acceleration of the ball released by the cart is $g \sin\theta$. Thus, the x component of the cart's acceleration is *smaller* than that of the ball by the factor $M/(M + 2m)$. Use this fact and kinematic equations to show that the ball overshoots the cart by an amount Δx, where

$$\Delta x = \left(\frac{4m}{M + 2m}\right)\left(\frac{\sin\theta}{\cos^2\theta}\right)\frac{v_{yi}^2}{g}$$

and v_{yi} is the initial speed of the ball imparted to it by the spring in the cart. (c) Show that the distance d that the ball travels measured along the incline is

$$d = \frac{2v_{yi}^2}{g}\frac{\sin\theta}{\cos^2\theta}$$

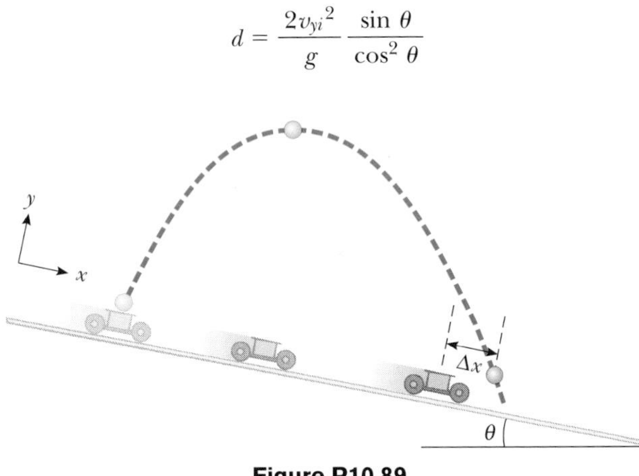

Figure P10.89

90. A spool of thread consists of a cylinder of radius R_1 with end caps of radius R_2 as in the end view shown in Figure P10.90. The mass of the spool, including the thread, is m and its moment of inertia about an axis through its center is I. The spool is placed on a rough horizontal surface so that it rolls without slipping when a force $\mathbf{T}$ acting to the right is applied to the free end of the thread. Show that the magnitude of the friction force exerted by the surface on the spool is given by

$$f = \left(\frac{I + mR_1R_2}{I + mR_2^2}\right) T$$

Determine the direction of the force of friction.

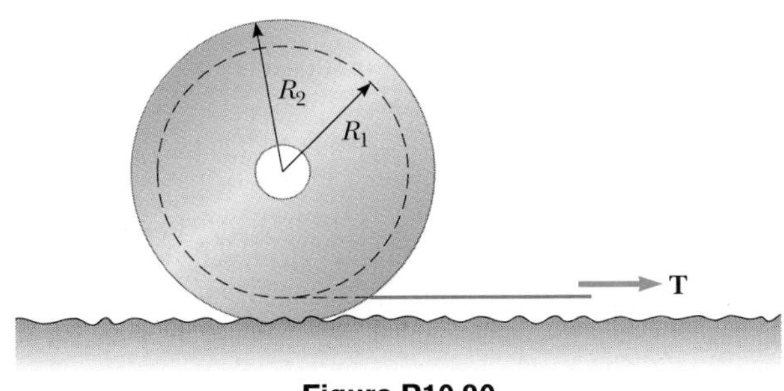

Figure P10.90

Answers to Quick Quizzes

10.1 (c). For a rotation of more than 180°, the angular displacement must be larger than $\pi = 3.14$ rad. The angular displacements in the three choices are (a) 6 rad − 3 rad = 3 rad (b) 1 rad − (−1) rad = 2 rad (c) 5 rad − 1 rad = 4 rad.

10.2 (b). Because all angular displacements occur in the same time interval, the displacement with the lowest value will be associated with the lowest average angular speed.

10.3 (b). The fact that ω is negative indicates that we are dealing with an object that is rotating in the clockwise direction. We also know that when $\boldsymbol{\omega}$ and $\boldsymbol{\alpha}$ are antiparallel, ω must be decreasing—the object is slowing down. Therefore, the object is spinning more and more slowly (with less and less angular speed) in the clockwise, or negative, direction.

10.4 (b). In Equation 10.8, both the initial and final angular speeds are the same in all three cases. As a result, the angular acceleration is inversely proportional to the angular displacement. Thus, the highest angular acceleration is associated with the lowest angular displacement.

10.5 (b). The system of the platform, Andy, and Charlie is a rigid object, so all points on the rigid object have the same angular speed.

10.6 (a). The tangential speed is proportional to the radial distance from the rotation axis.

10.7 (a). Almost all of the mass of the pipe is at the same distance from the rotation axis, so it has a larger moment of inertia than the solid cylinder.

10.8 (b). The fatter handle of the screwdriver gives you a larger moment arm and increases the torque that you can apply with a given force from your hand.

10.9 (a). The longer handle of the wrench gives you a larger moment arm and increases the torque that you can apply with a given force from your hand.

10.10 (b). With twice the moment of inertia and the same frictional torque, there is half the angular acceleration. With half the angular acceleration, it will require twice as long to change the speed to zero.

10.11 (d). When the rod is attached at its end, it offers four times as much moment of inertia as when attached in the center (see Table 10.2). Because the rotational kinetic energy of the rod depends on the square of the angular speed, the same work will result in half of the angular speed.

10.12 (b). All of the gravitational potential energy of the box–Earth system is transformed to kinetic energy of translation. For the ball, some of the gravitational potential energy of the ball–Earth system is transformed to rotational kinetic energy, leaving less for translational kinetic energy, so the ball moves downhill more slowly than the box does.

10.13 (c). In Equation 10.30, I_{CM} for a sphere is $\frac{2}{5}MR^2$. Thus, MR^2 will cancel and the remaining expression on the right-hand side of the equation is independent of mass and radius.

10.14 (a). The moment of inertia of the hollow sphere B is larger than that of sphere A. As a result, Equation 10.30 tells us that the center of mass of sphere B will have a smaller speed, so sphere A should arrive first.

Chapter 11

Angular Momentum

CHAPTER OUTLINE

▲ *Mark Ruiz undergoes a rotation during a dive at the U.S. Olympic trials in June 2000. He spins at a higher rate when he curls up and grabs his ankles due to the principle of conservation of angular momentum, as discussed in this chapter. (Otto Greule/Allsport/Getty)*

The central topic of this chapter is angular momentum, a quantity that plays a key role in rotational dynamics. In analogy to the principle of conservation of linear momentum, we find that the angular momentum of a system is conserved if no external torques act on the system. Like the law of conservation of linear momentum, the law of conservation of angular momentum is a fundamental law of physics, equally valid for relativistic and quantum systems.

11.1 The Vector Product and Torque

An important consideration in defining angular momentum is the process of multiplying two vectors by means of the operation called the *vector product.* We will introduce the vector product by considering torque as introduced in the preceding chapter.

Consider a force $\mathbf{F}$ acting on a rigid object at the vector position $\mathbf{r}$ (Fig. 11.1). As we saw in Section 10.6, the *magnitude* of the torque due to this force relative to the origin is $rF \sin \phi$, where ϕ is the angle between $\mathbf{r}$ and $\mathbf{F}$. The axis about which $\mathbf{F}$ tends to produce rotation is perpendicular to the plane formed by $\mathbf{r}$ and $\mathbf{F}$.

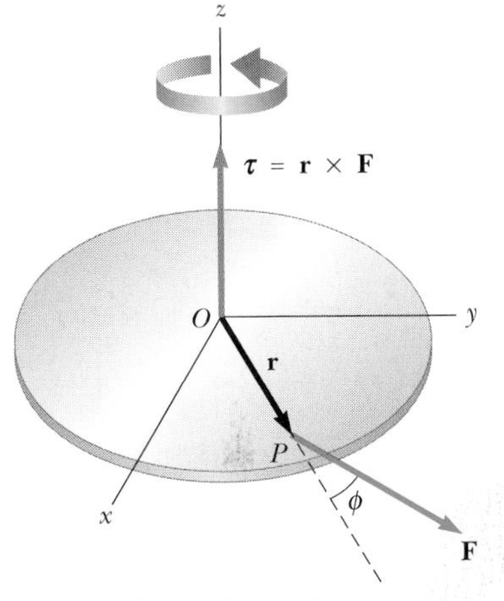

Active Figure 11.1 The torque vector $\boldsymbol{\tau}$ lies in a direction perpendicular to the plane formed by the position vector $\mathbf{r}$ and the applied force vector $\mathbf{F}$.

At the Active Figures link at http://www.pse6.com, you can move point P and change the force vector F to see the effect on the torque vector.

The torque vector $\boldsymbol{\tau}$ is related to the two vectors $\mathbf{r}$ and $\mathbf{F}$. We can establish a mathematical relationship between $\boldsymbol{\tau}$, $\mathbf{r}$, and $\mathbf{F}$ using a mathematical operation called the **vector product,** or **cross product:**

$$\boldsymbol{\tau} \equiv \mathbf{r} \times \mathbf{F} \tag{11.1}$$

We now give a formal definition of the vector product. Given any two vectors $\mathbf{A}$ and $\mathbf{B}$, the **vector product $\mathbf{A} \times \mathbf{B}$** is defined as a third vector $\mathbf{C}$, which has a magnitude of $AB \sin \theta$, where θ is the angle between $\mathbf{A}$ and $\mathbf{B}$. That is, if $\mathbf{C}$ is given by

$$\mathbf{C} = \mathbf{A} \times \mathbf{B} \tag{11.2}$$

then its magnitude is

$$C \equiv AB \sin \theta \tag{11.3}$$

The quantity $AB \sin \theta$ is equal to the area of the parallelogram formed by $\mathbf{A}$ and $\mathbf{B}$, as shown in Figure 11.2. The *direction* of $\mathbf{C}$ is perpendicular to the plane formed by $\mathbf{A}$ and $\mathbf{B}$, and the best way to determine this direction is to use the right-hand rule illustrated in Figure 11.2. The four fingers of the right hand are pointed along $\mathbf{A}$ and then "wrapped" into $\mathbf{B}$ through the angle θ. The direction of the upright thumb is the direction of $\mathbf{A} \times \mathbf{B} = \mathbf{C}$. Because of the notation, $\mathbf{A} \times \mathbf{B}$ is often read "$\mathbf{A}$ cross $\mathbf{B}$"; hence, the term *cross product.*

Some properties of the vector product that follow from its definition are as follows:

1. Unlike the scalar product, the vector product is *not* commutative. Instead, the order in which the two vectors are multiplied in a cross product is important:

$$\mathbf{A} \times \mathbf{B} = -\mathbf{B} \times \mathbf{A} \tag{11.4}$$

PITFALL PREVENTION

11.1 The Cross Product is a Vector

Remember that the result of taking a cross product between two vectors is *a third vector.* Equation 11.3 gives only the magnitude of this vector.

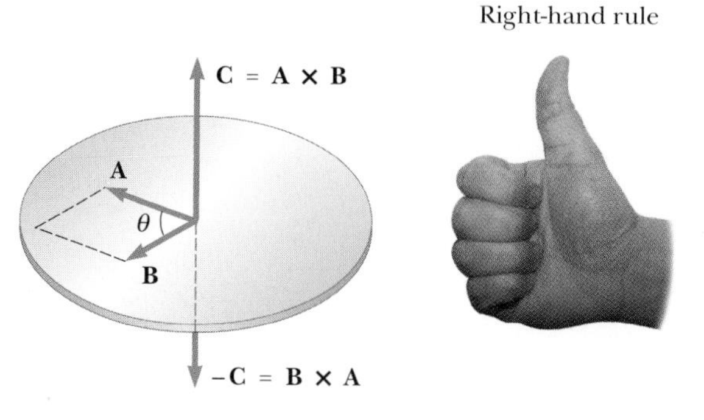

Right-hand rule

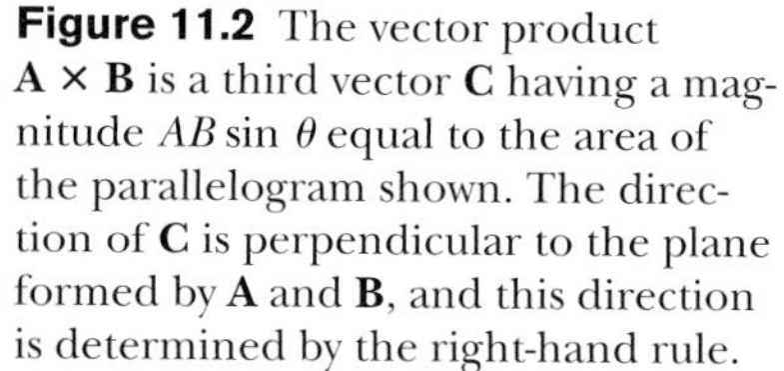

Figure 11.2 The vector product $\mathbf{A} \times \mathbf{B}$ is a third vector $\mathbf{C}$ having a magnitude $AB \sin \theta$ equal to the area of the parallelogram shown. The direction of $\mathbf{C}$ is perpendicular to the plane formed by $\mathbf{A}$ and $\mathbf{B}$, and this direction is determined by the right-hand rule.

Therefore, if you change the order of the vectors in a cross product, you must change the sign. You can easily verify this relationship with the right-hand rule.

2. If $\mathbf{A}$ is parallel to $\mathbf{B}$ ($\theta = 0°$ or $180°$), then $\mathbf{A} \times \mathbf{B} = 0$; therefore, it follows that $\mathbf{A} \times \mathbf{A} = 0$.
3. If $\mathbf{A}$ is perpendicular to $\mathbf{B}$, then $|\mathbf{A} \times \mathbf{B}| = AB$.
4. The vector product obeys the distributive law:

Properties of the vector product

$$\mathbf{A} \times (\mathbf{B} + \mathbf{C}) = \mathbf{A} \times \mathbf{B} + \mathbf{A} \times \mathbf{C} \tag{11.5}$$

5. The derivative of the cross product with respect to some variable such as t is

$$\frac{d}{dt}(\mathbf{A} \times \mathbf{B}) = \frac{d\mathbf{A}}{dt} \times \mathbf{B} + \mathbf{A} \times \frac{d\mathbf{B}}{dt} \tag{11.6}$$

where it is important to preserve the multiplicative order of $\mathbf{A}$ and $\mathbf{B}$, in view of Equation 11.4.

It is left as an exercise (Problem 10) to show from Equations 11.3 and 11.4 and from the definition of unit vectors that the cross products of the rectangular unit vectors $\hat{\mathbf{i}}$, $\hat{\mathbf{j}}$, and $\hat{\mathbf{k}}$ obey the following rules:

Cross products of unit vectors

$$\hat{\mathbf{i}} \times \hat{\mathbf{i}} = \hat{\mathbf{j}} \times \hat{\mathbf{j}} = \hat{\mathbf{k}} \times \hat{\mathbf{k}} = 0 \tag{11.7a}$$

$$\hat{\mathbf{i}} \times \hat{\mathbf{j}} = -\hat{\mathbf{j}} \times \hat{\mathbf{i}} = \hat{\mathbf{k}} \tag{11.7b}$$

$$\hat{\mathbf{j}} \times \hat{\mathbf{k}} = -\hat{\mathbf{k}} \times \hat{\mathbf{j}} = \hat{\mathbf{i}} \tag{11.7c}$$

$$\hat{\mathbf{k}} \times \hat{\mathbf{i}} = -\hat{\mathbf{i}} \times \hat{\mathbf{k}} = \hat{\mathbf{j}} \tag{11.7d}$$

Signs are interchangeable in cross products. For example, $\mathbf{A} \times (-\mathbf{B}) = -\mathbf{A} \times \mathbf{B}$ and $\hat{\mathbf{i}} \times (-\hat{\mathbf{j}}) = -\hat{\mathbf{i}} \times \hat{\mathbf{j}}$.

The cross product of any two vectors $\mathbf{A}$ and $\mathbf{B}$ can be expressed in the following determinant form:

$$\mathbf{A} \times \mathbf{B} = \begin{vmatrix} \hat{\mathbf{i}} & \hat{\mathbf{j}} & \hat{\mathbf{k}} \\ A_x & A_y & A_z \\ B_x & B_y & B_z \end{vmatrix} = \begin{vmatrix} A_y & A_z \\ B_y & B_z \end{vmatrix} \hat{\mathbf{i}} - \begin{vmatrix} A_x & A_z \\ B_x & B_z \end{vmatrix} \hat{\mathbf{j}} + \begin{vmatrix} A_x & A_y \\ B_x & B_y \end{vmatrix} \hat{\mathbf{k}}$$

Expanding these determinants gives the result

$$\mathbf{A} \times \mathbf{B} = (A_yB_z - A_zB_y)\hat{\mathbf{i}} - (A_xB_z - A_zB_x)\hat{\mathbf{j}} + (A_xB_y - A_yB_x)\hat{\mathbf{k}} \tag{11.8}$$

Given the definition of the cross product, we can now assign a direction to the torque vector. If the force lies in the xy plane, as in Figure 11.1, the torque $\boldsymbol{\tau}$ is represented by a vector parallel to the z axis. The force in Figure 11.1 creates a torque that tends to rotate the object counterclockwise about the z axis; thus the direction of $\boldsymbol{\tau}$ is toward increasing z, and $\boldsymbol{\tau}$ is therefore in the positive z direction. If we reversed the direction of $\mathbf{F}$ in Figure 11.1, then $\boldsymbol{\tau}$ would be in the negative z direction.

Quick Quiz 11.1 Which of the following is equivalent to the following *scalar* product: $(\mathbf{A} \times \mathbf{B}) \cdot (\mathbf{B} \times \mathbf{A})$? (a) $\mathbf{A} \cdot \mathbf{B} + \mathbf{B} \cdot \mathbf{A}$ (b) $(\mathbf{A} \times \mathbf{A}) \cdot (\mathbf{B} \times \mathbf{B})$ (c) $(\mathbf{A} \times \mathbf{B}) \cdot (\mathbf{A} \times \mathbf{B})$ (d) $-(\mathbf{A} \times \mathbf{B}) \cdot (\mathbf{A} \times \mathbf{B})$

Quick Quiz 11.2 Which of the following statements is true about the relationship between the magnitude of the cross product of two vectors and the product of the magnitudes of the vectors? (a) $|\mathbf{A} \times \mathbf{B}|$ is larger than AB; (b) $|\mathbf{A} \times \mathbf{B}|$ is smaller than AB; (c) $|\mathbf{A} \times \mathbf{B}|$ could be larger or smaller than AB, depending on the angle between the vectors; (d) $|\mathbf{A} \times \mathbf{B}|$ could be equal to AB.

Example 11.1 The Vector Product

Two vectors lying in the xy plane are given by the equations $\mathbf{A} = 2\hat{\mathbf{i}} + 3\hat{\mathbf{j}}$ and $\mathbf{B} = -\hat{\mathbf{i}} + 2\hat{\mathbf{j}}$. Find $\mathbf{A} \times \mathbf{B}$ and verify that $\mathbf{A} \times \mathbf{B} = -\mathbf{B} \times \mathbf{A}$.

Solution Using Equations 11.7a through 11.7d, we obtain

$$\mathbf{A} \times \mathbf{B} = (2\hat{\mathbf{i}} + 3\hat{\mathbf{j}}) \times (-\hat{\mathbf{i}} + 2\hat{\mathbf{j}})$$
$$= 2\hat{\mathbf{i}} \times 2\hat{\mathbf{j}} + 3\hat{\mathbf{j}} \times (-\hat{\mathbf{i}}) = 4\hat{\mathbf{k}} + 3\hat{\mathbf{k}} = 7\hat{\mathbf{k}}$$

(We have omitted the terms containing $\hat{\mathbf{i}} \times \hat{\mathbf{i}}$ and $\hat{\mathbf{j}} \times \hat{\mathbf{j}}$ because, as Equation 11.7a shows, they are equal to zero.)

We can show that $\mathbf{A} \times \mathbf{B} = -\mathbf{B} \times \mathbf{A}$, because

$$\mathbf{B} \times \mathbf{A} = (-\hat{\mathbf{i}} + 2\hat{\mathbf{j}}) \times (2\hat{\mathbf{i}} + 3\hat{\mathbf{j}})$$
$$= -\hat{\mathbf{i}} \times 3\hat{\mathbf{j}} + 2\hat{\mathbf{j}} \times 2\hat{\mathbf{i}} = -3\hat{\mathbf{k}} - 4\hat{\mathbf{k}} = -7\hat{\mathbf{k}}$$

Therefore, $\mathbf{A} \times \mathbf{B} = -\mathbf{B} \times \mathbf{A}$.

As an alternative method for finding $\mathbf{A} \times \mathbf{B}$, we could use Equation 11.8, with $A_x = 2$, $A_y = 3$, $A_z = 0$ and $B_x = -1$, $B_y = 2$, $B_z = 0$:

$$\mathbf{A} \times \mathbf{B} = (0)\hat{\mathbf{i}} - (0)\hat{\mathbf{j}} + [(2)(2) - (3)(-1)]\hat{\mathbf{k}} = 7\hat{\mathbf{k}}$$

Example 11.2 The Torque Vector

A force of $\mathbf{F} = (2.00\hat{\mathbf{i}} + 3.00\hat{\mathbf{j}})$ N is applied to an object that is pivoted about a fixed axis aligned along the z coordinate axis. If the force is applied at a point located at $\mathbf{r} = (4.00\hat{\mathbf{i}} + 5.00\hat{\mathbf{j}})$ m, find the torque vector $\boldsymbol{\tau}$.

Solution The torque vector is defined by means of a cross product in Equation 11.1:

$$\boldsymbol{\tau} = \mathbf{r} \times \mathbf{F} = [(4.00\hat{\mathbf{i}} + 5.00\hat{\mathbf{j}})\text{ m}] \times [(2.00\hat{\mathbf{i}} + 3.00\hat{\mathbf{j}})\text{ N}]$$
$$= [(4.00)(2.00)\hat{\mathbf{i}} \times \hat{\mathbf{i}} + (4.00)(3.00)\hat{\mathbf{i}} \times \hat{\mathbf{j}} + (5.00)(2.00)\hat{\mathbf{j}} \times \hat{\mathbf{i}} + (5.00)(3.00)\hat{\mathbf{j}} \times \hat{\mathbf{j}}]\text{ N}\cdot\text{m}$$
$$= [12.0\hat{\mathbf{i}} \times \hat{\mathbf{j}} + 10.0\hat{\mathbf{j}} \times \hat{\mathbf{i}}]\text{ N}\cdot\text{m}$$
$$= [12.0\hat{\mathbf{k}} - 10.0\hat{\mathbf{k}})\text{ N}\cdot\text{m}$$
$$= 2.0\hat{\mathbf{k}}\text{ N}\cdot\text{m}$$

Notice that both $\mathbf{r}$ and $\mathbf{F}$ are in the xy plane. As expected, the torque vector is perpendicular to this plane, having only a z component.

11.2 Angular Momentum

Imagine a rigid pole sticking up through the ice on a frozen pond (Fig. 11.3). A skater glides rapidly toward the pole, aiming a little to the side so that she does not hit it. As she approaches the pole, she reaches out and grabs it, an action that causes her to move in a circular path around the pole. Just as the idea of linear momentum helps us analyze translational motion, a rotational analog—*angular momentum*—helps us analyze the motion of this skater and other objects undergoing rotational motion.

Active Figure 11.3 As the skater passes the pole, she grabs hold of it. This causes her to swing around the pole rapidly in a circular path.

At the Active Figures link at http://www.pse6.com, you can change the speed of the skater and her distance to the pole and watch her spin when she grabs the pole.

In Chapter 9, we began by developing the mathematical form of linear momentum and then proceeded to show how this new quantity was valuable in problem-solving. We will follow a similar procedure for angular momentum.

Consider a particle of mass m located at the vector position $\mathbf{r}$ and moving with linear momentum $\mathbf{p}$ as in Figure 11.4. In describing linear motion, we found that the net force on the particle equals the time rate of change of its linear momentum, $\Sigma \mathbf{F} = d\mathbf{p}/dt$ (see Eq. 9.3). Let us take the cross product of each side of Equation 9.3 with $\mathbf{r}$, which gives us the net torque on the particle on the left side of the equation:

$$\mathbf{r} \times \sum \mathbf{F} = \sum \boldsymbol{\tau} = \mathbf{r} \times \frac{d\mathbf{p}}{dt}$$

Now let us add to the right-hand side the term $\frac{d\mathbf{r}}{dt} \times \mathbf{p}$, which is zero because $d\mathbf{r}/dt = \mathbf{v}$ and $\mathbf{v}$ and $\mathbf{p}$ are parallel. Thus,

$$\sum \boldsymbol{\tau} = \mathbf{r} \times \frac{d\mathbf{p}}{dt} + \frac{d\mathbf{r}}{dt} \times \mathbf{p}$$

We recognize the right-hand side of this equation as the derivative of $\mathbf{r} \times \mathbf{p}$ (see Equation 11.6). Therefore,

$$\sum \boldsymbol{\tau} = \frac{d(\mathbf{r} \times \mathbf{p})}{dt} \qquad (11.9)$$

This looks very similar in form to Equation 9.3, $\Sigma \mathbf{F} = d\mathbf{p}/dt$. This suggests that the combination $\mathbf{r} \times \mathbf{p}$ should play the same role in rotational motion that $\mathbf{p}$ plays in translational motion. We call this combination the *angular momentum* of the particle:

Angular momentum of a particle

The instantaneous **angular momentum L** of a particle relative to the origin O is defined by the cross product of the particle's instantaneous position vector $\mathbf{r}$ and its instantaneous linear momentum $\mathbf{p}$:

$$\mathbf{L} \equiv \mathbf{r} \times \mathbf{p} \qquad (11.10)$$

This allows us to write Equation 11.9 as

$$\sum \boldsymbol{\tau} = \frac{d\mathbf{L}}{dt} \qquad (11.11)$$

which is the rotational analog of Newton's second law, $\Sigma \mathbf{F} = d\mathbf{p}/dt$. Note that torque causes the angular momentum $\mathbf{L}$ to change just as force causes linear momentum $\mathbf{p}$ to change. Equation 11.11 states that **the torque acting on a particle is equal to the time rate of change of the particle's angular momentum.**

Note that Equation 11.11 is valid only if $\Sigma\boldsymbol{\tau}$ and $\mathbf{L}$ are measured about the same origin. (Of course, the same origin must be used in calculating all of the torques.) Furthermore, **the expression is valid for any origin fixed in an inertial frame.**

The SI unit of angular momentum is $\text{kg} \cdot \text{m}^2/\text{s}$. Note also that both the magnitude and the direction of $\mathbf{L}$ depend on the choice of origin. Following the right-hand rule, we see that the direction of $\mathbf{L}$ is perpendicular to the plane formed by $\mathbf{r}$ and $\mathbf{p}$. In Figure 11.4, $\mathbf{r}$ and $\mathbf{p}$ are in the xy plane, and so $\mathbf{L}$ points in the z direction. Because $\mathbf{p} = m\mathbf{v}$, the magnitude of $\mathbf{L}$ is

$$L = mvr \sin \phi \qquad (11.12)$$

where ϕ is the angle between $\mathbf{r}$ and $\mathbf{p}$. It follows that L is zero when $\mathbf{r}$ is parallel to $\mathbf{p}$ ($\phi = 0$ or $180°$). In other words, when the linear velocity of the particle is along a line that passes through the origin, the particle has zero angular momentum with respect to the origin. On the other hand, if $\mathbf{r}$ is perpendicular to $\mathbf{p}$ ($\phi = 90°$), then $L = mvr$. At that instant, the particle moves exactly as if it were on the rim of a wheel rotating about the origin in a plane defined by $\mathbf{r}$ and $\mathbf{p}$.

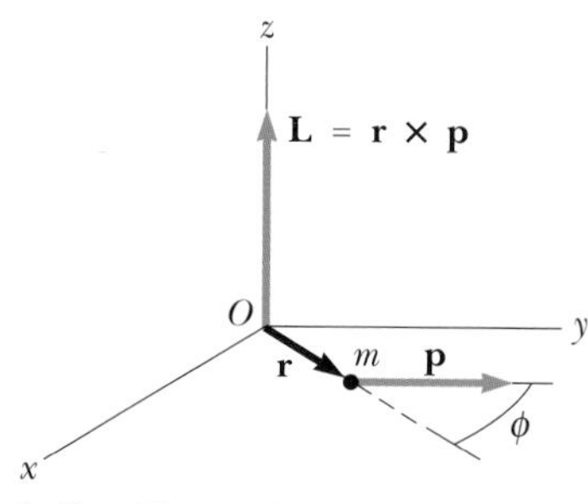

Active Figure 11.4 The angular momentum $\mathbf{L}$ of a particle of mass m and linear momentum $\mathbf{p}$ located at the vector position $\mathbf{r}$ is a vector given by $\mathbf{L} = \mathbf{r} \times \mathbf{p}$. The value of $\mathbf{L}$ depends on the origin about which it is measured and is a vector perpendicular to both $\mathbf{r}$ and $\mathbf{p}$.

At the Active Figures link at http://www.pse6.com, you can change the position vector r and the momentum vector p to see the effect on the angular momentum vector.

Quick Quiz 11.3 Recall the skater described at the beginning of this section. Let her mass be m. What would be her angular momentum relative to the pole at the instant she is a distance d from the pole if she were skating directly toward it at speed v? (a) zero (b) mvd (c) impossible to determine

Quick Quiz 11.4 Consider again the skater in Quick Quiz 11.3. What would be her angular momentum relative to the pole at the instant she is a distance d from the pole if she were skating at speed v along a straight line that would pass within a distance a from the pole? (a) zero (b) mvd (c) mva (d) impossible to determine

PITFALL PREVENTION

11.2 Is Rotation Necessary for Angular Momentum?

Notice that we can define angular momentum even if the particle is not moving in a circular path. Even a particle moving in a straight line has angular momentum about any axis displaced from the path of the particle.

Example 11.3 Angular Momentum of a Particle in Circular Motion

A particle moves in the xy plane in a circular path of radius r, as shown in Figure 11.5. Find the magnitude and direction of its angular momentum relative to O when its linear velocity is $\mathbf{v}$.

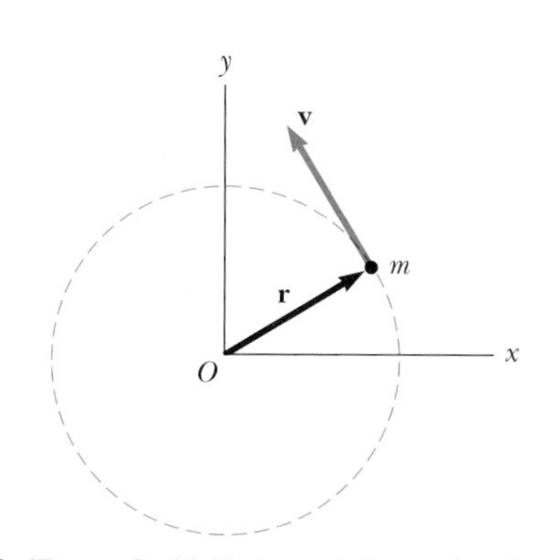

Figure 11.5 (Example 11.3) A particle moving in a circle of radius r has an angular momentum about O that has magnitude mvr. The vector $\mathbf{L} = \mathbf{r} \times \mathbf{p}$ points *out* of the diagram.

Solution The linear momentum of the particle is always changing (in direction, not magnitude). You might be tempted, therefore, to conclude that the angular momentum of the particle is always changing. In this situation, however, this is not the case—let us see why. From Equation 11.12, the magnitude of $\mathbf{L}$ is given by

$$L = mvr \sin 90° = mvr$$

where we have used $\phi = 90°$ because $\mathbf{v}$ is perpendicular to $\mathbf{r}$. This value of L is constant because all three factors on the right are constant.

The direction of $\mathbf{L}$ also is constant, even though the direction of $\mathbf{p} = m\mathbf{v}$ keeps changing. You can visualize this by applying the right-hand rule to find the direction of $\mathbf{L} = \mathbf{r} \times \mathbf{p} = m\mathbf{r} \times \mathbf{v}$ in Figure 11.5. Your thumb points upward and away from the page; this is the direction of $\mathbf{L}$ Hence, we can write the vector expression $\mathbf{L} = (mvr)\hat{\mathbf{k}}$. If the particle were to move clockwise, $\mathbf{L}$ would point downward and into the page. **A particle in uniform circular motion has a constant angular momentum about an axis through the center of its path.**

Angular Momentum of a System of Particles

In Section 9.6, we showed that Newton's second law for a particle could be extended to a system of particles, resulting in:

$$\sum \mathbf{F}_{\text{ext}} = \frac{d\mathbf{p}_{\text{tot}}}{dt}$$

This equation states that the net external force on a system of particles is equal to the time rate of change of the total linear momentum of the system. Let us see if there is a similar statement that can be made in rotational motion. The total angular momentum of a system of particles about some point is defined as the vector sum of the angular momenta of the individual particles:

$$\mathbf{L}_{\text{tot}} = \mathbf{L}_1 + \mathbf{L}_2 + \cdots + \mathbf{L}_n = \sum_i \mathbf{L}_i$$

where the vector sum is over all n particles in the system.

Let us differentiate this equation with respect to time:

$$\frac{d\mathbf{L}_{\text{tot}}}{dt} = \sum_i \frac{d\mathbf{L}_i}{dt} = \sum_i \boldsymbol{\tau}_i$$

where we have used Equation 11.11 to replace the time rate of change of the angular momentum of each particle with the net torque on the particle.

The torques acting on the particles of the system are those associated with internal forces between particles and those associated with external forces. However, the net torque associated with all internal forces is zero. To understand this, recall that Newton's third law tells us that internal forces between particles of the system are equal in magnitude and opposite in direction. If we assume that these forces lie along the line of separation of each pair of particles, then the total torque around some axis passing through an origin O due to each action–reaction force pair is zero. That is, the moment arm d from O to the line of action of the forces is equal for both particles and the forces are in opposite directions. In the summation, therefore, we see that the net internal torque vanishes. We conclude that the total angular momentum of a system can vary with time only if a net external torque is acting on the system, so that we have

The net external torque on a system equals the time rate of change of angular momentum of the system

$$\sum \boldsymbol{\tau}_{\text{ext}} = \frac{d\mathbf{L}_{\text{tot}}}{dt} \tag{11.13}$$

That is

> the net external torque acting on a system about some axis passing through an origin in an inertial frame equals the time rate of change of the total angular momentum of the system about that origin.

Note that Equation 11.13 is indeed the rotational analog of $\Sigma\mathbf{F}_{\text{ext}} = d\mathbf{p}_{\text{tot}}/dt$, for a system of particles.

Although we do not prove it here, the following statement is an important theorem concerning the angular momentum of a system relative to the system's center of mass:

> The resultant torque acting on a system about an axis through the center of mass equals the time rate of change of angular momentum of the system regardless of the motion of the center of mass.

This theorem applies even if the center of mass is accelerating, provided $\boldsymbol{\tau}$ and $\mathbf{L}$ are evaluated relative to the center of mass.

Example 11.4 Two Connected Objects

A sphere of mass m_1 and a block of mass m_2 are connected by a light cord that passes over a pulley, as shown in Figure 11.6. The radius of the pulley is R, and the mass of the rim is M. The spokes of the pulley have negligible mass. The block slides on a frictionless, horizontal surface. Find an expression for the linear acceleration of the two objects, using the concepts of angular momentum and torque.

Solution We need to determine the angular momentum of the system, which consists of the two objects and the pulley.

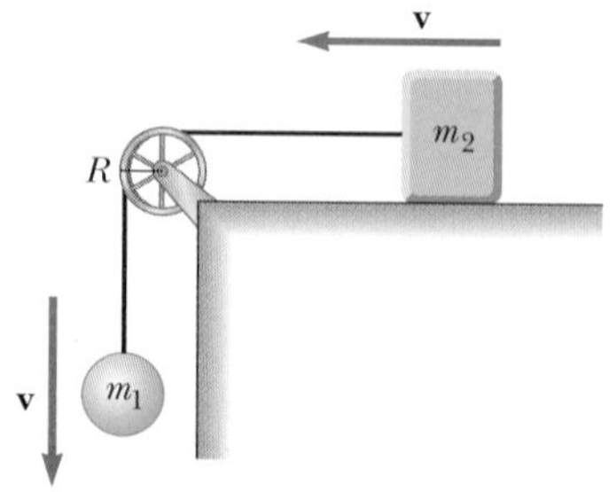

Figure 11.6 (Example 11.4) When the system is released, the sphere moves downward and the block moves to the left.

Let us calculate the angular momentum about an axis that coincides with the axle of the pulley. The angular momentum of the system includes that of two objects moving translationally (the sphere and the block) and one object undergoing pure rotation (the pulley).

At any instant of time, the sphere and the block have a common speed v, so the angular momentum of the sphere is m_1vR, and that of the block is m_2vR. At the same instant, all points on the rim of the pulley also move with speed v, so the angular momentum of the pulley is MvR. Hence, the total angular momentum of the system is

$$\text{(1)} \qquad L = m_1vR + m_2vR + MvR = (m_1 + m_2 + M)vR$$

Now let us evaluate the total external torque acting on the system about the pulley axle. Because it has a moment arm of zero, the force exerted by the axle on the pulley does not contribute to the torque. Furthermore, the normal force acting on the block is balanced by the gravitational force $m_2\mathbf{g}$, and so these forces do not contribute to the torque. The gravitational force $m_1\mathbf{g}$ acting on the sphere produces a torque about the axle equal in magnitude to m_1gR, where R is the moment arm of the force about the axle. This is the total external torque about the pulley axle;

that is, $\sum \tau_{ext} = m_1 gR$. Using this result, together with Equation (1) and Equation 11.13, we find

$$\sum \tau_{ext} = \frac{dL}{dt}$$

$$m_1 gR = \frac{d}{dt}[(m_1 + m_2 + M)vR]$$

$$(2) \qquad m_1 gR = (m_1 + m_2 + M)R\frac{dv}{dt}$$

Because $dv/dt = a$, we can solve this for a to obtain

$$a = \frac{m_1 g}{m_1 + m_2 + M}$$

You may wonder why we did not include the forces that the cord exerts on the objects in evaluating the net torque about the axle. The reason is that these forces are internal to the system under consideration, and we analyzed the system as a whole. Only *external* torques contribute to the change in the system's angular momentum.

11.3 Angular Momentum of a Rotating Rigid Object

In Example 11.4, we considered the angular momentum of a deformable system. Let us now restrict our attention to a nondeformable system—a rigid object. Consider a rigid object rotating about a fixed axis that coincides with the z axis of a coordinate system, as shown in Figure 11.7. Let us determine the angular momentum of this object. Each *particle* of the object rotates in the xy plane about the z axis with an angular speed ω. The magnitude of the angular momentum of a particle of mass m_i about the z axis is $m_i v_i r_i$. Because $v_i = r_i\omega$, we can express the magnitude of the angular momentum of this particle as

$$L_i = m_i r_i^2 \omega$$

The vector $\mathbf{L}_i$ is directed along the z axis, as is the vector $\boldsymbol{\omega}$.

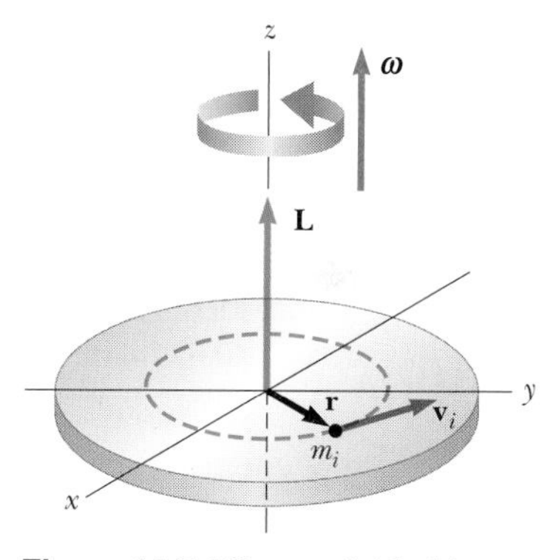

Figure 11.7 When a rigid object rotates about an axis, the angular momentum $\mathbf{L}$ is in the same direction as the angular velocity $\boldsymbol{\omega}$, according to the expression $\mathbf{L} = I\boldsymbol{\omega}$.

We can now find the angular momentum (which in this situation has only a z component) of the whole object by taking the sum of L_i over all particles:

$$L_z = \sum_i L_i = \sum_i m_i r_i^2 \omega = \left(\sum_i m_i r_i^2\right)\omega$$

$$L_z = I\omega \qquad (11.14)$$

where we have recognized $\sum_i m_i r_i^2$ as the moment of inertia I of the object about the z axis (Equation 10.15).

Now let us differentiate Equation 11.14 with respect to time, noting that I is constant for a rigid object:

$$\frac{dL_z}{dt} = I\frac{d\omega}{dt} = I\alpha \qquad (11.15)$$

where α is the angular acceleration relative to the axis of rotation. Because dL_z/dt is equal to the net external torque (see Eq. 11.13), we can express Equation 11.15 as

Rotational form of Newton's second law

$$\sum \tau_{ext} = I\alpha \qquad (11.16)$$

That is, the net external torque acting on a rigid object rotating about a fixed axis equals the moment of inertia about the rotation axis multiplied by the object's angular acceleration relative to that axis. This result is the same as Equation 10.21, which was derived using a force approach, but we derived Equation 11.16 using the concept of angular momentum. This equation is also valid for a rigid object rotating about a moving axis provided the moving axis (1) passes through the center of mass and (2) is a symmetry axis.

If a symmetrical object rotates about a fixed axis passing through its center of mass, you can write Equation 11.14 in vector form as $\mathbf{L} = I\boldsymbol{\omega}$, where $\mathbf{L}$ is the total angular momentum of the object measured with respect to the axis of rotation. Furthermore, the expression is valid for any object, regardless of its symmetry, if **L** stands for the component of angular momentum along the axis of rotation.[1]

Quick Quiz 11.5 A solid sphere and a hollow sphere have the same mass and radius. They are rotating with the same angular speed. The one with the higher angular momentum is (a) the solid sphere (b) the hollow sphere (c) they both have the same angular momentum (d) impossible to determine.

Example 11.5 Bowling Ball

Estimate the magnitude of the angular momentum of a bowling ball spinning at 10 rev/s, as shown in Figure 11.8.

Solution We start by making some estimates of the relevant physical parameters and model the ball as a uniform solid sphere. A typical bowling ball might have a mass of 6.0 kg and a radius of 12 cm. The moment of inertia of a solid sphere about an axis through its center is, from Table 10.2,

$$I = \tfrac{2}{5}MR^2 = \tfrac{2}{5}(6.0\text{ kg})(0.12\text{ m})^2 = 0.035\text{ kg}\cdot\text{m}^2$$

Therefore, the magnitude of the angular momentum is

$$\begin{aligned} L_z = I\omega &= (0.035\text{ kg}\cdot\text{m}^2)(10\text{ rev/s})(2\pi\text{ rad/rev}) \\ &= 2.2\text{ kg}\cdot\text{m}^2/\text{s} \end{aligned}$$

Because of the roughness of our estimates, we probably want to keep only one significant figure, and so $L_z \approx 2\text{ kg}\cdot\text{m}^2/\text{s}$.

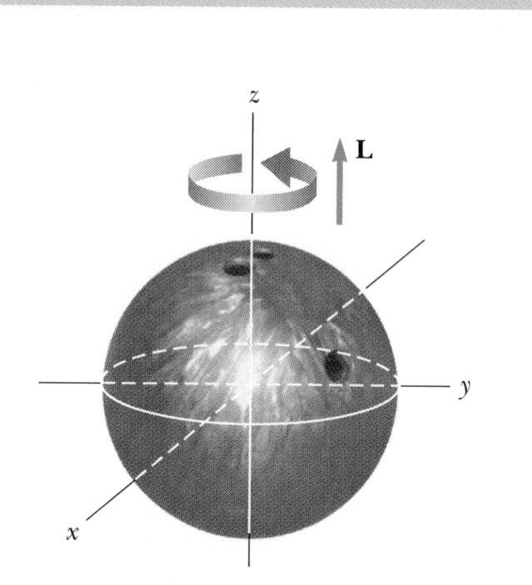

Figure 11.8 (Example 11.5) A bowling ball that rotates about the z axis in the direction shown has an angular momentum **L** in the positive z direction. If the direction of rotation is reversed, **L** points in the negative z direction.

Example 11.6 The Seesaw

Interactive

A father of mass m_f and his daughter of mass m_d sit on opposite ends of a seesaw at equal distances from the pivot at the center (Fig. 11.9). The seesaw is modeled as a rigid rod of mass M and length ℓ and is pivoted without friction. At a given moment, the combination rotates in a vertical plane with an angular speed ω.

(A) Find an expression for the magnitude of the system's angular momentum.

Solution The moment of inertia of the system equals the sum of the moments of inertia of the three components: the seesaw and the two individuals, whom we will model as particles. Referring to Table 10.2 to obtain the expression for the moment of inertia of the rod, and using the expression $I = mr^2$ for each person, we find that the total moment of inertia about the z axis through O is

$$I = \tfrac{1}{12}M\ell^2 + m_f\left(\frac{\ell}{2}\right)^2 + m_d\left(\frac{\ell}{2}\right)^2 = \frac{\ell^2}{4}\left(\frac{M}{3} + m_f + m_d\right)$$

Therefore, the magnitude of the angular momentum is

$$L = I\omega = \frac{\ell^2}{4}\left(\frac{M}{3} + m_f + m_d\right)\omega$$

(B) Find an expression for the magnitude of the angular acceleration of the system when the seesaw makes an angle θ with the horizontal.

Solution To find the angular acceleration of the system at any angle θ, we first calculate the net torque on the system and then use $\Sigma\tau_{\text{ext}} = I\alpha$ to obtain an expression for α.

The torque due to the force $m_f g$ about the pivot is

$$\tau_f = m_f g\frac{\ell}{2}\cos\theta \qquad (\boldsymbol{\tau}_f \text{ out of page})$$

The torque due to the force $m_d g$ about the pivot is

$$\tau_d = -m_d g\frac{\ell}{2}\cos\theta \qquad (\boldsymbol{\tau}_d \text{ into page})$$

[1] In general, the expression $\mathbf{L} = I\boldsymbol{\omega}$ is not always valid. If a rigid object rotates about an *arbitrary* axis, **L** and $\boldsymbol{\omega}$ may point in different directions. In this case, the moment of inertia cannot be treated as a scalar. Strictly speaking, $\mathbf{L} = I\boldsymbol{\omega}$ applies only to rigid objects of any shape that rotate about one of three mutually perpendicular axes (called *principal axes*) through the center of mass. This is discussed in more advanced texts on mechanics.

Hence, the net torque exerted on the system about O is

$$\sum \tau_{\text{ext}} = \tau_f + \tau_d = \tfrac{1}{2}(m_f - m_d)g\ell \cos\theta$$

To find α, we use $\Sigma\tau_{\text{ext}} = I\alpha$, where I was obtained in part (A):

$$\alpha = \frac{\sum \tau_{\text{ext}}}{I} = \frac{2(m_f - m_d)g\cos\theta}{\ell\left(\dfrac{M}{3} + m_f + m_d\right)}$$

Generally, fathers are more massive than daughters, so the angular acceleration is positive. If the seesaw begins in a horizontal orientation ($\theta = 0$) and is released, the rotation will be counterclockwise in Figure 11.9 and the father's end of the seesaw drops. This is consistent with everyday experience.

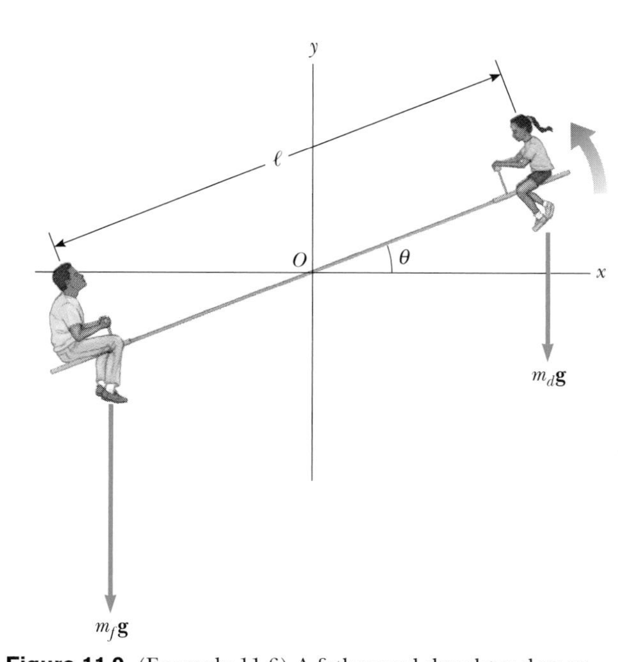

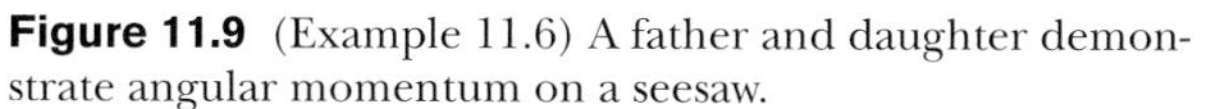

Figure 11.9 (Example 11.6) A father and daughter demonstrate angular momentum on a seesaw.

What If? **After several complaints from the daughter that she simply rises into the air rather than moving up and down as planned, the father moves inward on the seesaw to try to balance the two sides. He moves in to a position that is a distance *d* from the pivot. What is the angular acceleration of the system in this case when it is released from an arbitrary angle θ?**

Answer The angular acceleration of the system should decrease if the system is more balanced. As the father continues to slide inward, he should reach a point at which the seesaw is balanced and there is no angular acceleration of the system when released.

The total moment of inertia about the z axis through O for the modified system is

$$I = \tfrac{1}{12}M\ell^2 + m_f d^2 + m_d\left(\frac{\ell}{2}\right)^2$$

$$= \frac{\ell^2}{4}\left(\frac{M}{3} + m_d\right) + m_f d^2$$

The net torque exerted on the system about O is

$$\tau_{\text{net}} = \tau_f + \tau_d = m_f g d\cos\theta - \tfrac{1}{2}m_d g\ell\cos\theta$$

Now, the angular acceleration of the system is

$$\alpha = \frac{\tau_{\text{net}}}{I} = \frac{m_f g d\cos\theta - \tfrac{1}{2}m_d g\ell\cos\theta}{\dfrac{\ell^2}{4}[(M/3) + m_d] + m_f d^2}$$

The seesaw will be balanced when the angular acceleration is zero. In this situation, both father and daughter can push off the ground and rise to the highest possible point. We find the required position of the father by setting $\alpha = 0$:

$$\alpha = \frac{m_f g d\cos\theta - \tfrac{1}{2}m_d g\ell\cos\theta}{(\ell^2/4)[(M/3) + m_d] + m_f d^2} = 0$$

$$m_f g d\cos\theta - \tfrac{1}{2}m_d g\ell\cos\theta = 0$$

$$d = \left(\frac{m_d}{m_f}\right)\tfrac{1}{2}\ell$$

In the rare case that the father and daughter have the same mass, the father is located at the end of the seesaw, $d = \ell/2$.

At the Interactive Worked Example link at **http://www.pse6.com,** *you can move the father and daughter to see the effect on the motion of the system.*

11.4 Conservation of Angular Momentum

In Chapter 9 we found that the total linear momentum of a system of particles remains constant if the system is isolated, that is, if the resultant external force acting on the system is zero. We have an analogous conservation law in rotational motion:

Conservation of angular momentum

The total angular momentum of a system is constant in both magnitude and direction if the resultant external torque acting on the system is zero, that is, if the system is isolated.

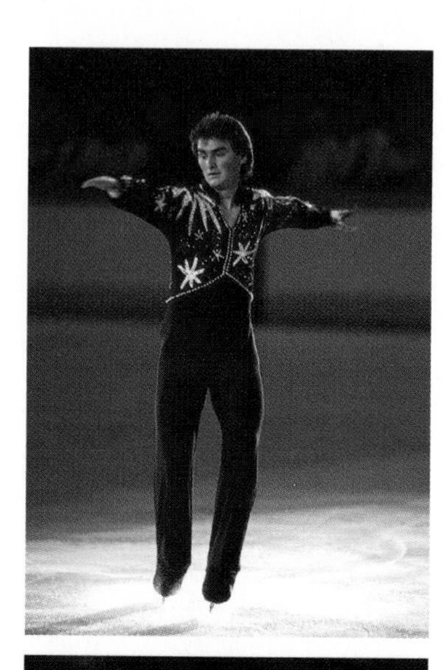

Figure 11.10 Figure skater Todd Eldridge is demonstrating angular momentum conservation. When he pulls his arms toward his body, he spins faster.

This follows directly from Equation 11.13, which indicates that if

$$\sum \boldsymbol{\tau}_{\text{ext}} = \frac{d\mathbf{L}_{\text{tot}}}{dt} = 0 \quad (11.17)$$

then

$$\mathbf{L}_{\text{tot}} = \text{constant} \quad \text{or} \quad \mathbf{L}_i = \mathbf{L}_f \quad (11.18)$$

For an isolated system consisting of a number of particles, we write this conservation law as $\mathbf{L}_{\text{tot}} = \Sigma\mathbf{L}_n = \text{constant}$, where the index n denotes the nth particle in the system.

If the mass of an isolated rotating system undergoes redistribution in some way, the system's moment of inertia changes. Because the magnitude of the angular momentum of the system is $L = I\omega$ (Eq. 11.14), conservation of angular momentum requires that the product of I and ω must remain constant. Thus, a change in I for an isolated system requires a change in ω. In this case, we can express the principle of conservation of angular momentum as

$$I_i\omega_i = I_f\omega_f = \text{constant} \quad (11.19)$$

This expression is valid both for rotation about a fixed axis and for rotation about an axis through the center of mass of a moving system as long as that axis remains fixed in direction. We require only that the net external torque be zero.

There are many examples that demonstrate conservation of angular momentum for a deformable system. You may have observed a figure skater spinning in the finale of a program (Fig. 11.10). The angular speed of the skater increases when the skater pulls his hands and feet close to his body, thereby decreasing I. Neglecting friction between skates and ice, no external torques act on the skater. Because the angular momentum of the skater is conserved, the product $I\omega$ remains constant, and a decrease in the moment of inertia of the skater causes an increase in the angular speed. Similarly, when divers or acrobats wish to make several somersaults, they pull their hands and feet close to their bodies to rotate at a higher rate, as in the opening photograph of this chapter. In these cases, the external force due to gravity acts through the center of mass and hence exerts no torque about this point. Therefore, the angular momentum about the center of mass must be conserved—that is, $I_i\omega_i = I_f\omega_f$. For example, when divers wish to double their angular speed, they must reduce their moment of inertia to half its initial value.

In Equation 11.18 we have a third conservation law to add to our list. We can now state that the energy, linear momentum, and angular momentum of an isolated system all remain constant:

$$\left.\begin{aligned} E_i &= E_f \\ \mathbf{p}_i &= \mathbf{p}_f \\ \mathbf{L}_i &= \mathbf{L}_f \end{aligned}\right\} \quad \text{For an isolated system}$$

Quick Quiz 11.6 A competitive diver leaves the diving board and falls toward the water with her body straight and rotating slowly. She pulls her arms and legs into a tight tuck position. Her angular speed (a) increases (b) decreases (c) stays the same (d) is impossible to determine.

Quick Quiz 11.7 Consider the competitive diver in Quick Quiz 11.6 again. When she goes into the tuck position, the rotational kinetic energy of her body (a) increases (b) decreases (c) stays the same (d) is impossible to determine.

Example 11.7 Formation of a Neutron Star

A star rotates with a period of 30 days about an axis through its center. After the star undergoes a supernova explosion, the stellar core, which had a radius of 1.0×10^4 km, collapses into a neutron star of radius 3.0 km. Determine the period of rotation of the neutron star.

Solution The same physics that makes a skater spin faster with his arms pulled in describes the motion of the neutron star. Let us assume that during the collapse of the stellar core, (1) no external torque acts on it, (2) it remains spherical with the same relative mass distribution, and (3) its mass remains constant. Also, let us use the symbol T for the period, with T_i being the initial period of the star and T_f being the period of the neutron star. The period is the length of time a point on the star's equator takes to make one complete circle around the axis of rotation. The angular speed of the star is given by $\omega = 2\pi/T$. Therefore, Equation 11.19 gives

$$I_i\omega_i = I_f\omega_f$$

$$I_i\left(\frac{2\pi}{T_i}\right) = I_f\left(\frac{2\pi}{T_f}\right)$$

We don't know the mass distribution of the star, but we have assumed that the distribution is symmetric, so that the moment of inertia can be expressed as kMR^2, where k is some numerical constant. (From Table 10.2, for example, we see that $k = 2/5$ for a solid sphere and $k = 2/3$ for a spherical shell.) Thus, we can rewrite the preceding equation as

$$kMR_i^2\left(\frac{2\pi}{T_i}\right) = kMR_f^2\left(\frac{2\pi}{T_f}\right)$$

$$T_f = \left(\frac{R_f^2}{R_i^2}\right)T_i$$

Substituting numerical values gives

$$T_f = (30 \text{ days})\left(\frac{3.0 \text{ km}}{1.0 \times 10^4 \text{ km}}\right)^2 = 2.7 \times 10^{-6} \text{ days}$$

$$= 0.23 \text{ s}$$

Thus, the neutron star rotates about four times each second.

Example 11.8 The Merry-Go-Round

A horizontal platform in the shape of a circular disk rotates freely in a horizontal plane about a frictionless vertical axle (Fig. 11.11). The platform has a mass $M = 100$ kg and a radius $R = 2.0$ m. A student whose mass is $m = 60$ kg walks slowly from the rim of the disk toward its center. If the angular speed of the system is 2.0 rad/s when the student is at the rim, what is the angular speed when he reaches a point $r = 0.50$ m from the center?

Solution The speed change here is similar to the increase in angular speed of the spinning skater when he pulls his arms inward. Let us denote the moment of inertia of the platform as I_p and that of the student as I_s. Modeling the student as a particle, we can write the initial moment of inertia I_i of the system (student plus platform) about the axis of rotation:

$$I_i = I_{pi} + I_{si} = \tfrac{1}{2}MR^2 + mR^2$$

When the student walks to the position $r < R$, the moment of inertia of the system reduces to

$$I_f = I_{pf} + I_{sf} = \tfrac{1}{2}MR^2 + mr^2$$

Note that we still use the greater radius R when calculating I_{pf} because the radius of the platform does not change. Because no external torques act on the system about the axis of rotation, we can apply the law of conservation of angular momentum:

$$I_i\omega_i = I_f\omega_f$$

$$(\tfrac{1}{2}MR^2 + mR^2)\omega_i = (\tfrac{1}{2}MR^2 + mr^2)\omega_f$$

$$\omega_f = \left(\frac{\frac{1}{2}MR^2 + mR^2}{\frac{1}{2}MR^2 + mr^2}\right)\omega_i$$

$$= \left(\frac{\frac{1}{2}(100 \text{ kg})(2.0 \text{ m})^2 + (60 \text{ kg})(2.0 \text{ m})^2}{\frac{1}{2}(100 \text{ kg})(2.0 \text{ m})^2 + (60 \text{ kg})(0.50 \text{ m})^2}\right)(2.0 \text{ rad/s})$$

$$= \left(\frac{440 \text{ kg}\cdot\text{m}^2}{215 \text{ kg}\cdot\text{m}^2}\right)(2.0 \text{ rad/s}) = 4.1 \text{ rad/s}$$

As expected, the angular speed increases.

What If? **What if we were to measure the kinetic energy of the system before and after the student walks inward? Are they the same?**

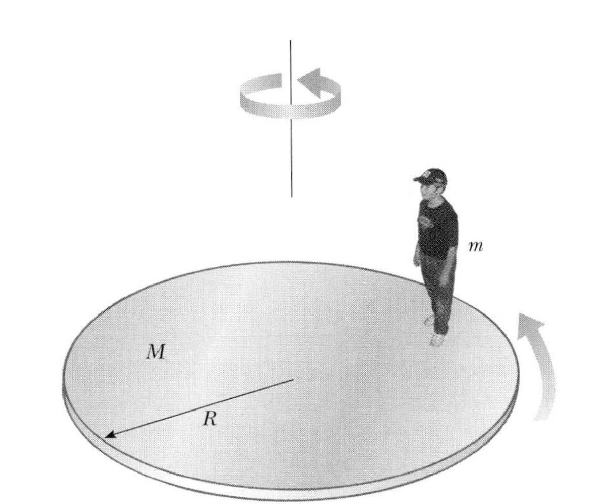

Figure 11.11 (Example 11.8) As the student walks toward the center of the rotating platform, the angular speed of the system increases because the angular momentum of the system remains constant.

Answer You may be tempted to say yes because the system is isolated. But remember that energy comes in several forms, so we have to handle an energy question carefully. The initial kinetic energy is

$$K_i = \frac{1}{2}I_i\omega_i^2 = \frac{1}{2}(440 \text{ kg}\cdot\text{m}^2)(2.0 \text{ rad/s})^2 = 880 \text{ J}$$

The final kinetic energy is

$$K_f = \frac{1}{2}I_f\omega_f^2 = \frac{1}{2}(215 \text{ kg}\cdot\text{m}^2)(4.1 \text{ rad/s})^2 = 1.81 \times 10^3 \text{ J}$$

Thus, the kinetic energy of the system *increases*. The student must do work to move himself closer to the center of rotation, so this extra kinetic energy comes from chemical potential energy in the body of the student.

Example 11.9 The Spinning Bicycle Wheel

In a favorite classroom demonstration, a student holds the axle of a spinning bicycle wheel while seated on a stool that is free to rotate (Fig. 11.12). The student and stool are initially at rest while the wheel is spinning in a horizontal plane with an initial angular momentum $\mathbf{L}_i$ that points upward. When the wheel is inverted about its center by 180°, the student and stool start rotating. In terms of $\mathbf{L}_i$, what are the magnitude and the direction of $\mathbf{L}$ for the student plus stool?

Figure 11.12 (Example 11.9) The wheel is initially spinning when the student is at rest. What happens when the wheel is inverted?

Solution The system consists of the student, the wheel, and the stool. Initially, the total angular momentum of the system $\mathbf{L}_i$ comes entirely from the spinning wheel. As the wheel is inverted, the student applies a torque to the wheel, but this torque is internal to the system. No external torque is acting on the system about the vertical axis. Therefore, the angular momentum of the system is conserved. Initially, we have

$$\mathbf{L}_{\text{system}} = \mathbf{L}_i = \mathbf{L}_{\text{wheel}} \quad \text{(upward)}$$

After the wheel is inverted, we have $\mathbf{L}_{\text{inverted wheel}} = -\mathbf{L}_i$. For angular momentum to be conserved, some other part of the system has to start rotating so that the total final angular momentum equals the initial angular momentum $\mathbf{L}_i$. That other part of the system is the student plus the stool she is sitting on. So, we can now state that

$$\mathbf{L}_f = \mathbf{L}_i = \mathbf{L}_{\text{student+stool}} - \mathbf{L}_i$$

$$\mathbf{L}_{\text{student+stool}} = 2\mathbf{L}_i$$

Example 11.10 Disk and Stick

Interactive

A 2.0-kg disk traveling at 3.0 m/s strikes a 1.0-kg stick of length 4.0 m that is lying flat on nearly frictionless ice, as shown in Figure 11.13. Assume that the collision is elastic and that the disk does not deviate from its original line of motion. Find the translational speed of the disk, the translational speed of the stick, and the angular speed of the stick after the collision. The moment of inertia of the stick about its center of mass is 1.33 kg · m^2.

Solution Conceptualize the situation by considering Figure 11.13 and imagining what happens after the disk hits the stick. Because the disk and stick form an isolated system, we can assume that total energy, linear momentum, and angular momentum are all conserved. Thus, we can categorize this as a problem in which all three conservation laws might play a part. To analyze the problem, first note that we have three unknowns, and so we need three equations to solve simultaneously. The first comes from the law of the conservation of linear momentum:

$$p_i = p_f$$

$$m_d v_{di} = m_d v_{df} + m_s v_s$$

$$(2.0 \text{ kg})(3.0 \text{ m/s}) = (2.0 \text{ kg})v_{df} + (1.0 \text{ kg})v_s$$

$$(1) \qquad 6.0 \text{ kg}\cdot\text{m/s} - (2.0 \text{ kg})v_{df} = (1.0 \text{ kg})v_s$$

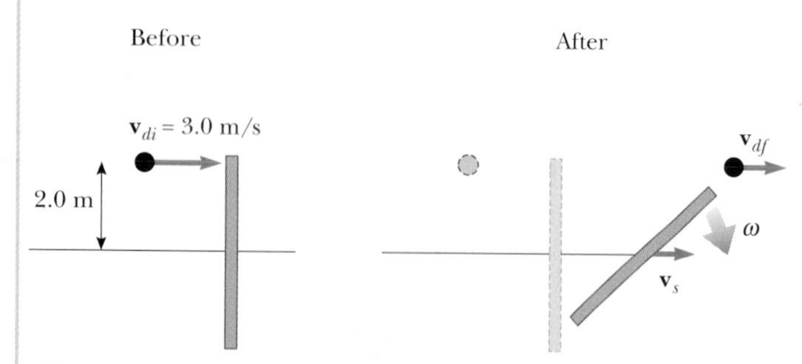

Figure 11.13 (Example 11.10) Overhead view of a disk striking a stick in an elastic collision, which causes the stick to rotate and move to the right.

Now we apply the law of conservation of angular momentum, using the initial position of the center of the stick as our reference point. We know that the component of angular momentum of the disk along the axis perpendicular to the plane of the ice is negative. (The right-hand rule shows that $\mathbf{L}_d$ points into the ice.) Applying conservation of angular momentum to the system gives

$$L_i = L_f$$

$$-rm_d v_{di} = -rm_d v_{df} + I\omega$$

$$-(2.0\text{ m})(2.0\text{ kg})(3.0\text{ m/s}) = -(2.0\text{ m})(2.0\text{ kg})v_{df} + (1.33\text{ kg}\cdot\text{m}^2)\omega$$

$$-12\text{ kg}\cdot\text{m}^2/\text{s} = -(4.0\text{ kg}\cdot\text{m})v_{df} + (1.33\text{ kg}\cdot\text{m}^2)\omega$$

$$(2)\quad -9.0\text{ rad/s} + (3.0\text{ rad/m})v_{df} = \omega$$

We use the fact that radians are dimensionless to ensure consistent units for each term.

Finally, the elastic nature of the collision tells us that kinetic energy is conserved; in this case, the kinetic energy consists of translational and rotational forms:

$$K_i = K_f$$

$$\tfrac{1}{2}m_d v_{di}^2 = \tfrac{1}{2}m_d v_{df}^2 + \tfrac{1}{2}m_s v_s^2 + \tfrac{1}{2}I\omega^2$$

$$\tfrac{1}{2}(2.0\text{ kg})(3.0\text{ m/s})^2 = \tfrac{1}{2}(2.0\text{ kg})v_{df}^2 + \tfrac{1}{2}(1.0\text{ kg})v_s^2 + \tfrac{1}{2}(1.33\text{ kg}\cdot\text{m}^2)\omega^2$$

$$(3)\quad 18\text{ m}^2/\text{s}^2 = 2.0\,v_{df}^2 + v_s^2 + (1.33\text{ m}^2)\omega^2$$

In solving Equations (1), (2), and (3) simultaneously, we find that $v_{df} = 2.3\text{ m/s}$, $v_s = 1.3\text{ m/s}$, and $\omega = -2.0\text{ rad/s}$.

To finalize the problem, note that these values seem reasonable. The disk is moving more slowly after the collision than it was before the collision, and the stick has a small translational speed. Table 11.1 summarizes the initial and final values of variables for the disk and the stick, and verifies the conservation of linear momentum, angular momentum, and kinetic energy.

What If? **What if the collision between the disk and the stick is perfectly inelastic? How does this change the analysis?**

Answer In this case, the disk adheres to the end of the stick upon collision. The conservation of linear momentum principle leading to Equation (1) would be altered to

$$p_i = p_f$$

$$m_d v_{di} = (m_d + m_s)v_{\text{CM}}$$

$$(2.0\text{ kg})(3.0\text{ m/s}) = (2.0\text{ kg} + 1.0\text{ kg})v_{\text{CM}}$$

$$v_{\text{CM}} = 2.0\text{ m/s}$$

For the rotational part of this question, we need to find the center of mass of the system of the disk and the stick. Choosing the center of the stick as the origin, the y position of the center of mass along the vertical stick is

$$y_{\text{CM}} = \frac{(2.0\text{ kg})(2.0\text{ m}) + (1.0\text{ kg})(0)}{(2.0\text{ kg} + 1.0\text{ kg})} = 1.33\text{ m}$$

Thus, the center of mass of the system is $2.0\text{ m} - 1.33\text{ m} = 0.67\text{ m}$ from the upper end of the stick.

The conservation of angular momentum principle leading to Equation (2) would be altered to the following, evaluating angular momenta around the center of mass of the system:

$$L_i = L_f$$

$$-rm_d v_{di} = I_d\omega + I_s\omega$$

$$(4)\quad -(0.67\text{ m})m_d v_{di} = [m_d(0.67\text{ m})^2]\omega + I_s\omega$$

The moment of inertia of the stick around the center of mass *of the system* is found from the parallel-axis theorem:

$$I_s = I_{\text{CM}} + MD^2 = 1.33\text{ kg}\cdot\text{m}^2 + (1.0\text{ kg})(1.33\text{ m})^2 = 3.1\text{ kg}\cdot\text{m}^2$$

Thus, Equation (4) becomes

$$-(0.67\text{ m})(2.0\text{ kg})(3.0\text{ m/s}) = [(2.0\text{ kg})(0.67\text{ m})^2]\omega + (3.1\text{ kg}\cdot\text{m}^2)\omega$$

$$-4.0\text{ kg}\cdot\text{m}^2/\text{s} = (4.0\text{ kg}\cdot\text{m}^2)\omega$$

$$\omega = \frac{-4.0\text{ kg}\cdot\text{m}^2/\text{s}}{4.0\text{ kg}\cdot\text{m}^2} = -1.0\text{ rad/s}$$

Evaluating the total kinetic energy of the system after the collision shows that it is less than that before the collision because kinetic energy is not conserved in an inelastic collision.

Table 11.1

Comparison of Values in Example 11.10 Before and After the Collision[a]

	v (m/s)	ω (rad/s)	p (kg·m/s)	L (kg·m²/s)	K_{trans} (J)	K_{rot} (J)
Before						
Disk	3.0	—	6.0	−12	9.0	—
Stick	0	0	0	0	0	0
Total for System	—	—	6.0	−12	9.0	0
After						
Disk	2.3	—	4.7	−9.3	5.4	—
Stick	1.3	−2.0	1.3	−2.7	0.9	2.7
Total for System	—	—	6.0	−12	6.3	2.7

[a] Notice that linear momentum, angular momentum, and total kinetic energy are conserved.

At the Interactive Worked Example link at **http://www.pse6.com,** *you can adjust the speed and position of the disk and observe the collision.*

11.5 The Motion of Gyroscopes and Tops

A very unusual and fascinating type of motion you probably have observed is that of a top spinning about its axis of symmetry, as shown in Figure 11.14a. If the top spins very rapidly, the symmetry axis rotates about the z axis, sweeping out a cone (see Fig. 11.14b). The motion of the symmetry axis about the vertical—known as **precessional motion**—is usually slow relative to the spinning motion of the top.

Precessional motion

It is quite natural to wonder why the top does not fall over. Because the center of mass is not directly above the pivot point O, a net torque is clearly acting on the top about O—a torque resulting from the gravitational force $M\mathbf{g}$. The top would certainly fall over if it were not spinning. Because it is spinning, however, it has an angular momentum $\mathbf{L}$ directed along its symmetry axis. We shall show that this symmetry axis moves about the z axis (precessional motion occurs) because the torque produces a change in the *direction* of the symmetry axis. This is an excellent example of the importance of the directional nature of angular momentum.

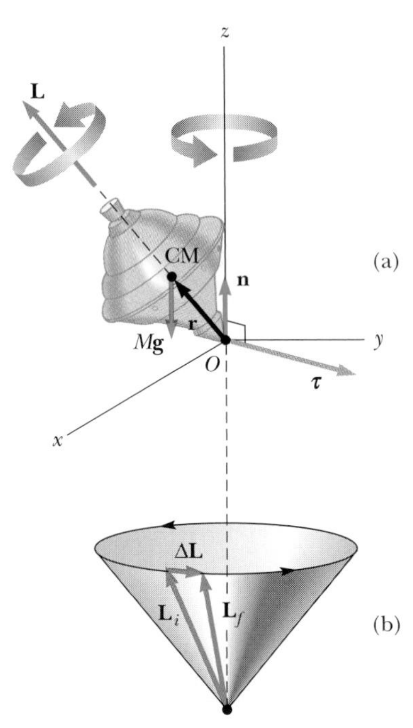

Figure 11.14 Precessional motion of a top spinning about its symmetry axis. (a) The only external forces acting on the top are the normal force $\mathbf{n}$ and the gravitational force $M\mathbf{g}$. The direction of the angular momentum $\mathbf{L}$ is along the axis of symmetry. The right-hand rule indicates that $\boldsymbol{\tau} = \mathbf{r} \times \mathbf{F} = \mathbf{r} \times M\mathbf{g}$ is in the xy plane. (b). The direction of $\Delta\mathbf{L}$ is parallel to that of $\boldsymbol{\tau}$ in part (a). The fact that $\mathbf{L}_f = \Delta\mathbf{L} + \mathbf{L}_i$ indicates that the top precesses about the z axis.

The essential features of precessional motion can be illustrated by considering the simple gyroscope shown in Figure 11.15a. The two forces acting on the top are the downward gravitational force $M\mathbf{g}$ and the normal force $\mathbf{n}$ acting upward at the pivot point O. The normal force produces no torque about the pivot because its moment arm through that point is zero. However, the gravitational force produces a torque $\boldsymbol{\tau} = \mathbf{r} \times M\mathbf{g}$ about O, where the direction of $\boldsymbol{\tau}$ is perpendicular to the plane formed by $\mathbf{r}$ and $M\mathbf{g}$. By necessity, the vector $\boldsymbol{\tau}$ lies in a horizontal xy plane perpendicular to the angular momentum vector. The net torque and angular momentum of the gyroscope are related through Equation 11.13:

$$\boldsymbol{\tau} = \frac{d\mathbf{L}}{dt}$$

From this expression, we see that the nonzero torque produces a change in angular momentum $d\mathbf{L}$—a change that is in the same direction as $\boldsymbol{\tau}$. Therefore, like the torque vector, $d\mathbf{L}$ must also be perpendicular to $\mathbf{L}$. Figure 11.15b illustrates the resulting precessional motion of the symmetry axis of the gyroscope. In a time interval dt, the change in angular momentum is $d\mathbf{L} = \mathbf{L}_f - \mathbf{L}_i = \boldsymbol{\tau}\,dt$. Because $d\mathbf{L}$ is perpendicular to $\mathbf{L}$, the magnitude of $\mathbf{L}$ does not change ($|\mathbf{L}_i| = |\mathbf{L}_f|$). Rather, what is changing is the *direction* of $\mathbf{L}$. Because the change in angular momentum $d\mathbf{L}$ is in the direction of $\boldsymbol{\tau}$, which lies in the xy plane, the gyroscope undergoes precessional motion.

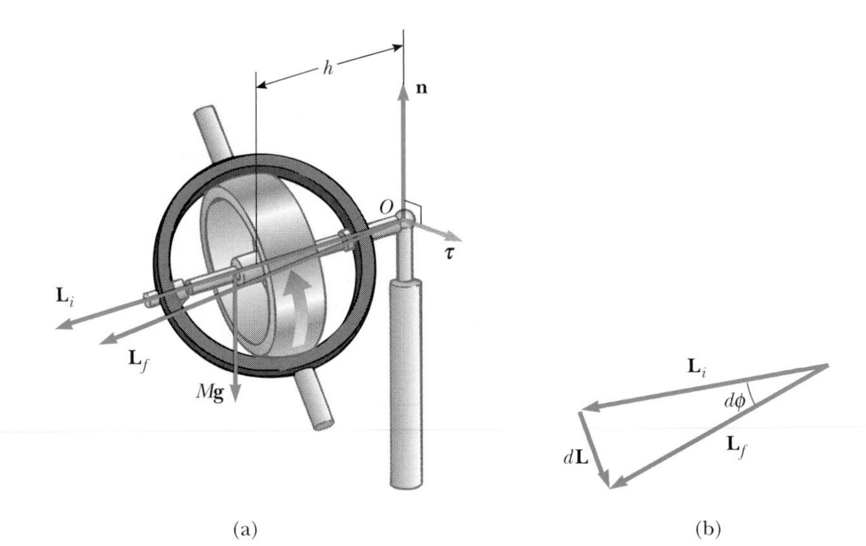

Figure 11.15 (a) The motion of a simple gyroscope pivoted a distance h from its center of mass. The gravitational force $M\mathbf{g}$ produces a torque about the pivot, and this torque is perpendicular to the axle. (b) Overhead view of the initial and final angular momentum vectors. The torque results in a change in angular momentum $d\mathbf{L}$ in a direction perpendicular to the axle. The axle sweeps out an angle $d\phi$ in a time interval dt.

To simplify the description of the system, we must make an assumption: The total angular momentum of the precessing wheel is the sum of the angular momentum $I\boldsymbol{\omega}$ due to the spinning and the angular momentum due to the motion of the center of mass about the pivot. In our treatment, we shall neglect the contribution from the center-of-mass motion and take the total angular momentum to be just $I\boldsymbol{\omega}$. In practice, this is a good approximation if $\boldsymbol{\omega}$ is made very large.

The vector diagram in Figure 11.15b shows that in the time interval dt, the angular momentum vector rotates through an angle $d\phi$, which is also the angle through which the axle rotates. From the vector triangle formed by the vectors $\mathbf{L}_i$, $\mathbf{L}_f$, and $d\mathbf{L}$, we see that

$$\sin(d\phi) \approx d\phi = \frac{dL}{L} = \frac{\tau\, dt}{L} = \frac{(Mgh)\, dt}{L}$$

where we have used the fact that, for small values of any angle θ, $\sin\theta \approx \theta$. Dividing through by dt and using the relationship $L = I\omega$, we find that the rate at which the axle rotates about the vertical axis is

$$\omega_p = \frac{d\phi}{dt} = \frac{Mgh}{I\omega} \tag{11.20}$$

The angular speed ω_p is called the **precessional frequency.** This result is valid only when $\omega_p \ll \omega$. Otherwise, a much more complicated motion is involved. As you can see from Equation 11.20, the condition $\omega_p \ll \omega$ is met when ω is large, that is, the wheel spins rapidly. Furthermore, note that the precessional frequency decreases as ω increases—that is, as the wheel spins faster about its axis of symmetry.

As an example of the usefulness of gyroscopes, suppose you are in a spacecraft in deep space and you need to alter your trajectory. You need to turn the spacecraft around in order to fire the engines in the correct direction. But how do you turn a spacecraft around in empty space? One way is to have small rocket engines that fire perpendicularly out the side of the spacecraft, providing a torque around its center of mass. This is desirable, and many spacecraft have such rockets.

Let us consider another method, however, that is related to angular momentum and does not require the consumption of rocket fuel. Suppose that the spacecraft carries a gyroscope that is not rotating, as in Figure 11.16a. In this case, the angular momentum of the spacecraft about its center of mass is zero. Suppose the gyroscope is set into rotation, giving the gyroscope a nonzero angular momentum. There is no external torque on the isolated system (spacecraft + gyroscope), so the angular momentum of this system must remain zero according to the principle of conservation of angular momentum. This principle can be satisfied if the spacecraft rotates in the direction opposite to that of the gyroscope, so that the angular momentum vectors of the gyroscope and the spacecraft cancel, resulting in no angular momentum of the system. The result of rotating the gyroscope, as in Figure 11.16b, is that the spacecraft turns around! By including three gyroscopes with mutually perpendicular axles, any desired rotation in space can be achieved.

This effect created an undesirable situation with the *Voyager 2* spacecraft during its flight. The spacecraft carried a tape recorder whose reels rotated at high speeds. Each time the tape recorder was turned on, the reels acted as gyroscopes, and the spacecraft started an undesirable rotation in the opposite direction. This had to be counteracted by Mission Control by using the sideward-firing jets to stop the rotation!

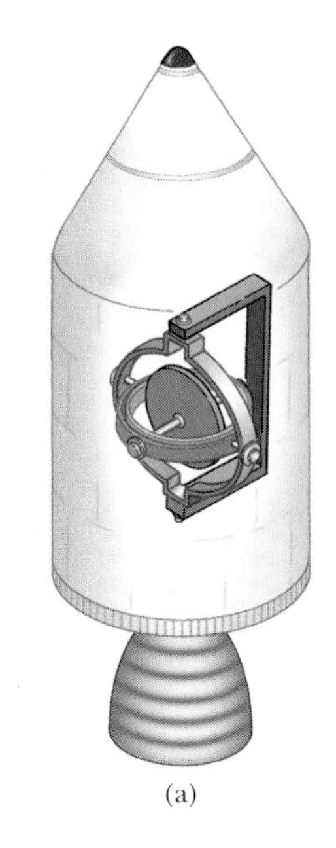

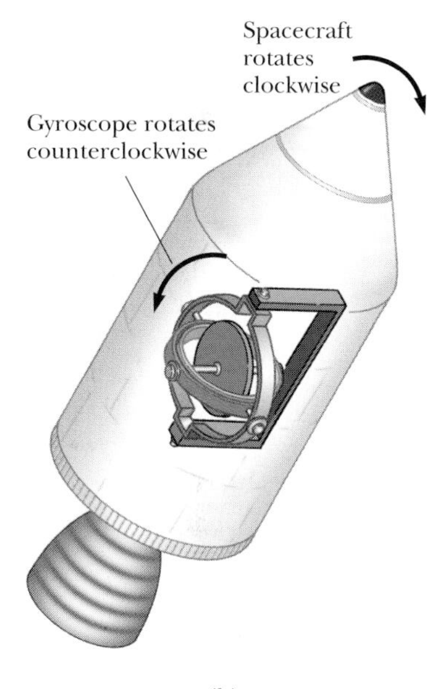

Figure 11.16 (a) A spacecraft carries a gyroscope that is not spinning. (b) When the gyroscope is set into rotation, the spacecraft turns the other way so that the angular momentum of the system is conserved.

11.6 Angular Momentum as a Fundamental Quantity

We have seen that the concept of angular momentum is very useful for describing the motion of macroscopic systems. However, the concept also is valid on a submicroscopic scale and has been used extensively in the development of modern theories of atomic, molecular, and nuclear physics. In these developments, it has been found that the angular momentum of a system is a fundamental quantity. The word *fundamental* in this context implies that angular momentum is an intrinsic property of atoms, molecules, and their constituents, a property that is a part of their very nature.

To explain the results of a variety of experiments on atomic and molecular systems, we rely on the fact that the angular momentum has discrete values. These discrete values are multiples of the fundamental unit of angular momentum $\hbar = h/2\pi$, where h is called Planck's constant:

$$\text{Fundamental unit of angular momentum} = \hbar = 1.054 \times 10^{-34}\ \text{kg} \cdot \text{m}^2/\text{s}$$

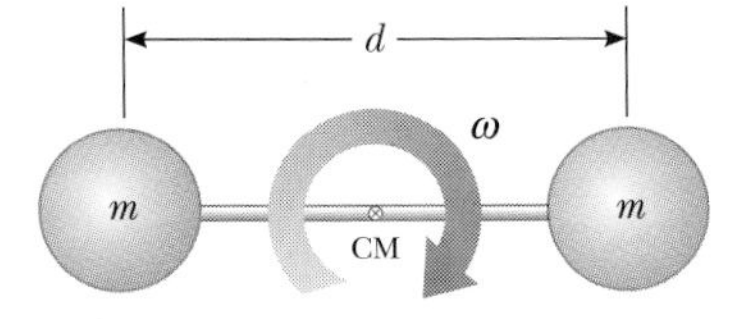

Figure 11.17 The rigid-rotor model of a diatomic molecule. The rotation occurs about the center of mass in the plane of the page.

Let us accept this postulate without proof for the time being and show how it can be used to estimate the angular speed of a diatomic molecule. Consider the O_2 molecule as a rigid rotor, that is, two atoms separated by a fixed distance d and rotating about the center of mass (Fig. 11.17). Equating the angular momentum to the fundamental unit $\hbar$, we can find the order of magnitude of the lowest angular speed:

$$I_{\text{CM}}\omega \approx \hbar \qquad \text{or} \qquad \omega \approx \frac{\hbar}{I_{\text{CM}}}$$

In Example 10.3, we found that the moment of inertia of the O_2 molecule about this axis of rotation is $1.95 \times 10^{-46}\ \text{kg} \cdot \text{m}^2$. Therefore,

$$\omega \approx \frac{\hbar}{I_{\text{CM}}} = \frac{1.054 \times 10^{-34}\ \text{kg} \cdot \text{m}^2/\text{s}}{1.95 \times 10^{-46}\ \text{kg} \cdot \text{m}^2} \sim 10^{12}\ \text{rad/s}$$

Actual angular speeds are found to be multiples of a number with this order of magnitude.

This simple example shows that certain classical concepts and models, when properly modified, are useful in describing some features of atomic and molecular systems. A wide variety of phenomena on the submicroscopic scale can be explained only if we assume discrete values of the angular momentum associated with a particular type of motion.

The Danish physicist Niels Bohr (1885–1962) accepted and adopted this radical idea of discrete angular momentum values in developing his theory of the hydrogen atom. Strictly classical models were unsuccessful in describing many of the hydrogen atom's properties. Bohr postulated that the electron could occupy only those circular orbits about the proton for which the orbital angular momentum was equal to $n\hbar$, where n is an integer. That is, he made the bold claim that orbital angular momentum is quantized. One can use this simple model to estimate the rotational frequencies of the electron in the various orbits (see Problem 42).

SUMMARY

Take a practice test for this chapter by clicking on the Practice Test link at http://www.pse6.com.

The **torque** $\boldsymbol{\tau}$ due to a force $\mathbf{F}$ about an origin in an inertial frame is defined to be

$$\boldsymbol{\tau} \equiv \mathbf{r} \times \mathbf{F} \tag{11.1}$$

Given two vectors $\mathbf{A}$ and $\mathbf{B}$, the **cross product $\mathbf{A} \times \mathbf{B}$** is a vector $\mathbf{C}$ having a magnitude

$$C \equiv AB \sin \theta \tag{11.3}$$

where θ is the angle between $\mathbf{A}$ and $\mathbf{B}$. The direction of the vector $\mathbf{C} = \mathbf{A} \times \mathbf{B}$ is perpendicular to the plane formed by $\mathbf{A}$ and $\mathbf{B}$, and this direction is determined by the right-hand rule.

The **angular momentum $\mathbf{L}$** of a particle having linear momentum $\mathbf{p} = m\mathbf{v}$ is

$$\mathbf{L} \equiv \mathbf{r} \times \mathbf{p} \tag{11.10}$$

where $\mathbf{r}$ is the vector position of the particle relative to an origin in an inertial frame.

The **net external torque** acting on a system is equal to the time rate of change of its angular momentum:

$$\sum \boldsymbol{\tau}_{\text{ext}} = \frac{d\mathbf{L}_{\text{tot}}}{dt} \tag{11.13}$$

The z component of **angular momentum** of a rigid object rotating about a fixed z axis is

$$L_z = I\omega \tag{11.14}$$

where I is the moment of inertia of the object about the axis of rotation and ω is its angular speed.

The **net external torque** acting on a rigid object equals the product of its moment of inertia about the axis of rotation and its angular acceleration:

$$\sum \tau_{\text{ext}} = I\alpha \tag{11.16}$$

If the net external torque acting on a system is zero, then the total angular momentum of the system is constant:

$$\mathbf{L}_i = \mathbf{L}_f \tag{11.18}$$

Applying this **law of conservation of angular momentum** to a system whose moment of inertia changes gives

$$I_i\omega_i = I_f\omega_f = \text{constant} \tag{11.19}$$

QUESTIONS

1. Is it possible to calculate the torque acting on a rigid object without specifying an axis of rotation? Is the torque independent of the location of the axis of rotation?

2. Is the triple product defined by $\mathbf{A} \cdot (\mathbf{B} \times \mathbf{C})$ a scalar or a vector quantity? Explain why the operation $(\mathbf{A} \cdot \mathbf{B}) \times \mathbf{C}$ has no meaning.

3. Vector $\mathbf{A}$ is in the negative y direction, and vector $\mathbf{B}$ is in the negative x direction. What are the directions of (a) $\mathbf{A} \times \mathbf{B}$ (b) $\mathbf{B} \times \mathbf{A}$?

4. If a single force acts on an object and the torque caused by the force is nonzero about some point, is there any other point about which the torque is zero?

5. Suppose that the vector velocity of a particle is completely specified. What can you conclude about the direction of its angular momentum vector with respect to the direction of motion?

6. If a system of particles is in motion, is it possible for the total angular momentum to be zero about some origin? Explain.

7. If the torque acting on a particle about a certain origin is zero, what can you say about its angular momentum about that origin?

8. A ball is thrown in such a way that it does not spin about its own axis. Does this mean that the angular momentum is zero about an arbitrary origin? Explain.

9. For a helicopter to be stable as it flies, it must have at least two propellers. Why?

10. A particle is moving in a circle with constant speed. Locate one point about which the particle's angular momentum is constant and another point about which it changes in time.

11. Why does a long pole help a tightrope walker stay balanced?

12. Often when a high diver wants to turn a flip in midair, she draws her legs up against her chest. Why does this make her rotate faster? What should she do when she wants to come out of her flip?

13. In some motorcycle races, the riders drive over small hills, and the motorcycle becomes airborne for a short time. If the motorcycle racer keeps the throttle open while leaving the hill and going into the air, the motorcycle tends to nose upward. Why does this happen?

14. Stars originate as large bodies of slowly rotating gas. Because of gravitation, these clumps of gas slowly decrease in size. What happens to the angular speed of a star as it shrinks? Explain.

15. If global warming occurs over the next century, it is likely that some polar ice will melt and the water will be distributed closer to the Equator. How would this change the moment of inertia of the Earth? Would the length of the day (one revolution) increase or decrease?

16. A mouse is initially at rest on a horizontal turntable mounted on a frictionless vertical axle. If the mouse

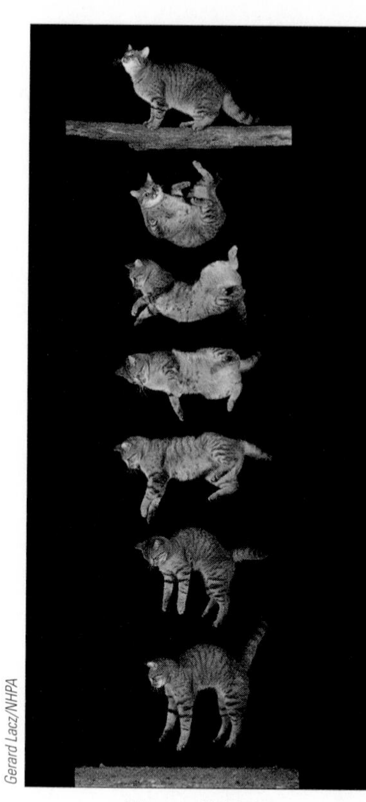
Gerard Lacz/NHPA

Figure Q11.17

begins to walk clockwise around the perimeter, what happens to the turntable? Explain.

17. A cat usually lands on its feet regardless of the position from which it is dropped. A slow-motion film of a cat falling shows that the upper half of its body twists in one direction while the lower half twists in the opposite direction. (See Figure Q11.17.) Why does this type of rotation occur?

18. As the cord holding a tether ball winds around a thin pole, what happens to the angular speed of the ball? Explain.

19. If you toss a textbook into the air, rotating it each time about one of the three axes perpendicular to the textbook, you will find that it will not rotate smoothly about one of these axis. (Try placing a strong rubber band around the book before the toss so it will stay closed.) Its rotation is stable about those axes having the largest and smallest moment of inertia but unstable about the axis of intermediate moment. Try this on your own to find the axis that has this intermediate moment.

20. A scientist arriving at a hotel asks a bellhop to carry a heavy suitcase. When the bellhop rounds a corner, the suitcase suddenly swings away from him for some unknown reason. The alarmed bellhop drops the suitcase and runs away. What might be in the suitcase?

PROBLEMS

1, 2, 3 = straightforward, intermediate, challenging □ = full solution available in the *Student Solutions Manual and Study Guide*

= coached solution with hints available at http://www.pse6.com = computer useful in solving problem

= paired numerical and symbolic problems

Section 11.1 The Vector Product and Torque

1. Given $\mathbf{M} = 6\hat{\mathbf{i}} + 2\hat{\mathbf{j}} - \hat{\mathbf{k}}$ and $\mathbf{N} = 2\hat{\mathbf{i}} - \hat{\mathbf{j}} - 3\hat{\mathbf{k}}$, calculate the vector product $\mathbf{M} \times \mathbf{N}$.

2. The vectors 42.0 cm at 15.0° and 23.0 cm at 65.0° both start from the origin. Both angles are measured counterclockwise from the x axis. The vectors form two sides of a parallelogram. (a) Find the area of the parallelogram. (b) Find the length of its longer diagonal.

3. Two vectors are given by $\mathbf{A} = -3\hat{\mathbf{i}} + 4\hat{\mathbf{j}}$ and $\mathbf{B} = 2\hat{\mathbf{i}} + 3\hat{\mathbf{j}}$. Find (a) $\mathbf{A} \times \mathbf{B}$ and (b) the angle between $\mathbf{A}$ and $\mathbf{B}$.

4. Two vectors are given by $\mathbf{A} = -3\hat{\mathbf{i}} + 7\hat{\mathbf{j}} - 4\hat{\mathbf{k}}$ and $\mathbf{B} = 6\hat{\mathbf{i}} - 10\hat{\mathbf{j}} + 9\hat{\mathbf{k}}$. Evaluate the quantities (a) $\cos^{-1}[\mathbf{A}\cdot\mathbf{B}/AB]$ and (b) $\sin^{-1}[|\mathbf{A} \times \mathbf{B}|/AB]$. (c) Which give(s) the angle between the vectors?

5. The wind exerts on a flower the force 0.785 N horizontally to the east. The stem of the flower is 0.450 m long and tilts toward the east, making an angle of 14.0° with the vertical. Find the vector torque of the wind force about the base of the stem.

6. A student claims that she has found a vector $\mathbf{A}$ such that $(2\hat{\mathbf{i}} - 3\hat{\mathbf{j}} + 4\hat{\mathbf{k}}) \times \mathbf{A} = (4\hat{\mathbf{i}} + 3\hat{\mathbf{j}} - \hat{\mathbf{k}})$. Do you believe this claim? Explain.

7. If $|\mathbf{A} \times \mathbf{B}| = \mathbf{A}\cdot\mathbf{B}$, what is the angle between $\mathbf{A}$ and $\mathbf{B}$?

8. A particle is located at the vector position $\mathbf{r} = (\hat{\mathbf{i}} + 3\hat{\mathbf{j}})$ m, and the force acting on it is $\mathbf{F} = (3\hat{\mathbf{i}} + 2\hat{\mathbf{j}})$ N. What is the torque about (a) the origin and (b) the point having coordinates (0, 6) m?

9. Two forces $\mathbf{F}_1$ and $\mathbf{F}_2$ act along the two sides of an equilateral triangle as shown in Figure P11.9. Point O is the intersection of the altitudes of the triangle. Find a third force $\mathbf{F}_3$ to be applied at B and along BC that will make the total torque

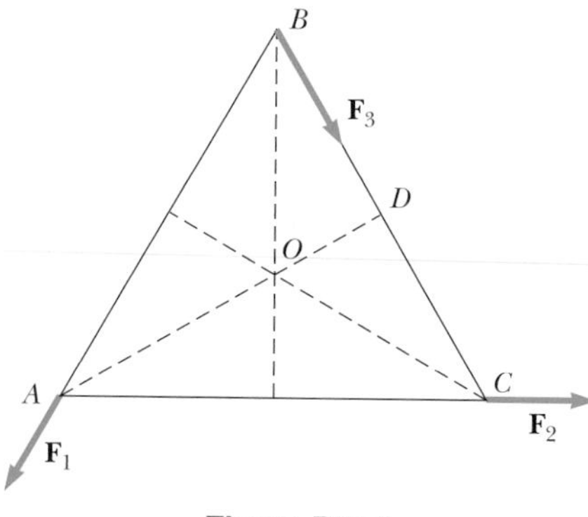

Figure P11.9

zero about the point O. **What If?** Will the total torque change if $\mathbf{F}_3$ is applied not at B but at any other point along BC?

10. Use the definition of the vector product and the definitions of the unit vectors $\hat{\mathbf{i}}$, $\hat{\mathbf{j}}$, and $\hat{\mathbf{k}}$ to prove Equations 11.7. You may assume that the x axis points to the right, the y axis up, and the z axis toward you (not away from you). This choice is said to make the coordinate system *right-handed.*

Section 11.2 Angular Momentum

11. A light rigid rod 1.00 m in length joins two particles, with masses 4.00 kg and 3.00 kg, at its ends. The combination rotates in the xy plane about a pivot through the center of the rod (Fig. P11.11). Determine the angular momentum of the system about the origin when the speed of each particle is 5.00 m/s.

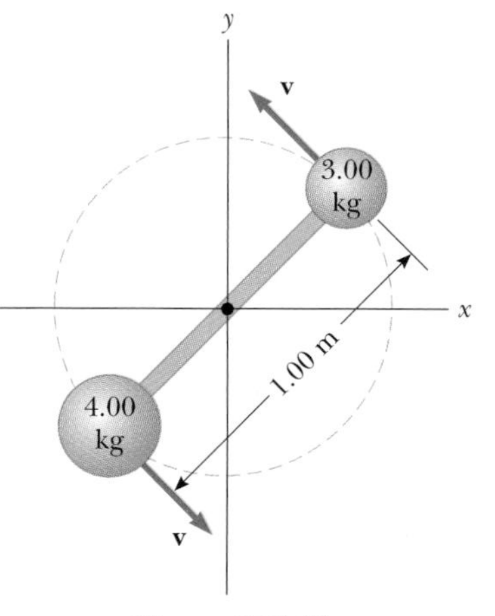

Figure P11.11

12. A 1.50-kg particle moves in the xy plane with a velocity of $\mathbf{v} = (4.20\hat{\mathbf{i}} - 3.60\hat{\mathbf{j}})$ m/s. Determine the angular momentum of the particle when its position vector is $\mathbf{r} = (1.50\hat{\mathbf{i}} + 2.20\hat{\mathbf{j}})$ m.

13. The position vector of a particle of mass 2.00 kg is given as a function of time by $\mathbf{r} = (6.00\hat{\mathbf{i}} + 5.00t\hat{\mathbf{j}})$ m. Determine the angular momentum of the particle about the origin, as a function of time.

14. A conical pendulum consists of a bob of mass m in motion in a circular path in a horizontal plane as shown in Figure P11.14. During the motion, the supporting wire of length ℓ maintains the constant angle θ with the vertical. Show that the magnitude of the angular momentum of the bob about the center of the circle is

$$L = \left(\frac{m^2 g\, \ell^3 \sin^4 \theta}{\cos \theta}\right)^{1/2}$$

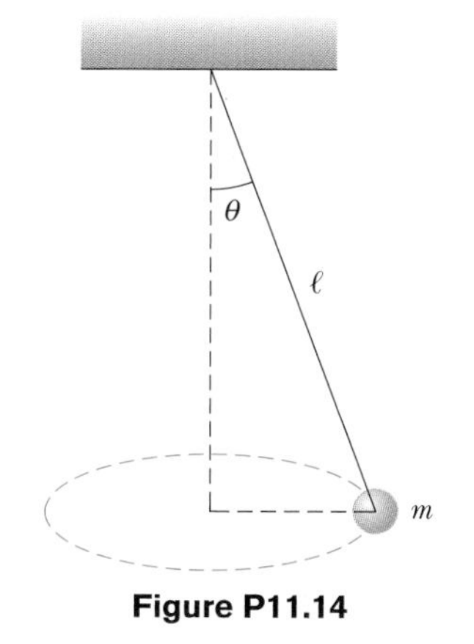

Figure P11.14

15. A particle of mass m moves in a circle of radius R at a constant speed v, as shown in Figure P11.15. If the motion begins at point Q at time $t = 0$, determine the angular momentum of the particle about point P as a function of time.

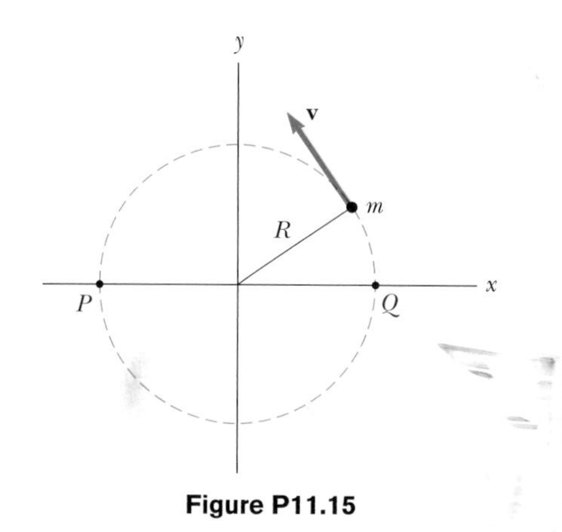

Figure P11.15

16. A 4.00-kg counterweight is attached to a light cord, which is wound around a spool (refer to Fig. 10.20). The spool is a uniform solid cylinder of radius 8.00 cm and mass 2.00 kg. (a) What is the net torque on the system about the point O? (b) When the counterweight has a speed v, the pulley has an angular speed $\omega = v/R$. Determine the total angular momentum of the system about O. (c) Using the fact that $\boldsymbol{\tau} = d\mathbf{L}/dt$ and your result from (b), calculate the acceleration of the counterweight.

17. A particle of mass m is shot with an initial velocity $\mathbf{v}_i$ making an angle θ with the horizontal as shown in Figure P11.17. The particle moves in the gravitational field of the Earth. Find the angular momentum of the particle about the origin when the particle is (a) at the origin, (b) at the highest point of its trajectory, and (c) just before it hits the ground. (d) What torque causes its angular momentum to change?

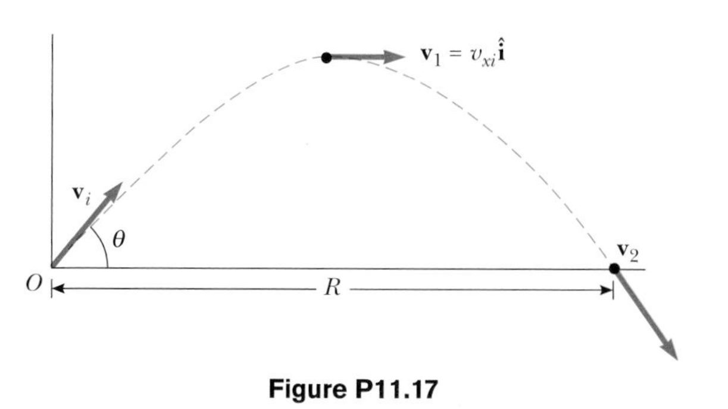

Figure P11.17

18. Heading straight toward the summit of Pike's Peak, an airplane of mass 12 000 kg flies over the plains of Kansas at nearly constant altitude 4.30 km, with a constant velocity of 175 m/s west. (a) What is the airplane's vector angular momentum relative to a wheat farmer on the ground directly below the airplane? (b) Does this value change as the airplane continues its motion along a straight line? (c) **What If?** What is its angular momentum relative to the summit of Pike's Peak?

19. A ball having mass m is fastened at the end of a flagpole that is connected to the side of a tall building at point P shown in Figure P11.19. The length of the flagpole is ℓ and it makes an angle θ with the horizontal. If the ball becomes loose and starts to fall, determine the angular momentum (as a function of time) of the ball about point P. Neglect air resistance.

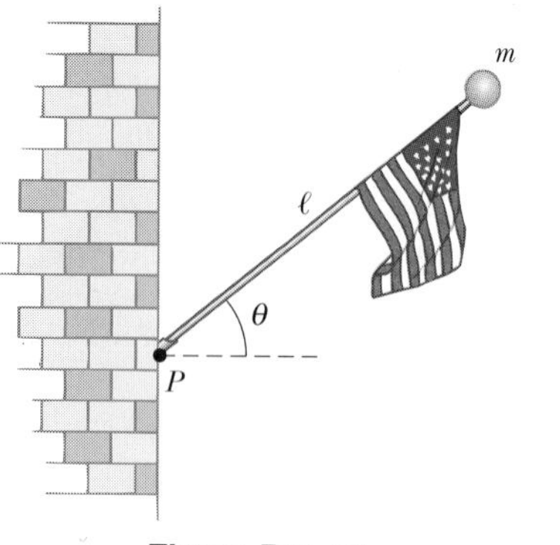

Figure P11.19

20. A fireman clings to a vertical ladder and directs the nozzle of a hose horizontally toward a burning building. The rate of water flow is 6.31 kg/s, and the nozzle speed is 12.5 m/s. The hose passes vertically between the fireman's feet, which are 1.30 m below the nozzle. Choose the origin to be inside the hose between the fireman's feet. What torque must the fireman exert on the hose? That is, what is the rate of change of the angular momentum of the water?

Section 11.3 Angular Momentum of a Rotating Rigid Object

21. Show that the kinetic energy of an object rotating about a fixed axis with angular momentum $L = I\omega$ can be written as $K = L^2/2I$.

22. A uniform solid sphere of radius 0.500 m and mass 15.0 kg turns counterclockwise about a vertical axis through its center. Find its vector angular momentum when its angular speed is 3.00 rad/s.

23. A uniform solid disk of mass 3.00 kg and radius 0.200 m rotates about a fixed axis perpendicular to its face. If the angular frequency of rotation is 6.00 rad/s, calculate the angular momentum of the disk when the axis of rotation (a) passes through its center of mass and (b) passes through a point midway between the center and the rim.

24. Big Ben (Figure P10.40), the Parliament Building tower clock in London, has hour and minute hands with lengths of 2.70 m and 4.50 m and masses of 60.0 kg and 100 kg, respectively. Calculate the total angular momentum of these hands about the center point. Treat the hands as long, thin uniform rods.

25. A particle of mass 0.400 kg is attached to the 100-cm mark of a meter stick of mass 0.100 kg. The meter stick rotates on a horizontal, frictionless table with an angular speed of 4.00 rad/s. Calculate the angular momentum of the system when the stick is pivoted about an axis (a) perpendicular to the table through the 50.0-cm mark and (b) perpendicular to the table through the 0-cm mark.

26. The distance between the centers of the wheels of a motorcycle is 155 cm. The center of mass of the motorcycle, including the biker, is 88.0 cm above the ground and halfway between the wheels. Assume the mass of each wheel is small compared to the body of the motorcycle. The engine drives the rear wheel only. What horizontal acceleration of the motorcycle will make the front wheel rise off the ground?

27. A space station is constructed in the shape of a hollow ring of mass 5.00×10^4 kg. Members of the crew walk on a deck formed by the inner surface of the outer cylindrical wall of the ring, with radius 100 m. At rest when constructed, the ring is set rotating about its axis so that the people inside experience an effective free-fall acceleration equal to g. (Figure P11.27 shows the ring together with some other parts that make a negligible contribution to the total moment of inertia.) The rotation is achieved by firing two small rockets attached tangentially to opposite points on the outside of the ring. (a) What angular momentum does the space station acquire? (b) How long must the rockets be fired if each exerts a thrust of 125 N? (c) Prove that the total torque on the ring, multiplied by the time interval found in part (b), is equal to the change in angular momentum, found in part (a). This equality represents the *angular impulse–angular momentum theorem.*

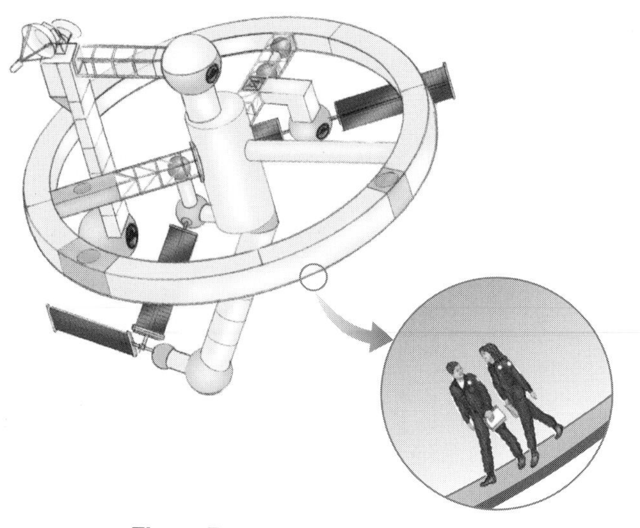
Figure P11.27 Problems 27 and 36.

Section 11.4 Conservation of Angular Momentum

28. A cylinder with moment of inertia I_1 rotates about a vertical, frictionless axle with angular speed ω_i. A second cylinder, this one having moment of inertia I_2 and initially not rotating, drops onto the first cylinder (Fig. P11.28). Because of friction between the surfaces, the two eventually reach the same angular speed ω_f. (a) Calculate ω_f. (b) Show that the kinetic energy of the system decreases in this interaction, and calculate the ratio of the final to the initial rotational energy.

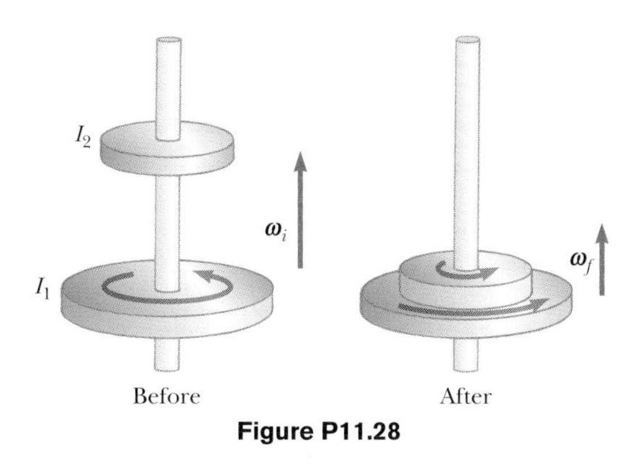

Figure P11.28

29. A playground merry-go-round of radius $R = 2.00$ m has a moment of inertia $I = 250$ kg · m^2 and is rotating at 10.0 rev/min about a frictionless vertical axle. Facing the axle, a 25.0-kg child hops onto the merry-go-round and manages to sit down on the edge. What is the new angular speed of the merry-go-round?

30. A student sits on a freely rotating stool holding two weights, each of mass 3.00 kg (Figure P11.30). When his arms are extended horizontally, the weights are 1.00 m from the axis of rotation and he rotates with an angular speed of 0.750 rad/s. The moment of inertia of the student plus stool is 3.00 kg · m^2 and is assumed to be constant. The student pulls the weights inward horizontally to a position 0.300 m from the rotation axis. (a) Find the new angular speed of the student. (b) Find the kinetic energy of the rotating system before and after he pulls the weights inward.

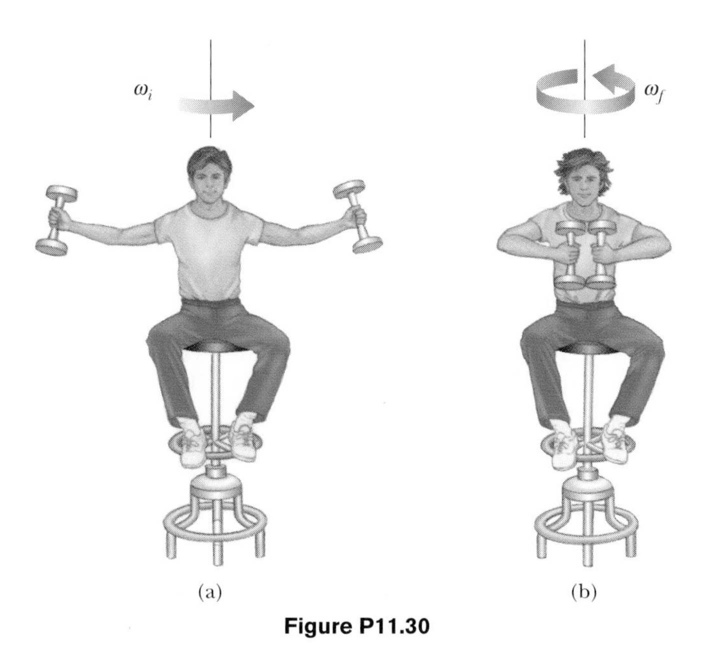

Figure P11.30

31. A uniform rod of mass 100 g and length 50.0 cm rotates in a horizontal plane about a fixed, vertical, frictionless pin through its center. Two small beads, each of mass 30.0 g, are mounted on the rod so that they are able to slide without friction along its length. Initially the beads are held by catches at positions 10.0 cm on each side of center, at which time the system rotates at an angular speed of 20.0 rad/s. Suddenly, the catches are released and the small beads slide outward along the rod. (a) Find the angular speed of the system at the instant the beads reach the ends of the rod. (b) **What if** the beads fly off the ends? What is the angular speed of the rod after this occurs?

32. An umbrella consists of a circle of cloth, a thin rod with the handle at one end and the center of the cloth at the other end, and several straight uniform ribs hinged to the top end of the rod and holding the cloth taut. With the ribs perpendicular to the rod, the umbrella is set rotating about the rod with an angular speed of 1.25 rad/s. The cloth is so light and the rod is so thin that they make negligible contributions to the moment of inertia, in comparison to the ribs. The spinning umbrella is balanced on its handle and keeps rotating without friction. Suddenly its latch breaks and the umbrella partly folds up, until each rib makes an angle of 22.5° with the rod. What is the final angular speed of the umbrella?

33. A 60.0-kg woman stands at the rim of a horizontal turntable having a moment of inertia of 500 kg · m^2 and a radius of 2.00 m. The turntable is initially at rest and is free to rotate about a frictionless, vertical axle through its center. The woman then starts walking around the rim clockwise (as viewed from above the system) at a constant speed of 1.50 m/s relative to the Earth. (a) In what direction and with what angular speed does the turntable rotate? (b) How much work does the woman do to set herself and the turntable into motion?

34. A puck of mass 80.0 g and radius 4.00 cm slides along an air table at a speed of 1.50 m/s as shown in Figure P11.34a. It makes a glancing collision with a second puck of radius 6.00 cm and mass 120 g (initially at rest) such that their rims just touch. Because their rims are coated with instant-acting glue, the pucks stick together and spin after the collision (Fig. P11.34b). (a) What is the angular momentum of the system relative to the center of mass? (b) What is the angular speed about the center of mass?

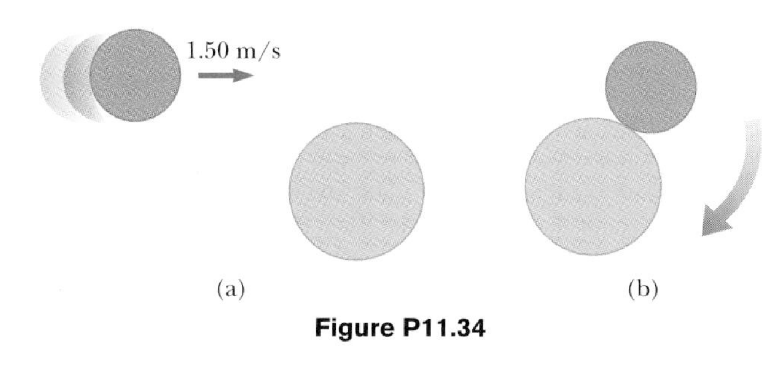

Figure P11.34

35. A wooden block of mass M resting on a frictionless horizontal surface is attached to a rigid rod of length ℓ and of negligible mass (Fig. P11.35). The rod is pivoted at the other end. A bullet of mass m traveling parallel to the horizontal surface and perpendicular to the rod with speed v hits the block and becomes embedded in it. (a) What is the angular momentum of the bullet–block system? (b) What fraction of the original kinetic energy is lost in the collision?

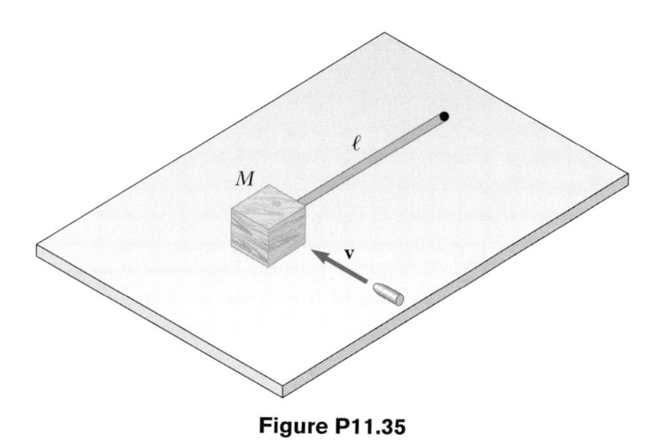

Figure P11.35

36. A space station shaped like a giant wheel has a radius of 100 m and a moment of inertia of $5.00 \times 10^8\ \text{kg}\cdot\text{m}^2$. A crew of 150 is living on the rim, and the station's rotation causes the crew to experience an apparent free-fall acceleration of g (Fig. P11.27). When 100 people move to the center of the station for a union meeting, the angular speed changes. What apparent free-fall acceleration is experienced by the managers remaining at the rim? Assume that the average mass for each inhabitant is 65.0 kg.

37. A wad of sticky clay with mass m and velocity $\mathbf{v}_i$ is fired at a solid cylinder of mass M and radius R (Figure P11.37). The cylinder is initially at rest and is mounted on a fixed horizontal axle that runs through its center of mass. The line of motion of the projectile is perpendicular to the axle and at a distance $d < R$ from the center. (a) Find the angular speed of the system just after the clay strikes and sticks to the surface of the cylinder. (b) Is mechanical energy of the clay–cylinder system conserved in this process? Explain your answer.

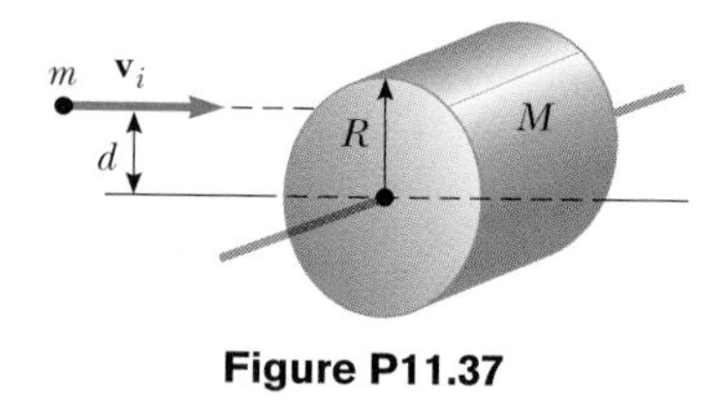

Figure P11.37

38. A thin uniform rectangular sign hangs vertically above the door of a shop. The sign is hinged to a stationary horizontal rod along its top edge. The mass of the sign is 2.40 kg and its vertical dimension is 50.0 cm. The sign is swinging without friction, becoming a tempting target for children armed with snowballs. The maximum angular displacement of the sign is 25.0° on both sides of the vertical. At a moment when the sign is vertical and moving to the left, a snowball of mass 400 g, traveling horizontally with a velocity of 160 cm/s to the right, strikes perpendicularly the lower edge of the sign and sticks there. (a) Calculate the angular speed of the sign immediately before the impact. (b) Calculate its angular speed immediately after the impact. (c) The spattered sign will swing up through what maximum angle?

39. Suppose a meteor of mass 3.00×10^{13} kg, moving at 30.0 km/s relative to the center of the Earth, strikes the Earth. What is the order of magnitude of the maximum possible decrease in the angular speed of the Earth due to this collision? Explain your answer.

Section 11.5 The Motion of Gyroscopes and Tops

40. A spacecraft is in empty space. It carries on board a gyroscope with a moment of inertia of $I_g = 20.0\ \text{kg}\cdot\text{m}^2$ about the axis of the gyroscope. The moment of inertia of the spacecraft around the same axis is $I_s = 5.00 \times 10^5\ \text{kg}\cdot\text{m}^2$. Neither the spacecraft nor the gyroscope is originally rotating. The gyroscope can be powered up in a negligible period of time to an angular speed of $100\ \text{s}^{-1}$. If the orientation of the spacecraft is to be changed by 30.0°, for how long should the gyroscope be operated?

41. The angular momentum vector of a precessing gyroscope sweeps out a cone, as in Figure 11.14b. Its angular speed, called its precessional frequency, is given by $\omega_p = \tau/L$, where τ is the magnitude of the torque on the gyroscope and L is the magnitude of its angular momentum. In the motion called *precession of the equinoxes*, the Earth's axis of rotation precesses about the perpendicular to its orbital plane with a period of 2.58×10^4 yr. Model the Earth as a uniform sphere and calculate the torque on the Earth that is causing this precession.

Section 11.6 Angular Momentum as a Fundamental Quantity

42. In the Bohr model of the hydrogen atom, the electron moves in a circular orbit of radius 0.529×10^{-10} m around the proton. Assuming the orbital angular momentum of the electron is equal to $h/2\pi$, calculate (a) the orbital speed of the electron, (b) the kinetic energy of the electron, and (c) the angular frequency of the electron's motion.

Additional Problems

43. We have all complained that there aren't enough hours in a day. In an attempt to change that, suppose that all the people in the world line up at the equator, and all start running east at 2.50 m/s relative to the surface of the Earth. By how much does the length of a day increase? Assume that the world population is 5.50×10^9 people with an average mass of 70.0 kg each, and that the Earth is a solid homogeneous sphere. In addition, you may use the approximation $1/(1 - x) \approx 1 + x$ for small x.

44. A skateboarder with his board can be modeled as a particle of mass 76.0 kg, located at his center of mass. As shown in Figure P8.67 on page 248, the skateboarder starts from rest in a crouching position at one lip of a half-pipe (point Ⓐ). The half-pipe forms one half of a cylinder of radius 6.80 m

with its axis horizontal. On his descent, the skateboarder moves without friction and maintains his crouch, so that his center of mass moves through one quarter of a circle of radius 6.30 m. (a) Find his speed at the bottom of the half-pipe (point Ⓑ). (b) Find his angular momentum about the center of curvature. (c) Immediately after passing point Ⓑ, he stands up and raises his arms, lifting his center of gravity from 0.500 m to 0.950 m above the concrete (point Ⓒ). Explain why his angular momentum is constant in this maneuver, while his linear momentum and his mechanical energy are not constant. (d) Find his speed immediately after he stands up, when his center of mass is moving in a quarter circle of radius 5.85 m. (e) What work did the skateboarder's legs do on his body as he stood up? Next, the skateboarder glides upward with his center of mass moving in a quarter circle of radius 5.85 m. His body is horizontal when he passes point Ⓓ, the far lip of the half-pipe. (f) Find his speed at this location. At last he goes ballistic, twisting around while his center of mass moves vertically. (g) How high above point Ⓓ does he rise? (h) Over what time interval is he airborne before he touches down, facing downward and again in a crouch, 2.34 m below the level of point Ⓓ? (i) Compare the solution to this problem with the solution to Problem 8.67. Which is more accurate? Why? (*Caution*: Do not try this yourself without the required skill and protective equipment, or in a drainage channel to which you do not have legal access.)

45. A rigid, massless rod has three particles with equal masses attached to it as shown in Figure P11.45. The rod is free to rotate in a vertical plane about a frictionless axle perpendicular to the rod through the point P, and is released from rest in the horizontal position at $t = 0$. Assuming m and d are known, find (a) the moment of inertia of the system (rod plus particles) about the pivot, (b) the torque acting on the system at $t = 0$, (c) the angular acceleration of the system at $t = 0$, (d) the linear acceleration of the particle labeled 3 at $t = 0$, (e) the maximum kinetic energy of the system, (f) the maximum angular speed reached by the rod, (g) the maximum angular momentum of the system, and (h) the maximum speed reached by the particle labeled 2.

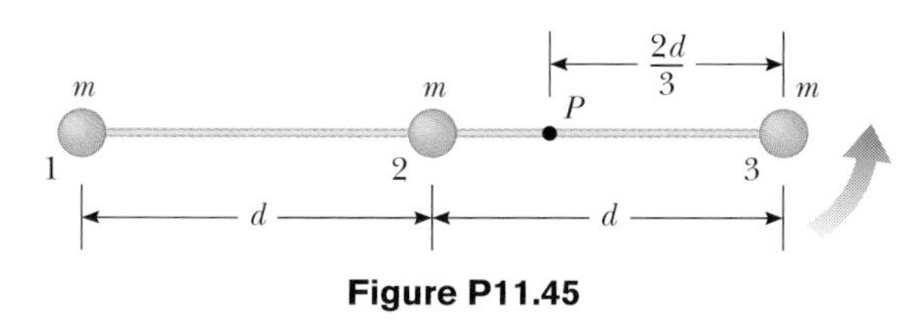

Figure P11.45

46. A 100-kg uniform horizontal disk of radius 5.50 m turns without friction at 2.50 rev/s on a vertical axis through its center, as in Figure P11.46. A feedback mechanism senses the angular speed of the disk, and a drive motor at A maintains the angular speed constant while a 1.20 kg block on top of the disk slides outward in a radial slot. The 1.20-kg block starts at the center of the disk at time $t = 0$ and moves outward with constant speed 1.25 cm/s relative to the disk until it reaches the edge at $t = 440$ s. The sliding block feels no friction. Its motion is constrained to have constant radial speed by a brake at B, producing tension in a light string tied to the block. (a) Find the torque that the drive motor must provide as a function of time, while the block is sliding. (b) Find the value of this torque at $t = 440$ s, just before the sliding block finishes its motion. (c) Find the power that the drive motor must deliver as a function of time. (d) Find the value of the power when the sliding block is just reaching the end of the slot. (e) Find the string tension as a function of time. (f) Find the work done by the drive motor during the 440-s motion. (g) Find the work done by the string brake on the sliding block. (h) Find the total work on the system consisting of the disk and the sliding block.

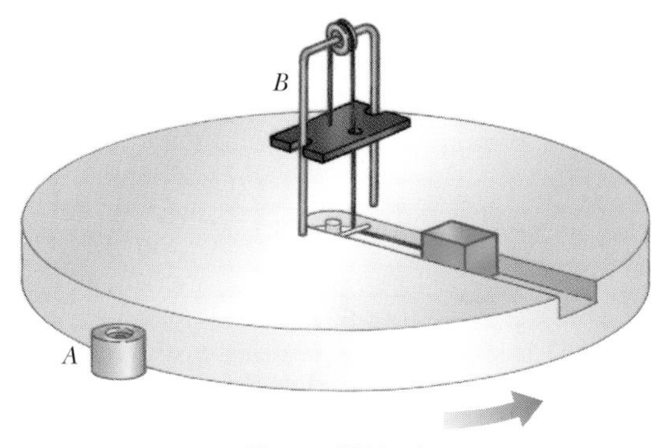

Figure P11.46

47. Comet Halley moves about the Sun in an elliptical orbit, with its closest approach to the Sun being about 0.590 AU and its greatest distance 35.0 AU (1 AU = the Earth–Sun distance). If the comet's speed at closest approach is 54.0 km/s, what is its speed when it is farthest from the Sun? The angular momentum of the comet about the Sun is conserved, because no torque acts on the comet. The gravitational force exerted by the Sun has zero moment arm.

48. A light rope passes over a light, frictionless pulley. One end is fastened to a bunch of bananas of mass M, and a monkey of mass M clings to the other end (Fig. P11.48). The mon-

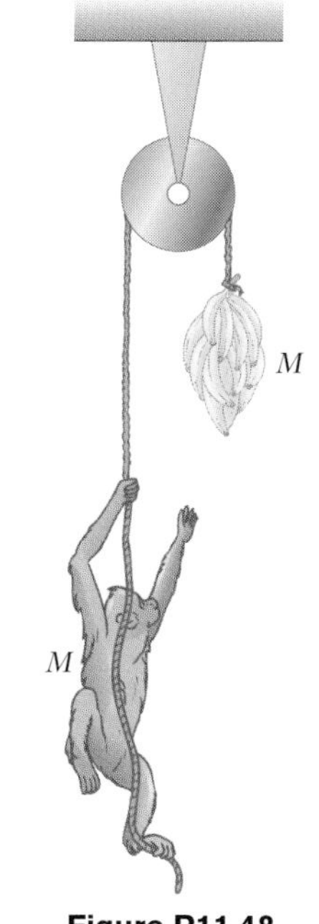

Figure P11.48

key climbs the rope in an attempt to reach the bananas. (a) Treating the system as consisting of the monkey, bananas, rope, and pulley, evaluate the net torque about the pulley axis. (b) Using the results of (a), determine the total angular momentum about the pulley axis and describe the motion of the system. Will the monkey reach the bananas?

49. A puck of mass m is attached to a cord passing through a small hole in a frictionless, horizontal surface (Fig. P11.49). The puck is initially orbiting with speed v_i in a circle of radius r_i. The cord is then slowly pulled from below, decreasing the radius of the circle to r. (a) What is the speed of the puck when the radius is r? (b) Find the tension in the cord as a function of r. (c) How much work W is done in moving m from r_i to r? (*Note:* The tension depends on r.) (d) Obtain numerical values for v, T, and W when $r = 0.100$ m, $m = 50.0$ g, $r_i = 0.300$ m, and $v_i = 1.50$ m/s.

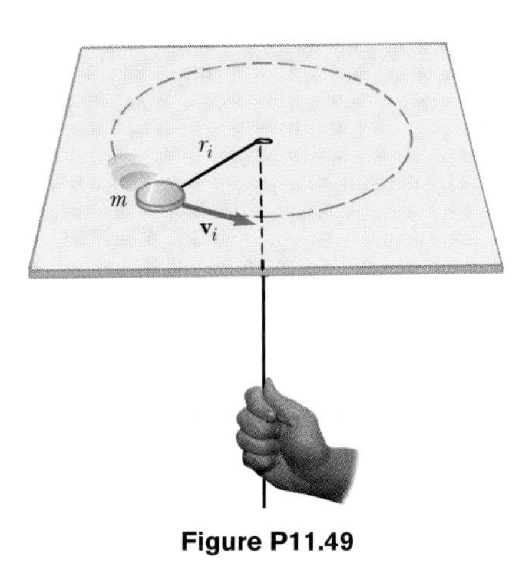

Figure P11.49

50. A projectile of mass m moves to the right with a speed v_i (Fig. P11.50a). The projectile strikes and sticks to the end of a stationary rod of mass M, length d, pivoted about a frictionless axle through its center (Fig. P11.50b). (a) Find the angular speed of the system right after the collision. (b) Determine the fractional loss in mechanical energy due to the collision.

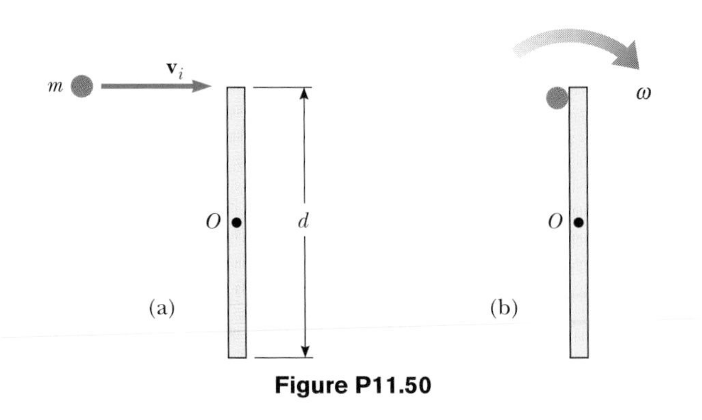

Figure P11.50

51. 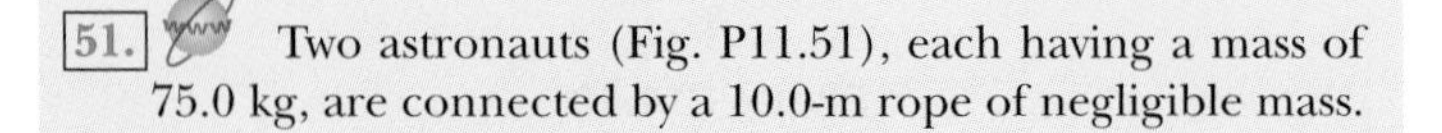 Two astronauts (Fig. P11.51), each having a mass of 75.0 kg, are connected by a 10.0-m rope of negligible mass. They are isolated in space, orbiting their center of mass at speeds of 5.00 m/s. Treating the astronauts as particles, calculate (a) the magnitude of the angular momentum of the system and (b) the rotational energy of the system. By pulling on the rope, one of the astronauts shortens the distance between them to 5.00 m. (c) What is the new angular momentum of the system? (d) What are the astronauts' new speeds? (e) What is the new rotational energy of the system? (f) How much work does the astronaut do in shortening the rope?

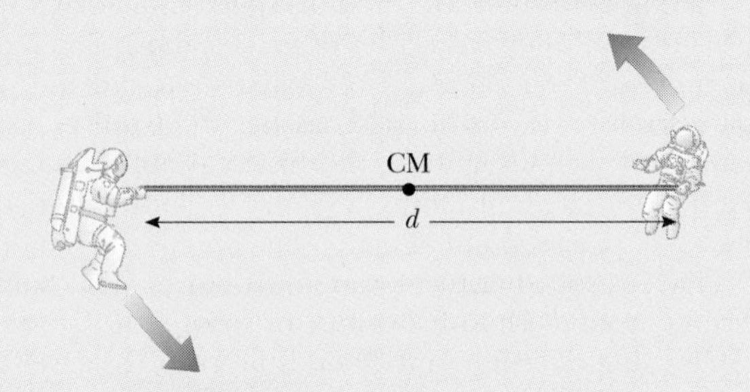

Figure P11.51 Problems 51 and 52.

52. Two astronauts (Fig. P11.51), each having a mass M, are connected by a rope of length d having negligible mass. They are isolated in space, orbiting their center of mass at speeds v. Treating the astronauts as particles, calculate (a) the magnitude of the angular momentum of the system and (b) the rotational energy of the system. By pulling on the rope, one of the astronauts shortens the distance between them to $d/2$. (c) What is the new angular momentum of the system? (d) What are the astronauts' new speeds? (e) What is the new rotational energy of the system? (f) How much work does the astronaut do in shortening the rope?

53. Global warming is a cause for concern because even small changes in the Earth's temperature can have significant consequences. For example, if the Earth's polar ice caps were to melt entirely, the resulting additional water in the oceans would flood many coastal cities. Would it appreciably change the length of a day? Calculate the resulting change in the duration of one day. Model the polar ice as having mass 2.30×10^{19} kg and forming two flat disks of radius 6.00×10^5 m. Assume the water spreads into an unbroken thin spherical shell after it melts.

54. A solid cube of wood of side $2a$ and mass M is resting on a horizontal surface. The cube is constrained to rotate about an axis AB (Fig. P11.54). A bullet of mass m and speed v is shot at the face opposite $ABCD$ at a height of $4a/3$. The bullet becomes embedded in the cube. Find the minimum value of v required to tip the cube so that it falls on face $ABCD$. Assume $m \ll M$.

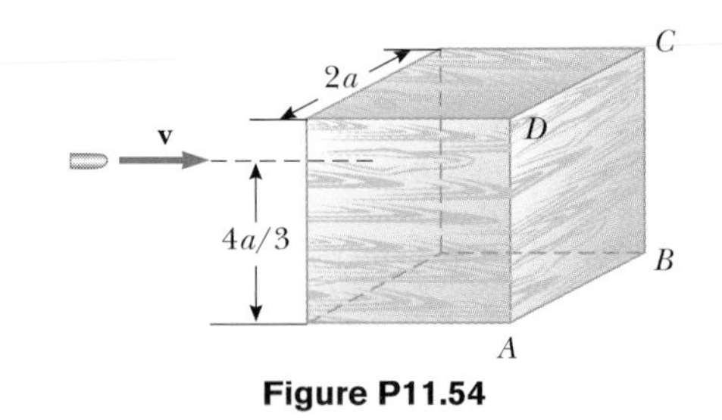

Figure P11.54

55. A solid cube of side $2a$ and mass M is sliding on a frictionless surface with uniform velocity **v** as in Figure P11.55a. It hits a small obstacle at the end of the table, which causes the cube to tilt as in Figure P11.55b. Find the minimum value of **v** such that the cube falls off the table. Note that the moment of inertia of the cube about an axis along one of its edges is $8Ma^2/3$. (*Hint:* The cube undergoes an inelastic collision at the edge.)

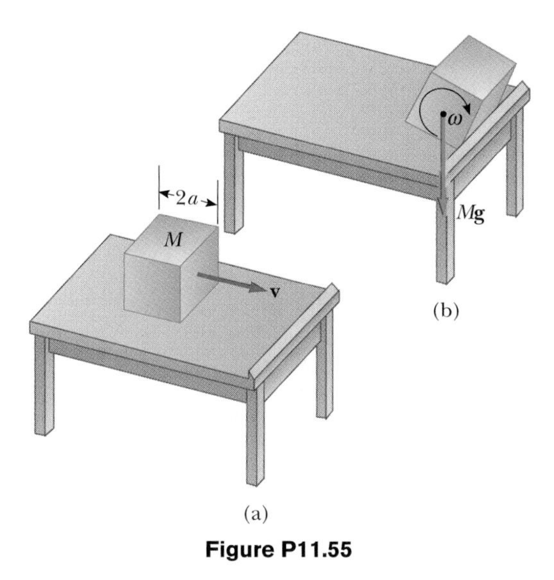

Figure P11.55

56. A uniform solid disk is set into rotation with an angular speed ω_i about an axis through its center. While still rotating at this speed, the disk is placed into contact with a horizontal surface and released as in Figure P11.56. (a) What is the angular speed of the disk once pure rolling takes place? (b) Find the fractional loss in kinetic energy from the time the disk is released until pure rolling occurs. (*Hint:* Consider torques about the center of mass.)

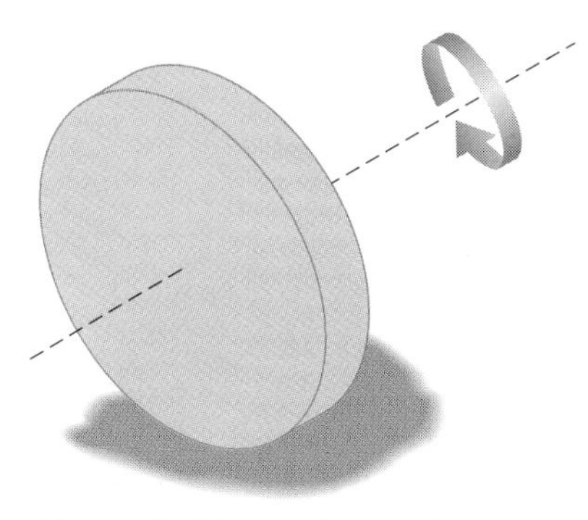

Figure P11.56 Problems 56 and 57.

57. Suppose a solid disk of radius R is given an angular speed ω_i about an axis through its center and then lowered to a horizontal surface and released, as in Problem 56 (Fig. P11.56). Furthermore, assume that the coefficient of friction between disk and surface is μ. (a) Show that the time interval before pure rolling motion occurs is $R\omega_i/3\mu g$. (b) Show that the distance the disk travels before pure rolling occurs is $R^2\omega_i^2/18\mu g$.

Answers to Quick Quizzes

11.1 (d). This result can be obtained by replacing $\mathbf{B} \times \mathbf{A}$ with $-(\mathbf{A} \times \mathbf{B})$, according to Equation 11.4.

11.2 (d). Because of the $\sin\theta$ function, $|\mathbf{A} \times \mathbf{B}|$ is either equal to or smaller than AB, depending on the angle θ.

11.3 (a). If **p** and **r** are parallel or antiparallel, the angular momentum is zero. For a nonzero angular momentum, the linear momentum vector must be offset from the rotation axis.

11.4 (c). The angular momentum is the product of the linear momentum and the perpendicular distance from the rotation axis to the line along which the linear momentum vector lies.

11.5 (b). The hollow sphere has a larger moment of inertia than the solid sphere.

11.6 (a). The diver is an isolated system, so the product $I\omega$ remains constant. Because her moment of inertia decreases, her angular speed increases.

11.7 (a). As the moment of inertia of the diver decreases, the angular speed increases by the same factor. For example, if I goes down by a factor of 2, ω goes up by a factor of 2. The rotational kinetic energy varies as the square of ω. If I is halved, ω^2 increases by a factor of 4 and the energy increases by a factor of 2.

Chapter 12

Static Equilibrium and Elasticity

CHAPTER OUTLINE

▲ *Balanced Rock in Arches National Park, Utah, is a 3 000 000-kg boulder that has been in stable equilibrium for several millennia. It had a smaller companion nearby, called "Chip Off the Old Block," which fell during the winter of 1975. Balanced Rock appeared in an early scene of the movie* Indiana Jones and the Last Crusade. *We will study the conditions under which an object is in equilibrium in this chapter. (John W. Jewett, Jr.)*

In Chapters 10 and 11 we studied the dynamics of rigid objects. Part of this current chapter addresses the conditions under which a rigid object is in equilibrium. The term *equilibrium* implies either that the object is at rest or that its center of mass moves with constant velocity relative to the observer. We deal here only with the former case, in which the object is in *static equilibrium.* Static equilibrium represents a common situation in engineering practice, and the principles it involves are of special interest to civil engineers, architects, and mechanical engineers. If you are an engineering student, you will undoubtedly take an advanced course in statics in the future.

The last section of this chapter deals with how objects deform under load conditions. An *elastic* object returns to its original shape when the deforming forces are removed. Several elastic constants are defined, each corresponding to a different type of deformation.

12.1 The Conditions for Equilibrium

In Chapter 5 we found that one necessary condition for equilibrium is that the net force acting on an object must be zero. If the object is modeled as a particle, then this is the only condition that must be satisfied for equilibrium. The situation with real (extended) objects is more complex, however, because these objects often cannot be modeled as particles. For an extended object to be in static equilibrium, a second condition must be satisfied. This second condition involves the net torque acting on the extended object.

Consider a single force **F** acting on a rigid object, as shown in Figure 12.1. The effect of the force depends on the location of its point of application P. If **r** is the position vector of this point relative to O, the torque associated with the force **F** about O is given by Equation 11.1:

$$\boldsymbol{\tau} = \mathbf{r} \times \mathbf{F}$$

Recall from the discussion of the vector product in Section 11.1 that the vector $\boldsymbol{\tau}$ is perpendicular to the plane formed by **r** and **F**. You can use the right-hand rule to determine the direction of $\boldsymbol{\tau}$ as shown in Figure 11.2. Hence, in Figure 12.1 $\boldsymbol{\tau}$ is directed toward you out of the page.

As you can see from Figure 12.1, the tendency of **F** to rotate the object about an axis through O depends on the moment arm d, as well as on the magnitude of **F**. Recall that the magnitude of $\boldsymbol{\tau}$ is Fd (see Eq. 10.19). According to Equation 10.21, the net torque on a rigid object will cause it to undergo an angular acceleration.

In the current discussion, we want to look at those rotational situations in which the angular acceleration of a rigid object is zero. Such an object is in **rotational equilibrium**. Because $\Sigma\boldsymbol{\tau} = I\alpha$ for rotation about a fixed axis, the necessary condition for rotational equilibrium is that **the net torque about any axis must be zero**. We now

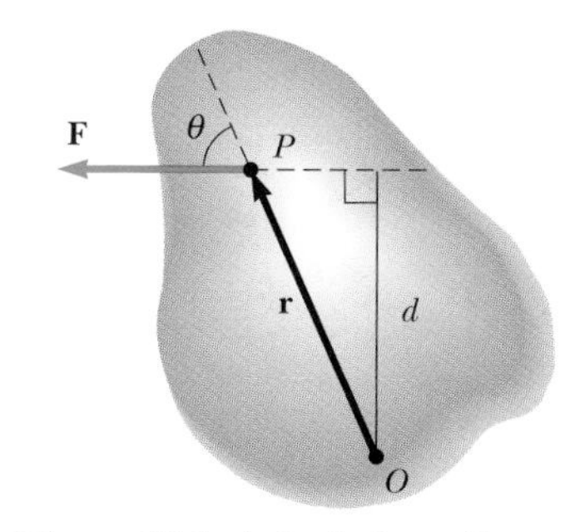

Figure 12.1 A single force **F** acts on a rigid object at the point P.

have two necessary conditions for equilibrium of an object:

1. The resultant external force must equal zero:

$$\sum \mathbf{F} = 0 \tag{12.1}$$

2. The resultant external torque about *any* axis must be zero:

$$\sum \boldsymbol{\tau} = 0 \tag{12.2}$$

The first condition is a statement of translational equilibrium; it tells us that the linear acceleration of the center of mass of the object must be zero when viewed from an inertial reference frame. The second condition is a statement of rotational equilibrium and tells us that the angular acceleration about any axis must be zero. In the special case of **static equilibrium**, which is the main subject of this chapter, the object is at rest relative to the observer and so has no linear or angular speed (that is, $v_{CM} = 0$ and $\omega = 0$).

PITFALL PREVENTION

12.1 Zero Torque

Zero net torque does not mean an absence of rotational motion. An object which is rotating at a constant angular speed can be under the influence of a net torque of zero. This is analogous to the translational situation—zero net force does not mean an absence of translational motion.

Quick Quiz 12.1 Consider the object subject to the two forces in Figure 12.2. Choose the correct statement with regard to this situation. (a) The object is in force equilibrium but not torque equilibrium. (b) The object is in torque equilibrium but not force equilibrium. (c) The object is in both force and torque equilibrium. (d) The object is in neither force nor torque equilibrium.

Quick Quiz 12.2 Consider the object subject to the three forces in Figure 12.3. Choose the correct statement with regard to this situation. (a) The object is in force equilibrium but not torque equilibrium. (b) The object is in torque equilibrium but not force equilibrium. (c) The object is in both force and torque equilibrium. (d) The object is in neither force nor torque equilibrium.

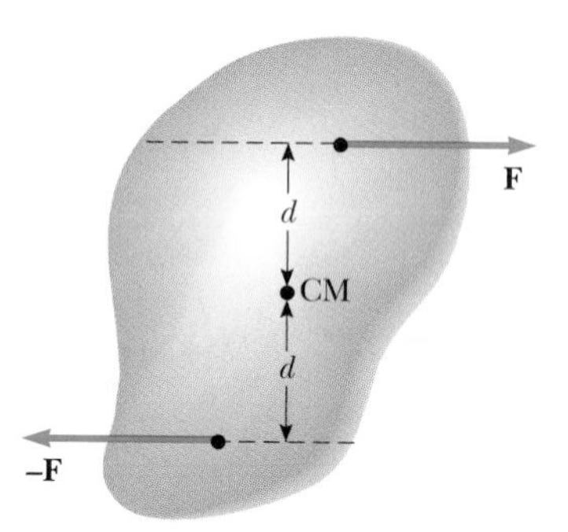

Figure 12.2 (Quick Quiz 12.1) Two forces of equal magnitude are applied at equal distances from the center of mass of a rigid object.

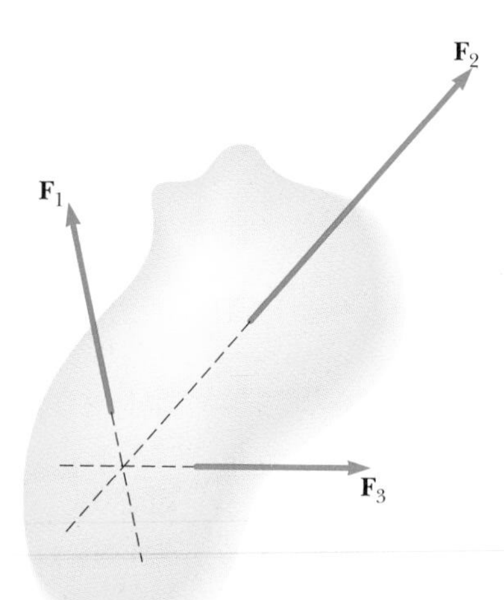

Figure 12.3 (Quick Quiz 12.2) Three forces act on an object. Notice that the lines of action of all three forces pass through a common point.

The two vector expressions given by Equations 12.1 and 12.2 are equivalent, in general, to six scalar equations: three from the first condition for equilibrium, and three from the second (corresponding to x, y, and z components). Hence, in a complex system involving several forces acting in various directions, you could be faced with solving a set of equations with many unknowns. Here, we restrict our discussion to situations in which all the forces lie in the xy plane. (Forces whose vector representations are in the same plane are said to be *coplanar.*) With this restriction, we must deal with only three scalar equations. Two of these come from balancing the forces in the x and y directions. The third comes from the torque equation—namely, that the net torque about a perpendicular axis through *any* point in the xy plane must be zero. Hence, the two conditions of equilibrium provide the equations

$$\sum F_x = 0 \qquad \sum F_y = 0 \qquad \sum \tau_z = 0 \tag{12.3}$$

where the location of the axis of the torque equation is arbitrary, as we now show.

Regardless of the number of forces that are acting, if an object is in translational equilibrium and if the net torque is zero about one axis, then the net torque must also be zero about any other axis. The axis can pass through a point that is inside or outside the boundaries of the object. Consider an object being acted on by several forces such that the resultant force $\sum \mathbf{F} = \mathbf{F}_1 + \mathbf{F}_2 + \mathbf{F}_3 + \cdots = 0$. Figure 12.4 describes this situation (for clarity, only four forces are shown). The point of application of $\mathbf{F}_1$ relative to O is specified by the position vector $\mathbf{r}_1$. Similarly, the points of application of $\mathbf{F}_2, \mathbf{F}_3, \ldots$ are specified by $\mathbf{r}_2, \mathbf{r}_3, \ldots$ (not shown). The net torque about an axis through O is

$$\sum \boldsymbol{\tau}_O = \mathbf{r}_1 \times \mathbf{F}_1 + \mathbf{r}_2 \times \mathbf{F}_2 + \mathbf{r}_3 \times \mathbf{F}_3 + \cdots$$

Now consider another arbitrary point O' having a position vector $\mathbf{r}'$ relative to O. The point of application of $\mathbf{F}_1$ relative to O' is identified by the vector $\mathbf{r}_1 - \mathbf{r}'$. Likewise, the point of application of $\mathbf{F}_2$ relative to O' is $\mathbf{r}_2 - \mathbf{r}'$, and so forth. Therefore, the torque about an axis through O' is

$$\begin{aligned}\sum \boldsymbol{\tau}_{O'} &= (\mathbf{r}_1 - \mathbf{r}') \times \mathbf{F}_1 + (\mathbf{r}_2 - \mathbf{r}') \times \mathbf{F}_2 + (\mathbf{r}_3 - \mathbf{r}') \times (\mathbf{F}_3 + \cdots \\ &= \mathbf{r}_1 \times \mathbf{F}_1 + \mathbf{r}_2 \times \mathbf{F}_2 + \mathbf{r}_3 \times \mathbf{F}_3 + \cdots - \mathbf{r}' \times (\mathbf{F}_1 + \mathbf{F}_2 + \mathbf{F}_3 + \cdots)\end{aligned}$$

Because the net force is assumed to be zero (given that the object is in translational equilibrium), the last term vanishes, and we see that the torque about an axis through O' is equal to the torque about an axis through O. Hence, **if an object is in translational equilibrium and the net torque is zero about one axis, then the net torque must be zero about any other axis.**

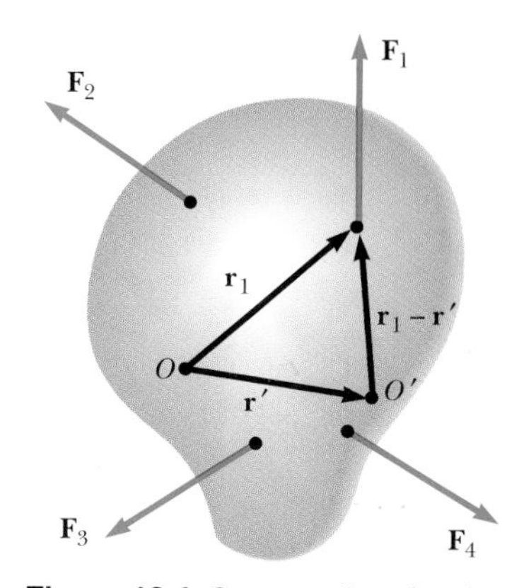

Figure 12.4 Construction showing that if the net torque is zero about origin O, it is also zero about any other origin, such as O'.

12.2 More on the Center of Gravity

We have seen that the point at which a force is applied can be critical in determining how an object responds to that force. For example, two equal-magnitude but oppositely directed forces result in equilibrium if they are applied at the same point on an object. However, if the point of application of one of the forces is moved, so that the two forces no longer act along the same line of action, then the object undergoes an angular acceleration.

Whenever we deal with a rigid object, one of the forces we must consider is the gravitational force acting on it, and we must know the point of application of this force. As we learned in Section 9.5, associated with every object is a special point called its center of gravity. All the various gravitational forces acting on all the various mass elements of the object are equivalent to a single gravitational force acting through this point. Thus, to compute the torque due to the gravitational force on an object of mass M, we need only consider the force $M\mathbf{g}$ acting at the center of gravity of the object.

How do we find this special point? As we mentioned in Section 9.5, if we assume that $\mathbf{g}$ is uniform over the object, then the center of gravity of the object coincides with its center of mass. To see that this is so, consider an object of arbitrary shape lying in the xy plane, as illustrated in Figure 12.5. Suppose the object is divided into a large number of particles of masses $m_1, m_2, m_3, \ldots$ having coordinates (x_1, y_1), (x_2, y_2), (x_3, y_3), ... In Equation 9.28 we defined the x coordinate of the center of mass of such an object to be

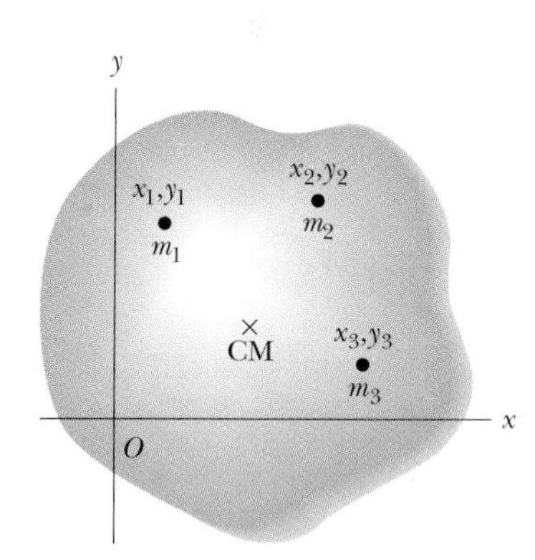

Figure 12.5 An object can be divided into many small particles each having a specific mass and specific coordinates. These particles can be used to locate the center of mass.

$$x_{\text{CM}} = \frac{m_1x_1 + m_2x_2 + m_3x_3 + \cdots}{m_1 + m_2 + m_3 + \cdots} = \frac{\sum_i m_ix_i}{\sum_i m_i}$$

We use a similar equation to define the y coordinate of the center of mass, replacing each x with its y counterpart.

Let us now examine the situation from another point of view by considering the gravitational force exerted on each particle, as shown in Figure 12.6. Each particle contributes a torque about the origin equal in magnitude to the particle's weight mg multiplied by its moment arm. For example, the magnitude of the torque due to the force $m_1\mathbf{g}_1$ is $m_1g_1x_1$, where g_1 is the value of the gravitational acceleration at the position of the particle of mass m_1. We wish to locate the center of gravity, the point at which application of the single gravitational force $M\mathbf{g}$ (where $M = m_1 + m_2 + m_3 + \cdots$ is the total mass of the object) has the same effect on rotation as does the combined effect of all the individual gravitational forces $m_i\mathbf{g}_i$. Equating the torque resulting from $M\mathbf{g}$ acting at the center of gravity to the sum of the torques acting on the individual particles gives

$$(m_1g_1 + m_2g_2 + m_3g_3 + \cdots)x_{\text{CG}} = m_1g_1x_1 + m_2g_2x_2 + m_3g_3x_3 + \cdots$$

This expression accounts for the fact that the value of g can in general vary over the object. If we assume uniform g over the object (as is usually the case), then the

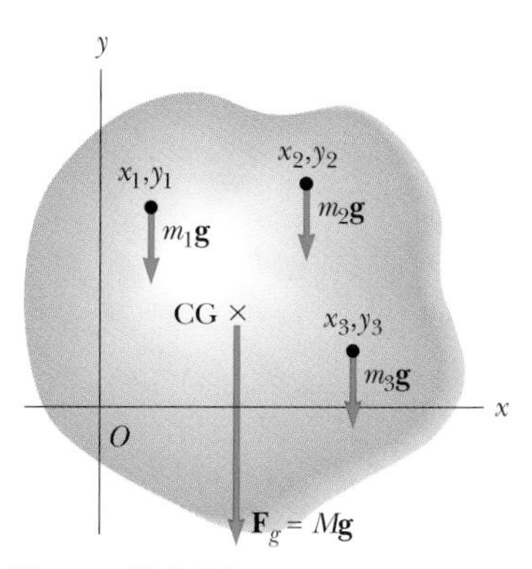

Figure 12.6 The center of gravity of an object is located at the center of mass if $\mathbf{g}$ is constant over the object.

g terms cancel and we obtain

$$x_{\text{CG}} = \frac{m_1x_1 + m_2x_2 + m_3x_3 + \cdots}{m_1 + m_2 + m_3 + \cdots} \qquad (12.4)$$

Comparing this result with Equation 9.28, we see that **the center of gravity is located at the center of mass as long as g is uniform over the entire object**. In several examples presented in the next section, we will deal with homogeneous, symmetric objects. The center of gravity for any such object coincides with its geometric center.

Quick Quiz 12.3 A meter stick is supported on a fulcrum at the 25-cm mark. A 0.50-kg object is hung from the zero end of the meter stick, and the stick is balanced horizontally. The mass of the meter stick is (a) 0.25 kg (b) 0.50 kg (c) 0.75 kg (d) 1.0 kg (e) 2.0 kg (f) impossible to determine.

12.3 Examples of Rigid Objects in Static Equilibrium

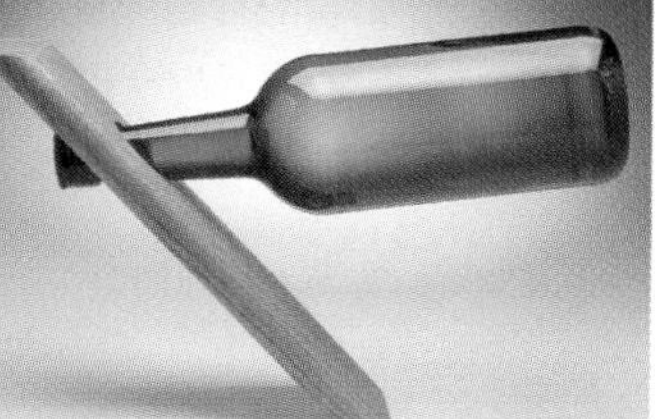

Charles D. Winters

Figure 12.7 This one-bottle wine holder is a surprising display of static equilibrium. The center of gravity of the system (bottle plus holder) is directly over the support point.

The photograph of the one-bottle wine holder in Figure 12.7 shows one example of a balanced mechanical system that seems to defy gravity. For the system (wine holder plus bottle) to be in equilibrium, the net external force must be zero (see Eq. 12.1) and the net external torque must be zero (see Eq. 12.2). The second condition can be satisfied only when the center of gravity of the system is directly over the support point.

When working static equilibrium problems, you must recognize all the external forces acting on the object. Failure to do so results in an incorrect analysis. When analyzing an object in equilibrium under the action of several external forces, use the following procedure.

PROBLEM-SOLVING STRATEGY

OBJECTS IN STATIC EQUILIBRIUM

- Draw a simple, neat diagram of the system.
- Isolate the object being analyzed. Draw a free-body diagram. Then show and label all external forces acting on the object, indicating where those forces are applied. Do not include forces exerted by the object on its surroundings. (For systems that contain more than one object, draw a *separate* free-body diagram for each one.) Try to guess the correct direction for each force.
- Establish a convenient coordinate system and find the components of the forces on the object along the two axes. Then apply the first condition for equilibrium. Remember to keep track of the signs of the various force components.
- Choose a convenient axis for calculating the net torque on the object. Remember that the choice of origin for the torque equation is arbitrary; therefore choose an origin that simplifies your calculation as much as possible. Note that a force that acts along a line passing through the point chosen as the origin gives zero contribution to the torque and so can be ignored.
- The first and second conditions for equilibrium give a set of linear equations containing several unknowns, and these equations can be solved simultaneously. If the direction you selected for a force leads to a negative value, do not be alarmed; this merely means that the direction of the force is the opposite of what you guessed.

Example 12.1 The Seesaw Revisited

A seesaw consisting of a uniform board of mass M and length ℓ supports a father and daughter with masses m_f and m_d, respectively, as shown in Figure 12.8. The support (called the *fulcrum*) is under the center of gravity of the board, the father is a distance d from the center, and the daughter is a distance $\ell/2$ from the center.

(A) Determine the magnitude of the upward force **n** exerted by the support on the board.

Solution First note that, in addition to **n**, the external forces acting on the board are the downward forces exerted by each person and the gravitational force acting on the board. We know that the board's center of gravity is at its geometric center because we are told that the board is uniform. Because the system is in static equilibrium, the net force on the board is zero. Thus, the upward force **n** must balance all the downward forces. From $\Sigma F_y = 0$, and defining upward as the positive y direction, we have

$$n - m_f g - m_d g - Mg = 0$$

$$n = m_f g + m_d g + Mg$$

(The equation $\Sigma F_x = 0$ also applies, but we do not need to consider it because no forces act horizontally on the board.)

(B) Determine where the father should sit to balance the system.

Solution To find this position, we must invoke the second condition for equilibrium. If we take an axis perpendicular to the page through the center of gravity of the board as the axis for our torque equation, the torques produced by **n** and the gravitational force acting on the board are zero. We see from $\Sigma\tau = 0$ that

$$(m_f g)(d) - (m_d g)\frac{\ell}{2} = 0$$

$$d = \left(\frac{m_d}{m_f}\right)\tfrac{1}{2}\ell$$

This is the same result that we obtained in Example 11.6 by evaluating the angular acceleration of the system and setting the angular acceleration equal to zero.

What If? Suppose we had chosen another point through which the rotation axis were to pass. For example, suppose the axis is perpendicular to the page and passes through the location of the father. Does this change the results to parts (A) and (B)?

Answer Part (A) is unaffected because the calculation of the net force does not involve a rotation axis. In part (B), we would conceptually expect there to be no change if a different rotation axis is chosen because the second condition of equilibrium claims that the torque is zero about any rotation axis.

Let us verify this mathematically. Recall that the sign of the torque associated with a force is positive if that force tends to rotate the system counterclockwise, while the sign of the torque is negative if the force tends to rotate the system clockwise. In the case of a rotation axis passing through the location of the father, $\Sigma\tau = 0$ yields

$$n(d) - (Mg)(d) - (m_d g)(d + \ell/2) = 0$$

From part (A) we know that $n = m_f g + m_d g + Mg$. Thus, we can substitute this expression for n and solve for d:

$$(m_f g + m_d g + Mg)(d) - (Mg)(d) - (m_d g)\left(d + \frac{\ell}{2}\right) = 0$$

$$(m_f g)(d) - (m_d g)\left(\frac{\ell}{2}\right) = 0$$

$$d = \left(\frac{m_d}{m_f}\right)\tfrac{1}{2}\ell$$

This result is in agreement with the one we obtained in part (B).

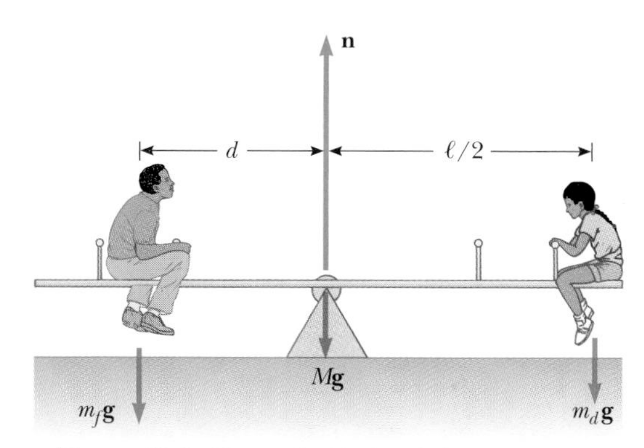

Figure 12.8 (Example 12.1) A balanced system.

Example 12.2 A Weighted Hand

A person holds a 50.0-N sphere in his hand. The forearm is horizontal, as shown in Figure 12.9a. The biceps muscle is attached 3.00 cm from the joint, and the sphere is 35.0 cm from the joint. Find the upward force exerted by the biceps on the forearm and the downward force exerted by the upper arm on the forearm and acting at the joint. Neglect the weight of the forearm.

Solution We simplify the situation by modeling the forearm as a bar as shown in Figure 12.9b, where **F** is the upward force exerted by the biceps and **R** is the downward force exerted by the upper arm at the joint. From the first condition for equilibrium, we have, with upward as the positive y direction,

$$(1) \qquad \sum F_y = F - R - 50.0\text{ N} = 0$$

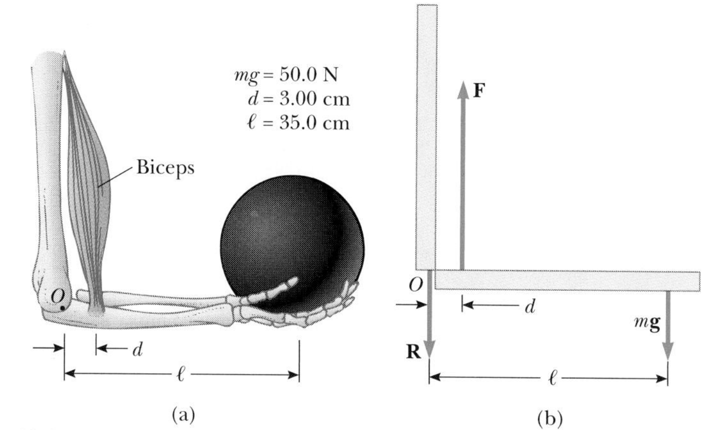

Figure 12.9 (Example 12.2) (a) The biceps muscle pulls upward with a force **F** that is essentially at a right angle to the forearm. (b) The mechanical model for the system described in part (a).

From the second condition for equilibrium, we know that the sum of the torques about any point must be zero. With the joint O as the axis, we have

$$\sum \tau = Fd - mg\ell = 0$$

$$F(3.00 \text{ cm}) - (50.0 \text{ N})(35.0 \text{ cm}) = 0$$

$$F = 583 \text{ N}$$

This value for F can be substituted into Equation (1) to give $R = 533$ N. As this example shows, the forces at joints and in muscles can be extremely large.

Example 12.3 Standing on a Horizontal Beam

Interactive

A uniform horizontal beam with a length of 8.00 m and a weight of 200 N is attached to a wall by a pin connection. Its far end is supported by a cable that makes an angle of 53.0° with the beam (Fig. 12.10a). If a 600-N person stands 2.00 m from the wall, find the tension in the cable as well as the magnitude and direction of the force exerted by the wall on the beam.

Solution Conceptualize this problem by imagining that the person in Figure 12.10 moves outward on the beam. It seems reasonable that the farther he moves outward, the larger the torque that he applies about the pivot and the larger the tension in the cable must be to balance this torque. Because the system is at rest, we categorize this as a static equilibrium problem. We begin to analyze the problem by identifying all the external forces acting on the beam: the 200-N gravitational force, the force **T** exerted by the cable, the force **R** exerted by the wall at the pivot, and the 600-N force that the person exerts on the beam. These forces are all indicated in the free-body diagram for the beam shown in Figure 12.10b. When we assign directions for forces, it is sometimes helpful to imagine what would happen if a force were suddenly removed. For example, if the wall were to vanish suddenly, the left end of the beam would move to the left as it begins to fall. This tells us that the wall is not only holding the beam up but is also pressing outward against it. Thus, we draw the vector **R** as shown in Figure 12.10b. If we resolve **T** and **R** into horizontal and vertical components, as shown in Figure 12.10c, and apply the first condition for equilibrium, we obtain

$$(1) \qquad \sum F_x = R\cos\theta - T\cos 53.0° = 0$$

$$(2) \qquad \sum F_y = R\sin\theta + T\sin 53.0° - 600 \text{ N} - 200 \text{ N} = 0$$

where we have chosen rightward and upward as our positive directions. Because R, T, and θ are all unknown, we cannot obtain a solution from these expressions alone. (The number of simultaneous equations must equal the number of unknowns for us to be able to solve for the unknowns.)

Now let us invoke the condition for rotational equilibrium. A convenient axis to choose for our torque equation is the one that passes through the pin connection. The feature that makes this point so convenient is that the force **R** and the horizontal component of **T** both have a moment arm of zero; hence, these forces provide no torque about this point. Recalling our counterclockwise-equals-positive convention for the sign of the torque about an axis and noting that the moment arms of the 600-N, 200-N, and $T \sin 53.0°$ forces are 2.00 m, 4.00 m, and 8.00 m, respectively, we obtain

$$(3) \qquad \sum \tau = (T\sin 53.0°)(8.00 \text{ m}) - (600 \text{ N})(2.00 \text{ m}) - (200 \text{ N})(4.00 \text{ m}) = 0$$

$$T = 313 \text{ N}$$

Thus, the torque equation with this axis gives us one of the unknowns directly! We now substitute this value into Equations (1) and (2) and find that

$$R\cos\theta = 188 \text{ N}$$

$$R\sin\theta = 550 \text{ N}$$

We divide the second equation by the first and, recalling the trigonometric identity $\sin\theta/\cos\theta = \tan\theta$, we obtain

$$\tan\theta = \frac{550 \text{ N}}{188 \text{ N}} = 2.93$$

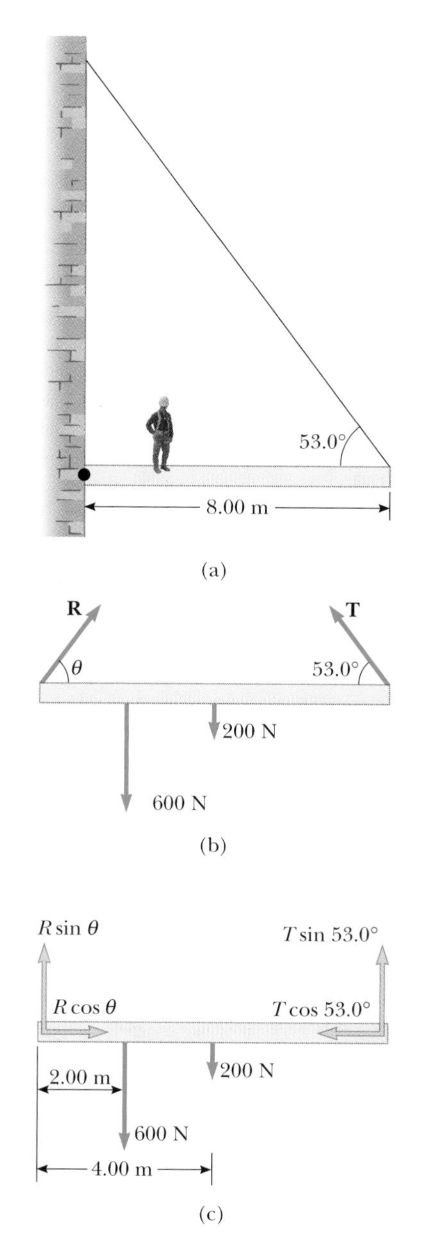

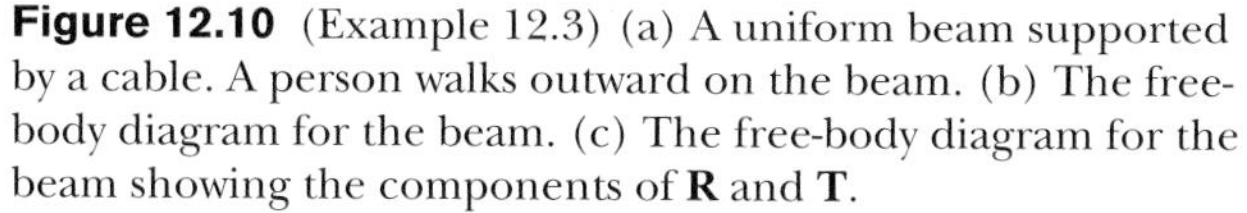

Figure 12.10 (Example 12.3) (a) A uniform beam supported by a cable. A person walks outward on the beam. (b) The free-body diagram for the beam. (c) The free-body diagram for the beam showing the components of **R** and **T**.

$$\theta = \boxed{71.1^\circ}$$

This positive value indicates that our estimate of the direction of **R** was accurate.

Finally,

$$R = \frac{188\text{ N}}{\cos\theta} = \frac{188\text{ N}}{\cos 71.1^\circ} = \boxed{580\text{ N}}$$

To finalize this problem, note that if we had selected some other axis for the torque equation, the solution might differ in the details, but the answers would be the same. For example, if we had chosen an axis through the center of gravity of the beam, the torque equation would involve both T and R. However, this equation, coupled with Equations (1) and (2), could still be solved for the unknowns. Try it!

When many forces are involved in a problem of this nature, it is convenient in your analysis to set up a table. For instance, for the example just given, we could construct the following table. Setting the sum of the terms in the last column equal to zero represents the condition of rotational equilibrium.

Force component	Moment arm relative to O (m)	Torque about O (N·m)
$T\sin 53.0^\circ$	8.00	$(8.00)T\sin 53.0^\circ$
$T\cos 53.0^\circ$	0	0
200 N	4.00	$-(4.00)(200)$
600 N	2.00	$-(2.00)(600)$
$R\sin\theta$	0	0
$R\cos\theta$	0	0

What If? **What if the person walks farther out on the beam? Does T change? Does R change? Does θ change?**

Answer T must increase because the weight of the person exerts a larger torque about the pin connection, which must be countered by a larger torque in the opposite direction due to an increased value of T. If T increases, the vertical component of **R** decreases to maintain force equilibrium in the vertical direction. But force equilibrium in the horizontal direction requires an increased horizontal component of **R** to balance the horizontal component of the increased **T**. This suggests that θ will become smaller, but it is hard to predict what will happen to R. Problem 26 allows you to explore the behavior of R.

At the Interactive Worked Example link at **http://www.pse6.com,** *you can adjust the position of the person and observe the effect on the forces.*

Example 12.4 The Leaning Ladder — Interactive

A uniform ladder of length ℓ rests against a smooth, vertical wall (Fig. 12.11a). If the mass of the ladder is m and the coefficient of static friction between the ladder and the ground is $\mu_s = 0.40$, find the minimum angle θ_{min} at which the ladder does not slip.

Solution The free-body diagram showing all the external forces acting on the ladder is illustrated in Figure 12.11b. The force exerted by the ground on the ladder is the vector sum of a normal force **n** and the force of static friction $\mathbf{f}_s$. The reaction force **P** exerted by the wall on the ladder is horizontal because the wall is frictionless. Notice how we have included only forces that act on the ladder. For example, the forces exerted by the ladder on the ground and on the wall are not part of the problem and thus do not appear in the free-body diagram. Applying the first condition for equilibrium to the ladder, we have

$$(1) \qquad \sum F_x = f_s - P = 0$$

$$(2) \qquad \sum F_y = n - mg = 0$$

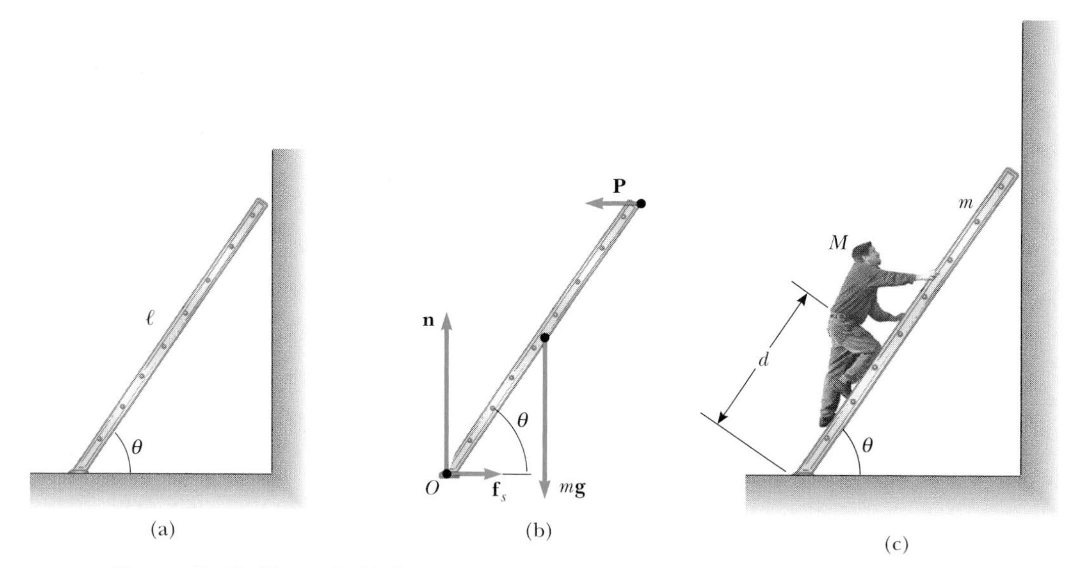

Figure 12.11 (Example 12.4) (a) A uniform ladder at rest, leaning against a smooth wall. The ground is rough. (b) The free-body diagram for the ladder. (c) A person of mass M begins to climb the ladder when it is at the minimum angle found in part (a) of the example. Will the ladder slip?

The first equation tells us that $P = f_s$. From the second equation we see that $n = mg$. Furthermore, when the ladder is on the verge of slipping, the force of friction must be a maximum, which is given by $f_{s,\,\max} = \mu_s n$. (Recall Eq. 5.8: $f_s \le \mu_s n$.) Thus, we must have $P = f_s = \mu_s n = \mu_s mg$.

To find $\theta_{\min}$, we must use the second condition for equilibrium. When we take the torques about an axis through the origin O at the bottom of the ladder, we have

$$(3) \qquad \sum \tau_O = P\ell \sin\theta - mg\frac{\ell}{2}\cos\theta = 0$$

This expression gives

$$\tan\theta_{\min} = \frac{mg}{2P} = \frac{mg}{2\mu_s mg} = \frac{1}{2\mu_s} = 1.25$$

$$\theta_{\min} = 51^\circ$$

What If? **What if a person begins to climb the ladder when the angle is 51°? Will the presence of a person on the ladder make it more or less likely to slip?**

Answer The presence of the additional weight of a person on the ladder will increase the clockwise torque about its base in Figure 12.11b. To maintain static equilibrium, the counterclockwise torque must increase, which can occur if P increases. Because equilibrium in the horizontal direction tells us that $P = f_s$, this would suggest that the friction force rises above the maximum value $f_{s,\,\max}$ and the ladder slips. However, the increased weight of the person also causes n to increase, which increases the maximum friction force $f_{s,\,\max}$! Thus, it is not clear conceptually whether the ladder is more or less likely to slip.

Imagine that the person of mass M is at a position d that is measured along the ladder from its base (Fig. 12.11c). Equations (1) and (2) can be rewritten

$$(4) \qquad \sum F_x = f_s - P = 0$$

$$(5) \qquad \sum F_y = n - (m + M)g = 0$$

Equation (3) can be rewritten

$$\sum \tau_O = P\ell \sin\theta - mg\frac{\ell}{2}\cos\theta - Mgd\cos\theta = 0$$

Solving this equation for $\tan\theta$, we find

$$\tan\theta = \frac{mg(\ell/2) + Mgd}{P\ell}$$

Incorporating Equations (4) and (5), and imposing the condition that the ladder is about to slip, this becomes

$$(6) \qquad \tan\theta_{\min} = \frac{m(\ell/2) + Md}{\mu_s \ell(m + M)}$$

When the person is at the bottom of the ladder, $d = 0$. In this case, there is no additional torque about the bottom of the ladder and the increased normal force causes the maximum static friction force to increase. Thus, the ladder is less likely to slip than in the absence of the person. As the person climbs and d becomes larger, however, the numerator in Equation (6) becomes larger. Thus, the minimum angle at which the ladder does not slip increases. Eventually, as the person climbs higher, the minimum angle becomes larger than 51° and the ladder slips. The particular value of d at which the ladder slips depends on the coefficient of friction and the masses of the person and the ladder.

At the Interactive Worked Example link at **http://www.pse6.com,** *you can adjust the angle of the ladder and watch what happens when it is released.*

Example 12.5 Negotiating a Curb

(A) Estimate the magnitude of the force **F** a person must apply to a wheelchair's main wheel to roll up over a sidewalk curb (Fig. 12.12a). This main wheel that comes in contact with the curb has a radius r, and the height of the curb is h.

Solution Normally, the person's hands supply the required force to a slightly smaller wheel that is concentric with the main wheel. For simplicity, we assume that the radius of the smaller wheel is the same as the radius of the main wheel. Let us estimate a combined weight of $mg = 1\,400$ N for the person and the wheelchair and choose a wheel radius of $r = 30$ cm. We also pick a curb height of $h = 10$ cm. We assume that the wheelchair and occupant are symmetric, and that each wheel supports a weight of 700 N. We then proceed to analyze only one of the wheels. Figure 12.12b shows the geometry for a single wheel.

When the wheel is just about to be raised from the street, the reaction force exerted by the ground on the wheel at point B goes to zero. Hence, at this time only three forces act on the wheel, as shown in the free-body diagram in Figure 12.12c. However, the force **R**, which is the force exerted by the curb on the wheel, acts at point A, and so if we choose to have our axis of rotation pass through point A, we do not need to include **R** in our torque equation. From the triangle OAC shown in Figure 12.12b, we see that the moment arm d of the gravitational force $m\mathbf{g}$ acting on the wheel relative to point A is

$$d = \sqrt{r^2 - (r - h)^2} = \sqrt{2rh - h^2}$$

The moment arm of **F** relative to point A is $2r - h$ (see Figure 12.12c). Therefore, the net torque acting on the wheel about point A is

$$mgd - F(2r - h) = 0$$

$$mg\sqrt{2rh - h^2} - F(2r - h) = 0$$

$$F = \frac{mg\sqrt{2rh - h^2}}{2r - h}$$

$$= \frac{(700 \text{ N})\sqrt{2(0.3 \text{ m})(0.1 \text{ m}) - (0.1 \text{ m})^2}}{2(0.3 \text{ m}) - 0.1 \text{ m}}$$

$$= 3 \times 10^2 \text{ N}$$

(Notice that we have kept only one digit as significant.) This result indicates that the force that must be applied to each wheel is substantial. You may want to estimate the force required to roll a wheelchair up a typical sidewalk accessibility ramp for comparison.

(B) Determine the magnitude and direction of **R**.

Solution We use the first condition for equilibrium to determine the direction:

$$\sum F_x = F - R\cos\theta = 0$$

$$\sum F_y = R\sin\theta - mg = 0$$

Dividing the second equation by the first gives

$$\tan\theta = \frac{mg}{F} = \frac{700 \text{ N}}{300 \text{ N}}$$

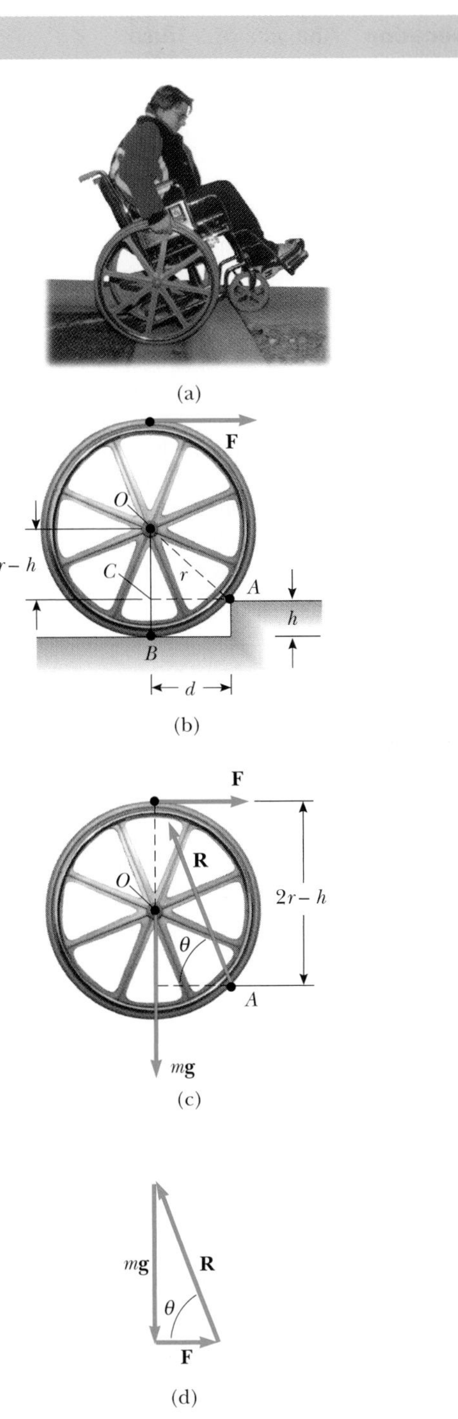

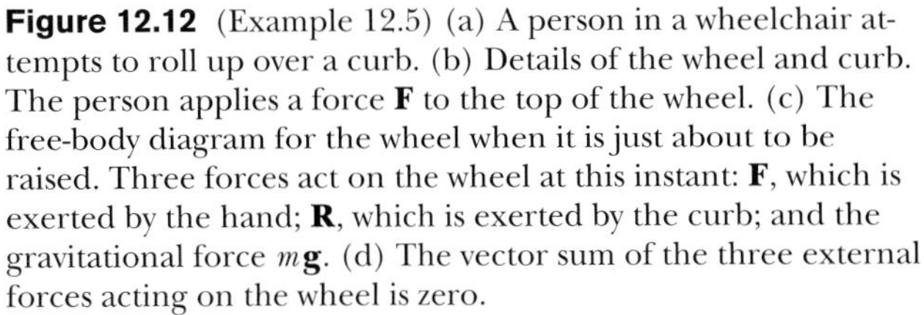

Figure 12.12 (Example 12.5) (a) A person in a wheelchair attempts to roll up over a curb. (b) Details of the wheel and curb. The person applies a force **F** to the top of the wheel. (c) The free-body diagram for the wheel when it is just about to be raised. Three forces act on the wheel at this instant: **F**, which is exerted by the hand; **R**, which is exerted by the curb; and the gravitational force $m\mathbf{g}$. (d) The vector sum of the three external forces acting on the wheel is zero.

$$\theta = 70°$$

We can use the right triangle shown in Figure 12.12d to obtain R:

$$R = \sqrt{(mg)^2 + F^2} = \sqrt{(700 \text{ N})^2 + (300 \text{ N})^2} = 800 \text{ N}$$

Application Analysis of a Truss

Roofs, bridges, and other structures that must be both strong and lightweight often are made of trusses similar to the one shown in Figure 12.13a. Imagine that this truss structure represents part of a bridge. To approach this problem, we assume that the structural components are connected by pin joints. We also assume that the entire structure is free to slide horizontally because it rests on "rockers" on each end, which allow it to move back and forth as it undergoes thermal expansion and contraction. We assume the mass of the bridge structure is negligible compared with the load. In this situation, the force exerted by each of the bars (struts) on the hinge pins is a force of tension or of compression and must be along the length of the bar. Let us calculate the force in each strut when the bridge is supporting a 7 200-N load at its center. We will do this by determining the forces that act at the pins.

The force notation that we use here is not of our usual format. Until now, we have used the notation F_{AB} to mean "the force exerted by A on B." For this application, however, the first letter in a double-letter subscript on F indicates the location of the pin on which the force is exerted. The combination of two letters identifies the strut exerting the force on the pin. For example, in Figure 12.13b, F_{AB} is the force exerted by strut AB on the pin at A. The subscripts are symmetric in that strut AB is the same as strut BA and $F_{AB} = F_{BA}$.

First, we apply Newton's second law to the truss as a whole in the vertical direction. Internal forces do not enter into this accounting. We balance the weight of the load with the normal forces exerted at the two ends by the supports on which the bridge rests:

$$\sum F_y = n_A + n_E - F_g = 0$$

$$n_A + n_E = 7\,200 \text{ N}$$

Next, we calculate the torque about A, noting that the overall length of the bridge structure is $L = 50$ m:

$$\sum \tau = Ln_E - (L/2)F_g = 0$$

$$n_E = F_g/2 = 3\,600 \text{ N}$$

Although we could repeat the torque calculation for the right end (point E), it should be clear from symmetry arguments that $n_A = 3\,600$ N.

Now let us balance the vertical forces acting on the pin at point A. If we assume that strut AB is in compression, then the force F_{AB} that the strut exerts on the pin at point A has a negative y component. (If the strut is actually in tension, our calculations will result in a negative value for the magnitude of the force, still of the correct size):

$$\sum F_y = n_A - F_{AB} \sin 30° = 0$$

$$F_{AB} = 7\,200 \text{ N}$$

The positive result shows that our assumption of compression was correct.

We can now find the force F_{AC} by considering the horizontal forces acting on the pin at point A. Because point A is not accelerating, we can safely assume that F_{AC} must point toward the right (Fig. 12.13b); this indicates that the bar between points A and C is under tension:

$$\sum F_x = F_{AC} - F_{AB} \cos 30° = 0$$

$$F_{AC} = (7\,200 \text{ N}) \cos 30° = 6\,200 \text{ N}$$

Now consider the vertical forces acting on the pin at point C. We shall assume that strut CB is in tension. (Imagine the subsequent motion of the pin at point C if strut CB were to

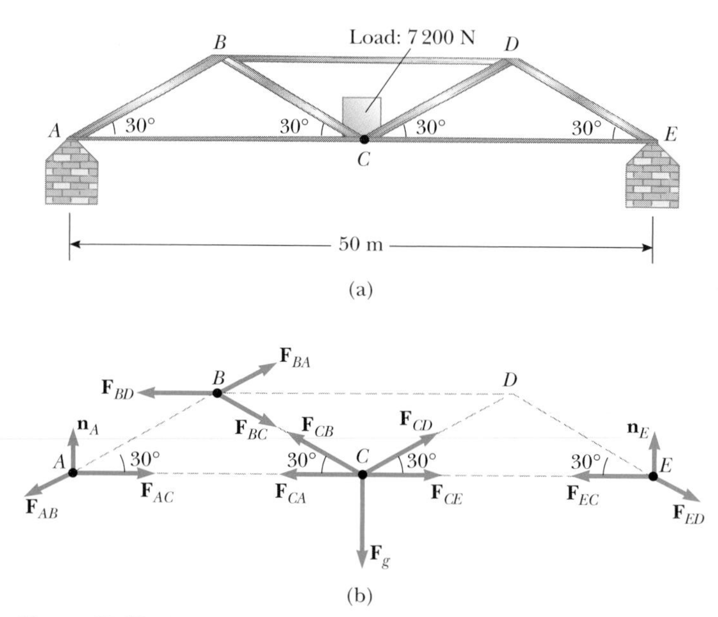

Figure 12.13 (a) Truss structure for a bridge. (b) The forces acting on the pins at points A, B, C, and E. Force vectors are not to scale.

break suddenly.) On the basis of symmetry, we assert that $F_{CB} = F_{CD}$ and $F_{CA} = F_{CE}$:

$$\sum F_y = 2\,F_{CB} \sin 30° - 7\,200 \text{ N} = 0$$

$$F_{CB} = 7\,200 \text{ N}$$

Finally, we balance the horizontal forces on B, assuming that strut BD is in compression:

$$\sum F_x = F_{BA} \cos 30° + F_{BC} \cos 30° - F_{BD} = 0$$

$$(7\,200 \text{ N}) \cos 30° + (7\,200 \text{ N}) \cos 30° - F_{BD} = 0$$

$$F_{BD} = 12\,000 \text{ N}$$

Thus, the top bar in a bridge of this design must be very strong.

12.4 Elastic Properties of Solids

Except for our discussion about springs in earlier chapters, we have assumed that objects remain rigid when external forces act on them. In reality, all objects are deformable. That is, it is possible to change the shape or the size (or both) of an object by applying external forces. As these changes take place, however, internal forces in the object resist the deformation.

We shall discuss the deformation of solids in terms of the concepts of *stress* and *strain*. **Stress** is a quantity that is proportional to the force causing a deformation; more specifically, stress is the external force acting on an object per unit cross-sectional area. The result of a stress is **strain**, which is a measure of the degree of deformation. It is found that, for sufficiently small stresses, **strain is proportional to stress;** the constant of proportionality depends on the material being deformed and on the nature of the deformation. We call this proportionality constant the **elastic modulus**. The elastic modulus is therefore defined as the ratio of the stress to the resulting strain:

$$\text{Elastic modulus} \equiv \frac{\text{stress}}{\text{strain}} \qquad (12.5)$$

The elastic modulus in general relates what is done to a solid object (a force is applied) to how that object responds (it deforms to some extent).

We consider three types of deformation and define an elastic modulus for each:

1. **Young's modulus**, which measures the resistance of a solid to a change in its length
2. **Shear modulus**, which measures the resistance to motion of the planes within a solid parallel to each other
3. **Bulk modulus**, which measures the resistance of solids or liquids to changes in their volume

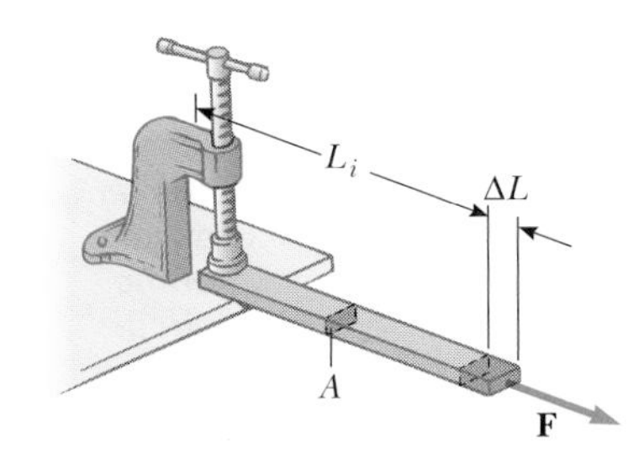

Active Figure 12.14 A long bar clamped at one end is stretched by an amount ΔL under the action of a force **F**.

At the Active Figures link at http://www.pse6.com, you can adjust the values of the applied force and Young's modulus to observe the change in length of the bar.

Young's Modulus: Elasticity in Length

Consider a long bar of cross-sectional area A and initial length L_i that is clamped at one end, as in Figure 12.14. When an external force is applied perpendicular to the cross section, internal forces in the bar resist distortion ("stretching"), but the bar reaches an equilibrium situation in which its final length L_f is greater than L_i and in which the external force is exactly balanced by internal forces. In such a situation, the bar is said to be stressed. We define the **tensile stress** as the ratio of the magnitude of the external force F to the cross-sectional area A. The **tensile strain** in this case is defined as the ratio of the change in length ΔL to the original length L_i. We define **Young's modulus** by a combination of these two ratios:

$$Y \equiv \frac{\text{tensile stress}}{\text{tensile strain}} = \frac{F/A}{\Delta L/L_i} \qquad (12.6)$$

Young's modulus

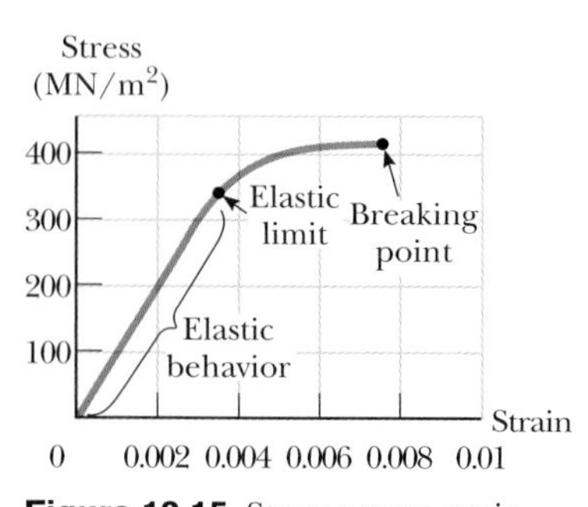

Figure 12.15 Stress-versus-strain curve for an elastic solid.

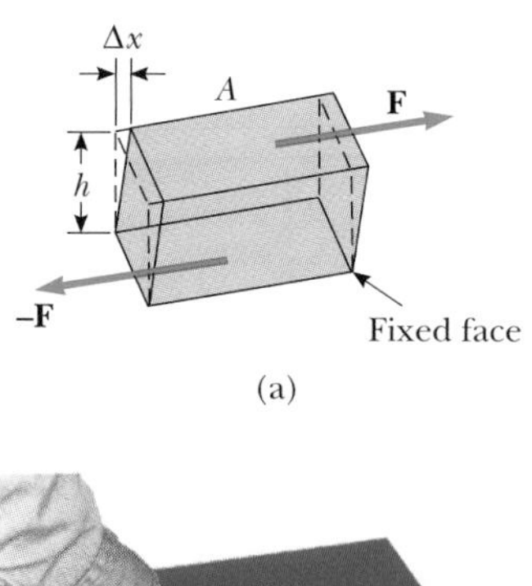

(a)

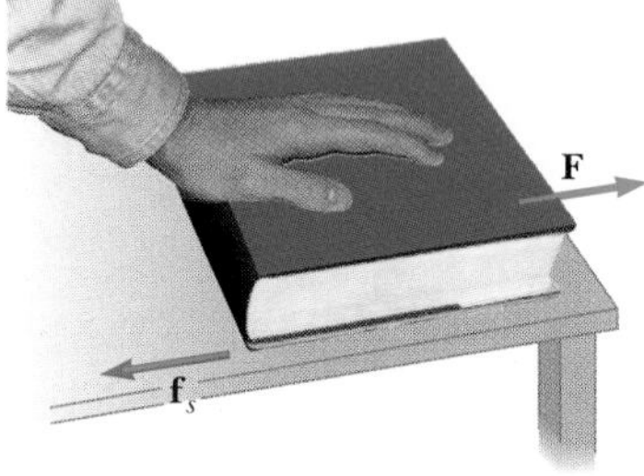

(b)

Active Figure 12.16 (a) A shear deformation in which a rectangular block is distorted by two forces of equal magnitude but opposite directions applied to two parallel faces. (b) A book under shear stress.

At the Active Figures link at http://www.pse6.com, you can adjust the values of the applied force and the shear modulus to observe the change in shape of the block in part (a).

Table 12.1

Typical Values for Elastic Moduli

Substance	Young's Modulus (N/m^2)	Shear Modulus (N/m^2)	Bulk Modulus (N/m^2)
Tungsten	35×10^{10}	14×10^{10}	20×10^{10}
Steel	20×10^{10}	8.4×10^{10}	6×10^{10}
Copper	11×10^{10}	4.2×10^{10}	14×10^{10}
Brass	9.1×10^{10}	3.5×10^{10}	6.1×10^{10}
Aluminum	7.0×10^{10}	2.5×10^{10}	7.0×10^{10}
Glass	$6.5–7.8 \times 10^{10}$	$2.6–3.2 \times 10^{10}$	$5.0–5.5 \times 10^{10}$
Quartz	5.6×10^{10}	2.6×10^{10}	2.7×10^{10}
Water	—	—	0.21×10^{10}
Mercury	—	—	2.8×10^{10}

Young's modulus is typically used to characterize a rod or wire stressed under either tension or compression. Note that because strain is a dimensionless quantity, Y has units of force per unit area. Typical values are given in Table 12.1. Experiments show (a) that for a fixed applied force, the change in length is proportional to the original length and (b) that the force necessary to produce a given strain is proportional to the cross-sectional area. Both of these observations are in accord with Equation 12.6.

For relatively small stresses, the bar will return to its initial length when the force is removed. The **elastic limit** of a substance is defined as the maximum stress that can be applied to the substance before it becomes permanently deformed and does not return to its initial length. It is possible to exceed the elastic limit of a substance by applying a sufficiently large stress, as seen in Figure 12.15. Initially, a stress-versus-strain curve is a straight line. As the stress increases, however, the curve is no longer a straight line. When the stress exceeds the elastic limit, the object is permanently distorted and does not return to its original shape after the stress is removed. As the stress is increased even further, the material ultimately breaks.

Shear Modulus: Elasticity of Shape

Another type of deformation occurs when an object is subjected to a force parallel to one of its faces while the opposite face is held fixed by another force (Fig. 12.16a). The stress in this case is called a shear stress. If the object is originally a rectangular block, a shear stress results in a shape whose cross section is a parallelogram. A book pushed sideways, as shown in Figure 12.16b, is an example of an object subjected to a shear stress. To a first approximation (for small distortions), no change in volume occurs with this deformation.

We define the **shear stress** as F/A, the ratio of the tangential force to the area A of the face being sheared. The **shear strain** is defined as the ratio $\Delta x/h$, where Δx is the horizontal distance that the sheared face moves and h is the height of the object. In terms of these quantities, the **shear modulus** is

Shear modulus

$$S \equiv \frac{\text{shear stress}}{\text{shear strain}} = \frac{F/A}{\Delta x/h} \tag{12.7}$$

Values of the shear modulus for some representative materials are given in Table 12.1. Like Young's modulus, the unit of shear modulus is the ratio of that for force to that for area.

Bulk Modulus: Volume Elasticity

Bulk modulus characterizes the response of an object to changes in a force of uniform magnitude applied perpendicularly over the entire surface of the object, as shown in Figure 12.17. (We assume here that the object is made of a single substance.) As we shall see in Chapter 14, such a uniform distribution of forces occurs when an object is immersed in a fluid. An object subject to this type of deformation undergoes a change in volume but no change in shape. The **volume stress** is defined as the ratio of the magnitude of the total force F exerted on a surface to the area A of the surface. The quantity $P = F/A$ is called **pressure**, which we will study in more detail in Chapter 14. If the pressure on an object changes by an amount $\Delta P = \Delta F/A$, then the object will experience a volume change ΔV. The **volume strain** is equal to the change in volume ΔV divided by the initial volume V_i. Thus, from Equation 12.5, we can characterize a volume ("bulk") compression in terms of the **bulk modulus**, which is defined as

$$B \equiv \frac{\text{volume stress}}{\text{volume strain}} = -\frac{\Delta F/A}{\Delta V/V_i} = -\frac{\Delta P}{\Delta V/V_i} \quad (12.8)$$

Bulk modulus

A negative sign is inserted in this defining equation so that B is a positive number. This maneuver is necessary because an increase in pressure (positive ΔP) causes a decrease in volume (negative ΔV) and vice versa.

Table 12.1 lists bulk moduli for some materials. If you look up such values in a different source, you often find that the reciprocal of the bulk modulus is listed. The reciprocal of the bulk modulus is called the **compressibility** of the material.

Note from Table 12.1 that both solids and liquids have a bulk modulus. However, no shear modulus and no Young's modulus are given for liquids because a liquid does not sustain a shearing stress or a tensile stress. If a shearing force or a tensile force is applied to a liquid, the liquid simply flows in response.

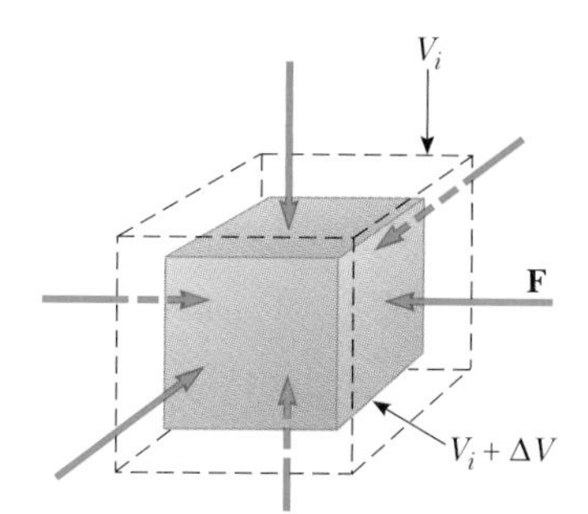

Active Figure 12.17 When a solid is under uniform pressure, it undergoes a change in volume but no change in shape. This cube is compressed on all sides by forces normal to its six faces.

At the Active Figures link at http://www.pse6.com, you can adjust the values of the applied force and the bulk modulus to observe the change in volume of the cube.

Quick Quiz 12.4 A block of iron is sliding across a horizontal floor. The friction force between the block and the floor causes the block to deform. To describe the relationship between stress and strain for the block, you would use (a) Young's modulus (b) shear modulus (c) bulk modulus (d) none of these.

Quick Quiz 12.5 A trapeze artist swings through a circular arc. At the bottom of the swing, the wires supporting the trapeze are longer than when the trapeze artist simply hangs from the trapeze, due to the increased tension in them. To describe the relationship between stress and strain for the wires, you would use (a) Young's modulus (b) shear modulus (c) bulk modulus (d) none of these.

Quick Quiz 12.6 A spacecraft carries a steel sphere to a planet on which atmospheric pressure is much higher than on the Earth. The higher pressure causes the radius of the sphere to decrease. To describe the relationship between stress and strain for the sphere, you would use (a) Young's modulus (b) shear modulus (c) bulk modulus (d) none of these.

Prestressed Concrete

If the stress on a solid object exceeds a certain value, the object fractures. The maximum stress that can be applied before fracture occurs depends on the nature of the material and on the type of applied stress. For example, concrete has a tensile strength of about 2×10^6 N/m^2, a compressive strength of 20×10^6 N/m^2, and a shear strength of 2×10^6 N/m^2. If the applied stress exceeds these values, the concrete fractures. It is common practice to use large safety factors to prevent failure in concrete structures.

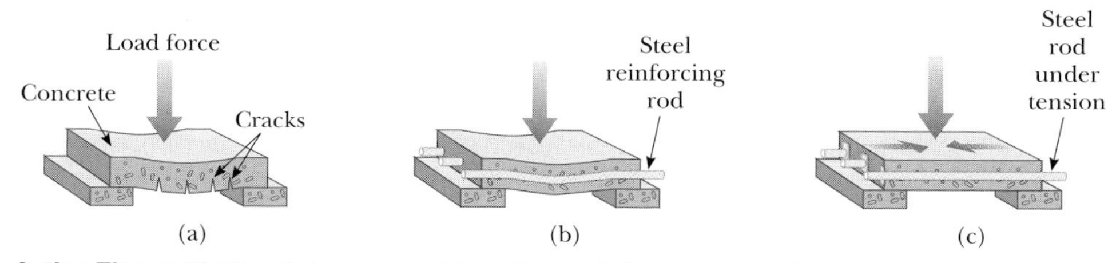

Active Figure 12.18 (a) A concrete slab with no reinforcement tends to crack under a heavy load. (b) The strength of the concrete is increased by using steel reinforcement rods. (c) The concrete is further strengthened by prestressing it with steel rods under tension.

Concrete is normally very brittle when it is cast in thin sections. Thus, concrete slabs tend to sag and crack at unsupported areas, as shown in Figure 12.18a. The slab can be strengthened by the use of steel rods to reinforce the concrete, as illustrated in Figure 12.18b. Because concrete is much stronger under compression (squeezing) than under tension (stretching) or shear, vertical columns of concrete can support very heavy loads, whereas horizontal beams of concrete tend to sag and crack. However, a significant increase in shear strength is achieved if the reinforced concrete is prestressed, as shown in Figure 12.18c. As the concrete is being poured, the steel rods are held under tension by external forces. The external forces are released after the concrete cures; this results in a permanent tension in the steel and hence a compressive stress on the concrete. This enables the concrete slab to support a much heavier load.

Example 12.6 Stage Design

Recall Example 8.4, in which we analyzed a cable used to support an actor as he swung onto the stage. Suppose that the tension in the cable is 940 N as the actor reaches the lowest point. What diameter should a 10-m-long steel wire have if we do not want it to stretch more than 0.5 cm under these conditions?

Solution From the definition of Young's modulus, we can solve for the required cross-sectional area. Assuming that the cross section is circular, we can determine the diameter of the wire. From Equation 12.6, we have

$$Y = \frac{F/A}{\Delta L/L_i}$$

$$A = \frac{FL_i}{Y\Delta L} = \frac{(940\text{ N})(10\text{ m})}{(20 \times 10^{10}\text{ N/m}^2)(0.005\text{ m})}$$

$$= 9.4 \times 10^{-6}\text{ m}^2$$

Because $A = \pi r^2$, the radius of the wire can be found from

$$r = \sqrt{\frac{A}{\pi}} = \sqrt{\frac{9.4 \times 10^{-6}\text{ m}^2}{\pi}} = 1.7 \times 10^{-3}\text{ m} = 1.7\text{ mm}$$

$$d = 2r = 2(1.7\text{ mm}) = 3.4\text{ mm}$$

To provide a large margin of safety, we would probably use a flexible cable made up of many smaller wires having a total cross-sectional area substantially greater than our calculated value.

Example 12.7 Squeezing a Brass Sphere

A solid brass sphere is initially surrounded by air, and the air pressure exerted on it is $1.0 \times 10^5\text{ N/m}^2$ (normal atmospheric pressure). The sphere is lowered into the ocean to a depth where the pressure is $2.0 \times 10^7\text{ N/m}^2$. The volume of the sphere in air is 0.50 m^3. By how much does this volume change once the sphere is submerged?

Solution From the definition of bulk modulus, we have

$$B = -\frac{\Delta P}{\Delta V/V_i}$$

$$\Delta V = -\frac{V_i \Delta P}{B}$$

Substituting the numerical values, we obtain

$$\Delta V = -\frac{(0.50\text{ m}^3)(2.0 \times 10^7\text{ N/m}^2 - 1.0 \times 10^5\text{ N/m}^2)}{6.1 \times 10^{10}\text{ N/m}^2}$$

$$= -1.6 \times 10^{-4}\text{ m}^3$$

The negative sign indicates that the volume of the sphere decreases.

SUMMARY

Take a practice test for this chapter by clicking on the Practice Test link at http://www.pse6.com.

A rigid object is in **equilibrium** if and only if **the resultant external force acting on it is zero and the resultant external torque on it is zero about any axis:**

$$\Sigma \mathbf{F} = 0 \qquad (12.1)$$

$$\Sigma \boldsymbol{\tau} = 0 \qquad (12.2)$$

The first condition is the condition for translational equilibrium, and the second is the condition for rotational equilibrium. These two equations allow you to analyze a great variety of problems. Make sure you can identify forces unambiguously, create a free-body diagram, and then apply Equations 12.1 and 12.2 and solve for the unknowns.

The gravitational force exerted on an object can be considered as acting at a single point called the **center of gravity**. The center of gravity of an object coincides with its center of mass if the object is in a uniform gravitational field.

We can describe the elastic properties of a substance using the concepts of stress and strain. **Stress** is a quantity proportional to the force producing a deformation; **strain** is a measure of the degree of deformation. Strain is proportional to stress, and the constant of proportionality is the **elastic modulus:**

$$\text{Elastic modulus} \equiv \frac{\text{stress}}{\text{strain}} \qquad (12.5)$$

Three common types of deformation are represented by (1) the resistance of a solid to elongation under a load, characterized by **Young's modulus** Y; (2) the resistance of a solid to the motion of internal planes sliding past each other, characterized by the **shear modulus** S; and (3) the resistance of a solid or fluid to a volume change, characterized by the **bulk modulus** B.

QUESTIONS

1. Stand with your back against a wall. Why can't you put your heels firmly against the wall and then bend forward without falling?

2. Can an object be in equilibrium if it is in motion? Explain.

3. Can an object be in equilibrium when only one force acts upon it? If you believe the answer is yes, give an example to support your conclusion.

4. (a) Give an example in which the net force acting on an object is zero and yet the net torque is nonzero. (b) Give an example in which the net torque acting on an object is zero and yet the net force is nonzero.

5. Can an object be in equilibrium if the only torques acting on it produce clockwise rotation?

6. If you measure the net force and the net torque on a system to be zero, (a) could the system still be rotating with respect to you? (b) Could it be translating with respect to you?

7. The center of gravity of an object may be located outside the object. Give a few examples for which this is the case.

8. Assume you are given an arbitrarily shaped piece of plywood, together with a hammer, nail, and plumb bob. How could you use these items to locate the center of gravity of the plywood? *Suggestion:* Use the nail to suspend the plywood.

9. For a chair to be balanced on one leg, where must the center of gravity of the chair be located?

10. A girl has a large, docile dog she wishes to weigh on a small bathroom scale. She reasons that she can determine her dog's weight with the following method: First she puts the dog's two front feet on the scale and records the scale reading. Then she places the dog's two back feet on the scale and records the reading. She thinks that the sum of the readings will be the dog's weight. Is she correct? Explain your answer.

11. A tall crate and a short crate of equal mass are placed side by side on an incline, without touching each other. As the incline angle is increased, which crate will topple first? Explain.

12. A ladder stands on the ground, leaning against a wall. Would you feel safer climbing up the ladder if you were told that the ground is frictionless but the wall is rough, or that the wall is frictionless but the ground is rough? Justify your answer.

13. When you are lifting a heavy object, it is recommended that you keep your back as nearly vertical as possible, lifting from your knees. Why is this better than bending over and lifting from your waist?

14. What kind of deformation does a cube of Jell-O exhibit when it jiggles?

15. Ruins of ancient Greek temples often have intact vertical columns, but few horizontal slabs of stone are still in place. Can you think of a reason why this is so?

PROBLEMS

1, 2, 3 = straightforward, intermediate, challenging □ = full solution available in the *Student Solutions Manual and Study Guide*

= coached solution with hints available at http://www.pse6.com = computer useful in solving problem

= paired numerical and symbolic problems

Section 12.1 The Conditions for Equilibrium of a Rigid Body

1. A baseball player holds a 36-oz bat (weight = 10.0 N) with one hand at the point O (Fig. P12.1). The bat is in equilibrium. The weight of the bat acts along a line 60.0 cm to the right of O. Determine the force and the torque exerted by the player on the bat around an axis through O.

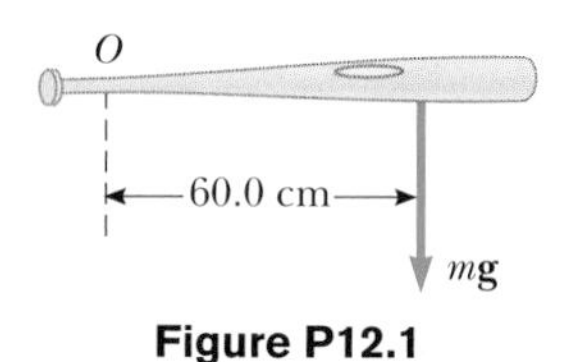

Figure P12.1

2. Write the necessary conditions for equilibrium of the object shown in Figure P12.2. Take the origin of the torque equation at the point O.

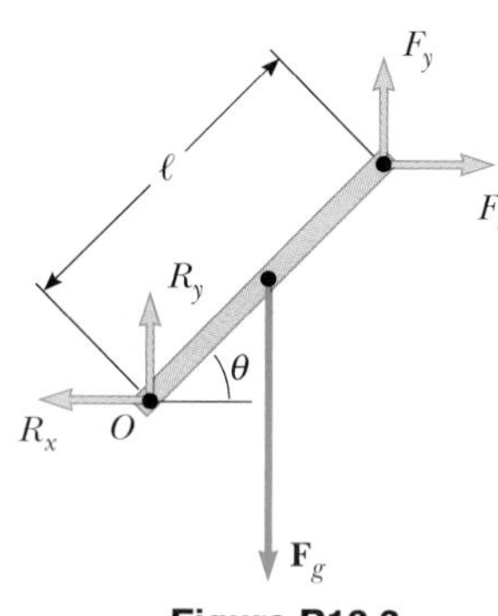

Figure P12.2

3. A uniform beam of mass m_b and length ℓ supports blocks with masses m_1 and m_2 at two positions, as in Figure P12.3. The beam rests on two knife edges. For what value of x will the beam be balanced at P such that the normal force at O is zero?

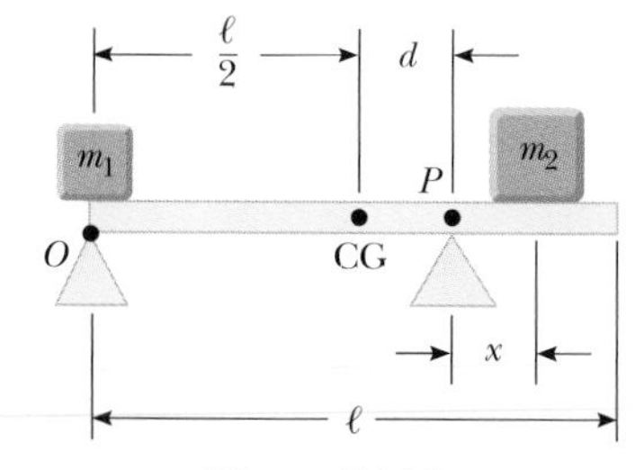

Figure P12.3

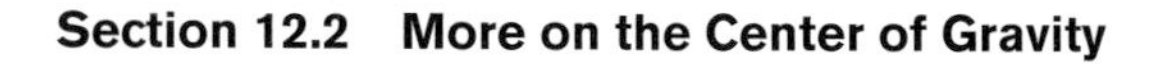

Section 12.2 More on the Center of Gravity

Problems 38, 39, 41, 43, and 44 in Chapter 9 can also be assigned with this section.

4. A circular pizza of radius R has a circular piece of radius $R/2$ removed from one side as shown in Figure P12.4. The center of gravity has moved from C to C' along the x axis. Show that the distance from C to C' is $R/6$. Assume the thickness and density of the pizza are uniform throughout.

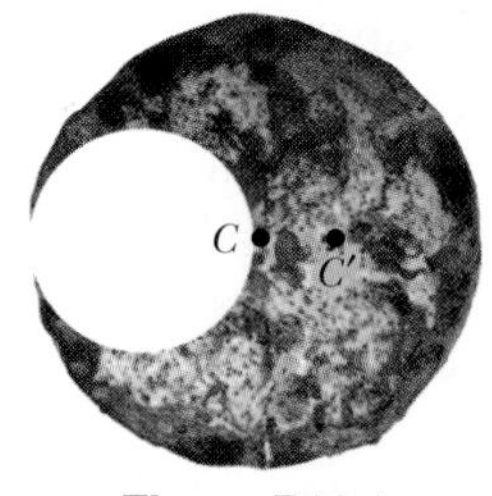

Figure P12.4

5. A carpenter's square has the shape of an L, as in Figure P12.5. Locate its center of gravity.

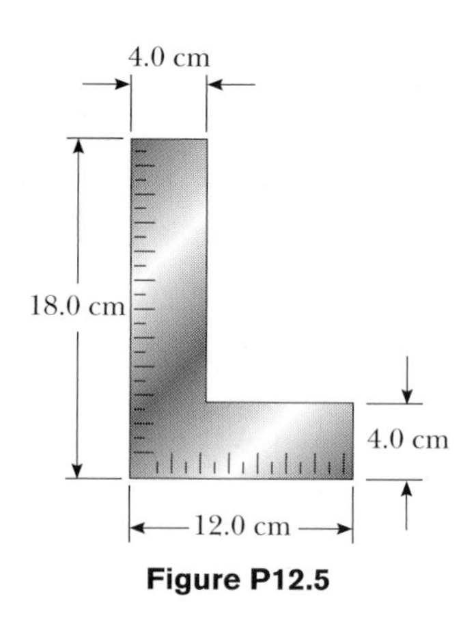

Figure P12.5

6. Pat builds a track for his model car out of wood, as in Figure P12.6. The track is 5.00 cm wide, 1.00 m high, and 3.00 m

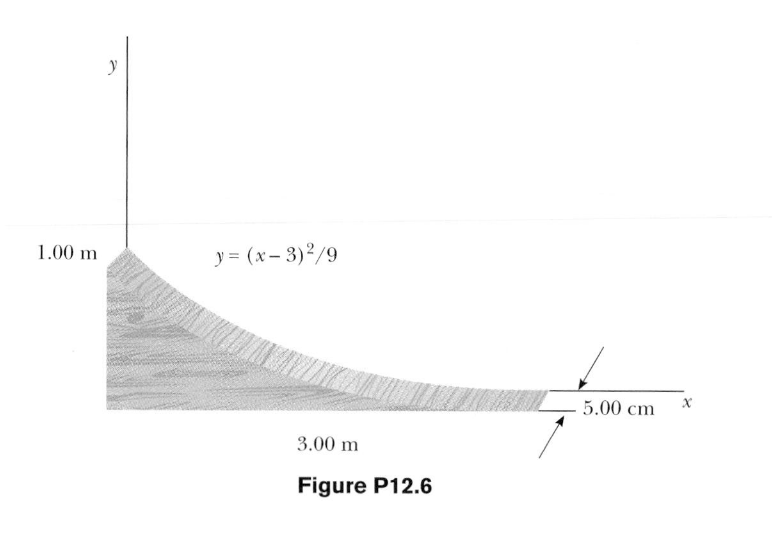

Figure P12.6

long, and is solid. The runway is cut such that it forms a parabola with the equation $y = (x - 3)^2/9$. Locate the horizontal coordinate of the center of gravity of this track.

7. Consider the following mass distribution: 5.00 kg at (0, 0) m, 3.00 kg at (0, 4.00) m, and 4.00 kg at (3.00, 0) m. Where should a fourth object of mass 8.00 kg be placed so that the center of gravity of the four-object arrangement will be at (0, 0)?

8. Figure P12.8 shows three uniform objects: a rod, a right triangle, and a square. Their masses and their coordinates in meters are given. Determine the center of gravity for the three-object system.

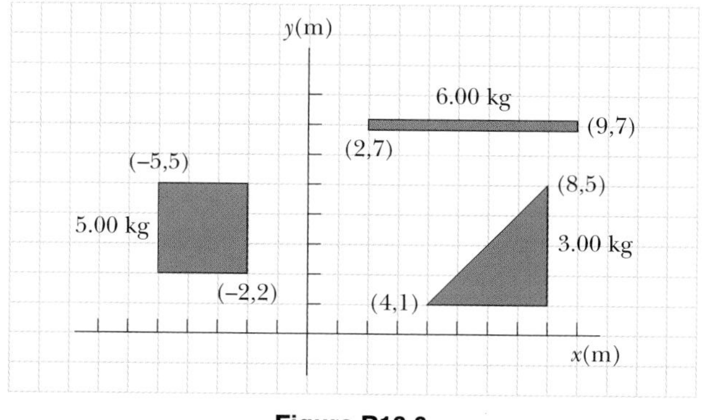

Figure P12.8

Section 12.3 Examples of Rigid Objects in Static Equilibrium

Problems 17, 18, 19, 20, 21, 27, 40, 46, 57, 59, and 73 in Chapter 5 can also be assigned with this section.

9. Find the mass m of the counterweight needed to balance the 1 500-kg truck on the incline shown in Figure P12.9. Assume all pulleys are frictionless and massless.

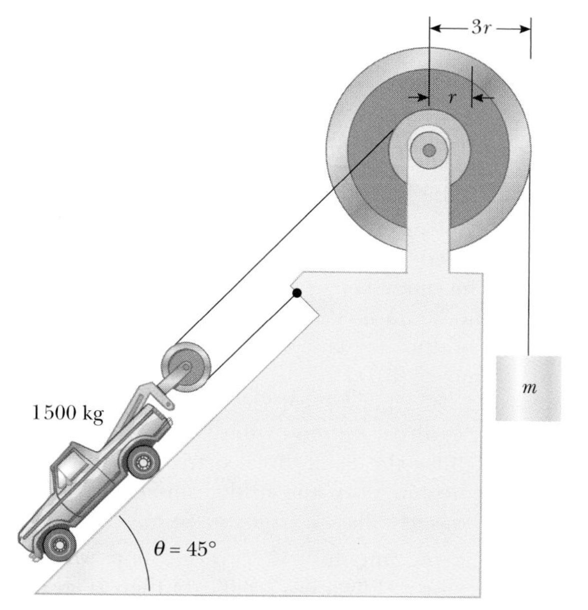

Figure P12.9

10. A mobile is constructed of light rods, light strings, and beach souvenirs, as shown in Figure P12.10. Determine the masses of the objects (a) m_1, (b) m_2, and (c) m_3.

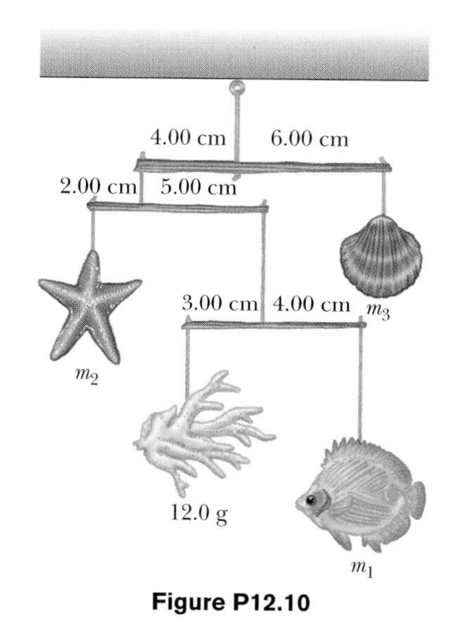

Figure P12.10

11. Two pans of a balance are 50.0 cm apart. The fulcrum of the balance has been shifted 1.00 cm away from the center by a dishonest shopkeeper. By what percentage is the true weight of the goods being marked up by the shopkeeper? (Assume the balance has negligible mass.)

12. A 20.0-kg floodlight in a park is supported at the end of a horizontal beam of negligible mass that is hinged to a pole, as shown in Figure P12.12. A cable at an angle of 30.0° with the beam helps to support the light. Find (a) the tension in the cable and (b) the horizontal and vertical forces exerted on the beam by the pole.

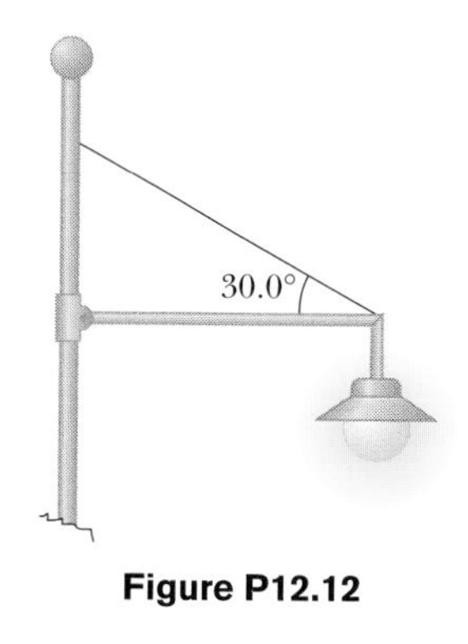

Figure P12.12

13. A 15.0-m uniform ladder weighing 500 N rests against a frictionless wall. The ladder makes a 60.0° angle with the horizontal. (a) Find the horizontal and vertical forces the ground exerts on the base of the ladder when an 800-N firefighter is 4.00 m from the bottom. (b) If the ladder is just on the verge of slipping when the firefighter is 9.00 m up, what is the coefficient of static friction between ladder and ground?

14. A uniform ladder of length L and mass m_1 rests against a frictionless wall. The ladder makes an angle θ with the horizontal. (a) Find the horizontal and vertical forces the ground exerts on the base of the ladder when a firefighter of mass m_2 is a distance x from the bottom. (b) If the ladder is just on the verge of slipping when the firefighter is a distance d from the bottom, what is the coefficient of static friction between ladder and ground?

15. Figure P12.15 shows a claw hammer as it is being used to pull a nail out of a horizontal board. If a force of 150 N is exerted horizontally as shown, find (a) the force exerted by the hammer claws on the nail and (b) the force exerted by the surface on the point of contact with the hammer head. Assume that the force the hammer exerts on the nail is parallel to the nail.

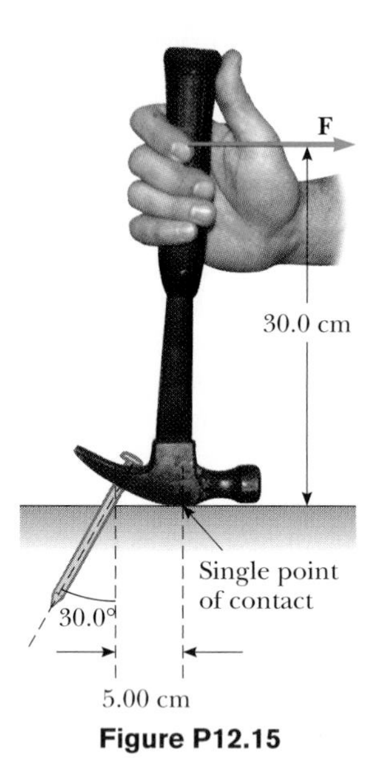

Figure P12.15

16. A uniform plank of length 6.00 m and mass 30.0 kg rests horizontally across two horizontal bars of a scaffold. The bars are 4.50 m apart, and 1.50 m of the plank hangs over one side of the scaffold. Draw a free-body diagram of the plank. How far can a painter of mass 70.0 kg walk on the overhanging part of the plank before it tips?

17. A 1 500-kg automobile has a wheel base (the distance between the axles) of 3.00 m. The center of mass of the automobile is on the center line at a point 1.20 m behind the front axle. Find the force exerted by the ground on each wheel.

18. A vertical post with a square cross section is 10.0 m tall. Its bottom end is encased in a base 1.50 m tall, which is precisely square but slightly loose. A force 5.50 N to the right acts on the top of the post. The base maintains the post in equilibrium. Find the force that the top of the right side wall of the base exerts on the post. Find the force that the bottom of the left side wall of the base exerts on the post.

19. A flexible chain weighing 40.0 N hangs between two hooks located at the same height (Fig. P12.19). At each hook, the tangent to the chain makes an angle $\theta = 42.0°$ with the horizontal. Find (a) the magnitude of the force each hook exerts on the chain and (b) the tension in the chain at its midpoint. (*Suggestion:* for part (b), make a free-body diagram for half of the chain.)

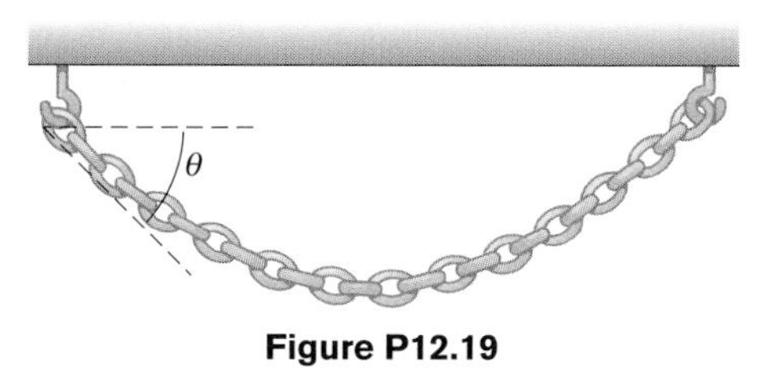

Figure P12.19

20. Sir Lost-a-Lot dons his armor and sets out from the castle on his trusty steed in his quest to improve communication between damsels and dragons (Fig. P12.20). Unfortunately his squire lowered the drawbridge too far and finally stopped it 20.0° below the horizontal. Lost-a-Lot and his horse stop when their combined center of mass is 1.00 m from the end of the bridge. The uniform bridge is 8.00 m long and has mass 2 000 kg. The lift cable is attached to the bridge 5.00 m from the hinge at the castle end, and to a point on the castle wall 12.0 m above the bridge. Lost-a-Lot's mass combined with his armor and steed is 1 000 kg. Determine (a) the tension in the cable and the (b) horizontal and (c) vertical force components acting on the bridge at the hinge.

Figure P12.20 Problems 20 and 21.

21. **Review problem.** In the situation described in Problem 20 and illustrated in Figure P12.20, the lift cable suddenly breaks! The hinge between the castle wall and the bridge is frictionless, and the bridge swings freely until it is vertical. (a) Find the angular acceleration of the bridge once it starts to move. (b) Find the angular speed of the bridge when it strikes the vertical castle wall below the hinge. (c) Find the force exerted by the hinge on the bridge immediately after the cable breaks. (d) Find the force exerted by the hinge on the bridge immediately before it strikes the castle wall.

22. Stephen is pushing his sister Joyce in a wheelbarrow when it is stopped by a brick 8.00 cm high (Fig. P12.22).

The handles make an angle of 15.0° below the horizontal. A downward force of 400 N is exerted on the wheel, which has a radius of 20.0 cm. (a) What force must Stephen apply along the handles in order to just start the wheel over the brick? (b) What is the force (magnitude and direction) that the brick exerts on the wheel just as the wheel begins to lift over the brick? Assume in both parts that the brick remains fixed and does not slide along the ground.

Figure P12.22

23. One end of a uniform 4.00-m-long rod of weight F_g is supported by a cable. The other end rests against the wall, where it is held by friction, as in Figure P12.23. The coefficient of static friction between the wall and the rod is $\mu_s = 0.500$. Determine the minimum distance x from point A at which an additional weight F_g (the same as the weight of the rod) can be hung without causing the rod to slip at point A.

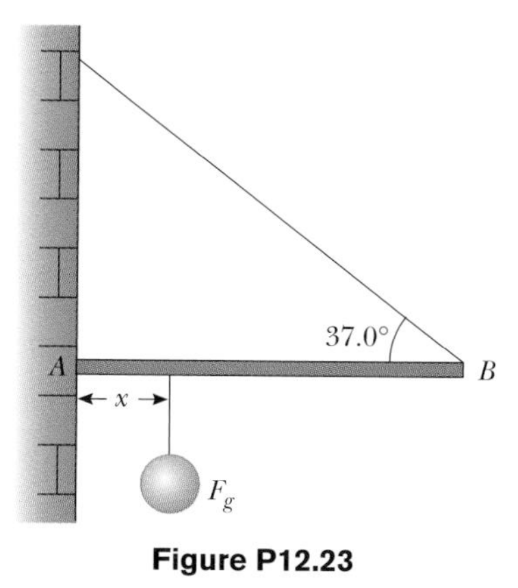

Figure P12.23

24. Two identical uniform bricks of length L are placed in a stack over the edge of a horizontal surface with the maximum overhang possible without falling, as in Figure P12.24. Find the distance x.

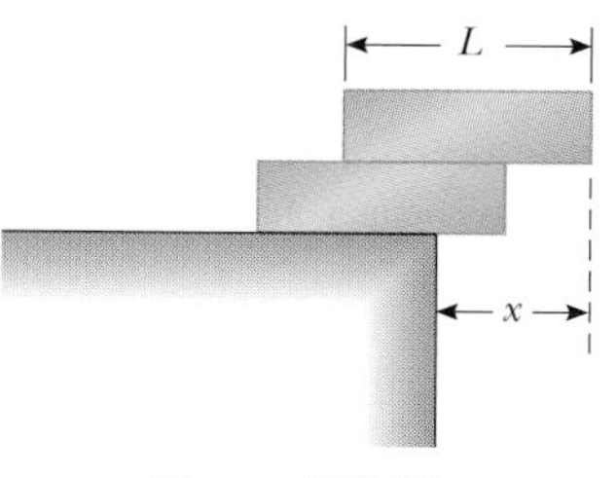

Figure P12.24

25. A vaulter holds a 29.4-N pole in equilibrium by exerting an upward force **U** with her leading hand and a downward force **D** with her trailing hand, as shown in Figure P12.25. Point C is the center of gravity of the pole. What are the magnitudes of **U** and **D**?

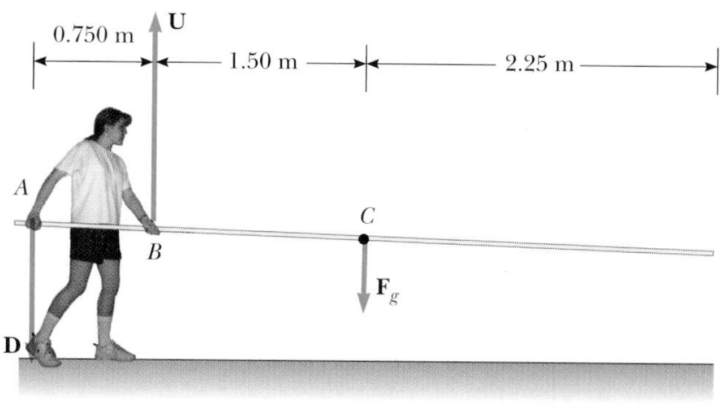

Figure P12.25

26. In the **What If?** section of Example 12.3, let x represent the distance in meters between the person and the hinge at the left end of the beam. (a) Show that the cable tension in newtons is given by $T = 93.9x + 125$. Argue that T increases as x increases. (b) Show that the direction angle θ of the hinge force is described by

$$\tan\theta = \left(\frac{32}{3x+4} - 1\right)\tan 53.0°$$

How does θ change as x increases? (c) Show that the magnitude of the hinge force is given by

$$R = \sqrt{8.82 \times 10^3 x^2 - 9.65 \times 10^4 x + 4.96 \times 10^5}$$

How does R change as x increases?

Section 12.4 Elastic Properties of Solids

27. A 200-kg load is hung on a wire having a length of 4.00 m, cross-sectional area 0.200×10^{-4} m^2, and Young's modulus 8.00×10^{10} N/m^2. What is its increase in length?

28. Assume that Young's modulus is 1.50×10^{10} N/m^2 for bone and that the bone will fracture if stress greater than 1.50×10^8 N/m^2 is imposed on it. (a) What is the maximum force that can be exerted on the femur bone in the leg if it has a minimum effective diameter of 2.50 cm? (b) If this much force is applied compressively, by how much does the 25.0-cm-long bone shorten?

29. Evaluate Young's modulus for the material whose stress-versus-strain curve is shown in Figure 12.15.

30. A steel wire of diameter 1 mm can support a tension of 0.2 kN. A cable to support a tension of 20 kN should have diameter of what order of magnitude?

31. A child slides across a floor in a pair of rubber-soled shoes. The friction force acting on each foot is 20.0 N.

The footprint area of each shoe sole is 14.0 cm^2, and the thickness of each sole is 5.00 mm. Find the horizontal distance by which the upper and lower surfaces of each sole are offset. The shear modulus of the rubber is 3.00 MN/m^2.

32. **Review problem.** A 30.0-kg hammer strikes a steel spike 2.30 cm in diameter while moving with speed 20.0 m/s. The hammer rebounds with speed 10.0 m/s after 0.110 s. What is the average strain in the spike during the impact?

33. If the shear stress in steel exceeds 4.00×10^8 N/m^2, the steel ruptures. Determine the shearing force necessary to (a) shear a steel bolt 1.00 cm in diameter and (b) punch a 1.00-cm-diameter hole in a steel plate 0.500 cm thick.

34. **Review problem.** A 2.00-m-long cylindrical steel wire with a cross-sectional diameter of 4.00 mm is placed over a light frictionless pulley, with one end of the wire connected to a 5.00-kg object and the other end connected to a 3.00-kg object. By how much does the wire stretch while the objects are in motion?

35. When water freezes, it expands by about 9.00%. What pressure increase would occur inside your automobile engine block if the water in it froze? (The bulk modulus of ice is 2.00×10^9 N/m^2.)

36. The deepest point in the ocean is in the Mariana Trench, about 11 km deep. The pressure at this depth is huge, about 1.13×10^8 N/m^2. (a) Calculate the change in volume of 1.00 m^3 of seawater carried from the surface to this deepest point in the Pacific ocean. (b) The density of seawater at the surface is 1.03×10^3 kg/m^3. Find its density at the bottom. (c) Is it a good approximation to think of water as incompressible?

37. A walkway suspended across a hotel lobby is supported at numerous points along its edges by a vertical cable above each point and a vertical column underneath. The steel cable is 1.27 cm in diameter and is 5.75 m long before loading. The aluminum column is a hollow cylinder with an inside diameter of 16.14 cm, an outside diameter of 16.24 cm, and unloaded length of 3.25 m. When the walkway exerts a load force of 8 500 N on one of the support points, how much does the point move down?

Additional Problems

38. A lightweight, rigid beam 10.0 m long is supported by a cable attached to a spring of force constant $k = 8.25$ kN/m as shown in Figure P12.38. When no load is hung on the beam ($F_g = 0$), the length L is equal to 5.00 m. (a) Find the angle θ in this situation. (b) Now a load of $F_g = 250$ N is hung on the end of the beam. Temporarily ignore the extension of the spring and the change in the angle θ. Calculate the tension in the cable with this approximation. (c) Use the answer to part (b) to calculate the spring elongation and a new value for the angle θ. (d) With the value of θ from part (c), find a second approximation for the tension in the cable. (e) Use the answer to part (d) to calculate more precise values for the spring elongation and

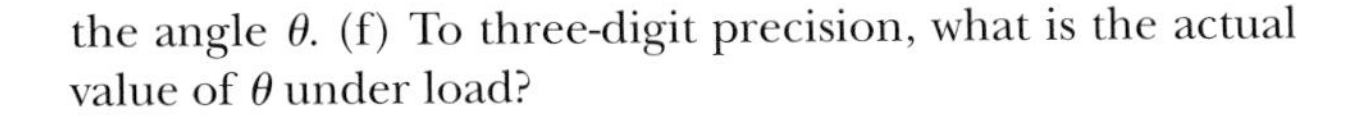
the angle θ. (f) To three-digit precision, what is the actual value of θ under load?

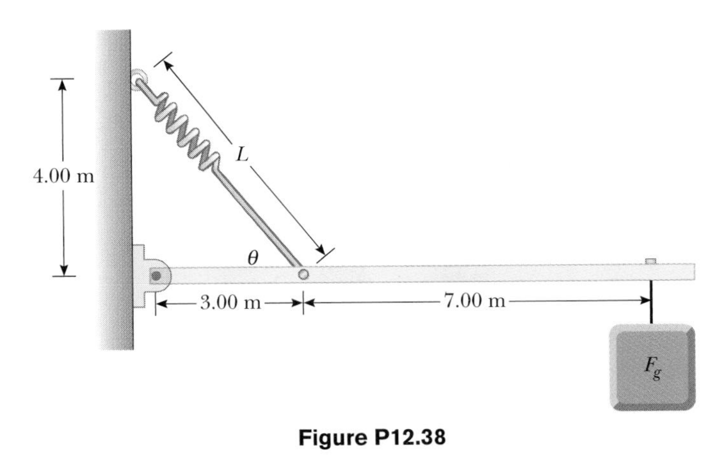

Figure P12.38

39. A bridge of length 50.0 m and mass 8.00×10^4 kg is supported on a smooth pier at each end as in Figure P12.39. A truck of mass 3.00×10^4 kg is located 15.0 m from one end. What are the forces on the bridge at the points of support?

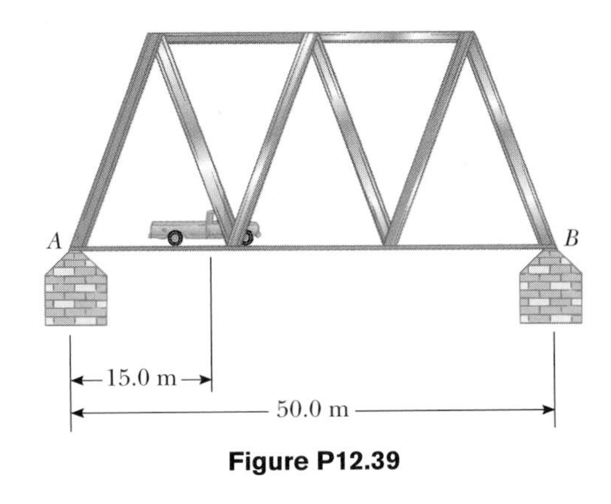

Figure P12.39

40. Refer to Figure 12.18(c). A lintel of prestressed reinforced concrete is 1.50 m long. The cross-sectional area of the concrete is 50.0 cm^2. The concrete encloses one steel reinforcing rod with cross-sectional area 1.50 cm^2. The rod joins two strong end plates. Young's modulus for the concrete is 30.0×10^9 N/m^2. After the concrete cures and the original tension T_1 in the rod is released, the concrete is to be under compressive stress 8.00×10^6 N/m^2. (a) By what distance will the rod compress the concrete when the original tension in the rod is released? (b) The rod will still be under what tension T_2? (c) The rod will then be how much longer than its unstressed length? (d) When the concrete was poured, the rod should have been stretched by what extension distance from its unstressed length? (e) Find the required original tension T_1 in the rod.

41. A uniform pole is propped between the floor and the ceiling of a room. The height of the room is 7.80 ft, and the coefficient of static friction between the pole and the ceiling is 0.576. The coefficient of static friction between the pole and the floor is greater than that. What is the length of the longest pole that can be propped between the floor and the ceiling?

42. A solid sphere of radius R and mass M is placed in a trough as shown in Figure P12.42. The inner surfaces of the trough are frictionless. Determine the forces exerted by the trough on the sphere at the two contact points.

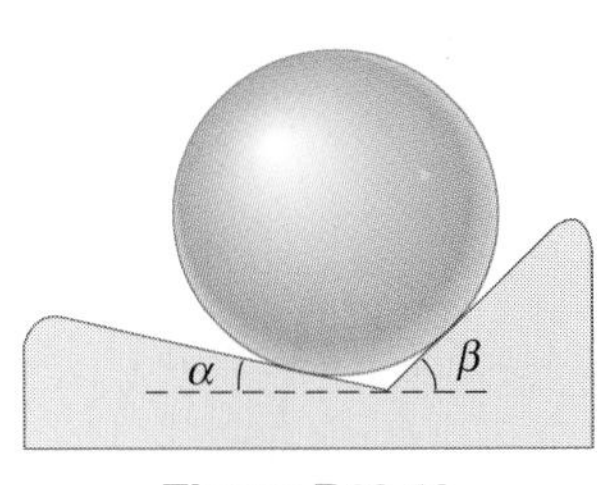

Figure P12.42

43. A hungry bear weighing 700 N walks out on a beam in an attempt to retrieve a basket of food hanging at the end of the beam (Fig. P12.43). The beam is uniform, weighs 200 N, and is 6.00 m long; the basket weighs 80.0 N. (a) Draw a free-body diagram for the beam. (b) When the bear is at $x = 1.00$ m, find the tension in the wire and the components of the force exerted by the wall on the left end of the beam. (c) **What If?** If the wire can withstand a maximum tension of 900 N, what is the maximum distance the bear can walk before the wire breaks?

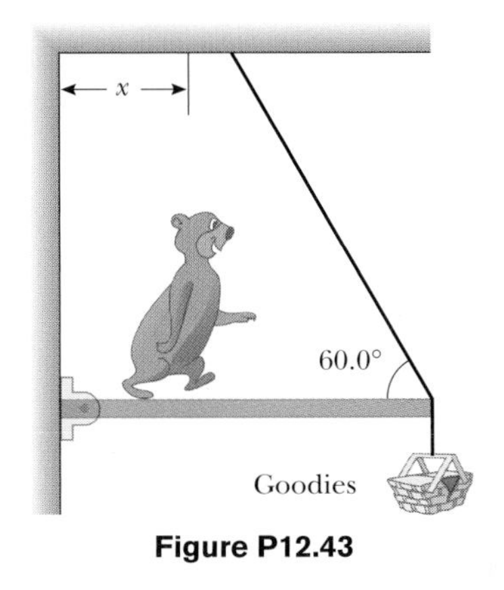

Figure P12.43

44. A farm gate (Fig. P12.44) is 3.00 m wide and 1.80 m high, with hinges attached to the top and bottom. The guy wire makes an angle of 30.0° with the top of the gate and is tightened by a turnbuckle to a tension of 200 N. The mass of the gate is 40.0 kg. (a) Determine the horizontal force exerted by the bottom hinge on the gate. (b) Find the horizontal force exerted by the upper hinge. (c) Determine the combined vertical force exerted by both hinges. (d) **What If?** What must be the tension in the guy wire so that the horizontal force exerted by the upper hinge is zero?

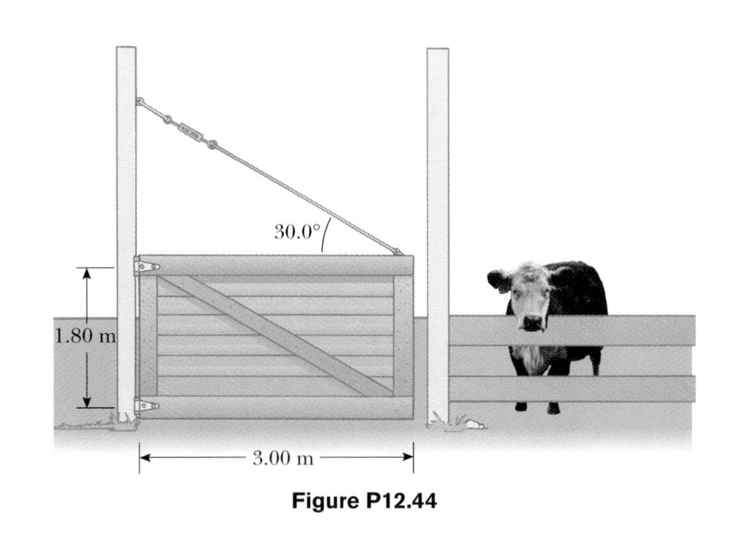

Figure P12.44

45. A uniform sign of weight F_g and width $2L$ hangs from a light, horizontal beam, hinged at the wall and supported by a cable (Fig. P12.45). Determine (a) the tension in the cable and (b) the components of the reaction force exerted by the wall on the beam, in terms of F_g, d, L, and θ.

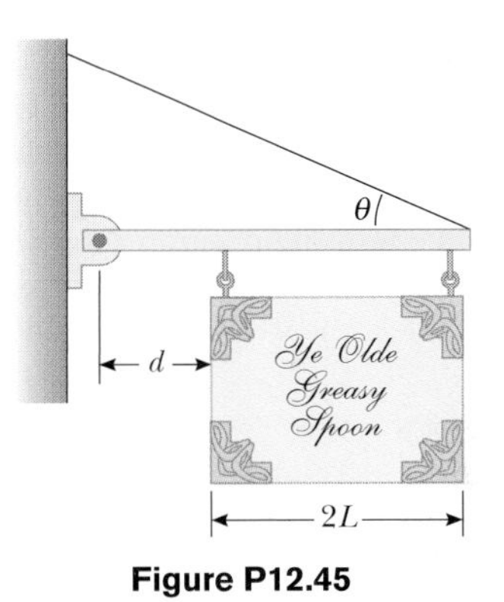

Figure P12.45

46. A 1 200-N uniform boom is supported by a cable as in Figure P12.46. The boom is pivoted at the bottom, and a 2 000-N object hangs from its top. Find the tension in the cable and the components of the reaction force exerted by the floor on the boom.

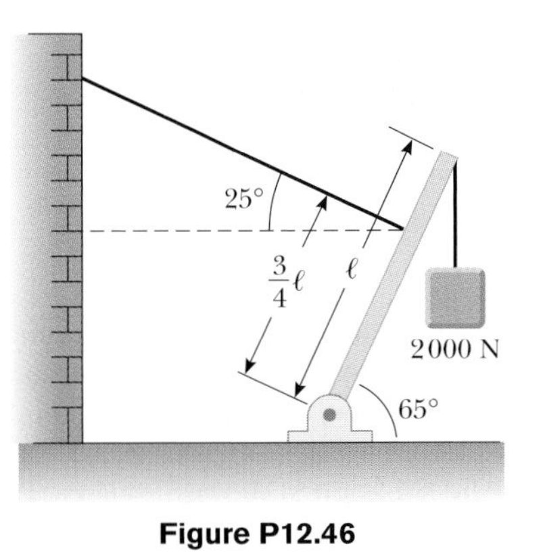

Figure P12.46

47. A crane of mass 3 000 kg supports a load of 10 000 kg as in Figure P12.47. The crane is pivoted with a frictionless pin at *A* and rests against a smooth support at *B*. Find the reaction forces at *A* and *B*.

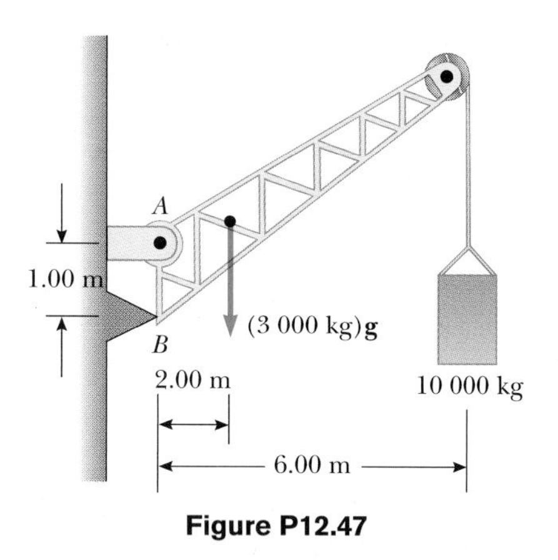

Figure P12.47

48. A ladder of uniform density and mass *m* rests against a frictionless vertical wall, making an angle of 60.0° with the horizontal. The lower end rests on a flat surface where the coefficient of static friction is $\mu_s = 0.400$. A window cleaner with mass $M = 2m$ attempts to climb the ladder. What fraction of the length *L* of the ladder will the worker have reached when the ladder begins to slip?

49. A 10 000-N shark is supported by a cable attached to a 4.00-m rod that can pivot at the base. Calculate the tension in the tie-rope between the rod and the wall if it is holding the system in the position shown in Figure P12.49. Find the horizontal and vertical forces exerted on the base of the rod. (Neglect the weight of the rod.)

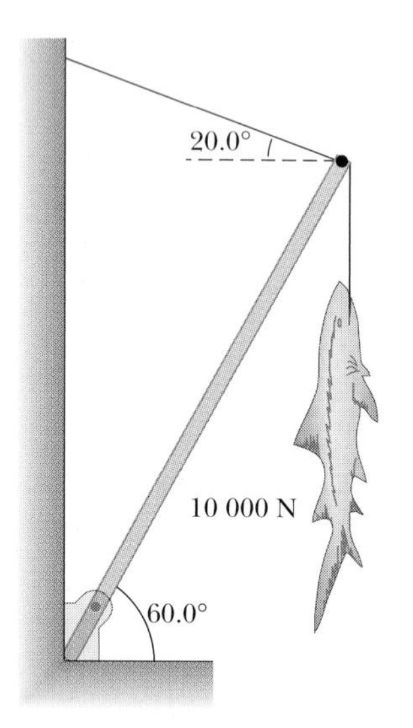

Figure P12.49

50. When a person stands on tiptoe (a strenuous position), the position of the foot is as shown in Figure P12.50a. The gravitational force on the body $\mathbf{F}_g$ is supported by the force **n** exerted by the floor on the toe. A mechanical model for the situation is shown in Figure P12.50b, where **T** is the force exerted by the Achilles tendon on the foot

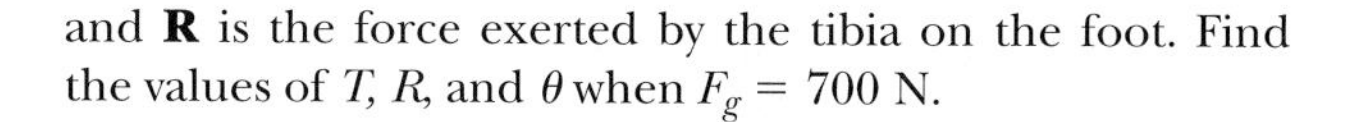
and **R** is the force exerted by the tibia on the foot. Find the values of *T*, *R*, and θ when $F_g = 700$ N.

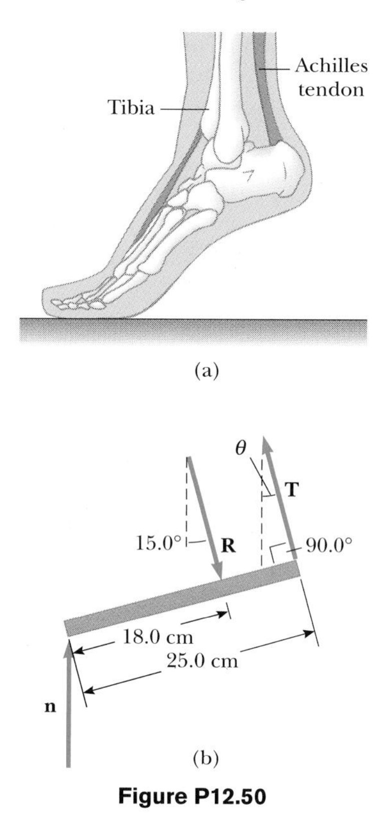

Figure P12.50

51. A person bending forward to lift a load "with his back" (Fig. P12.51a) rather than "with his knees" can be injured by large forces exerted on the muscles and vertebrae. The spine pivots mainly at the fifth lumbar vertebra, with the principal supporting force provided by the erector spinalis muscle in the back. To see the magnitude of the forces involved, and to understand why back problems are common among humans, consider the model shown in Figure P12.51b for a person bending forward to lift a 200-N object. The spine and upper body are represented as a uniform horizontal rod of weight 350 N, pivoted at the base of the spine. The erector spinalis muscle, attached at a

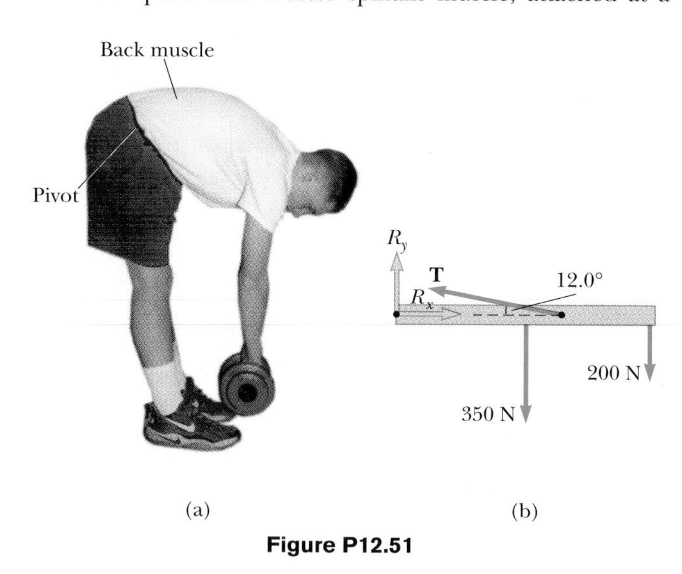

Figure P12.51

point two thirds of the way up the spine, maintains the position of the back. The angle between the spine and this muscle is 12.0°. Find the tension in the back muscle and the compressional force in the spine.

52. A uniform rod of weight F_g and length L is supported at its ends by a frictionless trough as shown in Figure P12.52. (a) Show that the center of gravity of the rod must be vertically over point O when the rod is in equilibrium. (b) Determine the equilibrium value of the angle θ.

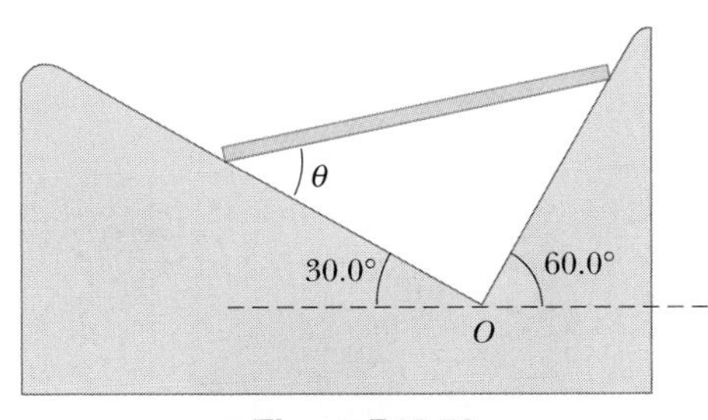

Figure P12.52

53. A force acts on a rectangular cabinet weighing 400 N, as in Figure P12.53. (a) If the cabinet slides with constant speed when $F = 200$ N and $h = 0.400$ m, find the coefficient of kinetic friction and the position of the resultant normal force. (b) If $F = 300$ N, find the value of h for which the cabinet just begins to tip.

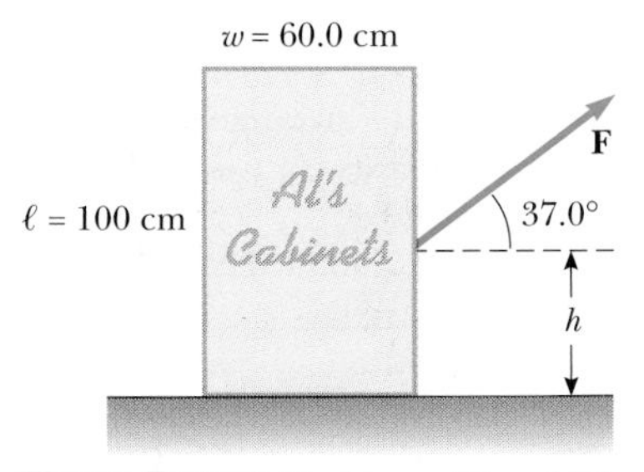

Figure P12.53 Problems 53 and 54.

54. Consider the rectangular cabinet of Problem 53, but with a force **F** applied horizontally at the upper edge. (a) What is the minimum force required to start to tip the cabinet? (b) What is the minimum coefficient of static friction required for the cabinet not to slide with the application of a force of this magnitude? (c) Find the magnitude and direction of the minimum force required to tip the cabinet if the point of application can be chosen anywhere on the cabinet.

55. A uniform beam of mass m is inclined at an angle θ to the horizontal. Its upper end produces a ninety-degree bend in a very rough rope tied to a wall, and its lower end rests on a rough floor (Fig. P12.55). (a) If the coefficient of static friction between beam and floor is μ_s, determine an expression for the maximum mass M that can be suspended from the top before the beam slips. (b) Determine the magnitude of the reaction force at the floor and the magnitude of the force exerted by the beam on the rope at P in terms of m, M, and μ_s.

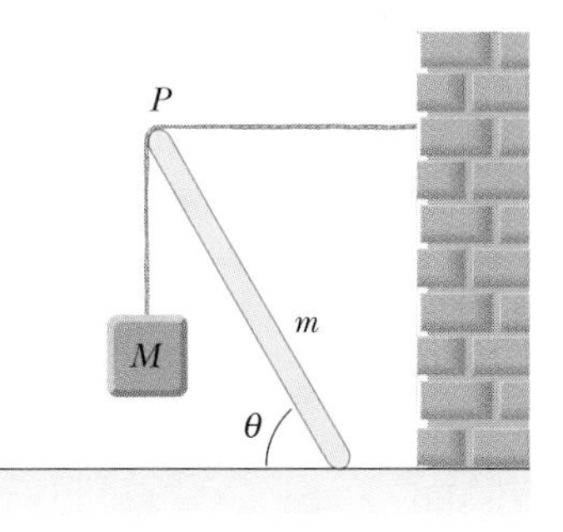

Figure P12.55

56. Figure P12.56 shows a truss that supports a downward force of 1 000 N applied at the point B. The truss has negligible weight. The piers at A and C are smooth. (a) Apply the conditions of equilibrium to prove that $n_A = 366$ N and $n_C = 634$ N. (b) Show that, because forces act on the light truss only at the hinge joints, each bar of the truss must exert on each hinge pin only a force along the length of that bar—a force of tension or compression. (c) Find the force of tension or of compression in each of the three bars.

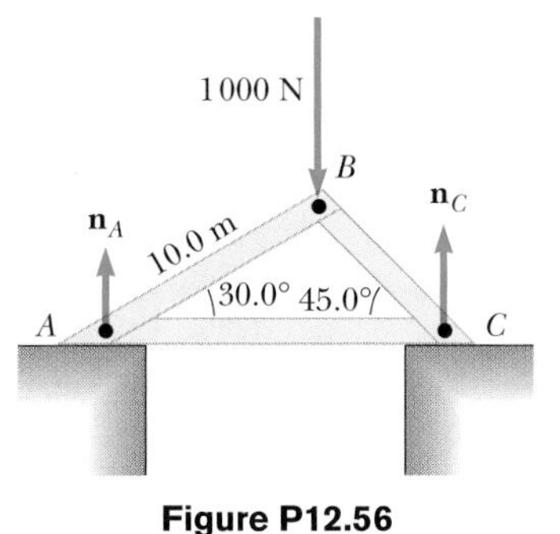

Figure P12.56

57. A stepladder of negligible weight is constructed as shown in Figure P12.57. A painter of mass 70.0 kg stands on the ladder 3.00 m from the bottom. Assuming the floor is frictionless, find (a) the tension in the horizontal bar connecting the two halves of the ladder, (b) the normal forces at A

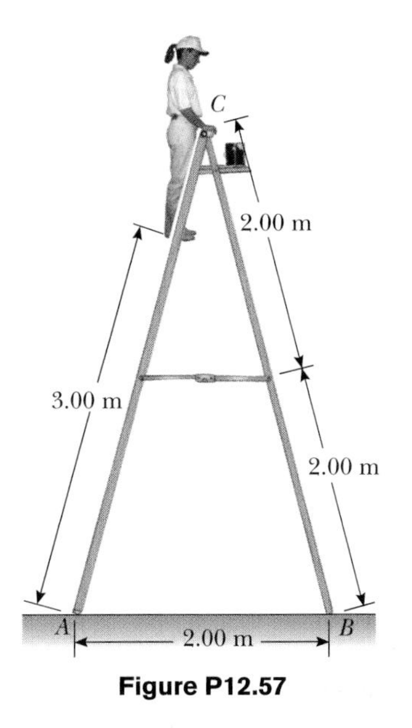

Figure P12.57

and *B*, and (c) the components of the reaction force at the single hinge *C* that the left half of the ladder exerts on the right half. (*Suggestion:* Treat the ladder as a single object, but also each half of the ladder separately.)

58. A flat dance floor of dimensions 20.0 m by 20.0 m has a mass of 1 000 kg. Three dance couples, each of mass 125 kg, start in the top left, top right, and bottom left corners. (a) Where is the initial center of gravity? (b) The couple in the bottom left corner moves 10.0 m to the right. Where is the new center of gravity? (c) What was the average velocity of the center of gravity if it took that couple 8.00 s to change positions?

59. A shelf bracket is mounted on a vertical wall by a single screw, as shown in Figure P12.59. Neglecting the weight of the bracket, find the horizontal component of the force that the screw exerts on the bracket when an 80.0 N vertical force is applied as shown. (*Hint:* Imagine that the bracket is slightly loose.)

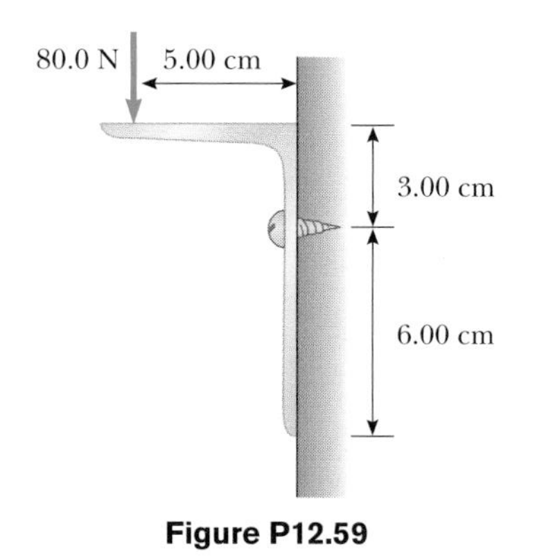

Figure P12.59

60. Figure P12.60 shows a vertical force applied tangentially to a uniform cylinder of weight F_g. The coefficient of static friction between the cylinder and all surfaces is 0.500. In terms of F_g, find the maximum force **P** that can be applied that does not cause the cylinder to rotate. (*Hint:* When the cylinder is on the verge of slipping, both friction forces are at their maximum values. Why?)

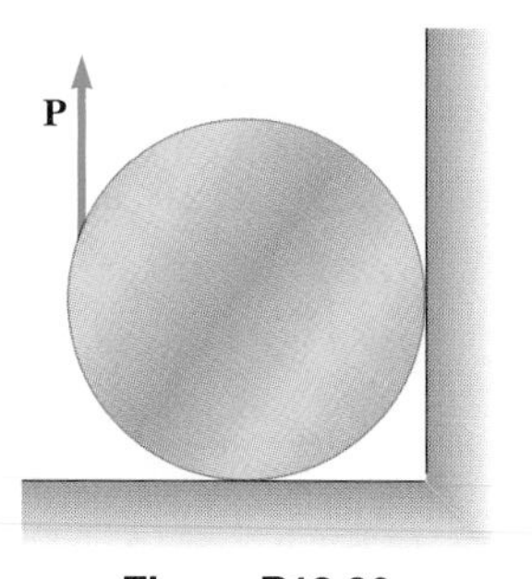

Figure P12.60

61. **Review problem.** A wire of length *L*, Young's modulus *Y*, and cross-sectional area *A* is stretched elastically by an amount ΔL. By Hooke's law (Section 7.4), the restoring force is $-k\,\Delta L$. (a) Show that $k = YA/L$. (b) Show that the work done in stretching the wire by an amount ΔL is

$$W = \tfrac{1}{2}YA(\Delta L)^2/L$$

62. Two racquetballs are placed in a glass jar, as shown in Figure P12.62. Their centers and the point *A* lie on a straight line. (a) Assume that the walls are frictionless, and determine P_1, P_2, and P_3. (b) Determine the magnitude of the force exerted by the left ball on the right ball. Assume each ball has a mass of 170 g.

Figure P12.62

63. In exercise physiology studies it is sometimes important to determine the location of a person's center of mass. This can be done with the arrangement shown in Figure P12.63. A light plank rests on two scales, which give readings of $F_{g1} = 380$ N and $F_{g2} = 320$ N. The scales are separated by a distance of 2.00 m. How far from the woman's feet is her center of mass?

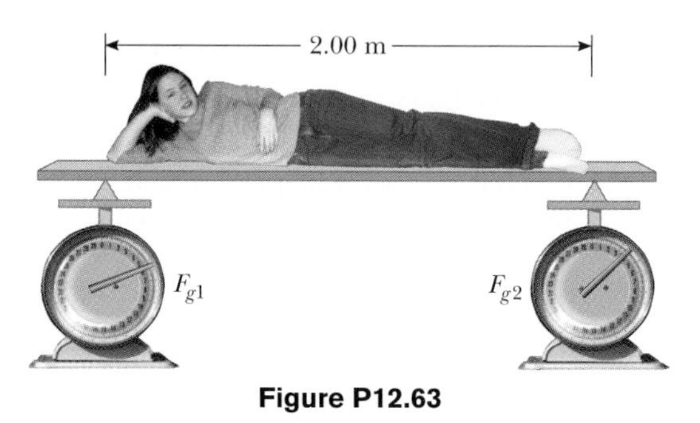

Figure P12.63

64. A steel cable 3.00 cm^2 in cross-sectional area has a mass of 2.40 kg per meter of length. If 500 m of the cable is hung over a vertical cliff, how much does the cable stretch under its own weight? $Y_{steel} = 2.00 \times 10^{11}$ N/m^2.

65. (a) Estimate the force with which a karate master strikes a board if the hand's speed at time of impact is 10.0 m/s, decreasing to 1.00 m/s during a 0.002 00-s time-of-contact with the board. The mass of his hand and arm is 1.00 kg. (b) Estimate the shear stress if this force is exerted on a 1.00-cm-thick pine board that is 10.0 cm wide. (c) If the maximum shear stress a pine board can support before breaking is 3.60×10^6 N/m^2, will the board break?

66. A bucket is made from thin sheet metal. The bottom and top of the bucket have radii of 25.0 cm and 35.0 cm, respectively. The bucket is 30.0 cm high and filled with water. Where is the center of gravity? (Ignore the weight of the bucket itself.)

67. **Review problem.** An aluminum wire is 0.850 m long and has a circular cross section of diameter 0.780 mm. Fixed at the top end, the wire supports a 1.20-kg object that swings in a horizontal circle. Determine the angular velocity required to produce a strain of 1.00×10^{-3}.

68. A bridge truss extends 200 m across a river (Fig. P12.68). The structure is free to slide horizontally to permit thermal expansion. The structural components are connected by pin joints, and the masses of the bars are small compared with the mass of a 1 360-kg car at the center. Calculate the force of tension or compression in each structural component.

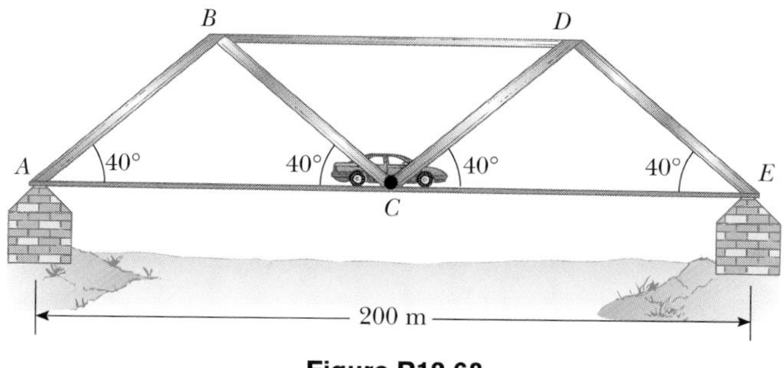

Figure P12.68

69. A bridge truss extends 100 m across a river (Fig. P12.69). The structure is free to slide horizontally to permit thermal expansion. The structural components are connected by pin joints, and the masses of the bars are small compared with the mass of a 1 500-kg car halfway between points A and C. Show that the weight of the car is in effect equally distributed between points A and C. Specify whether each structural component is under tension or compression and find the force in each.

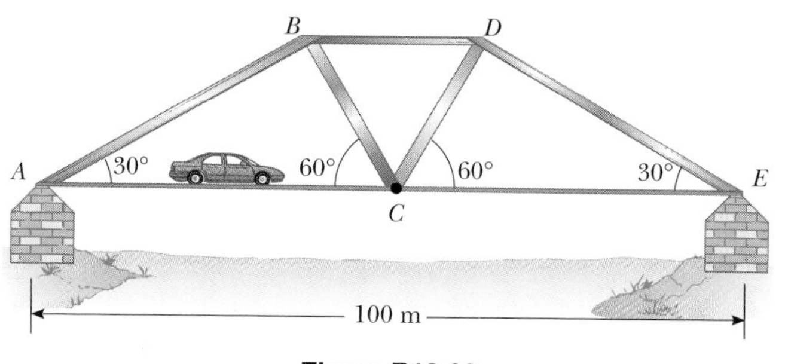

Figure P12.69

70. **Review problem.** A cue strikes a cue ball and delivers a horizontal impulse in such a way that the ball rolls without slipping as it starts to move. At what height above the ball's center (in terms of the radius of the ball) was the blow struck?

71. **Review problem.** A trailer with loaded weight F_g is being pulled by a vehicle with a force $\mathbf{P}$, as in Figure P12.71. The trailer is loaded such that its center of mass is located as shown. Neglect the force of rolling friction and let a represent the x component of the acceleration of the trailer. (a) Find the vertical component of $\mathbf{P}$ in terms of the given parameters. (b) If $a = 2.00\ \text{m/s}^2$ and $h = 1.50$ m, what must be the value of d in order that $P_y = 0$ (no vertical load on the vehicle)? (c) Find the values of P_x and P_y given that $F_g = 1\ 500$ N, $d = 0.800$ m, $L = 3.00$ m, $h = 1.50$ m, and $a = -2.00\ \text{m/s}^2$.

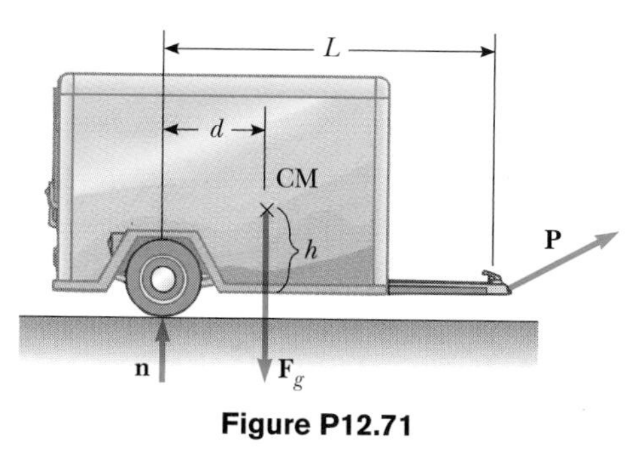

Figure P12.71

72. **Review problem.** A bicycle is traveling downhill at a high speed. Suddenly, the cyclist sees that a bridge ahead has collapsed, so she has to stop. What is the maximum magnitude of acceleration the bicycle can have if it is not to flip over its front wheel—in particular, if its rear wheel is not to leave the ground? The slope makes an angle of 20.0° with the horizontal. On level ground, the center of mass of the woman–bicycle system is at a point 1.05 m above the ground, 65.0 cm horizontally behind the axle of the front wheel, and 35.0 cm in front of the rear axle. Assume that the tires do not skid.

73. **Review problem.** A car moves with speed v on a horizontal circular track of radius R. A head-on view of the car is shown in Figure P12.73. The height of the car's center of mass above the ground is h, and the separation between its inner and outer wheels is d. The road is dry, and the car does not skid. Show that the maximum speed the car can

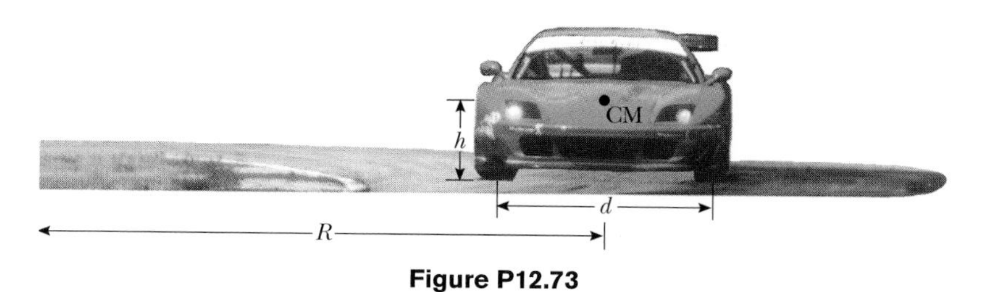

Figure P12.73

have without overturning is given by

$$v_{\max} = \sqrt{\frac{gRd}{2h}}$$

To reduce the risk of rollover, should one increase or decrease h? Should one increase or decrease the width d of the wheel base?

Answers to Quick Quizzes

12.1 (a). The unbalanced torques due to the forces in Figure 12.2 cause an angular acceleration even though the linear acceleration is zero.

12.2 (b). Notice that the lines of action of all the forces in Figure 12.3 intersect at a common point. Thus, the net torque about this point is zero. This zero value of the net torque is independent of the values of the forces. Because no force has a downward component, there is a net force and the object is not in force equilibrium.

12.3 (b). Both the object and the center of gravity of the meter stick are 25 cm from the pivot point. Thus, the meter stick and the object must have the same mass if the system is balanced.

12.4 (b). The friction force on the block as it slides along the surface is parallel to the lower surface and will cause the block to undergo a shear deformation.

12.5 (a). The stretching of the wire due to the increased tension is described by Young's modulus.

12.6 (c). The pressure of the atmosphere results in a force of uniform magnitude perpendicular at all points on the surface of the sphere.

Chapter 13

Universal Gravitation

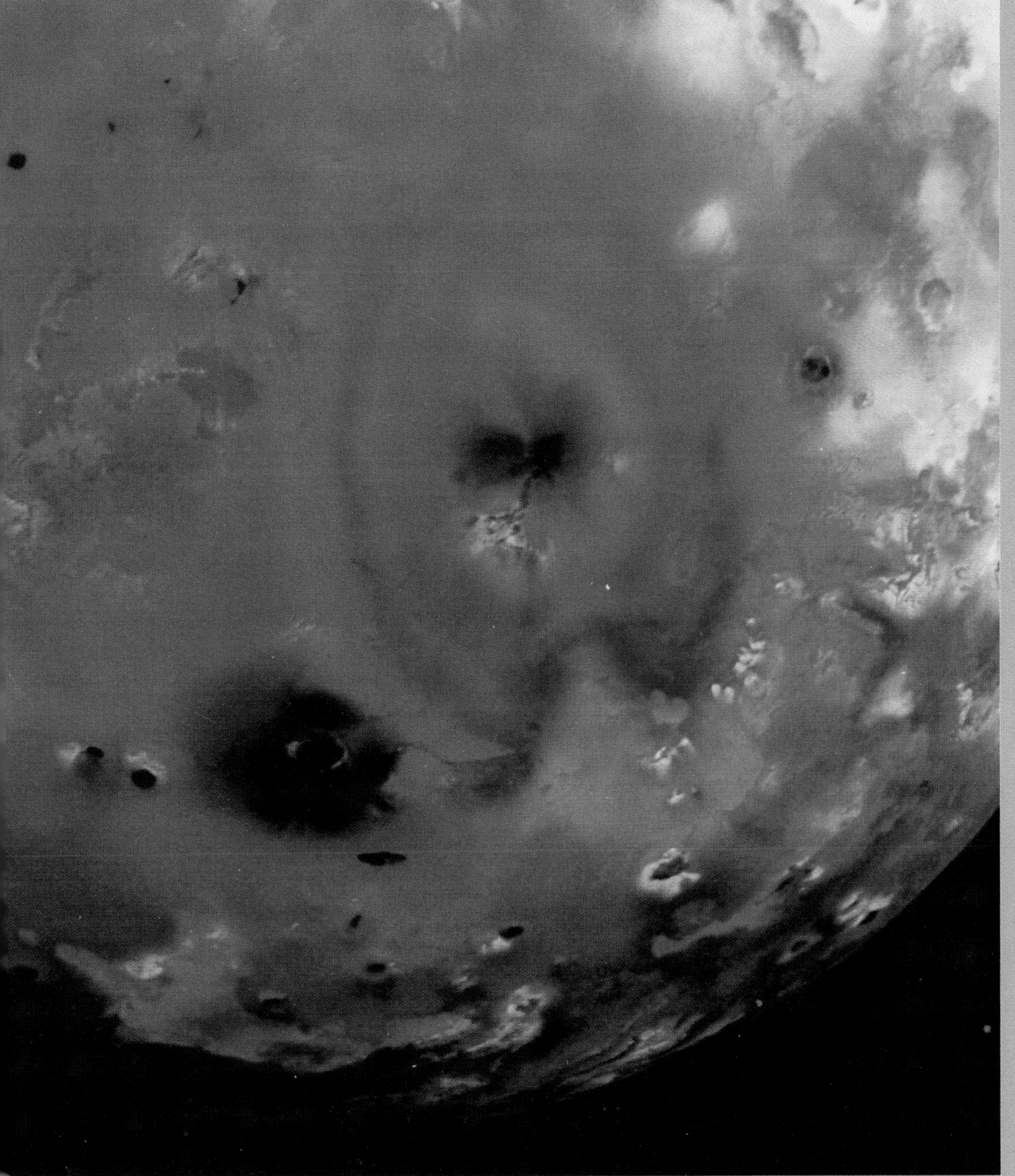

▲ *An understanding of the law of universal gravitation has allowed scientists to send spacecraft on impressively accurate journeys to other parts of our solar system. This photo of a volcano on Io, a moon of Jupiter, was taken by the Galileo spacecraft, which has been orbiting Jupiter since 1995. The red material has been vented from below the surface. (Univ. of Arizona/JPL/NASA)*

CHAPTER OUTLINE

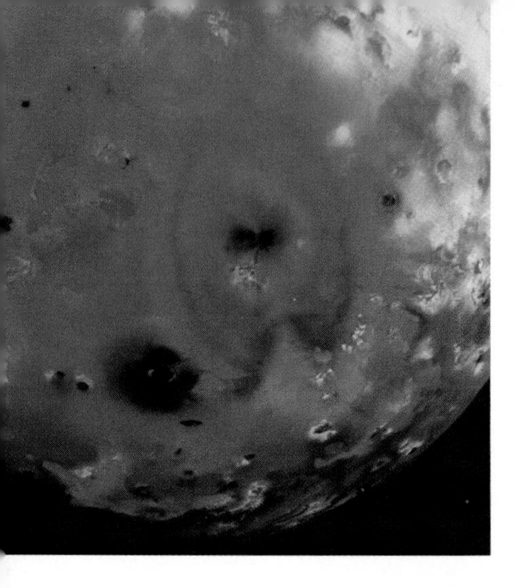

Before 1687, a large amount of data had been collected on the motions of the Moon and the planets, but a clear understanding of the forces related to these motions was not available. In that year, Isaac Newton provided the key that unlocked the secrets of the heavens. He knew, from his first law, that a net force had to be acting on the Moon because without such a force the Moon would move in a straight-line path rather than in its almost circular orbit. Newton reasoned that this force was the gravitational attraction exerted by the Earth on the Moon. He realized that the forces involved in the Earth–Moon attraction and in the Sun–planet attraction were not something special to those systems, but rather were particular cases of a general and universal attraction between objects. In other words, Newton saw that the same force of attraction that causes the Moon to follow its path around the Earth also causes an apple to fall from a tree. As he put it, "I deduced that the forces which keep the planets in their orbs must be reciprocally as the squares of their distances from the centers about which they revolve; and thereby compared the force requisite to keep the Moon in her orb with the force of gravity at the surface of the Earth; and found them answer pretty nearly."

In this chapter we study the law of universal gravitation. We emphasize a description of planetary motion because astronomical data provide an important test of this law's validity. We then show that the laws of planetary motion developed by Johannes Kepler follow from the law of universal gravitation and the concept of conservation of angular momentum. We conclude by deriving a general expression for gravitational potential energy and examining the energetics of planetary and satellite motion.

13.1 Newton's Law of Universal Gravitation

You may have heard the legend that Newton was struck on the head by a falling apple while napping under a tree. This alleged accident supposedly prompted him to imagine that perhaps all objects in the Universe were attracted to each other in the same way the apple was attracted to the Earth. Newton analyzed astronomical data on the motion of the Moon around the Earth. From that analysis, he made the bold assertion that the force law governing the motion of planets was the *same* as the force law that attracted a falling apple to the Earth. This was the first time that "earthly" and "heavenly" motions were unified. We shall look at the mathematical details of Newton's analysis in this section.

In 1687 Newton published his work on the law of gravity in his treatise *Mathematical Principles of Natural Philosophy.* **Newton's law of universal gravitation** states that

The law of universal gravitation

every particle in the Universe attracts every other particle with a force that is directly proportional to the product of their masses and inversely proportional to the square of the distance between them.

If the particles have masses m_1 and m_2 and are separated by a distance r, the magnitude of this gravitational force is

$$F_g = G\frac{m_1 m_2}{r^2} \tag{13.1}$$

where G is a constant, called the *universal gravitational constant,* that has been measured experimentally. Its value in SI units is

$$G = 6.673 \times 10^{-11}\ \text{N}\cdot\text{m}^2/\text{kg}^2 \tag{13.2}$$

The form of the force law given by Equation 13.1 is often referred to as an **inverse-square law** because the magnitude of the force varies as the inverse square of the separation of the particles.[1] We shall see other examples of this type of force law in subsequent chapters. We can express this force in vector form by defining a unit vector $\hat{\mathbf{r}}_{12}$ (Fig. 13.1). Because this unit vector is directed from particle 1 toward particle 2, the force exerted by particle 1 on particle 2 is

$$\mathbf{F}_{12} = -G\frac{m_1 m_2}{r^2}\,\hat{\mathbf{r}}_{12} \tag{13.3}$$

where the negative sign indicates that particle 2 is attracted to particle 1, and hence the force on particle 2 must be directed toward particle 1. By Newton's third law, the force exerted by particle 2 on particle 1, designated $\mathbf{F}_{21}$, is equal in magnitude to $\mathbf{F}_{12}$ and in the opposite direction. That is, these forces form an action–reaction pair, and $\mathbf{F}_{21} = -\mathbf{F}_{12}$.

Several features of Equation 13.3 deserve mention. The gravitational force is a field force that always exists between two particles, regardless of the medium that separates them. Because the force varies as the inverse square of the distance between the particles, it decreases rapidly with increasing separation.

Another important point that we can show from Equation 13.3 is that **the gravitational force exerted by a finite-size, spherically symmetric mass distribution on a particle outside the distribution is the same as if the entire mass of the distribution were concentrated at the center.** For example, the magnitude of the force exerted by the Earth on a particle of mass m near the Earth's surface is

$$F_g = G\frac{M_E m}{R_E^2} \tag{13.4}$$

where M_E is the Earth's mass and R_E its radius. This force is directed toward the center of the Earth.

In formulating his law of universal gravitation, Newton used the following reasoning, which supports the assumption that the gravitational force is proportional to the inverse square of the separation between the two interacting objects. He compared the acceleration of the Moon in its orbit with the acceleration of an object falling near the Earth's surface, such as the legendary apple (Fig. 13.2). Assuming that both accelerations had the same cause—namely, the gravitational attraction of the Earth—Newton used the inverse-square law to reason that the acceleration of the Moon toward the Earth (centripetal acceleration) should be proportional to $1/r_M^2$, where r_M is the distance between the centers of the Earth and the Moon. Furthermore, the acceleration of the apple toward the Earth should be proportional to $1/R_a^2$, where R_a is the distance between the centers of the Earth and the apple. Because the apple is located at the surface of the earth, $R_a = R_E$, the radius of the Earth. Using the values $r_M = 3.84 \times 10^8$ m and $R_E = 6.37 \times 10^6$ m, Newton predicted that the ratio of the Moon's acceleration a_M to the apple's acceleration g would be

$$\frac{a_M}{g} = \frac{(1/r_M)^2}{(1/R_E)^2} = \left(\frac{R_E}{r_M}\right)^2 = \left(\frac{6.37 \times 10^6\ \text{m}}{3.84 \times 10^8\ \text{m}}\right)^2 = 2.75 \times 10^{-4}$$

[1] An *inverse* proportionality between two quantities x and y is one in which $y = k/x$, where k is a constant. A *direct* proportion between x and y exists when $y = kx$.

PITFALL PREVENTION

13.1 Be Clear on *g* and *G*

The symbol g represents the magnitude of the free-fall acceleration near a planet. At the surface of the Earth, g has the value $9.80\ \text{m/s}^2$. On the other hand, G is a universal constant that has the same value everywhere in the Universe.

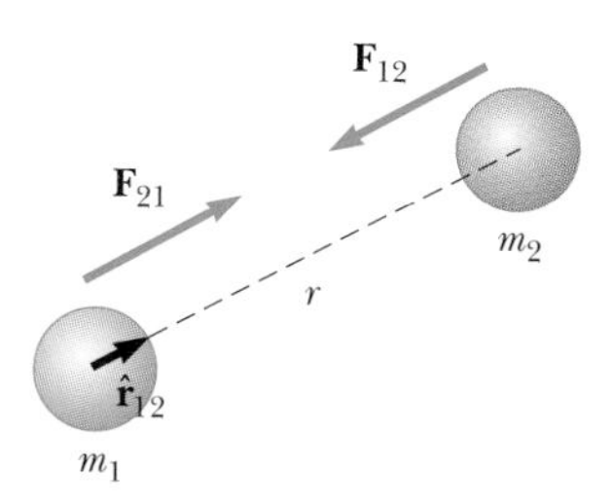

Active Figure 13.1 The gravitational force between two particles is attractive. The unit vector $\hat{\mathbf{r}}_{12}$ is directed from particle 1 toward particle 2. Note that $\mathbf{F}_{21} = -\mathbf{F}_{12}$.

At the Active Figures link at http://www/pse6.com, *you can change the masses of the particles and the separation distance between the particles to see the effect on the gravitational force.*

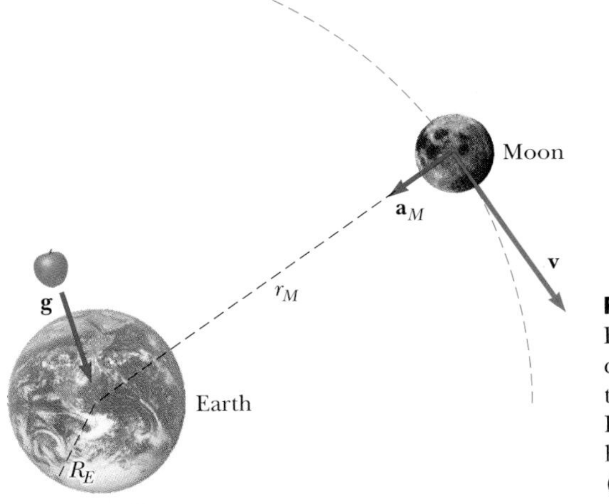

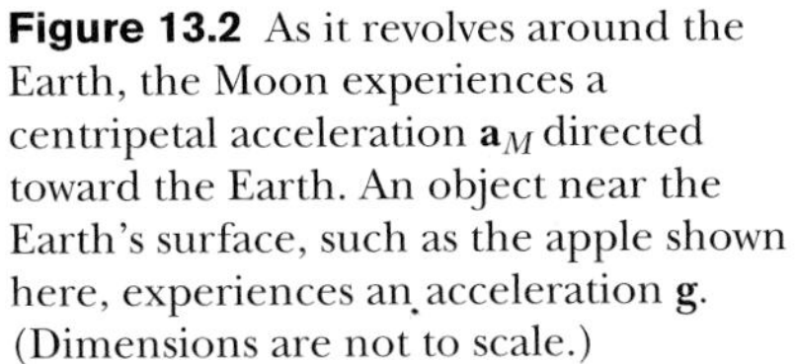

Figure 13.2 As it revolves around the Earth, the Moon experiences a centripetal acceleration $\mathbf{a}_M$ directed toward the Earth. An object near the Earth's surface, such as the apple shown here, experiences an acceleration $\mathbf{g}$. (Dimensions are not to scale.)

Therefore, the centripetal acceleration of the Moon is

$$a_M = (2.75 \times 10^{-4})(9.80 \text{ m/s}^2) = 2.70 \times 10^{-3} \text{ m/s}^2$$

Newton also calculated the centripetal acceleration of the Moon from a knowledge of its mean distance from the Earth and the known value of its orbital period, $T = 27.32$ days $= 2.36 \times 10^6$ s. In a time interval T, the Moon travels a distance $2\pi r_M$, which equals the circumference of its orbit. Therefore, its orbital speed is $2\pi r_M/T$ and its centripetal acceleration is

$$a_M = \frac{v^2}{r_M} = \frac{(2\pi r_M/T)^2}{r_M} = \frac{4\pi^2 r_M}{T^2} = \frac{4\pi^2(3.84 \times 10^8 \text{ m})}{(2.36 \times 10^6 \text{ s})^2}$$
$$= 2.72 \times 10^{-3} \text{ m/s}^2$$

The nearly perfect agreement between this value and the value Newton obtained using g provides strong evidence of the inverse-square nature of the gravitational force law.

Although these results must have been very encouraging to Newton, he was deeply troubled by an assumption he made in the analysis. To evaluate the acceleration of an object at the Earth's surface, Newton treated the Earth as if its mass were all concentrated at its center. That is, he assumed that the Earth acted as a particle as far as its influence on an exterior object was concerned. Several years later, in 1687, on the basis of his pioneering work in the development of calculus, Newton proved that this assumption was valid and was a natural consequence of the law of universal gravitation.

We have evidence that the gravitational force acting on an object is directly proportional to its mass from our observations of falling objects, discussed in Chapter 2. All objects, regardless of mass, fall in the absence of air resistance at the same acceleration g near the surface of the Earth. According to Newton's second law, this acceleration is given by $g = F_g/m$, where m is the mass of the falling object. If this ratio is to be the same for all falling objects, then F_g must be directly proportional to m, so that the mass cancels in the ratio. If we consider the more general situation of a gravitational force between any two objects with mass, such as two planets, this same argument can be applied to show that the gravitational force is proportional to one of the masses. We can choose *either* of the masses in the argument, however; thus, the gravitational force must be directly proportional to *both* masses, as can be seen in Equation 13.3.

Quick Quiz 13.1 The Moon remains in its orbit around the Earth rather than falling to the Earth because (a) it is outside of the gravitational influence of the Earth (b) it is in balance with the gravitational forces from the Sun and other planets (c) the net force on the Moon is zero (d) none of these (e) all of these.

Quick Quiz 13.2 A planet has two moons of equal mass. Moon 1 is in a circular orbit of radius r. Moon 2 is in a circular orbit of radius $2r$. The magnitude of the gravitational force exerted by the planet on moon 2 is (a) four times as large as that on moon 1 (b) twice as large as that on moon 1 (c) equal to that on moon 1 (d) half as large as that on moon 1 (e) one fourth as large as that on moon 1.

Example 13.1 Billiards, Anyone? **Interactive**

Three 0.300-kg billiard balls are placed on a table at the corners of a right triangle, as shown in Figure 13.3. Calculate the gravitational force on the cue ball (designated m_1) resulting from the other two balls.

Solution First we calculate separately the individual forces on the cue ball due to the other two balls, and then we find the vector sum to obtain the resultant force. We can see graphically that this force should point upward and toward the right. We locate our coordinate axes as shown in Figure 13.3, placing our origin at the position of the cue ball.

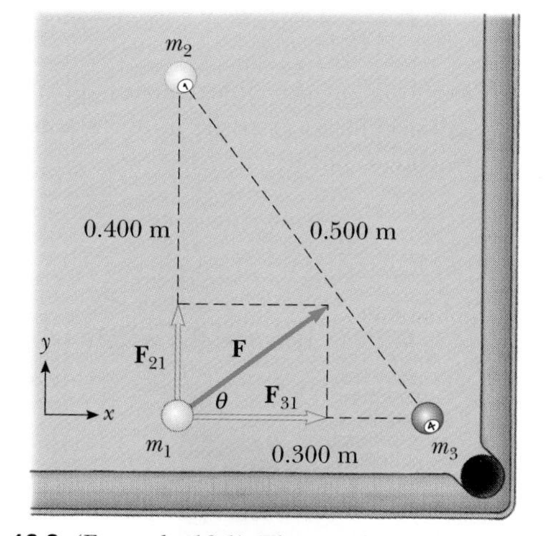

Figure 13.3 (Example 13.1) The resultant gravitational force acting on the cue ball is the vector sum $\mathbf{F}_{21} + \mathbf{F}_{31}$.

The force exerted by m_2 on the cue ball is directed upward and is given by

$$\mathbf{F}_{21} = G\frac{m_2 m_1}{{r_{21}}^2}\,\hat{\mathbf{j}}$$

$$= (6.67 \times 10^{-11}\ \text{N}\cdot\text{m}^2/\text{kg}^2)\,\frac{(0.300\ \text{kg})(0.300\ \text{kg})}{(0.400\ \text{m})^2}\,\hat{\mathbf{j}}$$

$$= 3.75 \times 10^{-11}\,\hat{\mathbf{j}}\ \text{N}$$

This result shows that the gravitational forces between everyday objects have extremely small magnitudes. The force exerted by m_3 on the cue ball is directed to the right:

$$\mathbf{F}_{31} = G\frac{m_3 m_1}{{r_{31}}^2}\,\hat{\mathbf{i}}$$

$$= (6.67 \times 10^{-11}\ \text{N}\cdot\text{m}^2/\text{kg}^2)\,\frac{(0.300\ \text{kg})(0.300\ \text{kg})}{(0.300\ \text{m})^2}\,\hat{\mathbf{i}}$$

$$= 6.67 \times 10^{-11}\,\hat{\mathbf{i}}\ \text{N}$$

Therefore, the net gravitational force on the cue ball is

$$\mathbf{F} = \mathbf{F}_{21} + \mathbf{F}_{31} = (6.67\hat{\mathbf{i}} + 3.75\hat{\mathbf{j}}) \times 10^{-11}\ \text{N}$$

and the magnitude of this force is

$$F = \sqrt{{F_{21}}^2 + {F_{31}}^2} = \sqrt{(3.75)^2 + (6.67)^2} \times 10^{-11}\ \text{N}$$
$$= 7.65 \times 10^{-11}\ \text{N}$$

From $\tan\theta = 3.75/6.67 = 0.562$, the direction of the net gravitational force is $\theta = 29.3°$ counterclockwise from the x axis.

At the Interactive Worked Example link at **http://www.pse6.com,** *you can move balls 2 and 3 to see the effect on the net gravitational force on ball 1.*

13.2 Measuring the Gravitational Constant

The universal gravitational constant G was measured in an important experiment by Henry Cavendish (1731–1810) in 1798. The Cavendish apparatus consists of two small spheres, each of mass m, fixed to the ends of a light horizontal rod suspended by a fine fiber or thin metal wire, as illustrated in Figure 13.4. When two large spheres, each of mass M, are placed near the smaller ones, the attractive force between smaller and larger spheres causes the rod to rotate and twist the wire suspension to a new equilibrium orientation. The angle of rotation is measured by the deflection of a light beam reflected from a mirror attached to the vertical suspension. The deflection of the light beam is an effective technique for amplifying the motion. The experiment is carefully repeated with different masses at various separations. In addition to providing a value

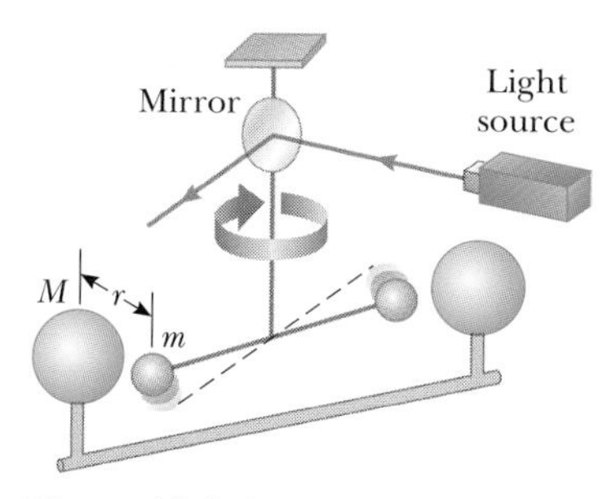

Figure 13.4 Cavendish apparatus for measuring G. The dashed line represents the original position of the rod.

for G, the results show experimentally that the force is attractive, proportional to the product mM, and inversely proportional to the square of the distance r.

13.3 Free-Fall Acceleration and the Gravitational Force

In Chapter 5, when defining mg as the weight of an object of mass m, we referred to g as the magnitude of the free-fall acceleration. Now we are in a position to obtain a more fundamental description of g. Because the magnitude of the force acting on a freely falling object of mass m near the Earth's surface is given by Equation 13.4, we can equate mg to this force to obtain

$$mg = G\frac{M_E m}{{R_E}^2}$$

$$g = G\frac{M_E}{{R_E}^2} \tag{13.5}$$

Now consider an object of mass m located a distance h above the Earth's surface or a distance r from the Earth's center, where $r = R_E + h$. The magnitude of the gravitational force acting on this object is

$$F_g = G\frac{M_E m}{r^2} = G\frac{M_E m}{(R_E + h)^2}$$

The magnitude of the gravitational force acting on the object at this position is also $F_g = mg$, where g is the value of the free-fall acceleration at the altitude h. Substituting this expression for F_g into the last equation shows that g is

Variation of g with altitude

$$g = \frac{GM_E}{r^2} = \frac{GM_E}{(R_E + h)^2} \tag{13.6}$$

Thus, it follows that g *decreases* with *increasing altitude*. Because the weight of an object is mg, we see that as $r \to \infty$, its weight approaches zero.

Courtesy NASA

Astronauts F. Story Musgrave and Jeffrey A. Hoffman, along with the Hubble Space Telescope and the space shuttle *Endeavor*, are all in free fall while orbiting the Earth.

Quick Quiz 13.3 Superman stands on top of a very tall mountain and throws a baseball horizontally with a speed such that the baseball goes into a circular orbit around the Earth. While the baseball is in orbit, the acceleration of the ball (a) depends on how fast the baseball is thrown (b) is zero because the ball does not fall to the ground (c) is slightly less than 9.80 m/s^2 (d) is equal to 9.80 m/s^2.

Example 13.2 Variation of g with Altitude h

The International Space Station operates at an altitude of 350 km. When final construction is completed, it will have a weight (measured at the Earth's surface) of 4.22×10^6 N. What is its weight when in orbit?

Solution We first find the mass of the space station from its weight at the surface of the Earth:

$$m = \frac{F_g}{g} = \frac{4.22 \times 10^6 \text{ N}}{9.80 \text{ m/s}^2} = 4.31 \times 10^5 \text{ kg}$$

This mass is fixed—it is independent of the location of the space station. Because the station is above the surface of the Earth, however, we expect its weight in orbit to be less than its weight on the Earth. Using Equation 13.6 with $h = 350$ km, we obtain

$$g = \frac{GM_E}{(R_E + h)^2}$$

$$= \frac{(6.67 \times 10^{-11} \text{ N}\cdot\text{m}^2/\text{kg}^2)(5.98 \times 10^{24} \text{ kg})}{(6.37 \times 10^6 \text{ m} + 0.350 \times 10^6 \text{ m})^2}$$

$$= 8.83 \text{ m/s}^2$$

Because this value is about 90% of the value of g at the Earth surface, we expect that the weight of the station at an altitude of 350 km is 90% of the value at the Earth's surface.

Using the value of g at the location of the station, the station's weight in orbit is

$$mg = (4.31 \times 10^5 \text{ kg})(8.83 \text{ m/s}^2) = 3.80 \times 10^6 \text{ N}$$

Values of g at other altitudes are listed in Table 13.1.

Table 13.1

Free-Fall Acceleration g at Various Altitudes Above the Earth's Surface

Altitude h (km)	g (m/s^2)
1 000	7.33
2 000	5.68
3 000	4.53
4 000	3.70
5 000	3.08
6 000	2.60
7 000	2.23
8 000	1.93
9 000	1.69
10 000	1.49
50 000	0.13
∞	0

Example 13.3 The Density of the Earth

Using the known radius of the Earth and the fact that $g = 9.80$ m/s^2 at the Earth's surface, find the average density of the Earth.

Solution From Eq. 1.1, we know that the average density is

$$\rho = \frac{M_E}{V_E}$$

where M_E is the mass of the Earth and V_E is its volume.

From Equation 13.5, we can relate the mass of the Earth to the value of g:

$$g = G\frac{M_E}{{R_E}^2} \longrightarrow M_E = \frac{g{R_E}^2}{G}$$

Substituting this into the definition of density, we obtain

$$\rho_E = \frac{M_E}{V_E} = \frac{(g{R_E}^2/G)}{\frac{4}{3}\pi {R_E}^3} = \frac{3}{4}\frac{g}{\pi G R_E}$$

$$= \frac{3}{4}\frac{9.80 \text{ m/s}^2}{\pi(6.67 \times 10^{-11} \text{ N}\cdot\text{m}^2/\text{kg}^2)(6.37 \times 10^6 \text{ m})}$$

$$= 5.51 \times 10^3 \text{ kg/m}^3$$

What If? **What if you were told that a typical density of granite at the Earth's surface were 2.75×10^3 kg/m^3—what would you conclude about the density of the material in the Earth's interior?**

Answer Because this value is about half the density that we calculated as an average for the entire Earth, we conclude that the inner core of the Earth has a density much higher than the average value. It is most amazing that the Cavendish experiment, which determines G and can be done on a tabletop, combined with simple free-fall measurements of g provides information about the core of the Earth!

13.4 Kepler's Laws and the Motion of Planets

Johannes Kepler
German astronomer
(1571–1630)

The German astronomer Kepler is best known for developing the laws of planetary motion based on the careful observations of Tycho Brahe. *(Art Resource)*

People have observed the movements of the planets, stars, and other celestial objects for thousands of years. In early history, scientists regarded the Earth as the center of the Universe. This so-called *geocentric model* was elaborated and formalized by the Greek astronomer Claudius Ptolemy (c. 100–c. 170) in the second century A.D. and was accepted for the next 1 400 years. In 1543 the Polish astronomer Nicolaus Copernicus (1473–1543) suggested that the Earth and the other planets revolved in circular orbits around the Sun (the *heliocentric model*).

The Danish astronomer Tycho Brahe (1546–1601) wanted to determine how the heavens were constructed, and thus he developed a program to determine the positions of both stars and planets. It is interesting to note that those observations of the planets and 777 stars visible to the naked eye were carried out with only a large sextant and a compass. (The telescope had not yet been invented.)

The German astronomer Johannes Kepler was Brahe's assistant for a short while before Brahe's death, whereupon he acquired his mentor's astronomical data and spent 16 years trying to deduce a mathematical model for the motion of the planets. Such data are difficult to sort out because the Earth is also in motion around the Sun. After many laborious calculations, Kepler found that Brahe's data on the revolution of Mars around the Sun provided the answer.

Kepler's complete analysis of planetary motion is summarized in three statements known as **Kepler's laws:**

Kepler's laws

1. All planets move in elliptical orbits with the Sun at one focus.
2. The radius vector drawn from the Sun to a planet sweeps out equal areas in equal time intervals.
3. The square of the orbital period of any planet is proportional to the cube of the semimajor axis of the elliptical orbit.

We discuss each of these laws below.

Kepler's First Law

We are familiar with circular orbits of objects around gravitational force centers from our discussions in this chapter. Kepler's first law indicates that the circular orbit is a very special case and elliptical orbits are the general situation. This was a difficult notion for scientists of the time to accept, because they felt that perfect circular orbits of the planets reflected the perfection of heaven.

Active Figure 13.5 Plot of an ellipse. The semimajor axis has length a, and the semiminor axis has length b. Each focus is located at a distance c from the center on each side of the center.

At the Active Figures link at http://www.pse6.com, you can move the focal points or enter values for a, b, c, and e to see the resulting elliptical shape.

Figure 13.5 shows the geometry of an ellipse, which serves as our model for the elliptical orbit of a planet. An ellipse is mathematically defined by choosing two points F_1 and F_2, each of which is a called a **focus,** and then drawing a curve through points for which the sum of the distances r_1 and r_2 from F_1 and F_2, respectively, is a constant. The longest distance through the center between points on the ellipse (and passing through both foci) is called the **major axis,** and this distance is $2a$. In Figure 13.5, the major axis is drawn along the x direction. The distance a is called the **semimajor axis.** Similarly, the shortest distance through the center between points on the ellipse is called the **minor axis** of length $2b$, where the distance b is the **semiminor axis.** Either focus of the ellipse is located at a distance c from the center of the ellipse, where $a^2 = b^2 + c^2$. In the elliptical orbit of a planet around the Sun, the Sun is at one focus of the ellipse. There is nothing at the other focus.

The **eccentricity** of an ellipse is defined as $e = c/a$ and describes the general shape of the ellipse. For a circle, $c = 0$, and the eccentricity is therefore zero. The smaller b is than a, the shorter the ellipse is along the y direction compared to its extent in the x direction in Figure 13.5. As b decreases, c increases, and the eccentricity e increases.

PITFALL PREVENTION

13.2 Where is the Sun?

The Sun is located at one focus of the elliptical orbit of a planet. It is *not* located at the center of the ellipse.

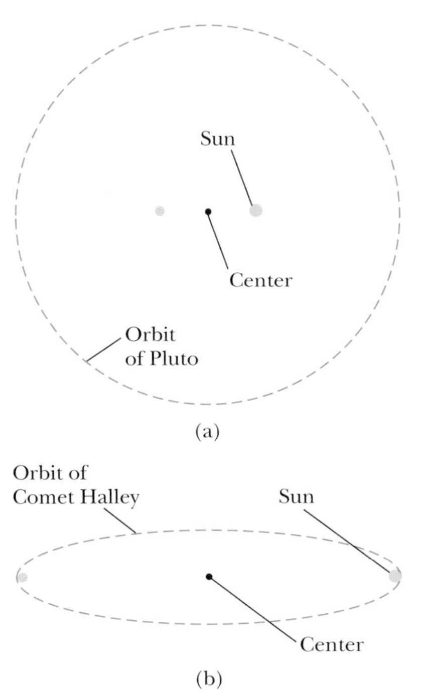

Figure 13.6 (a) The shape of the orbit of Pluto, which has the highest eccentricity ($e = 0.25$) among the planets in the solar system. The Sun is located at the large yellow dot, which is a focus of the ellipse. There is nothing physical located at the center (the small dot) or the other focus (the blue dot). (b) The shape of the orbit of Comet Halley.

Thus, higher values of eccentricity correspond to longer and thinner ellipses. The range of values of the eccentricity for an ellipse is $0 < e < 1$.

Eccentricities for planetary orbits vary widely in the solar system. The eccentricity of the Earth's orbit is 0.017, which makes it nearly circular. On the other hand, the eccentricity of Pluto's orbit is 0.25, the highest of all the nine planets. Figure 13.6a shows an ellipse with the eccentricity of that of Pluto's orbit. Notice that even this highest-eccentricity orbit is difficult to distinguish from a circle. This is why Kepler's first law is an admirable accomplishment. The eccentricity of the orbit of Comet Halley is 0.97, describing an orbit whose major axis is much longer than its minor axis, as shown in Figure 13.6b. As a result, Comet Halley spends much of its 76-year period far from the Sun and invisible from the Earth. It is only visible to the naked eye during a small part of its orbit when it is near the Sun.

Now imagine a planet in an elliptical orbit such as that shown in Figure 13.5, with the Sun at focus F_2. When the planet is at the far left in the diagram, the distance between the planet and the Sun is $a + c$. This point is called the *aphelion*, where the planet is the farthest away from the Sun that it can be in the orbit. (For an object in orbit around the Earth, this point is called the *apogee*). Conversely, when the planet is at the right end of the ellipse, the point is called the *perihelion* (for an Earth orbit, the *perigee*), and the distance between the planet and the Sun is $a - c$.

Kepler's first law is a direct result of the inverse square nature of the gravitational force. We have discussed circular and elliptical orbits. These are the allowed shapes of orbits for objects that are *bound* to the gravitational force center. These objects include planets, asteroids, and comets that move repeatedly around the Sun, as well as moons orbiting a planet. There could also be *unbound* objects, such as a meteoroid from deep space that might pass by the Sun once and then never return. The gravitational force between the Sun and these objects also varies as the inverse square of the separation distance, and the allowed paths for these objects include parabolas ($e = 1$) and hyperbolas ($e > 1$).

Kepler's Second Law

Kepler's second law can be shown to be a consequence of angular momentum conservation as follows. Consider a planet of mass M_P moving about the Sun in an elliptical orbit (Fig. 13.7a). Let us consider the planet as a system. We will model the Sun to be

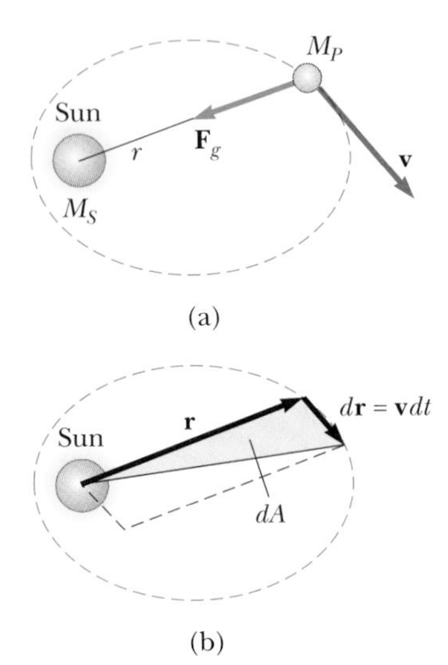

Active Figure 13.7 (a) The gravitational force acting on a planet is directed toward the Sun. (b) As a planet orbits the Sun, the area swept out by the radius vector in a time interval dt is equal to half the area of the parallelogram formed by the vectors $\mathbf{r}$ and $d\mathbf{r} = \mathbf{v}\,dt$.

At the Active Figures link at http://www.pse6.com, you can assign a value of the eccentricity and see the resulting motion of the planet around the Sun.

so much more massive than the planet that the Sun does not move. The gravitational force acting on the planet is a central force, always along the radius vector, directed toward the Sun (Fig. 13.7a). The torque on the planet due to this central force is clearly zero, because $\mathbf{F}$ is parallel to $\mathbf{r}$. That is

$$\boldsymbol{\tau} = \mathbf{r} \times \mathbf{F} = \mathbf{r} \times F(r)\hat{\mathbf{r}} = 0$$

Recall that the external net torque on a system equals the time rate of change of angular momentum of the system; that is, $\boldsymbol{\tau} = d\mathbf{L}/dt$. Therefore, because $\boldsymbol{\tau} = 0$, **the angular momentum L of the planet is a constant of the motion:**

$$\mathbf{L} = \mathbf{r} \times \mathbf{p} = M_P\mathbf{r} \times \mathbf{v} = \text{constant}$$

We can relate this result to the following geometric consideration. In a time interval dt, the radius vector $\mathbf{r}$ in Figure 13.7b sweeps out the area dA, which equals half the area $|\mathbf{r} \times d\mathbf{r}|$ of the parallelogram formed by the vectors $\mathbf{r}$ and $d\mathbf{r}$. Because the displacement of the planet in the time interval dt is given by $d\mathbf{r} = \mathbf{v}\,dt$, we have

$$dA = \tfrac{1}{2}|\mathbf{r} \times d\mathbf{r}| = \tfrac{1}{2}|\mathbf{r} \times \mathbf{v}\,dt| = \frac{L}{2M_P}\,dt$$

$$\frac{dA}{dt} = \frac{L}{2M_P} = \text{constant} \tag{13.7}$$

where L and M_P are both constants. Thus, we conclude that **the radius vector from the Sun to any planet sweeps out equal areas in equal times.**

It is important to recognize that this result is a consequence of the fact that the gravitational force is a central force, which in turn implies that angular momentum of the planet is constant. Therefore, the law applies to *any* situation that involves a central force, whether inverse-square or not.

Kepler's Third Law

It is informative to show that Kepler's third law can be predicted from the inverse-square law for circular orbits.[2] Consider a planet of mass M_P that is assumed to be moving about the Sun (mass M_S) in a circular orbit, as in Figure 13.8. Because the gravitational force provides the centripetal acceleration of the planet as it moves in a circle, we use Newton's second law for a particle in uniform circular motion,

$$\frac{GM_SM_P}{r^2} = \frac{M_Pv^2}{r}$$

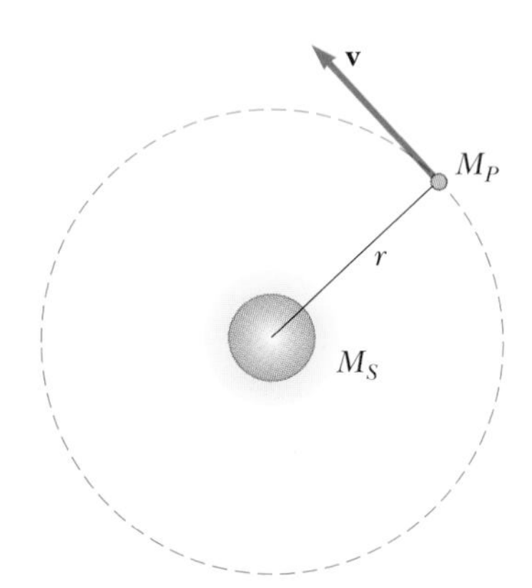

Figure 13.8 A planet of mass M_P moving in a circular orbit around the Sun. The orbits of all planets except Mercury and Pluto are nearly circular.

The orbital speed of the planet is $2\pi r/T$, where T is the period; therefore, the preceding expression becomes

$$\frac{GM_S}{r^2} = \frac{(2\pi r/T)^2}{r}$$

$$T^2 = \left(\frac{4\pi^2}{GM_S}\right)r^3 = K_Sr^3$$

where K_S is a constant given by

$$K_S = \frac{4\pi^2}{GM_S} = 2.97 \times 10^{-19}\ \text{s}^2/\text{m}^3$$

[2] The orbits of all planets except Mercury and Pluto are very close to being circular; hence, we do not introduce much error with this assumption. For example, the ratio of the semiminor axis to the semimajor axis for the Earth's orbit is $b/a = 0.999\ 86$.

Table 13.2

Useful Planetary Data

Body	Mass (kg)	Mean Radius (m)	Period of Revolution (s)	Mean Distance from Sun (m)	$\frac{T^2}{r^3}$ (s^2/m^3)
Mercury	3.18×10^{23}	2.43×10^{6}	7.60×10^{6}	5.79×10^{10}	2.97×10^{-19}
Venus	4.88×10^{24}	6.06×10^{6}	1.94×10^{7}	1.08×10^{11}	2.99×10^{-19}
Earth	5.98×10^{24}	6.37×10^{6}	3.156×10^{7}	1.496×10^{11}	2.97×10^{-19}
Mars	6.42×10^{23}	3.37×10^{6}	5.94×10^{7}	2.28×10^{11}	2.98×10^{-19}
Jupiter	1.90×10^{27}	6.99×10^{7}	3.74×10^{8}	7.78×10^{11}	2.97×10^{-19}
Saturn	5.68×10^{26}	5.85×10^{7}	9.35×10^{8}	1.43×10^{12}	2.99×10^{-19}
Uranus	8.68×10^{25}	2.33×10^{7}	2.64×10^{9}	2.87×10^{12}	2.95×10^{-19}
Neptune	1.03×10^{26}	2.21×10^{7}	5.22×10^{9}	4.50×10^{12}	2.99×10^{-19}
Pluto	$\approx 1.4 \times 10^{22}$	$\approx 1.5 \times 10^{6}$	7.82×10^{9}	5.91×10^{12}	2.96×10^{-19}
Moon	7.36×10^{22}	1.74×10^{6}	—	—	—
Sun	1.991×10^{30}	6.96×10^{8}	—	—	—

This equation is also valid for elliptical orbits if we replace r with the length a of the semimajor axis (Fig. 13.5):

$$T^2 = \left(\frac{4\pi^2}{GM_S}\right) a^3 = K_S a^3 \qquad (13.8)$$

Kepler's third law

Equation 13.8 is Kepler's third law. Because the semimajor axis of a circular orbit is its radius, Equation 13.8 is valid for both circular and elliptical orbits. Note that the constant of proportionality K_S is independent of the mass of the planet. Equation 13.8 is therefore valid for *any* planet.[3] If we were to consider the orbit of a satellite such as the Moon about the Earth, then the constant would have a different value, with the Sun's mass replaced by the Earth's mass, that is, $K_E = 4\pi^2/GM_E$.

Table 13.2 is a collection of useful planetary data. The last column verifies that the ratio T^2/r^3 is constant. The small variations in the values in this column are due to uncertainties in the data measured for the periods and semimajor axes of the planets.

Recent astronomical work has revealed the existence of a large number of solar system objects beyond the orbit of Neptune. In general, these lie in the *Kuiper belt,* a region that extends from about 30 AU (the orbital radius of Neptune) to 50 AU. (An AU is an *astronomical unit*—the radius of the Earth's orbit.) Current estimates identify at least 70 000 objects in this region with diameters larger than 100 km. The first KBO (Kuiper Belt Object) was discovered in 1992. Since then, many more have been detected and some have been given names, such as Varuna (diameter about 900–1 000 km, discovered in 2000), Ixion (diameter about 900–1 000 km, discovered in 2001), and Quaoar (diameter about 800 km, discovered in 2002).

A subset of about 1 400 KBOs are called "Plutinos" because, like Pluto, they exhibit a resonance phenomenon, orbiting the Sun two times in the same time interval as Neptune revolves three times. Some astronomers even claim that Pluto should not be considered a planet but should be identified as a KBO. The contemporary application of Kepler's laws and such exotic proposals as planetary angular momentum exchange and migrating planets[4] suggest the excitement of this active area of current research.

Quick Quiz 13.4 Pluto, the farthest planet from the Sun, has an orbital period that is (a) greater than a year (b) less than a year (c) equal to a year.

[3] Equation 13.8 is indeed a proportion because the ratio of the two quantities T^2 and a^3 is a constant. The variables in a proportion are not required to be limited to the first power only.

[4] Malhotra, R., "Migrating Planets," *Scientific American,* September 1999, volume 281, number 3.

Quick Quiz 13.5 An asteroid is in a highly eccentric elliptical orbit around the Sun. The period of the asteroid's orbit is 90 days. Which of the following statements is true about the possibility of a collision between this asteroid and the Earth? (a) There is no possible danger of a collision. (b) There is a possibility of a collision. (c) There is not enough information to determine whether there is danger of a collision.

Quick Quiz 13.6 A satellite moves in an elliptical orbit about the Earth such that, at perigee and apogee positions, its distances from the Earth's center are respectively D and $4D$. The relationship between the speeds at these two positions is (a) $v_p = v_a$ (b) $v_p = 4v_a$ (c) $v_a = 4v_p$ (d) $v_p = 2v_a$ (e) $v_a = 2v_p$.

Example 13.4 The Mass of the Sun

Calculate the mass of the Sun using the fact that the period of the Earth's orbit around the Sun is 3.156×10^7 s and its distance from the Sun is 1.496×10^{11} m.

Solution Using Equation 13.8, we find that

$$M_S = \frac{4\pi^2 r^3}{GT^2} = \frac{4\pi^2(1.496 \times 10^{11}\text{ m})^3}{(6.67 \times 10^{-11}\text{ N}\cdot\text{m}^2/\text{kg}^2)(3.156 \times 10^7\text{ s})^2}$$

$$= 1.99 \times 10^{30}\text{ kg}$$

In Example 13.3, an understanding of gravitational forces enabled us to find out something about the density of the Earth's core, and now we have used this understanding to determine the mass of the Sun!

What If? Suppose you were asked for the mass of Mars. How could you determine this value?

Answer Kepler's third law is valid for any system of objects in orbit around an object with a large mass. Mars has two moons, Phobos and Deimos. If we rewrite Equation 13.8 for these moons of Mars, we have

$$T^2 = \left(\frac{4\pi^2}{GM_M}\right)a^3$$

where M_M is the mass of Mars. Solving for this mass,

$$M_M = \left(\frac{4\pi^2}{G}\right)\frac{a^3}{T^2} = \left(\frac{4\pi^2}{6.67 \times 10^{-11}\text{ N}\cdot\text{m}^2/\text{kg}^2}\right)\frac{a^3}{T^2}$$

$$= (5.92 \times 10^{11}\text{ kg}\cdot\text{s}^2/\text{m}^3)\frac{a^3}{T^2}$$

Phobos has an orbital period of 0.32 days and an almost circular orbit of radius 9 380 km. The orbit of Deimos is even more circular, with a radius of 23 460 km and an orbital period of 1.26 days. Let us calculate the mass of Mars using each of these sets of data:

Phobos:

$$M_M = (5.92 \times 10^{11}\text{ kg}\cdot\text{s}^2/\text{m}^3) \times \frac{(9.380 \times 10^6\text{ m})^3}{(0.32\text{ d})^2}\left(\frac{1\text{ d}}{86\,400\text{ s}}\right)^2 = 6.39 \times 10^{23}\text{ kg}$$

Deimos:

$$M_M = (5.92 \times 10^{11}\text{ kg}\cdot\text{s}^2/\text{m}^3) \times \frac{(2.346 \times 10^7\text{ m})^3}{(1.26\text{ d})^2}\left(\frac{1\text{ d}}{86\,400\text{ s}}\right)^2 = 6.45 \times 10^{23}\text{ kg}$$

These two calculations are within 1% of each other and both are within 0.5% of the value of the mass of Mars given in Table 13.2.

Example 13.5 A Geosynchronous Satellite

Interactive

Consider a satellite of mass m moving in a circular orbit around the Earth at a constant speed v and at an altitude h above the Earth's surface, as illustrated in Figure 13.9.

(A) Determine the speed of the satellite in terms of G, h, R_E (the radius of the Earth), and M_E (the mass of the Earth).

Solution Conceptualize by imagining the satellite moving around the Earth in a circular orbit under the influence of the gravitational force. The satellite must have a centripetal acceleration. Thus, we categorize this problem as one involving Newton's second law, the law of universal gravitation, and circular motion. To analyze the problem, note that the only external force acting on the satellite is the gravitational force, which acts toward the center of the Earth and keeps the satellite in its circular orbit. Therefore, the net force on the satellite is the gravitational force

$$F_r = F_g = G\frac{M_E m}{r^2}$$

From Newton's second law and the fact that the acceleration of the satellite is centripetal, we obtain

$$G\frac{M_E m}{r^2} = m\frac{v^2}{r}$$

Solving for v and remembering that the distance r from the center of the Earth to the satellite is $r = R_E + h$, we obtain

$$(1) \qquad v = \sqrt{\frac{GM_E}{r}} = \sqrt{\frac{GM_E}{R_E + h}}$$

(B) If the satellite is to be *geosynchronous* (that is, appearing to remain over a fixed position on the Earth), how fast is it moving through space?

Solution In order to appear to remain over a fixed position on the Earth, the period of the satellite must be 24 h and the satellite must be in orbit directly over the equator. From Kepler's third law (Equation 13.8) with $a = r$ and $M_S \rightarrow M_E$, we find the radius of the orbit:

$$T^2 = \left(\frac{4\pi^2}{GM_E}\right) r^3$$

$$r = \sqrt[3]{\frac{GM_E T^2}{4\pi^2}}$$

Substituting numerical values and noting that the period is $T = 24\text{ h} = 86\,400\text{ s}$, we find

$$r = \sqrt[3]{\frac{(6.67 \times 10^{-11}\text{ N}\cdot\text{m}^2/\text{kg}^2)(5.98 \times 10^{24}\text{ kg})(86\,400\text{ s})^2}{4\pi^2}}$$

$$= 4.23 \times 10^7\text{ m}$$

To find the speed of the satellite, we use Equation (1):

$$v = \sqrt{\frac{GM_E}{r}}$$

$$= \sqrt{\frac{(6.67 \times 10^{-11}\text{ N}\cdot\text{m}^2/\text{kg}^2)(5.98 \times 10^{24}\text{ kg})}{4.23 \times 10^7\text{ m}}}$$

$$= 3.07 \times 10^3\text{ m/s}$$

To finalize this problem, it is interesting to note that the value of r calculated here translates to a height of the satellite above the surface of the Earth of almost 36 000 km. Thus, geosynchronous satellites have the advantage of allowing an earthbound antenna to be aimed in a fixed direction, but there is a disadvantage in that the signals between Earth and the satellite must travel a long distance. It is difficult to use geosynchronous satellites for optical observation of the Earth's surface because of their high altitude.

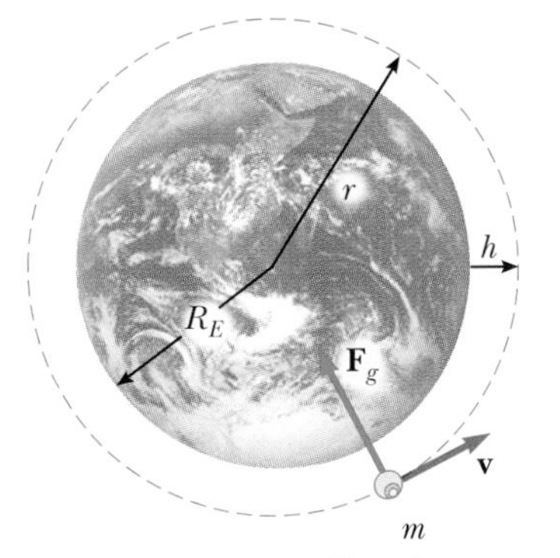

Figure 13.9 (Example 13.5) A satellite of mass m moving around the Earth in a circular orbit of radius r with constant speed v. The only force acting on the satellite is the gravitational force $\mathbf{F}_g$. (Not drawn to scale.)

What If? What if the satellite motion in part (A) were taking place at height *h* above the surface of another planet more massive than the Earth but of the same radius? Would the satellite be moving at a higher or a lower speed than it does around the Earth?

Answer If the planet pulls downward on the satellite with more gravitational force due to its larger mass, the satellite would have to move with a higher speed to avoid moving toward the surface. This is consistent with the predictions of Equation (1), which shows that because the speed v is proportional to the square root of the mass of the planet, as the mass increases, the speed also increases.

You can adjust the altitude of the satellite at the Interactive Worked Example link at **http://www.pse6.com.**

13.5 The Gravitational Field

When Newton published his theory of universal gravitation, it was considered a success because it satisfactorily explained the motion of the planets. Since 1687 the same theory has been used to account for the motions of comets, the deflection of a Cavendish balance, the orbits of binary stars, and the rotation of galaxies. Nevertheless, both Newton's contemporaries and his successors found it difficult to accept the concept of a force that acts at a distance, as mentioned in Section 5.1. They asked how it was possible for two objects to interact when they were not in contact with each other. Newton himself could not answer that question.

An approach to describing interactions between objects that are not in contact came well after Newton's death, and it enables us to look at the gravitational interaction in a different way, using the concept of a **gravitational field** that exists at every point in space. When a particle of mass m is placed at a point where the gravitational

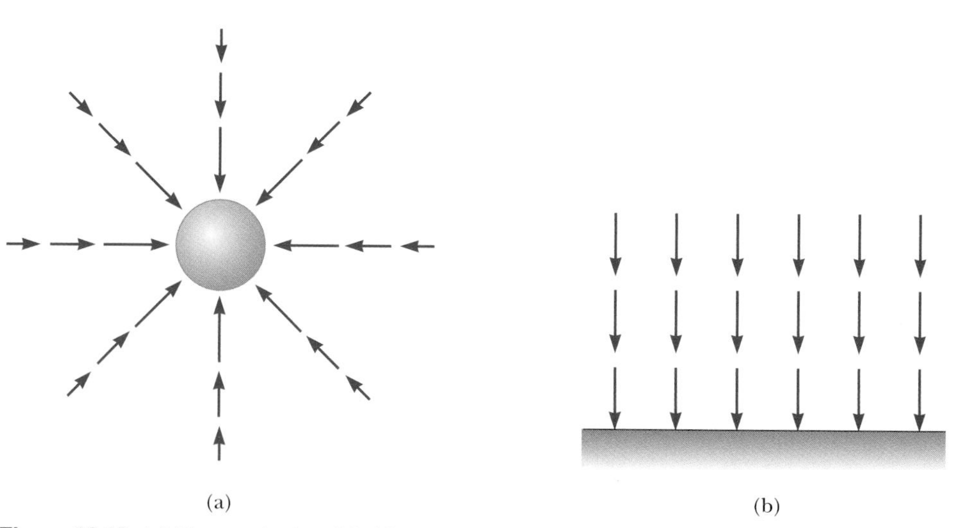

Figure 13.10 (a) The gravitational field vectors in the vicinity of a uniform spherical mass such as the Earth vary in both direction and magnitude. The vectors point in the direction of the acceleration a particle would experience if it were placed in the field. The magnitude of the field vector at any location is the magnitude of the free-fall acceleration at that location. (b) The gravitational field vectors in a small region near the Earth's surface are uniform in both direction and magnitude.

field is $\mathbf{g}$, the particle experiences a force $\mathbf{F}_g = m\mathbf{g}$. In other words, the field exerts a force on the particle. The gravitational field $\mathbf{g}$ is defined as

Gravitational field

$$\mathbf{g} \equiv \frac{\mathbf{F}_g}{m} \tag{13.9}$$

That is, the gravitational field at a point in space equals the gravitational force experienced by a *test particle* placed at that point divided by the mass of the test particle. Notice that the presence of the test particle is not necessary for the field to exist—the Earth creates the gravitational field. We call the object creating the field the *source particle.* (Although the Earth is clearly not a particle, it is possible to show that we can approximate the Earth as a particle for the purpose of finding the gravitational field that it creates.) We can detect the presence of the field and measure its strength by placing a test particle in the field and noting the force exerted on it.

Although the gravitational force is inherently an interaction between two objects, the concept of a gravitational field allows us to "factor out" the mass of one of the objects. In essence, we are describing the "effect" that any object (in this case, the Earth) has on the empty space around itself in terms of the force that *would* be present *if* a second object were somewhere in that space.[5]

As an example of how the field concept works, consider an object of mass m near the Earth's surface. Because the gravitational force acting on the object has a magnitude GM_Em/r^2 (see Eq. 13.4), the field $\mathbf{g}$ at a distance r from the center of the Earth is

$$\mathbf{g} = \frac{\mathbf{F}_g}{m} = -\frac{GM_E}{r^2}\,\hat{\mathbf{r}} \tag{13.10}$$

where $\hat{\mathbf{r}}$ is a unit vector pointing radially outward from the Earth and the negative sign indicates that the field points toward the center of the Earth, as illustrated in Figure 13.10a. Note that the field vectors at different points surrounding the Earth vary in both direction and magnitude. In a small region near the Earth's surface, the downward field $\mathbf{g}$ is approximately constant and uniform, as indicated in Figure 13.10b. Equation 13.10 is valid at all points *outside* the Earth's surface, assuming that the Earth is spherical. At the Earth's surface, where $r = R_E$, $\mathbf{g}$ has a magnitude of 9.80 N/kg. (The unit N/kg is the same as m/s^2.)

[5] We shall return to this idea of mass affecting the space around it when we discuss Einstein's theory of gravitation in Chapter 39.

13.6 Gravitational Potential Energy

In Chapter 8 we introduced the concept of gravitational potential energy, which is the energy associated with the configuration of a system of objects interacting via the gravitational force. We emphasized that the gravitational potential-energy function mgy for a particle–Earth system is valid only when the particle is near the Earth's surface, where the gravitational force is constant. Because the gravitational force between two particles varies as $1/r^2$, we expect that a more general potential-energy function—one that is valid without the restriction of having to be near the Earth's surface—will be significantly different from $U = mgy$.

Before we calculate this general form for the gravitational potential energy function, let us first verify that *the gravitational force is conservative.* (Recall from Section 8.3 that a force is conservative if the work it does on an object moving between any two points is independent of the path taken by the object.) To do this, we first note that the gravitational force is a central force. By definition, a central force is any force that is directed along a radial line to a fixed center and has a magnitude that depends only on the radial coordinate r. Hence, a central force can be represented by $F(r)\hat{\mathbf{r}}$ where $\hat{\mathbf{r}}$ is a unit vector directed from the origin toward the particle, as shown in Figure 13.11.

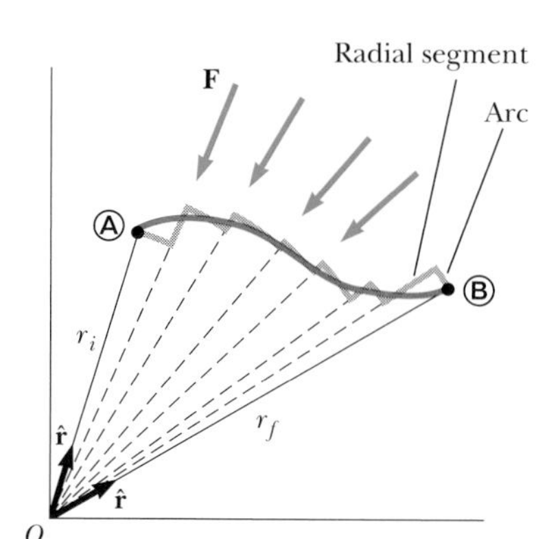

Figure 13.11 A particle moves from Ⓐ to Ⓑ while acted on by a central force **F**, which is directed radially. The path is broken into a series of radial segments and arcs. Because the work done along the arcs is zero, the work done is independent of the path and depends only on r_f and r_i.

Consider a central force acting on a particle moving along the general path Ⓐ to Ⓑ in Figure 13.11. The path from Ⓐ to Ⓑ can be approximated by a series of steps according to the following procedure. In Figure 13.11, we draw several thin wedges, which are shown as dashed lines. The outer boundary of our set of wedges is a path consisting of short radial line segments and arcs (gray in the figure). We select the length of the radial dimension of each wedge such that the short arc at the wedge's wide end intersects the actual path of the particle. Then we can approximate the actual path with a series of zigzag movements that alternate between moving along an arc and moving along a radial line.

By definition, a central force is always directed along one of the radial segments; therefore, the work done by **F** along any radial segment is

$$dW = \mathbf{F} \cdot d\mathbf{r} = F(r)\ dr$$

By definition, the work done by a force that is perpendicular to the displacement is zero. Hence, the work done in moving along any arc is zero because **F** is perpendicular to the displacement along these segments. Therefore, the total work done by **F** is the sum of the contributions along the radial segments:

Work done by a central force

$$W = \int_{r_i}^{r_f} F(r)\ dr$$

where the subscripts i and f refer to the initial and final positions. Because the integrand is a function only of the radial position, this integral depends only on the initial and final values of r. Thus, the work done is the same over *any* path from Ⓐ to Ⓑ. Because the work done is independent of the path and depends only on the end points, we conclude that *any central force is conservative.* We are now assured that a potential energy function can be obtained once the form of the central force is specified.

Recall from Equation 8.15 that the change in the gravitational potential energy of a system associated with a given displacement of a member of the system is defined as the negative of the work done by the gravitational force on that member during the displacement:

$$\Delta U = U_f - U_i = -\int_{r_i}^{r_f} F(r)\ dr \qquad (13.11)$$

We can use this result to evaluate the gravitational potential energy function. Consider a particle of mass m moving between two points Ⓐ and Ⓑ above the Earth's surface (Fig. 13.12). The particle is subject to the gravitational force given by Equation 13.1. We can express this force as

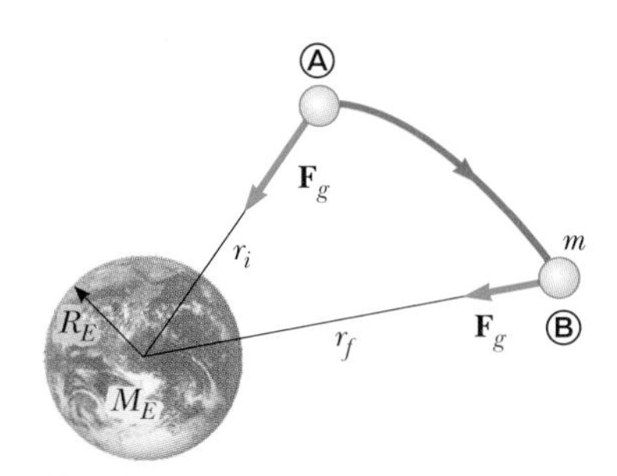

Figure 13.12 As a particle of mass m moves from Ⓐ to Ⓑ above the Earth's surface, the gravitational potential energy changes according to Equation 13.11.

$$F(r) = -\frac{GM_E m}{r^2}$$

where the negative sign indicates that the force is attractive. Substituting this expression for $F(r)$ into Equation 13.11, we can compute the change in the gravitational potential energy function:

$$U_f - U_i = GM_E m \int_{r_i}^{r_f} \frac{dr}{r^2} = GM_E m \left[-\frac{1}{r} \right]_{r_i}^{r_f}$$

$$U_f - U_i = -GM_E m \left(\frac{1}{r_f} - \frac{1}{r_i} \right) \tag{13.12}$$

As always, the choice of a reference configuration for the potential energy is completely arbitrary. It is customary to choose the reference configuration for zero potential energy to be the same as that for which the force is zero. Taking $U_i = 0$ at $r_i = \infty$, we obtain the important result

Gravitational potential energy of the Earth–particle system for $r > R_E$

$$U(r) = -\frac{GM_E m}{r} \tag{13.13}$$

This expression applies to the Earth–particle system where the particle is separated from the center of the Earth by a distance r, provided that $r \geq R_E$. The result is not valid for particles inside the Earth, where $r < R_E$. Because of our choice of U_i, the function U is always negative (Fig. 13.13).

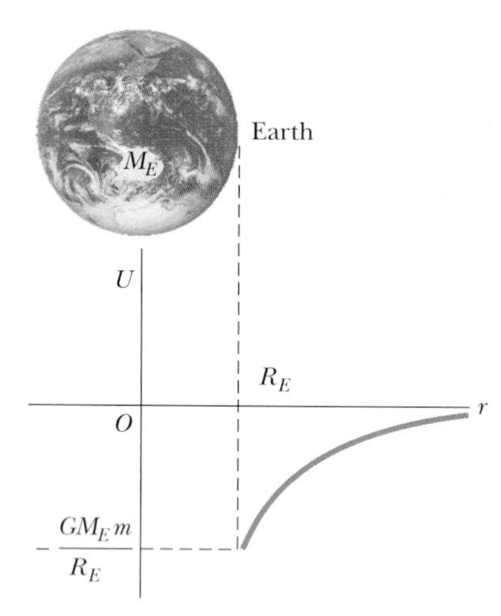

Figure 13.13 Graph of the gravitational potential energy U versus r for an object above the Earth's surface. The potential energy goes to zero as r approaches infinity.

Although Equation 13.13 was derived for the particle–Earth system, it can be applied to any two particles. That is, the gravitational potential energy associated with any pair of particles of masses m_1 and m_2 separated by a distance r is

$$U = -\frac{Gm_1 m_2}{r} \tag{13.14}$$

This expression shows that the gravitational potential energy for any pair of particles varies as $1/r$, whereas the force between them varies as $1/r^2$. Furthermore, the potential energy is negative because the force is attractive and we have taken the potential energy as zero when the particle separation is infinite. Because the force between the particles is attractive, we know that an external agent must do positive work to increase the separation between them. The work done by the external agent produces an increase in the potential energy as the two particles are separated. That is, U becomes less negative as r increases.

When two particles are at rest and separated by a distance r, an external agent has to supply an energy at least equal to $+Gm_1m_2/r$ in order to separate the particles to an infinite distance. It is therefore convenient to think of the absolute value of the potential energy as the *binding energy* of the system. If the external agent supplies an energy greater than the binding energy, the excess energy of the system will be in the form of kinetic energy of the particles when the particles are at an infinite separation.

We can extend this concept to three or more particles. In this case, the total potential energy of the system is the sum over all pairs of particles.[6] Each pair contributes a term of the form given by Equation 13.14. For example, if the system contains three particles, as in Figure 13.14, we find that

$$U_{\text{total}} = U_{12} + U_{13} + U_{23} = -G\left(\frac{m_1 m_2}{r_{12}} + \frac{m_1 m_3}{r_{13}} + \frac{m_2 m_3}{r_{23}} \right) \tag{13.15}$$

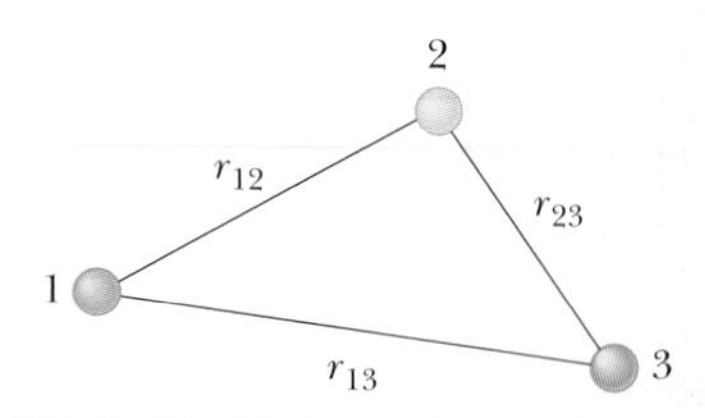

Figure 13.14 Three interacting particles.

The absolute value of U_{total} represents the work needed to separate the particles by an infinite distance.

[6] The fact that potential energy terms can be added for all pairs of particles stems from the experimental fact that gravitational forces obey the superposition principle.

Example 13.6 The Change in Potential Energy

A particle of mass m is displaced through a small vertical distance Δy near the Earth's surface. Show that in this situation the general expression for the change in gravitational potential energy given by Equation 13.12 reduces to the familiar relationship $\Delta U = mg\,\Delta y$.

Solution We can express Equation 13.12 in the form

$$\Delta U = -GM_E m\left(\frac{1}{r_f} - \frac{1}{r_i}\right) = GM_E m\left(\frac{r_f - r_i}{r_i r_f}\right)$$

If both the initial and final positions of the particle are close to the Earth's surface, then $r_f - r_i = \Delta y$ and $r_i r_f \approx R_E^2$. (Recall that r is measured from the center of the Earth.) Therefore, the change in potential energy becomes

$$\Delta U \approx \frac{GM_E m}{R_E^2}\Delta y = mg\,\Delta y$$

where we have used the fact that $g = GM_E/R_E^2$ (Eq. 13.5). Keep in mind that the reference configuration is arbitrary because it is the *change* in potential energy that is meaningful.

What If? Suppose you are performing upper-atmosphere studies and are asked by your supervisor to find the height in the Earth's atmosphere at which the "surface equation" $\Delta U = mg\,\Delta y$ gives a 1.0% error in the change in the potential energy. What is this height?

Answer Because the surface equation assumes a constant value for g, it will give a ΔU value that is larger than the value given by the general equation, Equation 13.12. Thus, a 1.0% error would be described by the ratio

$$\frac{\Delta U_{\text{surface}}}{\Delta U_{\text{general}}} = 1.010$$

Substituting the expressions for each of these changes ΔU, we have

$$\frac{mg\,\Delta y}{GM_E m(\Delta y/r_i r_f)} = \frac{g r_i r_f}{GM_E} = 1.010$$

where $r_i = R_E$ and $r_f = R_E + \Delta y$. Substituting for g from Equation 13.5, we find

$$\frac{(GM_E/R_E^2)R_E(R_E + \Delta y)}{GM_E} = \frac{R_E + \Delta y}{R_E} = 1 + \frac{\Delta y}{R_E} = 1.010$$

Thus,

$$\Delta y = 0.010R_E = 0.010(6.37 \times 10^6\text{ m})$$
$$= 6.37 \times 10^4\text{ m} = 63.7\text{ km}$$

13.7 Energy Considerations in Planetary and Satellite Motion

Consider an object of mass m moving with a speed v in the vicinity of a massive object of mass M, where $M \gg m$. The system might be a planet moving around the Sun, a satellite in orbit around the Earth, or a comet making a one-time flyby of the Sun. If we assume that the object of mass M is at rest in an inertial reference frame, then the total mechanical energy E of the two-object system when the objects are separated by a distance r is the sum of the kinetic energy of the object of mass m and the potential energy of the system, given by Equation 13.14:[7]

$$E = K + U$$

$$E = \tfrac{1}{2}mv^2 - \frac{GMm}{r} \tag{13.16}$$

[7] You might recognize that we have ignored the kinetic energy of the larger body. To see that this simplification is reasonable, consider an object of mass m falling toward the Earth. Because the center of mass of the object–Earth system is effectively stationary, it follows from conservation of momentum that $mv = M_E v_E$. Thus, the Earth acquires a kinetic energy equal to

$$\tfrac{1}{2}M_E v_E^2 = \tfrac{1}{2}\frac{m^2}{M_E}v^2 = \frac{m}{M_E}K$$

where K is the kinetic energy of the object. Because $M_E \gg m$, this result shows that the kinetic energy of the Earth is negligible.

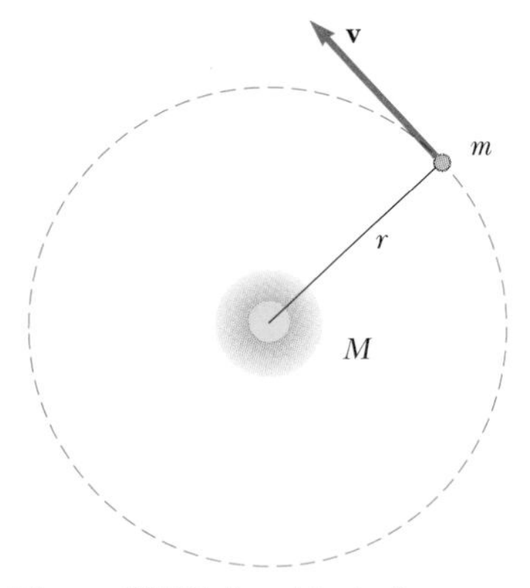

Figure 13.15 An object of mass m moving in a circular orbit about a much larger object of mass M.

This equation shows that E may be positive, negative, or zero, depending on the value of v. However, for a bound system,[8] such as the Earth–Sun system, E is necessarily *less than zero* because we have chosen the convention that $U \rightarrow 0$ as $r \rightarrow \infty$.

We can easily establish that $E < 0$ for the system consisting of an object of mass m moving in a circular orbit about an object of mass $M >> m$ (Fig. 13.15). Newton's second law applied to the object of mass m gives

$$\frac{GMm}{r^2} = ma = \frac{mv^2}{r}$$

Multiplying both sides by r and dividing by 2 gives

$$\frac{1}{2}mv^2 = \frac{GMm}{2r} \tag{13.17}$$

Substituting this into Equation 13.16, we obtain

$$E = \frac{GMm}{2r} - \frac{GMm}{r}$$

Total energy for circular orbits

$$E = -\frac{GMm}{2r} \quad \text{(circular orbits)} \tag{13.18}$$

This result clearly shows that **the total mechanical energy is negative in the case of circular orbits.** Note that **the kinetic energy is positive and equal to half the absolute value of the potential energy.** The absolute value of E is also equal to the binding energy of the system, because this amount of energy must be provided to the system to move the two objects infinitely far apart.

The total mechanical energy is also negative in the case of elliptical orbits. The expression for E for elliptical orbits is the same as Equation 13.18 with r replaced by the semimajor axis length a:

Total energy for elliptical orbits

$$E = -\frac{GMm}{2a} \quad \text{(elliptical orbits)} \tag{13.19}$$

Furthermore, the total energy is constant if we assume that the system is isolated. Therefore, as the object of mass m moves from Ⓐ to Ⓑ in Figure 13.12, the total energy remains constant and Equation 13.16 gives

$$E = \frac{1}{2}mv_i^2 - \frac{GMm}{r_i} = \frac{1}{2}mv_f^2 - \frac{GMm}{r_f} \tag{13.20}$$

Combining this statement of energy conservation with our earlier discussion of conservation of angular momentum, we see that **both the total energy and the total angular momentum of a gravitationally bound, two-object system are constants of the motion.**

Quick Quiz 13.7 A comet moves in an elliptical orbit around the Sun. Which point in its orbit (perihelion or aphelion) represents the highest value of (a) the speed of the comet (b) the potential energy of the comet–Sun system (c) the kinetic energy of the comet (d) the total energy of the comet–Sun system?

[8] Of the three examples provided at the beginning of this section, the planet moving around the Sun and a satellite in orbit around the Earth are bound systems—the Earth will always stay near the Sun, and the satellite will always stay near the Earth. The one-time comet flyby represents an unbound system—the comet interacts once with the Sun but is not bound to it. Thus, in theory the comet can move infinitely far away from the Sun.

Example 13.7 Changing the Orbit of a Satellite

The space shuttle releases a 470-kg communications satellite while in an orbit 280 km above the surface of the Earth. A rocket engine on the satellite boosts it into a geosynchronous orbit, which is an orbit in which the satellite stays directly over a single location on the Earth. How much energy does the engine have to provide?

Solution We first determine the initial radius (not the altitude above the Earth's surface) of the satellite's orbit when it is still in the shuttle's cargo bay. This is simply

$$R_E + 280 \text{ km} = 6.65 \times 10^6 \text{ m} = r_i$$

In Example 13.5, we found that the radius of the orbit of a geosynchronous satellite is $r_f = 4.23 \times 10^7$ m. Applying Equation 13.18, we obtain, for the total initial and final energies,

$$E_i = -\frac{GM_E m}{2r_i} \qquad E_f = -\frac{GM_E m}{2r_f}$$

The energy required from the engine to boost the satellite is

$$\Delta E = E_f - E_i = -\frac{GM_E m}{2}\left(\frac{1}{r_f} - \frac{1}{r_i}\right)$$

$$= -\frac{(6.67 \times 10^{-11}\text{ N}\cdot\text{m}^2/\text{kg}^2)(5.98 \times 10^{24}\text{ kg})(470\text{ kg})}{2}$$

$$\times \left(\frac{1}{4.23 \times 10^7\text{ m}} - \frac{1}{6.65 \times 10^6\text{ m}}\right)$$

$$= 1.19 \times 10^{10}\text{ J}$$

This is the energy equivalent of 89 gal of gasoline. NASA engineers must account for the changing mass of the spacecraft as it ejects burned fuel, something we have not done here. Would you expect the calculation that includes the effect of this changing mass to yield a greater or lesser amount of energy required from the engine?

If we wish to determine how the energy is distributed after the engine is fired, we find from Equation 13.17 that the change in kinetic energy is $\Delta K = (GM_E m/2)(1/r_f - 1/r_i) = -1.19 \times 10^{10}$ J (a decrease), and the corresponding change in potential energy is $\Delta U = -GM_E m(1/r_f - 1/r_i) = 2.38 \times 10^{10}$ J (an increase). Thus, the change in orbital energy of the system is $\Delta E = \Delta K + \Delta U = 1.19 \times 10^{10}$ J, as we already calculated. The firing of the engine results in a transformation of chemical potential energy in the fuel to orbital energy of the system. Because an increase in gravitational potential energy is accompanied by a decrease in kinetic energy, we conclude that the speed of an orbiting satellite decreases as its altitude increases.

Escape Speed

Suppose an object of mass m is projected vertically upward from the Earth's surface with an initial speed v_i, as illustrated in Figure 13.16. We can use energy considerations to find the minimum value of the initial speed needed to allow the object to move infinitely far away from the Earth. Equation 13.16 gives the total energy of the system at any point. At the surface of the Earth, $v = v_i$ and $r = r_i = R_E$. When the object reaches its maximum altitude, $v = v_f = 0$ and $r = r_f = r_{max}$. Because the total energy of the system is constant, substituting these conditions into Equation 13.20 gives

$$\tfrac{1}{2}mv_i^2 - \frac{GM_E m}{R_E} = -\frac{GM_E m}{r_{max}}$$

Solving for v_i^2 gives

$$v_i^2 = 2GM_E\left(\frac{1}{R_E} - \frac{1}{r_{max}}\right) \tag{13.21}$$

Therefore, if the initial speed is known, this expression can be used to calculate the maximum altitude h because we know that

$$h = r_{max} - R_E$$

We are now in a position to calculate **escape speed,** which is the minimum speed the object must have at the Earth's surface in order to approach an infinite separation distance from the Earth. Traveling at this minimum speed, the object continues to

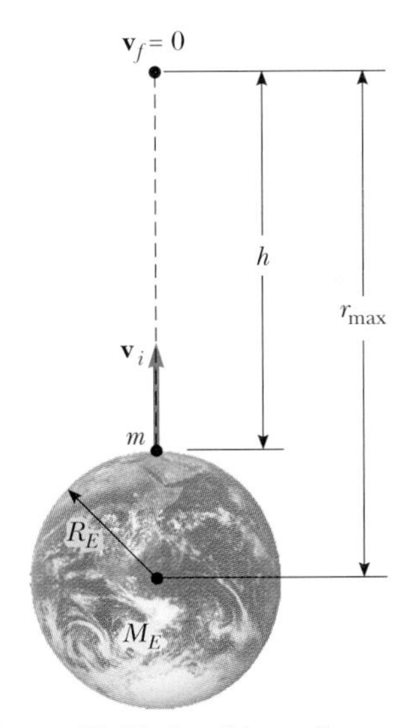

Figure 13.16 An object of mass m projected upward from the Earth's surface with an initial speed v_i reaches a maximum altitude h.

PITFALL PREVENTION

13.3 You Can't Really Escape

Although Equation 13.22 provides the "escape speed" from the Earth, *complete* escape from the Earth's gravitational influence is impossible because the gravitational force is of infinite range. No matter how far away you are, you will always feel some gravitational force due to the Earth.

move farther and farther away from the Earth as its speed asymptotically approaches zero. Letting $r_{max} \rightarrow \infty$ in Equation 13.21 and taking $v_i = v_{esc}$, we obtain

$$v_{esc} = \sqrt{\frac{2GM_E}{R_E}} \tag{13.22}$$

Note that this expression for v_{esc} is independent of the mass of the object. In other words, a spacecraft has the same escape speed as a molecule. Furthermore, the result is independent of the direction of the velocity and ignores air resistance.

If the object is given an initial speed equal to v_{esc}, the total energy of the system is equal to zero. This can be seen by noting that when $r \rightarrow \infty$, the object's kinetic energy and the potential energy of the system are both zero. If v_i is greater than v_{esc}, the total energy of the system is greater than zero and the object has some residual kinetic energy as $r \rightarrow \infty$.

Example 13.8 Escape Speed of a Rocket

Calculate the escape speed from the Earth for a 5 000-kg spacecraft, and determine the kinetic energy it must have at the Earth's surface in order to move infinitely far away from the Earth.

Solution Using Equation 13.22 gives

$$v_{esc} = \sqrt{\frac{2GM_E}{R_E}}$$

$$= \sqrt{\frac{2(6.67 \times 10^{-11}\ \text{N}\cdot\text{m}^2/\text{kg}^2)(5.98 \times 10^{24}\ \text{kg})}{6.37 \times 10^6\ \text{m}}}$$

$$= 1.12 \times 10^4\ \text{m/s}$$

This corresponds to about 25 000 mi/h.

The kinetic energy of the spacecraft is

$$K = \tfrac{1}{2}mv_{esc}^2 = \tfrac{1}{2}(5.00 \times 10^3\ \text{kg})(1.12 \times 10^4\ \text{m/s})^2$$

$$= 3.14 \times 10^{11}\ \text{J}$$

This is equivalent to about 2 300 gal of gasoline.

What If? What if we wish to launch a 1 000-kg spacecraft at the escape speed? How much energy does this require?

Answer In Equation 13.22, the mass of the object moving with the escape speed does not appear. Thus, the escape speed for the 1 000-kg spacecraft is the same as that for the 5 000-kg spacecraft. The only change in the kinetic energy is due to the mass, so the 1 000-kg spacecraft will require one fifth of the energy of the 5 000-kg spacecraft:

$$K = \tfrac{1}{5}(3.14 \times 10^{11}\ \text{J}) = 6.28 \times 10^{10}\ \text{J}$$

Equations 13.21 and 13.22 can be applied to objects projected from any planet. That is, in general, the escape speed from the surface of any planet of mass M and radius R is

$$v_{esc} = \sqrt{\frac{2GM}{R}} \tag{13.23}$$

Escape speeds for the planets, the Moon, and the Sun are provided in Table 13.3. Note that the values vary from 1.1 km/s for Pluto to about 618 km/s for the Sun. These results, together with some ideas from the kinetic theory of gases (see Chapter 21), explain why some planets have atmospheres and others do not. As we shall see later, at a given temperature the average kinetic energy of a gas molecule depends only on the mass of the molecule. Lighter molecules, such as hydrogen and helium, have a higher average speed than heavier molecules at the same temperature. When the average speed of the lighter molecules is not much less than the escape speed of a planet, a significant fraction of them have a chance to escape.

This mechanism also explains why the Earth does not retain hydrogen molecules and helium atoms in its atmosphere but does retain heavier molecules, such as oxygen and nitrogen. On the other hand, the very large escape speed for Jupiter enables that planet to retain hydrogen, the primary constituent of its atmosphere.

Table 13.3

Escape Speeds from the Surfaces of the Planets, Moon, and Sun

Planet	v_{esc} (km/s)
Mercury	4.3
Venus	10.3
Earth	11.2
Mars	5.0
Jupiter	60
Saturn	36
Uranus	22
Neptune	24
Pluto	1.1
Moon	2.3
Sun	618

Black Holes

In Example 11.7 we briefly described a rare event called a supernova—the catastrophic explosion of a very massive star. The material that remains in the central core of such an object continues to collapse, and the core's ultimate fate depends on its mass. If the core has a mass less than 1.4 times the mass of our Sun, it gradually cools down and ends its life as a white dwarf star. However, if the core's mass is greater than this, it may collapse further due to gravitational forces. What remains is a neutron star, discussed in Example 11.7, in which the mass of a star is compressed to a radius of about 10 km. (On Earth, a teaspoon of this material would weigh about 5 billion tons!)

An even more unusual star death may occur when the core has a mass greater than about three solar masses. The collapse may continue until the star becomes a very small object in space, commonly referred to as a **black hole.** In effect, black holes are remains of stars that have collapsed under their own gravitational force. If an object such as a spacecraft comes close to a black hole, it experiences an extremely strong gravitational force and is trapped forever.

The escape speed for a black hole is very high, due to the concentration of the mass of the star into a sphere of very small radius (see Eq. 13.23). If the escape speed exceeds the speed of light c, radiation from the object (such as visible light) cannot escape, and the object appears to be black; hence the origin of the terminology "black hole." The critical radius R_S at which the escape speed is c is called the **Schwarzschild radius** (Fig. 13.17). The imaginary surface of a sphere of this radius surrounding the black hole is called the **event horizon.** This is the limit of how close you can approach the black hole and hope to escape.

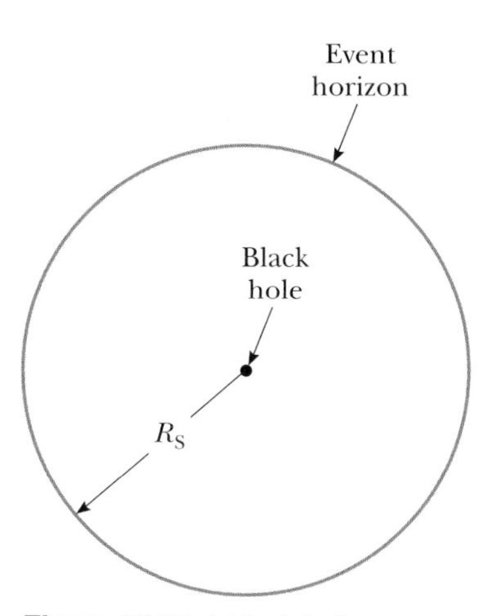

Figure 13.17 A black hole. The distance R_S equals the Schwarzschild radius. Any event occurring within the boundary of radius R_S, called the event horizon, is invisible to an outside observer.

Although light from a black hole cannot escape, light from events taking place near the black hole should be visible. For example, it is possible for a binary star system to consist of one normal star and one black hole. Material surrounding the ordinary star can be pulled into the black hole, forming an **accretion disk** around the black hole, as suggested in Figure 13.18. Friction among particles in the accretion disk results in transformation of mechanical energy into internal energy. As a result, the orbital height of the material above the event horizon decreases and the temperature rises. This high-temperature material emits a large amount of radiation, extending well into the x-ray region of the electromagnetic spectrum. These x-rays are characteristic of a black hole. Several possible candidates for black holes have been identified by observation of these x-rays.

Figure 13.18 A binary star system consisting of an ordinary star on the left and a black hole on the right. Matter pulled from the ordinary star forms an accretion disk around the black hole, in which matter is raised to very high temperatures, resulting in the emission of x-rays.

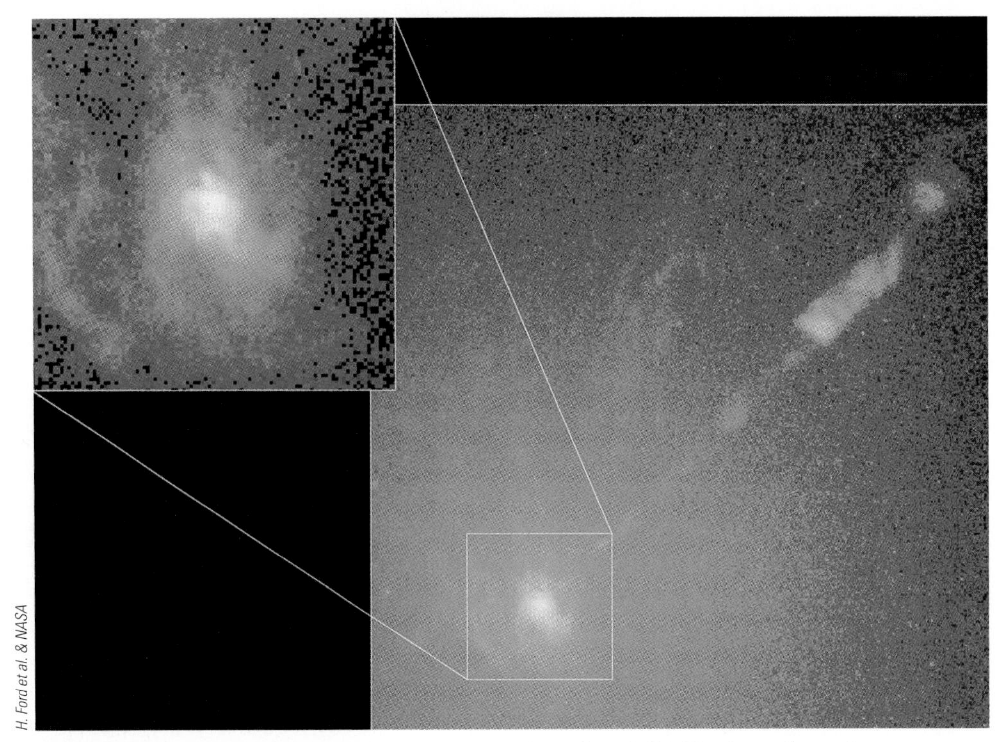

H. Ford et al. & NASA

Figure 13.19 Hubble Space Telescope images of the galaxy M87. The inset shows the center of the galaxy. The wider view shows a jet of material moving away from the center of the galaxy toward the upper right of the figure at about one tenth of the speed of light. Such jets are believed to be evidence of a supermassive black hole at the galaxy center.

There is also evidence that supermassive black holes exist at the centers of galaxies, with masses very much larger than the Sun. (There is strong evidence of a supermassive black hole of mass 2–3 million solar masses at the center of our galaxy.) Theoretical models for these bizarre objects predict that jets of material should be evident along the rotation axis of the black hole. Figure 13.19 shows a Hubble Space Telescope photograph of galaxy M87. The jet of material coming from this galaxy is believed to be evidence for a supermassive black hole at the center of the galaxy.

SUMMARY

Take a practice test for this chapter by clicking on the Practice Test link at http://www.pse6.com.

Newton's law of universal gravitation states that the gravitational force of attraction between any two particles of masses m_1 and m_2 separated by a distance r has the magnitude

$$F_g = G\frac{m_1 m_2}{r^2} \tag{13.1}$$

where $G = 6.673 \times 10^{-11}\ \text{N}\cdot\text{m}^2/\text{kg}^2$ is the **universal gravitational constant.** This equation enables us to calculate the force of attraction between masses under a wide variety of circumstances.

An object at a distance h above the Earth's surface experiences a gravitational force of magnitude mg, where g is the free-fall acceleration at that elevation:

$$g = \frac{GM_E}{r^2} = \frac{GM_E}{(R_E + h)^2} \tag{13.6}$$

In this expression, M_E is the mass of the Earth and R_E is its radius. Thus, the weight of an object decreases as the object moves away from the Earth's surface.

Kepler's laws of planetary motion state that

1. All planets move in elliptical orbits with the Sun at one focus.
2. The radius vector drawn from the Sun to a planet sweeps out equal areas in equal time intervals.
3. The square of the orbital period of any planet is proportional to the cube of the semimajor axis of the elliptical orbit.

Kepler's third law can be expressed as

$$T^2 = \left(\frac{4\pi^2}{GM_S}\right)a^3 \tag{13.8}$$

where M_S is the mass of the Sun and a is the semimajor axis. For a circular orbit, a can be replaced in Equation 13.8 by the radius r. Most planets have nearly circular orbits around the Sun.

The **gravitational field** at a point in space is defined as the gravitational force experienced by any test particle located at that point divided by the mass of the test particle:

$$\mathbf{g} \equiv \frac{\mathbf{F}_g}{m} \tag{13.9}$$

The gravitational force is conservative, and therefore a potential energy function can be defined for a system of two objects interacting gravitationally. The **gravitational potential energy** associated with two particles separated by a distance r is

$$U = -\frac{Gm_1m_2}{r} \tag{13.14}$$

where U is taken to be zero as $r \rightarrow \infty$. The total potential energy for a system of particles is the sum of energies for all pairs of particles, with each pair represented by a term of the form given by Equation 13.14.

If an isolated system consists of an object of mass m moving with a speed v in the vicinity of a massive object of mass M, the total energy E of the system is the sum of the kinetic and potential energies:

$$E = \tfrac{1}{2}mv^2 - \frac{GMm}{r} \tag{13.16}$$

The total energy is a constant of the motion. If the object moves in an elliptical orbit of semimajor axis a around the massive object and if $M \gg m$, the total energy of the system is

$$E = -\frac{GMm}{2a} \tag{13.19}$$

For a circular orbit, this same equation applies with $a = r$. The total energy is negative for any bound system.

The **escape speed** for an object projected from the surface of a planet of mass M and radius R is

$$v_{\text{esc}} = \sqrt{\frac{2GM}{R}} \tag{13.23}$$

QUESTIONS

1. If the gravitational force on an object is directly proportional to its mass, why don't objects with large masses fall with greater acceleration than small ones?

2. The gravitational force exerted by the Sun on you is downward into the Earth at night, and upward into the sky during the day. If you had a sensitive enough bathroom scale, would you expect to weigh more at night than during the day? Note also that you are farther away from the Sun at night than during the day. Would you expect to weigh less?

3. Use Kepler's second law to convince yourself that the Earth must move faster in its orbit during December, when it is closest to the Sun, than during June, when it is farthest from the Sun.

4. The gravitational force that the Sun exerts on the Moon is about twice as great as the gravitational force that the Earth exerts on the Moon. Why doesn't the Sun pull the Moon away from the Earth during a total eclipse of the Sun?

5. A satellite in orbit is not truly traveling through a vacuum. It is moving through very, very thin air. Does the resulting air friction cause the satellite to slow down?

6. How would you explain the fact that Jupiter and Saturn have periods much greater than one year?

7. If a system consists of five particles, how many terms appear in the expression for the total potential energy? How many terms appear if the system consists of N particles?

8. Does the escape speed of a rocket depend on its mass? Explain.

9. Compare the energies required to reach the Moon for a 10^5-kg spacecraft and a 10^3-kg satellite.

10. Explain why it takes more fuel for a spacecraft to travel from the Earth to the Moon than for the return trip. Estimate the difference.

11. A particular set of directions forms the *celestial equator.* If you live at 40° north latitude, these directions lie in an arc across your southern sky, including horizontally east, horizontally west, and south at 50° above the horizontal. In order to enjoy satellite TV, you need to install a dish with an unobstructed view to a particular point on the celestial equator. Why is this requirement so specific?

12. Why don't we put a geosynchronous weather satellite in orbit around the 45th parallel? Wouldn't this be more useful in the United States than one in orbit around the equator?

13. Is the absolute value of the potential energy associated with the Earth–Moon system greater than, less than, or equal to the kinetic energy of the Moon relative to the Earth?

14. Explain why no work is done on a planet as it moves in a circular orbit around the Sun, even though a gravitational force is acting on the planet. What is the net work done on a planet during each revolution as it moves around the Sun in an elliptical orbit?

15. Explain why the force exerted on a particle by a uniform sphere must be directed toward the center of the sphere. Would this be the case if the mass distribution of the sphere were not spherically symmetric?

16. At what position in its elliptical orbit is the speed of a planet a maximum? At what position is the speed a minimum?

17. If you are given the mass and radius of planet X, how would you calculate the free-fall acceleration on the surface of this planet?

18. If a hole could be dug to the center of the Earth, would the force on an object of mass m still obey Equation 13.1 there? What do you think the force on m would be at the center of the Earth?

19. In his 1798 experiment, Cavendish was said to have "weighed the Earth." Explain this statement.

20. The *Voyager* spacecraft was accelerated toward escape speed from the Sun by Jupiter's gravitational force exerted on the spacecraft. How is this possible?

21. How would you find the mass of the Moon?

22. The *Apollo 13* spacecraft developed trouble in the oxygen system about halfway to the Moon. Why did the mission continue on around the Moon, and then return home, rather than immediately turn back to Earth?

PROBLEMS

1, 2, 3 = straightforward, intermediate, challenging □ = full solution available in the *Student Solutions Manual and Study Guide*

= coached solution with hints available at http://www.pse6.com   = computer useful in solving problem

= paired numerical and symbolic problems

Section 13.1 Newton's Law of Universal Gravitation

Problem 17 in Chapter 1 can also be assigned with this section.

1. Determine the order of magnitude of the gravitational force that you exert on another person 2 m away. In your solution state the quantities you measure or estimate and their values.

2. Two ocean liners, each with a mass of 40 000 metric tons, are moving on parallel courses, 100 m apart. What is the magnitude of the acceleration of one of the liners toward the other due to their mutual gravitational attraction? Treat the ships as particles.

3. A 200-kg object and a 500-kg object are separated by 0.400 m. (a) Find the net gravitational force exerted by these objects on a 50.0-kg object placed midway between them. (b) At what position (other than an infinitely remote one) can the 50.0-kg object be placed so as to experience a net force of zero?

4. Two objects attract each other with a gravitational force of magnitude 1.00×10^{-8} N when separated by 20.0 cm. If the total mass of the two objects is 5.00 kg, what is the mass of each?

5. Three uniform spheres of mass 2.00 kg, 4.00 kg, and 6.00 kg are placed at the corners of a right triangle as in Figure P13.5. Calculate the resultant gravitational force on the 4.00-kg object, assuming the spheres are isolated from the rest of the Universe.

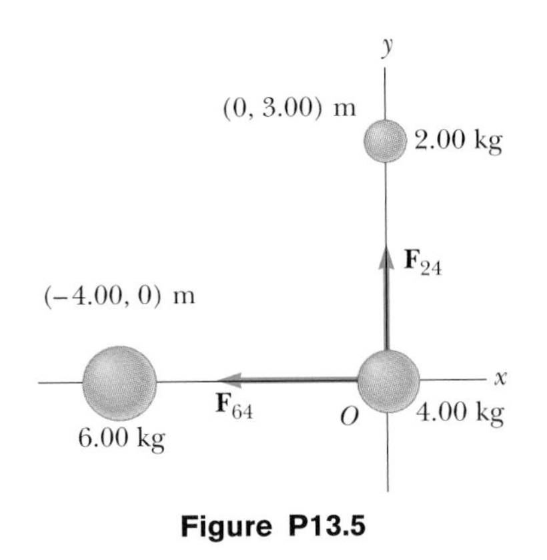

Figure P13.5

6. During a solar eclipse, the Moon, Earth, and Sun all lie on the same line, with the Moon between the Earth and the Sun. (a) What force is exerted by the Sun on the Moon? (b) What force is exerted by the Earth on the Moon? (c) What force is exerted by the Sun on the Earth?

Section 13.2 Measuring the Gravitational Constant

7. In introductory physics laboratories, a typical Cavendish balance for measuring the gravitational constant G uses lead spheres with masses of 1.50 kg and 15.0 g whose centers are separated by about 4.50 cm. Calculate the gravitational force between these spheres, treating each as a particle located at the center of the sphere.

8. A student proposes to measure the gravitational constant G by suspending two spherical objects from the ceiling of a tall cathedral and measuring the deflection of the cables from the vertical. Draw a free-body diagram of one of the objects. If two 100.0-kg objects are suspended at the lower ends of cables 45.00 m long and the cables are attached to the ceiling 1.000 m apart, what is the separation of the objects?

Section 13.3 Free-Fall Acceleration and the Gravitational Force

9. When a falling meteoroid is at a distance above the Earth's surface of 3.00 times the Earth's radius, what is its acceleration due to the Earth's gravitation?

10. The free-fall acceleration on the surface of the Moon is about one sixth of that on the surface of the Earth. If the radius of the Moon is about $0.250R_E$, find the ratio of their average densities, $\rho_{\text{Moon}}/\rho_{\text{Earth}}$.

11. On the way to the Moon the *Apollo* astronauts reached a point where the Moon's gravitational pull became stronger than the Earth's. (a) Determine the distance of this point from the center of the Earth. (b) What is the acceleration due to the Earth's gravitation at this point?

Section 13.4 Kepler's Laws and the Motion of Planets

12. The center-to-center distance between Earth and Moon is 384 400 km. The Moon completes an orbit in 27.3 days. (a) Determine the Moon's orbital speed. (b) If gravity were switched off, the Moon would move along a straight line tangent to its orbit, as described by Newton's first law. In its actual orbit in 1.00 s, how far does the Moon fall below the tangent line and toward the Earth?

13. Plaskett's binary system consists of two stars that revolve in a circular orbit about a center of mass midway between them. This means that the masses of the two stars are equal (Fig. P13.13). Assume the orbital speed of each star is 220 km/s and the orbital period of each is 14.4 days. Find the mass M of each star. (For comparison, the mass of our Sun is 1.99×10^{30} kg.)

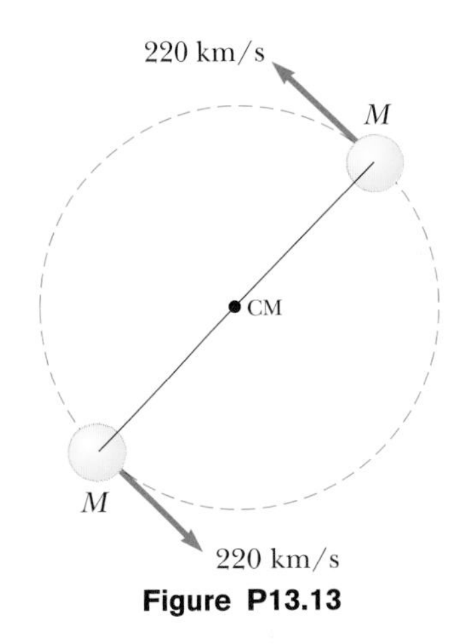

Figure P13.13

14. A particle of mass m moves along a straight line with constant speed in the x direction, a distance b from the x axis (Fig. P13.14). Show that Kepler's second law is satisfied by showing that the two shaded triangles in the figure have the same area when $t_4 - t_3 = t_2 - t_1$.

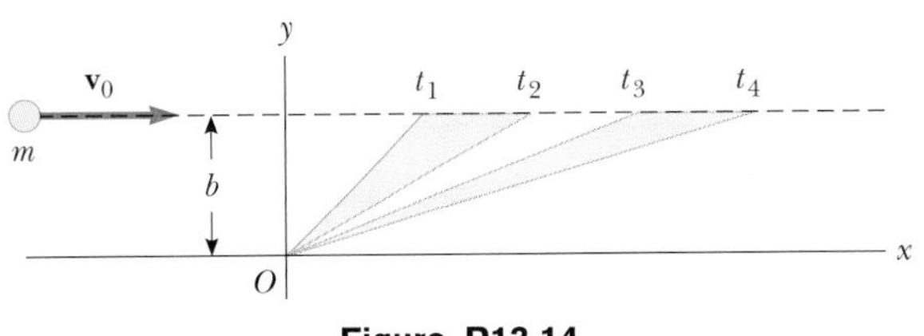

Figure P13.14

15. Io, a moon of Jupiter, has an orbital period of 1.77 days and an orbital radius of 4.22×10^5 km. From these data, determine the mass of Jupiter.

16. The *Explorer VIII* satellite, placed into orbit November 3, 1960, to investigate the ionosphere, had the following orbit parameters: perigee, 459 km; apogee, 2 289 km (both distances above the Earth's surface); period, 112.7 min. Find the ratio v_p/v_a of the speed at perigee to that at apogee.

17. Comet Halley (Figure P13.17) approaches the Sun to within 0.570 AU, and its orbital period is 75.6 years. (AU is the symbol for astronomical unit, where 1 AU = 1.50×10^{11} m is the mean Earth–Sun distance.) How far from the Sun will Halley's comet travel before it starts its return journey?

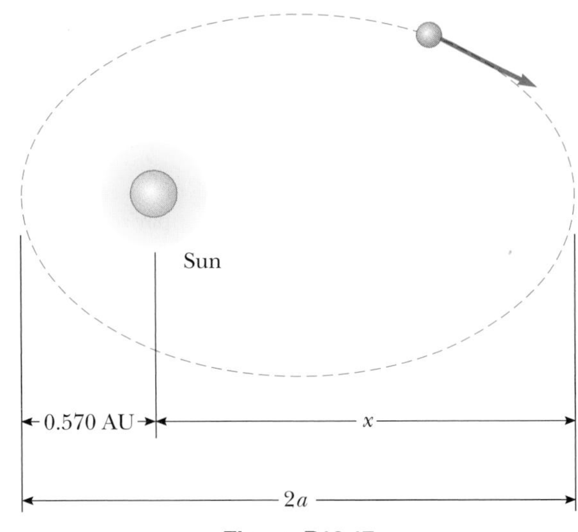

Figure P13.17

18. Two planets X and Y travel counterclockwise in circular orbits about a star as in Figure P13.18. The radii of their orbits are in the ratio 3:1. At some time, they are aligned as in Figure P13.18a, making a straight line with the star. During the next five years, the angular displacement of planet X is 90.0°, as in Figure P13.18b. Where is planet Y at this time?

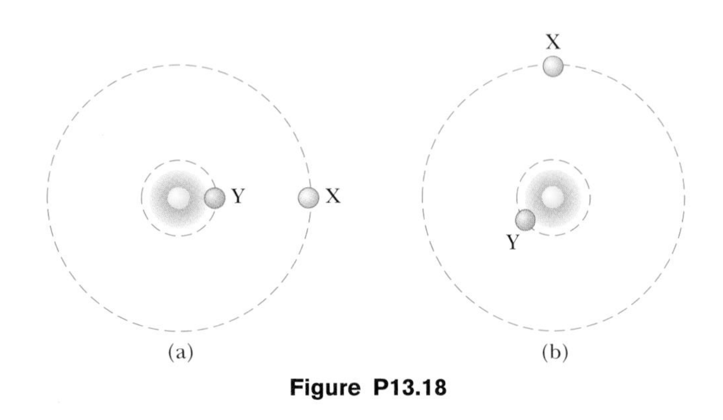

Figure P13.18

19. A synchronous satellite, which always remains above the same point on a planet's equator, is put in orbit around Jupiter to study the famous red spot. Jupiter rotates about its axis once every 9.84 h. Use the data of Table 13.2 to find the altitude of the satellite.

20. Neutron stars are extremely dense objects that are formed from the remnants of supernova explosions. Many rotate very rapidly. Suppose that the mass of a certain spherical neutron star is twice the mass of the Sun and its radius is 10.0 km. Determine the greatest possible angular speed it can have so that the matter at the surface of the star on its equator is just held in orbit by the gravitational force.

21. Suppose the Sun's gravity were switched off. The planets would leave their nearly circular orbits and fly away in straight lines, as described by Newton's first law. Would Mercury ever be farther from the Sun than Pluto? If so, find how long it would take for Mercury to achieve this passage. If not, give a convincing argument that Pluto is always farther from the Sun.

22. As thermonuclear fusion proceeds in its core, the Sun loses mass at a rate of 3.64×10^9 kg/s. During the 5 000-yr period of recorded history, by how much has the length of the year changed due to the loss of mass from the Sun? *Suggestions:* Assume the Earth's orbit is circular. No external torque acts on the Earth–Sun system, so its angular momentum is conserved. If x is small compared to 1, then $(1 + x)^n$ is nearly equal to $1 + nx$.

Section 13.5 The Gravitational Field

23. Three objects of equal mass are located at three corners of a square of edge length ℓ as in Figure P13.23. Find the gravitational field at the fourth corner due to these objects.

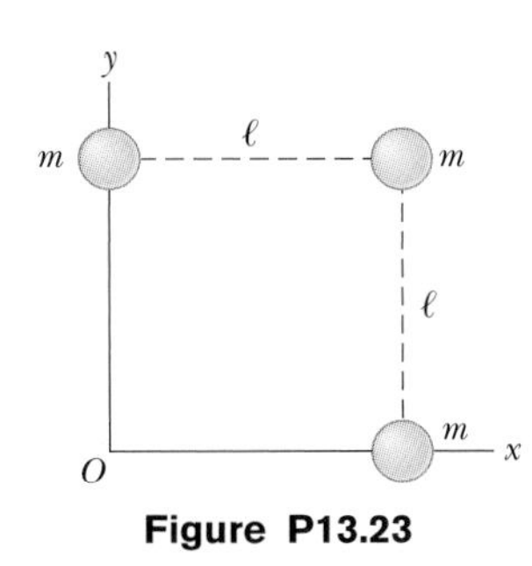

Figure P13.23

24. A spacecraft in the shape of a long cylinder has a length of 100 m, and its mass with occupants is 1 000 kg. It has strayed too close to a black hole having a mass 100 times that of the Sun (Fig. P13.24). The nose of the spacecraft points toward the black hole, and the distance between the nose and the center of the black hole is 10.0 km. (a) Determine the total force on the spacecraft. (b) What is the difference in the gravitational fields acting on the occupants in the nose of the ship and on those in the rear of the ship, farthest from the black hole? This difference in accelerations grows rapidly as the ship approaches the black hole. It puts the body of the ship under extreme tension and eventually tears it apart.

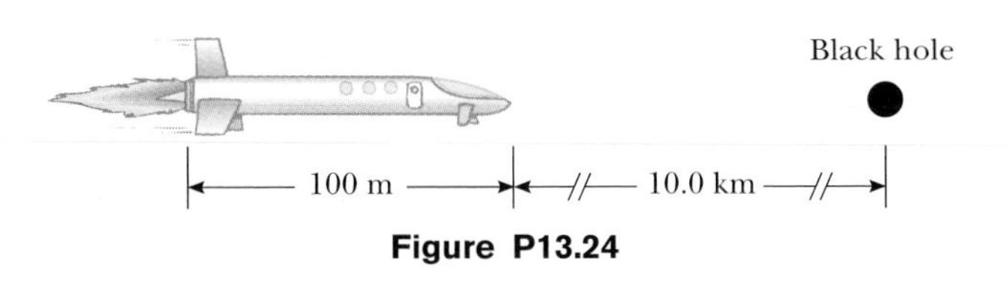

Figure P13.24

25. Compute the magnitude and direction of the gravitational field at a point P on the perpendicular bisector of the line joining two objects of equal mass separated by a distance $2a$ as shown in Figure P13.25.

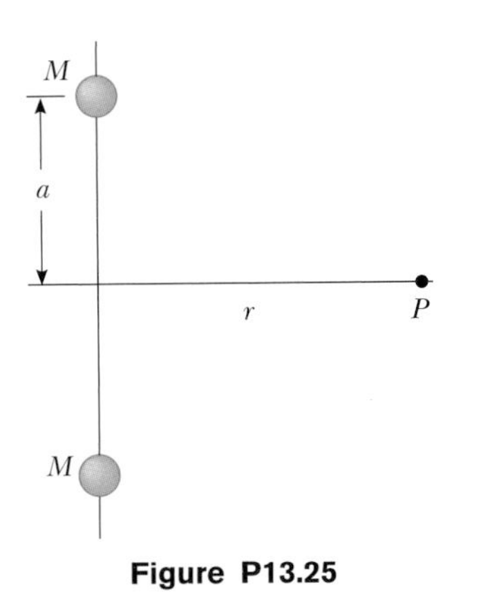

Figure P13.25

Section 13.6 Gravitational Potential Energy

Assume $U = 0$ at $r = \infty$.

26. A satellite of the Earth has a mass of 100 kg and is at an altitude of 2.00×10^6 m. (a) What is the potential energy of the satellite–Earth system? (b) What is the magnitude of the gravitational force exerted by the Earth on the satellite? (c) **What If?** What force does the satellite exert on the Earth?

27. How much energy is required to move a 1 000-kg object from the Earth's surface to an altitude twice the Earth's radius?

28. At the Earth's surface a projectile is launched straight up at a speed of 10.0 km/s. To what height will it rise? Ignore air resistance and the rotation of the Earth.

29. After our Sun exhausts its nuclear fuel, its ultimate fate may be to collapse to a *white dwarf* state, in which it has approximately the same mass as it has now, but a radius equal to the radius of the Earth. Calculate (a) the average density of the white dwarf, (b) the free-fall acceleration, and (c) the gravitational potential energy of a 1.00-kg object at its surface.

30. How much work is done by the Moon's gravitational field as a 1 000-kg meteor comes in from outer space and impacts on the Moon's surface?

31. A system consists of three particles, each of mass 5.00 g, located at the corners of an equilateral triangle with sides of 30.0 cm. (a) Calculate the potential energy of the system. (b) If the particles are released simultaneously, where will they collide?

32. An object is released from rest at an altitude h above the surface of the Earth. (a) Show that its speed at a distance r from the Earth's center, where $R_E \le r \le R_E + h$, is given by

$$v = \sqrt{2GM_E\left(\frac{1}{r} - \frac{1}{R_E + h}\right)}$$

(b) Assume the release altitude is 500 km. Perform the integral

$$\Delta t = \int_i^f dt = -\int_i^f \frac{dr}{v}$$

to find the time of fall as the object moves from the release point to the Earth's surface. The negative sign appears because the object is moving opposite to the radial direction, so its speed is $v = -dr/dt$. Perform the integral numerically.

Section 13.7 Energy Considerations in Planetary and Satellite Motion

33. A space probe is fired as a projectile from the Earth's surface with an initial speed of 2.00×10^4 m/s. What will its speed be when it is very far from the Earth? Ignore friction and the rotation of the Earth.

By permission of John Hart and Creators Syndicate, Inc.

Figure P13.35

34. (a) What is the minimum speed, relative to the Sun, necessary for a spacecraft to escape the solar system if it starts at the Earth's orbit? (b) *Voyager 1* achieved a maximum speed of 125 000 km/h on its way to photograph Jupiter. Beyond what distance from the Sun is this speed sufficient to escape the solar system?

35. A "treetop satellite" (Fig. P13.35) moves in a circular orbit just above the surface of a planet, assumed to offer no air resistance. Show that its orbital speed v and the escape speed from the planet are related by the expression

$$v_{\text{esc}} = \sqrt{2}\,v.$$

36. A 500-kg satellite is in a circular orbit at an altitude of 500 km above the Earth's surface. Because of air friction, the satellite eventually falls to the Earth's surface, where it hits the ground with a speed of 2.00 km/s. How much energy was transformed into internal energy by means of friction?

37. A satellite of mass 200 kg is placed in Earth orbit at a height of 200 km above the surface. (a) With a circular orbit, how long does the satellite take to complete one orbit? (b) What is the satellite's speed? (c) What is the minimum energy input necessary to place this satellite in orbit? Ignore air resistance but include the effect of the planet's daily rotation.

38. A satellite of mass m, originally on the surface of the Earth, is placed into Earth orbit at an altitude h. (a) With a circular orbit, how long does the satellite take to complete one orbit? (b) What is the satellite's speed? (c) What is the minimum energy input necessary to place this satellite in orbit? Ignore air resistance but include the effect of the planet's daily rotation. At what location on the Earth's surface and in what direction should the satellite be launched to minimize the required energy investment? Represent the mass and radius of the Earth as M_E and R_E.

39. A 1 000-kg satellite orbits the Earth at a constant altitude of 100 km. How much energy must be added to the system to move the satellite into a circular orbit with altitude 200 km?

40. The planet Uranus has a mass about 14 times the Earth's mass, and its radius is equal to about 3.7 Earth radii. (a) By setting up ratios with the corresponding Earth values, find the free-fall acceleration at the cloud tops of Uranus. (b) Ignoring the rotation of the planet, find the minimum escape speed from Uranus.

41. Determine the escape speed for a rocket on the far side of Ganymede, the largest of Jupiter's moons (Figure P13.41). The radius of Ganymede is 2.64×10^6 m, and its mass is 1.495×10^{23} kg. The mass of Jupiter is 1.90×10^{27} kg, and the distance between Jupiter and Ganymede is 1.071×10^9 m. Be sure to include the gravitational effect due to Jupiter, but you may ignore the motion of Jupiter and Ganymede as they revolve about their center of mass.

Figure P13.41

42. In Robert Heinlein's "The Moon is a Harsh Mistress," the colonial inhabitants of the Moon threaten to launch rocks down onto the Earth if they are not given independence (or at least representation). Assuming that a rail gun could launch a rock of mass m at twice the lunar escape speed, calculate the speed of the rock as it enters the Earth's atmosphere. (By *lunar escape speed* we mean the speed required to move infinitely far away from a stationary Moon alone in the Universe. Problem 61 in Chapter 30 describes a rail gun.)

43. An object is fired vertically upward from the surface of the Earth (of radius R_E) with an initial speed v_i that is comparable to but less than the escape speed v_{esc}. (a) Show that the object attains a maximum height h given by

$$h = \frac{R_E v_i^2}{v_{\text{esc}}^2 - v_i^2}$$

(b) A space vehicle is launched vertically upward from the Earth's surface with an initial speed of 8.76 km/s, which is less than the escape speed of 11.2 km/s. What maximum height does it attain? (c) A meteorite falls toward the Earth. It is essentially at rest with respect to the Earth when it is at a height of 2.51×10^7 m. With what speed does the meteorite strike the Earth? (d) **What If?** Assume that a baseball is tossed up with an initial speed that is very small compared to the escape speed. Show that the equation from part (a) is consistent with Equation 4.13.

44. Derive an expression for the work required to move an Earth satellite of mass m from a circular orbit of radius $2R_E$ to one of radius $3R_E$.

45. A comet of mass 1.20×10^{10} kg moves in an elliptical orbit around the Sun. Its distance from the Sun ranges between 0.500 AU and 50.0 AU. (a) What is the eccentricity of its orbit? (b) What is its period? (c) At aphelion what is the potential energy of the comet–Sun system? *Note:* 1 AU = one astronomical unit = the average distance from Sun to Earth = 1.496×10^{11} m.

46. A satellite moves around the Earth in a circular orbit of radius r. (a) What is the speed v_0 of the satellite? Suddenly, an explosion breaks the satellite into two pieces, with masses m and $4m$. Immediately after the explosion the smaller piece of mass m is stationary with respect to the Earth and falls directly toward the Earth. (b) What is the speed v_i of the larger piece immediately after the explosion? (c) Because of the increase in its speed, this larger piece now moves in a new elliptical orbit. Find its distance away from the center of the Earth when it reaches the other end of the ellipse.

Additional Problems

47. The Solar and Heliospheric Observatory (SOHO) spacecraft has a special orbit, chosen so that its view of the Sun is never eclipsed and it is always close enough to the Earth to

transmit data easily. It moves in a near-circle around the Sun that is smaller than the Earth's circular orbit. Its period, however, is just equal to 1 yr. It is always located between the Earth and the Sun along the line joining them. Both objects exert gravitational forces on the observatory. Show that its distance from the Earth must be between 1.47×10^9 m and 1.48×10^9 m. In 1772 Joseph Louis Lagrange determined theoretically the special location allowing this orbit. The SOHO spacecraft took this position on February 14, 1996. *Suggestion:* Use data that are precise to four digits. The mass of the Earth is 5.983×10^{24} kg.

48. Let Δg_M represent the difference in the gravitational fields produced by the Moon at the points on the Earth's surface nearest to and farthest from the Moon. Find the fraction $\Delta g_M/g$, where g is the Earth's gravitational field. (This difference is responsible for the occurrence of the *lunar tides* on the Earth.)

49. **Review problem.** Two identical hard spheres, each of mass m and radius r, are released from rest in otherwise empty space with their centers separated by the distance R. They are allowed to collide under the influence of their gravitational attraction. (a) Show that the magnitude of the impulse received by each sphere before they make contact is given by $[Gm^3(1/2r - 1/R)]^{1/2}$. (b) **What If?** Find the magnitude of the impulse each receives if they collide elastically.

50. Two spheres having masses M and $2M$ and radii R and $3R$, respectively, are released from rest when the distance between their centers is $12R$. How fast will each sphere be moving when they collide? Assume that the two spheres interact only with each other.

51. In Larry Niven's science-fiction novel *Ringworld,* a rigid ring of material rotates about a star (Fig. P13.51). The tangential speed of the ring is 1.25×10^6 m/s, and its radius is 1.53×10^{11} m. (a) Show that the centripetal acceleration of the inhabitants is 10.2 m/s^2. (b) The inhabitants of this ring world live on the starlit inner surface of the ring. Each person experiences a normal contact force **n**. Acting alone, this normal force would produce an inward acceleration of 9.90 m/s^2. Additionally, the star at the center of the ring exerts a gravitational force on the ring and its inhabitants. The difference between the total acceleration and the acceleration provided by the normal force is due to the gravitational attraction of the central star. Show that the mass of the star is approximately 10^{32} kg.

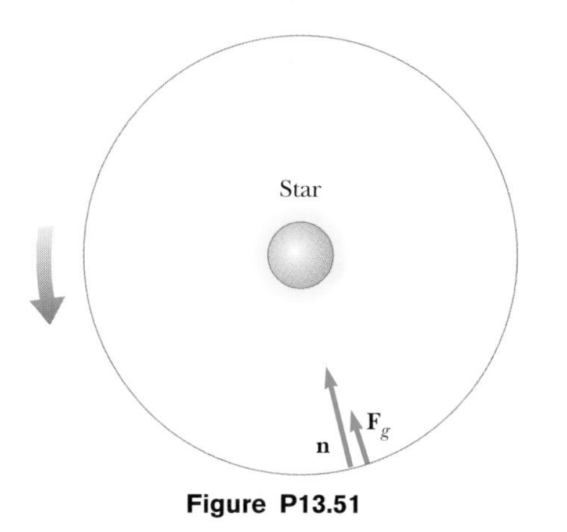

Figure P13.51

52. (a) Show that the rate of change of the free-fall acceleration with distance above the Earth's surface is

$$\frac{dg}{dr} = -\frac{2GM_E}{R_E^{\ 3}}$$

This rate of change over distance is called a *gradient.* (b) If h is small in comparison to the radius of the Earth, show that the difference in free-fall acceleration between two points separated by vertical distance h is

$$|\Delta g| = \frac{2GM_E h}{R_E^{\ 3}}$$

(c) Evaluate this difference for $h = 6.00$ m, a typical height for a two-story building.

53. A ring of matter is a familiar structure in planetary and stellar astronomy. Examples include Saturn's rings and a ring nebula. Consider a uniform ring of mass 2.36×10^{20} kg and radius 1.00×10^8 m. An object of mass 1 000 kg is placed at a point A on the axis of the ring, 2.00×10^8 m from the center of the ring (Figure P13.53). When the object is released, the attraction of the ring makes the object move along the axis toward the center of the ring (point B). (a) Calculate the gravitational potential energy of the object–ring system when the object is at A. (b) Calculate the gravitational potential energy of the system when the object is at B. (c) Calculate the speed of the object as it passes through B.

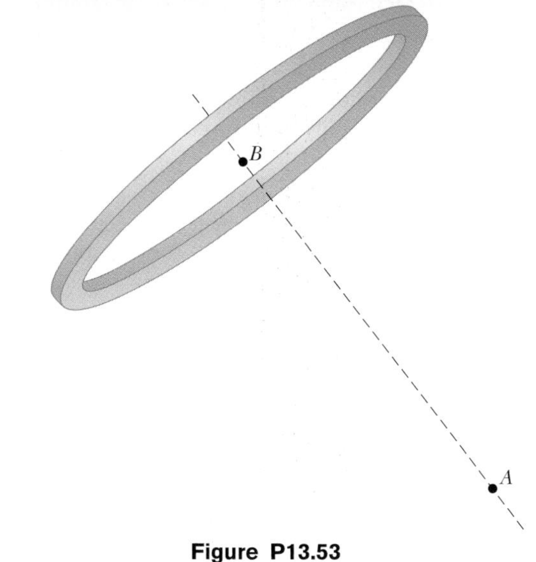

Figure P13.53

54. *Voyagers 1* and *2* surveyed the surface of Jupiter's moon Io and photographed active volcanoes spewing liquid sulfur to heights of 70 km above the surface of this moon. Find

the speed with which the liquid sulfur left the volcano. Io's mass is 8.9×10^{22} kg, and its radius is 1 820 km.

55. As an astronaut, you observe a small planet to be spherical. After landing on the planet, you set off, walking always straight ahead, and find yourself returning to your spacecraft from the opposite side after completing a lap of 25.0 km. You hold a hammer and a falcon feather at a height of 1.40 m, release them, and observe that they fall together to the surface in 29.2 s. Determine the mass of the planet.

56. A certain quaternary star system consists of three stars, each of mass m, moving in the same circular orbit of radius r about a central star of mass M. The stars orbit in the same sense, and are positioned one third of a revolution apart from each other. Show that the period of each of the three stars is given by

$$T = 2\pi\sqrt{\frac{r^3}{g\left(M + m/\sqrt{3}\right)}}$$

57. **Review problem.** A cylindrical habitat in space 6.00 km in diameter and 30 km long has been proposed (by G. K. O'Neill, 1974). Such a habitat would have cities, land, and lakes on the inside surface and air and clouds in the center. This would all be held in place by rotation of the cylinder about its long axis. How fast would the cylinder have to rotate to imitate the Earth's gravitational field at the walls of the cylinder?

58. Newton's law of universal gravitation is valid for distances covering an enormous range, but it is thought to fail for very small distances, where the structure of space itself is uncertain. Far smaller than an atomic nucleus, this crossover distance is called the Planck length. It is determined by a combination of the constants G, c, and h, where c is the speed of light in vacuum and h is Planck's constant (introduced in Chapter 11) with units of angular momentum. (a) Use dimensional analysis to find a combination of these three universal constants that has units of length. (b) Determine the order of magnitude of the Planck length. You will need to consider noninteger powers of the constants.

59. Show that the escape speed from the surface of a planet of uniform density is directly proportional to the radius of the planet.

60. Many people assume that air resistance acting on a moving object will always make the object slow down. It can actually be responsible for making the object speed up. Consider a 100-kg Earth satellite in a circular orbit at an altitude of 200 km. A small force of air resistance makes the satellite drop into a circular orbit with an altitude of 100 km. (a) Calculate its initial speed. (b) Calculate its final speed in this process. (c) Calculate the initial energy of the satellite–Earth system. (d) Calculate the final energy of the system. (e) Show that the system has lost mechanical energy and find the amount of the loss due to friction. (f) What force makes the satellite's speed increase? You will find a free-body diagram useful in explaining your answer.

61. Two hypothetical planets of masses m_1 and m_2 and radii r_1 and r_2, respectively, are nearly at rest when they are an infinite distance apart. Because of their gravitational attraction, they head toward each other on a collision course. (a) When their center-to-center separation is d, find expressions for the speed of each planet and for their relative speed. (b) Find the kinetic energy of each planet just before they collide, if $m_1 = 2.00 \times 10^{24}$ kg, $m_2 = 8.00 \times 10^{24}$ kg, $r_1 = 3.00 \times 10^6$ m, and $r_2 = 5.00 \times 10^6$ m. (*Note:* Both energy and momentum of the system are conserved.)

62. The maximum distance from the Earth to the Sun (at our aphelion) is 1.521×10^{11} m, and the distance of closest approach (at perihelion) is 1.471×10^{11} m. If the Earth's orbital speed at perihelion is 3.027×10^4 m/s, determine (a) the Earth's orbital speed at aphelion, (b) the kinetic and potential energies of the Earth–Sun system at perihelion, and (c) the kinetic and potential energies at aphelion. Is the total energy constant? (Ignore the effect of the Moon and other planets.)

63. (a) Determine the amount of work (in joules) that must be done on a 100-kg payload to elevate it to a height of 1 000 km above the Earth's surface. (b) Determine the amount of additional work that is required to put the payload into circular orbit at this elevation.

64. X-ray pulses from Cygnus X-1, a celestial x-ray source, have been recorded during high-altitude rocket flights. The signals can be interpreted as originating when a blob of ionized matter orbits a black hole with a period of 5.0 ms. If the blob were in a circular orbit about a black hole whose mass is $20M_{\text{Sun}}$, what is the orbit radius?

65. Studies of the relationship of the Sun to its galaxy—the Milky Way—have revealed that the Sun is located near the outer edge of the galactic disk, about 30 000 lightyears from the center. The Sun has an orbital speed of approximately 250 km/s around the galactic center. (a) What is the period of the Sun's galactic motion? (b) What is the order of magnitude of the mass of the Milky Way galaxy? Suppose that the galaxy is made mostly of stars of which the Sun is typical. What is the order of magnitude of the number of stars in the Milky Way?

66. The oldest artificial satellite in orbit is *Vanguard I*, launched March 3, 1958. Its mass is 1.60 kg. In its initial orbit, its minimum distance from the center of the Earth was 7.02 Mm, and its speed at this perigee point was 8.23 km/s. (a) Find the total energy of the satellite–Earth system. (b) Find the magnitude of the angular momentum of the satellite. (c) Find its speed at apogee and its maximum (apogee) distance from the center of the Earth. (d) Find the semimajor axis of its orbit. (e) Determine its period.

67. Astronomers detect a distant meteoroid moving along a straight line that, if extended, would pass at a distance $3R_E$ from the center of the Earth, where R_E is the radius of the Earth. What minimum speed must the meteoroid have if the Earth's gravitation is not to deflect the meteoroid to make it strike the Earth?

68. A spherical planet has uniform density ρ. Show that the minimum period for a satellite in orbit around it is

$$T_{\text{min}} = \sqrt{\frac{3\pi}{G\rho}}$$

independent of the radius of the planet.

69. Two stars of masses M and m, separated by a distance d, revolve in circular orbits about their center of mass (Fig. P13.69). Show that each star has a period given by

$$T^2 = \frac{4\pi^2 d^3}{G(M+m)}$$

Proceed as follows: Apply Newton's second law to each star. Note that the center-of-mass condition requires that $Mr_2 = mr_1$, where $r_1 + r_2 = d$.

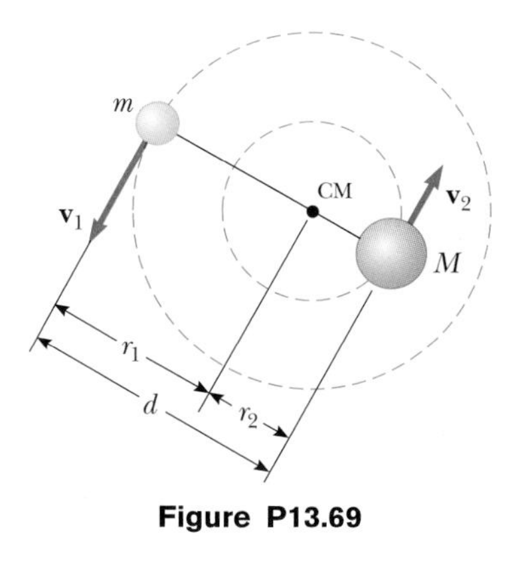

Figure P13.69

70. (a) A 5.00-kg object is released 1.20×10^7 m from the center of the Earth. It moves with what acceleration relative to the Earth? (b) **What If?** A 2.00×10^{24} kg object is released 1.20×10^7 m from the center of the Earth. It moves with what acceleration relative to the Earth? Assume that the objects behave as pairs of particles, isolated from the rest of the Universe.

71. The acceleration of an object moving in the gravitational field of the Earth is

$$\mathbf{a} = -\frac{GM_E\mathbf{r}}{r^3}$$

where $\mathbf{r}$ is the position vector directed from the center of the Earth toward the object. Choosing the origin at the center of the Earth and assuming that the small object is moving in the xy plane, we find that the rectangular (Cartesian) components of its acceleration are

$$a_x = -\frac{GM_E x}{(x^2 + y^2)^{3/2}} \qquad a_y = -\frac{GM_E y}{(x^2 + y^2)^{3/2}}$$

Use a computer to set up and carry out a numerical prediction of the motion of the object, according to Euler's method. Assume the initial position of the object is $x = 0$ and $y = 2R_E$, where R_E is the radius of the Earth. Give the object an initial velocity of 5 000 m/s in the x direction. The time increment should be made as small as practical. Try 5 s. Plot the x and y coordinates of the object as time goes on. Does the object hit the Earth? Vary the initial velocity until you find a circular orbit.

Answers to Quick Quizzes

13.1 (d). The gravitational force exerted by the Earth on the Moon provides a net force that causes the Moon's centripetal acceleration.

13.2 (e). The gravitational force follows an inverse-square behavior, so doubling the distance causes the force to be one fourth as large.

13.3 (c). An object in orbit is simply falling while it moves around the Earth. The acceleration of the object is that due to gravity. Because the object was launched from a very tall mountain, the value for g is slightly less than that at the surface.

13.4 (a). Kepler's third law (Eq. 13.8), which applies to all the planets, tells us that the period of a planet is proportional to $a^{3/2}$. Because Pluto is farther from the Sun than the Earth, it has a longer period. The Sun's gravitational field is much weaker at Pluto than it is at the Earth. Thus, this planet experiences much less centripetal acceleration than the Earth does, and it has a correspondingly longer period.

13.5 (a). From Kepler's third law and the given period, the major axis of the asteroid can be calculated. It is found to be 1.2×10^{11} m. Because this is smaller than the Earth–Sun distance, the asteroid cannot possibly collide with the Earth.

13.6 (b). From conservation of angular momentum, $mv_p r_p = mv_a r_a$, so that $v_p = (r_a/r_p)v_a = (4D/D)v_a = 4v_a$.

13.7 (a) Perihelion. Because of conservation of angular momentum, the speed of the comet is highest at its closest position to the Sun. (b) Aphelion. The potential energy of the comet–Sun system is highest when the comet is at its farthest distance from the Sun. (c) Perihelion. The kinetic energy is highest at the point at which the speed of the comet is highest. (d) All points. The total energy of the system is the same regardless of where the comet is in its orbit.

Chapter 14

Fluid Mechanics

CHAPTER OUTLINE

▲ *These hot-air balloons float because they are filled with air at high temperature and are surrounded by denser air at a lower temperature. In this chapter, we will explore the buoyant force that supports these balloons and other floating objects. (Richard Megna/Fundamental Photographs)*

Matter is normally classified as being in one of three states: solid, liquid, or gas. From everyday experience, we know that a solid has a definite volume and shape. A brick maintains its familiar shape and size day in and day out. We also know that a liquid has a definite volume but no definite shape. Finally, we know that an unconfined gas has neither a definite volume nor a definite shape. These descriptions help us picture the states of matter, but they are somewhat artificial. For example, asphalt and plastics are normally considered solids, but over long periods of time they tend to flow like liquids. Likewise, most substances can be a solid, a liquid, or a gas (or a combination of any of these), depending on the temperature and pressure. In general, the time it takes a particular substance to change its shape in response to an external force determines whether we treat the substance as a solid, a liquid, or a gas.

A **fluid** is a collection of molecules that are randomly arranged and held together by weak cohesive forces and by forces exerted by the walls of a container. Both liquids and gases are fluids.

In our treatment of the mechanics of fluids, we do not need to learn any new physical principles to explain such effects as the buoyant force acting on a submerged object and the dynamic lift acting on an airplane wing. First, we consider the mechanics of a fluid at rest—that is, *fluid statics.* We then treat the mechanics of fluids in motion—that is, *fluid dynamics.* We can describe a fluid in motion by using a model that is based upon certain simplifying assumptions.

Figure 14.1 At any point on the surface of a submerged object, the force exerted by the fluid is perpendicular to the surface of the object. The force exerted by the fluid on the walls of the container is perpendicular to the walls at all points.

14.1 Pressure

Fluids do not sustain shearing stresses or tensile stresses; thus, the only stress that can be exerted on an object submerged in a static fluid is one that tends to compress the object from all sides. In other words, the force exerted by a static fluid on an object is always perpendicular to the surfaces of the object, as shown in Figure 14.1.

The pressure in a fluid can be measured with the device pictured in Figure 14.2. The device consists of an evacuated cylinder that encloses a light piston connected to a spring. As the device is submerged in a fluid, the fluid presses on the top of the piston and compresses the spring until the inward force exerted by the fluid is balanced by the outward force exerted by the spring. The fluid pressure can be measured directly if the spring is calibrated in advance. If F is the magnitude of the force exerted on the piston and A is the surface area of the piston, then the **pressure** P of the fluid at the level to which the device has been submerged is defined as the ratio F/A:

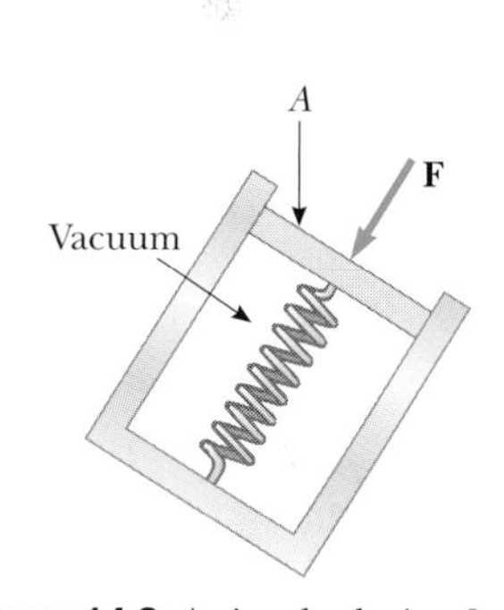

Figure 14.2 A simple device for measuring the pressure exerted by a fluid.

Definition of pressure

$$P \equiv \frac{F}{A} \tag{14.1}$$

Note that pressure is a scalar quantity because it is proportional to the magnitude of the force on the piston.

If the pressure varies over an area, we can evaluate the infinitesimal force dF on an infinitesimal surface element of area dA as

Earl Young/Getty Images

Snowshoes keep you from sinking into soft snow because they spread the downward force you exert on the snow over a large area, reducing the pressure on the snow surface.

PITFALL PREVENTION

14.1 Force and Pressure

Equations 14.1 and 14.2 make a clear distinction between force and pressure. Another important distinction is that *force is a vector* and *pressure is a scalar*. There is no direction associated with pressure, but the direction of the force associated with the pressure is perpendicular to the surface of interest.

$$dF = P\,dA \tag{14.2}$$

where P is the pressure at the location of the area dA. The pressure exerted by a fluid varies with depth. Therefore, to calculate the total force exerted on a flat vertical wall of a container, we must integrate Equation 14.2 over the surface area of the wall.

Because pressure is force per unit area, it has units of newtons per square meter (N/m^2) in the SI system. Another name for the SI unit of pressure is **pascal** (Pa):

$$1 \text{ Pa} \equiv 1 \text{ N/m}^2 \tag{14.3}$$

Quick Quiz 14.1 Suppose you are standing directly behind someone who steps back and accidentally stomps on your foot with the heel of one shoe. Would you be better off if that person were (a) a large professional basketball player wearing sneakers (b) a petite woman wearing spike-heeled shoes?

Example 14.1 The Water Bed

The mattress of a water bed is 2.00 m long by 2.00 m wide and 30.0 cm deep.

(A) Find the weight of the water in the mattress.

Solution The density of fresh water is 1 000 kg/m³ (see Table 14.1 on page 423), and the volume of the water filling the mattress is $V = (2.00 \text{ m})(2.00 \text{ m})(0.300 \text{ m}) = 1.20 \text{ m}^3$. Hence, using Equation 1.1, the mass of the water in the bed is

$$M = \rho V = (1\,000 \text{ kg/m}^3)(1.20 \text{ m}^3) = 1.20 \times 10^3 \text{ kg}$$

and its weight is

$$Mg = (1.20 \times 10^3 \text{ kg})(9.80 \text{ m/s}^2) = 1.18 \times 10^4 \text{ N}$$

This is approximately 2 650 lb. (A regular bed weighs approximately 300 lb.) Because this load is so great, such a water bed is best placed in the basement or on a sturdy, well-supported floor.

(B) Find the pressure exerted by the water on the floor when the bed rests in its normal position. Assume that the entire lower surface of the bed makes contact with the floor.

Solution When the bed is in its normal position, the area in contact with the floor is 4.00 m²; thus, from Equation 14.1, we find that

$$P = \frac{1.18 \times 10^4 \text{ N}}{4.00 \text{ m}^2} = 2.95 \times 10^3 \text{ Pa}$$

What If? What if the water bed is replaced by a 300-lb ordinary bed that is supported by four legs? Each leg has a circular cross section of radius 2.00 cm. What pressure does this bed exert on the floor?

Answer The weight of the bed is distributed over four circular cross sections at the bottom of the legs. Thus, the pressure is

$$P = \frac{F}{A} = \frac{mg}{4(\pi r^2)} = \frac{300 \text{ lb}}{4\pi(0.0200 \text{ m})^2}\left(\frac{1 \text{ N}}{0.225 \text{ lb}}\right)$$
$$= 2.65 \times 10^5 \text{ Pa}$$

Note that this is almost 100 times larger than the pressure due to the water bed! This is because the weight of the ordinary bed, even though it is much less than the weight of the water bed, is applied over the very small area of the four legs. The high pressure on the floor at the feet of an ordinary bed could cause denting of wood floors or permanently crush carpet pile. In contrast, a water bed requires a sturdy floor to support the very large weight.

Table 14.1

Densities of Some Common Substances at Standard Temperature (0°C) and Pressure (Atmospheric)

Substance	$\rho(\text{kg/m}^3)$	Substance	$\rho(\text{kg/m}^3)$
Air	1.29	Ice	0.917×10^3
Aluminum	2.70×10^3	Iron	7.86×10^3
Benzene	0.879×10^3	Lead	11.3×10^3
Copper	8.92×10^3	Mercury	13.6×10^3
Ethyl alcohol	0.806×10^3	Oak	0.710×10^3
Fresh water	1.00×10^3	Oxygen gas	1.43
Glycerin	1.26×10^3	Pine	0.373×10^3
Gold	19.3×10^3	Platinum	21.4×10^3
Helium gas	1.79×10^{-1}	Seawater	1.03×10^3
Hydrogen gas	8.99×10^{-2}	Silver	10.5×10^3

14.2 Variation of Pressure with Depth

As divers well know, water pressure increases with depth. Likewise, atmospheric pressure decreases with increasing altitude; for this reason, aircraft flying at high altitudes must have pressurized cabins.

We now show how the pressure in a liquid increases with depth. As Equation 1.1 describes, the *density* of a substance is defined as its mass per unit volume; Table 14.1 lists the densities of various substances. These values vary slightly with temperature because the volume of a substance is temperature-dependent (as shown in Chapter 19). Under standard conditions (at 0°C and at atmospheric pressure) the densities of gases are about 1/1 000 the densities of solids and liquids. This difference in densities implies that the average molecular spacing in a gas under these conditions is about ten times greater than that in a solid or liquid.

Now consider a liquid of density ρ at rest as shown in Figure 14.3. We assume that ρ is uniform throughout the liquid; this means that the liquid is incompressible. Let us select a sample of the liquid contained within an imaginary cylinder of cross-sectional area A extending from depth d to depth $d + h$. The liquid external to our sample exerts forces at all points on the surface of the sample, perpendicular to the surface. The pressure exerted by the liquid on the bottom face of the sample is P, and the pressure on the top face is P_0. Therefore, the upward force exerted by the outside fluid on the bottom of the cylinder has a magnitude PA, and the downward force exerted on the top has a magnitude P_0A. The mass of liquid in the cylinder is $M = \rho V = \rho Ah$; therefore, the weight of the liquid in the cylinder is $Mg = \rho Ahg$. Because the cylinder is in equilibrium, the net force acting on it must be zero. Choosing upward to be the positive y direction, we see that

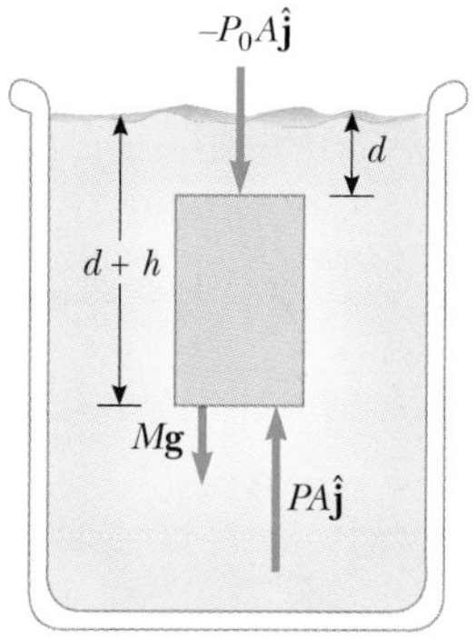

Figure 14.3 A parcel of fluid (darker region) in a larger volume of fluid is singled out. The net force exerted on the parcel of fluid must be zero because it is in equilibrium.

$$\sum \mathbf{F} = PA\hat{\mathbf{j}} - P_0A\hat{\mathbf{j}} - M\mathbf{g}\hat{\mathbf{j}} = 0$$

or

$$PA - P_0A - \rho Ahg = 0$$

$$PA - P_0A = \rho Ahg$$

Variation of pressure with depth

$$P = P_0 + \rho gh \qquad (14.4)$$

That is, **the pressure P at a depth h below a point in the liquid at which the pressure is P_0 is greater by an amount ρgh.** If the liquid is open to the atmosphere and P_0

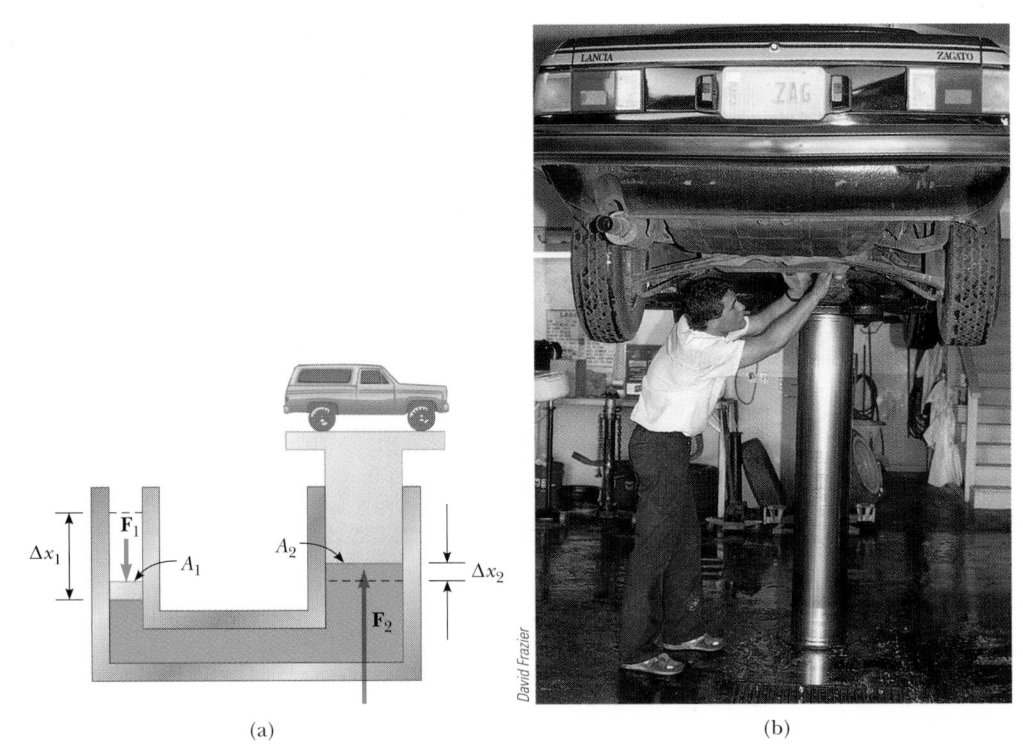

(a) (b)

Figure 14.4 (a) Diagram of a hydraulic press. Because the increase in pressure is the same on the two sides, a small force $\mathbf{F}_1$ at the left produces a much greater force $\mathbf{F}_2$ at the right. (b) A vehicle undergoing repair is supported by a hydraulic lift in a garage.

is the pressure at the surface of the liquid, then P_0 is atmospheric pressure. In our calculations and working of end-of-chapter problems, we usually take atmospheric pressure to be

$$P_0 = 1.00 \text{ atm} = 1.013 \times 10^5 \text{ Pa}$$

Equation 14.4 implies that the pressure is the same at all points having the same depth, independent of the shape of the container.

In view of the fact that the pressure in a fluid depends on depth and on the value of P_0, any increase in pressure at the surface must be transmitted to every other point in the fluid. This concept was first recognized by the French scientist Blaise Pascal (1623–1662) and is called **Pascal's law: a change in the pressure applied to a fluid is transmitted undiminished to every point of the fluid and to the walls of the container.**

Pascal's law

An important application of Pascal's law is the hydraulic press illustrated in Figure 14.4a. A force of magnitude F_1 is applied to a small piston of surface area A_1. The pressure is transmitted through an incompressible liquid to a larger piston of surface area A_2. Because the pressure must be the same on both sides, $P = F_1/A_1 = F_2/A_2$. Therefore, the force F_2 is greater than the force F_1 by a factor A_2/A_1. By designing a hydraulic press with appropriate areas A_1 and A_2, a large output force can be applied by means of a small input force. Hydraulic brakes, car lifts, hydraulic jacks, and forklifts all make use of this principle (Fig. 14.4b).

Because liquid is neither added nor removed from the system, the volume of liquid pushed down on the left in Figure 14.4a as the piston moves downward through a displacement Δx_1 equals the volume of liquid pushed up on the right as the right piston moves upward through a displacement Δx_2. That is, $A_1 \Delta x_1 = A_2 \Delta x_2$; thus, $A_2/A_1 = \Delta x_1/\Delta x_2$. We have already shown that $A_2/A_1 = F_2/F_1$. Thus, $F_2/F_1 = \Delta x_1/\Delta x_2$, so $F_1 \Delta x_1 = F_2 \Delta x_2$. Each side of this equation is the work done by the force. Thus, the work done by $\mathbf{F}_1$ on the input piston equals the work done by $\mathbf{F}_2$ on the output piston, as it must in order to conserve energy.

Quick Quiz 14.2 The pressure at the bottom of a filled glass of water (ρ = 1 000 kg/m^3) is P. The water is poured out and the glass is filled with ethyl alcohol (ρ = 806 kg/m^3). The pressure at the bottom of the glass is (a) smaller than P (b) equal to P (c) larger than P (d) indeterminate.

Example 14.2 The Car Lift

Interactive

In a car lift used in a service station, compressed air exerts a force on a small piston that has a circular cross section and a radius of 5.00 cm. This pressure is transmitted by a liquid to a piston that has a radius of 15.0 cm. What force must the compressed air exert to lift a car weighing 13 300 N? What air pressure produces this force?

Solution Because the pressure exerted by the compressed air is transmitted undiminished throughout the liquid, we have

$$F_1 = \left(\frac{A_1}{A_2}\right) F_2 = \frac{\pi(5.00 \times 10^{-2}\text{ m})^2}{\pi(15.0 \times 10^{-2}\text{ m})^2}(1.33 \times 10^4\text{ N})$$

$$= 1.48 \times 10^3\text{ N}$$

The air pressure that produces this force is

$$P = \frac{F_1}{A_1} = \frac{1.48 \times 10^3\text{ N}}{\pi(5.00 \times 10^{-2}\text{ m})^2}$$

$$= 1.88 \times 10^5\text{ Pa}$$

This pressure is approximately twice atmospheric pressure.

You can adjust the weight of the truck in Figure 14.4a at the Interactive Worked Example link at **http://www.pse6.com.**

Example 14.3 A Pain in Your Ear

Estimate the force exerted on your eardrum due to the water above when you are swimming at the bottom of a pool that is 5.0 m deep.

Solution First, we must find the unbalanced pressure on the eardrum; then, after estimating the eardrum's surface area, we can determine the force that the water exerts on it.

The air inside the middle ear is normally at atmospheric pressure P_0. Therefore, to find the net force on the eardrum, we must consider the difference between the total pressure at the bottom of the pool and atmospheric pressure:

$$\begin{aligned} P_{\text{bot}} - P_0 &= \rho g h \\ &= (1.00 \times 10^3\text{ kg/m}^3)(9.80\text{ m/s}^2)(5.0\text{ m}) \\ &= 4.9 \times 10^4\text{ Pa} \end{aligned}$$

We estimate the surface area of the eardrum to be approximately 1 cm^2 = 1×10^{-4} m^2. This means that the force on it is $F = (P_{\text{bot}} - P_0)A \approx 5$ N. Because a force on the eardrum of this magnitude is extremely uncomfortable, swimmers often "pop their ears" while under water, an action that pushes air from the lungs into the middle ear. Using this technique equalizes the pressure on the two sides of the eardrum and relieves the discomfort.

Example 14.4 The Force on a Dam

Water is filled to a height H behind a dam of width w (Fig. 14.5). Determine the resultant force exerted by the water on the dam.

Solution Because pressure varies with depth, we cannot calculate the force simply by multiplying the area by the pressure. We can solve the problem by using Equation 14.2 to find the force dF exerted on a narrow horizontal strip at depth h and then integrating the expression to find the total force. Let us imagine a vertical y axis, with $y = 0$ at the bottom of the dam and our strip a distance y above the bottom.

We can use Equation 14.4 to calculate the pressure at the depth h; we omit atmospheric pressure because it acts on both sides of the dam:

$$P = \rho g h = \rho g(H - y)$$

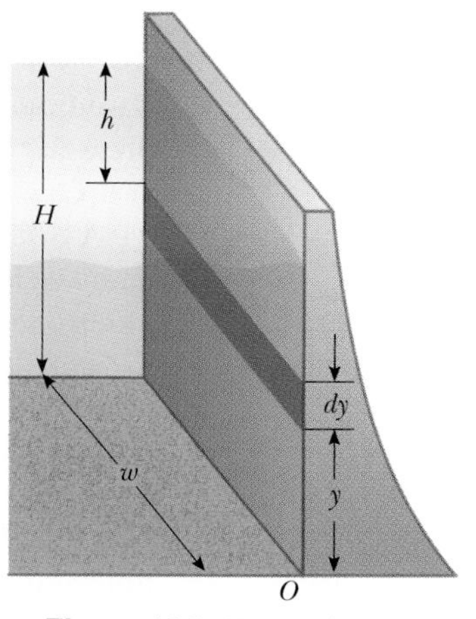

Figure 14.5 (Example 14.4)

Using Equation 14.2, we find that the force exerted on the shaded strip of area $dA = w\,dy$ is

$$dF = P\,dA = \rho g(H - y)\,w\,dy$$

Therefore, the total force on the dam is

$$F = \int P\,dA = \int_0^H \rho g(H - y)\,w\,dy = \tfrac{1}{2}\rho g w H^2$$

Note that the thickness of the dam shown in Figure 14.5 increases with depth. This design accounts for the greater and greater pressure that the water exerts on the dam at greater depths.

What If? **What if you were asked to find this force without using calculus? How could you determine its value?**

Answer We know from Equation 14.4 that the pressure varies linearly with depth. Thus, the average pressure due to the water over the face of the dam is the average of the pressure at the top and the pressure at the bottom:

$$P_{\text{av}} = \frac{P_{\text{top}} + P_{\text{bottom}}}{2} = \frac{0 + \rho g H}{2} = \tfrac{1}{2}\rho g H$$

Now, the total force is equal to the average pressure times the area of the face of the dam:

$$F = P_{\text{av}}A = (\tfrac{1}{2}\rho g H)(Hw) = \tfrac{1}{2}\rho g w H^2$$

which is the same result we obtained using calculus.

14.3 Pressure Measurements

During the weather report on a television news program, the *barometric pressure* is often provided. This is the current pressure of the atmosphere, which varies over a small range from the standard value provided earlier. How is this pressure measured?

One instrument used to measure atmospheric pressure is the common barometer, invented by Evangelista Torricelli (1608–1647). A long tube closed at one end is filled with mercury and then inverted into a dish of mercury (Fig. 14.6a). The closed end of the tube is nearly a vacuum, so the pressure at the top of the mercury column can be taken as zero. In Figure 14.6a, the pressure at point A, due to the column of mercury, must equal the pressure at point B, due to the atmosphere. If this were not the case, there would be a net force that would move mercury from one point to the other until equilibrium is established. Therefore, it follows that $P_0 = \rho_{\text{Hg}} g h$, where ρ_{Hg} is the density of the mercury and h is the height of the mercury column. As atmospheric pressure varies, the height of the mercury column varies, so the height can be calibrated to measure atmospheric pressure. Let us determine the height of a mercury column for one atmosphere of pressure, $P_0 = 1\text{ atm} = 1.013 \times 10^5\text{ Pa}$:

$$P_0 = \rho_{\text{Hg}} g h \quad \longrightarrow \quad h = \frac{P_0}{\rho_{\text{Hg}} g} = \frac{1.013 \times 10^5\text{ Pa}}{(13.6 \times 10^3\text{ kg/m}^3)(9.80\text{ m/s}^2)} = 0.760\text{ m}$$

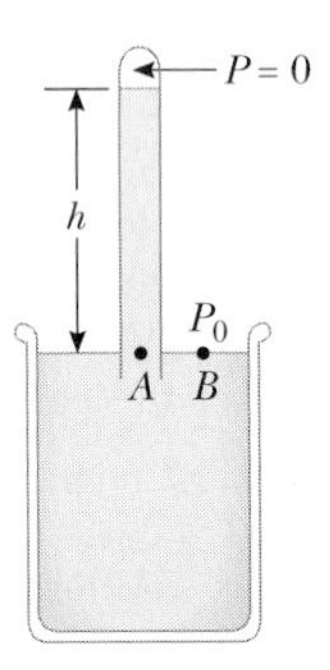

(a)

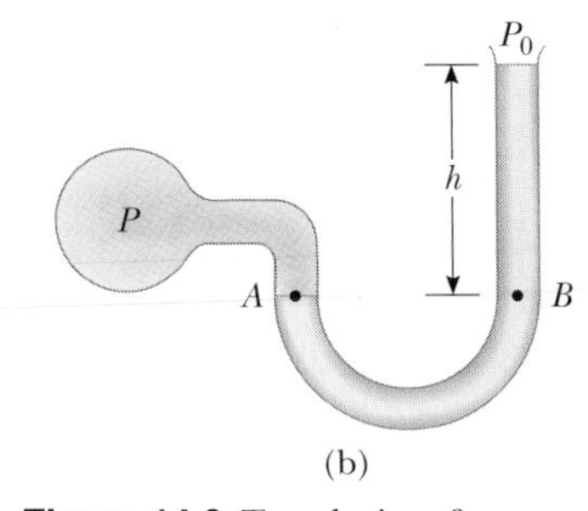

(b)

Figure 14.6 Two devices for measuring pressure: (a) a mercury barometer and (b) an open-tube manometer.

Based on a calculation such as this, one atmosphere of pressure is defined to be the pressure equivalent of a column of mercury that is exactly 0.760 0 m in height at 0°C.

A device for measuring the pressure of a gas contained in a vessel is the open-tube manometer illustrated in Figure 14.6b. One end of a U-shaped tube containing a liquid is open to the atmosphere, and the other end is connected to a system of unknown pressure P. The pressures at points A and B must be the same (otherwise, the curved portion of the liquid would experience a net force and would accelerate), and the pressure at A is the unknown pressure of the gas. Therefore, equating the unknown pressure P to the pressure at point B, we see that $P = P_0 + \rho g h$. The difference in pressure $P - P_0$ is equal to $\rho g h$. The pressure P is called the **absolute pressure,** while the difference $P - P_0$ is called the **gauge pressure.** For example, the pressure you measure in your bicycle tire is gauge pressure.

Quick Quiz 14.3 Several common barometers are built, with a variety of fluids. For which fluid will the column of fluid in the barometer be the highest? (a) mercury (b) water (c) ethyl alcohol (d) benzene

Quick Quiz 14.4 You have invented a spacesuit with a straw passing through the faceplate so that you can drink from a glass while on the surface of a planet. Out on the surface of the Moon, you attempt to drink through the straw from an open glass of water. The value of g on the Moon is about one sixth of that on Earth. Compared to the difficulty in drinking through a straw on Earth, you find drinking through a straw on the Moon to be (a) easier (b) equally difficult (c) harder (d) impossible.

14.4 Buoyant Forces and Archimedes's Principle

Have you ever tried to push a beach ball under water (Fig. 14.7a)? This is extremely difficult to do because of the large upward force exerted by the water on the ball. The upward force exerted by a fluid on any immersed object is called a **buoyant force.** We can determine the magnitude of a buoyant force by applying some logic. Imagine a beach ball–sized parcel of water beneath the water surface, as in Figure 14.7b. Because this parcel is in equilibrium, there must be an upward force that balances the downward gravitational force on the parcel. This upward force is the buoyant force, and its magnitude is equal to the weight of the water in the parcel. The buoyant force is the resultant force due to all forces applied by the fluid surrounding the parcel.

Now imagine replacing the beach ball–sized parcel of water with a beach ball of the same size. The resultant force applied by the fluid surrounding the beach ball is the same, regardless of whether it is applied to a beach ball or to a parcel of water. Consequently, we can claim that **the magnitude of the buoyant force always equals the weight of the fluid displaced by the object.** This statement is known as **Archimedes's principle.**

With the beach ball under water, the buoyant force, equal to the weight of a beach ball-sized parcel of water, is much larger than the weight of the beach ball. Thus, there is a net upward force of large magnitude—this is why it is so hard to hold the beach ball under the water. Note that Archimedes's principle does not refer to the makeup of the object experiencing the buoyant force. The object's composition is not a factor in the buoyant force because the buoyant force is exerted by the fluid.

Archimedes
(287–212 B.C.)

Archimedes, a Greek mathematician, physicist, and engineer, was perhaps the greatest scientist of antiquity. He was the first to compute accurately the ratio of a circle's circumference to its diameter, and he also showed how to calculate the volume and surface area of spheres, cylinders, and other geometric shapes. He is well known for discovering the nature of the buoyant force and was also a gifted inventor. One of his practical inventions, still in use today, is Archimedes's screw, an inclined, rotating, coiled tube used originally to lift water from the holds of ships. He also invented the catapult and devised systems of levers, pulleys, and weights for raising heavy loads. Such inventions were successfully used to defend his native city, Syracuse, during a two-year siege by Romans.

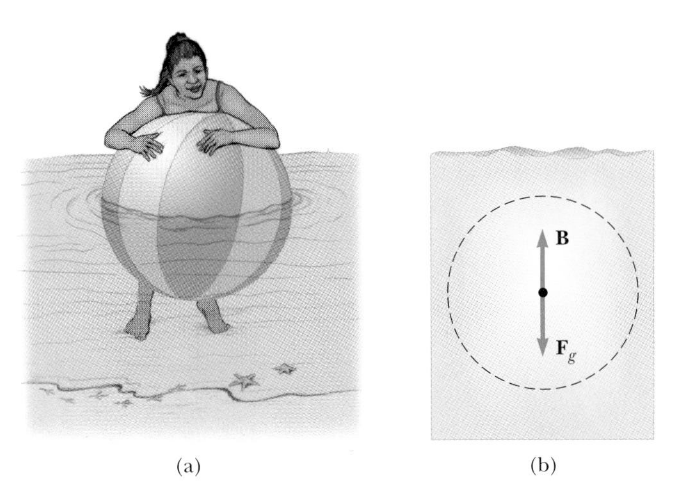

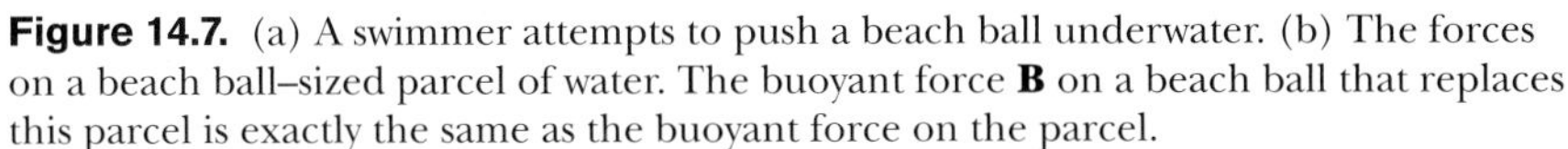

Figure 14.7. (a) A swimmer attempts to push a beach ball underwater. (b) The forces on a beach ball–sized parcel of water. The buoyant force **B** on a beach ball that replaces this parcel is exactly the same as the buoyant force on the parcel.

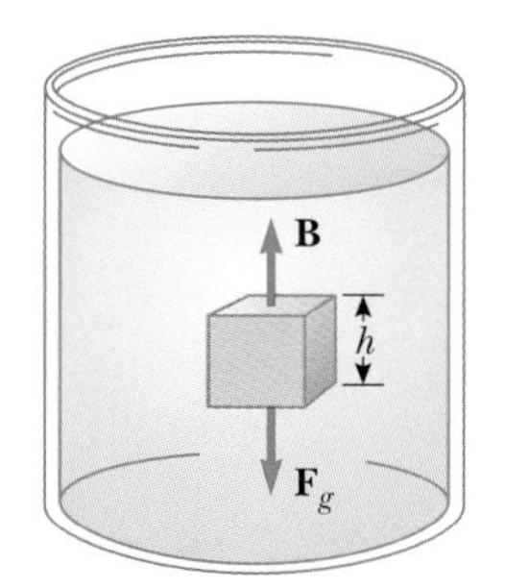

Figure 14.8 The external forces acting on the cube of liquid are the gravitational force $\mathbf{F}_g$ and the buoyant force $\mathbf{B}$. Under equilibrium conditions, $B = F_g$.

To understand the origin of the buoyant force, consider a cube immersed in a liquid as in Figure 14.8. The pressure P_b at the bottom of the cube is greater than the pressure P_t at the top by an amount $\rho_{\text{fluid}}gh$, where h is the height of the cube and ρ_{fluid} is the density of the fluid. The pressure at the bottom of the cube causes an *upward* force equal to P_bA, where A is the area of the bottom face. The pressure at the top of the cube causes a *downward* force equal to P_tA. The resultant of these two forces is the buoyant force $\mathbf{B}$:

$$B = (P_b - P_t)A = (\rho_{\text{fluid}}gh)A = \rho_{\text{fluid}}gV \quad (14.5)$$

where V is the volume of the fluid displaced by the cube. Because the product $\rho_{\text{fluid}}V$ is equal to the mass of fluid displaced by the object, we see that

$$B = Mg$$

where Mg is the weight of the fluid displaced by the cube. This is consistent with our initial statement about Archimedes's principle above, based on the discussion of the beach ball.

Before we proceed with a few examples, it is instructive for us to discuss two common situations—a totally submerged object and a floating (partly submerged) object.

PITFALL PREVENTION

14.2 Buoyant Force is Exerted by the Fluid

Remember that **the buoyant force is exerted by the fluid.** It is not determined by properties of the object, except for the amount of fluid displaced by the object. Thus, if several objects of different densities but the same volume are immersed in a fluid, they will all experience the same buoyant force. Whether they sink or float will be determined by the relationship between the buoyant force and the weight.

Case 1: Totally Submerged Object When an object is totally submerged in a fluid of density ρ_{fluid}, the magnitude of the upward buoyant force is $B = \rho_{\text{fluid}}gV = \rho_{\text{fluid}}gV_{\text{obj}}$, where V_{obj} is the volume of the object. If the object has a mass M and density ρ_{obj}, its weight is equal to $F_g = Mg = \rho_{\text{obj}}gV_{\text{obj}}$, and the net force on it is $B - F_g = (\rho_{\text{fluid}} - \rho_{\text{obj}})gV_{\text{obj}}$. Hence, if the density of the object is less than the density of the fluid, then the downward gravitational force is less than the buoyant force, and the unsupported object accelerates upward (Fig. 14.9a). If the density of the object is greater than the density of the fluid, then the upward buoyant force is less than the downward gravitational force, and the unsupported object sinks (Fig. 14.9b). If the density of the submerged object equals the density of the fluid, the net force on the object is zero and it remains in equilibrium. Thus, **the motion of an object submerged in a fluid is determined only by the densities of the object and the fluid.**

Case 2: Floating Object Now consider an object of volume V_{obj} and density $\rho_{\text{obj}} < \rho_{\text{fluid}}$ in static equilibrium floating on the surface of a fluid—that is, an object that is only *partially* submerged (Fig. 14.10). In this case, the upward buoyant force is balanced by the downward gravitational force acting on the object. If V_{fluid} is the volume of the fluid displaced by the object (this volume is the same as the volume of that part of the object that is beneath the surface of the fluid), the buoyant force has a magnitude $B = \rho_{\text{fluid}}gV_{\text{fluid}}$. Because the weight of the object is $F_g = Mg = \rho_{\text{obj}}gV_{\text{obj}}$, and because $F_g = B$, we see that $\rho_{\text{fluid}}gV_{\text{fluid}} = \rho_{\text{obj}}gV_{\text{obj}}$, or

$$\frac{V_{\text{fluid}}}{V_{\text{obj}}} = \frac{\rho_{\text{obj}}}{\rho_{\text{fluid}}} \quad (14.6)$$

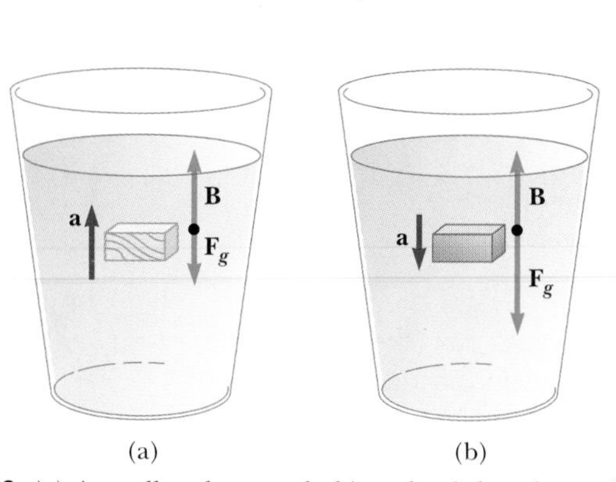

Active Figure 14.9 (a) A totally submerged object that is less dense than the fluid in which it is submerged experiences a net upward force. (b) A totally submerged object that is denser than the fluid experiences a net downward force and sinks.

At the Active Figures link at http://www.pse6.com, you can move the object to new positions as well as change the density of the object to see the results.

This equation tells us that **the fraction of the volume of a floating object that is below the fluid surface is equal to the ratio of the density of the object to that of the fluid.**

Under normal conditions, the weight of a fish is slightly greater than the buoyant force due to water. It follows that the fish would sink if it did not have some mechanism for adjusting the buoyant force. The fish accomplishes this by internally regulating the size of its air-filled swim bladder to increase its volume and the magnitude of the buoyant force acting on it. In this manner, fish are able to swim to various depths.

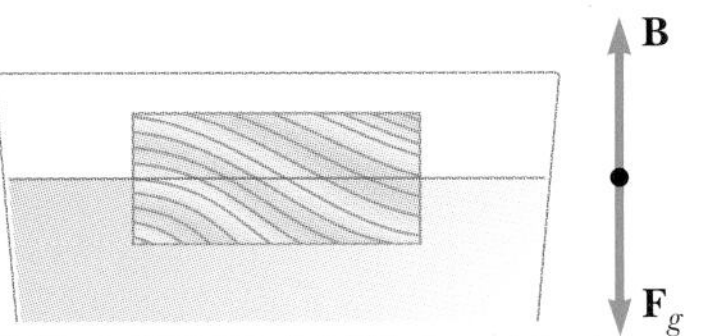

Active Figure 14.10 An object floating on the surface of a fluid experiences two forces, the gravitational force $\mathbf{F}_g$ and the buoyant force $\mathbf{B}$. Because the object floats in equilibrium, $B = F_g$.

At the Active Figures link at http://www.pse6.com, you can change the densities of the object and the fluid.

Quick Quiz 14.5 An apple is held completely submerged just below the surface of a container of water. The apple is then moved to a deeper point in the water. Compared to the force needed to hold the apple just below the surface, the force needed to hold it at a deeper point is (a) larger (b) the same (c) smaller (d) impossible to determine.

Quick Quiz 14.6 A glass of water contains a single floating ice cube (Fig. 14.11). When the ice melts, does the water level (a) go up (b) go down (c) remain the same?

Figure 14.11 (Quick Quiz 14.6) An ice cube floats on the surface of water. What happens to the water level as the ice cube melts?

Quick Quiz 14.7 You are shipwrecked and floating in the middle of the ocean on a raft. Your cargo on the raft includes a treasure chest full of gold that you found before your ship sank, and the raft is just barely afloat. To keep you floating as high as possible in the water, should you (a) leave the treasure chest on top of the raft (b) secure the treasure chest to the underside of the raft (c) hang the treasure chest in the water with a rope attached to the raft? (Assume that throwing the treasure chest overboard is not an option you wish to consider!)

Example 14.5 Eureka!

Archimedes supposedly was asked to determine whether a crown made for the king consisted of pure gold. Legend has it that he solved this problem by weighing the crown first in air and then in water, as shown in Figure 14.12. Suppose the scale read 7.84 N in air and 6.84 N in water. What should Archimedes have told the king?

Solution Figure 14.12 helps us to conceptualize the problem. Because of our understanding of the buoyant force, we realize that the scale reading will be smaller in Figure 14.12b than in Figure 14.12a. The scale reading is a measure of one of the forces on the crown, and we recognize that the crown is stationary. Thus, we can categorize this as a force equilibrium problem. To analyze the problem, note that when the crown is suspended in air, the scale reads the true weight $T_1 = F_g$ (neglecting the buoyancy of air). When it is immersed in water, the buoyant force $\mathbf{B}$ reduces the scale reading to an *apparent* weight of $T_2 = F_g - B$. Because the crown is in equilibrium, the net force on it is zero. When the crown is in water,

$$\sum F = B + T_2 - F_g = 0$$

so that

$$B = F_g - T_2 = 7.84 \text{ N} - 6.84 \text{ N} = 1.00 \text{ N}$$

Because this buoyant force is equal in magnitude to the weight of the displaced water, we have $\rho_w g V_w = 1.00$ N, where V_w is the volume of the displaced water and ρ_w is its

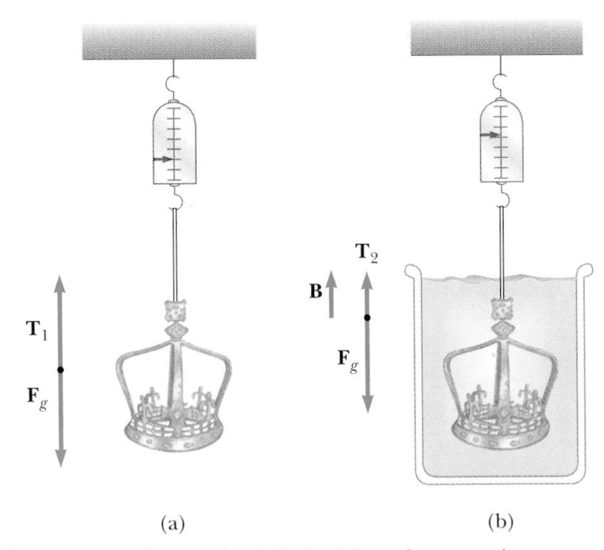

Figure 14.12 (Example 14.5) (a) When the crown is suspended in air, the scale reads its true weight because $T_1 = F_g$ (the buoyancy of air is negligible). (b) When the crown is immersed in water, the buoyant force **B** changes the scale reading to a lower value $T_2 = F_g - B$.

density. Also, the volume of the crown V_c is equal to the volume of the displaced water because the crown is completely submerged. Therefore,

$$V_c = V_w = \frac{1.00\text{ N}}{\rho_w g} = \frac{1.00\text{ N}}{(1\,000\text{ kg/m}^3)(9.80\text{ m/s}^2)}$$
$$= 1.02 \times 10^{-4}\text{ m}^3$$

Finally, the density of the crown is

$$\rho_c = \frac{m_c}{V_c} = \frac{m_c g}{V_c g} = \frac{7.84\text{ N}}{(1.02 \times 10^{-4}\text{ m}^3)(9.80\text{ m/s}^2)}$$
$$= 7.84 \times 10^3\text{ kg/m}^3$$

To finalize the problem, from Table 14.1 we see that the density of gold is $19.3 \times 10^3\text{ kg/m}^3$. Thus, Archimedes should have told the king that he had been cheated. Either the crown was hollow, or it was not made of pure gold.

What If? Suppose the crown has the same weight but were indeed pure gold and not hollow. What would the scale reading be when the crown is immersed in water?

Answer We first find the volume of the solid gold crown:

$$V_c = \frac{m_c}{\rho_c} = \frac{m_c g}{\rho_c g} = \frac{7.84\text{ N}}{(19.3 \times 10^3\text{ kg/m}^3)(9.80\text{ m/s}^2)}$$
$$= 4.15 \times 10^{-5}\text{ m}^3$$

Now, the buoyant force on the crown will be

$$B = \rho_w g V_w = \rho_w g V_c$$
$$= (1.00 \times 10^3\text{ kg/m}^3)(9.80\text{ m/s}^2)(4.15 \times 10^{-5}\text{ m}^3)$$
$$= 0.406\text{ N}$$

and the tension in the string hanging from the scale is

$$T_2 = F_g - B = 7.84\text{ N} - 0.406\text{ N} = 7.43\text{ N}$$

Example 14.6 A Titanic Surprise

An iceberg floating in seawater, as shown in Figure 14.13a, is extremely dangerous because most of the ice is below the surface. This hidden ice can damage a ship that is still a considerable distance from the visible ice. What fraction of the iceberg lies below the water level?

Solution This problem corresponds to Case 2. The weight of the iceberg is $F_g = \rho_i V_i g$, where $\rho_i = 917\text{ kg/m}^3$ and V_i is the volume of the whole iceberg. The magnitude of the upward buoyant force equals the weight of the displaced water: $B = \rho_w V_w g$, where V_w, the volume of the displaced water, is equal to the volume of the ice beneath the water (the shaded region in Fig. 14.13b) and ρ_w is the density of seawater, $\rho_w = 1\,030\text{ kg/m}^3$. Because $\rho_i V_i g = \rho_w V_w g$, the fraction of ice beneath the water's surface is

$$f = \frac{V_w}{V_i} = \frac{\rho_i}{\rho_w} = \frac{917\text{ kg/m}^3}{1\,030\text{ kg/m}^3} = 0.890 \text{ or } 89.0\%$$

Figure 14.13 (Example 14.6) (a) Much of the volume of this iceberg is beneath the water. (b) A ship can be damaged even when it is not near the visible ice.

Andy Sacks/Getty

Figure 14.14 Laminar flow around an automobile in a test wind tunnel.

14.5 Fluid Dynamics

Thus far, our study of fluids has been restricted to fluids at rest. We now turn our attention to fluids in motion. When fluid is in motion, its flow can be characterized as being one of two main types. The flow is said to be **steady, or laminar,** if each particle of the fluid follows a smooth path, such that the paths of different particles never cross each other, as shown in Figure 14.14. In steady flow, the velocity of fluid particles passing any point remains constant in time.

Above a certain critical speed, fluid flow becomes **turbulent;** turbulent flow is irregular flow characterized by small whirlpool-like regions, as shown in Figure 14.15.

Werner Wolff/Black Star

Figure 14.15 Hot gases from a cigarette made visible by smoke particles. The smoke first moves in laminar flow at the bottom and then in turbulent flow above.

The term **viscosity** is commonly used in the description of fluid flow to characterize the degree of internal friction in the fluid. This internal friction, or *viscous force,* is associated with the resistance that two adjacent layers of fluid have to moving relative to each other. Viscosity causes part of the kinetic energy of a fluid to be converted to internal energy. This mechanism is similar to the one by which an object sliding on a rough horizontal surface loses kinetic energy.

Because the motion of real fluids is very complex and not fully understood, we make some simplifying assumptions in our approach. In our model of **ideal fluid flow,** we make the following four assumptions:

1. **The fluid is nonviscous.** In a nonviscous fluid, internal friction is neglected. An object moving through the fluid experiences no viscous force.
2. **The flow is steady.** In steady (laminar) flow, the velocity of the fluid at each point remains constant.
3. **The fluid is incompressible.** The density of an incompressible fluid is constant.
4. **The flow is irrotational.** In irrotational flow, the fluid has no angular momentum about any point. If a small paddle wheel placed anywhere in the fluid does not rotate about the wheel's center of mass, then the flow is irrotational.

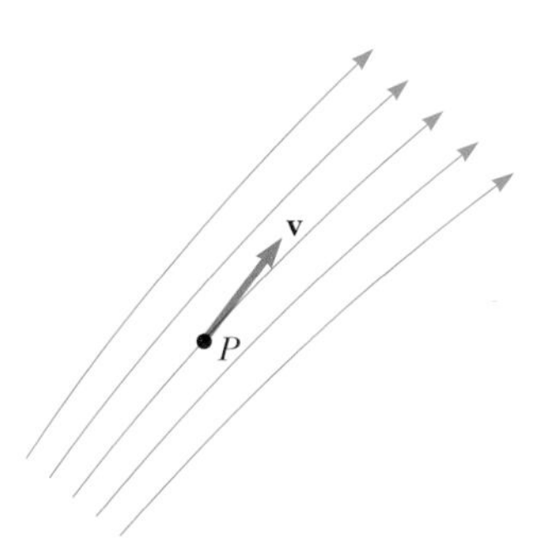

Figure 14.16 A particle in laminar flow follows a streamline, and at each point along its path the particle's velocity is tangent to the streamline.

The path taken by a fluid particle under steady flow is called a **streamline.** The velocity of the particle is always tangent to the streamline, as shown in Figure 14.16. A set of streamlines like the ones shown in Figure 14.16 form a *tube of flow.* Note that fluid

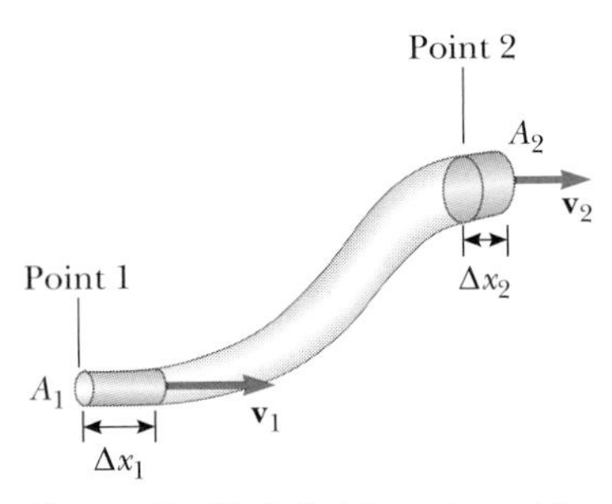

Figure 14.17 A fluid moving with steady flow through a pipe of varying cross-sectional area. The volume of fluid flowing through area A_1 in a time interval Δt must equal the volume flowing through area A_2 in the same time interval. Therefore, $A_1v_1 = A_2v_2$.

particles cannot flow into or out of the sides of this tube; if they could, then the streamlines would cross each other.

Consider an ideal fluid flowing through a pipe of nonuniform size, as illustrated in Figure 14.17. The particles in the fluid move along streamlines in steady flow. In a time interval Δt, the fluid at the bottom end of the pipe moves a distance $\Delta x_1 = v_1 \Delta t$. If A_1 is the cross-sectional area in this region, then the mass of fluid contained in the left shaded region in Figure 14.17 is $m_1 = \rho A_1 \Delta x_1 = \rho A_1 v_1 \Delta t$, where ρ is the (unchanging) density of the ideal fluid. Similarly, the fluid that moves through the upper end of the pipe in the time interval Δt has a mass $m_2 = \rho A_2 v_2 \Delta t$. However, because the fluid is incompressible and because the flow is steady, the mass that crosses A_1 in a time interval Δt must equal the mass that crosses A_2 in the same time interval. That is, $m_1 = m_2$, or $\rho A_1 v_1 = \rho A_2 v_2$; this means that

$$A_1 v_1 = A_2 v_2 = \text{constant} \tag{14.7}$$

This expression is called the **equation of continuity for fluids.** It states that

> the product of the area and the fluid speed at all points along a pipe is constant for an incompressible fluid.

Equation 14.7 tells us that the speed is high where the tube is constricted (small A) and low where the tube is wide (large A). The product Av, which has the dimensions of volume per unit time, is called either the *volume flux* or the *flow rate.* The condition $Av =$ constant is equivalent to the statement that the volume of fluid that enters one end of a tube in a given time interval equals the volume leaving the other end of the tube in the same time interval if no leaks are present.

You demonstrate the equation of continuity each time you water your garden with your thumb over the end of a garden hose as in Figure 14.18. By partially blocking the opening with your thumb, you reduce the cross-sectional area through which the water passes. As a result, the speed of the water increases as it exits the hose, and it can be sprayed over a long distance.

Quick Quiz 14.8 You tape two different soda straws together end-to-end to make a longer straw with no leaks. The two straws have radii of 3 mm and 5 mm. You drink a soda through your combination straw. In which straw is the speed of the liquid the highest? (a) whichever one is nearest your mouth (b) the one of radius 3 mm (c) the one of radius 5 mm (d) Neither—the speed is the same in both straws.

Figure 14.18 The speed of water spraying from the end of a garden hose increases as the size of the opening is decreased with the thumb.

Example 14.7 Niagara Falls

Each second, 5 525 m^3 of water flows over the 670-m-wide cliff of the Horseshoe Falls portion of Niagara Falls. The water is approximately 2 m deep as it reaches the cliff. What is its speed at that instant?

Solution The cross-sectional area of the water as it reaches the edge of the cliff is $A = (670\text{ m})(2\text{ m}) = 1\,340\text{ m}^2$. The flow rate of 5 525 m^3/s is equal to Av. This gives

$$v = \frac{5\,525\text{ m}^3/\text{s}}{A} = \frac{5\,525\text{ m}^3/\text{s}}{1\,340\text{ m}^2} = 4\text{ m/s}$$

Note that we have kept only one significant figure because our value for the depth has only one significant figure.

Example 14.8 Watering a Garden

A water hose 2.50 cm in diameter is used by a gardener to fill a 30.0-L bucket. The gardener notes that it takes 1.00 min to fill the bucket. A nozzle with an opening of cross-sectional area 0.500 cm^2 is then attached to the hose. The nozzle is held so that water is projected horizontally from a point 1.00 m above the ground. Over what horizontal distance can the water be projected?

Solution We identify point 1 within the hose and point 2 at the exit of the nozzle. We first find the speed of the water in the hose from the bucket-filling information. The cross-sectional area of the hose is

$$A_1 = \pi r^2 = \pi \frac{d^2}{4} = \pi\left(\frac{(2.50\text{ cm})^2}{4}\right) = 4.91\text{ cm}^2$$

According to the data given, the volume flow rate is equal to 30.0 L/min:

$$A_1 v_1 = 30.0\text{ L/min} = \frac{30.0 \times 10^3\text{ cm}^3}{60.0\text{ s}} = 500\text{ cm}^3/\text{s}$$

$$v_1 = \frac{500\text{ cm}^3/\text{s}}{A_1} = \frac{500\text{ cm}^3/\text{s}}{4.91\text{ cm}^2} = 102\text{ cm/s} = 1.02\text{ m/s}$$

Now we use the continuity equation for fluids to find the speed $v_2 = v_{xi}$ with which the water exits the nozzle. The subscript i anticipates that this will be the *initial* velocity component of the water projected from the hose, and the subscript x recognizes that the initial velocity vector of the projected water is in the horizontal direction.

$$A_1 v_1 = A_2 v_2 = A_2 v_{xi} \longrightarrow v_{xi} = \frac{A_1}{A_2} v_1$$

$$v_{xi} = \frac{4.91\text{ cm}^2}{0.500\text{ cm}^2}(1.02\text{ m/s})$$
$$= 10.0\text{ m/s}$$

We now shift our thinking away from fluids and to projectile motion because the water is in free fall once it exits the nozzle. A particle of the water falls through a vertical distance of 1.00 m starting from rest, and lands on the ground at a time that we find from Equation 2.12:

$$y_f = y_i + v_{yi}t - \tfrac{1}{2}gt^2$$

$$-1.00\text{ m} = 0 + 0 - \tfrac{1}{2}(9.80\text{ m/s}^2)t^2$$

$$t = \sqrt{\frac{2(1.00\text{ m})}{9.80\text{ m/s}^2}} = 0.452\text{ s}$$

In the horizontal direction, we apply Equation 2.12 with $a_x = 0$ to a particle of water to find the horizontal distance:

$$x_f = x_i + v_{xi}t = 0 + (10.0\text{ m/s})(0.452\text{ s}) = 4.52\text{ m}$$

14.6 Bernoulli's Equation

You have probably had the experience of driving on a highway and having a large truck pass you at high speed. In this situation, you may have had the frightening feeling that your car was being pulled in toward the truck as it passed. We will investigate the origin of this effect in this section.

As a fluid moves through a region where its speed and/or elevation above the Earth's surface changes, the pressure in the fluid varies with these changes. The relationship between fluid speed, pressure, and elevation was first derived in 1738 by the Swiss physicist Daniel Bernoulli. Consider the flow of a segment of an ideal fluid through a nonuniform pipe in a time interval Δt, as illustrated in Figure 14.19. At the beginning of the time interval, the segment of fluid consists of the blue shaded portion (portion 1) at the left and the unshaded portion. During the time interval, the left end of the segment moves to the right by a distance Δx_1, which is the length of the blue shaded portion at the left. Meanwhile, the right end of the segment moves to the right through a distance Δx_2, which is the length of the blue shaded portion (portion 2) at the upper right of Figure 14.19.

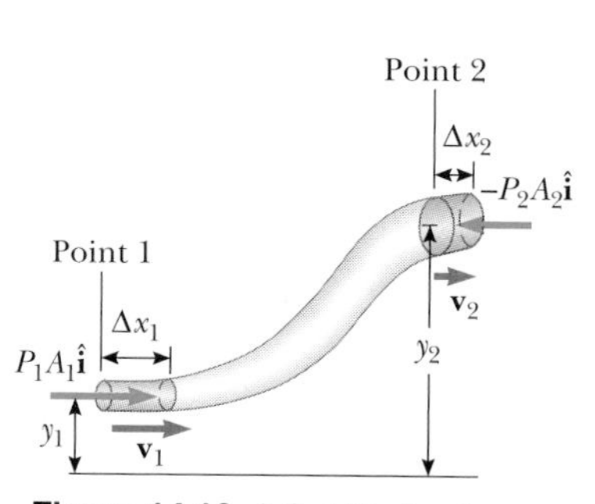

Figure 14.19 A fluid in laminar flow through a constricted pipe. The volume of the shaded portion on the left is equal to the volume of the shaded portion on the right.

Daniel Bernoulli
Swiss physicist (1700–1782)

Daniel Bernoulli made important discoveries in fluid dynamics. Born into a family of mathematicians, he was the only member of the family to make a mark in physics.

Bernoulli's most famous work, *Hydrodynamica,* was published in 1738; it is both a theoretical and a practical study of equilibrium, pressure, and speed in fluids. He showed that as the speed of a fluid increases, its pressure decreases. Referred to as "Bernoulli's principle," his work is used to produce a partial vacuum in chemical laboratories by connecting a vessel to a tube through which water is running rapidly.

In *Hydrodynamica,* Bernoulli also attempted the first explanation of the behavior of gases with changing pressure and temperature; this was the beginning of the kinetic theory of gases, a topic we study in Chapter 21. *(Corbis–Bettmann)*

Thus, at the end of the time interval, the segment of fluid consists of the unshaded portion and the blue shaded portion at the upper right.

Now consider forces exerted on this segment by fluid to the left and the right of the segment. The force exerted by the fluid on the left end has a magnitude P_1A_1. The work done by this force on the segment in a time interval Δt is $W_1 = F_1\,\Delta x_1 = P_1A_1\,\Delta x_1 = P_1V$, where V is the volume of portion 1. In a similar manner, the work done by the fluid to the right of the segment in the same time interval Δt is $W_2 = -P_2A_2\,\Delta x_2 = -P_2V$. (The volume of portion 1 equals the volume of portion 2.) This work is negative because the force on the segment of fluid is to the left and the displacement is to the right. Thus, the net work done on the segment by these forces in the time interval Δt is

$$W = (P_1 - P_2)V$$

Part of this work goes into changing the kinetic energy of the segment of fluid, and part goes into changing the gravitational potential energy of the segment–Earth system. Because we are assuming streamline flow, the kinetic energy of the unshaded portion of the segment in Figure 14.19 is unchanged during the time interval. The only change is as follows: before the time interval we have portion 1 traveling at v_1, whereas after the time interval, we have portion 2 traveling at v_2. Thus, the change in the kinetic energy of the segment of fluid is

$$\Delta K = \tfrac{1}{2}mv_2^2 - \tfrac{1}{2}mv_1^2$$

where m is the mass of both portion 1 and portion 2. (Because the volumes of both portions are the same, they also have the same mass.)

Considering the gravitational potential energy of the segment–Earth system, once again there is no change during the time interval for the unshaded portion of the fluid. The net change is that the mass of the fluid in portion 1 has effectively been moved to the location of portion 2. Consequently, the change in gravitational potential energy is

$$\Delta U = mgy_2 - mgy_1$$

The total work done on the system by the fluid outside the segment is equal to the change in mechanical energy of the system: $W = \Delta K + \Delta U$. Substituting for each of these terms, we obtain

$$(P_1 - P_2)V = \tfrac{1}{2}mv_2^2 - \tfrac{1}{2}mv_1^2 + mgy_2 - mgy_1$$

If we divide each term by the portion volume V and recall that $\rho = m/V$, this expression reduces to

$$P_1 - P_2 = \tfrac{1}{2}\rho v_2^2 - \tfrac{1}{2}\rho v_1^2 + \rho g y_2 - \rho g y_1$$

Rearranging terms, we obtain

$$P_1 + \tfrac{1}{2}\rho v_1^2 + \rho g y_1 = P_2 + \tfrac{1}{2}\rho v_2^2 + \rho g y_2 \qquad (14.8)$$

This is **Bernoulli's equation** as applied to an ideal fluid. It is often expressed as

Bernoulli's equation

$$P + \tfrac{1}{2}\rho v^2 + \rho g y = \text{constant} \qquad (14.9)$$

This expression shows that the pressure of a fluid decreases as the speed of the fluid increases. In addition, the pressure decreases as the elevation increases. This explains why water pressure from faucets on the upper floors of a tall building is weak unless measures are taken to provide higher pressure for these upper floors.

When the fluid is at rest, $v_1 = v_2 = 0$ and Equation 14.8 becomes

$$P_1 - P_2 = \rho g(y_2 - y_1) = \rho g h$$

This is in agreement with Equation 14.4.

While Equation 14.9 was derived for an incompressible fluid, the general behavior of pressure with speed is true even for gases—as the speed increases, the pressure decreases. This *Bernoulli effect* explains the experience with the truck on the highway at the opening of this section. As air passes between you and the truck, it must pass through a relatively narrow channel. According to the continuity equation, the speed of the air is higher. According to the Bernoulli effect, this higher speed air exerts less pressure on your car than the slower moving air on the other side of your car. Thus, there is a net force pushing you toward the truck!

Quick Quiz 14.9 You observe two helium balloons floating next to each other at the ends of strings secured to a table. The facing surfaces of the balloons are separated by 1–2 cm. You blow through the small space between the balloons. What happens to the balloons? (a) They move toward each other. (b) They move away from each other. (c) They are unaffected.

Example 14.9 The Venturi Tube

The horizontal constricted pipe illustrated in Figure 14.20, known as a *Venturi tube*, can be used to measure the flow speed of an incompressible fluid. Determine the flow speed at point 2 if the pressure difference $P_1 - P_2$ is known.

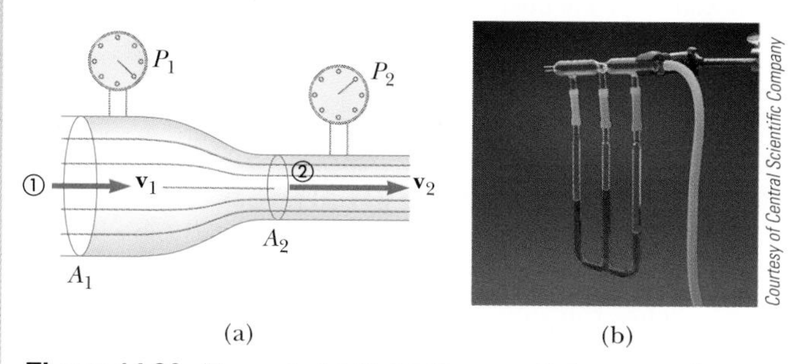

Figure 14.20 (Example 14.9) (a) Pressure P_1 is greater than pressure P_2 because $v_1 < v_2$. This device can be used to measure the speed of fluid flow. (b) A Venturi tube, located at the top of the photograph. The higher level of fluid in the middle column shows that the pressure at the top of the column, which is in the constricted region of the Venturi tube, is lower.

Solution Because the pipe is horizontal, $y_1 = y_2$, and applying Equation 14.8 to points 1 and 2 gives

$$(1) \qquad P_1 + \tfrac{1}{2}\rho v_1^2 = P_2 + \tfrac{1}{2}\rho v_2^2$$

From the equation of continuity, $A_1 v_1 = A_2 v_2$, we find that

$$(2) \qquad v_1 = \frac{A_2}{A_1} v_2$$

Substituting this expression into Equation (1) gives

$$P_1 + \tfrac{1}{2}\rho\left(\frac{A_2}{A_1}\right)^2 v_2^2 = P_2 + \tfrac{1}{2}\rho v_2^2$$

$$v_2 = A_1\sqrt{\frac{2(P_1 - P_2)}{\rho(A_1^2 - A_2^2)}}$$

We can use this result and the continuity equation to obtain an expression for v_1. Because $A_2 < A_1$, Equation (2) shows us that $v_2 > v_1$. This result, together with Equation (1), indicates that $P_1 > P_2$. In other words, the pressure is reduced in the constricted part of the pipe.

Example 14.10 Torricelli's Law

Interactive

An enclosed tank containing a liquid of density ρ has a hole in its side at a distance y_1 from the tank's bottom (Fig. 14.21). The hole is open to the atmosphere, and its diameter is much smaller than the diameter of the tank. The air above the liquid is maintained at a pressure P. Determine the speed of the liquid as it leaves the hole when the liquid's level is a distance h above the hole.

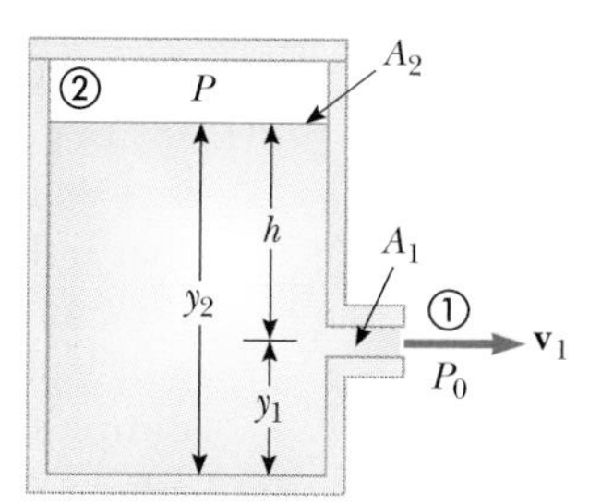

Figure 14.21 (Example 14.10) A liquid leaves a hole in a tank at speed v_1.

Solution Because $A_2 >> A_1$, the liquid is approximately at rest at the top of the tank, where the pressure is P. Applying Bernoulli's equation to points 1 and 2 and noting that at the hole P_1 is equal to atmospheric pressure P_0, we find that

$$P_0 + \tfrac{1}{2}\rho v_1^2 + \rho g y_1 = P + \rho g y_2$$

But $y_2 - y_1 = h$; thus, this expression reduces to

$$v_1 = \sqrt{\frac{2(P - P_0)}{\rho} + 2gh}$$

When P is much greater than P_0 (so that the term $2gh$ can be neglected), the exit speed of the water is mainly a function of P. If the tank is open to the atmosphere, then $P = P_0$ and $v_1 = \sqrt{2gh}$. In other words, for an open tank, the speed of liquid coming out through a hole a distance h below the surface is equal to that acquired by an object falling freely through a vertical distance h. This phenomenon is known as **Torricelli's law.**

What If? What if the position of the hole in Figure 14.21 could be adjusted vertically? If the tank is open to the atmosphere and sitting on a table, what position of the hole would cause the water to land on the table at the farthest distance from the tank?

Answer We model a parcel of water exiting the hole as a projectile. We find the time at which the parcel strikes the table from a hole at an arbitrary position:

$$y_f = y_i + v_{yi}t - \tfrac{1}{2}gt^2$$

$$0 = y_1 + 0 - \tfrac{1}{2}gt^2$$

$$t = \sqrt{\frac{2y_1}{g}}$$

Thus, the horizontal position of the parcel at the time it strikes the table is

$$x_f = x_i + v_{xi}t = 0 + \sqrt{2g(y_2 - y_1)}\,\sqrt{\frac{2y_1}{g}}$$

$$= 2\sqrt{(y_2y_1 - y_1^2)}$$

Now we maximize the horizontal position by taking the derivative of x_f with respect to y_1 (because y_1, the height of the hole, is the variable that can be adjusted) and setting it equal to zero:

$$\frac{dx_f}{dy_1} = \tfrac{1}{2}(2)(y_2y_1 - y_1^2)^{-1/2}(y_2 - 2y_1) = 0$$

This is satisfied if

$$y_1 = \tfrac{1}{2}y_2$$

Thus, the hole should be halfway between the bottom of the tank and the upper surface of the water to maximize the horizontal distance. Below this location, the water is projected at a higher speed, but falls for a short time interval, reducing the horizontal range. Above this point, the water is in the air for a longer time interval, but is projected with a smaller horizontal speed.

At the Interactive Worked Example link at http://www.pse6.com, *you can move the hole vertically to see where the water lands.*

14.7 Other Applications of Fluid Dynamics

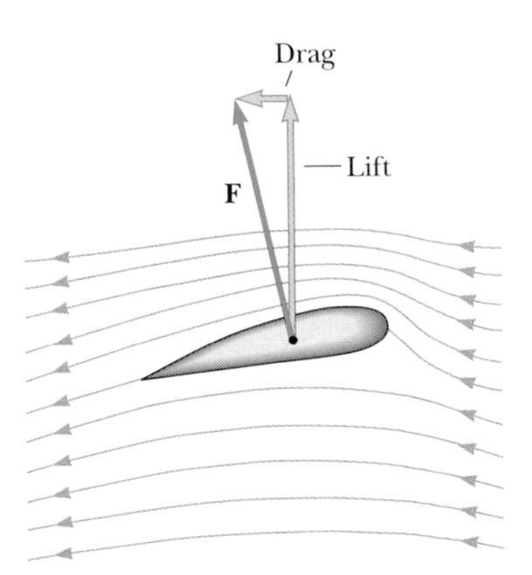

Figure 14.22 Streamline flow around a moving airplane wing. The air approaching from the right is deflected downward by the wing. By Newton's third law, this must coincide with an upward force on the wing from the air—*lift.* Because of air resistance, there is also a force opposite the velocity of the wing—*drag.*

Consider the streamlines that flow around an airplane wing as shown in Figure 14.22. Let us assume that the airstream approaches the wing horizontally from the right with a velocity $\mathbf{v}_1$. The tilt of the wing causes the airstream to be deflected downward with a velocity $\mathbf{v}_2$. Because the airstream is deflected by the wing, the wing must exert a force on the airstream. According to Newton's third law, the airstream exerts a force **F** on the wing that is equal in magnitude and opposite in direction. This force has a vertical component called the **lift** (or aerodynamic lift) and a horizontal component called **drag.** The lift depends on several factors, such as the speed of the airplane, the area of the wing, its curvature, and the angle between the wing and the horizontal. The curvature of the wing surfaces causes the pressure above the wing to be lower than that below the wing, due to the Bernoulli effect. This assists with the lift on the wing. As the angle between the wing and the horizontal increases, turbulent flow can set in above the wing to reduce the lift.

In general, an object moving through a fluid experiences lift as the result of any effect that causes the fluid to change its direction as it flows past the object. Some factors that influence lift are the shape of the object, its orientation with respect to the fluid flow, any spinning motion it might have, and the texture of its surface. For example, a golf ball struck with a club is given a rapid backspin due to the slant of the club. The dimples on the ball increase the friction force between the ball and the air so that air adheres to the ball's surface. This effect is most pronounced on the top half of the ball, where the ball's surface is moving in the same direction as the air flow. Figure 14.23 shows air adhering to the ball and being deflected downward as a result. Because the ball pushes the air down, the air must push up on the ball. Without the dimples, the friction force is lower, and the golf ball does not travel as far. It may seem counterintuitive to increase the range by increasing the friction force, but the lift gained by spinning the ball more than compensates for the loss of range due to the effect of friction

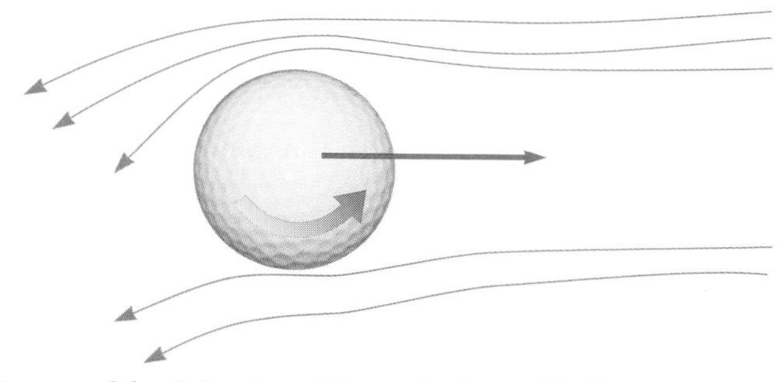

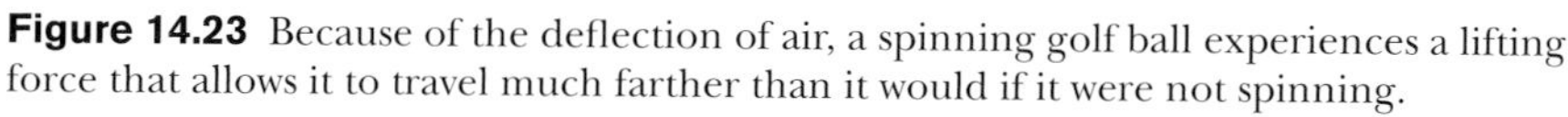

Figure 14.23 Because of the deflection of air, a spinning golf ball experiences a lifting force that allows it to travel much farther than it would if it were not spinning.

on the translational motion of the ball! For the same reason, a baseball's cover helps the spinning ball "grab" the air rushing by and helps to deflect it when a "curve ball" is thrown.

A number of devices operate by means of the pressure differentials that result from differences in a fluid's speed. For example, a stream of air passing over one end of an open tube, the other end of which is immersed in a liquid, reduces the pressure above the tube, as illustrated in Figure 14.24. This reduction in pressure causes the liquid to rise into the air stream. The liquid is then dispersed into a fine spray of droplets. You might recognize that this so-called atomizer is used in perfume bottles and paint sprayers.

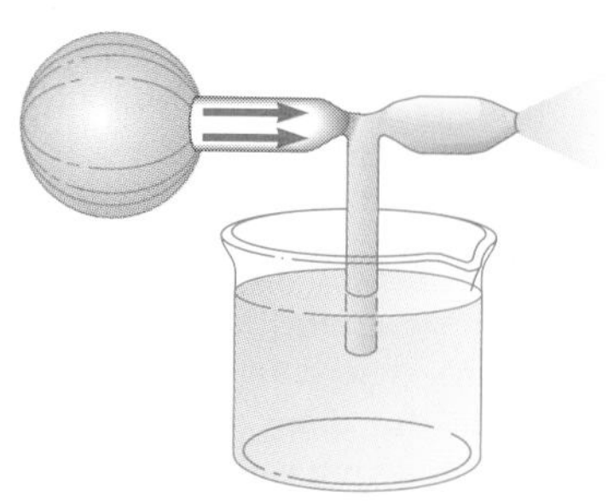

Figure 14.24 A stream of air passing over a tube dipped into a liquid causes the liquid to rise in the tube.

SUMMARY

Take a practice test for this chapter by clicking on the Practice Test link at **http://www.pse6.com.**

The **pressure** P in a fluid is the force per unit area exerted by the fluid on a surface:

$$P \equiv \frac{F}{A} \quad (14.1)$$

In the SI system, pressure has units of newtons per square meter (N/m^2), and 1 N/m^2 = 1 **pascal** (Pa).

The pressure in a fluid at rest varies with depth h in the fluid according to the expression

$$P = P_0 + \rho g h \quad (14.4)$$

where P_0 is the pressure at $h = 0$ and ρ is the density of the fluid, assumed uniform.

Pascal's law states that when pressure is applied to an enclosed fluid, the pressure is transmitted undiminished to every point in the fluid and to every point on the walls of the container.

When an object is partially or fully submerged in a fluid, the fluid exerts on the object an upward force called the **buoyant force.** According to **Archimedes's principle,** the magnitude of the buoyant force is equal to the weight of the fluid displaced by the object:

$$B = \rho_{\text{fluid}} g V \quad (14.5)$$

You can understand various aspects of a fluid's dynamics by assuming that the fluid is nonviscous and incompressible, and that the fluid's motion is a steady flow with no rotation.

Two important concepts regarding ideal fluid flow through a pipe of nonuniform size are as follows:

1. The flow rate (volume flux) through the pipe is constant; this is equivalent to stating that the product of the cross-sectional area A and the speed v at any point is a constant. This result is expressed in the **equation of continuity for fluids:**

$$A_1 v_1 = A_2 v_2 = \text{constant} \quad (14.7)$$

2. The sum of the pressure, kinetic energy per unit volume, and gravitational potential energy per unit volume has the same value at all points along a streamline. This result is summarized in **Bernoulli's equation:**

$$P + \tfrac{1}{2}\rho v^2 + \rho g y = \text{constant} \tag{14.9}$$

QUESTIONS

1. Two drinking glasses having equal weights but different shapes and different cross-sectional areas are filled to the same level with water. According to the expression $P = P_0 + \rho g h$, the pressure is the same at the bottom of both glasses. In view of this, why does one weigh more than the other?

2. Figure Q14.2 shows aerial views from directly above two dams. Both dams are equally wide (the vertical dimension in the diagram) and equally high (into the page in the diagram). The dam on the left holds back a very large lake, while the dam on the right holds back a narrow river. Which dam has to be built stronger?

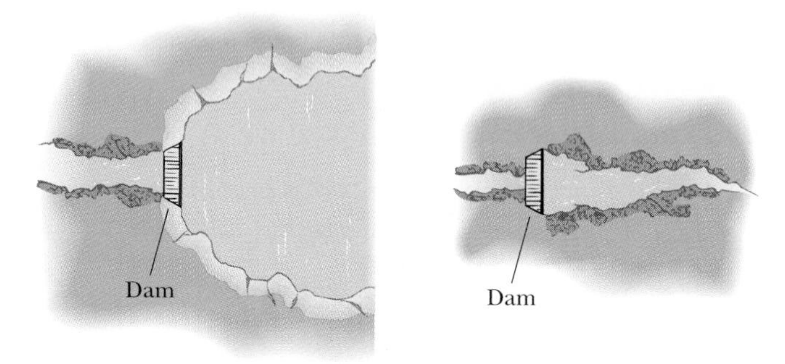

Figure Q14.2

3. Some physics students attach a long tube to the opening of a hot water bottle made of strong rubber. Leaving the hot water bottle on the ground, they hoist the other end of the tube to the roof of a multistory campus building. Students at the top of the building pour water into the tube. The students on the ground watch the bottle fill with water. On the roof, the students are surprised to see that the tube never seems to fill up—they can continue to pour more and more water down the tube. On the ground, the hot water bottle swells up like a balloon and bursts, drenching the students. Explain these observations.

4. If the top of your head has a surface area of 100 cm^2, what is the weight of the air above your head?

5. A helium-filled balloon rises until its density becomes the same as that of the surrounding air. If a sealed submarine begins to sink, will it go all the way to the bottom of the ocean or will it stop when its density becomes the same as that of the surrounding water?

6. A fish rests on the bottom of a bucket of water while the bucket is being weighed on a scale. When the fish begins to swim around, does the scale reading change?

7. Will a ship ride higher in the water of an inland lake or in the ocean? Why?

8. Suppose a damaged ship can just barely keep afloat in the ocean. It is towed toward shore and into a river, heading toward a dry dock for repair. As it is pulled up the river, it sinks. Why?

9. Lead has a greater density than iron, and both are denser than water. Is the buoyant force on a lead object greater than, less than, or equal to the buoyant force on an iron object of the same volume?

10. The water supply for a city is often provided from reservoirs built on high ground. Water flows from the reservoir, through pipes, and into your home when you turn the tap on your faucet. Why is the water flow more rapid out of a faucet on the first floor of a building than in an apartment on a higher floor?

11. Smoke rises in a chimney faster when a breeze is blowing. Use the Bernoulli effect to explain this phenomenon.

12. If the air stream from a hair dryer is directed over a Ping-Pong ball, the ball can be levitated. Explain.

13. When ski jumpers are airborne (Fig. Q14.13), why do they bend their bodies forward and keep their hands at their sides?

Figure Q14.13

14. When an object is immersed in a liquid at rest, why is the net force on the object in the horizontal direction equal to zero?

15. Explain why a sealed bottle partially filled with a liquid can float in a basin of the same liquid.

16. When is the buoyant force on a swimmer greater—after exhaling or after inhaling?

17. A barge is carrying a load of gravel along a river. It approaches a low bridge and the captain realizes that the top of the pile of gravel is not going to make it under the bridge. The captain orders the crew to quickly shovel gravel from the pile into the water. Is this a good decision?

18. A person in a boat floating in a small pond throws an anchor overboard. Does the level of the pond rise, fall, or remain the same?

19. An empty metal soap dish barely floats in water. A bar of Ivory soap floats in water. When the soap is stuck in the soap dish, the combination sinks. Explain why.

20. A piece of unpainted porous wood barely floats in a container partly filled with water. If the container is sealed and pressurized above atmospheric pressure, does the wood rise, fall, or remain at the same level?

21. A flat plate is immersed in a liquid at rest. For what orientation of the plate is the pressure on its flat surface uniform?

22. Because atmospheric pressure is about 10^5 N/m^2 and the area of a person's chest is about 0.13 m^2, the force of the atmosphere on one's chest is around 13 000 N. In view of this enormous force, why don't our bodies collapse?

23. How would you determine the density of an irregularly shaped rock?

24. Why do airplane pilots prefer to take off into the wind?

25. If you release a ball while inside a freely falling elevator, the ball remains in front of you rather than falling to the floor, because the ball, the elevator, and you all experience the same downward acceleration **g**. What happens if you repeat this experiment with a helium-filled balloon? (This one is tricky.)

26. Two identical ships set out to sea. One is loaded with a cargo of Styrofoam, and the other is empty. Which ship is more submerged?

27. A small piece of steel is tied to a block of wood. When the wood is placed in a tub of water with the steel on top, half of the block is submerged. If the block is inverted so that the steel is under water, does the amount of the block submerged increase, decrease, or remain the same? What happens to the water level in the tub when the block is inverted?

28. Prairie dogs (Fig. Q14.28) ventilate their burrows by building a mound around one entrance, which is open to a stream of air when wind blows from any direction. A second entrance at ground level is open to almost stagnant air. How does this construction create an air flow through the burrow?

Pamela Zilly/The Image Bank

Figure Q14.28

29. An unopened can of diet cola floats when placed in a tank of water, whereas a can of regular cola of the same brand sinks in the tank. What do you suppose could explain this behavior?

30. Figure Q14.30 shows a glass cylinder containing four liquids of different densities. From top to bottom, the liquids are oil (orange), water (yellow), salt water (green), and mercury (silver). The cylinder also contains, from top to bottom, a Ping-Pong ball, a piece of wood, an egg, and a steel ball. (a) Which of these liquids has the lowest density, and which has the greatest? (b) What can you conclude about the density of each object?

Henry Leap and Jim Lehman

Figure Q14.30

31. In Figure Q14.31, an air stream moves from right to left through a tube that is constricted at the middle. Three Ping-Pong balls are levitated in equilibrium above the vertical columns through which the air escapes. (a) Why is the ball at the right higher than the one in the middle? (b) Why is the ball at the left lower than the ball at the right even though the horizontal tube has the same dimensions at these two points?

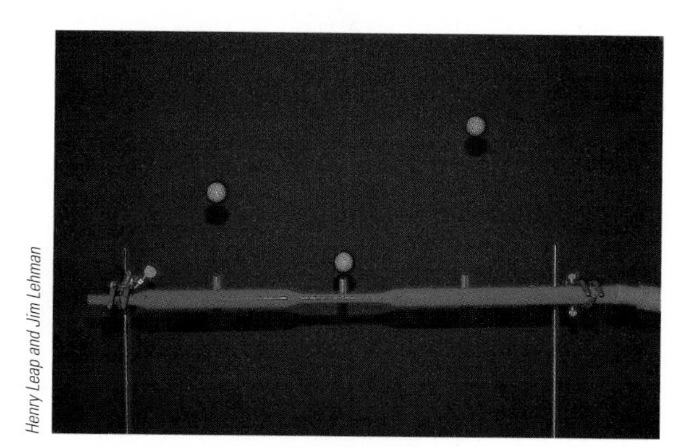

Henry Leap and Jim Lehman

Figure Q14.31

32. You are a passenger on a spacecraft. For your survival and comfort, the interior contains air just like that at the surface of the Earth. The craft is coasting through a very empty region of space. That is, a nearly perfect vacuum exists just outside the wall. Suddenly, a meteoroid pokes a hole, about the size of a large coin, right through the wall next to your seat. What will happen? Is there anything you can or should do about it?

PROBLEMS

1, 2, 3 = straightforward, intermediate, challenging ☐ = full solution available in the *Student Solutions Manual and Study Guide*

= coached solution with hints available at http://www.pse6.com = computer useful in solving problem

= paired numerical and symbolic problems

Section 14.1 Pressure

1. Calculate the mass of a solid iron sphere that has a diameter of 3.00 cm.

2. Find the order of magnitude of the density of the nucleus of an atom. What does this result suggest concerning the structure of matter? Model a nucleus as protons and neutrons closely packed together. Each has mass 1.67×10^{-27} kg and radius on the order of 10^{-15} m.

3. A 50.0-kg woman balances on one heel of a pair of high-heeled shoes. If the heel is circular and has a radius of 0.500 cm, what pressure does she exert on the floor?

4. The four tires of an automobile are inflated to a gauge pressure of 200 kPa. Each tire has an area of 0.024 0 m^2 in contact with the ground. Determine the weight of the automobile.

5. What is the total mass of the Earth's atmosphere? (The radius of the Earth is 6.37×10^6 m, and atmospheric pressure at the surface is 1.013×10^5 N/m^2.)

Section 14.2 Variation of Pressure with Depth

6. (a) Calculate the absolute pressure at an ocean depth of 1 000 m. Assume the density of seawater is 1 024 kg/m^3 and that the air above exerts a pressure of 101.3 kPa. (b) At this depth, what force must the frame around a circular submarine porthole having a diameter of 30.0 cm exert to counterbalance the force exerted by the water?

7. The spring of the pressure gauge shown in Figure 14.2 has a force constant of 1 000 N/m, and the piston has a diameter of 2.00 cm. As the gauge is lowered into water, what change in depth causes the piston to move in by 0.500 cm?

8. The small piston of a hydraulic lift has a cross-sectional area of 3.00 cm^2, and its large piston has a cross-sectional area of 200 cm^2 (Figure 14.4). What force must be applied to the small piston for the lift to raise a load of 15.0 kN? (In service stations, this force is usually exerted by compressed air.)

9. What must be the contact area between a suction cup (completely exhausted) and a ceiling if the cup is to support the weight of an 80.0-kg student?

10. (a) A very powerful vacuum cleaner has a hose 2.86 cm in diameter. With no nozzle on the hose, what is the weight of the heaviest brick that the cleaner can lift? (Fig. P14.10a) (b) **What If?** A very powerful octopus uses one sucker of diameter 2.86 cm on each of the two shells of a clam in an attempt to pull the shells apart (Fig. P14.10b). Find the greatest force the octopus can exert in salt water 32.3 m deep. *Caution:* Experimental verification can be interesting, but do not drop a brick on your foot. Do not overheat the motor of a vacuum cleaner. Do not get an octopus mad at you.

11. For the cellar of a new house, a hole is dug in the ground, with vertical sides going down 2.40 m. A concrete foundation wall is built all the way across the 9.60-m width of the

(a) (b)

Figure P14.10

excavation. This foundation wall is 0.183 m away from the front of the cellar hole. During a rainstorm, drainage from the street fills up the space in front of the concrete wall, but not the cellar behind the wall. The water does not soak into the clay soil. Find the force the water causes on the foundation wall. For comparison, the weight of the water is given by $2.40 \text{ m} \times 9.60 \text{ m} \times 0.183 \text{ m} \times 1\,000 \text{ kg/m}^3 \times 9.80 \text{ m/s}^2 = 41.3 \text{ kN}$.

12. A swimming pool has dimensions 30.0 m × 10.0 m and a flat bottom. When the pool is filled to a depth of 2.00 m with fresh water, what is the force caused by the water on the bottom? On each end? On each side?

13. A sealed spherical shell of diameter d is rigidly attached to a cart, which is moving horizontally with an acceleration a as in Figure P14.13. The sphere is nearly filled with a fluid having density ρ and also contains one small bubble of air at atmospheric pressure. Determine the pressure P at the center of the sphere.

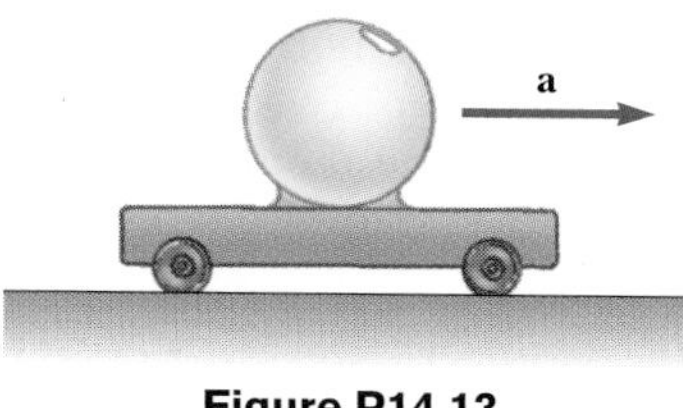

Figure P14.13

14. The tank in Figure P14.14 is filled with water 2.00 m deep. At the bottom of one side wall is a rectangular hatch 1.00 m high and 2.00 m wide, which is hinged at the top of the hatch. (a) Determine the force the water causes on the hatch. (b) Find the torque caused by the water about the hinges.

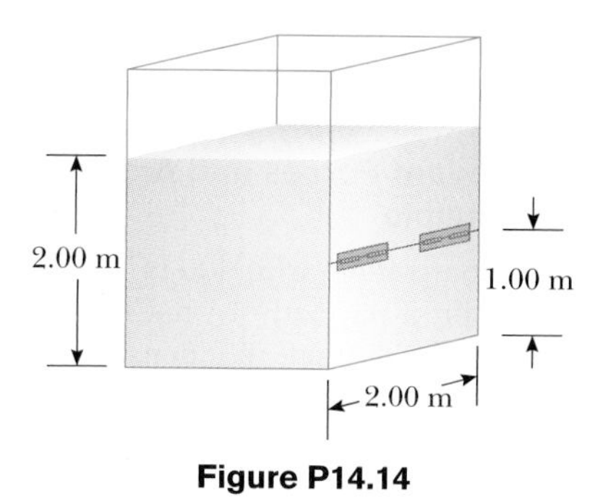

Figure P14.14

15. **Review problem.** The Abbott of Aberbrothock paid to have a bell moored to the Inchcape Rock to warn seamen of the hazard. Assume the bell was 3.00 m in diameter, cast from brass with a bulk modulus of $14.0 \times 10^{10} \text{ N/m}^2$. The pirate Ralph the Rover cut loose the warning bell and threw it into the ocean. By how much did the diameter of the bell decrease as it sank to a depth of 10.0 km? Years later, Ralph drowned when his ship collided with the rock. *Note*: The brass is compressed uniformly, so you may model the bell as a sphere of diameter 3.00 m.

Section 14.3 Pressure Measurements

16. Figure P14.16 shows Superman attempting to drink water through a very long straw. With his great strength he achieves maximum possible suction. The walls of the tubular straw do not collapse. (a) Find the maximum height through which he can lift the water. (b) **What If?** Still thirsty, the Man of Steel repeats his attempt on the Moon, which has no atmosphere. Find the difference between the water levels inside and outside the straw.

Figure P14.16

17. Blaise Pascal duplicated Torricelli's barometer using a red Bordeaux wine, of density 984 kg/m^3, as the working liquid (Fig. P14.17). What was the height h of the wine

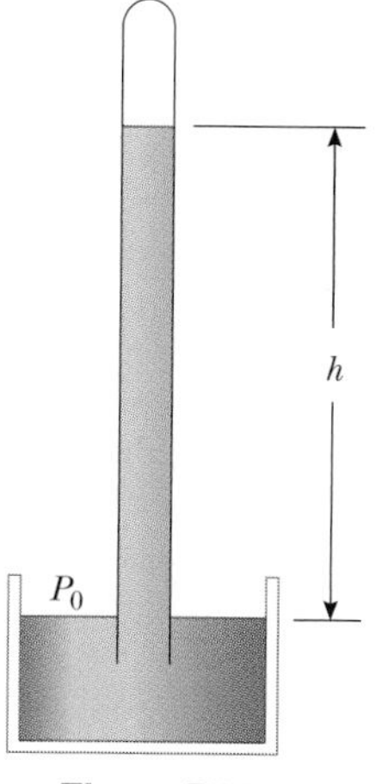

Figure P14.17

column for normal atmospheric pressure? Would you expect the vacuum above the column to be as good as for mercury?

18. Mercury is poured into a U-tube as in Figure P14.18a. The left arm of the tube has cross-sectional area A_1 of 10.0 cm^2, and the right arm has a cross-sectional area A_2 of 5.00 cm^2. One hundred grams of water are then poured into the right arm as in Figure P14.18b. (a) Determine the length of the water column in the right arm of the U-tube. (b) Given that the density of mercury is 13.6 g/cm^3, what distance h does the mercury rise in the left arm?

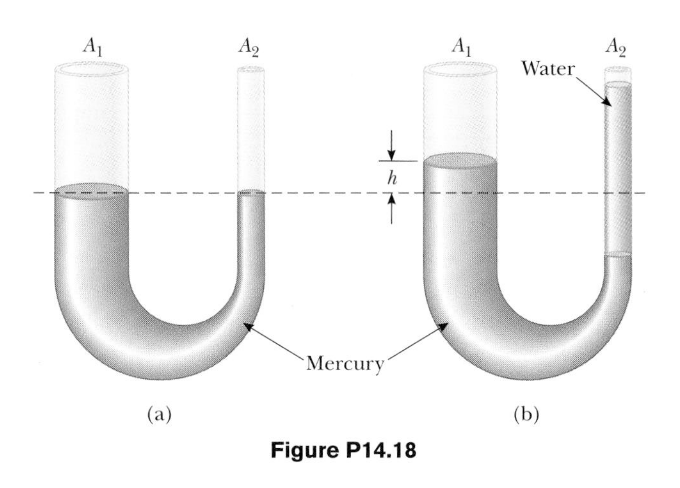

Figure P14.18

19. Normal atmospheric pressure is 1.013×10^5 Pa. The approach of a storm causes the height of a mercury barometer to drop by 20.0 mm from the normal height. What is the atmospheric pressure? (The density of mercury is 13.59 g/cm^3.)

20. A U-tube of uniform cross-sectional area, open to the atmosphere, is partially filled with mercury. Water is then poured into both arms. If the equilibrium configuration of the tube is as shown in Figure P14.20, with $h_2 = 1.00$ cm, determine the value of h_1.

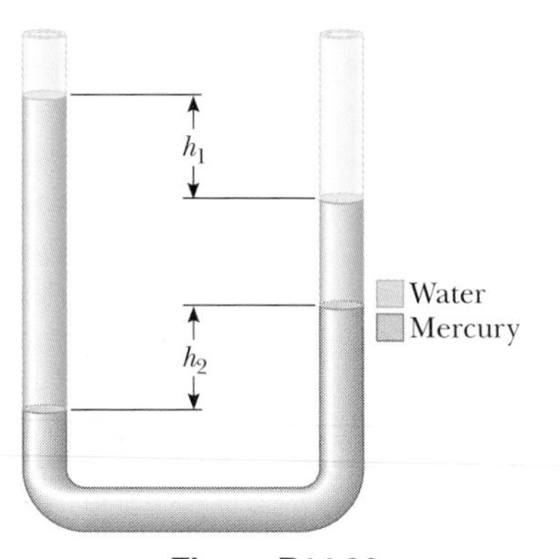

Figure P14.20

21. The human brain and spinal cord are immersed in the cerebrospinal fluid. The fluid is normally continuous between the cranial and spinal cavities. It normally exerts a pressure of 100 to 200 mm of H_2O above the prevailing atmospheric pressure. In medical work pressures are often measured in units of millimeters of H_2O because body fluids, including the cerebrospinal fluid, typically have the same density as water. The pressure of the cerebrospinal fluid can be measured by means of a *spinal tap*, as illustrated in Figure P14.21. A hollow tube is inserted into the spinal column, and the height to which the fluid rises is observed. If the fluid rises to a height of 160 mm, we write its gauge pressure as 160 mm H_2O. (a) Express this pressure in pascals, in atmospheres, and in millimeters of mercury. (b) Sometimes it is necessary to determine if an accident victim has suffered a crushed vertebra that is blocking flow of the cerebrospinal fluid in the spinal column. In other cases a physician may suspect a tumor or other growth is blocking the spinal column and inhibiting flow of cerebrospinal fluid. Such conditions can be investigated by means of the *Queckensted test.* In this procedure, the veins in the patient's neck are compressed, to make the blood pressure rise in the brain. The increase in pressure in the blood vessels is transmitted to the cerebrospinal fluid. What should be the normal effect on the height of the fluid in the spinal tap? (c) Suppose that compressing the veins had no effect on the fluid level. What might account for this?

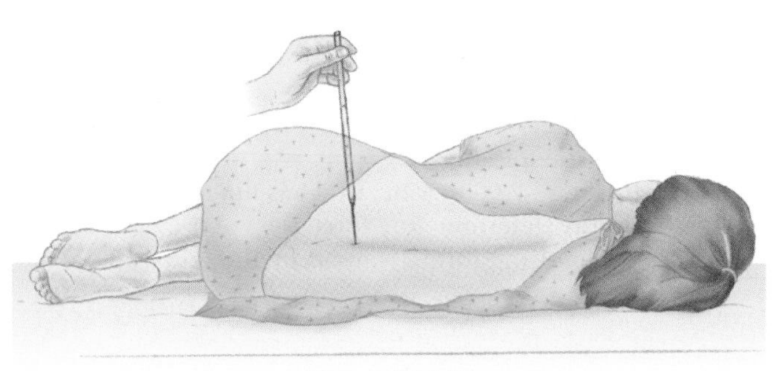

Figure P14.21

Section 14.4 Buoyant Forces and Archimedes's Principle

22. (a) A light balloon is filled with 400 m^3 of helium. At 0°C, the balloon can lift a payload of what mass? (b) **What If?** In Table 14.1, observe that the density of hydrogen is nearly half the density of helium. What load can the balloon lift if filled with hydrogen?

23. A Ping-Pong ball has a diameter of 3.80 cm and average density of 0.084 0 g/cm^3. What force is required to hold it completely submerged under water?

24. A Styrofoam slab has thickness h and density ρ_s. When a swimmer of mass m is resting on it, the slab floats in fresh water with its top at the same level as the water surface. Find the area of the slab.

25. A piece of aluminum with mass 1.00 kg and density 2 700 kg/m^3 is suspended from a string and then completely immersed in a container of water (Figure P14.25). Calculate the tension in the string (a) before and (b) after the metal is immersed.

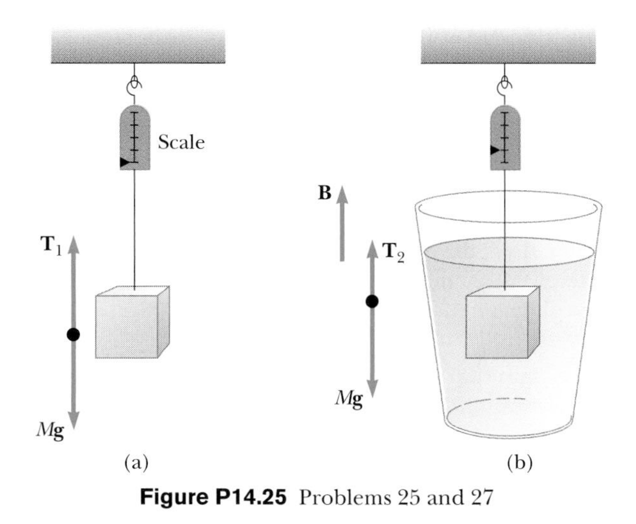

Figure P14.25 Problems 25 and 27

26. The weight of a rectangular block of low-density material is 15.0 N. With a thin string, the center of the horizontal bottom face of the block is tied to the bottom of a beaker partly filled with water. When 25.0% of the block's volume is submerged, the tension in the string is 10.0 N. (a) Sketch a free-body diagram for the block, showing all forces acting on it. (b) Find the buoyant force on the block. (c) Oil of density 800 kg/m^3 is now steadily added to the beaker, forming a layer above the water and surrounding the block. The oil exerts forces on each of the four side walls of the block that the oil touches. What are the directions of these forces? (d) What happens to the string tension as the oil is added? Explain how the oil has this effect on the string tension. (e) The string breaks when its tension reaches 60.0 N. At this moment, 25.0% of the block's volume is still below the water line; what additional fraction of the block's volume is below the top surface of the oil? (f) After the string breaks, the block comes to a new equilibrium position in the beaker. It is now in contact only with the oil. What fraction of the block's volume is submerged?

27. A 10.0-kg block of metal measuring 12.0 cm $\times$ 10.0 cm $\times$ 10.0 cm is suspended from a scale and immersed in water as in Figure P14.25b. The 12.0-cm dimension is vertical, and the top of the block is 5.00 cm below the surface of the water. (a) What are the forces acting on the top and on the bottom of the block? (Take $P_0 = 1.013\,0 \times 10^5$ N/m^2.) (b) What is the reading of the spring scale? (c) Show that the buoyant force equals the difference between the forces at the top and bottom of the block.

28. To an order of magnitude, how many helium-filled toy balloons would be required to lift you? Because helium is an irreplaceable resource, develop a theoretical answer rather than an experimental answer. In your solution state what physical quantities you take as data and the values you measure or estimate for them.

29. A cube of wood having an edge dimension of 20.0 cm and a density of 650 kg/m^3 floats on water. (a) What is the distance from the horizontal top surface of the cube to the water level? (b) How much lead weight has to be placed on top of the cube so that its top is just level with the water?

30. A spherical aluminum ball of mass 1.26 kg contains an empty spherical cavity that is concentric with the ball. The ball just barely floats in water. Calculate (a) the outer radius of the ball and (b) the radius of the cavity.

31. Determination of the density of a fluid has many important applications. A car battery contains sulfuric acid, for which density is a measure of concentration. For the battery to function properly, the density must be inside a range specified by the manufacturer. Similarly, the effectiveness of antifreeze in your car's engine coolant depends on the density of the mixture (usually ethylene glycol and water). When you donate blood to a blood bank, its screening includes determination of the density of the blood, since higher density correlates with higher hemoglobin content. A *hydrometer* is an instrument used to determine liquid density. A simple one is sketched in Figure P14.31. The bulb of a syringe is squeezed and released to let the atmosphere lift a sample of the liquid of interest into a tube containing a calibrated rod of known density. The rod, of length L and average density ρ_0, floats partially immersed in the liquid of density ρ. A length h of the rod protrudes above the surface of the liquid. Show that the density of the liquid is given by

$$\rho = \frac{\rho_0 L}{L - h}$$

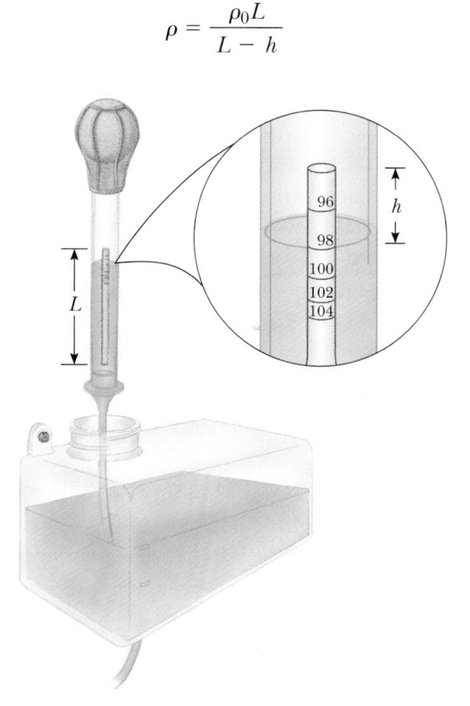

Figure P14.31 Problems 31 and 32

32. Refer to Problem 31 and Figure P14.31. A hydrometer is to be constructed with a cylindrical floating rod. Nine fiduciary marks are to be placed along the rod to indicate densities of 0.98 g/cm^3, 1.00 g/cm^3, 1.02 g/cm^3, 1.04 g/cm^3, . . . 1.14 g/cm^3. The row of marks is to start 0.200 cm from the top end of the rod and end 1.80 cm from the top end. (a) What is the required length of the rod? (b) What must be its average density? (c) Should the marks be equally spaced? Explain your answer.

33. How many cubic meters of helium are required to lift a balloon with a 400-kg payload to a height of 8 000 m? (Take $\rho_{He} = 0.180$ kg/m^3.) Assume that the balloon maintains a

constant volume and that the density of air decreases with the altitude z according to the expression $\rho_{air} = \rho_0 e^{-z/8\,000}$, where z is in meters and $\rho_0 = 1.25 \text{ kg/m}^3$ is the density of air at sea level.

34. A frog in a hemispherical pod (Fig. P14.34) just floats without sinking into a sea of blue-green ooze with density 1.35 g/cm^3. If the pod has radius 6.00 cm and negligible mass, what is the mass of the frog?

Figure P14.34

35. A plastic sphere floats in water with 50.0 percent of its volume submerged. This same sphere floats in glycerin with 40.0 percent of its volume submerged. Determine the densities of the glycerin and the sphere.

36. A bathysphere used for deep-sea exploration has a radius of 1.50 m and a mass of 1.20×10^4 kg. To dive, this submarine takes on mass in the form of seawater. Determine the amount of mass the submarine must take on if it is to descend at a constant speed of 1.20 m/s, when the resistive force on it is 1 100 N in the upward direction. The density of seawater is $1.03 \times 10^3 \text{ kg/m}^3$.

37. The United States possesses the eight largest warships in the world—aircraft carriers of the *Nimitz* class—and is building two more. Suppose one of the ships bobs up to float 11.0 cm higher in the water when 50 fighters take off from it in 25 min, at a location where the free-fall acceleration is 9.78 m/s^2. Bristling with bombs and missiles, the planes have average mass 29 000 kg. Find the horizontal area enclosed by the waterline of the $4-billion ship. By comparison, its flight deck has area 18 000 m^2. Below decks are passageways hundreds of meters long, so narrow that two large men cannot pass each other.

Section 14.5 Fluid Dynamics
Section 14.6 Bernoulli's Equation

38. A horizontal pipe 10.0 cm in diameter has a smooth reduction to a pipe 5.00 cm in diameter. If the pressure of the water in the larger pipe is 8.00×10^4 Pa and the pressure in the smaller pipe is 6.00×10^4 Pa, at what rate does water flow through the pipes?

39. A large storage tank, open at the top and filled with water, develops a small hole in its side at a point 16.0 m below the water level. If the rate of flow from the leak is equal to $2.50 \times 10^{-3} \text{ m}^3/\text{min}$, determine (a) the speed at which the water leaves the hole and (b) the diameter of the hole.

40. A village maintains a large tank with an open top, containing water for emergencies. The water can drain from the tank through a hose of diameter 6.60 cm. The hose ends with a nozzle of diameter 2.20 cm. A rubber stopper is inserted into the nozzle. The water level in the tank is kept 7.50 m above the nozzle. (a) Calculate the friction force exerted by the nozzle on the stopper. (b) The stopper is removed. What mass of water flows from the nozzle in 2.00 h? (c) Calculate the gauge pressure of the flowing water in the hose just behind the nozzle.

41. Water flows through a fire hose of diameter 6.35 cm at a rate of $0.0120 \text{ m}^3/\text{s}$. The fire hose ends in a nozzle of inner diameter 2.20 cm. What is the speed with which the water exits the nozzle?

42. Water falls over a dam of height h with a mass flow rate of R, in units of kg/s. (a) Show that the power available from the water is

$$\mathcal{P} = Rgh$$

where g is the free-fall acceleration. (b) Each hydroelectric unit at the Grand Coulee Dam takes in water at a rate of 8.50×10^5 kg/s from a height of 87.0 m. The power developed by the falling water is converted to electric power with an efficiency of 85.0%. How much electric power is produced by each hydroelectric unit?

43. Figure P14.43 shows a stream of water in steady flow from a kitchen faucet. At the faucet the diameter of the stream is 0.960 cm. The stream fills a 125-cm^3 container in 16.3 s. Find the diameter of the stream 13.0 cm below the opening of the faucet.

George Semple

Figure P14.43

44. A legendary Dutch boy saved Holland by plugging a hole in a dike with his finger, which is 1.20 cm in diameter. If the hole was 2.00 m below the surface of the North Sea (density 1 030 kg/m^3), (a) what was the force on his finger? (b) If he pulled his finger out of the hole, how long would it take the released water to fill 1 acre of land to a depth of 1 ft, assuming the hole remained constant in size? (A typical U.S. family of four uses 1 acre-foot of water, 1 234 m^3, in 1 year.)

45. Through a pipe 15.0 cm in diameter, water is pumped from the Colorado River up to Grand Canyon Village, located on the rim of the canyon. The river is at an elevation of 564 m, and the village is at an elevation of 2 096 m.

(a) What is the minimum pressure at which the water must be pumped if it is to arrive at the village? (b) If 4 500 m^3 are pumped per day, what is the speed of the water in the pipe? (c) What additional pressure is necessary to deliver this flow? *Note*: Assume that the free-fall acceleration and the density of air are constant over this range of elevations.

46. Old Faithful Geyser in Yellowstone Park (Fig. P14.46) erupts at approximately 1-h intervals, and the height of the water column reaches 40.0 m. (a) Model the rising stream as a series of separate drops. Analyze the free-fall motion of one of the drops to determine the speed at which the water leaves the ground. (b) **What If?** Model the rising stream as an ideal fluid in streamline flow. Use Bernoulli's equation to determine the speed of the water as it leaves ground level. (c) What is the pressure (above atmospheric) in the heated underground chamber if its depth is 175 m? You may assume that the chamber is large compared with the geyser's vent.

Figure P14.46

47. A Venturi tube may be used as a fluid flow meter (see Fig. 14.20). If the difference in pressure is $P_1 - P_2 = 21.0$ kPa, find the fluid flow rate in cubic meters per second, given that the radius of the outlet tube is 1.00 cm, the radius of the inlet tube is 2.00 cm, and the fluid is gasoline ($\rho = 700$ kg/m^3).

Section 14.7 Other Applications of Fluid Dynamics

48. An airplane has a mass of 1.60×10^4 kg, and each wing has an area of 40.0 m^2. During level flight, the pressure on the lower wing surface is 7.00×10^4 Pa. Determine the pressure on the upper wing surface.

49. A Pitot tube can be used to determine the velocity of air flow by measuring the difference between the total pressure and the static pressure (Fig. P14.49). If the fluid in the tube is mercury, density $\rho_{Hg} = 13\ 600$ kg/m^3, and $\Delta h = 5.00$ cm, find the speed of air flow. (Assume that the air is stagnant at point A, and take $\rho_{air} = 1.25$ kg/m^3.)

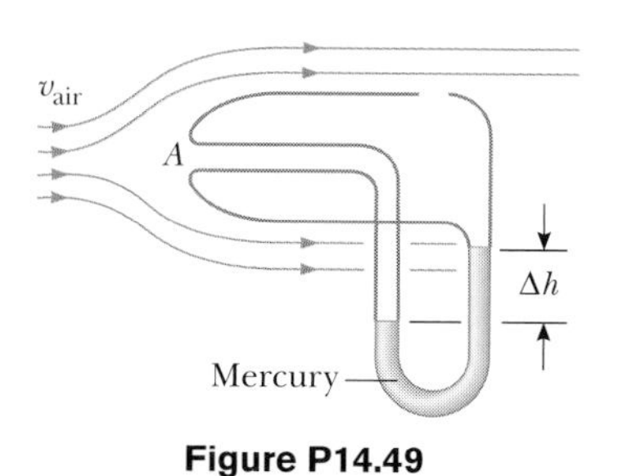

Figure P14.49

50. An airplane is cruising at an altitude of 10 km. The pressure outside the craft is 0.287 atm; within the passenger compartment the pressure is 1.00 atm and the temperature is 20°C. A small leak occurs in one of the window seals in the passenger compartment. Model the air as an ideal fluid to find the speed of the stream of air flowing through the leak.

51. A siphon is used to drain water from a tank, as illustrated in Figure P14.51. The siphon has a uniform diameter. Assume steady flow without friction. (a) If the distance $h = 1.00$ m, find the speed of outflow at the end of the siphon. (b) **What If?** What is the limitation on the height of the top of the siphon above the water surface? (For the flow of the liquid to be continuous, the pressure must not drop below the vapor pressure of the liquid.)

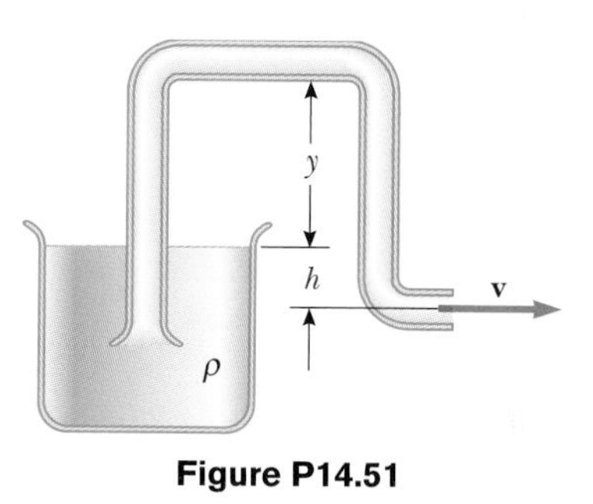

Figure P14.51

52. The Bernoulli effect can have important consequences for the design of buildings. For example, wind can blow around a skyscraper at remarkably high speed, creating low pressure. The higher atmospheric pressure in the still air inside the buildings can cause windows to pop out. As originally constructed, the John Hancock building in Boston popped window panes, which fell many stories to the sidewalk below. (a) Suppose that a horizontal wind blows in streamline flow with a speed of 11.2 m/s outside a large pane of plate glass with dimensions 4.00 m $\times$ 1.50 m. Assume the density of the air to be uniform at 1.30 kg/m^3. The air inside the building is at atmospheric pressure. What is the total force exerted by air on the window pane? (b) **What If?** If a second skyscraper is built nearby, the air speed can be especially high where wind passes through the narrow separation between the buildings. Solve part (a) again if the wind speed is 22.4 m/s, twice as high.

53. A hypodermic syringe contains a medicine with the density of water (Figure P14.53). The barrel of the syringe has a cross-sectional area $A = 2.50 \times 10^{-5}$ m^2, and the needle has a cross-sectional area $a = 1.00 \times 10^{-8}$ m^2. In the absence of a force on the plunger, the pressure everywhere is 1 atm. A force $\mathbf{F}$ of magnitude 2.00 N acts on the plunger, making medicine squirt horizontally from the needle. Determine the speed of the medicine as it leaves the needle's tip.

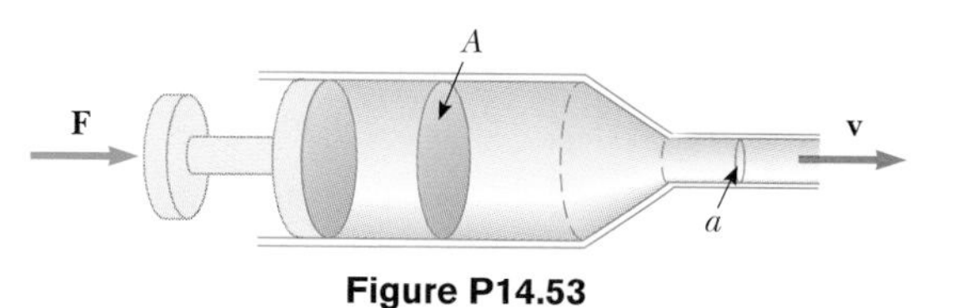

Figure P14.53

Additional Problems

54. Figure P14.54 shows a water tank with a valve at the bottom. If this valve is opened, what is the maximum height attained by the water stream coming out of the right side of the tank? Assume that h = 10.0 m, L = 2.00 m, and θ = 30.0°, and that the cross-sectional area at A is very large compared with that at B.

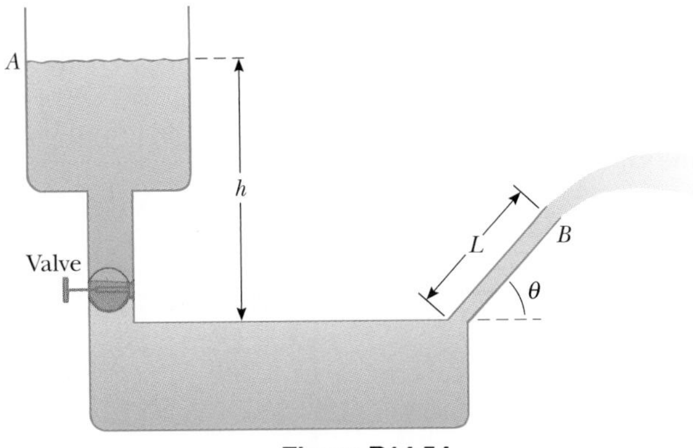

Figure P14.54

55. A helium-filled balloon is tied to a 2.00-m-long, 0.050 0-kg uniform string. The balloon is spherical with a radius of 0.400 m. When released, it lifts a length h of string and then remains in equilibrium, as in Figure P14.55. Determine the value of h. The envelope of the balloon has mass 0.250 kg.

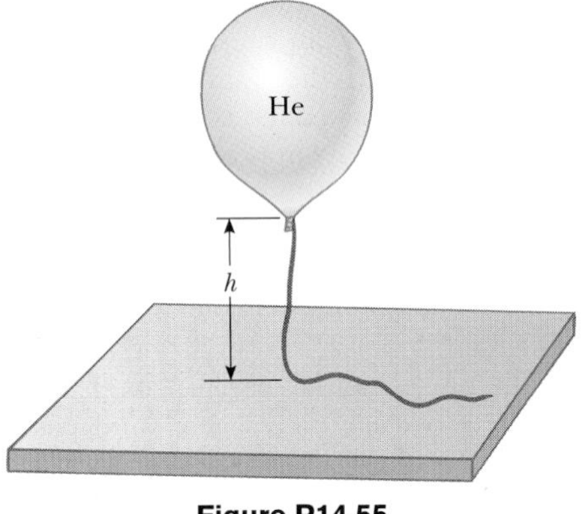

Figure P14.55

56. Water is forced out of a fire extinguisher by air pressure, as shown in Figure P14.56. How much gauge air pressure in the tank (above atmospheric) is required for the water jet to have a speed of 30.0 m/s when the water level in the tank is 0.500 m below the nozzle?

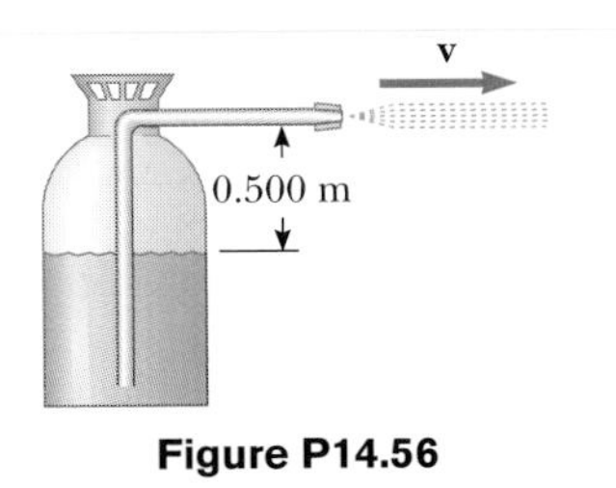

Figure P14.56

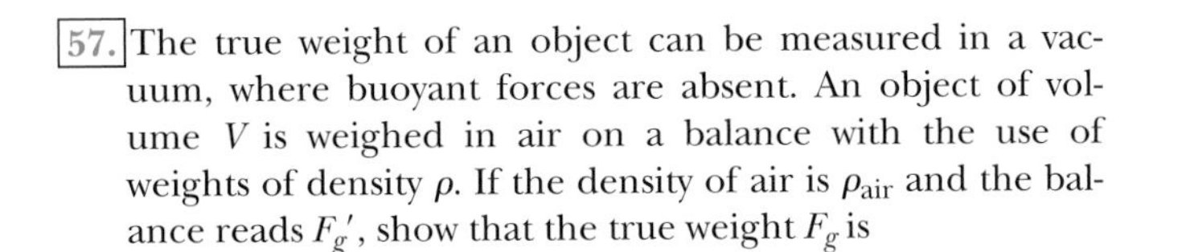

57. The true weight of an object can be measured in a vacuum, where buoyant forces are absent. An object of volume V is weighed in air on a balance with the use of weights of density ρ. If the density of air is ρ_{air} and the balance reads F_g', show that the true weight F_g is

$$F_g = F_g' + \left(V - \frac{F_g'}{\rho g}\right)\rho_{air}\, g$$

58. A wooden dowel has a diameter of 1.20 cm. It floats in water with 0.400 cm of its diameter above water (Fig. P14.58). Determine the density of the dowel.

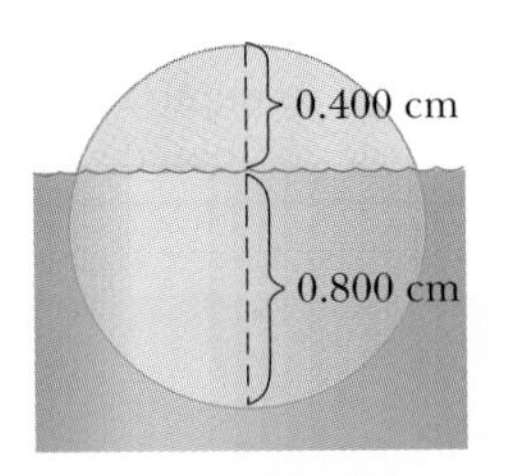

Figure P14.58

59. A light spring of constant k = 90.0 N/m is attached vertically to a table (Fig. P14.59a). A 2.00-g balloon is filled with helium (density = 0.180 kg/m^3) to a volume of 5.00 m^3 and is then connected to the spring, causing it to stretch as in Figure P14.59b. Determine the extension distance L when the balloon is in equilibrium.

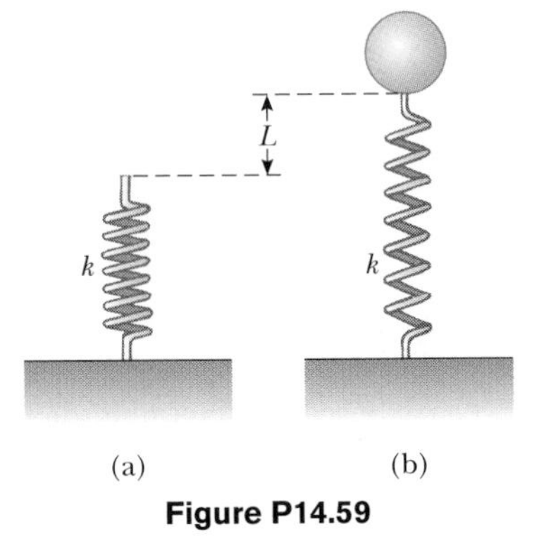

Figure P14.59

60. Evangelista Torricelli was the first person to realize that we live at the bottom of an ocean of air. He correctly surmised that the pressure of our atmosphere is attributable to the weight of the air. The density of air at 0°C at the Earth's surface is 1.29 kg/m^3. The density decreases with increasing altitude (as the atmosphere thins). On the other hand, if we assume that the density is constant at 1.29 kg/m^3 up to some altitude h, and zero above that altitude, then h would represent the depth of the ocean of air. Use this model to determine the value of h that gives a pressure of 1.00 atm at the surface of the Earth. Would the peak of

Mount Everest rise above the surface of such an atmosphere?

61. **Review problem**. With reference to Figure 14.5, show that the total torque exerted by the water behind the dam about a horizontal axis through O is $\frac{1}{6}\rho g w H^3$. Show that the effective line of action of the total force exerted by the water is at a distance $\frac{1}{3}H$ above O.

62. In about 1657 Otto von Guericke, inventor of the air pump, evacuated a sphere made of two brass hemispheres. Two teams of eight horses each could pull the hemispheres apart only on some trials, and then "with greatest difficulty," with the resulting sound likened to a cannon firing (Fig. P14.62). (a) Show that the force F required to pull the evacuated hemispheres apart is $\pi R^2(P_0 - P)$, where R is the radius of the hemispheres and P is the pressure inside the hemispheres, which is much less than P_0. (b) Determine the force if $P = 0.100P_0$ and $R = 0.300$ m.

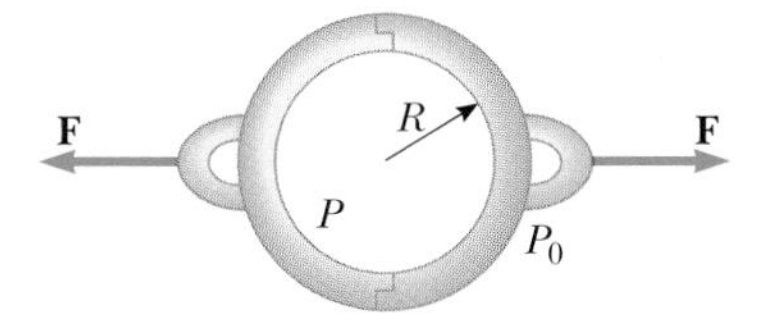

The Granger Collection

Figure P14.62 The colored engraving, dated 1672, illustrates Otto von Guericke's demonstration of the force due to air pressure as performed before Emperor Ferdinand III in 1657.

63. A 1.00-kg beaker containing 2.00 kg of oil (density = 916.0 kg/m^3) rests on a scale. A 2.00-kg block of iron is suspended from a spring scale and completely submerged in the oil as in Figure P14.63. Determine the equilibrium readings of both scales.

64. A beaker of mass m_{beaker} containing oil of mass m_{oil} (density = ρ_{oil}) rests on a scale. A block of iron of mass

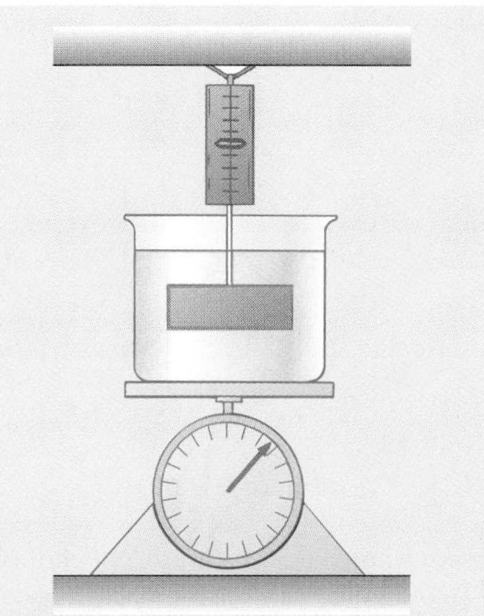

Figure P14.63 Problems 63 and 64

m_{iron} is suspended from a spring scale and completely submerged in the oil as in Figure P14.63. Determine the equilibrium readings of both scales.

65. In 1983, the United States began coining the cent piece out of copper-clad zinc rather than pure copper. The mass of the old copper penny is 3.083 g, while that of the new cent is 2.517 g. Calculate the percentage of zinc (by volume) in the new cent. The density of copper is 8.960 g/cm^3 and that of zinc is 7.133 g/cm^3. The new and old coins have the same volume.

66. A thin spherical shell of mass 4.00 kg and diameter 0.200 m is filled with helium (density = 0.180 kg/m^3). It is then released from rest on the bottom of a pool of water that is 4.00 m deep. (a) Neglecting frictional effects, show that the shell rises with constant acceleration and determine the value of that acceleration. (b) How long will it take for the top of the shell to reach the water surface?

67. **Review problem.** A uniform disk of mass 10.0 kg and radius 0.250 m spins at 300 rev/min on a low-friction axle. It must be brought to a stop in 1.00 min by a brake pad that makes contact with the disk at average distance 0.220 m from the axis. The coefficient of friction between pad and disk is 0.500. A piston in a cylinder of diameter 5.00 cm presses the brake pad against the disk. Find the pressure required for the brake fluid in the cylinder.

68. Show that the variation of atmospheric pressure with altitude is given by $P = P_0 e^{-\alpha y}$, where $\alpha = \rho_0 g/P_0$, P_0 is atmospheric pressure at some reference level $y = 0$, and ρ_0 is the atmospheric density at this level. Assume that the decrease in atmospheric pressure over an infinitesimal change in altitude (so that the density is approximately uniform) is given by $dP = -\rho g\, dy$, and that the density of air is proportional to the pressure.

69. An incompressible, nonviscous fluid is initially at rest in the vertical portion of the pipe shown in Figure P14.69a, where $L = 2.00$ m. When the valve is opened, the fluid flows into the horizontal section of the pipe. What is the speed of the fluid when all of it is in the horizontal section, as in Figure P14.69b? Assume the cross-sectional area of the entire pipe is constant.

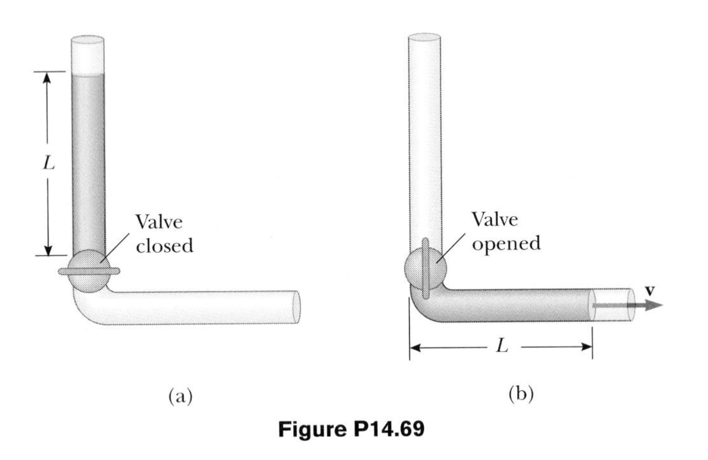

Figure P14.69

70. A cube of ice whose edges measure 20.0 mm is floating in a glass of ice-cold water with one of its faces parallel to the water's surface. (a) How far below the water surface is the bottom face of the block? (b) Ice-cold ethyl alcohol is gently poured onto the water surface to form a layer 5.00 mm thick above the water. The alcohol does not mix with the water. When the ice cube again attains hydrostatic equilibrium, what will be the distance from the top of the water to the bottom face of the block? (c) Additional cold ethyl alcohol is poured onto the water's surface until the top surface of the alcohol coincides with the top surface of the ice cube (in hydrostatic equilibrium). How thick is the required layer of ethyl alcohol?

71. A U-tube open at both ends is partially filled with water (Fig. P14.71a). Oil having a density of 750 kg/m^3 is then poured into the right arm and forms a column $L = 5.00$ cm high (Fig. P14.71b). (a) Determine the difference h in the heights of the two liquid surfaces. (b) The right arm is then shielded from any air motion while air is blown across the top of the left arm until the surfaces of the two liquids are at the same height (Fig. P14.71c). Determine the speed of the air being blown across the left arm. Take the density of air as 1.29 kg/m^3.

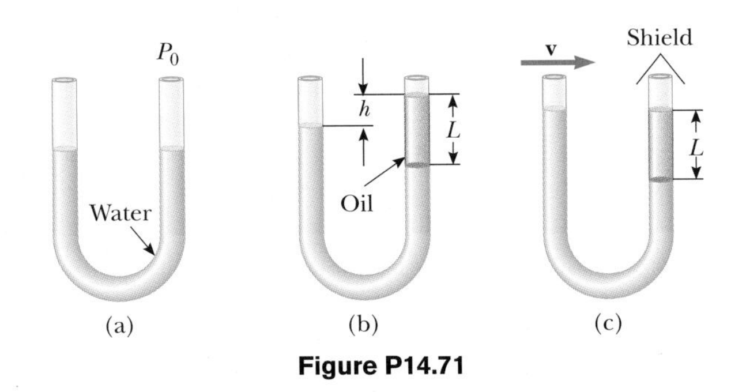

Figure P14.71

72. The water supply of a building is fed through a main pipe 6.00 cm in diameter. A 2.00-cm-diameter faucet tap, located 2.00 m above the main pipe, is observed to fill a 25.0-L container in 30.0 s. (a) What is the speed at which the water leaves the faucet? (b) What is the gauge pressure in the 6-cm main pipe? (Assume the faucet is the only "leak" in the building.)

73. The *spirit-in-glass thermometer*, invented in Florence, Italy, around 1654, consists of a tube of liquid (the spirit) containing a number of submerged glass spheres with slightly different masses (Fig. P14.73). At sufficiently low temperatures all the spheres float, but as the temperature rises, the spheres sink one after another. The device is a crude but interesting tool for measuring temperature. Suppose that the tube is filled with ethyl alcohol, whose density is 0.789 45 g/cm^3 at 20.0°C and decreases to 0.780 97 g/cm^3 at 30.0°C. (a) If one of the spheres has a radius of 1.000 cm and is in equilibrium halfway up the tube at 20.0°C, determine its mass. (b) When the temperature increases to 30.0°C, what mass must a second sphere of the same radius have in order to be in equilibrium at the halfway point? (c) At 30.0°C the first sphere has fallen to the bottom of the tube. What upward force does the bottom of the tube exert on this sphere?

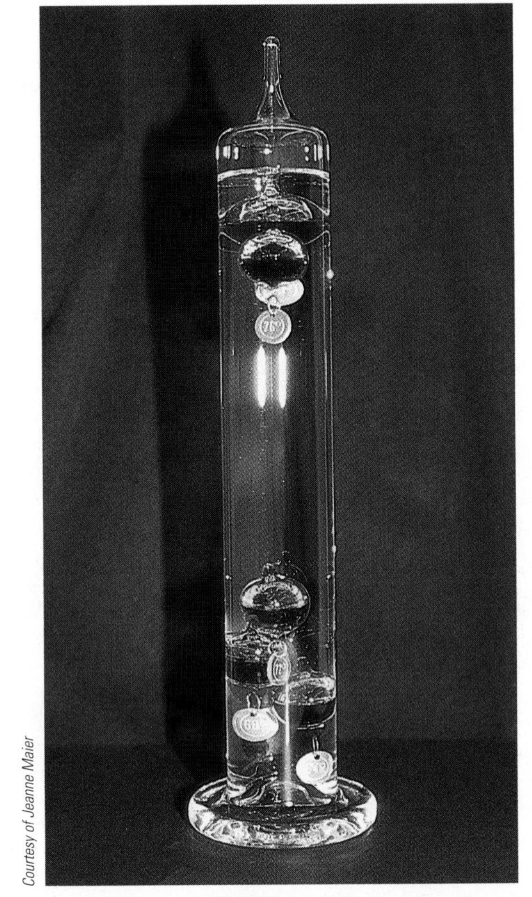

Figure P14.73

74. A woman is draining her fish tank by siphoning the water into an outdoor drain, as shown in Figure P14.74. The rectangular tank has footprint area A and depth h. The drain is located a distance d below the surface of the water in the tank, where $d \gg h$. The cross-sectional area of the siphon tube is A'. Model the water as flowing without friction. (a) Show that the time interval required to empty the tank is given by

$$\Delta t = \frac{Ah}{A'\sqrt{2gd}}$$

(b) Evaluate the time interval required to empty the tank if it is a cube 0.500 m on each edge, if $A' = 2.00$ cm^2, and $d = 10.0$ m.

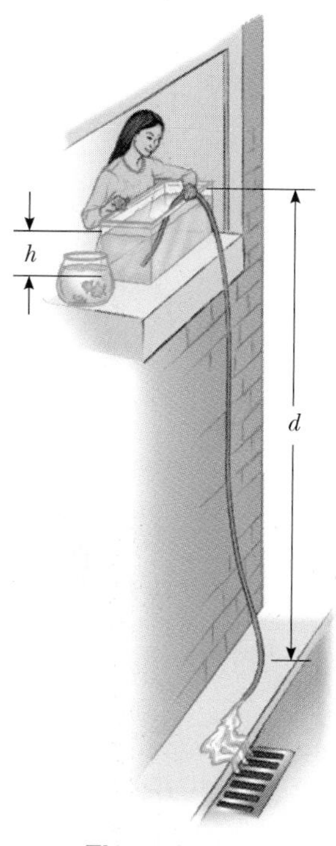

Figure P14.74

75. The hull of an experimental boat is to be lifted above the water by a hydrofoil mounted below its keel, as shown in Figure P14.75. The hydrofoil has a shape like that of an airplane wing. Its area projected onto a horizontal surface is A. When the boat is towed at sufficiently high speed, water of density ρ moves in streamline flow so that its average speed at the top of the hydrofoil is n times larger than its speed v_b below the hydrofoil. (a) Neglecting the buoyant force, show that the upward lift force exerted by the water on the hydrofoil has a magnitude given by

$$F \approx \frac{1}{2}(n^2 - 1)\rho v_b{}^2 A$$

(b) The boat has mass M. Show that the liftoff speed is given by

$$v \approx \sqrt{\frac{2Mg}{(n^2 - 1)A\rho}}$$

(c) Assume that an 800-kg boat is to lift off at 9.50 m/s. Evaluate the area A required for the hydrofoil if its design yields $n = 1.05$.

Figure P14.75

Answer to Quick Quizzes

14.1 (a). Because the basketball player's weight is distributed over the larger surface area of the shoe, the pressure (F/A) that he applies is relatively small. The woman's lesser weight is distributed over the very small cross-sectional area of the spiked heel, so the pressure is high.

14.2 (a). Because both fluids have the same depth, the one with the smaller density (alcohol) will exert the smaller pressure.

14.3 (c). All barometers will have the same pressure at the bottom of the column of fluid—atmospheric pressure. Thus, the barometer with the highest column will be the one with the fluid of lowest density.

14.4 (d). Because there is no atmosphere on the Moon, there is no atmospheric pressure to provide a force to push the water up the straw.

14.5 (b). For a totally submerged object, the buoyant force does not depend on the depth in an incompressible fluid.

14.6 (c). The ice cube displaces a volume of water that has a weight equal to that of the ice cube. When the ice cube melts, it becomes a parcel of water with the same weight and exactly the volume that was displaced by the ice cube before.

14.7 (b) or (c). In all three cases, the weight of the treasure chest causes a downward force on the raft that makes it sink into the water. In (b) and (c), however, the treasure chest also displaces water, which provides a buoyant force in the upward direction, reducing the effect of the chest's weight.

14.8 (b). The liquid moves at the highest speed in the straw with the smaller cross sectional area.

14.9 (a). The high-speed air between the balloons results in low pressure in this region. The higher pressure on the outer surfaces of the balloons pushes them toward each other.

Oscillations and Mechanical Waves

We begin this new part of the text by studying a special type of motion called *periodic* motion. This is a *repeating* motion of an object in which the object continues to return to a given position after a fixed time interval. Familiar objects that exhibit periodic motion include a pendulum and a beach ball floating on the waves at a beach. The back and forth movements of such an object are called *oscillations.* We will focus our attention on a special case of periodic motion called *simple harmonic motion.* We shall find that all periodic motions can be modeled as combinations of simple harmonic motions. Thus, simple harmonic motion forms a basic building block for more complicated periodic motion.

Simple harmonic motion also forms the basis for our understanding of *mechanical waves.* Sound waves, seismic waves, waves on stretched strings, and water waves are all produced by some source of oscillation. As a sound wave travels through the air, elements of the air oscillate back and forth; as a water wave travels across a pond, elements of the water oscillate up and down and backward and forward. In general, as waves travel through any medium, the elements of the medium move in repetitive cycles. Therefore, the motion of the elements of the medium bears a strong resemblance to the periodic motion of an oscillating pendulum or an object attached to a spring.

To explain many other phenomena in nature, we must understand the concepts of oscillations and waves. For instance, although skyscrapers and bridges appear to be rigid, they actually oscillate, a fact that the architects and engineers who design and build them must take into account. To understand how radio and television work, we must understand the origin and nature of electromagnetic waves and how they propagate through space. Finally, much of what scientists have learned about atomic structure has come from information carried by waves. Therefore, we must first study oscillations and waves if we are to understand the concepts and theories of atomic physics. ■

◀ *Drops of water fall from a leaf into a pond. The disturbance caused by the falling water causes the water surface to oscillate. These oscillations are associated with waves moving away from the point at which the water fell. In Part 2 of the text, we will explore the principles related to oscillations and waves. (Don Bonsey/Getty Images)*

Chapter 15

Oscillatory Motion

CHAPTER OUTLINE

▲ *In the Bay of Fundy, Nova Scotia, the tides undergo oscillations with very large amplitudes, such that boats often end up sitting on dry ground for part of the day. In this chapter, we will investigate the physics of oscillatory motion. (www.comstock.com)*

Periodic motion is motion of an object that regularly repeats—the object returns to a given position after a fixed time interval. With a little thought, we can identify several types of periodic motion in everyday life. Your car returns to the driveway each afternoon. You return to the dinner table each night to eat. A bumped chandelier swings back and forth, returning to the same position at a regular rate. The Earth returns to the same position in its orbit around the Sun each year, resulting in the variation among the four seasons. The Moon returns to the same relationship with the Earth and the Sun, resulting in a full Moon approximately once a month.

In addition to these everyday examples, numerous other systems exhibit periodic motion. For example, the molecules in a solid oscillate about their equilibrium positions; electromagnetic waves, such as light waves, radar, and radio waves, are characterized by oscillating electric and magnetic field vectors; and in alternating-current electrical circuits, voltage, current, and electric charge vary periodically with time.

A special kind of periodic motion occurs in mechanical systems when the force acting on an object is proportional to the position of the object relative to some equilibrium position. If this force is always directed toward the equilibrium position, the motion is called *simple harmonic motion,* which is the primary focus of this chapter.

Active Figure 15.1 A block attached to a spring moving on a frictionless surface. (a) When the block is displaced to the right of equilibrium ($x > 0$), the force exerted by the spring acts to the left. (b) When the block is at its equilibrium position ($x = 0$), the force exerted by the spring is zero. (c) When the block is displaced to the left of equilibrium ($x < 0$), the force exerted by the spring acts to the right.

At the Active Figures link, at http://www.pse6.com, you can choose the spring constant and the initial position and velocities of the block to see the resulting simple harmonic motion.

15.1 Motion of an Object Attached to a Spring

As a model for simple harmonic motion, consider a block of mass m attached to the end of a spring, with the block free to move on a horizontal, frictionless surface (Fig. 15.1). When the spring is neither stretched nor compressed, the block is at the position called the **equilibrium position** of the system, which we identify as $x = 0$. We know from experience that such a system oscillates back and forth if disturbed from its equilibrium position.

We can understand the motion in Figure 15.1 qualitatively by first recalling that when the block is displaced to a position x, the spring exerts on the block a force that is proportional to the position and given by **Hooke's law** (see Section 7.4):

Hooke's law

$$F_s = -kx \tag{15.1}$$

We call this a **restoring force** because it is always directed toward the equilibrium position and therefore *opposite* the displacement from equilibrium. That is, when the block is displaced to the right of $x = 0$ in Figure 15.1, then the position is positive and the restoring force is directed to the left. When the block is displaced to the left of $x = 0$, then the position is negative and the restoring force is directed to the right.

Applying Newton's second law $\Sigma F_x = ma_x$ to the motion of the block, with Equation 15.1 providing the net force in the x direction, we obtain

$$-kx = ma_x$$

$$a_x = -\frac{k}{m}x \tag{15.2}$$

PITFALL PREVENTION

15.1 The Orientation of the Spring

Figure 15.1 shows a *horizontal* spring, with an attached block sliding on a frictionless surface. Another possibility is a block hanging from a *vertical* spring. All of the results that we discuss for the horizontal spring will be the same for the vertical spring, except that when the block is placed on the vertical spring, its weight will cause the spring to extend. If the resting position of the block is defined as $x = 0$, the results of this chapter will apply to this vertical system also.

That is, the acceleration is proportional to the position of the block, and its direction is opposite the direction of the displacement from equilibrium. Systems that behave in this way are said to exhibit **simple harmonic motion. An object moves with simple harmonic motion whenever its acceleration is proportional to its position and is oppositely directed to the displacement from equilibrium.**

If the block in Figure 15.1 is displaced to a position $x = A$ and released from rest, its *initial* acceleration is $-kA/m$. When the block passes through the equilibrium position $x = 0$, its acceleration is zero. At this instant, its speed is a maximum because the acceleration changes sign. The block then continues to travel to the left of equilibrium with a positive acceleration and finally reaches $x = -A$, at which time its acceleration is $+kA/m$ and its speed is again zero, as discussed in Sections 7.4 and 8.6. The block completes a full cycle of its motion by returning to the original position, again passing through $x = 0$ with maximum speed. Thus, we see that the block oscillates between the turning points $x = \pm A$. In the absence of friction, because the force exerted by the spring is conservative, this idealized motion will continue forever. Real systems are generally subject to friction, so they do not oscillate forever. We explore the details of the situation with friction in Section 15.6.

As Pitfall Prevention 15.1 points out, the principles that we develop in this chapter are also valid for an object hanging from a vertical spring, as long as we recognize that the weight of the object will stretch the spring to a new equilibrium position $x = 0$. To prove this statement, let x_s represent the total extension of the spring from its equilibrium position *without* the hanging object. Then, $x_s = -(mg/k) + x$, where $-(mg/k)$ is the extension of the spring due to the weight of the hanging object and x is the instantaneous extension of the spring due to the simple harmonic motion. The magnitude of the net force on the object is then $F_s - F_g = -k(-(mg/k) + x) - mg = -kx$. The net force on the object is the same as that on a block connected to a horizontal spring as in Equation 15.1, so the same simple harmonic motion results.

Quick Quiz 15.1 A block on the end of a spring is pulled to position $x = A$ and released. In one full cycle of its motion, through what total distance does it travel? (a) $A/2$ (b) A (c) $2A$ (d) $4A$

15.2 Mathematical Representation of Simple Harmonic Motion

Let us now develop a mathematical representation of the motion we described in the preceding section. We model the block as a particle subject to the force in Equation 15.1. We will generally choose x as the axis along which the oscillation occurs; hence, we will drop the subscript-x notation in this discussion. Recall that, by definition, $a = dv/dt = d^2x/dt^2$, and so we can express Equation 15.2 as

$$\frac{d^2x}{dt^2} = -\frac{k}{m}x \tag{15.3}$$

If we denote the ratio k/m with the symbol ω^2 (we choose ω^2 rather than ω in order to make the solution that we develop below simpler in form), then

$$\omega^2 = \frac{k}{m} \tag{15.4}$$

and Equation 15.3 can be written in the form

$$\frac{d^2x}{dt^2} = -\omega^2 x \tag{15.5}$$

PITFALL PREVENTION

15.2 A Nonconstant Acceleration

Notice that the acceleration of the particle in simple harmonic motion is not constant. Equation 15.3 shows that it varies with position x. Thus, we *cannot* apply the kinematic equations of Chapter 2 in this situation.

What we now require is a mathematical solution to Equation 15.5—that is, a function $x(t)$ that satisfies this second-order differential equation. This is a mathematical representation of the position of the particle as a function of time. We seek a function $x(t)$ whose second derivative is the same as the original function with a negative sign and multiplied by ω^2. The trigonometric functions sine and cosine exhibit this behavior, so we can build a solution around one or both of these. The following cosine function is a solution to the differential equation:

$$x(t) = A\cos(\omega t + \phi) \tag{15.6}$$

Position versus time for an object in simple harmonic motion

where A, ω, and ϕ are constants. To see explicitly that this equation satisfies Equation 15.5, note that

$$\frac{dx}{dt} = A\frac{d}{dt}\cos(\omega t + \phi) = -\omega A\sin(\omega t + \phi) \tag{15.7}$$

$$\frac{d^2x}{dt^2} = -\omega A\frac{d}{dt}\sin(\omega t + \phi) = -\omega^2 A\cos(\omega t + \phi) \tag{15.8}$$

Comparing Equations 15.6 and 15.8, we see that $d^2x/dt^2 = -\omega^2 x$ and Equation 15.5 is satisfied.

The parameters A, ω, and ϕ are constants of the motion. In order to give physical significance to these constants, it is convenient to form a graphical representation of the motion by plotting x as a function of t, as in Figure 15.2a. First, note that A, called the **amplitude** of the motion, is simply **the maximum value of the position of the particle in either the positive or negative x direction.** The constant ω is called the **angular frequency,** and has units of rad/s.[1] It is a measure of how rapidly the oscillations are occurring—the more oscillations per unit time, the higher is the value of ω. From Equation 15.4, the angular frequency is

$$\omega = \sqrt{\frac{k}{m}} \tag{15.9}$$

The constant angle ϕ is called the **phase constant** (or initial phase angle) and, along with the amplitude A, is determined uniquely by the position and velocity of the particle at $t = 0$. If the particle is at its maximum position $x = A$ at $t = 0$, the phase constant is $\phi = 0$ and the graphical representation of the motion is shown in Figure 15.2b. The quantity $(\omega t + \phi)$ is called the **phase** of the motion. Note that the function $x(t)$ is periodic and its value is the same each time ωt increases by 2π radians.

Equations 15.1, 15.5, and 15.6 form the basis of the mathematical representation of simple harmonic motion. If we are analyzing a situation and find that the force on a particle is of the mathematical form of Equation 15.1, we know that the motion will be that of a simple harmonic oscillator and that the position of the particle is described by Equation 15.6. If we analyze a system and find that it is described by a differential equation of the form of Equation 15.5, the motion will be that of a simple harmonic oscillator. If we analyze a situation and find that the position of a particle is described by Equation 15.6, we know the particle is undergoing simple harmonic motion.

PITFALL PREVENTION

15.3 Where's the Triangle?

Equation 15.6 includes a trigonometric function, a *mathematical function* that can be used whether it refers to a triangle or not. In this case, the cosine function happens to have the correct behavior for representing the position of a particle in simple harmonic motion.

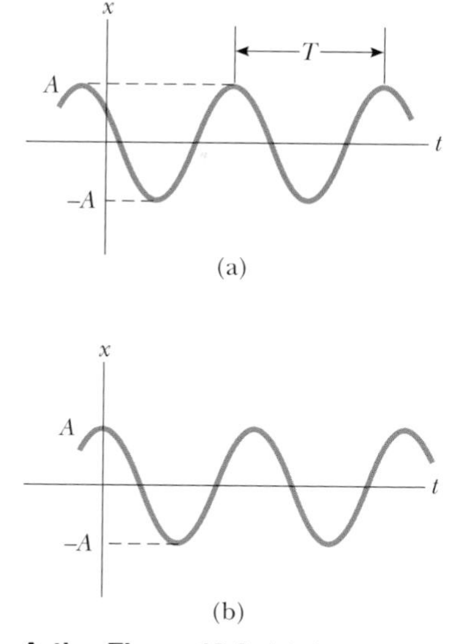

Active Figure 15.2 (a) An x-vs.-t graph for an object undergoing simple harmonic motion. The amplitude of the motion is A, the period (page 456) is T, and the phase constant is ϕ. (b) The x-vs.-t graph in the special case in which $x = A$ at $t = 0$ and hence $\phi = 0$.

At the Active Figures link at http://www.pse6.com, you can adjust the graphical representation and see the resulting simple harmonic motion of the block in Figure 15.1.

[1] We have seen many examples in earlier chapters in which we evaluate a trigonometric function of an angle. The argument of a trigonometric function, such as sine or cosine, *must* be a pure number. The radian is a pure number because it is a ratio of lengths. Angles in degrees are pure numbers simply because the degree is a completely artificial "unit"—it is not related to measurements of lengths. The notion of requiring a pure number for a trigonometric function is important in Equation 15.6, where the angle is expressed in terms of other measurements. Thus, ω *must* be expressed in rad/s (and not, for example, in revolutions per second) if t is expressed in seconds. Furthermore, other types of functions such as logarithms and exponential functions require arguments that are pure numbers.

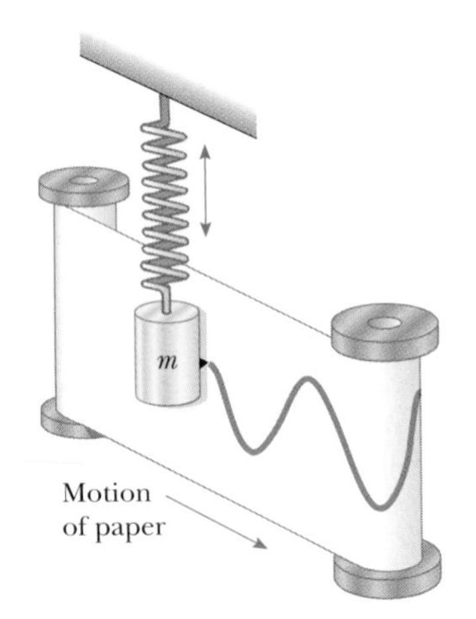

Figure 15.3 An experimental apparatus for demonstrating simple harmonic motion. A pen attached to the oscillating object traces out a sinusoidal pattern on the moving chart paper.

An experimental arrangement that exhibits simple harmonic motion is illustrated in Figure 15.3. An object oscillating vertically on a spring has a pen attached to it. While the object is oscillating, a sheet of paper is moved perpendicular to the direction of motion of the spring, and the pen traces out the cosine curve in Equation 15.6.

Quick Quiz 15.2 Consider a graphical representation (Fig. 15.4) of simple harmonic motion, as described mathematically in Equation 15.6. When the object is at point Ⓐ on the graph, its (a) position and velocity are both positive (b) position and velocity are both negative (c) position is positive and its velocity is zero (d) position is negative and its velocity is zero (e) position is positive and its velocity is negative (f) position is negative and its velocity is positive.

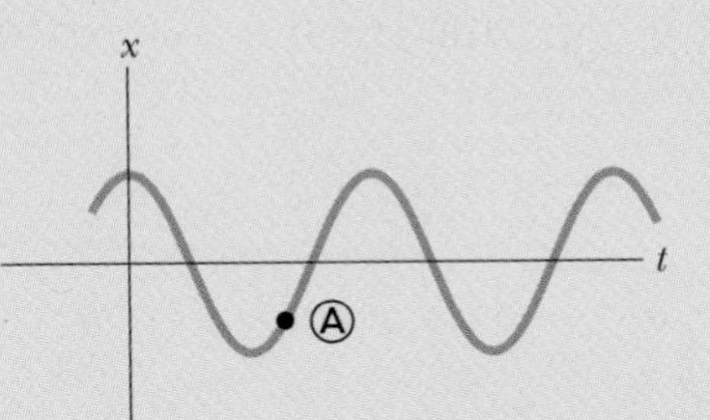

Figure 15.4 (Quick Quiz 15.2) An x-t graph for an object undergoing simple harmonic motion. At a particular time, the object's position is indicated by Ⓐ in the graph.

Quick Quiz 15.3 Figure 15.5 shows two curves representing objects undergoing simple harmonic motion. The correct description of these two motions is that the simple harmonic motion of object B is (a) of larger angular frequency and larger amplitude than that of object A (b) of larger angular frequency and smaller amplitude than that of object A (c) of smaller angular frequency and larger amplitude than that of object A (d) of smaller angular frequency and smaller amplitude than that of object A.

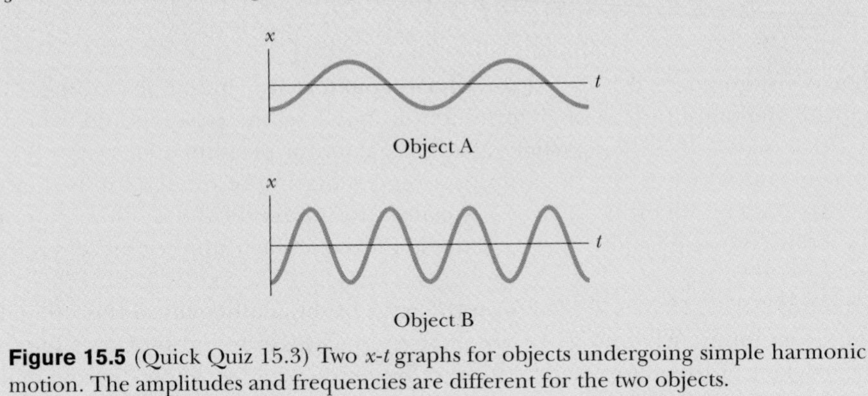

Figure 15.5 (Quick Quiz 15.3) Two x-t graphs for objects undergoing simple harmonic motion. The amplitudes and frequencies are different for the two objects.

Let us investigate further the mathematical description of simple harmonic motion. The **period** T of the motion is the time interval required for the particle to go through one full cycle of its motion (Fig. 15.2a). That is, the values of x and v for the particle at time t equal the values of x and v at time $t + T$. We can relate the period to the angular frequency by using the fact that the phase increases by 2π radians in a time interval of T:

$$[\omega(t + T) + \phi] - (\omega t + \phi) = 2\pi$$

Simplifying this expression, we see that $\omega T = 2\pi$, or

$$T = \frac{2\pi}{\omega} \tag{15.10}$$

The inverse of the period is called the **frequency** f of the motion. Whereas the period is the time interval per oscillation, the frequency represents the **number of oscillations that the particle undergoes per unit time interval:**

$$f = \frac{1}{T} = \frac{\omega}{2\pi} \tag{15.11}$$

The units of f are cycles per second, or **hertz** (Hz). Rearranging Equation 15.11 gives

$$\omega = 2\pi f = \frac{2\pi}{T} \tag{15.12}$$

We can use Equations 15.9, 15.10, and 15.11 to express the period and frequency of the motion for the particle–spring system in terms of the characteristics m and k of the system as

Period

$$T = \frac{2\pi}{\omega} = 2\pi\sqrt{\frac{m}{k}} \tag{15.13}$$

Frequency

$$f = \frac{1}{T} = \frac{1}{2\pi}\sqrt{\frac{k}{m}} \tag{15.14}$$

That is, the period and frequency depend *only* on the mass of the particle and the force constant of the spring, and *not* on the parameters of the motion, such as A or ϕ. As we might expect, the frequency is larger for a stiffer spring (larger value of k) and decreases with increasing mass of the particle.

We can obtain the velocity and acceleration[2] of a particle undergoing simple harmonic motion from Equations 15.7 and 15.8:

Velocity of an object in simple harmonic motion

$$v = \frac{dx}{dt} = -\omega A \sin(\omega t + \phi) \tag{15.15}$$

Acceleration of an object in simple harmonic motion

$$a = \frac{d^2x}{dt^2} = -\omega^2 A \cos(\omega t + \phi) \tag{15.16}$$

From Equation 15.15 we see that, because the sine and cosine functions oscillate between ± 1, the extreme values of the velocity v are $\pm\,\omega A$. Likewise, Equation 15.16 tells us that the extreme values of the acceleration a are $\pm\,\omega^2 A$. Therefore, the *maximum* values of the magnitudes of the velocity and acceleration are

Maximum magnitudes of speed and acceleration in simple harmonic motion

$$v_{\max} = \omega A = \sqrt{\frac{k}{m}}\,A \tag{15.17}$$

$$a_{\max} = \omega^2 A = \frac{k}{m}\,A \tag{15.18}$$

Figure 15.6a plots position versus time for an arbitrary value of the phase constant. The associated velocity–time and acceleration–time curves are illustrated in Figures 15.6b and 15.6c. They show that the phase of the velocity differs from the phase of the position by $\pi/2$ rad, or 90°. That is, when x is a maximum or a minimum, the velocity is zero. Likewise, when x is zero, the speed is a maximum. Furthermore, note that the

15.4 Two Kinds of Frequency

We identify two kinds of frequency for a simple harmonic oscillator—f, called simply the *frequency*, is measured in hertz, and ω, the *angular frequency*, is measured in radians per second. Be sure that you are clear about which frequency is being discussed or requested in a given problem. Equations 15.11 and 15.12 show the relationship between the two frequencies.

[2] Because the motion of a simple harmonic oscillator takes place in one dimension, we will denote velocity as v and acceleration as a, with the direction indicated by a positive or negative sign, as in Chapter 2.

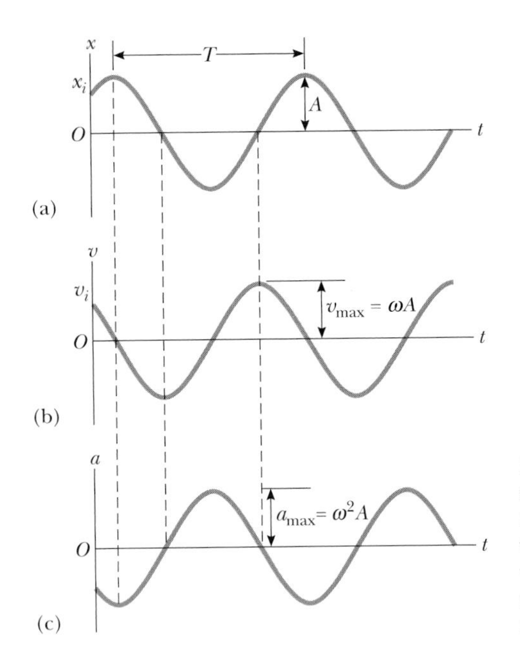

Figure 15.6 Graphical representation of simple harmonic motion. (a) Position versus time. (b) Velocity versus time. (c) Acceleration versus time. Note that at any specified time the velocity is 90° out of phase with the position and the acceleration is 180° out of phase with the position.

phase of the acceleration differs from the phase of the position by π radians, or 180°. For example, when x is a maximum, a has a maximum magnitude in the opposite direction.

Quick Quiz 15.4 Consider a graphical representation (Fig. 15.4) of simple harmonic motion, as described mathematically in Equation 15.6. When the object is at position Ⓐ on the graph, its (a) velocity and acceleration are both positive (b) velocity and acceleration are both negative (c) velocity is positive and its acceleration is zero (d) velocity is negative and its acceleration is zero (e) velocity is positive and its acceleration is negative (f) velocity is negative and its acceleration is positive.

Quick Quiz 15.5 An object of mass m is hung from a spring and set into oscillation. The period of the oscillation is measured and recorded as T. The object of mass m is removed and replaced with an object of mass $2m$. When this object is set into oscillation, the period of the motion is (a) $2T$ (b) $\sqrt{2}T$ (c) T (d) $T/\sqrt{2}$ (e) $T/2$.

Equation 15.6 describes simple harmonic motion of a particle in general. Let us now see how to evaluate the constants of the motion. The angular frequency ω is evaluated using Equation 15.9. The constants A and ϕ are evaluated from the initial conditions, that is, the state of the oscillator at $t = 0$.

Suppose we initiate the motion by pulling the particle from equilibrium by a distance A and releasing it from rest at $t = 0$, as in Figure 15.7. We must then require that

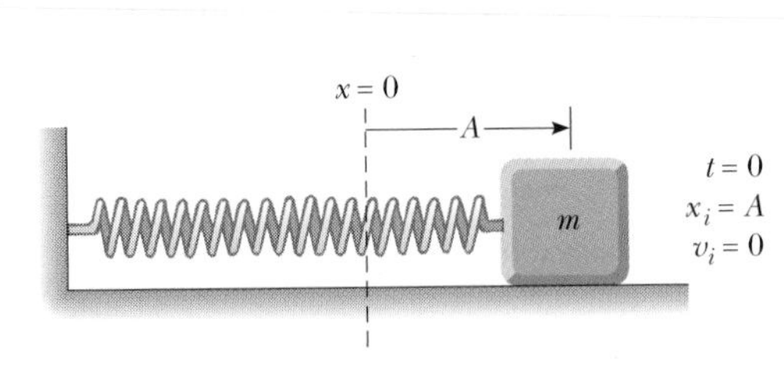

Active Figure 15.7 A block–spring system that begins its motion from rest with the block at $x = A$ at $t = 0$. In this case, $\phi = 0$ and thus $x = A \cos \omega t$.

At the Active Figures link at http://www.pse6.com, you can compare the oscillations of two blocks starting from different initial positions to see that the frequency is independent of the amplitude.

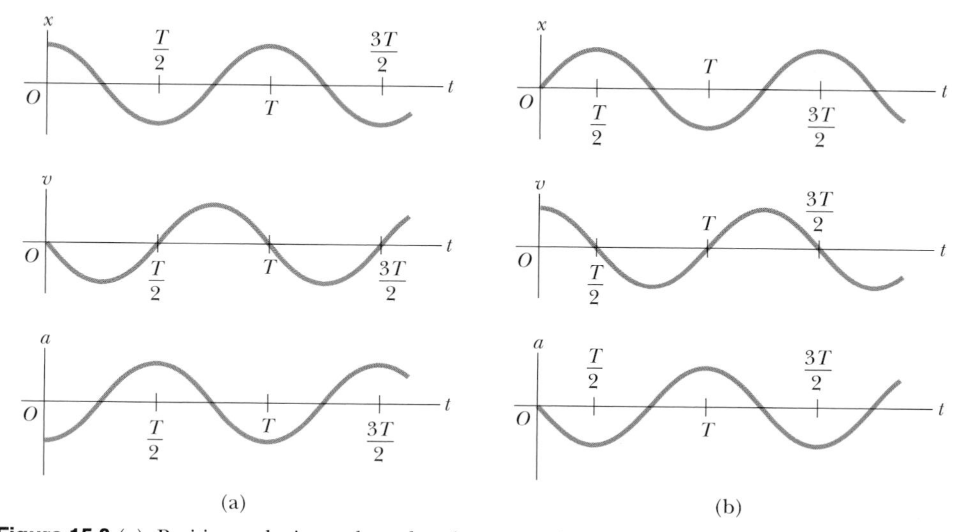

Figure 15.8 (a) Position, velocity, and acceleration versus time for a block undergoing simple harmonic motion under the initial conditions that at $t = 0$, $x(0) = A$ and $v(0) = 0$. (b) Position, velocity, and acceleration versus time for a block undergoing simple harmonic motion under the initial conditions that at $t = 0$, $x(0) = 0$ and $v(0) = v_i$.

our solutions for $x(t)$ and $v(t)$ (Eqs. 15.6 and 15.15) obey the initial conditions that $x(0) = A$ and $v(0) = 0$:

$$x(0) = A\cos\phi = A$$

$$v(0) = -\omega A\sin\phi = 0$$

These conditions are met if we choose $\phi = 0$, giving $x = A\cos\omega t$ as our solution. To check this solution, note that it satisfies the condition that $x(0) = A$, because $\cos 0 = 1$.

The position, velocity, and acceleration versus time are plotted in Figure 15.8a for this special case. The acceleration reaches extreme values of $\mp\omega^2 A$ when the position has extreme values of $\pm A$. Furthermore, the velocity has extreme values of $\pm\omega A$, which both occur at $x = 0$. Hence, the quantitative solution agrees with our qualitative description of this system.

Let us consider another possibility. Suppose that the system is oscillating and we define $t = 0$ as the instant that the particle passes through the unstretched position of the spring while moving to the right (Fig. 15.9). In this case we must require that our solutions for $x(t)$ and $v(t)$ obey the initial conditions that $x(0) = 0$ and $v(0) = v_i$:

$$x(0) = A\cos\phi = 0$$

$$v(0) = -\omega A\sin\phi = v_i$$

The first of these conditions tells us that $\phi = \pm\pi/2$. With these choices for ϕ, the second condition tells us that $A = \mp v_i/\omega$. Because the initial velocity is positive and the amplitude must be positive, we must have $\phi = -\pi/2$. Hence, the solution is given by

$$x = \frac{v_i}{\omega}\cos\left(\omega t - \frac{\pi}{2}\right)$$

The graphs of position, velocity, and acceleration versus time for this choice of $t = 0$ are shown in Figure 15.8b. Note that these curves are the same as those in Figure 15.8a, but shifted to the right by one fourth of a cycle. This is described mathematically by the phase constant $\phi = -\pi/2$, which is one fourth of a full cycle of 2π.

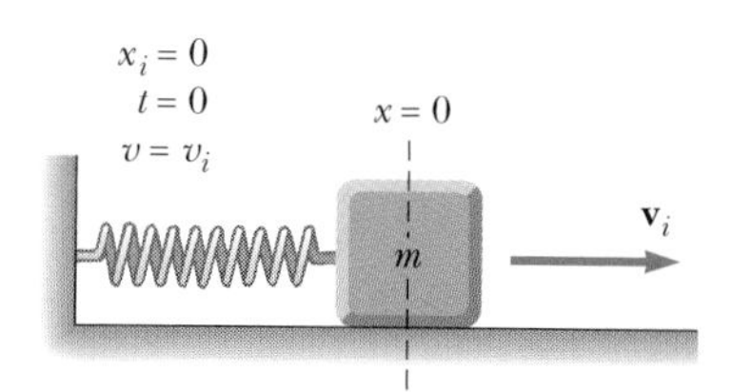

Active Figure 15.9 The block–spring system is undergoing oscillation, and $t = 0$ is defined at an instant when the block passes through the equilibrium position $x = 0$ and is moving to the right with speed v_i.

At the Active Figures link at http://www.pse6.com, you can compare the oscillations of two blocks with different velocities at t = 0 to see that the frequency is independent of the amplitude.

Example 15.1 An Oscillating Object

An object oscillates with simple harmonic motion along the x axis. Its position varies with time according to the equation

$$x = (4.00 \text{ m}) \cos\left(\pi t + \frac{\pi}{4}\right)$$

where t is in seconds and the angles in the parentheses are in radians.

(A) Determine the amplitude, frequency, and period of the motion.

Solution By comparing this equation with Equation 15.6, $x = A\cos(\omega t + \phi)$, we see that $A =$ 4.00 m and $\omega = \pi$ rad/s. Therefore, $f = \omega/2\pi = \pi/2\pi =$ 0.500 Hz and $T = 1/f =$ 2.00 s.

(B) Calculate the velocity and acceleration of the object at any time t.

Solution Differentiating x to find v, and v to find a, we obtain

$$v = \frac{dx}{dt} = -(4.00 \text{ m/s}) \sin\left(\pi t + \frac{\pi}{4}\right)\frac{d}{dt}(\pi t)$$

$$= -(4.00\pi \text{ m/s}) \sin\left(\pi t + \frac{\pi}{4}\right)$$

$$a = \frac{dv}{dt} = -(4.00\pi \text{ m/s}) \cos\left(\pi t + \frac{\pi}{4}\right)\frac{d}{dt}(\pi t)$$

$$= -(4.00\pi^2 \text{ m/s}^2) \cos\left(\pi t + \frac{\pi}{4}\right)$$

(C) Using the results of part (B), determine the position, velocity, and acceleration of the object at $t = 1.00$ s.

Solution Noting that the angles in the trigonometric functions are in radians, we obtain, at $t = 1.00$ s,

$$x = (4.00 \text{ m}) \cos\left(\pi + \frac{\pi}{4}\right) = (4.00 \text{ m}) \cos\left(\frac{5\pi}{4}\right)$$

$$= (4.00 \text{ m})(-0.707) = -2.83 \text{ m}$$

$$v = -(4.00\pi \text{ m/s}) \sin\left(\frac{5\pi}{4}\right)$$

$$= -(4.00\pi \text{ m/s})(-0.707) = 8.89 \text{ m/s}$$

$$a = -(4.00\pi^2 \text{ m/s}^2) \cos\left(\frac{5\pi}{4}\right)$$

$$= -(4.00\pi^2 \text{ m/s}^2)(-0.707) = 27.9 \text{ m/s}^2$$

(D) Determine the maximum speed and maximum acceleration of the object.

Solution In the general expressions for v and a found in part (B), we use the fact that the maximum values of the sine and cosine functions are unity. Therefore, v varies between $\pm 4.00\pi$ m/s, and a varies between $\pm 4.00\pi^2$ m/s^2. Thus,

$$v_{\max} = 4.00\pi \text{ m/s} = 12.6 \text{ m/s}$$

$$a_{\max} = 4.00\pi^2 \text{ m/s}^2 = 39.5 \text{ m/s}^2$$

We obtain the same results using the relations $v_{\max} = \omega A$ and $a_{\max} = \omega^2 A$, where $A = 4.00$ m and $\omega = \pi$ rad/s.

(E) Find the displacement of the object between $t = 0$ and $t = 1.00$ s.

Solution The position at $t = 0$ is

$$x_i = (4.00 \text{ m}) \cos\left(0 + \frac{\pi}{4}\right) = (4.00 \text{ m})(0.707) = 2.83 \text{ m}$$

In part (C), we found that the position at $t = 1.00$ s is -2.83 m; therefore, the displacement between $t = 0$ and $t = 1.00$ s is

$$\Delta x = x_f - x_i = -2.83 \text{ m} - 2.83 \text{ m} = -5.66 \text{ m}$$

Because the object's velocity changes sign during the first second, the magnitude of Δx is not the same as the distance traveled in the first second. (By the time the first second is over, the object has been through the point $x = -2.83$ m once, traveled to $x = -4.00$ m, and come back to $x = -2.83$ m.)

Example 15.2 Watch Out for Potholes!

A car with a mass of 1 300 kg is constructed so that its frame is supported by four springs. Each spring has a force constant of 20 000 N/m. If two people riding in the car have a combined mass of 160 kg, find the frequency of vibration of the car after it is driven over a pothole in the road.

Solution To conceptualize this problem, think about your experiences with automobiles. When you sit in a car, it moves downward a small distance because your weight is compressing the springs further. If you push down on the front bumper and release, the front of the car oscillates a couple of times. We can model the car as being supported by a single spring and categorize this as an oscillation problem based on our simple spring model. To analyze the problem, we first need to consider the effective spring constant of the four springs combined. For a given extension x of the springs, the combined force on the car is the sum of the forces from the individual springs:

$$F_{\text{total}} = \sum(-kx) = -\left(\sum k\right)x$$

where x has been factored from the sum because it is the

same for all four springs. We see that the effective spring constant for the combined springs is the sum of the individual spring constants:

$$k_{eff} = \sum k = 4 \times 20\ 000 \text{ N/m} = 80\ 000 \text{ N/m}$$

Hence, the frequency of vibration is, from Equation 15.14,

$$f = \frac{1}{2\pi}\sqrt{\frac{k_{eff}}{m}} = \frac{1}{2\pi}\sqrt{\frac{80\ 000 \text{ N/m}}{1\ 460 \text{ kg}}} = 1.18 \text{ Hz}$$

To finalize the problem, note that the mass we used here is that of the car plus the people, because this is the total mass that is oscillating. Also note that we have explored only up-and-down motion of the car. If an oscillation is established in which the car rocks back and forth such that the front end goes up when the back end goes down, the frequency will be different.

What If? Suppose the two people exit the car on the side of the road. One of them pushes downward on the car and releases it so that it oscillates vertically. Is the frequency of the oscillation the same as the value we just calculated?

Answer The suspension system of the car is the same, but the mass that is oscillating is smaller—it no longer includes the mass of the two people. Thus, the frequency should be higher. Let us calculate the new frequency:

$$f = \frac{1}{2\pi}\sqrt{\frac{k_{eff}}{m}} = \frac{1}{2\pi}\sqrt{\frac{80\ 000 \text{ N/m}}{1\ 300 \text{ kg}}} = 1.25 \text{ Hz}$$

As we predicted conceptually, the frequency is a bit higher.

Example 15.3 A Block–Spring System

A 200-g block connected to a light spring for which the force constant is 5.00 N/m is free to oscillate on a horizontal, frictionless surface. The block is displaced 5.00 cm from equilibrium and released from rest, as in Figure 15.7.

(A) Find the period of its motion.

Solution From Equations 15.9 and 15.10, we know that the angular frequency of a block–spring system is

$$\omega = \sqrt{\frac{k}{m}} = \sqrt{\frac{5.00 \text{ N/m}}{200 \times 10^{-3} \text{ kg}}} = 5.00 \text{ rad/s}$$

and the period is

$$T = \frac{2\pi}{\omega} = \frac{2\pi}{5.00 \text{ rad/s}} = 1.26 \text{ s}$$

(B) Determine the maximum speed of the block.

Solution We use Equation 15.17:

$$v_{max} = \omega A = (5.00 \text{ rad/s})(5.00 \times 10^{-2} \text{ m}) = 0.250 \text{ m/s}$$

(C) What is the maximum acceleration of the block?

Solution We use Equation 15.18:

$$a_{max} = \omega^2 A = (5.00 \text{ rad/s})^2(5.00 \times 10^{-2} \text{ m}) = 1.25 \text{ m/s}^2$$

(D) Express the position, speed, and acceleration as functions of time.

Solution We find the phase constant from the initial condition that $x = A$ at $t = 0$:

$$x(0) = A\cos\phi = A$$

which tells us that $\phi = 0$. Thus, our solution is $x = A\cos\omega t$. Using this expression and the results from (A), (B), and (C), we find that

$$x = A\cos\omega t = (0.050\ 0 \text{ m})\cos 5.00t$$

$$v = \omega A \sin\omega t = -(0.250 \text{ m/s})\sin 5.00t$$

$$a = -\omega^2 A\cos\omega t = -(1.25 \text{ m/s}^2)\cos 5.00t$$

What If? What if the block is released from the same initial position, $x_i = 5.00$ cm, but with an initial velocity of $v_i = -0.100$ m/s? Which parts of the solution change and what are the new answers for those that do change?

Answers Part (A) does not change—the period is independent of how the oscillator is set into motion. Parts (B), (C), and (D) will change. We begin by considering position and velocity expressions for the initial conditions:

$$(1) \qquad x(0) = A\cos\phi = x_i$$

$$(2) \qquad v(0) = -\omega A\sin\phi = v_i$$

Dividing Equation (2) by Equation (1) gives us the phase constant:

$$\frac{-\omega A\sin\phi}{A\cos\phi} = \frac{v_i}{x_i}$$

$$\tan\phi = -\frac{v_i}{\omega x_i} = -\frac{-0.100 \text{ m}}{(5.00 \text{ rad/s})(0.050\ 0 \text{ m})} = 0.400$$

$$\phi = 0.12\pi$$

Now, Equation (1) allows us to find A:

$$A = \frac{x_i}{\cos\phi} = \frac{0.050\ 0 \text{ m}}{\cos(0.12\pi)} = 0.053\ 9 \text{ m}$$

The new maximum speed is

$$v_{max} = \omega A = (5.00 \text{ rad/s})(5.39 \times 10^{-2} \text{ m}) = 0.269 \text{ m/s}$$

The new magnitude of the maximum acceleration is

$$a_{max} = \omega^2 A = (5.00 \text{ rad/s})^2(5.39 \times 10^{-2} \text{ m}) = 1.35 \text{ m/s}^2$$

The new expressions for position, velocity, and acceleration are

$$x = (0.053\ 9 \text{ m})\cos(5.00t + 0.12\pi)$$

$$v = -(0.269 \text{ m/s})\sin(5.00t + 0.12\pi)$$

$$a = -(1.35 \text{ m/s}^2)\cos(5.00t + 0.12\pi)$$

As we saw in Chapters 7 and 8, many problems are easier to solve with an energy approach rather than one based on variables of motion. This particular **What If?** is easier to solve from an energy approach. Therefore, in the next section we shall investigate the energy of the simple harmonic oscillator.

15.3 Energy of the Simple Harmonic Oscillator

Let us examine the mechanical energy of the block–spring system illustrated in Figure 15.1. Because the surface is frictionless, we expect the total mechanical energy of the system to be constant, as was shown in Chapter 8. We assume a massless spring, so the kinetic energy of the system corresponds only to that of the block. We can use Equation 15.15 to express the kinetic energy of the block as

Kinetic energy of a simple harmonic oscillator

$$K = \tfrac{1}{2}mv^2 = \tfrac{1}{2}m\omega^2 A^2 \sin^2(\omega t + \phi) \tag{15.19}$$

The elastic potential energy stored in the spring for any elongation x is given by $\frac{1}{2}kx^2$ (see Eq. 8.11). Using Equation 15.6, we obtain

Potential energy of a simple harmonic oscillator

$$U = \tfrac{1}{2}kx^2 = \tfrac{1}{2}kA^2 \cos^2(\omega t + \phi) \tag{15.20}$$

We see that K and U are *always* positive quantities. Because $\omega^2 = k/m$, we can express the total mechanical energy of the simple harmonic oscillator as

$$E = K + U = \tfrac{1}{2}kA^2[\sin^2(\omega t + \phi) + \cos^2(\omega t + \phi)]$$

From the identity $\sin^2\theta + \cos^2\theta = 1$, we see that the quantity in square brackets is unity. Therefore, this equation reduces to

Total energy of a simple harmonic oscillator

$$E = \tfrac{1}{2}kA^2 \tag{15.21}$$

That is, **the total mechanical energy of a simple harmonic oscillator is a constant of the motion and is proportional to the square of the amplitude.** Note that U is small when K is large, and vice versa, because the sum must be constant. In fact, the total mechanical energy is equal to the maximum potential energy stored in the spring when $x = \pm A$ because $v = 0$ at these points and thus there is no kinetic energy. At the equilibrium position, where $U = 0$ because $x = 0$, the total energy, all in the form of kinetic energy, is again $\frac{1}{2}kA^2$. That is,

$$E = \tfrac{1}{2}mv_{\text{max}}^2 = \tfrac{1}{2}m\omega^2 A^2 = \tfrac{1}{2}m\frac{k}{m}A^2 = \tfrac{1}{2}kA^2 \qquad (\text{at } x = 0)$$

Plots of the kinetic and potential energies versus time appear in Figure 15.10a, where we have taken $\phi = 0$. As already mentioned, both K and U are always positive, and at all times their sum is a constant equal to $\frac{1}{2}kA^2$, the total energy of the system. The variations of K and U with the position x of the block are plotted in Figure 15.10b.

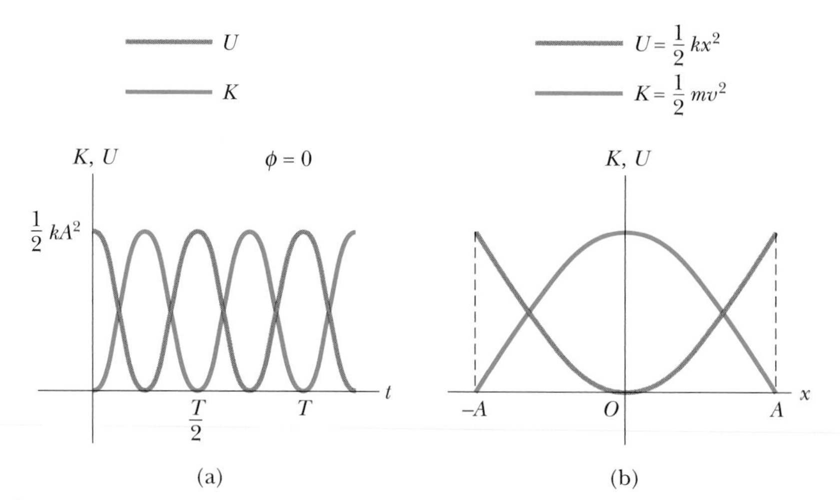

Active Figure 15.10 (a) Kinetic energy and potential energy versus time for a simple harmonic oscillator with $\phi = 0$. (b) Kinetic energy and potential energy versus position for a simple harmonic oscillator. In either plot, note that $K + U =$ constant.

At the Active Figures link at* http://www.pse6.com, *you can compare the physical oscillation of a block with energy graphs in this figure as well as with energy bar graphs.

Energy is continuously being transformed between potential energy stored in the spring and kinetic energy of the block.

Figure 15.11 illustrates the position, velocity, acceleration, kinetic energy, and potential energy of the block–spring system for one full period of the motion. Most of the ideas discussed so far are incorporated in this important figure. Study it carefully.

Finally, we can use the principle of conservation of energy to obtain the velocity for an arbitrary position by expressing the total energy at some arbitrary position x as

$$E = K + U = \frac{1}{2}mv^2 + \frac{1}{2}kx^2 = \frac{1}{2}kA^2$$

$$v = \pm\sqrt{\frac{k}{m}(A^2 - x^2)} = \pm\omega\sqrt{A^2 - x^2} \qquad (15.22)$$

Velocity as a function of position for a simple harmonic oscillator

When we check Equation 15.22 to see whether it agrees with known cases, we find that it verifies the fact that the speed is a maximum at $x = 0$ and is zero at the turning points $x = \pm A$.

You may wonder why we are spending so much time studying simple harmonic oscillators. We do so because they are good models of a wide variety of physical phenomena. For example, recall the Lennard–Jones potential discussed in Example 8.11. This complicated function describes the forces holding atoms together. Figure 15.12a shows that, for small displacements from the equilibrium position, the potential energy curve

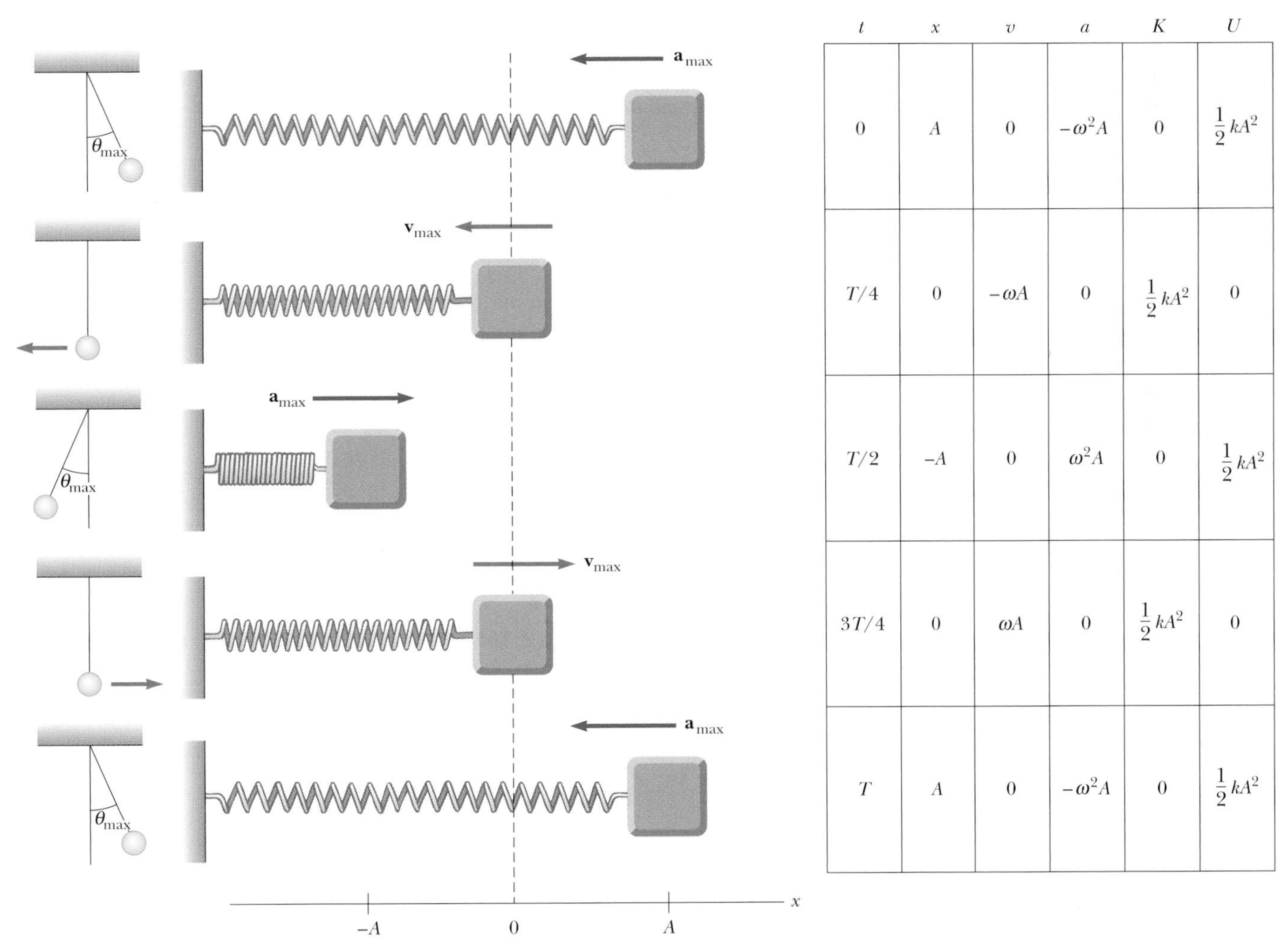

t	x	v	a	K	U
0	A	0	$-\omega^2 A$	0	$\frac{1}{2}kA^2$
$T/4$	0	$-\omega A$	0	$\frac{1}{2}kA^2$	0
$T/2$	$-A$	0	$\omega^2 A$	0	$\frac{1}{2}kA^2$
$3T/4$	0	ωA	0	$\frac{1}{2}kA^2$	0
T	A	0	$-\omega^2 A$	0	$\frac{1}{2}kA^2$

Active Figure 15.11 Simple harmonic motion for a block–spring system and its analogy to the motion of a simple pendulum (Section 15.5). The parameters in the table at the right refer to the block–spring system, assuming that at $t = 0$, $x = A$ so that $x = A \cos \omega t$.

At the Active Figures link at* http://www.pse6.com, *you can set the initial position of the block and see the block–spring system and the analogous pendulum in motion.

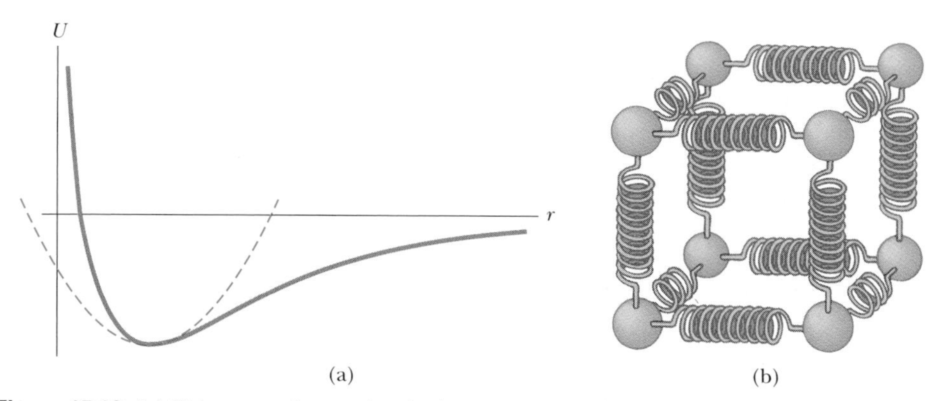

Figure 15.12 (a) If the atoms in a molecule do not move too far from their equilibrium positions, a graph of potential energy versus separation distance between atoms is similar to the graph of potential energy versus position for a simple harmonic oscillator (blue curve). (b) The forces between atoms in a solid can be modeled by imagining springs between neighboring atoms.

for this function approximates a parabola, which represents the potential energy function for a simple harmonic oscillator. Thus, we can model the complex atomic binding forces as being due to tiny springs, as depicted in Figure 15.12b.

The ideas presented in this chapter apply not only to block–spring systems and atoms, but also to a wide range of situations that include bungee jumping, tuning in a television station, and viewing the light emitted by a laser. You will see more examples of simple harmonic oscillators as you work through this book.

Example 15.4 Oscillations on a Horizontal Surface

A 0.500-kg cart connected to a light spring for which the force constant is 20.0 N/m oscillates on a horizontal, frictionless air track.

(A) Calculate the total energy of the system and the maximum speed of the cart if the amplitude of the motion is 3.00 cm.

Solution Using Equation 15.21, we obtain

$$E = \tfrac{1}{2}kA^2 = \tfrac{1}{2}(20.0\ \text{N/m})(3.00 \times 10^{-2}\ \text{m})^2$$

$$= 9.00 \times 10^{-3}\ \text{J}$$

When the cart is located at $x = 0$, we know that $U = 0$ and $E = \frac{1}{2}mv_{\text{max}}^2$; therefore,

$$\tfrac{1}{2}mv_{\text{max}}^2 = 9.00 \times 10^{-3}\ \text{J}$$

$$v_{\text{max}} = \sqrt{\frac{2(9.00 \times 10^{-3}\ \text{J})}{0.500\ \text{kg}}} = 0.190\ \text{m/s}$$

(B) What is the velocity of the cart when the position is 2.00 cm?

Solution We can apply Equation 15.22 directly:

$$v = \pm\sqrt{\frac{k}{m}(A^2 - x^2)}$$

$$= \pm\sqrt{\frac{20.0\ \text{N/m}}{0.500\ \text{kg}}[(0.0300\ \text{m})^2 - (0.0200\ \text{m})^2]}$$

$$= \pm 0.141\ \text{m/s}$$

The positive and negative signs indicate that the cart could be moving to either the right or the left at this instant.

(C) Compute the kinetic and potential energies of the system when the position is 2.00 cm.

Solution Using the result of (B), we find that

$$K = \tfrac{1}{2}mv^2 = \tfrac{1}{2}(0.500\ \text{kg})(0.141\ \text{m/s})^2 = 5.00 \times 10^{-3}\ \text{J}$$

$$U = \tfrac{1}{2}kx^2 = \tfrac{1}{2}(20.0\ \text{N/m})(0.0200\ \text{m})^2 = 4.00 \times 10^{-3}\ \text{J}$$

Note that $K + U = E$.

What If? The motion of the cart in this example could have been initiated by releasing the cart from rest at x = 3.00 cm. What if the cart were released from the same position, but with an initial velocity of v = −0.100 m/s? What are the new amplitude and maximum speed of the cart?

Answer This is the same type of question as we asked at the end of Example 15.3, but here we apply an energy approach. First let us calculate the total energy of the system at $t = 0$, which consists of both kinetic energy and potential energy:

$$E = \tfrac{1}{2}mv^2 + \tfrac{1}{2}kx^2$$

$$= \tfrac{1}{2}(0.500\ \text{kg})(-0.100\ \text{m/s})^2 + \tfrac{1}{2}(20.0\ \text{N/m})(0.030\,0\ \text{m})^2$$

$$= 1.15 \times 10^{-2}\ \text{J}$$

To find the new amplitude, we equate this total energy to the potential energy when the cart is at the end point of the motion:

$$E = \frac{1}{2}kA^2$$

$$A = \sqrt{\frac{2E}{k}} = \sqrt{\frac{2(1.15 \times 10^{-2}\text{ J})}{20.0\text{ N/m}}} = 0.033\,9\text{ m}$$

Note that this is larger than the previous amplitude of 0.030 0 m. To find the new maximum speed, we equate this total energy to the kinetic energy when the cart is at the equilibrium position:

$$E = \frac{1}{2}mv_{\text{max}}^2$$

$$v_{\text{max}} = \sqrt{\frac{2E}{m}} = \sqrt{\frac{2(1.15 \times 10^{-2}\text{ J})}{0.500\text{ kg}}} = 0.214\text{ m/s}$$

This is larger than the value found in part (a) as expected because the cart has an initial velocity at $t = 0$.

15.4 Comparing Simple Harmonic Motion with Uniform Circular Motion

Some common devices in our everyday life exhibit a relationship between oscillatory motion and circular motion. For example, the pistons in an automobile engine (Figure 15.13a) go up and down—oscillatory motion—yet the net result of this motion is circular motion of the wheels. In an old-fashioned locomotive (Figure 15.13b), the drive shaft goes back and forth in oscillatory motion, causing a circular motion of the wheels. In this section, we explore this interesting relationship between these two types of motion. We shall use this relationship again when we study electromagnetism and when we explore optics.

Figure 15.14 is an overhead view of an experimental arrangement that shows this relationship. A ball is attached to the rim of a turntable of radius A, which is illuminated from the side by a lamp. The ball casts a shadow on a screen. We find that **as the turntable rotates with constant angular speed, the shadow of the ball moves back and forth in simple harmonic motion.**

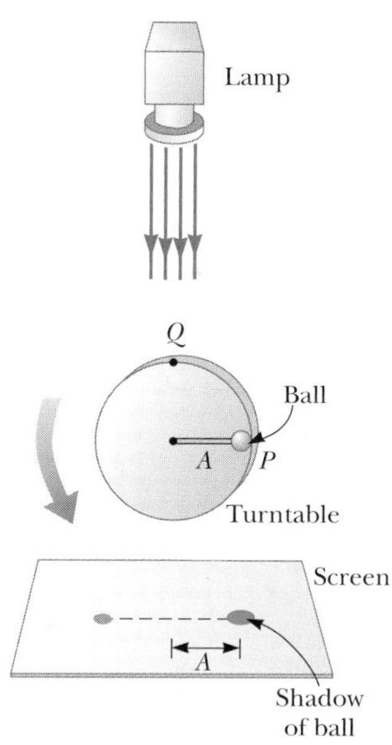

Active Figure 15.14 An experimental setup for demonstrating the connection between simple harmonic motion and uniform circular motion. As the ball rotates on the turntable with constant angular speed, its shadow on the screen moves back and forth in simple harmonic motion.

At the Active Figures link at http://www.pse6.com, you can adjust the frequency and radial position of the ball and see the resulting simple harmonic motion of the shadow.

Courtesy of Ford Motor Company

(a)

© Link / Visuals Unlimited

(b)

Figure 15.13 (a) The pistons of an automobile engine move in periodic motion along a single dimension. This photograph shows a cutaway view of two of these pistons. This motion is converted to circular motion of the crankshaft, at the lower right, and ultimately of the wheels of the automobile. (b) The back-and-forth motion of pistons (in the curved housing at the left) in an old-fashioned locomotive is converted to circular motion of the wheels.

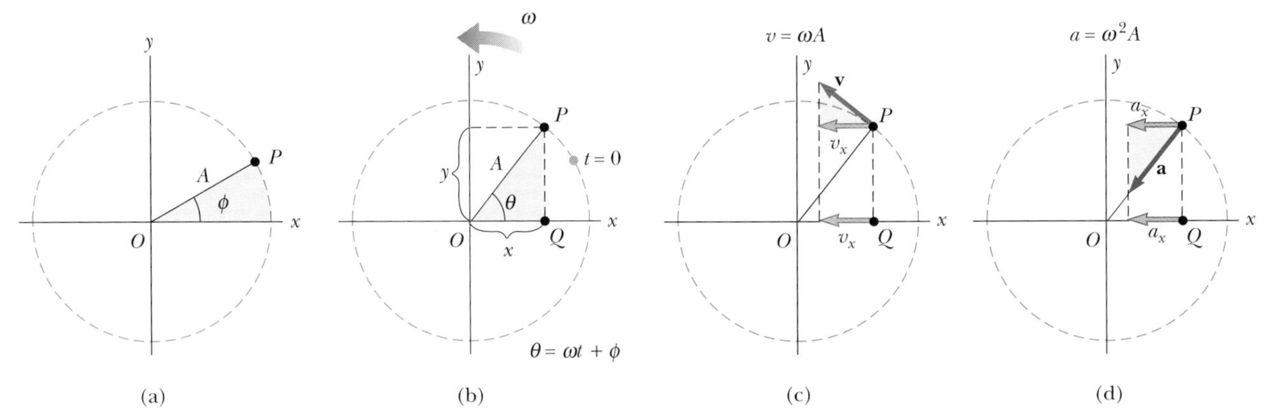

Figure 15.15 Relationship between the uniform circular motion of a point P and the simple harmonic motion of a point Q. A particle at P moves in a circle of radius A with constant angular speed ω. (a) A reference circle showing the position of P at $t = 0$. (b) The x coordinates of points P and Q are equal and vary in time according to the expression $x = A\cos(\omega t + \phi)$. (c) The x component of the velocity of P equals the velocity of Q. (d) The x component of the acceleration of P equals the acceleration of Q.

Consider a particle located at point P on the circumference of a circle of radius A, as in Figure 15.15a, with the line OP making an angle ϕ with the x axis at $t = 0$. We call this circle a *reference circle* for comparing simple harmonic motion with uniform circular motion, and we take the position of P at $t = 0$ as our reference position. If the particle moves along the circle with constant angular speed ω until OP makes an angle θ with the x axis, as in Figure 15.15b, then at some time $t > 0$, the angle between OP and the x axis is $\theta = \omega t + \phi$. As the particle moves along the circle, the projection of P on the x axis, labeled point Q, moves back and forth along the x axis between the limits $x = \pm A$.

Note that points P and Q always have the same x coordinate. From the right triangle OPQ, we see that this x coordinate is

$$x(t) = A\cos(\omega t + \phi) \tag{15.23}$$

This expression is the same as Equation 15.6 and shows that the point Q moves with simple harmonic motion along the x axis. Therefore, we conclude that

simple harmonic motion along a straight line can be represented by the projection of uniform circular motion along a diameter of a reference circle.

We can make a similar argument by noting from Figure 15.15b that the projection of P along the y axis also exhibits simple harmonic motion. Therefore, **uniform circular motion can be considered a combination of two simple harmonic motions, one along the x axis and one along the y axis, with the two differing in phase by 90°.**

This geometric interpretation shows that the time interval for one complete revolution of the point P on the reference circle is equal to the period of motion T for simple harmonic motion between $x = \pm A$. That is, the angular speed ω of P is the same as the angular frequency ω of simple harmonic motion along the x axis. (This is why we use the same symbol.) The phase constant ϕ for simple harmonic motion corresponds to the initial angle that OP makes with the x axis. The radius A of the reference circle equals the amplitude of the simple harmonic motion.

Because the relationship between linear and angular speed for circular motion is $v = r\omega$ (see Eq. 10.10), the particle moving on the reference circle of radius A has a velocity of magnitude ωA. From the geometry in Figure 15.15c, we see that the x component of this velocity is $-\omega A\sin(\omega t + \phi)$. By definition, point Q has a velocity given by dx/dt. Differentiating Equation 15.23 with respect to time, we find that the velocity of Q is the same as the x component of the velocity of P.

The acceleration of P on the reference circle is directed radially inward toward O and has a magnitude $v^2/A = \omega^2 A$. From the geometry in Figure 15.15d, we see that the x component of this acceleration is $-\omega^2 A \cos(\omega t + \phi)$. This value is also the acceleration of the projected point Q along the x axis, as you can verify by taking the second derivative of Equation 15.23.

Quick Quiz 15.6 Figure 15.16 shows the position of an object in uniform circular motion at $t = 0$. A light shines from above and projects a shadow of the object on a screen below the circular motion. The correct values for the *amplitude* and *phase constant* (relative to an x axis to the right) of the simple harmonic motion of the shadow are (a) 0.50 m and 0 (b) 1.00 m and 0 (c) 0.50 m and π (d) 1.00 m and π.

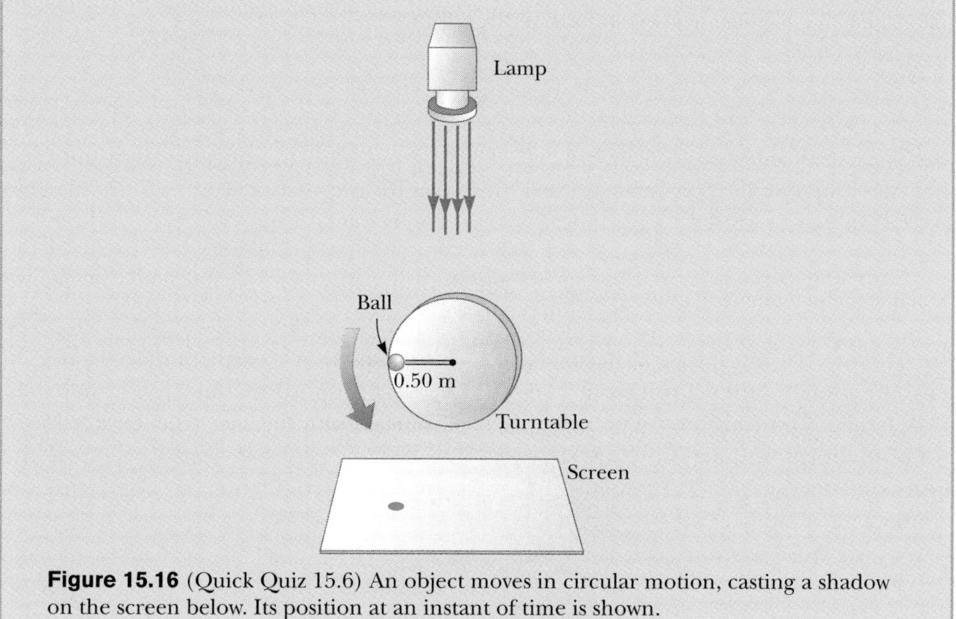

Figure 15.16 (Quick Quiz 15.6) An object moves in circular motion, casting a shadow on the screen below. Its position at an instant of time is shown.

Example 15.5 Circular Motion with Constant Angular Speed

A particle rotates counterclockwise in a circle of radius 3.00 m with a constant angular speed of 8.00 rad/s. At $t = 0$, the particle has an x coordinate of 2.00 m and is moving to the right.

(A) Determine the x coordinate as a function of time.

Solution Because the amplitude of the particle's motion equals the radius of the circle and $\omega = 8.00$ rad/s, we have

$$x = A\cos(\omega t + \phi) = (3.00 \text{ m})\cos(8.00t + \phi)$$

We can evaluate ϕ by using the initial condition that $x = 2.00$ m at $t = 0$:

$$2.00 \text{ m} = (3.00 \text{ m})\cos(0 + \phi)$$

$$\phi = \cos^{-1}\left(\frac{2.00 \text{ m}}{3.00 \text{ m}}\right)$$

If we were to take our answer as $\phi = 48.2° = 0.841$ rad, then the coordinate $x = (3.00 \text{ m})\cos(8.00t + 0.841)$ would be decreasing at time $t = 0$ (that is, moving to the left). Because our particle is first moving to the right, we must choose $\phi = -0.841$ rad. The x coordinate as a function of time is then

$$x = (3.00 \text{ m})\cos(8.00t - 0.841)$$

Note that the angle ϕ in the cosine function must be in radians.

(B) Find the x components of the particle's velocity and acceleration at any time t.

Solution

$$v_x = \frac{dx}{dt} = (-3.00 \text{ m})(8.00 \text{ rad/s})\sin(8.00t - 0.841)$$

$$= -(24.0 \text{ m/s})\sin(8.00t - 0.841)$$

$$a_x = \frac{dv}{dt} = (-24.0 \text{ m/s})(8.00 \text{ rad/s})\cos(8.00t - 0.841)$$

$$= -(192 \text{ m/s}^2)\cos(8.00t - 0.841)$$

From these results, we conclude that $v_{max} = 24.0$ m/s and that $a_{max} = 192 \text{ m/s}^2$.

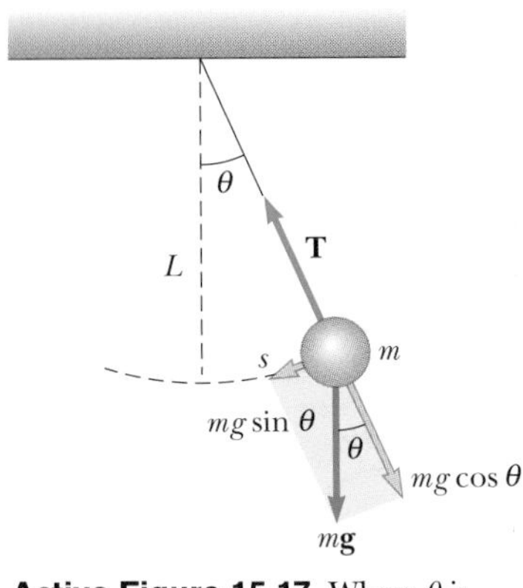

Active Figure 15.17 When θ is small, a simple pendulum oscillates in simple harmonic motion about the equilibrium position $\theta = 0$. The restoring force is $-mg\sin\theta$, the component of the gravitational force tangent to the arc.

At the Active Figures link at http://www.pse6.com, you can adjust the mass of the bob, the length of the string, and the initial angle and see the resulting oscillation of the pendulum.

15.5 The Pendulum

The **simple pendulum** is another mechanical system that exhibits periodic motion. It consists of a particle-like bob of mass m suspended by a light string of length L that is fixed at the upper end, as shown in Figure 15.17. The motion occurs in the vertical plane and is driven by the gravitational force. We shall show that, provided the angle θ is small (less than about 10°), the motion is very close to that of a simple harmonic oscillator.

The forces acting on the bob are the force **T** exerted by the string and the gravitational force $m\mathbf{g}$. The tangential component $mg\sin\theta$ of the gravitational force always acts toward $\theta = 0$, opposite the displacement of the bob from the lowest position. Therefore, the tangential component is a restoring force, and we can apply Newton's second law for motion in the tangential direction:

$$F_t = -mg\sin\theta = m\frac{d^2s}{dt^2}$$

where s is the bob's position measured along the arc and the negative sign indicates that the tangential force acts toward the equilibrium (vertical) position. Because $s = L\theta$ (Eq. 10.1a) and L is constant, this equation reduces to

$$\frac{d^2\theta}{dt^2} = -\frac{g}{L}\sin\theta$$

Considering θ as the position, let us compare this equation to Equation 15.3—does it have the same mathematical form? The right side is proportional to $\sin\theta$ rather than to θ; hence, we would not expect simple harmonic motion because this expression is not of the form of Equation 15.3. However, if we assume that θ is *small*, we can use the approximation $\sin\theta \approx \theta$; thus, in this approximation, the equation of motion for the simple pendulum becomes

PITFALL PREVENTION

15.5 Not True Simple Harmonic Motion

Remember that the pendulum *does not* exhibit true simple harmonic motion for *any* angle. If the angle is less than about 10°, the motion is close to and can be *modeled* as simple harmonic.

$$\frac{d^2\theta}{dt^2} = -\frac{g}{L}\theta \quad \text{(for small values of } \theta\text{)} \tag{15.24}$$

Now we have an expression that has the same form as Equation 15.3, and we conclude that the motion for small amplitudes of oscillation is simple harmonic motion. Therefore, the function θ can be written as $\theta = \theta_{\max}\cos(\omega t + \phi)$, where $\theta_{\max}$ is the *maximum angular position* and the angular frequency ω is

Angular frequency for a simple pendulum

$$\omega = \sqrt{\frac{g}{L}} \tag{15.25}$$

The period of the motion is

Period of a simple pendulum

$$T = \frac{2\pi}{\omega} = 2\pi\sqrt{\frac{L}{g}} \tag{15.26}$$

In other words, **the period and frequency of a simple pendulum depend only on the length of the string and the acceleration due to gravity.** Because the period is independent of the mass, we conclude that all simple pendula that are of equal length and are at the same location (so that g is constant) oscillate with the same period. The analogy between the motion of a simple pendulum and that of a block–spring system is illustrated in Figure 15.11.

The simple pendulum can be used as a timekeeper because its period depends only on its length and the local value of g. It is also a convenient device for making precise measurements of the free-fall acceleration. Such measurements are important because variations in local values of g can provide information on the location of oil and of other valuable underground resources.

Quick Quiz 15.7 A grandfather clock depends on the period of a pendulum to keep correct time. Suppose a grandfather clock is calibrated correctly and then a mischievous child slides the bob of the pendulum downward on the oscillating rod. Does the grandfather clock run (a) slow (b) fast (c) correctly?

Quick Quiz 15.8 Suppose a grandfather clock is calibrated correctly at sea level and is then taken to the top of a very tall mountain. Does the grandfather clock run (a) slow (b) fast (c) correctly?

Example 15.6 A Connection Between Length and Time

Christian Huygens (1629–1695), the greatest clockmaker in history, suggested that an international unit of length could be defined as the length of a simple pendulum having a period of exactly 1 s. How much shorter would our length unit be had his suggestion been followed?

Solution Solving Equation 15.26 for the length gives

$$L = \frac{T^2 g}{4\pi^2} = \frac{(1.00\text{ s})^2(9.80\text{ m/s}^2)}{4\pi^2} = 0.248\text{ m}$$

Thus, the meter's length would be slightly less than one fourth of its current length. Note that the number of significant digits depends only on how precisely we know g because the time has been defined to be exactly 1 s.

What If? **What if Huygens had been born on another planet? What would the value for g have to be on that planet such that the meter based on Huygens's pendulum would have the same value as our meter?**

Answer We solve Equation 15.26 for g:

$$g = \frac{4\pi^2 L}{T^2} = \frac{4\pi^2(1.00\text{ m})}{(1.00\text{ s})^2} = 4\pi^2\text{ m/s}^2 = 39.5\text{ m/s}^2$$

No planet in our solar system has an acceleration due to gravity that is this large.

Physical Pendulum

Suppose you balance a wire coat hanger so that the hook is supported by your extended index finger. When you give the hanger a small angular displacement (with your other hand) and then release it, it oscillates. If a hanging object oscillates about a fixed axis that does not pass through its center of mass and the object cannot be approximated as a point mass, we cannot treat the system as a simple pendulum. In this case the system is called a **physical pendulum.**

Consider a rigid object pivoted at a point O that is a distance d from the center of mass (Fig. 15.18). The gravitational force provides a torque about an axis through O, and the magnitude of that torque is $mgd \sin\theta$, where θ is as shown in Figure 15.18. Using the rotational form of Newton's second law, $\Sigma\tau = I\alpha$, where I is the moment of inertia about the axis through O, we obtain

$$-mgd\sin\theta = I\frac{d^2\theta}{dt^2}$$

The negative sign indicates that the torque about O tends to decrease θ. That is, the gravitational force produces a restoring torque. If we again assume that θ is small, the approximation $\sin\theta \approx \theta$ is valid, and the equation of motion reduces to

$$\frac{d^2\theta}{dt^2} = -\left(\frac{mgd}{I}\right)\theta = -\omega^2\theta \qquad (15.27)$$

Because this equation is of the same form as Equation 15.3, the motion is simple harmonic motion. That is, the solution of Equation 15.27 is $\theta = \theta_{max}\cos(\omega t + \phi)$, where θ_{max} is the maximum angular position and

$$\omega = \sqrt{\frac{mgd}{I}}$$

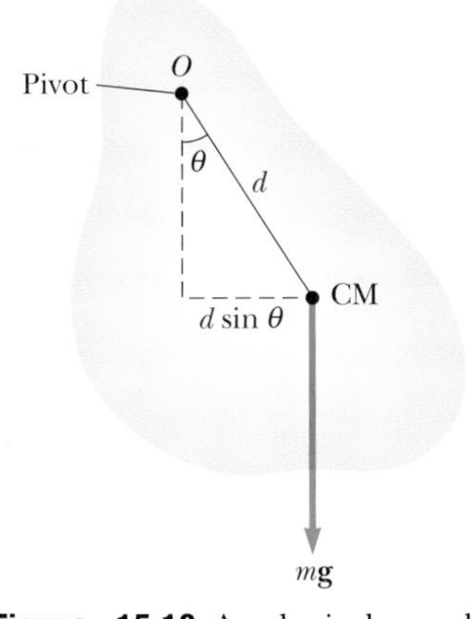

Figure 15.18 A physical pendulum pivoted at O.

The period is

Period of a physical pendulum

$$T = \frac{2\pi}{\omega} = 2\pi\sqrt{\frac{I}{mgd}} \tag{15.28}$$

One can use this result to measure the moment of inertia of a flat rigid object. If the location of the center of mass—and hence the value of d—is known, the moment of inertia can be obtained by measuring the period. Finally, note that Equation 15.28 reduces to the period of a simple pendulum (Eq. 15.26) when $I = md^2$—that is, when all the mass is concentrated at the center of mass.

Example 15.7 A Swinging Rod

A uniform rod of mass M and length L is pivoted about one end and oscillates in a vertical plane (Fig. 15.19). Find the period of oscillation if the amplitude of the motion is small.

Solution In Chapter 10 we found that the moment of inertia of a uniform rod about an axis through one end is $\frac{1}{3}ML^2$. The distance d from the pivot to the center of mass is $L/2$. Substituting these quantities into Equation 15.28 gives

$$T = 2\pi\sqrt{\frac{\frac{1}{3}ML^2}{Mg(L/2)}} = 2\pi\sqrt{\frac{2L}{3g}}$$

Comment In one of the Moon landings, an astronaut walking on the Moon's surface had a belt hanging from his space suit, and the belt oscillated as a physical pendulum. A scientist on the Earth observed this motion on television and used it to estimate the free-fall acceleration on the Moon. How did the scientist make this calculation?

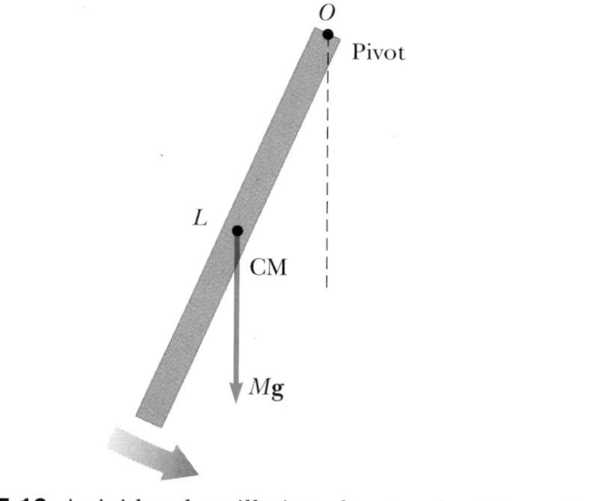

Figure 15.19 A rigid rod oscillating about a pivot through one end is a physical pendulum with $d = L/2$ and, from Table 10.2, $I = \frac{1}{3}ML^2$.

Torsional Pendulum

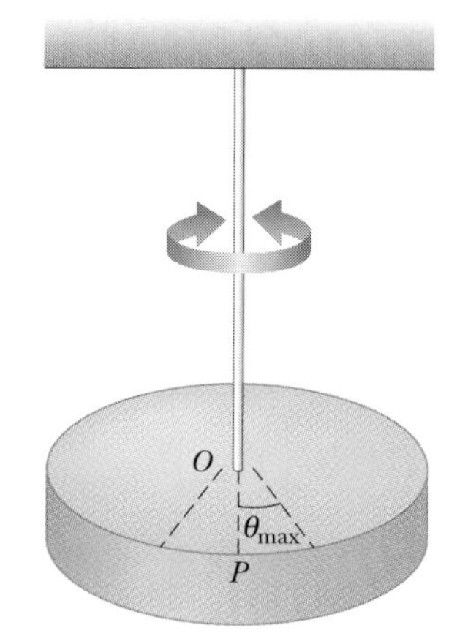

Figure 15.20 A torsional pendulum consists of a rigid object suspended by a wire attached to a rigid support. The object oscillates about the line OP with an amplitude θ_{max}.

Figure 15.20 shows a rigid object suspended by a wire attached at the top to a fixed support. When the object is twisted through some angle θ, the twisted wire exerts on the object a restoring torque that is proportional to the angular position. That is,

$$\tau = -\kappa\theta$$

where κ (kappa) is called the *torsion constant* of the support wire. The value of κ can be obtained by applying a known torque to twist the wire through a measurable angle θ. Applying Newton's second law for rotational motion, we find

$$\tau = -\kappa\theta = I\frac{d^2\theta}{dt^2}$$

$$\frac{d^2\theta}{dt^2} = -\frac{\kappa}{I}\theta \tag{15.29}$$

Again, this is the equation of motion for a simple harmonic oscillator, with $\omega = \sqrt{\kappa/I}$ and a period

Period of a torsional pendulum

$$T = 2\pi\sqrt{\frac{I}{\kappa}} \tag{15.30}$$

This system is called a *torsional pendulum*. There is no small-angle restriction in this situation as long as the elastic limit of the wire is not exceeded.

15.6 Damped Oscillations

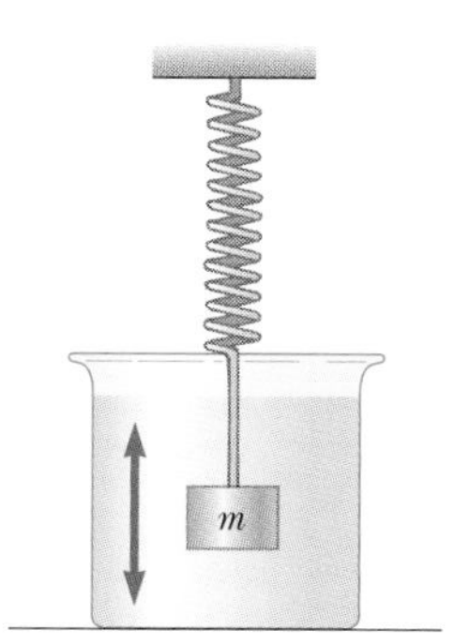

Figure 15.21 One example of a damped oscillator is an object attached to a spring and submersed in a viscous liquid.

The oscillatory motions we have considered so far have been for ideal systems—that is, systems that oscillate indefinitely under the action of only one force—a linear restoring force. In many real systems, nonconservative forces, such as friction, retard the motion. Consequently, the mechanical energy of the system diminishes in time, and the motion is said to be *damped.* Figure 15.21 depicts one such system: an object attached to a spring and submersed in a viscous liquid.

One common type of retarding force is the one discussed in Section 6.4, where the force is proportional to the speed of the moving object and acts in the direction opposite the motion. This retarding force is often observed when an object moves through air, for instance. Because the retarding force can be expressed as $\mathbf{R} = -b\mathbf{v}$ (where b is a constant called the *damping coefficient*) and the restoring force of the system is $-kx$, we can write Newton's second law as

$$\sum F_x = -kx - bv_x = ma_x$$

$$-kx - b\frac{dx}{dt} = m\frac{d^2x}{dt^2} \qquad (15.31)$$

The solution of this equation requires mathematics that may not be familiar to you; we simply state it here without proof. When the retarding force is small compared with the maximum restoring force—that is, when b is small—the solution to Equation 15.31 is

$$x = Ae^{-\frac{b}{2m}t}\cos(\omega t + \phi) \qquad (15.32)$$

where the angular frequency of oscillation is

$$\omega = \sqrt{\frac{k}{m} - \left(\frac{b}{2m}\right)^2} \qquad (15.33)$$

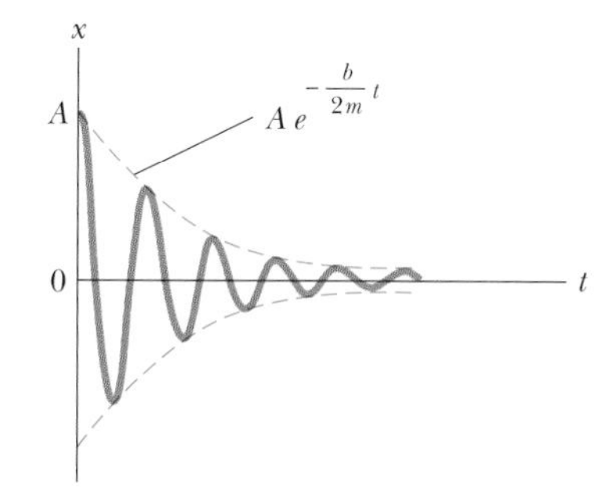

Active Figure 15.22 Graph of position versus time for a damped oscillator. Note the decrease in amplitude with time.

At the Active Figures link at http://www.pse6.com, you can adjust the spring constant, the mass of the object, and the damping constant and see the resulting damped oscillation of the object.

This result can be verified by substituting Equation 15.32 into Equation 15.31.

Figure 15.22 shows the position as a function of time for an object oscillating in the presence of a retarding force. We see that **when the retarding force is small, the oscillatory character of the motion is preserved but the amplitude decreases in time, with the result that the motion ultimately ceases.** Any system that behaves in this way is known as a **damped oscillator.** The dashed blue lines in Figure 15.22, which define the *envelope* of the oscillatory curve, represent the exponential factor in Equation 15.32. This envelope shows that **the amplitude decays exponentially with time.** For motion with a given spring constant and object mass, the oscillations dampen more rapidly as the maximum value of the retarding force approaches the maximum value of the restoring force.

It is convenient to express the angular frequency (Eq. 15.33) of a damped oscillator in the form

$$\omega = \sqrt{\omega_0^2 - \left(\frac{b}{2m}\right)^2}$$

where $\omega_0 = \sqrt{k/m}$ represents the angular frequency in the absence of a retarding force (the undamped oscillator) and is called the **natural frequency** of the system.

When the magnitude of the maximum retarding force $R_{max} = bv_{max} < kA$, the system is said to be **underdamped.** The resulting motion is represented by the blue curve in Figure 15.23. As the value of b increases, the amplitude of the oscillations decreases more and more rapidly. When b reaches a critical value b_c such that $b_c/2m = \omega_0$, the system does not oscillate and is said to be **critically damped.** In this case the system, once released from rest at some nonequilibrium position, approaches but does not pass through the equilibrium position. The graph of position versus time for this case is the red curve in Figure 15.23.

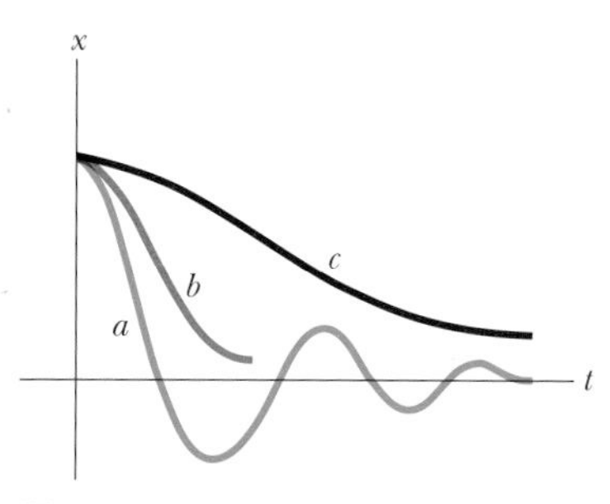

Figure 15.23 Graphs of position versus time for (a) an underdamped oscillator, (b) a critically damped oscillator, and (c) an overdamped oscillator.

If the medium is so viscous that the retarding force is greater than the restoring force—that is, if $R_{max} = bv_{max} > kA$ and $b/2m > \omega_0$—the system is **overdamped.** Again, the displaced system, when free to move, does not oscillate but simply returns to its equilibrium position. As the damping increases, the time interval required for the system to approach equilibrium also increases, as indicated by the black curve in Figure 15.23. For critically damped and overdamped systems, there is no angular frequency ω and the solution in Equation 15.32 is not valid.

Whenever friction is present in a system, whether the system is overdamped or underdamped, the energy of the oscillator eventually falls to zero. The lost mechanical energy is transformed into internal energy in the object and the retarding medium.

Quick Quiz 15.9 An automotive suspension system consists of a combination of springs and shock absorbers, as shown in Figure 15.24. If you were an automotive engineer, would you design a suspension system that was (a) underdamped (b) critically damped (c) overdamped?

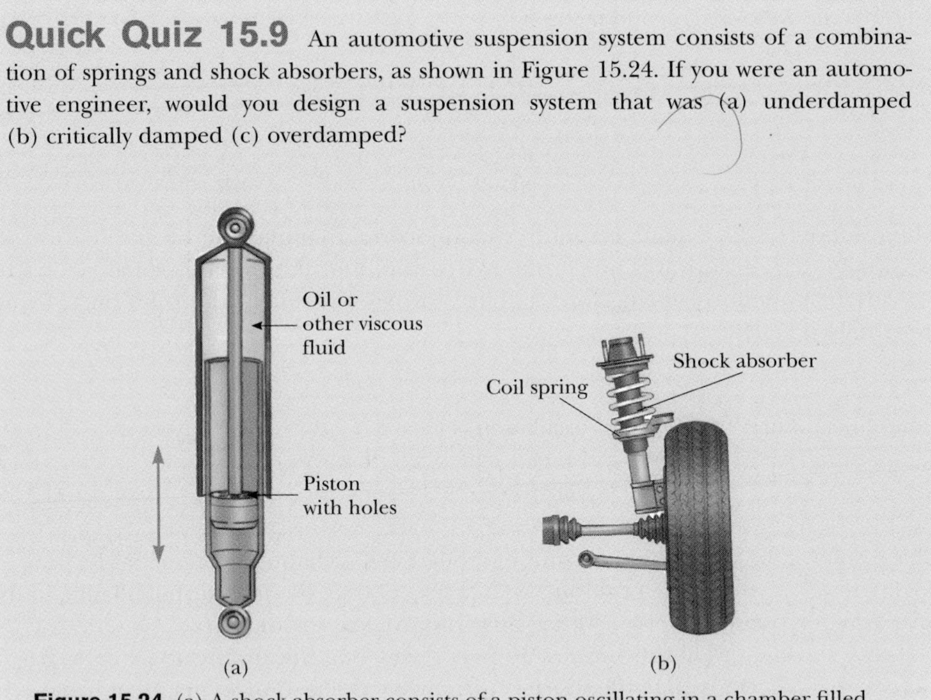

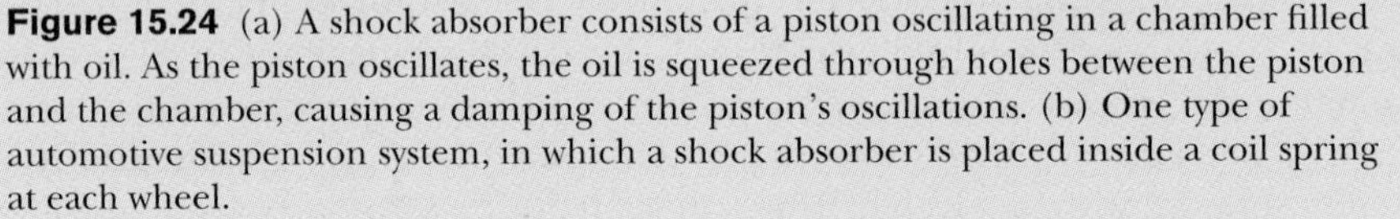

Figure 15.24 (a) A shock absorber consists of a piston oscillating in a chamber filled with oil. As the piston oscillates, the oil is squeezed through holes between the piston and the chamber, causing a damping of the piston's oscillations. (b) One type of automotive suspension system, in which a shock absorber is placed inside a coil spring at each wheel.

15.7 Forced Oscillations

We have seen that the mechanical energy of a damped oscillator decreases in time as a result of the resistive force. It is possible to compensate for this energy decrease by applying an external force that does positive work on the system. At any instant, energy can be transferred into the system by an applied force that acts in the direction of motion of the oscillator. For example, a child on a swing can be kept in motion by appropriately timed "pushes." The amplitude of motion remains constant if the energy input per cycle of motion exactly equals the decrease in mechanical energy in each cycle that results from resistive forces.

A common example of a forced oscillator is a damped oscillator driven by an external force that varies periodically, such as $F(t) = F_0 \sin \omega t$, where ω is the angular frequency of the driving force and F_0 is a constant. In general, the frequency ω of the driving force is variable while the natural frequency ω_0 of the oscillator is fixed by the values of k and m. Newton's second law in this situation gives

$$\sum F = ma \longrightarrow F_0 \sin \omega t - b\frac{dx}{dt} - kx = m\frac{d^2x}{dt^2} \qquad (15.34)$$

Again, the solution of this equation is rather lengthy and will not be presented. After the driving force on an initially stationary object begins to act, the amplitude of the oscillation will increase. After a sufficiently long period of time, when the energy input per cycle from the driving force equals the amount of mechanical energy transformed to internal energy for each cycle, a steady-state condition is reached in which the oscillations proceed with constant amplitude. In this situation, Equation 15.34 has the solution

$$x = A\cos(\omega t + \phi) \qquad (15.35)$$

where

$$A = \frac{F_0/m}{\sqrt{(\omega^2 - {\omega_0}^2)^2 + \left(\frac{b\omega}{m}\right)^2}} \qquad (15.36)$$

Amplitude of a driven oscillator

and where $\omega_0 = \sqrt{k/m}$ is the natural frequency of the undamped oscillator $(b = 0)$.

Equations 15.35 and 15.36 show that the forced oscillator vibrates at the frequency of the driving force and that the amplitude of the oscillator is constant for a given driving force because it is being driven in steady-state by an external force. For small damping, the amplitude is large when the frequency of the driving force is near the natural frequency of oscillation, or when $\omega \approx \omega_0$. The dramatic increase in amplitude near the natural frequency is called **resonance,** and the natural frequency ω_0 is also called the **resonance frequency** of the system.

The reason for large-amplitude oscillations at the resonance frequency is that energy is being transferred to the system under the most favorable conditions. We can better understand this by taking the first time derivative of x in Equation 15.35, which gives an expression for the velocity of the oscillator. We find that v is proportional to $\sin(\omega t + \phi)$, which is the same trigonometric function as that describing the driving force. Thus, the applied force $\mathbf{F}$ is in phase with the velocity. The rate at which work is done on the oscillator by $\mathbf{F}$ equals the dot product $\mathbf{F}\cdot\mathbf{v}$; this rate is the power delivered to the oscillator. Because the product $\mathbf{F}\cdot\mathbf{v}$ is a maximum when $\mathbf{F}$ and $\mathbf{v}$ are in phase, we conclude that **at resonance the applied force is in phase with the velocity and the power transferred to the oscillator is a maximum.**

Figure 15.25 is a graph of amplitude as a function of frequency for a forced oscillator with and without damping. Note that the amplitude increases with decreasing damping $(b \rightarrow 0)$ and that the resonance curve broadens as the damping increases. Under steady-state conditions and at any driving frequency, the energy transferred into the system equals the energy lost because of the damping force; hence, the average total energy of the oscillator remains constant. In the absence of a damping force $(b = 0)$, we see from Equation 15.36 that the steady-state amplitude approaches infinity as ω approaches ω_0. In other words, if there are no losses in the system and if we continue to drive an initially motionless oscillator with a periodic force that is in phase with the velocity, the amplitude of motion builds without limit (see the brown curve in Fig. 15.25). This limitless building does not occur in practice because some damping is always present in reality.

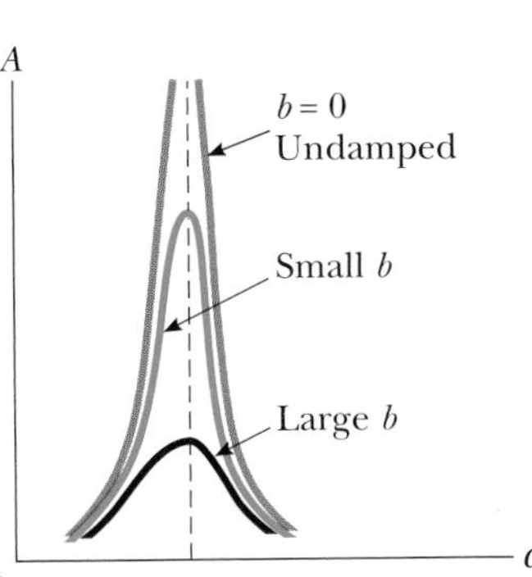

Figure 15.25 Graph of amplitude versus frequency for a damped oscillator when a periodic driving force is present. When the frequency ω of the driving force equals the natural frequency ω_0 of the oscillator, resonance occurs. Note that the shape of the resonance curve depends on the size of the damping coefficient b.

Later in this book we shall see that resonance appears in other areas of physics. For example, certain electric circuits have natural frequencies. A bridge has natural frequencies that can be set into resonance by an appropriate driving force. A dramatic example of such resonance occurred in 1940, when the Tacoma Narrows Bridge in the state of Washington was destroyed by resonant vibrations. Although the winds were not particularly strong on that occasion, the "flapping" of the wind across the roadway (think of the "flapping" of a flag in a strong wind) provided a periodic driving force whose frequency matched that of the bridge. The resulting oscillations of the bridge caused it to ultimately collapse (Fig. 15.26) because the bridge design had inadequate built-in safety features.

(a) (b)

Figure 15.26 (a) In 1940 turbulent winds set up torsional vibrations in the Tacoma Narrows Bridge, causing it to oscillate at a frequency near one of the natural frequencies of the bridge structure. (b) Once established, this resonance condition led to the bridge's collapse.

Many other examples of resonant vibrations can be cited. A resonant vibration that you may have experienced is the "singing" of telephone wires in the wind. Machines often break if one vibrating part is in resonance with some other moving part. Soldiers marching in cadence across a bridge have been known to set up resonant vibrations in the structure and thereby cause it to collapse. Whenever any real physical system is driven near its resonance frequency, you can expect oscillations of very large amplitudes.

SUMMARY

Take a practice test for this chapter by clicking on the Practice Test link at http://www.pse6.com.

When the acceleration of an object is proportional to its position and is in the direction opposite the displacement from equilibrium, the object moves with simple harmonic motion. The position x of a simple harmonic oscillator varies periodically in time according to the expression

$$x(t) = A\cos(\omega t + \phi) \tag{15.6}$$

where A is the **amplitude** of the motion, ω is the **angular frequency,** and ϕ is the **phase constant.** The value of ϕ depends on the initial position and initial velocity of the oscillator.

The time interval T needed for one complete oscillation is defined as the **period** of the motion:

$$T = \frac{2\pi}{\omega} \tag{15.10}$$

A block–spring system moves in simple harmonic motion on a frictionless surface with a period

$$T = \frac{2\pi}{\omega} = 2\pi\sqrt{\frac{m}{k}} \tag{15.13}$$

The inverse of the period is the **frequency** of the motion, which equals the number of oscillations per second.

The velocity and acceleration of a simple harmonic oscillator are

$$v = \frac{dx}{dt} = -\omega A\sin(\omega t + \phi) \tag{15.15}$$

$$a = \frac{d^2x}{dt^2} = -\omega^2 A\cos(\omega t + \phi) \tag{15.16}$$

$$v = \pm\,\omega\sqrt{A^2 - x^2} \tag{15.22}$$

Thus, the maximum speed is ωA, and the maximum acceleration is $\omega^2 A$. The speed is zero when the oscillator is at its turning points $x = \pm A$ and is a maximum when the oscillator is at the equilibrium position $x = 0$. The magnitude of the acceleration is a maximum at the turning points and zero at the equilibrium position.

The kinetic energy and potential energy for a simple harmonic oscillator vary with time and are given by

$$K = \frac{1}{2}mv^2 = \frac{1}{2}m\omega^2A^2 \sin^2(\omega t + \phi) \tag{15.19}$$

$$U = \frac{1}{2}kx^2 = \frac{1}{2}kA^2 \cos^2(\omega t + \phi) \tag{15.20}$$

The total energy of a simple harmonic oscillator is a constant of the motion and is given by

$$E = \frac{1}{2}kA^2 \tag{15.21}$$

The potential energy of the oscillator is a maximum when the oscillator is at its turning points and is zero when the oscillator is at the equilibrium position. The kinetic energy is zero at the turning points and a maximum at the equilibrium position.

A **simple pendulum** of length L moves in simple harmonic motion for small angular displacements from the vertical. Its period is

$$T = 2\pi\sqrt{\frac{L}{g}} \tag{15.26}$$

For small angular displacements from the vertical, a **physical pendulum** moves in simple harmonic motion about a pivot that does not go through the center of mass. The period of this motion is

$$T = 2\pi\sqrt{\frac{I}{mgd}} \tag{15.28}$$

where I is the moment of inertia about an axis through the pivot and d is the distance from the pivot to the center of mass.

If an oscillator experiences a damping force $\mathbf{R} = -b\mathbf{v}$, its position for small damping is described by

$$x = Ae^{-\frac{b}{2m}t}\cos(\omega t + \phi) \tag{15.32}$$

where

$$\omega = \sqrt{\frac{k}{m} - \left(\frac{b}{2m}\right)^2} \tag{15.33}$$

If an oscillator is subject to a sinusoidal driving force $F(t) = F_0 \sin \omega t$, it exhibits **resonance,** in which the amplitude is largest when the driving frequency matches the natural frequency of the oscillator.

QUESTIONS

1. Is a bouncing ball an example of simple harmonic motion? Is the daily movement of a student from home to school and back simple harmonic motion? Why or why not?

2. If the coordinate of a particle varies as $x = -A \cos \omega t$, what is the phase constant in Equation 15.6? At what position is the particle at $t = 0$?

3. Does the displacement of an oscillating particle between $t = 0$ and a later time t necessarily equal the position of the particle at time t? Explain.

4. Determine whether or not the following quantities can be in the same direction for a simple harmonic oscillator: (a) position and velocity, (b) velocity and acceleration, (c) position and acceleration.

5. Can the amplitude A and phase constant ϕ be determined for an oscillator if only the position is specified at $t = 0$? Explain.

6. Describe qualitatively the motion of a block–spring system when the mass of the spring is not neglected.

7. A block is hung on a spring, and the frequency f of the oscillation of the system is measured. The block, a second identical block, and the spring are carried in the Space Shuttle to space. The two blocks are attached to the ends of the spring, and the system is taken out into space on a space walk. The spring is extended, and the system is released to oscillate while floating in space. What is the frequency of oscillation for this system, in terms of f?

8. A block–spring system undergoes simple harmonic motion with amplitude A. Does the total energy change if the mass is doubled but the amplitude is not changed? Do the kinetic and potential energies depend on the mass? Explain.

9. The equations listed in Table 2.2 give position as a function of time, velocity as a function of time, and velocity as function of position for an object moving in a straight line with constant acceleration. The quantity v_{xi} appears in every equation. Do any of these equations apply to an object moving in a straight line with simple harmonic motion? Using a similar format, make a table of equations describing simple harmonic motion. Include equations giving acceleration as a function of time and acceleration as a function of position. State the equations in such a form that they apply equally to a block–spring system, to a pendulum, and to other vibrating systems. What quantity appears in every equation?

10. What happens to the period of a simple pendulum if the pendulum's length is doubled? What happens to the period if the mass of the suspended bob is doubled?

11. A simple pendulum is suspended from the ceiling of a stationary elevator, and the period is determined. Describe the changes, if any, in the period when the elevator (a) accelerates upward, (b) accelerates downward, and (c) moves with constant velocity.

12. Imagine that a pendulum is hanging from the ceiling of a car. As the car coasts freely down a hill, is the equilibrium position of the pendulum vertical? Does the period of oscillation differ from that in a stationary car?

13. A simple pendulum undergoes simple harmonic motion when θ is small. Is the motion periodic when θ is large? How does the period of motion change as θ increases?

14. If a grandfather clock were running slow, how could we adjust the length of the pendulum to correct the time?

15. Will damped oscillations occur for any values of b and k? Explain.

16. Is it possible to have damped oscillations when a system is at resonance? Explain.

17. At resonance, what does the phase constant ϕ equal in Equation 15.35? (*Suggestion:* Compare this equation with the expression for the driving force, which must be in phase with the velocity at resonance.)

18. You stand on the end of a diving board and bounce to set it into oscillation. You find a maximum response, in terms of the amplitude of oscillation of the end of the board, when you bounce at frequency f. You now move to the middle of the board and repeat the experiment. Is the resonance frequency for forced oscillations at this point higher, lower, or the same as f? Why?

19. Some parachutes have holes in them to allow air to move smoothly through the chute. Without the holes, the air gathered under the chute as the parachutist falls is sometimes released from under the edges of the chute alternately and periodically from one side and then the other. Why might this periodic release of air cause a problem?

20. You are looking at a small tree. You do not notice any breeze, and most of the leaves on the tree are motionless. However, one leaf is fluttering back and forth wildly. After you wait for a while, that leaf stops moving and you notice a different leaf moving much more than all the others. Explain what could cause the large motion of one particular leaf.

21. A pendulum bob is made with a sphere filled with water. What would happen to the frequency of vibration of this pendulum if there were a hole in the sphere that allowed the water to leak out slowly?

PROBLEMS

1, 2, 3 = straightforward, intermediate, challenging □ = full solution available in the *Student Solutions Manual and Study Guide*

= coached solution with hints available at http://www.pse6.com = computer useful in solving problem

= paired numerical and symbolic problems

Note: Neglect the mass of every spring, except in problems 66 and 68.

Section 15.1 Motion of an Object Attached to a Spring

Problems 15, 16, 19, 23, 56, and 62 in Chapter 7 can also be assigned with this section.

1. A ball dropped from a height of 4.00 m makes a perfectly elastic collision with the ground. Assuming no mechanical energy is lost due to air resistance, (a) show that the ensuing motion is periodic and (b) determine the period of the motion. (c) Is the motion simple harmonic? Explain.

Section 15.2 Mathematical Representation of Simple Harmonic Motion

2. In an engine, a piston oscillates with simple harmonic motion so that its position varies according to the expression

$$x = (5.00 \text{ cm})\cos(2t + \pi/6)$$

where x is in centimeters and t is in seconds. At $t = 0$, find (a) the position of the piston, (b) its velocity, and (c) its acceleration. (d) Find the period and amplitude of the motion.

3. The position of a particle is given by the expression $x = (4.00 \text{ m})\cos(3.00\,\pi t + \pi)$, where x is in meters and t is in seconds. Determine (a) the frequency and period of the motion, (b) the amplitude of the motion, (c) the phase constant, and (d) the position of the particle at $t = 0.250$ s.

4. (a) A hanging spring stretches by 35.0 cm when an object of mass 450 g is hung on it at rest. In this situation, we define its position as $x = 0$. The object is pulled down an additional 18.0 cm and released from rest to oscillate without friction. What is its position x at a time 84.4 s later? (b) **What If?** A hanging spring stretches by 35.5 cm when an object of mass 440 g is hung on it at rest. We define this new position as $x = 0$. This object is also pulled down an additional 18.0 cm and released from rest to oscillate without friction. Find its position 84.4 s later. (c) Why are the answers to (a) and (b) different by such a large percentage when the data are so similar? Does this circumstance reveal a fundamental difficulty in calculating the future? (d) Find the distance traveled by the vibrating object in part (a). (e) Find the distance traveled by the object in part (b).

5. A particle moving along the x axis in simple harmonic motion starts from its equilibrium position, the origin, at $t = 0$ and moves to the right. The amplitude of its motion is 2.00 cm, and the frequency is 1.50 Hz. (a) Show that the position of the particle is given by

$$x = (2.00\text{ cm})\sin(3.00\pi t)$$

Determine (b) the maximum speed and the earliest time ($t > 0$) at which the particle has this speed, (c) the maximum acceleration and the earliest time ($t > 0$) at which the particle has this acceleration, and (d) the total distance traveled between $t = 0$ and $t = 1.00$ s.

6. The initial position, velocity, and acceleration of an object moving in simple harmonic motion are x_i, v_i, and a_i; the angular frequency of oscillation is ω. (a) Show that the position and velocity of the object for all time can be written as

$$x(t) = x_i \cos \omega t + \left(\frac{v_i}{\omega}\right) \sin \omega t$$

$$v(t) = -x_i\omega \sin \omega t + v_i \cos \omega t$$

(b) If the amplitude of the motion is A, show that

$$v^2 - ax = v_i^{\,2} - a_i x_i = \omega^2 A^2$$

7. A simple harmonic oscillator takes 12.0 s to undergo five complete vibrations. Find (a) the period of its motion, (b) the frequency in hertz, and (c) the angular frequency in radians per second.

8. A vibration sensor, used in testing a washing machine, consists of a cube of aluminum 1.50 cm on edge mounted on one end of a strip of spring steel (like a hacksaw blade) that lies in a vertical plane. The mass of the strip is small compared to that of the cube, but the length of the strip is large compared to the size of the cube. The other end of the strip is clamped to the frame of the washing machine, which is not operating. A horizontal force of 1.43 N applied to the cube is required to hold it 2.75 cm away from its equilibrium position. If the cube is released, what is its frequency of vibration?

9. A 7.00-kg object is hung from the bottom end of a vertical spring fastened to an overhead beam. The object is set into vertical oscillations having a period of 2.60 s. Find the force constant of the spring.

10. A piston in a gasoline engine is in simple harmonic motion. If the extremes of its position relative to its center point are ± 5.00 cm, find the maximum velocity and acceleration of the piston when the engine is running at the rate of 3 600 rev/min.

11. A 0.500-kg object attached to a spring with a force constant of 8.00 N/m vibrates in simple harmonic motion with an amplitude of 10.0 cm. Calculate (a) the maximum value of its speed and acceleration, (b) the speed and acceleration when the object is 6.00 cm from the equilibrium position, and (c) the time interval required for the object to move from $x = 0$ to $x = 8.00$ cm.

12. A 1.00-kg glider attached to a spring with a force constant of 25.0 N/m oscillates on a horizontal, frictionless air track. At $t = 0$ the glider is released from rest at $x = -3.00$ cm. (That is, the spring is compressed by 3.00 cm.) Find (a) the period of its motion, (b) the maximum values of its speed and acceleration, and (c) the position, velocity, and acceleration as functions of time.

13. A 1.00-kg object is attached to a horizontal spring. The spring is initially stretched by 0.100 m, and the object is released from rest there. It proceeds to move without friction. The next time the speed of the object is zero is 0.500 s later. What is the maximum speed of the object?

14. A particle that hangs from a spring oscillates with an angular frequency ω. The spring is suspended from the ceiling of an elevator car and hangs motionless (relative to the elevator car) as the car descends at a constant speed v. The car then stops suddenly. (a) With what amplitude does the particle oscillate? (b) What is the equation of motion for the particle? (Choose the upward direction to be positive.)

Section 15.3 Energy of the Simple Harmonic Oscillator

15. A block of unknown mass is attached to a spring with a spring constant of 6.50 N/m and undergoes simple harmonic motion with an amplitude of 10.0 cm. When the block is halfway between its equilibrium position and the end point, its speed is measured to be 30.0 cm/s. Calculate (a) the mass of the block, (b) the period of the motion, and (c) the maximum acceleration of the block.

16. A 200-g block is attached to a horizontal spring and executes simple harmonic motion with a period of 0.250 s. If the total energy of the system is 2.00 J, find (a) the force constant of the spring and (b) the amplitude of the motion.

17. An automobile having a mass of 1 000 kg is driven into a brick wall in a safety test. The bumper behaves like a spring of force constant 5.00×10^6 N/m and compresses 3.16 cm as the car is brought to rest. What was the speed of the car before impact, assuming that no mechanical energy is lost during impact with the wall?

18. A block–spring system oscillates with an amplitude of 3.50 cm. If the spring constant is 250 N/m and the mass of the block is 0.500 kg, determine (a) the mechanical energy of the system, (b) the maximum speed of the block, and (c) the maximum acceleration.

19. A 50.0-g object connected to a spring with a force constant of 35.0 N/m oscillates on a horizontal, frictionless surface with an amplitude of 4.00 cm. Find (a) the total energy of the system and (b) the speed of the object when the position is 1.00 cm. Find (c) the kinetic energy and (d) the potential energy when the position is 3.00 cm.

20. A 2.00-kg object is attached to a spring and placed on a horizontal, smooth surface. A horizontal force of 20.0 N is required to hold the object at rest when it is pulled 0.200 m from its equilibrium position (the origin of the x axis). The object is now released from rest with an initial position of $x_i = 0.200$ m, and it subsequently undergoes simple harmonic oscillations. Find (a) the force constant of the spring, (b) the frequency of the oscillations, and (c) the maximum speed of the object. Where does this maximum speed occur? (d) Find the maximum acceleration of the object. Where does it occur? (e) Find the total energy of the oscillating system. Find (f) the speed and (g) the acceleration of the object when its position is equal to one third of the maximum value.

21. The amplitude of a system moving in simple harmonic motion is doubled. Determine the change in (a) the total energy, (b) the maximum speed, (c) the maximum acceleration, and (d) the period.

22. A 65.0-kg bungee jumper steps off a bridge with a light bungee cord tied to herself and to the bridge (Figure P15.22). The unstretched length of the cord is 11.0 m. She reaches the bottom of her motion 36.0 m below the bridge before bouncing back. Her motion can be separated into an 11.0-m free fall and a 25.0-m section of simple harmonic oscillation. (a) For what time interval is she in free fall? (b) Use the principle of conservation of energy to find the spring constant of the bungee cord. (c) What is the location of the equilibrium point where the spring force balances the gravitational force acting on the jumper? Note that this point is taken as the origin in our mathematical description of simple harmonic oscillation. (d) What is the angular frequency of the oscillation? (e) What time interval is required for the cord to stretch by 25.0 m? (f) What is the total time interval for the entire 36.0-m drop?

Figure P15.22 Problems 22 and 58.

23. A particle executes simple harmonic motion with an amplitude of 3.00 cm. At what position does its speed equal half its maximum speed?

24. A cart attached to a spring with constant 3.24 N/m vibrates with position given by $x = (5.00 \text{ cm}) \cos(3.60t \text{ rad/s})$. (a) During the first cycle, for $0 < t < 1.75$ s, just when is the system's potential energy changing most rapidly into kinetic energy? (b) What is the maximum rate of energy transformation?

Section 15.4 Comparing Simple Harmonic Motion with Uniform Circular Motion

25. While riding behind a car traveling at 3.00 m/s, you notice that one of the car's tires has a small hemispherical bump on its rim, as in Figure P15.25. (a) Explain why the bump, from your viewpoint behind the car, executes simple harmonic motion. (b) If the radii of the car's tires are 0.300 m, what is the bump's period of oscillation?

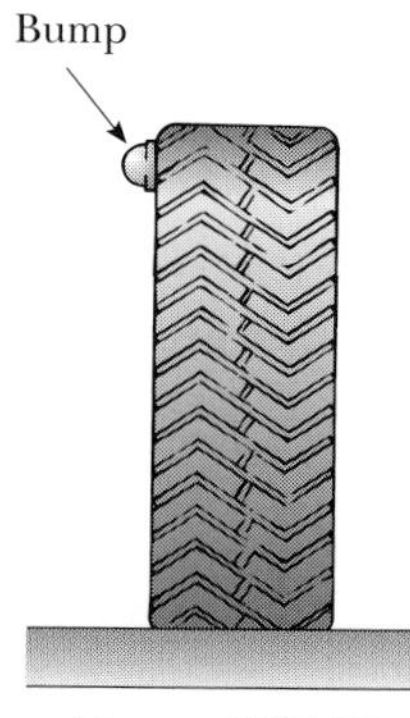

Figure P15.25

26. Consider the simplified single-piston engine in Figure P15.26. If the wheel rotates with constant angular speed, explain why the piston rod oscillates in simple harmonic motion.

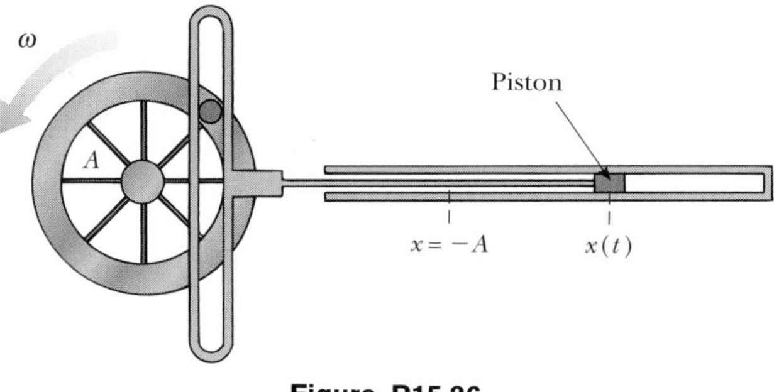

Figure P15.26

Section 15.5 The Pendulum

Problem 60 in Chapter 1 can also be assigned with this section.

27. A man enters a tall tower, needing to know its height. He notes that a long pendulum extends from the ceiling almost to the floor and that its period is 12.0 s. (a) How tall is the tower? (b) **What If?** If this pendulum is taken to the Moon, where the free-fall acceleration is 1.67 m/s^2, what is its period there?

28. A "seconds pendulum" is one that moves through its equilibrium position once each second. (The period of the pendulum is precisely 2 s.) The length of a seconds pendulum is 0.992 7 m at Tokyo, Japan and 0.994 2 m at Cambridge, England. What is the ratio of the free-fall accelerations at these two locations?

29. A rigid steel frame above a street intersection supports standard traffic lights, each of which is hinged to hang immediately below the frame. A gust of wind sets a light swinging in a vertical plane. Find the order of magnitude of its period. State the quantities you take as data and their values.

30. The angular position of a pendulum is represented by the equation $\theta = (0.320 \text{ rad})\cos \omega t$, where θ is in radians and $\omega = 4.43$ rad/s. Determine the period and length of the pendulum.

31. A simple pendulum has a mass of 0.250 kg and a length of 1.00 m. It is displaced through an angle of 15.0° and then released. What are (a) the maximum speed, (b) the maximum angular acceleration, and (c) the maximum restoring force? **What If?** Solve this problem by using the simple harmonic motion model for the motion of the pendulum, and then solve the problem more precisely by using more general principles.

32. **Review problem.** A simple pendulum is 5.00 m long. (a) What is the period of small oscillations for this pendulum if it is located in an elevator accelerating upward at 5.00 m/s^2? (b) What is its period if the elevator is accelerating downward at 5.00 m/s^2? (c) What is the period of this pendulum if it is placed in a truck that is accelerating horizontally at 5.00 m/s^2?

33. A particle of mass m slides without friction inside a hemispherical bowl of radius R. Show that, if it starts from rest with a small displacement from equilibrium, the particle moves in simple harmonic motion with an angular frequency equal to that of a simple pendulum of length R. That is, $\omega = \sqrt{g/R}$.

34. A small object is attached to the end of a string to form a simple pendulum. The period of its harmonic motion is measured for small angular displacements and three lengths, each time clocking the motion with a stopwatch for 50 oscillations. For lengths of 1.000 m, 0.750 m, and 0.500 m, total times of 99.8 s, 86.6 s, and 71.1 s are measured for 50 oscillations. (a) Determine the period of motion for each length. (b) Determine the mean value of g obtained from these three independent measurements, and compare it with the accepted value. (c) Plot T^2 versus L, and obtain a value for g from the slope of your best-fit straight-line graph. Compare this value with that obtained in part (b).

35. A physical pendulum in the form of a planar body moves in simple harmonic motion with a frequency of 0.450 Hz. If the pendulum has a mass of 2.20 kg and the pivot is located 0.350 m from the center of mass, determine the moment of inertia of the pendulum about the pivot point.

36. A very light rigid rod with a length of 0.500 m extends straight out from one end of a meter stick. The stick is suspended from a pivot at the far end of the rod and is set into oscillation. (a) Determine the period of oscillation. *Suggestion:* Use the parallel-axis theorem from Section 10.5. (b) By what percentage does the period differ from the period of a simple pendulum 1.00 m long?

37. Consider the physical pendulum of Figure 15.18. (a) If its moment of inertia about an axis passing through its center of mass and parallel to the axis passing through its pivot point is I_{CM}, show that its period is

$$T = 2\pi\sqrt{\frac{I_{CM} + md^2}{mgd}}$$

where d is the distance between the pivot point and center of mass. (b) Show that the period has a minimum value when d satisfies $md^2 = I_{CM}$.

38. A torsional pendulum is formed by taking a meter stick of mass 2.00 kg, and attaching to its center a wire. With its upper end clamped, the vertical wire supports the stick as the stick turns in a horizontal plane. If the resulting period is 3.00 minutes, what is the torsion constant for the wire?

39. A clock balance wheel (Fig. P15.39) has a period of oscillation of 0.250 s. The wheel is constructed so that its mass of 20.0 g is concentrated around a rim of radius 0.500 cm. What are (a) the wheel's moment of inertia and (b) the torsion constant of the attached spring?

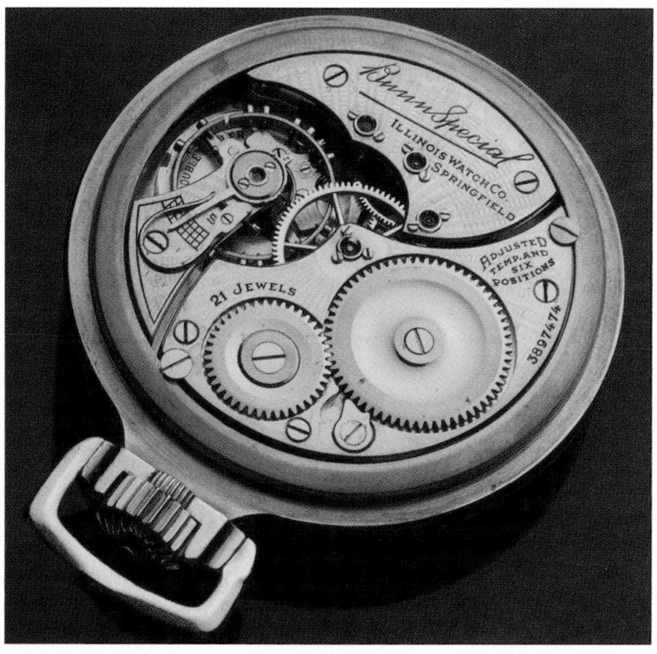

George Semple

Figure P15.39

Section 15.6 Damped Oscillations

40. Show that the time rate of change of mechanical energy for a damped, undriven oscillator is given by $dE/dt = -bv^2$ and hence is always negative. Proceed as follows: Differentiate the expression for the mechanical energy of an oscillator, $E = \frac{1}{2}mv^2 + \frac{1}{2}kx^2$, and use Equation 15.31.

41. A pendulum with a length of 1.00 m is released from an initial angle of 15.0°. After 1 000 s, its amplitude has been reduced by friction to 5.50°. What is the value of $b/2m$?

42. Show that Equation 15.32 is a solution of Equation 15.31 provided that $b^2 < 4mk$.

43. A 10.6-kg object oscillates at the end of a vertical spring that has a spring constant of 2.05×10^4 N/m. The effect of air resistance is represented by the damping coefficient $b = 3.00$ N·s/m. (a) Calculate the frequency of the damped oscillation. (b) By what percentage does the amplitude of the oscillation decrease in each cycle? (c) Find the time interval that elapses while the energy of the system drops to 5.00% of its initial value.

Section 15.7 Forced Oscillations

44. The front of her sleeper wet from teething, a baby rejoices in the day by crowing and bouncing up and down in her crib. Her mass is 12.5 kg, and the crib mattress can be modeled as a light spring with force constant 4.30 kN/m. (a) The baby soon learns to bounce with maximum amplitude and minimum effort by bending her knees at what frequency? (b) She learns to use the mattress as a trampoline—losing contact with it for part of each cycle—when her amplitude exceeds what value?

45. A 2.00-kg object attached to a spring moves without friction and is driven by an external force given by $F = (3.00\text{ N})\sin(2\pi t)$. If the force constant of the spring is 20.0 N/m, determine (a) the period and (b) the amplitude of the motion.

46. Considering an undamped, forced oscillator ($b = 0$), show that Equation 15.35 is a solution of Equation 15.34, with an amplitude given by Equation 15.36.

47. A weight of 40.0 N is suspended from a spring that has a force constant of 200 N/m. The system is undamped and is subjected to a harmonic driving force of frequency 10.0 Hz, resulting in a forced-motion amplitude of 2.00 cm. Determine the maximum value of the driving force.

48. Damping is negligible for a 0.150-kg object hanging from a light 6.30-N/m spring. A sinusoidal force with an amplitude of 1.70 N drives the system. At what frequency will the force make the object vibrate with an amplitude of 0.440 m?

49. You are a research biologist. You take your emergency pager along to a fine restaurant. You switch the small pager to vibrate instead of beep, and you put it into a side pocket of your suit coat. The arm of your chair presses the light cloth against your body at one spot. Fabric with a length of 8.21 cm hangs freely below that spot, with the pager at the bottom. A coworker urgently needs instructions and calls you from your laboratory. The motion of the pager makes the hanging part of your coat swing back and forth with remarkably large amplitude. The waiter and nearby diners notice immediately and fall silent. Your daughter pipes up and says, "Daddy, look! Your cockroaches must have gotten out again!" Find the frequency at which your pager vibrates.

50. Four people, each with a mass of 72.4 kg, are in a car with a mass of 1 130 kg. An earthquake strikes. The driver manages to pull off the road and stop, as the vertical oscillations of the ground surface make the car bounce up and down on its suspension springs. When the frequency of the shaking is 1.80 Hz, the car exhibits a maximum amplitude of vibration. The earthquake ends, and the four people leave the car as fast as they can. By what distance does the car's undamaged suspension lift the car body as the people get out?

Additional Problems

51. A small ball of mass M is attached to the end of a uniform rod of equal mass M and length L that is pivoted at the top (Fig. P15.51). (a) Determine the tensions in the rod at the pivot and at the point P when the system is stationary. (b) Calculate the period of oscillation for small displacements from equilibrium, and determine this period for $L = 2.00$ m. (*Suggestions*: Model the object at the end of the rod as a particle and use Eq. 15.28.)

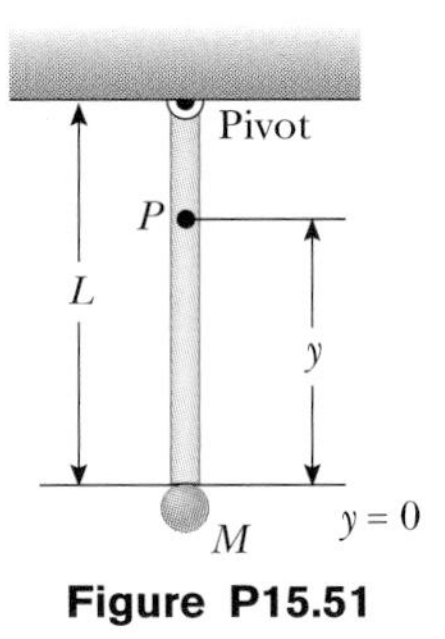

Figure P15.51

52. An object of mass $m_1 = 9.00$ kg is in equilibrium while connected to a light spring of constant $k = 100$ N/m that is fastened to a wall as shown in Figure P15.52a. A second object, $m_2 = 7.00$ kg, is slowly pushed up against m_1, compressing the spring by the amount $A = 0.200$ m, (see Figure P15.52b). The system is then released, and both objects start moving to the right on the frictionless surface. (a) When m_1 reaches the equilibrium point, m_2 loses contact with m_1 (see Fig. P15.5c) and moves to the right with speed v. Determine the value of v. (b) How far apart are the objects when the spring is fully stretched for the first time (D in Fig. P15.52d)? (*Suggestion:* First determine the period of oscillation and the amplitude of the m_1–spring system after m_2 loses contact with m_1.)

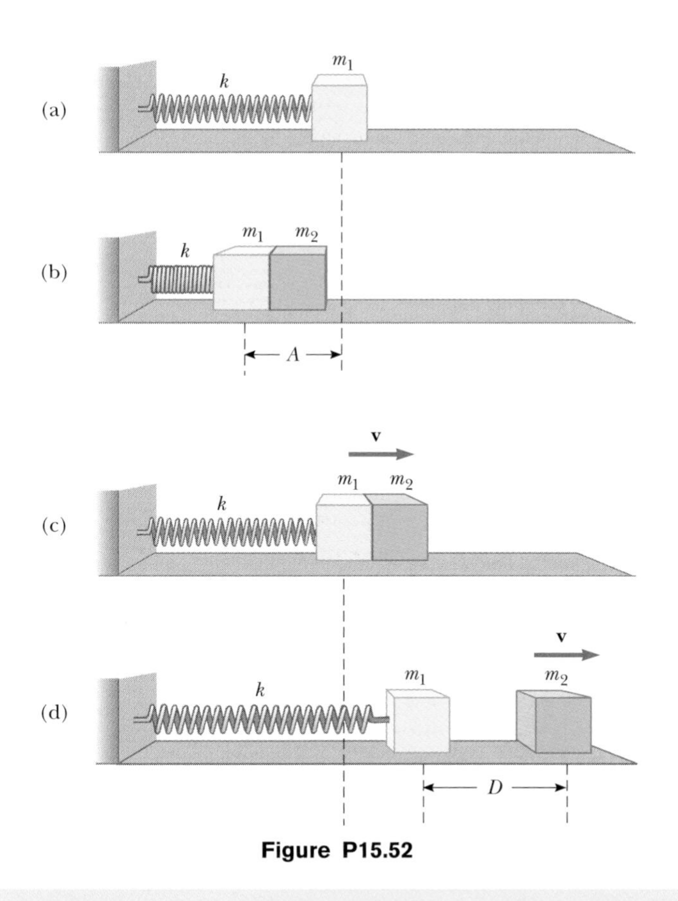

Figure P15.52

53. A large block P executes horizontal simple harmonic motion as it slides across a frictionless surface with a frequency $f = 1.50$ Hz. Block B rests on it, as shown in Figure P15.53, and the coefficient of static friction between the two is $\mu_s = 0.600$. What maximum amplitude of oscillation can the system have if block B is not to slip?

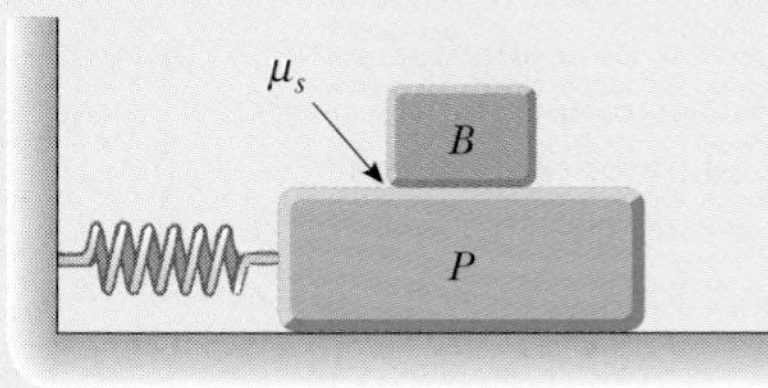

Figure P15.53 Problems 53 and 54.

54. A large block P executes horizontal simple harmonic motion as it slides across a frictionless surface with a frequency f. Block B rests on it, as shown in Figure P15.53, and the coefficient of static friction between the two is μ_s. What maximum amplitude of oscillation can the system have if the upper block is not to slip?

55. The mass of the deuterium molecule (D_2) is twice that of the hydrogen molecule (H_2). If the vibrational frequency of H_2 is 1.30×10^{14} Hz, what is the vibrational frequency of D_2? Assume that the "spring constant" of attracting forces is the same for the two molecules.

56. A solid sphere (radius = R) rolls without slipping in a cylindrical trough (radius = $5R$) as shown in Figure P15.56. Show that, for small displacements from equilibrium perpendicular to the length of the trough, the sphere executes simple harmonic motion with a period $T = 2\pi\sqrt{28R/5g}$.

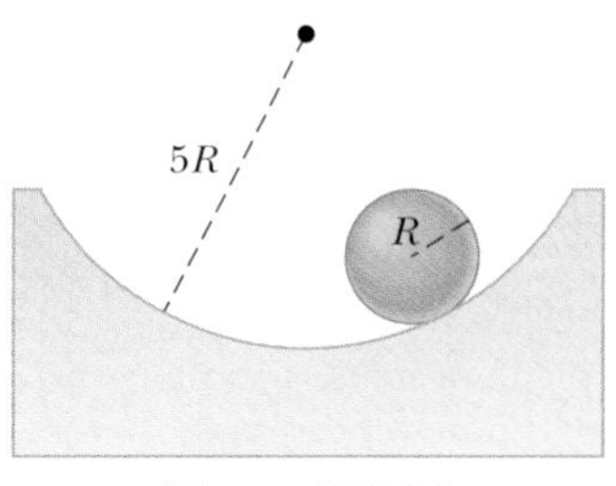

Figure P15.56

57. A light, cubical container of volume a^3 is initially filled with a liquid of mass density ρ. The cube is initially supported by a light string to form a simple pendulum of length L_i, measured from the center of mass of the filled container, where $L_i \gg a$. The liquid is allowed to flow from the bottom of the container at a constant rate (dM/dt). At any time t, the level of the fluid in the container is h and the length of the pendulum is L (measured relative to the instantaneous center of mass). (a) Sketch the apparatus and label the dimensions a, h, L_i, and L. (b) Find the time rate of change of the period as a function of time t. (c) Find the period as a function of time.

58. After a thrilling plunge, bungee-jumpers bounce freely on the bungee cord through many cycles (Fig. P15.22). After the first few cycles, the cord does not go slack. Your little brother can make a pest of himself by figuring out the mass of each person, using a proportion which you set up by solving this problem: An object of mass m is oscillating freely on a vertical spring with a period T. An object of unknown mass m' on the same spring oscillates with a period T'. Determine (a) the spring constant and (b) the unknown mass.

59. A pendulum of length L and mass M has a spring of force constant k connected to it at a distance h below its point of suspension (Fig. P15.59). Find the frequency of vibration

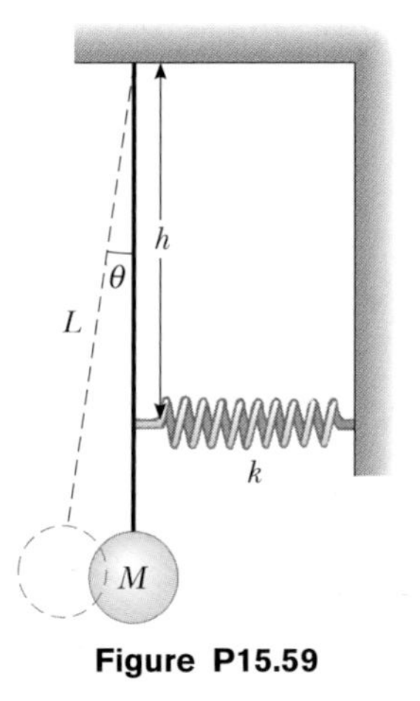

Figure P15.59

of the system for small values of the amplitude (small θ). Assume the vertical suspension of length L is rigid, but ignore its mass.

60. A particle with a mass of 0.500 kg is attached to a spring with a force constant of 50.0 N/m. At time $t = 0$ the particle has its maximum speed of 20.0 m/s and is moving to the left. (a) Determine the particle's equation of motion, specifying its position as a function of time. (b) Where in the motion is the potential energy three times the kinetic energy? (c) Find the length of a simple pendulum with the same period. (d) Find the minimum time interval required for the particle to move from $x = 0$ to $x = 1.00$ m.

61. A horizontal plank of mass m and length L is pivoted at one end. The plank's other end is supported by a spring of force constant k (Fig P15.61). The moment of inertia of the plank about the pivot is $\frac{1}{3}mL^2$. The plank is displaced by a small angle θ from its horizontal equilibrium position and released. (a) Show that it moves with simple harmonic motion with an angular frequency $\omega = \sqrt{3k/m}$. (b) Evaluate the frequency if the mass is 5.00 kg and the spring has a force constant of 100 N/m.

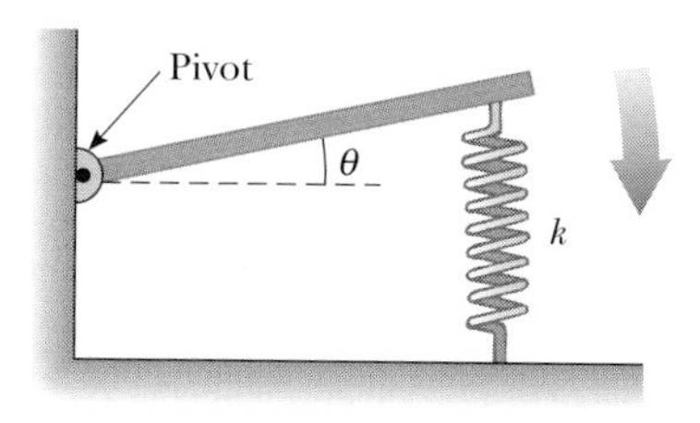

Figure P15.61

62. **Review problem.** A particle of mass 4.00 kg is attached to a spring with a force constant of 100 N/m. It is oscillating on a horizontal frictionless surface with an amplitude of 2.00 m. A 6.00-kg object is dropped vertically on top of the 4.00-kg object as it passes through its equilibrium point. The two objects stick together. (a) By how much does the amplitude of the vibrating system change as a result of the collision? (b) By how much does the period change? (c) By how much does the energy change? (d) Account for the change in energy.

63. A simple pendulum with a length of 2.23 m and a mass of 6.74 kg is given an initial speed of 2.06 m/s at its equilibrium position. Assume it undergoes simple harmonic motion, and determine its (a) period, (b) total energy, and (c) maximum angular displacement.

64. **Review problem.** One end of a light spring with force constant 100 N/m is attached to a vertical wall. A light string is tied to the other end of the horizontal spring. The string changes from horizontal to vertical as it passes over a solid pulley of diameter 4.00 cm. The pulley is free to turn on a fixed smooth axle. The vertical section of the string supports a 200-g object. The string does not slip at its contact with the pulley. Find the frequency of oscillation of the object if the mass of the pulley is (a) negligible, (b) 250 g, and (c) 750 g.

65. People who ride motorcycles and bicycles learn to look out for bumps in the road, and especially for *washboarding*, a condition in which many equally spaced ridges are worn into the road. What is so bad about washboarding? A motorcycle has several springs and shock absorbers in its suspension, but you can model it as a single spring supporting a block. You can estimate the force constant by thinking about how far the spring compresses when a big biker sits down on the seat. A motorcyclist traveling at highway speed must be particularly careful of washboard bumps that are a certain distance apart. What is the order of magnitude of their separation distance? State the quantities you take as data and the values you measure or estimate for them.

66. A block of mass M is connected to a spring of mass m and oscillates in simple harmonic motion on a horizontal, frictionless track (Fig. P15.66). The force constant of the spring is k and the equilibrium length is ℓ. Assume that all portions of the spring oscillate in phase and that the velocity of a segment dx is proportional to the distance x from the fixed end; that is, $v_x = (x/\ell)v$. Also, note that the mass of a segment of the spring is $dm = (m/\ell)dx$. Find (a) the kinetic energy of the system when the block has a speed v and (b) the period of oscillation.

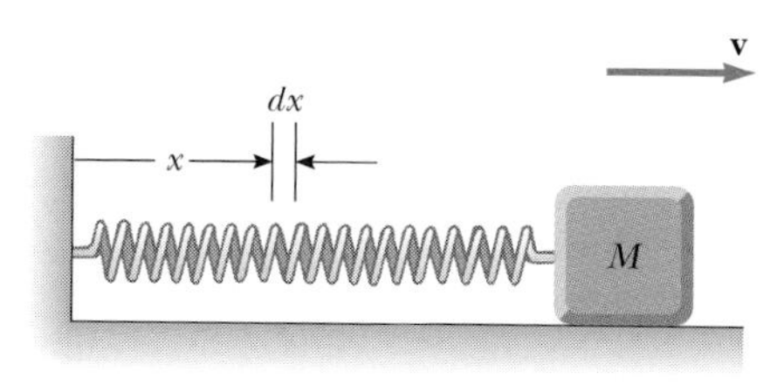

Figure P15.66

67. A ball of mass m is connected to two rubber bands of length L, each under tension T, as in Figure P15.67. The ball is displaced by a small distance y perpendicular to the length of the rubber bands. Assuming that the tension does not change, show that (a) the restoring force is $-(2T/L)y$ and (b) the system exhibits simple harmonic motion with an angular frequency $\omega = \sqrt{2T/mL}$.

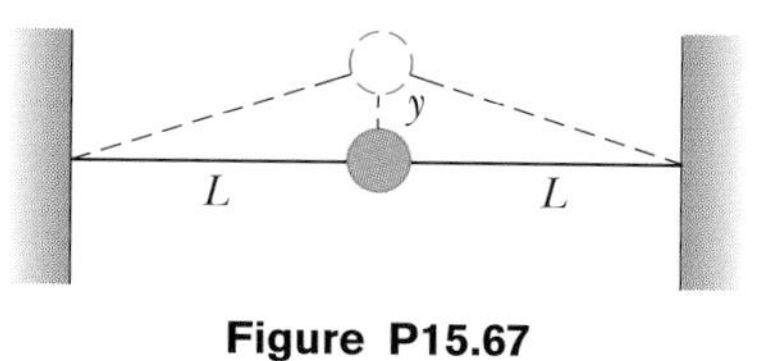

Figure P15.67

68. When a block of mass M, connected to the end of a spring of mass $m_s = 7.40$ g and force constant k, is set into simple harmonic motion, the period of its motion is

$$T = 2\pi\sqrt{\frac{M + (m_s/3)}{k}}$$

A two-part experiment is conducted with the use of blocks of various masses suspended vertically from the spring, as shown in Figure P15.68. (a) Static extensions of 17.0, 29.3, 35.3, 41.3, 47.1, and 49.3 cm are measured for M values of 20.0, 40.0, 50.0, 60.0, 70.0, and 80.0 g, respectively. Construct a graph of Mg versus x, and perform a linear least-squares fit to the data. From the slope of your graph, determine a value for k for this spring. (b) The system is now set into simple harmonic motion, and periods are measured with a stopwatch. With $M = 80.0$ g, the total time for 10 oscillations is measured to be 13.41 s. The experiment is repeated with M values of 70.0, 60.0, 50.0, 40.0, and 20.0 g, with corresponding times for 10 oscillations of 12.52, 11.67, 10.67, 9.62, and 7.03 s. Compute the experimental value for T from each of these measurements. Plot a graph of T^2 versus M, and determine a value for k from the slope of the linear least-squares fit through the data points. Compare this value of k with that obtained in part (a). (c) Obtain a value for m_s from your graph and compare it with the given value of 7.40 g.

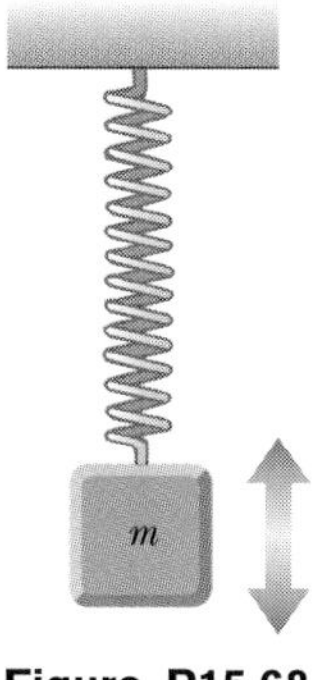

Figure P15.68

69. A smaller disk of radius r and mass m is attached rigidly to the face of a second larger disk of radius R and mass M as shown in Figure P15.69. The center of the small disk is located at the edge of the large disk. The large disk is mounted at its center on a frictionless axle. The assembly is rotated through a small angle θ from its equilibrium position and released. (a) Show that the speed of the center of the small disk as it passes through the equilibrium position is

$$v = 2\left[\frac{Rg(1 - \cos\theta)}{(M/m) + (r/R)^2 + 2}\right]^{1/2}$$

(b) Show that the period of the motion is

$$T = 2\pi\left[\frac{(M + 2m)R^2 + mr^2}{2mgR}\right]^{1/2}$$

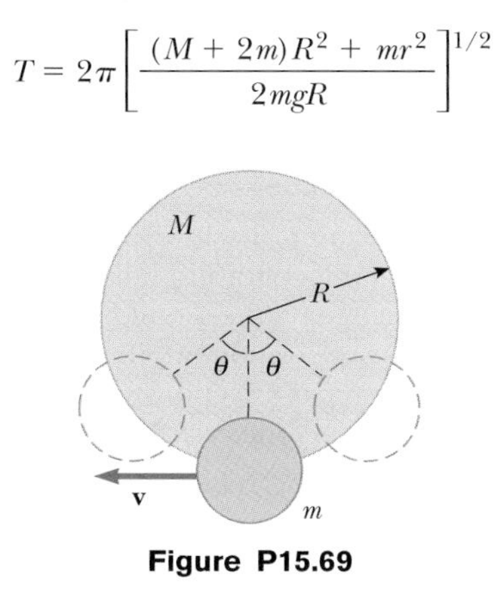

Figure P15.69

70. Consider a damped oscillator as illustrated in Figures 15.21 and 15.22. Assume the mass is 375 g, the spring constant is 100 N/m, and $b = 0.100$ N·s/m. (a) How long does it takes for the amplitude to drop to half its initial value? (b) **What If?** How long does it take for the mechanical energy to drop to half its initial value? (c) Show that, in general, the fractional rate at which the amplitude decreases in a damped harmonic oscillator is half the fractional rate at which the mechanical energy decreases.

71. A block of mass m is connected to two springs of force constants k_1 and k_2 as shown in Figures P15.71a and P15.71b. In each case, the block moves on a frictionless table after it is displaced from equilibrium and released. Show that in the two cases the block exhibits simple harmonic motion with periods

(a) $$T = 2\pi\sqrt{\frac{m(k_1 + k_2)}{k_1k_2}}$$

(b) $$T = 2\pi\sqrt{\frac{m}{k_1 + k_2}}$$

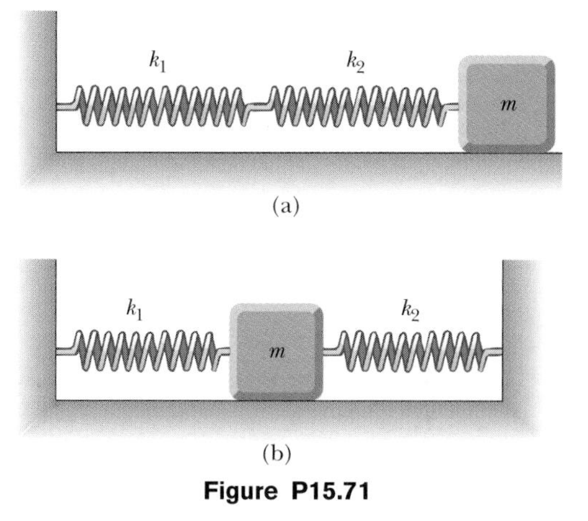

Figure P15.71

72. A lobsterman's buoy is a solid wooden cylinder of radius r and mass M. It is weighted at one end so that it floats upright in calm sea water, having density ρ. A passing shark tugs on the slack rope mooring the buoy to a lobster trap, pulling the buoy down a distance x from its equilibrium position and releasing it. Show that the buoy will execute simple harmonic motion if the resistive effects of the water are neglected, and determine the period of the oscillations.

73. Consider a bob on a light stiff rod, forming a simple pendulum of length $L = 1.20$ m. It is displaced from the vertical by an angle θ_{max} and then released. Predict the subsequent angular positions if θ_{max} is small or if it is large. Proceed as follows: Set up and carry out a numerical method to integrate the equation of motion for the simple pendulum:

$$\frac{d^2\theta}{dt^2} = -\frac{g}{L}\sin\theta$$

Take the initial conditions to be $\theta = \theta_{max}$ and $d\theta/dt = 0$ at $t = 0$. On one trial choose $\theta_{max} = 5.00°$, and on another trial take $\theta_{max} = 100°$. In each case find the position θ as a function of time. Using the same values of θ_{max}, compare your results for θ with those obtained from $\theta(t) = \theta_{max}\cos\omega t$. How does the period for the large value of θ_{max} compare with that for the small value of θ_{max}? *Note:* Using the Euler method to solve this differential equation, you may find that the amplitude tends to increase with time. The fourth-order Runge–Kutta method would be a better choice to solve the differential equation. However, if you choose Δt small enough, the solution using Euler's method can still be good.

74. Your thumb squeaks on a plate you have just washed. Your sneakers often squeak on the gym floor. Car tires squeal when you start or stop abruptly. You can make a goblet sing by wiping your moistened finger around its rim. As you slide it across the table, a Styrofoam cup may not make much sound, but it makes the surface of some water inside it dance in a complicated resonance vibration. When chalk squeaks on a blackboard, you can see that it makes a row of regularly spaced dashes. As these examples suggest, vibration commonly results when friction acts on a moving elastic object. The oscillation is not simple harmonic motion, but is called *stick-and-slip*. This problem models stick-and-slip motion.

A block of mass m is attached to a fixed support by a horizontal spring with force constant k and negligible mass (Fig. P15.74). Hooke's law describes the spring both in extension and in compression. The block sits on a long horizontal board, with which it has coefficient of static friction μ_s and a smaller coefficient of kinetic friction μ_k. The board moves to the right at constant speed v. Assume that the block spends most of its time sticking to the board and moving to the right, so that the speed v is small in comparison to the average speed the block has as it slips back toward the left. (a) Show that the maximum extension of the spring from its unstressed position is very nearly given by $\mu_s mg/k$. (b) Show that the block oscillates around an equilibrium position at which the spring is stretched by $\mu_k mg/k$. (c) Graph the block's position versus time. (d) Show that the amplitude of the block's motion is

$$A = \frac{(\mu_s - \mu_k)\,mg}{k}$$

(e) Show that the period of the block's motion is

$$T = \frac{2(\mu_s - \mu_k)\,mg}{vk} + \pi\sqrt{\frac{m}{k}}$$

(f) Evaluate the frequency of the motion if $\mu_s = 0.400$, $\mu_k = 0.250$, $m = 0.300$ kg, $k = 12.0$ N/m, and $v = 2.40$ cm/s. (g) **What If?** What happens to the frequency if the mass increases? (h) If the spring constant increases? (i) If the speed of the board increases? (j) If the coefficient of static friction increases relative to the coefficient of kinetic friction? Note that it is the excess of static over kinetic friction that is important for the vibration. "The squeaky wheel gets the grease" because even a viscous fluid cannot exert a force of static friction.

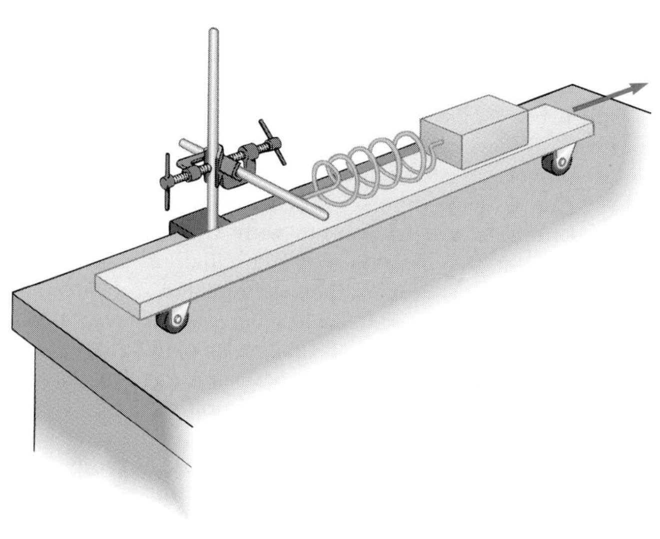

Figure P15.74

75. **Review problem.** Imagine that a hole is drilled through the center of the Earth to the other side. An object of mass m at a distance r from the center of the Earth is pulled toward the center of the Earth only by the mass within the sphere of radius r (the reddish region in Fig. P15.75). (a) Write Newton's law of gravitation for an object at the distance r from the center of the Earth, and show that the force on it is of Hooke's law form, $F = -kr$, where the effective force constant is $k = (4/3)\pi\rho Gm$. Here ρ is the density of the Earth, assumed uniform, and G is the gravitational constant. (b) Show that a sack of mail dropped into the hole will execute simple harmonic motion if it moves without friction. When will it arrive at the other side of the Earth?

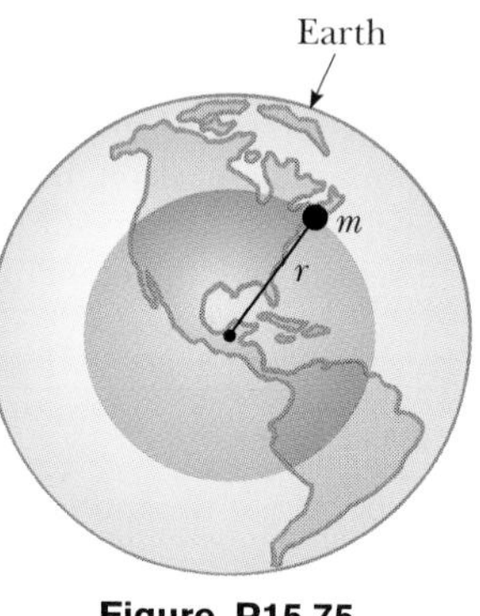

Figure P15.75

Answers to Quick Quizzes

15.1 (d). From its maximum positive position to the equilibrium position, the block travels a distance A. It then goes an equal distance past the equilibrium position to its maximum negative position. It then repeats these two motions in the reverse direction to return to its original position and complete one cycle.

15.2 (f). The object is in the region $x < 0$, so the position is negative. Because the object is moving back toward the origin in this region, the velocity is positive.

15.3 (a). The amplitude is larger because the curve for Object B shows that the displacement from the origin (the vertical axis on the graph) is larger. The frequency is larger for Object B because there are more oscillations per unit time interval.

15.4 (a). The velocity is positive, as in Quick Quiz 15.2. Because the spring is pulling the object toward equilibrium from the negative x region, the acceleration is also positive.

15.5 (b). According to Equation 15.13, the period is proportional to the square root of the mass.

15.6 (c). The amplitude of the simple harmonic motion is the same as the radius of the circular motion. The initial position of the object in its circular motion is π radians from the positive x axis.

15.7 (a). With a longer length, the period of the pendulum will increase. Thus, it will take longer to execute each swing, so that each second according to the clock will take longer than an actual second—the clock will run *slow.*

15.8 (a). At the top of the mountain, the value of g is less than that at sea level. As a result, the period of the pendulum will increase and the clock will run slow.

15.9 (a). If your goal is simply to stop the bounce from an absorbed shock as rapidly as possible, you should critically damp the suspension. Unfortunately, the stiffness of this design makes for an uncomfortable ride. If you underdamp the suspension, the ride is more comfortable but the car bounces. If you overdamp the suspension, the wheel is displaced from its equilibrium position longer than it should be. (For example, after hitting a bump, the spring stays compressed for a short time and the wheel does not quickly drop back down into contact with the road after the wheel is past the bump—a dangerous situation.) Because of all these considerations, automotive engineers usually design suspensions to be slightly underdamped. This allows the suspension to absorb a shock rapidly (minimizing the roughness of the ride) and then return to equilibrium after only one or two noticeable oscillations.

Chapter 16

Wave Motion

CHAPTER OUTLINE

16.1 Propagation of a Disturbance

16.2 Sinusoidal Waves

16.3 The Speed of Waves on Strings

16.4 Reflection and Transmission

16.5 Rate of Energy Transfer by Sinusoidal Waves on Strings

16.6 The Linear Wave Equation

▲ *The rich sound of a piano is due to waves on strings that are under tension. Many such strings can be seen in this photograph. Waves also travel on the soundboard, which is visible below the strings. In this chapter, we study the fundamental principles of wave phenomena. (Kathy Ferguson Johnson/PhotoEdit/PictureQuest)*

Most of us experienced waves as children when we dropped a pebble into a pond. At the point where the pebble hits the water's surface, waves are created. These waves move outward from the creation point in expanding circles until they reach the shore. If you were to examine carefully the motion of a beach ball floating on the disturbed water, you would see that the ball moves vertically and horizontally about its original position but does not undergo any net displacement away from or toward the point where the pebble hit the water. The small elements of water in contact with the beach ball, as well as all the other water elements on the pond's surface, behave in the same way. That is, the water *wave* moves from the point of origin to the shore, but the water is not carried with it.

The world is full of waves, the two main types being *mechanical* waves and *electromagnetic* waves. In the case of mechanical waves, some physical medium is being disturbed—in our pebble and beach ball example, elements of water are disturbed. Electromagnetic waves do not require a medium to propagate; some examples of electromagnetic waves are visible light, radio waves, television signals, and x-rays. Here, in this part of the book, we study only mechanical waves.

The wave concept is abstract. When we observe what we call a water wave, what we see is a rearrangement of the water's surface. Without the water, there would be no wave. A wave traveling on a string would not exist without the string. Sound waves could not travel from one point to another if there were no air molecules between the two points. With mechanical waves, what we interpret as a wave corresponds to the propagation of a disturbance through a medium.

Considering further the beach ball floating on the water, note that we have caused the ball to move at one point in the water by dropping a pebble at another location. The ball has gained kinetic energy from our action, so energy must have transferred from the point at which we drop the pebble to the position of the ball. This is a central feature of wave motion—*energy* is transferred over a distance, but *matter* is not.

All waves carry energy, but the amount of energy transmitted through a medium and the mechanism responsible for that transport of energy differ from case to case. For instance, the power of ocean waves during a storm is much greater than the power of sound waves generated by a single human voice.

16.1 Propagation of a Disturbance

In the introduction, we alluded to the essence of wave motion—the transfer of energy through space without the accompanying transfer of matter. In the list of energy transfer mechanisms in Chapter 7, two mechanisms depend on waves—mechanical waves and electromagnetic radiation. By contrast, in another mechanism—matter transfer—the energy transfer is accompanied by a movement of matter through space.

All mechanical waves require (1) some source of disturbance, (2) a medium that can be disturbed, and (3) some physical mechanism through which elements of the medium can influence each other. One way to demonstrate wave

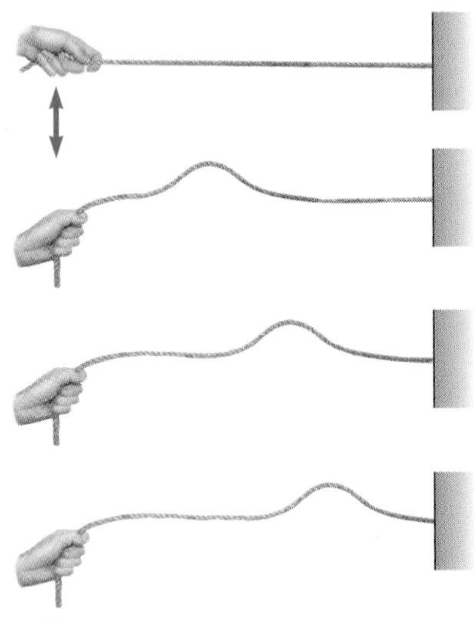

Figure 16.1 A pulse traveling down a stretched rope. The shape of the pulse is approximately unchanged as it travels along the rope.

motion is to flick one end of a long rope that is under tension and has its opposite end fixed, as shown in Figure 16.1. In this manner, a single bump (called a *pulse*) is formed and travels along the rope with a definite speed. Figure 16.1 represents four consecutive "snapshots" of the creation and propagation of the traveling pulse. The rope is the medium through which the pulse travels. The pulse has a definite height and a definite speed of propagation along the medium (the rope). As we shall see later, the properties of this particular medium that determine the speed of the disturbance are the tension in the rope and its mass per unit length. The shape of the pulse changes very little as it travels along the rope.[1]

We shall first focus our attention on a pulse traveling through a medium. Once we have explored the behavior of a pulse, we will then turn our attention to a *wave*, which is a *periodic* disturbance traveling through a medium. We created a pulse on our rope by flicking the end of the rope once, as in Figure 16.1. If we were to move the end of the rope up and down repeatedly, we would create a traveling wave, which has characteristics that a pulse does not have. We shall explore these characteristics in Section 16.2.

As the pulse in Figure 16.1 travels, each disturbed element of the rope moves in a direction *perpendicular* to the direction of propagation. Figure 16.2 illustrates this point for one particular element, labeled P. Note that no part of the rope ever moves in the direction of the propagation.

A traveling wave or pulse that causes the elements of the disturbed medium to move perpendicular to the direction of propagation is called a **transverse wave.**

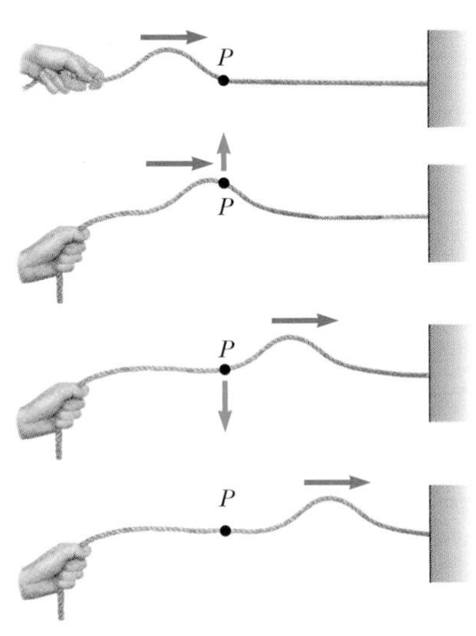

Figure 16.2 A transverse pulse traveling on a stretched rope. The direction of motion of any element P of the rope (blue arrows) is perpendicular to the direction of propagation (red arrows).

Compare this with another type of pulse—one moving down a long, stretched spring, as shown in Figure 16.3. The left end of the spring is pushed briefly to the right and then pulled briefly to the left. This movement creates a sudden compression of a region of the coils. The compressed region travels along the spring (to the right in Figure 16.3). The compressed region is followed by a region where the coils are extended. Notice that the direction of the displacement of the coils is *parallel* to the direction of propagation of the compressed region.

A traveling wave or pulse that causes the elements of the medium to move parallel to the direction of propagation is called a **longitudinal wave.**

Sound waves, which we shall discuss in Chapter 17, are another example of longitudinal waves. The disturbance in a sound wave is a series of high-pressure and low-pressure regions that travel through air.

Some waves in nature exhibit a combination of transverse and longitudinal displacements. Surface water waves are a good example. When a water wave travels on the surface of deep water, elements of water at the surface move in nearly circular paths, as shown in Figure 16.4. Note that the disturbance has both transverse and longitudinal

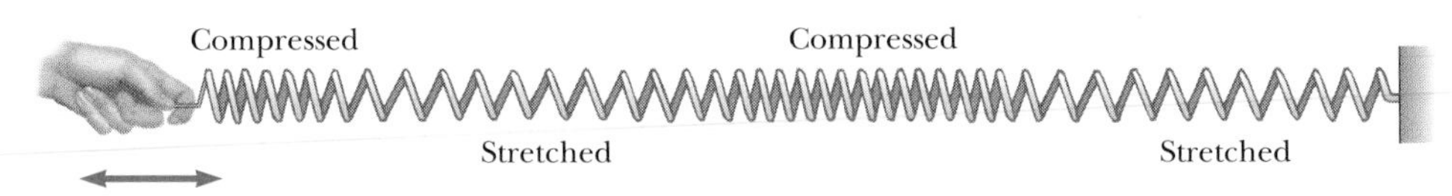

Figure 16.3 A longitudinal pulse along a stretched spring. The displacement of the coils is parallel to the direction of the propagation.

[1] In reality, the pulse changes shape and gradually spreads out during the motion. This effect is called *dispersion* and is common to many mechanical waves as well as to electromagnetic waves. We do not consider dispersion in this chapter.

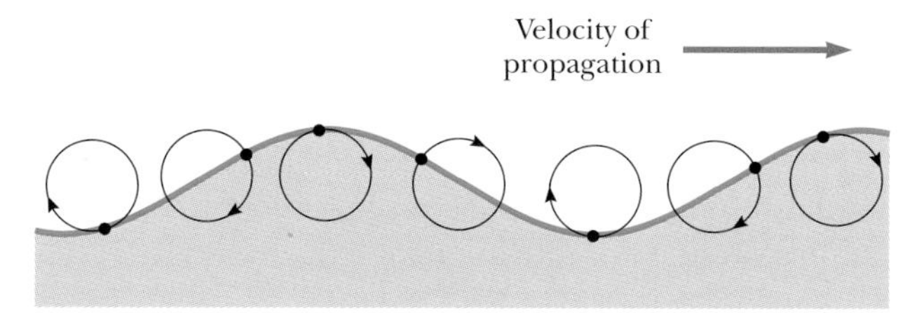

Active Figure 16.4 The motion of water elements on the surface of deep water in which a wave is propagating is a combination of transverse and longitudinal displacements, with the result that elements at the surface move in nearly circular paths. Each element is displaced both horizontally and vertically from its equilibrium position.

At the Active Figures link at* http://www.pse6.com, *you can observe the displacement of water elements at the surface of the moving waves.

components. The transverse displacements seen in Figure 16.4 represent the variations in vertical position of the water elements. The longitudinal displacement can be explained as follows: as the wave passes over the water's surface, water elements at the highest points move in the direction of propagation of the wave, whereas elements at the lowest points move in the direction opposite the propagation.

The three-dimensional waves that travel out from points under the Earth's surface along a fault at which an earthquake occurs are of both types—transverse and longitudinal. The longitudinal waves are the faster of the two, traveling at speeds in the range of 7 to 8 km/s near the surface. These are called **P waves** (with "P" standing for *primary*) because they travel faster than the transverse waves and arrive at a seismograph (a device used to detect waves due to earthquakes) first. The slower transverse waves, called **S waves** (with "S" standing for *secondary*), travel through the Earth at 4 to 5 km/s near the surface. By recording the time interval between the arrivals of these two types of waves at a seismograph, the distance from the seismograph to the point of origin of the waves can be determined. A single measurement establishes an imaginary sphere centered on the seismograph, with the radius of the sphere determined by the difference in arrival times of the P and S waves. The origin of the waves is located somewhere on that sphere. The imaginary spheres from three or more monitoring stations located far apart from each other intersect at one region of the Earth, and this region is where the earthquake occurred.

Consider a pulse traveling to the right on a long string, as shown in Figure 16.5. Figure 16.5a represents the shape and position of the pulse at time $t = 0$. At this time, the shape of the pulse, whatever it may be, can be represented by some mathematical function which we will write as $y(x, 0) = f(x)$. This function describes the transverse position y of the element of the string located at each value of x at time $t = 0$. Because the speed of the pulse is v, the pulse has traveled to the right a distance vt at the time t (Fig. 16.5b). We assume that the shape of the pulse does not change with time. Thus, at time t, the shape of the pulse is the same as it was at time $t = 0$, as in Figure 16.5a.

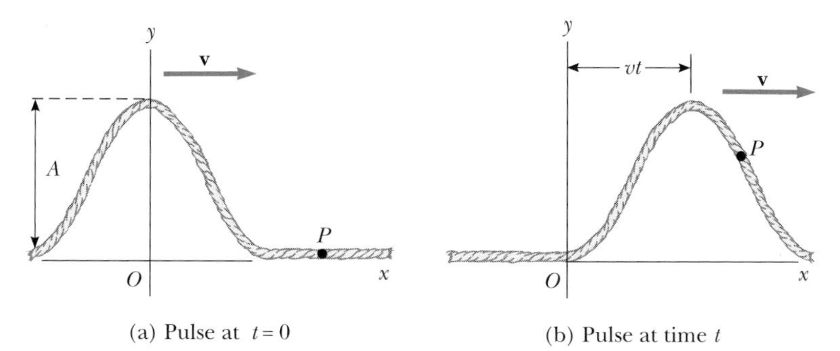

Figure 16.5 A one-dimensional pulse traveling to the right with a speed v. (a) At $t = 0$, the shape of the pulse is given by $y = f(x)$. (b) At some later time t, the shape remains unchanged and the vertical position of an element of the medium any point P is given by $y = f(x - vt)$.

Consequently, an element of the string at x at this time has the same y position as an element located at $x - vt$ had at time $t = 0$:

$$y(x, t) = y(x - vt, 0)$$

In general, then, we can represent the transverse position y for all positions and times, measured in a stationary frame with the origin at O, as

Pulse traveling to the right

$$y(x, t) = f(x - vt) \tag{16.1}$$

Similarly, if the pulse travels to the left, the transverse positions of elements of the string are described by

Pulse traveling to the left

$$y(x, t) = f(x + vt) \tag{16.2}$$

The function y, sometimes called the **wave function,** depends on the two variables x and t. For this reason, it is often written $y(x, t)$, which is read "y as a function of x and t."

It is important to understand the meaning of y. Consider an element of the string at point P, identified by a particular value of its x coordinate. As the pulse passes through P, the y coordinate of this element increases, reaches a maximum, and then decreases to zero. **The wave function $y(x, t)$ represents the y coordinate—the transverse position—of any element located at position x at any time t.** Furthermore, if t is fixed (as, for example, in the case of taking a snapshot of the pulse), then the wave function $y(x)$, sometimes called the **waveform,** defines a curve representing the actual geometric shape of the pulse at that time.

Quick Quiz 16.1 In a long line of people waiting to buy tickets, the first person leaves and a pulse of motion occurs as people step forward to fill the gap. As each person steps forward, the gap moves through the line. Is the propagation of this gap (a) transverse (b) longitudinal?

Quick Quiz 16.2 Consider the "wave" at a baseball game: people stand up and shout as the wave arrives at their location, and the resultant pulse moves around the stadium. Is this wave (a) transverse (b) longitudinal?

Example 16.1 A Pulse Moving to the Right

A pulse moving to the right along the x axis is represented by the wave function

$$y(x, t) = \frac{2}{(x - 3.0t)^2 + 1}$$

where x and y are measured in centimeters and t is measured in seconds. Plot the wave function at $t = 0$, $t = 1.0$ s, and $t = 2.0$ s.

Solution First, note that this function is of the form $y = f(x - vt)$. By inspection, we see that the wave speed is $v = 3.0$ cm/s. Furthermore, the maximum value of y is given by $A = 2.0$ cm. (We find the maximum value of the function representing y by letting $x - 3.0t = 0$.) The wave function expressions are

$$y(x, 0) = \frac{2}{x^2 + 1} \quad \text{at } t = 0$$

$$y(x, 1.0) = \frac{2}{(x - 3.0)^2 + 1} \quad \text{at } t = 1.0 \text{ s}$$

$$y(x, 2.0) = \frac{2}{(x - 6.0)^2 + 1} \quad \text{at } t = 2.0 \text{ s}$$

We now use these expressions to plot the wave function versus x at these times. For example, let us evaluate $y(x, 0)$ at $x = 0.50$ cm:

$$y(0.50, 0) = \frac{2}{(0.50)^2 + 1} = 1.6 \text{ cm}$$

Likewise, at $x = 1.0$ cm, $y(1.0, 0) = 1.0$ cm, and at $y = 2.0$ cm, $y(2.0, 0) = 0.40$ cm. Continuing this procedure for other values of x yields the wave function shown in Figure 16.6a. In a similar manner, we obtain the graphs of $y(x, 1.0)$ and $y(x, 2.0)$, shown in Figure 16.6b and c, respectively. These snapshots show that the pulse moves to the right without changing its shape and that it has a constant speed of 3.0 cm/s.

What If? (A) What if the wave function were

$$y(x, t) = \frac{2}{(x + 3.0t)^2 + 1}$$

How would this change the situation?

(B) What if the wave function were

$$y(x, t) = \frac{4}{(x - 3.0t)^2 + 1}$$

How would this change the situation?

Answer (A) The new feature in this expression is the plus sign in the denominator rather than the minus sign. This results in a pulse with the same shape as that in Figure 16.6, but moving to the left as time progresses.

(B) The new feature here is the numerator of 4 rather than 2. This results in a pulse moving to the right, but with twice the height of that in Figure 16.6.

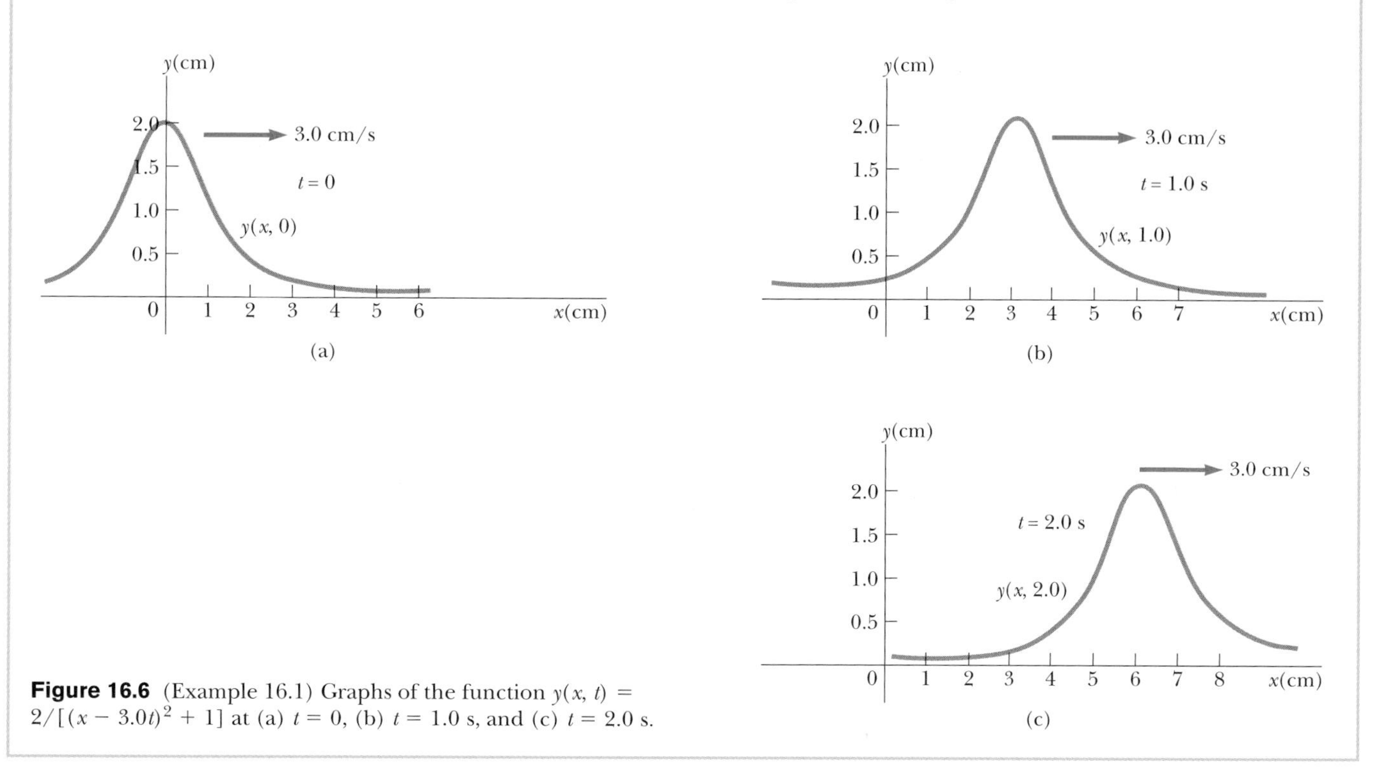

Figure 16.6 (Example 16.1) Graphs of the function $y(x, t) = 2/[(x - 3.0t)^2 + 1]$ at (a) $t = 0$, (b) $t = 1.0$ s, and (c) $t = 2.0$ s.

16.2 Sinusoidal Waves

In this section, we introduce an important wave function whose shape is shown in Figure 16.7. The wave represented by this curve is called a **sinusoidal wave** because the curve is the same as that of the function sin θ plotted against θ. On a rope, a sinusoidal wave could be established by shaking the end of the rope up and down in simple harmonic motion.

The sinusoidal wave is the simplest example of a periodic continuous wave and can be used to build more complex waves (see Section 18.8). The brown curve in Figure 16.7 represents a snapshot of a traveling sinusoidal wave at $t = 0$, and the blue curve represents a snapshot of the wave at some later time t. Notice two types of motion that can be seen in your mind. First, the entire waveform in Figure 16.7 moves to the right, so that the brown curve moves toward the right and eventually reaches the position of the blue curve. This is the motion of the *wave.* If we focus on one element of the medium, such as the element at $x = 0$, we see that each element moves up and down along the y axis in simple harmonic motion. This is the motion of the *elements of the medium.* It is important to differentiate between the motion of the wave and the motion of the elements of the medium.

Figure 16.8a shows a snapshot of a wave moving through a medium. Figure 16.8b shows a graph of the position of one element of the medium as a function of time. The

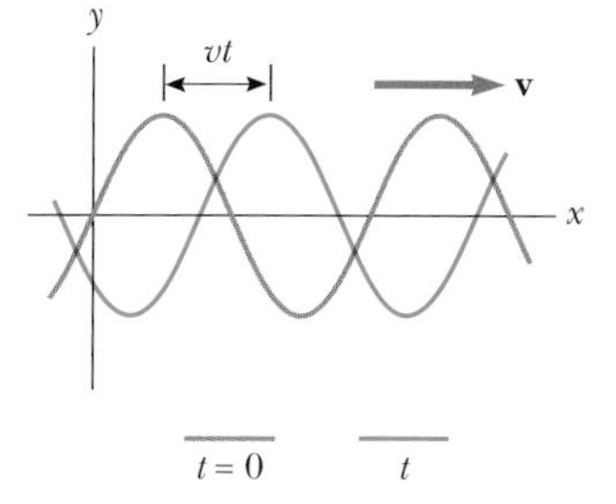

Active Figure 16.7 A one-dimensional sinusoidal wave traveling to the right with a speed v. The brown curve represents a snapshot of the wave at $t = 0$, and the blue curve represents a snapshot at some later time t.

At the Active Figures link at http://www.pse6.com, you can watch the wave move and take snapshots of it at various times.

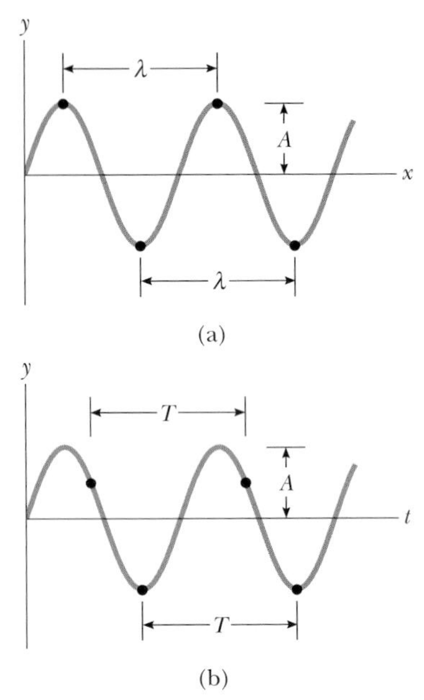

Active Figure 16.8 (a) The wavelength λ of a wave is the distance between adjacent crests or adjacent troughs. (b) The period T of a wave is the time interval required for the wave to travel one wavelength.

At the Active Figures link at http://www.pse6.com, you can change the parameters to see the effect on the wave function.

point at which the displacement of the element from its normal position is highest is called the **crest** of the wave. The distance from one crest to the next is called the **wavelength** λ (Greek lambda). More generally, **the wavelength is the minimum distance between any two identical points (such as the crests) on adjacent waves,** as shown in Figure 16.8a.

If you count the number of seconds between the arrivals of two adjacent crests at a given point in space, you are measuring the **period** T of the waves. In general, **the period is the time interval required for two identical points (such as the crests) of adjacent waves to pass by a point.** The period of the wave is the same as the period of the simple harmonic oscillation of one element of the medium.

The same information is more often given by the inverse of the period, which is called the **frequency** f. In general, **the frequency of a periodic wave is the number of crests (or troughs, or any other point on the wave) that pass a given point in a unit time interval.** The frequency of a sinusoidal wave is related to the period by the expression

$$f = \frac{1}{T} \tag{16.3}$$

The frequency of the wave is the same as the frequency of the simple harmonic oscillation of one element of the medium. The most common unit for frequency, as we learned in Chapter 15, is second^{-1}, or **hertz** (Hz). The corresponding unit for T is seconds.

The maximum displacement from equilibrium of an element of the medium is called the **amplitude** A of the wave.

Waves travel with a specific speed, and this speed depends on the properties of the medium being disturbed. For instance, sound waves travel through room-temperature air with a speed of about 343 m/s (781 mi/h), whereas they travel through most solids with a speed greater than 343 m/s.

Consider the sinusoidal wave in Figure 16.8a, which shows the position of the wave at $t = 0$. Because the wave is sinusoidal, we expect the wave function at this instant to be expressed as $y(x, 0) = A \sin ax$, where A is the amplitude and a is a constant to be determined. At $x = 0$, we see that $y(0, 0) = A \sin a(0) = 0$, consistent with Figure 16.8a. The next value of x for which y is zero is $x = \lambda/2$. Thus,

$$y\left(\frac{\lambda}{2}, 0\right) = A \sin a\left(\frac{\lambda}{2}\right) = 0$$

For this to be true, we must have $a(\lambda/2) = \pi$, or $a = 2\pi/\lambda$. Thus, the function describing the positions of the elements of the medium through which the sinusoidal wave is traveling can be written

$$y(x, 0) = A \sin\left(\frac{2\pi}{\lambda} x\right) \tag{16.4}$$

where the constant A represents the wave amplitude and the constant λ is the wavelength. We see that the vertical position of an element of the medium is the same whenever x is increased by an integral multiple of λ. If the wave moves to the right with a speed v, then the wave function at some later time t is

$$y(x, t) = A \sin\left[\frac{2\pi}{\lambda}(x - vt)\right] \tag{16.5}$$

That is, the traveling sinusoidal wave moves to the right a distance vt in the time t, as shown in Figure 16.7. Note that the wave function has the form $f(x - vt)$ (Eq. 16.1). If the wave were traveling to the left, the quantity $x - vt$ would be replaced by $x + vt$, as we learned when we developed Equations 16.1 and 16.2.

16.1 What's the Difference Between Figure 16.8a and 16.8b?

Notice the visual similarity between Figures 16.8a and 16.8b. The shapes are the same, but (a) is a graph of vertical position versus horizontal position while (b) is vertical position versus time. Figure 16.8a is a pictorial representation of the wave *for a series of particles of the medium*—this is what you would see at an instant of time. Figure 16.8b is a graphical representation of the position of *one element of the medium* as a function of time. The fact that both figures have the identical shape represents Equation 16.1—a wave is the *same* function of both x and t.

By definition, the wave travels a distance of one wavelength in one period T. Therefore, the wave speed, wavelength, and period are related by the expression

$$v = \frac{\lambda}{T} \tag{16.6}$$

Substituting this expression for v into Equation 16.5, we find that

$$y = A \sin\left[2\pi\left(\frac{x}{\lambda} - \frac{t}{T}\right)\right] \tag{16.7}$$

This form of the wave function shows the *periodic* nature of y. (We will often use y rather than $y(x, t)$ as a shorthand notation.) At any given time t, y has the *same* value at the positions x, $x + \lambda$, $x + 2\lambda$, and so on. Furthermore, at any given position x, the value of y is the same at times t, $t + T$, $t + 2T$, and so on.

We can express the wave function in a convenient form by defining two other quantities, the **angular wave number** k (usually called simply the **wave number**) and the **angular frequency** ω:

$$k \equiv \frac{2\pi}{\lambda} \tag{16.8}$$

Angular wave number

$$\omega \equiv \frac{2\pi}{T} \tag{16.9}$$

Angular frequency

Using these definitions, we see that Equation 16.7 can be written in the more compact form

$$y = A \sin(kx - \omega t) \tag{16.10}$$

Wave function for a sinusoidal wave

Using Equations 16.3, 16.8, and 16.9, we can express the wave speed v originally given in Equation 16.6 in the alternative forms

$$v = \frac{\omega}{k} \tag{16.11}$$

$$v = \lambda f \tag{16.12}$$

Speed of a sinusoidal wave

The wave function given by Equation 16.10 assumes that the vertical position y of an element of the medium is zero at $x = 0$ and $t = 0$. This need not be the case. If it is not, we generally express the wave function in the form

$$y = A \sin(kx - \omega t + \phi) \tag{16.13}$$

General expression for a sinusoidal wave

where ϕ is the **phase constant,** just as we learned in our study of periodic motion in Chapter 15. This constant can be determined from the initial conditions.

Quick Quiz 16.3 A sinusoidal wave of frequency f is traveling along a stretched string. The string is brought to rest, and a second traveling wave of frequency $2f$ is established on the string. The wave speed of the second wave is (a) twice that of the first wave (b) half that of the first wave (c) the same as that of the first wave (d) impossible to determine.

Quick Quiz 16.4 Consider the waves in Quick Quiz 16.3 again. The wavelength of the second wave is (a) twice that of the first wave (b) half that of the first wave (c) the same as that of the first wave (d) impossible to determine.

Quick Quiz 16.5 Consider the waves in Quick Quiz 16.3 again. The amplitude of the second wave is (a) twice that of the first wave (b) half that of the first wave (c) the same as that of the first wave (d) impossible to determine.

Example 16.2 A Traveling Sinusoidal Wave

A sinusoidal wave traveling in the positive x direction has an amplitude of 15.0 cm, a wavelength of 40.0 cm, and a frequency of 8.00 Hz. The vertical position of an element of the medium at $t = 0$ and $x = 0$ is also 15.0 cm, as shown in Figure 16.9.

(A) Find the wave number k, period T, angular frequency ω, and speed v of the wave.

Solution Using Equations 16.8, 16.3, 16.9, and 16.12, we find the following:

$$k = \frac{2\pi}{\lambda} = \frac{2\pi \text{ rad}}{40.0 \text{ cm}} = 0.157 \text{ rad/cm}$$

$$T = \frac{1}{f} = \frac{1}{8.00 \text{ s}^{-1}} = 0.125 \text{ s}$$

$$\omega = 2\pi f = 2\pi(8.00 \text{ s}^{-1}) = 50.3 \text{ rad/s}$$

$$v = \lambda f = (40.0 \text{ cm})(8.00 \text{ s}^{-1}) = 320 \text{ cm/s}$$

(B) Determine the phase constant ϕ, and write a general expression for the wave function.

Solution Because $A = 15.0$ cm and because $y = 15.0$ cm at $x = 0$ and $t = 0$, substitution into Equation 16.13 gives

$$15.0 = (15.0) \sin \phi \qquad \text{or} \qquad \sin \phi = 1$$

We may take the principal value $\phi = \pi/2$ rad (or 90°). Hence, the wave function is of the form

$$y = A \sin\left(kx - \omega t + \frac{\pi}{2}\right) = A \cos(kx - \omega t)$$

By inspection, we can see that the wave function must have this form, noting that the cosine function has the same shape as the sine function displaced by 90°. Substituting the values for A, k, and ω into this expression, we obtain

$$y = (15.0 \text{ cm}) \cos(0.157x - 50.3t)$$

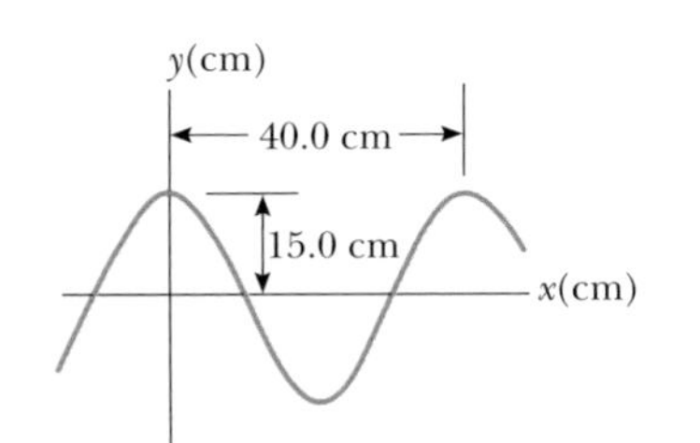

Figure 16.9 (Example 16.2) A sinusoidal wave of wavelength $\lambda = 40.0$ cm and amplitude $A = 15.0$ cm. The wave function can be written in the form $y = A \cos(kx - \omega t)$.

Sinusoidal Waves on Strings

In Figure 16.1, we demonstrated how to create a pulse by jerking a taut string up and down once. To create a series of such pulses—a wave—we can replace the hand with an oscillating blade. If the wave consists of a series of identical waveforms, whatever their shape, the relationships $f = 1/T$ and $v = f\lambda$ among speed, frequency, period, and wavelength hold true. We can make more definite statements about the wave function if the source of the waves vibrates in simple harmonic motion. Figure 16.10 represents snapshots of the wave created in this way at intervals of $T/4$. Because the end of the blade oscillates in simple harmonic motion, **each element of the string, such as that at *P*, also oscillates vertically with simple harmonic motion.** This must be the case because each element follows the simple harmonic motion of the blade. Therefore, every element of the string can be treated as a simple harmonic oscillator vibrating with a frequency equal to the frequency of oscillation of the blade.[2] Note that although each element oscillates in the y direction, the wave travels in the x direction with a speed v. Of course, this is the definition of a transverse wave.

If the wave at $t = 0$ is as described in Figure 16.10b, then the wave function can be written as

$$y = A \sin(kx - \omega t)$$

[2] In this arrangement, we are assuming that a string element always oscillates in a vertical line. The tension in the string would vary if an element were allowed to move sideways. Such motion would make the analysis very complex.

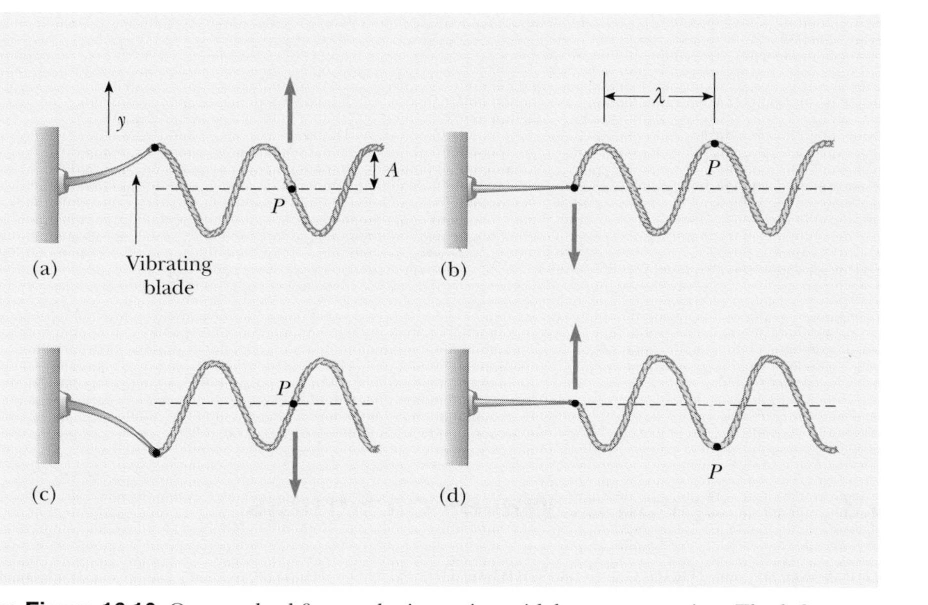

Active Figure 16.10 One method for producing a sinusoidal wave on a string. The left end of the string is connected to a blade that is set into oscillation. Every element of the string, such as that at point P, oscillates with simple harmonic motion in the vertical direction.

At the Active Figures link at http://www.pse6.com, you can adjust the frequency of the blade.

We can use this expression to describe the motion of any element of the string. An element at point P (or any other element of the string) moves only vertically, and so its x coordinate remains constant. Therefore, the **transverse speed** v_y (not to be confused with the wave speed v) and the **transverse acceleration** a_y of elements of the string are

$$v_y = \left.\frac{dy}{dt}\right]_{x = \text{constant}} = \frac{\partial y}{\partial t} = -\omega A \cos(kx - \omega t) \quad (16.14)$$

$$a_y = \left.\frac{dv_y}{dt}\right]_{x = \text{constant}} = \frac{\partial v_y}{\partial t} = -\omega^2 A \sin(kx - \omega t) \quad (16.15)$$

In these expressions, we must use partial derivatives (see Section 8.5) because y depends on both x and t. In the operation $\partial y/\partial t$, for example, we take a derivative with respect to t while holding x constant. The maximum values of the transverse speed and transverse acceleration are simply the absolute values of the coefficients of the cosine and sine functions:

$$v_{y,\,\max} = \omega A \quad (16.16)$$

$$a_{y,\,\max} = \omega^2 A \quad (16.17)$$

The transverse speed and transverse acceleration of elements of the string do not reach their maximum values simultaneously. The transverse speed reaches its maximum value (ωA) when $y = 0$, whereas the magnitude of the transverse acceleration reaches its maximum value ($\omega^2 A$) when $y = \pm A$. Finally, Equations 16.16 and 16.17 are identical in mathematical form to the corresponding equations for simple harmonic motion, Equations 15.17 and 15.18.

PITFALL PREVENTION

16.2 Two Kinds of Speed/Velocity

Do not confuse v, the speed of the wave as it propagates along the string, with v_y, the transverse velocity of a point on the string. The speed v is constant while v_y varies sinusoidally.

Quick Quiz 16.6 The amplitude of a wave is doubled, with no other changes made to the wave. As a result of this doubling, which of the following statements is correct? (a) The speed of the wave changes. (b) The frequency of the wave changes. (c) The maximum transverse speed of an element of the medium changes. (d) All of these are true. (e) None of these is true.

Example 16.3 A Sinusoidally Driven String

The string shown in Figure 16.10 is driven at a frequency of 5.00 Hz. The amplitude of the motion is 12.0 cm, and the wave speed is 20.0 m/s. Determine the angular frequency ω and wave number k for this wave, and write an expression for the wave function.

Solution Using Equations 16.3, 16.9, and 16.11, we find that

$$\omega = \frac{2\pi}{T} = 2\pi f = 2\pi(5.00 \text{ Hz}) = 31.4 \text{ rad/s}$$

$$k = \frac{\omega}{v} = \frac{31.4 \text{ rad/s}}{20.0 \text{ m/s}} = 1.57 \text{ rad/m}$$

Because $A = 12.0 \text{ cm} = 0.120 \text{ m}$, we have

$$y = A \sin(kx - \omega t)$$
$$= (0.120 \text{ m}) \sin(1.57x - 31.4t)$$

16.3 The Speed of Waves on Strings

In this section, we focus on determining the speed of a transverse pulse traveling on a taut string. Let us first conceptually predict the parameters that determine the speed. If a string under tension is pulled sideways and then released, the tension is responsible for accelerating a particular element of the string back toward its equilibrium position. According to Newton's second law, the acceleration of the element increases with increasing tension. If the element returns to equilibrium more rapidly due to this increased acceleration, we would intuitively argue that the wave speed is greater. Thus, we expect the wave speed to increase with increasing tension.

Likewise, the wave speed should decrease as the mass per unit length of the string increases. This is because it is more difficult to accelerate a massive element of the string than a light element. If the tension in the string is T and its mass per unit length is μ (Greek mu), then as we shall show, the wave speed is

Speed of a wave on a stretched string

$$v = \sqrt{\frac{T}{\mu}} \tag{16.18}$$

First, let us verify that this expression is dimensionally correct. The dimensions of T are ML/T^2, and the dimensions of μ are M/L. Therefore, the dimensions of T/μ are L^2/T^2; hence, the dimensions of $\sqrt{T/\mu}$ are L/T, the dimensions of speed. No other combination of T and μ is dimensionally correct, and if we assume that these are the only variables relevant to the situation, the speed must be proportional to $\sqrt{T/\mu}$.

Now let us use a mechanical analysis to derive Equation 16.18. Consider a pulse moving on a taut string to the right with a uniform speed v measured relative to a stationary frame of reference. Instead of staying in this reference frame, it is more convenient to choose as our reference frame one that moves along with the pulse with the same speed as the pulse, so that the pulse is at rest within the frame. This change of reference frame is permitted because Newton's laws are valid in either a stationary frame or one that moves with constant velocity. In our new reference frame, all elements of the string move to the left—a given element of the string initially to the right of the pulse moves to the left, rises up and follows the shape of the pulse, and then continues to move to the left. Figure 16.11a shows such an element at the instant it is located at the top of the pulse.

The small element of the string of length Δs shown in Figure 16.11a, and magnified in Figure 16.11b, forms an approximate arc of a circle of radius R. In our moving frame of reference (which is moving to the right at a speed v along with the pulse), the shaded element is moving to the left with a speed v. This element has a centripetal acceleration equal to v^2/R, which is supplied by components of the force **T** whose magnitude is the tension in the string. The force **T** acts on both sides of the element and is tangent to the arc, as shown in Figure 16.11b. The horizontal components of **T** cancel, and each vertical component $T \sin \theta$ acts radially toward the center of the arc. Hence, the total

PITFALL PREVENTION

16.3 Multiple *T*'s

Do not confuse the T in Equation 16.18 for the tension with the symbol T used in this chapter for the period of a wave. The context of the equation should help you to identify which quantity is meant. There simply aren't enough letters in the alphabet to assign a unique letter to each variable!

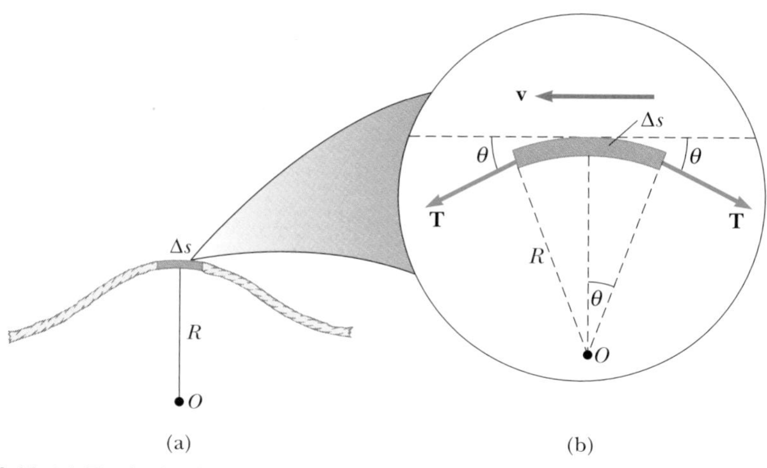

Figure 16.11 (a) To obtain the speed v of a wave on a stretched string, it is convenient to describe the motion of a small element of the string in a moving frame of reference. (b) In the moving frame of reference, the small element of length Δs moves to the left with speed v. The net force on the element is in the radial direction because the horizontal components of the tension force cancel.

radial force on the element is $2T \sin \theta$. Because the element is small, θ is small, and we can use the small-angle approximation $\sin \theta \approx \theta$. Therefore, the total radial force is

$$F_r = 2T \sin \theta \approx 2T\theta$$

The element has a mass $m = \mu \, \Delta s$. Because the element forms part of a circle and subtends an angle 2θ at the center, $\Delta s = R(2\theta)$, we find that

$$m = \mu \, \Delta s = 2\mu R\theta$$

If we apply Newton's second law to this element in the radial direction, we have

$$F_r = ma = \frac{mv^2}{R}$$

$$2T\theta = \frac{2\mu R\theta v^2}{R} \quad \longrightarrow \quad v = \sqrt{\frac{T}{\mu}}$$

This expression for v is Equation 16.18.

Notice that this derivation is based on the assumption that the pulse height is small relative to the length of the string. Using this assumption, we were able to use the approximation $\sin \theta \approx \theta$. Furthermore, the model assumes that the tension T is not affected by the presence of the pulse; thus, T is the same at all points on the string. Finally, this proof does *not* assume any particular shape for the pulse. Therefore, we conclude that a pulse of *any shape* travels along the string with speed $v = \sqrt{T/\mu}$ without any change in pulse shape.

Quick Quiz 16.7 Suppose you create a pulse by moving the free end of a taut string up and down once with your hand beginning at $t = 0$. The string is attached at its other end to a distant wall. The pulse reaches the wall at time t. Which of the following actions, taken by itself, decreases the time interval that it takes for the pulse to reach the wall? More than one choice may be correct. (a) moving your hand more quickly, but still only up and down once by the same amount (b) moving your hand more slowly, but still only up and down once by the same amount (c) moving your hand a greater distance up and down in the same amount of time (d) moving your hand a lesser distance up and down in the same amount of time (e) using a heavier string of the same length and under the same tension (f) using a lighter string of the same length and under the same tension (g) using a string of the same linear mass density but under decreased tension (h) using a string of the same linear mass density but under increased tension

Example 16.4 The Speed of a Pulse on a Cord

A uniform cord has a mass of 0.300 kg and a length of 6.00 m (Fig. 16.12). The cord passes over a pulley and supports a 2.00-kg object. Find the speed of a pulse traveling along this cord.

Solution The tension T in the cord is equal to the weight of the suspended 2.00-kg object:

$$T = mg = (2.00 \text{ kg})(9.80 \text{ m/s}^2) = 19.6 \text{ N}$$

(This calculation of the tension neglects the small mass of the cord. Strictly speaking, the cord can never be exactly horizontal, and therefore the tension is not uniform.) The mass per unit length μ of the cord is

$$\mu = \frac{m}{\ell} = \frac{0.300 \text{ kg}}{6.00 \text{ m}} = 0.050\,0 \text{ kg/m}$$

Therefore, the wave speed is

$$v = \sqrt{\frac{T}{\mu}} = \sqrt{\frac{19.6 \text{ N}}{0.050\,0 \text{ kg/m}}} = 19.8 \text{ m/s}$$

What If? What if the block were swinging back and forth between maximum angles of ± 20° with respect to the vertical? What range of wave speeds would this create on the horizontal cord?

Answer Figure 16.13 shows the swinging block at three positions—its highest position, its lowest position, and an arbitrary position. Summing the forces on the block in the radial direction when the block is at an arbitrary position, Newton's second law gives

$$(1) \qquad \sum F = T - mg\cos\theta = m\frac{v_{\text{block}}^2}{L}$$

where the acceleration of the block is centripetal, L is the length of the vertical piece of string, and v_{block} is the instantaneous speed of the block at the arbitrary position. Now consider conservation of mechanical energy for the block–Earth system. We define the zero of gravitational potential energy for the system when the block is at its lowest point, point Ⓒ in Figure 16.13. Equating the mechanical energy of the system when the block is at Ⓐ to the mechanical energy when the block is at an arbitrary position Ⓑ, we have,

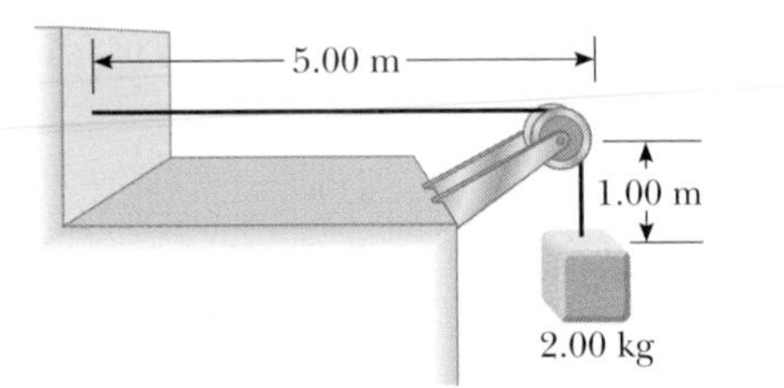

Figure 16.12 (Example 16.4) The tension T in the cord is maintained by the suspended object. The speed of any wave traveling along the cord is given by $v = \sqrt{T/\mu}$.

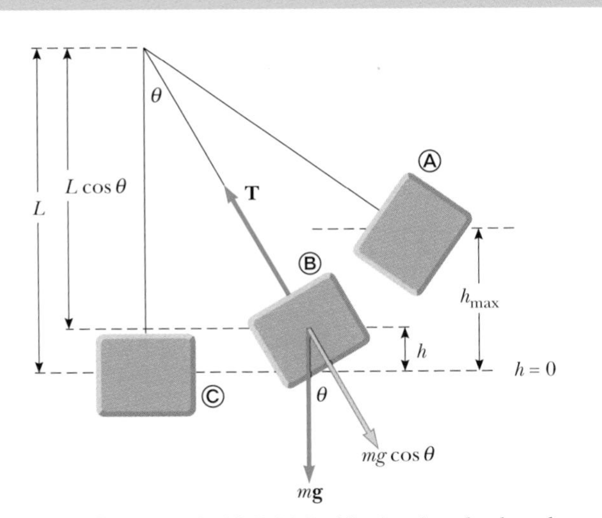

Figure 16.13 (Example 16.4) If the block swings back and forth, the tension in the cord changes, which causes a variation in the wave speed on the horizontal section of cord in Figure 16.12. The forces on the block when it is at arbitrary position Ⓑ are shown. Position Ⓐ is the highest position and Ⓒ is the lowest. (The maximum angle is exaggerated for clarity.)

$$E_A = E_B$$

$$mgh_{\text{max}} = mgh + \tfrac{1}{2}mv_{\text{block}}^2$$

$$mv_{\text{block}}^2 = 2mg(h_{\text{max}} - h)$$

Substituting this into Equation (1), we find an expression for T as a function of angle θ and height h:

$$T - mg\cos\theta = \frac{2mg(h_{\text{max}} - h)}{L}$$

$$T = mg\left[\cos\theta + \frac{2}{L}(h_{\text{max}} - h)\right]$$

The maximum value of T occurs when $\theta = 0$ and $h = 0$:

$$T_{\text{max}} = mg\left[\cos 0 + \frac{2}{L}(h_{\text{max}} - 0)\right] = mg\left(1 + \frac{2h_{\text{max}}}{L}\right)$$

The minimum value of T occurs when $h = h_{\text{max}}$ and $\theta = \theta_{\text{max}}$:

$$T_{\text{min}} = mg\left[\cos\theta_{\text{max}} + \frac{2}{L}(h_{\text{max}} - h_{\text{max}})\right] = mg\cos\theta_{\text{max}}$$

Now we find the maximum and minimum values of the wave speed v, using the fact that, as we see from Figure 16.13, h and θ are related by $h = L - L\cos\theta$:

$$v_{\text{max}} = \sqrt{\frac{T_{\text{max}}}{\mu}} = \sqrt{\frac{mg[1 + (2h_{\text{max}}/L)]}{\mu}}$$

$$= \sqrt{\frac{mg\{1 + [2(L - L\cos\theta_{\text{max}})/L]\}}{\mu}}$$

$$= \sqrt{\frac{mg(3 - 2\cos\theta_{max})}{\mu}}$$

$$= \sqrt{\frac{(2.00\text{ kg})(9.80\text{ m/s}^2)(3 - 2\cos 20°)}{0.050\,0\text{ kg/m}}} = 21.0\text{ m/s}$$

$$v_{min} = \sqrt{\frac{T_{min}}{\mu}} = \sqrt{\frac{mg\cos\theta_{max}}{\mu}}$$

$$= \sqrt{\frac{(2.00\text{ kg})(9.80\text{ m/s}^2)(\cos 20°)}{0.050\,0\text{ kg/m}}} = 19.2\text{ m/s}$$

Example 16.5 Rescuing the Hiker **Interactive**

An 80.0-kg hiker is trapped on a mountain ledge following a storm. A helicopter rescues the hiker by hovering above him and lowering a cable to him. The mass of the cable is 8.00 kg, and its length is 15.0 m. A chair of mass 70.0 kg is attached to the end of the cable. The hiker attaches himself to the chair, and the helicopter then accelerates upward. Terrified by hanging from the cable in midair, the hiker tries to signal the pilot by sending transverse pulses up the cable. A pulse takes 0.250 s to travel the length of the cable. What is the acceleration of the helicopter?

Solution To conceptualize this problem, imagine the effect of the acceleration of the helicopter on the cable. The higher the upward acceleration, the larger is the tension in the cable. In turn, the larger the tension, the higher is the speed of pulses on the cable. Thus, we categorize this problem as a combination of one involving Newton's laws and one involving the speed of pulses on a string. To analyze the problem, we use the time interval for the pulse to travel from the hiker to the helicopter to find the speed of the pulses on the cable:

$$v = \frac{\Delta x}{\Delta t} = \frac{15.0\text{ m}}{0.250\text{ s}} = 60.0\text{ m/s}$$

The speed of pulses on the cable is given by Equation 16.18, which allows us to find the tension in the cable:

$$v = \sqrt{\frac{T}{\mu}} \longrightarrow T = \mu v^2 = \left(\frac{8.00\text{ kg}}{15.0\text{ m}}\right)(60.0\text{ m/s})^2$$

$$T = 1.92 \times 10^3\text{ N}$$

Newton's second law relates the tension in the cable to the acceleration of the hiker and the chair, which is the same as the acceleration of the helicopter:

$$\sum F = ma \longrightarrow T - mg = ma$$

$$a = \frac{T}{m} - g = \frac{1.92 \times 10^3\text{ N}}{150.0\text{ kg}} - 9.80\text{ m/s}^2$$

$$= 3.00\text{ m/s}^2$$

To finalize this problem, note that a real cable has stiffness in addition to tension. Stiffness tends to return a wire to its original straight-line shape even when it is not under tension. For example, a piano wire straightens if released from a curved shape; package wrapping string does not.

Stiffness represents a restoring force in addition to tension, and increases the wave speed. Consequently, for a real cable, the speed of 60.0 m/s that we determined is most likely associated with a tension lower than 1.92×10^3 N and a correspondingly smaller acceleration of the helicopter.

Investigate this situation at the Interactive Worked Example link at **http://www.pse6.com.**

16.4 Reflection and Transmission

We have discussed waves traveling through a uniform medium. We now consider how a traveling wave is affected when it encounters a change in the medium. For example, consider a pulse traveling on a string that is rigidly attached to a support at one end as in Figure 16.14. When the pulse reaches the support, a severe change in the medium occurs—the string ends. The result of this change is that the pulse undergoes **reflection**—that is, the pulse moves back along the string in the opposite direction.

Note that the reflected pulse is *inverted.* This inversion can be explained as follows. When the pulse reaches the fixed end of the string, the string produces an upward force on the support. By Newton's third law, the support must exert an equal-magnitude and oppositely directed (downward) reaction force on the string. This downward force causes the pulse to invert upon reflection.

Now consider another case: this time, the pulse arrives at the end of a string that is free to move vertically, as in Figure 16.15. The tension at the free end is maintained because the string is tied to a ring of negligible mass that is free to slide vertically on a smooth post without friction. Again, the pulse is reflected, but this time it is not inverted. When it reaches the post, the pulse exerts a force on the free end of the string, causing the ring to accelerate upward. The ring rises as high as the incoming pulse,

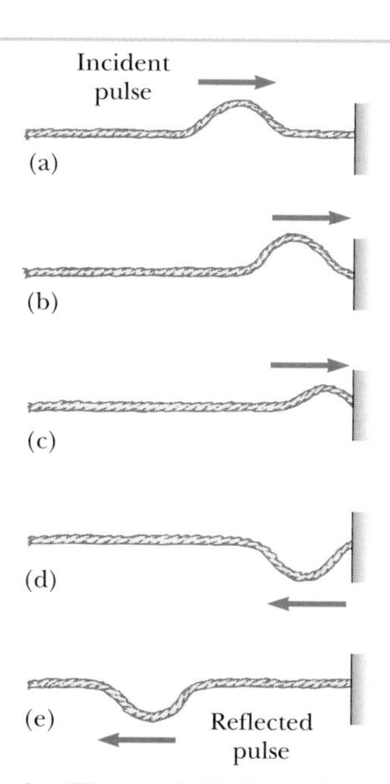

Active Figure 16.14 The reflection of a traveling pulse at the fixed end of a stretched string. The reflected pulse is inverted, but its shape is otherwise unchanged.

At the Active Figures link at http://www.pse6.com, *you can adjust the linear mass density of the string and the transverse direction of the initial pulse.*

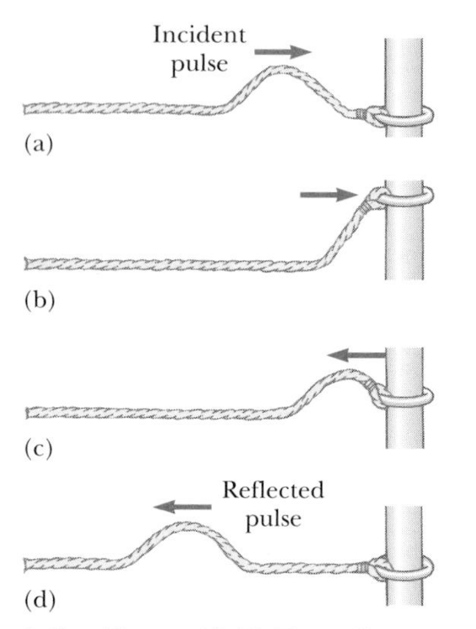

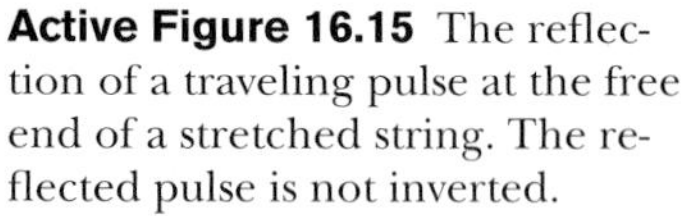

Active Figure 16.15 The reflection of a traveling pulse at the free end of a stretched string. The reflected pulse is not inverted.

At the Active Figures link at http://www.pse6.com, you can adjust the linear mass density of the string and the transverse direction of the initial pulse.

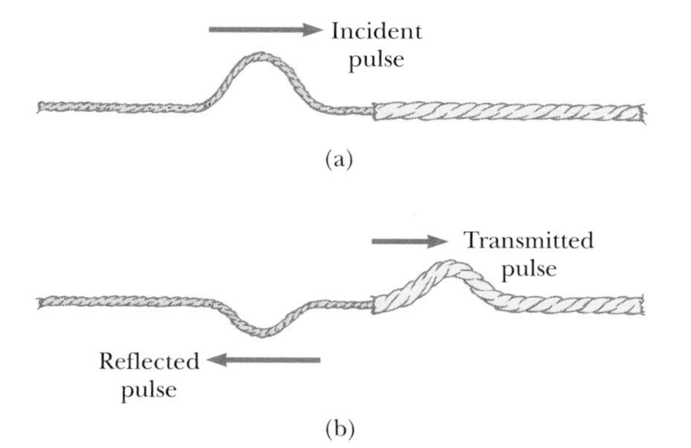

Figure 16.16 (a) A pulse traveling to the right on a light string attached to a heavier string. (b) Part of the incident pulse is reflected (and inverted), and part is transmitted to the heavier string. See Figure 16.17 for an animation available for both figures at the Active Figures link.

and then the downward component of the tension force pulls the ring back down. This movement of the ring produces a reflected pulse that is not inverted and that has the same amplitude as the incoming pulse.

Finally, we may have a situation in which the boundary is intermediate between these two extremes. In this case, part of the energy in the incident pulse is reflected and part undergoes **transmission**—that is, some of the energy passes through the boundary. For instance, suppose a light string is attached to a heavier string, as in Figure 16.16. When a pulse traveling on the light string reaches the boundary between the two, part of the pulse is reflected and inverted and part is transmitted to the heavier string. The reflected pulse is inverted for the same reasons described earlier in the case of the string rigidly attached to a support.

Note that the reflected pulse has a smaller amplitude than the incident pulse. In Section 16.5, we show that the energy carried by a wave is related to its amplitude. According to the principle of the conservation of energy, when the pulse breaks up into a reflected pulse and a transmitted pulse at the boundary, the sum of the energies of these two pulses must equal the energy of the incident pulse. Because the reflected pulse contains only part of the energy of the incident pulse, its amplitude must be smaller.

When a pulse traveling on a heavy string strikes the boundary between the heavy string and a lighter one, as in Figure 16.17, again part is reflected and part is transmitted. In this case, the reflected pulse is not inverted.

In either case, the relative heights of the reflected and transmitted pulses depend on the relative densities of the two strings. If the strings are identical, there is no discontinuity at the boundary and no reflection takes place.

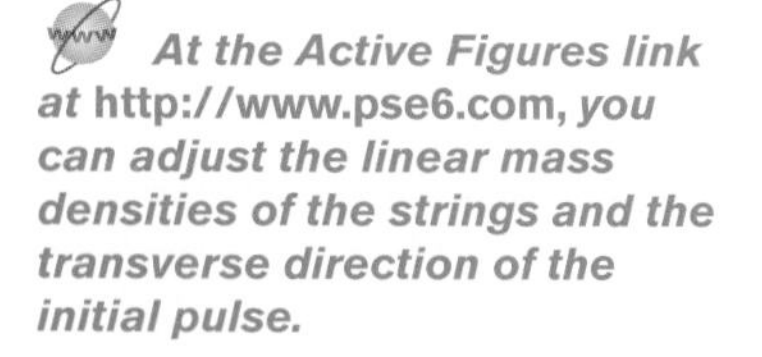

At the Active Figures link at http://www.pse6.com, you can adjust the linear mass densities of the strings and the transverse direction of the initial pulse.

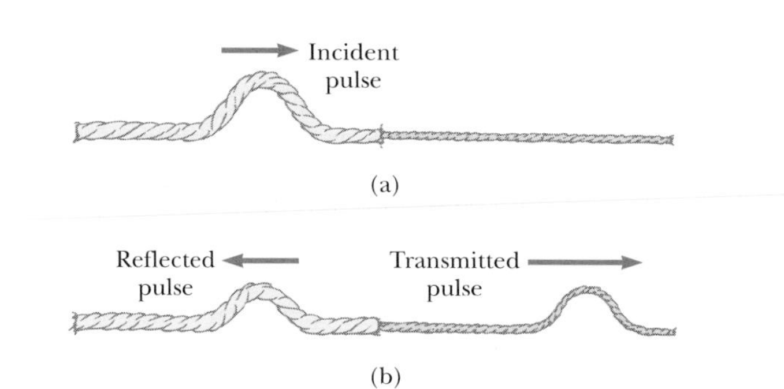

Active Figure 16.17 (a) A pulse traveling to the right on a heavy string attached to a lighter string. (b) The incident pulse is partially reflected and partially transmitted, and the reflected pulse is not inverted.

According to Equation 16.18, the speed of a wave on a string increases as the mass per unit length of the string decreases. In other words, a wave travels more slowly on a heavy string than on a light string if both are under the same tension. The following general rules apply to reflected waves: **when a wave or pulse travels from medium A to medium B and $v_A > v_B$ (that is, when B is denser than A), it is inverted upon reflection. When a wave or pulse travels from medium A to medium B and $v_A < v_B$ (that is, when A is denser than B), it is not inverted upon reflection.**

16.5 Rate of Energy Transfer by Sinusoidal Waves on Strings

Waves transport energy when they propagate through a medium. We can easily demonstrate this by hanging an object on a stretched string and then sending a pulse down the string, as in Figure 16.18a. When the pulse meets the suspended object, the object is momentarily displaced upward, as in Figure 16.18b. In the process, energy is transferred to the object and appears as an increase in the gravitational potential energy of the object–Earth system. This section examines the rate at which energy is transported along a string. We shall assume a one-dimensional sinusoidal wave in the calculation of the energy transferred.

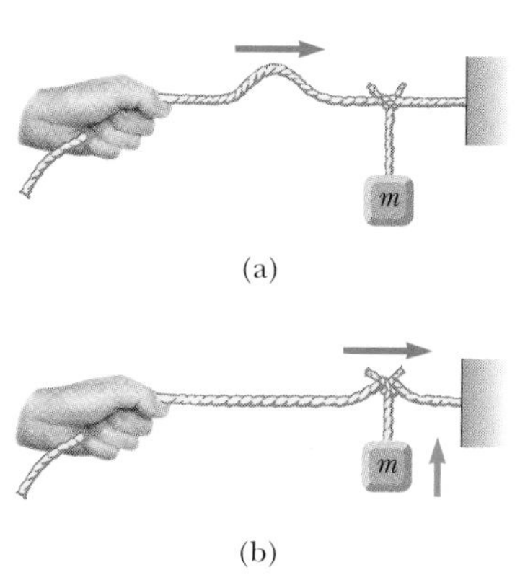

Figure 16.18 (a) A pulse traveling to the right on a stretched string that has an object suspended from it. (b) Energy is transmitted to the suspended object when the pulse arrives.

Consider a sinusoidal wave traveling on a string (Fig. 16.19). The source of the energy is some external agent at the left end of the string, which does work in producing the oscillations. We can consider the string to be a nonisolated system. As the external agent performs work on the end of the string, moving it up and down, energy enters the system of the string and propagates along its length. Let us focus our attention on an element of the string of length Δx and mass Δm. Each such element moves vertically with simple harmonic motion. Thus, we can model each element of the string as a simple harmonic oscillator, with the oscillation in the y direction. All elements have the same angular frequency ω and the same amplitude A. The kinetic energy K associated with a moving particle is $K = \frac{1}{2}mv^2$. If we apply this equation to an element of length Δx and mass Δm, we see that the kinetic energy ΔK of this element is

$$\Delta K = \tfrac{1}{2}(\Delta m)\, v_y^{\,2}$$

where v_y is the transverse speed of the element. If μ is the mass per unit length of the string, then the mass Δm of the element of length Δx is equal to $\mu\,\Delta x$. Hence, we can express the kinetic energy of an element of the string as

$$\Delta K = \tfrac{1}{2}(\mu\,\Delta x)\, v_y^{\,2} \qquad (16.19)$$

As the length of the element of the string shrinks to zero, this becomes a differential relationship:

$$dK = \tfrac{1}{2}(\mu\, dx)\, v_y^{\,2}$$

We substitute for the general transverse speed of a simple harmonic oscillator using Equation 16.14:

$$\begin{aligned} dK &= \tfrac{1}{2}\mu[\omega A \cos(kx - \omega t)]^2\, dx \\ &= \tfrac{1}{2}\mu\omega^2 A^2 \cos^2(kx - \omega t)\, dx \end{aligned}$$

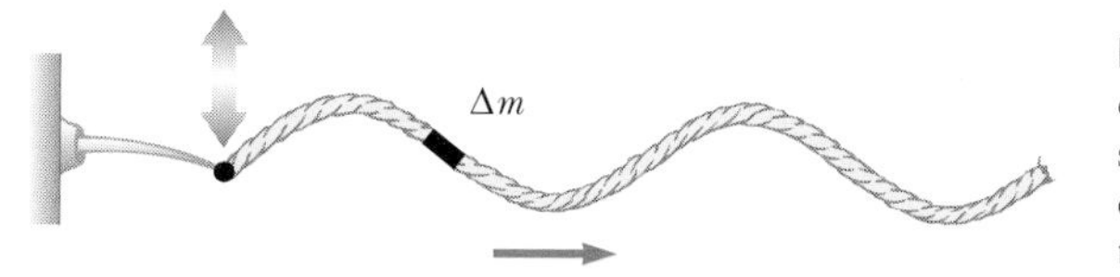

Figure 16.19 A sinusoidal wave traveling along the x axis on a stretched string. Every element moves vertically, and every element has the same total energy.

If we take a snapshot of the wave at time $t = 0$, then the kinetic energy of a given element is

$$dK = \tfrac{1}{2}\mu\omega^2 A^2 \cos^2 kx\, dx$$

Let us integrate this expression over all the string elements in a wavelength of the wave, which will give us the total kinetic energy K_λ in one wavelength:

$$K_\lambda = \int dK = \int_0^\lambda \tfrac{1}{2}\mu\omega^2 A^2 \cos^2 kx\, dx = \tfrac{1}{2}\mu\omega^2 A^2 \int_0^\lambda \cos^2 kx\, dx$$

$$= \tfrac{1}{2}\mu\omega^2 A^2 \left[\tfrac{1}{2}x + \frac{1}{4k}\sin 2kx\right]_0^\lambda = \tfrac{1}{2}\mu\omega^2 A^2 \left[\tfrac{1}{2}\lambda\right] = \tfrac{1}{4}\mu\omega^2 A^2 \lambda$$

In addition to kinetic energy, each element of the string has potential energy associated with it due to its displacement from the equilibrium position and the restoring forces from neighboring elements. A similar analysis to that above for the total potential energy U_λ in one wavelength will give exactly the same result:

$$U_\lambda = \tfrac{1}{4}\mu\omega^2 A^2 \lambda$$

The total energy in one wavelength of the wave is the sum of the potential and kinetic energies:

$$E_\lambda = U_\lambda + K_\lambda = \tfrac{1}{2}\mu\omega^2 A^2 \lambda \qquad (16.20)$$

As the wave moves along the string, this amount of energy passes by a given point on the string during a time interval of one period of the oscillation. Thus, the power, or rate of energy transfer, associated with the wave is

$$\mathcal{P} = \frac{\Delta E}{\Delta t} = \frac{E_\lambda}{T} = \frac{\tfrac{1}{2}\mu\omega^2 A^2 \lambda}{T} = \tfrac{1}{2}\mu\omega^2 A^2 \left(\frac{\lambda}{T}\right)$$

Power of a wave

$$\mathcal{P} = \tfrac{1}{2}\mu\omega^2 A^2 v \qquad (16.21)$$

This expression shows that the rate of energy transfer by a sinusoidal wave on a string is proportional to (a) the square of the frequency, (b) the square of the amplitude, and (c) the wave speed. In fact: **the rate of energy transfer in any sinusoidal wave is proportional to the square of the angular frequency and to the square of the amplitude.**

Quick Quiz 16.8 Which of the following, taken by itself, would be most effective in increasing the rate at which energy is transferred by a wave traveling along a string? (a) reducing the linear mass density of the string by one half (b) doubling the wavelength of the wave (c) doubling the tension in the string (d) doubling the amplitude of the wave

Example 16.6 Power Supplied to a Vibrating String

A taut string for which $\mu = 5.00 \times 10^{-2}$ kg/m is under a tension of 80.0 N. How much power must be supplied to the string to generate sinusoidal waves at a frequency of 60.0 Hz and an amplitude of 6.00 cm?

Solution The wave speed on the string is, from Equation 16.18,

$$v = \sqrt{\frac{T}{\mu}} = \sqrt{\frac{80.0\ \text{N}}{5.00 \times 10^{-2}\ \text{kg/m}}} = 40.0\ \text{m/s}$$

Because $f = 60.0$ Hz, the angular frequency ω of the sinusoidal waves on the string has the value

$$\omega = 2\pi f = 2\pi(60.0\ \text{Hz}) = 377\ \text{s}^{-1}$$

Using these values in Equation 16.21 for the power, with $A = 6.00 \times 10^{-2}$ m, we obtain

$$\mathcal{P} = \tfrac{1}{2}\mu\omega^2 A^2 v$$
$$= \tfrac{1}{2}(5.00 \times 10^{-2}\ \text{kg/m})(377\ \text{s}^{-1})^2 \times (6.00 \times 10^{-2}\ \text{m})^2(40.0\ \text{m/s})$$
$$= 512\ \text{W}$$

What If? **What if the string is to transfer energy at a rate of 1 000 W? What must be the required amplitude if all other parameters remain the same?**

Answer We set up a ratio of the new and old power, reflecting only a change in the amplitude:

$$\frac{\mathcal{P}_{\text{new}}}{\mathcal{P}_{\text{old}}} = \frac{\frac{1}{2}\mu\omega^2 A_{\text{new}}^2 v}{\frac{1}{2}\mu\omega^2 A_{\text{old}}^2 v} = \frac{A_{\text{new}}^2}{A_{\text{old}}^2}$$

Solving for the new amplitude,

$$A_{\text{new}} = A_{\text{old}}\sqrt{\frac{\mathcal{P}_{\text{new}}}{\mathcal{P}_{\text{old}}}} = (6.00\ \text{cm})\sqrt{\frac{1\,000\ \text{W}}{512\ \text{W}}}$$
$$= 8.39\ \text{cm}$$

16.6 The Linear Wave Equation

In Section 16.1 we introduced the concept of the wave function to represent waves traveling on a string. All wave functions $y(x, t)$ represent solutions of an equation called the *linear wave equation*. This equation gives a complete description of the wave motion, and from it one can derive an expression for the wave speed. Furthermore, the linear wave equation is basic to many forms of wave motion. In this section, we derive this equation as applied to waves on strings.

Suppose a traveling wave is propagating along a string that is under a tension T. Let us consider one small string element of length Δx (Fig. 16.20). The ends of the element make small angles θ_A and θ_B with the x axis. The net force acting on the element in the vertical direction is

$$\sum F_y = T\sin\theta_B - T\sin\theta_A = T(\sin\theta_B - \sin\theta_A)$$

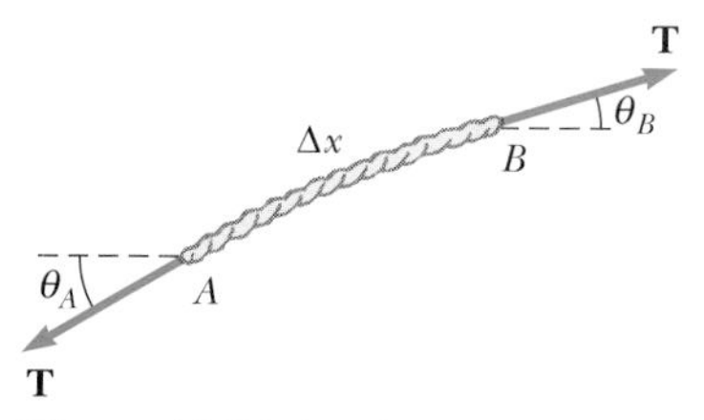

Figure 16.20 An element of a string under tension T.

Because the angles are small, we can use the small-angle approximation $\sin\theta \approx \tan\theta$ to express the net force as

$$\sum F_y \approx T(\tan\theta_B - \tan\theta_A) \tag{16.22}$$

Imagine undergoing an infinitesimal displacement outward from the end of the rope element in Figure 16.20 along the blue line representing the force $\mathbf{T}$. This displacement has infinitesimal x and y components and can be represented by the vector $dx\hat{\mathbf{i}} + dy\hat{\mathbf{j}}$. The tangent of the angle with respect to the x axis for this displacement is dy/dx. Because we are evaluating this tangent at a particular instant of time, we need to express this in partial form as $\partial y/\partial x$. Substituting for the tangents in Equation 16.22 gives

$$\sum F_y \approx T\left[\left(\frac{\partial y}{\partial x}\right)_B - \left(\frac{\partial y}{\partial x}\right)_A\right] \tag{16.23}$$

We now apply Newton's second law to the element, with the mass of the element given by $m = \mu\,\Delta x$:

$$\sum F_y = ma_y = \mu\,\Delta x\left(\frac{\partial^2 y}{\partial t^2}\right) \tag{16.24}$$

Combining Equation 16.23 with Equation 16.24, we obtain

$$\mu\,\Delta x\left(\frac{\partial^2 y}{\partial t^2}\right) = T\left[\left(\frac{\partial y}{\partial x}\right)_B - \left(\frac{\partial y}{\partial x}\right)_A\right]$$

$$\frac{\mu}{T}\frac{\partial^2 y}{\partial t^2} = \frac{(\partial y/\partial x)_B - (\partial y/dx)_A}{\Delta x} \tag{16.25}$$

The right side of this equation can be expressed in a different form if we note that the partial derivative of any function is defined as

$$\frac{\partial f}{\partial x} \equiv \lim_{\Delta x \to 0} \frac{f(x + \Delta x) - f(x)}{\Delta x}$$

If we associate $f(x + \Delta x)$ with $(\partial y/\partial x)_B$ and $f(x)$ with $(\partial y/\partial x)_A$, we see that, in the limit $\Delta x \to 0$, Equation 16.25 becomes

Linear wave equation for a string

$$\frac{\mu}{T}\frac{\partial^2 y}{\partial t^2} = \frac{\partial^2 y}{\partial x^2} \qquad (16.26)$$

This is the linear wave equation as it applies to waves on a string.

We now show that the sinusoidal wave function (Eq. 16.10) represents a solution of the linear wave equation. If we take the sinusoidal wave function to be of the form $y(x, t) = A \sin(kx - \omega t)$, then the appropriate derivatives are

$$\frac{\partial^2 y}{\partial t^2} = -\omega^2 A \sin(kx - \omega t)$$

$$\frac{\partial^2 y}{\partial x^2} = -k^2 A \sin(kx - \omega t)$$

Substituting these expressions into Equation 16.26, we obtain

$$-\frac{\mu\omega^2}{T}\sin(kx - \omega t) = -k^2 \sin(kx - \omega t)$$

This equation must be true for all values of the variables x and t in order for the sinusoidal wave function to be a solution of the wave equation. Both sides of the equation depend on x and t through the same function $\sin(kx - \omega t)$. Because this function divides out, we do indeed have an identity, provided that

$$k^2 = \frac{\mu}{T}\omega^2$$

Using the relationship $v = \omega/k$ (Eq. 16.11) in this expression, we see that

$$v^2 = \frac{\omega^2}{k^2} = \frac{T}{\mu}$$

$$v = \sqrt{\frac{T}{\mu}}$$

which is Equation 16.18. This derivation represents another proof of the expression for the wave speed on a taut string.

The linear wave equation (Eq. 16.26) is often written in the form

Linear wave equation in general

$$\frac{\partial^2 y}{\partial x^2} = \frac{1}{v^2}\frac{\partial^2 y}{\partial t^2} \qquad (16.27)$$

This expression applies in general to various types of traveling waves. For waves on strings, y represents the vertical position of elements of the string. For sound waves, y corresponds to longitudinal position of elements of air from equilibrium or variations in either the pressure or the density of the gas through which the sound waves are propagating. In the case of electromagnetic waves, y corresponds to electric or magnetic field components.

We have shown that the sinusoidal wave function (Eq. 16.10) is one solution of the linear wave equation (Eq. 16.27). Although we do not prove it here, the linear wave equation is satisfied by *any* wave function having the form $y = f(x \pm vt)$. Furthermore, we have seen that the linear wave equation is a direct consequence of Newton's second law applied to any element of a string carrying a traveling wave.

SUMMARY

A **transverse wave** is one in which the elements of the medium move in a direction *perpendicular* to the direction of propagation. An example is a wave on a taut string. A **longitudinal wave** is one in which the elements of the medium move in a direction *parallel* to the direction of propagation. Sound waves in fluids are longitudinal.

Any one-dimensional wave traveling with a speed v in the x direction can be represented by a wave function of the form

$$y(x, t) = f(x \pm vt) \quad (16.1, 16.2)$$

where the positive sign applies to a wave traveling in the negative x direction and the negative sign applies to a wave traveling in the positive x direction. The shape of the wave at any instant in time (a snapshot of the wave) is obtained by holding t constant.

The **wave function** for a one-dimensional sinusoidal wave traveling to the right can be expressed as

$$y = A \sin\left[\frac{2\pi}{\lambda}(x - vt)\right] = A \sin(kx - \omega t) \quad (16.5, 16.10)$$

where A is the **amplitude,** λ is the **wavelength,** k is the **angular wave number,** and ω is the **angular frequency.** If T is the **period** and f the **frequency,** v, k, and ω can be written

$$v = \frac{\lambda}{T} = \lambda f \quad (16.6, 16.12)$$

$$k \equiv \frac{2\pi}{\lambda} \quad (16.8)$$

$$\omega \equiv \frac{2\pi}{T} = 2\pi f \quad (16.3, 16.9)$$

The speed of a wave traveling on a taut string of mass per unit length μ and tension T is

$$v = \sqrt{\frac{T}{\mu}} \quad (16.18)$$

A wave is totally or partially reflected when it reaches the end of the medium in which it propagates or when it reaches a boundary where its speed changes discontinuously. If a wave traveling on a string meets a fixed end, the wave is reflected and inverted. If the wave reaches a free end, it is reflected but not inverted.

The **power** transmitted by a sinusoidal wave on a stretched string is

$$\mathcal{P} = \tfrac{1}{2}\mu\omega^2 A^2 v \quad (16.21)$$

Wave functions are solutions to a differential equation called the **linear wave equation:**

$$\frac{\partial^2 y}{\partial x^2} = \frac{1}{v^2}\frac{\partial^2 y}{\partial t^2} \quad (16.27)$$

Take a practice test for this chapter by clicking on the Practice Test link at http://www.pse6.com.

QUESTIONS

1. Why is a pulse on a string considered to be transverse?
2. How would you create a longitudinal wave in a stretched spring? Would it be possible to create a transverse wave in a spring?
3. By what factor would you have to multiply the tension in a stretched string in order to double the wave speed?
4. When traveling on a taut string, does a pulse always invert upon reflection? Explain.
5. Does the vertical speed of a segment of a horizontal taut string, through which a wave is traveling, depend on the wave speed?
6. If you shake one end of a taut rope steadily three times each second, what would be the period of the sinusoidal wave set up in the rope?
7. A vibrating source generates a sinusoidal wave on a string under constant tension. If the power delivered to the

string is doubled, by what factor does the amplitude change? Does the wave speed change under these circumstances?

8. Consider a wave traveling on a taut rope. What is the difference, if any, between the speed of the wave and the speed of a small segment of the rope?

9. If a long rope is hung from a ceiling and waves are sent up the rope from its lower end, they do not ascend with constant speed. Explain.

10. How do transverse waves differ from longitudinal waves?

11. When all the strings on a guitar are stretched to the same tension, will the speed of a wave along the most massive bass string be faster, slower, or the same as the speed of a wave on the lighter strings?

12. If one end of a heavy rope is attached to one end of a light rope, the speed of a wave will change as the wave goes from the heavy rope to the light one. Will it increase or decrease? What happens to the frequency? To the wavelength?

13. If you stretch a rubber hose and pluck it, you can observe a pulse traveling up and down the hose. What happens to the speed of the pulse if you stretch the hose more tightly? What happens to the speed if you fill the hose with water?

14. In a longitudinal wave in a spring, the coils move back and forth in the direction of wave motion. Does the speed of the wave depend on the maximum speed of each coil?

15. Both longitudinal and transverse waves can propagate through a solid. A wave on the surface of a liquid can involve both longitudinal and transverse motion of elements of the medium. On the other hand, a wave propagating through the volume of a fluid must be purely longitudinal, not transverse. Why?

16. In an earthquake both S (transverse) and P (longitudinal) waves propagate from the focus of the earthquake. The focus is in the ground below the epicenter on the surface. The S waves travel through the Earth more slowly than the P waves (at about 5 km/s versus 8 km/s). By detecting the time of arrival of the waves, how can one determine the distance to the focus of the quake? How many detection stations are necessary to locate the focus unambiguously?

17. In mechanics, massless strings are often assumed. Why is this not a good assumption when discussing waves on strings?

PROBLEMS

1, 2, 3 = straightforward, intermediate, challenging □ = full solution available in the *Student Solutions Manual and Study Guide*

= coached solution with hints available at http://www.pse6.com = computer useful in solving problem

= paired numerical and symbolic problems

Section 16.1 Propagation of a Disturbance

1. At $t = 0$, a transverse pulse in a wire is described by the function

$$y = \frac{6}{x^2 + 3}$$

where x and y are in meters. Write the function $y(x, t)$ that describes this pulse if it is traveling in the positive x direction with a speed of 4.50 m/s.

2. Ocean waves with a crest-to-crest distance of 10.0 m can be described by the wave function

$$y(x, t) = (0.800 \text{ m}) \sin[0.628(x - vt)]$$

where $v = 1.20$ m/s. (a) Sketch $y(x, t)$ at $t = 0$. (b) Sketch $y(x, t)$ at $t = 2.00$ s. Note that the entire wave form has shifted 2.40 m in the positive x direction in this time interval.

3. A pulse moving along the x axis is described by

$$y(x, t) = 5.00e^{-(x+5.00t)^2}$$

where x is in meters and t is in seconds. Determine (a) the direction of the wave motion, and (b) the speed of the pulse.

4. Two points A and B on the surface of the Earth are at the same longitude and 60.0° apart in latitude. Suppose that an earthquake at point A creates a P wave that reaches point B by traveling straight through the body of the Earth at a constant speed of 7.80 km/s. The earthquake also radiates a *Rayleigh wave,* which travels across the surface of the Earth in an analogous way to a surface wave on water, at 4.50 km/s. (a) Which of these two seismic waves arrives at B first? (b) What is the time difference between the arrivals of the two waves at B? Take the radius of the Earth to be 6 370 km.

5. S and P waves, simultaneously radiated from the hypocenter of an earthquake, are received at a seismographic station 17.3 s apart. Assume the waves have traveled over the same path at speeds of 4.50 km/s and 7.80 km/s. Find the distance from the seismograph to the hypocenter of the quake.

Section 16.2 Sinusoidal Waves

6. For a certain transverse wave, the distance between two successive crests is 1.20 m, and eight crests pass a given point along the direction of travel every 12.0 s. Calculate the wave speed.

7. A sinusoidal wave is traveling along a rope. The oscillator that generates the wave completes 40.0 vibrations in 30.0 s. Also, a given maximum travels 425 cm along the rope in 10.0 s. What is the wavelength?

8. When a particular wire is vibrating with a frequency of 4.00 Hz, a transverse wave of wavelength 60.0 cm is produced. Determine the speed of waves along the wire.

9. A wave is described by $y = (2.00 \text{ cm}) \sin(kx - \omega t)$, where $k = 2.11$ rad/m, $\omega = 3.62$ rad/s, x is in meters, and t is in seconds. Determine the amplitude, wavelength, frequency, and speed of the wave.

10. A sinusoidal wave on a string is described by

$$y = (0.51 \text{ cm}) \sin(kx - \omega t)$$

where $k = 3.10$ rad/cm and $\omega = 9.30$ rad/s. How far does a wave crest move in 10.0 s? Does it move in the positive or negative x direction?

11. Consider further the string shown in Figure 16.10 and treated in Example 16.3. Calculate (a) the maximum transverse speed and (b) the maximum transverse acceleration of a point on the string.

12. Consider the sinusoidal wave of Example 16.2, with the wave function

$$y = (15.0 \text{ cm}) \cos(0.157x - 50.3t).$$

At a certain instant, let point A be at the origin and point B be the first point along the x axis where the wave is 60.0° out of phase with point A. What is the coordinate of point B?

13. A sinusoidal wave is described by

$$y = (0.25 \text{ m}) \sin(0.30x - 40t)$$

where x and y are in meters and t is in seconds. Determine for this wave the (a) amplitude, (b) angular frequency, (c) angular wave number, (d) wavelength, (e) wave speed, and (f) direction of motion.

14. (a) Plot y versus t at $x = 0$ for a sinusoidal wave of the form $y = (15.0 \text{ cm}) \cos(0.157x - 50.3t)$, where x and y are in centimeters and t is in seconds. (b) Determine the period of vibration from this plot and compare your result with the value found in Example 16.2.

15. (a) Write the expression for y as a function of x and t for a sinusoidal wave traveling along a rope in the *negative* x direction with the following characteristics: $A = 8.00$ cm, $\lambda = 80.0$ cm, $f = 3.00$ Hz, and $y(0, t) = 0$ at $t = 0$. (b) **What If?** Write the expression for y as a function of x and t for the wave in part (a) assuming that $y(x, 0) = 0$ at the point $x = 10.0$ cm.

16. A sinusoidal wave traveling in the $-x$ direction (to the left) has an amplitude of 20.0 cm, a wavelength of 35.0 cm, and a frequency of 12.0 Hz. The transverse position of an element of the medium at $t = 0$, $x = 0$ is $y = -3.00$ cm, and the element has a positive velocity here. (a) Sketch the wave at $t = 0$. (b) Find the angular wave number, period, angular frequency, and wave speed of the wave. (c) Write an expression for the wave function $y(x, t)$.

17. A transverse wave on a string is described by the wave function

$$y = (0.120 \text{ m}) \sin[(\pi x/8) + 4\pi t]$$

(a) Determine the transverse speed and acceleration at $t = 0.200$ s for the point on the string located at $x = 1.60$ m. (b) What are the wavelength, period, and speed of propagation of this wave?

18. A transverse sinusoidal wave on a string has a period $T = 25.0$ ms and travels in the negative x direction with a speed of 30.0 m/s. At $t = 0$, a particle on the string at $x = 0$ has a transverse position of 2.00 cm and is traveling downward with a speed of 2.00 m/s. (a) What is the amplitude of the wave? (b) What is the initial phase angle? (c) What is the maximum transverse speed of the string? (d) Write the wave function for the wave.

19. A sinusoidal wave of wavelength 2.00 m and amplitude 0.100 m travels on a string with a speed of 1.00 m/s to the right. Initially, the left end of the string is at the origin. Find (a) the frequency and angular frequency, (b) the angular wave number, and (c) the wave function for this wave. Determine the equation of motion for (d) the left end of the string and (e) the point on the string at $x = 1.50$ m to the right of the left end. (f) What is the maximum speed of any point on the string?

20. A wave on a string is described by the wave function $y = (0.100 \text{ m}) \sin(0.50x - 20t)$. (a) Show that a particle in the string at $x = 2.00$ m executes simple harmonic motion. (b) Determine the frequency of oscillation of this particular point.

Section 16.3 The Speed of Waves on Strings

21. A telephone cord is 4.00 m long. The cord has a mass of 0.200 kg. A transverse pulse is produced by plucking one end of the taut cord. The pulse makes four trips down and back along the cord in 0.800 s. What is the tension in the cord?

22. Transverse waves with a speed of 50.0 m/s are to be produced in a taut string. A 5.00-m length of string with a total mass of 0.060 0 kg is used. What is the required tension?

23. A piano string having a mass per unit length equal to 5.00×10^{-3} kg/m is under a tension of 1 350 N. Find the speed of a wave traveling on this string.

24. A transverse traveling wave on a taut wire has an amplitude of 0.200 mm and a frequency of 500 Hz. It travels with a speed of 196 m/s. (a) Write an equation in SI units of the form $y = A \sin(kx - \omega t)$ for this wave. (b) The mass per unit length of this wire is 4.10 g/m. Find the tension in the wire.

25. An astronaut on the Moon wishes to measure the local value of the free-fall acceleration by timing pulses traveling down a wire that has an object of large mass suspended from it. Assume a wire has a mass of 4.00 g and a length of 1.60 m, and that a 3.00-kg object is suspended from it. A pulse requires 36.1 ms to traverse the length of the wire. Calculate g_{Moon} from these data. (You may ignore the mass of the wire when calculating the tension in it.)

26. Transverse pulses travel with a speed of 200 m/s along a taut copper wire whose diameter is 1.50 mm. What is the tension in the wire? (The density of copper is 8.92 g/cm^3.)

27. Transverse waves travel with a speed of 20.0 m/s in a string under a tension of 6.00 N. What tension is required for a wave speed of 30.0 m/s in the same string?

28. A simple pendulum consists of a ball of mass M hanging from a uniform string of mass m and length L, with $m \ll M$. If the period of oscillations for the pendulum is T, determine the speed of a transverse wave in the string when the pendulum hangs at rest.

29. The elastic limit of the steel forming a piece of wire is equal to 2.70×10^8 Pa. What is the maximum speed at which transverse wave pulses can propagate along this wire without exceeding this stress? (The density of steel is 7.86×10^3 km/m^3.)

30. **Review problem.** A light string with a mass per unit length of 8.00 g/m has its ends tied to two walls separated by a distance equal to three fourths of the length of the string (Fig. P16.30). An object of mass m is suspended from the center of the string, putting a tension in the string. (a) Find an expression for the transverse wave speed in the string as a function of the mass of the hanging object. (b) What should be the mass of the object suspended from the string in order to produce a wave speed of 60.0 m/s?

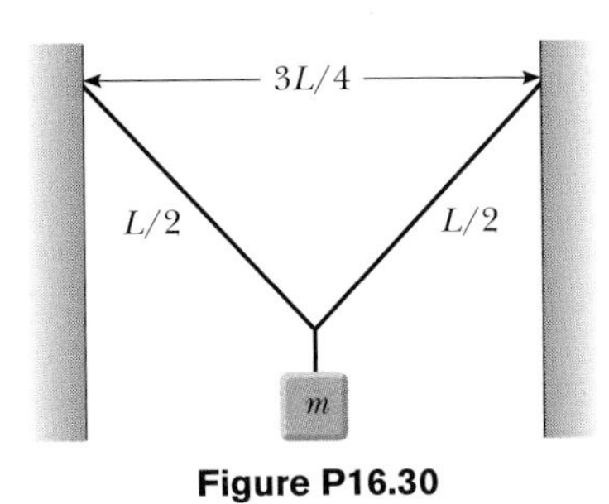

Figure P16.30

31. A 30.0-m steel wire and a 20.0-m copper wire, both with 1.00-mm diameters, are connected end to end and stretched to a tension of 150 N. How long does it take a transverse wave to travel the entire length of the two wires?

32. **Review problem.** A light string of mass m and length L has its ends tied to two walls that are separated by the distance D. Two objects, each of mass M, are suspended from the string as in Figure P16.32. If a wave pulse is sent from point A, how long does it take to travel to point B?

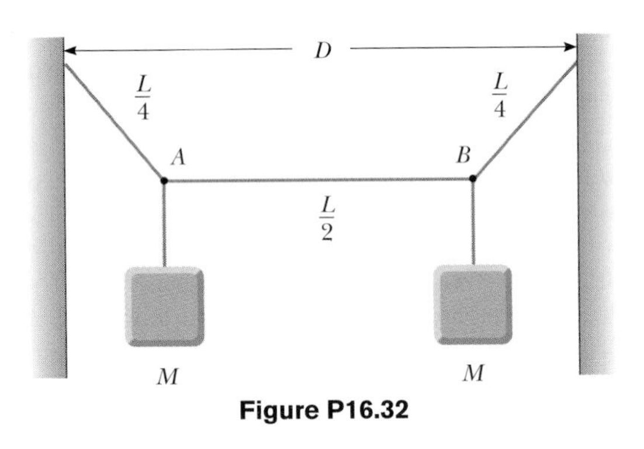

Figure P16.32

33. A student taking a quiz finds on a reference sheet the two equations

$$f = 1/T \quad \text{and} \quad v = \sqrt{T/\mu}$$

She has forgotten what T represents in each equation. (a) Use dimensional analysis to determine the units required for T in each equation. (b) Identify the physical quantity each T represents.

Section 16.5 Rate of Energy Transfer by Sinusoidal Waves on Strings

34. A taut rope has a mass of 0.180 kg and a length of 3.60 m. What power must be supplied to the rope in order to generate sinusoidal waves having an amplitude of 0.100 m and a wavelength of 0.500 m and traveling with a speed of 30.0 m/s?

35. A two-dimensional water wave spreads in circular ripples. Show that the amplitude A at a distance r from the initial disturbance is proportional to $1/\sqrt{r}$. (*Suggestion:* Consider the energy carried by one outward-moving ripple.)

36. Transverse waves are being generated on a rope under constant tension. By what factor is the required power increased or decreased if (a) the length of the rope is doubled and the angular frequency remains constant, (b) the amplitude is doubled and the angular frequency is halved, (c) both the wavelength and the amplitude are doubled, and (d) both the length of the rope and the wavelength are halved?

37. Sinusoidal waves 5.00 cm in amplitude are to be transmitted along a string that has a linear mass density of 4.00×10^{-2} kg/m. If the source can deliver a maximum power of 300 W and the string is under a tension of 100 N, what is the highest frequency at which the source can operate?

38. It is found that a 6.00-m segment of a long string contains four complete waves and has a mass of 180 g. The string is vibrating sinusoidally with a frequency of 50.0 Hz and a peak-to-valley distance of 15.0 cm. (The "peak-to-valley" distance is the vertical distance from the farthest positive position to the farthest negative position.) (a) Write the function that describes this wave traveling in the positive x direction. (b) Determine the power being supplied to the string.

39. A sinusoidal wave on a string is described by the equation

$$y = (0.15 \text{ m}) \sin(0.80x - 50t)$$

where x and y are in meters and t is in seconds. If the mass per unit length of this string is 12.0 g/m, determine (a) the speed of the wave, (b) the wavelength, (c) the frequency, and (d) the power transmitted to the wave.

40. The wave function for a wave on a taut string is

$$y(x, t) = (0.350 \text{ m}) \sin(10\pi t - 3\pi x + \pi/4)$$

where x is in meters and t in seconds. (a) What is the average rate at which energy is transmitted along the string if the linear mass density is 75.0 g/m? (b) What is the energy contained in each cycle of the wave?

41. A horizontal string can transmit a maximum power $\mathcal{P}_0$ (without breaking) if a wave with amplitude A and angular frequency ω is traveling along it. In order to increase this maximum power, a student folds the string and uses this "double string" as a medium. Determine the maximum power that can be transmitted along the "double string," assuming that the tension is constant.

42. In a region far from the epicenter of an earthquake, a seismic wave can be modeled as transporting energy in a single direction without absorption, just as a string wave does. Suppose the seismic wave moves from granite into mudfill with similar density but with a much lower bulk modulus. Assume the speed of the wave gradually drops by a factor of 25.0, with negligible reflection of the wave. Will the amplitude of the ground shaking increase or decrease? By

what factor? This phenomenon led to the collapse of part of the Nimitz Freeway in Oakland, California, during the Loma Prieta earthquake of 1989.

Section 16.6 The Linear Wave Equation

43. (a) Evaluate A in the scalar equality $(7 + 3)4 = A$. (b) Evaluate A, B, and C in the vector equality $7.00\hat{\mathbf{i}} + 3.00\hat{\mathbf{k}} = A\hat{\mathbf{i}} + B\hat{\mathbf{j}} + C\hat{\mathbf{k}}$. Explain how you arrive at the answers to convince a student who thinks that you cannot solve a single equation for three different unknowns. (c) **What If?** The functional equality or identity

$$A + B\cos(Cx + Dt + E) = (7.00 \text{ mm}) \cos(3x + 4t + 2)$$

is true for all values of the variables x and t, which are measured in meters and in seconds, respectively. Evaluate the constants A, B, C, D, and E. Explain how you arrive at the answers.

44. Show that the wave function $y = e^{b(x-vt)}$ is a solution of the linear wave equation (Eq. 16.27), where b is a constant.

45. Show that the wave function $y = \ln[b(x - vt)]$ is a solution to Equation 16.27, where b is a constant.

46. (a) Show that the function $y(x, t) = x^2 + v^2t^2$ is a solution to the wave equation. (b) Show that the function in part (a) can be written as $f(x + vt) + g(x - vt)$, and determine the functional forms for f and g. (c) **What If?** Repeat parts (a) and (b) for the function $y(x, t) = \sin(x)\cos(vt)$.

Additional Problems

47. "The wave" is a particular type of pulse that can propagate through a large crowd gathered at a sports arena to watch a soccer or American football match (Figure P16.47). The elements of the medium are the spectators, with zero position corresponding to their being seated and maximum position corresponding to their standing and raising their arms. When a large fraction of the spectators participate in the wave motion, a somewhat stable pulse shape can develop. The wave speed depends on people's reaction time, which is typically on the order of 0.1 s. Estimate the order of magnitude, in minutes, of the time required for such a pulse to make one circuit around a large sports stadium. State the quantities you measure or estimate and their values.

Figure P16.47

48. A traveling wave propagates according to the expression $y = (4.0 \text{ cm}) \sin(2.0x - 3.0t)$, where x is in centimeters and t is in seconds. Determine (a) the amplitude, (b) the wavelength, (c) the frequency, (d) the period, and (e) the direction of travel of the wave.

49. The wave function for a traveling wave on a taut string is (in SI units)

$$y(x, t) = (0.350 \text{ m}) \sin(10\pi t - 3\pi x + \pi/4)$$

(a) What are the speed and direction of travel of the wave? (b) What is the vertical position of an element of the string at $t = 0$, $x = 0.100$ m? (c) What are the wavelength and frequency of the wave? (d) What is the maximum magnitude of the transverse speed of the string?

50. A transverse wave on a string is described by the equation

$$y(x, t) = (0.350 \text{ m}) \sin[(1.25 \text{ rad/m})x + (99.6 \text{ rad/s})t]$$

Consider the element of the string at $x = 0$. (a) What is the time interval between the first two instants when this element has a position of $y = 0.175$ m? (b) What distance does the wave travel during this time interval?

51. Motion picture film is projected at 24.0 frames per second. Each frame is a photograph 19.0 mm high. At what constant speed does the film pass into the projector?

52. Review problem. A block of mass M, supported by a string, rests on a frictionless incline making an angle θ with the horizontal (Fig. P16.52). The length of the string is L, and its mass is $m \ll M$. Derive an expression for the time interval required for a transverse wave to travel from one end of the string to the other.

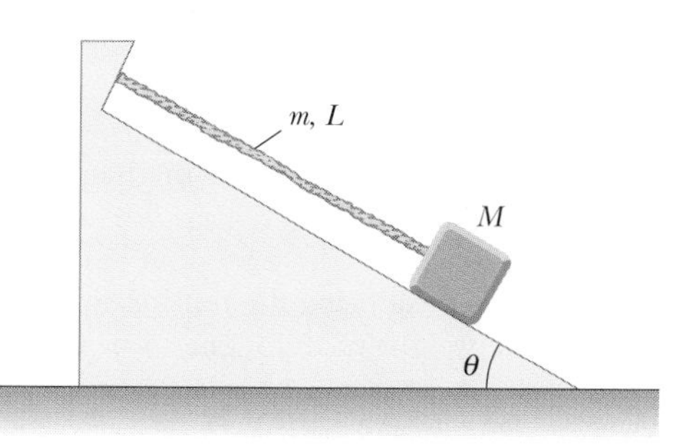

Figure P16.52

53. Review problem. A 2.00-kg block hangs from a rubber cord, being supported so that the cord is not stretched. The unstretched length of the cord is 0.500 m, and its mass is 5.00 g. The "spring constant" for the cord is 100 N/m. The block is released and stops at the lowest point. (a) Determine the tension in the cord when the block is at this lowest point. (b) What is the length of the cord in this "stretched" position? (c) Find the speed of a transverse wave in the cord if the block is held in this lowest position.

54. **Review problem.** A block of mass M hangs from a rubber cord. The block is supported so that the cord is not stretched. The unstretched length of the cord is L_0 and its mass is m, much less than M. The "spring constant" for the cord is k. The block is released and stops at the lowest point. (a) Determine the tension in the string when the block is at this lowest point. (b) What is the length of the cord in this "stretched" position? (c) Find the speed of a transverse wave in the cord if the block is held in this lowest position.

55. (a) Determine the speed of transverse waves on a string under a tension of 80.0 N if the string has a length of 2.00 m and a mass of 5.00 g. (b) Calculate the power required to generate these waves if they have a wavelength of 16.0 cm and an amplitude of 4.00 cm.

56. A sinusoidal wave in a rope is described by the wave function

$$y = (0.20 \text{ m}) \sin(0.75\pi x + 18\pi t)$$

where x and y are in meters and t is in seconds. The rope has a linear mass density of 0.250 kg/m. If the tension in the rope is provided by an arrangement like the one illustrated in Figure 16.12, what is the value of the suspended mass?

57. A block of mass 0.450 kg is attached to one end of a cord of mass 0.003 20 kg; the other end of the cord is attached to a fixed point. The block rotates with constant angular speed in a circle on a horizontal frictionless table. Through what angle does the block rotate in the time that a transverse wave takes to travel along the string from the center of the circle to the block?

58. A wire of density ρ is tapered so that its cross-sectional area varies with x according to

$$A = (1.0 \times 10^{-3}\, x + 0.010) \text{ cm}^2$$

(a) If the wire is subject to a tension T, derive a relationship for the speed of a wave as a function of position. (b) **What If?** If the wire is aluminum and is subject to a tension of 24.0 N, determine the speed at the origin and at $x = 10.0$ m.

59. A rope of total mass m and length L is suspended vertically. Show that a transverse pulse travels the length of the rope in a time interval $\Delta t = 2\sqrt{L/g}$. (*Suggestion:* First find an expression for the wave speed at any point a distance x from the lower end by considering the tension in the rope as resulting from the weight of the segment below that point.)

60. If an object of mass M is suspended from the bottom of the rope in Problem 59, (a) show that the time interval for a transverse pulse to travel the length of the rope is

$$\Delta t = 2\sqrt{\frac{L}{mg}}\left(\sqrt{M+m} - \sqrt{M}\right)$$

What If? (b) Show that this reduces to the result of Problem 59 when $M = 0$. (c) Show that for $m \ll M$, the expression in part (a) reduces to

$$\Delta t = \sqrt{\frac{mL}{Mg}}$$

61. It is stated in Problem 59 that a pulse travels from the bottom to the top of a hanging rope of length L in a time interval $\Delta t = 2\sqrt{L/g}$. Use this result to answer the following questions. (It is not necessary to set up any new integrations.) (a) How long does it take for a pulse to travel halfway up the rope? Give your answer as a fraction of the quantity $2\sqrt{L/g}$. (b) A pulse starts traveling up the rope. How far has it traveled after a time interval $\sqrt{L/g}$?

62. Determine the speed and direction of propagation of each of the following sinusoidal waves, assuming that x and y are measured in meters and t in seconds.

(a) $y = 0.60 \cos(3.0x - 15t + 2)$
(b) $y = 0.40 \cos(3.0x + 15t - 2)$
(c) $y = 1.2 \sin(15t + 2.0x)$
(d) $y = 0.20 \sin[12t - (x/2) + \pi]$

63. An aluminum wire is clamped at each end under zero tension at room temperature. The tension in the wire is increased by reducing the temperature, which results in a decrease in the wire's equilibrium length. What strain $(\Delta L/L)$ results in a transverse wave speed of 100 m/s? Take the cross-sectional area of the wire to be 5.00×10^{-6} m^2, the density to be 2.70×10^3 kg/m^3, and Young's modulus to be 7.00×10^{10} N/m^2.

64. If a loop of chain is spun at high speed, it can roll along the ground like a circular hoop without slipping or collapsing. Consider a chain of uniform linear mass density μ whose center of mass travels to the right at a high speed v_0. (a) Determine the tension in the chain in terms of μ and v_0. (b) If the loop rolls over a bump, the resulting deformation of the chain causes two transverse pulses to propagate along the chain, one moving clockwise and one moving counterclockwise. What is the speed of the pulses traveling along the chain? (c) Through what angle does each pulse travel during the time it takes the loop to make one revolution?

65. (a) Show that the speed of longitudinal waves along a spring of force constant k is $v = \sqrt{kL/\mu}$, where L is the unstretched length of the spring and μ is the mass per unit length. (b) A spring with a mass of 0.400 kg has an unstretched length of 2.00 m and a force constant of 100 N/m. Using the result you obtained in (a), determine the speed of longitudinal waves along this spring.

66. A string of length L consists of two sections. The left half has mass per unit length $\mu = \mu_0/2$, while the right has a mass per unit length $\mu' = 3\mu = 3\mu_0/2$. Tension in the string is T_0. Notice from the data given that this string has the same total mass as a uniform string of length L and mass per unit length μ_0. (a) Find the speeds v and v' at which transverse pulses travel in the two sections. Express the speeds in terms of T_0 and μ_0, and also as multiples of the speed $v_0 = (T_0/\mu_0)^{1/2}$. (b) Find the time interval required for a pulse to travel from one end of the string to the other. Give your result as a multiple of $\Delta t_0 = L/v_0$.

67. A pulse traveling along a string of linear mass density μ is described by the wave function

$$y = [A_0 e^{-bx}] \sin(kx - \omega t)$$

where the factor in brackets before the sine function is said to be the amplitude. (a) What is the power $\mathcal{P}(x)$ carried by this wave at a point x? (b) What is the power carried by this wave at the origin? (c) Compute the ratio $\mathcal{P}(x)/\mathcal{P}(0)$.

68. An earthquake on the ocean floor in the Gulf of Alaska produces a *tsunami* (sometimes incorrectly called a "tidal wave") that reaches Hilo, Hawaii, 4 450 km away, in a time interval of 9 h 30 min. Tsunamis have enormous wavelengths (100 to 200 km), and the propagation speed for these waves is $v \approx \sqrt{gd}$, where d is the average depth of the water. From the information given, find the average wave speed and the average ocean depth between Alaska and Hawaii. (This method was used in 1856 to estimate the average depth of the Pacific Ocean long before soundings were made to give a direct determination.)

69. A string on a musical instrument is held under tension T and extends from the point $x = 0$ to the point $x = L$. The string is overwound with wire in such a way that its mass per unit length $\mu(x)$ increases uniformly from μ_0 at $x = 0$ to μ_L at $x = L$. (a) Find an expression for $\mu(x)$ as a function of x over the range $0 \leq x \leq L$. (b) Show that the time interval required for a transverse pulse to travel the length of the string is given by

$$\Delta t = \frac{2L\left(\mu_L + \mu_0 + \sqrt{\mu_L \mu_0}\right)}{3\sqrt{T}\left(\sqrt{\mu_L} + \sqrt{\mu_0}\right)}$$

Answers to Quick Quizzes

16.1 (b). It is longitudinal because the disturbance (the shift of position of the people) is parallel to the direction in which the wave travels.

16.2 (a). It is transverse because the people stand up and sit down (vertical motion), whereas the wave moves either to the left or to the right.

16.3 (c). The wave speed is determined by the medium, so it is unaffected by changing the frequency.

16.4 (b). Because the wave speed remains the same, the result of doubling the frequency is that the wavelength is half as large.

16.5 (d). The amplitude of a wave is unrelated to the wave speed, so we cannot determine the new amplitude without further information.

16.6 (c). With a larger amplitude, an element of the string has more energy associated with its simple harmonic motion, so the element passes through the equilibrium position with a higher maximum transverse speed.

16.7 Only answers (f) and (h) are correct. (a) and (b) affect the transverse speed of a particle of the string, but not the wave speed along the string. (c) and (d) change the amplitude. (e) and (g) increase the time interval by decreasing the wave speed.

16.8 (d). Doubling the amplitude of the wave causes the power to be larger by a factor of 4. In (a), halving the linear mass density of the string causes the power to change by a factor of 0.71—the rate decreases. In (b), doubling the wavelength of the wave halves the frequency and causes the power to change by a factor of 0.25—the rate decreases. In (c), doubling the tension in the string changes the wave speed and causes the power to change by a factor of 1.4—not as large as in part (d).

Chapter 17

Sound Waves

CHAPTER OUTLINE

▲ *Human ears have evolved to detect sound waves and interpret them as music or speech. Some animals, such as this young bat-eared fox, have ears adapted for the detection of very weak sounds. (Getty Images)*

Sound waves are the most common example of longitudinal waves. They travel through any material medium with a speed that depends on the properties of the medium. As the waves travel through air, the elements of air vibrate to produce changes in density and pressure along the direction of motion of the wave. If the source of the sound waves vibrates sinusoidally, the pressure variations are also sinusoidal. The mathematical description of sinusoidal sound waves is very similar to that of sinusoidal string waves, which were discussed in the previous chapter.

Sound waves are divided into three categories that cover different frequency ranges. (1) *Audible waves* lie within the range of sensitivity of the human ear. They can be generated in a variety of ways, such as by musical instruments, human voices, or loudspeakers. (2) *Infrasonic waves* have frequencies below the audible range. Elephants can use infrasonic waves to communicate with each other, even when separated by many kilometers. (3) *Ultrasonic waves* have frequencies above the audible range. You may have used a "silent" whistle to retrieve your dog. The ultrasonic sound it emits is easily heard by dogs, although humans cannot detect it at all. Ultrasonic waves are also used in medical imaging.

We begin this chapter by discussing the speed of sound waves and then wave intensity, which is a function of wave amplitude. We then provide an alternative description of the intensity of sound waves that compresses the wide range of intensities to which the ear is sensitive into a smaller range for convenience. We investigate the effects of the motion of sources and/or listeners on the frequency of a sound. Finally, we explore digital reproduction of sound, focusing in particular on sound systems used in modern motion pictures.

17.1 Speed of Sound Waves

Let us describe pictorially the motion of a one-dimensional longitudinal pulse moving through a long tube containing a compressible gas (Fig. 17.1). A piston at the left end can be moved to the right to compress the gas and create the pulse. Before the piston is moved, the gas is undisturbed and of uniform density, as represented by the uniformly shaded region in Figure 17.1a. When the piston is suddenly pushed to the right (Fig. 17.1b), the gas just in front of it is compressed (as represented by the more heavily shaded region); the pressure and density in this region are now higher than they were before the piston moved. When the piston comes to rest (Fig. 17.1c), the compressed region of the gas continues to move to the right, corresponding to a longitudinal pulse traveling through the tube with speed v. Note that the piston speed does *not* equal v. Furthermore, the compressed region does not "stay with" the piston as the piston moves, because the speed of the wave is usually greater than the speed of the piston.

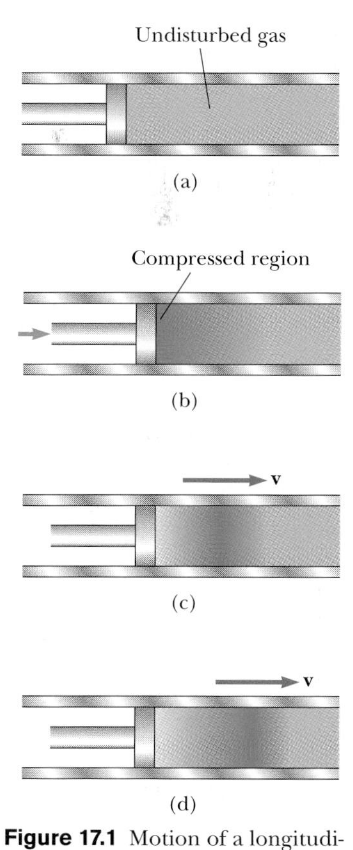

Figure 17.1 Motion of a longitudinal pulse through a compressible gas. The compression (darker region) is produced by the moving piston.

The speed of sound waves in a medium depends on the compressibility and density of the medium. If the medium is a liquid or a gas and has a bulk modulus B (see

Table 17.1

Speed of Sound in Various Media

Medium	v (m/s)
Gases	
Hydrogen (0°C)	1 286
Helium (0°C)	972
Air (20°C)	343
Air (0°C)	331
Oxygen (0°C)	317
Liquids at 25°C	
Glycerol	1 904
Seawater	1 533
Water	1 493
Mercury	1 450
Kerosene	1 324
Methyl alcohol	1 143
Carbon tetrachloride	926
Solids[a]	
Pyrex glass	5 640
Iron	5 950
Aluminum	6 420
Brass	4 700
Copper	5 010
Gold	3 240
Lucite	2 680
Lead	1 960
Rubber	1 600

[a] Values given are for propagation of longitudinal waves in bulk media. Speeds for longitudinal waves in thin rods are smaller, and speeds of transverse waves in bulk are smaller yet.

Section 12.4) and density ρ, the speed of sound waves in that medium is

$$v = \sqrt{\frac{B}{\rho}} \tag{17.1}$$

It is interesting to compare this expression with Equation 16.18 for the speed of transverse waves on a string, $v = \sqrt{T/\mu}$. In both cases, the wave speed depends on an elastic property of the medium—bulk modulus B or string tension T—and on an inertial property of the medium—ρ or μ. In fact, the *speed of all mechanical waves* follows an expression of the general form

$$v = \sqrt{\frac{\text{elastic property}}{\text{inertial property}}}$$

For longitudinal sound waves in a solid rod of material, for example, the speed of sound depends on Young's modulus Y and the density ρ. Table 17.1 provides the speed of sound in several different materials.

The speed of sound also depends on the temperature of the medium. For sound traveling through air, the relationship between wave speed and medium temperature is

$$v = (331 \text{ m/s})\sqrt{1 + \frac{T_C}{273°C}}$$

where 331 m/s is the speed of sound in air at 0°C, and T_C is the air temperature in degrees Celsius. Using this equation, one finds that at 20°C the speed of sound in air is approximately 343 m/s.

This information provides a convenient way to estimate the distance to a thunderstorm. You count the number of seconds between seeing the flash of lightning and hearing the thunder. Dividing this time by 3 gives the approximate distance to the lightning in kilometers, because 343 m/s is approximately $\frac{1}{3}$ km/s. Dividing the time in seconds by 5 gives the approximate distance to the lightning in miles, because the speed of sound in ft/s (1 125 ft/s) is approximately $\frac{1}{5}$ mi/s.

Quick Quiz 17.1 The speed of sound in air is a function of (a) wavelength (b) frequency (c) temperature (d) amplitude.

Example 17.1 Speed of Sound in a Liquid — Interactive

(A) Find the speed of sound in water, which has a bulk modulus of 2.1×10^9 N/m^2 at a temperature of 0°C and a density of 1.00×10^3 kg/m^3.

Solution Using Equation 17.1, we find that

$$v_{\text{water}} = \sqrt{\frac{B}{\rho}} = \sqrt{\frac{2.1 \times 10^9 \text{ N/m}^2}{1.00 \times 10^3 \text{ kg/m}^3}} = 1.4 \text{ km/s}$$

In general, sound waves travel more slowly in liquids than in solids because liquids are more compressible than solids. Note that the speed of sound in water is lower at 0°C than at 25°C (Table 17.1).

(B) Dolphins use sound waves to locate food. Experiments have shown that a dolphin can detect a 7.5-cm target 110 m away, even in murky water. For a bit of "dinner" at that distance, how much time passes between the moment the dolphin emits a sound pulse and the moment the dolphin hears its reflection and thereby detects the distant target?

Solution The total distance covered by the sound wave as it travels from dolphin to target and back is 2×110 m = 220 m. From Equation 2.2, we have, for 25°C water

$$\Delta t = \frac{\Delta x}{v_x} = \frac{220 \text{ m}}{1\,533 \text{ m/s}} = 0.14 \text{ s}$$

At the Interactive Worked Example link at **http://www.pse6.com,** *you can compare the speed of sound through the various media found in Table 17.1.*

17.2 Periodic Sound Waves

This section will help you better comprehend the nature of sound waves. An important fact for understanding how our ears work is that *pressure variations control what we hear.*

One can produce a one-dimensional periodic sound wave in a long, narrow tube containing a gas by means of an oscillating piston at one end, as shown in Figure 17.2. The darker parts of the colored areas in this figure represent regions where the gas is compressed and thus the density and pressure are above their equilibrium values. A compressed region is formed whenever the piston is pushed into the tube. This compressed region, called a **compression,** moves through the tube as a pulse, continuously compressing the region just in front of itself. When the piston is pulled back, the gas in front of it expands, and the pressure and density in this region fall below their equilibrium values (represented by the lighter parts of the colored areas in Fig. 17.2). These low-pressure regions, called **rarefactions,** also propagate along the tube, following the compressions. Both regions move with a speed equal to the speed of sound in the medium.

As the piston oscillates sinusoidally, regions of compression and rarefaction are continuously set up. The distance between two successive compressions (or two successive rarefactions) equals the wavelength λ. As these regions travel through the tube, any small element of the medium moves with simple harmonic motion parallel to the direction of the wave. If $s(x, t)$ is the position of a small element relative to its equilibrium position,[1] we can express this harmonic position function as

$$s(x, t) = s_{\max} \cos(kx - \omega t) \tag{17.2}$$

where **$s_{\max}$ is the maximum position of the element relative to equilibrium.** This is often called the **displacement amplitude** of the wave. The parameter k is the wave number and ω is the angular frequency of the piston. Note that the displacement of the element is along x, in the direction of propagation of the sound wave, which means we are describing a longitudinal wave.

The variation in the gas pressure ΔP measured from the equilibrium value is also periodic. For the position function in Equation 17.2, ΔP is given by

$$\Delta P = \Delta P_{\max} \sin(kx - \omega t) \tag{17.3}$$

where **the pressure amplitude $\Delta P_{\max}$**—which is the **maximum change in pressure from the equilibrium value**—is given by

$$\Delta P_{\max} = \rho v \omega s_{\max} \tag{17.4}$$

Thus, we see that a sound wave may be considered as either a displacement wave or a pressure wave. A comparison of Equations 17.2 and 17.3 shows that **the pressure wave is 90° out of phase with the displacement wave.** Graphs of these functions are shown in Figure 17.3. Note that the pressure variation is a maximum when the displacement from equilibrium is zero, and the displacement from equilibrium is a maximum when the pressure variation is zero.

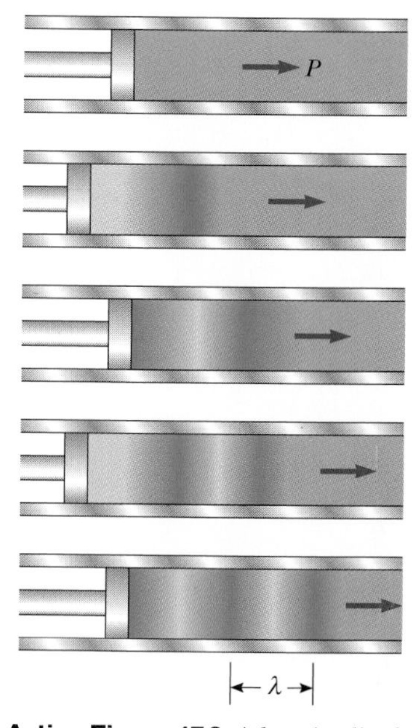

Active Figure 17.2 A longitudinal wave propagating through a gas-filled tube. The source of the wave is an oscillating piston at the left.

At the Active Figures link at http://www.pse6.com, you can adjust the frequency of the piston.

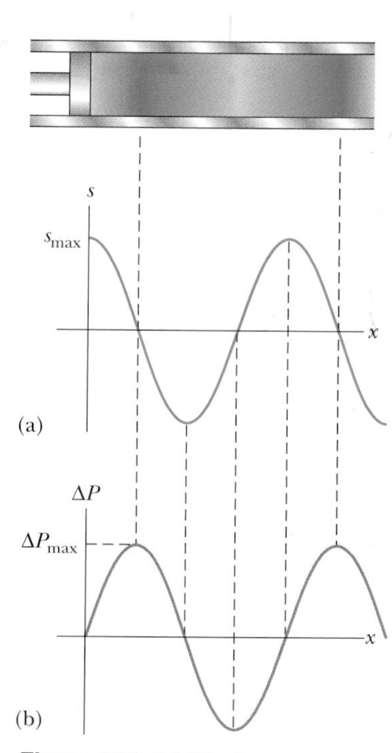

Figure 17.3 (a) Displacement amplitude and (b) pressure amplitude versus position for a sinusoidal longitudinal wave.

Quick Quiz 17.2 If you blow across the top of an empty soft-drink bottle, a pulse of sound travels down through the air in the bottle. At the moment the pulse reaches the bottom of the bottle, the correct descriptions of the displacement of elements of air from their equilibrium positions and the pressure of the air at this point are (a) the displacement and pressure are both at a maximum (b) the displacement and pressure are both at a minimum (c) the displacement is zero and the pressure is a maximum (d) the displacement is zero and the pressure is a minimum.

[1] We use $s(x, t)$ here instead of $y(x, t)$ because the displacement of elements of the medium is not perpendicular to the x direction.

Derivation of Equation 17.3

Consider a thin disk-shaped element of gas whose circular cross section is parallel to the piston in Figure 17.2. This element will undergo changes in position, pressure, and density as a sound wave propagates through the gas. From the definition of bulk modulus (see Eq. 12.8), the pressure variation in the gas is

$$\Delta P = -B\frac{\Delta V}{V_i}$$

The element has a thickness Δx in the horizontal direction and a cross-sectional area A, so its volume is $V_i = A\Delta x$. The change in volume ΔV accompanying the pressure change is equal to $A\Delta s$, where Δs is the difference between the value of s at $x + \Delta x$ and the value of s at x. Hence, we can express ΔP as

$$\Delta P = -B\frac{\Delta V}{V_i} = -B\frac{A\Delta s}{A\Delta x} = -B\frac{\Delta s}{\Delta x}$$

As Δx approaches zero, the ratio $\Delta s/\Delta x$ becomes $\partial s/\partial x$. (The partial derivative indicates that we are interested in the variation of s with position at a *fixed* time.) Therefore,

$$\Delta P = -B\frac{\partial s}{\partial x}$$

If the position function is the simple sinusoidal function given by Equation 17.2, we find that

$$\Delta P = -B\frac{\partial}{\partial x}[s_{\max}\cos(kx - \omega t)] = Bs_{\max}k\sin(kx - \omega t)$$

Because the bulk modulus is given by $B = \rho v^2$ (see Eq. 17.1), the pressure variation reduces to

$$\Delta P = \rho v^2 s_{\max}k\sin(kx - \omega t)$$

From Equation 16.11, we can write $k = \omega/v$; hence, ΔP can be expressed as

$$\Delta P = \rho v\omega s_{\max}\sin(kx - \omega t)$$

Because the sine function has a maximum value of 1, we see that the maximum value of the pressure variation is $\Delta P_{\max} = \rho v\omega s_{\max}$ (see Eq. 17.4), and we arrive at Equation 17.3:

$$\Delta P = \Delta P_{\max}\sin(kx - \omega t)$$

17.3 Intensity of Periodic Sound Waves

In the preceding chapter, we showed that a wave traveling on a taut string transports energy. The same concept applies to sound waves. Consider an element of air of mass Δm and width Δx in front of a piston oscillating with a frequency ω, as shown in Figure 17.4.

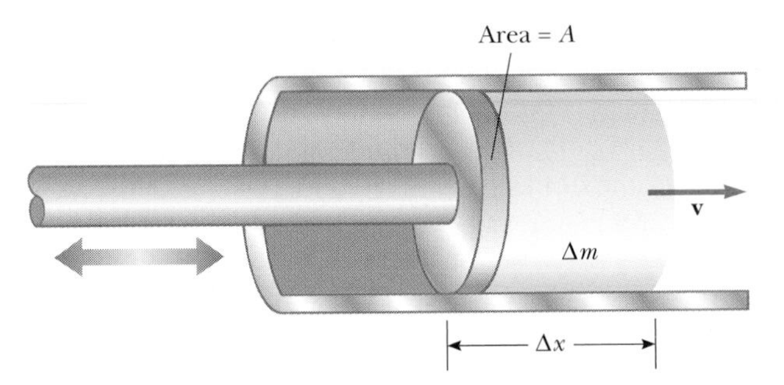

Figure 17.4 An oscillating piston transfers energy to the air in the tube, causing the element of air of width Δx and mass Δm to oscillate with an amplitude $s_{\max}$.

The piston transmits energy to this element of air in the tube, and the energy is propagated away from the piston by the sound wave. To evaluate the rate of energy transfer for the sound wave, we shall evaluate the kinetic energy of this element of air, which is undergoing simple harmonic motion. We shall follow a procedure similar to that in Section 16.5, in which we evaluated the rate of energy transfer for a wave on a string.

As the sound wave propagates away from the piston, the position of any element of air in front of the piston is given by Equation 17.2. To evaluate the kinetic energy of this element of air, we need to know its speed. We find the speed by taking the time derivative of Equation 17.2:

$$v(x, t) = \frac{\partial}{\partial t} s(x, t) = \frac{\partial}{\partial t}[s_{\max} \cos(kx - \omega t)] = -\omega s_{\max} \sin(kx - \omega t)$$

Imagine that we take a "snapshot" of the wave at $t = 0$. The kinetic energy of a given element of air at this time is

$$\Delta K = \tfrac{1}{2}\Delta m(v)^2 = \tfrac{1}{2}\Delta m(-\omega s_{\max} \sin kx)^2 = \tfrac{1}{2}\rho A\, \Delta x(-\omega s_{\max} \sin kx)^2$$
$$= \tfrac{1}{2}\rho A\, \Delta x(\omega s_{\max})^2 \sin^2 kx$$

where A is the cross-sectional area of the element and $A\Delta x$ is its volume. Now, as in Section 16.5, we integrate this expression over a full wavelength to find the total kinetic energy in one wavelength. Letting the element of air shrink to infinitesimal thickness, so that $\Delta x \rightarrow dx$, we have

$$K_\lambda = \int dK = \int_0^\lambda \tfrac{1}{2}\rho A(\omega s_{\max})^2 \sin^2 kx\, dx = \tfrac{1}{2}\rho A(\omega s_{\max})^2 \int_0^\lambda \sin^2 kx\, dx$$
$$= \tfrac{1}{2}\rho A(\omega s_{\max})^2(\tfrac{1}{2}\lambda) = \tfrac{1}{4}\rho A(\omega s_{\max})^2 \lambda$$

As in the case of the string wave in Section 16.5, the total potential energy for one wavelength has the same value as the total kinetic energy; thus, the total mechanical energy for one wavelength is

$$E_\lambda = K_\lambda + U_\lambda = \tfrac{1}{2}\rho A(\omega s_{\max})^2 \lambda$$

As the sound wave moves through the air, this amount of energy passes by a given point during one period of oscillation. Hence, the rate of energy transfer is

$$\mathcal{P} = \frac{\Delta E}{\Delta t} = \frac{E_\lambda}{T} = \frac{\tfrac{1}{2}\rho A(\omega s_{\max})^2 \lambda}{T} = \tfrac{1}{2}\rho A(\omega s_{\max})^2 \left(\frac{\lambda}{T}\right) = \tfrac{1}{2}\rho A v(\omega s_{\max})^2$$

where v is the speed of sound in air.

We define the **intensity** I of a wave, or the power per unit area, to be the rate at which the energy being transported by the wave transfers through a unit area A perpendicular to the direction of travel of the wave:

$$I \equiv \frac{\mathcal{P}}{A} \qquad (17.5)$$

In the present case, therefore, the intensity is

$$I = \frac{\mathcal{P}}{A} = \tfrac{1}{2}\rho v(\omega s_{\max})^2$$

Intensity of a sound wave

Thus, we see that the intensity of a periodic sound wave is proportional to the square of the displacement amplitude and to the square of the angular frequency (as in the case of a periodic string wave). This can also be written in terms of the pressure amplitude ΔP_{max}; in this case, we use Equation 17.4 to obtain

$$I = \frac{\Delta P_{max}^2}{2\rho v} \tag{17.6}$$

Now consider a point source emitting sound waves equally in all directions. From everyday experience, we know that the intensity of sound decreases as we move farther from the source. We identify an imaginary sphere of radius r centered on the source. When a source emits sound equally in all directions, we describe the result as a **spherical wave.** The average power $\mathcal{P}_{av}$ emitted by the source must be distributed uniformly over this spherical surface of area $4\pi r^2$. Hence, the wave intensity at a distance r from the source is

Inverse-square behavior of intensity for a point source

$$I = \frac{\mathcal{P}_{av}}{A} = \frac{\mathcal{P}_{av}}{4\pi r^2} \tag{17.7}$$

This inverse-square law, which is reminiscent of the behavior of gravity in Chapter 13, states that the intensity decreases in proportion to the square of the distance from the source.

Quick Quiz 17.3 An *ear trumpet* is a cone-shaped shell, like a megaphone, that was used before hearing aids were developed to help persons who were hard of hearing. The small end of the cone was held in the ear, and the large end was aimed toward the source of sound as in Figure 17.5. The ear trumpet increases the intensity of sound because (a) it increases the speed of sound (b) it reflects sound back toward the source (c) it gathers sound that would normally miss the ear and concentrates it into a smaller area (d) it increases the density of the air.

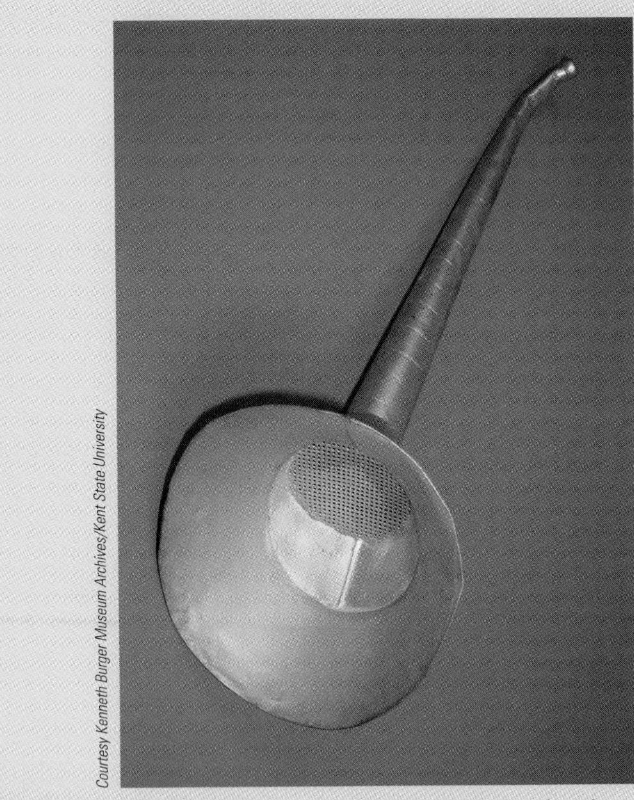

Courtesy Kenneth Burger Museum Archives/Kent State University

Figure 17.5 (Quick Quiz 17.3) An ear trumpet, used before hearing aids to make sounds intense enough for people who were hard of hearing. You can simulate the effect of an ear trumpet by cupping your hands behind your ears.

Quick Quiz 17.4 A vibrating guitar string makes very little sound if it is not mounted on the guitar. But if this vibrating string is attached to the guitar body, so that the body of the guitar vibrates, the sound is higher in intensity. This is because (a) the power of the vibration is spread out over a larger area (b) the energy leaves the guitar at a higher rate (c) the speed of sound is higher in the material of the guitar body (d) none of these.

Example 17.2 Hearing Limits

The faintest sounds the human ear can detect at a frequency of 1 000 Hz correspond to an intensity of about 1.00×10^{-12} W/m^2—the so-called *threshold of hearing*. The loudest sounds the ear can tolerate at this frequency correspond to an intensity of about 1.00 W/m^2—the *threshold of pain*. Determine the pressure amplitude and displacement amplitude associated with these two limits.

Solution First, consider the faintest sounds. Using Equation 17.6 and taking $v = 343$ m/s as the speed of sound waves in air and $\rho = 1.20$ kg/m^3 as the density of air, we obtain

$$\Delta P_{\text{max}} = \sqrt{2\rho v I}$$

$$= \sqrt{2(1.20 \text{ kg/m}^3)(343 \text{ m/s})(1.00 \times 10^{-12} \text{ W/m}^2)}$$

$$= 2.87 \times 10^{-5} \text{ N/m}^2$$

Because atmospheric pressure is about 10^5 N/m^2, this result tells us that the ear is sensitive to pressure fluctuations as small as 3 parts in 10^{10}!

We can calculate the corresponding displacement amplitude by using Equation 17.4, recalling that $\omega = 2\pi f$ (see Eqs. 16.3 and 16.9):

$$s_{\text{max}} = \frac{\Delta P_{\text{max}}}{\rho v \omega} = \frac{2.87 \times 10^{-5} \text{ N/m}^2}{(1.20 \text{ kg/m}^3)(343 \text{ m/s})(2\pi \times 1\,000 \text{ Hz})}$$

$$= 1.11 \times 10^{-11} \text{ m}$$

This is a remarkably small number! If we compare this result for s_{max} with the size of an atom (about 10^{-10} m), we see that the ear is an extremely sensitive detector of sound waves.

In a similar manner, one finds that the loudest sounds the human ear can tolerate correspond to a pressure amplitude of 28.7 N/m^2 and a displacement amplitude equal to 1.11×10^{-5} m.

Example 17.3 Intensity Variations of a Point Source

A point source emits sound waves with an average power output of 80.0 W.

(A) Find the intensity 3.00 m from the source.

Solution A point source emits energy in the form of spherical waves. Using Equation 17.7, we have

$$I = \frac{\mathcal{P}_{\text{av}}}{4\pi r^2} = \frac{80.0 \text{ W}}{4\pi (3.00 \text{ m})^2} = 0.707 \text{ W/m}^2$$

an intensity that is close to the threshold of pain.

(B) Find the distance at which the intensity of the sound is 1.00×10^{-8} W/m^2.

Solution Using this value for I in Equation 17.7 and solving for r, we obtain

$$r = \sqrt{\frac{\mathcal{P}_{\text{av}}}{4\pi I}} = \sqrt{\frac{80.0 \text{ W}}{4\pi(1.00 \times 10^{-8} \text{ W/m}^2)}}$$

$$= 2.52 \times 10^4 \text{ m}$$

which equals about 16 miles!

Sound Level in Decibels

Example 17.2 illustrates the wide range of intensities the human ear can detect. Because this range is so wide, it is convenient to use a logarithmic scale, where the **sound level** β (Greek beta) is defined by the equation

$$\beta \equiv 10 \log\left(\frac{I}{I_0}\right) \qquad (17.8)$$

Sound level in decibels

The constant I_0 is the *reference intensity*, taken to be at the threshold of hearing ($I_0 = 1.00 \times 10^{-12}$ W/m^2), and I is the intensity in watts per square meter to which the sound level β corresponds, where β is measured[2] in **decibels** (dB). On this scale,

[2] The unit *bel* is named after the inventor of the telephone, Alexander Graham Bell (1847–1922). The prefix *deci-* is the SI prefix that stands for 10^{-1}.

Table 17.2

Sound Levels

Source of Sound	β (dB)
Nearby jet airplane	150
Jackhammer; machine gun	130
Siren; rock concert	120
Subway; power mower	100
Busy traffic	80
Vacuum cleaner	70
Normal conversation	50
Mosquito buzzing	40
Whisper	30
Rustling leaves	10
Threshold of hearing	0

the threshold of pain ($I = 1.00\ \text{W/m}^2$) corresponds to a sound level of $\beta = 10\log[(1\ \text{W/m}^2)/(10^{-12}\ \text{W/m}^2)] = 10\ \log(10^{12}) = 120$ dB, and the threshold of hearing corresponds to $\beta = 10\log[(10^{-12}\ \text{W/m}^2)/(10^{-12}\ \text{W/m}^2)] = 0$ dB.

Prolonged exposure to high sound levels may seriously damage the ear. Ear plugs are recommended whenever sound levels exceed 90 dB. Recent evidence suggests that "noise pollution" may be a contributing factor to high blood pressure, anxiety, and nervousness. Table 17.2 gives some typical sound-level values.

Quick Quiz 17.5 A violin plays a melody line and is then joined by a second violin, playing at the same intensity as the first violin, in a repeat of the same melody. With both violins playing, what physical parameter has doubled compared to the situation with only one violin playing? (a) wavelength (b) frequency (c) intensity (d) sound level in dB (e) none of these.

Quick Quiz 17.6 Increasing the intensity of a sound by a factor of 100 causes the sound level to increase by (a) 100 dB (b) 20 dB (c) 10 dB (d) 2 dB.

Example 17.4 Sound Levels

Two identical machines are positioned the same distance from a worker. The intensity of sound delivered by each machine at the location of the worker is $2.0 \times 10^{-7}\ \text{W/m}^2$. Find the sound level heard by the worker

(A) when one machine is operating

(B) when both machines are operating.

Solution

(A) The sound level at the location of the worker with one machine operating is calculated from Equation 17.8:

$$\beta_1 = 10\log\left(\frac{2.0 \times 10^{-7}\ \text{W/m}^2}{1.00 \times 10^{-12}\ \text{W/m}^2}\right) = 10\log(2.0 \times 10^5)$$

$$= 53\ \text{dB}$$

(B) When both machines are operating, the intensity is doubled to $4.0 \times 10^{-7}\ \text{W/m}^2$; therefore, the sound level now is

$$\beta_2 = 10\log\left(\frac{4.0 \times 10^{-7}\ \text{W/m}^2}{1.00 \times 10^{-12}\ \text{W/m}^2}\right) = 10\log(4.0 \times 10^5)$$

$$= 56\ \text{dB}$$

From these results, we see that when the intensity is doubled, the sound level increases by only 3 dB.

What If? ***Loudness*** **is a psychological response to a sound and depends on both the intensity and the frequency of the sound. As a rule of thumb, a doubling in loudness is approximately associated with an increase in sound level of 10 dB. (Note that this rule of thumb is relatively inaccurate at very low or very high frequencies.) If the loudness of the**

machines in this example is to be doubled, how many machines must be running?

Answer Using the rule of thumb, a doubling of loudness corresponds to a sound level increase of 10 dB. Thus,

$$\beta_2 - \beta_1 = 10 \text{ dB} = 10 \log\left(\frac{I_2}{I_0}\right) - 10 \log\left(\frac{I_1}{I_0}\right) = 10 \log\left(\frac{I_2}{I_1}\right)$$

$$\log\left(\frac{I_2}{I_1}\right) = 1$$

$$I_2 = 10 I_1$$

Thus, ten machines must be operating to double the loudness.

Loudness and Frequency

The discussion of sound level in decibels relates to a *physical* measurement of the strength of a sound. Let us now consider how we describe the *psychological* "measurement" of the strength of a sound.

Of course, we don't have meters in our bodies that can read out numerical values of our reactions to stimuli. We have to "calibrate" our reactions somehow by comparing different sounds to a reference sound. However, this is not easy to accomplish. For example, earlier we mentioned that the threshold intensity is 10^{-12} W/m^2, corresponding to an intensity level of 0 dB. In reality, this value is the threshold only for a sound of frequency 1 000 Hz, which is a standard reference frequency in acoustics. If we perform an experiment to measure the threshold intensity at other frequencies, we find a distinct variation of this threshold as a function of frequency. For example, at 100 Hz, a sound must have an intensity level of about 30 dB in order to be just barely audible! Unfortunately, there is no simple relationship between physical measurements and psychological "measurements." The 100-Hz, 30-dB sound is psychologically "equal" to the 1 000-Hz, 0-dB sound (both are just barely audible) but they are not physically equal (30 dB $\neq$ 0 dB).

By using test subjects, the human response to sound has been studied, and the results are shown in Figure 17.6 (the white area), along with the approximate frequency and sound-level ranges of other sound sources. The lower curve of the white area corresponds to the threshold of hearing. Its variation with frequency is clear from this diagram. Note that humans are sensitive to frequencies ranging from about 20 Hz to about 20 000 Hz. The upper bound of the white area is the threshold of pain. Here the

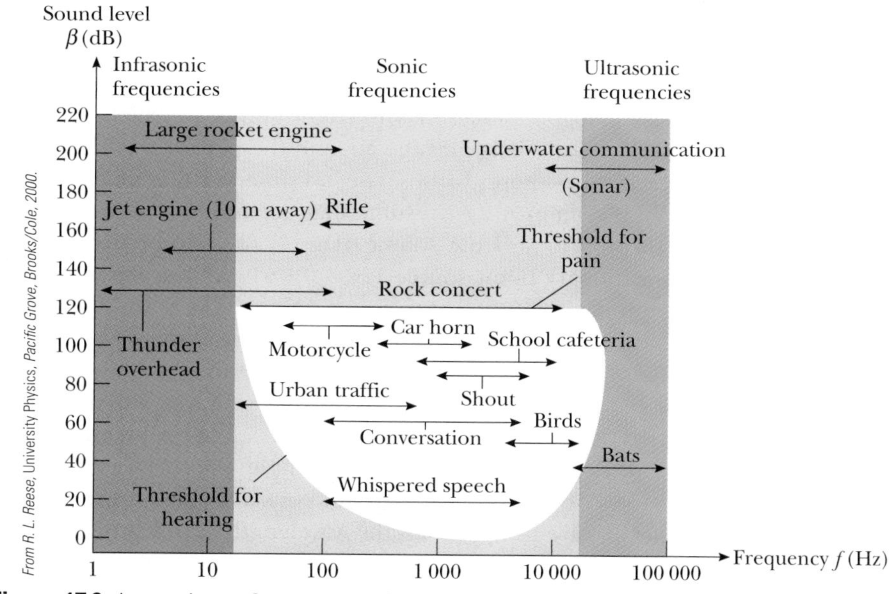

Figure 17.6 Approximate frequency and sound level ranges of various sources and that of normal human hearing, shown by the white area.

boundary of the white area is straight, because the psychological response is relatively independent of frequency at this high sound level.

The most dramatic change with frequency is in the lower left region of the white area, for low frequencies and low intensity levels. Our ears are particularly insensitive in this region. If you are listening to your stereo and the bass (low frequencies) and treble (high frequencies) sound balanced at a high volume, try turning the volume down and listening again. You will probably notice that the bass seems weak, which is due to the insensitivity of the ear to low frequencies at low sound levels, as shown in Figure 17.6.

17.4 The Doppler Effect

Perhaps you have noticed how the sound of a vehicle's horn changes as the vehicle moves past you. The frequency of the sound you hear as the vehicle approaches you is higher than the frequency you hear as it moves away from you. This is one example of the **Doppler effect.**[3]

To see what causes this apparent frequency change, imagine you are in a boat that is lying at anchor on a gentle sea where the waves have a period of $T = 3.0$ s. This means that every 3.0 s a crest hits your boat. Figure 17.7a shows this situation, with the water waves moving toward the left. If you set your watch to $t = 0$ just as one crest hits, the watch reads 3.0 s when the next crest hits, 6.0 s when the third crest hits, and so on. From these observations you conclude that the wave frequency is $f = 1/T = 1/(3.0 \text{ s}) = 0.33$ Hz. Now suppose you start your motor and head directly into the oncoming waves, as in Figure 17.7b. Again you set your watch to $t = 0$ as a crest hits the front of your boat. Now, however, because you are moving toward the next wave crest as it moves toward you, it hits you less than 3.0 s after the first hit. In other words, the period you observe is shorter than the 3.0-s period you observed when you were stationary. Because $f = 1/T$, you observe a higher wave frequency than when you were at rest.

If you turn around and move in the same direction as the waves (see Fig. 17.7c), you observe the opposite effect. You set your watch to $t = 0$ as a crest hits the back of the boat. Because you are now moving away from the next crest, more than 3.0 s has elapsed on your watch by the time that crest catches you. Thus, you observe a lower frequency than when you were at rest.

These effects occur because the *relative* speed between your boat and the waves depends on the direction of travel and on the speed of your boat. When you are moving toward the right in Figure 17.7b, this relative speed is higher than that of the wave speed, which leads to the observation of an increased frequency. When you turn around and move to the left, the relative speed is lower, as is the observed frequency of the water waves.

Let us now examine an analogous situation with sound waves, in which the water waves become sound waves, the water becomes the air, and the person on the boat becomes an observer listening to the sound. In this case, an observer O is moving and a sound source S is stationary. For simplicity, we assume that the air is also stationary and that the observer moves directly toward the source (Fig. 17.8). The observer moves with a speed v_O toward a stationary point source ($v_S = 0$), where *stationary* means at rest with respect to the medium, air.

If a point source emits sound waves and the medium is uniform, the waves move at the same speed in all directions radially away from the source; this is a spherical wave, as was mentioned in Section 17.3. It is useful to represent these waves with a series of circular arcs concentric with the source, as in Figure 17.8. Each arc represents a surface over which the phase of the wave is constant. For example, the surface could pass through the crests of all waves. We call such a surface of constant phase a **wave front.** The distance between adjacent wave fronts equals the wavelength λ. In Figure 17.8, the

[3] Named after the Austrian physicist Christian Johann Doppler (1803–1853), who in 1842 predicted the effect for both sound waves and light waves.

(a)

(b)

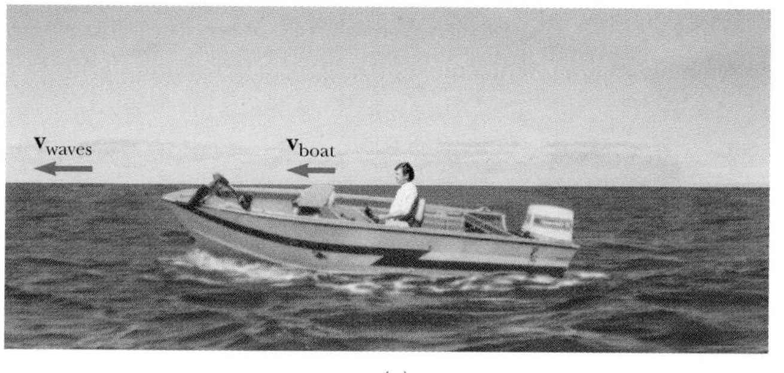

(c)

Figure 17.7 (a) Waves moving toward a stationary boat. The waves travel to the left, and their source is far to the right of the boat, out of the frame of the photograph. (b) The boat moving toward the wave source. (c) The boat moving away from the wave source.

circles are the intersections of these three-dimensional wave fronts with the two-dimensional paper.

We take the frequency of the source in Figure 17.8 to be f, the wavelength to be λ, and the speed of sound to be v. If the observer were also stationary, he or she would detect wave fronts at a rate f. (That is, when $v_O = 0$ and $v_S = 0$, the observed frequency equals the source frequency.) When the observer moves toward the source, the speed of the waves relative to the observer is $v' = v + v_O$, as in the case of the boat, but the

Active Figure 17.8 An observer O (the cyclist) moves with a speed v_O toward a stationary point source S, the horn of a parked truck. The observer hears a frequency f' that is greater than the source frequency.

At the Active Figures link at http://www.pse6.com, you can adjust the speed of the observer.

wavelength λ is unchanged. Hence, using Equation 16.12, $v = \lambda f$, we can say that the frequency f' heard by the observer is *increased* and is given by

$$f' = \frac{v'}{\lambda} = \frac{v + v_O}{\lambda}$$

Because $\lambda = v/f$, we can express f' as

$$f' = \left(\frac{v + v_O}{v}\right)f \qquad \text{(observer moving toward source)} \tag{17.9}$$

If the observer is moving away from the source, the speed of the wave relative to the observer is $v' = v - v_O$. The frequency heard by the observer in this case is *decreased* and is given by

$$f' = \left(\frac{v - v_O}{v}\right)f \qquad \text{(observer moving away from source)} \tag{17.10}$$

In general, whenever an observer moves with a speed v_O relative to a stationary source, the frequency heard by the observer is given by Equation 17.9, with a sign convention: a positive value is substituted for v_O when the observer moves toward the source and a negative value is substituted when the observer moves away from the source.

Now consider the situation in which the source is in motion and the observer is at rest. If the source moves directly toward observer A in Figure 17.9a, the wave fronts heard by the observer are closer together than they would be if the source were not moving. As a result, the wavelength λ' measured by observer A is shorter than the wavelength λ of the source. During each vibration, which lasts for a time interval T (the period), the source moves a distance $v_S T = v_S/f$ and the wavelength is *shortened* by this amount. Therefore, the observed wavelength λ' is

$$\lambda' = \lambda - \Delta\lambda = \lambda - \frac{v_S}{f}$$

Because $\lambda = v/f$, the frequency f' heard by observer A is

$$f' = \frac{v}{\lambda'} = \frac{v}{\lambda - (v_S/f)} = \frac{v}{(v/f) - (v_S/f)}$$

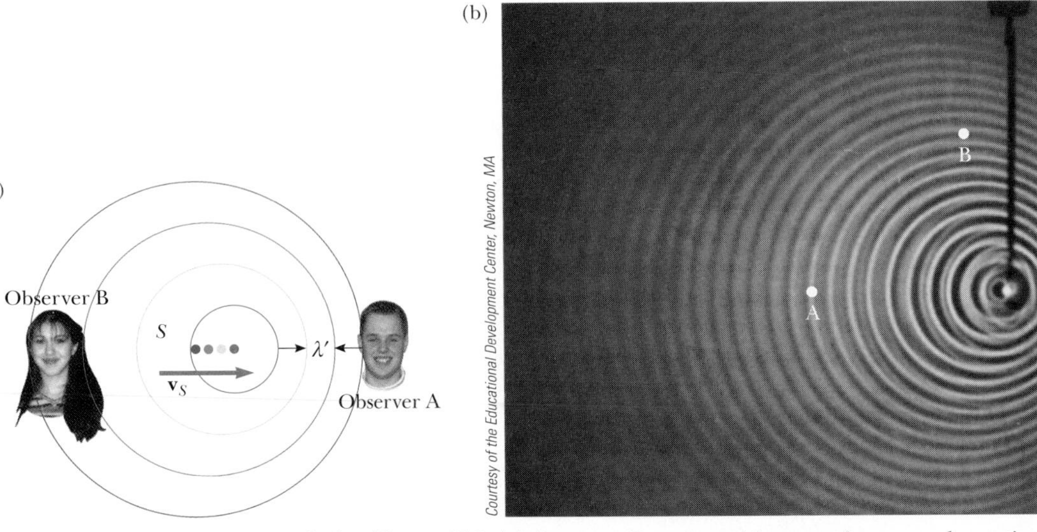

Active Figure 17.9 (a) A source S moving with a speed v_S toward a stationary observer A and away from a stationary observer B. Observer A hears an increased frequency, and observer B hears a decreased frequency. (b) The Doppler effect in water, observed in a ripple tank. A point source is moving to the right with speed v_S. Letters shown in the photo refer to Quick Quiz 17.7.

At the Active Figures link at http://www.pse6.com, you can adjust the speed of the source.

$$f' = \left(\frac{v}{v - v_S}\right)f \quad \text{(source moving toward observer)} \tag{17.11}$$

That is, the observed frequency is *increased* whenever the source is moving toward the observer.

When the source moves away from a stationary observer, as is the case for observer B in Figure 17.9a, the observer measures a wavelength λ' that is *greater* than λ and hears a *decreased* frequency:

$$f' = \left(\frac{v}{v + v_S}\right)f \quad \text{(source moving away from observer)} \tag{17.12}$$

We can express the general relationship for the observed frequency when a source is moving and an observer is at rest as Equation 17.11, with the same sign convention applied to v_S as was applied to v_O: a positive value is substituted for v_S when the source moves toward the observer and a negative value is substituted when the source moves away from the observer.

Finally, we find the following general relationship for the observed frequency:

General Doppler-shift expression

$$f' = \left(\frac{v + v_O}{v - v_S}\right)f \tag{17.13}$$

In this expression, the signs for the values substituted for v_O and v_S depend on the direction of the velocity. A positive value is used for motion of the observer or the source *toward* the other, and a negative sign for motion of one *away from* the other.

A convenient rule concerning signs for you to remember when working with all Doppler-effect problems is as follows:

The word *toward* is associated with an *increase* in observed frequency. The words *away from* are associated with a *decrease* in observed frequency.

Although the Doppler effect is most typically experienced with sound waves, it is a phenomenon that is common to all waves. For example, the relative motion of source and observer produces a frequency shift in light waves. The Doppler effect is used in police radar systems to measure the speeds of motor vehicles. Likewise, astronomers use the effect to determine the speeds of stars, galaxies, and other celestial objects relative to the Earth.

PITFALL PREVENTION

17.1 Doppler Effect Does Not Depend on Distance

Many people think that the Doppler effect depends on the distance between the source and the observer. While the intensity of a sound varies as the distance changes, the apparent frequency depends only on the relative speed of source and observer. As you listen to an approaching source, you will detect increasing intensity but constant frequency. As the source passes, you will hear the frequency suddenly drop to a new constant value and the intensity begin to decrease.

Quick Quiz 17.7 Consider detectors of water waves at three locations A, B, and C in Figure 17.9b. Which of the following statements is true? (a) The wave speed is highest at location A. (b) The wave speed is highest at location C. (c) The detected wavelength is largest at location B. (c) The detected wavelength is largest at location C. (e) The detected frequency is highest at location C. (f) The detected frequency is highest at location A.

Quick Quiz 17.8 You stand on a platform at a train station and listen to a train approaching the station at a constant velocity. While the train approaches, but before it arrives, you hear (a) the intensity and the frequency of the sound both increasing (b) the intensity and the frequency of the sound both decreasing (c) the intensity increasing and the frequency decreasing (d) the intensity decreasing and the frequency increasing (e) the intensity increasing and the frequency remaining the same (f) the intensity decreasing and the frequency remaining the same.

Example 17.5 The Broken Clock Radio

Your clock radio awakens you with a steady and irritating sound of frequency 600 Hz. One morning, it malfunctions and cannot be turned off. In frustration, you drop the clock radio out of your fourth-story dorm window, 15.0 m from the ground. Assume the speed of sound is 343 m/s.

(A) As you listen to the falling clock radio, what frequency do you hear just before you hear the radio striking the ground?

(B) At what rate does the frequency that you hear change with time just before you hear the radio striking the ground?

Solution

(A) In conceptualizing the problem, note that the speed of the radio increases as it falls. Thus, it is a source of sound moving away from you with an increasing speed. We categorize this problem as one in which we must combine our understanding of falling objects with that of the frequency shift due to the Doppler effect. To analyze the problem, we identify the clock radio as a moving source of sound for which the Doppler-shifted frequency is given by

$$f' = \left(\frac{v}{v - v_S}\right) f$$

The speed of the source of sound is given by Equation 2.9 for a falling object:

$$v_S = v_{yi} + a_y t = 0 - gt = -gt$$

Thus, the Doppler-shifted frequency of the falling clock radio is

$$(1) \qquad f' = \left(\frac{v}{v - (-gt)}\right) f = \left(\frac{v}{v + gt}\right) f$$

The time at which the radio strikes the ground is found from Equation 2.12:

$$y_f = y_i + v_{yi}t - \tfrac{1}{2}gt^2$$

$$-15.0 \text{ m} = 0 + 0 - \tfrac{1}{2}(9.80 \text{ m/s}^2)t^2$$

$$t = 1.75 \text{ s}$$

Thus, the Doppler-shifted frequency just as the radio strikes the ground is

$$f' = \left(\frac{v}{v + gt}\right) f$$

$$= \left(\frac{343 \text{ m/s}}{343 \text{ m/s} + (9.80 \text{ m/s}^2)(1.75 \text{ s})}\right)(600 \text{ Hz})$$

$$= 571 \text{ Hz}$$

(B) The rate at which the frequency changes is found by differentiating Equation (1) with respect to t:

$$\frac{df'}{dt} = \frac{d}{dt}\left(\frac{vf}{v + gt}\right) = \frac{-vg}{(v + gt)^2} f$$

$$= \frac{-(343 \text{ m/s})(9.80 \text{ m/s}^2)}{[343 \text{ m/s} + (9.80 \text{ m/s}^2)(1.75 \text{ s})]^2}(600 \text{ Hz})$$

$$= -15.5 \text{ Hz/s}$$

To finalize this problem, consider the following **What If?**

What If? Suppose you live on the eighth floor instead of the fourth floor. If you repeat the radio-dropping activity, does the frequency shift in part (A) and the rate of change of frequency in part (B) of this example double?

Answer The doubled height does not give a time at which the radio lands that is twice the time found in part (A). From Equation 2.12:

$$y_f = y_i + v_{yi}t - \tfrac{1}{2}gt^2$$

$$-30.0 \text{ m} = 0 + 0 - \tfrac{1}{2}(9.80 \text{ m/s}^2)t^2$$

$$t = 2.47 \text{ s}$$

The new frequency heard just before you hear the radio strike the ground is

$$f' = \left(\frac{v}{v + gt}\right) f$$

$$= \left(\frac{343 \text{ m/s}}{343 \text{ m/s} + (9.80 \text{ m/s}^2)(2.47 \text{ s})}\right)(600 \text{ Hz})$$

$$= 560 \text{ Hz}$$

The frequency shift heard on the fourth floor is 600 Hz − 571 Hz = 29 Hz, while the frequency shift heard from the eighth floor is 600 Hz − 560 Hz = 40 Hz, which is not twice as large.

The new rate of change of frequency is

$$\frac{df'}{dt} = \frac{-vg}{(v + gt)^2} f$$

$$= \frac{-(343 \text{ m/s})(9.80 \text{ m/s}^2)}{[343 \text{ m/s} + (9.80 \text{ m/s}^2)(2.47 \text{ s})]^2}(600 \text{ Hz})$$

$$= -15.0 \text{ Hz/s}$$

Note that this value is actually *smaller* in magnitude than the previous value of − 15.5 Hz/s!

Example 17.6 Doppler Submarines

Interactive

A submarine (sub A) travels through water at a speed of 8.00 m/s, emitting a sonar wave at a frequency of 1 400 Hz. The speed of sound in the water is 1 533 m/s. A second submarine (sub B) is located such that both submarines are traveling directly toward one another. The second submarine is moving at 9.00 m/s.

(A) What frequency is detected by an observer riding on sub B as the subs approach each other?

(B) The subs barely miss each other and pass. What frequency is detected by an observer riding on sub B as the subs recede from each other?

Solution

(A) We use Equation 17.13 to find the Doppler-shifted frequency. As the two submarines approach each other, the observer in sub B hears the frequency

$$f' = \left(\frac{v + v_O}{v - v_S}\right) f$$
$$= \left(\frac{1\,533\,\mathrm{m/s} + (+9.00\,\mathrm{m/s})}{1\,533\,\mathrm{m/s} - (+8.00\,\mathrm{m/s})}\right)(1\,400\,\mathrm{Hz}) = 1\,416\,\mathrm{Hz}$$

(B) As the two submarines recede from each other, the observer in sub B hears the frequency

$$f' = \left(\frac{v + v_O}{v - v_S}\right) f$$
$$= \left(\frac{1\,533\,\mathrm{m/s} + (-9.00\,\mathrm{m/s})}{1\,533\,\mathrm{m/s} - (-8.00\,\mathrm{m/s})}\right)(1\,400\,\mathrm{Hz}) = 1\,385\,\mathrm{Hz}$$

What If? While the subs are approaching each other, some of the sound from sub A will reflect from sub B and return to sub A. If this sound were to be detected by an observer on sub A, what is its frequency?

Answer The sound of apparent frequency 1 416 Hz found in part (A) will be reflected from a moving source (sub B) and then detected by a moving observer (sub A). Thus, the frequency detected by sub A is

$$f'' = \left(\frac{v + v_O}{v - v_S}\right) f'$$
$$= \left(\frac{1\,533\,\mathrm{m/s} + (+8.00\,\mathrm{m/s})}{1\,533\,\mathrm{m/s} - (+9.00\,\mathrm{m/s})}\right)(1\,416\,\mathrm{Hz}) = 1\,432\,\mathrm{Hz}$$

This technique is used by police officers to measure the speed of a moving car. Microwaves are emitted from the police car and reflected by the moving car. By detecting the Doppler-shifted frequency of the reflected microwaves, the police officer can determine the speed of the moving car.

At the Interactive Worked Example link at http://www.pse6.com, *you can alter the relative speeds of the submarines and observe the Doppler-shifted frequency.*

Shock Waves

Now consider what happens when the speed v_S of a source *exceeds* the wave speed v. This situation is depicted graphically in Figure 17.10a. The circles represent spherical wave fronts emitted by the source at various times during its motion. At $t = 0$, the source is at S_0, and at a later time t, the source is at S_n. At the time t, the wave front

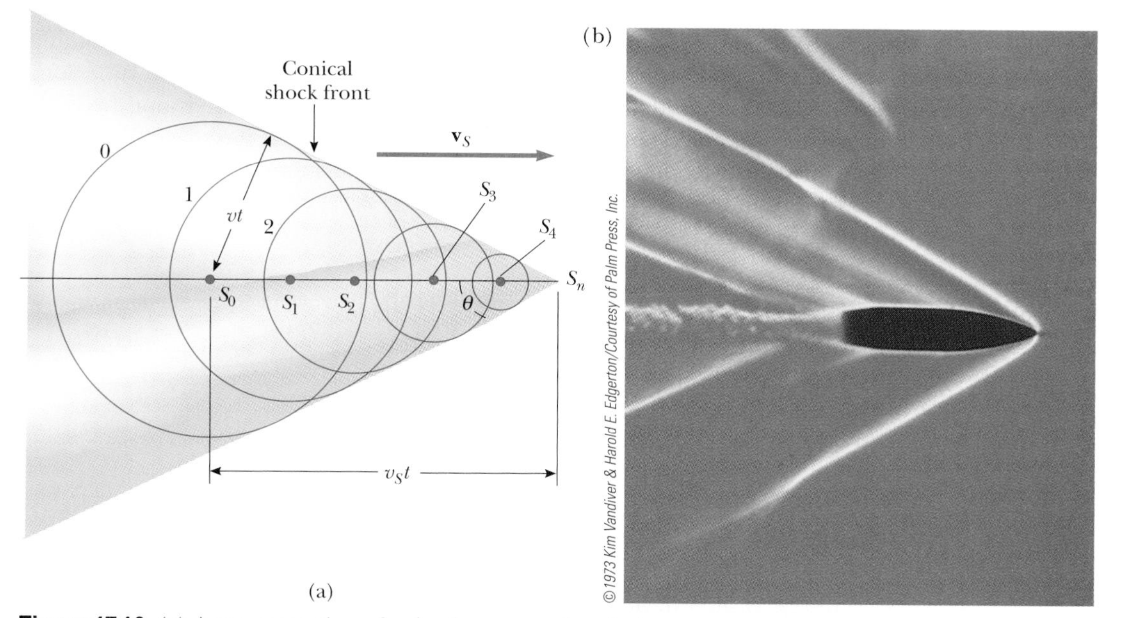

Figure 17.10 (a) A representation of a shock wave produced when a source moves from S_0 to S_n with a speed v_S, which is greater than the wave speed v in the medium. The envelope of the wave fronts forms a cone whose apex half-angle is given by $\sin\theta = v/v_S$. (b) A stroboscopic photograph of a bullet moving at supersonic speed through the hot air above a candle. Note the shock wave in the vicinity of the bullet.

Figure 17.11 The **V**-shaped bow wave of a boat is formed because the boat speed is greater than the speed of the water waves it generates. A bow wave is analogous to a shock wave formed by an airplane traveling faster than sound.

centered at S_0 reaches a radius of vt. In this same time interval, the source travels a distance $v_S t$ to S_n. At the instant the source is at S_n, waves are just beginning to be generated at this location, and hence the wave front has zero radius at this point. The tangent line drawn from S_n to the wave front centered on S_0 is tangent to all other wave fronts generated at intermediate times. Thus, we see that the envelope of these wave fronts is a cone whose apex half-angle θ (the "Mach angle") is given by

$$\sin \theta = \frac{vt}{v_S t} = \frac{v}{v_S}$$

The ratio v_S/v is referred to as the *Mach number*, and the conical wave front produced when $v_S > v$ (supersonic speeds) is known as a *shock wave.* An interesting analogy to shock waves is the **V**-shaped wave fronts produced by a boat (the bow wave) when the boat's speed exceeds the speed of the surface-water waves (Fig. 17.11).

Jet airplanes traveling at supersonic speeds produce shock waves, which are responsible for the loud "sonic boom" one hears. The shock wave carries a great deal of energy concentrated on the surface of the cone, with correspondingly great pressure variations. Such shock waves are unpleasant to hear and can cause damage to buildings when aircraft fly supersonically at low altitudes. In fact, an airplane flying at supersonic speeds produces a double boom because two shock waves are formed, one from the nose of the plane and one from the tail. People near the path of the space shuttle as it glides toward its landing point often report hearing what sounds like two very closely spaced cracks of thunder.

Quick Quiz 17.9 An airplane flying with a constant velocity moves from a cold air mass into a warm air mass. Does the Mach number (a) increase (b) decrease (c) stay the same?

17.5 Digital Sound Recording

The first sound recording device, the phonograph, was invented by Thomas Edison in the nineteenth century. Sound waves were recorded in early phonographs by encoding the sound waveforms as variations in the depth of a continuous groove cut in tin foil wrapped around a cylinder. During playback, as a needle followed along the groove of the rotating cylinder, the needle was pushed back and forth according to the sound

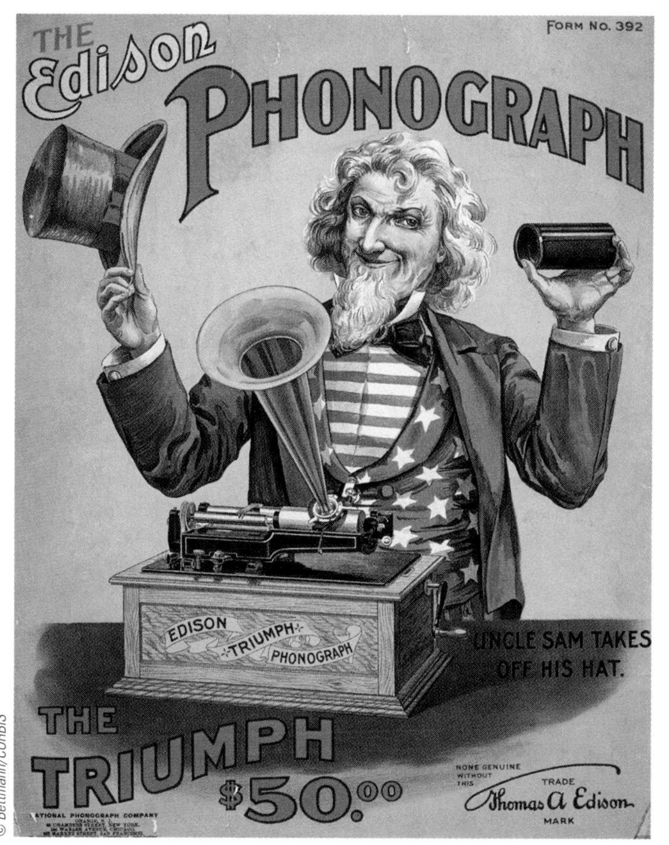

Figure 17.12 An Edison phonograph. Sound information is recorded in a groove on a rotating cylinder of wax. A needle follows the groove and vibrates according to the sound information. A diaphragm and a horn make the sound intense enough to hear.

waves encoded on the record. The needle was attached to a diaphragm and a horn (Fig. 17.12), which made the sound loud enough to be heard.

As the development of the phonograph continued, sound was recorded on cardboard cylinders coated with wax. During the last decade of the nineteenth century and the first half of the twentieth century, sound was recorded on disks made of shellac and clay. In 1948, the plastic phonograph disk was introduced and dominated the recording industry market until the advent of compact discs in the 1980s.

There are a number of problems with phonograph records. As the needle follows along the groove of the rotating phonograph record, the needle is pushed back and forth according to the sound waves encoded on the record. By Newton's third law, the needle also pushes on the plastic. As a result, the recording quality diminishes with each playing as small pieces of plastic break off and the record wears away.

Another problem occurs at high frequencies. The wavelength of the sound on the record is so small that natural bumps and graininess in the plastic create signals as loud as the sound signal, resulting in noise. The noise is especially noticeable during quiet passages in which high frequencies are being played. This is handled electronically by a process known as *pre-emphasis.* In this process, the high frequencies are recorded with more intensity than they actually have, which increases the amplitude of the vibrations and overshadows the sources of noise. Then, an *equalization circuit* in the playback system is used to reduce the intensity of the high-frequency sounds, which also reduces the intensity of the noise.

Example 17.7 Wavelengths on a Phonograph Record

Consider a 10 000-Hz sound recorded on a phonograph record which rotates at $33\frac{1}{3}$ rev/min. How far apart are the crests of the wave for this sound on the record

(A) at the outer edge of the record, 6.0 inches from the center?

(B) at the inner edge, 1.0 inch from the center?

Solution

(A) The linear speed v of a point at the outer edge of the record is $2\pi r/T$ where T is the period of the rotation and r

is the distance from the center. We first find T:

$$T = \frac{1}{f} = \frac{1}{33.33 \text{ rev/min}} = 0.030 \text{ min}\left(\frac{60 \text{ s}}{1 \text{ min}}\right) = 1.8 \text{ s}$$

Now, the linear speed at the outer edge is

$$v = \frac{2\pi r}{T} = \frac{2\pi(6.0 \text{ in.})}{1.8 \text{ s}} = 21 \text{ in./s}\left(\frac{2.54 \text{ cm}}{1 \text{ in.}}\right)$$
$$= 53 \text{ cm/s}$$

Thus, the wave on the record is moving past the needle at this speed. The wavelength is

$$\lambda = \frac{v}{f} = \frac{53 \text{ cm/s}}{10\,000 \text{ Hz}} = 5.3 \times 10^{-5} \text{ m}$$
$$= 53 \ \mu\text{m}$$

(B) The linear speed at the inner edge is

$$v = \frac{2\pi r}{T} = \frac{2\pi(1.0 \text{ in.})}{1.8 \text{ s}} = 3.5 \text{ in./s}\left(\frac{2.54 \text{ cm}}{1 \text{ in.}}\right)$$
$$= 8.9 \text{ cm/s}$$

The wavelength is

$$\lambda = \frac{v}{f} = \frac{8.9 \text{ cm/s}}{10\,000 \text{ Hz}} = 8.9 \times 10^{-6} \text{ m}$$
$$= 8.9 \ \mu\text{m}$$

Thus, the problem with noise interfering with the recorded sound is more severe at the inner edge of the disk than at the outer edge.

Digital Recording

In digital recording, information is converted to binary code (ones and zeroes), similar to the dots and dashes of Morse code. First, the waveform of the sound is *sampled,* typically at the rate of 44 100 times per second. Figure 17.13 illustrates this process. The sampling frequency is much higher than the upper range of hearing, about 20 000 Hz, so all frequencies of sound are sampled at this rate. During each sampling, the pressure of the wave is measured and converted to a voltage. Thus, there are 44 100 numbers associated with each second of the sound being sampled.

These measurements are then converted to *binary numbers,* which are numbers expressed using base 2 rather than base 10. Table 17.3 shows some sample binary numbers. Generally, voltage measurements are recorded in 16-bit "words," where each bit is a one or a zero. Thus, the number of different voltage levels that can be assigned codes is $2^{16} = 65\,536$. The number of bits in one second of sound is $16 \times 44\,100 = 705\,600$. It is these strings of ones and zeroes, in 16-bit words, that are recorded on the surface of a compact disc.

Figure 17.14 shows a magnification of the surface of a compact disc. There are two types of areas that are detected by the laser playback system—*lands* and *pits.* The lands are untouched regions of the disc surface that are highly reflective. The pits, which are areas burned into the surface, scatter light rather than reflecting it back to the detection system. The playback system samples the reflected light 705 600 times per second. When the laser moves from a pit to a flat or from a flat to a pit, the reflected light changes during the sampling and the bit is recorded as a one. If there is no change during the sampling, the bit is recorded as a zero.

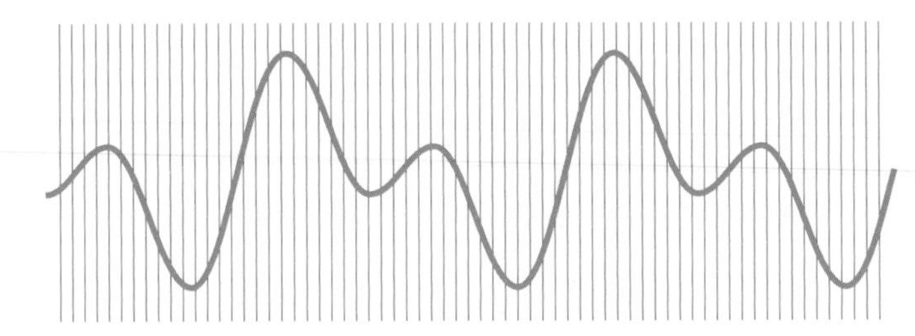

Figure 17.13 Sound is digitized by electronically sampling the sound waveform at periodic intervals. During each time interval between the blue lines, a number is recorded for the average voltage during the interval. The sampling rate shown here is much slower than the actual sampling rate of 44 100 samples per second.

Table 17.3

Sample Binary Numbers

Number in Base 10	Number in Binary	Sum
1	0000000000000001	1
2	0000000000000010	2 + 0
3	0000000000000011	2 + 1
10	0000000000001010	8 + 0 + 2 + 0
37	0000000000100101	32 + 0 + 0 + 4 + 0 + 1
275	0000000100010011	256 + 0 + 0 + 0 + 16 + 0 + 0 + 2 + 1

The binary numbers read from the CD are converted back to voltages, and the waveform is reconstructed, as shown in Figure 17.15. Because the sampling rate is so high—44 100 voltage readings each second—the fact that the waveform is constructed from step-wise discrete voltages is not evident in the sound.

The advantage of digital recording is in the high fidelity of the sound. With analog recording, any small imperfection in the record surface or the recording equipment can cause a distortion of the waveform. If all peaks of a maximum in a waveform are clipped off so as to be only 90% as high, for example, this will have a major effect on the spectrum of the sound in an analog recording. With digital recording, however, it takes a major imperfection to turn a one into a zero. If an imperfection causes the magnitude of a one to be 90% of the original value, it still registers as a one, and there is no distortion. Another advantage of digital recording is that the information is extracted optically, so that there is no mechanical wear on the disc.

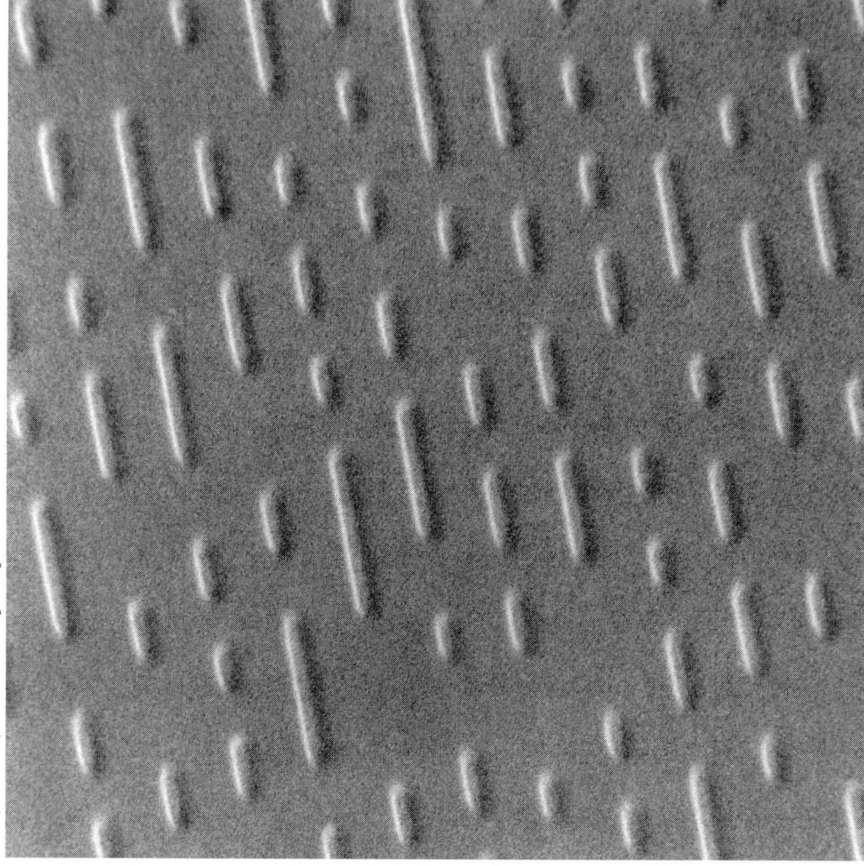

Courtesy of University of Miami, Music Engineering

Figure 17.14 The surface of a compact disc, showing the pits. Transitions between pits and lands correspond to ones. Regions without transitions correspond to zeroes.

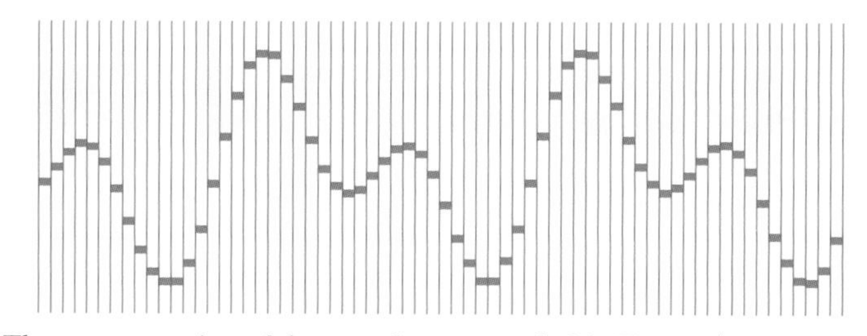

Figure 17.15 The reconstruction of the sound wave sampled in Figure 17.13. Notice that the reconstruction is step-wise, rather than the continuous waveform in Figure 17.13.

Example 17.8 How Big Are the Pits?

In Example 10.2, we mentioned that the speed with which the CD surface passes the laser is 1.3 m/s. What is the average length of the audio track on a CD associated with each bit of the audio information?

Solution In one second, a 1.3-m length of audio track passes by the laser. This length includes 705 600 bits of audio information. Thus, the average length per bit is

$$\frac{1.3 \text{ m}}{705\,600 \text{ bits}} = 1.8 \times 10^{-6} \text{ m/bit}$$

$$= 1.8\ \mu\text{m/bit}$$

The average length per bit of *total* information on the CD is smaller than this because there is additional information on the disc besides the audio information. This information includes error correction codes, song numbers, timing codes, etc. As a result, the shortest length per bit is actually about 0.8 μm.

Example 17.9 What's the Number?

Consider the photograph of the compact disc surface in Figure 17.14. Audio data undergoes complicated processing in order to reduce a variety of errors in reading the data. Thus, an audio "word" is not laid out linearly on the disc. Suppose that data has been read from the disc, the error encoding has been removed, and the resulting audio word is

1 0 1 1 1 0 1 1 1 0 1 1 1 0 1 1

What is the decimal number represented by this 16-bit word?

Solution We convert each of these bits to a power of 2 and add the results:

$1 \times 2^{15} = 32\,768$ $\quad 1 \times 2^{9} = 512$ $\quad 1 \times 2^{3} = 8$

$0 \times 2^{14} = 0$ $\quad 1 \times 2^{8} = 256$ $\quad 0 \times 2^{2} = 0$

$1 \times 2^{13} = 8\,192$ $\quad 1 \times 2^{7} = 128$ $\quad 1 \times 2^{1} = 2$

$1 \times 2^{12} = 4\,096$ $\quad 0 \times 2^{6} = 0$ $\quad 1 \times 2^{0} = 1$

$1 \times 2^{11} = 2\,048$ $\quad 1 \times 2^{5} = 32$

$0 \times 2^{10} = 0$ $\quad 1 \times 2^{4} = 16$ $\quad \text{sum} = 48\,059$

This number is converted by the CD player into a voltage, representing one of the 44 100 values that will be used to build one second of the electronic waveform that represents the recorded sound.

17.6 Motion Picture Sound

Another interesting application of digital sound is the soundtrack in a motion picture. Early twentieth-century movies recorded sound on phonograph records, which were synchronized with the action on the screen. Beginning with early newsreel films, the *variable-area optical soundtrack* process was introduced, in which sound was recorded on an optical track on the film. The width of the transparent portion of the track varied according to the sound wave that was recorded. A photocell detecting light passing through the track converted the varying light intensity to a sound wave. As with phonograph recording, there are a number of difficulties with this recording system. For example, dirt or fingerprints on the film cause fluctuations in intensity and loss of fidelity.

Digital recording on film first appeared with *Dick Tracy* (1990), using the Cinema Digital Sound (CDS) system. This system suffered from lack of an analog backup system in case of equipment failure and is no longer used in the film industry. It did, however, introduce the use of 5.1 channels of sound—Left, Center, Right, Right Surround, Left Surround, and Low Frequency Effects (LFE). The LFE channel, which is the "0.1

channel" of 5.1, carries very low frequencies for dramatic sound from explosions, earthquakes, and the like.

Current motion pictures are produced with three systems of digital sound recording:

Dolby Digital; In this format, 5.1 channels of digital sound are optically stored between the sprocket holes of the film. There is an analog optical backup in case the digital system fails. The first film to use this technique was *Batman Returns* (1992).

DTS (Digital Theater Sound); 5.1 channels of sound are stored on a separate CD-ROM which is synchronized to the film print by time codes on the film. There is an analog optical backup in case the digital system fails. The first film to use this technique was *Jurassic Park* (1993).

SDDS (Sony Dynamic Digital Sound); Eight full channels of digital sound are optically stored outside the sprocket holes on both sides of film. There is an analog optical backup in case the digital system fails. The first film to use this technique was *Last Action Hero* (1993). The existence of information on both sides of the tape is a system of redundancy—in case one side is damaged, the system will still operate. SDDS employs a full-spectrum LFE channel and two additional channels (left center and right center behind the screen). In Figure 17.16, showing a section of SDDS film, both the analog optical soundtrack and the dual digital soundtracks can be seen.

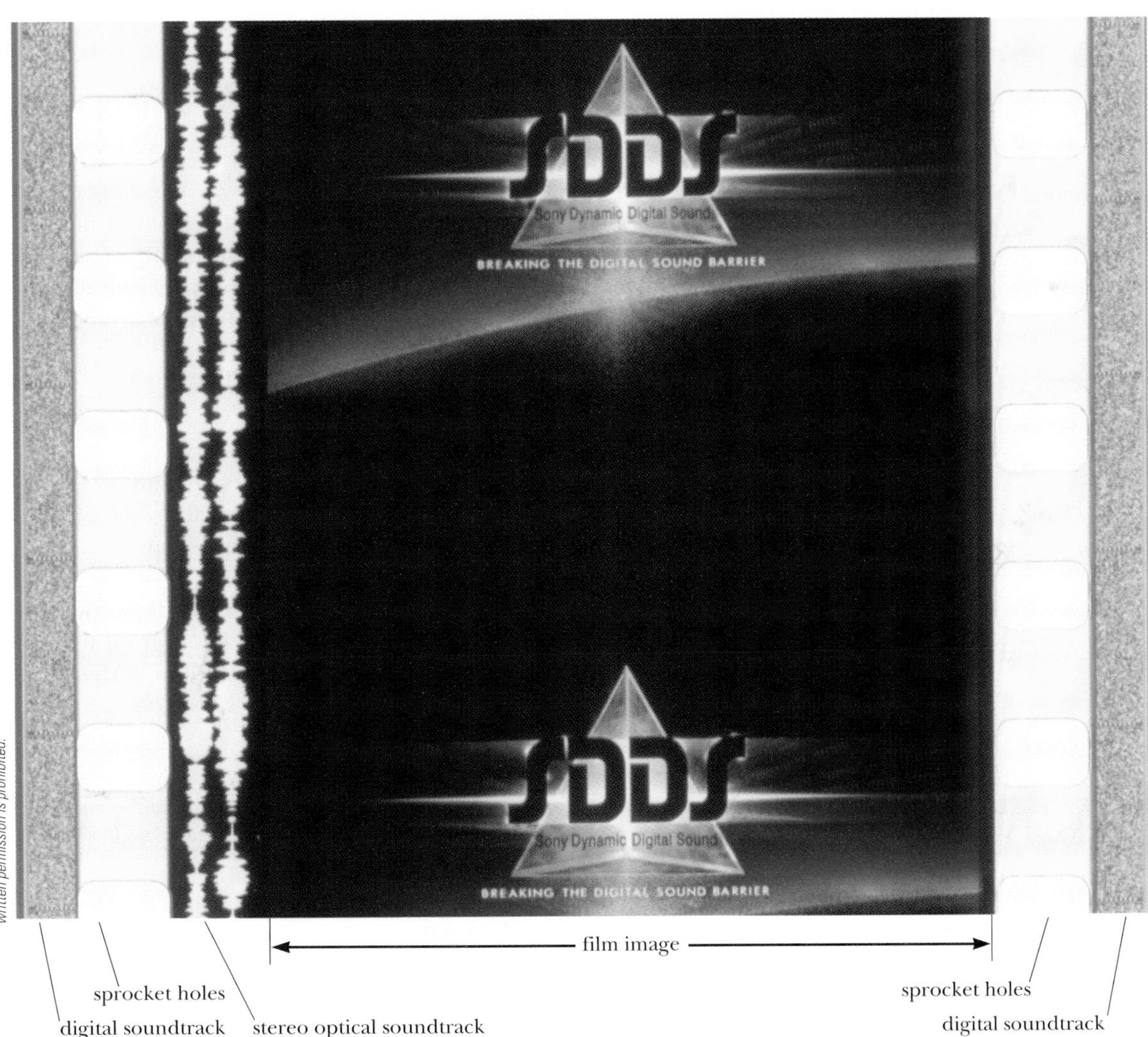

Figure 17.16 The layout of information on motion picture film using the SDDS digital sound system.

SUMMARY

Take a practice test for this chapter by clicking on the Practice Test link at **http://www.pse6.com.**

Sound waves are longitudinal and travel through a compressible medium with a speed that depends on the elastic and inertial properties of that medium. The speed of sound in a liquid or gas having a bulk modulus B and density ρ is

$$v = \sqrt{\frac{B}{\rho}} \tag{17.1}$$

For sinusoidal sound waves, the variation in the position of an element of the medium is given by

$$s(x, t) = s_{\max} \cos(kx - \omega t) \tag{17.2}$$

and the variation in pressure from the equilibrium value is

$$\Delta P = \Delta P_{\max} \sin(kx - \omega t) \tag{17.3}$$

where $\Delta P_{\max}$ is the **pressure amplitude.** The pressure wave is 90° out of phase with the displacement wave. The relationship between $s_{\max}$ and $\Delta P_{\max}$ is given by

$$\Delta P_{\max} = \rho v \omega s_{\max} \tag{17.4}$$

The intensity of a periodic sound wave, which is the power per unit area, is

$$I \equiv \frac{\mathcal{P}}{A} = \frac{\Delta P_{\max}^2}{2\rho v} \tag{17.5, 17.6}$$

The sound level of a sound wave, in decibels, is given by

$$\beta \equiv 10 \log\left(\frac{I}{I_0}\right) \tag{17.8}$$

The constant I_0 is a reference intensity, usually taken to be at the threshold of hearing (1.00×10^{-12} W/m^2), and I is the intensity of the sound wave in watts per square meter.

The change in frequency heard by an observer whenever there is relative motion between a source of sound waves and the observer is called the **Doppler effect.** The observed frequency is

$$f' = \left(\frac{v + v_O}{v - v_S}\right) f \tag{17.13}$$

In this expression, the signs for the values substituted for v_O and v_S depend on the direction of the velocity. A positive value for the velocity of the observer or source is substituted if the velocity of one is toward the other, while a negative value represents a velocity of one away from the other.

In digital recording of sound, the sound waveform is sampled 44 100 times per second. The pressure of the wave for each sampling is measured and converted to a binary number. In playback, these binary numbers are read and used to build the original waveform.

QUESTIONS

1. Why are sound waves characterized as longitudinal?

2. If an alarm clock is placed in a good vacuum and then activated, no sound is heard. Explain.

3. A sonic ranger is a device that determines the distance to an object by sending out an ultrasonic sound pulse and measuring how long it takes for the wave to return after it reflects from the object. Typically these devices cannot reliably detect an object that is less than half a meter from the sensor. Why is that?

4. A friend sitting in her car far down the road waves to you and beeps her horn at the same time. How far away must she be for you to calculate the speed of sound to two significant figures by measuring the time it takes for the sound to reach you?

5. If the wavelength of sound is reduced by a factor of 2, what happens to its frequency? Its speed?

6. By listening to a band or orchestra, how can you determine that the speed of sound is the same for all frequencies?

7. In Example 17.3 we found that a point source with a power output of 80 W produces sound with an intensity of 1.00×10^{-8} W/m^2, which corresponds to 40 dB, at a distance of about 16 miles. Why do you suppose you cannot normally hear a rock concert that is going on 16 miles away? (See Table 17.2.)

8. If the distance from a point source is tripled, by what factor does the intensity decrease?

9. *The Tunguska Event.* On June 30, 1908, a meteor burned up and exploded in the atmosphere above the Tunguska River valley in Siberia. It knocked down trees over thousands of square kilometers and started a forest fire, but apparently caused no human casualties. A witness sitting on his doorstep outside the zone of falling trees recalled events in the following sequence: He saw a moving light in the sky, brighter than the sun and descending at a low angle to the horizon. He felt his face become warm. He felt the ground shake. An invisible agent picked him up and immediately dropped him about a meter farther away from where the light had been. He heard a very loud protracted rumbling. Suggest an explanation for these observations and for the order in which they happened.

10. Explain how the Doppler effect with microwaves is used to determine the speed of an automobile.

11. Explain what happens to the frequency of the echo of your car horn as you move in a vehicle toward the wall of a canyon. What happens to the frequency as you move away from the wall?

12. Of the following sounds, which is most likely to have a sound level of 60 dB: a rock concert, the turning of a page in this textbook, normal conversation, or a cheering crowd at a football game?

13. Estimate the decibel level of each of the sounds in the previous question.

14. A binary star system consists of two stars revolving about their common center of mass. If we observe the light reaching us from one of these stars as it makes one complete revolution, what does the Doppler effect predict will happen to this light?

15. How can an object move with respect to an observer so that the sound from it is not shifted in frequency?

16. Suppose the wind blows. Does this cause a Doppler effect for sound propagating through the air? Is it like a moving source or a moving observer?

17. Why is it not possible to use sonar (sound waves) to determine the speed of an object traveling faster than the speed of sound?

18. Why is it so quiet after a snowfall?

19. Why is the intensity of an echo less than that of the original sound?

20. A loudspeaker built into the exterior wall of an airplane produces a large-amplitude burst of vibration at 200 Hz, then a burst at 300 Hz, and then a burst at 400 Hz (Boop . . . baap . . . beep), all while the plane is flying faster than the speed of sound. Describe qualitatively what an observer hears if she is in front of the plane, close to its flight path. **What If?** What will the observer hear if the pilot uses the loudspeaker to say, "How are you?"

21. In several cases, a nearby star has been found to have a large planet orbiting about it, although the planet could not be seen. Using the ideas of a system rotating about its center of mass and of the Doppler shift for light (which is in several ways similar to the Doppler effect for sound), explain how an astronomer could determine the presence of the invisible planet.

PROBLEMS

1, 2, 3 = straightforward, intermediate, challenging □ = full solution available in the *Student Solutions Manual and Study Guide*

= coached solution with hints available at http://www.pse6.com = computer useful in solving problem

= paired numerical and symbolic problems

Section 17.1 Speed of Sound Waves

1. Suppose that you hear a clap of thunder 16.2 s after seeing the associated lightning stroke. The speed of sound waves in air is 343 m/s, and the speed of light is 3.00×10^8 m/s. How far are you from the lightning stroke?

2. Find the speed of sound in mercury, which has a bulk modulus of approximately 2.80×10^{10} N/m^2 and a density of 13 600 kg/m^3.

3. A flowerpot is knocked off a balcony 20.0 m above the sidewalk and falls toward an unsuspecting 1.75-m-tall man who is standing below. How close to the sidewalk can the flower pot fall before it is too late for a warning shouted from the balcony to reach the man in time? Assume that the man below requires 0.300 s to respond to the warning.

4. The speed of sound in air (in m/s) depends on temperature according to the approximate expression

$$v = 331.5 + 0.607T_C$$

where T_C is the Celsius temperature. In dry air the temperature decreases about 1°C for every 150 m rise in altitude. (a) Assuming this change is constant up to an altitude of 9 000 m, how long will it take the sound from an airplane flying at 9 000 m to reach the ground on a day when the ground temperature is 30°C? (b) **What If?** Compare this to the time interval required if the air were a constant 30°C. Which time interval is longer?

5. A cowboy stands on horizontal ground between two parallel vertical cliffs. He is not midway between the cliffs. He fires a shot and hears its echoes. The second echo arrives 1.92 s after the first and 1.47 s before the third. Consider only the sound traveling parallel to the ground and reflecting from the cliffs. Take the speed of sound as 340 m/s. (a) What is the distance between the cliffs? (b) **What If?** If he can hear a fourth echo, how long after the third echo does it arrive?

6. A rescue plane flies horizontally at a constant speed searching for a disabled boat. When the plane is directly above the boat, the boat's crew blows a loud horn. By the time the plane's sound detector perceives the horn's sound, the plane has traveled a distance equal to half its altitude above the ocean. If it takes the sound 2.00 s to reach the plane, determine (a) the speed of the plane and (b) its altitude. Take the speed of sound to be 343 m/s.

Section 17.2 Periodic Sound Waves

Note: Use the following values as needed unless otherwise specified: the equilibrium density of air at 20°C is $\rho = 1.20\ \text{kg/m}^3$. The speed of sound in air is $v = 343\ \text{m/s}$. Pressure variations ΔP are measured relative to atmospheric pressure, $1.013 \times 10^5\ \text{N/m}^2$.
Problem 70 in Chapter 2 can also be assigned with this section.

7. A bat (Fig. P17.7) can detect very small objects, such as an insect whose length is approximately equal to one wavelength of the sound the bat makes. If a bat emits chirps at a frequency of 60.0 kHz, and if the speed of sound in air is 340 m/s, what is the smallest insect the bat can detect?

Figure P17.7 Problems 7 and 60.

8. An ultrasonic tape measure uses frequencies above 20 MHz to determine dimensions of structures such as buildings. It does this by emitting a pulse of ultrasound into air and then measuring the time for an echo to return from a reflecting surface whose distance away is to be measured. The distance is displayed as a digital read-out. For a tape measure that emits a pulse of ultrasound with a frequency of 22.0 MHz, (a) What is the distance to an object from which the echo pulse returns after 24.0 ms when the air temperature is 26°C? (b) What should be the duration of the emitted pulse if it is to include 10 cycles of the ultrasonic wave? (c) What is the spatial length of such a pulse?

9. Ultrasound is used in medicine both for diagnostic imaging and for therapy. For diagnosis, short pulses of ultrasound are passed through the patient's body. An echo reflected from a structure of interest is recorded, and from the time delay for the return of the echo the distance to the structure can be determined. A single transducer emits and detects the ultrasound. An image of the structure is obtained by reducing the data with a computer. With sound of low intensity, this technique is noninvasive and harmless. It is used to examine fetuses, tumors, aneurysms, gallstones, and many other structures. A Doppler ultrasound unit is used to study blood flow and functioning of the heart. To reveal detail, the wavelength of the reflected ultrasound must be small compared to the size of the object reflecting the wave. For this reason, frequencies in the range 1.00 to 20.0 MHz are used. What is the range of wavelengths corresponding to this range of frequencies? The speed of ultrasound in human tissue is about 1 500 m/s (nearly the same as the speed of sound in water).

10. A sound wave in air has a pressure amplitude equal to $4.00 \times 10^{-3}\ \text{N/m}^2$. Calculate the displacement amplitude of the wave at a frequency of 10.0 kHz.

11. A sinusoidal sound wave is described by the displacement wave function

$$s(x, t) = (2.00\ \mu\text{m}) \cos[(15.7\ \text{m}^{-1})x - (858\ \text{s}^{-1})t]$$

(a) Find the amplitude, wavelength, and speed of this wave. (b) Determine the instantaneous displacement from equilibrium of the elements of air at the position $x = 0.050\ 0$ m at $t = 3.00$ ms. (c) Determine the maximum speed of the element's oscillatory motion.

12. As a certain sound wave travels through the air, it produces pressure variations (above and below atmospheric pressure) given by $\Delta P = 1.27 \sin(\pi x - 340\pi t)$ in SI units. Find (a) the amplitude of the pressure variations, (b) the frequency, (c) the wavelength in air, and (d) the speed of the sound wave.

13. Write an expression that describes the pressure variation as a function of position and time for a sinusoidal sound wave in air, if $\lambda = 0.100$ m and $\Delta P_{max} = 0.200\ \text{N/m}^2$.

14. Write the function that describes the displacement wave corresponding to the pressure wave in Problem 13.

15. An experimenter wishes to generate in air a sound wave that has a displacement amplitude of 5.50×10^{-6} m. The pressure amplitude is to be limited to $0.840\ \text{N/m}^2$. What is the minimum wavelength the sound wave can have?

16. The tensile stress in a thick copper bar is 99.5% of its elastic breaking point of 13.0×10^{10} N/m^2. If a 500-Hz sound wave is transmitted through the material, (a) what displacement amplitude will cause the bar to break? (b) What is the maximum speed of the elements of copper at this moment? (c) What is the sound intensity in the bar?

17. Prove that sound waves propagate with a speed given by Equation 17.1. Proceed as follows. In Figure 17.3, consider a thin cylindrical layer of air in the cylinder, with face area A and thickness Δx. Draw a free-body diagram of this thin layer. Show that $\Sigma F_x = ma_x$ implies that $-[\partial(\Delta P)/\partial x]A\,\Delta x = \rho A\,\Delta x(\partial^2 s/\partial t^2)$. By substituting $\Delta P = -B(\partial s/\partial x)$, obtain the wave equation for sound, $(B/\rho)(\partial^2 s/\partial x^2) = (\partial^2 s/\partial t^2)$. To a mathematical physicist, this equation demonstrates the existence of sound waves and determines their speed. As a physics student, you must take another step or two. Substitute into the wave equation the trial solution $s(x, t) = s_{\max}\cos(kx - \omega t)$. Show that this function satisfies the wave equation provided that $\omega/k = \sqrt{B/\rho}$. This result reveals that sound waves exist provided that they move with the speed $v = f\lambda = (2\pi f)(\lambda/2\pi) = \omega/k = \sqrt{B/\rho}$.

Section 17.3 Intensity of Periodic Sound Waves

18. The area of a typical eardrum is about 5.00×10^{-5} m^2. Calculate the sound power incident on an eardrum at (a) the threshold of hearing and (b) the threshold of pain.

19. Calculate the sound level in decibels of a sound wave that has an intensity of 4.00 μW/m^2.

20. A vacuum cleaner produces sound with a measured sound level of 70.0 dB. (a) What is the intensity of this sound in W/m^2? (b) What is the pressure amplitude of the sound?

21. The intensity of a sound wave at a fixed distance from a speaker vibrating at 1.00 kHz is 0.600 W/m^2. (a) Determine the intensity if the frequency is increased to 2.50 kHz while a constant displacement amplitude is maintained. (b) Calculate the intensity if the frequency is reduced to 0.500 kHz and the displacement amplitude is doubled.

22. The intensity of a sound wave at a fixed distance from a speaker vibrating at a frequency f is I. (a) Determine the intensity if the frequency is increased to f' while a constant displacement amplitude is maintained. (b) Calculate the intensity if the frequency is reduced to $f/2$ and the displacement amplitude is doubled.

23. The most soaring vocal melody is in Johann Sebastian Bach's *Mass in B minor.* A portion of the score for the Credo section, number 9, bars 25 to 33, appears in Figure P17.23. The repeating syllable O in the phrase "resurrectionem mortuorum" (the resurrection of the dead) is seamlessly passed from basses to tenors to altos to first sopranos, like a baton in a relay. Each voice carries the melody up in a run of an octave or more. Together they carry it from D below middle C to A above a tenor's high C. In concert pitch, these notes are now assigned frequencies of 146.8 Hz and 880.0 Hz. (a) Find the wavelengths of the initial and final notes. (b) Assume that the choir sings the melody with a uniform sound level of 75.0 dB. Find the pressure amplitudes of the initial and final notes. (c) Find the displacement amplitudes of the initial and final notes. (d) **What If?** In Bach's time, before the invention of the tuning fork, frequencies were assigned to notes as a matter of immediate local convenience. Assume that the rising melody was sung starting from 134.3 Hz and ending at 804.9 Hz. How would the answers to parts (a) through (c) change?

24. The tube depicted in Figure 17.2 is filled with air at 20°C and equilibrium pressure 1 atm. The diameter of the tube is 8.00 cm. The piston is driven at a frequency of 600 Hz with an amplitude of 0.120 cm. What power must be supplied to maintain the oscillation of the piston?

25. A family ice show is held at an enclosed arena. The skaters perform to music with level 80.0 dB. This is too loud for your baby, who yells at 75.0 dB. (a) What total sound intensity engulfs you? (b) What is the combined sound level?

26. Consider sinusoidal sound waves propagating in these three different media: air at 0°C, water, and iron. Use densities and speeds from Tables 14.1 and 17.1. Each wave has the same intensity I_0 and the same angular frequency ω_0. (a) Compare the values of the wavelength in the three media. (b) Compare the values of the displacement amplitude in the three media. (c) Compare the values of the pressure amplitude in the three media. (d) For values of $\omega_0 = 2\,000\ \pi$ rad/s and $I_0 = 1.00 \times 10^{-6}$ W/m^2, evaluate the wavelength, displacement amplitude, and pressure amplitude in each of the three media.

27. The power output of a certain public address speaker is 6.00 W. Suppose it broadcasts equally in all directions. (a) Within what distance from the speaker would the sound be painful to the ear? (b) At what distance from the speaker would the sound be barely audible?

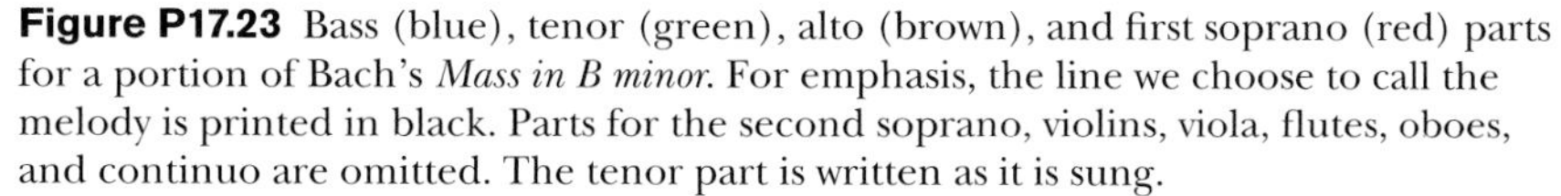

Figure P17.23 Bass (blue), tenor (green), alto (brown), and first soprano (red) parts for a portion of Bach's *Mass in B minor.* For emphasis, the line we choose to call the melody is printed in black. Parts for the second soprano, violins, viola, flutes, oboes, and continuo are omitted. The tenor part is written as it is sung.

28. Show that the difference between decibel levels β_1 and β_2 of a sound is related to the ratio of the distances r_1 and r_2 from the sound source by

$$\beta_2 - \beta_1 = 20\log\left(\frac{r_1}{r_2}\right)$$

29. A firework charge is detonated many meters above the ground. At a distance of 400 m from the explosion, the acoustic pressure reaches a maximum of 10.0 N/m^2. Assume that the speed of sound is constant at 343 m/s throughout the atmosphere over the region considered, that the ground absorbs all the sound falling on it, and that the air absorbs sound energy as described by the rate 7.00 dB/km. What is the sound level (in dB) at 4.00 km from the explosion?

30. A loudspeaker is placed between two observers who are 110 m apart, along the line connecting them. If one observer records a sound level of 60.0 dB and the other records a sound level of 80.0 dB, how far is the speaker from each observer?

31. Two small speakers emit sound waves of different frequencies. Speaker *A* has an output of 1.00 mW, and speaker *B* has an output of 1.50 mW. Determine the sound level (in dB) at point *C* (Fig. P17.31) if (a) only speaker *A* emits sound, (b) only speaker *B* emits sound, and (c) both speakers emit sound.

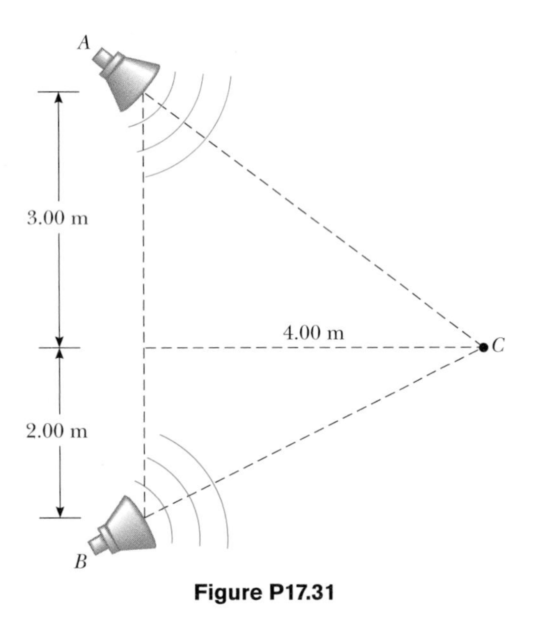

Figure P17.31

32. A jackhammer, operated continuously at a construction site, behaves as a point source of spherical sound waves. A construction supervisor stands 50.0 m due north of this sound source and begins to walk due west. How far does she have to walk in order for the amplitude of the wave function to drop by a factor of 2.00?

33. The sound level at a distance of 3.00 m from a source is 120 dB. At what distance will the sound level be (a) 100 dB and (b) 10.0 dB?

34. A fireworks rocket explodes at a height of 100 m above the ground. An observer on the ground directly under the explosion experiences an average sound intensity of 7.00×10^{-2} W/m^2 for 0.200 s. (a) What is the total sound energy of the explosion? (b) What is the sound level in decibels heard by the observer?

35. As the people sing in church, the sound level everywhere inside is 101 dB. No sound is transmitted through the massive walls, but all the windows and doors are open on a summer morning. Their total area is 22.0 m^2. (a) How much sound energy is radiated in 20.0 min? (b) Suppose the ground is a good reflector and sound radiates uniformly in all horizontal and upward directions. Find the sound level 1 km away.

36. The smallest change in sound level that a person can distinguish is approximately 1 dB. When you are standing next to your power lawnmower as it is running, can you hear the steady roar of your neighbor's lawnmower? Perform an order-of-magnitude calculation to substantiate your answer, stating the data you measure or estimate.

Section 17.4 The Doppler Effect

37. A train is moving parallel to a highway with a constant speed of 20.0 m/s. A car is traveling in the same direction as the train with a speed of 40.0 m/s. The car horn sounds at a frequency of 510 Hz, and the train whistle sounds at a frequency of 320 Hz. (a) When the car is behind the train, what frequency does an occupant of the car observe for the train whistle? (b) After the car passes and is in front of the train, what frequency does a train passenger observe for the car horn?

38. Expectant parents are thrilled to hear their unborn baby's heartbeat, revealed by an ultrasonic motion detector. Suppose the fetus's ventricular wall moves in simple harmonic motion with an amplitude of 1.80 mm and a frequency of 115 per minute. (a) Find the maximum linear speed of the heart wall. Suppose the motion detector in contact with the mother's abdomen produces sound at 2 000 000.0 Hz, which travels through tissue at 1.50 km/s. (b) Find the maximum frequency at which sound arrives at the wall of the baby's heart. (c) Find the maximum frequency at which reflected sound is received by the motion detector. By electronically "listening" for echoes at a frequency different from the broadcast frequency, the motion detector can produce beeps of audible sound in synchronization with the fetal heartbeat.

39. Standing at a crosswalk, you hear a frequency of 560 Hz from the siren of an approaching ambulance. After the ambulance passes, the observed frequency of the siren is 480 Hz. Determine the ambulance's speed from these observations.

40. A block with a speaker bolted to it is connected to a spring having spring constant $k = 20.0$ N/m as in Figure P17.40. The total mass of the block and speaker is 5.00 kg, and the amplitude of this unit's motion is 0.500 m. (a) If the speaker emits sound waves of frequency 440 Hz, determine the highest and lowest frequencies heard by the person to the right of the speaker. (b) If the maximum sound level heard by the person is 60.0 dB when he is closest to the

speaker, 1.00 m away, what is the minimum sound level heard by the observer? Assume that the speed of sound is 343 m/s.

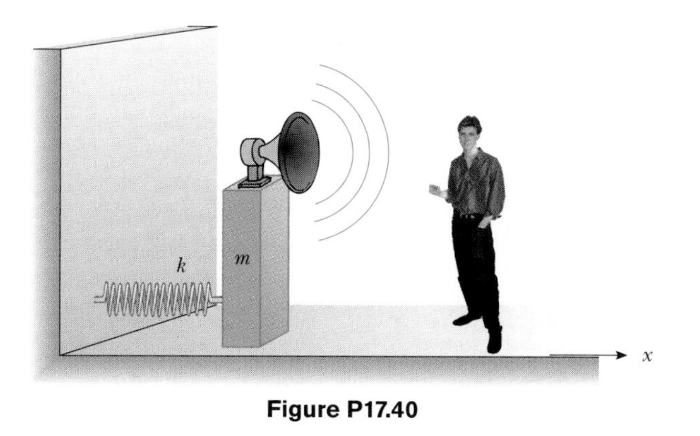

Figure P17.40

41. A tuning fork vibrating at 512 Hz falls from rest and accelerates at 9.80 m/s^2. How far below the point of release is the tuning fork when waves of frequency 485 Hz reach the release point? Take the speed of sound in air to be 340 m/s.

42. At the Winter Olympics, an athlete rides her luge down the track while a bell just above the wall of the chute rings continuously. When her sled passes the bell, she hears the frequency of the bell fall by the musical interval called a minor third. That is, the frequency she hears drops to five sixths of its original value. (a) Find the speed of sound in air at the ambient temperature -10.0°C. (b) Find the speed of the athlete.

43. A siren mounted on the roof of a firehouse emits sound at a frequency of 900 Hz. A steady wind is blowing with a speed of 15.0 m/s. Taking the speed of sound in calm air to be 343 m/s, find the wavelength of the sound (a) upwind of the siren and (b) downwind of the siren. Firefighters are approaching the siren from various directions at 15.0 m/s. What frequency does a firefighter hear (c) if he or she is approaching from an upwind position, so that he or she is moving in the direction in which the wind is blowing? (d) if he or she is approaching from a downwind position and moving against the wind?

44. The Concorde can fly at Mach 1.50, which means the speed of the plane is 1.50 times the speed of sound in air. What is the angle between the direction of propagation of the shock wave and the direction of the plane's velocity?

45. When high-energy charged particles move through a transparent medium with a speed greater than the speed of light in that medium, a shock wave, or bow wave, of light is produced. This phenomenon is called the *Cerenkov effect.* When a nuclear reactor is shielded by a large pool of water, Cerenkov radiation can be seen as a blue glow in the vicinity of the reactor core, due to high-speed electrons moving through the water. In a particular case, the Cerenkov radiation produces a wave front with an apex half-angle of 53.0°. Calculate the speed of the electrons in the water. (The speed of light in water is 2.25×10^8 m/s.)

46. The loop of a circus ringmaster's whip travels at Mach 1.38 (that is, $v_S/v = 1.38$). What angle does the shock wave make with the direction of the whip's motion?

47. A supersonic jet traveling at Mach 3.00 at an altitude of 20 000 m is directly over a person at time $t = 0$ as in Figure P17.47. (a) How long will it be before the person encounters the shock wave? (b) Where will the plane be when it is finally heard? (Assume the speed of sound in air is 335 m/s.)

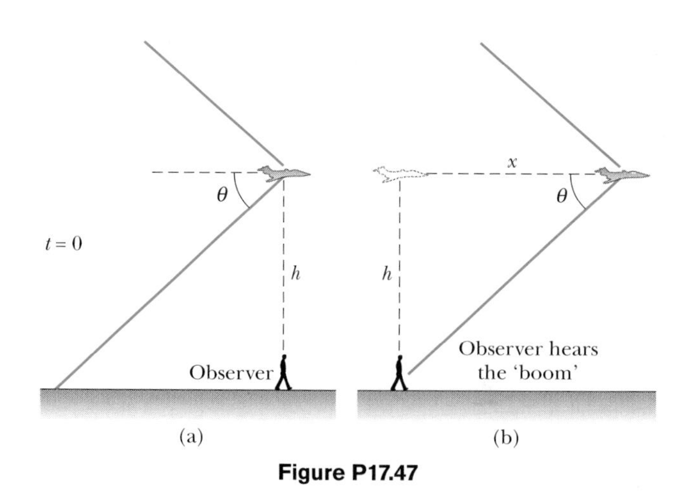

Figure P17.47

Section 17.5 Digital Sound Recording
Section 17.6 Motion Picture Sound

48. This problem represents a possible (but not recommended) way to code instantaneous pressures in a sound wave into 16-bit digital words. Example 17.2 mentions that the pressure amplitude of a 120-dB sound is 28.7 N/m^2. Let this pressure variation be represented by the digital code 65 536. Let zero pressure variation be represented on the recording by the digital word 0. Let other intermediate pressures be represented by digital words of intermediate size, in direct proportion to the pressure. (a) What digital word would represent the maximum pressure in a 40 dB sound? (b) Explain why this scheme works poorly for soft sounds. (c) Explain how this coding scheme would clip off half of the waveform of any sound, ignoring the actual shape of the wave and turning it into a string of zeros. By introducing sharp corners into every recorded waveform, this coding scheme would make everything sound like a buzzer or a kazoo.

49. Only two recording channels are required to give the illusion of sound coming from any point located between two speakers of a stereophonic sound system. If the same signal is recorded in both channels, a listener will hear it coming from a single direction halfway between the two speakers. This "phantom orchestra" illusion can be heard in the two-channel original Broadway cast recording of the song "Do-Re-Mi" from *The Sound of Music* (Columbia Records KOS 2020). Each of the eight singers can be heard at a different location between the loudspeakers. All listeners with normal hearing will agree on their locations. The brain can sense the direction of sound by noting how

much earlier a sound is heard in one ear than in the other. Model your ears as two sensors 19.0 cm apart in a flat screen. If a click from a distant source is heard 210 μs earlier in the left ear than in the right, from what direction does it appear to originate?

50. Assume that a loudspeaker broadcasts sound equally in all directions and produces sound with a level of 103 dB at a distance of 1.60 m from its center. (a) Find its sound power output. (b) If the salesperson claims to be giving you 150 W per channel, he is referring to the electrical power input to the speaker. Find the efficiency of the speaker—that is, the fraction of input power that is converted into useful output power.

Additional Problems

51. A large set of unoccupied football bleachers has solid seats and risers. You stand on the field in front of the bleachers and fire a starter's pistol or sharply clap two wooden boards together once. The sound pulse you produce has no definite frequency and no wavelength. The sound you hear reflected from the bleachers has an identifiable frequency and may remind you of a short toot on a trumpet, or of a buzzer or kazoo. Account for this sound. Compute order-of-magnitude estimates for its frequency, wavelength, and duration, on the basis of data you specify.

52. Many artists sing very high notes in *ad lib* ornaments and cadenzas. The highest note written for a singer in a published score was F-sharp above high C, 1.480 kHz, for Zerbinetta in the original version of Richard Strauss's opera *Ariadne auf Naxos.* (a) Find the wavelength of this sound in air. (b) Suppose people in the fourth row of seats hear this note with level 81.0 dB. Find the displacement amplitude of the sound. (c) **What If?** Because of complaints, Strauss later transposed the note down to F above high C, 1.397 kHz. By what increment did the wavelength change?

53. A sound wave in a cylinder is described by Equations 17.2 through 17.4. Show that $\Delta P = \pm \rho v \omega \sqrt{s_{\max}^2 - s^2}$.

54. On a Saturday morning, pickup trucks and sport utility vehicles carrying garbage to the town dump form a nearly steady procession on a country road, all traveling at 19.7 m/s. From one direction, two trucks arrive at the dump every 3 min. A bicyclist is also traveling toward the dump, at 4.47 m/s. (a) With what frequency do the trucks pass him? (b) **What If?** A hill does not slow down the trucks, but makes the out-of-shape cyclist's speed drop to 1.56 m/s. How often do noisy, smelly, inefficient, garbage-dripping, roadhogging trucks whiz past him now?

55. The ocean floor is underlain by a layer of basalt that constitutes the crust, or uppermost layer, of the Earth in that region. Below this crust is found denser periodotite rock, which forms the Earth's mantle. The boundary between these two layers is called the Mohorovicic discontinuity ("Moho" for short). If an explosive charge is set off at the surface of the basalt, it generates a seismic wave that is reflected back out at the Moho. If the speed of this wave in basalt is 6.50 km/s and the two-way travel time is 1.85 s, what is the thickness of this oceanic crust?

56. For a certain type of steel, stress is always proportional to strain with Young's modulus as shown in Table 12.1. The steel has the density listed for iron in Table 14.1. It will fail by bending permanently if subjected to compressive stress greater than its yield strength $\sigma_y = 400$ MPa. A rod 80.0 cm long, made of this steel, is fired at 12.0 m/s straight at a very hard wall, or at another identical rod moving in the opposite direction. (a) The speed of a one-dimensional compressional wave moving along the rod is given by $\sqrt{Y/\rho}$, where ρ is the density and Y is Young's modulus for the rod. Calculate this speed. (b) After the front end of the rod hits the wall and stops, the back end of the rod keeps moving, as described by Newton's first law, until it is stopped by excess pressure in a sound wave moving back through the rod. How much time elapses before the back end of the rod receives the message that it should stop? (c) How far has the back end of the rod moved in this time? Find (d) the strain in the rod and (e) the stress. (f) If it is not to fail, show that the maximum impact speed a rod can have is given by the expression $\sigma_y/\sqrt{\rho Y}$.

57. To permit measurement of her speed, a skydiver carries a buzzer emitting a steady tone at 1 800 Hz. A friend on the ground at the landing site directly below listens to the amplified sound he receives. Assume that the air is calm and that the sound speed is 343 m/s, independent of altitude. While the skydiver is falling at terminal speed, her friend on the ground receives waves of frequency 2 150 Hz. (a) What is the skydiver's speed of descent? (b) **What If?** Suppose the skydiver can hear the sound of the buzzer reflected from the ground. What frequency does she receive?

58. A train whistle ($f = 400$ Hz) sounds higher or lower in frequency depending on whether it approaches or recedes. (a) Prove that the difference in frequency between the approaching and receding train whistle is

$$\Delta f = \frac{2u/v}{1 - u^2/v^2} f$$

where u is the speed of the train and v is the speed of sound. (b) Calculate this difference for a train moving at a speed of 130 km/h. Take the speed of sound in air to be 340 m/s.

59. Two ships are moving along a line due east. The trailing vessel has a speed relative to a land-based observation point of 64.0 km/h, and the leading ship has a speed of 45.0 km/h relative to that point. The two ships are in a region of the ocean where the current is moving uniformly due west at 10.0 km/h. The trailing ship transmits a sonar signal at a frequency of 1 200.0 Hz. What frequency is monitored by the leading ship? (Use 1 520 m/s as the speed of sound in ocean water.)

60. A bat, moving at 5.00 m/s, is chasing a flying insect (Fig. P17.7). If the bat emits a 40.0 kHz chirp and receives back an echo at 40.4 kHz, at what speed is the insect moving toward or away from the bat? (Take the speed of sound in air to be $v = 340$ m/s.)

61. A supersonic aircraft is flying parallel to the ground. When the aircraft is directly overhead, an observer sees a rocket fired from the aircraft. Ten seconds later the observer hears the sonic boom, followed 2.80 s later by the sound of the rocket engine. What is the Mach number of the aircraft?

62. A police car is traveling east at 40.0 m/s along a straight road, overtaking a car ahead of it moving east at 30.0 m/s. The police car has a malfunctioning siren that is stuck at 1 000 Hz. (a) Sketch the appearance of the wave fronts of the sound produced by the siren. Show the wave fronts both to the east and to the west of the police car. (b) What would be the wavelength in air of the siren sound if the police car were at rest? (c) What is the wavelength in front of the police car? (d) What is it behind the police car? (e) What is the frequency heard by the driver being chased?

63. The speed of a one-dimensional compressional wave traveling along a thin copper rod is 3.56 km/s. A copper bar is given a sharp compressional blow at one end. The sound of the blow, traveling through air at 0°C, reaches the opposite end of the bar 6.40 ms later than the sound transmitted through the metal of the bar. What is the length of the bar?

64. A jet flies toward higher altitude at a constant speed of 1 963 m/s in a direction making an angle θ with the horizontal (Fig. P17.64). An observer on the ground hears the jet for the first time when it is directly overhead. Determine the value of θ if the speed of sound in air is 340 m/s.

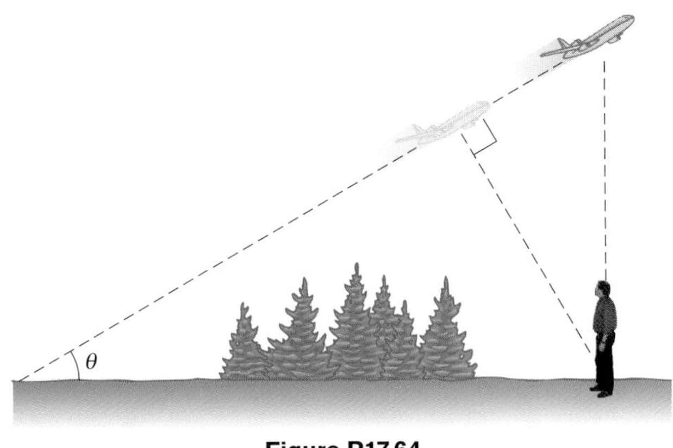

Figure P17.64

65. A meteoroid the size of a truck enters the earth's atmosphere at a speed of 20.0 km/s and is not significantly slowed before entering the ocean. (a) What is the Mach angle of the shock wave from the meteoroid in the atmosphere? (Use 331 m/s as the sound speed.) (b) Assuming that the meteoroid survives the impact with the ocean surface, what is the (initial) Mach angle of the shock wave that the meteoroid produces in the water? (Use the wave speed for seawater given in Table 17.1.)

66. An interstate highway has been built through a poor neighborhood in a city. In the afternoon, the sound level in a rented room is 80.0 dB, as 100 cars pass outside the window every minute. Late at night, when the tenant is working in a factory, the traffic flow is only five cars per minute. What is the average late-night sound level?

67. With particular experimental methods, it is possible to produce and observe in a long thin rod both a longitudinal wave and a transverse wave whose speed depends primarily on tension in the rod. The speed of the longitudinal wave is determined by the Young's modulus and the density of the material as $\sqrt{Y/\rho}$. The transverse wave can be modeled as a wave in a stretched string. A particular metal rod is 150 cm long and has a radius of 0.200 cm and a mass of 50.9 g. Young's modulus for the material is 6.80×10^{10} N/m^2. What must the tension in the rod be if the ratio of the speed of longitudinal waves to the speed of transverse waves is 8.00?

68. A siren creates sound with a level β at a distance d from the speaker. The siren is powered by a battery that delivers a total energy E. Let e represent the efficiency of the siren. (That is, e is equal to the output sound energy divided by the supplied energy). Determine the total time the siren can sound.

69. The Doppler equation presented in the text is valid when the motion between the observer and the source occurs on a straight line, so that the source and observer are moving either directly toward or directly away from each other. If this restriction is relaxed, one must use the more general Doppler equation

$$f' = \left(\frac{v + v_O \cos\theta_O}{v - v_S \cos\theta_S}\right) f$$

where θ_O and θ_S are defined in Figure P17.69a. (a) Show that if the observer and source are moving away from each other, the preceding equation reduces to Equation 17.13 with negative values for both v_O and v_S. (b) Use the preceding equation to solve the following problem. A train moves at a constant speed of 25.0 m/s toward the intersection shown in Figure P17.69b. A car is stopped near the intersection, 30.0 m from the tracks. If the train's horn emits a frequency of 500 Hz, what is the frequency heard by the passengers in the car when the train is 40.0 m from the intersection? Take the speed of sound to be 343 m/s.

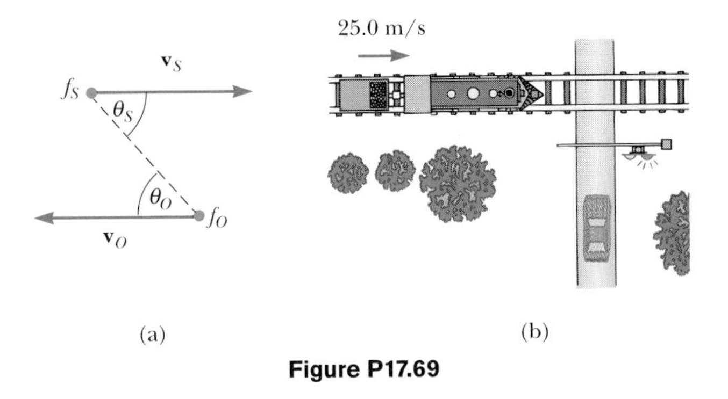

Figure P17.69

70. Equation 17.7 states that, at distance r away from a point source with power $\mathcal{P}_{av}$, the wave intensity is

$$I = \frac{\mathcal{P}_{av}}{4\pi r^2}$$

Study Figure 17.9 and prove that, at distance r straight in front of a point source with power $\mathcal{P}_{av}$ moving with constant speed v_S, the wave intensity is

$$I = \frac{\mathcal{P}_{\text{av}}}{4\pi r^2}\left(\frac{v - v_S}{v}\right)$$

71. Three metal rods are located relative to each other as shown in Figure P17.71, where $L_1 + L_2 = L_3$. The speed of sound in a rod is given by $v = \sqrt{Y/\rho}$, where ρ is the density and Y is Young's modulus for the rod. Values of density and Young's modulus for the three materials are $\rho_1 = 2.70 \times 10^3\ \text{kg/m}^3$, $Y_1 = 7.00 \times 10^{10}\ \text{N/m}^2$, $\rho_2 = 11.3 \times 10^3\ \text{kg/m}^3$, $Y_2 = 1.60 \times 10^{10}\ \text{N/m}^2$, $\rho_3 = 8.80 \times 10^3\ \text{kg/m}^3$, $Y_3 = 11.0 \times 10^{10}\ \text{N/m}^2$. (a) If $L_3 = 1.50$ m, what must the ratio L_1/L_2 be if a sound wave is to travel the length of rods 1 and 2 in the same time as it takes for the wave to travel the length of rod 3? (b) If the frequency of the source is 4.00 kHz, determine the phase difference between the wave traveling along rods 1 and 2 and the one traveling along rod 3.

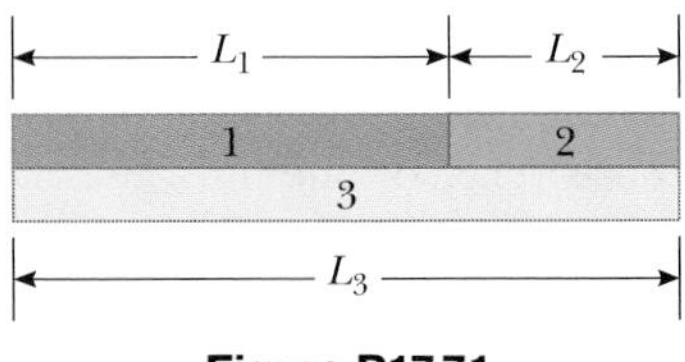

Figure P17.71

72. The smallest wavelength possible for a sound wave in air is on the order of the separation distance between air molecules. Find the order of magnitude of the highest-frequency sound wave possible in air, assuming a wave speed of 343 m/s, density $1.20\ \text{kg/m}^3$, and an average molecular mass of 4.82×10^{-26} kg.

Answers to Quick Quizzes

17.1 (c). Although the speed of a wave is given by the product of its wavelength (a) and frequency (b), it is not affected by changes in either one. The amplitude (d) of a sound wave determines the size of the oscillations of elements of air but does not affect the speed of the wave through the air.

17.2 (c). Because the bottom of the bottle is a rigid barrier, the displacement of elements of air at the bottom is zero. Because the pressure variation is a minimum or a maximum when the displacement is zero, and the pulse is moving downward, the pressure variation at the bottom is a maximum.

17.3 (c). The ear trumpet collects sound waves from the large area of its opening and directs it toward the ear. Most of the sound in this large area would miss the ear in the absence of the trumpet.

17.4 (b). The large area of the guitar body sets many elements of air into oscillation and allows the energy to leave the system by mechanical waves at a much larger rate than from the thin vibrating string.

17.5 (c). The only parameter that adds directly is intensity. Because of the logarithm function in the definition of sound level, sound levels cannot be added directly.

17.6 (b). The factor of 100 is two powers of ten. Thus, the logarithm of 100 is 2, which multiplied by 10 gives 20 dB.

17.7 (e). The wave speed cannot be changed by moving the source, so (a) and (b) are incorrect. The detected wavelength is largest at A, so (c) and (d) are incorrect. Choice (f) is incorrect because the detected frequency is lowest at location A.

17.8 (e). The intensity of the sound increases because the train is moving closer to you. Because the train moves at a constant velocity, the Doppler-shifted frequency remains fixed.

17.9 (b). The Mach number is the ratio of the plane's speed (which does not change) to the speed of sound, which is greater in the warm air than in the cold. The denominator of this ratio increases while the numerator stays constant. Therefore, the ratio as a whole—the Mach number—decreases.

Chapter 18

Superposition and Standing Waves

▲ *Guitarist Carlos Santana takes advantage of standing waves on strings. He changes to a higher note on the guitar by pushing the strings against the frets on the fingerboard, shortening the lengths of the portions of the strings that vibrate. (Bettmann/Corbis)*

CHAPTER OUTLINE

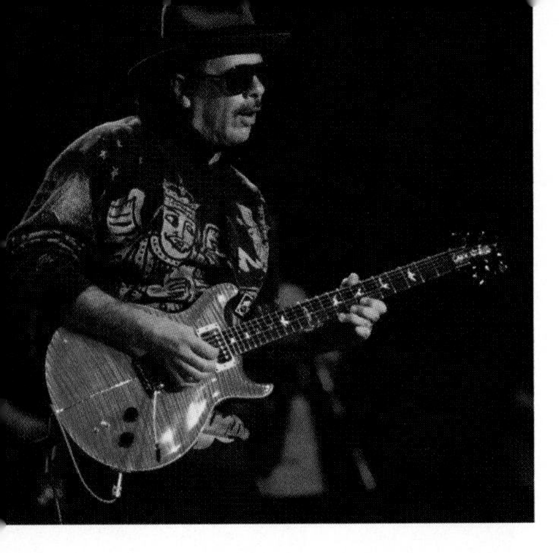

In the previous two chapters, we introduced the wave model. We have seen that waves are very different from particles. A particle is of zero size, while a wave has a characteristic size—the wavelength. Another important difference between waves and particles is that we can explore the possibility of two or more waves combining at one point in the same medium. We can combine particles to form extended objects, but the particles must be at *different* locations. In contrast, two waves can both be present at the same location, and the ramifications of this possibility are explored in this chapter.

When waves are combined, only certain allowed frequencies can exist on systems with boundary conditions—the frequencies are *quantized.* Quantization is a notion that is at the heart of quantum mechanics, a subject that we introduce formally in Chapter 40. There we show that waves under boundary conditions explain many of the quantum phenomena. For our present purposes in this chapter, quantization enables us to understand the behavior of the wide array of musical instruments that are based on strings and air columns.

We also consider the combination of waves having different frequencies and wavelengths. When two sound waves having nearly the same frequency interfere, we hear variations in the loudness called *beats.* The beat frequency corresponds to the rate of alternation between constructive and destructive interference. Finally, we discuss how any nonsinusoidal periodic wave can be described as a sum of sine and cosine functions.

18.1 Superposition and Interference

Many interesting wave phenomena in nature cannot be described by a single traveling wave. Instead, one must analyze complex waves in terms of a combination of traveling waves. To analyze such wave combinations, one can make use of the **superposition principle:**

Superposition principle

If two or more traveling waves are moving through a medium, the resultant value of the wave function at any point is the algebraic sum of the values of the wave functions of the individual waves.

Waves that obey this principle are called *linear waves.* In the case of mechanical waves, linear waves are generally characterized by having amplitudes much smaller than their wavelengths. Waves that violate the superposition principle are called *nonlinear waves* and are often characterized by large amplitudes. In this book, we deal only with linear waves.

One consequence of the superposition principle is that **two traveling waves can pass through each other without being destroyed or even altered.** For instance,

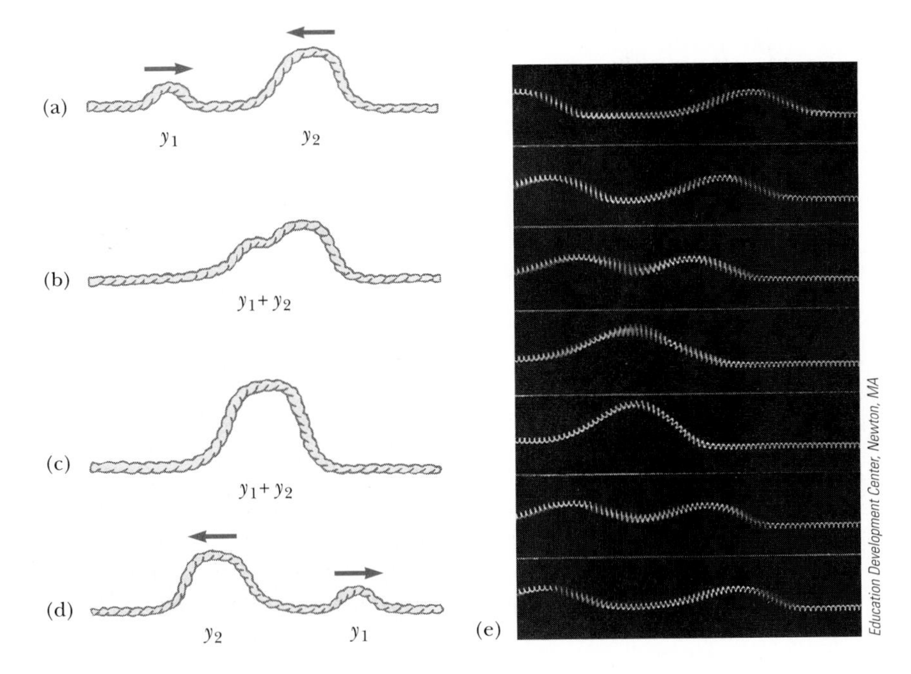

Active Figure 18.1 (a–d) Two pulses traveling on a stretched string in opposite directions pass through each other. When the pulses overlap, as shown in (b) and (c), the net displacement of the string equals the sum of the displacements produced by each pulse. Because each pulse produces positive displacements of the string, we refer to their superposition as *constructive interference.* (e) Photograph of the superposition of two equal, symmetric pulses traveling in opposite directions on a stretched spring.

At the Active Figures link at* http://www.pse6.com, *you can choose the amplitude and orientation of each of the pulses and study the interference between them as they pass each other.

when two pebbles are thrown into a pond and hit the surface at different places, the expanding circular surface waves do not destroy each other but rather pass through each other. The complex pattern that is observed can be viewed as two independent sets of expanding circles. Likewise, when sound waves from two sources move through air, they pass through each other.

Figure 18.1 is a pictorial representation of the superposition of two pulses. The wave function for the pulse moving to the right is y_1, and the wave function for the pulse moving to the left is y_2. The pulses have the same speed but different shapes, and the displacement of the elements of the medium is in the positive y direction for both pulses. When the waves begin to overlap (Fig. 18.1b), the wave function for the resulting complex wave is given by $y_1 + y_2$. When the crests of the pulses coincide (Fig. 18.1c), the resulting wave given by $y_1 + y_2$ has a larger amplitude than that of the individual pulses. The two pulses finally separate and continue moving in their original directions (Fig. 18.1d). Note that the pulse shapes remain unchanged after the interaction, as if the two pulses had never met!

The combination of separate waves in the same region of space to produce a resultant wave is called **interference.** For the two pulses shown in Figure 18.1, the displacement of the elements of the medium is in the positive y direction for both pulses, and the resultant pulse (created when the individual pulses overlap) exhibits an amplitude greater than that of either individual pulse. Because the displacements caused by the two pulses are in the same direction, we refer to their superposition as **constructive interference.**

Constructive interference

Now consider two pulses traveling in opposite directions on a taut string where one pulse is inverted relative to the other, as illustrated in Figure 18.2. In this case, when the pulses begin to overlap, the resultant pulse is given by $y_1 + y_2$, but the values of the function y_2 are negative. Again, the two pulses pass through each other; however, because the displacements caused by the two pulses are in opposite directions, we refer to their superposition as **destructive interference.**

Destructive interference

PITFALL PREVENTION

18.1 Do Waves Really *Interfere*?

In popular usage, the term *interfere* implies that an agent affects a situation in some way so as to preclude something from happening. For example, in American football, *pass interference* means that a defending player has affected the receiver so that he is unable to catch the ball. This is very different from its use in physics, where waves pass through each other and interfere, but do not affect each other in any way. In physics, interference is similar to the notion of *combination* as described in this chapter.

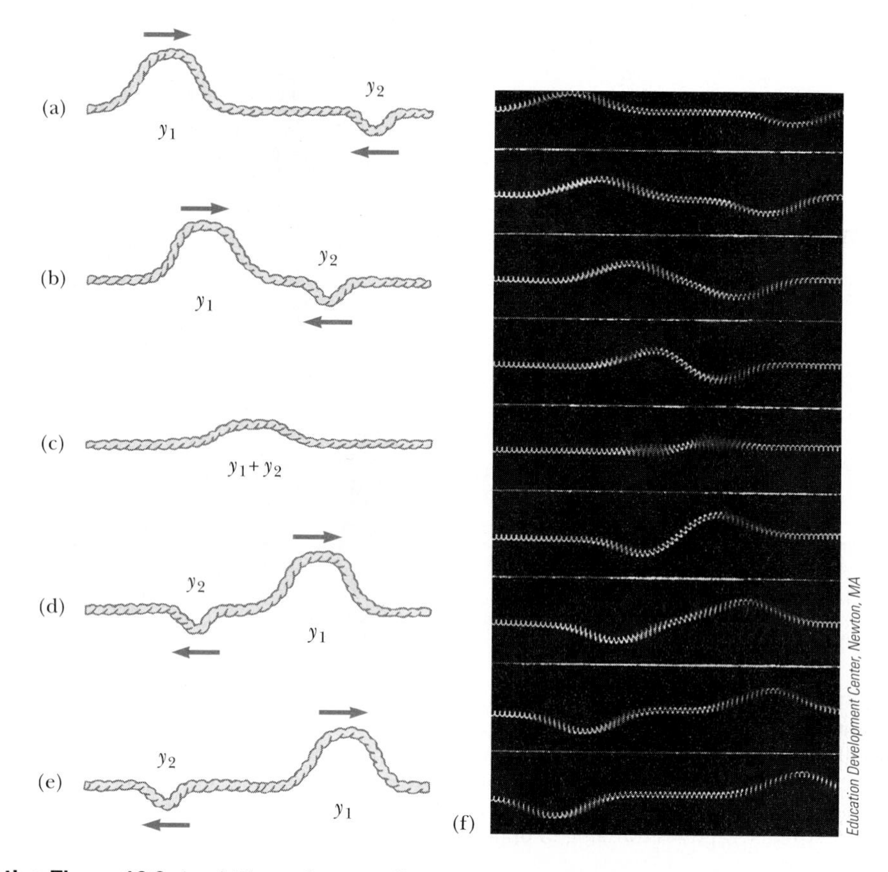

At the Active Figures link at http://www.pse6.com, you can choose the amplitude and orientation of each of the pulses and watch the interference as they pass each other.

Active Figure 18.2 (a–e) Two pulses traveling in opposite directions and having displacements that are inverted relative to each other. When the two overlap in (c), their displacements partially cancel each other. (f) Photograph of the superposition of two symmetric pulses traveling in opposite directions, where one is inverted relative to the other.

Quick Quiz 18.1 Two pulses are traveling toward each other, each at 10 cm/s on a long string, as shown in Figure 18.3. Sketch the shape of the string at $t = 0.6$ s.

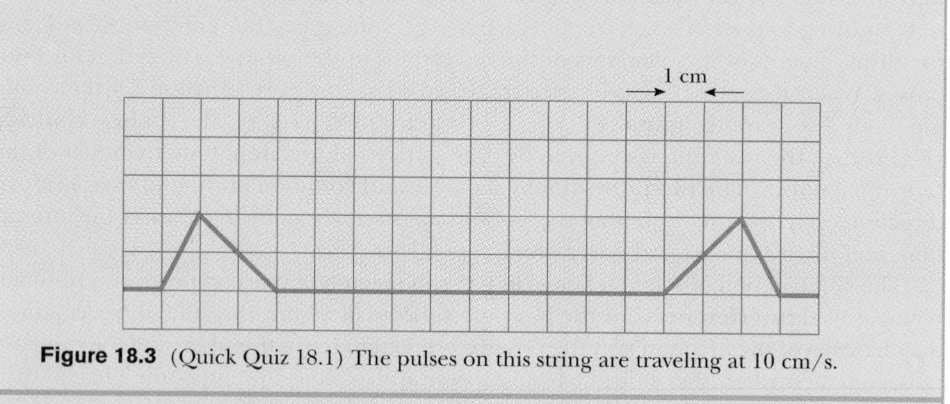

Figure 18.3 (Quick Quiz 18.1) The pulses on this string are traveling at 10 cm/s.

Quick Quiz 18.2 Two pulses move in opposite directions on a string and are identical in shape except that one has positive displacements of the elements of the string and the other has negative displacements. At the moment that the two pulses completely overlap on the string, (a) the energy associated with the pulses has disappeared (b) the string is not moving (c) the string forms a straight line (d) the pulses have vanished and will not reappear.

Superposition of Sinusoidal Waves

Let us now apply the principle of superposition to two sinusoidal waves traveling in the same direction in a linear medium. If the two waves are traveling to the right and have the same frequency, wavelength, and amplitude but differ in phase, we can express their individual wave functions as

$$y_1 = A\sin(kx - \omega t) \qquad y_2 = A\sin(kx - \omega t + \phi)$$

where, as usual, $k = 2\pi/\lambda$, $\omega = 2\pi f$, and ϕ is the phase constant, which we discussed in Section 16.2. Hence, the resultant wave function y is

$$y = y_1 + y_2 = A[\sin(kx - \omega t) + \sin(kx - \omega t + \phi)]$$

To simplify this expression, we use the trigonometric identity

$$\sin a + \sin b = 2\cos\left(\frac{a-b}{2}\right)\sin\left(\frac{a+b}{2}\right)$$

If we let $a = kx - \omega t$ and $b = kx - \omega t + \phi$, we find that the resultant wave function y reduces to

Resultant of two traveling sinusoidal waves

$$y = 2A\cos\left(\frac{\phi}{2}\right)\sin\left(kx - \omega t + \frac{\phi}{2}\right)$$

This result has several important features. The resultant wave function y also is sinusoidal and has the same frequency and wavelength as the individual waves because the sine function incorporates the same values of k and ω that appear in the original wave functions. The amplitude of the resultant wave is $2A\cos(\phi/2)$, and its phase is $\phi/2$. If the phase constant ϕ equals 0, then $\cos(\phi/2) = \cos 0 = 1$, and the amplitude of the resultant wave is $2A$—twice the amplitude of either individual wave. In this case the waves are said to be everywhere *in phase* and thus interfere constructively. That is, the crests and troughs of the individual waves y_1 and y_2 occur at the same positions and combine to form the red curve y of amplitude $2A$ shown in Figure 18.4a. Because the

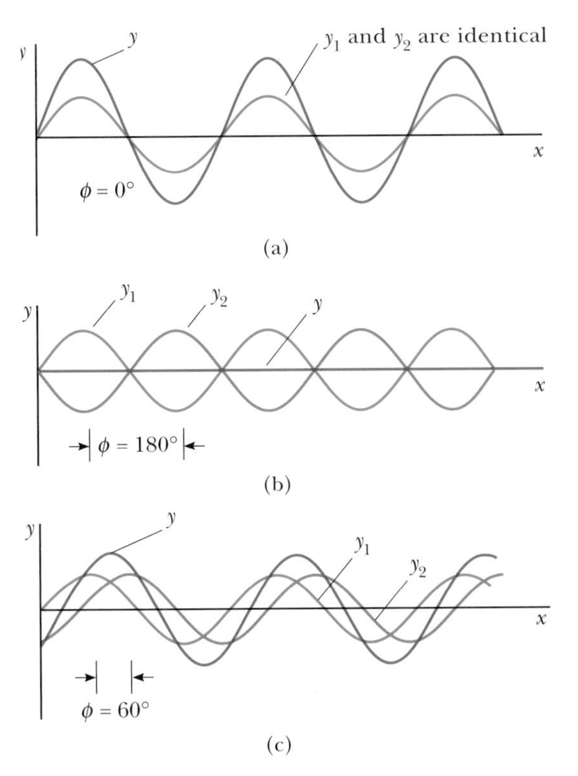

Active Figure 18.4 The superposition of two identical waves y_1 and y_2 (blue and green) to yield a resultant wave (red). (a) When y_1 and y_2 are in phase, the result is constructive interference. (b) When y_1 and y_2 are π rad out of phase, the result is destructive interference. (c) When the phase angle has a value other than 0 or π rad, the resultant wave y falls somewhere between the extremes shown in (a) and (b).

At the Active Figures link at http://www.pse6.com, you can change the phase relationship between the waves and observe the wave representing the superposition.

individual waves are in phase, they are indistinguishable in Figure 18.4a, in which they appear as a single blue curve. In general, constructive interference occurs when $\cos(\phi/2) = \pm 1$. This is true, for example, when $\phi = 0, 2\pi, 4\pi, \ldots$ rad—that is, when ϕ is an *even* multiple of π.

When ϕ is equal to π rad or to any *odd* multiple of π, then $\cos(\phi/2) = \cos(\pi/2) = 0$, and the crests of one wave occur at the same positions as the troughs of the second wave (Fig. 18.4b). Thus, the resultant wave has *zero* amplitude everywhere, as a consequence of destructive interference. Finally, when the phase constant has an arbitrary value other than 0 or an integer multiple of π rad (Fig. 18.4c), the resultant wave has an amplitude whose value is somewhere between 0 and $2A$.

Interference of Sound Waves

One simple device for demonstrating interference of sound waves is illustrated in Figure 18.5. Sound from a loudspeaker S is sent into a tube at point P, where there is a T-shaped junction. Half of the sound energy travels in one direction, and half travels in the opposite direction. Thus, the sound waves that reach the receiver R can travel along either of the two paths. The distance along any path from speaker to receiver is called the **path length** r. The lower path length r_1 is fixed, but the upper path length r_2 can be varied by sliding the U-shaped tube, which is similar to that on a slide trombone. When the difference in the path lengths $\Delta r = |r_2 - r_1|$ is either zero or some integer multiple of the wavelength λ (that is $\Delta r = n\lambda$, where $n = 0, 1, 2, 3, \ldots$), the two waves reaching the receiver at any instant are in phase and interfere constructively, as shown in Figure 18.4a. For this case, a maximum in the sound intensity is detected at the receiver. If the path length r_2 is adjusted such that the path difference $\Delta r = \lambda/2, 3\lambda/2, \ldots, n\lambda/2$ (for n odd), the two waves are exactly π rad, or 180°, out of phase at the receiver and hence cancel each other. In this case of destructive interference, no sound is detected at the receiver. This simple experiment demonstrates that a phase difference may arise between two waves generated by the same source when they travel along paths of unequal lengths. This important phenomenon will be indispensable in our investigation of the interference of light waves in Chapter 37.

It is often useful to express the path difference in terms of the phase angle ϕ between the two waves. Because a path difference of one wavelength corresponds to a phase angle of 2π rad, we obtain the ratio $\phi/2\pi = \Delta r/\lambda$ or

Relationship between path difference and phase angle

$$\Delta r = \frac{\phi}{2\pi}\lambda \tag{18.1}$$

Using the notion of path difference, we can express our conditions for constructive and destructive interference in a different way. If the path difference is any even multiple of $\lambda/2$, then the phase angle $\phi = 2n\pi$, where $n = 0, 1, 2, 3, \ldots$, and the interference is constructive. For path differences of odd multiples of $\lambda/2$, $\phi = (2n + 1)\pi$, where $n = 0, 1, 2, 3, \ldots$, and the interference is destructive. Thus, we have the conditions

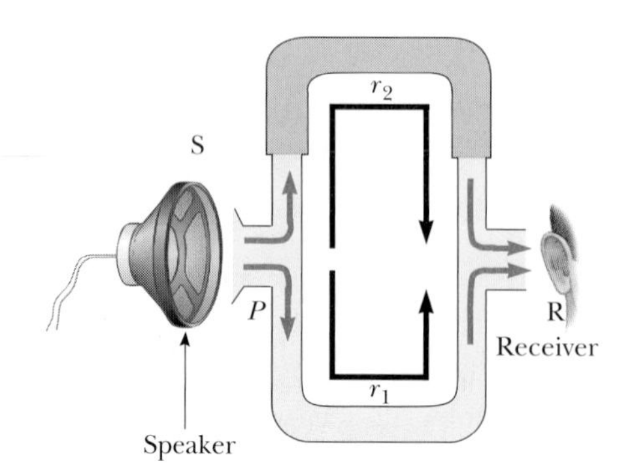

Figure 18.5 An acoustical system for demonstrating interference of sound waves. A sound wave from the speaker (S) propagates into the tube and splits into two parts at point P. The two waves, which combine at the opposite side, are detected at the receiver (R). The upper path length r_2 can be varied by sliding the upper section.

$$\Delta r = (2n)\frac{\lambda}{2} \quad \text{for constructive interference}$$

and (18.2)

$$\Delta r = (2n + 1)\frac{\lambda}{2} \quad \text{for destructive interference}$$

This discussion enables us to understand why the speaker wires in a stereo system should be connected properly. When connected the wrong way—that is, when the positive (or red) wire is connected to the negative (or black) terminal on one of the speakers and the other is correctly wired—the speakers are said to be "out of phase"—one speaker cone moves outward while the other moves inward. As a consequence, the sound wave coming from one speaker destructively interferes with the wave coming from the other—along a line midway between the two, a rarefaction region due to one speaker is superposed on a compression region from the other speaker. Although the two sounds probably do not completely cancel each other (because the left and right stereo signals are usually not identical), a substantial loss of sound quality occurs at points along this line.

Example 18.1 Two Speakers Driven by the Same Source

A pair of speakers placed 3.00 m apart are driven by the same oscillator (Fig. 18.6). A listener is originally at point O, which is located 8.00 m from the center of the line connecting the two speakers. The listener then walks to point P, which is a perpendicular distance 0.350 m from O, before reaching the *first minimum* in sound intensity. What is the frequency of the oscillator?

Solution To find the frequency, we must know the wavelength of the sound coming from the speakers. With this information, combined with our knowledge of the speed of sound, we can calculate the frequency. The wavelength can be determined from the interference information given. The first minimum occurs when the two waves reaching the listener at point P are 180° out of phase—in other words, when their path difference Δr equals $\lambda/2$. To calculate the path difference, we must first find the path lengths r_1 and r_2.

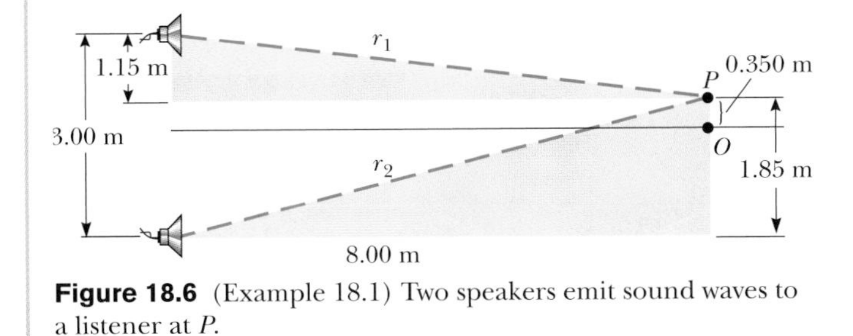

Figure 18.6 (Example 18.1) Two speakers emit sound waves to a listener at P.

Figure 18.6 shows the physical arrangement of the speakers, along with two shaded right triangles that can be drawn on the basis of the lengths described in the problem. From these triangles, we find that the path lengths are

$$r_1 = \sqrt{(8.00 \text{ m})^2 + (1.15 \text{ m})^2} = 8.08 \text{ m}$$

and

$$r_2 = \sqrt{(8.00 \text{ m})^2 + (1.85 \text{ m})^2} = 8.21 \text{ m}$$

Hence, the path difference is $r_2 - r_1 = 0.13$ m. Because we require that this path difference be equal to $\lambda/2$ for the first minimum, we find that $\lambda = 0.26$ m.

To obtain the oscillator frequency, we use Equation 16.12, $v = \lambda f$, where v is the speed of sound in air, 343 m/s:

$$f = \frac{v}{\lambda} = \frac{343 \text{ m/s}}{0.26 \text{ m}} = \boxed{1.3 \text{ kHz}}$$

What If? **What if the speakers were connected out of phase? What happens at point *P* in Figure 18.6?**

Answer In this situation, the path difference of $\lambda/2$ combines with a phase difference of $\lambda/2$ due to the incorrect wiring to give a full phase difference of λ. As a result, the waves are in phase and there is a *maximum* intensity at point P.

18.2 Standing Waves

The sound waves from the speakers in Example 18.1 leave the speakers in the forward direction, and we considered interference at a point in front of the speakers. Suppose that we turn the speakers so that they face each other and then have them emit sound of the same frequency and amplitude. In this situation, two identical waves travel in

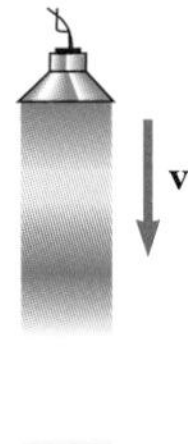

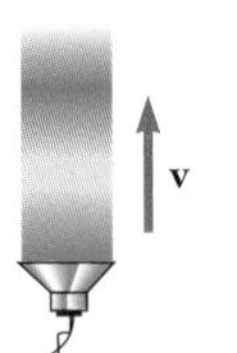

Figure 18.7 Two speakers emit sound waves toward each other. When they overlap, identical waves traveling in opposite directions will combine to form standing waves.

opposite directions in the same medium, as in Figure 18.7. These waves combine in accordance with the superposition principle.

We can analyze such a situation by considering wave functions for two transverse sinusoidal waves having the same amplitude, frequency, and wavelength but traveling in opposite directions in the same medium:

$$y_1 = A\sin(kx - \omega t) \qquad y_2 = A\sin(kx + \omega t)$$

where y_1 represents a wave traveling in the $+x$ direction and y_2 represents one traveling in the $-x$ direction. Adding these two functions gives the resultant wave function y:

$$y = y_1 + y_2 = A\sin(kx - \omega t) + A\sin(kx + \omega t)$$

When we use the trigonometric identity $\sin(a \pm b) = \sin(a)\cos(b) \pm \cos(a)\sin(b)$, this expression reduces to

$$y = (2A\sin kx)\cos \omega t \tag{18.3}$$

Equation 18.3 represents the wave function of a **standing wave.** A standing wave, such as the one shown in Figure 18.8, is an oscillation pattern *with a stationary outline* that results from the superposition of two identical waves traveling in opposite directions.

Notice that Equation 18.3 does not contain a function of $kx - \omega t$. Thus, it is not an expression for a traveling wave. If we observe a standing wave, we have no sense of motion in the direction of propagation of either of the original waves. If we compare this equation with Equation 15.6, we see that Equation 18.3 describes a special kind of simple harmonic motion. Every element of the medium oscillates in simple harmonic motion with the same frequency ω (according to the $\cos \omega t$ factor in the equation). However, the amplitude of the simple harmonic motion of a given element (given by the factor $2A\sin kx$, the coefficient of the cosine function) depends on the location x of the element in the medium.

The maximum amplitude of an element of the medium has a minimum value of zero when x satisfies the condition $\sin kx = 0$, that is, when

$$kx = \pi, 2\pi, 3\pi, \ldots$$

Because $k = 2\pi/\lambda$, these values for kx give

$$x = \frac{\lambda}{2}, \lambda, \frac{3\lambda}{2}, \ldots = \frac{n\lambda}{2} \qquad n = 0, 1, 2, 3, \ldots \tag{18.4}$$

These points of zero amplitude are called **nodes.**

PITFALL PREVENTION

18.2 Three Types of Amplitude

We need to distinguish carefully here between the **amplitude of the individual waves,** which is A, and the **amplitude of the simple harmonic motion of the elements of the medium,** which is $2A\sin kx$. A given element in a standing wave vibrates within the constraints of the *envelope* function $2A\sin kx$, where x is that element's position in the medium. This is in contrast to traveling sinusoidal waves, in which all elements oscillate with the same amplitude and the same frequency, and the amplitude A of the wave is the same as the amplitude A of the simple harmonic motion of the elements. Furthermore, we can identify the **amplitude of the standing wave** as $2A$.

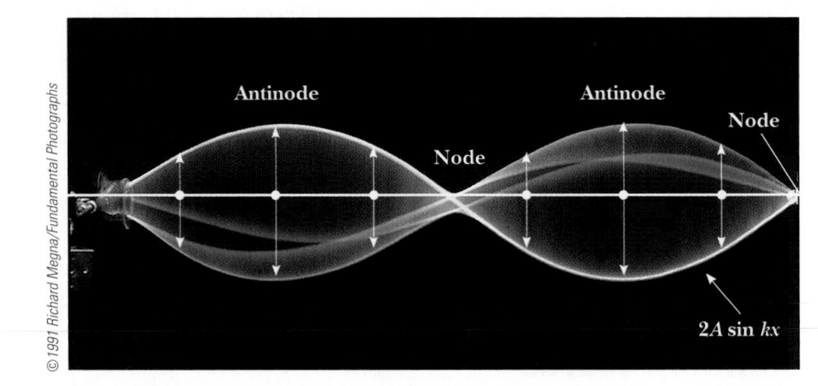

Figure 18.8 Multiflash photograph of a standing wave on a string. The time behavior of the vertical displacement from equilibrium of an individual element of the string is given by $\cos \omega t$. That is, each element vibrates at an angular frequency ω. The amplitude of the vertical oscillation of any elements of the string depends on the horizontal position of the element. Each element vibrates within the confines of the envelope function $2A\sin kx$.

The element with the *greatest* possible displacement from equilibrium has an amplitude of $2A$, and we define this as the amplitude of the standing wave. The positions in the medium at which this maximum displacement occurs are called **antinodes.** The antinodes are located at positions for which the coordinate x satisfies the condition $\sin kx = \pm 1$, that is, when

$$kx = \frac{\pi}{2}, \frac{3\pi}{2}, \frac{5\pi}{2}, \ldots$$

Thus, the positions of the antinodes are given by

$$x = \frac{\lambda}{4}, \frac{3\lambda}{4}, \frac{5\lambda}{4}, \ldots = \frac{n\lambda}{4} \qquad n = 1, 3, 5, \ldots \tag{18.5}$$

Position of antinodes

In examining Equations 18.4 and 18.5, we note the following important features of the locations of nodes and antinodes:

The distance between adjacent antinodes is equal to $\lambda/2$.

The distance between adjacent nodes is equal to $\lambda/2$.

The distance between a node and an adjacent antinode is $\lambda/4$.

Wave patterns of the elements of the medium produced at various times by two waves traveling in opposite directions are shown in Figure 18.9. The blue and green curves are the wave patterns for the individual traveling waves, and the red curves are the wave patterns for the resultant standing wave. At $t = 0$ (Fig. 18.9a), the two traveling waves are in phase, giving a wave pattern in which each element of the medium is experiencing its maximum displacement from equilibrium. One quarter of a period later, at $t = T/4$ (Fig. 18.9b), the traveling waves have moved one quarter of a wavelength (one to the right and the other to the left). At this time, the traveling waves are out of phase, and each element of the medium is passing through the equilibrium position in its simple harmonic motion. The result is zero displacement for elements at all values of x—that is, the wave pattern is a straight line. At $t = T/2$ (Fig. 18.9c), the traveling waves are again in phase, producing a wave pattern that is inverted relative to the $t = 0$ pattern. In the standing wave, the elements of the medium alternate in time between the extremes shown in Figure 18.9a and c.

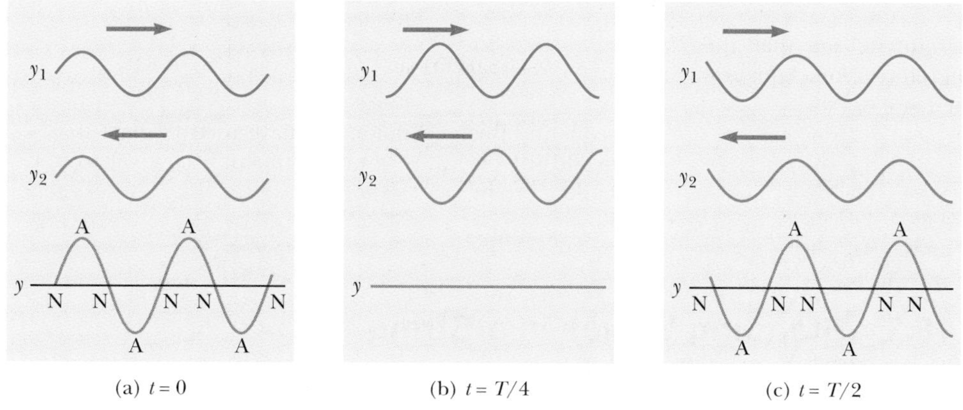

Active Figure 18.9 Standing-wave patterns produced at various times by two waves of equal amplitude traveling in opposite directions. For the resultant wave y, the nodes (N) are points of zero displacement, and the antinodes (A) are points of maximum displacement.

At the Active Figures link at http://www.pse6.com, you can choose the wavelength of the waves and see the standing wave that results.

Quick Quiz 18.3 Consider a standing wave on a string as shown in Figure 18.9. Define the velocity of elements of the string as positive if they are moving upward in the figure. At the moment the string has the shape shown by the red curve in Figure 18.9a, the instantaneous velocity of elements along the string (a) is zero for all elements (b) is positive for all elements (c) is negative for all elements (d) varies with the position of the element.

Quick Quiz 18.4 Continuing with the scenario in Quick Quiz 18.3, at the moment the string has the shape shown by the red curve in Figure 18.9b, the instantaneous velocity of elements along the string (a) is zero for all elements (b) is positive for all elements (c) is negative for all elements (d) varies with the position of the element.

Example 18.2 Formation of a Standing Wave

Two waves traveling in opposite directions produce a standing wave. The individual wave functions are

$$y_1 = (4.0 \text{ cm}) \sin(3.0x - 2.0t)$$
$$y_2 = (4.0 \text{ cm}) \sin(3.0x + 2.0t)$$

where x and y are measured in centimeters.

(A) Find the amplitude of the simple harmonic motion of the element of the medium located at $x = 2.3$ cm.

Solution The standing wave is described by Equation 18.3; in this problem, we have $A = 4.0$ cm, $k = 3.0$ rad/cm, and $\omega = 2.0$ rad/s. Thus,

$$y = (2A \sin kx) \cos \omega t = [(8.0 \text{ cm}) \sin 3.0x] \cos 2.0t$$

Thus, we obtain the amplitude of the simple harmonic motion of the element at the position $x = 2.3$ cm by evaluating the coefficient of the cosine function at this position:

$$y_{\text{max}} = (8.0 \text{ cm}) \sin 3.0x\big|_{x=2.3}$$
$$= (8.0 \text{ cm}) \sin (6.9 \text{ rad}) = 4.6 \text{ cm}$$

(B) Find the positions of the nodes and antinodes if one end of the string is at $x = 0$.

Solution With $k = 2\pi/\lambda = 3.0$ rad/cm, we see that the wavelength is $\lambda = (2\pi/3.0)$ cm. Therefore, from Equation 18.4 we find that the nodes are located at

$$x = n\frac{\lambda}{2} = n\left(\frac{\pi}{3}\right) \text{cm} \qquad n = 0, 1, 2, 3, \ldots$$

and from Equation 18.5 we find that the antinodes are located at

$$x = n\frac{\lambda}{4} = n\left(\frac{\pi}{6}\right) \text{cm} \qquad n = 1, 3, 5, \ldots$$

(C) What is the maximum value of the position in the simple harmonic motion of an element located at an antinode?

Solution According to Equation 18.3, the maximum position of an element at an antinode is the amplitude of the standing wave, which is twice the amplitude of the individual traveling waves:

$$y_{\text{max}} = 2A(\sin kx)_{\text{max}} = 2(4.0 \text{ cm})(\pm 1) = \pm 8.0 \text{ cm}$$

where we have used the fact that the maximum value of $\sin kx$ is ± 1. Let us check this result by evaluating the coefficient of our standing-wave function at the positions we found for the antinodes:

$$y_{\text{max}} = (8.0 \text{ cm}) \sin 3.0x\big|_{x=n(\pi/6)}$$
$$= (8.0 \text{ cm}) \sin\left[3.0n\left(\frac{\pi}{6}\right) \text{rad}\right]$$
$$= (8.0 \text{ cm}) \sin\left[n\left(\frac{\pi}{2}\right) \text{rad}\right] = \pm 8.0 \text{ cm}$$

In evaluating this expression, we have used the fact that n is an odd integer; thus, the sine function is equal to ± 1, depending on the value of n.

18.3 Standing Waves in a String Fixed at Both Ends

Consider a string of length L fixed at both ends, as shown in Figure 18.10. Standing waves are set up in the string by a continuous superposition of waves incident on and reflected from the ends. Note that there is a boundary condition for the waves on the

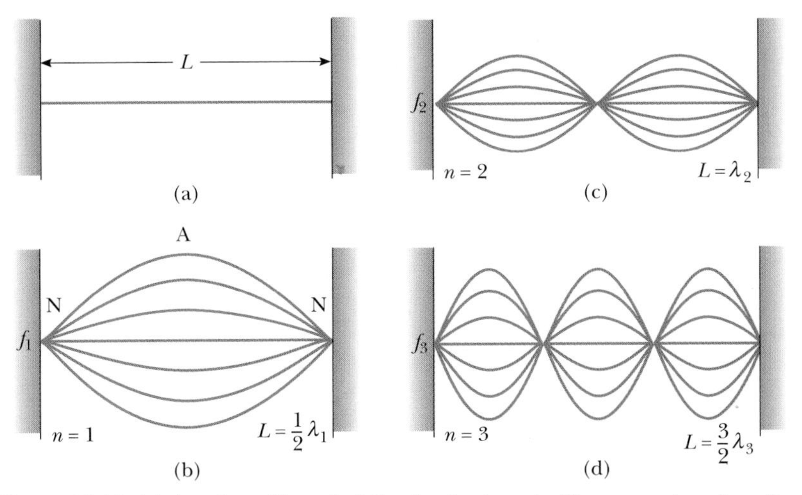

Active Figure 18.10 (a) A string of length L fixed at both ends. The normal modes of vibration form a harmonic series: (b) the fundamental, or first harmonic; (c) the second harmonic; (d) the third harmonic.

At the Active Figures link at http://www.pse6.com, you can choose the mode number and see the corresponding standing wave.

string. The ends of the string, because they are fixed, must necessarily have zero displacement and are, therefore, nodes by definition. The boundary condition results in the string having a number of natural patterns of oscillation, called **normal modes,** each of which has a characteristic frequency that is easily calculated. This situation in which only certain frequencies of oscillation are allowed is called **quantization.** Quantization is a common occurrence when waves are subject to boundary conditions and will be a central feature in our discussions of quantum physics in the extended version of this text.

Figure 18.11 shows one of the normal modes of oscillation of a string fixed at both ends. Except for the nodes, which are always stationary, all elements of the string oscillate vertically with the same frequency but with different amplitudes of simple harmonic motion. Figure 18.11 represents snapshots of the standing wave at various times over one half of a period. The red arrows show the velocities of various elements of the string at various times. As we found in Quick Quizzes 18.3 and 18.4,

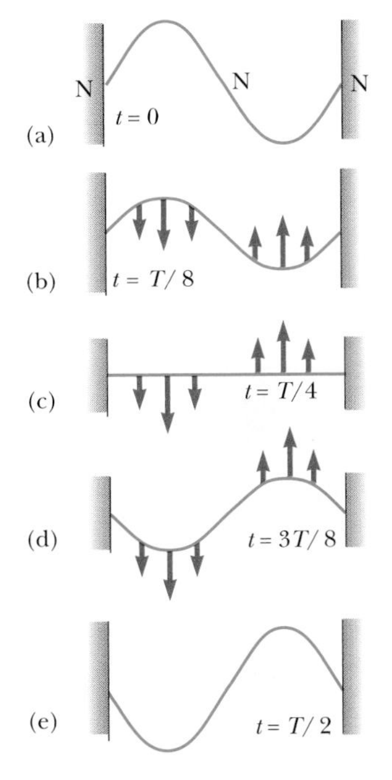

Figure 18.11 A standing-wave pattern in a taut string. The five "snapshots" were taken at intervals of one eighth of the period. (a) At $t = 0$, the string is momentarily at rest. (b) At $t = T/8$, the string is in motion, as indicated by the red arrows, and different parts of the string move in different directions with different speeds. (c) At $t = T/4$, the string is moving but horizontal (undeformed). (d) The motion continues as indicated. (e) At $t = T/2$, the string is again momentarily at rest, but the crests and troughs of (a) are reversed. The cycle continues until ultimately, when a time interval equal to T has passed, the configuration shown in (a) is repeated.

all elements of the string have zero velocity at the extreme positions (Figs. 18.11a and 18.11e) and elements have varying velocities at other positions (Figs. 18.11b through 18.11d).

The normal modes of oscillation for the string can be described by imposing the requirements that the ends be nodes and that the nodes and antinodes be separated by one fourth of a wavelength. The first normal mode that is consistent with the boundary conditions, shown in Figure 18.10b, has nodes at its ends and one antinode in the middle. This is the longest-wavelength mode that is consistent with our requirements. This first normal mode occurs when the length of the string is half the wavelength λ_1, as indicated in Figure 18.10b, or $\lambda_1 = 2L$. The next normal mode (see Fig. 18.10c) of wavelength λ_2 occurs when the wavelength equals the length of the string, that is, when $\lambda_2 = L$. The third normal mode (see Fig. 18.10d) corresponds to the case in which $\lambda_3 = 2L/3$. In general, the wavelengths of the various normal modes for a string of length L fixed at both ends are

Wavelengths of normal modes

$$\lambda_n = \frac{2L}{n} \qquad n = 1, 2, 3, \ldots \tag{18.6}$$

where the index n refers to the nth normal mode of oscillation. These are the *possible* modes of oscillation for the string. The *actual* modes that are excited on a string are discussed shortly.

The natural frequencies associated with these modes are obtained from the relationship $f = v/\lambda$, where the wave speed v is the same for all frequencies. Using Equation 18.6, we find that the natural frequencies f_n of the normal modes are

Frequencies of normal modes as functions of wave speed and length of string

$$f_n = \frac{v}{\lambda_n} = n\frac{v}{2L} \qquad n = 1, 2, 3, \ldots \tag{18.7}$$

These natural frequencies are also called the *quantized frequencies* associated with the vibrating string fixed at both ends.

Because $v = \sqrt{T/\mu}$ (see Eq. 16.18), where T is the tension in the string and μ is its linear mass density, we can also express the natural frequencies of a taut string as

Frequencies of normal modes as functions of string tension and linear mass density

$$f_n = \frac{n}{2L}\sqrt{\frac{T}{\mu}} \qquad n = 1, 2, 3, \ldots \tag{18.8}$$

The lowest frequency f_1, which corresponds to $n = 1$, is called either the **fundamental** or the **fundamental frequency** and is given by

Fundamental frequency of a taut string

$$f_1 = \frac{1}{2L}\sqrt{\frac{T}{\mu}} \tag{18.9}$$

The frequencies of the remaining normal modes are integer multiples of the fundamental frequency. Frequencies of normal modes that exhibit an integer-multiple relationship such as this form a **harmonic series,** and the normal modes are called

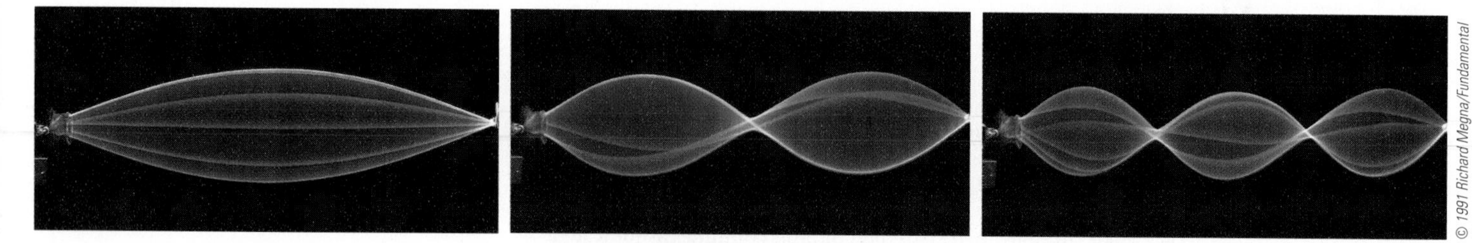

Multiflash photographs of standing-wave patterns in a cord driven by a vibrator at its left end. The single-loop pattern represents the first normal mode (the fundamental), $n = 1$). The double-loop pattern represents the second normal mode ($n = 2$), and the triple-loop pattern represents the third normal mode ($n = 3$).

harmonics. The fundamental frequency f_1 is the frequency of the first harmonic; the frequency $f_2 = 2f_1$ is the frequency of the second harmonic; and the frequency $f_n = nf_1$ is the frequency of the nth harmonic. Other oscillating systems, such as a drumhead, exhibit normal modes, but the frequencies are not related as integer multiples of a fundamental. Thus, we do not use the term *harmonic* in association with these types of systems.

In obtaining Equation 18.6, we used a technique based on the separation distance between nodes and antinodes. We can obtain this equation in an alternative manner. Because we require that the string be fixed at $x = 0$ and $x = L$, the wave function $y(x, t)$ given by Equation 18.3 must be zero at these points for all times. That is, the *boundary conditions* require that $y(0, t) = 0$ and $y(L, t) = 0$ for all values of t. Because the standing wave is described by $y = 2A(\sin kx) \cos \omega t$, the first boundary condition, $y(0, t) = 0$, is automatically satisfied because $\sin kx = 0$ at $x = 0$. To meet the second boundary condition, $y(L, t) = 0$, we require that $\sin kL = 0$. This condition is satisfied when the angle kL equals an integer multiple of π rad. Therefore, the allowed values of k are given by[1]

$$k_n L = n\pi \qquad n = 1, 2, 3, \ldots \tag{18.10}$$

Because $k_n = 2\pi/\lambda_n$, we find that

$$\left(\frac{2\pi}{\lambda_n}\right) L = n\pi \qquad \text{or} \qquad \lambda_n = \frac{2L}{n}$$

which is identical to Equation 18.6.

Let us examine further how these various harmonics are created in a string. If we wish to excite just a single harmonic, we must distort the string in such a way that its distorted shape corresponds to that of the desired harmonic. After being released, the string vibrates at the frequency of that harmonic. This maneuver is difficult to perform, however, and it is not how we excite a string of a musical instrument. If the string is distorted such that its distorted shape is not that of just one harmonic, the resulting vibration includes various harmonics. Such a distortion occurs in musical instruments when the string is plucked (as in a guitar), bowed (as in a cello), or struck (as in a piano). When the string is distorted into a nonsinusoidal shape, only waves that satisfy the boundary conditions can persist on the string. These are the harmonics.

The frequency of a string that defines the musical note that it plays is that of the fundamental. The frequency of the string can be varied by changing either the tension or the string's length. For example, the tension in guitar and violin strings is varied by a screw adjustment mechanism or by tuning pegs located on the neck of the instrument. As the tension is increased, the frequency of the normal modes increases in accordance with Equation 18.8. Once the instrument is "tuned," players vary the frequency by moving their fingers along the neck, thereby changing the length of the oscillating portion of the string. As the length is shortened, the frequency increases because, as Equation 18.8 specifies, the normal-mode frequencies are inversely proportional to string length.

Quick Quiz 18.5 When a standing wave is set up on a string fixed at both ends, (a) the number of nodes is equal to the number of antinodes (b) the wavelength is equal to the length of the string divided by an integer (c) the frequency is equal to the number of nodes times the fundamental frequency (d) the shape of the string at any time is symmetric about the midpoint of the string.

[1] We exclude $n = 0$ because this value corresponds to the trivial case in which no wave exists ($k = 0$).

Example 18.3 Give Me a C Note!

Middle C on a piano has a fundamental frequency of 262 Hz, and the first A above middle C has a fundamental frequency of 440 Hz.

(A) Calculate the frequencies of the next two harmonics of the C string.

Solution Knowing that the frequencies of higher harmonics are integer multiples of the fundamental frequency $f_1 = 262$ Hz, we find that

$$f_2 = 2f_1 = \boxed{524 \text{ Hz}}$$

$$f_3 = 3f_1 = \boxed{786 \text{ Hz}}$$

(B) If the A and C strings have the same linear mass density μ and length L, determine the ratio of tensions in the two strings.

Solution Using Equation 18.9 for the two strings vibrating at their fundamental frequencies gives

$$f_{1\text{A}} = \frac{1}{2L}\sqrt{\frac{T_\text{A}}{\mu}} \quad \text{and} \quad f_{1\text{C}} = \frac{1}{2L}\sqrt{\frac{T_\text{C}}{\mu}}$$

Setting up the ratio of these frequencies, we find that

$$\frac{f_{1\text{A}}}{f_{1\text{C}}} = \sqrt{\frac{T_\text{A}}{T_\text{C}}}$$

$$\frac{T_\text{A}}{T_\text{C}} = \left(\frac{f_{1\text{A}}}{f_{1\text{C}}}\right)^2 = \left(\frac{440}{262}\right)^2 = \boxed{2.82}$$

What If? What if we look inside a real piano? In this case, the assumption we made in part (B) is only partially true. The string densities are equal, but the length of the A string is only 64 percent of the length of the C string. What is the ratio of their tensions?

Answer Using Equation 18.8 again, we set up the ratio of frequencies:

$$\frac{f_{1\text{A}}}{f_{1\text{C}}} = \frac{L_\text{C}}{L_\text{A}}\sqrt{\frac{T_\text{A}}{T_\text{C}}} = \left(\frac{100}{64}\right)\sqrt{\frac{T_\text{A}}{T_\text{C}}}$$

$$\frac{T_\text{A}}{T_\text{C}} = (0.64)^2\left(\frac{440}{262}\right)^2 = 1.16$$

Example 18.4 Guitar Basics

Interactive

The high E string on a guitar measures 64.0 cm in length and has a fundamental frequency of 330 Hz. By pressing down so that the string is in contact with the first fret (Fig. 18.12), the string is shortened so that it plays an F note that has a frequency of 350 Hz. How far is the fret from the neck end of the string?

Solution Equation 18.7 relates the string's length to the fundamental frequency. With $n = 1$, we can solve for the speed of the wave on the string,

$$v = \frac{2L}{n} f_n = \frac{2(0.640 \text{ m})}{1}(330 \text{ Hz}) = 422 \text{ m/s}$$

Because we have not adjusted the tuning peg, the tension in the string, and hence the wave speed, remain constant. We can again use Equation 18.7, this time solving for L and substituting the new frequency to find the shortened string length:

$$L = n\frac{v}{2f_n} = (1)\frac{422 \text{ m/s}}{2(350 \text{ Hz})} = 0.603 \text{ m} = 60.3 \text{ cm}$$

The difference between this length and the measured length of 64.0 cm is the distance from the fret to the neck end of the string, or $\boxed{3.7 \text{ cm.}}$

Figure 18.12 (Example 18.4) Playing an F note on a guitar.

What If? What if we wish to play an F sharp, which we do by pressing down on the second fret from the neck in Figure 18.12? The frequency of F sharp is 370 Hz. Is this fret another 3.7 cm from the neck?

Answer If you inspect a guitar fingerboard, you will find that the frets are *not* equally spaced. They are far apart near the neck and close together near the opposite end. Consequently, from this observation, we would not expect the F sharp fret to be another 3.7 cm from the end.

Let us repeat the calculation of the string length, this time for the frequency of F sharp:

$$L = n\frac{v}{2f_n} = (1)\frac{422 \text{ m/s}}{2(370 \text{ Hz})} = 0.571 \text{ m}$$

This gives a distance of 0.640 m − 0.571 m = 0.069 m = 6.9 cm from the neck. Subtracting the distance from the neck to the first fret, the separation distance between the first and second frets is 6.9 cm − 3.7 cm = 3.2 cm.

Explore this situation at the Interactive Worked Example link at **http://www.pse6.com.**

Example 18.5 Changing String Vibration with Water Interactive

One end of a horizontal string is attached to a vibrating blade and the other end passes over a pulley as in Figure 18.13a. A sphere of mass 2.00 kg hangs on the end of the string. The string is vibrating in its second harmonic. A container of water is raised under the sphere so that the sphere is completely submerged. After this is done, the string vibrates in its fifth harmonic, as shown in Figure 18.13b. What is the radius of the sphere?

Solution To conceptualize the problem, imagine what happens when the sphere is immersed in the water. The buoyant force acts upward on the sphere, reducing the tension in the string. The change in tension causes a change in the speed of waves on the string, which in turn causes a change in the wavelength. This altered wavelength results in the string vibrating in its fifth normal mode rather than the second. We categorize the problem as one in which we will need to combine our understanding of Newton's second law, buoyant forces, and standing waves on strings. We begin to analyze the problem by studying Figure 18.13a. Newton's second law applied to the sphere tells us that the tension in the string is equal to the weight of the sphere:

$$\Sigma F = T_1 - mg = 0$$

$$T_1 = mg = (2.00 \text{ kg})(9.80 \text{ m/s}^2) = 19.6 \text{ N}$$

where the subscript 1 is used to indicate initial variables before we immerse the sphere in water. Once the sphere is immersed in water, the tension in the string decreases to T_2. Applying Newton's second law to the sphere again in this situation, we have

$$T_2 + B - mg = 0$$

$$(1) \qquad B = mg - T_2$$

The desired quantity, the radius of the sphere, will appear in the expression for the buoyant force B. Before proceeding in this direction, however, we must evaluate T_2. We do this from the standing wave information. We write the equation for the frequency of a standing wave on a string (Equation 18.8) twice, once before we immerse the sphere and once after, and divide the equations:

$$\left.\begin{aligned} f &= \frac{n_1}{2L}\sqrt{\frac{T_1}{\mu}} \\ f &= \frac{n_2}{2L}\sqrt{\frac{T_2}{\mu}} \end{aligned}\right. \longrightarrow \quad 1 = \frac{n_1}{n_2}\sqrt{\frac{T_1}{T_2}}$$

where the frequency f is the same in both cases, because it is determined by the vibrating blade. In addition, the linear mass density μ and the length L of the vibrating portion of the string are the same in both cases. Solving for T_2, we have

$$T_2 = \left(\frac{n_1}{n_2}\right)^2 T_1 = \left(\frac{2}{5}\right)^2 (19.6 \text{ N}) = 3.14 \text{ N}$$

Substituting this into Equation (1), we can evaluate the buoyant force on the sphere:

$$B = mg - T_2 = 19.6 \text{ N} - 3.14 \text{ N} = 16.5 \text{ N}$$

Finally, expressing the buoyant force (Eq. 14.5) in terms of the radius of the sphere, we solve for the radius:

$$B = \rho_{\text{water}} g V_{\text{sphere}} = \rho_{\text{water}} g \left(\tfrac{4}{3}\pi r^3\right)$$

$$r = \sqrt[3]{\frac{3B}{4\pi\rho_{\text{water}} g}} = \sqrt[3]{\frac{3(16.5 \text{ N})}{4\pi(1\,000 \text{ kg/m}^3)(9.80 \text{ m/s}^2)}}$$

$$= 7.38 \times 10^{-2} \text{ m} = \boxed{7.38 \text{ cm}}$$

To finalize this problem, note that only certain radii of the sphere will result in the string vibrating in a normal mode. This is because the speed of waves on the string must be changed to a value such that the length of the string is an integer multiple of half wavelengths. This is a feature of the *quantization* that we introduced earlier in this chapter—the sphere radii that cause the string to vibrate in a normal mode are *quantized*.

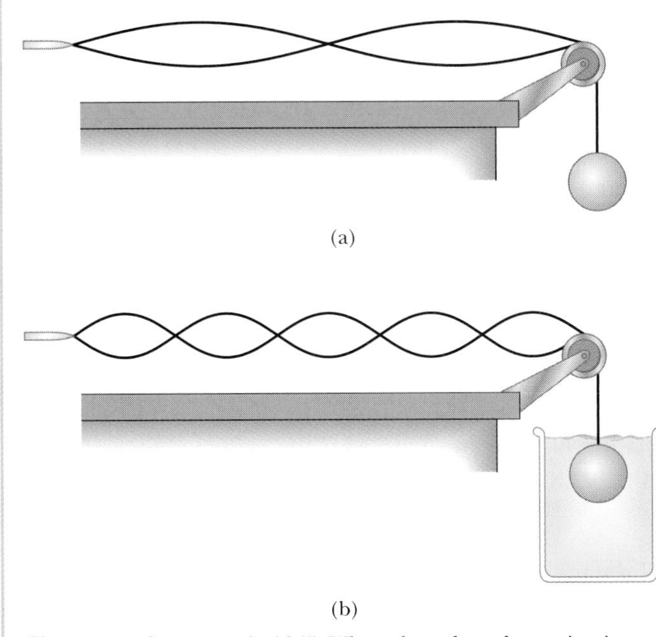

Figure 18.13 (Example 18.5) When the sphere hangs in air, the string vibrates in its second harmonic. When the sphere is immersed in water, the string vibrates in its fifth harmonic.

You can adjust the mass at the Interactive Worked Example link at **http://www.pse6.com.**

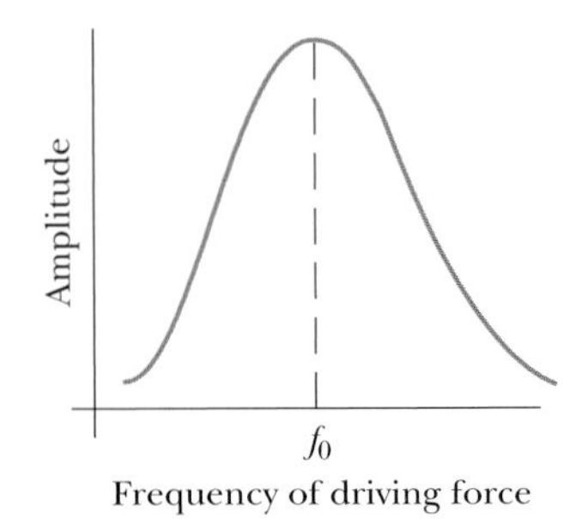

Figure 18.14 Graph of the amplitude (response) versus driving frequency for an oscillating system. The amplitude is a maximum at the resonance frequency f_0.

18.4 Resonance

We have seen that a system such as a taut string is capable of oscillating in one or more normal modes of oscillation. **If a periodic force is applied to such a system, the amplitude of the resulting motion is greatest when the frequency of the applied force is equal to one of the natural frequencies of the system.** We discussed this phenomenon, known as *resonance,* briefly in Section 15.7. Although a block–spring system or a simple pendulum has only one natural frequency, standing-wave systems have a whole set of natural frequencies, such as that given by Equation 18.7 for a string. Because an oscillating system exhibits a large amplitude when driven at any of its natural frequencies, these frequencies are often referred to as **resonance frequencies.**

Figure 18.14 shows the response of an oscillating system to various driving frequencies, where one of the resonance frequencies of the system is denoted by f_0. Note that the amplitude of oscillation of the system is greatest when the frequency of the driving force equals the resonance frequency. The maximum amplitude is limited by friction in the system. If a driving force does work on an oscillating system that is initially at rest, the input energy is used both to increase the amplitude of the oscillation and to overcome the friction force. Once maximum amplitude is reached, the work done by the driving force is used only to compensate for mechanical energy loss due to friction.

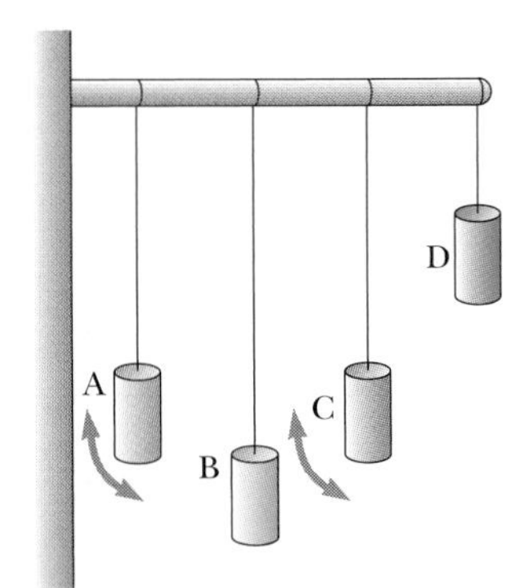

Figure 18.15 An example of resonance. If pendulum A is set into oscillation, only pendulum C, whose length matches that of A, eventually oscillates with large amplitude, or resonates. The arrows indicate motion in a plane perpendicular to the page.

Examples of Resonance

A playground swing is a pendulum having a natural frequency that depends on its length. Whenever we use a series of regular impulses to push a child in a swing, the swing goes higher if the frequency of the periodic force equals the natural frequency of the swing. We can demonstrate a similar effect by suspending pendulums of different lengths from a horizontal support, as shown in Figure 18.15. If pendulum A is set into oscillation, the other pendulums begin to oscillate as a result of waves transmitted along the beam. However, pendulum C, the length of which is close to the length of A, oscillates with a much greater amplitude than pendulums B and D, the lengths of which are much different from that of pendulum A. Pendulum C moves the way it does because its natural frequency is nearly the same as the driving frequency associated with pendulum A.

Next, consider a taut string fixed at one end and connected at the opposite end to an oscillating blade, as illustrated in Figure 18.16. The fixed end is a node, and the end connected to the blade is very nearly a node because the amplitude of the blade's motion is small compared with that of the elements of the string. As the blade oscillates, transverse waves sent down the string are reflected from the fixed end. As we learned in Section 18.3, the string has natural frequencies that are determined by its length, tension, and linear mass density (see Eq. 18.8). When the frequency of the blade equals one of the natural frequencies of the string, standing waves are produced and the string oscillates with a large amplitude. In this resonance case, the wave generated by the oscillating blade is in phase with the reflected wave, and the string absorbs energy from the blade. If the string is driven at a frequency that is not one of its natural frequencies, then the oscillations are of low amplitude and exhibit no stable pattern.

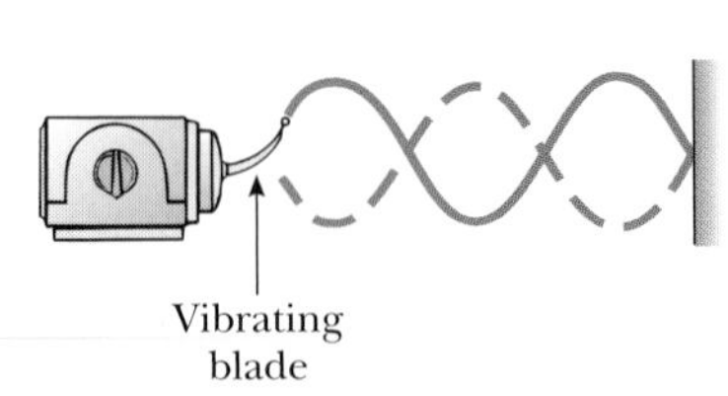

Figure 18.16 Standing waves are set up in a string when one end is connected to a vibrating blade. When the blade vibrates at one of the natural frequencies of the string, large-amplitude standing waves are created.

Once the amplitude of the standing-wave oscillations is a maximum, the mechanical energy delivered by the blade and absorbed by the system is transformed to internal energy because of the damping forces caused by friction in the system. If the applied frequency differs from one of the natural frequencies, energy is transferred to the string at first, but later the phase of the wave becomes such that it forces the blade to receive energy from the string, thereby reducing the energy in the string.

Resonance is very important in the excitation of musical instruments based on air columns. We shall discuss this application of resonance in Section 18.5.

Quick Quiz 18.6 A wine glass can be shattered through resonance by maintaining a certain frequency of a high-intensity sound wave. Figure 18.17a shows a side view of a wine glass vibrating in response to such a sound wave. Sketch the standing-wave pattern in the rim of the glass as seen from above. If an integral number of waves "fit" around the circumference of the vibrating rim, how many wavelengths fit around the rim in Figure 18.17a?

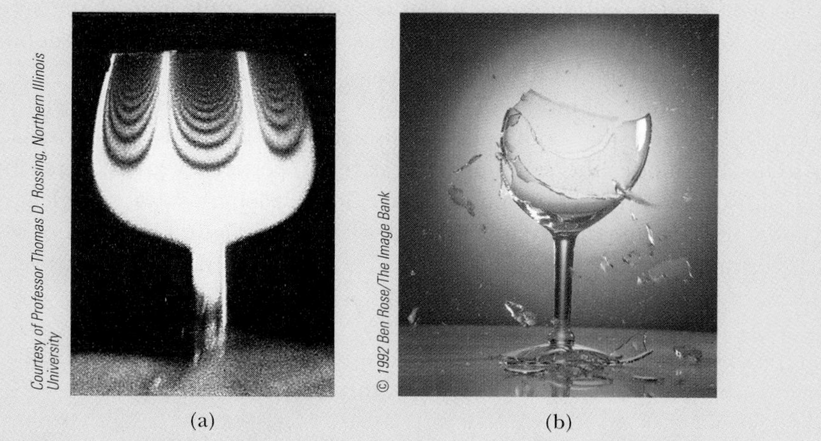

Figure 18.17 (Quick Quiz 18.6) (a) Standing-wave pattern in a vibrating wine glass. The glass shatters if the amplitude of vibration becomes too great. (b) A wine glass shattered by the amplified sound of a human voice.

18.5 Standing Waves in Air Columns

Standing waves can be set up in a tube of air, such as that inside an organ pipe, as the result of interference between longitudinal sound waves traveling in opposite directions. The phase relationship between the incident wave and the wave reflected from one end of the pipe depends on whether that end is open or closed. This relationship is analogous to the phase relationships between incident and reflected transverse waves at the end of a string when the end is either fixed or free to move (see Figs. 16.14 and 16.15).

In a pipe closed at one end, **the closed end is a displacement node because the wall at this end does not allow longitudinal motion of the air.** As a result, at a closed end of a pipe, the reflected sound wave is 180° out of phase with the incident wave. Furthermore, because the pressure wave is 90° out of phase with the displacement wave (see Section 17.2), **the closed end of an air column corresponds to a pressure antinode** (that is, a point of maximum pressure variation).

The open end of an air column is approximately a displacement antinode[2] **and a pressure node.** We can understand why no pressure variation occurs at an open end by noting that the end of the air column is open to the atmosphere; thus, the pressure at this end must remain constant at atmospheric pressure.

You may wonder how a sound wave can reflect from an open end, as there may not appear to be a change in the medium at this point. It is indeed true that the medium

[2] Strictly speaking, the open end of an air column is not exactly a displacement antinode. A compression reaching an open end does not reflect until it passes beyond the end. For a tube of circular cross section, an end correction equal to approximately $0.6R$, where R is the tube's radius, must be added to the length of the air column. Hence, the effective length of the air column is longer than the true length L. We ignore this end correction in this discussion.

through which the sound wave moves is air both inside and outside the pipe. However, sound is a pressure wave, and a compression region of the sound wave is constrained by the sides of the pipe as long as the region is inside the pipe. As the compression region exits at the open end of the pipe, the constraint of the pipe is removed and the compressed air is free to expand into the atmosphere. Thus, there is a change in the *character* of the medium between the inside of the pipe and the outside even though there is no change in the *material* of the medium. This change in character is sufficient to allow some reflection.

With the boundary conditions of nodes or antinodes at the ends of the air column, we have a set of normal modes of oscillation, as we do for the string fixed at both ends. Thus, the air column has quantized frequencies.

The first three normal modes of oscillation of a pipe open at both ends are shown in Figure 18.18a. Note that both ends are displacement antinodes (approximately). In the first normal mode, the standing wave extends between two adjacent antinodes, which is a distance of half a wavelength. Thus, the wavelength is twice the length of the pipe, and the fundamental frequency is $f_1 = v/2L$. As Figure 18.18a shows, the frequencies of the higher harmonics are $2f_1, 3f_1, \ldots$. Thus, we can say that

> In a pipe open at both ends, the natural frequencies of oscillation form a harmonic series that includes all integral multiples of the fundamental frequency.

18.3 Sound Waves in Air Are Longitudinal, not Transverse

Note that the standing longitudinal waves are drawn as transverse waves in Figure 18.18. This is because it is difficult to draw longitudinal displacements—they are in the same direction as the propagation. Thus, it is best to interpret the curves in Figure 18.18 as a graphical representation of the waves (our diagrams of string waves are pictorial representations), with the vertical axis representing horizontal displacement of the elements of the medium.

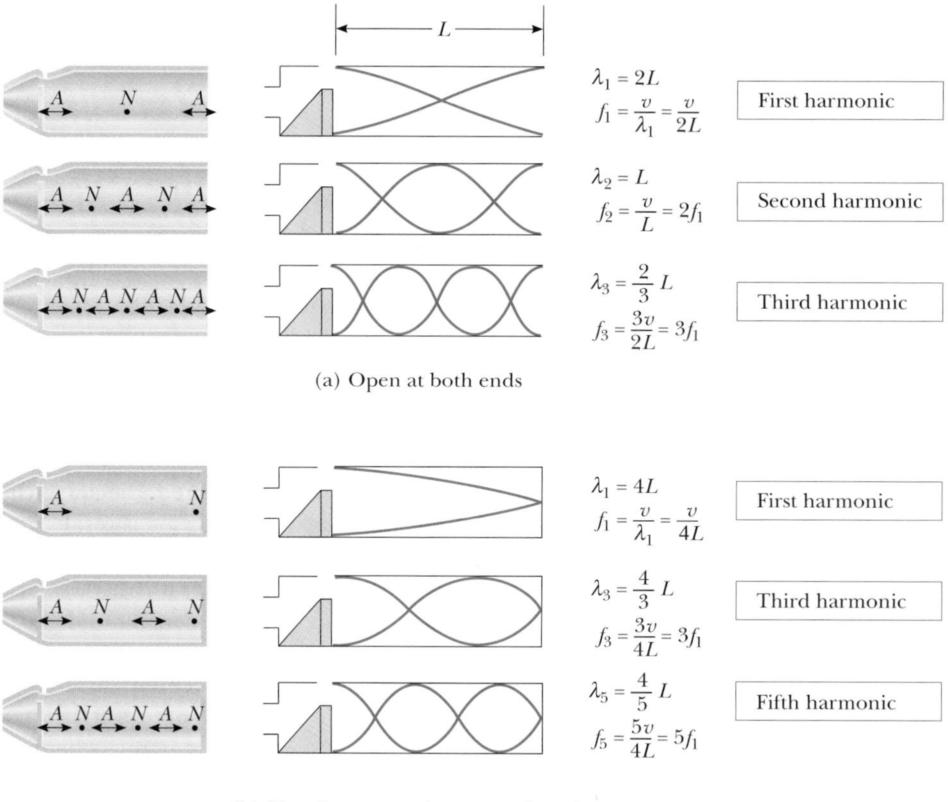

(a) Open at both ends

(b) Closed at one end, open at the other

Figure 18.18 Motion of elements of air in standing longitudinal waves in a pipe, along with schematic representations of the waves. In the schematic representations, the structure at the left end has the purpose of exciting the air column into a normal mode. The hole in the upper edge of the column assures that the left end acts as an open end. The graphs represent the displacement amplitudes, not the pressure amplitudes. (a) In a pipe open at both ends, the harmonic series created consists of all integer multiples of the fundamental frequency: $f_1, 2f_1, 3f_1, \ldots$. (b) In a pipe closed at one end and open at the other, the harmonic series created consists of only odd-integer multiples of the fundamental frequency: $f_1, 3f_1, 5f_1, \ldots$.

Because all harmonics are present, and because the fundamental frequency is given by the same expression as that for a string (see Eq. 18.7), we can express the natural frequencies of oscillation as

$$f_n = n\frac{v}{2L} \qquad n = 1, 2, 3, \ldots \tag{18.11}$$

Natural frequencies of a pipe open at both ends

Despite the similarity between Equations 18.7 and 18.11, you must remember that v in Equation 18.7 is the speed of waves on the string, whereas v in Equation 18.11 is the speed of sound in air.

If a pipe is closed at one end and open at the other, the closed end is a displacement node (see Fig. 18.18b). In this case, the standing wave for the fundamental mode extends from an antinode to the adjacent node, which is one fourth of a wavelength. Hence, the wavelength for the first normal mode is $4L$, and the fundamental frequency is $f_1 = v/4L$. As Figure 18.18b shows, the higher-frequency waves that satisfy our conditions are those that have a node at the closed end and an antinode at the open end; this means that the higher harmonics have frequencies $3f_1, 5f_1, \ldots$.

In a pipe closed at one end, the natural frequencies of oscillation form a harmonic series that includes only odd integral multiples of the fundamental frequency.

We express this result mathematically as

$$f_n = n\frac{v}{4L} \qquad n = 1, 3, 5, \ldots \tag{18.12}$$

Natural frequencies of a pipe closed at one end and open at the other

It is interesting to investigate what happens to the frequencies of instruments based on air columns and strings during a concert as the temperature rises. The sound emitted by a flute, for example, becomes sharp (increases in frequency) as it warms up because the speed of sound increases in the increasingly warmer air inside the flute (consider Eq. 18.11). The sound produced by a violin becomes flat (decreases in frequency) as the strings thermally expand because the expansion causes their tension to decrease (see Eq. 18.8).

Musical instruments based on air columns are generally excited by resonance. The air column is presented with a sound wave that is rich in many frequencies. The air column then responds with a large-amplitude oscillation to the frequencies that match the quantized frequencies in its set of harmonics. In many woodwind instruments, the initial rich sound is provided by a vibrating reed. In the brasses, this excitation is provided by the sound coming from the vibration of the player's lips. In a flute, the initial excitation comes from blowing over an edge at the mouthpiece of the instrument. This is similar to blowing across the opening of a bottle with a narrow neck. The sound of the air rushing across the edge has many frequencies, including one that sets the air cavity in the bottle into resonance.

Quick Quiz 18.7 A pipe open at both ends resonates at a fundamental frequency f_{open}. When one end is covered and the pipe is again made to resonate, the fundamental frequency is f_{closed}. Which of the following expressions describes how these two resonant frequencies compare? (a) $f_{\text{closed}} = f_{\text{open}}$ (b) $f_{\text{closed}} = \frac{1}{2}f_{\text{open}}$ (c) $f_{\text{closed}} = 2f_{\text{open}}$ (d) $f_{\text{closed}} = \frac{3}{2}f_{\text{open}}$

Quick Quiz 18.8 Balboa Park in San Diego has an outdoor organ. When the air temperature increases, the fundamental frequency of one of the organ pipes (a) stays the same (b) goes down (c) goes up (d) is impossible to determine.

Example 18.6 Wind in a Culvert

A section of drainage culvert 1.23 m in length makes a howling noise when the wind blows.

(A) Determine the frequencies of the first three harmonics of the culvert if it is cylindrical in shape and open at both ends. Take $v = 343$ m/s as the speed of sound in air.

Solution The frequency of the first harmonic of a pipe open at both ends is

$$f_1 = \frac{v}{2L} = \frac{343 \text{ m/s}}{2(1.23 \text{ m})} = \boxed{139 \text{ Hz}}$$

Because both ends are open, all harmonics are present; thus,

$$f_2 = 2f_1 = \boxed{278 \text{ Hz}} \quad \text{and} \quad f_3 = 3f_1 = \boxed{417 \text{ Hz}}$$

(B) What are the three lowest natural frequencies of the culvert if it is blocked at one end?

Solution The fundamental frequency of a pipe closed at one end is

$$f_1 = \frac{v}{4L} = \frac{343 \text{ m/s}}{4(1.23 \text{ m})} = \boxed{69.7 \text{ Hz}}$$

In this case, only odd harmonics are present; hence, the next two harmonics have frequencies $f_3 = 3f_1 = \boxed{209 \text{ Hz}}$ and $f_5 = 5f_1 = \boxed{349 \text{ Hz.}}$

(C) For the culvert open at both ends, how many of the harmonics present fall within the normal human hearing range (20 to 20 000 Hz)?

Solution Because all harmonics are present for a pipe open at both ends, we can express the frequency of the highest harmonic heard as $f_n = nf_1$ where n is the number of harmonics that we can hear. For $f_n = 20\ 000$ Hz, we find that the number of harmonics present in the audible range is

$$n = \frac{20\ 000 \text{ Hz}}{139 \text{ Hz}} = \boxed{143}$$

Only the first few harmonics are of sufficient amplitude to be heard.

Example 18.7 Measuring the Frequency of a Tuning Fork

A simple apparatus for demonstrating resonance in an air column is depicted in Figure 18.19. A vertical pipe open at both ends is partially submerged in water, and a tuning fork vibrating at an unknown frequency is placed near the top of the pipe. The length L of the air column can be adjusted by moving the pipe vertically. The sound waves generated by the fork are reinforced when L corresponds to one of the resonance frequencies of the pipe.

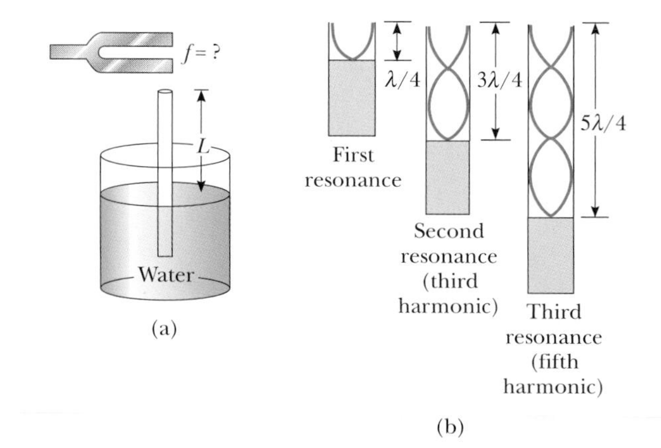

Figure 18.19 (Example 18.7) (a) Apparatus for demonstrating the resonance of sound waves in a pipe closed at one end. The length L of the air column is varied by moving the pipe vertically while it is partially submerged in water. (b) The first three normal modes of the system shown in part (a).

For a certain pipe, the smallest value of L for which a peak occurs in the sound intensity is 9.00 cm. What are

(A) the frequency of the tuning fork

(B) the values of L for the next two resonance frequencies?

Solution

(A) Although the pipe is open at its lower end to allow the water to enter, the water's surface acts like a wall at one end. Therefore, this setup can be modeled as an air column closed at one end, and so the fundamental frequency is given by $f_1 = v/4L$. Taking $v = 343$ m/s for the speed of sound in air and $L = 0.090\ 0$ m, we obtain

$$f_1 = \frac{v}{4L} = \frac{343 \text{ m/s}}{4(0.090\ 0 \text{ m})} = \boxed{953 \text{ Hz}}$$

Because the tuning fork causes the air column to resonate at this frequency, this must also be the frequency of the tuning fork.

(B) Because the pipe is closed at one end, we know from Figure 18.18b that the wavelength of the fundamental mode is $\lambda = 4L = 4(0.090\ 0 \text{ m}) = 0.360$ m. Because the frequency of the tuning fork is constant, the next two normal modes (see Fig. 18.19b) correspond to lengths of

$$L = 3\lambda/4 = \boxed{0.270 \text{ m}} \quad \text{and} \quad L = 5\lambda/4 = \boxed{0.450 \text{ m.}}$$

18.6 Standing Waves in Rods and Membranes

Standing waves can also be set up in rods and membranes. A rod clamped in the middle and stroked parallel to the rod at one end oscillates, as depicted in Figure 18.20a. The oscillations of the elements of the rod are longitudinal, and so the broken lines in Figure 18.20 represent *longitudinal* displacements of various parts of the rod. For clarity, we have drawn them in the transverse direction, just as we did for air columns. The midpoint is a displacement node because it is fixed by the clamp, whereas the ends are displacement antinodes because they are free to oscillate. The oscillations in this setup are analogous to those in a pipe open at both ends. The broken lines in Figure 18.20a represent the first normal mode, for which the wavelength is $2L$ and the frequency is $f = v/2L$, where v is the speed of longitudinal waves in the rod. Other normal modes may be excited by clamping the rod at different points. For example, the second normal mode (Fig. 18.20b) is excited by clamping the rod a distance $L/4$ away from one end.

Musical instruments that depend on standing waves in rods include triangles, marimbas, xylophones, glockenspiels, chimes, and vibraphones. Other devices that make sounds from bars include music boxes and wind chimes.

Two-dimensional oscillations can be set up in a flexible membrane stretched over a circular hoop, such as that in a drumhead. As the membrane is struck at some point, waves that arrive at the fixed boundary are reflected many times. The resulting sound is not harmonic because the standing waves have frequencies that are *not* related by integer multiples. Without this relationship, the sound may be more correctly described as *noise* than as music. This is in contrast to the situation in wind and stringed instruments, which produce sounds that we describe as musical.

Some possible normal modes of oscillation for a two-dimensional circular membrane are shown in Figure 18.21. While nodes are *points* in one-dimensional standing

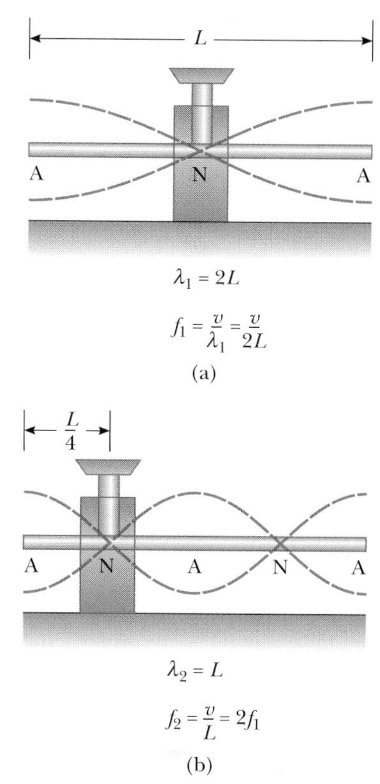

Figure 18.20 Normal-mode longitudinal vibrations of a rod of length L (a) clamped at the middle to produce the first normal mode and (b) clamped at a distance $L/4$ from one end to produce the second normal mode. Note that the broken lines represent oscillations parallel to the rod (longitudinal waves).

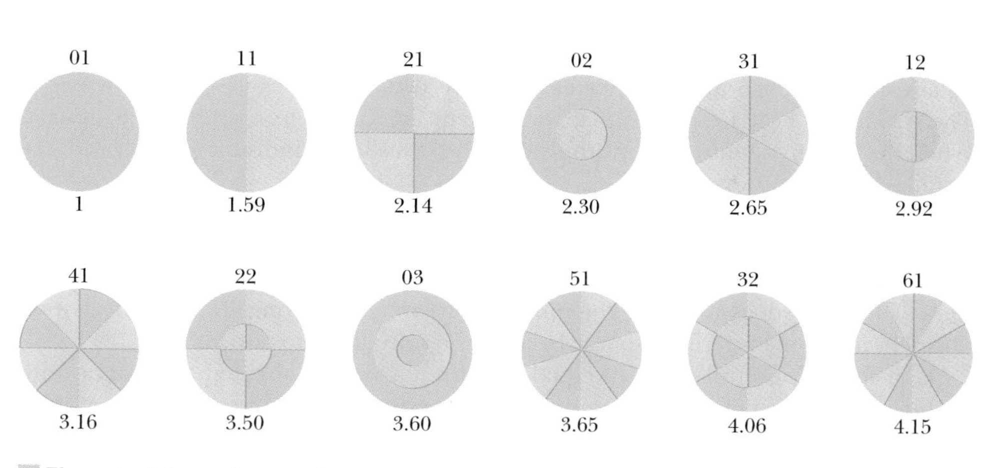

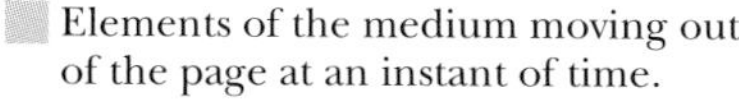

Elements of the medium moving out of the page at an instant of time.
Elements of the medium moving into the page at an instant of time.

Figure 18.21 Representation of some of the normal modes possible in a circular membrane fixed at its perimeter. The pair of numbers above each pattern corresponds to the number of radial nodes and the number of circular nodes. Below each pattern is a factor by which the frequency of the mode is larger than that of the 01 mode. The frequencies of oscillation do not form a harmonic series because these factors are not integers. In each diagram, elements of the membrane on either side of a nodal line move in opposite directions, as indicated by the colors. (*Adapted from T. D. Rossing,* The Science of Sound, 2nd ed, *Reading, Massachusetts, Addison-Wesley Publishing Co., 1990*)

waves on strings and in air columns, a two-dimensional oscillator has *curves* along which there is no displacement of the elements of the medium. The lowest normal mode, which has a frequency f_1, contains only one nodal curve; this curve runs around the outer edge of the membrane. The other possible normal modes show additional nodal curves that are circles and straight lines across the diameter of the membrane.

18.7 Beats: Interference in Time

The interference phenomena with which we have been dealing so far involve the superposition of two or more waves having the same frequency. Because the amplitude of the oscillation of elements of the medium varies with the position in space of the element, we refer to the phenomenon as *spatial interference.* Standing waves in strings and pipes are common examples of spatial interference.

We now consider another type of interference, one that results from the superposition of two waves having slightly *different* frequencies. In this case, when the two waves are observed at the point of superposition, they are periodically in and out of phase. That is, there is a *temporal* (time) alternation between constructive and destructive interference. As a consequence, we refer to this phenomenon as *interference in time* or *temporal interference.* For example, if two tuning forks of slightly different frequencies are struck, one hears a sound of periodically varying amplitude. This phenomenon is called **beating:**

Definition of beating

Beating is the periodic variation in amplitude at a given point due to the superposition of two waves having slightly different frequencies.

The number of amplitude maxima one hears per second, or the *beat frequency,* equals the difference in frequency between the two sources, as we shall show below. The maximum beat frequency that the human ear can detect is about 20 beats/s. When the beat frequency exceeds this value, the beats blend indistinguishably with the sounds producing them.

A piano tuner can use beats to tune a stringed instrument by "beating" a note against a reference tone of known frequency. The tuner can then adjust the string tension until the frequency of the sound it emits equals the frequency of the reference tone. The tuner does this by tightening or loosening the string until the beats produced by it and the reference source become too infrequent to notice.

Consider two sound waves of equal amplitude traveling through a medium with slightly different frequencies f_1 and f_2. We use equations similar to Equation 16.10 to represent the wave functions for these two waves at a point that we choose as $x = 0$:

$$y_1 = A \cos \omega_1 t = A \cos 2\pi f_1 t$$
$$y_2 = A \cos \omega_2 t = A \cos 2\pi f_2 t$$

Using the superposition principle, we find that the resultant wave function at this point is

$$y = y_1 + y_2 = A(\cos 2\pi f_1 t + \cos 2\pi f_2 t)$$

The trigonometric identity

$$\cos a + \cos b = 2 \cos\left(\frac{a - b}{2}\right) \cos\left(\frac{a + b}{2}\right)$$

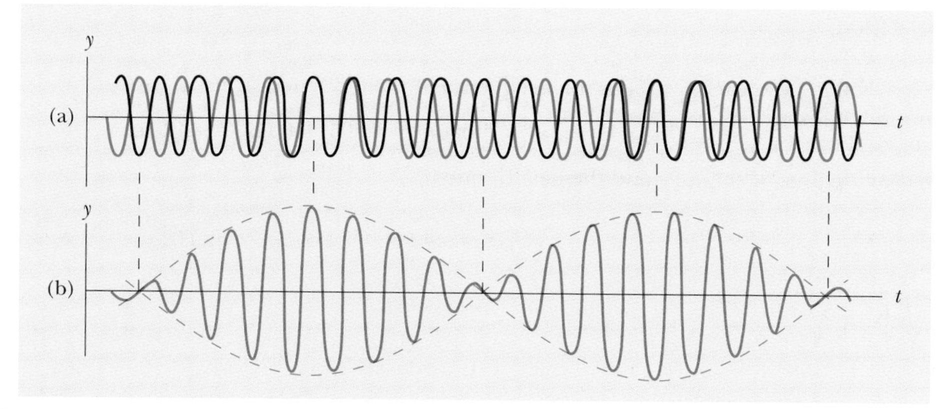

Active Figure 18.22 Beats are formed by the combination of two waves of slightly different frequencies. (a) The individual waves. (b) The combined wave has an amplitude (broken line) that oscillates in time.

At the Active Figures link at http://www.pse6.com, you can choose the two frequencies and see the corresponding beats.

allows us to write the expression for y as

$$y = \left[2A \cos 2\pi \left(\frac{f_1 - f_2}{2}\right) t\right] \cos 2\pi \left(\frac{f_1 + f_2}{2}\right) t \tag{18.13}$$

Resultant of two waves of different frequencies but equal amplitude

Graphs of the individual waves and the resultant wave are shown in Figure 18.22. From the factors in Equation 18.13, we see that the resultant sound for a listener standing at any given point has an effective frequency equal to the average frequency $(f_1 + f_2)/2$ and an amplitude given by the expression in the square brackets:

$$A_{\text{resultant}} = 2A \cos 2\pi \left(\frac{f_1 - f_2}{2}\right) t \tag{18.14}$$

That is, the **amplitude and therefore the intensity of the resultant sound vary in time.** The broken blue line in Figure 18.22b is a graphical representation of Equation 18.14 and is a sine wave varying with frequency $(f_1 - f_2)/2$.

Note that a maximum in the amplitude of the resultant sound wave is detected whenever

$$\cos 2\pi\left(\frac{f_1 - f_2}{2}\right)t = \pm 1$$

This means there are *two* maxima in each period of the resultant wave. Because the amplitude varies with frequency as $(f_1 - f_2)/2$, the number of beats per second, or the beat frequency f_{beat}, is twice this value. That is,

$$f_{\text{beat}} = |f_1 - f_2| \tag{18.15}$$

Beat frequency

For instance, if one tuning fork vibrates at 438 Hz and a second one vibrates at 442 Hz, the resultant sound wave of the combination has a frequency of 440 Hz (the musical note A) and a beat frequency of 4 Hz. A listener would hear a 440-Hz sound wave go through an intensity maximum four times every second.

Quick Quiz 18.9 You are tuning a guitar by comparing the sound of the string with that of a standard tuning fork. You notice a beat frequency of 5 Hz when both sounds are present. You tighten the guitar string and the beat frequency rises to 8 Hz. In order to tune the string exactly to the tuning fork, you should (a) continue to tighten the string (b) loosen the string (c) impossible to determine.

Example 18.8 The Mistuned Piano Strings

Two identical piano strings of length 0.750 m are each tuned exactly to 440 Hz. The tension in one of the strings is then increased by 1.0%. If they are now struck, what is the beat frequency between the fundamentals of the two strings?

Solution We find the ratio of frequencies if the tension in one string is 1.0% larger than the other:

$$\frac{f_2}{f_1} = \frac{(v_2/2L)}{(v_1/2L)} = \frac{v_2}{v_1} = \frac{\sqrt{T_2/\mu}}{\sqrt{T_1/\mu}} = \sqrt{\frac{T_2}{T_1}} = \sqrt{\frac{1.010T_1}{T_1}}$$
$$= 1.005$$

Thus, the frequency of the tightened string is

$$f_2 = 1.005 f_1 = 1.005(440 \text{ Hz}) = 442 \text{ Hz}$$

and the beat frequency is

$$f_{\text{beat}} = 442 \text{ Hz} - 440 \text{ Hz} = \boxed{2 \text{ Hz.}}$$

18.8 Nonsinusoidal Wave Patterns

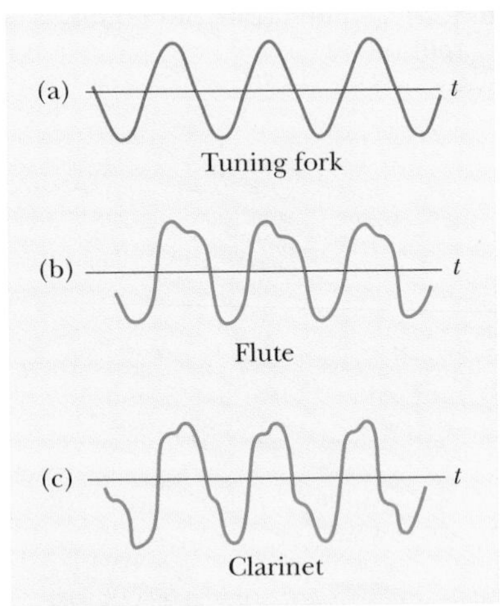

Figure 18.23 Sound wave patterns produced by (a) a tuning fork, (b) a flute, and (c) a clarinet, each at approximately the same frequency. *(Adapted from C. A. Culver,* Musical Acoustics, 4th ed., *New York, McGraw-Hill Book Company, 1956, p. 128.)*

The sound wave patterns produced by the majority of musical instruments are nonsinusoidal. Characteristic patterns produced by a tuning fork, a flute, and a clarinet, each playing the same note, are shown in Figure 18.23. Each instrument has its own characteristic pattern. Note, however, that despite the differences in the patterns, each pattern is periodic. This point is important for our analysis of these waves.

It is relatively easy to distinguish the sounds coming from a violin and a saxophone even when they are both playing the same note. On the other hand, an individual untrained in music may have difficulty distinguishing a note played on a clarinet from the same note played on an oboe. We can use the pattern of the sound waves from various sources to explain these effects.

This is in contrast to a musical instrument that makes a noise, such as the drum, in which the combination of frequencies do not form a harmonic series. When frequencies that are integer multiples of a fundamental frequency are combined, the result is a *musical* sound. A listener can assign a pitch to the sound, based on the fundamental frequency. Pitch is a psychological reaction to a sound that allows the listener to place the sound on a scale of low to high (bass to treble). Combinations of frequencies that are not integer multiples of a fundamental result in a *noise*, rather than a musical sound. It is much harder for a listener to assign a pitch to a noise than to a musical sound.

The wave patterns produced by a musical instrument are the result of the superposition of various harmonics. This superposition results in the corresponding richness of musical tones. The human perceptive response associated with various mixtures of harmonics is the *quality* or *timbre* of the sound. For instance, the sound of the trumpet is perceived to have a "brassy" quality (that is, we have learned to associate the adjective *brassy* with that sound); this quality enables us to distinguish the sound of the trumpet from that of the saxophone, whose quality is perceived as "reedy." The clarinet and oboe, however, both contain air columns excited by reeds; because of this similarity, it is more difficult for the ear to distinguish them on the basis of their sound quality.

The problem of analyzing nonsinusoidal wave patterns appears at first sight to be a formidable task. However, if the wave pattern is periodic, it can be represented as closely as desired by the combination of a sufficiently large number of sinusoidal waves that form a harmonic series. In fact, we can represent any periodic function as a series of sine and cosine terms by using a mathematical technique based on **Fourier's theorem.**[3] The corresponding sum of terms that represents the periodic wave pattern

[3] Developed by Jean Baptiste Joseph Fourier (1786–1830).

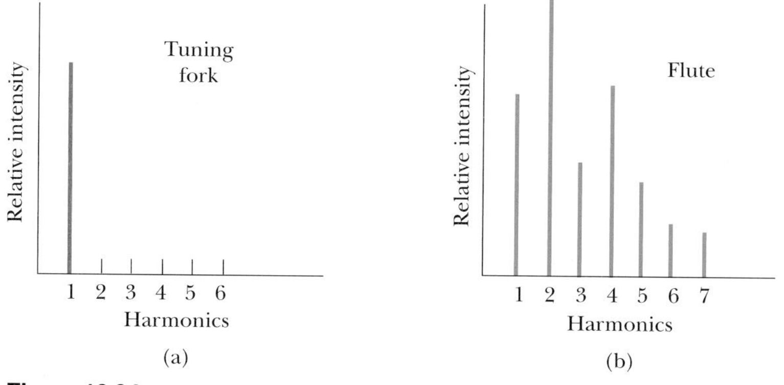

Figure 18.24 Harmonics of the wave patterns shown in Figure 18.23. Note the variations in intensity of the various harmonics. (*Adapted from C. A. Culver,* Musical Acoustics, 4th ed., *New York, McGraw-Hill Book Company, 1956.*)

is called a **Fourier series.** Let $y(t)$ be any function that is periodic in time with period T, such that $y(t + T) = y(t)$. Fourier's theorem states that this function can be written as

$$y(t) = \sum_n (A_n \sin 2\pi f_n t + B_n \cos 2\pi f_n t) \tag{18.16}$$

Fourier's theorem

where the lowest frequency is $f_1 = 1/T$. The higher frequencies are integer multiples of the fundamental, $f_n = nf_1$, and the coefficients A_n and B_n represent the amplitudes of the various waves. Figure 18.24 represents a harmonic analysis of the wave patterns shown in Figure 18.23. Note that a struck tuning fork produces only one harmonic (the first), whereas the flute and clarinet produce the first harmonic and many higher ones.

Note the variation in relative intensity of the various harmonics for the flute and the clarinet. In general, any musical sound consists of a fundamental frequency f plus other frequencies that are integer multiples of f, all having different intensities.

18.4 Pitch vs. Frequency

Do not confuse the term *pitch* with *frequency*. Frequency is the physical measurement of the number of oscillations per second. Pitch is a psychological reaction to sound that enables a person to place the sound on a scale from high to low, or from treble to bass. Thus, frequency is the stimulus and pitch is the response. Although pitch is related mostly (but not completely) to frequency, they are not the same. A phrase such as "the pitch of the sound" is incorrect because pitch is not a physical property of the sound.

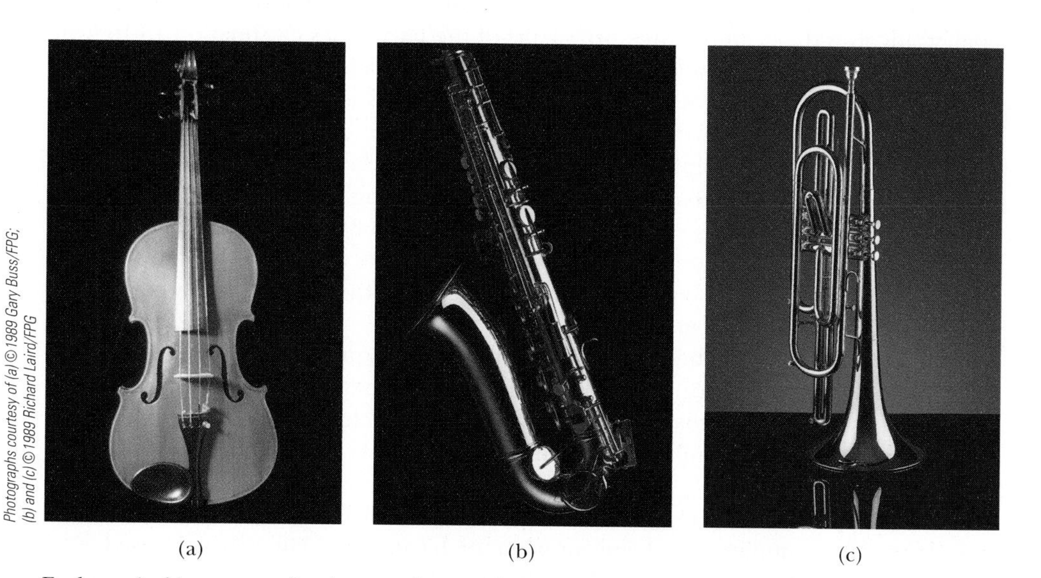

(a) (b) (c)

Each musical instrument has its own characteristic sound and mixture of harmonics. Instruments shown are (a) the violin, (b) the saxophone, and (c) the trumpet.

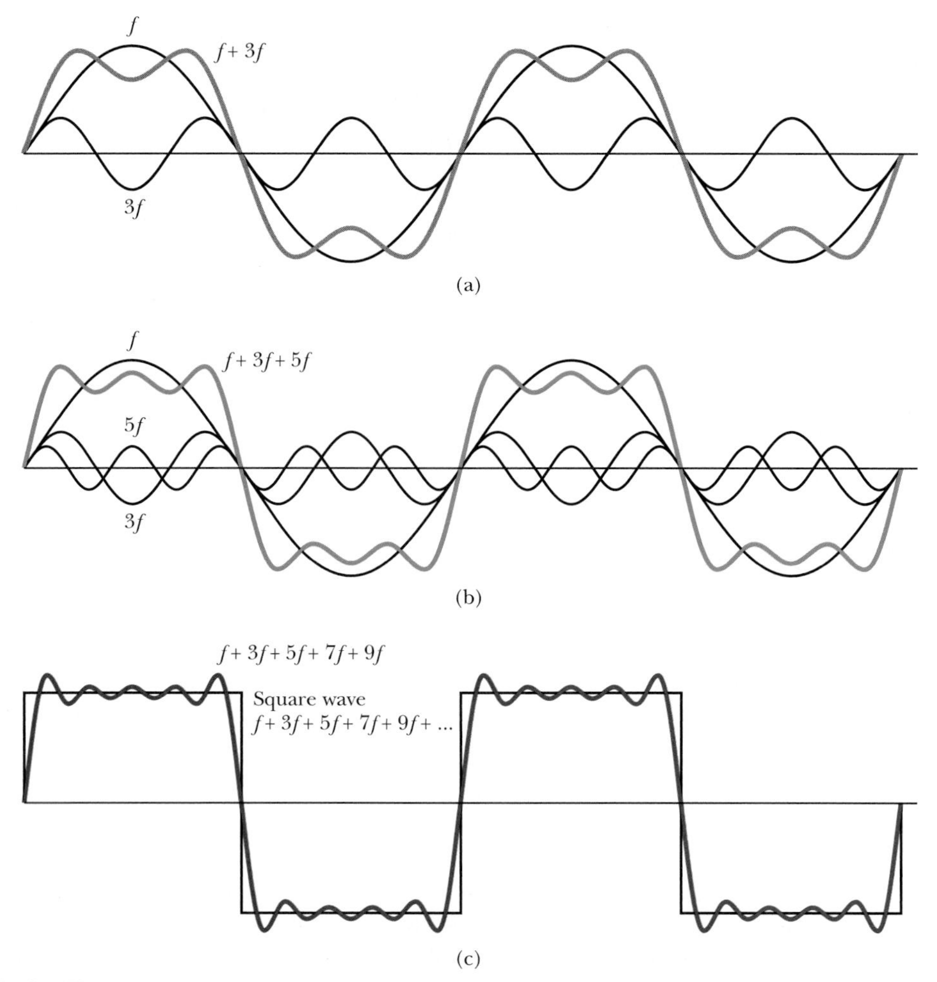

At the Active Figures link at http://www.pse6.com, *you can add in harmonics with frequencies higher than 9f to try to synthesize a square wave.*

Active Figure 18.25 Fourier synthesis of a square wave, which is represented by the sum of odd multiples of the first harmonic, which has frequency f. (a) Waves of frequency f and $3f$ are added. (b) One more odd harmonic of frequency $5f$ is added. (c) The synthesis curve approaches closer to the square wave when odd frequencies up to $9f$ are added.

We have discussed the *analysis* of a wave pattern using Fourier's theorem. The analysis involves determining the coefficients of the harmonics in Equation 18.16 from a knowledge of the wave pattern. The reverse process, called *Fourier synthesis,* can also be performed. In this process, the various harmonics are added together to form a resultant wave pattern. As an example of Fourier synthesis, consider the building of a square wave, as shown in Figure 18.25. The symmetry of the square wave results in only odd multiples of the fundamental frequency combining in its synthesis. In Figure 18.25a, the orange curve shows the combination of f and $3f$. In Figure 18.25b, we have added $5f$ to the combination and obtained the green curve. Notice how the general shape of the square wave is approximated, even though the upper and lower portions are not flat as they should be.

Figure 18.25c shows the result of adding odd frequencies up to $9f$. This approximation (purple curve) to the square wave is better than the approximations in parts a and b. To approximate the square wave as closely as possible, we would need to add all odd multiples of the fundamental frequency, up to infinite frequency.

Using modern technology, we can generate musical sounds electronically by mixing different amplitudes of any number of harmonics. These widely used electronic music synthesizers are capable of producing an infinite variety of musical tones.

SUMMARY

Take a practice test for this chapter by clicking on the Practice Test link at http://www.pse6.com.

The **superposition principle** specifies that when two or more waves move through a medium, the value of the resultant wave function equals the algebraic sum of the values of the individual wave functions.

When two traveling waves having equal amplitudes and frequencies superimpose, the resultant wave has an amplitude that depends on the phase angle ϕ between the two waves. **Constructive interference** occurs when the two waves are in phase, corresponding to $\phi = 0, 2\pi, 4\pi, \ldots$ rad. **Destructive interference** occurs when the two waves are 180° out of phase, corresponding to $\phi = \pi, 3\pi, 5\pi, \ldots$ rad.

Standing waves are formed from the superposition of two sinusoidal waves having the same frequency, amplitude, and wavelength but traveling in opposite directions. The resultant standing wave is described by the wave function

$$y = (2A \sin kx) \cos \omega t \qquad (18.3)$$

Hence, the amplitude of the standing wave is $2A$, and the amplitude of the simple harmonic motion of any particle of the medium varies according to its position as $2A \sin kx$. The points of zero amplitude (called **nodes**) occur at $x = n\lambda/2$ ($n = 0, 1, 2, 3, \ldots$). The maximum amplitude points (called **antinodes**) occur at $x = n\lambda/4$ ($n = 1, 3, 5, \ldots$). Adjacent antinodes are separated by a distance $\lambda/2$. Adjacent nodes also are separated by a distance $\lambda/2$.

The natural frequencies of vibration of a taut string of length L and fixed at both ends are quantized and are given by

$$f_n = \frac{n}{2L}\sqrt{\frac{T}{\mu}} \qquad n = 1, 2, 3, \ldots \qquad (18.8)$$

where T is the tension in the string and μ is its linear mass density. The natural frequencies of vibration $f_1, 2f_1, 3f_1, \ldots$ form a **harmonic series.**

An oscillating system is in **resonance** with some driving force whenever the frequency of the driving force matches one of the natural frequencies of the system. When the system is resonating, it responds by oscillating with a relatively large amplitude.

Standing waves can be produced in a column of air inside a pipe. If the pipe is open at both ends, all harmonics are present and the natural frequencies of oscillation are

$$f_n = n\frac{v}{2L} \qquad n = 1, 2, 3, \ldots \qquad (18.11)$$

If the pipe is open at one end and closed at the other, only the odd harmonics are present, and the natural frequencies of oscillation are

$$f_n = n\frac{v}{4L} \qquad n = 1, 3, 5, \ldots \qquad (18.12)$$

The phenomenon of **beating** is the periodic variation in intensity at a given point due to the superposition of two waves having slightly different frequencies.

QUESTIONS

1. Does the phenomenon of wave interference apply only to sinusoidal waves?

2. As oppositely moving pulses of the same shape (one upward, one downward) on a string pass through each other, there is one instant at which the string shows no displacement from the equilibrium position at any point. Has the energy carried by the pulses disappeared at this instant of time? If not, where is it?

3. Can two pulses traveling in opposite directions on the same string reflect from each other? Explain.

4. When two waves interfere, can the amplitude of the resultant wave be greater than either of the two original waves? Under what conditions?

5. For certain positions of the movable section shown in Figure 18.5, no sound is detected at the receiver—a situation corresponding to destructive interference. This suggests that energy is somehow lost. What happens to the energy transmitted by the speaker?

6. When two waves interfere constructively or destructively, is there any gain or loss in energy? Explain.

7. A standing wave is set up on a string, as shown in Figure 18.10. Explain why no energy is transmitted along the string.

8. What limits the amplitude of motion of a real vibrating system that is driven at one of its resonant frequencies?

9. Explain why your voice seems to sound better than usual when you sing in the shower.

10. What is the purpose of the slide on a trombone or of the valves on a trumpet?

11. Explain why all harmonics are present in an organ pipe open at both ends, but only odd harmonics are present in a pipe closed at one end.

12. Explain how a musical instrument such as a piano may be tuned by using the phenomenon of beats.

13. To keep animals away from their cars, some people mount short, thin pipes on the fenders. The pipes give out a high-pitched wail when the cars are moving. How do they create the sound?

14. When a bell is rung, standing waves are set up around the bell's circumference. What boundary conditions must be satisfied by the resonant wavelengths? How does a crack in the bell, such as in the Liberty Bell, affect the satisfying of the boundary conditions and the sound emanating from the bell?

15. An archer shoots an arrow from a bow. Does the string of the bow exhibit standing waves after the arrow leaves? If so, and if the bow is perfectly symmetric so that the arrow leaves from the center of the string, what harmonics are excited?

16. Despite a reasonably steady hand, a person often spills his coffee when carrying it to his seat. Discuss resonance as a possible cause of this difficulty, and devise a means for solving the problem.

17. An airplane mechanic notices that the sound from a twin-engine aircraft rapidly varies in loudness when both engines are running. What could be causing this variation from loud to soft?

18. When the base of a vibrating tuning fork is placed against a chalkboard, the sound that it emits becomes louder. This is because the vibrations of the tuning fork are transmitted to the chalkboard. Because it has a larger area than the tuning fork, the vibrating chalkboard sets more air into vibration. Thus, the chalkboard is a better radiator of sound than the tuning fork. How does this affect the length of time during which the fork vibrates? Does this agree with the principle of conservation of energy?

19. If you wet your finger and lightly run it around the rim of a fine wineglass, a high-frequency sound is heard. Why? How could you produce various musical notes with a set of wineglasses, each of which contains a different amount of water?

20. If you inhale helium from a balloon and do your best to speak normally, your voice will have a comical quacky quality. Explain why this "Donald Duck effect" happens. *Caution*: Helium is an asphyxiating gas and asphyxiation can cause panic. Helium can contain poisonous contaminants.

21. You have a standard tuning fork whose frequency is 262 Hz and a second tuning fork with an unknown frequency. When you tap both of them on the heel of one of your sneakers, you hear beats with a frequency of 4 per second. Thoughtfully chewing your gum, you wonder whether the unknown frequency is 258 Hz or 266 Hz. How can you decide?

PROBLEMS

1, 2, 3 = straightforward, intermediate, challenging ☐ = full solution available in the *Student Solutions Manual and Study Guide*

= coached solution with hints available at http://www.pse6.com = computer useful in solving problem

= paired numerical and symbolic problems

Section 18.1 Superposition and Interference

1. Two waves in one string are described by the wave functions

$$y_1 = 3.0 \cos(4.0x - 1.6t)$$

and

$$y_2 = 4.0 \sin(5.0x - 2.0t)$$

where y and x are in centimeters and t is in seconds. Find the superposition of the waves $y_1 + y_2$ at the points (a) $x = 1.00$, $t = 1.00$, (b) $x = 1.00$, $t = 0.500$, and (c) $x = 0.500$, $t = 0$. (Remember that the arguments of the trigonometric functions are in radians.)

2. Two pulses A and B are moving in opposite directions along a taut string with a speed of 2.00 cm/s. The amplitude of A is twice the amplitude of B. The pulses are shown in Figure P18.2 at $t = 0$. Sketch the shape of the string at $t = 1$, 1.5, 2, 2.5, and 3 s.

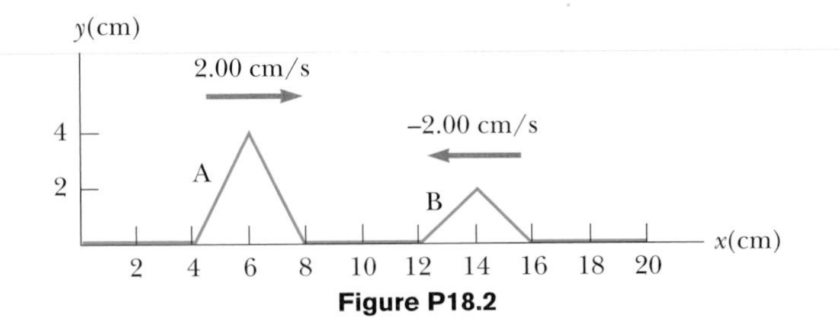

Figure P18.2

3. Two pulses traveling on the same string are described by

$$y_1 = \frac{5}{(3x - 4t)^2 + 2} \quad \text{and} \quad y_2 = \frac{-5}{(3x + 4t - 6)^2 + 2}$$

(a) In which direction does each pulse travel? (b) At what time do the two cancel everywhere? (c) At what point do the two pulses always cancel?

4. Two waves are traveling in the same direction along a stretched string. The waves are 90.0° out of phase. Each wave has an amplitude of 4.00 cm. Find the amplitude of the resultant wave.

5. Two traveling sinusoidal waves are described by the wave functions

$$y_1 = (5.00 \text{ m}) \sin[\pi(4.00x - 1\,200t)]$$

and

$$y_2 = (5.00 \text{ m}) \sin[\pi(4.00x - 1\,200t - 0.250)]$$

where x, y_1, and y_2 are in meters and t is in seconds. (a) What is the amplitude of the resultant wave? (b) What is the frequency of the resultant wave?

6. Two identical sinusoidal waves with wavelengths of 3.00 m travel in the same direction at a speed of 2.00 m/s. The second wave originates from the same point as the first, but at a later time. Determine the minimum possible time interval between the starting moments of the two waves if the amplitude of the resultant wave is the same as that of each of the two initial waves.

7. **Review problem.** A series of pulses, each of amplitude 0.150 m, is sent down a string that is attached to a post at one end. The pulses are reflected at the post and travel back along the string without loss of amplitude. What is the net displacement at a point on the string where two pulses are crossing, (a) if the string is rigidly attached to the post? (b) if the end at which reflection occurs is free to slide up and down?

8. Two loudspeakers are placed on a wall 2.00 m apart. A listener stands 3.00 m from the wall directly in front of one of the speakers. A single oscillator is driving the speakers at a frequency of 300 Hz. (a) What is the phase difference between the two waves when they reach the observer? (b) **What If?** What is the frequency closest to 300 Hz to which the oscillator may be adjusted such that the observer hears minimal sound?

9. Two speakers are driven by the same oscillator whose frequency is 200 Hz. They are located on a vertical pole a distance of 4.00 m from each other. A man walks straight toward the lower speaker in a direction perpendicular to the pole as shown in Figure P18.9. (a) How many times will he hear a minimum in sound intensity, and (b) how far is he from the pole at these moments? Take the speed of sound to be 330 m/s and ignore any sound reflections coming off the ground.

10. Two speakers are driven by the same oscillator whose frequency is f. They are located a distance d from each other on a vertical pole. A man walks straight toward the lower speaker in a direction perpendicular to the pole, as shown in Figure P18.9. (a) How many times will he hear a minimum in sound intensity? (b) How far is he from the pole at these moments? Let v represent the speed of sound, and assume that the ground does not reflect sound.

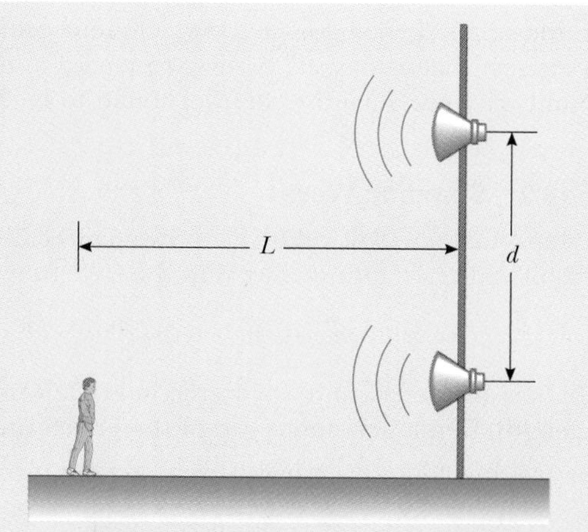

Figure P18.9 Problems 9 and 10.

11. Two sinusoidal waves in a string are defined by the functions

$$y_1 = (2.00 \text{ cm}) \sin(20.0x - 32.0t)$$

and

$$y_2 = (2.00 \text{ cm}) \sin(25.0x - 40.0t)$$

where y_1, y_2, and x are in centimeters and t is in seconds. (a) What is the phase difference between these two waves at the point $x = 5.00$ cm at $t = 2.00$ s? (b) What is the positive x value closest to the origin for which the two phases differ by $\pm\pi$ at $t = 2.00$ s? (This is where the two waves add to zero.)

12. Two identical speakers 10.0 m apart are driven by the same oscillator with a frequency of $f = 21.5$ Hz (Fig. P18.12). (a) Explain why a receiver at point A records a minimum in sound intensity from the two speakers. (b) If the receiver is moved in the plane of the speakers, what path should it take so that the intensity remains at a minimum? That is, deter-

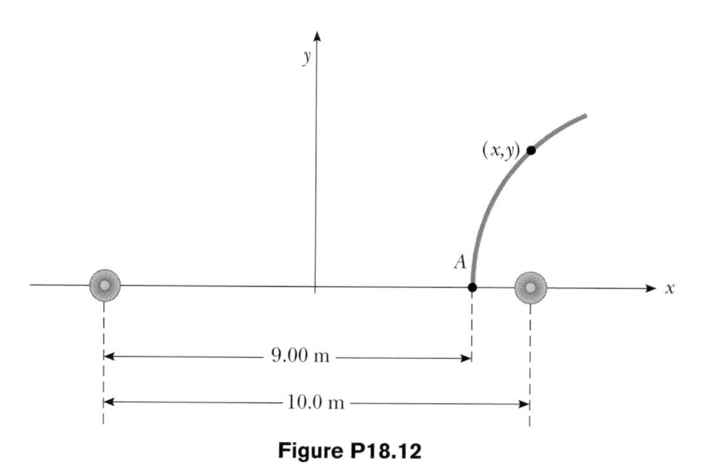

Figure P18.12

mine the relationship between x and y (the coordinates of the receiver) that causes the receiver to record a minimum in sound intensity. Take the speed of sound to be 344 m/s.

Section 18.2 Standing Waves

13. Two sinusoidal waves traveling in opposite directions interfere to produce a standing wave with the wave function

$$y = (1.50 \text{ m}) \sin(0.400x) \cos(200t)$$

where x is in meters and t is in seconds. Determine the wavelength, frequency, and speed of the interfering waves.

14. Two waves in a long string have wave functions given by

$$y_1 = (0.015\,0 \text{ m}) \cos\left(\frac{x}{2} - 40t\right)$$

and

$$y_2 = (0.015\,0 \text{ m}) \cos\left(\frac{x}{2} + 40t\right)$$

where y_1, y_2, and x are in meters and t is in seconds. (a) Determine the positions of the nodes of the resulting standing wave. (b) What is the maximum transverse position of an element of the string at the position $x = 0.400$ m?

15. Two speakers are driven in phase by a common oscillator at 800 Hz and face each other at a distance of 1.25 m. Locate the points along a line joining the two speakers where relative minima of sound pressure amplitude would be expected. (Use $v = 343$ m/s.)

16. Verify by direct substitution that the wave function for a standing wave given in Equation 18.3,

$$y = 2A \sin kx \cos \omega t$$

is a solution of the general linear wave equation, Equation 16.27:

$$\frac{\partial^2 y}{\partial x^2} = \frac{1}{v^2}\frac{\partial^2 y}{\partial t^2}$$

17. Two sinusoidal waves combining in a medium are described by the wave functions

$$y_1 = (3.0 \text{ cm}) \sin \pi(x + 0.60t)$$

and

$$y_2 = (3.0 \text{ cm}) \sin \pi(x - 0.60t)$$

where x is in centimeters and t is in seconds. Determine the *maximum* transverse position of an element of the medium at (a) $x = 0.250$ cm, (b) $x = 0.500$ cm, and (c) $x = 1.50$ cm. (d) Find the three smallest values of x corresponding to antinodes.

18. Two waves that set up a standing wave in a long string are given by the wave functions

$$y_1 = A \sin(kx - \omega t + \phi) \quad \text{and} \quad y_2 = A \sin(kx + \omega t)$$

Show (a) that the addition of the arbitrary phase constant ϕ changes only the position of the nodes and, in particular, (b) that the distance between nodes is still one half the wavelength.

Section 18.3 Standing Waves in a String Fixed at Both Ends

19. Find the fundamental frequency and the next three frequencies that could cause standing-wave patterns on a string that is 30.0 m long, has a mass per length of 9.00×10^{-3} kg/m, and is stretched to a tension of 20.0 N.

20. A string with a mass of 8.00 g and a length of 5.00 m has one end attached to a wall; the other end is draped over a pulley and attached to a hanging object with a mass of 4.00 kg. If the string is plucked, what is the fundamental frequency of vibration?

21. In the arrangement shown in Figure P18.21, an object can be hung from a string (with linear mass density $\mu = 0.002\,00$ kg/m) that passes over a light pulley. The string is connected to a vibrator (of constant frequency f), and the length of the string between point P and the pulley is $L = 2.00$ m. When the mass m of the object is either 16.0 kg or 25.0 kg, standing waves are observed; however, no standing waves are observed with any mass between these values. (a) What is the frequency of the vibrator? (*Note:* The greater the tension in the string, the smaller the number of nodes in the standing wave.) (b) What is the largest object mass for which standing waves could be observed?

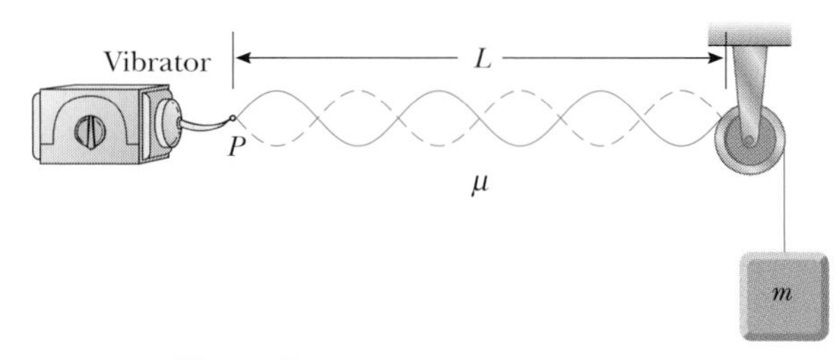

Figure P18.21 Problems 21 and 22.

22. A vibrator, pulley, and hanging object are arranged as in Figure P18.21, with a compound string, consisting of two strings of different masses and lengths fastened together end-to-end. The first string, which has a mass of 1.56 g and a length of 65.8 cm, runs from the vibrator to the junction of the two strings. The second string runs from the junction over the pulley to the suspended 6.93-kg object. The mass and length of the string from the junction to the pulley are, respectively, 6.75 g and 95.0 cm. (a) Find the lowest frequency for which standing waves are observed in both strings, with a node at the junction. The standing wave patterns in the two strings may have different numbers of nodes. (b) What is the total number of nodes observed along the compound string at this frequency, excluding the nodes at the vibrator and the pulley?

23. Example 18.4 tells you that the adjacent notes E, F, and F-sharp can be assigned frequencies of 330 Hz, 350 Hz, and 370 Hz. You might not guess how the pattern continues. The next notes, G, G-sharp, and A, have frequencies of 392 Hz, 416 Hz, and 440 Hz. On the equally tempered or chromatic scale used in Western music, the frequency of each higher note is obtained by multiplying the previous frequency by $\sqrt[12]{2}$. A standard guitar has strings 64.0 cm long and nineteen frets. In Example 18.4, we found the spacings of the first two frets. Calculate the distance between the last two frets.

24. The top string of a guitar has a fundamental frequency of 330 Hz when it is allowed to vibrate as a whole, along all of its 64.0-cm length from the neck to the bridge. A fret is provided for limiting vibration to just the lower two-thirds of the string. (a) If the string is pressed down at this fret and plucked, what is the new fundamental frequency? (b) **What If?** The guitarist can play a "natural harmonic" by gently touching the string at the location of this fret and plucking the string at about one sixth of the way along its length from the bridge. What frequency will be heard then?

25. A string of length L, mass per unit length μ, and tension T is vibrating at its fundamental frequency. What effect will the following have on the fundamental frequency? (a) The length of the string is doubled, with all other factors held constant. (b) The mass per unit length is doubled, with all other factors held constant. (c) The tension is doubled, with all other factors held constant.

26. A 60.000-cm guitar string under a tension of 50.000 N has a mass per unit length of 0.100 00 g/cm. What is the highest resonant frequency that can be heard by a person capable of hearing frequencies up to 20 000 Hz?

27. A cello A-string vibrates in its first normal mode with a frequency of 220 Hz. The vibrating segment is 70.0 cm long and has a mass of 1.20 g. (a) Find the tension in the string. (b) Determine the frequency of vibration when the string vibrates in three segments.

28. A violin string has a length of 0.350 m and is tuned to concert G, with $f_G = 392$ Hz. Where must the violinist place her finger to play concert A, with $f_A = 440$ Hz? If this position is to remain correct to half the width of a finger (that is, to within 0.600 cm), what is the maximum allowable percentage change in the string tension?

29. **Review problem.** A sphere of mass M is supported by a string that passes over a light horizontal rod of length L (Fig. P18.29). Given that the angle is θ and that f represents the fundamental frequency of standing waves in the portion of the string above the rod, determine the mass of this portion of the string.

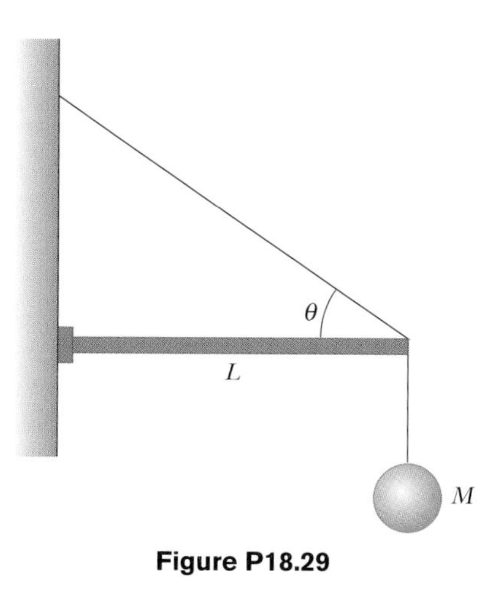

Figure P18.29

30. **Review problem.** A copper cylinder hangs at the bottom of a steel wire of negligible mass. The top end of the wire is fixed. When the wire is struck, it emits sound with a fundamental frequency of 300 Hz. If the copper cylinder is then submerged in water so that half its volume is below the water line, determine the new fundamental frequency.

31. A standing-wave pattern is observed in a thin wire with a length of 3.00 m. The equation of the wave is

$$y = (0.002 \text{ m}) \sin(\pi x) \cos(100\pi t)$$

where x is in meters and t is in seconds. (a) How many loops does this pattern exhibit? (b) What is the fundamental frequency of vibration of the wire? (c) **What If?** If the original frequency is held constant and the tension in the wire is increased by a factor of 9, how many loops are present in the new pattern?

Section 18.4 Resonance

32. The chains suspending a child's swing are 2.00 m long. At what frequency should a big brother push to make the child swing with largest amplitude?

33. An earthquake can produce a *seiche* in a lake, in which the water sloshes back and forth from end to end with remarkably large amplitude and long period. Consider a seiche produced in a rectangular farm pond, as in the cross-sectional view of Figure P18.33. (The figure is not drawn to scale.) Suppose that the pond is 9.15 m long and of uniform width and depth. You measure that a pulse produced at one end reaches the other end in 2.50 s. (a) What is the wave speed? (b) To produce the seiche, several people stand on the bank at one end and paddle together with snow shovels, moving them in simple harmonic motion. What should be the frequency of this motion?

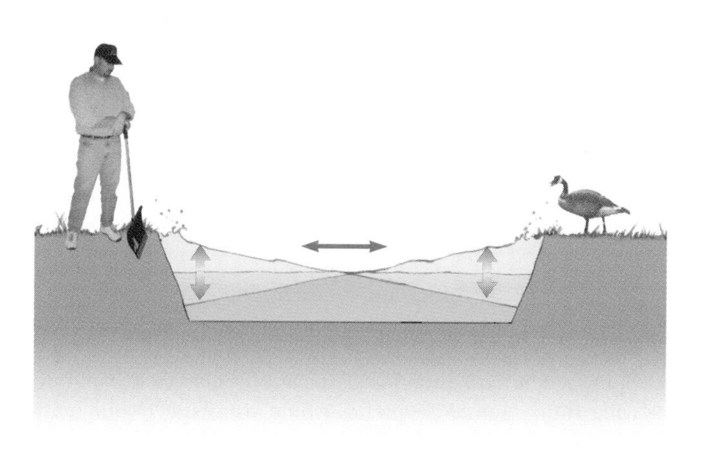

Figure P18.33

34. The Bay of Fundy, Nova Scotia, has the highest tides in the world, as suggested in the photographs on page 452. Assume that in mid-ocean and at the mouth of the bay, the Moon's gravity gradient and the Earth's rotation make the water surface oscillate with an amplitude of a few centimeters and a period of 12 h 24 min. At the head of the bay, the amplitude is several meters. Argue for or against the proposition that the tide is amplified by standing-wave resonance. Assume the bay has a length of 210 km and a uniform depth of 36.1 m. The speed of long-wavelength water waves is given by $\sqrt{gd}$, where d is the water's depth.

35. Standing-wave vibrations are set up in a crystal goblet with four nodes and four antinodes equally spaced around the 20.0-cm circumference of its rim. If transverse waves move around the glass at 900 m/s, an opera singer would have to produce a high harmonic with what frequency to shatter the glass with a resonant vibration?

Section 18.5 Standing Waves in Air Columns

Note: Unless otherwise specified, assume that the speed of sound in air is 343 m/s at 20°C, and is described by

$$v = (331 \text{ m/s})\sqrt{1 + \frac{T_C}{273°}}$$

at any Celsius temperature T_C.

36. The overall length of a piccolo is 32.0 cm. The resonating air column vibrates as in a pipe open at both ends. (a) Find the frequency of the lowest note that a piccolo can play, assuming that the speed of sound in air is 340 m/s. (b) Opening holes in the side effectively shortens the length of the resonant column. If the highest note a piccolo can sound is 4 000 Hz, find the distance between adjacent antinodes for this mode of vibration.

37. Calculate the length of a pipe that has a fundamental frequency of 240 Hz if the pipe is (a) closed at one end and (b) open at both ends.

38. The fundamental frequency of an open organ pipe corresponds to middle C (261.6 Hz on the chromatic musical scale). The third resonance of a closed organ pipe has the same frequency. What are the lengths of the two pipes?

39. The windpipe of one typical whooping crane is 5.00 ft long. What is the fundamental resonant frequency of the bird's trachea, modeled as a narrow pipe closed at one end? Assume a temperature of 37°C.

40. Do not stick anything into your ear! Estimate the length of your ear canal, from its opening at the external ear to the eardrum. If you regard the canal as a narrow tube that is open at one end and closed at the other, at approximately what fundamental frequency would you expect your hearing to be most sensitive? Explain why you can hear especially soft sounds just around this frequency.

41. A shower stall measures 86.0 cm × 86.0 cm × 210 cm. If you were singing in this shower, which frequencies would sound the richest (because of resonance)? Assume that the stall acts as a pipe closed at both ends, with nodes at opposite sides. Assume that the voices of various singers range from 130 Hz to 2 000 Hz. Let the speed of sound in the hot shower stall be 355 m/s.

42. As shown in Figure P18.42, water is pumped into a tall vertical cylinder at a volume flow rate R. The radius of the cylinder is r, and at the open top of the cylinder a tuning fork is vibrating with a frequency f. As the water rises, how much time elapses between successive resonances?

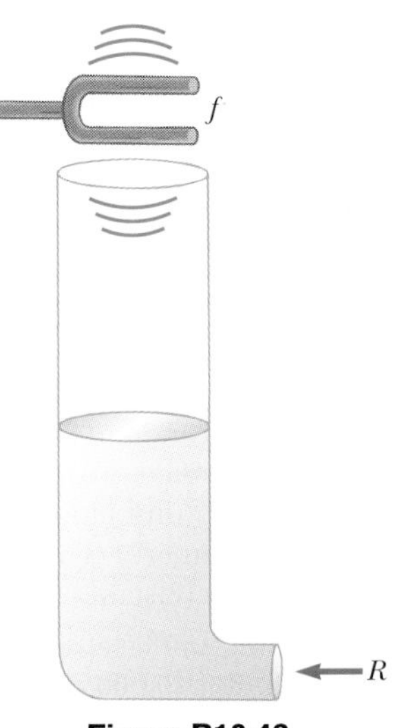

Figure P18.42

43. If two adjacent natural frequencies of an organ pipe are determined to be 550 Hz and 650 Hz, calculate the fundamental frequency and length of this pipe. (Use v = 340 m/s.)

44. A glass tube (open at both ends) of length L is positioned near an audio speaker of frequency f = 680 Hz. For what values of L will the tube resonate with the speaker?

45. An air column in a glass tube is open at one end and closed at the other by a movable piston. The air in the tube is warmed above room temperature, and a 384-Hz tuning fork is held at the open end. Resonance is heard when the piston is 22.8 cm from the open end and again when it is 68.3 cm from the open end. (a) What speed of sound is implied by these data? (b) How far from the open end will the piston be when the next resonance is heard?

46. A tuning fork with a frequency of 512 Hz is placed near the top of the pipe shown in Figure 18.19a. The water level is lowered so that the length L slowly increases from an initial value of 20.0 cm. Determine the next two values of L that correspond to resonant modes.

47. When an open metal pipe is cut into two pieces, the lowest resonance frequency for the air column in one piece is 256 Hz and that for the other is 440 Hz. (a) What resonant frequency would have been produced by the original length of pipe? (b) How long was the original pipe?

48. With a particular fingering, a flute plays a note with frequency 880 Hz at 20.0°C. The flute is open at both ends. (a) Find the air column length. (b) Find the frequency it produces at the beginning of the half-time performance at a late-season American football game, when the ambient temperature is −5.00°C and the musician has not had a chance to warm up the flute.

Section 18.6 Standing Waves in Rods and Membranes

49. An aluminum rod 1.60 m long is held at its center. It is stroked with a rosin-coated cloth to set up a longitudinal vibration. The speed of sound in a thin rod of aluminum is 5 100 m/s. (a) What is the fundamental frequency of the waves established in the rod? (b) What harmonics are set up in the rod held in this manner? (c) **What If?** What would be the fundamental frequency if the rod were made of copper, in which the speed of sound is 3 560 m/s?

50. An aluminum rod is clamped one quarter of the way along its length and set into longitudinal vibration by a variable-frequency driving source. The lowest frequency that produces resonance is 4 400 Hz. The speed of sound in an aluminum rod is 5 100 m/s. Find the length of the rod.

Section 18.7 Beats: Interference in Time

51. In certain ranges of a piano keyboard, more than one string is tuned to the same note to provide extra loudness. For example, the note at 110 Hz has two strings at this frequency. If one string slips from its normal tension of 600 N to 540 N, what beat frequency is heard when the hammer strikes the two strings simultaneously?

52. While attempting to tune the note C at 523 Hz, a piano tuner hears 2 beats/s between a reference oscillator and the string. (a) What are the possible frequencies of the string? (b) When she tightens the string slightly, she hears 3 beats/s. What is the frequency of the string now? (c) By what percentage should the piano tuner now change the tension in the string to bring it into tune?

53. A student holds a tuning fork oscillating at 256 Hz. He walks toward a wall at a constant speed of 1.33 m/s. (a) What beat frequency does he observe between the tuning fork and its echo? (b) How fast must he walk away from the wall to observe a beat frequency of 5.00 Hz?

54. When beats occur at a rate higher than about 20 per second, they are not heard individually but rather as a steady hum, called a *combination tone.* The player of a typical pipe organ can press a single key and make the organ produce sound with different fundamental frequencies. She can select and pull out different stops to make the same key for the note C produce sound at the following frequencies: 65.4 Hz from a so-called eight-foot pipe; $2 \times 65.4 =$ 131 Hz from a four-foot pipe; $3 \times 65.4 = 196$ Hz from a two-and-two-thirds-foot pipe; $4 \times 65.4 = 262$ Hz from a two-foot pipe; or any combination of these. With notes at low frequencies, she obtains sound with the richest quality by pulling out all the stops. When an air leak develops in one of the pipes, that pipe cannot be used. If a leak occurs in an eight-foot pipe, playing a combination of other pipes can create the sensation of sound at the frequency that the eight-foot pipe would produce. Which sets of stops, among those listed, could be pulled out to do this?

Section 18.8 Nonsinusoidal Wave Patterns

55. An A-major chord consists of the notes called A, C$^{\#}$, and E. It can be played on a piano by simultaneously striking strings with fundamental frequencies of 440.00 Hz, 554.37 Hz, and 659.26 Hz. The rich consonance of the chord is associated with near equality of the frequencies of some of the higher harmonics of the three tones. Consider the first five harmonics of each string and determine which harmonics show near equality.

56. Suppose that a flutist plays a 523-Hz C note with first harmonic displacement amplitude $A_1 = 100$ nm. From Figure 18.24b read, by proportion, the displacement amplitudes of harmonics 2 through 7. Take these as the values A_2 through A_7 in the Fourier analysis of the sound, and assume that $B_1 = B_2 = \cdots = B_7 = 0$. Construct a graph of the waveform of the sound. Your waveform will not look exactly like the flute waveform in Figure 18.23b because you simplify by ignoring cosine terms; nevertheless, it produces the same sensation to human hearing.

Additional Problems

57. On a marimba (Fig. P18.57), the wooden bar that sounds a tone when struck vibrates in a transverse standing wave having three antinodes and two nodes. The lowest frequency note is 87.0 Hz, produced by a bar 40.0 cm long. (a) Find the speed of transverse waves on the bar. (b) A resonant pipe suspended vertically below the center of the bar enhances the loudness of the emitted sound. If the pipe is open at the top end only and the speed of sound in air is 340 m/s, what is the length of the pipe required to resonate with the bar in part (a)?

Figure P18.57 Marimba players in Mexico City.

58. A loudspeaker at the front of a room and an identical loudspeaker at the rear of the room are being driven by the same oscillator at 456 Hz. A student walks at a uniform rate of 1.50 m/s along the length of the room. She hears a single tone, repeatedly becoming louder and softer. (a) Model these variations as beats between the Doppler-shifted sounds the student receives. Calculate the number of beats the student hears each second. (b) **What If?** Model the two speakers as producing a standing wave in the room and the student as walking between antinodes. Calculate the number of intensity maxima the student hears each second.

59. Two train whistles have identical frequencies of 180 Hz. When one train is at rest in the station and the other is moving nearby, a commuter standing on the station platform hears beats with a frequency of 2.00 beats/s when the whistles sound at the same time. What are the two possible speeds and directions that the moving train can have?

60. A string fixed at both ends and having a mass of 4.80 g, a length of 2.00 m, and a tension of 48.0 N vibrates in its second ($n = 2$) normal mode. What is the wavelength in air of the sound emitted by this vibrating string?

61. A student uses an audio oscillator of adjustable frequency to measure the depth of a water well. The student hears two successive resonances at 51.5 Hz and 60.0 Hz. How deep is the well?

62. A string has a mass per unit length of 9.00×10^{-3} kg/m and a length of 0.400 m. What must be the tension in the string if its second harmonic has the same frequency as the second resonance mode of a 1.75-m-long pipe open at one end?

63. Two wires are welded together end to end. The wires are made of the same material, but the diameter of one is twice that of the other. They are subjected to a tension of 4.60 N. The thin wire has a length of 40.0 cm and a linear mass density of 2.00 g/m. The combination is fixed at both ends and vibrated in such a way that two antinodes are present, with the node between them being right at the weld. (a) What is the frequency of vibration? (b) How long is the thick wire?

64. **Review problem.** For the arrangement shown in Figure P18.64, $\theta = 30.0°$, the inclined plane and the small pulley are frictionless, the string supports the object of mass M at the bottom of the plane, and the string has mass m that is small compared to M. The system is in equilibrium and the vertical part of the string has a length h. Standing waves are set up in the vertical section of the string. (a) Find the tension in the string. (b) Model the shape of the string as one leg and the hypotenuse of a right triangle. Find the whole length of the string. (c) Find the mass per unit length of the string. (d) Find the speed of waves on the string. (e) Find the lowest frequency for a standing wave. (f) Find the period of the standing wave having three nodes. (g) Find the wavelength of the standing wave having three nodes. (h) Find the frequency of the beats resulting from the interference of the sound wave of lowest frequency generated by the string with another sound wave having a frequency that is 2.00% greater.

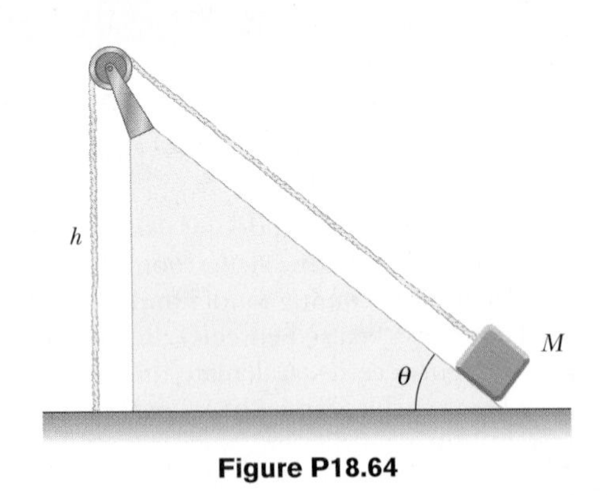

Figure P18.64

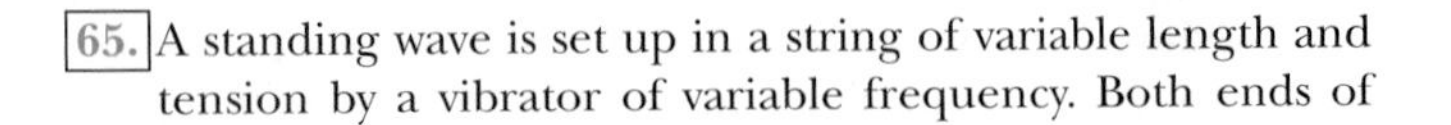

65. A standing wave is set up in a string of variable length and tension by a vibrator of variable frequency. Both ends of the string are fixed. When the vibrator has a frequency f, in a string of length L and under tension T, n antinodes are set up in the string. (a) If the length of the string is doubled, by what factor should the frequency be changed so that the same number of antinodes is produced? (b) If the frequency and length are held constant, what tension will produce $n + 1$ antinodes? (c) If the frequency is tripled and the length of the string is halved, by what factor should the tension be changed so that twice as many antinodes are produced?

66. A 0.010 0-kg wire, 2.00 m long, is fixed at both ends and vibrates in its simplest mode under a tension of 200 N. When a vibrating tuning fork is placed near the wire, a beat frequency of 5.00 Hz is heard. (a) What could be the frequency of the tuning fork? (b) What should the tension in the wire be if the beats are to disappear?

67. Two waves are described by the wave functions

$$y_1(x, t) = 5.0 \sin(2.0x - 10t)$$

and

$$y_2(x, t) = 10 \cos(2.0x - 10t)$$

where y_1, y_2, and x are in meters and t is in seconds. Show that the wave resulting from their superposition is also sinusoidal. Determine the amplitude and phase of this sinusoidal wave.

68. The wave function for a standing wave is given in Equation 18.3 as $y = 2A \sin kx \cos \omega t$. (a) Rewrite this wave function in terms of the wavelength λ and the wave speed v of the wave. (b) Write the wave function of the simplest standing-wave vibration of a stretched string of length L. (c) Write the wave function for the second harmonic. (d) Generalize these results and write the wave function for the nth resonance vibration.

69. **Review problem.** A 12.0-kg object hangs in equilibrium from a string with a total length of $L = 5.00$ m and a linear mass density of $\mu = 0.001\,00$ kg/m. The string is wrapped around two light, frictionless pulleys that are separated by a distance of $d = 2.00$ m (Fig. P18.69a). (a) Determine the tension in the string. (b) At what frequency must the string between the pulleys vibrate in order to form the standing wave pattern shown in Figure P18.69b?

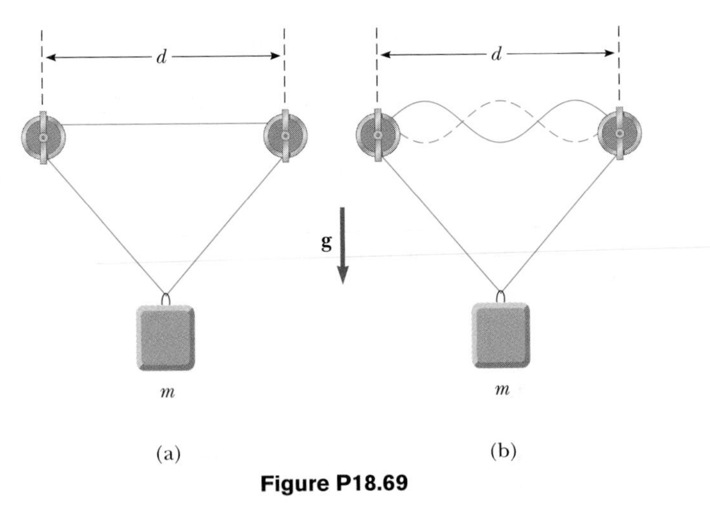

Figure P18.69

70. A quartz watch contains a crystal oscillator in the form of a block of quartz that vibrates by contracting and expanding. Two opposite faces of the block, 7.05 mm apart, are antinodes, moving alternately toward each other and away from each other. The plane halfway between these two faces is a node of the vibration. The speed of sound in quartz is 3.70 km/s. Find the frequency of the vibration. An oscillating electric voltage accompanies the mechanical oscillation—the quartz is described as *piezoelectric.* An electric circuit feeds in energy to maintain the oscillation and also counts the voltage pulses to keep time.

Answers to Quick Quizzes

18.1 The shape of the string at $t = 0.6$ s is shown below.

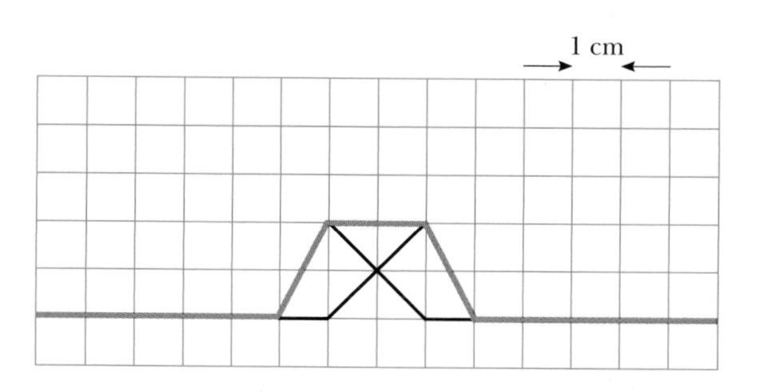

18.2 (c). The pulses completely cancel each other in terms of displacement of elements of the string from equilibrium, but the string is still moving. A short time later, the string will be displaced again and the pulses will have passed each other.

18.3 (a). The pattern shown at the bottom of Figure 18.9a corresponds to the extreme position of the string. All elements of the string have momentarily come to rest.

18.4 (d). Near a nodal point, elements on one side of the point are moving upward at this instant and elements on the other side are moving downward.

18.5 (d). Choice (a) is incorrect because the number of nodes is one greater than the number of antinodes. Choice (b) is only true for half of the modes; it is not true for any odd-numbered mode. Choice (c) would be correct if we replace the word *nodes* with *antinodes.*

18.6 For each natural frequency of the glass, the standing wave must "fit" exactly around the rim. In Figure 18.17a we see three antinodes on the near side of the glass, and thus there must be another three on the far side. This corresponds to three complete waves. In a top view, the wave pattern looks like this (although we have greatly exaggerated the amplitude):

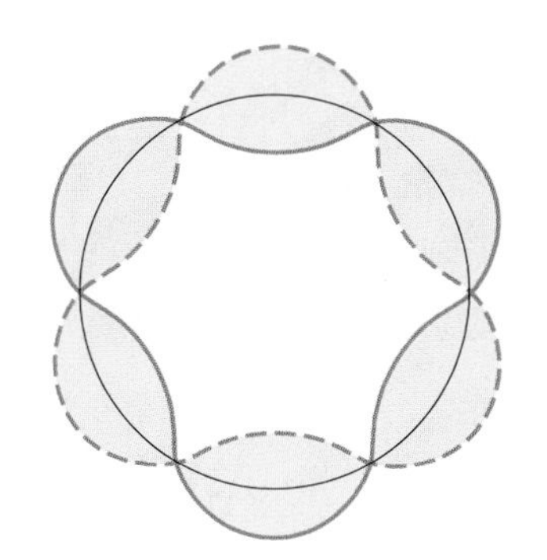

18.7 (b). With both ends open, the pipe has a fundamental frequency given by Equation 18.11: $f_{\text{open}} = v/2L$. With one end closed, the pipe has a fundamental frequency given by Equation 18.12:

$$f_{\text{closed}} = \frac{v}{4L} = \frac{1}{2}\frac{v}{2L} = \frac{1}{2}f_{\text{open}}$$

18.8 (c). The increase in temperature causes the speed of sound to go up. According to Equation 18.11, this will result in an increase in the fundamental frequency of a given organ pipe.

18.9 (b). Tightening the string has caused the frequencies to be farther apart, based on the increase in the beat frequency.

Thermodynamics

We now direct our attention to the study of thermodynamics, which involves situations in which the temperature or state (solid, liquid, gas) of a system changes due to energy transfers. As we shall see, thermodynamics is very successful in explaining the bulk properties of matter and the correlation between these properties and the mechanics of atoms and molecules.

Historically, the development of thermodynamics paralleled the development of the atomic theory of matter. By the 1820s, chemical experiments had provided solid evidence for the existence of atoms. At that time, scientists recognized that a connection between thermodynamics and the structure of matter must exist. In 1827, the botanist Robert Brown reported that grains of pollen suspended in a liquid move erratically from one place to another, as if under constant agitation. In 1905, Albert Einstein used kinetic theory to explain the cause of this erratic motion, which today is known as *Brownian motion*. Einstein explained this phenomenon by assuming that the grains are under constant bombardment by "invisible" molecules in the liquid, which themselves move erratically. This explanation gave scientists insight into the concept of molecular motion and gave credence to the idea that matter is made up of atoms. A connection was thus forged between the everyday world and the tiny, invisible building blocks that make up this world.

Thermodynamics also addresses more practical questions. Have you ever wondered how a refrigerator is able to cool its contents, what types of transformations occur in a power plant or in the engine of your automobile, or what happens to the kinetic energy of a moving object when the object comes to rest? The laws of thermodynamics can be used to provide explanations for these and other phenomena. ■

◀ *The Alyeska oil pipeline near the Tazlina River in Alaska. The oil in the pipeline is warm, and energy transferring from the pipeline could melt environmentally sensitive permafrost in the ground. The finned structures on top of the support posts are thermal radiators that allow the energy to be transferred into the air in order to protect the permafrost. (Topham Picturepoint/The Image Works)*

Chapter 19

Temperature

CHAPTER OUTLINE

▲ *Why would someone designing a pipeline include these strange loops? Pipelines carrying liquids often contain loops such as these to allow for expansion and contraction as the temperature changes. We will study thermal expansion in this chapter.* *(Lowell Georgia/CORBIS)*

In our study of mechanics, we carefully defined such concepts as *mass*, *force*, and *kinetic energy* to facilitate our quantitative approach. Likewise, a quantitative description of thermal phenomena requires careful definitions of such important terms as *temperature*, *heat*, and *internal energy*. This chapter begins with a discussion of temperature and with a description of one of the laws of thermodynamics (the so-called "zeroth law").

Next, we consider why an important factor when we are dealing with thermal phenomena is the particular substance we are investigating. For example, gases expand appreciably when heated, whereas liquids and solids expand only slightly.

This chapter concludes with a study of ideal gases on the macroscopic scale. Here, we are concerned with the relationships among such quantities as pressure, volume, and temperature. In Chapter 21, we shall examine gases on a microscopic scale, using a model that represents the components of a gas as small particles.

19.1 Temperature and the Zeroth Law of Thermodynamics

We often associate the concept of temperature with how hot or cold an object feels when we touch it. Thus, our senses provide us with a qualitative indication of temperature. However, our senses are unreliable and often mislead us. For example, if we remove a metal ice tray and a cardboard box of frozen vegetables from the freezer, the ice tray feels colder than the box *even though both are at the same temperature*. The two objects feel different because metal transfers energy by heat at a higher rate than cardboard does. What we need is a reliable and reproducible method for measuring the relative hotness or coldness of objects rather than the rate of energy transfer. Scientists have developed a variety of thermometers for making such quantitative measurements.

We are all familiar with the fact that two objects at different initial temperatures eventually reach some intermediate temperature when placed in contact with each other. For example, when hot water and cold water are mixed in a bathtub, the final temperature of the mixture is somewhere between the initial hot and cold temperatures. Likewise, when an ice cube is dropped into a cup of hot coffee, it melts and the coffee's temperature decreases.

To understand the concept of temperature, it is useful to define two often-used phrases: *thermal contact* and *thermal equilibrium*. To grasp the meaning of thermal contact, imagine that two objects are placed in an insulated container such that they interact with each other but not with the environment. If the objects are at different temperatures, energy is exchanged between them, even if they are initially not in physical contact with each other. The energy transfer mechanisms from Chapter 7 that we will focus on are heat and electromagnetic radiation. For purposes of the current discussion, we assume that two objects are in **thermal contact** with each other if energy can be exchanged between them by these processes due to a temperature difference.

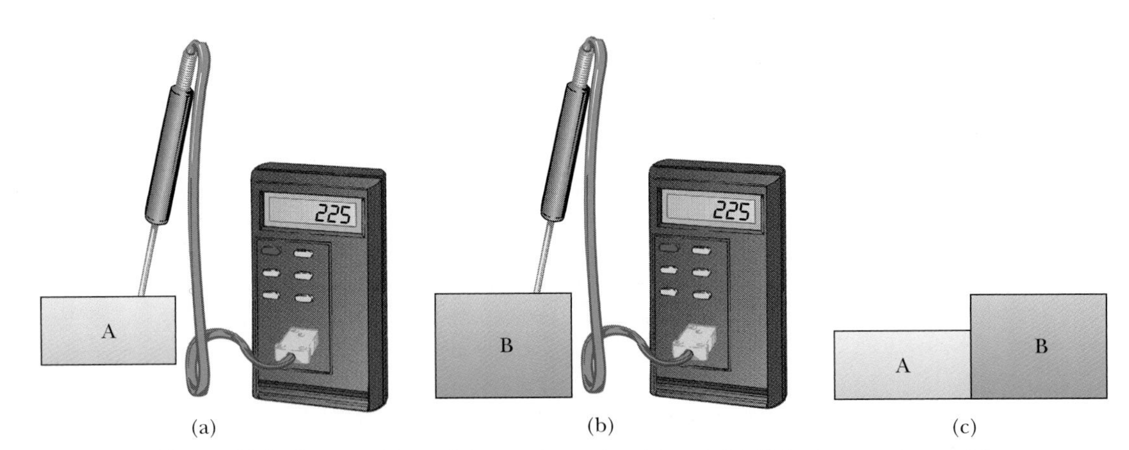

Figure 19.1 The zeroth law of thermodynamics. (a) and (b) If the temperatures of A and B are measured to be the same by placing them in thermal contact with a thermometer (object C), no energy will be exchanged between them when they are placed in thermal contact with each other (c).

Thermal equilibrium is a situation in which two objects would not exchange energy by heat or electromagnetic radiation if they were placed in thermal contact.

Let us consider two objects A and B, which are not in thermal contact, and a third object C, which is our thermometer. We wish to determine whether A and B are in thermal equilibrium with each other. The thermometer (object C) is first placed in thermal contact with object A until thermal equilibrium is reached,[1] as shown in Figure 19.1a. From that moment on, the thermometer's reading remains constant, and we record this reading. The thermometer is then removed from object A and placed in thermal contact with object B, as shown in Figure 19.1b. The reading is again recorded after thermal equilibrium is reached. If the two readings are the same, then object A and object B are in thermal equilibrium with each other. If they are placed in contact with each other as in Figure 19.1c, there is no exchange of energy between them.

We can summarize these results in a statement known as the **zeroth law of thermodynamics** (the law of equilibrium):

Zeroth law of thermodynamics

> If objects A and B are separately in thermal equilibrium with a third object C, then A and B are in thermal equilibrium with each other.

This statement can easily be proved experimentally and is very important because it enables us to define temperature. We can think of **temperature** as the property that determines whether an object is in thermal equilibrium with other objects. **Two objects in thermal equilibrium with each other are at the same temperature.** Conversely, if two objects have different temperatures, then they are not in thermal equilibrium with each other.

Quick Quiz 19.1 Two objects, with different sizes, masses, and temperatures, are placed in thermal contact. Energy travels (a) from the larger object to the smaller object (b) from the object with more mass to the one with less (c) from the object at higher temperature to the object at lower temperature.

[1] We assume that negligible energy transfers between the thermometer and object A during the equilibrium process. Without this assumption, which is also made for the thermometer and object B, the measurement of the temperature of an object disturbs the system so that the measured temperature is different from the initial temperature of the object. In practice, whenever you measure a temperature with a thermometer, you measure the disturbed system, not the original system.

19.2 Thermometers and the Celsius Temperature Scale

Thermometers are devices that are used to measure the temperature of a system. All thermometers are based on the principle that some physical property of a system changes as the system's temperature changes. Some physical properties that change with temperature are (1) the volume of a liquid, (2) the dimensions of a solid, (3) the pressure of a gas at constant volume, (4) the volume of a gas at constant pressure, (5) the electric resistance of a conductor, and (6) the color of an object. A temperature scale can be established on the basis of any one of these physical properties.

A common thermometer in everyday use consists of a mass of liquid—usually mercury or alcohol—that expands into a glass capillary tube when heated (Fig. 19.2). In this case the physical property that changes is the volume of a liquid. Any temperature change in the range of the thermometer can be defined as being proportional to the change in length of the liquid column. The thermometer can be calibrated by placing it in thermal contact with some natural systems that remain at constant temperature. One such system is a mixture of water and ice in thermal equilibrium at atmospheric pressure. On the **Celsius temperature scale**, this mixture is defined to have a temperature of zero degrees Celsius, which is written as 0°C; this temperature is called the *ice point* of water. Another commonly used system is a mixture of water and steam in thermal equilibrium at atmospheric pressure; its temperature is 100°C, which is the *steam point* of water. Once the liquid levels in the thermometer have been established at these two points, the length of the liquid column between the two points is divided into 100 equal segments to create the Celsius scale. Thus, each segment denotes a change in temperature of one Celsius degree.

Thermometers calibrated in this way present problems when extremely accurate readings are needed. For instance, the readings given by an alcohol thermometer calibrated at the ice and steam points of water might agree with those given by a mercury thermometer only at the calibration points. Because mercury and alcohol have different thermal expansion properties, when one thermometer reads a temperature of, for example, 50°C, the other may indicate a slightly different value. The discrepancies

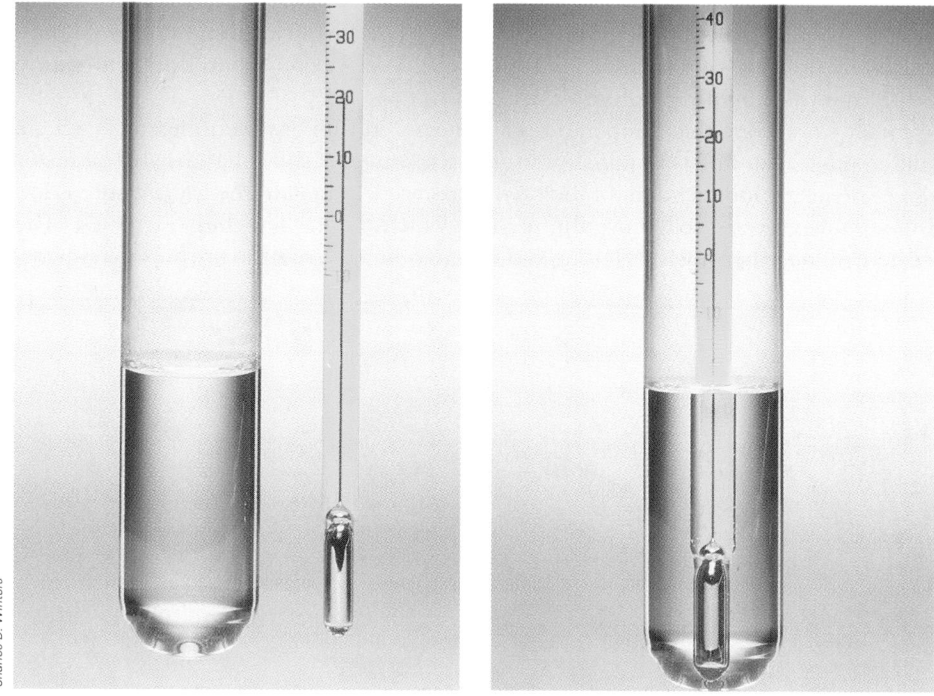

Charles D. Winters

Figure 19.2 As a result of thermal expansion, the level of the mercury in the thermometer rises as the mercury is heated by water in the test tube.

between thermometers are especially large when the temperatures to be measured are far from the calibration points.[2]

An additional practical problem of any thermometer is the limited range of temperatures over which it can be used. A mercury thermometer, for example, cannot be used below the freezing point of mercury, which is − 39°C, and an alcohol thermometer is not useful for measuring temperatures above 85°C, the boiling point of alcohol. To surmount this problem, we need a universal thermometer whose readings are independent of the substance used in it. The gas thermometer, discussed in the next section, approaches this requirement.

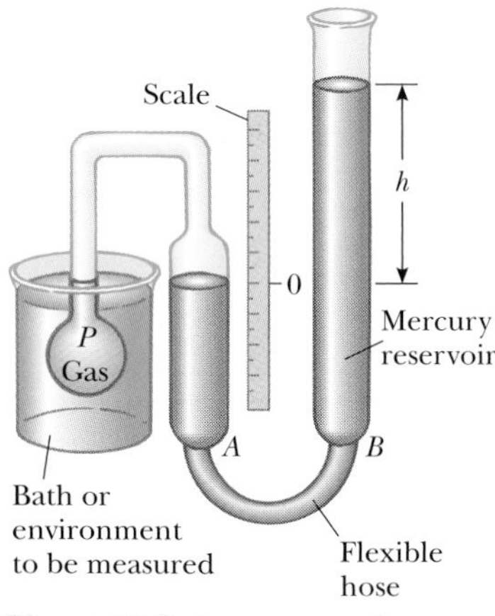

Figure 19.3 A constant-volume gas thermometer measures the pressure of the gas contained in the flask immersed in the bath. The volume of gas in the flask is kept constant by raising or lowering reservoir B to keep the mercury level in column A constant.

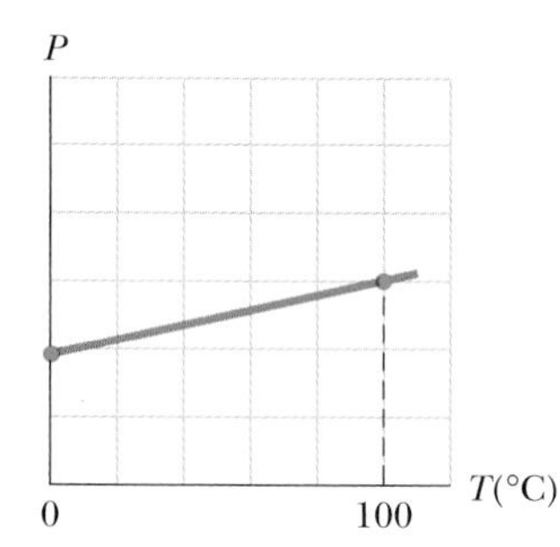

Figure 19.4 A typical graph of pressure versus temperature taken with a constant-volume gas thermometer. The two dots represent known reference temperatures (the ice and steam points of water).

19.3 The Constant-Volume Gas Thermometer and the Absolute Temperature Scale

One version of a gas thermometer is the constant-volume apparatus shown in Figure 19.3. The physical change exploited in this device is the variation of pressure of a fixed volume of gas with temperature. When the constant-volume gas thermometer was developed, it was calibrated by using the ice and steam points of water as follows. (A different calibration procedure, which we shall discuss shortly, is now used.) The flask was immersed in an ice-water bath, and mercury reservoir B was raised or lowered until the top of the mercury in column A was at the zero point on the scale. The height h, the difference between the mercury levels in reservoir B and column A, indicated the pressure in the flask at 0°C.

The flask was then immersed in water at the steam point, and reservoir B was readjusted until the top of the mercury in column A was again at zero on the scale; this ensured that the gas's volume was the same as it was when the flask was in the ice bath (hence, the designation "constant volume"). This adjustment of reservoir B gave a value for the gas pressure at 100°C. These two pressure and temperature values were then plotted, as shown in Figure 19.4. The line connecting the two points serves as a calibration curve for unknown temperatures. (Other experiments show that a linear relationship between pressure and temperature is a very good assumption.) If we wanted to measure the temperature of a substance, we would place the gas flask in thermal contact with the substance and adjust the height of reservoir B until the top of the mercury column in A is at zero on the scale. The height of the mercury column indicates the pressure of the gas; knowing the pressure, we could find the temperature of the substance using the graph in Figure 19.4.

Now let us suppose that temperatures are measured with gas thermometers containing different gases at different initial pressures. Experiments show that the thermometer readings are nearly independent of the type of gas used, as long as the gas pressure is low and the temperature is well above the point at which the gas liquefies (Fig. 19.5). The agreement among thermometers using various gases improves as the pressure is reduced.

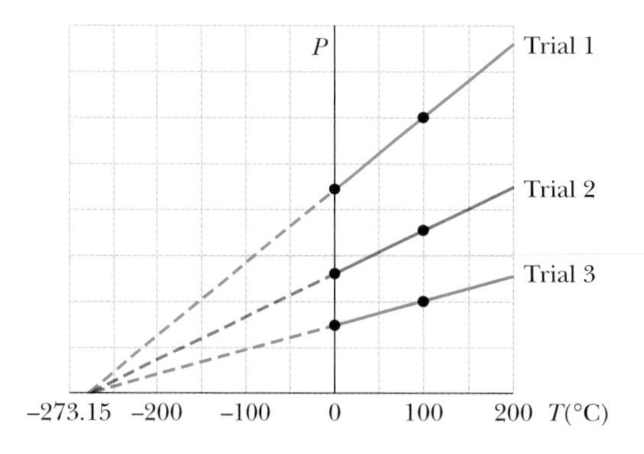

Figure 19.5 Pressure versus temperature for experimental trials in which gases have different pressures in a constant-volume gas thermometer. Note that, for all three trials, the pressure extrapolates to zero at the temperature − 273.15°C.

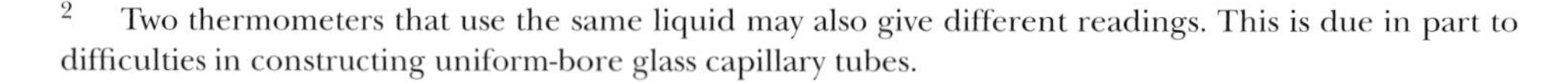
[2] Two thermometers that use the same liquid may also give different readings. This is due in part to difficulties in constructing uniform-bore glass capillary tubes.

If we extend the straight lines in Figure 19.5 toward negative temperatures, we find a remarkable result—**in every case, the pressure is zero when the temperature is − 273.15°C!** This suggests some special role that this particular temperature must play. It is used as the basis for the **absolute temperature scale**, which sets − 273.15°C as its zero point. This temperature is often referred to as **absolute zero**. The size of a degree on the absolute temperature scale is chosen to be identical to the size of a degree on the Celsius scale. Thus, the conversion between these temperatures is

$$T_C = T - 273.15 \tag{19.1}$$

where T_C is the Celsius temperature and T is the absolute temperature.

Because the ice and steam points are experimentally difficult to duplicate, an absolute temperature scale based on two new fixed points was adopted in 1954 by the International Committee on Weights and Measures. The first point is absolute zero. The second reference temperature for this new scale was chosen as the **triple point of water**, which is the single combination of temperature and pressure at which liquid water, gaseous water, and ice (solid water) coexist in equilibrium. This triple point occurs at a temperature of 0.01°C and a pressure of 4.58 mm of mercury. On the new scale, which uses the unit *kelvin*, the temperature of water at the triple point was set at 273.16 kelvins, abbreviated 273.16 K. This choice was made so that the old absolute temperature scale based on the ice and steam points would agree closely with the new scale based on the triple point. This new absolute temperature scale (also called the **Kelvin scale**) employs the SI unit of absolute temperature, the **kelvin**, which is defined to be **1/273.16 of the difference between absolute zero and the temperature of the triple point of water.**

19.1 A Matter of Degree

Note that notations for temperatures in the Kelvin scale do not use the degree sign. The unit for a Kelvin temperature is simply "kelvins" and not "degrees Kelvin."

Figure 19.6 shows the absolute temperature for various physical processes and structures. The temperature of absolute zero (0 K) cannot be achieved, although laboratory experiments incorporating the laser cooling of atoms have come very close.

What would happen to a gas if its temperature could reach 0 K (and it did not liquefy or solidify)? As Figure 19.5 indicates, the pressure it exerts on the walls of its container would be zero. In Chapter 21 we shall show that the pressure of a gas is proportional to the average kinetic energy of its molecules. Thus, according to classical physics, the kinetic energy of the gas molecules would become zero at absolute zero, and molecular motion would cease; hence, the molecules would settle out on the bottom of the container. Quantum theory modifies this prediction and shows that some residual energy, called the *zero-point energy*, would remain at this low temperature.

The Celsius, Fahrenheit, and Kelvin Temperature Scales[3]

Equation 19.1 shows that the Celsius temperature T_C is shifted from the absolute (Kelvin) temperature T by 273.15°. Because the size of a degree is the same on the two scales, a temperature difference of 5°C is equal to a temperature difference of 5 K. The two scales differ only in the choice of the zero point. Thus, the ice-point temperature on the Kelvin scale, 273.15 K, corresponds to 0.00°C, and the Kelvin-scale steam point, 373.15 K, is equivalent to 100.00°C.

A common temperature scale in everyday use in the United States is the **Fahrenheit scale**. This scale sets the temperature of the ice point at 32°F and the temperature of the steam point at 212°F. The relationship between the Celsius and Fahrenheit temperature scales is

$$T_F = \tfrac{9}{5}T_C + 32°\text{F} \tag{19.2}$$

We can use Equations 19.1 and 19.2 to find a relationship between changes in temperature on the Celsius, Kelvin, and Fahrenheit scales:

$$\Delta T_C = \Delta T = \tfrac{5}{9}\,\Delta T_F \tag{19.3}$$

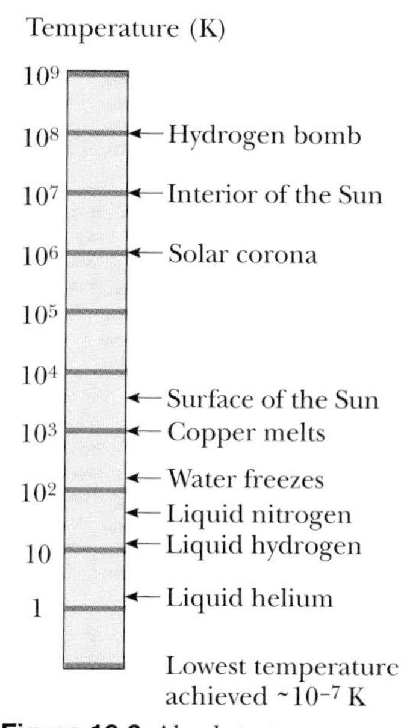

Figure 19.6 Absolute temperatures at which various physical processes occur. Note that the scale is logarithmic.

[3] Named after Anders Celsius (1701–1744), Daniel Gabriel Fahrenheit (1686–1736), and William Thomson, Lord Kelvin (1824–1907), respectively.

Of the three temperature scales that we have discussed, only the Kelvin scale is based on a true zero value of temperature. The Celsius and Fahrenheit scales are based on an arbitrary zero associated with one particular substance—water—on one particular planet—Earth. Thus, if you encounter an equation that calls for a temperature T or involves a ratio of temperatures, you *must* convert all temperatures to kelvins. If the equation contains a change in temperature ΔT, using Celsius temperatures will give you the correct answer, in light of Equation 19.3, but it is always *safest* to convert temperatures to the Kelvin scale.

Quick Quiz 19.2 Consider the following pairs of materials. Which pair represents two materials, one of which is twice as hot as the other? (a) boiling water at 100°C, a glass of water at 50°C (b) boiling water at 100°C, frozen methane at −50°C (c) an ice cube at −20°C, flames from a circus fire-eater at 233°C (d) No pair represents materials one of which is twice as hot as the other

Example 19.1 Converting Temperatures

On a day when the temperature reaches 50°F, what is the temperature in degrees Celsius and in kelvins?

Solution Substituting into Equation 19.2, we obtain

$$T_C = \frac{5}{9}(T_F - 32) = \frac{5}{9}(50 - 32)$$

$$= 10°\text{C}$$

From Equation 19.1, we find that

$$T = T_C + 273.15 = 10°\text{C} + 273.15 = 283\text{ K}$$

A convenient set of weather-related temperature equivalents to keep in mind is that 0°C is (literally) freezing at 32°F, 10°C is cool at 50°F, 20°C is room temperature, 30°C is warm at 86°F, and 40°C is a hot day at 104°F.

Example 19.2 Heating a Pan of Water

A pan of water is heated from 25°C to 80°C. What is the change in its temperature on the Kelvin scale and on the Fahrenheit scale?

Solution From Equation 19.3, we see that the change in temperature on the Celsius scale equals the change on the Kelvin scale. Therefore,

$$\Delta T = \Delta T_C = 80°\text{C} - 25°\text{C} = 55°\text{C} = 55\text{ K}$$

From Equation 19.3, we also find that

$$\Delta T_F = \frac{9}{5}\Delta T_C = \frac{9}{5}(55°\text{C}) = 99°\text{F}$$

19.4 Thermal Expansion of Solids and Liquids

Our discussion of the liquid thermometer makes use of one of the best-known changes in a substance: as its temperature increases, its volume increases. This phenomenon, known as **thermal expansion**, has an important role in numerous engineering applications. For example, thermal-expansion joints, such as those shown in Figure 19.7, must be included in buildings, concrete highways, railroad tracks, brick walls, and bridges to compensate for dimensional changes that occur as the temperature changes.

Thermal expansion is a consequence of the change in the *average* separation between the atoms in an object. To understand this, model the atoms as being connected by stiff springs, as discussed in Section 15.3 and shown in Figure 15.12b. At ordinary temperatures, the atoms in a solid oscillate about their equilibrium positions with an amplitude of approximately 10^{-11} m and a frequency of approximately 10^{13} Hz. The average spacing between the atoms is about 10^{-10} m. As the temperature of the solid

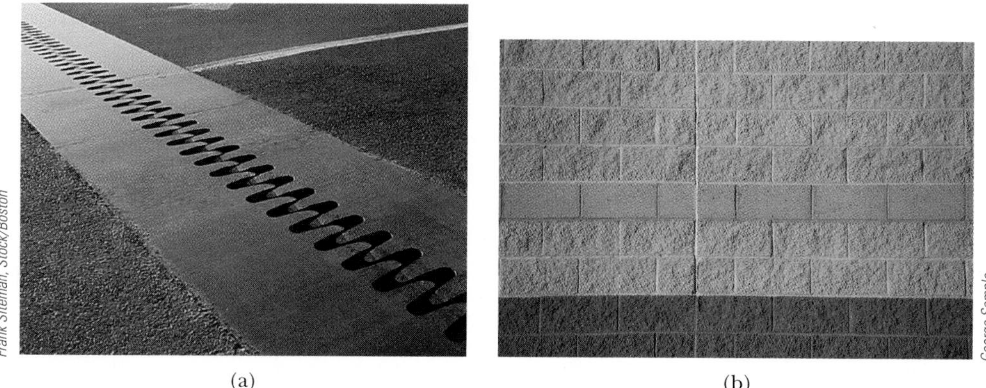

Figure 19.7 (a) Thermal-expansion joints are used to separate sections of roadways on bridges. Without these joints, the surfaces would buckle due to thermal expansion on very hot days or crack due to contraction on very cold days. (b) The long, vertical joint is filled with a soft material that allows the wall to expand and contract as the temperature of the bricks changes.

increases, the atoms oscillate with greater amplitudes; as a result, the average separation between them increases.[4] Consequently, the object expands.

If thermal expansion is sufficiently small relative to an object's initial dimensions, the change in any dimension is, to a good approximation, proportional to the first power of the temperature change. Suppose that an object has an initial length L_i along some direction at some temperature and that the length increases by an amount ΔL for a change in temperature ΔT. Because it is convenient to consider the fractional change in length per degree of temperature change, we define the **average coefficient of linear expansion** as

$$\alpha \equiv \frac{\Delta L / L_i}{\Delta T}$$

Experiments show that α is constant for small changes in temperature. For purposes of calculation, this equation is usually rewritten as

$$\Delta L = \alpha L_i \, \Delta T \tag{19.4}$$

or as

$$L_f - L_i = \alpha L_i (T_f - T_i) \tag{19.5}$$

where L_f is the final length, T_i and T_f are the initial and final temperatures, and the proportionality constant α is the average coefficient of linear expansion for a given material and has units of $(°C)^{-1}$.

It may be helpful to think of thermal expansion as an effective magnification or as a photographic enlargement of an object. For example, as a metal washer is heated (Fig. 19.8), all dimensions, including the radius of the hole, increase according to Equation 19.4. Notice that this is equivalent to saying that **a cavity in a piece of material expands in the same way as if the cavity were filled with the material.**

Table 19.1 lists the average coefficient of linear expansion for various materials. Note that for these materials α is positive, indicating an increase in length with increasing temperature. This is not always the case. Some substances—calcite ($CaCO_3$) is one example—expand along one dimension (positive α) and contract along another (negative α) as their temperatures are increased.

[4] More precisely, thermal expansion arises from the *asymmetrical* nature of the potential-energy curve for the atoms in a solid, as shown in Figure 15.12a. If the oscillators were truly harmonic, the average atomic separations would not change regardless of the amplitude of vibration.

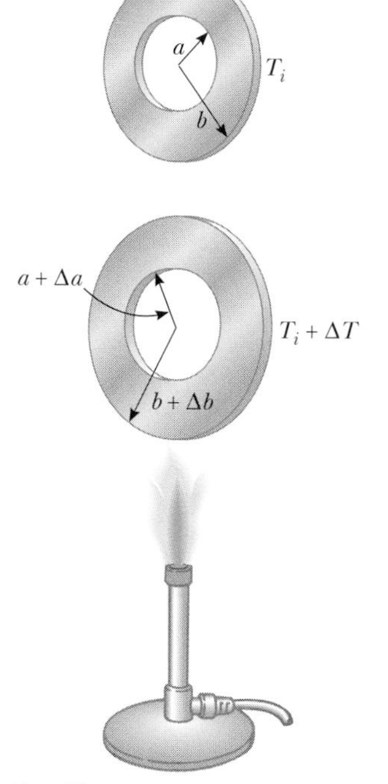

Active Figure 19.8 Thermal expansion of a homogeneous metal washer. As the washer is heated, all dimensions increase. (The expansion is exaggerated in this figure.)

At the Active Figures link at http://www.pse6.com, you can compare expansions for various temperatures of the burner and materials from which the washer is made.

PITFALL PREVENTION

19.2 Do Holes Become Larger or Smaller?

When an object's temperature is raised, every linear dimension increases in size. This includes any holes in the material, which expand in the same way as if the hole were filled with the material, as shown in Figure 19.8. Keep in mind the notion of thermal expansion as being similar to a photographic enlargement.

Table 19.1

Average Expansion Coefficients for Some Materials Near Room Temperature

Material	Average Linear Expansion Coefficient $(\alpha)(°C)^{-1}$	Material	Average Volume Expansion Coefficient $(\beta)(°C)^{-1}$
Aluminum	24×10^{-6}	Alcohol, ethyl	1.12×10^{-4}
Brass and bronze	19×10^{-6}	Benzene	1.24×10^{-4}
Copper	17×10^{-6}	Acetone	1.5×10^{-4}
Glass (ordinary)	9×10^{-6}	Glycerin	4.85×10^{-4}
Glass (Pyrex)	3.2×10^{-6}	Mercury	1.82×10^{-4}
Lead	29×10^{-6}	Turpentine	9.0×10^{-4}
Steel	11×10^{-6}	Gasoline	9.6×10^{-4}
Invar (Ni–Fe alloy)	0.9×10^{-6}	Air[a] at 0°C	3.67×10^{-3}
Concrete	12×10^{-6}	Helium[a]	3.665×10^{-3}

[a] Gases do not have a specific value for the volume expansion coefficient because the amount of expansion depends on the type of process through which the gas is taken. The values given here assume that the gas undergoes an expansion at constant pressure.

Because the linear dimensions of an object change with temperature, it follows that surface area and volume change as well. The change in volume is proportional to the initial volume V_i and to the change in temperature according to the relationship

$$\Delta V = \beta V_i \, \Delta T \tag{19.6}$$

where β is the **average coefficient of volume expansion**. For a solid, the average coefficient of volume expansion is three times the average linear expansion coefficient: $\beta = 3\alpha$. (This assumes that the average coefficient of linear expansion of the solid is the same in all directions—that is, the material is *isotropic.*)

To see that $\beta = 3\alpha$ for a solid, consider a solid box of dimensions ℓ, w, and h. Its volume at some temperature T_i is $V_i = \ell wh$. If the temperature changes to $T_i + \Delta T$, its volume changes to $V_i + \Delta V$, where each dimension changes according to Equation 19.4. Therefore,

$$\begin{aligned} V_i + \Delta V &= (\ell + \Delta\ell)(w + \Delta w)(h + \Delta h) \\ &= (\ell + \alpha\ell\,\Delta T)(w + \alpha w\,\Delta T)(h + \alpha h\,\Delta T) \\ &= \ell wh(1 + \alpha\,\Delta T)^3 \\ &= V_i[1 + 3\alpha\,\Delta T + 3(\alpha\,\Delta T)^2 + (\alpha\,\Delta T)^3] \end{aligned}$$

If we now divide both sides by V_i and isolate the term $\Delta V/V_i$, we obtain the fractional change in volume:

$$\frac{\Delta V}{V_i} = 3\alpha\,\Delta T + 3(\alpha\,\Delta T)^2 + (\alpha\,\Delta T)^3$$

Because $\alpha\Delta T \ll 1$ for typical values of ΔT ($<\sim 100°C$), we can neglect the terms $3(\alpha\,\Delta T)^2$ and $(\alpha\,\Delta T)^3$. Upon making this approximation, we see that

$$\frac{\Delta V}{V_i} = 3\alpha\,\Delta T$$

$$3\alpha = \frac{1}{V_i}\frac{\Delta V}{\Delta T}$$

Equation 19.6 shows that the right side of this expression is equal to β, and so we have $3\alpha = \beta$, the relationship we set out to prove. In a similar way, you can show that the change in area of a rectangular plate is given by $\Delta A = 2\alpha A_i\,\Delta T$ (see Problem 55).

As Table 19.1 indicates, each substance has its own characteristic average coefficient of expansion. For example, when the temperatures of a brass rod and a steel rod of

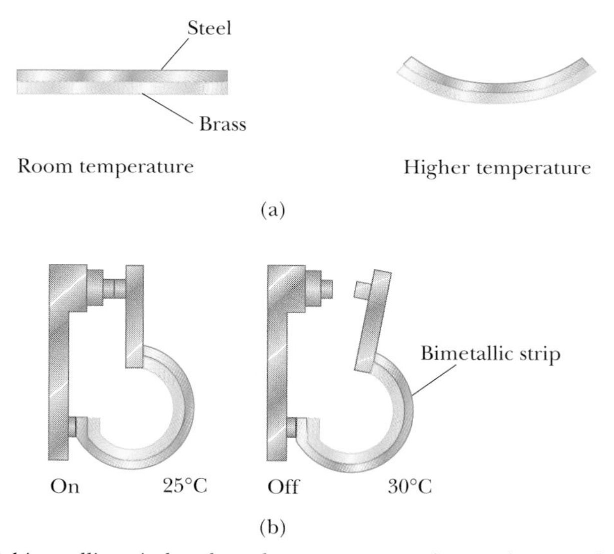

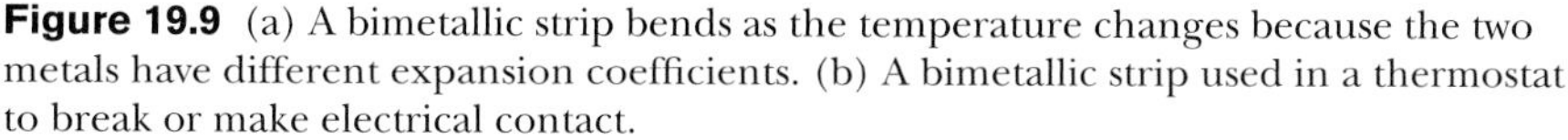

Figure 19.9 (a) A bimetallic strip bends as the temperature changes because the two metals have different expansion coefficients. (b) A bimetallic strip used in a thermostat to break or make electrical contact.

equal length are raised by the same amount from some common initial value, the brass rod expands more than the steel rod does because brass has a greater average coefficient of expansion than steel does. A simple mechanism called a *bimetallic strip* utilizes this principle and is found in practical devices such as thermostats. It consists of two thin strips of dissimilar metals bonded together. As the temperature of the strip increases, the two metals expand by different amounts and the strip bends, as shown in Figure 19.9.

Quick Quiz 19.3 If you are asked to make a very sensitive glass thermometer, which of the following working liquids would you choose? (a) mercury (b) alcohol (c) gasoline (d) glycerin

Quick Quiz 19.4 Two spheres are made of the same metal and have the same radius, but one is hollow and the other is solid. The spheres are taken through the same temperature increase. Which sphere expands more? (a) solid sphere (b) hollow sphere (c) They expand by the same amount. (d) not enough information to say

Example 19.3 Expansion of a Railroad Track

A segment of steel railroad track has a length of 30.000 m when the temperature is 0.0°C.

(A) What is its length when the temperature is 40.0°C?

Solution Making use of Table 19.1 and noting that the change in temperature is 40.0°C, we find that the increase in length is

$$\Delta L = \alpha L_i \, \Delta T = [11 \times 10^{-6} (°\text{C})^{-1}](30.000 \text{ m})(40.0°\text{C})$$
$$= 0.013 \text{ m}$$

If the track is 30.000 m long at 0.0°C, its length at 40.0°C is

30.013 m.

(B) Suppose that the ends of the rail are rigidly clamped at 0.0°C so that expansion is prevented. What is the thermal stress set up in the rail if its temperature is raised to 40.0°C?

Solution The thermal stress will be the same as that in the situation in which we allow the rail to expand freely and then compress it with a mechanical force F back to its original length. From the definition of Young's modulus for a solid (see Eq. 12.6), we have

$$\text{Tensile stress} = \frac{F}{A} = Y \frac{\Delta L}{L_i}$$

Because Y for steel is 20×10^{10} N/m^2 (see Table 12.1), we have

$$\frac{F}{A} = (20 \times 10^{10}\ \text{N/m}^2)\left(\frac{0.013\ \text{m}}{30.000\ \text{m}}\right) = 8.7 \times 10^7\ \text{N/m}^2$$

What If? **What if the temperature drops to −40.0°C? What is the length of the unclamped segment?**

The expression for the change in length in Equation 19.4 is the same whether the temperature increases or decreases. Thus, if there is an increase in length of 0.013 m when the temperature increases by 40°C, then there is a decrease in length of 0.013 m when the temperature decreases by 40°C. (We assume that α is constant over the entire range of temperatures.) The new length at the colder temperature is 30.000 m − 0.013 m = 29.987 m.

Example 19.4 The Thermal Electrical Short

An electronic device has been poorly designed so that two bolts attached to different parts of the device almost touch each other in its interior, as in Figure 19.10. The steel and brass bolts are at different electric potentials and if they touch, a short circuit will develop, damaging the device. (We will study electric potential in Chapter 25.) If the initial gap between the ends of the bolts is 5.0 μm at 27°C, at what temperature will the bolts touch?

Solution We can conceptualize the situation by imagining that the ends of both bolts expand into the gap between them as the temperature rises. We categorize this as a thermal expansion problem, in which the *sum* of the changes in length of the two bolts must equal the length of the initial gap between the ends. To analyze the problem, we write this condition mathematically:

$$\Delta L_{\text{br}} + \Delta L_{\text{st}} = \alpha_{\text{br}} L_{i,\text{br}} \Delta T + \alpha_{\text{st}} L_{i,\text{st}} \Delta T = 5.0 \times 10^{-6}\ \text{m}$$

Solving for ΔT, we find

$$\Delta T = \frac{5.0 \times 10^{-6}\ \text{m}}{\alpha_{\text{br}} L_{i,\,\text{br}} + \alpha_{\text{st}} L_{i,\,\text{st}}}$$

$$= \frac{5.0 \times 10^{-6}\ \text{m}}{(19 \times 10^{-6}\,°\text{C}^{-1})(0.030\ \text{m}) + (11 \times 10^{-6}\,°\text{C}^{-1})(0.010\ \text{m})}$$

$$= 7.4°\text{C}$$

Thus, the temperature at which the bolts touch is 27°C + 7.4°C = 34°C. To finalize this problem, note that this temperature is possible if the air conditioning in the building housing the device fails for a long period on a very hot summer day.

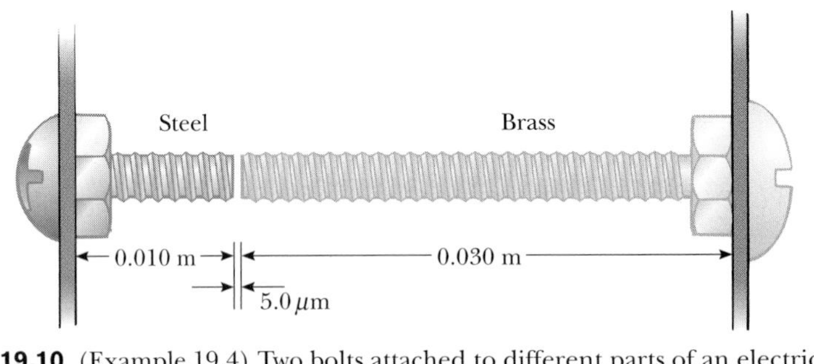

Figure 19.10 (Example 19.4) Two bolts attached to different parts of an electrical device are almost touching when the temperature is 27°C. As the temperature increases, the ends of the bolts move toward each other.

The Unusual Behavior of Water

Liquids generally increase in volume with increasing temperature and have average coefficients of volume expansion about ten times greater than those of solids. Cold water is an exception to this rule, as we can see from its density-versus-temperature curve, shown in Figure 19.11. As the temperature increases from 0°C to 4°C, water contracts and thus its density increases. Above 4°C, water expands with increasing temperature, and so its density decreases. Thus, the density of water reaches a maximum value of 1.000 g/cm^3 at 4°C.

We can use this unusual thermal-expansion behavior of water to explain why a pond begins freezing at the surface rather than at the bottom. When the atmospheric temperature drops from, for example, 7°C to 6°C, the surface water also cools and consequently decreases in volume. This means that the surface water is denser than the water below it, which has not cooled and decreased in volume. As a result, the surface water sinks, and warmer water from below is forced to the surface to be cooled. When the atmospheric temperature is between 4°C and 0°C, however, the surface water expands as it cools, becoming less dense than the water below it. The mixing process

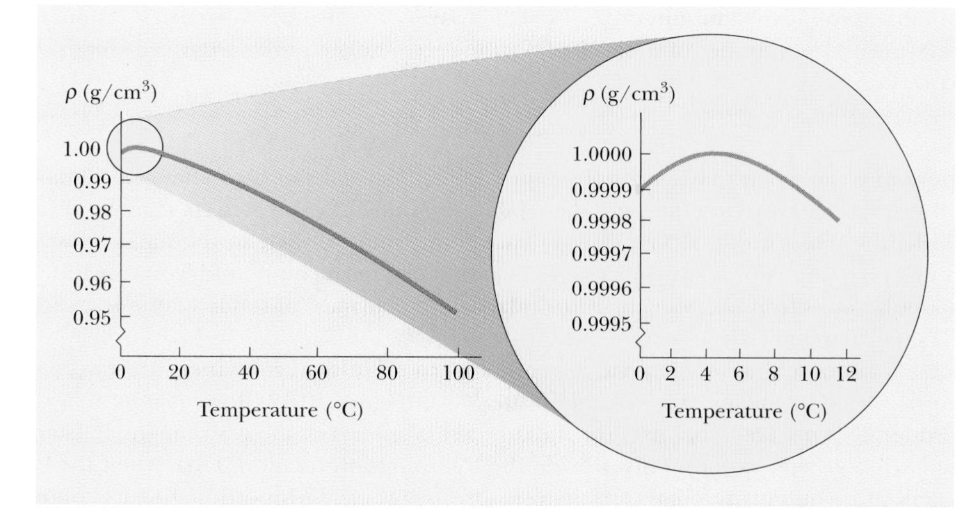

Figure 19.11 The variation in the density of water at atmospheric pressure with temperature. The inset at the right shows that the maximum density of water occurs at 4°C.

stops, and eventually the surface water freezes. As the water freezes, the ice remains on the surface because ice is less dense than water. The ice continues to build up at the surface, while water near the bottom remains at 4°C. If this were not the case, then fish and other forms of marine life would not survive.

19.5 Macroscopic Description of an Ideal Gas

The volume expansion equation $\Delta V = \beta V_i \, \Delta T$ is based on the assumption that the material has an initial volume V_i before the temperature change occurs. This is the case for solids and liquids because they have a fixed volume at a given temperature.

The case for gases is completely different. The interatomic forces within gases are very weak, and, in many cases, we can imagine these forces to be nonexistent and still make very good approximations. Note that *there is no equilibrium separation* for the atoms and, thus, no "standard" volume at a given temperature. As a result, we cannot express changes in volume ΔV in a process on a gas with Equation 19.6 because we have no defined volume V_i at the beginning of the process. For a gas, the volume is entirely determined by the container holding the gas. Thus, equations involving gases will contain the volume V as a *variable*, rather than focusing on a *change* in the volume from an initial value.

For a gas, it is useful to know how the quantities volume V, pressure P, and temperature T are related for a sample of gas of mass m. In general, the equation that interrelates these quantities, called the *equation of state*, is very complicated. However, if the gas is maintained at a very low pressure (or low density), the equation of state is quite simple and can be found experimentally. Such a low-density gas is commonly referred to as an *ideal gas*.[5]

It is convenient to express the amount of gas in a given volume in terms of the number of moles n. One **mole** of any substance is that amount of the substance that

[5] To be more specific, the assumption here is that the temperature of the gas must not be too low (the gas must not condense into a liquid) or too high, and that the pressure must be low. The concept of an ideal gas implies that the gas molecules do not interact except upon collision, and that the molecular volume is negligible compared with the volume of the container. In reality, an ideal gas does not exist. However, the concept of an ideal gas is very useful because real gases at low pressures behave as ideal gases do.

contains **Avogadro's number** $N_A = 6.022 \times 10^{23}$ of constituent particles (atoms or molecules). The number of moles n of a substance is related to its mass m through the expression

$$n = \frac{m}{M} \tag{19.7}$$

where M is the molar mass of the substance. The molar mass of each chemical element is the atomic mass (from the periodic table, Appendix C) expressed in g/mol. For example, the mass of one He atom is 4.00 u (atomic mass units), so the molar mass of He is 4.00 g/mol. For a molecular substance or a chemical compound, you can add up the molar mass from its molecular formula. The molar mass of stable diatomic oxygen (O_2) is 32.0 g/mol.

Now suppose that an ideal gas is confined to a cylindrical container whose volume can be varied by means of a movable piston, as in Figure 19.12. If we assume that the cylinder does not leak, the mass (or the number of moles) of the gas remains constant. For such a system, experiments provide the following information. First, when the gas is kept at a constant temperature, its pressure is inversely proportional to its volume (Boyle's law). Second, when the pressure of the gas is kept constant, its volume is directly proportional to its temperature (the law of Charles and Gay-Lussac). These observations are summarized by the **equation of state for an ideal gas:**

Equation of state for an ideal gas

$$PV = nRT \tag{19.8}$$

In this expression, known as the **ideal gas law**, R is a constant and n is the number of moles of gas in the sample. Experiments on numerous gases show that as the pressure approaches zero, the quantity PV/nT approaches the same value R for all gases. For this reason, R is called the **universal gas constant**. In SI units, in which pressure is expressed in pascals ($1 \text{ Pa} = 1 \text{ N/m}^2$) and volume in cubic meters, the product PV has units of newton·meters, or joules, and R has the value

$$R = 8.314 \text{ J/mol} \cdot \text{K} \tag{19.9}$$

If the pressure is expressed in atmospheres and the volume in liters ($1 \text{ L} = 10^3 \text{ cm}^3 = 10^{-3} \text{ m}^3$), then R has the value

$$R = 0.082\,14 \text{ L} \cdot \text{atm/mol} \cdot \text{K}$$

Using this value of R and Equation 19.8, we find that the volume occupied by 1 mol of any gas at atmospheric pressure and at 0°C (273 K) is 22.4 L.

The ideal gas law states that if the volume and temperature of a fixed amount of gas do not change, then the pressure also remains constant. Consider a bottle of champagne that is shaken and then spews liquid when opened, as shown in Figure 19.13.

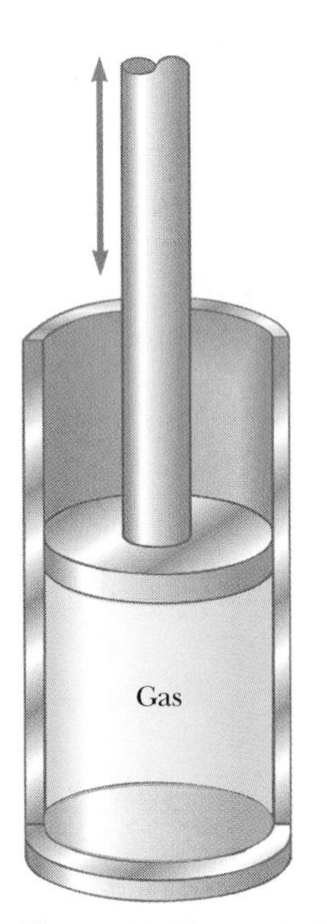

Active Figure 19.12 An ideal gas confined to a cylinder whose volume can be varied by means of a movable piston.

At the Active Figures link at http://www.pse6.com, you can choose to keep either the temperature or the pressure constant and verify Boyle's law and the law of Charles and Gay–Lussac.

Figure 19.13 A bottle of champagne is shaken and opened. Liquid spews out of the opening. A common misconception is that the pressure inside the bottle is increased due to the shaking.

A common misconception is that the pressure inside the bottle is increased when the bottle is shaken. On the contrary, because the temperature of the bottle and its contents remains constant as long as the bottle is sealed, so does the pressure, as can be shown by replacing the cork with a pressure gauge. The correct explanation is as follows. Carbon dioxide gas resides in the volume between the liquid surface and the cork. Shaking the bottle displaces some of this carbon dioxide gas into the liquid, where it forms bubbles, and these bubbles become attached to the inside of the bottle. (No new gas is generated by shaking.) When the bottle is opened, the pressure is reduced; this causes the volume of the bubbles to increase suddenly. If the bubbles are attached to the bottle (beneath the liquid surface), their rapid expansion expels liquid from the bottle. If the sides and bottom of the bottle are first tapped until no bubbles remain beneath the surface, then when the champagne is opened, the drop in pressure will not force liquid from the bottle.

The ideal gas law is often expressed in terms of the total number of molecules N. Because the total number of molecules equals the product of the number of moles n and Avogadro's number N_A, we can write Equation 19.8 as

$$PV = nRT = \frac{N}{N_A} RT$$

$$PV = Nk_BT \tag{19.10}$$

where k_B is **Boltzmann's constant**, which has the value

$$k_B = \frac{R}{N_A} = 1.38 \times 10^{-23} \text{ J/K} \tag{19.11}$$

Boltzmann's constant

It is common to call quantities such as P, V, and T the **thermodynamic variables** of an ideal gas. If the equation of state is known, then one of the variables can always be expressed as some function of the other two.

19.3 So Many *k*'s

There are a variety of physical quantities for which the letter k is used—we have seen two previously, the force constant for a spring (Chapter 15) and the wave number for a mechanical wave (Chapter 16). Boltzmann's constant is another k, and we will see k used for thermal conductivity in Chapter 20 and for an electrical constant in Chapter 23. In order to make some sense of this confusing state of affairs, we will use a subscript for Boltzmann's constant to help us recognize it. In this book, we will see Boltzmann's constant as k_B, but keep in mind that you may see Boltzmann's constant in other resources as simply k.

Quick Quiz 19.5 A common material for cushioning objects in packages is made by trapping bubbles of air between sheets of plastic. This material is more effective at keeping the contents of the package from moving around inside the package on (a) a hot day (b) a cold day (c) either hot or cold days.

Quick Quiz 19.6 A helium-filled rubber balloon is left in a car on a cold winter night. Compared to its size when it was in the warm car the afternoon before, the size the next morning is (a) larger (b) smaller (c) unchanged.

Quick Quiz 19.7 On a winter day, you turn on your furnace and the temperature of the air inside your home increases. Assuming that your home has the normal amount of leakage between inside air and outside air, the number of moles of air in your room at the higher temperature is (a) larger than before (b) smaller than before (c) the same as before.

Example 19.5 How Many Moles of Gas in a Container?

An ideal gas occupies a volume of 100 cm^3 at 20°C and 100 Pa. Find the number of moles of gas in the container.

Solution The quantities given are volume, pressure, and temperature: $V = 100 \text{ cm}^3 = 1.00 \times 10^{-4} \text{ m}^3$, $P = 100$ Pa, and $T = 20°\text{C} = 293$ K. Using Equation 19.8, we find that

$$n = \frac{PV}{RT} = \frac{(100 \text{ Pa})(1.00 \times 10^{-4} \text{ m}^3)}{(8.314 \text{ J/mol} \cdot \text{K})(293 \text{ K})}$$

$$= 4.11 \times 10^{-6} \text{ mol}$$

Example 19.6 Filling a Scuba Tank

A certain scuba tank is designed to hold 66.0 ft^3 of air when it is at atmospheric pressure at 22°C. When this volume of air is compressed to an absolute pressure of 3 000 lb/in.^{2} and stored in a 10.0-L (0.350-ft^3) tank, the air becomes so hot that the tank must be allowed to cool before it can be used. Before the air cools, what is its temperature? (Assume that the air behaves like an ideal gas.)

Solution If no air escapes during the compression, then the number of moles n of air remains constant; therefore, using $PV = nRT$, with n and R constant, we obtain a relationship between the initial and final values:

$$\frac{P_iV_i}{T_i} = \frac{P_fV_f}{T_f}$$

The initial pressure of the air is 14.7 lb/in.^{2}, its final pressure is 3 000 lb/in.^{2}, and the air is compressed from an initial volume of 66.0 ft^3 to a final volume of 0.350 ft^3. The initial temperature, converted to SI units, is 295 K. Solving for T_f, we obtain

$$T_f = \left(\frac{P_fV_f}{P_iV_i}\right)T_i = \frac{(3\,000 \text{ lb/in.}^2)(0.350 \text{ ft}^3)}{(14.7 \text{ lb/in.}^2)(66.0 \text{ ft}^3)}(295 \text{ K})$$

$$= 319 \text{ K}$$

Example 19.7 Heating a Spray Can

Interactive

A spray can containing a propellant gas at twice atmospheric pressure (202 kPa) and having a volume of 125.00 cm^3 is at 22°C. It is then tossed into an open fire. When the temperature of the gas in the can reaches 195°C, what is the pressure inside the can? Assume any change in the volume of the can is negligible.

Solution We employ the same approach we used in Example 19.6, starting with the expression

$$(1) \qquad \frac{P_iV_i}{T_i} = \frac{P_fV_f}{T_f}$$

Because the initial and final volumes of the gas are assumed to be equal, this expression reduces to

$$\frac{P_i}{T_i} = \frac{P_f}{T_f}$$

Solving for P_f gives

$$(2) \qquad P_f = \left(\frac{T_f}{T_i}\right)P_i = \left(\frac{468 \text{ K}}{295 \text{ K}}\right)(202 \text{ kPa}) = 320 \text{ kPa}$$

Obviously, the higher the temperature, the higher the pressure exerted by the trapped gas. Of course, if the pressure increases sufficiently, the can will explode. Because of this possibility, you should never dispose of spray cans in a fire.

What If? Suppose we include a volume change due to thermal expansion of the steel can as the temperature increases. Does this alter our answer for the final pressure significantly?

Because the thermal expansion coefficient of steel is very small, we do not expect much of an effect on our final answer. The change in the volume of the can is found using Equation 19.6 and the value for α for steel from Table 19.1:

$$\Delta V = \beta V_i\,\Delta T = 3\alpha V_i\,\Delta T$$
$$= 3(11 \times 10^{-6}\,°\text{C}^{-1})(125.00 \text{ cm}^3)(173°\text{C})$$
$$= 0.71 \text{ cm}^3$$

So the final volume of the can is 125.71 cm^3. Starting from Equation (1) again, the equation for the final pressure becomes

$$P_f = \left(\frac{T_f}{T_i}\right)\left(\frac{V_i}{V_f}\right)P_i$$

This differs from Equation (2) only in the factor V_i/V_f. Let us evaluate this factor:

$$\frac{V_i}{V_f} = \frac{125.00 \text{ cm}^3}{125.71 \text{ cm}^3} = 0.994 = 99.4\%$$

Thus, the final pressure will differ by only 0.6% from the value we calculated without considering the thermal expansion of the can. Taking 99.4% of the previous final pressure, the final pressure including thermal expansion is 318 kPa.

Explore this situation at the Interactive Worked Example link at **http://www.pse6.com.**

SUMMARY

Take a practice test for this chapter by clicking on the Practice Test link at **http://www.pse6.com.**

Two objects are in **thermal equilibrium** with each other if they do not exchange energy when in thermal contact.

The **zeroth law of thermodynamics** states that if objects A and B are separately in thermal equilibrium with a third object C, then objects A and B are in thermal equilibrium with each other.

Temperature is the property that determines whether an object is in thermal equilibrium with other objects. **Two objects in thermal equilibrium with each other are at the same temperature.**

The SI unit of absolute temperature is the **kelvin**, which is defined to be the fraction 1/273.16 of the temperature of the triple point of water.

When the temperature of an object is changed by an amount ΔT, its length changes by an amount ΔL that is proportional to ΔT and to its initial length L_i:

$$\Delta L = \alpha L_i \Delta T \tag{19.4}$$

where the constant α is the **average coefficient of linear expansion**. The **average coefficient of volume expansion** β for a solid is approximately equal to 3α.

An **ideal gas** is one for which PV/nT is constant. An ideal gas is described by the **equation of state,**

$$PV = nRT \tag{19.8}$$

where n equals the number of moles of the gas, V is its volume, R is the universal gas constant (8.314 J/mol · K), and T is the absolute temperature. A real gas behaves approximately as an ideal gas if it has a low density.

QUESTIONS

1. Is it possible for two objects to be in thermal equilibrium if they are not in contact with each other? Explain.

2. A piece of copper is dropped into a beaker of water. If the water's temperature rises, what happens to the temperature of the copper? Under what conditions are the water and copper in thermal equilibrium?

3. In describing his upcoming trip to the Moon, and as portrayed in the movie *Apollo 13* (Universal, 1995), astronaut Jim Lovell said, "I'll be walking in a place where there's a 400-degree difference between sunlight and shadow." What is it that is hot in sunlight and cold in shadow? Suppose an astronaut standing on the Moon holds a thermometer in his gloved hand. Is it reading the temperature of the vacuum at the Moon's surface? Does it read any temperature? If so, what object or substance has that temperature?

4. Rubber has a negative average coefficient of linear expansion. What happens to the size of a piece of rubber as it is warmed?

5. Explain why a column of mercury in a thermometer first descends slightly and then rises when the thermometer is placed into hot water.

6. Why should the amalgam used in dental fillings have the same average coefficient of expansion as a tooth? What would occur if they were mismatched?

7. Markings to indicate length are placed on a steel tape in a room that has a temperature of 22°C. Are measurements made with the tape on a day when the temperature is 27°C too long, too short, or accurate? Defend your answer.

8. Determine the number of grams in a mole of the following gases: (a) hydrogen (b) helium (c) carbon monoxide.

9. What does the ideal gas law predict about the volume of a sample of gas at absolute zero? Why is this prediction incorrect?

10. An inflated rubber balloon filled with air is immersed in a flask of liquid nitrogen that is at 77 K. Describe what happens to the balloon, assuming that it remains flexible while being cooled.

11. Two identical cylinders at the same temperature each contain the same kind of gas and the same number of moles of gas. If the volume of cylinder A is three times greater than the volume of cylinder B, what can you say about the relative pressures in the cylinders?

12. After food is cooked in a pressure cooker, why is it very important to cool off the container with cold water before attempting to remove the lid?

13. The shore of the ocean is very rocky at a particular place. The rocks form a cave sloping upward from an underwater opening, as shown in Figure Q19.13a. (a) Inside the cave is

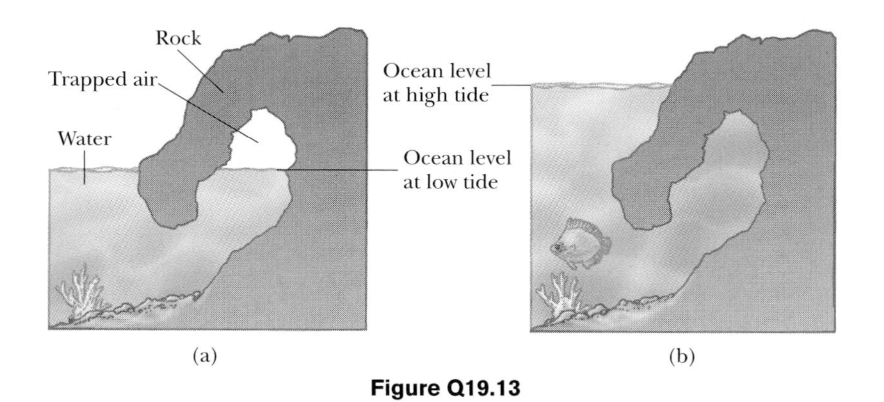

Figure Q19.13

a pocket of trapped air. As the level of the ocean rises and falls with the tides, will the level of water in the cave rise and fall? If so, will it have the same amplitude as that of the ocean? (b) **What If?** Now suppose that the cave is deeper in the water, so that it is completely submerged and filled with water at high tide, as shown in Figure Q19.13b. At low tide, will the level of the water in the cave be the same as that of the ocean?

14. In *Colonization: Second Contact* (Harry Turtledove, Ballantine Publishing Group, 1999), the Earth has been partially settled by aliens from another planet, whom humans call Lizards. Laboratory study by humans of Lizard science requires "shifting back and forth between the metric system and the one the Lizards used, which was also based on powers of ten but used different basic quantities for everything but temperature." Why might temperature be an exception?

15. The pendulum of a certain pendulum clock is made of brass. When the temperature increases, does the period of the clock increase, decrease, or remain the same? Explain.

16. An automobile radiator is filled to the brim with water while the engine is cool. What happens to the water when the engine is running and the water is heated? What do modern automobiles have in their cooling systems to prevent the loss of coolants?

17. Metal lids on glass jars can often be loosened by running hot water over them. How is this possible?

18. When the metal ring and metal sphere in Figure Q19.18 are both at room temperature, the sphere can just be passed through the ring. After the sphere is heated, it cannot be passed through the ring. Explain. **What If?** What if the ring is heated and the sphere is left at room temperature? Does the sphere pass through the ring?

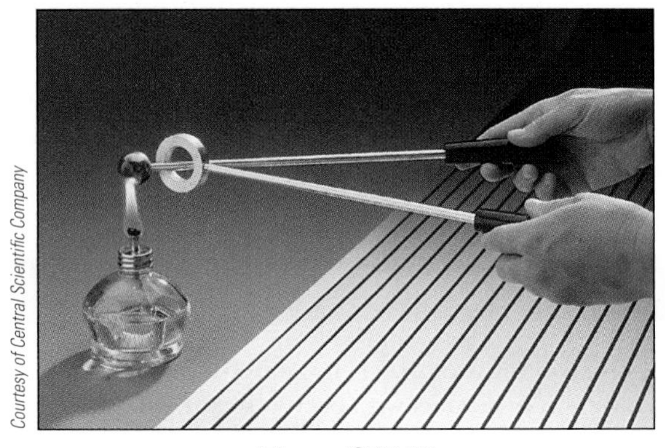

Courtesy of Central Scientific Company

Figure Q19.18

PROBLEMS

1, 2, 3 = straightforward, intermediate, challenging □ = full solution available in the *Student Solutions Manual and Study Guide*

= coached solution with hints available at http://www.pse6.com = computer useful in solving problem

= paired numerical and symbolic problems

Section 19.2 Thermometers and the Celsius Temperature Scale

Section 19.3 The Constant-Volume Gas Thermometer and the Absolute Temperature Scale

1. A constant-volume gas thermometer is calibrated in dry ice (that is, carbon dioxide in the solid state, which has a temperature of $-80.0°C$) and in boiling ethyl alcohol (78.0°C). The two pressures are 0.900 atm and 1.635 atm. (a) What Celsius value of absolute zero does the calibration yield? What is the pressure at (b) the freezing point of water and (c) the boiling point of water?

2. In a constant-volume gas thermometer, the pressure at 20.0°C is 0.980 atm. (a) What is the pressure at 45.0°C? (b) What is the temperature if the pressure is 0.500 atm?

3. Liquid nitrogen has a boiling point of $-195.81°C$ at atmospheric pressure. Express this temperature (a) in degrees Fahrenheit and (b) in kelvins.

4. Convert the following to equivalent temperatures on the Celsius and Kelvin scales: (a) the normal human body temperature, 98.6°F; (b) the air temperature on a cold day, $-5.00°F$.

5. The temperature difference between the inside and the outside of an automobile engine is 450°C. Express this temperature difference on (a) the Fahrenheit scale and (b) the Kelvin scale.

6. On a Strange temperature scale, the freezing point of water is $-15.0°S$ and the boiling point is $+60.0°S$. Develop a *linear* conversion equation between this temperature scale and the Celsius scale.

7. The melting point of gold is 1 064°C, and the boiling point is 2 660°C. (a) Express these temperatures in kelvins. (b) Compute the difference between these temperatures in Celsius degrees and kelvins.

Section 19.4 Thermal Expansion of Solids and Liquids

Note: Table 19.1 is available for use in solving problems in this section.

8. The New River Gorge bridge in West Virginia is a steel arch bridge 518 m in length. How much does the total length of the roadway decking change between temperature extremes of $-20.0°C$ and 35.0°C? The result indicates the size of the expansion joints that must be built into the structure.

9. A copper telephone wire has essentially no sag between poles 35.0 m apart on a winter day when the temperature is $-20.0°C$. How much longer is the wire on a summer day when $T_C = 35.0°C$?

10. The concrete sections of a certain superhighway are designed to have a length of 25.0 m. The sections are poured and cured at 10.0°C. What minimum spacing should the engineer leave between the sections to eliminate buckling if the concrete is to reach a temperature of 50.0°C?

11. A pair of eyeglass frames is made of epoxy plastic. At room temperature (20.0°C), the frames have circular lens holes 2.20 cm in radius. To what temperature must the frames be heated if lenses 2.21 cm in radius are to be inserted in them? The average coefficient of linear expansion for epoxy is $1.30 \times 10^{-4}\ (°C)^{-1}$.

12. Each year thousands of children are badly burned by hot tap water. Figure P19.12 shows a cross-sectional view of an antiscalding faucet attachment designed to prevent such accidents. Within the device, a spring made of material with a high coefficient of thermal expansion controls a movable plunger. When the water temperature rises above a preset safe value, the expansion of the spring causes the plunger to shut off the water flow. If the initial length L of the unstressed spring is 2.40 cm and its coefficient of linear expansion is $22.0 \times 10^{-6}\ (°C)^{-1}$, determine the increase in length of the spring when the water temperature rises by 30.0°C. (You will find the increase in length to be small. For this reason actual devices have a more complicated mechanical design, to provide a greater variation in valve opening for the temperature change anticipated.)

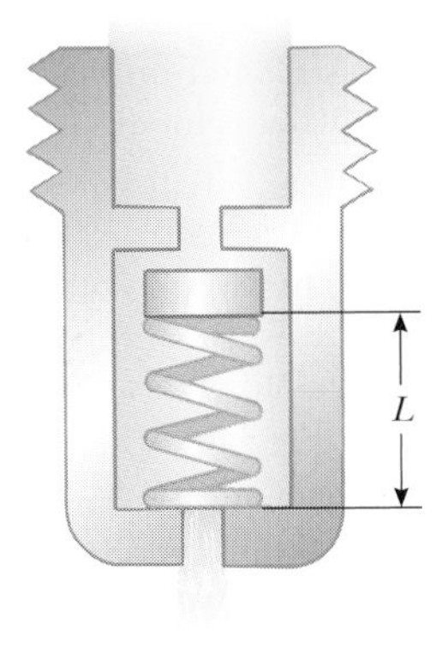

Figure P19.12

13. The active element of a certain laser is made of a glass rod 30.0 cm long by 1.50 cm in diameter. If the temperature of the rod increases by 65.0°C, what is the increase in (a) its length, (b) its diameter, and (c) its volume? Assume that the average coefficient of linear expansion of the glass is $9.00 \times 10^{-6}\ (°C)^{-1}$.

14. **Review problem.** Inside the wall of a house, an L-shaped section of hot-water pipe consists of a straight horizontal piece 28.0 cm long, an elbow, and a straight vertical piece 134 cm long (Figure P19.14). A stud and a second-story floorboard hold stationary the ends of this section of copper pipe. Find the magnitude and direction of the displacement of the pipe elbow when the water flow is turned on, raising the temperature of the pipe from 18.0°C to 46.5°C.

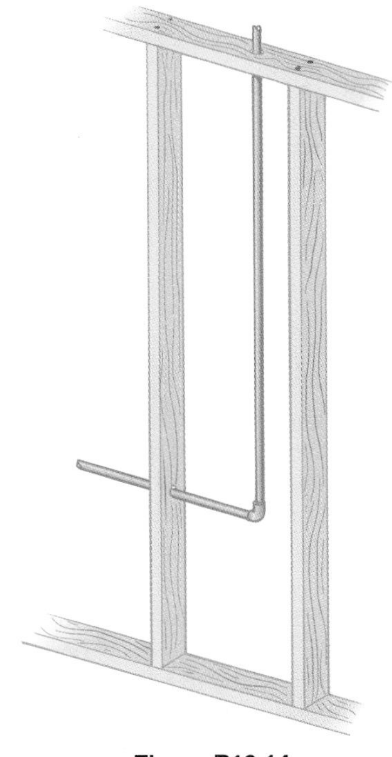

Figure P19.14

15. A brass ring of diameter 10.00 cm at 20.0°C is heated and slipped over an aluminum rod of diameter 10.01 cm at 20.0°C. Assuming the average coefficients of linear expansion are constant, (a) to what temperature must this combination be cooled to separate them? Is this attainable? (b) **What If?** What if the aluminum rod were 10.02 cm in diameter?

16. A square hole 8.00 cm along each side is cut in a sheet of copper. (a) Calculate the change in the area of this hole if the temperature of the sheet is increased by 50.0 K. (b) Does this change represent an increase or a decrease in the area enclosed by the hole?

17. The average coefficient of volume expansion for carbon tetrachloride is $5.81 \times 10^{-4}\ (°C)^{-1}$. If a 50.0-gal steel container is filled completely with carbon tetrachloride when the temperature is 10.0°C, how much will spill over when the temperature rises to 30.0°C?

18. At 20.0°C, an aluminum ring has an inner diameter of 5.000 0 cm and a brass rod has a diameter of 5.050 0 cm. (a) If only the ring is heated, what temperature must it reach so that it will just slip over the rod? (b) **What If?** If both are heated together, what temperature must they both reach so that the ring just slips over the rod? Would this latter process work?

19. A volumetric flask made of Pyrex is calibrated at 20.0°C. It is filled to the 100-mL mark with 35.0°C acetone. (a) What is the volume of the acetone when it cools to 20.0°C? (b) How significant is the change in volume of the flask?

20. A concrete walk is poured on a day when the temperature is 20.0°C in such a way that the ends are unable to move. (a) What is the stress in the cement on a hot day of 50.0°C? (b) Does the concrete fracture? Take Young's modulus for concrete to be 7.00×10^9 N/m^2 and the compressive strength to be 2.00×10^9 N/m^2.

21. A hollow aluminum cylinder 20.0 cm deep has an internal capacity of 2.000 L at 20.0°C. It is completely filled with turpentine and then slowly warmed to 80.0°C. (a) How much turpentine overflows? (b) If the cylinder is then cooled back to 20.0°C, how far below the cylinder's rim does the turpentine's surface recede?

22. A beaker made of ordinary glass contains a lead sphere of diameter 4.00 cm firmly attached to its bottom. At a uniform temperature of -10.0°C, the beaker is filled to the brim with 118 cm^3 of mercury, which completely covers the sphere. How much mercury overflows from the beaker if the temperature is raised to 30.0°C?

23. A steel rod undergoes a stretching force of 500 N. Its cross-sectional area is 2.00 cm^2. Find the change in temperature that would elongate the rod by the same amount as the 500-N force does. Tables 12.1 and 19.1 are available to you.

24. The Golden Gate Bridge in San Francisco has a main span of length 1.28 km—one of the longest in the world. Imagine that a taut steel wire with this length and a cross-sectional area of 4.00×10^{-6} m^2 is laid on the bridge deck with its ends attached to the towers of the bridge, on a summer day when the temperature of the wire is 35.0°C. (a) When winter arrives, the towers stay the same distance apart and the bridge deck keeps the same shape as its expansion joints open. When the temperature drops to -10.0°C, what is the tension in the wire? Take Young's modulus for steel to be 20.0×10^{10} N/m^2. (b) Permanent deformation occurs if the stress in the steel exceeds its elastic limit of 3.00×10^8 N/m^2. At what temperature would this happen? (c) **What If?** How would your answers to (a) and (b) change if the Golden Gate Bridge were twice as long?

25. A certain telescope forms an image of part of a cluster of stars on a square silicon charge-coupled detector (CCD) chip 2.00 cm on each side. A star field is focused on the CCD chip when it is first turned on and its temperature is 20.0°C. The star field contains 5 342 stars scattered uniformly. To make the detector more sensitive, it is cooled to -100°C. How many star images then fit onto the chip? The average coefficient of linear expansion of silicon is 4.68×10^{-6} (°C)$^{-1}$.

Section 19.5 Macroscopic Description of an Ideal Gas

Note: Problem 8 in Chapter 1 can be assigned with this section.

26. Gas is contained in an 8.00-L vessel at a temperature of 20.0°C and a pressure of 9.00 atm. (a) Determine the number of moles of gas in the vessel. (b) How many molecules are there in the vessel?

27. An automobile tire is inflated with air originally at 10.0°C and normal atmospheric pressure. During the process, the air is compressed to 28.0% of its original volume and the temperature is increased to 40.0°C. (a) What is the tire pressure? (b) After the car is driven at high speed, the tire air temperature rises to 85.0°C and the interior volume of the tire increases by 2.00%. What is the new tire pressure (absolute) in pascals?

28. A tank having a volume of 0.100 m^3 contains helium gas at 150 atm. How many balloons can the tank blow up if each filled balloon is a sphere 0.300 m in diameter at an absolute pressure of 1.20 atm?

29. An auditorium has dimensions 10.0 m × 20.0 m × 30.0 m. How many molecules of air fill the auditorium at 20.0°C and a pressure of 101 kPa?

30. Imagine a baby alien playing with a spherical balloon the size of the Earth in the outer solar system. Helium gas inside the balloon has a uniform temperature of 50.0 K due to radiation from the Sun. The uniform pressure of the helium is equal to normal atmospheric pressure on Earth. (a) Find the mass of the gas in the balloon. (b) The baby blows an additional mass of 8.00×10^{20} kg of helium into the balloon. At the same time, she wanders closer to the Sun and the pressure in the balloon doubles. Find the new temperature inside the balloon, whose volume remains constant.

31. Just 9.00 g of water is placed in a 2.00-L pressure cooker and heated to 500°C. What is the pressure inside the container?

32. One mole of oxygen gas is at a pressure of 6.00 atm and a temperature of 27.0°C. (a) If the gas is heated at constant volume until the pressure triples, what is the final temperature? (b) If the gas is heated until both the pressure and volume are doubled, what is the final temperature?

33. The mass of a hot-air balloon and its cargo (not including the air inside) is 200 kg. The air outside is at 10.0°C and 101 kPa. The volume of the balloon is 400 m^3. To what temperature must the air in the balloon be heated before the balloon will lift off? (Air density at 10.0°C is 1.25 kg/m^3.)

34. Your father and your little brother are confronted with the same puzzle. Your father's garden sprayer and your brother's water cannon both have tanks with a capacity of 5.00 L (Figure P19.34). Your father inserts a negligible amount of concentrated insecticide into his tank. They both pour in 4.00 L of water and seal up their tanks, so that they also contain air at atmospheric pressure. Next, each uses a hand-operated piston pump to inject more air, until the absolute pressure in the tank reaches 2.40 atm and it becomes too difficult to move the pump handle. Now each uses his device to spray out water—not air—until the stream becomes feeble, as it does when the pressure in the tank reaches 1.20 atm. Then he must pump it up again, spray again, and so on. In order to spray out all the water, each finds that he must pump up the tank three times. This is the puzzle: most of the water sprays out as a result of the second pumping. The first and the third pumping-up processes seem just as difficult, but result in a disappointingly small amount of water coming out. Account for this phenomenon.

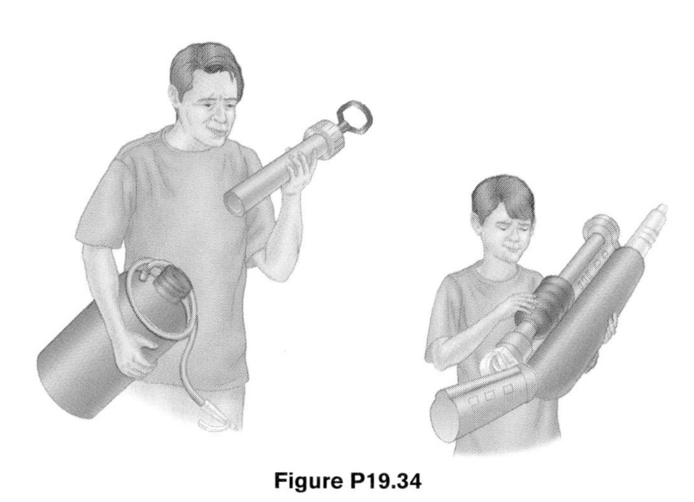

Figure P19.34

35. (a) Find the number of moles in one cubic meter of an ideal gas at 20.0°C and atmospheric pressure. (b) For air, Avogadro's number of molecules has mass 28.9 g. Calculate the mass of one cubic meter of air. Compare the result with the tabulated density of air.

36. The *void fraction* of a porous medium is the ratio of the void volume to the total volume of the material. The void is the hollow space within the material; it may be filled with a fluid. A cylindrical canister of diameter 2.54 cm and height 20.0 cm is filled with activated carbon having a void fraction of 0.765. Then it is flushed with an ideal gas at 25.0°C and pressure 12.5 atm. How many moles of gas are contained in the cylinder at the end of this process?

37. A cube 10.0 cm on each edge contains air (with equivalent molar mass 28.9 g/mol) at atmospheric pressure and temperature 300 K. Find (a) the mass of the gas, (b) its weight, and (c) the force it exerts on each face of the cube. (d) Comment on the physical reason why such a small sample can exert such a great force.

38. At 25.0 m below the surface of the sea (ρ = 1 025 kg/m^3), where the temperature is 5.00°C, a diver exhales an air bubble having a volume of 1.00 cm^3. If the surface temperature of the sea is 20.0°C, what is the volume of the bubble just before it breaks the surface?

39. The pressure gauge on a tank registers the gauge pressure, which is the difference between the interior and exterior pressure. When the tank is full of oxygen (O_2), it contains 12.0 kg of the gas at a gauge pressure of 40.0 atm. Determine the mass of oxygen that has been withdrawn from the tank when the pressure reading is 25.0 atm. Assume that the temperature of the tank remains constant.

40. Estimate the mass of the air in your bedroom. State the quantities you take as data and the value you measure or estimate for each.

41. A popular brand of cola contains 6.50 g of carbon dioxide dissolved in 1.00 L of soft drink. If the evaporating carbon dioxide is trapped in a cylinder at 1.00 atm and 20.0°C, what volume does the gas occupy?

42. In state-of-the-art vacuum systems, pressures as low as 10^{-9} Pa are being attained. Calculate the number of molecules in a 1.00-m^3 vessel at this pressure if the temperature is 27.0°C.

43. A room of volume V contains air having equivalent molar mass M (in g/mol). If the temperature of the room is raised from T_1 to T_2, what mass of air will leave the room? Assume that the air pressure in the room is maintained at P_0.

44. A diving bell in the shape of a cylinder with a height of 2.50 m is closed at the upper end and open at the lower end. The bell is lowered from air into sea water (ρ = 1.025 g/cm^3). The air in the bell is initially at 20.0°C. The bell is lowered to a depth (measured to the bottom of the bell) of 45.0 fathoms or 82.3 m. At this depth the water temperature is 4.0°C, and the bell is in thermal equilibrium with the water. (a) How high does sea water rise in the bell? (b) To what minimum pressure must the air in the bell be raised to expel the water that entered?

Additional Problems

45. A student measures the length of a brass rod with a steel tape at 20.0°C. The reading is 95.00 cm. What will the tape indicate for the length of the rod when the rod and the tape are at (a) − 15.0°C and (b) 55.0°C?

46. The density of gasoline is 730 kg/m^3 at 0°C. Its average coefficient of volume expansion is 9.60×10^{-4}/°C. If 1.00 gal of gasoline occupies 0.003 80 m^3, how many extra kilograms of gasoline would you get if you bought 10.0 gal of gasoline at 0°C rather than at 20.0°C from a pump that is not temperature compensated?

47. A mercury thermometer is constructed as shown in Figure P19.47. The capillary tube has a diameter of 0.004 00 cm,

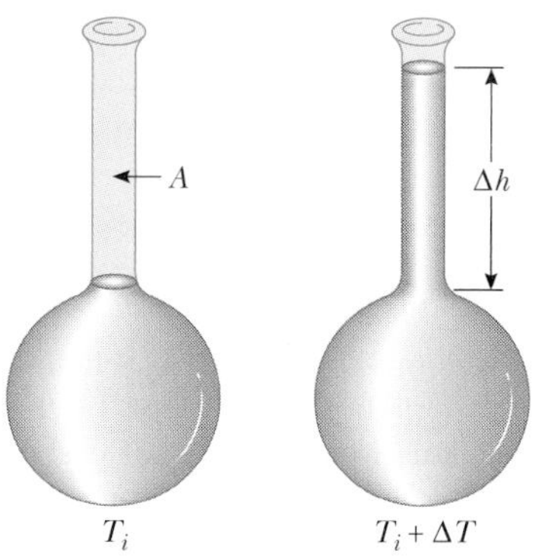

Figure P19.47 Problems 47 and 48.

and the bulb has a diameter of 0.250 cm. Neglecting the expansion of the glass, find the change in height of the mercury column that occurs with a temperature change of 30.0°C.

48. A liquid with a coefficient of volume expansion β just fills a spherical shell of volume V_i at a temperature of T_i (see Fig. P19.47). The shell is made of a material that has an average coefficient of linear expansion α. The liquid is free to expand into an open capillary of area A projecting from the top of the sphere. (a) If the temperature increases by ΔT, show that the liquid rises in the capillary by the amount Δh given by $\Delta h = (V_i/A)(\beta - 3\alpha)\Delta T$. (b) For a typical system, such as a mercury thermometer, why is it a good approximation to neglect the expansion of the shell?

49. **Review problem.** An aluminum pipe, 0.655 m long at 20.0°C and open at both ends, is used as a flute. The pipe is cooled to a low temperature but then is filled with air at 20.0°C as soon as you start to play it. After that, by how much does its fundamental frequency change as the metal rises in temperature from 5.00°C to 20.0°C?

50. A cylinder is closed by a piston connected to a spring of constant 2.00×10^3 N/m (see Fig. P19.50). With the spring relaxed, the cylinder is filled with 5.00 L of gas at a pressure of 1.00 atm and a temperature of 20.0°C. (a) If the piston has a cross-sectional area of 0.010 0 m^2 and negligible mass, how high will it rise when the temperature is raised to 250°C? (b) What is the pressure of the gas at 250°C?

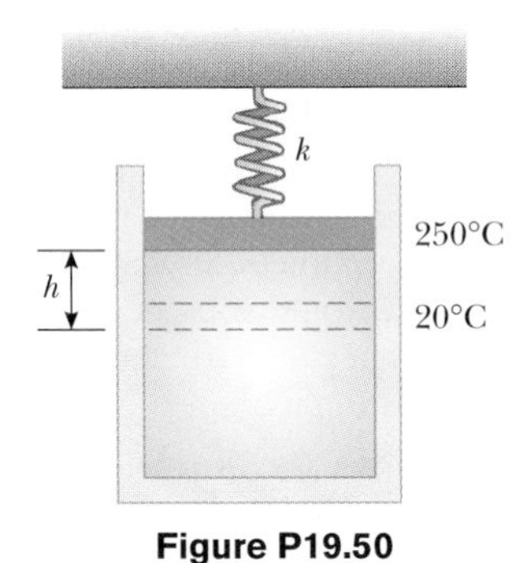

Figure P19.50

51. A liquid has a density ρ. (a) Show that the fractional change in density for a change in temperature ΔT is $\Delta\rho/\rho = -\beta\,\Delta T$. What does the negative sign signify? (b) Fresh water has a maximum density of 1.000 0 g/cm^3 at 4.0°C. At 10.0°C, its density is 0.999 7 g/cm^3. What is β for water over this temperature interval?

52. Long-term space missions require reclamation of the oxygen in the carbon dioxide exhaled by the crew. In one method of reclamation, 1.00 mol of carbon dioxide produces 1.00 mol of oxygen and 1.00 mol of methane as a byproduct. The methane is stored in a tank under pressure and is available to control the attitude of the spacecraft by controlled venting. A single astronaut exhales 1.09 kg of carbon dioxide each day. If the methane generated in the respiration recycling of three astronauts during one week of flight is stored in an originally empty 150-L tank at −45.0°C, what is the final pressure in the tank?

53. A vertical cylinder of cross-sectional area A is fitted with a tight-fitting, frictionless piston of mass m (Fig. P19.53). (a) If n moles of an ideal gas are in the cylinder at a temperature of T, what is the height h at which the piston is in equilibrium under its own weight? (b) What is the value for h if $n = 0.200$ mol, $T = 400$ K, $A = 0.008\,00$ m^2, and $m = 20.0$ kg?

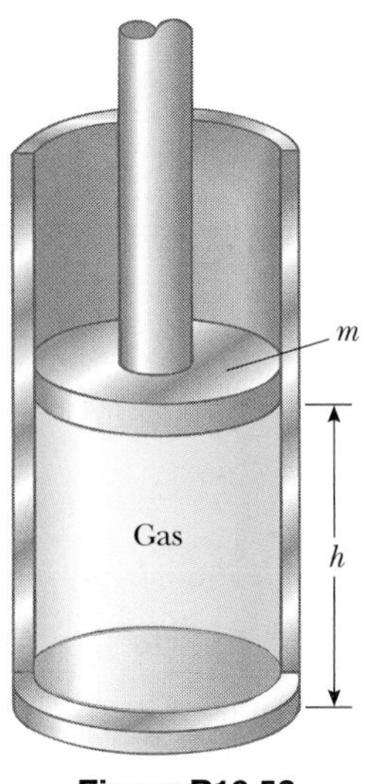

Figure P19.53

54. A bimetallic strip is made of two ribbons of dissimilar metals bonded together. (a) First assume the strip is originally straight. As they are heated, the metal with the greater average coefficient of expansion expands more than the other, forcing the strip into an arc, with the outer radius having a greater circumference (Fig. P19.54a). Derive an expression for the angle of bending θ as a function of the initial length of the strips, their average coefficients of linear expansion, the change in temperature, and the separation of the centers of the strips ($\Delta r = r_2 - r_1$). (b) Show that the angle of bending decreases to zero when ΔT decreases to zero and also when the two average coefficients of expansion become equal. (c) **What If?** What happens if the strip is cooled? (d) Figure P19.54b shows a compact spiral bimetallic strip in a home thermostat. The equation from part (a) applies to it as well, if θ is interpreted as the angle of additional bending caused by a change in temperature. The inner end of the spiral strip is fixed, and the outer end is free to move. Assume the metals are bronze and invar, the thickness of the strip is $2\,\Delta r = 0.500$ mm, and the overall length of the spiral strip is 20.0 cm. Find the angle through which the free end of the strip turns when the temperature changes by one Celsius degree. The free end of the strip supports a capsule partly filled with mercury, visible above the strip in Figure P19.54b. When the capsule tilts, the mercury shifts from one end to the other, to make or break an electrical contact switching the furnace on or off.

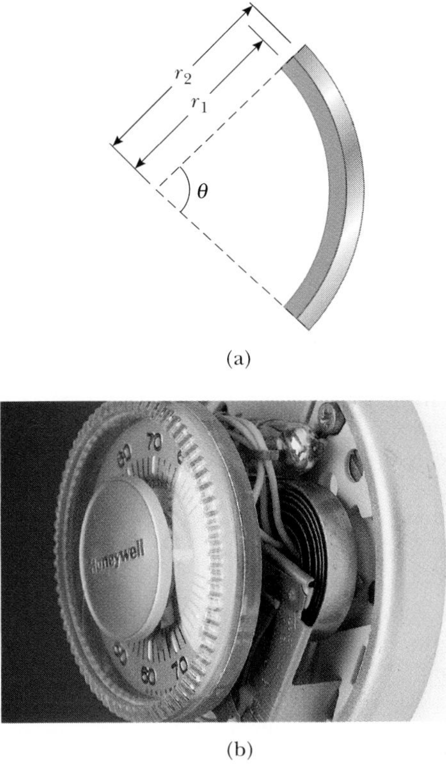

(b)

Figure P19.54

55. The rectangular plate shown in Figure P19.55 has an area A_i equal to ℓw. If the temperature increases by ΔT, each dimension increases according to the equation $\Delta L = \alpha L_i \Delta T$, where α is the average coefficient of linear expansion. Show that the increase in area is $\Delta A = 2\alpha A_i \Delta T$. What approximation does this expression assume?

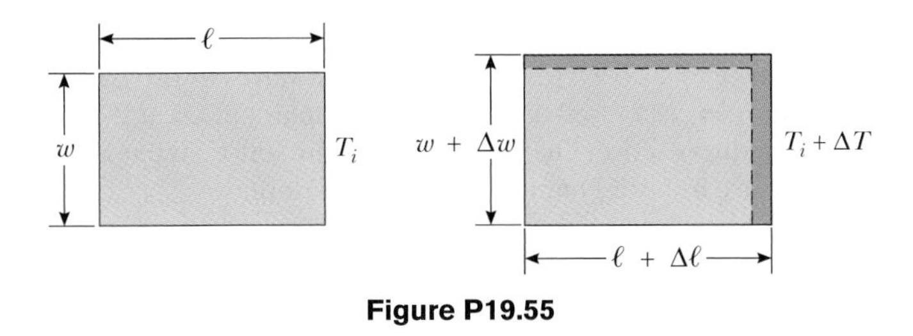

Figure P19.55

56. **Review problem.** A clock with a brass pendulum has a period of 1.000 s at 20.0°C. If the temperature increases to 30.0°C, (a) by how much does the period change, and (b) how much time does the clock gain or lose in one week?

57. **Review problem.** Consider an object with any one of the shapes displayed in Table 10.2. What is the percentage increase in the moment of inertia of the object when it is heated from 0°C to 100°C if it is composed of (a) copper or (b) aluminum? Assume that the average linear expansion coefficients shown in Table 19.1 do not vary between 0°C and 100°C.

58. (a) Derive an expression for the buoyant force on a spherical balloon, submerged in water, as a function of the depth below the surface, the volume of the balloon at the surface, the pressure at the surface, and the density of the water. (Assume water temperature does not change with depth.) (b) Does the buoyant force increase or decrease as the balloon is submerged? (c) At what depth is the buoyant force half the surface value?

59. A copper wire and a lead wire are joined together, end to end. The compound wire has an effective coefficient of linear expansion of $20.0 \times 10^{-6}\ (°\text{C})^{-1}$. What fraction of the length of the compound wire is copper?

60. **Review problem.** Following a collision in outer space, a copper disk at 850°C is rotating about its axis with an angular speed of 25.0 rad/s. As the disk radiates infrared light, its temperature falls to 20.0°C. No external torque acts on the disk. (a) Does the angular speed change as the disk cools off? Explain why. (b) What is its angular speed at the lower temperature?

61. Two concrete spans of a 250-m-long bridge are placed end to end so that no room is allowed for expansion (Fig. P19.61a). If a temperature increase of 20.0°C occurs, what is the height y to which the spans rise when they buckle (Fig. P19.61b)?

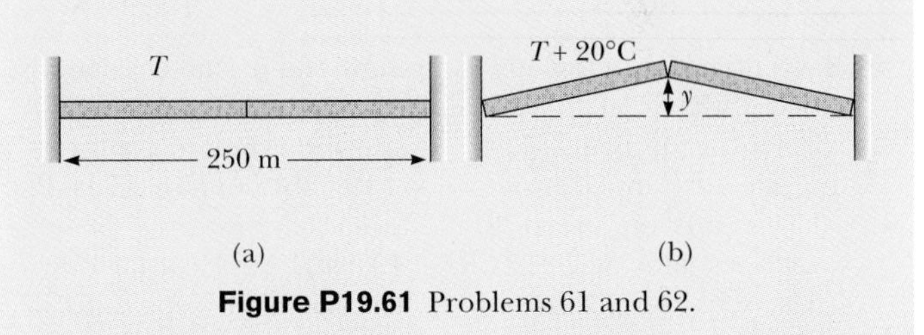

Figure P19.61 Problems 61 and 62.

62. Two concrete spans of a bridge of length L are placed end to end so that no room is allowed for expansion (Fig. P19.61a). If a temperature increase of ΔT occurs, what is the height y to which the spans rise when they buckle (Fig. P19.61b)?

63. (a) Show that the density of an ideal gas occupying a volume V is given by $\rho = PM/RT$, where M is the molar mass. (b) Determine the density of oxygen gas at atmospheric pressure and 20.0°C.

64. (a) Use the equation of state for an ideal gas and the definition of the coefficient of volume expansion, in the form $\beta = (1/V)\ dV/dT$, to show that the coefficient of volume expansion for an ideal gas at constant pressure is given by $\beta = 1/T$, where T is the absolute temperature. (b) What value does this expression predict for β at 0°C? Compare this result with the experimental values for helium and air in Table 19.1. Note that these are much larger than the coefficients of volume expansion for most liquids and solids.

65. Starting with Equation 19.10, show that the total pressure P in a container filled with a mixture of several ideal gases is $P = P_1 + P_2 + P_3 + \cdots$, where $P_1, P_2, \ldots$, are the pressures that each gas would exert if it alone filled the container (these individual pressures are called the *partial pres-*

sures of the respective gases). This result is known as *Dalton's law of partial pressures.*

66. A sample of dry air that has a mass of 100.00 g, collected at sea level, is analyzed and found to consist of the following gases:

    nitrogen (N_2) = 75.52 g
    oxygen (O_2) = 23.15 g
    argon (Ar) = 1.28 g
    carbon dioxide (CO_2) = 0.05 g

    plus trace amounts of neon, helium, methane, and other gases. (a) Calculate the partial pressure (see Problem 65) of each gas when the pressure is 1.013×10^5 Pa. (b) Determine the volume occupied by the 100-g sample at a temperature of 15.00°C and a pressure of 1.00 atm. What is the density of the air for these conditions? (c) What is the effective molar mass of the air sample?

67. Helium gas is sold in steel tanks. If the helium is used to inflate a balloon, could the balloon lift the spherical tank the helium came in? Justify your answer. Steel will rupture if subjected to tensile stress greater than its yield strength of 5×10^8 N/m^2. *Suggestion:* You may consider a steel shell of radius r and thickness t containing helium at high pressure and on the verge of breaking apart into two hemispheres.

68. A cylinder that has a 40.0-cm radius and is 50.0 cm deep is filled with air at 20.0°C and 1.00 atm (Fig. P19.68a). A 20.0-kg piston is now lowered into the cylinder, compressing the air trapped inside (Fig. P19.68b). Finally, a 75.0-kg man stands on the piston, further compressing the air, which remains at 20°C (Fig. P19.68c). (a) How far down (Δh) does the piston move when the man steps onto it? (b) To what temperature should the gas be heated to raise the piston and man back to h_i?

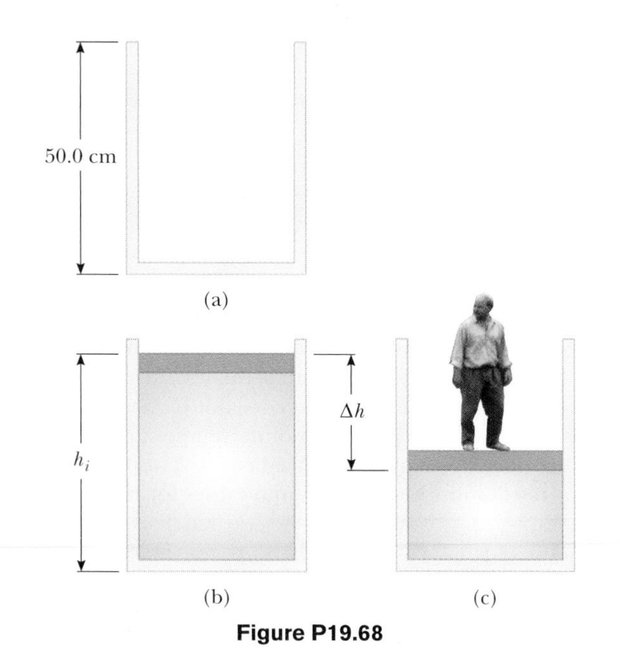

Figure P19.68

69. The relationship $L_f = L_i(1 + \alpha\Delta T)$ is an approximation that works when the average coefficient of expansion is small. If α is large, one must integrate the relationship $dL/dT = \alpha L$ to determine the final length. (a) Assuming that the coefficient of linear expansion is constant as L varies, determine a general expression for the final length. (b) Given a rod of length 1.00 m and a temperature change of 100.0°C, determine the error caused by the approximation when $\alpha = 2.00 \times 10^{-5}$ (°C)$^{-1}$ (a typical value for a metal) and when $\alpha = 0.020\,0$ (°C)$^{-1}$ (an unrealistically large value for comparison).

70. A steel wire and a copper wire, each of diameter 2.000 mm, are joined end to end. At 40.0°C, each has an unstretched length of 2.000 m; they are connected between two fixed supports 4.000 m apart on a tabletop, so that the steel wire extends from $x = -2.000$ m to $x = 0$, the copper wire extends from $x = 0$ to $x = 2.000$ m, and the tension is negligible. The temperature is then lowered to 20.0°C. At this lower temperature, find the tension in the wire and the x coordinate of the junction between the wires. (Refer to Tables 12.1 and 19.1.)

71. **Review problem.** A steel guitar string with a diameter of 1.00 mm is stretched between supports 80.0 cm apart. The temperature is 0.0°C. (a) Find the mass per unit length of this string. (Use the value 7.86×10^3 kg/m^3 for the density.) (b) The fundamental frequency of transverse oscillations of the string is 200 Hz. What is the tension in the string? (c) If the temperature is raised to 30.0°C, find the resulting values of the tension and the fundamental frequency. Assume that both the Young's modulus (Table 12.1) and the average coefficient of expansion (Table 19.1) have constant values between 0.0°C and 30.0°C.

72. In a chemical processing plant, a reaction chamber of fixed volume V_0 is connected to a reservoir chamber of fixed volume $4V_0$ by a passage containing a thermally insulating porous plug. The plug permits the chambers to be at different temperatures. The plug allows gas to pass from either chamber to the other, ensuring that the pressure is the same in both. At one point in the processing, both chambers contain gas at a pressure of 1.00 atm and a temperature of 27.0°C. Intake and exhaust valves to the pair of chambers are closed. The reservoir is maintained at 27.0°C while the reaction chamber is heated to 400°C. What is the pressure in both chambers after this is done?

73. A 1.00-km steel railroad rail is fastened securely at both ends when the temperature is 20.0°C. As the temperature increases, the rail begins to buckle. If its shape is an arc of a vertical circle, find the height h of the center of the rail when the temperature is 25.0°C. You will need to solve a transcendental equation.

74. **Review problem.** A perfectly plane house roof makes an angle θ with the horizontal. When its temperature changes, between T_c before dawn each day to T_h in the middle of each afternoon, the roof expands and contracts uniformly with a coefficient of thermal expansion α_1. Resting on the roof is a flat rectangular metal plate with expansion coefficient α_2, greater than α_1. The length of the plate is L, measured up the slope of the roof. The component of the plate's weight perpendicular to the roof is supported by a normal force uniformly distributed over the area of the plate. The coefficient of kinetic friction between the plate and the roof is μ_k. The plate is always at the same tempera-

ture as the roof, so we assume its temperature is continuously changing. Because of the difference in expansion coefficients, each bit of the plate is moving relative to the roof below it, except for points along a certain horizontal line running across the plate. We call this the stationary line. If the temperature is rising, parts of the plate below the stationary line are moving down relative to the roof and feel a force of kinetic friction acting up the roof. Elements of area above the stationary line are sliding up the roof and on them kinetic friction acts downward parallel to the roof. The stationary line occupies no area, so we assume no force of static friction acts on the plate while the temperature is changing. The plate as a whole is very nearly in equilibrium, so the net friction force on it must be equal to the component of its weight acting down the incline. (a) Prove that the stationary line is at a distance of

$$\frac{L}{2}\left(1 - \frac{\tan\theta}{\mu_k}\right)$$

below the top edge of the plate. (b) Analyze the forces that act on the plate when the temperature is falling, and prove that the stationary line is at that same distance above the bottom edge of the plate. (c) Show that the plate steps down the roof like an inchworm, moving each day by the distance

$$\frac{L(\alpha_2 - \alpha_1)(T_h - T_c)\tan\theta}{\mu_k}$$

(d) Evaluate the distance an aluminum plate moves each day if its length is 1.20 m, if the temperature cycles between 4.00°C and 36.0°C, and if the roof has slope 18.5°, coefficient of linear expansion 1.50×10^{-5} $(°C)^{-1}$, and coefficient of friction 0.420 with the plate. (e) **What If?** What if the expansion coefficient of the plate is less than that of the roof? Will the plate creep up the roof?

Answers to Quick Quizzes

19.1 (c). The direction of the transfer of energy depends only on temperature and not on the size of the object or on which object has more mass.

19.2 (c). The phrase "twice as hot" refers to a ratio of temperatures. When the given temperatures are converted to kelvins, only those in part (c) are in the correct ratio.

19.3 (c). Gasoline has the largest average coefficient of volume expansion.

19.4 (c). A cavity in a material expands in the same way as if it were filled with material.

19.5 (a). On a cold day, the trapped air in the bubbles is reduced in pressure, according to the ideal gas law. Thus, the volume of the bubbles may be smaller than on a hot day, and the package contents can shift more.

19.6 (b). Because of the decreased temperature of the helium, the pressure in the balloon is reduced. The atmospheric pressure around the balloon then compresses it to a smaller size until the pressure in the balloon reaches the atmospheric pressure.

19.7 (b). Because of the increased temperature, the air expands. Consequently, some of the air leaks to the outside, leaving less air in the house.

Chapter 20

Heat and the First Law of Thermodynamics

CHAPTER OUTLINE

▲ *In this photograph of Bow Lake in Banff National Park, Alberta, we see evidence of water in all three phases. In the lake is liquid water, and solid water in the form of snow appears on the ground. The clouds in the sky consist of liquid water droplets that have condensed from the gaseous water vapor in the air. Changes of a substance from one phase to another are a result of energy transfer. (Jacob Taposchaner/Getty Images)*

Until about 1850, the fields of thermodynamics and mechanics were considered to be two distinct branches of science, and the law of conservation of energy seemed to describe only certain kinds of mechanical systems. However, mid-nineteenth-century experiments performed by the Englishman James Joule and others showed that there was a strong connection between the transfer of energy by heat in thermal processes and the transfer of energy by work in mechanical processes. Today we know that internal energy, which we formally define in this chapter, can be transformed to mechanical energy. Once the concept of energy was generalized from mechanics to include internal energy, the law of conservation of energy emerged as a universal law of nature.

This chapter focuses on the concept of internal energy, the processes by which energy is transferred, the first law of thermodynamics, and some of the important applications of the first law. The first law of thermodynamics is a statement of conservation of energy. It describes systems in which the only energy change is that of internal energy and the transfers of energy are by heat and work. Furthermore, the first law makes no distinction between the results of heat and the results of work. According to the first law, a system's internal energy can be changed by an energy transfer to or from the system either by heat or by work. A major difference in our discussion of work in this chapter from that in the chapters on mechanics is that we will consider work done on *deformable* systems.

20.1 Heat and Internal Energy

At the outset, it is important that we make a major distinction between internal energy and heat. **Internal energy is all the energy of a system that is associated with its microscopic components—atoms and molecules—when viewed from a reference frame at rest with respect to the center of mass of the system.** The last part of this sentence ensures that any bulk kinetic energy of the system due to its motion through space is not included in internal energy. Internal energy includes kinetic energy of random translational, rotational, and vibrational motion of molecules, potential energy within molecules, and potential energy between molecules. It is useful to relate internal energy to the temperature of an object, but this relationship is limited—we show in Section 20.3 that internal energy changes can also occur in the absence of temperature changes.

Heat is defined as the transfer of energy across the boundary of a system due to a temperature difference between the system and its surroundings. When you *heat* a substance, you are transferring energy into it by placing it in contact with surroundings that have a higher temperature. This is the case, for example, when you place a pan of cold water on a stove burner—the burner is at a higher temperature than the water, and so the water gains energy. We shall also use the term *heat* to represent the amount of energy transferred by this method.

Scientists used to think of heat as a fluid called *caloric,* which they believed was transferred between objects; thus, they defined heat in terms of the temperature

PITFALL PREVENTION

20.1 Internal Energy, Thermal Energy, and Bond Energy

In reading other physics books, you may see terms such as *thermal energy* and *bond energy.* Thermal energy can be interpreted as that part of the internal energy associated with random motion of molecules and, therefore, related to temperature. Bond energy is the intermolecular potential energy. Thus,

internal energy = thermal energy + bond energy

While this breakdown is presented here for clarification with regard to other texts, we will not use these terms, because there is no need for them.

PITFALL PREVENTION

20.2 Heat, Temperature, and Internal Energy Are Different

As you read the newspaper or listen to the radio, be alert for incorrectly used phrases including the word *heat,* and think about the proper word to be used in place of *heat.* Incorrect examples include "As the truck braked to a stop, a large amount of heat was generated by friction" and "The heat of a hot summer day . . ."

changes produced in an object during heating. Today we recognize the distinct difference between internal energy and heat. Nevertheless, we refer to quantities using names that do not quite correctly define the quantities but which have become entrenched in physics tradition based on these early ideas. Examples of such quantities are *heat capacity* and *latent heat* (Sections 20.2 and 20.3).

As an analogy to the distinction between heat and internal energy, consider the distinction between work and mechanical energy discussed in Chapter 7. The work done on a system is a measure of the amount of energy transferred to the system from its surroundings, whereas the mechanical energy of the system (kinetic plus potential) is a consequence of the motion and configuration of the system. Thus, when a person does work on a system, energy is transferred from the person to the system. It makes no sense to talk about the work *of* a system—one can refer only to the work done *on* or *by* a system when some process has occurred in which energy has been transferred to or from the system. Likewise, it makes no sense to talk about the heat *of* a system—one can refer to *heat* only when energy has been transferred as a result of a temperature difference. Both heat and work are ways of changing the energy of a system.

It is also important to recognize that the internal energy of a system can be changed even when no energy is transferred by heat. For example, when a gas in an insulated container is compressed by a piston, the temperature of the gas and its internal energy increase, but no transfer of energy by heat from the surroundings to the gas has occurred. If the gas then expands rapidly, it cools and its internal energy decreases, but no transfer of energy by heat from it to the surroundings has taken place. The temperature changes in the gas are due not to a difference in temperature between the gas and its surroundings but rather to the compression and the expansion. In each case, energy is transferred to or from the gas by *work.* The changes in internal energy in these examples are evidenced by corresponding changes in the temperature of the gas.

James Prescott Joule
British physicist (1818–1889)

Joule received some formal education in mathematics, philosophy, and chemistry from John Dalton but was in large part self-educated. Joule's research led to the establishment of the principle of conservation of energy. His study of the quantitative relationship among electrical, mechanical, and chemical effects of heat culminated in his announcement in 1843 of the amount of work required to produce a unit of energy, called the mechanical equivalent of heat. *(By kind permission of the President and Council of the Royal Society)*

Units of Heat

As we have mentioned, early studies of heat focused on the resultant increase in temperature of a substance, which was often water. The early notions of heat based on caloric suggested that the flow of this fluid from one substance to another caused changes in temperature. From the name of this mythical fluid, we have an energy unit related to thermal processes, the **calorie (cal),** which is defined as **the amount of energy transfer necessary to raise the temperature of 1 g of water from 14.5°C to 15.5°C.**[1] (Note that the "Calorie," written with a capital "C" and used in describing the energy content of foods, is actually a kilocalorie.) The unit of energy in the U.S. customary system is the **British thermal unit (Btu),** which is defined as **the amount of energy transfer required to raise the temperature of 1 lb of water from 63°F to 64°F.**

Scientists are increasingly using the SI unit of energy, the *joule,* when describing thermal processes. In this textbook, heat, work, and internal energy are usually measured in joules. (Note that both heat and work are measured in energy units. Do not confuse these two means of energy *transfer* with energy itself, which is also measured in joules.)

The Mechanical Equivalent of Heat

In Chapters 7 and 8, we found that whenever friction is present in a mechanical system, some mechanical energy is lost—in other words, mechanical energy is not conserved in the presence of nonconservative forces. Various experiments show that this lost mechanical energy does not simply disappear but is transformed into internal

[1] Originally, the calorie was defined as the "heat" necessary to raise the temperature of 1 g of water by 1°C. However, careful measurements showed that the amount of energy required to produce a 1°C change depends somewhat on the initial temperature; hence, a more precise definition evolved.

energy. We can perform such an experiment at home by simply hammering a nail into a scrap piece of wood. What happens to all the kinetic energy of the hammer once we have finished? Some of it is now in the nail as internal energy, as demonstrated by the fact that the nail is measurably warmer. Although this connection between mechanical and internal energy was first suggested by Benjamin Thompson, it was Joule who established the equivalence of these two forms of energy.

A schematic diagram of Joule's most famous experiment is shown in Figure 20.1. The system of interest is the water in a thermally insulated container. Work is done on the water by a rotating paddle wheel, which is driven by heavy blocks falling at a constant speed. The temperature of the stirred water increases due to the friction between it and the paddles. If the energy lost in the bearings and through the walls is neglected, then the loss in potential energy associated with the blocks equals the work done by the paddle wheel on the water. If the two blocks fall through a distance h, the loss in potential energy is $2mgh$, where m is the mass of one block; this energy causes the temperature of the water to increase. By varying the conditions of the experiment, Joule found that the loss in mechanical energy $2mgh$ is proportional to the increase in water temperature ΔT. The proportionality constant was found to be approximately $4.18\text{ J/g}\cdot°\text{C}$. Hence, 4.18 J of mechanical energy raises the temperature of 1 g of water by 1°C. More precise measurements taken later demonstrated the proportionality to be $4.186\text{ J/g}\cdot°\text{C}$ when the temperature of the water was raised from 14.5°C to 15.5°C. We adopt this "15-degree calorie" value:

$$1\text{ cal} \equiv 4.186\text{ J} \tag{20.1}$$

This equality is known, for purely historical reasons, as the **mechanical equivalent of heat.**

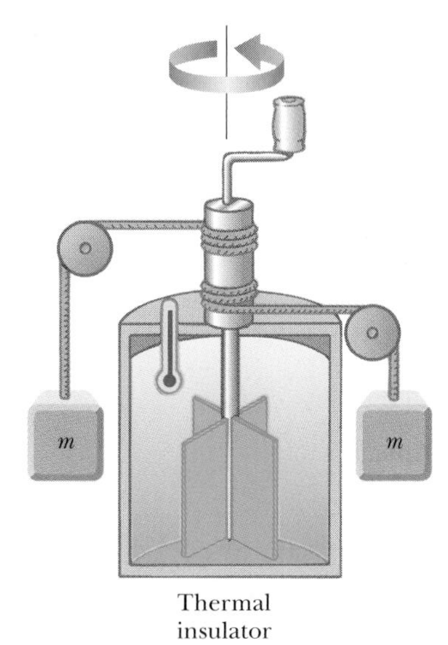

Figure 20.1 Joule's experiment for determining the mechanical equivalent of heat. The falling blocks rotate the paddles, causing the temperature of the water to increase.

Example 20.1 Losing Weight the Hard Way

A student eats a dinner rated at 2 000 Calories. He wishes to do an equivalent amount of work in the gymnasium by lifting a 50.0-kg barbell. How many times must he raise the barbell to expend this much energy? Assume that he raises the barbell 2.00 m each time he lifts it and that he regains no energy when he lowers the barbell.

Solution Because 1 Calorie = 1.00×10^3 cal, the total amount of work required to be done on the barbell–Earth system is 2.00×10^6 cal. Converting this value to joules, we have

$$W = (2.00 \times 10^6\text{ cal})(4.186\text{ J/cal}) = 8.37 \times 10^6\text{ J}$$

The work done in lifting the barbell a distance h is equal to mgh, and the work done in lifting it n times is $nmgh$. We equate this to the total work required:

$$W = nmgh = 8.37 \times 10^6\text{ J}$$

$$n = \frac{W}{mgh} = \frac{8.37 \times 10^6\text{ J}}{(50.0\text{ kg})(9.80\text{ m/s}^2)(2.00\text{ m})}$$

$$= 8.54 \times 10^3\text{ times}$$

If the student is in good shape and lifts the barbell once every 5 s, it will take him about 12 h to perform this feat. Clearly, it is much easier for this student to lose weight by dieting.

In reality, the human body is not 100% efficient. Thus, not all of the energy transformed within the body from the dinner transfers out of the body by work done on the barbell. Some of this energy is used to pump blood and perform other functions within the body. Thus, the 2 000 Calories can be worked off in less time than 12 h when these other energy requirements are included.

20.2 Specific Heat and Calorimetry

When energy is added to a system and there is no change in the kinetic or potential energy of the system, the temperature of the system usually rises. (An exception to this statement is the case in which a system undergoes a change of state—also called a *phase transition*—as discussed in the next section.) If the system consists of a sample of a substance, we find that the quantity of energy required to raise the temperature of a given mass of the substance by some amount varies from one substance to another. For example, the quantity of energy required to raise the temperature of 1 kg of water by 1°C is 4 186 J, but the quantity of energy required to raise the temperature of 1 kg of copper by 1°C is only 387 J. In the discussion that follows, we shall use heat as our example of energy transfer, but keep in mind that we could change the temperature of our system by means of any method of energy transfer.

The **heat capacity** C of a particular sample of a substance is defined as the amount of energy needed to raise the temperature of that sample by 1°C. From this definition, we see that if energy Q produces a change ΔT in the temperature of a sample, then

$$Q = C\Delta T \tag{20.2}$$

The **specific heat** c of a substance is the heat capacity per unit mass. Thus, if energy Q transfers to a sample of a substance with mass m and the temperature of the sample changes by ΔT, then the specific heat of the substance is

Specific heat

$$c \equiv \frac{Q}{m\,\Delta T} \tag{20.3}$$

Specific heat is essentially a measure of how thermally insensitive a substance is to the addition of energy. The greater a material's specific heat, the more energy must be added to a given mass of the material to cause a particular temperature change. Table 20.1 lists representative specific heats.

From this definition, we can relate the energy Q transferred between a sample of mass m of a material and its surroundings to a temperature change ΔT as

$$Q = mc\,\Delta T \tag{20.4}$$

PITFALL PREVENTION

20.3 An Unfortunate Choice of Terminology

The name *specific heat* is an unfortunate holdover from the days when thermodynamics and mechanics developed separately. A better name would be *specific energy transfer,* but the existing term is too entrenched to be replaced.

Table 20.1

Specific Heats of Some Substances at 25°C and Atmospheric Pressure

	Specific heat c	
Substance	**J/kg·°C**	**cal/g·°C**
Elemental solids		
Aluminum	900	0.215
Beryllium	1 830	0.436
Cadmium	230	0.055
Copper	387	0.092 4
Germanium	322	0.077
Gold	129	0.030 8
Iron	448	0.107
Lead	128	0.030 5
Silicon	703	0.168
Silver	234	0.056
Other solids		
Brass	380	0.092
Glass	837	0.200
Ice (−5°C)	2 090	0.50
Marble	860	0.21
Wood	1 700	0.41
Liquids		
Alcohol (ethyl)	2 400	0.58
Mercury	140	0.033
Water (15°C)	4 186	1.00
Gas		
Steam (100°C)	2 010	0.48

For example, the energy required to raise the temperature of 0.500 kg of water by 3.00°C is $(0.500 \text{ kg})(4\,186 \text{ J/kg}\cdot{}^\circ\text{C})(3.00^\circ\text{C}) = 6.28 \times 10^3$ J. Note that when the temperature increases, Q and ΔT are taken to be positive, and energy transfers into the system. When the temperature decreases, Q and ΔT are negative, and energy transfers out of the system.

Specific heat varies with temperature. However, if temperature intervals are not too great, the temperature variation can be ignored and c can be treated as a constant.[2] For example, the specific heat of water varies by only about 1% from 0°C to 100°C at atmospheric pressure. Unless stated otherwise, we shall neglect such variations.

Measured values of specific heats are found to depend on the conditions of the experiment. In general, measurements made in a constant-pressure process are different from those made in a constant-volume process. For solids and liquids, the difference between the two values is usually no greater than a few percent and is often neglected. Most of the values given in Table 20.1 were measured at atmospheric pressure and room temperature. The specific heats for gases measured at constant pressure are quite different from values measured at constant volume (see Chapter 21).

PITFALL PREVENTION

20.4 Energy Can Be Transferred by Any Method

We will use Q to represent the amount of energy transferred, but keep in mind that the energy transfer in Equation 20.4 could be by *any* of the methods introduced in Chapter 7; it does not have to be heat. For example, repeatedly bending a coat hanger wire raises the temperature at the bending point by *work*.

Quick Quiz 20.1 Imagine you have 1 kg each of iron, glass, and water, and that all three samples are at 10°C. Rank the samples from lowest to highest temperature after 100 J of energy is added to each sample.

Quick Quiz 20.2 Considering the same samples as in Quick Quiz 20.1, rank them from least to greatest amount of energy transferred by heat if each sample increases in temperature by 20°C.

It is interesting to note from Table 20.1 that water has the highest specific heat of common materials. This high specific heat is responsible, in part, for the moderate temperatures found near large bodies of water. As the temperature of a body of water decreases during the winter, energy is transferred from the cooling water to the air by heat, increasing the internal energy of the air. Because of the high specific heat of water, a relatively large amount of energy is transferred to the air for even modest temperature changes of the water. The air carries this internal energy landward when prevailing winds are favorable. For example, the prevailing winds on the West Coast of the United States are toward the land (eastward). Hence, the energy liberated by the Pacific Ocean as it cools keeps coastal areas much warmer than they would otherwise be. This explains why the western coastal states generally have more favorable winter weather than the eastern coastal states, where the prevailing winds do not tend to carry the energy toward land.

Conservation of Energy: Calorimetry

One technique for measuring specific heat involves heating a sample to some known temperature T_x, placing it in a vessel containing water of known mass and temperature $T_w < T_x$, and measuring the temperature of the water after equilibrium has been reached. This technique is called **calorimetry**, and devices in which this energy transfer occurs are called **calorimeters**. If the system of the sample and the water is isolated, the law of the conservation of energy requires that the amount of energy that leaves the sample (of unknown specific heat) equal the amount of energy that enters the water.[3]

[2] The definition given by Equation 20.3 assumes that the specific heat does not vary with temperature over the interval $\Delta T = T_f - T_i$. In general, if c varies with temperature over the interval, then the correct expression for Q is $Q = m\int_{T_i}^{T_f} c\, dT$.

[3] For precise measurements, the water container should be included in our calculations because it also exchanges energy with the sample. Doing so would require a knowledge of its mass and composition, however. If the mass of the water is much greater than that of the container, we can neglect the effects of the container.

PITFALL PREVENTION

20.5 Remember the Negative Sign

It is *critical* to include the negative sign in Equation 20.5. The negative sign in the equation is necessary for consistency with our sign convention for energy transfer. The energy transfer Q_{hot} has a negative value because energy is leaving the hot substance. The negative sign in the equation assures that the right-hand side is a positive number, consistent with the left-hand side, which is positive because energy is entering the cold water.

Conservation of energy allows us to write the mathematical representation of this energy statement as

$$Q_{\text{cold}} = -Q_{\text{hot}} \tag{20.5}$$

The negative sign in the equation is necessary to maintain consistency with our sign convention for heat.

Suppose m_x is the mass of a sample of some substance whose specific heat we wish to determine. Let us call its specific heat c_x and its initial temperature T_x. Likewise, let m_w, c_w, and T_w represent corresponding values for the water. If T_f is the final equilibrium temperature after everything is mixed, then from Equation 20.4, we find that the energy transfer for the water is $m_w c_w(T_f - T_w)$, which is positive because $T_f > T_w$, and that the energy transfer for the sample of unknown specific heat is $m_x c_x(T_f - T_x)$, which is negative. Substituting these expressions into Equation 20.5 gives

$$m_w c_w(T_f - T_w) = -m_x c_x(T_f - T_x)$$

Solving for c_x gives

$$c_x = \frac{m_w c_w(T_f - T_w)}{m_x(T_x - T_f)}$$

Example 20.2 Cooling a Hot Ingot

A 0.050 0-kg ingot of metal is heated to 200.0°C and then dropped into a beaker containing 0.400 kg of water initially at 20.0°C. If the final equilibrium temperature of the mixed system is 22.4°C, find the specific heat of the metal.

Solution According to Equation 20.5, we can write

$$m_w c_w(T_f - T_w) = -m_x c_x(T_f - T_x)$$

$$(0.400\text{ kg})(4\,186\text{ J/kg}\cdot{}^\circ\text{C})(22.4^\circ\text{C} - 20.0^\circ\text{C}) = -(0.050\,0\text{ kg})(c_x)(22.4^\circ\text{C} - 200.0^\circ\text{C})$$

From this we find that

$$c_x = 453\text{ J/kg}\cdot{}^\circ\text{C}$$

The ingot is most likely iron, as we can see by comparing this result with the data given in Table 20.1. Note that the temperature of the ingot is initially above the steam point. Thus, some of the water may vaporize when we drop the ingot into the water. We assume that we have a sealed system and that this steam cannot escape. Because the final equilibrium temperature is lower than the steam point, any steam that does result recondenses back into water.

What If? **Suppose you are performing an experiment in the laboratory that uses this technique to determine the specific heat of a sample and you wish to decrease the overall uncertainty in your final result for c_x. Of the data given in the text of this example, changing which value would be most effective in decreasing the uncertainty?**

Answer The largest experimental uncertainty is associated with the small temperature difference of 2.4°C for $T_f - T_w$. For example, an uncertainty of 0.1°C in each of these two temperature readings leads to an 8% uncertainty in their difference. In order for this temperature difference to be larger experimentally, the most effective change is to *decrease the amount of water.*

Example 20.3 Fun Time for a Cowboy

A cowboy fires a silver bullet with a muzzle speed of 200 m/s into the pine wall of a saloon. Assume that all the internal energy generated by the impact remains with the bullet. What is the temperature change of the bullet?

Solution The kinetic energy of the bullet is

$$K = \tfrac{1}{2}mv^2$$

Because nothing in the environment is hotter than the bullet, the bullet gains no energy by heat. Its temperature increases because the kinetic energy is transformed to extra internal energy when the bullet is stopped by the wall. The temperature change is the same as that which would take place if energy $Q = K$ were transferred by heat from a stove to the bullet. If we imagine this latter process taking place, we can calculate ΔT from Equation 20.4. Using 234 J/kg · °C as the specific heat of silver (see Table 20.1), we obtain

$$(1) \qquad \Delta T = \frac{Q}{mc} = \frac{K}{mc} = \frac{\frac{1}{2}\cancel{m}(200\text{ m/s})^2}{\cancel{m}(234\text{ J/kg}\cdot{}^\circ\text{C})} = 85.5^\circ\text{C}$$

Note that the result does not depend on the mass of the bullet.

What If? **Suppose that the cowboy runs out of silver bullets and fires a lead bullet at the same speed into the wall. Will the temperature change of the bullet be larger or smaller?**

Answer Consulting Table 20.1, we find that the specific heat of lead is 128 J/kg · °C, which is smaller than that for silver. Thus, a given amount of energy input will raise lead to a higher temperature than silver and the final temperature of the lead bullet will be larger. In Equation (1), we substitute the new value for the specific heat:

$$\Delta T = \frac{Q}{mc} = \frac{K}{mc} = \frac{\frac{1}{2}\cancel{m}(200\text{ m/s})^2}{\cancel{m}(128\text{ J/kg}\cdot{}^\circ\text{C})} = 156^\circ\text{C}$$

Note that there is no requirement that the silver and lead bullets have the same mass to determine this temperature. The only requirement is that they have the same speed.

20.3 Latent Heat

A substance often undergoes a change in temperature when energy is transferred between it and its surroundings. There are situations, however, in which the transfer of energy does not result in a change in temperature. This is the case whenever the physical characteristics of the substance change from one form to another; such a change is commonly referred to as a **phase change**. Two common phase changes are from solid to liquid (melting) and from liquid to gas (boiling); another is a change in the crystalline structure of a solid. All such phase changes involve a change in internal energy but no change in temperature. The increase in internal energy in boiling, for example, is represented by the breaking of bonds between molecules in the liquid state; this bond breaking allows the molecules to move farther apart in the gaseous state, with a corresponding increase in intermolecular potential energy.

20.6 Signs Are Critical

Sign errors occur very often when students apply calorimetry equations, so we will make this point once again. For phase changes, use the correct explicit sign in Equation 20.6, depending on whether you are adding or removing energy from the substance. In Equation 20.4, there is no explicit sign to consider, but be sure that your ΔT is *always* the final temperature minus the initial temperature. In addition, make sure that you *always* include the negative sign on the right-hand side of Equation 20.5.

As you might expect, different substances respond differently to the addition or removal of energy as they change phase because their internal molecular arrangements vary. Also, the amount of energy transferred during a phase change depends on the amount of substance involved. (It takes less energy to melt an ice cube than it does to thaw a frozen lake.) If a quantity Q of energy transfer is required to change the phase of a mass m of a substance, the ratio $L \equiv Q/m$ characterizes an important thermal property of that substance. Because this added or removed energy does not result in a temperature change, the quantity L is called the **latent heat** (literally, the "hidden" heat) of the substance. The value of L for a substance depends on the nature of the phase change, as well as on the properties of the substance.

From the definition of latent heat, and again choosing heat as our energy transfer mechanism, we find that the energy required to change the phase of a given mass m of a pure substance is

Latent heat

$$Q = \pm mL \qquad (20.6)$$

Latent heat of fusion L_f is the term used when the phase change is from solid to liquid (*to fuse* means "to combine by melting"), and **latent heat of vaporization** L_v is the term used when the phase change is from liquid to gas (the liquid "vaporizes").[4] The latent heats of various substances vary considerably, as data in Table 20.2 show. The positive sign in Equation 20.6 is used when energy enters a system, causing melting or vaporization. The negative sign corresponds to energy leaving a system, such that the system freezes or condenses.

To understand the role of latent heat in phase changes, consider the energy required to convert a 1.00-g cube of ice at -30.0°C to steam at 120.0°C. Figure 20.2 indicates the experimental results obtained when energy is gradually added to the ice. Let us examine each portion of the red curve.

[4] When a gas cools, it eventually *condenses*—that is, it returns to the liquid phase. The energy given up per unit mass is called the *latent heat of condensation* and is numerically equal to the latent heat of vaporization. Likewise, when a liquid cools, it eventually solidifies, and the *latent heat of solidification* is numerically equal to the latent heat of fusion.

Table 20.2

Latent Heats of Fusion and Vaporization

Substance	Melting Point (°C)	Latent Heat of Fusion (J/kg)	Boiling Point (°C)	Latent Heat of Vaporization (J/kg)
Helium	−269.65	5.23×10^3	−268.93	2.09×10^4
Nitrogen	−209.97	2.55×10^4	−195.81	2.01×10^5
Oxygen	−218.79	1.38×10^4	−182.97	2.13×10^5
Ethyl alcohol	−114	1.04×10^5	78	8.54×10^5
Water	0.00	3.33×10^5	100.00	2.26×10^6
Sulfur	119	3.81×10^4	444.60	3.26×10^5
Lead	327.3	2.45×10^4	1 750	8.70×10^5
Aluminum	660	3.97×10^5	2 450	1.14×10^7
Silver	960.80	8.82×10^4	2 193	2.33×10^6
Gold	1 063.00	6.44×10^4	2 660	1.58×10^6
Copper	1 083	1.34×10^5	1 187	5.06×10^6

Part A. On this portion of the curve, the temperature of the ice changes from −30.0°C to 0.0°C. Because the specific heat of ice is 2 090 J/kg · °C, we can calculate the amount of energy added by using Equation 20.4:

$$Q = m_i c_i \, \Delta T = (1.00 \times 10^{-3} \text{ kg})(2\,090 \text{ J/kg} \cdot {}^\circ\text{C})(30.0^\circ\text{C}) = 62.7 \text{ J}$$

Part B. When the temperature of the ice reaches 0.0°C, the ice–water mixture remains at this temperature—even though energy is being added—until all the ice melts. The energy required to melt 1.00 g of ice at 0.0°C is, from Equation 20.6,

$$Q = m_i L_f = (1.00 \times 10^{-3} \text{ kg})(3.33 \times 10^5 \text{ J/kg}) = 333 \text{ J}$$

Thus, we have moved to the 396 J (= 62.7 J + 333 J) mark on the energy axis in Figure 20.2.

Part C. Between 0.0°C and 100.0°C, nothing surprising happens. No phase change occurs, and so all energy added to the water is used to increase its temperature. The amount of energy necessary to increase the temperature from 0.0°C to 100.0°C is

$$Q = m_w c_w \, \Delta T = (1.00 \times 10^{-3} \text{ kg})(4.19 \times 10^3 \text{ J/kg} \cdot {}^\circ\text{C})(100.0^\circ\text{C}) = 419 \text{ J}$$

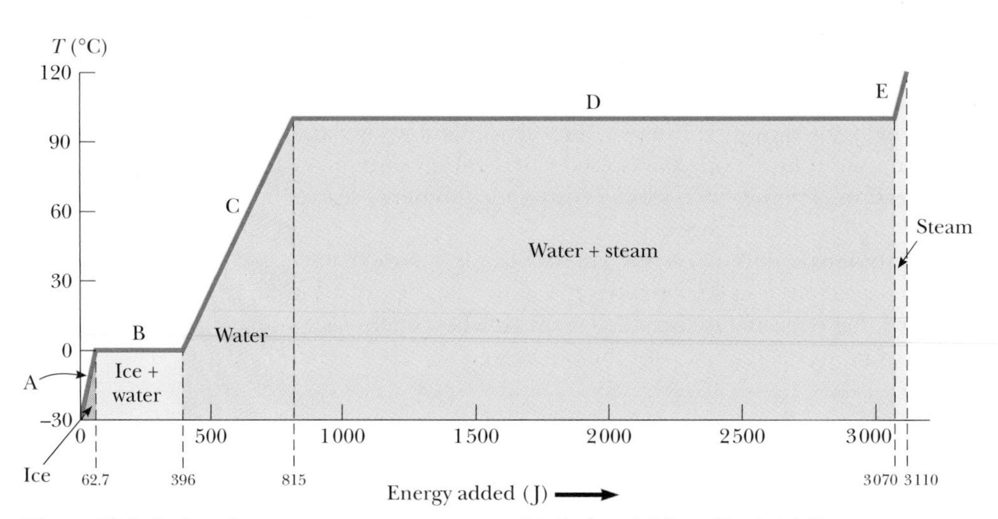

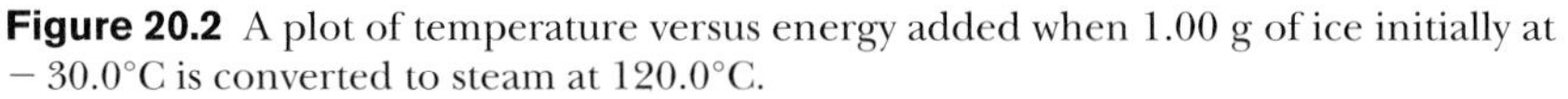

Figure 20.2 A plot of temperature versus energy added when 1.00 g of ice initially at −30.0°C is converted to steam at 120.0°C.

Part D. At 100.0°C, another phase change occurs as the water changes from water at 100.0°C to steam at 100.0°C. Similar to the ice–water mixture in part B, the water–steam mixture remains at 100.0°C—even though energy is being added—until all of the liquid has been converted to steam. The energy required to convert 1.00 g of water to steam at 100.0°C is

$$Q = m_w L_v = (1.00 \times 10^{-3}\ \text{kg})(2.26 \times 10^6\ \text{J/kg}) = 2.26 \times 10^3\ \text{J}$$

Part E. On this portion of the curve, as in parts A and C, no phase change occurs; thus, all energy added is used to increase the temperature of the steam. The energy that must be added to raise the temperature of the steam from 100.0°C to 120.0°C is

$$Q = m_s c_s\, \Delta T = (1.00 \times 10^{-3}\ \text{kg})(2.01 \times 10^3\ \text{J/kg}\cdot{}^\circ\text{C})(20.0^\circ\text{C}) = 40.2\ \text{J}$$

The total amount of energy that must be added to change 1 g of ice at −30.0°C to steam at 120.0°C is the sum of the results from all five parts of the curve, which is 3.11×10^3 J. Conversely, to cool 1 g of steam at 120.0°C to ice at −30.0°C, we must remove 3.11×10^3 J of energy.

Note in Figure 20.2 the relatively large amount of energy that is transferred into the water to vaporize it to steam. Imagine reversing this process—there is a large amount of energy transferred out of steam to condense it into water. This is why a burn to your skin from steam at 100°C is much more damaging than exposure of your skin to water at 100°C. A very large amount of energy enters your skin from the steam and the steam remains at 100°C for a long time while it condenses. Conversely, when your skin makes contact with water at 100°C, the water immediately begins to drop in temperature as energy transfers from the water to your skin.

We can describe phase changes in terms of a rearrangement of molecules when energy is added to or removed from a substance. (For elemental substances in which the atoms do not combine to form molecules, the following discussion should be interpreted in terms of atoms. We use the general term *molecules* to refer to both chemical compounds and elemental substances.) Consider first the liquid-to-gas phase change. The molecules in a liquid are close together, and the forces between them are stronger than those between the more widely separated molecules of a gas. Therefore, work must be done on the liquid against these attractive molecular forces if the molecules are to separate. The latent heat of vaporization is the amount of energy per unit mass that must be added to the liquid to accomplish this separation.

Similarly, for a solid, we imagine that the addition of energy causes the amplitude of vibration of the molecules about their equilibrium positions to become greater as the temperature increases. At the melting point of the solid, the amplitude is great enough to break the bonds between molecules and to allow molecules to move to new positions. The molecules in the liquid also are bound to each other, but less strongly than those in the solid phase. The latent heat of fusion is equal to the energy required per unit mass to transform the bonds among all molecules from the solid-type bond to the liquid-type bond.

As you can see from Table 20.2, the latent heat of vaporization for a given substance is usually somewhat higher than the latent heat of fusion. This is not surprising if we consider that the average distance between molecules in the gas phase is much greater than that in either the liquid or the solid phase. In the solid-to-liquid phase change, we transform solid-type bonds between molecules into liquid-type bonds between molecules, which are only slightly less strong. In the liquid-to-gas phase change, however, we break liquid-type bonds and create a situation in which the molecules of the gas essentially are not bonded to each other. Therefore, it is not surprising that more energy is required to vaporize a given mass of substance than is required to melt it.

Quick Quiz 20.3 Suppose the same process of adding energy to the ice cube is performed as discussed above, but we graph the internal energy of the system as a function of energy input. What would this graph look like?

PITFALL PREVENTION

20.7 Celsius vs. Kelvin

In equations in which T appears—for example, the ideal gas law—the Kelvin temperature *must* be used. In equations involving ΔT, such as calorimetry equations, it is *possible* to use Celsius temperatures, because a change in temperature is the same on both scales. It is *safest*, however, to *consistently* use Kelvin temperatures in all equations involving T or ΔT.

Quick Quiz 20.4 Calculate the slopes for the A, C, and E portions of Figure 20.2. Rank the slopes from least to greatest and explain what this ordering means.

PROBLEM-SOLVING HINTS

Calorimetry Problems

If you have difficulty in solving calorimetry problems, be sure to consider the following points:

- Units of measure must be consistent. For instance, if you are using specific heats measured in J/kg · °C, be sure that masses are in kilograms and temperatures are in Celsius degrees.
- Transfers of energy are given by the equation $Q = mc\,\Delta T$ only for those processes in which no phase changes occur. Use the equations $Q = \pm mL_f$ and $Q = \pm mL_v$ only when phase changes *are* taking place; be sure to select the proper sign for these equations depending on the direction of energy transfer.
- Often, errors in sign are made when the equation $Q_{\text{cold}} = -Q_{\text{hot}}$ is used. Make sure that you use the negative sign in the equation, and remember that ΔT is always the final temperature minus the initial temperature.

Example 20.4 Cooling the Steam

What mass of steam initially at 130°C is needed to warm 200 g of water in a 100-g glass container from 20.0°C to 50.0°C?

Solution The steam loses energy in three stages. In the first stage, the steam is cooled to 100°C. The energy transfer in the process is

$$Q_1 = m_s c_s\,\Delta T = m_s(2.01 \times 10^3\ \text{J/kg}\cdot{}^\circ\text{C})(-30.0^\circ\text{C}) = -m_s(6.03 \times 10^4\ \text{J/kg})$$

where m_s is the unknown mass of the steam.

In the second stage, the steam is converted to water. To find the energy transfer during this phase change, we use $Q = -mL_v$, where the negative sign indicates that energy is leaving the steam:

$$Q_2 = -m_s(2.26 \times 10^6\ \text{J/kg})$$

In the third stage, the temperature of the water created from the steam is reduced to 50.0°C. This change requires an energy transfer of

$$Q_3 = m_s c_w \Delta T = m_s(4.19 \times 10^3\ \text{J/kg}\cdot{}^\circ\text{C})(-50.0^\circ\text{C}) = -m_s(2.09 \times 10^5\ \text{J/kg})$$

Adding the energy transfers in these three stages, we obtain

$$\begin{aligned} Q_{\text{hot}} &= Q_1 + Q_2 + Q_3 \\ &= -m_s[6.03 \times 10^4\ \text{J/kg} + 2.26 \times 10^6\ \text{J/kg} + 2.09 \times 10^5\ \text{J/kg}] \\ &= -m_s(2.53 \times 10^6\ \text{J/kg}) \end{aligned}$$

Now, we turn our attention to the temperature increase of the water and the glass. Using Equation 20.4, we find that

$$\begin{aligned} Q_{\text{cold}} &= (0.200\ \text{kg})(4.19 \times 10^3\ \text{J/kg}\cdot{}^\circ\text{C})(30.0^\circ\text{C}) \\ &\quad + (0.100\ \text{kg})(837\ \text{J/kg}\cdot{}^\circ\text{C})(30.0^\circ\text{C}) \\ &= 2.77 \times 10^4\ \text{J} \end{aligned}$$

Using Equation 20.5, we can solve for the unknown mass:

$$Q_{\text{cold}} = -Q_{\text{hot}}$$

$$2.77 \times 10^4\ \text{J} = -[-m_s(2.53 \times 10^6\ \text{J/kg})]$$

$$m_s = 1.09 \times 10^{-2}\ \text{kg} = 10.9\ \text{g}$$

What If? **What if the final state of the system is water at 100°C? Would we need more or less steam? How would the analysis above change?**

Answer More steam would be needed to raise the temperature of the water and glass to 100°C instead of 50.0°C. There would be two major changes in the analysis. First, we would not have a term Q_3 for the steam because the water that condenses from the steam does not cool below 100°C. Second, in Q_{cold}, the temperature change would be 80.0°C instead of 30.0°C. Thus, Q_{hot} becomes

$$\begin{aligned} Q_{\text{hot}} &= Q_1 + Q_2 \\ &= -m_s(6.03 \times 10^4\ \text{J/kg} + 2.26 \times 10^6\ \text{J/kg}) \\ &= -m_s(2.32 \times 10^6\ \text{J/kg}) \end{aligned}$$

and Q_{cold} becomes

$$\begin{aligned} Q_{\text{cold}} &= (0.200\ \text{kg})(4.19 \times 10^3\ \text{J/kg}\cdot{}^\circ\text{C})(80.0^\circ\text{C}) \\ &\quad + (0.100\ \text{kg})(837\ \text{J/kg}\cdot{}^\circ\text{C})(80.0^\circ\text{C}) \\ &= 7.37 \times 10^4\ \text{J} \end{aligned}$$

leading to $m_s = 3.18 \times 10^{-2}\ \text{kg} = 31.8\ \text{g}$.

Example 20.5 Boiling Liquid Helium

Liquid helium has a very low boiling point, 4.2 K, and a very low latent heat of vaporization, 2.09×10^4 J/kg. If energy is transferred to a container of boiling liquid helium from an immersed electric heater at a rate of 10.0 W, how long does it take to boil away 1.00 kg of the liquid?

Solution Because $L_v = 2.09 \times 10^4$ J/kg, we must supply 2.09×10^4 J of energy to boil away 1.00 kg. Because 10.0 W = 10.0 J/s, 10.0 J of energy is transferred to the helium each second. From $\mathcal{P} = \Delta E/\Delta t$, the time interval required to transfer 2.09×10^4 J of energy is

$$\Delta t = \frac{\Delta E}{\mathcal{P}} = \frac{2.09 \times 10^4 \text{ J}}{10.0 \text{ J/s}} = 2.09 \times 10^3 \text{ s} \approx 35 \text{ min}$$

20.4 Work and Heat in Thermodynamic Processes

In the macroscopic approach to thermodynamics, we describe the *state* of a system using such variables as pressure, volume, temperature, and internal energy. As a result, these quantities belong to a category called **state variables**. For any given configuration of the system, we can identify values of the state variables. It is important to note that a *macroscopic state* of an isolated system can be specified only if the system is in thermal equilibrium internally. In the case of a gas in a container, internal thermal equilibrium requires that every part of the gas be at the same pressure and temperature.

A second category of variables in situations involving energy is **transfer variables**. These variables are zero *unless* a process occurs in which energy is transferred across the boundary of the system. Because a transfer of energy across the boundary represents a change in the system, transfer variables are not associated with a given state of the system, but with a *change* in the state of the system. In the previous sections, we discussed heat as a transfer variable. For a given set of conditions of a system, there is no defined value for the heat. We can only assign a value of the heat if energy crosses the boundary by heat, resulting in a change in the system. State variables are characteristic of a system in thermal equilibrium. Transfer variables are characteristic of a process in which energy is transferred between a system and its environment.

In this section, we study another important transfer variable for thermodynamic systems—work. Work performed on particles was studied extensively in Chapter 7, and here we investigate the work done on a deformable system—a gas. Consider a gas contained in a cylinder fitted with a movable piston (Fig. 20.3). At equilibrium, the gas oc-

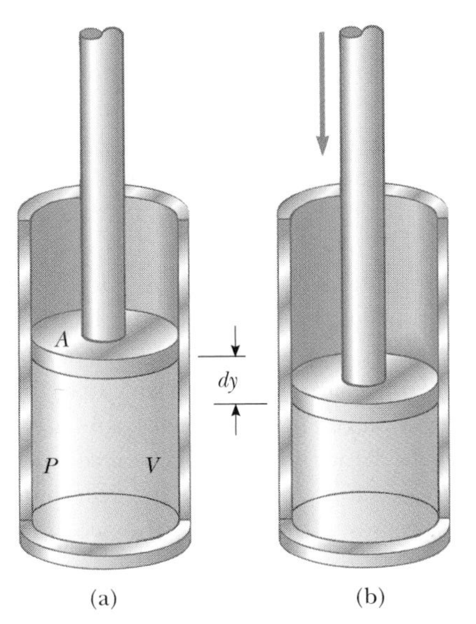

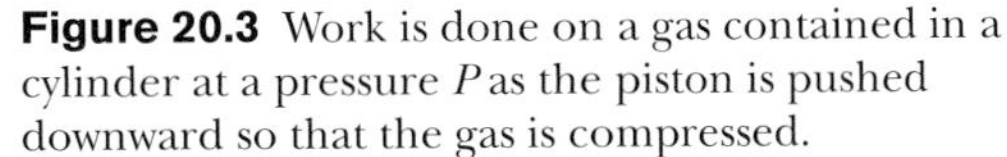

Figure 20.3 Work is done on a gas contained in a cylinder at a pressure P as the piston is pushed downward so that the gas is compressed.

cupies a volume V and exerts a uniform pressure P on the cylinder's walls and on the piston. If the piston has a cross-sectional area A, the force exerted by the gas on the piston is $F = PA$. Now let us assume that we push the piston inward and compress the gas **quasi-statically**, that is, slowly enough to allow the system to remain essentially in thermal equilibrium at all times. As the piston is pushed downward by an external force $\mathbf{F} = -F\hat{\mathbf{j}}$ through a displacement of $d\mathbf{r} = dy\hat{\mathbf{j}}$ (Fig. 20.3b), the work done on the gas is, according to our definition of work in Chapter 7,

$$dW = \mathbf{F} \cdot d\mathbf{r} = -F\hat{\mathbf{j}} \cdot dy\hat{\mathbf{j}} = -F\,dy = -PA\,dy$$

where we have set the magnitude F of the external force equal to PA because the piston is always in equilibrium between the external force and the force from the gas. For this discussion, we assume the mass of the piston is negligible. Because $A\,dy$ is the change in volume of the gas dV, we can express the work done on the gas as

$$dW = -P\,dV \tag{20.7}$$

If the gas is compressed, dV is negative and the work done on the gas is positive. If the gas expands, dV is positive and the work done on the gas is negative. If the volume remains constant, the work done on the gas is zero. The total work done on the gas as its volume changes from V_i to V_f is given by the integral of Equation 20.7:

Work done on a gas

$$W = -\int_{V_i}^{V_f} P\,dV \tag{20.8}$$

To evaluate this integral, one must know how the pressure varies with volume during the process.

In general, the pressure is not constant during a process followed by a gas, but depends on the volume and temperature. If the pressure and volume are known at each step of the process, the state of the gas at each step can be plotted on a graph called a ***PV* diagram**, as in Figure 20.4. This type of diagram allows us to visualize a process through which a gas is progressing. The curve on a PV diagram is called the *path* taken between the initial and final states.

Note that the integral in Equation 20.8 is equal to the area under a curve on a PV diagram. Thus, we can identify an important use for PV diagrams:

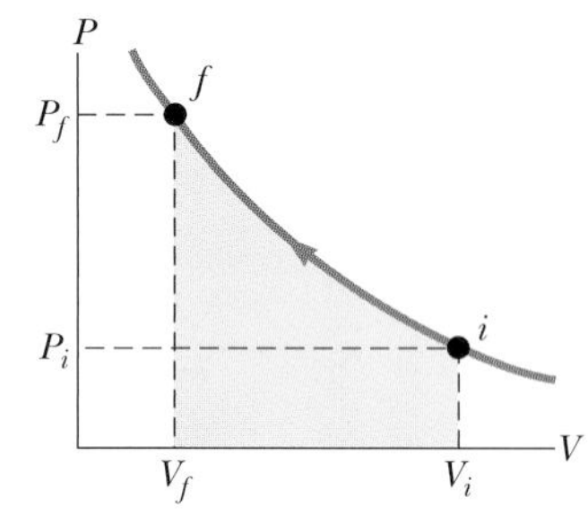

Active Figure 20.4 A gas is compressed quasi-statically (slowly) from state i to state f. The work done on the gas equals the negative of the area under the PV curve.

At the Active Figures link at http://www.pse6.com, you can compress the piston in Figure 20.3 and see the result on the PV diagram in this figure.

> The work done on a gas in a quasi-static process that takes the gas from an initial state to a final state is the negative of the area under the curve on a PV diagram, evaluated between the initial and final states.

As Figure 20.4 suggests, for our process of compressing a gas in the cylinder, the work done depends on the particular path taken between the initial and final states. To illustrate this important point, consider several different paths connecting i and f (Fig. 20.5). In the process depicted in Figure 20.5a, the volume of the gas is first reduced from V_i to V_f at constant pressure P_i and the pressure of the gas then increases from P_i to P_f by heating at constant volume V_f. The work done on the gas along this path is $-P_i(V_f - V_i)$. In Figure 20.5b, the pressure of the gas is increased from P_i to P_f at constant volume V_i and then the volume of the gas is reduced from V_i to V_f at constant pressure P_f. The work done on the gas is $-P_f(V_f - V_i)$, which is greater than that for the process described in Figure 20.5a. It is greater because the piston is moved through the same displacement by a larger force than for the situation in Figure 20.5a. Finally, for the process described in Figure 20.5c, where both P and V change continuously, the work done on the gas has some value intermediate between the values obtained in the first two processes. To evaluate the work in this case, the function $P(V)$ must be known, so that we can evaluate the integral in Equation 20.8.

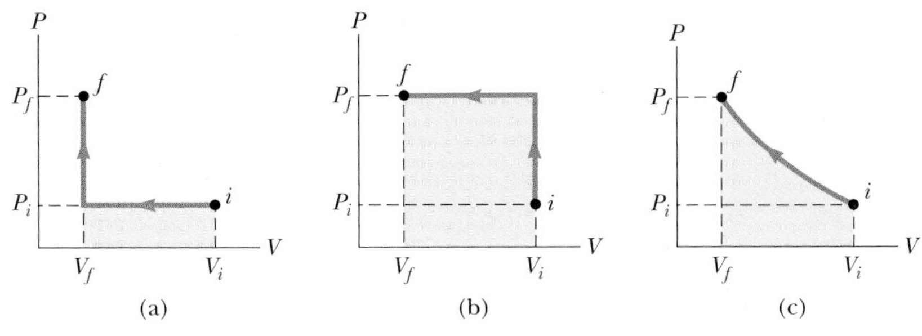

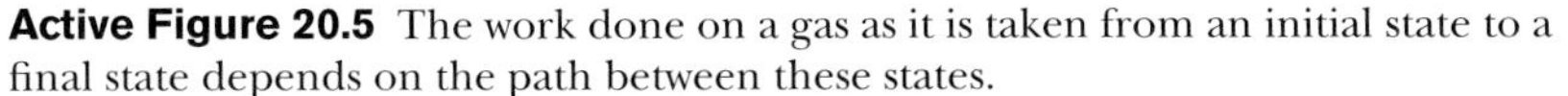
Active Figure 20.5 The work done on a gas as it is taken from an initial state to a final state depends on the path between these states.

At the Active Figures link at http://www.pse6.com, you can choose one of the three paths and see the movement of the piston in Figure 20.3 and of a point on the PV diagram in this figure.

The energy transfer Q into or out of a system by heat also depends on the process. Consider the situations depicted in Figure 20.6. In each case, the gas has the same initial volume, temperature, and pressure, and is assumed to be ideal. In Figure 20.6a, the gas is thermally insulated from its surroundings except at the bottom of the gas-filled region, where it is in thermal contact with an energy reservoir. An *energy reservoir* is a source of energy that is considered to be so great that a finite transfer of energy to or from the reservoir does not change its temperature. The piston is held at its initial position by an external agent—a hand, for instance. When the force holding the piston is reduced slightly, the piston rises very slowly to its final position. Because the piston is moving upward, the gas is doing work on the piston. During this expansion to the final volume V_f, just enough energy is transferred by heat from the reservoir to the gas to maintain a constant temperature T_i.

Now consider the completely thermally insulated system shown in Figure 20.6b. When the membrane is broken, the gas expands rapidly into the vacuum until it occupies a volume V_f and is at a pressure P_f. In this case, the gas does no work because it does not apply a force—no force is required to expand into a vacuum. Furthermore, no energy is transferred by heat through the insulating wall.

The initial and final states of the ideal gas in Figure 20.6a are identical to the initial and final states in Figure 20.6b, but the paths are different. In the first case, the gas does work on the piston, and energy is transferred slowly to the gas by heat. In the second case, no energy is transferred by heat, and the value of the work done is zero. Therefore, we conclude that **energy transfer by heat, like work done, depends on the initial, final, and intermediate states of the system.** In other words, because heat and work depend on the path, neither quantity is determined solely by the end points of a thermodynamic process.

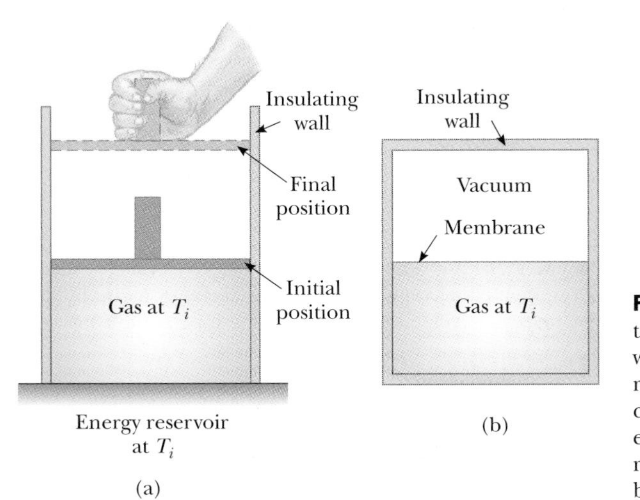

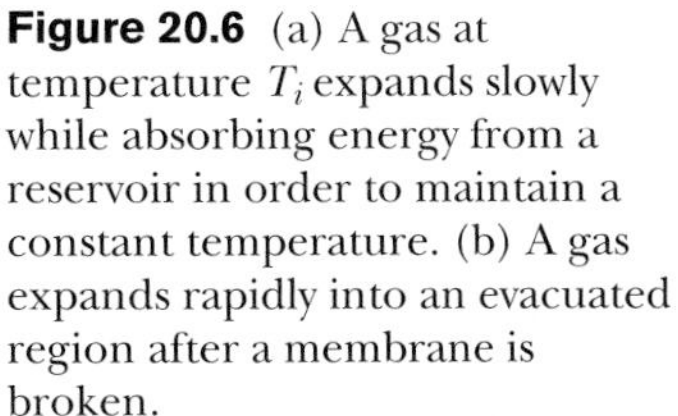
Figure 20.6 (a) A gas at temperature T_i expands slowly while absorbing energy from a reservoir in order to maintain a constant temperature. (b) A gas expands rapidly into an evacuated region after a membrane is broken.

20.5 The First Law of Thermodynamics

When we introduced the law of conservation of energy in Chapter 7, we stated that the change in the energy of a system is equal to the sum of all transfers of energy across the boundary of the system. The first law of thermodynamics is a special case of the law of conservation of energy that encompasses changes in internal energy and energy transfer by heat and work. It is a law that can be applied to many processes and provides a connection between the microscopic and macroscopic worlds.

We have discussed two ways in which energy can be transferred between a system and its surroundings. One is work done on the system, which requires that there be a macroscopic displacement of the point of application of a force. The other is heat, which occurs on a molecular level whenever a temperature difference exists across the boundary of the system. Both mechanisms result in a change in the internal energy of the system and therefore usually result in measurable changes in the macroscopic variables of the system, such as the pressure, temperature, and volume of a gas.

To better understand these ideas on a quantitative basis, suppose that a system undergoes a change from an initial state to a final state. During this change, energy transfer by heat Q to the system occurs, and work W is done on the system. As an example, suppose that the system is a gas in which the pressure and volume change from P_i and V_i to P_f and V_f. If the quantity $Q + W$ is measured for various paths connecting the initial and final equilibrium states, we find that it is the same for all paths connecting the two states. We conclude that the quantity $Q + W$ is determined completely by the initial and final states of the system, and we call this quantity the **change in the internal energy** of the system. Although Q and W both depend on the path, **the quantity $Q + W$ is independent of the path.** If we use the symbol E_{int} to represent the internal energy, then the *change* in internal energy ΔE_{int} can be expressed as[5]

First law of thermodynamics

$$\Delta E_{\text{int}} = Q + W \qquad (20.9)$$

where all quantities must have the same units of measure for energy. Equation 20.9 is known as the **first law of thermodynamics.** One of the important consequences of the first law of thermodynamics is that there exists a quantity known as internal energy whose value is determined by the state of the system. The internal energy is therefore a state variable like pressure, volume, and temperature.

When a system undergoes an infinitesimal change in state in which a small amount of energy dQ is transferred by heat and a small amount of work dW is done, the internal energy changes by a small amount dE_{int}. Thus, for infinitesimal processes we can express the first law as[6]

$$dE_{\text{int}} = dQ + dW$$

The first law of thermodynamics is an energy conservation equation specifying that the only type of energy that changes in the system is the internal energy E_{int}. Let us investigate some special cases in which this condition exists.

First, consider an *isolated system*—that is, one that does not interact with its surroundings. In this case, no energy transfer by heat takes place and the work done on

PITFALL PREVENTION

20.8 Dual Sign Conventions

Some physics and engineering textbooks present the first law as $\Delta E_{\text{int}} = Q - W$, with a minus sign between the heat and work. The reason for this is that work is defined in these treatments as the work done *by* the gas rather than *on* the gas, as in our treatment. The equivalent equation to Equation 20.8 in these treatments defines work as $W = \int_{V_i}^{V_f} P\,dV$. Thus, if positive work is done by the gas, energy is leaving the system, leading to the negative sign in the first law.

In your studies in other chemistry or engineering courses, or in your reading of other physics textbooks, be sure to note which sign convention is being used for the first law.

[5] It is an unfortunate accident of history that the traditional symbol for internal energy is U, which is also the traditional symbol for potential energy, as introduced in Chapter 8. To avoid confusion between potential energy and internal energy, we use the symbol E_{int} for internal energy in this book. If you take an advanced course in thermodynamics, however, be prepared to see U used as the symbol for internal energy.

[6] Note that dQ and dW are not true differential quantities because Q and W are not state variables; however, dE_{int} is. Because dQ and dW are *inexact differentials*, they are often represented by the symbols $đQ$ and $đW$. For further details on this point, see an advanced text on thermodynamics, such as R. P. Bauman, *Modern Thermodynamics and Statistical Mechanics*, New York, Macmillan Publishing Co., 1992.

the system is zero; hence, the internal energy remains constant. That is, because $Q = W = 0$, it follows that $\Delta E_{\text{int}} = 0$, and thus $E_{\text{int},\,i} = E_{\text{int},\,f}$. We conclude that **the internal energy E_{int} of an isolated system remains constant.**

Next, consider the case of a system (one not isolated from its surroundings) that is taken through a **cyclic process**—that is, a process that starts and ends at the same state. In this case, the change in the internal energy must again be zero, because E_{int} is a state variable, and therefore the energy Q added to the system must equal the negative of the work W done on the system during the cycle. That is, in a cyclic process,

$$\Delta E_{\text{int}} = 0 \quad \text{and} \quad Q = -W \quad \text{(cyclic process)}$$

On a PV diagram, a cyclic process appears as a closed curve. (The processes described in Figure 20.5 are represented by open curves because the initial and final states differ.) It can be shown that **in a cyclic process, the net work done on the system per cycle equals the area enclosed by the path representing the process on a PV diagram.**

PITFALL PREVENTION

20.9 The First Law

With our approach to energy in this book, the first law of thermodynamics is a special case of Equation 7.17. Some physicists argue that the first law is the general equation for energy conservation, equivalent to Equation 7.17. In this approach, the first law is applied to a closed system (so that there is no matter transfer), heat is interpreted so as to include electromagnetic radiation, and work is interpreted so as to include electrical transmission ("electrical work") and mechanical waves ("molecular work"). Keep this in mind if you run across the first law in your reading of other physics books.

20.6 Some Applications of the First Law of Thermodynamics

The first law of thermodynamics that we discussed in the preceding section relates the changes in internal energy of a system to transfers of energy by work or heat. In this section, we consider applications of the first law to processes through which a gas is taken. As a model, we consider the sample of gas contained in the piston–cylinder apparatus in Figure 20.7. This figure shows work being done on the gas and energy transferring in by heat, so the internal energy of the gas is rising. In the following discussion of various processes, refer back to this figure and mentally alter the directions of the transfer of energy so as to reflect what is happening in the process.

Before we apply the first law of thermodynamics to specific systems, it is useful to first define some idealized thermodynamic processes. An **adiabatic process** is one during which no energy enters or leaves the system by heat—that is, $Q = 0$. An adiabatic process can be achieved either by thermally insulating the walls of the system, such as the cylinder in Figure 20.7, or by performing the process rapidly, so that there is negligible time for energy to transfer by heat. Applying the first law of thermodynamics to an adiabatic process, we see that

$$\Delta E_{\text{int}} = W \quad \text{(adiabatic process)} \tag{20.10}$$

From this result, we see that if a gas is compressed adiabatically such that W is positive, then ΔE_{int} is positive and the temperature of the gas increases. Conversely, the temperature of a gas decreases when the gas expands adiabatically.

Adiabatic processes are very important in engineering practice. Some common examples are the expansion of hot gases in an internal combustion engine, the liquefaction of gases in a cooling system, and the compression stroke in a diesel engine.

The process described in Figure 20.6b, called an **adiabatic free expansion,** is unique. The process is adiabatic because it takes place in an insulated container. Because the gas expands into a vacuum, it does not apply a force on a piston as was depicted in Figure 20.6a, so no work is done on or by the gas. Thus, in this adiabatic process, both $Q = 0$ and $W = 0$. As a result, $\Delta E_{\text{int}} = 0$ for this process, as we can see from the first law. That is, **the initial and final internal energies of a gas are equal in an adiabatic free expansion.** As we shall see in the next chapter, the internal energy of an ideal gas depends only on its temperature. Thus, we expect no change in temperature during an adiabatic free expansion. This prediction is in accord with the results of experiments performed at low pressures. (Experiments performed at high pressures for real gases show a slight change in temperature after the expansion. This change is due to intermolecular interactions, which represent a deviation from the model of an ideal gas.)

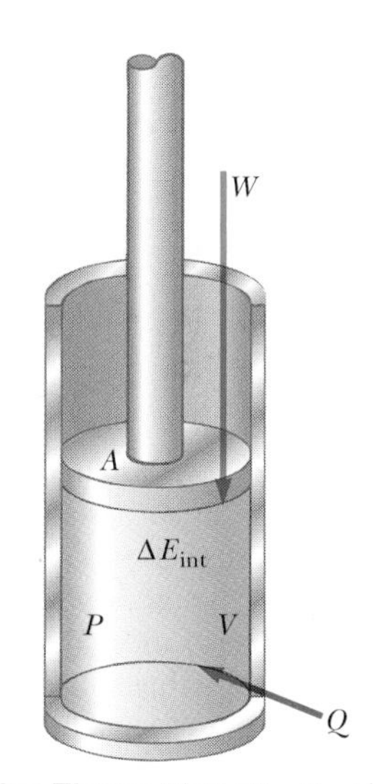

Active Figure 20.7 The first law of thermodynamics equates the change in internal energy E_{int} in a system to the net energy transfer to the system by heat Q and work W. In the situation shown here, the internal energy of the gas increases.

At the Active Figures link at http://www.pse6.com, you can choose one of the four processes for the gas discussed in this section and see the movement of the piston and of a point on a PV diagram.

A process that occurs at constant pressure is called an **isobaric process.** In Figure 20.7, an isobaric process could be established by allowing the piston to move freely so that it is always in equilibrium between the net force from the gas pushing upward and the weight of the piston plus the force due to atmospheric pressure pushing downward. In Figure 20.5, the first process in part (a) and the second process in part (b) are isobaric.

In such a process, the values of the heat and the work are both usually nonzero. The work done on the gas in an isobaric process is simply

Isobaric process

$$W = -P(V_f - V_i) \qquad \text{(isobaric process)} \qquad (20.11)$$

where P is the constant pressure.

A process that takes place at constant volume is called an **isovolumetric process.** In Figure 20.7, clamping the piston at a fixed position would ensure an isovolumetric process. In Figure 20.5, the second process in part (a) and the first process in part (b) are isovolumetric.

In such a process, the value of the work done is zero because the volume does not change. Hence, from the first law we see that in an isovolumetric process, because $W = 0$,

Isovolumetric process

$$\Delta E_{\text{int}} = Q \qquad \text{(isovolumetric process)} \qquad (20.12)$$

This expression specifies that **if energy is added by heat to a system kept at constant volume, then all of the transferred energy remains in the system as an increase in its internal energy.** For example, when a can of spray paint is thrown into a fire, energy enters the system (the gas in the can) by heat through the metal walls of the can. Consequently, the temperature, and thus the pressure, in the can increases until the can possibly explodes.

Isothermal process

A process that occurs at constant temperature is called an **isothermal process**. In Figure 20.7, this process can be established by immersing the cylinder in Figure 20.7 in an ice-water bath or by putting the cylinder in contact with some other constant-temperature reservoir. A plot of P versus V at constant temperature for an ideal gas yields a hyperbolic curve called an *isotherm*. The internal energy of an ideal gas is a function of temperature only. Hence, in an isothermal process involving an ideal gas, $\Delta E_{\text{int}} = 0$. For an isothermal process, then, we conclude from the first law that the energy transfer Q must be equal to the negative of the work done on the gas—that is, $Q = -W$. Any energy that enters the system by heat is transferred out of the system by work; as a result, no change in the internal energy of the system occurs in an isothermal process.

PITFALL PREVENTION

20.10 $Q \neq 0$ in an Isothermal Process

Do not fall into the common trap of thinking that there must be no transfer of energy by heat if the temperature does not change, as is the case in an isothermal process. Because the cause of temperature change can be either heat *or* work, the temperature can remain constant even if energy enters the gas by heat. This can only happen if the energy entering the gas by heat leaves by work.

Quick Quiz 20.5 In the last three columns of the following table, fill in the boxes with −, +, or 0. For each situation, the system to be considered is identified.

Situation	System	Q	W	ΔE_{int}
(a) Rapidly pumping up a bicycle tire	Air in the pump			
(b) Pan of room-temperature water sitting on a hot stove	Water in the pan			
(c) Air quickly leaking out of a balloon	Air originally in the balloon			

Isothermal Expansion of an Ideal Gas

Suppose that an ideal gas is allowed to expand quasi-statically at constant temperature. This process is described by the PV diagram shown in Figure 20.8. The curve is a hyperbola (see Appendix B, Eq. B.23), and the ideal gas law with T constant indicates that the equation of this curve is $PV =$ constant.

Let us calculate the work done on the gas in the expansion from state i to state f. The work done on the gas is given by Equation 20.8. Because the gas is ideal and the process is quasi-static, we can use the expression $PV = nRT$ for each point on the path. Therefore, we have

$$W = -\int_{V_i}^{V_f} P\,dV = -\int_{V_i}^{V_f} \frac{nRT}{V}\,dV$$

Because T is constant in this case, it can be removed from the integral along with n and R:

$$W = -nRT\int_{V_i}^{V_f} \frac{dV}{V} = -nRT\ln V\Big|_{V_i}^{V_f}$$

To evaluate the integral, we used $\int(dx/x) = \ln x$. Evaluating this at the initial and final volumes, we have

$$W = nRT\ln\left(\frac{V_i}{V_f}\right) \tag{20.13}$$

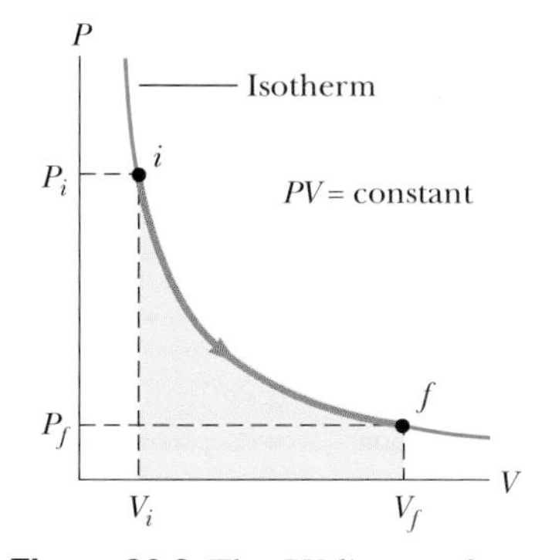

Figure 20.8 The PV diagram for an isothermal expansion of an ideal gas from an initial state to a final state. The curve is a hyperbola.

Numerically, this work W equals the negative of the shaded area under the PV curve shown in Figure 20.8. Because the gas expands, $V_f > V_i$ and the value for the work done on the gas is negative, as we expect. If the gas is compressed, then $V_f < V_i$ and the work done on the gas is positive.

Quick Quiz 20.6 Characterize the paths in Figure 20.9 as isobaric, isovolumetric, isothermal, or adiabatic. Note that $Q = 0$ for path B.

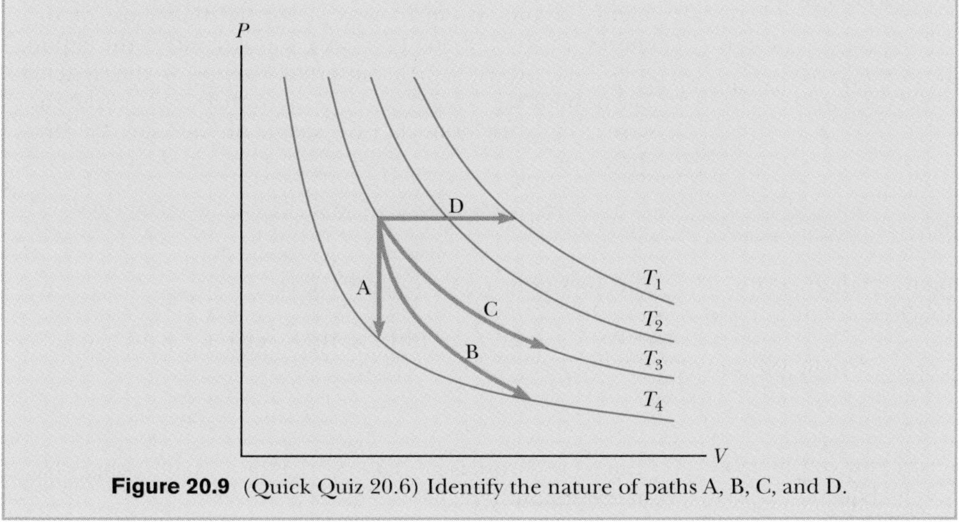

Figure 20.9 (Quick Quiz 20.6) Identify the nature of paths A, B, C, and D.

Example 20.6 An Isothermal Expansion

A 1.0-mol sample of an ideal gas is kept at 0.0°C during an expansion from 3.0 L to 10.0 L.

(A) How much work is done on the gas during the expansion?

Solution Substituting the values into Equation 20.13, we have

$$W = nRT\ln\left(\frac{V_i}{V_f}\right)$$

$$= (1.0\text{ mol})(8.31\text{ J/mol}\cdot\text{K})(273\text{ K})\ln\left(\frac{3.0\text{ L}}{10.0\text{ L}}\right)$$

$$= -2.7 \times 10^3\text{ J}$$

(B) How much energy transfer by heat occurs with the surroundings in this process?

Solution From the first law, we find that

$$\Delta E_{\text{int}} = Q + W$$
$$0 = Q + W$$
$$Q = -W = 2.7 \times 10^3 \text{ J}$$

(C) If the gas is returned to the original volume by means of an isobaric process, how much work is done on the gas?

Solution The work done in an isobaric process is given by Equation 20.11. In this case, the initial volume is 10.0 L and the final volume is 3.0 L, the reverse of the situation in part (A). We are not given the pressure, so we need to incorporate the ideal gas law:

$$W = -P(V_f - V_i) = -\frac{nRT_i}{V_i}(V_f - V_i)$$
$$= -\frac{(1.0 \text{ mol})(8.31 \text{ J/mol}\cdot\text{K})(273 \text{ K})}{10.0 \times 10^{-3} \text{ m}^3}$$
$$\times (3.0 \times 10^{-3} \text{ m}^3 - 10.0 \times 10^{-3} \text{ m}^3)$$
$$= 1.6 \times 10^3 \text{ J}$$

Notice that we use the initial temperature and volume to determine the value of the constant pressure because we do not know the final temperature. The work done on the gas is positive because the gas is being compressed.

Example 20.7 Boiling Water

Suppose 1.00 g of water vaporizes isobarically at atmospheric pressure (1.013×10^5 Pa). Its volume in the liquid state is $V_i = V_{\text{liquid}} = 1.00 \text{ cm}^3$, and its volume in the vapor state is $V_f = V_{\text{vapor}} = 1\,671 \text{ cm}^3$. Find the work done in the expansion and the change in internal energy of the system. Ignore any mixing of the steam and the surrounding air—imagine that the steam simply pushes the surrounding air out of the way.

Solution Because the expansion takes place at constant pressure, the work done on the system (the vaporizing water) as it pushes away the surrounding air is, from Equation 20.11,

$$W = -P(V_f - V_i)$$
$$= -(1.013 \times 10^5 \text{ Pa})(1\,671 \times 10^{-6} \text{ m}^3 - 1.00 \times 10^{-6} \text{ m}^3)$$
$$= -169 \text{ J}$$

To determine the change in internal energy, we must know the energy transfer Q needed to vaporize the water. Using Equation 20.6 and the latent heat of vaporization for water, we have

$$Q = mL_v = (1.00 \times 10^{-3} \text{ kg})(2.26 \times 10^6 \text{ J/kg}) = 2\,260 \text{ J}$$

Hence, from the first law, the change in internal energy is

$$\Delta E_{\text{int}} = Q + W = 2\,260 \text{ J} + (-169 \text{ J}) = 2.09 \text{ kJ}$$

The positive value for ΔE_{int} indicates that the internal energy of the system increases. We see that most of the energy ($2\,090 \text{ J}/2\,260 \text{ J} = 93\%$) transferred to the liquid goes into increasing the internal energy of the system. The remaining 7% of the energy transferred leaves the system by work done by the steam on the surrounding atmosphere.

Example 20.8 Heating a Solid

A 1.0-kg bar of copper is heated at atmospheric pressure. If its temperature increases from 20°C to 50°C,

(A) what is the work done on the copper bar by the surrounding atmosphere?

Solution Because the process is isobaric, we can find the work done on the copper bar using Equation 20.11, $W = -P(V_f - V_i)$. We can calculate the change in volume of the copper bar using Equation 19.6. Using the average linear expansion coefficient for copper given in Table 19.1, and remembering that $\beta = 3\alpha$, we obtain

$$\Delta V = \beta V_i \, \Delta T$$
$$= [5.1 \times 10^{-5} (^\circ\text{C})^{-1}](50^\circ\text{C} - 20^\circ\text{C})V_i = 1.5 \times 10^{-3} V_i$$

The volume V_i is equal to m/ρ, and Table 14.1 indicates that the density of copper is $8.92 \times 10^3 \text{ kg/m}^3$. Hence,

$$\Delta V = (1.5 \times 10^{-3})\left(\frac{1.0 \text{ kg}}{8.92 \times 10^3 \text{ kg/m}^3}\right)$$
$$= 1.7 \times 10^{-7} \text{ m}^3$$

The work done on the copper bar is

$$W = -P\,\Delta V = -(1.013 \times 10^5 \text{ N/m}^2)(1.7 \times 10^{-7} \text{ m}^3)$$
$$= -1.7 \times 10^{-2} \text{ J}$$

Because this work is negative, work is done *by* the copper bar on the atmosphere.

(B) What quantity of energy is transferred to the copper bar by heat?

Solution Taking the specific heat of copper from Table 20.1 and using Equation 20.4, we find that the energy transferred by heat is

$$Q = mc\,\Delta T = (1.0 \text{ kg})(387 \text{ J/kg}\cdot{}^\circ\text{C})(30^\circ\text{C})$$
$$= 1.2 \times 10^4 \text{ J}$$

(C) What is the increase in internal energy of the copper bar?

Solution From the first law of thermodynamics, we have

$$\Delta E_{\text{int}} = Q + W = 1.2 \times 10^4\,\text{J} + (-1.7 \times 10^{-2}\,\text{J})$$
$$= 1.2 \times 10^4\,\text{J}$$

Note that almost all of the energy transferred into the system by heat goes into increasing the internal energy of the copper bar. The fraction of energy used to do work on the surrounding atmosphere is only about 10^{-6}! Hence, when the thermal expansion of a solid or a liquid is analyzed, the small amount of work done on or by the system is usually ignored.

20.7 Energy Transfer Mechanisms

In Chapter 7, we introduced a global approach to energy analysis of physical processes through Equation 7.17, $\Delta E_{\text{system}} = \Sigma T$, where T represents energy transfer. Earlier in this chapter, we discussed two of the terms on the right-hand side of this equation, work and heat. In this section, we explore more details about heat as a means of energy transfer and consider two other energy transfer methods that are often related to temperature changes—convection (a form of matter transfer) and electromagnetic radiation.

Thermal Conduction

The process of energy transfer by heat can also be called **conduction** or **thermal conduction.** In this process, the transfer can be represented on an atomic scale as an exchange of kinetic energy between microscopic particles—molecules, atoms, and free electrons—in which less-energetic particles gain energy in collisions with more energetic particles. For example, if you hold one end of a long metal bar and insert the other end into a flame, you will find that the temperature of the metal in your hand soon increases. The energy reaches your hand by means of conduction. We can understand the process of conduction by examining what is happening to the microscopic particles in the metal. Initially, before the rod is inserted into the flame, the microscopic particles are vibrating about their equilibrium positions. As the flame heats the rod, the particles near the flame begin to vibrate with greater and greater amplitudes. These particles, in turn, collide with their neighbors and transfer some of their energy in the collisions. Slowly, the amplitudes of vibration of metal atoms and electrons farther and farther from the flame increase until, eventually, those in the metal near your hand are affected. This increased vibration is detected by an increase in the temperature of the metal and of your potentially burned hand.

A pan of boiling water sits on a stove burner. Energy enters the water through the bottom of the pan by thermal conduction.

The rate of thermal conduction depends on the properties of the substance being heated. For example, it is possible to hold a piece of asbestos in a flame indefinitely. This implies that very little energy is conducted through the asbestos. In general, metals are good thermal conductors, and materials such as asbestos, cork, paper, and fiberglass are poor conductors. Gases also are poor conductors because the separation distance between the particles is so great. Metals are good thermal conductors because they contain large numbers of electrons that are relatively free to move through the metal and so can transport energy over large distances. Thus, in a good conductor, such as copper, conduction takes place by means of both the vibration of atoms and the motion of free electrons.

Conduction occurs only if there is a difference in temperature between two parts of the conducting medium. Consider a slab of material of thickness Δx and cross-sectional area A. One face of the slab is at a temperature T_c, and the other face is at a temperature $T_h > T_c$ (Fig. 20.10). Experimentally, it is found that the energy Q transfers in a time interval Δt from the hotter face to the colder one. The rate $\mathcal{P} = Q/\Delta t$ at which this energy transfer occurs is found to be proportional to the cross-sectional area and the temperature difference $\Delta T = T_h - T_c$, and inversely proportional to the thickness:

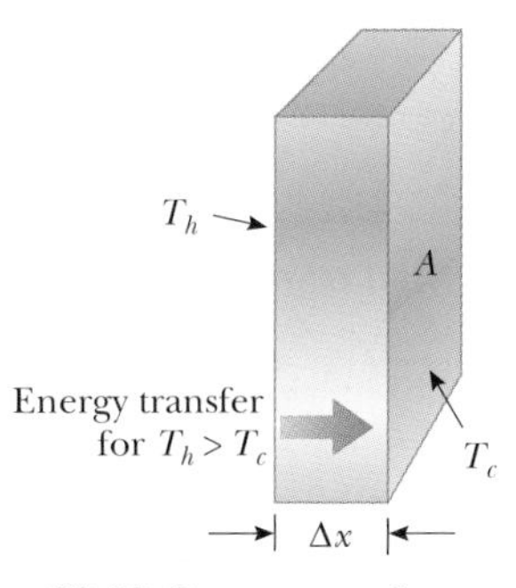

Figure 20.10 Energy transfer through a conducting slab with a cross-sectional area A and a thickness Δx. The opposite faces are at different temperatures T_c and T_h.

$$\mathcal{P} = \frac{Q}{\Delta t} \propto A \frac{\Delta T}{\Delta x}$$

Note that $\mathcal{P}$ has units of watts when Q is in joules and Δt is in seconds. This is not surprising because $\mathcal{P}$ is *power*—the rate of energy transfer by heat. For a slab of infinitesimal thickness dx and temperature difference dT, we can write the **law of thermal conduction** as

Law of thermal conduction

$$\mathcal{P} = kA\left|\frac{dT}{dx}\right| \qquad (20.14)$$

where the proportionality constant k is the **thermal conductivity** of the material and $|dT/dx|$ is the **temperature gradient** (the rate at which temperature varies with position).

Suppose that a long, uniform rod of length L is thermally insulated so that energy cannot escape by heat from its surface except at the ends, as shown in Figure 20.11. One end is in thermal contact with an energy reservoir at temperature T_c, and the other end is in thermal contact with a reservoir at temperature $T_h > T_c$. When a steady state has been reached, the temperature at each point along the rod is constant in time. In this case if we assume that k is not a function of temperature, the temperature gradient is the same everywhere along the rod and is

$$\left|\frac{dT}{dx}\right| = \frac{T_h - T_c}{L}$$

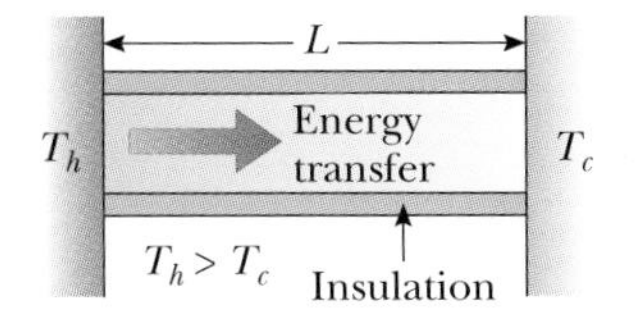

Figure 20.11 Conduction of energy through a uniform, insulated rod of length L. The opposite ends are in thermal contact with energy reservoirs at different temperatures.

Table 20.3

Thermal Conductivities

Substance	Thermal Conductivity (W/m · °C)
Metals (at 25°C)	
Aluminum	238
Copper	397
Gold	314
Iron	79.5
Lead	34.7
Silver	427
Nonmetals (approximate values)	
Asbestos	0.08
Concrete	0.8
Diamond	2 300
Glass	0.8
Ice	2
Rubber	0.2
Water	0.6
Wood	0.08
Gases (at 20°C)	
Air	0.023 4
Helium	0.138
Hydrogen	0.172
Nitrogen	0.023 4
Oxygen	0.023 8

Thus the rate of energy transfer by conduction through the rod is

$$\mathcal{P} = kA\left(\frac{T_h - T_c}{L}\right) \tag{20.15}$$

Substances that are good thermal conductors have large thermal conductivity values, whereas good thermal insulators have low thermal conductivity values. Table 20.3 lists thermal conductivities for various substances. Note that metals are generally better thermal conductors than nonmetals.

For a compound slab containing several materials of thicknesses $L_1, L_2, \ldots$ and thermal conductivities $k_1, k_2, \ldots$, the rate of energy transfer through the slab at steady state is

$$\mathcal{P} = \frac{A(T_h - T_c)}{\sum_i (L_i/k_i)} \tag{20.16}$$

where T_c and T_h are the temperatures of the outer surfaces (which are held constant) and the summation is over all slabs. Example 20.9 shows how this equation results from a consideration of two thicknesses of materials.

Example 20.9 Energy Transfer Through Two Slabs

Two slabs of thickness L_1 and L_2 and thermal conductivities k_1 and k_2 are in thermal contact with each other, as shown in Figure 20.12. The temperatures of their outer surfaces are T_c and T_h, respectively, and $T_h > T_c$. Determine the temperature at the interface and the rate of energy transfer by conduction through the slabs in the steady-state condition.

Solution To conceptualize this problem, notice the phrase "in the steady-state condition." We interpret this to mean that energy transfers through the compound slab at the same rate at all points. Otherwise, energy would be building up or disappearing at some point. Furthermore, the temperature will vary with position in the two slabs, most likely at different rates in each part of the compound slab. Thus, there will be some fixed temperature T at the interface when the system is in steady state. We categorize this as a thermal conduction problem and impose the condition that the power is the same in both slabs of material. To analyze the problem, we use Equation 20.15 to express the rate at which energy is transferred through slab 1:

$$(1) \qquad \mathcal{P}_1 = k_1 A\left(\frac{T - T_c}{L_1}\right)$$

The rate at which energy is transferred through slab 2 is

$$(2) \qquad \mathcal{P}_2 = k_2 A\left(\frac{T_h - T}{L_2}\right)$$

When a steady state is reached, these two rates must be equal; hence,

$$k_1 A\left(\frac{T - T_c}{L_1}\right) = k_2 A\left(\frac{T_h - T}{L_2}\right)$$

Solving for T gives

$$(3) \qquad T = \frac{k_1 L_2 T_c + k_2 L_1 T_h}{k_1 L_2 + k_2 L_1}$$

Substituting Equation (3) into either Equation (1) or Equation (2), we obtain

$$(4) \qquad \mathcal{P} = \frac{A(T_h - T_c)}{(L_1/k_1) + (L_2/k_2)}$$

To finalize this problem, note that extension of this procedure to several slabs of materials leads to Equation 20.16.

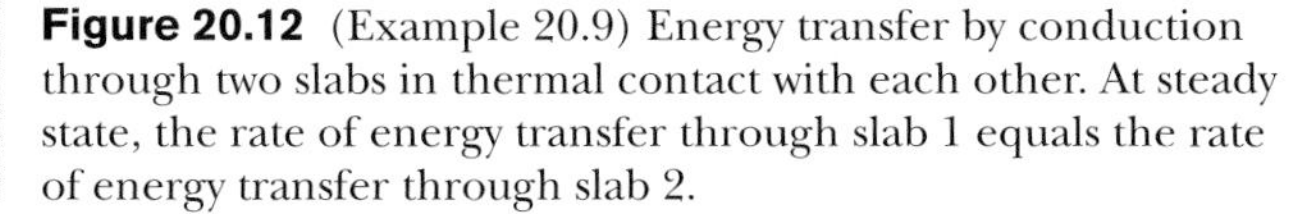

Figure 20.12 (Example 20.9) Energy transfer by conduction through two slabs in thermal contact with each other. At steady state, the rate of energy transfer through slab 1 equals the rate of energy transfer through slab 2.

What If? Suppose you are building an insulated container with two layers of insulation and the rate of energy transfer determined by Equation (4) turns out to be too high. You have enough room to increase the thickness of one of the two layers by 20%. How would you decide which layer to choose?

Answer To decrease the power as much as possible, you must increase the denominator in Equation (4) as much as possible. Whichever thickness you choose to increase, L_1 or L_2, you will increase the corresponding term L/k in the denominator by 20%. In order for this percentage change to represent the largest absolute change, you want to take 20% of the larger term. Thus, you should increase the thickness of the layer that has the larger value of L/k.

Quick Quiz 20.7 Will an ice cube wrapped in a wool blanket remain frozen for (a) a shorter length of time (b) the same length of time (c) a longer length of time than an identical ice cube exposed to air at room temperature?

Quick Quiz 20.8 You have two rods of the same length and diameter but they are formed from different materials. The rods will be used to connect two regions of different temperature such that energy will transfer through the rods by heat. They can be connected in series, as in Figure 20.13a, or in parallel, as in Figure 20.13b. In which case is the rate of energy transfer by heat larger? (a) when the rods are in series (b) when the rods are in parallel (c) The rate is the same in both cases.

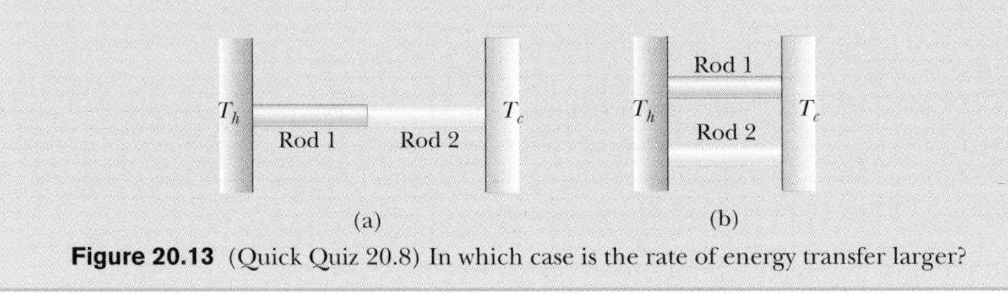

Figure 20.13 (Quick Quiz 20.8) In which case is the rate of energy transfer larger?

Courtesy of Dr. Albert A. Bartlett, University of Colorado, Boulder

Energy is conducted from the inside to the exterior more rapidly on the part of the roof where the snow has melted. The dormer appears to have been added and insulated. The main roof does not appear to be well insulated.

Home Insulation

In engineering practice, the term L/k for a particular substance is referred to as the ***R* value** of the material. Thus, Equation 20.16 reduces to

$$\mathcal{P} = \frac{A(T_h - T_c)}{\sum_i R_i} \qquad (20.17)$$

where $R_i = L_i/k_i$. The R values for a few common building materials are given in Table 20.4. In the United States, the insulating properties of materials used in buildings are usually expressed in U.S. customary units, not SI units. Thus, in Table 20.4, measurements of R values are given as a combination of British thermal units, feet, hours, and degrees Fahrenheit.

At any vertical surface open to the air, a very thin stagnant layer of air adheres to the surface. One must consider this layer when determining the R value for a wall. The thickness of this stagnant layer on an outside wall depends on the speed of the wind. Energy loss from a house on a windy day is greater than the loss on a day when the air is calm. A representative R value for this stagnant layer of air is given in Table 20.4.

Table 20.4

***R* Values for Some Common Building Materials**

Material	*R* value ($ft^2 \cdot °F \cdot h/Btu$)
Hardwood siding (1 in. thick)	0.91
Wood shingles (lapped)	0.87
Brick (4 in. thick)	4.00
Concrete block (filled cores)	1.93
Fiberglass insulation (3.5 in. thick)	10.90
Fiberglass insulation (6 in. thick)	18.80
Fiberglass board (1 in. thick)	4.35
Cellulose fiber (1 in. thick)	3.70
Flat glass (0.125 in. thick)	0.89
Insulating glass (0.25-in. space)	1.54
Air space (3.5 in. thick)	1.01
Stagnant air layer	0.17
Drywall (0.5 in. thick)	0.45
Sheathing (0.5 in. thick)	1.32

Example 20.10 The *R* Value of a Typical Wall — Interactive

Calculate the total R value for a wall constructed as shown in Figure 20.14a. Starting outside the house (toward the front in the figure) and moving inward, the wall consists of 4 in. of brick, 0.5 in. of sheathing, an air space 3.5 in. thick, and 0.5 in. of drywall. Do not forget the stagnant air layers inside and outside the house.

Solution Referring to Table 20.4, we find that

$$\begin{aligned}
R_1 \text{ (outside stagnant air layer)} &= 0.17\ ft^2 \cdot °F \cdot h/Btu \\
R_2 \text{ (brick)} &= 4.00\ ft^2 \cdot °F \cdot h/Btu \\
R_3 \text{ (sheathing)} &= 1.32\ ft^2 \cdot °F \cdot h/Btu \\
R_4 \text{ (air space)} &= 1.01\ ft^2 \cdot °F \cdot h/Btu \\
R_5 \text{ (drywall)} &= 0.45\ ft^2 \cdot °F \cdot h/Btu \\
R_6 \text{ (inside stagnant air layer)} &= 0.17\ ft^2 \cdot °F \cdot h/Btu \\
\hline
R_{total} &= 7.12\ ft^2 \cdot °F \cdot h/Btu
\end{aligned}$$

What If? You are not happy with this total *R* value for the wall. You cannot change the overall structure, but you can fill the air space as in Figure 20.14b. What material should you choose to fill the air space in order to *maximize* the total *R* value?

Answer Looking at Table 20.4, we see that 3.5 in. of fiberglass insulation is over ten times as effective at insulating the wall as 3.5 in. of air. Thus, we could fill the air space with fiberglass insulation. The result is that we add 10.90 $ft^2 \cdot °F \cdot h/Btu$ of R value and we lose 1.01 $ft^2 \cdot °F \cdot h/Btu$ due to the air space we have replaced, for a total change of 10.90 $ft^2 \cdot °F \cdot h/Btu - 1.01\ ft^2 \cdot °F \cdot h/Btu = 9.89\ ft^2 \cdot °F \cdot h/Btu$. The new total R value is $7.12\ ft^2 \cdot °F \cdot h/Btu + 9.89\ ft^2 \cdot °F \cdot h/Btu = 17.01\ ft^2 \cdot °F \cdot h/Btu$.

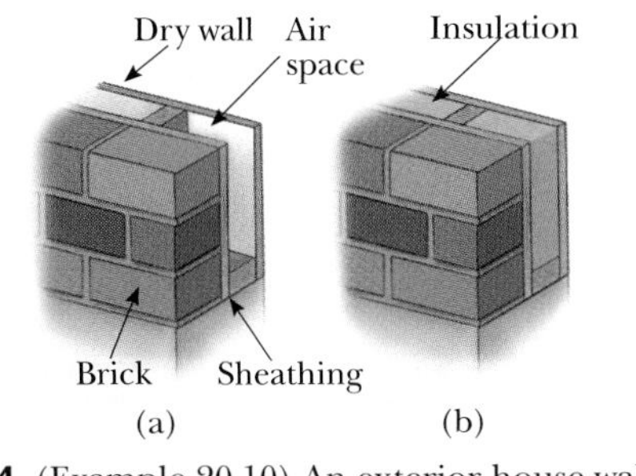

Figure 20.14 (Example 20.10) An exterior house wall containing (a) an air space and (b) insulation.

Study the R values of various types of common building materials at the Interactive Worked Example link at **http://www.pse6.com.**

Convection

At one time or another, you probably have warmed your hands by holding them over an open flame. In this situation, the air directly above the flame is heated and expands. As a result, the density of this air decreases and the air rises. This hot air warms your

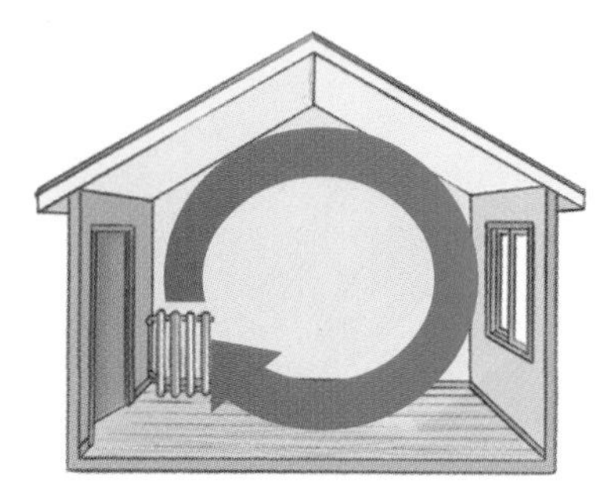

Figure 20.15 Convection currents are set up in a room warmed by a radiator.

hands as it flows by. **Energy transferred by the movement of a warm substance is said to have been transferred by convection.** When the movement results from differences in density, as with air around a fire, it is referred to as *natural convection.* Air flow at a beach is an example of natural convection, as is the mixing that occurs as surface water in a lake cools and sinks (see Section 19.4). When the heated substance is forced to move by a fan or pump, as in some hot-air and hot-water heating systems, the process is called *forced convection.*

If it were not for convection currents, it would be very difficult to boil water. As water is heated in a teakettle, the lower layers are warmed first. This water expands and rises to the top because its density is lowered. At the same time, the denser, cool water at the surface sinks to the bottom of the kettle and is heated.

The same process occurs when a room is heated by a radiator. The hot radiator warms the air in the lower regions of the room. The warm air expands and rises to the ceiling because of its lower density. The denser, cooler air from above sinks, and the continuous air current pattern shown in Figure 20.15 is established.

Radiation

The third means of energy transfer that we shall discuss is **radiation**. All objects radiate energy continuously in the form of electromagnetic waves (see Chapter 34) produced by thermal vibrations of the molecules. You are likely familiar with electromagnetic radiation in the form of the orange glow from an electric stove burner, an electric space heater, or the coils of a toaster.

The rate at which an object radiates energy is proportional to the fourth power of its absolute temperature. This is known as **Stefan's law** and is expressed in equation form as

Stefan's law

$$\mathcal{P} = \sigma A e T^4 \qquad (20.18)$$

where $\mathcal{P}$ is the power in watts radiated from the surface of the object, σ is a constant equal to $5.669\,6 \times 10^{-8}\ \text{W/m}^2 \cdot \text{K}^4$, A is the surface area of the object in square meters, e is the **emissivity**, and T is the surface temperature in kelvins. The value of e can vary between zero and unity, depending on the properties of the surface of the object. The emissivity is equal to the **absorptivity**, which is the fraction of the incoming radiation that the surface absorbs.

Approximately 1 340 J of electromagnetic radiation from the Sun passes perpendicularly through each 1 m^2 at the top of the Earth's atmosphere every second. This radiation is primarily visible and infrared light accompanied by a significant amount of ultraviolet radiation. We shall study these types of radiation in detail in Chapter 34. Some of this energy is reflected back into space, and some is absorbed by the atmosphere. However, enough energy arrives at the surface of the Earth each day to supply all our energy needs on this planet hundreds of times over—if only it could be captured and used efficiently. The growth in the number of solar energy–powered houses built in this country reflects the increasing efforts being made to use this abundant energy. Radiant energy from the Sun affects our day-to-day existence in a number of ways. For example, it influences the Earth's average temperature, ocean currents, agriculture, and rain patterns.

What happens to the atmospheric temperature at night is another example of the effects of energy transfer by radiation. If there is a cloud cover above the Earth, the water vapor in the clouds absorbs part of the infrared radiation emitted by the Earth and re-emits it back to the surface. Consequently, temperature levels at the surface remain moderate. In the absence of this cloud cover, there is less in the way to prevent this radiation from escaping into space; thus the temperature decreases more on a clear night than on a cloudy one.

As an object radiates energy at a rate given by Equation 20.18, it also absorbs electromagnetic radiation. If the latter process did not occur, an object would eventually radiate all its energy, and its temperature would reach absolute zero. The energy an object absorbs comes from its surroundings, which consist of other objects that radiate energy. If an object is at a temperature T and its surroundings are at an average temperature T_0, then the net rate of energy gained or lost by the object as a result of radiation is

$$\mathcal{P}_{\text{net}} = \sigma A e(T^4 - T_0^4) \tag{20.19}$$

When an object is in equilibrium with its surroundings, it radiates and absorbs energy at the same rate, and its temperature remains constant. When an object is hotter than its surroundings, it radiates more energy than it absorbs, and its temperature decreases.

An **ideal absorber** is defined as an object that absorbs all the energy incident on it, and for such an object, $e = 1$. An object for which $e = 1$ is often referred to as a **black body.** We shall investigate experimental and theoretical approaches to radiation from a black body in Chapter 40. An ideal absorber is also an ideal radiator of energy. In contrast, an object for which $e = 0$ absorbs none of the energy incident on it. Such an object reflects all the incident energy, and thus is an **ideal reflector.**

The Dewar Flask

The *Dewar flask*[7] is a container designed to minimize energy losses by conduction, convection, and radiation. Such a container is used to store either cold or hot liquids for long periods of time. (A Thermos bottle is a common household equivalent of a Dewar flask.) The standard construction (Fig. 20.16) consists of a double-walled Pyrex glass vessel with silvered walls. The space between the walls is evacuated to minimize energy transfer by conduction and convection. The silvered surfaces minimize energy transfer by radiation because silver is a very good reflector and has very low emissivity. A further reduction in energy loss is obtained by reducing the size of the neck. Dewar flasks are commonly used to store liquid nitrogen (boiling point: 77 K) and liquid oxygen (boiling point: 90 K).

To confine liquid helium (boiling point: 4.2 K), which has a very low heat of vaporization, it is often necessary to use a double Dewar system, in which the Dewar flask containing the liquid is surrounded by a second Dewar flask. The space between the two flasks is filled with liquid nitrogen.

Newer designs of storage containers use "super insulation" that consists of many layers of reflecting material separated by fiberglass. All of this is in a vacuum, and no liquid nitrogen is needed with this design.

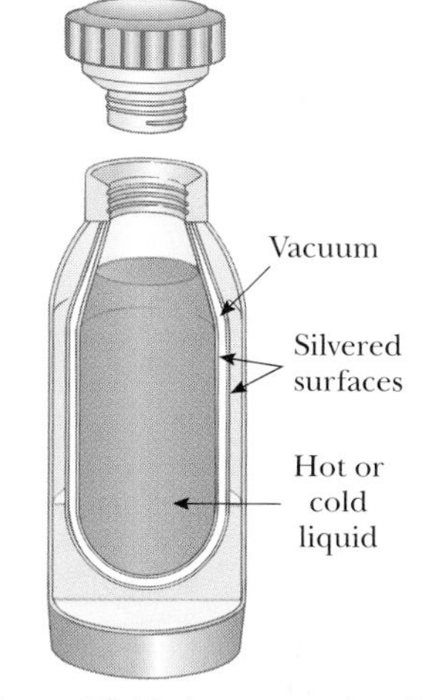

Figure 20.16 A cross-sectional view of a Dewar flask, which is used to store hot or cold substances.

Example 20.11 Who Turned Down the Thermostat?

A student is trying to decide what to wear. The surroundings (his bedroom) are at 20.0°C. If the skin temperature of the unclothed student is 35°C, what is the net energy loss from his body in 10.0 min by radiation? Assume that the emissivity of skin is 0.900 and that the surface area of the student is 1.50 m^2.

Solution Using Equation 20.19, we find that the net rate of energy loss from the skin is

$$\mathcal{P}_{\text{net}} = \sigma A e(T^4 - T_0^4)$$

$$= (5.67 \times 10^{-8}\ \text{W/m}^2\cdot\text{K}^4)(1.50\ \text{m}^2)$$
$$\times (0.900)[(308\ \text{K})^4 - (293\ \text{K})^4] = 125\ \text{W}$$

At this rate, the total energy lost by the skin in 10 min is

$$Q = \mathcal{P}_{\text{net}}\,\Delta t = (125\ \text{W})(600\ \text{s}) = \boxed{7.5 \times 10^4\ \text{J}}$$

Note that the energy radiated by the student is roughly equivalent to that produced by two 60-W light bulbs!

[7] Invented by Sir James Dewar (1842–1923).

SUMMARY

Take a practice test for this chapter by clicking on the Practice Test link at **http://www.pse6.com.**

Internal energy is all of a system's energy that is associated with the system's microscopic components. Internal energy includes kinetic energy of random translation, rotation, and vibration of molecules, potential energy within molecules, and potential energy between molecules.

Heat is the transfer of energy across the boundary of a system resulting from a temperature difference between the system and its surroundings. We use the symbol Q for the amount of energy transferred by this process.

The **calorie** is the amount of energy necessary to raise the temperature of 1 g of water from 14.5°C to 15.5°C. The **mechanical equivalent of heat** is 1 cal = 4.186 J.

The **heat capacity** C of any sample is the amount of energy needed to raise the temperature of the sample by 1°C. The energy Q required to change the temperature of a mass m of a substance by an amount ΔT is

$$Q = mc\,\Delta T \qquad (20.4)$$

where c is the **specific heat** of the substance.

The energy required to change the phase of a pure substance of mass m is

$$Q = \pm mL \qquad (20.6)$$

where L is the **latent heat** of the substance and depends on the nature of the phase change and the properties of the substance. The positive sign is used if energy is entering the system, and the negative sign is used if energy is leaving.

The **work done** on a gas as its volume changes from some initial value V_i to some final value V_f is

$$W = -\int_{V_i}^{V_f} P\,dV \qquad (20.8)$$

where P is the pressure, which may vary during the process. In order to evaluate W, the process must be fully specified—that is, P and V must be known during each step. In other words, the work done depends on the path taken between the initial and final states.

The **first law of thermodynamics** states that when a system undergoes a change from one state to another, the change in its internal energy is

$$\Delta E_{\text{int}} = Q + W \qquad (20.9)$$

where Q is the energy transferred into the system by heat and W is the work done on the system. Although Q and W both depend on the path taken from the initial state to the final state, the quantity ΔE_{int} is path-independent.

In a **cyclic process** (one that originates and terminates at the same state), $\Delta E_{\text{int}} = 0$ and, therefore, $Q = -W$. That is, the energy transferred into the system by heat equals the negative of the work done on the system during the process.

In an **adiabatic process**, no energy is transferred by heat between the system and its surroundings ($Q = 0$). In this case, the first law gives $\Delta E_{\text{int}} = W$. That is, the internal energy changes as a consequence of work being done on the system. In the **adiabatic free expansion** of a gas $Q = 0$ and $W = 0$, and so $\Delta E_{\text{int}} = 0$. That is, the internal energy of the gas does not change in such a process.

An **isobaric process** is one that occurs at constant pressure. The work done on a gas in such a process is $W = -P(V_f - V_i)$.

An **isovolumetric process** is one that occurs at constant volume. No work is done in such a process, so $\Delta E_{\text{int}} = Q$.

An **isothermal process** is one that occurs at constant temperature. The work done on an ideal gas during an isothermal process is

$$W = nRT \ln\left(\frac{V_i}{V_f}\right) \qquad (20.13)$$

Energy may be transferred by work, which we addressed in Chapter 7, and by conduction, convection, or radiation. **Conduction** can be viewed as an exchange of kinetic energy between colliding molecules or electrons. The rate of energy transfer by conduction through a slab of area A is

$$\mathcal{P} = kA\left|\frac{dT}{dx}\right| \tag{20.14}$$

where k is the **thermal conductivity** of the material from which the slab is made and $|dT/dx|$ is the **temperature gradient.** This equation can be used in many situations in which the rate of transfer of energy through materials is important.

In **convection,** a warm substance transfers energy from one location to another.

All objects emit **radiation** in the form of electromagnetic waves at the rate

$$\mathcal{P} = \sigma AeT^4 \tag{20.18}$$

An object that is hotter than its surroundings radiates more energy than it absorbs, whereas an object that is cooler than its surroundings absorbs more energy than it radiates.

QUESTIONS

1. Clearly distinguish among temperature, heat, and internal energy.

2. Ethyl alcohol has about half the specific heat of water. If equal-mass samples of alcohol and water in separate beakers are supplied with the same amount of energy, compare the temperature increases of the two liquids.

3. A small metal crucible is taken from a 200°C oven and immersed in a tub full of water at room temperature (this process is often referred to as *quenching*). What is the approximate final equilibrium temperature?

4. What is a major problem that arises in measuring specific heats if a sample with a temperature above 100°C is placed in water?

5. In a daring lecture demonstration, an instructor dips his wetted fingers into molten lead (327°C) and withdraws them quickly, without getting burned. How is this possible? (This is a dangerous experiment, which you should *NOT* attempt.)

6. What is wrong with the following statement? "Given any two objects, the one with the higher temperature contains more heat."

7. Why is a person able to remove a piece of dry aluminum foil from a hot oven with bare fingers, while a burn results if there is moisture on the foil?

8. The air temperature above coastal areas is profoundly influenced by the large specific heat of water. One reason is that the energy released when 1 m^3 of water cools by 1°C will raise the temperature of a much larger volume of air by 1°C. Find this volume of air. The specific heat of air is approximately 1 kJ/kg · °C. Take the density of air to be 1.3 kg/m^3.

9. Concrete has a higher specific heat than soil. Use this fact to explain (partially) why cities have a higher average nighttime temperature than the surrounding countryside. If a city is hotter than the surrounding countryside, would you expect breezes to blow from city to country or from country to city? Explain.

10. Using the first law of thermodynamics, explain why the *total* energy of an isolated system is always constant.

11. When a sealed Thermos bottle full of hot coffee is shaken, what are the changes, if any, in (a) the temperature of the coffee (b) the internal energy of the coffee?

12. Is it possible to convert internal energy to mechanical energy? Explain with examples.

13. The U.S. penny was formerly made mostly of copper and is now made of copper-coated zinc. Can a calorimetric experiment be devised to test for the metal content in a collection of pennies? If so, describe the procedure you would use.

14. Figure Q20.14 shows a pattern formed by snow on the roof of a barn. What causes the alternating pattern of snow-covered and exposed roof?

Courtesy of Dr. Albert A. Bartlett, University of Colorado, Boulder, CO

Figure Q20.14 Alternating patterns on a snow-covered roof.

15. A tile floor in a bathroom may feel uncomfortably cold to your bare feet, but a carpeted floor in an adjoining room at the same temperature will feel warm. Why?

16. Why can potatoes be baked more quickly when a metal skewer has been inserted through them?

17. A piece of paper is wrapped around a rod made half of wood and half of copper. When held over a flame, the paper in contact with the wood burns but the half in contact with the metal does not. Explain.

18. Why do heavy draperies over the windows help keep a home cool in the summer, as well as warm in the winter?

19. If you wish to cook a piece of meat thoroughly on an open fire, why should you not use a high flame? (Note that carbon is a good thermal insulator.)

20. In an experimental house, Styrofoam beads were pumped into the air space between the panes of glass in double windows at night in the winter, and pumped out to holding bins during the day. How would this assist in conserving energy in the house?

21. Pioneers stored fruits and vegetables in underground cellars. Discuss the advantages of this choice for a storage site.

22. The pioneers referred to in the last question found that a large tub of water placed in a storage cellar would prevent their food from freezing on really cold nights. Explain why this is so.

23. When camping in a canyon on a still night, one notices that as soon as the sun strikes the surrounding peaks, a breeze begins to stir. What causes the breeze?

24. Updrafts of air are familiar to all pilots and are used to keep nonmotorized gliders aloft. What causes these currents?

25. If water is a poor thermal conductor, why can its temperature be raised quickly when it is placed over a flame?

26. Why is it more comfortable to hold a cup of hot tea by the handle rather than by wrapping your hands around the cup itself?

27. If you hold water in a paper cup over a flame, you can bring the water to a boil without burning the cup. How is this possible?

28. You need to pick up a very hot cooking pot in your kitchen. You have a pair of hot pads. Should you soak them in cold water or keep them dry, to be able to pick up the pot most comfortably?

29. Suppose you pour hot coffee for your guests, and one of them wants to drink it with cream, several minutes later, and then as warm as possible. In order to have the warmest coffee, should the person add the cream just after the coffee is poured or just before drinking? Explain.

30. Two identical cups both at room temperature are filled with the same amount of hot coffee. One cup contains a metal spoon, while the other does not. If you wait for several minutes, which of the two will have the warmer coffee? Which energy transfer process explains your answer?

31. A warning sign often seen on highways just before a bridge is "Caution—Bridge surface freezes before road surface." Which of the three energy transfer processes discussed in Section 20.7 is most important in causing a bridge surface to freeze before a road surface on very cold days?

32. A professional physics teacher drops one marshmallow into a flask of liquid nitrogen, waits for the most energetic boiling to stop, fishes it out with tongs, shakes it off, pops it into his mouth, chews it up, and swallows it. Clouds of ice crystals issue from his mouth as he crunches noisily and comments on the sweet taste. How can he do this without injury? *Caution*: Liquid nitrogen can be a dangerous substance and you should *not* try this yourself. The teacher might be badly injured if he did not shake it off, if he touched the tongs to a tooth, or if he did not start with a mouthful of saliva.

33. In 1801 Humphry Davy rubbed together pieces of ice inside an ice-house. He took care that nothing in their environment was at a higher temperature than the rubbed pieces. He observed the production of drops of liquid water. Make a table listing this and other experiments or processes, to illustrate each of the following. (a) A system can absorb energy by heat, increase in internal energy, and increase in temperature. (b) A system can absorb energy by heat and increase in internal energy, without an increase in temperature. (c) A system can absorb energy by heat without increasing in temperature or in internal energy. (d) A system can increase in internal energy and in temperature, without absorbing energy by heat. (e) A system can increase in internal energy without absorbing energy by heat or increasing in temperature. (f) **What If?** If a system's temperature increases, is it necessarily true that its internal energy increases?

34. Consider the opening photograph for Part 3 on page 578. Discuss the roles of conduction, convection, and radiation in the operation of the cooling fins on the support posts of the Alaskan oil pipeline.

PROBLEMS

1, 2, 3 = straightforward, intermediate, challenging □ = full solution available in the *Student Solutions Manual and Study Guide*

= coached solution with hints available at http://www.pse6.com = computer useful in solving problem

= paired numerical and symbolic problems

Section 20.1 Heat and Internal Energy

1. On his honeymoon James Joule traveled from England to Switzerland. He attempted to verify his idea of the interconvertibility of mechanical energy and internal energy by measuring the increase in temperature of water that fell in a waterfall. If water at the top of an alpine waterfall has a temperature of 10.0°C and then falls 50.0 m (as at Niagara Falls), what maximum temperature at the bottom of the falls could Joule expect? He did not succeed in measuring the temperature change, partly because evaporation cooled the falling water, and also because his thermometer was not sufficiently sensitive.

2. Consider Joule's apparatus described in Figure 20.1. The mass of each of the two blocks is 1.50 kg, and the insulated tank is filled with 200 g of water. What is the increase in the temperature of the water after the blocks fall through a distance of 3.00 m?

Section 20.2 Specific Heat and Calorimetry

3. The temperature of a silver bar rises by 10.0°C when it absorbs 1.23 kJ of energy by heat. The mass of the bar is 525 g. Determine the specific heat of silver.

4. A 50.0-g sample of copper is at 25.0°C. If 1 200 J of energy is added to it by heat, what is the final temperature of the copper?

5. Systematic use of solar energy can yield a large saving in the cost of winter space heating for a typical house in the north central United States. If the house has good insulation, you may model it as losing energy by heat steadily at the rate 6 000 W on a day in April when the average exterior temperature is 4°C, and when the conventional heating system is not used at all. The passive solar energy collector can consist simply of very large windows in a room facing south. Sunlight shining in during the daytime is absorbed by the floor, interior walls, and objects in the room, raising their temperature to 38°C. As the sun goes down, insulating draperies or shutters are closed over the windows. During the period between 5:00 P.M. and 7:00 A.M. the temperature of the house will drop, and a sufficiently large "thermal mass" is required to keep it from dropping too far. The thermal mass can be a large quantity of stone (with specific heat 850 J/kg · °C) in the floor and the interior walls exposed to sunlight. What mass of stone is required if the temperature is not to drop below 18°C overnight?

6. The *Nova* laser at Lawrence Livermore National Laboratory in California is used in studies of initiating controlled nuclear fusion (Section 45.4). It can deliver a power of 1.60×10^{13} W over a time interval of 2.50 ns. Compare its energy output in one such time interval to the energy required to make a pot of tea by warming 0.800 kg of water from 20.0°C to 100°C.

7. A 1.50-kg iron horseshoe initially at 600°C is dropped into a bucket containing 20.0 kg of water at 25.0°C. What is the final temperature? (Ignore the heat capacity of the container, and assume that a negligible amount of water boils away.)

8. An aluminum cup of mass 200 g contains 800 g of water in thermal equilibrium at 80.0°C. The combination of cup and water is cooled uniformly so that the temperature decreases by 1.50°C per minute. At what rate is energy being removed by heat? Express your answer in watts.

9. An aluminum calorimeter with a mass of 100 g contains 250 g of water. The calorimeter and water are in thermal equilibrium at 10.0°C. Two metallic blocks are placed into the water. One is a 50.0-g piece of copper at 80.0°C. The other block has a mass of 70.0 g and is originally at a temperature of 100°C. The entire system stabilizes at a final temperature of 20.0°C. (a) Determine the specific heat of the unknown sample. (b) Guess the material of the unknown, using the data in Table 20.1.

10. A 3.00-g copper penny at 25.0°C drops 50.0 m to the ground. (a) Assuming that 60.0% of the change in potential energy of the penny–Earth system goes into increasing the internal energy of the penny, determine its final temperature. (b) **What If?** Does the result depend on the mass of the penny? Explain.

11. A combination of 0.250 kg of water at 20.0°C, 0.400 kg of aluminum at 26.0°C, and 0.100 kg of copper at 100°C is mixed in an insulated container and allowed to come to thermal equilibrium. Ignore any energy transfer to or from the container and determine the final temperature of the mixture.

12. If water with a mass m_h at temperature T_h is poured into an aluminum cup of mass m_{Al} containing mass m_c of water at T_c, where $T_h > T_c$, what is the equilibrium temperature of the system?

13. A water heater is operated by solar power. If the solar collector has an area of 6.00 m^2 and the intensity delivered by sunlight is 550 W/m^2, how long does it take to increase the temperature of 1.00 m^3 of water from 20.0°C to 60.0°C?

14. Two thermally insulated vessels are connected by a narrow tube fitted with a valve that is initially closed. One vessel, of volume 16.8 L, contains oxygen at a temperature of 300 K and a pressure of 1.75 atm. The other vessel, of volume 22.4 L, contains oxygen at a temperature of 450 K and a pressure of 2.25 atm. When the valve is opened, the gases in the two vessels mix, and the temperature and pressure become uniform throughout. (a) What is the final temperature? (b) What is the final pressure?

Section 20.3 Latent Heat

15. How much energy is required to change a 40.0-g ice cube from ice at −10.0°C to steam at 110°C?

16. A 50.0-g copper calorimeter contains 250 g of water at 20.0°C. How much steam must be condensed into the water if the final temperature of the system is to reach 50.0°C?

17. A 3.00-g lead bullet at 30.0°C is fired at a speed of 240 m/s into a large block of ice at 0°C, in which it becomes embedded. What quantity of ice melts?

18. Steam at 100°C is added to ice at 0°C. (a) Find the amount of ice melted and the final temperature when the mass of steam is 10.0 g and the mass of ice is 50.0 g. (b) **What If?** Repeat when the mass of steam is 1.00 g and the mass of ice is 50.0 g.

19. A 1.00-kg block of copper at 20.0°C is dropped into a large vessel of liquid nitrogen at 77.3 K. How many kilograms of nitrogen boil away by the time the copper reaches 77.3 K? (The specific heat of copper is 0.092 0 cal/g · °C. The latent heat of vaporization of nitrogen is 48.0 cal/g.)

20. Assume that a hailstone at 0°C falls through air at a uniform temperature of 0°C and lands on a sidewalk also at this temperature. From what initial height must the hailstone fall in order to entirely melt on impact?

21. In an insulated vessel, 250 g of ice at 0°C is added to 600 g of water at 18.0°C. (a) What is the final temperature

of the system? (b) How much ice remains when the system reaches equilibrium?

22. **Review problem.** Two speeding lead bullets, each of mass 5.00 g, and at temperature 20.0°C, collide head-on at speeds of 500 m/s each. Assuming a perfectly inelastic collision and no loss of energy by heat to the atmosphere, describe the final state of the two-bullet system.

Section 20.4 Work and Heat in Thermodynamic Processes

23. A sample of ideal gas is expanded to twice its original volume of 1.00 m^3 in a quasi-static process for which $P = \alpha V^2$, with $\alpha = 5.00\ \text{atm/m}^6$, as shown in Figure P20.23. How much work is done on the expanding gas?

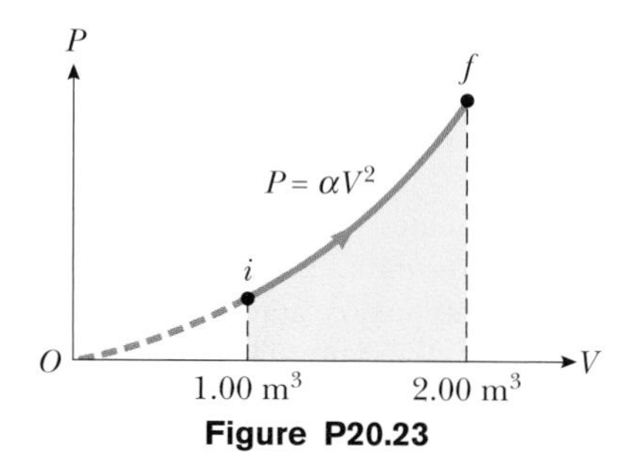

Figure P20.23

24. (a) Determine the work done on a fluid that expands from i to f as indicated in Figure P20.24. (b) **What If?** How much work is performed on the fluid if it is compressed from f to i along the same path?

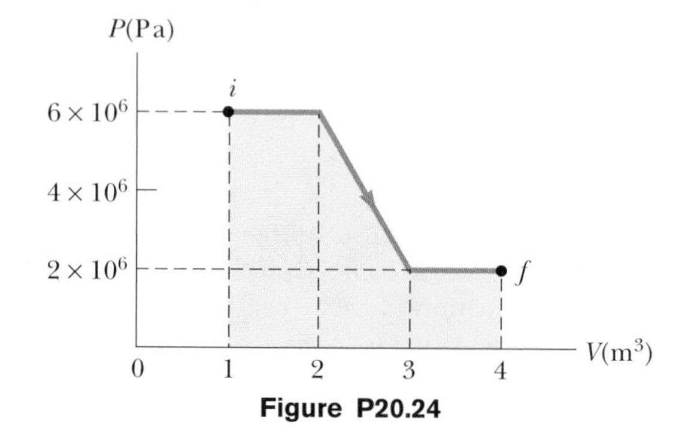

Figure P20.24

25. An ideal gas is enclosed in a cylinder with a movable piston on top of it. The piston has a mass of 8 000 g and an area of 5.00 cm^2 and is free to slide up and down, keeping the pressure of the gas constant. How much work is done on the gas as the temperature of 0.200 mol of the gas is raised from 20.0°C to 300°C?

26. An ideal gas is enclosed in a cylinder that has a movable piston on top. The piston has a mass m and an area A and is free to slide up and down, keeping the pressure of the gas constant. How much work is done on the gas as the temperature of n mol of the gas is raised from T_1 to T_2?

27. One mole of an ideal gas is heated slowly so that it goes from the PV state (P_i, V_i) to $(3P_i, 3V_i)$ in such a way that the pressure is directly proportional to the volume. (a) How much work is done on the gas in the process? (b) How is the temperature of the gas related to its volume during this process?

Section 20.5 The First Law of Thermodynamics

28. A gas is compressed at a constant pressure of 0.800 atm from 9.00 L to 2.00 L. In the process, 400 J of energy leaves the gas by heat. (a) What is the work done on the gas? (b) What is the change in its internal energy?

29. A thermodynamic system undergoes a process in which its internal energy decreases by 500 J. At the same time, 220 J of work is done on the system. Find the energy transferred to or from it by heat.

30. A gas is taken through the cyclic process described in Figure P20.30. (a) Find the net energy transferred to the system by heat during one complete cycle. (b) **What If?** If the cycle is reversed—that is, the process follows the path $ACBA$—what is the net energy input per cycle by heat?

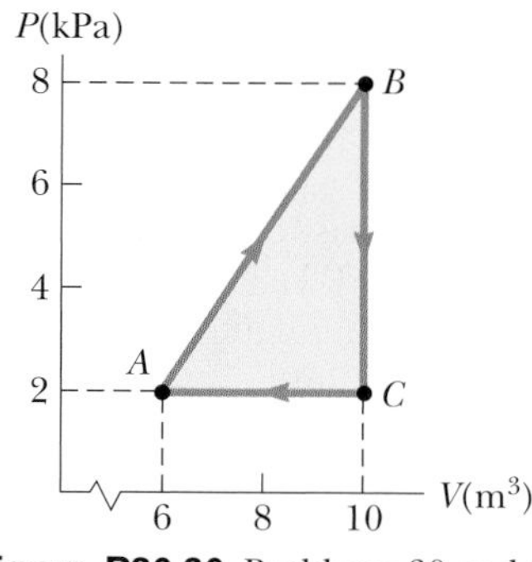

Figure P20.30 Problems 30 and 31.

31. Consider the cyclic process depicted in Figure P20.30. If Q is negative for the process BC and ΔE_{int} is negative for the process CA, what are the signs of Q, W, and ΔE_{int} that are associated with each process?

32. A sample of an ideal gas goes through the process shown in Figure P20.32. From A to B, the process is adiabatic; from B to C, it is isobaric with 100 kJ of energy entering the system by heat. From C to D, the process is isothermal; from D to A, it is isobaric with 150 kJ of energy leaving the system by heat. Determine the difference in internal energy $E_{\text{int},B} - E_{\text{int},A}$.

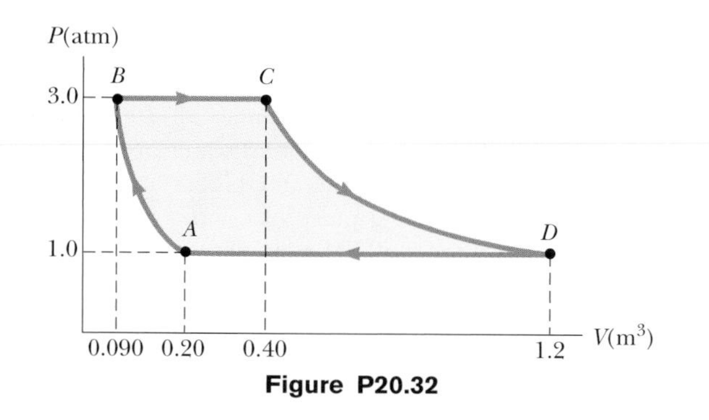

Figure P20.32

33. A sample of an ideal gas is in a vertical cylinder fitted with a piston. As 5.79 kJ of energy is transferred to the gas by heat to raise its temperature, the weight on the piston is adjusted so that the state of the gas changes from point A to point B along the semicircle shown in Figure P20.33. Find the change in internal energy of the gas.

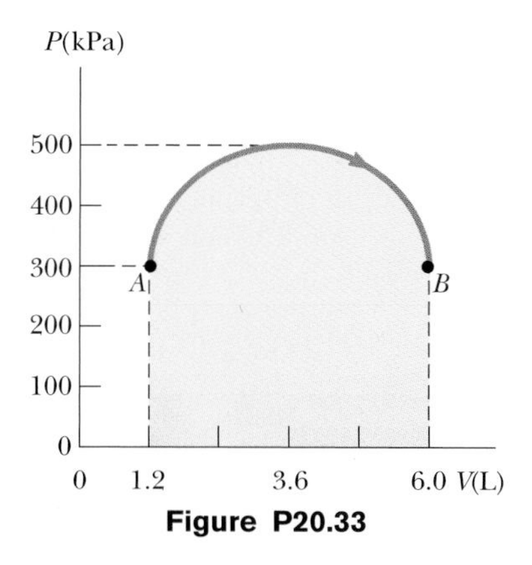

Figure P20.33

Section 20.6 Some Applications of the First Law of Thermodynamics

34. One mole of an ideal gas does 3 000 J of work on its surroundings as it expands isothermally to a final pressure of 1.00 atm and volume of 25.0 L. Determine (a) the initial volume and (b) the temperature of the gas.

35. An ideal gas initially at 300 K undergoes an isobaric expansion at 2.50 kPa. If the volume increases from 1.00 m^3 to 3.00 m^3 and 12.5 kJ is transferred to the gas by heat, what are (a) the change in its internal energy and (b) its final temperature?

36. A 1.00-kg block of aluminum is heated at atmospheric pressure so that its temperature increases from 22.0°C to 40.0°C. Find (a) the work done on the aluminum, (b) the energy added to it by heat, and (c) the change in its internal energy.

37. How much work is done on the steam when 1.00 mol of water at 100°C boils and becomes 1.00 mol of steam at 100°C at 1.00 atm pressure? Assuming the steam to behave as an ideal gas, determine the change in internal energy of the material as it vaporizes.

38. An ideal gas initially at P_i, V_i, and T_i is taken through a cycle as in Figure P20.38. (a) Find the net work done on the gas per cycle. (b) What is the net energy added by heat to the system per cycle? (c) Obtain a numerical value for the net work done per cycle for 1.00 mol of gas initially at 0°C.

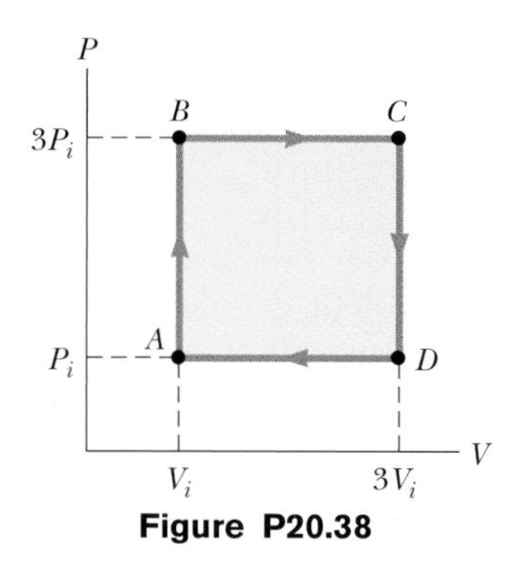

Figure P20.38

39. A 2.00-mol sample of helium gas initially at 300 K and 0.400 atm is compressed isothermally to 1.20 atm. Noting that the helium behaves as an ideal gas, find (a) the final volume of the gas, (b) the work done on the gas, and (c) the energy transferred by heat.

40. In Figure P20.40, the change in internal energy of a gas that is taken from A to C is + 800 J. The work done on the gas along path ABC is −500 J. (a) How much energy must be added to the system by heat as it goes from A through B to C? (b) If the pressure at point A is five times that of point C, what is the work done on the system in going from C to D? (c) What is the energy exchanged with the surroundings by heat as the cycle goes from C to A along the green path? (d) If the change in internal energy in going from point D to point A is + 500 J, how much energy must be added to the system by heat as it goes from point C to point D?

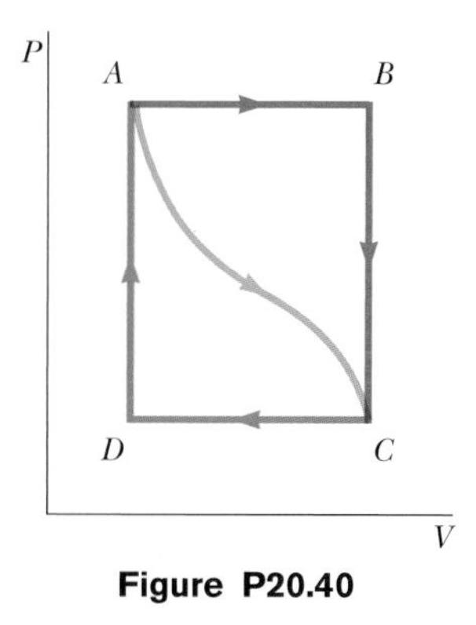

Figure P20.40

Section 20.7 Energy-Transfer Mechanisms

41. A box with a total surface area of 1.20 m^2 and a wall thickness of 4.00 cm is made of an insulating material. A 10.0-W electric heater inside the box maintains the inside temperature at 15.0°C above the outside temperature. Find the thermal conductivity k of the insulating material.

42. A glass window pane has an area of 3.00 m^2 and a thickness of 0.600 cm. If the temperature difference between its faces is 25.0°C, what is the rate of energy transfer by conduction through the window?

43. A bar of gold is in thermal contact with a bar of silver of the same length and area (Fig. P20.43). One end of the compound bar is maintained at 80.0°C while the opposite end is at 30.0°C. When the energy transfer reaches steady state, what is the temperature at the junction?

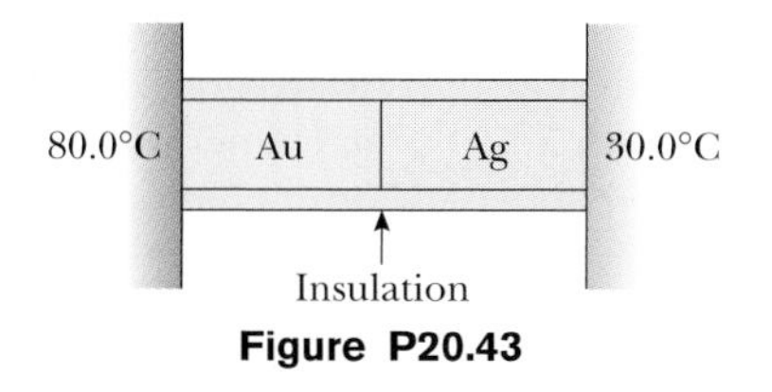

Figure P20.43

44. A thermal window with an area of 6.00 m^2 is constructed of two layers of glass, each 4.00 mm thick, and separated from each other by an air space of 5.00 mm. If the inside surface is at 20.0°C and the outside is at −30.0°C, what is the rate of energy transfer by conduction through the window?

45. A power transistor is a solid-state electronic device. Assume that energy entering the device at the rate of 1.50 W by electrical transmission causes the internal energy of the device to increase. The surface area of the transistor is so small that it tends to overheat. To prevent overheating, the transistor is attached to a larger metal heat sink with fins. The temperature of the heat sink remains constant at 35.0°C under steady-state conditions. The transistor is electrically insulated from the heat sink by a rectangular sheet of mica measuring 8.25 mm by 6.25 mm, and 0.085 2 mm thick. The thermal conductivity of mica is equal to 0.075 3 W/m · °C. What is the operating temperature of the transistor?

46. Calculate the *R* value of (a) a window made of a single pane of flat glass $\frac{1}{8}$ in. thick, and (b) a thermal window made of two single panes each $\frac{1}{8}$ in. thick and separated by a $\frac{1}{4}$-in. air space. (c) By what factor is the transfer of energy by heat through the window reduced by using the thermal window instead of the single pane window?

47. The surface of the Sun has a temperature of about 5 800 K. The radius of the Sun is 6.96×10^8 m. Calculate the total energy radiated by the Sun each second. Assume that the emissivity of the Sun is 0.965.

48. A large hot pizza floats in outer space. What is the order of magnitude of (a) its rate of energy loss? (b) its rate of temperature change? List the quantities you estimate and the value you estimate for each.

49. The tungsten filament of a certain 100-W light bulb radiates 2.00 W of light. (The other 98 W is carried away by convection and conduction.) The filament has a surface area of 0.250 mm^2 and an emissivity of 0.950. Find the filament's temperature. (The melting point of tungsten is 3 683 K.)

50. At high noon, the Sun delivers 1 000 W to each square meter of a blacktop road. If the hot asphalt loses energy only by radiation, what is its equilibrium temperature?

51. The intensity of solar radiation reaching the top of the Earth's atmosphere is 1 340 W/m^2. The temperature of the Earth is affected by the so-called greenhouse effect of the atmosphere. That effect makes our planet's emissivity for visible light higher than its emissivity for infrared light. For comparison, consider a spherical object with no atmosphere, at the same distance from the Sun as the Earth. Assume that its emissivity is the same for all kinds of electromagnetic waves and that its temperature is uniform over its surface. Identify the projected area over which it absorbs sunlight and the surface area over which it radiates. Compute its equilibrium temperature. Chilly, isn't it? Your calculation applies to (a) the average temperature of the Moon, (b) astronauts in mortal danger aboard the crippled *Apollo 13* spacecraft, and (c) global catastrophe on the Earth if widespread fires should cause a layer of soot to accumulate throughout the upper atmosphere, so that most of the radiation from the Sun were absorbed there rather than at the surface below the atmosphere.

Additional Problems

52. Liquid nitrogen with a mass of 100 g at 77.3 K is stirred into a beaker containing 200 g of 5.00°C water. If the nitrogen leaves the solution as soon as it turns to gas, how much water freezes? (The latent heat of vaporization of nitrogen is 48.0 cal/g, and the latent heat of fusion of water is 79.6 cal/g.)

53. A 75.0-kg cross-country skier moves across the snow (Fig. P20.53). The coefficient of friction between the skis and the snow is 0.200. Assume that all the snow beneath his skis is at 0°C and that all the internal energy generated by friction is added to the snow, which sticks to his skis until it melts. How far would he have to ski to melt 1.00 kg of snow?

Nathan Bilow/Leo de Wys, Inc.

Figure P20.53

54. On a cold winter day you buy roasted chestnuts from a street vendor. Into the pocket of your down parka you put the change he gives you—coins constituting 9.00 g of copper at −12.0°C. Your pocket already contains 14.0 g of silver coins at 30.0°C. A short time later the temperature of the copper coins is 4.00°C and is increasing at a rate of 0.500°C/s. At this time, (a) what is the temperature of the silver coins, and (b) at what rate is it changing?

55. An aluminum rod 0.500 m in length and with a cross-sectional area of 2.50 cm^2 is inserted into a thermally insulated vessel containing liquid helium at 4.20 K. The rod is initially at 300 K. (a) If half of the rod is inserted into the helium, how many liters of helium boil off by the time the inserted half cools to 4.20 K? (Assume the upper half does not yet cool.) (b) If the upper end of the rod is maintained at 300 K, what is the approximate boil-off rate of liquid helium after the lower half has reached 4.20 K? (Aluminum has thermal conductivity of 31.0 J/s · cm · K at 4.2 K; ignore its temperature variation. Aluminum has a specific heat of 0.210 cal/g · °C and density of 2.70 g/cm^3. The density of liquid helium is 0.125 g/cm^3.)

56. A copper ring (with mass of 25.0 g, coefficient of linear expansion of 1.70×10^{-5} (°C)$^{-1}$, and specific heat of 9.24×10^{-2} cal/g · °C) has a diameter of 5.00 cm at its temperature of 15.0°C. A spherical aluminum shell (with mass 10.9 g, coefficient of linear expansion 2.40×10^{-5} (°C)$^{-1}$, and specific heat 0.215 cal/g·°C) has a diameter of 5.01 cm at a temperature higher than 15.0°C. The sphere is placed on top of the horizontal ring, and the two are allowed to come to thermal equilibrium without any exchange of energy with the surroundings. As soon as the sphere and ring reach thermal equilibrium, the sphere barely falls through the ring. Find (a) the equilibrium temperature, and (b) the initial temperature of the sphere.

57. A *flow calorimeter* is an apparatus used to measure the specific heat of a liquid. The technique of flow calorimetry involves measuring the temperature difference between the input and output points of a flowing stream of the liquid while energy is added by heat at a known rate. A liquid of density ρ flows through the calorimeter with volume flow rate R. At steady state, a temperature difference ΔT is established between the input and output points when energy is supplied at the rate $\mathcal{P}$. What is the specific heat of the liquid?

58. One mole of an ideal gas is contained in a cylinder with a movable piston. The initial pressure, volume, and temperature are P_i, V_i, and T_i, respectively. Find the work done on the gas for the following processes and show each process on a PV diagram: (a) An isobaric compression in which the final volume is half the initial volume. (b) An isothermal compression in which the final pressure is four times the initial pressure. (c) An isovolumetric process in which the final pressure is three times the initial pressure.

59. One mole of an ideal gas, initially at 300 K, is cooled at constant volume so that the final pressure is one fourth of the initial pressure. Then the gas expands at constant pressure until it reaches the initial temperature. Determine the work done on the gas.

60. **Review problem.** Continue the analysis of Problem 60 in Chapter 19. Following a collision between a large spacecraft and an asteroid, a copper disk of radius 28.0 m and thickness 1.20 m, at a temperature of 850°C, is floating in space, rotating about its axis with an angular speed of 25.0 rad/s. As the disk radiates infrared light, its temperature falls to 20.0°C. No external torque acts on the disk. (a) Find the change in kinetic energy of the disk. (b) Find the change in internal energy of the disk. (b) Find the amount of energy it radiates.

61. **Review problem.** A 670-kg meteorite happens to be composed of aluminum. When it is far from the Earth, its temperature is − 15°C and it moves with a speed of 14.0 km/s relative to the Earth. As it crashes into the planet, assume that the resulting additional internal energy is shared equally between the meteor and the planet, and that all of the material of the meteor rises momentarily to the same final temperature. Find this temperature. Assume that the specific heat of liquid and of gaseous aluminum is 1 170 J/kg · °C.

62. An iron plate is held against an iron wheel so that a kinetic friction force of 50.0 N acts between the two pieces of metal. The relative speed at which the two surfaces slide over each other is 40.0 m/s. (a) Calculate the rate at which mechanical energy is converted to internal energy. (b) The plate and the wheel each have a mass of 5.00 kg, and each receives 50.0% of the internal energy. If the system is run as described for 10.0 s and each object is then allowed to reach a uniform internal temperature, what is the resultant temperature increase?

63. A solar cooker consists of a curved reflecting surface that concentrates sunlight onto the object to be warmed (Fig. P20.63). The solar power per unit area reaching the Earth's surface at the location is 600 W/m^2. The cooker faces the Sun and has a diameter of 0.600 m. Assume that 40.0% of the incident energy is transferred to 0.500 L of water in an open container, initially at 20.0°C. How long does it take to completely boil away the water? (Ignore the heat capacity of the container.)

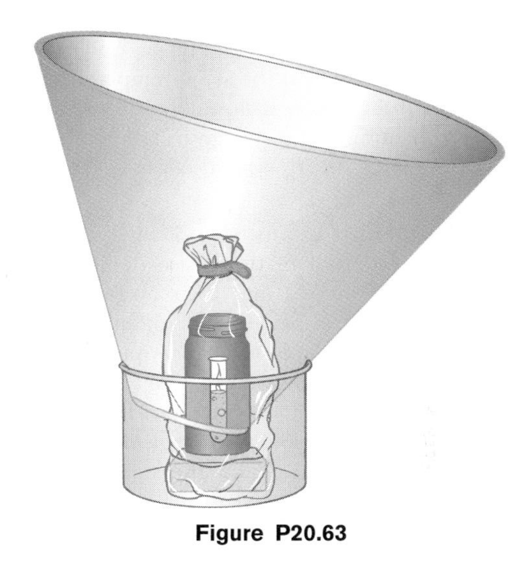

Figure P20.63

64. Water in an electric teakettle is boiling. The power absorbed by the water is 1.00 kW. Assuming that the pressure of vapor in the kettle equals atmospheric pressure, determine the speed of effusion of vapor from the kettle's spout, if the spout has a cross-sectional area of 2.00 cm^2.

65. A cooking vessel on a slow burner contains 10.0 kg of water and an unknown mass of ice in equilibrium at 0°C at time $t = 0$. The temperature of the mixture is measured at various times, and the result is plotted in Figure P20.65. During the first 50.0 min, the mixture remains at 0°C. From 50.0 min to 60.0 min, the temperature increases to 2.00°C. Ignoring the heat capacity of the vessel, determine the initial mass of ice.

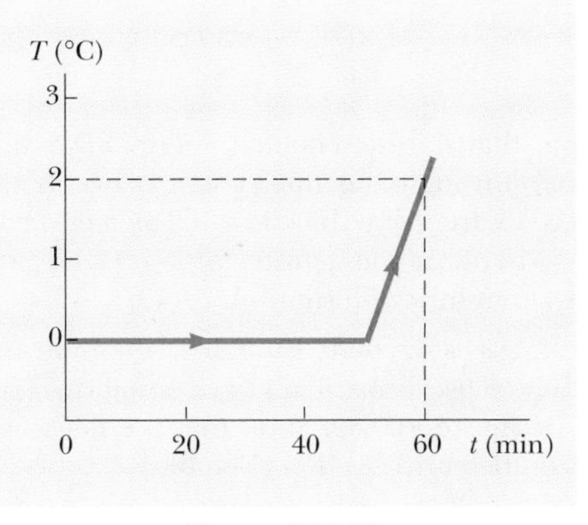

Figure P20.65

66. (a) In air at 0°C, a 1.60-kg copper block at 0°C is set sliding at 2.50 m/s over a sheet of ice at 0°C. Friction brings the block to rest. Find the mass of the ice that melts. To describe the process of slowing down, identify the energy input Q, the work input W, the change in internal energy ΔE_{int}, and the change in mechanical energy ΔK for the block and also for the ice. (b) A 1.60-kg block of ice at 0°C is set sliding at 2.50 m/s over a sheet of copper at 0°C. Friction brings the block to rest. Find the mass of the ice that melts. Identify Q, W, ΔE_{int}, and ΔK for the block and for the metal sheet during the process. (c) A thin 1.60-kg slab of copper at 20°C is set sliding at 2.50 m/s over an identical stationary slab at the same temperature. Friction quickly stops the motion. If no energy is lost to the environment by heat, find the change in temperature of both objects. Identify Q, W, ΔE_{int}, and ΔK for each object during the process.

67. The average thermal conductivity of the walls (including the windows) and roof of the house depicted in Figure P20.67 is 0.480 W/m · °C, and their average thickness is 21.0 cm. The house is heated with natural gas having a heat of combustion (that is, the energy provided per cubic meter of gas burned) of 9 300 kcal/m^3. How many cubic meters of gas must be burned each day to maintain an inside temperature of 25.0°C if the outside temperature is 0.0°C? Disregard radiation and the energy lost by heat through the ground.

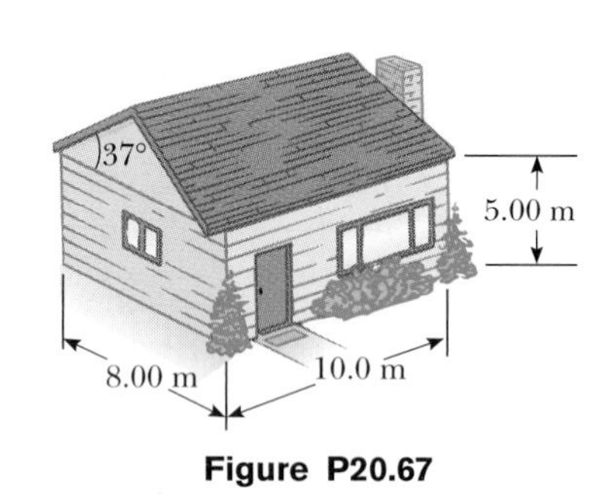

Figure P20.67

68. A pond of water at 0°C is covered with a layer of ice 4.00 cm thick. If the air temperature stays constant at −10.0°C, how long does it take for the ice thickness to increase to 8.00 cm? *Suggestion:* Utilize Equation 20.15 in the form

$$\frac{dQ}{dt} = kA\frac{\Delta T}{x}$$

and note that the incremental energy dQ extracted from the water through the thickness x of ice is the amount required to freeze a thickness dx of ice. That is, $dQ = L\rho A\, dx$, where ρ is the density of the ice, A is the area, and L is the latent heat of fusion.

69. An ideal gas is carried through a thermodynamic cycle consisting of two isobaric and two isothermal processes as shown in Figure P20.69. Show that the net work done on the gas in the entire cycle is given by

$$W_{net} = -P_1(V_2 - V_1)\ln\frac{P_2}{P_1}$$

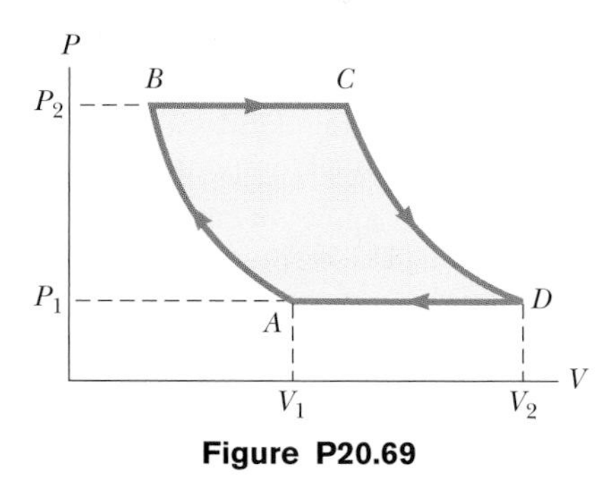

Figure P20.69

70. The inside of a hollow cylinder is maintained at a temperature T_a while the outside is at a lower temperature, T_b (Fig. P20.70). The wall of the cylinder has a thermal conductivity k. Ignoring end effects, show that the rate of energy conduction from the inner to the outer surface in the radial direction is

$$\frac{dQ}{dt} = 2\pi Lk\left[\frac{T_a - T_b}{\ln(b/a)}\right]$$

(*Suggestions:* The temperature gradient is dT/dr. Note that a radial energy current passes through a concentric cylinder of area $2\pi rL$.)

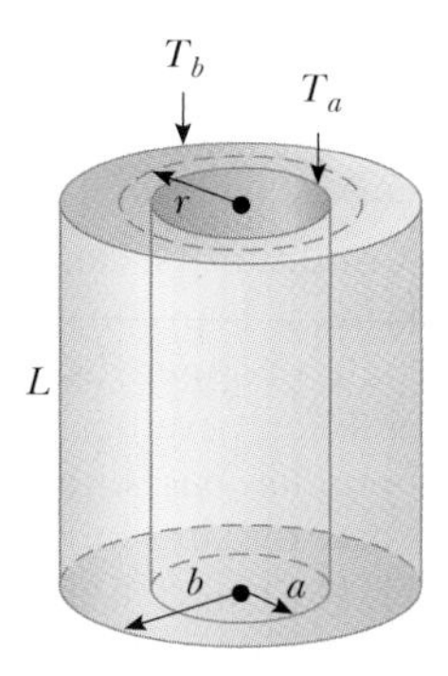

Figure P20.70

71. The passenger section of a jet airliner is in the shape of a cylindrical tube with a length of 35.0 m and an inner radius of 2.50 m. Its walls are lined with an insulating material 6.00 cm in thickness and having a thermal conductivity of 4.00×10^{-5} cal/s · cm · °C. A heater must maintain the interior temperature at 25.0°C while the outside temperature is −35.0°C. What power must be supplied to the heater? (Use the result of Problem 70.)

72. A student obtains the following data in a calorimetry experiment designed to measure the specific heat of aluminum:

Initial temperature of water and calorimeter:	70°C
Mass of water:	0.400 kg
Mass of calorimeter:	0.040 kg
Specific heat of calorimeter:	0.63 kJ/kg · °C
Initial temperature of aluminum:	27°C
Mass of aluminum:	0.200 kg
Final temperature of mixture:	66.3°C

Use these data to determine the specific heat of aluminum. Your result should be within 15% of the value listed in Table 20.1.

73. During periods of high activity, the Sun has more sunspots than usual. Sunspots are cooler than the rest of the luminous layer of the Sun's atmosphere (the photosphere). Paradoxically, the total power output of the active Sun is not lower than average but is the same or slightly higher than average. Work out the details of the following crude model of this phenomenon. Consider a patch of the photosphere with an area of $5.10 \times 10^{14}\ \text{m}^2$. Its emissivity is 0.965. (a) Find the power it radiates if its temperature is uniformly 5 800 K, corresponding to the quiet Sun. (b) To represent a sunspot, assume that 10.0% of the area is at 4 800 K and the other 90.0% is at 5 890 K. That is, a section with the surface area of the Earth is 1 000 K cooler than before and a section nine times as large is 90 K warmer. Find the average temperature of the patch. (c) Find the power output of the patch. Compare it with the answer to part (a). (The next sunspot maximum is expected around the year 2012.)

Answers to Quick Quizzes

20.1 Water, glass, iron. Because water has the highest specific heat (4 186 J/kg · °C), it has the smallest change in temperature. Glass is next (837 J/kg · °C), and iron is last (448 J/kg · °C).

20.2 Iron, glass, water. For a given temperature increase, the energy transfer by heat is proportional to the specific heat.

20.3 The figure below shows a graphical representation of the internal energy of the ice in parts A to E as a function of energy added. Notice that this graph looks quite different from Figure 20.2—it doesn't have the flat portions during the phase changes. Regardless of how the temperature is varying in Figure 20.2, the internal energy of the system simply increases linearly with energy input.

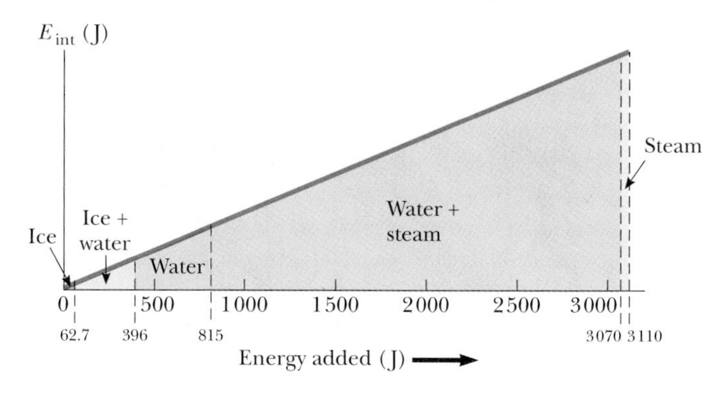

20.4 C, A, E. The slope is the ratio of the temperature change to the amount of energy input. Thus, the slope is proportional to the reciprocal of the specific heat. Water, which has the highest specific heat, has the smallest slope.

20.5

Situation	System	Q	W	ΔE_{int}
(a) Rapidly pumping up a bicycle tire	Air in the pump	0	+	+
(b) Pan of room-temperature water sitting on a hot stove	Water in the pan	+	0	+
(c) Air quickly leaking out of a balloon	Air originally in the balloon	0	−	−

(a) Because the pumping is rapid, no energy enters or leaves the system by heat. Because $W > 0$ when work is done *on* the system, it is positive here. Thus, we see that $\Delta E_{int} = Q + W$ must be positive. The air in the pump is warmer. (b) There is no work done either on or by the system, but energy transfers into the water by heat from the hot burner, making both Q and ΔE_{int} positive. (c) Again no energy transfers into or out of the system by heat, but the air molecules escaping from the balloon do work on the surrounding air molecules as they push them out of the way. Thus W is negative and ΔE_{int} is negative. The decrease in internal energy is evidenced by the fact that the escaping air becomes cooler.

20.6 A is isovolumetric, B is adiabatic, C is isothermal, and D is isobaric.

20.7 (c). The blanket acts as a thermal insulator, slowing the transfer of energy by heat from the air into the cube.

20.8 (b). In parallel, the rods present a larger area through which energy can transfer and a smaller length.

Chapter 21

The Kinetic Theory of Gases

CHAPTER OUTLINE

▲ *Dogs do not have sweat glands like humans. In hot weather, dogs pant to promote evaporation from the tongue. In this chapter, we show that evaporation is a cooling process based on the removal of molecules with high kinetic energy from a liquid. (Frank Oberle/Getty Images)*

In Chapter 19 we discussed the properties of an ideal gas, using such macroscopic variables as pressure, volume, and temperature. We shall now show that such large-scale properties can be related to a description on a microscopic scale, where matter is treated as a collection of molecules. Newton's laws of motion applied in a statistical manner to a collection of particles provide a reasonable description of thermodynamic processes. To keep the mathematics relatively simple, we shall consider primarily the behavior of gases, because in gases the interactions between molecules are much weaker than they are in liquids or solids. In our model of gas behavior, called **kinetic theory**, gas molecules move about in a random fashion, colliding with the walls of their container and with each other. Kinetic theory provides us with a physical basis for our understanding of the concept of temperature.

21.1 Molecular Model of an Ideal Gas

We begin this chapter by developing a microscopic model of an ideal gas. The model shows that the pressure that a gas exerts on the walls of its container is a consequence of the collisions of the gas molecules with the walls and is consistent with the macroscopic description of Chapter 19. In developing this model, we make the following assumptions:

Assumptions of the molecular model of an ideal gas

1. **The number of molecules in the gas is large, and the average separation between them is large compared with their dimensions.** This means that the molecules occupy a negligible volume in the container. This is consistent with the ideal gas model, in which we imagine the molecules to be point-like.
2. **The molecules obey Newton's laws of motion, but as a whole they move randomly.** By "randomly" we mean that any molecule can move in any direction with any speed. At any given moment, a certain percentage of molecules move at high speeds, and a certain percentage move at low speeds.
3. **The molecules interact only by short-range forces during elastic collisions.** This is consistent with the ideal gas model, in which the molecules exert no long-range forces on each other.
4. **The molecules make elastic collisions with the walls.**
5. **The gas under consideration is a pure substance; that is, all molecules are identical.**

Although we often picture an ideal gas as consisting of single atoms, we can assume that the behavior of molecular gases approximates that of ideal gases rather well at low pressures. Molecular rotations or vibrations have no effect, on the average, on the motions that we consider here.

For our first application of kinetic theory, let us derive an expression for the pressure of N molecules of an ideal gas in a container of volume V in terms of microscopic quantities. The container is a cube with edges of length d (Fig. 21.1). We shall first

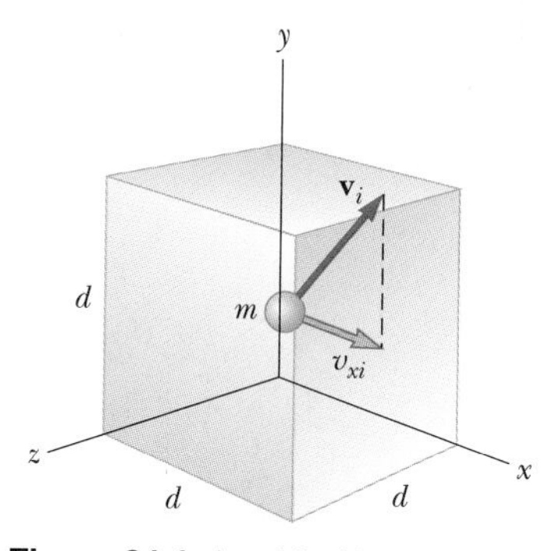

Figure 21.1 A cubical box with sides of length d containing an ideal gas. The molecule shown moves with velocity $\mathbf{v}_i$.

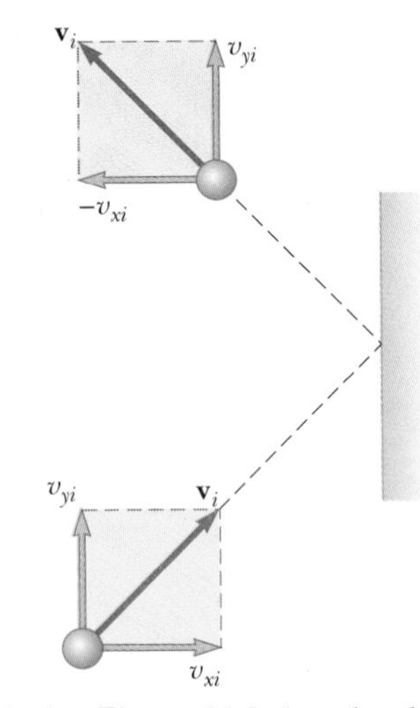

Active Figure 21.2 A molecule makes an elastic collision with the wall of the container. Its x component of momentum is reversed, while its y component remains unchanged. In this construction, we assume that the molecule moves in the xy plane.

At the Active Figures link at* http://www.pse6.com, *you can observe molecules within a container making collisions with the walls of the container and with each other.

focus our attention on one of these molecules of mass m, and assume that it is moving so that its component of velocity in the x direction is v_{xi} as in Figure 21.2. (The subscript i here refers to the ith molecule, not to an initial value. We will combine the effects of all of the molecules shortly.) As the molecule collides elastically with any wall (assumption 4), its velocity component perpendicular to the wall is reversed because the mass of the wall is far greater than the mass of the molecule. Because the momentum component p_{xi} of the molecule is mv_{xi} before the collision and $-mv_{xi}$ after the collision, the change in the x component of the momentum of the molecule is

$$\Delta p_{xi} = -mv_{xi} - (mv_{xi}) = -2mv_{xi}$$

Because the molecules obey Newton's laws (assumption 2), we can apply the impulse-momentum theorem (Eq. 9.8) to the molecule to give us

$$\overline{F}_{i,\text{ on molecule}} \Delta t_{\text{collision}} = \Delta p_{xi} = -2mv_{xi}$$

where $\overline{F}_{i,\text{on molecule}}$ is the x component of the average force that the wall exerts on the molecule during the collision and $\Delta t_{\text{collision}}$ is the duration of the collision. In order for the molecule to make another collision with the same wall after this first collision, it must travel a distance of $2d$ in the x direction (across the container and back). Therefore, the time interval between two collisions with the same wall is

$$\Delta t = \frac{2d}{v_{xi}}$$

The force that causes the change in momentum of the molecule in the collision with the wall occurs only during the collision. However, we can average the force over the time interval for the molecule to move across the cube and back. Sometime during this time interval, the collision occurs, so that the change in momentum for this time interval is the same as that for the short duration of the collision. Thus, we can rewrite the impulse-momentum theorem as

$$\overline{F}_i \, \Delta t = -2mv_{xi}$$

where $\overline{F}_i$ is the average force component over the time for the molecule to move across the cube and back. Because exactly one collision occurs for each such time interval, this is also the long-term average force on the molecule, over long time intervals containing any number of multiples of Δt.

This equation and the preceding one enable us to express the x component of the long-term average force exerted by the wall on the molecule as

$$\overline{F}_i = \frac{-2mv_{xi}}{\Delta t} = \frac{-2mv_{xi}^2}{2d} = \frac{-mv_{xi}^2}{d}$$

Now, by Newton's third law, the average x component of the force exerted by the molecule on the wall is equal in magnitude and opposite in direction:

$$\overline{F}_{i,\text{ on wall}} = -\overline{F}_i = -\left(\frac{-mv_{xi}^2}{d}\right) = \frac{mv_{xi}^2}{d}$$

The total average force $\overline{F}$ exerted by the gas on the wall is found by adding the average forces exerted by the individual molecules. We add terms such as that above for all molecules:

$$\overline{F} = \sum_{i=1}^{N} \frac{mv_{xi}^2}{d} = \frac{m}{d} \sum_{i=1}^{N} v_{xi}^2$$

where we have factored out the length of the box and the mass m, because assumption 5 tells us that all of the molecules are the same. We now impose assumption 1, that the number of molecules is large. For a small number of molecules, the actual force on the wall would vary with time. It would be nonzero during the short interval of a collision of a molecule with the wall and zero when no molecule happens to be hitting the wall. For a very large number of molecules, however, such as Avogadro's number, these variations in force are smoothed out, so that the average force given above is the same over *any* time interval. Thus, the *constant* force F on the wall due to the molecular collisions is

$$F = \frac{m}{d} \sum_{i=1}^{N} v_{xi}^2$$

To proceed further, let us consider how to express the average value of the square of the x component of the velocity for N molecules. The traditional average of a set of values is the sum of the values over the number of values:

$$\overline{v_x^2} = \frac{\sum_{i=1}^{N} v_{xi}^2}{N}$$

The numerator of this expression is contained in the right-hand side of the preceding equation. Thus, combining the two expressions, the total force on the wall can be written

$$F = \frac{m}{d} N\overline{v_x^2} \qquad (21.1)$$

Now let us focus again on one molecule with velocity components v_{xi}, v_{yi}, and v_{zi}. The Pythagorean theorem relates the square of the speed of the molecule to the squares of the velocity components:

$$v_i^2 = v_{xi}^2 + v_{yi}^2 + v_{zi}^2$$

Hence, the average value of v^2 for all the molecules in the container is related to the average values of v_x^2, v_y^2, and v_z^2 according to the expression

$$\overline{v^2} = \overline{v_x^2} + \overline{v_y^2} + \overline{v_z^2}$$

Because the motion is completely random (assumption 2), the average values $\overline{v_x^2}$, $\overline{v_y^2}$, and $\overline{v_z^2}$ are equal to each other. Using this fact and the preceding equation, we find that

$$\overline{v^2} = 3\overline{v_x^2}$$

Thus, from Equation 21.1, the total force exerted on the wall is

$$F = \frac{N}{3}\left(\frac{m\overline{v^2}}{d}\right)$$

Using this expression, we can find the total pressure exerted on the wall:

$$P = \frac{F}{A} = \frac{F}{d^2} = \frac{1}{3}\left(\frac{N}{d^3}m\overline{v^2}\right) = \frac{1}{3}\left(\frac{N}{V}\right)m\overline{v^2}$$

$$P = \frac{2}{3}\left(\frac{N}{V}\right)\left(\frac{1}{2}m\overline{v^2}\right) \qquad (21.2)$$

Relationship between pressure and molecular kinetic energy

This result indicates that **the pressure of a gas is proportional to the number of molecules per unit volume and to the average translational kinetic energy of the molecules, $\frac{1}{2}m\overline{v^2}$.** In analyzing this simplified model of an ideal gas, we obtain an important result that relates the macroscopic quantity of pressure to a microscopic quantity—the average value of the square of the molecular speed. Thus, we have established a key link between the molecular world and the large-scale world.

You should note that Equation 21.2 verifies some features of pressure with which you are probably familiar. One way to increase the pressure inside a container is to increase the number of molecules per unit volume N/V in the container. This is what you do when you add air to a tire. The pressure in the tire can also be increased by increasing the average translational kinetic energy of the air molecules in the tire.

This can be accomplished by increasing the temperature of that air, as we shall soon show mathematically. This is why the pressure inside a tire increases as the tire warms up during long trips. The continuous flexing of the tire as it moves along the road surface results in work done as parts of the tire distort, causing an increase in internal energy of the rubber. The increased temperature of the rubber results in the transfer of energy by heat into the air inside the tire. This transfer increases the air's temperature, and this increase in temperature in turn produces an increase in pressure.

Molecular Interpretation of Temperature

We can gain some insight into the meaning of temperature by first writing Equation 21.2 in the form

$$PV = \tfrac{2}{3}N(\tfrac{1}{2}m\overline{v^2})$$

Let us now compare this with the equation of state for an ideal gas (Eq. 19.10):

$$PV = Nk_BT$$

Recall that the equation of state is based on experimental facts concerning the macroscopic behavior of gases. Equating the right sides of these expressions, we find that

Temperature is proportional to average kinetic energy

$$T = \frac{2}{3k_B}(\tfrac{1}{2}m\overline{v^2}) \tag{21.3}$$

This result tells us that temperature is a direct measure of average molecular kinetic energy. By rearranging Equation 21.3, we can relate the translational molecular kinetic energy to the temperature:

Average kinetic energy per molecule

$$\tfrac{1}{2}m\overline{v^2} = \tfrac{3}{2}k_BT \tag{21.4}$$

That is, the average translational kinetic energy per molecule is $\frac{3}{2}k_BT$. Because $\overline{v_x^2} = \frac{1}{3}\overline{v^2}$, it follows that

$$\tfrac{1}{2}m\overline{v_x^2} = \tfrac{1}{2}k_BT \tag{21.5}$$

In a similar manner, it follows that the motions in the y and z directions give us

$$\tfrac{1}{2}m\overline{v_y^2} = \tfrac{1}{2}k_BT \qquad \text{and} \qquad \tfrac{1}{2}m\overline{v_z^2} = \tfrac{1}{2}k_BT$$

Thus, each translational degree of freedom contributes an equal amount of energy, $\frac{1}{2}k_BT$, to the gas. (In general, a "degree of freedom" refers to an independent means by which a molecule can possess energy.) A generalization of this result, known as the **theorem of equipartition of energy**, states that

Theorem of equipartition of energy

each degree of freedom contributes $\frac{1}{2}k_BT$ to the energy of a system, where possible degrees of freedom in addition to those associated with translation arise from rotation and vibration of molecules.

The total translational kinetic energy of N molecules of gas is simply N times the average energy per molecule, which is given by Equation 21.4:

Total translational kinetic energy of N molecules

$$K_{\text{tot trans}} = N(\tfrac{1}{2}m\overline{v^2}) = \tfrac{3}{2}Nk_BT = \tfrac{3}{2}nRT \tag{21.6}$$

where we have used $k_B = R/N_A$ for Boltzmann's constant and $n = N/N_A$ for the number of moles of gas. If we consider a gas in which molecules possess only translational kinetic energy, Equation 21.6 represents the internal energy of the gas. This result implies that **the internal energy of an ideal gas depends only on the temperature.** We will follow up on this point in Section 21.2.

The square root of $\overline{v^2}$ is called the *root-mean-square* (rms) *speed* of the molecules. From Equation 21.4 we find that the rms speed is

$$v_{\text{rms}} = \sqrt{\overline{v^2}} = \sqrt{\frac{3k_BT}{m}} = \sqrt{\frac{3RT}{M}} \qquad (21.7)$$

Root-mean-square speed

where M is the molar mass in kilograms per mole and is equal to mN_A. This expression shows that, at a given temperature, lighter molecules move faster, on the average, than do heavier molecules. For example, at a given temperature, hydrogen molecules, whose molar mass is 2.02×10^{-3} kg/mol, have an average speed approximately four times that of oxygen molecules, whose molar mass is 32.0×10^{-3} kg/mol. Table 21.1 lists the rms speeds for various molecules at 20°C.

21.1 The Square Root of the Square?

Notice that taking the square root of $\overline{v^2}$ does not "undo" the square because we have taken an average *between* squaring and taking the square root. While the square root of $(\overline{v})^2$ is $\overline{v}$ because the squaring is done after the averaging, the square root of $\overline{v^2}$ is *not* $\overline{v}$, but rather v_{rms}.

Table 21.1

Some rms Speeds

Gas	Molar mass (g/mol)	v_{rms} at 20°C(m/s)
H_2	2.02	1 902
He	4.00	1 352
H_2O	18.0	637
Ne	20.2	602
N_2 or CO	28.0	511
NO	30.0	494
O_2	32.0	478
CO_2	44.0	408
SO_2	64.1	338

Example 21.1 A Tank of Helium

A tank used for filling helium balloons has a volume of 0.300 m^3 and contains 2.00 mol of helium gas at 20.0°C. Assume that the helium behaves like an ideal gas.

(A) What is the total translational kinetic energy of the gas molecules?

Solution Using Equation 21.6 with $n = 2.00$ mol and $T = 293$ K, we find that

$$K_{\text{tot trans}} = \tfrac{3}{2}nRT = \tfrac{3}{2}(2.00 \text{ mol})(8.31 \text{ J/mol}\cdot\text{K})(293 \text{ K})$$

$$= 7.30 \times 10^3 \text{ J}$$

(B) What is the average kinetic energy per molecule?

Solution Using Equation 21.4, we find that the average kinetic energy per molecule is

$$\tfrac{1}{2}m\overline{v^2} = \tfrac{3}{2}k_BT = \tfrac{3}{2}(1.38 \times 10^{-23} \text{ J/K})(293 \text{ K})$$

$$= 6.07 \times 10^{-21} \text{ J}$$

What If? What if the temperature is raised from 20.0°C to 40.0°C? Because 40.0 is twice as large as 20.0, is the total translational energy of the molecules of the gas twice as large at the higher temperature?

Answer The expression for the total translational energy depends on the temperature, and the value for the temperature must be expressed in kelvins, not in degrees Celsius. Thus, the ratio of 40.0 to 20.0 is *not* the appropriate ratio. Converting the Celsius temperatures to kelvins, 20.0°C is 293 K and 40.0°C is 313 K. Thus, the total translational energy increases by a factor of 313 K/293 K = 1.07.

Quick Quiz 21.1 Two containers hold an ideal gas at the same temperature and pressure. Both containers hold the same type of gas but container B has twice the volume of container A. The average translational kinetic energy per molecule in container B is (a) twice that for container A (b) the same as that for container A (c) half that for container A (d) impossible to determine.

Quick Quiz 21.2 Consider again the situation in Quick Quiz 21.1. The internal energy of the gas in container B is (a) twice that for container A (b) the same as that for container A (c) half that for container A (d) impossible to determine.

Quick Quiz 21.3 Consider again the situation in Quick Quiz 21.1. The rms speed of the gas molecules in container B is (a) twice that for container A (b) the same as that for container A (c) half that for container A (d) impossible to determine.

21.2 Molar Specific Heat of an Ideal Gas

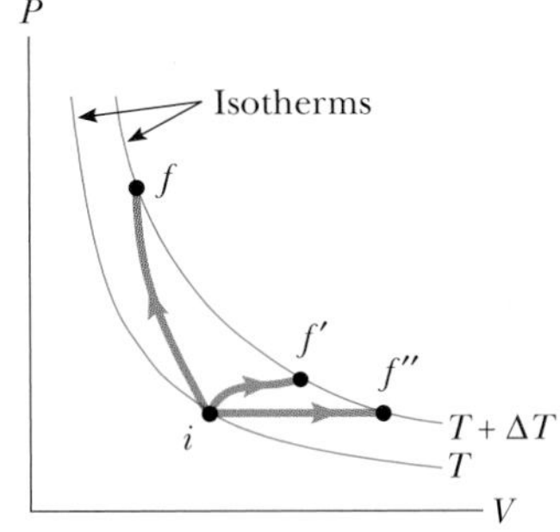

Figure 21.3 An ideal gas is taken from one isotherm at temperature T to another at temperature $T + \Delta T$ along three different paths.

Consider an ideal gas undergoing several processes such that the change in temperature is $\Delta T = T_f - T_i$ for all processes. The temperature change can be achieved by taking a variety of paths from one isotherm to another, as shown in Figure 21.3. Because ΔT is the same for each path, the change in internal energy ΔE_{int} is the same for all paths. However, we know from the first law, $Q = \Delta E_{\text{int}} - W$, that the heat Q is different for each path because W (the negative of the area under the curves) is different for each path. Thus, the heat associated with a given change in temperature does *not* have a unique value.

We can address this difficulty by defining specific heats for two processes that frequently occur: changes at constant volume and changes at constant pressure. Because the number of moles is a convenient measure of the amount of gas, we define the **molar specific heats** associated with these processes with the following equations:

$$Q = nC_V \Delta T \qquad \text{(constant volume)} \tag{21.8}$$

$$Q = nC_P \Delta T \qquad \text{(constant pressure)} \tag{21.9}$$

where C_V is the **molar specific heat at constant volume** and C_P is the **molar specific heat at constant pressure.** When we add energy to a gas by heat at constant pressure, not only does the internal energy of the gas increase, but work is done on the gas because of the change in volume. Therefore, the heat $Q_{\text{constant } P}$ must account for both the increase in internal energy and the transfer of energy out of the system by work. For this reason, $Q_{\text{constant } P}$ is greater than $Q_{\text{constant } V}$ for given values of n and ΔT. Thus, C_P is greater than C_V.

In the previous section, we found that the temperature of a gas is a measure of the average translational kinetic energy of the gas molecules. This kinetic energy is associated with the motion of the center of mass of each molecule. It does not include the energy associated with the internal motion of the molecule—namely, vibrations and rotations about the center of mass. This should not be surprising because the simple kinetic theory model assumes a structureless molecule.

In view of this, let us first consider the simplest case of an ideal monatomic gas, that is, a gas containing one atom per molecule, such as helium, neon, or argon. When energy is added to a monatomic gas in a container of fixed volume, all of the added energy goes into increasing the translational kinetic energy of the atoms. There is no other way to store the energy in a monatomic gas. Therefore, from Equation 21.6, we see that the internal energy E_{int} of N molecules (or n mol) of an ideal monatomic gas is

Internal energy of an ideal monatomic gas

$$E_{\text{int}} = K_{\text{tot trans}} = \tfrac{3}{2}Nk_BT = \tfrac{3}{2}nRT \tag{21.10}$$

Note that for a monatomic ideal gas, E_{int} is a function of T only, and the functional relationship is given by Equation 21.10. In general, the internal energy of an ideal gas is a function of T only, and the exact relationship depends on the type of gas.

If energy is transferred by heat to a system at *constant volume,* then no work is done on the system. That is, $W = -\int P\,dV = 0$ for a constant-volume process. Hence, from the first law of thermodynamics, we see that

$$Q = \Delta E_{\text{int}} \tag{21.11}$$

In other words, all of the energy transferred by heat goes into increasing the internal energy of the system. A constant-volume process from i to f for an ideal gas is described in Figure 21.4, where ΔT is the temperature difference between the two isotherms. Substituting the expression for Q given by Equation 21.8 into Equation 21.11, we obtain

$$\Delta E_{\text{int}} = nC_V\,\Delta T \tag{21.12}$$

If the molar specific heat is constant, we can express the internal energy of a gas as

$$E_{\text{int}} = nC_V T$$

This equation applies to all ideal gases—to gases having more than one atom per molecule as well as to monatomic ideal gases. In the limit of infinitesimal changes, we can use Equation 21.12 to express the molar specific heat at constant volume as

$$C_V = \frac{1}{n}\frac{dE_{\text{int}}}{dT} \tag{21.13}$$

Let us now apply the results of this discussion to the monatomic gas that we have been studying. Substituting the internal energy from Equation 21.10 into Equation 21.13, we find that

$$C_V = \tfrac{3}{2}R \tag{21.14}$$

This expression predicts a value of $C_V = \frac{3}{2}R = 12.5\text{ J/mol}\cdot\text{K}$ for *all* monatomic gases. This prediction is in excellent agreement with measured values of molar specific heats for such gases as helium, neon, argon, and xenon over a wide range of temperatures (Table 21.2). Small variations in Table 21.2 from the predicted values are due to the

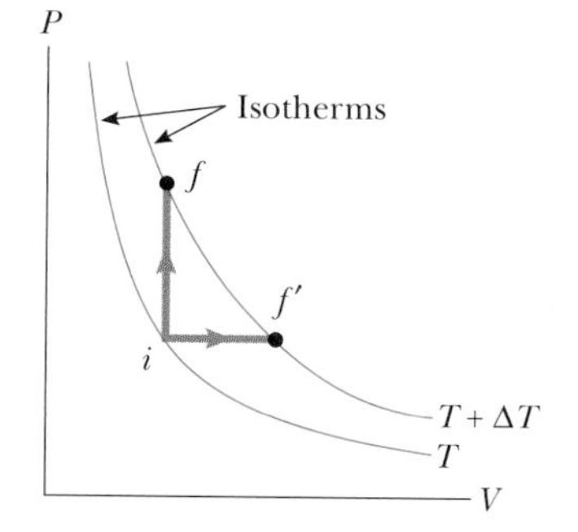

Active Figure 21.4 Energy is transferred by heat to an ideal gas in two ways. For the constant-volume path $i \to f$, all the energy goes into increasing the internal energy of the gas because no work is done. Along the constant-pressure path $i \to f'$, part of the energy transferred in by heat is transferred out by work.

At the Active Figures link at http://www.pse6.com, you can choose initial and final temperatures for one mole of an ideal gas undergoing constant-volume and constant pressure processes and measure Q, W, ΔE_{int}, C_V, and C_P.

Table 21.2

Molar Specific Heats of Various Gases

	Molar Specific Heat (J/mol·K)[a]			
Gas	C_P	C_V	$C_P - C_V$	$\gamma = C_P/C_V$
Monatomic Gases				
He	20.8	12.5	8.33	1.67
Ar	20.8	12.5	8.33	1.67
Ne	20.8	12.7	8.12	1.64
Kr	20.8	12.3	8.49	1.69
Diatomic Gases				
H_2	28.8	20.4	8.33	1.41
N_2	29.1	20.8	8.33	1.40
O_2	29.4	21.1	8.33	1.40
CO	29.3	21.0	8.33	1.40
Cl_2	34.7	25.7	8.96	1.35
Polyatomic Gases				
CO_2	37.0	28.5	8.50	1.30
SO_2	40.4	31.4	9.00	1.29
H_2O	35.4	27.0	8.37	1.30
CH_4	35.5	27.1	8.41	1.31

[a] All values except that for water were obtained at 300 K.

fact that real gases are not ideal gases. In real gases, weak intermolecular interactions occur, which are not addressed in our ideal gas model.

Now suppose that the gas is taken along the constant-pressure path $i \rightarrow f'$ shown in Figure 21.4. Along this path, the temperature again increases by ΔT. The energy that must be transferred by heat to the gas in this process is $Q = nC_P \Delta T$. Because the volume changes in this process, the work done on the gas is $W = -P \Delta V$ where P is the constant pressure at which the process occurs. Applying the first law of thermodynamics to this process, we have

$$\Delta E_{\text{int}} = Q + W = nC_P \Delta T + (-P \Delta V) \tag{21.15}$$

In this case, the energy added to the gas by heat is channeled as follows: Part of it leaves the system by work (that is, the gas moves a piston through a displacement), and the remainder appears as an increase in the internal energy of the gas. But the change in internal energy for the process $i \rightarrow f'$ is equal to that for the process $i \rightarrow f$ because E_{int} depends only on temperature for an ideal gas and because ΔT is the same for both processes. In addition, because $PV = nRT$, we note that for a constant-pressure process, $P \Delta V = nR \Delta T$. Substituting this value for $P \Delta V$ into Equation 21.15 with $\Delta E_{\text{int}} = nC_V \Delta T$ (Eq. 21.12) gives

$$nC_V \Delta T = nC_P \Delta T - nR \Delta T$$

$$C_P - C_V = R \tag{21.16}$$

This expression applies to *any* ideal gas. It predicts that the molar specific heat of an ideal gas at constant pressure is greater than the molar specific heat at constant volume by an amount R, the universal gas constant (which has the value 8.31 J/mol·K). This expression is applicable to real gases, as the data in Table 21.2 show.

Because $C_V = \frac{3}{2}R$ for a monatomic ideal gas, Equation 21.16 predicts a value $C_P = \frac{5}{2}R = 20.8$ J/mol·K for the molar specific heat of a monatomic gas at constant pressure. The ratio of these molar specific heats is a dimensionless quantity γ (Greek gamma):

Ratio of molar specific heats for a monatomic ideal gas

$$\gamma = \frac{C_P}{C_V} = \frac{5R/2}{3R/2} = \frac{5}{3} = 1.67 \tag{21.17}$$

Theoretical values of C_V, C_P and γ are in excellent agreement with experimental values obtained for monatomic gases, but they are in serious disagreement with the values for the more complex gases (see Table 21.2). This is not surprising because the value $C_V = \frac{3}{2}R$ was derived for a monatomic ideal gas and we expect some additional contribution to the molar specific heat from the internal structure of the more complex molecules. In Section 21.4, we describe the effect of molecular structure on the molar specific heat of a gas. The internal energy—and, hence, the molar specific heat—of a complex gas must include contributions from the rotational and the vibrational motions of the molecule.

In the case of solids and liquids heated at constant pressure, very little work is done because the thermal expansion is small. Consequently, C_P and C_V are approximately equal for solids and liquids.

Quick Quiz 21.4 How does the internal energy of an ideal gas change as it follows path $i \rightarrow f$ in Figure 21.4? (a) E_{int} increases. (b) E_{int} decreases. (c) E_{int} stays the same. (d) There is not enough information to determine how E_{int} changes.

Quick Quiz 21.5 How does the internal energy of an ideal gas change as it follows path $f \rightarrow f'$ along the isotherm labeled $T + \Delta T$ in Figure 21.4? (a) E_{int} increases. (b) E_{int} decreases. (c) E_{int} stays the same. (d) There is not enough information to determine how E_{int} changes.

Example 21.2 Heating a Cylinder of Helium

A cylinder contains 3.00 mol of helium gas at a temperature of 300 K.

(A) If the gas is heated at constant volume, how much energy must be transferred by heat to the gas for its temperature to increase to 500 K?

Solution For the constant-volume process, we have

$$Q_1 = nC_V \Delta T$$

Because $C_V = 12.5\ \text{J/mol}\cdot\text{K}$ for helium and $\Delta T = 200$ K, we obtain

$$Q_1 = (3.00\ \text{mol})(12.5\ \text{J/mol}\cdot\text{K})(200\ \text{K})$$

$$= \boxed{7.50 \times 10^3\ \text{J}}$$

(B) How much energy must be transferred by heat to the gas at constant pressure to raise the temperature to 500 K?

Solution Making use of Table 21.2, we obtain

$$Q_2 = nC_P \Delta T$$

$$= (3.00\ \text{mol})(20.8\ \text{J/mol}\cdot\text{K})(200\ \text{K})$$

$$= \boxed{12.5 \times 10^3\ \text{J}}$$

Note that this is larger than Q_1, due to the transfer of energy out of the gas by work in the constant pressure process.

21.3 Adiabatic Processes for an Ideal Gas

As we noted in Section 20.6, an **adiabatic process** is one in which no energy is transferred by heat between a system and its surroundings. For example, if a gas is compressed (or expanded) very rapidly, very little energy is transferred out of (or into) the system by heat, and so the process is nearly adiabatic. Such processes occur in the cycle of a gasoline engine, which we discuss in detail in the next chapter. Another example of an adiabatic process is the very slow expansion of a gas that is thermally insulated from its surroundings.

Suppose that an ideal gas undergoes an adiabatic expansion. At any time during the process, we assume that the gas is in an equilibrium state, so that the equation of state $PV = nRT$ is valid. As we show below, the pressure and volume of an ideal gas at any time during an adiabatic process are related by the expression

$$\boxed{PV^{\gamma} = \text{constant}} \tag{21.18}$$

Relationship between *P* and *V* for an adiabatic process involving an ideal gas

where $\gamma = C_P/C_V$ is assumed to be constant during the process. Thus, we see that all three variables in the ideal gas law—P, V, and T—change during an adiabatic process.

Proof That PV^{γ} = Constant for an Adiabatic Process

When a gas is compressed adiabatically in a thermally insulated cylinder, no energy is transferred by heat between the gas and its surroundings; thus, $Q = 0$. Let us imagine an infinitesimal change in volume dV and an accompanying infinitesimal change in temperature dT. The work done on the gas is $-P\,dV$. Because the internal energy of an ideal gas depends only on temperature, the change in the internal energy in an adiabatic process is the same as that for an isovolumetric process between the same temperatures, $dE_{\text{int}} = nC_V\,dT$ (Eq. 21.12). Hence, the first law of thermodynamics, $\Delta E_{\text{int}} = Q + W$, with $Q = 0$ becomes

$$dE_{\text{int}} = nC_V\,dT = -P\,dV$$

Taking the total differential of the equation of state of an ideal gas, $PV = nRT$, we see that

$$P\,dV + V\,dP = nR\,dT$$

Eliminating dT from these two equations, we find that

$$P\,dV + V\,dP = -\frac{R}{C_V}\,P\,dV$$

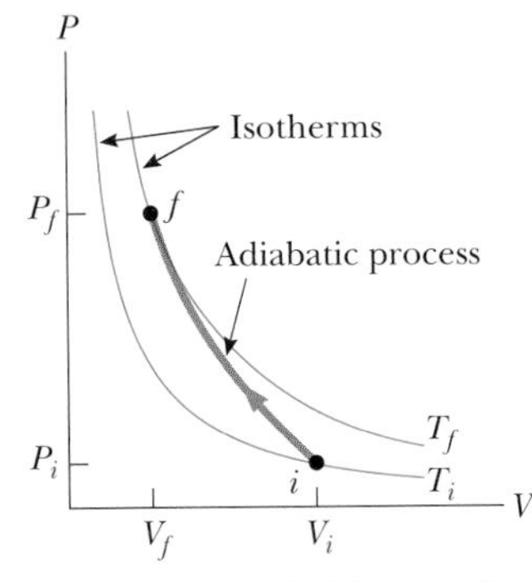

Figure 21.5 The PV diagram for an adiabatic compression. Note that $T_f > T_i$ in this process, so the temperature of the gas increases.

Substituting $R = C_P - C_V$ and dividing by PV, we obtain

$$\frac{dV}{V} + \frac{dP}{P} = -\left(\frac{C_P - C_V}{C_V}\right)\frac{dV}{V} = (1 - \gamma)\frac{dV}{V}$$

$$\frac{dP}{P} + \gamma\frac{dV}{V} = 0$$

Integrating this expression, we have

$$\ln P + \gamma \ln V = \text{constant}$$

which is equivalent to Equation 21.18:

$$PV^{\gamma} = \text{constant}$$

The PV diagram for an adiabatic compression is shown in Figure 21.5. Because $\gamma > 1$, the PV curve is steeper than it would be for an isothermal compression. By the definition of an adiabatic process, no energy is transferred by heat into or out of the system. Hence, from the first law, we see that ΔE_{int} is positive (work is done on the gas, so its internal energy increases) and so ΔT also is positive. Thus, the temperature of the gas increases ($T_f > T_i$) during an adiabatic compression. Conversely, the temperature decreases if the gas expands adiabatically.[1] Applying Equation 21.18 to the initial and final states, we see that

$$P_iV_i^{\gamma} = P_fV_f^{\gamma} \tag{21.19}$$

Relationship between T and V for an adiabatic process involving an ideal gas

Using the ideal gas law, we can express Equation 21.19 as

$$T_iV_i^{\gamma-1} = T_fV_f^{\gamma-1} \tag{21.20}$$

Example 21.3 A Diesel Engine Cylinder

Air at 20.0°C in the cylinder of a diesel engine is compressed from an initial pressure of 1.00 atm and volume of 800.0 cm^3 to a volume of 60.0 cm^3. Assume that air behaves as an ideal gas with $\gamma = 1.40$ and that the compression is adiabatic. Find the final pressure and temperature of the air.

Solution Conceptualize by imagining what happens if we compress a gas into a smaller volume. Our discussion above and Figure 21.5 tell us that the pressure and temperature both increase. We categorize this as a problem involving an adiabatic compression. To analyze the problem, we use Equation 21.19 to find the final pressure:

$$P_f = P_i\left(\frac{V_i}{V_f}\right)^{\gamma} = (1.00\text{ atm})\left(\frac{800.0\text{ cm}^3}{60.0\text{ cm}^3}\right)^{1.40}$$

$$= 37.6\text{ atm}$$

Because $PV = nRT$ is valid throughout an ideal gas process and because no gas escapes from the cylinder,

$$\frac{P_iV_i}{T_i} = \frac{P_fV_f}{T_f}$$

$$T_f = \frac{P_fV_f}{P_iV_i}T_i = \frac{(37.6\text{ atm})(60.0\text{ cm}^3)}{(1.00\text{ atm})(800.0\text{ cm}^3)}(293\text{ K})$$

$$= 826\text{ K} = 553°\text{C}$$

To finalize the problem, note that the temperature of the gas has increased by a factor of 2.82. The high compression in a diesel engine raises the temperature of the fuel enough to cause its combustion without the use of spark plugs.

21.4 The Equipartition of Energy

We have found that predictions based on our model for molar specific heat agree quite well with the behavior of monatomic gases but not with the behavior of complex gases (see Table 21.2). The value predicted by the model for the quantity $C_P - C_V = R$, however, is the same for all gases. This is not surprising because this difference is the result of the work done on the gas, which is independent of its molecular structure.

[1] In the adiabatic free expansion discussed in Section 20.6, the temperature remains constant. This is a special process in which no work is done because the gas expands into a vacuum. In general, the temperature decreases in an adiabatic expansion in which work is done.

To clarify the variations in C_V and C_P in gases more complex than monatomic gases, let us explore further the origin of molar specific heat. So far, we have assumed that the sole contribution to the internal energy of a gas is the translational kinetic energy of the molecules. However, the internal energy of a gas includes contributions from the translational, vibrational, and rotational motion of the molecules. The rotational and vibrational motions of molecules can be activated by collisions and therefore are "coupled" to the translational motion of the molecules. The branch of physics known as *statistical mechanics* has shown that, for a large number of particles obeying the laws of Newtonian mechanics, the available energy is, on the average, shared equally by each independent degree of freedom. Recall from Section 21.1 that the equipartition theorem states that, at equilibrium, each degree of freedom contributes $\frac{1}{2}k_BT$ of energy per molecule.

Let us consider a diatomic gas whose molecules have the shape of a dumbbell (Fig. 21.6). In this model, the center of mass of the molecule can translate in the x, y, and z directions (Fig. 21.6a). In addition, the molecule can rotate about three mutually perpendicular axes (Fig. 21.6b). We can neglect the rotation about the y axis because the molecule's moment of inertia I_y and its rotational energy $\frac{1}{2}I_y\omega^2$ about this axis are negligible compared with those associated with the x and z axes. (If the two atoms are taken to be point masses, then I_y is identically zero.) Thus, there are five degrees of freedom for translation and rotation: three associated with the translational motion and two associated with the rotational motion. Because each degree of freedom contributes, on the average, $\frac{1}{2}k_BT$ of energy per molecule, the internal energy for a system of N molecules, ignoring vibration for now, is

$$E_{\text{int}} = 3N(\tfrac{1}{2}k_BT) + 2N(\tfrac{1}{2}k_BT) = \tfrac{5}{2}Nk_BT = \tfrac{5}{2}nRT$$

We can use this result and Equation 21.13 to find the molar specific heat at constant volume:

$$C_V = \frac{1}{n}\frac{dE_{\text{int}}}{dT} = \frac{1}{n}\frac{d}{dT}(\tfrac{5}{2}nRT) = \tfrac{5}{2}R \qquad (21.21)$$

From Equations 21.16 and 21.17, we find that

$$C_P = C_V + R = \tfrac{7}{2}R$$

$$\gamma = \frac{C_P}{C_V} = \frac{\frac{7}{2}R}{\frac{5}{2}R} = \frac{7}{5} = 1.40$$

These results agree quite well with most of the data for diatomic molecules given in Table 21.2. This is rather surprising because we have not yet accounted for the possible vibrations of the molecule.

In the model for vibration, the two atoms are joined by an imaginary spring (see Fig. 21.6c). The vibrational motion adds two more degrees of freedom, which correspond to the kinetic energy and the potential energy associated with vibrations along the length of the molecule. Hence, classical physics and the equipartition theorem in a model that includes all three types of motion predict a total internal energy of

$$E_{\text{int}} = 3N(\tfrac{1}{2}k_BT) + 2N(\tfrac{1}{2}k_BT) + 2N(\tfrac{1}{2}k_BT) = \tfrac{7}{2}Nk_BT = \tfrac{7}{2}nRT$$

and a molar specific heat at constant volume of

$$C_V = \frac{1}{n}\frac{dE_{\text{int}}}{dT} = \frac{1}{n}\frac{d}{dT}(\tfrac{7}{2}nRT) = \tfrac{7}{2}R \qquad (21.22)$$

This value is inconsistent with experimental data for molecules such as H_2 and N_2 (see Table 21.2) and suggests a breakdown of our model based on classical physics.

It might seem that our model is a failure for predicting molar specific heats for diatomic gases. We can claim some success for our model, however, if measurements of molar specific heat are made over a wide temperature range, rather than at the

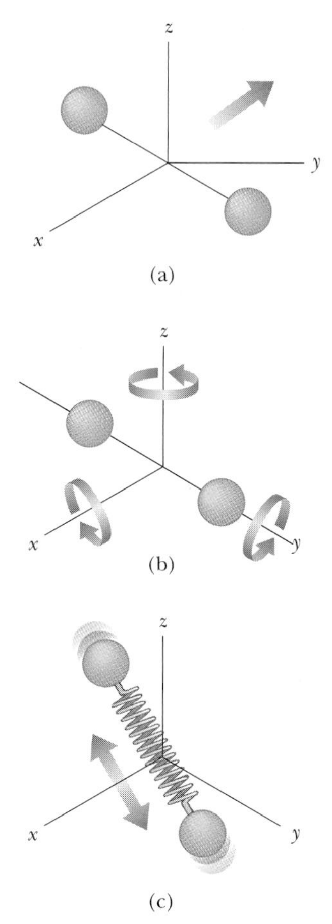

Figure 21.6 Possible motions of a diatomic molecule: (a) translational motion of the center of mass, (b) rotational motion about the various axes, and (c) vibrational motion along the molecular axis.

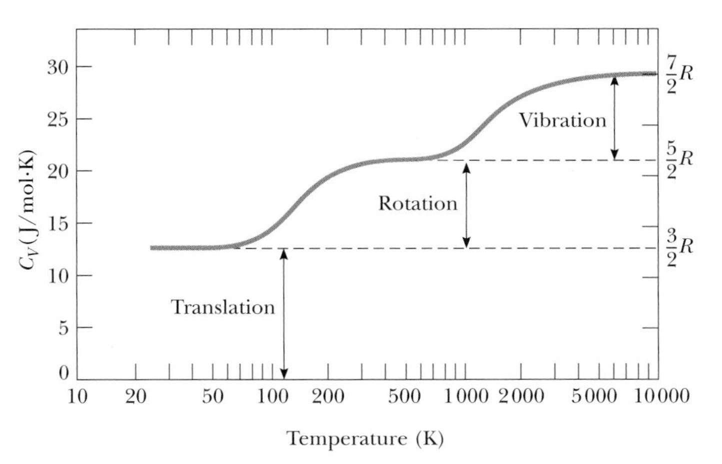

Figure 21.7 The molar specific heat of hydrogen as a function of temperature. The horizontal scale is logarithmic. Note that hydrogen liquefies at 20 K.

single temperature that gives us the values in Table 21.2. Figure 21.7 shows the molar specific heat of hydrogen as a function of temperature. There are three plateaus in the curve. The remarkable feature of these plateaus is that they are at the values of the molar specific heat predicted by Equations 21.14, 21.21, and 21.22! For low temperatures, the diatomic hydrogen gas behaves like a monatomic gas. As the temperature rises to room temperature, its molar specific heat rises to a value for a diatomic gas, consistent with the inclusion of rotation but not vibration. For high temperatures, the molar specific heat is consistent with a model including all types of motion.

Before addressing the reason for this mysterious behavior, let us make a brief remark about polyatomic gases. For molecules with more than two atoms, the vibrations are more complex than for diatomic molecules and the number of degrees of freedom is even larger. This results in an even higher predicted molar specific heat, which is in qualitative agreement with experiment. For the polyatomic gases shown in Table 21.2 we see that the molar specific heats are higher than those for diatomic gases. The more degrees of freedom available to a molecule, the more "ways" there are to store energy, resulting in a higher molar specific heat.

A Hint of Energy Quantization

Our model for molar specific heats has been based so far on purely classical notions. It predicts a value of the specific heat for a diatomic gas that, according to Figure 21.7, only agrees with experimental measurements made at high temperatures. In order to explain why this value is only true at high temperatures and why the plateaus exist in Figure 21.7, we must go beyond classical physics and introduce some quantum physics into the model. In Chapter 18, we discussed quantization of frequency for vibrating strings and air columns. This is a natural result whenever waves are subject to boundary conditions.

Quantum physics (Chapters 40 to 43) shows that atoms and molecules can be described by the physics of waves under boundary conditions. Consequently, these waves have quantized frequencies. Furthermore, in quantum physics, the energy of a system is proportional to the frequency of the wave representing the system. Hence, **the energies of atoms and molecules are quantized.**

For a molecule, quantum physics tells us that the rotational and vibrational energies are quantized. Figure 21.8 shows an **energy-level diagram** for the rotational and vibrational quantum states of a diatomic molecule. The lowest allowed state is called the **ground state**. Notice that vibrational states are separated by larger energy gaps than are rotational states.

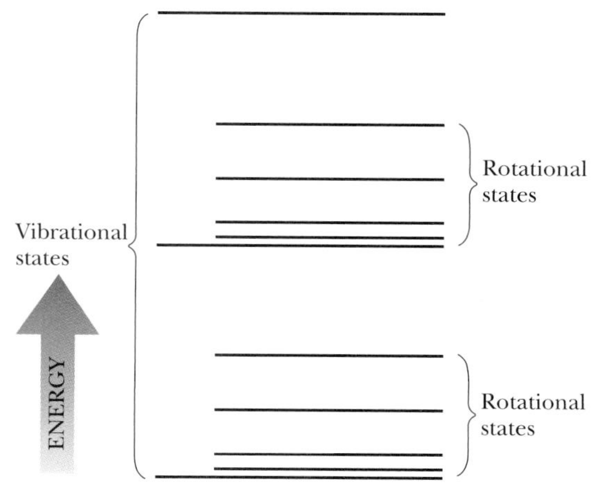

Figure 21.8 An energy-level diagram for vibrational and rotational states of a diatomic molecule. Note that the rotational states lie closer together in energy than the vibrational states.

At low temperatures, the energy that a molecule gains in collisions with its neighbors is generally not large enough to raise it to the first excited state of either rotation or vibration. Thus, even though rotation and vibration are classically allowed, they do not occur at low temperatures. All molecules are in the ground state for rotation and vibration. Thus, the only contribution to the molecules' average energy is from translation, and the specific heat is that predicted by Equation 21.14.

As the temperature is raised, the average energy of the molecules increases. In some collisions, a molecule may have enough energy transferred to it from another molecule to excite the first rotational state. As the temperature is raised further, more molecules can be excited to this state. The result is that rotation begins to contribute to the internal energy and the molar specific heat rises. At about room temperature in Figure 21.7, the second plateau has been reached and rotation contributes fully to the molar specific heat. The molar specific heat is now equal to the value predicted by Equation 21.21.

There is no contribution at room temperature from vibration, because the molecules are still in the ground vibrational state. The temperature must be raised even further to excite the first vibrational state. This happens in Figure 21.7 between 1 000 K and 10 000 K. At 10 000 K on the right side of the figure, vibration is contributing fully to the internal energy and the molar specific heat has the value predicted by Equation 21.22.

The predictions of this model are supportive of the theorem of equipartition of energy. In addition, the inclusion in the model of energy quantization from quantum physics allows a full understanding of Figure 21.7.

Quick Quiz 21.6 The molar specific heat of a diatomic gas is measured at constant volume and found to be 29.1 J/mol · K. The types of energy that are contributing to the molar specific heat are (a) translation only (b) translation and rotation only (c) translation and vibration only (d) translation, rotation, and vibration.

Quick Quiz 21.7 The molar specific heat of a gas is measured at constant volume and found to be $11R/2$. The gas is most likely to be (a) monatomic (b) diatomic (c) polyatomic.

The Molar Specific Heat of Solids

The molar specific heats of solids also demonstrate a marked temperature dependence. Solids have molar specific heats that generally decrease in a nonlinear manner with decreasing temperature and approach zero as the temperature approaches

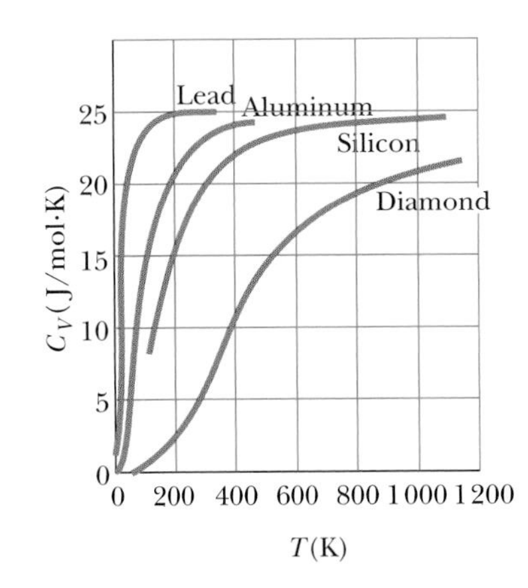

Figure 21.9 Molar specific heat of four solids. As T approaches zero, the molar specific heat also approaches zero.

absolute zero. At high temperatures (usually above 300 K), the molar specific heats approach the value of $3R \approx 25\ \mathrm{J/mol \cdot K}$, a result known as the *DuLong–Petit law.* The typical data shown in Figure 21.9 demonstrate the temperature dependence of the molar specific heats for several solids.

We can explain the molar specific heat of a solid at high temperatures using the equipartition theorem. For small displacements of an atom from its equilibrium position, each atom executes simple harmonic motion in the x, y, and z directions. The energy associated with vibrational motion in the x direction is

$$E = \tfrac{1}{2}mv_x^2 + \tfrac{1}{2}kx^2$$

The expressions for vibrational motions in the y and z directions are analogous. Therefore, each atom of the solid has six degrees of freedom. According to the equipartition theorem, this corresponds to an average vibrational energy of $6(\tfrac{1}{2}k_BT) = 3k_BT$ per atom. Therefore, the internal energy of a solid consisting of N atoms is

Total internal energy of a solid

$$E_{\text{int}} = 3Nk_BT = 3nRT \tag{21.23}$$

From this result, we find that the molar specific heat of a solid at constant volume is

Molar specific heat of a solid at constant volume

$$C_V = \frac{1}{n}\frac{dE_{\text{int}}}{dT} = 3R \tag{21.24}$$

This result is in agreement with the empirical DuLong–Petit law. The discrepancies between this model and the experimental data at low temperatures are again due to the inadequacy of classical physics in describing the world at the atomic level.

21.5 The Boltzmann Distribution Law

PITFALL PREVENTION

21.2 The Distribution Function

Notice that the distribution function $n_V(E)$ is defined in terms of the number of molecules with energy in the range E to $E + dE$ rather than in terms of the number of molecules with energy E. Because the number of molecules is finite and the number of possible values of the energy is infinite, the number of molecules with an *exact* energy E may be zero.

Thus far we have considered only average values of the energies of molecules in a gas and have not addressed the distribution of energies among molecules. In reality, the motion of the molecules is extremely chaotic. Any individual molecule is colliding with others at an enormous rate—typically, a billion times per second. Each collision results in a change in the speed and direction of motion of each of the participant molecules. Equation 21.7 shows that rms molecular speeds increase with increasing temperature. What is the relative number of molecules that possess some characteristic, such as energy within a certain range?

We shall address this question by considering the **number density** $n_V(E)$. This quantity, called a *distribution function*, is defined so that $n_V(E)\ dE$ is the number of molecules per unit volume with energy between E and $E + dE$. (Note that the ratio of the number of molecules that have the desired characteristic to the total number of molecules is the probability that a particular molecule has that characteristic.) In general, the number density is found from statistical mechanics to be

$$n_V(E) = n_0 e^{-E/k_BT} \qquad (21.25)$$

Boltzmann distribution law

where n_0 is defined such that $n_0\, dE$ is the number of molecules per unit volume having energy between $E = 0$ and $E = dE$. This equation, known as the **Boltzmann distribution law,** is important in describing the statistical mechanics of a large number of molecules. It states that **the probability of finding the molecules in a particular energy state varies exponentially as the negative of the energy divided by k_BT.** All the molecules would fall into the lowest energy level if the thermal agitation at a temperature T did not excite the molecules to higher energy levels.

Example 21.4 Thermal Excitation of Atomic Energy Levels

Interactive

As we discussed in Section 21.4, atoms can occupy only certain discrete energy levels. Consider a gas at a temperature of 2 500 K whose atoms can occupy only two energy levels separated by 1.50 eV, where 1 eV (electron volt) is an energy unit equal to 1.60×10^{-19} J (Fig. 21.10). Determine the ratio of the number of atoms in the higher energy level to the number in the lower energy level.

Solution Equation 21.25 gives the relative number of atoms in a given energy level. In this case, the atom has two possible energies, E_1 and E_2, where E_1 is the lower energy level. Hence, the ratio of the number of atoms in the higher energy level to the number in the lower energy level is

$$(1) \qquad \frac{n_V(E_2)}{n_V(E_1)} = \frac{n_0 e^{-E_2/k_BT}}{n_0 e^{-E_1/k_BT}} = e^{-(E_2 - E_1)/k_BT}$$

In this problem, $E_2 - E_1 = 1.50$ eV, and the denominator of the exponent is

$$k_BT = (1.38 \times 10^{-23}\ \text{J/K})(2\,500\ \text{K})\left(\frac{1\ \text{eV}}{1.60 \times 10^{-19}\ \text{J}}\right)$$

$$= 0.216\ \text{eV}$$

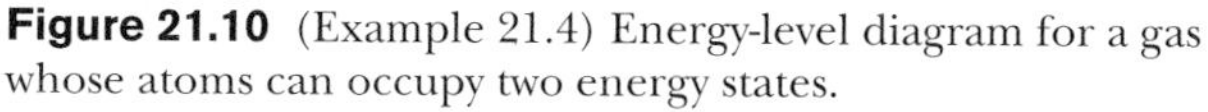

Figure 21.10 (Example 21.4) Energy-level diagram for a gas whose atoms can occupy two energy states.

Therefore, the required ratio is

$$\frac{n_V(E_2)}{n_V(E_1)} = e^{-1.50\ \text{eV}/0.216\ \text{eV}} = e^{-6.94}$$

$$= 9.64 \times 10^{-4}$$

This result indicates that at $T = 2\,500$ K, only a small fraction of the atoms are in the higher energy level. In fact, for every atom in the higher energy level, there are about 1 000 atoms in the lower level. The number of atoms in the higher level increases at even higher temperatures, but the distribution law specifies that at equilibrium there are always more atoms in the lower level than in the higher level.

What If? What if the energy levels in Figure 21.10 were closer together in energy? Would this increase or decrease the fraction of the atoms in the upper energy level?

Answer If the excited level is lower in energy than that in Figure 21.10, it would be easier for thermal agitation to excite atoms to this level, and the fraction of atoms in this energy level would be larger. Let us see this mathematically by expressing Equation (1) as

$$r_2 = e^{-(E_2 - E_1)/k_BT}$$

where r_2 is the ratio of atoms having energy E_2 to those with energy E_1. Differentiating with respect to E_2, we find

$$\frac{dr_2}{dE_2} = \frac{d}{dE_2}\left(e^{-(E_2 - E_1)/k_BT}\right) = -\frac{1}{k_BT} e^{-(E_2 - E_1)/k_BT} < 0$$

Because the derivative has a negative value, we see that as E_2 decreases, r_2 increases.

At the Interactive Worked Example link at **http://www.pse6.com,** *you can investigate the effects of changing the temperature and the energy difference between the states.*

21.6 Distribution of Molecular Speeds

In 1860 James Clerk Maxwell (1831–1879) derived an expression that describes the distribution of molecular speeds in a very definite manner. His work and subsequent developments by other scientists were highly controversial because direct detection of molecules could not be achieved experimentally at that time. However, about 60 years later, experiments were devised that confirmed Maxwell's predictions.

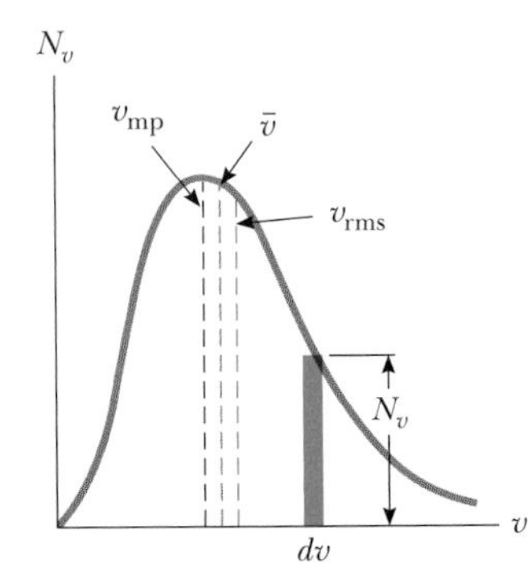

Active Figure 21.11 The speed distribution of gas molecules at some temperature. The number of molecules having speeds in the range v to $v + dv$ is equal to the area of the shaded rectangle, $N_v dv$. The function N_v approaches zero as v approaches infinity.

At the Active Figures link at http://www.pse6.com, you can move the blue triangle and measure the number of molecules with speeds within a small range.

Let us consider a container of gas whose molecules have some distribution of speeds. Suppose we want to determine how many gas molecules have a speed in the range from, for example, 400 to 410 m/s. Intuitively, we expect that the speed distribution depends on temperature. Furthermore, we expect that the distribution peaks in the vicinity of v_{rms}. That is, few molecules are expected to have speeds much less than or much greater than v_{rms} because these extreme speeds result only from an unlikely chain of collisions.

The observed speed distribution of gas molecules in thermal equilibrium is shown in Figure 21.11. The quantity N_v, called the **Maxwell–Boltzmann speed distribution function,** is defined as follows. If N is the total number of molecules, then the number of molecules with speeds between v and $v + dv$ is $dN = N_v\,dv$. This number is also equal to the area of the shaded rectangle in Figure 21.11. Furthermore, the fraction of molecules with speeds between v and $v + dv$ is $(N_v\,dv)/N$. This fraction is also equal to the probability that a molecule has a speed in the range v to $v + dv$.

The fundamental expression that describes the distribution of speeds of N gas molecules is

$$N_v = 4\pi N\left(\frac{m}{2\pi k_B T}\right)^{3/2} v^2 e^{-mv^2/2k_B T} \tag{21.26}$$

where m is the mass of a gas molecule, k_B is Boltzmann's constant, and T is the absolute temperature.[2] Observe the appearance of the Boltzmann factor $e^{-E/k_B T}$ with $E = \frac{1}{2}mv^2$.

As indicated in Figure 21.11, the average speed is somewhat lower than the rms speed. The *most probable speed* v_{mp} is the speed at which the distribution curve reaches a peak. Using Equation 21.26, one finds that

$$v_{rms} = \sqrt{\overline{v^2}} = \sqrt{\frac{3k_B T}{m}} = 1.73\sqrt{\frac{k_B T}{m}} \tag{21.27}$$

$$\bar{v} = \sqrt{\frac{8k_B T}{\pi m}} = 1.60\sqrt{\frac{k_B T}{m}} \tag{21.28}$$

$$v_{mp} = \sqrt{\frac{2k_B T}{m}} = 1.41\sqrt{\frac{k_B T}{m}} \tag{21.29}$$

Equation 21.27 has previously appeared as Equation 21.7. The details of the derivations of these equations from Equation 21.26 are left for the student (see Problems 39 and 65). From these equations, we see that

$$v_{rms} > \bar{v} > v_{mp}$$

Figure 21.12 represents speed distribution curves for nitrogen, N_2. The curves were obtained by using Equation 21.26 to evaluate the distribution function at various speeds and at two temperatures. Note that the peak in the curve shifts to the right as T increases, indicating that the average speed increases with increasing temperature, as expected. The asymmetric shape of the curves is due to the fact that the lowest speed possible is zero while the upper classical limit of the speed is infinity. (In Chapter 39, we will show that the actual upper limit is the speed of light.)

Equation 21.26 shows that the distribution of molecular speeds in a gas depends both on mass and on temperature. At a given temperature, the fraction of molecules with speeds exceeding a fixed value increases as the mass decreases. This explains why lighter molecules, such as H_2 and He, escape more readily from the Earth's atmosphere than do heavier molecules, such as N_2 and O_2. (See the discussion of escape speed in Chapter 13. Gas molecules escape even more readily from the Moon's surface than from the Earth's because the escape speed on the Moon is lower than that on the Earth.)

The speed distribution curves for molecules in a liquid are similar to those shown in Figure 21.12. We can understand the phenomenon of evaporation of a liquid from this distribution in speeds, using the fact that some molecules in the liquid are more

Ludwig Boltzmann

Austrian physicist (1844–1906)

Boltzmann made many important contributions to the development of the kinetic theory of gases, electromagnetism, and thermodynamics. His pioneering work in the field of kinetic theory led to the branch of physics known as statistical mechanics. *(Courtesy of AIP Niels Bohr Library, Lande Collection)*

[2] For the derivation of this expression, see an advanced textbook on thermodynamics, such as that by R. P. Bauman, *Modern Thermodynamics with Statistical Mechanics*, New York, Macmillan Publishing Co., 1992.

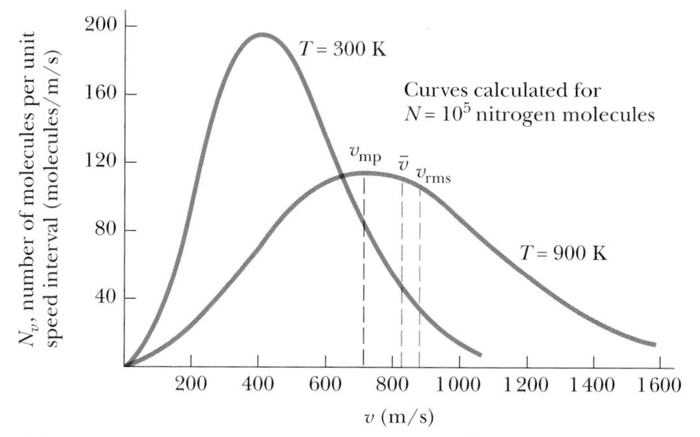

Active Figure 21.12 The speed distribution function for 10^5 nitrogen molecules at 300 K and 900 K. The total area under either curve is equal to the total number of molecules, which in this case equals 10^5. Note that $v_{rms} > \bar{v} > v_{mp}$.

At the Active Figures link at **http://www.pse6.com**, *you can set the desired temperature and see the effect on the distribution curve.*

energetic than others. Some of the faster-moving molecules in the liquid penetrate the surface and leave the liquid even at temperatures well below the boiling point. The molecules that escape the liquid by evaporation are those that have sufficient energy to overcome the attractive forces of the molecules in the liquid phase. Consequently, the molecules left behind in the liquid phase have a lower average kinetic energy; as a result, the temperature of the liquid decreases. Hence, evaporation is a cooling process. For example, an alcohol-soaked cloth often is placed on a feverish head to cool and comfort a patient.

Quick Quiz 21.8 Consider the qualitative shapes of the two curves in Figure 21.12, without regard for the numerical values or labels in the graph. Suppose you have two containers of gas *at the same temperature.* Container A has 10^5 nitrogen molecules and container B has 10^5 hydrogen molecules. The correct qualitative matching between the containers and the two curves in Figure 21.12 is (a) container A corresponds to the blue curve and container B to the brown curve (b) container B corresponds to the blue curve and container A to the brown curve (c) both containers correspond to the same curve.

Example 21.5 A System of Nine Particles

Nine particles have speeds of 5.00, 8.00, 12.0, 12.0, 12.0, 14.0, 14.0, 17.0, and 20.0 m/s.

(A) Find the particles' average speed.

Solution The average speed of the particles is the sum of the speeds divided by the total number of particles:

$$\bar{v} = \frac{(5.00 + 8.00 + 12.0 + 12.0 + 12.0 + 14.0 + 14.0 + 17.0 + 20.0)\ \text{m/s}}{9}$$

$$= 12.7\ \text{m/s}$$

(B) What is the rms speed of the particles?

Solution The average value of the square of the speed is

$$\overline{v^2} = \frac{(5.00^2 + 8.00^2 + 12.0^2 + 12.0^2 + 12.0^2 + 14.0^2 + 14.0^2 + 17.0^2 + 20.0^2)\ \text{m}^2/\text{s}^2}{9}$$

$$= 178\ \text{m}^2/\text{s}^2$$

Hence, the rms speed of the particles is

$$v_{rms} = \sqrt{\overline{v^2}} = \sqrt{178\ \text{m}^2/\text{s}^2} = 13.3\ \text{m/s}$$

(C) What is the most probable speed of the particles?

Solution Three of the particles have a speed of 12.0 m/s, two have a speed of 14.0 m/s, and the remaining have different speeds. Hence, we see that the most probable speed v_{mp} is

12.0 m/s.

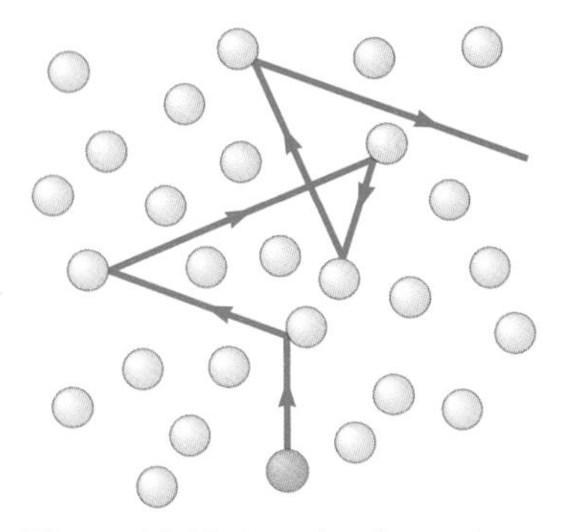

Figure 21.13 A molecule moving through a gas collides with other molecules in a random fashion. This behavior is sometimes referred to as a *random-walk process.* The mean free path increases as the number of molecules per unit volume decreases. Note that the motion is not limited to the plane of the paper.

21.7 Mean Free Path

Most of us are familiar with the fact that the strong odor associated with a gas such as ammonia may take a fraction of a minute to diffuse throughout a room. However, because average molecular speeds are typically several hundred meters per second at room temperature, we might expect a diffusion time of much less than one second. The reason for this difference is that molecules collide with one other because they are not geometrical points. Therefore, they do not travel from one side of a room to the other in a straight line. Between collisions, the molecules move with constant speed along straight lines. The average distance between collisions is called the **mean free path**. The path of an individual molecule is random and resembles that shown in Figure 21.13. As we would expect from this description, the mean free path is related to the diameter of the molecules and the density of the gas.

We now describe how to estimate the mean free path for a gas molecule. For this calculation, we assume that the molecules are spheres of diameter d. We see from Figure 21.14a that no two molecules collide unless their paths, assumed perpendicular to the page in Figure 21.14a are less than a distance d apart as the molecules approach each other. An equivalent way to describe the collisions is to imagine that one of the molecules has a diameter $2d$ and that the rest are geometrical points (Fig. 21.14b). Let us choose the large molecule to be one moving with the average speed $\overline{v}$. In a time interval Δt, this molecule travels a distance $\overline{v}\,\Delta t$. In this time interval, the molecule sweeps out a cylinder having a cross-sectional area πd^2 and a length $\overline{v}\,\Delta t$ (Fig. 21.15). Hence, the volume of the cylinder is $\pi d^2 \overline{v}\,\Delta t$. If n_V is the number of molecules per unit volume, then the number of point-size molecules in the cylinder is $(\pi d^2 \overline{v}\,\Delta t)\,n_V$. The molecule of equivalent diameter $2d$ collides with every molecule in this cylinder in the time interval Δt. Hence, the number of collisions in the time interval Δt is equal to the number of molecules in the cylinder, $(\pi d^2 \overline{v}\,\Delta t)\,n_V$.

The mean free path ℓ equals the average distance $\overline{v}\,\Delta t$ traveled in a time interval Δt divided by the number of collisions that occur in that time interval:

$$\ell = \frac{\overline{v}\,\Delta t}{(\pi d^2 \overline{v}\,\Delta t)\,n_V} = \frac{1}{\pi d^2 n_V}$$

Because the number of collisions in a time interval Δt is $(\pi d^2 \overline{v}\,\Delta t)\,n_V$, the number of collisions per unit time interval, or **collision frequency** f, is

$$f = \pi d^2 \overline{v} n_V$$

The inverse of the collision frequency is the average time interval between collisions, known as the **mean free time.**

Our analysis has assumed that molecules in the cylinder are stationary. When the motion of these molecules is included in the calculation, the correct results are

Mean free path

$$\ell = \frac{1}{\sqrt{2}\,\pi d^2 n_V} \qquad (21.30)$$

Collision frequency

$$f = \sqrt{2}\,\pi d^2 \overline{v} n_V = \frac{\overline{v}}{\ell} \qquad (21.31)$$

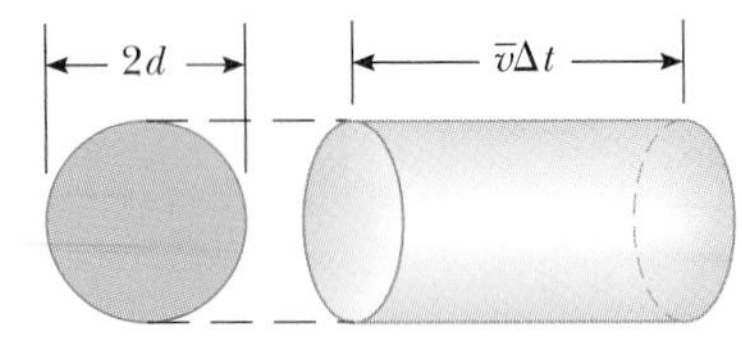

Figure 21.15 In a time interval Δt, a molecule of effective diameter $2d$ and moving to the right sweeps out a cylinder of length $\overline{v}\Delta t$ where $\overline{v}$ is its average speed. In this time interval, it collides with every point molecule within this cylinder.

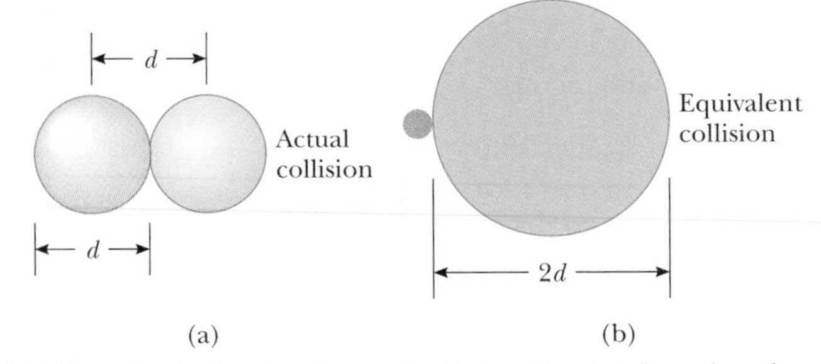

Figure 21.14 (a) Two spherical molecules, each of diameter d and moving along paths perpendicular to the page, collide if their paths are within a distance d of each other. (b) The collision between the two molecules is equivalent to a point molecule colliding with a molecule having an effective diameter of $2d$.

Example 21.6 Bouncing Around in the Air

Approximate the air around you as a collection of nitrogen molecules, each having a diameter of 2.00×10^{-10} m.

(A) How far does a typical molecule move before it collides with another molecule?

Solution Assuming that the gas is ideal, we can use the equation $PV = Nk_BT$ to obtain the number of molecules per unit volume under typical room conditions:

$$n_V = \frac{N}{V} = \frac{P}{k_BT} = \frac{1.01 \times 10^5 \text{ N/m}^2}{(1.38 \times 10^{-23} \text{ J/K})(293 \text{ K})}$$
$$= 2.50 \times 10^{25} \text{ molecules/m}^3$$

Hence, the mean free path is

$$\ell = \frac{1}{\sqrt{2}\pi d^2 n_V}$$
$$= \frac{1}{\sqrt{2}\pi(2.00 \times 10^{-10} \text{ m})^2(2.50 \times 10^{25} \text{ molecules/m}^3)}$$
$$= 2.25 \times 10^{-7} \text{ m}$$

This value is about 10^3 times greater than the molecular diameter.

(B) On average, how frequently does one molecule collide with another?

Solution Because the rms speed of a nitrogen molecule at 20.0°C is 511 m/s (see Table 21.1), we know from Equations 21.27 and 21.28 that $\bar{v} = (1.60/1.73)(511 \text{ m/s}) = 473 \text{ m/s}$. Therefore, the collision frequency is

$$f = \frac{\bar{v}}{\ell} = \frac{473 \text{ m/s}}{2.25 \times 10^{-7} \text{ m}} = 2.10 \times 10^9\text{/s}$$

The molecule collides with other molecules at the average rate of about two billion times each second!

The mean free path ℓ is *not* the same as the average separation between particles. In fact, the average separation d between particles is approximately $n_V^{-1/3}$. In this example, the average molecular separation is

$$d = \frac{1}{n_V^{1/3}} = \frac{1}{(2.5 \times 10^{25})^{1/3}} = 3.4 \times 10^{-9} \text{ m}$$

SUMMARY

Take a practice test for this chapter by clicking on the Practice Test link at **http://www.pse6.com.**

The pressure of N molecules of an ideal gas contained in a volume V is

$$P = \frac{2}{3}\frac{N}{V}\left(\tfrac{1}{2}m\overline{v^2}\right) \qquad (21.2)$$

The average translational kinetic energy per molecule of a gas, $\frac{1}{2}m\overline{v^2}$, is related to the temperature T of the gas through the expression

$$\tfrac{1}{2}m\overline{v^2} = \tfrac{3}{2}k_BT \qquad (21.4)$$

where k_B is Boltzmann's constant. Each translational degree of freedom (x, y, or z) has $\frac{1}{2}k_BT$ of energy associated with it.

The **theorem of equipartition of energy** states that the energy of a system in thermal equilibrium is equally divided among all degrees of freedom.

The internal energy of N molecules (or n mol) of an ideal monatomic gas is

$$E_{\text{int}} = \tfrac{3}{2}Nk_BT = \tfrac{3}{2}nRT \qquad (21.10)$$

The change in internal energy for n mol of any ideal gas that undergoes a change in temperature ΔT is

$$\Delta E_{\text{int}} = nC_V\Delta T \qquad (21.12)$$

where C_V is the molar specific heat at constant volume.

The molar specific heat of an ideal monatomic gas at constant volume is $C_V = \frac{3}{2}R$; the molar specific heat at constant pressure is $C_P = \frac{5}{2}R$. The ratio of specific heats is given by $\gamma = C_P/C_V = \frac{5}{3}$.

If an ideal gas undergoes an adiabatic expansion or compression, the first law of thermodynamics, together with the equation of state, shows that

$$PV^\gamma = \text{constant} \qquad (21.18)$$

The **Boltzmann distribution law** describes the distribution of particles among available energy states. The relative number of particles having energy between E and $E + dE$ is $n_V(E)\, dE$, where

$$n_V(E) = n_0 e^{-E/k_B T} \tag{21.25}$$

The **Maxwell–Boltzmann speed distribution function** describes the distribution of speeds of molecules in a gas:

$$N_v = 4\pi N \left(\frac{m}{2\pi k_B T}\right)^{3/2} v^2 e^{-mv^2/2k_B T} \tag{21.26}$$

This expression enables us to calculate the **root-mean-square speed,** the **average speed,** and **the most probable speed:**

$$v_{\text{rms}} = \sqrt{\overline{v^2}} = \sqrt{\frac{3k_B T}{m}} = 1.73\sqrt{\frac{k_B T}{m}} \tag{21.27}$$

$$\overline{v} = \sqrt{\frac{8k_B T}{\pi m}} = 1.60\sqrt{\frac{k_B T}{m}} \tag{21.28}$$

$$v_{\text{mp}} = \sqrt{\frac{2k_B T}{m}} = 1.41\sqrt{\frac{k_B T}{m}} \tag{21.29}$$

QUESTIONS

1. Dalton's law of partial pressures states that the total pressure of a mixture of gases is equal to the sum of the partial pressures of gases making up the mixture. Give a convincing argument for this law based on the kinetic theory of gases.
2. One container is filled with helium gas and another with argon gas. If both containers are at the same temperature, which molecules have the higher rms speed? Explain.
3. A gas consists of a mixture of He and N_2 molecules. Do the lighter He molecules travel faster than the N_2 molecules? Explain.
4. Although the average speed of gas molecules in thermal equilibrium at some temperature is greater than zero, the average velocity is zero. Explain why this statement must be true.
5. When alcohol is rubbed on your body, it lowers your skin temperature. Explain this effect.
6. A liquid partially fills a container. Explain why the temperature of the liquid decreases if the container is then partially evacuated. (Using this technique, it is possible to freeze water at temperatures above 0°C.)
7. A vessel containing a fixed volume of gas is cooled. Does the mean free path of the molecules increase, decrease, or remain constant in the cooling process? What about the collision frequency?
8. A gas is compressed at a constant temperature. What happens to the mean free path of the molecules in this process?
9. If a helium-filled balloon initially at room temperature is placed in a freezer, will its volume increase, decrease, or remain the same?
10. Which is denser, dry air or air saturated with water vapor? Explain.
11. What happens to a helium-filled balloon released into the air? Will it expand or contract? Will it stop rising at some height?
12. Why does a diatomic gas have a greater energy content per mole than a monatomic gas at the same temperature?
13. An ideal gas is contained in a vessel at 300 K. If the temperature is increased to 900 K, by what factor does each one of the following change? (a) The average kinetic energy of the molecules. (b) The rms molecular speed. (c) The average momentum change of one molecule in a collision with a wall. (d) The rate of collisions of molecules with walls. (e) The pressure of the gas.
14. A vessel is filled with gas at some equilibrium pressure and temperature. Can all gas molecules in the vessel have the same speed?
15. In our model of the kinetic theory of gases, molecules were viewed as hard spheres colliding elastically with the walls of the container. Is this model realistic?
16. In view of the fact that hot air rises, why does it generally become cooler as you climb a mountain? (Note that air has low thermal conductivity.)
17. Inspecting the magnitudes of C_V and C_P for the diatomic and polyatomic gases in Table 21.2, we find that the values increase with increasing molecular mass. Give a qualitative explanation of this observation.

PROBLEMS

1, 2, 3 = straightforward, intermediate, challenging □ = full solution available in the *Student Solutions Manual and Study Guide*
= coached solution with hints available at http://www.pse6.com = computer useful in solving problem
= paired numerical and symbolic problems

Section 21.1 Molecular Model of an Ideal Gas

1. In a 30.0-s interval, 500 hailstones strike a glass window of area $0.600\ m^2$ at an angle of $45.0°$ to the window surface. Each hailstone has a mass of 5.00 g and moves with a speed of 8.00 m/s. Assuming the collisions are elastic, find the average force and pressure on the window.

2. In a period of 1.00 s, 5.00×10^{23} nitrogen molecules strike a wall with an area of $8.00\ cm^2$. If the molecules move with a speed of 300 m/s and strike the wall head-on in elastic collisions, what is the pressure exerted on the wall? (The mass of one N_2 molecule is 4.68×10^{-26} kg.)

3. A sealed cubical container 20.0 cm on a side contains three times Avogadro's number of molecules at a temperature of 20.0°C. Find the force exerted by the gas on one of the walls of the container.

4. A 2.00-mol sample of oxygen gas is confined to a 5.00-L vessel at a pressure of 8.00 atm. Find the average translational kinetic energy of an oxygen molecule under these conditions.

5. A spherical balloon of volume $4\ 000\ cm^3$ contains helium at an (inside) pressure of 1.20×10^5 Pa. How many moles of helium are in the balloon if the average kinetic energy of the helium atoms is 3.60×10^{-22} J?

6. Use the definition of Avogadro's number to find the mass of a helium atom.

7. (a) How many atoms of helium gas fill a balloon having a diameter of 30.0 cm at 20.0°C and 1.00 atm? (b) What is the average kinetic energy of the helium atoms? (c) What is the root-mean-square speed of the helium atoms?

8. Given that the rms speed of a helium atom at a certain temperature is 1 350 m/s, find by proportion the rms speed of an oxygen (O_2) molecule at this temperature. The molar mass of O_2 is 32.0 g/mol, and the molar mass of He is 4.00 g/mol.

9. A cylinder contains a mixture of helium and argon gas in equilibrium at 150°C. (a) What is the average kinetic energy for each type of gas molecule? (b) What is the root-mean-square speed of each type of molecule?

10. A 5.00-L vessel contains nitrogen gas at 27.0°C and a pressure of 3.00 atm. Find (a) the total translational kinetic energy of the gas molecules and (b) the average kinetic energy per molecule.

11. (a) Show that $1\ Pa = 1\ J/m^3$. (b) Show that the density in space of the translational kinetic energy of an ideal gas is $3P/2$.

Section 21.2 Molar Specific Heat of an Ideal Gas

Note: You may use data in Table 21.2 about particular gases. Here we define a "monatomic ideal gas" to have molar specific heats $C_V = 3R/2$ and $C_P = 5R/2$, and a "diatomic ideal gas" to have $C_V = 5R/2$ and $C_P = 7R/2$.

12. Calculate the change in internal energy of 3.00 mol of helium gas when its temperature is increased by 2.00 K.

13. A 1.00-mol sample of hydrogen gas is heated at constant pressure from 300 K to 420 K. Calculate (a) the energy transferred to the gas by heat, (b) the increase in its internal energy, and (c) the work done on the gas.

14. A 1.00-mol sample of air (a diatomic ideal gas) at 300 K, confined in a cylinder under a heavy piston, occupies a volume of 5.00 L. Determine the final volume of the gas after 4.40 kJ of energy is transferred to the air by heat.

15. In a constant-volume process, 209 J of energy is transferred by heat to 1.00 mol of an ideal monatomic gas initially at 300 K. Find (a) the increase in internal energy of the gas, (b) the work done on it, and (c) its final temperature.

16. A house has well-insulated walls. It contains a volume of $100\ m^3$ of air at 300 K. (a) Calculate the energy required to increase the temperature of this diatomic ideal gas by 1.00°C. (b) **What If?** If this energy could be used to lift an object of mass m through a height of 2.00 m, what is the value of m?

17. An incandescent lightbulb contains a volume V of argon at pressure P_i. The bulb is switched on and constant power $\mathcal{P}$ is transferred to the argon for a time interval Δt. (a) Show that the pressure P_f in the bulb at the end of this process is $P_f = P_i[1 + (\mathcal{P}\Delta t R)/(P_i V C_V)]$. (b) Find the pressure in a spherical light bulb 10.0 cm in diameter 4.00 s after it is switched on, given that it has initial pressure 1.00 atm and that 3.60 W of power is transferred to the gas.

18. A vertical cylinder with a heavy piston contains air at a temperature of 300 K. The initial pressure is 200 kPa, and the initial volume is $0.350\ m^3$. Take the molar mass of air as 28.9 g/mol and assume that $C_V = 5R/2$. (a) Find the specific heat of air at constant volume in units of J/kg · °C. (b) Calculate the mass of the air in the cylinder. (c) Suppose the piston is held fixed. Find the energy input required to raise the temperature of the air to 700 K. (d) **What If?** Assume again the conditions of the initial state and that the heavy piston is free to move. Find the energy input required to raise the temperature to 700 K.

19. A 1-L Thermos bottle is full of tea at 90°C. You pour out one cup and immediately screw the stopper back on. Make an order-of-magnitude estimate of the change in temperature of the tea remaining in the flask that results from the admission of air at room temperature. State the quantities you take as data and the values you measure or estimate for them.

20. A 1.00-mol sample of a diatomic ideal gas has pressure P and volume V. When the gas is heated, its pressure triples and its volume doubles. This heating process includes two steps, the first at constant pressure and the second at constant volume. Determine the amount of energy transferred to the gas by heat.

21. A 1.00-mol sample of an ideal monatomic gas is at an initial temperature of 300 K. The gas undergoes an isovolumetric process acquiring 500 J of energy by heat. It then undergoes an isobaric process losing this same amount of energy by heat. Determine (a) the new temperature of the gas and (b) the work done on the gas.

22. A vertical cylinder with a movable piston contains 1.00 mol of a diatomic ideal gas. The volume of the gas is V_i, and its temperature is T_i. Then the cylinder is set on a stove and additional weights are piled onto the piston as it moves up, in such a way that the pressure is proportional to the volume and the final volume is $2V_i$. (a) What is the final temperature? (b) How much energy is transferred to the gas by heat?

23. A container has a mixture of two gases: n_1 mol of gas 1 having molar specific heat C_1 and n_2 mol of gas 2 of molar specific heat C_2. (a) Find the molar specific heat of the mixture. (b) **What If?** What is the molar specific heat if the mixture has m gases in the amounts $n_1, n_2, n_3, \ldots, n_m$, with molar specific heats $C_1, C_2, C_3, \ldots, C_m$, respectively?

Section 21.3 Adiabatic Processes for an Ideal Gas

24. During the compression stroke of a certain gasoline engine, the pressure increases from 1.00 atm to 20.0 atm. If the process is adiabatic and the fuel–air mixture behaves as a diatomic ideal gas, (a) by what factor does the volume change and (b) by what factor does the temperature change? (c) Assuming that the compression starts with 0.016 0 mol of gas at 27.0°C, find the values of Q, W, and ΔE_{int} that characterize the process.

25. A 2.00-mol sample of a diatomic ideal gas expands slowly and adiabatically from a pressure of 5.00 atm and a volume of 12.0 L to a final volume of 30.0 L. (a) What is the final pressure of the gas? (b) What are the initial and final temperatures? (c) Find Q, W, and ΔE_{int}.

26. Air (a diatomic ideal gas) at 27.0°C and atmospheric pressure is drawn into a bicycle pump that has a cylinder with an inner diameter of 2.50 cm and length 50.0 cm. The down stroke adiabatically compresses the air, which reaches a gauge pressure of 800 kPa before entering the tire (Fig. P21.26). Determine (a) the volume of the compressed air and (b) the temperature of the compressed air. (c) **What If?** The pump is made of steel and has an inner wall that is 2.00 mm thick. Assume that 4.00 cm of the cylinder's length is allowed to come to thermal equilibrium with the air. What will be the increase in wall temperature?

27. Air in a thundercloud expands as it rises. If its initial temperature is 300 K and no energy is lost by thermal conduction on expansion, what is its temperature when the initial volume has doubled?

28. The largest bottle ever made by blowing glass has a volume of about 0.720 m^3. Imagine that this bottle is filled with air that behaves as an ideal diatomic gas. The bottle is held with its opening at the bottom and rapidly submerged into the ocean. No air escapes or mixes with the water. No energy is exchanged with the ocean by heat. (a) If the final volume of the air is 0.240 m^3, by what factor does the internal energy of the air increase? (b) If the bottle is submerged so that the air temperature doubles, how much volume is occupied by air?

George Semple

Figure P21.26

29. A 4.00-L sample of a diatomic ideal gas with specific heat ratio 1.40, confined to a cylinder, is carried through a closed cycle. The gas is initially at 1.00 atm and at 300 K. First, its pressure is tripled under constant volume. Then, it expands adiabatically to its original pressure. Finally, the gas is compressed isobarically to its original volume. (a) Draw a PV diagram of this cycle. (b) Determine the volume of the gas at the end of the adiabatic expansion. (c) Find the temperature of the gas at the start of the adiabatic expansion. (d) Find the temperature at the end of the cycle. (e) What was the net work done on the gas for this cycle?

30. A diatomic ideal gas ($\gamma = 1.40$) confined to a cylinder is put through a closed cycle. Initially the gas is at P_i, V_i, and T_i. First, its pressure is tripled under constant volume. It then expands adiabatically to its original pressure and finally is compressed isobarically to its original volume. (a) Draw a PV diagram of this cycle. (b) Determine the volume at the end of the adiabatic expansion. Find (c) the temperature of the gas at the start of the adiabatic expansion and (d) the temperature at the end of the cycle. (e) What was the net work done on the gas for this cycle?

31. How much work is required to compress 5.00 mol of air at 20.0°C and 1.00 atm to one tenth of the original volume (a) by an isothermal process? (b) by an adiabatic process? (c) What is the final pressure in each of these two cases?

32. During the power stroke in a four-stroke automobile engine, the piston is forced down as the mixture of combustion products and air undergoes an adiabatic expansion (Fig. P21.32). Assume that (1) the engine is running at 2 500 cycles/min, (2) the gauge pressure right before the expansion is 20.0 atm, (3) the volumes of the mixture right before and after the expansion are 50.0 and 400 cm^3, respectively, (4) the time involved in the expansion is one-fourth that of the total cycle, and (5) the mixture behaves like an ideal gas with specific heat ratio 1.40. Find the average power generated during the expansion.

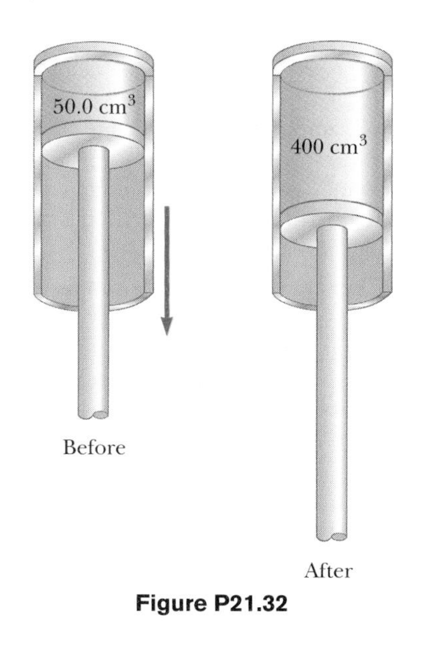

Figure P21.32

Section 21.4 The Equipartition of Energy

33. Consider 2.00 mol of an ideal diatomic gas. (a) Find the total heat capacity of the gas at constant volume and at constant pressure assuming the molecules rotate but do not vibrate. (b) **What If?** Repeat, assuming the molecules both rotate and vibrate.

34. A certain molecule has f degrees of freedom. Show that an ideal gas consisting of such molecules has the following properties: (1) its total internal energy is $fnRT/2$; (2) its molar specific heat at constant volume is $fR/2$; (3) its molar specific heat at constant pressure is $(f + 2)R/2$; (4) its specific heat ratio is $\gamma = C_P/C_V = (f + 2)/f$.

35. In a crude model (Fig. P21.35) of a rotating diatomic molecule of chlorine (Cl_2), the two Cl atoms are 2.00×10^{-10} m apart and rotate about their center of mass with angular speed $\omega = 2.00 \times 10^{12}$ rad/s. What is the rotational kinetic energy of one molecule of Cl_2, which has a molar mass of 70.0 g/mol?

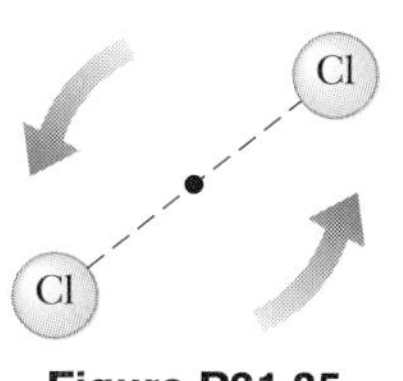

Figure P21.35

Section 21.5 The Boltzmann Distribution Law

Section 21.6 Distribution of Molecular Speeds

36. One cubic meter of atomic hydrogen at 0°C and atmospheric pressure contains approximately 2.70×10^{25} atoms. The first excited state of the hydrogen atom has an energy of 10.2 eV above the lowest energy level, called the ground state. Use the Boltzmann factor to find the number of atoms in the first excited state at 0°C and at 10 000°C.

37. Fifteen identical particles have various speeds: one has a speed of 2.00 m/s; two have speeds of 3.00 m/s; three have speeds of 5.00 m/s; four have speeds of 7.00 m/s; three have speeds of 9.00 m/s; and two have speeds of 12.0 m/s. Find (a) the average speed, (b) the rms speed, and (c) the most probable speed of these particles.

38. Two gases in a mixture diffuse through a filter at rates proportional to the gases' rms speeds. (a) Find the ratio of speeds for the two isotopes of chlorine, ^{35}Cl and ^{37}Cl, as they diffuse through the air. (b) Which isotope moves faster?

39. From the Maxwell–Boltzmann speed distribution, show that the most probable speed of a gas molecule is given by Equation 21.29. Note that the most probable speed corresponds to the point at which the slope of the speed distribution curve dN_v/dv is zero.

40. Helium gas is in thermal equilibrium with liquid helium at 4.20 K. Even though it is on the point of condensation, model the gas as ideal and determine the most probable speed of a helium atom (mass $= 6.64 \times 10^{-27}$ kg) in it.

41. **Review problem.** At what temperature would the average speed of helium atoms equal (a) the escape speed from Earth, 1.12×10^4 m/s and (b) the escape speed from the Moon, 2.37×10^3 m/s? (See Chapter 13 for a discussion of escape speed, and note that the mass of a helium atom is 6.64×10^{-27} kg.)

42. A gas is at 0°C. If we wish to double the rms speed of its molecules, to what temperature must the gas be brought?

43. Assume that the Earth's atmosphere has a uniform temperature of 20°C and uniform composition, with an effective molar mass of 28.9 g/mol. (a) Show that the number density of molecules depends on height according to

$$n_V(y) = n_0 e^{-mgy/k_BT}$$

where n_0 is the number density at sea level, where $y = 0$. This result is called the *law of atmospheres*. (b) Commercial jetliners typically cruise at an altitude of 11.0 km. Find the ratio of the atmospheric density there to the density at sea level.

44. *If you can't walk to outer space, can you at least walk halfway?* Using the law of atmospheres from Problem 43, we find that the average height of a molecule in the Earth's atmosphere is given by

$$\bar{y} = \frac{\int_0^\infty y n_V(y)\, dy}{\int_0^\infty n_V(y)\, dy} = \frac{\int_0^\infty y e^{-mgy/k_BT}\, dy}{\int_0^\infty e^{-mgy/k_BT}\, dy}$$

(a) Prove that this average height is equal to k_BT/mg. (b) Evaluate the average height, assuming the temperature is 10°C and the molecular mass is 28.9 u.

Section 21.7 Mean Free Path

45. In an ultra-high-vacuum system, the pressure is measured to be 1.00×10^{-10} torr (where 1 torr = 133 Pa). Assuming the molecular diameter is 3.00×10^{-10} m, the average molecular speed is 500 m/s, and the temperature is 300 K, find (a) the number of molecules in a volume of 1.00 m^3, (b) the mean free path of the molecules, and (c) the collision frequency.

46. In deep space the number density of particles can be one particle per cubic meter. Using the average temperature of 3.00 K and assuming the particle is H_2 with a diameter of 0.200 nm, (a) determine the mean free path of the particle and the average time between collisions. (b) **What If?** Repeat part (a) assuming a density of one particle per cubic centimeter.

47. Show that the mean free path for the molecules of an ideal gas is

$$\ell = \frac{k_B T}{\sqrt{2}\pi d^2 P}$$

where d is the molecular diameter.

48. In a tank full of oxygen, how many molecular diameters d (on average) does an oxygen molecule travel (at 1.00 atm and 20.0°C) before colliding with another O_2 molecule? (The diameter of the O_2 molecule is approximately 3.60×10^{-10} m.)

49. Argon gas at atmospheric pressure and 20.0°C is confined in a 1.00-m^3 vessel. The effective hard-sphere diameter of the argon atom is 3.10×10^{-10} m. (a) Determine the mean free path ℓ. (b) Find the pressure when $\ell = 1.00$ m. (c) Find the pressure when $\ell = 3.10 \times 10^{-10}$ m.

Additional Problems

50. The dimensions of a room are 4.20 m × 3.00 m × 2.50 m. (a) Find the number of molecules of air in the room at atmospheric pressure and 20.0°C. (b) Find the mass of this air, assuming that the air consists of diatomic molecules with molar mass 28.9 g/mol. (c) Find the average kinetic energy of one molecule. (d) Find the root-mean-square molecular speed. (e) On the assumption that the molar specific heat is a constant independent of temperature, we have $E_{\text{int}} = 5nRT/2$. Find the internal energy in the air. (f) **What If?** Find the internal energy of the air in the room at 25.0°C.

51. The function $E_{\text{int}} = 3.50nRT$ describes the internal energy of a certain ideal gas. A sample comprising 2.00 mol of the gas always starts at pressure 100 kPa and temperature 300 K. For each one of the following processes, determine the final pressure, volume, and temperature; the change in internal energy of the gas; the energy added to the gas by heat; and the work done on the gas. (a) The gas is heated at constant pressure to 400 K. (b) The gas is heated at constant volume to 400 K. (c) The gas is compressed at constant temperature to 120 kPa. (d) The gas is compressed adiabatically to 120 kPa.

52. Twenty particles, each of mass m and confined to a volume V, have various speeds: two have speed v; three have speed $2v$; five have speed $3v$; four have speed $4v$; three have speed $5v$; two have speed $6v$; one has speed $7v$. Find (a) the average speed, (b) the rms speed, (c) the most probable speed, (d) the pressure the particles exert on the walls of the vessel, and (e) the average kinetic energy per particle.

53. A cylinder containing n mol of an ideal gas undergoes an adiabatic process. (a) Starting with the expression $W = -\int P dV$ and using the condition PV^γ = constant, show that the work done on the gas is

$$W = \left(\frac{1}{\gamma - 1}\right)(P_f V_f - P_i V_i)$$

(b) Starting with the first law of thermodynamics in differential form, prove that the work done on the gas is also equal to $nC_V(T_f - T_i)$. Show that this result is consistent with the equation in part (a).

54. As a 1.00-mol sample of a monatomic ideal gas expands adiabatically, the work done on it is −2 500 J. The initial temperature and pressure of the gas are 500 K and 3.60 atm. Calculate (a) the final temperature and (b) the final pressure. You may use the result of Problem 53.

55. A cylinder is closed at both ends and has insulating walls. It is divided into two compartments by a perfectly insulating partition that is perpendicular to the axis of the cylinder. Each compartment contains 1.00 mol of oxygen, which behaves as an ideal gas with $\gamma = 7/5$. Initially the two compartments have equal volumes, and their temperatures are 550 K and 250 K. The partition is then allowed to move slowly until the pressures on its two sides are equal. Find the final temperatures in the two compartments. You may use the result of Problem 53.

56. An air rifle shoots a lead pellet by allowing high-pressure air to expand, propelling the pellet down the rifle barrel. Because this process happens very quickly, no appreciable thermal conduction occurs, and the expansion is essentially adiabatic. Suppose that the rifle starts by admitting to the barrel 12.0 cm^3 of compressed air, which behaves as an ideal gas with $\gamma = 1.40$. The air expands behind a 1.10-g pellet and pushes on it as a piston with cross-sectional area 0.030 0 cm^2, as the pellet moves 50.0 cm along the gun barrel. The pellet emerges with muzzle speed 120 m/s. Use the result of problem 53 to find the initial pressure required.

57. **Review problem.** Oxygen at pressures much greater than 1 atm is toxic to lung cells. Assume that a deep-sea diver breathes a mixture of oxygen (O_2) and helium (He). By weight, what ratio of helium to oxygen must be used if the diver is at an ocean depth of 50.0 m?

58. A vessel contains 1.00×10^4 oxygen molecules at 500 K. (a) Make an accurate graph of the Maxwell–Boltzmann speed distribution function versus speed with points at speed intervals of 100 m/s. (b) Determine the most probable speed from this graph. (c) Calculate the average and rms speeds for the molecules and label these points on your graph. (d) From the graph, estimate the fraction of molecules with speeds in the range 300 m/s to 600 m/s.

59. The compressibility κ of a substance is defined as the fractional change in volume of that substance for a given change in pressure:

$$\kappa = -\frac{1}{V}\frac{dV}{dP}$$

(a) Explain why the negative sign in this expression ensures that κ is always positive. (b) Show that if an ideal gas is compressed isothermally, its compressibility is given by $\kappa_1 = 1/P$. (c) **What If?** Show that if an ideal gas is compressed adiabatically, its compressibility is given by $\kappa_2 = 1/\gamma P$. (d) Determine values for κ_1 and κ_2 for a monatomic ideal gas at a pressure of 2.00 atm.

60. **Review problem.** (a) Show that the speed of sound in an ideal gas is

$$v = \sqrt{\frac{\gamma RT}{M}}$$

where M is the molar mass. Use the general expression for the speed of sound in a fluid from Section 17.1, the definition of the bulk modulus from Section 12.4, and the result of Problem 59 in this chapter. As a sound wave passes through a gas, the compressions are either so rapid or so far apart that thermal conduction is prevented by a negligible time interval or by effective thickness of insulation. The compressions and rarefactions are adiabatic. (b) Compute the theoretical speed of sound in air at 20°C and compare it with the value in Table 17.1. Take $M = 28.9$ g/mol. (c) Show that the speed of sound in an ideal gas is

$$v = \sqrt{\frac{\gamma k_B T}{m}}$$

where m is the mass of one molecule. Compare it with the most probable, average, and rms molecular speeds.

61. Model air as a diatomic ideal gas with $M = 28.9$ g/mol. A cylinder with a piston contains 1.20 kg of air at 25.0°C and 200 kPa. Energy is transferred by heat into the system as it is allowed to expand, with the pressure rising to 400 kPa. Throughout the expansion, the relationship between pressure and volume is given by

$$P = CV^{1/2}$$

where C is a constant. (a) Find the initial volume. (b) Find the final volume. (c) Find the final temperature. (d) Find the work done on the air. (e) Find the energy transferred by heat.

62. *Smokin'!* A pitcher throws a 0.142-kg baseball at 47.2 m/s (Fig. P21.62). As it travels 19.4 m, the ball slows to a speed of 42.5 m/s because of air resistance. Find the change in temperature of the air through which it passes. To find the greatest possible temperature change, you may make the following assumptions: Air has a molar specific heat of $C_P = 7R/2$ and an equivalent molar mass of 28.9 g/mol. The process is so rapid that the cover of the baseball acts as thermal insulation, and the temperature of the ball itself does not change. A change in temperature happens initially only for the air in a cylinder 19.4 m in length and 3.70 cm in radius. This air is initially at 20.0°C.

63. For a Maxwellian gas, use a computer or programmable calculator to find the numerical value of the ratio $N_v(v)/N_v(v_{mp})$ for the following values of v: $v = (v_{mp}/50)$, $(v_{mp}/10)$, $(v_{mp}/2)$, v_{mp}, $2v_{mp}$, $10v_{mp}$, and $50v_{mp}$. Give your results to three significant figures.

64. Consider the particles in a gas centrifuge, a device used to separate particles of different mass by whirling them in a circular path of radius r at angular speed ω. The force acting toward the center of the circular path on a given particle is $m\omega^2 r$. (a) Discuss how a gas centrifuge can be used to separate particles of different mass. (b) Show that the density of the particles as a function of r is

$$n(r) = n_0 e^{mr^2\omega^2/2k_B T}$$

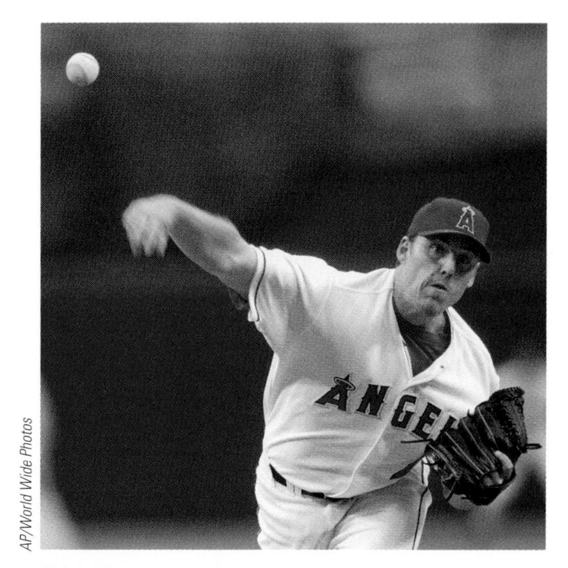

Figure P21.62 John Lackey, the first rookie to win a World Series game 7 in 93 years, pitches for the Anaheim Angels during the final game of the 2002 World Series.

65. Verify Equations 21.27 and 21.28 for the rms and average speed of the molecules of a gas at a temperature T. Note that the average value of v^n is

$$\overline{v^n} = \frac{1}{N}\int_0^\infty v^n N_v\, dv$$

Use the table of definite integrals in Appendix B (Table B.6).

66. On the PV diagram for an ideal gas, one isothermal curve and one adiabatic curve pass through each point. Prove that the slope of the adiabat is steeper than the slope of the isotherm by the factor γ.

67. A sample of monatomic ideal gas occupies 5.00 L at atmospheric pressure and 300 K (point A in Figure P21.67). It is heated at constant volume to 3.00 atm (point B). Then it is allowed to expand isothermally to 1.00 atm (point C) and at last compressed isobarically to its original state. (a) Find the number of moles in the sample. (b) Find the temperature at points B and C and the volume at point C. (c) Assuming that the molar specific heat does not depend on temperature, so that $E_{int} = 3nRT/2$, find the internal energy at points A, B, and C. (d) Tabulate P, V, T, and E_{int} for the states at points A, B, and C. (e) Now consider the processes $A \rightarrow B$, $B \rightarrow C$, and $C \rightarrow A$. Describe just how to carry out each process experimentally. (f) Find Q, W, and ΔE_{int} for each of the processes. (g) For the whole cycle $A \rightarrow B \rightarrow C \rightarrow A$ find Q, W, and ΔE_{int}.

68. This problem can help you to think about the size of molecules. In the city of Beijing a restaurant keeps a pot of chicken broth simmering continuously. Every morning it is topped up to contain 10.0 L of water, along with a fresh chicken, vegetables, and spices. The soup is thoroughly stirred. The molar mass of water is 18.0 g/mol. (a) Find

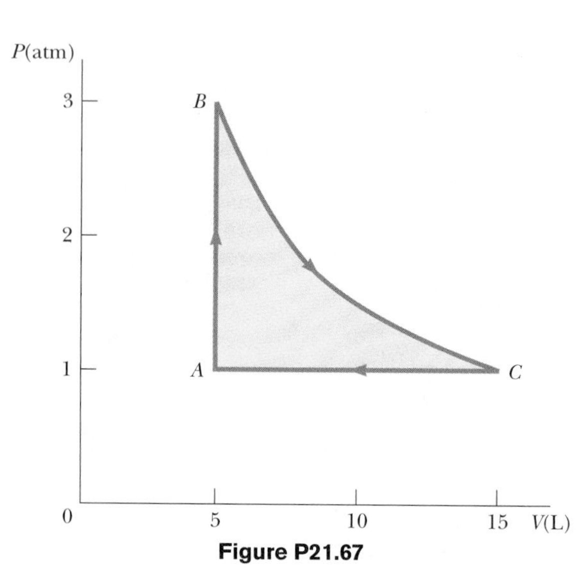

Figure P21.67

the number of molecules of water in the pot. (b) During a certain month, 90.0% of the broth was served each day to people who then emigrated immediately. Of the water molecules in the pot on the first day of the month, when was the last one likely to have been ladled out of the pot? (c) The broth has been simmering for centuries, through wars, earthquakes, and stove repairs. Suppose the water that was in the pot long ago has thoroughly mixed into the Earth's hydrosphere, of mass 1.32×10^{21} kg. How many of the water molecules originally in the pot are likely to be present in it again today?

69. **Review problem.** (a) If it has enough kinetic energy, a molecule at the surface of the Earth can "escape the Earth's gravitation," in the sense that it can continue to move away from the Earth forever, as discussed in Section 13.7. Using the principle of conservation of energy, show that the minimum kinetic energy needed for "escape" is mgR_E, where m is the mass of the molecule, g is the free-fall acceleration at the surface, and R_E is the radius of the Earth. (b) Calculate the temperature for which the minimum escape kinetic energy is ten times the average kinetic energy of an oxygen molecule.

70. Using multiple laser beams, physicists have been able to cool and trap sodium atoms in a small region. In one experiment the temperature of the atoms was reduced to 0.240 mK. (a) Determine the rms speed of the sodium atoms at this temperature. The atoms can be trapped for about 1.00 s. The trap has a linear dimension of roughly 1.00 cm. (b) Approximately how long would it take an atom to wander out of the trap region if there were no trapping action?

Answers to Quick Quizzes

21.1 (b). The average translational kinetic energy per molecule is a function only of temperature.

21.2 (a). Because there are twice as many molecules and the temperature of both containers is the same, the total energy in B is twice that in A.

21.3 (b). Because both containers hold the same type of gas, the rms speed is a function only of temperature.

21.4 (a). According to Equation 21.10, E_{int} is a function of temperature only. Because the temperature increases, the internal energy increases.

21.5 (c). Along an isotherm, T is constant by definition. Therefore, the internal energy of the gas does not change.

21.6 (d). The value of 29.1 J/mol·K is $7R/2$. According to Figure 21.7, this suggests that all three types of motion are occurring.

21.7 (c). The highest possible value of C_V for a diatomic gas is $7R/2$, so the gas must be polyatomic.

21.8 (a). Because the hydrogen atoms are lighter than the nitrogen molecules, they move with a higher average speed and the distribution curve is stretched out more along the horizontal axis. See Equation 21.26 for a mathematical statement of the dependence of N_v on m.

Chapter 22

Heat Engines, Entropy, and the Second Law of Thermodynamics

▲ *This cutaway image of an automobile engine shows two pistons that have work done on them by an explosive mixture of air and fuel, ultimately leading to the motion of the automobile. This apparatus can be modeled as a heat engine, which we study in this chapter. (Courtesy of Ford Motor Company)*

CHAPTER OUTLINE

The first law of thermodynamics, which we studied in Chapter 20, is a statement of conservation of energy. This law states that a change in internal energy in a system can occur as a result of energy transfer by heat or by work, or by both. As was stated in Chapter 20, the law makes no distinction between the results of heat and the results of work—either heat or work can cause a change in internal energy. However, there is an important distinction between heat and work that is not evident from the first law. One manifestation of this distinction is that it is impossible to design a device that, operating in a cyclic fashion, takes in energy by heat and expels an *equal* amount of energy by work. A cyclic device that takes in energy by heat and expels a *fraction* of this energy by work is possible and is called a *heat engine.*

Although the first law of thermodynamics is very important, it makes no distinction between processes that occur spontaneously and those that do not. However, only certain types of energy-conversion and energy-transfer processes actually take place in nature. The *second law of thermodynamics,* the major topic in this chapter, establishes which processes do and which do not occur. The following are examples of processes that do not violate the principle of conservation of energy if they proceed in either direction, but are observed to proceed in only one direction, governed by the second law:

- When two objects at different temperatures are placed in thermal contact with each other, the net transfer of energy by heat is always from the warmer object to the cooler object, never from the cooler to the warmer.
- A rubber ball dropped to the ground bounces several times and eventually comes to rest, but a ball lying on the ground never gathers internal energy from the ground and begins bouncing on its own.
- An oscillating pendulum eventually comes to rest because of collisions with air molecules and friction at the point of suspension. The mechanical energy of the system is converted to internal energy in the air, the pendulum, and the suspension; the reverse conversion of energy never occurs.

All these processes are *irreversible*—that is, they are processes that occur naturally in one direction only. No irreversible process has ever been observed to run backward—if it were to do so, it would violate the second law of thermodynamics.[1]

From an engineering standpoint, perhaps the most important implication of the second law is the limited efficiency of heat engines. The second law states that a machine that operates in a cycle, taking in energy by heat and expelling an equal amount of energy by work, cannot be constructed.

Lord Kelvin
British physicist and mathematician (1824–1907)

Born William Thomson in Belfast, Kelvin was the first to propose the use of an absolute scale of temperature. The Kelvin temperature scale is named in his honor. Kelvin's work in thermodynamics led to the idea that energy cannot pass spontaneously from a colder object to a hotter object.
(J. L. Charmet/SPL/Photo Researchers, Inc.)

[1] Although we have never *observed* a process occurring in the time-reversed sense, it is *possible* for it to occur. As we shall see later in the chapter, however, the probability of such a process occurring is infinitesimally small. From this viewpoint, we say that processes occur with a vastly greater probability in one direction than in the opposite direction.

22.1 Heat Engines and the Second Law of Thermodynamics

A **heat engine** is a device that takes in energy by heat[2] and, operating in a cyclic process, expels a fraction of that energy by means of work. For instance, in a typical process by which a power plant produces electricity, coal or some other fuel is burned, and the high-temperature gases produced are used to convert liquid water to steam. This steam is directed at the blades of a turbine, setting it into rotation. The mechanical energy associated with this rotation is used to drive an electric generator. Another device that can be modeled as a heat engine—the internal combustion engine in an automobile—uses energy from a burning fuel to perform work on pistons that results in the motion of the automobile.

A heat engine carries some working substance through a cyclic process during which (1) the working substance absorbs energy by heat from a high-temperature energy reservoir, (2) work is done by the engine, and (3) energy is expelled by heat to a lower-temperature reservoir. As an example, consider the operation of a steam engine (Fig. 22.1), which uses water as the working substance. The water in a boiler absorbs energy from burning fuel and evaporates to steam, which then does work by expanding against a piston. After the steam cools and condenses, the liquid water produced returns to the boiler and the cycle repeats.

It is useful to represent a heat engine schematically as in Figure 22.2. The engine absorbs a quantity of energy $|Q_h|$ from the hot reservoir. For this discussion of heat engines, we will use absolute values to make all energy transfers positive and will indicate the direction of transfer with an explicit positive or negative sign. The engine does work W_{eng} (so that *negative* work $W = -W_{\text{eng}}$ is done *on* the engine), and then gives up a quantity of energy $|Q_c|$ to the cold reservoir. Because the working substance goes

Figure 22.1 This steam-driven locomotive runs from Durango to Silverton, Colorado. It obtains its energy by burning wood or coal. The generated energy vaporizes water into steam, which powers the locomotive. (This locomotive must take on water from tanks located along the route to replace steam lost through the funnel.) Modern locomotives use diesel fuel instead of wood or coal. Whether old-fashioned or modern, such locomotives can be modeled as heat engines, which extract energy from a burning fuel and convert a fraction of it to mechanical energy.

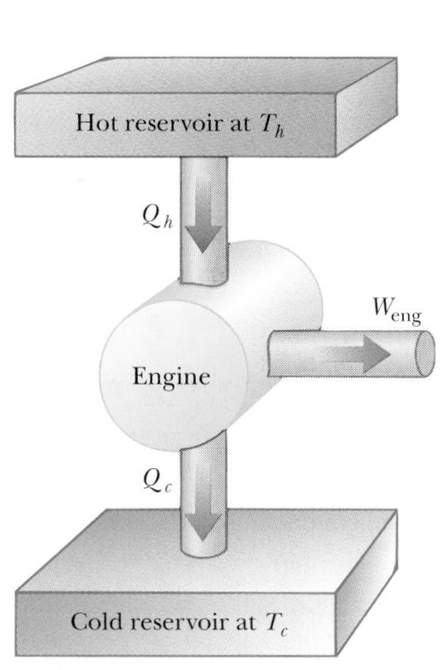

Active Figure 22.2 Schematic representation of a heat engine. The engine does work W_{eng}. The arrow at the top represents energy $Q_h > 0$ entering the engine. At the bottom, $Q_c < 0$ represents energy leaving the engine.

At the Active Figures link at http://www.pse6.com, you can select the efficiency of the engine and observe the transfer of energy.

[2] We will use heat as our model for energy transfer into a heat engine. Other methods of energy transfer are also possible in the model of a heat engine, however. For example, the Earth's atmosphere can be modeled as a heat engine, in which the input energy transfer is by means of electromagnetic radiation from the Sun. The output of the atmospheric heat engine causes the wind structure in the atmosphere.

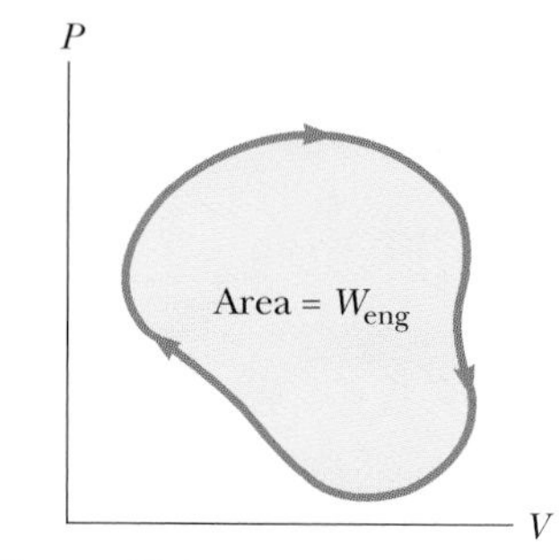

Figure 22.3 *PV* diagram for an arbitrary cyclic process taking place in an engine. The value of the net work done by the engine in one cycle equals the area enclosed by the curve.

through a cycle, its initial and final internal energies are equal, and so $\Delta E_{int} = 0$. Hence, from the first law of thermodynamics, $\Delta E_{int} = Q + W = Q - W_{eng}$, and with no change in internal energy, **the net work W_{eng} done by a heat engine is equal to the net energy Q_{net} transferred to it.** As we can see from Figure 22.2, $Q_{net} = |Q_h| - |Q_c|$; therefore,

$$W_{eng} = |Q_h| - |Q_c| \quad (22.1)$$

If the working substance is a gas, **the net work done in a cyclic process is the area enclosed by the curve representing the process on a *PV* diagram.** This is shown for an arbitrary cyclic process in Figure 22.3.

The **thermal efficiency** e of a heat engine is defined as the ratio of the net work done by the engine during one cycle to the energy input at the higher temperature during the cycle:

Thermal efficiency of a heat engine

$$e = \frac{W_{eng}}{|Q_h|} = \frac{|Q_h| - |Q_c|}{|Q_h|} = 1 - \frac{|Q_c|}{|Q_h|} \quad (22.2)$$

We can think of the efficiency as the ratio of what you gain (work) to what you give (energy transfer at the higher temperature). In practice, all heat engines expel only a fraction of the input energy Q_h by mechanical work and consequently their efficiency is always less than 100%. For example, a good automobile engine has an efficiency of about 20%, and diesel engines have efficiencies ranging from 35% to 40%.

Equation 22.2 shows that a heat engine has 100% efficiency ($e = 1$) only if $|Q_c| = 0$—that is, if no energy is expelled to the cold reservoir. In other words, a heat engine with perfect efficiency would have to expel all of the input energy by work. On the basis of the fact that efficiencies of real engines are well below 100%, the **Kelvin–Planck form of the second law of thermodynamics** states the following:

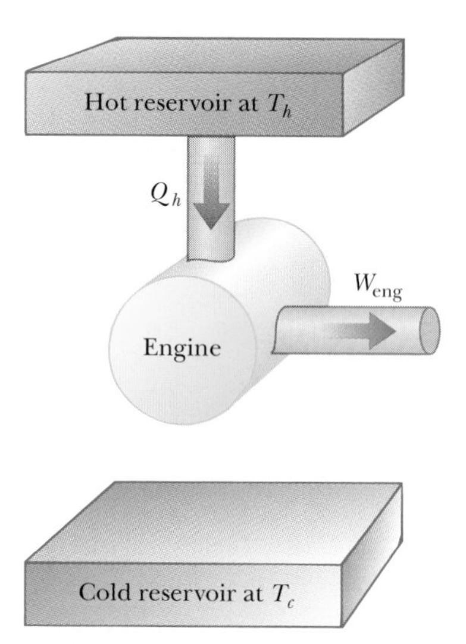

The impossible engine

Figure 22.4 Schematic diagram of a heat engine that takes in energy from a hot reservoir and does an equivalent amount of work. It is impossible to construct such a perfect engine.

> It is impossible to construct a heat engine that, operating in a cycle, produces no effect other than the input of energy by heat from a reservoir and the performance of an equal amount of work.

This statement of the second law means that, during the operation of a heat engine, W_{eng} can never be equal to $|Q_h|$, or, alternatively, that some energy $|Q_c|$ must be rejected to the environment. Figure 22.4 is a schematic diagram of the impossible "perfect" heat engine.

Quick Quiz 22.1 The energy input to an engine is 3.00 times greater than the work it performs. What is its thermal efficiency? (a) 3.00 (b) 1.00 (c) 0.333 (d) impossible to determine

Quick Quiz 22.2 For the engine of Quick Quiz 22.1, what fraction of the energy input is expelled to the cold reservoir? (a) 0.333 (b) 0.667 (c) 1.00 (d) impossible to determine

Example 22.1 The Efficiency of an Engine

An engine transfers 2.00×10^3 J of energy from a hot reservoir during a cycle and transfers 1.50×10^3 J as exhaust to a cold reservoir.

(A) Find the efficiency of the engine.

Solution The efficiency of the engine is given by Equation 22.2 as

$$e = 1 - \frac{|Q_c|}{|Q_h|} = 1 - \frac{1.50 \times 10^3 \text{ J}}{2.00 \times 10^3 \text{ J}} = 0.250, \text{ or } 25.0\%$$

(B) How much work does this engine do in one cycle?

Solution The work done is the difference between the input and output energies:

$$W_{\text{eng}} = |Q_h| - |Q_c| = 2.00 \times 10^3\text{ J} - 1.50 \times 10^3\text{ J}$$

$$= 5.0 \times 10^2\text{ J}$$

What If? **Suppose you were asked for the power output of this engine? Do you have sufficient information to answer this question?**

Answer No, you do not have enough information. The power of an engine is the *rate* at which work is done by the engine. You know how much work is done per cycle but you have no information about the time interval associated with one cycle. However, if you were told that the engine operates at 2 000 rpm (revolutions per minute), you could relate this rate to the period of rotation T of the mechanism of the engine. If we assume that there is one thermodynamic cycle per revolution, then the power is

$$\mathcal{P} = \frac{W_{\text{eng}}}{T} = \frac{5.0 \times 10^2\text{ J}}{\left(\frac{1}{2000}\text{ min}\right)}\left(\frac{1\text{ min}}{60\text{ s}}\right) = 1.7 \times 10^4\text{ W}$$

22.2 Heat Pumps and Refrigerators

In a heat engine, the direction of energy transfer is from the hot reservoir to the cold reservoir, which is the natural direction. The role of the heat engine is to process the energy from the hot reservoir so as to do useful work. What if we wanted to transfer energy from the cold reservoir to the hot reservoir? Because this is not the natural direction of energy transfer, we must put some energy into a device in order to accomplish this. Devices that perform this task are called **heat pumps** or **refrigerators.** For example, we cool homes in summer using heat pumps called *air conditioners.* The air conditioner transfers energy from the cool room in the home to the warm air outside.

In a refrigerator or heat pump, the engine takes in energy $|Q_c|$ from a cold reservoir and expels energy $|Q_h|$ to a hot reservoir (Fig. 22.5). This can be accomplished only if work is done *on* the engine. From the first law, we know that the energy given up to the hot reservoir must equal the sum of the work done and the energy taken in from the cold reservoir. Therefore, the refrigerator or heat pump transfers energy from a colder body (for example, the contents of a kitchen refrigerator or the winter air outside a building) to a hotter body (the air in the kitchen or a room in the building). In practice, it is desirable to carry out this process with a minimum of work. If it could be accomplished without doing any work, then the refrigerator or heat pump would be "perfect" (Fig. 22.6). Again, the existence of such a device would be in violation of the second law of thermodynamics, which in the form of the **Clausius statement**[3] states:

PITFALL PREVENTION

22.1 The First and Second Laws

Notice the distinction between the first and second laws of thermodynamics. If a gas undergoes a *one-time isothermal process* $\Delta E_{\text{int}} = Q + W = 0$. Therefore, the first law allows *all* energy input by heat to be expelled by work. In a heat engine, however, in which a substance undergoes a cyclic process, only a *portion* of the energy input by heat can be expelled by work according to the second law.

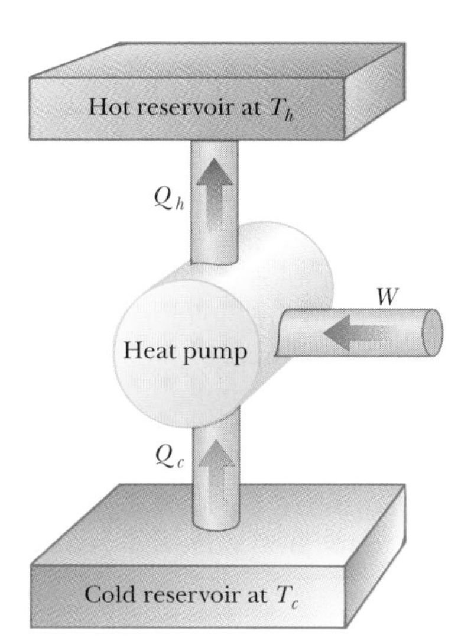

Active Figure 22.5 Schematic diagram of a heat pump, which takes in energy $Q_c > 0$ from a cold reservoir and expels energy $Q_h < 0$ to a hot reservoir. Work W is done *on* the heat pump. A refrigerator works the same way.

At the Active Figures link at http://www.pse6.com, you can select the COP of the heat pump and observe the transfer of energy.

[3] First expressed by Rudolf Clausius (1822–1888).

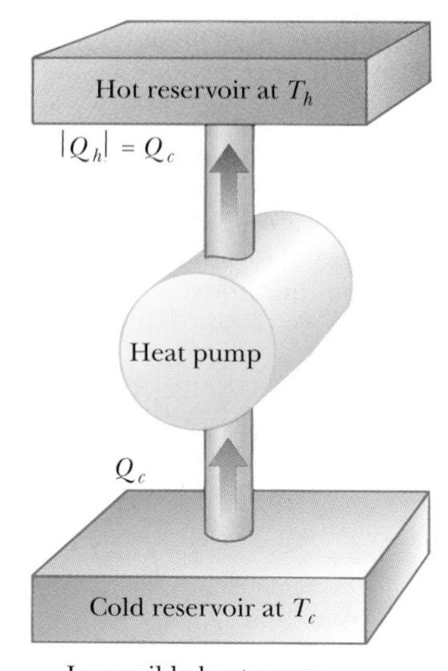

Impossible heat pump

Figure 22.6 Schematic diagram of an impossible heat pump or refrigerator—that is, one that takes in energy from a cold reservoir and expels an equivalent amount of energy to a hot reservoir without the input of energy by work.

It is impossible to construct a cyclical machine whose sole effect is to transfer energy continuously by heat from one object to another object at a higher temperature without the input of energy by work.

In simpler terms, **energy does not transfer spontaneously by heat from a cold object to a hot object.** This direction of energy transfer requires an input of energy to a heat pump, which is often supplied by means of electricity.

The Clausius and Kelvin–Planck statements of the second law of thermodynamics appear, at first sight, to be unrelated, but in fact they are equivalent in all respects. Although we do not prove so here, if either statement is false, then so is the other.[4]

Heat pumps have long been used for cooling homes and buildings, and they are now becoming increasingly popular for heating them as well. The heat pump contains two sets of metal coils that can exchange energy by heat with the surroundings: one set on the outside of the building, in contact with the air or buried in the ground, and the other set in the interior of the building. In the heating mode, a circulating fluid flowing through the coils absorbs energy from the outside and releases it to the interior of the building from the interior coils. The fluid is cold and at low pressure when it is in the external coils, where it absorbs energy by heat from either the air or the ground. The resulting warm fluid is then compressed and enters the interior coils as a hot, high-pressure fluid, where it releases its stored energy to the interior air.

An air conditioner is simply a heat pump with its exterior and interior coils interchanged, so that it operates in the cooling mode. Energy is absorbed into the circulating fluid in the interior coils; then, after the fluid is compressed, energy leaves the fluid through the external coils. The air conditioner must have a way to release energy to the outside. Otherwise, the work done on the air conditioner would represent energy added to the air inside the house, and the temperature would increase. In the same manner, a refrigerator cannot cool the kitchen if the refrigerator door is left open. The amount of energy leaving the external coils (Fig. 22.7) behind or underneath the refrigerator is greater than the amount of energy removed from the food. The difference between the energy out and the energy in is the work done by the electricity supplied to the refrigerator.

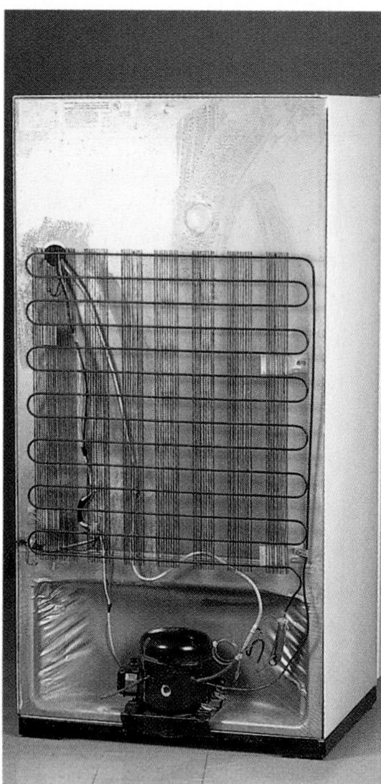

Charles D. Winters

Figure 22.7 The coils on the back of a refrigerator transfer energy by heat to the air. The second law of thermodynamics states that this amount of energy must be greater than the amount of energy removed from the contents of the refrigerator, due to the input of energy by work.

The effectiveness of a heat pump is described in terms of a number called the **coefficient of performance** (COP). In the heating mode, the COP is defined as the ratio of the energy transferred to the hot reservoir to the work required to transfer that energy:

$$\text{COP (heating mode)} \equiv \frac{\text{energy transferred at high temperature}}{\text{work done on heat pump}} = \frac{|Q_h|}{W} \qquad (22.3)$$

Note that the COP is similar to the thermal efficiency for a heat engine in that it is a ratio of what you gain (energy delivered to the interior of the building) to what you give (work input). Because $|Q_h|$ is generally greater than W, typical values for the COP are greater than unity. It is desirable for the COP to be as high as possible, just as it is desirable for the thermal efficiency of an engine to be as high as possible.

If the outside temperature is 25°F (−4°C) or higher, a typical value of the COP for a heat pump is about 4. That is, the amount of energy transferred to the building is about four times greater than the work done by the motor in the heat pump. However, as the outside temperature decreases, it becomes more difficult for the heat pump to extract sufficient energy from the air, and so the COP decreases. In fact, the COP can fall below unity for temperatures below about 15°F (−9°C). Thus, the use of heat pumps that extract energy from the air, while satisfactory in moderate climates, is not appropriate in areas where winter temperatures are very low. It is possible to use heat pumps in colder

[4] See, for example, R. P. Bauman, *Modern Thermodynamics and Statistical Mechanics*, New York, Macmillan Publishing Co., 1992.

areas by burying the external coils deep in the ground. In this case, the energy is extracted from the ground, which tends to be warmer than the air in the winter.

For a heat pump operating in the cooling mode, "what you gain" is energy removed from the cold reservoir. The most effective refrigerator or air conditioner is one that removes the greatest amount of energy from the cold reservoir in exchange for the least amount of work. Thus, for these devices we define the COP in terms of $|Q_c|$:

$$\text{COP (cooling mode)} = \frac{|Q_c|}{W} \tag{22.4}$$

A good refrigerator should have a high COP, typically 5 or 6.

Quick Quiz 22.3 The energy entering an electric heater by electrical transmission can be converted to internal energy with an efficiency of 100%. By what factor does the cost of heating your home change when you replace your electric heating system with an electric heat pump that has a COP of 4.00? Assume that the motor running the heat pump is 100% efficient. (a) 4.00 (b) 2.00 (c) 0.500 (d) 0.250

Example 22.2 Freezing Water

A certain refrigerator has a COP of 5.00. When the refrigerator is running, its power input is 500 W. A sample of water of mass 500 g and temperature 20.0°C is placed in the freezer compartment. How long does it take to freeze the water to ice at 0°C? Assume that all other parts of the refrigerator stay at the same temperature and there is no leakage of energy from the exterior, so that the operation of the refrigerator results only in energy being extracted from the water.

Solution Conceptualize this problem by realizing that energy leaves the water, reducing its temperature and then freezing it into ice. The time interval required for this entire process is related to the rate at which energy is withdrawn from the water, which, in turn is related to the power input of the refrigerator. We categorize this problem as one in which we will need to combine our understanding of temperature changes and phase changes from Chapter 20 with our understanding of heat pumps from the current chapter. To analyze the problem, we first find the amount of energy that we must extract from 500 g of water at 20°C to turn it into ice at 0°C. Using Equations 20.4 and 20.6,

$$|Q_c| = |mc\,\Delta T + mL_f| = m\,|c\,\Delta T + L_f|$$
$$= (0.500\text{ kg})[(4\,186\text{ J/kg}\cdot{}^\circ\text{C})(20.0^\circ\text{C}) + 3.33 \times 10^5\text{ J/kg}]$$
$$= 2.08 \times 10^5\text{ J}$$

Now we use Equation 22.4 to find out how much energy we need to provide to the refrigerator to extract this much energy from the water:

$$\text{COP} = \frac{|Q_c|}{W} \longrightarrow W = \frac{|Q_c|}{\text{COP}} = \frac{2.08 \times 10^5\text{ J}}{5.00}$$
$$W = 4.17 \times 10^4\text{ J}$$

Using the power rating of the refrigerator, we find out the time interval required for the freezing process to occur:

$$\mathcal{P} = \frac{W}{\Delta t} \longrightarrow \Delta t = \frac{W}{\mathcal{P}} = \frac{4.17 \times 10^4\text{ J}}{500\text{ W}} = 83.3\text{ s}$$

To finalize this problem, note that this time interval is very different from that of our everyday experience; this suggests the difficulties with our assumptions. Only a small part of the energy extracted from the refrigerator interior in a given time interval will come from the water. Energy must also be extracted from the container in which the water is placed, and energy that continuously leaks into the interior from the exterior must be continuously extracted. In reality, the time interval for the water to freeze is much longer than 83.3 s.

22.3 Reversible and Irreversible Processes

In the next section we discuss a theoretical heat engine that is the most efficient possible. To understand its nature, we must first examine the meaning of reversible and irreversible processes. In a **reversible** process, the system undergoing the process can be

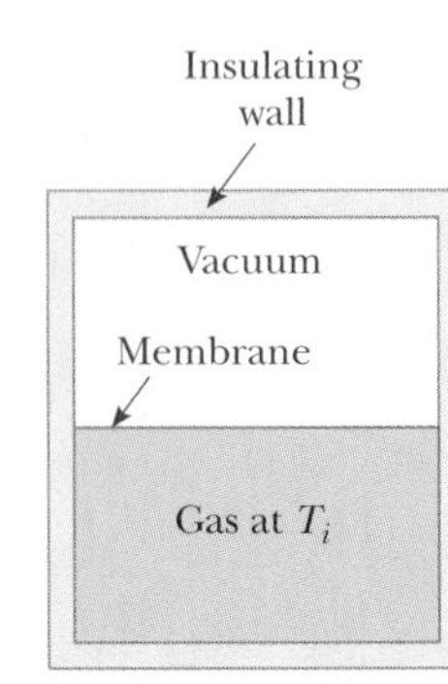

Figure 22.8 Adiabatic free expansion of a gas.

returned to its initial conditions along the same path on a *PV* diagram, and every point along this path is an equilibrium state. A process that does not satisfy these requirements is **irreversible.**

All natural processes are known to be irreversible. From the endless number of examples that could be selected, let us examine the adiabatic free expansion of a gas, which was already discussed in Section 20.6, and show that it cannot be reversible. Consider a gas in a thermally insulated container, as shown in Figure 22.8. A membrane separates the gas from a vacuum. When the membrane is punctured, the gas expands freely into the vacuum. As a result of the puncture, the system has changed because it occupies a greater volume after the expansion. Because the gas does not exert a force through a displacement, it does no work on the surroundings as it expands. In addition, no energy is transferred to or from the gas by heat because the container is insulated from its surroundings. Thus, in this adiabatic process, the system has changed but the surroundings have not.

For this process to be reversible, we need to be able to return the gas to its original volume and temperature without changing the surroundings. Imagine that we try to reverse the process by compressing the gas to its original volume. To do so, we fit the container with a piston and use an engine to force the piston inward. During this process, the surroundings change because work is being done by an outside agent on the system. In addition, the system changes because the compression increases the temperature of the gas. We can lower the temperature of the gas by allowing it to come into contact with an external energy reservoir. Although this step returns the gas to its original conditions, the surroundings are again affected because energy is being added to the surroundings from the gas. If this energy could somehow be used to drive the engine that compressed the gas, then the net energy transfer to the surroundings would be zero. In this way, the system and its surroundings could be returned to their initial conditions, and we could identify the process as reversible. However, the Kelvin–Planck statement of the second law specifies that the energy removed from the gas to return the temperature to its original value cannot be completely converted to mechanical energy in the form of the work done by the engine in compressing the gas. Thus, we must conclude that the process is irreversible.

22.2 All Real Processes Are Irreversible

The reversible process is an idealization—all real processes on Earth are irreversible.

We could also argue that the adiabatic free expansion is irreversible by relying on the portion of the definition of a reversible process that refers to equilibrium states. For example, during the expansion, significant variations in pressure occur throughout the gas. Thus, there is no well-defined value of the pressure for the entire system at any time between the initial and final states. In fact, the process cannot even be represented as a path on a *PV* diagram. The *PV* diagram for an adiabatic free expansion would show the initial and final conditions as points, but these points would not be connected by a path. Thus, because the intermediate conditions between the initial and final states are not equilibrium states, the process is irreversible.

Although all real processes are irreversible, some are almost reversible. If a real process occurs very slowly such that the system is always very nearly in an equilibrium state, then the process can be approximated as being reversible. Suppose that a gas is compressed isothermally in a piston–cylinder arrangement in which the gas is in thermal contact with an energy reservoir, and we continuously transfer just enough energy from the gas to the reservoir during the process to keep the temperature constant. For example, imagine that the gas is compressed very slowly by dropping grains of sand onto a frictionless piston, as shown in Figure 22.9. As each grain lands on the piston and compresses the gas a bit, the system deviates from an equilibrium state, but is so close to one that it achieves a new equilibrium state in a relatively short time interval. Each grain added represents a change to a new equilibrium state but the differences between states are so small that we can approximate the entire process as occurring through continuous equilibrium states. We can reverse the process by slowly removing grains from the piston.

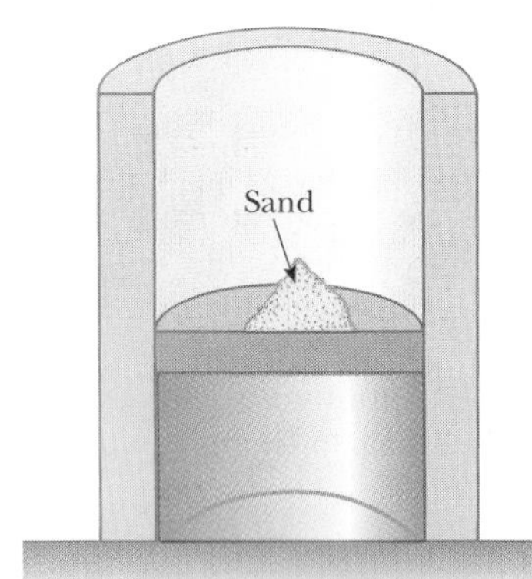

Figure 22.9 A gas in thermal contact with an energy reservoir is compressed slowly as individual grains of sand drop onto the piston. The compression is isothermal and reversible.

A general characteristic of a reversible process is that no dissipative effects (such as turbulence or friction) that convert mechanical energy to internal energy can be present. Such effects can be impossible to eliminate completely. Hence, it is not surprising that real processes in nature are irreversible.

22.4 The Carnot Engine

In 1824 a French engineer named Sadi Carnot described a theoretical engine, now called a **Carnot engine,** which is of great importance from both practical and theoretical viewpoints. He showed that a heat engine operating in an ideal, reversible cycle—called a **Carnot cycle**—between two energy reservoirs is the most efficient engine possible. Such an ideal engine establishes an upper limit on the efficiencies of all other engines. That is, the net work done by a working substance taken through the Carnot cycle is the greatest amount of work possible for a given amount of energy supplied to the substance at the higher temperature. **Carnot's theorem** can be stated as follows:

> No real heat engine operating between two energy reservoirs can be more efficient than a Carnot engine operating between the same two reservoirs.

Sadi Carnot
French engineer (1796–1832)

Carnot was the first to show the quantitative relationship between work and heat. In 1824 he published his only work–*Reflections on the Motive Power of Heat*–which reviewed the industrial, political, and economic importance of the steam engine. In it, he defined work as "weight lifted through a height." *(J.-L. Charmet/Science Photo Library/Photo Researchers, Inc.)*

To argue the validity of this theorem, imagine two heat engines operating between the *same* energy reservoirs. One is a Carnot engine with efficiency e_C, and the other is an engine with efficiency e, where we assume that $e > e_C$. The more efficient engine is used to drive the Carnot engine as a Carnot refrigerator. The output by work of the more efficient engine is matched to the input by work of the Carnot refrigerator. For the *combination* of the engine and refrigerator, no exchange by work with the surroundings occurs. Because we have assumed that the engine is more efficient than the refrigerator, the net result of the combination is a transfer of energy from the cold to the hot reservoir without work being done on the combination. According to the Clausius statement of the second law, this is impossible. Hence, the assumption that $e > e_C$ must be false. **All real engines are less efficient than the Carnot engine because they do not operate through a reversible cycle.** The efficiency of a real engine is further reduced by such practical difficulties as friction and energy losses by conduction.

To describe the Carnot cycle taking place between temperatures T_c and T_h, we assume that the working substance is an ideal gas contained in a cylinder fitted with a movable piston at one end. The cylinder's walls and the piston are thermally nonconducting. Four stages of the Carnot cycle are shown in Figure 22.10, and the PV diagram for the cycle is shown in Figure 22.11. The Carnot cycle consists of two adiabatic processes and two isothermal processes, all reversible:

PITFALL PREVENTION

22.3 Don't Shop for a Carnot Engine

The Carnot engine is an idealization—do not expect a Carnot engine to be developed for commercial use. We explore the Carnot engine only for theoretical considerations.

1. Process $A \rightarrow B$ (Fig. 22.10a) is an isothermal expansion at temperature T_h. The gas is placed in thermal contact with an energy reservoir at temperature T_h. During the expansion, the gas absorbs energy $|Q_h|$ from the reservoir through the base of the cylinder and does work W_{AB} in raising the piston.
2. In process $B \rightarrow C$ (Fig. 22.10b), the base of the cylinder is replaced by a thermally nonconducting wall, and the gas expands adiabatically—that is, no energy enters or leaves the system by heat. During the expansion, the temperature of the gas decreases from T_h to T_c and the gas does work W_{BC} in raising the piston.
3. In process $C \rightarrow D$ (Fig. 22.10c), the gas is placed in thermal contact with an energy reservoir at temperature T_c and is compressed isothermally at temperature T_c. During this time, the gas expels energy $|Q_c|$ to the reservoir, and the work done by the piston on the gas is W_{CD}.
4. In the final process $D \rightarrow A$ (Fig. 22.10d), the base of the cylinder is replaced by a nonconducting wall, and the gas is compressed adiabatically. The temperature of the gas increases to T_h, and the work done by the piston on the gas is W_{DA}.

At the Active Figures link at http://www.pse6.com, you can observe the motion of the piston in the Carnot cycle while you also observe the cycle on the PV diagram of Figure 22.11.

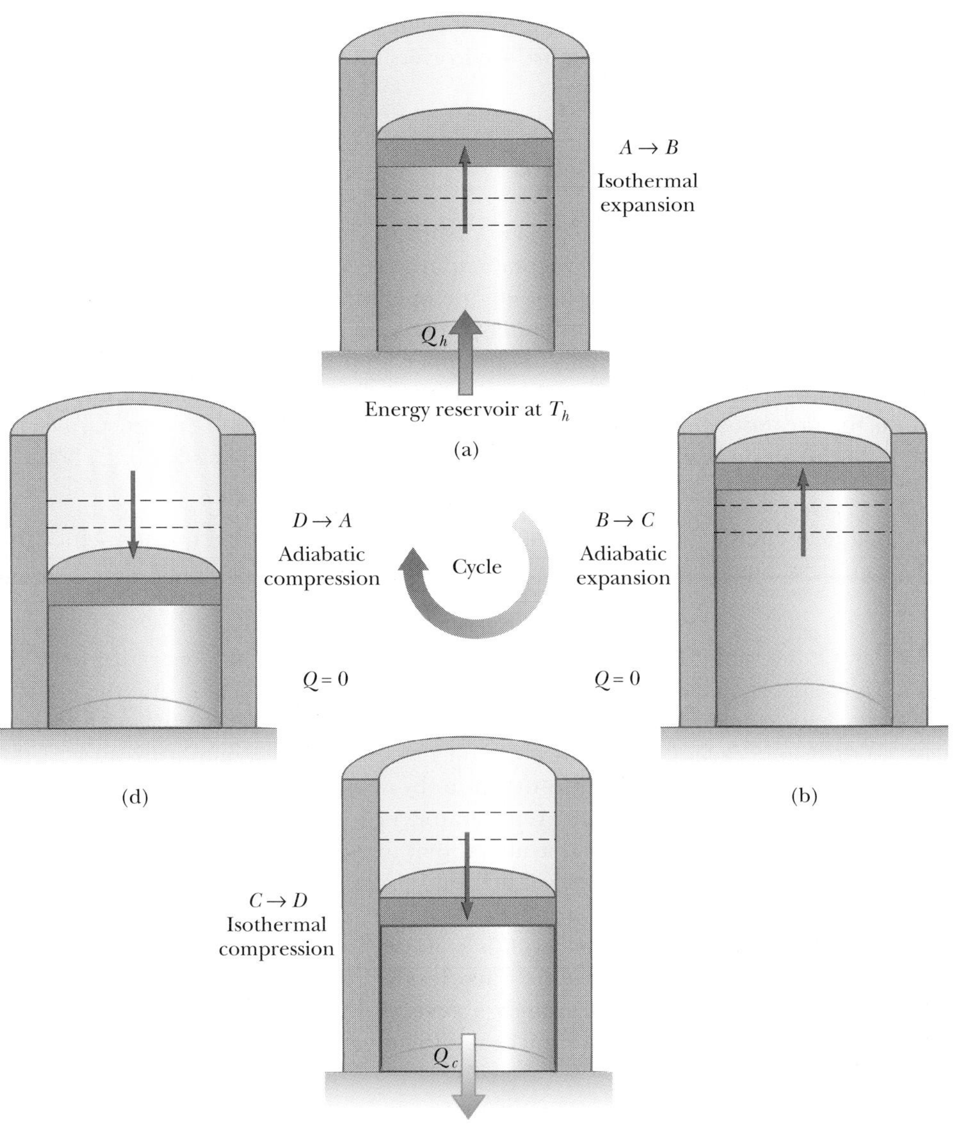

Active Figure 22.10 The Carnot cycle. (a) In process $A \to B$, the gas expands isothermally while in contact with a reservoir at T_h. (b) In process $B \to C$, the gas expands adiabatically ($Q = 0$). (c) In process $C \to D$, the gas is compressed isothermally while in contact with a reservoir at $T_c < T_h$. (d) In process $D \to A$, the gas is compressed adiabatically. The arrows on the piston indicate the direction of its motion during each process.

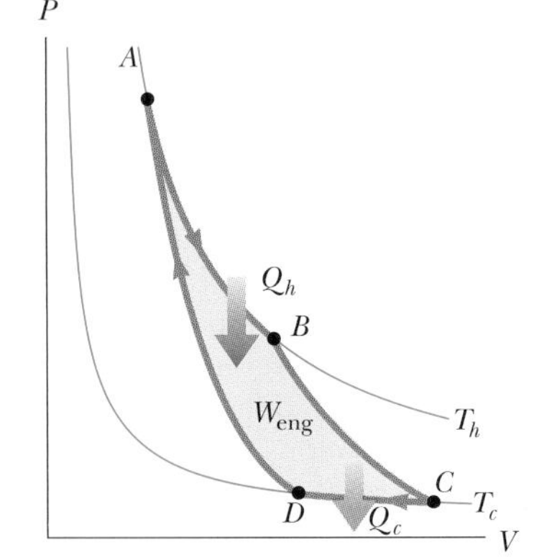

Active Figure 22.11 *PV* diagram for the Carnot cycle. The net work done W_{eng} equals the net energy transferred into the Carnot engine in one cycle, $|Q_h| - |Q_c|$. Note that $\Delta E_{int} = 0$ for the cycle.

At the Active Figures link at http://www.pse6.com, you can observe the Carnot cycle on the PV diagram while you also observe the motion of the piston in Figure 22.10.

The net work done in this reversible, cyclic process is equal to the area enclosed by the path *ABCDA* in Figure 22.11. As we demonstrated in Section 22.1, because the change in internal energy is zero, the net work W_{eng} done by the gas in one cycle equals the net energy transferred into the system, $|Q_h| - |Q_c|$. The thermal efficiency of the engine is given by Equation 22.2:

$$e = \frac{W_{eng}}{|Q_h|} = \frac{|Q_h| - |Q_c|}{|Q_h|} = 1 - \frac{|Q_c|}{|Q_h|}$$

In Example 22.3, we show that for a Carnot cycle

$$\frac{|Q_c|}{|Q_h|} = \frac{T_c}{T_h} \tag{22.5}$$

Hence, the thermal efficiency of a Carnot engine is

$$e_C = 1 - \frac{T_c}{T_h} \qquad (22.6)$$

Efficiency of a Carnot engine

This result indicates that **all Carnot engines operating between the same two temperatures have the same efficiency.**[5]

Equation 22.6 can be applied to any working substance operating in a Carnot cycle between two energy reservoirs. According to this equation, the efficiency is zero if $T_c = T_h$, as one would expect. The efficiency increases as T_c is lowered and as T_h is raised. However, the efficiency can be unity (100%) only if $T_c = 0$ K. Such reservoirs are not available; thus, the maximum efficiency is always less than 100%. In most practical cases, T_c is near room temperature, which is about 300 K. Therefore, one usually strives to increase the efficiency by raising T_h. Theoretically, a Carnot-cycle heat engine run in reverse constitutes the most effective heat pump possible, and it determines the maximum COP for a given combination of hot and cold reservoir temperatures. Using Equations 22.1 and 22.3, we see that the maximum COP for a heat pump in its heating mode is

$$\text{COP}_C \text{ (heating mode)} = \frac{|Q_h|}{W}$$

$$= \frac{|Q_h|}{|Q_h| - |Q_c|} = \frac{1}{1 - \dfrac{|Q_c|}{|Q_h|}} = \frac{1}{1 - \dfrac{T_c}{T_h}} = \frac{T_h}{T_h - T_c}$$

The Carnot COP for a heat pump in the cooling mode is

$$\text{COP}_C \text{ (cooling mode)} = \frac{T_c}{T_h - T_c}$$

As the difference between the temperatures of the two reservoirs approaches zero in this expression, the theoretical COP approaches infinity. In practice, the low temperature of the cooling coils and the high temperature at the compressor limit the COP to values below 10.

Quick Quiz 22.4 Three engines operate between reservoirs separated in temperature by 300 K. The reservoir temperatures are as follows: Engine A: $T_h = 1\,000$ K, $T_c = 700$ K; Engine B: $T_h = 800$ K, $T_c = 500$ K; Engine C: $T_h = 600$ K, $T_c = 300$ K. Rank the engines in order of theoretically possible efficiency, from highest to lowest.

[5] In order for the processes in the Carnot cycle to be reversible, they must be carried out infinitesimally slowly. Thus, although the Carnot engine is the most efficient engine possible, it has zero power output, because it takes an infinite time interval to complete one cycle! For a real engine, the short time interval for each cycle results in the working substance reaching a high temperature lower than that of the hot reservoir and a low temperature higher than that of the cold reservoir. An engine undergoing a Carnot cycle between this narrower temperature range was analyzed by Curzon and Ahlborn (*Am. J. Phys.*, **43**(1), 22, 1975), who found that the efficiency at maximum power output depends only on the reservoir temperatures T_c and T_h, and is given by $e_{C\text{-}A} = 1 - (T_c/T_h)^{1/2}$. The Curzon–Ahlborn efficiency $e_{C\text{-}A}$ provides a closer approximation to the efficiencies of real engines than does the Carnot efficiency.

Example 22.3 Efficiency of the Carnot Engine

Show that the efficiency of a heat engine operating in a Carnot cycle using an ideal gas is given by Equation 22.6.

Solution During the isothermal expansion (process $A \rightarrow B$ in Fig. 22.10), the temperature of the gas does not change. Thus, its internal energy remains constant. The work done on a gas during an isothermal process is given by Equation 20.13. According to the first law,

$$|Q_h| = |-W_{AB}| = nRT_h \ln \frac{V_B}{V_A}$$

In a similar manner, the energy transferred to the cold reservoir during the isothermal compression $C \rightarrow D$ is

$$|Q_c| = |-W_{CD}| = nRT_c \ln \frac{V_C}{V_D}$$

Dividing the second expression by the first, we find that

$$(1) \qquad \frac{|Q_c|}{|Q_h|} = \frac{T_c}{T_h} \frac{\ln(V_C/V_D)}{\ln(V_B/V_A)}$$

We now show that the ratio of the logarithmic quantities is unity by establishing a relationship between the ratio of volumes. For any quasi-static, adiabatic process, the temperature and volume are related by Equation 21.20:

$$T_iV_i^{\gamma-1} = T_fV_f^{\gamma-1}$$

Applying this result to the adiabatic processes $B \rightarrow C$ and $D \rightarrow A$, we obtain

$$T_hV_B^{\gamma-1} = T_cV_C^{\gamma-1}$$

$$T_hV_A^{\gamma-1} = T_cV_D^{\gamma-1}$$

Dividing the first equation by the second, we obtain

$$(V_B/V_A)^{\gamma-1} = (V_C/V_D)^{\gamma-1}$$

$$(2) \qquad \frac{V_B}{V_A} = \frac{V_C}{V_D}$$

Substituting Equation (2) into Equation (1), we find that the logarithmic terms cancel, and we obtain the relationship

$$\frac{|Q_c|}{|Q_h|} = \frac{T_c}{T_h}$$

Using this result and Equation 22.2, we see that the thermal efficiency of the Carnot engine is

$$e_C = 1 - \frac{|Q_c|}{|Q_h|} = 1 - \frac{T_c}{T_h}$$

which is Equation 22.6, the one we set out to prove.

Example 22.4 The Steam Engine

A steam engine has a boiler that operates at 500 K. The energy from the burning fuel changes water to steam, and this steam then drives a piston. The cold reservoir's temperature is that of the outside air, approximately 300 K. What is the maximum thermal efficiency of this steam engine?

Solution Using Equation 22.6, we find that the maximum thermal efficiency for any engine operating between these temperatures is

$$e_C = 1 - \frac{T_c}{T_h} = 1 - \frac{300 \text{ K}}{500 \text{ K}} = 0.400 \quad \text{or} \quad 40.0\%$$

You should note that this is the highest *theoretical* efficiency of the engine. In practice, the efficiency is considerably lower.

What If? Suppose we wished to increase the theoretical efficiency of this engine and we could do so by increasing T_h by ΔT or by decreasing T_c by the same ΔT. Which would be more effective?

Answer A given ΔT would have a larger fractional effect on a smaller temperature, so we would expect a larger change in efficiency if we alter T_c by ΔT. Let us test this numerically. Increasing T_h by 50 K, corresponding to $T_h = 550$ K, would give a maximum efficiency of

$$e_C = 1 - \frac{T_c}{T_h} = 1 - \frac{300 \text{ K}}{550 \text{ K}} = 0.455$$

Decreasing T_c by 50 K, corresponding to $T_c = 250$ K, would give a maximum efficiency of

$$e_C = 1 - \frac{T_c}{T_h} = 1 - \frac{250 \text{ K}}{500 \text{ K}} = 0.500$$

While changing T_c is *mathematically* more effective, often changing T_h is *practically* more feasible.

Example 22.5 The Carnot Efficiency

The highest theoretical efficiency of a certain engine is 30.0%. If this engine uses the atmosphere, which has a temperature of 300 K, as its cold reservoir, what is the temperature of its hot reservoir?

Solution We use the Carnot efficiency to find T_h:

$$e_C = 1 - \frac{T_c}{T_h}$$

$$T_h = \frac{T_c}{1 - e_C} = \frac{300 \text{ K}}{1 - 0.300} = 429 \text{ K}$$

22.5 Gasoline and Diesel Engines

In a gasoline engine, six processes occur in each cycle; five of these are illustrated in Figure 22.12. In this discussion, we consider the interior of the cylinder above the piston to be the system that is taken through repeated cycles in the operation of the engine. For a given cycle, the piston moves up and down twice. This represents a four-stroke cycle consisting of two upstrokes and two downstrokes. The processes in the cycle can be approximated by the **Otto cycle**, shown in the *PV* diagram in Figure 22.13. In the following discussion, refer to Figure 22.12 for the pictorial representation of the strokes and to Figure 22.13 for the significance on the *PV* diagram of the letter designations below:

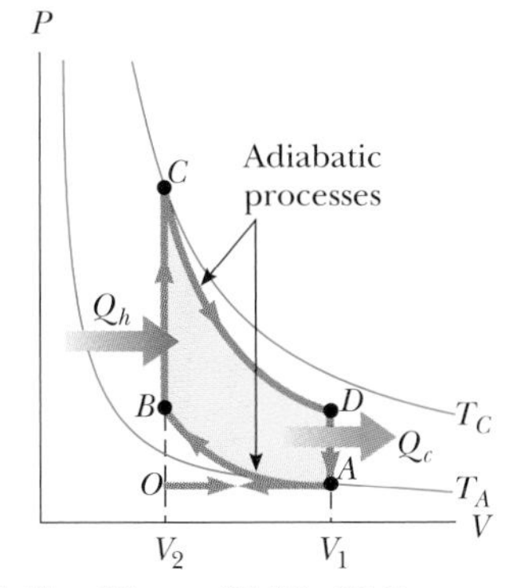

Active Figure 22.13 *PV* diagram for the Otto cycle, which approximately represents the processes occurring in an internal combustion engine.

At the Active Figures link at http://www.pse6.com, you can observe the Otto cycle on the PV diagram while you observe the motion of the piston and crankshaft in Figure 22.12.

1. During the *intake stroke* $O \rightarrow A$ (Fig. 22.12a), the piston moves downward, and a gaseous mixture of air and fuel is drawn into the cylinder at atmospheric pressure. In this process, the volume increases from V_2 to V_1. This is the energy input part of the cycle—energy enters the system (the interior of the cylinder) as potential energy stored in the fuel.
2. During the *compression stroke* $A \rightarrow B$ (Fig. 22.12b), the piston moves upward, the air–fuel mixture is compressed adiabatically from volume V_1 to volume V_2, and the temperature increases from T_A to T_B. The work done on the gas is positive, and its value is equal to the negative of the area under the curve *AB* in Figure 22.13.
3. In process $B \rightarrow C$, combustion occurs when the spark plug fires (Fig. 22.12c). This is not one of the strokes of the cycle because it occurs in a very short period of time while the piston is at its highest position. The combustion represents a rapid transformation from potential energy stored in chemical bonds in the fuel to internal energy associated with molecular motion, which is related to temperature. During this time, the pressure and temperature in the cylinder increase rapidly, with the temperature rising from T_B to T_C. The volume, however, remains approximately constant because of the short time interval. As a result, approximately no work is done on or by the gas. We can model this process in the *PV* diagram (Fig. 22.13) as

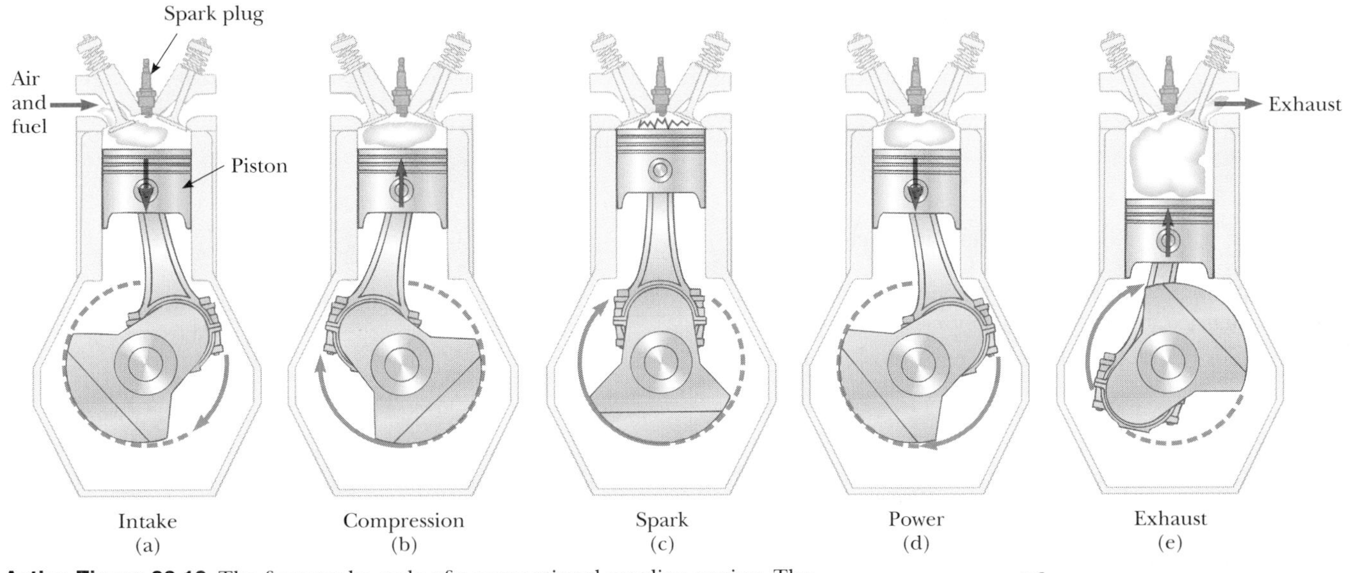

Active Figure 22.12 The four-stroke cycle of a conventional gasoline engine. The arrows on the piston indicate the direction of its motion during each process. (a) In the intake stroke, air and fuel enter the cylinder. (b) The intake valve is then closed, and the air–fuel mixture is compressed by the piston. (c) The mixture is ignited by the spark plug, with the result that the temperature of the mixture increases at essentially constant volume. (d) In the power stroke, the gas expands against the piston. (e) Finally, the residual gases are expelled, and the cycle repeats.

At the Active Figures link at http://www.pse6.com, you can observe the motion of the piston and crankshaft while you also observe the cycle on the PV diagram of Figure 22.13.

that process in which the energy $|Q_h|$ enters the system. (However, in reality this process is a *conversion* of energy already in the cylinder from process $O \rightarrow A$.)

4. In the *power stroke* $C \rightarrow D$ (Fig. 22.12d), the gas expands adiabatically from V_2 to V_1. This expansion causes the temperature to drop from T_C to T_D. Work is done by the gas in pushing the piston downward, and the value of this work is equal to the area under the curve CD.
5. In the process $D \rightarrow A$ (not shown in Fig. 22.12), an exhaust valve is opened as the piston reaches the bottom of its travel, and the pressure suddenly drops for a short time interval. During this interval, the piston is almost stationary and the volume is approximately constant. Energy is expelled from the interior of the cylinder and continues to be expelled during the next process.
6. In the final process, the *exhaust stroke* $A \rightarrow O$ (Fig. 22.12e), the piston moves upward while the exhaust valve remains open. Residual gases are exhausted at atmospheric pressure, and the volume decreases from V_1 to V_2. The cycle then repeats.

If the air–fuel mixture is assumed to be an ideal gas, then the efficiency of the Otto cycle is

$$e = 1 - \frac{1}{(V_1/V_2)^{\gamma-1}} \quad \text{(Otto cycle)} \tag{22.7}$$

where γ is the ratio of the molar specific heats C_P/C_V for the fuel–air mixture and V_1/V_2 is the **compression ratio.** Equation 22.7, which we derive in Example 22.6, shows that the efficiency increases as the compression ratio increases. For a typical compression ratio of 8 and with $\gamma = 1.4$, we predict a theoretical efficiency of 56% for an engine operating in the idealized Otto cycle. This value is much greater than that achieved in real engines (15% to 20%) because of such effects as friction, energy transfer by conduction through the cylinder walls, and incomplete combustion of the air–fuel mixture.

Diesel engines operate on a cycle similar to the Otto cycle but do not employ a spark plug. The compression ratio for a diesel engine is much greater than that for a gasoline engine. Air in the cylinder is compressed to a very small volume, and, as a consequence, the cylinder temperature at the end of the compression stroke is very high. At this point, fuel is injected into the cylinder. The temperature is high enough for the fuel–air mixture to ignite without the assistance of a spark plug. Diesel engines are more efficient than gasoline engines because of their greater compression ratios and resulting higher combustion temperatures.

Example 22.6 Efficiency of the Otto Cycle

Show that the thermal efficiency of an engine operating in an idealized Otto cycle (see Figs. 22.12 and 22.13) is given by Equation 22.7. Treat the working substance as an ideal gas.

Solution First, let us calculate the work done on the gas during each cycle. No work is done during processes $B \rightarrow C$ and $D \rightarrow A$. The work done on the gas during the adiabatic compression $A \rightarrow B$ is positive, and the work done on the gas during the adiabatic expansion $C \rightarrow D$ is negative. The value of the net work done equals the area of the shaded region bounded by the closed curve in Figure 22.13. Because the change in internal energy for one cycle is zero, we see from the first law that the net work done during one cycle equals the net energy transfer to the system:

$$W_{\text{eng}} = |Q_h| - |Q_c|$$

Because processes $B \rightarrow C$ and $D \rightarrow A$ take place at constant volume, and because the gas is ideal, we find from the definition of molar specific heat (Eq. 21.8) that

$$|Q_h| = nC_V(T_C - T_B) \quad \text{and} \quad |Q_c| = nC_V(T_D - T_A)$$

Using these expressions together with Equation 22.2, we obtain for the thermal efficiency

$$\text{(1)} \quad e = \frac{W_{\text{eng}}}{|Q_h|} = 1 - \frac{|Q_c|}{|Q_h|} = 1 - \frac{T_D - T_A}{T_C - T_B}$$

We can simplify this expression by noting that processes $A \rightarrow B$ and $C \rightarrow D$ are adiabatic and hence obey Equation 21.20. For the two adiabatic processes, then,

$$A \rightarrow B: \quad T_A V_A^{\gamma-1} = T_B V_B^{\gamma-1}$$

$$C \rightarrow D: \quad T_C V_C^{\gamma-1} = T_D V_D^{\gamma-1}$$

Using these equations and relying on the fact that $V_A = V_D = V_1$ and $V_B = V_C = V_2$, we find that

$$T_A V_1^{\gamma-1} = T_B V_2^{\gamma-1}$$

$$(2) \qquad T_A = T_B\left(\frac{V_2}{V_1}\right)^{\gamma-1}$$

$$T_D V_1^{\gamma-1} = T_C V_2^{\gamma-1}$$

$$(3) \qquad T_D = T_C\left(\frac{V_2}{V_1}\right)^{\gamma-1}$$

Subtracting Equation (2) from Equation (3) and rearranging, we find that

$$(4) \qquad \frac{T_D - T_A}{T_C - T_B} = \left(\frac{V_2}{V_1}\right)^{\gamma-1}$$

Substituting Equation (4) into Equation (1), we obtain for the thermal efficiency

$$(5) \qquad e = 1 - \frac{1}{(V_1/V_2)^{\gamma-1}}$$

which is Equation 22.7.

We can also express this efficiency in terms of temperatures by noting from Equations (2) and (3) that

$$\left(\frac{V_2}{V_1}\right)^{\gamma-1} = \frac{T_A}{T_B} = \frac{T_D}{T_C}$$

Therefore, Equation (5) becomes

$$(6) \qquad e = 1 - \frac{T_A}{T_B} = 1 - \frac{T_D}{T_C}$$

During the Otto cycle, the lowest temperature is T_A and the highest temperature is T_C. Therefore, the efficiency of a Carnot engine operating between reservoirs at these two temperatures, which is given by the expression $e_C = 1 - (T_A/T_C)$, is *greater* than the efficiency of the Otto cycle given by Equation (6), as expected.

Application Models of Gasoline and Diesel Engines

We can use the thermodynamic principles discussed in this and earlier chapters to model the performance of gasoline and diesel engines. In both types of engine, a gas is first compressed in the cylinders of the engine and then the fuel–air mixture is ignited. Work is done on the gas during compression, but significantly more work is done on the piston by the mixture as the products of combustion expand in the cylinder. The power of the engine is transferred from the piston to the crankshaft by the connecting rod.

Two important quantities of either engine are the **displacement volume,** which is the volume displaced by the piston as it moves from the bottom to the top of the cylinder, and the compression ratio r, which is the ratio of the maximum and minimum volumes of the cylinder, as discussed earlier. Most gasoline and diesel engines operate with a four-stroke cycle (intake, compression, power, exhaust), in which the net work of the intake and exhaust strokes can be considered negligible. Therefore, power is developed only once for every two revolutions of the crankshaft (see Fig. 22.12).

In a diesel engine, only air (and no fuel) is present in the cylinder at the beginning of the compression. In the idealized diesel cycle of Figure 22.14, air in the cylinder undergoes an adiabatic compression from A to B. Starting at B, fuel is injected into the cylinder. The high temperature of the mixture causes combustion of the fuel–air mixture. Fuel continues to be injected in such a way that during the time interval while the fuel is being injected, the fuel–air mixture undergoes a constant-pressure expansion to an intermediate volume V_C ($B \rightarrow C$). At C, the fuel injection is cut off and the power stroke is an adiabatic expansion back to $V_D = V_A$ ($C \rightarrow D$). The exhaust valve is opened, and a constant-volume output of energy occurs ($D \rightarrow A$) as the cylinder empties.

To simplify our calculations, we assume that the mixture in the cylinder is air modeled as an ideal gas. We use specific heats c instead of molar specific heats

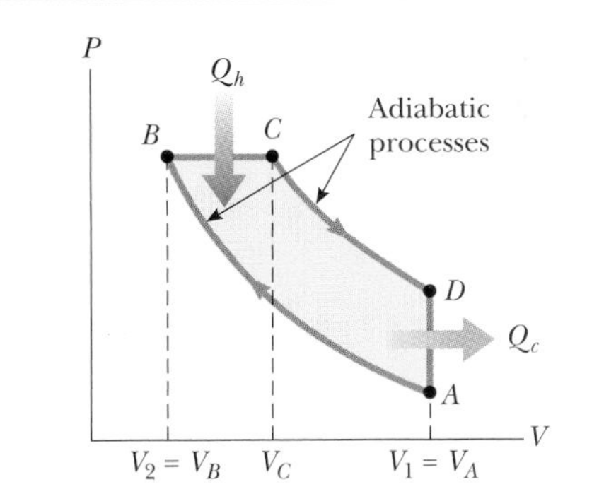

Figure 22.14 *PV* diagram for an ideal diesel engine.

C and assume constant values for air at 300 K. We express the specific heats and the universal gas constant in terms of unit masses rather than moles. Thus, $c_V = 0.718$ kJ/kg·K, $c_P = 1.005$ kJ/kg·K, $\gamma = c_P/c_V = 1.40$, and $R = c_P - c_V = 0.287$ kJ/kg·K $= 0.287$ kPa·m^3/kg·K.

A 3.00-L Gasoline Engine

Let us calculate the power delivered by a six-cylinder gasoline engine that has a displacement volume of 3.00 L operating at 4 000 rpm and having a compression ratio of $r = 9.50$. The air–fuel mixture enters a cylinder at atmospheric pressure and an ambient temperature of 27°C. During combustion, the mixture reaches a temperature of 1 350°C.

First, let us calculate the work done in an individual cylinder. Using the initial pressure $P_A = 100$ kPa, and the initial temperature $T_A = 300$ K, we calculate the initial volume and the mass of the air–fuel mixture. We know that the ratio of the initial and final volumes is the compression ratio,

$$\frac{V_A}{V_B} = r = 9.50$$

We also know that the difference in volumes is the displacement volume. The 3.00-L rating of the engine is the total displacement volume for all six cylinders. Thus, for one cylinder,

$$V_A - V_B = \frac{3.00 \text{ L}}{6} = 0.500 \times 10^{-3} \text{ m}^3$$

Solving these two equations simultaneously, we find the initial and final volumes:

$$V_A = 0.559 \times 10^{-3} \text{ m}^3 \qquad V_B = 0.588 \times 10^{-4} \text{ m}^3$$

Using the ideal gas law (in the form $PV = mRT$, because we are using the universal gas constant in terms of mass rather than moles), we can find the mass of the air–fuel mixture:

$$m = \frac{P_A V_A}{R T_A} = \frac{(100 \text{ kPa})(0.559 \times 10^{-3} \text{ m}^3)}{(0.287 \text{ kPa}\cdot\text{m}^3/\text{kg}\cdot\text{K})(300 \text{ K})}$$
$$= 6.49 \times 10^{-4} \text{ kg}$$

Process $A \rightarrow B$ (see Fig. 22.13) is an adiabatic compression, and this means that $PV^\gamma =$ constant; hence,

$$P_B V_B{}^\gamma = P_A V_A{}^\gamma$$

$$P_B = P_A\left(\frac{V_A}{V_B}\right)^\gamma = P_A(r)^\gamma = (100 \text{ kPa})(9.50)^{1.40}$$
$$= 2.34 \times 10^3 \text{ kPa}$$

Using the ideal gas law, we find that the temperature after the compression is

$$T_B = \frac{P_B V_B}{mR} = \frac{(2.34 \times 10^3 \text{ kPa})(0.588 \times 10^{-4} \text{ m}^3)}{(6.49 \times 10^{-4} \text{ kg})\ (0.287 \text{ kPa}\cdot\text{m}^3/\text{kg}\cdot\text{K})}$$
$$= 739 \text{ K}$$

In process $B \rightarrow C$, the combustion that transforms the potential energy in chemical bonds into internal energy of molecular motion occurs at constant volume; thus, $V_C = V_B$. Combustion causes the temperature to increase to $T_C = 1\,350°\text{C} = 1\,623$ K. Using this value and the ideal gas law, we can calculate P_C:

$$P_C = \frac{mRT_C}{V_C}$$
$$= \frac{(6.49 \times 10^{-4} \text{ kg})(0.287 \text{ kPa}\cdot\text{m}^3/\text{kg}\cdot\text{K})(1\,623 \text{ K})}{(0.588 \times 10^{-4} \text{ m}^3)}$$
$$= 5.14 \times 10^3 \text{ kPa}$$

Process $C \rightarrow D$ is an adiabatic expansion; the pressure after the expansion is

$$P_D = P_C\left(\frac{V_C}{V_D}\right)^\gamma = P_C\left(\frac{V_B}{V_A}\right)^\gamma = P_C\left(\frac{1}{r}\right)^\gamma$$
$$= (5.14 \times 10^3 \text{ kPa})\left(\frac{1}{9.50}\right)^{1.40} = 220 \text{ kPa}$$

Using the ideal gas law again, we find the final temperature:

$$T_D = \frac{P_D V_D}{mR} = \frac{(220 \text{ kPa})(0.559 \times 10^{-3} \text{ m}^3)}{(6.49 \times 10^{-4} \text{ kg})(0.287 \text{ kPa}\cdot\text{m}^3/\text{kg}\cdot\text{K})}$$
$$= 660 \text{ K}$$

Now that we have the temperatures at the beginning and end of each process of the cycle, we can calculate the net energy transfer and net work done in each cylinder every two cycles:

$$|Q_h| = |Q_{\text{in}}| = mc_V(T_C - T_B)$$
$$= (6.49 \times 10^{-4} \text{ kg})(0.718 \text{ kJ/kg}\cdot\text{K})(1\,623 - 739 \text{ K})$$
$$= 0.412 \text{ kJ}$$
$$|Q_c| = |Q_{\text{out}}| = mc_V(T_D - T_A)$$
$$= (6.49 \times 10^{-4} \text{ kg})(0.718 \text{ kJ/kg}\cdot\text{K})(660 \text{ K} - 300 \text{ K})$$
$$= 0.168 \text{ kJ}$$
$$W_{\text{net}} = |Q_{\text{in}}| - |Q_{\text{out}}| = 0.244 \text{ kJ}$$

From Equation 22.2, the efficiency is $e = W_{\text{net}}/|Q_{\text{in}}| = 59\%$. (We can also use Equation 22.7 to calculate the efficiency directly from the compression ratio.)

Recalling that power is delivered every other revolution of the crankshaft, we find that the net power for the six-cylinder engine operating at 4 000 rpm is

$$\mathcal{P}_{\text{net}} = 6(\tfrac{1}{2} \text{ rev})[(4\,000 \text{ rev/min})(1 \text{ min}/60 \text{ s})](0.244 \text{ kJ})$$
$$= 48.8 \text{ kW} = 65 \text{ hp}$$

A 2.00-L Diesel Engine

Let us calculate the power delivered by a four-cylinder diesel engine that has a displacement volume of 2.00 L and is operating at 3 000 rpm. The compression ratio is $r = V_A/V_B = 22.0$, and the **cutoff ratio,** which is the ratio of the volume change during the constant-pressure process $B \rightarrow C$ in Figure 22.14, is $r_c = V_C/V_B = 2.00$. The air enters each cylinder at the beginning of the compression cycle at atmospheric pressure and at an ambient temperature of 27°C.

Our model of the diesel engine is similar to our model of the gasoline engine except that now the fuel is injected at point B and the mixture self-ignites near the end of the compression cycle $A \rightarrow B$, when the temperature reaches the ignition temperature. We assume that the energy input occurs in the constant-pressure process $B \rightarrow C$, and that the expansion process continues from C to D with no further energy transfer by heat.

Let us calculate the work done in an individual cylinder that has an initial volume of $V_A = (2.00 \times 10^{-3} \text{ m}^3)/4 = 0.500 \times 10^{-3} \text{ m}^3$. Because the compression ratio is quite high, we approximate the maximum cylinder volume to be the displacement volume. Using the initial pressure $P_A = 100$ kPa and initial temperature $T_A = 300$ K, we can calculate the mass of the air in the cylinder using the ideal gas law:

$$m = \frac{P_A V_A}{R T_A} = \frac{(100 \text{ kPa})(0.500 \times 10^{-3} \text{ m}^3)}{(0.287 \text{ kPa}\cdot\text{m}^3/\text{kg}\cdot\text{K})(300 \text{ K})}$$
$$= 5.81 \times 10^{-4} \text{ kg}$$

Process $A \rightarrow B$ is an adiabatic compression, so $PV^\gamma =$ constant; thus,

$$P_B V_B{}^\gamma = P_A V_A{}^\gamma$$

$$P_B = P_A\left(\frac{V_A}{V_B}\right)^\gamma = (100 \text{ kPa})(22.0)^{1.40} = 7.58 \times 10^3 \text{ kPa}$$

Using the ideal gas law, we find that the temperature of the air after the compression is

$$T_B = \frac{P_B V_B}{mR} = \frac{(7.58 \times 10^3 \text{ kPa})(0.500 \times 10^{-3} \text{ m}^3)(1/22.0)}{(5.81 \times 10^{-4} \text{ kg})(0.287 \text{ kPa}\cdot\text{m}^3/\text{kg}\cdot\text{K})}$$
$$= 1.03 \times 10^3 \text{ K}$$

Process $B \rightarrow C$ is a constant-pressure expansion; thus, $P_C = P_B$. We know from the cutoff ratio of 2.00 that the volume doubles in this process. According to the ideal gas law, a doubling of volume in an isobaric process results in a doubling of the temperature, so

$$T_C = 2T_B = 2.06 \times 10^3\ \text{K}$$

Process $C \rightarrow D$ is an adiabatic expansion; therefore,

$$P_D = P_C\left(\frac{V_C}{V_D}\right)^{\gamma} = P_C\left(\frac{V_C}{V_B}\frac{V_B}{V_D}\right)^{\gamma} = P_C\left(r_c\frac{1}{r}\right)^{\gamma}$$

$$= (7.57 \times 10^3\ \text{kPa})\left(\frac{2.00}{22.0}\right)^{1.40}$$

$$= 264\ \text{kPa}$$

We find the temperature at D from the ideal gas law:

$$T_D = \frac{P_D V_D}{mR} = \frac{(264\ \text{kPa})(0.500 \times 10^{-3}\ \text{m}^3)}{(5.81 \times 10^{-4}\ \text{kg})(0.287\ \text{kPa}\cdot\text{m}^3/\text{kg}\cdot\text{K})}$$

$$= 792\ \text{K}$$

Now that we have the temperatures at the beginning and the end of each process, we can calculate the net energy transfer by heat and the net work done in each cylinder every two cycles:

$$|Q_h| = |Q_{\text{in}}| = mc_P(T_C - T_B) = 0.601\ \text{kJ}$$

$$|Q_c| = |Q_{\text{out}}| = mc_V(T_D - T_A) = 0.205\ \text{kJ}$$

$$W_{\text{net}} = |Q_{\text{in}}| - |Q_{\text{out}}| = 0.396\ \text{kJ}$$

The efficiency is $e = W_{\text{net}}/|Q_{\text{in}}| = 66\%$.

The net power for the four-cylinder engine operating at 3 000 rpm is

$$\mathcal{P}_{\text{net}} = 4(\tfrac{1}{2}\ \text{rev})[(3\ 000\ \text{rev/min})(1\ \text{min}/60\ \text{s})](0.396\ \text{kJ})$$

$$= 39.6\ \text{kW} = 53\ \text{hp}$$

Modern engine design goes beyond this very simple thermodynamic treatment, which uses idealized cycles.

22.6 Entropy

The zeroth law of thermodynamics involves the concept of temperature, and the first law involves the concept of internal energy. Temperature and internal energy are both state variables—that is, they can be used to describe the thermodynamic state of a system. Another state variable—this one related to the second law of thermodynamics—is **entropy** S. In this section we define entropy on a macroscopic scale as it was first expressed by Clausius in 1865.

Entropy was originally formulated as a useful concept in thermodynamics; however, its importance grew as the field of statistical mechanics developed because the analytical techniques of statistical mechanics provide an alternative means of interpreting entropy and a more global significance to the concept. In statistical mechanics, the behavior of a substance is described in terms of the statistical behavior of its atoms and molecules. One of the main results of this treatment is that **isolated systems tend toward disorder and that entropy is a measure of this disorder.** For example, consider the molecules of a gas in the air in your room. If half of the gas molecules had velocity vectors of equal magnitude directed toward the left and the other half had velocity vectors of the same magnitude directed toward the right, the situation would be very ordered. However, such a situation is extremely unlikely. If you could actually view the molecules, you would see that they move haphazardly in all directions, bumping into one another, changing speed upon collision, some going fast and others going slowly. This situation is highly disordered.

The cause of the tendency of an isolated system toward disorder is easily explained. To do so, we distinguish between *microstates* and *macrostates* of a system. A **microstate** is a particular configuration of the individual constituents of the system. For example, the description of the ordered velocity vectors of the air molecules in your room refers to a particular microstate, and the more likely haphazard motion is another microstate—one that represents disorder. A **macrostate** is a description of the conditions of the system from a macroscopic point of view and makes use of macroscopic variables such as pressure, density, and temperature for gases.

For any given macrostate of the system, a number of microstates are possible. For example, the macrostate of a four on a pair of dice can be formed from the possible microstates 1-3, 2-2, and 3-1. It is assumed that all microstates are equally probable. However, when all possible macrostates are examined, it is found that macrostates

PITFALL PREVENTION

22.4 Entropy Is Abstract

Entropy is one of the most abstract notions in physics, so follow the discussion in this and the subsequent sections very carefully. Do not confuse energy with entropy—even though the names sound similar, they are very different concepts.

(a)

(b)

Figure 22.15 (a) A royal flush is a highly ordered poker hand with low probability of occurring. (b) A disordered and worthless poker hand. The probability of this *particular* hand occurring is the same as that of the royal flush. There are so many worthless hands, however, that the probability of being dealt a worthless hand is much higher than that of a royal flush.

associated with disorder have far more possible microstates than those associated with order. For example, there is only one microstate associated with the macrostate of a royal flush in a poker hand of five spades, laid out in order from ten to ace (Fig. 22.15a). This is a highly ordered hand. However, there are many microstates (the set of five individual cards in a poker hand) associated with a worthless hand in poker (Fig. 22.15b).

The probability of being dealt the royal flush in spades is exactly the same as the probability of being dealt any *particular* worthless hand. Because there are so many worthless hands, however, the probability of a macrostate of a worthless hand is far larger than the probability of a macrostate of a royal flush in spades.

Quick Quiz 22.5 Suppose that you select four cards at random from a standard deck of playing cards and end up with a macrostate of four deuces. How many microstates are associated with this macrostate?

Quick Quiz 22.6 Suppose you pick up two cards at random from a standard deck of playing cards and end up with a macrostate of two aces. How many microstates are associated with this macrostate?

We can also imagine ordered macrostates and disordered macrostates in physical processes, not just in games of dice and poker. The probability of a system moving in time from an ordered macrostate to a disordered macrostate is far greater than the probability of the reverse, because there are more microstates in a disordered macrostate.

If we consider a system and its surroundings to include the entire Universe, then the Universe is always moving toward a macrostate corresponding to greater disorder. Because entropy is a measure of disorder, an alternative way of stating this is **the entropy of the Universe increases in all real processes.** This is yet another statement of the second law of thermodynamics that can be shown to be equivalent to the Kelvin–Planck and Clausius statements.

The original formulation of entropy in thermodynamics involves the transfer of energy by heat during a reversible process. Consider any infinitesimal process in which a system changes from one equilibrium state to another. If dQ_r is the amount of energy transferred by heat when the system follows a reversible path between the states, then the change in entropy dS is equal to this amount of energy for the reversible process divided by the absolute temperature of the system:

$$dS = \frac{dQ_r}{T} \tag{22.8}$$

We have assumed that the temperature is constant because the process is infinitesimal. Because we have claimed that entropy is a state variable, **the change in entropy during a process depends only on the end points and therefore is independent of the actual path followed. Consequently, the entropy change for an irreversible process can be determined by calculating the entropy change for a reversible process that connects the same initial and final states.**

The subscript r on the quantity dQ_r is a reminder that the transferred energy is to be measured along a reversible path, even though the system may actually have followed some irreversible path. When energy is absorbed by the system, dQ_r is positive and the entropy of the system increases. When energy is expelled by the system, dQ_r is negative and the entropy of the system decreases. Note that Equation 22.8 defines not entropy but rather the *change* in entropy. Hence, the meaningful quantity in describing a process is the *change* in entropy.

To calculate the change in entropy for a *finite* process, we must recognize that T is generally not constant. If dQ_r is the energy transferred by heat when the system follows an arbitrary reversible process between the same initial and final states as the irreversible process, then

$$\Delta S = \int_i^f dS = \int_i^f \frac{dQ_r}{T} \tag{22.9}$$

Change in entropy for a finite process

As with an infinitesimal process, the change in entropy ΔS of a system going from one state to another has the same value for *all* paths connecting the two states. That is, the finite change in entropy ΔS of a system depends only on the properties of the initial and final equilibrium states. Thus, we are free to choose a particular reversible path over which to evaluate the entropy in place of the actual path, as long as the initial and final states are the same for both paths. This point is explored further in Section 22.7.

Quick Quiz 22.7 Which of the following is true for the entropy change of a system that undergoes a reversible, adiabatic process? (a) $\Delta S < 0$ (b) $\Delta S = 0$ (c) $\Delta S > 0$

Quick Quiz 22.8 An ideal gas is taken from an initial temperature T_i to a higher final temperature T_f along two different reversible paths: Path A is at constant pressure; Path B is at constant volume. The relation between the entropy changes of the gas for these paths is (a) $\Delta S_A > \Delta S_B$ (b) $\Delta S_A = \Delta S_B$ (c) $\Delta S_A < \Delta S_B$.

Let us consider the changes in entropy that occur in a Carnot heat engine that operates between the temperatures T_c and T_h. In one cycle, the engine takes in energy Q_h from the hot reservoir and expels energy Q_c to the cold reservoir. These energy transfers occur only during the isothermal portions of the Carnot cycle; thus, the constant temperature can be brought out in front of the integral sign in Equation 22.9. The integral then simply has the value of the total amount of energy transferred by heat. Thus, the total change in entropy for one cycle is

$$\Delta S = \frac{|Q_h|}{T_h} - \frac{|Q_c|}{T_c}$$

where the negative sign represents the fact that $|Q_c|$ is positive, but this term must represent energy leaving the engine. In Example 22.3 we showed that, for a Carnot engine,

$$\frac{|Q_c|}{|Q_h|} = \frac{T_c}{T_h}$$

Using this result in the previous expression for ΔS, we find that the total change in entropy for a Carnot engine operating in a cycle is *zero*:

$$\Delta S = 0$$

Now consider a system taken through an arbitrary (non-Carnot) reversible cycle. Because entropy is a state variable—and hence depends only on the properties of a given equilibrium state—we conclude that $\Delta S = 0$ for *any* reversible cycle. In general, we can write this condition in the mathematical form

$$\oint \frac{dQ_r}{T} = 0 \tag{22.10}$$

where the symbol $\oint$ indicates that the integration is over a closed path.

Quasi-Static, Reversible Process for an Ideal Gas

Suppose that an ideal gas undergoes a quasi-static, reversible process from an initial state having temperature T_i and volume V_i to a final state described by T_f and V_f. Let us calculate the change in entropy of the gas for this process.

Writing the first law of thermodynamics in differential form and rearranging the terms, we have $dQ_r = dE_{\text{int}} - dW$, where $dW = -P\,dV$. For an ideal gas, recall that $dE_{\text{int}} = nC_V\,dT$ (Eq. 21.12), and from the ideal gas law, we have $P = nRT/V$. Therefore, we can express the energy transferred by heat in the process as

$$dQ_r = dE_{\text{int}} + P\,dV = nC_V\,dT + nRT\frac{dV}{V}$$

We cannot integrate this expression as it stands because the last term contains two variables, T and V. However, if we divide all terms by T, each of the terms on the right-hand side depends on only one variable:

$$\frac{dQ_r}{T} = nC_V\frac{dT}{T} + nR\frac{dV}{V} \tag{22.11}$$

Assuming that C_V is constant over the process, and integrating Equation 22.11 from the initial state to the final state, we obtain

$$\Delta S = \int_i^f \frac{dQ_r}{T} = nC_V \ln\frac{T_f}{T_i} + nR\ln\frac{V_f}{V_i} \tag{22.12}$$

This expression demonstrates mathematically what we argued earlier—ΔS depends only on the initial and final states and is independent of the path between the states. We can claim this because we have not specified the path taken between the initial and final states. We have only required that the path be reversible. Also, note in Equation 22.12 that ΔS can be positive or negative, depending on the values of the initial and final volumes and temperatures. Finally, for a cyclic process ($T_i = T_f$ and $V_i = V_f$), we see from Equation 22.12 that $\Delta S = 0$. This is further evidence that entropy is a state variable.

Example 22.7 Change in Entropy–Melting

A solid that has a latent heat of fusion L_f melts at a temperature T_m.

(A) Calculate the change in entropy of this substance when a mass m of the substance melts.

Solution Let us assume that the melting occurs so slowly that it can be considered a reversible process. In this case the temperature can be regarded as constant and equal to T_m. Making use of Equations 22.9 and that for the latent heat of fusion $Q = mL_f$ (Eq. 20.6, choosing the positive sign because energy is entering the ice), we find that

$$\Delta S = \int \frac{dQ_r}{T} = \frac{1}{T_m}\int dQ = \frac{Q}{T_m} = \frac{mL_f}{T_m}$$

Note that we are able to remove T_m from the integral because the process is modeled as isothermal. Note also that ΔS is positive.

(B) Estimate the value of the change in entropy of an ice cube when it melts.

Solution Let us assume an ice tray makes cubes that are about 3 cm on a side. The volume per cube is then (very roughly) 30 cm^3. This much liquid water has a mass of 30 g. From Table 20.2 we find that the latent heat of fusion of ice is 3.33×10^5 J/kg. Substituting these values into our answer for part (A), we find that

$$\Delta S = \frac{mL_f}{T_m} = \frac{(0.03\text{ kg})(3.33 \times 10^5\text{ J/kg})}{273\text{ K}} = 4 \times 10^1\text{ J/K}$$

We retain only one significant figure, in keeping with the nature of our estimations.

What If? Suppose you did not have Equation 22.9 available so that you could not calculate an entropy change. How could you argue from the statistical description of entropy that the changes in entropy for parts (A) and (B) should be positive?

Answer When a solid melts, its entropy increases because the molecules are much more disordered in the liquid state than they are in the solid state. The positive value for ΔS also means that the substance in its liquid state does not spontaneously transfer energy from itself to the surroundings and freeze because to do so would involve a spontaneous increase in order and a decrease in entropy.

22.7 Entropy Changes in Irreversible Processes

By definition, a calculation of the change in entropy for a system requires information about a reversible path connecting the initial and final equilibrium states. To calculate changes in entropy for real (irreversible) processes, we must remember that entropy (like internal energy) depends only on the *state* of the system. That is, entropy is a state variable. Hence, the change in entropy when a system moves between any two equilibrium states depends only on the initial and final states.

We can calculate the entropy change in some irreversible process between two equilibrium states by devising a reversible process (or series of reversible processes) between the same two states and computing $\Delta S = \int dQ_r/T$ for the reversible process. In irreversible processes, it is critically important that we distinguish between Q, the actual energy transfer in the process, and Q_r, the energy that would have been transferred by heat along a reversible path. Only Q_r is the correct value to be used in calculating the entropy change.

As we show in the following examples, the change in entropy for a system and its surroundings is always positive for an irreversible process. In general, the total entropy—and therefore the disorder—always increases in an irreversible process. Keeping these considerations in mind, we can state the second law of thermodynamics as follows:

The total entropy of an isolated system that undergoes a change cannot decrease.

Furthermore, **if the process is irreversible, then the total entropy of an isolated system always increases. In a reversible process, the total entropy of an isolated system remains constant.**

When dealing with a system that is not isolated from its surroundings, remember that the increase in entropy described in the second law is that of the system *and* its surroundings. When a system and its surroundings interact in an irreversible process, the increase in entropy of one is greater than the decrease in entropy of the other. Hence, we conclude that **the change in entropy of the Universe must be greater than zero for an irreversible process and equal to zero for a reversible process.** Ultimately, the entropy of the Universe should reach a maximum value. At this value, the Universe will be in a state of uniform temperature and density. All physical, chemical, and biological processes will cease because a state of perfect disorder implies that no energy is available for doing work. This gloomy state of affairs is sometimes referred to as the heat death of the Universe.

Quick Quiz 22.9 True or false: The entropy change in an adiabatic process must be zero because $Q = 0$.

Entropy Change in Thermal Conduction

Let us now consider a system consisting of a hot reservoir and a cold reservoir that are in thermal contact with each other and isolated from the rest of the Universe. A process occurs during which energy Q is transferred by heat from the hot reservoir at temperature T_h to the cold reservoir at temperature T_c. The process as described is irreversible, and so we must find an equivalent reversible process. Let us assume that the objects are connected by a poor thermal conductor whose temperature spans the range from T_c to T_h. This conductor transfers energy slowly, and its state does not change during the process. Under this assumption, the energy transfer to or from each object is reversible, and we may set $Q = Q_r$.

Because the cold reservoir absorbs energy Q, its entropy increases by Q/T_c. At the same time, the hot reservoir loses energy Q, and so its entropy change is $-Q/T_h$. Because $T_h > T_c$, the increase in entropy of the cold reservoir is greater than the decrease in entropy of the hot reservoir. Therefore, the change in entropy of the system (and of the Universe) is greater than zero:

$$\Delta S_U = \frac{Q}{T_c} + \frac{-Q}{T_h} > 0$$

Example 22.8 Which Way Does the Energy Go?

A large, cold object is at 273 K, and a second large, hot object is at 373 K. Show that it is impossible for a small amount of energy—for example, 8.00 J—to be transferred spontaneously by heat from the cold object to the hot one without a decrease in the entropy of the Universe and therefore a violation of the second law.

Solution We assume that, during the energy transfer, the two objects do not undergo a temperature change. This is not a necessary assumption; we make it only to avoid complicating the situation by having to use integral calculus in our calculations. The entropy change of the hot object is

$$\Delta S_h = \frac{Q_r}{T_h} = \frac{8.00 \text{ J}}{373 \text{ K}} = 0.021\,4 \text{ J/K}$$

The cold object loses energy, and its entropy change is

$$\Delta S_c = \frac{Q_r}{T_c} = \frac{-8.00 \text{ J}}{273 \text{ K}} = -0.029\,3 \text{ J/K}$$

We consider the two objects to be isolated from the rest of the Universe. Thus, the entropy change of the Universe is just that of our two-object system, which is

$$\Delta S_U = \Delta S_c + \Delta S_h = -0.007\,9 \text{ J/K}$$

This decrease in entropy of the Universe is in violation of the second law. That is, **the spontaneous transfer of energy by heat from a cold to a hot object cannot occur.**

Suppose energy were to continue to transfer spontaneously from a cold object to a hot object, in violation of the second law. We can describe this impossible energy transfer in terms of disorder. Before the transfer, a certain degree of order is associated with the different temperatures of the objects. The hot object's molecules have a higher average energy than the cold object's molecules. If energy spontaneously transfers from the cold object to the hot object, then, over a period of time, the cold object will become colder and the hot object will become hotter. The difference in average molecular energy will become even greater; this would represent an increase in order for the system and a violation of the second law.

In comparison, the process that does occur naturally is the transfer of energy from the hot object to the cold object. In this process, the difference in average molecular energy decreases; this represents a more random distribution of energy and an increase in disorder.

Entropy Change in a Free Expansion

Let us again consider the adiabatic free expansion of a gas occupying an initial volume V_i (Fig. 22.16). In this situation, a membrane separating the gas from an evacuated region is broken, and the gas expands (irreversibly) to a volume V_f. What are the changes in entropy of the gas and of the Universe during this process?

The process is neither reversible nor quasi-static. The work done by the gas against the vacuum is zero, and because the walls are insulating, no energy is transferred by heat during the expansion. That is, $W = 0$ and $Q = 0$. Using the first law, we see that the change in internal energy is zero. Because the gas is ideal, E_{int} depends on temperature only, and we conclude that $\Delta T = 0$ or $T_i = T_f$.

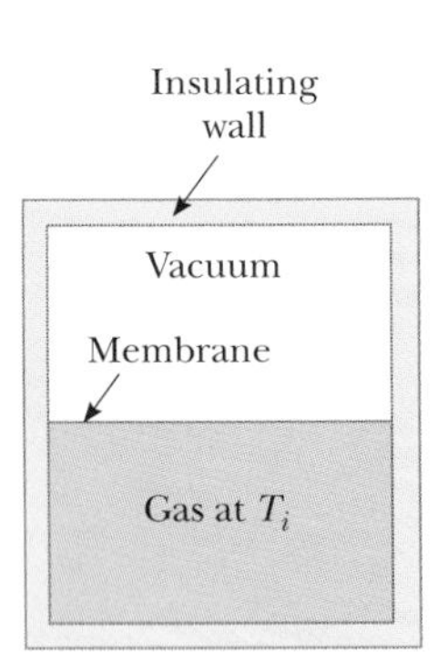

Figure 22.16 Adiabatic free expansion of a gas. When the membrane separating the gas from the evacuated region is ruptured, the gas expands freely and irreversibly. As a result, it occupies a greater final volume. The container is thermally insulated from its surroundings; thus, $Q = 0$.

To apply Equation 22.9, we cannot use $Q = 0$, the value for the irreversible process, but must instead find Q_r; that is, we must find an equivalent reversible path that shares the same initial and final states. A simple choice is an isothermal, reversible expansion in which the gas pushes slowly against a piston while energy enters the gas by heat from a reservoir to hold the temperature constant. Because T is constant in this process, Equation 22.9 gives

$$\Delta S = \int_i^f \frac{dQ_r}{T} = \frac{1}{T}\int_i^f dQ_r$$

For an isothermal process, the first law of thermodynamics specifies that $\int_i^f dQ_r$ is equal to the negative of the work done on the gas during the expansion from V_i to V_f, which is given by Equation 20.13. Using this result, we find that the entropy change for the gas is

$$\Delta S = nR \ln \frac{V_f}{V_i} \tag{22.13}$$

Because $V_f > V_i$, we conclude that ΔS is positive. This positive result indicates that both the entropy and the disorder of the gas *increase* as a result of the irreversible, adiabatic expansion.

It is easy to see that the gas is more disordered after the expansion. Instead of being concentrated in a relatively small space, the molecules are scattered over a larger region.

Because the free expansion takes place in an insulated container, no energy is transferred by heat from the surroundings. (Remember that the isothermal, reversible expansion is only a *replacement* process that we use to calculate the entropy change for the gas; it is not the *actual* process.) Thus, the free expansion has no effect on the surroundings, and the entropy change of the surroundings is zero. Thus, the entropy change for the Universe is positive; this is consistent with the second law.

Entropy Change in Calorimetric Processes

A substance of mass m_1, specific heat c_1, and initial temperature T_c is placed in thermal contact with a second substance of mass m_2, specific heat c_2, and initial temperature $T_h > T_c$. The two substances are contained in a calorimeter so that no energy is lost to the surroundings. The system of the two substances is allowed to reach thermal equilibrium. What is the total entropy change for the system?

First, let us calculate the final equilibrium temperature T_f. Using the techniques of Section 20.2—namely, Equation 20.5, $Q_{\text{cold}} = -Q_{\text{hot}}$, and Equation 20.4, $Q = mc\,\Delta T$, we obtain

$$m_1c_1\,\Delta T_c = -m_2c_2\,\Delta T_h$$

$$m_1c_1(T_f - T_c) = -m_2c_2(T_f - T_h)$$

Solving for T_f, we have

$$T_f = \frac{m_1c_1T_c + m_2c_2T_h}{m_1c_1 + m_2c_2} \tag{22.14}$$

The process is irreversible because the system goes through a series of nonequilibrium states. During such a transformation, the temperature of the system at any time is not well defined because different parts of the system have different temperatures. However, we can imagine that the hot substance at the initial temperature T_h is slowly cooled to the temperature T_f as it comes into contact with a series of reservoirs differing infinitesimally in temperature, the first reservoir being at T_h and the last being at T_f. Such a series of very small changes in temperature would approximate a reversible process. We imagine doing the same thing for the cold substance. Applying Equation 22.9 and noting that $dQ = mc\,dT$ for an infinitesimal change, we have

$$\Delta S = \int_1 \frac{dQ_{\text{cold}}}{T} + \int_2 \frac{dQ_{\text{hot}}}{T} = m_1c_1\int_{T_c}^{T_f}\frac{dT}{T} + m_2c_2\int_{T_h}^{T_f}\frac{dT}{T}$$

where we have assumed that the specific heats remain constant. Integrating, we find that

$$\Delta S = m_1c_1\ln\frac{T_f}{T_c} + m_2c_2\ln\frac{T_f}{T_h} \tag{22.15}$$

where T_f is given by Equation 22.14. If Equation 22.14 is substituted into Equation 22.15, we can show that one of the terms in Equation 22.15 is always positive and the other is always negative. (You may want to verify this for yourself.) The positive term is always greater than the negative term, and this results in a positive value for ΔS. Thus, we conclude that the entropy of the Universe increases in this irreversible process.

Finally, you should note that Equation 22.15 is valid only when no mixing of different substances occurs, because a further entropy increase is associated with the increase in disorder during the mixing. If the substances are liquids or gases and mixing occurs, the result applies only if the two fluids are identical, as in the following example.

Example 22.9 Calculating ΔS for a Calorimetric Process

Suppose that 1.00 kg of water at 0.00°C is mixed with an equal mass of water at 100°C. After equilibrium is reached, the mixture has a uniform temperature of 50.0°C. What is the change in entropy of the system?

Solution We can calculate the change in entropy from Equation 22.15 using the given values $m_1 = m_2 = 1.00$ kg, $c_1 = c_2 = 4\,186\text{ J/kg}\cdot\text{K}$, $T_1 = 273$ K, $T_2 = 373$ K, and $T_f = 323$ K:

$$\Delta S = m_1c_1 \ln \frac{T_f}{T_1} + m_2c_2 \ln \frac{T_f}{T_2}$$

$$\Delta S = (1.00\text{ kg})(4\,186\text{ J/kg}\cdot\text{K}) \ln\left(\frac{323\text{ K}}{273\text{ K}}\right)$$

$$+ (1.00\text{ kg})(4\,186\text{ J/kg}\cdot\text{K}) \ln\left(\frac{323\text{ K}}{373\text{ K}}\right)$$

$$= 704\text{ J/K} - 602\text{ J/K} = 102\text{ J/K}$$

That is, as a result of this irreversible process, the increase in entropy of the cold water is greater than the decrease in entropy of the warm water. Consequently, the increase in entropy of the system is 102 J/K.

22.8 Entropy on a Microscopic Scale[6]

As we have seen, we can approach entropy by relying on macroscopic concepts. We can also treat entropy from a microscopic viewpoint through statistical analysis of molecular motions. We now use a microscopic model to investigate once again the free expansion of an ideal gas, which was discussed from a macroscopic point of view in the preceding section.

In the kinetic theory of gases, gas molecules are represented as particles moving randomly. Let us suppose that the gas is initially confined to a volume V_i, as shown in Figure 22.17a. When the partition separating V_i from a larger container is removed, the molecules eventually are distributed throughout the greater volume V_f (Fig. 22.17b). For a given uniform distribution of gas in the volume, there are a large number of equivalent microstates, and we can relate the entropy of the gas to the number of microstates corresponding to a given macrostate.

We count the number of microstates by considering the variety of molecular locations involved in the free expansion. The instant after the partition is removed (and before the molecules have had a chance to rush into the other half of the container), all the molecules are in the initial volume. We assume that each molecule occupies some microscopic volume V_m. The total number of possible locations of a single molecule in a macroscopic initial volume V_i is the ratio $w_i = V_i/V_m$, which is a huge number. We use w_i here to represent the number of *ways* that the molecule can be placed in the volume, or the number of microstates, which is equivalent to the number of available locations. We assume that the probabilities of a molecule occupying any of these locations are equal.

As more molecules are added to the system, the number of possible ways that the molecules can be positioned in the volume multiplies. For example, if we consider two molecules, for every possible placement of the first, all possible placements of the second are available. Thus, there are w_1 ways of locating the first molecule, and for each of these, there are w_2 ways of locating the second molecule. The total number of ways of locating the two molecules is w_1w_2.

Neglecting the very small probability of having two molecules occupy the same location, each molecule may go into any of the V_i/V_m locations, and so the number of ways of locating N molecules in the volume becomes $W_i = w_i^{\,N} = (V_i/V_m)^N$. ($W_i$ is not to be confused with work.) Similarly, when the volume is increased to V_f, the number of ways of locating N molecules increases to $W_f = w_f^{\,N} = (V_f/V_m)^N$. The ratio of the number of ways of placing the molecules in the volume for the initial and final configurations is

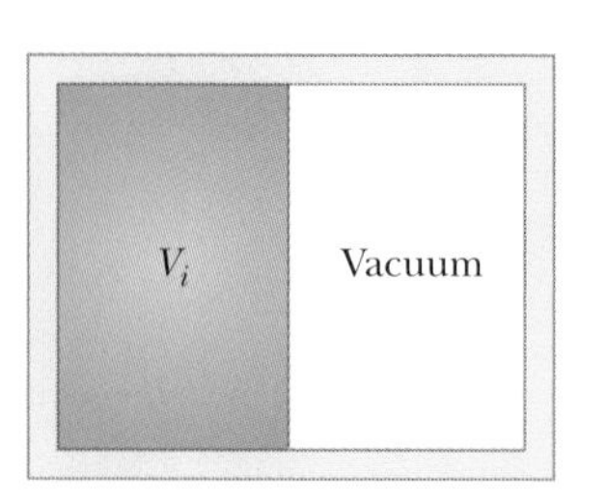

(a)

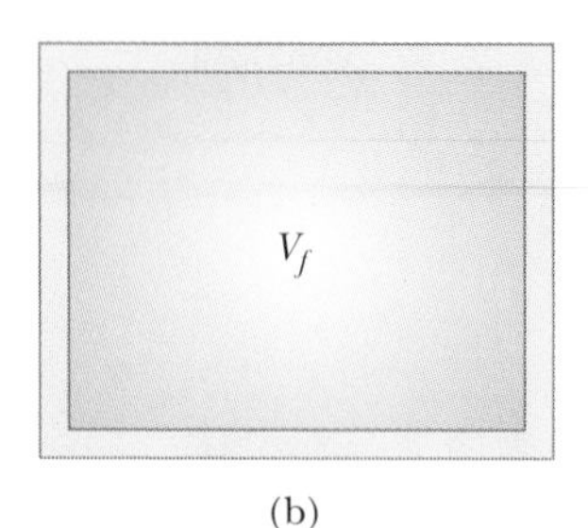

(b)

Figure 22.17 In a free expansion, the gas is allowed to expand into a region that was previously evacuated.

[6] This section was adapted from A. Hudson and R. Nelson, *University Physics,* Philadelphia, Saunders College Publishing, 1990.

$$\frac{W_f}{W_i} = \frac{(V_f/V_m)^N}{(V_i/V_m)^N} = \left(\frac{V_f}{V_i}\right)^N$$

If we now take the natural logarithm of this equation and multiply by Boltzmann's constant, we find that

$$k_B \ln\left(\frac{W_f}{W_i}\right) = nN_A k_B \ln\left(\frac{V_f}{V_i}\right)$$

where we have used the equality $N = nN_A$. We know from Equation 19.11 that $N_A k_B$ is the universal gas constant R; thus, we can write this equation as

$$k_B \ln W_f - k_B \ln W_i = nR \ln\left(\frac{V_f}{V_i}\right) \quad (22.16)$$

From Equation 22.13 we know that when n mol of a gas undergoes a free expansion from V_i to V_f, the change in entropy is

$$S_f - S_i = nR \ln\left(\frac{V_f}{V_i}\right) \quad (22.17)$$

Note that the right-hand sides of Equations 22.16 and 22.17 are identical. Thus, from the left-hand sides, we make the following important connection between entropy and the number of microstates for a given macrostate:

$$S \equiv k_B \ln W \quad (22.18)$$

Entropy (microscopic definition)

The more microstates there are that correspond to a given macrostate, the greater is the entropy of that macrostate. As we have discussed previously, there are many more microstates associated with disordered macrostates than with ordered macrostates. Thus, Equation 22.18 indicates mathematically that **entropy is a measure of disorder.** Although in our discussion we used the specific example of the free expansion of an ideal gas, a more rigorous development of the statistical interpretation of entropy would lead us to the same conclusion.

We have stated that individual microstates are equally probable. However, because there are far more microstates associated with a disordered macrostate than with an ordered microstate, a disordered macrostate is much more probable than an ordered one.

Figure 22.18 shows a real-world example of this concept. There are two possible macrostates for the carnival game—winning a goldfish and winning a black fish. Because only one jar in the array of jars contains a black fish, only one possible microstate corresponds to the macrostate of winning a black fish. A large number of microstates are described by the coin's falling into a jar containing a goldfish. Thus, for the macrostate of winning a goldfish, there are many equivalent microstates. As a result, the probability of winning a goldfish is much greater than the probability of winning a black fish. If there are 24 goldfish and 1 black fish, the probability of winning the black fish is 1 in 25. This assumes that all microstates have the same probability, a situation

Figure 22.18 By tossing a coin into a jar, the carnival-goer can win the fish in the jar. It is more likely that the coin will land in a jar containing a goldfish than in the one containing the black fish.

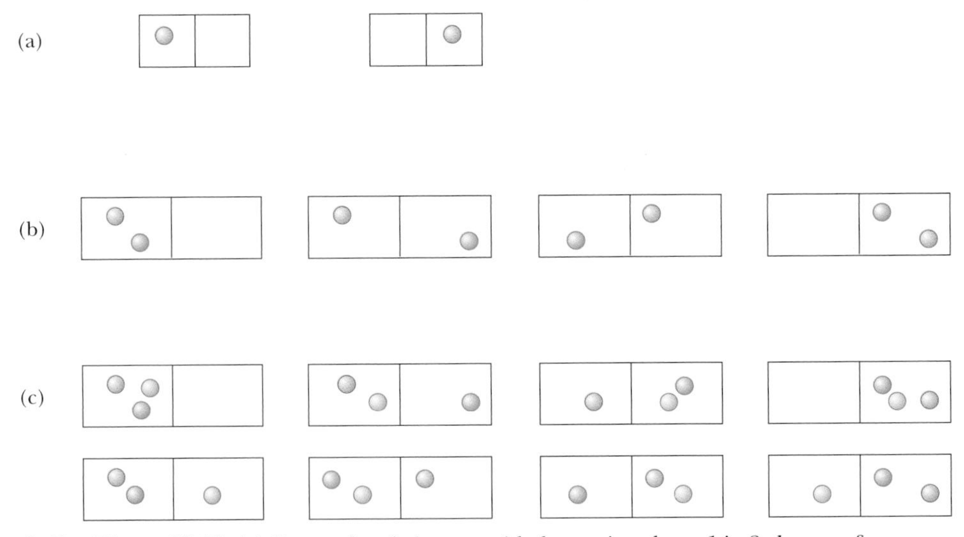

At the Active Figures link at http://www.pse6.com, you can choose the number of molecules to put in the container and measure the probability of all of them being in the left hand side.

Active Figure 22.19 (a) One molecule in a two-sided container has a 1-in-2 chance of being on the left side. (b) Two molecules have a 1-in-4 chance of being on the left side at the same time. (c) Three molecules have a 1-in-8 chance of being on the left side at the same time.

that may not be quite true for the situation shown in Figure 22.18. For example, if you are an accurate coin tosser and you are aiming for the edge of the array of jars, then the probability of the coin's landing in a jar near the edge is likely to be greater than the probability of its landing in a jar near the center.

Let us consider a similar type of probability problem for 100 molecules in a container. At any given moment, the probability of one molecule being in the left part of the container shown in Figure 22.19a as a result of random motion is $\frac{1}{2}$. If there are two molecules, as shown in Figure 22.19b, the probability of both being in the left part is $\left(\frac{1}{2}\right)^2$ or 1 in 4. If there are three molecules (Fig. 22.19c), the probability of all of them being in the left portion at the same moment is $\left(\frac{1}{2}\right)^3$, or 1 in 8. For 100 independently moving molecules, the probability that the 50 fastest ones will be found in the left part at any moment is $\left(\frac{1}{2}\right)^{50}$. Likewise, the probability that the remaining 50 slower molecules will be found in the right part at any moment is $\left(\frac{1}{2}\right)^{50}$. Therefore, the probability of finding this fast-slow separation as a result of random motion is the product $\left(\frac{1}{2}\right)^{50}\left(\frac{1}{2}\right)^{50} = \left(\frac{1}{2}\right)^{100}$, which corresponds to about 1 in 10^{30}. When this calculation is extrapolated from 100 molecules to the number in 1 mol of gas (6.02×10^{23}), the ordered arrangement is found to be *extremely* improbable!

Conceptual Example 22.10 Let's Play Marbles! — Interactive

Suppose you have a bag of 100 marbles. Fifty of the marbles are red, and 50 are green. You are allowed to draw four marbles from the bag according to the following rules. Draw one marble, record its color, and return it to the bag. Shake the bag and then draw another marble. Continue this process until you have drawn and returned four marbles. What are the possible macrostates for this set of events? What is the most likely macrostate? What is the least likely macrostate?

Solution Because each marble is returned to the bag before the next one is drawn, and the bag is shaken, the probability of drawing a red marble is always the same as the probability of drawing a green one. All the possible microstates and macrostates are shown in Table 22.1. As this table indicates, there is only one way to draw a macrostate of four red marbles, and so there is only one microstate for that macrostate. However, there are four possible microstates that correspond to the macrostate of one green marble and three red marbles; six microstates that correspond to two green marbles and two red marbles; four microstates that correspond to three green marbles and one red marble; and one microstate that corresponds to four green marbles. The most likely, and most disordered, macrostate—two red marbles and two green marbles—corresponds to the largest number of microstates. The least likely, most ordered macrostates—four red marbles or four green marbles—correspond to the smallest number of microstates.

Table 22.1

Possible Results of Drawing Four Marbles from a Bag

Macrostate	Possible Microstates	Total Number of Microstates
All R	RRRR	1
1G, 3R	RRRG, RRGR, RGRR, GRRR	4
2G, 2R	RRGG, RGRG, GRRG, RGGR, GRGR, GGRR	6
3G, 1R	GGGR, GGRG, GRGG, RGGG	4
All G	GGGG	1

Explore the generation of microstates and macrostates at the Interactive Worked Example link at **http://www.pse6.com.**

Example 22.11 Adiabatic Free Expansion–One Last Time

Let us verify that the macroscopic and microscopic approaches to the calculation of entropy lead to the same conclusion for the adiabatic free expansion of an ideal gas. Suppose that an ideal gas expands to four times its initial volume. As we have seen for this process, the initial and final temperatures are the same.

(A) Using a macroscopic approach, calculate the entropy change for the gas.

(B) Using statistical considerations, calculate the change in entropy for the gas and show that it agrees with the answer you obtained in part (A).

Solution

(A) Using Equation 22.13, we have

$$\Delta S = nR\ln\left(\frac{V_f}{V_i}\right) = nR\ln\left(\frac{4V_i}{V_i}\right) = nR\ln 4$$

(B) The number of microstates available to a single molecule in the initial volume V_i is $w_i = V_i/V_m$. For N molecules, the number of available microstates is

$$W_i = w_i^N = \left(\frac{V_i}{V_m}\right)^N$$

The number of microstates for all N molecules in the final volume $V_f = 4V_i$ is

$$W_f = \left(\frac{V_f}{V_m}\right)^N = \left(\frac{4V_i}{V_m}\right)^N$$

Thus, the ratio of the number of final microstates to initial microstates is

$$\frac{W_f}{W_i} = 4^N$$

Using Equation 22.18, we obtain

$$\Delta S = k_B\ln W_f - k_B\ln W_i = k_B\ln\left(\frac{W_f}{W_i}\right)$$

$$= k_B\ln(4^N) = Nk_B\ln 4 = nR\ln 4$$

The answer is the same as that for part (A), which dealt with macroscopic parameters.

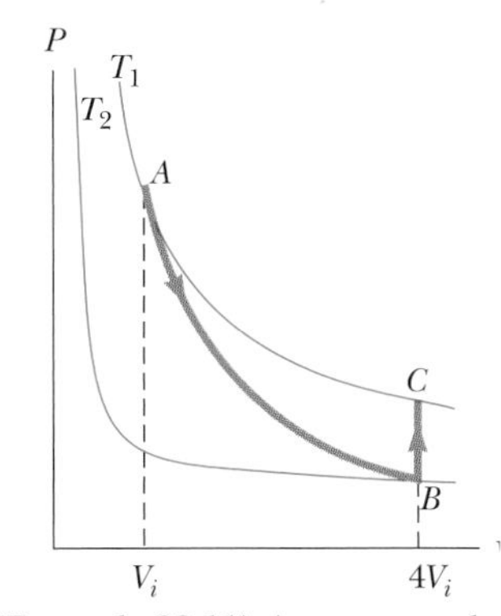

Figure 22.20 (Example 22.11) A gas expands to four times its initial volume and back to the initial temperature by means of a two-step process.

What If? **In part (A) we used Equation 22.13, which was based on a reversible isothermal process connecting the initial and final states. What if we were to choose a different reversible process? Would we arrive at the same result?**

Answer We *must* arrive at the same result because entropy is a state variable. For example, consider the two-step process in Figure 22.20—a reversible adiabatic expansion from V_i to $4V_i$, ($A \rightarrow B$) during which the temperature drops from T_1 to T_2, and a reversible isovolumetric process ($B \rightarrow C$) that takes the gas back to the initial temperature T_1.

During the reversible adiabatic process, $\Delta S = 0$ because $Q_r = 0$. During the reversible isovolumetric process ($B \rightarrow C$), we have from Equation 22.9,

$$\Delta S = \int_i^f \frac{dQ_r}{T} = \int_B^C \frac{nC_V\, dT}{T} = nC_V \ln\left(\frac{T_1}{T_2}\right)$$

Now, we can find the relationship of temperature T_2 to T_1 from Equation 21.20 for the adiabatic process:

$$\frac{T_1}{T_2} = \left(\frac{4V_i}{V_i}\right)^{\gamma-1} = (4)^{\gamma-1}$$

Thus,

$$\Delta S = nC_V \ln (4)^{\gamma-1} = nC_V(\gamma - 1) \ln 4$$

$$= nC_V\left(\frac{C_P}{C_V} - 1\right)\ln 4 = n(C_P - C_V) \ln 4 = nR \ln 4$$

and we do indeed obtain the exact same result for the entropy change.

SUMMARY

Take a practice test for this chapter by clicking on the Practice Test link at http://www.pse6.com.

A **heat engine** is a device that takes in energy by heat and, operating in a cyclic process, expels a fraction of that energy by means of work. The net work done by a heat engine in carrying a working substance through a cyclic process ($\Delta E_{\text{int}} = 0$) is

$$W_{\text{eng}} = |Q_h| - |Q_c| \tag{22.1}$$

where $|Q_h|$ is the energy taken in from a hot reservoir and $|Q_c|$ is the energy expelled to a cold reservoir.

The **thermal efficiency** e of a heat engine is

$$e = \frac{W_{\text{eng}}}{|Q_h|} = 1 - \frac{|Q_c|}{|Q_h|} \tag{22.2}$$

The **second law of thermodynamics** can be stated in the following two ways:

- It is impossible to construct a heat engine that, operating in a cycle, produces no effect other than the input of energy by heat from a reservoir and the performance of an equal amount of work (the Kelvin–Planck statement).
- It is impossible to construct a cyclical machine whose sole effect is to transfer energy continuously by heat from one object to another object at a higher temperature without the input of energy by work (the Clausius statement).

In a **reversible** process, the system can be returned to its initial conditions along the same path on a *PV* diagram, and every point along this path is an equilibrium state. A process that does not satisfy these requirements is **irreversible. Carnot's theorem** states that no real heat engine operating (irreversibly) between the temperatures T_c and T_h can be more efficient than an engine operating reversibly in a Carnot cycle between the same two temperatures.

The **thermal efficiency** of a heat engine operating in the Carnot cycle is

$$e_C = 1 - \frac{T_c}{T_h} \tag{22.6}$$

The second law of thermodynamics states that when real (irreversible) processes occur, the degree of disorder in the system plus the surroundings increases. When a process occurs in an isolated system, the state of the system becomes more disordered. The measure of disorder in a system is called **entropy** S. Thus, another way in which the second law can be stated is

- The entropy of the Universe increases in all real processes.

The **change in entropy** dS of a system during a process between two infinitesimally separated equilibrium states is

$$dS = \frac{dQ_r}{T} \tag{22.8}$$

where dQ_r is the energy transfer by heat for a reversible process that connects the initial and final states. The change in entropy of a system during an arbitrary process between an initial state and a final state is

$$\Delta S = \int_i^f \frac{dQ_r}{T} \qquad (22.9)$$

The value of ΔS for the system is the same for all paths connecting the initial and final states. The change in entropy for a system undergoing any reversible, cyclic process is zero, and when such a process occurs, the entropy of the Universe remains constant.

From a microscopic viewpoint, the entropy of a given macrostate is defined as

$$S \equiv k_B \ln W \qquad (22.18)$$

where k_B is Boltzmann's constant and W is the number of microstates of the system corresponding to the macrostate.

QUESTIONS

1. What are some factors that affect the efficiency of automobile engines?

2. In practical heat engines, which are we better able to control: the temperature of the hot reservoir or the temperature of the cold reservoir? Explain.

3. A steam-driven turbine is one major component of an electric power plant. Why is it advantageous to have the temperature of the steam as high as possible?

4. Is it possible to construct a heat engine that creates no thermal pollution? What does this tell us about environmental considerations for an industrialized society?

5. Does the second law of thermodynamics contradict or correct the first law? Argue for your answer.

6. "The first law of thermodynamics says you can't really win, and the second law says you can't even break even." Explain how this statement applies to a particular device or process; alternatively, argue against the statement.

7. In solar ponds constructed in Israel, the Sun's energy is concentrated near the bottom of a salty pond. With the proper layering of salt in the water, convection is prevented, and temperatures of 100°C may be reached. Can you estimate the maximum efficiency with which useful energy can be extracted from the pond?

8. Can a heat pump have a coefficient of performance less than unity? Explain.

9. Give various examples of irreversible processes that occur in nature. Give an example of a process in nature that is nearly reversible.

10. A heat pump is to be installed in a region where the average outdoor temperature in the winter months is −20°C. In view of this, why would it be advisable to place the outdoor compressor unit deep in the ground? Why are heat pumps not commonly used for heating in cold climates?

11. The device shown in Figure Q22.11, called a thermoelectric converter, uses a series of semiconductor cells to convert internal energy to electric potential energy, which we will study in Chapter 25. In the photograph at the left,

Figure Q22.11

both legs of the device are at the same temperature, and no electric potential energy is produced. However, when one leg is at a higher temperature than the other, as in the photograph on the right, electric potential energy is produced as the device extracts energy from the hot reservoir and drives a small electric motor. (a) Why does the temperature differential produce electric potential energy in this demonstration? (b) In what sense does this intriguing experiment demonstrate the second law of thermodynamics?

12. Discuss three common examples of natural processes that involve an increase in entropy. Be sure to account for all parts of each system under consideration.

13. Discuss the change in entropy of a gas that expands (a) at constant temperature and (b) adiabatically.

14. A thermodynamic process occurs in which the entropy of a system changes by -8.0 J/K. According to the second law of thermodynamics, what can you conclude about the entropy change of the environment?

15. If a supersaturated sugar solution is allowed to evaporate slowly, sugar crystals form in the container. Hence, sugar molecules go from a disordered form (in solution) to a highly ordered crystalline form. Does this process violate the second law of thermodynamics? Explain.

16. How could you increase the entropy of 1 mol of a metal that is at room temperature? How could you decrease its entropy?

17. Suppose your roommate is "Mr. Clean" and tidies up your messy room after a big party. Because your roommate is creating more order, does this represent a violation of the second law of thermodynamics?

18. Discuss the entropy changes that occur when you (a) bake a loaf of bread and (b) consume the bread.

19. "Energy is the mistress of the Universe and entropy is her shadow." Writing for an audience of general readers, argue for this statement with examples. Alternatively, argue for the view that entropy is like a decisive hands-on executive instantly determining what will happen, while energy is like a wretched back-office bookkeeper telling us how little we can afford.

20. A classmate tells you that it is just as likely for all the air molecules in the room you are both in to be concentrated in one corner (with the rest of the room being a vacuum) as it is for the air molecules to be distributed uniformly about the room in their current state. Is this a true statement? Why doesn't the situation he describes actually happen?

21. If you shake a jar full of jellybeans of different sizes, the larger beans tend to appear near the top, and the smaller ones tend to fall to the bottom. Why? Does this process violate the second law of thermodynamics?

PROBLEMS

1, 2, 3 = straightforward, intermediate, challenging ☐ = full solution available in the *Student Solutions Manual and Study Guide*

= coached solution with hints available at http://www.pse6.com = computer useful in solving problem

= paired numerical and symbolic problems

Section 22.1 Heat Engines and the Second Law of Thermodynamics

1. A heat engine takes in 360 J of energy from a hot reservoir and performs 25.0 J of work in each cycle. Find (a) the efficiency of the engine and (b) the energy expelled to the cold reservoir in each cycle.

2. A heat engine performs 200 J of work in each cycle and has an efficiency of 30.0%. For each cycle, how much energy is (a) taken in and (b) expelled by heat?

3. A particular heat engine has a useful power output of 5.00 kW and an efficiency of 25.0%. The engine expels 8 000 J of exhaust energy in each cycle. Find (a) the energy taken in during each cycle and (b) the time interval for each cycle.

4. Heat engine *X* takes in four times more energy by heat from the hot reservoir than heat engine *Y*. Engine *X* delivers two times more work, and it rejects seven times more energy by heat to the cold reservoir than heat engine *Y*. Find the efficiency of (a) heat engine *X* and (b) heat engine *Y*.

5. A multicylinder gasoline engine in an airplane, operating at 2 500 rev/min, takes in energy 7.89×10^3 J and exhausts 4.58×10^3 J for each revolution of the crankshaft. (a) How many liters of fuel does it consume in 1.00 h of operation if the heat of combustion is 4.03×10^7 J/L? (b) What is the mechanical power output of the engine? Ignore friction and express the answer in horsepower. (c) What is the torque exerted by the crankshaft on the load? (d) What power must the exhaust and cooling system transfer out of the engine?

6. Suppose a heat engine is connected to two energy reservoirs, one a pool of molten aluminum (660°C) and the other a block of solid mercury (−38.9°C). The engine runs by freezing 1.00 g of aluminum and melting 15.0 g of mercury during each cycle. The heat of fusion of aluminum is 3.97×10^5 J/kg; the heat of fusion of mercury is 1.18×10^4 J/kg. What is the efficiency of this engine?

Section 22.2 Heat Pumps and Refrigerators

7. A refrigerator has a coefficient of performance equal to 5.00. The refrigerator takes in 120 J of energy from a cold reservoir in each cycle. Find (a) the work required in each cycle and (b) the energy expelled to the hot reservoir.

8. A refrigerator has a coefficient of performance of 3.00. The ice tray compartment is at −20.0°C, and the room temperature is 22.0°C. The refrigerator can convert 30.0 g of water at 22.0°C to 30.0 g of ice at − 20.0°C each minute. What input power is required? Give your answer in watts.

9. In 1993 the federal government instituted a requirement that all room air conditioners sold in the United States must have an energy efficiency ratio (EER) of 10 or higher. The EER is defined as the ratio of the cooling capacity of the air conditioner, measured in Btu/h, to its electrical power requirement in watts. (a) Convert the EER of 10.0 to dimensionless form, using the conversion 1 Btu = 1 055 J. (b) What is the appropriate name for this dimensionless quantity? (c) In the 1970s it was common to find room air conditioners with EERs of 5 or lower. Compare the operating costs for 10 000-Btu/h air conditioners with EERs of 5.00 and 10.0. Assume that each air conditioner operates for 1 500 h during the summer in a city where electricity costs 10.0¢ per kWh.

Section 22.3 Reversible and Irreversible Processes
Section 22.4 The Carnot Engine

10. A Carnot engine has a power output of 150 kW. The engine operates between two reservoirs at 20.0°C and 500°C. (a) How much energy does it take in per hour? (b) How much energy is lost per hour in its exhaust?

11. One of the most efficient heat engines ever built is a steam turbine in the Ohio valley, operating between 430°C and 1 870°C on energy from West Virginia coal to produce electricity for the Midwest. (a) What is its maximum theoretical efficiency? (b) The actual efficiency of the engine is 42.0%. How much useful power does the engine deliver if it takes in 1.40×10^5 J of energy each second from its hot reservoir?

12. A heat engine operating between 200°C and 80.0°C achieves 20.0% of the maximum possible efficiency. What energy input will enable the engine to perform 10.0 kJ of work?

13. An ideal gas is taken through a Carnot cycle. The isothermal expansion occurs at 250°C, and the isothermal compression takes place at 50.0°C. The gas takes in 1 200 J of energy from the hot reservoir during the isothermal expansion. Find (a) the energy expelled to the cold reservoir in each cycle and (b) the net work done by the gas in each cycle.

14. The exhaust temperature of a Carnot heat engine is 300°C. What is the intake temperature if the efficiency of the engine is 30.0%?

15. A Carnot heat engine uses a steam boiler at 100°C as the high-temperature reservoir. The low-temperature reservoir is the outside environment at 20.0°C. Energy is exhausted to the low-temperature reservoir at the rate of 15.4 W. (a) Determine the useful power output of the heat engine. (b) How much steam will it cause to condense in the high-temperature reservoir in 1.00 h?

16. A power plant operates at a 32.0% efficiency during the summer when the sea water used for cooling is at 20.0°C. The plant uses 350°C steam to drive turbines. If the plant's efficiency changes in the same proportion as the ideal efficiency, what would be the plant's efficiency in the winter, when the sea water is 10.0°C?

17. Argon enters a turbine at a rate of 80.0 kg/min, a temperature of 800°C and a pressure of 1.50 MPa. It expands adiabatically as it pushes on the turbine blades and exits at pressure 300 kPa. (a) Calculate its temperature at exit. (b) Calculate the (maximum) power output of the turning turbine. (c) The turbine is one component of a model closed-cycle gas turbine engine. Calculate the maximum efficiency of the engine.

18. An electric power plant that would make use of the temperature gradient in the ocean has been proposed. The system is to operate between 20.0°C (surface water temperature) and 5.00°C (water temperature at a depth of about 1 km). (a) What is the maximum efficiency of such a system? (b) If the useful power output of the plant is 75.0 MW, how much energy is taken in from the warm reservoir per hour? (c) In view of your answer to part (a), do you think such a system is worthwhile? Note that the "fuel" is free.

19. Here is a clever idea. Suppose you build a two-engine device such that the exhaust energy output from one heat engine is the input energy for a second heat engine. We say that the two engines are running *in series*. Let e_1 and e_2 represent the efficiencies of the two engines. (a) The overall efficiency of the two-engine device is defined as the total work output divided by the energy put into the first engine by heat. Show that the overall efficiency is given by

$$e = e_1 + e_2 - e_1 e_2$$

(b) **What If?** Assume the two engines are Carnot engines. Engine 1 operates between temperatures T_h and T_i. The gas in engine 2 varies in temperature between T_i and T_c. In terms of the temperatures, what is the efficiency of the combination engine? (c) What value of the intermediate temperature T_i will result in equal work being done by each of the two engines in series? (d) What value of T_i will result in each of the two engines in series having the same efficiency?

20. A 20.0%-efficient real engine is used to speed up a train from rest to 5.00 m/s. It is known that an ideal (Carnot) engine using the same cold and hot reservoirs would accelerate the same train from rest to a speed of 6.50 m/s using the same amount of fuel. The engines use air at 300 K as a cold reservoir. Find the temperature of the steam serving as the hot reservoir.

21. A firebox is at 750 K, and the ambient temperature is 300 K. The efficiency of a Carnot engine doing 150 J of work as it transports energy between these constant-temperature baths is 60.0%. The Carnot engine must take in energy 150 J/0.600 = 250 J from the hot reservoir and must put out 100 J of energy by heat into the environment. To follow Carnot's reasoning, suppose that some other heat engine S could have efficiency 70.0%. (a) Find the energy input and wasted energy output of engine S as it does 150 J of work. (b) Let engine S operate as in part (a) and run the Carnot engine in reverse. Find the total energy the firebox puts out as both engines operate together, and the total energy trans-

ferred to the environment. Show that the Clausius statement of the second law of thermodynamics is violated. (c) Find the energy input and work output of engine S as it puts out exhaust energy of 100 J. (d) Let engine S operate as in (c) and contribute 150 J of its work output to running the Carnot engine in reverse. Find the total energy the firebox puts out as both engines operate together, the total work output, and the total energy transferred to the environment. Show that the Kelvin–Planck statement of the second law is violated. Thus our assumption about the efficiency of engine S must be false. (e) Let the engines operate together through one cycle as in part (d). Find the change in entropy of the Universe. Show that the entropy statement of the second law is violated.

22. At point A in a Carnot cycle, 2.34 mol of a monatomic ideal gas has a pressure of 1 400 kPa, a volume of 10.0 L, and a temperature of 720 K. It expands isothermally to point B, and then expands adiabatically to point C where its volume is 24.0 L. An isothermal compression brings it to point D, where its volume is 15.0 L. An adiabatic process returns the gas to point A. (a) Determine all the unknown pressures, volumes and temperatures as you fill in the following table:

	P	V	T
A	1 400 kPa	10.0 L	720 K
B			
C		24.0 L	
D		15.0 L	

(b) Find the energy added by heat, the work done by the engine, and the change in internal energy for each of the steps $A \rightarrow B$, $B \rightarrow C$, $C \rightarrow D$, and $D \rightarrow A$. (c) Calculate the efficiency W_{net}/Q_h. Show that it is equal to $1 - T_C/T_A$, the Carnot efficiency.

23. What is the coefficient of performance of a refrigerator that operates with Carnot efficiency between temperatures $-3.00°\text{C}$ and $+27.0°\text{C}$?

24. What is the maximum possible coefficient of performance of a heat pump that brings energy from outdoors at $-3.00°\text{C}$ into a 22.0°C house? Note that the work done to run the heat pump is also available to warm up the house.

25. An ideal refrigerator or ideal heat pump is equivalent to a Carnot engine running in reverse. That is, energy Q_c is taken in from a cold reservoir and energy Q_h is rejected to a hot reservoir. (a) Show that the work that must be supplied to run the refrigerator or heat pump is

$$W = \frac{T_h - T_c}{T_c} Q_c$$

(b) Show that the coefficient of performance of the ideal refrigerator is

$$\text{COP} = \frac{T_c}{T_h - T_c}$$

26. A heat pump, shown in Figure P22.26, is essentially an air conditioner installed backward. It extracts energy from colder air outside and deposits it in a warmer room. Suppose that the ratio of the actual energy entering the room to the work done by the device's motor is 10.0% of the theoretical maximum ratio. Determine the energy entering the room per joule of work done by the motor, given that the inside temperature is 20.0°C and the outside temperature is $-5.00°\text{C}$.

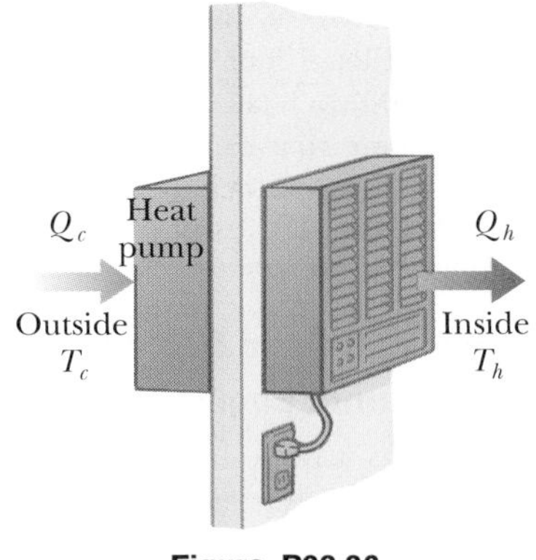

Figure P22.26

27. How much work does an ideal Carnot refrigerator require to remove 1.00 J of energy from helium at 4.00 K and reject this energy to a room-temperature (293-K) environment?

28. A refrigerator maintains a temperature of 0°C in the cold compartment with a room temperature of 25.0°C. It removes energy from the cold compartment at the rate of 8 000 kJ/h. (a) What minimum power is required to operate the refrigerator? (b) The refrigerator exhausts energy into the room at what rate?

29. If a 35.0%-efficient Carnot heat engine (Fig. 22.2) is run in reverse so as to form a refrigerator (Fig. 22.5), what would be this refrigerator's coefficient of performance?

30. Two Carnot engines have the same efficiency. One engine runs in reverse as a heat pump, and the other runs in reverse as a refrigerator. The coefficient of performance of the heat pump is 1.50 times the coefficient of performance of the refrigerator. Find (a) the coefficient of performance of the refrigerator, (b) the coefficient of performance of the heat pump, and (c) the efficiency of each heat engine.

Section 22.5 Gasoline and Diesel Engines

31. In a cylinder of an automobile engine, just after combustion, the gas is confined to a volume of 50.0 cm^3 and has an initial pressure of 3.00×10^6 Pa. The piston moves outward to a final volume of 300 cm^3, and the gas expands without energy loss by heat. (a) If $\gamma = 1.40$ for the gas, what is the final pressure? (b) How much work is done by the gas in expanding?

32. A gasoline engine has a compression ratio of 6.00 and uses a gas for which $\gamma = 1.40$. (a) What is the efficiency of the engine if it operates in an idealized Otto cycle? (b) **What If?** If the actual efficiency is 15.0%, what fraction of the fuel is wasted as a result of friction and energy losses by heat that could by avoided in a reversible engine? (Assume complete combustion of the air–fuel mixture.)

33. A 1.60-L gasoline engine with a compression ratio of 6.20 has a useful power output of 102 hp. Assuming the engine operates in an idealized Otto cycle, find the energy taken in and the energy exhausted each second. Assume the fuel–air mixture behaves like an ideal gas with $\gamma = 1.40$.

34. The compression ratio of an Otto cycle, as shown in Figure 22.13, is $V_A/V_B = 8.00$. At the beginning A of the compression process, 500 cm^3 of gas is at 100 kPa and 20.0°C. At the beginning of the adiabatic expansion the temperature is $T_C = 750$°C. Model the working fluid as an ideal gas with $E_{\text{int}} = nC_VT = 2.50nRT$ and $\gamma = 1.40$. (a) Fill in the table below to follow the states of the gas:

	T (K)	P (kPa)	V (cm^3)	E_{int}
A	293	100	500	
B				
C	1 023			
D				
A				

(b) Fill in the table below to follow the processes:

	Q (input)	W (output)	ΔE_{int}
$A \rightarrow B$			
$B \rightarrow C$			
$C \rightarrow D$			
$D \rightarrow A$			
$ABCDA$			

(c) Identify the energy input Q_h, the energy exhaust Q_c, and the net output work W_{eng}. (d) Calculate the thermal efficiency. (e) Find the number of crankshaft revolutions per minute required for a one-cylinder engine to have an output power of 1.00 kW = 1.34 hp. Note that the thermodynamic cycle involves four piston strokes.

Section 22.6 Entropy

35. An ice tray contains 500 g of liquid water at 0°C. Calculate the change in entropy of the water as it freezes slowly and completely at 0°C.

36. At a pressure of 1 atm, liquid helium boils at 4.20 K. The latent heat of vaporization is 20.5 kJ/kg. Determine the entropy change (per kilogram) of the helium resulting from vaporization.

37. Calculate the change in entropy of 250 g of water heated slowly from 20.0°C to 80.0°C. (*Suggestion:* Note that $dQ = mc\,dT$.)

38. In making raspberry jelly, 900 g of raspberry juice is combined with 930 g of sugar. The mixture starts at room temperature, 23.0°C, and is slowly heated on a stove until it reaches 220°F. It is then poured into heated jars and allowed to cool. Assume that the juice has the same specific heat as water. The specific heat of sucrose is 0.299 cal/g · °C. Consider the heating process. (a) Which of the following terms describe(s) this process: adiabatic, isobaric, isothermal, isovolumetric, cyclic, reversible, isentropic? (b) How much energy does the mixture absorb? (c) What is the minimum change in entropy of the jelly while it is heated?

39. What change in entropy occurs when a 27.9-g ice cube at −12°C is transformed into steam at 115°C?

Section 22.7 Entropy Changes in Irreversible Processes

40. The temperature at the surface of the Sun is approximately 5 700 K, and the temperature at the surface of the Earth is approximately 290 K. What entropy change occurs when 1 000 J of energy is transferred by radiation from the Sun to the Earth?

41. A 1 500-kg car is moving at 20.0 m/s. The driver brakes to a stop. The brakes cool off to the temperature of the surrounding air, which is nearly constant at 20.0°C. What is the total entropy change?

42. A 1.00-kg iron horseshoe is taken from a forge at 900°C and dropped into 4.00 kg of water at 10.0°C. Assuming that no energy is lost by heat to the surroundings, determine the total entropy change of the horseshoe-plus-water system.

43. How fast are you personally making the entropy of the Universe increase right now? Compute an order-of-magnitude estimate, stating what quantities you take as data and the values you measure or estimate for them.

44. A rigid tank of small mass contains 40.0 g of argon, initially at 200°C and 100 kPa. The tank is placed into a reservoir at 0°C and allowed to cool to thermal equilibrium. (a) Calculate the volume of the tank. (b) Calculate the change in internal energy of the argon. (c) Calculate the energy transferred by heat. (d) Calculate the change in entropy of the argon. (e) Calculate the change in entropy of the constant-temperature bath.

45. A 1.00-mol sample of H_2 gas is contained in the left-hand side of the container shown in Figure P22.45, which has equal volumes left and right. The right-hand side is evacuated. When the valve is opened, the gas streams into the right-hand side. What is the final entropy change of the gas? Does the temperature of the gas change?

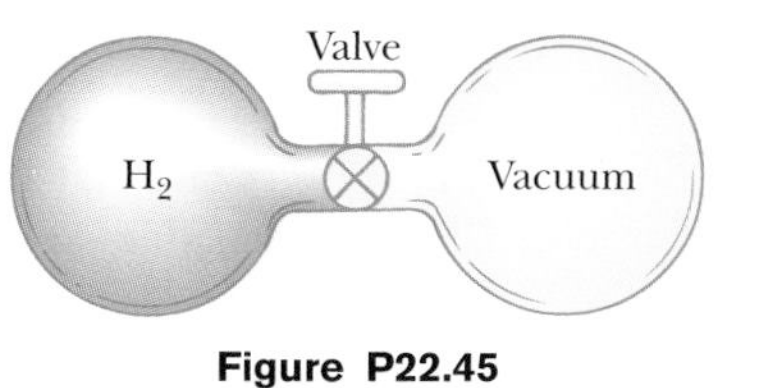

Figure P22.45

46. A 2.00-L container has a center partition that divides it into two equal parts, as shown in Figure P22.46. The left side contains H_2 gas, and the right side contains O_2 gas. Both gases are at room temperature and at atmospheric pressure. The partition is removed, and the gases are allowed to mix. What is the entropy increase of the system?

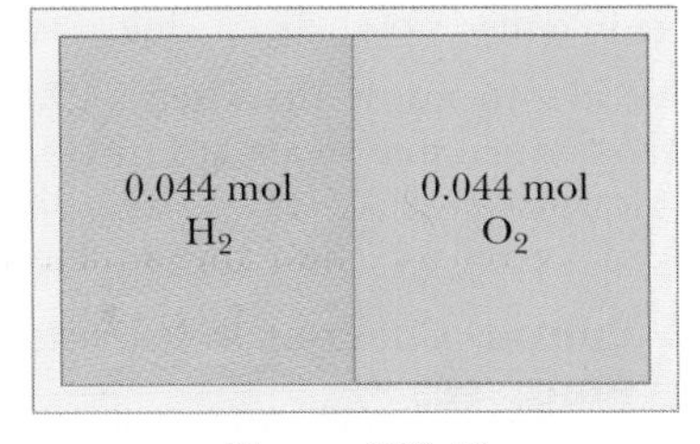

Figure P22.46

47. A 1.00-mol sample of an ideal monatomic gas, initially at a pressure of 1.00 atm and a volume of 0.025 0 m^3, is heated to a final state with a pressure of 2.00 atm and a volume of 0.040 0 m^3. Determine the change in entropy of the gas in this process.

48. A 1.00-mol sample of a diatomic ideal gas, initially having pressure P and volume V, expands so as to have pressure $2P$ and volume $2V$. Determine the entropy change of the gas in the process.

Section 22.8 Entropy on a Microscopic Scale

49. If you toss two dice, what is the total number of ways in which you can obtain (a) a 12 and (b) a 7?

50. Prepare a table like Table 22.1 for the following occurrence. You toss four coins into the air simultaneously and then record the results of your tosses in terms of the numbers of heads and tails that result. For example, HHTH and HTHH are two possible ways in which three heads and one tail can be achieved. (a) On the basis of your table, what is the most probable result of a toss? In terms of entropy, (b) what is the most ordered state and (c) what is the most disordered state?

51. Repeat the procedure used to construct Table 22.1 (a) for the case in which you draw three marbles from your bag rather than four and (b) for the case in which you draw five rather than four.

Additional Problems

52. Every second at Niagara Falls (Fig. P22.52), some 5 000 m^3 of water falls a distance of 50.0 m. What is the increase in entropy per second due to the falling water? Assume that the mass of the surroundings is so great that its temperature and that of the water stay nearly constant at 20.0°C. Suppose that a negligible amount of water evaporates.

CORBIS/Stock Market

Figure P22.52 Niagara Falls, a popular tourist attraction.

53. A house loses energy through the exterior walls and roof at a rate of 5 000 J/s = 5.00 kW when the interior temperature is 22.0°C and the outside temperature is − 5.00°C. Calculate the electric power required to maintain the interior temperature at 22.0°C for the following two cases. (a) The electric power is used in electric resistance heaters (which convert all of the energy transferred in by electrical transmission into internal energy). (b) **What If?** The electric power is used to drive an electric motor that operates the compressor of a heat pump, which has a coefficient of performance equal to 60.0% of the Carnot-cycle value.

54. How much work is required, using an ideal Carnot refrigerator, to change 0.500 kg of tap water at 10.0°C into ice at − 20.0°C? Assume the temperature of the freezer compartment is held at − 20.0°C and the refrigerator exhausts energy into a room at 20.0°C.

55. A heat engine operates between two reservoirs at T_2 = 600 K and T_1 = 350 K. It takes in 1 000 J of energy from the higher-temperature reservoir and performs 250 J of work. Find (a) the entropy change of the Universe ΔS_U for this process and (b) the work W that could have been done by an ideal Carnot engine operating between these two reservoirs. (c) Show that the difference between the amounts of work done in parts (a) and (b) is $T_1 \Delta S_U$.

56. Two identically constructed objects, surrounded by thermal insulation, are used as energy reservoirs for a Carnot engine. The finite reservoirs both have mass m and specific heat c. They start out at temperatures T_h and T_c, where $T_h > T_c$. (a) Show that the engine will stop working when the final temperature of each object is $(T_h T_c)^{1/2}$. (b) Show that the total work done by the

Carnot engine is

$$W_{eng} = mc(T_h^{1/2} - T_c^{1/2})^2$$

57. In 1816 Robert Stirling, a Scottish clergyman, patented the *Stirling engine*, which has found a wide variety of applications ever since. Fuel is burned externally to warm one of the engine's two cylinders. A fixed quantity of inert gas moves cyclically between the cylinders, expanding in the hot one and contracting in the cold one. Figure P22.57 represents a model for its thermodynamic cycle. Consider n mol of an ideal monatomic gas being taken once through the cycle, consisting of two isothermal processes at temperatures $3T_i$ and T_i and two constant-volume processes. Determine, in terms of n, R, and T_i, (a) the net energy transferred by heat to the gas and (b) the efficiency of the engine. A Stirling engine is easier to manufacture than an internal combustion engine or a turbine. It can run on burning garbage. It can run on the energy of sunlight and produce no material exhaust.

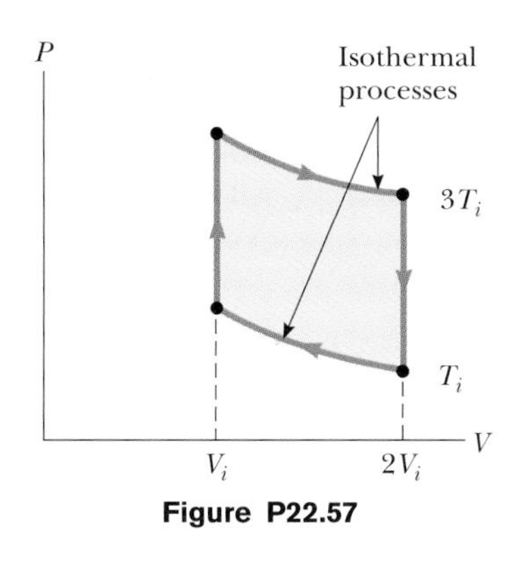

Figure P22.57

58. An electric power plant has an overall efficiency of 15.0%. The plant is to deliver 150 MW of power to a city, and its turbines use coal as the fuel. The burning coal produces steam that drives the turbines. This steam is then condensed to water at 25.0°C by passing it through cooling coils in contact with river water. (a) How many metric tons of coal does the plant consume each day (1 metric ton = 10^3 kg)? (b) What is the total cost of the fuel per year if the delivered price is \$8.00/metric ton? (c) If the river water is delivered at 20.0°C, at what minimum rate must it flow over the cooling coils in order that its temperature not exceed 25.0°C? (*Note:* The heat of combustion of coal is 33.0 kJ/g.)

59. A power plant, having a Carnot efficiency, produces 1 000 MW of electrical power from turbines that take in steam at 500 K and reject water at 300 K into a flowing river. The water downstream is 6.00 K warmer due to the output of the power plant. Determine the flow rate of the river.

60. A power plant, having a Carnot efficiency, produces electric power $\mathcal{P}$ from turbines that take in energy from steam at temperature T_h and discharge energy at temperature T_c through a heat exchanger into a flowing river. The water downstream is warmer by ΔT due to the output of the power plant. Determine the flow rate of the river.

61. An athlete whose mass is 70.0 kg drinks 16 oz (453.6 g) of refrigerated water. The water is at a temperature of 35.0°F. (a) Ignoring the temperature change of the body that results from the water intake (so that the body is regarded as a reservoir always at 98.6°F), find the entropy increase of the entire system. (b) **What If?** Assume that the entire body is cooled by the drink and that the average specific heat of a person is equal to the specific heat of liquid water. Ignoring any other energy transfers by heat and any metabolic energy release, find the athlete's temperature after she drinks the cold water, given an initial body temperature of 98.6°F. Under these assumptions, what is the entropy increase of the entire system? Compare this result with the one you obtained in part (a).

62. A 1.00-mol sample of an ideal monatomic gas is taken through the cycle shown in Figure P22.62. The process $A \rightarrow B$ is a reversible isothermal expansion. Calculate (a) the net work done by the gas, (b) the energy added to the gas by heat, (c) the energy exhausted from the gas by heat, and (d) the efficiency of the cycle.

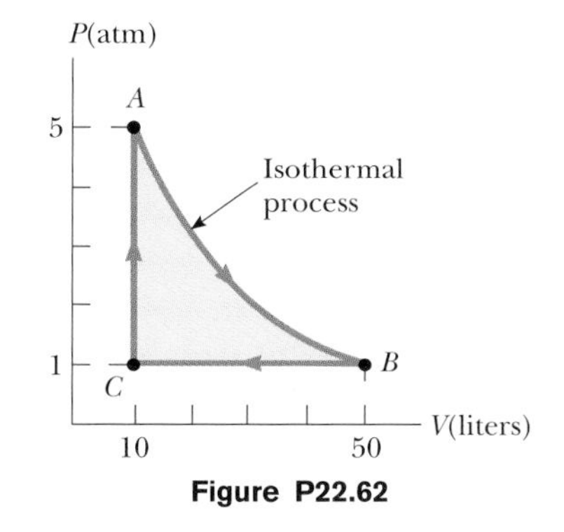

Figure P22.62

63. A biology laboratory is maintained at a constant temperature of 7.00°C by an air conditioner, which is vented to the air outside. On a typical hot summer day the outside temperature is 27.0°C and the air conditioning unit emits energy to the outside at a rate of 10.0 kW. Model the unit as having a coefficient of performance equal to 40.0% of the coefficient of performance of an ideal Carnot device. (a) At what rate does the air conditioner remove energy from the laboratory? (b) Calculate the power required for the work input. (c) Find the change in entropy produced by the air conditioner in 1.00 h. (d) **What If?** The outside temperature increases to 32.0°C. Find the fractional change in the coefficient of performance of the air conditioner.

64. A 1.00-mol sample of an ideal gas expands isothermally, doubling in volume. (a) Show that the work it does in ex-

panding is $W = RT \ln 2$. (b) Because the internal energy E_{int} of an ideal gas depends solely on its temperature, the change in internal energy is zero during the expansion. It follows from the first law that the energy input to the gas by heat during the expansion is equal to the energy output by work. Why does this conversion *not* violate the second law?

65. A 1.00-mol sample of a monatomic ideal gas is taken through the cycle shown in Figure P22.65. At point A, the pressure, volume, and temperature are P_i, V_i, and T_i, respectively. In terms of R and T_i, find (a) the total energy entering the system by heat per cycle, (b) the total energy leaving the system by heat per cycle, (c) the efficiency of an engine operating in this cycle, and (d) the efficiency of an engine operating in a Carnot cycle between the same temperature extremes.

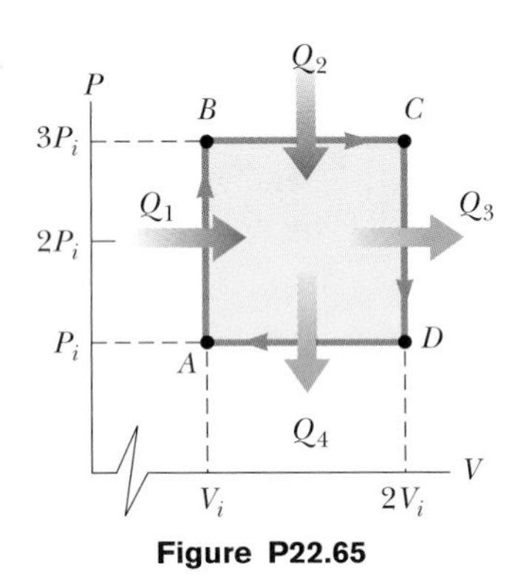

Figure P22.65

66. A sample consisting of n mol of an ideal gas undergoes a reversible isobaric expansion from volume V_i to volume $3V_i$. Find the change in entropy of the gas by calculating $\int_i^f dQ/T$ where $dQ = nC_P\,dT$.

67. A system consisting of n mol of an ideal gas undergoes two reversible processes. It starts with pressure P_i and volume V_i, expands isothermally, and then contracts adiabatically to reach a final state with pressure P_i and volume $3V_i$. (a) Find its change in entropy in the isothermal process. The entropy does not change in the adiabatic process. (b) **What If?** Explain why the answer to part (a) must be the same as the answer to Problem 66.

68. Suppose you are working in a patent office, and an inventor comes to you with the claim that her heat engine, which employs water as a working substance, has a thermodynamic efficiency of 0.61. She explains that it operates between energy reservoirs at 4°C and 0°C. It is a very complicated device, with many pistons, gears, and pulleys, and the cycle involves freezing and melting. Does her claim that $e = 0.61$ warrant serious consideration? Explain.

69. An idealized diesel engine operates in a cycle known as the *air-standard diesel cycle,* shown in Figure 22.14. Fuel is sprayed into the cylinder at the point of maximum compression, B. Combustion occurs during the expansion $B \to C$, which is modeled as an isobaric process. Show that the efficiency of an engine operating in this idealized diesel cycle is

$$e = 1 - \frac{1}{\gamma}\left(\frac{T_D - T_A}{T_C - T_B}\right)$$

70. A 1.00-mol sample of an ideal gas ($\gamma = 1.40$) is carried through the Carnot cycle described in Figure 22.11. At point A, the pressure is 25.0 atm and the temperature is 600 K. At point C, the pressure is 1.00 atm and the temperature is 400 K. (a) Determine the pressures and volumes at points A, B, C, and D. (b) Calculate the net work done per cycle. (c) Determine the efficiency of an engine operating in this cycle.

71. Suppose 1.00 kg of water at 10.0°C is mixed with 1.00 kg of water at 30.0°C at constant pressure. When the mixture has reached equilibrium, (a) what is the final temperature? (b) Take $c_P = 4.19$ kJ/kg · K for water and show that the entropy of the system increases by

$$\Delta S = 4.19 \ln\left[\left(\frac{293}{283}\right)\left(\frac{293}{303}\right)\right] \text{kJ/K}$$

(c) Verify numerically that $\Delta S > 0$. (d) Is the mixing an irreversible process?

Answers to Quick Quizzes

22.1 (c). Equation 22.2 gives this result directly.

22.2 (b). The work represents one third of the input energy. The remainder, two thirds, must be expelled to the cold reservoir.

22.3 (d). The COP of 4.00 for the heat pump means that you are receiving four times as much energy as the energy entering by electrical transmission. With four times as much energy per unit of energy from electricity, you need only one fourth as much electricity.

22.4 C, B, A. Although all three engines operate over a 300-K temperature difference, the efficiency depends on the ratio of temperatures, not the difference.

22.5 One microstate—all four deuces.

22.6 Six microstates—club–diamond, club–heart, club–spade, diamond–heart, diamond–spade, heart–spade. The macrostate of two aces is more probable than that of four deuces in Quick Quiz 22.5 because there are six times as many microstates for this particular macrostate compared to the macrostate of four deuces. Thus, in a hand of poker, two of a kind is less valuable than four of a kind.

22.7 (b). Because the process is reversible and adiabatic, $Q_r = 0$; therefore, $\Delta S = 0$.

22.8 (a). From the first law of thermodynamics, for these two reversible processes, $Q_r = \Delta E_{\text{int}} - W$. During the constant-volume process, $W = 0$, while the work W is nonzero and negative during the constant-pressure expansion. Thus, Q_r is larger for the constant-pressure process, leading to a larger value for the change in entropy. In terms of entropy as disorder, during the constant-pressure process, the gas must expand. The increase in volume results in more ways of locating the molecules of the gas in a container, resulting in a larger increase in entropy.

22.9 False. The determining factor for the entropy change is Q_r, not Q. If the adiabatic process is not reversible, the entropy change is not necessarily zero because a reversible path between the same initial and final states may involve energy transfer by heat.

Appendix A • Tables

Table A.1

Conversion Factors

Length

	m	cm	km	in.	ft	mi
1 meter	1	10^2	10^{-3}	39.37	3.281	6.214×10^{-4}
1 centimeter	10^{-2}	1	10^{-5}	0.393 7	3.281×10^{-2}	6.214×10^{-6}
1 kilometer	10^3	10^5	1	3.937×10^4	3.281×10^3	0.621 4
1 inch	2.540×10^{-2}	2.540	2.540×10^{-5}	1	8.333×10^{-2}	1.578×10^{-5}
1 foot	0.304 8	30.48	3.048×10^{-4}	12	1	1.894×10^{-4}
1 mile	1 609	1.609×10^5	1.609	6.336×10^4	5 280	1

Mass

	kg	g	slug	u
1 kilogram	1	10^3	6.852×10^{-2}	6.024×10^{26}
1 gram	10^{-3}	1	6.852×10^{-5}	6.024×10^{23}
1 slug	14.59	1.459×10^4	1	8.789×10^{27}
1 atomic mass unit	1.660×10^{-27}	1.660×10^{-24}	1.137×10^{-28}	1

Note: 1 metric ton = 1 000 kg.

Time

	s	min	h	day	yr
1 second	1	1.667×10^{-2}	2.778×10^{-4}	1.157×10^{-5}	3.169×10^{-8}
1 minute	60	1	1.667×10^{-2}	6.994×10^{-4}	1.901×10^{-6}
1 hour	3 600	60	1	4.167×10^{-2}	1.141×10^{-4}
1 day	8.640×10^4	1 440	24	1	2.738×10^{-5}
1 year	3.156×10^7	5.259×10^5	8.766×10^3	365.2	1

Speed

	m/s	cm/s	ft/s	mi/h
1 meter per second	1	10^2	3.281	2.237
1 centimeter per second	10^{-2}	1	3.281×10^{-2}	2.237×10^{-2}
1 foot per second	0.304 8	30.48	1	0.681 8
1 mile per hour	0.447 0	44.70	1.467	1

Note: 1 mi/min = 60 mi/h = 88 ft/s.

Force

	N	lb
1 newton	1	0.224 8
1 pound	4.448	1

continued

Table A.1

Conversion Factors *continued*

Work, Energy, Heat

	J	ft · lb	eV
1 joule	1	0.737 6	6.242×10^{18}
1 foot-pound	1.356	1	8.464×10^{18}
1 electron volt	1.602×10^{-19}	1.182×10^{-19}	1
1 calorie	4.186	3.087	2.613×10^{19}
1 British thermal unit	1.055×10^{3}	7.779×10^{2}	6.585×10^{21}
1 kilowatt hour	3.600×10^{6}	2.655×10^{6}	2.247×10^{25}

	cal	Btu	kWh
1 joule	0.238 9	9.481×10^{-4}	2.778×10^{-7}
1 foot-pound	0.323 9	1.285×10^{-3}	3.766×10^{-7}
1 electron volt	3.827×10^{-20}	1.519×10^{-22}	4.450×10^{-26}
1 calorie	1	3.968×10^{-3}	1.163×10^{-6}
1 British thermal unit	2.520×10^{2}	1	2.930×10^{-4}
1 kilowatt hour	8.601×10^{5}	3.413×10^{2}	1

Pressure

	Pa	atm
1 pascal	1	9.869×10^{-6}
1 atmosphere	1.013×10^{5}	1
1 centimeter mercury[a]	1.333×10^{3}	1.316×10^{-2}
1 pound per square inch	6.895×10^{3}	6.805×10^{-2}
1 pound per square foot	47.88	4.725×10^{-4}

	cm Hg	lb/in.^{2}	lb/ft^2
1 pascal	7.501×10^{-4}	1.450×10^{-4}	2.089×10^{-2}
1 atmosphere	76	14.70	2.116×10^{3}
1 centimeter mercury[a]	1	0.194 3	27.85
1 pound per square inch	5.171	1	144
1 pound per square foot	3.591×10^{-2}	6.944×10^{-3}	1

[a] At 0°C and at a location where the free-fall acceleration has its "standard" value, 9.806 65 m/s^2.

Table A.2

Symbols, Dimensions, and Units of Physical Quantities

Quantity	Common Symbol	Unit[a]	Dimensions[b]	Unit in Terms of Base SI Units
Acceleration	$\mathbf{a}$	m/s^2	L/T^2	m/s^2
Amount of substance	n	MOLE		mol
Angle	θ, ϕ	radian (rad)	1	
Angular acceleration	$\boldsymbol{\alpha}$	rad/s^2	T^{-2}	s^{-2}
Angular frequency	ω	rad/s	T^{-1}	s^{-1}
Angular momentum	$\mathbf{L}$	kg · m^2/s	ML^2/T	kg · m^2/s
Angular velocity	$\boldsymbol{\omega}$	rad/s	T^{-1}	s^{-1}
Area	A	m^2	L^2	m^2
Atomic number	Z			

continued

Table A.2

Symbols, Dimensions, and Units of Physical Quantities *continued*

Quantity	Common Symbol	Unit[a]	Dimensions[b]	Unit in Terms of Base SI Units
Capacitance	C	farad (F)	Q^2T^2/ML^2	$A^2 \cdot s^4/kg \cdot m^2$
Charge	q, Q, e	coulomb (C)	Q	$A \cdot s$
Charge density				
Line	λ	C/m	Q/L	$A \cdot s/m$
Surface	σ	C/m^2	Q/L^2	$A \cdot s/m^2$
Volume	ρ	C/m^3	Q/L^3	$A \cdot s/m^3$
Conductivity	σ	$1/\Omega \cdot m$	Q^2T/ML^3	$A^2 \cdot s^3/kg \cdot m^3$
Current	I	AMPERE	Q/T	A
Current density	$\mathbf{J}$	A/m^2	Q/T^2	A/m^2
Density	ρ	kg/m^3	M/L^3	kg/m^3
Dielectric constant	κ			
Length	ℓ, L	METER	L	m
Position	$x, y, z, \mathbf{r}$			
Displacement	$\Delta x, \Delta \mathbf{r}$			
Distance	d, h			
Electric dipole moment	$\mathbf{p}$	$C \cdot m$	QL	$A \cdot s \cdot m$
Electric field	$\mathbf{E}$	V/m	ML/QT^2	$kg \cdot m/A \cdot s^3$
Electric flux	Φ_E	$V \cdot m$	ML^3/QT^2	$kg \cdot m^3/A \cdot s^3$
Electromotive force	$\mathcal{E}$	volt (V)	ML^2/QT^2	$kg \cdot m^2/A \cdot s^3$
Energy	E, U, K	joule (J)	ML^2/T^2	$kg \cdot m^2/s^2$
Entropy	S	J/K	$ML^2/T^2 \cdot K$	$kg \cdot m^2/s^2 \cdot K$
Force	$\mathbf{F}$	newton (N)	ML/T^2	$kg \cdot m/s^2$
Frequency	f	hertz (Hz)	T^{-1}	s^{-1}
Heat	Q	joule (J)	ML^2/T^2	$kg \cdot m^2/s^2$
Inductance	L	henry (H)	ML^2/Q^2	$kg \cdot m^2/A^2 \cdot s^2$
Magnetic dipole moment	$\boldsymbol{\mu}$	$N \cdot m/T$	QL^2/T	$A \cdot m^2$
Magnetic field	$\mathbf{B}$	tesla (T) (= Wb/m^2)	M/QT	$kg/A \cdot s^2$
Magnetic flux	Φ_B	weber (Wb)	ML^2/QT	$kg \cdot m^2/A \cdot s^2$
Mass	m, M	KILOGRAM	M	kg
Molar specific heat	C	$J/mol \cdot K$		$kg \cdot m^2/s^2 \cdot mol \cdot K$
Moment of inertia	I	$kg \cdot m^2$	ML^2	$kg \cdot m^2$
Momentum	$\mathbf{p}$	$kg \cdot m/s$	ML/T	$kg \cdot m/s$
Period	T	s	T	s
Permeability of free space	μ_0	N/A^2 (= H/m)	ML/Q^2T	$kg \cdot m/A^2 \cdot s^2$
Permittivity of free space	ϵ_0	$C^2/N \cdot m^2$ (= F/m)	Q^2T^2/ML^3	$A^2 \cdot s^4/kg \cdot m^3$
Potential	V	volt (V) (= J/C)	ML^2/QT^2	$kg \cdot m^2/A \cdot s^3$
Power	$\mathcal{P}$	watt (W) (= J/s)	ML^2/T^3	$kg \cdot m^2/s^3$
Pressure	P	pascal (Pa) (= N/m^2)	M/LT^2	$kg/m \cdot s^2$
Resistance	R	ohm (Ω) (= V/A)	ML^2/Q^2T	$kg \cdot m^2/A^2 \cdot s^3$
Specific heat	c	$J/kg \cdot K$	$L^2/T^2 \cdot K$	$m^2/s^2 \cdot K$
Speed	v	m/s	L/T	m/s
Temperature	T	KELVIN	K	K
Time	t	SECOND	T	s
Torque	τ	$N \cdot m$	ML^2/T^2	$kg \cdot m^2/s^2$
Velocity	$\mathbf{v}$	m/s	L/T	m/s
Volume	V	m^3	L^3	m^3
Wavelength	λ	m	L	m
Work	W	joule (J) (= $N \cdot m$)	ML^2/T^2	$kg \cdot m^2/s^2$

[a] The base SI units are given in uppercase letters.

[b] The symbols M, L, T, and Q denote mass, length, time, and charge, respectively.

Table A.3

Table of Atomic Masses[a]

Atomic Number Z	Element	Symbol	Chemical Atomic Mass (u)	Mass Number (*Indicates Radioactive) *A*	Atomic Mass (u)	Percent Abundance	Half-Life (If Radioactive) $T_{1/2}$
0	(Neutron)	n		1*	1.008 665		10.4 min
1	Hydrogen	H	1.007 94	1	1.007 825	99.988 5	
	Deuterium	D		2	2.014 102	0.011 5	
	Tritium	T		3*	3.016 049		12.33 yr
2	Helium	He	4.002 602	3	3.016 029	0.000 137	
				4	4.002 603	99.999 863	
				6*	6.018 888		0.81 s
3	Lithium	Li	6.941	6	6.015 122	7.5	
				7	7.016 004	92.5	
				8*	8.022 487		0.84 s
4	Beryllium	Be	9.012 182	7*	7.016 929		53.3 days
				9	9.012 182	100	
				10*	10.013 534		1.5×10^6 yr
5	Boron	B	10.811	10	10.012 937	19.9	
				11	11.009 306	80.1	
				12*	12.014 352		0.020 2 s
6	Carbon	C	12.010 7	10*	10.016 853		19.3 s
				11*	11.011 434		20.4 min
				12	12.000 000	98.93	
				13	13.003 355	1.07	
				14*	14.003 242		5 730 yr
				15*	15.010 599		2.45 s
7	Nitrogen	N	14.006 7	12*	12.018 613		0.011 0 s
				13*	13.005 739		9.96 min
				14	14.003 074	99.632	
				15	15.000 109	0.368	
				16*	16.006 101		7.13 s
				17*	17.008 450		4.17 s
8	Oxygen	O	15.999 4	14*	14.008 595		70.6 s
				15*	15.003 065		122 s
				16	15.994 915	99.757	
				17	16.999 132	0.038	
				18	17.999 160	0.205	
				19*	19.003 579		26.9 s
9	Fluorine	F	18.998 403 2	17*	17.002 095		64.5 s
				18*	18.000 938		109.8 min
				19	18.998 403	100	
				20*	19.999 981		11.0 s
				21*	20.999 949		4.2 s
10	Neon	Ne	20.179 7	18*	18.005 697		1.67 s
				19*	19.001 880		17.2 s
				20	19.992 440	90.48	
				21	20.993 847	0.27	
				22	21.991 385	9.25	
				23*	22.994 467		37.2 s
11	Sodium	Na	22.989 77	21*	20.997 655		22.5 s
				22*	21.994 437		2.61 yr

continued

Table A.3

Table of Atomic Masses[a] *continued*

Atomic Number Z	Element	Symbol	Chemical Atomic Mass (u)	Mass Number (*Indicates Radioactive) A	Atomic Mass (u)	Percent Abundance	Half-Life (If Radioactive) $T_{1/2}$
(11)	Sodium			23	22.989 770	100	
				24*	23.990 963		14.96 h
12	Magnesium	Mg	24.305 0	23*	22.994 125		11.3 s
				24	23.985 042	78.99	
				25	24.985 837	10.00	
				26	25.982 593	11.01	
				27*	26.984 341		9.46 min
13	Aluminum	Al	26.981 538	26*	25.986 892		7.4×10^5 yr
				27	26.981 539	100	
				28*	27.981 910		2.24 min
14	Silicon	Si	28.085 5	28	27.976 926	92.229 7	
				29	28.976 495	4.683 2	
				30	29.973 770	3.087 2	
				31*	30.975 363		2.62 h
				32*	31.974 148		172 yr
15	Phosphorus	P	30.973 761	30*	29.978 314		2.50 min
				31	30.973 762	100	
				32*	31.973 907		14.26 days
				33*	32.971 725		25.3 days
16	Sulfur	S	32.066	32	31.972 071	94.93	
				33	32.971 458	0.76	
				34	33.967 869	4.29	
				35*	34.969 032		87.5 days
				36	35.967 081	0.02	
17	Chlorine	Cl	35.452 7	35	34.968 853	75.78	
				36*	35.968 307		3.0×10^5 yr
				37	36.965 903	24.22	
18	Argon	Ar	39.948	36	35.967 546	0.336 5	
				37*	36.966 776		35.04 days
				38	37.962 732	0.063 2	
				39*	38.964 313		269 yr
				40	39.962 383	99.600 3	
				42*	41.963 046		33 yr
19	Potassium	K	39.098 3	39	38.963 707	93.258 1	
				40*	39.963 999	0.011 7	1.28×10^9 yr
				41	40.961 826	6.730 2	
20	Calcium	Ca	40.078	40	39.962 591	96.941	
				41*	40.962 278		1.0×10^5 yr
				42	41.958 618	0.647	
				43	42.958 767	0.135	
				44	43.955 481	2.086	
				46	45.953 693	0.004	
				48	47.952 534	0.187	
21	Scandium	Sc	44.955 910	41*	40.969 251		0.596 s
				45	44.955 910	100	
22	Titanium	Ti	47.867	44*	43.959 690		49 yr
				46	45.952 630	8.25	

continued

Table A.3

Table of Atomic Masses[a] *continued*

Atomic Number Z	Element	Symbol	Chemical Atomic Mass (u)	Mass Number (*Indicates Radioactive) *A*	Atomic Mass (u)	Percent Abundance	Half-Life (If Radioactive) $T_{1/2}$
(22)	Titanium			47	46.951 764	7.44	
				48	47.947 947	73.72	
				49	48.947 871	5.41	
				50	49.944 792	5.18	
23	Vanadium	V	50.941 5	48*	47.952 254		15.97 days
				50*	49.947 163	0.250	1.5×10^{17} yr
				51	50.943 964	99.750	
24	Chromium	Cr	51.996 1	48*	47.954 036		21.6 h
				50	49.946 050	4.345	
				52	51.940 512	83.789	
				53	52.940 654	9.501	
				54	53.938 885	2.365	
25	Manganese	Mn	54.938 049	54*	53.940 363		312.1 days
				55	54.938 050	100	
26	Iron	Fe	55.845	54	53.939 615	5.845	
				55*	54.938 298		2.7 yr
				56	55.934 942	91.754	
				57	56.935 399	2.119	
				58	57.933 280	0.282	
				60*	59.934 077		1.5×10^{6} yr
27	Cobalt	Co	58.933 200	59	58.933 200	100	
				60*	59.933 822		5.27 yr
28	Nickel	Ni	58.693 4	58	57.935 348	68.076 9	
				59*	58.934 351		7.5×10^{4} yr
				60	59.930 790	26.223 1	
				61	60.931 060	1.139 9	
				62	61.928 349	3.634 5	
				63*	62.929 673		100 yr
				64	63.927 970	0.925 6	
29	Copper	Cu	63.546	63	62.929 601	69.17	
				65	64.927 794	30.83	
30	Zinc	Zn	65.39	64	63.929 147	48.63	
				66	65.926 037	27.90	
				67	66.927 131	4.10	
				68	67.924 848	18.75	
				70	69.925 325	0.62	
31	Gallium	Ga	69.723	69	68.925 581	60.108	
				71	70.924 705	39.892	
32	Germanium	Ge	72.61	70	69.924 250	20.84	
				72	71.922 076	27.54	
				73	72.923 459	7.73	
				74	73.921 178	36.28	
				76	75.921 403	7.61	
33	Arsenic	As	74.921 60	75	74.921 596	100	
34	Selenium	Se	78.96	74	73.922 477	0.89	
				76	75.919 214	9.37	
				77	76.919 915	7.63	

continued

Table A.3

Table of Atomic Masses[a] *continued*

Atomic Number Z	Element	Symbol	Chemical Atomic Mass (u)	Mass Number (*Indicates Radioactive) *A*	Atomic Mass (u)	Percent Abundance	Half-Life (If Radioactive) $T_{1/2}$
(34)	Selenium			78	77.917 310	23.77	
				79*	78.918 500		$\leq 6.5 \times 10^4$ yr
				80	79.916 522	49.61	
				82*	81.916 700	8.73	1.4×10^{20} yr
35	Bromine	Br	79.904	79	78.918 338	50.69	
				81	80.916 291	49.31	
36	Krypton	Kr	83.80	78	77.920 386	0.35	
				80	79.916 378	2.28	
				81*	80.916 592		2.1×10^5 yr
				82	81.913 485	11.58	
				83	82.914 136	11.49	
				84	83.911 507	57.00	
				85*	84.912 527		10.76 yr
				86	85.910 610	17.30	
37	Rubidium	Rb	85.467 8	85	84.911 789	72.17	
				87*	86.909 184	27.83	4.75×10^{10} yr
38	Strontium	Sr	87.62	84	83.913 425	0.56	
				86	85.909 262	9.86	
				87	86.908 880	7.00	
				88	87.905 614	82.58	
				90*	89.907 738		29.1 yr
39	Yttrium	Y	88.905 85	89	88.905 848	100	
40	Zirconium	Zr	91.224	90	89.904 704	51.45	
				91	90.905 645	11.22	
				92	91.905 040	17.15	
				93*	92.906 476		1.5×10^6 yr
				94	93.906 316	17.38	
				96	95.908 276	2.80	
41	Niobium	Nb	92.906 38	91*	90.906 990		6.8×10^2 yr
				92*	91.907 193		3.5×10^7 yr
				93	92.906 378	100	
				94*	93.907 284		2×10^4 yr
42	Molybdenum	Mo	95.94	92	91.906 810	14.84	
				93*	92.906 812		3.5×10^3 yr
				94	93.905 088	9.25	
				95	94.905 842	15.92	
				96	95.904 679	16.68	
				97	96.906 021	9.55	
				98	97.905 408	24.13	
				100	99.907 477	9.63	
43	Technetium	Tc		97*	96.906 365		2.6×10^6 yr
				98*	97.907 216		4.2×10^6 yr
				99*	98.906 255		2.1×10^5 yr
44	Ruthenium	Ru	101.07	96	95.907 598	5.54	
				98	97.905 287	1.87	
				99	98.905 939	12.76	
				100	99.904 220	12.60	

continued

Table A.3

Table of Atomic Masses[a] *continued*

Atomic Number Z	Element	Symbol	Chemical Atomic Mass (u)	Mass Number (*Indicates Radioactive) A	Atomic Mass (u)	Percent Abundance	Half-Life (If Radioactive) $T_{1/2}$
(44)	Ruthenium			101	100.905 582	17.06	
				102	101.904 350	31.55	
				104	103.905 430	18.62	
45	Rhodium	Rh	102.905 50	103	102.905 504	100	
46	Palladium	Pd	106.42	102	101.905 608	1.02	
				104	103.904 035	11.14	
				105	104.905 084	22.33	
				106	105.903 483	27.33	
				107*	106.905 128		6.5×10^6 yr
				108	107.903 894	26.46	
				110	109.905 152	11.72	
47	Silver	Ag	107.868 2	107	106.905 093	51.839	
				109	108.904 756	48.161	
48	Cadmium	Cd	112.411	106	105.906 458	1.25	
				108	107.904 183	0.89	
				109*	108.904 986		462 days
				110	109.903 006	12.49	
				111	110.904 182	12.80	
				112	111.902 757	24.13	
				113*	112.904 401	12.22	9.3×10^{15} yr
				114	113.903 358	28.73	
				116	115.904 755	7.49	
49	Indium	In	114.818	113	112.904 061	4.29	
				115*	114.903 878	95.71	4.4×10^{14} yr
50	Tin	Sn	118.710	112	111.904 821	0.97	
				114	113.902 782	0.66	
				115	114.903 346	0.34	
				116	115.901 744	14.54	
				117	116.902 954	7.68	
				118	117.901 606	24.22	
				119	118.903 309	8.59	
				120	119.902 197	32.58	
				121*	120.904 237		55 yr
				122	121.903 440	4.63	
				124	123.905 275	5.79	
51	Antimony	Sb	121.760	121	120.903 818	57.21	
				123	122.904 216	42.79	
				125*	124.905 248		2.7 yr
52	Tellurium	Te	127.60	120	119.904 020	0.09	
				122	121.903 047	2.55	
				123*	122.904 273	0.89	1.3×10^{13} yr
				124	123.902 820	4.74	
				125	124.904 425	7.07	
				126	125.903 306	18.84	
				128*	127.904 461	31.74	$>8 \times 10^{24}$ yr
				130*	129.906 223	34.08	$\leq 1.25 \times 10^{21}$ yr

continued

Table A.3

Table of Atomic Masses[a] *continued*

Atomic Number Z	Element	Symbol	Chemical Atomic Mass (u)	Mass Number (*Indicates Radioactive) A	Atomic Mass (u)	Percent Abundance	Half-Life (If Radioactive) $T_{1/2}$
53	Iodine	I	126.904 47	127	126.904 468	100	
				129*	128.904 988		1.6×10^{7} yr
54	Xenon	Xe	131.29	124	123.905 896	0.09	
				126	125.904 269	0.09	
				128	127.903 530	1.92	
				129	128.904 780	26.44	
				130	129.903 508	4.08	
				131	130.905 082	21.18	
				132	131.904 145	26.89	
				134	133.905 394	10.44	
				136*	135.907 220	8.87	$\geq 2.36 \times 10^{21}$ yr
55	Cesium	Cs	132.905 45	133	132.905 447	100	
				134*	133.906 713		2.1 yr
				135*	134.905 972		2×10^{6} yr
				137*	136.907 074		30 yr
56	Barium	Ba	137.327	130	129.906 310	0.106	
				132	131.905 056	0.101	
				133*	132.906 002		10.5 yr
				134	133.904 503	2.417	
				135	134.905 683	6.592	
				136	135.904 570	7.854	
				137	136.905 821	11.232	
				138	137.905 241	71.698	
57	Lanthanum	La	138.905 5	137*	136.906 466		6×10^{4} yr
				138*	137.907 107	0.090	1.05×10^{11} yr
				139	138.906 349	99.910	
58	Cerium	Ce	140.116	136	135.907 144	0.185	
				138	137.905 986	0.251	
				140	139.905 434	88.450	
				142*	141.909 240	11.114	$> 5 \times 10^{16}$ yr
59	Praseodymium	Pr	140.907 65	141	140.907 648	100	
60	Neodymium	Nd	144.24	142	141.907 719	27.2	
				143	142.909 810	12.2	
				144*	143.910 083	23.8	2.3×10^{15} yr
				145	144.912 569	8.3	
				146	145.913 112	17.2	
				148	147.916 888	5.7	
				150*	149.920 887	5.6	$> 1 \times 10^{18}$ yr
61	Promethium	Pm		143*	142.910 928		265 days
				145*	144.912 744		17.7 yr
				146*	145.914 692		5.5 yr
				147*	146.915 134		2.623 yr
62	Samarium	Sm	150.36	144	143.911 995	3.07	
				146*	145.913 037		1.0×10^{8} yr
				147*	146.914 893	14.99	1.06×10^{11} yr
				148*	147.914 818	11.24	7×10^{15} yr

continued

Table A.3

Table of Atomic Masses[a] *continued*

Atomic Number Z	Element	Symbol	Chemical Atomic Mass (u)	Mass Number (*Indicates Radioactive) A	Atomic Mass (u)	Percent Abundance	Half-Life (If Radioactive) $T_{1/2}$
(62)	Samarium			149*	148.917 180	13.82	$>2 \times 10^{15}$ yr
				150	149.917 272	7.38	
				151*	150.919 928		90 yr
				152	151.919 728	26.75	
				154	153.922 205	22.75	
63	Europium	Eu	151.964	151	150.919 846	47.81	
				152*	151.921 740		13.5 yr
				153	152.921 226	52.19	
				154*	153.922 975		8.59 yr
				155*	154.922 889		4.7 yr
64	Gadolinium	Gd	157.25	148*	147.918 110		75 yr
				150*	149.918 656		1.8×10^{6} yr
				152*	151.919 788	0.20	1.1×10^{14} yr
				154	153.920 862	2.18	
				155	154.922 619	14.80	
				156	155.922 120	20.47	
				157	156.923 957	15.65	
				158	157.924 100	24.84	
				160	159.927 051	21.86	
65	Terbium	Tb	158.925 34	159	158.925 343	100	
66	Dysprosium	Dy	162.50	156	155.924 278	0.06	
				158	157.924 405	0.10	
				160	159.925 194	2.34	
				161	160.926 930	18.91	
				162	161.926 795	25.51	
				163	162.928 728	24.90	
				164	163.929 171	28.18	
67	Holmium	Ho	164.930 32	165	164.930 320	100	
				166*	165.932 281		1.2×10^{3} yr
68	Erbium	Er	167.6	162	161.928 775	0.14	
				164	163.929 197	1.61	
				166	165.930 290	33.61	
				167	166.932 045	22.93	
				168	167.932 368	26.78	
				170	169.935 460	14.93	
69	Thulium	Tm	168.934 21	169	168.934 211	100	
				171*	170.936 426		1.92 yr
70	Ytterbium	Yb	173.04	168	167.933 894	0.13	
				170	169.934 759	3.04	
				171	170.936 322	14.28	
				172	171.936 378	21.83	
				173	172.938 207	16.13	
				174	173.938 858	31.83	
				176	175.942 568	12.76	
71	Lutecium	Lu	174.967	173*	172.938 927		1.37 yr
				175	174.940 768	97.41	
				176*	175.942 682	2.59	3.78×10^{10} yr

continued

Table A.3

Table of Atomic Masses[a] *continued*

Atomic Number Z	Element	Symbol	Chemical Atomic Mass (u)	Mass Number (*Indicates Radioactive) *A*	Atomic Mass (u)	Percent Abundance	Half-Life (If Radioactive) $T_{1/2}$
72	Hafnium	Hf	178.49	174*	173.940 040	0.16	2.0×10^{15} yr
				176	175.941 402	5.26	
				177	176.943 220	18.60	
				178	177.943 698	27.28	
				179	178.945 815	13.62	
				180	179.946 549	35.08	
73	Tantalum	Ta	180.947 9	180*	179.947 466	0.012	8.152 h
				181	180.947 996	99.988	
74	Tungsten (Wolfram)	W	183.84	180	179.946 706	0.12	
				182	181.948 206	26.50	
				183	182.950 224	14.31	
				184*	183.950 933	30.64	$>3 \times 10^{17}$ yr
				186	185.954 362	28.43	
75	Rhenium	Re	186.207	185	184.952 956	37.40	
				187*	186.955 751	62.60	4.4×10^{10} yr
76	Osmium	Os	190.23	184	183.952 491	0.02	
				186*	185.953 838	1.59	2.0×10^{15} yr
				187	186.955 748	1.96	
				188	187.955 836	13.24	
				189	188.958 145	16.15	
				190	189.958 445	26.26	
				192	191.961 479	40.78	
				194*	193.965 179		6.0 yr
77	Iridium	Ir	192.217	191	190.960 591	37.3	
				193	192.962 924	62.7	
78	Platinum	Pt	195.078	190*	189.959 930	0.014	6.5×10^{11} yr
				192	191.961 035	0.782	
				194	193.962 664	32.967	
				195	194.964 774	33.832	
				196	195.964 935	25.242	
				198	197.967 876	7.163	
79	Gold	Au	196.966 55	197	196.966 552	100	
80	Mercury	Hg	200.59	196	195.965 815	0.15	
				198	197.966 752	9.97	
				199	198.968 262	16.87	
				200	199.968 309	23.10	
				201	200.970 285	13.18	
				202	201.970 626	29.86	
				204	203.973 476	6.87	
81	Thallium	Tl	204.383 3	203	202.972 329	29.524	
				204*	203.973 849		3.78 yr
				205	204.974 412	70.476	
		(Ra E″)		206*	205.976 095		4.2 min
		(Ac C″)		207*	206.977 408		4.77 min
		(Th C″)		208*	207.982 005		3.053 min
		(Ra C″)		210*	209.990 066		1.30 min

continued

Table A.3

Table of Atomic Masses[a] *continued*

Atomic Number Z	Element	Symbol	Chemical Atomic Mass (u)	Mass Number (*Indicates Radioactive) A	Atomic Mass (u)	Percent Abundance	Half-Life (If Radioactive) $T_{1/2}$
82	Lead	Pb	207.2	202*	201.972 144		5×10^4 yr
				204*	203.973 029	1.4	$\geq 1.4 \times 10^{17}$ yr
				205*	204.974 467		1.5×10^7 yr
				206	205.974 449	24.1	
				207	206.975 881	22.1	
				208	207.976 636	52.4	
		(Ra D)		210*	209.984 173		22.3 yr
		(Ac B)		211*	210.988 732		36.1 min
		(Th B)		212*	211.991 888		10.64 h
		(Ra B)		214*	213.999 798		26.8 min
83	Bismuth	Bi	208.980 38	207*	206.978 455		32.2 yr
				208*	207.979 727		3.7×10^5 yr
				209	208.980 383	100	
		(Ra E)		210*	209.984 105		5.01 days
		(Th C)		211*	210.987 258		2.14 min
				212*	211.991 272		60.6 min
		(Ra C)		214*	213.998 699		19.9 min
				215*	215.001 832		7.4 min
84	Polonium	Po		209*	208.982 416		102 yr
		(Ra F)		210*	209.982 857		138.38 days
		(Ac C′)		211*	210.986 637		0.52 s
		(Th C′)		212*	211.988 852		0.30 μs
		(Ra C′)		214*	213.995 186		164 μs
		(Ac A)		215*	214.999 415		0.001 8 s
		(Th A)		216*	216.001 905		0.145 s
		(Ra A)		218*	218.008 966		3.10 min
85	Astatine	At		215*	214.998 641		$\approx$ 100 μs
				218*	218.008 682		1.6 s
				219*	219.011 297		0.9 min
86	Radon	Rn					
		(An)		219*	219.009 475		3.96 s
		(Tn)		220*	220.011 384		55.6 s
		(Rn)		222*	222.017 570		3.823 days
87	Francium	Fr					
		(Ac K)		223*	223.019 731		22 min
88	Radium	Ra					
		(Ac X)		223*	223.018 497		11.43 days
		(Th X)		224*	224.020 202		3.66 days
		(Ra)		226*	226.025 403		1 600 yr
		(Ms Th_1)		228*	228.031 064		5.75 yr
89	Actinium	Ac		227*	227.027 747		21.77 yr
		(Ms Th_2)		228*	228.031 015		6.15 h
90	Thorium	Th	232.038 1				
		(Rd Ac)		227*	227.027 699		18.72 days
		(Rd Th)		228*	228.028 731		1.913 yr
				229*	229.031 755		7 300 yr
		(Io)		230*	230.033 127		75.000 yr

continued

Table A.3

Table of Atomic Masses[a] *continued*

Atomic Number Z	Element	Symbol	Chemical Atomic Mass (u)	Mass Number (*Indicates Radioactive) *A*	Atomic Mass (u)	Percent Abundance	Half-Life (If Radioactive) $T_{1/2}$
(90)	Thorium	(UY)		231*	231.036 297		25.52 h
		(Th)		232*	232.038 050	100	1.40×10^{10} yr
		(UX_1)		234*	234.043 596		24.1 days
91	Protactinium	Pa	231.035 88	231*	231.035 879		32.760 yr
		(Uz)		234*	234.043 302		6.7 h
92	Uranium	U	238.028 9	232*	232.037 146		69 yr
				233*	233.039 628		1.59×10^5 yr
				234*	234.040 946	0.005 5	2.45×10^5 yr
		(Ac U)		235*	235.043 923	0.720 0	7.04×10^8 yr
				236*	236.045 562		2.34×10^7 yr
		(UI)		238*	238.050 783	99.274 5	4.47×10^9 yr
93	Neptunium	Np		235*	235.044 056		396 days
				236*	236.046 560		1.15×10^5 yr
				237*	237.048 167		2.14×10^6 yr
94	Plutonium	Pu		236*	236.046 048		2.87 yr
				238*	238.049 553		87.7 yr
				239*	239.052 156		2.412×10^4 yr
				240*	240.053 808		6 560 yr
				241*	241.056 845		14.4 yr
				242*	242.058 737		3.73×10^6 yr
				244*	244.064 198		8.1×10^7 yr

[a] Chemical atomic masses are from T. B. Coplen, "Atomic Weights of the Elements 1999," a technical report to the International Union of Pure and Applied Chemistry, and published in *Pure and Applied Chemistry*, 73(4), 667–683, 2001. Atomic masses of the isotopes are from G. Audi and A. H. Wapstra, "The 1995 Update to the Atomic Mass Evaluation," *Nuclear Physics*, A595, vol. 4, 409–480, December 25, 1995. Percent abundance values are from K. J. R. Rosman and P. D. P. Taylor, "Isotopic Compositions of the Elements 1999", a technical report to the International Union of Pure and Applied Chemistry, and published in *Pure and Applied Chemistry*, 70(1), 217–236, 1998.

Appendix B • Mathematics Review

These appendices in mathematics are intended as a brief review of operations and methods. Early in this course, you should be totally familiar with basic algebraic techniques, analytic geometry, and trigonometry. The appendices on differential and integral calculus are more detailed and are intended for those students who have difficulty applying calculus concepts to physical situations.

B.1 Scientific Notation

Many quantities that scientists deal with often have very large or very small values. For example, the speed of light is about 300 000 000 m/s, and the ink required to make the dot over an *i* in this textbook has a mass of about 0.000 000 001 kg. Obviously, it is very cumbersome to read, write, and keep track of numbers such as these. We avoid this problem by using a method dealing with powers of the number 10:

$$10^0 = 1$$

$$10^1 = 10$$

$$10^2 = 10 \times 10 = 100$$

$$10^3 = 10 \times 10 \times 10 = 1000$$

$$10^4 = 10 \times 10 \times 10 \times 10 = 10\ 000$$

$$10^5 = 10 \times 10 \times 10 \times 10 \times 10 = 100\ 000$$

and so on. The number of zeros corresponds to the power to which 10 is raised, called the **exponent** of 10. For example, the speed of light, 300 000 000 m/s, can be expressed as 3×10^8 m/s.

In this method, some representative numbers smaller than unity are

$$10^{-1} = \frac{1}{10} = 0.1$$

$$10^{-2} = \frac{1}{10 \times 10} = 0.01$$

$$10^{-3} = \frac{1}{10 \times 10 \times 10} = 0.001$$

$$10^{-4} = \frac{1}{10 \times 10 \times 10 \times 10} = 0.000\ 1$$

$$10^{-5} = \frac{1}{10 \times 10 \times 10 \times 10 \times 10} = 0.000\ 01$$

In these cases, the number of places the decimal point is to the left of the digit 1 equals the value of the (negative) exponent. Numbers expressed as some power of 10 multiplied by another number between 1 and 10 are said to be in **scientific notation.** For example, the scientific notation for 5 943 000 000 is 5.943×10^9 and that for 0.000 083 2 is 8.32×10^{-5}.

When numbers expressed in scientific notation are being multiplied, the following general rule is very useful:

$$10^n \times 10^m = 10^{n+m} \qquad \text{(B.1)}$$

where n and m can be *any* numbers (not necessarily integers). For example, $10^2 \times 10^5 = 10^7$. The rule also applies if one of the exponents is negative: $10^3 \times 10^{-8} = 10^{-5}$.

When dividing numbers expressed in scientific notation, note that

$$\frac{10^n}{10^m} = 10^n \times 10^{-m} = 10^{n-m} \tag{B.2}$$

Exercises

With help from the above rules, verify the answers to the following:

1. $86\,400 = 8.64 \times 10^4$
2. $9\,816\,762.5 = 9.816\,762\,5 \times 10^6$
3. $0.000\,000\,039\,8 = 3.98 \times 10^{-8}$
4. $(4 \times 10^8)\,(9 \times 10^9) = 3.6 \times 10^{18}$
5. $(3 \times 10^7)\,(6 \times 10^{-12}) = 1.8 \times 10^{-4}$
6. $\dfrac{75 \times 10^{-11}}{5 \times 10^{-3}} = 1.5 \times 10^{-7}$
7. $\dfrac{(3 \times 10^6)(8 \times 10^{-2})}{(2 \times 10^{17})(6 \times 10^5)} = 2 \times 10^{-18}$

B.2 Algebra

Some Basic Rules

When algebraic operations are performed, the laws of arithmetic apply. Symbols such as x, y, and z are usually used to represent quantities that are not specified, what are called the **unknowns.**

First, consider the equation

$$8x = 32$$

If we wish to solve for x, we can divide (or multiply) each side of the equation by the same factor without destroying the equality. In this case, if we divide both sides by 8, we have

$$\frac{8x}{8} = \frac{32}{8}$$

$$x = 4$$

Next consider the equation

$$x + 2 = 8$$

In this type of expression, we can add or subtract the same quantity from each side. If we subtract 2 from each side, we obtain

$$x + 2 - 2 = 8 - 2$$

$$x = 6$$

In general, if $x + a = b$, then $x = b - a$.

Now consider the equation

$$\frac{x}{5} = 9$$

If we multiply each side by 5, we are left with x on the left by itself and 45 on the right:

$$\left(\frac{x}{5}\right)(5) = 9 \times 5$$

$$x = 45$$

In all cases, *whatever operation is performed on the left side of the equality must also be performed on the right side.*

The following rules for multiplying, dividing, adding, and subtracting fractions should be recalled, where a, b, and c are three numbers:

	Rule	Example
Multiplying	$\left(\frac{a}{b}\right)\left(\frac{c}{d}\right) = \frac{ac}{bd}$	$\left(\frac{2}{3}\right)\left(\frac{4}{5}\right) = \frac{8}{15}$
Dividing	$\frac{(a/b)}{(c/d)} = \frac{ad}{bc}$	$\frac{2/3}{4/5} = \frac{(2)(5)}{(4)(3)} = \frac{10}{12}$
Adding	$\frac{a}{b} \pm \frac{c}{d} = \frac{ad \pm bc}{bd}$	$\frac{2}{3} - \frac{4}{5} = \frac{(2)(5) - (4)(3)}{(3)(5)} = -\frac{2}{15}$

Exercises

In the following exercises, solve for x:

	Answers
1. $a = \frac{1}{1+x}$	$x = \frac{1-a}{a}$
2. $3x - 5 = 13$	$x = 6$
3. $ax - 5 = bx + 2$	$x = \frac{7}{a-b}$
4. $\frac{5}{2x+6} = \frac{3}{4x+8}$	$x = -\frac{11}{7}$

Powers

When powers of a given quantity x are multiplied, the following rule applies:

$$x^n x^m = x^{n+m} \tag{B.3}$$

For example, $x^2x^4 = x^{2+4} = x^6$.

When dividing the powers of a given quantity, the rule is

$$\frac{x^n}{x^m} = x^{n-m} \tag{B.4}$$

For example, $x^8/x^2 = x^{8-2} = x^6$.

A power that is a fraction, such as $\frac{1}{3}$, corresponds to a root as follows:

$$x^{1/n} = \sqrt[n]{x} \tag{B.5}$$

For example, $4^{1/3} = \sqrt[3]{4} = 1.5874$. (A scientific calculator is useful for such calculations.)

Finally, any quantity x^n raised to the mth power is

$$(x^n)^m = x^{nm} \tag{B.6}$$

Table B.1 summarizes the rules of exponents.

Table B.1

Rules of Exponents
$x^0 = 1$
$x^1 = x$
$x^n x^m = x^{n+m}$
$x^n/x^m = x^{n-m}$
$x^{1/n} = \sqrt[n]{x}$
$(x^n)^m = x^{nm}$

Exercises

Verify the following:

1. $3^2 \times 3^3 = 243$
2. $x^5x^{-8} = x^{-3}$

3. $x^{10}/x^{-5} = x^{15}$
4. $5^{1/3} = 1.709\ 975$ (Use your calculator.)
5. $60^{1/4} = 2.783\ 158$ (Use your calculator.)
6. $(x^4)^3 = x^{12}$

Factoring

Some useful formulas for factoring an equation are

$$ax + ay + az = a(x + y + z) \qquad \text{common factor}$$

$$a^2 + 2ab + b^2 = (a + b)^2 \qquad \text{perfect square}$$

$$a^2 - b^2 = (a + b)(a - b) \qquad \text{differences of squares}$$

Quadratic Equations

The general form of a quadratic equation is

$$ax^2 + bx + c = 0 \tag{B.7}$$

where x is the unknown quantity and a, b, and c are numerical factors referred to as **coefficients** of the equation. This equation has two roots, given by

$$x = \frac{-b \pm \sqrt{b^2 - 4ac}}{2a} \tag{B.8}$$

If $b^2 \geq 4ac$, the roots are real.

Example 1

The equation $x^2 + 5x + 4 = 0$ has the following roots corresponding to the two signs of the square-root term:

$$x = \frac{-5 \pm \sqrt{5^2 - (4)(1)(4)}}{2(1)} = \frac{-5 \pm \sqrt{9}}{2} = \frac{-5 \pm 3}{2}$$

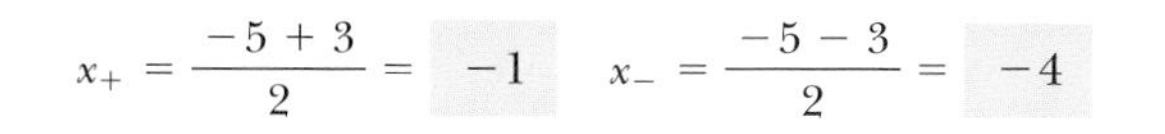

$$x_+ = \frac{-5 + 3}{2} = -1 \qquad x_- = \frac{-5 - 3}{2} = -4$$

where x_+ refers to the root corresponding to the positive sign and x_- refers to the root corresponding to the negative sign.

Exercises

Solve the following quadratic equations:

	Answers	
1. $x^2 + 2x - 3 = 0$	$x_+ = 1$	$x_- = -3$
2. $2x^2 - 5x + 2 = 0$	$x_+ = 2$	$x_- = \frac{1}{2}$
3. $2x^2 - 4x - 9 = 0$	$x_+ = 1 + \sqrt{22}/2$	$x_- = 1 - \sqrt{22}/2$

Linear Equations

A linear equation has the general form

$$y = mx + b \tag{B.9}$$

where m and b are constants. This equation is referred to as being linear because the graph of y versus x is a straight line, as shown in Figure B.1. The constant b, called the **y-intercept,** represents the value of y at which the straight line intersects the y axis. The constant m is equal to the **slope** of the straight line. If any two points on the straight line are specified by the coordinates (x_1, y_1) and (x_2, y_2), as in Figure B.1, then

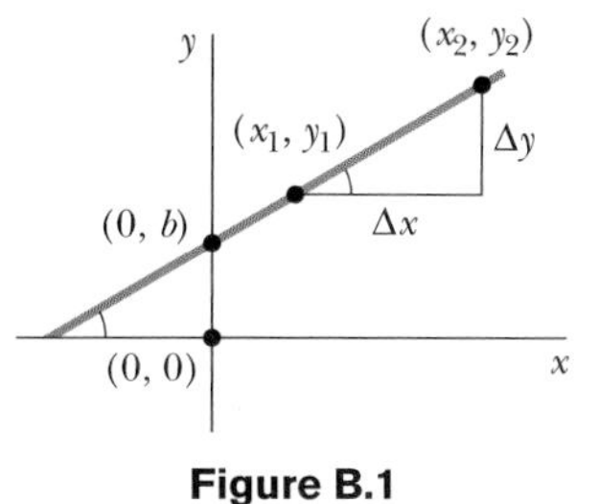

Figure B.1

the slope of the straight line can be expressed as

$$\text{Slope} = \frac{y_2 - y_1}{x_2 - x_1} = \frac{\Delta y}{\Delta x} \tag{B.10}$$

Note that m and b can have either positive or negative values. If $m > 0$, the straight line has a *positive* slope, as in Figure B.1. If $m < 0$, the straight line has a *negative* slope. In Figure B.1, both m and b are positive. Three other possible situations are shown in Figure B.2.

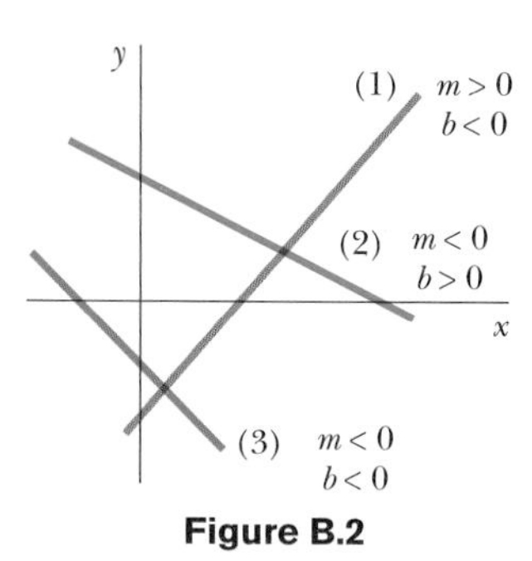

Figure B.2

Exercises

1. Draw graphs of the following straight lines:
   (a) $y = 5x + 3$ (b) $y = -2x + 4$ (c) $y = -3x - 6$
2. Find the slopes of the straight lines described in Exercise 1.

Answers (a) 5 (b) -2 (c) -3

3. Find the slopes of the straight lines that pass through the following sets of points:
   (a) $(0, -4)$ and $(4, 2)$ (b) $(0, 0)$ and $(2, -5)$ (c) $(-5, 2)$ and $(4, -2)$

Answers (a) 3/2 (b) $-5/2$ (c) $-4/9$

Solving Simultaneous Linear Equations

Consider the equation $3x + 5y = 15$, which has two unknowns, x and y. Such an equation does not have a unique solution. For example, note that $(x = 0, y = 3)$, $(x = 5, y = 0)$, and $(x = 2, y = 9/5)$ are all solutions to this equation.

If a problem has two unknowns, a unique solution is possible only if we have *two* equations. In general, if a problem has n unknowns, its solution requires n equations. In order to solve two simultaneous equations involving two unknowns, x and y, we solve one of the equations for x in terms of y and substitute this expression into the other equation.

Example 2

Solve the following two simultaneous equations:

$$(1) \qquad 5x + y = -8$$
$$(2) \qquad 2x - 2y = 4$$

Solution From Equation (2), $x = y + 2$. Substitution of this into Equation (1) gives

$$5(y + 2) + y = -8$$
$$6y = -18$$
$$y = -3$$
$$x = y + 2 = -1$$

Alternate Solution Multiply each term in Equation (1) by the factor 2 and add the result to Equation (2):

$$10x + 2y = -16$$
$$2x - 2y = 4$$
$$12x = -12$$
$$x = -1$$
$$y = x - 2 = -3$$

Two linear equations containing two unknowns can also be solved by a graphical method. If the straight lines corresponding to the two equations are plotted in a conventional coordinate system, the intersection of the two lines represents the solution. For example, consider the two equations

$$x - y = 2$$
$$x - 2y = -1$$

These are plotted in Figure B.3. The intersection of the two lines has the coordinates $x = 5$, $y = 3$. This represents the solution to the equations. You should check this solution by the analytical technique discussed above.

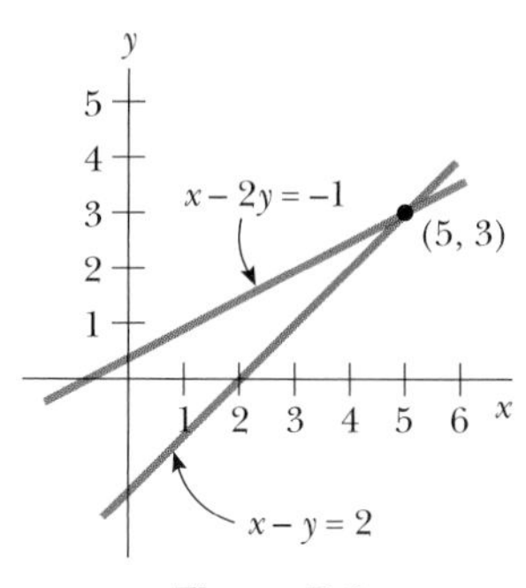

Figure B.3

Exercises

Solve the following pairs of simultaneous equations involving two unknowns:

	Answers
1. $x + y = 8$ $x - y = 2$	$x = 5, y = 3$
2. $98 - T = 10a$ $T - 49 = 5a$	$T = 65, a = 3.3$
3. $6x + 2y = 6$ $8x - 4y = 28$	$x = 2, y = -3$

Logarithms

Suppose that a quantity x is expressed as a power of some quantity a:

$$x = a^y \tag{B.11}$$

The number a is called the **base** number. The **logarithm** of x with respect to the base a is equal to the exponent to which the base must be raised in order to satisfy the expression $x = a^y$:

$$y = \log_a x \tag{B.12}$$

Conversely, the **antilogarithm** of y is the number x:

$$x = \text{antilog}_a\, y \tag{B.13}$$

In practice, the two bases most often used are base 10, called the *common* logarithm base, and base $e = 2.718\,282$, called Euler's constant or the *natural* logarithm base. When common logarithms are used,

$$y = \log_{10} x \qquad (\text{or } x = 10^y) \tag{B.14}$$

When natural logarithms are used,

$$y = \ln x \qquad (\text{or } x = e^y) \tag{B.15}$$

For example, $\log_{10} 52 = 1.716$, so that $\text{antilog}_{10}\, 1.716 = 10^{1.716} = 52$. Likewise, $\ln 52 = 3.951$, so $\text{antiln}\, 3.951 = e^{3.951} = 52$.

In general, note that you can convert between base 10 and base e with the equality

$$\ln x = (2.302\,585) \log_{10} x \tag{B.16}$$

Finally, some useful properties of logarithms are

$$\left.\begin{aligned} \log(ab) &= \log a + \log b \\ \log(a/b) &= \log a - \log b \\ \log(a^n) &= n \log a \end{aligned}\right\} \text{any base}$$

$$\ln e = 1$$

$$\ln e^a = a$$

$$\ln\left(\frac{1}{a}\right) = -\ln a$$

B.3 Geometry

The **distance** d between two points having coordinates (x_1, y_1) and (x_2, y_2) is

$$d = \sqrt{(x_2 - x_1)^2 + (y_2 - y_1)^2} \tag{B.17}$$

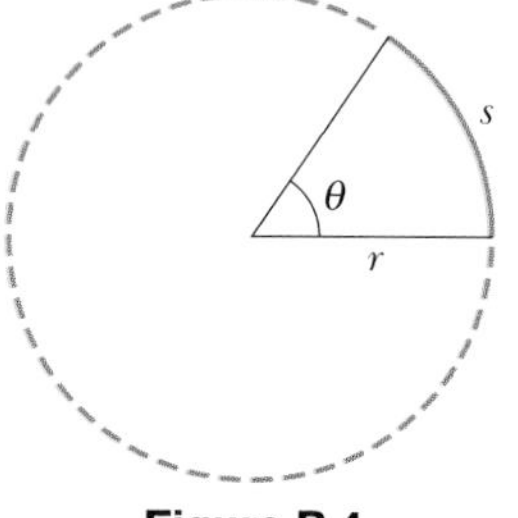

Figure B.4

Radian measure: The arc length s of a circular arc (Fig. B.4) is proportional to the radius r for a fixed value of θ (in radians):

$$s = r\theta$$
$$\theta = \frac{s}{r} \tag{B.18}$$

Table B.2 gives the areas and volumes for several geometric shapes used throughout this text:

Table B.2

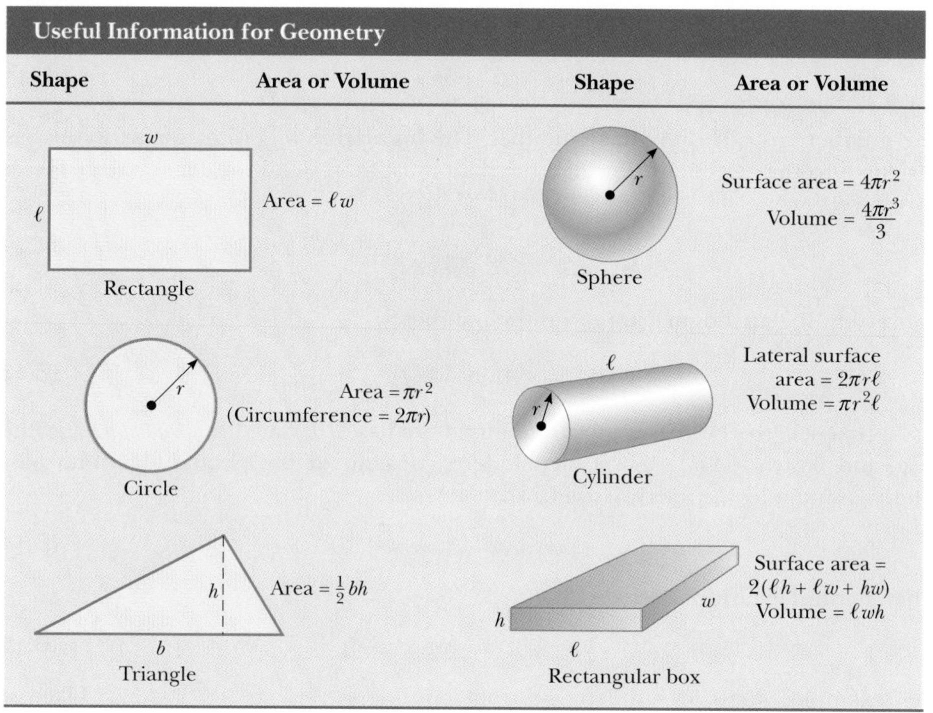

Useful Information for Geometry

Shape	Area or Volume	Shape	Area or Volume
Rectangle (ℓ, w)	Area = ℓw	Sphere (r)	Surface area = $4\pi r^2$ Volume = $\frac{4\pi r^3}{3}$
Circle (r)	Area = πr^2 (Circumference = $2\pi r$)	Cylinder (r, ℓ)	Lateral surface area = $2\pi r\ell$ Volume = $\pi r^2\ell$
Triangle (b, h)	Area = $\frac{1}{2}bh$	Rectangular box (ℓ, w, h)	Surface area = $2(\ell h + \ell w + hw)$ Volume = ℓwh

The equation of a **straight line** (Fig. B.5) is

$$y = mx + b \tag{B.19}$$

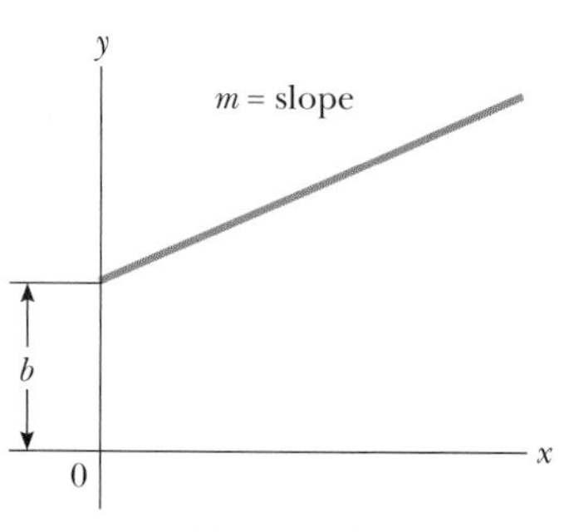

Figure B.5

where b is the y intercept and m is the slope of the line.

The equation of a **circle** of radius R centered at the origin is

$$x^2 + y^2 = R^2 \tag{B.20}$$

The equation of an **ellipse** having the origin at its center (Fig. B.6) is

$$\frac{x^2}{a^2} + \frac{y^2}{b^2} = 1 \tag{B.21}$$

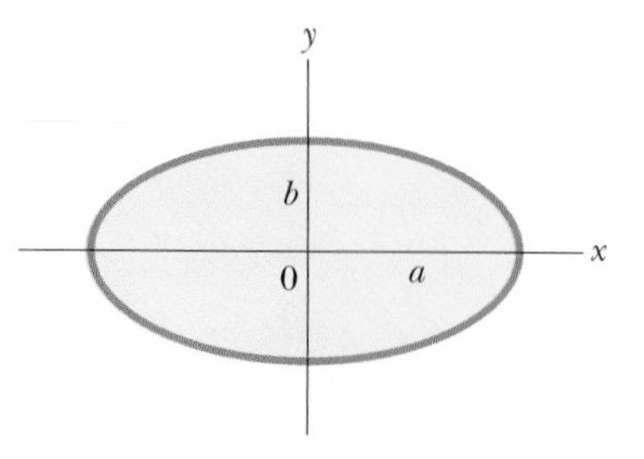

Figure B.6

where a is the length of the semimajor axis (the longer one) and b is the length of the semiminor axis (the shorter one).

The equation of a **parabola** the vertex of which is at $y = b$ (Fig. B.7) is

$$y = ax^2 + b \tag{B.22}$$

The equation of a **rectangular hyperbola** (Fig. B.8) is

$$xy = \text{constant} \tag{B.23}$$

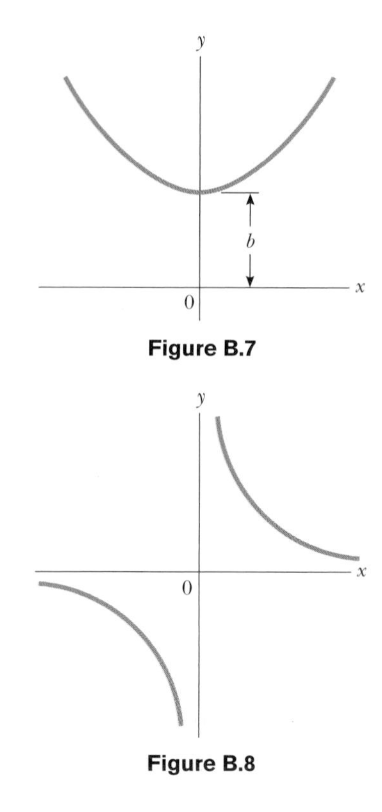

Figure B.7

Figure B.8

B.4 Trigonometry

That portion of mathematics based on the special properties of the right triangle is called trigonometry. By definition, a right triangle is one containing a 90° angle. Consider the right triangle shown in Figure B.9, where side a is opposite the angle θ, side b is adjacent to the angle θ, and side c is the hypotenuse of the triangle. The three basic trigonometric functions defined by such a triangle are the sine (sin), cosine (cos), and tangent (tan) functions. In terms of the angle θ, these functions are defined by

$$\sin\theta \equiv \frac{\text{side opposite } \theta}{\text{hypotenuse}} = \frac{a}{c} \tag{B.24}$$

$$\cos\theta \equiv \frac{\text{side adjacent to } \theta}{\text{hypotenuse}} = \frac{b}{c} \tag{B.25}$$

$$\tan\theta \equiv \frac{\text{side opposite } \theta}{\text{side adjacent to } \theta} = \frac{a}{b} \tag{B.26}$$

The Pythagorean theorem provides the following relationship among the sides of a right triangle:

$$c^2 = a^2 + b^2 \tag{B.27}$$

From the above definitions and the Pythagorean theorem, it follows that

$$\sin^2\theta + \cos^2\theta = 1$$

$$\tan\theta = \frac{\sin\theta}{\cos\theta}$$

The cosecant, secant, and cotangent functions are defined by

$$\csc\theta \equiv \frac{1}{\sin\theta} \qquad \sec\theta \equiv \frac{1}{\cos\theta} \qquad \cot\theta \equiv \frac{1}{\tan\theta}$$

The relationships below follow directly from the right triangle shown in Figure B.9:

$$\sin\theta = \cos(90° - \theta)$$

$$\cos\theta = \sin(90° - \theta)$$

$$\cot\theta = \tan(90° - \theta)$$

Some properties of trigonometric functions are

$$\sin(-\theta) = -\sin\theta$$

$$\cos(-\theta) = \cos\theta$$

$$\tan(-\theta) = -\tan\theta$$

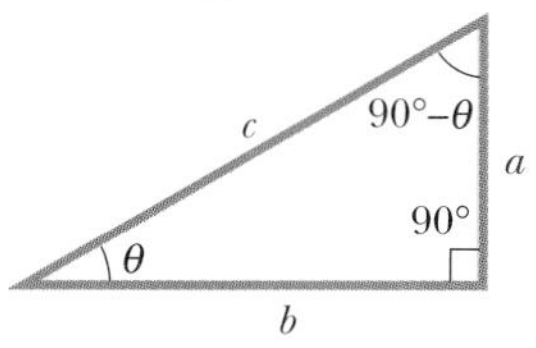

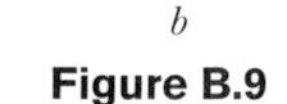

Figure B.9

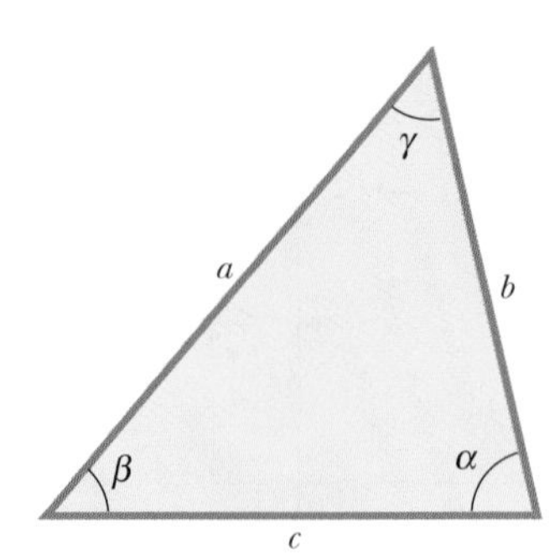

Figure B.10

The following relationships apply to *any* triangle, as shown in Figure B.10:

$$\alpha + \beta + \gamma = 180°$$

Law of cosines

$$a^2 = b^2 + c^2 - 2bc \cos \alpha$$
$$b^2 = a^2 + c^2 - 2ac \cos \beta$$
$$c^2 = a^2 + b^2 - 2ab \cos \gamma$$

Law of sines

$$\frac{a}{\sin \alpha} = \frac{b}{\sin \beta} = \frac{c}{\sin \gamma}$$

Table B.3 lists a number of useful trigonometric identities.

Table B.3

Some Trigonometric Identities	
$\sin^2 \theta + \cos^2 \theta = 1$	$\csc^2 \theta = 1 + \cot^2 \theta$
$\sec^2 \theta = 1 + \tan^2 \theta$	$\sin^2 \frac{\theta}{2} = \frac{1}{2}(1 - \cos \theta)$
$\sin 2\theta = 2 \sin \theta \cos \theta$	$\cos^2 \frac{\theta}{2} = \frac{1}{2}(1 + \cos \theta)$
$\cos 2\theta = \cos^2 \theta - \sin^2 \theta$	$1 - \cos \theta = 2 \sin^2 \frac{\theta}{2}$
$\tan 2\theta = \frac{2 \tan \theta}{1 - \tan^2 \theta}$	$\tan \frac{\theta}{2} = \sqrt{\frac{1 - \cos \theta}{1 + \cos \theta}}$
$\sin(A \pm B) = \sin A \cos B \pm \cos A \sin B$	
$\cos(A \pm B) = \cos A \cos B \mp \sin A \sin B$	
$\sin A \pm \sin B = 2 \sin[\frac{1}{2}(A \pm B)]\cos[\frac{1}{2}(A \mp B)]$	
$\cos A + \cos B = 2 \cos[\frac{1}{2}(A + B)]\cos[\frac{1}{2}(A - B)]$	
$\cos A - \cos B = 2 \sin[\frac{1}{2}(A + B)]\sin[\frac{1}{2}(B - A)]$	

Example 3

Consider the right triangle in Figure B.11, in which $a = 2$, $b = 5$, and c is unknown. From the Pythagorean theorem, we have

$$c^2 = a^2 + b^2 = 2^2 + 5^2 = 4 + 25 = 29$$

$$c = \sqrt{29} = 5.39$$

To find the angle θ, note that

$$\tan \theta = \frac{a}{b} = \frac{2}{5} = 0.400$$

From a table of functions or from a calculator, we have

$$\theta = \tan^{-1}(0.400) = 21.8°$$

where $\tan^{-1}$ (0.400) is the notation for "angle whose tangent is 0.400," sometimes written as arctan (0.400).

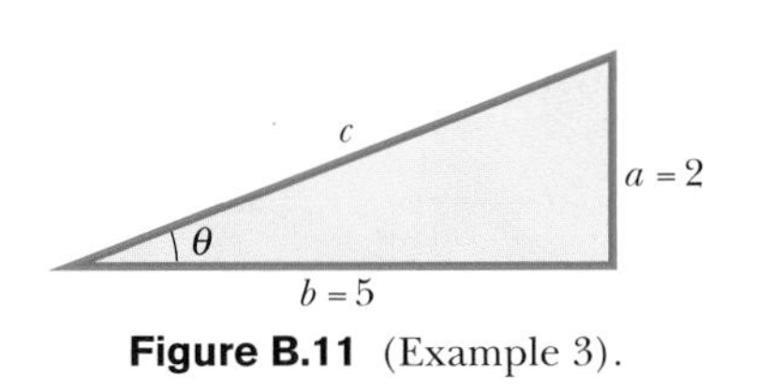

Figure B.11 (Example 3).

Exercises

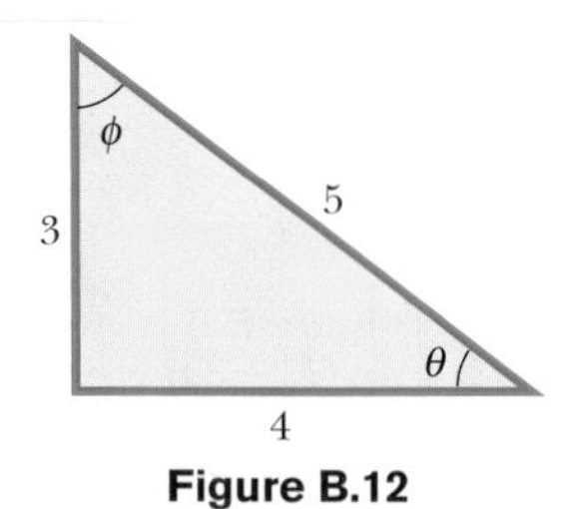

Figure B.12

1. In Figure B.12, identify (a) the side opposite θ (b) the side adjacent to ϕ. Then find (c) $\cos \theta$ (d) $\sin \phi$ (e) $\tan \phi$.

Answers (a) 3 (b) 3 (c) $\frac{4}{5}$ (d) $\frac{4}{5}$ (e) $\frac{4}{3}$

2. In a certain right triangle, the two sides that are perpendicular to each other are 5 m and 7 m long. What is the length of the third side?

Answer 8.60 m

3. A right triangle has a hypotenuse of length 3 m, and one of its angles is 30°. What is the length of (a) the side opposite the 30° angle (b) the side adjacent to the 30° angle?

Answers (a) 1.5 m (b) 2.60 m

B.5 Series Expansions

$$(a+b)^n = a^n + \frac{n}{1!}a^{n-1}b + \frac{n(n-1)}{2!}a^{n-2}b^2 + \cdots$$

$$(1+x)^n = 1 + nx + \frac{n(n-1)}{2!}x^2 + \cdots$$

$$e^x = 1 + x + \frac{x^2}{2!} + \frac{x^3}{3!} + \cdots$$

$$\ln(1 \pm x) = \pm x - \tfrac{1}{2}x^2 \pm \tfrac{1}{3}x^3 - \cdots$$

$$\left.\begin{aligned} \sin x &= x - \frac{x^3}{3!} + \frac{x^5}{5!} - \cdots \\ \cos x &= 1 - \frac{x^2}{2!} + \frac{x^4}{4!} - \cdots \\ \tan x &= x + \frac{x^3}{3} + \frac{2x^5}{15} + \cdots \qquad |x| < \pi/2 \end{aligned}\right\} x \text{ in radians}$$

For $x \ll 1$, the following approximations can be used[1]:

$$(1+x)^n \approx 1 + nx \qquad \sin x \approx x$$

$$e^x \approx 1 + x \qquad \cos x \approx 1$$

$$\ln(1 \pm x) \approx \pm x \qquad \tan x \approx x$$

B.6 Differential Calculus

In various branches of science, it is sometimes necessary to use the basic tools of calculus, invented by Newton, to describe physical phenomena. The use of calculus is fundamental in the treatment of various problems in Newtonian mechanics, electricity, and magnetism. In this section, we simply state some basic properties and "rules of thumb" that should be a useful review to the student.

First, a **function** must be specified that relates one variable to another (such as a coordinate as a function of time). Suppose one of the variables is called y (the dependent variable), the other x (the independent variable). We might have a function relationship such as

$$y(x) = ax^3 + bx^2 + cx + d$$

If a, b, c, and d are specified constants, then y can be calculated for any value of x. We usually deal with continuous functions, that is, those for which y varies "smoothly" with x.

The **derivative** of y with respect to x is defined as the limit, as Δx approaches zero, of the slopes of chords drawn between two points on the y versus x curve. Mathematically, we write this definition as

$$\frac{dy}{dx} = \lim_{\Delta x \to 0} \frac{\Delta y}{\Delta x} = \lim_{\Delta x \to 0} \frac{y(x+\Delta x) - y(x)}{\Delta x} \qquad \text{(B.28)}$$

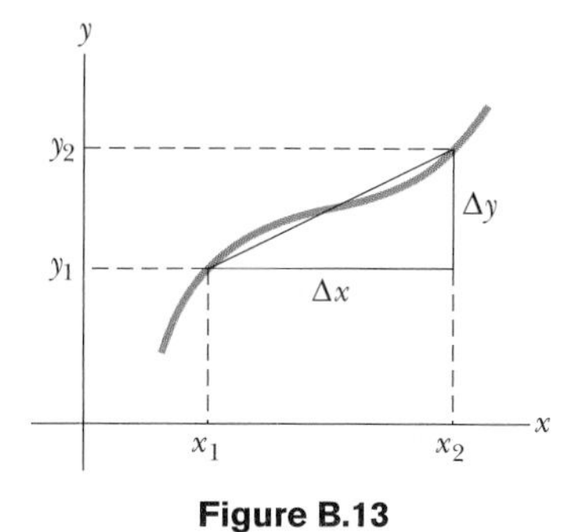

Figure B.13

where Δy and Δx are defined as $\Delta x = x_2 - x_1$ and $\Delta y = y_2 - y_1$ (Fig. B.13). It is important to note that dy/dx *does not* mean dy divided by dx, but is simply a notation of the limiting process of the derivative as defined by Equation B.28.

[1] The approximations for the functions sin x, cos x, and tan x are for $x \leq 0.1$ rad.

A useful expression to remember when $y(x) = ax^n$, where a is a *constant* and n is *any* positive or negative number (integer or fraction), is

$$\frac{dy}{dx} = nax^{n-1} \tag{B.29}$$

If $y(x)$ is a polynomial or algebraic function of x, we apply Equation B.29 to *each* term in the polynomial and take $d[\text{constant}]/dx = 0$. In Examples 4 through 7, we evaluate the derivatives of several functions.

Special Properties of the Derivative

A. Derivative of the product of two functions If a function $f(x)$ is given by the product of two functions, say, $g(x)$ and $h(x)$, then the derivative of $f(x)$ is defined as

$$\frac{d}{dx} f(x) = \frac{d}{dx}[g(x)h(x)] = g\frac{dh}{dx} + h\frac{dg}{dx} \tag{B.30}$$

B. Derivative of the sum of two functions If a function $f(x)$ is equal to the sum of two functions, then the derivative of the sum is equal to the sum of the derivatives:

$$\frac{d}{dx} f(x) = \frac{d}{dx}[g(x) + h(x)] = \frac{dg}{dx} + \frac{dh}{dx} \tag{B.31}$$

C. Chain rule of differential calculus If $y = f(x)$ and $x = g(z)$, then dy/dz can be written as the product of two derivatives:

$$\frac{dy}{dz} = \frac{dy}{dx}\frac{dx}{dz} \tag{B.32}$$

D. The second derivative The second derivative of y with respect to x is defined as the derivative of the function dy/dx (the derivative of the derivative). It is usually written

$$\frac{d^2y}{dx^2} = \frac{d}{dx}\left(\frac{dy}{dx}\right) \tag{B.33}$$

Example 4

Suppose $y(x)$ (that is, y as a function of x) is given by

$$y(x) = ax^3 + bx + c$$

where a and b are constants. Then it follows that

$$y(x + \Delta x) = a(x + \Delta x)^3 + b(x + \Delta x) + c$$
$$y(x + \Delta x) = a(x^3 + 3x^2\Delta x + 3x\Delta x^2 + \Delta x^3) + b(x + \Delta x) + c$$

so

$$\Delta y = y(x + \Delta x) - y(x) = a(3x^2\Delta x + 3x\Delta x^2 + \Delta x^3) + b\Delta x$$

Substituting this into Equation B.28 gives

$$\frac{dy}{dx} = \lim_{\Delta x \to 0} \frac{\Delta y}{\Delta x} = \lim_{\Delta x \to 0} [3ax^2 + 3x\Delta x + \Delta x^2] + b$$

$$\frac{dy}{dx} = 3ax^2 + b$$

Example 5

Find the derivative of

$$y(x) = 8x^5 + 4x^3 + 2x + 7$$

Solution Applying Equation B.29 to each term independently, and remembering that d/dx (constant) $= 0$, we have

$$\frac{dy}{dx} = 8(5)x^4 + 4(3)x^2 + 2(1)x^0 + 0$$

$$\frac{dy}{dx} = 40x^4 + 12x^2 + 2$$

Example 6

Find the derivative of $y(x) = x^3/(x+1)^2$ with respect to x.

Solution We can rewrite this function as $y(x) = x^3(x+1)^{-2}$ and apply Equation B.30:

$$\frac{dy}{dx} = (x+1)^{-2}\frac{d}{dx}(x^3) + x^3\frac{d}{dx}(x+1)^{-2}$$

$$= (x+1)^{-2}3x^2 + x^3(-2)(x+1)^{-3}$$

$$\frac{dy}{dx} = \frac{3x^2}{(x+1)^2} - \frac{2x^3}{(x+1)^3}$$

Example 7

A useful formula that follows from Equation B.30 is the derivative of the quotient of two functions. Show that

$$\frac{d}{dx}\left[\frac{g(x)}{h(x)}\right] = \frac{h\frac{dg}{dx} - g\frac{dh}{dx}}{h^2}$$

Solution We can write the quotient as gh^{-1} and then apply Equations B.29 and B.30:

$$\frac{d}{dx}\left(\frac{g}{h}\right) = \frac{d}{dx}(gh^{-1}) = g\frac{d}{dx}(h^{-1}) + h^{-1}\frac{d}{dx}(g)$$

$$= -gh^{-2}\frac{dh}{dx} + h^{-1}\frac{dg}{dx}$$

$$= \frac{h\frac{dg}{dx} - g\frac{dh}{dx}}{h^2}$$

Some of the more commonly used derivatives of functions are listed in Table B.4.

Table B.4

Derivative for Several Functions
$\frac{d}{dx}(a) = 0$
$\frac{d}{dx}(ax^n) = nax^{n-1}$
$\frac{d}{dx}(e^{ax}) = ae^{ax}$
$\frac{d}{dx}(\sin ax) = a\cos ax$
$\frac{d}{dx}(\cos ax) = -a\sin ax$
$\frac{d}{dx}(\tan ax) = a\sec^2 ax$
$\frac{d}{dx}(\cot ax) = -a\csc^2 dx$
$\frac{d}{dx}(\sec x) = \tan x\sec x$
$\frac{d}{dx}(\csc x) = -\cot x\csc x$
$\frac{d}{dx}(\ln ax) = \frac{1}{x}$

Note: The symbols a and n represent constants.

B.7 Integral Calculus

We think of integration as the inverse of differentiation. As an example, consider the expression

$$f(x) = \frac{dy}{dx} = 3ax^2 + b \tag{B.34}$$

which was the result of differentiating the function

$$y(x) = ax^3 + bx + c$$

in Example 4. We can write Equation B.34 as $dy = f(x)\,dx = (3ax^2 + b)\,dx$ and obtain $y(x)$ by "summing" over all values of x. Mathematically, we write this inverse operation

$$y(x) = \int f(x)\,dx$$

For the function $f(x)$ given by Equation B.34, we have

$$y(x) = \int (3ax^2 + b)\,dx = ax^3 + bx + c$$

where c is a constant of the integration. This type of integral is called an *indefinite integral* because its value depends on the choice of c.

A general **indefinite integral** $I(x)$ is defined as

$$I(x) = \int f(x)\,dx \tag{B.35}$$

where $f(x)$ is called the *integrand* and $f(x) = dI(x)/dx$.

For a *general continuous* function $f(x)$, the integral can be described as the area under the curve bounded by $f(x)$ and the x axis, between two specified values of x, say, x_1 and x_2, as in Figure B.14.

The area of the blue element is approximately $f(x_i)\,\Delta x_i$. If we sum all these area elements from x_1 and x_2 and take the limit of this sum as $\Delta x_i \rightarrow 0$, we obtain the *true*

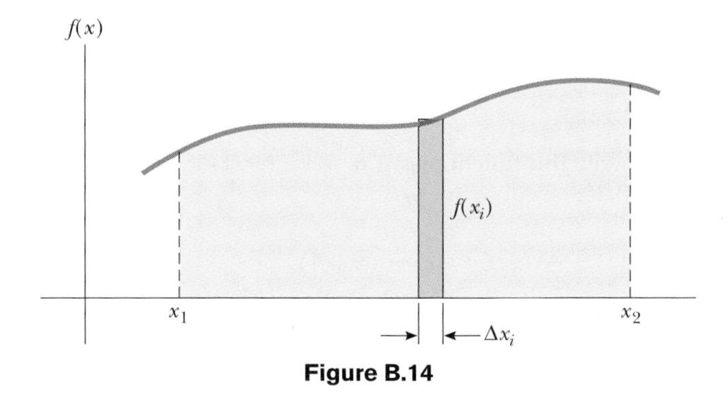

Figure B.14

area under the curve bounded by $f(x)$ and x, between the limits x_1 and x_2:

$$\text{Area} = \lim_{\Delta x \to 0} \sum_i f(x_i)\, \Delta x_i = \int_{x_1}^{x_2} f(x)\, dx \tag{B.36}$$

Integrals of the type defined by Equation B.36 are called **definite integrals.**

One common integral that arises in practical situations has the form

$$\int x^n\, dx = \frac{x^{n+1}}{n+1} + c \qquad (n \neq -1) \tag{B.37}$$

This result is obvious, being that differentiation of the right-hand side with respect to x gives $f(x) = x^n$ directly. If the limits of the integration are known, this integral becomes a *definite integral* and is written

$$\int_{x_1}^{x_2} x^n\, dx = \frac{x^{n+1}}{n+1}\bigg|_{x_1}^{x_2} = \frac{x_2^{\,n+1} - x_1^{\,n+1}}{n+1} \qquad (n \neq -1) \tag{B.38}$$

Examples

1. $\displaystyle \int_0^a x^2\, dx = \frac{x^3}{3}\bigg]_0^a = \frac{a^3}{3}$

2. $\displaystyle \int_0^b x^{3/2}\, dx = \frac{x^{5/2}}{5/2}\bigg]_0^b = \tfrac{2}{5} b^{5/2}$

3. $\displaystyle \int_3^5 x\, dx = \frac{x^2}{2}\bigg]_3^5 = \frac{5^2 - 3^2}{2} = 8$

Partial Integration

Sometimes it is useful to apply the method of *partial integration* (also called "integrating by parts") to evaluate certain integrals. The method uses the property that

$$\int u\, dv = uv - \int v\, du \tag{B.39}$$

where u and v are *carefully* chosen so as to reduce a complex integral to a simpler one. In many cases, several reductions have to be made. Consider the function

$$I(x) = \int x^2 e^x\, dx$$

This can be evaluated by integrating by parts twice. First, if we choose $u = x^2$, $v = e^x$, we obtain

$$\int x^2 e^x\, dx = \int x^2\, d(e^x) = x^2 e^x - 2\int e^x x\, dx + c_1$$

Now, in the second term, choose $u = x$, $v = e^x$, which gives

$$\int x^2 e^x\, dx = x^2 e^x - 2xe^x + 2\int e^x\, dx + c_1$$

or

$$\int x^2 e^x dx = x^2 e^x - 2xe^x + 2e^x + c_2$$

The Perfect Differential

Another useful method to remember is the use of the *perfect differential,* in which we look for a change of variable such that the differential of the function is the differential of the independent variable appearing in the integrand. For example, consider the integral

$$I(x) = \int \cos^2 x \sin x\, dx$$

This becomes easy to evaluate if we rewrite the differential as $d(\cos x) = -\sin x\, dx$. The integral then becomes

$$\int \cos^2 x \sin x\, dx = -\int \cos^2 x\, d(\cos x)$$

If we now change variables, letting $y = \cos x$, we obtain

$$\int \cos^2 x \sin x\, dx = -\int y^2 dy = -\frac{y^3}{3} + c = -\frac{\cos^3 x}{3} + c$$

Table B.5 lists some useful indefinite integrals. Table B.6 gives Gauss's probability integral and other definite integrals. A more complete list can be found in various handbooks, such as *The Handbook of Chemistry and Physics,* CRC Press.

Table B.5

Some Indefinite Integrals (An arbitrary constant should be added to each of these integrals.)

$$\int x^n\, dx = \frac{x^{n+1}}{n+1} \quad \text{(provided } n \neq -1)$$

$$\int \frac{dx}{x} = \int x^{-1} dx = \ln x$$

$$\int \frac{dx}{a+bx} = \frac{1}{b}\ln(a+bx)$$

$$\int \frac{xdx}{a+bx} = \frac{x}{b} - \frac{a}{b^2}\ln(a+bx)$$

$$\int \frac{dx}{x(x+a)} = -\frac{1}{a}\ln\frac{x+a}{x}$$

$$\int \frac{dx}{(a+bx)^2} = -\frac{1}{b(a+bx)}$$

$$\int \frac{dx}{a^2+x^2} = \frac{1}{a}\tan^{-1}\frac{x}{a}$$

$$\int \frac{dx}{a^2-x^2} = \frac{1}{2a}\ln\frac{a+x}{a-x} \quad (a^2 - x^2 > 0)$$

$$\int \frac{dx}{x^2-a^2} = \frac{1}{2a}\ln\frac{x-a}{x+a} \quad (x^2 - a^2 > 0)$$

$$\int \frac{x\, dx}{a^2 \pm x^2} = \pm\tfrac{1}{2}\ln(a^2 \pm x^2)$$

$$\int \frac{dx}{\sqrt{a^2-x^2}} = \sin^{-1}\frac{x}{a} = -\cos^{-1}\frac{x}{a} \quad (a^2 - x^2 > 0)$$

$$\int \frac{dx}{\sqrt{x^2 \pm a^2}} = \ln(x + \sqrt{x^2 \pm a^2})$$

$$\int \frac{x\, dx}{\sqrt{a^2-x^2}} = -\sqrt{a^2-x^2}$$

$$\int \frac{x\, dx}{\sqrt{x^2 \pm a^2}} = \sqrt{x^2 \pm a^2}$$

$$\int \sqrt{a^2-x^2}\, dx = \tfrac{1}{2}\left(x\sqrt{a^2-x^2} + a^2\sin^{-1}\frac{x}{a}\right)$$

$$\int x\sqrt{a^2-x^2}\, dx = -\tfrac{1}{3}(a^2-x^2)^{3/2}$$

$$\int \sqrt{x^2 \pm a^2}\, dx = \tfrac{1}{2}[x\sqrt{x^2 \pm a^2} \pm a^2\ln(x + \sqrt{x^2 \pm a^2})]$$

$$\int x(\sqrt{x^2 \pm a^2})\, dx = \tfrac{1}{3}(x^2 \pm a^2)^{3/2}$$

$$\int e^{ax}\, dx = \frac{1}{a}e^{ax}$$

$$\int \ln ax\, dx = (x\ln ax) - x$$

continued

Table B.5

Some Indefinite Integrals (An arbitrary constant should be added to each of these integrals.) *continued*

$$\int xe^{ax}\,dx = \frac{e^{ax}}{a^2}(ax - 1)$$

$$\int \frac{dx}{a + be^{cx}} = \frac{x}{a} - \frac{1}{ac}\ln(a + be^{cx})$$

$$\int \sin ax\,dx = -\frac{1}{a}\cos ax$$

$$\int \cos ax\,dx = \frac{1}{a}\sin ax$$

$$\int \tan ax\,dx = -\frac{1}{a}\ln(\cos ax) = \frac{1}{a}\ln(\sec ax)$$

$$\int \cot ax\,dx = \frac{1}{a}\ln(\sin ax)$$

$$\int \sec ax\,dx = \frac{1}{a}\ln(\sec ax + \tan ax) = \frac{1}{a}\ln\left[\tan\left(\frac{ax}{2} + \frac{\pi}{4}\right)\right]$$

$$\int \csc ax\,dx = \frac{1}{a}\ln(\csc ax - \cot ax) = \frac{1}{a}\ln\left(\tan\frac{ax}{2}\right)$$

$$\int \sin^2 ax\,dx = \frac{x}{2} - \frac{\sin 2ax}{4a}$$

$$\int \cos^2 ax\,dx = \frac{x}{2} + \frac{\sin 2ax}{4a}$$

$$\int \frac{dx}{\sin^2 ax} = -\frac{1}{a}\cot ax$$

$$\int \frac{dx}{\cos^2 ax} = \frac{1}{a}\tan ax$$

$$\int \tan^2 ax\,dx = \frac{1}{a}(\tan ax) - x$$

$$\int \cot^2 ax\,dx = -\frac{1}{a}(\cot ax) - x$$

$$\int \sin^{-1} ax\,dx = x(\sin^{-1} ax) + \frac{\sqrt{1 - a^2x^2}}{a}$$

$$\int \cos^{-1} ax\,dx = x(\cos^{-1} ax) - \frac{\sqrt{1 - a^2x^2}}{a}$$

$$\int \frac{dx}{(x^2 + a^2)^{3/2}} = \frac{x}{a^2\sqrt{x^2 + a^2}}$$

$$\int \frac{x\,dx}{(x^2 + a^2)^{3/2}} = -\frac{1}{\sqrt{x^2 + a^2}}$$

Table B.6

Gauss's Probability Integral and Other Definite Integrals

$$\int_0^\infty x^n e^{-ax}\,dx = \frac{n!}{a^{n+1}}$$

$$I_0 = \int_0^\infty e^{-ax^2}\,dx = \frac{1}{2}\sqrt{\frac{\pi}{a}} \quad \text{(Gauss's probability integral)}$$

$$I_1 = \int_0^\infty xe^{-ax^2}\,dx = \frac{1}{2a}$$

$$I_2 = \int_0^\infty x^2e^{-ax^2}\,dx = -\frac{dI_0}{da} = \frac{1}{4}\sqrt{\frac{\pi}{a^3}}$$

$$I_3 = \int_0^\infty x^3e^{-ax^2}\,dx = -\frac{dI_1}{da} = \frac{1}{2a^2}$$

$$I_4 = \int_0^\infty x^4e^{-ax^2}\,dx = \frac{d^2I_0}{da^2} = \frac{3}{8}\sqrt{\frac{\pi}{a^5}}$$

$$I_5 = \int_0^\infty x^5e^{-ax^2}\,dx = \frac{d^2I_1}{da^2} = \frac{1}{a^3}$$

$$\vdots$$

$$I_{2n} = (-1)^n\frac{d^n}{da^n}I_0$$

$$I_{2n+1} = (-1)^n\frac{d^n}{da^n}I_1$$

B.8 Propagation of Uncertainty

In laboratory experiments, a common activity is to take measurements that act as raw data. These measurements are of several types—length, time interval, temperature, voltage, etc.—and are taken by a variety of instruments. Regardless of the measurement and the quality of the instrumentation, **there is always uncertainty associated with a physical measurement.** This uncertainty is a combination of that associated with the instrument and that related to the system being measured. An example of the former is the inability to exactly determine the position of a length measurement between the lines on a meter stick. An example of uncertainty related to the system being measured is the variation of temperature within a sample of water so that a single temperature for the sample is difficult to determine.

Uncertainties can be expressed in two ways. **Absolute uncertainty** refers to an uncertainty expressed in the same units as the measurement. Thus, a length might be expressed as (5.5 ± 0.1) cm, as was the length of the computer disk label in Section 1.7. The uncertainty of ± 0.1 cm by itself is not descriptive enough for some purposes, however. This is a large uncertainty if the measurement is 1.0 cm, but it is a small uncertainty if the measurement is 100 m. To give a more descriptive account of the uncertainty, **fractional uncertainty** or **percent uncertainty** is used. In this type of description, the uncertainty is divided by the actual measurement. Thus, the length of the computer disk label could be expressed as

$$\ell = 5.5 \text{ cm} \pm \frac{0.1 \text{ cm}}{5.5 \text{ cm}} = 5.5 \text{ cm} \pm 0.018 \qquad \text{(fractional uncertainty)}$$

or as

$$\ell = 5.5 \text{ cm} \pm 1.8\% \qquad \text{(percent uncertainty)}$$

When combining measurements in a calculation, the uncertainty in the final result is larger than the uncertainty in the individual measurements. This is called **propagation of uncertainty** and is one of the challenges of experimental physics. As a calculation becomes more complicated, there is increased propagation of uncertainty and the uncertainty in the value of the final result can grow to be quite large.

There are simple rules that can provide a reasonable estimate of the uncertainty in a calculated result:

Multiplication and division: When measurements with uncertainties are multiplied or divided, add the *percent uncertainties* to obtain the percent uncertainty in the result.

Example: The Area of a Rectangular Plate

$$\begin{aligned} A = \ell w &= (5.5 \text{ cm} \pm 1.8\%) \times (6.4 \text{ cm} \pm 1.6\%) = 35 \text{ cm}^2 \pm 3.4\% \\ &= (35 \pm 1) \text{ cm}^2 \end{aligned}$$

Addition and subtraction: When measurements with uncertainties are added or subtracted, add the *absolute uncertainties* to obtain the absolute uncertainty in the result.

Example: A Change in Temperature

$$\begin{aligned} \Delta T = T_2 - T_1 &= (99.2 \pm 1.5)^\circ\text{C} - (27.6 \pm 1.5)^\circ\text{C} = (71.6 \pm 3.0)^\circ\text{C} \\ &= 71.6^\circ\text{C} \pm 4.2\% \end{aligned}$$

Powers: If a measurement is taken to a power, the percent uncertainty is multiplied by that power to obtain the percent uncertainty in the result.

Example: The Volume of a Sphere

$$\begin{aligned} V = \tfrac{4}{3}\pi r^3 &= \tfrac{4}{3}\pi(6.20 \text{ cm} \pm 2.0\%)^3 = 998 \text{ cm}^3 \pm 6.0\% \\ &= (998 \pm 60) \text{ cm}^3 \end{aligned}$$

Notice that uncertainties in a calculation always add. As a result, an experiment involving a subtraction should be avoided if possible. This is especially true if the measurements being subtracted are close together. The result of such a calculation is a small difference in the measurements and uncertainties that add together. It is possible that the uncertainty in the result could be larger than the result itself!

Appendix C • Periodic Table of the Elements

Symbol — **Ca** 20 — Atomic number
Atomic mass † — 40.078
$4s^2$ — Electron configuration

Group I	Group II	Transition elements						
H 1 1.007 9 $1s$								
Li 3 6.941 $2s^1$	**Be** 4 9.0122 $2s^2$							
Na 11 22.990 $3s^1$	**Mg** 12 24.305 $3s^2$							
K 19 39.098 $4s^1$	**Ca** 20 40.078 $4s^2$	**Sc** 21 44.956 $3d^14s^2$	**Ti** 22 47.867 $3d^24s^2$	**V** 23 50.942 $3d^34s^2$	**Cr** 24 51.996 $3d^54s^1$	**Mn** 25 54.938 $3d^54s^2$	**Fe** 26 55.845 $3d^64s^2$	**Co** 27 58.933 $3d^74s^2$
Rb 37 85.468 $5s^1$	**Sr** 38 87.62 $5s^2$	**Y** 39 88.906 $4d^15s^2$	**Zr** 40 91.224 $4d^25s^2$	**Nb** 41 92.906 $4d^45s^1$	**Mo** 42 95.94 $4d^55s^1$	**Tc** 43 (98) $4d^55s^2$	**Ru** 44 101.07 $4d^75s^1$	**Rh** 45 102.91 $4d^85s^1$
Cs 55 132.91 $6s^1$	**Ba** 56 137.33 $6s^2$	57-71*	**Hf** 72 178.49 $5d^26s^2$	**Ta** 73 180.95 $5d^36s^2$	**W** 74 183.84 $5d^46s^2$	**Re** 75 186.21 $5d^56s^2$	**Os** 76 190.23 $5d^66s^2$	**Ir** 77 192.2 $5d^76s^2$
Fr 87 (223) $7s^1$	**Ra** 88 (226) $7s^2$	89-103**	**Rf** 104 (261) $6d^27s^2$	**Db** 105 (262) $6d^37s^2$	**Sg** 106 (266)	**Bh** 107 (264)	**Hs** 108 (269)	**Mt** 109 (268)

*Lanthanide series	**La** 57 138.91 $5d^16s^2$	**Ce** 58 140.12 $5d^14f^16s^2$	**Pr** 59 140.91 $4f^36s^2$	**Nd** 60 144.24 $4f^46s^2$	**Pm** 61 (145) $4f^56s^2$	**Sm** 62 150.36 $4f^66s^2$
**Actinide series	**Ac** 89 (227) $6d^17s^2$	**Th** 90 232.04 $6d^27s^2$	**Pa** 91 231.04 $5f^26d^17s^2$	**U** 92 238.03 $5f^36d^17s^2$	**Np** 93 (237) $5f^46d^17s^2$	**Pu** 94 (244) $5f^66d^07s^2$

□ Atomic mass values given are averaged over isotopes in the percentages in which they exist in nature.
† For an unstable element, mass number of the most stable known isotope is given in parentheses.
†† Elements 110, 111, 112, and 114 have not yet been named.
††† For a description of the atomic data, visit *physics.nist.gov/atomic*

			Group III	Group IV	Group V	Group VI	Group VII	Group 0
							H 1 1.007 9 $1s^1$	**He** 2 4.002 6 $1s^2$
			B 5 10.811 $2p^1$	**C** 6 12.011 $2p^2$	**N** 7 14.007 $2p^3$	**O** 8 15.999 $2p^4$	**F** 9 18.998 $2p^5$	**Ne** 10 20.180 $2p^6$
			Al 13 26.982 $3p^1$	**Si** 14 28.086 $3p^2$	**P** 15 30.974 $3p^3$	**S** 16 32.066 $3p^4$	**Cl** 17 35.453 $3p^5$	**Ar** 18 39.948 $3p^6$
Ni 28 58.693 $3d^8 4s^2$	**Cu** 29 63.546 $3d^{10} 4s^1$	**Zn** 30 65.39 $3d^{10} 4s^2$	**Ga** 31 69.723 $4p^1$	**Ge** 32 72.61 $4p^2$	**As** 33 74.922 $4p^3$	**Se** 34 78.96 $4p^4$	**Br** 35 79.904 $4p^5$	**Kr** 36 83.80 $4p^6$
Pd 46 106.42 $4d^{10}$	**Ag** 47 107.87 $4d^{10} 5s^1$	**Cd** 48 112.41 $4d^{10} 5s^2$	**In** 49 114.82 $5p^1$	**Sn** 50 118.71 $5p^2$	**Sb** 51 121.76 $5p^3$	**Te** 52 127.60 $5p^4$	**I** 53 126.90 $5p^5$	**Xe** 54 131.29 $5p^6$
Pt 78 195.08 $5d^9 6s^1$	**Au** 79 196.97 $5d^{10} 6s^1$	**Hg** 80 200.59 $5d^{10} 6s^2$	**Tl** 81 204.38 $6p^1$	**Pb** 82 207.2 $6p^2$	**Bi** 83 208.98 $6p^3$	**Po** 84 (209) $6p^4$	**At** 85 (210) $6p^5$	**Rn** 86 (222) $6p^6$
110†† (271)	111†† (272)	112†† (285)		114†† (289)				

Eu 63 151.96 $4f^7 6s^2$	**Gd** 64 157.25 $5d^1 4f^7 6s^2$	**Tb** 65 158.93 $5d^1 4f^8 6s^2$	**Dy** 66 162.50 $4f^{10} 6s^2$	**Ho** 67 164.93 $4f^{11} 6s^2$	**Er** 68 167.26 $4f^{12} 6s^2$	**Tm** 69 168.93 $4f^{13} 6s^2$	**Yb** 70 173.04 $4f^{14} 6s^2$	**Lu** 71 174.97 $5d^1 4f^{14} 6s^2$
Am 95 (243) $5f^7 6d^0 7s^2$	**Cm** 96 (247) $5f^7 6d^1 7s^2$	**Bk** 97 (247) $5f^8 6d^1 7s^2$	**Cf** 98 (251) $5f^{10} 6d^0 7s^2$	**Es** 99 (252) $5f^{11} 6d^0 7s^2$	**Fm** 100 (257) $5f^{12} 6d^0 7s^2$	**Md** 101 (258) $5f^{13} 6d^0 7s^2$	**No** 102 (259) $6d^0 7s^2$	**Lr** 103 (262) $6d^1 7s^2$

Appendix D • SI Units

Table D.1

SI Units

Base Quantity	SI Base Unit Name	Symbol
Length	Meter	m
Mass	Kilogram	kg
Time	Second	s
Electric current	Ampere	A
Temperature	Kelvin	K
Amount of substance	Mole	mol
Luminous intensity	Candela	cd

Table D.2

Some Derived SI Units

Quantity	Name	Symbol	Expression in Terms of Base Units	Expression in Terms of Other SI Units
Plane angle	radian	rad	m/m	
Frequency	hertz	Hz	s^{-1}	
Force	newton	N	$kg \cdot m/s^2$	J/m
Pressure	pascal	Pa	$kg/m \cdot s^2$	N/m^2
Energy; work	joule	J	$kg \cdot m^2/s^2$	$N \cdot m$
Power	watt	W	$kg \cdot m^2/s^3$	J/s
Electric charge	coulomb	C	$A \cdot s$	
Electric potential	volt	V	$kg \cdot m^2/A \cdot s^3$	W/A
Capacitance	farad	F	$A^2 \cdot s^4/kg \cdot m^2$	C/V
Electric resistance	ohm	Ω	$kg \cdot m^2/A^2 \cdot s^3$	V/A
Magnetic flux	weber	Wb	$kg \cdot m^2/A \cdot s^2$	$V \cdot s$
Magnetic field	tesla	T	$kg/A \cdot s^2$	
Inductance	henry	H	$kg \cdot m^2/A^2 \cdot s^2$	$T \cdot m^2/A$

Appendix E • Nobel Prizes

All Nobel Prizes in physics are listed (and marked with a P), as well as relevant Nobel Prizes in Chemistry (C). The key dates for some of the scientific work are supplied; they often antedate the prize considerably.

1901 (P) *Wilhelm Roentgen* for discovering x-rays (1895).
1902 (P) *Hendrik A. Lorentz* for predicting the Zeeman effect and *Pieter Zeeman* for discovering the Zeeman effect, the splitting of spectral lines in magnetic fields.
1903 (P) *Antoine-Henri Becquerel* for discovering radioactivity (1896) and *Pierre* and *Marie Curie* for studying radioactivity.
1904 (P) *Lord Rayleigh* for studying the density of gases and discovering argon.
(C) *William Ramsay* for discovering the inert gas elements helium, neon, xenon, and krypton, and placing them in the periodic table.
1905 (P) *Philipp Lenard* for studying cathode rays, electrons (1898–1899).
1906 (P) *J. J. Thomson* for studying electrical discharge through gases and discovering the electron (1897).
1907 (P) *Albert A. Michelson* for inventing optical instruments and measuring the speed of light (1880s).
1908 (P) *Gabriel Lippmann* for making the first color photographic plate, using interference methods (1891).
(C) *Ernest Rutherford* for discovering that atoms can be broken apart by alpha rays and for studying radioactivity.
1909 (P) *Guglielmo Marconi* and *Carl Ferdinand Braun* for developing wireless telegraphy.
1910 (P) *Johannes D. van der Waals* for studying the equation of state for gases and liquids (1881).
1911 (P) *Wilhelm Wien* for discovering Wien's law giving the peak of a blackbody spectrum (1893).
(C) *Marie Curie* for discovering radium and polonium (1898) and isolating radium.
1912 (P) *Nils Dalén* for inventing automatic gas regulators for lighthouses.
1913 (P) *Heike Kamerlingh Onnes* for the discovery of superconductivity and liquefying helium (1908).
1914 (P) *Max T. F. von Laue* for studying x-rays from their diffraction by crystals, showing that x-rays are electromagnetic waves (1912).
(C) *Theodore W. Richards* for determining the atomic weights of sixty elements, indicating the existence of isotopes.
1915 (P) *William Henry Bragg* and *William Lawrence Bragg*, his son, for studying the diffraction of x-rays in crystals.
1917 (P) *Charles Barkla* for studying atoms by x-ray scattering (1906).
1918 (P) *Max Planck* for discovering energy quanta (1900).
1919 (P) *Johannes Stark*, for discovering the Stark effect, the splitting of spectral lines in electric fields (1913).
1920 (P) *Charles-Édouard Guillaume* for discovering invar, a nickel–steel alloy with low coefficient of expansion.
(C) *Walther Nernst* for studying heat changes in chemical reactions and formulating the third law of thermodynamics (1918).
1921 (P) *Albert Einstein* for explaining the photoelectric effect and for his services to theoretical physics (1905).
(C) *Frederick Soddy* for studying the chemistry of radioactive substances and discovering isotopes (1912).

1922 (P) *Niels Bohr* for his model of the atom and its radiation (1913).
(C) *Francis W. Aston* for using the mass spectrograph to study atomic weights, thus discovering 212 of the 287 naturally occurring isotopes.

1923 (P) *Robert A. Millikan* for measuring the charge on an electron (1911) and for studying the photoelectric effect experimentally (1914).

1924 (P) *Karl M. G. Siegbahn* for his work in x-ray spectroscopy.

1925 (P) *James Franck* and *Gustav Hertz* for discovering the Franck–Hertz effect in electron–atom collisions.

1926 (P) *Jean-Baptiste Perrin* for studying Brownian motion to validate the discontinuous structure of matter and measure the size of atoms.

1927 (P) *Arthur Holly Compton* for discovering the Compton effect on x-rays, their change in wavelength when they collide with matter (1922), and *Charles T. R. Wilson* for inventing the cloud chamber, used to study charged particles (1906).

1928 (P) *Owen W. Richardson* for studying the thermionic effect and electrons emitted by hot metals (1911).

1929 (P) *Louis Victor de Broglie* for discovering the wave nature of electrons (1923).

1930 (P) *Chandrasekhara Venkata Raman* for studying Raman scattering, the scattering of light by atoms and molecules with a change in wavelength (1928).

1932 (P) *Werner Heisenberg* for creating quantum mechanics (1925).

1933 (P) *Erwin Schrödinger* and *Paul A. M. Dirac* for developing wave mechanics (1925) and relativistic quantum mechanics (1927).
(C) *Harold Urey* for discovering heavy hydrogen, deuterium (1931).

1935 (P) *James Chadwick* for discovering the neutron (1932).
(C) *Irène* and *Frédéric Joliot-Curie* for synthesizing new radioactive elements.

1936 (P) *Carl D. Anderson* for discovering the positron in particular and antimatter in general (1932) and *Victor F. Hess* for discovering cosmic rays.
(C) *Peter J. W. Debye* for studying dipole moments and diffraction of x-rays and electrons in gases.

1937 (P) *Clinton Davisson* and *George Thomson* for discovering the diffraction of electrons by crystals, confirming de Broglie's hypothesis (1927).

1938 (P) *Enrico Fermi* for producing the transuranic radioactive elements by neutron irradiation (1934–1937).

1939 (P) *Ernest O. Lawrence* for inventing the cyclotron.

1943 (P) *Otto Stern* for developing molecular-beam studies (1923) and using them to discover the magnetic moment of the proton (1933).

1944 (P) *Isidor I. Rabi* for discovering nuclear magnetic resonance in atomic and molecular beams.
(C) *Otto Hahn* for discovering nuclear fission (1938).

1945 (P) *Wolfgang Pauli* for discovering the exclusion principle (1924).

1946 (P) *Percy W. Bridgman* for studying physics at high pressures.

1947 (P) *Edward V. Appleton* for studying the ionosphere.

1948 (P) *Patrick M. S. Blackett* for studying nuclear physics with cloud-chamber photographs of cosmic-ray interactions.

1949 (P) *Hideki Yukawa* for predicting the existence of mesons (1935).

1950 (P) *Cecil F. Powell* for developing the method of studying cosmic rays with photographic emulsions and discovering new mesons.

1951 (P) *John D. Cockcroft* and *Ernest T. S. Walton* for transmuting nuclei in an accelerator (1932).
(C) *Edwin M. McMillan* for producing neptunium (1940) and *Glenn T. Seaborg* for producing plutonium (1941) and further transuranic elements.

1952 (P) *Felix Bloch* and *Edward Mills Purcell* for discovering nuclear magnetic resonance in liquids and gases (1946).

1953 (P) *Frits Zernike* for inventing the phase-contrast microscope, which uses interference to provide high contrast.

1954 (P) *Max Born* for interpreting the wave function as a probability (1926) and other quantum-mechanical discoveries and *Walther Bothe* for developing the coincidence method to study subatomic particles (1930–1931), producing, in particular, the particle interpreted by Chadwick as the neutron.

1955 (P) *Willis E. Lamb, Jr.*, for discovering the Lamb shift in the hydrogen spectrum (1947) and *Polykarp Kusch* for determining the magnetic moment of the electron (1947).

1956 (P) *John Bardeen, Walter H. Brattain,* and *William Shockley* for inventing the transistor (1956).

1957 (P) *T.-D. Lee* and *C.-N. Yang* for predicting that parity is not conserved in beta decay (1956).

1958 (P) *Pavel A. Čerenkov* for discovering Čerenkov radiation (1935) and *Ilya M. Frank* and *Igor Tamm* for interpreting it (1937).

1959 (P) *Emilio G. Segrè* and *Owen Chamberlain* for discovering the antiproton (1955).

1960 (P) *Donald A. Glaser* for inventing the bubble chamber to study elementary particles (1952).

(C) *Willard Libby* for developing radiocarbon dating (1947).

1961 (P) *Robert Hofstadter* for discovering internal structure in protons and neutrons and *Rudolf L. Mössbauer* for discovering the Mössbauer effect of recoilless gamma-ray emission (1957).

1962 (P) *Lev Davidovich Landau* for studying liquid helium and other condensed matter theoretically.

1963 (P) *Eugene P. Wigner* for applying symmetry principles to elementary-particle theory and *Maria Goeppert Mayer* and *J. Hans D. Jensen* for studying the shell model of nuclei (1947).

1964 (P) *Charles H. Townes, Nikolai G. Basov,* and *Alexandr M. Prokhorov* for developing masers (1951–1952) and lasers.

1965 (P) *Sin-itiro Tomonaga, Julian S. Schwinger,* and *Richard P. Feynman* for developing quantum electrodynamics (1948).

1966 (P) *Alfred Kastler* for his optical methods of studying atomic energy levels.

1967 (P) *Hans Albrecht Bethe* for discovering the routes of energy production in stars (1939).

1968 (P) *Luis W. Alvarez* for discovering resonance states of elementary particles.

1969 (P) *Murray Gell-Mann* for classifying elementary particles (1963).

1970 (P) *Hannes Alfvén* for developing magnetohydrodynamic theory and *Louis Eugène Félix Néel* for discovering antiferromagnetism and ferrimagnetism (1930s).

1971 (P) *Dennis Gabor* for developing holography (1947).

(C) *Gerhard Herzberg* for studying the structure of molecules spectroscopically.

1972 (P) *John Bardeen, Leon N. Cooper,* and *John Robert Schrieffer* for explaining superconductivity (1957).

1973 (P) *Leo Esaki* for discovering tunneling in semiconductors, *Ivar Giaever* for discovering tunneling in superconductors, and *Brian D. Josephson* for predicting the Josephson effect, which involves tunneling of paired electrons (1958–1962).

1974 (P) *Anthony Hewish* for discovering pulsars and *Martin Ryle* for developing radio interferometry.

1975 (P) *Aage N. Bohr, Ben R. Mottelson,* and *James Rainwater* for discovering why some nuclei take asymmetric shapes.

1976 (P) *Burton Richter* and *Samuel C. C. Ting* for discovering the J/psi particle, the first charmed particle (1974).

1977 (P) *John H. Van Vleck, Nevill F. Mott,* and *Philip W. Anderson* for studying solids quantum-mechanically.

(C) *Ilya Prigogine* for extending thermodynamics to show how life could arise in the face of the second law.

1978 (P) *Arno A. Penzias* and *Robert W. Wilson* for discovering the cosmic background radiation (1965) and *Pyotr Kapitsa* for his studies of liquid helium.

1979 (P) *Sheldon L. Glashow, Abdus Salam,* and *Steven Weinberg* for developing the theory that unified the weak and electromagnetic forces (1958–1971).

1980 (P) *Val Fitch* and *James W. Cronin* for discovering CP (charge-parity) violation (1964), which possibly explains the cosmological dominance of matter over antimatter.

1981 (P) *Nicolaas Bloembergen* and *Arthur L. Schawlow* for developing laser spectroscopy and *Kai M. Siegbahn* for developing high-resolution electron spectroscopy (1958).

1982 (P) *Kenneth G. Wilson* for developing a method of constructing theories of phase transitions to analyze critical phenomena.

1983 (P) *William A. Fowler* for theoretical studies of astrophysical nucleosynthesis and *Subramanyan Chandrasekhar* for studying physical processes of importance to stellar structure and evolution, including the prediction of white dwarf stars (1930).

1984 (P) *Carlo Rubbia* for discovering the W and Z particles, verifying the electroweak unification, and *Simon van der Meer,* for developing the method of stochastic cooling of the CERN beam that allowed the discovery (1982–1983).

1985 (P) *Klaus von Klitzing* for the quantized Hall effect, relating to conductivity in the presence of a magnetic field (1980).

1986 (P) *Ernst Ruska* for inventing the electron microscope (1931), and *Gerd Binnig* and *Heinrich Rohrer* for inventing the scanning-tunneling electron microscope (1981).

1987 (P) *J. Georg Bednorz* and *Karl Alex Müller* for the discovery of high-temperature superconductivity (1986).

1988 (P) *Leon M. Lederman, Melvin Schwartz,* and *Jack Steinberger* for a collaborative experiment that led to the development of a new tool for studying the weak nuclear force, which affects the radioactive decay of atoms.

1989 (P) *Norman Ramsay* for various techniques in atomic physics; and *Hans Dehmelt* and *Wolfgang Paul* for the development of techniques for trapping single-charge particles.

1990 (P) *Jerome Friedman, Henry Kendall* and *Richard Taylor* for experiments important to the development of the quark model.

1991 (P) *Pierre-Gilles de Gennes* for discovering that methods developed for studying order phenomena in simple systems can be generalized to more complex forms of matter, in particular to liquid crystals and polymers.

1992 (P) *George Charpak* for developing detectors that trace the paths of evanescent subatomic particles produced in particle accelerators.

1993 (P) *Russell Hulse* and *Joseph Taylor* for discovering evidence of gravitational waves.

1994 (P) *Bertram N. Brockhouse* and *Clifford G. Shull* for pioneering work in neutron scattering.

1995 (P) *Martin L. Perl* and *Frederick Reines* for discovering the tau particle and the neutrino, respectively.

1996 (P) *David M. Lee, Douglas C. Osheroff,* and *Robert C. Richardson* for developing a superfluid using helium-3.

1997 (P) *Steven Chu, Claude Cohen-Tannoudji,* and *William D. Phillips* for developing methods to cool and trap atoms with laser light.

1998 (P) *Robert B. Laughlin, Horst L. Störmer,* and *Daniel C. Tsui* for discovering a new form of quantum fluid with fractionally charged excitations.

1999 (P) *Gerardus 'T Hooft* and *Martinus J. G. Veltman* for studies in the quantum structure of electroweak interactions in physics.

2000 (P) *Zhores I. Alferov* and *Herbert Kroemer* for developing semiconductor heterostructures used in high-speed electronics and optoelectronics and *Jack St. Clair Kilby* for participating in the invention of the integrated circuit.

2001 (P) *Eric A. Cornell, Wolfgang Ketterle,* and *Carl E. Wieman* for the achievement of Bose–Einstein condensation in dilute gases of alkali atoms.

2002 (P) *Raymond Davis Jr.* and *Masatoshi Koshiba* for the detection of cosmic neutrinos and *Riccardo Giacconi* for contributions to astrophysics that led to the discovery of cosmic x-ray sources.

Answers to Odd-Numbered Problems

CHAPTER 1

1. 0.141 nm
3. 2.15×10^4 kg/m^3
5. $4\pi\rho(r_2{}^3 - r_1{}^3)/3$
7. (a) 4.00 u = 6.64×10^{-24} g (b) 55.9 u = 9.28×10^{-23} g (c) 207 u = 3.44×10^{-22} g
9. 8.72×10^{11} atom/s
11. (a) 72.6 kg (b) 7.82×10^{26} atoms
13. No.
15. (b) only
17. The units of G are m^3/kg · s^2
19. 9.19 nm/s
21. 1.39×10^3 m^2
23. (a) 0.071 4 gal/s (b) 2.70×10^{-4} m^3/s (c) 1.03 h
25. 11.4×10^3 kg/m^3
27. 667 lb/s
29. (a) 190 yr (b) 2.32×10^4 times
31. 151 μm
33. 1.00×10^{10} lb
35. (a) 2.07 mm (b) 8.62×10^{13} times as large
37. 5.0 m
39. 2.86 cm
41. $\sim 10^6$ balls
43. $\sim 10^7$
45. $\sim 10^2$ kg; $\sim 10^3$ kg
47. $\sim 10^2$ tuners
49. (a) (346 ± 13) m^2 (b) (66.0 ± 1.3) m
51. $(1.61 \pm 0.17) \times 10^3$ kg/m^3
53. 31 556 926.0 s
55. 5.2 m^3, 3%
57. 2.57×10^{-10} m
59. $0.579\,t$ ft^3/s + $1.19 \times 10^{-9}\,t^2$ ft^3/s^2
61. 3.41 m
63. 0.449%
65. (a) 0.529 cm/s (b) 11.5 cm/s
67. 1×10^{10} gal/yr
69. $\sim 10^{11}$ stars
71. (a) 3.16×10^7 s/yr (b) 6.05×10^{10} yr

CHAPTER 2

1. (a) 2.30 m/s (b) 16.1 m/s (c) 11.5 m/s
3. (a) 5 m/s (b) 1.2 m/s (c) -2.5 m/s (d) -3.3 m/s (e) 0
5. (a) 3.75 m/s (b) 0
7. (a) -2.4 m/s (b) -3.8 m/s (c) 4.0 s
9. (a) 5.0 m/s (b) -2.5 m/s (c) 0 (d) 5.0 m/s
11. 1.34×10^4 m/s^2
13. (a) 52.4 ft/s, 55.0 ft/s, 55.5 ft/s, 57.4 ft/s (b) 0.598 ft/s^2
15. (a) 2.00 m (b) -3.00 m/s (c) -2.00 m/s^2
17. (a) 1.3 m/s^2 (b) 2.0 m/s^2 at 3 s (c) at $t = 6$ s and for $t > 10$ s (d) -1.5 m/s^2 at 8 s
19. 2.74×10^5 m/s^2, which is 2.79×10^4 g
21. -16.0 cm/s^2
23. (a) 4.53 s (b) 14.1 m/s
25. (a) 2.56 m (b) -3.00 m/s
27. (a) 20.0 s (b) no
29. 3.10 m/s
31. (a) -202 m/s^2 (b) 198 m
33. (a) 4.98×10^{-9} s (b) 1.20×10^{15} m/s^2
35. (a) v_c/t_m (c) $v_c t_0/2$ (d) $v_c t_0$ (e) yes, no
37. (a) 3.00 m/s (b) 6.00 s (c) -0.300 m/s^2 (d) 2.05 m/s
39. 31 s
41. $99.3/h
43. (a) 10.0 m/s up (b) 4.68 m/s down
45. (a) 2.17 s (b) -21.2 m/s (c) 2.23 s
47. (a) 29.4 m/s (b) 44.1 m
49. (a) 7.82 m (b) 0.782 s
51. 7.96 s
53. (a) $a_x(t) = a_{xi} + Jt$, $v_x(t) = v_{xi} + a_{xi}t + (1/2)Jt^2$, $x(t) = x_i + v_{xi}t + (1/2)a_{xi}t^2 + (1/6)Jt^3$
55. (a) $a = -(10.0 \times 10^7\ \text{m/s}^3)\,t + 3.00 \times 10^5\ \text{m/s}^2$; $x = -(1.67 \times 10^7\ \text{m/s}^3)\,t^3 + (1.50 \times 10^5\ \text{m/s}^2)\,t^2$ (b) 3.00×10^{-3} s (c) 450 m/s (d) 0.900 m
59. (a) Acela steadily cruises out of the city center at 45 mi/h. In less than a minute it smoothly speeds up to 150 mi/h; then its speed is nudged up to 170 mi/h. Next it smoothly slows to a very low speed, which it maintains as it rolls into a railroad yard. When it stops, it immediately begins backing up and smoothly speeds up to 50 mi/h in reverse, all in less than seven minutes after it started. (b) 2.2 mi/h/s = 0.98 m/s^2 (c) 6.7 mi
61. 48.0 mm
63. (a) 15.0 s (b) 30.0 m/s (c) 225 m
65. (a) 5.43 m/s^2 and 3.83 m/s^2 (b) 10.9 m/s and 11.5 m/s (c) Maggie by 2.62 m
67. $\sim 10^3$ m/s^2

69. (a) 3.00 s (b) − 15.3 m/s (c) 31.4 m/s down and 34.8 m/s down

71. (c) v_{boy}^2/h, 0 (d) v_{boy}, 0

73. (a) 5.46 s (b) 73.0 m (c) $v_{\text{Stan}} = 22.6$ m/s, $v_{\text{Kathy}} = 26.7$ m/s

75. 0.577 v

CHAPTER 3

1. (− 2.75, − 4.76) m

3. (a) 2.24 m (b) 2.24 m at 26.6°

5. $y = 1.15$; $r = 2.31$

7. 70.0 m

9. 310 km at 57° S of W

11. (a) 10.0 m (b) 15.7 m (c) 0

13. (a) $\sim 10^5$ m vertically upward (b) $\sim 10^3$ m vertically upward

15. (a) 5.2 m at 60° (b) 3.0 m at 330° (c) 3.0 m at 150° (d) 5.2 m at 300°

17. approximately 420 ft at − 3°

19. 47.2 units at 122°

21. (a) $(-11.1\hat{\mathbf{i}} + 6.40\hat{\mathbf{j}})$ m (b) $(1.65\hat{\mathbf{i}} + 2.86\hat{\mathbf{j}})$ cm (c) $(-18.0\hat{\mathbf{i}} - 12.6\hat{\mathbf{j}})$ in.

23. (a) 5.00 blocks at 53.1° N of E (b) 13.0 blocks

25. 358 m at 2.00° S of E

27. (a)

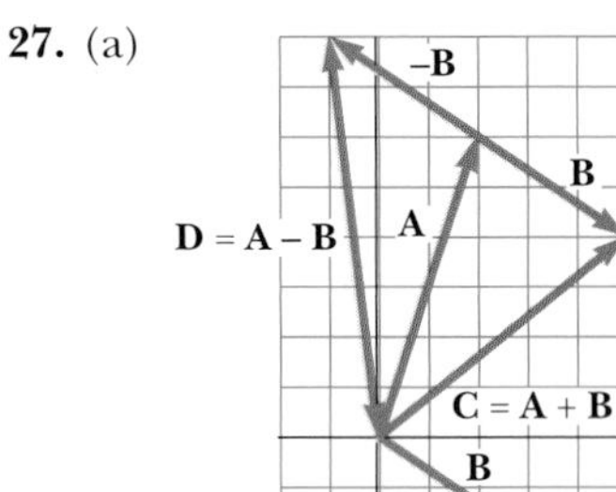

(b) $\mathbf{C} = 5.00\hat{\mathbf{i}} + 4.00\hat{\mathbf{j}}$ or 6.40 at 38.7°; $\mathbf{D} = -1.00\hat{\mathbf{i}} + 8.00\hat{\mathbf{j}}$ or 8.06 at 97.2°

29. 196 cm at 345°

31. (a) $2.00\hat{\mathbf{i}} - 6.00\hat{\mathbf{j}}$ (b) $4.00\hat{\mathbf{i}} + 2.00\hat{\mathbf{j}}$ (c) 6.32 (d) 4.47 (e) 288°; 26.6°

33. 9.48 m at 166°

35. (a) 185 N at 77.8° from the + x axis (b) $(-39.3\hat{\mathbf{i}} - 181\hat{\mathbf{j}})$ N

37. $\mathbf{A} + \mathbf{B} = (2.60\hat{\mathbf{i}} + 4.50\hat{\mathbf{j}})$ m

39. $|\mathbf{B}| = 7.81$, $\theta_x = 59.2°$, $\theta_y = 39.8°$, $\theta_z = 67.4°$

41. (a) $8.00\hat{\mathbf{i}} + 12.0\hat{\mathbf{j}} - 4.00\hat{\mathbf{k}}$ (b) $2.00\hat{\mathbf{i}} + 3.00\hat{\mathbf{j}} - 1.00\hat{\mathbf{k}}$ (c) $-24.0\hat{\mathbf{i}} - 36.0\hat{\mathbf{j}} + 12.0\hat{\mathbf{k}}$

43. (a) 5.92 m is the magnitude of $(5.00\hat{\mathbf{i}} - 1.00\hat{\mathbf{j}} - 3.00\hat{\mathbf{k}})$ m (b) 19.0 m is the magnitude of $(4.00\hat{\mathbf{i}} - 11.0\hat{\mathbf{j}} - 15.0\hat{\mathbf{k}})$ m

45. 157 km

47. (a) $-3.00\hat{\mathbf{i}} + 2.00\hat{\mathbf{j}}$ (b) 3.61 at 146° (c) $3.00\hat{\mathbf{i}} - 6.00\hat{\mathbf{j}}$

49. (a) $49.5\hat{\mathbf{i}} + 27.1\hat{\mathbf{j}}$ (b) 56.4 units at 28.7°

51. 1.15°

53. (a) 2.00, 1.00, 3.00 (b) 3.74 (c) $\theta_x = 57.7°$, $\theta_y = 74.5°$, $\theta_z = 36.7°$

55. 2.29 km

57. (a) 11.2 m (b) 12.9 m at 36.4°

59. 240 m at 237°

61. 390 mi/h at 7.37° north of east

63. (a) zero (b) zero

65. 106°

CHAPTER 4

1. (a) 4.87 km at 209° from east (b) 23.3 m/s (c) 13.5 m/s at 209°

3. 2.50 m/s

5. (a) $(2.00\hat{\mathbf{i}} + 3.00\hat{\mathbf{j}})$ m/s^2 (b) $(3.00t + t^2)\hat{\mathbf{i}}$ m + $(1.50t^2 - 2.00t)\hat{\mathbf{j}}$ m

7. (a) $(0.800\hat{\mathbf{i}} - 0.300\hat{\mathbf{j}})$ m/s^2 (b) 339° (c) $(360\hat{\mathbf{i}} - 72.7\hat{\mathbf{j}})$ m, −15.2°

9. (a) $x = 0.010\,0$ m, $y = 2.41 \times 10^{-4}$ m (b) $\mathbf{v} = (1.84 \times 10^7\hat{\mathbf{i}} + 8.78 \times 10^5\hat{\mathbf{j}})$ m/s (c) $v = 1.85 \times 10^7$ m/s (d) $\theta = 2.73°$

11. (a) $3.34\hat{\mathbf{i}}$ m/s (b) −50.9°

13. (a) 20.0° (b) 3.05 s

15. 53.1°

17. (a) 22.6 m (b) 52.3 m (c) 1.18 s

19. (a) The ball clears by 0.889 m while (b) descending

21. (a) 18.1 m/s (b) 1.13 m (c) 2.79 m

23. 9.91 m/s

25. (a) 30.3 m/s (b) 2.09 s

27. 377 m/s^2

29. 10.5 m/s, 219 m/s^2 inward

31. (a) 6.00 rev/s (b) 1.52 km/s^2 (c) 1.28 km/s^2

33. 1.48 m/s^2 inward and 29.9° backward

35. (a) 13.0 m/s^2 (b) 5.70 m/s (c) 7.50 m/s^2

37. $\theta = \tan^{-1}(1/4\pi) = 4.55°$

39. (a) 57.7 km/h at 60.0° west of vertical (b) 28.9 km/h downward

41. 2.02×10^3 s; 21.0% longer

43. $t_{\text{Alan}} = \dfrac{2L/c}{1 - v^2/c^2}$, $t_{\text{Beth}} = \dfrac{2L/c}{\sqrt{1 - v^2/c^2}}$. Beth returns first.

45. 15.3 m

47. (a) 101 m/s (b) 32 700 ft (c) 20.6 s (d) 180 m/s

49. 54.4 m/s^2

51. (a) 41.7 m/s (b) 3.81 s (c) $(34.1\hat{\mathbf{i}} - 13.4\hat{\mathbf{j}})$ m/s; 36.7 m/s

53. (a) 25.0 m/s^2; 9.80 m/s^2

(b)

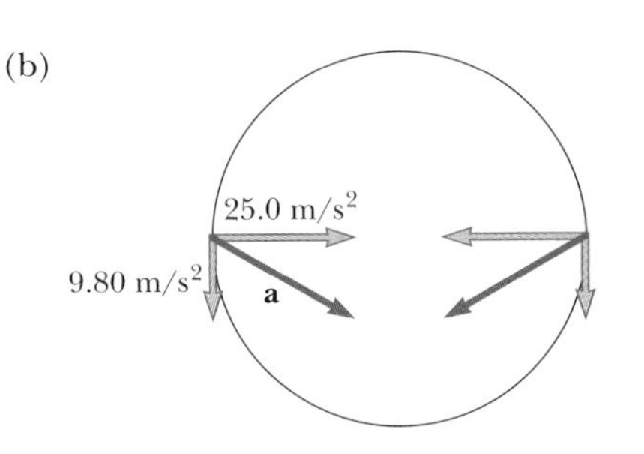

(c) 26.8 m/s^2 inward at 21.4° below the horizontal

55. (a) 26.6° (b) 0.949

57. (a) 0.600 m (b) 0.402 m (c) 1.87 m/s^2 toward center (d) 9.80 m/s^2 down

59. (a) 6.80 km (b) 3.00 km vertically above the impact point (c) 66.2°

61. (a) 46.5 m/s (b) −77.6° (c) 6.34 s

63. (a) 20.0 m/s, 5.00 s (b) $(16.0\hat{\mathbf{i}} - 27.1\hat{\mathbf{j}})$ m/s (c) 6.53 s (d) $24.5\hat{\mathbf{i}}$ m

65. (a) 22.9 m/s (b) 360 m from the base of the cliff (c) $\mathbf{v} = (114\hat{\mathbf{i}} - 44.3\hat{\mathbf{j}})$ m/s

67. (a) 43.2 m (b) $(9.66\hat{\mathbf{i}} - 25.5\hat{\mathbf{j}})$ m/s

69. (a) 4.00 km/h (b) 4.00 km/h

71. Safe distances are less than 270 m or greater than 3.48×10^3 m from the western shore.

CHAPTER 5

1. (a) 1/3 (b) 0.750 m/s^2

3. $(6.00\hat{\mathbf{i}} + 15.0\hat{\mathbf{j}})$ N; 16.2 N

5. (a) $(2.50\hat{\mathbf{i}} + 5.00\hat{\mathbf{j}})$ N (b) 5.59 N

7. (a) 3.64×10^{-18} N (b) 8.93×10^{-30} N is 408 billion times smaller

9. 2.38 kN

11. (a) 5.00 m/s^2 at 36.9° (b) 6.08 m/s^2 at 25.3°

13. (a) $\sim 10^{-22}$ m/s^2 (b) $\sim 10^{-23}$ m

15. (a) 15.0 lb up (b) 5.00 lb up (c) 0

17. 613 N

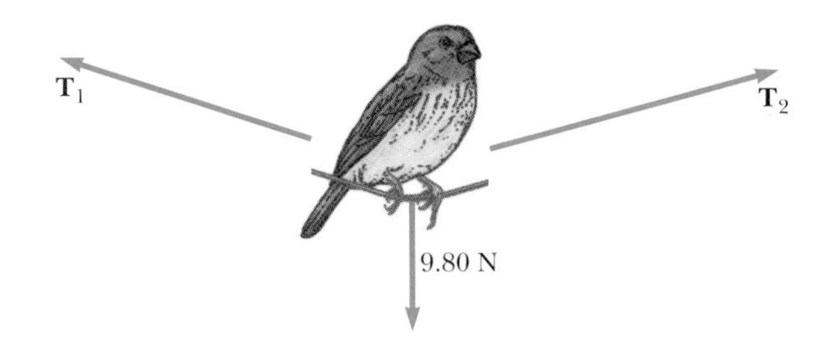

21. (a) 49.0 N (b) 98.0 N (c) 24.5 N

23. 8.66 N east

25. 3.73 m

27. A is in compression 3.83 kN and B is in tension 3.37 kN

29. 950 N

31. (a) $F_x > 19.6$ N (b) $F_x \leq -78.4$ N

(c)

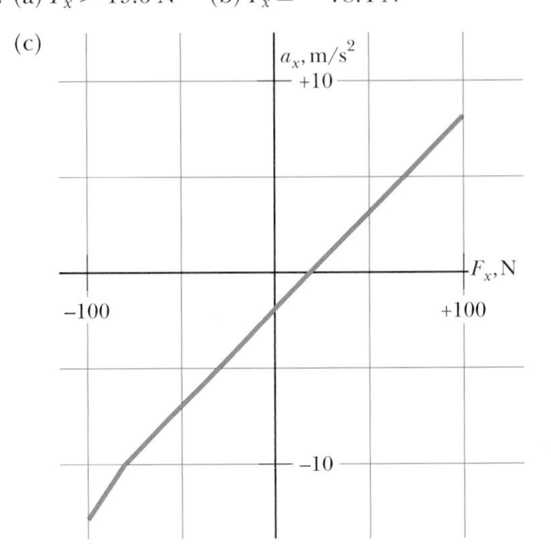

33. (a) 706 N (b) 814 N (c) 706 N (d) 648 N

35. (a) 0.404 (b) 45.8 lb

37. (a) 256 m (b) 42.7 m

39. (a) 1.10 s (b) 0.875 s

41. (a) 1.78 m/s^2 (b) 0.368 (c) 9.37 N (d) 2.67 m/s

43. 37.8 N

45. (a)

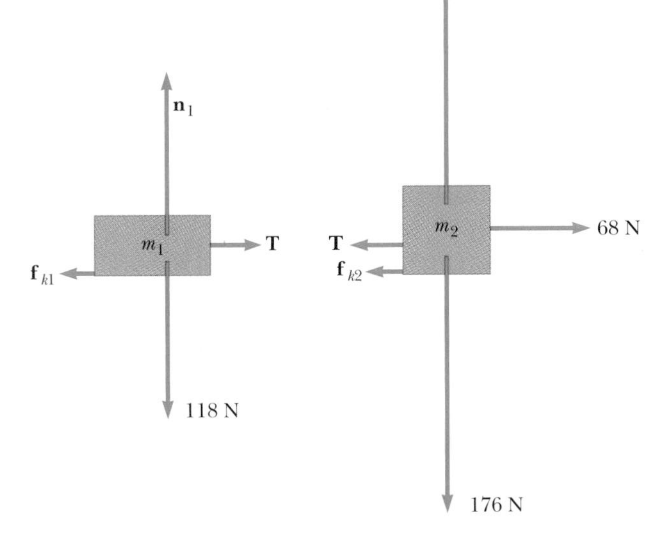

(b) 27.2 N, 1.29 m/s^2

47. $\mu_k = (3/5)\tan\theta$

49. (a) 8.05 N (b) 53.2 N (c) 42.0 N

51. (a)

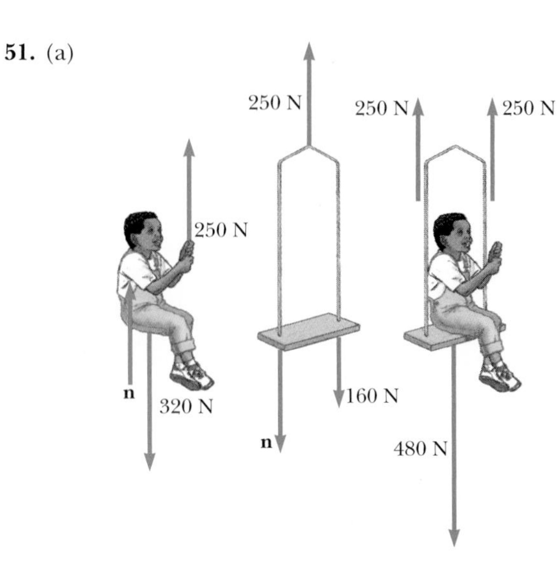

(b) 0.408 m/s^2 (c) 83.3 N

53. (a) $F_A = mg(\sin\theta - \mu_s \cos\theta)$
(b) $F_B = mg(\sin\theta - \mu_s \cos\theta)/(\cos\theta + \mu_s \sin\theta)$
(c) A's job is easier
(d) B's job is easier

55. (a) $Mg/2$, $Mg/2$, $Mg/2$, $3Mg/2$, Mg (b) $Mg/2$

57. (b)

θ	0	15°	30°	45°	60°
P(N)	40.0	46.4	60.1	94.3	260

59. (a) 19.3° (b) 4.21 N

61. $(M + m_1 + m_2)(m_2 g/m_1)$

63. (a) $m_2 g\left[\dfrac{m_1 M}{m_1 M + m_2(m_1 + M)}\right]$

(b) $\dfrac{m_2 g(M + m_1)}{m_1 M + m_2(m_1 + M)}$

(c) $\dfrac{m_1 m_2 g}{m_1 M + m_2(m_1 + M)}$

(d) $\dfrac{M m_2 g}{m_1 M + m_2(m_1 + M)}$

65. (c) 3.56 N

67. (a) $T = f/(2\sin\theta)$ (b) 410 N

69. (a) 30.7° (b) 0.843 N

71. 0.060 0 m

73. (a) $T_1 = \dfrac{2mg}{\sin\theta_1}$

$T_2 = \dfrac{mg}{\sin\theta_2} = \dfrac{mg}{\sin[\tan^{-1}(\frac{1}{2}\tan\theta_1)]}$

$T_3 = 2mg/\tan\theta_1$

(b) $\theta_2 = \tan^{-1}\left(\dfrac{\tan\theta_1}{2}\right)$

CHAPTER 6

1. Any speed up to 8.08 m/s

3. (a) 8.32×10^{-8} N toward the nucleus
(b) $9.13 \times 10^{22} \text{ m/s}^2$ inward

5. (a) static friction (b) 0.085 0

7. $v \leq 14.3$ m/s

9. (a) 68.6 N toward the center of the circle and 784 N up
(b) 0.857 m/s^2

11. (a) 108 N (b) 56.2 N

13. (a) 4.81 m/s (b) 700 N up

15. No. The jungle lord needs a vine of tensile strength 1.38 kN.

17. 3.13 m/s

19. (a) 2.49×10^4 N up (b) 12.1 m/s

21. (a) 3.60 m/s^2 (b) zero (c) An observer in the car (a noninertial frame) claims an 18.0-N force toward the left and an 18.0-N force toward the right. An inertial observer (outside the car) claims only an 18.0-N force toward the right.

23. (a) 17.0° (b) 5.12 N

25. (a) 491 N (b) 50.1 kg (c) 2.00 m/s^2

27. (a) $v = [2(a - \mu_k g)\ell]^{1/2}$; (b) $v' = (2\mu_k g\ell/v)$, where $v = [2(a - \mu_k g)\ell]^{1/2}$

29. 93.8 N

31. 0.092 7°

33. (a) 32.7 s^{-1} (b) 9.80 m/s^2 down (c) 4.90 m/s^2 down

35. 3.01 N up

37. (a) 1.47 N·s/m (b) 2.04×10^{-3} s (c) 2.94×10^{-2} N

39. (a) $0.034\,7 \text{ s}^{-1}$ (b) 2.50 m/s (c) $a = -cv$

41. (a) $x = k^{-1}\ln(1 + kv_0 t)$ (b) $v = v_0 e^{-kx}$

43. $\sim 10^1$ N

45. (a) 13.7 m/s down

(b)

	t (s)	x (m)	v (m/s)
	0	0	0
	0.2	0	−1.96
	0.4	−0.392	−3.88
. . .	1.0	−3.77	−8.71
. . .	2.0	−14.4	−12.56
. . .	4.0	−41.0	−13.67

47. (a) 49.5 m/s down and 4.95 m/s down

(b)

	t (s)	y (m)	v (m/s)
	0	1 000	0
. . .	1	995	−9.7
. . .	2	980	−18.6
. . .	10	674	−47.7
. . .	10.1	671	−16.7
. . .	12	659	−4.95
. . .	145	0	−4.95

49. (a) 2.33×10^{-4} kg/m (b) 57.0 m/s (c) 44.9 m/s. The second trajectory is higher and shorter than the first. In both cases, the ball attains maximum height when it has covered 56% of its horizontal range, and attains minimum speed a little later. The impact speeds are also similar, 30 m/s and 29 m/s.

51. (a) 11.5 kN (b) 14.1 m/s = 50.9 km/h

53. (a) 0.016 2 kg/m (b) $\frac{1}{2}D\rho A$ (c) 0.778 (d) 1.5% (e) For stacked coffee filters falling in air at terminal speed, the graph of resistive force as a function of squared speed demonstrates that the force is proportional to the speed squared, within the experimental uncertainty, estimated as ±2%. This proportionality agrees with that predicted by the theoretical equation $R = \frac{1}{2}D\rho Av^2$. The value of the constant slope of the graph implies that the drag coefficient for coffee filters is $D = 0.78 \pm 2\%$.

55. $g(\cos\phi \tan\theta - \sin\phi)$

57. (b) 732 N down at the equator and 735 N down at the poles

59. (a) 967 lb (b) −647 lb (pilot must be strapped in) (c) Speed and radius of path can be adjusted so that $v^2 = gR$.

61. (a) 1.58 m/s² (b) 455 N (c) 329 N (d) 397 N upward and 9.15° inward

63. (a) 5.19 m/s (b) T = 555 N

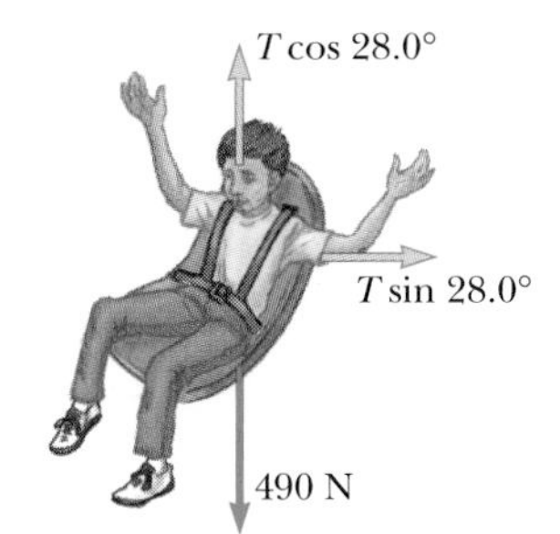

65. (b) 2.54 s; 23.6 rev/min

67. (a) $v_{\text{min}} = \sqrt{\dfrac{Rg(\tan\theta - \mu_s)}{1 + \mu_s \tan\theta}}$, $v_{\text{max}} = \sqrt{\dfrac{Rg(\tan\theta + \mu_s)}{1 - \mu_s \tan\theta}}$
(b) $\mu_s = \tan\theta$ (c) 8.57 m/s ≤ v ≤ 16.6 m/s

69. (a) 0.013 2 m/s (b) 1.03 m/s (c) 6.87 m/s

71. 12.8 N

73. $\Sigma\mathbf{F} = -km\mathbf{v}$

CHAPTER 7

1. (a) 31.9 J (b) 0 (c) 0 (d) 31.9 J

3. −4.70 kJ

5. 28.9

7. (a) 16.0 J (b) 36.9°

9. (a) 11.3° (b) 156° (c) 82.3°

11. (a) 24.0 J (b) −3.00 J (c) 21.0 J

13. (a) 7.50 J (b) 15.0 J (c) 7.50 J (d) 30.0 J

15. (a) 0.938 cm (b) 1.25 J

17. (a) 0.768 m (b) 1.68 × 10⁵ J

19. 12.0 J

21. (a) 0.020 4 m (b) 720 N/m

23. kg/s²

25. (a) 33.8 J (b) 135 J

27. 878 kN up

29. (a) 4.56 kJ (b) 6.34 kN (c) 422 km/s² (d) 6.34 kN

31. (a) 650 J (b) 588 J (c) 0 (d) 0 (e) 62.0 J (f) 1.76 m/s

33. (a) −168 J (b) 184 J (c) 500 J (d) 148 J (e) 5.65 m/s

35. 2.04 m

37. 875 W

39. (a) 20.6 kJ (b) 686 W

41. $46.2

43. (a) 423 mi/gal (b) 776 mi/gal

45. (a) 0.013 5 gal (b) 73.8 (c) 8.08 kW

47. 2.92 m/s

49. (a) $(2 + 24t^2 + 72t^4)$ J (b) $12t$ m/s²; $48t$ N (c) $(48t + 288t^3)$ W (d) 1 250 J

51. $k_1x_{\text{max}}^2/2 + k_2x_{\text{max}}^3/3$

53. (a) $\sqrt{2W/m}$ (b) W/d

55. (b) 240 W

57. (a) 1.38 × 10⁴ J (b) 3.02 × 10⁴ W

59. (a) $\mathcal{P} = 2Mgv_T$ (b) $\mathcal{P} = 24Mgv_T$

61. (a) 4.12 m (b) 3.35 m

63. 1.68 m/s

65. -1.37×10^{-21} J

67. 0.799 J

69. (b) For a block of weight w pushed over a rough horizontal surface at constant velocity, $b = \mu_k$. For a load pulled vertically upward at constant velocity, $b = 1$.

CHAPTER 8

1. (a) 259 kJ, 0, −259 kJ (b) 0, −259 kJ, −259 kJ

3. 22.0 kW

5. (a) $v = (3gR)^{1/2}$ (b) 0.098 0 N down

7. (a) 1.47 m/s (b) 1.35 m/s

9. (a) 2.29 m/s (b) 1.98 m/s

11. 10.2 m

13. (a) 4.43 m/s (b) 5.00 m

15. 5.49 m/s

17. (a) 18.5 km, 51.0 km (b) 10.0 MJ

19. (a) 25.8 m (b) 27.1 m/s²

21. (a) −196 J (b) −196 J (c) −196 J. The force is conservative.

23. (a) 125 J (b) 50.0 J (c) 66.7 J (d) Nonconservative. The results differ.

25. (a) −9.00 J; no; the force is conservative. (b) 3.39 m/s (c) 9.00 J

27. 26.5 m/s

29. 6.92 m/s

31. 3.74 m/s

33. (a) -160 J (b) 73.5 J (c) 28.8 N (d) 0.679

35. (a) 1.40 m/s (b) 4.60 cm after release (c) 1.79 m/s

37. (a) 0.381 m (b) 0.143 m (c) 0.371 m

39. (a) $a_x = -\mu_k gx/L$ (b) $v = (\mu_k gL)^{1/2}$

41. (a) 40.0 J (b) -40.0 J (c) 62.5 J

43. (A/r^2) away from the other particle

45. (a) + at Ⓑ, − at Ⓓ, 0 at Ⓐ, Ⓒ, and Ⓔ (b) Ⓒ stable; Ⓐ and Ⓔ unstable

(c)

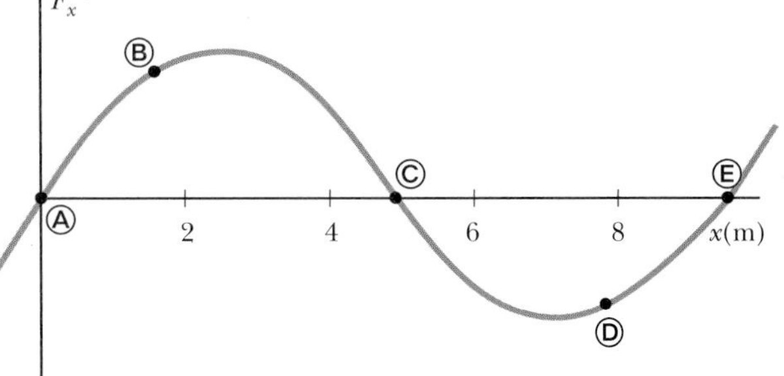

47. (b)

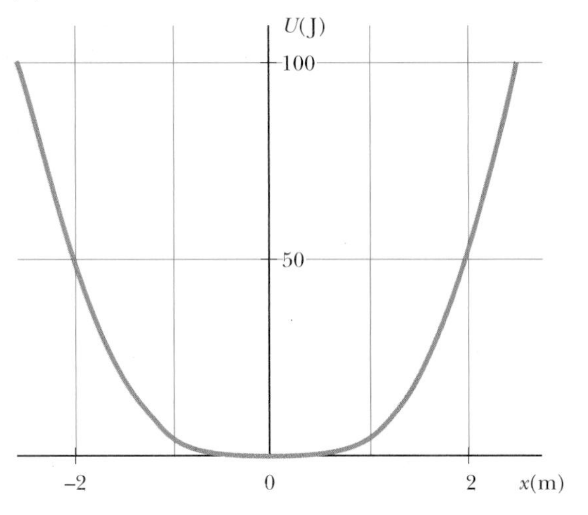

Equilibrium at $x = 0$

(c) 0.823 m/s

49. ~10^3 W peak or ~10^2 W sustainable

51. 48.2°

53. (a) 0.225 J (b) $\Delta E_{\text{mech}} = -0.363$ J (c) No; the normal force changes in a complicated way.

55. (a) 23.6 cm (b) 5.90 m/s^2 up the incline; no. (c) Gravitational potential energy is transformed into kinetic energy plus elastic potential energy and then entirely into elastic potential energy.

57. 0.328

59. 1.24 m/s

61. (a) 0.400 m (b) 4.10 m/s (c) The block stays on the track.

63. $(h/5)(4\sin^2\theta + 1)$

65. (a) 6.15 m/s (b) 9.87 m/s

67. (a) 11.1 m/s (b) 19.6 m/s^2 upward (c) 2.23×10^3 N upward (d) 1.01×10^3 J (e) 5.14 m/s (f) 1.35 m (g) 1.39 s

69. (b) 1.44 m (c) 0.400 m (d) No. A very strong wind pulls the string out horizontally. The largest possible equilibrium height is equal to L.

73. (a) $2.5R$

CHAPTER 9

1. (a) $(9.00\hat{\mathbf{i}} - 12.0\hat{\mathbf{j}})$ kg·m/s (b) 15.0 kg·m/s at 307°

3. ~10^{-23} m/s

5. (b) $p = \sqrt{2mK}$

7. (a) 13.5 N·s (b) 9.00 kN (c) 18.0 kN

9. 260 N normal to the wall

11. (a) $(9.05\hat{\mathbf{i}} + 6.12\hat{\mathbf{j}})$ N·s (b) $(377\hat{\mathbf{i}} + 255\hat{\mathbf{j}})$ N

13. 15.0 N in the direction of the initial velocity of the exiting water stream

15. 65.2 m/s

17. 301 m/s

19. (a) 2.50 m/s (b) 37.5 kJ (c) Each process is the time-reversal of the other. The same momentum conservation equation describes both.

21. (a) $v_{gx} = 1.15$ m/s (b) $v_{px} = -0.346$ m/s

23. (a) 0.284 (b) 115 fJ and 45.4 fJ

25. 91.2 m/s

27. (a) 2.24 m/s to the right (b) No. Coupling order makes no difference.

29. $v_{\text{orange}} = 3.99$ m/s, $v_{\text{yellow}} = 3.01$ m/s

31. $v_{\text{green}} = 7.07$ m/s, $v_{\text{blue}} = 5.89$ m/s

33. 2.50 m/s at $-60.0°$

35. $(3.00\hat{\mathbf{i}} - 1.20\hat{\mathbf{j}})$ m/s

37. (a) $(-9.33\hat{\mathbf{i}} - 8.33\hat{\mathbf{j}})$ Mm/s (b) 439 fJ

39. 0.006 73 nm from the oxygen nucleus along the bisector of the angle

41. $\mathbf{r}_{\text{CM}} = (11.7\hat{\mathbf{i}} + 13.3\hat{\mathbf{j}})$ cm

43. (a) 15.9 g (b) 0.153 m

45. (a) $(1.40\hat{\mathbf{i}} + 2.40\hat{\mathbf{j}})$ m/s (b) $(7.00\hat{\mathbf{i}} + 12.0\hat{\mathbf{j}})$ kg·m/s

47. 0.700 m

49. (a) 39.0 MN (b) 3.20 m/s^2 up

51. (a) 442 metric tons (b) 19.2 metric tons

53. 4.41 kg

55. (a) $1.33\hat{\mathbf{i}}$ m/s (b) $-235\hat{\mathbf{i}}$ N (c) 0.680 s (d) $-160\hat{\mathbf{i}}$ N·s and $+160\hat{\mathbf{i}}$ N·s (e) 1.81 m (f) 0.454 m (g) -427 J (h) $+107$ J (i) Equal friction forces act through different distances on person and cart, to do different amounts of work on them. The total work on both together, -320 J, becomes $+320$ J of extra internal energy in this perfectly inelastic collision.

57. 240 s

59. (a) 0; inelastic (b) $(-0.250\hat{\mathbf{i}} + 0.750\hat{\mathbf{j}} - 2.00\hat{\mathbf{k}})$ m/s; perfectly inelastic (c) either $a = -6.74$ with $\mathbf{v} = -0.419\hat{\mathbf{k}}$ m/s or $a = 2.74$ with $\mathbf{v} = -3.58\hat{\mathbf{k}}$ m/s

61. (a) $v_i = v(m + \rho V)/m$ (b) The cart slows with constant acceleration and eventually comes to rest.

63. (a) $m/M = 0.403$ (b) No changes; no difference.

65. (a) 6.29 m/s (b) 6.16 m/s

67. (a) 100 m/s (b) 374 J

69. (a) $(20.0\hat{\mathbf{i}} + 7.00\hat{\mathbf{j}})$ m/s (b) $4.00\hat{\mathbf{i}}$ m/s^2 (c) $4.00\hat{\mathbf{i}}$ m/s^2 (d) $(50.0\hat{\mathbf{i}} + 35.0\hat{\mathbf{j}})$ m (e) 600 J (f) 674 J (g) 674 J

71. $(3Mgx/L)\hat{\mathbf{j}}$

73. $\dfrac{m_1(R + \ell/2)}{(m_1 + m_2)}$

CHAPTER 10

1. (a) 5.00 rad, 10.0 rad/s, 4.00 rad/s^2 (b) 53.0 rad, 22.0 rad/s, 4.00 rad/s^2

3. (a) 4.00 rad/s^2 (b) 18.0 rad

5. (a) 5.24 s (b) 27.4 rad

7. 50.0 rev

9. (a) 7.27×10^{-5} rad/s (b) 2.57×10^4 s = 428 min

11. $\sim 10^7$ rev

13. (a) 8.00 rad/s (b) 8.00 m/s, $a_r = -64.0$ m/s^2, $a_t = 4.00$ m/s^2 (c) 9.00 rad

15. (a) 25.0 rad/s (b) 39.8 rad/s^2 (c) 0.628 s

17. (a) 126 rad/s (b) 3.77 m/s (c) 1.26 km/s^2 (d) 20.1 m

19. (a) $\omega(2h^3/g)^{1/2}$ (b) 0.011 6 m (c) Yes; the deflection is only 0.02% of the original height.

21. (a) 143 kg · m^2 (b) 2.57 kJ

23. $11mL^2/12$

25. 5.80 kg · m^2; the height makes no difference

29. $(23/48)MR^2\omega^2$

31. -3.55 N · m

33. 8.02×10^3 N

35. (a) 24.0 N · m (b) 0.035 6 rad/s^2 (c) 1.07 m/s^2

37. (a) 0.309 m/s^2 (b) 7.67 N and 9.22 N

39. 21.5 N

41. 24.5 km

43. (a) 1.59 m/s (b) 53.1 rad/s

45. (a) 11.4 N, 7.57 m/s^2, 9.53 m/s down (b) 9.53 m/s

49. (a) $2(Rg/3)^{1/2}$ (b) $4(Rg/3)^{1/2}$ (c) $(Rg)^{1/2}$

51. (a) 500 J (b) 250 J (c) 750 J

53. (a) $\frac{2}{3}g \sin\theta$ for the disk, larger than $\frac{1}{2}g\sin\theta$ for the hoop (b) $\frac{1}{3}\tan\theta$

55. 1.21×10^{-4} kg · m^2; height is unnecessary

57. $\frac{1}{3}\ell$

59. (a) 4.00 J (b) 1.60 s (c) yes

61. (a) $(3g/L)^{1/2}$ (b) $3g/2L$ (c) $-\frac{3}{2}g\hat{\mathbf{i}} - \frac{3}{4}g\hat{\mathbf{j}}$ (d) $-\frac{3}{2}Mg\hat{\mathbf{i}} + \frac{1}{4}Mg\hat{\mathbf{j}}$

63. -0.322 rad/s^2

65. (b) $2gM(\sin\theta - \mu\cos\theta)(m + 2M)^{-1}$

67. (a) $\sim -10^{-22}$ s^{-2} (b) $\sim -10^{16}$ N · m (c) $\sim 10^{13}$ m

71. (a) 118 N and 156 N (b) 1.17 kg · m^2

73. (a) $\alpha = -0.176$ rad/s^2 (b) 1.29 rev (c) 9.26 rev

75. (a) 61.2 J (b) 50.8 J

79. (a) $2.70R$ (b) $F_x = -20mg/7$, $F_y = -mg$

81. $\sim 10^1$ m

83. (a) $(3gh/4)^{1/2}$ (b) $(3gh/4)^{1/2}$

85. (c) $(8Fd/3M)^{1/2}$

87. $\mathbf{F}_1$ to right, $\mathbf{F}_2$ no rolling, $\mathbf{F}_3$ and $\mathbf{F}_4$ to left

CHAPTER 11

1. $-7.00\hat{\mathbf{i}} + 16.0\hat{\mathbf{j}} - 10.0\hat{\mathbf{k}}$

3. (a) $-17.0\hat{\mathbf{k}}$ (b) 70.6°

5. 0.343 N · m horizontally north

7. 45.0°

9. $F_3 = F_1 + F_2$; no

11. $(17.5\hat{\mathbf{k}})$ kg · m^2/s

13. $(60.0\hat{\mathbf{k}})$ kg · m^2/s

15. $mvR[\cos(vt/R) + 1]\hat{\mathbf{k}}$

17. (a) zero (b) $[-mv_i^3\sin^2\theta\cos\theta/2g]\hat{\mathbf{k}}$ (c) $[-2mv_i^3\sin^2\theta\cos\theta/g]\hat{\mathbf{k}}$ (d) The downward gravitational force exerts a torque in the $-z$ direction.

19. $-m\ell gt\cos\theta\hat{\mathbf{k}}$

23. (a) 0.360 kg · m^2/s (b) 0.540 kg · m^2/s

25. (a) 0.433 kg · m^2/s (b) 1.73 kg · m^2/s

27. (a) 1.57×10^8 kg · m^2/s (b) 6.26×10^3 s = 1.74 h

29. 7.14 rev/min

31. (a) 9.20 rad/s (b) 9.20 rad/s

33. (a) 0.360 rad/s counterclockwise (b) 99.9 J

35. (a) $mv\ell$ down (b) $M/(M + m)$

37. (a) $\omega = 2mv_id/(M + 2m)R^2$ (b) No; some mechanical energy changes into internal energy

39. $\sim 10^{-13}$ rad/s

41. 5.45×10^{22} N · m

43. 7.50×10^{-11} s

45. (a) $7md^2/3$ (b) $mgd\,\hat{\mathbf{k}}$ (c) $3g/7d$ counterclockwise (d) $2g/7$ upward (e) mgd (f) $\sqrt{6g/7d}$ (g) $m\sqrt{14gd^3/3}$ (h) $\sqrt{2gd/21}$

47. 0.910 km/s

49. (a) v_ir_i/r (b) $T = (mv_i^2r_i^2)r^{-3}$ (c) $\frac{1}{2}mv_i^2(r_i^2/r^2 - 1)$ (d) 4.50 m/s, 10.1 N, 0.450 J

51. (a) 3 750 kg · m^2/s (b) 1.88 kJ (c) 3 750 kg · m^2/s (d) 10.0 m/s (e) 7.50 kJ (f) 5.62 kJ

53. An increase of 0.550 s

55. $4[ga(\sqrt{2} - 1)/3]^{1/2}$

CHAPTER 12

1. 10.0 N up; 6.00 N · m counterclockwise

3. $[(m_1 + m_b)d + (m_1\ell/2)]/m_2$

5. (3.85 cm, 6.85 cm)

7. (− 1.50 m, − 1.50 m)

9. 177 kg

11. 8.33%

13. (a) $f_s = 268$ N, $n = 1\,300$ N (b) 0.324

15. (a) 1.04 kN at 60.0° (b) $(370\hat{\mathbf{i}} + 900\hat{\mathbf{j}})$ N

17. 2.94 kN on each rear wheel and 4.41 kN on each front wheel

19. (a) 29.9 N (b) 22.2 N

21. (a) 1.73 rad/s² (b) 1.56 rad/s (c) $(-4.72\hat{\mathbf{i}} + 6.62\hat{\mathbf{j}})$ kN (d) $38.9\hat{\mathbf{j}}$ kN

23. 2.82 m

25. 88.2 N and 58.8 N

27. 4.90 mm

29. 10×10^{10} N/m²

31. 23.8 μm

33. (a) 3.14×10^4 N (b) 6.28×10^4 N

35. 1.65×10^8 N/m²

37. 0.860 mm

39. $n_A = 5.98 \times 10^5$ N, $n_B = 4.80 \times 10^5$ N

41. 9.00 ft

43. (a)

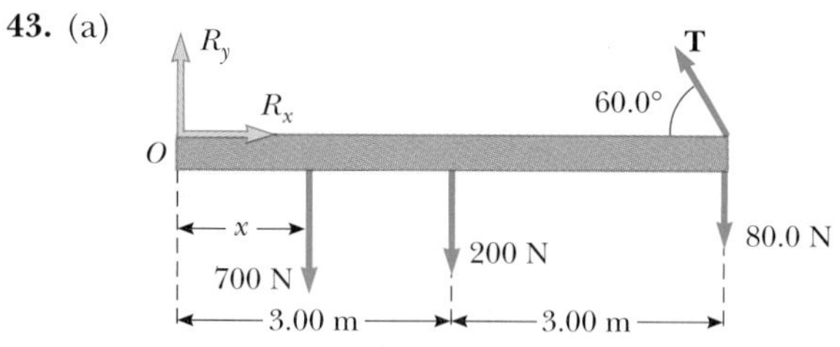

(b) $T = 343$ N; $R_x = 171$ N to the right, $R_y = 683$ N up (c) 5.13 m

45. (a) $T = F_g(L + d)/\sin\theta\,(2L + d)$ (b) $R_x = F_g(L + d)\cot\theta/(2L + d)$; $R_y = F_gL/(2L + d)$

47. $\mathbf{F}_A = (-6.47 \times 10^5\hat{\mathbf{i}} + 1.27 \times 10^5\hat{\mathbf{j}})$ N, $\mathbf{F}_B = 6.47 \times 10^5\hat{\mathbf{i}}$ N

49. 5.08 kN; $R_x = 4.77$ kN, $R_y = 8.26$ kN

51. $T = 2.71$ kN, $R_x = 2.65$ kN

53. (a) 20.1 cm to the left of the front edge; $\mu_k = 0.571$ (b) 0.501 m

55. (a) $M = (m/2)(2\mu_s \sin\theta - \cos\theta)(\cos\theta - \mu_s \sin\theta)^{-1}$ (b) $R = (m + M)g\,(1 + \mu_s^2)^{1/2}$; $F = g[M^2 + \mu_s^2(m + M)^2]^{1/2}$

57. (a) 133 N (b) $n_A = 429$ N and $n_B = 257$ N (c) $R_x = 133$ N and $R_y = -257$ N

59. 66.7 N

63. 1.09 m

65. (a) 4 500 N (b) 4.50×10^6 N/m² (c) The board will break.

67. 5.73 rad/s

69. $n_A = 11.0$ kN, $n_E = 3.67$ kN; $F_{AB} = F_{DE} = 7.35$ kN compression; $F_{AC} = F_{CE} = 6.37$ kN compression; $F_{BC} = F_{CD} = 4.24$ kN tension; $F_{BD} = 8.49$ kN compression

71. (a) $P_y = (F_g/L)(d - ah/g)$ (b) 0.306 m (c) $(-306\hat{\mathbf{i}} + 5.53\hat{\mathbf{j}})$ N

73. Decrease h, increase d

CHAPTER 13

1. ~10^{-7} N toward you

3. (a) 2.50×10^{-5} N toward the 500-kg object (b) between the objects and 0.245 m from the 500-kg object

5. $(-100\hat{\mathbf{i}} + 59.3\hat{\mathbf{j}})$pN

7. 7.41×10^{-10} N

9. 0.613 m/s² toward the Earth

11. (a) 3.46×10^8 m (b) 3.34×10^{-3} m/s² toward the Earth

13. 1.26×10^{32} kg

15. 1.90×10^{27} kg

17. 35.2 AU

19. 8.92×10^7 m

21. After 393 yr, Mercury would be farther from the Sun than Pluto

23. $\mathbf{g} = (Gm/\ell^2)(\frac{1}{2} + \sqrt{2})$ toward the opposite corner

25. $\mathbf{g} = 2MGr(r^2 + a^2)^{-3/2}$ toward the center of mass

27. 4.17×10^{10} J

29. (a) 1.84×10^9 kg/m³ (b) 3.27×10^6 m/s² (c) -2.08×10^{13} J

31. (a) -1.67×10^{-14} J (b) at the center

33. 1.66×10^4 m/s

37. (a) 5.30×10^3 s (b) 7.79 km/s (c) 6.43×10^9 J

39. 469 MJ

41. 15.6 km/s

43. (b) 1.00×10^7 m (c) 1.00×10^4 m/s

45. (a) 0.980 (b) 127 yr (c) -2.13×10^{17} J

49. (b) $2[Gm^3(1/2r - 1/R)]^{1/2}$

51. (b) 1.10×10^{32} kg

53. (a) -7.04×10^4 J (b) -1.57×10^5 J (c) 13.2 m/s

55. 7.79×10^{14} kg

57. $\omega = 0.057\,2$ rad/s or 1 rev in 110 s

59. $v_{esc} = (8\pi G\rho/3)^{1/2}R$

61. (a) $m_2(2G/d)^{1/2}(m_1 + m_2)^{-1/2}$ and $m_1(2G/d)^{1/2}(m_1 + m_2)^{-1/2}$; relative speed $(2G/d)^{1/2}(m_1 + m_2)^{1/2}$ (b) 1.07×10^{32} J and 2.67×10^{31} J

63. (a) 8.50×10^8 J (b) 2.71×10^9 J

65. (a) 200 Myr (b) ~10^{41} kg; ~10^{11} stars

67. $(GM_E/4R_E)^{1/2}$

71.

t (s)	x (m)	y (m)	v_x (m/s)	v_y (m/s)
0	0	12 740 000	5 000	0
10	50 000	12 740 000	4 999.9	− 24.6
20	99 999	12 739 754	4 999.7	− 49.1
30	149 996	12 739 263	4 999.4	− 73.7…

The object does not hit the Earth; its minimum radius is $1.33R_E$ as shown in the diagram below. Its period is 1.09×10^4 s. A circular orbit would require speed 5.60 km/s.

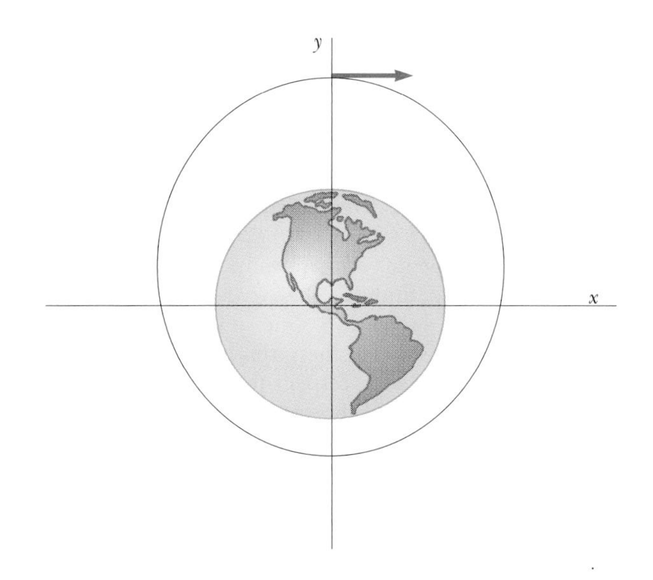

CHAPTER 14

1. 0.111 kg

3. 6.24 MPa

5. 5.27×10^{18} kg

7. 1.62 m

9. 7.74×10^{-3} m^2

11. 271 kN horizontally backward

13. $P_0 + \frac{1}{2}\rho d\sqrt{g^2 + a^2}$

15. 0.722 mm

17. 10.5 m; no; some alcohol and water evaporate

19. 98.6 kPa

21. (a) 1.57 Pa, 1.55×10^{-2} atm, 11.8 mm Hg (b) The fluid level in the tap should rise. (c) Blockage of flow of the cerebrospinal fluid

23. 0.258 N

25. (a) 9.80 N (b) 6.17 N

27. (a) $1.017\,9 \times 10^3$ N down, $1.029\,7 \times 10^3$ N up (b) 86.2 N

29. (a) 7.00 cm (b) 2.80 kg

33. 1 430 m^3

35. 1 250 kg/m^3 and 500 kg/m^3

37. 1.28×10^4 m^2

39. (a) 17.7 m/s (b) 1.73 mm

41. 31.6 m/s

43. 0.247 cm

45. (a) 1 atm + 15.0 MPa (b) 2.95 m/s (c) 4.34 kPa

47. 2.51×10^{-3} m^3/s

49. 103 m/s

51. (a) 4.43 m/s
(b) The siphon can be no higher than 10.3 m.

53. 12.6 m/s

55. 1.91 m

59. 0.604 m

63. 17.3 N and 31.7 N

65. 90.04%

67. 758 Pa

69. 4.43 m/s

71. (a) 1.25 cm (b) 13.8 m/s

73. (a) 3.307 g (b) 3.271 g (c) 3.48×10^{-4} N

75. (c) 1.70 m^2

CHAPTER 15

1. (a) The motion repeats precisely. (b) 1.82 s
(c) No, the force is not in the form of Hooke's law.

3. (a) 1.50 Hz, 0.667 s (b) 4.00 m (c) π rad
(d) 2.83 m

5. (b) 18.8 cm/s, 0.333 s (c) 178 cm/s^2, 0.500 s
(d) 12.0 cm

7. (a) 2.40 s (b) 0.417 Hz (c) 2.62 rad/s

9. 40.9 N/m

11. (a) 40.0 cm/s, 160 cm/s^2 (b) 32.0 cm/s, $-$96.0 cm/s^2
(c) 0.232 s

13. 0.628 m/s

15. (a) 0.542 kg (b) 1.81 s (c) 1.20 m/s^2

17. 2.23 m/s

19. (a) 28.0 mJ (b) 1.02 m/s (c) 12.2 mJ (d) 15.8 mJ

21. (a) E increases by a factor of 4. (b) v_{max} is doubled.
(c) a_{max} is doubled. (d) Period is unchanged.

23. 2.60 cm and $-$2.60 cm

25. (b) 0.628 s

27. (a) 35.7 m (b) 29.1 s

29. $\sim 10^0$ s

31. Assuming simple harmonic motion, (a) 0.820 m/s
(b) 2.57 rad/s^2 (c) 0.641 N. More precisely,
(a) 0.817 m/s (b) 2.54 rad/s^2 (c) 0.634 N

35. 0.944 kg·m^2

39. (a) 5.00×10^{-7} kg·m^2 (b) 3.16×10^{-4} N·m/rad

41. 1.00×10^{-3} s^{-1}

43. (a) 7.00 Hz (b) 2.00% (c) 10.6 s

45. (a) 1.00 s (b) 5.09 cm

47. 318 N

49. 1.74 Hz

51. (a) $2Mg$; $Mg(1 + y/L)$
(b) $T = (4\pi/3)(2L/g)^{1/2}$; 2.68 s

53. 6.62 cm

55. 9.19×10^{13} Hz

57. (a)

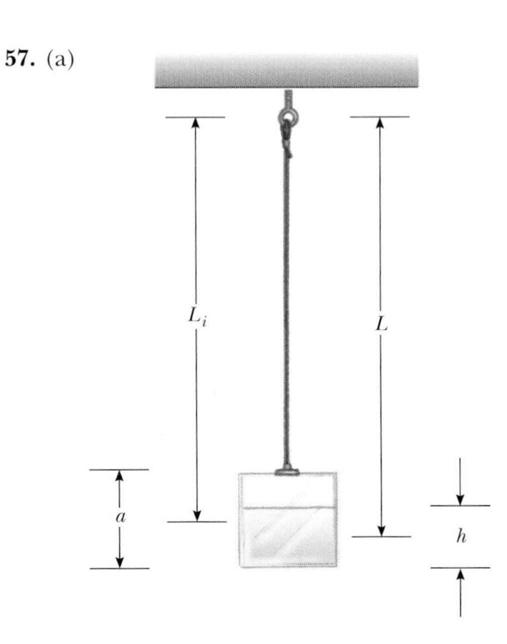

(b) $\dfrac{dT}{dt} = \dfrac{\pi(dM/dt)}{2\rho a^2 g^{1/2}[L_i + (dM/dt)t/2\rho a^2]^{1/2}}$

(c) $T = 2\pi g^{-1/2}\,[L_i + (dM/dt)t/2\rho a^2]^{1/2}$

59. $f = (2\pi L)^{-1}\,(gL + kh^2/M)^{1/2}$

61. (b) 1.23 Hz

63. (a) 3.00 s (b) 14.3 J (c) 25.5°

65. If the cyclist goes over them at one certain speed, the washboard bumps can excite a resonance vibration of the bike, so large in amplitude as to make the rider lose control. $\sim 10^1$ m

73. For $\theta_{max} = 5.00°$ there is precise agreement. For $\theta_{max} = 100°$ there are large differences, and the period is 23% greater than small-angle period.

75. (b) after 42.1 min

CHAPTER 16

1. $y = 6\,[(x - 4.5t)^2 + 3]^{-1}$

3. (a) left (b) 5.00 m/s

5. 184 km

7. 0.319 m

9. 2.00 cm, 2.98 m, 0.576 Hz, 1.72 m/s

11. (a) 3.77 m/s (b) 118 m/s^2

13. (a) 0.250 m (b) 40.0 rad/s (c) 0.300 rad/m (d) 20.9 m (e) 133 m/s (f) $+x$

15. (a) $y = (8.00\text{ cm})\sin(7.85x + 6\pi t)$
(b) $y = (8.00\text{ cm})\sin(7.85x + 6\pi t - 0.785)$

17. (a) -1.51 m/s, 0 (b) 16.0 m, 0.500 s, 32.0 m/s

19. (a) 0.500 Hz, 3.14 rad/s (b) 3.14 rad/m
(c) $(0.100\text{ m})\sin(3.14x/\text{m} - 3.14t/\text{s})$
(d) $(0.100\text{ m})\sin(-3.14t/\text{s})$
(e) $(0.100\text{ m})\sin(4.71\text{ rad} - 3.14t/\text{s})$ (f) 0.314 m/s

21. 80.0 N

23. 520 m/s

25. 1.64 m/s^2

27. 13.5 N

29. 185 m/s

31. 0.329 s

33. (a) s and kg·m/s^2 (b) time interval (period) and force (tension)

37. 55.1 Hz

39. (a) 62.5 m/s (b) 7.85 m (c) 7.96 Hz (d) 21.1 W

41. $\sqrt{2}\mathcal{P}_0$

43. (a) $A = 40$ (b) $A = 7.00$, $B = 0$, $C = 3.00$. One can take the dot product of the given equation with each one of $\hat{\mathbf{i}}, \hat{\mathbf{j}}$, and $\hat{\mathbf{k}}$. (c) $A = 0$, $B = 7.00$ mm, $C = 3.00$/m, $D = 4.00$/s, $E = 2.00$. Consider the average value of both sides of the given equation to find A. Then consider the maximum value of both sides to find B. One can evaluate the partial derivative of both sides of the given equation with respect to x and separately with respect to t to obtain equations yielding C and D upon chosen substitutions for x and t. Then substitute $x = 0$ and $t = 0$ to obtain E.

47. ~ 1 min

49. (a) $(3.33\hat{\mathbf{i}})$ m/s (b) -5.48 cm (c) 0.667 m, 5.00 Hz (d) 11.0 m/s

51. 0.456 m/s

53. (a) 39.2 N (b) 0.892 m (c) 83.6 m/s

55. (a) 179 m/s (b) 17.7 kW

57. 0.084 3 rad

61. (a) $(0.707)2(L/g)^{1/2}$ (b) $L/4$

63. 3.86×10^{-4}

65. (b) 31.6 m/s

67. (a) $\dfrac{\mu\omega^3}{2k}A_0^2e^{-2bx}$ (b) $\dfrac{\mu\omega^3}{2k}A_0^2$ (c) e^{-2bx}

69. (a) $\mu_0 + (\mu_L - \mu_0)x/L$

CHAPTER 17

1. 5.56 km

3. 7.82 m

5. (a) 826 m (b) 1.47 s

7. 5.67 mm

9. 1.50 mm to 75.0 μm

11. (a) 2.00 μm, 40.0 cm, 54.6 m/s (b) -0.433 μm (c) 1.72 mm/s

13. $\Delta P = (0.200\text{ N/m}^2)\sin(62.8\,x/\text{m} - 2.16 \times 10^4\,t/\text{s})$

15. 5.81 m

19. 66.0 dB

21. (a) 3.75 W/m^2 (b) 0.600 W/m^2

23. (a) 2.34 m and 0.390 m (b) 0.161 N/m^2 for both notes (c) 4.25×10^{-7} m and 7.09×10^{-8} m (d) The wavelengths and displacement amplitudes would be larger by a factor of 1.09. The answer to (b) is unchanged.

25. (a) 1.32×10^{-4} W/m^2 (b) 81.2 dB

27. (a) 0.691 m (b) 691 km

29. 65.6 dB

31. (a) 65.0 dB (b) 67.8 dB (c) 69.6 dB

33. (a) 30.0 m (b) 9.49×10^5 m

35. (a) 332 J (b) 46.4 dB

37. (a) 338 Hz (b) 483 Hz

39. 26.4 m/s

41. 19.3 m

43. (a) 0.364 m (b) 0.398 m (c) 941 Hz (d) 938 Hz

45. 2.82×10^8 m/s

47. (a) 56.3 s (b) 56.6 km farther along

49. 22.3° left of center

51. $f \sim 300$ Hz, $\lambda \sim 10^0$ m, duration $\sim 10^{-1}$ s

55. 6.01 km

57. (a) 55.8 m/s (b) 2 500 Hz

59. 1 204.2 Hz

61. 1.60

63. 2.34 m

65. (a) 0.948° (b) 4.40°

67. 1.34×10^4 N

69. (b) 531 Hz

71. (a) 6.45 (b) 0

CHAPTER 18

1. (a) −1.65 cm (b) −6.02 cm (c) 1.15 cm

3. (a) $+x$, $-x$ (b) 0.750 s (c) 1.00 m

5. (a) 9.24 m (b) 600 Hz

7. (a) zero (b) 0.300 m

9. (a) 2 (b) 9.28 m and 1.99 m

11. (a) 156° (b) 0.058 4 cm

13. 15.7 m, 31.8 Hz, 500 m/s

15. At 0.089 1 m, 0.303 m, 0.518 m, 0.732 m, 0.947 m, 1.16 m from one speaker

17. (a) 4.24 cm (b) 6.00 cm (c) 6.00 cm (d) 0.500 cm, 1.50 cm, 2.50 cm

19. 0.786 Hz, 1.57 Hz, 2.36 Hz, 3.14 Hz

21. (a) 350 Hz (b) 400 kg

23. 1.27 cm

25. (a) reduced by 1/2 (b) reduced by $1/\sqrt{2}$ (c) increased by $\sqrt{2}$

27. (a) 163 N (b) 660 Hz

29. $\dfrac{Mg}{4Lf^2 \tan\theta}$

31. (a) 3 loops (b) 16.7 Hz (c) 1 loop

33. (a) 3.66 m/s (b) 0.200 Hz

35. 9.00 kHz

37. (a) 0.357 m (b) 0.715 m

39. 57.6 Hz

41. n(206 Hz) for $n = 1$ to 9 and n(84.5 Hz) for $n = 2$ to 23

43. 50.0 Hz, 1.70 m

45. (a) 350 m/s (b) 1.14 m

47. (a) 162 Hz (b) 1.06 m

49. (a) 1.59 kHz (b) odd-numbered harmonics (c) 1.11 kHz

51. 5.64 beats/s

53. (a) 1.99 beats/s (b) 3.38 m/s

55. The second harmonic of E is close to the third harmonic of A, and the fourth harmonic of C# is close to the fifth harmonic of A.

57. (a) 34.8 m/s (b) 0.977 m

59. 3.85 m/s away from the station or 3.77 m/s toward the station

61. 21.5 m

63. (a) 59.9 Hz (b) 20.0 cm

65. (a) 1/2 (b) $[n/(n+1)]^2 T$ (c) 9/16

67. $y_1 + y_2 = 11.2 \sin(2.00x - 10.0t + 63.4°)$

69. (a) 78.9 N (b) 211 Hz

CHAPTER 19

1. (a) − 274°C (b) 1.27 atm (c) 1.74 atm

3. (a) − 320°F (b) 77.3 K

5. (a) 810°F (b) 450 K

7. (a) 1 337 K, 2 993 K (b) 1 596°C = 1 596 K

9. 3.27 cm

11. 55.0°C

13. (a) 0.176 mm (b) 8.78 μm (c) 0.093 0 cm^3

15. (a) − 179°C (attainable) (b) − 376°C (below 0 K, unattainable)

17. 0.548 gal

19. (a) 99.8 mL (b) about 6% of the volume change of the acetone

21. (a) 99.4 cm^3 (b) 0.943 cm

23. 1.14°C

25. 5 336 images

27. (a) 400 kPa (b) 449 kPa

29. 1.50×10^{29} molecules

31. 1.61 MPa = 15.9 atm

33. 472 K

35. (a) 41.6 mol (b) 1.20 kg, nearly in agreement with the tabulated density

37. (a) 1.17 g (b) 11.5 mN (c) 1.01 kN (d) The molecules must be moving very fast.

39. 4.39 kg

41. 3.55 L

43. $m_1 - m_2 = \dfrac{P_0 VM}{R}\left(\dfrac{1}{T_1} - \dfrac{1}{T_2}\right)$

45. (a) 94.97 cm (b) 95.03 cm

47. 3.55 cm

49. It falls by 0.094 3 Hz

51. (a) Expansion makes density drop. (b) $5 \times 10^{-5}/°C$

53. (a) $h = nRT/(mg + P_0A)$ (b) 0.661 m

55. We assume that $\alpha\Delta T$ is much less than 1.

57. (a) 0.340% (b) 0.480%

59. 0.750

61. 2.74 m

63. (b) 1.33 kg/m^3

67. No. Steel is not strong enough.

69. (a) $L_f = L_i e^{\alpha \Delta T}$ (b) $2.00 \times 10^{-4}\%$; 59.4%

71. (a) 6.17×10^{-3} kg/m (b) 632 N (c) 580 N; 192 Hz

73. 4.54 m

CHAPTER 20

1. (10.0 + 0.117)°C

3. 0.234 kJ/kg · °C

5. 1.78×10^4 kg

7. 29.6°C

9. (a) 0.435 cal/g · °C (b) beryllium

11. 23.6°C

13. 50.7 ks

15. 1.22×10^5 J

17. 0.294 g

19. 0.414 kg

21. (a) 0°C (b) 114 g

23. − 1.18 MJ

25. − 466 J

27. (a) $-4P_iV_i$ (b) It is proportional to the square of the volume, according to $T = (P_i/nRV_i)V^2$

29. $Q = -720$ J

31.

	Q	W	ΔE_{int}
BC	−	0	−
CA	−	+	−
AB	+	−	+

33. 3.60 kJ

35. (a) 7.50 kJ (b) 900 K

37. −3.10 kJ; 37.6 kJ

39. (a) 0.041 0 m^3 (b) + 5.48 kJ (c) − 5.48 kJ

41. 2.22×10^{-2} W/m · °C

43. 51.2°C

45. 67.9°C

47. 3.77×10^{26} J/s

49. 3.49×10^3 K

51. 277 K = 4°C

53. 2.27 km

55. (a) 16.8 L (b) 0.351 L/s

57. $c = \mathcal{P}/\rho R \Delta T$

59. − 1.87 kJ

61. 5.87×10^4 °C

63. 5.31 h

65. 1.44 kg

67. 38.6 m^3/d

71. 9.32 kW

73. (a) 3.16×10^{22} W (b) 5.78×10^3 K, 0.327% less than 5 800 K (c) 3.17×10^{22} W, 0.408% larger

CHAPTER 21

1. 0.943 N; 1.57 Pa

3. 3.65×10^4 N

5. 3.32 mol

7. (a) 3.54×10^{23} atoms (b) 6.07×10^{-21} J (c) 1.35 km/s

9. (a) 8.76×10^{-21} J for both (b) 1.62 km/s for helium and 514 m/s for argon

13. (a) 3.46 kJ (b) 2.45 kJ (c) − 1.01 kJ

15. (a) 209 J (b) zero (c) 317 K

17. 1.18 atm

19. Between 10^{-2} and 10^{-3} °C

21. (a) 316 K (b) 200 J

23. (a) $C = \dfrac{n_1C_1 + n_2C_2}{n_1 + n_2}$ (b) $C = \dfrac{\sum_{i=1}^{m} n_iC_i}{\sum_{i=1}^{m} n_i}$

25. (a) 1.39 atm (b) 366 K, 253 K (c) 0, − 4.66 kJ, − 4.66 kJ

27. 227 K

29. (a)

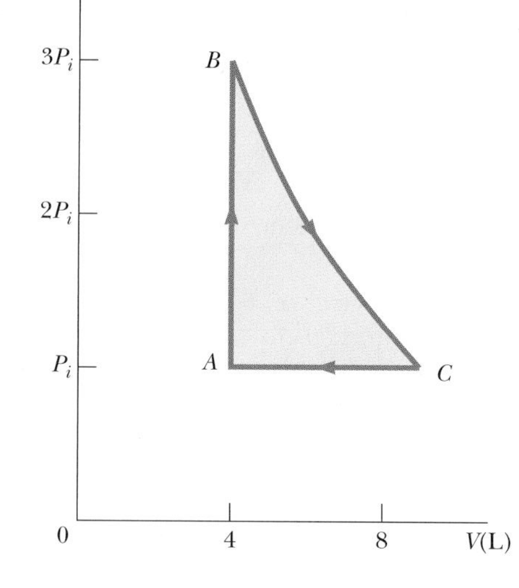

(b) 8.77 L (c) 900 K (d) 300 K (e) − 336 J

31. (a) 28.0 kJ (b) 46.1 kJ (c) Isothermal process, $P_f = 10.0$ atm; adiabatic process, $P_f = 25.1$ atm

33. (a) 9.95 cal/K, 13.9 cal/K (b) 13.9 cal/K, 17.9 cal/K

35. 2.33×10^{-21} J

37. (a) 6.80 m/s (b) 7.41 m/s (c) 7.00 m/s

41. (a) 2.37×10^4 K (b) 1.06×10^3 K

43. (b) 0.278

45. (a) 3.21×10^{12} molecules (b) 779 km
(c) 6.42×10^{-4} s^{-1}

49. (a) 9.36×10^{-8} m (b) 9.36×10^{-8} atm (c) 302 atm

51. (a) 100 kPa, 66.5 L, 400 K, 5.82 kJ, 7.48 kJ, − 1.66 kJ
(b) 133 kPa, 49.9 L, 400 K, 5.82 kJ, 5.82 kJ, 0
(c) 120 kPa, 41.6 L, 300 K, 0, − 910 J, + 910 J
(d) 120 kPa, 43.3 L, 312 K, 722 J, 0, + 722 J

55. 510 K and 290 K

57. 0.623

59. (a) Pressure increases as volume decreases
(d) 0.500 atm^{-1}, 0.300 atm^{-1}

61. (a) 0.514 m^3 (b) 2.06 m^3 (c) 2.38×10^3 K
(d) − 480 kJ (e) 2.28 MJ

63. 1.09×10^{-3}; 2.69×10^{-2}; 0.529; 1.00; 0.199; 1.01×10^{-41}; 1.25×10^{-1082}

67. (a) 0.203 mol (b) $T_B = T_C = 900$ K, $V_C = 15.0$ L

(c, d)	P, atm	V, L	T, K	E_{int}, kJ
A	1.00	5.00	300	0.760
B	3.00	5.00	900	2.28
C	1.00	15.0	900	2.28
A	1.00	5.00	300	0.760

(e) Lock the piston in place and put the cylinder into an oven at 900 K. Keep the gas in the oven while gradually letting the gas expand to lift a load on the piston as far as it can. Move the cylinder from the oven back to the 300-K room and let the gas cool and contract.

(f, g)	Q, kJ	W, kJ	ΔE_{int}, kJ
AB	1.52	0	1.52
BC	1.67	− 1.67	0
CD	− 2.53	+ 1.01	− 1.52
$ABCA$	0.656	− 0.656	0

69. 1.60×10^4 K

CHAPTER 22

1. (a) 6.94% (b) 335 J

3. (a) 10.7 kJ (b) 0.533 s

5. (a) 29.4 L/h (b) 185 hp (c) 527 N · m
(d) 1.91×10^5 W

7. (a) 24.0 J (b) 144 J

9. (a) 2.93 (b) coefficient of performance for a refrigerator
(c) $300 is twice as large as $150

11. (a) 67.2% (b) 58.8 kW

13. (a) 741 J (b) 459 J

15. (a) 4.20 W (b) 31.2 g

17. (a) 564 K (b) 212 kW (c) 47.5%

19. (b) $1 - T_c/T_h$ (c) $(T_c + T_h)/2$ (d) $(T_hT_c)^{1/2}$

21. (a) 214 J, 64.3 J
(b) − 35.7 J, − 35.7 J. The net effect is the transport of energy by heat from the cold to the hot reservoir without expenditure of external work. (c) 333 J, 233 J
(d) 83.3 J, 83.3 J, 0. The net effect is converting energy, taken in by heat, entirely into energy output by work in a cyclic process.
(e) − 0.111 J/K. The entropy of the Universe has decreased.

23. 9.00

27. 72.2 J

29. 1.86

31. (a) 244 kPa (b) 192 J

33. 146 kW, 70.8 kW

35. − 610 J/K

37. 195 J/K

39. 236 J/K

41. 1.02 kJ/K

43. ~ 10^0 W/K from metabolism; much more if you are using high-power electric appliances or an automobile, or if your taxes are paying for a war.

45. 5.76 J/K; temperature is constant if the gas is ideal

47. 18.4 J/K

49. (a) 1 (b) 6

51. (a)

Result	Number of Ways to Draw
All R	1
2 R, 1 G	3
1R, 2 G	3
All G	1

(b)

Result	Number of Ways to Draw
All R	1
4R, 1G	5
3R, 2G	10
2R, 3G	10
1R, 4G	5
All G	1

53. (a) 5.00 kW (b) 763 W

55. (a) 0.476 J/K (b) 417 J (c) $W_{net} = T_1 \Delta S_U = 167$ J

57. (a) $2nRT_i \ln 2$ (b) 0.273

59. 5.97×10^4 kg/s

61. (a) 3.19 cal/K (b) 98.19°F, 2.59 cal/K

63. (a) 8.48 kW (b) 1.52 kW (c) 1.09×10^4 J/K
(d) COP drops by 20.0%

65. (a) $10.5nRT_i$ (b) $8.50nRT_i$ (c) 0.190 (d) 0.833

67. (a) $nC_P \ln 3$
(b) Both ask for the change in entropy between the same two states of the same system. Entropy is a function of state. The change in entropy does not depend on path, but only on original and final states.

71. (a) 20.0°C (c) $\Delta S = +4.88$ J/K (d) Yes

Credits

Photographs

Chapter 1. **xxiv:** Courtesy of NASA **2:** elektraVision/Index Stock Imagery **6:** top left, Courtesy of National Institute of Standards and Technology, U.S. Dept. of Commerce; top right, Courtesy of National Institute of Standards and Technology, U.S. Dept. of Commerce **13:** Phil Boorman/Getty Images **19:** Sylvain Grandadam/Photo Researchers, Inc.

Chapter 2. **23:** George Lepp/Getty Images **26:** Ken White/Allsport/Getty Images **46:** North Wind Picture Archive **52:** bottom left, Courtesy U.S. Air Force; bottom right, Photri, Inc. **53:** George Semple **55:** Courtesy Amtrak Nec Media Relations

Chapter 3. **58:** Mark Wagner/Getty Images

Chapter 4. **77:** © Arndt/Premium Stock/PictureQuest **79:** Mark C. Burnett/Photo Researchers, Inc. **84:** The Telegraph Colour Library/Getty Images **86:** Mike Powell/Allsport/Getty Images **88:** Central Scientific Company **103:** bottom left, McKinney/Getty Images **104:** top left, Jed Jacobsohn/Allsport/Getty Images; top right, Bill Lee/Dembinsky Photo Associates; bottom left, Sam Sargent/Liaison International; bottom right, Courtesy of NASA **106:** Courtesy of NASA

Chapter 5. **111:** © Steve Raymer/CORBIS **114:** Giraudon/Art Resource **119:** Courtesy of NASA **120:** John Gillmoure/corbisstockmarket.com **122:** © John Elk III/Stock, Boston Inc./PictureQuest **139:** Roger Violet, Mill Valley, CA, University Science Books, 1992

Chapter 6. **150:** © Paul Hardy/CORBIS **151:** Mike Powell/Allsport/Getty Images **152:** © Tom Carroll/Index Stock Imagery/PictureQuest **157:** Robin Smith/Getty Images **165:** Jump Run Productions/Getty Images **166:** Charles D. Winters **174:** Frank Cezus/Getty Images **178:** Color Box/Getty Images

Chapter 7. **181:** Billy Hustace/Getty Images **183:** all, Charles D. Winters **184:** Gerard Vandystandt/Photo Researchers, Inc. **198:** all except top right, George Semple **198:** top right, Digital Vision/Getty Images **213:** Ron Chapple/Getty Images

Chapter 8. **217:** Harold E. Edgerton/Courtesy of Palm Press, Inc. **242:** Gamma **250:** Engraving from Scientific American, July 1888

Chapter 9. **251:** Mark Cooper/corbisstockmarket.com **257:** Courtesy of Saab **259:** top, Harold & Esther Edgerton Foundation 2002, courtesy of Palm Press, Inc; bottom, © Tim Wright/CORBIS **263:** No credit available **264:** Courtesy of Central Scientific Company **276:** Richard Megna/Fundamental Photographs **278:** Courtesy of NASA **279:** Courtesy of NASA **281:** Bill Stormont/corbisstockmarket.com **286:** Eye Ubiquitous/CORBIS

Chapter 10. **292:** Courtesy Tourism Malaysia **299:** George Semple **316:** Henry Leap and Jim Lehman **323:** Bruce Ayers/Getty Images **327:** John Lawrence/Getty Images **329:** Jerry Wachter/Photo Researchers, Inc.

Chapter 11. **336:** Otto Gruele/Getty Images **346:** both, © 1998 David Madison **354:** Gerard Lacz/NHPA

Chapter 12. **362:** John W. Jewett, Jr. **366:** Charles D. Winters

Chapter 13. **389:** University of Arizona/JPL/NASA **394:** Courtesy of NASA **410:** Courtesy H. Ford, et al., & NASA **417:** Courtesy of NASA

Chapter 14. **420:** Austin MacRae **422:** Earl Young/Getty Images **424:** right, David Frazier **427:** © Hulton-Deutsch Collection/CORBIS **430:** Geraldine Prentice/Getty Images **431:** Andy Sachs/Getty Images **431:** Werner Wolff/stockphoto.com **432:** George Semple **434:** Bettmann/CORBIS **435:** Courtesy of Central Scientific Company **438:** Galen Rowell/Peter Arnold, Inc. **439:** bottom left, Pamela Zilly/Getty Images **439:** top right, Henry Leap and Jim Lehman **439:** bottom right, Henry Leap and Jim Lehman **444:** George Semple **445:** Stan Osolinski/Dembinsky Photo Associates **447:** The Granger Collection **448:** Courtesy of Jeanne Maier

Chapter 15. **450:** Don Bonsey/Getty Images **452:** both, www.comstock.com **465:** top left, Courtesy of Ford Motor Company; bottom left, © Link/Visuals Unlimited **474:** both, UPI-Bettmann/CORBIS **478:** Telegraph Colour Library/Getty Images **479:** George Semple

Chapter 16. **486:** Kathy Ferguson Johnson/PhotoEdit/PictureQuest **509:** Gregg Adams/Getty Images

Chapter 17. **512:** Getty Images **518:** Courtesy Kenneth Burger Museum Archives/Kent State University **524:** Courtesy of the Educational Development Center, Newton, MA **527:** © 1973 Kim Vandiver and Harold E. Edgerton/Courtesy of Palm Press, Inc. **528:** www.comstock.com

529: © Bettmann/CORBIS **531:** Courtesy of University of Miami, Music Engineering **533:** Courtesy of Sony Cinema Products Corporation © 2003 Sony Cinema Products Corporation. All Rights Reserved. Reproduction in whole or in part without written permission is prohibited. **536:** Joe McDonald/Visuals Unlimited

Chapter 18. **543:** Bettmann/CORBIS **545:** Courtesy of the Educational Development Center, Newton, MA **546:** Courtesy of the Educational Development Center, Newton, MA **550:** © 1991 Richard Megna/Fundamental Photographs **554:** © 1991 Richard Megna/Fundamental Photographs **556:** Charles D. Winters **559:** a) Courtesy of Professor Thomas D. Rossing, Northern Illinois University; b) © 1992 Ben Rose/Getty Images **567:** a) © 1989 Gary Buss/Getty Images; b, c) © 1989 Richard Laird/Getty Images **575:** Murray Greenberg

Chapter 19. **578:** Topham Picturepoint/The Image Works **580:** Lowell Georgia/CORBIS **583:** both, Charles D. Winters **587:** Frank Siteman/Stock, Boston **587:** George Semple **592:** Steve Niedorf/Getty Images **596:** Courtesy of Central Scientific Company **601:** Charles D. Winters

Chapter 20. **604:** Jacob Taposchaner/Getty Images **606:** By kind permission of the President and Council of the Royal Society **623:** Charles D. Winters **626:** Courtesy of Dr. Albert A. Barlett, University of Colorado, Boulder, CO **631:** Courtesy of Dr. Albert A. Barlett, University of Colorado, Boulder, CO **636:** Nathan Bilow/Leo de Wys, Inc.

Chapter 21. **640:** Frank Oberle/Getty Images **654:** Courtesy of AIP Niels Bohr Library, Lande Collection **662:** George Semple **666:** AP/Wide World Photos

Chapter 22. **668:** Courtesy of Ford Motor Company **670:** © Phil Degginger/Getty Images **670:** J. L. Charmet/SPL/Photo Researchers, Inc. **673:** Charles D. Winters **676:** J. L. Charmet/SPL/Photo Researchers, Inc. **686:** both, George Semple **697:** both, Courtesy of PASCO Scientific Company **702:** corbisstockmarket.com

Tables and Illustrations

This page constitutes an extension of the copyright page. We have made every effort to trace the ownership of all copyrighted material and to secure permission from copyright holders. In the event of any question arising as to the use of any material, we will be pleased to make the necessary corrections in future printings. Thanks are due to the following authors, publishers, and agents for permission to use the material indicated.

Chapter 2. **56:** By permission of John Hart and Creators Syndicate, Inc.

Chapter 13. **415:** By permission of John Hart and Creators Syndicate, Inc.

Chapter 14. **449:** Calvin & Hobbes © 1992 Watterson. Reprinted with permission of Universal Press Syndicate. All rights reserved.

Chapter 21. **666:** By permission of John Hart and Creators Syndicate, Inc.

Index

Locator notes: **boldface** indicates a definition; *italics* indicates a figure; *t* indicates a table

Standard Abbreviations and Symbols for Units

Symbol	Unit	Symbol	Unit
A	ampere	K	kelvin
u	atomic mass unit	kg	kilogram
atm	atmosphere	kmol	kilomole
Btu	British thermal unit	L	liter
C	coulomb	lb	pound
°C	degree Celsius	ly	lightyear
cal	calorie	m	meter
d	day	min	minute
eV	electron volt	mol	mole
°F	degree Fahrenheit	N	newton
F	farad	Pa	pascal
ft	foot	rad	radian
G	gauss	rev	revolution
g	gram	s	second
H	henry	T	tesla
h	hour	V	volt
hp	horsepower	W	watt
Hz	hertz	Wb	weber
in.	inch	yr	year
J	joule	Ω	ohm

Mathematical Symbols Used in the Text and Their Meaning

Symbol	Meaning
$=$	is equal to
$\equiv$	is defined as
$\neq$	is not equal to
$\propto$	is proportional to
$\sim$	is on the order of
$>$	is greater than
$<$	is less than
$\gg (\ll)$	is much greater (less) than
$\approx$	is approximately equal to
Δx	the change in x
$\sum_{i=1}^{N} x_i$	the sum of all quantities x_i from $i = 1$ to $i = N$
$\lvert x \rvert$	the magnitude of x (always a nonnegative quantity)
$\Delta x \rightarrow 0$	Δx approaches zero
$\frac{dx}{dt}$	the derivative of x with respect to t
$\frac{\partial x}{\partial t}$	the partial derivative of x with respect to t
$\int$	integral